高山景行為人師表

學貫東西道播五洲

李正名先生雅正

陳洪渊書

癸巳年中秋

题词人简介

程津培院士

男，1948年生于天津，江苏连云港人。美国西北大学博士，美国杜克大学博士后。2001年当选为中国科学院院士和发展中国家科学院（TWAS）院士。1988年起在南开大学化学系任教，历任副教授、教授、物理有机化学研究室主任。曾任南开大学副校长、天津市政协副主席、科学技术部副部长。现任中国科学院学部主席团成员、国家奖励委员会委员、国家教育咨询委员会委员、教育部“2011计划”专家咨询委员会委员等职。

李正名　院士

题 词 人 简 介

陈洪渊院士

男，1937 年生于浙江三门县。1961 年毕业于南京大学化学系，德国 Mainz 大学访问学者（1981—1984）。2001 年当选为中国科学院院士。南京大学教授、博士生导师。后兼任：南京大学教学委员会副主任、南京大学分析科学研究所和化学生物学研究所所长；教育部科技委委员、化学化工学部主任，科技委学风建设委员会副主任。曾荣获国家自然科学二等奖（2007）和三等奖（1982）各 1 项，教育部自然科学一等奖（2001、2006）2 项，全国科学大会奖（1978）1 项，何梁何利科技进步奖（2006）1 项，中国侨联科技进步奖（2002）1 项等。2004 年被评为全国模范教师；2005 年被评为全国先进工作者，荣获国家“五一”劳动奖章。

耕耘六秩
春华秋实
科学人生
农药传人

程津培敬贺
2013年8月

题 词 人 简 介

张礼和院士

男，1937 年生，江苏扬州市人。1958 年毕业于北京医学院药学系，1967 年北京医学院药学系（有机化学）研究生毕业。美国弗吉尼亚大学化学系访问学者（1981—1983）。1995 年当选为中国科学院院士。任北京医科大学药学院副研究员，教授（1983—1999），北京大学医学部药学院教授（1999 至今），*Eur.J.Med.Chem* 副主编。曾任北京大学天然药物及仿生药物国家重点实验室主任，国家自然科学基金委化学部主任 (1999—2006)，IUPAC 有机和生物分子化学委员会常务委员（Titular Member, 2006—2009），英国皇家化学会 Fellow（FRSC），*Organic & Biomolecular Chemistry*、*ChemMedChem*、*Medicinal Research Review* 和 *Current Topics of Medicinal Chemistry* 国际编委。

李正名先生在发展元素有机化学学科和新农药研究等方面都取得了很多成果、做出了杰出贡献，在李正名院士文集出版之际特此祝贺并表敬意。

张礼和

2013年9月26日

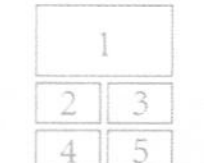

1. 在恩师杨石先先生家
2. 师生合影（一）
3. 师生合影（二）
4. 在波士顿会见指导过的博士生
5. 新年全家团聚照

中国工程院 院士文集

Collections from Members of the Chinese Academy of Engineering

李正名文集

A Collection from Li Zhengming

本书编委会 编

北 京
冶金工业出版社
2014

内 容 提 要

《李正名文集》将李正名院士及其科研团队自1957年至今所研究发表的论文，按照时间顺序进行了整合编排，内容涉及农药与环境安全、农药合成、分析、毒理、抗性、应用等诸多领域。由于论文数量较多，本书从中选取了106篇有代表性的论文进行了全文收录，其他论文则只对题目、收录期刊以及发表时间进行整理。

图书在版编目(CIP)数据

李正名文集/《李正名文集》编委会编.—北京：冶金工业出版社，2014.5

(中国工程院院士文集)

ISBN 978-7-5024-6470-7

Ⅰ.①李… Ⅱ.①李… Ⅲ.①李正名—文集 ②农药—文集 Ⅳ.①TQ45-53

中国版本图书馆CIP数据核字(2014)第055531号

出 版 人 谭学余
地　　址 北京北河沿大街嵩祝院北巷39号，邮编100009
电　　话 (010)64027926 电子信箱 yjcbs@cnmip.com.cn
策　　划 任静波 责任编辑 谢冠伦 任静波 美术编辑 彭子赫 罗栋青
版式设计 孙跃红 责任校对 卿文春 刘 倩 责任印制 牛晓波
ISBN 978-7-5024-6470-7
冶金工业出版社出版发行；各地新华书店经销；三河市双峰印刷装订有限公司印刷
2014年5月第1版，2014年5月第1次印刷
787mm×1092mm 1/16；58.5印张；4彩页；1426千字；916页
360.00元

冶金工业出版社投稿电话：(010)64027932 投稿信箱：tougao@cnmip.com.cn
冶金工业出版社发行部 电话：(010)64044283 传真：(010)64027893
冶金书店 地址：北京东四西大街46号(100010) 电话：(010)65289081(兼传真)

《中国工程院院士文集》总序

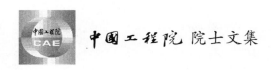

2012年暮秋，中国工程院开始组织并陆续出版《中国工程院院士文集》系列丛书。《中国工程院院士文集》收录了院士的传略、学术论著、中外论文及其目录、讲话文稿与科普作品等。其中，既有院士们早年初涉工程科技领域的学术论文，亦有其成为学科领军人物后，学术观点日趋成熟的思想硕果。卷卷文集在手，众多院士数十载辛勤耕耘的学术人生跃然纸上，透过严谨的工程科技论文，院士笑谈宏论的生动形象历历在目。

中国工程院是中国工程科学技术界的最高荣誉性、咨询性学术机构，由院士组成，致力于促进工程科学技术事业的发展。作为工程科学技术方面的领军人物，院士们在各自的研究领域具有极高的学术造诣，为我国工程科技事业发展做出了重大的、创造性的成就和贡献。《中国工程院院士文集》既是院士们一生事业成果的凝炼，也是他们高尚人格情操的写照。工程院出版史上能够留下这样丰富深刻的一笔，余有荣焉。

我向来认为，为中国工程院院士们组织出版院士文集之意义，贵在“真、善、美”三字。他们脚踏实地，放眼未来，自朴实的工程技术升华至引领学术前沿的至高境界，此谓其“真”；他们热爱祖国，提携后进，具有坚定的理想信念和高尚的人格魅力，此谓其“善”；他们治学严谨，著作等身，求真务实，科学创新，此谓其“美”。《中国工程院院士文集》集真、善、美于一体，辩而不华，质而不俚，既有“居高声自远”之澹泊意蕴，又有“大济于苍生”之战略胸怀，斯人斯事，斯情斯志，令人阅后难忘。

读一本文集，犹如阅读一段院士的“攀登”高峰的人生。让我们翻开《中国工程院院士文集》，进入院士们的学术世界。愿后之览者，亦有感于斯文，体味院士们的学术历程。

2012年7月

自　序

1948年，我在东吴大学高中毕业后得到一个获奖学金赴美留学的机会，所去的Erskine大学在美国南卡州，规模不大但历史悠久（成立于1839年）。由于1951年国际局势剧变，我经过反复考虑，克服种种困难于1953年回到了祖国。教育部派我来到南开大学。当时，我作为一个对祖国的未来充满憧憬的22岁青年只身来到南开园，至今弹指一挥间61年过去了，自己已进入耄耋之年。值得庆幸的是，我在这个时期亲眼目睹了祖国的飞速发展与南开大学日新月异的变化。新中国成立之前的100多年来，我国屡受列强侵略，多数战争都以割地赔款告终，这段屈辱的历史一直深刻地留在我的脑海之中。我在国外读书时，每当听到对中国人的歧视性言论时，常奋起据理反驳。当今我国发展飞速，已成为世界大国，赢得了各国应有的尊重。近年来，我国倡导的科技兴国、教育优先、人才强国、自主创新等各项政策日益深入人心。我们正在进入中华民族5000年历史长河中最为辉煌的时代，怎能不让我们感到无比自豪和兴奋呢！

最近接到中国工程院化工冶金材料学部的通知，为了庆祝2014年工程院成立20周年，决定资助我出版文集。我对工程院各位领导、学部院士以及工作人员的一贯关怀和帮助常存感激之心。这本文集收录了我的部分论文和发明专利，记载了多年来我课题组的一些踪迹。

回忆在不同时期由于工作需要，自己长期担任基层科技管理工作，虽感能力有限但不敢有所懈怠。在杨石先校长指导下，我一直在第一线坚持参加了有机化学和农药化学的科教实践。

这里收集的部分论文和发明专利，反映了在不同历史时期我课题组的科技活动。回想最早在1953—1956年，在杨老指导下开展新植物生长素研究的情景，当时实验室条件十分简单，缺乏基本的化学试剂。由于没有任何分析仪器，所合成的新化学结构迟迟不能确定。杨老通过中国科学院有机化学所黄鸣龙先生从国外带回来的一台元素分析仪才确定了结构。杨老又请刚到南开大学的崔徵教授协助测定生物活性，几经周折才成文发表。1962年，根据中苏科学院合作协议，杨老创建了我国高校中的第一个专职研究所——元素有机化学研究所。60年代初，苏联科学院陆续派来了4位专家，带来了一套当时在国内尚属空白的半微量磨口合成仪。经我所王琴荪教授参考设计交给天津玻璃仪器厂试制成功推广到全国，结束了多年来我们打橡皮塞孔的历史。当时实验条件虽然困难，但挡不住我们向科研进军的决心。

受到“文革”影响，元素所科研活动一度停滞。“文革”后杨校长根据当时国家改革开放政策决定派送一批中青年科技人员出国进修。1980年，我被选送到美国国家农业研究中心访问两年，参与该中心舞毒蛾信息素和澳洲原始小蜂信息素等研究课题。1982年回国后曾先后开展昆虫信息素、杂环化学、手性化学等课题研究并多次参加国家农药重要科技攻关项目，有些成果由于客观原因没有发表。《国家中长期科技发展规划纲要（2006—2020）》将农药创制列入优先主题，为了继承杨老的学术思想和优秀传统，我组工作逐步转到了这个方向。

农药对我国国民经济的重要性不言而喻。2012年5月18日，我国农业部公开声明：“农作物病虫草害引起的损失最多可达70%，通过正确使用农药可以挽回40%左右（粮食与果蔬等）的损失。我国是一个人口众多、耕地紧张的国家，粮食增产和农民增收始终是农业生产的主要目标。农药对植物来说，犹如医药对人类一样重要，必不可少。如果不用农药，我国肯定会出现饥荒。”

农药创制关键技术的掌握与否标志着一个国家农药科技有无核心技术和创新能力。由于历史条件所限，我国农药工业长期以仿制为主。农药创制有其特殊的跨学科、跨专业的交叉性，因此具有复杂性、长期性和风险性。现在国际上仅有美国、日本、德国、瑞士等国家有完整的新农药自主创制体系。

2002年，在有关农药创制的香山会议上，根据当时国际发展趋势我在报告中坚决支持我国今后农药创制目标应为绿色农药，并提出其具体的定义为：

（1）对人类健康安全无害（包括食品、水的安全）；

（2）对环境生态友好（包括对大气、土壤、有益生物无影响）；

（3）超低用量；

（4）高选择性；

（5）作用方式和代谢途径清楚；

（6）采用绿色工艺流程生产。

当时虽然提出了要注意对环境生态影响的观点，但没有达到现在认识的深度。最近，党的十八大文件中提出的要把我国建设成为生态文明社会的战略性部署及时指出了我们今后科技发展的努力方向，这对我们化学化工科技工作者来说，有着特别重要和特殊的指导意义。

回顾我的教学和科研生涯，要特别感谢杨石先先生对我的长期指导和培养，他的爱国情怀、重视人才和敬业精神对我的学术成长影响至深。我还要感谢范思涝副校长的殷切教导，他要我在工作中学好方针政策，注意团结同志和处理好各类矛盾。由于在化学院工作时间较长，还需感谢很多老同志多年来的热情帮助和支持：如金桂玉、李云峰、黄润秋、杨华铮、彭永冰、刘秀珍、杨光明、郝邦增、王玲秀、尚稚珍、赖城明、张岳军、战胜、廖仁安、方建新等。还要感谢先后参加我组学习的145名博士、硕士们。这本文集所记载的很多科学实验记录应归功于他们的刻苦努力和创新精神。毕业后，他们在不同岗位上勤恳工作，成绩显著，充分发扬了南开大学"允公允能、日新月异"的优秀传统，我从他们身上看到了祖国科技事业的光辉前景。

这里还要特别感谢，工程院老院长徐匡迪院士为《院士文集》作了总序，我国著名科学家陈洪渊、程津培、张礼和三位先生为本文集亲自题词，无疑是对我巨大的勉励和鞭策。在多年工作中，我曾得到天津市和南开大学主要领导的殷切教导和帮助以及各位老师和科技人员的指导与支持，这些都不断地激励自己的成长和进步。

学习永无止境，创新是科研的永恒主题。我们一定要以自主创新的开拓精神把各项工作搞好，一定要在平凡的岗位上做出不平凡的业绩，一定要在世界科技界的前沿阵地占有一席之地。

最后介绍一下本文集的编辑过程。在90年代，计算机的应用尚未普及。我的同事王惠林、叶春芝、岳铭秀、王素华、付丽娅、钱颖、玄镇爱等同志曾将我组有关论文编成软盘和讲义以便学生查阅。最近化学院农药国家工程研究中心和图书馆的李芳工程师、郝晋清博士、王宝雷副教授、赵毓高工等都投入大量精力和时间进一步整理资料，冶金工业出版社的任静波总编、谢冠伦编辑也多次给予专业性指导，在此对以上所有付出辛勤劳动的同志们表示深切谢意。

最后，我要表示对我父母的深切怀念，感谢他们对我的养育和教育之恩。由于工作繁重，在他们晚年身患重病时期未能长期伺伴身旁，深感内疚。感谢他们对我的长期教诲和鼓励，他们的善良、宽容和理解，永远激励自己在人生的道路上不断前进。半个世纪以来得到了夫人宋青雯在工作和生活上的一贯支持和照料，解除了我的后顾之忧，使我能全身心投入工作，特此表示由衷感谢。

再次衷心地感谢所有关心和帮助我的家人、领导、老师、同学、学生和朋友们。敬祝大家身体健康，平安幸福！

南开大学讲席教授 李正名

2014年3月18日于南开园

编者说明

中国工程院院士李正名是一位有机化学和农药化学家，在国内外学术界与产业界享有盛誉。在半个多世纪的教学科研生涯中，他为我国有机化学、农药化学的学科建设贡献了毕生的力量。

李正名重视理论研究的源头创新，多年来他以生物活性物质分子设计与有机合成为主要研究方向，参加了国家重大研究计划（“863”、“973”计划）、国家自然科学基金、国家新农药创制与产业化及天津市自然科学基金等项目，取得了一批研究成果。20世纪60年代在杨石先教授领导下，研究的“有机磷化合物设计合成与构效关系”于1987年获国家自然科学二等奖；70年代开展了天然生物调控物质研究、昆虫信息素化学、杂环化学与有机立体化学研究；80年代开展农药化学基础研究、新农药创制与开发等。

李正名曾主持和参加了国家“六五”到“十二五”期间的国家科技攻关项目，均出色完成了任务。所主要参加的“六五”国家科技攻关项目——“高效杀菌剂FXN新工艺研究”于1993年获国家科技进步一等奖。2003年，所主持的国家自然科学重点基金“新农药创制基础研究”项目被评为“特优”。参加的“十五”国家科技攻关项目，创制了具有自主知识产权的新型超高效除草剂单嘧磺隆。经过长期的工艺开发，药效鉴定和大量毒理、环境、生态评价的完成，成为我国第一个获得国家新农药正式登记的创制的除草剂，已经过谷田除草大规模田间示范推广。另一个创制品种小麦专用的单嘧磺酯也获得国家新农药正式登记证。这些具有自主知识产权的创新研发成果于2007年获国家技术发明二等奖。2014年获天津市科技重大成就奖。据统计共获国家级奖励7项、省部级奖励11项、个人奖励21项。

李正名曾指导博士生和硕士生共145名，发表论文550篇、编写专著2部，发明专利37项（其中授权17项）。

为追溯先生的科教人生轨迹，弘扬先生的学术思想，展示先生坚持不懈的创新精神，鼓励青年学子，我们按照时间顺序将先生的部分论文集编成《李正名文集》以供学术交流与参考。

谨以此书向中国工程院20周年院庆表示祝贺！

本书编委会
李　芳（主编）
郝晋清　王宝雷　赵　毓
王素华　战　胜　彭丽娜
2014年3月

目 录

承师创新

论文选登

1984 年

1987 年

1989 年

1990 年

1991 年

1992 年

1993 年

1994 年

1995 年

1996 年

1997 年

1998 年

1999 年

2000 年

2001 年

2002 年

2003 年

2004 年

2005 年

2006 年

2007 年

2008 年

2009 年

2010年

2011年

2012 年

2013 年

附　录

承师创新

李正名简历

1931 年	出生于上海市
1942—1945 年	上海震旦大学附属初中毕业
1945—1948 年	东吴大学附属高中毕业
1949—1953 年	美国埃斯金大学化学系毕业
1953—1956 年	南开大学化学系研究生毕业
1959—1979 年	南开大学讲师、副教授
1980—1982 年	美国国家农业研究中心访问学者
1983—1996 年	南开大学元素有机化学研究所所长、教授
1983 年—	国际纯粹与应用化学联合会（IUPAC）应用化学部中国代表（1983—1997），资深代表（2004— ）
1987—1996 年	南开大学元素有机化学国家重点实验室主任
1987—1990 年	南开大学元素有机化学国家重点实验室学术委员会主任
1990 年—	中国科学院上海有机化学研究所生命有机化学国家重点实验室学术委员
1992—1996 年	国务院学位委员会学科评议组成员
1992—1998 年	中国化学会常务理事兼副秘书长
1993—1996 年	国家自然科学基金委员会化学部有机化学组评委
1994—1998 年	教育部长江学者化学化工评审组组长
1994—1998 年	南开大学化学院副院长
1995 年	当选为中国工程院院士
1995 年—	“Pesticide Management Science” 编委、执行编委
1997 年—	南开大学农药国家工程研究中心主任
1997—2002 年	天津市科学技术协会副主席
1998—2002 年	中国工程院化工冶金材料学部常委
1999—2003 年	农业部农药化学与应用重点开放实验室学术委员会主任
1998—2000 年	国家自然科学基金委员会化学部有机化学级评委

2000—2004 年	国家自然科学基金委员会化学部有机化学评审组组长
2001—2004 年	国家自然科学基金委员会国家杰出青年评审组成员
2001 年—	大连理工大学精细化工国家重点实验室第三、四届学术委员，第五届学术委员会指导委员
2002—2008 年	中国科学院上海有机化学研究所学术委员会顾问委员
2002—2006 年	国家干细胞工程技术研究中心工程技术委员会委员
2003—2008 年	联合国南通农药新剂型工程技术中心技术委员会主任
2004 年—	南开大学元素有机化学国家重点实验室学术委员会副主任
2004—2006 年	福建农林大学农药与化学生物学教育部重点实验室学术委员
2006 年—	华中师范大学农药与化学生物学教育部重点实验室学术委员会主任
2006 年—	贵州大学绿色农药与农业生物工程教育部重点实验室学术委员，客座教授
2005—2008 年	农业部农药检定所专家咨询顾问委员会委员
2005 年—	清华大学生命有机磷化学及化学生物学教育部重点实验室学术委员
2006 年—	湖南化工研究院国家农药创制工程技术研究中心技术委员会高级顾问
2007 年 —	中国农药工业协会第八届理事会高级顾问
2009—2013 年	南开大学植物保护学位评定分委员会主席
2008 年—	南开大学科学技术协会主席
2011 年—	江苏丰山集团有限公司工程技术院士工作站管委会名誉主任
2014 年—	中国科学院上海有机化学研究所“一三五”战略咨询委员会委员

• 李正名自述

我出生在上海的一个知识分子家庭中。上世纪30年代，家父李中道在上海东吴大学法学院毕业后赴美国密歇根大学获法学博士学位，家母陈女英在美国伊利诺伊大学获文学硕士学位。他们学成回国后分别在上海东吴大学和复旦大学执教。后来父亲还参与了一些法律事务所的工作。我在上海的小学、初中毕业后去了苏州的东吴大学念完高中。记得在解放前国内物价飞涨，寄一封平信竟要法币500万元。家里每月汇到苏州的生活费必须马上去观前街换成银元，不然过几天就贬得不值钱了。在1948年我考取了美国私立大学联合奖学金，但当时办赴美签证要去香港美国大使馆办理。经过繁杂的手续，办成了留学签证。我去Erskine大学后，根据从小的志愿选择了化学专业。本来想按计划读下去，但当时国际局势骤变，在大三时，美国麦卡锡主义掀起反华高潮，促使我考虑是否要中断学习回国。当时我的系主任E. S. Sloan教授对我十分关心，他告诉我毕业后将推荐我去南卡州立大学研究生院，将获全额奖学金。一些亲友对当时国内局势存有疑虑，劝我慎重考虑。当时美国移民局不准中国理工科留学生回国，我通过大学校长R. C. Grier博士向有关管理部门反映，因家祖母病重而获准回国。1953年，我大学毕业后随第一批中国留学生集体坐船回国，来北京教育部报到，被分配来到南开大学。有人说不到外国不知道什么叫爱国，当时我国经济落后，科技不发达，在国际上往往被人瞧不起。作为一名知识分子，都会产生这种强烈愿望希望祖国早日现代化，中华民族尽快富强起来，这种朴素的爱国思想也成了我后来工作的动力。

1953年，来南开大学后，学校分配我做杨石先校长的助手。后来教育部开始建立研究生制度，杨校长让我做他的研究生兼科研助手，开始做新植物激素的创制工作。并将研究结果刊登在南开大学刚创刊的《自然科学》学报上。杨老德高望重，学识渊博，十分关心青年教师的成长。我有幸跟随杨老三十余年，在杨老和其他老同志的关心和帮助下，先后参与了有机二室、元素有机化学研究所、中苏科技合作项目、国家重点实验室、亚非拉研究中心、国家工程研究中心等项目的筹建和建设工作。在前辈对我国教育战线的热爱和对科学事业的责任心的言传身教下受到很大的教育。在60年代初由于工作需要，我曾负责过200多名同学的大有机化学实验班，亲自编写实验教材，设计各实验内容等，从大量教学实践中学到很多的经验和知识，较好地完成了任务。1962年元素有机化学研究所成立后，我始终坚持参加一线工作，对所里的人员和环境都怀有深厚的感情。

由于当时形势要求，我们经常下乡，后来由于参加国家科技攻关等项目也经常下厂。对于我来讲，这是了解国情、直接了解工农大众的大好机会。对于从小在大城市中成长的我来说，通过和广大劳动人民一起工作，克服原先严重脱离实际的“优越感”，打破了我身上的骄娇两气，劳动人民勤劳朴素的优良传统对我思想形成了深远的影响，从而使我对从农村来就读的研究生怀有一种亲切感和责任感。70年代末我国迎来了“改革、开放”的大好形势，杨老抓紧时机启动我校派遣教师出国进修的重要举措。杨老通过任之恭、左天觉等著名华人的帮助，通过教育部支持送我到美国联邦政府的国家农业研究中心，成为新中国派去该单位做访问学者的第一人。该中心是国家级研究部门，有1600名科技人员（含500多名博士）。在这种现代化的大型研究机构的运行模式下工作，使我对美国立国以来对农业科技的重视有了新的体会。

1982年9月赴美学术访问期告一段落，当时有些亲友介绍我去 Walter Reed 国家抗疟研究中心和加州西北大学生物化学所作博士后，由于杨老的召唤，我立即回国。1982年12月被学校任命为元素有机化学研究所所长，在此岗位上一直工作了13年。1985年，在杨老的领导下，我和同志们一起努力争取，成功地申请到元素有机化学国家重点实验室，曾主持并参与建设全过程。1993—1997年，我被任命为化学院副院长。1995年我校申请农药国家工程研究中心获得批准，我被任命为主任，和同志们一起艰苦创业，从当时一片空白到初具规模。回顾30年的科技管理工作，根据工作需要在不同时期承担不同性质的工作，虚心向领导和同事学习，恪守职责、不敢懈怠，基本完成了各个时段的任务。杨老的敬业精神、尊重他人的作风和其他老同志给我鼓励和帮助的情景犹历历在目，我对之十分珍惜。在大家努力下，不断开拓创新，所在单位曾先后获得“全国高校科研工作先进集体”、“优秀国家重点实验室”等称号。2004年国家验收我工程研究中心的评价：“出色地完成了项目建设任务，形成了较强技术创新能力，取得了较好成果转化经验，为高校中建立国家级工程研究中心提供了一个机制和体制上的范例”。这些鼓励给了我们克服困难的勇气和信心。

光阴荏苒，时光飞驰，我在南开大学工作至今已58年了。随着国家经济和科技的飞速发展，我国已进入历史上千载难逢的黄金时代。我感到十分幸运能亲身参与到这些实践活动中去。回想过去一直想当一名纯粹的科学工作者或教育工作者。1982年杨老看到我访问回国后表示不愿承担任何行政工作时，曾严肃地对我说：“我为了学校和元素所的发展，呕心沥血，全心投入，现在年纪大了身体不行了，让你来承担一些责任，你不能光考虑自己而不考虑大局呀”。当时杨老的严厉批评给了我很大的震动。此后在我承担领导所分配的各项任务中，虽然有时遇到的困难很多，责任很重，但当想起杨老的叮咛，把这些岗位看作自己为国家和集体服务的一种责任时就有了巨大的动力。我决不辜负各级领导对自己的期望，要和大家一起为学校、为社会多做贡献。

人到老年阶段回顾一下自己的一生，总结一些经验，或有所裨益：

（1）我在青年留学生涯中由于国际局势的骤变中断了深造回国，对此曾有不同的评论：有人认为没有读完学位回来太感情用事了，在待遇方面吃亏了，还遭到个别人的误解等。我虽然没能按原计划读完博士学位，但回国后我的专业知识能和祖国科教事业的建设紧密地结合起来，在不同的岗位上做出一定的业绩，看到所参加的科教事业的成果和培养人才的成长，感到自己的人生过得很有价值。当时如留在美国读完学位后找一待遇好的工作或有可能，但那将是另外一种人生轨迹了。人生的意义不能仅为满足个人的物质利益，在精神上也应有所追求。为了建设现代化的祖国，能和同志们一起团结拼搏，共同分担遇到挫折时的忧虑和分享事业成功时的喜悦，将青春年华的汗水撒在祖国的大地上是个人生涯中最大的幸福。根据当时的历史条件，选择回国是一个明智的抉择、理性的回归。

（2）杨老以高尚的品格对青年潜移默化的教育是他对人才培养的重要贡献。他提出的“繁荣经济、发展学科”的思想现在看来仍有很大的指导意义。他多次提出“只有高尚的思想才能产生巨大的动力”的观点使我印象深刻。我们为了建设现代化的祖国，一定要在思想上严格要求自己，培养崇高的理想才能有坚强的事业心和持续的进取精神，将科研工作结合国家的需求与科技前沿方向是我们的努力方向，积极承担建设国家的任务、坚持开拓创新是我们的历史使命。

(3) 家祖父李维格出身贫寒，在清末期间去英国半工半读，后来在英、美、日等使馆供职。回国后和梁启超一起创办“时务报”与“时务学堂”(湖南大学前身)。又按清朝大臣盛宣怀的委派，参加南洋公学（交通大学前身）和汉冶萍钢铁公司的筹建工作，克服无数艰辛，解决了技术难题，和全体职工一起奋斗，炼出了我国历史上第一炉优质钢（中国科学技术专家传略（冶金卷)，朱光亚总编、陆达主编）。我从小钦佩家祖父的事迹，后来通过一些史料和殷瑞钰院士的热情介绍，进一步了解到在贫穷落后、历经屈辱的清末时期，家祖父立志“教育救国”、“科技兴国”的事迹，对我有深刻的教育意义。在遇到困难时，我常想现在各种条件比100年前的半封建半殖民地社会的旧中国要强多了，为什么还要被这些困难所吓倒呢？实际上，历史上很多先进知识分子的远大理想和良好的愿望也只有在新中国才能真正得到实现，怎能不珍惜今天中华民族崛起的大好时机呢？

最后，要借此机会衷心感谢在南开大学工作和生活的半个多世纪中，中央各部委（教育部、科技部、原国家计委、原化工部、农业部)、中国工程院、国家自然科学基金委、天津市委、天津市科委、学校等各位领导、同事和同学们所给予我长期的关怀与鼓励。我感到十分幸运，得到了不少前辈对我的教育和帮助，一直鼓励自己不断进步。深切感谢组织多年来对自己的培养和教育，使我明确了正确的努力方向。感谢众多朴素勤劳的劳动人民，正是他们提供了粮食、建设了住房和实验大楼，让我们有发挥知识与才能的基础。我要感谢父母和家人的支持，使我全身心投入工作。我要在有生之年继续发挥余热，和大家一起为祖国的科技教育事业添砖加瓦，一起见证我国几代人向往的伟大“中国梦”早日实现。

饮水思源——恩师杨石先先生对农药化学学科的重要贡献及其学术思想*

李正名

（南开大学元素有机化学研究所，天津　300071）

摘　要　杨石先先生一生献身于我国的教育事业与化学学科的发展，在62年中为我国培养了无数高质量的科教人才。他除了长期担任南开大学校长之外，还创建了我国大学第一个专职研究所，即元素有机化学研究所。他率先开展了我国元素有机化学与农药化学的科学研究，领导了元素有机化学国家重点实验室的建立，是我国元素有机化学和农药化学的奠基人和开拓者。他倡导用有机化学的专业知识，科学和系统地开展农药化学研究，组建队伍获得20项科研成果，发表上百篇科学与论述性论文，为我国开展自主创新农药研究事业做出重要贡献。在农药化学学科的学术思想中，他强调要弄清该学科的交叉性、系统性和内在规律性，倡导要学习国际先进经验，要结合国情自主创新，要为国家经济服务，要对世界农药科技做出贡献。他毕生对人才培养给予了特别的重视，为我国科技事业持续发展做出了重大贡献。

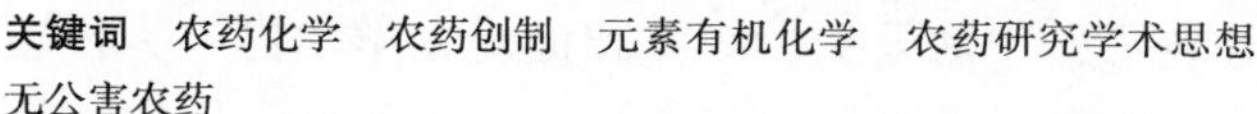

关键词　农药化学　农药创制　元素有机化学　农药研究学术思想　无公害农药

Important Contribution and Academic Thoughts on the Pesticide Chemistry Discipline by Yang Shixian

Zhengming Li

(The Research Institute of Elemento - Organic Chemistry, Nankai University, Tianjin, 300071, China)

Abstract　Professor Yang Shixian devoted his whole lifetime to the advancement of China education and chemistry discipline. During 62 years, he fostered numerous high - quality specialists in the field of science and education. He was the president of Nankai University for 28 years, during which he founded the first research institute in Chinese universities, the Research Institute of Elemento - Organic Chemistry. He was the first one to establish the research field of Elemento - Organic Chemistry and pesticide chemistry in China and was acknowledged as the founder and explorer in these disciplines. He advocated to utilize the knowledge of organic chemistry to explore scientifically and systematically the research in pesticide chemistry. He organized a research team to accomplish 20 projects, published more than a hundred papers. He made important contribution in the innovation of new agrochemical bio - active candidates. He emphasized to learn international experience, to incorporate into our national background, to serve our economy and to contribute to the world pesticide science. His extraordinary devotion during his entire career to rear young qualified specialists to sustain our science and education in China has been

*　原发表于《化学进展》，2011，23（1）：14－18。

highly appraised.

Key words elemento - organic chemistry; academic thoughts on pesticide research; pesticide innovation; environmentally benign pesticides

Contents

1 业绩简介

杨石先先生（1897—1985）是我国著名教育家和杰出化学家，曾先后担任南开大学理学院院长（1928—1937）、西南联大教务长和化学系主任（1938—1947），任南开大学校长达28年之久（1957—1985），跨越我国各个重大历史时期。不论时局怎么艰苦辛难，杨老始终孜孜不倦地为我国的高等教育事业和化学学科的发展呕心沥血，辛勤耕耘。他对祖国的衷心热爱，在工作中谦虚谨慎、兢兢业业、严于律己、客观公正的崇高品质令人肃然起敬。他对晚辈人才的爱护、培养和提携在学术界广为传颂。他培养的几代优秀学生遍及国内外，桃李满天下。他被誉为中国元素有机化学和农药化学的奠基人和开拓者[1]。

杨老历任中国科学院数理化学部委员，化学组组长。先后于1956年、1962年两次参加制定我国重大科学技术长远规划，为我国科技决策献计献策做出了重要贡献。1957年参加中国科学院赴苏科技代表团。1962年创建我国高校中第一个专职化学研究机构——南开大学元素有机化学研究所（简称元素所），开拓了我国元素有机化学和农药化学的研究领域。1985年在他的关怀下，创建南开大学元素有机化学国家重点实验室的申报被批准。1995年根据杨老关于“经济繁荣，学术发展”和搞农药研发要“小配套、大协作”的遗训，国家批准南开大学成立农药国家工程研究中心。由于杨老高瞻远瞩、德高望重，他对我国的科技事业和对南开大学的诸多重大贡献业绩永存。杨老一生的杰出贡献很难在此短文中予以概括。他对我国早期元素有机化学研究进展曾有总结[2]。以下仅就他对南开大学农药化学学科方面的贡献作一介绍。

2 创立我国农药化学学科

杨老在西南联大工作时，所承担的教学和行政任务十分繁重，在当时十分艰难的条件下坚持开展有关抗疟药物常山化学合成的研究。据杨老谈，他从小对植物学很感兴趣，早在解放初期考虑到我国是一个农业大国，经过慎重思考，根据国家的需要决定将自己的药物化学研究方向转到农药化学。解放初期，在当时的历史条件下，我国农药化学研究一片空白，我国连抗美援朝期间急需的666，DDT都不会生产。1953年为了探讨我国粮食增产的科学方法，在十分简陋的实验条件下杨老开展植物生长刺激剂研究，1957年他的第一篇研究论文发表在南开大学自然科学学报（创刊号）上[3]。

1958 年，杨老为了满足新建的天津农药厂急需上马的重点项目，组织了一批年轻教师战斗了 40 个日日夜夜，在国内首先完成了我国第一个有机磷杀虫剂“对硫磷”的合成工艺，并交付该厂使用。当时杨老领导在南开大学创建的“敌百虫”和“马拉硫磷”两个车间也首次生产出我国当时急需的杀虫剂产品，填补了我国的技术空白，并于 1958 年 8 月 13 日光荣地接受了毛主席的亲莅视察。60 年代初为了筹备元素所的成立，杨老先后邀请了一批苏联专家来所讲学，其中有 Kabachnik、Mastrukova 等院士与专家，对我国有机磷化学的发展起了重要的推动作用。后来还有 Martinov、Gefter 等有机硅化学、有机氟化学等专家来所讲学，当时在教育部的统一布置下，全国大学不少进修教师来到南开大学听课，对我国元素有机化学的学科建设和传播起了推动作用。在建设元素有机化学研究所的当年，还专门成立了第一研究室（农药化学），为我校的农药学科起了奠基作用。1966 年文革开始后元素所被迫摘牌解散，形势十分危急！1971 年杨老不顾自身安危，在南开大学贴出第一张大字报《请求保留元素所》，同时亲自写信给周总理，使得该所得以幸存下来。1977 年杨老出席全国科学大会，他所主持的 10 项农药科研成果荣获全国科学大会奖。1978 年根据中央的部署，杨老开始分批选派中青年教师出国深造，为我国的化学和农药学科培养了一批人才，后来这些人才都成为各单位的领导和科技骨干。在杨老领导下，南开大学有关农药研究成果，如有机磷 32 号及 47 号、灭锈一号、除草剂一号、大豆激素、矮健素、螟蛉威、燕麦敌、叶枯净、多霉净、久效磷新工艺、氯氰转位技术等课题曾先后获得全国科技大会奖、国家自然科学奖、国家科技进步奖、国家发明奖等部级以上的奖项 20 余项[4]，为南开大学有机化学学科与农药学科于 20 世纪 90 年代申请全国重点学科和博士点打下了扎实的基础。

杨老一生对化学学科发展十分关心，他曾写道“我从事化学工作已 60 年了，我对化学还有深厚的感情。——希望化学还是不断有新的发明创造，在赶超世界水平上作出我国独特的贡献”[5]。他在有机化学和农药学科方面的学术思想和观点也很精辟[6~8]，在今天看来还有很大的前瞻性。本文试将杨老在农药化学学科范畴内的重要贡献与其学术思想概述如下。

3　农药化学学科的学术思想

3.1　农药创制是一个典型的交叉学科和系统工程

在杨老启动农药化学研究的 20 世纪 50 年代，北京农业大学黄瑞伦教授在国内早已开展了卓有成就的农药研究，但侧重点是从农学的角度研究农药在实际应用中的一些重要基础理论问题。杨老作为一位在有机化学和药物学方面有很高造诣的专家，他所开展的农药化学研究，从发挥有机化学专业创制我国具有自己知识产权的视角来开展工作。在 1962 年成立元素有机化学研究所时，他在“谈谈农药问题”一文[9]中强调“农药不能脱离现代科学技术水平而单一发展，它是一个综合性极强的学科，涉及的面很广，如有机化学、农业化学、植物、动物、昆虫、真菌、细菌、病毒、土壤、医药、农林渔牧、生化、化工等”。因此在建所初期他坚持要有搞生物的人才参加，先后设立了生物测定室、有机分析室、剂型室、毒理室和中试车间等。诚然在这里有机合成追求的目标是寻找有新的具有特殊生物活性的新结构分子，在农药创新中起了关键的作用，但杨老高瞻远瞩从一开始就把农药研究看作是一个系统工程来开展，在“搭架子”的阶段即能注意农药学科的跨学科、跨专业的特点，说明他对我国开展农药科学的正确认识。这对后来南开大学解决了很多国家农药重大攻关中的技术难题和为启动农药创制工作打下扎实的基础。在当时国家召开的科技会议上，杨老还多次呼吁由政府成立一个跨部门、跨专业的国家级农药管理委员会来协调生产和研究中出现的很多交叉性的复杂问题，在今天看来也有很好的指导意义。

多年来国际公认“农药创制”是一个风险高（分子设计命中率小于几万分之一）、投资大（1 亿~2 亿美元/每个创制产品），周期长（十年以上），难度与创新医药相比有过之而无不及。国际上新农药创新一直由美、日、德和瑞士等少数发达国家垄断。长期以来我国农药创制工作的基础十分薄弱，缺少自己主打的创制产品。自 50 年代起杨老很清晰地看到我国和国际水平存在着的这种巨大差距，但他对我国的农药创制前景十分憧憬，1962 年杨老号召我们：“走中国自己的农药发展道路——赶上和超过国际先进水平，创制我国所需要的更多更好的新农药”。他在《全国十年科学规划会议报告》中指出：“我们一方面走中国自己发展农药的道路，另一方面还要吸收外国先进的经验——力争在最短期间内赶上和超过国际先进水平，创制我国所需要的更多更好的新农药，促进我国农业现代化的早日实现”。1964 年在北京举行的首届亚非拉科学大会上杨老在《中国农药化学的研究》中写道[9]：“我们完全有信心在化学农药的研究上作出有益的成果，不仅在解决国内病虫害防治的主要问题上，并且将对提高世界农药水平作出一定贡献”。

1982 年杨老在任中国化学会名誉理事长时给全国化学会的题字为“化学要为中国的经济繁荣、学术进展作出更大的贡献”，近年中央号召科技工作者为国家重大需求和国家科技前沿发展服务中更加科学地明确了这个指导思想，并已将“农药创制”项目列入国家科技中长期规划纲要（2006—2020），近年在国家的大力支持下开始出现一批志士仁人，特别是青年科技人员，开始作出了不少新农药创制的科技成果，杨老毕生为之奋斗的我国农药创制事业正在逐步成为现实。

3.2 关于农药研究的学术思想

1964 年，杨老在我国当时的历史条件下提出了农药科研 4 条原则[9]：

（1）在国内外已有的科学研究成果的基础上，密切结合我国实际，使研究成果尽快地服务于农业生产。

（2）贯彻理论联系实际的方针，使科学研究在踏实可靠的基础上稳步发展。

（3）充分利用我国资源，使我国农药科学研究工作具有本国的特色。

（4）组织分工合作，充分调动业务部门的力量，共同完成国家所制定的农药任务。

杨老还针对农药科研提出三项任务：

（1）选择国际上成效卓著而又适合我国情况的若干农药品种迅速投入生产。

（2）整理国内的“土农药”，加以鉴定、分析和找出有效成分以便测定结构和进行合成改造之用。

（3）在现有的基础上结合我国的具体情况，进行创新性的探索工作，试图制出崭新的具有优良性能，特别是高效低毒的农药。

在众多场合中杨老一再呼吁要从战略上看问题，创制自己的农药是我们要坚持的方向。但要考虑国情和我国的实际情况，要创仿结合、抓紧开发新产品的新工艺，对国际上有苗头的研究成果要提前开发，要考虑使用后对环境和天敌生态的影响，要重视天然植物源的活性结构优化，要向“无公害农药”、“第三代农药”的“崭新领域”发展[10,11,15]。

在具体工作中，他积极鼓励开拓在当时国内还很少开展的除虫菊素、昆虫信息素、非抗胆碱酯酶性杀虫剂、内吸性杀菌剂、异噻唑杂环等新课题，还亲自部署了甘肃地区的天然植物骆驼蓬草中具有植物激素活性结构的鉴定和研发[12,13]，支持了内蒙古地区的娃尔藤碱天然植物活性结构的研究。

杨老通过和当时苏联科学院涅斯米扬诺夫院长签订了在元素有机化学领域的全面合作协议，先后开展了有机磷化学、有机氟化学、有机硅化学、有机硼化学和有机金属化学等合作课题。杨老鉴于当时国内对磷化学研究基础少，而磷化学和生命科学关系密切，首先亲自指导了磷化学的研究并解决了其分离提纯的关键难题，重点是阐明有机磷化学农药的杀虫活性和分子结构之间的构效关系，研究了当时国内外大量使用的高毒高效杀虫剂3911的结构改造问题。杨老带领助手先后合成了上百个新结构，经过生物筛选，找到了P32和P47的杀虫活性（对棉红蜘蛛、棉蚜等有很高活性，药效与当时进口的1059相当，但对人畜的毒性要低5~10倍）。此成果曾获1978年全国科技大会科技新产品奖。在有机磷化学的基础研究中经过严格基本功锻炼的科研人员，后来承担了新杀虫剂“久效磷”的新工艺攻关任务，很快地掌握了关键技术，协助国内农药厂建立了生产车间，供应全国。在杨老鼓励下，有机硅和有机锡等研究也研发了一些很有开发前景的新的农药结构分子。在杨老学术思想的引导下，元素所勇于承担国家重大攻关项目，如高难课题溴氰菊酯、高效氯氰菊酯转位技术，叶枯净、螟蛉畏等，均获得同行好评，推动了我国农药生产水平的提高，产生了数亿元的经济效益。1988年杨老带领的团队在有机磷化学的基础理论成果也获得了国家自然科学奖（二等奖）[14~19]。

杨老对国际农药研究的最新动向十分关注，1971年在评论当时美国卡尔逊写的《寂寞的春天》时提出“要用一分为二的观点来观察事物，——综合防治的道路是植保农药今后发展必然的道路”。同年在《批判现代植物保护与农药发展中的一些形而上学的观点》中提出“我国是社会主义国家，应是关心人民群众的生命安危，毫无疑问今后农药发展的方向必须坚持高效低毒（包括低残毒）这一点，使植保农药永远沿着健康正确的路线向前发展”。1979年在《国外农药发展》一书的前言中写道[11]“除采取必要的防止污染的措施外，人们还希望合成一些无公害的农药。通过不断的研究探索，近年来出现了继无机农药、有机农药之后的所谓第三代农药——第三代农药的出现显示出很强的生命力。可以预料今后会有较快的发展，从而开辟出农药的一个崭新领域”。这些学术观点对当时我们科研工作方向都有很重要的指导意义。

据统计，自1959年到1979年期间，杨老先后写了“国外农药发展的趋势”、“谈谈国外农药发展的趋势”、“加快农药科研的步伐”等20多篇有关农药科研的指导性文章，平均每年一篇，对我国农药科研的同行起了重要的引领和借鉴作用。综述内容涉及有机磷化学、元素有机化学、杂环化学、天然产物等领域，覆盖了植物激素、杀虫剂、昆虫不育剂、杀菌剂、杀病毒剂、除草剂等研究方向。杨老在农药化学领域发表了84篇论文，论述性文章22篇，编著（译）作6部。在担任校长和其他16项兼职的情况下，还能以如此的热情和精力进行研究和编写工作，他对科学事业的不懈追求和敬业精神是我们永远学习的榜样。

3.3 十分重视科技队伍的建设

杨老不仅是一名出色的科学家，还是我国一位十分著名的教育家。杨老较早提出

“科学研究人员的培养和提高是战略任务，提高科学研究人员的水平是提高科学研究工作的关键”。敬爱的聂荣臻元帅曾经评价杨老为“学者楷模，人之师表”。中科院原院长卢嘉锡院士称杨老为“勤於育才、善於相才”。杨老在长期的教学工作中，以他高尚的品德、渊博的学识，严谨的治学精神在我国不同的历史时期，以无比的热情，培养了一代又一代的青年学子，使他们成长为我国各条战线上的骨干和栋梁，如较早期的著名学者唐敖庆、蒋明谦、邹承鲁、殷宏章、钮经义、胡秉芳、申泮文、何炳林、陈茹玉等院士。申泮文院士在谈到杨老对他一生的培养和影响时无比激动，他回忆杨老在西南联大的讲课“极为精彩，引人入胜，——更突出地显示他的精湛学识和高超讲课才能”。著名物理学家任之恭教授在美国曾对作者详细叙述了杨老在西南联大十分艰苦的环境中任劳任怨，精心培养我国未来科技人才的动人情景，一贯作风正派，办事公道，他强调说杨老是他一生中最为钦佩的几位学者之一。定居在国外的老化学家蔡麟博士，情不自禁谈起杨老在西南联大时对学生谆谆引导，严格要求的往事，使他终生受益。解放后，南开大学刚从昆明迁回天津，杨老积极从美国分批动员了一批学者回国充实化学系的师资力量，大大增强了南开化学学科的教学实力。在筹建元素所的初期，杨老从各重点大学调来应届大学生，还从全国重点职校招聘了一批毕业生，对后来元素所生测室、分析室、中试车间的建立起了十分重要的作用。在60年代初杨老还从教育部动员当时从苏联、民主德国回国的留学生来元素所工作，积极争取名额送元素所两名青年学者到苏联科学院进修。元素所科技人员在“文革”中被迫改行分散到全国各地，都被杨老陆续找回。在“文革”后期他不顾自身的安危，积极为在“文革”中受到不公正待遇的有机磷专家陈天池教授平反。70年代末，在邓小平同志创导的“改革、开放”的大好形势下，积极推动往国外派送访问学者，为元素所的持续发展呕尽心血。杨老一生（包括他的学生们）培养了无数高质量的各种专门人才，为祖国的现代化作出了重要的贡献。

4　学习楷模

至今杨老离开我们已有26个年头了。自从1978年我国改革开放以来，我国农药科技的综合实力和其他战线一样已发生了翻天覆地的变化。我国的农药产量已达世界第一，年产值已达1400亿元，其中约有一半出口到世界各地。在各级领导大力支持下，我国的农药科技已从长期的“以仿制为主”逐步转到启动“以创制为主”，从南开大学校园毕业后走向全国的一代代优秀人才，正在各条战线大显身手，一展风采，为我国农药科技赶上世界水平不断取得可喜成果。杨老预言的“走中国自己的农药发展道路，——赶上和超过国际先进水平，创制我国所需要的更多更好的新农药”的时代已经到来。杨老自1923年受聘为南开大学教授到1985年去世，在南开大学辛勤工作62年之久，他的一生是全心全意，鞠躬尽瘁，勤恳敬业，为我国的教育和科学事业无私贡献的一生。他培养的不同时期的学生正在祖国大地为中华民族的崛起努力工作。我们要牢记杨老的光辉业绩，以他为榜样，克服一切前进中的困难，为我国的科技事业和工农业现代化做出更大的贡献。

参 考 文 献

[1] 中国科学技术专家传略，理学编（化学卷Ⅰ）（Brief Biography of Chinese Specialists in Science and Technology（Chemistry Volume 1））．北京：中国科学技术出版社（Beijing：China Science and Technology Press），1993，150－163.

[2] 杨石先（Yang S X），陈天池（Chen T C），王积涛（Wang J T），林一（Lin Y）．元素有机化合物化学．十年来的中国科学－化学（1949—1959）．北京：科学出版社（Beijing：Science Press），1963：421－467.

[3] 杨石先（Yang S X），李正名（Li Z M），崔澂（Cui Z），姚珍（Yao Z），叶超然（Ye C R）．南开大学学报（自然科学）（Acta Scientiarum Naturalium Universitatis Nankaiensis）：1967（4）：4－10.

[4] 杨光伟（Yang G W）．杨石先传（The Biography of Yang Shixian）．天津：南开大学出版社（Tianjin：Nankai University Press），1991.

[5] 现代化（Modernization），1979（1）：1.

[6] 李正名（Li Z M）．农药学学报（Chinese Journal of Pesticide Science），2009，（11）（增刊）：72－74.

[7] 李正名（Li Z M）．杨石先纪念文集（A Collection of Memorial Articles to Yang Shixian）．天津：南开大学出版社（Tianjin：Nankai University Press），1999：104－107.

[8] 南开大学．杨石先纪念文集．天津：南开大学出版社，1999.

[9] 杨石先（Yang S X）．中国农药化学的研究（The Research on Pesticide Chemistry in China），亚非拉科学学术讨论会（Asia，Africa and Latin－America Scientific Workshop）（北京，1964）．

[10] 杨石先（Yang S X）．农药工业译丛（Translated Collection of Pesticide Industry），1979（1）．

[11] 杨石先（Yang S X）．前言（Introduction），国外农药进展（二）（World Progress in Pesticide Science）．北京：化工出版社（Beijing：Chemical Industry Press），1979.

[12] 杨石先（Yang S X），陈茹玉（Chen R Y），武振亮（Wu Z L），郑巧兰（Zheng Q L），史延年（Shi Y N），刘准（Liu Z），张春香（Zhang C X）．植物生理学通讯（Plant Physiology Communications），1987，（1）：18－21.

[13] 杨石先（Yang S X），陈茹玉（Chen R Y），史延年（Shi Y N）．第五届国际农药化学会议（The 5th IUPAC International Pesticide Chemistry Congress）（日本京都（Kyoto），1982年8月）．

[14] 杨石先（Yang S X），陈天池（Chen T C），李正名（Li Z M），李玉桂（Li Y G），王琴荪（Wang Q S），颜茂恭（Yan M G），董希阳（Dong X Y）．中国科学（Scientia Sinica），1960，9（7）：897－906.

[15] 杨石先（Yang S X），陈茹玉（Chen R Y），黎万琳（Li W L），曹慧芳（Cao H F），何闵章（He M Z），胡声闻（Hu S W），曹善耆（Cao S Q）．化学学报（Acta Chimica Sinica），1960，26：49－52.

[16] 杨石先（Yang S X），陈天池（Chen T C），李正名（Li Z M），李毓桂（Li Y G），董希阳（Dong X Y），高绍仪（Gao S Y），董松琦（Dong S Q）．化学学报（Acta Chimica Sinica），1962，28（3）：187－190.

[17] 杨石先（Yang S X），陈天池（Chen T C），李正名（Li Z M）．化学学报（Acta Chimica Sinica），1963，29（3）：153－158.

[18] 杨石先（Yang S X），陈天池（Chen T C），李正名（Li Z M），王惠林（Wang H L），黄润秋（Huang R Q），唐除痴（Tang C C），刘天麟（Liu T L），张金培（Zhang J P）．化学学报（Acta Chimica Sinica），1965，31（5）：399－412.

[19] 杨石先（Yang S X），陈天池（Chen T C），王琴荪（Wang Q S），金桂玉（Jin G Y），邵瑞莲（Shao R L），刘论祖（Liu L Z）．化学学报（Acta Chimica Sinica），1965，（31）：406－412.

业绩与奖项

一、国家级

序号	获奖时间	奖项名称	奖励等级及排名	授奖部门
1	1964 年	新产品（磷 32，磷 47）	国家科委新产品二等奖（集体）	国家科学技术委员会
2	1978 年	新杀菌剂叶枯净	全国科技大会奖（1）	国家科学技术委员会
3	1987 年	高效杀菌剂粉锈宁小试合成	国家“六五”科技攻关奖（集体）	国家计委、经委、科委、财政部联合颁发
4	1987 年	有机磷生物活性物质与有机磷化学	国家自然科学奖（二等）（4）	国家科学技术委员会
5	1992 年	有机化学农药创制	国家“七五”科技攻关重大成果奖（1）	国家计委、科委和财政部联合颁发
6	1993 年	粉锈宁新技术开发	国家科技进步奖（一等）（2）	国家科学技术委员会
7	2007 年	对环境友好的超高效除草剂的创制和开发研究	国家技术发明奖（二等）（1）	国务院

二、省部级

序号	获奖时间	奖项名称	奖励等级及排名	授奖部门
1	1989	无公害农药——昆虫信息素	天津化工学会一等优秀学术论文奖/第一	天津化工学会
2	1990	农药化学基础研究	科技进步二等奖/第一	国家教育委员会
3	1991	高效杀菌剂粉锈宁	科技进步一等奖/第二	中华人民共和国化学工业部
4	1992	新磺酰脲类化合物的合成及构效关系研究	年会优秀论文奖/第一	中国农药学会
5	1992	新农药的创制与农药新品种合成技术研究	科技进步二等奖/第二	贵州省
6	1999	新超高效除草剂#92825的创制研究	科技进步二等奖/第一	国家教育委员会
7	2004	磺酰脲类化合物及其除草用途	天津市专利金奖/第一	天津市人民政府
8	2005	对环境友好超高效除草的创制和开发研究	天津市科学技术奖一等奖/第一	天津市人民政府
9	2006	超高效绿色除草剂创制开发	全国发明创业奖（个人奖）	科技部中国发明协会

续表

序号	获奖时间	奖项名称	奖励等级及排名	授奖部门
10	2013	发明专利 93101976.1	天津市最有价值发明专利（第一）	天津市知识产权局
11	2013		天津市科技重大成就奖（个人奖）	天津市人民政府

三、个人奖

1. 1986 年天津市劳动模范
2. 1990 年国家重点实验室先进个人
3. 1990 年国家计委、国家科委、国家教委、中国科学院、卫生部、农业部、国家基金委联合授予国家重点实验室重大贡献“金牛奖”
4. 1990 年国家中青年有突出贡献专家的称号
5. 1991 年国家特殊津贴
6. 1992 年中国农药学会年会优秀论文奖
7. 1995 年日本农药学会外国科学家荣誉奖
8. 1995 年天津市政府特别津贴
9. 1998 年天津市科普工作特别奖
10. 1999 年南开大学自然科学优秀成果奖
11. 1999 年南开大学优秀教师一等奖
12. 2006 年全国杰出专利工程技术评价
13. 2006 年南开大学科技成果优秀奖
14. 2007 年中国农药工业协会中国农药工业杰出成就奖
15. 2007 年南开大学国家重点学科建设工作先进个人奖
16. 2009 年中国农药工业协会建国 60 周年中国农药工业突出贡献奖
17. 2009 年南开大学第三届“良师益友”优秀导师奖
18. 2011 年南开大学第四届“良师益友”优秀导师提名奖
19. 2011 年天津市教工委先进个人
20. 2012 年天津市教育工会“教工先锋岗”先进个人荣誉称号
21. 2013 年南开大学第五届“良师益友”优秀导师奖

主要科研项目

序号	项目名称	起止时间	项目来源	合作单位
国家级课题（共21项，课题负责人）				
1	基于特殊酶结构的植保药物的分子设计	2002.01－2003.12	863计划	华东理工大学
2	基于靶标结构的超高效除草先导及靶标与先导间相互作用	2004.01－2008.12	973计划	
3	新分子靶标导向的杀虫候选药物研究	2010.01－2014.08	973计划	华东理工大学
4	茶尺蠖和槐尺蠖性信息素的分离、鉴定、化学合成及其应用	1986.06－1990.12	国家攻关	
5	有机化学农药的创制	1986.08－1990.12	国家攻关	
6	创制化学新农药	1991.09－1995.12	国家攻关	
7	药用化合物的筛选及计算机辅助系统研究－化合物的筛选	1991.06－1995.12	国家攻关	中科院北医大医科院
8	新磺酰脲除草剂	1995.01－1997.04	国家攻关	
9	功能药物分子工程研究	1997－2000	教育部天南大合作项目	与天大王静康院士团队合作
10	创制除草剂单嘧磺隆的产业化	2002.01－2003.10	国家攻关	沈阳化工院
11	创制除草剂NK#94827的开发研究	2002.01－2003.10	国家攻关	江苏农药所
12	昆虫信息素和拟信息素的人工合成及其结构和生理活性关系研究	1983.01－1985.12	国家自然科学基金项目	
13	使用金属有机试剂在生物活性分子中定向引入双键及转化	1988.01－1990.12	国家自然科学基金项目	
14	西格玛迁移重排反应中的化学	1990.01－1992.12	国家自然科学基金项目	
15	农药化学基础研究	1993.01－1995.12	国家自然科学基金项目	
16	α－唑基烯酮二硫缩醛的合成、反应与生物活性研究	1994.01－1996.12	国家自然科学基金项目	
17	新磺酰脲除草剂92825	1994.01－1996.12	国家自然科学基金项目	
18	新磺酰脲类的合成、生物活性和三维构效关系	1998.01－2000.12	国家自然科学基金项目	
19	新农药创制基础研究	1999.01－2002.12	国家自然科学基金项目	中国农业大学，上海有机化学研究所

续表

序号	项目名称	起止时间	项目来源	合作单位
20	基于鱼尼丁受体新靶标的绿色杀虫剂的设计合成、生物活性及构效关系研究	2009.01-2011.12	国家自然科学基金项目	
21	新型磺酰脲分子多功能生物活性的基础研究	2013.01-2016.12	国家自然科学基金项目	
省部级课题（共17项，课题负责人）				
1	高效、低残毒农药新品种——粉锈宁	1983.01-1985.12	化工部攻关项目	
2	单嘧磺隆与AHAS酶的复合物研究及绿色农药分子设计	2004.12-2007.12	科技部国际合作	
3	有机磷化学反应及有关化合物在农业中的应用	1981.01-1983.12	教育部	
4	具有生物活性手性化合物的分析、结构测定及其不对称合成	1984.01-1987.12	教育部	
5	第三代农药的创制与开发	1989.04-1995.12	教育部	
6	具有生物活性物质的不对称和立体有择合成	1990.12-1991.12	教育部	
7	新农药创制体系	1991.11-1994.12	教育部	
8	新细胞分裂因子类植物生长调节剂的合成及构效关系	1993.09-1995.12	教育部	
9	超高效除草剂92825的开发研究	1997.06-2000.06	教育部	
10	新农药创制与开发研究	1998.12-2002.12	教育部	
11	苯环5-取代的超高效单取代磺酰脲类除草剂的设计合成	2012.01-2014.12	教育部	
12	棉蚜虫警戒信息素及其电生理行为学的研究	1984.01-1985.12	天津市攻关项目	
13	新化合物农用活性筛选	1996.06-1998.12	天津市科技发展计划项目	
14	绿色除草剂#92825的开发研究	1998.10-1999.03	天津市科技发展计划项目	
15	超高效磺酰脲除草剂作用机制及农药分子设计研究	2003.04-2005.12	天津市科技发展计划项目	
16	创制超高效绿色除草剂的应用研究	2005.04-2008.03	天津市科技发展计划项目	天津市绿保农用化学科技开发有限公司
17	抗癌细胞活性先导化合物的发现，优化及构效关系研究	2009.04-2012.03	天津市科技发展计划项目	
横向课题（共6项，课题负责人）				
1	化合物农药活性CA201011010	2011.03-	巴斯夫（中国）有限公司	

续表

序号	项目名称	起止时间	项目来源	合作单位
2	Research Services Agreement	2008.09－2009.09	Vertex Pharmaceuticals Incorporated	
3	超高效除草剂单嘧磺隆的生产技术	2007.04－2014.12	天津市绿保农用化学科技开发有限公司	
4	超高效除草剂单嘧磺酯的生产技术	2007.04－2014.12	天津市绿保农用化学科技开发有限公司	
5	2，5－二氯吡啶的生产技术转让	2007.08－2012.08	天津合诺达化工贸易有限公司	
6	博士生奖学金	2013－2015	瑞士 Syngenta 公司	

论文选登

萘和苯的衍生物类生长素对植物插枝生根作用的初步报告*

杨石先　姚　珍　叶超然　李正名　崔　徵

一、引言

插枝为园艺上常用的无性繁殖方法，有些植物，例如菊花、杨树和柳树等，插枝时很容易生根成活。也有些植物，例如松柏科植物则很难进行插枝繁殖。自从1934年Thimann和Went[7]发现生长素有促进生根的作用后，许多人相继在这方面进行了研究并获得很大成效[1,6,8]，这些工作在农业实践上具有重大的意义。虽然已经发现有很多生长素具有促进生根的作用，但是其效果较好的只有少数几种。根据Hitchcock和Zimmerman[2]的报告，他们曾比较研究过吲哚丁酸及萘乙酸的一些衍生物，这些化合物都有一定程度的作用，有一个卤族元素的衍生物的作用比吲哚丁酸和萘乙酸差些，有2，4二氯或2，4二溴的衍生物作用的大小与共侧链的性质有关。Hitchcock和Zimmerman[3]，Osborne，Wain和Wolker[4]都曾指出2，4，5-三氯苯氧乙酸，2，4，5-三氯苯氧丁酸，2，4-二氯苯氧丙酸等的作用大于吲哚丙酸及萘乙酸。但是这些苯氧酸类的化合物有以下的缺点，就是引起根部粗短，微高于适于生根的浓度即有毒害作用。不同化合物对生根的促进作用不但在根的数量上有差异而且影响根的质量，例如吲哚乙酸引起的根数目少而长，比较正常，而2，4-二氯代苯氧乙酸则产生像毛刷一样的粗短根系。从经济的观点而言，后者较为便宜。在这种情况下进一步研究对插枝生根有效的生长素仍是十分必要的。本实验主要目的是想在苯及萘衍生物中找出能有效和正常地促进生根的物质。

二、材料和方法

实验所用植物有下列几种：女贞（*Ligustrum lucidum*），黄刺梅（*Rosa xanthina*），槐树（*Sophora japonica*），山桃（*Prunus Davidiana*），榆叶梅（*Prunus trilota*），木槿（*Hibiscus syriacus*），杨树（*Populus tomentosa*），柳树（*Salix Matsudana*）。过去许多人用过这些植物进行生长素引起插枝生根的研究[6]。作者将这些树的当年生枝条剪成长约18厘米，分别用下列几种生长素（表1）处理：萘乙酸，1，5-萘二乙酸，1，4-及1，5-萘二乙酸，β-萘氧乙酸，2，6-萘二氧乙酸，对苯二氧二乙酸，间苯二氧二乙酸，1，3，5-间苯三氧三乙酸，2，4-二氯代苯氧乙酸和4，6-二氯代间苯二氧乙酸。处理的方法分为两种，一种是用生长素羊毛脂软膏，另一种是用生长素水溶液。处理的部位一般是枝条的形态下端，只有个别实验处理过上端，为的是与处理下端者比较其

* 原发表于《南开大学学报》（自然科学），1957（04）：132-148。本文插枝实验部分系叶超然、姚珍的毕业论文。

效果有何不同。处理的枝条有些实验留叶，有些去叶。处理的时间用生长素羊毛脂软膏者就是一次，生长素水溶液则浸泡 24 小时，生长素羊毛脂软膏的浓度为 0.1%，0.5%，1% 三种，生长素水溶液的浓度为 0.001%，0.005% 及 0.01% 三种。

表 1　生长素的化学名称和化学构造①

英文名称	中文名称	化学构造
α – naphthalene acetic acid	α – 萘乙酸	CH_2COOH
1，5 – naphthalene diacetic acid	1，5 – 萘二乙酸	CH_2COOH $HOOCCH_2$
1，4 – naphthalene diacetic acid	1，4 – 萘二乙酸	CH_2COOH CH_2COOH
β – naphthoxy acetic acid	β – 萘氧乙酸	$-OCH_2COOH$
2，6 – naphthalene dioxy diacetic acid	2，6 – 萘二氧二乙酸	$HOOCCH_2O-$　$-OCH_2COOH$
phenoxy – acetic acid	苯氧乙酸	OCH_2COOH
p – phenyl – dioxy – diacetic acid	对苯二氧二乙酸	OCH_2COOH OCH_2COOH
m – phenyl – dioxy – diacetic acid	间苯二氧二乙酸	$-OCH_2COOH$ OCH_2COOH
1，3，5 – pheny triacetic acid	1，3，5 – 间苯三氧三乙酸	OCH_2COOH $HOOCCH_2O-$　$-OCH_2COOH$
2，4 – dichlorophenoxy – acetic acid	2，4 – 二氯代苯氧乙酸	OCH_2COOH $-Cl$ Cl

续表2

英文名称	中文名称	化学构造
4, 6 – dichloro – m – phenyl – dioxydiacetic acid	4, 6 – 二氯代间苯二氧二乙酸	(Cl, Cl, OCH_2COOH, OCH_2COOH 取代苯环)

①药剂中除萘乙酸 β – 萘氧乙酸和 2, 4 – 二氯代苯氧乙酸外，均为杨石先、李正名所合成，制备方法尚未发表。

三、结果和讨论

实验 1：生长素羊毛脂软膏对女贞和黄刺梅生根的作用

过去很多人利用混有生长素的羊毛脂软膏处理植物枝条引起生根，近来由于此法大量处理枝条时不方便，采用者渐少。但是这个方法也有可使药效持久的好处，因此对研究新药剂的效用比较适宜。本实验用的植物是女贞和黄刺梅，分为有叶及无叶两种枝条，处理部位分为形态的上端及下端，生长素在羊毛脂中的浓度分为 0.1%，0.5% 和 1% 三种，以无生长素的羊毛脂为对照。

实验结果记在表 2 里，插枝成活率有叶枝条高于无叶枝条，在无叶枝条中处理下端的成活率又高于处理上端。这可能是由于无叶枝条中缺乏有机营养物质，而生长素把原有的物质又积累在上端。以致下端缺乏物质而不能生根。生长素促进生根的作用与留叶与否，处理部位以及生长素的种类和浓度都有关系。0.1% 萘乙酸和 1% 的间苯二氧二乙酸处理女贞留叶枝条上端有显著促进生根的作用，0.1% 2, 4 – 二氯代苯氧乙酸处理下端也有很好的作用，但用它处理上端则效果很坏。关于留叶与插枝生根的关系，van Overbeek 等[5]早就指出当木槿插枝留叶时吲哚丁酸促进生根的作用比不带叶的枝条大，他们还证明用蔗糖与某些有机或无机氮化合物可以代替叶子的作用。此外还可以看出处理留叶枝条上端一般比处理下端好些，对无叶枝条则处理下端较上端好，这可能是由于羊毛脂阻碍了切面对水分的吸收，而使体内水分失掉平衡。生长素对黄刺梅生根的作用与女贞不同，处理上端与下端的结果无大差异，萘乙酸的效果很坏，1% 间苯二氧二乙酸对生根有很好的作用，超过对照 11 倍以上。

表 2　生长素羊毛脂软膏促进插枝生根的作用和留叶及处理部位的关系

（实验在 1955 年 11 月 4 日开始，1956 年 2 月 22 日结束）

植物	留叶或去叶	处理部位	生长素	浓度(%)	枝条数目	成活数目	生根枝数	每株根鲜重(克)
女贞	留叶	枝条下端	萘乙酸	0.1	5	4		
				0.5	5	5		
				1	5	5	2	0.05
			1, 5 – 萘二乙酸	0.1	5	4		
				0.5	5	5	1	0.15
				1	5	5	1	0.10
			间苯二氧二乙酸	0.1	5	5		
				0.5	5	5	1	0.60
				1	5	5	1	0.20

续表2

植物	留叶或去叶	处理部位	生长素	浓度(%)	枝条数目	成活数目	生根枝数	每株根鲜重(克)
女贞	留叶	枝条下端	2,4-二氯代苯氧乙酸	0.1	5	4	3	2.27
				0.5	5	5		
				1	—	—		
			对照		5	5	3	0.10
		枝条上端	萘乙酸	0.1	5	5	3	2.27
				0.5	5	5	3	1.20
				1	5	5	2	0.05
			1,5-萘二乙酸	0.1	5	5	3	1.27
				0.5	5	5	3	1.03
				1	5	5	4	0.47
			间苯二氧二乙酸	0.1	5	5	4	1.17
				0.5	5	4	3	1.50
				1	5	5	4	1.75
			2,4-二氯代苯氧乙酸	0.1	5	4	4	0.87
				0.5	5	3	5	0.26
				1	5	1	5	0.24
			对照		5	5	5	1.56
	去叶	枝条下端	萘乙酸	0.1	5	5	4	0.11
				0.5	5	4	2	0.10
				1	5	5	3	0.97
			1,5-萘二乙酸	0.1	5	4	2	0.05
				0.5	5	5	5	0.03
				1	5	5	3	0.10
			间苯二氧二乙酸	0.1	5	3	1	0.10
				0.5	5	4	4	0.25
				1	5	5	4	0.37
			2,4-二氯代苯氧乙酸	0.1	5	3	1	1.30
				0.5	5	0		
				1	5	0		
			对照		5	4	5	0.62
		枝条上端	萘乙酸	0.1	5	2	1	0.60
				0.5	5	0		
				1	5	0		
			1,5-萘二乙酸	0.1	5	0		
				0.5	5	0		
				1	5	0		
			间苯二氧二乙酸	0.1	5	0		
				0.5	5	0	3	
				1	5	3		0.87
			2,4-二氯代苯氧乙酸	0.1	5	0		
				0.5	5	0		
				1	5	0		
			对照		5	2		

续表 2

植　物	留叶或去叶	处理部位	生　长　素	浓度(%)	枝条数目	成活数目	生根枝数	每株根鲜重(克)
黄刺梅	去　叶（原枝无叶）	枝条下端	萘　乙　酸	0.1	5	4	0	
				0.5	5	5	0	
				1	5	1	0	
			1，5－萘二乙酸	0.1	5	4	1	0.20
				0.5	5	4	1	0.15
				1	5	4	1	0.05
			间苯二氧二乙酸	0.1	5	4	1	0.20
				0.5	5	5	2	0.10
				1	5	5	1	1.15
			2，4－二氯代苯氧乙酸	0.1	5	5	2	0.17
				0.5	5	2	1	0.50
				1	5	2	0	—
			对　照		5	3	1	0.10
		枝条上端	萘　乙　酸	0.1	5	1		
				0.5	5	3		
				1	5	4	1	0.10
			1，5－萘二乙酸	0.1	5	3	2	0.35
				0.5	5	4	3	0.33
				1	5	4		
			间苯二氧二乙酸	0.1	5	5	2	0.10
				0.5	5	5	2	0.25
				1	5	4	1	0.45
			2，4－二氯代苯氧乙酸	0.1	5	4	1	0.10
				0.5	5	4		
				1	5	4	1	0.50
			对　照		5	4	1	1.1

从上面的结果可以知道生长素促进生根的作用是很复杂的，同一生长素对同一植物而采用不同的处理方法，就可以得到不同的结果，因此在农业实践上应用生长素于插枝时必须特别注意，不能机械地硬搬。

实验 2：生长素水溶液对女贞和黄刺梅插枝生根的作用

这一实验的材料与处理方法除采用生长素水溶液代替羊毛脂软膏外，其他均与前一实验相同。生长素水溶液的浓度为 0.001%，0.005% 及 0.01% 三种，用女贞为材料的结果证明留叶的枝条不论处理上端或下端，生根数量大大地超过了无叶枝条（表 3）。不论留叶与否，处理枝条下端的效果又比处理上端好些，同时在去叶枝条的实验中，对照枝条都没有生根。但用生长素处理者，除个别处理由于浓度过高者外，都生了相当多的根，其中萘二乙酸的效果最好。0.001% 2，4－二氯代苯氧乙酸处理枝条下端可以促进生根，处理上端反而有抑制作用，高浓度时甚至引起枝条死亡。Hitchcock 和 Zimmerman[2,3] 早已证明了这一事实。

用黄刺梅为材料的实验，各种处理的成活率都很低，萘二乙酸对促进生根的作用较其他生长素好些。从表 3 的结果可以看出生长素水溶液处理枝条有明显的促进生根的作用，但效果因植物或处理方法而异。一般而论，用合适的浓度处理留叶枝条下端的结果最好，各种生长素都能促进女贞插枝生根，但对黄刺梅多数都没有作用甚至引起死亡。

表 3 生长素水溶液促进插枝生根的作用和留叶及处理部位的关系

（实验在 1955 年 12 月 4 日开始，1956 年 3 月 8 日结束）

植 物	留叶或去叶	处理部位	生 长 素	浓度(%)	枝条数目	成活数目	生根株数	每株根鲜重(克)
女 贞	留 叶	枝条下端	萘 乙 酸	0.001	5	5	5	2.10
				0.005	5	4	4	1.32
				0.01	5	3	1	1.30
			1，5－萘二乙酸	0.001	5	5	5	0.40
				0.005	5	4	5	0.90
				0.01	5	4	5	0.46
			间苯二氧二乙酸	0.001	5	5	5	0.14
				0.005	5	5	5	0.62
				0.01	5	5	5	0.30
			2，4－二氯代苯氧乙酸	0.001	5	2	2	1.05
				0.005	5	0		
				0.01	5	0		
			对 照		5	5	5	0.24
		枝条上端	萘 乙 酸	0.001	5	5	5	0.56
				0.005	5	5	5	0.40
				0.01	5	5		
			1，5－萘二乙酸	0.001	5	5	5	0.40
				0.005	5	5	5	0.61
				0.01	5	5	4	0.56
			间苯二氧二乙酸	0.001	5	5	4	0.42
				0.005	5	5	5	0.66
				0.01	5	5	5	0.86
			2，4－二氯代苯氧乙酸	0.001	5	3		
				0.005	5	0		
				0.01	5	0		
			对 照		5	5	5	0.14
	去 叶	枝条下端	萘 乙 酸	0.001	5	3	3	0.40
				0.005	5	4	3	0.17
				0.01	5	2		
			1，5－萘二乙酸	0.001	5	2	1	0.60
				0.005	5	1	1	1.70
				0.01	5	5	5	0.24
			间苯二氧二乙酸	0.001	5	3	2	0.35
				0.005	5	1	1	0.15
				0.01	5	3	3	0.13
			2，4－二氯代苯氧乙酸	0.001	5	3	3	0.47
				0.005	5	0		
				0.01	5	0		
			对 照		5	2		

续表3

植　物	留叶或去叶	处理部位	生　长　素	浓度（%）	枝条数目	成活数目	生根株数	每株根鲜重（克）
女　贞	去　叶	枝条上端	萘　乙　酸	0.001	5	1		
				0.005	5	0		
				0.01	5	3	1	0.7
			1，5－萘二乙酸	0.001	5	0		
				0.005	5	2	1	0.2
				0.01	5	2	1	0.3
			间苯二氧二乙酸	0.001	5	1		
				0.005	5	1	1	0.25
				0.01	5	3	3	0.33
			2，4－二氯代苯氧乙酸	0.001	5	2		
				0.005	5	0		
				0.01	5	0		
			对　照		5	2		
黄刺梅	去　叶（原无叶）	枝条下端	萘　乙　酸	0.001	5	0		
				0.005	5	0		
				0.01	5	0		
			1，5－萘二乙酸	0.001	5	1		
				0.005	5	2	2	0.30
				0.01	5	0		
			间苯二氧二乙酸	0.001	5	1		
				0.005	5	2		
				0.01	5	0		
			2，4－二氯代苯氧乙酸	0.001	5	2		
				0.005	5	2		
				0.01	5	1		
			对　照		5	2		
		枝条上端	萘　乙　酸	0.001	5	0		
				0.005	5	0		
				0.01	5	0		
			1，5－萘二乙酸	0.001	5	1	1	0.30
				0.005	5	2	2	0.30
				0.01	5	0		
			间苯二氧二乙酸	0.001	5	1		
				0.005	5	2		
				0.01	5	0		
			2，4－二氯代苯氧乙酸	0.001	5	0		
				0.005	5	2	1	0.60
				0.01	5	0		
			对　照		5	1	1	0.20

实验3：萘乙酸，萘二乙酸，间苯二氧二乙酸及2，4－二氯代苯氧乙酸对女贞插枝生根的作用

根据前两个实验的结果，女贞是一个很好的实验材料，用生长素水溶液处理留叶枝条的下端得到最好的结果，同时发现生长素的作用因浓度而异。我们为了比较一下前面试过的生长素的作用，就选择萘乙酸，1，5－萘二乙酸，间苯二氧二乙酸及2，4－二氯代苯氧乙酸的最适浓度处理留叶枝条下端24小时。结果从表4可以看出萘乙酸的效果最好，间苯二氧二乙酸次之，萘二乙酸和0.001%的2，4－二氯代苯氧乙酸略低于对照，0.005%的2，4－二氯代苯氧乙酸效果很坏。

表4　不同生长素水溶液对女贞插枝生根的作用

（1956年3月14日开始，1956年6月6日结束）

生　长　素	浓度(%)	枝条数目	成活株数	生根株数	每株根鲜重(克)
萘　乙　酸	0.001	30	19	19	2.19
1，5－萘二乙酸	0.005	30	19	19	1.18
间苯二氧二乙酸	0.005	30	18	18	1.56
2，4－二氯代苯氧乙酸	0.001	30	13	13	1.15
	0.005	30	6	6	0.16
对　照		30	13	13	1.41

实验4：2，4－二氯代苯氧乙酸钠盐，间苯二氧二乙酸，2，4－二氯代间苯二氧二乙酸和苯氧乙酸对木槿，杨树，槐树，榆叶梅和桃树插枝生根的作用

前面的实验已经证明生长素对插枝生根的作用因植物种类而异，为了探讨生长素对其他植物生根的作用，本实验采用了木槿，杨树，槐树，榆叶梅，桃树等为材料。生长素是2，4－二氯代苯氧乙酸钠盐，间苯二氧二乙酸，4，6－二氯代间苯二氧二乙酸及苯氧乙酸四种，水溶液的浓度分为0.001%，0.005%及0.01%三种。用木槿的实验结果（表5）证明间苯二氧二乙酸，二氯代间苯二氧二乙酸和苯氧乙酸在各种浓度下生根的数量均大于对照，其中苯氧乙酸最好，二氯代间苯二氧二乙酸次之，间苯二氧二乙酸较差。这个结果说明这类生长素在化学构造上增加两个氯对生根更有好处，但是在苯环上有两个乙酸不及一个乙酸。至于2，4－二氯代苯氧乙酸钠盐，浓度为0.001%时有非常好的效果，当浓度略高时就有抑制作用。用杨树为实验材料的结果，2，4－二氯代苯氧乙酸钠盐各种浓度处理的枝条全部死亡，其他三种生长素对杨树生根都有不同程度的促进作用。生长素对槐树的生根作用极不明显，榆叶梅插枝的成活率极低，桃树全部死亡。

表5　各种生长素对不同植物插枝生根的作用

（1956年5月4日开始，1956年7月10日结束）

植　物	生　长　素	浓度(%)	枝条数目	成活株数	生根株数	每株根鲜重(克)
木　槿	2，4－二氯代苯氧乙酸钠盐	0.001 0.005 0.01	8 8 8	8 8 0	8 8	1.42 0.07 1.00

续表 5

植 物	生长素	浓度(%)	枝条数目	成活株数	生根株数	每株根鲜重(克)
木 槿	间苯二氧二乙酸	0.001	8	8	8	0.71
		0.005	8	8	8	0.67
		0.01	8	8	8	0.80
	4，6－二氯代间苯二氧二乙酸	0.001	8	8	8	1.07
		0.005	8	8	8	1.12
		0.01	8	8	8	1.10
	苯氧乙酸	0.001	8	8	8	0.97
		0.005	8	8	8	1.23
		0.01	8	8	8	1.27
	对 照		8	8	8	0.68
杨 树	2，4－二氯代苯氧乙酸钠盐	0.001	8	0		
		0.005	8	0		
		0.01	8	0		
	间苯二氧二乙酸	0.001	8	8	8	0.25
		0.005	8	8	8	0.26
		0.01	8	8	8	0.38
	4，6－二氯代间苯二氧二乙酸	0.001	8	8	8	0.56
		0.005	8	8	8	0.77
		0.01	8	8	8	0.23
	苯氧乙酸	0.001	8	8	8	0.23
		0.005	8	8	8	0.21
		0.01	8	8	8	0.54
	对 照		8	8	8	0.21
槐 树	2，4－二氯代苯氧乙酸钠盐	0.001	10	2	2	0.40
		0.005	10	2		
		0.01	10	2		
	间苯二氧二乙酸	0.001	10	1		
		0.005	10	3		
		0.01	10	5	2	0.15
	4，6－二氯代间苯二氧二乙酸	0.001	10	4	3	
		0.005	10	4		
		0.01	10	5		
	苯氧乙酸	0.001	10	3		0.20
		0.005	10	2		
		0.01	10	2		
	对 照		10	5	3	0.26

实验 5：1，5－萘二乙酸和对苯二氧二乙酸水溶液对木槿及杨树插枝生根的作用

从表 5 的结果已知间苯二氧二乙酸对生根有一定的促进作用，本实验目的是研究对苯二氧二乙酸的效果并用 1，5－萘二乙酸互相比较。每种生长素用了四种浓度，实

验材料是木槿和杨树。结果载于表6，用木槿的实验，以上两种生长素在低浓度时略有促进生根的作用，高浓度则有抑制作用。用杨树的实验，各种浓度的1，5－萘二乙酸均表现了抑制生长的作用。但是对苯二氧二乙酸浓度愈高，促进生根的作用愈大。由此可知生长素对生根的作用与植物特性有密切关系。

表6　1，5－萘二乙酸和对苯二氧二乙酸水溶液对木槿、杨树插枝生根的作用

（1956年5月20日开始，1956年6月26日结束）

植　物	生　长　素	浓度（%）	枝条数目	成活株数	生根株数	每株根鲜重（克）
木　槿	1，5－萘二乙酸	0.0005	10	10	10	0.61
		0.001	10	10	10	0.67
		0.005	10	10	10	0.59
		0.01	10	10	10	0.30
	对苯二氧二乙酸	0.0005	10	10	10	0.72
		0.001	10	10	10	0.69
		0.005	10	10	10	0.67
		0.05	10	10	10	0.47
	对　照		10	10	10	0.57
杨　树	1，5－萘二乙酸	0.0005	10	4	4	0.32
		0.001	10	10	10	0.48
		0.005	10	10	10	0.31
		0.01	10	10	10	0.43
	对苯二氧二乙酸	0.0005	10	5	5	0.32
		0.001	10	9	9	0.62
		0.005	10	10	10	0.62
		0.01	10	10	10	1.02
	对　照		10	10	10	0.65

实验6：1，5－萘二乙酸，2，6－萘二氧二乙酸，β－萘氧乙酸和间苯三氧三乙酸对木槿插枝生根的作用

为了比较生长素侧链上有氧无氧以及一个和两个乙酸的效果，选用了1，5－萘二乙酸，2，6－萘二氧二乙酸，β－萘氧乙酸和间苯三氧三乙酸四种生长素，每种生长素有0.001%，0.005%和0.01%三种浓度。实验材料是木槿，表7是所得的结果，四种生长素的各种浓度对生根均表现了不同程度的促进作用，2，6－萘二氧二乙酸较1，5－萘二乙酸和β－萘氧乙酸略好些。

表7　1，5－萘二乙酸，2，6－萘二氧二乙酸，β－萘氧乙酸，间苯三氧三乙酸对木槿插枝生根的作用

（1956年5月4日开始，1956年7月10日结束）

植　物	生长素	浓度（%）	枝条数目	成活株数	生根株数	每株根鲜重（克）
木　槿	1，5－萘二乙酸	0.001	9	9	9	0.59
		0.005	9	9	9	0.64
		0.01	9	9	9	0.62

续表7

植　物	生长素	浓度(%)	枝条数目	成活株数	生根株数	每株根鲜重(克)
木　槿	2，6-萘二氧二乙酸	0.001	9	9	9	0.66
		0.005	9	9	9	0.82
		0.01	9	9	9	0.61
	β-萘氧乙酸	0.001	9	9	9	0.62
		0.005	9	9	9	0.66
		0.01	9	9	9	0.61
	间苯三氧三乙酸	0.0005	9	9	8	0.65
		0.001	9	9	8	0.56
		0.005	9	9	9	0.58
	对　照		9	9	9	0.50

四、摘要

实验材料共用了女贞，黄刺梅，槐树，山桃，榆叶梅，木槿，杨树和柳树八种植物。生长素共有11种（表1），分为羊毛脂软膏及水溶液两种处理，软膏只涂于枝条上端或下端一次，水溶液处理24小时，处理完毕后即插于砂床中使其生根。结果摘要如下：

1. 不同植物的生根能力及成活率不同，其中以木槿、女贞和杨树的生根能力较强，成活率也较高。黄刺梅和槐树较差，山桃全部死亡。

2. 带叶枝条较去叶枝条生根多些，但因处理部位及方法而有不同的结果。留叶的女贞用生长素羊毛脂软膏处理上端较下端好些，而去叶的枝条处理下端较上端好些。用水溶液所得结果与羊毛脂相反，留叶的枝条处理下端者较上端好。

3. 各种生长素对插枝生根有不同的作用，有的作用大，有的作用小。但因植物种类，处理方法及浓度而异。女贞的留叶枝条，用生长素羊毛脂处理下端，以0.1%的2，4-二氯代苯氧乙酸为最好（表2）。如果处理上端，则以萘乙酸为最好。用水溶液处理女贞枝条的下端者，各种浓度的萘乙酸均表现有良好的作用（表3）。

4. 用萘乙酸，1，5-萘二乙酸，间苯二氧二乙酸和2，4-二氯代苯氧乙酸水溶液的最适浓度处理女贞的实验，萘乙酸最好，间苯二氧二乙酸次之，其他均不及对照（表4）。

5. 用2，4-二氯代苯氧乙酸钠盐，间苯二氧二乙酸及苯氧乙酸各种浓度的水溶液处理木槿，杨树及槐树，结果因植物种类及浓度而异。以木槿而论，各种生长素除间苯二氧二乙酸外对生根均有显著促进作用。2，4-二氯代苯氧乙酸只有在低浓度时（0.001%）有好的作用，而苯氧乙酸及二氯代间苯二氧二乙酸则在各种浓度中都有促进生根的作用（表5）。其中应该注意的是4，6-二氯代间苯二氧二乙酸较间苯二氧二乙酸有较好的表现。

6. 1，5-萘二乙酸与对苯二氧二乙酸的作用，用木槿的实验证明后者较前者略好，一般是因浓度增加而效果减少。但用杨树的实验，浓度愈大效果愈好，对苯二氧二乙酸表现尤为显著（表6）。

7. 1，5－萘二乙酸，2，6－萘二氧二乙酸，β－萘氧乙酸和间苯三氧三乙酸的作用，用木槿进行了比较实验，这四种生长素在各种浓度下均有促进生长的作用，互相之间的差异不大（表7）。

参考文献

[1] Avery，G. S. Jr.，and. Johnson，E. B.（1947）Hormones and Horticulture McGraw－Hill Book Co.

[2] Hitchcock，A. E. and Zimmerman，P. W.（1942）Contr. Boyce Thompson Inst. 12：497－580.

[3] Hitchcock，A. E. and Zimmerman，P. W.（1945）Contr. Boyce Thompson Inst. 14：14－38.

[4] Osborne，D. J.，Wain，R. L. and Walker，R. D.（1952）Jour. Hort. Sci. 27：44－52.

[5] Overbeek，J. van，Gordon，S. A. and Gregory，L. E.（1946）Amer. Jour. Bot. 33：100－107.

[6] Thimann，K. V. and Janc Behnke.（1947）The use of anxins in the rooting of woody cuttings. Petersham，Mass，U. S. A.

[7] Thimann K. V. and P. W. Went.（1934）Proc. Kon. Nederl. Akad. Wetensch. Amsterdam，37：8456－8459.

[8] Tukcy. H. B.（1954）Plant regulators in agriculture. John Wiley and Sons. Inc. U. S. A.

The Effect of Auxins of Phenoxy and Naphthoxy Compounds on the Rooting of Cuttings

Shixian Yang，Zhen Yao，Chaoran Ye，Zhengming Li，Zhi Cui

Abstract Eleven synthetic auxins of phenoxy and napthoxy compounds were used to study their effect on the rooting of cuttings；the chemical name of these compounds are α－naphthalene acetic acid，1，5－naphthalene diacetic acid，1，4－and 1，5－naphthalene diacetic acid，β－naphthoxy acetic acid，2，6－naphthalene－dioxy－diacetic acid，phenoxy－acetic acid，p－phenylene－dioxy－diacetic acid，m－phenylene dioxy－diacetic acid，1，3，5－bensene－trioxy－triacetic acid，2，4－dichlorophenoxy－acetic acid and m－phenylene－dioxy－diacetic acid. The materials used for cuttings were taken from six species of plants：they were Ligustrum lucidum Ait.，Rosa xanthina Lindl，Sophora japonica L.，Prunus trilota Lindl.，Hibiscus sxriacus L.，Populus tomentosa Carr. and Salix Matsudane. Koidz. The auxins were either mixed with lanolin or dissolved in water in various concentrations. The lanolin paste was applied once on the cutting surface and the water solution was used to immerse one end of the cuttings for 24 hours. After treatment the cuttings were placed in moist sand of a propagating frame.

The results showed that root formation of the cuttings was different in different plant matcrials. It was much better in Ligustrum lucidum and Hibiscus syriacus than in the other plants. The cuttings with leaves formed more roots than these without. Most compounds used for the experiments in optimum concentration gave better root formation than the control. 2，4－D stimulated root formation only at lower concentration，but phenoxy－acetic acid and 4，6 dichloro－m－phenyl－one dioxydiacetic acid were effective at various concentrations used in the experiments.

有机磷杀虫剂的研究 I*

——O，O-二烷基 S-烃基(取代烃基)硫(氧)甲基(取代甲基)二硫代磷酸酯的合成

杨石先　陈天池　李正名　李毓桂
王琴荪　颜茂恭　董希阳

（南开大学化学系；中国科学院河北分院元素有机化学研究所）

二硫代磷酸酯衍生物是一类新兴的有机磷杀虫剂，虽研究工作做得不够多，但已出现一些性能较好的药剂。Schrader[1]认为一硫代或二硫代磷酸酯中，凡具下列通式的均可能是有效的杀虫剂。

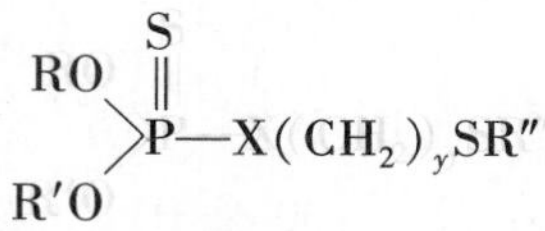

$$\begin{matrix} RO \\ R'O \end{matrix}\!\!>\overset{\overset{\displaystyle S}{\|}}{P}-X(CH_2)_ySR''$$

其中，R，R′，R″均为烷烃基，X 为氧或硫，y 为 1 或 2。它们对棉花和其他农业害虫呈不同程度的触杀和内吸性能。近年来，Кабачник[2]，Мастрюкова[3]等曾对通式 $\begin{matrix} RO \\ RO \end{matrix}\!\!>\overset{\overset{S}{\|}}{P}-S\,CH_2CH_2\,SR''$ 的化合物做过较系统的研究，发现 M-74（$R=R''=C_2H_5$）、M-81（$R=CH_3$；$R''=C_2H_5$）及 M-82（$R=R''=CH_3$）是极有效的内吸杀虫剂，并可作处理种子的内吸杀虫剂。关于 $\begin{matrix} RO \\ RO \end{matrix}\!\!>\overset{\overset{S}{\|}}{P}-S\,\underset{\underset{R'}{|}}{CH}XR''$ 型化合物的研究（X=硫或氧），最近亦有所开展。特别对 R=初级烷烃，R′=氢或甲基，R″=烷烃的一类化合物。Hook 等[4]利用二硫代磷酸酯、甲醛或乙醛和硫醇进行类似 Mannich 反应的缩合：

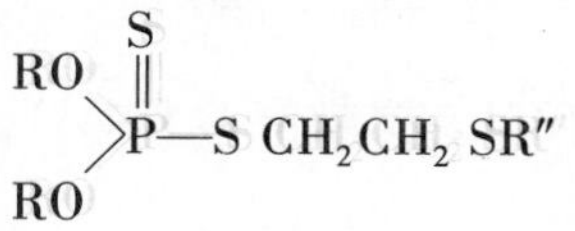

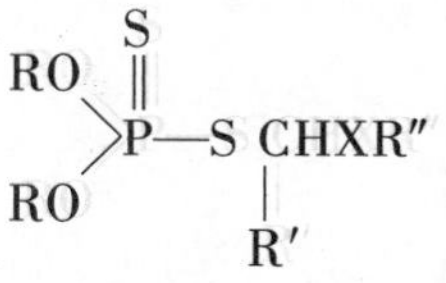

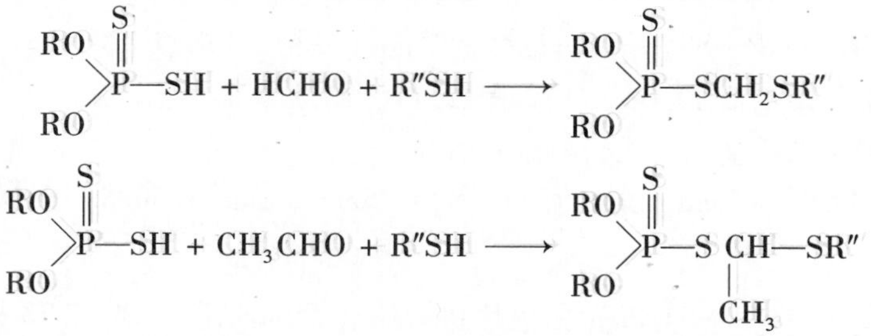

$$\begin{matrix} RO \\ RO \end{matrix}\!\!>\overset{\overset{\displaystyle S}{\|}}{P}-SH + HCHO + R''SH \longrightarrow \begin{matrix} RO \\ RO \end{matrix}\!\!>\overset{\overset{\displaystyle S}{\|}}{P}-SCH_2SR''$$

$$\begin{matrix} RO \\ RO \end{matrix}\!\!>\overset{\overset{\displaystyle S}{\|}}{P}-SH + CH_3CHO + R''SH \longrightarrow \begin{matrix} RO \\ RO \end{matrix}\!\!>\overset{\overset{\displaystyle S}{\|}}{P}-S\,\underset{\underset{\displaystyle CH_3}{|}}{CH}-SR''$$

而仅获得粗产品。此后各国学者对同类化合物陆续有所报告[5~9]。惟大部分为专利文

* 原发表于《化学学报》，1959，25（6）：402-408。

献，无具体的合成步骤，亦缺乏较完整的有关常数。本文将报告下列三种类型化合物的合成方法及其物理常数，并初步进行昆虫试验。

$$(RO)_2P(=S)—SCH(R')—XR'' \quad (X = 硫或氧)$$

Ⅰ. R = 甲基或乙基，R′ = 氢，R″ = 脂烃基，共十三种（见表1）

Ⅱ. R = 甲基或乙基，R′ = 氢，R″ = 烷氧代乙基或烷硫代乙基，共六种（见表2）

Ⅲ. R = 甲基或乙基，R′ = 甲基，R″ = 脂烃基或烷氧代乙基，共八种（见表3）

化合物Ⅰ类中有少数为已知化合物，本文补充一些新的常数，余为新化合物。

初步昆虫试验结果说明 P_2，P_7，P_{11}，P_{20}，P_{27}，P_{154} 等对四龄蚊幼虫有显著触杀作用，并对刺吸口器害虫有内吸毒效。此外 P_1，P_5，P_{32}，P_{47} 等虽触杀效率较差，但均具良好的内吸效率，其中大部分化合物的内吸效能相当于目前常用的内吸剂 *E* 1059。其药效测定将另文报告。

实验部分

O，O－二烷基二硫代磷酸酯　$(RO)_2P(=S)—SH$

在一装有回流装置的三颈瓶中置苯 100 毫升及五硫化二磷粉末（事先用二硫化碳处理）0.5 克分子。在 45℃时滴加无水乙醇 2 克分子。令反应液在 60℃左右保持 1 小时，搅拌回流半小时，冷却过滤，减压蒸馏。本实验中所用的醇为甲醇及乙醇。

（1）O，O－二甲基二硫代磷酸酯：

沸点 72～73.5℃/10mm，文献值 62～63℃/5mm[3]；n_D^{20}1.5331，文献值 1.5343[3]。

镍盐 $[(CH_3O)_2P(=S)—S]_2Ni$，紫色片状结晶（乙醚），熔点 124～126℃，文献值 113℃[10]，124～125℃[3]。

铅盐 $[(CH_3O)_2P(=S)—S]_2Pb$，白色针状结晶（酒精），熔点 128～129℃。

分析，P%，理论值 11.92，实验值 11.59，11.87。

汞盐 $[(CH_3O)_2P(=S)—S]_2Hg$，白色针状结晶（酒精），熔点 120～121℃。

分析，P%，理论值 12.08，实验值 12.11，12.12。

（2）O，O－二乙基二硫代磷酸酯：

沸点 79～80.5℃/5mm，文献值 81～82℃/5mm[3]；n_D^{20}1.5088，文献值 1.5070[3]，

铅盐 $[(C_2H_5O)_2P(=S)—S]_2Pb$，白色片状结晶（无水乙醇），熔点 73～74℃，文献值 74℃[11]，75～76℃[3]。

硫醇　R″SH

（1）脂烃基硫醇：本实验所用的硫醇是按照 Vogel[12] 的方法制备。

表1　O,O－二烷基 S－烃基硫甲基二硫代磷酸酯

$$(RO)_2P(=S)—SCH_2SR''$$

编号	分子式	产率(粗,%)	n_D^{20}(粗)	沸点(℃/mm 汞柱)	产率(%)	n_D^{20}	d_4^{20}	MR_D		P(%)		文献沸点(℃/mm 汞柱)
								计算值	实验值	计算值	实验值	
Y_1	$(CH_3O)_2P(=S)—S—CH_2SC_2H_5$	—	—	118～123/2.5	34.2	1.5522	1.2208	60.15	60.73	13.41	13.05，13.99	78/0.01[7] 114～116/1.5[6]
Y_2	$(C_2H_5O)_2P(=S)—SCH_2SC_2H_5$	—	—	127.5～130/2	44.7	1.5349	1.1615	69.38	69.68	12.00	11.93，11.98	75～78/0.01[7] 125～127/2[6]
P_5	$(CH_3O)_2P(=S)—SCH_2SC_3H_{7-i}$	57.7	1.5481	74～76/1×10^{-5}	56.1	1.5447	1.1970	64.76	64.94	12.60	12.99，13.07	
P_{130}	$(C_2H_5O)_2P(=S)—SCH_2SC_3H_{7-i}$	—	—	128.5～129/0.5	38.7	1.5278	1.1454	74.00	73.65	11.31	10.56，10.47	133～134/4[8]
P_{20}	$(CH_3O)_2P(=S)—SCH_2SC_3H_{7-n}$	57.3	1.5492	85～90/1×10^{-5}	49.4	1.5477	1.2045	65.04	64.87	12.60	12.77，12.89	
P_{154}	$(C_2H_5O)_2P(=S)—SCH_2SC_3H_{7-n}$	86.1	1.5254	84～87/1×10^{-5}	80.8	1.5275	1.1425	74.28	73.82	11.31	11.53，11.70	145～146/4[3]

续表1

编号	分子式	产率（粗,%）	n_D^{20}（粗）	沸点（℃/mm汞柱）	产率（%）	n_D^{20}	d_4^{20}	MR_D 计算值	MR_D 实验值	P（%）计算值	P（%）实验值	文献沸点（℃/mm汞柱）
P_7	$(CH_3O)_2P(=S)—SCH_2SC_4H_{9-n}$	36.5	1.5420	$94\sim98/2\times10^{-6}$	28.0	1.5397	1.1905	68.87	68.22	11.92	11.46，11.69	
P_6	$(C_2H_5O)_2P(=S)—SCH_2SC_4H_{9-n}$	30.6	1.5135	$89\sim91.5/1\times10^{-6}$	28.1	1.5250	1.1220	79.01	78.66	10.76	10.54，10.90	95/0.01[7]
P_9	$(CH_3O)_2P(=S)—SCH_2SC_4H_{9-i}$	63.2	1.5393	$76\sim78/1\times10^{-5}$	56.3	1.5393	1.1811	69.66	69.03	11.92	12.09，12.14	
P_{11}	$(CH_3O)_2P(=S)—SCH_2SC_4H_{9-i}$	53.4	1.5360	$82.5\sim84/1\times10^{-4}$	48.3	1.5378①	1.1640①	69.38	69.62	11.92	12.24，12.47	
P_{10}	$(C_2H_5O)_2P(=S)—SCH_2SC_4H_{9-i}$	53.8	1.5254	$96.5\sim98/2.5\times10^{-8}$	28.6	1.5255	1.1125	78.90	79.43	10.76	11.41，11.54	
P_{24}	$(CH_3O)_2P(=S)—SCH_2SCH_2CH=CH_2$	45.5	1.5640	$84\sim87/1\times10^{-5}$	31.9	1.5598	1.2265	64.58	64.30	12.74	12.26，12.50	
P_{41}	$(C_2H_5O)_2P(=S)—SCH_2SCH_2CH=CH_2$	41.2	1.5468	$71\sim77/1\times10^{-6}$	27.2	1.5399	1.1583	73.81	73.66	11.39	10.97，11.18	

①在25°时之测定值。

表2 O,O－二烷基S－烷氧(硫)乙基硫(氧)甲基二硫代磷酸酯

$$(RO)_2P(=S)—SCH_2XR''$$

编号	分子式	产率(粗,%)	n_D^{20}(粗)	沸点(℃/mm汞柱)	产率(%)	n_D^{20}	d_4^{20}	MR_D		P(%)	
								计算值	实验值	计算值	实验值
P_{27}	$(CH_3O)_2P(=S)—SCH_2SC_2H_4OCH_3$	66.5	1.5490	$114 \sim 118/1\times10^{-6}$	47.8	1.5472	1.2409	66.69	66.97	11.83	11.27, 11.42
P_{28}	$(C_2H_5O)_2P(=S)—SCH_2SC_2H_4OCH_3$	65.9	1.5278	$88 \sim 91/7\times10^{-5}$	54.0	1.5247	1.1675	75.29	76.09	10.69	10.22, 10.25
P_2	$(CH_3O)_2P(=S)—SCH_2SC_2H_4OC_2H_5$①	56.9	1.5440	$118/5\times10^{-5}$分解	—	—	—	71.31	71.31	11.22	11.14, 11.16
P_1	$(C_2H_5O)_2P(=S)—SCH_2SC_2H_4OC_2H_6$	53.0	1.5212	$98 \sim 100/5\times10^{-6}$	49.1	1.5200	1.1402	78.14	78.15	10.19	9.49, 9.56
P_{32}	$(CH_3O)_2P(=S)—SCH_2OC_3H_4SC_2H_5$	25.4	1.5455	$99 \sim 103/4\times10^{-5}$	17.4	1.5260	1.1861	71.31	71.44	11.22	10.77, 10.81
P_{47}	$(C_2H_5O)_2P(=S)—SCH_2OC_3H_4SC_2H_5$	77	1.5218	$110 \sim 114/5\times10^{-6}$	27.7	1.5278	1.1590	80.54	80.76	10.19	10.08, 10.09

①由于 P_2 在扩散蒸馏时分解，其 MR_D 及 P 分析值系自粗产品得到。

表3　O,O－二烷基S－烃基(取代烃基)α－硫乙基二硫代磷酸酯

$$(RO)_2P(=S)-S\,CH(CH_3)-XR''$$

编号	分子式	产率(粗,%)	n_D^{20}(粗)	沸点(℃/mm汞柱)	产率(%)	n_D^{20}	d_4^{20}	MR_D		P(%)	
								计算值	实验值	计算值	实验值
P_{97}	$(C_2H_5O)_2P(=S)-SCH(CH_3)-SCH_2CH=CH_3$	54.4	1.5443	$80\sim82/2\times10^{-6}$	40.3	1.5356	1.1402	78.43	78.16	10.83	10.85，11.02
P_{153}	$(CH_3O)_2P(=S)-SCH(CH_3)-SC_3H_{7-n}$	44.2	1.5541	$81\sim82/1\times10^{-6}$	25.8	1.5322	1.1525	69.66	69.90	11.92	11.77，11.90
P_{168}	$(C_2H_5O)_2P(=S)-SCH(CH_3)-SC_3H_{7-n}$	79.8	1.5213	$82\sim84/1\times10^{-5}$	60.0	1.5228	1.1085	78.90	79.82	10.75	10.27，10.29
P_{142}	$(C_2H_5O)_2P(=S)-SCH(CH_3)-SC_3H_{7-i}$	30.8	1.5431	$74\sim77.8/1\times10^{-5}$	21.8	1.5378	1.1769	69.66	69.09	11.92	12.41，12.57

续表 3

编号	分子式	产率(粗,%)	n_D^{20}(粗)	沸点(℃/mm 汞柱)	产率(%)	n_D^{20}	d_4^{20}	MR_D		P(%)	
								计算值	实验值	计算值	实验值
P_{161}	$(C_2H_5O)_2P(=S)—SCH(CH_3)—SC_4H_{9-n}$	45.6	1.5200	$84\sim86/1\times10^{-6}$	39.2	1.5192	1.1021	83.52	83.19	10.26	10.74, 10.90
P_{187}	$(C_2H_5O)_2P(=S)—SCH(CH_3)—SC_4H_{9-i}$	72.8	1.5198	$97\sim101/1\times10^{-6}$	68.7	1.5190	1.0980	83.52	83.49	10.26	10.69, 10.96
P_{146}	$(CH_3O)_2P(=S)—SCH(CH_3)—SC_2H_4OCH_3$	75.4	1.5401	$98\sim100/3\times10^{-6}$	59.5	1.5386	1.2121	71.31	71.28	11.22	11.08, 11.29
P_{138}	$(C_2H_5O)_2P(=S)—SCH(CH_3)—SC_2H_4OCH_3$	77.3	1.5228	$98\sim105/9\times10^{-6}$	69.6	1.5251	1.1570	79.91	80.84	10.19	10.22, 10.40

（2）β－烷氧代乙硫醇 $ROCH_2CH_2SH$：在三颈瓶中，加入半克分子硫脲及35毫升水，搅拌成糊状物，然后加热。在100℃滴加约0.4克分子β－烷氧基溴代乙烷（$ROCH_2CH_2Br$）加毕回流3小时，再在室温搅拌4小时。加入氢氧化钠溶液（30克氢氧化钠溶于28.3毫升水），继续搅拌回流1小时，有油层出现。冷却至室温，在冰浴中缓缓用盐酸（10毫升浓盐酸加入20毫升水中）酸化至溶液呈中性。分出油状物，母液用乙醚萃取三次（每次用乙醚100毫升）。乙醚萃取液合并，干燥，过滤，减压蒸馏。

O，O－二烷基S－烃基硫甲基二硫代磷酸酯 $(RO)_2P(=S)-SCH_2SR''$

置O，O－二烷基二硫代磷酸酯0.1克分子及硫醇R″SH 0.1克分子于三颈瓶中，冷却并不停搅拌。慢慢滴加甲醛溶液（36%）0.1克分子。保持反应液在3～10℃约7小时，然后移至水浴上加热至40～50℃，继续搅拌6小时，冷却。用乙醚萃取之，继用5%碳酸氢钠或碳酸钠溶液洗萃取液，后用水洗至中性，干燥。在氮气保护下减压抽去低沸点产物，进行减压或扩散蒸馏。

O，O－二烷基S－烷氧（硫）乙基硫（氧）甲基二硫代磷酸酯

$(RO)_2P(=S)-SCH_2XR''$　（X＝硫或氧，R″＝烷氧代乙基或烷硫代乙基）

取等克分子的二硫代磷酸酯，硫醇或醇及甲醛进行反应，步骤同上。

O，O－二烷基S－烃基（取代烃基）α－硫乙基二硫代磷酸酯

$(RO)_2P(=S)-SCH(CH_3)XR''$

将0.1克分子二硫代磷酸酯和0.1克分子硫醇置于三颈瓶中。用冰冷却至3℃，滴加乙醛0.1克分子，约需8分钟滴完。反应温度上升，最高达到28℃，然后在室温（14～16℃）搅拌6小时，移到水浴上加热至40～50℃，继续搅拌6小时。用乙醚萃取，先用5%碳酸氢钠或碳酸钠溶液洗涤，后用水洗至中性。干燥，过滤。在氮气保护下减压抽去低沸点产物，进行扩散蒸馏。

摘要

本文报告下列三类型化合物的合成：

$(RO)_2P(=S)-SCH(R')XR''$　（X＝硫或氧）

Ⅰ. R＝甲基或乙基，R′＝氢，R″＝脂烃基，共十三种

Ⅱ. R＝甲基或乙基，R′＝氢，R″＝烷氧代乙基或烷硫代乙基，共六种

Ⅲ. R＝甲基或乙基，R′＝甲基，R″＝脂烃基或烷氧代乙基，共八种

这些化合物是用相当的O，O－二烷基二硫代磷酸酯，醛，与硫醇或醇缩合制得。并测定了各化合物的一些物理常数。

初步昆虫试验结果说明 P_2，P_7，P_{11}，P_{20}，P_{27}，P_{154} 等对四龄蚊幼虫有显著的触杀

作用，并对刺吸口器害虫有内吸毒效。此外，P_1，P_5，P_{32}，P_{47}等虽触杀效率较差，但均具良好的内吸效率，其中大部分化合物的内吸效能相当于目前常用的内吸剂 *E* 1059。

致谢：本文承刘增勋同志参加部分工作；刘燕华，翟宝英二位同志代为作磷的分析；又蒙中国科学院昆虫研究所协助解决药效测定工作，谨致谢意。

参考文献

[1] G. Schrader，Ger. 850，677，C. A. 47. 4034(1953).

[2] М. И. Кабачник и Т. А. Мастрюкова，Д. А. Н. 109，974(1956).

[3] Т. А. Мастрюкова，"Хнмия и Применение Фосфорорганическях Соединений，" строе. 148，1957.

[4] E. O. Hook and P. H. Moss，(a) U. S. 2,565,920，C. A. 46，3067(1952)；(b) U. S. 2,596,076，C. A. 46，8322(1952)；(c) U. S. 2,614,988，C. A. 47，4079(1953)；(d) Brit. 676，776，C. A. 47，9995(1953).

[5] Б. А. Арбузов Изв. А. Н. СССР. 4，672 (1952).

[6] W. Lorenz and G. Schrader，(a) Ger. 918. 688，C. A. 49，12，528 (1955)；(b) U. S. 2,759,010，C. A. 51，2849(1957).

[7] G. Schrader，Ger. 947，369；C. A. 51，4425(1957).

[8] J. R. Goigy，Brit. 772，213；C. A. 51，10，830(1957).

[9] G. Schrader，Angew. Chem，69，90(1957).

[10] O. Foss，Acta Chem. Scand. 1，8(1947)；Kgl. Norske Videnskab selskab；Forh. 15，119(1942).

[11] T. W. Mastin，et al.，I. Am. Chem. Soc. 67，1662(1945).

[12] A. I. Vogel. "Practical organic chemistry" 3rd. ed.，p. 497，Longmans，Green，London，1956.

Researches on Organophosphorus Compounds Ⅰ.

O,O – Dialkyl S – alkyl (or Substituted Alkyl) Thio (or Oxy) Methyl (or Substituted Methyl) Phosphorodithioate

Siixian Yang，Tianchi Chen，Zhengming Li，Yugui Li，Qinsun Wang，Maogong Yan，Xiyang Dong

(Department of Chemistry，Nankai University)

Abstract The present communication deals with the preparation of three groups of compounds represented by the general formula：

$$(RO)_2P(=S)-SCH(R')XR'' \quad (X = S \text{ or } O)$$

Ⅰ. R = methyl of ethyl，R′ = hydrogen，R″ = alkyl，thirteen compounds.

Ⅱ. R = methyl of ethyl，R′ = hydrogen，R″ = alkoxycthyl or alkylthiacthyl，six compounds.

Ⅲ. R = methyl of ethyl，R′ = methyl，R″ = alkyl or alkoxyethyl，eight compounds.

These compounds were prepared by condensing O，O – dialkyl dithiophosphate，mercaptan (alcohol) with an aldehyde. Some physical constants of these compounds were also recorded.

The preliminary biological screening test gives interesting results. Compounds P_2，P_7，P_{11}，P_{20}，P_{27}，P_{154} exhibit remarkably high contact effect against 4th instar larva of *Calex pipiens*，and also systemic action to some sucking insect species. Compounds P_1，P_5，P_{32}，P_{47} also show marked systemic，though with weaker contact activity. The systemic properties of most of these compounds are equivalent to the well – known systemic insecticide *E* 1059.

1962年

有机磷杀虫剂的研究Ⅱ*

——O,O－二烷基S－烃基(取代烃基)硫甲基二硫代磷酸酯的合成

杨石先　陈天池　李正名　李毓桂
董希阳　高绍仪　董松琦

（南开大学化学系）

前文[1]报告 $(RO)_2P(=S)—SCHR'XR''$ 型二硫代磷酸酯的合成，发现当R为甲基或乙基，R′为H，R″为丙基、甲氧（硫）乙基或乙氧（硫）乙基时，具有良好的杀虫性能。本文继续报道具上述通式的一系列化合物，但R基团适当增大，如丙基、异丙基、丁基、氯乙基等，R′为H，R″不变，以探讨R增大对化合物毒性的影响。

我们得到十种新化合物，其中有一半即使在压力为 1×10^{-5} 毫米汞柱的高真空度时，仍发生分解。我们采用了在扩散泵压力下抽去低沸点杂质的方法以提纯产物，效果良好，结果见表1。

初步昆虫试验结果说明R基团增大降低各化合物的毒效。

实验部分

O,O－二烷基二硫代磷酸酯 $(RO)_2P(=S)—SH$

这类化合物是按前文[1]所载的方法制备的。不同之点是滴醇后，回流温度需提高到80～90℃，回流时间延长为两三小时。化合物的常数见表2。

O,O－二烷基（取代烷基）S－烃基（取代烃基）硫甲基二硫代磷酸酯 $(RO)_2P(=S)—SCH_2SR'$

置O,O－二烷基二硫代磷酸酯0.2克分子及硫醇R′SH 0.22克分子于三颈瓶中，冷却并不断搅拌。慢慢滴加甲醛溶液（36%）0.22克分子。在室温下搅拌7小时，然后移至水浴上加热至40～50℃，继续搅拌6小时，冷却。用乙醚萃取，5%碳酸氢钠或碳酸钠溶液洗萃取液，再用水洗至中性，干燥。在氮气流下减压抽去低沸点物质，用扩散蒸馏法蒸出产品。若产品在蒸馏温度有分解现象，则可适当降低温度，利用扩散泵抽去低沸点物质，仍可得到纯品。

* 原发表于《化学学报》，1962，28（3）：187－190。

表1　O,O－二烷基S－烃基(取代烃基)硫甲基二硫代磷酸酯

编号	分子式	产率（粗,%）	n_D^{25}（粗）	d_4^{25}（粗）	沸点（℃/毫米）	产率（精,%）	n_D^{25}（精）	d_4^{25}（精）	MR_D		P(%)	
									理论值	实验值	理论值	实验值
P_{182}	$(n-C_3H_7O)_2P(=S)SCH_2SC_3H_7\ (n)$	75	—	—	$98\sim100/6\times10^{-6}$	39.6	1.5184	1.1090	83.24	82.58	10.27	10.16，10.34
P_{174}	$(n-C_3H_7O)_2P(=S)SCH_2SC_3H_7\ (i)$	87	—	—	$90\sim92/1\times10^{-5}$	50.9	1.5140	1.1090	83.24	84.00	10.27	10.07，10.30
P_{211}	$(n-C_3H_7O)_2P(=S)SCH_3SC_2H_4OCH_3$	51	—	—	$114\sim117/8\times10^{-6}$	24.8	1.5139	1.1297	84.86	84.76	9.74	9.51，9.59
P_{203}	$(i-C_3H_7O)_2P(=S)SCH_2SC_3H_7\ (n)$	59.6	—	—	$98\sim100/3\times10^{-6}$	40.1	1.5140	1.0880	83.24	83.56	10.27	10.20，9.77
P_{206}	$(n-C_4H_9O)_2P(=S)SCH_2SC_3H_7\ (n)$	70.61	—	—	$110\sim112/5\times10^{-6}$	57.2	1.5072	1.0615	93.14	92.68	9.40	9.32，9.47
P_{207}	$(n-C_4H_9O)_2P(=S)SCH_2SC_2H_4OCH_3$	52.6	1.5163	1.1154	$93/1\times10^{-5}$ ①	—	—	—	94.78	93.88	8.95	8.28，8.61
P_{208}	$(n-C_4H_9O)_2P(=S)SCH_2SC_2H_4OC_2H_5$	52.6	1.5090	1.0876	$104/6\times10^{-5}$ ①	—	—	—	99.50	99.00	8.60	8.56，8.58
P_{192}	$(ClCH_2CH_2O)_2P(=S)SCH_2SC_3H_7\ (n)$	72.6	1.5526	1.3231	$117/7\times10^{-6}$ ①	—	—	—	84.09	83.13	9.07	8.63，9.18
P_{196}	$(ClCH_2CH_2O)_2P(=S)SCH_2SC_2H_4OCH_3$	50	1.5556	1.3632	$119/1\times10^{-5}$ ①	—	—	—	85.79	84.71	8.64	8.28，9.04
P_{195}	$(ClCH_2CH_2O)_2P(=S)SCH_2SC_2H_4OC_2H_5$	80	1.5416	1.3013	$117/8\times10^{-6}$ ①	—	—	—	90.41	90.22	8.30	8.37，8.85

①是抽低沸点的最高温度。

表2 O,O－二烷基二硫代磷酸酯

编号	分子式	相对分子质量	沸点(℃/毫米)	n_D^{25}	d_4^{25}	产率(%)	MR_D		P(%)		汞盐			备注
							理论值	实验值	理论值	实验值	熔点(℃)	P(%) 理论值	P(%) 实验值	
P_{161}	$(n-C_2H_7O)_2P(=S)SH$	214	102～104/0.9	1.4976	1.0899	57.0	57.5	56.7	14.48	14.46	59～61[①]			
P_{170}	$(i-C_2H_7O)_2P(=S)SH$	214	84～87/2	1.4906	1.0779	53.9	56.8	57.4	14.48	14.71	92～93			文献值 92－93℃
P_{164}	$(n-C_4H_9O)_2P(=S)SH'$	242	117.5～118/1.2	1.4916	1.0605	12.5	66.0	66.1	12.86	12.76	63～64			文献值 61～62℃
P_{169}	$(ClCH_2CH_2O)_2P(=S)SH$	255	117/5×10^{-5}	1.5503	1.4407	41.5	56.3	56.4	12.55	12.59	92～93	8.76	8.76	

①为镍盐。

摘要

本文报告 $(RO)_2P(=S)—SCH_2XR'$ 型二硫代磷酸酯十种，和它们的一些物理常数。初步昆虫试验结果说明它们对四龄蚊幼虫的触杀作用及对刺吸口器害虫的内吸毒效均较低。

参考文献

[1] 杨石先，等. 化学学报 25，402（1959）。

Researches on Organophosphorus Insecticides. Ⅱ.

O,O – Dialkyl S – alkyl (Substituted Alkyl) Thio Methyl Phosphorodithioate

Shixian Yang, Tianchi Chen, Zhengming Li, Yugui Li, Xiyang Dong, Shaoyi Gao, Songqi Dong

(Department of Chemistry, Nankai University)

Abstract The present communication deals with the synthesis of ten new compounds represented by the general formula:

$$(RO)_2P(=S)—SCH_2SR'$$

some physical constants of these compounds have been recorded.

The preliminary biological screening test showed that both the systemic action to some sucking insect species and contact action against fourth instar larva of *Culex pipiens* are comparatively low.

有机磷杀虫剂的研究Ⅲ*

O-乙基N，N-二乙氨基硫代磷酰氯的合成及其与硫氢化钠的反应

杨石先　陈天池　王琴荪　李正名

（南开大学化学系）

近年来对含P—N键杀虫剂的研究曾屡见于文献[1~6]，因而引起了我们对中间体（Ⅰ）的兴趣，并企图用Ⅰ以合成另一重要中间体（Ⅱ）

$$\underset{(\text{I})}{(C_2H_5O)((C_2H_5)_2N)P(=S)-Cl} \qquad \underset{(\text{II})}{(C_2H_5O)((C_2H_5)_2N)P(=S)-SH}$$

我们先用以下两法合成Ⅰ：

方法A：

$$PSCl_3 + C_2H_5OH + C_5H_5N \xrightarrow{0\sim5℃} \underset{(\text{III})}{C_2H_5OP(=S)Cl_2} + C_5H_5N\cdot HCl \tag{1}$$

$$C_2H_5OP(=S)Cl_2 + 2(C_2H_5)_2NH \xrightarrow{-5\sim-10℃} (C_2H_5O)((C_2H_5)_2N)P(=S)Cl + (C_2H_5)_2NH\cdot HCl \tag{2}$$

方法B：

$$PSCl_3 + 2(C_2H_5)_2NH \xrightarrow{50℃} \underset{(\text{IV})}{(C_2H_5)_2NP(=S)Cl_2} + (C_2H_5)_2NH\cdot HCl \tag{3}$$

$$(C_2H_5)_2NP(=S)Cl_2 + C_2H_5OH + C_2H_5ONa \xrightarrow{<30℃} (C_2H_5O)((C_2H_5)_2N)P(=S)Cl + NaCl \tag{4}$$

由上述两法所得到的Ⅰ具相同的折射率，沸点也一致**。

关于Ⅰ与硫氢化钠的反应尚未见报道。我们在进行这反应时，曾试用不同的溶剂。发现在石油醚、乙醚、苯、氯仿中均不起反应。在石油醚-丙酮混合溶剂中，或不用溶剂直接将Ⅰ与粉状硫氢化钠经长时间搅拌，皆得不到Ⅱ。在丙酮中反应很迟缓，产

---

* 原发表于《化学学报》，1963，29（3）：153-158。

** 上述工作在1959年4月即已完成，后来看到Кабачник[7]等也用方法A得到了Ⅰ，其折射率与我们的相同。

物经检定亦不是Ⅱ。在无水乙醇中进行反应，除有大量氯化钠析出外，还分离出一化合物。定性检验表明，它不使高锰酸钾溶液褪色，不溶于稀或浓氢氧化钠溶液，不与苯胺或氯化汞－酒精溶液形成相应的盐（证明无—SH），Beilstein 试验证明不含卤素，pH 接近中性。同时在反应过程中有硫化氢气不断逸出，经硝酸银溶液吸收后得到的沉淀经检定为硫化银。根据以上现象，可以确定发生下列反应：

$$\begin{matrix} C_2H_5O \\ (C_2H_5)_2N \end{matrix}\!\!>\!\overset{S}{\overset{\|}{P}}\text{—}Cl + NaSH + C_2H_5OH \xrightarrow{75℃} \underset{(V)}{\begin{matrix} C_2H_5O \\ (C_2H_5)_2N \end{matrix}\!\!>\!\overset{S}{\overset{\|}{P}}OC_2H_5} + NaCl + H_2S \quad (5)$$

在这里硫氢化钠作为碱促使化合物Ⅰ与乙醇分子间脱去氯化氢：

欲证明上述产物的结构，我们进行以下的合成工作：

（1）在氢氧化钠存在下，Ⅰ与无水乙醇进行反应

$$\begin{matrix} C_2H_5O \\ (C_2H_5)_2N \end{matrix}\!\!>\!\overset{S}{\overset{\|}{P}}\text{—}Cl + C_2H_5OH + NaOH \xrightarrow{75℃} V + NaCl \quad (6)$$

（2）在乙醇中令Ⅳ与乙醇钠反应

$$\underset{(IV)}{(C_2H_5)_2N\overset{S}{\overset{\|}{P}}Cl_2} + 2C_2H_5ONa \xrightarrow{C_2H_5OH} V + 2NaCl \quad (7)$$

经一系列物理常数的测定，证明反应（5），（6），（7）所得的均属同一化合物（见表1），因而也证明Ⅴ的结构。

反应（5）可以用于同一类型的合成，如丁酯Ⅵ、Ⅶ、Ⅷ等，方法简便，产率也较满意。

$$\underset{(VI)}{\begin{matrix} C_2H_5O \\ (C_2H_5)_2N \end{matrix}\!\!>\!\overset{S}{\overset{\|}{P}}OC_4H_9} \qquad \underset{(VII)}{\begin{matrix} C_4H_9O \\ (C_2H_5)_2N \end{matrix}\!\!>\!\overset{S}{\overset{\|}{P}}OC_2H_5} \qquad \underset{(VIII)}{\begin{matrix} C_4H_9O \\ (C_2H_5)_2N \end{matrix}\!\!>\!\overset{S}{\overset{\|}{P}}OC_4H_9}$$

实验部分

（一）O－乙基－N，N－二乙氨基硫代磷酰氯（Ⅰ）

方法 A：

（1）O－乙基硫代磷酰二氯（Ⅲ）：

置 169.5 克（1 克分子）三氯硫磷及 150 毫升无水乙醚于 500 毫升三颈瓶中，用冰浴冷却，在 0℃搅拌，同时滴加 46 克（1 克分子）无水乙醇 79 克（1 克分子）无水吡啶及 200 毫升无水乙醚的混合液。反应温度随即升至 5～6℃，并不断析出白色固体。滴加约 4 小时完成，然后继续搅拌 1 小时，继放置过夜。滤去吡啶盐酸盐，用无水乙醚洗涤。滤液用水泵在 10℃抽去乙醚后，减压蒸出产品。沸点 43～44℃/8.5 毫米。n_D^{75} 1.5041，产率 70.2%，文献值[3]：沸点 68℃/20 毫米。

分析：$C_2H_5Cl_2OPS$

计算值/%：P，17.32

实验值/%：P，18.20

（2）O－乙基N,N－二乙氨基硫代磷酰氯（Ⅰ）：

置116克（0.65克分子）Ⅲ与150毫升无水乙醚于三颈瓶中，用冰浴冷却到－10℃以下。搅拌滴加95克（1.3克分子）二乙胺（经氢氧化钠干燥）及100毫升无水乙醚的混合溶液。白色固体立即析出。经6小时后，滴加完成，继续在－8℃搅拌2小时，然后放置过夜。滤去二乙胺盐酸盐。滤液用水泵抽去乙醚，在氮气流中减压蒸馏产物。沸点71～73℃/3毫米，n_D^{20}1.4932，产率77.2%。Кабачник[7]报告的沸点为105～106℃/10毫米，n_D^{20}1.4931，产率48.0%。

分析：$C_6H_{15}ClNOPS$

计算值/%：N，6.50；P，14.39

实验值/%：N，5.91；P，15.06

在本反应中曾用二乙胺与三乙胺各0.65克分子以替代1.3克分子二乙胺也得到Ⅰ，但产率仅65%。

方法B：

（1）N,N－二乙氨基硫代磷酰二氯（Ⅳ）：

置212.5克（1.25克分子）三氯硫磷及350克无水苯于三颈瓶中，强烈搅拌。在室温滴入干燥的二乙胺82.5克（2.5克分子），令反应温度保持在50℃以下。滴加完毕后继续搅拌3小时，静置过夜。在氮气保护下在45℃/35毫米（水浴65℃）抽去过量的苯，然后仍在氮气流中进行减压蒸馏。沸点77℃/0.7毫米，n_D^{25}1.5240，产率71.0%。

分析：$C_4H_{10}Cl_2NPS$

计算值/%：P，15.07；S，15.53

实验值/%：P，15.64；S，15.99

（2）O－乙基N,N－二乙氨基硫代磷酰氯（Ⅰ）

溶21克（0.1克分子）Ⅳ于15毫升无水乙醇中，在搅拌下缓缓滴入相当于金属钠0.1克分子的乙醇钠－乙醇溶液（约60毫升）。反应温度保持在30℃以下，析出大量白色固体。继续激烈搅拌4.5小时，然后在60℃加热0.5小时。冷却过滤，减压蒸去过量乙醇（水浴不超过30℃）。在氮气流下减压蒸出产物。沸点79～80℃/2.5毫米，n_D^{20}1.4930，产率73%。

分析：$C_6H_{15}ClNOPS$

计算值/%：N，6.50；P，14.39

实验值/%：N，5.69；P，14.91

（二）O,O－二乙基N,N－二乙基硫代磷酰胺（Ⅴ）（反应5，6，7）

（1）溶43.1克（0.2克分子）Ⅰ于40毫升无水乙醇，在室温及搅拌下滴入含11.2克（0.2克分子）硫氢化钠的200毫升无水乙醇溶液。温度升到27℃。滴加2小时后，外面水浴升到80℃，继续搅拌2小时，白色固体逐渐增多，并有硫化氢逸出。反应完成后，用冰浴充分冷却，过滤。从滤液抽去过量乙醇后，在氮气流下减压蒸出产品，沸点90.5～91℃/1毫米。n_D^{20}1.4680，n_D^{25}1.4663，d_4^{25}1.0373，产率77.0%。文献值[9]：沸点110℃/20毫米。

分析：$C_8H_{20}NO_2PS$

计算值/%：N，6.54；P，13.78；S，14.22

实验值/%：N，6.14；P，13.93；S，15.40

（2）溶16.16克（0.075克分子）的Ⅰ于50毫升无水酒精，滴入含3克氢氧化钠的无水乙醇120毫升，反应物的温度升到26℃，白色固体开始析出。在室温搅拌2小时，继在75℃又2小时。冷却过滤，抽去溶剂，减压蒸馏。沸点97.5～98℃/1毫米，n_D^{20}1.4680，n_D^{25}1.4663，d_4^{25}1.0351，产率75.2%。

分析：$C_8H_{20}NO_2PS$

计算值/%：P，13.78；S，14.22

实验值/%：P，14.33；S，14.73

（3）溶20.6克（0.1克分子）Ⅰ于100毫升无水乙醇，滴加相当于0.2克分子金属钠的乙醇钠－乙醇溶液50毫升。在室温搅拌2小时。继在75℃又2小时。冷却过滤，抽去过量溶剂，减压蒸出产物。沸点88～89℃/1毫米，n_D^{20}1.4680，n_D^{25}1.4662，d_4^{25}1.0370，产率66.6%。

分析：$C_8H_{20}NO_2PS$

计算值/%：P，13.78；S，14.22

实验值/%：P，13.88；S，14.86

表1　由上述三反应所得到的产物Ⅴ的比较

反应	沸点（℃/毫米）	n_D^{20}	n_D^{25}	d_4^{25}	产率（%）	P分析值（%）		S分析值（%）		MR_D	
						计算	实验	计算	实验	计算	实验
（5）	90～90.5/1	1.4680	1.4663	1.0373	77.0	13.78	13.93 14.23	14.22	15.40① 15.68	60.75	60.37
（6）	97～98/1.1	1.4680	1.4663	1.0351	75.0	13.78	14.33 14.77	14.22	14.73 14.98	60.75	60.47
（7）	88～89/1	1.4680	1.4663	1.0370	66.6	13.78	13.88 14.08	14.22	14.86 15.07	60.75	60.35

①S%分析值稍大，可能由于产物中含少量硫化氢。

（三）O－乙基O－丁基N,N－二乙基硫代磷酰胺（Ⅵ）

制备方法同（二）（反应（5）），不同之处是硫氢化钠溶液滴加完毕后，需在水浴温度85℃搅拌5.5小时。沸点103～104℃/1.8毫米，n_D^{25}1.4665，产率54%。

分析：$C_{10}H_{24}NO_2PS$

计算值/%：P，12.21

实验值/%：P，12.17

（四）O－丁基N,N－二乙氨基磷酰氯（Ⅶ）

于74.12克（1克分子）的丁醇中分次加入金属钠5.75克（0.25克分子），继在油浴上回流1小时。冷却后，再加入约30毫升丁醇使金属钠全部溶解。将此丁醇钠－丁醇溶液慢慢滴加到含54克（0.25克分子）Ⅳ的丁醇溶液（20毫升）中。反应温度保持22～35℃，滴加时间约2小时。滴加完毕后继续搅拌2小时，静置6小时，滤去固体。减压抽去溶剂，减压蒸馏。沸点 101 ~ 103℃/1 毫米，n_D^{25}1.4862，产率 48.4%。

分析：$C_8H_{19}ClNOPS$

计算值/%：P，12.75

实验值/%：P，13.25

（五）O,O－二丁基 N,N－二乙氨基硫代磷酰胺（Ⅷ）

置 16 克（0.06 克分子）Ⅶ及 10 毫升丁醇于三颈瓶中，缓缓滴加含 3.4 克（0.06 克分子）硫氢化钠的丁醇溶液（40 毫升）。加完后温度略上升。在油浴上保持 100 ~ 110℃4 小时，冷却过滤。滤液减压抽去丁醇及低沸点物质，再经扩散蒸馏蒸出产物。沸点 89 ~ 90℃/1 × 10^{-1}毫米，n_D^{20}1.4703，产率 70.2%。

分析：$C_{12}H_{28}NO_2PS$

计算值/%：P，11.05

实验值/%：P，11.56

摘要

利用下列两法合成 O－乙基 N，N－二乙氨基硫代磷酰氯：

方法 A：

$$PSCl_3 + C_2H_5OH + C_5H_5N \xrightarrow{0\sim5℃} C_2H_5O\overset{S}{\overset{\|}{P}}Cl_2 + C_5H_5N \cdot HCl$$

$$C_2H_5O\overset{S}{\overset{\|}{P}}Cl_2 + 2(C_2H_5)_2NH \xrightarrow{-5\sim-10℃} \begin{matrix}C_2H_5O\\(C_2H_5)_2N\end{matrix}\!\!>\overset{S}{\overset{\|}{P}}Cl + (C_2H_5)_2NH \cdot HCl$$

方法 B：

$$PSCl_3 + 2(C_2H_5)_2NH \xrightarrow{50℃} (C_2H_5)_2N\overset{S}{\overset{\|}{P}}Cl_2 + (C_2H_5)_2NH \cdot HCl$$

$$(C_2H_5)_2N\overset{S}{\overset{\|}{P}}Cl_2 + C_2H_5OH + C_2H_5ONa \xrightarrow{<30℃} \begin{matrix}C_2H_5O\\(C_2H_5)_2N\end{matrix}\!\!>\overset{S}{\overset{\|}{P}}Cl + NaCl$$

并试验它在不同溶剂中与硫氢化钠的反应。在无水乙醇中所得到的产物确定为 O，O－二乙基 N,N－二乙基硫代磷酰胺，产率在 70% 以上。

致谢： 承刘燕华、翟宝英二位同志代作元素分析，谨致谢意。

参考文献

［1］Б. А. Арбузов，*Изв. АН СССР* 6，1038（1954）.

［2］Н. Н. Мелвников，*Ж. обш. хцм.* 25，828（1955）.

［3］L. F. Audrieth，*J. prakt. Chem.* 8，117（1959）.

［4］P. C. Coe，*J. Agr. and Food Chem.* 4，251（1959）.

［5］Dow Chem. Co，Brit. pat. 673，877；*C. A.* 43，3332（1953）.

［6］杨石先、陈天池等，化学学报 25，402（1959）.

［7］М. И. Кабачник，*Ж. обш. хцм.* 29，2182（1959）.

[8] P. C. Pishchimuka, *Ber.* 41, 3854 (1908).
[9] A. Michaelis *Ann.* 326, 129 (1903).

Researches on Organophosphorus Insecticides Ⅲ.

The Synthesis of O－ethyl N,N－Diethyl Phosphoroamido－Thionochloridate and Its Reaction with Sodium Hydrosulfide

Shixian Yang, Tianchi Chen, Qinsun Wang, Zhengming Li

(Department of Chemistry, Nankai University)

Abstract The preparation of O－ethyl N,N－diethyl phosphoro－amidothionochloridate by means of the following two methods was described:

Method A:

$$PSCl_3 + C_2H_5OH + C_5H_5N \xrightarrow{0\sim5℃} C_2H_5O\overset{S}{\overset{\|}{P}}Cl_2 + C_5H_5N\cdot HCl$$

$$C_2H_5O\overset{S}{\overset{\|}{P}}Cl_2 + 2(C_2H_5)_2NH \xrightarrow{-5\sim-10℃} (C_2H_5O)((C_2H_5)_2N)\overset{S}{\overset{\|}{P}}Cl + (C_2H_5)_2NH\cdot HCl$$

Method B:

$$PSCl_3 + 2(C_2H_5)_2NH \xrightarrow{50℃} (C_2H_5)_2N\overset{S}{\overset{\|}{P}}Cl_2 + (C_2H_5)_2NH\cdot HCl$$

$$(C_2H_5)_2N\overset{S}{\overset{\|}{P}}Cl_2 + C_2H_5OH + C_2H_5ONa \xrightarrow{<30℃} (C_2H_5O)((C_2H_5)_2N)\overset{S}{\overset{\|}{P}}Cl + NaCl$$

Its reaction with sodium hydrosulfide in different solvents has also been studied. In anhydrous ethyl alcohol, the reaction proceeded smoothly and the product was identified as O,O－diethyl N,N－diethyl phosphorothionoamidate (yield >70%).

1965年

有机磷杀虫剂的研究Ⅶ*

——某些苯基对位取代硫代膦酸酯的合成

杨石先　陈天池　李正名　王惠林
黄润秋　唐除痴　刘天麟　张金碚

（南开大学元素有机化学研究所）

在以通式 $\begin{matrix}R'\\R''\end{matrix}\!>\!P\!\!\begin{matrix}\nearrow S\\ \searrow OR'''\end{matrix}$ （Ⅰ）表示的有机磷杀虫剂中，R′、R″为甲（乙）氧基的各类化合物曾被较详细地研究。R′为烃基或芳基，R″为烷氧基，R‴为有关酸性基团如 $-C_6H_3(Cl)-NO_2$，$-C_6H_3(CH_3)-SCH_3$ 等，大多数均系专利文献[1~11]。至于R′为苯基，R″为烃基、烷硫基、仲胺基等衍生物则报道很少[12~16]。

本文中，作者合成（Ⅰ）型的硫代膦酸酯类化合物，其中R′为 C_6H_5—，R″分别为 C_2H_5O—，$(C_2H_5)_2N$—，C_2H_5S—，C_2H_5—等基团，R‴为 $-C_6H_4-NO_2$，$-$(4-甲基香豆素-7-基)（$C=O$，CH_3）等。为便于比较，作者亦合成已知的高效杀虫剂E-605（$R'=R''=C_2H_5O$，$R'''=-C_6H_4-NO_2$）和扑打杀（$R'=R''=C_2H_5O$，$R'''=$ 4-甲基香豆素-7-基），以探求R′、R″基团的变化与杀虫性能之间的关系。

这类化合物（Ⅰ）是按下列反应合成的：

$$\underset{(\text{Ⅱ})}{\begin{matrix}R'\\R''\end{matrix}\!>\!P\!\!\begin{matrix}\nearrow S\\ \searrow Cl\end{matrix}} + R'''OH \xrightarrow{\text{碱}} \underset{(\text{Ⅰ})}{\begin{matrix}R'\\R''\end{matrix}\!>\!P\!\!\begin{matrix}\nearrow S\\ \searrow OR'''\end{matrix}} + \text{碱}\cdot HCl$$

中间体 $\begin{matrix}R'\\R''\end{matrix}\!>\!P\!\!\begin{matrix}\nearrow S\\ \searrow Cl\end{matrix}$ （Ⅱ）的合成曾采取两种方法：

第一：苯基乙基硫代膦酰氯（Ⅲ）（Ⅱ，$R'=C_6H_5$，$R''=C_2H_5$）是以苯基二氯化

* 原发表于《化学学报》，1965，31（5）：399-406。

膦（Ⅳ）为原料（Ⅳ是按 Buchner 的方法[17]制得，我们对该法作若干改进），先参照 Камай 法[18]与四乙基铅（Ⅴ）反应，得苯基乙基氯化膦（Ⅵ），然后加硫得Ⅲ。

$$\underset{(\mathrm{V})}{(C_2H_5)_4Pb} + \underset{(\mathrm{IV})}{C_6H_5PCl_2} \longrightarrow (C_6H_5)(C_2H_5)PCl \quad (\mathrm{VI})$$

$$(C_6H_5)(C_2H_5)PCl + S \xrightarrow{AlCl_3} (C_6H_5)(C_2H_5)P(=S)Cl \quad (\mathrm{III})$$

第二：苯基 p－取代硫代膦酰氯［$R'=C_6H_5$，$R''=C_2H_5O$（Ⅶ），$R'=C_6H_5$，$R''=C_2H_5S$（Ⅷ），$R'=C_6H_5$，$R''=(CH_3)_2N$（Ⅸ），$R'=C_6H_5$，$R''=(C_2H_5)_2N$（Ⅹ）］的合成先制得苯基硫代膦酰二氯（Ⅺ）[19]，然后与乙醇、乙硫醇、二甲胺、二乙胺等在有机碱存在下缩合得到相应的中间体。

$$\underset{(\mathrm{XI})}{C_6H_5P(=S)Cl_2} + XH \xrightarrow{B:} (C_6H_5)(X)P(=S)Cl$$

（Ⅶ：$X=C_2H_5O$；Ⅷ：$X=C_2H_5S$；Ⅸ：$X=(CH_3)_2N$；Ⅹ：$X=(C_2H_5)_2N$）

以上化合物，除Ⅶ外，均未见诸文献（表1）。上述中间体Ⅲ、Ⅶ－Ⅹ与对硝基酚及4－甲基－7－羟基香豆素等缩合所得到的产物分别列于表2～表4。这些化合物的生物测定正在进行中。

表1　苯基 p－取代硫代膦酰氯

化合物	沸点(℃)（熔点）	折光率 n_D^{20}	产率(%)	元素分析					
				Cl(%)		P(%)		S(%)	
				计算值	实验值	计算值	实验值	计算值	实验值
$(C_6H_5)(C_2H_5O)P(=S)Cl$ ①	101～102/0.5 毫米	1.5712	88	—	—	—	—	—	—
$(C_6H_5)(C_2H_5S)P(=S)Cl$	122～124/0.2 毫米	1.6367	50	—	—	13.11	13.19 13.24	27.05	26.67 26.42
$(C_6H_5)((CH_3)_2N)P(=S)Cl$ ②③	126～127/0.4 毫米 (33～34)	(白色菱形晶体)	65	16.17	16.62 16.73	14.15	13.50 13.27	14.58	14.31 14.42
$(C_6H_5)((C_2H_5)_2N)P(=S)Cl$ ②	138～141/0.1 毫米 (43.5～44.5)	(白色菱形晶体)	55	14.34	14.03 14.12	12.53	12.41 12.45	—	—

续表1

化合物	沸点/(℃)(熔点)	折光率 n_D^{20}	产率(%)	元素分析 Cl(%) 计算值	Cl(%) 实验值	P(%) 计算值	P(%) 实验值	S(%) 计算值	S(%) 实验值
$(C_6H_5)(C_2H_5)P(=S)Cl$	104～106/1 毫米	1.6062	93	17.36	17.37 17.45	15.16	15.11 15.28	—	—

①文献值[20]：沸点 90℃/0.32 毫米，$n_D^{21.5}$ 1.5700。

②经减压蒸出的纯产品，冷冻后即成晶体。

③N 的分析数据：计算值：6.37；实验值：6.05，6.12。

表2　苯基 *p*－取代 *o*－对硝基苯基硫代膦酸酯

化合物	沸点/(℃)(熔点)	结晶形状	产率(%)	元素分析 N(%) 计算值	N(%) 实验值	P(%) 计算值	P(%) 实验值	S(%) 计算值	S(%) 实验值	比移值 R_f
$(C_6H_5)(C_2H_5O)P(=S)-O-C_6H_4-NO_2$ ①	184～186/ 1×10^{-8}毫米	淡黄色液体 n_D^{25} 1.5980	54	4.33	4.24 4.00	—	—	9.91	9.57 9.72	0.51
$(C_6H_5)(C_2H_5S)P(=S)-O-C_6H_4-NO_2$	96.5～98	白色针状	87	4.13	4.23 4.40	9.14	8.97 8.87	18.88	18.22 18.17	0.55
$(C_6H_5)((CH_3)_2N)P(=S)-O-C_6H_4-NO_2$	97.5～99	白色针状	85	8.69	8.38 8.19	9.62	9.45 9.36	9.93	9.75 9.70	0.33
$(C_6H_5)((C_2H_5)_2N)P(=S)-O-C_6H_4-NO_2$	99～100	白色针状	89	8.00	8.09 7.80	8.85	9.01 9.02	9.14	9.01 9.05	0.39
$(C_6H_5)(C_2H_5)P(=S)-O-C_6H_4-NO_2$	60～61	白色针状	51	4.56	4.51 4.50	10.10	10.16 10.10	10.43	10.21 10.36	0.42
$(C_2H_5O)_2P(=S)-O-C_6H_4-NO_2$ ②	164～165 0.1 毫米	淡黄色液体 $n_D^{25.5}$ 1.5361	52	—	—	—	—	—	—	0.49

①文献上报道的绝大部分为粗产品，曾参照专利文献方法[28]制备，未能得到所称的固体产物。

②文献值[21]沸点 157～162℃/0.6 毫米，n_D^{25}1.5370。

表3　苯基 *p* - 取代 *o* -4 - 甲基香豆素 -7 - 硫代膦酸酯

化合物	熔点(℃)	结晶形状	产率(%)	元素分析					
				N(%)		P(%)		S(%)	
				计算值	实验值	计算值	实验值	计算值	实验值
$(C_6H_5)(C_2H_5O)P(=S)-O-$(4-$CH_3$-香豆素-7-基) ①	116 ~ 117.5	白色针状	55	—	—	8.61	8.43 8.52	8.88	8.80 8.81
$(C_6H_5)(C_2H_5S)P(=S)-O-$(4-$CH_3$-香豆素-7-基)	95.5 ~ 97.5	白色菱形	68	—	—	8.24	8.17 8.15	17.03	16.84 16.81
$(C_6H_5)((CH_3)_2N)P(=S)-O-$(4-$CH_3$-香豆素-7-基)	164 ~ 165	白色片状	78	3.89	3.62 3.44	8.63	8.63 8.47	8.91	8.71 8.87
$(C_6H_5)((C_2H_5)_2N)P(=S)-O-$(4-$CH_3$-香豆素-7-基)	101 ~ 102	白色片状	53	3.61	3.66 3.66	8.01	8.04 7.89	8.27	8.22 8.35
$(C_6H_5)(C_2H_5)P(=S)-O-$(4-CH_3-香豆素-7-基)	88 ~ 89	白色菱形	65	—	—	8.98	8.73 8.83	9.30	9.37 9.40
$(C_2H_5O)_2P(=S)-O-$(4-CH_3-香豆素-7-基) ②	36 ~ 37	白色针状	59	—	—	9.45	9.01 9.00	—	—

①在得到该化合物后，文献[10]也报告同一化合物，熔点 116℃，产率 35%。

②文献值熔点 38℃[27]。

表4 苯基N，N－二乙基o－取代芳基硫代膦酰胺

化合物	熔点(℃)	结晶形状	产率(%)	元素分析					
				N(%)		P(%)		S(%)	
				计算值	实验值	计算值	实验值	计算值	实验值
C_6H_5, S, P—O, CH_3, NO_2, $(C_2H_5)_2N$	70～71	白色针状	48	7.70	7.64 7.93	8.51	8.39 8.36	8.79	8.70 8.77
C_6H_5, S, P—O, Cl, Cl, $(C_2H_5)_2N$ ①	64～65	白色片状	70.5	3.74	3.98 4.20	8.27	8.58 8.37	8.80	8.48 8.61
C_6H_5, S, P—O, NO_2, NO_2, $(C_2H_5)_2N$	109～111	浅黄色针状	57	9.44	9.42 9.62	—	—	—	—
C_6H_5, S, P—O, $(C_2H_5)_2N$, NO_2	106～107	黄色针状	82	7.00	7.15 7.19	7.75	8.02 8.06	8.00	8.09 8.12

①Cl 的分析计算值：18.95；实验值：18.95，18.72。

实验部分

四乙基铅*（Ⅴ）

参照文献[22]方法制备。由35克（1.45克分子）镁，300毫升无水乙醚及152.6克（1.4克分子）溴乙烷制得的格氏试剂于2.5小时滴入111.2克（0.4克分子）的二氯化铅，300毫升无水乙醚，100克（0.64克分子）碘乙烷及0.5克碘片的溶液中。在氮气流保护下回流7.5小时，有灰白色及黑色颗粒物析出，放置后溶液分为两层。然后将反应液倒入30毫升浓盐酸及500克碎冰的混合物中。有机层用少量冰水洗两次，水层用乙醚提取一次。合并溶液经硫酸镁干燥后除去乙醚，析出黄色沉淀。过滤，滤液于氮气流下减压蒸馏，得无色液体74.5克，产率86.7%。沸点62～63℃/4.5毫米，n_D^{22}1.5198（文献值[23]沸点78℃/10毫米，n_D^{20}1.5195）。

苯基二氯化膦（Ⅳ）

参考Buchner[17]法。最后一步的石油醚用量减去四分之三以上，得到同样的产率。

在三颈瓶中置330克（2.4克分子）三氯化磷及46.8克（0.6克分子）苯，然

* 四乙基铅为剧毒物质，据称在减压蒸馏其同系物四甲基铅时曾发生爆炸[22]，因此上述提纯操作需有相应的防护措施。减压蒸馏时，外浴不能超过110℃，以免发生意外事故。

后加入 106 克（0.8 克分子）无水三氯化铝。在搅拌下将此混合物缓缓加热并回流 8 小时，趁热在半小时内分批滴入 124 克（0.8 克分子）膦酰氯，发热，并有大量胶状固体物析出。滴完后再搅拌 15 分钟，加入 200 毫升无水石油醚（沸点 60 ~ 90℃）并回流搅拌约半小时。冷却，滤去颗粒状淡黄色复合物（$AlCl_3 \cdot POCl_3$），再用石油醚洗涤复合物 3 次（每次 50 毫升）。滤液减压蒸去溶剂及未反应的三氯化磷，然后减压蒸馏，得无色透明液体 78 克，产率 72.6%。沸点 62℃/0.6 毫米，n_D^{20}1.5836（文献值沸点 68 ~ 70℃/1 毫米，n_D^{24}1.5919）。

按 Weil 法[24]得到的苯基二氯化膦，沸点 67 ~ 69℃/0.5 毫米，$n_D^{31.5}$1.5946，产率32%。

苯基乙基氯化膦（Ⅵ）

取苯基二氯化膦与上述四乙基铅在 145 ~ 150℃反应 2 小时[13]，得到无色液体，产率65.5%。沸点 70 ~ 72℃/2.5 毫米，n_D^{20}1.5728（文献值[18]沸点 100 ~ 101℃/11 毫米，n_D^{20}1.5707）。

苯基乙基硫代膦酰氯（Ⅲ）*

置 17.3 克（0.1 克分子）苯基乙基氯化膦及 0.81 克（0.006 克分子）三氯化铝于装有搅拌器，回流冷凝器，氮气导管，温度计及固体加料装置的 50 毫升四口瓶中。逐渐升温到 80℃时，在氮气流下分批加入 3.5 克（0.11 克分子）硫粉。保持反应温度在 80 ~ 90℃之间并搅拌 3 小时。过滤，滤液减压蒸馏得无色液体，产率 93.3%。沸点 104 ~ 106℃/0.2 毫米，n_D^{20}1.6062，d_4^{20}1.2299。

分析 $C_8H_{10}ClPS$

计算值/%：Cl，17.36；P，15.16；MR_D 57.08

实验值/%：Cl，17.37，17.45；P，15.11，15.28；MR_D 57.36

苯基硫代膦酰二氯（Ⅺ）

方法 1　取上述苯基二氯化膦与硫黄在三氯化铝的催化下加热而得[26]。沸点 94 ~ 96℃/0.5 毫米，n_D^{20}1.6217（文献值[26]沸点 130 ~ 135℃/13 毫米，n_D^{21}1.6225）。产率 83.1%。

方法 2　系参照文献[17]制得苯基二氯化膦·三氯化铝复合物后，再采用下列操作提纯之。

在三颈瓶中，加入经升华并研成粉状的 240 克（1.8 克分子）三氯化铝，在搅拌下滴加 744 克（5.4 克分子）三氯化磷，加热回流。然后滴加 138 克（1.77 克分子）苯，搅拌回流 8 小时。蒸去过量的三氯化磷，冷至 30℃，分批加入 60 克（1.87 克分子）硫粉，温度逐渐自行上升到 50 ~ 60℃，维持半小时。冷却，加入 276 克（1.8 克分子）膦酰氯，即有白色固体物析出。加完后搅拌冷至室温，加入 800 毫升无水石油醚（60 ~ 90℃），分三次萃取。减压过滤，滤液用水泵减压除去溶剂，残液减压蒸馏，收集沸点 94 ~ 95℃/0.25 毫米部分。n_D^{20}1.6231。产量 234.6 克，产率 63%。

方法 3　参照 Jensen 一步法[19]制得，收集沸点 107℃/1 毫米，120 ~ 122℃/3 ~ 4 毫米馏分，n_D^{20}1.6234。产率 68.5%。

苯基乙氧基硫代膦酰氯（Ⅶ）

---

* 最近文献[25]曾简略述及此化合物，但无详细的物理数据。

以苯基硫代膦酰二氯与乙醇在吡啶存在下缩合而得[20]，收集沸点 110 ~ 112℃/0.5 毫米部分，n_D^{20}1.5712（文献值[20]：沸点 90℃/0.32 毫米，$n_D^{21.5}$1.5700）。产率 88%。

苯基乙硫基硫代膦酰氯（Ⅷ）

取含 45.4 克（0.7 克分子）乙硫醇及 70.7 克（0.7 克分子）三乙胺的 200 毫升石油醚溶液，在 -10 ~ -12℃间滴加到 147.7 克（0.7 克分子）苯基硫代膦酰二氯的 300 毫升石油醚溶液中。保持此温度 1 小时，在室温继续搅拌 2 小时。反应物倾入水中，分出醚层，用冰水洗濯 3 次，无水硫酸镁干燥。收集沸点 122 ~ 124℃/0.2 毫米部分，n_D^{20}1.6363，产量 83 克。产率 50%。

分析 $C_8H_{10}ClPS_2$

计算值/%：P，13.11；S，27.05

实验值/%：P，13.19，13.24；S，26.67，26.42

苯基二甲氨基硫代膦酰氯（Ⅸ）

在三颈瓶中置 16.4 克（0.0775 克分子）苯基硫代膦酰二氯及 80 毫升无水乙醚，冷至 -14℃时开始滴加含 7 克（0.155 克分子）二甲胺的 80 毫升无水乙醚溶液。控制滴加速度，保持温度在 -12 ~ -14℃，于 1 小时半内滴完。继续搅拌半小时，在室温再搅拌 1 小时。抽滤，滤液用含 1 ~ 2 滴 6*N* 盐酸的水洗涤，分出醚层。干燥，蒸去乙醚后，收集沸点 126 ~ 127℃/0.4 毫米馏分，产量 8.4 克，产率 65%。冷却即成白色针状固体。经石油醚重结晶，得白色菱形晶体，熔点 33 ~ 34℃，重 7.2 克。

分析 $C_8H_{11}ClNPS$

计算值/%：N，6.37；S，14.58

实验值/%：N，6.05，6.12；S，14.31，14.42

若用过量的二甲胺与之反应，得到 N，N′-四甲基苯基硫代膦酰二胺。经石油醚重结晶，得白色菱形结晶，熔点 45.5 ~ 46.5℃。

分析 $C_{10}H_{17}N_2PS$

计算值/%：N，12.27

实验值/%：N，12.20，12.26

苯基二乙胺基硫代膦酰氯（Ⅹ）

按类似于制备Ⅸ的方法。15.5 克（0.073 克分子）苯基硫代膦酰二氯与 11.5 克（0.15 克分子）二乙胺在氮气流下，-10℃反应 2 小时。滤去二乙胺盐酸盐，收集沸点 136 ~ 141℃/0.1 毫米产物。经石油醚重结晶得白色菱形结晶 9.9 克，熔点 43.5 ~ 44.5℃，产率 55%。粗产品不经减压蒸馏，直接冰冻也可得熔点为 43.5 ~ 44.5℃的白色结晶。

分析 $C_{10}H_{15}ClNPS$

计算值/%：Cl，14.34；P，12.53

实验值/%：Cl，14.03，14.12；P，12.41，12.45

缩合反应

表 2 ~ 表 4 中所列化合物采用的缩合条件参照以下两例：

O-(2,4-二氯苯基)-N，N-二乙氨基苯基硫代膦酸酯

在 100 毫升三颈瓶中溶 1.31 克（0.008 克分子）二氯酚于 25 毫升无水丙酮中，加入 2.21 克（0.016 克分子）无水碳酸钾及少量铜粉，搅拌回流。然后滴加含 2 克

(0.008 克分子）苯基二乙胺基硫代膦酰氯的 20 毫升无水丙酮溶液，继续回流 6 小时。冷却，过滤，减压除去丙酮。残留黏稠液体冷冻后析出结晶，经乙醇重结晶二次，得白色片状结晶 2.2 克，熔点 64 ~65℃，产率 70.5%。

苯基乙基 O－对硝基苯基硫代膦酸酯

取等克分子的乙醇钠和对硝基酚在无水乙醇中反应，减压除去乙醇即得对硝基酚钠盐。在三颈瓶中溶 1.61 克（0.01 克分子）钠盐于 40 毫升丁酮中，于回流下滴入 2.05 克（0.01 克分子）苯基乙基硫代膦酰氯的 10 毫升丁酮溶液，继续回流 4 小时。冷却，过滤，减压蒸去溶剂。剩余物用乙醇重结晶，得 1.5 克白色菱形结晶，熔点 60 ~61℃，产率 51%。

化合物 I 的薄层层析

将表 2 中所列诸化合物进行薄层层析。取活性氧化铝（E. Merck 层析规格）在 120℃活化 2 小时，在（18×24）厘米玻璃板上涂成厚度约 0.5 毫米层。取各样品的丙酮溶液（浓度 0.5 毫克/毫升）1 滴，点在离薄层底边 2.5 厘米处，各点相距 2 厘米。用四氯化碳显层，显层时间为 16 分钟，溶剂前缘为 13.6 厘米。晾干，在碘蒸气中显色，用紫外光照射。在白色背底上呈黄色斑点。所测定各化合物的比移值为二次实验的平均值（偏差 ±0.03）（表 2）。

致谢：本文承郑人祺同志参加部分实验，中国科学院化学研究所郁向荣、段惠、李允阁及本所分析室左育民、王菊仙、刘燕华、翟宝英、黄熙亮诸同志协助元素分析，特此一并致以谢忱。

摘　要

1. 制备一些新型的硫代膦酰氯 $\mathrm{R'R''P(=S)Cl}$，其中 R′为 C_6H_5，R″为 C_2H_5O，C_2H_5，$(CH_3)_2N$，$(C_2H_5)_2N$，C_2H_5S 等。

2. 合成十二种具不同程度的杀虫能力的有机磷化合物 $\mathrm{R'R''P(=S)OR'''}$，其中 R‴为 对硝基苯基（$-C_6H_4NO_2$），4-甲基香豆素-7-基 等，并列出它们的若干物理常数。

参 考 文 献

[1] R. Coelln. G. Schrader, Ger. Pat. 1, 132, 132; *C. A.* 58, 1491 (1963).

[2] G. Schrader, Belg. Pat. 614, 005; *C. A.* 58, 2472 (1963).

[3] E. Schegk. G. Schrader, Ger. Pat. 1, 078, 124; *C. A.* 58, 1492 (1963).

[4] P. E. Newallis, J. W. Baker. J. P. Chupp, U. S. Pat. 3, 070, 489; *C. A.* 58, 9144 (1963).

[5] J. W. Baker. J. P. Chupp. P. E. Newallis, U. S. Pat. 3, 096, 238; *C. A.* 59, 14023 (1963).

[6] G. Schrader, Ger. Pat. 1, 139, 494; *C. A.* 58, 12601 (1963).

[7] G. Schrader, Ger. Pat. 1, 139, 119; *C. A.* 58, 12601 (1963).

[8] G. Schrader, Ger. Pat. 1, 142, 606; *C. A.* 59, 1683 (1963).

[9] G. Schrader, Belg. Pat. 617, 721; *C. A.* 59, 5198 (1963).

[10] G. Schrader, U. S. Pat. 3, 067, 210; *C. A.* 58, 9141 (1963).

[11] Y. Ura, S. Sato, N. Shindo, I. Otsubo, Y. Takahashi, M. Hayakawa, K. Sakata, 农药生产技术 8, 7 (1963); *C. A.* 60, 6159 (1964).

[12] A. J. Floyd, R. C. Hinton. Belg. Pat. 616, 760; *C. A.* 59, 14024 (1963).

[13] Farbenfabriken Bayer, Brit. Pat. 900, 590; *C. A.* 58, 6864 (1963).

[14] G. Schrader, R. Coelln, Ger. Pat. 1, 139, 493; *C. A.* 58, 12601 (1963).

[15] G. Schrader, Ger. Pat. 1, 142, 605; *C, A.* 58, 12468 (1963).

[16] R. Coelln, G. Schrader, Ger. Pat. 1, 141, 990; *C. A.* 59, 1684 (1963).

[17] B. Buchner, L. B. Lockhart, *Org. Syn.* 31. 88 (1951).

[18] Г. Камай, Г. М. Русецкая. *Ж. общ. хим.* 32, 2848 (1962).

[19] W. L. Jensen, U. S. Pat. 2, 662, 917; *C. A.* 48, 13711 (1954).

[20] M. F. Hersman, L. F. Audrieth, *J. Org. Chem.* 23, 1891 (1958).

[21] J. H. Fletcher. J. C. Hamilton. I. Hechenbleikner, E. I. Hoegberg, B. J. Sertl, J. T. Cassaday. *J. Am. Chem. Soc.* 70, 3943 (1948).

[22] H. Gilman, R. G. Jones, *ibid.* 72, 1760 (1950).

[23] R. W. Leeper. L. Summers, H. Gilman, *Chem. Rev.* 54, 106 (1954).

[24] Th. Weil, B. Prijs, H. Erlenmeyer, *Helv. Chim. Acta* 35, 1412 (1952).

[25] M. Green, R. F. Hudson, *Proc. Chem. Soc.* 145 (1961).

[26] F. Seel, K. Bullreich, R. Schmutzler, *Ber.* 95, 199 (1961).

[27] T. F. West, J. E. Hardy, "*Chemical Control of Insects*", 2nd ed., p. 112, Chapman & Hall, London, 1961.

[28] A. G. Jelinek, U. S. Pat. 2, 503, 390 (1948): *C. A.* 44, 6435 (1950).

UV – Ozonation and Land Disposal of Aqueous Pesticide Wastes*

Philip C. Kearney, Jack R. Plimmer, Zhengming Li

(Agricultural Research Service(USDA) BARC. West. Beltsville Maryland, 20705, USA)

Abstract Land disposal of pesticides offers many advantages over incineration and other forms of destruction of aqueous liquid wastes. One of the major advantages is cost. Many chlorinated pesticides, however, are not readily degraded by soil processes, and consequently land disposal is not a feasible option. Any pretreatment process that makes these compounds more biodegradable would enhance their disposal on land. We have experimental evidence that pretreatment of aqueous solution of the isooctyl ester of 2,4,5 – trichlorophenoxyacetic acid(2,4,5 – T), 2,5,2′,5′ – tetrachlorobiphenyl(PCB), 2,3,7,8 – tetrachlorobenzo – p – dioxin (TCDD), and pentachlorophenol (PCP) by ultraviolet (UV) ozonation in a laboratory size 450 W mercury vapor lamp causes dechlorination and some ring cleavage. In fresh field soils ^{14}C – 2,4,5 – T for 1 hr in the presence of oxygen. ^{14}C – TCDD and PCB were degraded slower under the same conditions.

Concentration of the pesticide in aqueous solution, UV intensity, irradiation time, ozonoation, and chemical structure were found to be important variables affecting the total degradation process. Preliminary experiments with a 66 lamp commercial water purified (~ 2. 244W) suggest that UV ozonation of some pesticide and industrial wastes can be destroyed in sufficient volumes to make land disposal a viable option.

Introduction

A need exists for safe, simple waste disposal techniques to deal with an ever increasing number of compounds designated as toxic substances. Current options for disposing of hazardous materials include biological degradation, adsorption, oxidation, incineration, precipitation and filtration, and land disposal. Dissposal via landfills is limited, in part, by the ability of the soil microorganisms to metabolize some of the more refractory chlorinated compounds. Chemical pretreatment offers some promise for rendering such molecules more amenable to soil disposal, if the pretreatment does not disrupt the ecological systems responsible for metabolism. Ultraviolet (UV) ozonation offer several advantages as a pretreatment step since it leads to products that appear to be more biodegradable and less likely to adversely alter the soil ecology. Generally, the effect of UV irradiation of chlorinated organic compounds is to liberate chlorine, which may be replaced by hydrogen, oxygen or a nucleophile. The resultant molecule may be more susceptible to dedgradation than the parent. An extensive review of ozonation and UV ozonation for the treatment of hazardous materials in waste water has prepared by Rice[1].

UV ozonation has been examined as a disposal technique for the degradation of 2,3,7,8 – tetra – chlorodibenzo – p – dioxin(TCDD) by Wong et al. [2], where 1 ppb was completely de-

* Reprinted from IUPAC Pesticide Chemistry Monographs published by Pergamon Press, 1983, 397 – 400.

graded. Prengle et al. [3] reported 100% destruction of PCP. as measured by loss in total organiccarbon levels, in a 70ppm solution. after 45 minutes by UV ozonation. These same authors reported a slower rate of decomposition for the polychlorinated biphenyls (PCBs) at 0.5 ppm in a saturated a queous solution. Arisman and Musick[4] examined the efficiency and cost efficiency of a commercial UV – ozone system to destroy PCB and found the cost of a full – scale treatment compared favorably with other current disposal methods, Commercially available high energy UV sources make photochemical pretreatment of agricultural and industrial wastes pssible.

The objective of the uresent study was to determine the extent of degradation in soil of aqueous solutions of four chlorinated compounds pretreated by UV – ozonation. Primary emphasis is directed toward studies on the isooctyl ester of 2,4,5 – trichlorophenoxy acetic acid [2,4, –5 – T], other compounds invest igated include pentachlorophenol (PCP), TCDD, and a PCB isomer, 2,5,2′,5′ – tetrachlorobiphenyl.

P. C. Kearney, J. R. Plimmer and Z. – M. Li

Materials and methods

Procedure

To study the effect of UV – ozonation pretreatment on soil metabolism, the four test compounds were spiked with their ^{14}C – labeled counterparts, irradiated for various time periods, and then the irradiated solution added to soil in a biometer flask[5]. Ten mL of the irradiated 1 ppm solution of 2,4,5 – T, PCP and PCB's were added to 50 g soil which represents a soil concentration of 0.2 ppm or about 0.14 ppm of the free 2,4,5 – T acid. For TCDD, the soil concentration was about 4 ppb. Metabolism was determined by measuring $^{14}CO_2$ evolution.

Soli

Soli selected for this study was Matapeake loam (Typic hapluldts) with the following properties: 38.4% sand, 49.4% silt, 12.2% clay, 1.5% organic matter, and pH 5.5. The soil was adjusted to pH 6.8 with $CaCO_3$ and amended with 1 g/kg glucose to stimulate microbial activity. Soils were maintained at 24 ± 2℃ for about 2 weeks before use. Just prior to use, soils were dried to 10% ~15% moisture, added to the biometer flask in triplicate and treated with 10 mL of the irradiated solution added uniformly over the soil surface. $^{14}CO_2$ was trapped in 10 mL of 0.2 N NaOH and 1 mL samples were removed for scintillation counting.

Chemicals

Formulated isooctyl ester of 2,4,5 – T(63% 2,4,5 – T) was obtained from the Pesticide Chemistry Laboratory, Environmental Protection Agency, Beltsville, Maryland. This sample was part of a 1979 survey for dioxins and contained 20 ppb TCDD. Pentachlorophenol was a technical grace sample containing 86% PCP, 10% other chlorinated phenols, and 4% inert ingredients. The labeled PCB was added to Arochlor 1242. Technical grade chemicals were selected to simulate the conditions most likely to be encountered in a waste disposal situation. Use of formulated 2,4,5 – T also permitted greater flexibility in examining concentration effects during photolysis,

All solutions were prepared in H_2O, with the exception of the PCB, which was made up in 10% methanol due to its low solubility.

The ^{14}C compounds added to the aqueous solutions of the four chlorinted test compounds were isooctyl ester of [U - ring]2,4,5 - T(11.86mCi/mM, 97% pure by TLC) purchased from New England Nuclear, Boston Massachusetts; [U - ring] PCP(14.05mCi/mM, > 98% pure by TLC) purchased from Mallinkrodt, St. Louis, Missouri; [U - ring] PCB(9.87mCi/mM, > 97% pure by TLC) purchased from (Mallinkrodt) and [U - ring]2,3,7,8 - TCDD(107mCi/ mM, 98% pure by TLC) purchased from, KOR Isotopes, Cambridge, Massachusetts.

UV - Ozonation

Laboratory studies employed a 450 W quartz mercury vapor lamp (Hanovia Catalog No. 679 - A - 36) housed in a water cooled double - walled quartz immersion well and a 300 mL reaction vessel fitted with a gas inlet tube. Ozone was generated in situ by slowly bubbling oxygen into the reaction vessel during irradiation. For comparison, nitrogen was bubbled into the system prior to and during irradiation to reduce the oxygen tension of the solution; more elaborate methods of removing oxygen were not undertaken. For ^{14}C - 2,4,5 - T and PCP, gases emanating from the reaction vessel during photolysis were trapped in 0.2 N NaOH and $^{14}CO_2$ determined by scintillation counting. To determine whether any volatile compounds were produced during UV ozonation, a small polyurethane foam plug[6] was placed in the glass inlet tube of the CO_2 trap. After irradiation the plug was extracted with ethyl acetate and assayed for ^{14}C.

A time course study was conducted on a 10 ppm 2,4,5 - T solution submitted to UV - ozonation for 0, 2, 5, 10, 30, and 60 min. The aqueous solutions were extracted with hexane and 2,4,5 - T lose measured by GLC. Samples were analyzed by ^{63} Ni electron - capture GLC. The column was 1.8 × 4 mm i. d. glass packed with 3% OV - 17 Gas Chorm. Q. Column emperature was 205℃ and the gas was CH_4—Ar(5:95) 50mL/min flow rate. Relative retention times against standards were used for qualitative analyses and peak heights for quantitative analyses. After 23 days iniubation time, soils were extracted with 150mL $CHCl_3$ - CH_3OH(1:1) and nonextractable residues measured in soil by combustion.

Large scale UV - ozonation was conducted in a Ultra - violet Purifier manufactured by Pure Waters Systems, Inc., 4 Edison Place, Fairfield. N. J. 07006. The unit consists of 66 low - pressure mercury vapor lamps with a maximum energy output at 2437Å. Each lamp is encapsulated in long quartz tubes, and the tubes are arranged so that each lamp is located at 1.27 cm from each adjacent lamp. The lamps are housed in a stainless steel cylinder approximately 40 cm in diameter and liquid is delivered to the lamp unit by a pump at a flow rate of 8 to 40 L/min. A large stainless steel holding tank (ca 210 L) is connected to the pump and the lamp unit is connected to the holding tank so that liquid can be recycled through the lamp unit. The total output of the lamp unit is about 2.244W.

The 450 W lamp was used to irradiate 2,4,5 - T at:ppm for 1 h in the presence of O_2 and N_2 (Fig. 1). Losses of 15% and 20% of $^{14}CO_2$ occurred during irradiation with solutions purged with N_2 and O_2 respectively. The pH of the solutions decreased from 6.0 to 4.8 after irradiation. No ^{14}C was detected in the polyurethane foam plugs. Rapid soil degradation of ring labeled 2,4,5 – T was observed during the first three days following UV – ozonation which amounted to 82.7% after 23 days. This rapid generation of $^{14}CO_2$ in soil suggested either metabolism of some very liable compounds or chemical breakdown of some unstable species on the soil surface. A subsequent study with sterile soil controls showed a 10.9% chemical degradation of 1 ppm 2,4,5 – T to $^{14}CO_2$ compared to 52.5% biodegradation from nonsterile soils after 4 days. Struif et al. [7] and Weil et al. [8] reported ozonation of 2,4,5 – T was very rapid at pH 8.0 and yielded glycolic acid, oxalic acid, glyoxylic acid, dichloromaleic acid, and Cl^- as oxidation products. These organic acids would be rapidly metabolized in soil. Partial exclusion of O_2 by the use of N_2 lowered the apparent rate of ring opening and subsequent soil metabolism which amounted to 64.1% degradation after 28 days.

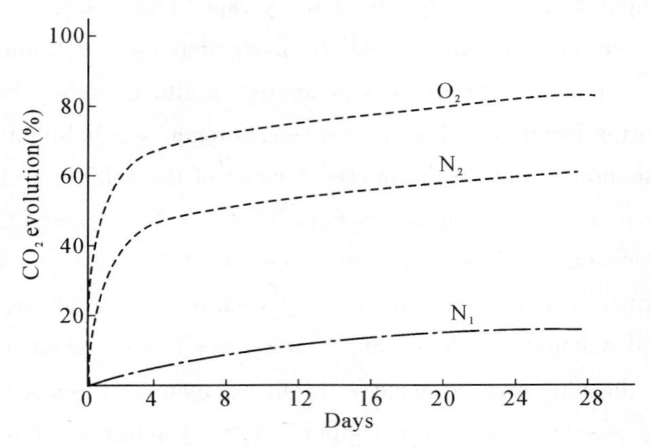

Fig. 1. Soil degradation of a 1 ppm aqueous solution of form ulated isooctyl exter of 2,4,5 – T subjected to UV photolysis (N_2) UV ozonation (O_2) and no irradiation (N_1) for 1 h.

A similar pattern of metabolism occurs when PCP at 1 ppm is irradiated for 1 h and then incubated in soil. Losses of 25% and 25% $^{14}CO_2$ occurred when PCP irradiated solutions were purged with O_2 and N_2, respectively. The pH after irradiation was 4.05 for the N_2 Parged solution and 3.75 for the O_2 Purged solution. No volacile ^{14}C compounds were detected. After 28 days in soil, the % of $^{14}CO_2$ crapped for nonirradiaced PCP was 20.7, O_2 73.1 and N_2 60.0.

Soil degradation of UV – irradiated 2,5,2′,5′ – tecrachlorobiphenyl in Arochlor 1242 was considarably less than 2,4,5 – T and PCP under identical conditions. Essentially no metabolism occurred in the nonirradiated (0.25%) or N_2 Purged solutions (1.5%) after 14 days. Only UV ozonation resulted in a significant loss of ^{14}C (27% after 14 days), and this occurred within the first 3 days. A similar pattern occurred with TCDD; after 28 days, the conversion to $^{14}CO_2$ was 1.5% nonirradiated, 3.1% N_2 and 12.7% O_2.

A time course study of 10 ppm 2,4,5 – T during the UV – ozonation phase revealed a rapid loss of the parent compound by GLC. Over 30% of the isooctyl ester of 2,4,5 – T nad disappeared after 10 min. The ring portion of the molecule degraded more slowly during UV – ozonation, and about 40% of the ring ^{14}C was lost after 1 h. Most of this ring ^{14}C was trapped as $^{14}CO_2$.

Table 1 shows the distribution of ^{14}C from irradiated ^{14}C – 2,4,5 – T in soil after 23 days. 4

progressive increase in $^{14}CO_2$ occurred from soil as the pretreatment time of UV – ozonation increased . The nonextractable residues in soil determined by combustion were larger when compared to unirradiated 2,4,5 – T solutions. The extractable products decreased with time of prior irradiation

Table 1　Distribution of ^{14}C after UV – ozonation of 10 ppm isooctyl ester of [U – ring]2,4,5 – T added to soil after 23 days

Irradiation time/(min)	Evolved $^{14}CO_2$	Extracted	Nonextractable Residues	Total
0	12.3	—	15.2	—
2	16.3	27.3	27.8	71.4
5	25.1	11.8	35.1	72.0
10	40.9	0.5	32.8	74.2
30	51.3	0.0	33.2	84.4
60	67.1	0.0	28.1	95.2

Results of a preliminary run using the large scale UV unit are shown in Table 2. A 10 ppm 2,4,5 – T solution (80L) was passed through the 66 lamp chamber at a rate of 5 – 22L/min, Oxygen was piped directly into the lamp chamber at a rate of 100 mL/min. Substantial decomposition of the ring structure of 2,4,5 – T occurred during the UV – ozonation stage, and amounted to 90% in 60 min. The extent of soil degradation, measured by $^{14}CO_2$ evolution , was dependent on pretreatment time and soil incubation time. An increase in both resulted in more soil degradation.

Table 2　Large scale (80L) UV – ozonation of 10 ppm isooctyl ester of [U – ring]2,4,5 – T in a 55 lamp unit

Dwell time(min)	^{14}C Recovered in solution(%)	Soil Degradation Day 4	$^{14}CO_2$ Evolution Day 8
0	100	1.0	2.3
20	43	16.9	26.6
40	26	33.7	57.0
60	20	66.6	103.0

Much remains to be done with the large system. Nevertheless the economics of operation, i. e, the energy use is estimated to be 1.5kW · h and if this hourly cost is 7cents (US)/hr. (based on 5cents/kW · h), then the total cost for I week's operation (7days at 24h/day) would be $12, make this option attractive from a cost standpoint.

REFERENCES

[1] R. G. Rice. Ozone for the Treatment of Hazardous Materials in Water. 1981. G. F. Bennett, Ed, A. I. Ch. E. N. Y. N. Y. In press(1981).

[2] A. S. Wong. M. W. Orbanosky, and W. J. Luksemburg, 178th Amer. Chem. Soc. Natl Mtg, Pest – 83, (1979).

[3] W. J. Prengle, Jr. , C. E. Mauk, and J. E. Payne, in Forum on Ozone Disinfection, E. G. Fochtman, R. E. Rice, and M. E. Browning. eds. Intl. Ozone Assoc. , eveland, Ohio. p 286 – 295(1977).

[4] R. K. Arisman, and R. C. Musick. Demonstration of Waste Treatment processes for the Destruction of PCB's and PCB Substitutes in an Industrial Effluent. Draft Report, Grant No. s – 804901, US EPA Research Triangle Park, N. C. 59 pp . 979.
[5] R. Bartha, and D. Pramer, Soil Sci. 100. 68. (1965).
[6] P. C. Kearney, and A. Kontson. J. Agric. Food Chem. 24. 424(1976).
[7] B. Struif, L. Weil and K. E. Quentin. Z. Wasser – und Abwasser – Forschung. 11. 118. (1978).
[8] L. Weil, B. Struif , and K. E. Quentin. wasser Berlin' 77. AMK Berlin. p. 294 – 307. (1978).

Mention of a proprietary product does not constitute endorsement by the United Sates Department of Agriculture.

内吸杀菌剂三唑醇立体异构体研究初报*

邵瑞链　李正名　陈宗庭　王笃祜　谢龙观

（南开大学元素有机化学研究所）

三唑醇（1－（4－氯苯氧基）－3，3二甲基－1－（1H，1，2，4－三唑－1－基）丁醇－2）是一个优良的内吸广谱杀菌剂，特别是用于谷类作物的种子处理，防治效果尤为突出[1~6]。它的分子式为

```
            H(a)      OH
             |         |
Cl—⟨苯环⟩—O—C1*——C2*— C(CH3)3
             |         |
             N        H(b)
           /   \
          ‖     N
          N_____‖
```

此化合物中有两个不相似的不对称碳原子，故存在四个立体异构体，即1S、2S、1R、2R；1S、2R，1R、2S；其中一对对映异构体为苏式（Threo form），另一对对映异构体为赤式（Erythro form），苏式与赤式互为非对映异构体。由于尚未测定异构体的绝对构型，我们把其中一对称为A体，另一对称为B体。

据文献报道[7~9]，上述非对映异构体对大麦散黑穗病、腥黑穗病和大麦白粉病在杀菌活性上表现很大差异。A体的亲水性强，在氢核磁共振谱上$H_{(a)}$和$H_{(b)}$之间的偶合常数小。A体具有较高的杀菌活性，而B体活性较低。所见专利中未有分离方法和B体制备方法的报道。

为研究杀菌剂分子中空间因素对药效的影响，进一步探讨化学结构与活性的关系，我们对三唑醇的非对映异构体采用薄层、离心薄层层析、柱层析和重结晶等法分离提纯，并用核磁共振谱进行分析鉴定，所得非对映异构体的纯度在98%以上。

一、实验方法

1. 样品来源

（1）由三唑酮经硼氢化钠还原[1]所得产物，为微黄色固体，熔点108~112℃。

（2）德国进口样品DS15% Baytan用丙酮萃取后浓缩，红色固体。

（注：红色染料鉴定为颜料，系警戒色，推想为安全目的而用）

2. 薄层层析法

吸附剂：硅胶GF（青岛海洋化工厂）

展开剂：乙醚或乙酸乙酯

紫外灯显色

由三唑醇薄层展开结果看来，无论是德国进口样品和我们所合成的三唑醇在薄层层析

* 原发表于《农药》，1983，5：2－3，14。

上都展示出 R_f 值相近的两个斑点。亲水性的 A 体 R_f 值较小，极性较小的 B 体具有较大的 R_f 值。

3. 离心薄层层析法

仪器：北京青云仪器厂生产 LBC－1 离心薄层层析仪

薄层板制备：20 克硅胶 GF 254 与适当比例水混合，涂成 1 毫米厚度的转子硅胶板，放置过夜、晾干，在 60～90℃活化 3 小时。

分离条件	进样量	100 毫克
	洗脱剂	乙酸乙酯或乙醚
	洗脱剂流速	3.5 毫升/分
	分离时间	30～45 分钟
	洗脱剂用量	150 毫升

分离结果，紫外灯观察硅胶 GF 层上呈现两条色谱环，依次接收组分，分别得到 A、B 异构体纯品和部分混合体。

4. 柱层析：长 60 厘米直径 3 厘米玻璃柱，内装 120 克（100～200 目）硅胶，湿法装柱，样品 4～5 克。

淋洗剂先用正己烷，逐渐增大极性，改用正己烷和乙酸乙酯（或乙醚）的混合液，比例从 4∶1，2∶1，1∶1，0∶1 分段接收若干组分，各组分用薄层层析鉴定，而后根据 A、B 异构体组分合并浓缩，所得 A、B 体尚需进一步用异丙醇重结晶后方可得到纯品。

二、结果和分析

上述层析法分离所得非对映异构体经多次重结晶至熔点不变，经元素分析和核磁共振谱鉴定，证明它们是所推测的结构，结果如下：

1. A 体：白色结晶，熔点 138～139℃

元素分析 $C_{14}H_{18}ClN_3O_4$（相对分子质量 295.8）

Cl(%)	理论值 12.01	实验值 12.05	12.14
N(%)	理论值 14.18	实验值 14.02	14.31

核磁分析

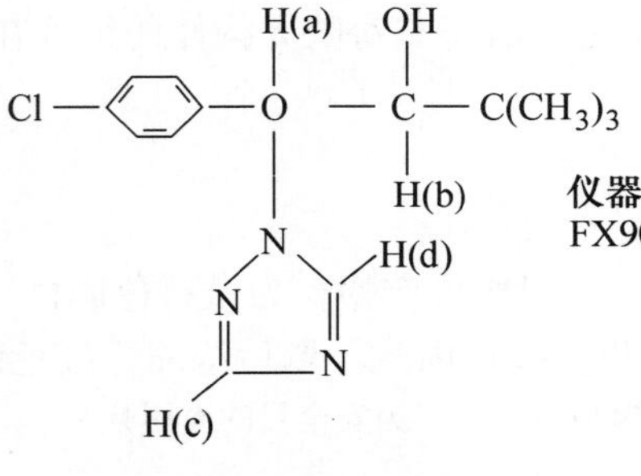

H(a) OH

Cl—⟨苯环⟩—O — C — C(CH$_3$)$_3$

H(b)

N H(d)

N N

H(c)

仪器JEOL—FX90QNMR

δ(ppm) CD_3OD

H(a) 6.47 (J_{ab}=2Hz)

H(b) 3.60

H(t) (Bu-t)1.02

H(c) 8.07

H(d) 8.67

2. B 体：白色结晶，熔点 132～133℃

元素分析

Cl(%)　理论值　12.01　　实验值　12.18　12.03
N(%)　　　　14.18　　　　　14.20　14.23

核磁分析　δ（ppm）CD_3OD
H（a）6.36
H（b）3.94
H（t）0.92　　（J_{ab} = 4.8Hz）
H（c）8.04
H（d）8.75

比较A、B体的氢核磁共振谱可看出，H(a)和H(b)之间的耦合常数在两个异构体中不同，A体为2赫兹，B体为4.8赫兹。我们知道相邻的碳原子上氢的偶合常数J_{ab}的大小主要是和两个氢之间的两面角（立体角）有关，根据交叉构象为最稳定的构象可作出稳定的苏式和赤式异构体。

由karplus半经验方程不难推断赤式构象的J_{ab}值将高于苏式构象，故我们初步认为A体为苏式，1S、2R；1R、2S对映异构混合体。B体为赤式，1R、2R；1S、2S对映异构混合体。

3. 三唑醇中A体和B体的相对含量分析

由于A体和B体中对应氢原子在NMR谱上化学位移不同，故我们可根据其积分曲线求出两对非对映异构体在三唑醇中的相对含量。下面列举由积分曲线计算所得的几个样品分析结果。

样品编号	来　源	相对含量（%）	
		A体	B体
BT_A	柱层析后重结晶二次所得样品熔点139℃	>98	<2
BT_B	柱层析后重结晶二次所得样品熔点132～133℃	<2	>98
$BT_{进口}$	德国进口样品DS15%丙酮萃取液浓缩	64	35
BT_1	国内样品1号	66	34

我们将A、B异构体分别进行了生物测定，初步试验表明，A体对小麦锈病的药效要比B体高得多，研究工作正在进行中。

（**附：**本实验完成后看到1982年国外进行了三唑醇非对映体的分离工作[10]，但实验方法和条件与我们不同）。

致谢：核磁共振谱由李国玮、张殿坤、陆秀菁同志协助完成；在薄层分析工作中，钱宝英同志提了宝贵意见，生测结果由张祖新同志提供，在此一并致谢。

参考文献

[1] Ger. Offen. 2，324，010（1975）；CA 82 156325（1975）.
[2] Ger. Offen. 2，632，603（1978）；CA 88 170152（1978）.
[3] Moere. M. S.，et al. *Proc. N. Z. Weed Post Control Conf.* 32nd 267～71 1979；CA 91 205529（1979）.
[4] Moere. M. S.，et al. *Proc. N. Z. Weed Post Control Conf.* 32nd 272～7（1979）；CA 91 205530（1979）.
[5] Tracyner－Born，J. et al. *Pflanzehscl. ulz－Nackr.* 31（1）25～38（1978）；CA 92 70903（1980）.

[6] Wainwright, A. et al. *Proc. Br. Crop. Prot. Coaf – Pests* Dis. （2）565 – 74（1979）；江燕敏《农药译丛》6 46（1982）.
[7] Ger. Offen. 2, 743, 767（1979）；CA 91 20518（1979）.
[8] Gasztonyi, *Maya Postic. sci.* 12（4）433 ~ 8（1981）.
[9] Wolfgang Krämer, et al.. U. S. P. 4, 232, 033（1980）.
[10] A. H. B. Deas, et al.. *Postic. Biochem. Physiol*, 17（2）, 120 ~ 33（1982）；CA 96 175967（1982）.

棉铃虫拟信息素正十四烷基甲酸酯及有关化合物的触角电位反应*

任自立　王银淑　尚稚珍　刘天麟　李正名　郭虎森

（南开大学元素有机化学研究所）

棉铃虫是我国和世界许多地区棉花的主要害虫。分离、鉴定棉铃虫性信息素组分并寻找拟性信息素作为对棉铃虫的一种综合防治方法已引起人们的重视。1974 年，Roelofs 等从棉铃虫（H. virencens）雌体中分离鉴定出(Z) - 9 - 十四碳烯醛和(Z) - 11 - 十六碳烯醛作为其性诱组分。并指出这两种组分的混合物在试验室和田间诱引同种雄蛾，次年，Sekul A. A 等从棉铃虫（H. zea）中分离鉴定出(Z) - 11 - 十六碳烯醛。后来，Nesbitt 对棉铃虫（H. amigera）性信息素组分作了系统的研究，分离鉴定出(Z) - 11 - 十六碳烯醛和(Z) - 11 - 十六碳烯醇两个化合物，并对此化合物作了电生理和田间诱蛾试验，发现有一定的生物活性。最近，Kydonieus 和 Kitterman 合成了正十四烷基甲酸酯，并报道了这个化合物作为拟性信息素能使雌蛾迷向，打断棉铃虫的交尾活动，进而破坏棉铃虫的生殖循环，因为这两个化合物结构简单、性质稳定、制备简便，值得进一步研究。我们按照 Kydonieus 的方法将正十四碳醇和甲酸在苯中回流脱水，制备了这个化合物，使用类似的方法还制备了另外八个有关的化合物（见表 1）对上述化合物作了触角电位的生理测定，现将试验结果报告如下。

表 1　正十四烷基甲酸酯及有关化合物

编号	化　合　物	沸点(℃/mm)	收率(%)	文献值沸点(℃/mm)
82501	$C_{14}H_{29}—O—C(=O)—H$	127 ~ 8/0.3	58.2	102 ~ 108/0.25[4]
82502	$C_{14}H_{29}—O—C(=O)—CH_3$	116 ~ 118/0.3	70.0	175.5/15 Beilstein，2，136
82503	$C_{14}H_{29}—O—C(=O)—C_2H_5$	151 ~ 3/0.3	63.0	C(%)计算值 75.48，实验值 75.35 H(%)计算值 12.69，实验值 12.75
82504	$C_{12}H_{25}—O—C(=O)—H$	145 ~ 7/1.5	70.0	145 ~ 146/15 Beilstein，2，Ⅱ，32
82505	$C_{12}H_{25}—O—C(=O)—CH_3$	127 ~ 9/3	54.3	150.5 ~ 151.5/15 Beilstein，2，136

* 原发表于《昆虫激素》，1984，1：43 - 46。本课题（科基金化 457 号）得到中国科学院科学基金委员会的资助和支持。

续表 1

编号	化合物	沸点(℃/mm)	收率(%)	文献值沸点(℃/mm)
82506	$C_{12}H_{25}—O—C(=O)—C_2H_5$	125~7/3	46.0	166~168/20 Beilstein, 2, Ⅱ, 222
82507	$C_{16}H_{33}—O—C(=O)—H$	138~140/0.3	72.0	188/17 Beilstein, 2, Ⅱ, 32
82508	$C_{16}H_{33}—O—C(=O)—CH_3$	144/0.3	67.7	199.5~200.5/15 Beilstein, 2, Ⅱ, 146
82509	$C_{13}H_{27}—C(=O)—H$	柱层分离	84.9	155/10 Beilstein, 1, Ⅱ, 770

材料和方法

试验用虫是中国农科院植保所提供的室内饲养棉铃虫，然后将蛹逐个鉴别雌雄，分别置于小玻璃瓶内，保持一定温度，让其在室温下羽化，每天收集羽化雄蛾，用 1~2 日龄作触角电位测试用。

将供试的九种化合物分别用 CH_2C1_2 配成 10^{-2}、10^{-1}、10^0、10^1、10^2、10^3（μg）六个浓度，每剂量取 10μL 加到样品管的滤纸片上，以 CH_2Cl_2 作空白对照，待溶剂挥发后，即可进行触角电位试验。从雄蛾头部取下带有部分头部组织的触角，剪去顶端 2~3 节，把记录电极的玻璃管套进触角端部，用 Ag－AgCl 玻璃电极作记录，形成闭合电路，电极接放大器，信号经放大器放大，引进示波器和记录仪，记下不同化合物不同剂量所引起的触角电位。

结果和讨论

一、九种化合物的 EAG 反应

试验结果表明，通过九种化合物九次反复试验，能诱发雄蛾 EAG 反应，剂量都是用 10^3 μg（即 1mg），其活性以 82504、82505、82506、82501 四个化合物为好，如图 1 所示，由于不同个体雄蛾的生理状态差异，对同一化合物同一剂量触角电位反应也不相同，但几种化合物触角电位活性的相对大小是一致的。

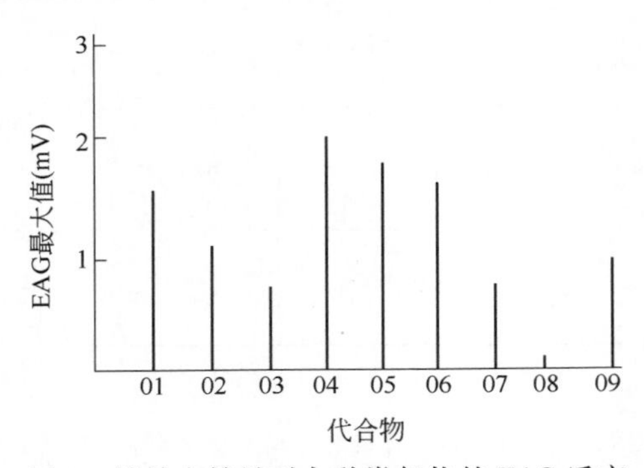

图 1　棉铃虫雄蛾对九种类似物的 EAG 反应

二、对不同剂量性信息素的 EAG 反应

我们在初试中活性较好的四个化合物，又用六个不同剂量进一步测试，其结果 82504 的反应最低阈值为 100μg，而 82505、82506 的为 10^1μg，82501 则是 10^2μg，如图 2 所示，四个化合物的活性大小的相对关系，82504 为其相同的最高浓度或最低浓度都是最佳的，而 20FE 纯性信息素量以 1000ng（假设每个雌蛾含纯性信息素 50ng，20 头蛾子含 $50 \times 20ng = 1000ng$ 即 1μg）来算，则 82504 的 10μg 的活性相当 1μg 性信息素的活性，说明 82504 的活性比粗提物低 10 倍。

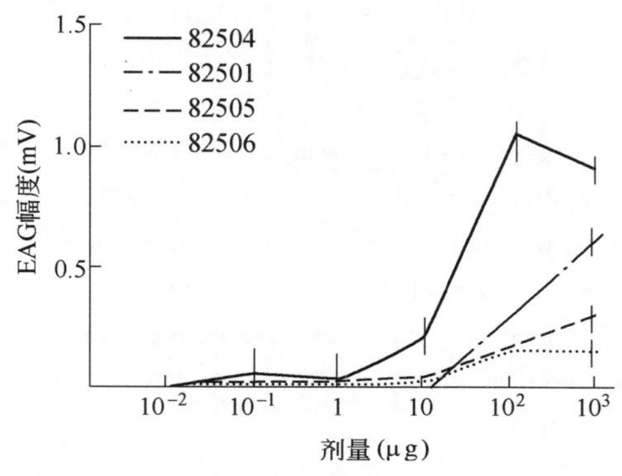

图 2　棉铃虫雄蛾对四种有活性的化合物不同剂量性信息素的 EAG 反应

从化学结构看 82504、82505 和 82506 分别为十二碳醇的甲酸酯，和 n – tetradecyl formate（即 82507）相比较，则易于挥发，在同样电生理测试条件下，触角接触到 82504 的分子要相对多些，因此电位反应较大，而这种化合物在田间是否引起比 82507 大的打断交尾作用，尚待田间试验证明。

另外，从上述试验中可以初步看出 EAG 反应与化合物结构有一定关系，为了进一步探索结构与 EAG 活性的规律性，需要对更多的类似物进行试验。

参 考 文 献

[1] Roelofs，Wendell L et al.，Life Sci.，14（8），1552 – 62（1974）；C. A.，82，40888.

[2] Sekul，A. A. et al.，J. Econ. Entomol.，68（5），603 – 4（1975）.

[3] Nesbitt，B. E.，J. Insect Physiol，6，535（1979）.

[4] Agis F Kydonieus，Roger L Kitterman U. S. 4，272 – 520（1981）.

A Convenient Synthetic Route for the Sex Pheromone of the Asian Corn Borer Moth (*Ostrinia furnacalis Guenée*) *

Zhengming Li[1] (李正名), Meyer Schwarz[2]

([1]The Research Institute of Elemento - Organic Chemistry, Nankai University, Tianjin;
[2]OCSL, AEQI, BARC, USDA, USA)

Abstract By hydroboration of 10 - undecen - 1 - ol acetate, a new synthetic route for the sex pheromone of the Asian corn. borer (*Ostrinia furnacalis Guenée*) was devised to shorten the original 5 steps to a 2 steps reaction, The E - and Z - 12 - tetradecen - 1 - ol acetates were obtained separately with a purity greater than 98%. The location of the double bond position was ascertained by capillary GC and GC - MS.

The important intermediate 12 tetradecyn - 1 - ol was also obtained from 2 tridecyn - 1 - ol by using a zipper reaction, then by methylation and acetylation. The intermediate obtained from 3 different routes were compared by GC and GC - MS spectra to be identical.

INTRODUCTION

The Asian corn borer (*Ostrinia furnacalis Guenée*) is one of the major pests reducing the corn production in the Far East. Klun et al. [1] identified the structure of the sex pheromone of the female species as 1:1 blend of (*E*) and (*Z*) - 12 - tetradecen - 1 - ol acetates. These compounds have the same carbon skeleton as the sex pheromone of the European corn borer, *Ostrinia nubilalis* (*Hübner*), except that the latter has a double bond situated at carbon 11. In the same period, Ando et al. [2] in Japan using a different approach identified the sex pheromone of the Asian corn borer as the same compounds: (*Z*) and (*E*) - 12 - tetradecen - 1 - ol acetates in a ratio of 3:2. Although both reports recorded their biological response in the laboratory bioassay, the reports on field tests differed strikingly. Ando et al. stated that the 2 synthetic compounds of (*Z* + *E*) tetradecen - 1 - ol did not attract the male moths, while Klun et al. reported that through preliminary field test in Honan Province, the People's Republic of China, an average of 9.8 Asina corn borer males/trap was caught in 1979[1]. Further tests were being undertaken in 1981 in different provinces of the People's Republic of China[3]. After the experiment in this paper was concluded, we noticed that the sex pheromone components from the Asian corn borer have also been identified independently by Cheng Zhiqing et al. [4] in the People's Republic of China to have the same structures given above.

Ⅰ. METHODS AND MATERIALS

1. Instrumentation

1) Capillary GC: Hewlett - Packard 5880a equipped with a fused silica open tubular capillary

* Peprinted from *Scientia Sinica* (*Serie B*), 1984. 1:38 - 43.

column (SP – 1000 or Durabond, 60M × 0.25mm). H_2 pressure 25psi.

2) HPLC: Waters Associates M – 6000.

3) Computerized GC – MS: a Finnigan model 4000 with 6000 data system, electron impact mass spectra obtained at 70 eV with a source temperature at 270℃.

2. Synthetic Methods

It is interesting to note that among the sex pheromones of lepidoptera the monounsaturated compounds with unsaturation at even – numbered position in the carbon chain are extremely rare[1]. To stereospecifically synthesize the (Z) and (E) – 12 – tetradecen – 1 – ol acetates separately, Klun et al. employed a 5 steps route starting from 6 – heptyn – 1 – ol[1]. Ando et al. used a 6 steps route starting from propargyl alcohol[2]. In view of the recent progress in hydroboration reactions [5] and the report by Brown and Wang of a one – pot reaction to synthesize olefins via organoboranes[6], we adopted a 2 steps route, starting from 10 – undecen – 1 – ol acetate (reactions a, b, c were carried out in a one – pot operation, Fig. 1).

one–pot reaction:

(a) $CH_2{=}CH{-}(CH_2)_2{-}OAc + BH_3/THF \xrightarrow{0℃} [AcO(CH_2)_{11}]_3B$ — I → II

(b) $II + LiC{\equiv}C{-}CH_3 \rightarrow [(AcO(CH_2)_{11})_3B{-}C{\equiv}C{-}CH_3]Li$ — III

$III + I_2 \rightarrow \left[\begin{matrix} AcO(CH_2)_{11} & & CH_3 \\ & C{=}C & \\ [(AcO(CH_2)_{11}]_2B & & I \end{matrix} \right]$

(c) ↓

$AcO(CH_2)_{11}{-}C{\equiv}C{-}CH_3$ (IV)

(d) P-2 nickel ↙ (e) Na/NH$_3$ ↘

$AcO(CH_2)_{11}{-}\underset{H}{C}{=}\underset{H}{C}{-}CH_3$ Z–12–tda (V)

$AcO(CH_2)_{11}{-}\underset{H}{C}{=}\overset{H}{C}{-}CH_3$ E–12–tda (VI)

Fig. 1 2 steps synthetic route

In the reaction, an intramolecular 1,2 – migration occurred, resulting an alkyl substituent migrated from boron to the neighboring carbon atom. Reaction c possibly involved an electrophilic attack of iodine on the triple bond followed by a subsequent dehaloboration to form the acetylene intermediate – 12 – tetradecyn – 1 – ol acetate (Ⅳ)[7]. Ⅳ was further hydrogenated with P – 2 nickel catalyst[8] to form the (Z) – tetradecen – 1 – ol, or reduced stereospecifically by sodium and liquid ammonia to the (E) – isomer[9]. Both isomers were further purified by HPLC, monitored by capillary GC. (SP – 1000), to obtain seqarately pure (Z) and (E) isomers with a purity greater than 98%. In neither case was the other geometrical isomer present as a contaminant.

3. Structural Identification of the Products

On capillary GC (SP – 1000 and Durabond, all 60 meters, i. d. 0.25 mm), (Z) and (E) – tetradecen – I – ol acetates (12 – tda) have the identical retention time as the authentic samples supplied by OCSL. AEQI, BARC, USDA (Fig. 2 and Table 1). The (*E*) – isomer elutes earlier than the (*Z*) – isomer as reported earlier[1,2] and the GC – MS spectra of both isomers obtained via 5 steps and 2 steps routes are identical too. Aliphatic acetates usually have typical peaks which were also observed here at $m/z = 61$ ($CH_3COOH_2^+$) and 43 ($C_2H_3O^+$). According to the linear relationship established between the ratio of the abundance *m/z* 55/54 and the different location of double bond in mono – unsaturated acetates[10], we compared our products with the earlier products via a 5 steps route and the following results were obtained (Table 2).

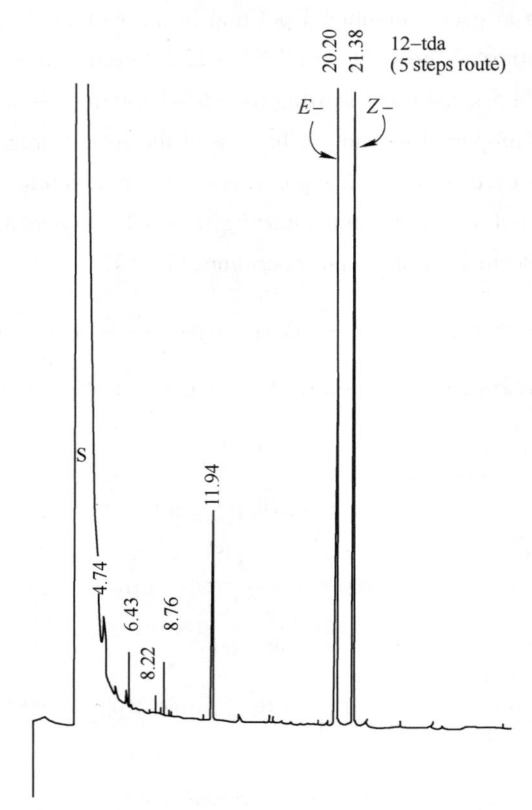

Fig. 2 RT of *E* and *Z* – 12 – tda obtained via 5 steps route on a SP – 1000 capillary column

Table 1 Comparison of Retention Time of Both Isomers via Different Routes

12 – tda	*E* –	*Z* –	Lit.
5 steps route	20. 20	21. 38	[1]
2 steps route	20. 18	21. 36	

RT on SP – 1000 (60M × 0. 25mm) in minutes.

Table 2 Mass Fragments Ratio of *Z* – and *E* – Isomers

m/z ratio / isomer / 2 – tda	55/54		Lit.
	E –	*Z* –	
5 steps reaction	2. 49	2. 58	[10]
2 steps reaction	2. 48	2. 57	

The conformity of the above data suggested that the double bonds in our products are also in the 12th position.

In order to unequivocally ascertain the structure of our products the (*E*) and (*Z*) – 12 – tetradecen – 1 – ol acetates were processed by the recent analytical technique developed by Bierl – Leonhardt; Devilbiss and Plimmer[11]. Each of 200 ng. of the synthesized (*E*) and (*Z*)12 – tetradecen – 1 – ol acetate was treated with m – chloroperbenzoic acid to form the corresponding epoxide, which was then injected into a combined gas chromatograph/mass spectrometer for analysis. The preferential cleavage alpha to the epoxy group produced the characteristic mass fragments $m/z = 55$ and 255, which are the same as obtained from 5 steps route product (Fig. 3). Further oxidation of these epoxides by periodic acid (HIO_3, predried at 100℃ for 12hr) yielded the expected aldehyde acetates which have the same fragments as previously reported[11] (Table 3).

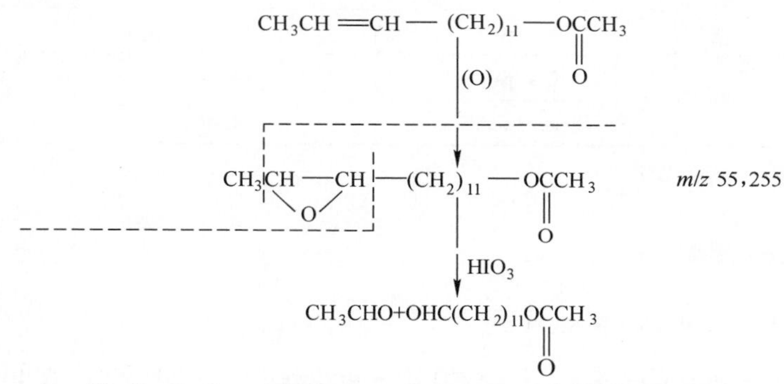

Fig. 3 Preferential cleavage of epoxy compounds of 12 – tda in GC – MS (1980 Bierl – Leonhardt, Devilbiss, Plimmer)

Table 3 $OHC(CH_2)_{11}OAc$ Fragments; m/z (% Abundance)

Epoxide of	M^+	$M^+ - 43$	$M^+ - 78$	$M^+ - 88$	$M^+ - 103$	Lit.
E – 12 – tda (5 steps route)	242	199 (4)	164 (2)	154 (5)	139 (11)	[11]
E – 12 – tda (2 steps route)	242	199 (2.56)	164 (1.25)	154 (3.57)	139 (8.22)	
Z – 12 – tda (2 steps route)	242	199 (0.62)	164 (0.54)	154 (0.57)	139 (2.09)	

These data further ascertained that our products are identical to the 2 isomers of 12 – tetradecen – 1 – ol acetates synthesized before by a 5 steps route. Referring to the procedure modified recently by Zoecon Co. [12], the intermediate 12 – tetradecyn – 1 – ol acetate (Ⅳ) was also successively synthesized by using a zipper reaction to obtain 12 – tridecyn – 1 – ol from 2 – tridecyn – 1 – ol (obtained from lithio dodecyne – 1 and formalde – hyde[13]), then further methylated and acetylated (Fig. 4).

By capillary GC (Durabond, 60M × 0.25mm i. d.), the retention time of 12 – tridecyn – 1 – ol acetate obtained via zipper reaction was compared with the intermediate obtained via 5 steps and 2 steps routes to be identical (Table 4). The GC – MS spectra from 3 different routes are identical too. Nevertheless, the zipper reaction offered a product with higher purity than the other 2 routes. The relating details involved in the zipper reaction will be discussed later.

$$CH_3(CH_2)_9C\equiv C-CH_2OH + NaH \xrightarrow{H_2NCH_2CH_2CH_2NH_2} HC\equiv C-(CH_2)_{11}-OH \quad (VII)\ 72\%$$

$$HC\equiv C-(CH_2)_{11}-OH + \text{(dihydropyran)} \xrightarrow{H^+} HC\equiv C-(CH_2)_{11}OTHP \quad (VIII)\ 68\%$$

$$HC\equiv C-(CH_2)_{11}-OTHP + CH_3I \xrightarrow[HMPA]{CH_3Li/THF} CH_3C\equiv C(CH_2)_{11}OTHP \quad (IX)\ 70\%$$

$$CH_3C\equiv C(CH_2)_{11}OTHP + (CH_3CO)_2O + CH_3COOH \longrightarrow CH_3C\equiv C-(CH_2)_{11}OC(=O)CH_3 \ (IV) \quad 63\%$$

Fig. 4 Synthetic route of 12 – tetradecyn – 1 – ol acetate via zipper reaction

Table 4 Comparison of RT and Purity of 12 – tetradecyn – 1 – ol Acetate From 3 Different Routes

Sample No.	Route	RT(min)	Purity(%)
1	5 steps	14. 15	89. 08
2	2 steps	14. 14	93. 19
3	zipper reaction	14. 01	94. 70

$H_3C\cdot C\equiv C(CH_2)_{11}OAc$(Ⅳ).

Ⅱ. EXPERIMENTAL

1. 10 – undecen – 1 – ol Acetate(Ⅰ)

In a 250mL flask, 40mL (34. 4 gm, 0. 2 M) 10 – undecen – 1 – ol(99% Aldrich Co.) 50mL acetic anhydride and 2mL pyridine were stirred at room temperature. The reaction was continuously monitored by GC(hp 5730A), and the reaction was completed after 12h. Excess acetic anhydride was evaporated *in vacuo*, and the residue was poured into a 500mL separatory funnel, then 150mL water and 100mL ether were added. The ether solution was washed with 3 × 100mL distilled water and dried over anhydrous magnesium sulphate. After removal of the solvent, the residue was vacuum distilled under nitrogen protection : b. p. 100℃/0. 7 mm or 118℃/2 mm. , lit. b. p. 107 – 108℃/2 mm[14]. Sample purity checked by GC > 99%. Yield 35. 2gm(82. 1%).

2. 12 – tetradecyn – 1 – ol Acetate(Ⅳ)

In a 4 – necked 250mL flask, equipped with nitrogen inlet, outlet, silicon rubber septum and overhead stirrer, the reaction was cooled in ice – bath under nitrogen protection.

12. 7gm (0. 06M) I and 30mL tetrahydrofuran (freshly distilled from $LiAlH_4$) were added. Under N_2 protection, 20mL BH_3/THF solution (1M solution, Aldrich Chemical Co.) was added gradually by a syringe. The reactants were stirred at room temperature for 1. 5h and then cooled to 0℃. The solution remained clear.

In another dry nitrogen – flushed 100mL 4 necked flask, under N_2 protection 12. 5mL (0. 02M) of 1. 6M butyl lithium/hexane solution (Aldrich Chemical Co.) was injected by a syringe. A very small piece of triphenylmethane was added as indicator. Then propyne gas was bubbled in and the solution turned from red to white within 1 min. The contents of this flask were carefully transferred to the 250mL flask under N_2 protection. The whole reaction was cooled in dry ice – acetone bath. A 30mL tetrahy – drofuran solution containing 5. 1 gm iodine (0. 02 M) was gradually delivered via syringe into the solution. The whole solution was stirred at −70℃ for 1. 5h and remained red – tinted. Finally it was allowed to warm up to room temperature for 2h. It was then stored at 0℃ for 48h. The volume of the reaction mixture was reduced to 1/3 under vacuum. Thirty mL methanol and 60mL heptane were added to dissolve the residue. The organic phase was washed with 60mL dilute sodium thiosulphate solution (1%) to remove the residual iodine (red color disappeared completely). The solution was dried over anhydrous magnesium sulphate. The crude product of 12 – tetradecyn – 1 – ol acetate 14. 88 gm, from which 5. 9 gm was taken for vacuum distillation : b. p. 154℃/0. 5 – 0. 7 mm, 2. 0 gm. Yield 43. 5%. There was a residue (3. 2 gm) which could not be distilled above 300℃. The product has identical GC and GC – MS spectra with the authentic sample. GC – MS date: *m/z* 252 (0. 09%) [M^+], 192 (0. 45%) [M^+—CH_3COOH], 164 (0. 65%) [M^+—(CH_3COOH) + C_2H_4], 163 (1. 5%) [M^+—(CH_3COOH + C_2H_5)], 150 (1. 45%) [M^+—(CH_3COOH + C_3H_6)], 149 (2. 65%) [M^+—(CH_3COOH + C_3H_7)], 68 (100%), 61 (9. 0%) [$CH_3COOH_2^+$], 43 (94. 7%) [$C_2H_3O^+$].

3. (*Z*) −12 – tetradecen –1 – ol Acetate (Ⅴ)

A two – necked 50mL flask was first thoroughly flushed with hydrogen gas. Powdered 0. 031gm (0. 125mmol) nickel (Ⅱ) acetate – tetrahydrate (Fisher certified grade, 99. 6%) was added along with 15mL 95% ethanol, then sodium borohydride solution (0. 125mL of 1 M ethanolic solution, 0. 125mmol) and 0. 015gm (0. 25mmol) of ethylene diamine were added sequentially. Finally 0. 24gm (1mmol) 12 tetradecyn – 1 – ol acetate (Ⅳ) was injected by syringe. Hydrogenation was carried out under the pressure of one atmosphere for 55 min, the process being monitored by GC. Upon completion of the hydrogen uptake. 25mL each of water and hexane were added. The aqueous solution was repeatedly extracted with ether and hexane, which were combined and washed with brine and water, dried over anhydrous magnesium sulphate. After evaporation *in vacuo*, a light yellow – tinted liquid was obtained, 0. 2798gm. GC analysis indicated it contained *Z* – isomer 81. 8%, *E* – isomer 5. 1%. The product was further purified by HPLC on silver nitrate treated silica column (16. 0gm spherisorb 10μm. 20% $AgNO_3$ column 10in × 0. 5in). Eluant: toluene; flow rate 3. 0mL/min. Each run carried 20μL sample. The contents of 12 runs were combined and the solvent was removed in a rotary evaporator. A colorless residue was obtained, 0. 13370gm. On capillary GC (SP – 1000, 60M × 0. 25mm, H_2 = 20psi, T_1 = 90℃, Time 1 = 0. 35min, rate 30℃/min, T_2 = 180℃, Time 2 = 30min), its retention time is 21. 38min, which coincides with that of the authentic sample (21. 36 min). Analysis by capillary GC showed its concentration 98. 6% (Fig. 6), yield 52. 26%. GC – MS data: *m/z* = 254 (0. 2%) [M^+], 194 (7. 58%) [M^+—CH_3COOH], 138 (4. 04%) [M^+—(CH_3COOH + C_4H_8)], 137 (3. 32%) [M^+—(CH_3COOH + C_4H_7)], 124 (5. 98%) [M^+—(CH_3COOH + C_5H_{10})], 123 (5. 57%) [M^+—(CH_3COOH + C_5H_9)], 61 (13. 8%) [$CH_3COOH_2^+$], 43 (100%) [$C_2H_3O^+$].

4. (E) – 12 – tetradecene – 1 – ol Acetate(Ⅵ)

A 4 – necked 250mL flask, equipped with a dry – ice trap and overhead stirrer (made of Ni – Cr alloy), was immersed in a dry ice – acetone mixture. Ammonia gas was passed in and condensed to give about 20mL of liquid ammonia. Sodium pellets (0.5gm) was added. The solution turned blue instantly. 300mg (1.25mmol) 12 – tetradecynol acetate diluted with 20mL tetrahydrofuran (freshly distilled over $LiAlH_4$) was added. The solution was stirred under reflux for 2h, then allowed to rise to room temperature; excess ammonia evaporated, while the solution turned from blue to colorless. 20mL saturated ammonium chloride solution was cautiously added. 2 layers appeared. The upper layer was separated, diluted with hexane and further washed with saturated sodium chloride solution and dried over anhydrous magnesium sulphate. Then the product was evaporated in vacuum, the residue dissolved in 25mL hexane and transferred to a 100mL round – bottom flask for reacetylation. Two hundred mg acetic anhydride and 50 mg pyridine were added to the hexane solution. It was refluxed 3h, washed with 2 × 15mL of water, the hexane solution dried over anhydrous magnesium sulphate. After evaporation of solvent, a slightly yellow – tinted liquid was obtained. 0.22883gm. GC analysis showed it contained about 5% undesired Z – isomer.

The desired E – isomer was separated from the Z – isomer by HPLC as described above. All 14 runs on HPLC (each run represented 20μL of sample) were combined, and evaporated. A colorless liquid was obtained. 104 mg. Capillary GC (SP – 1000, program profile described above) showed the retention time = 20.18 min, while the authentic sample has a retention time as 20.20 min. The purity is 98.5%, the yield 50.75%. Its GC – MS spectra is identical with the *Z – isomer* mentioned above.

We are especially grateful to Dr. J. R. Plimmer, Dr. Jerone Klun and Dr. Barbara Bierl – Leonhardt for the helpful assistance and suggestions for this project, and also to Mr. Everett D. Devilbiss who obtained all the mass spectra for us.

REFERENCES

[1] Klun, J. A. et al., *Life Sciences*, 27(1980), 17:1603 – 1606.
[2] Ando, T. et al., *Agrio. Biol. Chem.*, 44(1980), 11:2643 – 2649.
[3] Private communication with Dr. J. A. Klun, BARC, USDA.
[4] Cheng Zhiqing et al., *J. Chem. Ecol.*, 7(1981), 5:841.
[5] Brown, H, C. et al., "*Organic syntheses. via Boranes*", Wiley Interscience, N. Y., 1975.
[6] Private communication with H. C. Brown and K. K. Wang. Purdue University, 1981.
[7] Suzuki, A. et al., *J. Am. Chem. Soc.*, 95(1973), 9:3080 – 3081.
[8] Brown, C. A. & Aluiza, A. K., *J. Org. Chem.*, 38(1973), 2226.
[9] Schwarz, M. & Waters, R. M., *Synthesis*, 10(1972), 567.
[10] Bierl – Leonhardt, B. A. & Devilbiss, E. D., *BARC, USDA*, 1981, to be published.
[11] Bierl – Leonhardt, B. A. et al., *Journal of Chromatographic Science*, 18(1980), 364.
[12] Specific cooperative research projcect between USDA and Zoecon Corporation.
[13] Brandsma, Li, *Preparative Acetylene Chemistry*, p. 69, Elsevier Publishing Co., Amsterdam, 1971.
[14] Colonge, J. et al., *Bull. Soc. Chim. Er.*, 3(1963), 551.

内吸性杀菌剂三唑醇（酮）立体异构体的研究（Ⅱ）*

李正名[1] 董丽雯[1] 李国炜[1] 张祖新[1]
曹秋文[1] 王素华[1] 窦士琦[2]

（[1] 南开大学元素有机化学研究所；[2] 中国科学院生物物理研究所）

摘　要　确定了拆分三唑醇 A、B 两非对称异构体所形成的非对映中间体的结构，并得到 100% 光学纯的三唑醇右旋 A 体。经 X – 射线法等确定其绝对构型为（+）1R，2S，由 Jones 氧化反应得到 100% 光学纯的三唑酮右旋体，利用格氏氢转移反应对右旋三唑酮进行不对称还原得三唑醇的不同光学异构体。对小麦叶锈病的生物活性测定表明三唑醇中(－)－A－1S，2R 的生物活性最高；三唑酮中，左右旋体的活性相当。

三唑醇（Triadimenol）又称 Baytan，是 1975 年 Büchel 等发现的一种优良的内吸性广谱性杀菌剂[1]。用于各类作物的种子处理，防病效果突出[2~4]，可增产 20% 左右。三唑醇分子中含有两个手性碳原子，应有四种光学异构体。由于杀菌剂和病原菌受体的结合有其立体选择性，藉受体学说可阐明光学活性药剂的立体结构、原子间的距离、电荷分布等因素与受体构型之间的相互关系。1982 年我们分离出三唑醇的两个非对映异构体，它们的生物活性有明显差异[2]。国外也进行了分离工作，但方法及条件不同[5]，有效 A 体的两个光学异构体的分离工作及生物活性尚待阐明。

三唑酮（Triadimefon）又称粉锈宁或 Bayleton 是三唑醇的氧化产物。它是一种高效内吸性杀菌剂，对大麦、小麦白粉病[6]，小麦锈病[7]，玉米丝黑穗病[8]等防治效果显著，可增产 10% ~30%，已广泛应用。最近报道某些菌类能将三唑酮代谢为三唑醇[9,10]。此分子中有一手性碳原子，应有两种光学异构体，但未见其立体化学研究报道。由于三唑酮在植物保护中的重要性，从立体化学阐明生物活性关系有实际意义。

根据^{1}H NMR 中偶合常数⌡，我们曾按 Karplus 公式推测三唑醇的 A 体应为苏式即 1S，2R 或 1R，2S 对映异构体，B 体为赤式即 1S，2S 或 1R，2R 对映异构体，并曾用柱层析法得到毫克量的 A 体和 B 体[2]。参考王笃祜等实验方法[11]，我们采用Meerwein－Ponndorf－Verley 反应定向得到富集 A 体或 B 体的三唑醇，得到克数量级的纯 A 或纯 B 体，为进一步的光学体拆分打下基础。

我们采用（+）－樟脑－β－磺酸为拆分剂，对三唑醇 A 体拆分（反应式 1）。所得对映异构中间体经重结晶，氨解得到 100% 光学纯度的三唑醇右旋 A 体。又利用 A 体为一个手性第二醇的特点采用 Horeau 反应[12]（反应式 2）对（+）－A 进行测定，其绝对构型应为（+）－1R，2S；后经单晶 X－射线衍射法加以证实。由此推理另一对映体的绝对构型为（－）－1S，2R。

* 原发表于《高等学校化学学报》，1987，8（3）：235－239。国家教育委员会科学基金资助课题。

$$Cl-C_6H_4-O-\overset{H}{C^*}(N\text{-triazolyl})-\overset{OH}{\underset{H}{C^*}}-CMe_3 + \text{(camphor-10-sulfonic acid)}^* CH_2SO_3H \longrightarrow Cl-C_6H_4-O-\overset{H}{C^*}(N\text{-triazolyl}, H^{\oplus})-\overset{OH}{\underset{H}{C^*}}-CMe_3 \quad {}^{\ominus}O_3SCH_2\text{(camphor)}^*$$

反应式 1

*手性碳原子

$$Cl-C_6H_4-O-\overset{H}{C}(N\text{-triazolyl})-\overset{H}{\underset{OH}{C}}-CMe_3 + d,l-(C_2H_5\overset{C_2H_5}{C}HCO)_2O \longrightarrow Cl-C_6H_4-O-\overset{H}{C}(N\text{-triazolyl})-\overset{H}{C}(O-\underset{O}{\overset{\|}{C}}-\underset{C_2H_5}{CH}-C_6H_5)-CMe_3 + C_2H_5\overset{C_2H_5}{C^*H}-COOH$$

d-光学体

反应式2

将三唑醇右旋 A 体在温和条件下进行 Jones 氧化反应[13]（反应式 3），首次得到右旋三唑酮（右旋粉锈宁），其光学纯度为 100%，构型（+）－1R。

$$Cl-C_6H_4-O-\overset{H}{C}(N\text{-triazolyl})-\overset{H}{\underset{OH}{C}}-CMe_3 \xrightarrow[\text{丙酮}]{CrO_3-H_2SO_4} Cl-C_6H_4-O-\overset{H}{C}(N\text{-triazolyl})-\underset{O}{\overset{\|}{C}}-CMe_3$$

反应式3

利用带有空间障碍基团的 Grignard 试剂对右旋三唑酮进行还原反应（氢负离子转移机理），得到一个旋光性产物。经薄层层析和^{1}H NMR 测定为三唑醇 A 体及 B 体的混合物，根据 Cramm 规律对其相应的绝对构型的归属进行了讨论。

将三唑醇（酮）各右旋体和各对对映体对小麦叶锈病（Puccinia recondita Rob. ct. Desm. f. sp. fritici Erikss）混合菌种的小麦孢子进行生物活性对比测定，结果表明对三唑醇A 体来讲，药效较高的是（－）－A（1S，2R），而对三唑酮来讲，两个光学体的药效相当。

实验方法

（一）仪器

Autopal－Ⅲ自动旋光仪，WZZ 自动指示旋光仪（上海光学仪器修理厂）；Perkin－Elmer 241 MC 旋光仪（美国）；SP－2305 气相色谱仪（北分）与色谱数据处理机 C－R1B（日本岛津）；JNM－4H－100（日本 JEOL）核磁共振仪；DS－301 型（日本）红外分光光度计；JOBIN－YVON CD－5（法国）圆二色散仪。

（二）化合物的合成及拆分

1. 三唑醇 A、B 体的定向合成

由 Meerwein－Ponndorf－Verly 反应得到的富集 A 体的粗产品[11]，以乙醇（95%）：水＝3∶5 的溶液重结晶（含量 99.80%）或苯进行重结晶（含量 99.99%）。

由羧酸法得到的富集 B 体的粗产品[11]，以苯∶石油醚(60～90℃)＝6∶7 的混合溶剂重结晶（含量 99.93%）。

^{1}H NMR（TMS，$CDCl_3$），δ（ppm）：A 体 6.290（双峰，1H），3.582（双峰，1H），7.866（单峰，1H），8.430（单峰，1H），1.064（单峰，9H），Jab；1.76ppm。

2. 三唑醇（±）-A 体的拆分

置（±）-A 体 6.747g（22.8mmol，含量 99.4%）及 D-樟脑-β-磺酸 5.343g（23.0mmol）（上海试剂一厂，98.0%）于反应瓶中，加入乙酸乙酯与95%乙醇（2:1）混合溶剂 30mL，再加入石油醚（60~90℃）与乙酸乙酯的比例为 1:1 的混合溶剂 120mL，搅拌 1.5h，抽滤，得白色固体状盐（A-CSA）11.5g（收率 95.51%）。此即非对映异构中间体，元素分析：C% 54.55（计算值）；54.51、54.69（实验值）。H% 6.44（计算值）；6.40、6.38（实验值）。Cl% 6.72（计算值）；6.73、6.55（实验值）。^{1}H NMR（TMS，$CDCl_3$），δ(ppm)；8.41（单峰，1H），10.02（单峰，1H）。

所得非对映异构中间体（A-CSA）以乙腈重结晶（g: mL 为 1:4）。重结晶七次后取 2.09g 溶于 25mL 水中，加入氨水（25%~28%）10mL，搅拌 2h，抽滤得白色固体 1.2g。在乙醇: 水 =3:5 的溶剂中重结晶七次，得到白色片状晶体，熔点 115.8~116℃，$[a]_D^{13℃}$ = +99.34°。经手性 NMR 位移试剂（三-（3-t 氟丙基羟甲叉基-d-樟脑））铕（Ⅲ）即 Eu（hfc）$_3$ 测定其光学纯度（o. p.）为 100%。

3.（+）-A 体绝对构型的测定（Horeau 方法）[12]

称取（+）-A 体 70.11mg 置于反应瓶中，与 231.62mg 的 2-苯基丁酸酐混合，加入无水吡啶 3mL，于 24℃搅拌 4h，加水 0.5mL，再搅拌 0.5h。然后加入苯 0.5mL，酚酞一滴，以 0.0968N 的氢氧化钠溶液滴定至微红色，转入分液漏斗中，以 3mL 氯仿萃取三次（每次 1mL）。水层以 20% 盐酸酸化至 pH=2，以 6mL 苯萃取三次（每次 2mL），干燥，浓缩，得无色透明油状液体，加苯 2mL，测定其旋光度 a = +0.025。

4. 三唑醇右旋 A 体的氧化反应

置（+）-A 体 2.92g 及 25mL 丙酮于反应瓶中。冰浴下搅拌滴加 Jones 试剂[13,14] 30mL，在室温反应至呈深棕色。在冰浴下加入氨水（25%~28%）至碱性，搅拌 2h。抽滤得白色固体。干燥后用乙醚提取，蒸去乙醚后得白色固体 2.44g（收率 84.03%）；在甲醇中重结晶得白色片状晶体，熔点 80~82℃。文献值（三唑酮消旋体）82.3℃[15]。经气相色谱分析，保留时间和消旋体一致，其旋光度 $[a]_D^{29℃}$ = +15.139°。光学纯度为 100%（手性 NMR 位移试剂法）。（+）-三唑酮的 CD 谱为负性谱线，其 λ_{max}在 300nm 附近。

5. 右旋三唑酮的不对称还原反应

称取金属镁 0.96g（0.040mol）置于干燥的反应瓶中，加入无水乙醚 30mL。将溴代异丙烷 4.2g（0.034mol）与 5mL 无水乙醚混合，将其四分之一的溶液先加入上述金属镁中，搅拌引发，然后滴加其余的溴代异丙烷溶液，维持回流温度。

置（+）-三唑酮 0.410g（1.38mmol）及无水乙醚 5mL 于反应瓶中，加入上述格氏试剂 20mL，回流搅拌 4~5h，倒入水中，加 10% 盐酸 10mL 水解。用 60mL 乙醚分三次提取，干燥过滤，蒸去溶剂，得黄色黏稠物。以苯-石油醚重结晶，经^{1}H NMR 分析和比旋光度测定结果：$[a]_D^{21℃}$ = +102.22°，产物组成 A（1R，2S）: B（1R，2R）=1:1.23。

结果与讨论

1. 三唑醇（+）-A 体及（+）-三唑酮的光学纯度是通过手性 NMR 位移试剂

Eu（hfc），进行测定的。比较图 1、图 2，可看到分别加 Eu（hfc）至三唑醇 A 体的（+）和（±）体后，在（±）体裂分为双峰处，在（+）－A 体中相应的峰均呈单峰（图 1、图 2）。因此（+）－A 体的光学纯度为 100%，同样测定三唑酮的光学纯度为 100%。

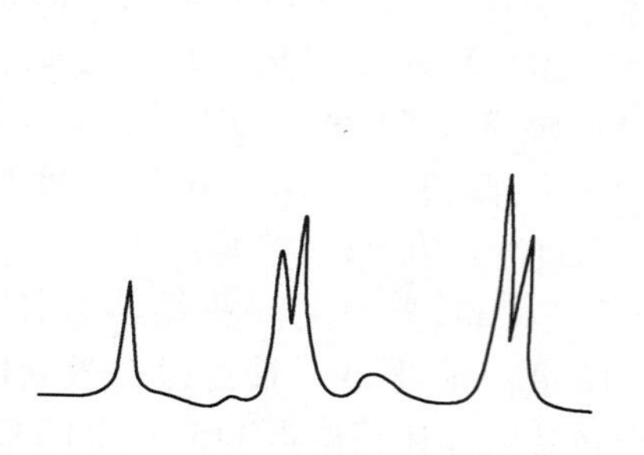

图 1　（+）－三唑醇 A 体加手性位移试剂 NMR 谱图

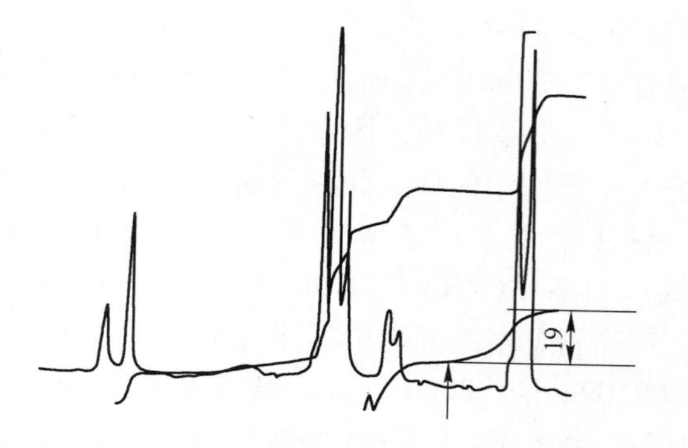

图 2　d，l－三唑醇 A 体：（+）－A＝2∶1 加手性位移试剂 NMR 谱图

2. 三唑醇（+）－A 体的绝对构型测定是采用 Horeau 方法[12]及其单晶的 X－射线衍射法进行的，利用三唑醇（+）－A 体分子中 C_2 原子是第二醇的特点使消旋的 2－苯基丁酸酐与之按动力学原理进行酯化反应，测得酯化后游离出来的 2－苯基丁酸为右旋。因此推论三唑醇（+）－A 体中 C_2 原子的绝对构型为 S。前文已报道 A 体为苏式[2]所以（+）－A 体的绝对构型应为 1R，2S。

应该指出（+）－A 体的单晶 X－射线衍射结构分析结果（图 3）与 Horeau 方法推测的结果完全一致，这就验证了 A 体为苏式结构的推测是正确的。

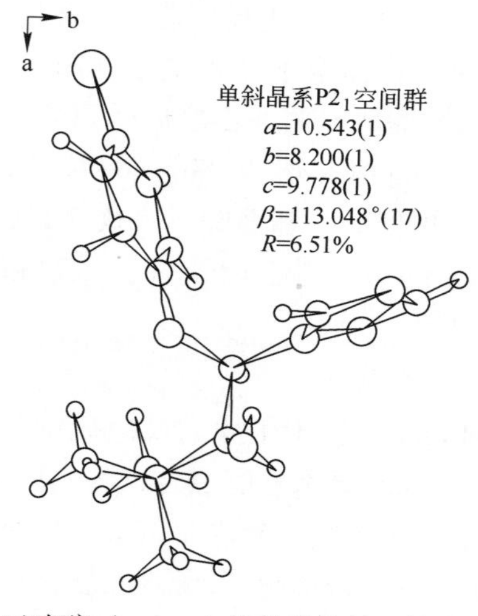

图 3　三唑醇（+）－A 体单晶的 X－射线衍射图

3. 对于右旋三唑酮的还原反应，一般格氏试剂与羰基化合物的反应机理应是以 2∶1 的分子数形成六元环过渡态中间体，形成新碳—碳键的产物，考虑到（+）－三唑酮中羰基邻位有一位阻大的叔丁基，因而反应经氢转移历程而进行还原反应，不形成新的碳—碳键。因此（+）－三唑酮与溴代异丙烷格氏试剂反应得到三唑醇。产物经 TLC 分离（展开剂为乙醚和乙酸乙酯），^{1}H NMR 测定证明为 A、B 体混合物，且以 A 体为主。按照

Cram 规则，可认为产物应以（+）-A（1R，2S）为主，B（1R，2R）为次（反应式4）。将产物以石油醚-苯的混合溶剂重结晶，发现其旋光度及 A（1R，2S）与 B（1R，2R）的比例均发生明显变化，由此推测 B（1R，2R）为右旋体，其结构有待进一步的测定。

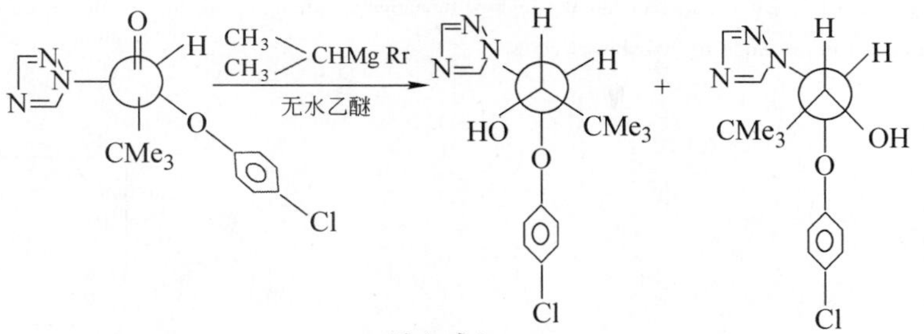

反应式4

上海有机化学研究所周维善教授、王中其和林国强同志热情协助，测定了 CD 谱，在此表示感谢。

参 考 文 献

[1] Büchel，K. H.，*J*，*Pest. Sci.*，（Japan），2，576（1977）.

[2] 邵瑞链，李正名，陈宗庭，王笃祜，谢龙观，农药，5，2（1983）.

[3] Ger. Offen. 2. 324，010 *O. A.*，32，156325，（1975）.

[4] Ger. Offen. 2. 632，603 *O. A.*，88，170152，（1978）.

[5] Deas. A. H. B.，*et al.*，*Pestio*，*Bioohem. Physiol.*，17(2)，120（1982）.

[6] Kolbe. W.，*Pflanzenschulz-Nachr.*，29（3）.310 *O. A.*，89 85651y.（1976）.

[7] 陈杨林，谢水仙，孙永厚，秦海阔. 植物保护学报，9(4).265（1982）.

[8] 吴新兰，庞志超，田立民，陈宗庭，张法庆，植物保护，6(2).381（1980）.

[9] Gasztonyi，M.，Josepovits，G.，*Pestio. Soi.*，10，57（1979）.

[10] Gasztonyi. M.，*Pestio. Sci.*，12，433（1981）.

[11] 王笃祜，彭永冰，辛泽欣，谢龙观. 农药，4，9（1985）.

[12] Horeau A. Stereochemistry Fundamentals and Methods，3，51. Edit. Kayan，H. B.，Stuttgart. Georg Thieme，1977.

[13] Bowers. A.，*et al.*，*J. Chem. Soo.* 2555（1953）.

[14] Haslarger，M.，Lawton. R. G.，*Syn. Commun.* 4（1），155（1974）.

[15] Kolbe. W.，*Pflanzenschutz Nachrichten Bayer* 29（3），312（1976）.

Studies on Steromers of Triadimenol（Ⅱ）

Zhengming Li[1]，Li wen Dong[1]，Guowei Li[1]，
Zuxin Zhang[1]，Qiuwen Cao[1]，Suhua Wang[1]，Shiqi Dou[2]

（[1]Institute of Elemento-Organio Chemistry，Nankai University，Tianjin，300071
[2]Institute of Biophysics，Academic Sinica，Beijing）

Abstract The two diastereomers（designated as A and B）of Triadimenol were separated and the enantiomers of the active triadimenol A were further resolved. The absolute configuration of（+）-triadimenol A was determined both by Horeau method and X – ray diffraction method. Its oxidation to the corresponding (+) – triadimefon was successfully carried out for the first time by Jones' oxidation. The asymmetric reduction of (+) triadimefon was carried out by using an hydrogen transfer mechanism of Grignard reaction to obtain different optical isomers of triadimenol.

The preliminary bioassay on mixed spores of wheat rust indicated that among the four optical isomers of triadimenol, (–) – A – 1S, 2R enantiomer has the highest bioactivity, whereas both levo – and dextro – isomers of triadimefon do not differ much in their bioactivities.

1-(4氯苯氧基)-3,3-二甲基-1-(1,2,4-三氮唑-1基)-丁醇2右旋A体的晶体结构和绝对构型*

窦士琦[1]　姚家星[2]　李正名[3]　董丽雯[3]

([1]中国科学院生物物理研究所，北京；[2]中国科学院物理研究所，北京；
[3]南开大学元素有机化学研究所，天津)

三唑醇（Triadimenol）是一种优良内吸性的广谱杀菌剂，用于处理各类作物种子时，防病效果显著。该化合物的分子式为 $C_{14}H_{18}N_3OCl$，化学结构式如右。其中含有两个手性碳原子，应具有四种光学异构体。据文献报道[1]，其中的两个非对映体之间的药效有明显的差别，有效体称为A体。A体的两个对映体之间的药效以（-）-A体为最佳有效体[2]。鉴于了解A体的一对对映体的构型，对探讨其生物活性的作用机理及其与立体构型关系方面有重要意义，本文采用X-射线单晶衍射法测定经拆分后得到的（+）-A体的晶体结构和分子的绝对构型。这也就提供了（-）-A体的立体化学的结构参数。

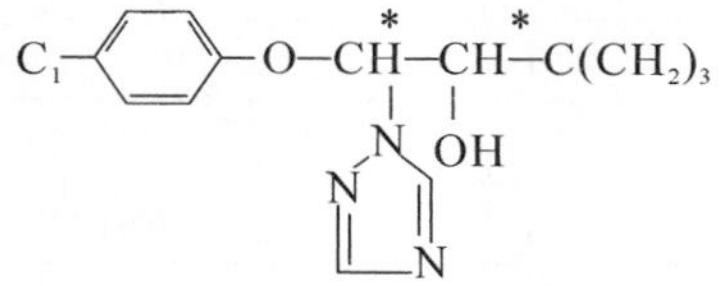

实验

三唑醇（+）-A体是通过对（±）-A体的拆分而得到的，拆分剂为 D-樟脑-β-磺酸[2]。在甲醇中结晶，晶体为无色透明柱状体，无辐射损伤。相对分子质量为295.7，$[a]_D^{13℃}$（甲醇）=+99.34℃。熔点：115.8~116.0℃。

在PW1100四圆衍射仪上收集了晶体的衍射强度数据，由实验数据确定晶体的空间群为 $C_2^2-P_{21}$，晶胞参数为 $a=10.543$（2）Å，$b=8.202$（1）Å，$c=9.778$（1）A，$\beta=113.048°$（17），$v=777.857Å^3$。计算密度 $D_2=1.261g\cdot cm^{-3}$，晶胞内分子数 $Z=2$，F（000）=312.00。选用的X射线为Cu*K*a，经石墨单色器单色化。采用 $\omega-2\theta$ 步进扫描方式，扫描速度为0.04°/sec，扫描宽度为1.08°，最大ω角为68°。数据收集过程中（422）和（$\bar{5}23$）反射为监测点。在1610个独立衍射点中，$I>3\sigma(I)$ 的可观察反射点有1524个。其中

$$\sigma(I)=\{N_P+N_B+[0.03(N_P-N_B)]^2\}^{\frac{1}{2}}$$

N_P 和 N_B 分别为每个衍射点的峰值和背景值。

晶体结构和绝对构型测定

结构用随机多解RANTAN法[3]测出全部非氢原子。然后用SHELX-76[4]程序进行

* 原发表于《物理化学学报》，1987，3（1）：74-77。

全矩阵最小二乘法修正结构参数，非氢原子都进行各向异性温度因子修正。甲基上的氢原子由程序给定，其余氢原子由差值电子密度图上获得，全部氢原子都进行了各向同性温度因子修正。最后的 R 因子为0.065，原子坐标参数，分子内键长、键角见表1～表3。

表1 非氢原子坐标

Table 1 Atomic coordinates for nonhydrogen atoms

atom	$x/a\times10^4$	$y/b\times10^4$	$z/c\times10^4$	$Ueq\times10^3$ (Å^2)
C1	−725(1)	−2383(1)	−1285(2)	104(1)
O_1	4933(3)	−797(5)	2782(3)	54(1)
O_2	7358(3)	685(5)	5050(3)	60(1)
N_1	5070(4)	2032(5)	2693(4)	54(2)
N_2	4911(6)	3315(7)	1774(5)	77(2)
N_3	4177(4)	3923(7)	3546(5)	73(2)
C_1	953(5)	−1917(7)	−112(6)	65(2)
C_2	1414(5)	−2234(9)	1400(6)	75(3)
C_3	2709(4)	−1785(8)	2325(5)	64(2)
C_4	3592(3)	−1061(5)	1761(4)	47(2)
C_5	3150(5)	−759(8)	264(5)	69(2)
C_6	1799(5)	−1162(9)	−670(6)	72(3)
C_7	5695(4)	504(7)	2500(4)	54(2)
C_8	7205(4)	395(6)	3568(4)	54(2)
C_9	7973(4)	−1138(8)	3406(5)	64(2)
C_{10}	9519(7)	−847(12)	4290(8)	84(4)
C_{11}	7765(9)	−1365(14)	1776(8)	94(5)
C_{12}	7599(6)	−2698(10)	4023(9)	85(3)
C_{13}	4350(7)	4407(9)	2335(7)	78(3)
C_{14}	4568(6)	2382(8)	3726(6)	69(3)

表2 键角（°）

Table 2 Bond angles（°）

$C_7-O_1-C_4$	118.0(3)	$C_3-C_2-C_1$	120.0(6)	$C_9-C_8-C_2$	112.9(3)
$C_7-N_1-N_2$	121.8(4)	$C_4-C_3-C_2$	120.4(5)	$C_7-C_8-O_2$	111.2(4)
$C_{14}-N_1-N_2$	110.6(5)	$C_5-C_4-O_1$	124.2(4)	$C_9-C_8-C_7$	115.3(4)
$C_{14}-N_1-C_7$	127.7(5)	$C_5-C_4-C_3$	120.0(3)	$C_8-C_9-C_{10}$	107.5(5)
$C_{13}-N_2-N_1$	102.3(5)	$C_3-C_4-O_1$	115.6(3)	$C_8-C_9-C_{11}$	109.9(5)
$C_{14}-N_3-C_{13}$	102.4(6)	$C_6-C_5-C_4$	119.2(5)	$C_8-C_9-C_{12}$	114.4(5)
C_2-C_1-Cl	120.3(5)	$C_5-C_6-C_1$	120.3(5)	$C_{10}-C_9-C_{11}$	107.0(6)
C_6-C_1-Cl	119.6(4)	$C_8-C_7-O_1$	109.9(4)	$C_{10}-C_9-C_{12}$	106.8(5)
$C_6-C_1-C_2$	120.0(5)	$C_8-C_7-N_1$	111.3(4)	$C_{11}-C_9-C_{12}$	110.9(6)

表3 键长（Å）

Table 3 Bond length（Å）

$Cl-C_1$	1.734(5)	N_2-C_{13}	1.306(10)	C_4-C_5	1.374(6)
O_1-C_4	1.391(4)	N_3-C_{13}	1.359(10)	C_5-C_6	1.398(6)
O_1-C_7	1.425(6)	N_3-C_{14}	1.337(9)	C_7-C_8	1.527(5)
O_2-C_8	1.421(5)	C_1-C_2	1.388(9)	C_8-C_9	1.537(8)
N_1-N_2	1.351(7)	C_1-C_6	1.363(9)	C_9-C_{10}	1.536(8)
N_1-C_7	1.462(7)	C_2-C_3	1.360(6)	C_9-C_{11}	1.533(9)
N_1-C_{14}	1.341(8)	C_3-C_4	1.388(7)	C_9-C_{12}	1.525(10)

应用氯原子对X射线反常散射效应，测定（+）-A三唑醇分子的绝对构型。首先计算Bijvoet点对的结构因子值 F_C（hkl）和 F_C（$\bar{h}\ \bar{k}\ \bar{l}$），选取Bijvoet点对衍射强度相对变化量较大的18对反射进行实验测量。为了克服测量中的偶然误差，采用了等精度测量法，累积衍射强度值达 10^5 数量级，这样可使测量实验误差处于千分之几的等精度范围之内。在每个Bijvoet点对的邻近，选择反常散射效应小的最邻近的点对，以同样的方式测量，用以校正晶体的吸收误差。最后计算出 ΔI 的符号，与理论计算值 ΔF_c 的符号进行比较，结果表明两者符号完全相反，由此唯一确定该分子的绝对构型。即分子的绝对构型是与从电子密度图上得到的结构模型相反，如图1所示。

讨论

X射线结构分析表明，三唑醇（+）-A体的绝对构型为 $1R\cdot 2S$，如图1和图2所示。由三唑环内各原子之键长及各原子与唑环标准平面的偏离，均表明三唑环具有芳香性。三唑环内各原子间键长为：$N_1-N_2=1.351$（7）Å；$N_1-C_{14}=1.341$（8）Å；$N_3-N_4=1.337$（9）Å；$N_3-C_{13}=1.395$（10）Å；$N_2-C_{13}=1.306$（10）Å。三唑环最小二乘平面方程为

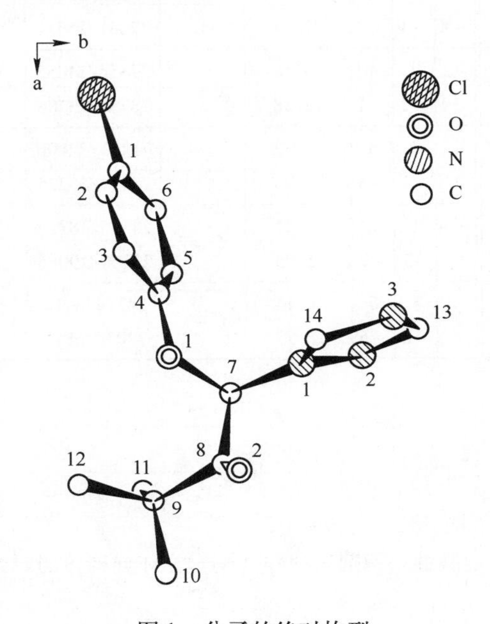

图1 分子的绝对构型

Fig. 1 The Configuration of moiecule

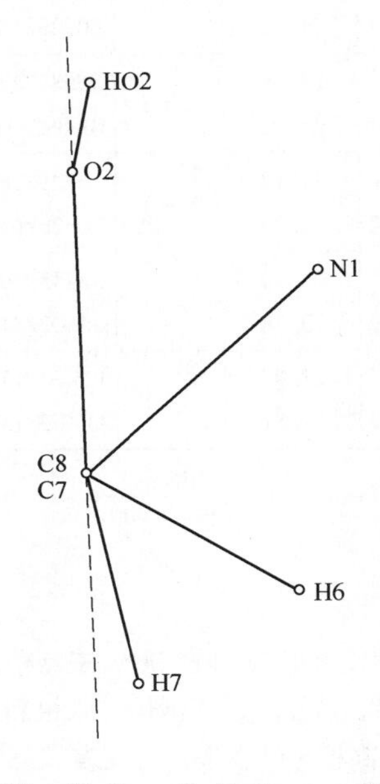

图2 沿 $C_8\rightarrow C_7$ 的Newman投影

Fig. 2 Newman projection along $C_8\rightarrow C_7$

$$0.74068X + 0.31773Y + 0.59198Z - 5.14186 = 0$$

各原子与最小二乘平面的偏离是，N_1：0.0174Å；N_2：-0.0013Å；N_3：0.0252Å；C_{13}：-0.0152Å；C_{14}：-0.0261Å。

三唑环内三个氮原子的键角相应地为：$C_{14}-N_1-N_2=110.6\ (5)°$；$C_{13}-N_2-N_1=102.3\ (5)°$；$C_{14}-N_3-C_{13}=102.4\ (6)°$。三唑环要保持其芳香性，则环上相应的各原子必须采取 sp^2 型杂化轨道来成键，由五元环的性质所知，五元环内角和应为540°，正常情况下，相应的各角应为108°，原子 N_1 形成 sp^2 杂化轨道后，p 轨道上有一对弧对电子参与了五元环的共轭 π 键的成键，$\angle C_{14}-N_1-N_2=110.6\ (5)°$与正常的108°相近。$N_2$，$N_3$ 所形成的杂化轨道为 sp^2 不等性杂化轨道。一对孤对电子占据一个 sp^2 轨道。由于电子云密度的影响，使得相应的各角度要小于正常的108°，致使$\angle C_{13}-N_2-N_1=102.4°$；$\angle C_{14}-N_3-C_{13}=102.4°$。

表4　分子绝对构型测定

Table 4　Determination of molecular absolute configuration

h	k	l	F_C(count)	S (ΔF_C)	I_0(exp.)	S (ΔI_0)	h	k	l	F_C(count)	S (ΔF_C)	I_0(exp.)	S (ΔI_0)
-5 5	1 -1	1 -1	22.41 21.57	(+)	56328/22903 60886/19611	(-)	-1 1	2 -2	2 -2	17.70 18.49	(-)	73615/126954 78116/150720	(+)
-3 3	1 -1	1 -1	11.51 12.10	(-)	26942/14929 25657/14402	(+)	-3 3	1 -1	3 -3	27.67 26.66	(+)	135792/32806 156260/35225	(-)
3 -3	1 -1	1 -1	11.63 12.95	(-)	25635/58769 21934/63392	(+)	-4 4	1 -1	3 -3	18.51 18.04	(+)	58421/11580 56975/10306	(-)
1 -1	2 -2	1 -1	20.41 20.06	(+)	94362/17298 116150/18199	(-)	-1 1	1 -1	3 -3	12.34 11.67	(+)	28726/32806 37062/35225	(-)
3 -3	2 -2	1 -1	20.65 21.12	(-)	61899/17298 60923/18199	(+)	1 -1	1 -1	3 -3	21.98 21.28	(+)	73856/25113 77651/25421	(-)
4 -4	1 -1	1 -1	28.79 27.97	(+)	95584/17298 100892/18199	(-)	-2 2	2 -2	4 -4	13.18 12.38	(+)	27354/14052 28751/13788	(-)
2 -2	1 -1	2 -2	11.17 12.23	(-)	25729/58769 20626/63392	(+)	-3 3	2 -2	3 -3	22.37 21.73	(+)	71463/32806 77431/35225	(-)
4 -4	2 -2	2 -2	22.70 23.78	(-)	52245/39863 48602/41916	(+)	-5 5	1 -1	2 -2	14.51 15.25	(-)	24545/18316 23260/19086	(+)
0 0	1 -1	2 -2	21.97 22.51	(-)	125431/151590 117872/152460	(+)	-1 1	1 -6	6 -6	16.14 16.77	(-)	24714/6010 22039/6812	(+)

注：$I_0\ (\text{exp.}) = \dfrac{I\ (\text{Bijvoet})}{I\ (\text{neightour})}$。

参考文献

[1] 邵瑞链，李正名，陈宗庭，王笃祜，谢龙观. 农药，5，2（1983）.

[2] 李正名，董丽雯，李国炜，张祖新，窦士琦. 内吸性杀菌剂三唑醇（酮）立体异构体的研究Ⅱ[J]. 高校化学学报，将发表。

[3] Yao Jia-xing. *Acta Cryst.* A37，642-644，（1981）.

[4] Sheldrick，G M. Program Crystal Structure Determination，University of Cambridge，England，（1976）.

A Stereoselective Synthesis of Pear Ester via Arsenic Ylide*

Zhengming Li, Tiansheng Wang, Diankun Zhang, Zhenheng Gao

(Institute of Elemento - Organic Chemistry, Nankai University, Tianjin, China)

Abstract The paper describes a four - step synthesis of ethyl(2E,4Z) -2,4 - decadienoate(pear ester) from propargyl alcohol with a 50% total yield. It also gives the synthesis of ethyl (2E,4E) - 2,4 - decadienoate. In both cases arsenic ylides were used to give the satisfactory results.

Ethyl(2E,4Z) -2,4 - decadienoate(1), named as pear ester, was a component found in Bartlett Pears[1]. Several previous syntheses[2-5] of 1 gave comparatively low stereoselectivity. Herein we would like to present a highly stereoselective synthesis of 1. In our route (see SCHEME), easily available propargyl alcohol(2) was used as the starting material. Alkylation of the dianion of 2 gave 2 - octyn - 1 - ol(3). PCC oxidation of 3 yielded 2 - octynal (4). Condensation of arsenic ylide 5 with 4 gave ethyl E - 2 - decen - 4 - ynoate(6). Lindlar reduction of acetylenic bond of 6 afforded pear ester(1). It is worthy to note that in our synthesis arsenic ylide(5) was used to condense with aldehyde 4 to give pure E - form of 6, no Z - isomer of 6 was detectable by GC analysis. The final - product thus obtained contained 96% of 2E,4Z - isomer 3% of 2E,4E - isomer and 1% of 6.

SCHEME

OH → OH → (2, 3, 4)

→ COOEt → (6, 1)

Ph3As=CHCOOEt 5

Ph3X=CHCH=CHCOOEt X=As(7), P(8)

In order to get 2E,4E - form of 1, 3 - ethoxycarbonylallylidenetriphenylarsorane (7) condensed with n - hexanal to form a product which consisted of 90% of ethyl (2E,4E), -2,4 - decadienoate and 10% of its 2E,4Z - isomer. If, however, the corresponding phosphorous ylide (8) was used, a mixture of its four isomers was obtained (2Z,4Z: 2Z,4E: 2E,4Z: 2E,4E = 9.5: 23.7: 27.6: 39.2) (GC - MS analysis).

---

* Reprinted from *Synthetic Communication*, 1989, 19: 91 - 96.

EXPERIMENTAL

All IR spectra were recorded as films on a Shimadzu Model IR – 435 instrument. NMR spectra were recorded on either Varian Model EM – 360 or JEOL FX – 90Q spectrometer. GC analysis was carried out on a Hewlett Packard Model 5890A with column OV – 1.

2 – Octyn – 1 – ol(3) In a 1L rounded flask 700mL of ammonia was condensed and catalytic amount of ferric nitrate added. 10min. later pieces of lithium (11. 1g, 1. 6mol) were added and the resulting gray mixture stirred for 1h. Fresh distilled propargyl alcohol(44. 8g, 0. 8mol) in 100mL of dry THF was added over 20min. After 2h, 60. 4g(0. 4mol) of n – pentyl bromide in 60mL of dry THE was added dropwise and the reaction continued overnight. To the resulting mixture was added ammonium chloride and the ammonia evaporated. The reaction was acidified with concentrated hydrochloric acid and then extracted with ether (3×200mL). The combined ether extract was washed with brine and dried (Na_2SO_4). Evaporation of volatile material gave a residul which after distillation gave 45. 8g of 2 – octyn – 1 – ol(90. 7%). b. p. 91 ~ 92℃/9mmHg. GC purity 99. 9%. IR: 3300, 2900, 2840, 2280, 2100, 1460, 1130, 1005 and 720 cm^{-1}, ^1H – NMR(CCl_4, $Me_3SiOSiMe_3$): δ3. 99 (t, J = 3Hz, 2H, CH_2O), 2. 57 (s, 1H, OH), 1. 99 – 2. 18(m, 2H, $CH_2C \equiv C$), 1. 23(m, 6H, $3 \times CH_2$), 0. 80(t, J = 6Hz, CH_3) ppm.

PCC oxidation of 2 – octyn – 1 – ol was accomplished similar to reported method[6]. 3. 97g of 2 – octynal(4) was obtained from 6. 3g 2 – octyn – 1 – ol(64%). b. p. 65 – 68℃ 17mmHg. GC purity 96%. IR: 2920, 2850, 2270, 2100, 1665 and 1130cm^{-1}; ^1H – NMR (CCl_4, $Me_3SiOSiMe_3$): δ8. 97(1H, s, CHO), 2. 28(t, J = 6Hz, 2H, $CH_2C \equiv C$), 1. 13 – 1. 63(m, 6H, $3 \times CH_2$), 0. 83(t, J = 6Hz, 3H, CH_3) ppm.

Ethyl E – 2 – decen – 4 – ynoate(6) Sodium (0. 23g, 10mmol) was dissolved in 10mL absolute ethanol. To the solution was added ethoxycarbonylmethyltriphenylarsonium bromide[7] (5. 20g, 11mmol) in 25mL of absolute ethanol. After 10min., 1. 24g(10mmol) of 2 – octynal in 15mL of ethanol was slowly added and the reaction continued for 1. 5h. The mixture was diluted with water (100mL), extracted with ether (3×50mL). The extract was washed with saturated sodium bisulphate solution remove unreacted aldehyde, brine and dried (Na_2SO_4). Removal of ether gave the crude product which after silica gel column chromatography (ether – light ptroleum) gave 1. 746g of ethyl E – 2 – decen – 4 – ynoate(89. 9%). GC purity 99. 8% (100% E – form). IR: 2950, 2850, 2200, 1715, 1620, 1320, 1300, 1260, 1175, 1150, 1030, 960, 840, and 726cm^{-1}, ^1H – NMR($CDCl_3$, TMS): 6. 76(d, t, J = 16 and 2Hz1H, CHC—C), 6. 12(d, J = 16Hz, 1H, CHCOO), 4. 20(q, J = 7Hz, 2H, $COOCH_2$), 2. 36(t, d, J = 6 and 2Hz, 2H, CH_2C—C), 1. 20 – 1. 50(m, 6H, $3 \times CH_2$), 0. 90(t, J = 7Hz, 3H, CH_3) ppm. ^{13}C – NMR: δ 166. 1(C – 1), 126. 2(C – 2), 129. 2(C – 3), 78. 0(C – 4), 100. 8(C – 5), 19. 8(C – 6), 28. 1(C – 7), 31. 1(C – 8), 22. 3(C – 9), 13. 9(C – 10), 60. 6(C – 1'), 14. 2(C – 2') ppm.

Ethyl (2E, 4Z) – 2, 4 – decadienoate(1) A 100mL rounded flask charged with 0. 1g Lindlar's catalyst[8], 3 drops of quinoline, 5mL of hexane, 5mL of ethyl acetate and 0. 777g (2mmol) of ethyl E – 2 – decen – 4 – cynoate began to absorb hydrogen. The process was monitored by GC. The absorption was stopped, 1% of the starting ester being left unreacted. The catalyst was filtered off and the filtrate washed with diluted hydrochloric acid, brine and driec

(Na_2SO_4). Removal of the solvent gave the residue which after silica gel column chromatography(ether – light petroleum) gave 0.75g of ethyl(2E,4Z) –2,4 – decadienoate (95.5%). (Found: C, 72.96; H, 10.58; $C_{12}H_{20}O_2$ required: C, 73.43; H, 10.27). GC – MS analysis showed that the product contained 96% of 1,3% of 2E,4E – isomer and 1% of the starting ester. IR: 2940, 2850, 1710, 1635, 1605, 992 and $705cm^{-1}$, ^1H – NMR ($CDCl_3$, TMS): δ 7.65 (d,d,J = 15 and 11Hz, 1H, C═C—CH═C), 5.90(d, J = 15Hz, 1H, CHCOO), 5.68 – 6.23 (m, 2H, CH═CH), 4.27(q, J = 6Hz, 2H, OCH_2) 2.17 – 2.32(m, 2H, CH_2C═C), 1.20 – 1.36(m, 9H, 3 × CH_2 + CH_3), 0.88(t, J = 6Hz, 3H, CH_3) ppm. 13C – NMR: δ167.0(C – 1), 121.1(C – 2), 139.2(C – 3), 126.3(C – 4), 141.3(C – 5), 28.1(C – 6), 28.9(C – 7), 31.3(C – 8), 22.4(C – 9), 13.8(C – 10), 60.0(C – 1'), 14.1(C – 2') ppm. MS: m/e 196, 167, 151, 125, 97, 81(100), 66, 55, 41.

Ethyl (2E,4E) –2,4 – decadienoate n – Hexanal(0.5g, 5mmol) and 3 – ethoxy – carbonylallyidenetriphenylarsorane[9] (2.1g, 5mmol) were mixed in 20mL absolute ether and stirred at room tempraure overnight. The solid was filtered off and the filtrate concentrated. Silica gel column chromatography of the residue gave 0.81g of product(82.5%) which contained 90% of ethyl(2E,4E) –2,4 – decadienoate and 10% of 2E,4Z – isomer (by GC – MS).

The preparation of phosphorane 8 (without isolated) and its condensation with n – hexanal were accomplished similar to the known procedure[10].

Acknowledgment

We are grateful to prof. Y. Z. Huang, L. L. Shi and Dr. J. H. Yang (Shanghai Institute of Organic Chemistry, Academia Sinica) for their kindly assistance.

References and note

[1] Heinz, D. e., Jennings, W. G., J. Food Sci., 1996, 31, 69.
[2] Ohloff, G., Pawlak, M., Helv. Chim Acta, 1973, 56, 1176.
[3] Devos, M. J. Heces, L., Bayet, P., Krief, A., Tetrahedron Lett., 1976, 3911.
[4] Bestmann, H. J., Suss, J., Liebigs Ann. Chem., 1982, 363.
[5] Byrne, B., Lawter, L. M. L., Wengenroth, K. J., J. Org. Chem., 1986, 51, 2607.
[6] Corey, E. J., Suggs, J. W., Tetrahedron Lett., 1975, 2647.
[7] Ethoxycarbonylmethyltriphenylarsonium bromide was prepared according to: Huang, Y. Z., Tai, H. I., Ting, W. Y., Chen, L., Tu, H. M., Wang, C. W., Acta Chimica Sinica, 1978, 36, 215. m. p. 148 – 50℃, ^1H – NMR($CDCl_3$), HMDS: 7.57(m, 15H, ArH), 5.42(s, 1H, CH_2CO), 3.93(q, 2H, J = 7Hz, CH_3) ppm.
[8] Lindlar, H., Dubuls, R, "Organic Syntheses", coll. Vol. V. Wiley, New York, N. Y. 1973, p880.
[9] Huang, Y. Z., Shen. Y. C., Zhang, S. X., Synthesis, 1985, 57.
[10] Howe, R. K., J. Am. Chem. Soc., 1971, 93, 1519.

The Chemical Characterization of the Cephalic Secretion of the Australian Colletid Bee, *Hylaeus albonitens* (Gnathoprosopls Cockerell)*

Zhengming Li[1], BATRA S. W. T. [2], PLIMMER J. R. [2]

([1]Research Institute of Elemento – Orgunic Chemistry, Nankai University, Tianjin, 300071;
[2]BARC, USDA – ARS, Beltsville, Maryland 20705, USA)

Abstract The cephalic extract of Australian colletid bee, *Hylaeus albonitens*, compriscs 10 components: linalool. citronellal, neral, geranial, nerolic acid, gcranic acid, heptadecane, pentacosane, Z, Z – and E, E – 9, 12 – octadccadienoic acid. Quantification of these major components suggested it to be an aggregation pheromone composition. The fairly large amounts of 9, 12 – octadecadienoic acids have not been reported before in the cephalic extract of bees. The chromatographic pattern is quite different from the previous reports on the constituents of mandibular glands of American *Hylaeus modestus* and the related European and North American Colletes bees.

Introduction

Citral was first reported in the 60's by Boch and Shearer[1] as a major constituent of the Nassanoff gland of the honey bee. Subsequently, there have been reports that predominant constituents in the mandibular glands of several bee species were citra[2-6] linalool[7] Some aldehydes and long chain hydrocarbons[8,9] were also reported to be present in the mandibular glands of Colletes.

Comparatively very few publications have dealt with the small primitive bees of the genus *Hylaeus*. The major constituents in the mandibular glands of American *Hylaeus modestus* are Z – and E – citral. [10]

On the biological evolution scale. the colletid bees may be an intermediate stage between the original Sphecoidca (wasps) and Apoidea (bees). Taxonomic comparison may be of great interest in view of the constituents in their respective mandibular glands. This paper describes for the first time the identification and measurement of the different constituents from the cephalic secretion of *Hylaeus* (Gnathoprosopis) *albonitens* (Cockerell) which were recently collected in Australia.

Methods and materials

Collection of cephalic secretion of Hylaeus albonitens

The male and female bees were collected from *Melaleuca* and Eusalyptus flowers at Dimbulah

* Reprinted from *Chinese Journal of Chemistry*, 1990, 2: 160 – 168. This is part of the research work complcted in BARC – West. USDA. Beltsville, Maryland. U. S. A. During 1980 – 1982.

and Mount Molloy in northern Queensland, Australia in August 1980. These metallically dark blue and sparsely haired bees are about one – eighth as large as honey bees. The heads were removed immediately. The malar space was slit to expose the scented mandibular glands and the heads were placed in 2mL vial containing 1mL of heptane(Burdick and Jackson, G. C. grade). Living bees of both sexes orally release a distinct lemon like odor when they were handled.

Instrumentation

The analyse were conducted using Hewlett Packard 5830A and 5840A gas chromatograph equipped with splitless capillary injectors and flame ionization detectors as well as Hewlett Packard 18850 GC terminal. For this project, 4 capillary columns were used:

Column Ⅰ, fused silica capillary column SE – 52(Quadrex), 50 meters, i. d. 0. 230mm. Theoretical plates/meter 3490 (it had been used 6 months prior to this project).

Column Ⅱ, fused silica capillary column SE – 52(Quadrex), 50 meters, i. d. 0. 235mm. Theoretical plates/meter 3200.

Column Ⅲ, fused silica glass capillary column Silar 10 (Quadrex), i. d. 0. 25mm, 76 meters. Theoretical plates/meter 5315.

Column Ⅳ, fused silica capillary column Durabond DB – 1 (J&W, Scientific), 60 meters. i. d. 0. 25mm.

Samples were run on a program as Temp. 1 = 90℃, Time 7min, Rate = 2℃/min, Temp. 2 = 170℃. Injector temp. = 225℃. Purge delay time = 2. 0min, Detector temp. (FID) = 275℃.

The computerized gas chromatograph – mass spectrometer (GC – MS) used was a Finnigan model 4000 with 6000 data system, electron impact mass spectra were obtained at 70 eV with a source temperature at 270℃. The capillary columns used for GC – MS were: a) SP – 2100 (Supelco) 60 meters; b) SP – 1000 (Supclco) 60 meters. The stereoisomers of methyl nerolate and geranate were scparated on a Nester/Faust Spinning Band Distillation Annular Still NFA – 200 and methyl nerolate was further purified by preparative GC (varian Aerograph Model 10) equipped with a glass column (i. d. 4mm. 2 meters) packing 5% SE – 30 on Chrom G DMCS. NMR spectra were determined by Varian XL 100 NMR spectrometer. 100. 0MHz(^{1}H)、255. 2MHz(^{13}C). and Varian EM 360A NMR spectrometer 60 MHz(^{1}H).

Results

The cephalic extract of female *Hylaeus albonitens* was chromatographed separatively on nonpolar capillary columns Ⅰ, Ⅱ, Ⅳ and on polar column Ⅲ. By comparison, it was noted that peaks 5 and 6 showed up strongly on column Ⅰ, Ⅳ and less strongly on column Ⅱ. did not show up at all on column Ⅲ. Peaks 9 and 10 showed up on column Ⅰ and Ⅳ, did not show up on column Ⅱ or Ⅲ. The typical glc spectrum on column Ⅰ is shown in Fig. 1.

Identification of peaks 1, 2, 7 and 11

Peaks were first detected by glc. structures assinged by MS spectra, finally verified by comparison with authentic samples which had identical retention times.

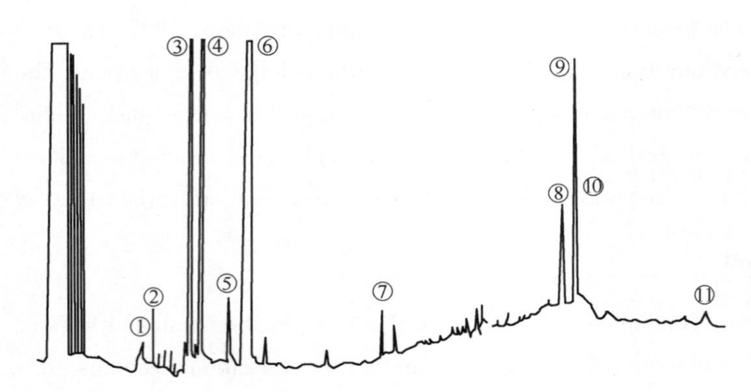

Fig. 1 The capillary gas chromatography spectrum of cephalic secretion of *Hylaeus ablonitens* (column Ⅰ, splitless injection mode)

Table 1 Retention times of peaks 1,2,7 and 11 on different columns in comparison with authentic samples

Component	Retention time (in min)			m/z
	col. Ⅰ	col. Ⅱ	col. Ⅲ	
peak 1	16.01	8.35	15.91	—
linalool	16.02	8.47	15.90	—
peak 2	—	10.66	15.38	154
citronellal	—	10.44	15.42	154
peak 7	53.58	—	17.83	240
heptadecane	53.62	—	17.73	240
peak 11	108.60	—	45.73	352
pentacosane	108.57	—	46.08	352

Identification of peaks 3 and 4

Peaks 3 and 4 showed up strongly on columns Ⅰ, Ⅱ and Ⅲ. GC – MC demonstrated that they both have the molecular ion at m/z 152. The strong fragment ion at m/z 69 suggested a terpenoid structure. By comparison of the data from columns Ⅱ and Ⅲ, there was a considerable shift of peaks 3 and 4 in relation to the retention time of the saturated hydrocarbons from C_{12} to C_{18}. This suggested that the two molecules have somewhat polar structures. The molecular formula $C_{10}H_{16}O$ was tentatively assigned. Comparison with the authentic sample of citral (Aldrich) proved that peaks 3 and 4 were 2 citral isomers. We confirmed Bergstrom's observation[3] that the fragment ions of these isomers above m/z 100 differed slightly in their respective intensities (Table 2).

Table 2 Intensities of individual peaks of citral isomers in GC – MC①

compound	m/z 119	m/z 123	m/z 134	m/z 137
Z – citral	9.63 (w)	2.86 (vw)	4.26 (w)	4.55 (vw)
E – citral	4.96 (vw)	17.14 (w)	2.78 (vw)	15.09 (w)

①w = weak, vw = very weak.

Thus peak 3 and peak 4 were identified as Z – citral (neral) and E – citral(geranial) respectively. The observation that peak 3 elutes earlier than the other isomer is also in accordance with the previous literature[10-12]

Identification of peaks 5 and 6

Peaks 5 and 6 showed up clearly on columns Ⅰ and Ⅳ, but were relatively reduced in size on column Ⅱ and completely disappeared on the polar column Ⅲ. Two noteworthy characteristics were observed:

a) When the mandibular extract was diluted stepwise, the intensity of peaks 5, 6, 9 and 10 on column Ⅰ dropped more sharply than would be expected by comparison with the peak intensity of the citral isomers.

b) On columns Ⅰ, Ⅱ and Ⅳ, the peaks 5 and 6 appeared to tail, an indication that these peaks might represent materials that have strongly polar functional groups. Furthermore. When the mandibular extract was washed with dilute sodium bicarbonate solution, peaks 5, 6, 9 and 10 disappeared, and on acidification of the aqueous layer with dilute hydrochloric acid, peak 6 appeared again. This suggested that acids were present in the mandibular extraction.

Peak 6 has a molecular ion of m/z 168. The fragments at m/z 123 and m/z 69 suggested a possible aliphatic(C_{10}) terpenic acid with a tentative molecular formula $C_{10}H_{16}O_2$. which might be closely related to citral isomers $C_{10}H_{16}O$. We oxidized citral by silver hydroxide[13] to obtain a mixture of nerolic and geranic acids, which was shown to have the same retention time (29.15 and 31.59 min) as that of peaks 5 and 6 (29.91 and 32.15 min).

Therefore it was postulated that peaks 5 and 6 corresponded to nerolic and geranic acids, or vice versa.

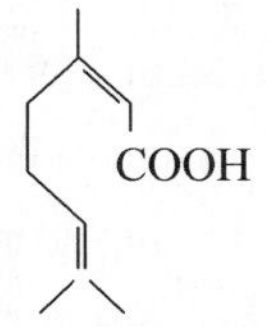

COOH COOH

Ncrolic acid 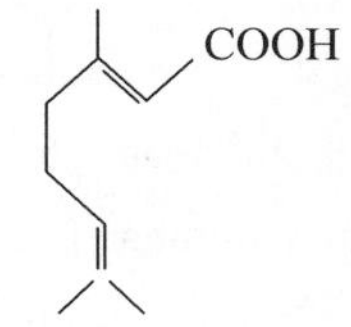 Geranic acid

We then used similar reaction conditions to those described by Wakabashi *et al.* [14] by reacting 6 – methyl – 5 – heptenone – 2 with trimethyl phosphonoacetate to prepare a mixture of methyl nerolate and geranate in satisfactory yield (76%):

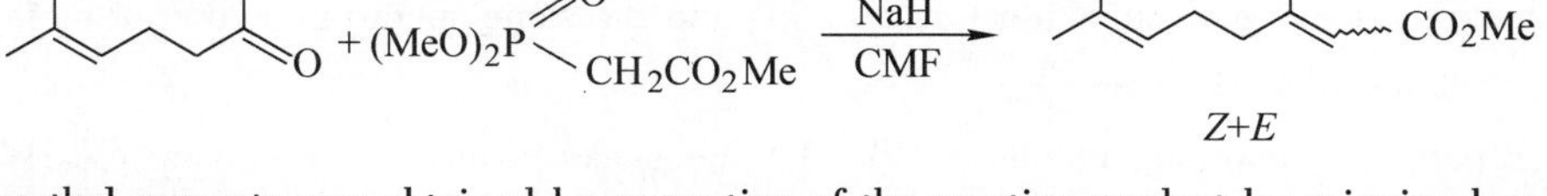

O + $(MeO)_2P$(=O)CH_2CO_2Me —NaH / CMF→ CO_2Me

Z+E

Pure methyl geranate was obtained by separation of the reaction product by spinning band distillation, while methyl nerolate was further purified by preparative GC. The configuration of both purified isomers was examined by ^1H NMR and ^{13}C NMR; therefore peaks 5 and 6 were conclusively identified as nerolic and geranic acids respectively with geranic acid being the predominant one.

Identification of peaks 9 and 10

On column 1, we observed a large peak 9, sometimes as a doublet (peaks 9 and 10), with a retention time between hydrocarbon C_{20} to C_{22}. On column Ⅱ, this peak was hardly observable for unknown reason. On column Ⅲ, this peak was not observed at all. Under the conditions used in our chromatographic program on both non - polar columns Ⅰ and Ⅱ, the retention time of the peak (or peaks) was very long (60 min). These observations suggest that it might be an acid of high molecular weight. The disappearance of the peak on extraction with dilute alkali indicated the presence of a carboxylic group. Through GC - MS, its molecular ion is estimated at *m/z* 280. The fragments at *m/z* 235 (M - 45) and *m/z* 263 corresponded to the loss of carboxyl and hydroxyl groups. Therefore it possibly could be another aliphatic acid C_{18}. A careful GC - MS data analysis indicated that it might be a 9,12 - octadecadienoic acid.

Difficulties were encountered in checking the retention time of an authentic 9,12 - octadecadienoic acid on the glc spectrum, because only in rare cases was a peak due to 9,12 - octadecadienoic observed. To investigate the possible presence of its isomers. the mandibular extract was further methylated with diazomethane. The product was observed on column Ⅱ as a triplet peak around the retention time of 63 min. Possibly the new peaks (16 and 17) as a doublet superimposed on the unknown peak 8. The phenomenon suggested that there were 2 isomeric C_{18} acids present. Derivatization with diazoethane was also performed. After comparison with authentic samples, it was found that the retention times of new peaks 16 (62.51), 17 (62.77), 18 (66.24) and 19 (66.59) on column Ⅱ were coincidental to those of methyl (9*Z*,12*Z*) -9,12 - octadecadienoate (methyl linoleate) (62.48), (9*E*,12*E*) -9,12 - octadecadienoate (methyl linolelaidate) (62.99), ethyl (9*Z*,12*Z*) -9,12 - octadecadienoate (ethyl linoleate) (66.03) and ethyl (9*E*,12*E*) -9,12 - octadecadienoate (ethyl linolelaidate) (66.66min) respectively. On column Ⅲ, the new peaks 16 and 17 were also identical in their retention times (49.80 and 50.15 min) to the authentic samples of methyl (9*Z*,12*Z*) -9,12 - octadecadienoate (50.24 min) and methyl (9*E*,12*E*) -9,12 - octadecadienoate (49.80 min).

The following observations were also made:

a) By peak enhancement method with authentic samples, it was established that methyl (9*E*, 12*E*) -9,12 - octadecadienoate cluted earlier on column Ⅲ than the (9*Z*,12*Z*) - isomer This sequence is contraty to the results obtained from non - polar columns Ⅰ and Ⅱ. This nevertheless is in accordance with the observation by Tamaki *et al* [15] and J. Klun[16] that the elution sequence of non - conjugated dienes on a polar column is *E*,*E*;*Z*,*E*;*E*,*Z*;*Z*,*Z*.

b) By GC - MS. it was proved that in the methylation product of cephalic extract. the peaks 16 and 17 have the same molecular ion (*m/z* = 294) and the same spectrum as that of methyl ester of 9,12 - octadecadienoic acids.

c) On a particular capillary column (DB - 1) the peaks 9 and 10 showed up clearly, which exactly have the same retention times (65.72 and 66.18) as those of the authentic samples (65.79 and 65.80 min). (Authentic samples required a concentration greater than 100 ng/μL to show up their peaks.)

From the above results, it can be concluded that peaks 9 and 10 are octadecadienoic acids with two double bonds located at C_9 and C_{12}, Peak 9 is the *Z*,*Z* - isomer and peak 10 is the *E*, *E* - isomer. From the calculation of the peak areas of the two isomers. it is apparent that the *E*, *E* - isomer is not the predominant one. This is the reason why on a non - polar column the larger peak, the *Z*,*Z* - isomer, elutes first and on a polar column the larger peak elutes last.

The identification of peak 8

Both on non – polar and polar columns, an unknown peak 8 was observed with a retention time above that of hydrocarbon C_{20}. By GC – MS, its molecular ion is at 286 and it has an unusually strong peak at m/z = 230.

The relative retention values of peak 8 and citrals 3 and 4 in relation to standard hydrocarbons on non – polar and polar columns were compared in Table 3.

Table 3 Comparison of retention times of peaks 3, 4 and 8 on different columns with respect to standard hydrocarbons

Compound	Non – polar column SE – 52(60M)	Polar column SP – 1000(60M)	Shift
Z – citral(peak 3)	$C_{12.3}$	$C_{17.1}$	$C_{4.8}$
E – citral (peak 4)	$C_{12.5}$	$C_{17.8}$	$C_{4.1}$
peak 8	$C_{22.7}$	$C_{24.4}$	$C_{1.7}$

The shift of peak 8(a hypothetical $C_{1,7}$) is much less than that of an aldehyde($C_{1.8}$ to $C_{5.1}$). So the possible formula range was narrowed down to the less polar $C_{21}H_{34}$ or $C_{20}H_{30}O$. the latter probably an internal ether. Peak 8 is not likely to be an alcohol in consideration of its polarity. It was noticed that a few sesquiterpenes as elemene($C_{15}H_{24}$) have some similarity in losing a mass fragment of (M – 203) in their mass spectra. Considering the characteristic peak at m/z 69. It could be a diterpene or substituted sesquiterpene.

Quantification of the major components present in the different sexes of Hylaeus albonitens

The cehalic secretion of female species collected in different locations in Australia were examined on capillary gic.

a) Collected at Dimbula. Queensland. Aug. 8. 1980.

b) Collected at Mount Molloy. Queensland. Aug. 6, 1980.

The chromatographic pattern on glc of the samples from these different locations remains the same. Analysis of the methylation products of cephalic extract of both sexes revealed that all the major peaks are present in each extract. but with some differences in the amount of individual components. The velative amount of each component from each sex were summarized in Table 4. The absolute quantity of the major components in male species were listed in Table 5.

Table 4 Relative amount of major components in the cephalic extract from each sex

	Neral	Geranial	Nerolic acid	Geranic acid	Linoleic acid	Linolelaidic acid
Male 10 heads, Mt. Molloy	100.0	151.5	0.23	2.69	133.0	30.6
Female 56 heads, Dimbula	100.0	140.0	0.53	8.67	75.7	5.0

Table 5 Absolute quantity of the major components in the cephalic extract from male species collected in Mt. Molloy(microgram/head)

Neral	Geranial	Nerolic acid	Geranic acid	Linoleic acio	Linolelaidic acid
3.4	5.6	0.01	0.10	4.4①	1.02①

① Calculated through the corresponding methyl esters.

The composition of cephalic extract of *Hylaeus albonitens* was summarized in Table 6.

Table 6 composition of the cephalic extract of *hylaeus albonitens*

Peak No.	Compound	Formula	M. W.	Structure	Analytical methods
1	3. 7 – Dimethyl – 1,6 – octadien – 3 – o (linalool)	$C_{10}H_{15}O$	154	1	CGC[①], GC – MS
2	3,7 – Dimethyl – 6 – octenal (citronellal)	$C_{10}H_{16}O$	154	2	CGC, GC – MS peak enhancement
3	(2*Z*) – 3,7 – Dimethyl – 2,6 – octadienal (neral)	$C_{10}H_{16}O$	152	3	CGC, GC – MS
4	(2*E*) – 3,7 – Dimethyl – 2,6 – octadienal (geranial)	$C_{10}H_{16}O$	152	4	CGC, GC – MS
5	(2*Z*) – 3,7 – Dimethyl – 2,6 – octadienoic acid (nerolic acid)	$C_{10}H_{16}O_2$	168	5	CGC, GC – MS, alkaline extraction, derivatization, hydrogenation
6	(2*E*) – 3,7 – Dimethyl – 2,6 – octadienoic acid (geranic acid)	$C_{10}H_{16}O_2$	168	6	*ibid.*
7	Heptadecane	$C_{17}H_{38}$	240	$C_{17}H_{36}$	CGC, GC – MS
8	Unknown	$C_{21}H_{34}$	286		CGC, GC – MS
9	(9*Z*,12*Z*) – 9,12 – Octadecadienoic acid (linoleic acid)	$C_{16}H_{32}O_2$	280	7	CGC, GC – MS, alkaline, extraction, derivatization
10	(9E,12E) – 9,12 – Octadecadienoic acid (linolelaidic acid)	$C_{18}O_{32}O_2$	280	8	*ibid.*
11	Pentacosane	$C_{25}H_{32}$	352	$C_{25}H_{32}$	CGC, GC – MS

①CGC = Capillary gas chromatography.

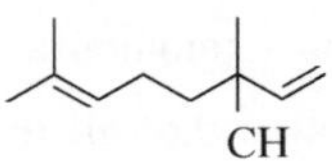

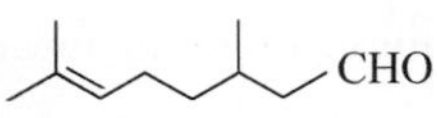

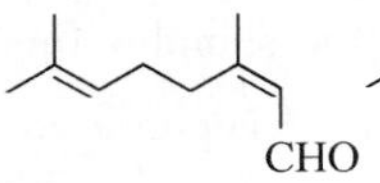

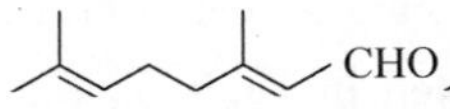

Discussion

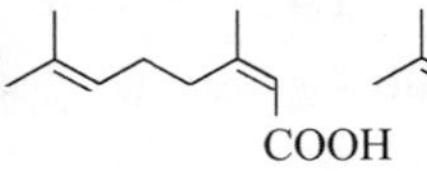

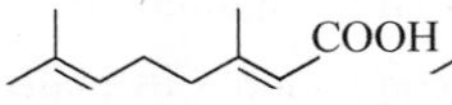

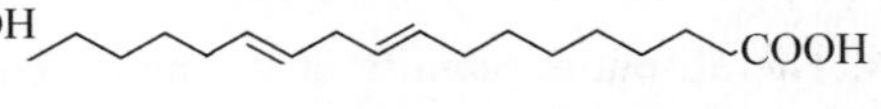

This is the first report of the chemical composition of the Australian Colletid bee, *Hylacus albonitens*. Of the 10 components identified in the cephalic extract, the 4 aliphatic acids (nerolie, geranic, linoleic, linolelaidic) posed some problems during the process of glc identification (Table 6) GC – MS often failed here in differentiating the structure of each isomer. Observation of their characteristic unsymmetrical peaks, treatment, with dilute alkali solution. derivatization and peak enhancement methods could bring satisfactory results in identifying these acids as well as their isomers. The elution scquence of the major peaks on SE52 and DB - 1 is neral. geranial, nerolic acid, geranic acid. linoleic acid and linolelaidic acid; while on Silar 10c the sequence is neral, geranial, nerolic acid, geranic acid, linolelaidic acid and linoleic acid. , Geranic acid was previously found in honey bees[17] and small carpenter bee[18] Besides, it exists widely in nature: in rose oil. black tea. gardenia flower, in fragrance raw material etc. Linoleic acid was also found in apple, grape, citrus, cottonseed oil, plant seed oil. peanut and peanut oil, corn, cocoa. rice bran oil, olive oil etc.

The wide spread of these acids in nature probably accounts for their presence in the cephalic extract of *Hylaeus albonitens*, it is appropriate to mention here that according to our quantitative calculation the geranic and linoleic acids are also the predominant isomer in comparison to their respective counterparts—nerolic and linolelaidic acids. The difference of the ratio of isomers in the cephalic extract composition is somehow, elected in the sex of *Hylaeus albonitens*: *the ratio of citral isomers* $Z:E = 1:1.5$ (male) and $1:14$ (female); for 3,7 - dimethyl - 2,6 - octadienoic acid, $Z:E = 1:11.5$ (male) and $1:16.4$ (female); for 9,12 - octadecadienoic acids, $9Z$、$12Z : 9E$、$12E = 4.41:1$ (male) and $15.1:1$ (female). The not too striking difference between sexes suggests that this is probably and aggregation pheromone composition instead of a sex pheromone one. This pattern is quite. different from the results. obtained from the mandibular glands of the American species *Hylaeus mondestus* which were reported recently to contain Z - and E - citral only, [10] and of the closely related *Colletes* bees, which were reported to contain linalool, neral and geranial in a ratio of 3: 1: 1[6].

Though ethyl linoleate has been reported once as the component of aggregation pheromone from the extract of the whole bodies of Khapra beetles *Trogoderma granarium*, [19] the 9,12 - octadecadienoic acids recorded here have not been reported previously as a natural product form gland extracts of bees.

Due to the relatively large amount of these C_{18} acids present in the cephalic extract, their function as possible pheromone constituents will be tested by bioassay. The comparison of these constituents from the mandibualr glands of closely related species of bees would be worthwhile to elucidate their taxonomic relationship in view of the progress of chemical evolution. The primitive colletid bees that have evolutionally radiated in Australia and South America are of particular interest in this respect.

Acknowledgments

We are deeply grateful to our colleagues in BARC - W, USDA: Dr. Jerone Klun, Dr. Barbara Bierl - Leonhardt, Dr. Meyer Schwarz and Dr. T. P. McGovern who offered kindly assistance and helpful suggestions. Mr. Everett D. Devilbiss obtained the related mass spectra for the project. We are also indebted to Dr. Yiu Fai Lam, University of Maryland, for his NMR spectrometer and Dr. Elizabeth Exley, University of Queensland. Australia for taxonomic determination.

References

[1] Both, R. ; Shearer, D. A. , *Nature*. 202. 320(1964).

[2] Blum, M. S. ; Crewe, R. M. ; Kerr, W. E. ; Keith, L. H. ; Garrison, A. W. ; Walker, M. M. , J, *Jnsect physiol.* ,

16,1637(1970).

[3] Bergstrom,G. ;Tengo,J. ,*ZOON Suppl.* 1,55(1973).

[4] Wheeler, J. W. ; Blum, M. S. ; Daly, H. V. ; Lislow, C. J. ; Brand, J. M. , *Ann. Entomol. Soc. Am.* , 70, 635 (1977).

[5] Crewe,R. M. ;Flelcher,D. J. C. ;*South African Journal of Science*,72,119(1976).

[6] Hefetz,A. ;Batra,S. W. T. ;Blum,M. S. ,*Experientia*,315,319(1979).

[7] Bergstrom,G. ;Tengo,J. ,*J. Chem. Ecol.* ,4,737(1978).

[8] Hefetz,A. ;Batra. S. W. T. ;Blum,M. S. ,*J. Chem. Ecol.* ,5,753(1979).

[9] Hefetz. A. ;Batra,S. W. T. ,*Experientia*,35,1138(1979).

[10] Duffield, R. M. ; Fernades, A. ; Mckay, S. ; Wheeler, J. W. ; Snelling. R. , *Comparative Biochemistry and physiology*, part B,67B,159(1980).

[11] *Applications of Naclear Magnetic Resonance Spectroscopy in Organic Chemistry* , p. 223, Jackman, L. M. ; Sternhell,S. ,Pergaman press Ltd,1959.

[12] Pickett,J. A. ;Williams,I. H. ;Martin,A. P. ;Smith,M. C. ,*J. Chem. Ecol.* ,6,425(1980).

[13] Cardillo. G. ;Contento. M. ;Sandri,S. ,*Tetrahedron Lett.* ,25,2215(1974).

[14] Wakabayashi,N. ;Sonnet,P. E. ;Law. M. W. ,*J. Medicinal Chemistry*,12. 911(1969).

[15] J. Klun. OCSL,AEQI. BARC – West. USDA(unpublished data).

[16] Hill,H. C. ,*J. Chem. Soc.* ,C 1,93(1953).

[17] Shearer,D. A. ;Boch,R. ,*J. Insect. Physiol.* ,12,1513(1966).

[18] Wheeler. J. W. , Blum, M. S. ; Daly, H. V. ; Kislow, C. J. ; Brand, J. M. , *Ann*, *Entomol. Soc. Am.* , 70, 635 (1977).

[19] Ikan,R. ;Bergmann,E. D. ;Yinov. U. ;Shulov. S. ;*Nature*,223,317(1969).

The Wittig Sigmatropic Rearrangement in A Conjugated Diene System I *

Zhengming Li[1], Tiansheng Wang[1], Enyun Yao[1], Zhengheng Gao[2]

([1]Institute of Elemento – Organic Chemistry, Nankai University, Tianjin, 300071;
[2]Department of Chemistry, Nankai University, Tianjin, 300071)

Abstract A conjugated dienic benzyl ether was shown to undergo Wittig sigmatropic rearrangement to give a mixture of [1,2] and [1,4] besides the expected [2,3] rearrangement product. The solvent effect as well as the reaction pathways were discussed.

In 1981, Nakai *et al* [1] reported that when allylic ether underwent a [2,3] Wittig sigmatropic rearrengement, the stereoselectivity relied heavily on the geometric configuration of the original substrate. Since then, [2,3] Wittig sigmatropic rearrangement has become one of the important methods in carbon – carbon bond synthesis. It is also applicable in some of the important total synthesis of natural products.

Nevertheless, none of the literature has yet been related to an allylic ether which is in corporated in a conjugated diene system. It was recorded here a Wittig sigmatropic rearrangement which occured in conjugated dienic phenyl ether. The behaviour of Wittig rearrangement in a dienic environment is very interesting both from a practical as well as theoretical viewpoint.

1 —(BuLi, −78℃)→ 2[1,2] + 3[2,3] + 4[1,4]

The starting material methyl(2*E*,4*E*) –2,4 – hexadienoate was reduced by lithium aluminium hydride to (2*E*,4*E*) –2,4 – hexadien – 1 – ol with much higher yield and purity than the reduction from sorbic acid as reported[2]. The product could be obtained up to 50 grams in one pot. The hexadienol was further proceeded to (2*E*,4*E*) –2,4 – hexadicnyl benzyl ether 1 with GC purity >99%. The Wittig sigmatropic rearrangement of 1 was carried out with butyl lithium at –78℃ in THF or THF – HMPA. The products were separated by GC and identified by GC – IR, GC – MS, ^{1}H NMR and ^{13}C NMR to be 3 compounds: 1 – phenyl – (3*E*,5*E*) –3,5 – heptadien –2 – ol(2), 1 – phenyl –2 – vinyl – (3*E*) –3 – penten –1 – ol(3) and 3 – benzyl – (4*E*) –4 – hexenal(4).

* Reprinted from *Chinese Journal of Chemistry*, 1990, 3: 265 – 270. The research work was supported by the National Natural Science Foundation of China.

Table 1　Effect of solvents on yields of rearrangement product

Solvent	Compound		
	2	**3**	**4**
THF	65.6%	15.9%	18.5%
THF - HMPA	71.6%	0	28.4%

Structural identification

Compound **2** Its structure was established by 1R, ^{1}H NMR and ^{13}C NMR. In MS data, it was observed to have a molecular ion of *m/z* 188, and also *m/z* 97 ascribed to bond breakage α - to the hydroxy functional group. The cleavage pathways of mass fragments was described as following:

Ph]+ OH *m/z* 188 —($-H_2O$)→ Ph]+ *m/z* 170 —($-CH_3$)→ *m/z* 155

m/z 188 → $PhCH_2$]+ *m/z* 91 —(-CH=OH)→ *m/z* 65

m/z 188 —($-PhCH_2$)→ OH]+ *m/z* 97 (base Deak) —($-H_2O$)→ *m/z* 79

The structure of **2** was further confirmed by stardard sample synthesized in the following way:

OH —PCC→ CHO —$PhCH_2MgCl$→ Ph, OH

All the spectra and GC retention time of **2** and the synthetic sample are identical.

Compound 3 IR data showed peaks at 3604cm^{-1} for hydroxy group, 701cm^{-1} for monosubstituted benzene ring, 920 and 999cm^{-1} for terminal alkene and 968cm^{-1} for CH ═CH. Miginiac - Groizeleau *et al.* [3] reported the same IR data for the identical structure synthesized by zinc bromide reagent. Their v_{OH} was at 3410cm^{-1}, which usually is mobile according to the degree of association of hydroxy group in different solvents.

The cleavage pathways of mass fragments was described as following:

The absence of molecular ion explained the instability of the compound under electron bombardment. There were **2** base peaks: *m/z* 107 corresponding to the α - breakage of sigmabond and *m/z* 82 ascribed to the rearrangement through intermediate cyclic structure. The structure of **3** was thus assigned.

Compound **4** IR spectra denoted an aliphatic aldehyde at 1740(strong) and 2710 cm^{-1}. There was no hydroxy absorption. The peak at 697 cm^{-1} was for monosubstituted benzene and at 965 cm^{-1} for *E* - olefine ^{1}H NMR showed a characteristic 9.22 ppm for aldehyde hydrogen. The integral curve at 5 - 6 ppm showed only two olefinic hydrogens. ^{13}C NMR spectra showed eleven peaks with a characteristic carbonyl carbon peak at 202.45 ppm.

Ph OH

PhCH=OH ⁺
m/z 107 (base)

−CO

C_6H_7
m/z 79

Ph O–H⁻

Ph O⁺

−PhCHO

m/z 82 (base)

−CH_3

m/z 67

The cleavage pathways of mass fragments of **4** was proposed as following:

m/z 170 ← −H_2O

m/z 159 ← −CHO

Ph CHO⁺

m/z 183

O H

CHO

m/z 97

−CO

m/z 69

$PhCH_2^+$
m/z 91

−CH=CH

m/z 65

OH
H +

m/z 144 (base)

The mechanism of the Wittig sigmatropic rearrangement was postulated as rollowing: Namely **2** was formed through [1,2] sigmatropic rearrangement, **3** through [2,3] and **4** through [1,4] arrangement.

It is very interesting to note that the Wittig sigmatropic rearrangement in THF resulted in **3** rearrangement products simultaneously. Besides the normally expected [2,3] product, there also appeared [1,2] and [1,4] products. As contrast to the allyl benzyl ether which gave predominant [2,3] product as reported earlier[4], an unexpected high yield of **2** was obtained. One reason why the yield of [2,3] product was low in THF and nil in THF – HMPA might due to the presence of the conjugated double bond in the molecule, which seems to favour the negative ion stabilizing on 2 – C in **6** instead of the α – position of 2 – C in **5**. The delocalization of the negative ion might explain a drastic shift from [2,3] to [1,2] and [1,4] pathways as well as the simultaneous [1,2] and [1,4] formation. The competition among these rearrangements enhanced when the reaction solvent was shifted to THF – HMPA, in which HMPA as a very polar solvent with its negatively charged oxygen in P=O severely suppressed the intermediate formation of 2 – C in **5**. As a general feature of Wittig [2,3] rearrangement, **3** was assigned to an *E* – configuration. The chirality of the products will be investigated later.

Experimental

IR: Shimadzu IR – 435. NMR: JEOL FX – 90Q. GC: HP 5890 OV – 1, capillary column (25 ×

0.25 mm). GC – IR: Nicholet 170 FT – IR. GC – MS: HP 5890A EI.

(2E,4E) –2,4 – Hexadien –1 – ol

At –30℃ to –40℃, 88.3 g (0.7 mol) of methyl sorbate in 150mL absolute ether was added into 1000mL absolute ether containing 24 g (0.653 mol) lithium aluminium hydride. Then the reaction temperature was raised to –10℃ to 0℃ for 1 h. 10 g ethyl acetate was added to decompose excess lithium aluminium hydride, followed by 130mL of saturated ammonium chloride solution. After filtration, the filtrate was concentrated to 300mL, then washed with dilute hydrochloric acid and saturated NaCl solution, dried with anhydrous sodium sulfate. The solvent was removed and 56.9 g (yield 83%) colorless liquid was obtained by vacuum distillation. bp 68 ~ 70℃/8 mmHg. GC purity, 99.5%. v_{max}: 3300 (s, OH), 1600 (w, C=C), 980 (s, *E* – CH = CH) cm^{-1}. δ_H: 6.02 – 5.37 (4H, m, 2 × CH=CH), 3.96 (2H, d, *J* = 6Hz, CH_2O), 2.60 (1H, s, OH), 1.61 (3H, d, *J* = 6Hz, CH_3) ppm. δ_c:

10.03 131.69 129.85 OH
129.30 130.38 83.18

(2E,4E) –2,4 – Hexadienyl benzyl ether (1)

3.8 g (80%) sodium hydride (127 mmol) was added into 100mL flask, followed by 20mL anhydrous THF and 1.96g (20 mmol) of (2*E*,4*E*) –2,4 – hexadien – 1 – ol. After reaction for 10 min 5.0g (39.5 mol) benzyl chloride was added and refluxed for 12h. After cooling, 100mL water was added and extracted with 3 × 50mL ether. The solvent was removed and the crude product was column chromatographed. Eluting solvents sequence: petroleum ether, pethroleum ether – ether (20∶1). A colorless liquid was obtained, 3.28g, yield 87.1%. Further purified by spinning band distillation to remove a small quantity of byproduct, the dibenzyl derivatives. GC purity > 99%. Anal. $C_{13}H_{16}O$, Calcd: C, 82.74; H, 8.75. Found: C, 82.88; H, 8.68. v_{max}: 1660 (w, CH), 1450, 1360, 1105, 985 (s, *E*—CH=CH), 694 (s, Ar) cm^{-1}. δ_H: 7.29 (5H, s, Ar – H), 6.37 – 5.48 (4H, m, 2 × CH=CH), 4.47 (2H, s, CH_2Ar), 4.01 (2H, d, *J* = 6 Hz, OCH_2C), 1.74 (3H, d, *J* = 6 Hz, CH_2) ppm. δ_C:

127.74 128.34
10.09 129.96 126.71 O 130.41 127.52
133.10 130.98 70.53 71.89

The Wittig sigmatropic rearrangement of (2*E*,4*E*) –2,4 – hexadienyl benzyl ether

(2*E*,4*E*) –2,4 – Hexadienyl benzyl ether 0.75g (3.98 mmol) was dissolved in 4mL anhydrous THF under N_2 protection at –78℃. 3.5mL (5.50 mmol) of *n* – butyl lithium/ether solution (1.57*N*) was added dropwise, and allowed to react for 8h, temperature gradually raised to 0℃ for 1h. The reaction mixture was acidified with dilute hydrochloric acid, extracted with ether (3 × 30mL), washed with saturated brine, dried with anhydrous sodium sulfate. Solvent was stripped off to give a colorless liquid, 0.685 g. GC analysis showed the crude product to be a mixture with retention time **2** 6. 42, **3** 4. 22 and **4** 4. 57 min respectively. The components were separated by preparative GC: 65. 6%, **3** 15. 9% and **4** 18. 5%. In THF – HMPA: Same procedure as above. Product 1. 2 g Purified by column chromatography (cluting solvent: petroleum ether – ether), GC analysis: 2 71. 6%, 4 28. 4%.

2 v_{max} (GC – IR): 3643 (w, OH), 3072, 3032 (=C—H), 1604 (C=C), 984 (s, *E*—CH=CH), 699 (s, monosubstituted benzent) cm^{-1}, δ_H: 7. 20 (5H, s, Ar—H), 6. 30 – 5. 43 (4H, m, 2 × CH=CH), 4. 29 (1H, q, *J* = 6Hz, C=C—CHO), 2. 17 (2H, d, *J* = 7Hz, CH_2Ar), 2. 09 (1H, s, OH), 1. 71 (3H, d, *J* = 6Hz, CH_3) ppm. δ_c: 137. 93, 132. 24, 130. 83, 129. 85, 129. 53, 128. 34, 128. 01, 126. 39, 73. 13, 44. 10, 18. 09ppm. *m/z*: 188 (M), 170 (M—H_2O), 155 (M—H_2O—CH_2) 141, 128, 115. 97, 97 (base, M—$PhCH_2$), 91 ($PhCH_2$), 79 (C_6H_7), 69. 43.

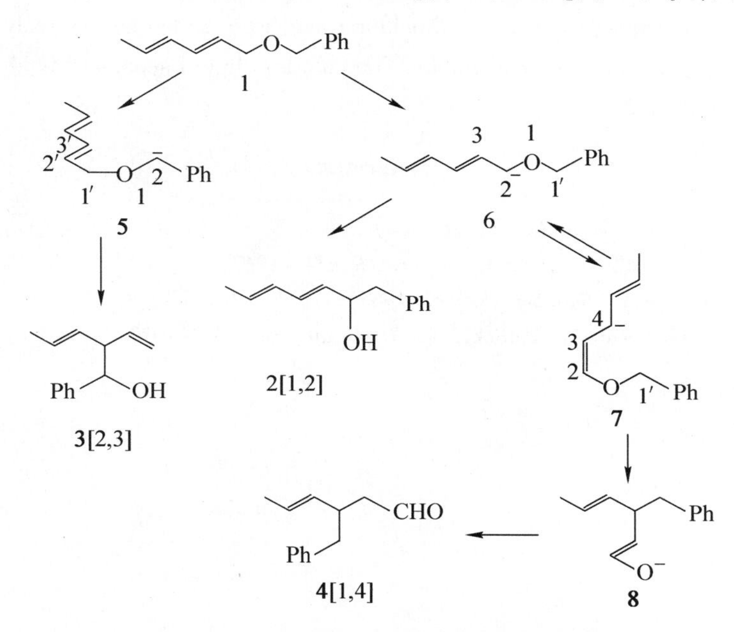

Fig. 1 The reaction mechanism of Wittig sigmatropic rearrangement of (2*E*,4*E*) –2,4 – hexadienyl benzyl ether

3 v_{max} (GC – IR): 3604 (w, OH), 1639 (w, C=C), 999 (m), 920 (s, CH=CH_2), 701 (s, monosubstituted benzene) cm^{-1}. *m/z*: 107 (base, M – C_6H_9), 82 (M – PhCHO), 79 (C_6H_7), 67 (C_5H_7), 57. 41 (C_3H_5), 29. 27.

4 v_{max} (GC – IR): 2711 (m), 1740 (s, CHO), 965 (m, *E*—CH=CH), 698 (s, monosubstituted benzene). δ_H: 9. 22 (1H, CHO), 7. 12 (5H, ArH), 5. 38 – 5. 29 (2H, m, CH=CH), 3. 58 (3H, m, $PhCH_2CH$), 2. 25 (2H, d, d, CH_2CO), 1. 58 (3H, d, CH_2) ppm. δ_c: 202. 45, 139. 43, 133. 00, 129. 31, 128. 28, 128. 01, 126. 12, 47. 78, 41. 71, 39. 01, 17. 88ppm. *m/z*: 188 (M), 170 (M—H_2O), 159 (M—CHO), 144 (M—CH_2=CHOH), 129, 115, 102, 97 (M—$PhCH_2$), 91 ($PhCH_2$), 82. 69 (C_5H_9), 65 (C_5H_5), 41 (C_3H_5).

2E, 4E – Hexadienal

16. 5 g (76. 6mmole) chromic anhydride pyridine hydrochloride (PCC) and 100mL anhydrous dichloromethane and 5 g (50. 9mmol) 2*E*, 4*E* – 2,4 – hexadien – 1 – ol in 10mL dichlorometh-

ane was allowed to react for 1. 5h, then extracted with ether. The product was purified by column chromatography. The colorless liquid was purified further by vacuum distillation. bp 62 – 64℃/20mmHg. 2. 35g (48%), GC purity >94%. v_{max}:2720(w), 1675(s, CHO), 1640(s, C=C), 985(s, E—CH=CH) cm^{-1}. δ_H:9. 06(1H, d, J = 7 Hz, CHO), 7. 20 – 5. 90(4H, m, 2 × CH = CH), 1. 88(3H, d, J = 7 Hz, CH_2) ppm. δ_c:193. 83(1 – C), 129. 64(2 – C), 152. 71 (3 – C), 129. 96(4 – C), 141. 93(5 – C), 18. 74(6 – C) ppm.

1 – Phenyl – (3E,5E) – 3,5 – heptadien – 2 – ol(2)

Excess magnesium powder with 10 mmole benzyl chloride was allowed to react in ether to form corresponding Grignard reagent. 0. 3g(2*E*, 4*E*) – 2, 4 – hexadienal was then added. The mixture was stirred overnight, water and hydrochloric acid were added successively, and the product was purified by column chromatography. The colorless liquid separated is identical with the rearrangement product **2**.

References

[1] Nakai, T.; Midame, K., *Chem. Rev.*, 86, 885(1986).
[2] Nystron, R. F.; Brown, W. C., *J. Am. Chem. Soc.*, 69, 2548(1947).
[3] Miginiac – Groizeleau, L., *Bull. Soc. Fr.* 3563(1965).
[4] Baldwin J. E.; DeBermardis, J.; Patrick, J. E., *Tetrahedron Lett.*, 353(1970).

山梨酸的合成及其生物活性研究*

李正名[1]　王天生[1]　张祖新[1]　张殿坤[1]　么恩云[1]　高振衡[2]

（[1]南开大学元素有机化学研究所，天津　300071；
[2]南开大学化学系，天津　300071）

关键词　山梨酸　2，4－己二烯酸　立体选择性合成

在不少生物活性分子中，如维生素A，白三烯（leukotriene）[1]和某些昆虫信息素等[2]含有共轭烯键结构。它们对立体构型有特定要求。为了研究立体定向合成共轭双键的方法，本文选择山梨酸（2，4－己二烯酸sorbic acid，1）作为模型化合物，试图通过山梨酸的四个异构体来研究共轭双键与生物活性之间的关系。山梨酸是含二个共轭烯键的不饱和羧酸，1953年美国正式批准作为食品添加剂。

$$CH_3CH{=}CHCH{=}CHCOOH$$

由于其毒性极低（与食盐差不多），已广泛用于各类食品中，但目前工业生产的山梨酸主要是其中2*E*，4*E*体，其他三个异构体的生物活性未见文献报道。因此，研究其构型与活性关系无论从理论上或应用上都有重要意义。

1955年，Allan[3]合成了山梨酸的三个异构体，但合成路线复杂，收率低，限于当时仪器的限制，没有对异构体的纯度和反应的立体选择性进行专门研究。工业上采用的合成工艺如山梨醛氧化法、巴豆醛与乙烯酮气相法、丙二酸缩合法、γ－丁烯－γ－丁酯开环法所得到的都是热力学稳定的2*E*，4*E*异构体，对各反应的立体化学均未涉及。

为了得到四个纯异构体，我们从易得的巴豆醛和炔丙醇为起始原料，合成山梨酸（2）及其乙酯（3）八个异构体。

下面分述八个异构体的合成方法：

E－巴豆醛与O，O－二乙基膦酰乙酸三甲基硅酯（a）在丁基锂作用下，反应产物水解得到2*E*，4*E*－山梨酸（2）。

1. （2*E*，4*E*）－2，4－Hexadienoic acid

CHO —a, 68.6%→ COOH

*m.p.*131～133℃
2*E*,4*E*–2

2. （2*Z*，4*E*）－2，4－Hexadienoic acid

CHO —b, 93%→ Br, Br —c→ [≡—Li]

* 原发表于《有机化学》，1990，10：117－125。

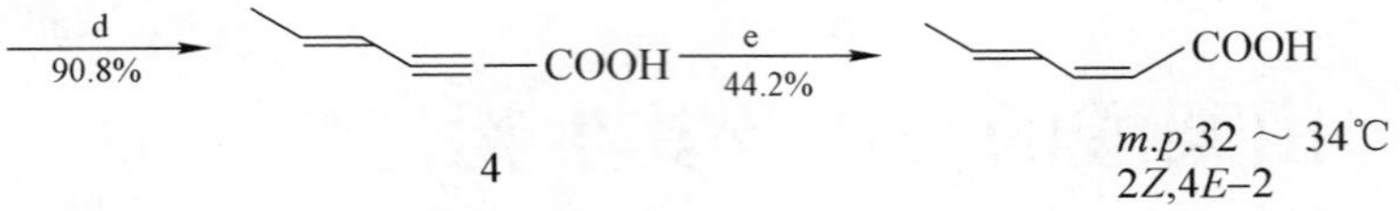

$\xrightarrow[90.8\%]{d}$ (4) $\xrightarrow[44.2\%]{e}$ COOH

m.p.32 ~ 34℃
2Z,4E-2

3. (2E, 4Z) -2, 4-Hexadienoic acid

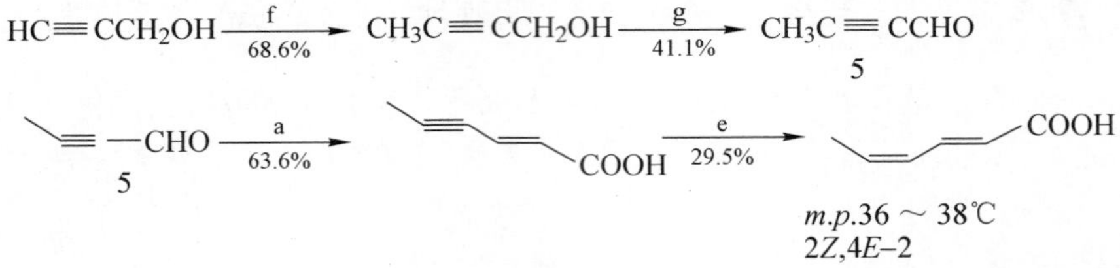

$HC \equiv CCH_2OH \xrightarrow[68.6\%]{f} CH_3C \equiv CCH_2OH \xrightarrow[41.1\%]{g} CH_3C \equiv CCHO$ (5)

—CHO (5) $\xrightarrow[63.6\%]{a}$ COOH $\xrightarrow[29.5\%]{e}$ COOH

m.p.36 ~ 38℃
2Z,4E-2

4. (2Z, 4Z) -2, 4-Hexadienoic acid

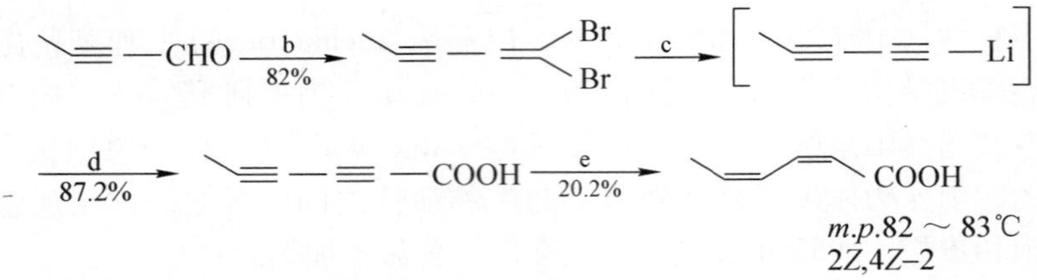

—CHO $\xrightarrow[82\%]{b}$ Br, Br $\xrightarrow{c}$ [—Li]

$\xrightarrow[87.2\%]{d}$ —COOH $\xrightarrow[20.2\%]{e}$ COOH

m.p.82 ~ 83℃
2Z,4Z-2

5. Ethyl (2E, 4E) -2, 4-Hexadienoate

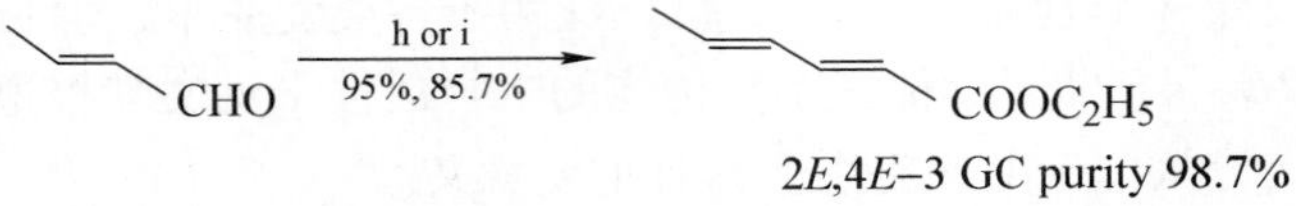

CHO $\xrightarrow[95\%,\ 85.7\%]{h\ or\ i}$ $COOC_2H_5$

2E,4E-3 GC purity 98.7%

6. Ethyl (2Z, 4E) -2, 4-Hexadienoate

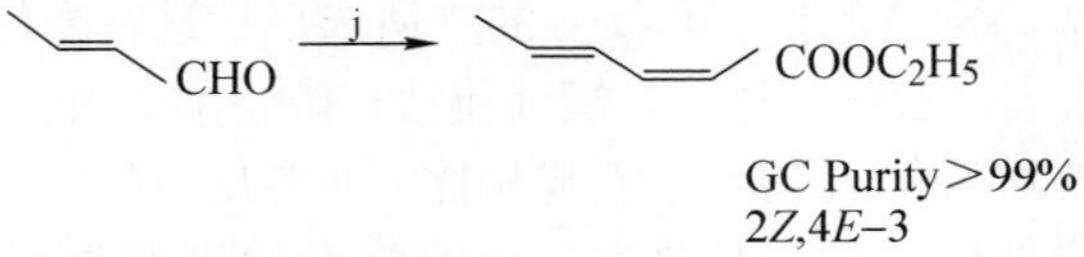

CHO $\xrightarrow{j}$ $COOC_2H_5$

GC Purity>99%
2Z,4E-3

7. Ethyl (2E, 4Z) -2, 4-Hexadienoate

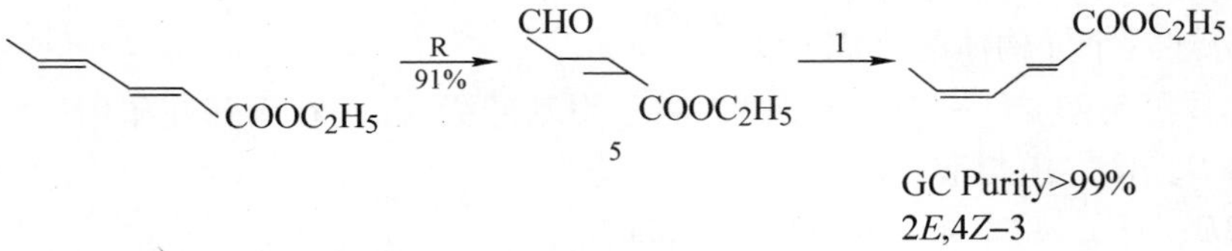

$COOC_2H_5$ $\xrightarrow[91\%]{R}$ CHO, $COOC_2H_5$ (5) $\xrightarrow{l}$ $COOC_2H_5$

GC Purity>99%
2E,4Z-3

8. (2Z, 4Z) -2, 4-Hexadienoic acetate

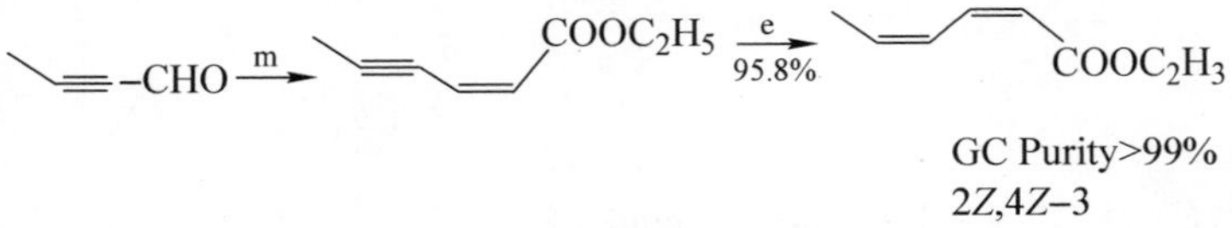

-CHO $\xrightarrow{m}$ $COOC_2H_5$ $\xrightarrow[95.8\%]{e}$ $COOC_2H_3$

GC Purity>99%
2Z,4Z-3

图1 山梨酸及其乙酯各异构体的立体有择合成

Fig. 1 The stereoselective synthesis of the isomers of sorbic acid and their ethyl esters

a. $(EtO)_2P(O)CH_2COOSiMe_3$, BuLi/THF, CO_{21}; b. $Ph_3P-CBr_4-Zn/CHCl_2$; c. 2 BuLi/THF; d. CO_7, H_2O_3; e. H_2, Lindlar catalyst/quinoline; f. NH_2Li/liq. NH_3, CH_31; g. PCC/CH_2Cl_2; h. $Ph_3AsCH_2COOC_2H_5Br^-$, EtONa/EtOH; i. $Ph_3AsCH_2COOC_2H_5Br^-$, K_2CO_3 - acetonitrile; J. $Ph_3P=CHCOOC_2H_5/MeOH_3$, revalving fractional; k. O_3/EtOH, -78℃ Me_2S; 1. $Ph_3PCH_2CH_3Br^-/NaN(SiMe_3)_2$, revalving fractional; m. $Ph_3P-CHCOOC_2H_5$/EtOH resolution of flash chromatogram

E－巴豆醛与三苯膦－四溴化碳－锌（b）生成的二溴代甲叉三苯基膦烷反应，得到E－1，1－二溴代－1，3－戊二烯，后者与两分子正丁基锂反应得一炔基锂，再与干冰反应，得到E－4－己烯－2－炔酸（4），经 Lindlar 催化加氢，得2Z，4E－山梨酸（2）。

从丁炔醇氧化，经丁炔醛（5）与上述反应相似，可得 2E，4Z－山梨酸（2）。从丙炔醛开始可得2Z，4Z－山梨酸（2）。

为了进一步验证山梨酸各异构体的构型，还合成四种构型的山梨酸乙酯。

巴豆醛与溴化乙氧羰基亚甲基三苯基胂（h）反应得到纯度 >98.7% 的 2E，4E－山梨酸乙酯（3）。反应中用乙醇钠或固体碳酸钾都有立体专一性。

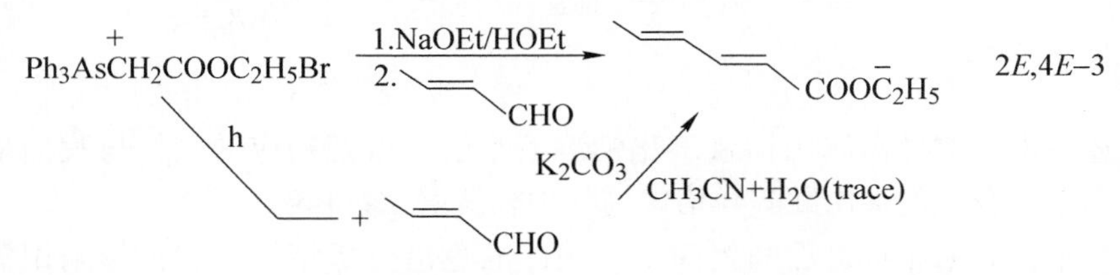

为了得到2Z，4E－山梨酰乙酯，用巴豆醛与乙氧羰基甲叉三苯基膦烷（j）反应，在不同溶剂中所得的异构体用硅胶柱层析或旋转带分馏，得到纯度 97% ~99% 的 2Z，6E－3（表1）。

$Ph_3P{=}CHCOOC_2H_5$ + CHO ⟶ $COOC_2H_5$ + $COOC_2H_5$

j

2Z,4E–3(34%)　　2E,4E–3(86%)

为了得到2E，4Z－山梨酸乙酯，将2E，4E－3 进行选择性臭氧化[4]，再用二甲硫醚还原得到E－4－氧代丁烯酸乙酯（6）然后再与溴代乙基三苯基膦/六甲基二硅胺钠[1]反应，得到比例约为 1∶1 的 2E，4Z 和 2E，4E 山梨酸乙酯的混合物。最后，用旋带分馏分离得到纯度为 98% 的 2E，4Z－（3）。

表1　巴豆醛与乙氧羰基甲叉三苯基膦烷反应条件

Tab 1　The reaction condition of crotonaldehyde with ethoxycarbonyl methylidene triphenyl phosphorane

Reaction condition				Yield（%）	Isomeric Ratio（°）		Purity①（%）
solvent	additive	t（℃）	h		2Z，4E	2E，4E	
benzene	—	78	6	84.2	13.4	86.6	99.8
DMF	—	25	12	76.3	10.6	89.4	99
DMF	H_2O	25	12	76.1	13.8	86.2	99.4
DMF	$LiNO_3$	25	12	79.5	15.4	84.6	99.4
methanol	—	25	12	84.7	34.3	65.7	97

①determined by GC.

$COOC_2H_5$ —(1.O_2; 2.Me_aS)→ CHO / $COOC_2H_5$

5

$Ph_3\overset{+}{P}CH_2CH_5Br^-$ —(1. $NaN(SiMe_3)_2$; 2. 5)→ $COOC_2H_5$ + $COOC_2H_5$

2E,4Z–3(53%)　　2E,4E–3(47%)

由 PCC 氧化2－丁炔醇生成的2－丁炔醛与膦烷（m）进行 Wittig 反应，得到了 *Z* 体和 *E* 体为 1∶1 的2－己烯－4－炔酸乙酯（7），经柱层析分别得到纯度97%以上的 *Z* 体和 *E* 体。

CHO + $Ph_3P=CHCOOC_2H_5$ ⟶ $COOC_2H_5$ + $COOC_2H_5$

Z－7(48%)　　*E*－7(52%)

将 *Z*－7 进行 Lindlar 氧化，可选择性得到2*Z*，4*Z*－山梨酸乙酯。

$COOC_2H_5$ $\xrightarrow[\text{Lindlar}]{H_2}$ $COOC_2H_5$

Z－7　　2*Z*,4*Z*－3

将上述各山梨酸异构体用稀氢氧化钠溶液水解，得到相应的酸，与我们从巴豆醛和炔丙醇为原料合成的各山梨酸异构体的物理和波谱常数相吻合。

迄今为止，尚未见报道乙酯类四个异构体的色谱分析结果。本工作采用极性大的聚乙烯醇毛细管柱分离四个异构体可得满意的结果（图2）。

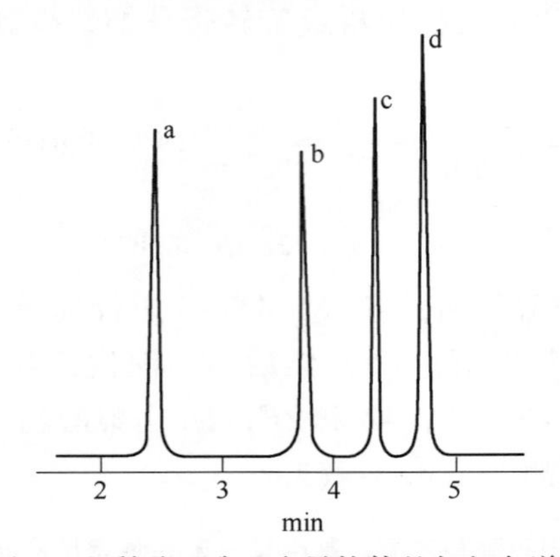

图2　山梨酸乙酯四个异构体的气相色谱分析

Fig. 2　GC analysis of 4 stereoisomers of ethyl esters of sorbic acid

色谱柱：WCOT PEG－20 M，毛细管柱（25m×0. 25mm）

分离条件：柱温110℃，进样及鉴定器温度200℃。柱前压1. 5kg/m^2

保留时间：a）2*Z*，4*Z*－体（2. 56min）b）2*Z*，4*E*－体（3. 84min）c）2*E*，4*Z*－体（4. 35min）d）2*E*，4*E*－体（4. 79min）

实验

分析仪器：岛津 RI－435 型红外光谱仪。JEOL FX－90Q 核磁共振仪。HP－5890A 毛细管气相色谱仪。PE－251 旋带分馏仪。VG－7070E 质谱仪。所有试剂重蒸，溶剂均经无水处理。

（2*E*，4*E*）－2. 4－己二烯酸（2*E*，4*E*－2）

在250mL 反应瓶中，氮气保护下，加入5. 4g（20. 1mmol）磷试剂 a 和100mL 无水四氢呋喃，在水浴冷却下，滴入1. 16mol/L 的正丁基锂－乙醚溶液 17mL（19. 7mmol）室温搅拌0. 5h，一次加入1. 25g（17. 8mmol）巴豆醛的10mL 无水四氢呋喃溶液。室温反应1 h，反应液浓缩到50mL，倒入300mL15%的氢氧化钠溶液中，100mL 乙醚提取，分出水层，用3mol/L盐酸酸化，乙醚提取（3×150mL），水洗，干燥，浓缩所得固体在水中结晶，得1.37 g白色晶体，收率68.6%，m. p. 131～133℃（文献值：134～135℃）[5]。$C_6H_8O_2$，计算值：C，64.27，H，7.19；实测值：C，64.41，H，7.43。v_{max}：1690（s，C＝O），1635（s，C＝C），1610（s，C＝C），995（s，E—C＝CH）cm^{-1}. δ_H：11.63（1 H，s，COOH），7.33（1 H，d，t，J＝15，5 Hz，β－H），6.21～6.13（2 H，m，CH_3—CH＝CH），5.73（1H，d，J＝15 Hz，α－H），1.85（3 H，d，J＝5Hz，CH_3）ppm。δ_{13C}：172.93，147.29，140.74，129.79，118.36，18.79ppm。m/z：113（M^+＋1，6），112（M^+，72），111（M^+－1，9）。

E－1，1－二溴代－1，3－戊二烯

在氮气保护下，用7 g锌粉（90%），26.2 g（0.1 mol）三苯基膦及100mL经P_2O_5干燥的二氯甲烷，置于250mL反应瓶中，在20℃下，滴入33.2 g（0.1 mol）四溴化碳的30mL二氯甲烷溶液。在20℃，反应24 h。同样温度下，加入3.5 g（0.05 mol）巴豆醛的20mL二氯甲烷溶液，搅拌2 h，加入石油醚，过滤，滤液浓缩后得10.5 g二溴化物。收率93%，用于下步反应。

E－4－己烯－2－炔烯（4）

在250mL反应瓶中，加入8.8g（39mmol）*E*－1，1－二溴代－1，3－戊二烯和100mL无水四氢呋喃，氮气保护下，于－78℃滴入57mL（74mmol）1.30mol/L正丁基锂－乙醚溶液，搅拌1h，升温到25℃反应1h。在－70℃下加入固体干冰20g，自然升温到25℃。1h后，浓缩到80mL，倒入含3g氢氧化钠的400mL水中，乙醚提取，水层酸化至pH＝1，再次乙醚提取，干燥，浓缩，石油醚重结晶，得3.7g白色絮状固体，收率90.8%，m. p. 128～132℃（文献值130.5～132.5[3]）。

(2*Z*，4*Z*)－2，4－己二烯酸（2*Z*，4*E*－2）

在反应瓶中，放入400mg（3.63mmol）*E*－4－己烯－2－炔酸，45mL正己烷、7mL乙酸乙酯、100mg Lindlar催化剂及2滴喹啉，搅拌下吸氢，1 h吸氢92mL。加入20mL乙醚，过滤，滤液用5%盐酸洗二次，饱和食盐水洗到中性，干燥氮气下浓缩，残余物用正戊烷在－78℃结晶二次，得180mg白色针状晶体，收率44.2%，m. p. 32～34℃（文献值32～35℃[6]），$C_6H_8O_2$，计算值：C，64.27；H，7.19。实测值：C，63.83；H，7.23。v_{max}：1690（s，C＝O），1637（s，C＝C），1600（s，C＝C），997.960（s，E—C＝CH），835（s，Z－C＝CH）cm^{-1}。δ_H：9.71（1H，s，COOH），7.28（1H，m，γ－H），6.59（1H，t，J＝11Hz，β－H），6.08（1H，d，q，J＝15.7 Hz，δ－H），5.51（1H，d，J＝11 Hz，α－H），1.84（3H，d，d，J＝7，1 Hz，CH_3）ppm。δ_{13C}：172.27，147.56，141.77，128.55，114.63，18.79ppm。m/z：113（M^+＋1，5），112（M^+，66），111（M^+－1，8）。

***E*－2－己烯－4－炔酸**

在反应瓶中，加入5.70g（21.2mmol）磷试剂a和100mL无水四氢呋喃，水浴冷却，氮气保护下加入17mL（21mmol）的1.23mol/L正丁基锂－乙醚溶液，室温搅拌0.5h，加入含1g（14mmol）2－丁炔醛的10mL无水四氢呋喃溶液。反应1h，浓缩至50mL，倾入300mL 15%氢氧化钠溶液中，乙醚提取。水层用稀盐酸酸化至pH＝1，乙醚提取，水洗，干燥，蒸去溶剂后，残留物用氯仿－石油醚结晶，得1.0g白色固体，收率63.6%，m.p.165～170℃。v_{max}：2210（m，C≡C），1680（s，C═O），1615（s，C═C），962（s，*E*—C═CH）cm^{-1}。δ_H：11.4（1 H，s，COOH），6.79（1H，d，q，*J*＝16，1 Hz，≡CCH），6.10（1 H，d，*J*＝16 Hz，CHCOO），2.04（3H，d，*J*＝1Hz，CH_3）ppm。δ_{13C}：171.30，128.61，128.61，98.16，77.03，4.88ppm。*m/z*：111（M^+＋1，7），110（M^+，100）。

(2*E*，4*Z*)－2，4－己二烯酸（2*E*，4*Z*－2）

反应瓶中加入500mg（4.54mmol）上述炔酸、60mL正己烷、5mL乙酸乙酯、150mg Lindlar催化剂和3滴喹啉。操作同2*Z*，4*E*－2，得150mg白色结晶，收率29.5%，m.p.36～38℃（文献值35～38℃[3]）。$C_6H_8O_2$，计算值：C，64，27；H，7.19，实测值：C，64.69，H，7.46。v_{max}：1685（s，C═O），1627（s，C═C），1605（s，C═C），990，945，870（m），760（w），685（m）cm^{-1}。δ_H：12.0（1H，s，COOH），7.14（1H，d，d，*J*＝15，11 Hz，β－H），6.32～5.88（2 H，m，CH_3CH═CH），5.85（1H，d，*J*＝15Hz，α－H），1.92（3 H，d，*J*＝7Hz CH_3C═C）ppm。δ_{13C}：173.13，141.55，137.22，127.25，120.21，14.13ppm。*m/z*：113（M^+＋1，5%），112（M^+，67%），111（M^+－1，12%）。

1，1－二溴代－1－戊烯－3－炔

操作同1，1－二溴代－1，3－戊二烯。由1.16g（17mmol）2－丁炔醛，得3.13g此二溴化合物，收率80%，直接用于下步反应。

2，4－己二炔酸

氮气保护下，加28mmol丁基锂－乙醚溶液和50mL无水四氢呋喃，在－78℃下，滴入3.13g（14mmol）上述二溴物溶于20mL无水四氢呋喃的溶液，反应1h，升温，再在25℃反应1h。参用前述*E*－4－己烯－2－炔酸的实验操作，得1.32g浅黄色固体，收率87.2%，m.p.116～117℃（文献值：118～120℃[3]）。$C_6H_4O_2$，计算值：C，66.67；H，3.73）；实测值C，67.21；H，3.57。

(2*Z*，4*Z*)－2，4－己二烯酸（2*Z*，4*Z*－2）

500mg（4.62mmol）2，4－己二炔酸、40mL乙酸乙酯、300mg Lindlar催化剂及0.3mL喹啉，氢化操作参照2*Z*，4*E*－2，经60～90℃石油醚结晶，得105mg白色晶体，收率20.2%，m.p.82～83℃（文献值：81.5～82.5℃[3]）。$C_6H_8O_2$，计算值：C，64.27；H，7.19。实测值；C，64.27，H，7.28。v_{max}：1690（s，C═O），1620（s，C═C），1590（s，C═C），825（m），780（m，Z—C═CH）cm^{-1}。δ_H：11.94（1H，s，COOH），7.26（1H，t. q. *J*＝11.2Hz，α－H），7.03（t，1 H，*J*＝11 Hz，β－H），6.01（m，1 H，δ－H），5.65（1 H，d，*J*＝11 Hz，α－H），1.85（3 H，d，d，*J*＝7，2 Hz，CH_3）ppm。δ_{13C}：166.53，138.47，135.54，125.46，117.37，59.81，

14.30，13.22ppm。m/z：113（M^+ +1，4），112（M^+，61），111（M^+ −1，8）。

(2*E*，4*E*) −2，4−己二烯酸乙酯 (2*E*，4*E*−3)

氮气保护下，将28mg（1.2mmol）金属钠溶于3mL无水乙醇，在20℃，滴入含0.71g（1.5mmol）溴代乙氧羰基亚甲基三苯基胂（h）的5mL无水乙醇溶液。搅拌10min后，缓加含70mg（1.0mmol）巴豆醛的4mL无水乙醇溶液，反应3h。加20mL水稀释，乙醚提取，干燥，蒸去溶剂后柱层析分离。先用石油醚淋洗痕量的三苯基胂，再改用石油醚∶乙醚＝10∶1淋洗，TLC检测，浓缩后得产品，收率96%，GC纯度大于99%，其中2*E*，4*E*体98.7%，2*Z*，4*E*体1.3%。v_{max}：1710（s，C═O），1642（s，C═C），1615（m，C═C），995（s，*E*—C═CH），945（w，*E*—C═CH）cm^{-1}。δ_H：7.25（1 H，d，d，d，*J*＝15，7，4 Hz，β−H），6.29～6.10（2H，m，CH_3CH═CH），5.74（1 H，d，*J*＝15 Hz，α−H），4.18（2 H，q，*J*＝7 Hz，CH_2O），1.84（3 H，d，*J*＝5 Hz，CH_3C═C），1.29（3 H，t，*J*＝7 Hz，CH_3CH_2）。δ_{13C}：166.91，144.59，138.79，129.64，118.91，59.86，18.31，14.09ppm. m/z：141（M^+ +1，12），140（M^+，100）125（M^+—CH_3，53%）。

(2*Z*，4*E*) −2，4−己二烯酸乙酯 (2*Z*，4*E*−3)

氮气保护下，将34.8g（0.1mol）乙氧羰基甲叉三苯基膦烷（j）溶于100mL无水甲醇，搅拌，滴加8g（0.11mol）巴豆醛。室温搅拌过夜，加水稀释，乙醚提取，干燥，真空赶去溶剂后，滤去三苯氧膦，母液经浓缩，减压蒸馏，得10.5g，收率74.9%，b.p.101～106℃/20mmHg。GC纯度95%。2*Z*，4*E*体含量34.1%，2E，4E体含量65.9%，将此混合物进行旋带分馏分离，收集纯度为99%的2*Z*，4*E*体2g。第二馏分为纯度99%的2*E*，4*E*体3g。v_{max}：1710（s，C═O），1640（s，C═C），1600（m，C═C），995（m），960（m，E—C═CH），835（m，Z—C═CH）cm^{-1}。δ_H：7.33（1H，d，d，q，*J*＝15，11，1.5Hz，γ−H），6.47（1H，t，*J*＝11Hz，β−H），6.00（1H，d，q，*J*＝15，7Hz，γ−H），5.47（1H，d，*J*＝11Hz，α−H），4.11（2H，q，*J*＝7Hz，CH_2O），1.81（3H，d，d，*J*＝7，1.5Hz，CH_3C═C），1.23（3H，t，*J*＝7Hz，CH_3CH_2O）ppm。δ_{13C}：166.15，144.86，139.71，128.23，115.17，59.48，18.36，14.09ppm。m/z：141（M^+ +1，8），140（M^+，77），125（M^+—CH_3，57）。

_E_−4−氧代丁烯酸乙酯 (6)

在臭氧反应瓶中，加入7g（50mmol）山梨酸乙酯和150mL无水乙醇，在−78℃下臭氧化，TLC监测，反应液倾入250mL反应瓶中，加5mL二甲硫醚，渐升至室温，搅拌过夜，真空浓缩，残余物用石油醚∶乙醚＝7∶3作淋洗剂进行柱层析，分离浓缩，减压蒸馏，收集b.p.86～88℃/23mmHg 5.6g，收率91%。GC纯度为100%。文献值：b.p.70～75℃/15mmHg[4]。δ_H：9.58（1H，d，d，*J*＝6，2Hz，CHO），6.72～6.57（2H，m，CH═CH），4.11（2H，q，*J*＝7Hz，CH_2）1.25（3H，t，*J*＝7Hz，OCH_3）ppm。

(2E，4Z) −2，4−己二烯酸乙酯 (2*E*，4*Z*−3)

在反应瓶中，放入13g（35mmol）溴化乙基三苯基膦[7]和90mL无水四氢呋喃，在氮气保护下，加入6.4g（35mmol）自制的六甲基二硅胺钠[8]，室温搅拌30min，回流2h，反应液呈橙红色，在-78℃滴入4.5g（35.1mmol）*E*-4-氧代丁烯酸乙酯5mL无水四氢呋喃溶液，在-78℃反应3h，逐渐升至室温，搅拌过夜，减压蒸去溶剂，残余物用石油醚∶乙醚=10∶1洗涤，过滤，滤液用饱和亚硫酸氢钠溶液除去未反应的醛，水洗，干燥，蒸除溶剂后得4.965g粗品，减压收集b.p. 98~102℃/20mm Hg 2.955g，收率60.2%。用PEG-20毛细管气相色谱分析，纯度98%两异构体比例2*E*，4*Z*∶2*E*，4*E*=53∶47。GC-MS检测还有2%的2*Z*，4*Z*和2*Z*，4*E*体。在20mmHg真空度下取2.7g产品，旋带分馏分离。气相色谱跟踪，收集1g纯2*E*，4*Z*-2，4-己二烯酸乙酯。v_{max}：1705（s，C=O），1635（s，C=C），1608（m，C=C），995（s），972（m），865（m），690（m）cm^{-1}. δ_H：12.0（1H，s，COOH），7.74（1H，d，d，*J*=15，11Hz，β-H），6.32~5.88（2H，m，$CH_3CH=CH$），5.85（1H，d，*J*=15Hz，α-H），1.92（3H，d，*J*=7Hz，$CH_3C=C$），ppm。δ_{13C}：166.96，138.85，135.27，127.25，120.97，59.97，14.09，13.71ppm。*m/z*：141（M^++1，4），140（M^+，46），125（M^+-CH_3，36）。

Z和E-2-己烯-4-炔酸乙酯（Z-7+E-7）

20.7g（59.4mmol）乙氧羰基甲叉三苯膦烷和120mL无水乙醚，氮气保护下，滴入含7g丁炔醛（含量76%）的5mL无水乙醇溶液，室温反应6h，放置过夜，倒入水中，乙醚提取，水洗，分层，干燥，蒸溶剂，滤去三苯基氧膦。并用石油醚∶乙醚=10∶1洗，滤液浓缩得残留物9.7g，TLC分析R_f=0.36，0.74（石油醚∶乙醚=9∶1展开）。毛细管气相色谱分析（OV-1柱），配合NMR分析收集1.02g*E*体，纯度97.3%。1.895g*Z*体，纯度97.7%。

E-2-己烯-4-炔酸乙酯：v_{max}：2200（s，C≡C），1710（s，C=O），1620（s，C=C），960（s，*E*—C=CH）cm^{-1}，δ_H：6.71（1H，d，q，*J*=15，2Hz，C≡CH），6.09（1H，d，*J*=15Hz，CHCOO），4.17（2H，q，*J*=7Hz，OCH_2），1.99（3H，d，*J*=2Hz，$CH_3C\equiv C$），1.26（3H，t，*J*=7Hz，CH_3CH_2O）ppm，δ_{13C}：165.93，129.30，125.89，96.01，77.03，60.45，14.08，4.49ppm，*m/z*：139（M^++1，2），138（M^+，24）。

Z-2-己烯-4-炔酸乙酯：v_{max}：2250（w），2200（m，C=C），1718（s），1700（s，C=O），1608（s，C=C），815（s，Z—C=CH）cm^{-1}. δ_H：6.04（2H，m，CH=CH），4.20（q），4.16（2H，q，*J*=7Hz，CH_2O），2.09（3H，d，*J*=2Hz，$CH_3C=C$），1.30（t），1.26（3H，t，*J*=7Hz，CH_3CH_2O）ppm。δ_{13C}：164.68，127.36，123.84，99.40，76.76，60.02，14.06，4.82ppm。*m/z*：139（M^++1，3），138（M^+，38）。

（2Z，4Z）-2，4-己二烯酸乙酯（2Z，4Z-3）

制备同（2*Z*，4*E*）-2，4-己二烯酸，最后柱层析用石油醚∶乙醚=20∶1展开。旋带分馏，得1.0g无色液体，收率95.8%，GC纯度96.2%。v_{max}：1712（s，C=O），1630（m，C=C），1592（m，C=C），820，800，775（m，Z—C=CH）cm^{-1}，δ_H：7.24（1H，t，q，*J*=11，2Hz，γ-H），6.94（1H，t，*J*=11Hz，β-H），6.02（m，

1H, δ－H), 5.68 (d, 1H, $J=11Hz$, α－H), 4.18 (q, 2H, $J=7Hz$, CH_2O), 1.84 (3H, d, d, $J=7$, 2Hz, $CH_3C=C$), 1.29 (3H, t, J = 7Hz, CH_3CH_2) ppm。δ_{13C}: 166.53, 138.47, 135.54, 125.46, 117.37, 59.81, 14.30, 13.22ppm。m/z: 141 ($M^+ +1$, 4), 140 (M^+, 39)。

山梨酸四个异构体杀菌活性比较

分别将上述四个纯异构体对五种植物病原真菌：小麦赤霉病菌 *Gibberella Saubinetii* (*Mont.*) *Sacc*, *Fusarium spp. G. Zeae* (*Schu.*) *Petch*；芦笋茎枯病 *Phoma asparagi Sacc*；苹果轮纹病 *Physalospora Piricola Nose*；棉花立枯病 *Rhizoctonia Solani kuhn*, *Pellicularila filamentosa* (*Pat.*) *Rogers*, *Corticium* centrifugum (Lev.) Bres；甜菜褐斑病 *Cercospora beticola Sacc* 和其他三种霉菌；桔青霉 *Pennicillum Citrinum*，黄曲霉 *Aspergillus flavas*；黑曲霉 *Aspergillum niger*；进行室内生物活性测定。各异构体使用浓度 300，500，700，1000ppm。

实验结果指出，当 2 位烯键为 Z 体时，活性明显增高。山梨酸四个异构体的生物活性与其两个双键的构型密切相关。其活性比较次序为：$2Z$, $4E > 2Z$, $4Z > 2E$, $4E > 2E$, $4Z$。详细生物测定数据将另文报道。

致谢：本工作曾得到中国科学院上海有机化学研究所黄耀曾教授，施莉兰教授和杨建华同志的热情指导和帮助，光谱和元素分析由本所分析室同志测定，在此一并谨致谢意。

参 考 文 献

[1] Greger, H., Zdero, C., Bohlmann, F., *Phytochemistry*, 1984, 23, 1503.

[2] Butenandt, A., Hecker, E., *Angew. Chem.*, 1961, 73.349.

[3] Allan, J. L. H., Jones, E. R. H., Whiting, M. C., J. *Chem. Soc.*, 1955, 1862.

[4] Veysoglu, T., Mitscher, L. A., Swayze, J. K., *Synthesis*, 1980, 807.

[5] Robert, C. W., Melvin, J. A., "*C. R. C. Handbook* of *Chemistry and Physics*" Press, 1982－1983.

[6] Eisner, U, Elvidge, J. A., Linstead, R. P., J. *Chem. Soc.*, 1953, 1372.

[7] 王葆仁，"有机合成反应"，北京：科学出版社，1980.

[8] Holtzclaw, H. F. Jr., *Inorganic Synthesis*, 1977, 8, 11.

A Study on the Synthesis and Bioassay of Sorbic Acid

Zhengming Li[1], Tiansheng Wang[1], Zuxin Zhang[1], Diankun Zhang[1], Enyun Yao[1], Zhenheng Gao[2]

([1]Institute of Elemento－Organic Chemistry, Nankai University, Tianjin, 300071; [2]Department of Chemistry, Nankai University, Tianjin, 300071)

Abstract All theroretically possible isomers of 2, 4－hexadienoic acid (sorbic acid) and their ethyl esters were synthesized stereotelectively with readily available crotonaldehyde and propargyl alcohol as starting materials. The $2Z$, $4E$ and $2Z$, $4Z$－isomers. showed better bioassay results than the commercially available $2E$, $4E$－isomer. The IR, [1]H－NMR, [13]C－NMR and MS spectroscopic properties were described for respectire geometric isomers.

Key words sorbic acid; 2, 4－hexadienoic acid; stereo－selecive synthesis

1991 年

茶尺蠖性信息素化学结构研究初报*

么恩云[1] 李正名[1] 罗志强[1] 尚稚珍[1] 殷坤山[2] 洪北边[2]

([1]南开大学元素有机化学国家重点实验室，南开大学元素有机化学研究所，天津 300071；
[2]中国农业科学院茶叶研究所，杭州 310024)

关键词 茶尺蠖 性信息素 化学结构 色质联用分析

茶尺蠖（*Ectropis Obliqua*）是我国浙、苏、皖茶区重要的食叶性茶树害虫。长期使用化学农药治虫，严重地污染了自然环境，并使茶叶中农药残留量上升，迫切需要寻找某些高效低毒、无公害、无残留的新防治措施。1978 – 1982 年，国内有关单位曾对茶尺蠖性信息素做过探索性研究，证明茶尺蠖性信息素为多组分，但未确定化学结构[1]。茶尺蠖属于鳞翅目尺蛾科昆虫，而尺蛾科昆虫性信息素不同于鳞翅目其他科的昆虫。由于多双键体系的化学结构增加了化学结构鉴定的难度致使研究报道较少，在国际上直到 80 年代初始有少量报道[2-7]。但茶尺蠖性信息素化学结构未见报道，因此我们对茶尺蠖性信息素化学结构进行了分析研究。

首先进行采样和分离，分别以两种不同方式采样：（1）溶剂萃取精剪尾和性腺体；（2）气体吸附，以树脂 Porapak Q 吸附雌虫释放出的性信息素气体。（1）和（2）样品经净化后分别以制备气相色谱分离。采用 Shimadzu GC9A 气相色谱仪，氢焰离子化（FID）检测，不锈钢柱（3m × 3mmID）OV – 1，柱温 190℃ $\xrightarrow{7℃/min}$ 240℃程序升温，氮作载气，分流比 20∶1 分离出的各馏分分别以干冰冷却下的玻璃毛细管收集，同时以生物测定跟踪找出具有诱引活性的馏分。（1）和（2）样品分离后活性组分均集中于 GC，保留时间 5 ~ 15min 之间馏出，且生测试验和 GC 测定表明活性馏分为多组分构成。

对分离出的活性馏分进行色质联用（GC/MS）分析（包括电子轰击电离 EI 和化学电离 CI）。

EI GC/MS 分析采用 HP 5890/5970B MSD 色质联用仪。电离电压 70eV，源温 200℃，气相色谱仪装 OV – 17 熔硅弹性毛细管柱（25m × 0.2mmID），柱温 35℃ $\xrightarrow{15℃/min}$ 220℃程序升温，载气氦。

CI GC/MS 分析采用 HP5890GC/5988MS 色质联用仪。电离电压 150eV，源温 134℃，源压 133.332 × 0.8Pa，发射电流 300μA。色谱柱同前，柱温 35℃ $\xrightarrow{20℃/min}$ 210℃程序升温，载气氦，反应气甲烷。

分析结果发现活性馏分中存在（z，z，z）–3，6，9 – 十八碳三烯，（z，z，z）– 3，6，9 – 十九碳三烯以及更高级同系物。在 GC/MS 总离子流色谱图上有 – RT（保留

---

* 原发表于《自然科学进展》，1991，5：452 – 454。

时间）=12. 58min 的目标峰，其 EI 质谱数据 m/z（相对离子丰度）：

41（97. 77），55（48. 53），67（65. 84），79（100），80（54. 51），93（28. 92），95（23. 34），108（32. 40），121（4. 89），135（2. 95），149（0. 91），192（3. 42），248（0. 26）。从质谱数据分析：m/z 55（$C_4H_7^+$），67（$C_5H_7^+$），79（$C_6H_7^+$），93（$C_7H_9^+$），95（$C_7H_{11}^+$）；$CH_3(CH_2)_n(CH{=}CH)_3^+$ 系列离子 107，121，135，149 以及特征的 108，192 使（3，6，9）三烯结构得以证实。

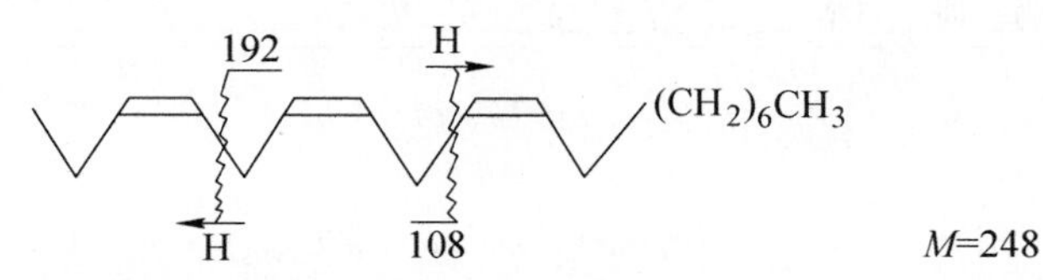

这种亚甲基间隔不饱和结构可由质谱判断[8,9]。另外从两组离子碎片 m/z：108/95，108/93 的丰度比均大于 1，也可判断它是三烯烃而不是带有其他官能团的同类型三烯化合物。CI GC/MS 质谱中 m/z 247（M－1）和 249（M＋1）峰的出现确证分子量为 248。推测为（z，z，z）－3，6，9－十八碳三烯，最后经化学合成标样对照核实（见图 1）。

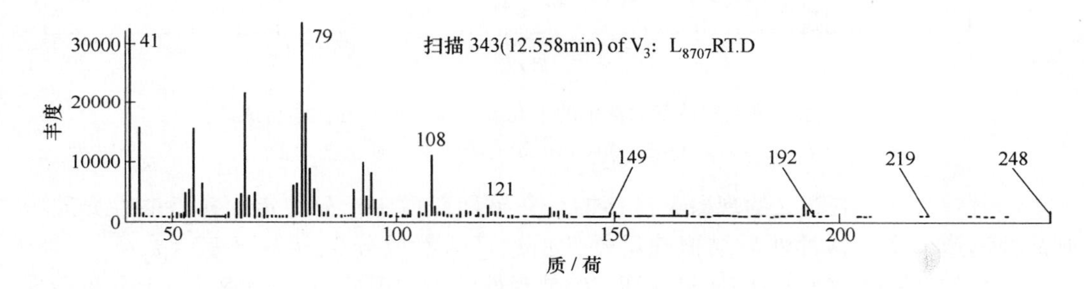

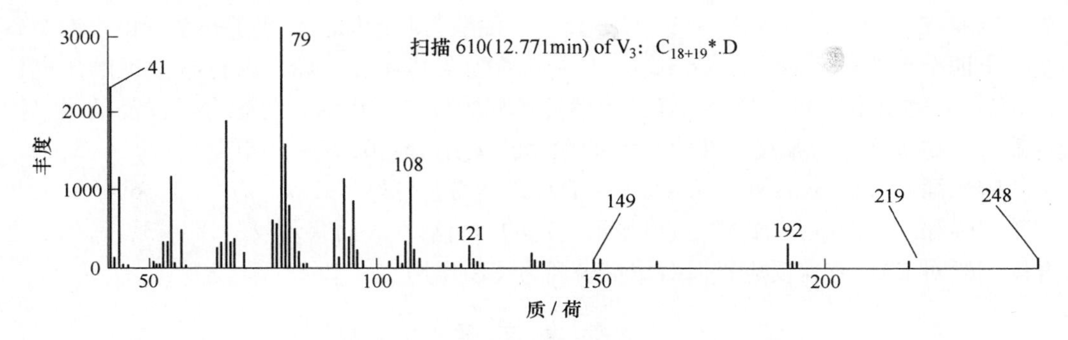

图 1　茶尺蠖性腺体萃取液中的十八碳三烯（上）和化学合成的（z，z，z）－3，6，9－十八碳三烯（下）质谱图对照

另一目标峰 RT＝13. 30min 其质谱与（z，z，z）－3，6，9－十八碳三烯非常类似，其特征差别仅在于 m/z 262（M^+）和 206，这与（z，z，z）－3，6，9－十八碳三烯中相应的 m/z 248（M^+）和 192 均差 14Amu，因此推断此峰为（z，z，z）－3，6，

9－十九碳三烯。后经化学合成标样对照核实（见图2）。

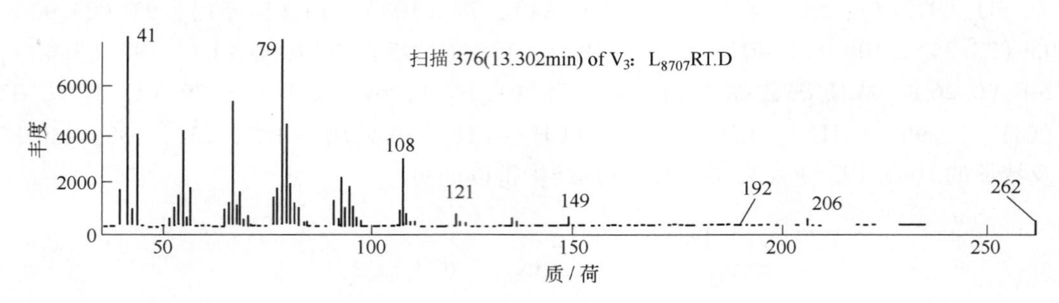

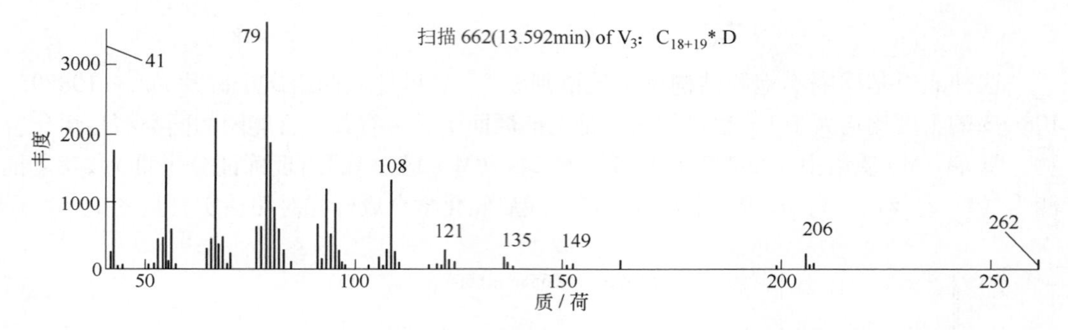

图2　茶尺蠖性腺体萃取液中的十九碳三烯（上）和化学合成的（z，z，z）－3，6，9－十九碳三烯（下）质谱图对照

同时观察到一有意义的现象：活性馏分在碱性介质中十分稳定而在微酸性介质中则活性迅速下降。这种性质酷似槐尺蠖性信息素活性成分3，4－环氧（z，z）－6，9－十七碳二烯在酸性介质中环氧开环，致使活性消失的现象[10]。为确证以上分析和现象，以亚麻油酸（z，z，z）－9，12，15－三烯酸为起始原料合成了一系列由十八个碳至二十四个碳的（z，z，z）－3，6，9－三烯烃及其环氧二烯，进行电生理触角电位（EAG）和雄虫生物测定筛选。通过天然提取样的分离分析结合化学合成样品的生物测定筛选，初步确定了茶尺蠖性信息素的两个重要组分：6，7－环氧（z，z）－3，9－十八碳二烯和（z，z，z）－3，6，9－十八碳三烯。另外（z，z，z）－3，6，9－二十二碳三烯和二十四碳三烯对雄虫也显示一定的诱引活性。而（z，z，z）－3，6，9－十九碳三烯对EAG有强反应，但对雄虫表现为抑制作用。

参考文献

[1] 杜家纬，等，昆虫激素，1981，1：44.
[2] Becker，D. et al.，*Tettra. Lett.*，24（1983），5505－5508.
[3] Underhill，E. W. et al.，*J. Chem. Ecol.*，9（1983），1413－1423.
[4] Wong，J. W. et al.，*ibid.*，10（1984a），463－473.
[5] Wong，J. W. et al.，*ibid.*，10（1984b），1579－1596.
[6] Wong，J. W. et al.，*ibid.*，11（1985），727－756.
[7] Miller，J. G. et al.，*ibid.*，13（1987），1371－1383.
[8] Lee. R. F. et al.，*Biochem. Biophys. Acta.*，202（1970），386－388.
[9] Karunen，P.，*Photochemistry.* 13（1974），2209－2213.
[10] Li，Z. M. et al.，*China Japan Seminar on Insect Semiochemicals Abstracts*，Beijing，China. 1988，29－32.

新磺酰脲类化合物的合成、结构及构效关系研究(Ⅰ)*

——N-(2′-嘧啶基)-2-甲酸乙酯-苯磺酰脲的晶体及分子结构

李正名[1] 贾国锋[1] 王玲秀[1] 赖城明[2] 王如骥[3] 王宏根[3]

([1]南开大学元素有机化学国家重点实验室，南开大学元素有机化学研究所，天津 300071；[2]南开大学化学系；[3]南开大学中心实验室)

摘 要 本文在合成标题化合物的基础上，测定了其晶体结构。晶体属单斜晶系，空间群为 $P2_1/a$，晶胞参数 $a=0.7348(2)\,nm$，$b=2.1496(4)\,nm$，$c=0.9973(2)\,nm$，$\beta=90.19(2)^\circ$，$V=1.576\,nm^3$，$Z=4$。晶体结构由直接法解出，进行全矩阵最小二乘法修正，$R=0.059$。分子中的嘧啶磺酰脲、苯环及酯基三部分分别形成3个独立的平面共轭体系。S-N 原子间形成 $d-p\pi$ 键并参加到第一个共轭体系中。分子中还存在着 N 与 H 原子间的分子内氢键。

关键词 磺酰脲类化合物 晶体结构 分子结构

磺酰脲类农药是近年来发展起来的一类极为重要的超高效低毒除草剂，研究该类化合物的结构及其与性能的关系，对于了解农药分子的作用机理，改进和提高药效、设计和探索合成新型农药都有重要的理论意义和实际应用价值。

样品的合成

标题化合物的合成反应为[1,2]

$$C_6H_4(COOC_2H_5)SO_2NH_2\ (Ⅱ) + (COCl)_2 \xrightarrow{甲苯} C_6H_4(COOC_2H_5)SO_2NCO$$

$$C_6H_4(COOC_2H_5)SO_2NCO + H_2N\text{-}C_4H_3N_2 \xrightarrow{乙腈} C_6H_4(COOC_2H_5)SO_2NHCONH\text{-}C_4H_3N_2\ (Ⅰ)$$

向反应瓶中加入5.0g（Ⅱ）、甲苯50mL和草酰氯11.0g，加热搅拌，回流32h，减压蒸去草酰氯及甲苯，残留物放入含1.5g 2-氨基嘧啶的25mL乙腈中，反应3h，滤出沉淀，用1，2-二氯乙烷重结晶，得产物（Ⅰ）5.0g，m.p.170.5~171.6℃，产率66%。$C_{14}H_{14}N_4O_5S$ 的元素分析实测值（%，计算值）：C 48.26(48.00)，H 3.86

* 原发表于《高等学校化学学报》,1992,13(11):1411-1414。国家教育委员会博士点基金重点资助课题。

(4.03), N 15.70 (15.9)。IR (cm^{-1}): 1723.5 (C ═O), 1710 (—NHCONH—), 1357.9、1160.8 (—SO_2—)。^{1}H NMR (δ_{ppm}, Acetone - d_6): 8.1 ~ 8.34(2H, br, —NHCONH—), 7.52 ~ 7.84(4H, br, Ar—), 8.64(2H, d, —$C_4N_2H_2$—), 4.29(2H, q, —CH_2—), 1.24(3H, t, —CH_3)。由此确定了化合物结构。

晶体结构测定

从丙酮中培养出 $C_{14}H_{14}N_4O_5S$ 无色透明晶体，选取约为 0.2mm × 0.2mm × 0.3mm 的单晶用于衍射实验。在 Enraf - Nonius CAD_4 衍射仪上，用经石墨单色器单色化的 Mo K_α 射线，在 $2° \leq \theta \leq 25°$ 范围内，以 $\omega - 2\theta$ 扫描方式，共收集 3019 个独立反射点。其中 1421 个为可观测反射［$I \geq 3\sigma(I)$］。强度数据均经 LP 因子校正及经验吸收校正。晶体属单斜晶系，空间群为 $P2_1/a$，$a = 0.7348(2)$ nm，$b = 2.1496(4)$ nm，$c = 0.9973(2)$ nm，$\beta = 90.19(2)°$，$V = 1.576 nm^3$，$M = 350.36$，$Z = 4$，$D_x = 1.48 g/cm^3$，$\mu = 2.27 cm^{-1}$，$F(000) = 728$。晶体结构由直接法（MULTAN82）解出。根据 E - 图确定了大部分非氢原子位置，剩余非氢原子坐标在以后的数轮差值 Fourier 合成中陆续确定。对全部非氢原子坐标及各向异性热参数进行全矩阵最小二乘法修正。最终一致性因子 $R = 0.059$。在最终差值 Fourier 图上，最高电子密度峰值为 450e/nm^3。全部计算在 PDP 11/44 计算机上用 SDP - PLUS 程序完成。

结果与讨论

由晶体结构分析得到标题分子的立体结构（不包括氢原子）见图 1。分子中全部非氢原子的分数坐标、等价各向同性热参量见表 1，分子中的键长、键角和扭角值见表 2 和表 3。

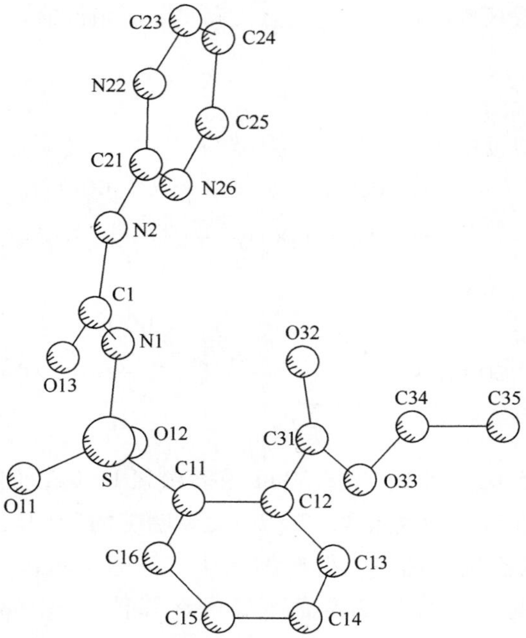

图 1　分子立体图

表1　原子坐标及热参量值（$10^{-2}nm^2$）

原　子	x	y	z	B_{eq}
C(11)	0.1279(8)	0.1836(3)	0.2986(7)	3.1(1)
C(12)	0.0768(9)	0.2086(3)	0.1734(7)	3.4(2)
C(13)	0.065(1)	0.2734(3)	0.1600(8)	3.8(2)
C(14)	0.098(1)	0.3122(4)	0.2695(8)	4.2(2)
C(15)	0.144(1)	0.2870(4)	0.3938(9)	4.5(2)
C(16)	0.161(1)	0.2215(4)	0.4078(8)	3.9(2)
S	0.1831(3)	0.10381(9)	0.3181(2)	3.74(4)
O(11)	0.2800(7)	0.0977(3)	0.4412(5)	4.9(1)
O(12)	0.2641(7)	0.0830(2)	0.1953(5)	4.5(1)
N(1)	−0.0100(8)	0.0647(3)	0.3250(6)	4.0(1)
C(1)	−0.124(1)	0.0665(3)	0.4334(7)	3.7(2)
O(13)	−0.0955(7)	0.0951(2)	0.5357(5)	4.8(1)
N(2)	−0.2841(8)	0.0328(3)	0.4207(6)	3.6(1)
C(21)	−0.343(1)	−0.0026(3)	0.3144(7)	3.6(2)
N(22)	−0.5023(8)	−0.0320(3)	0.3359(6)	3.9(1)
C(23)	−0.571(1)	−0.0657(4)	0.2343(8)	4.6(2)
C(24)	−0.483(1)	−0.0706(4)	0.1133(9)	5.2(2)
C(25)	−0.316(1)	−0.0388(4)	0.0993(8)	5.1(2)
N(26)	−0.2492(9)	−0.0044(3)	0.2007(6)	4.4(1)
C(31)	0.026(1)	0.1697(3)	0.0548(7)	3.9(2)
O(32)	−0.0783(8)	0.1266(2)	0.0573(5)	5.3(1)
O(33)	0.1086(8)	0.1905(3)	−0.0556(5)	5.2(1)
C(34)	0.049(1)	0.1648(4)	−0.1849(8)	6.1(2)
C(35)	−0.085(1)	0.2108(5)	−0.243(1)	7.3(3)

表2　主要键长（$10^{-1}nm$）及键角（°）

C(11)—C(12)	1.411(7)	S—O(12)	1.434(4)	N(22)—C(23)	1.344(7)
C(11)—C(16)	1.382(7)	S—N(1)	1.650(5)	C(23)—C(24)	1.373(9)
C(11)—S	1.773(5)	N(1)—C(1)	1.370(7)	C(24)—C(25)	1.415(9)
C(12)—C(13)	1.402(8)	C(1)—O(13)	1.211(6)	C(25)—N(26)	1.345(7)
C(12)—C(31)	1.498(8)	C(1)—N(2)	1.385(7)	C(31)—O(32)	1.200(7)
C(13)—C(14)	1.395(9)	N(2)—C(21)	1.376(7)	C(31)—O(33)	1.337(7)
C(14)—C(15)	1.395(8)	C(21)—N(22)	1.346(7)	O(33)—C(34)	1.470(8)
C(15)—C(16)	1.421(8)	C(21)—N(26)	1.328(8)	C(34)—C(35)	1.52(2)
S—O(11)	1.426(5)				
C(12)—C(11)—C(16)	121.3(6)	C(11)—S—N(1)	107.6(2)	N(22)—C(21)—N(26)	125.0(5)
C(12)—C(11)—S	121.8(4)	O(11)—S—O(12)	120.0(3)	C(21)—N(22)—C(23)	117.3(5)
C(16)—C(11)—S	116.4(4)	O(11)—S—N(1)	110.3(3)	N(22)—C(23)—C(24)	121.9(6)

续表 2

C(11)—C(12)—C(13)	118.7(5)	O(12)—S—N(1)	103.5(2)	C(23)—C(24)—C(25)	117.2(6)
C(11)—C(12)—C(31)	123.6(6)	S—N(1)—C(1)	123.1(4)	C(24)—C(25)—N(26)	120.6(6)
C(13)—C(12)—C(31)	117.7(5)	N(1)—C(1)—O(13)	124.8(6)	C(21)—N(26)—C(25)	117.9(6)
C(12)—C(13)—C(14)	120.5(6)	N(1)—C(1)—N(2)	115.7(5)	C(12)—C(31)—O(32)	125.2(6)
C(13)—C(14)—C(15)	120.5(5)	O(13)—C(1)—N(2)	119.5(6)	C(12)—C(31)—O(33)	110.4(5)
C(14)—C(15)—C(16)	119.5(6)	C(1)—N(2)—C(21)	128.8(5)	O(32)—C(31)—O(33)	124.5(6)
C(11)—C(16)—C(15)	119.5(6)	N(2)—C(21)—N(22)	114.3(5)	C(31)—O(33)—C(34)	117.4(5)
C(11)—S—O(11)	107.3(3)	N(2)—C(21)—N(26)	120.7(5)	O(33)—C(34)—C(35)	106.7(6)
C(11)—S—O(12)	107.7(2)				

表 3　扭角（°）

C(16)—C(11)—C(12)—C(13)	1.77(0.98)	C(11)—S—N(1)—C(1)	-72.17(0.63)
C(16)—C(11)—C(12)—C(31)	-175.08(0.64)	O(11)—S—N(1)—C(1)	44.54(0.65)
S—C(11)—C(12)—C(13)	-169.49(0.52)	O(12)—S—N(1)—C(1)	174.06(0.56)
S—C(11)—C(12)—C(31)	13.65(0.92)	S—N(1)—C(1)—O(13)	-1.81(1.03)
C(12)—C(11)—C(16)—C(15)	-0.38(1.02)	S—N(1)—C(1)—N(2)	177.63(0.48)
S—C(11)—C(16)—C(15)	171.34(0.54)	N(1)—C(1)—N(2)—C(21)	0.54(1.02)
C(12)—C(11)—S—O(11)	162.47(0.53)	O(13)—C(1)—N(2)—C(21)	-179.98(0.52)
C(12)—C(11)—S—O(12)	32.06(0.62)	C(1)—N(2)—C(21)—N(22)	177.15(0.63)
C(12)—C(11)—S—N(1)	-78.89(0.60)	C(1)—N(2)—C(21)—N(26)	-4.97(1.08)
C(16)—C(11)—S—O(11)	-9.20(0.61)	N(2)—C(21)—N(22)—C(23)	177.84(0.61)
C(16)—C(11)—S—O(12)	-139.60(0.53)	N(26)—C(21)—N(22)—C(23)	0.07(1.14)
C(16)—C(11)—S—N(1)	109.44(0.55)	N(2)—C(21)—N(26)—C(25)	-178.56(0.66)
C(11)—C(12)—C(13)—C(14)	-1.66(1.01)	N(22)—C(21)—N(26)—C(25)	-0.92(1.07)
C(31)—C(12)—C(13)—C(14)	175.38(0.64)	C(21)—N(22)—C(23)—C(24)	0.26(1.08)
C(11)—C(12)—C(31)—O(32)	47.46(1.08)	N(22)—C(23)—C(24)—C(25)	0.25(1.18)
C(11)—C(12)—C(31)—O(33)	-134.06(0.68)	C(23)—C(24)—C(25)—N(26)	-1.12(1.20)
C(13)—C(12)—C(31)—O(32)	-129.43(0.80)	C(24)—C(25)—N(26)—C(21)	1.43(1.12)
C(13)—C(12)—C(31)—O(33)	49.06(0.85)	C(12)—C(31)—O(33)—C(34)	-169.34(0.63)
C(12)—C(13)—C(14)—C(15)	0.17(1.06)	O(32)—C(31)—O(33)—C(34)	9.16(1.09)
C(13)—C(14)—C(15)—C(16)	1.24(1.08)	C(31)—O(33)—C(34)—C(35)	97.36(0.81)
C(14)—C(15)—C(16)—C(11)	-1.13(1.05)		

考察分子的几何结构可以发现，分子中的原子主要分布在嘧啶磺酰脲平面Ⅰ（不包括磺酰的氧），苯环平面Ⅱ，酯基中的 COOC 平面Ⅲ3 个独立的平面上，见表 4。

表4　最小二乘平面及其二面角（°）

平　面	平 面 方 程	原子到平面距离
Ⅰ	$-0.476(1)x+0.8055y-0.353(2)z=0.055(7)$	C(21) -0.0088, N(22) -0.0368, C(23) -0.0237, C(24) 0.0113, C(25) 0.0277, N(26) 0.0320, N(2) 0.0229, C(1) 0.0027, N(1) -0.0455, O(13) 0.0373, S -0.0190
Ⅱ	$0.9657(8)x+0.050(3)y-0.255(3)z=0.34(2)$	C(11) 0.0066, C(12) -0.0113, C(13) 0.0057, C(14) 0.0048, C(15) -0.0095, C(16) 0.0037
Ⅲ	$-0.767(2)x+0.616(3)y-0.184(4)z=2.0(1)$	C(31) -0.0078, C(12) 0.0022, O(32) 0.0031, O(33) 0.0024
平面间二面角		Ⅰ—Ⅱ:109.21°, Ⅰ—Ⅲ:22.30°, Ⅱ—Ⅲ:131.51°

研究农药分子中的原子平面的意义就在于这些平面通常均形成共轭体系，它们对于电子的传递、反应性能，进而对农药的活性都会有重要的影响。从表2可见，分子中所有C—N键长均短于C—N单键（0.147nm）而大于C=N双键（0.127nm），所有含N原子键的键角∠CNC及∠SNC均在120°±3°（仅∠C（1）N（2）C（21）为128.8°）。S—N键长为0.165nm，小于二者共价半径之和（0.178nm）。这说明S—N键具有部分双键成分。很显然，其π键部分是由S原子以空d_{xz}轨道与N的$2p_z$轨道间同相位重叠而成。由此推知，嘧啶磺酰脲骨架形成包括S原子在内的$Ⅱ_{11}^{12}$多电子共轭π键。由于苯甲酸乙酯中苯环与羧基不共面，故分别形成$Ⅱ_6^6$及$Ⅱ_3^4$共轭π键。由红外光谱（3395cm^{-1}，变宽，3265cm^{-1}，肩部）说明分子中存在N与H间的分子内氢键。为更好地了解氢键的形成，用分子图形学方法加上H原子，估测N（1）上的H与N（26）原子间距约为0.186nm，可形成氢键并组成六元环。这一结构对保持嘧啶磺酰脲骨架的平面性，从而保证共轭体系的稳定存在，都有重要的意义。

参 考 文 献

[1] Franz J. F. et al.; J. Org. Chem., 1964, 29: 2592.

[2] EP, 162 723.

Sytheses, Structure and SAR Study on New Sulfonylurea Compounds (Ⅰ)

——Crystal and Molecular Structure of N－(2′－pyrimidinyl)－2－ethyl Carboxylate－benzene Sulfonylurea

Zhengming Li[1], Guofeng Jia[1], Lingxiu Wang[1], Chengming Lai[2], Ruji Wang[3], Honggen Wang[3]

([1]National Key Laboratory of Elemento－Organic Chemistry, Elemento－Organic Chemistry Institute, Nankai University, Tianjin, 300071; [2]Department of Chemistry, Nankai University; [3]Central Laboratory of Nankai University)

Abstract　Synthesis and crystal structure of title molecule was reported here. The crystal is monoclinic, belong-

ing to space group $P2_1/a$ with unit cell parameter $a = 0.7384$ (2) nm, $b = 2.1496(4)$ nm, $c = 0.9973(2)$ nm, $\beta = 90.12(2)°$, $V = 1.576$ nm^3, $Z = 4$, $R = 0.059$ for 1421 reflections. Molecular structure was discussed. There are three different planes in molecule, in each of which a conjugated system was formed. The empty $3d_{xz}$ orbital of atom S enters the formation of conjugated bond. The intra – molecule hydrogen bond between atom N and H was found.

Key words crystal structure; molecular structure; new sulfonylurea compound

新磺酰脲类化合物的合成、结构及构效关系研究(Ⅱ)*

——N－[2－(4－甲基)嘧啶基]－2－甲酸乙酯－苯磺酰脲的晶体及分子结构

李正名[1]　贾国锋[1]　王玲秀[1]　赖城明[2]　王宏根[3]　王如骥[3]

([1]南开大学元素有机化学国家重点实验室，元素有机化学研究所，天津　300071；
[2]南开大学化学系；[3]南开大学中心实验室)

摘　要　本文报道了标题化合物的合成和晶体结构。该晶体由标题分子与溶剂分子 CH_2Cl_2 组成，属三斜晶系，空间群为 $P\bar{1}$，晶胞参数 $a=0.7413(2)$ nm，$b=0.9166(5)$ nm，$c=1.6289(5)$ nm，$\alpha=75.97(4)°$，$\beta=82.28(2)°$，$\gamma=76.22(3)°$，$V=1.039\text{nm}^3$，$Z=2$。晶体结构由直接法解出，进行全矩阵最小二乘法修正，$R=0.067$。标题分子中存在 3 个独立的平面共轭体系，还存在 N、H 原子间的分子内及分子间氢键。

关键词　磺酰脲类化合物　晶体结构　分子结构

为了系统地研究超高效、低毒除草剂磺酰脲类化合物的分子结构和构象对农药活性的影响，在前文[1]工作的基础上，我们研究了 N－[2－(4－甲基) 嘧啶基]－2－甲酸乙酯－苯磺酰脲的晶体及分子结构。结果表明，与嘧啶环上不带甲基的同类化合物相比，二者构象有很大的差异。这一差异，对于这 2 个化合物的农药活性，可能会有着较重要的影响。

样品的合成与表征

标题化合物的制备反应，参考文献[2,3]改进。

$$\text{(Ⅱ)}\ \ C_6H_4(COOC_2H_5)\text{—}SO_2NH_2 + (COCl)_2 \xrightarrow{\text{甲苯}} C_6H_4(COOC_2H_5)\text{—}SO_2NCO$$

$$C_6H_4(COOC_2H_5)\text{—}SO_2NCO + H_2N\text{—}C_4N_2H_2(CH_3) \xrightarrow{\text{乙腈}} C_6H_4(COOC_2H_5)\text{—}SO_2NHCONH\text{—}C_4N_2H_2(CH_3)\ \ \text{(Ⅰ)}$$

向反应瓶中加入（Ⅰ）3.5g、甲苯 25mL，然后一次加入草酰氯 6.7g，加热慢慢升温至回流 32h。冷却、过滤后将滤液减压蒸去草酰氯及甲苯，将残留物加入含 1.1g 的

* 原发表于《高等学校化学学报》,1993,14(3):349－352。国家自然科学基金重点资助课题。

4-甲基-2-氨基嘧啶的20mL乙腈中，反应3h，滤出沉淀，以二氯甲烷/石油醚重结晶，得产物（Ⅰ）3.0g，溶点 147～149℃，总收率 58%。元素分析（$C_{15}H_{16}N_4O_5S$,%）计算值：C 49.44，H 4.42，N 15.38；实测值：C 49.22，H 4.64，N 15.23。IR（cm^{-1}）：3400（N—H），1712（C ═O 及—NHCONH 合），1352，1163（—SO_2—）。^1H NMR（$CDCl_3$，ppm）：1.28（3H，t，CH_3），2.56（3H，s，—CH_3），4.28（2H，q，—CH_2—），6.76（1H，d，N N —H），7.36～8.32（4H，m，Ar），8.44（1H，d，N N H），9.52（1H，Br，—CONH－Het），12.64（1H，Br，—SO_2NH—CO—）。从而确定了其结构。

晶体结构的测定

将产物（Ⅰ）溶于体积比为 1/4 的二氯甲烷/石油醚的混合液中，在室温（10～20℃）下自然挥发，得到无色透明的 $C_{16}H_{13}Cl_2N_4O_5S$ 晶体。由于晶体容易风化，测定前不能将晶体从母液中取出。从中选取一粒大小约为 0.2mm×0.3mm×0.3mm 的单晶用于 X 射线衍射实验。在 Enraf－Nonius CAD4 四圆衍射仪上，使用经过石墨单色器单色化的 Mo $K\alpha$ 入射线，在 $2°\leqslant\theta\leqslant24°$的范围内，以 $\omega-2\theta$ 扫描方式，共收集到 3278 个独立反射点，其中 1914 个为可观测的反射点［$I\geqslant3\sigma(I)$］。全部强度数据均经过 Lp 因子校正及经验吸收校正。该晶体属三斜晶系，空间群为 $P\bar{1}$，$a=0.7413(2)$ nm，$b=0.9166(5)$ nm，$c=1.6289(5)$ nm，$\alpha=75.97(4)°$，$\beta=82.28(2)°$，$\gamma=76.22(3)°$，$V=1.039nm^3$，Mr＝449.32，$Z=2$，$D_x=1.44g/cm^3$，$\mu=4.4cm^{-1}$，$F(000)=464$。晶体结构由直接法（MULTAN 82）解出。根据 E－图确定了大部分非氢原子的位置，剩余的非氢原子坐标是在以后数轮差值 Fourier 函数中陆续确定的。对全部非氢原子坐标及各向异性热参数进行全矩阵最小二乘修正。最终的一致性因子为 $R=0.067$。在最终的差值 Fourier 图上，最高的电子密度峰高为 $350e/nm^3$。全部计算在 PDP 11/44 计算机上用 SDP－PLUS 程序包完成。

结果与讨论

晶体结构分析表明，该晶体是由标题分子与溶剂分子 CH_2Cl_2 组成。图 1 为分子在晶体中的排列，画斜线部分为溶剂分子。图 2 为标题分子的立体结构（均未标出氢原子）。表 1 给出了全部非氢原子的分数坐标以及等价各向同性热参量，表 2 列出了分子的键长及键角值，表 3 列出了部分扭角值。

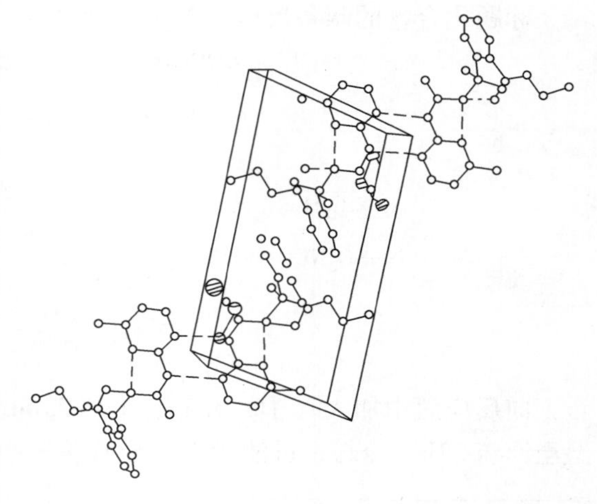

Fig. 1　Arrangement of molecules in crystal cell

分子的几何结构说明，分子中的原子主要分布在 3 个独立平面上，即嘧啶磺酰脲平面Ⅰ包括磺酰基上的氧，苯环平面Ⅰ，羧基平面Ⅱ。这 3 个平面间的两面角

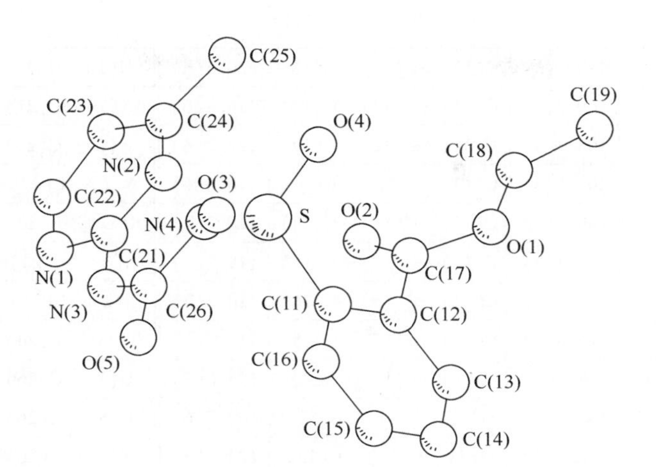

Fig. 2 Stereo structure of title molecule

Table 1 Fractional coordinates and equivalent isotropic thermal parameters for non – hydrogen atoms

Atom	x	y	z	B_{eq} (10^{-2} nm^2)	Atom	x	y	z	B_{eq} (10^{-2} nm^2)
S	0. 5948(3)	0. 4694(2)	0. 6970(1)	4. 25(4)	C(15)	0. 493(1)	0. 2602(8)	0. 5247(5)	5. 6(2)
O(1)	0. 0707(7)	0. 6808(5)	0. 6362(3)	4. 9(1)	C(16)	0. 580(1)	0. 3082(8)	0. 5804(5)	4. 7(2)
O(2)	0. 1456(7)	0. 5163(6)	0. 7598(3)	5. 7(1)	C(17)	0. 1619(9)	0. 5522(8)	0. 6836(5)	4. 3(2)
O(3)	0. 7905(6)	0. 4228(6)	0. 6761(3)	5. 8(1)	C(18)	−0. 060(1)	0. 7844(9)	0. 6858(5)	6. 5(2)
O(4)	0. 5148(8)	0. 6288(5)	0. 6945(3)	5. 6(1)	C(19)	−0. 113(2)	0. 934(1)	0. 6244(7)	10. 3(4)
O(5)	0. 6741(7)	0. 1382(5)	0. 7804(3)	5. 2(1)	C(21)	0. 4177(9)	0. 2504(7)	0. 9648(4)	3. 9(2)
N(1)	0. 3878(9)	0. 1620(6)	1. 0407(4)	4. 8(2)	C(22)	0. 295(1)	0. 2338(9)	1. 1018(5)	5. 8(2)
N(2)	0. 3510(8)	0. 4025(6)	0. 9428(4)	4. 3(1)	C(23)	0. 232(1)	0. 3912(9)	1. 0863(5)	5. 6(2)
N(3)	0. 5217(8)	0. 1716(6)	0. 9058(3)	4. 3(1)	C(24)	0. 265(1)	0. 4741(8)	1. 0042(5)	4. 7(2)
N(4)	0. 5434(8)	0. 3855(6)	0. 7952(4)	4. 3(1)	C(25)	0. 200(1)	0. 6467(8)	0. 9772(6)	6. 3(2)
C(11)	0. 4738(9)	0. 4031(7)	0. 6327(4)	3. 7(2)	C(26)	0. 5847(9)	0. 2254(7)	0. 8240(4)	4. 2(2)
C(12)	0. 2812(9)	0. 4522(7)	0. 6279(4)	3. 9(2)	Cl(1)	0. 0802(5)	0. 0668(4)	0. 8935(2)	10. 4(1)
C(13)	0. 194(1)	0. 4030(9)	0. 5730(4)	5. 0(2)	Cl(2)	0. 2941(4)	−0. 0235(4)	0. 7472(2)	11. 5(1)
C(14)	0. 297(1)	0. 3073(8)	0. 5203(5)	5. 5(2)	C	0. 127(1)	0. 122(1)	0. 7840(7)	7. 8(3)

Table 2 Bond distances (nm) and bond angles (°) between non – hydrogen atoms

S—O(3)	0. 1431(4)	O(5)—C(25)	0. 1215(9)	N(4)—C(25)	0. 1400(7)	C(15)—C(16)	0. 138(1)
S—O(4)	0. 1433(5)	N(1)—C(21)	0. 1328(7)	C(11)—C(12)	0. 1398(8)	C(18)—C(19)	0. 150(2)
S—N(4)	0. 1635(5)	N(1)—C(22)	0. 1347(8)	C(11)—C(16)	0. 1399(8)	C(22)—C(23)	0. 1378(9)
S—C(11)	0. 1759(7)	N(2)—C(21)	0. 1330(8)	C(12)—C(13)	0. 138(1)	C(23)—C(24)	0. 1391(9)
O(1)—C(17)	0. 1328(7)	N(2)—C(24)	0. 1357(8)	C(12)—C(17)	0. 1498(9)	C(24)—C(25)	0. 1512(9)
O(1)—C(18)	0. 1489(8)	N(3)—C(21)	0. 1382(7)	C(13)—C(14)	0. 1397(9)	Cl(1)—C	0. 1742(9)
O(2)—C(17)	0. 1202(7)	N(3)—C(26)	0. 1364(7)	C(14)—C(15)	0. 142(2)	Cl(2)—C	0. 1748(9)

续表 2

O(3)—S—O(4)	119.3(3)	C(12)—C(11)—C(16)	120.6(6)	N(1)—C(21)—N(2)	126.3(6)
O(3)—S—N(4)	109.2(3)	C(11)—C(12)—C(13)	119.8(6)	N(1)—C(21)—N(3)	114.3(5)
O(3)—S—C(11)	108.7(4)	C(11)—C(12)—C(17)	122.2(6)	N(2)—C(21)—N(3)	119.4(5)
O(4)—S—N(4)	103.4(3)	C(13)—C(12)—C(17)	118.0(6)	N(1)—C(22)—C(23)	121.9(7)
O(4)—S—C(11)	109.0(3)	C(12)—C(13)—C(14)	120.7(7)	C(22)—C(23)—C(24)	117.3(7)
N(4)—S—C(11)	106.7(3)	C(13)—C(14)—C(15)	119.2(7)	N(2)—C(24)—C(23)	121.0(6)
C(17)—O(1)—C(18)	114.1(6)	C(14)—C(15)—C(16)	120.2(7)	N(2)—C(24)—C(25)	115.3(6)
C(21)—N(1)—C(22)	116.8(6)	C(11)—C(16)—C(15)	119.7(7)	C(23)—C(24)—C(25)	123.3(7)
C(21)—N(2)—C(24)	116.8(5)	O(1)—C(17)—O(2)	125.6(7)	O(5)—C(26)—N(3)	121.3(6)
C(21)—N(3)—C(26)	130.2(5)	O(1)—C(17)—C(12)	109.9(6)	O(5)—C(26)—N(4)	123.0(5)
S—N(4)—C(26)	122.4(4)	O(2)—C(17)—C(12)	124.3(6)	N(3)—C(26)—N(4)	115.7(6)
S—C(11)—C(12)	122.2(5)	O(1)—C(18)—C(19)	105.5(6)	Cl(1)—C—Cl(2)	110.1(5)
S—C(11)—C(16)	117.2(5)				

Table 3　A part values of torsion angles (°)

Atom 1	Atom 2	Atom 3	Atom 4	Angle	Atom 1	Atom 2	Atom 3	Atom 4	Angle
O(3)	S	N(4)	C(26)	-56.79(0.63)	O(4)	S	C(11)	C(12)	-37.40(0.65)
O(4)	S	N(4)	C(26)	175.32(0.56)	O(4)	S	C(11)	C(16)	137.24(0.53)
O(11)	S	N(4)	C(26)	60.57(0.62)	N(4)	S	C(11)	C(12)	73.41(0.60)
O(3)	S	C(11)	C(12)	-168.92(0.54)	N(4)	S	C(11)	C(16)	-111.86(0.55)
O(3)	S	C(11)	C(16)	5.81(0.63)					

分别为：Ⅰ—Ⅱ95°，Ⅰ—Ⅲ23.06°，Ⅱ—Ⅲ119.64°。各原子到平面的距离（nm）为：

平面Ⅰ：S(—0.00020)，N(4)(-0.00681)，C(26)(-0.00057)，N(3)(0.00241)，C(21)(0.00419)，N(1)(0.00044)，C(22)(-0.00476)，C(23)(-0.00446)，C(24)(0.00319)，N (0.00656)，O (4)(0.00157)，C (25)(0.00809)，O (5)(-0.01880)；

平面Ⅱ：C(11)(-0.00079)，C(12)(0.00093)，C(13)(-0.00053)，C(14)(-0.00001)，C(15)(0.00016)，C(16)(0.00024)；

平面Ⅲ：C(17)(-0.00176)，O(1)(0.00056)，O(2)(0.00071)，C(12)(0.00049)。

从分子的键长、键角及扭角值可推知，分子中的3个平面分别形成3组离域π键。该化合与嘧啶环上不带甲基的同类化合物[1]的键结构相似，但构象上有很大差别。比较—S—N—C的扭角值，本文为60.57°，而后者为-72.17°。说明苯环位置相对于磺酰脲平面而言，在2个分子中的取向正好相反。这一构象的差异，反映在农药活性上，便呈现出很大的不同。

参考文献

[1] Li Zheng-Ming（李正名），JIA Guo-Feng（贾国锋），LAI Cheng-ming（赖城明）*et al.*；Chem. J. Chinese Univ.（高等学校化学学报），1992，13（11）；1411.

[2] Franz J. F. *et al.*；J. Org. Chem.，1964，29：2592.

[3] E. P.；162 723.

Synthesis, Structure and SAR Study on Sulfonylurea Compounds(Ⅱ)

——Crystal and Molecular Structure of N - [2 - (4 - methyl) pyrimidinyl] - 2 - ethyl carboxylate - benzene sulfonylurea

Zhengming Li[1], Guofeng Jia[1], Lingxiu Wang[1],
Chengming Lai[2], Honggen Wang[3], Ruji Wang[3]

([1]National Key Laboratory of Elemento - Organic Chemistry, Elemento - Organic Chemistry Institute, Nankai University, Tianjin, 300071;
[2]Department of Chemistry, Nankai University;
[3]Central Laboratory of Nankai University)

Abstract Synthesis and crystal structure of title molecule was reported here. This crystal consists of title molecules with solvent molecules CH_2Cl_2. It is triclinic, belonging to space group $p\bar{1}$ with unit cell parameter a = 0.7413(2) nm, b = 0.9166 (5) nm, c = 1.6289(5) nm, α = 75.97 (4)°, β = 82.28(2)°, γ = 76.22 (3)°, V = 1.039 nm^3, Z = 2, R = 0.067 for 1914 reflections. Molecular structure was discussed. There are three different planes in the molecule, in each of which a conjugated system was formed. The intra - molecule and inter - molecule hydrogen bonds between atoms N and H have been found.

Key words sulfonylurea compounds; crystal structure; molecular structure

Structural Elucidation of Sex Pheromone Components of the Geometridae *Semiothisa cinerearia* (Bremer *et* Grey) in China*

Zhengming Li[1], Enyun Yao[1], Tianlin Liu[1], Ziping Liu[1], Suhua Wang[1], Haiqing Zhu[2], Gang Zhao[2], Zili Ren[2]

([1]Elemento - Organic Chemistry Institute, Nankai University, Tianjin, 300071;
[2]Department of Biology, Nankai University, Tianjin, 300071)

Abstract An extract from the female sex gland of *Semiothisa cincreraria* attracted conspecific males in field tests. A major active component was isolated from the extract and identified by GO - MS, GC - IR and microchemical reactions as cis - 3,4 - epoxy - (*Z*,*Z*) - 6,9 - heptadecadiene, which showed strong EAG response. Another minor yet important component was identified as (*Z*,*Z*,*Z*) - 3,6,9 - heptadecatriene.

Introduction

Chinese scholartree (*Sophora japonica L.*) which has considerable economical value, grows in China and Korea. Its bark, twigs, leaves, biossoms and seeds are used as traditional Chinese medicines. It is also an excellent source of bee honey. In recent years Chinese scholartree has been seriously damaged by *Semiothisa cinerearia*, a Geometridae insect. Insecticides have been applied which has brought some unfavourable impact on the ecosystem. In order to devise a pheromone bioregulator to control the pest, a project was devised to undertake the structural elucidation of the sex pheromone of *Semiothisa cinerearia*.

In literature, there were a few reports on the elucidation of sex pheromone of Geometridae insects: winter moth, [1,2] giant looper, [3] fall cankerworm, [4] Bormia selenaria[5] and *Semiothisa signaria dispuneta*. [6] Yet no research on *Semiothisa cinererea* Bremer *et* Grey has been reported.

Methods and materials

Insect rearing and pheromone extraction

S. *cinerearia* larvae and pupae were collected locally from the field and were fed by fresh scholartree leaves. The pupae were segregated by different sex and kept in a thermotank (26 + 2℃) with relative humidity of 70%.

The adults emerged in the first and second days after eclosion, the sex glands which were drawn from the 8 - 9th segments on the abdomen of females at 10pm in the evenings were dipped into or washed by redistilled hexane (each gland with 10 drops of the solvent). The dipping solution after purification by silica gel (120 - 160mesh) column chromatography was ana-

* Reprinted from *Chinese Journal of Chemistry*, 1993, 11(3): 251 - 256.

lyzed or the washings were sent directly to preparative GC.

Separation by column chromatography

150 FE(femalc equivalent) glands extracts preliminarily purified fractionated by preparative gas chromatography(Shimadzu GC 9A, FID, N_2 3mL/min) using an OV - 17 stainless steel column (2m ×0.25cm i.d.) programmed from 35℃ to 210℃ at 20℃/min. The components were collected in dry - ice cooled glass capillaries(40cm ×0.2cm i.d.) by intervals of appearance of peaks. Each peak collected starting from the following retention time[minutes(′) and seconds (″)]: peak 0—up to 7′00″, peak Ⅰ—7′40″, peak Ⅱ—19′30″, peak Ⅲ—20′40″, peak Ⅳ—26′50″. Peaks were monitored by EAG evaluation. Peak Ⅱ was concluded to be the bioactive component.

Structural analysis by GC/MS

Chemical ionization(CI) and electron impact(EI) GC/MS analyses were obtained by using Hewleft - Packard 5890/5988 and Hewleft - Packard 5980/5970 GC/MS respectively equipped with OV - 17 or DB - 5 capillary columns(25m ×0.20mm i.d.) with splitless mode (helium):

GC/EIMS: Ionization voltage 70 eV. Ion source temperature 200℃. The columns programmed from 35℃ to 220℃ at 15℃/min.

GC/CIMS: Ionization voltage 110 eV. Ion source temperature 134℃, emission current 300μA. The columns programmed from 35℃ to 210℃ at 20℃/min. Methane was used as reagent gas.

The washing of 20 FE sex glands was concentrated to a volume of 5μL under a stream of N_2. For each time 1μL was injected into the GC capillary column with splitless mode for analysis.

Structural analysis by GC/IR

The IR spectra were obtained by using Hewlett - Packard GC 5890/5965/5970B, a special GC/IR/MS tandem apparatus to elucidate micro - organic subtances. The capillary column OV - 17 was programmed from 50℃ to 240℃ at 20℃/min. Transfer line: 260℃, flow cell 250℃ with splitless mode. A concentrated washes of 170 FE sex glands was used for infrared spectra analysis.

Microchemical reactions

1. Catalytic hydrogenation reaction

The washing of 60 FE sex glands by hexane was concentrated under N_2 in a micro - reaction vessel(0.5μL), methanol was added as solvent and 3 mg PtO_2 was added as catalyst. Hydrogen was introduced into the solution with stirring at room temperature for 22h. The hydrogenated product was filtered through glass wool to remove catalyst and further concentrated for GC - MS analysis. The unhydrogenated and the hydrogenated samples were compared in GC/CIMS under identical conditions.

2. Trimethylsilylation reaction

(a) As a reference sample, into 2μL of pentadecanol solution (1μL of 10% pentadecanol/1mL of hexane) in a micro - reaction vessel, chloroform was added as solvent, then 1μL of bis - (trimethylsilyl) - trifluoro - acetamide (Aldrich)/1mL of chloroform was added, stirred at 40℃ for 3h.

(b) With concentrated washes of 20 FE sex glands, the same reaction with silylanizing reagent was carried out under the identical conditions.

The products from reactions (a) and (b) were checked by GC/MS separately.

Results and discussion

1. Field tests established that washings or extracts (dipping solution) of female sex glands of *S. cinerearia* were attractive to conspecific males. The extract was fractionated by preparative GC and the EAG active component was found in the second fraction as shown above.

2. By GC anaylsis (analytical condition was the same as GC/EIMS mentioned above) of washings of female sex glanda, a peak at $t_R = 13$ min had a sharp EAG activity. The washings of abdominal tips of conspecific males of *S. cinerearia* did not appear this specific peak.

3. GC/MS Analysis

The EI mass spectrum of this major active component (t_R = 13min) was shown as following. *m/z* (relative intensity): 250 (M^+, 3.52), 232 (2.99), 221 (4.93), 192 (5.49), 178 (20.07), 164 (3.52), 163 (1.28), 147 (6.51), 135 (6.69), 124 (9.16), 121 (10.74), 119 (7.75), 109 (11.3), 108 (13.56), 107 (18.31), 105 (7.22), 95 (22.36), 94 (28.17), 93 (44.54), 91 (27.11), 80 (86.80), 79 (100), 69 (22.36), 67 (51.06), 59 (16.37), 57 (23.59), 55 (29.75), 43 (29.75), 41 (55.81).

EIMS data analysis: the molecular ion is 250, with the base peak 79, the fragmentation ion 232 (M - 18) losing a water molecule indicated the possible existence of an alcohol, an aldehyde or an epoxide. Besides, ions of the series (C_nH_{2n-5}) indicated a long chain unsaturated hydrocarbon structure. The result of library search (NBS MS database) proposed three candidate compounds: heptadecatrienol, heptadecadienal and 3,6,9 - heptadecatriene. The CIMS spectra indicated that its molecular weight is 250 and 232 ($M - H_2O$) ion was present. Ions 251 (M + 1) and 233 [(M + 1) - 18] also appeared.

4. Microchemical reaction

The functional groups and the number of the double bonds in the structure of the active component are identified by using microchemical techniques and GC/IR.

After catalytic hydrogenation reaction, the product was analyzed by GC/CIMS. The ions 255 (M + 1) and 237 [(M + 1) - 18] were found which means that the molecular weight increased from 250 to 254. There must exist two double bonds in the molecule.

When the active component reacted with bis - (trimethylsilyl) - trifluoracetamide, no derivatization product appeared. Thus, and alcohol moiety, was precluded. Its stability in acidic or basic medium was further checked by GC/MS. The results showed that the active component' was very stable in basic and was easily decomposed in acidic medium which conformed to the distinct property of an epoxide moiety.

5. GC/IR spectra

GC/IR spectra:3020,2934,2867,1653,1460,1387,1274,912,817cm^{-1}.

The IR data indicated that:

(a) Hydroxy and carbonyl functional groups were absent in the molecule.

(b) 1653 cm^{-1} for *cis*—CH ═CH— stretching was present but 940 ~ 980 cm^{-1} for *trans*—CH ═CH— out of plane bending(strong) was absent. A ring vibration of 817 cm^{-1} of *cis* – epoxide was present(literature reported:[7] *cis* – epoxides 865 ~ 785 cm^{-1} *trans* – epoxides 950 ~ 860cm^{-1}). This indicated that the active component contained *cis* – olefin and *cis* – epoyio structural features and precluded other two candidate structures.

6. The location of the epoxide moiety

In conclusion, the unknown active component is a *cis* – epoxide. To locate the position of the epoxide group, the EIMS spectra of its hydrogenated product gave a clear picture. The EIMS data 25(12.52),194(1.48),266(1.28),166(1.28),155(1.84),139(1.43),138(2.57),125(4.41),123(3.67),111(12.59),109(7.41),97(35.41),83(71.10),82(48.36),71(34.78),70(41.44),69(80.62),68(57.14),67(31.32),59(46.04),57(100),56(39.77),55(77.69),43(89.67),41(85.24).

The saturated hydrocarbon epoxide possessed both characteristic 225 and 71 fragment ions for *a* – split of epoxide functionality(Fig. 1). It denoted that the epoxy group was situated between C_3 and C_4.

Usually, epoxide exhibits ion character analogous to a double bond as oxygen atom is lost during GC/MS operation. The MS library search showed up that 3,6,9 – triene actually contains an artifact of one double bond instead of an epoxy, moiety. The literature[8-10] denoted that "skipped" dienes usually show up MS fragments 124 and 178(Fig. 2), which also appeared in our MS data recorded in part 3 of this paper.

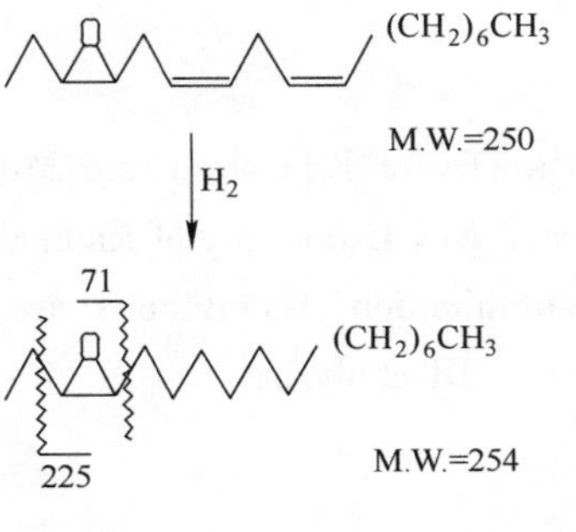

$(CH_2)_6CH_3$ M.W.=250 H_2 71 $(CH_2)_6CH_3$ 225 M.W.=254

Fig. 1 MS fragments for *a* – split of epoxide of the hydrogenated product

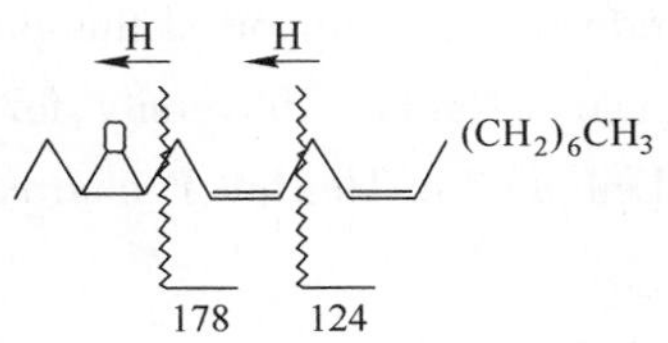

H H $(CH_2)_6CH_3$ 178 124

Fig. 2 A typical "skipped diene" MS fragments shown in the major active component

Summarizing the analyzing results mentioned above, the active component of *S. cinerearia* was identified as *cis* –3,4 – epoxy – (Z,Z) –6,9 – heptadecadiene. The MS data of the synthetic sample prepared in this laboratory are entirely identical. The EIMS data of synthetic sample *m/z*(relative intensity):250(M$^+$,0.6),221(0.3),203(1.1),192(2.0),178(8.1),163(0.2),147(1.1),135(2.6),124(3.3),121(4.8),109(9.0),108(9.8),107(11),105(6.6),95(20.4),94(20.6),93(36),91(25.2),80(72.6),79(100),69(19.7),67

(51.3),59(14.7),57(38.2),55(43.6),43(41.2).

The above data are also identical to the published data by Miller.[6]

7. Synthetic route of the target sample[11,12]

Starting from propargyl alcohol,3,6,9 – heptadecatriyne was hydrogenated with Lindlar catalyst to form(*Z*,*Z*,*Z*) –3,6,9 – heptadecatriene. The triene was then subjected to epoxidation by perbenzoic acid to form *cis* –3,4 – epoxy – (*Z*,*Z*) –6,9 – heptadecadiene which was separated and purified by column chromatography, yield 22.4%. By similar approach, (*Z*,*Z*,*Z*) – 2,5,8 – hexadecatrienol obtained was oxidated with *D* – (–) – tartaric diethyl ester(Sharpless reaction), *cis* –2*R*,3*S* – epoxy – (*Z*,*Z*) –5,8 – hexadecadienol – I was obtained with a yield of 76%, $[a]_D^{20}$ = +9.8. By further tosylation, reaction with lithium dimethyl cuprate, 3*R*,4*S* – epoxy – (*Z*,*Z*) –6,9 – heptadecadiene was synthesized with a yield 35%. All target samples were checked with a characteristic MS fragment 178 denoting the C_3 – C_4 epoxy functionality.

8. Minor active component

A minor active component was found in dipping solution(CH_2Cl_2) of abdominal tips which was also separated by GC(t_R = 11.79min) and checked by EAG. It existed in a ratio of 1∶6 to the major active component.

The EIMS spectrum of the minor active component is as following: 234 (M^+, 2.11), 178 (18.21), 135 (8.05), 121 (15.96), 108 (42.28), 95 (29.02), 93 (37.86), 79 (100), 67 (62.93), 55(46.04), 43(45.91), 41(71.90).

NBS library search denoted that the structure of minor active component was identified as(*Z*,*Z*,*Z*) –3,6,9 – heptadecatriene.

9. Bioassay results

The experimental results of electroantennogram(EAG) assay and flight tunnel behaviour assay of the synthesized samples of *cis* –3,4 – epoxy – (*Z*,*Z*) –6,9 – heptadecadiene(EHD) and(*Z*,*Z*,*Z*) –3,6,9 – heptadecatriene(HDT) were published.[13]

Acknowledgement

Sincere thanks are due to State Planning Commission, Ministry of Education and Ministry of Chemical Industries for support of this project and to National Key Laboratory of Elemento – Organic Chemistry, Nankai University for the spectral determination. Also thanks are due to Mr. Roger Leibramd of Hewlett Packard Co., U. S. A. for GC/IR analysis.

References

[1] Roelofs, W. L.; Hill, A. S.; Linn. C. E.; Meinwold, J.; Jain, S. C.; Herebert, H. J.; Smith, R. F., *Science*, 217, 657(1982).

[2] Bestmann, H. J.; Brosche, T.; Koschatzky. K. H.; Mich Clis, K.; Platz, H.; Roth. K.; Suss, J.; Vostrowsky, O.; Knant, W., *Tetrahedron Lett.*, 23, 4007(1982).

[3] Becker, D.; Kimmel, T.; Cyjon, R.; Moore, L.; Wysoki, M.; Bestmann, H. J.; Platz, H.; Roth, K.; Vostrowsky, O., *Tetrahedron Left.*, 24, 5505(1983).

[4] a) Wong, J. W.; Palaniswamy, P.; Underhill, E. W.; Steek, W. F.; Chisholm, M. D., *J. Chem. Eeol.*, 10, 463 (1984).

b) Wong, J. W.; Palaniswamy, P.; Underhill. E. W.; Steek, F.; Chisholm, M. D., *J. Chem. Ecol.* 10. 1579(1984).

[5] Wysoki, M. ; Scheepens, M. H. M. ; Moore, I. ; Becker, D. ; Cyjon, R. , *Ann. Entomol. Soc. Am.* , 78, 446 (1985).

[6] Miller, J. G. ; Underhill, E. W. ; Giblin, M. ; Barton, D. , *J. Chem. Ecol.* , 13, 1371(1987).

[7] Socrates, G. , *Infrared Characteristic Group Frequencies*, John Wiley, New York, 1980.

[8] Lee, R. R. ; Nevenzei, J. C. ; Paffenhofer, G. A. ; Benson, A. A. ; Patton, S. ; Kavanagh, T. E. , *Biochem. Biophys. Acta*, 202, 386(1970).

[9] Karunen, P. , *Photochemistry*, 13, 2209(1974).

[10] Yao, E. -Y. ; Li, Z. -M. ; Luo, Z. -Q. ; Shang, Z. -Z. ; Yin, K. -S. ; Hong, B. -B. , *Progress in Natural Science.* 1, 566(1991).

[11] Liu, T. -L. ; Liu, Z. -P. ; Li, Z. -M. , *Pesticides*, 30, 12(1991).

[12] Wang, S. -X. ; Yao, E. -Y. ; Li, Z. -M. ; *Huaxue Tongbao* 10, 22(1992).

[13] Ren, Z. -L. ; Zhao, G. ; Xu, J. -H. ; Zhu, H. -Q. , *Acta Entomol. Sin.* , 34, 2711(1991).

新磺酰脲类化合物的合成、结构及构效关系研究（Ⅲ）*

——N－［2′－（4，6－二甲基）嘧啶基］－2－甲酸乙酯－苯磺酰脲的晶体及分子结构

李正名[1]　贾国锋[1]　王玲秀[1]　赖城明[2]

（[1]南开大学元素有机化学国家重点实验室，南开大学元素有机化学研究所，天津 300071；
[2]南开大学化学系，天津 300071）

关键词　磺酰脲类化合物　晶体结构　分子结构

农药分子的几何结构对其活性有着重要的影响。为了系统研究这一问题，在前文[1~4]报道的合成 N－［2′－嘧啶基］－2－甲酸乙酯－苯磺酰脲、N－［2′－（4－甲基）嘧啶基］－2－甲酸乙酯－苯磺酰脲并测定其晶体结构的基础上，本文应用相似方法合成了标题化合物并测定了晶体结构。

组成为 $C_{16}H_{18}N_4O_5S$ 的标题化合物的晶体无色透明，经元素分析及 IR、NMR 确认并表征了结构。元素分析（%）计算值（实测值）：C 50.79（50.82），H 4.79（4.75），N 14.81（14.60）。IR（cm^{-1}）：3406（N—H），1693（C═O），1354，1176（—SO_2—）。^1H NMR（$CDCl_3$，δ，ppm）：1.32（3H，t，CH_3），2.48（6H，s，CH_3），4.32（2H，q，CH_2），6.64（1H，s，嘧啶环质子），7.50～8.30（4H，m，苯），8.54（1H，s，—NH—），13.08（1H，s，—SO_3—NH—）。

选取粒径线度约为 0.3mm×0.3mm×0.3mm 的单晶，在 Enraf－Nonius CAD4 四圆衍射仪上，用经单色化的 Mo $K\alpha$，入射线，在 $2°\leqslant\theta\leqslant24°$范围内，以 $\omega-2\theta$ 扫描方式，共收集到 3057 个独立反射点，其中 1818 个为可观测衍射点［$I\geqslant3\sigma(I)$］，全部强度数据均经过 L_P 因子校正及经验吸收校正。该晶体属三斜晶系，空间群为 $P\bar{1}$，$a=0.9134$（2）nm，$b=0.9567$（4）nm，$c=1.2547$（3）nm，$\alpha=73.88$（3）°，$\beta=70.93$（2）°，$\gamma=64.05$（4）°，$V=0.9273\ nm^3$，$M_r=378.41$，$Z=2$，$D_r=1.36g/cm^3$，$\mu=1.99cm^{-1}$，F（000）＝396。晶体结构由直接法（MULTAN 82）解出，根据 E－图确定了全部非氢原子坐标，用理论方法计算出碳上氢原子位置，在差值 Fourier 图上确定了氮上的两个氢原子位置，氢原子的各向同性热参数均被指定为 5.0，全部氢原子未参与最小二乘法修正，仅参与了结构因子的计算。对全部非氢原子的坐标及各向异性热参数进行全矩阵最小二乘修正，最终的一致性因子 $R=0.063$。在最终的 Fourier 图上，最高的电子密度峰的高度为 $560e/nm^3$，全部计算在 PDP 11/44 计算机上、用 SDP－PLUS 程序包完成。

＊　原发表于《高等学校化学学报》，1994，15（2）：227－229。国家自然科学基金重点资助课题。

图 1 为分子立体结构图。表 1 为全部非氢原子的分数坐标及等价各向同性热参量，表 2 列出了分子的键长及键角值，表 3 为部分扭角值。

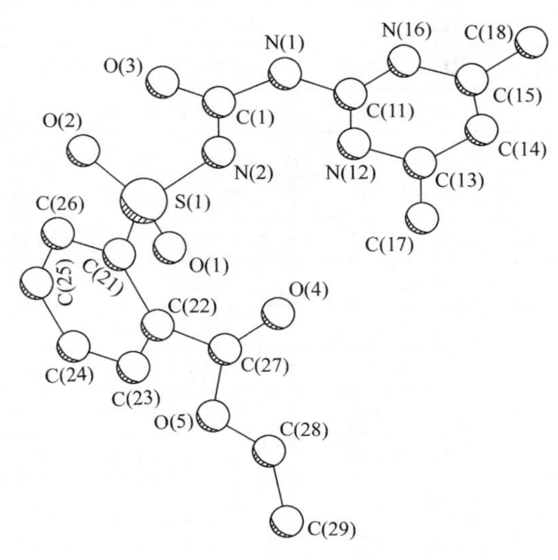

Fig. 1 Stereo structure of title molecule

Table 1 Fractional coordinates and equivalent isotropic thermal parameters for non - hydrogen atoms

Atom	x	y	z	B_{eq} ($10^{-2}nm^2$)	Atom	x	y	z	B_{eq} ($10^{-2}nm^2$)
S(1)	0.8547(2)	0.1231(2)	0.2677(1)	4.23(4)	C(14)	0.2722(7)	0.7298(7)	0.0897(6)	4.9(2)
O(1)	0.7178(5)	0.0748(4)	0.3266(4)	5.2(1)	C(15)	0.3807(7)	0.8032(6)	0.0208(5)	4.5(2)
O(2)	1.0008(6)	0.0181(5)	0.2052(4)	5.0(1)	C(17)	0.2314(7)	0.4935(8)	0.2243(5)	5.7(2)
O(3)	1.0125(4)	0.3334(5)	0.0960(4)	5.2(1)	C(18)	0.3219(9)	0.9634(8)	-0.0443(7)	6.8(2)
O(4)	0.5288(5)	0.3809(5)	0.4100(4)	5.7(1)	C(21)	0.9131(6)	0.1798(6)	0.3641(5)	3.7(1)
O(5)	0.5528(5)	0.2191(5)	0.5748(4)	5.5(1)	C(22)	0.7939(6)	0.2477(6)	0.4571(5)	3.9(2)
N(1)	0.7702(5)	0.5153(5)	0.0561(4)	3.6(1)	C(23)	0.8485(7)	0.2788(7)	0.5331(6)	5.1(2)
N(2)	0.7736(5)	0.2851(5)	0.1830(4)	3.7(1)	C(24)	1.0136(8)	0.2488(8)	0.5202(6)	6.1(2)
N(12)	0.5018(5)	0.5081(5)	0.1378(4)	3.6(1)	C(25)	1.1286(7)	0.1813(8)	0.4299(6)	6.0(2)
N(16)	0.5465(5)	0.7291(5)	0.0104(4)	4.1(1)	C(26)	1.0806(7)	0.1466(7)	0.3499(6)	4.9(2)
C(1)	0.8614(6)	0.3747(6)	0.1110(5)	3.8(2)	C(27)	0.6099(7)	0.2906(6)	0.4750(5)	4.1(2)
C(11)	0.5968(6)	0.5865(6)	0.0697(4)	3.4(1)	C(28)	0.3748(8)	0.2529(9)	0.6086(7)	7.5(2)
C(13)	0.3351(7)	0.5824(6)	0.1484(5)	4.1(2)	C(29)	0.303(1)	0.311(1)	0.7134(8)	9.0(3)

Table 2 Bond distances (nm) and angles (°) between non - hydrogen atoms

S(1)—O(1)	0.1427(5)	O(5)—C(27)	0.1317(7)	N(12)—C(13)	0.1347(7)
S(1)—O(2)	0.1412(5)	O(5)—C(28)	0.1447(8)	N(16)—C(11)	0.1321(6)
S(1)—N(2)	0.1640(5)	N(1)—C(1)	0.1366(6)	N(16)—C(15)	0.1339(8)
S(1)—C(21)	0.1765(8)	N(1)—C(11)	0.1394(6)	C(13)—C(14)	0.1363(7)
O(3)—C(1)	0.1220(7)	N(2)—C(1)	0.1377(7)	C(13)—C(17)	0.1484(9)
O(4)—C(27)	0.1189(8)	N(12)—C(11)	0.1335(7)	C(14)—C(15)	0.1394(9)

续表2

C(15)—C(18)	0.1478(8)	C(22)—C(23)	0.137(1)	C(24)—C(25)	0.1363(9)
C(21)—C(22)	0.1399(7)	C(22)—C(27)	0.1498(8)	C(25)—C(26)	0.139(1)
C(21)—C(26)	0.1379(9)	C(23)—C(24)	0.137(1)	C(28)—C(29)	0.142(1)

O(1)—S(1)—O(2)	118.8(3)	C(14)—C(13)—C(17)	124.3(6)
O(1)—S(1)—N(2)	104.2(2)	C(13)—C(14)—C(15)	119.7(5)
O(1)—S(1)—C(21)	109.2(3)	N(16)—C(15)—C(14)	120.6(5)
O(2)—S(1)—N(2)	109.9(2)	N(16)—C(15)—C(18)	116.5(5)
O(2)—S(1)—C(21)	108.2(3)	C(14)—C(15)—C(18)	122.9(5)
N(2)—S(1)—C(21)	105.9(3)	S(1)—C(21)—C(22)	120.9(5)
C(27)—O(5)—C(28)	117.1(5)	S(1)—C(21)—C(26)	118.4(5)
C(1)—N(1)—C(11)	128.3(5)	C(22)—C(21)—C(26)	120.5(7)
S(1)—N(2)—C(1)	124.2(4)	C(21)—C(22)—C(23)	118.1(5)
C(11)—N(12)—C(13)	116.8(5)	C(21)—C(22)—C(27)	122.9(7)
C(11)—N(16)—C(15)	115.9(5)	C(23)—C(22)—C(27)	118.9(5)
O(3)—C(1)—N(1)	120.6(5)	C(22)—C(23)—C(24)	122.4(6)
O(3)—C(1)—N(2)	122.9(4)	C(23)—C(24)—C(25)	118.9(8)
N(1)—C(1)—N(2)	116.4(5)	C(24)—C(25)—C(26)	121.4(6)
N(1)—C(11)—N(12)	119.1(4)	C(21)—C(26)—C(25)	118.7(6)
N(1)—C(11)—N(16)	113.4(5)	O(4)—C(27)—O(5)	126.2(5)
N(12)—C(11)—N(16)	127.5(5)	O(4)—C(27)—C(22)	124.2(6)
N(12)—C(13)—C(14)	119.7(5)	O(5)—C(27)—C(22)	109.5(5)
N(12)—C(13)—C(17)	116.0(4)	O(5)—C(28)—C(29)	110.2(9)

Table 3 Selected values of torsion angles

Atom 1	Atom 2	Atom 3	Atom 4	Angle (°)
O (1)	S (1)	N (2)	C (1)	178.67 (0.44)
O (2)	S (1)	N (2)	C (1)	-53.03 (0.54)
C (21)	S (1)	N (2)	C (1)	63.63 (0.50)
O (1)	S (1)	C (21)	C (22)	-31.84 (0.51)
O (1)	S (1)	C (21)	C (26)	144.15 (0.46)
O (2)	S (1)	C (21)	C (22)	-162.50 (0.44)
O (2)	S (1)	C (21)	C (26)	13.50 (0.55)
N (2)	S (1)	C (21)	C (22)	79.73 (0.50)
N (2)	S (1)	C (21)	C (26)	-104.26 (0.49)

考察分子的结构可见，分子中存在着3个独立的原子平面，即嘧啶磺酰脲平面、苯环平面、羧基平面，每个平面均各形成独立的共轭离域π键。分子中还存在N…H分子内氢键。

本工作得到南开大学中心实验室王如骥、王宏根二位老师的帮助，谨致谢意。

参 考 文 献

[1] Franz J. F. *et al.* J. Org. Chem, 1964, 29: 2592.

[2] Hillemann C. Lee; E. P., 162 723, 1985.

[3] LI Zheng – Ming (李正名), JIA Guo – Feng (贾国锋), LAI Cheng – Ming (赖城明) *et al.*; Chem. J. Chinese Univ. (高等学校化学学报). 1992, 13: 1411.

[4] LI Zheng – Ming (李正名), JIA Guo – Feng (贾国锋), LAI Cheng – Ming (赖城明) *et al*,; Chem. J. Chinese Univ. (高等学校化学学报). 1993, 14: 349.

Syntheses, Structure and SAR Study on Sulfonylurea Compounds (Ⅲ)

——Crystal and Molecular Structure of N – [2′ – (4,6 – dimethyl) pyrimidinyl] – 2 – ethyl Carboxylate – benzene Sulfonylurea

Zhengming Li[1], Guofeng Jia[1], Lingxiu Wang[1], Chengming Lai[2]

([1]National Key Lab. of Elemento – Organic Chem., Elemento – Organic Chem. Institute, Nankai University, Tianjin, 300071;
[2]Department of Chemistry, Nankai University, Tianjin, 300071)

Abstract Here we reported the crystal structure of title molecule. The crystal is monoclinic, belonging to space group $P\bar{1}$ with unit cell parameter $a = 0.9134$ (2) nm, $b = 0.9567$ (4) nm, $c = 1.2647$ (3) nm, $\alpha = 73.88$ (3)°, $\beta = 70.93$ (2)°, $\gamma = 64.05$ (4)°, $V = 0.9273$ nm^3, $D_x = 1.36$ g/cm^3, $\mu = 1.99$ cm^{-1}, $F(000) = 396$ for 1818 reflections. $R = 0.063$. Molecular structure has been discussed. There are three different planes in the molecule, in each of which a conjugated system was formed. The intra – moleucle hydrogen bond between N and H atom has been found.

Key words sulfonylurea compounds, crystal structure, molecular structure

The Reaction of α – oxo – α – Triazolylketene Dithioacetal with Amines, Hydrazine and Guanidine*

Zhengming Li, Zhengnian Huang

(Institute of Elemento – Organic Chemistry, Nankai University, Tianjin, 300071)

Abstract Triazolylbenzoylketene S,S – acetal(1) reacted with aniline to S,N – acetal(2), with ethylenediamine, o – phenylenediamine and ethanolamine to the corresponding cyclic N,N – acetalg(3), (5) and N,O – acetal(4). Treatment of (1), (2) with guanidine and hydrazine led to the biheterocycles(6), (7), (8) respectively.

The doubly activated ketene dithioacetals are useful intermediates in organic synthesis. In recent years, a variety of α – oxoketene dithioacetals were prepared and studied, most of which were briefly covered in recent reviews[1,2]. However, we surprisingly found that no reports on studies of α – oxo – α – azolylketene dithioacetals appeared in the literatures, although there were a few reports on studies of α – oxo – α – heterocyclylketene dithioacetals [2,3]. Therefore in a systematic work of exploring biologically active triazole compounds for new agrochemicals, the reaction of α – oxo – α – triazolylketene dithioacetal with amines, hydrazine and guanidine was described in this paper.

Referring to Gompper and Topfl's addition of amines to conjugated ketene dithioacetals to afford the corresponding conjugated ketene N,S – acetals[4], here the reaction of α – oxo – α – triazolylketene dithioacetal with amines proceeded smoothly to yield the corresponding α – triazolylketene N,S – acetal(2), cyclic N,N – acetals (3), (5) and N,O – acetal(4) in various yields as illustrated in scheme I. Besides, biheterocycles containing 1,2,4 – triazole were readily prepared by reacting α – oxo – α – triazolylketene S,S – acetals further with bifunctional nucleophiles such as hydrazine and guanidine in high yields (scheme Ⅱ). These novel compounds thus obtained were confirmed by ^{1}HNMR, IR, elemental analysis. The yields, melting points and reactiontimes were listed in the table 1.

Starting material: α – triazolyl – α – benzoylketene S,S – acetal was prepared according to the literature[5].

1 General procedure for the gynthesis of triazolylbenzoylketene S,N – acetal(2), cyclic N,N – acetals(3), (5) and N,O – acetal(4)

The mixture of (1) (0.87g. 0.006mol) and amines (0.003mol) in ethanol(20ml) was refluxed under stirring for 0.5 ~ 3 h and then the solvent removed by vacuum. In some cases products could be obtained directly, in other cases the viscous residues obtained were purified

* Reprinted from *Chinese Chemical Letters*, 1994, 5(1):31 – 34. This work was supported by the National Natural Science Foundation of China.

Table 1

Product No.	Reaction Time(h)	Yield(%)	M. P. (℃)	Molecular Formula
2	3	82	141 ~ 142	$C_{18}H_{18}N_4OS$
3	0.5	75	219 ~ 220	$C_{18}H_{18}N_6O$
4	0.5	55	199 ~ 200	$C_{13}H_{13}N_5$
5	3	87	278 ~ 279	$C_{17}H_{13}N_6O$
6	0.5	93	218 ~ 219	$C_{13}H_{11}N_5S$
7	0.5	96	244 ~ 245	$C_{17}H_{14}N_6$
8	1	76	201 ~ 203	$C_{14}H_{14}N_6O$

by column chromatography on silica gel using petroleum/ethyl acetate as eluant. The yields of products varied from moderate to high.

Scheme Ⅰ

Scheme Ⅱ

2 Preparation of (6),(7)

To a solution of (1) (0.87g,0.003mol) was added 2 ~ 3 equivalents of hydrazine, and the mixture was refluxed for 0.5h. Removal of solvent under reduced pressure gave the crude (6), which was further purified by crystallization. (7) was prepared from (2) under the identical condition.

3 Preparation of (8)

(0.87g,0.003mol) was added to a stirred suspension of a sodium ethoxy [prepared by dissolving sodium (0.003mol) in 20mL ethanol] and guanidine nitrate then the mixture was refluxed for 4h. The solvent was evaporated under reduced pressure and the residue poured into ice water 1000mL. With the pH of the mixture being adjusted to 6 - 7, the crude product was isolated by filtration and purified by crystallization using ethanol in 76% yield.

References

[1] R. K. Dieter, Tetrahedron, 1986, 42. 3029 - 96.
[2] H. Junjappa: H. Ha: C. V. Asokan Tetrahedron, 1990, 46, 5423 - 5503.
[3] M. Augustin: W. Doelling, Z. Chem, 1976, 16(10): 389 - 9.
[4] R. Gompper: W. Topfl. Chem. Ber, 1976, 95, 2871.
[5] Ger, offen. DE 3, 145, 890, 1963.

Studies on Sex Pheromone of *Ectropis obliqua* Prout*

Yin Kunshan(殷坤山)[1], Hong Beibian(洪北边)[1], Shang Zhizhen(尚稚珍)[2],
Yao Enyun(幺恩云)[2], Li Zhengming(李正名)[2]

([1]Institute of Tea, Chinese Academy of Agricultural Sciences, Hangzhou, 310008, China;
[2]Institute of Elemento-organic Chemistry, Nankai University, Tianjin, 300071, China)

Abstract All the evidence from the morphology, histology and scanning electron microscopy studies revealed that the sex pheromone gland of *Ectropis obliqua* prout was situated at the intersegmental membrane between the 8th and 9+10th abdominal segments. The methods used for collecting sex pheromone were: (i) excising abdominal tips of 2-d-old virgin females, and then keeping them in dark cycle for 6-8 h at 15-20℃ during their optimal calling period; (ii) rinsing and washing the abdominal tips with small quantities of dichloromethane and hexane; or (iii) adsorbing the sex pheromone with porapak Q. All of these methods were successful for quantitative analysis by HPLC and GC/MS. EAG response profile and laboratory bioassay with a wind tunnel were used to monitor and isolate the bioactive fractions, followed by GC/MS for identification of pheromone of *E. obliqua*. (z,z,z)-3,6,9-octadecatrine, (z,z,z)-3,6,9-docos trine, (z,z)-3,9-*cis*-6,7-epoxyocotadecadiene and (z,z,z)-3,6,9-tetracostriene were likely the main components of this system.

Key words *E. obliqua*; sex pheromone

Ectropis obliqua prout is an important tea pest in south China. From 1977 to 1979, the mating behavior and crude extracts from ovipositor were investigated by Du[1], but the components of sex pheromone (SP) have not been identified yet. Recently, the investigations of the gland location, the regularity of SP release, and the extraction and isolation of the active fractions, followed by bioassay, have been carried out[2,3]. The bioassay of artificial SP has been investigated too.

1 Materials and Methods

1.1 Insect Sources

Collected from tea plantations, and propagated in laboratory at 20℃. The virgin female adult was selected for investigation.

1.2 Synthesis of SP

Synthesized and supplied by Nankai University

1.3 Identification of Gland Location[4-6]

(1) Morphology. Fix the female adult on a wax flask with a little bio-salt water, incise its abdomen and ovipositor vertically and investigate the specific structure of the SP gland.

(2) Extraction. Excise the apophysis posterioris, the front and back semi-part of the 9+10th

* Reprinted from *Progress in Natural Science*, 1994, 4(6): 732-740.

abdominal segment, the intersegmental membrane between the 8th and 9 + 10th abdomen, the 8th abdomen segment with apophysis anterioris and the 7th abdomen segment individually, rinse 3 times with distilled water prior to extracting 10 female equivalences (FE) for 16 h with 0.5mL dichlormethane.

(3) Electron microscope scanning. Incise the intersegmental membrane between the 8th and 9 + 10th abdomen segments to monitor the micro – structure under the electron microscope.

1.4 Investigation of SP Release

(1) Relationship between the quantity of SP release and the age of female adult. Put 15 heads of emerging female adults into a glass bottle at 22:00 every night to obtain a series of emerging of 1 – 7 – d – old female adults, select 5 heads from each bottle to extract SP for bioassay and check the response rate of the male adults to the extracts.

(2) Relationship between the quantity of SP release and the light cycle. Put 10 heads of emerging female adults into a glass bottle which contained a round cubic brass sieve, keep the adults staying outside the sieve and inside the bottle, when two days old, put a piece of filter paper into the sieve to adsorb the SP every other hour before 2 h prior to starting the dark cycle. The quantity of SP release was monitored by the calling rate of the male adult to the filter paper.

1.5 Methods of SP Collecting

Keep the emerging female adults at 15 – 20℃, collect the SP after 6h during dark cycle from 2 – d – old adults. The collection methods were:

(1) Gland extracting. Merge the glands of 10 female adults in 0.2mL dichloromethane for 2h.

(2) Gland rinsing. Pull out the ovipositors, and rinse each ovipositor with 10 – 12 drops of n – hexane, and then concentrate 10 FE of the crude extract to 0.2mL.

(3) Porapak Q adsorbing. Put 3 filter paper bags (1.5cm × 11cm × 0.1cm) containing porapak Q and 30 emerging female adults into a glass bottle, and cover it well. Take the bags out after 48h, and put the adsorbent porapak Q into a glass column. The column was eluted with dichloromethane. 2mL of the elution was collected and 10 FE of the elution was concentrated to about 0.2mL.

(4) Ovipositor extracting. Press the abdomen to force the ovipositor outside, excise the abdominal segments from the 8th one, put them into a glass tube containing dichloromethane, and extract it for 14h. 10heads of female adults were used.

1.6 Bioassay Methods

(1) Apparatus and conditions. The apparatus is shown in Fig. 1.

The testing conditions were: at 15 – 25℃; 1 – 4 – d – old male adults; 2h after being in dark cycle; under a faint reflecting light.

(2) Methods. Transfer the crude extracts or artificial SP onto a piece of 3cm × 5cm large filter paper, put the paper into a tube after the resolvent evaporated, and put live females into another tube as CK. Connect the spherical suction tube and the sample tube to the aeration chamber

containing 5 heads of males, record the number of calling males in 1 min. The attraction rate testing was carried out by connecting an attraction tube to the other end of the wind tube as shown in Fig. 1, and record the number of males flying toward the wind.

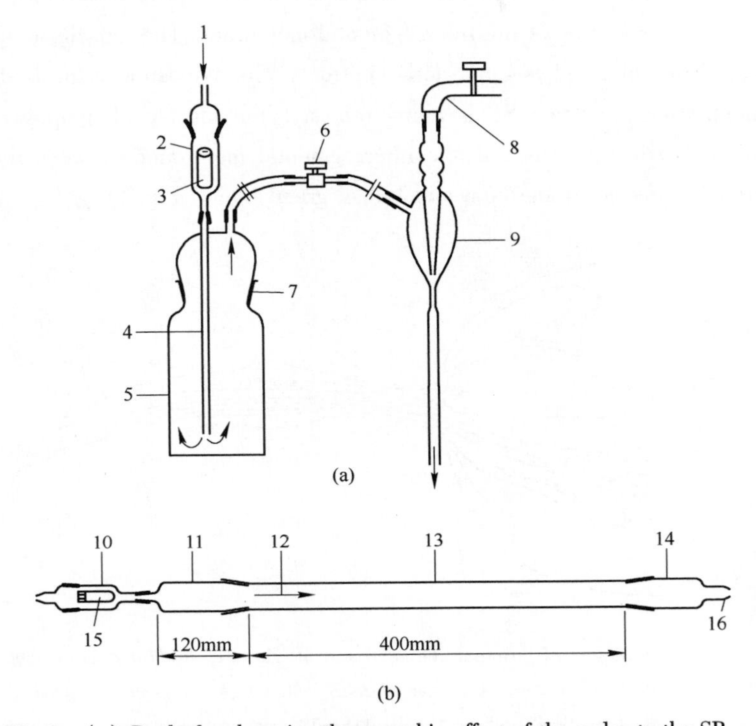

Fig. 1 (a) Bottle for observing the courtship effect of the males to the SP, (b) tube for observing the attracting effect to males

1—Airflow, 2—tube with sample (ϕ20mm. glass tube); 3—filter paper with sample; 4—ϕ6mm. glass tube; 5—aeration chamber for males (250mL glass bottle); 6—airflow controller; 7—normal joint; 8—tap; 9—spherical suction tube; 10—tube with sample; 11—tube trap; 12—airflow; 13—wind - tube (ϕ40mm); 14—tuble for storing males; 15—filter paper with sample; 16—to link spherical suction tube

1.7 Field testing

Put several bowls full of water containing 1% detergent among and above the tea plants. The bowls were about 20m apart, being hung in a 5cm × 5cm × 3cm bottomless wooden box which contained artificial SP above the middle of the bowl every evening. Record the number of the attracted males next morning. Set a 50 FE of crude extracts and empty box as CK.

2 Results

2.1 SP Gland Location of *E. obliqua* Prout

The abdomen of *E. obliqua* prout is composed of 10 segments. The 1st – 7th segments are larger and covered with many small scales. The 8th segment is smaller and is composed of ovipositor basis, several intersegmentalia and a membrane – like capsule inserted in the 7th abdomen segment. The capsule contains a pair of apophysis anterioris and much tomenta inside which are used to extrude ovum. Parts of the 8th and all the 9 + 10th segments make up a tube – like ovipositor. The intersegmental membrane between the 8th and 9 + 10th is a little longer, thin and semi – transparent, and can be folded and overturned. The ovipositor can move to and fro by the function of the contraction of the muscoid muscle between apophysis anterioris and apophysis posterioris and the overturning of the intersegmental membrane. The muscle of apophysis posterioris relaxes, and that of apophysis anterioris contracts. The ovipositor is inserted in the 7th abdominal segment. However, when SP is being released, the muscle of apophysis posterioris is contracted; the ovipositor is extended; the intersegmental membrane between the 8th and 9 + 10th abdomen segments is intumescent and full of green blood (Fig. 2).

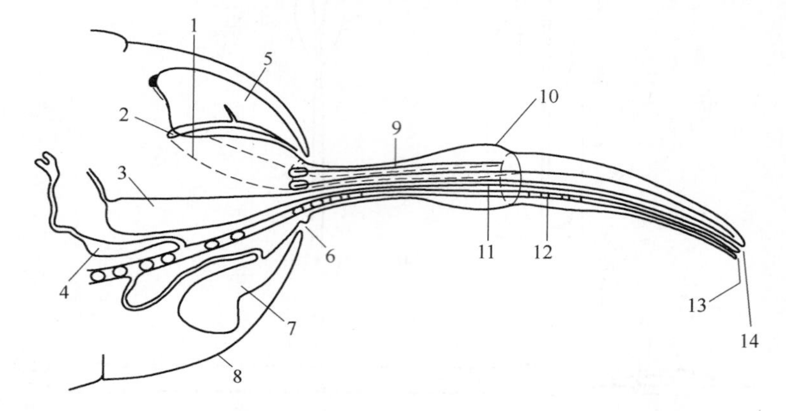

Fig. 2 The structure of terminal abdomen and SP gland of *E obliqua* prout females
1—Muscoid muscle; 2—apophysis anterioris; 3—intestine; 4—sex accessory gland; 5—hair bursa; 6—ostium; 7—bursa copulatrix; 8—7th segment; 9—apophysis posterioris; 10—sex pheromone gland; 11—intestine; 12—oviduct; 13—ovipore; 14—anales

There are many folds with a few spines but no hole was found under the electron microscope on the surface of the intersegmental membrane between the 8th and 9 + 10th segments (Fig. 3).

Fig. 3 Photograph of electron microscope scanning of SP gland

The bioassay results are shown in Table 1.

Table 1 The Bioactivity of Crude SP Extract

Extracting parts	Ovipositor	Intersegmental membrane between 8th and 9 + 10th Segment	Front semipart of 9 + 10th Segment	Behind semipart of 9 + 10th Segment	Pophysis posterioris	8th Segment with pophysis anterioris	7th Segment	CK
Calling rate for the males (%)	83	73	62	27	17	10	10	10
Significance of difference	a	ab	b	c	cd	de	de	e

2.2 Regularity of SP Release

At 20℃, the females would release SP as soon as they emerge. The amount of SP released increased rapidly with the time and reaches the peak after 2 d. The virgin female oviposited too, and the amount of SP released decreased gradually after 2 d, and the first oviposition completed in 4 d. If they mated again, the amount of SP released increased again and reached another peak after 6 – 7 d. (Fig. 4(a)). But most females died after 5 d.

The other factor affecting SP release is L – D cycle (Fig. 4(b)). Within 12 – 12/L – D cycle, no SP is released during light period, but SP releases when dark cycle starts, and reaches the peak at the 6th – 8th hour, and then decreases until the release stops. Therefore, the optimal period for extracting SP is at the 6th – 8th hour during the dark period for 2 – d – old females.

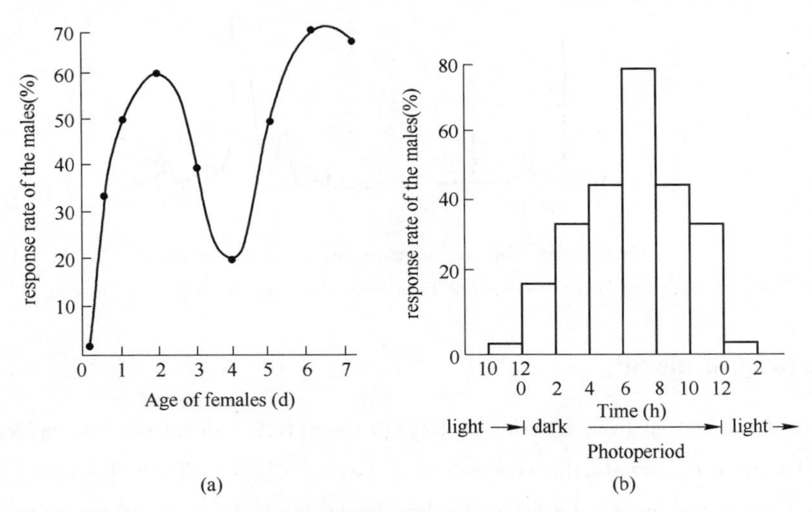

Fig. 4 (a) Regularity of SP release of females during a life; (b) regularity of SP release of females during a photoperiod

2.3 Effect of Extraction Methods on the Bioactivity of SP Crude Extracts

All the extracts from 4 different extraction methods show high bioactivity (Table 2). The respond rates for males to the crude extracts from gland rinsing and Porapak Q adsorbing are 74% – 84%. There is no significant difference from that of the ovipositor extracting. The attracting rates for males in wind tunnel to crude extracts from gland rinsing, gland extracting and Porapak Q adsorbing are 40% – 60%. There is also no significant difference from that of ovipositor extracting. However, the bioactivity of all the crude exrtacts from 4 methods are lower than that of the live females.

GC analysis results show that the extracts from gland rinsing, gland extracting and Porapak Q adsorbing are much clearer than that from ovipositor exrtacting(Fig. 5).

Table 2 Comparison of the Bioactivity of Extracts

Extraction methods	Females (CK)	Ovipositor extracting	Gland rinsing	Porapak Q adsorbing	Gland extracting
Calling rate of males(%)	100	84	82	77	74
Significance of difference	a	b	b	bc	c
Attracting rate to males(%)	93	60	50	50	40
Significance of difference	a	b	b	b	b
Coefficient	$r = 0.9679$				

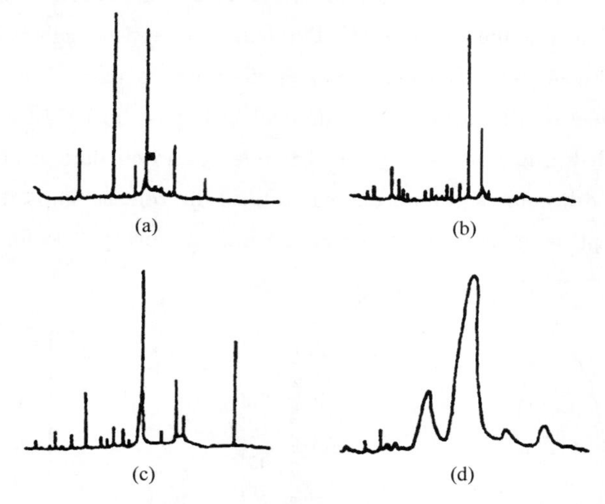

Fig. 5 Cas – liquid chromatogram of crude extracts

(a) Gland rinsing; (b) Gland extracting; (c) Porapak Q adsorbing; (d) Ovipositor extracting.

2.4 Bioactivity of the SP

According to the selection from a series of (z,z,z) – 3,6,9 – decatrine and epoxy decadiene with 18 – 24 carbon chemicals, the component of (z,z) – 3,9 – *cis* – 6,7 – epoxy – octadecad – iene has higher responce rate to males but lower than those of crude extracts and live males. The mixture of the components shows no significant difference in bioactivity from individual ones(Table 3).

Table 3 Calling Rates(%) for Males to Artificial sh

Dosage (μg)	Samples						
	octade – catriene	epoxyocta – decadiene	docostriene	tetracostriene	octadeca – dienal	crude extracts of gland rinsing(CK1)	live females (CK2)
10	0	44	0	0	0	—	—
1	0	30	13	0	0	—	—

Table 3 (continued)

Dosage (μg)	Samples						
	octade - catriene	epoxyocta - decadiene	docostriene	tetracostriene	octadeca - dienal	crude extracts of gland rinsing (CK1)	live females (CK2)
0.1	15	47	7	7	10	—	—
0.01	0	30	0	27	10	—	—
10 (FE)	—	—	—	—	—	82	—
1 (head)	—	—	—	—	—	—	100

The attracting rates of components of (z,z,z) -3,6,9 - docostriene, (z,z,z) -3,6,9 - tetracostriene and (z,z) -9,12 - octadecadienal are 12% - 22%, and that of the mixture of the above 3 components (0.1 μg equal dosage) is 27%. There are also some differences among dosages (Table 4).

Table 4 Attraction Rates (%) for Males to Artificial SP with Wind Tunnel in Laboratory

Dosage (μg)	Samples						
	octade - catriene	epoxyocta - decadiene	docostriene	tetracostriene	octadeca - dienal	crude extracts of gland rinsing (CK1)	live females (CK2)
10	0	0	8	0	0	—	—
1	0	0	13	12	9	—	—
0.1	0	0	18	22	12	—	—
0.01	0	0	5	0	0	—	—
10 (FE)	—	—	—	—	—	50	—
1 (head)	—	—	—	—	—	—	93

The field tests show that when 5μg dosage of every component is equally mixed, the attracting rate for males to the mixture of docostriene, tetracostriene and octadecadiene is 0.75 head/bowl, much lower than that of 50 FE of crude extracts (shown in Table 5).

Table 5 Attracted Males of Artificial SP in Field

Mixed dosage (μg)					Average number of attracted males per bowl
octadecatriene	epozyoctadecadiene	docostriene	tetracostriene	octadecadianal	
				5	0.50
		5	5	5	0.75
5	5	5	5	5	0.67
Crude extracts (50FE)					2.00
Empty (CK)					0

3 Discussion and Conclusions

(1) The SP gland is located at the intersegmental membrane between the 8th and 9 + 10th abdomen segments. The process of SP release is completed by the way of relaxing the muscle of

apophysis anterioris, contracting the muscle of apophysis posterioris, extruding ovipositor to force the intersegmental membrane overturned, and ultimately releasing the SP into the air through a great many vertical folds.

(2) The peak of SP release for virgin females is at the 6th – 8th hour during the dark cycle with 2 – d – old adults. The environmental conditions are 15 – 20℃ and 12/12 L – D cycle. That is the best period to extract SP.

(3) The gland rinsing, the gland extracting and the Porapak Q adsorbing are effective, and the crude extracts are purer than that from ovipositor extracting, which is successful in isolating the SP. However, the gland extracting is inconvenient, and the Porapak Q adsorbing is expensive. Only the gland rinsing is economical and convenient.

(4) SP of *E. obliqua* prout is composed of many components. (z, z) – 3, 9 – *cis* – 6, 7 – epoxyoctadecadient is the main one to induce calling rate for males, (z, z, z) – 3, 6, 9 – docostriene, (z, z, z) – 3, 6, 9 – tetracostriene and (z, z) – 9, 12 – octadecadienal can attract some males both in laboratory and field. The sex ratio of *E. obliqua* is 1: 1, and the attracting rate of artificial SP is much lower than that of the live females. There may be some other factors affecting the bioactivity of artificial SP, which should be studied further.

The authors thank Liu Tianlin and Luo Zhiqiang for synthesizing and providing synthetic sex pheromone.

References

[1] Du Jiawei, Zhao Qimin, Xu Shaopu *et al.*, *Insect Pheromone* (in Chinese), 1981(1): 41.

[2] Yin Kunshan, Hong Beibian, Hong Zhihua, *China Tea* (in Chinese). 1989(3): 20.

[3] Sei T., Seli H. & Matsnmoto, Y., *The Investigation Methods* (Study Section, in Japanese), Publishing Center of Society, 1980, pp. 133 – 140.

[4] John, W. W., *Motl. Journal Chemical Ecology*, 1985, 11(6): 725.

[5] Yao En – yun, Li Zheng – ming & Luo Zhi – qiang *Progress in Natural Science*, 1991, 1(6): 565.

[6] Ren Zili, *Journal of Entomology* (in Chinese), 1991, 34(3): 271.

Synthesis of 5 – Mercaptoalkylamino – and 5 – Anilinoalkylthiopyrazolyl – 1,2,4 – Triazoles by Ring Chain Transformation*

Zhennian Huang, Zhengming Li

(Institute of Elemento – Organic Chemistry, Nankai University, Tianjin, 300071, China)

Abstract Cyclic α – oxo – α – (1,2,4 – triazol – 1 – yl) ketene *N*, *S* – acetals (**1a – d**) and (**4a – d**) react with hydrazine by a ring chain transformation affording title compounds (**3a – d**, **6a – d** and **6'a – d**).

α – Oxoketene *N*, *S*; *S*, *S* – acetals are of great interests as synthetic intermediates in heterocyclic syntheses. [1,2] The reaction of these acetals with hydrazine and hydroxylamine can be applied to the preparation of substituted pyrazoles and isoxazoles. [3-8] In contrast to the open chain α – oxoketene *N*, *S*; *S*, *S* – acetals, the cyclic α – oxoketene *N*; *S*; *S* – acetals have not been paid attention yet. In this paper, therefore, the reaction of cyclic α – oxoketene *N*, *S* – acetals with hydrazine is first investigated. The cyclic α – oxo – α – (1,2,4 – triazol – 1 – yl) ketene *N*, *S* – acetals (**1a – d**) and (**4a – d**) are chosen as the starting materials in connection with our systematic work of probing bioactive triazole compounds. [9] The acetals (**1a – d**) and (**4a – d**) are easily available by condensation of α-(1,2,4 – triazol – 1 – yl) substituted acetophenone with phenyl isothiocyanate in the presence of powder potassium hydroxide and subsequent alkylation by 1,2 – dibromoethane and 1,3 – dibromopropane. [10,11]

Scheme I

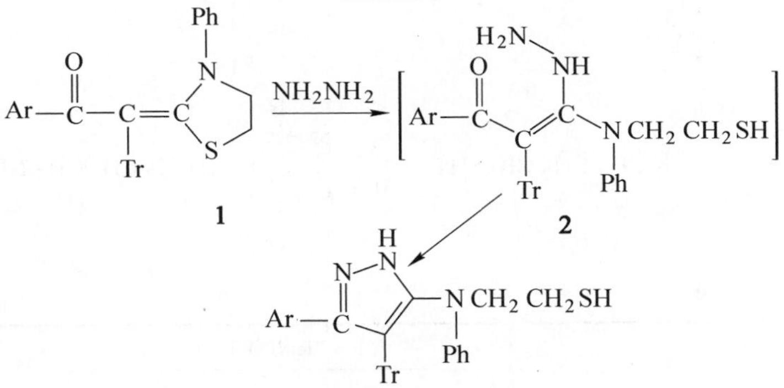

Ar: **a**, C_6H_5; **b**, 4 – ClC_6H_4; **c**, 4 – $CH_3OC_6H_4$; **d**, 4 – $CH_3C_6H_4$

Tr: 1,2,4 – triazol – 1 – yl

* Reprinted from *Heterocycles*, 1995, 41(8): 1633 – 1638.

When the mixture of **1** and **3 – 5** fold excess of 85% hydrazine hydrate in ethanol was refluxed under a nitrogen atmosphere for an hour, the final products(**3**) were obtained in moderate yields.

The formation of **3** can be explained by a initial attack of hydrazine at the ring – C atom in **1** leading to the assumed intermediate(**2**) and the final cyclization. Thus the whole reaction sequence indeed shows a process of ring chain transformation in which a pyrazole ring is formed by condensation while a thiazolidine ring of the substrate is opened to give an ω – substituted alkylheteroatomic side chain possessing two heteroatoms at its ends(Scheme Ⅰ). A few of such ring transformations have been studied. [12 – 17]

Further, the similar reaction of acetals(**4a – d**) with hydrazine was also investigated. To our surprise, the reaction of **4** with hydrazine in refluxing ethanol gave aminopyrazolyltriazoles(**6a – d**) and thio – iso – mers(**6′a – d**), respectively(Scheme Ⅱ).

Scheme II

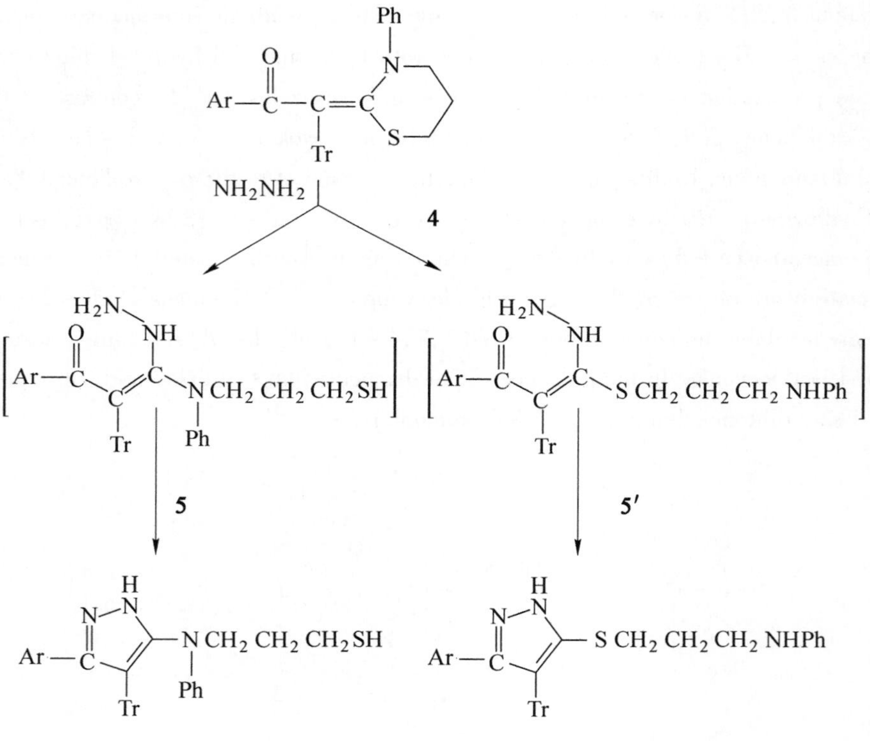

No	Ar	Yield(%)		6:6′
		6	6′	
a	C_6H_5	23	44	34:66
b	4 – ClC_6H_4	33	47	41:59
c	4 – $CH_3OC_6H_4$	26	38	41:59
d	4 – $CH_3C_6H_4$	18	55	25:75

Tr: 1, 2, 4 – triazol – 1 – yl

Obviously the C—N cleavage of **4a – d** under the attack of hydrazine affords **6′a – d**. It is worthy of note that the C—N cleavage predominated in the course of ring chain transformation as compared to only the C—S cleavage depicted in Scheme Ⅰ. Such preferential C—N cleavage in the process of ring chain transformation has been reported by Patzel and Iwata.[16,17]

These products have been confirmed by elemental analyses and spectral methods, particularly by the mass spectra. In the mass spectra, the mercaptoalkylamino substituent of **3** and **6** gives the dominant fragments due to an α – and β – cleavage, whereas **6′** presents its characteristic base peak (m/z = 106, $PhNHCH_2^+$) according to an α – cleavage of anilinopropylthio side chain; meanwhile, the McLafferty rearrangement typical for the anilinopropylthio substituent of **6′** is observed in the way as follows.

Ar, N—N(H), S, CH_2, CH—CH_2NHPh, H, Tr ⟶ Ar, N—N(H), =S + CH_2= CH — CH_2 NHPh, Tr

6′	Ar	m/z(%)	m/z = 133 (%)
6′a	C_6H_5	243(32)	(13)
6′b	4 – ClC_6H_4	277(14)	(9)
6′c	4 – $CH_3OC_6H_4$	273(23)	(18)
6′d	4 – $CH_3C_6H_4$	257(28)	(12)

In addition, the identical fragment derived from the characteristic cleavage of 3 – arylpyrazole of **3**, **6** and **6′** to arylcarbonitrile (Ar—C ≡ N^+) appears in the mass spectra. Melting points, yields and spectral data of these novel compounds thus obtained are described in the experimental section.

In conclusion, the forgoing results preliminarily demonstrate cyclic α – oxo – α – (1,2,4 – triazol – 1 – yl) ketene *N*,*S* – acetals can be successfully employed in the synthesis of heterocycles like the open chain α – oxo – α – (1,2,4 – triazol – 1 – yl) ketene *N*,*S* – acetals[9,18] based on the concept of ring chain transformation. In this way, it is probably assumed that other types of cyclic α – oxoketene *N*,*S* – acetals would possibly be applied to the preparation of heterocycles. Besides, this approach provides a convenient method for the preparation of ω – functionalized alkylheteroatomic substituted pyrazolyl – 1,2,4 – triazoles which are otherwise difficult to prepare.

EXPERIMENTAL

Melting points were determined with a Yanaco MT – 500 apparatus without correction. ^{1}H – Nmr spectra were taken with a Bruker Ac – P200 spectrometer using TMS as an internal reference and mass spectra on a VG – 7070E spectrometer. IR spectra were recorded on a Shimadzu – IR 435 spectrophotometer. Elemental analyses were performed by a Yanaco CHN RDER MT – 3 analyzer.

Starting materials (1a – d and 3a – d)

Starting materials (**1a – d** and **3a – d**) were prepared according to the literatures.[10,11]

General procedure for the preparation of 3a – d

A mixture of **1**(2mmol) and 85% hydrazine hydrate(300mg, 8 mmol) in ethanol(25 mL) was refluxed under nitrogen for an hour. After evaporation of some solvent and cooling, the resultant precipitates were collected by suction and purified by recrystallization.

1 – [5 – (*N* – Phenyl – *N* – mercaptoethyl) amino – 3 – phenyl – 1*H* – pyrazol – 4 – yl] – 1,2,4 – triazole. (**3a**)

mp 207 – 209℃ (DMF/H_2O), yield 55%; ^1H NMR(DMSO – d_6): δ 3.02(m, 2H, SCH_2), 3.94(m, 2H, NCH_2), 6.60 – 7.35(m, 10H, ArH), 7.88, 8.20(2s, 2H, Tr), 13.40(br, 1H, NH); ms: (*m/z*) 362(M^+, 8%), 360(28), 315(92), 104(49), 103(16), 77(100). IR: 3119 cm^{-1}(NH). Anal. Calcd for $C_{19}H_{18}N_6S$: C, 62.96; H, 5.01; N, 23.19. Found: C, 63.35; H, 5.05; N, 23.20.

1 – [5 – (*N* – Phenyl – *N* – mercaptoethyl) amino – 3 – *p* – chlorophenyl – 1*H* – pyrazol – 4 – yl] – 1,2,4 – triazole. (**3b**)

mp 240 – 242℃ (DMF/H_2O), yield 70%; ^1H NMR(DMSO – d_6): δ 2.94(m, 2H, SCH_2), 3.90(m, 2H, NCH_2), 6.64 – 7.34(m, 9H, ArH), 8.00, 8.40(2s, 2H, Tr), 13.60(br, 1H, NH); ms: (*m/z*) 396(M^+, 21%), 394(30), 363(37), 362(31), 349(82), 138(19), 137(13), 91(36), 77(100). IR: 3115 cm^{-1}(NH). Anal. Calcd for $C_{19}H_{17}N_6ClS$: C, 57.50; H, 4.32; N, 21.17. Found: C, 57.79; H, 3.93; N, 21.50.

1 – [5 – (*N* – Phenyl – *N* – mercaptoethyl) amino – 3 – *p* – methoxyphenyl – 1*H* – pyrazol – 4 – yl] – 1,2,4 – triazole. (**3c**)

mp 141 – 142℃ (ethanol), yield 61%; ^1H NMR(DMSO – d_6): δ 2.90(m, 2H, SCH_2), 3.75(s, 3H, OCH_3), 3.94(m, 2H, NCH_2), 6.76 – 6.28(m, 9H, ArH), 7.84(s, 2H, Tr), 13.34(br, 1H, NH); ms: (*m/z*) 392(M^+, 13%), 290(36), 359(41), 358(35), 345(100), 134(26), 133(10), 77(78), 28(37). IR: 3145cm^{-1}(NH). Anal. Calcd for $C_{20}H_{20}N_6OS$: C, 61.21; H, 5.14; N, 21.41. Found: C, 60.95; H, 4.80; N, 20.97.

1 – [5 – (*N* – Phenyl – *N* – mercaptoethyl) amino – 3 – *p* – methylphenyl – 1*H* – pyrazol – 4 – yl] – 1,2,4 – triazole. (**3d**)

mp 215 – 216℃ (ethanol), yield 73%; ^1H NMR(DMSO – d_6): δ 2.24(s, 3H, CH_3), 2.88(m, 2H, SCH_2), 3.90(m, 2H, NCH_2), 6.70 – 7.20(m, 9H, ArH), 7.80, 7.90(2s, 2H, Tr), 13.20(br, 1H, NH); ms: (*m/z*) 376(M^+, 8%), 374(28), 343(23), 342(17), 329(63), 118(25), 117(16), 91(31), 77(74), 28(100). IR: 3105 cm^{-1}(NH). Anal. Calcd For $C_{20}H_{20}N_6S$: C, 63.81; H, 5.35; N, 22.32. Found: C, 64.00; H, 4.94; N, 22.45.

General procedure for the preparation of 6a – d/6′a – d

A mixture of **4**(3mmol) and 85% hydrazine hydrate(450mg, 12mmol) in ethanol(30mL) was refluxed under nitrogen for 2 h. After evaporation of solvent and cooling, the residue was diluted with water(20mL) and extracted with ethyl acetate(3 × 12mL). The organic layer was dried with sodium sulfate and evaportued to give a mixture of **6a – d** and **6′a – d**, which were seperated by column chromatography on silica gel with petroleum/ethyl actate (1∶1) as eluents.

1 - [5 - (*N* - Phenyl - *N* - mercaptopropyl) amino - 3 - phenyl - 1*H* - pyrazol - 4 - yl] - 1,2,4 - triazole. (**6a**)

mp 152 - 154℃, yield 23%; ^{1}H NMR(CDCl$_3$): δ 2.09(m, 2H, NCH$_2$C$\underline{H}_2$CH$_2$SH), 2.79(t, 2H, J = 7.0Hz, SCH$_2$), 3.81(t, 2H, *J* = 7.3Hz, NCH$_2$), 6.86 - 7.26(m, 10H, ArH), 8.50 (br, 1H, NH); ms: (*m/z*) 376 (M$^+$, 37%), 374 (59), 342 (18), 315 (90), 301 (10), 104 (41), 103 (19), 77 (100), 47 (10). IR: 3084cm^{-1} (NH). Anal. Calcd for $C_{20}H_{20}N_6S$: C, 63.81; H, 5.35; N, 22.32. Found: C, 63.70; H, 5.23; H, 22.58.

1 - [5 - (3 - Anilinopropylthio) - 3 - phenyl - 1*H* - pyrazol - 4 - yl] - 1,2,4 - triazole. (**6'a**)

mp 137 - 138℃, yield 44%; ^{1}H NMR(CDCl$_3$): δ 1.94(m, 2H, SCH$_2$C$\underline{H}_2$CH$_2$N), 2.93(t, 2H, J = 6.8Hz, SCH$_2$), 3.27(t, 2H, J = 6.3Hz, NCH$_2$), 6.79 - 7.35(m, 10H, ArH), 8.15, 8.18(2s, 2H, Tr); ms: (*m/z*) 376(M$^+$, 18%), 243(32), 242(11), 133(13), 132(22), 106 (100), 104(15), 103(6). IR: 3127, 3115cm^{-1} (PhNH, NH). Anal. Calcd for $C_{20}H_{20}N_6S$: C, 63.81; H, 5.35; N, 22.32. Found: C, 63.86; H, 5.16; N, 22.07.

1 - [5 - (*N* - Phenyl - *N* - mercaptopropyl) amino - 3 - *p* - chlorophenyl - 1*H* - pyrazol - 4 - yl] - 1,2, 4 - triazole. (**6b**)

mp 145.5 - 147.5℃, yield 33%; ^{1}H NMR(CDCl$_3$): δ 2.08(m, 2H, NCH$_2$C$\underline{H}_2$CH$_2$SH), 2.79(t, 2H, J = 7.0Hz, SCH$_2$), 3.82(t, 2H, J = 6.8Hz, NCH$_2$), 6.88 - 7.26(m, 9H, ArH), 7.70, 7.79(2s, 2H, Tr); ms: (*m/z*) 410(M$^+$, 40%), 408(34), 377(12), 376(37), 349(53), 138 (18), 137(21), 106(31), 104(35), 77(100), 47(18). IR: 3105cm^{-1} (NH). Anal. Calcd for $C_{20}H_{19}N_6ClS$: C, 58.45; H, 4.66; N, 20.45. Found: C, 58.21; H, 4.62; N, 20.19.

1 - [5 - (3 - Anilinopropylthio) 3 - *p* - chlorophenyl - 1*H* - pyrazol - 4 - yl] - 1,2,4 - triazole. (**6'b**)

mp 165 - 166℃, yield 47%; ^{1}H NMR(CDCl$_3$): δ 1.85(m, 2H, SCH$_2$C$\underline{H}_2$CH$_2$N), 2.86(t, 2H, J = 7.0Hz, SCH$_2$), 3.17(t, 2H, J = 6.3Hz, NCH$_2$), 8.60 - 7.29(m, 9H, ArH), 8.18, 8.19(2s, 2H, Tr); ms: (*m/z*) 410 (M$^+$, 10%), 277 (14), 138 (4), 137 (4), 133 (9), 132 (18), 106(100), 77(21). IR: 3163, 3110cm^{-1} (PhNH, NH). Anal. Calcd for $C_{20}H_{19}N_6ClS$: C, 58.45; H, 4.66; N, 20.45. Found: C, 58.62; H, 4.75; N, 20.63.

1 - [5 - (*N* - Phenyl - *N* - mercaptopropyl) amino - 3 - *p* - methoxyphenyl - 1*H* - pyrazol - 4 - yl] - 1, 2,4 - triazole. (**6c**).

mp 117 - 118℃, yield 26%; ^{1}H NMR(CDCl$_3$): δ 2.05(m, 2H, NCH$_2$C$\underline{H}_2$CH$_2$SH), 2.25(t, 2H, J = 7.1Hz, SCH$_2$), 2.78(t, 2H, J = 6.8Hz, NCH$_2$), 3.74(s, 3H, CH$_3$O), 6.72 - 7.17 (m, 9H, ArH), 7.75, 7.83(2s, 2H, Tr); ms: (*m/z*) 406(M$^+$, 12%), 345(72), 373(12), 372 (11), 345 (63), 134 (11), 133 (13), 107 (9), 77 (100), 47 (24). IR: 3112cm^{-1} (NH). Anal. Calcd for $C_{21}H_{22}N_5OS$: C, 62.05; H, 5.46; N, 20.67. Found: C, 61.88; H, 5.29; N, 20.33.

1 - [5 - (3 - Anilinopropylthio) 3 - *p* - methoxyphenyl - 1*H* - pyrazol - 4 - yl] - 1,2,4 - triazole. (**6'c**)

mp 138 - 140℃, yield 38%; ^{1}H NMR(CDCl$_3$): δ 1.91(m, 2H, SCH$_2$C$\underline{H}_2$CH$_2$N), 2.93(t, 2H, J = 6.9Hz, SCH$_2$), 3.23(t, 2H, J = 6.3Hz, NCH$_2$), 3.74(s, 3H, OCH$_3$), 6.68 - 6.86 (m, 4H, ArH), 7.15 - 7.27(m, 5H, ArH), 8.15, 8.18(2s, 2H, Tr); ms: (*m/z*) 406(M$^+$, 13%), 273(23), 272(21), 134(13), 133(18), 132(25), 107(10), 106(100), 77(31). IR:

3145, 3119cm^{-1} (PhNH, NH). Anal. Calcd for $C_{21}H_{22}N_6OS$: C, 62.05; H, 5.46; N, 20.67. Found: C, 61.88; H, 5.29; N, 20.33.

1-[5-(*N*-Phenyl-*N*-mercaptopropyl)amino-3-*p*-methylphenyl-1*H*-pyrazol-4-yl]-1,2,4-triazole. (**6d**).

mp 128-130℃, yield 18%; ^{1}H NMR ($CDCl_3$): δ 1.90 (m, 2H, $NCH_2\underline{CH_2}CH_2SH$), 2.28 (s, 3H, CH_3), 2.54 (t, 2H, J = 6.8Hz, SCH_2), 3.73 (t, 2H, J = 7.1Hz, NCH_2), 6.83-7.25 (m, 9H, ArH), 7.69, 7.82 (2s, 2H, Tr), 9.00 (br, 2H, NH); ms: (*m/z*) 390 (M^+, 45%), 329 (100), 144 (12), 118 (39), 117 (18), 104 (20), 91 (54), 77 (65), 47 (42), 41 (42). IR: 3087cm^{-1} (NH). Anal. Calcd for $C_{21}H_{22}N_6S$: C, 64.59; H, 5.68; N, 21.52. Found: C, 64.31; H, 5.45; N, 21.34.

1-[5-(3-Anilinopropylthio)3-*p*-methylphenyl-1*H*-pyrazol-4-yl]-1,2,4-triazole. (**6′d**)

mp 101-103℃, yield 55%; ^{1}H NMR ($CDCl_3$): δ 1.88 (m, 2H, $SCH_2\underline{CH_2}CH_2N$), 2.31 (s, 3H, CH_3), 2.90 (t, 2H, J = 6.4Hz, SCH_2), 3.19 (t, 2H, J = 6.2Hz, NCH_2), 6.66-7.25 (m, 9H, ArH), 8.13, 8.18 (2s, 2H, Tr); ms: (m/z) 390 (M^+, 12%), 329 (15), 257 (28), 133 (12), 132 (23), 118 (13), 117 (9), 106 (100), 91 (13), 77 (30). IR: 3147, 3105cm^{-1} (PhNH, NH). Anal. Calcd for $C_{21}H_{22}N_6S$: C, 64.59; H, 5.68; N, 21.52. Found: C, 64.20; H, 5.29; N, 21.34.

REFERENCES AND NOTES

[1] R. K. Dieter, *Tetrahedron*, 1986, 42, 3029.

[2] H. Junjappa, H. Ila, and C. V. Asokan, *Tetrahedron*, 1990, 46, 5423.

[3] R. Gompper and W. Topfl, *Chem, Ber.*, 1962, 95, 2881.

[4] S. M. S. Chauhan and H. Junjappa, *Synthesis*, 1975, 798.

[5] W.-D. Rudorf and M. Augustin, *J. Prakt. Chem.*, 1978, 320, 585.

[6] G. Singh, H. Ila, and H. Junjappa, *Synthesis*, 1985, 165.

[7] G. Singh, B. Deb, H. Ila, and H. Junjappa, *Synthesis*, 1987, 286.

[8] M. L. Purkayastha, H. Ila, and H. Junjappa, *Synthesis*, 1989, 20.

[9] Z. N. Huang, *Dissertation*, Nankai University, Tianjin, (1994), P. R. China.

[10] Jpn. Kokai Tokkyo Koho Jp 6064965 (*Chem. Abstr.*, 1985, 103, 123486r).

[11] Z. M. Li and Z. N. Huang, *Chin. Chem. Lett.*, 1993, 4, 763.

[12] G. Dannhardt, Y. Geyer, K. K. Meyer, and R. Obergrusberger, *Arch. Pharm.*, 1988, 321, 17.

[13] G. Dannhardt, A. Grobe, S. Guβmann, R. Obergrusberger, and K. Ziereis, *Arch. Pharm.*, 1988, 321, 163.

[14] J. Liebscher, M. Patzel, and Y. F. Kelboro, *Synthesis*, 1989, 672.

[15] M. Patzel and J. Liebscher, *J. Heterocycl. Chem.*, 1991, 28, 1257.

[16] M. Patzel, A. Schulz, J. Liebscher, W. Richter, and M. Richter, *J. Heterocycl. Chem.*, 1992, 29, 1209.

[17] C. Iwata, M. Fujimoto, S. Okamoto, C. Nishihara, M. Sakae, M. Katsurada, M. Watanabe, T. Kawakami, T. Tanaka, and T. Imanishi, *Heterocycles*, 1990, 31, 1601.

[18] Z. M. Li and Z. N. Huang, *Chin. Chem. Lett.*, 1994, 5, 31.

Synthesis and base – catalyzed Protodesilylation of 5 – (Silylmethylthio) – 3(2H) – Pyridazinone Derivatives*

Zhengjie He, Zhengming Li

(Institute of Elemento – Organic Chemistry, Nankai University, Tianjin, 300071, China)

Abstract A series of 2 – tert – butyl – 4 – chloro – 5 – (silylmethylthio) – 3(2H) – pyridazinone derivatives **1** was synthesized. Their structures were confirmed by ^1H NMR, IR, MS and elemental analysis. Under mild conditions, base – catalyzed protodesilylation of **1** occurred extremely easily. Substituent effects on this reaction were discussed.

Key words Organosilicon compound; synthesis; 3(2H) – pyridazinone derivative; protodesilylation

INTRODUCTION

2 – tert – butyl – 4 – chloro – 5 – alkylthio – 3(2H) – pyridazinone derivatives possess excellent pesticidal activity[1]. In order to search for novel lead compounds with insecticidal and acaricidal activity. We synthesized a series of silicon – containing 3(2H) – pyridazinone derivatives **1** with bioisosterism in mind. Their structures were confirmed by ^1H NMR. IR. MS spectra and elemental analysis. The synthetic methods of **1** and their MS spectra were also discussed.

As known protodesilylation at an sp^3 carbon atom has a considerable significance in organic synthesis, especially as a method to remove the protecting groups or to introduce deuterium. Recently this type of reaction has been carried out mainly by fluoride ion catalysis[2]. KF. CsF or. n – Bu_4NF(TBAF) may be used as fluoride ion sources. t – BuOK[3] is also an effective catalyst for the protodesilylation with the cleavage of sp^3 hybridized carbon – silicon bond. During the course of the study concerned with pesticidal activity of **1**, we occasionally found that protodesilylation of **1** proceeds smoothly in the presence of catalytic amount of sodium hydroxide at room temperature (25℃) giving **2** and **3** in high yields (eq. 1).

$$\text{t-Bu–N(pyridazinone, 4-Cl, 5-SCH}_2\text{SiMeRR}^1\text{)}\ (\mathbf{1}) + H_2O \xrightarrow[\text{THF.25℃}]{\text{NaOH}} \text{t-Bu–N(pyridazinone, 4-Cl, 5-SCH}_3\text{)}\ (\mathbf{2}) + O\text{–}(SiMeRR^1)_2\ (\mathbf{3}) \quad (1)$$

When H_2O was replaced with D_2O, the deuterated product **2** was obtained. To the best of our knowledge, this is a first example of such a reaction. In this paper substituent effect on this reaction is discussed.

---

* Reprinted from *Phosphorus, Sulfur and Silicon*, 1996, 117(1): 1 – 9. Supported by the National Nature Science Foundation of China.

Results and Discussion

Synhesis The synthetic pathway for compound **1** is shown in Scheme 1. As the chlorine atom of chloromethylsilane **4** was difficult to be substituted. The attempt to synthesize **1** by treatment of 2 – tert – butyl – 4 – chloro – 5 – mercapto – 3(2H) – pyridazinone with **4** in the presence of inorganic or organic base was unsuccessful. **4** was converted into silylmethanethiol **5** by its reaction with thiourea. **5** reacted with 2 – tert – butyl – 4,5 – dichloro – 3(2H) – pyridazinone to give the desired product **1** in satisfactory yields (Table Ⅰ and Table Ⅱ). When R or R^1 is aryl group in **5**, method A for preparation of **1** was preferred. When R. R^1 is alkyl, method B was more suitable.

$$\text{a. } RR^1MeSiCH_2Cl + H_2N\overset{\overset{S}{\|}}{C}NH_2 \xrightarrow[\text{reflux}]{\text{ethanol}} \xrightarrow{\text{aq.NaOH}} RR^1MeSiCH_2SH$$

4 5

b. t–Bu–N (2-tert-butyl-4,5-dichloro-3(2H)-pyridazinone: O, Cl, Cl) + 5 $\xrightarrow[\text{or method B}]{\text{method A}}$ t–Bu–N (O, Cl, $SCH_2SiMeRR^1$) 1

method A: aqueous NaOH as HCl acceptor and CH_2Cl_2 as solvent.
method B: anhydrous K_2CO_3 as HCl acceptor and absolute ethanol as solvent.

I	*R*	R^1	*I*	*R*	R^1	*I*	*R*	R^1
a	C_6H_5	CH_3	f	m – $CH_3C_6H_4$	CH_3	k	C_6H_5	C_6H_5
b	p – ClC_6H_4	CH_3	g	p – $CH_3OC_6H_4$	CH_3	l	CH_3	CH_3
c	m – ClC_6H_4	CH_3	h	p – $PhOC_6H_4$	CH_3	m	C_2H_5	CH_3
d	p – FC_6H_4	CH_3	i	2 – thienyl	CH_3	n	C_2H_5	C_2H_5
e	p – $CH_3C_6H_4$	CH_3	j	2 – furyl	CH_3	o	C_4H_{9-n}	C_4H_{9-n}

Scheme 1 The synthetic pathway for preparation of 1

MS spectroscopy Main MS data of representative compounds **1** were listed in Table Ⅲ. All recorded compound **1** produced a molecular ion peak. The base peak ion is always RR^1MeSi^- when R equals aryl group in **1**. But it is R^1MeSi^+H resulting from β – elimination of the fragment RR^1MeSi^- when R is ethyl or butyl (Compound **1**m. **1**o in Table Ⅲ). There is a m/e(Ar + 14) ion peak in MS spectrum of **1**(R = Aryl). This ion peak suggests there is a rearrangement breakdown pattern. When R is aryl group in **1**. fragment ion $RR^1MeSiCH_2^-$ (A) probably tends to rearrange into $(RCH_2)R^1MeSi^+$ (B) which in turn undergoes a cleavage of Si – C bond to produce m/e(Ar + 14) ion RCH_2^- (Scheme 2). According to the rational calculations by Stang and coworkers[4] it is deduced that there is a substantial driving force for the above rearrangement.

Table Ⅰ　Data for compound 1

I	stale	*m. p.* (℃) or n_D^{25}	yield (%)	elemental analysis found (Calc)		
				C(%)	*H*(%)	*N*(%)
a	white crystal	88 – 89	58.5	56.07(55.64)	6.31(6.32)	7.59(7.63)
b	white crystal	94 – 95	74.8	50.64(50.86)	5.56(5.52)	7.44(6.98)
c	white crystal	90 – 92	59.9	51.13(50.86)	5.45(5.52)	7.06(6.98)
d	white crystal	114 – 115	78.0	52.88(53.04)	6.01(5.76)	7.16(7.28)
e	white crystal	116 – 118	90.8	56.65(56.74)	6.59(6.61)	7.41(7.35)
f	white crystal	86 – 87	31.5	56.96(56.74)	6.71(6.61)	7.38(7.35)
g	white crystal	130 – 131	70.0	54.45(54.46)	6.36(6.35)	6.95(7.06)
h	colorless viscous oil		87.1	60.23(60.17)	5.91(5.93)	6.07(6.10)
i	colorless crystal	119 – 120	64.3	48.71(48.30)	5.74(5.67)	7.50(7.51)
j	colorless crystal	115 – 116	70.0	50.42(50.47)	5.92(5.93)	7.83(7.85)
k	white solid	99 – 101	56.0	61.37(61.59)	6.12(5.87)	6.85(6.53)
l	white solid	48 – 50	45.9	47.20(47.27)	6.91(6.94)	9.30(9.19)
m	yellow oil	1.5518	62.7	48.69(48.96)	7.38(7.27)	8.46(8.78)
n	colorless oil	1.5508	55.3	50.53(50.50)	7.69(7.57)	8.86(8.41)
o	colorless oil	1.5358	72.0	55.68(55.57)	8.43(8.55)	7.21(7.20)

Table Ⅱ　IR and ^{1}H NMR data of compounds 1

I	*IR*(cm^{-1})			^{1}H NMR δ(ppm)
	γ_{C-O}	$\delta^{S}_{Si-CH_3}$	γ_{Si-CH_3}	
a	1635(s)	1246(s)	808(s)	0.47(s,611,$SiCH_3$),1.63(s,9H,C_4H_9 – t),2.39(s,2H,SCH_2),7.34 – 7.66 (m,5H,C_6H_5),7.66(s,1H,CH)
e	1633(s)	1247(m)	802(s)	0.48(s,6H,$SiCH_3$),1.63(s,9H,C_4H_9 – t),2.36(s,2H,SCH_2),7.37 – 7.50 (dd,4H,C_6H_4),7.66(s,1H,CH)
c	1633(s)	1251(m)	811(s)	0.48(s,6H,$SiCH_3$),1.60(s,9H,C_4H_9 – t),2.36(s,2H,SCH_2),7.32 – 7.52 (m,4H,C_6H_4),7.68(s,1H,CH)
d	1633(s)	1255(s)	805(s)	0.46(s,6H,$SiCH_3$),1.60(s,9H,C_4H_9 – t),2.36(s,2H,SCH_2),7.07,7.48 – 7.51(m,4H,C_6H_4),7.64(s,1H,CH)
e	1633(s)	1247(s)	807(s)	0.46(s,6H,$SiCH_3$),1.64(s,9H,C_4H_9 – t),2.38(d,5H,SCH_2,CH_3),7.16 – 7.52(dd,4H,C_6H_4),7.70(s,1H,CH)
f	1637(s)	1247(s)	815(s)	0.42(s,6H,$SiCH_3$),1.58(s,9H,C_4H_9 – t),2.32(d,5H,SCH_2,CH_3),7.20 – 7.40(m,4H,C_6H_4),7.70(s,1H,CH)
g	1620(s)	1244(s)	809(s)	0.40(s,6H,$SiCH_3$),1.56(s,9H,C_4H_9 – t),2.30(s,2H,SCH_2),3.76(s,3H, CH_3O),6.88,7.43(dd,4H,C_6H_4),7.62(s,1H,CH)
h	1630(s)	1245(s)	815(s)	0.48(s,6H,$SiCH_3$),1.63(s,9H,C_4H_9 – t),2.39(s,2H,SCH_2),7.00 – 7.53 (m,9H,C_6H_5,C_6H_4),7.69(s,1H,CH)
i	1635(s)	1248(s)	817(s)	0.53(s,6H,$SiCH_3$),1.63(s,9H,C_4H_9 – t),2.43(s,2H,SCH_2),7.24(m), 7.41(d),7.67(d,3H,C_4H_3S),7.70(s,1H,CH)

I	IR(cm^{-1})			^1H NMR δ(ppm)
	$\gamma_{C—O}$	$\delta^{S}_{Si-CH_3}$	γ_{Si-CH_3}	
j	1637(s)	1250(s)	815(s)	0.46(s,6H, $SiCH_3$), 1.62(s,9H, C_4H_9 – t), 2.42(s,2H, SCH_2), 6.41(m), 6.77(d), 7.60(d,3H, C_4H_3O), 7.67(s,1H,CH)
k	1638(s)	1253(m)	801(s)	0.74(s,6H, $SiCH_3$), 1.60(s,9H, C_4H_9 – t), 2.70(s,2H, SCH_2), 7.36 – 7.68 (m,10H, C_6H_5), 7.72(s,1H,CH)
l	1635(s)	1243(m)	845(s)	0.19(s,9H, $SiCH_3$), 1.62(s,9H, C_4H_9 – t), 2.19(s,2H, SCH_2), 7.70(s,1H, CH)
m	1633(s)	1234(m)	828(s)	0.17(s,6H, $SiCH_3$), 0.90(q,2H, CH_2), 0.97(t,3H, CH_3), 1.63(s,9H, C_4H_9 – t), 2.20(s,2H, SCH_2), 7.72(s,1H,CH)
n	1637(s)	1248(s)	795(s)	0.13(s,3H, $SiCH_3$), 0.67(q), 0.98(t,10H,2 × C_2H_5), 1.62(s,9H, C_4H_9 – t), 2.20(s,2H, SCH_2), 7.73(s,1H,CH)
o	1635(s)	1250(m)	799(s)	0.15(s,3H, $SiCH_3$), 0.67 – 0.72, 0.86 – 0.92, 1.29 – 1.36(m,18H, C_4H_9 – n), 1.63(s, 9H, C_4H_9 – t), 2.19(s,2H, SCH_2), 7.73(s,1H,CH)

Table Ⅲ Mass spectra data of representative compounds 1①

I	M′	base ion	RR′MeSiCH₂′ m/e	rearrangement ion m/e	other fragment ion m/e
a	366(16)	$PhMe_2Si'$	149(4)	91(8)	311(14), 310(7), 232(19), 57(27)
b	400(3)	p – $ClC_6H_4Me_2Si'$	183(8)	125(7)	345(18), 344(4), 232(33), 139(18)
g	396(1)	p – $CH_4C_6H_4Me_2Si'$	179(4)	121(12)	197(39), 135(12), 341(2), 340(2)
h	458(1)	p – $PhOC_6H_4Me_2Si'$	241(3)	183(4)	403(3), 402(2), 197(38), 57(69)
i	372(1)	$C_4H_3SMe_2Si'$	155(6)	97(13)	317(10), 316(3), 337(21), 232(24)
j	356(1)	$C_4H_3OMe_2Si'$	139(8)	81(4)	301(10), 321(55), 232(20), 57(52)
l	304(3)	Me_3Si'	87(3)		249(8), 248(4), 57(36), 45(22)
m	318(35)	$Me_2Si'H$	101(3)		263(54), 262(25), 87(78), 57(70)
o	388(43)	BuMeSi′H	171(0.5)		333(43), 332(26), 157(39), 57(91)

①In parentheses is relative intensity of fragment ion.

Prolodesilyation Ten compounds **1** were selected for protodesilylation and results are summarized in Table Ⅳ. Data listed in Table Ⅳ show that substituent R on silicon atom plays an important role for the reaction. The electron – withdrawing R groups facilitate the protodesilylation. For example, protodesily – lations of compounds **1a – d, i – j** were complete within 12 minutes to give product **2** in 80% ~95% yields. For other compounds with relatively weak electron withdrawing R. e. g. **1e – f.** substantially longer reaction times were required.

$$M^{+} \longrightarrow \underset{A}{RR^{1}MeSiCH_2^{+}} \xrightarrow{\text{rearrangement}} \underset{B}{(RCH_2)R^{1}MeSi^{+}} \longrightarrow \underset{m/e(Ar+14)}{RCH_2^{+}}$$

Scheme 2 A rearrangement breakdown pattem of 1(R = aryl)

On the other hand, the presence of 3(2H) – pyridaznon – 5 – yl group on sulfur atom also plays a critical role in the occurence of this protodesilylation. After compound **6**. which is analogous with compound **1a**. had been stirred continuously for 60 h under the same reaction conditions as compound **1a**, no reaction occurred and **6** was recovered completely unchanged (eq. 2). This result presumably attributes to the weaker electron – withdrawing ability of phenyl group compared to that of 3(2H) – pyridazinon – 5 – yl group for the strong electron – withdrawing group benefits sulfur atom stabilizing the carbanion intermediate from the cleavage of carbon – silicon bond. Theoretical calculations about net charge for sulfur atoms in compounds **1a** and **6** by CNDO/2 method are also in agreement with the above conclusion. The value of net charge on sulfur atom in compound **1a** is positive 0.0006081, whereas that in compound 6 is negative 0.014999.

EXPERIMENTAL

All temperatures were uncorrected. Melting points were determined with Yanaco MP – 500 apparatus. IR spectra were recorded on Shimadzu IR – 435 spectrophotometer as thin films or KBr tablet. ^{1}H NMR spectra were measured on a JEOL – FX – 90Q and a Bruker AC – 200 instruments using TMS as an internal standard and $CDCl_3$ as solvent. Mass spectra were recorded on an HP – 5988A instrument at 70eV. Elemental analyses were determined on an MT – 3 elemental analyzer.

Table Ⅳ Protodesilylation of compounds 1 catalyzed by NaOH①

1	Reaction time(min)	Yield of 2(%)	*1*	Reaction time(min)	Yield of 2(%)
a	10	95	f	50	87
b	6	95	g	80	84
c	8	90	i	11	80
d	11	92	j	12	82
e	40	88	l	12	94

①Reaction conditions: NaOH/1 molar ratio 1:10; THF – H_2O(10:1v/v) as solvent; Reaction temperature 25℃.

1. Silylmethanethiol 5

Silylmethanethiol **5** were prepared from chloromethylsilanes **4**. which were prepared from the reaction of $ClCH_2SiCl_2CH_3$ or $ClCH_2(CH_3)_2SiCl$ with organometallic reagents. as previously described in the literature[5]. The crude products **5** were purified by distillation at reduced pressure or column chromatography on silica gel using petroleum ether – CH_2Cl_2 (10:1 v/v) as an eluant (Table Ⅴ).

2. 2 – tert – Butyl – 4 – chloro – 5 – silymethylthio – pyridazinone 1 (Typical procedure)

Method A Sodium hydroxide 0.23 g (5mmol) was dissolved in **5** mL of water, and thereto were added 10 ml of dichloromethane. 1.1g(5mmol) of 2 – tert – butyl – 4.5 – dichloro – 3(2H) – pyridazinone and 0.05 g triethylbenzylammonium chloride. The resulting solution was incorporated with 0.91 g (5mmol) of dimethylphenylsilyl – methanethiol **5a** and then stirred at room temperature for 2h. After completion of the reaction, the organic layer was separated, washed with water and dried over anhydrous sodium sulfate. After filtration, solvent was distilled off under reduced pressure and the resulting oily residue was incorporated with 10 mL of petroleum ether (90 – 120℃) to give 1. 1 g of compound **1a**, m. p. 88℃.

$$C_6H_5SCH_2SiMe_2Ph\ (6) + H_2O \xrightarrow[\text{THF.r.t. 60h}]{\text{NaOH}} \text{no reaction} \quad (2)$$

Table V Data of silylmethanethiols 5

5	State	b. p(℃/mm)	n_D^{25}	yield(%)
a	Colorless liquid	79 ~ 81/6	1. 5432	76. 2
b	Colorless liquid	94 ~ 98/0. 4	1. 5573	60. 0
c	Colorless liquid	104 ~ 105/1	1. 5576	69. 4
d	Colorless liquid	96 ~ 97/9	1. 5270	67. 5
e	Colorless liquid	110 ~ 112/9	1. 5406	79. 1
f	Colorless liquid	104 ~ 108/9	1. 5400	81. 6
g	Colorless liquid	108 ~ 110/1	1. 5488	61. 3
h	Colorless liquid		1. 5758	85. 2
i	Colorless liquid	91 ~ 92/6	1. 5528	76. 6
j	Colorless liquid	60 ~ 61/6	1. 5094	66. 9
k	Colorless liquid	134 ~ 138/0. 3	1. 6050	53. 0
l	Colorless liquid	118 ~ 120/760	1. 4564	62. 2
m	Colorless liquid	46 ~ 48/27	1. 4586	55. 2
n	Colorless liquid	92 – 94/80	1. 4772	77. 7
o	Colorless liquid	84 ~ 88/6	1. 4632	52. 0

Compounds **1b** – 1 were prepared by a similar procedure (Table Ⅰ).

Method B To a solution of 1. 1g (5mmol) of 2 – tert – butyl – 4,5 – dichloro – 3(2H) – pyridazinone in 20mL of absolute ethanol were added 0. 74g (5mmol) of 5n and 0. 83g (6mmol) of anhydrous potassium carbonate. The resulting mixture was stirred at room temperature for 2h and refluxed for 4h. After removal of ethanol, the residue was dissolved in 20mL of ether, washed and dried over anhydrous sodium sulfate. Removal of the solvent gave a crude product which was purified by cloumn chromatography on silica gel(10 – 40μ) with ether/petroleum ether(1:10 v/v) as the eluent to give 0. 92g of **1**n as a colorless oil, n_D^{25} 1. 5508.

By method B compounds **1**l. m, o were also synthesized(Table Ⅰ).

3. Protodesilylation of 1(Typical procedure)

To a solution of compound 1(0. 25 mmol) in 10 mL of THF, 1. 0 mL of 0. 1% NaOH aqueous solution(0. 025 mmol) was added with stirring at 25℃. The reaction mixture was stirred continuously and monitored by TLC until the reaction was complete. The desired products **2** and **3** were isolated by silica gel column chromatography in high yields using petroleum ether – CH_2Cl_2 (10: 1) as an eluent.

Acknowledgements

We faithfully thank professor Cheng - Ming Lai (Department of chemistry, Nankai University) for some calculations about net charge.

References

[1] a) M. Taniguchi. M. Hirose. M. Babe. K. Hirata and Y. Ochiai. U. S. Patent 4877787: (1989) b) Y. Nakajima. Y. Kawamura. T. Ogura. T. Makabe. K. Hirata. et al.. European Patent 232825 (1987); C. A. ,108,37854(1988) c) K. Hirata. M. Kudo. T. Miyake Y. Kawamura and T. Ogura, Brighton Crop Prot. Conf. —Pesis Dis. ,(1),41(1988).

[2] a) S. F. Chen and P. S. Mariano. *Tetrahedron Lett.*. 26. 47(1985) b) G. Maier, M. Hoppe and H. P. Reisenauer, Angew. Chem. Int. Ed.. 22. 990(1983) c) R. M. Williams, J. S. Dung. J. Jasey. R. W. Armstrong and H. Meyers. *J. Am. Chem. Soc.* ,105,3214(1983) d) I. Hasan and Y. Kishi. *Tetrahedron Lett.* ,21,4229 (1980) e) T. H. Chan. B. S. Ong and W. Mychajlowski *Tetrahedron Lett.* ,17. 3253 (1976) f) R. J. Mills and V. Snieckus. *Tetrahedron Lett.* , 25. 479 (1984) g) H. G. Koser. G. E. Renzoni and W. T. Borden, *J. Am. Chem. Soc.* ,105,6359(1983) h) P. G. McDougal and B. D. Condon. *Tetrahedron Lett.* , 30. 789 (1989).

[3] a) P. F. Hudrlik, A, M. Hudrlik and A. K. Kulkami. *J. Am. Chem. Soc.*. 104,6809(1982) b) P. F. Hudrlik, P. E. Holmes and A. M. Hudrlik. *Tetrahedron Lett.*. 29, 6395 (1988) c) C. C. Price and J. R. Sowa. *J. Org. Chem.* ,32. 4126(1967).

[4] P. J. Stang. M. Ladika. Y. Apeloig. A. Stanger. M. D. Schiavelli and M. R. Hughey. *J. Am. Chem. Soc.*. 104. 6852 (1982).

[5] E. Block and J. A. Laffitte. *J. Org. Chem.* 51(18),3428(1986).

Synthesis of 5 – mercaptoalkylthiopyrazolyl – and 3 – mercaptoalkyl thioisoxazolyl – 1,2,4 – triazoles by Ring Chain Transformation*

Zhennian Huang[1], Zhengming Li[2]

([1]Department of Chemistry, Peking University, Beijing, 100871 China;
[2]Institute of Elemento – Organic Chemistry, Nankai University, Tianjin, 300071, China)

Abstract Cyclic α – oxo – α – (1,2,4 – triazolyl) ketene *S*,*S* – acetals **la – e** react with hydrazine affording the substituted pyrazolyl – 1,2,4 – triazoles **3a – e**. With hydroxylamine hydrochloride under basic conditions the substituted isoxazolyl – 1,2,4 – triazoles **5a – c** are obtained.

In recent decades, α – oxoketenc *N*,*S*;*S*,*S* – acetals came to considerable interesis as synthetic intermediates in heterocyclic synthesis due to their multiple intrinsic reactivity properties. However, in contrast to the huge body of literature dealing with the synthetic application of open chain α – oxoketene *N*,*S*;*S*,*S* – acetals[1-2], little is reported on cyclic α – oxoketene *N*,*S*;*S*,*S* – acetals. We had reported earlier that the cyclic α – oxo – α – (1,2,4 – triazolyl) ketene *N*,*S*;*S*,*S* and *N*,*O* – acetals reacl readily with hydrazine to give (ω – functionalized alkylheteroatomic pyrazolyl) – 1,2,4 – triazoles[3-4]. In addition, the reaction of cyclic α – oxo – α – (1,2,4 – triazolyl) ketene *N*,*S* with hydroxylamine was also investigated in our work[5]. We believe the mechanism of those reaction involve a process of ring chain transformation.

Here, we wish to report the reaction of α – oxo – α – (1,2,4 – triazolyl) ketene *S*,*S* – acetals **1a – e** with hydrazine and hydroxylamine.

Starting materials **1a – e** were prepared by condensation of α – (1,2,4 – triazolyl) substituted acetophenones with carbon disulfide in the presence of a base and subsequent alkylation by 1, 3 – dibromopropane.

When **1a** was treated with hydrazine at reflux in ethanol for 2h, a white product was isolated, identified as (5 – mercaptopropylthiopyrazolyl) – 1,2,4 – triazole (**3a**) (see Scheme 1). The reaction was found to be general. Other cyclic α – oxoketene *S*,*S* – acetals **1b – e** afforded the corresponding substituted pyrazolyl – 1,2,4 – triazoles **3b – e** in 45% – 61% overall yield.

When the cyclic *S*,*S* – acetals **1a** was treated with hydroxylamine hydrochloride in the presence of sodium methylate, the product isolated was identified as (3 – mercaptopropylthioisoxazolyl) – 1,2,4 – triazole (**5a**) (63%). Under the same conditions other substituted isoxazolyl – 1,2,4 – triazoles **5b – c** were obtained in 36% – 47% overall yield. In addition, a trace amount of disulfide **5c′** was separated from the reaction of **1c** with hydroxylamine. The analytical

* Reprinted from *Synthetic Communications*, 1996, 26 (16): 3115 – 3120. Research supported by the National Natural Science foundation of China.

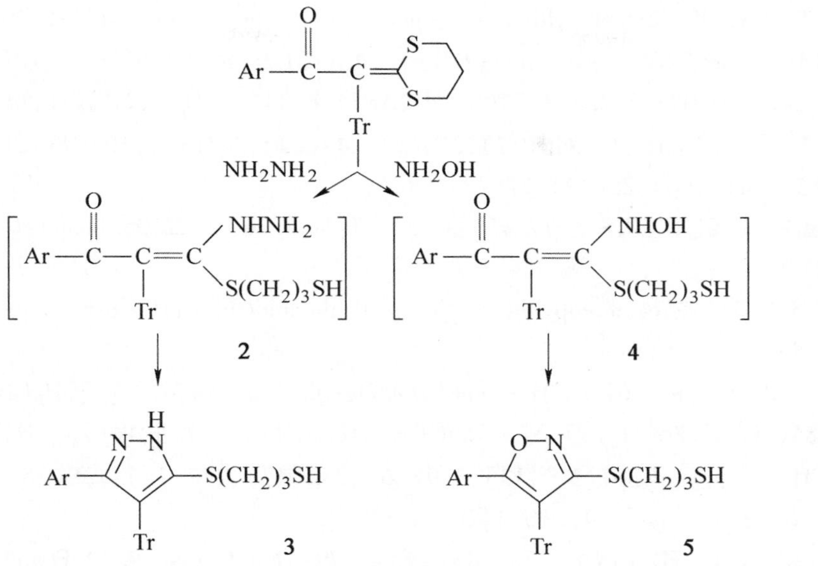

Ar : a, C_6H_5; b, 4-ClC_6H_4; c, 4-$CH_3OC_6H_4$; d, 4-$CH_3C_6H_4$; e, 4-BrC_6H_4

Tr : 1,2,4-triazolyl

Scheme 1

data of disulfide **5c′** is rather similar to **5c**. The formation of **5c′** can be rationalized from the oxidation of **5c**. Indeed, heating **5c** in ethanol in an oxygen atmosphere for half an hour provides **5c′**.

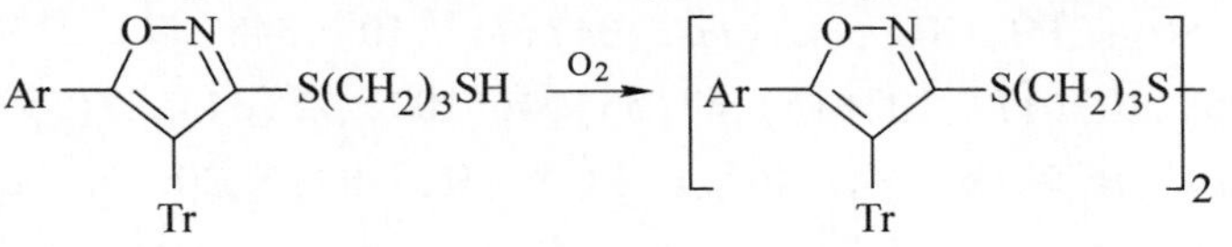

$$\text{5c} \xrightarrow{O_2} \text{5c}'$$

In our experiments the pure **5d – e** could not be obtained by recrystallization or chromatography from a rather complicated crude reaction mixtures owing to their bad solubility in usefull solvents.

Experimental

Melting points were observed with a Yanaco MT – 500 apparatus without correction. ^1H – nmr spectra were measured with a Bruker AC – P200 spectrometer using TMS as internal standard and mass spectra on a VG – 7070E spectrometer with 70eV. Elemental analyses were performered on a Perking – Elemer240 – C instrument. starting materials: **1a – e** were prepared according to the literature[6].

1. General procedure for the preparation of 3a – d

A mixture of 1 (3 mmol) and 85% hydrazine hydrate (15 mmol) in ethanol (30mL) was refluxed for an hour. After evaporation of solvent the residue was dissolved in chloroform (20mL), washed with water, dried with sodium sulfate and evaporated to give the reaction mixture. The pure **3** was isolated via column chromatography on silica gel using petroleum/ethyl acetate as eluents.

3a. 1 – [5 – (3 – Mercaptopropylthio) – 3 – phenyl – 1H – pyrazol – 4 – yl] – 1,2,4 – triazole mp153 – 155℃, yield 46%; ^{1}H – nmr (DMSO – d_6): δ1.9 (m, 2H, $SCH_2CH_2CH_2SH$), 2.71 – 2.89 (m, 4H, 2SCH_2), 7.25 – 7.40 (m, 5H, ArH), 8.28, 8.77 (2s, 2H, Tr), 13.98 (s, 1H, NH): ms: (*m/z*) 317 (M^+, 3), 315 (17), 270 (7), 248 (34), 243 (11), 104 (43), 103 (37), 77 (84), 47 (50), 41 (100), 28 (42), 27 (22).

Anal. Calcd. for $C_{14}H_{15}N_5S_2$ (317.43): C, 52.97: H, 4.76; N, 22.06. Found: C, 53.09; H, 4.39; N, 21.94.

3b. 1 – [5 – (3 – Mercaptopropylthio) – 3 – p – chlorophenyl – 1H – pyrazol – 4 – yl] – 1,2,4 – triazole.

mp 224 – 266℃, yield 61%; ^{1}H – nmr (DMSO – d_6): δ1.8 (m, 2H, $SCH_2CH_2CH_2SH$), 2.60 – 2.84 (m, 4H, 2SCH_2), 7.24 – 7.56 (m, 4H, ArH), 8.32, 8.40 (2s, 2H, Tr). 14.20 (br, 1H NH); ms: (*m/z*) 351 (M^+, 13), 349 (26), 282 (47), 277 (14), 138 (25), 137 (36), 111 (33), 106 (25), 47 (81), 41 (100), 28 (51), 27 (28).

Anal. Calcd. for $C_{14}H_{14}ClN_5S_2$ (351.87): C. 47.79: H. 4.01; N. 19.90. Found: C. 47.55; H. 4.05; N. 19.95.

3c. 1 – [5 – (3 – Mercaptoplropylthio) – 5 – p – methoxyphenyl – 1H – pyrazol – 4 – yl] – 1,2,4 – triazole

mp 150 – 153℃, yield 45%; ^{1}H – nmr (DMSO – d_6): δ 1.8 (m, 2H, $SCH_2CH_2CH_2SH$), 2.84 – 2.87 (m, 4H, 2SCH_2), 3.73 (s, 3H, CH_3O), 6.93 – 7.20 (m, 4H, ArH). 8.25, 8.72 (2s, 2H, Tr), 13.80 (s, 1H, NH); ms: (*m/z*) 347 (M^+, 10), 345 (47), 273 (10), 272 (12), 134 (34), 133 (26), 103 (17), 102 (15), 47 (55), 41 (100), 28 (34), 27 (37).

Anal. Calcd. for $C_{15}H_{17}N_5OS_2$ (347.46): C, 51.85: H, 4.93; N, 20.16. Found: C, 51.81: H, 4.97: N. 19.89.

3d. 1 – [5 – (3 – Mercaptopropylthio) – 3 – p – methylphenyl – 1H – pyrazol – 4 – yl] – 1,2,4 – triazole

mp 211 – 213℃, yield 52%, ^{1}H – nmr (DMSO – d_6): δ1.8 (m, 2H, $SCH_2CH_2CH_2SH$). 2.24 (s, 3H, CH_3), 2.60 – 2.96 (m, 4H, 2SCH_2), 7.16 (s, 4H, ArH), 8.24, 8.76 (2s, 2H, Tr), 13.94 (s, 1H, NH), ms: (*m/z*) 331 (M^+, 15), 329 (65), 284 (30), 262 (100), 257 (26), 256 (33), 118 (50), 117 (32), 91 (58), 47 (46), 41 (71), 28 (20).

Anal. Calcd. for $C_{15}H_{17}N_5S_2$ (331.46): C, 54.35; H, 5.17; N, 21.13. Found: C, 53.94, H. 4.76: N. 20.69.

3e. 1 – [5 – (3 – Mercaptopropylthio) – 3 – p – bromophenyl – 1H – pyrazol – 4 – yl] – 1,2,4 – triazole

mp 234 – 236℃, yield 59%; ^{1}H – nmr (DMSO – d_6): δ 1.8 (m, 2H, $SCH_2CH_2CH_2SH$), 2.30 – 2.80 (m, 4H, 2SCH_2), 7.08 – 7.48 (dd, 4H ArH), 8.24, 8.70 (2s, 2H, Tr), 14.00 (br, 1H, NH), ms: (*m/z*) 395 (M^+, 22), 393 (20), 348 (8), 326 (37), 321 (9), 320 (11), 182 (17), 181 (14), 74 (32), 47 (73), 41 (98), 28 (100), 27 (32).

Anal. Calcd. for $C_{14}H_{14}BrN_5S_2$ (396.33): C. 42.43; H. 3.56; N, 17.67. Found: C, 42.55: H, 3.55: N. 17.49.

General procedure for the preparation of **5a – c**

Hydroxylamine hydrochloride (15mmol, 1.05g) was added to a stirred solution of NaOMe [available by dissolving Na(20mmol 0.46g) in absolute methanol (30mL)]. 5 – 10 min later, **1**(3mmol) was added. and then the mixture was heated at retlux for two hours. After removal of methanol under vaccum. the residue was diluted by water (30mL), While the mixture was adjusted to pH = 6 – 7, the precipitatc appeared and was isolated by suction, the pure product **5** was obtained by column chromatography using petroleum/ ethyl acetate as eluant.

5a. 1 – [[3 – (3 – Mercaptopropylthio) – 5 – Pheny] isoxazol – 4 – yl] – 1,2,4 – triazole

mp 115 – 117℃, yield 63%; ^1H – nmr($CDCl_3$): δ2.2(m 2H, $SCH_2CH_2CH_2SH$), 2.80(m, 2H, CH_2SH), 3.28(t, 2H, SCH_2), 2.52(s, 4H, ArH), 8.36(s, 2H, Tr), ms: (*m/z*)318(M^+, 1), 316 (5), 211(19), 105(100), 77(49), 41(41).

Anal. Calcd. for $C_{14}H_{14}N_4OS_2$ (318.41): C, 52.81; H, 4.43; N, 17.60. Found: C, 52.16, H, 4.36, N, 17.44.

5b. 1 – [[3 – (3 – Mercaptopropylthio) – 5 – p – Chlorophenyl] isoxazol – 4 – yl] – 1,2,4 – triazole

mp 87 – 88℃, yield 52%, ^1H – nmr ($CDCl_3$): δ1.24 (t, 1H SH), 2.1 (m 2H, $SCH_2CH_2CH_2SH$), 2.7(m, 2H, CH_2SH), 3.30(t, 2H, SCH_2), 7.40(s, 4H, Ar). 8.26(s, 2H, Tr). ms: (*m/z*)352(M^+, 1). 246(22), 139(100), 111(38), 106(13), 75(25), 41(23), 28(41).

Anal. Calcd. for $C_{14}H_{13}ClN_4OS_2$ (352.78): C, 47.67; H, 3.77; N, 15.88. Found: C, 47.74, H, 3.63; N, 15.52.

5c. 1 – [[3 – (3 – Mercaptopropylthio) – 5 – p – mercaptophenyl] isoxazol – 4 – yl] – 1,2,4 – triazole

mp 84 – 85℃, yield 47%: ^1H – nmr ($CDCl_3$): δ1.40 (t, 1H, SH), 2.1 (m, 2H, $SCH_2CH_2CH_2SH$): 2.63(m, 2H, CH_2SH), 3.26(t, 2H, SCH_2), 3.82(s, 3H, CH_3O): 6.88 – 7.36(dd, 4H, Ar). 8.23, 8.24(2s, 2H, Tr); ms: (*m/z*)348(M^+, 1), 242(9). 135(100), 106 (79). 77(24), 47(23), 41(12).

Anal. Calcd. for $C_{15}H_{16}N_4O_2S_2$ (348.44): C, 51.71; H, 4.62; N, 16.08. Found: C, 51.47; H, 4.52; N, 16.03.

5c′. Disulfide

mp 157 – 159℃, yield 6%; ^1H – nmr(DMSO – d_6): δ 2.1(m, 2H, $SCH_2CH_2CH_2SH$). 2.80(m, 2H, CH_2SH), 3.20(t, 2H, SCH_2), 3.80(s, 3H, CH_3O), 7.04 – 7.44(dd, 4H, Ar), 8.48, 9.00(2s, 2H, Tr). ms: (*m/z*)347(1/2M^+, 1), 242(9), 135(100), 106(25), 107(8), 77(20), 41(22), 28(74).

Anal. Calcd. for $C_{30}H_{30}N_8O_4S_2$ (694.87): C, 51.06; H, 4.35; N, 16.12. Found: C, 51.92; H, 4.46; N, 15.98.

References

[1] Dieter, R. K. *Tetrahedron*, 1986, 42, 3029.

[2] Junjappa, H. and Ⅱa, H. Asokan, C, V. , *Tetrahedron* 1990, 46, 5423.

[3] Huang, Z. N. and Li, Z. M. *Synth. Commun.* 1995, 25(22): 3603 – 3609.

[4] Huang, Z. N. and Li, Z. M. *Heterocycles*. 1995, 41, 1653 – 1658.

[5] Huang, Z. N. and Li, Z. M. *Synth. Commun.* 1995, 25(20): 3219 – 3224.

[6] Ger. Offen. DE 3, 145, 890(1983). *Chem. Abstr.*, 99, 105256k.

1997年

新磺酰脲类化合物的合成、结构及构效关系研究（Ⅴ）*

——N－［2－（4－乙基）三嗪基］－2－硝基－苯磺酰脲的晶体及分子结构

李正名[1]　刘　洁[1]　王　霞[2]　袁满雪[2]　赖城明[2]

（[1]南开大学元素有机化学研究所，元素有机化学国家重点实验室，天津　300071；
[2]南开大学化学系，天津　300071）

关键词　磺酰脲类化合物　晶体结构　分子结构

为了进一步研究超高效、低毒除草剂磺酰脲类化合物的分子结构和构象对活性的影响，特别是考察其分子中杂环部分为三嗪基时的结构与活性关系，我们研究了N－［2－（4－乙基）三嗪基］－2－硝基－苯磺酰脲的晶体及分子结构，比较了它与磺酰脲分子中杂环为嘧啶环的构象差异。

1　样品的合成与表征

标题化合物的合成反应路线如下[1,2]：

NO_2-C_6H_4-SO_2NH_2 + $(COCl)_2$ (Ⅱ) ⟶ NO_2-C_6H_4-SO_2NCO

NO_2-C_6H_4-SO_2NCO + NH_2-(4-CH_2CH_3-三嗪基) ⟶ NO_2-C_6H_4-$SO_2NHCONH$-(4-CH_2CH_3-三嗪基) (Ⅰ)

产物Ⅰ的熔点为182～183℃。元素分析（$C_{12}H_{12}N_6O_5S$,%）实验值：C 41.11，H 3.37，N 24.24；计算值：C 40.90，H 3.67，N 23.85. IR，$\bar{\nu}/cm^{-1}$：3411（N－H），1724（C═O及—NHCONH），1 366.2，1163.5（—SO_2—）。^1H NMR，δ（丙酮－d_6）：2.90（2H，q，—CH_2），1.32（3H，t，—CH_3），8.04～8.38（4H，m，Ar），8.94（1H，s，三嗪环H）。从而确定了其结构。

2　晶体结构的测定

从丙酮中培养出$C_{12}H_{12}N_6O_5S$无色透明晶体，选取约为0.2mm×0.4mm×0.4mm

* 原发表于《高等学校化学学报》，1997，18（5）：750－752。国家自然科学基金资助课题。

的单晶用于衍射实验。在 Enraf - Nonius CAD4 四圆衍射仪上，用经单色化的 Mo$K\alpha$ 射线，在 $0° \leqslant \theta \leqslant 23°$范围内，以 $\omega/2\theta$ 扫描方式，共收集 2278 个独立衍射点，其中 1084 个为可观测衍射点［$I \geqslant 3\sigma$（I）］，全部强度数据均经过 LP 因子校正及经验吸收校正。该晶体属单斜晶系，空间群为 $P2_1/n$，$a = 0.7861(3)$ nm，$b = 0.5809(2)$ nm，$c = 3.2729(6)$ nm，$\beta = 93.25(2)°$，$V = 1.492(1)$ nm^3，$M_r = 352.33$，$Z = 4$，$D_x = 1.568$ g/cm^3，$\mu = 2.441$ cm^{-1}，$F(000) = 728$。晶体结构由直接法（MULTAN 82）解出，根据 E - 图确定了大部分非氢原子位置，剩余非氢原子坐标在以后的数轮差值 Fourier 合成中陆续确定。对全部非氢原子坐标及各向异性热参数进行全矩阵最小二乘法修正。最终一致性因子 $R = 0.046$，$R_W = 0.046$。在最终差值 Fourier 图上，最高电子密度峰值为 190e/nm^3，全部计算在 PDP11/44 计算机上用 SDP - PLUS 程序完成。

3　结果与讨论

由晶体结构分析得到标题分子的立体结构（见图 1），分子中全部非氢原子的分数坐标以及等价各向同性热参量见表 1，分子中的键长、键角和部分扭角值见表 2 和表 3。

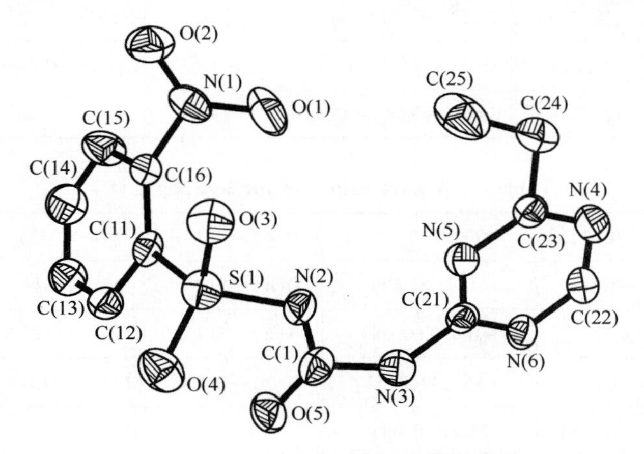

Fig. 1　Stereo structure of the title molecule

Table 1　Fractional coordinates and thermal parameters

Atom	x	y	z	$B_{eq} \times 10^2$ /nm^2	Atom	x	y	z	$B_{eq} 10^2 \times$ /nm^2
S(1)	0.7099(2)	0.3955(3)	0.1348(5)	4.05(3)	C(1)	0.5805(8)	0.629(1)	0.0701(2)	4.2(2)
O(1)	0.8362(7)	0.859(1)	0.1809(2)	7.2(1)	C(11)	0.5648(7)	0.502(1)	0.1689(2)	3.6(1)
O(2)	0.8526(6)	0.705(1)	0.2405(2)	6.1(1)	C(12)	0.3995(8)	0.424(1)	0.1642(2)	4.4(2)
O(3)	0.8756(5)	0.3990(9)	0.1532(1)	5.1(1)	C(13)	0.2797(8)	0.494(1)	0.1904(2)	5.0(2)
O(4)	0.6386(6)	0.1912(8)	0.1175(2)	5.4(1)	C(14)	0.3246(9)	0.636(1)	0.2216(2)	5.8(2)
O(5)	0.4499(5)	0.5227(9)	0.0687(1)	5.4(1)	C(15)	0.4876(9)	0.715(1)	0.2276(2)	5.5(2)
N(1)	0.7799(7)	0.743(1)	0.2081(2)	4.8(1)	C(16)	0.6050(7)	0.652(1)	0.2006(2)	3.8(2)
N(2)	0.7152(6)	0.588(1)	0.0978(2)	4.2(1)	C(21)	0.7347(8)	0.957(1)	0.0386(2)	3.8(1)
N(3)	0.6044(6)	0.802(1)	0.0418(2)	4.2(1)	C(22)	0.8309(8)	1.268(1)	0.0085(2)	4.6(2)
N(4)	0.9778(7)	1.268(1)	0.0311(2)	4.9(1)	C(23)	0.9935(8)	1.098(1)	0.0583(2)	3.9(1)
N(5)	0.8765(6)	0.9358(9)	0.0628(1)	3.6(1)	C(24)	1.1514(8)	1.094(1)	0.0843(2)	5.2(2)
N(6)	0.7068(5)	1.1207(9)	0.0112(1)	3.1(1)	C(25)	1.1739(9)	0.905(2)	0.1139(3)	6.9(2)

Table 2 Bond distances(nm) and angles(°) between some atoms

S(1)—O(3)	0.1413(4)	O(5)—C(1)	0.1198(6)	N(5)—C(21)	0.1337(6)	C(23)—C(24)	0.1464(7)
S(1)—O(4)	0.1409(4)	N(3)—C(1)	0.1384(8)	N(5)—C(23)	0.1329(7)	C(24)—C(25)	0.1469(9)
O(1)—N(1)	0.1222(7)	N(3)—C(21)	0.1374(7)	N(6)—C(21)	0.1316(7)	N(6)—C(22)	0.1306(7)
O(2)—N(1)	0.1196(6)						

O(3)—S(1)—O(4)	121.3(3)	O(3)—S(1)—C(11)	108.7(2)	C(21)—N(5)—C(23)	115.3(5)
O(3)—S(1)—N(2)	104.6(2)	N(2)—C(1)—N(3)	116.0(5)	C(21)—N(6)—C(22)	115.0(4)
O(4)—S(1)—N(2)	108.9(3)	S(1)—C(11)—C(12)	116.8(4)	O(5)—C(1)—N(2)	124.2(6)
O(4)—S(1)—C(11)	106.6(3)	N(1)—C(16)—C(11)	121.3(5)	O(5)—C(1)—N(3)	119.8(6)
N(2)—S(1)—C(11)	105.7(3)	N(1)—C(16)—C(15)	117.0(5)	N(4)—C(22)—N(6)	126.0(6)
O(1)—N(1)—O(2)	125.0(6)	N(3)—C(21)—N(5)	119.6(5)	N(4)—C(23)—N(5)	124.1(6)
O(1)—N(1)—C(16)	116.6(5)	N(3)—C(21)—N(6)	115.4(5)	N(4)—C(23)—C(24)	116.3(6)
O(2)—N(1)—C(16)	118.3(5)	N(5)—C(21)—N(6)	125.0(5)	N(5)—C(23)—C(24)	119.6(6)
S(1)—N(2)—C(1)	123.2(5)	C(1)—N(3)—C(21)	131.0(5)	C(23)—C(24)—C(25)	117.4(6)
S(1)—C(11)—C(16)	125.3(4)	C(22)—N(4)—C(23)	114.6(5)		

Table 3 A part values of torsion angles(°)

O(3)—S(1)—N(2)—C(1)	175.99(0.52)	O(4)—S(1)—C(11)—C(12)	-19.98(0.61)
O(4)—S(1)—N(2)—C(1)	44.93(0.60)	O(4)—S(1)—C(11)—C(16)	158.37(0.57)
C(11)—S(1)—N(2)—C(1)	-69.31(0.58)	N(2)—S(1)—C(11)—C(12)	95.81(0.56)
O(3)—S(1)—C(11)—C(12)	-152.39(0.52)	N(2)—S(1)—C(11)—C(16)	-85.85(0.61)
O(3)—S(1)—C(11)—C(16)	25.96(0.68)		

分子的几何结构说明，分子中的原子主要分布在3个平面上，即三嗪基磺酰脲平面（不包括磺酰基上的氧）、苯环平面和硝基平面。从分子的键长、键角及扭角可推知，这3个平面分别形成3组离域π键，分子中还存在着N…H分子内氢键。

比较我们曾研究的杂环基为嘧啶基的苯磺酰脲晶体结构[3]，两者在构象上有很大差异。由于引入三嗪环，以N取代C，N—C键长（0.1332nm，0.1337nm）明显短于C—C键长（0.1378nm，0.1391nm）。再比较C—S—N—C的扭角值，本文为-69.31°，而前者[3]为60.57°，说明硝基取代酯基后使苯环发生了较大的扭转，可能对其生物活性有所影响。

参考文献

[1] Joshua C. P., Rajan V. P.. Aust J. Chem., 1974, 27: 6627.

[2] Franz J. E.. J. Org. Chem., 1964, 29: 2595.

[3] LI Zheng - Ming（李正名），JIA Guo - Feng（贾国锋），LAI Cheng - Ming（赖城明）*et al.*. Chem. J. Chinese Universities（高等学校化学学报），1993，14（3）：349.

Syntheses, Structure and SAR Study on Sulfonylurea Compounds (V)

——Crystal and Molecular Structure of *N* – [2 – (4 – Ethyl) triazinyl] – 2 – nitrobenzene Sulfonylurea

Zhengming Li[1], Jie Liu[1], Xia Wang[2], Manxue Yuan[2], Chengming Lai[2]

([1]National Key Laboratory of Elemento – Organic Chemistry, Nankai University, Tianjin, 300071;
[2]Department of Chemistry, Nankai University, Tianjin, 300071)

Abstract Synthesis and crystal structure of the title molecule was reported here. The crystal is monoclinic, belonging to space group $P2_1/n$, with unit cell parameters $a = 0.7861$ (3) nm, $b = 0.5809$ (2) nm, $c = 3.2729$ (6) nm, $\beta = 93.25°$, $V = 1.492$ (1) nm^3, $D_x = 1.568 g/cm^3$, $\mu = 2.441 cm^{-1}$, $F(0\,0\,0) = 728$ for 1084 reflections. $R = 0.046$, $R_W = 0.046$. Molecular structure was discussed. There are three different planes in the molecule, with a conjugated system in each. The conformation difference between this molecule and other sulfonylurea molecule which has a different bioactivity was compared.

Key words sulfonylurea compounds; crystal structure; molecular structure

新型杀菌剂β-甲氧基丙烯酸甲酯类化合物的合成方法概述*

王忠文　李正名　刘天麟

（南开大学元素有机化学研究所，天津　300071）

摘　要　对新型杀菌剂β-甲氧基丙烯酸甲酯类化合物的合成方法进行了总结和概述。参考文献17篇。

关键词　β-甲氧基丙烯酸甲酯　杀菌剂　合成

β-甲氧基丙烯酸甲酯类化合物是近年来开发的一类优秀杀菌剂[1]，它们是以毒蕈中发现的 Strobilurin A（**1**）[2]为母体衍生的一系列化合物。Strobilurin A 对光不稳定。因此，人们合成了芪类似物（**2**）和二苯醚类似物（**3**），以提高其稳定性，由此衍生出了一类β-甲氧基丙烯酸甲酯的类似物。

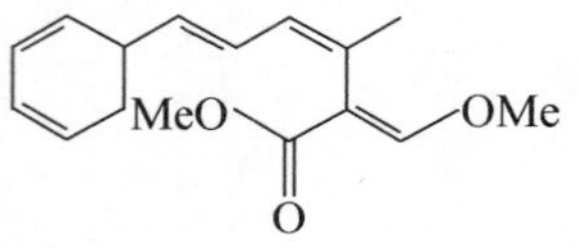

1 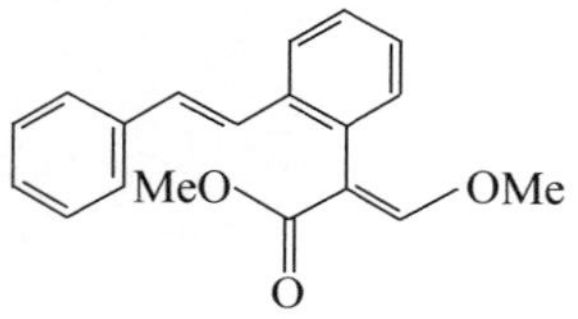　　**2** 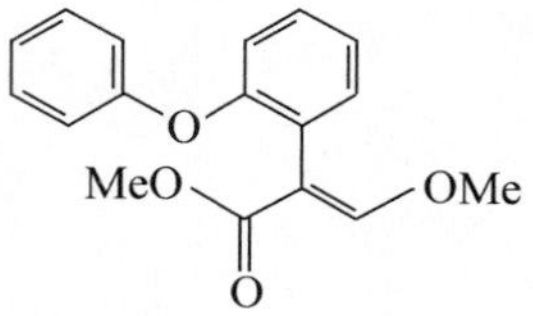　　**3**

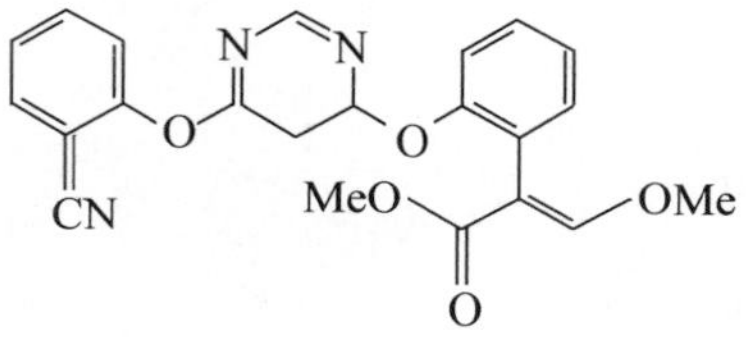

4

该类化合物具有超高效的杀菌活性。杀菌谱非常广，作用机制非常独特，是继三唑类杀菌剂之后的又一类重要的杀菌剂。目前已开发出了许多杀菌性能优良的品种，如 Zeneca 公司开发的 ICIA 5504（**4**）[3,4]。

我们将该类杀菌剂分成三个类型（Ⅰ、Ⅱ、Ⅲ），以便对其合成方法进行清晰的介绍。

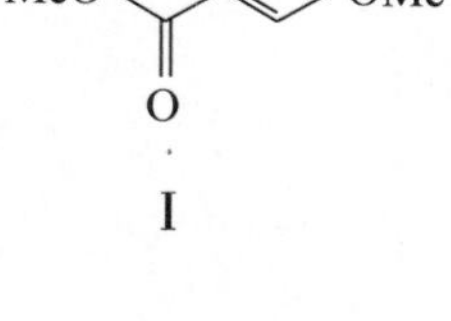

Ⅰ 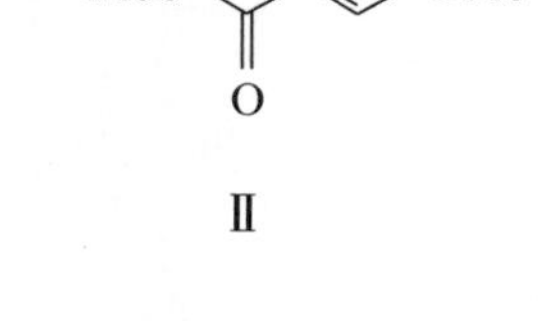　　Ⅱ 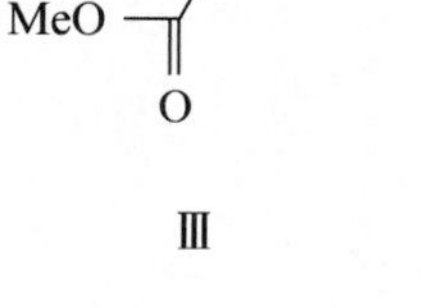　　Ⅲ

* 原发表于《合成化学》，1997，5（3）：241－245。

β－甲氧基丙烯酸甲酯类化合物结构变化较大，但其合成方法的共同点是双键的引入。双键的引入方法主要有两种，第一种是由取代的苯乙酸甲酯为原料进行合成。这是目前普遍采用的方法。

$$PhCH_2CO_2Me \xrightarrow[NaH]{HC(O)OMe} HOCH{=}C(Ph)CO_2Me \xrightarrow{Me_2SO_4} MeOCH{=}C(Ph)CO_2Me$$

另一种方法是用 α－酮酸酯作原料进行合成。

$$PhC(O)CO_2Me \xrightarrow[t\text{-}BuOK]{Ph_3\overset{+}{P}CH_2OMe\ Cl^-} MeOCH{=}C(Ph)CO_2Me$$

下面我们对Ⅰ、Ⅱ、Ⅲ类化合物的合成方法进行具体的介绍。

1　Ⅰ类化合物的合成（Ar＝芳环或芳杂环）

该类化合物的合成方法主要有以下两种。

1.1　先引入双键法

Ⅰ类化合物主要是通过此法进行合成，其中中间体（**5**）的合成是关键，合成方法如下[5]：

$$2\text{-}HOC_6H_4CH_2CO_2H \xrightarrow[2)\ PhCH_2Br]{1)\ MeOH} 2\text{-}PhCH_2OC_6H_4CH_2CO_2Me \xrightarrow[2)\ Me_2SO_4]{1)\ HC(O)OMe} 2\text{-}PhCH_2OC_6H_4C(CO_2Me){=}CHOMe \xrightarrow[H_2]{Pd/C} 2\text{-}HOC_6H_4C(CO_2Me){=}CHOMe\ (\mathbf{5})$$

由中间体（**5**）可衍生出大部分Ⅰ类化合物，如 IC IA 5504（**4**）。

另外，Beautement 等[6]报道了用邻溴苯乙酸甲酯为原料合成该类化合物的方法。

$$2\text{-}BrC_6H_4CH_2CO_2Me \xrightarrow[2)\ Me_2SO_4]{1)\ HC(O)OMe} 2\text{-}BrC_6H_4C(CO_2Me){=}CHOMe \xrightarrow{3\text{-}MeC_6H_4OH} \mathbf{6}$$

当 Ar 为芳杂环时，其合成方法也基本相同。Oda 等[7]报道了吡唑类似物的合成。

$$\text{1,3-}Me_2\text{-4-}OH\text{-5-}CO_2Et\text{-pyrazole} \longrightarrow \text{4-}OCH_2Ph\text{, 5-}CO_2Et \xrightarrow[3)\ NaCN]{1)\ LiAlH_4;\ 2)\ SOCl_2} \text{4-}OCH_2Ph\text{, 5-}CH_2CN$$

$$\longrightarrow \text{4-}OCH_2Ph\text{, 5-}CH_2CO_2Me \xrightarrow[2)\ Me_2SO_4]{1)\ HC(O)OMe} \text{4-}OCH_2Ph\text{, 5-}C(CO_2Me){=}CHOMe$$

$\xrightarrow[H_2]{Pd/C}$ Me, OH, N, N, Me, CO_2Me, MeO $\xrightarrow{COCl, Cl, Cl}$ Me, O, OC, Cl, Cl, N, N, Me, CO_2Me, MeO

7

1.2 后引入双键法

上述方法是先引入β-甲氧基丙烯酸甲酯的双键，再在芳（杂）环的邻位进行衍生化。而本法正相反。Schnetz 等[8]报道了化合物（**8**）的合成。

HO, CH_2, CO_2Me $\xrightarrow{PhO\text{~~~~}Br}$ PhO, O, CH_2, CO_2Me $\xrightarrow[2)\ Me_2SO_4]{1)\ HCOMe}$ PhO, O, MeO, OMe, O

8

以α-酮酸酯为原料经 Wittig 反应进行合成的方法，所得产物为顺反异构体的混合物，其应用没有上述方法普遍。

2 Ⅱ类化合物的合成（Ar = 芳环或芳杂环）

该类化合物主要通过芳环邻位甲基的溴化而进行衍生化。依照衍生化的顺序将该类化合物的合成分为两种：先溴化衍生法和后溴化衍生法。

2.1 先溴化衍生法

Clough 等由化合物（**9**）出发合成了化合物（**10**）[9]。

Me, CO_2Me **9** → Br, CO_2Me → Cl, O, CO_2Me →

Cl, O, CHO $\xrightarrow{MeSCH_2SMe\ (O)}$ Cl, O, O, SMe, SMe $\xrightarrow{MeOH/HCl}$

Cl, O, CO_2Me $\xrightarrow[2)\ Me_2SO_4]{1)\ HCOMe}$ Cl, O, MeO, OMe, O

10

2.2 后溴化衍生法

该法应用范围较广，大部分化合物的合成均采用此法。该方法的特点是易于衍生化。

Tsubata[10]、Schirmer 等[11]和 Cliff 等[12]由邻甲基苯甲醛出发合成了中间体（**11**），再分别与肟、酚或巯基苯并噻唑反应合成了化合物（**12 ~ 14**）。

1) $MeSCH_2S(O)Me$ 2) MeOH HCl；1) HCOMe 2) Me_2SO_4；NBS

Me CHO；Me CO_2Me；Me MeO OMe O

R O MeO OMe O **12**

RR′C ═NO MeO OMe O **13**

Br MeO OMe O **11**

N S S MeO OMe O **14**

3 Ⅲ类化合物的合成（X = O，S，NR 等）

该类化合物的合成路线较简单。

Clough 等[13]合成了化合物（**15**）。

OH Ph → OCH_2CO_2Me Ph → 1) HCOMe 2) Me_2SO_4 → MeO O CO_2Me Ph **15**

Klausener 等[14]合成了化合物（**16**）。

Br Br N S NMe H → Br Br N S NCH_2CO_2Me Me

1) HCOMe (O=C)
2) Me_2SO_4

Br, Br, N, S, N, MeO, CO_2Me, Me

16

由于Ⅰ类和Ⅱ类β－甲氧基丙烯酸甲酯类杀菌剂的活性体为E型，而部分反应（如经由Wittig反应的路线）得到的产物为Z型和E型的混合物，因此，对Z型异构体转化为E型异构体的研究引起了人们的重视。Clough等[15~17]系统地研究了该类方法。Z型异构体在Lewis酸、亲核试剂或自由基催化下可转化为E型异构体。

OMe, CO_2Me —Cat→ MeO, CO_2Me

Z型　　E型

β－甲氧基丙烯酸甲酯类杀菌剂是从80年代末开始大量系统地进行合成的，相信在不久的将来会有许多新类型和新合成方法出现。

参考文献

[1] 张一宾：农药译丛，1995，17（3），25.

[2] Anke，T.；Oberwinkler，F.；Steglic，W.：J. Antibiot.，1977，30（3），221.

[3] Clough，J. M.；Godfrey，C. R. A.；Streeting，I. T.：EP 382 375.

[4] Clough，J. M.；Godfrey，C. R. A.；Streeting，I. T.：EP 468 684.

[5] Godfrey，C. R. A.；Clough，J. M.；Streeting，I. T.：GB 2 193 495.

[6] Beautement，K.；Clough，J. M.；Godfrey，C. R. A.：EP 307 101.

[7] Oda，M.；Shike，T.；Miura，Y.；Kikutake，K.；Sekine，M.：EP 483 851.

[8] Schnetz，F.；Brand，S.；Wenderoth，B.；Sauter，H.；Ammermann，E.；Lorenz，G.：EP 336 211.

[9] Clough，J. M.；Godfrey，C. R. A.；DeFraine，P.；Hutchings，M. G.；Anthony，V. M.：EP 278 595.

[10] Tsubata，K.；Niino，N.；Endo，K.；Yamamoto，Y.；Kanno，H.：EP 414 153.

[11] Schirmer，U.；Karbach，S.；Pommer，E. H.；Ammermann，E.；Steglich，W.；Schwalge，B. A. M.；Anke，T.：DE 3 545 319.

[12] Cliff，G. R.；Richards，I. C.：EP 299 694.

[13] Clough，J. M.；Godfrey，C. R. A：EP 212 859.

[14] Klausener，A.；Kleefeld，G.；Branks，W.；Dutzmann，S.；Haenssler，G.：DD 298 394.

[15] Clough，J. M.；Eshelby，J. J.：GB 2 248 613.

[16] Clough，J. M.；Godfrey，C. R. A.；Eshelby，J. J.：GB 2 248 615.

[17] Clough，J. M.：GB 2 248 614.

An Introduction to the Synthetic Methods for Methyl β – methoxylacrylates and Related Compounds as Fungicides

Zhongwen Wang, Zhengming Li, Tianlin Liu

(Institute of Elemento – Organic Chemistry, Nankai University, Tianjin, 300071)

Abstract This paper summarized the main synthetic methods for methyl β – methoxylacry – lates and related compounds as fungicides with 17 references.

Key words methyl β – methoxylacrylates; fungicides; synthesis

3-吡唑基-6-取代均三唑并[3，4-b]-1，3，4-噻二唑的合成及生物活性*

陈寒松　李正名

（南开大学元素有机化学研究所，天津　300071）

摘　要　以1-甲基-3-乙基（-4-氯）-5-吡唑甲酰肼作原料，经两步得到4-氨基-3-（1′-甲基-3′-乙基（-4-氯）-5-吡唑基）-1，2，4-三唑-5-硫酮（**3**），**3**再与取代羧酸反应，得到一系列3-（1′-甲基-3′-乙基（-4′-氯）-5′-吡唑基）-6-取代均三唑并［3，4-b］-1，3，4-噻二唑（**4**、**5**、**6**）。元素分析、^{1}H NMR、IR和MS确定了它们的结构。初步生测结果表明：**3**具有植物生长调节活性，**4b**、**4d**、**6**具有杀菌活性。

关键词　吡唑　均三唑并［3，4-b］-1，3，4-噻二唑　合成　生物活性

自从Kanaoka将三唑和噻二唑稠合在一个分子中，制得均三唑并［3，4-b］-1，3，4-噻二唑杂环化合物[1]以来，一些文献相继报道这类稠杂环具有多种生物活性，如：消炎、抗肿瘤[2~4]、杀虫[5]、杀微生物[6]、杀菌[7,8]、除草[9,10]等，至今对该稠杂环的研究仍很活跃。

关于三唑并噻二唑3，6位引入烷基、芳基以及3位引入杂环衍生物的研究已多见文献[8,11,12]，但3位引入吡唑杂环的化合物尚未见报道。我们以5-吡唑酰肼为原料，合成了4-氨基-3-（1′-甲基-3′-乙基（-4′-氯）-5′-吡唑基）-1，2，4-三唑-5-硫酮（**3**），再与取代苯甲酸反应，制备了标题化合物，并研究了其生物活性。

1　实验部分

1.1　仪器与试剂

Yanaco CHN CORDER MT-3型自动元素分析仪；BRUKER AC-P200型核磁共振仪，溶剂为$CDCl_3$或氘代丙酮，TMS为内标；VG-ZAB型质谱仪；Shimadzu IR-435型红外光谱仪，KBr压片；Yanaco M P-500型熔点仪，温度计未经校正。柱层析硅胶（青岛海洋化工厂）100~140目。所有试剂均为国产分析纯或化学纯。

化合物**1**按文献［14］方法制备。

1.2　中间体及产物的合成

5-吡唑酰肼二硫代甲酸钾盐（**2**）的合成[9]：将15mmol **1**溶于60mL无水乙醇中，然后加入22.5mmol氢氧化钾，搅拌溶解后，再加入22.5mmol二硫化碳。室温下反应14h。加入200 mL无水乙醚稀释后，搅拌1h。滤出固体，烘干。无需提纯即可进行下

* 原发表于《应用化学》，1998，15（3）：59-62。

一步反应。

4－氨基－3－（1′－甲基－2′－乙基（－4′－氯）－5′－吡唑基）－1，2，4－三唑－5－硫酮（**3**）的合成[9]：将10mmol **2** 和2mL 85% 水合肼混合，加热回流4h，冷却，倾入冰水中，用稀盐酸溶液中和至 pH＝3，放置过夜。滤出固体，用水反复洗涤，烘干，乙醇重结晶得纯品（化合物数据见表1）。

标题化合物（**4**、**5**、**6**）的合成[10]方法：将2mmol **3** 和2.2mmol **1** 羧酸放在5mL $POCl_3$ 中回流6h。减压蒸除过量的 $POCl_3$，剩余物倾入冰水中，放置过夜。滤出固体，分别用稀氢氧化钠溶液和水洗涤，烘干。快速柱层析、梯度洗脱分离出产物（淋洗剂 $V_{石油醚}/V_{乙酸乙酯}=1.5/1$），物理数据见表1。

1a.X＝Cl；b. X＝H　　CS_2,KOH　　**2**a.X＝Cl；b. X＝H　　$NH_2NH_2·H_2O$

3a.X＝Cl；b. X＝H　　HOOC–C₆H₄–R, $POCl_3$

4a，b. X＝Cl；R＝4－OMe，4－F，
4c～g. X＝H；R＝4－F，4－OMe，4－Br，4－H，4－Me

$HOOCCH_2O$–(2,4-二氯苯基), $POCl_3$　　$POCl_3$, 2-萘氧乙酸（OCH_2COOH）

6X＝H　　**5**X＝H

表1　化合物3～6的有关数据

化合物	分子式	收率（%）	熔点（℃）	外观	元素分析值（计算值，%） C	H	N
3a	$C_8H_{11}ClN_6S$	63	176～177	白色固体	37.04（37.14）	4.01（4.28）	32.48（32.47）
3b	$C_8H_{12}N_6S$	60	172～173	白色固体	42.44（42.84）	5.41（5.39）	36.99（37.47）
4a	$C_{16}H_{15}ClN_6OS$	53	183～184	白色固体	51.45（51.27）	4.31（4.03）	22.01（22.42）
4b	$C_{15}H_{12}ClFN_6S$	74	185～186	白色固体	49.73（49.66）	3.62（3.33）	23.31（23.16）
4c	$C_{15}H_{13}FN_6S$	62	207～208	白色固体	55.14（54.87）	4.07（3.99）	25.17（25.59）
4d	$C_{16}H_{16}N_6OS$	29	171～172	白色固体	56.57（56.46）	4.76（4.74）	24.88（24.69）
4e	$C_{15}H_{13}BrN_6S$	26	210～211	白色固体	46.47（46.28）	3.75（3.37）	21.53（21.59）

续表 1

化合物	分子式	收率（%）	熔点（℃）	外观	元素分析值（计算值,%）		
					C	H	N
4f	$C_{15}H_{14}N_6S$	40	211 ~ 212	白色固体	57.89（58.05）	4.29（4.55）	26.87（27.08）
4g	$C_{16}H_{16}N_6S$	31	160 ~ 161	绿色固体	59.30（59.24）	4.87（4.97）	25.91（25.91）
5	$C_{20}H_{18}N_6OS$	13	209 ~ 210	棕色固体	61.30（61.52）	4.61（4.65）	21.26（21.52）
6	$C_{16}H_{14}Cl_2N_6OS$	41	184 ~ 185	白色固体	47.17（46.95）	3.28（3.45）	20.32（20.53）

2 结果与讨论

2.1 合成

制备关键中间体吡唑酰肼二硫代甲酸钾盐 **2b** 时，在室温反应 14h 后，必须加入大量无水乙醚，**2b** 才会析出。

3－取代－4－氨基－1，2，4－三唑－5－硫酮（**3**）同有机羧酸反应，得到稠杂环均三唑并［3，4－b］－1，3，4－噻二唑，产率偏低，原因是：**3** 为一共轭分子，3 位吡唑环的－C 和－I 效应，引起 4 位氨基孤对电子沿共轭链向吡唑环转移，降低了电子密度，致使其亲核活性降低，不利于和羧基缩合[13]。**4b**、**4c** 的产率较高，原因是对氟苯甲酸，氟的强吸电子效应使得羧基亲电子能力增强。

2.2 化合物结构与图谱解析

所有化合物的 ^1H NMR 数据列于表 2，部分化合物 MS 和 IR 数据列于表 3。

表 2 化合物 3 ~ 6 的 ^1H NMR 数据

化合物	溶剂	^1H NMR（δ）
3a	DMCO	1.24（t，3H，CH_3），2.31（q，2H，CH_2），3.86（s，3H，NCH_3），5.36（s，2H，NH_2），13.07（s，1H，NH）
3b	DMSO	1.18（t，3H，CH_3），2.50（q，2H，CH_2），3.92（s，3H，NCH_3），3.93（s，2H，NH_2），6.88（s，1H，H_4－吡唑），13.07（s，1H，NH）
4a	$CDCl_3$	1.30（t，3H，CH_3），2.71（q，2H，CH_2），3.87（s，3H，NCH_3），4.00（s，3H，OCH_3），7.00，7.81（2d，4H，Ph，J=8.5Hz）
4b	$CDCl_3$	1.31（t，3H，CH_3），2.72（q，2H，CH_2），3.87（s，3H，NCH_3），7.24（d，2H，Ph）7.90（q，2H，Ph，J_{HH}=9.2Hz，J_{FH} = 5.1Hz）
4c	$CDCl_3$	1.31（t，3H，CH_3），2.73（q，2H，CH_2），4.32（s，3H，NCH_3），6.98（s，1H，H_4－吡唑），7.25（d，2H，Ph），7.91（q，2H，Ph，J_{HH}=8.5Hz，J_{FH} = 5.2Hz）
4d	$CDCl_3$	1.31（t，3H，CH_3），2.72（q，2H，CH_2），3.88（s，3H，NCH_3），4.30（s，3H，OCH_3），6.97（s，1H，H_4－吡唑），7.00，7.83（2d，4H，Ph，J_{HH}=8.5Hz）
4e	$CDCl_3$	1.30（t，3H，CH_3），2.72（q，2H，CH_2），4.30（s，3H，NCH_3），6.96（s，1H，H_4－吡唑），7.77，7.68（2d，4H，Ph，J_{HH}=8.7Hz）
4f	$CDCl_3$	1.29（t，3H，CH_3），2.71（q，2H，CH_2），4.30（s，3H，NCH_3），6.98（s，1H，H_4－吡唑），7.52~7.91（m，5H，Ph）

化合物	溶剂	^{1}H NMR (δ)
4g	$CDCl_3$	1.31（t，3H，CH_3），2.43（s，3H，$PhCH_3$），2.72（q，2H，CH_2），4.30（s，3H，NCH_3），6.98（s，1H，H_4－吡唑），7.30，7.70（2d，4H，Ph，J_{HH}＝7.8Hz）
5	$CDCl_3$	1.31（t，3H，CH_3），2.72（q，2H，CH_2），4.31（s，3H，NCH_3），5.54（s，2H，OCH_2），6.92（s，1H，H_4－吡唑），6.66，8.25（m，7H，Napth）
6	$CDCl_3$	1.27（t，3H，CH_3），2.78（q，2H，CH_2），4.28（s，3H，NCH_3），5.40（s，2H，OCH_2），6.85（s，1H，H_4－吡唑），6.91～7.41（m，3H，Ph）

表3　3a、4a、4e、4g、5 和 6 的 IR 及 MS 数据

化合物	IR (cm^{-1})，MS (m/e)
3a	3406，3254（s，NH），1620，1532，1497（s，C═C），1296（s，C═S）
4a	3012（m，ph－H），1602，1489，1451（s，C═N 和 C═C），1224，1023（s，Ar—O—C），824（s，对二取代苯）
4g	3125（w，Ar—H），1607，1486，1453（s，C═N 和 C═C），815（s，对二取代苯）
5	3120（s，Ar—H），1629，1593，1505，1457（s，C═N 和 C═C），1244，1102（s，Ar—O—C），870，805（s，1，2，4－三取代苯），737（s，邻二取代苯）
6	3129（m，Ar—H），1588，1484，1457（s，C═N 和 C═C），1265，1070（s，Ar—O—C），870，805（s，1，2，4－三取代苯）
4e	328（M，69%），327（M－1，100%），207（9%），179（2.2%），165（1.3%），139（38.4%），135（16.4%），121（12.1%）

在^{1}H NMR谱图中，**4b** 和 **4c** 的苯环氢分为两组，远离 F 的两氢为二重峰，与 F 相邻的两氢为四重峰是由于 F 的耦合造成的。相应的 **4a**、**4d** 苯环氢为两组二重峰。

3a 的 IR 谱图有 1296 cm^{-1}吸收峰，说明分子中有 C═S 存在，它应以酮式存在。

2.3　生物活性

我们对化合物 **3a**、**3b**、**4a**～f 和 **6** 进行了杀菌活性测试。对番茄早疫和棉花立枯运用的是离体平皿法，浓度为 0.005%；对油菜菌核运用的是活体小株法，浓度为 0.05%，结果见表4。从表4 可以看出，**4b** 和 **4d** 对棉花立枯的抑制活性较好，**6** 对油菜菌核的抑制活性较好。

表4　化合物 3、4a～ f 和 6 的杀菌活性

化合物	**3a**	**3b**	**4a**	**4b**	**4c**	**4d**	**4e**	**4f**	**6**
番茄早疫	33.3	6.6	25	29.1	0	45	6.6	20.0	13.3
棉花立枯	14.3	12.2	33.9	63.5	32.1	69.4	2.0	20.4	18.4
油菜菌核	3.7	43.7	15.7	18.4	7.8	5.2	0	45.7	70.0

我们还对化合物 **3a**、**3b** 和 **4b** 进行了植物生长调节活性测试，每项浓度均为 0.001%，结果见表5。结果表明：**3b** 对芽鞘伸长有促进作用，**3a** 和 **3b** 对子叶扩张都有促进作用。

表5 3a、3b和4b的植物激素活性实验结果

化合物	3a	3b	4b	化合物	3a	3b	4b
芽鞘伸长	-3.0	11.0+	6.0	子叶扩张	18.4+	21+	5.2
子叶生根	-23.9	0	34.7	下胚轴抑制	4.2	7.1	3.6

参考文献

[1] Kanaoka. *J Pharm Soc J p n*, 1956, 76: 1133.

[2] Mody M K, Prasad A R, Ramalingam T *et al. J Indian Chem Soc*, 1982, 59: 769.

[3] Deshmukh A A, Mody M K, Ramalingam T *et al. Indian J Chem*, 1984, 23B: 793.

[4] Prasad A R, Ramalingam T, Rao A B *et al. Indian J Chem*, 1986, 25B: 566.

[5] Bandana C, Nirupama T, Nizamuddin. *Agric Boil Chem*, 1988, 52 (5): 1229.

[6] Mohan J, Anjaneyulu G S R, Kiran. *Indian J Chem*, 1988, 27B: 128.

[7] Patel H V, Fernandes P S, V yas K A. *Indian J Chem*, 1990, 29B: 135.

[8] Pant H K, Durgapal R, Joshi P. *Indian J Chem*, 1983, 22B: 712.

[9] 张自义，李明，赵岚．有机化学，1993，4：397.

[10] 张自义，李明，赵岚．高等学校化学学报，1994，15（2）：220.

[11] Eweiss N F, Bahajaj A A. *J Heterocy clic Chem*, 1987, 24: 1173.

[12] Potts K T, Huseby R M. *J Org Chem*, 1966, 31 (11): 3528.

[13] 张自义，陈新．化学学报，1991，49：513.

[14] 李正名，陈寒松，赵卫光，等．高等学校化学学报，1997，18（11）：1794.

Synthesis and Biological Activities of 3 – Pyrazolyl – 6 – substituted – s – triazolo [3, 4 – b] – 1, 3, 4 – thiadiazoles

Hansong Chen, Zhengming Li

(Institute of Elemento – Organic Chemistry, Nankai University, Tianjin, 300071)

Abstract In view of the biological activity of s – triazo lo [3, 4 – b] – 1, 3, 4 – thiadiazole fused heterocyclic derivatives, nine novel compounds 3 – ((4′ – chloro) – 3′ – ethyl – 1′ – methyl – 5′ – pyrazolyl) – 6 – substituted – s – triazolo [3, 4 – b] – 1, 3, 4 – thiadiazoles (**4**a ~ g, **5**, **6**) were synthesized by condensation of carboxylic acid with 4 – amino – 3 – ((4′ – chloro) – 3′ – ethyl – 1′ – methyl – 5′ – pyrazolyl) – 1, 2, 4 – triazole – 5 – thione. Their structures were characterized by elem ental analysis, ^{1}H NMR, IR and M S spectroscopy. The preliminary bioassay tests showed that compound **3** exhibited plant grow thregulation activity and **4**b, **4**d and **6** possessed moderate fungicidal activity.

Key words pyrazolyl – s – triazolo [3, 4 – b] – 1, 3, 4 – thiadiazole; synthesis; bioassay

Syntheses of Di – heterocyclic Compounds; Pyrazolylimidazoles and Isoxazolylimidazoles*

Zhengming Li, Jimin Li, Guofeng Jia

(Elemento – Organic Chemistry Institute, Nankai University, Tianjin, 300071, China)

Abstract Fifteen substituted di – heterocyclic imidazoles were prepared. Ten substituted pyrazolylimidazoles were obtained by cyclocondensation of α – oxo – α – imidazolylketene dithioacetal or N, S – acetals with hydrazine and by ring – chain transformation of cyclic α – oxo – α – imidazolylketene N, O – acetals. Five isoxazolylimidazoles were obtained by reaction of α – oxo – α – imidazolylketene dithioacetal with hydroxylamine.

RESULTS AND DISCUSSION

Substituted di – heterocyclic compounds, such as pyrazolylimidazoles and isoxazolylimidazoles (Table 1) were obtained by three methods. First, substituted pyrazolylimidazole **2**, isoxazolylimidazole **3**, and **4** were synthesized by cyclocondensation of α – oxo – α – imidazolylketene dithioacetal **1** with bifunctional nucleophiles[1,2] such as hydrazine and hydroxylamine (Scheme 1).

Table 1 Preparation of Pyrazolylimidazoles and Isoxazolylimidazoles

No.	Formula	Appearance	Mp(℃)	Yield (%)	Analysis(%) Calcd/Found C	H	N
2a	$C_{13}H_{12}N_4S$	white needle	235.5 – 236.5	98	60.92/60.98	4.72/4.74	21.86/21.69
2b	$C_{13}H_{11}ClN_4S$	white needle	228.0 – 229.0	95	53.70/53.77	3.80/3.85	19.27/19.57
2c	$C_{13}H_{11}BrN_4S$	white needle	224.0 – 225.0	91	46.60/46.89	3.30/3.41	16.72/16.76
2d	$C_{14}H_{14}N_4OS$	white needle	196.0 – 198.0	90	58.71/59.14	4.93/5.17	19.56/20.00
2e	$C_{14}H_{14}N_4S$	white needle	206.5 – 207.5	90	62.22/61.95	5.22/5.30	20.72/20.48
3b	$C_{13}H_{10}ClN_3OS$	white needle	124.5 – 125.5	56	53.53/53.62	3.46/3.48	14.41/14.14
4b	$C_{13}H_{10}ClN_3OS$	yellow needle	178.0 – 179.0	22	53.53/53.22	3.46/3.58	14.41/14.23
3c/4c	$C_{13}H_{10}BrN_3OS$	yellow needle	157.0 – 158.5	40	46.44/46.39	3.00/2.96	12.50/12.50
4d	$C_{14}H_{13}N_3O_2S$	yellow needle	130.0 – 131.0	44	58.53/58.25	4.56/4.59	14.63/14.54
6a	$C_{14}H_{15}N_5O$	white needle	216.5 – 217.0	90	62.44/62.14	5.61/5.65	25.99/25.56
6b	$C_{14}H_{14}ClN_5O$	white needle	232.0 – 233.0	83	55.36/55.51	4.65/4.95	23.04/22.77
6c	$C_{14}H_{14}BrN_5O$	white needle	235.0 – 236.0	93	48.29/48.30	4.05/4.05	20.10/20.13
9b	$C_{18}H_{14}ClN_5$	white powder	272.0 – 273.0	70	64.38/64.48	4.20/4.51	20.85/20.77
9c	$C_{18}H_{14}BrN_5$	white powder	282.0 – 284.0	69	56.86/57.13	3.71/4.00	18.42/18.20

* Reprinted from *Heteroatom Chemistry*, 1998, 9(3): 317 – 320. This project was founded by National Natural Science Foundation of China.

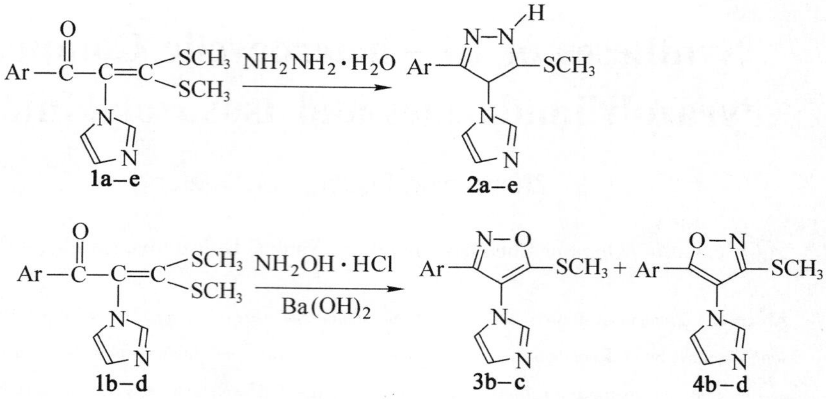

a,Ar=C_6H_5; b,Ar=p-ClC_6H_4; c, Ar=p-BrC_6H_4; d, Ar=p-$CH_3OC_6H_4$; e, Ar=p-$CH_3C_6H_4$

Scheme 1

Their structures were confirmed by spectroscopic data. The structures of **3** and **4** are isomers as determined by ^1H NMR and mass spectroscopy. In the high field of the proton spectrum, the protons of **4** resonate in a higher field than those of **3** because the inductive effect of O is a little stronger than that of N. For the same reason, in the low field of the proton spectrum, the protons of **4** resonate in a lower field than those of **3** (Table 2). The characteristic mass spectroscopy peaks of **3b** and **4b** are listed in Scheme 2.

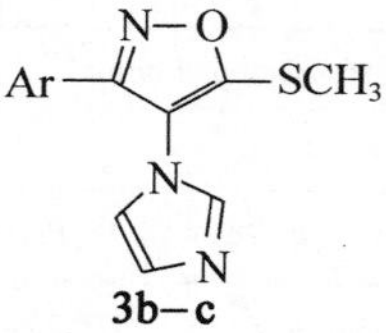

3b–c

No.	Ar	Ar—C≡N⌉ +	Ar—C=NH⌉ +
3b	p-ClC_6H_4	137	138
3c	p-BrC_6H_4	181	182

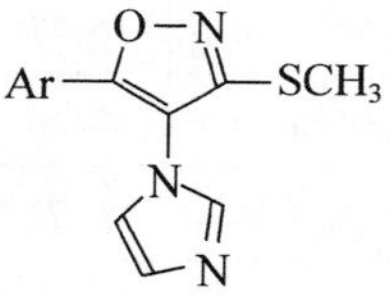

4b–d

No.	Ar	Ar—C=O⌉ +	Ar-C=OH⌉ +
4b	p-ClC_6H_4	139	140
4c	p-BrC_6H_4	183	—
4b	p-$CH_3OC_6H_4$	135	—

Scheme 2

Talbe 2 ^1H NMR Data

No.	^1H NMR, δ
2a	(1)2.40(s,3H,SCH_3),7.16-7.60(m,8H,C_6H_5 & $C_3H_3N_2$),13.68(s,NH)
2b	(1)2.38(s,3H,SCH_3),7.16-7.66(m,7H,C_6H_4 & $C_3H_3N_2$),13.80(s,NH)
2c	(1)2.36(s,3H,SCH_3),7.12-7.56(m,7H,C_6H_4 & $C_3H_3N_2$),13.72(s,NH)
2d	(1)2.40(s,3H,SCH_3),3.76(s,3H,OCH_3),6.88-7.68(m,7H,C_6H_4 & $C_3H_3N_2$),13.58(s,NH)
2e	(2)2.32(s,3H,CH_3),2.37(s,3H,OCH_3),7.16-7.68(m,7H,C_6H_4 & $C_3H_3N_2$)
3b	(3)2.62(s,3H,SCH_3),6.95-7.54(m,7H,C_6H_4 & $C_3H_3N_2$)
4b	(3)2.59(s,3H,SCH_3),7.00-7.57(m,7H,C_6H_4 & $C_3H_3N_2$)
3c	(3)2.60(s,3H,SCH_3),6.93-7.52(m,7H,C_6H_4 & $C_3H_3N_2$)

Table 2 (continued)

No.	^1H NMR, δ
4c	(3) 2.58 (s, 3H, SCH_3), 6.99 – 7.57 (m, 7H, C_6H_4 & $C_3H_3N_2$)
4d	(3) 2.68 (s, 3H, SCH_3), 3.80 (s, 3H, OCH_3), 6.90 – 8.01 (m, 7H, C_6H_4 & $C_3H_3N_2$)
6a	(1) 3.24 (t, 2H, NCH_2), 3.60 (t, 2H, OCH_2), 4.64 (s, OH), 7.20 – 7.50 (m, 8H, C_6H_5 & $C_3H_3N_2$), 8.12 [s, NH (branched)]
6b	(1) 3.20 (t, 2H, NCH_2), 3.60 (t, 2H, OCH_2), 4.60 (s, OH), 7.20 – 7.68 (m, 7H, C_6H_4 & $C_3H_3N_2$), 8.24 [s, NH (branched)]
6c	(1) 3.30 (t, 2H, NCH_2), 3.48 (t, 2H, OCH_2), 4.62 (s, OH), 7.10 – 7.63 (m, 7H, C_6H_4 & $C_3H_3N_2$), 7.50 [s, NH (branched)]
9b	(1) 6.50 – 8.00 [m, C_6H_4, $C_3H_3N_2$ & NH (branched)], 13.00 [s, NH (pyrazole)]
9c	(1) 6.72 – 7.89 [m, C_6H_4, $C_3H_3N_2$ & NH (branched)], 13.06 [s, NH (pyrazole)]

(1) DMSO – d_6.

(2) CD_3OD.

(3) $CDCl_3$.

Second, the ring – chain transformation of cyclic α – oxo – α – imidazolylketene N, O – acetals **5a – c** was investigated. The concept of ring – chain transformation is based on the opening of a saturated heterocyclic ring in the starting material while immediately afterward a new heteroaromatic ring is formed by condensation[3]. As starting intermediates, 1, 3 – dicarbonyl heteroanalogs were used. These substrates were "ring – chain" transformed by reaction with binucleophiles. We applied **5** as C – C – C building blocks and hydrazine as the binucleophile[4-7]. Pyrazolylimidazole compounds **6a – c** were obtained (Scheme 3).

Ar—C(=O)—C(=oxazolidine)—N-imidazolyl (**5a–c**) —[$NH_2NH_2 \cdot H_2O$, EtOH, reflux]→ [Ar—C(=O)—C(N-imidazolyl)=C(NHNH$_2$)(NHCH$_2$CH$_2$OH)] → Ar-pyrazole(N–N–H)-NHCH$_2$CH$_2$OH, N-imidazolyl (**6a–c**)

a, Ar = C_6H_5; **b**, Ar = p–ClC_6H_4; **c**, Ar = p– BrC_6H_4

Scheme 3

Finally, we focused on the reaction of α – oxo – α – imidazole substituted acetophenones **7b – c** with phenyl isothiocyanate, followed by alkylation. α – oxo – α – imidazolylketene N, S – acetals **8b – c** were obtained. On further reaction with hydrazine, new pyrazolylimidazole compounds **9b – c** were obtained (Scheme 4).

Ar—C(=O)—CH$_2$—N-imidazolyl (**7b–c**) —[PhNCS, KOH]→ [Ar—C(=O)—C(N-imidazolyl)=C(NHPh)(SK)] —[CH_3I]→ Ar—C(=O)—C(N-imidazolyl)=C(NHPh)(SCH$_3$) (**8b–c**) —[$NH_2NH_2 \cdot H_2O$, reflux]→ Ar-pyrazole(N–N–H)-NHPh, N-imidazolyl (**9b–c**)

b, Ar = p–ClC_6H_4; **c**, Ar = p– BrC_6H_4

Scheme 4

EXPERIMENTAL

Instruments

Melting points were determined with a Yanaco MP – 500 apparatus without correction. ^{1}H NMR spectra were measured with a Jeol FX – 90Q spectrometer using TMS as internal reference and mass spectra on a HP5988A spectrometer with 70eV. Elemental analyses were performed on a MT – 3CHN instrument.

Table 3 EI – MS Data

No.	m/z(abundant)
2b	292(39),290(M$^+$,100),264(7),262(13),248(27),250(18),179(13),140(19),138(46),113(13),111(37),102(21)
3b	293(4),291(M$^+$,10),246(5),244(16),218(9),216(28),140(5),139(8),138(14),137(11),113(9),111(40),102(16),79(100)
4b	293(42),291(M$^+$,110),246(47),244(137),142(28),141(345),140(85),139(1000),113(156),111(522)
3c/4c	337(47),335(M$^+$,45),290(45),288(44),262(48),260(48),242(25),240(24),228(77),201(57),199(56),185(40),184(23),183(52),182(20),181(12),157(23),155(24),111(36),79(100),52(72)
4d	287(M$^+$,51),272(10),240(30),135(100),111(13),92(16),79(25)
6a	269(M$^+$,41),238(100),211(36),104(26),77(30)
6b	305(15),303(M$^+$,42),274(38),272(100),247(14),245(44),140(12),138(22),113(8),111(16)
6c	349(45),347(M$^+$,44),318(99),316(100),291(30),289(33),237(15),184(23),182(33),157(16),155(25),76(33),75(21)
9b	337(37),335(M$^+$,100),309(15),307(43),140(5),138(11),113(2),111(5),77(20)
9c	381(95),379(M$^+$,100),353(44),351(46),184(13),182(16),157(9),155(10),93(13),91(12)

General Procedure for the Preparation of **2a – c**

A mixture of **1**(2mmol) and 85% hydrazine hydrate(360mg,6mmol) in ethanol(20mL) was refluxed for 2 hours. After evaporation of part of the solvent and cooling, precipitates formed that were collected and recrystallized from anhydrous ethanol.

General Procedure for the Preparation of **3b – c** and **4b – d**

Hydroxylamine hydrochloride(0. 66 g,9. 3mmol) was added to a stirred mixture of $Ba(OH)_2$ (4. 7mmol) in 20mL of ethanol. Ten minutes later, **1**(3. 1mmol) was added, and then the mixture was heated at reflux for 0. 5 hour. The solvent was evaporated under vacuum and the residue was diluted with water(40mL). Acetic acid(36%) was added to the solution until pH = 5. Products were extracted by ethyl acetate(10mL × 3) and dried with anhydrous sodium sulfate. The solvent was evaporated, and the products **3** and **4** were separated by column chromatography using petroleum ether/ethyl acetate as eluant.

General Procedure for the Preparation of **6a – c**

A mixture of 5(0. 78mmol), 85% hydrazine hydrate(0. 23g, 3. 9mmol) and 20mL of ethanol was refluxed for 4 hours. Part of the solvent was evaporated under vacuum. After cooling of the solution, the products **6** precipitated and were filtered off, then recrystallized from anhydrous ethanol.

General Procedure for the Preparation of 9b – c

Phenyl isothiocyanate (1.2mL, 10mmol) and powdered KOH (0.8g, 14mmol) were added to a solution of **5** (10mmol) in 20mL of DMSO. The mixture was stirred at room temperature for 2 hours. Methyl iodide (1.4g, 10mmol) was gradually added over 4 hours. Addition of water (100mL) caused a yellow precipitate **8b – c** to form. This was collected by filtration and then recrystallized from anhydrous ether.

Excess hydrazine hydrate (85%) was added to a mixture of **8** (3mmol) and ethanol (40mL). After the mixture had been refluxed for 3 hours, part of the solvent was evaporated under vacuum. The products **9b – c** were collected by filtration and recrystallized from DMF/H_2O.

REFERENCES

[1] G. Singh, H. Ila, H. Junjappa, *Synthesis*, 1987, 286.

[2] W. – D. Rudorf, M. Augustin, *J. Prakt. Chem.*, 320, 1978, 585.

[3] M. Patzel, A. Ushmajev, J. Liebscher, *Synthesis*, 5, 1993, 525.

[4] J. Bohrish, H. Fattz, et al., *Tetrahedron*, 50(36), 1994, 10701.

[5] Z. N. Huang, Z. M. Li, *Synth. Commun.*, 25(20), 1995, 3219.

[6] Z. N. Huang, Z. M. Li, *Synth. Commun.*, 25(20), 1995, 3603.

[7] A. K. Mukerjee, R. Ashare, *Chem. Rev.*, 1, 1991, 1.

3－芳基磺酰氧基(取代)异噻唑的合成及除草活性*

杨小平　李正名　王玲秀　李永红

（南开大学元素有机化学研究所、元素有机化学国家重点实验室，天津　300071）

摘　要　通过3－羟基异噻唑和4－氰基－5－甲硫基－3－羟基异噻唑与芳基磺酰氯在碱性条件下缩合，合成了3－芳基磺酰氧基异噻唑和3－芳基磺酰氧基－4－氰基－5－甲硫基－异噻唑，对所合成的化合物进行了除草活性的测定。

关键词　3－芳基磺酰氧基（取代）异噻唑　合成　除草活性

异噻唑（单核）杂环在杂环化合物中属较新的一族[1]，人们大多集中研究其合成方法和反应，而关于其生物活性方面的应用基础研究则较少。随着有关含异噻唑类化合物良好的杀菌活性[2]、杀虫活性[3]及改善血液循环[4]、抗癌[5]等生物活性的报道，对这类化合物进行结构修饰和生物活性研究已成为90年代杂环化学领域的又一个研究热点。为了寻找结构新颖、具有生物活性的化合物，我们通过3－羟基异噻唑（**1**）和4－氰基－5－甲硫基－3－羟基异噻唑（**2**）与芳基磺酰氯在碱性条件下缩合，得到了目标化合物3－芳基磺酰氧基异噻唑（**3**）和3－芳基磺酰氧基－4－氰基－5－甲硫基－异噻唑（**4**）。合成路线如下：

(1) + X-C$_6$H$_4$-SO$_2$Cl $\xrightarrow[CHCl_3]{Et_3N}$ (3)

(2) + X-C$_6$H$_4$-SO$_2$Cl $\xrightarrow[CHCl_3]{Et_3N}$ (4)

X: **3a** H; **3b** 4－Me; **3c** 2－NO_2; **3d** 4－NO_2; **3e** 4－Br; **3f** 2－CO_2Et;
4a H; **4b** 4－Me; **4c** 2－NO_2; **4d** 4－Br; **4e** 4－NO_2; **4f** 2－CO_2Et

以上产物的结构经元素分析、^{1}H NMR谱及部分化合物的红外吸收光谱和质谱证实。初步生物活性测定表明，这类化合物具有一定的除草活性，并且具有一定的活性规律。

1　实验部分

1.1　仪器和试剂

所用溶剂均为分析纯，用前经纯化处理。试剂为化学纯硅胶（青岛海洋化工厂）。

* 原发表于《高等学校化学学报》，1998，19（2）：228－231。

元素分析用 Yanaco CHN CORDER MT－3 型自动分析仪测定；^{1}H NMR 用 BRU KER AC－P200 型仪器测定（TMS 作内标）；质谱用 HP－5988A 型质谱仪（EI，70eV）；红外光谱用 SHMADZU－IR 435 型仪器测定，KBr 压片；熔点用 Yanaco 熔点仪测定；折光率用 WZS－I 型折光仪（上海光学仪器厂）测定；所用温度计未经校正。

1.2 化合物1～4的合成

1.2.1 化合物**1**的合成 参考文献［6］方法进行。产物为稍带黄色的晶体，总收率（按丙烯酰胺计）71%，熔点73.5～75.0℃（文献值：73～74℃）。

1.2.2 化合物**2**的合成 参考文献［7］方法进行。产物为土黄色针状晶体，总收率（按丙二腈计）75%，熔点239～240℃（文献值：239～241℃）。

1.2.3 化合物**3**的合成 在25mL三口瓶中，加入5mmol芳基磺酰氯和10mL经P_2O_5干燥的氯仿，搅拌使芳基磺酰氯溶解，分3次加入5mmol（0.51g）化合物**1**，搅拌溶解后，用冰水冷至0℃，慢慢滴加5mmol（0.51g）Et_3N/5mL $CHCl_3$的混合溶液，使反应液温度不超过5℃，滴毕，在室温下继续搅拌12h，用TLC检测至终点。将反应液用水（3×20mL）洗涤，无水硫酸镁干燥，脱溶剂并留下1～2mL溶液后，进行快速柱层析，用V（石油醚）∶V（乙酸乙酯）＝10∶3作淋洗液，收集所需的点，减压脱溶剂后得产物。

1.2.4 化合物**4**的合成 步骤同**3**的合成。滴加三乙胺时，保持反应温度在15～20℃。室温搅拌20h以上至终点。化合物**3**和**4**的收率、物理常数、元素分析和^{1}H NMR见表1。

Table 1 Physical properties and data of elemental analysis and ^{1}H NMR for title compounds①

No.	Molecular formula	Molecular weight	n_D or m. p. (℃)	Physical character	Yield (%)	Elemental anal. (found/calcd, %) C	H	N
3a	$C_9H_7NO_3S_2$	241.09	1.6642	Colorless liquid	45.2	44.78/44.80	2.86/2.90	5.79/5.79
3b	$C_{10}H_9NO_3S_2$	255.10	1.6300	Colorless liquid	53.3	46.93/47.08	3.63/3.53	5.29/5.49
3c	$C_9H_6N_2O_5S_2$	286.09	83.5～84.5	White powder	44.2	37.68/37.76	2.07/2.10	9.50/9.75
3d	$C_9H_6N_2O_5S_2$	286.09	126.0～126.5	White rhombus	58.6	37.62/37.76	2.00/2.10	9.82/9.75
3e	$C_9H_6BrNO_3S_2$	320.02	63.5～64.5	White needle	60.0	33.80/33.78	1.75/1.87	4.10/4.38
3f	$C_{12}H_{11}NO_5S_2$	313.23	105.0～106.0	Yellow powder	76.6	46.05/46.01	3.42/3.51	4.48/4.47
4a	$C_{11}H_8N_2O_3S_3$	312.11	122.0～123.5	Yellow rhombus	51.0	42.22/42.33	2.54/2.56	8.86/8.97
4b	$C_{12}H_{10}N_2O_3S_3$	326.12	111.5～112.5	White rhombus	45.5	44.19/44.19	3.02/3.07	8.48/8.58
4c	$C_{11}H_7N_3O_5S_3$	357.11	141.0～142.0	White rhombus	44.8	36.68/36.97	1.96/1.96	11.46/11.36
4d	$C_{11}H_7BrN_2O_3S_3$	391.11	122.0～123.0	White needle	59.0	33.83/33.76	1.75/1.79	7.16/7.16
4e	$C_{11}H_7N_3O_5S_3$	357.11	105.0～106.0	White needle	82.2	37.18/36.87	2.26/1.96	11.62/11.36
4f	$C_{14}H_{12}N_2O_5S_3$	384.32	176.5～178.0	Brown powder	80.0	43.69/43.75	3.12/3.12	7.30/7.29

No.	^{1}H NMR ($CDCl_3$), δ
3a	6.90 (d, 1H, J=4.6Hz), 7.51～7.95 (m, 5H, Ar—H), 8.57 (d, 1H, J=4.6Hz)
3b	2.40 (s, 3H, Ar—CH_3), 6.89 (d, 1H, J=4.8Hz), 7.30 (d, 2H, Ar—H), 7.80 (d, 2H, Ar—H), 8.57 (d, 1H, J=4.8Hz)
3c	7.02 (d, 1H, J=4.8Hz), 7.81～8.24 (m, 4H, Ar—H), 8.63 (d, 1H, J=4.8Hz)

续表 1

No.	^{1}H NMR ($CDCl_3$), δ
3d	6.99 (d, 1H, J=4.8Hz), 8.21 (d, 2H, Ar—H), 8.39 (d, 2H, Ar—H), 8.63 (d, 1H, J=4.8Hz)
3e	6.96 (d, 1H, J=4.8Hz), 7.68 (d, 2H, Ar—H), 7.83 (d, 2H, Ar—H), 8.61 (d, 1H, J=4.8Hz)
3f	1.02 (t, 3H, —CH_3), 4.02 (q, 2H, —CH_2—), 6.90 (d, 1H, J=4.8Hz), 7.32 ~ 8.00 (m, 4H, Ar—H), 8.60 (d, 1H, J=4.8Hz)
4a	2.65 (s, 3H, —SCH_3), 7.59 ~ 8.07 (m, 5H, Ar—H)
4b	2.46 (s, 3H, Ar—CH_3), 2.66 (s, 3H, —SCH_3), 7.37 (d, 2H, Ar—H), 7.92 (d, 2H, Ar—H)
4c	2.69 (s, 3H, —SCH_3), 7.85 ~ 8.29 (m, 4H, Ar—H)
4d	2.67 (s, 3H, —SCH_3), 7.73 (d, 2H, Ar—H), 7.90 (d, 2H, Ar—H)
4e	2.68 (s, 3H, —SCH_3), 8.27 (d, 2H, Ar—H), 8.43 (d, 2H, Ar—H)
4f	1.08 (t, 3H, —CH_3), 2.67 (s, 3H, —SCH_3), 4.05 (q, 2H, —CH_2—), 7.22 ~ 7.99 (m, 4H, Ar—H)

①Purified by column chromagraphy.

1.3 除草活性测定

对合成的所有化合物用“油菜根长法”进行了除草活性测定。将适量药样溶于适当溶剂和少量乳化剂中，用蒸馏水逐级稀释至所需浓度配成乳剂。吸取已配制好的乳剂 2.5mL 放入垫有两张直径 5.6cm 滤纸的培养皿中，然后把经浸泡的优选油菜种子 20 粒播于皿中，置于 (26±1)℃暗培养室中生长，48h 后测量油菜根长（与空白对照）。每个药样分 4～5 个剂量测定，然后计算出 I_{C50}，每个剂量重复 2 次，空白对照重复 3 次。测定结果见表 2。

Table 2 The I_{C50} of the title compounds

No.	I_{C50}	r①	No.	I_{C50}	r①
3a	1.2411×10^{-4}	0.9968	4a	2.7877×10^{-4}	0.9981
3b	2.0311×10^{-4}	0.9967	4b	2.4541×10^{-4}	0.9976
3c	7.8392×10^{-5}	0.9950	4c	3.6759×10^{-4}	0.9973
3d	5.3000×10^{-4}	0.9922	4d	1.8214×10^{-4}	0.9953
3e	9.7334×10^{-5}	0.9910	4e	1.6407×10^{-4}	0.9998
3f	6.0703×10^{-5}	0.9900	4f	4.1380×10^{-4}	0.9951

①Regression parameter.

2 结果与讨论

2.1 化合物 2 及 3，4 的合成特点

化合物 **2** 的合成参考文献[7]方法做了如下改进，直接使用含 4 个结晶水的二（钠硫基）甲叉丙二腈作原料，使用前不需在苯回流减压下用 P_2O_5 干燥脱结晶水，同样得到了比较理想的产率。这样大大节省了时间，减少了能耗。

在目标产物的合成中，以氯仿作溶剂及三乙胺作缚酸剂条件下，较理想地得到所期望的缩合产物（产率见表 1）。化合物 **3** 比化合物 **4** 反应所需温度要低，反应时间也

较短，说明前者比较容易，这可能是化合物**3**的空间效应和电子效应都对反应有利之故。

2.2 产物的波谱特性

在化合物**3**的^{1}H NMR中，可以明显地观察到5－位氢与4－位氢的相互裂分，分别形成两重峰，其耦合常数是4.8～5.0Hz。文献[8]认为：3－羟基异噻唑与酰氯发生酰基化时可能发生酰基的转移，形成*N*－酰化产物，此时，4－位氢与5－位氢之间的耦合常数为6.2～7.3Hz；如果没有发生酰基迁移，即酰化发生在氧原子上，则耦合常数为4.2～5.3Hz。我们的^{1}H NMR结果似乎表明磺酰化发生在氧原子上，但由于酰基和磺酰基的不同可能导致耦合常数的差异。因此，并不能从^{1}H NMR完全断定磺酰化位置发生在氧原子上。对于4－氰基－5－甲硫基－3－羟基异噻唑的磺酰化，由于异噻唑杂环的4位和5位均被取代，因此更不能从^{1}H NMR谱中断定磺酰化的位置。

为了确证目标化合物的结构，我们对化合物**3**c和**4**b的单晶进行了X射线衍射分析，结果清楚地表明磺酰化均发生在氧原子上，亦即产物的结构与目标化合物的结构一致[9]。

IR测定主要特征峰为（$\tilde{\nu}$/cm^{-1}）：3100～3000（s，C—H）；1660（s，杂环C＝C）；1600，1580（s，芳环C＝C）；1400～1350，1200～1195（s，SO_2—O）。对化合物**4**，在2220cm^{-1}处有1个中等强度吸收，为氰基的特征吸收峰，证实了杂环的存在。由于在IR中没有出现特征的羰基吸收峰，这也证实了反应过程中没有发生磺酰基迁移。

化合物的质谱特点是：能出现明显的分子离子峰，对其碎片分析归属表明，在质谱条件下，极易脱去1个分子的SO_2而成为基峰。

2.3 除草活性的规律

化合物**3**和**4**的除草活性呈现以下规律：具有未取代异噻唑环结构的目标产物比取代异噻唑的活性均要高，当其芳基磺酰基的邻位有硝基或乙氧羰基取代时，活性最高；但杂环被取代时，芳基磺酰基部分的邻位有取代基时活性最低。这个规律正好与赖城明教授等[10]提出的ALS酶抑制剂的“卡口模型”一致，说明这类化合物可能是ALS酶抑制剂。为了证实这一新的发现，我们对这类化合物进行了抑制ALS酶活性的测定。结果表明，这类化合物对ALS酶抑制效果明显，从而确证这类化合物是ALS酶抑制剂。

参考文献

[1] Adams A., Slack R.. Chem. & Ind. (London), 1956: 1232.
[2] Yagi M., Nakane K., Hiyane Y.. Eur. Pat. Appl. EP, 648415, 1995.
[3] Davis R. H., Krummel G.. Eur. Pat. Appl. EP, 832282, 1994.
[4] Burak K., Machon Z.. Pharmazie, 1992, 47 (7): 492.
[5] Lipnicka U., Regiec A., Machon Z.. Pharmazie, 1994, 49 (9): 652.
[6] FANG Ren－Ci（方仁慈），CHENG Jun－Ran（成俊然），LI Zheng－Ming（李正名）. CN－85－1－03016A, 1986.
[7] Hatchard W. R.. J. Org. Chem., 1963, 28 (8): 2163.

[8] Chan A. W. K. , Crow W. D. . Aust. J. Chem. , 1968, 21: 2967.
[9] YANG Xiao - Ping (杨小平), LI Zheng - Ming (李正名), WANG Hong - Gen (王宏根) et al. . Chinese Journal of Structural Chemistry (结构化学) (to be submitted) .
[10] LAI Cheng - Ming (赖城明), YUAN Man - Xue (袁满雪), LI Zheng - Ming (李正名) et al. . Chemical Journal of Chinese Universities (高等学校化学学报), 1994, 15 (5): 693.

Synthes is and Herbicidal Activity of 3 - Arylsulfonyloxy (substituted) isothiazoles

Xiaoping Yang, Zhengming Li, Lingxiu Wang, Yonghong Li

(Research Institute of Elemento - Organic Chemistry , Key Laboratory of Elemento - organic Chemistry , Nankai University , Tianjin, 300071)

Abstract To look for novel compounds with biological activity, 3 - arylsulfonyloxy (substituted) isothiazoles were synthesized by the condensation of 3 - hydroxy (substituted) isothiazoles with arylsulfonylch lorides in the presence of base. The strctures of target compounds were confirmed by elemental analysis, ^1H NMR, IR, MS. Also, the single crystal structures of compounds **3**c and **4**b were determined. All the facts ascertained that the sulfonylization position was oxygen atom of isothiazole heterocycle. The herbicidal activities of title compounds were determined and the result showed that it was possible that they were ALS inhibitors.

Key words 3 - Arylsulfonyloxy (substituted) isothiazole; synthesis; herbicidal activity

应用 CoMFA 研究磺酰脲类化合物的三维构效关系*

刘 洁 李正名 王 霞 马 翼 赖城明 贾国锋 王玲秀

（南开大学元素有机化学研究所，元素有机化学国家重点实验室，天津 300071）

摘 要 用比较分子力场分析（CoMFA）方法研究了磺酰脲类除草剂结构与活性的三维定量构效关系（3D－QSAR），并在此基础上对原有分子结构进行修饰改造，以获得较高活性的化合物。并在三维等值线图的基础上得到了此类除草剂作用靶标 ALS 酶的模拟作用模型，为进一步合成出全新结构的 ALS 酶抑制剂的先导化合物提供了有益的启示和帮助。

关键词 CoMFA 磺酰脲 3D－QSAR ALS 酶

磺酰脲除草剂由开发至今已有20年的历史，它作为超高效低毒的新型除草剂受到了人们的欢迎。它通过抑制体内 ALS 酶的活性，破坏支链氨基酸的合成，达到除草作用。但由于 ALS 酶的具体三维结构尚未得到，无法确切了解药物与受体的作用机制，因而对此类分子结构的改进和研究就受到了一定限制[1]。

本文从三维角度出发，利用美国 Tripos 公司开发的先进的三维分子模拟软件包 SYBYL/6.22，计算方法则采用处理3D－QSAR 问题中典型的比较分子力场分析（CoMFA）方法，对本室新近合成的共38个不同活性的磺酰脲分子进行三维构效关系的研究，得到其三维等值线图，可清晰观测出与受体分子可能的结合点，并据此对所改造的分子结构进行活性预测，获得了活性较高的新磺酰脲分子结构。并在等值线图的基础上继续利用 Leapfrog 软件，得到了 ALS 酶模拟作用模型，为今后设计出全新结构的 ALS 酶抑制剂的先导化合物提供了有益的帮助。

1 计算方法

比较分子力场分析（CoMFA）方法是近年来兴起的一种研究药物－受体三维定量构效关系（3D－QSAR）的新方法。它的核心内容是：在分子水平上，药物分子与受体之间的相互作用主要是通过非共价作用力如 Van der waals 力、静电相互作用实现的。作用于同一受体的一系列药物分子，它们与受体之间的这两种作用力场应该有一定的相似性。因此，在不了解受体三维结构的情况下，研究这些药物分子周围两种作用力场的分布，把它们与药物分子的生物活性定量地联系起来，既可以推测受体的某些性质，又可依此建立一个模型来设计新的化合物，并定量地预测新分子的药效强度[2]。具体的实现过程是：将同一系列分子确定药效构象后，在三维网格点中叠合定位，计算网格点上探针原子与它们之间的静电能及立体场相互作用能，得到的数据用偏最小二乘法（PLS）分析，经交叉验证和非交叉验证得到预测模型。其3D－QSAR 的结果以三维等值线图表示，据此可指导新分子的设计，并可用所得模型预测所设计分子的活性。

* 原发表于《中国科学》（B辑），1998，28（1）：60－64。

2　结果与讨论

2.1　磺酰脲分子的合成及活性测定

我们参照文献［3，4］，合成了系列新磺酰脲化合物，其化学结构经元素分析和谱图测定，其通式为：

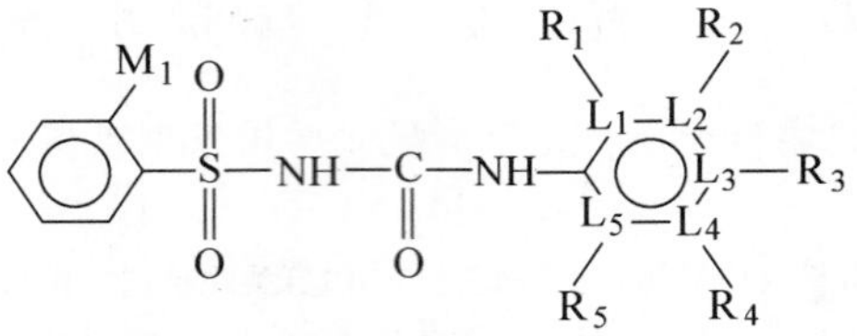

L_i=N or C, M_1, R_i=H, CH_3, Cl···

各分子的生物活性数据由本所生测室提供，即用“油菜根长法”测定了它们的 IC_{50}（抑制油菜根生长50%所需药剂的摩尔浓度）[5,6]，据此得到 $PI_{50} = -\lg IC_{50}$。通过计算发现，用 CoMFA 法研究系列化合物的构效关系时，所选分子的活性应该有恰当的分布。当选取化合物的活性集中于某一区域时，不容易得到好的计算结果，即模型的预报能力不佳。故在本文中我们所选择化合物的活性范围较宽，其 PI_{50} 落于 3.08 ~ 7.58 之间的 38 个化合物作为研究对象。

2.2　药效构象的获得

药效构象就是药物分子与受体结合时所采取的空间构象，它是获得正确的药物分子与受体作用模型的基础。选择合理药效构象也是 CoMFA 分析的关键步骤。药效构象的确定有各种方法，最合理的方法是参用已知药物分子与受体结合的 X 射线衍射结构数据。我们曾对此系列分子中较具代表性的一个化合物培养了单晶[7]，以 X 射线测定了结构，考虑到磺酰脲分子最终是进入溶液中以自由分子形式与受体相互作用，所以我们将其用 Tripos 力场优化后（相当于自由分子）作为药效构象应该更合理一些，这个分子也就是进行 CoMFA 分析的模板分子。其他没有晶体数据的同类磺酰脲分子则以上述药效构象为基础进行局部改造，然后再经优化而得。用这种方法所获得的全部 38 个磺酰脲分子的药效构象，是比较接近实际情况的。

2.3　PLS 分析结果

PLS 分析中，首先进行交叉验证（leave - one - out），得到 $r^2 = 0.809$，（r^2 用来衡量模型的预测能力，当 $r^2 > 0.5$ 时通常认为模型可以接受），说明所建立的模型体系有较好的预测可靠性。然后用所有分子，5 个主成分数进行非交叉验证回归，最终结果为：$r^2 = 0.959$，$F = 155.393$（F 用来判断统计回归的显著性），$s = 0.263$（s 为标准偏差）。以上结果表明此模型对同系列分子的活性有较高预报能力。总的结果见表 1。

表 1　所研究化合物的生物活性

No.	$PIC_{50\ obsd}$ ①	$PIC_{50\ calc}$ ②	$\Delta PIC_{50\ calc}$	$PIC_{50\ pred}$ ③	$\Delta PIC_{50\ pred}$
1	7.58	7.36	0.22	6.94	0.64
2	7.07	7.13	−0.06	6.78	0.29

续表 1

No.	$PIC_{50\ obsd}$ ①	$PIC_{50\ calc}$ ②	$\Delta PIC_{50\ calc}$	$PIC_{50\ pred}$ ③	$\Delta PIC_{50\ pred}$
3	6.56	6.28	0.28	5.59	0.97
4	6.44	6.36	0.08	6.65	-0.21
5	6.42	6.50	-0.08	6.07	0.35
6	6.38	6.18	0.20	5.68	0.70
7	6.27	6.01	0.26	5.39	0.88
8	6.10	6.55	-0.45	6.62	-0.52
9	6.10	5.88	0.22	5.00	1.10
10	5.90	5.89	0.01	5.40	0.50
11	5.39	5.14	0.25	4.67	0.72
12	5.15	5.40	-0.25	5.20	-0.05
13	5.13	5.26	-0.13	5.83	-0.70
14	4.39	4.19	0.20	4.49	-0.10
15	4.33	3.98	0.35	3.91	0.42
16	4.31	4.61	-0.30	5.00	-0.69
17	4.29	4.56	-0.27	4.30	-0.01
18	4.28	4.29	-0.01	4.38	-0.10
19	4.24	4.13	0.11	4.45	-0.21
20	4.13	4.11	0.02	4.12	0.01
21	4.07	3.84	0.23	3.60	0.47
22	4.05	3.93	0.12	3.46	0.59
23	3.97	3.86	0.11	4.33	-0.36
24	3.97	3.86	0.11	3.74	0.23
25	3.95	4.17	-0.22	4.30	-0.35
26	3.90	4.34	-0.45	4.64	-0.74
27	3.87	3.53	0.34	3.46	0.41
28	3.84	3.92	-0.08	3.85	-0.01
29	3.83	4.30	-0.47	4.46	-0.63
30	3.78	3.52	0.26	3.52	0.26
31	3.77	3.42	0.35	4.21	-0.44
32	3.70	3.88	-0.18	4.03	-0.33
33	3.44	3.55	-0.11	4.15	-0.71
34	3.43	3.85	-0.42	4.39	-0.96
35	3.39	3.34	0.05	3.54	-0.15
36	3.35	3.34	0.01	3.61	-0.26
37	3.13	3.52	-0.39	3.85	-0.72
38	3.08	2.87	0.21	3.19	-0.11

①生物活性实验值；②非交叉验证计算值；③交叉验证计算值。

2.4 三维等值线图显示 QSAR 结果

经 CoMFA 法分析，我们可以得到一个彩色的三维轮廓图形式，它用不同颜色来表示此系列分子主要空间和静电性能（图 1 和图 2），从图上可以看出：（1）图 1 表示围绕在模板分子周围的立体相互作用（黄色和绿色区域），增加绿色区域或减少黄色区域的空间体积有助于增加化合物的除草活性。因此，分子中苯环邻位取代基稍远处黄色区域表明该位取代基的立体位阻不要超过此位置。（2）图 2 表示围绕在模板分子周围的静电相互作用（红色和蓝色区域），在蓝色区域增加正电荷，红色区域增加负电荷有助于增加化合物的除草活性。因此，分子中苯环邻位取代基附近的红色区域表明该位取代基的第 1 个或第 2 个原子应为电负性较强的原子，杂环附近蓝色区域提示我们在分子、杂环的间位取代基选择带正电的原子，这将有利于提高分子的生物活性。

根据上述结果，我们有选择地对部分分子进行结构修饰和改造，经预测，其活性较高，目前正在合成之中。

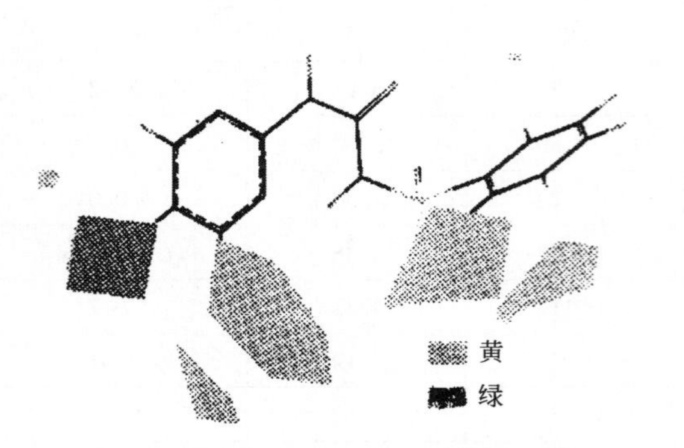

图 1　CoMFA 模型立体场三维等值线图

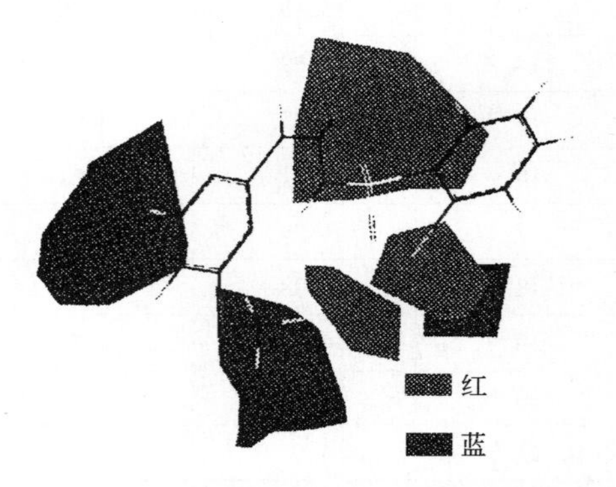

图 2　CoMFA 模型静电场三维等值线图

根据上述结果，我们有选择地对部分分子进行结构修饰和改造，经预测，其活性较高，目前正在合成之中。

2.5 靶标 ALS 酶模拟作用模型的获得

我们在三维等值线图的基础上，继续运用 Leapfrog 软件进行处理，得到了此类除草剂作用靶标 ALS 酶的模拟作用模型（图 3），它以轮廓图形式表示受体模型的立体场和静电场特征，然后根据一定的受体和配体之间的作用条件，合理地筛选出符合条件的结构碎片连接起来，据此可设计出全新结构的 ALS 酶抑制剂的先导化合物。

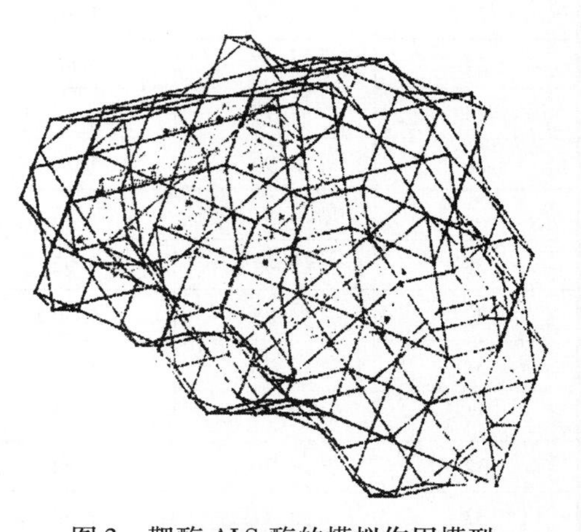

图 3　靶酶 ALS 酶的模拟作用模型

参 考 文 献

［1］ Li Zhengming. Chemistry of novel bio - regulating substances. J Pesticide Science，1996，21（1）：124.

［2］ Cramer R D Ⅲ，Patterson D E，Bunce J D. Comparative molecular field analysis（CoMFA）. 1. Effect of shape on binding of steroids to carrier proteins. J Am Chem Soc，1988，110：5959.

[3] Franz J E , Osuch C. The reactions of sulfonamides with oxalyl chloride. J Org Chem, 1964, 29: 2592.
[4] Levitt G. Herbicidal sulfonamides. US patent , 4127405 , 1978 - 11 - 28.
[5] Swanson C P. A simple bioassay method for the determination of low concentration of 2 , 4 - D in aqueous solutions. Bot Gaz , 1946 , 107: 507.
[6] Ready D , Gramt V Q. A rapid sensitive method for determination of low concentration of 2 , 4 - D in aqueous solution. Bot Gaz , 1947 , 109: 39.
[7] 李正名，贾国锋，赖城明，等．新磺酰脲类化合物的合成、结构及构效关系研究（I），N - (2′ - 嘧啶基) -2 - 甲酸乙酯 - 苯磺酰脲的晶体及分子结构．高等学校化学学报，1992 ，13 (11)：1411.

2，5－二吡唑基－1，3，4－噁二唑的一锅煮法合成及生物活性*

陈寒松　李正名

（南开大学元素有机化学研究所，天津　300071）

关键词　1，3，4－噁二唑　一锅煮法合成　生物活性

1，3，4－噁二唑类化合物具有广泛的生物生理活性，如杀菌[1]、除草[2]、杀虫[3]和消炎[4]等，而2，5－二芳基－1，3，4－噁二唑类化合物一般具有昆虫生长调节活性，1980年美国Dow公司报道了2，5－双（2，4－二氯苯基）－1，3，4－噁二唑具有良好的昆虫生长调节活性[5]。1，3，4－噁二唑类化合物有多种合成方法，花文廷等[6]曾对其作过详细总结：（1）先由甲酰肼和酰氯反应得双酰肼，然后由双酰肼在脱水剂的作用下关环可得，常用的脱水剂有硫酸、乙酸酐、三氯氧磷、多聚磷酸、多聚磷酸酯等；（2）先由甲酰肼得到酰氨基硫脲类化合物，然后在氧化剂（$Hg(OAc)_2$或KI/I_2等）的作用下合环而成；（3）先由甲酰肼得到酰腙类化合物，然后由氧化剂（$Pb(OAc)_4$等）氧化合环而成，最常用的方法是第一种。合成1，3，4－噁二唑的方法虽多，但步骤都较长，如果以甲酰肼作为起始物，以上3种方法至少需要2步或2步以上才能得到目标产物。我们考虑到三氯氧磷既可作酰化试剂，又可作脱水剂，可以将酰肼和羧酸放在三氯氧磷溶剂中回流，用一锅煮的方法得到1，3，4－噁二唑。我们用这种方法，从5－吡唑甲酸和5－吡唑甲酰肼直接得到了具有新颖结构的2，5－二吡唑基－1，3，4－噁二唑：

Et, X, N, N, Me, $CONHNH_2$ (1) + Et, X′, N, N, Me, COOH (2) $\xrightarrow[\text{Reflux}]{POCl_3}$ Et, X, N, N, Me, N—N, O, X′, Et, N, N, Me (3)

3a.X=Cl，X′=Cl；**3**b.X=Cl，X′=H；**3**c. X=H，X′=H

1　实验部分

Yanaco CHN CORDER MT－3型自动元素分析仪；BRU KER AC－P200型核磁共振仪，溶剂为$CDCl_3$，TMS为内标；Yanaco MP－500型熔点仪，温度计未经校正。

柱层析硅胶（青岛海洋化工厂）100～140目。三氯氧磷经重蒸处理。

化合物**1**、**2**按文献[7]方法合成。

* 原发表于《高等学校化学学报》，1998，19（4）：572－573。国家自然科学基金（批准号：29232010）资助课题。

将4mmol 5－吡唑甲酰肼、4.2mmol 5－吡唑甲酸和6mL三氯氧磷放入装有干燥管的三口瓶中，回流6h，减压抽除过量的三氯氧磷，剩余物倾入冰水混合物中，放置过夜。将滤出的固体先用稀氢氧化钠溶液洗涤，再用水洗至中性，烘干，减压柱层析（淋洗剂：石油醚/乙酸乙酯）分离出纯品。

2 结果与讨论

化合物**3a**－**3c**均经元素分析和^{1}H NMR验证，其物理常数列在表1中，^{1}H NMR数据列在表2中。实验结果表明，3个化合物的产率基本稳定在60%左右。在这个反应中，三氯氧磷既是酰化剂，又是脱水剂，同时还是反应溶剂，对其纯度要求较高，必须现蒸现用，若放置时间较长，则产率明显降低。该方法的后处理很容易进行，先用稀氢氧化钠溶液洗涤以除去酸，再用水洗除去氢氧化钠，烘干，即可进行柱层析。

Table 1 The physical data of compounds 3a－3c

Compd	Formula	m. p.（℃）	Yield（%）	Elementary anal（%，Cacld）		
				C	H	N
3a	$C_{14}H_{16}Cl_2N_6O$	151～152	58	47.30（47.34）	4.32（4.54）	23.63（23.66）
3b	$C_{14}H_{17}Cl_2N_6O$	81～82	60	52.22（52.42）	5.19（5.34）	26.01（26.20）
3c	$C_{14}H_{18}N_6O$	51～52	55	58.42（58.72）	6.25（6.34）	29.17（29.35）

Table 2 ^{1}H NMR data of compounds 3a－3c（$CDCl_3$）

Compd	^{1}H NMR（δ）
3a	1.27（6H，t，CH_3），2.70（4H，q，CH_2），4.26（6H，s，N－CH_3）
3b	1.35（6H，m，CH_3），2.77（4H，m，CH_2），4.35（6H，2s，N－CH_3），6.83（1H，s，pyrazole－H_4）
3c	1.25（6H，t，CH_3），2.66（4H，q，CH_2），4.25（6H，s，N－CH_3），6.68（1H，s，pyrazole－H_4）

对化合物**3a**－**3c**进行了初步的生物活性测试。用活体小株法作杀菌试验，结果表明：**3b**在0.05%浓度下对水稻纹枯病菌具有抑制作用，抑制率为70%；用盆栽法进行除草试验，结果表明：**3b**在1500g/ha浓度下，无论是茎叶处理还是土壤处理，对苋菜生长均有抑制作用，抑制率分别为60.7%和57.9%。化合物**3a**和**3c**也有一定的杀菌和除草活性。

参考文献

[1] Farbenfabriken Bayer A G. Ger. Patent，1956510，1971.
[2] Uniroyal Inc.. Canadian Patent，986123，1976.
[3] Qian X H，Zhang R J. Chem. Tech. Biotechnol.，1996，67：124.
[4] Raman K，Singh H K，Salzman S K，*et al.* J. Pharm. Sci.，1993，82（2）：167.
[5] Arrington J P，Wade L L. U S Patent，4215129，1980.
[6] YANG Jun（杨骏），HUA Wen－Ting（花文廷）. Chemistry（化学通报），1996，9：18.
[7] LI Zheng－Ming（李正名），CHEN Han－Song（陈寒松），ZHAO Wei－Guang（赵卫光）*et al.* Chem. J. Chinese Universities（高等学校化学学报），1997，18（11）：1794.

One Pot Synthes is and Biological Activity of 2, 5 – Dipyrazolyl – 1, 3, 4 – oxadiazoles

Hansong Chen, Zhengming Li

(Institute of Elemento – Organic Chemistry, Nankai University, Tianjin, 300071)

Abstract 2, 5 – Dipyrazolyl – 1, 3, 4 – oxadiazoles were synthesized from 5 – pyrazole form ic acid and its hydrazide by one pot synthetic method in the presence of phosphorus oxychloride. The method shortened synthetic route and simplified experimental operation. The preliminary biological activity tests showed that compound **3**b has fungicidal and herbicidal activity.

Key words 1, 3, 4 – oxadiazole; one pot synthesis; biological activity

2－芳甲酰氨基硫羰基－3－异噻唑酮及4－氰基－5－甲硫基－2－芳甲酰氨基硫羰基－3－异噻唑酮的合成与生物活性*

杨小平　李正名　陈寒松　刘　洁　李树正

（南开大学元素有机化学国家重点实验室，南开大学元素有机化学研究所，天津　300071）

摘　要　通过3－羟基异噻唑（**4**）和4－氰基－5－甲硫基－3－羟基异噻唑（**5**）分别与芳酰基异硫氰酸酯（**3**）反应，合成标题化合物2－芳甲酰氨基硫羰基－3－异噻唑酮（**6**a－**6**f）和4－氰基－5－甲硫基－2－芳甲酰氨基硫羰基－3－异噻唑酮（**7**a－**7**e），对此反应及产物的结构特点进行了探讨。对部分化合物进行了杀菌活性测定，结果表明个别化合物具有一定活性。

关键词　异噻唑　酰基硫脲　合成　杀菌活性

在寻找结构新颖具有生物活性的化合物中，设计和合成各种杂环化合物是非常重要的途径之一，业已成为相当活跃的领域[1]。近年来，含吡啶、三唑、吡唑、嘧啶、咪唑、噻唑以及稠杂环的优越的新农药品种不断问世[2]。相比之下，1956 年第一次出现的新型杂环化合物[3]异噻唑（单核）杂环，受关注的程度比其他杂环要低得多。80年代末以来，人们对低毒、高效的药用和农用化学品更加重视，由于异噻唑杂环具有对人体低毒的特点[4]及含异噻唑杂环化合物具有比较好的生物活性，因此，近年来对于异噻唑杂环化合物进行结构修饰，以期找到比较理想的生物活性化合物引起人们的重视，出现了良好的发展势头[5]。这一发展趋势引起了我们的极大兴趣，我们比较系统地开展了这方面的研究，前文[6]报道了含异噻唑杂环磺酸酯的研究结果，本文报道标题化合物的合成和生物活性的研究结果。

已有许多研究表明，酰基硫脲类化合物具有较好的生物活性[7,8]，为此，我们通过如下步骤比较巧妙地把酰基硫脲官能团引入异噻唑杂环中，得到了 11 个结构新颖的化合物：

R—C₆H₄—COOH (**1**) —$SOCl_2$→ R—C₆H₄—C(O)—Cl (**2**) —KSCN / 无水乙腈→ R—C₆H₄—C(O)—NCS (**3**)

3 + 3－羟基异噻唑 (**4**) → **6a－6f**；**3** + 4－氰基－5－甲硫基－3－羟基异噻唑 (**5**) → **7a－7e**

R = H(**6**a,**7**a)；4－MeO(**6**b,**7**b)；2－Cl(**6**c,**7**c)；4－F(**6**d,**7**d)；

4－NO_2(**6**e,**7**e)；2－NO_2(**6**f)

* 原发表于《高等学校化学学报》，1999，20（3）：395－398。国家自然科学基金（批准号：29232010）资助课题。

产物的结构经元素分析、^{1}H NMR 谱及红外吸收光谱证实，产物的^{1}H NMR 很有特点。初步生物活性测定表明，这类化合物具有一定的杀菌活性。

1 实验部分

1.1 仪器和试剂

溶剂均为分析纯，用前经纯化处理。试剂为化学纯，硅胶（青岛海洋化工厂生产）。

元素分析用 Yanaco CHN corder MT－3 型自动分析仪测定，^{1}H NMR 用 Bruker AC－P200 型仪器测定（TMS 作内标）；红外光谱用 Shimadzu－IR 435 型仪器测定，KBr 压片；熔点用 Yanaco 熔点仪测定，所用温度计未经校正。

1.2 化合物6 的合成

化合物 **4** 和 **5** 按文献［5］方法合成。芳酰基异硫氰酸酯 **3** 按文献［9］方法合成，按计量得到含 3mmol 化合物 **3** 的无水乙腈滤液后（无水乙腈用前经五氧化二磷干燥处理），不经脱溶和分离，直接往滤液中一次性加入含 0.3g（3mmol）3－羟基异噻唑 **4** 的无水乙腈混合溶液 5mL，在室温下搅拌 1h 后，85℃回流 2h。冷却后，有少量橙红色固体生成。过滤，滤液在不断搅拌下慢慢倾入 100mL 水中。有黄色沉淀析出，放置过夜，抽滤，得固体。固体用丙酮/石油醚（体积比 2∶5）混合溶剂进行重结晶，得 **6a**－**6f**。化合物 **6a**－**6f** 的物理常数、收率及元素分析值见表 1，^{1}H NMR 数据见表 2。

1.3 化合物7 的合成

7a－**7e** 的合成一般步骤同 **6a**－**6f**，只是因为 4－氰基－5－甲硫基－3－羟基异噻唑 **5** 在无水乙腈中溶解性不好，改用 DMF 作溶剂。化合物 **7a**－**7e** 的物理常数、收率及元素分析值见表 1，^{1}H NMR 数据见表 2。

Table 1 Physical properties and data of elemental analysis for the title compounds

No.	Molecular formula	Molecular weight	m. p. (℃)	Physical character	Yield (%)	Elemental anal. (Calcd., %) C	H	N
6a	$C_{11}H_8N_2O_2S_2$	264.24	214.0～215.5	Yellow needle	56.8	50.28 (50.02)	3.24 (3.03)	10.41 (10.60)
6b	$C_{12}H_{10}N_2O_3S_2$	294.12	230.0～232.0	Yellow rhombus	47.7	49.10 (49.00)	3.06 (3.40)	9.40 (9.52)
6c	$C_{11}H_7ClN_2O_2S_2$	298.69	176.5～178.0	Yellow powder	49.2	44.40 (44.24)	2.62 (2.34)	9.18 (9.38)
6d	$C_{11}H_7FN_2O_2S_2$	282.24	205.0～207.0	Yellow powder	52.3	46.80 (46.81)	2.32 (2.48)	9.98 (9.92)
6e	$C_{11}H_7N_3O_4S_2$	309.24	198.0～199.5	Yellow rhombus	64.1	42.70 (42.72)	2.30 (2.26)	13.60 (13.58)
6f	$C_{11}H_7N_3O_4S_2$	309.24	210.0～212.0	Yellow rhombus	68.1	42.71 (42.72)	2.26 (2.26)	13.69 (13.58)

续表 1

No.	Molecular formula	Molecular weight	m. p. (℃)	Physical character	Yield (%)	Elemental anal. (Calcd. ,%)		
						C	H	N
7a	$C_{13}H_9N_3O_2S_3$	335. 19	139. 0 ~ 141. 0	Yellow powder	75. 2	46. 58 (46. 58)	2. 69 (2. 68)	12. 44 (12. 53)
7b	$C_{14}H_{11}N_3O_3S_3$	365. 33	212. 0 ~ 213. 5	Brown powder	72. 6	46. 42 (46. 02)	2. 98 (3. 01)	11. 48 (11. 50)
7c	$C_{13}H_8ClN_3O_2S_3$	369. 77	170. 0 ~ 172. 0	Brown powder	68. 6	42. 32 (42. 22)	2. 08 (2. 16)	11. 30 (11. 36)
7d	$C_{13}H_8FN_3O_2S_3$	353. 32	165. 0 ~ 167. 0	Yellow powder	52. 6	44. 08 (44. 19)	2. 28 (2. 26)	11. 78 (11. 89)
7e	$C_{13}H_8N_4O_4S_3$	380. 32	218. 0 ~ 219. 5	Brown powder	65. 5	41. 08 (41. 05)	2. 06 (2. 10)	14. 58 (14. 73)

Table 2 ^{1}H NMR (δ, DM SO, d_6) for the title compounds

No.	δ
6a	6. 63 (d, 1H, J = 10. 6Hz), 7. 51 ~ 8. 11 (m, 6H, —ArH, C=C—H)
6b	3. 94 (s, 3H, —OCH_3), 6. 62 (d, 1H, J = 10. 2Hz), 7. 30 ~ 8. 17 (m, 5H, —ArH, C=C—H)
6c	6. 67 (d, 1H, J = 10. 3Hz), 7. 53 ~ 7. 68 (m, 4H, —ArH), 8. 07 (d, 1H, J = 10. 3Hz)
6d	6. 64 (d, 1H, J = 10. 5Hz), 7. 60 ~ 8. 30 (m, 5H, —ArH, C=C—H)
6e	6. 63 (d, 1H, J = 10. 6Hz), 7. 60 ~ 8. 30 (m, 5H, —ArH, C=C—H)
6f	6. 63 (d, 1H, J = 10. 6Hz), 7. 80 ~ 8. 30 (m, 4H, —ArH, C=C—H)
7a	2. 71 (s, 3H, —SCH_3), 7. 62 ~ 8. 17 (m, 5H, —ArH)
7b	2. 74 (s, 3H, —SCH_3), 3. 94 (s, 3H, —OCH_3), 7. 30 ~ 8. 20 (m, 4H, —ArH)
7c	2. 75 (s, 3H, —SCH_3), 7. 50 ~ 7. 66 (m, 4H, —ArH)
7d	2. 75 (s, 3H, —SCH_3), 7. 00 ~ 8. 20 (m, 4H, —ArH)
7e	2. 74 (s, 3H, —SCH_3), 7. 80 ~ 8. 03 (m, 4H, —ArH)

2 结果与讨论

2.1 合成

按文献［9］的合成方法可以顺利地得到芳酰基异硫氰酸酯。芳酰基异硫氰酸酯与3-羟基异噻唑反应，需以无水乙腈为溶剂，在回流条件下进行。从TLC检测原料消失的快慢看，芳酰基环上吸电子基的存在对此加成反应有利，这表明反应快速步骤是3-羟基异噻唑对芳酰基异硫氰酸酯的亲核加成。

2.2 结构特征

在化合物**6**的^{1}H NMR中，4位氢与5位氢之间的耦合常数在10Hz左右，这一结果有力地支持结构**6**，否则，此值应在5. 3Hz以内[6]。

化合物**6a**-**6f**的4位氢和5位氢的耦合常数达10Hz，而异噻唑环上氮原子连有羰基时相应值为 6.2 ~ 7.3Hz[10,11]，因而上述值异常高。我们认为当异噻唑环的氮原子与硫羰基（而非羰基）相连时，由于硫原子电子云较发散，故与异噻唑环共平面程度低，导致异噻唑环的刚性较小，从而导致偶合常数增大[12]。在化合物 **6**a – **6**f 和 **7**a – **7**e 的 IR 谱中，在 $1670cm^{-1}$ 和 $1603cm^{-1}$ 左右有两个强吸收峰存在，这是目标化合物中的两个羰基伸缩振动吸收位置，因而 IR 谱进一步证实了化合物 **6**a – **6**f，**7**a – **7**e 的结构。

2.3 生物活性

我们对大部分化合物用离体平皿法和活体小株法进行了杀菌活性的测定，结果见表 3。

Table 3 The assay results of fungic idal activity of the title compounds

	In vivo (5.0×10^{-5})					In vivo (500×10^{-6})		
No.	Gibberella zeae	A lternaria solani	Rhizoctonia solani	Physalo sporapiricola	Cercospora arachidicola	Puccinia graminis	Botrytis cinerea	Sclertinia sclertiorum
6a	23.8	0	52.8	40.0	9.1	20	16.6	0
7a	19.0	0	41.7	20.0	18.2	10	16.6	26.4
6b	19.0	0	33.3	20.0	9.1	10	40.5	52.9
6c	14.3	0	44.4	26.6	0	10	9.5	0
6d	4.8	0	8.3	13.3	0	0	9.5	0
7c	23.8	0	38.9	40.0	27.3	20	59.5	76.0

从测定结果看，这些化合物均具有一定的杀菌活性，此结果与我们的预想一致。在离体平皿法的测定结果中，在质量分数为 0.005% 下，对棉花立枯的效果为 40% 左右，但对番茄早疫根本无效，这表明这类化合物抑菌活性具有一定的选择性。在活体小株法的测试中，化合物 **7**c 在质量分数为 0.05% 下对油菜菌核的抑制率达 76%，对黄瓜灰霉抑制率为 59%。化合物 **6**b 在质量分数为 0.05% 下对油菜菌核的抑制率约为 52.9%，对黄瓜灰霉抑制率为 40.5%。活性测试结果表明，异噻唑环上的取代基对活性影响不大。

参 考 文 献

[1] CHEN Wan – Yi（陈万义）. Research and Development of New Pesticides（新农药研究与开发），Beijing：Chem. Industry Publishing House，1996：24.

[2] WANG Da – Xiang（王大翔），BO Zai – Su（柏再苏）. Reviews of New Pesticides（新农药论丛），Pesticide Special Committee of Academic Society of China Chemical Industry，1995：1.

[3] Adams A.，Slack R.. Chem. & Ind.（London），1956：1232.

[4] Hubennett，Flock F. H.，Hansel W. *et al.*. Angew. Chem.，Intern. E. Engl.，1963，2：714.

[5] YANG Xiao – Ping（杨小平）. Doctoral Thesis of Nankai University（南开大学博士论文），1997：3.

[6] YANG Xiao – Ping（杨小平），LI Zheng – Ming（李正名），WANG Ling – Xiu（王玲秀）*et al.*. Chem. J. Chinese Universities（高等学校化学学报），1998，19（2）：228.

[7] Goswami B. N.. J. Heterocyclic Chem.，1984，21：182.

[8] Mishra U. K.. J. Indian Chem. Soc.，1983，60（9）：867.

[9] FENG Xiao - Ming（冯小明）. Chem. J. Chinese Universities（高等学校化学学报），1992，13（2）：187.
[10] Chan A. W. K.，Grow W. D.，Gosney I.. Tetrahedron，1970，26：2497.
[11] Chan A. W. K.，Grow W. D.. Aust. J. Chem.，1968，21：2967.
[12] CHEN Yao - Zu（陈耀祖）. Organic Analyses（有机分析），Beijing：Advanced Educational Publishing House，1981：664.

Syn theses and Biolog ical Activities of 2 - Arylamidothioacyl - 3 - isothiozolones and 4 - Cyano - 5 - methylth io - 2 - arylam idothioacyl - 3 - isothiozolones

Xiaoping Yang，Zhengming Li，Hansong Chen，Jie Liu，Shuzheng Li

(Research Institute of Elemento - organic Chemistry，Key Laboratory of Elemento - Organic Chemistry，Nankai University，Tianjin，300071)

Abstract Novel compounds，2 - arylam idothioacyl - 3 - isothiozolones **6** and 4 - cyano - 5 - methylthio - 2 - arylamidothioacyl - 3 - isothiozolones **7** were synthesized by reacting of 3 - hydrox - y isothiozo les（**4**，**5**）with aryl isothiocyanates **3**. The unique coupling con stants of 4 - H and 5 - H of isothiazo lering of compounds **6** were discovered and explained according to the coplanarity of isothiazolering with thioacyl group. The preliminary test of fungicidal activity showed that some of the title compounds possessed promising activity.

Key words isothiozole；acylthiourea；synthesis；fungicidalactivity

The Design and Synthesis of ALS Inhibitors from Pharmacophore Models*

Jie Liu, Zhengming Li, Han Yan, Lingxiu Wang, Junpeng Chen

(Elemento – Organic Chemistry Institute, State Key Laboratory of Elemento – Organic Chemistry, Nankai University, Tianjin, 300071, China)

Abstract In search of new ALS inhibitors without the previous knowledge of receptor crystal structure, DISCO module was applied to produce 3D – pharmacophore models, which provided information to design novel molecules by 3D – database searching. Then a number of molecules were synthesized. Several of them have some ALS inhibitory activities.

Introduction

Following with the rapid development of computer technology, Computer Aided Molecular Design, (CAMD) has become a focus of attention in assisting the molecular design of some novel agrochemicals.

The research on the ALS enzyme inhibitors has been carried out consistently in our Laboratory. Since it is difficult to obtain ALS receptor crystal to clarify the relative mode of action, which restrain further development of new inhibitors. So a series of ligand molecules were calculated by using CoMFA method, and the 3D – QSAR results were obtained[1].

In order to explore new ALS inhibitors, another new DISCO module was adopted which translated the information of two – dimensional structures of active compounds into that of the three – dimension. Then the possible pharmacophore models could be displaced, which provide further information to design new molecules by 3D – database searching[2]. Afterwards, some molecules could be modified, synthesized and bio – assayed. In this way, the cycle of seeking potential leads can be found by using *Ligand Based Drug Design*, which is a promising method to contrive novel agrochemicals efficiently.

Method

The 3D – pharmacophore models were built by using the molecular software package SYBYL/6.22[3] on Silicon Graphics Worksations. The 3D – database searching was carried by using ISIS/3D chemical information management software.

Aiming at the ALS enzyme inhibitors, we selected 9 compounds, which are different in structures and all known act on ALS enzyme with high activities[4] (Scheme 1). These compounds constituted a set of molecular database. After the DISCO module ran, molecule d6 was selected

* Reprinted from *Bioorganic & Medicinal Chemistry Letters*, 1999, 9: 1927 – 1932. This project was supported by the National Natural Science Foundation of China (Granted Number: 29772679) and the State Key Laboratory of Elemento – Organic Chemistry.

d1 d4 d7

d2 d5 d8

d3 d6 d9

Scheme 1

as the reference compound by system. In addition, we also selected molecule dl, which was synthesized in our Laboratory, as the reference. Then two classes 3D – pharmacophore models were obtained. The process followed as: Firstly, two kinds of preferential models were chosen (Figure 1, 2). Secondly, their information was changed into 3D – query[5] (Figure 3, 4). And last, based on 3D – query, relevant molecules were searched from the ACD – 3D molecule database. Thus we obtained some molecules (Scheme 2), which have similar chemical features but are different in structures. Then molecule Ⅱ and I were chosen to modify and synthesize.

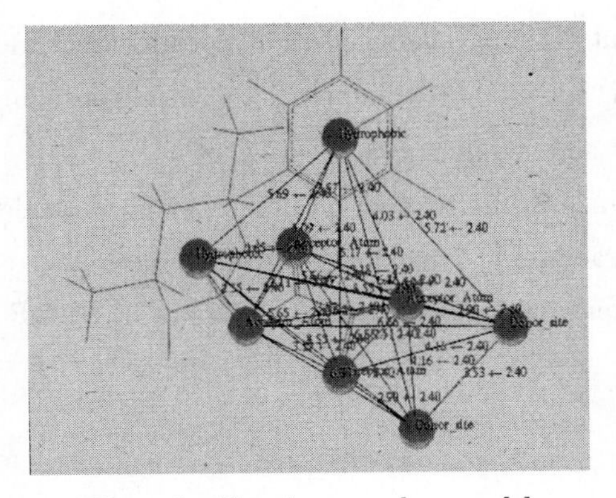

Figure 1 The pharmacophore model (d6 as reference)

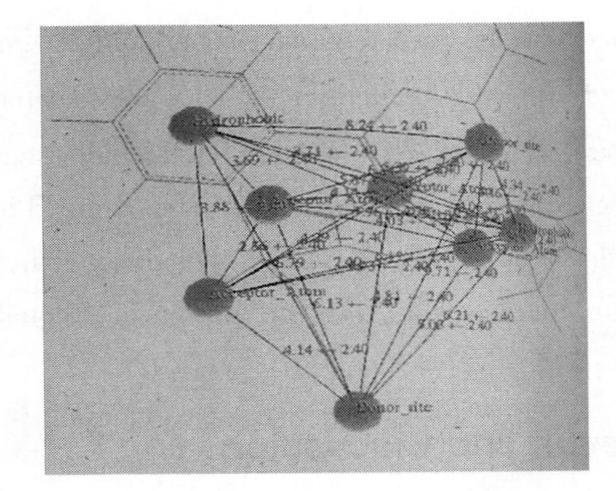

Figure 2 The pharmacophore model (dl as reference)

Synthesis

Since we have obtained the structure A from search result, we synthesized nine benzene sulfonic amide compounds I to screen for herbicidal lead (Scheme 3). The reaction of benzene sulfonyl chloride with different heterocyclic amines in pyridine gave the corresponding compounds I.

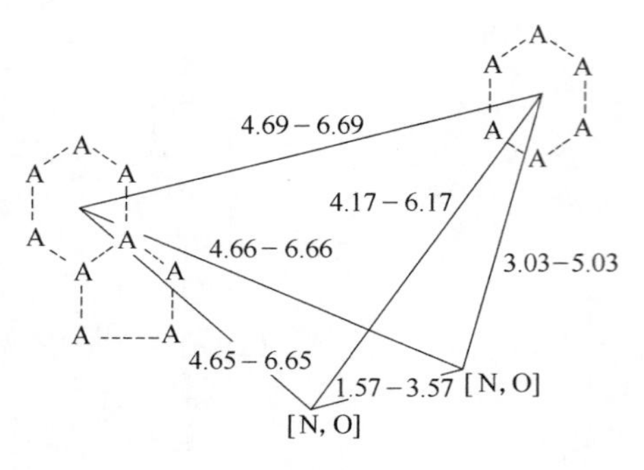

Figure 3 The 3D - query from the pharmacophore model(d6 as reference)

Figure 4 The 3D - query from the pharmacophore model(d1 as reference)

(A means any atom;(u) marks the atom with unsaturated bonds;⋯means any type of bond)

A B

Scheme 2

Scheme 3

In addition, we also selected structure B, which belongs to Uracil derivative. Some N^3 substituted uracils were reported as herbicides[6,7]. But N^1 long chain substituted compounds were rarely reported as herbicides. Thus we selected N^1 substituted structure Ⅱ as staring structure, modified group NH to atom O as bioisoster[8] and synthesized a series of compounds Ⅱ (Scheme 4). The synthesis of compounds Ⅱ began with the reaction of uracil(a) and formaldehyde at 60℃ to give 1,3 - bishydroxymethyl uracil(b). Then the substituted benzoic acids were coupled with(b) in the presence of equal equiv DCC and a catalytic amount of DAMP in dry nitrile[9].

Result and Discussion

Preliminary bioassays were carried out on two category compounds. Their biological activity data IC_{50} were calculated by using the rape - root growth method[10,11] (Table 1,2).

The results of biological tests indicated that some compounds synthesized have some extent of ALS inhibiting activities. That implied a starting point of designing novel structures of potential ALS inhibitors using CAMD without the previous knowledge of exact structure of the target enzyme. The exploring work still needs to be optimized in future.

HN, O, H, O, N, H, O, H_2C, H N, R ⟹ HN, O, H, O, N, H, O, CH_2 O, R

HN, O, H, O, N H, H + HCHO ⟶ $HOCH_2$ N, O, H, O, N, H, CH_2OH —R-C6H4-COOH, DCC, DMAP⟶ HN, O, H, O, N, H, CH_2—O—C(=O), R

(a) (b) II

Scheme4

Table 1 The biological activities of compounds I

No.	(Sub) Aryl		IC_{50} (mol/L)
	R	(Sub) Het	
I_a	o – NO_2	N, N, CH_3	2.87×10^{-4}
I_b	p – Br	N, N, CH_3	5.89×10^{-5}
I_c	p – F	N, N, CH_3	3.89×10^{-5}
I_d	2,5 – 2Cl	N, N, CH_3	9.72×10^{-5}
I_e	2,5 – 2Cl	N, N, CH_3, CH_3	3.30×10^{-5}
I_f	2,5 – 2Cl	N, S, CH_3	$>3.09 \times 10^{-4}$
I_g	p – F	N, CH_3	1.52×10^{-4}
l_h	2.5 – 2Cl	N, S	8.47×10^{-5}
I_i	p – F	N, N, Cl, OCH_3	1.11×10^{-4}

Table 2 The biological activities of compounds Ⅱ

No.	R	IC_{50} (ppm)
$Ⅱ_a$	m - CH_3	16.04
$Ⅱ_b$	p - F	55.86
$Ⅱ_c$	H	55.36
$Ⅱ_d$	p - NO_2	>100
$Ⅱ_e$	p - I	>100
$Ⅱ_f$	p - Cl	51.29
$Ⅱ_g$	o - OCH_3	>100
$Ⅱ_h$	p - Br	55.38
$Ⅱ_i$	2,6 - 2Cl,4 - CF_3	>100
$Ⅱ_j$	p - OCH_3	>100

Acknowledgements

Thanks to the assistance to this project from Prof. Kaixian CHEN and Dr. Jianzhong CHEN, Shanghai Institute of Materia Medica.

References and notes

[1] Liu Jie; Li Zhengming; Wang Xia et al. *Science in China (Series B)* 1998, 41(1), 50.

[2] Yvonne C. Martin; Mark G. Bures; Elizabeth A. Danaher, Jerry DeLazzer *Journal of Computer - Aided Molecular Design* 1993, 7, 83.

[3] Tripos Inc., 1699 South Hanley Road, St. Louis, Missouri 63144, USA.

[4] Shimizu, T.; Nakayama, I.; Wada, N. and Abe, H. *J. Pest. Sci.* 1994, 19, 257.

[5] Douglas R. Henry and Osman F. G. http://www. ch. ic. ac. uk/ectoc/papers/guner/.

[6] Lutz Albert W. U. S. Patent 3,635,977, 1972; *Chem. Abstr.* 1972, 76, 113244v.

[7] Enomoto Masayuki; Takemura Susumu et al. Eur. Pat. Appl. EP 540,023, 1993; *Chem. Abstr.* 1993, 119, 139259k.

[8] Toshio Fujita *Biosci. Biotech. Biochem.* 1996, 60(4), 557.

[9] Suhail Ahmad; Shoichiro Ozaki; Toshio Nagase et al. *Chem. Pharm. Bull.* 1987, 35(10), 4137.

[10] Swanson, C. P. *Bot. Gaz.* 1946, 107, 507.

[11] Ready, D.; Gramt, V. Q. *Bot. Gaz.* 1947, 109, 39.

新磺酰脲类化合物的合成、结构及构效关系研究（Ⅵ）*

——*N*－［2－（4－甲基）嘧啶基］－2－甲酸甲酯－苄基磺酰脲的晶体及分子结构

姜　林　李正名　翁林红　冷雪冰

（南开大学元素有机化学研究所，天津　300071）

摘　要　合成了标题化合物 C_6H_4（CO_2CH_3）$CH_2SO_2NHCONHC_4N_2H_2CH_3$（$C_{15}H_{16}N_4O_5S$，$Mr$ = 364.38），用X射线晶体衍射法测定了其晶体结构。它属单斜晶系，空间群为 $P2_1/c$，a = 10.220（1），b = 9.938（1），c = 17.246（2）Å，β = 106.26（2）°，V = 1681.6（3）Å^3，Z = 4，D_c = 1.439Mg/m^3，μ = 0.227mm^{-1}，F（000）=760。结构由直接法解出，全矩阵最小二乘法修正，最终偏离因子 R = 0.0498，wR = 0.1327（$I > 2\sigma(I)$），独立可观测点数为3448。分子中的嘧啶磺酰脲、苯环及酯基分别形成3个独立的平面共轭体系。

关键词　磺酰脲　晶体结构　分子结构

磺酰脲类化合物是一类超高效、低毒除草剂，自70年代末发现以来，大约有20种该类除草剂投入市场。在已商品化的磺酰脲除草剂中，绝大多数化合物的分子中苯（杂）环与磺酰基直接相连，如氯磺隆、豆草隆和氟嘧磺隆，而苄嘧磺隆与其不同，在苯环和磺酰基之间插入了1个亚甲基[1]。为研究苄嘧磺隆及其类似物的结构和构象对除草活性的影响，合成了标题化合物，测定了它的晶体结构，并比较了它与有关化合物分子结构的差异。

1　实验

1.1　标题化合物的合成[2,3]

向反应瓶中加入1.8g 2－甲氧羰基苄基磺酰胺，15mL无水甲苯和3.3g草酰氯，缓慢加热至回流，反应32h。冷却，过滤除去固体，滤液减压蒸去草酰氯及甲苯，残留物滴加至0.6g 4－甲基－2－氨基嘧啶的15mL乙腈中，搅拌3h，滤出棕黄色沉淀，用1，2－二氯乙烷重结晶，得产物1.64g，产率57.2%，m.p. 201～203℃。元素分析，实测值：C，49.18；H，4.89；N，15.77。计算值：C，49.44；H，4.45；N，15.48. IR（cm^{-1}）：3399.0（N—H），1712.1（C═O），1343.4、1146.9（—SO_2—）。^{1}H NMR（$CDCl_3$）：δ 2.28（3H，s，—CH_3），3.91（3H，s，—CO_2CH_3），4.86（2H，s，—CH_2—），6.48（1H，d，[嘧啶环5-H]），8.12（1H，d，[嘧啶环6-H]），7.58～

* 原发表于《结构化学》，2000，19（2）：149－152。国家自然科学基金资助课题。

7.91（4H，m，Ar）。MS：m/z 364（M^+，5%）。

1.2 晶体结构的测定

将标题化合物溶于乙腈，在室温下自然挥发，得到 $C_{15}H_{16}N_4O_5S$ 棕黄色晶体。选取 0.30mm×0.25mm×0.20mm 的单晶用于 X 射线衍射实验，实验温度为 293（2）K。在 BRUKER SMART 1000 衍射仪上，用经石墨单色器单色化的 Mo$K\alpha$ 射线（λ = 0.71073Å），以 $\omega-2\theta$ 扫描方式在 $2.08° < \theta < 26.41°$ 范围内，收集 3448 个独立衍射点（R_{int} = 0.0523），其中 2424 个可观察的衍射点。全部强度数据用 SADABS 程序校正，晶体结构由直接法（SHELXS-97）解出，根据 E 图确定了大部分非氢原子的位置，剩余非氢原子坐标用数轮差值 Fourier 合成陆续确定。对全部非氢原子坐标及各向异性热参数进行全矩阵最小二乘法修正。最终结果为，R = 0.0498，wR = 0.1327（$I > 2\sigma(I)$），$W = 1/[S^2(F_o^2) + (0.1000P)^2 + 0.0000P]$，其中 $P = (F_o^2 + 2F_c^2)/3$，$(\Delta/\rho)_{max}$ = 0.0001，S = 0.991。在最终差电子云图中，$(\Delta\rho)_{max}$ = 0.347 e·Å^{-3}，$(\Delta\rho)_{min}$ = −0.353 e·Å^{-3}。全部计算用 BRUKER SAINT 程序完成。

2 结果与讨论

标题化合物的立体结构（不包括氢原子）见图 1，晶胞堆积图见图 2，全部非氢原子坐标及热参数列于表 1，分子键长及键角、扭转角分列于表 2～表 4。

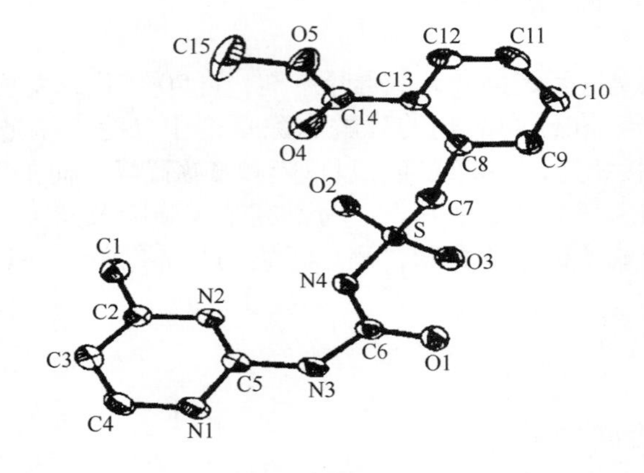

Fig. 1 Molecular structure of title compound

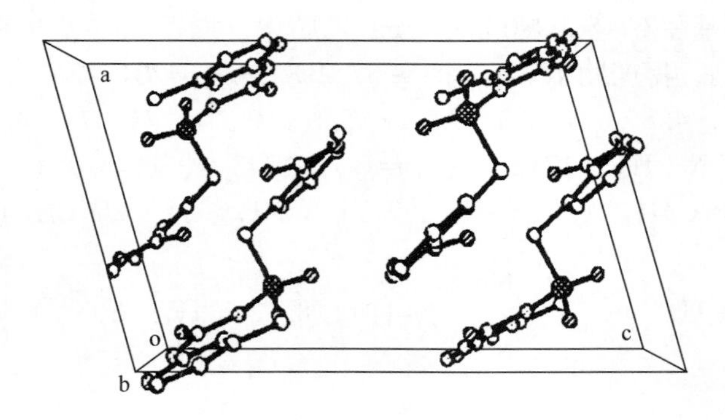

Fig. 2 Packing of the molecules in the unit cell

Table 1　Atomic Coordinates and Thermal Parameters (Å^2)

Atom	x	y	z	U_{eq}	Atom	x	y	z	U_{eq}
S	0.7407(1)	0.7896(1)	0.7025(1)	0.041	C(4)	0.9714(3)	0.2366(3)	0.9096(2)	0.056
O	0.8703(2)	0.8059(2)	0.8836(1)	0.062	C(5)	0.9198(2)	0.4518(2)	0.8714(1)	0.042
O(2)	0.7062(2)	0.7162(2)	0.6286(1)	0.055	C(6)	0.8676(2)	0.6983(2)	0.8514(1)	0.046
O(3)	0.8258(2)	0.9047(2)	0.7097(1)	0.057	C(7)	0.5886(2)	0.8357(3)	0.7269(1)	0.045
O(4)	0.4083(2)	0.6215(2)	0.6496(1)	0.084	C(8)	0.4949(2)	0.9080(2)	0.6556(1)	0.041
O(5)	0.3322(3)	0.6442(2)	0.5179(1)	0.095	C(9)	0.5052(3)	1.0467(2)	0.6523(2)	0.052
N(1)	0.9744(2)	0.3646(2)	0.9313(1)	0.051	C(10)	0.4290(3)	1.1196(3)	0.5878(2)	0.062
N(2)	0.8657(2)	0.4193(2)	0.7941(1)	0.044	C(11)	0.3389(3)	1.0537(3)	0.5243(2)	0.063
N(3)	0.9224(2)	0.5838(2)	0.8948(1)	0.046	C(12)	0.3272(2)	0.9158(3)	0.5262(1)	0.055
N(4)	0.8118(2)	0.6764(2)	0.7700(1)	0.049	C(13)	0.4040(2)	0.8407(3)	0.5913(1)	0.044
C(1)	0.8123(3)	0.2578(3)	0.6865(2)	0.067	C(14)	0.3847(3)	0.6925(3)	0.5915(2)	0.054
C(2)	0.8666(2)	0.2890(2)	0.7738(2)	0.048	C(15)	0.3028(5)	0.5020(4)	0.5111(3)	0.140
C(3)	0.9192(3)	0.1927(3)	0.8319(2)	0.058					

注：$U_{eq} = (1/3)\sum_i \sum_j U_{ij} a_i^* a_j^* a_i \cdot a_j$

Table 2　Selected Bond Lengths (Å)

Bond	Dist.	Bond	Dist.	Bond	Dist.	Bond	Dist.
S-O(2)	1.425(2)	O(5)-C(14)	1.323(3)	N(3)-C(5)	1.371(3)	C(7)-C(8)	1.512(3)
S-O(3)	1.422(2)	O(5)-C(15)	1.443(4)	N(3)-C(6)	1.391(3)	C(8)-C(9)	1.384(3)
S-N(4)	1.635(2)	N(1)-C(4)	1.324(3)	N(4)-C(6)	1.376(3)	C(8)-C(13)	1.401(3)
S-C(7)	1.780(2)	N(1)-C(5)	1.343(3)	C(1)-C(2)	1.484(4)	C(13)-C(14)	1.486(4)
O(1)-C(6)	1.202(3)	N(2)-C(2)	1.341(3)	C(2)-C(3)	1.382(4)		
O(4)-C(14)	1.194(3)	N(2)-C(5)	1.333(3)	C(3)-C(4)	1.367(4)		

Table 3　Selected Bond Angles (°)

Angle	(°)	Angle	(°)	Angle	(°)
O(1)-C(6)-N(3)	121.7(2)	N(1)-C(5)-N(3)	114.9(2)	C(7)-C(8)-C(9)	117.9(2)
O(1)-C(6)-N(4)	124.2(2)	N(2)-C(2)-C(1)	116.0(2)	C(7)-C(8)-C(13)	123.0(2)
O(2)-S-O(3)	118.8(1)	N(2)-C(2)-C(3)	120.6(2)	C(8)-C(7)-S	109.1(2)
O(2)-S-N(4)	103.6(1)	N(2)-C(5)-N(3)	119.7(2)	C(8)-C(9)-C(10)	121.8(2)
O(2)-S-C(7)	109.2(1)	N(3)-C(6)-N(4)	114.1(2)	C(8)-C(13)-C(12)	118.8(2)
O(3)-S-N(4)	110.7(1)	N(4)-S-C(7)	104.6(1)	C(8)-C(13)-C(14)	122.0(2)
O(3)-S-C(7)	109.0(1)	C(1)-C(2)-C(3)	123.4(2)	C(9)-C(8)-C(13)	118.9(2)
O(4)-C(14)-O(5)	121.6(3)	C(2)-C(3)-C(4)	117.1(2)	C(9)-C(10)-C(11)	119.5(2)
O(4)-C(14)-C(13)	126.3(2)	C(2)-N(2)-C(5)	117.6(2)	C(10)-C(11)-C(12)	119.7(2)
O(5)-C(14)-C(13)	112.1(2)	C(4)-N(1)-C(5)	115.5(2)	C(11)-C(12)-C(13)	121.3(3)
N(1)-C(4)-C(3)	123.7(2)	C(5)-N(3)-C(6)	130.8(2)	C(12)-C(13)-C(14)	119.2(2)
N(1)-C(5)-N(2)	125.4(2)	C(6)-N(4)-S	126.3(2)	C(14)-O(5)-C(15)	116.9(3)

Table 4 A Part Values of Torsion Angles (°)

O (2) -S-N (4) -C (6)	117.5 (2)	C (8) -C (7) -S-O (3)	78.0 (2)
O (3) -S-N (4) -C (6)	49.2 (2)	C (8) -C (7) -S-N (4)	-163.6 (2)
C (7) -S-N (4) -C (6)	-68.1 (2)	C (9) -C (8) -C (7) -S	-90.8 (2)
C (8) -C (7) -S-O (2)	-53.3 (2)	C (13) -C (8) -C (7) -S	85.8 (2)

由表5中最小二乘平面Ⅰ的有关数据可以看出，嘧啶环和磺酰脲基部分基本在一个平面上（不包括磺酰基上的氧原子）。由表2可知，磺酰脲基上的C—N键短于C—N单键（1.470 Å）而长于C＝N双键（1.270 Å），C（5）—N（3）—C（6），C（6）—N（4）—S夹角分别为130.8°和126.3°，均远大于有机脂肪胺中C—N—C夹角（约108°，N原子 sp^3 杂化），说明N原子已具有某些 sp^2 杂化的特征；S—N键长为1.635Å，小于二者的共价半径之和（1.780 Å），这表明S—N键具有部分双键成分。由此推知，嘧啶磺酰脲骨架形成包括S原子在内的多电子π键（π_{11}^{12}）。此外，分子中还存在苯环平面Ⅱ和酯基平面Ⅲ，见表5。这三个平面分别形成3组离域π键，它们对于电子的传递、反应性能，进而对农药的活性均产生重要的影响[4]。

Table 5 Least - squares Planes and Deviations from Them (Å)

Plane	Equation	Rms deviation	Deviation distance
Ⅰ	9.974 (1) x + 1.385 (4) y - 7.488 (8) z = 3.244 (9)	0.0356	C (2) 0.0065, C (3) -0.0380, C (4) -0.0385, C (5) 0.0075, C (5) 0.0307, C (6) 0.0015, N (1) 0.0075, N (2) 0.0260, N (3) 0.0650, N (4) 0.0245, O (1) -0.0630, S -0.0224
Ⅱ	-8.989 (4) x + 0.871 (9) y + 11.990 (9) z = 4.171 (13)	0.0179	C (8) 0.0255, C (9) 0.0158, C (10) -0.0059C (11) -0.0136, C (12) -0.0072, C (13) 0.0146
Ⅲ	-10.025 (6) x + 1.751 (33) y + 6.096 (27) z = 0.958 (24)	0.0031	C (14) 0.0041, C (15) 0.0017, O (4) -0.0022, O (5) -0.0036
Angle	Ⅰ—Ⅱ: 22.35° Ⅱ—Ⅲ: 23.43° Ⅰ—Ⅲ: 18.63°		

本文用分子图形学方法加上H原子，估测N（4）上的H与N（2）之间的距离约为1.905Å，可形成分子内氢键N（2）…H—N（4）。IR谱中3399cm^{-1}变宽，3226cm^{-1}肩部，也说明了氢键的存在[5]。该氢键的3个原子与C（5）、N（3）和C（6）一起组成六元环。这一结构对保持嘧啶磺酰脲骨架的平面性，从而保证共轭体系的稳定存在，具有非常重要的意义[4]。

比较标题化合物与*N*－［2′－（4－甲基）嘧啶基］－2－甲酸乙酯－苯基磺酰脲的分子结构，发现它们的构象有很大不同，C—S—N—C的扭角值前者为－68.1°，而后者为60.6°，说明标题化合物引入亚甲基后，苯环发生了反方向的扭转。此外，还可看出它们的二面角也有较大差异，标题化合物分子3个平面间的二面角均较小，而后者Ⅰ—Ⅱ、Ⅱ—Ⅲ二面角较大，分别为96.9°和119.6°[6]。这些结构上的差异，可影响分子主平面（平面Ⅰ）与羰基氧之间形成的凹隙（即“卡口”）的大小，从而影响药物分子与受体的结合[7]，因此二者表现出不同的除草活性。

参 考 文 献

[1] 化工部农药信息总站. 国外农药品种手册（新版合订本），867～890.

[2] Franz J E, Osuch C. The Reactions of Sulfonam ides with Oxalyl Chloride. J Org. Chem., 1964, 29: 2592～2595.

[3] Sauers R F, Del H. Herbicidal Sulfonam ides. U S 4 420 325 (1983).

[4] 李正名，贾国锋，王玲秀等. 新磺酰脲类化合物的合成、结构及构效关系研究（Ⅰ）－N－(2′－嘧啶基)－2－甲酸乙酯－苯基磺酰脲的晶体及分子结构. 高等学校化学学报，1992，13 (11):1411～1414.

[5] 西尔弗斯坦 R M et al. 著，姚海文译. 有机化合物光谱鉴定（第二版）. 北京：科学出版社，1988. 116～117.

[6] 李正名，贾国锋，王玲秀，等. 新磺酰脲类化合物的合成、结构及构效关系研究（Ⅱ）－N－[2′－（4－甲基)嘧啶基] －2－甲酸乙酯－苯基磺酰脲的晶体及分子结构. 高等学校化学学报，1993，14 (3): 349～352.

[7] 赖城明，袁满雪，李正名，等. 磺酰脲除草剂分子与受体作用的初级模型. 高等学校化学学报，1994: 15 (5): 693～694.

Syntheses, Structures and SAR Study of Sulfonylurea Compounds (Ⅵ)

——Crystal and Molecular Structure of N－[2－(4－methyl) pyrim idinyl] －2－Methylcarboxylate－benzyl sulfonylurea

Lin Jiang, Zhengming Li, Linhong Weng, Xuebing Leng

(Institute of Elemento－Organic Chemistry, Nankai University, Tianjin, 300071)

Abstract The synthesis and crystal structure of the title molecule are reported, the crystal is monoclinic, space group $P\,2_1/c$, with unit cell parameters $a = 10.220$ (1), $b = 9.938$ (1), $c = 17.246$ (2) Å, $\beta = 106.26$ (2)°. $V = 1681.6$ (3) $Å^3$, $Z = 4$, $D_c = 1.439 mg/m^3$, $R = 0.0498$, $wR = 0.1327$ for 2424 reflections. The molecular structure is discussed. There are three different planes in themolecule, in each of which a conjugated system is formed. The conformation difference between this molecule and related sulfonylurea molecule is illustrated.

Key words sulfonylurea; crystal structure; molecular structure

Synthesis and Fungicidal Activity against *Rhizoctonia solani* of 2 – Alkyl(Alkylthio) – 5 – pyrazolyl – 1,3,4 – oxadiazoles(Thiadiazoles) *

Hansong Chen, Zhengming Li, Yufeng Han

(State Key Laboratory of Elemento – Organic Chemistry, Institute of Elemento – Organic Chemistry, Nankai University, Tianjin, 300071, China)

Abstract Some series of 2 – alkyl(alkythio) – 5 – ((4 – chloro) – 3 – ethyl – 1 – methyl – 1H – pyrazole – 5 – yl) – 1,3,4 – oxadiazoles(thiadiazoles) were prepared as potential fungicides. Their fungicidal activity was evaluated against rice sheath blight, which is a major disease of rice in China. Structure – activity relationships for the screened compounds were evaluated and discussed. It was found that 5 – (4 – chloro – 3 – ethyl – 1 – methyl – 1H – pyrazole – 5 – yl) – 1,3,4 – thiadiazole – 2 – thione has the higher fungicidal activity.

Key words pyrazolyl – 1,3,4 – oxadiazole; pyrazolyl – 1,3,4 – thiadiazole; fungicide; rice sheath blight; *Rhizoctonia solani*

1 Introduction

It is well – known that the study of pyrazole derivatives is significant in pesticide chemistry. Among the pyrazole derivatives some compounds are broadly used as insecticides, herbicides and fungicides, such as fripronil(MB46030) (Colliot et al. ,1992), tebufenpyrad(Okada et al. ,1988), pyrazosulfuron – ethyl(NC – 311) (Nissan Chemical Industries, Ltd. 1982), and so on.

However, a survey of the literature revealed that linked biheterocyclic compounds containing pyrazole to possess biological activity are seldom reported. 1,3,4 – Oxadiazole, 1,3,4 – thiadiazole derivatives all being fungicidal(Ashour et al. , 1994; Funatsukuri and Ueda, 1966), if they are linked to pyrazole ring respectively, the biheterocydic compounds obtained could have better fungicidal activity. As part of our program aimed at developing a new class of agrochemicals we synthesized the title compounds and studied their fungicidal activity.

2 Materials and methods

2.1 Synthetic procedures

All melting points were determined with Yanaco MP – 300 micro melting points apparatus, and the values are uncorrected. ^{1}H NMR spectra were recorded on a Bruker AC – P200 spectrometer, using tetramethylsilane(TMS) as an internal reference and $CDCl_3$ or CD_3COCD_3 as the solvent. Elemental analyses were performed on a Yanaco CHN CORDER MT – 3 instrument.

5 – (4 – Chloro – 3 – ethyl – 1 – methyl – 1H – pyrazole – 5 – yl) – 1,3,4 – oxadiazole – 2 – thione (**2a**). To a solution of compound **1a**(2. 35g, 12. 5mmol) in ethanol(50mL) at 0℃ were

---

* Reprinted from *Journal of Agriculture and Food Chemistry*, 2000, 48: 5312 – 5315. This project was supported by the National Natural Science Foundation of China(project No. 29832050).

added carbon disulfide(3.8g,50mmol) and potassium hydroxide(0.86g,12.5mmol). After addition, the mixture was refluxed for 7 h. The solvent was evaporated. The residue was dissolved in water and acidified with dilute hydrochloric acid. The precipitate was filtered and recrystallized from ethanol to give 2.4g (yield 70%) of **2a** as a white solid. Yield, 70%; mp, 199℃ (d). ^1H NMR($CDCl_3$)δ: 1.28(t,3H), 2.68(q,2H), 4.12(s,3H), 11.39(s,1H). Analysis found: C, 39.63; H, 3.78; N, 23.20. Calcd. for $C_8H_9ClN_4OS$: C, 39.26; H, 3.71; N, 22.90.

5-(3-Ethyl-1-methyl-1H-pyrazole-5-yl)-1,3,4-oxadiazole-2-thione(**2b**). Compound **2b** was prepared by the same method as **2a**. Yield, 76%; mp, 199-200 ℃. ^1H NMR (CD_3COCD_3)δ: 1.69(t,3H), 3.09(q,2H), 4.53(s,3H), 7.19(s,1H), 13.80(s,1H). Analysis found: C, 46.03; H, 4.64; N, 26.26. Calcd. for $C_8H_{10}N_4OS$: C, 45.70; H, 4.79; N, 26.65.

2-(4-Chloro-3-ethyl-1-methyl-1H-pyrazole-5-yl)-5-methylthio-1,3,4-oxadiazole(**3a**). A mixture of compound **2a**(0.61g, 2.5mmol), 1N aqueous sodium hydroxide solution(2.5 mL, 2.5mmol), and tetrabutylammonium bromide(0.1g) was stirred for several minutes, and then methyl iodide(0.35g, 2.5mmol) and toluene(20mL) were added. After the mixture was stirred for 24h at room temperature, the organic layer was separated and dried over magnesium sulfate. The solvent was removed in vacuo, and the residue was subjected to silica gel column chromatography to give 0.5g of **3a** as a white powder. Yield, 77%; mp, 85-86℃. ^1H NMR($CDCl_3$)δ: 1.25(t,3H), 2.66(q,2H), 2.77(s,3H), 4.16(s,3H). Analysis found: C, 41.81; H, 4.32; N, 21.53. Calcd. for $C_9H_{11}ClN_4OS$: C, 45.78; H, 4.25; N, 21.65.

Compounds **3b-3l** were prepared in the same method as **3a** by using the corresponding alkyl halide instead of methyl iodide.

5-(4-Chloro-3-ethyl-1-methyl-1H-pyrazole-5-yl)-2-propylthio-1,3,4-oxadiazole(**3b**). Yield, 80%; mp, 37-38℃. ^1H NMR($CDCl_3$)δ: 1.05(t,3H), 1.24(t,3H), 1.86(m,2H), 2.64(q,2H), 3.27(t,2H), 4.15(s,3H). Analysis found: C, 45.99; H, 5.16; N, 19.58. Calcd. for $C_{11}H_{15}ClN_4OS$: C, 46.07; H, 5.27; N, 19.54.

2-(4-Chloro-3-ethyl-1-methyl-1H-pyrazole-5-yl)-5-(2-methylpropylthio)-1,3,4-oxadiazole(**3c**). Yield, 17%; an oil. ^1H NMR($CDCl_3$)δ: 1.03(t,6H), 1.21(t,3H), 2.08(m,1H), 2.62(q,2H), 3.15(d,2H), 4.13(s,3H). Analysis found: C, 47.83; H, 5.54; N, 18.48. Calcd. for $C_{12}H_{17}ClN_4OS$: C, 47.91; H, 5.70; N, 18.63.

2-Amylthio-5-(4-chloro-3-ethyl-1-methyl-1H-pyrazole-5-yl)-1,3,4-oxadiazole(**3d**). Yield, 79%; an oil. ^1H NMR($CDCl_3$)δ: 0.87(t,3H), 1.20(t,3H), 1.36(m,4H), 1.86(m,2H), 2.61(q,2H), 3.23(t,2H), 4.12(s,3H). Analysis found: C, 49.31; H, 5.92; N, 17.57. Calcd. for $C_{13}H_{19}ClN_4OS$: C, 49.60; H, 6.08; N, 17.80.

2-(4-Chloro-3-ethyl-1-methyl-1H-pyrazole-5-yl)-5-heptylthio-1,3,4-oxadiazole(**3e**). Yield, 73%; an oil. ^1H NMR($CDCl_3$)δ: 0.84(t,3H), 1.08-1.46(m,11H), 1.81(m,2H), 2.63(q,2H), 3.25(t,2H), 4.14(s,3H). Analysis found: C, 52.68; H, 6.74; N, 16.18. Calcd. for $C_{15}H_{23}ClN_4OS$: C, 52.54; H, 6.76; N, 16.34.

2-(4-Chloro-3-ethyl-1-methyl-1H-pyrazole-5-yl)-5-octylthio-1,3,4-oxadiazole(**3f**). Yield, 76%; mp, <30℃. ^1H NMR($CDCl_3$)δ: 0.85(t,3H), 1.20-1.49(m,13H), 1.87(m,2H), 2.63(q,2H), 3.26(t,2H), 4.15(s,3H). Analysis found: C, 53.52; H,

6.96;N,15.38. Calcd. for $C_{16}H_{25}ClN_4OS$:C,53.84;H,7.06;N,15.70.

2-(4-Chloro-3-ethyl-1-methyl-1H-pyrazole-5-yl)-5-(1-ethoxycarbonylethylthio)-1,3,4-oxadiazole (**3g**). Yield,87%;an oil. ^1H NMR($CDCl_3$)δ:1.22(m,6H),1.69(d,3H),2.64(q,2H),4.12(s,3H),4.20(q,2H),4.45(q,1H). Analysis found:C,45.20;H,4.78;N,15.96. Calcd. for $C_{13}H_{17}ClNO_3S$:C,45.28;H,4.97;N,16.25.

2-(3-Ethyl-1-methyl-1H-pyrazole-5-yl)-5-methylthio-1,3,4-oxadiazole (**3h**). Yield,77%;mp,68-69℃. ^1H NMR($CDCl_3$)δ:1.21(t,3H),2.62(q,2H),2.72(s,3H),4.16(s,3H),6.55(s,1H). Analysis found:C,48.30;H,5.28;N,24.93. Calcd. for $C_9H_{12}N_4OS$:C,48.20;H,5.39;N,24.98.

2-(3-Ethyl-1-methyl-1H-pyrazole-5-yl)-5-ethylthio-1,3,4-oxadiazole (**3i**). Yield,73%;mp,34-35℃. ^1H NMR($CDCl_3$)δ:1.21(t,3H),1.47(t,3H),2.63(q,2H),3.26(q,2H),4.16(s,3H),6.55(s,1H). Analysis found:C,50.35;H,5.93;N,23.17. Calcd. for $C_{10}H_{14}N_4OS$:C,50.40;H,5.92;N,23.51.

2-(3-Ethyl-1-methyl-1H-pyrazole-5-yl)-5-propylthio-1,3,4-oxadiazole (**3j**). Yield,80%;an oil. ^1H NMR($CDCl_3$)δ:1.02(t,3H),1.21(t,3H),1.84(m,2H),2.62(q,2H),3.22(t,2H),4.16(s,3H),6.54(s,1H). Analysis found:C,52.20;H,6.19;N,22.47. Calcd. for $C_{11}H_{16}N_4OS$:C,52.36;H,6.39;N,22.20.

2-Amylthio-5-(3-ethyl-1-methyl-1H-pyrazole-5-yl)-1,3,4-oxadiazole (**3k**). Yield,71%;an oil. ^1H NMR($CDCl_3$)δ:0.89(t,3H),1.22(t,3H),1.38(m,4H),1.79(m,2H),2.64(q,2H),3.25(t,2H),4.17(s,3H),8.56(s,1H). Analysis found:C,55.49;H,6.96;N,20.08. Calcd. for $C_{13}H_{20}N_4OS$:C,55.69;H,7.19;N,20.08.

2-(3-Ethyl-1-methyl-1H-pyrazole-5-yl)-5-(1-ethoxycarbonylethylthio)-1,3,4-oxadiazole (**3l**). Yield,84%;an oil. ^1H NMR($CDCl_3$)δ:1.16(m,6H),1.64(d,3H),2.5(q,2H),4.16(s,3H),4.19(q,2H),4.38(q,1H),6.52(s,1H). Analysis found:C,49.96;H,5.62;N,17.84. Calcd. for $C_{13}H_{18}NO_3S$:C,50.03;H,5.84;N,18.05.

5-(4-Chloro-3-ethyl-1-methyl-1H-pyrazole-5-yl)-1,3,4-oxadiazole-2-one (**4**). To a solution of compound **3b** (0.59g,2.2mmol) in formic acid (98%,15mL) were added 30% hydrogen peroxide (1.0g,8.8mmol,1.1mL) and water (1mL). The mixture was stirred for 24 h at room temperature. The solvent was removed in vacuo, and the solid obtained was washed with water then recrystallized from ethanol to give 0.45g of **4** as a white solid. Yield,75%;mp,175-176℃. ^1H NMR($CDCl_3$)δ:1.25(t,3H),2.66(q,2H),4.07(s,3H),9.75(s,1H). IR:1780cm^{-1}. Analysis found:C,41.70;H,4.16;N,24.20. Calcd. for $C_8H_9ClN_4O_2$:C,42.03;H,3.97;N,24.50.

5-(4-Chloro-3-ethyl-1-methyl-1H-pyrazole-5-yl)-3-methyl-1,3,4-oxadiazole-2-one (**5**). To a solution of compound **4** (0.40g,1.7mmol) in THF (15mL) was added 1N aqueous sodium hydroxide (1.7mL,1.7mmol). The mixture was stirred for several minutes and then methyl iodide (0.5mL) was added. After stirring for 5h at room temperature, the solvent was concentrated under reduced pressure. The residue was dissolved in chloroform (20mL), washed with water two times (20×2), and then dried over magnesium sulfate. The solvent was removed in vacuo, and the residue was subjected to silica gel column chromatography to give 0.3g of **5** as a white powder. Yield, 73%; mp, 114 – 115℃. ^1H NMR($CDCl_3$) δ: 1.22(t,3H), 2.61(q,2H), 3.50(s,3H), 4.01(s,3H). IR: 1784cm^{-1}. Analysis found: C, 44.40; H, 4.75; N, 22.93. Calcd. for $C_9H_{11}ClN_4O_2$: C, 44.55; H, 4.57; N, 23.09.

2 – (4 – Chloro – 3 – ethyl – 1 – methyl – 1H – pyrazole – 5 – yl) – 5 – chloromethyl – 1,3,4 – oxadiazole(**6a**). To a solution of **1a**(0.72g, 3.5mmol) in $POCl_3$ (7mL) was added dry chloroacetic acid(0.33g, 3.5mmol). The mixture was heated under reflux for 7h. The excess $POCl_3$ was removed under reduced pressure, and the residue was poured into ice water and allowed to stand overnight. The solid obtained was washed with dilute aqueous sodium hydroxide and water respectively, dried, and purified by silica gel column chromatography to give 0.52g of **6a** as a white solid. Yield, 56%; mp, 100 – 101℃. ^1H NMR($CDCl_3$) δ: 1.25(t,3H), 2.67(q,2H), 4.20(s,3H), 4.79(s,2H). Analysis found: C, 41.53; H, 3.65; N, 21.44. Calcd. for $C_9H_{10}Cl_2N_4O$: C, 41.40; H, 3.86; N, 21.46.

Compounds **6b – 6d** were prepared in the same method as **6a** using the corresponding aliphatic acid instead of chloroacetic acid.

2 – (4 – Chloro – 3 – ethyl – 1 – methyl – 1H – pyrazole – 5 – yl) – 5 – trifluoromethyl – 1,3,4 – oxadiazole(**6b**). Yield, 71%; mp, 52 – 53℃. ^1H NMR($CDCl_3$) δ: 1.25(t,3H), 2.68(q,2H), 4.23(s,3H). Analysis found: C, 38.81; H, 2.68; N, 20.14. Calcd. for $C_9H_8ClF_3N_4O$: C, 38.52; H, 2.87; N, 19.96.

2 – (4 – Chloro – 3 – ethyl – 1 – methyl – 1H – pyrazole – 5 – yl) – 2 – ethyl – 1,3,4 – oxadiazole(**6c**). Yield, 57%; mp, 62 – 63℃. ^1H NMR($CDCl_3$) δ: 1.24(t,3H), 1.42(t,3H), 2.65(q, 2H), 2.96(q, 2H), 4.17(s, 3H). Analysis found: C, 49.94; H, 5.50; N, 23.36. Calcd. for $C_{10}H_{13}ClN_4O$: C, 49.90; H, 5.44; N, 23.28.

2 – Butyl – 5 – (4 – chloro – 3 – ethyl – 1 – methyl – 1H – pyrazole – 5 – yl) – 1,3,4 – oxadiazole(**6d**). Yield, 64%; mp, 43 – 44℃. ^1H NMR($CDCl_3$) δ: 0.95(t,3H), 1.23(t,3H), 1.46(m,2H), 1.83(m,2H), 2.64(q,2H), 2.91(t,2H), 4.16(s,3H). Analysis found: C, 53.34; H, 6.13; N, 20.82. Calcd. for $C_{12}H_{17}ClN_4O$: C, 53.63; H, 6.38; N, 20.85.

Potassium 3 – (4 – chloro – 3 – ethyl – 1 – methyl – 1H – pyrazole – 5 – carbonyl) dithiocarbazate(**7**). A solution of potassium hydroxide(2.06g, 0.03mol), absolute ethanol(50mL), and **1a**(6.10g, 0.03mL) was treated to the addition of carbon disulfide(3.43g, 0.06mol). The mixture was diluted with absolute ethanol(70mL) and stirred for 16h. It was then diluted with dry ether(100mL) and vacuum – dried at 65℃. This salt, prepared as described above, was obtained in nearly quantitative yield and was used in the next reaction without further purification.

5 – (4 – Chloro – 3 – ethyl – 1 – methyl – 1H – pyrazole – 5 – yl) – 1,3,4 – thiadiazole – 2 – thione(**8**). Potassium dithiocarbazate **7**(8.5g, 27mmol) was added in portions to concentrated H_2SO_4(35mL) at 0℃ over 40min. The mixture was stirred for 1h at room temperature and poured over crushed ice and allowed to stand overnight. The separated solid was dissolved in dilute aqueous sodium hydroxide. The insoluble solid was filtered and the filtrate was acidified with dilute hydrochloric acid. The solid obtained was washed with water, dried, and recrystallized from ethanol to give 4.5g of **8** as a light green solid. Yield, 64%; mp, 182 – 183℃. ^1H NMR(CD_3COCD_3) δ: 1.20(t,3H), 2.59(q,2H), 3.21(s,1H), 4.06(s,3H). Analysis found: C, 37.00; H, 3.21; N, 21.88. Calcd. for $C_8H_9ClN_4S_2$: C, 36.85; H, 3.48; N, 21.49.

5-(4-Chloro-3-ethyl-1-methyl-1H-pyrazole-5-yl)-2-methylthio-1,3,4-thiadiazole (**9a**). A mixture of compound **8** (0.52g, 2mmol), 1N aqueous sodium hydroxide solution (2mL, 2mmol), and tetrabutylammonium bromide (0.1g) was stirred for several minutes, and then methyl iodide (0.29g, 2mmol) and toluene (20mL) were added. After stirring for 24h at room temperature, the organic layer was separated and dried over magnesium sulfate. The solvent was removed in vacuo and the residue was subjected to silica gel column chromatography to give 0.42g of **9a** as a white powder. Yield, 83%; mp, 106-107℃. ^1H NMR ($CDCl_3$) δ: 1.25 (t, 3H), 2.65 (q, 2H), 2.63 (s, 3H), 4.22 (s, 3H). Analysis found: C, 39.18; H, 4.03; N, 20.15. Calcd. for $C_9H_{11}ClN_4S_2$: C, 39.34; H, 4.04; N, 20.39.

Compounds **9b-9e** were prepared in the same method as **9a** by using the corresponding alkyl halide instead of methyl iodide.

2-(4-Chloro-3-ethyl-1-methyl-1H-pyrazole-5-yl)-5-ethylthio-1,3,4-thiadiazole (**9b**). Yield, 85%; mp, 93-94℃. ^1H NMR ($CDCl_3$) δ: 1.25 (t, 3H), 1.49 (t, 3H), 2.65 (q, 2H), 3.39 (q, 2H), 4.23 (s, 3H). Analysis found: C, 41.53; H, 4.52; N, 19.26. Calcd. for $C_{10}H_{13}ClN_4S_2$: C, 41.59; H, 4.54; N, 19.40.

2-(4-Chloro-3-ethyl-1-methyl-1H-pyrazole-5-yl)-5-propylthio-1,3,4-thiadiazole (**9c**). Yield, 71%; mp, 59-60℃. ^1H NMR ($CDCl_3$) δ: 1.05 (t, 3H), 1.24 (t, 3H), 1.83 (m, 2H), 2.65 (q, 2H), 3.34 (t, 2H), 4.22 (s, 3H). Analysis found: C, 43.70; H, 4.91; N, 18.27. Calcd. for $C_{11}H_{15}ClN_4S_2$: C, 43.60; H, 4.99; N, 18.50.

2-Allylthio-5-(4-chloro-3-ethyl-1-methyl-1H-pyrazole-5-yl)-1,3,4-thiadiazole (**9d**). Yield, 83%; mp, 66-67℃. ^1H NMR ($CDCl_3$) δ: 1.25 (t, 3H), 2.67 (q, 2H), 3.99 (d, 2H), 4.23 (s, 3H), 5.28 (s×4, 2H), 5.99 (m, 1H). Analysis found: C, 43.77; H, 4.23; N, 18.41. Calcd. for $C_{11}H_{13}ClN_4S_2$: C, 43.92; H, 4.36; N, 18.62.

2-Amylthio-5-(4-chloro-3-ethyl-1-methyl-1H-pyrazole-5-yl)-1,3,4-thiadiazole (**9e**). Yield, 83%; mp, 54-55℃. ^1H NMR ($CDCl_3$) δ: 0.92 (t, 3H), 1.25 (t, 3H), 1.38 (m, 4H), 1.82 (m, 2H), 2.65 (q, 2H), 3.34 (t, 2H), 4.22 (s, 3H). Analysis found: C, 47.47; H, 5.72; N, 16.82. Calcd. for $C_{13}H_{19}ClN_4S_2$: C, 47.19; H, 5.79; N, 16.93.

2.2 Biological assay

Most of compounds were tested for control of rice sheath blight pathogen, *Rhizoctonia solani*, on rice seedlings at the fifth-leaf stage. The compounds were formulated in water and DMF (5+1 by volume) (containing 2.5g liter^{-1} Tween 80) to 500- and 100-mg-liter^{-1} solutions which were applied to the rice seedlings as foliar sprays using a hand-held spray gun. The next day the seedlings were inoculated with the chaff medium within *Rhizoctonia solani* (the causal fungus of the rice sheath blight). Then the plants were immediately placed in a temperature- and humidity-controlled chamber at 28℃ for 4 days. After treatment, percentage of disease control in the treated seedlings was compared to that of seedlings with a treatment in the absence of the experimental compounds, and fungicidal activity was estimated. Four replicates were included in the evaluation. For comparative purposes, the commercial fungicide Carbendazim was tested under the same conditions as the title compounds.

3 RESULTS AND DISCUSSION

3.1 Synthesis

In our previous paper, we reported the synthesis of 5 – pyrazole formic acid hydrazide(**1**) (Li et al., 1997). 5 – pyrazolyl – 1,3,4 – oxadiazole – 2 – thione compounds(**2**) were obtained via compounds **1** reacted with carbon disulfide and potassium hydroxide in refluxing ethanol; then alkylation of **2** afforded 2 – alkylthio – 5 – pyrazolyl – 1,3,4 – oxadiazoles(**3**) (Scheme 1).

Oxidation of 2 – ethylthio – 5 – pyrazolyl – 1,3,4 – oxadiazole by hydrogen peroxide in formic acid afforded 5 – pyrazolyl – 1,3,4 – oxadiazole – 2 – one(**4**), which was alkylated by methyl iodide to give 3 – methyl – 5 – pyrazolyl – 1,3,4 – oxadiazole(**5**) (Scheme 2).

The chemical literature records several synthetic routes leading to the formation of 3,5 – disubstituted – 1,3,4 – oxadiazole (Yang and Hua, 1996). In this paper, 2 – alkyl – 5 – pyrazolyl – 1,3,4 – oxadiazoles were synthesized from compound **1a** and aliphatic acid via a one – pot synthetic method in the presence of phosphorus oxychloride (Scheme 3).

The synthesis of 2 – alkylthio – 5 – pyrazolyl – 1,3,4 – thiadiazoles(**9**) is outlined in Scheme 4. 5 – pyrazole hydrazide(**1a**) reacted with carbon disulfide in ethanolic potassium hydroxide at room temperature to yield the potassium 1 – (4 – chloro – 3 – ethyl – 1 – ethyl – 1H – pyrazole – 5 – carbonyl) dithiocarbazates(**7**), which were cyclized in concentrated sulfuric acid to afford 5 – pyrazolyl – 1,3,4 – thiadiazole – 2 – thiones(**8**), which, when followed by alkylation, gave compounds **9**.

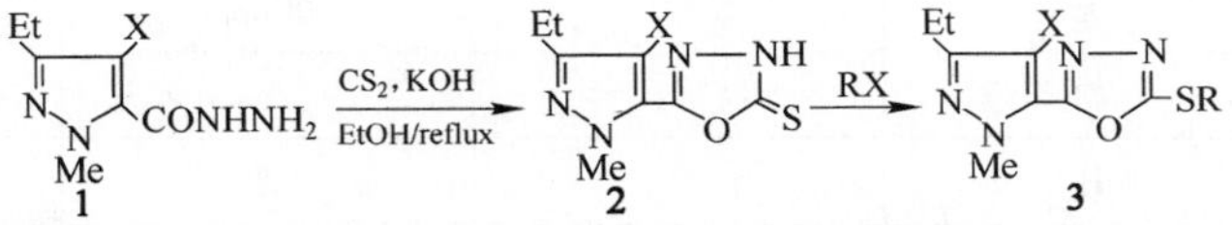

1a–b X = Cl, H; **2a–b** X = Cl, H; **3a–g** X = Cl, R = CH_3, n–C_3H_7, $CH_2CH(CH_3)_2$, n–C_5H_{11}, n–C_7H_{11}, n–C_8H_{17}, $CH(CH_3)CO_2C_2H_5$; **3h–l** X = H, R = CH_3, C_2H_5, n–C_3H_7, n–C_5H_{11}, $CH(CH_3)CO_2C_2H_5$

Scheme 1

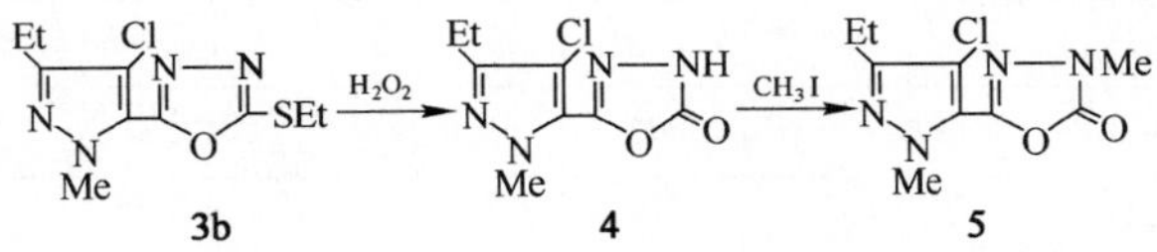

Scheme 2

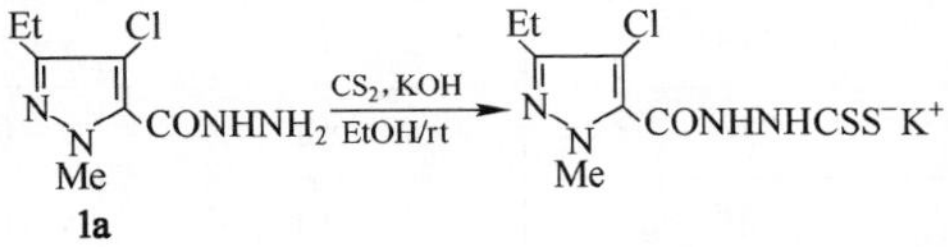

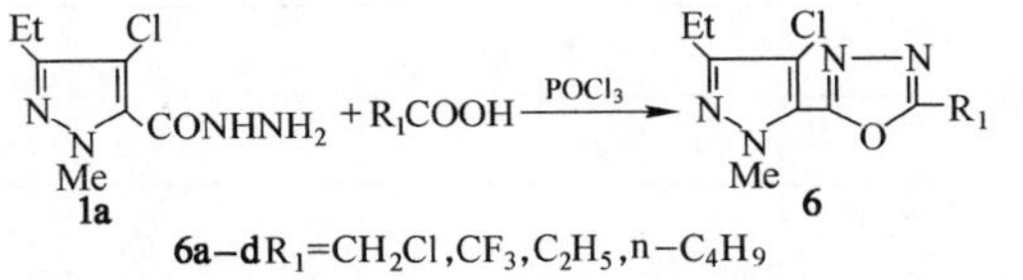

6a–d R_1 = CH_2Cl, CF_3, C_2H_5, n–C_4H_9

Scheme 3

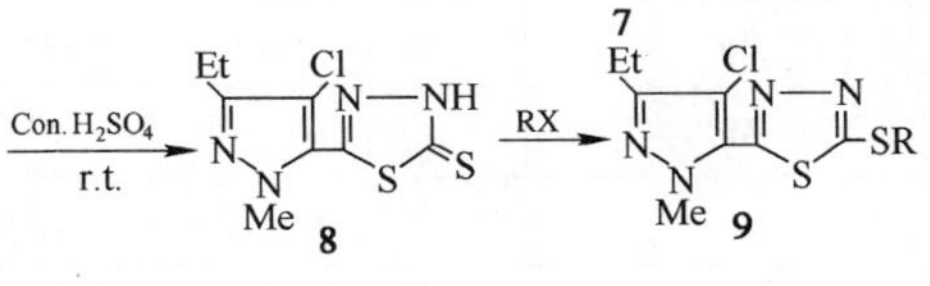

9a–b R = CH_3, C_2H_5, n–C_3H_7, $CH_2CH{=}CH_2$, n–C_5H_{11}

Scheme 4

Oxidation of **3b** by hydrogen peroxide gave **4** instead of the corresponding compound 2 – ethysulfonyl – 5 – pyrazolyl – 1, 3, 4 – oxadiazole. A strong oxidant is the cause of the result. Compound **4** exists in "one" form, which was confirmed by its infrared spectra. A strong absorption was observed in the region of 1780cm^{-1}, which is indicative of a carbonyl group.

Alkylation of **4** by methyl iodide yielded **5**, and infrared spectra confirmed its structure as well. A methyl group is linked to the N atom at the 3 – position of the 1,3,4 – oxadiazole ring instead of the O – atom of the 2 – position. This result is different from the methylation of compound **2**. The theory of hard soft acid base can explain this phenomenon. The S atom of **2** is a soft base, the N atom is a medium base, and the C atom of 4 is a hard acid. Methyl iodide as soft acid preferred to S rather than N when it reacted with **2**. But when it reacted with **4**, it preferred N to O.

3.2 Structure – Activity Relationship (SAR)

Tables 1 – 3 summarize the fungicidal screening results of the study compounds. The results indicated that compounds **3** had significant potency against *Rhizoctonia* solani, and compounds **2** had hardly any inhibition against *R. solani*, that is to say, the activity of compounds that were substituted by alkylthio at the 2 – position of the 1,3,4 – oxadiazole ring was higher than the activity of those compounds that were substituted by mercapto. The activity was found to fall off with increasing size of the alkyl group (*R*). The preferred substituent for *R* was found to be methyl. Loss of biological activity was observed if the chloride atom at the 4 – position of the pyrazole ring was substituted by hydrogen. Oxidation of the sulfide group of compound 3 to 2 – one analogue 5 almost eliminated the activity.

Tab. 1 Fungicidal Screening Results of Compounds2, 3, 4, and 5

(structure: Et, X, N—N, SR, N, O, N, Me)			inhibition of rice sheath blight①	
compd.	X	R	500ppm	100ppm
2a	Cl	H	1	0
2b	H	H	2	1
3a	Cl	CH_3	4	3
3b	Cl	$n-C_3H_7$	4	2
3d	Cl	$n-C_5H_{11}$	3	1
3e	Cl	$n-C_7H_{15}$	2	2
3f	Cl	$n-C_8H_{17}$	2	2
3g	Cl	$CH(CH_3)CO_2Et$	0	0
3h	H	CH_3	4	2
3i	H	C_2H_5	4	2
3j	H	$n-C_3H_7$	3	1
3k	H	$n-C_5H_{11}$	2	0
3l	H	$CH(CH_3)CO_2Et$	1	0
4			0	0
5			1	0
carbendazim			5	5

①On a scale of 0 – 5, where 0 = 0 – 24% control; **1** = 25% – 49% control; **2** = 50% – 69% control; **3** = 70% – 89% control; **4** = 90% – 99% control, and **5** = complete control.

Tab. 2 Fungicidal Screening Results of Compounds 6

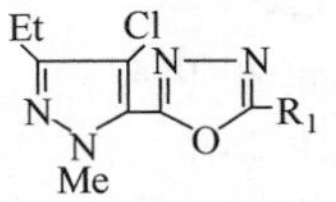		inhibition of rice sheath blight①	
compd.	R_1	500ppm	100ppm
6a	CH_2Cl	4	2
6b	CF_3	3	1
6c	C_2H_5	3	1
6d	$n-C_4H_9$	3	1

① On a scale of 0 – 5, where 0 = 0% – 24% control; 1 = 25% – 49% control; 2 = 50% – 69% control; 3 = 70% – 89% control; 4 = 90% – 99% control, and 5 = complete control.

Tab. 3 Fungicidal Screening Results of Compounds 8 and 9

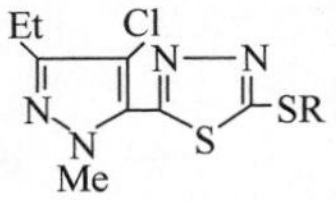		inhibition of rice sheath blight①	
compd.	R	500ppm	100ppm
8	H	5	4
9a	CH_3	4	4
9b	C_2H_5	4	4
9c	$n-C_3H_7$	3	3
9d	$CH_2CH{=}CH_2$	2	1
9e	$n-C_5H_{11}$	3	1

① On a scale of 0 – 5, where 0 = 0 – 24% control; 1 = 25% – 49% control; 2 = 50% – 69% control; 3 = 70% – 89% control; 4 = 90% – 99% control, and 5 = complete control.

Removal of the sulfur atom, by replacement of the alkylthio group with an alkyl group, resulted in some decrease in activity. A sulfur atom at the 2 – position of 1,3,4 – oxadiazole is necessary for fungicidal activity to occur.

The idea of bioisosterism is one of the most successful techniques of bioactive compound design (Lipinski, 1986). The substitution of oxygen for sulfur in the heterocyclic ring represents an example of an approach that is commonly known as bioisosterim. The 1,3,4 – thiadiazole ring is a bioisosteric analogue of the 1,3,4 – oxadiazole. The bioassay indicated that replacing O for S appears to retain fungicidal activity. As with compounds **3**, the activity of compounds **9** was found to increase with decreasing size of the alkyl group *R* at the 3 – position of the 1,3,4 – thiadiazole. Compound **8** has the same activity as compounds **9**, which is different from compounds **2** and **3**.

References

Ashour F. A.; El – Hawash S. A. M.; Mahran, M. A.; et al. Synthesis, antibacterial and antifungal activity of some new 1,3,4 – oxadiazoles and 2 – substituted amino – 1,3,4 – oxadizole derivatives containing benzimidazole moiety. *Bull. Pharm. Sci.*, *Assiut Univ.* 1994, 17(1), 17.

Colliot F. ; Kukorowski K. A. ; Hawkins D. W. , et al. Fipronil: a new soil and foliar broad spectrum insecticide. Proceedings of *Brighton Crop Prot. Conf. – Pests & Disease* 1992, 1, 29.

Funatsukuri, G. ; Ueda, M. 5 – Amino – 1, 3, 4 – thiadiazoles. Japanese Patent 20, 944.

Lipinski, C. A. Bioisosterism in drug design. *Annu. Rep. Med. Chem.* 1986, 21, 283.

Li, Z. ; Chen, H, ; Zhao, W. , et al. Synthesis and biological activity of pyrazole derivatives. *Chem. J. Chin. Univ.* 1997, 18(11), 1794.

Nissan chemical industries, Ltd. Pyrazolesulfonylurea derivatives. Japanese Patent 122, 488.

Okada I. ; Okui S. ; Tsakahshi Y. *et al.* Preparation and testing of pyrazolecarboxamides as insecticides and acaricides. European Patent 289, 879.

Yang, J. ; Hua, W. . Review: Synthesis of 2, 5 – substituted – 1, 3, 4 – oxadizole derivatives. *Chemistry* 1996, 9, 18.

N－（取代嘧啶－2′－基）－2－三氟乙酰氨基苯磺酰脲的合成及除草活性*

姜 林 刘 洁 高发旺 李正名

（南开大学元素有机化学国家重点实验室，南开大学元素有机化学研究所，天津 300071）

关键词 磺酰脲 合成 除草活性

磺酰脲类除草剂是近20年来开发出的超高效、广谱、低毒和高选择性除草剂。在已研究的磺酰脲分子中，芳环的邻位取代基多为酯基、卤素、取代烷氧基、三氟甲基等[1,2]，而邻位取代基为酰胺基的甚少。为寻找高活性的磺酰脲化合物，本文将活性基团三氟乙酰氨基引入磺酰脲分子中，合成了10种 *N*－（取代嘧啶－2′－基）－2－三氟乙酰氨基苯磺酰脲（其中9种为新化合物，**4b**～**4j**），并改进了实验方法，使产率明显提高。同时还测定了它们的除草活性，其中一些化合物具有良好的活性。合成路线如下：

1 (NO_2, $SO_2NHCO_2C_2H_5$) $\xrightarrow[HCl]{SnCl_2}$ **2** (NH_2, $SO_2NHCO_2C_2H_5$) $\xrightarrow{(CF_3CO)_2O}$ **3** ($NHCOCF_3$, $SO_2NHCO_2C_2H_5$)

$\xrightarrow{H_2N-Het}$ **4** ($NHCOCF_3$, $SO_2NHCONH$–嘧啶-2-基, R^1, R^2)

Compd.	R^1	R^2	Compd.	R^1	R^2
4a	H	CH_3	**4f**	CH_3	CH_3
4b	H	OCH_3	**4g**	OCH_3	OCH_3
4c	H	OC_2H_5	**4h**	CH_3	OCH_3
4d	H	SCH_3	**4i**	CH_3	Cl
4e	H	Cl	**4j**	OCH_3	Cl

X－4数字显示显微熔点仪，温度计未校正 Yanaco CHN CORDER MT－3型元素分析仪。BRU KER－ACP 200 核磁共振仪，DM SO－d_6 或 $CDCl_3$ 为溶剂，TMS 为内标。VG－ZAB 型质谱仪。Shimadzu IR－435 型红外光谱仪，KBr 压片。

化合物**3**的制备参照文献［3］方法，将17.5g（78mmol）二水氯化亚锡溶于100mL浓盐酸，搅拌下加入6.9g（25mmol）2－硝基苯磺酰基氨基甲酸乙酯（**1**），缓

* 原发表于《应用化学》，2001，18（3）：225－227。国家自然科学重点基金资助课题（29832050）。

慢升温，在60℃左右反应0.5h，冷至室温，抽滤，10mL冷水洗涤，干燥，得白色粉末5.2g，即为2－氨基苯磺酰基氨基甲酸乙酯（**2**），产率82.5%，mp 130～132℃。

2.70g（15mmol）**2**溶于25mL二氯甲烷，冰水浴下加入9.5g（45mmol）三氟乙酸酐，在此温度下搅拌0.5h，然后升至室温搅拌2h，混合物用15mL×3冷水洗涤，有机层用无水硫酸镁干燥，减压脱去溶剂，得白色粉末3.58g，即为2－三氟乙酰基氨基苯磺酰胺（**3**），收率70.2%，mp 120～122℃。

4的合成参照文献［3，4］方法：将0.43g（1.34mmol）**3**置于装有短分馏柱的烧瓶中，加10mL无水甲苯和等摩尔的2－氨基－4－取代嘧啶或2－氨基－4，6－二取代嘧啶，加热至回流，反应5h，反应过程中不断将生成的乙醇蒸出。冷至室温，抽滤析出的固体，丙酮（或丙酮－石油醚）洗涤，干燥，得产物**4a～4j**，化合物的物理常数见表1，^{1}H NMR数据见表2，化合物**4b**的IR、MS数据见表3。

表1　化合物4a～4j的熔点、元素分析和收率

Tab. 1　mp，elemental analysis data and yield of compounds 4a～4j

Compd.	Formula	mp（℃）	Elemental analysis，（Calcd，%）			Yield（%）
			C	H	N	
4a	$C_{14}H_{12}F_3N_5O_4S$	209～210	41.90（41.69）	2.59（3.00）	17.11（17.36）	79.1
4b	$C_{14}H_{12}F_3N_5O_5S$	198～200	40.10（40.12）	2.91（2.89）	16.98（16.70）	82.1
4c	$C_{15}H_{14}F_3N_5O_5S$	196～198	41.40（41.60）	3.30（3.26）	16.35（16.17）	78.6
4d	$C_{14}H_{12}F_3N_5O_4S_2$	210～212	38.43（38.65）	2.93（2.78）	15.75（16.09）	87.6
4e	$C_{13}H_9ClF_3N_5O_4S$	190～191	36.95（36.85）	3.35（3.38）	16.34（16.52）	70.6
4f	$C_{15}H_{14}F_3N_5O_4S$	180～182	43.38（43.20）	2.27（2.14）	16.70（16.78）	65.9
4g	$C_{15}H_{14}F_3N_5O_6S$	200～201	39.92（40.12）	3.15（3.14）	15.60（15.61）	84.7
4h	$C_{15}H_{14}F_3N_5O_5S$	179～181	41.85（41.60）	3.40（3.26）	16.10（16.17）	64.3
4i	$C_{14}H_{11}ClF_3N_5O_4S$	180～182	38.47（38.43）	2.34（2.53）	16.08（16.00）	66.8
4j	$C_{14}H_{11}ClF_3N_5O_5S$	194～196	37.04（37.08）	2.49（2.44）	15.35（15.44）	69.0

表2　化合物4a～4j的^{1}H NMR数据

Tab. 2　^{1}H NMR data of compounds 4a～4j

Compd.	Solvent	^{1}H NMR，δ
4a	DMSO	2.43（s，3H，CH_3），7.12（d，1H，J=5.2Hz，Pyrim－H_5），7.60～8.10（m，4H，Ar—H），8.48（d，1H，J=5.2Hz，Pyrim－H_6），10.85（bs，1H，NH），11.40（bs，1H，NH）
4b	DMSO	3.94（s，3H，OCH_3），6.68（d，1H，J=6.6Hz，Pyrim－H_5），7.43～7.99（m，4H，Ar—H），8.24（d，1H，J=6.6Hz，Pyrim－H_6），11.15（bs，1H，NH），11.40（bs，1H，NH），13.21（bs，1H，NH）
4c	DMSO	1.30（t，3H，J=7.0Hz，OCH_2—CH_3），4.39（q，2H，J=7.0Hz，OCH_2—），6.65（d，1H，J=6.6Hz，Pyrim－H_5），7.42～7.98（m，4H，Ar—H），8.23（d，1H，J=6.6Hz，Pyrim－H_6），11.23（bs，1H，NH），11.45（bs，1H，NH），13.35（bs，1H，NH）
4d	DMSO	2.55（s，3H，SCH_3），7.15（d，1H，J=6.4Hz，Pyrim－H_5），7.50～8.00（m，4H，Ar—H），8.21（d，1H，J=6.4Hz，Pyrim－H_6），10.95（bs，1H，NH），11.35（bs，1H，NH），13.30（bs，1H，NH）

续表 2

Compd.	Solvent	^{1}H NMR, δ
4e	DM SO	7.38 (d, 1H, J=5.4Hz, Pyrim-H_5), 7.62~8.14 (m, 4H, Ar—H), 8.60 (d, 1H, J=5.4Hz, Pyrim-H_6), 11.02 (bs, 1H, NH), 11.20 (bs, 1H, NH), 13.35 (bs, 1H, NH)
4f	$CDCl_3$	2.44 (s, 6H, CH_3), 6.76 (s, 1H, Pyrim-H_5), 7.35~8.41 (m, 4H, Ar—H), 10.73 (bs, 1H, NH), 13.25 (bs, 1H, NH)
4g	DM SO	3.88 (s, 6H, OCH_3), 6.00 (s, 1H, Pyrim-H_5), 7.65~8.20 (m, 4H, Ar—H), 7.95 (bs, 1H, NH)
4h	DM SO	2.36 (s, 3H, CH_3), 3.91 (s, 3H, OCH_3), 6.59 (s, 1H, Pyrim-H_5), 7.50~8.10 (m, 4H, Ar—H), 10.90 (bs, 1H, NH), 11.40 (bs, 1H, NH)
4i	$CDCl_3$	2.52 (s, 3H, CH_3), 6.94 (s, 1H, Pyrim-H_5), 7.36~8.41 (m, 4H, Ar—H), 10.63 (bs, 1H, NH), 12.45 (bs, 1H, NH)
4j	$CDCl_3$	3.98 (s, 3H, OCH_3), 6.49 (s, 1H, Pyrim-H_5), 7.34~8.45 (m, 4H, Ar—H), 10.66 (bs, 1H, NH), 12.27 (bs, 1H, NH)

表 3 化合物 4b 的 IR 和 MS 数据

Tab. 3 IR and MS data of compound 4b

IR, v (cm^{-1})	3390 (v_N—H), 1711 (v_C═O), 1580 (v_N—H+v_C—N), 1360 ($v_{as}SO_2$), 1155 (v_sSO_2)
MS, m/z (%)	421 (M+2, 0.2), 420 (M+1, 0.6), 419 (M+, 2.7), 350 (28), 188 (100), 168 (22), 152 (46), 125 (77), 124 (57), 118 (30), 90 (78), 78 (4.1), 77 (1.8), 69 (48)

将目标化合物 5.0mg，用少量二甲基甲酰胺溶解，再加入少量乳化剂和 100mL 水，制得浓度为 50μg/g 待测样品溶液。取此溶液 3mL 加至垫有 2 张滤纸的培养皿中，再加 30 粒油菜种子，将培养皿放入 28℃恒温暗室 48h，然后测量油菜幼苗初生根长度，分别计算其抑制百分率。实验中用清水+乳化剂作空白对照，氯嘧磺隆（chlorimuron）**4**s 作对照药剂，结果见表 4。

表 4 化合物 4a~4j 的除草活性

Tab. 4 Herbic idal activity of 4a~4j

Compd.	**4a**	**4b**	**4c**	**4d**	**4e**	**4f**	**4g**	**4h**	**4i**	**4j**	**4s**
Inhibitioni rate (%)	60.5	56.3	36.7	27.2	26.9	71.0	74.7	71.9	61.8	69.8	78.8

结果与讨论

制备三氟乙酰氨基苯磺酰氨基甲酸乙酯时，文献［1］直接用 2-氨基苯磺酰基氨基甲酸乙酯与三氟乙酰酐反应，后者既是酰化剂又作溶剂，这样，三氟乙酸酐过量 5 倍之多。本文将反应物先用二氯甲烷溶解，再滴加 3 倍于 2-氨基苯磺酰基氨基甲酸酯的三氟乙酸酐，因而节省了三氟乙酸酐的用量。在产物的合成中，适当延长反应时间，并采取用分馏柱分馏出生成醇的方法，使该可逆反应向生成产物方向移动，收率有较大提高，如 **4a** 产率文献值为 44.4%，本文为 79.1%。

在该系列化合物中，4-取代嘧啶磺酰脲（**4a**~**4e**）以 **4a** 除草活性最好（见表4），

4b 次之，其余 3 种化合物活性较低，表明嘧啶环上连有 CH_3、OCH_3 时化合物的活性较好。4，6－二取代嘧啶磺酰脲（**4**f～**4**j），除 **4**i（$R^1 = CH_3$，$R^2 = Cl$）活性稍低外，其余 4 种化合物活性均较高，而且与对照药剂氯嘧磺隆比较接近。此外，我们还用油菜根长法测定了化合物 **4**f 和 **4**g 的 IC_{50}，分别为 4.57×10^{-6}、1.09×10^{-6} mol/L，表明它们有良好的除草活性。

参考文献

[1] General Station of Pesticide Information of Ministry of Chemical Industry（化工部农药信息总站）. Handbook of Foreign Pesticide Species)（国外农药品种手册）(New Collectiveed)（新版合订本）. Shenyang（沈阳）: General Station Press of Pesticide Information of Ministry of Chemical Industry（化工部农药信息总站出版社），1996：867.

[2] LIU Chang－Ling（刘长令）. Nongyao（农药），1999，38（9）：40.

[3] LIU Jie（刘洁）. Doctor Dissertation（［博士学位论文］）. Tianjin（天津）：Elemento－organic Chemistry Institute，Nankai University（南开大学元素有机化学研究所），1998.

[4] Stephens J A. U S 3 577 375，1968.

Synthesis and Herbicidal Activity of *N*－(Substituted Pyrimidin－2′－yl)－2－Trifluoroacetylamino Benzenesulfonylurea

Lin Jiang，Jie Liu，Fawang Gao，Zhengming Li

(State Key Laboratory of Elemento－Organic Chemistry，Institute of Elemento－Organic Chemistry，Nankai Univeristy，Tianjin，300071)

Abstract Ten *N*－（4′－substituted pyrimidin－2′－yl）－2－trifluoroacetylamino benzenesulfonylureas were synthesized. Their structures were confirmed by elemental analysis and ^1H NMR. Some of the compounds showed good herbicidal activity.

Key words sulfonylurea；synthesis；herbicidal activity

新磺酰脲类除草活性构效关系的研究*

李正名　赖城明

（南开大学化学学院，天津　300071）

摘　要　磺酰脲类除草剂具有对环境友好和超高效的特点。本文采用X－衍射谱对其绝对构型进行分析，首次发现分子内氢键的存在。采用各种理论和软件计算，活性结构应符合三点要求：（a）分子内氢键使杂环和脲之间形成一个共轭体系；（b）羰基氧、磺酰氧和杂环氮形成分子中三个负电中心；（c）在磺酰胺与苯邻位取代基之间形成一个空穴。根据以上结论，构建了一个卡口模型，较合理地解释了磺酰脲类除草活性的构效关系。建立了一个虚拟靶酶ALS的模拟作用模型，供进一步分子设计ALS抑制剂，包括一些非磺酰脲类先导化合物时参考。

关键词　磺酰脲类除草剂　分子内氢键　构效关系　分子设计　卡口模型　虚拟ALS作用模型

杜邦公司Levilt等于80年代发现某些磺酰脲化合物具有超高效除草活性后引起了一股研究热潮，二十年来各国已合成了六万多个不同结构的新磺酰脲类化合物，申请了380篇专利，美、日、瑞士等国家已有十个商品问世，为此Levitt本人于1991年获美国化学会创造发明奖，在授奖仪式报告中他介绍了磺酰脲研究经过和他的构效关系（图1）[1]。

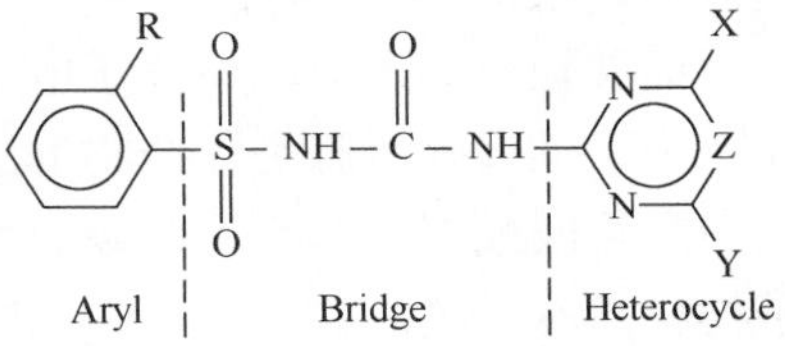

R　O　O　X

S－NH－C－NH

N　Z　N

O　Y

Aryl　Bridge　Heterocycle

R═COOCH$_3$，NO$_2$，F，Cl，Br，etc
X，Y═OCH$_3$，CH$_3$，Cl，etc
Z═C，N

图1　磺酰脲的分子结构

Fig. 1　General formula of bioactive sulfonyl ureas

自1990年开始，我们已合成了新型磺酰脲类化合物近500个，均进行了室内除草活性测定工作[2]，并在此基础上进行了各晶体的X－射线衍射图的测定，首次发现在活性结构中的N（2）—H与N（12）之间存在着一个分子内氢键（图2）。并通过其红外光谱［3395cm^{-1}，3215cm^{-1}（肩峰）］旁证了此氢键的存在。被测定的五个新磺酰脲类化合物晶体为[3－6,27]：*N*－（2′－嘧啶基）－2－甲酸乙酯－苯磺酰脲（Ⅰ），*N*－［2′－（4－甲基）－嘧啶基］－2－甲酸乙酯－苯磺酰脲（Ⅱ），*N*－［2′－（4，6－二甲基）－嘧啶基］－2－甲酸乙酯－苯磺酰脲（Ⅲ），*N*－［2′－（4－乙基）三嗪基］－2－硝基－苯磺酰脲（Ⅲ），*N*－［2－（4－甲基）－嘧啶基］－2－甲酸乙酯－苄基磺酰脲（Ⅴ）。

其分子结构具有以下特征：分子中的原子主要分布在嘧啶磺酰脲平面（不含磺酰基）、苯环平面、酰基平面等三个相互独立的平面上。含嘧啶、三嗪等杂环通过其杂氮原子与酰胺中的氢原子的氢键形成的平面导致一个共轭体系的形成，从而对活性结构中的电子传递、反应性能、生物活性产生重要影响。

应用分子图形学、分子力学和量子化学方法，研究了磺酰脲化合物的电子结构及化学键[7]，计算指出在分子中形成三个独立平面的体系中，整体平面形成了较大的共

＊　原发表于《有机化学》，2001，21（11）：810－815。国家自然科学重点基金（Noe. 29232010，29832050）和国家自然科学基金（Noe. 29322001，29772019）资助项目。

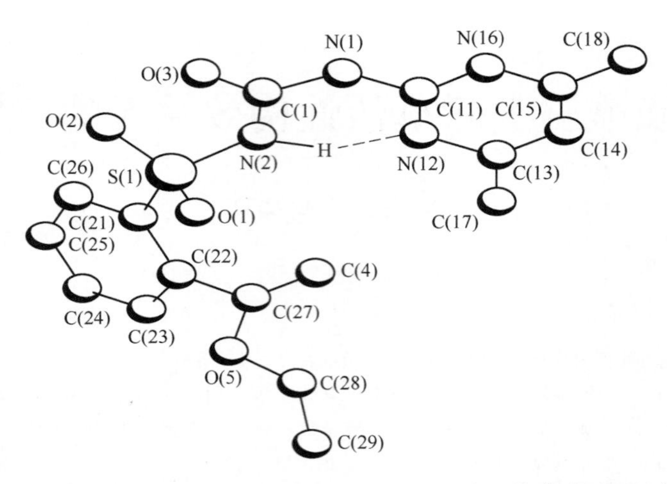

图 2　*N* –［2′–（4,6 – 二甲基）– 嘧啶基］– 2 – 甲酯乙酯 – 苯磺酰脲（Ⅲ）的晶体结构

Fig. 2　Molecular Structure of *N* –［2′–（4′，6′– dimethyl）– pyrimidinyl］– 2 – ethylcarboxylate – benzene sulfonylurea（Ⅲ）

轭体系，根据 S—N 键长测定值小于单键长，以及集居数分析，S 原子的 d 空轨道参与了平面共轭键的形成，保证了电子的有效传递。

为了估计磺酰脲类化合物的活性构象结构，应用多构象重叠方法[8-10]研究了 12 种已商品化磺酰脲除草剂的构象，研究指出这些分子均有相似的构象特征，与晶体中分子构象基本一致。

从静电势计算[11]，分子中存在三个独立的负电势区、即磺酰基氧区，脲基上羰基氧区和杂环上氮原子区。此外领位取代基与磺酰基之间共同形成了一个空穴（我们称为“卡口”）区域。根据药物（授体）与受体作用的三点作用模型，这些区域可能与生活活性密切相关（图 3，图 4）。

后来的研究证明了磺酰脲的靶酶是乙酰乳酸合成酶（ALS），但其结构不清。Schloss[12]认为磺酰脲类活性结构的作用机理可能是抑制了植物体内 ALS 靶酶中的辅酶——硫胺焦磷酸素（thiamine pyrophosphate，TPP）。根据这个假设，我们建立了磺酰脲分子与受体中 TPP 之间弱相互作用的初级作用模型[13]。根据轨道能量及电荷计算，两个分子的前线轨道能级相似，电荷分布具有适配性。即两者氮杂环之间，TPP 中噻唑环正电性与磺酰脲中羰基负势区之间，TPP 中的羟基与磺酰脲邻位取代基所形成凹隙之间的相互作用力使得授/受体能相互吸引和结合，这个假设较好地解释了不同磺酰脲生物活性的差异性。

上述化合物的晶体结构的空间群指出，晶胞中分别存在着对称元素（对称面或对称中心），对映体的出现应是由于阻碍内旋转而产生的构象异构体，两者之间，可通过 N—S，S—C 以及苯环与酰基相连的 C—C 键三者协同旋转而转换[14,15]，应用 Tripos 力场计算最低能量约为 20 ~ 30kJ/mol，因此若仅一种构象有活性，那么在溶液中分子会容易地转到活性构象而与受体结合。

在构效关系方面，研究了结构参数及计算方法的选择对提高磺酰脲类除草活性预报准确性的影响[16,17]

应用多元回归方法，以活性 Pl_{50}值处在 7. 575 ~ 3. 075 之间的 23 个新合成磺酰脲化合物为研究对象，结果表明，引入反映分子表面积大小的 AREA 值，以及反映分子构

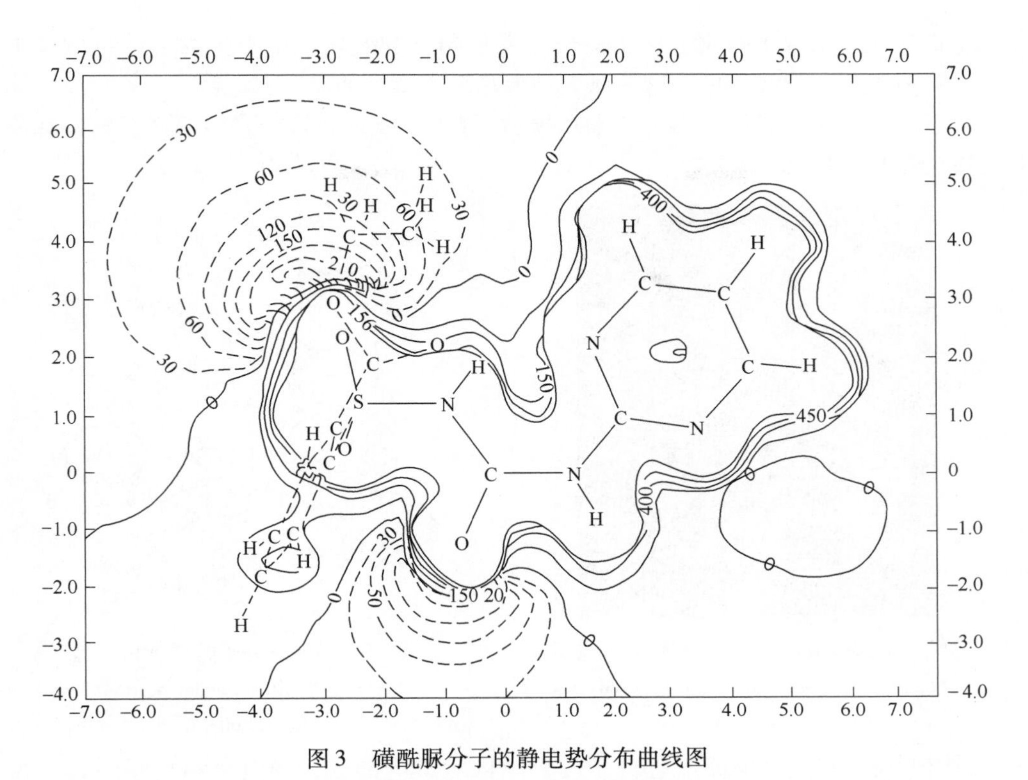

图 3　磺酰脲分子的静电势分布曲线图

Fig. 3　Electro - static potential（kJ/mol）in the sulfonylurea molecule

象差异的 d（RMS）值（d_1，d_2 为局部差异，d 为整体差异）明显可以改善预报结果。将上述构象差异值，表面积 AREA 值引入人工神经网络方法（ANN）计算时，所得预测值与实验值吻合较好[17]，预测均方误差 $mse=0.073$，选优于多元回归方法 $mse=0.82$。在探讨杂环为嘧啶环时，其单取代的药效与双取代基相似，而杂环改为三嗪环时，单取代的药效明显低于双取代基的药效这种现象时，我们认为分子与受体作用位点之间的几何相互作用存在一个临界区域范围，而单取代三嗪环的空间几何参数可能恰好低于这个最低临界域值，所以活性较低[26]。

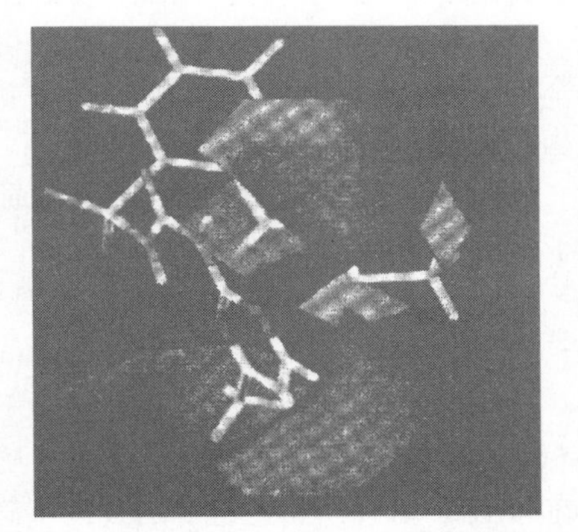

图 4　静电效应的 CoMFA 三维等值线图

Fig. 4　A 3D - contour map by CoMFA to show electrostatic effects

应用模式识别方法[18,19]，对具有不同活性的磺酰脲类化合物进行分类研究，选择杂环上负电性原子的电荷值作为特征参数组成模式空间，成功地将含嘧啶环及三嗪环单取代磺酰脲类化合物的活性进行分类判别。应用比较分子力场 CoMFA 方法研究了磺酰脲类构效关系 3D - QSAR[20,21]，从立体场图可知增加绿色区域或减少黄色区域的空间体积将有助于提高其除草活性，而从静电场的蓝色区域增加正电荷，红色区域减少负电荷将有助于提高活性，这样就为进一步优化结构提供参数（图 5）。

应用分子动力学模拟（MD）方法[22,23]研究了磺酰脲分子在不同极性溶剂如 H_2O，$CHCl_3$，CCl_4 中的构象变化，分子在水溶液中有较大几率发生弯曲，溶剂极性愈大，弯曲几率也愈大。水分子的氢键作用使构象发生明显的变化。由于生物活性物质基本上是在生物体中水相中起作用，这一研究结果是较有意义的。

基于上述工作，我们提出了“卡口模型”(caliper model)较好地解释了磺酰脲分子对受体的相互作用（图6）[13]。

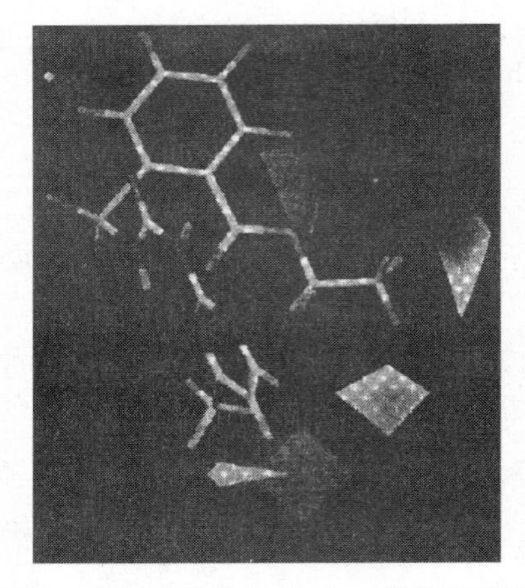

图5 空间效应的CoMFA三维等值线图

Fig. 5 A 3D-contour map by CoMFA to show steric effects

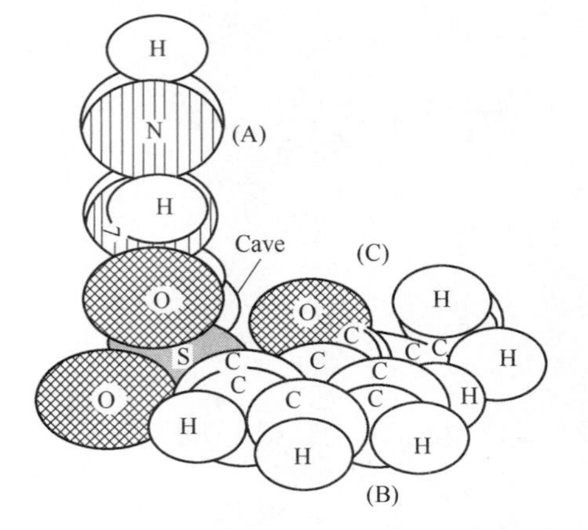

图6 磺酰脲分子的“卡口模型”

Fig. 6 Caliper model postulated for sulfonylureas

根据上述研究的有关磺酰脲分子的分子内氢键的形成，静电势和空间因素，我们首次提出了磺酰脲分子除草活性三要素[27]：

（a）由于分子内氢键的存在，使得酰脲基团与杂环形成一个共平面的共轭体系，整个分子由以一定角度相交的三个非共平面性结构所组成。

（b）羰基氧、磺酰基氧和杂环氮原子形成分子中三个电负中心。

（c）磺酰胺基和邻位取代基之间形成一个空穴。

虽经很多科学家的艰苦努力，由于得不到ALS酶的晶体，其绝对结构至今尚未确立，我们在运用CoMFA所得三维等值线的基础上继续运用Leapfrog软件进行处理，从轮廓形式受体模型的立体场和静电场特征，构建了一个虚拟的ALS酶的模拟作用模型（图7）[21]。

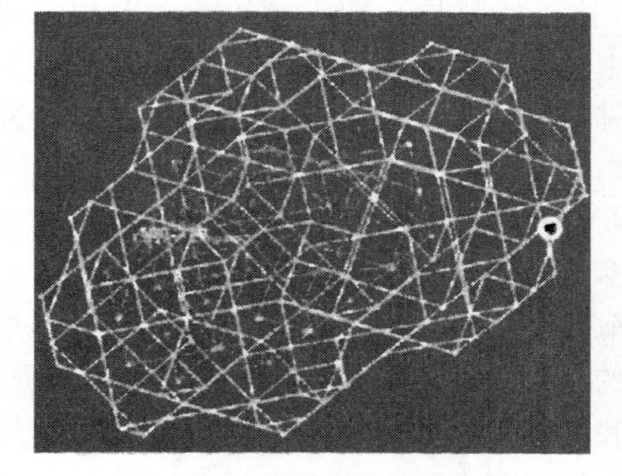

图7 ALS酶的模拟作用模型

Fig. 7 Virtual model of ALS

选取了四个作用点作为提问结构，采用ACD-3D进行三维结构搜索，得到一组符合条件的分子结构信息，选取了部分结构作为合成新ALS酶抑制剂的参考（图8）[24]

可将在靶酶ALS结构不清的前提下我们对新磺酰脲活性构效关系研究的过程用图9所示流程来表述之。

致谢：本基础研究工作曾多次得到国家自然科学基金重点基金及面上基金资助，在此表示衷心的感谢。本项目参加者有王玲秀，贾国锋，刘洁，范传文，姜林，袁满雪，马翼，钱宝英，王素华，王霞等同志，在此表示深切的谢意。

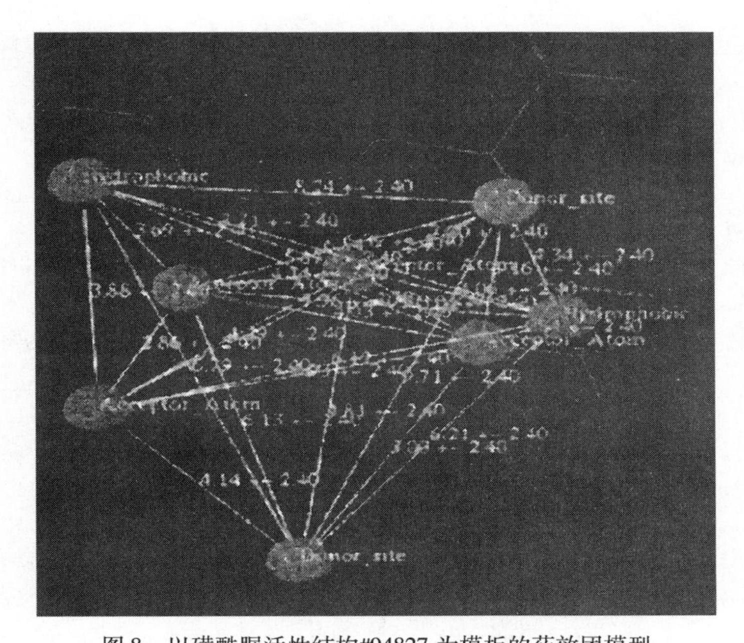

图 8 以磺酰脲活性结构#94827 为模板的药效团模型

Fig. 8 Pharmacophore model for bio - active sulfonylurea #94827 by DISCO

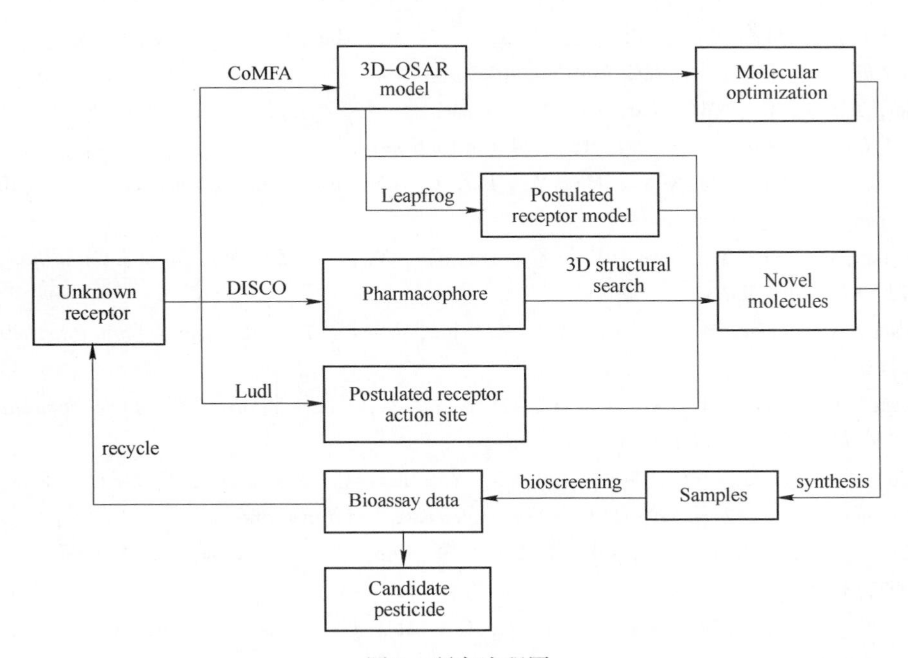

图 9 研究流程图

Fig. 9 Research flowchart

References

[1] Levitt, G. In *Synthesis and Chemistry of Agrochemicals*, Eds: Baker, D. R.; Fenyes, J. G.; Mobery, W. K., ACS Symposium Series 443, 1991, pp. 17 – 47.

[2] Li, Z. – M.; Jia, G – F.; Wang, L. – X.; Lai, C. – M. *Chem. J. Chin. Univ*, 1994, 15, 391 (in Chinese).

[3] Li, Z. –M.; Jia, G. –F.; Wang, L. –X.; Lai, C. –M.; Wang, R. –J.; Wang, H. –G. *Chem. J. Chin. Univ.* 1992, 13, 1411 (in Chinese).

[4] Li, Z. –M.; Jia, G. –F.; Wang, L. –X.; Lai, C. –M.; Wang, H. –G.; Wang, R. –J. *Chem. J. Chin. Univ.* 1993, 14, 349 (in Chinese).

[5] Li, Z. –M.; Jia, G. –F.; Wang, L. –X.; Lai, C. –M, *Chem. J. Chin. Univ.* 1994, 15, 227 (in Chinese).

[6] Li, Z. –M,; Liu, J.; Wang, X.; Yuan, M. –X.; Lai, C. –M. *Chem. J. Chin. Univ.* 1997, 18, 750 (in Chinese).

[7] Lai, C. –M.; Yuan, M. –X.; Li, Z. –M.; Jia. G. –F. *Chem. J. Chin. Univ.* 1994, 15. 1004 (in Chinese).

[8] Yan, B.; Lai, C. –M.; Lin, S. –F.; Li, Z. –M. *Chem. J. Chin. Univ.* 1992, 13. 1555 (in Chinese).

[9] Yan, B.; Lai, C. –M.; Lin, S. –F.; Li, Z. –M. *Chem. J. Chin. Univ.* 1993, 14, 1534 (in Chinese).

[10] Yan, B.; Lai, C. –M.; Lin, S. –F.; Zheng, J. –B.; Li, Z. –M. *Computer Chemistry Monograph Seriers* 3, 1992 ~ 1993, Science Press, China, 1994, p. 102.

[11] Lai. C. –M.; Luo, H. –H.; Yuan, M. –X., Li, Z. –M. *Acta Scientiarum Naturatium Universitatis Nankaiensis*, 1993. (1), 56 (in Chinese).

[12] Schloes, J. V. *Pest. Sci.* 1990, 29, 283.

[13] Lai, C. – M.; Yuan, M. – X.; Li, Z. – M.; Jia, G. – F.; Wang, L. – X. *Chem, J. Chin. Univ.* 1994, 15, 693 (in Chinese).

[14] Liu, A. –L; Cao, W.; Lai, C. –M.; Yuan, M. –X. Zhang, J. –P.; Lin, S. –F.; Li, Z. –M. *Chem, J. Chin. Univ.* 1997, 18, 574 (in Chinese).

[15] Wang, X.; Yuan, M. –X.; Ma, Y.; Lai, C. –M. *Comput. Appl. Chem*, 1997, 13 (10) supplement, 176 (in Chinese).

[16] Wang, X.; Sun, Y.; Yuan, M. –X.; Lai, C. –M.; Li, Z. –M, *Chem. J. Chin. Univ*, 1996, 17, 1874 (in Chinese).

[17] Wang, X.; Yuan, M. –X; Lai, C. –M.; Liu, J.; Li, Z. –M. *Chem. J. Chin. Univ.* 1997, 18, 60 (in Chinese).

[18] Wang, X.; Yuan, M. –X.; Lai, C. –M.; Liu, J.; Li. Z – M. *Acta Scientianum Naturatium Universties Nankaiensts.* 2000. 33. 11 (in Chinese).

[19] Liu, J.; Li, Z. – M.; Wang. X.; Ma, Y.; Lai, C. – M.; Jia, G. – F.; Wang, L – X. *Comput, Appl. Chem.* 1997, 13 (10) supplement, 155 (in Chinese).

[20] Liu, J.; Wang, X.; Li, Z. –M.; Lai, C. –M.; Jia, G. –F.; Wang, L. –X. *Chin. Chem. Lett.* 1997, 8, 503.

[21] Liu, J.; Li, Z. – M.; Wang, X.; Lai, C. – M., Jia, G. – F.; Wang, L. – X. *Sci. China, Ser. B* 1998, 41, 50.

[22] Shen, R. – X.; Fang, Y. – Y.; Ma, Y.; Sun, H – W.; Lai, C. – M.; Li, Z. – M. *Chem. J. Chin. Univ.* 2001, 22, 952 (in Chinese).

[23] Ma, Y.; Liu, J.; Li, Z. –M. *Chem. J. Chin. Univ.* 2000, 21, 85.

[24] Liu, J.; Li, Z. – M.; Yan, Y.; Wang, L. – X.; Chen, J. – P. *Bio – organic & Medicinal Chemistry Letters.* 1999, 9. 1927.

[25] Liu, J. *Ph. D Thesis* 1998, p. 44.

[26] Jiang, L.; Li, Z. –M.; Weng, L. –H.; Leng, X. –B. *Chin. J. Struc. Chem.* 2000, 19, 149 (in Chinese).

[27] Li. Z. – M. *UNIDO International Conference on Crop Protection Chemicals*, 1999. *Nantong*, *China*, China Agricultural Press, Beijing, 2000, pp. 30 – 36.

Research on the Structure/Activity Relationship of Herbicidal Sulfonylureas

Zhengming Li, Chengming Lai

(College of Chemistry, Nankai University, Tianjin, 300071)

Abstract: Some sulfonylureas are an important class of environmental benign, ultra – low dosage herbicides. X – ray diffraction spectra of bio – active sulfonylureas were carefully examined and an intra – molecular hydrogen bond was first reported here. By theoretical calculation. three essential structural requirements for the bio – activity were concluded as follows: (a) all intramolecular hydrogen bonds cause a coplanar conjugated system between the heterocycle and urea moiety; (b) carbonyl oxygen, sulfonyl oxygen and heterocyclic nitrogen form three negative cemlers; (c) between sulfonylamido and phenyl ortho substituent, there is a cavity.

A Caliper model thus deduced reasonably explains the structure/activity relationship of bioactive sulfonylureas. A 3D contour map of a pseudo – ALS model was set up with which could lead to further design of novel ALS inhibitors, including some possible non – sulfonylureas leads.

Key words sulfonylurea herbicides; intra – molecular hydrogen bond; structural/bioactivity relationship; molecular design; pseudo ALS model

卤代-2-(3-甲基-5-取代-4H-1，2，4-三唑-4-基)-苯甲酸的合成*

罗铁军[1]　李正名[2]　王素华[2]　廖仁安[2]　李之春[2]

([1] 湖南化工研究院，长沙；[2] 南开大学元素有机化学国家重点实验室，天津　300071)

关键词　卤代（甲基取代-4H-三唑基）苯甲酸　邻氨基苯甲酸　三唑　缩合反应

杂环化合物的结构变化繁多，具有广泛的生物活性，已受到化学研究者的广泛关注。在含氮杂环化合物中，三唑类化合物的研究引人注目，如已商品化的杀菌剂三唑酮[1]；植物生长调节剂多效唑[2]；杀虫剂唑蚜威[3]以及除草剂唑草胺（CH-900）[4]等。为了寻找新型高生物活性三唑化合物或其先导化合物，我们以邻氨基苯甲酸为起始原料，经过卤化得到卤代邻氨基苯甲酸（**1**）；**1** 和乙酸酐缩合得到卤代-3-甲基-1H-2，4-苯并噁嗪-1-酮（**2**）；**2** 与取代酰肼（**3**）缩合得到一类新型的卤代-2-（3-甲基-5-取代-4H-1，2，4-三唑-4-基）苯甲酸（**4**）目标产物。合成路线如下：

$\mathbf{4}$: X_n=4-Cl, 5-Br, 3,5-Br_2; R=H, Me, $CH_2OC_6H_5$, CH_2CN

仪器和试剂：Bruker AC-200 型核磁共振仪，TMS 为内标，MT-3 元素分析仪，X-4数字显示显微熔点仪，温度计未校正。

5-溴邻氨基苯甲酸和3，5-二溴邻氨基苯甲酸按文献［5］方法合成，4-氯邻氨基苯甲酸购自 Acros 公司。甲酰肼、乙酰肼、氰基乙酰肼、苯氧乙酰肼分别按文献［6~9］方法合成。其他试剂均为市售分析纯试剂。

化合物 **2** 的合成[5,10]：6-氯-3-甲基-1H-2，4-苯并噁嗪-1-酮（**2a**）的合成：在4-氯-2-氨基苯甲酸（17.2g，0.1mol）中加入 170mL 新蒸馏的乙酸酐，回流 1h，冷却至室温析出絮状的晶体，过滤，用少量乙酸酐和石油醚洗涤，干燥得到 **2a** 12.85g，mp 154~155℃（对母液浓缩处理后可得到纯度稍低的 **2a** 5.7g），产率 94%。

7-溴-3-甲基-1H-2，4-苯并噁嗪-1-酮（**2b**）的合成：在5-溴-2-氨基苯甲酸（4g，0.0185mol）中加入 40mL 新蒸馏的乙酸酐，回流 15min，冷却至室温

* 原发表于《应用化学》，2002，19（6）：594-596。国家自然科学重点基金（29832050）和高等学校国家重点实验室访问学者基金资助课题。

析出无色晶体，过滤，用少量乙酸酐和石油醚洗涤，干燥，得到**2b** 2.74g，mp 134～135 ℃（对母液浓缩处理可得到纯度稍低的**2b** 1.5g），产率96%。

5，7－二溴－3－甲基－3H－2，4－苯并噁嗪－1－酮（**2c**）的合成：在3，5－二溴－2－氨基苯甲酸（4g，0.0136mol）中加入40mL新蒸馏的乙酸酐，回流15min，冷却至室温析出无色晶体，过滤，用少量乙酸酐和石油醚洗涤，干燥得到**2c** 3.71g，mp 176～177 ℃（对母液浓缩处理可得到纯度稍低的**2c** 0.4g），产率95%。

卤代－2－（3－甲基－5－取代－4H－1，2，4－三唑－4－基）－苯甲酸（**4**）的合成通法[9,11]：将卤代－3－甲基－1H－2，4－苯并噁嗪－1－酮（**2**）（0.01mol）和取代酰肼**3**（0.0105mol）置于50mL的三口反应瓶中，加入25mL无水乙醇，通入N_2气，回流110min。减压脱除溶剂后，用适当的溶剂（表1）重结晶，过滤，用少量乙酸乙酯和石油醚洗涤，干燥，得到白色晶体**4**。

结果与讨论

系列目标产物**4**的合成收率、组成分析及物理化学常数列入表1。

表1　化合物4a～4l的组成及物理化学常数

Tab. 1　Compositions and physical constants of compounds 4a～4l

No.	X_n	R	recryst. solvent	mp（℃）	Yield（%）	Mol. formula	Elemental analysis（calcd.，%）		
							C	H	N
4a	4－Cl	H	MeOH	270～272	65	$C_{10}H_8N_3O_2Cl$	50.52（50.52）	3.31（3.39）	17.54（17.69）
4b	5－Br	H	EtOH	230～232	58	$C_{10}H_8N_3O_2Br$	42.35（42.5）	2.61（2.86）	14.76（14.90）
4c	3，5－Br_2	H	EtOH	215～217	70	$C_{10}H_7N_3O_2Br_2$	33.23（33.25）	1.90（1.96）	11.40（11.64）
4d	4－Cl	CH_3	DMSO	316～318	40	$C_{11}H_{10}N_3O_2Cl$	52.35（52.47）	4.00（4.01）	16.88（16.70）
4e	5－Br	CH_3	EtOH	277～279	49	$C_{11}H_{10}N_3O_2Br$	44.45（44.59）	3.17（3.41）	14.31（14.20）
4f	3，5－Br_2	CH_3	EtOH	266～268	81	$C_{11}H_9N_3O_2Br_2$	34.98（35.21）	2.40（2.42）	11.18（11.21）
4g	4－Cl	$CH_2OC_6H_5$	EtOH	258～260	76	$C_{17}H_{14}N_3O_3Cl$	59.13（59.37）	4.13（4.11）	12.12（12.23）
4h	5－Br	$CH_2OC_6H_5$	CH_3COOEt	112～114	77	$C_{17}H_{14}N_3O_3Br$	52.45（52.57）	3.42（3.64）	10.71（10.83）
4i	3，5－Br_2	$CH_2OC_6H_5$	EtOH	232～236	89	$C_{17}H_{13}N_3O_3Br_2$	43.58（43.69）	2.70（2.81）	8.95（9.00）
4j	4－Cl	CH_2CN	EtOH + DMSO	259～261	64	$C_{12}H_9N_4O_2Cl$	51.63（52.07）	3.32（3.28）	20.70（20.26）
4k	5－Br	CH_2CN	EtOH	＞310	59	$C_{12}H_9N_4O_2Br$	44.70（44.86）	2.78（2.83）	17.47（17.45）
4l	3，5－Br_2	CH_2CN	EtOH	242～244	48	$C_{12}H_8N_4O_2Br$	35.92（36.01）	1.94（2.02）	13.95（14.01）

关于目标产物**4**的形成机制，可以设想在反应过程中形成了中间体**5**[9]，我们在合成一系列**4**的反应中发现部分反应经由透明至混浊再至透明的过程，说明此反应可能经过某种特定中间体。但中间体**5**的结构有待进一步确证。

目标产物**4**的^{1}H NMR数据列于表2。

表2　化合物的4a～4l的^{1}H NMR数据

Tab. 2　^{1}H NMR data for compounds 4a～4l

No.	^{1}H NMR（DMSO－d_6，TMS），δ
4a	2.14（s，3H，CH_3），7.75～8.08（m，3H，Ar—H），8.53（s，1H，H— N—N N ）
4b	2.12（s，3H，CH_3），7.49～8.15（m，3H，Ar—H），8.50（s，1H，H— N—N N ）
4c	2.10（s，3H，CH_3），8.15～8.43（m，2H，Ar—H），8.51（s，1H，H— N—N N ）
4d	2.07（s，6H，CH_3），7.78～8.10（m，3H，Ar—H）
4e	2.03（s，6H，CH_3），7.48～8.17（m，3H，Ar—H）
4f	2.02（s，6H，CH_3），8.18～8.48（m，2H，Ar—H）
4g	2.12（s，3H，CH_3），4.89～5.03（q，2H，CH_2，J=13Hz），6.79～8.08（m，8H，Ar—H）
4h①	2.23（s，3H，CH_3），4.67～5.12（q，2H，CH_2，J=12.7Hz），6.72～8.37（m，8H，Ar—H），10.17～11.01（br，1H，COOH）
4i	2.13（s，3H，CH_3），4.80～5.04（q，2H，CH_2，J=12.6Hz），6.81～8.43（m，7H，Ar—H）
4j	2.05（s，3H，CH_3），4.07（s，2H，CH_2），7.81～8.16（m，3H，Ar—H）
4k	2.05（s，3H，CH_3），4.01（s，2H，CH_2），7.53～8.24（m，3H，Ar—H）
4l	2.05（s，3H，CH_3），3.97（s，2H，CH_2），8.21～8.51（m，2H，Ar—H）

①solvent $CDCl_3$。

其中**4**g，**4**h，**4**i结构中的—CH_2单元的^{1}H NMR谱属于AB体系，这可能是因为分子的空间位阻所引起的，计算了它们的偶合常数J。除采用$CDCl_3$为溶剂的4h的—COOH结构单元在低场出现一个很宽的δ外，其余的几种化合物**4**用DM SO－d_6作溶剂测定的结构单元—COOH均未见其化学位移δ，可能与DM SO－d_6溶剂本身带有少量的HOD有关。

X_n O O N CH_3 **2** —$H_2NNHCOR$→ [X_n $CO_2^{\ominus}$ NH⋯$\overset{\oplus}{C}$—NH—NHCOR CH_3 **5**] —→ X_n N CH_3 N N N R COOH **4**

参 考 文 献

[1] Werner M, Wolfgang K, Karl H B. DE 2247 186 [P], 1974.
[2] Sugavanam B. *Pestic Sci* [J], 1984, 15 (3): 296.
[3] Martin J R, Tu NL, Muthuvelu T. EP 337 815 [P], 1989.
[4] Couderchet M W M, Schmalfub J, Boger P. Pestic Sci [J], 1998, 52 (4): 381.
[5] Wheeler A S, OatsW M. *J Am Chem Soc* [J], 1910, 32: 770.
[6] Drake N L. *Org Synth* [J], 1956, 24: 12.
[7] Reiseregger H. Ber 16 662 [P], 1883.
[8] Hillers D. *J Prakt Chem* [J], 1915, 92 (2): 313.
[9] Ried W, Peters B. *Liebigs Ann Chem* [J], 1969, 729: 124.
[10] Coppola G M. *J Heterocyclic Chem* [J], 1999, 36: 563.
[11] Hester Jr T B, Von Voigtlander P, Evenson G N. *J Med Chem* [J], 1980, 23: 873.

Synthesis of Halogen Substituted –2 –(3 –methyl –5 –substitued –4H –1,2,4 –triazol –4 –yl) benzoic Acid

Tiejun Luo[1], Zhengming Li[2], Suhua Wang[2], Ren'an Liao[2], Zhichun Li[2]

([1]Hunan Institute of Chemical Industry, Changsha; [2]State Key Laburatory of Elemento – Organic Chemistry, Nankai University, Tianjin, 300071)

Abstract The halogen substituted anthranilic acid (**1**) was condensed with acetic anhydride to give halogen substituted – 3 – methyl – 1H – 2, 4 – benoxazin – 1 – one (**2**), which was further condensed with hydrazide (**3**) to give twelve novel halogen substituted – 2 – (3 – methyl – 5 – substitued – 4H – 1, 2, 4 – triazol – 4 – yl) benzoic acid. All compounds synthesized were identified by ^{1}H NMR and elemental analyses.

Key words halogen substitued (methyl substitued – 4H – triazol – yl) benzoic acid; anthranilic acid; triazole; condensation reaction

5-依维菌素 B_{1a} 酯的合成和生物活性*

廖联安[1]　李正名[2]　方红云[1]　赵卫光[2]　范志金[2]　刘桂龙[2]

([1] 厦门大学化学系，厦门　361005；
[2]南开大学元素有机化学研究所　元素有机化学重点实验室，天津　300071)

摘　要　依维菌素 B_{1a} 与羧酸在 DMAP/DCC 体系中直接酯化，得到 10 个 5-IV B_{1a} 酯衍生物，产率 66%～82%；它们的化学结构得到 IR，^{1}H NMR，^{13}CNMR 和 MS 谱的确证。它们具有良好的杀虫、杀螨活性。

关键词　酯化　5-依维菌素 B_{1a} 酯　生物活性　DMAP/DCC

依维菌素是阿维菌素[1～3]的二氢化物，属于十六元环内酯衍生物。依维菌素的主要成分为 22，23-二氢 AVB_{1a}，根据测试，B_{1a} 杀虫活性最高，其他异构体杀虫活性较低而且毒性高。依维菌素对家畜的肠内寄生虫具有极其有效的驱虫活性，并具有作用机制独特、有效剂量低、安全性高等特点，广泛用于畜牧业和宠物保健；在医药上依维菌素用于治疗人类由于螨虫或线虫引起的疾病，极为有效[4]。

为进一步提高其活性和稳定性、扩展（改变）其活性谱、降低毒性，许多化学家对它的结构进行了各种改造，合成了许多衍生物，并从中推出了几个比母体化合物活性更好的化合物，如 4″-C 羟基的改造产物 Emamectin[5] 和 Eprinomectin[6]，其活性比母体化合物提高了 1～2 个数量级，正在商品化进程中。本文依据生物活性基团拼接原理，在母体分子中引入生物活性片段、杂环和杂原子，设计合成了 10 个 5-依维菌素 B_{1a} 酯，并初筛了它们的杀虫、杀螨活性。

1　结果与讨论

在早期文献[7,8]中 5-依维菌素 B_{1a} 酯的制备方法是用大过量的酸酐或酰氯与依维菌素反应，产率仅 30%～50%，而且选择性不理想，通常得到 5-酯和 4″-、5-双酯。DMAP 对酯（酰）化反应是优良的催化剂，与脱水剂 DCC 合用，羧酸可以在很温和的条件下直接与醇反应生成高产率的酯[9]。我们将这一体系应用于依维菌素 B_{1a} 酯的合成，发现用 1.1 当量的羧酸可以高效、高选择性地进行反应，产率良好。所制备的 10 个 5-依维菌素 B_{1a} 酯均系新化合物。化合物的化学结构得到 IR，^{1}H NMR，^{13}CNMR 和 MS 谱的确证。

由于依维菌素系列化合物对酸碱均不稳定，所以整个合成和后处理过程中，必须尽可能避免长时间与酸碱接触或处于高温状态，否则这些化合物都会发生异构化、分子内脱水和开环降解等副反应。因此，反应物羧酸必须先与 DCC 反应，生成物混合酸酐可以溶解在二氯甲烷中，然后再加入依维菌素原料和催化剂 DMAP，这样反应就可以顺利进行，反应中伴有大约 3%～5% 的 4″-位和 5-位双酯生成。大过量（1∶1.5 或更

* 原发表于《化学学报》，2002，60（3）：468-474。国家自然科学基金（No. 29832050）资助项目。

高）的羧酸会导致4″-位和5-位双酯的大量生成而影响分离。反应在室温下进行，当提高反应温度至50℃以上，会使C(3)═C(4)双键移位成C(2)═C(3)双键和7-位羟基的脱水。

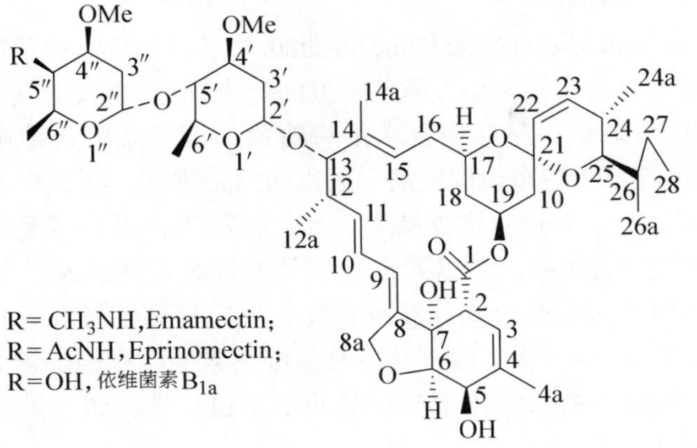

R= CH_3NH,Emamectin；
R= AcNH,Eprinomectin；
R= OH,依维菌素 B_{1a}

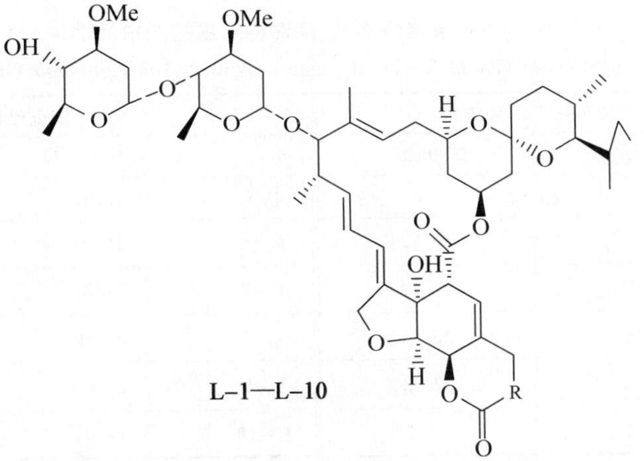

L-1—L-10

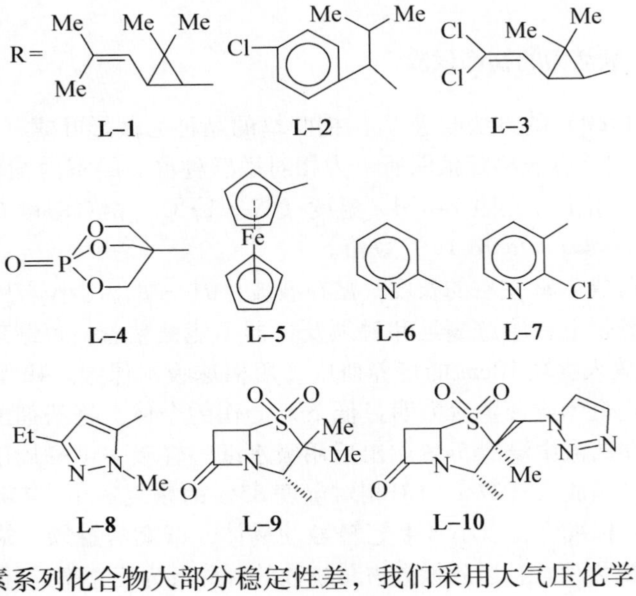

L-1　L-2　L-3

L-4　L-5　L-6　L-7

L-8　L-9　L-10

由于依维菌素系列化合物大部分稳定性差，我们采用大气压化学电离（APCI）和快原子轰击（FAB）来进行它们的质谱实验。

2 杀虫杀螨活性

2.1 对棉蚜杀虫活性的初筛试验

棉蚜（*Aphis gossypii* Glover）采自北京东北旺肖家河菜地的白瓜叶上。取干净新鲜的瓜叶一小片，用小号毛笔将大小、体色一致的蚜虫挑选30头于其上，将叶片连试虫一起浸入药液中2s后取出，立即用吸水纸将蚜虫周围多余的药液吸净，将叶片放入50mL的塑料杯中，杯口用纱布以橡皮筋套紧以防蚜虫逃逸。将试虫放入（27±1）℃相对湿度85%的恢复室中，16h后检查结果。死亡标准为，以小号毛笔轻触动虫体，试虫不动且无任何反应者视为死亡，以死虫数占总虫数的百分数为死亡率[10]。同时设清水对照，以清水对照按Abbott公式计算校正死亡率。每药剂设0.50mg/L和2.00mg/L两个浓度，各浓度重复2~3次。由于在商品依维菌素中含有的少量B_{1b}组分与B_{1a}组分的杀虫活性相近，纯化后的依维菌素与商品药剂杀虫活性一致，因此我们选用商品依维菌素为对照药剂（表1）。

表1 5－依维菌素B_{1a}酯对棉蚜虫的杀虫活性

Tab.1 Bioactivity of 5－IV B_{1a} esters against *Aphis gossypii* Glover

Compd.	校正死亡率（%）		Compd.	校正死亡率（%）	
	0.50mg/L	2.00mg/L		0.50mg/L	2.00mg/L
CK①	10.12		IV②	26.47	47.78
L－1	8.73	56.72	**L－6**	15.93	36.41
L－2	25.01	42.19	**L－7**	32.81	49.61
L－3	21.95	30.20	**L－8**	52.59	72.15
L－4	35.09	42.23	**L－9**	61.60	62.87
L－5	6.35	6.55	**L－10**	19.94	38.09

①CK为空白对照，26次平均值；②IV为依维菌素药剂对照，13次平均值。

2.2 对红蜘蛛杀螨活力的初筛试验

参照FAO（1980）的方法改进[11]。棉叶螨的品种为截形叶螨（*Tetranychus truncatus* Ehara），采自河北省农科院植保所，为相对敏感种群。经室内饲养纯化后使用。试螨饲养在光照培养箱中，光照14h/d，温度（27±1）℃，相对湿度60%~75%。饲料为新鲜的蚕（*Phaseolus vulgaris* L.）豆苗。

将双面胶带剪成2cm左右的长度，贴在载玻片的一端，用小号毛笔将3~4日龄的雌成螨背部粘在胶带上，注意螨足和触须及口器不能被粘着，每载玻片粘两行，每行10~15头。将其放入直径10cm的培养皿中，用棉球吸水保湿。4h后在双目解剖镜下用解剖针挑除死亡个体或者生活力明显低下及受伤的个体。将药剂配成一定浓度的水溶液后，将试螨在药液中浸泡5s，取出后用吸水纸轻轻吸去试螨周围多余的药液，避免损伤螨体。将试螨放入（27±1）℃相对湿度85%的恢复室中，24h后在双目解剖镜下检查结果。死亡标准为，以小号毛笔轻触动螨体，试螨的触须、螨足和口器无任何反应（不动）者视为死亡，以死螨数占总螨数的百分数为死亡率。同时设清水为对照，以清水对照按 Abbott 公式计算校正死亡率。药剂浓度设为 1.00mg/L，各浓度重复 2 次（表 2）。

表 2　依维菌素 B_{1a} 酯对红蜘蛛的杀螨活性

Tab. 2　Bioactivity of 5 – IV B_{1a} esters against *Tetranychus truncatus* Ehara①

Compd.	CK	IV	**L – 1**	**L – 2**	**L – 3**	**L – 4**	**L – 5**	**L – 6**	**L – 7**	**L – 8**	**L – 9**	**L – 10**
死亡率（%）	0	52.4	83.3	5.7	47.6	73.5	0	76.0	61.4	70.0	58.8	2.8

①10 次实验的平均值。

Mrozik 等[7]曾经对阿维菌素 5 – 位羟基进行了酯化研究，结果发现 5 – 位酯化产物的生物活性明显下降。然而从上述两表可以看出，不同 5 – 位酯化产物对活性有不同的影响。特别是化合物 **L – 1** 和 **L – 8** 对棉蚜和棉叶螨都具有较好的杀虫活性，在相同剂量下比商品依维菌素的杀虫活性提高了 0.3 到 1.3 倍。

3　实验部分

3.1　仪器和试剂

所用的试剂为化学纯或分析纯，溶剂经过重蒸后使用。山东齐鲁制药厂提供了依维菌素［w（B_1）=98.1%，w(B_{1a})=94.2%，w(B_{1b})=3.9%］样品。

X – 4 型数字显示显微熔点测定仪，温度计未校正。Alltech 426 液相色谱仪。Perkin Elmer Polarimeter 341 自动旋光仪（589nm，$CHCl_3$）。Shimadzu IR – 435 红外光谱仪，KBr 压片。Bruker ACP – 200 核磁共振仪、JEOL AL – 300 核磁共振仪和 Varian UNITY – plus 400 磁共振仪（$CDCl_3$，TMS 内标）。Fisons ZAB – HS 质谱仪（FAB）和 Finnigan MAT LCQ 质谱仪（APCI）。

3.2　依维菌素 B_{1a}的分离和纯化

从依维菌素商品中分离出各个组分，常用的方法是采用制备性液相色谱进行分离；但只能够得到很少量的样品，远远不能够满足合成上的需要。我们经过实验摸索，发现采用减压柱层析方法可以简便、大量地分离依维菌素商品中的 B_{1a}和 B_{1b}组分，回收率达到 92%，高于文献[8]的 75%，操作极为简便，并且层析柱可以重复使用。分离方法如下：

依维菌素［w(B_1)=98.1%，w(B_{1a})=94.2%，w(B_{1b})=3.9%］商品 3.00g 溶解在少量的二氯甲烷中，加入 1.0g 硅胶，减压旋干；取出放在一根装有 20g 硅胶的层析柱上，减压柱层析，用乙酸乙酯（EA）– 石油醚（PE，60 ~ 90℃）进行梯度淋洗，先用 1:2EA – PE 淋洗，洗脱出 AVB_{2a}和 AVB_{2b}；再用 1:1 EA – PE 淋洗，得白色固体粉末 IVB_{1a}2.60g，收率 87%，m. p. 155 ~ 157℃（154.5 ~ 157℃[2]）；^{1}H NMR δ：5.87（d，J =10.1Hz，1H，9 – H），5.74（m，2H，10 – H & 11 – H），5.34 ~ 5.43（m，3H，3 – H，1″ – H & 19 – H），4.98（d，J =6.8Hz，1H，15 – H），4.78（s，1H，1′ – H），4.69（s，2H，8a – H_2），4.30（d，J = 6.1Hz，1H，5 – H），3.98（d，J = 6.5Hz，1H，6 – H），3.95（d，J =3.6Hz，2H，17 – H & 13 – H），3.60 ~ 3.80（m，4H，5″ – H，5′ – H，3″ – H & 3′ – H），3.40（s，3H，OCH_3），3.39（s，3H，OCH_3），3.10 ~ 3.24（m，4H，4″ – H，4′ – H，2 – H & 25 – H），2.60 ~ 2.80（br，3H，3 × OH），

2.53 (m, 1H, 12 − H), 2.20 ~ 2.35 (m, 5H, 16 − H_2, 24 − H, 2′ − H_2), 1.78 ~ 2.03 (m, 5H, 4 − Me & 18 − H_2), 1.28 ~ 1.68 (m, 14H, 14 − Me, 20 − H_2, 26 − H, 27 − H_2, 2″ − H_2, 22 − H_2, 23H_2), 1.26 (d, J = 6.5Hz, 3H, 5″ − Me), 1.21 ~ 1.24 (d, J = 6.5Hz, 3H, 5′ − Me), 1.15 (d, J = 7.0Hz, 3H, 12 − Me), 0.90 (t, J = 7.0Hz, 3H, 28 − Me), 0.80 ~ 0.84 (d, J = 6.8Hz, 3H, 26 − Me), 0.76 (d, J = 7.2Hz, 3H, 24 − Me); ^{13}CNMR δ: 173.7 (1 − C), 139.5 (8 − C), 137.9 (11 − C), 137.8 (4 − C), 134.9 (14 − C), 124.6 (10 − C), 120.3 (9 − C), 118.2 (3 − C), 118.0 (15 − C), 98.4 (1″ − C), 97.4 (21 − C), 94.7 (1′ − C), 8.17 (13 − C), 80.3 (4′ − C), 79.3 (7 − C), 79.1 (6 − C), 78.1 (3′ − C), 77.4 (3″ − C), 77.0 (25 − C), 76.6 (4″ − C), 75.9 (5″ − C), 68.6 (17 − C), 68.3 (8a − C), 68.1 (5 − C), 67.6 (5′ − C), 67.1 (19 − C), 56.4 (OCH_3), 56.3 (OCH_3), 45.6 (2 − C), 41.1 (20 − C), 39.6 (12 − C), 36.8 (16 − C), 35.9 (26 − C), 35.4 (18 − C), 34.2 (2″ − C), 34.1 (2′ − C), 34.0 (24 − C), 31.1 (22 − C), 28.0 (23 − C), 27.2 (27 − C), 20.2 (12a − C), 19.9 (4a − C), 18.3 (6″ − C), 17.6 (6′ − C), 17.4 (24a − C), 15.1 (14a − C), 12.4 (26a − C), 12.0 (28 − C); IR (KBr) ν: 3443, 2966, 1733, 1457, 1381, 1160, 1050, 987cm^{-1}; MS (FAB) m/z (%): 891 (M^+ + Li, 100), 874 (M^+), 566 (M^+ − Ole − Ole − H_2O)。

3.3 5 − 依维菌素 B_{1a}酯的通用合成方法

在 50mL 锥形瓶中，加入 TAZOBACTAM 酸 132mg (0.44mmol)，N，N' − 二环己基碳二亚胺 (DCC) 91mg (0.44mmol)、4 − 二甲胺基吡啶 (DMAP) 10mg 和 25mL 二氯甲烷，电磁搅拌溶解，20min 后，加入依维菌素 B_{1a}350mg (0.40mmol)，室温下搅拌反应 6h；过滤除去生成的 N，N' − 二环己基脲 (DCU)，滤液用 3 × 10mL 水洗，无水硫酸钠干燥，旋转蒸发至干，得淡黄色固体。减压柱层析，用 EA − PE 进行梯度淋洗，得到 366mg 淡黄色固体，m. p. 138 ~ 140℃，$[\alpha]_D^{25}$ +49.9 (c 1.06, $CHCl_3$)，波谱数据证明其化学结构为 5 − TZ − IVB_{1a} (**L − 10**)。

按同样方法合成了 **L − 1** 至 **L − 9**，其理化数据和波谱数据如下：

L − 1, ^{1}H NMR δ: 5.82 ~ 5.84 (d, J = 9.6Hz, 1H, 9 − H), 5.72 ~ 5.75 (m, 2H, 10 − H &11 − H), 5.50 ~ 5.54 (d, J = 7.3Hz, 1H, 3 − H), 5.30 ~ 5.42 (m, 3H, Me_2C ═CH, 1″ − H & 19 − H), 4.98 ~ 5.02 (d, 1H, J = 6.8Hz, 15 − H), 4.87 ~ 4.89 (d, J = 3.5Hz, 1H, 1′ − H), 4.78 (d, J = 2.6Hz, 1H, 8a − H), 4.65 ~ 4.69 (m, 1H, 5 − H), 4.56 ~ 4.58 (d, J = 3.5Hz, 1H, 8a − H), 4.13 (br, 2H, 2 × OH), 3.84 ~ 4.04 (m, 3H, 6 − H, 17 − H & 13 − H), 3.62 ~ 3.69 (m, 4H, 5″ − H, 5′ − H, 3″ − H & 3′ − H), 3.34 ~ 3.43 (m, 2H, 25 − H & 2 − H), 3.18 ~ 3.22 (m, 2H, 4′ − H & 4″ − H), 2.52 (m, 1H, 12 − H), 2.16 ~ 2.32 (m, 7H, 16 − H_2, 24 − H, 环丙烷片段的 2 个 H & 2′ − H_2), 2.02 ~ 2.08 (m, 2H, 18 − H_2), 1.93 (s, 9H, 4 − Me & ═CMe_2), 1.57 ~ 1.82 (m, 14H, 20 − H_2, 14 − Me, 26 − H, 27 − H_2, 2″ − H_2, 22 − H_2 & 23 − H_2), 1.12 ~ 1.40 (m, 15H, 5′ − Me, 5″ − Me, 12 − Me & 环丙烷片段的 2 个 Me), 0.79 ~ 0.98 (m, 9H, 28 − Me, 26 − Me & 24 − Me); IR (KBr) ν: 3386, 2915, 1725, 1689, 1645, 1559, 1445, 1218, 1050, 996cm^{-1}; MS (APCI) m/z (%): 1024.5 (M^+, 16; 计算值, 1024.6123), 1047 (M^+ + Na, 8), 1006

（$M^+ - H_2O$，13）。

L－2，^1H NMR δ：7.265－7.313（m，4H，4－Cl－C_6H_4），5.793－5.817（m，1H，9－H），5.655～5.712（m，2H，10－H & 11－H），5.541（s，1H，3－H），5.486（d，*J*＝4.0Hz，1H，1″－H），5.352（m，1H，19－H），4.988（d，*J*＝7.0Hz，1H，15－H），4.763（s，1H，1′－H），4.607～4.653（m，2H，8a－H_2），4.415～4.445（dd，*J*＝9.6Hz，3.5Hz，1H，5－H），4.124（brs，2H，2×OH），3.941～3.961（m，1H，6－H），3.708～3.821（s，1H，13－H），3.119～3.591（m，8H，17－H，2×OCH_3 & 4″－H），2.920（d，*J*＝7.3Hz，1H，Ar－CH），2.513（m，1H，12－H），2.238～2.369（m，6H，16－H_2，24－H，2′－H_2 & Me_2CH），2.014（d，*J*＝6.5Hz，2H，18－H_2），1.620～1.745（m，18H，4－Me，14－Me，20－H_2，26－H，27－H_2，2″－H_2，22－H_2 & 23－H_2），1.030～1.502（m，15H，5′－Me，5″－Me，12－Me & CMe_2），0.652～0.919（m，9H，28－Me，26－Me & 24－Me）；^{13}C NMR δ：173.646（1－C），154.039（New，CO_2R），139.761（8－C），138.010（11－C），137.895（4－C），135.249（14－C），133.215，130.605，130.118，128.780，128.711，128.543（6×Ar－C），125.010（10－C），121.385（9－C），120.452（3－C），118.739（15－C），98.703（1″－C），97.732（21－C），95.163（1′－C），82.208（13－C），80.809（7－C），80.732（6－C），79.440（3′－C），78.326（3″－C），70.600（25－C），69.040（4″－C），67.480（4′－C），67.396（17－C），66.616（8a－C），60.154（5－C），59.833（5′－C），59.703（5″－C），58.624（19－C），56.835（OMe），56.705（OMe），50.534（Me_2－C），46.037（2－C），41.510（20－C），39.988（12－C），37.143（16－C），36.012（Ar－C），35.736（26－C），35.308（18－C），33.465（2′－C），33.090（2″－C），32.907（24－C），31.454（23－C），31.079（22－C），26.307（27－C），25.642（New Me），24.977（New Me），21.826（12a－C），20.442（4a－C），18.599（6″－C），17.643（6′－C），17.574（24a－C），15.356（14a－C），12.596（26a－C），12.305（28－C）；IR（KBr）ν：3395，2945，1732（br），1628，1510，1447，1113，1043，981cm^{-1}；MS（APCI）*m/z*（%）：1069.8（M^+，10；计算值，1069.7798），1093（M^+＋Na，18），1051（$M^+ - H_2O$，43）.

L－3，^1H NMR δ：6.26（m，1H，$Cl_2C=CH$），5.83（m，1H，9－H），5.71（m，2H，10－H & 11－H），5.28～5.60（m，3H，3－H，1″－H，19－H），4.97（d，*J*＝6.7Hz，1H，15－H），4.76（d，*J*＝3.5Hz，1H，1′－H），4.58～4.69（m，2H，8a－H_2），4.28（d，*J*＝6.1Hz，1H，5－H），3.60～3.98（m，3H，13－H，6－H & 17－H），3.73（brs，2H，2×OH），3.42（s，3H，OMe），3.38（s，3H，OMe），3.36～3.53（m，4H，5″－H，5′－H，7H，3″－H & 3′－H），3.10～3.26（m，4H，4″－H，4′－H，2－H & 25－H），2.52（m，1H，12－H），2.23～2.35（m，16－H，24－H，2′－H & 环丙烷片段的两个 H），1.78～1.96（m，5H，4－Me & 18－H_2），1.30～1.68（m，14H，14－Me，20－H_2，26－H，27－H_2，2″－H_2，22－H_2 & 23－H_2），1.13～1.28（m，15H，5″－Me，5′－Me，12－Me & 环丙烷片段的两个 Me），0.81～0.90（m，9H，28－Me，26－Me & 24－Me）；IR（KBr）ν：3408，2916，1690，1647，1527，1446，1245，1047，993cm^{-1}；MS（APCI）*m/z*（%）：

1066.3（M^+，13；计算值，1066.1838），1089（$M^+ + Na$，8），1048（$M^+ - H_2O$，26）.

L-4，^1H NMR δ：6.03～6.10（d，1H，J=8.7Hz，9-H），5.70～5.73（m，2H，10-H & 11-H），5.52（m，1H，3-H），5.34～5.41（m，2H，1″-H & 19-H），4.98（d，J=7.0Hz，1H，15-H），4.81～4.86（d，J=8.7Hz，6H，PO（OCH_2）$_3$C），4.70～4.78（m，3H，1′-H & 8a-H_2），4.26～4.32（d，J=6.1Hz，1H，5-H），4.07（brs，2H，2×OH），3.86～3.99（m，4H，13-H，6-H，17-H & 5″-H），3.62～3.71（m，3H，5′-H，3″-H & 3′-H），3.40（s，3H，O-Me），3.39（s，3H，O-Me），3.10～3.18（m，4H，4′-H，4″-H，2-H & 25-H），2.52（m，1H，12-H），2.23（m，5H，16-H_2，24-H & 2′-H_2），1.87～2.03（m，5H，4-Me & 18-H_2），1.53～1.74（m，14H，20-H_2，26-H，27-H_2，14-Me，2″-H_2，22-H_2 & 23-H_2），1.12～1.40（m，9H，5′-Me，5″-Me & 12-Me），0.79～0.98（m，9H，28-Me，26-Me & 24-Me）；IR（KBr）ν：3413，2926，1715，1583，1368，1115，1050，981cm^{-1}；MS（APCI）m/z（%）：1051.3（M^+，43；计算值，1051.1759），1074（$M^+ + Na$，18），1033（$M^+ - H_2O$，12）.

L-5，^1H NMR δ：5.856（d，J=9.8Hz，1H，9-H），5.753（m，2H，10-H & 11-H），5.600（d，J=4.5Hz，1H，3-H），5.390～5.425（m，1H，1″-H），5.297（m，1H，19-H），5.031（d，J=7.0Hz，1H，15-H），4.630～4.864（m，3H，1′-H & 8a-H_2），4.408（brs，2H，2×OH），4.242（s，10H，二茂铁环上的10个H），4.159（m，1H，5-H），3.942（s，1H，6-H），3.658～3.903（m，6H，13-H，17-H，5″-H，5′-H，3″-H & 3′-H），3.428（s，6H，2×O-Me），3.164～3.239（m，4H，4″-H，4′-H，2-H & 25-H），2.513（m，1H，12-H），2.257～2.323（m，5H，16-H_2，24-H & 2″-H_2），2.039（d，2H，18-H_2），1.869（s，3H，4-Me），1.296～1.735（m，14H，20-H_2，26-H，27-H_2，14-Me，2″-H_2，22-H_2 & 23-H_2），1.148～1.273（m，9H，5′-Me，5″-Me & 12-Me），0.792～1.024（m，9H，28-Me，26-Me & 24-Me）；^{13}C NMR δ：173.701（1-C），171.162（New，CO_2R），139.671（8-C），137.981（11-C），137.825（4-C），135.023（14-C），124.754（10-C），123.847（二茂铁环上的10个C），120.394（9-C），118.358（3-C），118.086（15-C），98.504（1″-C），97.507（21-C），94.820（1′-C），81.806（13-C），80.405（7-C），79.375（6-C），79.260（3′-C），78.246（3″-C），77.125（25-C），76.705（4″-C），76.029（4′-C），68.645（5″-C），68.389（17-C），68.175（8a-C），67.730（5-C），67.243（5′-C），60.386（19-C），56.505（O-Me），56.414（O-Me），45.716（2-C），41.216（20-C），39.766（12-C），36.906（16-C），35.752（26-C），35.464（18-C），34.508（2″-C），34.269（2′-C），34.096（24-C），31.219（22-C），28.079（23-C），27.288（27-C），20.225（12a-C），19.920（4a-C），18.404（6″-C），17.703（6′-C），15.140（14a-C），14.184（24a-C），12.428（26a-C），12.099（28-C）；IR（KBr）ν：3393，2916，1705，1625，1596，1451，1235，1045，890cm^{-1}；MS（FAB）m/z（%）：1087（M^+，63），1100（$M^+ + Na$，100）.

L-6，^1H NMR δ：8.56～8.57（d，J=4.0Hz，1H，Py-H），7.79～7.81（m，

1H, Py－H), 7.67～7.69 (d, J=8.0Hz, 1H, Py－H), 7.35～7.38 (m, 1H, Py－H), 5.90 (brd, J=12.5Hz, 1H, 9－H), 5.71～5.78 (m, 2H, 10－H & 11－H), 5.36～5.43 (m, 3H, 3－H, 1″－H, 19－H), 4.96 (d, J=7.1Hz, 1H, 15－H), 4.66～4.80 (m, 3H, 1′－H & 8a－H_2), 4.27 (d, J=6.0Hz, 1H, 5－H), 4.18～4.21 (br, 2H, 2×OH), 4.02 (d, J=3.5Hz, 1H, 6－H), 3.93～3.96 (d, J=6.1Hz, 2H, 17－H & 13－H), 3.65～3.83 (m, 4H, 5″－H, 5′－H, 3″－H & 3′－H), 3.42 (s, 6H, 2×O－Me), 3.12～3.26 (m, 4H, 4″－H, 4′－H, 2－H & 25－H), 2.55 (m, 1H, 12－H), 2.22～2.36 (m, 5H, 16－H_2, 24－H, 2″－H_2), 1.83～2.01 (m, 5H, 4－Me & 18－H_2), 1.30～1.69 (m, 14H, 14－Me, 20－H_2, 26－H, 27－H_2, 2′－H_2, 22－H_2 & 23－H_2), 1.05～1.26 (m, 9H, 5′－Me, 5″－Me & 12－Me), 0.79～0.96 (m, 9H, 28－Me, 26－Me & 24－Me); IR (KBr) ν: 3455, 2918, 1677, 1627, 1528, 1382, 1249, 1081, 1001cm^{-1}; MS (FAB) m/z (%): 980 (M^+, 26), 1003 (M^++Na, 13).

L－7, ^1H NMR δ: 8.42～8.44 (q, 1H, Py－H), 7.63～7.65 (q, 1H, Py－H), 7.28～7.31 (q, 1H, Py－H), 5.86～5.90 (d, J=10.3Hz, 1H, 9－H), 5.74～5.78 (m, 2H, 10H & 11－H), 5.34～5.42 (m, 3H, 3－H, 1″－H & 19－H), 4.92～5.00 (d, J=7.3Hz, 1H, 15－H), 4.72 (s, 1H, 1′－H), 4.64 (s, 2H, 8a－H_2), 4.28～4.32 (d, J=6.1Hz, 1H, 5－H), 4.12 (m, 3H, 6－H, 17－H & 13－H), 3.96 (brs, 2H, 2×OH), 3.60～3.70 (m, 4H, 5′－H, 5″－H, 3′－H & 3″－H), 3.43 (s, 3H, O－Me), 3.42 (s, 3H, O－Me), 3.24～3.35 (m, 4H, 4′－H, 4″－H, 2－H & 25－H), 2.59 (m, 1H, 12－H), 1.78～2.03 (m, 5H, 4－Me & 18－H_2), 1.49～1.67 (m, 10H, 14－Me, 20－H_2, 26－H, 27－H_2 & 2″－H_2), 1.01～1.32 (m, 13H, 22－H_2, 23－H_2, 5″－Me, 5′－Me & 12－Me), 0.80～0.94 (m, 9H, 28－Me, 26－Me & 24－Me); IR (KBr) ν: 3416, 2946, 1732, 1621, 1578, 1447, 1115, 1045, 981cm^{-1}; MS (FAB) m/z (%): 1014 (M^+, 6), 1037 (M^++Na, 23), 996 (M^+－H_2O).

L－8, ^1H NMR δ: 6.704 (s, 1H, Pyrazole－H), 5.842～5.869 (d, J=9.5Hz, 1H, 9－H), 5.655～5.782 (m, 2H, 10－H & 11－H), 5.602 (s, 1H, 3－H), 5.390～5.425 (d, J=3.6Hz, 1H, 1″－H), 5.321～5.349 (m, 1H, 19－H), 4.968 (d, J=7.1Hz, 1H, 15－H), 4.777 (m, 1H, 1′－H), 4.543～4.672 (m, 2H, 8a－H_2), 4.288 (d, J=6.0Hz, 1H, 5－H), 4.124 (brs, 2H, 2×OH), 3.961 (d, J=6.6Hz, 1H, 6－H), 3.933 (s, 1H, 13－H), 3.62～3.822 (m, 4H, 5′－H, 5″－H, 3′－H & 3″－H), 3.421 (d, 6H, 2×O－Me), 3.356 (s, 1H, 2－H), 3.163～3.283 (m, 3H, 4′－H, 4″－H & 25－H), 2.600～2.672 (m, 2H, Pz－CH_2), 2.512 (m, 1H, 12－H), 2.206～2.398 (m, 5H, 16－H_2, 24－H & 2′－H_2), 1.872～2.010 (m, 5H, 20－H_2 & 4－Me), 1.808 (s, 3H, PzNMe), 1.642～1.778 (m, 5H, 18－H_2 & 14－Me), 1.425～1.546 (m, 9H, 26－H, 27－H_2, 2″－H_2, 22－H_2, 23－H_2), 1.166～1.389 (m, 12H, 5′－Me, 5″－Me, 12－Me & Pz－CMe), 0.932 (t, 3H, 28－Me), 0.847 (d, J=6.8Hz, 3H, 26－Me), 0.778 (d, J=6.8Hz, 3H, 24－Me); ^{13}C NMR δ: 173.707 (1－C), 159.789 (New CO_2R), 139.578 (8－C), 138.324 (11－C), 138.171 (4－C), 135.280 (Pz－5－C),

135.171 (Pz-3-C), 133.483 (14-C), 124.918 (10-C), 120.582 (9-C), 118.747 (3-C), 118.318 (15-C), 109.662 (Pz-4-C), 98.787 (1″-C), 97.740 (21-C), 95.097 (1′-C), 82.071 (13-C), 81.467 (4′-C), 80.855 (7-C), 80.656 (6-C), 79.616 (3′-C), 79.471 (3″-C), 78.476 (25-C), 76.442 (4″-C), 76.045 (5″-C), 70.921 (PzNMe), 69.132 (17-C), 68.512 (8a-C), 68.405 (5-C), 67.985 (5′-C), 67.510 (19-C), 56.667 (OMe), 56.567 (OMe), 49.578 (PzCH), 46.037 (2-C), 41.510 (20-C), 40.049 (12-C), 37.182 (16-C), 36.027 (26-C), 35.767 (18-C), 34.520 (2″-C), 34.337 (2′-C), 34.054 (24-C), 31.469 (22-C), 28.334 (23-C), 27.546 (27-C), 20.365 (12a-C), 19.799 (4a-C), 18.614 (6″-C), 17.865 (6′-C), 17.628 (24a-C), 15.349 (14a-C), 13.903 (PzCMe), 12.580 (26a-C), 12.313 (28-C); IR (KBr) ν: 3415, 2943, 1716, 1613, 1571, 1450, 1112, 1045, 981cm^{-1}; MS (APCI) m/z (%): 1011.5 (M^+, 100, 计算值, 1011.264), 1034 (M^+ + Na, 8).

L-9, ^1H NMR δ: 5.946 (m, 1H, 9-H), 5.712~5.753 (m, 2H, 10-H & 11-H), 5.410 (s, 1H, 3-H), 5.326 (m, 2H, 1″-H & 19-H), 5.201 (m, 1H, $NCHSO_2$), 4.987 (d, J=7.3Hz, 1H, 15-H), 4.779 (d, J=3.6Hz, 1H, 1′-H), 4.692 (s, 1H, $NCHCO_2$), 4.625~4.723 (m, 2H, 8a-H_2), 4.301 (d, J=6.1Hz, 1H, 5-H), 4.143 (brs, 2H, 2×OH), 3.921~3.986 (m, 3H, 6-H, 17-H & 13-H), 3.601 (m, 4H, 5′-H, 5″-H, 3′-H & 3″-H), 3.472~3.490 (d, J=12.3Hz, 2H, CH_2CO), 3.432 (s, 3H, OMe), 3.421 (s, 3H, OMe), 3.142~3.264 (m, 4H, 4′-H, 4″-H, 2-H & 25-H), 2.638~2.677 (m, 5H, 12-H, 16-H_2 & 2′-H_2), 2.233 (m, 1H, 24-H), 2.007 (d, J=3.5Hz, 2H, 18-H_2), 1.932~1.956 (m, 3H, 4-Me), 1.715~1.830 (m, 2H, 20-H_2), 1.607~1.626 (3H, m, 14-Me), 1.450~1.500 (d, J=6.5Hz, 6H, CMe_2), 1.338~1.427 (m, 9H, 26-H, 27-H_2, 2″-H_2, 22-H_2 & 23-H_2), 1.097~1.275 (m, 5′-Me, 5″-Me & 9H, 12-Me), 0.813~0.942 (m, 9H, 28-Me, 26-Me & 24-Me); IR (KBr) ν: 3318, 2911, 1703, 1621, 1566, 1445, 1261, 1045, 983cm^{-1}. MS (APCI) m/z (%): 1090.1 (M^+, 46; 计算值, 1090.339), 1097 (M^+ + Li, 100), 1072 (M^+ - H_2O, 16).

L-10, ^1H NMR δ: 7.834 (s, 1H, =NCH=), 7.734 (s, 1H, NCH=), 5.857~5.881 (d, J=9.6Hz, 1H, 9-H), 5.741~5.793 (m, 2H, 10-H & 11-H), 5.307~5.411 (m, 3H, 3-H, 1″-H & 19-H), 5.149 (m, 1H, $NCHSO_2$), 4.969~4.993 (d, J=7.3Hz, 1H, 15-H), 4.898 (d, J=6.5Hz, 2H, $TrCH_2$), 4.785 (s, 2H, 1′-H & $NCHCO_2$), 4.685 (s, 2H, 8a-H_2), 4.297 (d, J=6.0Hz, 1H, 5-H), 4.203 (brs, 2H, 2×OH), 3.946~3.986 (m, 2H, 6-H & 13-H), 3.754~3.867 (m, 3H, 17-H, 5′-H & 5″-H), 3.589~3.650 (m, 2H, 3′-H & 3″-H), 3.430 (s, 6H, 2×O-Me), 3.378~3.458 (m, 2H, CH_2CON), 3.153~3.287 (m, 4H, 4′-H, 4″-H, 2-H & 25-H), 2.524 (m, 1H, 12-H), 2.219~2.353 (m, 5H, 16-H_2, 24-H & 2′-H_2), 1.831~1.993 (m, 5H, 4-Me & 18-H_2), 1.648~1.770 (5H, 14-Me & 20-H_2), 1.361~1.569 (m, 9H, 26-H, 27-

H_2, 2″ – H_2, 22 – H_2 & 23 – H_2), 1.126 ~ 1.334 (m, 12H, New Me, 5′ – Me, 5″ – Me & 12 – Me), 0.934 (t, 3H, 28 – Me), 0.848 (d, J = 7.3Hz, 3H, 26 – Me), 0.797 (d, J = 7.3Hz, 3H, 24 – Me); IR (KBr) ν: 3332, 2919, 1706, 1623, 1569, 1450, 1266, 1045, 981cm^{-1}; MS (APCI) m/z (%): 1156.6 (M^+, 16; 计算值, 1156.5501), 1163 (M^+ + Li, 45), 1174 (M^+ + H_2O, 100), 1180 (M^+ + Na), 1138 (M^+ – H_2O, 23).

参考文献

[1] Campbell, W C.; Clark, J. N. *New Zealand Veterinary Journal* 1981, 29, 174.

[2] Fisher, M. H. *Pure Appl. Chem.* 1990, 62, 1231.

[3] Daoquan, W. *Agricultruae Universitatis Pekinensis* 1994, 20 (4), 431 (in Chinese). (王道全, 中国农业大学学报, 1994, 20 (4), 431.)

[4] Meinke, P. T. *New Engl. J. Med.* 1995, 333 (1), 26.

[5] Loewe, M. F.; Ccvetovich, R. J.; DiMichele, L. M.; Shuman, R. F.; Grabowski, E. J. *J. Org. Chem.* 1994, 59 (25), 7870.

[6] Dorny, P.; Demeulenaere, D.; Smets, K.; Vercruysse, J. *Vet . Parasitol.* 2000, 89, 279.

[7] Mrozik, H. H.; Fisher, M. H; Kulsa, P. *US* 4201861, 1980 [*Chem. Abstr.* 1980, 92, 22780].

[8] Lian' an, L.; Zhengming, L. *Chem. Eng. Prog.* 1998, 17 (6), 43 (in Chinese). (廖联安, 李正名, 化工进展, 1998, 17, (6), 43.)

[9] Smith, A. B.; Thompson, A. S. *Tetrahedron Lett.* 1985, 26 (36), 4279.

[10] Mu, L. – Y. *The Research Method of Plant Chemical Protection*, Agricultural Press of China, Beijing, 1994, p. 56 (in Chinese). (慕立义主编, 植物化学保护研究方法, 中国农业出版社, 北京, 1994, p. 56.)

[11] *FAO Plant Production and Protection Paper* 21, 1980, pp. 49253.

Synthesis and Bioactivity of 5 – Ivermectin B_{1a} Esters

Lian' an Liao[1], Zhengming Li[2], Hongyun Fang[1], Weiguang Zhao[2], Zhijin Fan[2], Guilong Liu[2]

([1]Department of Chemistry, Xiamen University, Xiamen, 361005;
[2]Institute of Elemento – Organic Chemistry, Nankai University, Tianjin, 300071)

Abstract Ten 5 – Ivermectin B_{1a} ester derivatives were prepared from Ivermectin B_{1a} and carboxylic acids in the systems of DCC/DMAP. Their chemical structures were determined by IR, ^{1}H NMR, ^{13}C NMR and MS spectra. The preliminary results of bioassay showed that **L – 1 ~ L – 10** have good insecticidal activities.

Key words esterification; 5 – Ivermectin B_{1a}; bioactivity; DMAP/DCC

Synthesis and Bioactivity of 5 – Ivermectin B_{1a} Esters

OMe OMe OH H O O O O O O O O O OH O H O R O

L–1—L–10

R = Me Me Me Me Me Me Cl Me Me Cl Cl Me Me

L–1 L–2 L–3

O=P O O O Fe N N Cl

L–4 L–5 L–6 L–7

Et N–N Me O S O Me Me N O O S O N N=N N Me O N

L–8 L–9 L–10

LIAO, Lian – An; LI, Zheng – Ming; FANG, Hong – Yun; ZHAO, Wei – Guang; FAN, Zhi – Jin; LIU, Gui – Long

Acta Chimica Sinica **2002**, 60 (3), 468

The Conformation of α – Monosubstituted Cyclododecanone

O R H ⇌ O R H

WANG, Dao – Quan; YANG, Xiao – Liang; WANG, Ming – An; LIANG, Xiao – Mei; YOU, Tian – Ba

Acta Chimica Sinica **2002**, 60 (3), 475

The ring skeleton of the preferred conformations of α – monosubstituted cyclododecanones is still [3333], with the carbonyl group present in the 2 – C position. These molecules may adopt two different conformations in solution, the predominant α – side – *exo* – substituted and α – corner – *syn* – substituted conformation, which exist in dynamic equilibrium. The preferred conformation in the crystal isα – corner – *syn* – substituted conformation.

M (OTf)$_3$ Catalyzed Novel Mannich Reaction of N – Alkyoxycarbonylpyrrole or Thiophene, Formaldehyde and Primary Amine Hydrochlo – ride

$$\text{pyrrole(N-R)} + HCHO + HCl \cdot H_2NR^1 \xrightarrow[V(THF):V(H_2O) = 10:1]{M(OTf)_3(10\%mmol)} \text{pyrrole(N-R)}-CH_2NHR^1$$

ZHANG, Chuan – Xin; CHENG, Tie – Ming; LI, Run – Tao

Acta Chimica Sinica **2002**, 60 (3), 481

The novel Mannich of *N* – alkoxycarbonylpyrrole or thiophene, formaldehyde, primary amines hydrochlorides is catalyzed by M (OTf)$_3$ in aqueous media, Y (OTf)$_3$ was the best catalyst. The products are obtained in good yield under mild conditions.

Design, Synthesis and Biological Activity Determination of New Type Photosystem Ⅱ Inhibitors

LIU, Xiao – Lan; YANG, Xia; SUN, Ming; LIU, Xiao – Hong; ZHAO, Ru; MIAO, Fang – Ming

Acta Chimica Sinica **2002**, 60 (3), 487

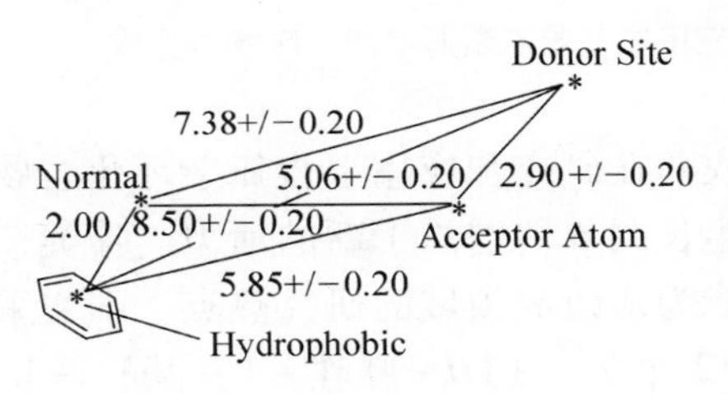

According to pharmacophore model, a series of new type photosystem Ⅱ inhibitors were designed, synthe – sized and characterized with element analysis, UV, IR and ^{1}H NMR. All compounds present the biological activities by Hill reaction.

2-(1*H*-咪唑-1-基)-1-(2, 3, 4-三甲氧基)苯乙酮肟酯新化合物合成与生物活性研究*

杨　松[1]，宋宝安[1]，李正名[2]，廖仁安[3]，刘　刚[1]，胡德禹[1]

([1]贵州大学精细化工研究开发中心，贵阳　550025；[2]南开大学国家农药工程中心，天津　300071；[3]南开大学元素有机国家重点实验室，天津　300071)

摘　要　首次合成了12个2-（1*H*-咪唑-1-基）-1-（2，3，4-三甲氧基）苯乙酮肟酯新化合物，并经元素分析、红外光谱、核磁共振氢谱对其结构进行了表征，初步生物活性测试表明，有些化合物具有杀菌活性。研究了这些新化合物结构与光谱特征之间的关系。

关键词　2，3，4-三甲氧基苯乙酮肟　酯　咪唑　合成

近年来，国内外农药在研究和应用上含氮杂环化合物的应用使原有的杀虫剂、杀菌剂、除草剂和植物生长调节剂增添了新的活力。唑类杂环化合物具有良好的生物活性且易生物降解，已成为杀菌剂领域的研究热点。本文采用焦性没食子酸为先导化合物，衍生设计和合成12个2-（1*H*-咪唑-1-基）-1-（2，3，4-三甲氧基）苯乙酮肟酯新化合物，利用元素分析、红外光谱、核磁共振氢谱表征了这些化合物结构。经文献查阅，该类工作未见文献报道。生物活性测定表明，该类化合物有良好的杀菌活性。研究了其结构和光谱特征之间的关系。化学反应式如Scheme1。

MeO, MeO, MeO, CCH_2Br (C=O) + H—N(咪唑) —DMF, 0～5℃→ MeO, MeO, MeO, CCH_2—N(咪唑) (C=O)

1 + $HONH_2\cdot HCl$ —回流, NH_3/CH_3CN→ **2** (C=N—OH, CH_2—N(咪唑))

* 原发表于《有机化学》，2002，22（5）：345-349。国家自然科学基金（No. 19789201）、元素有机国家重点实验室（2002）和贵州省优秀人才省长基金（20013）资助项目。

MeO MeO MeO CH_2—N (imidazole) C=N—OH **2** + RCOCl **3a～3l** —吡啶/$CHCl_3$, 20～30℃→ MeO MeO MeO CH_2—N (imidazole) C=N—OC(=O)—R **4a～4l**

	a	b	c	d	e	f
R	CH_3	C_6H_5	$p-CH_3OC_6H_4$	$o-ClC_6H_4$	$p-ClC_6H_4$	$m-ClC_6H_4$
	g	**h**	**i**	**j**	**k**	**l**
R	$m-CH_3OC_6H_4$	$p-FC_6H_4$	$m-CH_3C_6H_4$	$o-CH_3OC_6H_4$	$o-FC_6H_4$	2 - furyl

Scheme 1

1 实验

1.1 仪器与试剂

元素分析用日本柳本 MT - 3 型元素分析仪；IR 用日本岛津 IR - 435 红外光谱仪测定，KBr 压片；^{1}H NMR 用 Bruker AC - P200 型核磁共振仪测定，$CDCl_3$ 为溶剂，TMS 为内标；质谱用 HP6890A 质谱仪；熔点用 Yanaco MT - 500 熔点仪，温度计未校正。实验用的试剂均为国产分析纯或化学纯试剂。

1.2 中间体制备

溴 - 1 - （2，3，4 - 三甲氧基）苯乙酮按文献［1］方法合成，2 - （1*H* - 咪唑 - 1 - 基） - 1 - （2，3，4 - 三甲氧基）苯乙酮按文献［2］方法合成，酰氯（**3a～3l**）按文献［3］方法合成。

1.3 2 - (1*H* - 咪唑 - 1 - 基) - 1 - (2，3，4 - 三甲氧基)苯乙酮肟(2)的制备

将盐酸羟胺（4mmol，2.8g）加入到盛有 15mL 无水乙腈的 50mL 三颈瓶中，控制温度在 10～20℃，通入氨气（4.2mmol）；控制温度在 15～25℃，滴加 **1**（4mmol，1.1g）和 5mL 无水乙腈混合液，15min 滴加完毕，在 10～15℃反应 1h，加热升温到 80℃，回流 6h，负压脱溶回收乙腈溶剂，残留物加入到 40mL 碎冰水中，析出固体，过滤，硅胶柱层析纯化（硅胶 100～200 目，洗脱剂正己烷: 四氢呋喃 = 1: 1，体积比），分离得 0.85g 无色晶体 **2**，产率 72.8%，m.p. 163～164℃；^{1}H NMR δ：3.68～4.22（m，9H，3 × MeO），5.30（d，2H，$CH_2N=$），6.55～6.99（m，2H，ArH），7.20（s，1H，=NCH=N），7.90（s，1H，=NCH=C），8.20（s，1H，C=CHN=），12.20（s，1H，C=NOH）；Anal. calcd for $C_{14}H_{17}N_3O_4$：C 57.73，H 5.88，N 14.43；found C 57.58，H 5.72，N 14.50。

1.4 化合物4a~4l的合成

在100mL三口瓶中，加入11.6g（4mmol）**2**，用40mL氯仿溶解，加入0.4g（5mmol）无水吡啶，控制温度0~5℃，滴加0.32g（4mmol）乙酰氯和10mL无水氯仿混合液，反应1h后，升温到20~30℃，保温搅拌16h，负压洗脱回收苯，残留物注入60mL碎冰水中，析出固体，过滤，硅胶柱层析纯化（硅胶100~200目，洗脱剂丙酮: 石油醚（60~90℃）=1∶1，体积比），分离可得1.1g白色固体**4a**，产率81.2%（以**2**计算）。化合物**4b~4l**的合成方法[4,5]与**4a**相同。**4a~4l**物理数据见表1，IR和^{1}H NMR数据见表2。

表1 化合物4a~4l的物理数据和实验数据

Tab.1 Physical data and elemental analysis results of 4a~4l

Compd.	R	Formula	m.p. (℃)	Appearance	Yield (%)	Elemental analysis found(calcd.,%) C	H	N
4a	CH_3	$C_{16}H_{19}N_3O_5$	110~111	White solid	78.0	57.42(57.65)	5.48(5.75)	12.9(12.61)
4b		$C_{21}H_{21}N_3O_5$	120~121	White solid	62.4	63.58(63.74)	5.13(5.35)	10.49(10.63)
4c	OMe	$C_{22}H_{23}N_3O_6$	131~132	White solid	84.7	62.58(62.11)	5.17(5.45)	9.74(9.85)
4d	Cl	$C_{21}H_{20}ClN_3O_5$	134~135	White solid	82.2	58.29(58.68)	4.52(4.69)	9.36(9.78)
4e	Cl	$C_{21}H_{20}ClN_3O_5$	138~139	White solid	85.3	58.45(58.68)	4.48(4.69)	9.45(9.78)
4f	Cl	$C_{21}H_{20}ClN_3O_5$	140~141	White solid	83.7	58.51(58.68)	4.52(4.69)	9.62(9.78)
4g	OMe	$C_{22}H_{23}N_3O_6$	128~129	White solid	80.7	62.09(62.11)	5.32(5.45)	9.69(9.85)
4h	F	$C_{21}H_{20}FN_3O_5$	129~130	White solid	76.2	59.91(61.02)	4.62(4.84)	10.01(10.17)
4i	Me	$C_{22}H_{23}N_3O_5$	127~129	White solid	92.8	64.07(64.54)	5.53(5.66)	10.11(10.26)
4j	MeO	$C_{22}H_{23}N_3O_6$	110~112	White solid	92.4	62.15(62.11)	5.53(5.45)	9.78(9.88)
4k	F	$C_{21}H_{20}FN_3O_5$	111~113	White solid	90.5	60.99(61.01)	4.78(4.88)	9.79(10.16)
4l	O	$C_{19}H_{19}N_3O_6$	134~136	White solid	83.1	58.73(58.92)	4.70(4.97)	10.90(10.90)

表 2　化合物 4a ~ 4l 的 IR 和 ^{1}H NMR 数据

Tab. 2　^{1}H NMR and IR spectral data of compounds 4a ~ 4l

Compd.	IRν/ (cm^{-1})		^{1}H NMR, δ
	C ═N	O ═C—O	
4a	1470	1711	2.25 (s, 3H, CH_3), 3.77 (s, 3H, MeO), 3.82 (s, 3H, MeO), 3.88 (s, 3H , MeO), 5.47 (s, 2H, CH_2N), 6.57 ~ 7.03 (m, 2H, ArH), 7.69 (s, 1H, NCH ═N), 7.92 (s, 1H, NCH ═C), 8.26 (s, 1H, C ═CHN ═)
4b	1485	1725	3.76 (s, 3H, MeO), 3.83 (s, 3H, MeO), 3.92 (s, 3H, MeO), 5.70 (s, 2H, CH_2N), 6.59 ~ 7.48 (m, 7H, ArH), 7.72 (s, 1H, NCH ═N), 8.01 (s, 1H, NCH ═C), 8.31 (s, 1H, CCHN ═)
4c	1481	1718	3.76 (s, 3H, MeO), 3.79 (s, 3H, MeO), 3.85 (s, 3H, MeO), 3.88 (s, 3H, MeO), 5.38 (s, 2H, CH_2N), 6.56 ~ 7.01 (m, 6H, ArH), 7.59 (s, 1H, NCH ═N), 7.96 (s, 1H, NCH ═C), 8.28 (s, 1H, C ═CHN ═)
4d	1476	1720	3.79 (s, 3H, MeO), 3.82 (s, 3H, MeO), 3.91 (s, 3H, MeO), 5.42 (s, 1H, CH_2N), 6.56 ~ 7.53 (m, 6H, ArH), 7.67 (s, 1H, NCH ═N), 7.92 (s, 1H, NCH ═C), 8.20 (s, 1H, C ═CHN ═)
4e	1492	1730	3.79 (s, 3H, MeO), 3.82 (s, 3H, MeO), 3.92 (s, 3H, MeO), 5.60 ~ 5.75 (d, 2H, CH_2N), 6.62 ~ 7.45 (m, 6H, ArH), 7.61 (s, 1H, NCH ═ N), 7.98 (s, 1H, NCH ═C), 1.21 (s, 1H, C ═CHN ═)
4f	1482	1727	3.87 (s, 3H, MeO), 3.92 (s, 3H, MeO), 4.01 (s, 3H, MeO), 5.28 (s, 1H, CH_2N), 6.62 ~ 7.71 (m, 6H, ArH) 7.81 (s, 1H, NCH ═N), 8.03 (s, 1H, NCH ═C), 8.26 (s, 1H, C ═CHN ═)
4g	1472	1718	3.72 (s, 3H, MeO), 3.84 (s, 3H, MeO), 3.92 (s, 3H, MeO), 5.21 (s, 1H, CH_2N), 6.51 ~ 7.71 (m, 6H, ArH), 7.79 (s, 1H, NCH ═N), 7.91 (s, 1H, NCH ═C), 8.21 (s, 1H, C ═CHN ═)
4h	1482	1729	3.78 (s, 3H, MeO), 3.82 (s, 3H, MeO), 3.88 (s, 3H, MeO), 3.91 (s, 3H, MeO), 5.33 (s, 1H, CH_2N), 6.39 ~ 7.41 (m, 6H, ArH), 7.76 (s, 1H, NCH ═N), 7.90 (s, 1H, NCH ═C), 8.19 (s, 1H, C ═CHN ═)
4i	1473	1721	2.41 (s, 3H, CH_3), 3.78 (s, 3H, MeO), 3.79 (s, 3H, MeO), 3.81 (s, 3H, MeO), 3.89 (s, 3H, MeO), 5.52 (s, 1H, CH_2N), 6.61 ~ 7.40 (m, 6H, ArH), 7.70 (s, 1H, NCH ═N), 7.86 (s, 1H, NCH ═C), 8.09 (s, 1H, C ═CHN ═)
4j	1480	1728	3.79 (s, 3H, MeO), 3.83 (s, 3H, MeO), 3.87 (s, 3H, MeO), 5.39 (s, 1H, CH_2N), 6.52 ~ 7.40 (m, 6H, ArH), 7.43 (s, 1H, NCH ═N), 7.81 (s, 1H, NCH ═C), 7.91 (s, 1H, C ═CHN ═)
4k	1487	1739	3.78 (s, 3H, MeO), 3.80 (s, 3H, MeO), 3.87 (s, 3H, MeO), 5.44 (s, 1H, CH_2N), 6.54 ~ 7.23 (m, 6H, ArH), 7.50 (s, 1H, NCH ═N), 7.92 (s, 1H, NCH ═C), 8.12 (s, 1H, C ═CHN ═)
4l	1421	1710	3.77 (s, 3H, MeO), 3.81 (s, 3H, MeO), 3.90 (s, 3H, MeO), 5.36 (s, 2H, CH_2N), 6.53 ~ 7.42 (m, 5H, ArH + FuH), 7.60 (s, 1H, NCH ═N), 7.79 (s, 1H, NCH ═C), 8.09 (s, 1H, C ═CHN)

2 结果与讨论

2.1 波谱性质和结构

肟酯系列化合物选了代表性的化合物**4k**做MS测试，有明显的分子离子峰（M^+ 413）及分子离子峰失去甲氧自由基MeO·的裂解形成 *m/z* 382的碎片；形成碎片Ⅰ（*m/z* 193），以及 *m/z* 193碎片失去一个甲基后形成的 *m/z* 178碎片Ⅱ，除此之外也形成 *m/z* 81的碎片即碎片Ⅲ，以及Ⅲ失去HCN形成的 *m/z* 54碎片Ⅳ，另外，还有强度很高的 *m/z* 123碎片Ⅴ生成以及随后的特征的苯环系列的裂解即形成 *m/z*95和 *m/z*75的碎片（Ⅵ和Ⅶ）见Scheme 2。

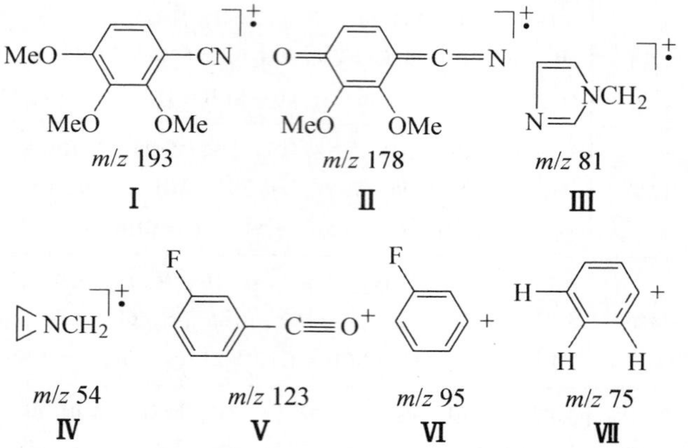

m/z 193 Ⅰ　*m/z* 178 Ⅱ　*m/z* 81 Ⅲ

m/z 54 Ⅳ　*m/z* 123 Ⅴ　*m/z* 95 Ⅵ　*m/z* 75 Ⅶ

Scheme 2

2.2 生物活性

在离体条件下，用含毒介质法对合成的12个新化合物进行生物活性测定，测定菌种为棉花枯萎病菌（*Fusarium vasinfectum Atk*）、棉花立枯病菌（*Rhizoctonia solani Kuehn*）。结果表明，在测定浓度为 50×10^{-6} 下，该新化合物有良好抑菌活性，其中**4e**对棉花枯萎病的抑制率为100%，**4h**对棉花立枯病的抑制率为100%，药剂浓度为 5×10^{-6} 下，**4h**对棉花立枯病的抑制率为100%。

References

[1] Horton, W. J.; Jach, T. S. *J. Am. Chem. Soc.* 1955, 77, 2894.

[2] Nardii, D.; Tajanu, A.; Leonardi, A. *J. Med. Chem.* 1981, 24, 724.

[3] Li, S. W. *The Preparation Handbook of Applied Organic Chemistry*, Shanghai Science Publishing House, Shanghai, 1981, p. 343 (in Chinese).

[4] Song, B. A.; Li, Z. M.; Li, S. Z. *Prog. Nat. Sci.* 1992, 2, 532.

[5] Song, B. A.; Hu, D. Y.; Zeng, S.; Huang, R. M.; Yang, S.; Hung, J. *Chin. J. Org. Chem.* 2001, 21, 524 (in Chinese).
(宋宝安，胡德禹，曾松，黄荣茂，杨松，黄剑，有机化学，2001，21，524.)

Synthesis and Bioactivity of 2 –(1*H* – Imidazol – 1 – yl) – 1 –(2,3,4 – trimethoxy) acetophenoxime Ester Derivatives

Song Yang[1], Bao'an Song[1], Zhengming Li[2], Ren'an Liao[3], Gang Liu[1], Deyu Hu[1]

([1]Research and Development Centre of Fine Chemicals, Guizhou University, Guiyang, 550025;
[2]National Pesticide Engineering Research Centre, Tianjin, 300071;
[3]State Key Laboratory of Elemento – Organic Chemistry, Nankai University, Tianjin, 300071)

Abstract Twelve new 2 – (1*H* – imidazol – 1 – yl) – 1 – (2, 3, 4 – trimethoxy) acetophenoxime esters were synthesized and characterized by elemental analysis, IR and ^1H NMR spectra. The bioassay of some compounds showed that they had good fungicidal activity. The relationship between the structure of these new compounds and their spectral characters is discussed.

Key words 2, 3, 4 – trimethoxyacetophenoxime; ester; imidazole; synthesis

1，3－二甲基－5－甲硫基－4－苯腙基羰基吡唑的合成及抑菌活性*

范志金　钟　滨　王素华　李正名

（南开大学元素有机化学研究所，元素有机化学国家重点实验室，天津　300071）

关键词　吡唑衍生物　合成　抑菌活性

吡唑衍生物具有广泛的生物活性，已成功开发出多种医药和农药新品种[1~3]。近年来，吡唑类衍生物已成为农药研究的热点之一。考虑到酰腙是一类很好的活性亚结构[4]，本文以1，3－二甲基－5－甲硫基－4－肼基羰基吡唑为原料按以下反应设计合成了8种新型的1，3－二甲基－5－甲硫基－4－苯腙基羰基吡唑衍生物，并测定了抑制农作物病菌活性，以期筛选出具有较高抑菌活性的化合物。

H_3C　$COOC_2H_5$ … N—N(H)，SCH_3 (**1**) $\xrightarrow{(CH_3)_2SO_4/NaOH}$ H_3C　$COOC_2H_5$ … N—N(CH_3)，SCH_3 (**2**) $\xrightarrow{NH_2NH_2\cdot H_2O(85\%)}$ H_3C　$CONHNH_2$ … N—N(CH_3)，SCH_3 (**3**) $\xrightarrow{RCHO}$ H_3C　$CONHN{=}CHR$ … N—N(CH_3)，SCH_3 (**4**)

仪器和试剂：Yanaco CHN CORDER MT－3 型自动元素分析仪；BRUKER ACP－200 型核磁共振仪，TMS 为内标；NICOLET 5DX 型红外光谱仪；Yanaco MP－500 型熔点仪，温度计未经校正。所用试剂均为国产分析纯或化学纯。原料 **1** 参照文献［5，6］方法合成，mp 130～131 ℃（文献值：129～130℃）。

1，3－二甲基－5－甲硫基－4－乙氧羰基吡唑（**2**）的合成：将 15.0g（75mmol）**1** 溶于 105mL 丙酮，冰水浴冷却到10℃左右，分批加入 6g（150mmol）NaOH 粉末，在搅拌下慢慢滴加 9.75g（78mmol）硫酸二甲酯，滴毕，室温搅拌 4h。抽滤除去固体，滤液脱掉丙酮得粗品，乙醇重结晶得白色晶体 210.8g，产率 67%，mp 83～84℃。^{1}H NMR，δ_H（$CDCl_3$）：1.28（t，3H，CH_3）；2.42（s，3H，S—CH_3）；2.45（s，3H，3－CH_3）；3.71（s，3H，N—CH_3）；4.22（q，2H，OCH_2）。

1，3－二甲基－5－甲硫基－4－肼基羰基吡唑（**3**）的合成：将 10.0g（46.7mmol）**2** 加到 20mL 无水乙醇中，在搅拌下加入 32mL（28mmol）85% 的水合肼，回流反应 5h，冷却后过滤收集固体，用乙醇重结晶得白色晶体 **3** 为 8.6g，产率 92%，mp 137～138℃。^{1}H NMR，δ_H（DMSO－d_6）：2.40（s，3H，SCH_3）；2.49（s，3H，3－CH_3）；3.71（s，3H，N—CH_3）；3.42（s，2H，NH_2）；7.84（d，1H，NH）。

* 原发表于《应用化学》，2003，20（4）：365－367。国家自然科学基金资助课题（30270883）。

目标化合物**4**的合成：将5mmol（0.93g）**3**，30mL无水乙醇，5mmol取代苯甲醛和催化剂量的乙酸加到反应瓶中，搅拌回流，TLC跟踪反应（展开剂V（石油醚）：V（乙酸乙酯）=3:2）至完全，冷却析出固体、过滤，用乙醇洗涤固体，并由乙醇或DMF/乙醇混合溶剂重结晶，得到目标化合物**4**。

结果与讨论

所得化合物的结构和物理化学常数见表1，^{1}H NMR和IR数据见表2。

表1　化合物4a～4h的熔点、收率和元素分析数据

Tab. 1　Melting point, yield and elemental analysis data of 4a～4h

Compd.	R	Formula	Appearance	mp(℃)	Yield (%)	Elemental analysis(calcd.,%) C	H	N
4a	C_6H_5	$C_{14}H_{16}N_4OS$	white crystal	190～191	86	58.21(58.31)	5.62(5.59)	19.64(19.43)
4b	3-$NO_2C_6H_5$	$C_{14}H_{15}N_5O_3$	yellow crystal	236～238	87	50.65(50.44)	4.77(4.54)	21.32(21.01)
4c	4-$NMe_2C_6H_5$	$C_{16}H_{21}N_5OS$	yellow crystal	190～191	81	58.18(57.98)	6.51(6.39)	21.34(21.13)
4d	C_6H_5—CH═CH—	$C_{16}H_{18}N_4OS$	white crystal	204～206	79	61.42(61.12)	5.99(5.77)	18.03(17.82)
4e	4-ClC_6H_5	$C_{14}H_{15}ClN_4S$	white crystal	196～197	84	52.01(52.09)	4.83(4.68)	17.57(17.36)
4f	4-HOC_6H_5	$C_{14}H_{16}N_4O_2$	white crystal	261～263	80	55.47(55.25)	5.55(5.30)	18.63(18.41)
4g	3-CH_3O-4-C_6H_5	$C_{15}H_{18}N_4O_3$	white crystal	206～208	83	53.71(53.88)	5.62(5.43)	16.89(16.75)
4h	4-$CH_3C_6H_5$	$C_{15}H_{18}N_4OS$	white crystal	170～171	85	59.79(59.58)	6.29(6.00)	18.71(18.53)

表2　化合物4a～4h的^{1}H NMR和IR数据

Tab. 2　^{1}H NMR and IR data of compounds 4a～4h

Compd.	Solvent	^{1}H NMR, δ	IR, σ/cm^{-1}
4a	$CDCl_3$	2.25（s，3H，SCH_3）；2.36（s，3H，3-CH_3）；3.58（s，3H，N—CH_3）；7.32～7.73（m，5H，Ph—H）；7.92（s，1H，N═CH）；9.72（s，1H，NH）	3245（ν_{N-H}）；3114，2977（$\nu_{=C-H}$，Ph）；1624（$\nu_{C=O}$）；1597（$\nu_{C=N}$）；1541，1407（$\nu_{=C-H}$，Ph）；[964.5]
4b	DMSO-d_6	2.35（s，3H，SCH_3）；2.40（s，3H，3-CH_3）；3.74（s，3H，N—CH_3）；7.73～8.31（m，4H，Ph—H）；8.62（s，1H，N═CH）；9.83（s，H，NH）	3393（ν_{N-H}）；3 115，2974（$\nu_{=C-H}$，Ph）；1632（$\nu_{C=O}$）；1599（$\nu_{C=N}$）；1544，1498（$\nu_{=C-H}$，Ph）；[984.0]
4c	$CDCl_3$	2.47（s，3H，SCH_3）；2.54（s，6H，N—CH_3）；2.60（s，3H，3-CH_3）；3.79（s，3H，N—CH_3）；7.62～7.94（m，4H，Ph—H）；8.19（s，1H，N═CH）；9.96（s，1H，NH）	3393（ν_{N-H}）；3130，3020（$\nu_{=C-H}$，Ph）；2910，（ν_{C-H}，CH_3）；1640（$\nu_{C=O}$）；1552（$\nu_{C=N}$）；1612，1521（$\nu_{=C-H}$，Ph）；[966.5]
4d	DMSO-d_6	2.48（s，3H，SCH_3）；2.59（s，3H，3-CH_3）；3.80（s，3H，N—CH_3）；6.85～7.17（m，2H，CH═CH）；7.42～7.49（m，5H，Ph—H）；7.91（s，1H，N═CH）；9.79（s，1H，NH）	3390（ν_{N-H}）；3181，3020（$\nu_{=C-H}$，Ph）；2912，（ν_{C-H}，CH_3）；1641（$\nu_{C=O}$）；1621（$\nu_{C=C}$）；1489，1579（$\nu_{=C-H}$，Ph）；1542（$\nu_{C=N}$）；[990.8]

续表 2

Compd.	Solvent	^{1}H NMR, δ	IR, σ/cm^{-1}
4e	DMSO - d_6	2.34 (s, 3H, SCH_3); 2.50 (s, 3H, 3 - CH_3); 3.73 (s, 3H, N—CH_3); 7.48 ~ 7.67 (dd, 4H, Ph—H); 8.15 (s, 1H, N=CH); 9.82 (s, 1H, NH)	3392 (ν_{N-H}); 3240, 3044 ($\nu_{=C-H}$, Ph); 2913, (ν_{C-H}, CH_3); 1647 ($\nu_{C=O}$); 1603 ($\nu_{C=N}$); 1566, 1488 ($\nu_{=C-H}$, Ph); [965.4]
4f	DMSO - d_6	2.33 (s, 3H, SCH_3); 2.39 (s, 3H, 3 - CH_3); 3.72 (s, 3H, N—CH_3); 6.79 ~ 7.48 (dd, 4H, Ph); 8.08 (s, 1H, N=CH); 9.92 (s, 1H, NH)	3646 ~ 3143 (ν_{O-H}, ν_{N-H}); 3143, 2882 (ν_{C-H}, CH_3); 1640 ($\nu_{C=O}$); 1595, 1509 ($\nu_{=C-H}$, Ph); 1558 ($\nu_{C=N}$); [973.8]
4g	DMSO - d_6	2.39 (s, 3H, SCH_3); 2.47 (s, 3H, 3 - CH_3); 2.49 (s, 3H, OCH_3); 3.78 (s, 3H, N—CH_3); 6.86 ~ 7.32 (m, 3H, Ph—H); 8.12 (s, 1H, N=CH); 9.41 (s, 1H, OH), 9.71 (s, 1H, NH)	3374 (ν_{N-H}); 3137, 3020 ($\nu_{=C-H}$, Ph); 1636 ($\nu_{C=O}$); 1509, 1593 ($\nu_{=C-H}$, Ph); 1558 ($\nu_{C=N}$); [969.9]
4h	$CDCl_3$	2.45 (s, 3H, SCH_3); 2.51 (s, 3H, Ph—CH_3); 2.64 (s, 3H, 3 - CH_3); 3.89 (s, 3H, N—CH_3); 7.31 ~ 7.76 (dd, 4H, Ph—H); 7.85 (s, 1H, N=CH); 9.58 (s, 1H, NH)	3399 (ν_{N-H}); 3202, 3051 ($\nu_{=C-H}$, Ph); 1639, ($\nu_{C=O}$); 1507, 1602 ($\nu_{=C-H}$, Ph); 1578 ($\nu_{C=N}$) [962.0]

注：IR data were obtained using KBr pellet, data in [ ] was infrared data of *E* isomer.

化合物**1**在甲基化时通常是在强碱条件下用碘甲烷做甲基化试剂，此法虽然收率高，但碘甲烷价格昂贵且容易挥发，改用硫酸二甲酯做甲基化试剂，在 2 倍摩尔量碱的存在下，反应速度较快，化合物**2**的收率也较高。利用中间体**3**与芳香醛在加热条件下进行亲核加成消除反应合成了 8 种目标化合物。实验中发现，当苯甲醛的苯环上连有吸电子取代基（如 NO_2、Cl 等）时，不需催化剂可在短时间内反应完全，若连有供电子取代基（如 CH_3、OCH_3、N（CH_3）$_2$ 等）时，需要加入少量冰醋酸作催化剂才能反应完全，不加冰醋酸作催化剂，长时间也难反应完全。

目标化合物的^{1}H NMR 谱中，除苯环上的 H 外其余质子谱均为单峰。因此，目标化合物在溶剂中只以一种几何异构体存在。IR 谱图（KBr 压片法）在 960 ~ 980cm^{-1}处有特征吸收证明其为反式 *E* 式异构体（表 2 IR 中的数据）。

利用菌体生长速率测定法[7]对上述化合物进行了离体筛选，结果见表 3。由表 3 可见，在质量浓度为 50mg/L 条件下大部分化合物具有一定抑菌活性，所有化合物对芦笋茎枯病菌（*Phoma asparagi*）的抑制作用较差，部分化合物对番茄早疫病菌（*Alternaria solani*）和西瓜炭疽病菌（*Colletotrichum lagenarium*）的抑菌作用较好。其中化合物**4e**对番茄早疫病菌（*Alternaria solani*）的抑制率为 58.3%，化合物**4a**对小麦赤霉病菌（*Gibberella zeae*），化合物**4f**和**4g**对苹果轮纹病菌（*Physalospora piricola*）有较好的抑制作用。苯环上有强吸电子取代基和强供电子取代基时，化合物对上述 5 种菌体的生长影响较小，苯环上取代基的差异与抑制菌体生长之间无明显规律性。

表3 化合物4a～4h的抑菌活性（%）

Tab. 3 Fungicidal activity of compound 4a～4h（%）

Compd.	*Phoma asparagi*	*Alternaria solani*	*Gibberella zeae*	*Physalospora piricola*	*Colletotrichum lagenarium*
4a	20. 6	33. 3	46. 5	27. 1	40. 0
4b	33. 3	0	25. 0	18. 2	42. 9
4c	36. 5	41. 7	21. 8	15. 0	20. 0
4d	23. 1	42. 9	21. 1	29. 4	22. 2
4e	20. 6	58. 3	30. 8	31. 2	30. 0
4f	4. 8	33. 3	25. 9	47. 4	20. 0
4g	4. 8	41. 7	21. 8	47. 4	40. 0
4h	4. 8	0	5. 3	0	10

注：**4**a～**4**h：50mg/L。

参 考 文 献

[1] Takao Hisashi, Wakisaka Seiichi, Murai Keizaburo. Jpn Kokai, Tokyo Koho, JP 06 329 633, 1994; CA 122: 160 634a.

[2] Delany J J. US 4 999 368, 1991; CA 115: 29 321j.

[3] Ohvchi Seigo, Okada Soji. JP 10 338 583, 1998; CA 130: 120 912y.

[4] Labouta I M, Hasson A M M, Aboulwafa O M, *et al. Monatshefte fur Chim* [J], 1989, 120: 571.

[5] Jensen L, Dalgaard L, Lawesson S D. *Tetrahedron* [J], 1974, 30: 2413.

[6] Okajima Nobuyuki, Aoki Isai, Okada Yoshiyuki, *et al.* JP 62 242 668, 1987; CA 108: 1 674 639.

[7] LI Shu－Zheng（李树正），WANGDu－Ku（王笃枯），JIAO Shu－Mei（焦书梅）Trans（译）. Method of Pesticide Experiment ——Fungicide（农药实验法——杀菌剂篇）1st Edn（第1版）[M]. Beijing（北京）：Agicultural Press（农业出版社），1991：35.

Synthesis and Fungicidal Activity of Derivatives of 1, 3－Dimethyl－5－methylthio－4－phenylhydrazonocarboxyl Pyrazole

Zhijin Fan, Bin Zhong, Suhua Wang, Zhengming Li

(State Key Laboratory of Elemento Organic Chemistry, Institute of Elemento Organic Chemistry, Nankai University, Tianjin, 300071)

Abstract Eight derivatives of 1, 3－dimethyl－5－methylthio－4－phenylhydrazonocarboxyl pyrazole were synthesized from 1, 3－dimethyl－5－methylthio－4－hydrazinocarboxyl pyrazole. All compounds synthesized were identified by ^{1}H NMR and elemental analysis. Some of them exhibited certain growth inhibition action to *Phoma asparagi*, *Alternaria solani*, *Gibberella zeae*, *Physalospora piricola*, *Colletotrichum lagenarium as tested in vitro* at dosage of 50mg/L.

Key words pyrazole compound; synthesis; fungicidal activity

Metal Ion Interactions with Sugars. The Crystal Structure and FT – IR Study of the $NdCl_3$ – ribose Complex*

Yan Lu[1], Guocai Deng[1], Fangming Miao[2], Zhengming Li[1]

([1]College of Chemistry, Nankai University, Tianjin, 300071, China;
[2]Institute of Crystal Chemistry, Tianjin Normal University, Tianjin, 300074, China)

Abstract The single – crystal structure of neodymium chloride – ribopyranose pentahydrate, $NdCl_3 \cdot C_5H_{10}O_5 \cdot 5H_2O$ was determined to have $M_r = 490.80$, $a = 9.138(11)$, $b = 8.830(10)$, $c = 9.811(11)$ Å, $\beta = 94.087(18)°$, $V = 789.7(16)$ Å^3, $P2_1$, $Z = 2$, $\mu = 0.71073$Å and $R = 0.0198$ for 2075 observed reflections. The ligand of the title complex was observed in a disordered state and two molecular configurations of $NdCl_3 \cdot C_5H_{10}O_5 \cdot 5H_2O$ were found in the single crystal as a pair of isomers. Both ligand moieties of the two molecules are ribopyranose forms, providing three hydroxyl groups in *ax – eq – ax* orientation for coordination. One ligand of the pair of isomers is β – D – ribopyranose in the $^1C_4$ conformation, and the other is α – D – ribopyranose in the $^4C_1$ conformation. The Nd^{3+} ion is nine – coordinated with five Nd—O bonds from water molecules, three Nd—O bonds from hydroxyl groups of the ribopyranose and one Nd—Cl bond from chloride ion. The hydroxyl groups, water molecules, chloride ions form an extensive hydrogen – bond network. The IR spectral C—C, O—H, C—O and C—O—H vibrations were observed to be shifted in the complex and the IR results are in accordance with those of X – ray spectroscopy.

Key words D – Ribose; complexation; neodymium chloride; crystal structure; FT – IR

1 Introduction

Recent research has shown that the metal – binding properties of carbohydrates have fundamental importance in many biochemical processes, such as the transference and storage of metal ions[1,2], the action of metal – containing pharmaceuticals, toxic – metal metabolism[3,4], and Ca^{2+} mediated carbohydrate—protein binding[5,6]. Although coordination chemistry plays a central role in these processes, information regarding the interactions of saccharides with metal cations is rather limited in the literature, especially the detailed threedimensional structures of the resultant complexes. Most previous work concerning metal—carbohydrate complexes is focused on their aqueous solution chemistry, and relatively few well – characterized solid complexes have been reported, especially for those sugars containing only alcoholic oxygen donor atoms. Only a few crystal structures of such sugar—metal complexes have been determined by X – ray diffraction; these include the D – xylose complex of Mo^{6+}[7], the D – fructose complex of Ca^{2+}[8-10], the D – galactose complex of Ca^{2+}[11], the lactose complex of Ca^{2+}[12], the mannose complex of Ca^{2+}[13], and the D – ribose complex of Pr^{3+}[14], and the roles of metal ions in determining and regulating these structures still remain obscure. Study of the metal ion binding properties of simple carbohydrates should aid in understanding the action of polysaccharides ac-

* Reprinted from *Carbohydrate Research*, 2003, 338(24): 2913 – 2919.

tion in the biological processes.

D – Ribose, as a component of nucleic acids, is an important pentose that exists in all organisms, and its metal – ion – binding properties may have biological significance. Solution studies show that D – ribopyranose, having an *ax – eq – ax* sequence of three adjacent hydroxyl groups is readily coordinated with metal cations and forms 1∶1 complexes in hydrophilic solvents[15]. The interactions of Ca^{2+}, Sr^{2+}, Ba^{2+}, La^{3+}, Ce^{3+}, Pr^{3+}, Nd^{3+}, Sm^{3+}, Eu^{3+}, Gd^{3+}, and Tb^{3+} with D – ribose in neutral solutions have been studied by NMR and calorimetric methods[16-18]. The ribose—metal complexes of Ti(Ⅳ), VO(Ⅱ), Cr(Ⅱ), Mn(Ⅱ), Fe(Ⅲ), Co(Ⅱ), Ni(Ⅱ), Cu(Ⅱ), Zn(Ⅱ), Ce(Ⅲ), Pr(Ⅲ), and Nd(Ⅲ) were prepared in strongly alkaline solutions and appropriate structures assigned for the complexes[19-29]. A review on the lanthanide—saccharide complexes has been published by C. P. Rao[30]. However, only one detailed three – dimensional structure of a ribose—metal ion complex, a single crystal of $PrCl_3 \cdot C_5H_{10}O_5 \cdot 5H_2O$, has been reported[14]. In this study, two configurations of $NdCl_3 \cdot C_5H_{10}O_5 \cdot 5H_2O$ are here observed in a single crystal, and are compared with $PrCl_3 \cdot C_5H_{10}O_5 \cdot 5H_2O$. The vibration spectrum of $NdCl_3 \cdot C_5H_{10}O_5 \cdot 5H_2O$ is assigned and interpreted in correlation with the crystal structure.

2 Experimental

2.1 Materials

$NdCl_3$ was prepared from the corresponding rare earth oxide of high purity (99.99%) and crystallized[31]. D – Ribose was purchased from Acros, and was used without further purification.

2.2 Preparation of $NdCl_3$ · ribose · $5H_2O$

D – Ribose (0.45g, 3mmol) and equivalent amounts of neodymium chloride were dissolved in distilled water, and the solution was evaporated slowly until crystallization occurred; yield of purified complex: 71.33%. Anal. Calcd for $NdCl_3 \cdot C_5H_{10}O_5 \cdot 5H_2O$: C, 12.22; H, 4.07; Nd, 29.38. Found: C, 12.04; H, 4.00; Nd, 29.21.

2.3 Physical measurements

The mid – IR spectrum was measured on a Nicolet Magna – IR 750 spectrometer using the micro – IR method, 128 scans at 4 cm^{-1} resolution.

A single crystal (0.30 × 0.25 × 0.20mm) of $NdCl_3$ · ribopyranose · $5H_2O$ was mounted on a glass capillary, and data collection was made on a Bruker Smart 1000 diffractometer using monochromatic MoK_α radiation (λ = 0.71073Å) in the θ range from 2.08 to 25.03° at 293 K. The final cycle of full – matrix, least – squares refinement was based on 2075 observed reflections. Calculations were completed with the SHELX – 97 program.

3 Results and discussion

3.1 X – Ray crystal structure

Two molecular configurations of $NdCl_3 \cdot C_5H_{10}O_5 \cdot 5H_2O$ were observed in the single crystal and these structures are shown together in Fig. 1. Fig. 2 is the projection of the crystal cell in the unit structure of $NdCl_3 \cdot C_5H_{10}O_5 \cdot 5H_2O$ with the ${}^1C_4$ conformation. The crystal data and structure refinements are listed in Table 1; atomic coordinates and equivalent isotropic displacement parameters in Table 2; selected bond lengths and angles in Table 3.

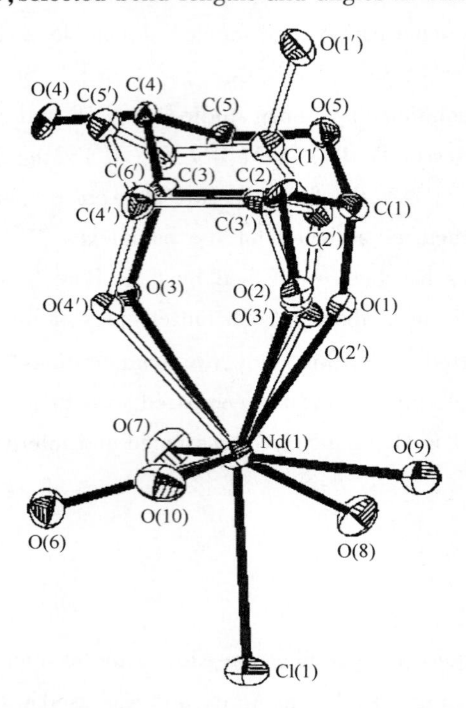

Fig. 1 The structure and atom numbering scheme of $NdCl_3 \cdot C_5H_{10}O_5 \cdot 5H_2O$. The ligand of the complex is in a disordered state, which indicates that there are two configurations of $NdCl_3 \cdot C_5H_{10}O_5 \cdot 5H_2O$ molecules in the single crystal. The one shown by solid bands is α-D-ribopyranose in the ${}^4C_1$ conformation, and the other shown by hollow bands is β-D-ribopyranose in the ${}^1C_4$ conformation (labeled by numbers with commas)

Table 1 Crystal data and structure refinement parameters for $NdCl_3 \cdot C_5H_{10}O_5 \cdot 5H_2O$

Formula	$NdCl_3 \cdot C_5H_{10}O_5 \cdot 5H_2O$
Formula weight	490.80
Crystal system, space group	monoclinic, $P2(1)$
a(Å)	9.138(11)
b(Å)	8.830(10)
c(Å)	9.811(11)
β(°)	94.087(18)
V(Å^3)	789.7(16)
Z	2
D_{calcd}(g/cm^3)	2.064
Absorption coefficient(mm^{-1})	3.832
F(000)	482
Crystal size(mm)	0.30 × 0.25 × 0.20
θ Range for data collection(°)	2.08 – 25.03

Table 1 (continued)

Index ranges	$-10 \leqslant h \leqslant 10$, $-7 \leqslant k \leqslant 10$, $-11 \leqslant l \leqslant 11$
Reflections collected/unique	3237/2075 [R_{int} = 0.0207]
Completeness to θ = 25.03 (%)	99.7
Absorption correction	Semi – empirical from equivalents
Max/min transmission	0.5145 and 0.3927
Refinement method	Full – matrix least – squares on F^2
S	1.039
Final R indices [$I > 2\sigma(I)$]	R_1 = 0.0198, wR_2 = 0.0422
R indices (all data)	R_1 = 0.0228, wR_2 = 0.0429
Absolute structure parameter	0.035(15)
Largest difference peak and hole (e/Å^3)	0.372 and −0.350
Data/restraints/parameters	2075/579/265

Table 2 Atomic coordinates ($\times 10^4$) and equivalent isotropic displacement parameters ($A^2 \times 10^3$) for $NdCl_3$ · ribopyranose · $5H_2O$

	x	y	z	U_{eq}
Nd(1)	2776(1)	1569(1)	7999(1)	25(1)
Cl(1)	5637(1)	1735(4)	9167(1)	40(1)
Cl(2)	9638(1)	1576(4)	1641(1)	41(1)
Cl(3)	5452(1)	6553(4)	6764(1)	43(1)
O(6)	3469(4)	−1133(5)	8306(4)	41(1)
O(7)	3194(5)	271(5)	5803(4)	53(1)
O(8)	4142(4)	3177(5)	6392(4)	47(1)
O(9)	2928(4)	4179(5)	8888(4)	46(1)
O(10)	2577(4)	842(4)	10421(3)	42(1)
O(1)	889(10)	3025(11)	6392(11)	25(2)
O(2)	268(19)	2570(30)	8908(15)	20(3)
O(3)	476(13)	42(14)	7439(12)	21(3)
O(4)	−1904(7)	−1296(8)	6008(6)	28(2)
O(5)	−1597(12)	2794(13)	5505(10)	30(2)
C(1)	−606(12)	3236(13)	6663(12)	26(3)
C(2)	−944(19)	2316(17)	7927(13)	25(3)
C(3)	−955(16)	618(16)	7642(12)	24(3)
C(4)	−1988(15)	252(10)	6385(12)	21(3)
C(5)	−1556(17)	1198(13)	5160(10)	26(3)
O(1′)	−2870(8)	2410(9)	5197(7)	43(2)
O(2′)	1004(10)	2483(11)	6201(10)	27(2)
O(3′)	398(19)	2540(30)	8669(15)	23(3)
O(4′)	628(14)	−221(15)	7741(12)	27(3)

Table 2(continued)

	x	y	z	U_{eq}
O(5′)	-1347(12)	299(13)	5363(10)	33(3)
C(1′)	-1404(14)	1904(13)	5217(10)	30(3)
C(2′)	-509(11)	2707(13)	6386(12)	25(3)
C(3′)	-811(18)	2020(15)	7765(11)	20(3)
C(4′)	-835(16)	302(15)	7749(15)	31(3)
C(5′)	-1863(17)	-329(13)	6597(13)	36(3)

U_{eq} is defined as one third of the trace of the orthogonalized U_{ij} tensor.

Table 3 Selected bond lengths(Å) and angles(°) for $NdCl_3 \cdot C_5H_{10}O_5 \cdot 5H_2O$

Bond lengths(Å)			
Nd(1)-O(2′)	2.445(10)	O(5)-C(5)	1.450(11)
Nd(1)-O(9)	2.464(5)	O(5)-C(1)	1.455(10)
Nd(1)-O(3′)	2.47(2)	C(1)-C(2)	1.532(9)
Nd(1)-O(10)	2.480(4)	C(2)-C(3)	1.525(8)
Nd(1)-O(6)	2.481(5)	C(3)-C(4)	1.533(8)
Nd(1)-O(7)	2.494(4)	C(4)-C(5)	1.539(9)
Nd(1)-O(8)	2.518(4)	O(1′)-C(1′)	1.411(12)
Nd(1)-O(4′)	2.520(14)	O(2′)-C(2′)	1.422(10)
Nd(1)-O(3)	2.525(13)	O(3′)-C(3′)	1.442(11)
Nd(1)-O(1)	2.593(10)	O(4′)-C(4′)	1.415(11)
Nd(1)-O(2)	2.67(2)	O(5′)-C(1′)	1.425(11)
Nd(1)-Cl(1)	2.782(3)	O(5′)-C(5′)	1.441(11)
O(1)-C(1)	1.423(11)	C(1′)-C(2′)	1.534(9)
O(2)-C(2)	1.432(11)	C(2′)-C(3′)	1.525(8)
O(3)-C(3)	1.431(11)	C(3′)-C(4′)	1.517(8)
O(4)-C(4)	1.420(10)	C(4′)-C(5′)	1.523(8)
Bond angles(°)			
O(2′)-Nd(1)-O(9)	88.0(3)	O(7)-Nd(1)-O(3)	75.1(3)
O(2′)-Nd(1)-O(3′)	61.8(4)	O(8)-Nd(1)-O(3)	127.4(3)
O(9)-Nd(1)-O(3′)	67.0(5)	O(9)-Nd(1)-O(1)	76.7(2)
O(2′)-Nd(1)-O(10)	133.4(2)	O(10)-Nd(1)-O(1)	128.3(2)
O(9)-Nd(1)-O(10)	84.80(15)	O(6)-Nd(1)-O(1)	134.7(2)
O(3′)-Nd(1)-O(10)	73.1(4)	O(7)-Nd(1)-O(1)	80.7(3)
O(2′)-Nd(1)-O(6)	123.7(3)	O(8)-Nd(1)-O(1)	71.2(2)
O(9)-Nd(1)-O(6)	148.08(12)	O(3)-Nd(1)-O(1)	68.0(3)
O(3′)-Nd(1)-O(6)	121.6(5)	O(9)-Nd(1)-O(2)	66.4(4)
O(10)-Nd(1)-O(6)	70.74(12)	O(10)-Nd(1)-O(2)	69.0(4)
O(2′)-Nd(1)-O(7)	70.3(3)	O(6)-Nd(1)-O(2)	119.7(5)

Table 3 (continued)

Bond angles(°)			
O(9) – Nd(1) – O(7)	136.67(15)	O(7) – Nd(1) – O(2)	129.0(3)
O(3′) – Nd(1) – O(7)	125.6(3)	O(8) – Nd(1) – O(2)	119.9(5)
O(10) – Nd(1) – O(7)	137.28(15)	O(3) – Nd(1) – O(2)	61.8(5)
O(6) – Nd(1) – O(7)	67.00(14)	O(1) – Nd(1) – O(2)	59.3(4)
O(2′) – Nd(1) – O(8)	72.2(3)	O(2′) – Nd(1) – Cl(1)	146.5(2)
O(9) – Nd(1) – O(8)	71.13(15)	O(9) – Nd(1) – Cl(1)	77.14(11)
O(3′) – Nd(1) – O(8)	117.1(5)	O(3′) – Nd(1) – Cl(1)	133.7(3)
O(10) – Nd(1) – O(8)	145.02(13)	O(10) – Nd(1) – Cl(1)	75.53(9)
O(6) – Nd(1) – O(8)	118.98(14)	O(6) – Nd(1) – Cl(1)	77.10(12)
O(7) – Nd(1) – O(8)	66.72(15)	O(7) – Nd(1) – Cl(1)	100.53(12)
O(2′) – Nd(1) – O(4′)	70.3(4)	O(8) – Nd(1) – Cl(1)	74.63(12)
O(9) – Nd(1) – O(4′)	130.3(3)	O(4′) – Nd(1) – Cl(1)	141.0(3)
O(3′) – Nd(1) – O(4′)	63.3(6)	O(3) – Nd(1) – Cl(1)	149.4(3)
O(10) – Nd(1) – O(4′)	79.9(3)	O(1) – Nd(1) – Cl(1)	142.1(2)
O(6) – Nd(1) – O(4′)	66.4(3)	O(2) – Nd(1) – Cl(1)	130.5(3)
O(7) – Nd(1) – O(4′)	77.9(3)	C(2) – O(2) – Nd(1)	111.4(11)
O(8) – Nd(1) – O(4′)	135.1(3)	C(2′) – O(2′) – Nd(1)	123.6(7)
O(9) – Nd(1) – O(3)	127.0(3)	C(3′) – O(3′) – Nd(1)	112.1(12)
O(10) – Nd(1) – O(3)	87.3(3)	C(4′) – O(4′) – Nd(1)	121.6(9)
O(6) – Nd(1) – O(3)	73.4(3)	C(1′) – O(5′) – C(5′)	117.1(10)
C(1) – O(1) – Nd(1)	123.9(8)	O(1′) – C(1′) – O(5′)	110.1(9)
C(3) – O(3) – Nd(1)	122.3(9)	O(1′) – C(1′) – C(2′)	108.6(9)
C(5) – O(5) – C(1)	114.7(9)	O(5′) – C(1′) – C(2′)	111.8(7)
O(1) – C(1) – O(5)	111.8(9)	O(2′) – C(2′) – C(3′)	107.3(10)
O(1) – C(1) – C(2)	109.5(11)	O(2′) – C(2′) – C(1′)	108.2(9)
O(5) – C(1) – C(2)	109.8(8)	C(3′) – C(2′) – C(1′)	111.0(7)
O(2) – C(2) – C(3)	105.8(15)	O(3′) – C(3′) – C(4′)	109.6(13)
O(2) – C(2) – C(1)	105.7(12)	O(3′) – C(3′) – C(2′)	103.9(11)
C(3) – C(2) – C(1)	111.8(8)	C(4′) – C(3′) – C(2′)	113.1(7)
O(3) – C(3) – C(2)	112.4(12)	O(4′) – C(4′) – C(3′)	108.3(11)
O(3) – C(3) – C(4)	109.2(11)	O(4′) – C(4′) – C(5′)	114.2(13)
C(2) – C(3) – C(4)	110.5(7)	C(3′) – C(4′) – C(5′)	112.4(7)

The D – ribose moiety of the $NdCl_3$ · ribopyranose · $5H_2O$ complex is in a disordered state (Fig. 1). This indicates that there are two kinds of $NdCl_3 \cdot C_5H_{10}O_5 \cdot 5H_2O$ molecules in the single crystal, having different configurations. The ribose moiety of one of the molecules is α – pyranose in the $^4C_1$ conformation (shown by solid bands in Fig. 1), and thus resembles $PrCl_3$ · ribose · $5H_2O$[14]. In the other $NdCl_3 \cdot C_5H_{10}O_5 \cdot 5H_2O$ molecule, the ribose moiety is the β –

pyranose in the $^{1}C_4$ conformation (shown by hollow bands in Fig. 1 and the atoms labeled by numbers with commas), which has not been observed before. The sugar moieties in the two $NdCl_3 \cdot C_5H_{10}O_5 \cdot 5H_2O$ molecules are thus a pair of configurational isomers in 1∶1 ratio. In both of the $NdCl_3$ · ribopyranose · $5H_2O$ molecules, the Nd ion is nine – coordinated and binds to three hydroxyl groups of one D – ribopyranose molecule, five water molecules, and a chloride ion. The other two Cl^- ions in the molecule are free. The three adjacent hydroxyl groups labeled $O_{(1)}H$, $O_{(2)}H$ and $O_{(3)}H$ or $O_{(2')}H$, $O_{(3')}H$ and $O_{(4')}H$ which containing the *ax* – *eq* – *ax* sequence are the coordination sites of the ribose, and the Nd – O distances at the coordination sites (from 2.445 to 2.520Å in the β – pyranose $^{1}C_4$ conformation or from 2.525 to 2.67Å in the α – pyranose $^{4}C_1$ conformation) are comparable to those of Pr – O (from 2.535 to 2.589Å)[14]. The ring oxygen of D – ribose does not coordinate with Nd^{3+} in either molecule. All water molecules in the crystal are coordinated in the two structures.

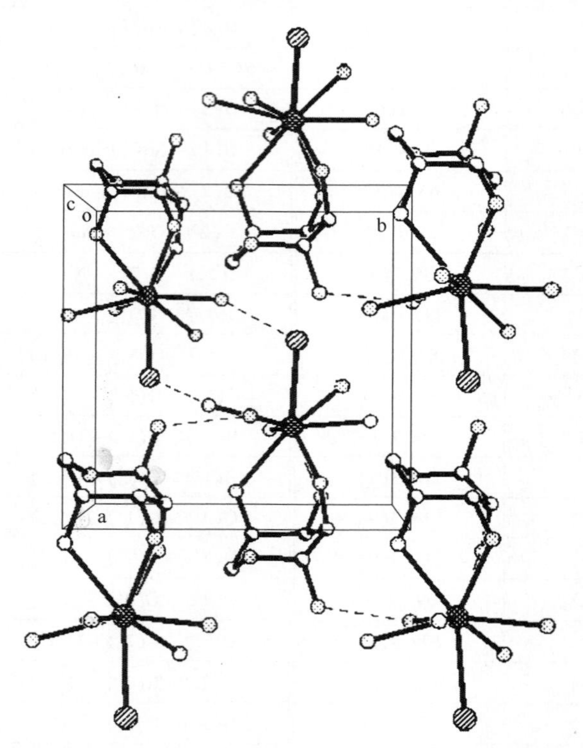

Fig. 2 Projection of the crystal cell in the structure of $NdCl_3 \cdot C_5H_{10}O_5 \cdot 5H_2O$ with the β – ribopyranose $^{1}C_4$ conformation

As expected, there is an extensive network of hydrogen bonds that involve all hydroxyl groups, water molecules, and chloride ions in the crystal structure of $NdCl_3$ · ribopyranose · $5H_2O$ (Table 4). The Nd – ribose molecules are organized by these hydrogen bonds and thus form a layer parallel to the plane of (10 – 1). These layers are then also held together by hydrogen bonds with regular spaces between them. Free Cl ions distributed in the layers are responsible for the connections in the layers, and the bonding between the layers is by hydrogen bonds. The network of hydrogen bonds thus forms the packed structure of the whole crystal. The Cl^- ions thus play important roles not only as counterions, but also as the predominant feature in the network of hydrogen bonds.

Table 4 Hydrogen bonds for $NdCl_3 \cdot C_5H_{10}O_5 \cdot 5H_2O$ with H···A <r(A) +2.000 Å and <DHA >110°

D – H···A	d(D – H)	d(H···A)	d(D···A)	∠(DHA)
O6 – H6A···Cl3#1	0.840	2.529	3.185	135.77
O6 – H6B···Cl2#2	0.844	2.689	3.490	158.87
O7 – H7B···O5#3	0.839	2.065	2.880	163.64
O7 – H7B···O1′#3	0.839	2.104	2.719	129.83
O8 – H8A···Cl3	0.838	2.390	3.223	172.62
O8 – H8B···O4#4	0.837	2.328	3.042	143.60
O8 – H8B···Cl3#2	0.837	2.782	3.456	138.78
O9 – H9A···Cl1#5	0.840	2.435	3.178	147.92
O9 – H9B···Cl2#6	0.840	2.340	3.174	172.63
O10 – H10A···Cl3#7	0.845	2.644	3.249	129.61
O10 – H10A···Cl1	0.845	2.837	3.232	110.47
O10 – H10B···Cl2#8	0.844	2.290	3.086	157.52
O1 – H1···O4#4	0.930	1.751	2.660	164.99
O2 – H2···Cl2#8	0.930	2.335	2.917	120.38
O3 – H3···Cl2#2	0.930	2.544	3.195	127.32
O4 – H4···Cl3#9	0.820	2.397	3.201	167.09
O2′ – H2′···O5′#4	0.930	2.403	2.951	117.54
O3′ – H3′···Cl2#8	0.930	2.700	3.162	111.56
O4′ – H4′···Cl2#2	0.930	2.093	2.906	145.30

Symmetry transformations used to generate equivalent atoms: #1 $x, y-1, z$; #2 $-x+1, y-1/2, -z+1$; #3 $-x, y-1/2, -z+1$; #4 $-x, y+1/2, -z+1$; #5 $-x+1, y+1/2, -z+2$; #6 $-x+1, y+1/2, -z+1$; #7 $-x+1, y-1/2, -z+2$; #8 $x-1, y, z+1$; #9 $x-1, y-1, z$.

3.2 IR spectroscopic study of $NdCl_3 \cdot C_5H_{10}O_5 \cdot 5H_2O$

The FT – IR spectra of D – ribose and the Nd^{3+} salt are shown in Fig. 3. The absorption bands and tentative assignments are given in Table 5. The broad absorption band at around 3400 cm^{-1} in the spectrum of D – ribose can be assigned to the hydrogen – bonded OH groups. This band appears broader in the spectrum of the metal complex (see Fig. 3). This observed spectral change is due to the rearrangement of the strong hydrogenbonding network observed in the crystal structure of the complex (see Table 4). The C – H stretching vibration bands (2947, 2852, 2670, 2538 cm^{-1}) in the spectrum of the complex, corresponding to the band of 2928 cm^{-1} in the spectrum of free D – ribose, are masked by the broaden v OH bands in the salt spectrum, and the band at 1627 cm^{-1} can be assigned to the δ_{HOH} vibration of the coordinated water. The bands at 1452, 1417, 1340, 1250 cm^{-1} in the spectrum of free D – ribose, which are assigned to the multiple bending vibrations of O—C—H, C—C—H, C—O—H and CH_2, are observed to be shifted to bands of 1456, 1408, 1315, and 1244 cm^{-1} in the spectrum of the complex; the intensities become weaker upon salt formation. The characteristic vibration of the pyranose is observed in both of the spectra of D – ribose and the complex (1153cm^{-1} for D – ribose, 1152cm^{-1} for Nd – ribose). In the 1200 – 970 cm^{-1} region, the C—O, C—C stretching vibrations and the C—O—H, C—C—O bending vibrations of D – ribose are observed to be shifted and split in the spectrum of the salt (see Table 5). Such observed splitting and shifting is indicative of the participation of the sugar hydroxyl groups in metal – ligand bonding, which therefore affect the C—O, C—C stretching vibrations and the C—O—H, C—C—O bending vibrations of the sugar moiety.

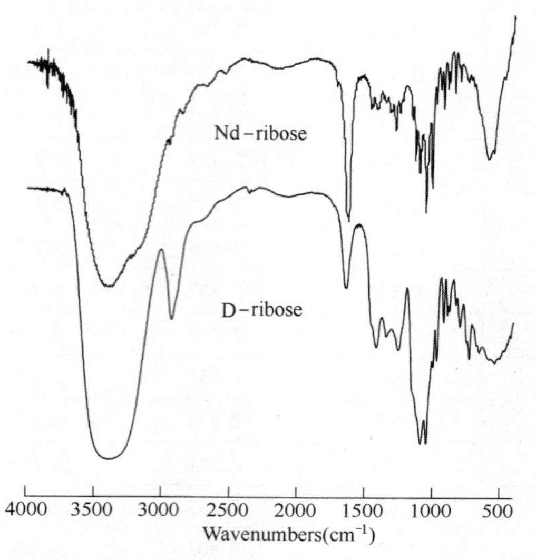

Fig. 3 The mid – IR spectra of D – ribose and Nd – ribose

Table 5 IR data for D – ribose, Nd – ribose and Pr – ribose complex (1500 – 700cm^{-1})

D – ribose	Nd – ribose	Pr – ribose[14]	Possible assignment[32 – 37]
1452	1456	1458	δOCH + δCCH + δCH_2
1417	1408	1409	δOCH + δCCH
1340	1315	1318	δOCH + δCCH + δCOH
1250	1244	1244	δCCH + δCOH + δOCH
1153	1153	1152	vC—O + vC—C + δCOH (β – pyranose)
1117	1128	1128	vC—O + vC—C + δCOH
1085	1091	1095	vC—O + vC—C + δCOH
1042	1047	1048	vC—O + vC—C + δCCO
	1003	1004	vC—O + vC—C + δCCO
987	971	972	vC—O + vC—C + δCCO
914	918	918	vC—O + δ CCH + $v$asy (ring of pyranose)
887	888	887	vC—O + vC—C + δCCH
872	874	874	δCH (β – pyranose)
	836	836	δCH (α – pyranose)
826			δCH
795	796	795	τC—O + δCCO + δCCH + $v$sy (ring of pyranose)
746	734	736	τC—O + δCCO + δCCH

δ, bending mode; v, stretching mode; τ, twisting.

The ring skeletal deformation bands(δ C—C—O and δ C—C—C) of free D – ribose, mainly in the 1000 – 400cm^{-1} region, show considerable changes on complex formation (see Fig. 3 and Table 5). This may be attributed to distortion of the sugar ring upon metalation, although there are no published crystal data for free D – ribose for comparison with those of the complex. The spectral changes observed in this region may be interpreted as showing that metalation of the sugar perturbs the electron distribution within the sugar ring system where the vibrations are mostly localized, and causes ring distortion, resulting in the alterations in the spectrum. The 914cm^{-1} and 795 cm^{-1} bands in the ribose spectrum are attributed respectively to the asymmetric and symmetric ring – breathing modes of the pyranose, and they are observed as bands at 918 and 796cm^{-1} in the spectrum of the title complex, indicating that the six – membered ring is retained in the complex. The absorption bands at about 870 and 840cm^{-1} in the pyranose spectra are generally assigned to the presence of the β and α anomers, respectively[32,33]. In relation to the spectrum of the Nd – ribose complex, the coexistence of absorption bands at 874 and 836cm^{-1} is indicative of the presence of both ribopyranose anomers in the complex. These anomers in the complex indicated by the IR data are proved by the X – ray spectroscopy.

The IR results indicate that the hydroxyl groups of D – ribose take part in the metal – oxygen interaction; the hydrogen – bond network rearranges upon metalation; the sugar moiety of the complex is D – ribopyranose in two anomeric forms and the skeleton is deformed as a result of salt formation. Since the spectrum of Nd – ribose is similar to that of Pr – ribose[14], it may be concluded that the constitutions of the two complexes are in fact similar. The IR results are in accordance with those of X – ray spectroscopy, and the FT – IR technique is thus a useful method for detecting the formation of such complexes.

4 Supplementary material

Crystallographic data (without structure factors) for the structure reported in this paper have been deposited with the Cambridge Crystallographic Data Centre, CCDC No. 195540. Copies of this information may be obtained free of charge from The Director, CCDC, 12 Union Road, Cambridge CB2 1EZ, UK (Fax: +44 – 1223 – 336 – 033; e – mail: deposit@ ccdc. cam. ac. uk or www: http://www. ccdc. cam. ac. uk).

References

[1] Sauchelli, V. *Trace Elements in Agriculture*; Van Nostrand: New York, 1969; p 248.
[2] Holm, R. P.; Berg, J. M. *Pure Appl. Chem.* 1984, 56, 1645 – 1657.
[3] Predki, P. F.; Whitfield, D. M.; Sarkar, B. *Biochem. J.* 1992, 281, 835 – 841.
[4] Templeton, D. M.; Sarkar, B. *Biochem. J.* 1985, 230, 35 – 42.
[5] Weis, W. I.; Drickamer, K.; Hendrickson, W. A. *Nature* 1992, 360, 127 – 134.
[6] Drickamer, K. *Nature* 1992, 360, 183 – 186.
[7] Taylor, G. E.; Waters, J. M. *Tetrahedron Lett.* 1981, 22, 1277 – 1278.
[8] Craig, D. C.; Stephenson, N. C.; Stevens, J. D. *Cryst. Struct. Commun.* 1974, 3, 277 – 281.
[9] Craig, D. C.; Stephenson, N. C.; Stevens, J. D. *Cryst. Struct. Commun.* 1974, 3, 195 – 199.
[10] Cook, W. J.; Bugg, C. E. *Acta Crystallogr.* 1976, 32, 656 – 659.
[11] Cook, W. J.; Bugg, C. E. *J. Am. Chem. Soc.* 1973, 95, 6442 – 6446.

[12] Bugg, C. E. *J. Am. Chem. Soc.* 1973, 95, 908 - 913.
[13] Craig, D. C.; Stephenson, N. C.; Stevens, J. D. *Carbohydr. Res.* 1972, 22, 494 - 495.
[14] Yang, L. M.; Zhao, Y.; Xu, Y. Z.; et al. *Carbohydr. Res.* 2001, 334, 91 - 95.
[15] Angyal, S. J. *Pure Appl. Chem.* 1973, 35, 131 - 146.
[16] Alvarez, A. M.; Desrosiers, N. M.; Morel, J. P. *Can. J. Chem.* 1987, 65, 2656 - 2660.
[17] Desrosiers, N. M.; Lhermet, C.; Morel, J. P. *J. Chem. Soc., Faraday Trans.* 1991, 87, 2173 - 2177.
[18] Desrosiers, N. M.; Lhermet, C.; Morel, J. P. *J. Chem. Soc., Faraday Trans.* 1993, 89, 1223 - 1228.
[19] (a) Rao, C. P.; Kaiwar, S. P. *Inorg. Chim. Acta* 1991, 186, 11 - 12;
(b) Kaiwar, S. P.; Rao, C. P. *Carbohydr. Res.* 1992, 237, 203 - 210.
[20] (a) Sreedhara, A.; Raghavan, M. S. S.; Rao, C. P. *Carbohydr. Res.* 1994, 264, 227 - 235;
(b) Sreedhara, A.; Rao, C. P.; Rao, B. J. *Carbohydr. Res.* 1996, 289, 39 - 52.
[21] (a) Rao, C. P.; Kaiwar, S. P.; Raghavan, M. S. S. *Polyhedron* 1994, 13, 1895 - 1906;
(b) Kaiwar, S. P.; Raghavan, M. S. S.; Rao, C. P. *J. Chem. Soc., Dalton Trans.* 1995, 10, 1569 - 1576.
[22] Bandwar, R. P.; Rao, C. P. *Carbohydr. Res.* 1996, 287, 157 - 168.
[23] (a) Rao, C. P.; Geetha, K.; Bandwar, R. P. *Bioorg. Med. Chem. Lett.* 1992, 2, 997 - 1002;
(b) Geetha, K.; Kulshreshtha, S. K.; Rao, C. P.; et al. *Carbohydr. Res.* 1995, 271, 163 - 175.
[24] Bandwar, R. P.; Sastry, M. D.; Rao, C. P.; et al. *Carbohydr. Res.* 1997, 297, 337 - 339.
[25] Bandwar, R. P.; Rao, C. P. *Carbohydr. Res.* 1996, 297, 341 - 346.
[26] Bandwar, R. P.; Rao, C. P.; Giralt, M.; et al. *J. Inorg. Biochem.* 1997, 67, 172.
[27] Bandwar, R. P.; Giralt, M.; Rao, C. P. *Carbohydr. Res.* 1996, 284, 73 - 84.
[28] Mukhopadhyay, A.; Kolehmainen, E.; Rao, C. P. *Carbohydr. Res.* 2000, 324, 30 - 37.
[29] Mukhopadhyay, A.; Kolehmainen, E.; Rao, C. P. *Carbohydr. Res.* 2000, 328, 103 - 113.
[30] Rao, C. P.; Das, T. M. *Indian J. Chem.* 2003, 42A, 227 - 239.
[31] Chandrasekhar, A. *J. Imaging Technol.* 1990, 16, 158 - 161.
[32] Tajmir - Riahi, A. H. *Carbohydr. Res.* 1983, 122, 241 - 248.
[33] Tajmir - Riahi, A. H. *Carbohydr. Res.* 1984, 127, 1 - 8.
[34] Zhang, W. J. *Biochemical Technology of Complexes of Carbohydrate*; Zhejiang University Press: Hangzhou, 1999; pp 193 - 198.
[35] Mathlouthi, M.; Seuvre, A. M.; Koenig, J. L. *Carbohydr. Res.* 1983, 122, 31 - 47.
[36] Cael, J. J.; Koenig, J. L.; Blackwell, J. *Carbohydr. Res.* 1974, 32, 79 - 91.
[37] Tajmir - Riahi, A. H. *Biophys. Chem.* 1986, 23, 223 - 228.

3 – Methylthio – pyrano[4,3 – *c*]pyrazol – 4(2*H*) – ones from 3 – (Bis – methylthio)methylene – 2*H* – pyran – 2,4 – diones and Hydrazines*

Yuxin Li[2], Youming Wang[1], Xiaoping Yang[2], Suhua Wang[1], Zhengming Li[1]

([1]State Key Laboratory of Elemento – Organic Chemistry, Institute of Elemento – Organic Chemistry, Nankai University, Tianjin, 300071, China;
[2]Department of Chemistry, Xiangtan Normal University, Xiangtan, Hunan Province, 411201, China)

Abstract A useful synthesis of 3 – methylthio – 6 – methyl – pyrano[4,3 – c]pyrazol – 4(2H) – ones via 3 – (bismethylthio) methylene – 5,6 – dihydro – 6 – alkyl(aryl) – 2H – pyran – 2,4 – dione with hydrazine as well as methyl and phenyl hydrazines is described and the mechanism of the formation is discussed.

Introduction

Being a structural subunit in numerous natural bioactivited products, pyrone exhibits a broad range of biological activity[1]. In addition, it is a useful intermediate in the synthesis of a variety of important heterocyclic molecules, which exhibit antimicrobial, antifungal, antiviral, or phytotoxic activity [2-4]. It is reported that β – keto – δ – valerolactone derivatives inhibit the activity of HIV proteases[5-7]. Recently we have reported that 3 – anilinomethylene – 5,6 – dihydro – 6 – alkyl(aryl) – 2*H* – pyran – 2,4 – diones have interesting fungicidal, tobacco virucidal activities[8,9].

As part of our program aimed at developing a new class of pesticides, we have been interested in synthesizing nitrogen heterocyclic compounds such as isothiazoles and pyrazoles because nitrogen heterocycles play an important role among useful herbicides[10]. Combining β – keto – δ – valerolactone and pyrazole, we expect to find better lead compounds of herbicides. With these considerations in mind, we synthesized substituted pyrano[4,3 – *c*]pyrazol – 4(2*H*) – ones.

Results and discussion

6 – Methyl – 5,6 – dihydro – 2*H* – pyran – 2,4 – dione(**1a**) was prepared from dehydroacetic acid in two steps[8]. 6 – Methyl – 6 – benzyl and 6 – methyl – 6 – piperonyl – 5,6 – dihydro – 2*H* – pyran – 2,4 – dione(**1b**,**c**) were prepared as described[8]. It is generally believed that 6 – alkyl(aryl) – 5,6 – dihydro – 2*H* – pyran – 2,4 – diones react with carbon disulfide and methyl iodide to give low yields of 3 – (bis – methylthio)methylene – 5,6 – dihydro – 6 – alkyl (aryl) – 2*H* – pyran – 2,4 – diones(38%)[11]. We found that **1a – c** react with carbon disulfide in the presence of weak base such as anhydrous potassium carbonate and methyl iodide to

* Reprinted from *Heteroatom Chemistry*, 2003, 14(4): 342 – 344. Contract grant number: NNSFC 29832050.

give compounds **2a – c** in reasonable yield(70%)(Scheme 1). Butyl bromide and phenyl iodide did not give analogous compounds.

K_2CO_3 / DMF; CH_3I

1a R^1=H ;
1b R^1=$PhCH_2$;
1c R^1=(3,4-methylenedioxyphenyl)CH_2

2a–c

Scheme 1

3 – Methylthio – 6 – methyl – pyrano[4,3 – *c*]pyrazol – 4(2*H*) – ones **4a – h** were synthesized by the process described in Scheme 2. Generally, the reactions proceeded quickly with excellent yields as shown in Table 1. However, for the coupling reaction with phenylhydrazine at room temperature, 3h were needed(monitored by TLC).

NH_2NHR^2 / $-H_2O$; $-CH_3SH$

2a–c → **3** → **4a–h**

Scheme 2

Table 1 Data and Elemental Analyses of Compounds 4

	R^1	R^2	Yield(%)	mp(℃)	Elemental Analysis(Calcd) C	H	N
4a	H	CH_3	98.1	102 – 103	50.90(50.94)	5.72(5.66)	13.20(13.20)
4b	H	Ph	94.3	167 – 169	61.20(61.30)	5.12(5.11)	10.18(10.22)
4c	$PhCH_2$	H	68.7	145 – 146	62.55(62.50)	5.54(5.56)	10.00(9.72)
4d	$PhCH_2$	CH_3	63.2	oil	63.60(63.58)	6.02(5.96)	9.20(9.27)
4e	$PhCH_2$	Ph	73.5	140 – 141	69.08(69.23)	5.54(5.49)	7.69(7.69)
4f	$CH_2O_2C_6H_3CH_2$	H	88.3	165 – 166	57.90(57.83)	5.06(4.82)	8.32(8.43)
4g	$CH_2O_2C_6H_3CH_2$	CH_3	80.2	140 – 141	58.79(58.96)	5.19(5.20)	7.90(8.09)
4h	$CH_2O_2C_6H_3CH_2$	Ph	85.6	169 – 170	64.75(64.70)	4.93(4.90)	6.82(6.86)

Interestingly, although the reaction of 1c with 2,4 – dinitrophenylhydrazine failed to gain a compound **4**, we successfully separated an oily substance. ^1H NMR ($CDCl_3$): 1.49(s,3H, CH_3), 2.54(s,2H,CH_2), 2.84(s,6H,2CH_3), 2.94(dd,2H,J = 6.8Hz,CH_2), 5.92(s,2H, CH_2), 6.70(m,3H), 7.92(m,3H), 11.68(s,1H,NH), and(EI) m/z = 546 of the compound show it is an uncyclized product **3**(R^1 = piperonyl, R^2 = 2,4 – dinitrophenyl). The results indicated that the formation of the transition state **3** is reasonable.

Experimental

Melting points were conducted on a Yanaco MP – 500 micromelting point apparatus. ^1H NMR spectra were recorded in $CDCl_3$ as solvent on AC – 200 instrument, using TMS as internal standard. Elemental analyses were performed on MF – 3 automatic elemental analyzer.

*6 – Methyl – 5,6 – dihydro – 3 – (bis – methylthio) – methylene – 2H – pyran – 2,4 – diones **2a – c***

2a: Compound **1a** (1.28g, 0.01mol) was dissolved in 20 mL DMF. With constant agitation, 3.18g (0.03mol) of potassium carbonate anhydrous was added in portions to the mixture, which was agitated for 0.5h at room temperature. The mixture was cooled to 0℃ and 0.76g (0.01mol) carbon disulfide was added dropwise. After the addition, the reaction mixture was agitated for 0.5h at 0℃, and then 2.84g (0.02mol) methyl iodide was added dropwise at 0℃. The brown suspension was agitated for 4 h. The suspension was poured into 200mL ice – cooled water and extracted with 100mL dichloromethane twice. The organic phase was separated and dried over magnesium sulfate. After the filtration, the solvent was evaporated in a vacuum. The crude product was purified with silicon gel column using ethyl acetate/petroleum ether (v/v, 1:3) as eluent and **2a** was obtained. Yield: 70.0%. mp 148 – 149℃. ^1H NMR ($CDCl_3$): 1.40 (d, 3H, J = 6.8, CH_3), 2.56 (m, 2H, CH_2), 3.70 (s, 6H, SCH_3), 4.45 (m, 1H, CH).

2b: Following the above method and using 2.18g 1b, 2.07g **2b** was obtained. Yield: 64.3%. mp 112 – 113℃. ^1H NMR ($CDCl_3$): 1.40 (s, 3H, CH_3), 2.53 (s, 6H, SCH_3), 2.64 (s, 2H, CH_2), 2.97 (dd, 2H, J = 6.8, CH_2), 7.23 (m, 5H).

2c: Following the above method and using 2.62g **1c**, 2.49g **2c** was obtained. Yield: 68.2%. mp 129 – 131℃. ^1H NMR ($CDCl_3$): 1.39 (s, 3H, CH_3), 2.53 (s, 6H, SCH_3), 2.64 (s, 2H, CH_2), 2.93 (dd, 2H, J = 6.8, CH_2), 5.91 (s, 2H, CH_2), 6.70 (m, 3H).

*3 – Methylthio – 6 – methyl – 5,6 – dihydro – 2H – pyrano[4,3 – c]pyrazole – 4(2H) – ones **4a – h***

4a: Methylhydrazine (0.28g, 0.006mol) was added dropwise to a mixture of 1.16g (0.005mol) **2a** in 30mL ethanol. After agitation for 1h at room temperature, the solvent was evaporated in vacuum and gave 1.03g **4a**, which was purified with silicon gel column using ethyl acetate/petroleum ether (1:2) as eluent. ^1H NMR ($CDCl_3$): 1.49 (d, 3H, J = 6.4, CH_3), 2.57 (s, 3H, SCH_3), 2.85 (m, 2H, CH_2), 3.87 (s, 3H, NCH_3), 4.70 (m, 1H, CH).

4b: Following the above method and using 1.16g **2a** and 0.65g phenylhydrazine, 1.3g **4b** was obtained. ^1H NMR ($CDCl_3$): 1.53 (d, 3H, J = 6.2, CH_3), 2.61 (s, 3H, SCH_3), 2.97 (m, 2H, CH_2), 4.65 (m, 1H), 7.44 (m, 5H).

4c: Following the above method and using 1.61g **2b** and 0.3g 85% hydrazine hydrate solution, 0.97g **4c** was obtained. ^1H NMR ($CDCl_3$): 1.48 (s, 3H, CH_3), 2.68 (s, 3H, SCH_3), 2.94 (s, 2H, CH_2), 3.08 (dd, 2H, J = 7.0, CH_2), 7.20 (m, 5H), 8.18 (bs, 1H, NH).

4d: Following the above method and using 1.61g **2b** and 0.28g methylhydrazine, 0.95g **4d** was obtained. ^1H NMR ($CDCl_3$): 1.43 (s, 3H, CH_3), 2.62 (s, 3H, SCH_3), 2.86 (s, 2H, CH_2), 3.02 (dd, 2H, J = 5.6, CH_2), 3.92 (s, 3H, NCH_3), 7.24 (m, 5H).

4e: Following the above method and using 1.61g **2b** and 0.65g phenylhydrazine, 1.32g **4e**

was obtained. ^{1}H NMR($CDCl_3$):1.46(s,3H,CH_3),2.59(s,3H,SCH_3),2.95(s,2H,CH_2),3.03(dd,2H,J=8.0,CH_2),7.16(m,5H),7.45(m,5H).

4f: Following the above method and using 1.83g **2c** and 0.3g 85% hydrazine hydrate solution, 1.46g **4f** was obtained. ^{1}H NMR($CDCl_3$):1.51(s,3H,CH_3),2.56(s,3H,SCH_3),2.90(s,2H,CH_2),3.09(dd,2H,J=8.2,CH_2),5.90(s,2H,CH_2),6.65(m,3H),8.38(bs,1H,NH).

4g: Following the above method and using 1.83g **2c** and 0.28g methylhydrazine, 1.38g **4g** was obtained. ^{1}H NMR($CDCl_3$):1.43(s,3H,CH_3),2.52(s,3H,SCH_3),2.85(s,2H,CH_2),2.89(dd,2H,J=8.0,CH_2),3.72(s,3H,NCH_3),5.89(s,2H,CH_2),6.63(m,3H).

4h: Following the above method and using 1.83g **2c** and 0.65g phenylhydrazine, 1.73g **4h** was obtained. ^{1}H NMR($CDCl_3$):1.46(s,3H,CH_3),2.59(s,3H,SCH_3),2.88(s,2H,CH_2),2.97(dd,2H,J=8.2,CH_2),5.88(s,2H,CH_2),6.61(m,3H),7.45(m,5H).

References

[1] Brian, P. W.; Curtis, P. J.; Hemming, H. G.; Unwin, C. H.; Wright, J. M. Nature 1949, 164, 534.

[2] Knudsen, C. G.; Michaely, W. J.; James, D. R.; Chin, H. L. M. EP Patent 249812, 1987.

[3] Oishi, H.; Ueda, A.; Tomita, K.; Hosaka, H. JP Patent 2111577, 1989.

[4] Adachi, H.; Aihara, T.; Toshio, T.; Tanaka, K. WO Patent 01171, 1993.

[5] Thaisrivongs, S.; Yang, C. P.; Strohbach, J. W.; Turner, S. R. WO Patent 11361, 1994.

[6] Ellsworch, E. L.; Lunney, E.; Tait, B. D. WO Patent 14011, 1995.

[7] Tait, B. D.; Hagen, S.; Domagala, J.; Ellsworth, E.; Gajda, C.; Hamilton, H.; Vara Prasad, J. V. N.; Ferguson, D.; Graham, N.; Hupe, D.; Nouhan, C.; Tummino, P. J.; Humblet, C.; Lunney, E. A.; Pavlovsky, A.; Rubin, J.; Baldwin, E. T.; Bhat, T. N.; Erickson, J. W.; Gulnik, S. V.; Liu, B. J Med Chem 1997, 40, 3781.

[8] Wang, Y. M.; Li, Z. M.; Li, J. F.; Li, S. Z.; Zhang, S. H. Chem J Chinese Univ 1999, 20(5), 1559.

[9] Wang, Y. M.; Li, Z. M.; Han, Y. F. J Appl Chem 2001, 12(6), 475(in Chinese).

[10] Yang, X. P.; Li, Z. M.; Chen, H. S.; Liu, J.; Li, S. Z. Chem J Chinese Univ 1999, 20, 393.

[11] Wexler, B. A. US Patent 4622062, 1986.

新单取代苯磺酰脲衍生物的合成及生物活性*

马 宁[1,2] 李鹏飞[1] 李永红[1] 李正名[1] 王玲秀[1] 王素华[1]

([1]南开大学元素有机化学研究所，元素有机化学国家重点实验室，天津 300071；
[2]天津大学理学院化学系，天津 300072)

摘 要 以较高活性的单取代苯磺酰脲为基础，设计合成了14个新的含烷硫基和烷胺基嘧啶环的单取代苯磺酰脲化合物，其结构经^{1}H NMR及元素分析确证。用油菜根长法测定IC_{50}，并进行盆栽除草活性测试，结果表明，嘧啶环4位取代基的变化对分子除草活性影响较大，活性大致按烷氧基、烷硫基和烷胺基的顺序递减。

关键词 磺酰脲 除草剂 合成 嘧啶

磺酰脲类除草剂是一类高效、低毒、高选择性、环境友好的农药，自 Levitt[1]发现第一个品种氯磺隆之后，已有20多个品种商品化。在商品化的品种中，杂环部分通常为4位和6位双取代的嘧啶环或三嗪环。本课题组的研究发现[2~5]，某些含单取代嘧啶环的磺酰脲化合物，其除草活性与嘧啶环为双取代的磺酰脲相当。近年来商品化的品种中，嘧啶环上的取代基不再是甲基或甲氧基，而是甲胺基和三氟甲氧基等，中国专利[6,7]也报道了嘧啶环为烷硫基取代的磺酰脲。这些结构修饰使得某些磺酰脲分子在保持除草活性的同时，也可能改变了对杂草和作物的选择性。为发现具有新性能的磺酰脲除草剂，本文合成了含烷硫基和烷胺基嘧啶环的单取代苯磺酰脲化合物，测试了它们的除草活性，并对这类化合物进行了构效关系探讨。合成路线如下：

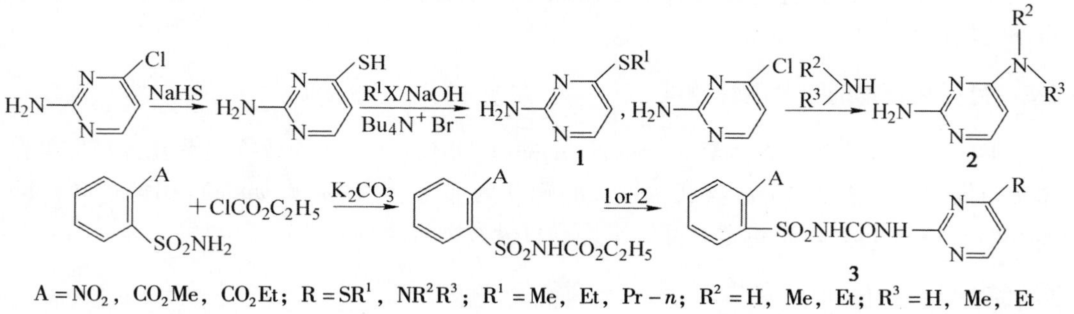

$A = NO_2$，CO_2Me，CO_2Et；$R = SR^1$，NR^2R^3；$R^1 = Me$，Et，Pr－*n*；$R^2 = H$，Me，Et；$R^3 = H$，Me，Et

1 实验部分

1.1 试剂与仪器

所有试剂均为市售分析纯或化学纯。X－4数字显示显微熔点测定仪（北京泰克仪

* 原发表于《高等学校化学学报》，2004，25（12）：2259－2262。基金项目：国家“973”计划项目（批准号：2003CB114406）、教育部南开大学、天津大学科技合作基金和天津市自然科学基金重点基金（批准号：033803411）资助。

器有限公司)，Yanaco CHN CORDER MT－3 元素分析仪；BRUKER ACP－200，Varian Mercury Vx300 核磁共振仪，TMS 为内标，$CDCl_3$ 或 DMSO－d_6 为溶剂。

1.2 中间体2－氨基－4－烷硫基嘧啶（1）的合成

将1.28g（0.01mol）2－氨基－4－巯基嘧啶[8]溶于20mL质量分数为4%的NaOH溶液中，加入1.42g（0.01mol）碘甲烷和0.1g四丁基溴化铵，于室温搅拌0.5h，滤出固体，水洗，少量乙醚洗，甲苯重结晶，得2－氨基－4－甲硫基嘧啶，收率72%，熔点150～152℃（文献值150～153℃[8]）。用类似方法制备2－氨基－4－乙硫基嘧啶，收率78%，熔点152～153℃（文献值155℃[8]）；2－氨基－4－丙硫基嘧啶（**1**a）未见文献报道。

1.3 中间体2－氨基－4－烷胺基嘧啶（2）的合成

在50mL锥形瓶中，加入15mL质量分数为25%～30%的甲胺水溶液及1.3g（0.01mol）2－氨基－4－氯嘧啶[9]，塞好塞子，于室温搅拌过夜。滤去不溶物，将溶液旋转蒸发至干，用体积分数为20%的盐酸溶解后，经质量分数为30%的NaOH中和至pH＝8～9。用冷水洗涤，析出固体，用乙醇重结晶，得2－氨基－4－甲胺基嘧啶，收率67%，熔点161～163℃（文献值161～163.5℃[10]）。用类似方法得到2－氨基－4－二甲胺基嘧啶，收率47%，熔点152～154℃（文献值155～156℃[11]）；2－氨基－4－乙胺基嘧啶（**2**a）未见文献报道。

1.4 1－（4－取代嘧啶－2－基）－3－（2－取代苯磺酰基）脲（3）的合成

以1－（4－甲硫基嘧啶－2－基）－3－（2－硝基苯磺酰基）脲（**3**a）的合成为例。将0.55g（2mmol）邻硝基苯磺酰氨基甲酸乙酯[12]和0.28g（2mmol）2－氨基－4－甲硫基嘧啶加入到20mL甲苯中，加热回流8h。蒸出10mL甲苯，再补加10mL甲苯，回流3h，重复此操作直至TLC检测至终点。冷却后析出固体，过滤，用丙酮、丙酮－石油醚洗涤，得0.70g **3**a。

1.5 目标化合物的除草活性测试

平皿法：试验靶标为油菜（Brassica napus），用最小二乘法计算油菜胚芽长度抑制50%药剂的摩尔浓度IC_{50}。盆栽法：试验靶标为油菜（Brassica napus）和稗草（Echinochloa Crusgalli），以植株地上部鲜重抑制百分数来表示药效。

2 结果与讨论

化合物的物理化学常数及^{1}H NMR数据见表1和表2。

Table 1 Physico－chemical data and elemental analysis data of compounds 1a，2a and 3a－3n

Compd.	Substituent		m. p. (℃)	Yield(%)	Appearance	Elemental analysis(%, Calcd.)		
	A	R				C	H	N
1a	—	—	140～142	75	Yellow crystal	49.59(49.68)	6.63(6.55)	24.78(24.83)
2a	—	—	185～186	63	White crystal	51.97(52.16)	7.24(7.29)	40.37(40.55)

续表 1

Compd.	Substituent		m. p. (℃)	Yield(%)	Appearance	Elemental analysis(%, Calcd.)		
	A	R				C	H	N
3a	NO_2	SMe	190 ~ 192	95	White powder	39.04(39.02)	3.01(3.00)	18.79(18.96)
3b	NO_2	SPr - *n*	184 ~ 185	91	White crystal	42.15(42.31)	3.82(3.80)	17.62(17.62)
3c	NO_2	NHMe	194 ~ 196	91	Yellow powder	40.71(40.91)	3.23(3.43)	23.58(23.85)
3d	CO_2Me	SMe	187 ~ 188	80	White crystal	43.72(43.97)	3.86(3.69)	14.41(14.65)
3e	CO_2Me	SEt	208 ~ 210	94	White powder	45.32(45.45)	4.04(4.07)	13.94(14.13)
3f	CO_2Me	SPr - *n*	184 ~ 186	83	White powder	46.80(46.82)	4.36(4.42)	13.73(13.65)
3g	CO_2Me	NHMe	213 ~ 215	51	White powder	45.80(46.02)	4.14(4.14)	18.96(19.17)
3h	CO_2Me	NMe_2	203 ~ 204	89	White powder	47.31(47.49)	4.49(4.52)	18.48(18.46)
3i	CO_2Me	NHEt	186 ~ 188	87	White powder	47.36(47.49)	4.68(4.52)	18.28(18.46)
3j	CO_2Et	SMe	180 ~ 181	63	White crystal	45.50(45.45)	4.06(4.07)	13.97(14.13)
3k	CO_2Et	SEt	174 ~ 176	61	White crystal	46.70(46.82)	4.20(4.42)	13.70(13.65)
3l	CO_2Et	SPr - *n*	174 ~ 176	57	White powder	47.90(48.10)	4.57(4.75)	13.35(13.20)
3m	CO_2Et	NHMe	188 ~ 190	46	White powder	47.21(47.49)	4.26(4.52)	18.27(18.46)
3n	CO_2Et	NMe_2	213 ~ 214	57	White powder	48.74(48.85)	5.04(4.87)	17.80(17.80)
3o①	CO_2Me	OMe						

①Known compound.

Table 2 ^{1}H NMR data (δ, DMSO - d_6) of compounds 1a, 2a, 3a - 3n

Compd.	Data
1a	0.97 ~ 1.05 (t, 3H, J = 7.4Hz, $SCH_2CH_2C\mathbf{H}_3$), 1.79 ~ 1.61 (m, 2H, J = 7.4Hz, $SCH_2C\mathbf{H}_2CH_3$), 3.03 ~ 3.10 (t, 2H, J = 7.4Hz, $SC\mathbf{H}_2CH_2CH_3$), 5.49 (s, 2H, NH_2), 6.48 ~ 8.51 (d, 1H, J = 6.0Hz, Pyrim - H_5), 7.82 ~ 7.85 (d, 1H, J = 6.0Hz, Pyrim - H_6)
2a	1.08 ~ 1.15 (s, 3H, $NHCH_2C\mathbf{H}_3$), 3.21 ~ 3.33 (q, 2H, J = 7.4Hz, $NHC\mathbf{H}_2CH_3$), 5.75 ~ 5.78 (d, 1H, J = 6.0Hz, Pyrim - H_5), 7.33 (s, 1H, $N\mathbf{H}CH_2CH_3$), 7.59 ~ 7.62 (1H, d, J = 6.0Hz, Pyrim - H_6)
3a	2.56 (s, 3H, SCH_3), 7.16 ~ 7.19, (d, 1H, J = 6.0Hz, Pyrim - H_5), 7.89 ~ 7.93 (m, 3H, Ar—H), 8.24 ~ 8.31 (m, 2H, Ar, Pyrim - H_6)
3b	0.94 ~ 1.01 (t, 3H, J = 7.0Hz, $SCH_2CH_2C\mathbf{H}_3$), 1.59 ~ 1.72 (m, 2H, J = 7.0Hz, $SCH_2C\mathbf{H}_2CH_3$), 3.11 ~ 3.18 (t, 2H, J = 7.0Hz, $SC\mathbf{H}_2CH_2CH_3$), 7.13 ~ 7.17 (d, 1H, J = 6.6Hz, Pyrim - H_5), 7.73 ~ 7.88 (m, 3H, Ar—H), 8.18 ~ 8.25 (m, 2H, Ar—H, Pyrim - H_6), 11.05 (s, 1H, CONH - Pyrim), 13.30 (s, 1H, SO_2NH)
3c	2.86 (s, 3H, $NHC\mathbf{H}_3$), 6.21 ~ 6.24 (d, 1H, J = 6.0Hz, Pyrim - H_5), 7.66 ~ 7.69 (m, 3H, Ar—H), 7.85 ~ 7.88 (d, 1H, J = 6.0Hz, Pyrim - H_6), 8.04 ~ 8.07 (m, 1H, Ar—H), 8.86 (s, 1H, $\mathbf{NH}$—CH_3)
3d	2.57 (s, 3H, SCH_3), 3.81 (s, 3H, CO_2CH_3), 7.15 ~ 7.18 (d, 1H, J = 5.4Hz, Pyrim - H_5), 7.75 ~ 7.83 (m, 3H, Ar—H), 8.12 ~ 8.15 (m, 1H, Ar—H), 8.35 ~ 8.38 (d, 1H, J = 5.4Hz, Pyrim - H_6), 10.75 (s, 1H, CONH - Pyrim), 12.81 (s, 1H, SO_2NH)
3e	1.21 ~ 1.33 (t, 3H, J = 7.4Hz, $SCH_2C\mathbf{H}_3$), 2.98 ~ 3.09 (q, 2H, J = 7.4Hz, $SC\mathbf{H}_2CH_3$), 3.82 (s, 3H, CO_2CH_3), 7.10 ~ 7.12 (d, 1H, J = 5.2Hz, Pyrim - H_5), 7.66 ~ 7.79 (m, 3H, Ar—H), 7.89 ~ 7.96 (m, 1H, Ar—H), 8.33 ~ 8.35 (d, 1H, J = 5.2Hz, Pyrim - H_6), 10.73 (s, 1H, CONH - Pyrim), 12.62 (s, 1H, SO_2NHCO)

续表 2

3f	0.93～1.01（t，3H，J=7.4Hz，$SCH_2CH_2CH_3$），1.56～1.75（m，2H，$SCH_2CH_2CH_3$），3.11～3.18（t，2H，J=7.2Hz，$SCH_2CH_2CH_3$），3.81（s，3H，CO_2CH_3），7.12～7.15（d，1H，J=5.8Hz，Pyrim－H_5），7.67～7.84（m，3H，Ar—H），8.15～8.19（m，1H，Ar—H），8.33～8.36（d，1H，J=5.8Hz，Pyrim－H_6）
3g	2.86（s，3H，$NHCH_3$），3.75（s，3H，CO_2CH_3），6.20～6.23（d，1H，J=7.0Hz，Pyrim－H_5），7.40～7.43（m，1H，Ar—H），7.52～7.57（m，2H，Ar—H），7.70～7.73（d，1H，J=7.0Hz，Pyrim－H_6），7.96～8.00（m，1H，Ar—H），8.74（s，1H，NH—CH_3）
3h	3.13（s，3H，N—CH_3），3.30（s，3H，N—CH_3），3.74（s，3H，CO_2CH_3），6.51～6.55（d，1H，J=7.0Hz，Pyrim－H_5），7.38～7.44（m，1H，Ar—H），7.54～7.58（m，2H，Ar—H），7.79～7.82（d，1H，J=7.0Hz，Pyrim－H_6），7.96～8.00（m，1H，Ar—H）
3i①	1.08～1.13（t，3H，J=7.2Hz，$NHCH_2CH_3$），3.31～3.38（q，2H，J=7.2Hz，$NHCH_2CH_3$），3.73（s，3H，CO_2CH_3），6.17～6.20（d，1H，J=7.5Hz，Pyrim－H_5），7.38～7.41（m，1H，Ar—H），7.48～7.57（m，2H，Ar—H），7.69～7.71（d，1H，J=7.5Hz，Pyrim－H_6），7.95～7.98（m，1H，Ar—H），10.23（s，1H，CONH－Pyrim），13.30（s，1H，SO_2NHCO）
3j	1.21～1.28（t，3H，J=7.0Hz，$CO_2CH_2CH_3$），2.57（s，3H，SCH_3），4.23～4.33（q，2H，J=7.0Hz，$CO_2CH_2CH_3$），7.14～7.17（d，1H，J=5.6Hz，Pyrim－H_5），7.69～7.82（m，3H，Ar—H），8.14～8.18（m，1H，Ar—H），8.32～8.35（d，1H，J=5.6Hz，Pyrim－H_6）
3k	1.22～1.34（m，6H，$CO_2CH_2CH_3$，SCH_2CH_3），3.08～3.20（q，2H，J=7.4Hz，SCH_2CH_3），4.23～4.34（q，2H，J=7.4Hz，$CO_2CH_2CH_3$），7.11～7.14（d，1H，J=6.0Hz，Pyrim－H_5），7.69～7.82（m，3H，Ar—H），8.15～8.19（m，1H，Ar—H），8.31～8.34（d，1H，J=6.0Hz，Pyrim－H_6），10.73（s，1H，CONH－Pyrim），12.61（s，1H，SO_2NHCO）
3l	0.93～1.00（t，3H，J=7.4Hz，$SCH_2CH_2CH_3$），1.20～1.28（t，3H，J=7.4Hz，$CO_2CH_2CH_3$），1.56～1.74（m，2H，$SCH_2CH_2CH_3$），3.11～3.18（t，2H，J=7.2Hz，$SCH_2CH_2CH_3$），4.26～4.33（q，2H，J=7.4Hz，$CO_2CH_2CH_3$），7.11～7.13（d，1H，J=6.0Hz，Pyrim－H_5），7.69～7.84（m，3H，Ar—H），8.15～8.18（m，1H，Ar—H），8.29～8.32（d，1H，J=6.0Hz，Pyrim－H_6），10.72（s，1H，CONH－Pyrim），12.61（s，1H，SO_2NHCO）
3m	1.20～1.27（t，3H，J=7.2Hz，$CO_2CH_2CH_3$），2.86（s，3H，$NHCH_3$），4.16～4.25（q，2H，J=7.2Hz，$CO_2CH_2CH_3$），6.19～6.23（d，1H，J=7.2Hz，Pyrim－H_5），7.38～7.42（m，1H，Ar—H），7.52～7.56（m，2H，Ar—H），7.69～7.72（1H，d，J=7.2Hz，Pyrim－H_6），7.95～8.00（m，1H，Ar—H），8.72（s，1H，NH—CH_3）
3n	0.74～0.82（t，3H，J=7.2Hz，$CO_2CH_2CH_3$），2.67（s，3H，N—CH_3），2.90（s，3H，N—CH_3），3.71～3.82（q，2H，J=7.2Hz，$CO_2CH_2CH_3$），6.05～6.08（d，1H，J=7.2Hz，Pyrim－H_5），6.94～6.99（m，1H，Ar—H），7.08～7.12（m，1H，Ar—H），7.23～7.27（m，1H，Ar—H），7.38～7.42（d，1H，J=7.2Hz，Pyrim－H_6），7.52～7.56（m，1H，Ar—H）

①300MHz NMR.

按文献［8］中的2－氨基－4－巯基嘧啶与相应卤代烃回流的方法合成2－氨基－4－烷硫基嘧啶，产品纯度较差。本文采用相转移催化剂于室温反应，反应时间可适当缩短，产品质量较好，可不经提纯直接进行下一步反应。反应在质量分数为50%的氢氧化钠溶液中进行时，相同反应时间所得产品的收率略有提高，但影响不大，因此采用稀氢氧化钠作反应介质。在文献［10，11］中，2－氨基－4－烷胺基嘧啶的合成一般采用相应卤代嘧啶与胺的水溶液封管反应来进行，考虑到4位氯很容易被亲核取代，本文尝试在室温常压下和胺的水溶液中进行反应，发现很容易得到相应产物，且收率与文献值相近。

从生物活性测试结果（表3）可看出，在新化合物**3**a～**3**n中，嘧啶环4位为甲硫基取代（**3**d，**3**j）时保持了一定的除草活性，其他烷硫基或烷胺基取代后所得化合物的除草活性均低于甲氧基取代化合物（**3**o），其活性由高到低的顺序为烷氧基取代＞烷硫基取代＞烷胺基取代。另外，同为4位烷硫基取代，随烷基增大除草活性随之降低。总体来看，对嘧啶环为单取代的磺酰脲化合物，嘧啶环4位取代基的变化对整个分子的除草活性影响很大，这与含双取代嘧啶环的磺酰脲类有较大区别，后者在两个取代基中保持一个高活性取代基（如甲基、甲氧基），仅变化另一个取代基，整个分子除草活性的变化并不明显[6,7]。

Table 3　Herbicidal activity of the target compounds

Compd.	IC_{50} (mol/L)	Pot experiment of *Echinochloa Crusgalli* L (%, 0.3kg/hm^2)		Pot experiment of *Brassica napus* L (%, 0.3kg/hm^2)	
		Foliage spray	Soil treatment	Foliage spray	Soil treatment
3a	4.253×10^{-5}	11.4	0	25.1	39.2
3b	3.976×10^{-4}	8.2	0	0	14.7
3c	$>5.677\times10^{-4}$	35.0	0	0	62.6
3d	6.199×10^{-6}	38.2	0	1.0	56.7
3e	$>5.045\times10^{-4}$	14.5	3.4	13.9	29.2
3f	$>4.873\times10^{-4}$	27.3	0	0	16.7
3g	$>5.474\times10^{-4}$	27.3	0	0	36.8
3h	2.345×10^{-4}	0	0	0	5.8
3i	$>2.650\times10^{-4}$	20.5	0	32.6	15.4
3j	4.295×10^{-6}	47.3	9.1	10.6	56.1
3k	1.625×10^{-4}	3.6	0	7.4	29.4
3l	$>4.712\times10^{-4}$	11.4	0	0	2.2
3m	7.052×10^{-4}	9.5	0	21.9	33.5
3n	$>4.484\times10^{-4}$	16.8	0	0	0
3o	2.358×10^{-7}	83.2	87.2	96.8	98.3
Chlorosulfuron	2.242×10^{-7}				

参 考 文 献

［1］Levitt G.. Synthesis and Chemistry of Agrochemicals Ⅱ, ACS Symposium Series No. 443［M］, Washington DC：American Chemistry Society，1991：16－31.

［2］LI Zheng－Ming（李正名），JIA Guo－Feng（贾国锋），WANG Ling－Xiu（王玲秀）*et al.*. CN 1080 116A［P］，1994.

[3] LI Zheng - Ming（李正名），JIA Guo - Feng（贾国锋），WANG Ling - Xiu（王玲秀）*et al.*. CN 1038 679C [P]，1998.

[4] LI Zheng - Ming（李正名），LAI Cheng - Ming（赖城明）. Chinese Journal of Organic Chemistry（有机化学）[J]，2001，21（11）：810 - 815.

[5] YE Guo - Zhong（野国中），FAN Zhi - Jin（范志金），LI Zheng - Ming（李正名）*et al.*. Chem. J. Chinese Universities（高等学校化学学报）[J]，2003，24（9）：1599 - 1603.

[6] HUANG Ming - Zhi（黄明智），HUANG Lu（黄路），CHEN Can（陈灿）*et al.*. CN 1323 789A [P]，2001.

[7] ZHAO Feng - Ge（赵凤革），CAO Guang - Hong（曹广宏），XIE Long - Guan（谢龙观）*et al.*. CN 1337 158A [P]，2002.

[8] Koppd H. C.，Springer R. H.，Robins R. K. *et al.*. J. Org. Chem. [J]，1961，26（3）：792 - 803.

[9] Kuh E.，Chapper T. W.. US 2425 248 [P]，1947.

[10] Fidler W. E.，Wood H. C. S.. J. Chem. Soc. [J]，1957，(10)：4157 - 4162.

[11] Brown D. J.，Harper J. S.. J. Chem. Soc. [J]，1963，(2)：1276 - 1284.

[12] Demosthene C. G.，Aspisi C. R.. US 4226 995 [P]，1980.

Synthesis and Biological Activity of New Phenylsulfonylurea Derivatives

Ning Ma[1,2]，Pengfei Li[1]，Yonghong Li[1]，Zhengming Li[1]，
Lingxiu Wang[1]，Suhua Wang[1]

([1]State Key Laboratory of Elemento - Organic Chemistry，Institute of Elemento - Organic Chemistry，Nankai University，Tianjin，300071，China;
[2]Department of Chemistry，School of Science，Tianjin University，Tianjin，300072，China)

Abstract In order to develop new herbicidal active structures - and research the structure - activity relationship, on the basis of the known monosubstituted phenylsulfonylureas with a high activity we designed and synthesized 14 novel sulfonylurea compounds with a single substituting group，such as alkoxy，alkylthio and alkylamino in the pyrimidine ring. The structures of all compounds were confirmed by ^1H NMR spectra data and elemental analysis. Herbicidal activitiy including IC_{50} was determined by rape root length experiment and pot experiment. The results show that the variation of 4 - substituting group of the pyrimidine ring affects the activity of the compounds respectively. Herbicidal activity decreases in the sequence of alkoxy，alkylthio and alkylamino substituent.

Key words sulfonylurea；herbicide；synthesis；pyrimidine

N－（4－取代嘧啶－2－基）苄基磺酰脲和苯氧基磺酰脲的 3D－QSAR 研究*

马 翼 姜 林 李正名 赖城明

（南开大学化学学院，元素有机化学国家重点实验室，
农药国家工程研究中心，天津 300071）

摘 要 采用比较分子力场分析（CoMFA）方法，对两类单取代嘧啶类似物、6 个 N－（4－取代嘧啶－2－基）－2－甲氧羰基苄基磺酰脲（**1**a～**1**f）和 14 个 N－（4－取代嘧啶－2－基）－2－取代苯氧基磺酰脲（**2**a～**2**n）进行三维定量构效关系（3D－QSAR）研究，建立了一个较为可靠的预测模型。结果表明，分子中苯环邻位、嘧啶环形成氢键的 N 原子处以及嘧啶环 4 位和 6 位附近负电荷增加；苯环邻位乙氧基的 CH_2CH_3 附近选择带正电的原子；苯环邻位乙氧基附近空间体积增加，而嘧啶环 4 位甲氧基稍远处取代基的立体位阻不超过此位置，将有利于提高活性。最后解释了修饰磺酰脲的除草剂仍具有较高活性的原因。

关键词 比较分子力场分析 三维定量构效关系 磺酰脲化合物 除草活性

磺酰脲类除草剂是近 20 年来开发出的一类超高效、广谱、低毒和高选择性的除草剂，它通过抑制植物体内乙酰乳酸合成酶（ALS）的活性来阻止支链氨基酸的生物合成，从而使杂草死亡。

迄今为止，已开发出 20 多种此类除草剂，其中苄嘧磺隆（Bensulfuronmethyl）和乙氧嘧磺隆（Ethoxysulfuron）是磺酰脲桥经过修饰的除草剂，其桥中分别插入了氧原子和亚甲基。

为进一步寻找高活性的磺酰脲化合物，研究其构效关系，我们合成了这两种化合物的单取代嘧啶类似物[1]、6 个 N－（4－取代嘧啶－2－基）－2－甲氧羰基苄基磺酰脲（**1**a～**1**f）和 14 个 N－（4－取代嘧啶－2－基）－2－取代苯氧基磺酰脲（**2**a～**2**n），并测定了其除草活性。

鉴于这两类化合物具有相似的结构，本文采用比较分子力场分析（CoMFA）方法对其进行三维定量构效关系（3D－QSAR）研究。

1 计算方法

CoMFA 方法是研究药物－受体三维定量构效关系的一种新方法。该方法认为，药物分子与受体之间的相互作用主要是通过非共价作用力（如范德华力和静电相互作用）而实现的，作用于同一受体的一系列药物分子与受体之间的非共价作用应该有一定的相似性。通过计算，将药物分子周围的作用力场与药物分子的生物活性定量地联系起来，建立一个模型来设计新化合物，并可以预测新物质的活性大小。

本文使用美国 Tripos 公司的 SYBYL/6.5 软件进行计算，在 SGI 图形工作站上完成。

* 原发表于《高等学校化学学报》，2004，25（11）：2031－2033。基金项目：国家“973”计划（批准号：s2003CB114406）和天津大学与南开大学联合研究项目资助。

化合物的除草活性用油菜根长法测定，并以苄嘧磺隆（**1s**）和乙氧嘧磺隆（**2s**）为对照，对油菜根长的抑制率（50μ/g）见表1。抑制率在34.1%～76.6%范围内，化合物的活性有恰当的分布。

Table 1　Herbicidal activities of compounds 1 and 2①

Compd.	M	Z	R^1	R^2	Inhibition ratio（%）	Compd.	M	Z	R^1	R^2	Inhibition ratio（%）
1a	CO_2CH_3	CH_2	CH_3	H	55.8	**2**e	OCH_3	O	CH_3	H	65.6
1b	CO_2CH_3	CH_2	OCH_3	H	61.2	**2**f	OCH_3	O	OCH_3	H	66.3
1c	CO_2CH_3	CH_2	OC_2H_5	H	51.1	**2**g	OCH_3	O	OC_2H_5	H	53.2
1d	CO_2CH_3	CH_2	$NHCH_3$	H	48.2	**2**h	OCH_3	O	SCH_3	H	34.1
1e	CO_2CH_3	CH_2	SCH_3	H	47.0	**2**i	NO_2	O	CH_3	H	60.9
1f	CO_2CH_3	CH_2	Cl	H	47.6	**2**j	NO_2	O	OCH_3	H	57.7
1s	CO_2CH_3	CH_2	OCH_3	OCH_3	70.8	**2**k	Cl	O	CH_3	H	66.8
2a	CO_2CH_3	O	CH_3	H	62.9	**2**l	Cl	O	OCH_3	H	53.8
2b	CO_2CH_3	O	OCH_3	H	62.1	**2**m	CH_3	O	CH_3	H	67.7
2c	CO_2CH_3	O	OC_2H_5	H	55.1	**2**n	CH_3	O	OCH_3	H	68.4
2d	CO_2CH_3	O	SCH_3	H	45.6	**2**s	OC_2H_5	O	OCH_3	OCH_3	76.6

① (structure: benzene ring bearing M, –Z–$SO_2NHCONH$–pyrimidinyl bearing R^1, R^2) 1a—1f，2a—2n。

2　计算与结果讨论

2.1　分子的药效构象及叠合

计算时，既参考化合物**1a**的X射线衍射晶体结构数据[2]，又考虑到磺酰脲分子最终进入溶液中以自由分子形式与受体相互作用，因而将化合物**1a**的构象用Tripos力场优化，得到该分子的低能药效构象。以活性最高的化合物**2s**为模板，将分子中有共同原子特征的骨架结构（嘧啶磺酰脲部分）进行叠合。计算过程中，三维网格边距为1nm×1nm×1nm，以带一个单位正电荷的sp^3杂化C原子为探针，步长为0.2nm，计算分子与网格点上探针原子的相互作用。

2.2　偏最小二乘法（PLS）分析结果

首先进行交叉验证（Leave - one - out）。$R^2_{cross}=0.732$（R^2_{cross}用来衡量模型的预测能力），最佳主成分数（Optimal component）为5，说明所建立的模型具有较好的预测可靠性。然后进行非交叉验证（No - validation），$R^2=0.951$，$S_{error}=3.583$，$F_{ratio}=85.111$。以上数据表明，用CoMFA参数与化合物分子的除草活性定量联系，可建立一个较为可靠的预测模型。

2.3　3D－QSAR 结果

用 CoMFA 方法得到的 3D－QSAR 结果以三维等值线图表示，用不同颜色表示此系列分子的主要空间和静电性能。红色和蓝色区域代表静电效应，绿色和黄色区域代表立体效应。图 1 为模板分子 **2**s 的三维等值线图。

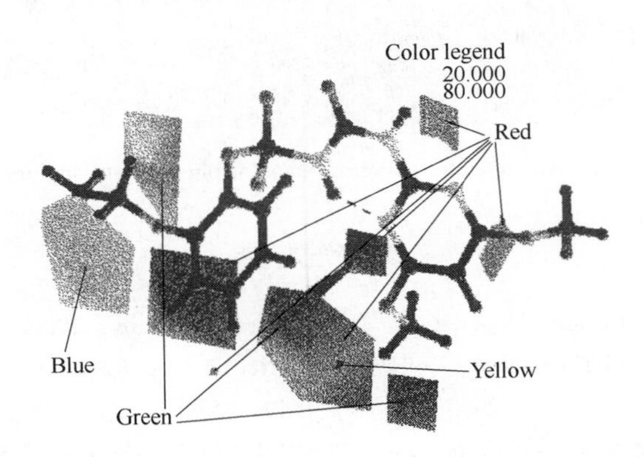

Fig. 1　3D－QSAR result of compound **2**s obtained by using CoMFA

由图 1 可以看出：（1）分子中苯环邻位、嘧啶环形成氢键的 N 原子处以及嘧啶环 4 位，6 位附近的红色区表明，增加负电荷有助于增加化合物的除草活性，即这几个部位应有电负性较强的原子；苯环邻位乙氧基的 CH_2CH_3 附近的蓝色区说明，在此处选择带正电的原子将有利于提高活性。（2）分子中苯环邻位乙氧基附近的绿色区表明，增加空间体积有利于提高除草活性，而嘧啶环 4 位甲氧基稍远处的黄色区说明，取代基的立体位阻不要超过此位置，否则，分子的活性会降低。

2.4　修饰磺酰脲的除草剂仍具有较高除草活性的原因

我们曾提出了能较合理地解释磺酰脲分子的构效关系的磺酰脲分子与受体 ALS 酶中的磺胺焦磷酸素作用的“卡口模型”[3]。根据这一模型，磺酰脲分子中需要存在 3 个负电势较强的区域与受体相互作用。比较一般的磺酰脲分子及磺酰脲桥上修饰—CH_2 或—O—的磺酰脲分子，其构象见图 2，其磺酰脲基，即分子主平面部分的构象差异不大，说明这些分子仍能满足与受体结合的几何条件，进一步比较其静电势图（图 3），其负电势区位置变化不大，因而仍能很好地与受体进行三点相互作用，从而呈现出较好的除草活性。

Conformational Comparison of classical sulfonylurea with modified bridge sulfonylurea

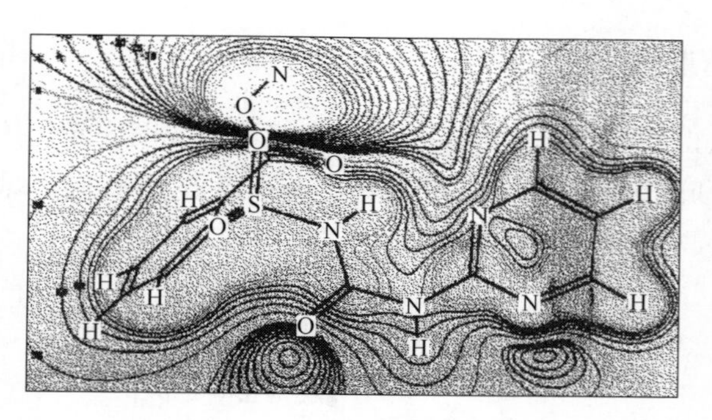

Fig. 3 Electrostatic potential maps of phenoxy sulfonylurea

参考文献

[1] JIANG Lin（姜林）. Doctoral Dissertation [D], Nankai University, 2000.

[2] JIANG Lin（姜林）, LI Zhengming（李正名）, WENG Linhong（翁林红）*et al.*. Chinese J. of Structural Chem.（结构化学）[J], 2000, 19 (2): 149 - 152.

[3] LAI Chengming（赖城明）, YUAN Manxue（袁满雪）, LI Zhengming（李正名）*et al.*. Chem. J. Chinese Universities（高等学校化学学报）[J], 1994, 15 (5): 693 - 696.

3D - QSAR Study on *N* -（4 - Substituted Pyrimidin - 2 - yl）Benzyl Sulfonylurea and Phenoxy Sulfonylurea

Yi Ma, Lin Jiang, Zhengming Li, Chengming Lai

(State Key Laboratory of Elemento - Organic Chemistry, National Pesticide Engineering Resarch Center, College of Chemistry, Nankai University, Tianjin, 300071, China)

Abstract Comparative molecular field analysis (CoMFA) method was applied to the study of the three - dimensional quantitative structure activity relationship (3D - QSAR) on two series of *N* - (4 - substituted pyrimidin - 2 - yl) - 2 - methyl carboxylate - benzyl sulfonylureas and *N* - (4 - substituted pyrimidin - 2 - yl) - 2 - substituted phenoxy sulfonylureas. A reasonable model with predictive ability was obtained from the investigation. According to the resulting contour map of electrostatic field, compounds with enhanced activity could be designed by introduction of more negative charged group into the red region, and positive charged substituent in the blue region will lead to more herbicidal activity. For the steric field, the introduction of smaller groups into the yellow region gives a compound with a higher activity and larger groups in the green region will improve the activity. The models were used to explain why the sulfonylureas with amodified bridge still remain a high herbicidal activity and will help us in our further study.

Key words CoMFA; 3D - QSAR; sulfonylurea compound; herbicidal activity

基于酮醇酸还原异构酶 KARI 复合物晶体结构的三维数据库搜寻*

王宝雷[1]　李正名[1]　马　翼[1]　王建国[1]　罗小民[2]　左之利[2]

（[1] 南开大学元素有机化学研究所，元素有机化学国家重点实验室，天津　300071；
[2] 中国科学院上海药物研究所，上海　200031）

摘　要　以酮醇酸还原异构酶 KARI 复合物0.165nm 高分辨率晶体结构为基础，采用 DOCK 4.0 分子对接程序通过 MDL/ACD 三维数据库搜寻，找到了279 个与 KARI 结合能较低的小分子，讨论了能量打分较高分子同靶酶的作用模式。这些分子作为潜在的 KARI 抑制剂为基于 KARI 的农药分子设计提供了有利指导。

关键词　KARI　分子对接　三维数据库

酮醇酸还原异构酶（Ketol - acid reductoisomerase，简称 KARI）作为植物体内支链氨基酸——缬氨酸、亮氨酸及异亮氨酸生物合成过程中的第二个酶，是继乙酰乳酸合成酶 ALS（acetolactate synthase）之后的另一个关键酶[1]，通过抑制 KARI 的活性，可以阻止支链氨基酸的合成，从而使植物死亡，将其应用于杂草体内则可生物合理设计靶向 KARI 的除草剂。

迄今为止，所报道的 KARI 抑制剂种类和数量稀少，主要为 KARI 所催化的天然底物类似物2 - 二甲基膦酰基 - 2 - 羟基乙酸盐（Hoe 704）[2]和 N - 羟基 - N - 异丙基草酰胺盐（IpOHA）[3]，以及1，2，3 - 噻二唑类化合物[4]。这些化合物虽在抑制 KARI 活性上（invitro）具有很好的效果，但在活体测试中（invivo）结果却不理想，到目前为止尚没有一个商品化的靶向 KARI 除草剂问世，因此有关 KARI 抑制剂设计方面还有很大的潜力。可喜的是，法国的 Biou 等[5] 1997 年首次报道了 KARI - IpOHA 复合物的0.165nm 高分辨率的晶体结构，这为基于结构的药物设计提供了新的机遇。目前有关基于 KARI 的计算化学方面的研究较为罕见。

因此，我们在 KARI 复合物结构基础上利用分子对接方法进行计算机辅助药物设计，以期获得结构新颖，种类较多的 KARI 潜在抑制剂，是一项较为有意义的工作。

1　研究方法

分子对接法是基于生物受体三维结构的药物设计方法之一，是将小分子配体放置于受体的活性位点处，并寻找合理的取向和构象使得配体与受体的形状和相互作用能够最佳匹配。配体和受体的结合强弱取决于结合过程中自由能的变化，如方程（1）所示[6]：

$$\Delta G_{\text{binding}} = RT\ \ln k_i \qquad (1)$$

其中，k_i 是药物与受体的结合常数。配体与受体的相互作用包括静电作用（$E_{\text{electrostatic}}$）、范德华作用（E_{vdw}）和氢键相互作用（$E_{\text{H\_ bond}}$），其中氢键作用部分可以包含在静电作

* 原发表于《有机化学》，2004，24（8）：973 - 976。国家“973”计划（No. 2003CB114406）、天津市科委（No. 033803411）以及天津大学/南开大学联合研究项目“功能药物的分子工程研究”资助项目。

用能中。不同的程序处理配体与受体结合时的刚柔性问题不同，本工作采用 DOCK 4.0 版本，较之以前的几个版本，考虑了配体的柔性，这不但更加接近真实情况，而且在进行分子对接时将大大改善对接结果[6,7]。

网格文件的计算及分子对接工作在中国科学院上海药物研究所 SGI Indigo R10000 图形工作站上进行，其余工作均在南开大国家重点实验室 SGI Indy 工作站上完成。KARI－IpOHA 复合物晶体结构在蛋白质数据库（PDB）中的序号是 1yve[5]，基于这个复合物结构，我们运用分子对接程序 DOCK 4.0 对其进行了三维数据库搜寻，以期寻找可能的 KARI 抑制剂。所搜寻的数据库为 MDL/ACD 3D 小分子数据库，约有 25 万个化合物。

1.1 KARI 活性位点表征（site characterization）

1yve 是一个双二聚体，即有四个单体，分别有 I，J，K，L 四条链[5]，为处理方便，删除了其中的 J，K，L 三条链及其包含的抑制剂 IpOHA，删除所有 NADPH，Mg^{2+} 及 H_2O，进而给分子加氢（all）和加电荷（kollman all）。将配体 IpOHA 及其周围 0.5nm 范围内的氨基酸残基定义为活性区域。利用 autoMS 程序计算 1yve 活性部位的分子表面，然后用 sphgen 程序产生描述活性位点的形状球集（sphere cluster）。分子对接程序利用这些球集加快对接时小分子与大分子活性部位的匹配。

1.2 计算网格文件

利用 grid 程序计算盒子（box）内静电作用以及范德华作用能的网格文件。

1.3 分子对接

以上准备工作完成后，便可以利用 DOCK 4.0 程序进行 MDL/ACD 3D 三维数据库搜寻。一些重要的参数设置如下：考虑小分子柔性（flexible ligand），并用多个“锚”（multiple anchors）搜寻和扭转优化（torsion minimize）来优化小分子的取向（position）和构象（conformation）；采用“锚”优先算法（anchorfirst），使每个分子每轮优化 100 次（configurations per cycle 为 100），并保留最好的构型，每个分子允许最大的“锚”方位数为 5000（maximum orientations 为 5000）；所有对接过的分子构型用 100 个迭代次数（maximum iterations 为 100）进行能量优化。另外在数据库搜寻中还使用了几个滤过参数（filter）：超过 10 根柔性键、少于 5 个重原子以及超过 40 个重原子的小分子不予考虑，排除了柔性太强、太小或太大的分子；打分排序按照计算静电作用能和范德华作用能的总和进行；只保留得分相对较高的前 300 个分子进行研究。

2 结果和讨论

通过基于 KARI 复合物晶体结构的三维数据库搜寻，得到了 300 个能量较低，打分较高的小分子，其中有 21 个分子由于含有同位素或金属元素，Sybyl 无法识别和处理这些原子，故而将其舍弃，仅保留 279 个分子。这 279 个分子中，分属各种类型，约计 14 大类，有胺类及醇类、（氨基）羧酸类、肽类、糖苷类、哌嗪、哌啶及吗啉类、酰胺类、（氨基酸）酯类、脲，胍及脒类、酮类、磺酰胺（肼）类、砜类、schiff 碱类、杂环类等。图 1 中列出了部分化合物的结构式及其与 KARI 的结合能。我们发现得分较高的分子结构中都含有 NH，OH 或 COOH 一些带有活泼氢的基团，还有的则是肽类化合物，这些化合物在水溶液中都较容易电离，因此和 KARI 可能会有较为有利的静电相互作用。

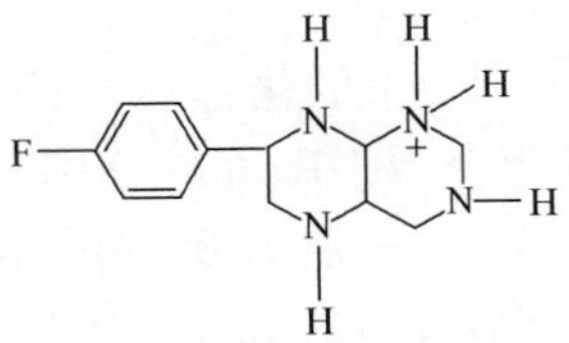

1—312.75kJ/mol

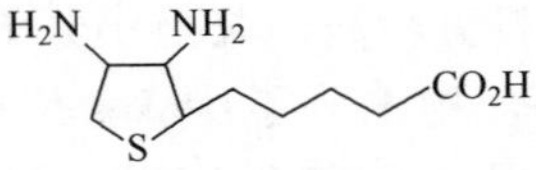

2—290.80kJ/mol

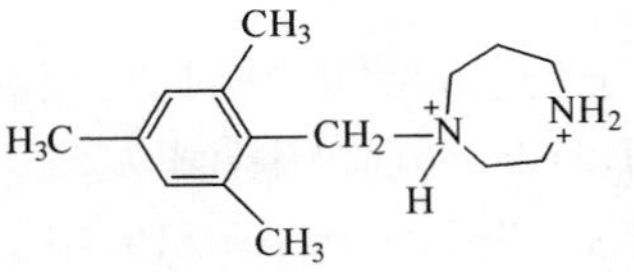

3—265.85kJ/mol

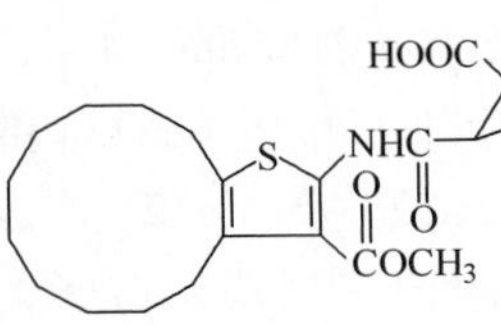

4—245.62kJ/mol

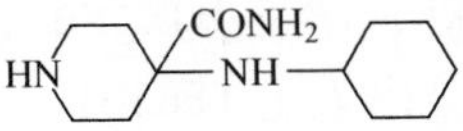

5—245.62kJ/mol

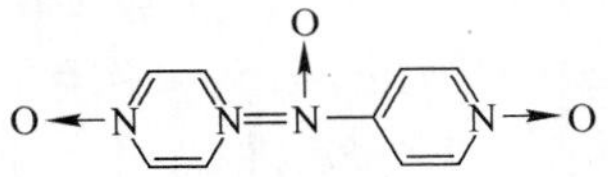

6—240.18kJ/mol

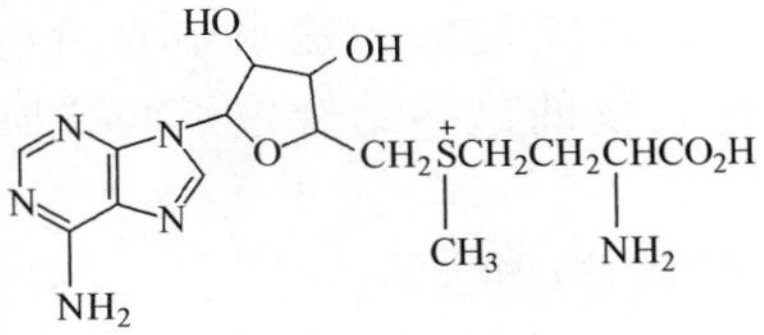

7—239.05kJ/mol

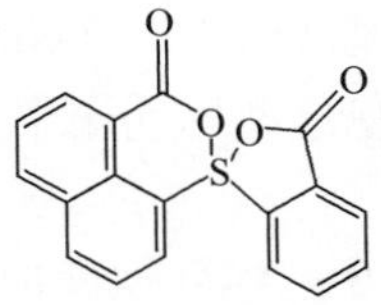

8—235.21kJ/mol

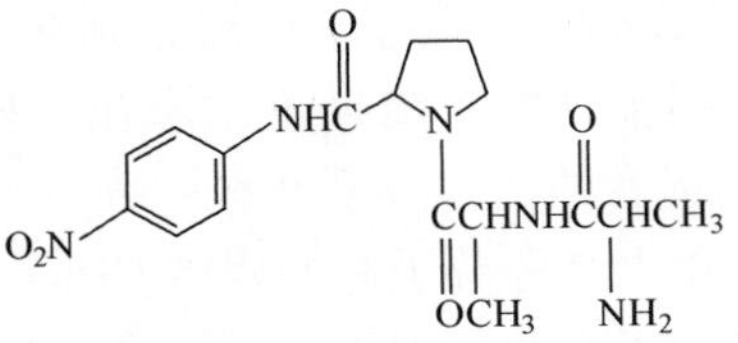

9—230.544kJ/mol

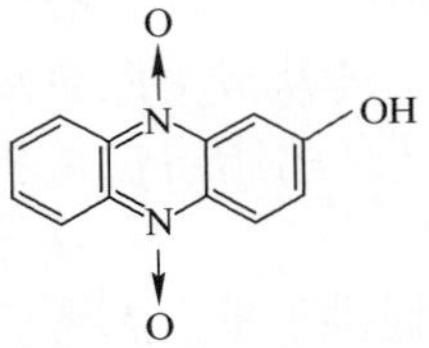

10—227.02kJ/mol

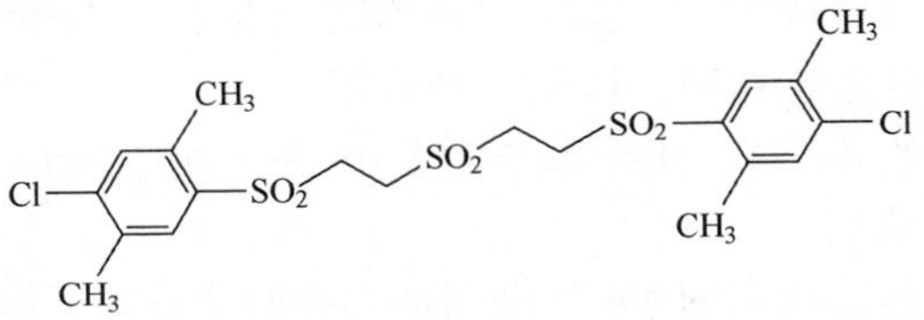

11—227.02kJ/mol

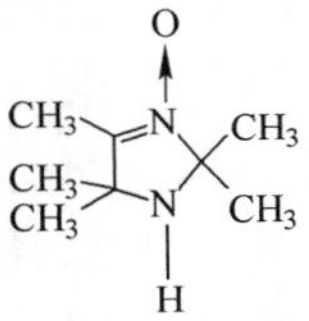

12—223.05kJ/mol

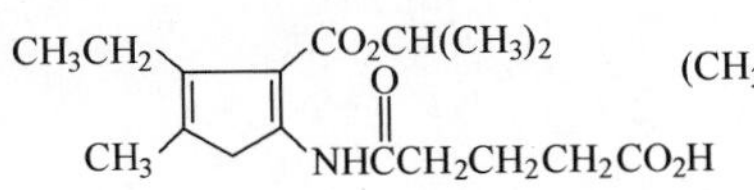

13—221.12kJ/mol

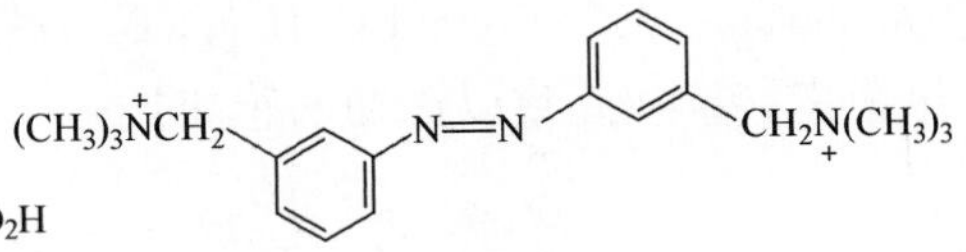

14—220.33kJ/mol

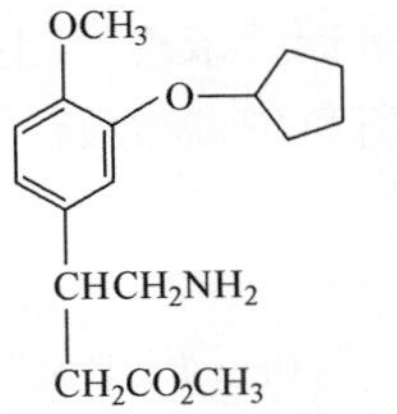

15—212.85kJ/mol

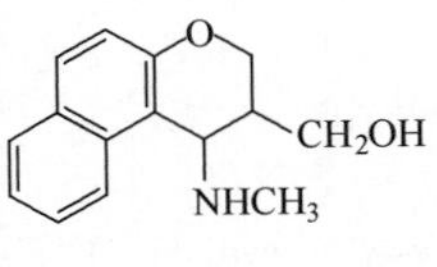

16—212.09kJ/mol

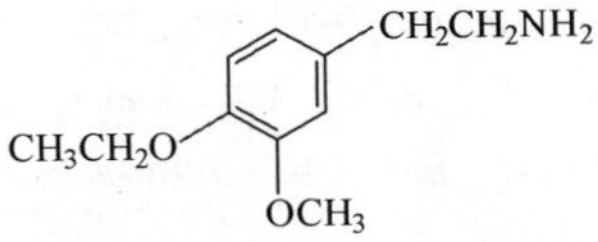

17—210.00kJ/mol

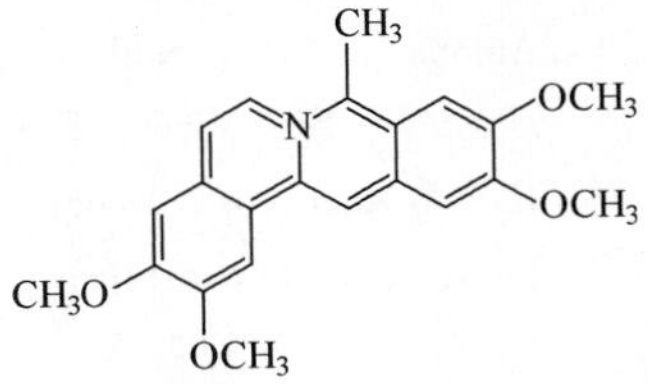

18—205.91kJ/mol

图 1　MDL/ACD 三维数据库搜寻结果中的部分化合物的结构以及它们与 KARI 的结合能

Figure 1　Partial listed structures and their binding energies with KARI

前 279 个分子平均结合能为 -220. 54kJ/mol，其中范德华作用对总能量的贡献平均为 -145. 09kJ/mol，静电作用的贡献为 -75. 45kJ/mol，从总的统计结果来看，范德华作用能明显大于静电作用能，但就局部来看，如前 50 个分子，主要还是静电作用能的贡献要多。另外，前 100 个分子与受体结合较强，100 以后的分子结合差别不是较为明显。

我们通过分析结合能相对较低、打分较高的几个小分子（如图 1 中的化合物**1～5**）与受体 KARI 相互作用模式发现，化合物**1** 分子中**3** - 位季铵两个 N—H 可以分别与 KARI 残基 Glu319 和 Asp315 中的羧基氧原子作用形成氢键，键长分别为 0. 1912 和 0. 2037nm，**4** - 位 N—H 也与残基 Glu319 作用形成了氢键，键长为 0. 2778nm，而 5 - 位苯环则与辅酶 NADPH 平行靠近；化合物**2** 中两个氨基也具有类似的结合情况；化合物**3** 中 1 - 位 N—H 与残基 Asp315 羧基氧原子作用形成氢键，键长为 0. 2582nm，4 - 位 N—H 与残基 Glu496 羧基作用形成氢键，键长为 0. 2614nm，苯环上甲基接近 Leu324 的疏水中心；化合物**4** 除了十二元环部分落入由 Leu501，Leu324，Leu323 及 Ala521 形成的大的疏水口袋中，噻吩环 2 - 位酰胺 N—H 还与 3 - 位甲氧羰基氧原子形成了分子内氢键；化合物 **5** 中酰胺 N - H 和残基 Asp315 羧基氧原子作用形成氢键，键长为 0. 2443nm，哌啶环中 N—H 也与 Asp315 中羧基形成了两个氢键，键长分别为 0. 2134 和 0. 2612nm，分子的疏水部分则被残基 Leu501，Leu323，Leu324 形成的疏水中心所包围。初步分析的结果可以看出，所有这些打分较高的化合物的这些结合位点与文献[5] 提及的活性残基基本一致。

以上是经分子对接程序 DOCK 4. 0 三维数据库搜寻得到的初步结果。因为工作的目的是为了在此基础之上寻找除草剂，而作为除草剂其 lg*P* 是一项较为重要的参数，因而我们通过杜邦公司 SRC 预测软件[8]还对这 279 个分子进行了 lg*P* 预测，lg*P*≤ 5 的[9]有 230 个分子，例如图 1 中化合物**15**，**17** 的 lg*P* 分别为 2. 94 和 1. 47，若从成为除草剂来看，是较好的，而化合物**4**，**18**，由于 lg*P* 为 7. 01 和 5. 06，会因其高的脂溶性蓄积于脂质膜中而无望成为除草剂。

搜寻得到的小分子作为潜在的 KARI 抑制剂为我们今后的工作提供了重要参考。目前我们正在合成部分化合物以验证对 KARI 的抑制活性。

致谢：承澳大利亚 Queensland 大学 Ronald G. Duggleby 教授和中国科学院上海药物研究所蒋华良教授给予热情指导，在此致以深切的感谢！

参考文献

[1] Garault，P.；Letort，C.；Juilard，V. *Appl . Environ. Microbiol.* 2000，66 ，5128.

[2] Schulz，A.；Sponemann，P.；Kocher，H.；Wengenmayer，F. *FEBS LETTERS* 1988，238，375.

[3] Aulabaugh，A.；Schloss，J. V. *Biochemistry* 1990，29，2824.

[4] Halgand，F.；Vives，F.；Dumas，R.；Biou，V.；Andersen，J.；Andrieu，J. - P.；Cantegril，R.；Gagnon，J.；Douce，R.；Forest，E.；Job，D. *Biochemistry* 1998，37 ，4773.

[5] Biou，V.；Dumas，R.；Cohen - Addad，C.；Douce，R.；Job，D.；Pebay - Peyroula，E. *EMBOJ* . 1997，16 ，3405.

[6] Ewing ，T. J. A.；Kuntz，I. D. *J . Comput. Chem.* 1997，18，1175.

[7] *DOCK* 4 . 0 *Manual*，Ed.：Todd Ewing，Regents of the University of California，1998，p. 17.

[8] Meylan，W. M.；Howard，P. H. *J. Pharm. Sci.* 1995，84，83.

[9] Zhou，W. - P. *World Pest.* 2001，23，6 (in Chinese) .（周卫平，世界农药，2001，23 ，6.）

3D – Database Searching Based on the Crystal Structure of Ketol – acid Reductoisomerase (KARI) Complex

Baolei Wang[1], Zhengming Li[1], Yi Ma[1], Jianguo Wang[1]
Xiaomin Luo[2], Zhili Zuo[2]

([1]Institute of Elemento – Organic Chemistry, National Laboratory of Elemento – Organic Chemistry, Nankai University, Tianjin, 300071;
[2]Shanghai Institute of Materia Medica, Chinese Academy of Sciences, Shanghai, 200031)

Abstract Based on the reported crystal structure of complexes of the enzyme ketol – acid reductoisomerase (KARI), 279 molecules were obtained with predicted high affinity for KARI from MDL/ACD 3D – database searching, using program DOCK 4. 0. The interaction pattern between some top – ranked molecules and KARI was described. These structures provide information for further design of new potential herbicidal molecules targeted at KARI.

Key words KARI; DOCK; 3D – database

N－杂环－3－N′－苄氧羰基－β－氨基丁酰胺的合成和结构表征*

臧洪俊　李正名　韩　亮　王宝雷　赵卫光

（南开大学元素有机化学研究所，元素有机化学国家重点实验室，天津　300071）

摘　要　以具有诱导抗性的β－氨基丁酸（BABA）为先导化合物，合成了8个新的N－杂环－3－N′－苄氧羰基－β－氨基丁酰胺类化合物，所有新化合物经元素分析、^{1}H NMR确证，讨论了目标化合物的合成方法。

关键词　β－氨基丁酸　3－N′－苄氧羰基－β－氨基丁酰胺　杂环

目前，我国农作物和蔬菜的病害防治仍以化学农药为主。加入WTO后出口扩大，农产品的农药污染问题，已成为广泛关注的热点。植物系统获得诱导抗性（Systemic Acquired Resistance，SAR）是指经外界因子诱导后，植物体内产生的对病原菌的抗性现象。能够激发作物产生SAR的化学物质被称作植物激活剂或诱导剂，其在农药中代表着一种新型化学调控手段，实现了由治病到防病，提供了一种全新的植保策略[1]。β－氨基丁酸（BABA）是一种具有诱导抗性的新型抗病激活剂，它是一种由番茄根系分泌的非蛋白质氨基酸[2]。它能激活植物的防卫系统，同时作为一种氨基酸，不污染环境。有文献报道β－氨基丁酸（BABA）可诱导番茄、马铃薯、棉花、花生、西瓜、向日葵、辣椒等作物获得抗病性[3]，以及可诱导拟南芥和烟草等植物产生诱导抗性[4]，是一种对环境安全、具有高效诱抗作用的非蛋白质氨基酸。

作者在以β－氨基丁酸为母体结构的基础上，把具有杀菌活性的杂环化合物引入分子中，设计并合成了一系列新N－杂环－3－N′－苄氧羰基－β－氨基丁酰胺化合物（Schemes 1，2），并对其结构进行了表征。目标化合物的活性在进一步研究之中。

1　实验部分

1.1　仪器及药品

Yanaco CHN CORDER MT－3型自动分析仪；BRUKER ACP－200型[1]H NMR仪（TMS为内标）；Yanaco熔点仪；Perkin Elmer 241MC型旋光仪。所有溶剂皆为分析纯，用前经重蒸，试剂为分析纯或化学纯。

1.2　中间体1的合成

按照文献［5］的方法合成，白色固体，用乙醇重结晶，得到白色片状固体，产率38%，为消旋的化合物 $[\alpha]_D=0$（c4，H_2O），熔点216～217℃（文献值[5]为217～218℃）。

* 原发表于《有机化学》，2004，24（6）：669－672。国家“863”（No. 2001AA235011）及天津大学/南开大学联合研究院资助项目。

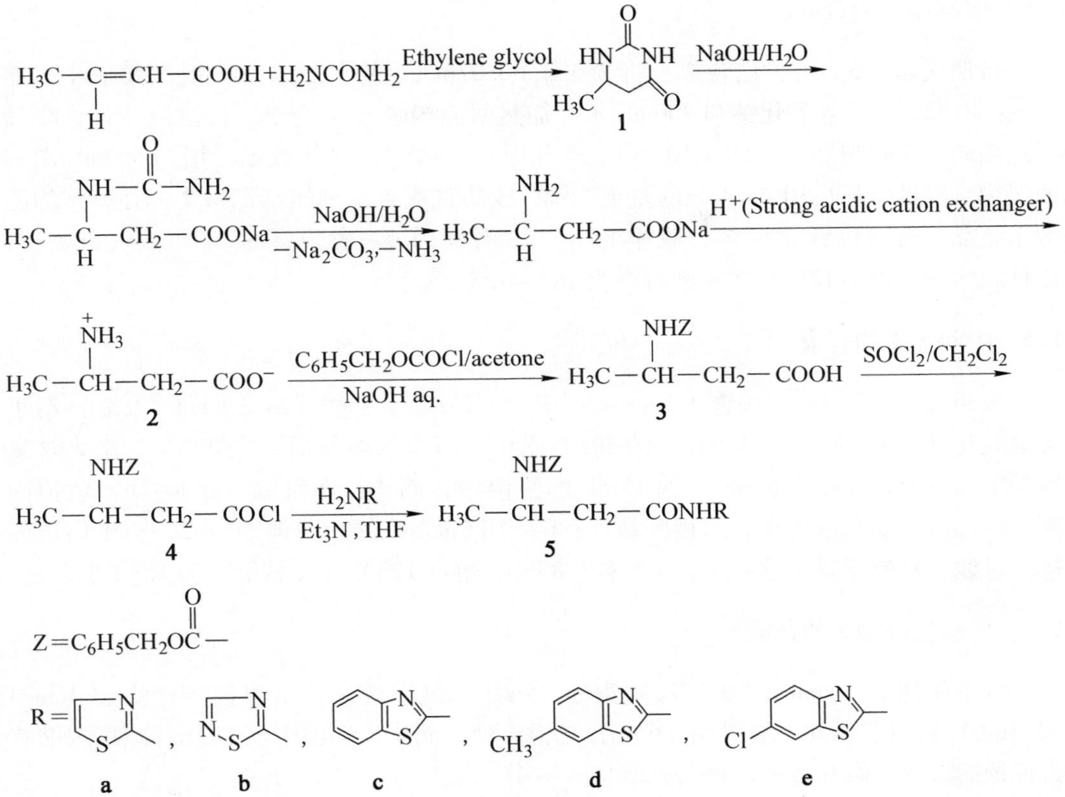

$H_3C-CH=CH-COOH + H_2NCONH_2 \xrightarrow{\text{Ethylene glycol}}$ 1 $\xrightarrow{NaOH/H_2O}$

$H_3C-CH(NHCONH_2)-CH_2-COONa \xrightarrow[-Na_2CO_3, -NH_3]{NaOH/H_2O} H_3C-CH(NH_2)-CH_2-COONa \xrightarrow{H^+\text{(Strong acidic cation exchanger)}}$

$H_3C-CH(\overset{+}{N}H_3)-CH_2-COO^-$ (2) $\xrightarrow[NaOH\ aq.]{C_6H_5CH_2OCOCl/acetone} H_3C-CH(NHZ)-CH_2-COOH$ (3) $\xrightarrow{SOCl_2/CH_2Cl_2}$

$H_3C-CH(NHZ)-CH_2-COCl$ (4) $\xrightarrow[Et_3N, THF]{H_2NR} H_3C-CH(NHZ)-CH_2-CONHR$ (5)

$Z=C_6H_5CH_2OC(=O)-$

R = a, b, c, d, e

Scheme 1

4 + f,g (site 2: H_2N; site 1: NH) $\xrightarrow{NEt_3, THF}$ 5f,5g + 6f,6g

f: X=C $R^1=CH_3$

g: X=N $R^1=Ph$

4 + h (site 2: NH_2; site 1: NH) $\xrightarrow{NEt_3, THF}$ 5h + 6h

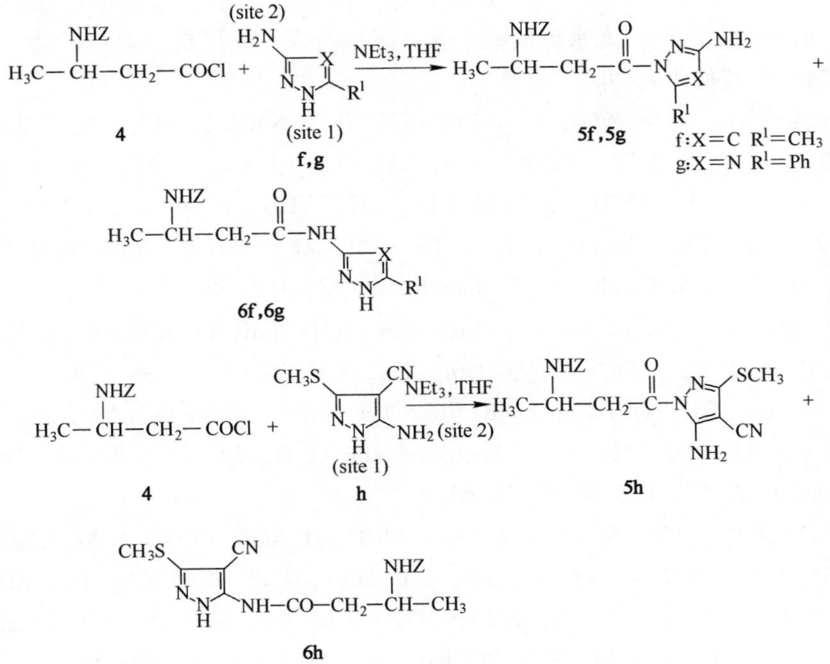

Scheme 2

1.3 中间体2的合成

按照文献［6］的方法合成。将2.56g（0.02mol）的**1**加入500mL三颈瓶中，加入8g（0.02mol）氢氧化钠和200mL水，加热回流8h。反应完毕，待溶液冷却，将其通过阳离子交换树脂柱（50～10目），开始时有二氧化碳气体放出，用约800mL左右的蒸馏水淋洗，然后用1mol/L的氨水洗脱，收集洗脱液，减压蒸馏得**2**。用95%的乙醇重结晶，得白色针状固体，收率87%，为消旋的β－氨基丁酸［α］$_D$＝0（c2.5，H_2O），熔点184～185℃（文献值[6]为184～185.5℃）。

1.4 中间体3的合成

按照文献［7］的方法合成，将（S，R）－3－氨基丁酸（24.3mmol，2.5g）溶于含氢氧化钠（38.5mmol，1.94g）的20mL水中，当全部溶解后，在0℃下，滴入含氯甲酸苄酯（25mmol，3.57mL）的15mL丙酮溶液，滴毕，在室温下反应1h，蒸掉丙酮，用乙酸乙酯萃取两次，水相冷却到0℃，用6mol/L的盐酸调pH＝2，析出白色固体。过滤，真空干燥，重4.99g，产率为83%，熔点122℃（文献值[7]为123℃）。

1.5 目标化合物5的合成

将中间体**3**（1mmol，0.237g）溶于15mL二氯甲烷中，室温搅拌，滴加4mmol（0.3mL）二氯亚砜和二氯甲烷（6mL）的混合液，滴毕，室温下反应1h，减压浓缩蒸去过量的二氯亚砜和溶剂，得白色蜡状固体**4**[7]。

将1.5mmol的芳香胺溶于8mL四氢呋喃中，加入1.65mmol（0.23mL）三乙胺，冰水浴冷却，在0℃下，滴入含**4**的6mL四氢呋喃溶液，有白色不溶物生成，为三乙胺盐酸盐，TLC监测，反应0.5～1h，过滤，蒸掉四氢呋喃，加入15mL三氯甲烷溶解，用1mol/L盐酸洗、水洗，无水硫酸钠干燥，减压浓缩，柱层析分离得产物。用甲苯重结晶。同法可得其他化合物。

5a：白色固体，产率76%，m.p.156～157℃；^{1}H NMR（$CDCl_3$）δ：1.32（d，J＝6.94Hz，3H，CH_3），2.71～2.83（m，2H，CH_2CHN），4.11～4.24（m，1H，CHN），5.06（s，2H，OCH_2），5.40（bs，1H，NH），6.96（s，1H，噻唑—H），7.24～7.28（m，5H，Ph—H），7.30（s，1H，噻唑—H）。Anal. calcd for $C_{15}H_{17}N_3O_3S$：C 56.43，H 5.33，N 13.17；found C 56.43，H 5.35，N 13.28。

5b：白色固体，产率81%，m.p.196～198℃；^{1}H NMR（$CDCl_3$）δ：1.32（d，J＝6.78Hz，3H，CH_3），2.78～2.93（m，2H，CH_2CHN），4.26～4.38（m，1H，CHN），4.99（s，2H，OCH_2），5.48（bs，1H，NH），7.24～7.26（m，5H，Ph—H），8.85（s，1H，噻二唑—H）。Anal. calcd for $C_{14}H_{16}N_4O_3S$：C 52.50，H 5.00，N 17.50；found C 52.55，H 5.00，N 17.45。

5c：白色固体，产率61%，m.p.190～191℃；^{1}H NMR（$CDCl_3$）δ：1.32（d，J＝6.44Hz，3H，CH_3），2.78～2.93（m，2H，CH_2CHN），4.18～4.25（m，1H，CHN），5.07（s，2H，OCH_2），5.23（bs，1H，NH），7.29～7.63（m，9H，Ph—H）。Anal. calcd for $C_{19}H_{19}N_3O_3S$：C 61.79，H 5.15，N 11.38；found C 61.66，H 5.15，N 11.19。

5d：白色固体，产率57%，m.p.190～191℃；^{1}H NMR（$CDCl_3$）δ：1.27（d，J＝

6.68Hz，3H，CH_3），2.43（s，3H，CH_3Ph），2.73～2.76（m，2H，CH_2CHN），4.13～4.20（m，1H，CHN），5.03（s，2H，OCH_2），5.20（bs，1H，NH），7.23～7.63（m，8H，Ph—H）。Anal. calcd for $C_{20}H_{21}N_3O_3S$：C 62.66，H 5.48，N 10.97；found C 62.59，H 5.55，N 11.05。

5e：白色固体，产率53%，m.p. 178～179℃；^1H NMR（$CDCl_3$）δ：1.31（d，J=6.46Hz，3H，CH_3），2.74～2.76（m，2H，CH_2CHN），4.16～4.20（m，1H，CHN），5.06（s，2H，OCH_2），5.25（bs，1H，NH），7.35～7.74（m，8H，Ph—H）。Anal. calcd for $C_{19}H_{18}N_3O_3SCl$：C 56.51，H 4.46，N 10.41；found C 56.55，H 4.55，N 10.50。

5f：白色针状固体，产率67%，m.p. 119～120℃；^1H NMR（$CDCl_3$）δ：1.29（d，J=6.42Hz，3H，CH_3），2.13（s，3H，5—CH_3），3.13～3.33（m，2H，CH_2CHN），4.22～4.35（m，1H，CHN），5.05（s，2H，OCH_2），5.21（s，1H，Py—H），5.44（bs，1H，NH），7.24～7.32（m，5H，Ph—H）。Anal. calcd for $C_{16}H_{20}N_4O_3$：C 60.75，H 6.33，N 17.72；found C 60.75，H 6.30，N 17.87。

5g：白色针状固体，产率63%，m.p. 189～190℃；^1H NMR（$CDCl_3$）δ：1.34（d，J=6.84Hz，3H，CH_3），3.19～3.35（m，2H，CH_2CHN），4.26～4.39（m，1H，CHN），5.05（s，2H，OCH_2），5.12（bs，1H，NH），6.55（s，2H，NH_2），7.30～8.01（m，10H，Ph—H）。Anal. calcd for $C_{20}H_{21}N_5O_3$：C 63.32，H 5.54，N 18.47；found C 63.30，H 5.54，N 18.40。

5h：白色絮状固体，产率74%，m.p. 241～242℃；^1H NMR（$CDCl_3$）δ：1.30（d，J=6.78Hz，3H，CH_3），2.51（s，3H，SCH_3），3.15～3.21（m，2H，CH_2CHN），4.20～4.35（m，1H，CHN），5.06（s，2H，OCH_2），5.20（bs，1H，NH），6.38（s，2H，NH_2），7.25～7.32（m，5H，Ph—H）。Anal. calcd for $C_{17}H_{19}N_5O_3S$：C 54.69，H 5.09，N 18.77；found C 54.74，H 4.99，N 18.55。

2　结果与讨论

2.1　化合物2的合成

按文献［5］的方法制备，采用200目的阳离子交换树脂，作为柱层析的吸附剂，分离时间比较长，且该树脂比较昂贵，费用较高。我们采用50目以下的阳离子交换树脂，作为柱层析的吸附剂，大大缩短分离时间，且得到很好的收率。

2.2　中间体化合物3的合成

按文献［7］的方法制备，在后处理中，水相用乙酸乙酯萃取两次，无水硫酸钠干燥，减压浓缩得产品。但在实验中我们将水相冷却到0℃，用6mol/L的盐酸调pH=2，即可析出白色固体，过滤，真空干燥，产率为83%。

2.3　讨论目标化合物5的合成

在**f**，**g**，**h**这三个反应物中，有两个位置可以和化合物**4**发生反应，应该得到两种产物，如Scheme 2所示。但实验结果发现，只得到一种产物分别为**5f**，**5g**，**5h**；**6f**，**6g**，

6h 几乎为痕量，此结果说明位置 1 和 2 有不同的反应活性。此外，在 $CHCl_3$，$CHCl_2$，CH_3CN 不同溶剂中，进行该反应，实验结果发现，也只得到一种产物分别为 **5f**，**5g**，**5h**；**6f**，**6g**，**6h** 几乎为痕量，该结果说明，位置 1 和 2 的反应活性不受溶剂的影响。

2.4 化合物 5 的结构表征

化合物所有非活泼氢均得到合理归属。*N* – 杂环 – 3 – *N′* – 苄氧羰基 – *β* – 氨基丁酰胺的甲基质子化学位移 *δ* 值在 1.25 ~ 1.31（双峰），亚甲基质子化学位移 *δ* 值在 2.59 ~ 2.64（多峰），主要是由于旁边连接了一个不对称 C 原子，使 CH_2 上的两个氢表现为磁不等价，同碳偶合裂分为四重峰，同时又受到手性 C 上氢的偶合，因此在谱图上表现为多重峰；次甲基质子化学位移 *δ* 值在 4.01 ~ 4.12（多峰），苄氧羰基上的亚甲基质子化学位移 *δ* 值在 5.04 ~ 5.09（单峰），芳环上氢的化学位移值 *δ* 在 7.25 ~ 7.34（多峰）。化合物 **5a ~ 5d**，**5e**，**5g**，只出现一个活泼 H 的化学位移，为 $NHCOOCH_2Ph$ 的 NH 质子，化学位移 *δ* 值在 5.12 ~ 5.48 之间，峰型均为馒头状；化合物 **5f**，**5h** 的谱图上出现了两处活泼氢的化学位移，化学位移 *δ* 值在 5.12 ~ 5.48 之间，峰型均为馒头状，为 $NHCOOCH_2Ph$ 的 NH 质子，杂环上的 NH_2 质子化学位移 *δ* 值在 6.38 ~ 6.55 之间，峰型为尖峰，积分面积为 2 个氢。

参考文献

[1] Fan, H. Y.; Li, B. J.; Lu, C. M.; Li, T. L.; Zhou, B. L. *Plant Prot.* 2003, 29, 14 (in Chinese).
(范海延，李宝聚，吕春茂，李天来，周宝利，植物保护，2003，29，14.)

[2] Gamliel, A.; Katan, J. *Phytopathology* 1992, 82, 320.

[3] Cohen, Y.; Reuweni, M.; Baider, A. *Eur. J. Plant Pathol.* 1999, 105, 351.

[4] (a) Zimmer, L.; Metranx, J. P.; Mauch – Mani, B. *Plant Physiol.* 2001, 126, 517.
(b) Zimmer, L.; Jakab, G.; Metranx, J. P., Mauch – Mani, B. *Proc. Natl. Acad. Sci. U. S. A.* 2000, 97, 12920.

[5] Zee – Cheng, K. Y.; Robins, R. K.; Cheng, C. C. *J. Org. Chem.* 1961, 26, 1877.

[6] Rachina, V. *Synthesis* 1982, 967.

[7] Amoroso, R.; Cardillo, G.; Mobbili, G.; Tomasini, C. *Tetrahedron: Asymmetry* 1993, 4 (10), 2241.

Synthesis and Characterization of *N* – Heterocycle – 3 – *N′* – benzyloxycarbonyl – *β* – aminobutanamide

Hongjun Zang, Zhengming Li, Liang Han, Baolei Wang, Weiguang Zhao

(State Key Laboratory of Elemento – Organic Chemistry, Institute of Elemento – Organic Chemistry, Nankai University, Tianjin, 300071)

Abstract *β* – Aminobutyric acid was used as a leading compound for its induced plant resistance activity. Eight new *N* – heterocycle – 3 – *N′* – benzyloxycarbonyl – *β* – aminobutanamides were prepared. The structures were confirmed by elemental analyses and ^{1}H NMR spectra.

Key words *β* – aminobutyric acid; 3 – *N′* – benzyloxycarbonyl – *β* – aminobutanamide; heterocycle

Synthesis and Characterization of N – Heterocycle – 3 – N′ – benzyloxycarbonyl – β – aminobutanamide

ZANG, Hong – Jun;
LI, Zheng – Ming*;
HAN, Liang;
WANG, Bao – Lei;
ZHAO, Wei – Guang
Chin. J. Org. Chem. **2004**, 24 (6), 669

β – Aminobutyric acid was used as a leading compound for its induced plant resistance activity. Eight new *N* – heterocyclic – 3 – *N′* – benzyloxycarbonyl – β – aminobutanamides were prepared.

Synthesis of 1 – Hydroxy – 4 – (3 – pyridyl) butan – 2 – one Using Heck Reaction

CAI, Chun*;
LÜ, Chun – Xu
Chin. J. Org. Chem. **2004**, 24 (6), 673

Under the Heck reaction condition with Ph_3P as ligand, 1 – hydroxy – 4 – (3 – pyridyl) butan – 2 – one can be synthesized utilizing cross – coupling of 3 – halopyridine and 3 – butene – 1, 2 – diol.

Synthesis of 4 – Oxo – 2 – thioxohexahydropyrimidines under Ultrasound Irradiation

LI, Ji – Tai*;
HAN, Jun – Fen;
LIN, Zhi – Ping;
LI, Tong – Shuang
Chin. J. Org. Chem. **2004**, 24 (6), 675

Under ultrasound irradiation, the cyclocondensation of ethyl α – cyanocinnamates with thiourea activated by potassium carbonate at room temperature in ethanol resulted in 4 – oxo – 2 – thioxohexahydropyrimidines in good yields.

Cyanating Addition to the Derivatives of Isopropylidene Methylenemalonate

LI, Jing – Hua*;
SHI, Jie – Hua;
SHENG, Hua – Feng
Chin. J. Org. Chem. **2004**, 24 (6), 680

Isopropylidene 5 – (1 – cyanoalkyl) malonates were synthesized by Michael addition of cyanide to the derivatives of isopropylidene methylenemalonate.

Synthesis and Biological Activity of Novel 2 – Methyl – 4 – trifluoromethyl-thiazole – 5 – carboxamide Derivatives*

Changling Liu[1,2], Zhengming Li[1], Bin Zhong[1]

([1]Institute of Elemento – Organic Chemistry, State Key Laboratory of Elemento – Organic Chemistry, Nankai University, Tianjin, 300071, China;
[2]Shenyang Research Institute of Chemical Industry, Shenyang, 110021, China)

Abstract Nine novel 2 – methyl – 4 – trifluoromethylthiazole – 5 – carboxamide derivatives were designed and synthesized utilizing ethyl 4,4,4 – trifluoroacetoacetate as a starting material. Subsequently, the biological activity of the compounds was evaluated in the greenhouse. Results indicated that all of the compounds have some fungicidal and insecticidal activity but no herbicidal activity. Compound **1** has fungicidal activity with 90% control of tomato late blight at 375g ai/ha, while two compounds **2F** and **2H** show insecticidal activity with 80 and 100% control, respectively, against potato leafhopper at 600 g ai/ha.

Key words synthesis; biological activity; 2 – methyl – 5 – trifluoromethyl – thiazolecarboxamides

1 Introduction

The thiazole nucleus plays a vital role in many biological activities making it one of the extensively studied heterocycles[1-9]. For example 2,4 – dimethylthiazole – 5 – carboxamide and 2 – methyl – 4 – trifluoromethylthiazole – 5 – carboxamide derivatives such as metsulfovax[10] and thifluzamide[11] are known as agricultural fungicides where the 4 – trifluoromethylthiazole – 5 – carboxamide derivatives are usually better than the 4 – methylthiazole – 5 – carboxamides[3]. Propamocarb [12] is also a known agricultural fungicide and with the goal of discovering new fungicides we incorporated the two active moieties of thifuzamide and propamocarb into a single compound, **1**. The hydrochloride salt (**2A**), and additional analogs (**2B** – **2D**) and comparative compounds (**2E** – **2H**) were synthesized, and their biological activities were evaluated in the greenhouse.

2 Results and discussion

2.1 Synthesis

Scheme 1 illustrates the synthetic route used to prepare the derivatives.

Compound **6** was synthesized from ethyl 4,4,4 – trifluoroacetoacetate[13] and reacted easily with amines in acetonitrile to afford **1** and **2** in high purity and yield. Compound **2A** was prepared from **1** with hydrochloride acid in methanol.

---

* Reprinted from *Journal of Fluorine Chemistry*, 2004, 125: 1287 – 1290. The project is funded by the National Key Project for Basic Research (2003CB114400).

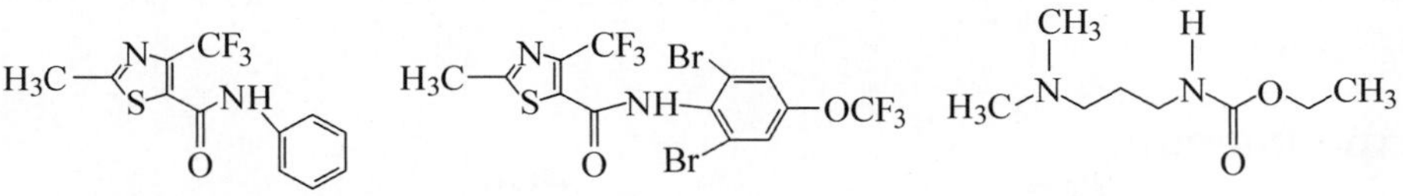

metsulfovax thifuzamide propamocarb

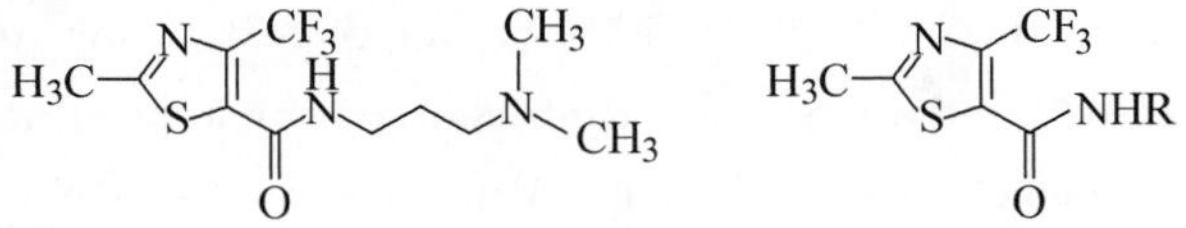

1 2

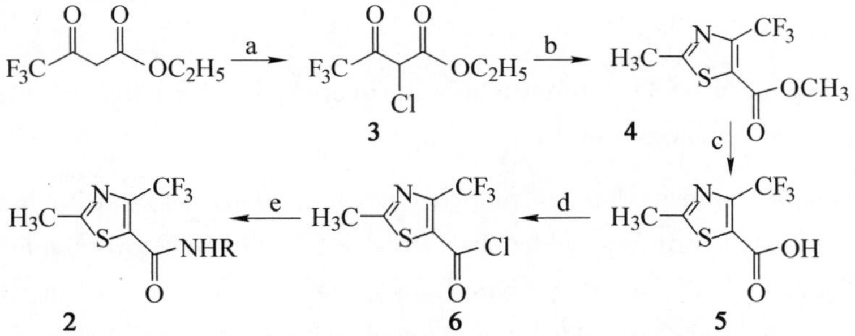

3 4 5 6 2

Scheme 1 Synthesis of 2 – methyl – 4 – trifluoromethyl – thiazole – 5 – carboxamide derivatives.
(a) $SO_2Cl_2CCl_4$; (b) thioacetamide/DMF; (c) $NaOH/H_2O$, HCl(aq); (d) $SOCl_2$; (e) amine/Et_3N/CH_3CN. R: **2A**, 3 – (dimethylamino) propyl HCl salt; **2B**, 3 – (diethylamino) propyl; **2C**, 3 – (1H – imidazol – 1 – yl) propyl; **2D**, 3 – (2 – oxo – pyrrolidin – 1 – yl) propyl; **2E**, 1, 1 – dimethyl – prop – 2 – ynyl; **2F**, 1 – ethyl – 1 – methyl – prop – 2 – ynyl; **2G**, tert – butyl; **2H**, (N – tert – butyl) amino.

The synthesized compounds **1** and **2** were characterized by IR, ^1H NMR, mass spectra and elemental analyses were consistent with the assigned structures. The IR spectra of compounds showed NH and C =O stretching bands at 3240 – 3440 and 1640 – 1675cm^{-1}, respectively. The ^1H NMR of spectra of compounds **1** and **2** showed signals at δ 5.98 – 8.45 ppm attributed to NH, while the CH_2 of compound **2F** was affected by CH_3 and chiral carbon simultaneously, and showed two groups of quadruple peak 2.11 – 2.15 (m, 1H, J = 7.5Hz, CH_aMe), and 1.85 – 1.92 (m, 1H, J = 7.5Hz, CH_bMe).

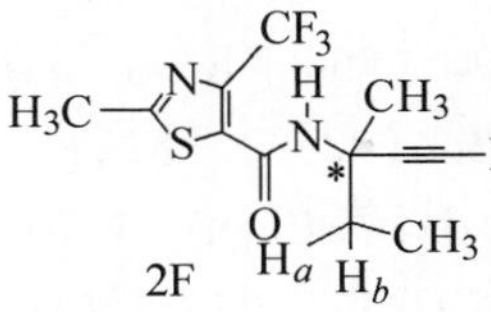

2F

2.2 Biological activities

The biological activities were evaluated at Rohm and Haas Co. using a previously reported procedure[14-17]. All of the compounds have some flingicidal and insecticidal activity with no herbicidal activity, where 100% is complete control of the flingus. Compound **1** showed good activity of tomato late blight providing 90% at 375g ai/ha. Additionally, two compounds, **2F** and

2H, have insecticidal activity with 80 and 100% control, respectively, against potato leafhopper at 600g ai/ha.

3 Experimental

Melting points were determined on a Büchi melting point apparatus and are uncorrected. ^1H NMR spectra were recorded with Mercury 300 (Varian, 300MHz) spectrometer with $CDCl_3$ as the solvent and TMS as the internal standard. Infrared spectra were measured on KBr disks using a PF-983G instrument (Perkin-Elmer). Mass spectra (GC-MS) were obtained on an MD-800 (Fisons) instrument. Combustion analyses for elemental composition were made with an EA 1106 analyzer (Fisons). All chemicals or reagents were purchased from standard commercial suppliers.

3.1 Preparation of N-(3-(dimethylamino)propyl)-2-methyl-4-trifluoromethylthiazole-5-carboxamide (1)

To the solution of 3-(dimethylaniino)propylamine (1.1g, 0.01mol) and triethylamine (1.0g, 0.01mol) in 10mL acetonitrile was added 2-methyl-4-trifluoromethylthiazole-5-carbonyl chloride (**6**)[13] (2.3g, 0.01mol), the mixture was heated to reflux for 10 min, rotovaped to remove the acetonitrile. Water was added and extracted with ethyl acetate, washed with saturated aqueous $NaHCO_3$, water and brine, dried over magnesium sulfate and rotovaped to give 2.36g of the target compound (**1**) as an oil, yield 80%. IR (KBr) v: 3300 (N—H), 2960, 2840, 1660, 1565, 1495, 1465, 1360, 1295, 1175, 1140, 910, 730cm^{-1}; GC-MS, m/z (%): 295 (M^+, 15%), 273 (25%), 254, 245 (10%), 232 (8%), 222 (88%), 206 (10%), 194 (75%), 181 (18%), 166 (70%), 147, 125 (40%), 106 (10%), 97 (100%), 84 (20%), 71 (25%), 58 (25%), 42 (80%); ^1H NMR (300MHz, $CDCl_3$) δ: 8.45 (bs, 1H, NH), 3.53 (d, 2H, J = 5.4Hz, CH_2), 2.73 (s, 3H, CH_3), 2.60 (t, 2H, J = 6Hz, CH_2), 2.34 (s, 6H, NMe_2), 1.82 (t, 2H, J = 6Hz, CH_2); Anal. Calc. (%) for $C_{11}H_{16}F_3N_3OS$: C, 44.73; H, 5.46; N, 14.24. Found: C, 44.49; H, 5.56; N, 14.28.

3.2 Preparation of N-(3-(dimethylamino)propyl)-2-methyl-4-trifluoromethylthiazole-5-carboxamide hydrochloride salt (2A)

Yield 90%, 143-144℃. IR (KBr) v: 3440 (N—H), 2980, 2960, 1640, 1520, 1505, 1475, 1440, 1320, 1250, 1160, 1135, 1080, 905, 850, 710cm^{-1}; GC-MS, m/z (%): 331 (M^+), 295 (13%), 280, 275, 251, 194 (15%), 166 (22%), 151, 125 (22%), 106 (5%), 84 (10%), 72 (20%), 58 (100%); ^1H NMR (300MHz, $CDCl_3$) δ: 11.76 (bs, 1H, HCl), 8.22 (bs, 1H, NH), 3.60 (bs, 2H, CH_2), 3.15 (bs, 2H, CH_2), 2.84 (d, 6H, J = 3.6Hz, NMe_2), 2.73 (s, 3H, CH_3), 2.18 (bs, 2H, CH_2).

3.3 Preparation of N-(3-(diethylamino)propyl)-2-methyl-4-trifluoromethylthiazole-5-carboxamide (2B)

Oil, yield 86%. IR (KBr) v: 3280 (N—H), 2980, 2820, 1660, 1565, 1490, 1360, 1295, 1200, 1175, 1135, 910, 730cm^{-1}; GC-MS, m/z (%): 323 (M^+, 15%), 308 (68%), 294 (45%), 231 (5%), 223 (8%), 183 (3%), 166 (65%), 125 (50%), 112 (20%), 100 (22%), 86

(100%), 72 (75%), 58 (65%), 42 (38%); ^{1}H NMR (300MHz, $CDCl_3$) δ: 8.44 (bs, 1H, NH), 3.56 (d, 2H, J = 6Hz, CH_2), 2.70 – 2.81 (m, 9H, CH_3, $3CH_2$), 1.92 (t, 2H, J = 6Hz, CH_2), 1.13 – 1.18 (m, 6H, $2CH_3$); Anal. Calc. (%) for $C_{13}H_{20}F_3N_3OS$: C, 48.29; H, 6.23; N 12.99. Found: C, 48.40; H, 6.18; N 12.78.

3.4 Preparation of N – (3 – (1H – imidazol – 1 – yl) propyl) – 2 – methyl – 4 – trifluoromethylthiazole – 5 – carboxamide (2C)

Oil, yield 78%. IR (KBr) v: 3240 (N—H), 2950, 2880, 1660, 1570, 1505, 1495, 1440, 1360, 1295, 1230, 1205, 1175, 1140, 1090, 910, 820, 730, 670 cm^{-1}; GC – MS, m/z (%): 318 (M^+, 5%), 300, 275, 341, 259 (6%), 251, 203 (3%), 194 (80%), 183 (15%), 175 (18%), 166 (90%), 146 (8%), 125 (65%), 107 (42%), 95 (100%), 82 (85%), 69 (28%), 55 (45%), 41 (35%); ^{1}H NMR (300MHz, $CDCl_3$) δ: 7.62 (bs, 1H, NH), 7.45 (s, 1H, CH), 6.95 (s, 2H, CH=CH), 4.04 (t, 2H, J = 6.6Hz, CH_2), 3.38 (d, 2H, J = 6.3Hz, CH_2), 2.73 (s, 3H, CH_3), 2.10 (t, 2H, J = 6.6Hz, CH_2); Anal. Calc. (%) for $C_{12}H_{13}F_3N_4OS$: C, 45.27; H, 4.12; N 17.61. Found: C, 45.08; H, 4.25; N 17.45.

3.5 Preparation of N – (3 – (2 – oxo – pyrrolidin – 1 – yl) propyl) – 2 – methyl – 4 – trifluoromethylthiazole – 5 – carboxamide (2D)

Oil, yield 85%. IR (KBr) v: 3265 (N—H), 2940, 2880, 1675, 1655, 1560, 1500, 1475, 1440, 1365, 1295, 1205, 1175, 1140, 905, 820, 730 cm^{-1}; GC – MS, m/z (%): 335 (M^+, 30%), 315, 302, 287 (5%), 266 (10%), 251 (5%), 237 (15%), 224 (23%), 211, 194 (62%), 183 (15%), 175 (12%), 166 (62%), 146 (3%), 141 (12%), 125 (64%), 112 (78%), 98 (100%), 84 (38%), 70 (64%), 56 (52%), 41 (49%); ^{1}H NMR (300MHz, $CDCl_3$) δ: 7.66 (bs, 1H, NH), 3.33 – 3.46 (m, 6H, $3CH_2$), 2.73 (s, 3H, CH_3), 2.43 (t, 2H, J = 8.1Hz, CH_2), 2.11 (m, 2H, CH_2), 1.80 (m, 2H, CH_2); Anal. Calc. (%) for $C_{13}H_{16}F_3N_3O_2S$: C, 46.55; H, 4.81; N, 12.54. Found: C, 46.72; H, 4.67; N, 12.62.

3.6 Preparation of N – (1,1 – dimethyl – prop – 2 – ynyl) – 2 – methyl – 4 – trifluoromethylthiazole – 5 – carboxamide (2E)

Yield 95%, 99 – 100℃. IR (KBr) v: 3320 (C≡CH), 3290 (N—H), 2980, 2920, 2180 (C≡C), 1660, 1560, 1530, 1485, 1360, 1295, 1205, 1175, 1135, 1000, 905, 850, 730, 665, 640 cm^{-1}; GC – MS, m/z (%): 276 (M^+, 50%), 261 (60%), 256 (78%), 241 (15%), 228 (12%), 220 (8%), 207 (15%), 194 (100%), 187 (60%), 171 (15%), 166 (88%), 146 (15%), 125 (78%), 106 (50%), 96 (15%), 84 (55%), 67 (48%), 52 (39%), 41 (65%); ^{1}H NMR (300MHz, $CDCl_3$) δ: 6.21 (bs, 1H, NH), 2.73 (s, 3H, CH_3), 2.42 (s, 1H, ≡CH), 1.73 (s, 6H, $2CH_3$); Anal. Calc. (%) for $C_{11}H_{11}F_3N_2OS$: C, 47.82; H, 4.02; N, 10.15. Found: C, 47.68; H, 3.90; N, 10.28.

3.7 Preparation of N – (1 – ethyl – 1 – methyl – prop – 2 – ynyl) – 2 – methyl – 4 – trifluoromethylthiazole – 5 – carboxamide (2F)

Yield 95%, 79 – 80℃. IR (KBr) v: 3310 (C≡CH), 3280 (N—H), 2980, 2940, 2880, 2180 (C≡C), 1665, 1575, 1540, 1485, 1485, 1465, 1360, 1320, 1295, 1205, 1185, 1135, 1000,

910, 840, 730, 675, 635cm^{-1}; GC－MS, m/z(%): 291(M^+, 2%), 275(10%), 261(100%), 255(15%), 242(70%), 235, 221, 194(90%), 173(10%), 166(80%), 146(15%), 125(65%), 106(32%), 96(15%), 79(52%), 65(12%), 53(25%), 42(20%); ^{1}H NMR (300MHz, $CDCl_3$) δ: 6.20(bs, 1H, NH), 2.73(s, 3H, CH_3), 2.11－2.15(m, J = 7.5, 1H, CH_aMe), 1.85－1.92(m, J = 7.5, 1H, CH_bMe), 2.43(s, 1H, C≡CH), 1.71(s, 3H, CH_3), 1.58(t, 3H, J = 7.2Hz, CH_3); Anal. Calc. (%) for $C_{12}H_{13}F_3N_2OS$: C, 49.64; H, 4.52; N, 9.65. Found: C, 49.45; H, 4.42; N, 9.78.

3.8 Preparation of N－tert－butyl－2－methyl－4－trifluoromethylthiazole－5－carboxamide(2G)

Yield 95%, 110－112℃. IR(KBr) v: 3300(N—H), 2980, 2940, 1650, 1575, 1540, 1495, 1375, 1305, 1200, 1175, 1140, 910, 850, 730, 695cm^{-1}; GC－MS, m/z(%): 266(M^+, 35%), 251(88%), 231, 223, 211(60%), 194(100%), 171(44%), 166(88%), 151(25%), 146(10%), 125(75%), 118(20%), 106(30%), 96(8%), 87(5%), 70(8%), 56(48%), 42(41%); ^{1}H NMR(300MHz, $CDCl_3$) δ: 5.98(bs, 1H, NH), 2.72(s, 3H, CH_3), 1.44(s, 9H, 3CH_3); Anal. Calc. (%) for $C_{10}H_{13}F_3N_2OS$: C, 45.10; H, 4.92; N, 10.53. Found: C, 45.29; H, 4.78; N, 10.37.

3.9 Preparation of N－((N－tert－butyl)amino)－2－methyl－4－trifluoromethylthiazole－5－carboxamide(2H)

Yield 85%, 123－124℃. IR(KBr) v: 3260(N—H), 3205(N—NH), 2980, 1675, 1530, 1460, 1375, 1340, 1210, 1155, 910, 865, 780, 715cm^{-1}; GC－MS, m/z(%): 281(M^+, 20%), 266(52%), 246(15%), 225(75%), 205(25%), 194(100%), 186(10%), 166(45%), 148(2%), 146(3%), 125(30%), 106(10%), 87(15%), 75(4%), 576(95%), 41(55%); ^{1}H NMR(300MHz, $CDCl_3$) δ: 7.64(bs, 1H, NH), 3.56(bs, 1H, NH), 2.75(s, 3H, CH_3), 1.14(s, 9H, 3CH_3); Anal. Calc. (%) for $C_{10}H_{14}F_3N_3OS$: C, 42.69; H, 5.02; N, 14.95. Found: C, 42.88; H, 5.21; N, 14.78.

Acknowledgements

Thanks to Dr. R. M. Jacobson, Dr. S. H. Shaber, Dr. Y. M. Zhu, and Dr. J. M. Renga for their help in the project and paper preparation.

References

[1] L. F. Lee, F. M. Schleppnik, R. K. Howe, J. Heterocyclic Chem. 22(1985)1621－1630.

[2] G. A. White, Pestic. Biochem. Physiol. 34(1989)255－276.

[3] W. G. Phillips, M. Rejda－Heath, Pestic. Sci. 38(1993)1－7.

[4] H. Dolman, J. Kuipers, US Patent 4 889 863(1989).

[5] S. Berg, S. Hellberg, PCT Application WO 03 089 419(2003).

[6] E. Harrington, J. Wang, J. Cochran, S. Nanthakumar, US Patent Application 2003119856(2003).

[7] F. J. Gellibert, C. D. Hartley, J. M. Woolven, N. Mathews, European Patent Application EP1 355 892A1 (2003).

[8] G. Seifert, T. Rapold, M. Senn, European Patent Application EP1330446A1(2003).

[9] M. Kulka, W. Harrison, US Patent 3 7254 27(1973).
[10] C. D. S. Tomlin, The Pesticide Manual, 12th ed., BCPC, 2000, p. 1008.
[11] C. D. S. Tomlin, The Pesticide Manual, 12th ed., BCPC, 2000, p. 901.
[12] C. D. S. Tomlin, The Pesticide Manual, 12th ed., BCPC, 2000, pp. 769 - 770.
[13] G. H. Alt, J. K. Pratt, W. G. Phillips, G. H. Srouji, US Patent 5045554(1991).
[14] C. L. Liu, X. N. Liu, R. M. Jacobson, M. J. Mulvihill, PR China Patent CN1118465c(2003).
[15] R. Ross, T. T. Fujimoto, S. H. Shaber, European Patent Application EP0811614A1(1997).
[16] R. M. Jacobson, L. T. Nguyen, US Patent 6 147 062(2000).
[17] R. M. Jacobson, L. T. Nguyen, M. Thirugnanam, US Patent 4 970 224(1990).

Metal – ion Interactions with Sugars. Crystal Structures and FT – IR Studies of the $LaCl_3$ – ribopyranose and $CeCl_3$ – ribopyranose Complexes*

Yan Lu[1], Guocai Deng[1], Fangming Miao[2], Zhengming Li[1]

([1]College of Chemistry, Nankai University, Tianjin, 300071, China;
[2]Institute of Crystal Chemistry, Tianjin Normal University, Tianjin, 300074, China)

Abstract Single crystals of $LaCl_3 \cdot C_5H_{10}O_5 \cdot 5H_2O$(**1**) and $CeCl_3 \cdot C_5H_{10}O_5 \cdot 5H_2O$(**2**) were obtained from ethanol – water solutions and their structures determined by X – ray. The two complexes are isomorphous. Two configurations of complex **1** or complex **2**, as a pair of isomers, were found in each single crystal in a disordered state. The ligand of one of the isomer is α – D – ribopyranose in the $^4C_1$ conformation, the ligand of the other is β – D – ribopyranose in the $^1C_4$ conformation. For complex **1**, the α: β anomeric ratio is 51:49, and for complex **2**, the ratio is 52:48. Both ligands of the two isomers provide three hydroxyl groups in *ax – eq – ax* orientation for coordination. The Ln^{3+} (Ln = La or Ce) ion is nine – coordinated with five Ln – O bonds from water molecules, three Ln – O bonds from hydroxyl groups of the D – ribopyranose, and one Ln – Cl bond from chloride ion. The hydroxyl groups, water molecules, and chloride ions form an extensive hydrogen – bond network. The IR spectral C – C, O – H, C – O, and C – O – H vibrations were observed to be shifted in both the two complexes and the IR results are in accord with those of X – ray diffraction.

Key words La – D – ribose complex; Ce – D – ribose complex; crystal structure; FT – IR

1 Introduction

Saccharides are widely distributed in the biosphere, along with proteins and nucleic acids.[1] The multiple oxygen donor atoms in sugar molecules and the fact that metal cations coexist in biological fluids, suggest that coordination between saccharides may have biological relevance. Thus the formation of a Ni^{2+} – carbohydrate complex in human kidney was demonstrated in 1985.[2] Such interactions may have importance in such biochemical processes as the transfer and storage of metal ions,[3,4] the action of metal – containing pharmaceuticals, toxic – metal metabolism,[5,6] and Ca^{2+} – mediated carbohydrate – protein binding.[7,8]

Information in the literature, on the interactions of saccharides with metal cations is rather limited, especially for well – characterized solid complexes, with most previous work being focused on aqueous solution chemistry. Thus far, only the following neutral sugar – metal complexes have been determined by X – ray diffraction: the complexes of Mo^{6+} – D – xylose,[9] Ca^{2+} – D – fructose,[10-12] Ca^{2+} – D – galactose,[13] Ca^{2+} – lactose,[14,15] Na^+ – sucrose,[16,17] Na^+ – cellobiose,[18] Ca^{2+} – D – mannose,[19] Ca^{2+} – trehalose,[20] Ca^{2+} – D – xylose,[21] Ca^{2+} – L – arabinose,[22] Pr^{3+} – D – ribose,[23] Nd^{3+} – D – ribose,[24,25] and Ca^{2+} – D – ribose.[26] The detailed three – dimensional structures of the saccharides and the role of metal ions in determi-

* Reprinted from *Carbohydrate Research*, 2004, 339(10): 1689 – 1696.

ning and regulating these structures remains obscure.

D – Ribose($C_5H_{10}O_5$), is important as a component of nucleic acids, and its metal – ion – binding properties may have biological significance. D – Ribose exists in aqueous solution as an equilibrium mixture of six tautomer. Among these, those having an *ax* – *eq* – *ax* sequence of three adjacent hydroxyl groups coordinated readily with metal cations to form 1:1 complexes in hydrophilic solvents. It has been reported that the "complexing" isomers constitute altogether 43% of the total, being less only for those of talose.[27] The interactions of Ca^{2+}, Sr^{2+}, Ba^{2+}, La^{3+}, Ce^{3+}, Pr^{3+}, Nd^{3+}, Sm^{3+}, Eu^{3+}, Gd^{3+}, and Tb^{3+} with D – ribose in neutral solutions have been studied by NMR and calorimetric methods,[28–30] and the stability constants have been calculated($< 10M^{-1}$ in most cases). However, the structures of most of these complexes are still undefined, because solid complexes are difficult to obtain. Since the crystal structure of $PrCl_3 \cdot C_5H_{10}O_5 \cdot 5H_2O$ was reported in 2001, only four such ribose – metal complexes, from Ca^{2+}, La^{3+},[31] Pr^{3+}, and Nd^{3+} have so far been obtained as solids. No crystal data for the La^{3+} – ribose complex have been reported.

We have obtain single crystals of the La^{3+} and Ce^{3+} complexes with D – ribose, and crystal structures of $LaCl_3 \cdot C_5H_{10}O_5 \cdot 5H_2O$(**1**) and $CeCl_3 \cdot C_5H_{10}O_5 \cdot 5H_2O$(**2**) were determined by X – ray. In contrast to the single configuration observed in the single crystals of $PrCl_3 \cdot C_5H_{10}O_5 \cdot 5H_2O$[23] and $NdCl_3 \cdot C_5H_{10}O_5 \cdot 5H_2O$,[24] as reported by Yang et al., two configurations were observed in the single crystals of both **1** and **2**. The configurations reported in this article are similar to those observed in single crystals of $NdCl_3 \cdot C_5H_{10}O_5 \cdot 5H_2O$ as reported by our group.[25] The vibration spectra of complexes **1** and **2** were assigned and interpreted in correlation with the crystal structures.

2 Experimental

2.1 Materials

$LaCl_3$ and $CeCl_3$ were prepared from corresponding rare earth oxide of high purity (99.99%).[32] D – Ribose was purchased from Acros, and was used without further purification.

2.2 Preparation of $LaCl_3 \cdot C_5H_{10}O_5 \cdot 5H_2O$ and $CeCl_3 \cdot C_5H_{10}O_5 \cdot 5H_2O$

D – Ribose(0.45g, 3mmol) and equivalent amounts of metal chlorides were dissolved in H_2O – EtOH, and the solutions were evaporated slowly until crystallization occurred. *Anal.* Calcd for $LaCl_3 \cdot C_5H_{10}O_5 \cdot 5H_2O$: C, 12.37; H, 4.15. Found: C, 12.14; H, 4.23. *Anal.* Calcd for $CeCl_3 \cdot C_5H_{10}O_5 \cdot 5H_2O$: C, 12.34; H, 4.14. Found: C, 12.22; H, 4.21.

2.3 Physical measurements

The mid – IR spectra were measured on a Nicolet Magna – IR 750 spectrometer using the micro – IR method, with 128 scans at 4 cm^{-1} resolution.

The structures of $LaCl_3$ · ribopyranose · $5H_2O$(**1**) and $CeCl_3$ · ribopyranose · $5H_2O$(**2**) were determined on a Bruker Smart 1000 diffractometer using monochromatic Mo Kα radiation(λ = 0.71073Å) in the θ ranges from 2.07° to 26.42°(**1**) and from 2.07° to 26.41°(**2**) at 293 K, respectively. The final cycle of full – matrix least – squares refinements were based on 2649 observed reflections for (**1**) and 2906 observed reflections for (**2**). Calculations were completed with the SHELX – 97 program.

Crystallographic data (without structure factors) for the structures reported in this paper have been deposited with the Cambridge Crystallographic Data Centre as supplementary publication No. CCDC – 199655 (**1**) and No. CCDC – 220665 (**2**). Copies of the data can be obtained free of charge on application to the CCDC, 12 Union Road, Cambridge CB2 1EZ, UK. [Fax: (internat.) +44 – 1223/336033; E – mail: deposit@ ccdc. cam. ac. uk].

3 Results and discussion

3.1 X – Ray crystal structures

The crystal structures of the title complexes are isomorphous and are similar to that of $NdCl_3$ · ribose · $5H_2O$.[25] The single crystals of the two complexes are both in disordered state in that two configurations, as a pair of anomers, were observed in each single crystal. The structures of complex **1** are shown in Figures 1 and 2; Figures 3 and 4 are of the two structures of **2**; Figures 5 and 6 are the projections of the crystal cells in the unit structures of **1** in which the ligand is the α – pyranose in $^4C_1$ configuration and **2** with which the ligand is the β – pyranose in $^1C_4$ configuration, respectively. The crystal data and structure refinements of the two complexes are listed in Table 1; selected bond lengths and angles of **1** are listed in Table 2, and those of **2** in Table 3.

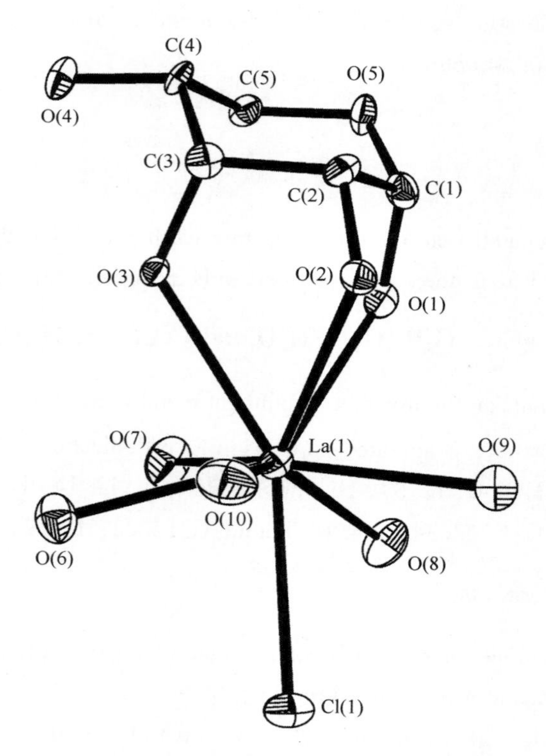

Figure 1 The structure and atom numbering scheme of $LaCl_3$ · $C_5H_{10}O_5$ · $5H_2O$ with which ligand in the αP^4C_1 configuration

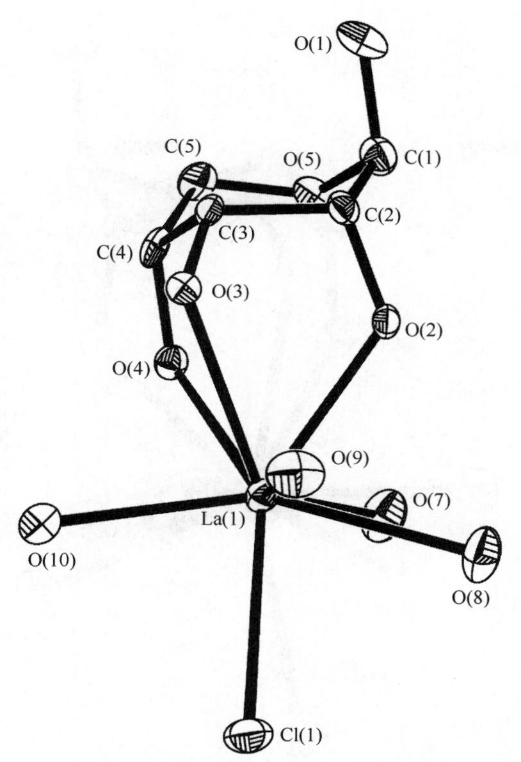

Figure 2 The structure and atom numbering scheme of

$LaCl_3 \cdot C_5H_{10}O_5 \cdot 5H_2O$ with which ligand in the $\beta P^1 C_4$ configuration

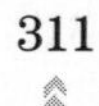

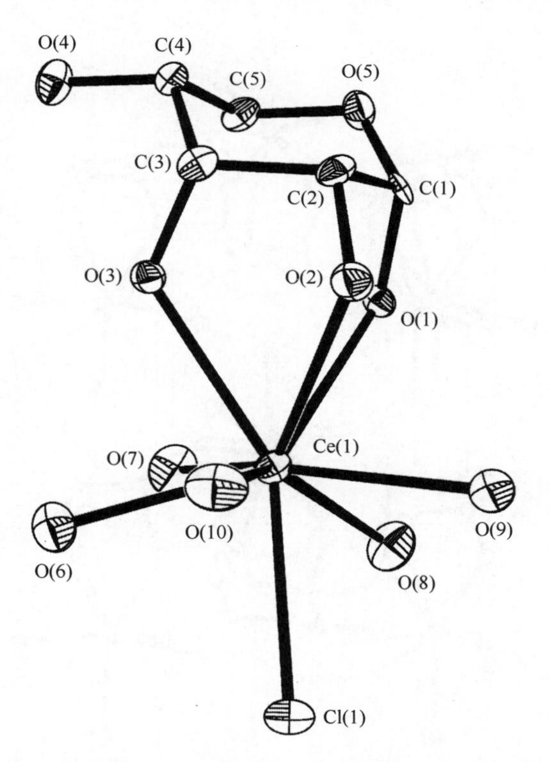

Figure 3 The structure and atom numbering scheme of

$CeCl_3 \cdot C_5H_{10}O_5 \cdot 5H_2O$ with which ligand in the $\alpha P^4 C_1$ configuration

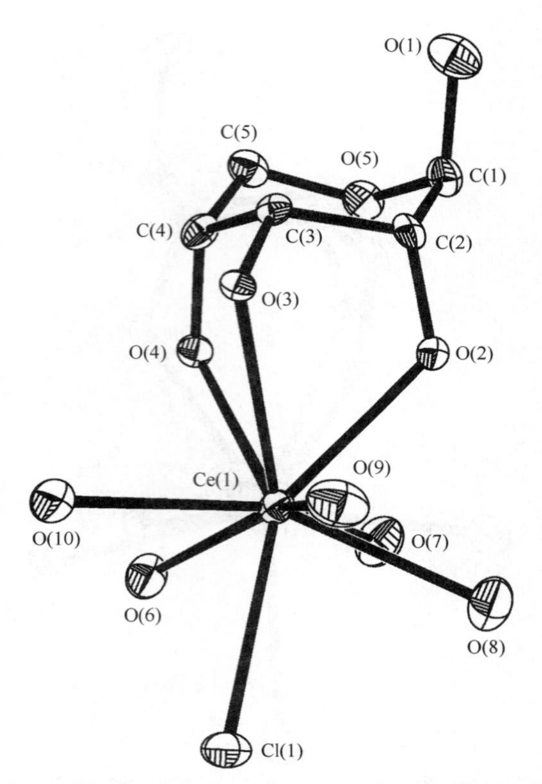

Figure 4　The structure and atom numbering scheme of $CeCl_3 \cdot C_5H_{10}O_5 \cdot 5H_2O$ with which ligand in the βP^1C_4 configuration

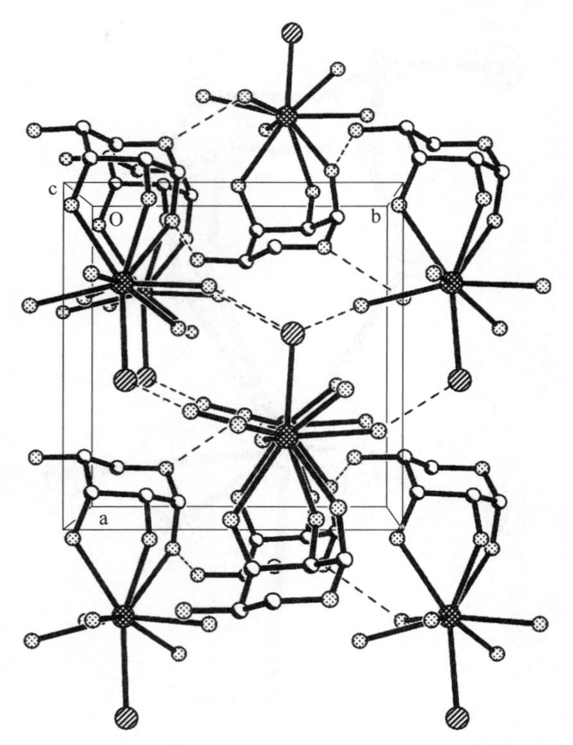

Figure 5　Projection of the crystal cell in the structure of $LaCl_3 \cdot C_5H_{10}O_5 \cdot 5H_2O$ with which ligand in the αP^4C_1 configuration

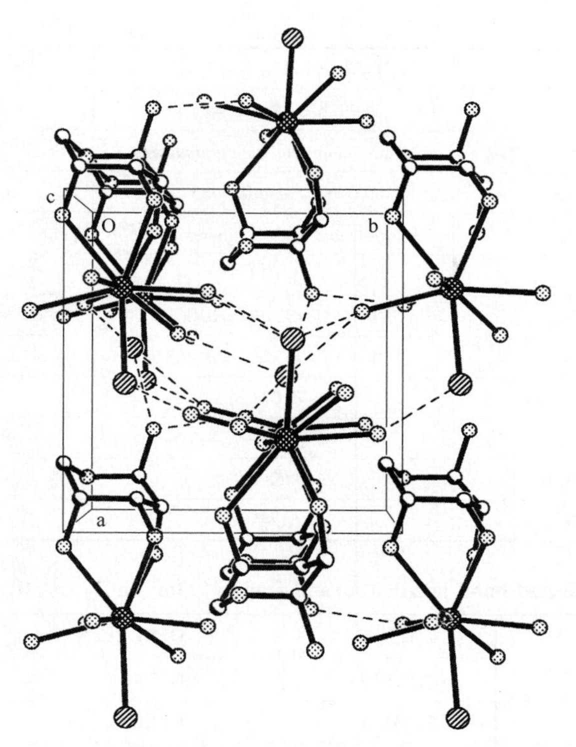

Figure 6 Projection of the crystal cell in the structure of $CeCl_3 \cdot C_5H_{10}O_5 \cdot 5H_2O$ with which ligand in the $\beta P^1 C_4$ configuration

Table 1 Crystal data and structure refinement parameters for $LaCl_3 \cdot C_5H_{10}O_5 \cdot 5H_2O$ and $CeCl_3 \cdot C_5H_{10}O_5 \cdot 5H_2O$

Formula	$LaCl_3 \cdot C_5H_{10}O_5 \cdot 5H_2O$	$CeCl_3 \cdot C_5H_{10}O_5 \cdot 5H_2O$
Formula weight	485. 47	486. 68
Crystal system, Space group	Monoclinic, $P2(1)$	Monoclinic, $P2(1)$
a(Å)	9. 281(3)	9. 230(5)
b(Å)	8. 847(3)	8. 840(5)
c(Å)	9. 849(3)	9. 850(6)
β(°)	94. 028(5)	93. 996(9)
V(Å^3)	806. 6(4)	801. 6(8)
Z	2	2
D_{calcd}(Mg/m^3)	1. 999	2. 016
Absorption coefficient(mm^{-1})	3. 181	3. 375
F(000)	476	478
Crystal size(mm)	0. 25 × 0. 18 × 0. 16	0. 25 × 0. 18 × 0. 16
θ Range for data collection(°)	2. 07 – 26. 42	2. 07 – 26. 41
Index ranges	$-11 \leqslant h \leqslant 8$, $-9 \leqslant k \leqslant 11$, $-12 \leqslant l \leqslant 12$	$-8 \leqslant h \leqslant 11$, $-11 \leqslant k \leqslant 10$, $-12 \leqslant l \leqslant 6$
Reflections collected/unique	4636/2649 [R(int) = 0. 0345]	4282/2906 [R(int) = 0. 0362]

Table 1 (continued)

Formula	$LaCl_3 \cdot C_5H_{10}O_5 \cdot 5H_2O$	$CeCl_3 \cdot C_5H_{10}O_5 \cdot 5H_2O$
Completeness to θ	99.8% (θ = 26.42)	98.9% (θ = 26.41)
Absorption correction	Semi – empirical from equivalents	Semi – empirical from equivalents
Max/min transmission	0.6301 and 0.5035	0.6142 and 0.4858
Refinement method	Full – matrix least – squares on F^2	Full – matrix least – squares on F^2
S	1.013	0.958
Final R indices [$I > 2\sigma(I)$]	$R1$ = 0.0348, $wR2$ = 0.0612	$R1$ = 0.0347, $wR2$ = 0.0790
R indices (all data)	$R1$ = 0.0496, $wR2$ = 0.0652	$R1$ = 0.0429, $wR2$ = 0.0815
Absolute structure parameter	−0.02(2)	0.04(3)
Largest difference peak and hole (e/Å^3)	0.483 and −0.446	0.876 and −0.681
Data/restraints/parameters	2649/585/265	2906/579/265
α,β anomer ratio	51/49	52/48

Table 2 Selected bond lengths (Å) and angles (°) for $LaCl_3 \cdot C_5H_{10}O_5 \cdot 5H_2O$①

La(1) – O(2′)	2.513(13)	O(5) – C(5)	1.444(12)
La(1) – O(9)	2.518(5)	O(5) – C(1)	1.442(12)
La(1) – O(6)	2.531(5)	C(1) – C(2)	1.536(9)
La(1) – O(10)	2.539(5)	C(2) – C(3)	1.531(8)
La(1) – O(7)	2.542(5)	C(3) – C(4)	1.524(8)
La(1) – O(4′)	2.559(17)	C(4) – C(5)	1.535(9)
La(1) – O(2)	2.56(2)	O(1′) – C(1′)	1.420(13)
La(1) – O(8)	2.572(5)	O(2′) – C(2′)	1.417(11)
La(1) – O(3)	2.593(15)	O(3′) – C(3′)	1.440(12)
La(1) – O(1)	2.643(13)	O(4′) – C(4′)	1.430(12)
La(1) – O(3′)	2.69(2)	O(5′) – C(1′)	1.442(12)
La(1) – Cl(1)	2.834(18)	O(5′) – C(5′)	1.439(13)
O(1) – C(1)	1.418(11)	C(1′) – C(2′)	1.530(9)
O(2) – C(2)	1.428(12)	C(2′) – C(3′)	1.528(9)
O(3) – C(3)	1.437(12)	C(3′) – C(4′)	1.525(8)
O(4) – C(4)	1.421(11)	C(4′) – C(5′)	1.526(9)
O(2′) – La(1) – O(9)	87.2(3)	O(7) – La(1) – Cl(1)	100.76(15)
O(2′) – La(1) – O(6)	123.4(3)	O(4′) – La(1) – Cl(1)	141.8(4)
O(9) – La(1) – O(6)	149.06(16)	O(2) – La(1) – Cl(1)	132.0(4)
O(2′) – La(1) – O(10)	132.1(3)	O(8) – La(1) – Cl(1)	74.77(14)
O(9) – La(1) – O(10)	84.64(18)	O(3) – La(1) – Cl(1)	149.5(3)
O(6) – La(1) – O(10)	71.34(16)	O(1) – La(1) – Cl(1)	142.2(3)
O(2′) – La(1) – O(7)	70.8(3)	O(3′) – La(1) – Cl(1)	134.0(4)
O(9) – La(1) – O(7)	136.80(18)	C(1) – O(1) – La(1)	122.7(10)

Table 2(continued)

O(6) - La(1) - O(7)	66.72(17)	C(2) - O(2) - La(1)	116.4(16)
O(10) - La(1) - O(7)	137.57(18)	C(3) - O(3) - La(1)	121.3(11)
O(2′) - La(1) - O(4′)	69.2(4)	C(1) - O(5) - C(5)	116.4(11)
O(9) - La(1) - O(4′)	128.7(4)	O(1) - C(1) - O(5)	111.5(12)
O(6) - La(1) - O(4′)	67.0(4)	O(1) - C(1) - C(2)	109.0(13)
O(10) - La(1) - O(4′)	79.9(3)	O(5) - C(1) - C(2)	111.5(10)
O(7) - La(1) - O(4′)	78.2(4)	O(2) - C(2) - C(3)	104.5(17)
O(9) - La(1) - O(2)	67.8(4)	O(2) - C(2) - C(1)	107.1(16)
O(6) - La(1) - O(2)	118.0(5)	C(3) - C(2) - C(1)	111.2(8)
O(10) - La(1) - O(2)	68.7(5)	O(3) - C(3) - C(4)	110.2(13)
O(7) - La(1) - O(2)	127.3(4)	O(3) - C(3) - C(2)	111.1(14)
O(2′) - La(1) - O(8)	72.8(3)	C(4) - C(3) - C(2)	111.3(7)
O(9) - La(1) - O(8)	71.19(18)	O(4) - C(4) - C(3)	112.0(12)
O(6) - La(1) - O(8)	119.23(16)	O(4) - C(4) - C(5)	107.0(10)
O(10) - La(1) - O(8)	145.08(18)	C(3) - C(4) - C(5)	111.1(8)
O(7) - La(1) - O(8)	66.99(16)	O(5) - C(5) - C(4)	108.8(10)
O(4′) - La(1) - O(8)	135.0(3)	C(2′) - O(2′) - La(1)	123.9(10)
O(2) - La(1) - O(8)	120.8(5)	C(3′) - O(3′) - La(1)	106.5(15)
O(9) - La(1) - O(3)	126.8(3)	C(4′) - O(4′) - La(1)	123.7(12)
O(6) - La(1) - O(3)	72.9(3)	C(5′) - O(5′) - C(1′)	116.3(12)
O(10) - La(1) - O(3)	88.1(3)	O(1′) - C(1′) - O(5′)	107.8(11)
O(7) - La(1) - O(3)	74.0(3)	O(1′) - C(1′) - C(2′)	107.3(12)
O(2) - La(1) - O(3)	60.6(6)	O(5′) - C(1′) - C(2′)	113.0(9)
O(8) - La(1) - O(3)	126.5(3)	O(2′) - C(2′) - C(3′)	109.1(13)
O(9) - La(1) - O(1)	76.6(3)	O(2′) - C(2′) - C(1′)	106.7(12)
O(6) - La(1) - O(1)	133.8(3)	C(3′) - C(2′) - C(1′)	111.6(8)
O(10) - La(1) - O(1)	128.3(3)	O(3′) - C(3′) - C(4′)	113.6(16)
O(7) - La(1) - O(1)	80.1(3)	O(3′) - C(3′) - C(2′)	104.1(15)
O(2) - La(1) - O(1)	59.5(5)	C(4′) - C(3′) - C(2′)	112.5(8)
O(8) - La(1) - O(1)	70.9(3)	O(4′) - C(4′) - C(3′)	108.0(14)
O(3) - La(1) - O(1)	67.5(4)	O(4′) - C(4′) - C(5′)	115.2(16)
O(2′) - La(1) - O(3′)	60.2(5)	C(3′) - C(4′) - C(5′)	111.8(8)
O(9) - La(1) - O(3′)	66.0(4)	O(5′) - C(5′) - C(4′)	106.6(11)
O(6) - La(1) - O(3′)	122.1(5)	O(2′) - La(1) - Cl(1)	147.2(3)
O(10) - La(1) - O(3′)	73.5(5)	O(9) - La(1) - Cl(1)	77.87(15)
O(7) - La(1) - O(3′)	125.0(4)	O(6) - La(1) - Cl(1)	77.52(16)
O(4′) - La(1) - O(3′)	62.6(6)	O(10) - La(1) - Cl(1)	75.77(12)
O(8) - La(1) - O(3′)	116.1(5)		

①To distinguish atoms of the ribose moieties between the two isomers, the atoms in the $\rho P^1 C_4$ configuration are numbered with a prime, and so are the atoms in the below tables.

Table 3 Selected bond lengths(Å) and angles(°) for $CeCl_3 \cdot C_5H_{10}O_5 \cdot 5H_2O$

Ce(1) - O(2′)	2.484(13)	O(5) - C(5)	1.442(12)
Ce(1) - O(9)	2.488(6)	O(5) - C(1)	1.437(11)
Ce(1) - O(6)	2.511(6)	C(1) - C(2)	1.529(9)
Ce(1) - O(10)	2.516(6)	C(2) - C(3)	1.527(8)
Ce(1) - O(2)	2.530(3)	C(3) - C(4)	1.536(8)
Ce(1) - O(7)	2.528(6)	C(4) - C(5)	1.537(9)
Ce(1) - O(4′)	2.543(17)	O(1′) - C(1′)	1.421(14)
Ce(1) - O(8)	2.557(6)	O(2′) - C(2′)	1.418(12)
Ce(1) - O(3)	2.583(14)	O(3′) - C(3′)	1.433(12)
Ce(1) - O(1)	2.624(11)	O(4′) - C(4′)	1.424(12)
Ce(1) - O(3′)	2.690(3)	O(5′) - C(1′)	1.444(13)
Ce(1) - Cl(1)	2.813(2)	O(5′) - C(5′)	1.437(13)
O(1) - C(1)	1.414(11)	C(1′) - C(2′)	1.532(9)
O(2) - C(2)	1.429(12)	C(2′) - C(3′)	1.524(9)
O(3) - C(3)	1.441(12)	C(3′) - C(4′)	1.527(9)
O(4) - C(4)	1.420(12)	C(4′) - C(5′)	1.531(9)
O(2′) - Ce(1) - O(9)	88.0(3)	O(8) - Ce(1) - O(3)	127.0(3)
O(2′) - Ce(1) - O(6)	123.1(3)	O(9) - Ce(1) - O(1)	77.1(3)
O(9) - Ce(1) - O(6)	148.62(18)	O(6) - Ce(1) - O(1)	133.7(3)
O(2′) - Ce(1) - O(10)	133.3(3)	O(10) - Ce(1) - O(1)	127.6(3)
O(9) - Ce(1) - O(10)	84.3(2)	O(2) - Ce(1) - O(1)	58.0(7)
O(6) - Ce(1) - O(10)	71.68(19)	O(7) - Ce(1) - O(1)	80.6(3)
O(9) - Ce(1) - O(2)	67.3(6)	O(8) - Ce(1) - O(1)	71.8(3)
O(6) - Ce(1) - O(2)	119.6(8)	O(3) - Ce(1) - O(1)	67.0(4)
O(10) - Ce(1) - O(2)	69.6(7)	O(2′) - Ce(1) - O(3′)	63.0(6)
O(2′) - Ce(1) - O(7)	70.0(3)	O(9) - Ce(1) - O(3′)	66.7(6)
O(9) - Ce(1) - O(7)	136.3(2)	O(6) - Ce(1) - O(3′)	121.3(7)
O(6) - Ce(1) - O(7)	66.8(2)	O(10) - Ce(1) - O(3′)	71.7(6)
O(10) - Ce(1) - O(7)	138.1(2)	O(7) - Ce(1) - O(3′)	126.6(5)
O(2) - Ce(1) - O(7)	127.7(5)	O(4′) - Ce(1) - O(3′)	63.1(7)
O(2′) - Ce(1) - O(4′)	70.2(5)	O(8) - Ce(1) - O(3′)	117.6(7)
O(9) - Ce(1) - O(4′)	129.8(4)	O(2′) - Ce(1) - Cl(1)	146.5(3)
O(6) - Ce(1) - O(4′)	66.5(4)	O(9) - Ce(1) - Cl(1)	77.39(15)
O(10) - Ce(1) - O(4′)	80.0(4)	O(6) - Ce(1) - Cl(1)	77.26(17)
O(7) - Ce(1) - O(4′)	78.5(4)	O(10) - Ce(1) - Cl(1)	75.61(12)
O(2′) - Ce(1) - O(8)	72.0(3)	O(2) - Ce(1) - Cl(1)	132.0(5)
O(9) - Ce(1) - O(8)	71.0(2)	O(7) - Ce(1) - Cl(1)	100.31(16)
O(6) - Ce(1) - O(8)	118.94(19)	O(4′) - Ce(1) - Cl(1)	141.1(4)
O(10) - Ce(1) - O(8)	144.8(2)	O(8) - Ce(1) - Cl(1)	74.77(15)

Table 3 (continued)

O(2) – Ce(1) – O(8)	119.6(8)	O(3) – Ce(1) – Cl(1)	149.6(3)
O(7) – Ce(1) – O(8)	66.5(2)	O(1) – Ce(1) – Cl(1)	142.9(3)
O(4′) – Ce(1) – O(8)	135.2(4)	O(3′) – Ce(1) – Cl(1)	133.0(4)
C(3′) – O(3′) – Ce(1)	104.2(18)	C(1) – O(1) – Ce(1)	125.9(9)
O(9) – Ce(1) – O(3)	126.9(3)	C(2) – O(2) – Ce(1)	118(2)
O(6) – Ce(1) – O(3)	73.3(3)	C(3) – O(3) – Ce(1)	120.4(10)
O(10) – Ce(1) – O(3)	87.8(3)	C(1) – O(5) – C(5)	116.6(10)
O(2) – Ce(1) – O(3)	60.8(7)	O(1) – C(1) – O(5)	112.3(11)
O(7) – Ce(1) – O(3)	75.1(3)	O(1) – C(1) – C(2)	105.9(12)
C(4′) – O(4′) – Ce(1)	122.6(12)	O(5) – C(1) – C(2)	110.2(9)
C(2′) – O(2′) – Ce(1)	120.9(10)	O(2) – C(2) – C(3)	103.2(18)
O(3′) – C(3′) – C(4′)	112.8(18)	O(2) – C(2) – C(1)	106.5(19)
C(2′) – C(3′) – C(4′)	112.6(8)	C(3) – C(2) – C(1)	112.1(8)
O(4′) – C(4′) – C(3′)	107.8(13)	O(3) – C(3) – C(2)	112.2(14)
C(5′) – O(5′) – C(1′)	116.5(12)	O(3) – C(3) – C(4)	109.0(12)
O(4′) – C(4′) – C(5′)	113.9(15)	C(2) – C(3) – C(4)	111.0(7)
C(3′) – C(4′) – C(5′)	111.4(8)	O(4) – C(4) – C(3)	112.6(11)
O(5′) – C(5′) – C(4′)	107.5(11)	O(4) – C(4) – C(5)	105.4(10)
O(5) – C(5) – C(4)	106.4(10)	C(3) – C(4) – C(5)	110.0(7)
C(3′) – C(2′) – C(1′)	111.5(8)	O(2′) – C(2′) – C(1′)	106.8(12)
O(1′) – C(1′) – O(5′)	107.4(12)	O(2′) – C(2′) – C(3′)	111.3(13)
O(3′) – C(3′) – C(2′)	106.3(19)	O(1′) – C(1′) – C(2′)	111.2(12)
O(5′) – C(1′) – C(2′)	112.0(10)		

Figures 1 and 3 show that the ribose moieties of complexes **1** and **2** are both α – pyranose in the $^4C_1$ conformation (abbreviated to αP^4C_1), which has been observed in $PrCl_3$ · ribose · $5H_2O$[23] and $NdCl_3$ · ribose · $5H_2O$.[24,25] Figures 2 and 4 show that the ribose moieties of the two complexes are both β – pyranose in $^1C_4$ conformation (abbreviated to βP^1C_4), which has only been observed in $NdCl_3 \cdot C_5H_{10}O_5 \cdot 5H_2O$.[25] For complex **1**, the α:β anomeric ratio is 51:49. For complex 2, the anomeric ratio is 52:48. In both of the title complexes, the Ln ion (Ln = La or Ce) is nine coordinated and binds to three hydroxyl groups of one D – ribopyranose molecule, five water molecules, and a chloride ion. The other two Cl^- ions in the complex are free. The three adjacent hydroxyl groups, HO – 1, 2, and 3 in the αP^4C_1 or HO – 2, 3, and 4 in the βP^1C_4 both have the *ax – eq – ax* sequence as the coordination sites of D – Ribose. The Ln – O distances determined in this study (for **1**: from 2.513 to 2.690Å; for **2**: from 2.484 to 2.690Å) are comparable to those of Pr – O (from 2.480 to 2.594Å)[23] and Nd – O (from 2.445 to 2.670Å[25] or from 2.463 to 2.681Å[24]). The torsion angles of the ribose moiety in **1**, **2**, and $NdCl_3 \cdot C_5H_{10}O_5 \cdot 5H_2O$[25] are similar (see Table 4), indicating that the sugar ring is not much changed by the metal – ion coordination. As no data have been reported on the crystal structure of D - ribose in the literature, and so we cannot calculate changes in the torsion angles of the sugar ring upon metalation. The ring oxygen atoms of D - ribose does not coordinate with Ln^{3+} in either of the complexes. All water molecules are coordinated in the two complexes. The coordination behavior of La^{3+}, Ce^{3+}, Pr^{3+}, and Nd^{3+} are shown to be similar from the X - ray results.

Table 4 Selected torsion angles(°) for $LaCl_3 \cdot C_5H_{10}O_5 \cdot 5H_2O$, $CeCl_3 \cdot C_5H_{10}O_5 \cdot 5H_2O$, and $NdCl_3 \cdot C_5H_{10}O_5 \cdot 5H_2O$①

Configuration and conformation of the ribopyranose moiety	Torsion angles	Ribose - La	Ribose - Ce	Ribose - Nd①
$\alpha P^4 C_1$	C(5) - O(5) - C(1) - C(2)	-55.5(17)	-58.2(16)	-57.6(13)
	O(5) - C(1) - C(2) - C(3)	49.5(17)	48.5(16)	52.7(14)
	C(1) - C(2) - C(3) - C(4)	-50.3(17)	-49.4(17)	-53.1(15)
	C(2) - C(3) - C(4) - C(5)	54.1(15)	55.3(14)	54.9(13)
	C(1) - O(5) - C(5) - C(4)	58.0(17)	63.2(16)	60.4(14)
	C(3) - C(4) - C(5) - O(5)	-55.8(16)	-59.2(15)	-57.3(13)
$\beta P^1 C_4$	C(5) - O(5) - C(1) - C(2)	55.6(18)	56.2(18)	58.6(13)
	O(5) - C(1) - C(2) - C(3)	-44.2(17)	-46.4(17)	-45.3(13)
	C(1) - C(2) - C(3) - C(4)	44.4(17)	46.4(17)	44.1(13)
	C(2) - C(3) - C(4) - C(5)	-53.0(17)	-52.6(17)	-52.2(15)
	C(1) - O(5) - C(5) - C(4)	-61.1(18)	-60.4(18)	-62.8(13)
	C(3) - C(4) - C(5) - O(5)	58.4(17)	56.8(17)	57.3(14)

①From our unpublished data.

As expected, there is an extensive network of hydrogen bonds involving all hydroxyl groups, water molecules, and chloride ions in the crystals of **1** and **2** (see Tables 5 and 6). The Ln - ribose complexes are organized by these hydrogen bonds and thus form layers parallel to the (10 - 1) plane. These layers are then also held together by hydrogen bonds with regular spaces between them. Free Cl ions distributed in these layers are responsible for not only the formation of a layer but also the bonding between the layers by hydrogen bonds. This network of hydrogen bonds forms the packed structure of the whole crystal. In the title complexes, the Cl^- ions play important roles not only as counterions but also as the predominant feature in the network of hydrogen bonds.

Table 5 Hydrogen bonds for $LaCl_3 \cdot C_5H_{10}O_5 \cdot 5H_2O$ with $H \cdots A < r(A) + 2.000Å$ and $< DHA > 110°$

D - H···A	*d*(D - H)	*d*(H···A)	*d*(D···A)	<(DHA)
O6 - H6A···Cl3#1	0.851	2.505	3.170	135.56
O6 - H6B···Cl2#2	0.850	2.720	3.522	157.85
O7 - H7B···O5#3	0.846	2.059	2.883	164.42
O7 - H7B···O1′#3	0.846	2.104	2.723	129.56
O8 - H8A···Cl3	0.842	2.388	3.224	172.09

Table 5 (continued)

D – H···A	d(D – H)	d(H···A)	d(D···A)	<(DHA)
O8 – H8B···O4#4	0.842	2.282	2.998	143.09
O8 – H8B···Cl3#2	0.842	2.742	3.409	137.33
O9 – H9A···Cl1#5	0.848	2.422	3.172	147.76
O9 – H9B···Cl2#6	0.848	2.325	3.170	173.46
O10 – H10A···Cl3#7	0.852	2.632	3.245	129.95
O10 – H10A···Cl1	0.852	2.916	3.308	110.18
O10 – H10B···Cl2#8	0.853	2.296	3.101	157.32
O1 – H1···O4#4	0.930	1.736	2.646	165.14
O2 – H2···Cl2#8	0.930	2.328	2.979	126.74
O3 – H3···Cl2#2	0.930	2.516	3.162	126.85
O4 – H4···Cl3#9	0.820	2.413	3.219	167.66
O1′ – H1′···O7#4	0.820	1.945	2.723	158.14
O2′ – H2′···O5′#4	0.930	2.365	2.934	119.23
O3′ – H3′···Cl2#8	0.930	2.729	3.127	106.80
O4′ – H4′···Cl2#2	0.930	2.088	2.914	147.28

Symmetry transformations used to generate equivalent atoms: #1 $x, y-1, z$; #2 $-x+1, y-1/2, -z+1$; #3 $-x, y-1/2, -z+1$; #4 $-x, y+1/2, -z+1$; #5 $-x+1, y+1/2, -z+2$; #6 $-x+1, y+1/2, -z+1$; #7 $-x+1, y-1/2, -z+2$; #8 $x-1, y, z+1$; #9 $x-1, y-1, z$.

Table 6 Hydrogen bonds for $CeCl_3 \cdot C_5H_{10}O_5 \cdot 5H_2O$ with H···A < r(A) + 2.000 Å and <DHA > 110°

D – H···A	d(D – H)	d(H···A)	d(D···A)	<(DHA)
O6 – H6A···Cl3#1	0.847	2.514	3.173	135.35
O6 – H6B···Cl2#2	0.848	2.718	3.519	158.25
O7 – H7B···O5#3	0.844	2.092	2.913	164.06
O7 – H7B···O1′#3	0.844	2.114	2.737	130.31
O8 – H8A···Cl3	0.841	2.381	3.216	172.19
O8 – H8B···O4#4	0.841	2.300	3.015	142.96
O8 – H8B···Cl3#2	0.841	2.759	3.430	137.93
O9 – H9A···Cl1#5	0.846	2.437	3.184	147.68
O9 – H9B···Cl2#6	0.845	2.338	3.179	173.06
O10 – H10A···Cl3#7	0.850	2.643	3.250	129.47
O10 – H10A···Cl1	0.850	2.885	3.275	110.01
O10 – H10B···Cl2#8	0.850	2.285	3.089	157.87
O1 – H1···O4#4	0.930	1.771	2.689	168.97
O2 – H2···Cl2#8	0.930	2.412	3.036	124.38
O3 – H3···Cl2#2	0.930	2.509	3.169	128.18
O4 – H4···Cl3#9	0.820	2.541	3.221	141.20
O1′ – H1′···O8#10	0.820	2.768	3.160	159.10

Table 6 (continued)

D－H···A	d(D－H)	d(H···A)	d(D···A)	<(DHA)
O2′－H2′···O5′#4	0.930	2.390	2.936	117.41
O3′－H3′···Cl2#8	0.930	2.617	3.059	109.70
O4′－H4′···Cl2#2	0.930	2.087	2.900	145.38

Symmetry transformations used to generate equivalent atoms: #1 $x, y-1, z$; #2 $-x+1, y-1/2, -z+1$; #3 $-x, y-1/2, -z+1$; #4 $-x, y+1/2, -z+1$; #5 $-x+1, y+1/2, -z+2$; #6 $-x+1, y+1/2, -z+1$; #7 $-x+1, y-1/2, -z+2$; #8 $x-1, y, z+1$; #9 $x-1, y-1, z$; #10 $x-1, y, z$.

3.2 IR spectroscopy studies of La－ribose and Ce－ribose complexes

The FT－IR spectra of D－ribose and the La^{3+}, Ce^{3+}, and Nd^{3+} complexes are shown in Figure 7. The absorption bands and tentative assignments are given in Table 7.

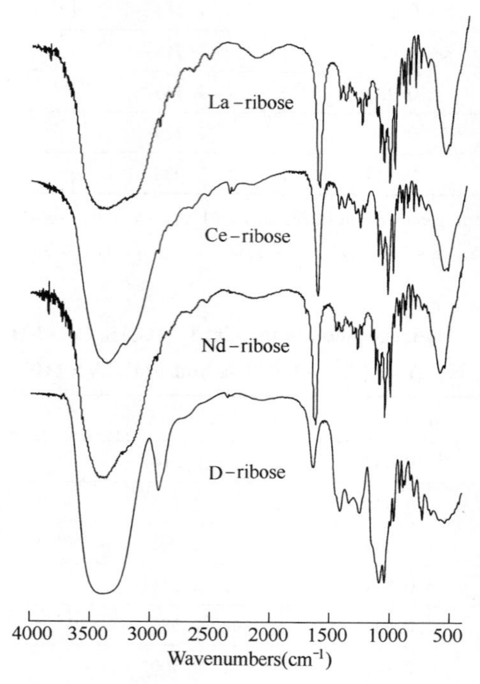

Figure 7 The mid－IR spectra of D－ribose, La－, Ce－, and Nd－D－ribose

The IR data of La－, Ce－, and Nd－ribose are similar, indicating that La^{3+}, Ce^{3+}, and Nd^{3+} have similar coordination modes. The broad absorption band around 3400cm^{-1} in the spectrum of D－ribose can be assigned to the hydrogen－bonded OH groups. This band appears broader in the spectra of the metal complexes (see Fig. 7). The observed spectral changes are due to the metalation of the sugar and the rearrangement of the strong hydrogen－bonding network in the crystal structures of the complexes (see Tables 5 and 6). The C－H stretching vibration bands (for La－ribose: 2946, 2853, 2673, 2539cm^{-1}, for Ce－ribose: 2946, 2853, 2672, 2541cm^{-1}) in the spectra of the complexes, corresponding to the band of 2928 cm^{-1} in the spectrum of free D－ribose, are observed weaker and are masked by the broadened vOH bands. The bands at about 1622cm $^{-1}$ in the spectra of the complexes can be assigned to the δHOH vibration of the coordinated water. The bands at 1452, 1417, 1340, 1250 cm $^{-1}$ in the spectrum of free D – ribose, assigned to the multiple bending vibrations of O – C – H, C – C – H, C – O – H, and CH_2, are shifted in the spectra of the complexes, and their intensities become weaker upon salt formation (see Table 7). The band at about 1150cm $^{-1}$ is the characteristic vibration of a pyranose,[33] and is observed in the spectra of the complexes and D – ribose itself. In the 1200 – 970cm $^{-1}$ region, the C – O, C – C stretching vibrations, and the C – O – H, C – C – O bending vibrations, of D – ribose are observed to be shifted and split in the spectra of the complexes (see Table 7). Such observed splitting and shifting is indicative of the participation of the sugar hydroxyl groups in metal – ligand bonding, which therefore affects the C – O, C – C stretching vibrations and the C – O – H, C – C – O bending vibrations of the sugar moiety.

Table 7 IR data for D – ribose, La – ribose, Ce – ribose, Pr – ribose, and Nd – ribose complexes (1500 – 700cm $^{-1}$)

D – Ribose	La – ribose	Ce – ribose	Pr – ribose[23]	Nd – ribose[25]	Possible assignment[33 – 38]
1452	1457	1458	1458	1456	δOCH + δCCH + $δCH_2$
1417	1410	1417	1409	1408	δOCH + δCCH
1340	1348	1348	1350	1350	δOCH + δCCH + δCOH
	1319	1318	1318	1315	δOCH + δCCH + δCOH
1250	1244	1244	1244	1244	δCCH + δCOH + δOCH
1153	1152	1153	1152	1153	vC – O + vC – C + δCOH (pyranose)
1117	1128	1127	1128	1128	vC – O + vC – C + δCOH
1085	1093	1094	1095	1091	vC – O + vC – C + δCOH
1042	1047	1047	1048	1047	vC – O + vC – C + δCCO
	1004	1004	1004	1003	vC – O + vC – C + δCCO
987	971	971	972	971	vC – O + vC – C + δCCO
914	917	917	918	918	vC – O + δCCH + vasy (ring of pyranose)
887	885	886	887	888	vC – O + vC – C + δCCH
872	874	874	874	874	δCH (β – pyranose)
	835	835	836	836	δCH (α – pyranose)
826					δCH
795	793	794	795	796	τC – O + δCCO + δCCH + vsy (ring of pyranose)
746	734	735	736	734	τC – O + δCCO + δCCH

δ, bending mode; v, stretching mode; τ, twisting.

The ring skeletal deformation bands (δC – C – O and δC – C – C) of free D – ribose, mainly in the 1000 – 400cm $^{-1}$ region, show considerable changes on complex formation (see Fig. 7 and Table 7). This may be attributed to distortion of the sugar ring upon metalation. However, no crystal data for free D – ribose have been reported to permit comparison with those of the complexes. The spectral changes observed in this region may be interpreted to indicate that metalation of the sugar perturbs the electron distribution within the sugar ring system, where the vibrations are mostly localized, and finally causes ring distortion, resulting in the alterations in the spectra. The 914 and 795cm^{-1} bands in the D – ribose spectrum are attributed to the asymmetric and symmetric ring – breathing modes of the pyranose, respectively. They are observed as bands at around 917 and 794cm^{-1} in the spectrum of each complex (see Table 7), indicating that the six – membered sugar ring is retained in each complex. The absorption bands at about 870 and 840cm^{-1} in a pyranose spectrum are generally assigned to the presence of the β and α anomers, respectively.[34,35] In relation to the spectra of the Ln – D – ribose complex, the coexistence of the absorption bands at about 874 and 835cm^{-1} in each spectrum indicates that the complex is, in fact, a mixture with both α – ribopyranose and β – ribopyranose as ligands. The two configurations of the complex indicated by the IR data are in accord with the X – ray results.

The IR results indicate that the hydroxyl groups of D – ribose take part in the metal – oxygen interaction; the hydrogen – bond network rearranges upon metalation; and there are two isomers coexisting in the complex with both α – ribopyranose and β – ribopyranose as ligands. The IR results are in accord with those of X – ray diffraction, and the FT – IR technique is thus a useful method for detecting the formation of such complexes.

References

[1] Luo, J. S. *Biochemistry*; Huadong Normal University Press: Shanghai, 1997.

[2] Templeton, D. M.; Sarkar, B. *Biochem. J.* 1985, 230, 35 – 42.

[3] Sauchelli, V. *Trace Elements in Agriculture*; Van Nostrand: New York, 1969. p 248.

[4] Holm, R. P.; Berg, J. M. *Pure Appl. Chem.* 1984, 56, 1645 – 1657.

[5] Predki, P. F.; Whitfield, D. M.; Sarkar, B. *Biochem. J.* 1992, 281, 835 – 841.

[6] Templeton, D. M.; Sarkar, B. *Biochem. J.* 1985, 230, 35 – 42.

[7] Weis, W. I.; Drickamer, K.; Hendrickson, W. A. *Nature* 1992, 360, 127 – 134.

[8] Drickamer, K. *Nature* 1992, 360, 183 – 186.

[9] Taylor, G. E.; Waters, J. M. *Tetrahedron Lett.* 1981, 22, 1277 – 1278.

[10] Craig, D. C.; Stephenson, N. C.; Stevens, J. D. *Cryst. Struct. Commun.* 1974, 3, 277 – 281.

[11] Craig, D. C.; Stephenson, N. C.; Stevens, J. D. *Cryst. Struct. Commun.* 1974, 3, 195 – 199.

[12] Cook, W. J.; Bugg, C. E. *Acta Cryst.* 1976, 32, 656 – 659.

[13] Cook, W. J.; Bugg, C. E. *J. Am. Chem. Soc.* 1973, 95, 6442 – 6446.

[14] Bugg, C. E. *J. Am. Chem. Soc.* 1973, 95, 908 – 913.

[15] Cook, W. J.; Bugg, C. E. *Acta Crystallogr., Sect. B.* 1973, 29, 907 – 909.

[16] Accorsi, C. A.; Bertolasi, V.; Ferretti, V.; Gastone, G. *Carbohydr. Res.* 1989, 191, 91 – 104.

[17] Accorsi, C. A.; Bellucci, F.; Bertolasi, V.; Valeria, F.; Gastone, G. *Carbohydr. Res.* 1989, 191, 105 – 116.

[18] Peralta – Inga, Z.; Johnson, G. P.; Dowd, M. K., Rendleman, J. A.; Stevens, E. D.; French, A. D. *Carbohydr. Res.* 2002, 337, 851 – 861.

[19] Craig, D. C.; Stephenson, N. C.; Stevens, J. D. *Carbohydr. Res.* 1972, 22, 494 – 495.

[20] Cook, W. J.; Bugg, C. E. *Carbohydr. Res.* 1973, 31, 265 – 275.

[21] Richards, G. F. *Carbohydr. Res.* 1973, 26, 448 – 449.

[22] Terzis, A. *Cryst. Struct. Commun.* 1978, 7, 95 – 99.

[23] Yang, L. M.; Zhao, Y.; Xu, Y. Z.; Jin, X. L.; Weng, S. F.; Wen, T.; Wu, J. G.; Xu, G. X. *Carbohydr. Res.* 2001, 334, 91 – 95.

[24] Yang, L. M.; Wu, J. G.; Weng, S. F.; Jin, X. L. *J. Mol. Struct.* 2002, 612, 49 – 57.

[25] Lu, Y.; Deng, G. C.; Miao, F. M.; Li, Z. M. *Carbohydr. Res.* 2003, 338, 2913 – 2919.

[26] Lu, Y.; Deng, G. C.; Miao, F. M.; Li, Z. M. *J. Inorg. Biochem.* 2003, 96, 487 - 492.
[27] Angyal, S. J. *Pure Appl. Chem.* 1973, 35, 131 - 146.
[28] Alvarez, A. M.; Desrosiers, N. M.; Morel, J. P. *Can. J. Chem.* 1987, 65, 2656 - 2660.
[29] Desrosiers, N. M.; Lhermet, C.; Morel, J. P. *J. Chem. Soc., Faraday Trans.* 1991, 87, 2173 - 2177.
[30] Desrosiers, N. M.; Lhermet, C.; Morel, J. P. *J. Chem. Soc., Faraday Trans.* 1993, 89, 1223 - 1228.
[31] Su, Y. L.; Yang, L. M.; Weng, S. F.; Wu, J. G. *J. Rare Earths.* 2002, 20, 339 - 342.
[32] Chandrasekhar, A. *J. Imaging Technol.* 1990, 16, 158 - 161.
[33] Cael, J. J.; Koenig, J. L.; Blackwell, J. *Carbohydr. Res.* 1974, 32, 79 - 91.
[34] Tajmir - Riahi, A. H. *Carbohydr. Res.* 1983, 122, 241 - 248.
[35] Tajmir - Riahi, A. H. *Carbohydr. Res.* 1984, 127, 1 - 8.
[36] Zhang, W. J. *Biochemical Technology of Complexes of Carbohydrate*; Zhejiang University Press: Hangzhou, 1999. pp 193 - 198.
[37] Mathlouthi, M.; Seuvre, A. M.; Koenig, J. L. *Carbohydr. Res.* 1983, 122, 31 - 47.
[38] Tajmir - Riahi, A. H. *Biophys. Chem.* 1986, 23, 223 - 228.

一类新型嘧啶苯氧(硫)醚的合成及生物活性*

袁德凯　李正名　赵卫光　范志金　王素华

（南开大学元素有机化学研究所，元素有机化学国家重点实验室，天津　300071）

摘　要　由 *N*，*N*－二乙基硝酸胍(**1**)与双乙烯酮制备了农药中间体嘧啶醇(**2**)，经硝化、氯化，得到4－氯－5－硝基嘧啶（**4**）。通过化合物**4**与苯(硫)酚反应，制备了新型嘧啶苯氧(硫)醚类化合物(**5**)。并发现由化合物**4**的巯基乙酸甲酯衍生物(**6**)与水合肼的新反应，得到嘧啶肼(**8**)。化合物结构均经过元素分析、^{1}H NMR、IR及MS确证。生物活性测试表明，所合成的化合物具有一定的除草、杀菌及抗烟草花叶病毒活性。

关键词　双乙烯酮　氯硝基嘧啶　苯氧(硫)醚　嘧啶肼　生物活性

嘧啶杂环在农药研究与应用中占据重要的地位，多种商品化的杀虫剂[1]、杀菌剂[2]、除草剂[3]中均含有嘧啶结构。例如具有杀虫活性的芳氧基嘧啶[4]和嘧啶胺[5]，以抗菌素Strobilurin为先导得到的新型杀菌剂ICIA 5504[6]、具有杀菌活性的芳基嘧啶[7]，具有除草活性的嘧啶水杨酸类[8]、芳氧嘧啶类[9]和脲嘧啶衍生物[10]等都是近期农药研究较为活跃的领域，人们对嘧啶类农药的作用机制[11]做了深入的探讨。

随着环保意识的增强，高毒有机磷农药将被禁用，寻找高效、低毒、环境友好的新农药已成为农药研究的主要方向之一。本文用久效磷的基本原料双乙烯酮与硝酸二乙胍反应，获得嘧啶磷中间体6－甲基－2－（*N*，*N*－二乙胺基）嘧啶－4－醇（**2**），通过对其硝化、氯化，得到重要中间体4－氯－5－硝基嘧啶（**4**），参照芳氧嘧啶结构，设计合成了嘧啶苯氧（硫）醚（**5**）。在研究化合物**4**的巯基乙酸甲酯衍生物**6**时，发现水合肼不是使酯基肼解生成预期的酰肼（**7**），而是在其母环上发生亲核取代，使硫代乙酸甲酯基团离去而得到嘧啶肼（**8**），该反应尚未见报道。化合物活性筛选结果表明，某些化合物具有除草、杀菌及抗烟草花叶病毒活性。

目标化合物的合成路线如Scheme 1所示。

1　实验部分

1.1　仪器和试剂

AC－200Q型核磁共振仪（Bruker公司，瑞士），$CDCl_3$，TMS；EQU INOX55型红外光谱仪（Bruker公司，德国），KBr压片；MT－3型CHN元素分析仪（Yanaco公司，日本）；300型熔点仪（Yanaco公司，日本），温度计未经校正；VG ZAB－HS型质谱仪（VG Scientific公司，英国）；柱层析采用硅胶H（青岛海洋化工厂）。

双乙烯酮由南通江山农化股份有限公司提供，工业品，含量≥98%；石灰氮，工

*　原发表于《应用化学》，2005，22（10）：1045－1049。国家自然科学基金重点项目（20432010），天津市自然科学重点基金资助（033803411）。

5a～5g: X= O; R= H, p-NO_2, o-MeO, m-CHO, p-Me, p-Cl, p-Br

5h～5j: X= S; R= H, p-iPr, p-Cl

Scheme 1 Synthetic route for title compounds

业品，含量≥50%；其它试剂均为分析纯。N，N－二乙基硝酸胍（**1**）和嘧啶醇（**2**）参照文献［12］方法制备。

1.2 中间体化合物3和化合物4的合成

在5℃下将4.0g（0.0017mol）化合物**2**分批加入10mL混酸（$V(HNO_3):V(H_2SO_4)=4:6$）中。加毕，室温搅拌12h，得浅黄色透明液体。倾入200g冰水中，析出少量浅黄色固体[13]。浓氨水中和至pH=6，析出大量黄色沉淀。过滤后无水乙醇重结晶，得2.6g黄色片状晶体（**3**）。

将2.26g(0.01mol)化合物**3**溶于10mL $POCl_3$中，60℃下搅拌至其溶解，加入4mL N，N－二甲基苯胺，搅拌回流4h，减压蒸除溶剂，残留物以100g冰水水解，得黄色固体。过滤，滤饼以浓盐酸和水依次洗涤，真空干燥，得2.21g黄色固体（**4**）。化合物**3**和化合物**4**的物理化学数据见表1和表2。

表1 化合物的收率、熔点和元素分析数据

Table 1 Yield, mp and elemental analysis data of the compounds

No.	mp/℃	Yield/%	Elemental analysis（calcd.）/%		
			C	H	N
3	215～218	67.7	47.80（47.80）	6.19（6.19）	24.53（24.78）
4	58～60	83.2	43.96（44.10）	5.26（5.32）	22.88（22.90）
5a*	85	63.2	59.59（59.60）	6.09（5.96）	17.46（18.54）
5b	92～95	58.6	51.62（51.87）	4.89（4.90）	20.06（20.17）
5c	96～97	78.3	57.83（57.83）	6.07（6.02）	16.89（16.87）
5d	92	66.8	58.08（58.18）	5.40（5.45）	17.00（16.97）
5e	58～60	77.2	60.79（60.76）	6.18（6.33）	17.90（17.72）
5f	65～68	87.0	53.50（53.49）	4.96（5.05）	16.46（16.64）
5g	65～67	74.7	47.22（47.24）	4.78（4.46）	14.60（14.70）
5h	104～105	89.5	56.70（56.60）	5.65（5.66）	17.48（17.61）

续表 1

No.	mp/℃	Yield/%	Elemental analysis (calcd.) /%		
			C	H	N
5i	80 ~ 82	89.8	60.02 (60.00)	6.45 (6.67)	15.48 (15.56)
5j	58 ~ 60	94.6	51.08 (51.06)	4.70 (4.82)	15.85 (15.89)
6	100 ~ 101	83.2	45.77 (45.86)	5.77 (5.73)	17.87 (17.83)
9	195 ~ 196	95	53.07 (52.97)	5.16 (5.24)	23.19 (23.17)

表 2 化合物的 ^{1}H NMR 的数据

Table 2 ^{1}H NMR data of the compounds

No.	^{1}H NMR (200MHz in $CDCl_3$), δ
3	1.23 (t, J=7.0Hz, 6H, $2\times CH_2CH_3$), 2.57 (s, 3H, pyrimidyl − CH_3), 3.69 (m, J=7.0Hz, 5H, $2\times CH_2CH_3$); 1.23 (t, J=7.0Hz, 6H, $2\times CH_2CH_3$), 2.57 (s, 3H, pyrimidyl − CH_3), 3.69 (q, J=7.0Hz, 4H, $2\times CH_2CH_3$) (D_2O)
4	1.25 (t, J=6.8Hz, 6H, $2\times CH_2CH_3$), 2.60 (s, 3H, pyrimidyl − CH_3), 3.58 (q, J=6.8Hz, 4H, $2\times CH_2CH_3$)
5a	0.85 (t, J=7.1Hz, 3H, CH_2CH_3), 1.13 (t, J=7.1Hz, 3H, CH_2CH_3), 2.55 (s, 3H, pyrimidyl − CH_3), 3.15 (q, J=7.1Hz, 2H, CH_2CH_3), 3.60 (q, J=7.1Hz, 2H, CH_2CH_3), 7.10 ~ 7.36 (m, 5H, Ar—H)
5b	0.90 (t, J=6.9Hz, 3H, CH_2CH_3), 1.15 (t, J=6.9Hz, 3H, CH_2CH_3), 2.58 (s, 3H, pyrimidyl − CH_3), 3.22 (q, J=6.9Hz, 2H, CH_2CH_3), 3.64 (q, J=6.9Hz, 2H, CH_2CH_3), 7.80 (q, 4H, Ar—H)
5c	0.800 ~ 1.14 (m, 6H, $2\times CH_2CH_3$), 2.61 (s, 3H, pyrimidyl − CH_3), 3.75 (s, 3H, OCH_3), 3.10 ~ 3.75 (m, 4H, $2\times CH_2CH_3$), 6.91 ~ 7.24 (m, 4H, Ar—H)
5d	0.85 (t, J=6.7Hz, 3H, CH_2CH_3), 1.15 (t, J=6.7Hz, 3H, CH_2CH_3), 2.56 (s, 3H, pyrimidyl − CH_3), 3.10 (q, J=6.7Hz, 2H, CH_2CH_3), 3.60 (q, J=6.7Hz, 2H, CH_2CH_3), 7.55 ~ 7.73 (m, 4H, Ar—H), 10.00 (s, 1H, —CHO)
5e	0.87 (t, J=7.1Hz, 3H, CH_2CH_3), 1.08 (t, J=7.1Hz, 3H, CH_2CH_3), 2.33 (s, 3H, pyrimidyl − CH_3), 2.51 (s, 3H, Ar—CH_3), 3.20 (q, J=7.1Hz, 2H, CH_2CH_3), 3.50 (q, J=7.1Hz, 2H, CH_2CH_3), 7.06 (q, 4H, At—H)
5f	0.95 (t, J=7.3Hz, 3H, CH_2CH_3), 1.15 (t, J=7.3Hz, 3H, CH_2CH_3), 2.57 (s, 3H, pyrimidyl − CH_3), 3.20 (q, J=7.3Hz, 2H, CH_2CH_3), 3.70 (q, J=7.3Hz, 2H, CH_2CH_3), 7.21 (q, 4H, Ar—H)
5g	0.89 (t, J=7.1Hz, 3H, CH_2CH_3), 1.13 (t, J=7.1Hz, 3H, CH_2CH_3), 2.52 (s, 3H, pyrimidyl − CH_3), 3.18 (q, J=7.1Hz, 2H, CH_2CH_3), 3.58 (q, J=7.1Hz, 2H, CH_2CH_3), 7.25 (q, 4H, Ar—H)
5h	0.64 (t, J=7.2Hz, 3H, CH_2CH_3), 1.10 (t, J=7.2Hz, 3H, CH_2CH_3), 2.69 (s, 3H, pyrimidyl − CH_3), 3.00 (q, J=7.2Hz, 2H, CH_2CH_3), 3.55 (q, J=7.2Hz, 2H, —CH_2CH_3), 7.39 ~ 7.47 (m, 5H, Ar—H)

续表2

No.	^1H NMR (200MHz in $CDCl_3$), δ
5i	0.64 (t, J=6.9Hz, 3H, CH_2CH_3), 1.10 (t, J=6.9Hz, 3H, CH_2CH_3), 1.14~1.26 (m, 7H, *i*-Pr—H), 2.72 (s, 3H, pyrimidyl-CH_3), 2.92 (q, J=6.9Hz, 2H, CH_2CH_3), 3.60 (q, J=6.9Hz, 2H, CH_2CH_3), 7.28 (q, 4H, Ar—H)
5j	0.700 (t, J=7.0Hz, 3H, CH_2CH_3), 1.11 (t, J=7.0Hz, 3H, CH_2CH_3), 2.67 (s, 3H, pyrimidyl-CH_3), 3.00 (q, J=7.0Hz, 2H, CH_2CH_3), 3.54 (q, J=7.0Hz, 2H, CH_2CH_3), 7.40 (q, 4H, Ar—H)
6	1.19 (t, J=7.1Hz, 6H, 2×CH_2CH_3), 2.70 (s, 3H, pyrimidyl-CH_2), 3.71 (s, 2H, SCH_2), 3.71 (q, J=7.1Hz, 4H, 2×CH_2CH_3), 3.79 (s, 3H, CO_2CH_3)
9	1.28 (s, 6H, 2×CH_2CH_3), 2.87 (s, 3H, pyrimidyl-CH_3), 3.76~3.90 (m, 4H, 2×CH_2CH_3), 7.36~7.67 (q, 4H, Ar—H), 8.09 (s, 1H, PhCH═), 11.65 (s, 1H, ═NNH—)

1.3 目标化合物的合成

1.3.1 2-*N*, *N*-二乙胺基-6-甲基-5-硝基嘧啶-4-芳氧(硫)醚(**5a**~**5j**)的制备

取0.0011mol酚或硫酚和0.06g研碎的KOH，加入5mL DMF中，加热至KOH溶解。冷却至室温，取0.27g（0.0011mol）化合物**4**分批加入DMF中，加毕后65℃反应8h。减压脱溶后减压柱层析（*V*(石油醚):*V*(乙酸乙酯)=10:1），得浅黄色固体化合物**5**[9,14]。化合物物理化学数据见表1和表2。

1.3.2 化合物6的合成及其相关反应

将1.14g（0.011mol）巯基乙酸甲酯溶于15mLDMF中，0℃下加入0.64g（0.13mol）50%的NaH，待其溶解后分批加入2.18g（0.009mol）化合物**4**，室温反应8h[15]。倾入50mL水中，析出黄色沉淀化合物（**6**），无水甲醇重结晶；将1g化合物6溶于10mL乙醇，冰水浴下滴加5mL乙醇与1g水合肼（85%）的混合液，滴毕室温反应0.5h，后回流至TLC检测原料点消失，冷却后析出浅黄色固体。蒸除溶剂，无水乙醇重结晶，获0.71g黄色片状结晶，但^1H NMR显示并非预期的酰肼（**7**），由化合物**6**的结构特点，生成物可能为嘧啶肼（**8**）。取0.43g化合物**8**和0.20g对氯苯甲醛投入20mL乙醇中，滴加2滴冰醋酸，回流至TLC检测原料点消失。冷却，析出黄色固体，无水乙醇重结晶。^1H NMR和元素分析结果证明，化合物**9**为嘧啶腙，从而验证了嘧啶肼（**8**）的结构。

2 结果与讨论

2.1 嘧啶醇的制备

我们主要研究了以双乙烯酮代替三乙法制备2-*N*, *N*-二乙胺基-6甲基嘧啶-4-醇（**2**）的方法。文献中嘧啶醇（**2**）多采用三乙法制备[12]。如能从双乙烯酮直接制备化合物**2**，将大大降低能耗和成本。参照三乙法[12]，我们对双乙烯酮制备嘧啶醇的反应条件进行了研究，从反应结果和原料成本考虑，选定*n*(双乙烯酮):*n*(硝酸胍):*n*(NaOH)=1:1:1的物料比来制备嘧啶醇。

2.2 目标化合物的合成

2.2.1 中间体合成

有关嘧啶硝化的文献较多，有报道[13]使用 HNO_3/HOAc 体系可对 4，6－嘧啶二醇进行 C_4 位硝化。但采用该体系对化合物 **2** 硝化时，60℃下反应仍不完全；采用浓硝酸/浓硫酸体系在 60℃反应，虽能得到化合物 **3**，但收率只有 30%，而常温反应 5h 则硝化不完全，延长反应时间至 12h 则可完全反应，收率较高。后处理中使用氨水中和过量的酸可使产物析出。参照文献［13］，我们选用 $POCl_3/Me_2NPh$ 对化合物 **3** 进行氯化，结果高收率获得化合物 **4**。

2.2.2 目标化合物的合成

由于硝基的存在，化合物 **4** 的 C4 氯非常活泼，常温下即可与酚钠或巯基作用得到芳氧（硫）醚衍生物 **5a～5j** 及巯基乙酸甲酯衍生物 **6**。但化合物 **6** 的肼解不能生成其酰肼衍生物 **7**，而是发生嘧啶环上的取代反应生成多取代嘧啶肼 **8**，表明在嘧啶环中引入硝基后化合物 **6** 中 C4 的巯基可在肼的作用下脱除，因此该反应可用于制备多取代嘧啶肼。

2.3 结构解析

在化合物 **2** 的 ^1H NMR 谱图中，δ 3.69 处的吸收无裂分，氢数为 5。重水交换后，变为四重峰，氢数减少 1 个，表明化合物 **2** 中羟基氢化学位移与亚甲基的接近。**5a～5j** 的谱图则显示结构中 N2 上乙基的吸收呈现裂分状态，这是由结构中硝基的阻碍作用使苯环不能自由旋转，而导致 2 个乙基所处化学环境不同而引起；另外我们通过对化合物 **8** 的衍生物 **9** 的谱图分析证明了化合物 **8** 为嘧啶肼。化合物的 ^1H NMR 数据见表 2。

我们还对化合物 **5h** 的 IR 和 MS 进行了测定，在红外光谱（IR）中有：3071，2980，2933，1561，1527，1460，846，755，708 等特征吸收（cm^{-1}）；质谱（MS）中有：318.4（M^+，100），303.4（100），289.4（60），242.3（44），177.3（30）等主要碎片峰（m/e,%）。

2.4 生物活性

部分化合物在 5×10^{-5} 质量分数浓度下的除草活性测试结果表明，该类化合物除 **5e**、**5f**、**6** 具有一定的除草活性以外，其它化合物的除草活性普遍偏低。表明要提高此类化合物的除草活性需对其结构进行改造。

表 3　部分化合物的除草活性（质量分数 5×10^{-5}）

Table 3　Herbicidal activities of some compounds（mass fraction 5×10^{-5}）

		3	4	5a	5b	5c	5e	5f	5h	5i	5j	6	9
Echinochla Crusgulli	Preemergent	0	23	5.8	4.5	23.2	8.9	6.2	16.1	18.6	20.9	25	27.9
	Postergent	0	6.4	0	19.2	0	19.2	44.1	19.2	0	0	37.9	0
B rassica N apus	Preemergent	23.2	7.4	19.1	16.9	4.5	0	20.9	0	12.7	5.4	0	1.8
	Postergent	7.9	12.2	0	0	0	44	21	24	0	0	12	0

采用离体平皿法对一些化合物的杀菌活性测试表明，化合物 **4** 对所有致病菌都显

示一定的杀菌活性。其中对苹果轮纹病的抑制活性达 74.3% ，对芦笋茎枯、花生褐斑、小麦赤霉病的抑制率在 40% 左右，对番茄早疫病也有一定的效果。由于该化合物的结构简单，值得进行深入研究。其它化合物的杀菌活性总体来讲都不是很高，故其结构需改造。对部分化合物的抗烟草花叶病毒活性测试表明该杂环的衍生物具有一定的抗病毒活性。

表 4 部分化合物的杀菌及抗烟草花叶病毒活性

Table 4 Fungcidal and an ti - TMV activities of some compounds

	Mass fraction	3	4	5a	5b	5e	5f	5h	6	9
Gibberella zeae	5×10^{-5}	19.0	38.1	—	0	15	0	—	—	0
A ltemaria solani		20.2	26.7	—	0	0	0	—	—	15.4
Pham a asparagi		0	44.4	—	0	25	8.3	—	—	30
Physalaspora piricola		0	74.3	—	0	0	0	—	—	7.1
Cercosporsa arachidicola		0	42.1	—	0	0	20	—	—	0
TMV（Tobacco Mosaic Virus）	5×10^{-4}	0	—	12.5	—	—	0	25	33.3	25

参 考 文 献

[1] ZHANG Min - Heng（张敏恒）Chief - Edr（主编）. Handbook of Commercial Pesticides（农药商品手册），1st Edn（第 1 版）.［M］. Shenyang（沈阳）：Shenyang Press（沈阳出版社），1999：92.

[2] LIU Chang - Ling（刘长令）. *Pest Sci Man*（农药科学与管理）［J］，2000，21（3）：20.

[3] LI Zong - Cheng（李宗成）. *Pesticides*（农药）［J］，1998，37（1）：1.

[4] Eberle M，Schaub F，Craig G W. EP 0 667 343A1［P］，1995.

[5] a. Krautstrunk G，Jakobi H，Maerkl M，*et al.* WO 0 007 998［P］，2000.
b. Brown M E，Bauser J W，Cullen T G，*et al.* WO 9 820 878［P］，1998.

[6] a. LIU Chang - Ling（刘长令），ZHANG Li - Xin（张立新），WANG Can - Ming（汪灿明），*et al. Pesticides*（农药）［J］，1998，37（3）：1.
b. Krueger B W，Mauler - Machnik A，Dunkel R，*et al.* DE 10 006 210［P］，2000.

[7] a. Yokoyama S，YuguchiM，Janaky T，*et al.* WO 0 037 460［P］，2000.
b. Shimozono T，Umeda T，Tachino H，*et al.* JP 10 147 584［P］，1998.

[8] a. LI Zong - Cheng（李宗成）. *Pesticides*（农药）［J］，2001，40（3）：43.
b. Bieringer H，Bauer K，Stark H，*et al.* WO 9 801 430［P］，1998.
c. Han S D，Lu L，Mao L S，*et al.* WO 0 234 724［P］，2002.

[9] FEIXue - Ning（费学宁），SONG Hong - Hai（宋洪海），WANG Xiang（王翔），*et al. Chem J Chin Univ*（高等学校化学学报）［J］，2000，21（2）：237.

[10] a. Tohyama Y，Gotou T，Senemitisu Y. EP 1 106 607［P］，2001.
b. Tsukamoto M，Gupta S，Ying B P，*et al.* WO 9 921 837［P］，1999.

[11] a. XIA Yu（夏禹）Edr and Trans（编译）. *World Pesticides*（世界农药）［J］，2000，22（5）：22.
b. TANG Chun - Feng（唐春风）Edr and Trans（编译），ZHANG Yi - Bing（张亦冰）Proof（校）. *World Pesticides*（世界农药）［J］，2000，22（5）：26.

[12] HUANG Yong - Ming（黄永明），LÜ Yin - Xiang（吕银祥），HUANG Yue - Fang（黄月芳），*et al. Chin J Appl Chem*（应用化学）［J］，2001，18（2）：171.

[13] MEIHe - Shan（梅和珊），WU Yin - Wen（武引文），LÜ Xin - Ying（吕新英）. *Hebei Chem Ind*

(河北化工) [J], 2000, 2: 34.

[14] WANG Xiang (王翔), REN Kang - Tai (任康太), YANG Hua - Zheng (杨华铮). *J Xiamen Univ* (Nat Sci) (厦门大学学报 (自然科学版)), Plus (增刊), 1999, 38: 505.

[15] Hornin E C Chief - Edr (主编). Education & Research Group in Organic Chemistry ofNanjingUniversity (南京大学有机化学教研室) Trans (译). Organic Synthesis (有机合成), Vol. 3 (第3集) [M]. Beijing (北京): Science Press (科学出版社), 1981: 446.

Synthesis and Biological Activities of a Novel Series of 4 - (Substituted - phenoxy or Phenylthio) Pyrimidines

Dekai Yuan, Zhengming Li, Weiguang Zhao, Zhijin Fan, Suhua Wang

(State Key Laboratory of Elemento - Organic Chemistry, Institute of Elemento - Organic Chemistry, Nankai University, Tianjin, 300071)

Abstract The synthesis of 6 - methyl - 2 - *N*, *N* - diethylaminopyrimidinol (**2**) from diethylgunidine nitrate (**1**) and diketene was studied. After nitration and then chlorization of **2**, 4 - chloro - 5 - nitropyrimidine (**4**) was obtained, and a novel series of 5 - nitro - 4 - (substituted - phenoxy or phenylthio) pyrimidines (**5**) were prepared *via* the reaction of **4** and substituted phenols or phenylthiols. Furthermore, a new reaction between the methylmercap - toacetate derivative of **4** (**6**) and hydrazine, which yields pyrimidinyl hydrazine (**8**) in which the thio - acetoacetic acid methyl ester behaves as a leaving group under the nucleophinlic attack. The structures of all of the compounds were confirmed by ^{1}H NMR, IR, MS, and elemental analysis. The preliminary bioassay indicated that some of the compounds showed moderate herbicidal, fungicidal and anti - TMV activities.

Key words diketene; chloro - nitropyrimidine; pyrimidinyl hydrazine; (substituted - phenoxy or phenylthio) pyrimidine; biological activity

4－（取代嘧啶－2－基）－1－芳磺酰基氨基脲化合物的合成与表征*

李鹏飞　王宝雷　马　宁　王素华　李正名

（南开大学元素有机化学研究所，元素有机化学国家重点实验室，天津　300071）

摘　要　以取代的苯磺酰肼与（取代嘧啶－2－基）－氨基甲酸苯酯在非亲核性碱存在下合成了12个4－（取代嘧啶－2－基）－1－芳磺酰基氨基脲化合物**5**和3个**5**与碱形成的有机盐**5as**，**5bs**和**5cs**。所有合成化合物均经过元素分析和^{1}HNMR的结构确证。^{1}HNMR数据证明，在**5as～5cs**中，磺酰氨基脲起到提供质子的作用。

关键词　磺酰氨基脲　合成　有机盐

具有磺酰脲结构的化合物以其独特的生物活性而著名，比如作为医用降血糖药物[1]和农用超高效除草剂等。一些氨基脲衍生物则具有抗惊厥活性[2]、昆虫生长调节活性[3]、杀菌活性[4]。而将磺酰基与氨基脲结构连接而成的磺酰氨基脲化合物尽管研究相对较少，但也有用作降血糖[5]、抗血栓[6]等功效。如Tolazamide[7]就是一个商品化的治疗糖尿病的药物。嘧啶杂环衍生物已经报道具有多种生物活性，在医药学上用作抗癌[8]、抗菌[9]等的药物，农业上作为杀菌剂、除草剂、杀虫剂[10]。为了筛选具有良好生物活性的新结构类型，本文将4，6－二取代或4－取代嘧啶杂环引入磺酰氨基脲结构中，合成了一系列新型4－（取代嘧啶－2－基）－1－苯磺酰基氨基脲化合物**5**。合成路线如下：

1 (SO_2Cl) $\xrightarrow{N_2H_4 \cdot H_2O}$ **2** (SO_2NHNH_2)

3 (NH_2) $\xrightarrow{ClCOOC_6H_5}$ **4** ($NHCOOC_6H_5$)

2 + **4** $\xrightarrow{base}$ **5**

R^1=H, CH_3; R^2=H, CH_3, F, Br, NO_2; R^3=H, NO_2;
R^4=CH_3, OCH_3; R^5=H, CH_3, OCH_3

* 原发表于《有机化学》，2005，25（19）：1057－1061。国家“973”项目（绿色化学农药先导结构及作用靶标的发现与研究，No. 2003CB114406）和天津市科委（超高效磺酰脲除草剂作用机制及农药分子设计研究，No. 033803411）资助项目。

1 实验部分

1.1 仪器和试剂

X－4 数字显示显微熔点测定仪（北京泰克仪器有限公司），温度计未校正；BRUKER AVANCE－300MHz 核磁共振仪，DMSO－d_6 或 $CDCl_3$ 为溶剂，TMS 为内标。Yanaco CHN CORDER MT－3 型元素分析仪。BRUKER EQUINOX 55 红外光谱仪，KBr 压片。

对氟苯磺酰氯[11]、间硝基苯磺酰氯[12]、2－甲基－5－硝基苯磺酰氯[13]和（取代嘧啶－2－基）氨基甲酸苯酯（**4**）[14]参照文献方法合成。其余所用试剂均为市售分析纯或化学纯。

1.2 化合物合成步骤和结构表征

1.2.1 （取代）苯磺酰肼（**2**）的制备

在 100mL 三口烧瓶中加入 9mL 85% 的水合肼和 16mL 水，搅匀，冷却至适当温度，保持此温度缓慢加入 50mmol（取代）苯磺酰氯，加入过程中反应有放热现象，加毕，在室温下搅拌反应 2～4h，将混合物倾倒入 50mL 冰水中，抽滤，用冷水洗涤滤饼，抽干，收集固体，红外灯下烘干，用甲醇或乙醇重结晶可得纯品。

苯磺酰肼（**2a**）：苯磺酰氯加入温度低于－5℃，产物为白色晶体，产率 89%，m. p. 101～102℃。

对甲苯磺酰肼（**2b**）：对甲苯磺酰氯加入时温度低于 10℃，产物为白色晶体，产率 93%，m. p. 106～108℃。

2－甲基－5－硝基苯磺酰肼（**2c**）：2－甲基－5－硝基苯磺酰氯加入时保持温度低于 20℃，产物为棕色粉末，产率 96%，m. p. 142～143 ℃。

对氟苯磺酰肼（**2d**）：对氟苯磺酰氯加入时保持温度低于 5℃，产物为白色晶体，产率 92%，m. p. 80～82℃。

对溴苯磺酰肼（**2e**）：对溴苯磺酰氯加入时保持温度低于 5℃，产物为淡黄色晶体，产率 85%，m. p. 110～111℃。

对硝基苯磺酰肼（**2f**）：对硝基苯磺酰氯加入时保持温度低于 20℃，产物为灰色粉末，产率 87%，m. p. 145～146℃。

间硝基苯磺酰肼（**2g**）：间硝基苯磺酰氯加入时保持温度低于 10℃，产物为灰色粉末，产率 86%，m. p. 123～125℃。

1.2.2 4－（取代嘧啶－2－基）－1－苯磺酰基氨基脲（**5**）的制备

在 50mL 烧瓶中加入 1.5mmol（取代）苯磺酰肼，20mL 干燥乙腈，搅拌溶解，再加入相应的（取代嘧啶－2－基）氨基甲酸苯酯 1.5mmol，再溶解。搅拌下加入催化剂 DBU（1，8－二氮杂双环［5，4，0］十一－7－烯，1，8－Diazabicyclo－［5.4.0］undec－7－ene）或 DABCO（11，4－二氮杂双环［2.2.2］辛烷，1，4－Diazabicyclo［2.2.2］octane）1.7mmol，然后在室温下搅拌反应 3～7h。（Ⅰ）如果有大量的沉淀生成，则抽滤收集固体，用乙腈、乙醚洗涤，干燥。（Ⅱ）如果反应中没有大量沉淀生成，则减压将混合物浓缩至约 5mL，冷却后用 5% 的盐酸酸化至 pH≈5，将形成白色沉淀，待沉淀充分析出，抽滤，用水、丙酮洗涤，干燥。产品用丙酮和二甲基亚砜的混合溶剂重结晶，可得纯品。

4 - （4，6 - 二甲基嘧啶 - 2 - 基） - 1 - 苯磺酰基氨基脲（**5a**）：DBU 催化，分离方法Ⅱ，白色晶体，产率 40.5%，m. p. 203 ~ 205℃。^{1}H NMR（DMSO - d_6，300MHz）δ：2.24（s，6H，2 × CH_3），6.89（s，1H，PyrimH - 5），7.68（t，J = 7.4Hz，2H，ArH - 3 and H - 5），7.71（t，J = 5.4Hz，1H，ArH - 4），7.84（d，J = 7.5Hz，2H，ArH - 2 and H - 6），9.91（br，1H，CO**NH**Het），10.01（br，1H，NH**NH**CO），10.81（br，1H，SO_2**NH**NH）。Anal. calcd for $C_{13}H_{15}N_5O_3S$：C 48.59，H 4.70，N 21.79；found C 48.38，H 4.85，N 21.49。

4 - （4 - 甲基嘧啶 - 2 - 基） - 1 - 苯磺酰基氨基脲（**5b**）：DBU 催化，分离方法Ⅱ，白色粉末，产率 62.2%，m. p. 215 ~ 216℃。^{1}H NMR（DMSO - d_6，300MHz）δ：2.23（s，3H，HetC**H**$_3$），6.95（d，J = 5.1Hz，1H，PyrimH - 5），7.61（t，J = 7.7Hz，2H，ArH - 3 and H - 5），7.71（t，J = 6.8Hz，1H，ArH - 4），7.84（d，J = 7.8Hz，2H，ArH - 2 and H - 6），8.34（d，J = 5.1Hz，1H，PyrimH - 6），9.88（br，1H，CO**NH**Het），10.03（br，1H，NH**NH**CO），10.61（br，1H，SO_2**NH**NH）。Anal. calcd for $C_{12}H_{13}N_5O_3S$：C 46.90，H 4.26，N 22.79；found C 46.87，H 4.29，N 22.68。

4 - （4，6 - 二甲基嘧啶 - 2 - 基） - 1 - （4 - 甲基苯磺酰基） - 氨基脲（**5c**）：分离方法Ⅱ，白色晶体，DBU 催化，产率 50.3%；DABCO 催化，产率 55.0%，m. p. 206 ~ 207℃。^{1}H NMR（DMSO - d_6，300MHz）δ：2.17（s，6H，2 × CH_3），2.39（s，3H，ArCH_3），6.82（s，1H，PyrimH - 5），7.42（d，J = 8.1Hz，2H，ArH - 3 and H - 5），7.72（d，J = 8.1Hz，2H，ArH - 2 and H - 6），9.74（br，1H，CO**NH**Het），9.93（br，1H，NH**NH**CO），10.71（br，1H，SO_2**NH**NH）。Anal. calcd for $C_{14}H_{17}N_5O_3S$：C 50.14，H 5.11，N 20.88；found C 50.10，H 4.98，N 21.08。

4 - （4 - 甲基嘧啶 - 2 - 基） - 1 - （4 - 甲基苯磺酰基） - 氨基脲（**5d**）：DBU 催化，分离方法Ⅱ，白色晶体，产率 58.2%，m. p. 222 ~ 223℃. ^{1}H NMR（DMSO - d_6，300MHz）δ：2.23（s，3H，HetCH_3），2.40（s，3H，ArCH_3），6.96（d，J = 5.1Hz，1H，PyrimH - 5），7.41（d，J = 8.1Hz，2H，ArH - 3 and H - 5），7.72（d，J = 8.1Hz，2H，ArH - 2 and H - 6），8.35（d，J = 5.1Hz，1H，PyrimH - 6），9.78（br，1H，CO**NH**Het），10.03（br，1H，NH**NH**CO），10.56（br，1H，SO_2**NH**NH）。Anal. calcd for $C_{13}H_{15}N_5O_3S$：C 48.59，H 4.70，N 21.79；found C 48.35，H 4.75，N 21.83。

4 - （4，6 - 二甲基嘧啶 - 2 - 基） - 1 - （4 - 氟苯磺酰基） - 氨基脲（**5e**）：分离方法Ⅱ，白色晶体，DBU 催化，产率 54.5%，m. p. 215℃。^{1}H NMR（DMSO - d_6，300MHz）δ：2.22（s，6H，2 × CH_3），6.84（s，1H，PyrimH - 5），7.46（t，J = 9.0Hz，2H，ArH - 3 and H - 5），7.90（q，2H，ArH - 2 and H - 6），9.92（s，1H，NH**NH**CO），9.94（s，1H，CO**NH**Het），10.78（br，1H，SO_2**NH**NH）。Anal. calcd for $C_{13}H_{14}FN_5O_3S$：C 46.01，H 4.16，N 20.64；found C 45.99，H 4.17，N 20.71。

4 - （4 - 甲基嘧啶 - 2 - 基） - 1 - （4 - 氟苯磺酰基） - 氨基脲（**5f**）：DBU 催化，分离方法Ⅱ，白色晶体，产率 72.9%，m. p. 212 ~ 213℃。^{1}H NMR（DMSO - d_6，300MHz）δ：2.29（s，3H，HetCH_3），6.98（d，1H，PyrimH - 5），7.45（t，2H，ArH - 3 and H - 5），7.91（q，2H，ArH - 2 and H - 6），9.98（br，1H，CO**NH**Het），

10.05（br，1H，NH**NH**CO），10.67（br，1H，SO_2**NH**NH）。Anal. calcd for $C_{12}H_{12}FN_5O_3S$：C 44.30，H 3.72，N 21.53；found C 44.20，H 3.65，N 21.58。

4－（4，6－二甲基嘧啶－2－基）－1－（4－溴苯磺酰基）－氨基脲（**5g**）：分离方法Ⅱ，白色晶体，DBU 催化，产率 62.4%，m. p. 214～215 ℃。^{1}H NMR（DMSO－d_6，300MHz）δ：2.23（s，6H，2×CH_3），6.85（s，1H，PyrimH－5），7.79（dd，J＝5.4Hz，4H，ArH－3 and H－5，ArH－2 and H－6），9.98（d，2H，CO**NH**Het and NH**NH**CO），10.79（br，1H，SO_2**NH**NH）。Anal. calcd for $C_{13}H_{14}BrN_5O_3S$：C 39.01，H 3.53，N 17.50；found C 38.92，H 3.78，N 17.14。

4－（4－甲氧基嘧啶－2－基）－1－（4－溴苯磺酰基）－氨基脲（**5h**）：DABCO 催化，分离方法Ⅰ，白色晶体，产率 68.2%，m. p. 166～168℃。^{1}H NMR（DMSO－d_6，300MHz）δ：3.83（s，3H，HetOCH_3），6.53（d，J＝6.0Hz，1H，PyrimH－5），7.75（d，J＝9.0Hz，2H，ArH－3 and H－5），7.80（d，J＝9.0Hz，2H，ArH－2 and H－6），8.24（d，J＝6.0Hz，1H，PyrimH－6），10.04（br，1H，CO**NH**Het），10.63（br，1H，SO_2**NH**NH）. Anal. calcd for $C_{12}H_{12}BrN_5O_4S$：C 45.49，H 5.09，N 17.68；found C 45.29，H 5.15，N 17.75。

4－（4，6－二甲氧基嘧啶－2－基）－1－（2－甲基－5－硝基苯磺酰基）－氨基脲（**5i**）：分离方法Ⅰ，白色粉末，DABCO 催化，产率 88.6%，m. p. 212～213℃。^{1}H NMR（DMSO－d_6，300MHz）δ：2.81（s，3H，ArCH_3），3.81（s，6H，2×OCH_3），5.88（s，1H，PyrimH－5），7.73（d，J＝7.4Hz，1H，ArH－3），8.39（dd，J_3＝7.4Hz，J_4＝2.4Hz，1H，ArH－4），8.55（d，J_4＝2.4Hz，1H，ArH－6），9.94（br，1H，CO**NH**Het），10.47（br，1H，NH**NH**CO）。Anal. calcd for $C_{14}H_{16}N_6O_7S$：C 40.78，H 3.91，N 20.38；found C 40.80，H 3.99，N 20.39。

4－（4，6－二甲基嘧啶－2－基）－1－（4－硝基苯磺酰基）－氨基脲（**5j**）：分离方法Ⅰ，灰色粉末，DABCO 催化，产率 63.9%，m. p. 213～214 ℃。^{1}H NMR（DMSO－d_6，300MHz）δ：2.26（s，6H，2×CH_3），6.86（s，1H，PyrimH－5），8.10（d，J＝8.7Hz，2H，ArH－2 and H－6），8.43（d，J＝8.7Hz，2H，ArH－3 and H－5），9.98（br，1H，CO**NH**Het），10.37（br，1H，NH**NH**CO），10.90（br，1H，SO_2**NH**NH）。Anal. calcd for $C_{13}H_{14}N_6O_5S$：C 42.62，H 3.85，N 22.94；found C 42.56，H 3.89，N 22.86。

4－（4－甲基嘧啶－2－基）－1－（4－硝基苯磺酰基）－氨基脲（**5k**）：分离方法Ⅰ，白色晶体，DABCO 催化，产率 71.4%，m. p. 222～224 ℃. ^{1}H NMR（DMSO－d_6，300MHz）δ：2.33（s，3H，HetCH_3），6.99（d，J＝5.1Hz，1H，PyrimH－5），8.10（d，J＝8.7Hz，2H，ArH－3 and H－5），8.42（t，3H，ArH－2 and H－6，PyrimH－6），10.07（br，1H，CO**NH**Het），10.37（br，1H，NH**NH**CO），10.77（br，1H，SO_2**NH**NH）。Anal. calcd for $C_{12}H_{12}N_6O_5S$：C 40.91，H 3.43，N 23.85；found C 40.71，H 3.50，N 23.79。

4－（4－甲基嘧啶－2－基）－1－（3－硝基苯磺酰基）－氨基脲（**5l**）：分离方法Ⅰ，白色晶体，DABCO 催化，产率 65.1%，m. p. 216～217 ℃。^{1}H NMR（DMSO－d_6，300MHz）δ：2.33（s，3H，HetCH_3），6.98（d，J＝5.1Hz，1H，PyrimH－5），7.90（t，J＝8.0Hz，1H，ArH－3），8.25（d，J＝7.8Hz，1H，ArH－2），8.41（d，J＝5.1Hz，1H，PyrimH－6），8.53（d，J＝8.4Hz，1H，ArH－4），8.58（s，1H，

ArH－6），10.05（br，1H，CO**NH**Het），10.37（br，1H，NH**NH**CO），10.77（br，1H，SO_2**NH**NH）。Anal. calcd for $C_{12}H_{12}N_6O_5S$：C 40.91，H 3.43，N 23.85；found C 40.80，H 3.55，N 23.78。

用与合成4－（取代嘧啶－2－基）－1－苯磺酰基氨基脲**5**相似的方法，有三个反应中分离得到了目标产物**5**与作催化剂的碱形成的盐**5as～5cs**。

4－（4－甲氧基嘧啶－2－基）－1－（4－溴苯磺酰基）－氨基脲与DBU形成的盐（**5as**）：分离方法Ⅱ，白色晶体，DBU催化，产率62.2%，m. p. 144～145 ℃。^{1}H NMR（DMSO－d_6，300MHz）δ：1.60～1.63（m，6H，DBUH－3，H－4 and H－5），1.89（m，2H，DBUH－10），2.71（d，2H，DBUH－6），3.26（t，2H，DBUH－9），3.46（t，2H，DBUH－11），3.53（d，2H，DBUH－2），3.86（s，3H，HetOCH_3），6.39（d，J=6Hz，1H，PyrimH－5），7.53（d，J=8.1Hz，2H，ArH－2 and H－6），7.65（d，J=8.1Hz，2H，ArH－3 and H－5），8.20（d，J=6Hz，1H，PyrimH－6）。Anal. calcd for $C_{21}H_{28}BrN_7O4S$：C 45.49，H 5.09，N 17.68；found C 45.29，H 5.15，N 17.75。

4－（4－甲基嘧啶－2－基）－1－（2－甲基－5－硝基苯磺酰基）－氨基脲与DABCO形成的盐（**5bs**）：分离方法Ⅰ，白色晶体，DABCO催化，产率70.9%，m. p. 218～219 ℃. ^{1}H NMR（DMSO－d_6，300MHz）δ：2.34（s，3H，HetCH_3），2.69（s，12H，DABCO），2.80（s，3H，ArCH_3），6.96（d，J=5.1Hz，1H，PyrimH－5），7.68（d，J=8.4Hz，1H，ArH－3），8.33（d，J=6.9Hz，1H，ArH－4），8.40（d，J=5.1Hz，1H，PyrimH－6），8.54（d，J=2.1Hz，1H，ArH－6），9.90（br，1H，CO**NH**Het），10.67（br，1H，SO_2**NH**NH）。Anal. calcd for $C_{19}H_{26}N_8O_5S$：C 47.69，H 5.48，N 23.42；found C 47.59，H 5.55，N 23.34。

4－（4，6－二甲基嘧啶－2－基）－1－（3－硝基苯磺酰基）－氨基脲与DABCO形成的盐（**5cs**）：分离方法Ⅰ，白色晶体，DABCO催化，产率68.4%，m. p. 160～161℃。^{1}H NMR（DMSO－d_6，300MHz）δ：2.27（s，6H，2×CH_3），2.70（s，12H，DABCO），6.84（s，1H，PyrimH－6），7.86（t，J=8.0Hz，1H，ArH－3），8.23（d，J=8.1Hz，1H，ArH－2），8.48（d，J=8.1Hz，1H，ArH－4），8.55（s，1H，ArH－6），9.85（br，1H，CO**NH**Het），10.85（br，1H，SO_2**NH**NH）。Anal. calcd for $C_{19}H_{26}N_8O_5S$：C 47.69，H 5.48，N 23.42；found C 47.66，H 5.67，N 23.81。

2 结果与讨论

本工作采用的磺酰氨基脲合成路线是参考磺酰脲类除草剂的经典合成路线。在磺酰脲的合成中通常需要用等摩尔或略过量的DBU作催化剂，磺酰胺与氨基甲酸苯酯反应，才能得到良好的效果，但是DBU在反应中所起的直接作用并不清楚。我们推测，具有强的碱性、弱的亲核性和高位阻的DBU将与弱酸性产物磺酰脲（pK_a 3.3～5.2）结合成盐，使反应的平衡向右移动。本文部分化合物用另一非亲核性碱DABCO代替DBU催化磺酰肼与氨基甲酸苯酯的反应，结果发现DABCO也能起到较好的催化效果，而在化合物**5c**的合成中，我们还同时分别使用两种催化剂，发现二者在反应中催化效果相当，而且在相同反应时间用DABCO得到略高的产率。DABCO价格较DBU便宜，又因为是固体而DBU是易被氧化的液体，所以在储存运输中更为有利。鉴于此，本文在用DBU催化合成了部分目标产物后，另一部分化合物的合成改用DABCO作催化剂。

DBU 或 DABCO 与磺酰氨基脲形成的盐有时在处理过程中仍稳定存在，因而本文制得了三个有机盐产物，并经过元素分析和^{1}H NMR 的鉴定。化合物 **5as** 的^{1}H NMR 谱图中，与磺酰氨基脲结合的 DBU 环上八个亚甲基氢的化学位移不同程度地移向低场。根据文献［14］结论，DBU 与三氟乙酸结合后各亚甲基氢的化学位移呈现规律性变化，与此相似，我们对 **5as** 的^{1}H NMR 谱图中高场 DBU 上质子的吸收作了详细归属，见表 1。而且，根据与自由 DBU 的^{1}H NMR 谱比较，可以得出结论：与三氟乙酸相似，在 **5as** 中，磺酰氨基脲起到提供质子的作用。

与 **5as** 相比，其余两个有机盐 **5bs**，**5cs** 的核磁谱图就简单得多。自由的 DABCO 的核磁谱图（溶剂为 DMSO－d_6）上只有 δ2.61 处的单峰，而 **5bs** 和 **5cs** 的谱图上 DABCO 的 12 个氢分别表现为 δ 2.69 和 δ 2.70 处的单峰，$\Delta\delta$ 值分别为 0.08 和 0.09。

DBU　　　　Protonized DBU

表 1　DBU 结合酸前后的化学位移变化

Table. 1　The difference of H chemical shifts of DBU with or without binding proton

H	Free DBU in CD_3CN	Complex with 1 equiv. of TFA in CD_3CN		Free DBU in DMSO－d_6	Complex with 1 equiv. of sulfonyl semicarbazide 5h in DMSO－d_6	
	δ	δ	$\Delta\delta$	δ	δ	$\Delta\delta$
H（11）	3.19	3.39	0.20	3.14	3.43	0.32
H（10）	1.70	1.91	0.21	1.65	1.89	0.24
H（9）	3.12	3.24	0.12	3.07	3.26	0.19
H（6）	2.29	2.61	0.32	2.24	2.71	0.47
H（5）	1.60	1.68	0.08	1.54	1.62	0.08
H（4）	1.62	1.73	0.11	1.57	1.63	0.06
H（3）	1.54	1.66	0.12	1.49	1.60	0.11
H（2）	3.17	3.44	0.27	3.13	3.53	0.40

References

［1］Scheen，A. J. *Foreign Med. Sci. Sect. Pharm.* 1998，25（3），148（in Chinese）.
（Scheen，A. J. 国外医学药学分册，1998，25（3），148.）

［2］Shridhar，V.；Andurkar，C. B.；Stables，J. P.；Kohn，H. *J. Med. Chem.* 2001，44（9），1475.

［3］Xu，H. X.；Chen，W. D.；Qian，X. H. *Chin. J. Pest. Sci.* 1999，1（3），20（in Chinese）.
（徐环昕，陈卫东，钱旭红，农药学学报，1999，1（3），20.）

［4］Chen，W. B.；Jin，G. Y. *Heteroatom Chem.* 2001，12，151.

［5］Heerdt，R.；Huebner，M.；Schmidt，F. H.；Stach，K.；Aumueller，W. S. *African* 6800662，1968［*Chem. Abstr.* 1969，70，87843u］.

［6］Lima，L. M.；Ormelli，C. B.；Fraga，C. A. M.；Miranda，A. L. P.；Barreiro，E. J. *J. Braz. Chem. Soc.* 1999，10，421.

［7］Thomas，R. C.；Duchamp，D. J.；Judy，R. W.；Ikeda，G. J. *J. Med. Chem.* 1978，21，725.

[8] Li, Y. S. *Foreign Med. Sci. Oncol. Sec.* 1995, 22, 101 (in Chinese).
(李玉升，国外医学，肿瘤学分册，1995，22，101.)
[9] Lin, M. R.; Liao, L. A. *Chin. J. New Drugs* 1998, 7, 345 (in Chinese).
(林木荣，廖联安，中国新药杂志，1998，7，345.)
[10] Shang, E. C.; Liu, C. L.; Du, Y. J. *Chem. Ind. Eng. Prog.* 1995, 8 (in Chinese).
(尚尔才，刘长令，杜英娟，化工进展，1995，8.)
[11] Li, Z. G.; Wang, Q. M.; Huang, J. M. *Preparation of Organic Intermediates*, Chemical Industry Press, Beijing 2001, p. 93 (in Chinese).
(李在国，汪清民，黄君珉，有机中间体制备，北京：化学工业出版社，2001，p. 93.)
[12] Han, G. D.; Zhao, S. W.; Li, S. W. *Handbook of Preparative Organic Chemistry*, Chemical Industry Press, Beijing, 1980, p. 208 (in Chinese).
(韩广甸，赵树纬，李述文，有机化学制备手册（上），北京：化学工业出版社，1980，p. 208.)
[13] Kamogawa, H.; Yamamoto, S.; Nanasawa, M. *Bull. Chem. Soc. Jpn.* 1982, 55, 3824.
[14] Wiench, J. W.; Stefaniak, L.; Grech, E.; Bednarek, E. *J. Chem. Soc., Perkin Trans.* 2 1999, 885.

Synthesis and Characterization of New 4-(substituted pyrimidin-2-yl)-1-(substituted benzenesulfonyl) Semicarbazides

Pengfei Li, Baolei Wang, Ning Ma, Suhua Wang, Zhengming Li

(State Key Laboratory of Elemento-Organic Chemistry, Institute of Elemento-Organic Chemistry, Nankai University, Tianjin, 300071)

Abstract Nucleophilic reaction of phenyl (substituted pyrimidin-2-yl)-carbamate with benzenesulfonyl hydrazine in the presence of DBU or DABCO afforded the title compounds 4-(substituted pyrimidin-2-yl)-1-(substituted benzenesulfonyl) semicarbazides **5**. Three organic salts **5as**~**5cs** formed from **5** and DBU or DABCO were also obtained. Structures of **5** and their salts were determined by elemental analysis and ^{1}H NMR spectra. The characteristic chemical shifts of ^{1}H NMR of such salts were discussed.

Key words sulfonyl semicarbazide; synthesis; organic salt

Microwave Syntheses of Disulfonate Salt-type Cleavable Surfactants with a 1, 3-Dioxane Ring

YUAN, Xian-You;
YANG, Nian-Fa*;
LUO, Huo-An;
LIU Yue-Jin

Chin. J. Org. Chem. 2005, 25 (9), 1049

$RCHO + C(CH_2OH)_4 \xrightarrow[MWI]{DMF, HPW}$ R(H)-1,3-dioxane with HOH_2C / CH_2OH $\xrightarrow[benzene, NaH]{\text{propane sultone } (SO_2)}$ R(H)-1,3-dioxane with $CH_2O(CH_2)_3SO_3Na$ / $CH_2O(CH_2)_3SO_3Na$

(a) $R = n-C_5H_{11}$; (b) $R = n-C_7H_{15}$;
(c) $R = n-C_9H_{19}$; (d) $R = n-C_{11}H_{23}$

Synthesis and Biological Activity of *N′* – (substituted pyrimidin – 2 – yl) – *N* – chrysanthemoylthiourea Derivatives

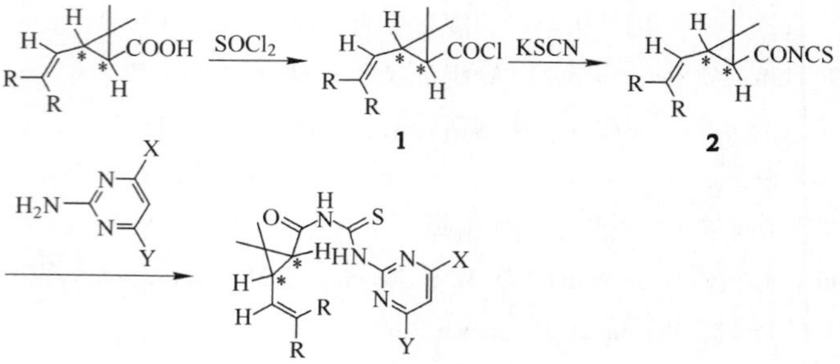

XU, Dong – Fang;
LI, Jing – Zhi;
GUO, Yan – Ling;
KE, Shao – Yong;
XUE, Si – Jia*
Chin. J. Org. Chem. 2005, 25(9), 1053

Five new *N′* – (substituted pyrimidin – 2 – yl) – *N* – chrysanthemoylthiourea derivatives and three new *N′* – (substituted pyrimidin – 2 – yl) – *N* – dichlorochrysanthemoylthiourea derivatives have been synthesized and characterized. The structures of these compounds were conformed by ^{1}H NMR, IR spectra and elemental analysis. The preliminary biological tests showed that some of these target compounds have excellent biological activity.

Synthesis and Characterization of New 4 – (Substituted pyrimidin – 2 – yl) – 1 – (substituted benzenesulfonyl) semicarbazides

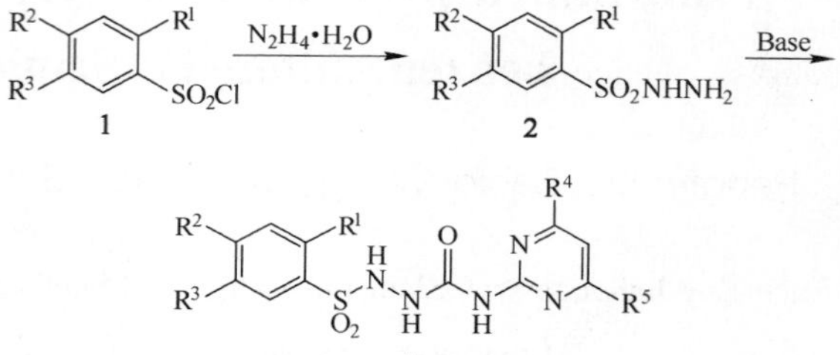

LI, Peng – Fei;
WANG, Bao – Lei;
MA, Ning;
WANG, Su – Hua;
LI, Zheng – Ming*
Chin. J. Org. Chem. 2005, 25(9), 1057

Nucleophilic reaction of phenyl (substituted pyrimidin – 2 – yl) – carbamate with benzenesulfonyl hydrazine in presence of DBU or DABCO afforded the title compounds 4 – (substituted pyrimidin – 2 – yl) – 1 – (substituted benzenesulfonyl) semicarbazides 5. Three organic salts **5as** ~ **5cs** formed from **5** and DBU or DABCO were also obtained. Structures of 5 and their salts were determined by elemental analysis and ^{1}H NMR spectra.

One – pot Synthesis of 3, 4 – Dihydropyrimidin – 2 (1*H*) – ones Catalyzed by Nontoxic Ionic Liquid

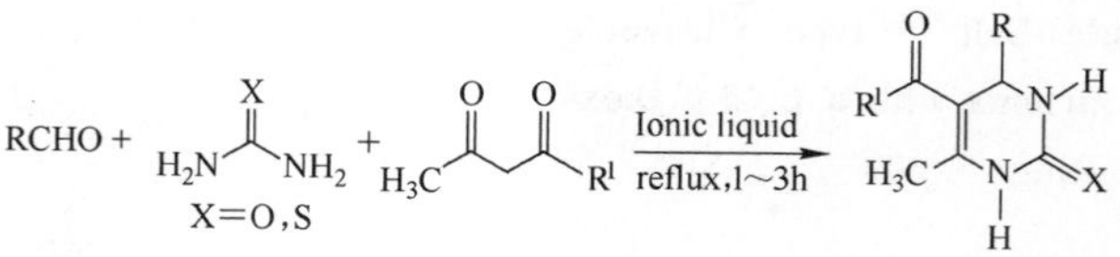

LI, Ming*; GUO, Wei – Si;
WEN, Li –Rong; ZHANG, Xiu – Li
Chin. J. Org. Chem. 2005, 25(9), 1062

3, 4 – Dihydropyrimidin – 2 (1*H*) – ones were synthesized via one – pot reaction of aldehydes, β – dicarbonyl compounds and urea or thiourea using nontoxic ionic liquid BMImSac as catalyst at room temperature. Compared to the classical Biginelli reaction, this is a simple, highly – yielding, time – saving and environmentally friendly method.

Synthesis and Fungicidal Activity of Novel 2 - Oxocycloalkylsulfonylureas*

Xinghai Li[1], Xinling Yang[1], Yun Ling[1], Zhijin Fan[2], Xiaomei Liang[1], Daoquan Wang[1], Fuheng Chen[1], Zhengming Li[2]

([1]Key Laboratory of Pesticide Chemistry and Application Technology, Department of Applied Chemistry, China Agricultural University, Beijing, 100094, China;
[2]The Research Institute of Elemento - Organic Chemistry, Nankai University, Tianjin, 300071, China)

Abstract A series of 2 - oxocycloalkylsulfonylureas (**2**) have been synthesized in a six - step, three - pot reaction sequence from readily available cyclododecanone, cycloheptanone, and cyclohexanone. Their structures were confirmed by IR, ^1H NMR, and elemental analysis. The bioassay indicated that some of them possess certain fungicidal activity against *Gibberella zeae* Petch. In general, compounds containing a 12 - membered ring (**2A**) are more active than those containing a 6 - or 7 - membered ring (**2B**, **2C**). In the series **2A**, the compounds in which R is a disubstituted phenyl or pyrimidyl showed better activity than those in which R is a monosubstituted phenyl or pyrimidyl, and aryl - substituted compounds have somewhat higher activity than those substituted by pyrimidyl. The further bioassay showed that the representative of **2A**, $\mathbf{2A_{15}}$, has good fungicidal activities against not only *G. zeae* Petch but also *Botrytis cinerea* Pers, *Colletotrichum orbiculare* Arx, *Pythium aphanidermatum* Fitzp, *Fusarium oxysporum* Schl. f. sp. Vasinfectum, etc.

Key words 2 - Oxocycloalkylsulfonylurea; synthesis; fungicidal activity

1 INTRODUCTION

In the searching for potential pesticides, more than 10 series of cyclododecanone derivatives have been synthesized and their biological activities evaluated. Among them, 2 - oxocyclododecylsulfonamides (**1**) were found to be active against *Gibberella zeae* Petch[1]. To our surprise, QSAR study (CoMFA) [2] showed that 2 - oxocyclodode - cylsulfonylureas (**2A**) may have higher predicted activity. It is known that some types of sulfonylureas are highly efficient chemical herbicides and hypoglycemic agents such as glimepiride[3]. However, very little of their fungicidal activities was studied. It was of interest for us to synthesize and evaluate their fungicidal activity. Meanwhile, to study the relationship between the size of the ring and their activity, 2 - oxocycloheptylsulfonamides (**2B**) and 2 - oxocyclohexylsulfonamides (**2C**) were also synthesized. The synthetic route of the title compounds (**2**) is shown in Scheme 1. Potassium 2 - oxocycloalkylsulfonates (**3**), prepared from readily available cycloalkanones by sulfonation with a sulfur trioxide - dioxane adduct and neutralization with potassium hydroxide, were allowed to react with oxalyl chloride to give corresponding sulfonyl chlorides, which were converted into sul-

* Reprinted from *Journal of Agriculture and Food Chemistry*, 2005, 53: 2202 - 2206. This work was supported by the National Key Project for Basic Research (2003CB114400), People's Republic of China, and the Foundation of State - Key Laboratory of Elemento - Organic Chemistry, Nankai University, People's Republic of China.

fonamides (**4**) using NH_3. The reaction of **4** with phenyl chloroformate and amines successively afforded title compounds 2 – oxocycloalkylsulfonylureas (**2**).

$SO_2NR^1R^2$ **1** $SO_2NHCONHR$ **2A**

1. $SO_3 \cdot$ dioxane 2. KOH/H_2O → SO_3K **3** → 3. $(COCl)_2$ 4. NH_3

SO_2NH_2 **4** → 5. ClCOOPh 6. $R\text{-}NH_2$ → $SO_2NHCONHR$ **2**

A, n=10; B, n=5; C, n=4

Scheme 1 Synthetic Route to the Title Compounds 2

Fungicidal activities of compounds **2** against some economically important fungus species were evaluated, and one of the compounds **1**, *N* – (4 – chlorophenyl) – 2 – oxocyclododecylsulfonamide (**1c**), was synthesized and used as a control in the bioassay.

2 MATERIALS AND METHODS

2.1 General

Infrared spectra were recorded in potassium bromide disks on a Shimadzu IR – 435 spectrophotometer; NMR spectra were recorded in $CDCl_3$, CD_3COCD_3, or DMSO – d_6 unless otherwise indicated with a Bruker DPX300 spectrometer, using TMS as internal standard; elemental analysis was performed by the analytical center at the Institute of Chemistry (Beijing), Chinese Academy of Science; melting points were measured on a Yanagimoto melting – point apparatus and are uncorrected. The solvents and reagents were used as received or were dried prior to use as needed.

2.2 Chemical Synthesis

2.2.1 N – (4 – Chlorophenyl) – 2 – oxocyclododecylsulfonamide (1c)

Compound **1c** was prepared according to the method given in ref 1.

2.2.2 Potassium 2 – Oxocyclododecylsulfonate (3A)

Compound **3A** was prepared according to the method given in ref 1.

2.2.3 Potassium 2 – Oxocycloheptylsulfonate (3B)

To a solution of cycloheptanone (11.5g, 0.10mol) in 1,2 – dichloroethane (40mL) at 5℃ under a nitrogen atmosphere was added sulfur trioxide – dioxane adduct (20.7g, 0.12mol) portion by portion within 15 min. The mixture was stirred for 3 h in an ice – water bath. Water (40mL) was poured into the mixture. The aqueous layer was separated, and the organic layer was extracted with water (10mL ×3). The combined aqueous layer was treated with $Ba(OH)_2 \cdot 8H_2O$ until no $BaSO_4$ precipitate could be observed, then filtered, and the filtrate was neutralized to pH 7 – 8 with potassium hydroxide. The aqueous solution was concentrated to dryness at reduced pressure. The resulting yellow solid was recrystallized in methanol to give 15g of white solid (66%): mp 268 – 270℃; ^1H NMR (D_2O) δ 1.07 – 1.39 (m, 3H), 1.75 – 1.98 (m, 4H), 2.22 – 2.31 (m, 1H), 2.38 – 2.45 (m, 1H), 2.72 – 2.81 (m, 1H), 3.80 (dd, 1H, $J_{\alpha,\beta1}$ = 11.5Hz, $J_{\alpha,\beta2}$ = 5.0Hz).

2.2.4 Potassium 2 – Oxocyclohexylsulfonate (**3C**)

The reaction was run similarly to that used to synthesize **3B**. After the reaction was completed, water was poured into the reaction mixture. The aqueous layer was separated, and the organic layer was extracted with water. The combined aqueous layer was treated with $Ba(OH)_2 \cdot 8H_2O$ until no $BaSO_4$ precipitate could be observed, then filtered, and the filtrate was neutralized to pH 7 – 8 with potassium hydroxide. Concentration of the aqueous solution at reduced pressure gave a yellow oil, which was cooled in an ice – water bath to give the first portion of yellow solid (D). To the filtrate was added 4 equivalent volumes of methanol. It was heated and then filtered to give the second portion of yellow solid (E), which was identified as disulfonated product [^1H NMR(D_2O) δ 1.87 – 1.95 (m, 2H), 2.08 – 2.17 (m, 2H), 2.21 – 2.37 (m, 2H), 3.98 – 4.03 (m, 2H)]. The filtrate was concentrated again at reduced pressure and then methanol added. After 24 h of standing, filtration gave the third portion of yellow solid (F). All of D and F were **3C** as identified by ^1H NMR, which could be recrystallized from methanol: yield, 56%; mp 235 – 238℃; ^1H NMR(D_2O) δ 1.52 – 2.03 (m, 5H), 2.22 – 2.46 (m, 3H), 3.76 (dd, 1H, $J_{\alpha,\beta1}$ = 6.3Hz, $J_{\alpha,\beta2}$ = 8.5Hz).

2.2.5 2 – Oxocyclododecylsulfonamide (**4A**)

To the slurry of **3A** (12g, 0.04mol) and DMF (0.1mL) in methylene chloride (60mL) was added oxalyl chloride (3.4mL, 0.04mol); the mixture was stirred under reflux for 1.5h. After the slurry had been cooled in an ice – water bath, it was filtered at reduced pressure before NH_3 (g) was introduced into the ice – cold filtrate. Ammonia was added until the pH rose to 7 – 8. The mixture was treated with water and extracted with methylene chloride. The combined organic layer was washed with water, dried over sodium sulfate, and evaporated to give a white solid, which was recrystallized from benzene to give 6.8g of **4A** (65%): mp 150 – 151℃; ^1H NMR($CDCl_3$) δ 1.22 – 1.42 (m, 14H), 1.75 – 1.81 (m, 2H,), 2.02 – 2.06 (m, 1H), 2.23 – 2.33 (m, 1H), 2.62 – 2.71 (m, 1H), 2.94 – 3.03 (m, 1H), 4.34 (dd, 1H, $J_{\alpha,\beta1}$ = 11.2Hz, $J_{\alpha,\beta2}$ = 3.1Hz), 4.67 (s, 2H); IR ν 3350, 3250, 2920, 2850, 1705, 1315, 1150cm^{-1}.

2.2.6 2 – Oxocycloheptylsulfonamide (**4B**)

The reaction was run similarly to that used to synthesize **4A**. A white solid was obtained, and it was recrystallized from benzene/petroleum ether to give **4B** in 61% yield: mp 75 – 76℃ [lit. (4) 76 – 78℃].

2.2.7 2 – *Oxocyclohexylsulfonamide* (**4C**)

The reaction was run similarly to that used to synthesize **4A**. The crude product was recrystallized from water to give **4C** in 52% yield: mp 114 - 116℃ [lit. [4] 118 - 119℃].

2.2.8 General Synthetic Procedure for 2 - Oxocycloalkylsulfonylureas (**2**)

To a solution of **4** (5.7mmol) and triethylamine (1.7mL, 12.0mmol) in acetonitrile (20mL) at 20℃ under a nitrogen atmosphere was added dropwise phenyl chloroformate (0.86mL, 6.9mmol) within 2 min. The mixture was further stirred at 20 - 25℃ for 30min, and to the suspension were added methanesulfonic acid (0.38mL, 6.0mmol) and then amine (6.9 mmol). The resulting mixture was stirred at 60℃ for 20min. After cooling, the title compounds were obtained by filtration. The physical and elemental data of the title compounds are listed in Table 1 and the ^{1}H NMR and IR data in Table 2.

Table 1 Physical and Elemental Data of Compounds 2

compd.	R	yield (%)	mp (℃)	elemental analysis (%)		
				C (calcd)	H (calcd)	N (calcd)
2A$_1$	4 - methyl - 2 - pyrimidyl	47	147 - 148	54.03 (54.52)	7.04 (7.12)	13.73 (14.13)
2A$_2$	4 - chloro - 6 - methoxy - 2 - pyrimidyl	58	134 - 136	47.89 (48.37)	6.34 (6.09)	12.76 (12.54)
2A$_3$	4,6 - dimethoxy - 2 - pyrimidyl	74	164 - 166	51.20 (51.57)	7.28 (6.83)	12.68 (12.66)
2A$_4$	4,6 - dimethyl - 2 - pyrimidyl	47	131 - 132	55.58 (55.59)	7.40 (7.37)	13.62 (13.65)
2A$_5$	4,6 - diethyl - 2 - pyrimidyl	67	136 - 138	53.32 (53.60)	7.46 (7.28)	11.47 (11.91)
2A$_6$	4,6 - dipropyl - 2 - pyrimidyl	58	120 - 122	55.23 (55.40)	7.88 (7.68)	11.10 (11.24)
2A$_7$	4,6 - diisopropyl - 2 - pyrimidyl	60	158 - 160	54.97 (55.40)	7.75 (7.68)	10.79 (11.24)
2A$_8$	*p* - tolyl	51	134 - 136	60.49 (60.89)	8.19 (7.66)	7.21 (7.10)
2A$_9$	*p* - methoxyphenyl	52	102 - 104	58.52 (58.51)	7.40 (7.37)	6.82 (6.82)
2A$_{10}$	*p* - chlorophenyl	56	148 - 149	54.80 (55.00)	6.71 (6.56)	6.86 (6.75)
2A$_{11}$	*o* - chlorophenyl	50	142 - 144	54.96 (55.00)	7.03 (6.56)	6.77 (6.75)
2A$_{12}$	5 - oxa - 6 - oxocyclohexadec - 1 - yl	36	150 - 152	61.71 (61.96)	9.64 (9.29)	5.53 (5.16)
2A$_{13}$	*p* - fluorophenyl	43	147 - 149	57.12 (57.27)	7.09 (6.83)	6.83 (7.03)
2A$_{14}$	3,4 - dichlorophenyl	63	163 - 164	50.70 (50.78)	5.83 (5.83)	6.16 (6.23)
2A$_{15}$	2,5 - dichlorophenyl	67	150 - 152	50.65 (50.78)	6.23 (5.83)	6.56 (6.23)
2A$_{16}$	*m* - tolyl	33	144 - 146	60.84 (60.89)	7.62 (7.66)	7.08 (7.10)
2A$_{17}$	3,4 - xylyl	58	148 - 150	61.75 (61.74)	8.00 (7.89)	6.90 (6.86)
2A$_{18}$	*m* - nitrophenyl	60	162 - 164	53.69 (53.63)	6.42 (6.40)	9.88 (9.88)
2A$_{19}$	*m* - chlorophenyl	54	146 - 148	55.46 (55.00)	6.47 (6.56)	6.58 (6.75)
2A$_{20}$	*p* - acetylphenyl	46	137 - 139	59.91 (59.69)	7.51 (7.16)	6.82 (6.63)
2B$_1$	2,5 - dichlorophenyl	57	152 - 153	44.28 (44.34)	4.27 (4.25)	7.28 (7.39)
2B$_2$	2,4 - xylyl	50	120 - 121	56.85 (56.78)	6.61 (6.55)	8.24 (8.28)
2B$_3$	4,6 - dimethoxy - 2 - pyrimidyl	62	147 - 148	45.20 (45.15)	5.49 (5.41)	15.06 (15.04)
2B$_4$	*p* - chlorophenyl	55	161 - 162	49.24 (48.77)	5.07 (4.97)	8.10 (8.12)
2B$_5$	*p* - tolyl	52	138 - 139	55.54 (55.54)	6.27 (6.21)	8.67 (8.64)
2B$_6$	3,4 - dichlorophenyl	43	150 - 151	44.40 (44.34)	4.29 (4.25)	7.28 (7.39)
2C$_1$	2,5 - dichlorophenyl	30	145 - 146	43.05 (42.75)	3.91 (3.86)	7.59 (7.67)
2C$_2$	2,4 - xylyl	35	140 - 141	55.85 (55.54)	6.27 (6.21)	8.66 (8.64)
2C$_3$	4,6 - dimethoxy - 2 - pyrimidyl	60	141 - 143	43.66 (43.57)	5.10 (5.06)	15.62 (15.63)

Table 1 (continued)

compd.	R	yield (%)	mp (℃)	elemental analysis (%)		
				C (calcd)	H (calcd)	N (calcd)
2C₄	4 - chlorophenyl	43	154 - 155	47.10 (47.20)	4.59 (4.57)	8.38 (8.47)
2C₅	*p* - tolyl	33	146 - 147	54.22 (54.18)	5.91 (5.85)	8.96 (9.03)
2C₆	3,4 - dichlorophenyl	36	161 - 162	42.99 (42.75)	3.89 (3.86)	7.53 (7.67)

Table 2 ^{1}H NMR and IR Data of Compounds 2

compd.	^{1}H NMR, δ	IR (ν, cm^{-1})
$2A_1$	1.20 - 1.40 (m, 14H), 1.62 - 1.69 (m, 2H), 1.86 - 2.05 (m, 2H), 2.53 (s, 3H), 2.81 - 2.98 (m, 2H), 4.88 (dd, 1H, $J_{\alpha,\beta1}$ = 11.9Hz, $J_{\alpha,\beta2}$ = 2.7Hz), 6.93 (d, 1H, J = 5.2Hz), 8.48 (d, 1H, J = 5.0Hz), 8.50 (s, 1H), 12.53 (s, 1H)	3400, 3150, 3080, 2920, 2860, 1720, 1600, 1340, 1155
$2A_2$	1.21 - 1.40 (m, 14H), 1.65 - 1.71 (m, 2H), 1.86 - 2.07 (m, 2H), 2.81 - 2.92 (m, 2H), 3.98 (s, 3H), 4.82 (dd, 1H, $J_{\alpha,\beta1}$ = 11.7Hz, $J_{\alpha,\beta2}$ = 2.9Hz), 6.50 (s, 1H), 7.60 (s, 1H), 12.53 (s, 1H)	3250, 3150, 3080, 2920, 2860, 1715, 1590, 1360, 1160
$2A_3$	1.21 - 1.42 (m, 14H), 1.63 - 1.85 (m, 2H), 1.99 - 2.44 (m, 2H), 2.77 - 2.97 (m, 2H), 3.91 (s, 6H), 4.85 (dd, 1H, $J_{\alpha,\beta1}$ = 11.6Hz, $J_{\alpha,\beta2}$ = 2.9Hz), 5.79 (s, 1H), 7.42 (s, 1H), 12.51 (s, 1H)	3200, 3150, 3030, 2920, 2860, 1720, 1610, 1360, 1150
$2A_4$	1.22 - 1.43 (m, 14H), 1.63 - 1.90 (m, 2H), 2.00 - 2.49 (m, 2H), 2.45 (s, 6H), 2.80 - 2.98 (m, 2H), 4.88 (dd, 1H, $J_{\alpha,\beta1}$ = 11.9Hz, $J_{\alpha,\beta2}$ = 2.9Hz), 6.78 (s, 1H), 7.97 (s, 1H), 12.40 (s, 1H)	3250, 3180, 2920, 2860, 1710, 1600, 1340, 1155
$2A_5$	1.33 - 1.41 (m, 20H), 1.66 - 1.87 (m, 2H), 1.98 - 2.44 (m, 2H), 2.77 - 2.98 (m, 2H), 4.29 (q, 4H, J = 7.1Hz), 4.87 (dd, 1H, $J_{\alpha,\beta1}$ = 11.7Hz, $J_{\alpha,\beta2}$ = 2.9Hz), 5.73 (s, 1H), 7.33 (s, 1H), 12.49 (s, 1H)	3350, 3220, 3150, 2920, 2860, 1720, 1700, 1610, 1340, 1160
$2A_6$	1.02 (t, 6H, J = 7.5Hz), 1.21 - 1.41 (m, 14H), 1.65 - 1.87 (m, 6H), 2.00 - 2.44 (m, 2H), 2.78 - 2.93 (m, 2H), 4.18 (t, 4H, J = 6.6Hz), 4.86 (dd, 1H, $J_{\alpha,\beta1}$ = 11.8Hz, $J_{\alpha,\beta2}$ = 2.9Hz), 5.75 (s, 1H), 7.32 (s, 1H), 12.51 (s, 1H)	3150, 3090, 2920, 2860, 1735, 1710, 1605, 1350, 1150
$2A_7$	1.20 - 1.41 (m, 26H), 1.66 - 1.89 (m, 2H), 1.97 - 2.44 (m, 2H), 2.72 - 2.99 (m, 2H), 4.89 (dd, 1H, $J_{\alpha,\beta1}$ = 11.8Hz, $J_{\alpha,\beta2}$ = 2.9Hz), 5.05 - 5.11 (m, 2H), 5.66 (s, 1H), 7.26 (s, 1H), 12.44 (s, 1H)	3240, 3180, 3060, 2920, 2860, 1720, 1700, 1610, 1360, 1155
$2A_8$	1.21 - 1.41 (m, 14H), 1.73 - 1.75 (m, 2H), 2.04 - 2.46 (m, 2H), 2.31 (s, 3H), 2.65 - 2.98 (m, 2H), 4.58 (dd, 1H, $J_{\alpha,\beta1}$ = 11.2Hz, $J_{\alpha,\beta2}$ = 2.9Hz), 7.12 (d, 2H, J = 8.3Hz), 7.27 (d, 2H, J = 8.3Hz), 8.19 (s, 1H), 8.48 (s, 1H)	3350, 3150, 3030, 2920, 2860, 1710, 1695, 1610, 1345, 1140
$2A_9$	1.21 - 1.40 (m, 14H), 1.74 - 1.76 (m, 2H), 2.00 - 2.44 (m, 2H), 2.65 - 2.98 (m, 2H), 3.79 (s, 3H), 4.57 (d, 1H, J = 9.4Hz), 6.82 - 6.87 (m, 2H), 7.25 - 7.30 (m, 2H), 8.12 (s, 1H), 8.69 (s, 1H)	3350, 3160, 3060, 2920, 2860, 1710, 1690, 1610, 1350, 1140
$2A_{10}$	1.24 - 1.43 (m, 14H), 1.72 - 1.76 (m, 2H), 2.03 - 2.42 (m, 2H), 2.63 - 2.98 (m, 2H), 4.55 (dd, 1H, $J_{\alpha,\beta1}$ = 11.1Hz, $J_{\alpha,\beta2}$ = 3.1Hz), 7.26 - 7.37 (m, 4H), 8.24 (s, 1H), 8.33 (s, 1H)	3350, 3210, 2920, 2850, 1725, 1705, 1600, 1335, 1150

Table 2 (continued)

compd.	^{1}H NMR, δ	IR (ν, cm^{-1})
$2A_{11}$	1.22 – 1.42 (m, 14H), 1.73 – 1.79 (m, 2H), 2.05 – 2.46 (m, 2H), 2.66 – 3.01 (m, 2H), 4.53 (dd, 1H, $J_{\alpha,\beta1}$ = 11.3Hz, $J_{\alpha,\beta2}$ = 3.2Hz), 7.05 – 7.11 (m, 1H), 7.28 – 7.31 (m, 1H), 7.37 – 7.40 (m, 1H), 8.13 – 8.16 (m, 1H), 7.95 (s, 1H), 8.88 (s, 1H)	3300, 3150, 3050, 2920, 2860, 1720, 1700, 1600, 1355, 1140
$2A_{12}$	1.23 – 1.77 (m, 38H), 1.97 – 2.01 (m, 1H), 2.31 – 2.40 (m, 3H), 2.71 – 2.75 (m, 1H), 2.86 – 2.94 (m, 1H), 3.86 (s, 1H), 4.08 – 4.21 (m, 2H), 4.41 (dd, 1H, $J_{\alpha,\beta1}$ = 11.2Hz, $J_{\alpha,\beta2}$ = 2.0Hz), 6.32 (d, 1H, J = 8.6Hz), 8.31 (s, 1H)	3300, 3250, 2920, 2860, 1730, 1705, 1690, 1530, 1340, 1155
$2A_{13}$	1.22 – 1.43 (m, 14H), 1.70 – 1.76 (m, 2H), 2.03 – 2.43 (m, 2H), 2.64 – 2.98 (m, 2H), 4.56 (dd, 1H, $J_{\alpha,\beta1}$ = 11.1Hz, $J_{\alpha,\beta2}$ = 3.2Hz), 6.99 – 7.06 (m, 2H), 7.33 – 7.38 (m, 2H), 8.27 (s, 1H), 8.37 (s, 1H)	3350, 3250, 3080, 2920, 2860, 1720, 1710, 1620, 1330, 1150
$2A_{14}$	1.21 – 1.41 (m, 14H), 1.65 – 1.89 (m, 2H), 1.99 – 2.36 (m, 2H), 2.65 – 3.03 (m, 2H), 5.04 (dd, 1H, $J_{\alpha,\beta1}$ = 11.7Hz, $J_{\alpha,\beta2}$ = 3.0Hz), 7.43 (dd, 1H, J = 8.8Hz, 2.5Hz), 7.53 (d, 1H, J = 8.8Hz), 7.86 (d, 1H, J = 2.5Hz), 8.66 (s, 1H), 9.70 (s, 1H)	3290, 3200, 2920, 2860, 1708, 1580, 1340, 1148
$2A_{15}$	1.22 – 1.42 (m, 14H), 1.73 – 1.80 (m, 2H), 2.02 – 2.46 (m, 2H), 2.65 – 3.02 (m, 2H), 4.54 (dd, 1H, $J_{\alpha,\beta1}$ = 11.2Hz, $J_{\alpha,\beta2}$ = 3.2Hz), 7.06 (dd, 1H, J = 8.6, 2.5Hz), 7.31 (d, 1H, J = 8.6Hz), 8.26 (d, 1H, J = 2.5Hz), 8.14 (s, 1H), 8.96 (s, 1H)	3300, 3150, 3080, 2920, 2860, 1715, 1695, 1595, 1360, 1140
$2A_{16}$	1.22 – 1.43 (m, 14H), 1.74 – 1.76 (m, 2H), 2.04 – 2.40 (m, 2H), 2.34 (s, 3H), 2.68 – 2.97 (m, 2H), 4.55 (dd, 1H, $J_{\alpha,\beta1}$ = 11.1Hz, $J_{\alpha,\beta2}$ = 2.8Hz), 6.95 – 6.98 (m, 1H), 7.21 – 7.22 (m, 3H), 8.10 (s, 1H), 8.24 (s, 1H)	3350, 3180, 2920, 2860, 1710, 1670, 1615, 1340, 1150
$2A_{17}$	1.21 – 1.42 (m, 14H), 1.70 – 1.77 (m, 2H), 1.99 – 2.07 (m, 1H), 2.20 (s, 3H), 2.29 (s, 3H), 2.33 – 2.45 (m, 1H), 2.66 – 2.97 (m, 2H), 4.55 (dd, 1H, $J_{\alpha,\beta1}$ = 11.4Hz, $J_{\alpha,\beta2}$ = 2.4Hz), 7.01 (d, 2H, J = 5.5Hz), 7.53 (s, 1H), 8.11 (s, 1H), 8.50 (s, 1H)	3350, 3150, 3050, 2920, 2860, 1710, 1695, 1605, 1345, 1140
$2A_{18}$	1.22 – 1.42 (m, 14H), 1.68 – 2.03 (m, 2H), 2.04 – 2.37 (m, 2H), 2.67 – 3.05 (m, 2H), 5.04 (dd, 1H, $J_{\alpha,\beta1}$ = 11.7Hz, $J_{\alpha,\beta2}$ = 3.0Hz), 7.61 – 7.66 (m, 1H), 7.83 – 7.87 (m, 1H), 7.95 – 7.99 (m, 1H), 8.54 – 8.56 (m, 1H), 8.89 (s, 1H), 9.80 (s, 1H)	3300, 3260, 2920, 2860, 1710, 1605, 1350, 1160
$2A_{19}$	1.22 – 1.42 (m, 14H), 1.67 – 2.03 (m, 2H), 2.04 – 2.36 (m, 2H), 2.66 – 3.04 (m, 2H), 5.05 (dd, 1H, $J_{\alpha,\beta1}$ = 11.7Hz, $J_{\alpha,\beta2}$ = 3.0Hz), 7.11 – 7.72 (m, 4H), 8.63 (s, 1H), 9.68 (s, 1H)	3350, 3250, 3080, 2920, 2860, 1720, 1710, 1600, 1335, 1150
$2A_{20}$	1.23 – 1.43 (m, 14H), 1.73 – 1.76 (m, 2H), 2.06 – 2.39 (m, 2H), 2.59 (s, 3H), 2.66 – 2.99 (m, 2H), 4.61 (dd, 1H, $J_{\alpha,\beta1}$ = 11.1Hz, $J_{\alpha,\beta2}$ = 3.1Hz), 7.54 (d, 2H, J = 8.7Hz), 7.95 (d, 2H, J = 8.7Hz), 8.39 (s, 1H), 8.56 (s, 1H)	3370, 3250, 3080, 2920, 2860, 1720, 1680, 1600, 1340, 1150
$2B_{1}$	1.46 – 1.53 (m, 3H), 1.89 – 1.93 (m, 2H), 2.03 – 2.12 (m, 2H), 2.50 – 2.58 (m, 2H), 2.76 – 2.86 (m, 1H), 4.60 (dd, 1H, $J_{\alpha,\beta1}$ = 11.2Hz, $J_{\alpha,\beta2}$ = 4.8Hz), 7.17 (dd, 1H, J = 8.6, 2.2Hz), 7.50 (d, 1H, J = 8.6Hz), 8.33 (d, 1H, J = 2.5Hz), 8.58 (s, 1H), 9.89 (s, 1H)	3200, 3150, 2800, 1700, 1680, 1580, 1345, 1150

Table 2 (continued)

compd.	^{1}H NMR, δ	IR (ν, cm^{-1})
2B$_2$	1.39 – 1.59 (m, 3H), 1.91 – 1.98 (m, 2H), 2.02 – 2.10 (m, 2H), 2.20 (s, 3H), 2.28 (s, 3H), 2.47 – 2.56 (m, 1H), 2.60 – 2.76 (m, 2H), 4.38 – 4.40 (d, 1H, J = 7.4Hz), 6.99 (s, 2H), 7.55 (s, 1H), 8.04 (s, 1H), 8.36 (s, 1H)	3320, 3200, 3080, 2920, 2880, 1700, 1665, 1600, 1330, 1140
2B$_3$	1.35 – 1.52 (m, 3H), 1.93 – 1.97 (m, 2H), 2.07 – 2.19 (m, 2H), 2.53 – 2.61 (m, 2H), 2.77 – 2.85 (m, 1H), 3.90 (s, 6H), 4.56 (dd, 1H, $J_{\alpha,\beta1}$ = 11.5Hz, $J_{\alpha,\beta2}$ = 4.7Hz), 5.78 (s, 1H), 7.51 (s, 1H), 12.48 (s, 1H)	3250, 3200, 3100, 3020, 2920, 2850, 1720, 1620, 1340, 1150
2B$_4$	1.27 – 1.45 (m, 3H), 1.79 – 2.03 (m, 4H), 2.29 – 2.37 (m, 1H), 2.46 – 2.52 (m, 1H), 2.67 – 2.76 (m, 1H), 4.56 (dd, 1H, $J_{\alpha,\beta1}$ = 11.2Hz, $J_{\alpha,\beta2}$ = 4.9Hz), 7.34 – 7.39 (m, 2H), 7.43 – 7.48 (m, 2H), 9.00 (s, 1H), 9.98 (s, 1H)	3380, 3200, 2920, 2880, 1710, 1600, 1320, 1140
2B$_5$	1.41 – 1.52 (m, 3H), 1.89 – 2.09 (m, 4H), 2.30 (s, 3H), 2.49 – 2.51 (m, 1H), 2.64 – 2.71 (m, 2H), 4.43 (dd, 1H, $J_{\alpha,\beta1}$ = 10.8Hz, $J_{\alpha,\beta2}$ = 3.9Hz), 7.10 (d, 2H, J = 8.3Hz), 7.27 (d, 2H, J = 8.1Hz), 8.14 (s, 1H), 8.34 (s, 1H)	3270, 3320, 2920, 2850, 1710, 1600, 1330, 1140
2B$_6$	1.24 – 1.45 (m, 3H), 1.79 – 2.03 (m, 4H), 2.29 – 2.38 (m, 1H), 2.46 – 2.52 (m, 1H), 2.67 – 2.76 (m, 1H), 4.56 (dd, 1H, $J_{\alpha,\beta1}$ = 11.2Hz, $J_{\alpha,\beta2}$ = 4.9Hz), 7.37 (dd, 1H, J = 8.9, 2.6Hz), 7.57 (d, 1H, J = 8.8Hz), 7.81 (d, 1H, J = 2.5Hz), 9.18 (s, 1H), 10.79 (s, 1H)	3350, 3200, 3110, 2920, 2850, 1710, 1610, 1580, 1370, 1130
2C$_1$	1.72 – 1.88 (m, 2H), 2.05 – 2.23 (m, 3H), 2.37 – 2.47 (m, 1H), 2.47 – 2.74 (m, 2H), 4.24 (dd, 1H, $J_{\alpha,\beta1}$ = 11.8Hz, $J_{\alpha,\beta2}$ = 5.4Hz), 7.04 (dd, 1H, J = 8.6, 2.4Hz), 7.30 (d, 1H, J = 8.6Hz), 8.26 (d, 1H, J = 2.4Hz), 8.04 (s, 1H), 8.98 (s, 1H)	3320, 3205, 2920, 2860, 1720, 1680, 1580, 1345, 1150
2C$_2$	1.82 – 1.88 (m, 2H), 1.96 – 2.06 (m, 3H), 2.21 – 2.29 (m, 7H), 2.48 – 2.54 (m, 2H), 4.63 (dd, 1H, $J_{\alpha,\beta1}$ = 9.3Hz, $J_{\alpha,\beta2}$ = 5.8Hz), 6.97 – 7.01 (m, 2H), 7.60 – 7.64 (m, 1H), 8.00 (s, 1H), 9.30 (s, 1H)	3380, 3300, 3160, 2920, 2860, 1705, 1680, 1600, 1355, 1160
2C$_3$	1.85 – 2.06 (m, 4H), 2.29 (m, 1H), 2.42 – 2.56 (m, 3H), 3.95 (s, 6H), 4.57 (dd, 1H, $J_{\alpha,\beta1}$ = 9.4Hz, $J_{\alpha,\beta2}$ = 5.9Hz), 5.86 (s, 1H), 9.42 (s, 1H), 12.49 (s, 1H)	3250, 3200, 3100, 3030, 2950, 2880, 1710, 1615, 1355, 1155
2C$_4$	1.82 – 2.08 (m, 4H), 2.23 – 2.28 (m, 1H), 2.48 – 2.55 (m, 3H), 4.65 (dd, 1H, $J_{\alpha,\beta1}$ = 9.5Hz, $J_{\alpha,\beta2}$ = 5.8Hz), 7.31 – 7.36 (m, 2H), 7.50 – 7.55 (m, 2H), 8.57 (s, 1H), 8.29 (s, 1H)	3360, 3240, 3080, 2950, 2880, 1710, 1600, 1320, 1140
2C$_5$	1.83 – 2.08 (m, 4H), 2.25 – 2.28 (m, 4H), 2.48 – 2.54 (m, 3H), 4.64 (dd, 1H, $J_{\alpha,\beta1}$ = 9.40Hz, $J_{\alpha,\beta2}$ = 5.9Hz), 7.10 – 7.14 (m, 2H), 7.34 – 7.38 (m, 2H), 8.37 (s, 1H), 9.15 (s, 1H)	3350, 3250, 2950, 2880, 1710, 1600, 1320, 1135
2C$_6$	1.68 – 1.96 (m, 4H), 2.11 – 2.16 (m, 1H), 2.33 – 2.58 (m, 3H), 4.64 (dd, 1H, $J_{\alpha,\beta1}$ = 8.8Hz, $J_{\alpha,\beta2}$ = 6.0Hz), 7.35 (dd, 1H, J = 8.8, 2.5Hz), 7.57 (d, 1H, J = 8.8Hz), 7.81 (d, 1H, J = 2.46Hz), 9.16 (s, 1H), 10.70 (s, 1H)	3360, 3200, 2920, 2860, 1715, 1695, 1530, 1320, 1140

2.3 Bioassay of Fungicidal Activities

2.3.1 Method

Fungicidal activities of the title compounds against *G. zeae* Petch, *B. cinerea* Pers, *C. orbiculare* Arx, *P. aphanidermatum* Fitzp, *F. oxysporum* Schl. f. sp. Vasinfectum, *R. solani* Kuhn, and *Verticillium dahliae* Kled were evaluated using the mycelium growth rate test [5]. The culture media, with known concentration of the test compounds, were obtained by mixing the solution of compounds **2** in acetone with potato dextrose agar (PDA), on which fungus cakes were placed. The blank test was made using acetone. The culture was carried out at 24 ± 0.5℃. Three replicates were performed. After the mycelia grew completely, the diameter of the mycelia was measured and the inhibition rate calculated according to the formula in which I is the inhibition rate, $\overline{D}_1$ is the average diameter of mycelia in the blank test, and $\overline{D}_0$ is the average diameter of mycelia in the presence of compounds **2**.

$$I = \frac{\overline{D}_1^2 - \overline{D}_0^2}{\overline{D}_1^2} \times 100\%$$

2.3.2 Fungicidal ActiVities of Compounds 2

The inhibition rate of compounds **2** against *G. zeae* Petch at 50μg/mL was determined first, and the results are given in Table 3. The most active compounds were **2A$_{15}$** and **2A$_{17}$**, and their inhibition rates against seven fungi were further determined at the concentrations of 100, 50, 25, 12.5, and 6.25μg/mL, respectively. EC_{50} and EC_{90} values were estimated using logit analysis[6]. The results were shown in Table 4. Compound **1c** and commercial fungicides carbendazim or procymidone were used as a control in the above bioassay.

Table 3 Inhibition Rate of Compounds 2 against *G. zeae*

compd.	inhibition rate (%)	compd.	inhibition rate (%)	compd.	inhibition rate (%)	compd.	inhibition rate (%)
2A$_1$	18.1	**2A$_{10}$**	49.0	**2A$_{19}$**	42.0	**2C$_1$**	26.5
2A$_2$	34.5	**2A$_{11}$**	49.0	**2A$_{20}$**	9.3	**2C$_2$**	26.5
2A$_3$	26.5	**2A$_{12}$**	26.1	**1c**	72.6	**2C$_3$**	18.1
2A$_4$	18.1	**2A$_{13}$**	18.1	**2B$_1$**	18.1	**2C$_4$**	18.1
2A$_5$	55.6	**2A$_{14}$**	55.6	**2B$_2$**	9.3	**2C$_5$**	18.1
2A$_6$	34.5	**2A$_{15}$**	81.6	**2B$_3$**	18.1	**2C$_6$**	18.1
2A$_7$	55.6	**2A$_{16}$**	34.5	**2B$_4$**	18.1		
2A$_8$	34.5	**2A$_{17}$**	77.3	**2B$_5$**	18.1		
2A$_9$	18.1	**2A$_{18}$**	18.1	**2B$_6$**	18.1		

3 RESULTS AND DISCUSSION

3.1 Synthesis

The preparation of 2-oxocyclododecylsulfonamides by reaction of cyclododecanone with phosphorus pentachloride and amines successively was reported in our previous paper[1]. In this paper we explored the use of oxalyl chloride as chlorinating agent for the preparation of 2-oxocycloalkylsulfonyl chloride and discovered that oxalyl chloride plus DMF as a catalyst gave a somewhat higher yield of 2 – oxocycloalkylsulfonyl chloride[7], which was converted into 2 – oxocycloalkylsulfonamides by ammoniation.

Irie et al[8] reported that Lewis acids, such as boron trifluoride – diethyl ether, catalyzed the synthesis of arylsulfonylureas by the reaction of arylsulfonamides with alkyl isocy – anates. However, no product can be obtained in the case of 2 – oxocycloalkylsulfonamides instead of arylsulfonamides, perhaps owing to its lower chemical reactivity. Finally, the title compounds were obtained in acceptable yield by the reaction of **4** with phenyl chloroformate and amines successively[9].

Table 4 Fungicidal Activity of Compounds $2A_{15}$ and $2A_{17}$ against Seven Fungus Species

fungus	compd.	regressioneq	r	EC_{50} (μg/mL)	EC_{90} (μg/mL)
Botrytis cinerea	$\mathbf{2A_{15}}$	$Y=3.20+2.02x$	0.9884	7.8	33.75
	$\mathbf{2A_{17}}$	$Y=2.81+2.23x$	0.9967	9.57	35.88
	1c	$Y=3.41+1.81x$	0.9559	7.62	39.05
	procymidone	$Y=3.46x+3.66$	0.9920	2.45	5.75
Colletotrichum orbiculare	$\mathbf{2A_{15}}$	$Y=2.86+1.89x$	0.9882	13.52	64.32
	$\mathbf{2A_{17}}$	$Y=1.45+2.39x$	0.9943	30.45	104.53
	$\mathbf{1c_1}$	$Y=2.57+2.14x$	0.9975	13.76	54.80
	carbendazim	$Y=1.81x+2.11$	0.9873	39.81	199.52
Pythium aphanidermatum	$\mathbf{2A_{15}}$	$Y=1.71+2.69x$	0.9910	16.66	49.88
	$\mathbf{2A_{17}}$	$Y=1.56+2.26x$	0.9900	33.53	123.84
	1c	$Y=1.09+2.42x$	0.9956	17.60	59.57
	carbendazim	$Y=2.08+2.24x$	0.9755	20.42	75.86
Fusarium oxysporum	$\mathbf{2A_{15}}$	$Y=1.24+2.91x$	0.9989	19.54	53.82
	$\mathbf{2A_{17}}$	$Y=2.10+2.10x$	0.9999	24.01	97.98
	1c	$Y=3.34+1.19x$	0.9992	24.60	293.1
	carbendazim	$Y=5.46+1.72x$	0.9436	0.54	2.98
Rhizoctonia solani	$\mathbf{2A_{15}}$	$Y=0.31+2.95x$	0.9973	38.93	105.80
	$\mathbf{2A_{17}}$	$Y=1.23+1.84x$	1.0000	111.79	556.10
	1c	$Y=3.84+1.39x$	0.9995	6.82	57.00
	carbendazim	$Y=4.40+4.16x$	0.9737	1.41	2.82
Verticillium dahliae	$\mathbf{2A_{15}}$	$Y=2.87+1.58x$	0.9937	22.03	142.17
	$\mathbf{2A_{17}}$	$Y=3.31+1.11x$	0.9995	33.69	483.93
	1c	$Y=3.24+1.75x$	0.9903	10.21	55.28
	carbendazim	$Y=2.72+1.80x$	0.9580	18.20	93.33
Gibberella zeae	$\mathbf{2A_{15}}$	$Y=2.37+1.35x$	0.9934	31.47	169.20
	$\mathbf{2A_{17}}$	$Y=1.88+2.00x$	0.9963	39.01	176.08
	1c	$Y=3.34+1.13x$	0.9990	30.05	413.54
	carbendazim	$Y=3.65x+3.93$	0.9767	1.95	4.46

3.2 Biological Assay

As shown in Table 3, title compounds **2** exhibited some fungicidal activity against *G. zeae*. Among them, compounds containing a 12 – membered ring (**2A**) are more active than those containing a 6 - or 7 - membered ring (**2B**,**2C**), which indicated that 2 - oxocyclododecyl may be an active group showing pesticidal activities and merits our attention in the research and development of novel pesticides. On the other hand, for the structure - activity of **2A**, it can be seen in Table 3 that the ureas in which R is a disubstituted phenyl or pyrimidyl showed better activity than those in which R is a monosubstituted phenyl or pyrimidyl, and aryl - substituted ureas have somewhat higher activity than those substituted by pyrimidyl. Further study on their QSAR has been planned.

Precise bioassay (Table 4) showed that among compounds **2A**, $\mathbf{2A_{15}}$ has better fungicidal activity against *F. oxysporum* and *G. zeae* than **1c**, almost the same fungicidal activities against *B. cinerea*, *C. orbiculare*, and *P. aphanidermatum* as **1c**, and lower activity against *R. solani* and *V. dahliae* than **1c**; these results do not fully agree with the prediction of CoMFA. The result indicates that the process should be further perfected for using CoMFA to predict the activities of pesticides. In addition, although the activity of $\mathbf{2A_{15}}$ against most of the seven fungal species is lower than that of the commercial fungicides carbendazim or procymidone, the activity against *C. orbiculare* and *P. aphanidermatum* is better than that of carbendazim. All of the results in this paper will be very useful for later research.

ACKNOWLEDGMENT

We thank Associate Professor Qi Shuhua and laboratory assistant Liu Xiufeng for help with biological assay.

LITERATURE CITED

[1] Wang, X. - p.; Wang, D. - q. Synthesis and antifungal activity of N - substituted - α - oxocyclododecylsulphonamides. *Chem. J. Chin. Univ.* 1997, 18 (6), 889 - 893.

[2] Xie, G. - r.; Wang, X. - p.; Wang, D. - q.; Su, Y.; Zhou, J. - ju. QSAR study of α - oxocyclododecylsulphonamides by CoMFA. *Chin. J. Pestic. Sci.* 1999, 1 (2), 17 - 24.

[3] Langtry, H. D.; Balfour, J. A. *Glimepiride*, a review of its use in the management of type 2 diabetes mellitus. *Drugs* 1998, 55(4), 563 - 584.

[4] Bender, A.; Güntner, D.; Willms, L.; Wingen, R. Eine einfache synthese von 2 - oxoalkansulfon - amiden. *Synthesis* 1985, 66 - 70.

[5] Chen, N. - c. *Bioassay of Pesticides*; Beijing Agricultural University Press: Beijing, China, 1991; pp 161 - 162.

[6] Berkson, J. A statistically precise and relatively simple method of estimating the bioassay with quantal response, based on the logistic function. *J. Am. Stat. Assoc.* 1953, 48, 565 - 599.

[7] Fujita, S. A convenient preparation of arenesulfonyl chlorides from the sodium sulfonates and phosphoryl chloride/sulfolane. *Synthesis* 1982, 423 - 424.

[8] Irie, H.; Nishimura, M.; Yoshida, M. Lewis acid - catalyzed preparation of carbamates and sulphonylureas. Application to the determination of enantiomeric purity of chiral alcohols. *J. Chem. Soc.*, *Perkin Trans.* 1 1989, 1209 - 1210.

[9] Hideki, M.; Yasuo, I. Process of producing sulfonylurea derivatives as herbicides. Eur. Patent Appl. EP 0305939, 1989.

Synthesis, Crystal Structure and Herbicidal Activity of Mimics of Intermediates of the KARI Reaction*

Baolei Wang[1], Ronald G Duggleby[2], Zhengming Li[1], Jianguo Wang[1], Yonghong Li[1], Suhua Wang[1], Haibin Song[1]

([1]Elemento - Organic Chemistry Institute, State - Key Laboratory of Elemento - Organic Chemistry, Nankai University, Tianjin, 300071, China;
[2]Department of Biochemistry & Molecular Biology, University of Queensland, Brisbane QLD 4072, Australia)

Abstract Two mimics of the intermediate in the reaction catalyzed by ketol - acid reductoisomerase (KARI) were synthesized. Their structures were established on the basis of elemental analyses, IR, ^{1}H NMR and GC/mass detector. The crystal structure of compound 2 was found to be a substituted dioxane, formed by the condensation of two molecules. The two compounds showed some herbicidal activity on the basis of tests using rape root and barnyard grass growth inhibition. However, the herbicidal effect was weaker in greenhouse tests.

Key words KARI; metabolite mimics; crystal structure; herbicidal activity

1 INTRODUCTION

Plants have a more extensive biosynthetic capacity than animals, because the latter rely on their diet to supply many biochemical compounds and their precursors. Because of this difference, plants contain numerous enzymes that are potential targets for herbicides. The enzymes involved in the biosynthesis of the branched - chain amino acids leucine, isoleucine and valine are an example of such a pathway. Valine and isoleucine are synthesized in a parallel set of four reactions while leucine synthesis is an extension of the valine pathway.

The first successful herbicides to target this pathway were the sulfonylureas[1] and the imidazolinones,[2] both of which inhibit the first enzyme, acetohydroxyacid synthase (AHAS). Since the discovery of these herbicides a variety of other compounds targeting this enzyme have been found.[3] The success of these herbicides has stimulated research into inhibitors of other enzymes in the pathway, including two of those in the leucine branch[4,5] and the second enzyme in the common pathway, ketol - acid reductoisomerase (KARI). The reaction catalyzed by KARI is shown in Fig. 1.[6]

The KARI reaction proceeds in two steps.[7,8] First, there is a magnesium - ion - dependent isomerization reaction that consists of an alkyl migration between carbon C2 and carbon C3 of the substrate and gives a methylhydroxyketol - acid. Second, the intermediate is transformed by an NADPH - dependent reduction of the ketone moiety to give the final product, and the reac-

* Reprinted from *Pest Management Science*, 2005, 61: 407 - 412. Contract/grant sponsor: China 973 Program; contract/grant number: 2003 CB114406. Contract/grant sponsor: Australian Research Council, Linkage - International Program; contract/grant number: LX0349233. Contract/grant sponsor: Tianjin S&T Commission Project; contract/grant number: 033803411.

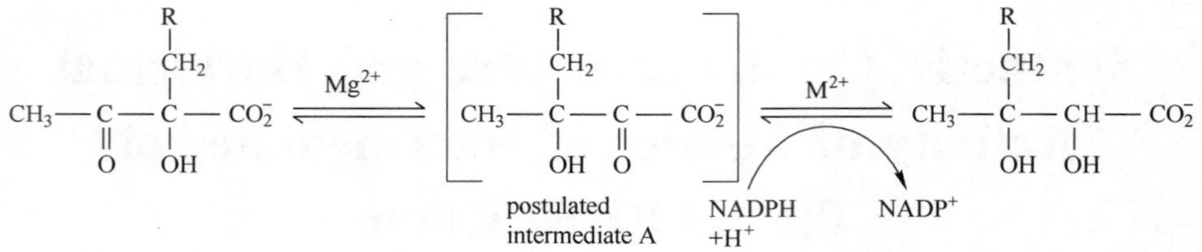

Fig. 1 Reaction catalyzed by KARI

tion requires a divalent metal ion, such as Mg^{2+}, Mn^{2+} or Co^{2+}.

Inhibitors of KARI have been synthesized and tested as herbicides. HOE 704[9] and IpO-HA[10] (Fig. 2) are potent competitive inhibitors of the enzyme *in vitro*, but their activity as herbicides is weak. Dumas *et al*[11] suggested that this is due to their slow binding to the enzyme rather than KARI being an intrinsically poor herbicide target. Accumulation of the substrate *in vivo* would reverse the inhibition faster than it could develop, a phenomenon that has been described as "metabolic resistance".[12]

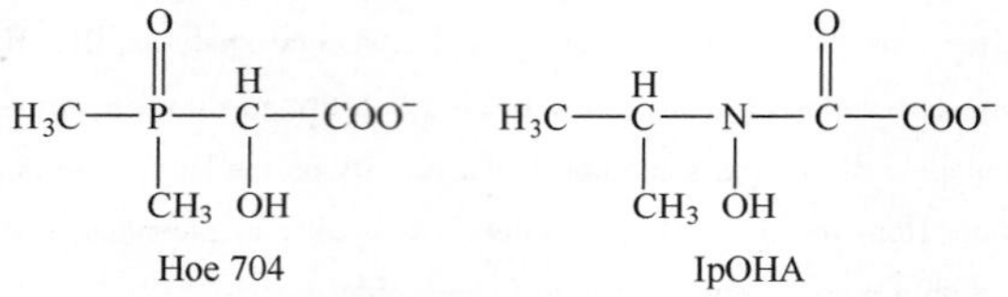

Fig. 2 Structures of Hoe 704 and IpOHA

New inhibitors of KARI remain as a potential source of novel herbicidal compounds. With this in mind, we designed and synthesized two compounds (Fig. 3; **1** and **2**) that mimic the structure of the β - hydroxy - α - keto - acid intermediate **A** shown as the bracketed structure in Fig. 1. We have characterized their structures and assayed their herbicidal activity. The route of synthesis of the compounds is shown in Fig. 3.

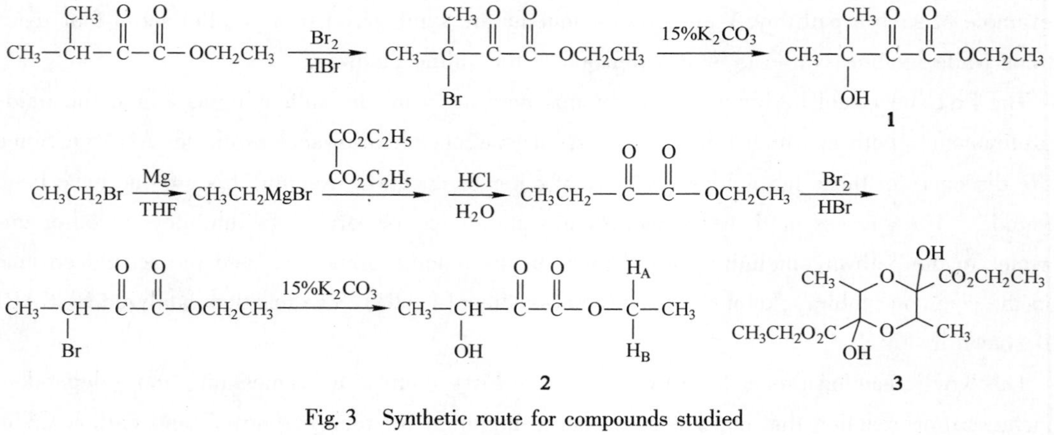

Fig. 3 Synthetic route for compounds studied

2 EXPERIMENTAL

2.1 Methods and materials

Melting points were determined using a Yanaco MP - 241 apparatus and are uncorrected. Infrared spectra were recorded on a Shimadzu IR - 435 spectrophotometer as thin films or potassium bromide tablets. ^1H NMR spectra were measured on a Bruker AC – P500 instrument (300MHz) using tetramethylsilane as an internal standard and deuterochloroform as solvent. Mass spectra were recorded on a Hewlett – Packard G1800A GC/mass detector instrument. Elemental analyses were performed on a Yanaco MT – 3CHN elemental analyzer. The single crystal structure of compound **3** was determined on a Bruker SMART 1000 CCD diffractometer. Ethyl 3 – methyl – 2 – oxobutanoate was purchased from Aldrich.

2. 2 Syntheses

2. 2. 1 Ethyl 3 – methyl – 3 – hydroxy – 2 – oxobutanoate(1)

The method is based on that described previously. [8] Ethyl 3 – methyl – 2 – oxobutanoate (3. 60g, 0. 025mol) was placed in a 50 – mL three – necked flask and a small amount of anhydrous hydrogen bromide passed through. The system was heated to 55℃ and then treated drop – wise with 1. 25mL(0. 025mol) of bromine while stirring. The red colour following the addition of one drop of bromine was allowed to dissipate before the next drop was added. After stirring at 55 – 60℃ for 2h, during which time large amounts of hydrogen bromide gas were released, the mixture was cooled to room temperature to give a pale orange solution. The resulting crude ethyl 3 – bromo – 3 – methyl – 2 – oxobutyrate was treated drop – wise with 3. 45g(0. 025mol) of potassium carbonate in a 150g litre^{-1} aqueous solution and reacted for 1. 5h at 25℃. The aqueous mixture was saturated with sodium chloride and extracted with ethyl acetate(5 × 15ml). The organic layer was dried(anhydrous sodium sulfate) and the ethyl acetate removed under reduced pressure. The residual clear oil was distilled to give 3. 10g of compound 1 as a pale yellow liquid, bp 108 – 109℃ (27 mm Hg); ^1H NMR: δ: 4. 33 – 4. 40(q, J = 7. 2Hz, 2H, CH_2), 3. 13(s, 1H, OH), 1. 52(s, 6H, 2CH_3), 1. 37 – 1. 41(t, J = 7. 2Hz, 3H, CH_3); IR, ν: 3514(O—H), 2986, 2940(CH_3, CH_2), 1732(C ═O); MS, *m/z*: 160(M^+); Anal: calcd for $C_7H_{12}O_4$; C 52. 49, H 7. 55; found C 52. 41, H 7. 50. The total yield of the two steps was 77. 5%.

2. 2. 2 Ethyl 3 – hydroxy – 2 – oxobutanoate(2)

Ethyl 2 – oxobutanoate(5g, 0. 038mol), prepared according to the literature[13] was placed in a 50 – mL three – necked flask and a small amount of anhydrous hydrogen bromide passed through. The system was heated to 40℃ and then treated drop – wise with 2mL(0. 04mole) of bromine while stirring. The red colour following the addition of one drop of bromine was allowed to dissipate before the next drop was added. After stirring at 40 – 45℃ for 2h, the mixture was cooled to room temperature to give a pale orange solution. The resulting crude ethyl 3 – bromo – 2 – oxobutyrate was treated drop – wise with a 15% aqueous solution containing 5. 52g(0. 04 mole) of potassium carbonate as a 150g litre^{-1} aqueous solution and reacted for 1. 5h at 35℃. The aqueous mixture was saturated with sodium chloride and extracted with ethyl acetate(5 × 20ml). There was white layer between the organic layer and the water layer and this was combined with the ethyl acetate layer. After standing, a white solid appeared and was removed. The filtrate was then dried(anhydrous sodium sulfate) and the ethyl acetate removed under reduced pressure to give compound **2**. This material combined with the solid separated earlier had a total weight of 2. 4g, yield 43%; mp 127 – 129℃; ^1H NMR; δ: 4. 47 – 4. 68(q, J = 6. 6Hz, 1H,

CH), 4.20 – 4.31 (dq, J_{AB} = 10.8Hz, J = 7.0Hz, 1H, CH—H_A), 4.36 – 4.47 (dq, J_{AB} = 10.8Hz, J = 7.0Hz, 1H, CH—H_B), 4.01 (s, 1H, OH), 1.32 – 1.36 (t, J = 7.0Hz, 3H, CH_3), 1.09 – 1.11 (d, J = 6.6Hz, 3H, CH_3); IR, ν: 3486 (OH), 2988, 2937 (CH_3, CH_2), 1735 (C = O); MS, m/z: 146 (M^+); Anal: calcd for $C_6H_{10}O_4$; C 49.31, H 6.90; found C 49.19, H 6.95.

2.3 Crystal structure determination

Compound **2** was dissolved in hot alcohol and diethyl ether, and the resulting colourless solution allowed to stand in air at room temperature to give single crystals of **3**. A colourless single crystal of **3** suitable for X – ray diffraction with dimensions of 0.24 × 0.24 × 0.18mm wasmounted on a Bruker SMART 1000 CCD diffractometer with Mo – Kα radiation (λ = 0.71073 Å) for data collection. A total of 2001 reflections were collected in the range of 2.57 < θ < 26.44 by the $\omega - 2\theta$ scan technique at 293 (2) K, of which 1413 were independent with R_{int} = 0.0188. The empirical absorption correction was applied by the SADABS program. The structure was solved by direct methods (SHELXS – 97) and refined by the full – matrix least – squares techniques on F^2. Most of the nonhydrogen atoms were located from an E – map, and the others, except the hydrogen atoms, were determined with successive difference Fourier syntheses. The compound crystallized in space group $P-1$ of the triclinic system with cell parameters: a = 6.733(2) Å, b = 6.980(3) Å, c = 8.078(3) Å, α = 85.684(6)°, β = 78.637(6)°, γ = 69.305(6)°, V = 348.2(2) $Å^3$, Z = 1, D_c = 1.394g · cm^{-3}, μ = 0.118mm^{-1} and $F(000)$ = 156. The final refinement converged at R = 0.0373, wR = 0.0884 for 1045 observed reflections with $I > 2\sigma(I)$, where $W = 1/[\sigma^2(F_0^2) + (0.0451P)^2 + 0.04P]$ with $P = (\max(F_0^2, 0) + 2F_C^2)/3$, S = 1.021, $(\Delta/\delta)_{max}$ = 0.000, $(\Delta\rho)_{max}$ = 0.211e/$Å^3$ and $(\Delta\rho)_{min}$ = –0.165e/$Å^3$.

2.4 Herbicidal activity tests

2.4.1 Inhibition of the root – growth of rape (Brassica campestris L)

The compounds to be tested were made into emulsions to aid dissolution. Rape seeds were soaked in distilled water for 4h before being placed on a filter paper in a 6 – cm Petri plate, to which 2 ml of inhibitor solution had been added in advance. Usually, 15 seeds were used on each plate. The plate was placed in a dark room and allowed to germinate for 65 h at 28 (± 1) ℃. The lengths of 10 rape roots selected from each plate were measured and the means were calculated. The percentage inhibition was calculated relative to controls using distilled water instead of the inhibitor solution.

2.4.2 Inhibition of the seedling growth of barnyard grass (*Echinochloa crus – galli*(L) *Beauv*)

The compounds to be evaluated were made into emulsions to aid dissolution. Ten *E crus – galli* seeds were placed into a 50 – mL cup covered with a layer of glass beads and a piece of filter paper at the bottom, to which 5 mL of inhibitor solution had been added in advance. The cup was placed in a bright room and the seeds allowed to germinate for 65h at 28 (± 1) ℃. The heights of the above – ground parts of the seedlings in each cup were measured and the means calculated. The percentage inhibition was calculated relative to controls using distilled water instead of the inhibitor solution.

2.4.3 Glasshouse tests

2.4.3.1 Pre - emergence

Sandy clay(100g) in a plastic box(11 × 7.5 × 6cm) was wetted with water. Sprouting seeds (15) of the weed under test were planted in fine earth(0.6cm depth) in the glasshouse and sprayed with the test compound dissolved in a suitable solvent at 1500g · ha^{-1}.

2.4.3.2 Post - emergence

Seedlings(one leaf and one stem) of the weed were sprayed with the test compounds at the same rate as used for the preemergence test.

For both methods, the fresh weights were determined 15 days later, and the percentage inhibition relative to water - sprayed controls was calculated.

3 RESULTS AND DISCUSSION

3.1 Synthesis

The ethyl 3 - hydroxy - 2 - oxo - (substituted) - butanoate products can be synthesized by bromination and hydrolysis of the corresponding α - keto esters. Several methods reported for the synthesis of α - keto esters are based on the alcoholysis of acyl cyanides,[14] the addition of alkyl lithium reagents to triethoxyacetonitrile[15] and the peracid oxidation of diazoesters.[16] Weinstock *et al*[13] reported a one - pot synthesis of α - keto esters using diethyl oxalate and Grignard reagent. We preferred Weinstock's method to prepare ethyl 2 - oxobutanoate for its less expensive reagents and greater convenience.

3.2 Spectroscopic properties

The IR spectra of the compounds tested shows absorption bands at 3500cm^{-1} originating from the stretching vibration of O—H. The strong band at 1735cm^{-1} can be assigned to the C=O stretching vibration. In the ^1H NMR spectrum of compound **2**, the methylene proton of ethoxyl was split into two octets of peaks instead of quadruple peaks by coupling with the adjacent methyl. However, the spectrum was normal for compound **1**. We suggest that the asymmetry of the molecular structure of **2** and the surrounding influence on CH_2 leads to the splitting of H_A and H_B. These protons are on the same carbon atom, but may be affected differently by the chiral carbon atom in the molecule. The two protons are split into double peaks that are affected by each other with $J = 10.8Hz$.

3.3 Crystal structure

Because of the unusual ^1H NMR spectrum of **2**, we investigated this compound further by determining its crystal structure. This analysis revealed that compound **3** was present. Possibly this is formed after standing in solvent for several weeks by condensation of two molecules of **2** (Fig. 4a). The β - hydroxy and α - carbonyl of one molecule of **2** reacted simultaneously with the corresponding α - carbonyl and β - hydroxy of another molecule of **2** to give two hemiketals, thus forming a dioxane structure. The similar dimerization of glycolaldehyde is well established. In the dioxane six - membered ring, the ethoxycarbonyl and methyl are in the a - bond position of the chair conformation and the hydroxyls are in the e – bond position. From the cell packing diagram (Fig. 4b) of compound **3**, it can be seen that there are intermolecular hydrogen bonds between H(3) and O(2) of two molecules, with bond lengths of 2.128Å for H(3) – O(2) and 2.278Å for O(2) – H(3). An eight – membered ring is formed through hydrogen bonds from the hydroxyl and ethoxycarbonyl of two dioxane structures. All atomic coordinates and equivalent isotropic displacement parameters are listed in Table 1, and selected bond lengths and bond angles are shown in Table 2.

Table 1 Atomic coordinates ($\times 10^4$) and equivalent isotropic displacement parameters ($Å^2 \times 10^3$) of compound 3

Atom	X	Y	Z	U_{eq} ①
O(1)	2860(2)	859(2)	1784(1)	39(1)
O(2)	1734(2)	2768(2)	4101(1)	50(1)
O(3)	1353(2)	5603(2)	2691(1)	38(1)
O(4)	1751(2)	5032(2)	641(1)	33(1)
C(1)	3134(4)	-2074(3)	3658(3)	59(1)
C(2)	4311(3)	-828(2)	2631(2)	42(1)
C(3)	1697(2)	2514(2)	2662(2)	31(1)
C(4)	267(2)	4177(2)	1625(2)	30(1)
C(5)	752(2)	6657(2)	-454(2)	31(1)
C(6)	2465(3)	7469(3)	-1357(2)	46(1)

①$U_{eq} = (1/3) \sum_i \sum_j U_{ij} \alpha_i * \alpha_j * \alpha_i \alpha_j$.

Table 2 Selected bond lengths (Å) and bond angles (°) for compound 3①

Bond length		Bond angle	
O(1) – C(3)	1.3104(18)	C(3) – O(1) – C(2)	117.59(12)
O(1) – C(2)	1.4627(18)	C(4) – O(4) – C(5)	112.73(11)
O(2) – C(3)	1.1953(18)	O(1) – C(2) – C(1)	111.15(15)
O(3) – C(4)	1.3832(17)	O(2) – C(3) – O(1)	125.33(14)
O(3) – H(3)	0.8200	O(2) – C(3) – C(4)	122.12(13)
O(4) – C(4)	1.4226(18)	O(1) – C(3) – C(4)	112.53(12)
O(4) – C(5)	1.4376(17)	O(3) – C(4) – O(4)	112.08(12)
C(1) – C(2)	1.481(3)	O(3) – C(4) – C(5)#1	108.92(13)
C(3) – C(4)	1.535(2)	O(4) – C(4) – C(5)#1	109.31(12)
C(4) – C(5)#	11.525(2)	O(3) – C(4) – C(3)	110.04(12)
C(5) – C(6)	1.501(2)	O(4) – C(4) – C(3)	102.75(12)
C(5) – C(4)#	11.525(2)	C(5)#1 – C(4) – C(3)	113.69(12)
		O(4) – C(5) – C(6)	107.13(13)
		O(4) – C(5) – C(4)#1	108.77(12)
		C(6) – C(5) – C(4)#1	113.99(13)

①Symmetry code: #1 – x, – y + 1, – z.

3.4 Herbicidal activity of compounds

Compound **1** is the ethyl ester of the normal intermediate in valine and leucine synthesis. Analysis of the three – dimensional structure of spinach KARI[17] suggests that this compound could fit into the active site of the enzyme with the ethyl group protruding towards the solvent. The smaller compound **2** could also fit into the active site. In this way they could probably inhibit KARI and possess herbicidal activity. Consequently, the herbicidal activities of compounds **1** and **2** were tested and the results are shown in Table 3. Both compounds showed some herbicidal activity in the rape root and barnyard grass cup tests and compound **1** showed 82.8% inhibition of rape root growth at a concentration of 100μg · mL. In addition, the herbicidal activity of these compounds was bioassayed in the glasshouse on five herbs representative of monocotyledonous and dicotyledonous plants. The ethyl esters were further hydrolyzed into the carboxylic form (compounds **4** and **5** in Fig. 5) using aqueous potassium hydroxide[8] and their herbicidal activities were also tested. The results of the greenhouse tests are shown in Table 4.

Table 3 Herbicidal activity of compounds (% inhibition)

Compound	Rape root test			Barnyardgrass cup test	
	$100\mu g \cdot mL^{-1}$	$10\mu g \cdot mL^{-1}$	$1\mu g \cdot mL^{-1}$	$100\mu g \cdot mL^{-1}$	$10\mu g \cdot mL^{-1}$
1	82.8	23.6	—	34.3	13.9
2	9.1	2.6	—	24.4	12.8
Chlorsulfuron	—	—	64.2	—	—
Metsulfuron – methyl	—	—	81.0	—	—

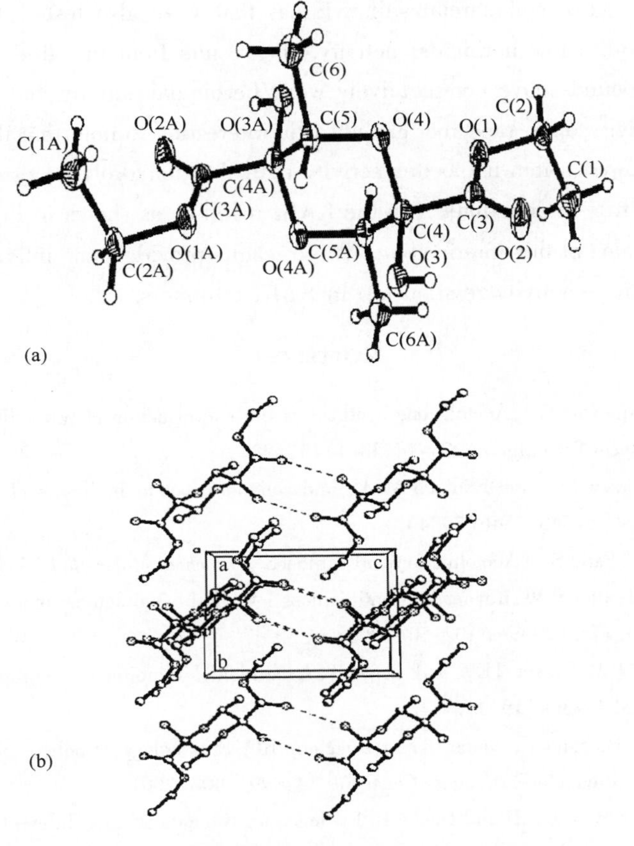

Fig. 4 (a) Molecular structure of compound **3** and (b) its packing diagram in the unit cell

$CH_3-C(CH_3)(OH)-C(=O)-COO^-$ (4) $CH_3-CH(OH)-C(=O)-COO^-$ (5)

Fig. 5 Structures of compounds **4** and **5**

Table 4 Herbicidal activity of compounds(% inhibition)①

Compound②	Echinochloa crus – galli		Digitaria adscendens		Brassica campestris		Amaranthus retroflexus		Lucerne	
	Post	Pre	Post	Pre	Post	Pre	Post	Pre	Post	Pre
1	14. 9	0	27. 9	0	0	0	17. 9	0	6. 8	0
2	9. 4	11. 7	21. 3	14. 5	0	0	0	0	0	6. 7
4	18. 2	5. 1	42. 6	17. 4	25. 8	0	31. 1	14. 3	34. 6	10. 0
5	15. 6	3. 0	25. 2	0	5. 9	0	11. 5	0	13. 6	13. 3
Chlorsulfuron	100	68. 0	30. 0	0	100	98. 0	100	92. 6	100	81. 0
Metsulfuron – methyl	100	88. 2	100	89. 0	100	98. 1	100	81. 0	100	92. 0

①Post: post – emergence; Pre: pre – emergence.

②Rates: **1**, **2**, **4**, **5** 1500g ha^{-1}, chlorsulfuron 30g · ha^{-1}, metsulfuron – methyl 15g · ha^{-1}.

In comparison with two commercial sulfonylureas that were also tested, the compounds described here have quite low herbicidal activity. The results from the greenhouse tests showed that all of our compounds have comparatively weak herbicidal activity, but compound **1** shows clear inhibitory activity upon rape root growth. It is interesting to note that the hydrolyzed compounds **4** and **5** showed somewhat better activity than the unhydrolyzed esters. Compound 4 is supposed to be the true intermediate A in the KARI reaction, as shown in Fig. 1, and compound **5** is a simplified analog of this intermediate. These results provide some interesting clues for further study of structure – activity relationship in KARI inhibitors.

References

[1] Chaleff R S and Mauvais C J, Acetolactate synthase is the site of action of two sulfonylurea herbicides in higher plants. *Science(Washington)* 224:1443 – 1445(1984).

[2] Shaner D L, Anderson P C and Stidham M A, Imidazolinones: potent inhibitors of acetohydroxyacid synthase. *Plant Physiol* 76:545 – 546(1984).

[3] Duggleby R G and Pang S S, Acetohydroxyacid synthase. *J Biochem Molec Biol* 33:1 – 36(2000).

[4] Wittenbach V A, Teaney P W, Rayner D R and Schloss J V, Herbicidal activity of an isopropylmalate dehydrogenase inhibitor. *Plant Physiol* 102:50(1993).

[5] Hawkes T R, Cox J M, Fraser TEM and Lewis T, A herbicidal inhibitor of isopropylmalate isomerase. *Z Naturforsch* C 48:364 – 368(1993).

[6] Dumas R, Biou V, Halgand F, Douce R and Duggleby RG, Enzymology, structure and dynamics of acetohydroxy acid isomeroreductase. *Accounts Chem Res* 34:399 – 408(2001).

[7] Dumas R, Butikofer M C, Job D and Douce R, Evidence for two catalytically different magnesium – binding sites in acetohydroxy acid isomeroreductase by site – directed mutagenesis. *Biochemistry* 34:6026 – 6036 (1995).

[8] Chunduru S K, Mrachko G T and Calvo K C, Mechanism of ketol acid reductoisomerase—steady - state analysis and metal ion requirement. *Biochemistry* 28:486 - 493(1989).

[9] Schulz A, Sponemann P, Kocher H and Wengenmayer F, The herbicidally active experimental compound Hoe 704 is a potent inhibitor of the enzyme acetolactate reductoisomerase. *FEBS Lett* 238: 375 - 378 (1988).

[10] Aulabaugh A and Schloss J V, Oxalyl hydroxamates as reactionintermediate analogues for ketol - acid reductoisomerase. *Biochemistry* 29:2824 - 2830(1990).

[11] Dumas R, Vives F, Job D, Douce R, Biou V, Pebay - Peyroula E and Cohen - Addad C, Inhibition of acetohydroxy acid isomeroreductase by reaction intermediate analogues: herbicidal effect, kinetic analysis and 3 - D structural studies, in *Proc Brighton Crop Prot Conf—Weeds*, BCPC, Farnham, Surrey, UK, pp 833 - 842(1993).

[12] Christopherson R I and Duggleby R G, Metabolic resistance: the protection of enzymes against drugs which are tight - binding inhibitors by the accumulation of substrate. *Eur J Biochem* 134:331 - 335(1983).

[13] Weinstock LM, Currie RB and Lovell AV, A general, one - step synthesis of α - keto esters. *Synth* Comm 11:943 - 946(1981).

[14] Photis J M, Halide - directed nitrile hydrolysis. *Tetrahedron Lett*, 21:3539 - 3540(1980).

[15] Axiotis GP, Reaction of organo - metallic reagents with triethoxyacetonitrile. A new and short synthesis of α - ketoesters. *Tetrahedron Lett* 22:1509 - 1510(1981).

[16] Thorsett E D, Conversion of α - aminoesters to α - ketoesters. *Tetrahedron Lett* 23:1875 - 1876(1982).

[17] Thomazeau K, Dumas R, Halgand F, Forest E, Douce R and Biou V, Structure of spinach acetohydroxyacid isomeroreductase complexed with its reaction product dihydroxymethylvalerate, manganese and (phospho) - ADP - ribose. *Acta Crystallogr D* 56:389 - 397(2000).

Synthesis and Herbicidal Activity of 2 - Alkyl(aryl) - 3 - methylsulfonyl(sulfinyl) pyrano - [4,3 - *c*] pyrazol - 4(2*H*) - ones*

Yuxin Li, Youming Wang, Bin Liu, Suhua Wang, Zhengming Li

(State Key Laboratory of Elemento - Organic Chemistry,
Nankai University, Tianjin, 300071, China)

Abstract Useful oxidation reaction of 2 - alkyl - (aryl) - 3 - methylthiopyrano[4,3 - *c*] pyrazol - 4(2*H*) - ones, leading to either the corresponding sulfoxides or sulfones, using hydrogen peroxide and acetic acid in 1,2 - dichloroethane, is described. Bioassay results showed that the products have some herbicidal activity.

INTRODUCTION

We previously found that 3 - (bis - methylthio) methylene - 5,6 - dihydro - 6 - methyl - 2*H* - pyran - 2,4 - diones have interesting herbicidal activity. Considering good herbicidal activity of some pyrazole compounds, we combined β - keto - δ - valerolactone and pyrazole and hoped to find better lead compounds of herbicide. We have reported that the reaction of 5,6 - dihydro - 2*H* - pyran - 2,4 - dione - 3 - dithioacetals with (un) substituted hydrazines affords 2 - alkyl (aryl) - 3 - methylthiopyrano[4,3 - *c*] pyrazol - 4(2*H*) - ones[1,2] (**1**), and we were interested in oxidation of 2 - alkyl(aryl) - 3 - methylthiopyrano[4,3 - *c*] pyrazol - 4(2*H*) - ones and testing the herbicidal activity of title compounds.

Various oxidative methods have been used for the preparation of sulfoxides or sulfones from the corresponding sulfides[3-7]. However, only few of them permit the oxidation in a selective manner. The simplest procedure for oxidation of sulfides to sulfoxides or sulfones involves hydrogen peroxide and acetic acid as the oxidative reagent. It was reported that it is necessary to use them in equivalent amounts with respect to sulfide in order to avoid the overoxidation. In our experiments, we used five equivalent amounts of hydrogen peroxide and acetic acid in 1,2 - dichloroethane to oxidize the corresponding sulfides, and sulfoxides or sulfones were obtained respectively by controlling the oxidation temperature.

RESULTS AND DISCUSSION

2 - Alkyl(aryl) - 3 - methylsulfinylpyrano[4,3 - *c*] pyrazol - 4(2*H*) - ones (**2a - i**) were prepared from 2 - alkyl(aryl) - 3 - methylthiopyrano[4,3 - *c*] pyrazol - 4(2*H*) - ones (**1a - i**)[1]. The process is displayed in Scheme 1. At first, we failed to separate 2 - alkyl(aryl) - 3 - methylsulfinylpyrano[4,3 - *c*] pyrazol - 4(2*H*) - ones by flash column chromatography because sulfoxides and sulfones were obtained at the same time and they have similar polarity, but sulfox-

* Reprinted from *Heteroatom Chemistry*, 2005, 16(4): 255 - 258. Contract grant number: 20302003.

ides or sulfones were obtained respectively by controlling the reaction condition and using 1,2 – dichloroethane as the solvent.

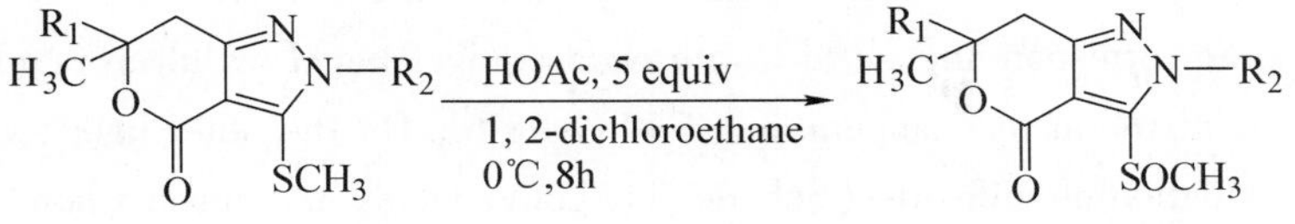

R_1, H_3C, O, O, N, N—R_2, SCH_3 → (HOAc, 5 equiv; 1, 2-dichloroethane; 0℃,8h) → R_1, H_3C, O, O, N, N—R_2, $SOCH_3$

1a-i → **2a-i**

R_1=H, Ph, Piperonyl, R_2=H, CH_3, Ph, SO_2Ph

Scheme 1

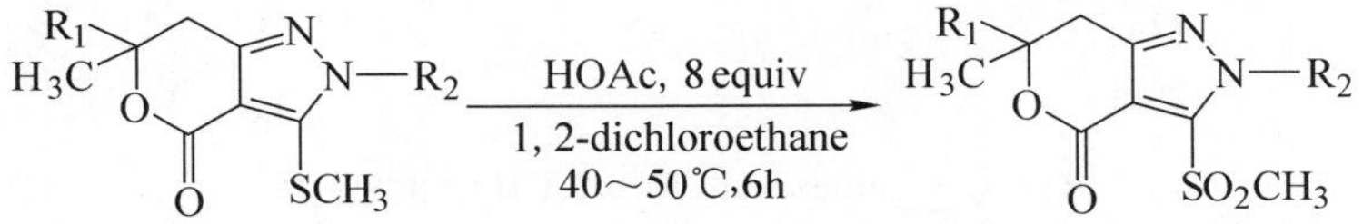

R_1, H_3C, O, O, N, N—R_2, SCH_3 → (HOAc, 8 equiv; 1, 2-dichloroethane; 40～50℃,6h) → R_1, H_3C, O, O, N, N—R_2, SO_2CH_3

1a-i → **3a-i**

R_1=H, Ph, Piperonyl, R_2=H, CH_3, Ph, SO_2Ph

Scheme 2

2 – Alkyl(aryl) – 3 – methylsulfinylpyrano[4,3 – *c*]pyrazol – 4(2*H*) – ones(**2a – i**) were obtained in high yields(80% – 96%)(Table 1) by oxidizing the corresponding sulfides with five equivalents of hydrogen peroxideacetic acid in 1,2 – dichloroethane at a temperature of 0℃ for 8h. In order to avoid overoxidation, low temperature is necessary.

Table 1 Physical Data of Title Compounds 2a – i and Elemental Analysis

Compound 2	R_1	R_2	Yield(%)	mp(℃)	Elemental analysis %(Calcd,%)		
					C	H	N
a	H	CH_3	82	185 – 187	47.19(47.36)	5.75(5.30)	12.00(12.27)
b	H	Ph	96	101 – 103	57.26(57.92)	4.99(4.86)	10.01(9.65)
c	$PhCH_2$	H	86	123 – 125	59.12(59.19)	4.90(5.30)	9.45(9.20)
d	$PhCH_2$	Ph	81	92.5 – 94	66.30(66.30)	5.52(5.30)	7.04(7.36)
e	$PhCH_2$	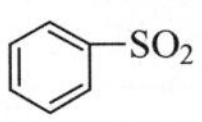 $-SO_2$	93	85.5 – 86	56.52(56.74)	4.58(4.53)	6.35(6.30)
f	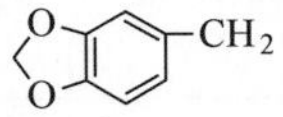 O, O, $-CH_2$	H	88	204 – 206	55.23(55.16)	4.62(4.63)	8.09(8.05)
g	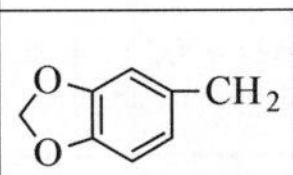 O, O, $-CH_2$	CH_3	83	124 – 125	56.42(56.34)	5.42(5.01)	7.55(7.73)
h	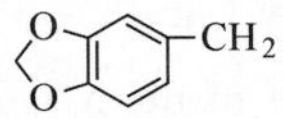 O, O, $-CH_2$	Ph	84	170 – 172	61.91(62.25)	4.52(4.75)	6.33(6.60)
i	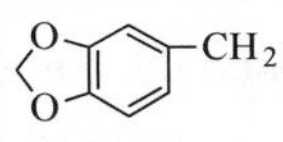 O, O, $-CH_2$	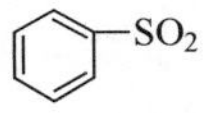 $-SO_2$	92	131 – 132	54.58(54.09)	4.18(4.13)	5.86(5.74)

2 – Alkyl(aryl) – 3 – methylsulfonyl – pyrano[4,3 – *c*]pyrazol – 4(2*H*) – ones(**3a – i**) were prepared by the processes described in Schemes 2 and 3. Two different routes could be expected to lead to the required sulfones. Sulfones can be obtained in better yields(65% – 96%)(Table 2) by oxidizing the corresponding sulfides with eight equivalents of hydrogen peroxideacetic acid in 1,2 – dichloroethane at the temperature of 40℃ for 6h. On the other hand, we also got sulfones by over oxidation of sulfoxides(Scheme 3). Lower yields may result when water solubility of the product is higher.

2a-i $\xrightarrow[\text{1, 2-dichloroethane, r.t.}]{\text{HOAc}}$ **3a-i**

(2a-i: R_1, H_3C, O, N, N—R_2, O, $SOCH_3$; 3a-i: R_1, H_3C, O, N, N—R_2, O, SO_2CH_3)

R_1=H, Ph, Piperonyl, R_2=H, CH_3, Ph, SO_2Ph

Scheme 3

Table 2 Physical Data of Title Compounds 3a – i and Elemental Analysis

Elemental Analysis %(Calcd, %)

Compound 2	R_1	R_2	Yield(%)	mp(℃)	Elemental analysis %(Calcd, %)		
					C	H	N
a	H	CH_3	96	155 – 157	44.34(44.25)	4.93(4.95)	11.3(11.47)
b	H	Ph	92	148 – 150	54.78(54.89)	4.71(4.61)	10.10(9.14)
c	$PhCH_2$	H	65	101 – 103	56.42(56.24)	4.85(5.03)	8.39(8.74)
d	$PhCH_2$	Ph	83	99 – 101	63.58(63.62)	5.37(5.08)	7.12(7.07)
e	$PhCH_2$	C_6H_5–SO_2	92	180 – 182	54.50(54.77)	4.56(4.38)	5.48(6.08)
f	$(OCH_2O)C_6H_3$–CH_2	H	77	119 – 120	52.23(52.74)	4.55(4.43)	8.06(7.69)
g	$(OCH_2O)C_6H_3$–CH_2	CH_3	68	126 – 127	53.41(53.96)	4.84(4.80)	7.54(7.41)
h	$(OCH_2O)C_6H_3$–CH_2	Ph	89	165 – 167	60.05(59.99)	4.33(4.58)	6.51(6.36)
i	$(OCH_2O)C_6H_3$–CH_2	C_6H_5–SO_2	93	186 – 188	52.67(52.37)	4.32(4.00)	5.56(5.56)

The ^{1}H NMR spectra were consistent with the structure of new 2 – alkyl(aryl) – 3 – methylsulfonyl – (sulfinyl) – pyrano[4,3 – *c*]pyrazol – 4(2*H*) – ones. A singlet of methylsulfinyl was shifted from higher field to lower field when the oxidation products were sulfones. When considering the advantages of hydrogen peroxide – acetic acid in 1,2 – dichloroethane, overoxidation could be avoided and mild conditions could be employed.

Compounds **2** were tested in soil treatment against many herbs such as *Polygonum tataricum*, *Digitaria sanguinalis*, *Portulaca oleracea*, *Medicago sativa*, and Rape at 1. 5kg/ha. The bioassay results showed that they have some herbicidal activity. Especially, the inhibitory rate of **2g** toward *Portulaca oleracea* was 81. 1% (Table 3). Considering the lowherbicidal activity of compound **3b**, we did not test the herbicidal activity of other compounds.

Table 3 Inhibitory Rate(%) of Compounds 2 or 3 Against Many Herbs at 1. 5kg/ha(Soil Treatment)

Compound	Polygonum tataricum	Digitaria sanguinalis	Portulaca oleracea	Medicago sativa	Rape
2a	17. 2	11. 1	2. 7	0	0
2c	0	0	0	12. 6	0
2d	0	44. 4	0	26. 2	0
2f	15. 6	11. 1	0	0	17. 6
2g	50	66. 7	81. 1	26. 2	33. 2
2h	1. 6	0	5. 4	9. 8	0
2i	20. 3	22. 2	2. 7	1. 6	0
3b	11. 0	0	0	7. 1	0

EXPERIMENTAL

Melting points were conducted on a Yanaco MP – 500 micromelting point apparatus. ^1H NMR spectra were recorded in $CDCl_3$ as solvent on AC – 200 instrument using TMS as internal standard. Elemental analyses were performed on a Bruker MF – 3 automatic elemental analyzer.

2 – Alkyl(aryl) – 3 – methylsulfinyl – pyrano[4,3 – c] – pyrazol – 4(2H) – ones ***2a – i***

2a: Compound **1a**(1. 06g, 0. 005mol) was dissolved in 20mL 1,2 – dichloroethane. With constant agitation, a solution of 2. 9g(0. 025mol, 30%) hydrogen peroxide and 1. 5g(0. 025mol) acetic acid was added dropwise to the mixture, which was agitated for 8h at a temperature of 0℃. After the agitation (monitored by TLC), the organic phase was separated and dried over anhydrous magnesium sulfate. After filtration, the solvent was evaporated in vacuo. The raw product was purified using silica gel column with ethyl acetate/petroleum ether(v/v, 1∶1) as eluent and 0. 93g **2a** was obtained. Yield: 82%. mp 185 – 187℃. ^1H NMR($CDCl_3$): 1. 54(d, 3H, J = 6. 2, CH_3), 2. 83(m, 2H), 3. 09(s, 3H, $SOCH_3$), 4. 22(s, 3H, NCH_3), 4. 75(m, 1H).

2b: Following the above method and using 1. 37g **1b**, 1. 39g **2b** was obtained. Yield: 96%. mp 101 – 103℃. ^1H NMR($CDCl_3$): 1. 56(d, 3H, J = 6. 2, CH_3), 2. 86(m, 2H), 3. 22(s, 3H, $SOCH_3$), 4. 75(m, 1H), 7. 51(m, 5H).

2c: Following the above method and using 1. 44g **1c**, 1. 30g **2c** was obtained. Yield: 86%. mp 123 – 125℃. ^1H NMR($CDCl_3$): 1. 46(s, 3H, CH_3), 3. 03(s, 2H, CH_2), 3. 07(m, 2H, CH_2), 3. 11(s, 3H, $SOCH_3$), 7. 18(m, 5H), 11. 32(bs, 1H, NH).

2d: Following the above method and using 1. 82g **1d**, 1. 54g **2d** was obtained. Yield: 81%. mp

92.5 – 94℃. ^1H NMR($CDCl_3$): 1.53(s,3H,CH_3), 2.98(s,2H,CH_2), 3.05(m,2H,CH_2), 3.10(s,3H,$SOCH_3$), 7.20(m,5H), 7.49(m,5H).

2e: Following the above method and using 2.14g **1e**, 2.06 g **2e** was obtained. Yield: 93%. mp 85.5 – 86℃. ^1H NMR($CDCl_3$): 1.55(s,3H,CH_3), 2.91(s,2H,CH_2), 3.01(m,2H,CH_2), 3.33(s,3H,$SOCH_3$), 7.18(m,5H), 8.05(m,5H).

2f: Following the above method and using 1.66g **1f**, 1.52g **2f** was obtained. Yield: 88%. mp 204 – 206℃. ^1H NMR($CDCl_3$): 1.45(s,3H,CH_3), 2.94(s,2H,CH_2), 3.06(m,2H,CH_2), 3.20(s,3H,$SOCH_3$), 5.88(s,2H,CH_2), 6.66(m,3H), 11.45(bs,1H).

2g: Following the above method and using 1.73g **1g**, 1.49g **2g** was obtained. Yield: 83%. mp 124 – 125℃. ^1H NMR($CDCl_3$): 1.43(s,3H,CH_3), 2.88(s,2H,CH_2), 2.96(m,2H,CH_2), 3.08(s,3H,$SOCH_3$), 4.19(s,3H,NCH_3), 5.90(s,2H,CH_2), 6.68(m,3H).

2h: Following the above method and using 2.04g **1h**, 1.78g **2h** was obtained. Yield: 84%. mp 170 – 172℃. ^1H NMR($CDCl_3$): 1.51(s,3H,CH_3), 2.88(s,2H,CH_2), 3.05(m,2H,CH_2), 3.14(s,3H,$SOCH_3$), 5.90(s,2H,CH_2), 6.62(m,3H), 7.48(m,5H).

2i: Following the above method and using 2.34 g **1i**, 2.22 g **2i** was obtained. Yield: 92%. mp131 – 132℃. ^1H NMR($CDCl_3$): 1.45(s,3H,CH_3), 2.88(s,2H,CH_2), 2.96(s,3H,$SOCH_3$), 3.25(m,2H,CH_2), 5.82(s,2H,CH_2), 6.52(m,3H), 7.68(m,5H).

2 – Alkyl(aryl) – 3 – methylsulfonyl – pyrano[4,3 – c] – pyrazol – 4(2H) – ones **3a – i**

3a: Compound **1a**(1.06g, 0.005mol) was dissolved in 20mL 1,2 – dichloroethane. With constant agitation, a solution of 4.6g(0.04mol, 30%) hydrogen peroxide and 2.4g(0.04 mol) acetic acid was added dropwise to the mixture, which was agitated for 6 h at a temperature of 40℃. After the agitation, the organic phase was separated off and dried over anhydrous magnesium sulfate. After filtration, the solvent was evaporated in vacuo. The raw product was purified using *silica* gel column with ethyl acetate/petroleum ether(v/v, 1∶2) as eluent and 1.17g **3a** was obtained. Yield: 96%. mp 155 – 157℃. ^1H NMR($CDCl_3$): 1.53(d,3H,J = 6.2, CH_3), 2.82(m,2H), 3.54(s,3H,SO_2CH_3), 4.21(s,3H,NCH_3), 4.80(m,1H).

3b: Following the above method and using 1.37g **1b**, 1.40g **3b** was obtained. Yield: 92%. mp 148 – 150℃. ^1H NMR($CDCl_3$): 1.56(d,3H,J = 6.2, CH_3), 3.05(m,2H), 3.46(s,3H,SO_2CH_3), 4.73(m,1H), 7.52(m,5H).

3c: Following the above method and using 1.44g **1c**, 1.04g **3c** was obtained. Yield: 65%. mp 101 – 103℃. ^1H NMR($CDCl_3$): 1.50(s,3H,CH_3), 3.01(m,2H), 3.15(s,2H), 3.42(s,3H,SO_2CH_3), 6.43(bs,1H), 7.22(m,5H).

3d: Following the above method and using 1.82g **1d**, 1.64g **3d** was obtained. Yield: 83%. mp 99 – 101℃. ^1H NMR($CDCl_3$): 1.56(s,3H,CH_3), 3.02(s,2H,CH_2), 3.12(m,2H,CH_2), 3.45(s,3H,SO_2CH_3), 7.12(m,5H), 7.55(m,5H).

3e: Following the above method and using 2.14g **1e**, 2.11g **3e** was obtained. Yield: 92%. mp 180 – 182℃. ^1H NMR($CDCl_3$): 1.54(s,3H,CH_3), 3.00(m,2H,CH_2), 3.10(s,2H,CH_2), 3.33(s,3H,SO_2CH_3), 7.17(m,5H), 7.66(m,5H).

3f: Following the above method and using 1.66g **1f**, 1.40g **3f** was obtained. Yield: 77%. mp 119 – 120℃. ^1H NMR($CDCl_3$): 1.49(s,3H,CH_3), 2.94(s,2H,CH_2), 3.15(m,2H,CH_2), 3.43(s,3H,SO_2CH_3), 5.86(s,2H,CH_2), 6.64(m,3H), 13.17(bs,1H,NH).

3g: Following the above method and using 1. 73g **1g**, 1. 28g **3g** was obtained. Yield: 68%. 126 – 127℃. ^1H NMR($CDCl_3$): 1. 40(s, 3H, CH_3), 2. 79(s, 2H, CH_2), 2. 93(m, 2H, CH_2), 3. 49(s, 3H, SO_2CH_3), 4. 15(s, 3H, NCH_3), 5. 87(s, 2H, CH_2), 6. 63(m, 3H).

3h: Following the above method and using 2. 04g **1h**, 1. 95g **3h** was obtained. Yield: 89%. mp 165 – 167℃. ^1H NMR($CDCl_3$): 1. 45(s, 3H, CH_3), 2. 72(s, 2H, CH_2), 2. 97(m, 2H, CH_2), 3. 34(s, 3H, SO_2CH_3), 5. 82(s, 2H, CH_2), 6. 48(m, 3H), 7. 40(m, 5H).

3i: Following the above method and using 2. 34g **1i**, 2. 32 g **3i** was obtained. Yield: 93%. mp 186 – 188℃. ^1H NMR($CDCl_3$): 1. 54(s, 3H, CH_3), 2. 90(s, 2H, CH_2), 3. 01(s, 2H, CH_2), 3. 33(s, 3H, SO_2CH_3), 5. 90(s, 2H, CH_2), 6. 51(m, 3H), 7. 78(m, 5H).

References

[1] Li, Y. X.; Wang, Y. M.; Yang, X. P.; Wang, S. H.; Li, Z. M. Heteroatom Chem 2003, 14(4), 342.

[2] Li, Y. X.; Wang, Y. M.; Yang, X. P.; Wang, S. H.; Li, Z. M. Chin Chem Lett 2004, 15(1), 14.

[3] Drabowicz, J.; Mikoajczyk, M. Synth Commun 1981, 11(12), 1025.

[4] Tarbell, D. S.; Weaver, C. J Am Chem Soc 1941, 63, 2939.

[5] Overberger, C. G.; Cummins, R. W. J Am Chem Soc 1953, 75, 4250.

[6] Curci, R.; Giovine, A.; Modena, G. Tetrahedron 1966, 22, 1235.

[7] Carpion, L. A.; Chen, H. W. J Am Chem Soc 1979, 101, 390.

2006年

单嘧磺隆除草剂的晶体构象-活性构象转换的密度泛函理论研究*

陈沛全[1,2]　孙宏伟[3,4]　李正名[1,2,4]　王建国[1,2]　马　翼[1,2]　赖城明[2,3]

([1]南开大学元素有机化学研究所,天津　300071;
[2]南开大学元素有机化学国家重点实验室,天津　300071;
[3]南开大学化学系,天津　300071;
[4]南开大学科学计算研究所,天津　300071)

摘　要　应用密度泛函理论在B3LYP/6-31G(d)水平上对新型除草剂单嘧磺隆绕脲桥部分两个C—N键的内旋转势能面进行计算,然后对势能面上的驻点进行构型优化和过渡态搜索,得到单嘧磺隆4种稳定构象和构象转换过程所涉及的8个过渡态结构。研究结果表明,单嘧磺隆晶体构象-活性构象转换过程中涉及4种稳定构象和8条转换途径,脲桥部分NH基团与嘧啶环上N原子所形成的分子内氢键对于构象的稳定性及转换过程起着十分重要的作用。应用极化连续介质溶剂模型(PCM)在B3LYP/6-31++G(d,p)水平下进行溶剂化效应计算,结果表明单嘧磺隆从晶体构象转换成活性构象主要是在水相中进行的。

关键词　单嘧磺隆　构象　内旋转　密度泛函理论　氢键

磺酰脲是一类十分重要的低毒超高效除草剂,有着十分重要的用途[1]。Levitt等[2]认为磺酰脲类化合物的杂环上面必须有两个取代基才能具有较好的活性,而我们多年新农药创新研究中发现的杂环单取代基磺酰脲化合物单嘧磺隆[3](图1)也可具有很好的除草活性,且已经取得国家新农药的三证。

研究分子的空间结构对于了解农药分子的构效关系及其作用机制,设计合成新型农药都具有重要理论意义和实际应用价值。药物分子的活性构象往往与它们的最低能量构象或晶体构象不同。近来的研究表明[4~7],磺酰脲类化合物的除草活性主要是通过抑制植物体内乙酰羟基酸合成酶(AHAS,EC 2.2.2.6),从而阻断生物体内支链氨基酸合成,达到除草目的的。最近我们[8]分别对单嘧磺隆和单嘧磺隆-拟南芥AHAS复合物进行结晶,并初步测定其晶体结构。晶体测定结果表明,单嘧磺隆的晶体构象与其活性构象在与嘧啶环相连的脲桥部分存在着较大的差异(见图1)。在晶体构象中,与我们所测定的一系列磺酰脲分子的晶体结构[9~14]类似,单嘧磺隆中与苯磺酰基相连的NH基团和嘧啶环中一个N原子形成分子内氢键;单嘧磺隆的活性构象与Duggleby等[5]所测定的氯嘧磺隆活性构象相似。与晶体构象相比,单嘧磺隆的活性构象为了更好地与靶酶作用而破坏了原先的分子内氢键绕C—N键旋转而得到活性构象。

研究农药分子如何从晶体构象转换为活性构象的过程对于了解药物药效性及药物如何与靶酶分子作用有着重要的意义。在前文[2,15~17]的基础上,本文进一步应用密度泛

---

* 原发表于《化学学报》,2006,64(13):1341-1348。国家"973"计划(No. 2003CB114406)、国家自然科学基金重点项目(No. 20432010)及天津市科委高性能计算(No. 043185111-5)资助项目。

函理论方法计算了单嘧磺隆分子中与晶体构象 - 活性构象转换过程有关的势能曲面，并在此基础上应用构型优化、过渡态搜索、频率分析及内禀反应坐标(IRC)计算等对构象转换过程的转换途径进行研究，从分子结构的水平上理解单嘧磺隆晶体构象转化为活性构象具体过程，从而为设计新型除草剂提供理论依据。

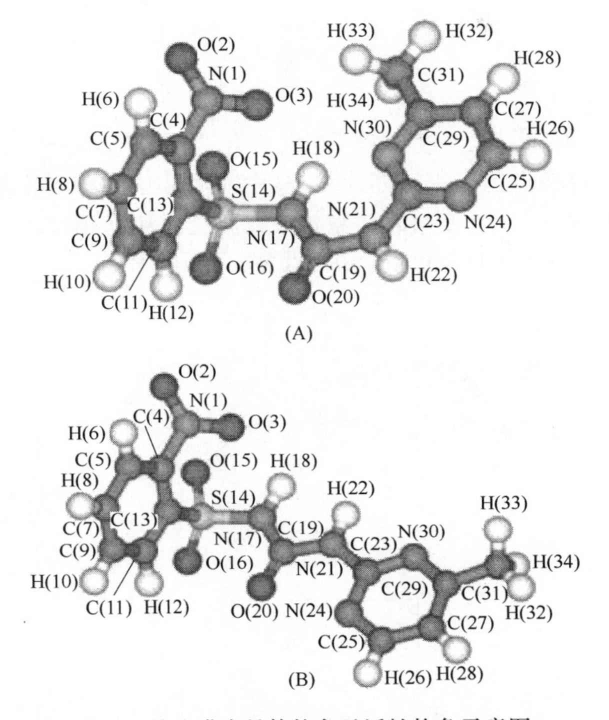

图 1　单嘧磺隆晶体构象及活性构象示意图

Figure 1　Crystal conformation(A) and active conformation(B) of monosulfuron

1　计算方法

对比单嘧磺隆的晶体构象和活性构象可见，构象转换过程中主要涉及到分子围绕键 C(19)—N(21) 和键 N(21)—C(23) 的旋转，因此在本文的计算中我们主要探讨分子绕这两个键旋转的势能面。

首先根据单嘧磺隆的晶体结构构建分子的初始构型，然后应用密度泛函理论在 B3LYP/6 - 31G(d) 水平上进行优化及频率分析。根据上述优化所得到的构型，定义二面角 θ_1[O(20)—C(19)—N(21)—H(22)] 及 θ_2[H(22)—N(21)—C(23)—N(24)]，在 0° ~360°范围内以步长 10°作 B3LYP/6 - 31G(d) 水平上的刚性势能面扫描，这样就得到了 1296(36×36) 组不同(θ_1,θ_2) 以及其对应的能量 E。以 θ_1 和 θ_2 为自变量，能量 E 为因变量则可构筑出单嘧磺隆绕 C(19)—N(21) 和 N(21)—C(23) 内旋转的势能曲面。

根据势能曲面上各驻点的性质，对不同类型的驻点采取不同的处理策略：对势能面上的极小点，取其构型作为初始构型，在 B3LYP/6 - 31G(d) 水平上进行构型全优化和频率分析计算，得到体系的稳定构象；对势能面上的一阶鞍点，取连接该鞍点的两个极小点及该鞍点的构型作为初始构型，应用 QST3 算法进行过渡态搜索并进行频率分析，并在此基础上从过渡态出发计算 IRC。

为了更好地描述体系的氢键作用和考虑溶剂对构象转换的影响，还在 B3LYP/6 -

31++G(d,p)水平上进行了单嘧磺隆在真空、水和正庚烷三种环境中的单点能计算，溶剂模型采用极化连续介质模型(polarizable continuum model,PCM)。此外，对某些构象还进行了NBO分析。

所有计算均采用Gaussian 03程序[18]在“南开之星”及“深腾6800”超级计算机上完成。

2 结果与讨论

2.1 内旋转势能面

图2(A)展示了单嘧磺隆分子绕单键C(19)—N(21)和N(21)—C(23)旋转在B3LYP/6-31G(d)水平上计算所得的三维势能曲线，其中能量数值为相对于最低能量构象(θ_1 = 0°, θ_2 = 0°)能量的相对值。为了更好地考察构象转换过程，还根据二面角的周期性特征，作出θ_1及θ_2在-270°~270°范围内的能量等高线图，见图2(B)。

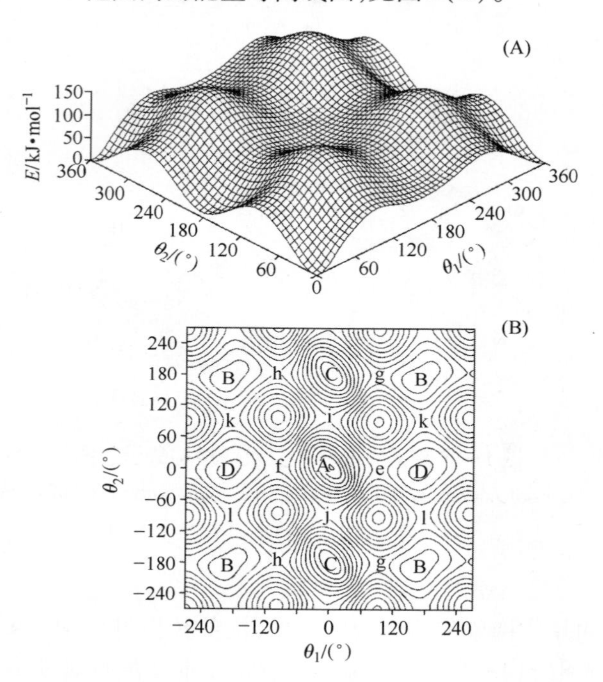

图2 单嘧磺隆分子围绕脲桥部分单键C(19)—N(21)和N(21)—C(23)内旋转的势能面

Figure 2 Potential energy surface of rotating C(19)—N(21) and N(21)—C(23) in monosulfuron (A) 3D potential energy surface; (B) contour map of potential energy surface

从图2可以看出，势能面上存在4类能量极小点，在0°~360°范围内对应的四组[θ_1, θ_2]分别是[0°,0°]，[180°,180°]，[0°,180°]和[180°,0°]，其对应的单嘧磺隆构象分别标志为A~D；势能面在[90°,0°]，[-90°,0°]，[90°,180°]，[-90°,180°]，[0,90°]，[0°,-90°]，[180°,90°]和[180°,-90°]处存在一级鞍点，对应于势能面上8个过渡态构象，标志为e~l。根据二面角的周期性特征，势能面上某些点的[θ_1, θ_2]虽不同，但其对应构象相同，如[90°,180°]和[90°,-180°]都对应构象g。

计算结果表明，构象A~D在B3LYP/6-31G(d)水平下的相对能量分别为0.00，66.45，10.92和58.10kJ·mol^{-1}，由此可见，构象的稳定性为A≈C>D≈B。从构型上看，构象A与晶体构象相似，构象B与活性构象相似。分析内旋转所产生的势能等高线图可见，单嘧磺隆从构象A转换成活性构象B与靶酶作用的最低能量途径并不是通过绕C(19)—N(21)和N(21)—C(23)作协同内旋转一次完成的，而是通过分别绕这两个单键作内旋转而变化到中间构象C或D进而转化为活性构象B的。总的说来，构象A可分别通过8条转换途径经过中间构象C或D转换成构象B，这8条途径分别为(其中e～l为过渡态构象)：

A→i→C→g→B(1)　　A→e→D→k→B(5)

A→i→C→h→B(2)　　A→f→D→k→B(6)

A→j→C→g→B(3)　　A→e→D→l→B(7)

A→j→C→h→B(4)　　A→f→D→l→B(8)

2.2　构象转换过程的构象分析

分子绕C(19)—N(21)和N(21)—C(23)作刚性势能面扫描计算在一定程度上反映了各个构象之间转换的过程。由于具体计算中，分子的内旋转是刚性的，而实际过程分子的内旋转并非是刚性的，而是在分子绕C(19)—N(21)和N(21)—C(23)作内旋转的过程中分子的其他部分会进行相应的调整。因此在分子绕C(19)—N(21)和N(21)—C(23)过程中考虑分子其他部分的柔性对于了解分子内旋转的实际过程相当重要。在如下的论述中我们将在刚性势能面扫描计算的基础上应用构型优化及过渡搜索等对构象转换过程进行探讨。

对势能面上4类极小点进行构型全优化得到4个稳定的构象，分别标记为MS(A)，MS(B)，MS(C)和MS(D)；对势能面上的8类鞍点进行过渡态搜索得到8个构象转换过程的过渡态构象，分别标记为TS(e)～TS(l)；这些构象的性质均通过频率分析确认，对于过渡态构象还进行IRC计算确认。

计算得到的4个稳定构象和8个过渡态构象如图3所示，表1分别列出了在6－31G(d)，6－31＋＋G(d,p)以及两种不同溶剂下各个构象的相对能量和虚频振动频率。表2列出了这些构象的特征结构参数。

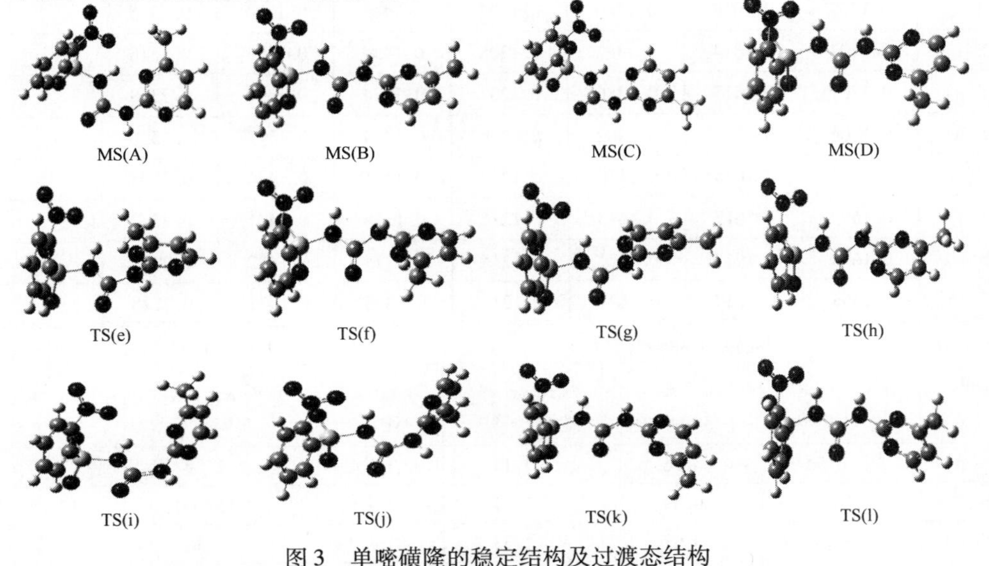

图3　单嘧磺隆的稳定结构及过渡态结构

Figure 3　Stable conformations and transition states of monosulfuron

表 1 单嘧磺隆的稳定结构及过渡态结构的相对能量

Table 1 Relative energies of stationary conformations and transition states of monosulfuron

Conf.	6 - 31G(d) /kJ · mol^{-1}	6 - 31 + + G(d,p) /kJ · mol^{-1}	v/cm^{-1}	PCM(Water) /kJ · mol^{-1}	PCM(Heptane) /kJ · mol^{-1}
MS(A)	0.00	0.00		0.00	0.00
MS(B)	37.78	37.75		19.63	33.95
MS(C)	2.95	1.68		-0.51	0.44
MS(D)	37.14	36.87		20.97	34.74
TS(e)	61.08	60.56	-50.80	50.98	58.38
TS(f)	64.04	62.22	-46.95	45.86	58.80
TS(g)	65.40	63.71	-55.78	38.06	57.80
TS(h)	66.01	64.28	-54.43	37.65	58.09
TS(i)	64.86	65.00	-73.56	50.99	62.37
TS(j)	64.86	61.46	-75.12	46.29	58.17
TS(k)	61.03	59.98	-122.31	50.53	58.27
TS(l)	64.86	63.28	-115.34	45.03	59.02

表 2 单嘧磺隆构象的部分结构参数

Table 2 Selected structural parameters of monosulfuron conformations

Conf.	Bond distance/nm							
	S(14)—N(17)	N(17)—H(18)	N(17)—C(19)	C(19)—O(20)	C(19)—N(21)	N(21)—C(23)	O(3)···H(18)	H(18)—N
MS(A)	0.1686	0.1030	0.1386	0.1217	0.1401	0.1388	0.2508	0.1880②
MS(B)	0.1697	0.1017	0.1422	0.1209	0.1389	0.1398	0.2208	
MS(C)	0.1689	0.1028	0.1383	0.1218	0.1400	0.1387	0.2589	0.1914①
MS(D)	0.1697	0.1017	0.1422	0.1209	0.1389	0.1399	0.2225	
TS(e)	0.1707	0.1015	0.1386	0.1207	0.1443	0.1390	0.2355	
TS(f)	0.1704	0.1014	0.1382	0.1208	0.1444	0.1393	0.2516	
TS(g)	0.1707	0.1015	0.1386	0.1207	0.1444	0.1390	0.2338	
TS(h)	0.1704	0.1014	0.1382	0.1208	0.1444	0.1393	0.2520	
TS(i)	0.1694	0.1015	0.1410	0.1217	0.1381	0.1430	0.2304	
TS(j)	0.1693	0.1017	0.1413	0.1216	0.1378	0.1431	0.2272	
TS(k)	0.1696	0.1017	0.1421	0.1215	0.1375	0.1432	0.2211	
TS(l)	0.1696	0.1017	0.1421	0.1214	0.1375	0.1432	0.2218	

Conf.	Improper torsion/(°)			Torsion/(°)		
	$\theta_{N(17)—S(14)—H(18)—C(19)}$	$\theta_{C(19)—N(17)—N(21)—O(20)}$	$\theta_{N(21)—C(19)—H(22)—C(23)}$	$\theta_{H(18)—N(17)—C(19)—O(20)}$	$\theta_{O(20)—C(19)—N(21)—H(22)}$	$\theta_{H(22)—N(21)—C(23)—N(24)}$
MS(A)	-10.54	0.26	-4.14	167.29	2.84	3.19
MS(B)	-21.18	0.49	-8.27	138.83	158.00	175.70
MS(C)	-8.27	0.02	-1.33	168.53	2.36	-177.50
MS(D)	-20.77	0.50	8.41	139.05	158.33	-2.29

续表 2

Conf.	Improper torsion/(°)			Torsion/(°)		
	$\theta_{N(17)—S(14)—H(18)—C(19)}$	$\theta_{C(19)—N(17)—N(21)—O(20)}$	$\theta_{N(21)—C(19)—H(22)—C(23)}$	$\theta_{H(18)—N(17)—C(19)—O(20)}$	$\theta_{O(20)—C(19)—N(21)—H(22)}$	$\theta_{H(22)—N(21)—C(23)—N(24)}$
TS(e)	-14.58	1.01	24.68	159.33	74.94	13.74
TS(f)	-8.52	-1.20	-25.92	168.47	-74.20	-14.65
TS(g)	-14.98	0.98	24.83	159.74	74.90	-168.55
TS(h)	-8.46	-1.27	-25.78	168.76	-74.59	168.16
TS(i)	-19.03	0.78	16.27	154.10	-13.54	110.21
TS(j)	-19.36	-1.26	-15.51	141.82	2.75	-94.79
TS(k)	-21.76	-0.65	14.69	140.07	162.42	95.40
TS(l)	-21.70	-0.69	14.98	139.79	162.38	-83.73

①Distance between H(18) and N(24); ②Distance between H(18) and N(30).

2.2.1 稳定构象

能量计算结果表明，在稳定构象中，与晶体构象相似的构象 MS(A) 的构象能最低，为最稳定构象，与活性构象相似的构象 MS(B) 的构象能最高；构象能 MS(A) 与 MS(C) 相近，MS(B) 与 MS(D) 相近，MS(A) 与 MS(C) 和 MS(B) 与 MS(D) 其构象差别主要体现在嘧啶环上甲基的取向上，这表明嘧啶环上甲基的取向对于单嘧磺隆的构象稳定性影响不大。4 个稳定构象的稳定性与前面势能面计算所得的结果一致，稳定性为 MS(A) ≈ MS(C) > MS(D) ≈ MS(B)。由此可见，在常温条件下，MS(A)及 MS(C) 构象存在的几率较 MS(B) 及 MS(D) 高，单嘧磺隆在与靶酶 AHAS 作用的过程中必须经历构象转换。

分析 4 个稳定构象可见，能量较低构象 MS(A) 和 MS(C) 中的 N(17) —H(18) 能与嘧啶环上的 N(24) 或 N(30) 形成分子内氢键，在脲桥部分与嘧啶环一起构造一个稳定的六元环，形成嘧啶基磺酰脲平面[16,17]。氢键的形成得到结构参数和 6 - 31 + + G(d, p) 水平下 NBO 计算的证实。分子的结构参数表明，H(18) 与 N(24)[或 N(30)] 之间的距离约为 0.19nm，两者满足形成氢键条件，能形成氢键；NBO 计算结果更表明，H(18) 与 N(24)[或 N(30)] 之间的 Wiberg 键级约为 0.06，是 N(17) —H(18)键级(约 0.6) 的十分之一，氢键的形成是 N(24) [或 N(30)] 上的孤对电子与 N(17) —H(18) 反键轨道相互作用的结果，其相互作用能约为 $70kJ \cdot mol^{-1}$。在构象 MS(B) 和 MS(D) 中，由于分子绕 C(19) —N(21) 旋转了约 180°，导致了N(17)—H(18) 与 N(24)[或 N(30)] 不能有效地形成分子内氢键，因而这两个构象的构象能较 MS(A) 和 MS(C) 高。

MS(A) 与 MS(C) 和 MS(B) 与 MS(D) 在脲桥部分的结构参数大致相同，进一步说明嘧啶环上甲基的空间取向对单嘧磺隆分子的空间结构及稳定性影响不大。分子构象从 MS(A) 或 MS(C) 转换成 MS(B) 或 MS(D)，S(14) —N(17)，N(17) —C (19)和 N(21) —C(23) 的键长有所增长，N(17) —H(18)，C(19) —O(20)，C(19)—N(21) 和 O(3) —H(18) 的键长有所缩短。二面角特征结构参数表明，MS (A) 和 MS(C) 的嘧啶环平面和磺酰脲平面基本上共面，形成一个大的共轭大 π 键便于电子的传递，而构象 MS(B) 及 MS(D) 中与 N(17) 相连的原子不在同一个平面内。从 MS(A) 或 MS(C) 转换成 MS(B) 或 MS(D) 的特征结构参数变化可见，

N(17)原子的杂化类型有从 sp^2 到 sp^3 转变的趋势，N(17)—H(18) 与 N(24) 或 N(30)不能形成分子内氢键而使分子原来的共轭大 π 键被破坏，从而影响了构象的稳定性。

值得一提的是，在 MS(B) 和 MS(D) 构象中 O(3) 与 H(18) 之间的距离大约为 0.221nm，两者间能形成分子内氢键，这个氢键形成也得到 NBO 计算的证实。这个氢键形成使得与 N(17) 相连的原子进一步偏离同一个平面，破坏大 π 键的形成。

2.2.2　过渡态构象

能量计算结果表明，气态下 8 个过渡态的能量较最稳定构象 MS(A) 相比高 61 ~ 66kJ · mol^{-1}。虚频振动频率基本上在 -122 ~ -45cm^{-1}范围内，与化学反应过渡态相比在数值上较小，这表明这些构象转换过渡态在虚频振动方向上的能量变化较小，是较为平缓的势能面。TS(e) ~ TS(h) 是分子绕键 C(19)—N(21) 内旋转产生的过渡态，TS(i) ~ TS(l) 是分子绕键 N(21)—C(23) 内旋转产生的过渡态，计算表明分子绕键 C(19)—N(21) 旋转所产生过渡态的虚频频率在数值上比绕 N(21)—C(23) 小。

表 2 的特征结构参数及分子的空间构型表明，构象转换过渡态中 N(17)—H(18) 都不能与嘧啶环上 N(24) [或 N(30)] 形成氢键，且嘧啶环平面与由 C(19)，N(17)，O(20) 及 N(21) 组成的平面基本上相互垂直，不能形成共轭大 π 键。构象转换过渡态的 S(14)—N(17) 键长比稳定构象 MS(A) 和 MS(C) 增长 0.01 ~ 0.02nm，且接近 MS(B) 和 MS(D) 的键长。由于没有形成分子内氢键，N(17)—H(18)不与嘧啶环上 N 原子作用，因而键长比 MS (A) 或 MS (C) 缩短了约 0.015nm。二面角结构参数表明，与 N(17)，C(19) 及 N(21) 相连的原子组成的平面在过渡态构象中不再共面，偏离约 15°。以上特征结构参数的变化反映了由于分子内氢键的破坏，共轭大 π 键难以形成，本来离域在嘧啶磺酰脲平面上的 π 电子变得更为定域化，N(17) 和 N(21) 原子的杂化类型有从 sp^2 向 sp^3 转变的趋势。结构参数表明，与稳定构象 MS(B) 和 MS(D) 类似，过渡态中 O(3) 与 H(18) 之间可形成另一分子内氢键。

由上述分析可见，单嘧磺隆分子从晶体构象 MS(A) 通过内旋转转换成 MS(B) 过程中涉及到分子内氢键和嘧啶环脲桥平面部分的共轭大 π 键的破坏，且可形成新的 O(3)…H(18)—N(17) 分子内氢键。这进一步说明了 N(17)—H(18) 与嘧啶环 N 原子形成的分子内氢键对于分子结构的稳定性及形成嘧啶磺酰脲平面大 π 键有着重要的作用。

2.3　构象转换途径分析

上述的讨论表明，单嘧磺隆从晶体构象 MS(A) 转换为活性构象 MS(B) 可沿 8 条不同的转换途径，每一条途径都经历一个转换中间态构象 MS(C) 或 MS(D)。图 4 所示为 6-31++G(d, p) 水平下晶体构象 - 活性构象转换过程中的能量关系图。这 8 条构象转换途径一般可分为两类：一类是晶体构象 MS(A) 先通过绕单键 N(21)—C(23)旋转大致 180°而形成中间构象 MS(C)，然后再绕单键 C(19)—N(21) 旋转大致 180°转换成活性构象 MS(B)，此类转换途径对应于 2.1 部分中的途径(1) ~ (4)；另一类是构象 MS(A) 先绕 C(19)—N(21) 旋转约 180°形成中间构象 MS(D)，然后绕 N(21)—C(23) 旋转约 180°转换成活性构象 MS(B)，对应于 2.1 部分中的途径 (5) ~ (8)。

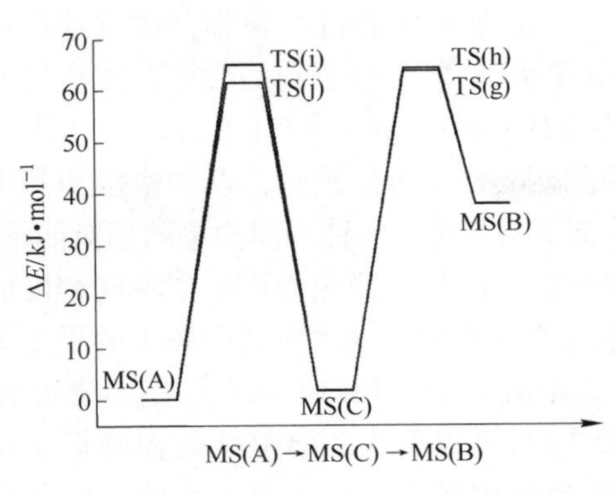

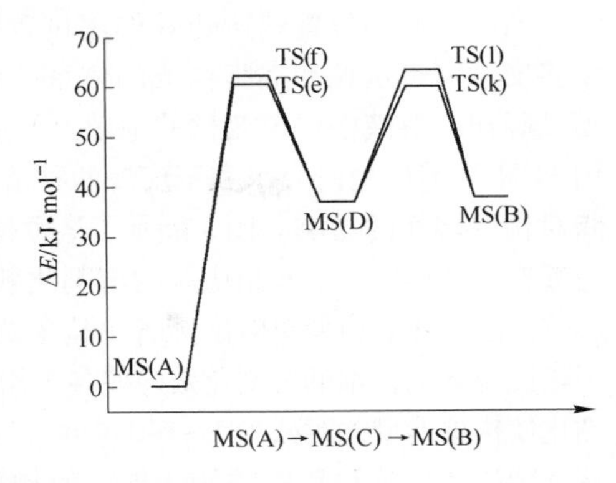

图 4　单嘧磺隆稳定构象转换中的能量关系

Figure 4　Energy relationship of conformational conversion of monosulfuron

对比两类转换途径可见，两类转换途径的第一步都需经历一个活化能为 60 ~ 65kJ · mol^{-1}的过渡态，两类不同的转换途径差别在于转换的第二步，第二步需要的活化能第一类为 60 ~ 65kJ · mol^{-1}，而第二类为 23 ~ 29kJ · mol^{-1}。分析转换途径经历的分子构象可见，第一类途径中构象 MS（A） 绕 N（21） —C（23） 旋转转换成中间构象 MS（C），N（17） —H（18） 与嘧啶环 N 原子所形成的氢键在转换过程中被破坏，但在构象 MS（C） 中氢键再次形成，不同的是形成氢键的受体原子从 N（30） 变成 N（24）；在第二类转换途径中，构象 MS（A） 的分子内氢键在转换成中间构象 MS（D） 过程中被破坏，且在构象 MS（D） 中不能形成这种类型的分子内氢键。两类转换途径中第二步的起始构象在是否形成 N—H…N 型氢键存在差异，从而导致了第二步所需的活化能不同，第二类转换途径第二步的活化能比第一类低。从动力学角度看来，由于两类转换途径第一步转换所需的活化能相差不大，而在第二步转换中，第二类所需活化能比第一类少得多，因而第二类转换在动力学上是有利的。由此可见，分子内氢键 N（17） —H（18） …N 的形成对于构象的稳定性及构象转换过程有着重要的作用。

过渡态构象 TS（e） 与 TS（f），TS（g） 与 TS（j），TS（i） 与 TS（j） 和 TS（k） 与 TS（l）之间的能量差距不大（ <4kJ · mol^{-1}），表明分子绕 C（19） —N（21） 和N（21）—C（23） 内旋转甲基取代基的空间取向对活化能的影响不大。这个结论的意义在于，若磺酰脲类化合物的药效跟构象转换途径有关，嘧啶环上取代基的引入对构象转换不产生较大的影响，取代基的改造主要是从引入取代基是否能增加药物分子与靶酶分子之间的相互作用考虑。

2.4　溶剂化对构象转换过程的影响

一般情况下，药物分子是在溶液条件下与靶酶作用的，溶液中分子一般以多种构象形式存在，各种构象的分布服从 Boltzmann 规律。单嘧磺隆分子从晶体构象转换为活性构象过程中涉及到分子绕 C（19） —N（21） 和 N（21） —C（23） 内旋转，由于共轭作用一般认为这两个键具有部分双键的性质。Bain 等[19]曾对 N 取代乙酰吡咯的 C（O） —N 酰胺键的内旋转进行 NMR 和理论计算表明，分子能绕 C（O） —N 键自由旋转，转换过渡态的活化能约为 55kJ · mol^{-1}，而我们的计算表明，单嘧磺隆构象转换所需的活化能为 60 ~ 65kJ · mol^{-1}和 23 ~ 29kJ · mol^{-1}，这说明了单嘧磺隆各种构象能在溶液条件下转换成活性构象与靶酶发生作用。

药物从给药位置到达药效作用位置的过程中，需要多次透过细胞膜，也就是说药物要交替穿过水相和类脂构成的体系。在应用分子力学[20]、分子动力学[2,15]方法研究磺酰脲在水溶液中构象变化的基础上，进一步在 B3LYP/6 - 31 + + G(d, p) 水平下应用 PCM 溶剂模型研究水和正庚烷两种溶剂对单嘧磺隆的溶剂化效应，以考察水相和类脂对构象转换的影响。图 5 展示了这两种不同溶剂下构象转换各种不同构象的相对能量。与气态条件相比，水和正庚烷的溶剂化效应对单嘧磺隆的构象稳定性产生了不同程度的影响。溶剂化效应使单嘧磺隆各个构象之间的能量差距减少，且水溶剂中减少的程度大于正庚烷溶剂，水的溶剂化效应使各个构象能量差距减少 10 ~ 18kJ · mol^{-1}，正庚烷的溶剂化使构象能量差距减少 1 ~ 6kJ · mol^{-1}。从图 5 同时可以看出，溶剂化效应降低了单嘧磺隆构象转换过程中的活化能，使构象转换在溶液中较真空中容易，水的溶剂化效应使构象转换活化能降低的程度大于正庚烷，因此可见单嘧磺隆在水溶剂中的构象转换比在正庚烷溶剂中容易，从而可以推断单嘧磺隆在与靶酶作用的过程中转换成活性构象主要是在水相中完成的。

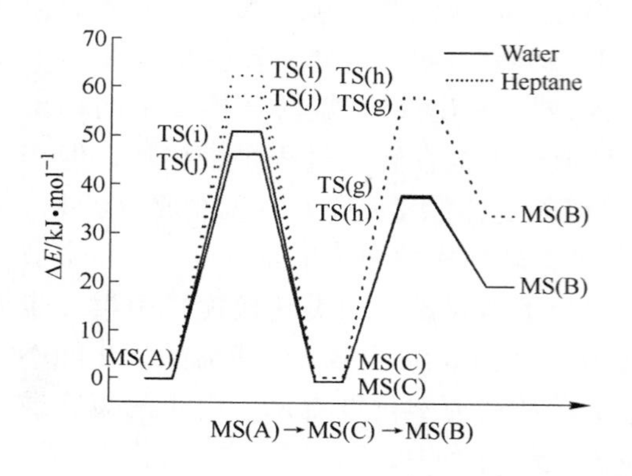

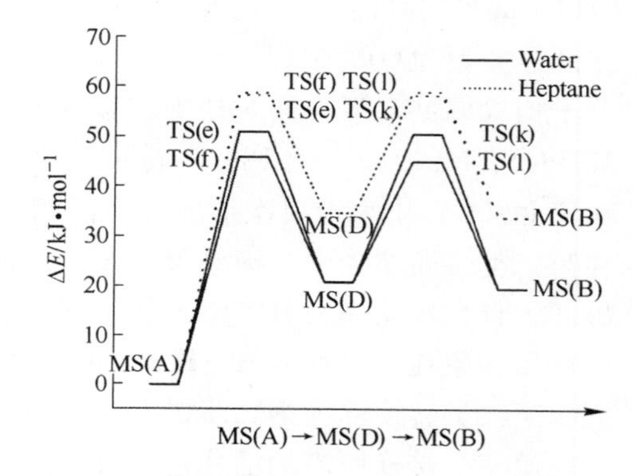

图 5 水和正庚烷中构象转换的能量关系

Figure 5 Energy relationship of conformational conversion in water and heptane

为探讨不同溶剂对构象能量差距造成影响的本质，对转换过程中涉及的 12 个构象的溶剂化能及偶极矩进行计算，表 3 列出了 6 - 31 + + G（d，p）水平下的计算结果。表中数据表明，不同构象在水中的溶剂化能大于在正庚烷中的溶剂化能，这说明了水对单嘧磺隆的稳定作用大于正庚烷的稳定作用，使构象转换在水中较易进行。通过分析气态、水和正庚烷三种不同状态下不同偶极矩的变化可见，溶剂化效应通过诱导作用使不同构象的偶极矩增加，且水对偶极矩的影响大于正庚烷，使单嘧磺隆分子与水的相互作用大于与正庚烷的相互作用，从而解释了单嘧磺隆在水中的溶剂化能大于在正庚烷中的溶剂化能。

表 3 不同构象的溶剂化能及不同状态下的偶极矩

Table 3 Solvation free energy and dipole moment of different conformations

Conf.	ΔG_{sol}^{Water} /kJ · mol^{-1}	$\Delta G_{sol}^{Heptane}$ /kJ · mol^{-1}	D^{Vacuum} /Debye	D^{Water} /Debye	$D^{Heptane}$ /Debye
MS（A）	-104.52	-30.84	5.2423	7.2595	5.8898
MS（B）	-127.32	-34.89	6.4846	8.3990	7.1939
MS（C）	-106.65	-32.13	5.7741	7.3764	6.3204

续表 3

Conf.	ΔG_{sol}^{Water} /kJ·mol^{-1}	$\Delta G_{sol}^{Heptane}$ /kJ·mol^{-1}	D^{Vacuum} /Debye	D^{Water} /Debye	$D^{Heptane}$ /Debye
MS (D)	-124.10	-33.30	7.0053	8.9216	7.7460
TS (e)	-114.52	-33.01	6.7453	8.8939	7.5251
TS (f)	-123.76	-34.31	4.7635	6.4998	5.3640
TS (g)	-137.49	-37.24	7.4418	9.6069	8.2275
TS (h)	-138.74	-37.57	7.8750	10.0849	8.6773
TS (i)	-121.67	-33.72	6.5744	8.8177	7.2935
TS (j)	-122.59	-34.43	3.7575	4.8340	4.0897
TS (k)	-114.47	-32.51	6.7019	8.8007	7.4518
TS (l) -	126.61	-35.27	5.4807	7.2967	6.0944

需要指出的是，溶剂化计算的 PCM 模型只描述溶剂的统计平均性质，而不采用可见的溶剂分子，因此上述计算的结果仅考虑了溶剂极性的影响，而没有考虑溶剂分子与药物分子间的氢键相互作用。实际上，单嘧磺隆的活性构象中并不存在其晶体中的 N(17)—H(18)…N（嘧啶环上）分子内氢键，而可能与带氢键接受体的溶剂分子形成分子间氢键，这使单嘧磺隆在水溶液中的构象转换变得更容易，即构象转换更主要在水相中完成。

若采用显示溶剂来考虑药物与溶剂分子间的氢键相互作用，对构象转换过程能有更全面的了解，有关这方面的研究我们将另文发表。

3 结论

本文应用密度泛函理论对新型除草剂单嘧磺隆的晶体构象 - 活性构象转换过程进行研究。计算结果表明，单嘧磺隆晶体构象 - 活性构象的转换过程涉及到 4 个稳定构象及 8 个转换过程的过渡态，单嘧磺隆的 4 个稳定构象之间可通过 8 条不同的转换途径进行构象转换。在构象转换的过程中，脲桥部分 NH 基团与嘧啶环上 N 原子所形成的分子内氢键对于构象的稳定性及转换过程起着十分重要的作用。溶剂化效应的计算结果表明，单嘧磺隆从晶体构象转换成活性构象主要是在水相中进行的。

致谢 作者感谢中国科学院计算机网络信息中心“计算化学虚拟实验室”的沈斌助理工程师对本计算工作的支持。

References

[1] Li, Z. -M.; Lai, C. -M. *Chin. J. Org. Chem.* 2001, 21, 810 (in Chinese).
(李正名，赖城明，有机化学，2001，21，810.)

[2] Ma, Y.; Li, Z. -M.; Lai, C. -M. *Chin. J. Pestic. Sci.* 2004, 6, 71 (in Chinese).
(马翼，李正名，赖城明，农药学学报，2004，6，71.)

[3] Li, Z. -M.; Jia, G. -F.; Wang, L. -X.; Fan, C. -W.; Yang, Z. *CN* 93 - 101976, 1994 [*Chem. Abstr.* 1994, 121, 29277] (in Chinese).
(李正名，贾国锋，王玲秀，范传文，杨炤，中国专利 94118793.4，1994.)

[4] Duggleby, R. G.; Pang, S. S. *J. Biochem. Mol. Biol.* 2000, 33, 1.

[5] Pang, S. S.; Duggleby, R. G.; Guddat, L. W. *J. Mol. Biol.* 2002, 317, 249.

[6] Pang, S. S.; Guddat, L. W.; Duggleby, R. G. *J. Biol. Chem.* 2003, 278, 7639.

[7] Chaleff, R. S.; Mauvais, C. J. *Science* 1984, 224, 1443.

[8] Wang, J. – G. *Ph. D. Dissertation*, Nankai University, Tianjin, 2004 (in Chinese).
(王建国，博士论文，南开大学，天津，2004.)

[9] Li, Z. – M.; Jia, G. – F.; Wang, L. – X.; Lai, C. – M.; Wang, R. – J.; Wang, H. – G. *Chem. J. Chin. Univ.* 1992, 13, 1411 (in Chinese).
(李正名，贾国锋，王玲秀，赖城明，王如骥，王宏根，高等学校化学学报，1992，13，1411.)

[10] Li, Z. – M.; Jia, G. – F.; Wang, L. – X.; Lai, C. – M.; Wang, H. – G.; Wang, R. – J. *Chem. J. Chin. Univ.* 1993, 14, 349 (in Chinese).
(李正名，贾国锋，王玲秀，赖城明，王宏根，王如骥，高等学校化学学报，1993，14，349.)

[11] Li, Z. – M.; Jia, G. – F.; Wang, L. – X.; Lai, C. – M. *Chem. J. Chin. Univ.* 1994, 15, 227 (in Chinese).
(李正名，贾国锋，王玲秀，赖城明，高等学校化学学报，1994，15，227.)

[12] Li, Z. – M.; Liu, J.; Wang, X.; Yuan, M. – X.; Lai, C. – M. *Chem. J. Chin. Univ.* 1997, 18, 750 (in Chinese).
(李正名，刘洁，王霞，袁满雪，赖城明，高等学校化学学报，1997，18，750.)

[13] Jiang, L.; Li, Z. – M.; Weng, L. – H.; Leng, X. – B. *Chin. J. Struct. Chem.* 2000, 19, 149 (in Chinese).
(姜林，李正名，翁林红，冷雪冰，结构化学，2000，19，149.)

[14] Wang, B. – L.; Ma, L.; Wang, J. – G.; Ma, Y.; Li, Z. – M.; Leng, X. – B. *Chin. J. Struct. Chem.* 2004, 23, 783 (in Chinese).
(王宝雷，马宁，王建国，马翼，李正名，冷雪冰，结构化学，2004，23，783.)

[15] Shen, R. – X.; Fang, Y. – Y.; Ma, Y.; Sun, H. – W.; Lai, C. – M.; Li, Z. – M. *Chem. J. Chin. Univ.* 2001, 22, 952 (in Chinese).
(沈荣欣，方亚寅，马翼，孙宏伟，赖城明，李正名，高等学校化学学报，2001，22，952.)

[16] Liu, A. – L.; Cao, W.; Lai, C. – M.; Yuan, M. – X.; Zhang, J. – P.; Lin, S. – F.; Li, Z. – M. *Chem. J. Chin. Univ.* 1997, 18, 574 (in Chinese).
(刘艾林，曹炜，赖城明，袁满雪，张金碚，林少凡，李正名，高等学校化学学报，1997，18，574.).

[17] Lai, C. – M.; Yuan, M. – X.; Li, Z. – M.; Jia, G. – F. *Chem. J. Chin. Univ.* 1994, 15, 1004 (in Chinese).
(赖城明，袁满雪，李正名，贾国锋，高等学校化学学报，1994，15，1004.)

[18] Frisch, M. J.; Trucks, G. W.; Schlegel, H. B.; Scuseria, G. E.; Robb, M. A.; Cheeseman, J. R.; Cheeseman, J. R.; Montgomery, J. A. Jr.; Vreven, T.; Kudin, K. N.; Burant, J. C.; Millam, J. M.; Iyengar, S. S.; Tomasi, J.; Barone, V.; Mennucci, B.; Cossi, M.; Rega, G. M. N.; Petersson, G. A.; Nakatsuji, H.; Hada, M.; Ehara, M.; Toyota, K.; Fukuda, R.; Hasegawa, J.; Ishida, M.; Nakajima, T.; Honda, Y.; Kitao, O.; Nakai, H.; Klene, M.; Li, X.; Knox, J. E.; Hratchian, H. P.; Cross, J. B.; Adamo, C.; Jaramillo, J.; Gomperts, R.; Stratmann, R. E.; Yazyev, O.; Austin, A. J.; Cammi, R.; Pomelli, C.; Ochterski, J. W.; Ayala, P. Y.; Morokuma, K.; Voth, G. A.; Salvador, P.; Dannenberg, J. J.; Zakrzewski, V. G.; Dapprich, S.; Daniels, A. D.; Strain, M. C.; Farkas, O.; Malick, D. K.; Rabuck, A. D.; Raghavachari, K.; Foresman, J. B.; Ortiz, J. V.; Cui, Q.; Baboul, A. G.; Clifford, S.; Cioslowski, J.; Stefanov, B. B.; Liu, G.; Liashenko, A.; Piskorz, P.; Komaromi, I.; Martin, R. L.; Fox, D. J.; Keith, T.; Al – Laham, M. A.; Peng, C. Y.; Nanayakkara, A.; Challacombe, M.; Gill, P. M. W.; Johnson, B.; Chen, W.; Wong, M. W.; Gonzal-

ez, C. ; Pople, J. A. *Gaussian* 03, Revision C. 01, Gaussian, Inc. , Wallingford CT, 2004.

[19] Bain, A. D. ; Duns, G. J. ; Ternieden, S. ; Ma, J. ; Werstiuk, N. H. J. *Phys. Chem.* 1994, 98, 7458.

[20] Ma, Y. ; Liu, J. ; Li, Z. – M. *Chem. J. Chin. Univ.* 2000, 21, 85 (in Chinese) .
(马翼，刘洁，李正名，高等学校化学学报，2000，21，85.）

Density Functional Theory Study on Conformational Conversion between Crystal Conformation and Active Conformation of Herbicidal Monosulfuron

Peiquan Chen[1,2], Hongwei Sun[3,4], Zhengming Li[1,2,4]
Jianguo Wang[1,2], Yi Ma[1,2], Chengming Lai[2,3]

([1]Institute of Elemento – Organic Chemistry, Nankai University, Tianjin, 300071;
[2]State Key Laboratory of Elemento – Organic Chemistry, Nankai University, Tianjin, 300071;
[3]Department of Chemistry, Nankai University, Tianjin, 300071;
[4]Institute of Scientific Computing, Nankai University, Tianjin, 300071)

Abstract To understand how the crystal conformer converts to the active conformer, here the internal rotation potential energy surface around the two C—N bonds in the sulfonylurea bridge of monosulfuron has been studied by density functional B3LYP method and 6 – 31G (d) basis set. For the stationary points along the internal rotation energy surface, full geometry optimization or QST3 transition structure search and frequency analysis calculations were also performed. The calculated results indicated that there were four stable conformers and eight transition state structures corresponding to the conformational conversion of monosulfuron. The results also implied that the four stable conformers could be converted mutually through eight conversional paths. During the conversion, the internal hydrogen bond in monosulfuron can play a key role in the conformational stability and conversional process. For the obtained conformers, B3LYP/6 – 31 + + G (d, p) method was also used to do single point calculation in vacuum and two PCM model solvents of water and heptane. The calculated results implied that the conversion mainly occurred in the aqueous solution.

Key words monosulfuron; conformation; internal rotation; density functional theory; hydrogen bond

O－取代苯基－*O*－（2－硬脂酰胺基）乙基－*N*，*N*－二（2－氯乙基）磷酰胺的合成和生物活性*

韩　亮[1]　李正名[1]　张　云[2]　郭维明[2]

（[1]南开大学元素有机化学研究所，元素有机化学国家重点实验室，天津　300071；
[2]南京农业大学园艺学院，南京　210095）

摘　要　*N*－脂肪酰基乙醇胺（NAE）作为植物体内的一种内源物质，在调节植物生长方面起着重要作用。为了弥补其分子结构中长的脂肪链所带来的溶解性能以及在植物体内传导性能的缺陷，我们将*N*－硬脂酰乙醇胺（NAE18）引入氮芥磷酸酯中，合成了一系列标题化合物。在合成工作中发现：在NAE18与氮芥芳基磷酰氯的反应过程中，4－二甲氨基吡啶（DMAP）起着关键性的催化作用。在不加DMAP的相同实验条件下，反应不能进行。对所合成的标题化合物进行了植物生长调节和杀菌活性的测定，初步生测结果表明：经过结构修饰后，大多数目标化合物的活性相对NAE18有所增强，但有关生物活性仍有待进一步研究。

关键词　*N*－脂肪酰基乙醇胺　氮芥磷酸酯　合成　生物活性

20世纪50年代，*N*－脂肪酰基乙醇胺（NAE）作为大豆卵磷脂和花生饼粉的组成成分首次被报道[1]。这些脂类在哺乳动物中起着重要的生理作用，但是NAE在植物体内的广泛存在、代谢以及生理重要性直至最近才得以认识[2]。近年来对NAE越来越多的研究表明，NAE可能参与了引发植物防卫反应的信号传导，例如0.1μmol·L^{-1}的肉豆蔻酰乙醇胺能激活烟草细胞中苯基丙氨酸－氨裂解酶表达（PAL2表达，对烟草中一系列胁迫所产生的反应）[3]，而且NAE还起着抑制在植物衰老过程中调节膜老化的磷脂酶α（PLDα）的作用[4]，其抑制作用是溶血磷脂酰乙醇胺（LPE）的100倍，后者对与花和水果收割后衰老相关的生理症状具有显著的作用[5]，由此可见，NAE在植物生长调节方面具有新的应用价值。实验表明，外源的NAE能抑制木聚糖酶引发的碱化效应[3]，对幼苗中初生根伸长、侧根形成和根毛形成具有显著的生理/形态影响[6]，并且能有效地延长花卉的保鲜时间[7]。当NAE作为外源物质对植物作用时，分子结构中长的脂肪链决定了其亲脂性高、亲水性差，因而NAE对植物表皮穿透能力好，而在植物体内传导能力差，从而影响了其活性的发挥。我们试图通过对其进行结构修饰，改善其溶解性能，使其对植物进行作用时亲水性和亲脂性达到平衡，并进一步提高活性。鉴于NAE在生物体内的前体是以磷酸酯的形式存在的[8]，而有机磷化合物在植物生长调节方面有着广泛的应用，如乙烯利[9]、乙二膦酸[10]、氮芥磷酸衍生物[11,12]等，因此我们选择将*N*－硬脂酰乙醇胺（NAE18）引入氮芥磷酸酯中，合成了一系列标题化合物，并对其生物活性进行了测定。合成路线见Scheme 1。

* 原发表于《有机化学》，2006，26（2）：242－246。国家自然科学基金（No. 20432010）、南开大学元素有机化学国家重点实验室（No. 0204）资助项目。

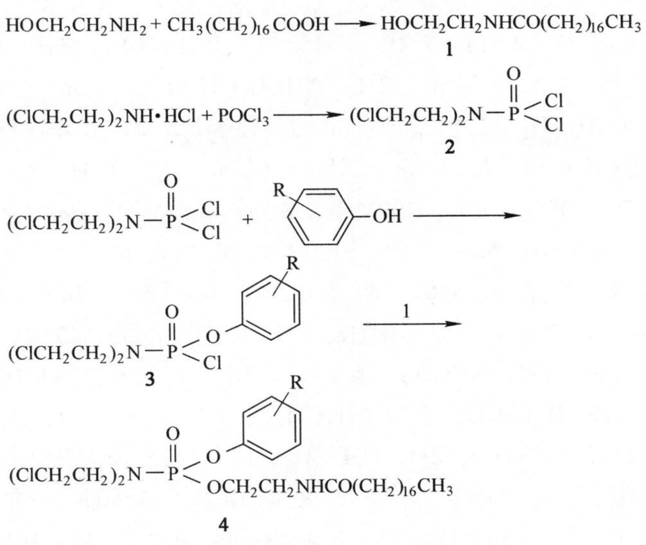

$$HOCH_2CH_2NH_2 + CH_3(CH_2)_{16}COOH \longrightarrow HOCH_2CH_2NHCO(CH_2)_{16}CH_3 \ (\mathbf{1})$$

$$(ClCH_2CH_2)_2NH \cdot HCl + POCl_3 \longrightarrow (ClCH_2CH_2)_2N-P(=O)Cl_2 \ (\mathbf{2})$$

$$(ClCH_2CH_2)_2N-P(=O)Cl_2 + R\text{-}C_6H_4\text{-}OH \longrightarrow (ClCH_2CH_2)_2N-P(=O)(OC_6H_4R)Cl \ (\mathbf{3}) \xrightarrow{\mathbf{1}} (ClCH_2CH_2)_2N-P(=O)(OC_6H_4R)OCH_2CH_2NHCO(CH_2)_{16}CH_3 \ (\mathbf{4})$$

R=H, *p*-F, *p*-Cl, *p*-Br, *o*-CH_3, *m*-CH_3, *m*-MeO, *p*-*t*-Bu, 3, 4-Me_2

Scheme 1

1 实验部分

Yanaco CHN CORDER MT－3 型自动元素分析仪；BRUKER AVANCE 300 型核磁共振仪（TMS 为内标，$CDCl_3$ 为溶剂）；Polaris－Q 型气质联用仪（EI 离子源）；EQUINOX 55 红外光谱仪；Yanaco 熔点仪。所用试剂均为国产分析纯，无水溶剂用常规方法干燥。

1.1 *N*－硬脂酰乙醇胺（NAE18）（1）的合成

参考文献［13］合成，粗产物用 $CHCl_3$－丙酮重结晶，收率96%，m. p. 99～100℃（文献值 m. p. 102℃）。

1.2 氮芥磷酰二氯（2）的合成

参照文献［14］合成，收率 85%，m. p. 54～55℃（文献值[14]：收率 75%，m. p. 54～56℃）。

1.3 *O*－取代苯基－*O*－（2－硬脂酰胺）乙基－*N*，*N*－二（2－氯乙基）磷酰胺（4）的合成

于 100mL 圆底烧瓶中加入氮芥磷酰二氯（1g，3.9mmol）和取代酚（3.9mmol），用 10mL 无水苯溶解，室温搅拌下缓慢滴加 Et_3N（0.47g，4.7mmol）的无水苯（5mL）溶液，滴毕，加热回流 3.5h，TLC［展开剂为 *V*（石油醚）：*V*（乙醚）＝3:1］监测，酚消失后停止反应，冷却，过滤，少量苯洗涤 $Et_3N \cdot HCl$，滤液浓缩至干，得黄色黏稠液体，加入 NAE18（0.84g，2.6mmol），DMAP（0.1g，0.8mmol），Et_3N（0.47g，4.7mmol），25mL $CHCl_3$ 溶解，室温搅拌，TLC［展开剂为 *V*（石油醚）：*V*（乙酸乙酯）＝1:1］监测反应完全后，停止反应，水洗 3 次，无水 $MgSO_4$ 干燥后过滤，浓缩至干，残余物柱层析分离，*V*（石油醚）：*V*（乙酸乙酯）＝2:1 和 1:1 淋洗得到产物。

4a：R=H，乳白色固体，收率41%，m. p. 32~33℃；^1H NMR（$CDCl_3$，300MHz）δ：7.34~7.30（m，2H，ArH），7.19~7.15（t，J=8.60Hz，3H，ArH），6.33（bs，1H，NHCO），4.19~4.12（m，2H，$NHCH_2C\mathbf{H}_2O$），3.55~3.44［m，10H，$NHC\mathbf{H}_2CH_2O$，N（$CH_2CH_2Cl)_2$］，2.08（t，J=7.04Hz，2H，CH_2CO），1.54（t，J=6.26Hz，2H，$C\mathbf{H}_2CH_2CO$），1.21［s，28H，（$CH_2)_{14}$］，0.84［t，J=7.04Hz，3H，$C\mathbf{H}_3$（$CH_2)_{14}$］；^{31}P NMR（$CDCl_3$，300MHz）δ：6.41。Anal. calcd for $C_{30}H_{53}Cl_2N_2O_4P$：C 59.30，H 8.79，N 4.61；found C 59.20，H 8.49，N 4.60。

4b：R=p－F，乳白色固体，收率52%，m. p. 36~38℃；^1H NMR（$CDCl_3$，300MHz）δ：7.20~7.17（d，J=9.04Hz，2H，ArH），7.06~7.01（d，J=9.04Hz，2H，ArH），6.23（bs，1H，NHCO），4.23~4.16（m，2H，$NHCH_2C\mathbf{H}_2O$），3.63~3.47［m，10H，$NHC\mathbf{H}_2CH_2O$，N（$CH_2C\mathbf{H}_2Cl)_2$］，2.13（t，J=7.54Hz，2H，CH_2CO），1.58（t，J=7.54Hz，2H，$C\mathbf{H}_2CH_2CO$），1.25［s，28H，（$CH_2)_{14}$］，0.88［t，J=6.03Hz，3H，$C\mathbf{H}_3(CH_2)_{14}$］；^{31}P NMR（$CDCl_3$，300MHz）δ：5.70。Anal. calcd for $C_{30}H_{52}Cl_2FN_2O_4P$：C 57.59，H 8.38，N 4.48；found C 57.33，H 8.11，N 4.45。

4c：R=p－Cl，乳白色固体，收率53%，m. p. 33~34℃；^1H NMR（$CDCl_3$，300MHz）δ：7.34~7.31（d，J=9.04Hz，2H，ArH），7.18~7.16（d，J=9.04Hz，2H，ArH），6.25（bs，1H，NHCO），4.24~4.17（m，2H，$NHCH_2C\mathbf{H}_2O$），3.63~3.45［m，10H，$NHC\mathbf{H}_2CH_2O$，N（$CH_2CH_2Cl)_2$］，2.14（t，J=7.54Hz，2H，CH_2CO），1.58（t，J=6.78Hz，2H，$C\mathbf{H}_2CH_2CO$），1.25［s，28H，（$CH_2)_{14}$］，0.88［t，J=6.78Hz，3H，$C\mathbf{H}_3$（$CH_2)_{14}$］；^{31}P NMR（$CDCl_3$，300MHz）δ：5.56。Anal. calcd for $C_{30}H_{52}Cl_3N_2O_4P$：C 56.12，H 8.16，N 4.36；found C 56.16，H 8.16，N 4.26。

4d：R=p－Br，乳白色固体，收率41%，m. p. 37~39℃；^1H NMR（$CDCl_3$，300MHz）δ：7.48~7.45（d，J=9.04Hz，2H，ArH），7.13~7.10（d，J=9.04Hz，2H，ArH），6.18（bs，1H，NHCO），4.23~4.17（m，2H，$NHCH_2C\mathbf{H}_2O$），3.63~3.45［m，10H，$NHC\mathbf{H}_2CH_2O$，N（$CH_2CH_2Cl)_2$］，2.13（t，J=7.54Hz，2H，CH_2CO），1.58（t，J=6.78Hz，2H，$C\mathbf{H}_2CH_2CO$），1.25［s，28H，（$CH_2)_{14}$］，0.88［t，J=6.78Hz，3H，$C\mathbf{H}_3$（$CH_2)_{14}$］；^{31}P NMR（$CDCl_3$，300MHz）δ：5.40。Anal. calcd for $C_{30}H_{52}BrCl_2N_2O_4P$：C 52.48，H 7.63，N 4.08；found C 52.39，H 7.77，N 4.01。

4e：R=o－CH_3，淡黄色固体，收率46%，m. p. <30℃；^1H NMR（$CDCl_3$，300MHz）δ：7.33~7.30（d，J=9.04Hz，1H，ArH），7.23~7.17（t，J=9.04Hz，2H，ArH），7.12~7.09（t，J=7.54Hz，1H，ArH），6.10（bs，1H，NHCO），4.24~4.09（m，2H，$NHCH_2C\mathbf{H}_2O$），3.67~3.47［m，10H，$NHC\mathbf{H}_2CH_2O$，N（$CH_2CH_2Cl)_2$］，2.31（s，3H，$ArCH_3$），2.07（t，J=8.29Hz，2H，CH_2CO），1.55（t，J=7.54Hz，2H，$C\mathbf{H}_2CH_2CO$），1.25［s，28H，（$CH_2)_{14}$］，0.88［t，J=6.78Hz，3H，$C\mathbf{H}_3$（$CH_2)_{14}$］；^{31}P NMR（$CDCl_3$，300MHz）δ：5.65。Anal. calcd for $C_{31}H_{55}Cl_2N_2O_4P$：C 59.89，H 8.81，N 4.51；found C 59.75，H 8.92，N 4.75。

4f：R=m－CH_3，淡黄色固体，收率55%，m. p. 31~35℃；^1H NMR（$CDCl_3$，300MHz）δ：7.25~7.20（t，J=7.54Hz，1H，ArH），7.02~6.99（d，J=9.04Hz，3H，ArH），6.23（bs，1H，NHCO），4.22~4.16（m，2H，$NHCH_2C\mathbf{H}_2O$），3.62~3.45［m，10H，$NHC\mathbf{H}_2CH_2O$，N（$CH_2CH_2Cl)_2$］，2.36（s，3H，$ArCH_3$），2.11（t，

$J = 7.54$Hz, 2H, CH_2CO), 1.58 (t, $J = 7.54$Hz, 2H, $C\mathbf{H}_2CH_2CO$), 1.25 [s, 28H, $(CH_2)_{14}$], 0.88 [t, $J = 7.54$Hz, 3H, $C\mathbf{H}_3(CH_2)_{14}$]; ^{31}P NMR ($CDCl_3$, 300MHz) δ: 5.41。Anal. calcd for $C_{31}H_{55}Cl_2N_2O_4P$: C 59.89, H 8.92, N 4.51; found C 59.70, H 9.05, N 4.59。

4g: R = $m-CH_3O$，白色固体，收率 55%，m.p. < 30℃; ^1H NMR ($CDCl_3$, 300MHz) δ: 7.35~7.32 (d, $J = 8.29$Hz, 1H, ArH), 7.19~7.13 (t, $J = 8.29$Hz, 1H, ArH), 6.98~6.89 (t, $J = 9.04$Hz, 2H, ArH), 6.25 (bs, 1H, NHCO), 4.25~4.16 (m, 2H, $NHCH_2C\mathbf{H}_2O$), 3.88 (s, 3H, $ArOCH_3$), 3.63~3.46 [m, 10H, $NHC\mathbf{H}_2CH_2O$, N($CH_2-CH_2Cl)_2$], 2.12 (t, $J = 7.54$Hz, 2H, CH_2CO), 1.58 (t, $J = 7.54$Hz, 2H, $C\mathbf{H}_2CH_2CO$), 1.25 [s, 28H, $(CH_2)_{14}$], 0.88 [t, $J = 7.54$Hz, 3H, $C\mathbf{H}_3(CH_2)_{14}$]; ^{31}P NMR ($CDCl_3$, 300MHz) δ: 5.80. Anal. calcd for $C_{31}H_{55}Cl_2N_2O_5P$: C 58.39, H 8.69, N 4.39; found C 58.27, H 8.70, N 4.48。

4h: R = $p-t-Bu$，白色固体，收率 45%，m.p. 33~35℃; ^1H NMR ($CDCl_3$, 300MHz) δ: 7.37~7.34 (d, $J = 8.29$Hz, 2H, ArH), 7.13~7.11 (d, $J = 9.04$Hz, 2H, ArH), 6.34 (bs, 1H, NHCO), 4.22~4.17 (m, 2H, $NHCH_2C\mathbf{H}_2O$), 3.61~3.44 [m, 10H, $NHC\mathbf{H}_2CH_2O$, N($CH_2CH_2Cl)_2$], 2.13 (t, $J = 8.29$Hz, 2H, CH_2CO), 1.59 (t, $J = 7.54$Hz, 2H, $C\mathbf{H}_2CH_2CO$), 1.31 (s, 9H, $t-$BuH), 1.25 [s, 28H, $(CH_2)_{14}$], 0.88 [t, $J = 6.78$Hz, 3H, $C\mathbf{H}_3(CH_2)_{14}$]; ^{31}P NMR ($CDCl_3$, 300MHz) δ: 5.55。Anal. calcd for $C_{34}H_{61}Cl_2N_2O_4P$: C 61.52, H 9.26, N 4.22; found C 61.31, H 9.11, N 4.27。

4i: R = 3, 4 $-(CH_3)_2$，白色固体，收率 71%，m.p. < 30℃; ^1H NMR ($CDCl_3$, 300MHz) δ: 7.09~7.07 (d, $J = 8.29$Hz, 1H, ArH), 6.98 (s, 1H, ArH), 6.94~6.92 (d, $J = 8.29$Hz, 1H, ArH), 6.27 (bs, 1H, NHCO), 4.22~4.13 (m, 2H, $NHCH_2C\mathbf{H}_2O$), 3.62~3.44 [m, 10H, $NHC\mathbf{H}_2CH_2O$, N($CH_2CH_2Cl)_2$], 2.25 (s, 3H, $ArCH_3$), 2.23 (s, 3H, $ArCH_3$), 2.11 (t, $J = 8.29$Hz, 2H, CH_2CO), 1.58 (t, $J = 7.54$Hz, 2H, $C\mathbf{H}_2CH_2CO$), 1.25 [s, 28H, $(CH_2)_{14}$], 0.88 [t, $J = 6.78$Hz, 3H, $C\mathbf{H}_3(CH_2)_{14}$]; ^{31}P NMR ($CDCl_3$, 300MHz) δ: 5.61。Anal. calcd for $C_{32}H_{57}Cl_2N_2O_4P$: C 60.46, H 9.04, N 4.41; found C 60.59, H 8.99, N 4.39。

2 结果和讨论

2.1 目标化合物的合成

实验中先用氮芥磷酰二氯和取代苯酚反应后，过滤除去三乙胺盐酸盐，浓缩至干，得到的黄色液体即为氮芥芳基磷酰氯 **3**。**3** 未经分离，直接在 DMAP 作催化剂、三乙胺作缚酸剂的条件下与 NAE18 进行反应，合成得到目标化合物 **4**。实验表明，在 **3** 与 NAE18 的反应过程中，DMAP 起着关键性的催化作用。同样的反应条件，不加 DMAP 时，反应几乎不发生。为提高产率，我们将反应温度提高，在回流的条件下进行反应，结果对产率并没有帮助。当使用 DMAP 同时作为缚酸剂和催化剂时，产率反而降低，副产物大大增加。

在柱层析过程中，我们分离到了一极性较小的白色固体，经核磁以及 EI-MS 证实为 NAE 的氯代产物，即 *N*-（2-氯乙基）硬脂酰胺 **D**。推测 DMAP 催化反应进行以及副产物生成机理见 Scheme 2。DMAP 先与氮芥芳基磷酰氯作用得到中间体 **A**，接着 NAE 与其亲核加成，生成磷的五配位体 **B**。在缓解中间体 **B** 张力以及形成稳定的磷氧双键的驱动下，DMAP 离去，生成中间体 **C**。反应过程中生成的氯负离子进攻 **C** 中 NAE 羟基上的氢，形成目标产物磷酸酯；反之，如果进攻 NAE 中与羟基相连的亚甲基，则生成副产物 **D**。

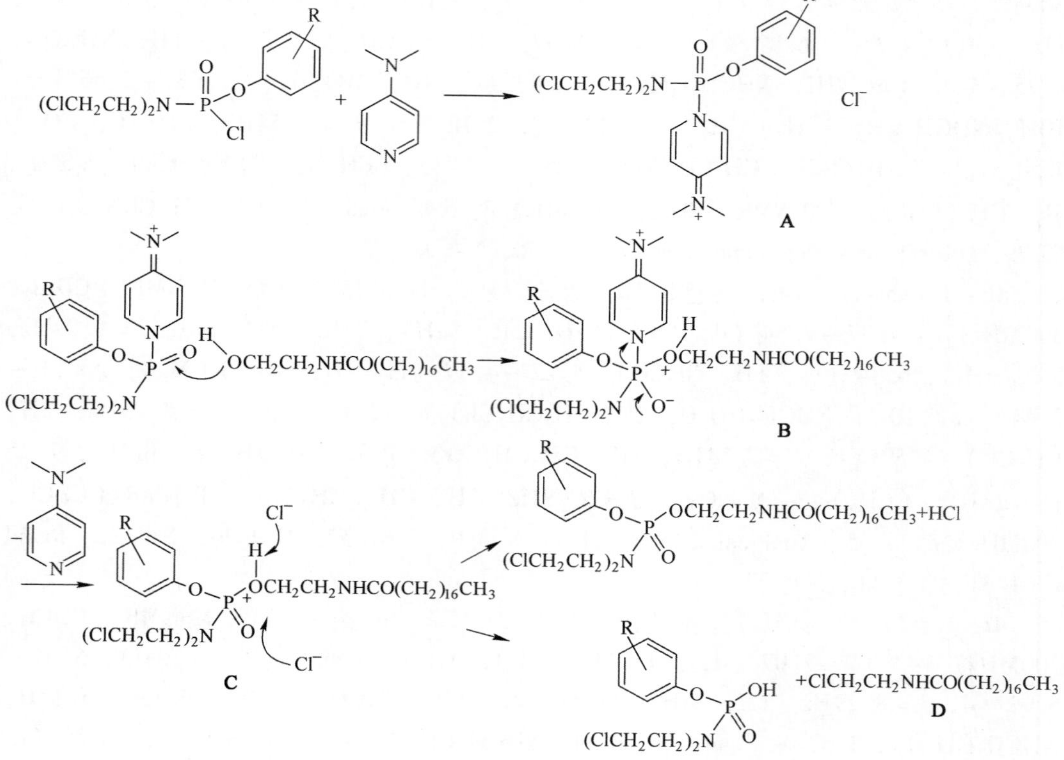

Scheme 2

2.2 目标化合物的谱学特征

从目标产物的核磁数据可以看出，产物氢的吸收峰主要分为三个区域，最低场为苯环氢的吸收峰，随取代基的变化以及取代位置的变化而有所不同；中间区域主要是与电负性较大的氮、氧等相连的亚甲基氢，其中与氧相连的亚甲基的氢的吸收峰偏向低场，而与氯和氮相连的亚甲基氢的吸收峰互相重叠，于 δ 3.6 ~ 3.4 之间表现为多重峰；脂肪酰基长链氢的吸收峰处于最高场，其中甲基氢的吸收峰于 δ 0.9 ~ 0.86 处表现为三重峰，紧挨着的是多个亚甲基的吸收峰互相重叠所表现出来的一个宽的单峰，与羰基相连的亚甲基由于受羰基的去屏蔽作用，其吸收峰位于此区域的最低场，同样受羰基的影响，羰基 β 位的亚甲基的吸收峰也相对其他亚甲基偏向低场，在 δ 1.6 ~ 1.5 之间表现为一较宽的三重峰。此类化合物的 ^{31}P NMR 吸收峰主要分布在 δ 4 ~ 6 之间。

在目标化合物的红外谱图中，于 1218cm^{-1} 处有一强的 P ═O 基团的特征吸收峰，酰胺中 C ═O 在 1649cm^{-1} 处有一强的吸收峰；同时，在 3304cm^{-1} 处出现酰胺中 N—H 的伸缩振动吸收峰，在 2925 和 2854cm^{-1} 处出现 CH_3 和 CH_2 的伸缩振动吸收峰。目标化合物的质谱图表明，此类化合物一般得不到分子离子峰，*m/z* 85 是其丰度最大的离子，其次则为 m/z 98。推测其碎裂机理见 Scheme 3。

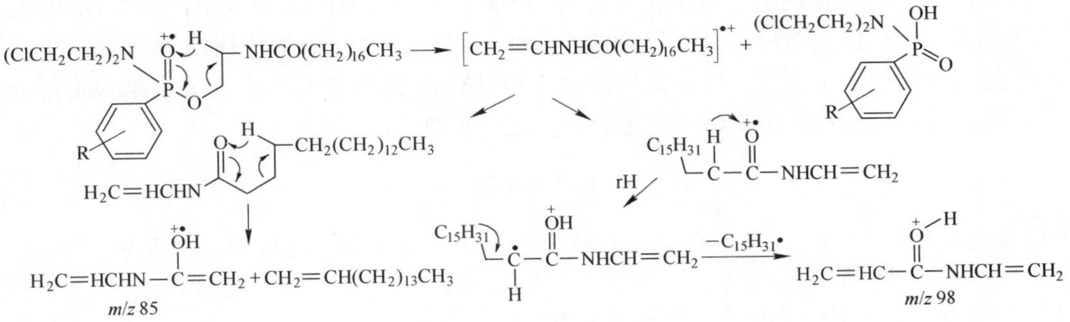

Scheme 3

2.3 目标化合物的生物活性初探

文献表明外源 NAE 不仅能激活防卫基因的表达，而且在秧苗根细胞成长中起着重要作用。例如，用 $\mu mol \cdot L^{-1}$浓度 NAE 处理生长了三天的秧苗 180min 后，根长度显著下降[6]。另外，在细胞悬浮液中加入 $100\mu mol \cdot L^{-1}$浓度的 NAE，对病原诱导剂所引发的碱化有着 65% 的抑制作用[3]。因此，除了化合物 **4h**，所有目标化合物均进行了植物生长调节活性（黄瓜子叶扩张法）和杀菌活性测定，并与 NAE18 的生物活性结果进行了比较，结果见表 1。

表 1　目标化合物的生物活性

Table 1　Biological activity of target compounds

化合物	抑菌活性(%)									促生活性(%)
	离体平皿法（50mg/L）					活体小株法（500mg/L）				黄瓜子叶扩张法（10mg/L）
	小麦赤霉	番茄早疫	花生褐斑	苹果轮纹	芦笋茎枯	小麦锈病	烟草病毒	黄瓜灰霉	油菜菌核	黄瓜
4a	0	0	0	21.9	0	6	11	0	0	5.9
4b	0	10.7	0	19.5	0	0	20	0	13.3	1.5
4c	0	0	0	19.5	0	11	8	0	18.7	7.0
4d	9.1	0	0	41.5	0	6	7	0	0	4.5
4e	12.1	0	0	26.8	0	0	14	0	18.7	6.2
4f	0	0	15.0	19.5	0	0	8	0	0	9.1
4g	12.1	0	0	24.4	0	11	8	0	12.5	4.2
4i	0	0	0	24.4	0	0	15	0	0	10.3
NAE18	0	14.3	0	24.4	0	6	13	0	0	6.2

从表中可以看出，当浓度为 10mg/L 时，目标化合物对黄瓜子叶扩张具有一定的促进作用。相对母体化合物 NAE18 而言，化合物 **4c**，**4f**，**4i** 的活性有所加强，其中 **4i** 对黄瓜子叶扩张的促进作用达到了 10.3%，**4e** 与 NAE 活性相当。目标化合物杀菌活性测定结果表明，部分化合物对相应病害的防治作用相对 NAE18 有所增强。例如，**4d**，**4e** 用于防治苹果轮纹、**4c**，**4g** 用于防治小麦锈病以及 **4b**，**4e** 和 **4i** 用于防治烟草病毒时，它们的抑菌活性均高于 NAE18，特别是 **4d** 对苹果轮纹的抑制活性达到了 41.5%。另外

4d，4e 和 4g 用于防治小麦赤霉病，4f 用于防治花生褐斑病，4b，4c，4e 和 4g 用于防治油菜菌核病时均表现出一定的活性，而 NAE18 对这三种病菌则完全没有抑制作用。总的说来，经结构修饰后，目标化合物的活性相对母体化合物 NAE18 有所加强，但所得到的活性数据并不理想。即使是母体化合物 NAE18 的活性结果也不如文献报道的显著，有关目标化合物的生物活性测定仍有待进一步研究。

References

[1] Kuehl, F. F.; Jacob, T. A.; Ganley, O. H.; Ormond, R. E.; Meisinger, M. A. P. *J. Am. Chem. Soc.* 1957, 79, 5577.

[2] Chapman, K. D. *Prog. Lipid Res.* 2004, 43, 302.

[3] Tripathy, S.; Venables, B. J.; Chapman, K. D. *Plant Physiol.* 1999, 121, 1299.

[4] Austin – Brown, S.; Chapman, K. D. *Plant Physiol.* 2002, 129, 1892.

[5] Palta, J. P.; Ryu S. B. *US* 6426105B1, 1999 [*Chem. Abstr.* 1999, 130, 321890].

[6] Blancaflor, E. B.; Hou, G.; Chapman, K. D. *Planta* 2003, 217, 206.

[7] Chapman, K. D. *WO* 01/30143, 2001 [*Chem. Abstr.* 2001, 134, 337133].

[8] Chapman, K. D. *Chem. Phys. Lipids* 2000, 108, 221.

[9] Randall, D. I.; Stahl, C. R. *DE* 1815999, 1969 [*Chem. Abstr.* 1969, 71, 124668].

[10] Sommer, K. *DE* 2158765, 1973 [*Chem. Abstr.* 1973, 79, 53558a].

[11] Chen, R. – Y.; Wang, H. – L.; Zhou, *J. Heteroat. Chem.* 1994, 5, 497.

[12] Kuang, Y. – M.; Li, Z. – H. *Chemistry* 2004, 67, 449 (in Chinese).
(匡宇明，李中华，化学通报，2004，67，449.)

[13] Roe, E. T.; Miles, T. D.; Swern, D. *J. Am. Chem. Soc.* 1952, 74, 3442.

[14] Li Z. – G. *The Preparation of Organic Intermediates*, Chemical Industry Press, Beijing, 2001, p. 203 (in Chinese).
(李在国，有机中间体制备，北京：化学工业出版社，2001，p. 203.)

Synthesis and Biological Activity of *O* – Substituted Phenyl *O* – 2 – Stearamidoethyl *N*, *N* – Bis(2 – chloroethyl) – phosphoramidate

Liang Han[1], Zhengming Li[1], Yun Zhang[2], Weiming Guo[2]

([1]National Key Laboratory of Elemento – organic Chemistry, Research Institute of Elemento – Organic Chemistry, Nankai University, Tianjin, 300071;
[2]College of Horticulture, Nanjing Agricultural University, Nanjing, 210095)

Abstract *N* – Acylethanolamine (NAE), an endogenous substance of plants, plays important role in regulation of plant growth. To make up for its deficiency in solubility and transportation *in vivo* that is led by its long fatty chain, NAE18 was introduced into nitrogen mustard phosphorate and a series of title compounds were prepared. In the reaction of NAE18 with nitrogen mustard aryl phosphorochloridate, 4 – dimethylaminopyridine (DMAP) played critical catalytic action. Under the same condition without DMAP, no reaction happened. The plant growth and antifungal bioactivities of title compounds were tested and the data suggested that after the structural modification most title compounds have better bioactivity than NAE18.

Key words *N* – acylethanolamine; nitrogen mustard phosphorate; synthesis; biological activity

3－N－苄氧羰基－β－氨基丁酸水杨酸酯类化合物的合成及其生物活性*

臧洪俊[1]　李正名[2]　倪长春[3]　沈　宙[3]　范志金[2]　刘秀峰[2]

（[1]天津工业大学材料科学与化学工程学院，天津　300160；
[2]南开大学元素有机化学研究所，元素有机化学国家重点实验室，天津　300071；
[3]国家南方农药创制中心上海基地，上海　200032）

摘　要　利用活性分子组合的原理，把两种具有诱导活性的水杨酸酯类化合物和 D，L－β－氨基丁酸连接在一个分子内，合成了 9 个新化合物，所有新化合物经元素分析及 ^{1}H NMR 确认。生物活性初步测定结果表明，该类化合物具有一定的诱导活性。

关键词　D，L－β－氨基丁酸　水杨酸酯　植物诱导抗性　生物活性

β－氨基丁酸（BABA）是一种具有诱导抗性的新型抗病激活剂，是由番茄根系分泌的非蛋白质氨基酸[1]，可诱导番茄、马铃薯、棉花、花生、西瓜、向日葵、辣椒等作物对卵菌或真菌病害蛋白获得抗病性[2]，拟南芥和烟草等植物产生诱导抗病性[3,4]，对环境安全、具有高效诱抗作用的非质氨基酸。水杨酸是植物体内普遍存在的一种酚类化合物，已经证实它具有诱导植物提高抗病性、引起天南星科植物佛焰花序生热、调节植物光周期等作用，因此，Raskin[5]提出水杨酸是一种新的植物激素。有关水杨酸对植物抗病诱导的报道所涉及的植物还包括番茄、黄瓜、水稻、大蒜、大豆、甜菜、拟南芥等[6]。因此，本文利用活性分子组合的原理[7]，把两种具有诱导活性的水杨酸酯类化合物和 D，L－β－氨基丁酸组合在一个分子内，共合成了 9 个新化合物，合成路线如下：

$H_3C-CH{=}CH-COOH + H_2NCONH_2 \xrightarrow{\text{Ethylene glycol}}$ **1** $\xrightarrow{NaOH/H_2O}\ \xrightarrow{H^+\text{(strong acidic cation exchanger)}}$

$H_3C-CH(\overset{+}{N}H_3)-CH_2-COO^-$ (**2**) $\xrightarrow[\text{NaOH(aq.)}]{C_6H_6CH_2OCOCl/\text{acetone}}$ $H_3C-CH(NHCO_2CH_2Ph)-CH_2-COOH$ (**3**)

$H_3C-CH(NHCOCH_2Ph)-CH_2-COOH$ (**3**) + ROOC(HO)C$_6$H$_4$ (**4**) $\xrightarrow[CHCl_3]{DCC,\ DMAP}$ C$_6$H$_4$(COOR)(OCOCH$_2$CHCH$_3$(NHCOCH$_2$Ph)) (**5**)

R＝CH_3—（a），C_2H_5—（b），n－C_3H_7—（c），n－C_4H_9—（d），$(CH_3)_2CH$—（e），CH_3CH（Cl）（f），$(CH_3)_2CHCH_2CH_2$—（g），$CH_3CH_2CH_2CH_2CH(CH_2CH_3)CH_2$（h），$(CH_3)_2CHCH_2$—（i）

* 原发表于《高等学校化学学报》，2006，27（3）：468－471。基金项目：国家自然科学基金重点项目（批准号：20432010）和国家自然科学基金（批准号：30270883）资助。

1 结果与讨论

1.1 化合物2的合成

按文献[8]方法，采用0.074mm的阳离子交换树脂为柱层析的吸附剂，分离时间比较长，需要8h完成，且该树脂比较昂贵。本文采用0.297mm以下的阳离子交换树脂为柱层析的吸附剂，在1h内分离得到产品，且得到很好的收率。

1.2 中间体3和化合物5的合成

按文献[9]方法，在后处理中，水相用乙酸乙酯萃取2次，无水硫酸钠干燥，减压浓缩得产品。我们将水相冷却到0℃，用6mol/L盐酸调pH = 2，即可析出白色固体，过滤，真空干燥，得化合物**5**，产率为83%，物理数据和元素分析结果见表1。

Table 1 Physical data and elemental analysis of compounds 5

Comp.	R	m. p. (℃)	Yield(%)	Elemental analysis(%, Calc.)		
				C	H	N
5a	CH_3	102~103	57.5	6.61(64.69)	5.79(5.66)	3.78(3.77)
5b	C_2H_5	100~101	45.6	65.35(65.45)	5.99(6.02)	3.55(3.64)
5c	C_3H_7	64~62	58.2	66.11(66.17)	6.46(6.27)	3.57(3.51)
5d	C_4H_9	78~79	54.0	66.82(66.83)	6.42(6.54)	3.29(3.39)
5e	$(CH_3)_2CH$	70~71	53.9	66.05(66.17)	6.20(6.27)	3.41(3.51)
5f	CH_3CHCl	74~75	56.7	60.01(60.07)	5.15(5.24)	3.16(3.34)
5g	$(CH_3)_2CHCH_2CH_2$	84~85	52.6	67.43(67.45)	6.68(6.79)	3.44(3.28)
5h	$CH_3(CH_2)_3CH(CH_2CH_3)CH_2$	74~75	47.8	69.11(69.08)	7.55(7.46)	2.94(2.99)
5i	$(CH_3)_2CHCH_2$	68~69	53.2	66.88(66.83)	6.47(6.54)	3.51(3.39)

1.3 ^1H NMR谱解析

化合物的^1H HMR谱的数据见表2。在^1H NMR中，3-*N*-苄氧羰基-β-氨基丁酸水杨酸酯的甲基质子化学位移δ1.24~1.41（双峰），归于甲基与一个次甲基（CH）相连接；亚甲基质子化学位移δ2.77~2.91（多峰），这是由于亚甲基连接了一个不对称C原子，使亚甲基上的两个氢表现为磁不等价，同碳耦合裂分为四重峰，同时又受到手性C上氢的耦合，因此在谱图上表现为多重峰；次甲基质子化学位移δ4.14~4.26（多峰），苄氧羰基上的亚甲基质子化学位移δ5.04~5.22（单峰）；$NHCOOCH_2Ph$的活泼NH的质子化学位移δ5.44~5.68（宽峰）；芳环上氢的化学位移值在δ7.09~8.04（多峰）。

Table 2 ^1H NMR data of compounds 5

Comp.	^1H NMR ($CDCl_3$), δ
5a	1.40 (d, 3H, CH_3), 2.89 (m, 2H, CH_2 CHN), 3.83 (s, 3H, $COOCH_3$), 4.25~4.29 (m, 1H, CHN), 5.13 (s, 2H, OCH_2 Ph), 5.48~5.52 (br s, 1H, NH), 7.09~8.04 (m, 9H, PhH)

Comp.	^1H NMR ($CDCl_3$), δ
5b	1.33～1.41 (m, 6H, $2CH_3$), 2.86～2.91 (m, 2H, CH_2 CHN), 4.29～4.31 (m, 3H, $COOCH_2$, CHN), 5.13 (s, 2H, OCH_2 Ph), 5.54～5.56 (b r. s, 1H, NH), 7.08～8.04 (m, 9H, PhH)
5c	0.99～1.02 (m, 3H, CH_2 **CH_3**), 1.38～1.41 (d, 3H, CH_3), 1.60～1.75 (m, 2H, **CH_2** CH_3), 2.86～2.91 (m, 2H, CH_2 CHN), 4.20～4.22 (m, 3H, $COOCH_2$, CHN), 5.13 (s, 2H, OCH_2 Ph), 5.52～5.56 (br. s, 1H, NH), 7.08～8.04 (m, 9H, PhH)
5d	0.88 (t, 3H, CH_2 CH_3), 1.31 (d, 2H, CH_3), 1.32～1.64 (m, 4H, CH_3 **CH_2** **CH_2**), 2.77～2.79 (m, 2H, CH_2 CHN), 4.14～4.16 (m, 3H, COOC**H_2** CH_2, C**H**N), 5.04 (s, 2H, OCH_2 Ph), 5.46～5.48 (b r. s, 1H, NH), 7.00～7.94 (m, 9H, PhH)
5e	1.29～1.41 (m, 9H, $3CH_3$), 2.86～2.87 (m, 2H, CH_2 CHN), 4.26～4.30 (m, 1H, CHN), 5.12～5.22 [m, 3H, OCH_2 Ph, CH (CH_3)$_2$], 5.61～5.63 (br. s, 1H, NH), 7.11～8.02 (m, 9H, PhH)
5f	1.39 (d, 3H, CH_3), 2.87～2.92 (m, 2H, CH_2 CHN), 3.72～3.74 (m, 3H, CH_3), 4.27～4.29 (m, 1H, CHN), 4.45～4.48 (m, 1H, ClCH CH_3), 5.12～5.22 [m, 3H, OCH_2Ph, **CH** (CH_3)$_2$], 5.49 (b r. s, 1H, NH), 7.10～8.08 (m, 9H, PhH)
5g	0.91～0.96 (m, 6H, $2CH_3$), 1.24～1.62 (m, 6H, CH_2 CH, CH_3), 2.85～2.91 (m, 2H, CH_2 CHN), 4.10～4.15 (m, 1H, CHN), 4.25～4.30 (m, 2H, OC**H_2**CH_2), 5.13 (m, 2H, OCH_2 Ph), 5.55～5.57 (br. s, 1H, NH), 7.11～8.02 (m, 9H, PhH)
5h	0.91～0.96 (m, 6H, $2CH_3$), 1.24～1.62 (m, 6H, CH_2 CH, CH_3), 2.85～2.91 (m, 2H, CH_2 CHN), 4.10～4.15 (m, 1H, CHN), 4.25～4.30 (m, 2H, OC**H_2**CH_2), 5.13 (m, 2H, OCH_2 Ph), 5.55～5.57 (br. s, 1H, NH), 7.11～8.02 (m, 9H, PhH)
5i	0.98～1.00 [d, 6H, CH (CH_3)$_2$], 1.38～1.41 (d, 3H, CH_3), 1.99～2.08 [m, 1H, **CH** (CH_3)$_2$], 2.85～2.87 (m, 2H, CH_2 CHN), 4.02～4.04 (d, 2H, OC**H_2** CH), 4.20～4.30 (m, 1H, CHN), 5.13 (m, 2H, OCH_2 Ph), 5.51～5.62 (br. s, 1H, NH), 7.08～8.02 (m, 9H, PhH)

1.4 生物活性

1.4.1 离体抗菌活性测定结果

在化合物质量分数为0.05%的条件下，采用平皿法对部分化合物的杀菌活性进行了测试（表3），结果显示，这9种化合物对供试的20种真菌菌体生长中部分菌有较弱活性（用BABA作对照）。

Table 3 Inhibition ratio of compounds 5 towards fungi *in vitro* at 0.05% (mass fraction)

Fungi	5a	5b	5d	5f	5g	5h	5i	5j	BABA
Cercospora beticola Sacc.	0	0	0	0	0	0	0	0	0
Fusarium moniliforme	0	0	0	0	0	0	0	0	0
Fusarium oxysporum	0	0	0	0	0	0	0	0	0

续表 3

Fungi	5a	5b	5d	5f	5g	5h	5i	5j	BABA
Cercosporium arachidicola	0	0	0	0	0	0	0	0	0
Alternaria solani	0	0	0	0	0	0	0	0	0
Gibberellazeae	0	0	0	0	29. 4	0	0	0	0
Oniothyrium diplodiella (*Speg.*) *Sacc*	0	0	0	0	0	0	0	0	0
Physalospora p iricola Nose	36. 2	42. 5	44. 6	31. 9	36. 2	0	34. 0	31. 9	0
Cladosporium cucumerinum	0	0	0	0	0	0	0	0	0
A. kikuchiana	0	0	0	0	0	0	0	0	0
Botrytis c inerea Pers	14. 5	27. 1	0	20. 8	16. 7	27. 1	20. 8	27. 1	0
Alternariamali	0	0	0	0	0	0	0	0	0
Valsa mali	0	23. 1	26. 9	23. 1	23. 1	23. 1	23. 1	23. 1	0
Glorosprium musarum Cookeet Mass	31. 0	0	0	0	34. 5	31. 0	34. 5	34. 5	0
Rhizoctonia solani	24. 6	0	24. 6	20. 0	44. 6	46. 1	23. 1	44. 6	0
Colletotrichum arbiculara	0	0	0	0	0	0	0	0	0
Glomerella cingulata	0	0	0	0	0	0	25. 0	0	0
Bipolaris sorokinianum	0	0	0	0	0	0	0	0	0
Rhizoctonia solani Ktihn	29. 8	12. 3	29. 8	12. 3	45. 6	47. 3	12. 3	45. 6	0
Colletrichum lagenarium	0	0	0	0	0	0	0	0	0

1. 4. 2 盆栽法测定结果

在化合物质量分数为0. 05% 的条件下，采用叶面喷雾方法对部分化合物的抑菌活性进行实验（表4），分4个间隔期，即接种前7d，5d，3d和1d对叶面定量喷雾处理，用水杨酸（ SA ）作对照测试。结果表明，该类化合物对细菌白叶枯病、真菌性稻瘟病、瓜类白粉病、炭疽病均有一定的防治效果，活性谱较广。化合物**5**a在喷药7d后，对水稻稻瘟病、瓜类白粉病的抑制率分别为72. 8% 和71. 6% ；化合物**5**b在喷药7d后，水稻白叶枯病、瓜类白粉病的抑制率分别为74% 和60. 5%；化合物**5**h在喷药7d后，对水稻稻瘟病的抑制率为72. 8%；化合物**5**a，**5**b，**5**d，**5**f对瓜类炭疽病在喷药3d或5d后，抑制率为62. 9% ~81. 4%，且其诱导活性高于水杨酸。说明该类化合物符合植物激活剂的特点，叶面喷洒后激活了黄瓜体内的诱导抗病性，部分化合物显示出较好的活性。

Table 4 Inhibition ratio（%）of compounds 5 towards fungi *in vivo* at 0. 05%（mass fraction)

Comp.	*Xanthomonasoryzea*				*Pyriricurariaoryzea*				*Helminthosporium maydis*				*Sphaerotheca fuliginea*				*Cucurbitsanthracnose*			
	1d	3d	5d	7d	1d	3d	5d	7d	1d	3d	5d	7d	1d	3d	5d	7d	1d	3d	5d	7d
5a	34. 9	0. 0	74. 3	21. 0	2. 6	23. 6	40. 0	72. 8	28. 6	0. 0	71. 9	37. 5	50. 7	73. 8	35. 4	71. 6	26. 3	80. 3	71. 9	37. 5
5b	14. 7	0. 0	69. 5	74. 0	84. 2	35. 3	80. 0	18. 3	48. 9	0. 0	80. 6	62. 2	30. 8	55. 2	52. 2	60. 5	35. 1	62. 9	80. 6	62. 2
5d	23. 6	0. 0	75. 4	45. 0	52. 6	35. 3	66. 7	54. 5	31. 4	0. 0	77. 2	33. 6	53. 5	46. 7	27. 6	47. 0	26. 3	56. 8	77. 2	33. 6
5f	68. 2	0. 0	61. 0	64. 0	73. 6	11. 8	53. 4	18. 3	59. 4	6. 7	81. 4	29. 7	47. 5	42. 3	44. 5	54. 4	14. 0	75. 1	81. 4	29. 7
5h	51. 1	0. 0	42. 6	10. 1	31. 6	23. 8	40. 0	72. 8	31. 6	12. 3	46. 8	30. 6	50. 5	56. 0	36. 8	51. 6	3. 5	61. 0	46. 8	30. 6

续表 3

Comp.	*Xanthomonasoryzea*				*Pyriricurariaoryzea*				*Helminthosporium maydis*				*Sphaerotheca fuliginea*				*Cucurbitsanthracnose*			
	1d	3d	5d	7d	1d	3d	5d	7d	1d	3d	5d	7d	1d	3d	5d	7d	1d	3d	5d	7d
5i	51.8	50.7	52.9	4.9	63.2	58.9	40.0	18.3	39.3	0.0	74.5	44.4	33.8	44.6	54.1	24.7	24.6	65.7	74.5	44.4
SA	48.4	0.0	51.5	32.3	42.0	0.0	13.4	0.0	50.1	0.0	68.1	47.7	29.6	50.8	39.1	37.7	56.1	77.0	68.1	47.7

2 实验部分

2.1 仪器及试剂

Yanaco C H N CORDER MT－3 型元素自动分析仪；BrukerAC－P300 型核磁共振仪，TMS 为内标；Yanaco MP－500 熔点仪；薄层层析硅胶 GF254，TLC 紫外检测或质量分数为 5% 的水合磷钼酸乙醇溶液显色，柱层析硅胶 H（60 型）为青岛海洋化工厂产品。所有溶剂均为分析纯，用前经重蒸处理；试剂为分析纯或化学纯。

2.2 中间体 1，2，3 及水杨酸酯 4 的合成

中间体 **1** 按文献［10］的方法合成；中间体 **2** 按文献［8］的方法合成。中间体 3 参照文献［9］的方法合成。熔点 122℃（文献值为 123℃）。水杨酸甲酯、水杨酸乙酯为市售产品，其他水杨酸酯参照文献［11］方法制备。

2.3 目标化合物 5 的合成

在 100mL 圆底烧瓶中，加入中间体化合物 **3**（2.5mmol，0.6g）和 DCC（2.5mmol，0.5g），再加入 20mL 三氯甲烷，于室温搅拌 1h，加入 DMAP（0.2g）和水杨酸酯（2.1mmol），于室温搅拌 6h，TLC 跟踪反应，反应完毕，过滤除去反应生成的二环己基脲，用 5mL 1mol/L HCl 洗涤 2 次，饱和 Na_2CO_3 溶液洗涤 1 次，然后用水、饱和氯化钠洗涤，无水 $MgSO_4$ 干燥。过滤，脱去溶剂，粗产品经减压柱层析分离，用乙酸乙酯和石油醚混合溶剂（体积比 1:2）重结晶，得到目标化合物 **5**。

参 考 文 献

［1］Gamliel A.，Katan J. Phytop athology［J］，1992，82：320～327.
［2］Cohen Y.，Reuweni M.，Baider A.. Eur. J. Plant Patho. l［J］，1999，105：351～361.
［3］Zimmer L.，Metranx J. P.，Mauch－Mani B.. Plant Physiology［J］，2001，126：517～523.
［4］Zimmer L.，Jakab G.，Metranx J. P. *et al.*. Proc. Nat. l A cad. Sci USA［J］，2000，97：12920～12925.
［5］Raskin I.. Plant Physiol［J］，1992，99（3）：799～803.
［6］Klessingdf M.. Plant Mo Biol［J］，1994，26：1439～1458.
［7］CHEN Kai－Xian（陈凯先），LUO Xiao－Min（罗小民），JIANG Hua－Liang（蒋华良）. Bulletin of Chinese A cademy of Sciences（中国科学院院刊）［J］，2000，（4）：265～269.
［8］Rachina V.. Synthesis［J］，1982：967～970.
［9］Amoroso R.，Cardillo G.，Mobbili G. *et al.*. Tetrahed ron Asymmetry［J］，1993，4（10）：2241～2254.
［10］Zee－Cheng K. Y.，Robins R. K.，Cheng C. C.. J. Org. Chem.［J］，1961，26：1877～1880.
［11］LI Zai－Guo（李在国），WANG Qing－Min（汪清民）. Synthesis of Organic Intermediate，2nd Ed.（有机中间体的制备，第二版）［M］，Beijing：Chemical Industry Press，2001：77.

Synthesis and Bioactivity of Salicylic Acid 3 – *N'* – Benzyloxycarbonyl – β – aminobutyric Ester

Hongjun Zang[1], Zhengming Li[2], Changchun Ni[3], Zhou Shen[3], Zhijin Fan[2], Xiufeng Liu[2]

([1]School of Material Science and Chemistry Engineering, TianjinPoly – technic University, Tianjin, 300160, China;
[2]Research Institute of Elemento – organic Chemistry, State Key Labortary of Elem – ento – Organic Chemisty, Nankai University, Tianjin, 300071, China;
[3]Shanghai Branch of National Pesticide R & D South Center, Shanghai, 200032, China)

Abstract The plant activators such as salicylic acid and *D*, *L* – β – aminobutyric acid can induce systemic resiatance acquired in the treated plants against following pathogen attack. Utilizing coordination of activemoieties, the chemical activators were designed to incorporate some structural moieties from β – aminobutyric and salicylic acid ester. Nine new compounds were prepared, whose structures were confirmed by elemental analysis and ^1H NMR. The bioactivity of them was tested and the results indicate that several compounds show a certain induced activity against pathogen infection.

Key words β – Aminobutyric acid; salicylic acid ester; plant – induced resistance; bioactivity

Synthesis and Crystal Structure of 5 – *N* – *i* – Propyl – 2 – (2′ – nitrobenzenesulfonyl) – glutamine*

Yongjun Xiao, Jianguo Wang, Baolei Wang, Zhengming Li, Haibin Song

(Elemento – Organic Chemistry Institute, State Key Laboratory of Elemento – Organic Chemistry, Nankai University, Tianjin, 300071, China)

Abstract The title compound, 5 – *N* – *i* – propyl – 2 – (2′ – nitrobenzenesulfonyl) – glutamine, was synthesized and its structure was confirmed by IR, MS, ^1H NMR, and elemental analysis. The single crystal structure of the title compound was determined by X – ray diffraction. The crystal belongs to Monoclinic, space group *P*2(1), with $a = 0.69281(11)$ nm, $b = 0.76508(12)$, $c = 1.5843(3)$ nm, $\alpha = 90°$, $\beta = 90.941(3)°$, $\gamma = 90°$, $V = 0.8397(2)$ nm^3, $Z = 2$, $D_C = 1.477$ g/cm^3, $\mu = 0.236$ mm^{-1}, $F(000) = 392$, $R = 0.0297$, and $wR = 0.0664$.

Key words 5 – *N* – *i* – Propyl – 2(2′ – nitrobenzenesulfonyl) – glutamine; synthesis; crystal structure

Introduction

Acetohydroxy acid synthase (AHAS) is a perfect target for the design of environmentally benign herbicides because it is a key enzyme that is absent in animals but is important for the biosynthesis of branchedchain amino acids in plants[1,2]. Since the synthesis of commercial herbicides such as sulfonylureas and imidazolinones, a variety of other compounds targeting this enzyme have been synthesized. An understanding of the structure of AHAS is very important for the further design of novel inhibitors as it can elucidate the exact binding site and mode. Recently, Duggleby *et al.* [3] first reported only the crystal structure of yeast AHAS and then its complex with chlorimuron ethyl, a commonly used sulfonylurea herbicide. These achievements have opened new vistas for the synthesis of novel inhibitors.

On the basis of the reported 0.28 – nm high – resolution yeast AHAS complex structure, molecular docking was carried out to screen the MDL/ACD – 3D database to obtain 300 novel molecules with a low binding energy *via* the program DOCK 4.0, which provides new structural information for designing and synthesizing new AHAS herbicidal inhibitors.

5 – *N* – *i* – Propyl – 2 – (2′ – nitrobenzenesulfonyl) – glutamine, one among the obtained 300 molecules, was synthesized and its structure was determined by singlecrystal X – ray diffraction analysis. The elucidation of the crystal structure will provide more information that will help to better understand the structure – activity relationship of potential AHAS inhibitors.

Experimental

1 Materials and Instrument

Melting points were determined by using a Yanaco MP – 241 apparatus, and the values thus de-

* Reprinted from *Chemical Research in Chinese Universities*, 2006, 22(6): 760 – 762. Supported by the National Natural Science Foundation of China (Nos. 20432010, 20602021).

termined were uncorrected. Infrared spectra were recorded on a Bruker Equinox 55 spectrophotometer *via* potassium bromide tablet technique. ^1H NMR spectra were measured on a Bruker AC – P500 instrument(300MHz) by using DMSO – d_6 as the solvent and TMS as the internal standard. Mass spectra were recorded by using a Thermo Finnigan LCQ Advantage LC/mass detector instrument. Elemental analyses were carried out with a Yanaco MT – 3CHN elemental analyzer. The single – crystal structure of the title compound was determined on a Bruker SMART 1000 CCD diffractometer. All the reagents wete of analytical reagent grade.

2 Synthesis of the Title Compound

The synthesis route of the title compound is described in Scheme 1. Intermediates **1** and **2** were prepared according to the methods reported previously[4,5]. Applying Schotten – Bauman reaction[6], intermediate **1** was prepared by one – step condensation of 2 – nitrobenzene – 1 – sulfonyl chloride with *L* – glutamic acid. Reaction of intermediate **1** with acetyl chloride afforded cyclized acid intermediate **2.**

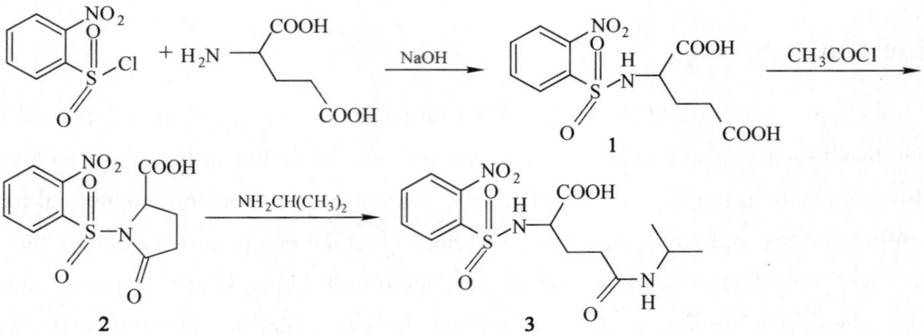

Scheme 1 Synthesis route of the title compound

The title compound was synthesized *via* the following procedure. In a 50 – mL loosely stoppered comical flask, intermediate **2** (0.1mol) was suspended in 15mL of water. To this suspension, excess of iso – propylamine(0.025mol) was added under continuous stirring and the mixture was allowed to stand overnight. The reaction mixture was concentrated to dryness on a steam bath, cooled, and acidified with dilute hydrochloric acid. The resultant mass was filtered, washed with cold water, and recrystallized from ethanol and water.

Yield: 75%, m.p.: 158 – 160℃. ^1H NMR(300MHz, DMSO – d_6), δ: 8.04 – 7.81(m, 4H, $C_6\mathbf{H}_4$), 3.90(q, 1H, NHC**H**CH_2, $^3J_{H-H}$ = 4.52Hz), 3.84 – 3.74[m, 1H, NHC**H**$(CH_3)_2$], 2.12(q, 2H, COC$\mathbf{H}_2$, $^3J_{H-H}$ = 7.91Hz), 2.04 – 1.91(m, 1H, NHCHC$\mathbf{H}_2$), 1.85 – 1.70(m, 1H, NHCHC$\mathbf{H}_2$), 1.01[d, 6H, CH(C$\mathbf{H}_3$)$_2$, $^3J_{H-H}$ = 6.03Hz]. IR(KBr), $\tilde{\nu}$/cm^{-1}: 3397, 3304 (N—H), 1702(C=O), 1550, 1398(N=O), 1365, and 1174(S=O). ESI – MS(negative ion), *m/z*(%): 372(M – 1, 12), 324(53), 297(71), 185(65), 157(100), 138 (26). Elemental anal. (%), calc. for $C_{14}H_{19}N_3O_7S$(M_r = 373.38): C 45.03, H 5.13, N 11.25; found: C 45.01, H 4.88, N 11.48.

3 X – ray Crystallographic Analysis

A crystal of the title compound was obtained from an ethanol – water(volume ratio is 4:1) solution by slow evaporation at room temperature. The colorless crystal with a dimension of 0. 28mm × 0. 22mm × 0. 20mm was selected for X – ray diffraction analysis. The data were collected by using a Bruker SMART CCD areadetector diffractometer under graphite monochromated Mo $K\alpha$ radiation ($\lambda = 0.071073$nm). In the range of $2.94° < \theta < 26.30°$, 3276 independent reflections were obtained.

The structure was solved by direct methods by using the SHELXS – 97 program[7,8]. All the nonhydrogen atoms were refined on F^2 anisotropically by using the full – matrix least squares method. The hydrogen atoms were located from the difference Fourier map and added to the structure calculation, but their positions were not refined. The contributions of these hydrogen atoms were included in structure factor calculations. The final least – square cycle gave $R_1 = 0.0297$ and $wR_2 = 0.0664$ for 3276 reflections with $I > 2\sigma(I)$; the weighing scheme is $w = 1/[\sigma^2(F_0^2) + (0.0306P)^2 + 0.1092P]$, where $P = (F_o^2 + 2F_c^2)/3$. The difference of maximum and minimum peaks and holes are 0. 180 and – 213 e/nm^3. $S = 1.061$ and $(\Delta/\delta)_{max} = 0.001$. Atomic scattering factors and anomalous dispersion correction were taken from the International Table for X – ray Crystallography.

Results and Discussion

Figs. 1 and 2 show the molecular structure and the perspective view of the crystal packing in a unit cell of the title compound. The atomic coordinates and equivalent isotropic displacement parameters are shown in Table 1. The selected bond lengths, bond angles, torsional angles, and hydrogen – binding geometry of the title compound are listed in Tables 2 and 3, respectively.

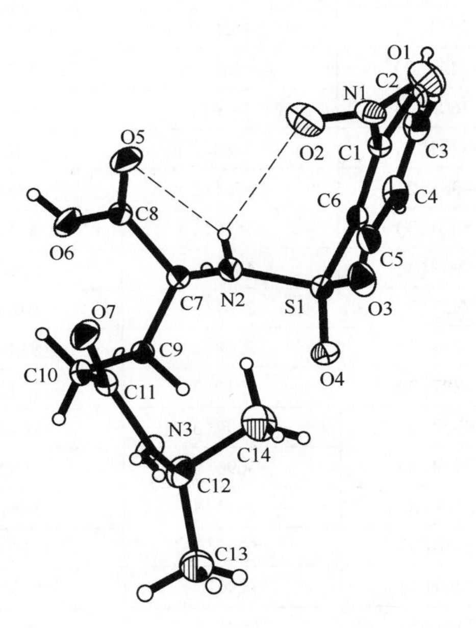

Fig. 1 Molecular structure of the title compound with atomic numbering scheme

Displacement ellipsoids are drawn at the 30% probability level.

Dashed lines indicate intramolecular hydrogen bonds.

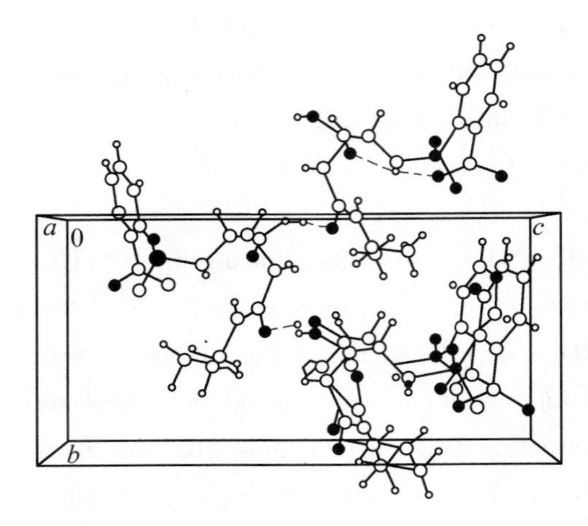

Fig. 2 A view of the crystal packing along c axis for the title compound

Dashed lines indicate hydrogen bonds.

Table 1 Atomic coordinates($\times 10^4$) and equivalent isotropic displacement parameters($\times 10 nm^2$) of the title compound

Atom	x	y	z	U_{aq}
S1	3261(1)	6751(1)	8205(1)	32(1)
O1	8408(2)	8052(2)	9418(1)	62(1)
O2	7469(2)	8031(2)	8112(1)	61(1)
O3	3554(2)	8360(2)	8639(1)	47(1)
O4	1430(2)	5903(2)	8211(1)	45(1)
O5	6566(2)	6748(2)	6120(1)	51(1)
O6	4953(2)	4825(2)	5325(1)	45(1)
O7	2450(2)	10187(2)	5781(1)	43(1)
N1	7608(2)	7369(2)	8814(1)	40(1)
N2	3764(2)	7080(2)	7223(1)	31(1)
N3	-386(2)	9515(2)	6354(1)	33(1)
C1	6804(3)	5619(3)	8923(1)	33(1)
C2	7979(3)	4394(3)	9314(1)	41(1)
C3	7325(3)	2707(3)	9405(1)	45(1)
C4	5494(4)	2268(3)	9136(1)	47(1)
C5	4297(3)	3510(3)	8771(1)	39(1)
C6	4932(3)	5199(3)	8642(1)	30(1)
C7	3509(2)	5696(2)	6592(1)	28(1)
C8	5188(3)	5830(3)	5981(1)	32(1)
C9	1530(3)	5766(3)	6149(1)	30(1)
C10	1204(3)	7290(3)	5536(1)	33(1)
C11	1148(3)	9099(3)	5910(1)	30(1)
C12	-672(3)	11241(2)	6734(1)	36(1)
C13	-2809(3)	11571(3)	6855(1)	49(1)
C14	480(3)	11431(3)	7545(1)	51(1)

Table 2　Selected bond lengths(nm), bond angles and torsion angles(°) of the title compound

S1—N2	0. 1619(2)	O2—N1	0. 1223(2)
S1—C6	0. 1790(2)	N1—C1	0. 1461(3)
O5—C8	0. 1202(2)	N2—C7	0. 1465(2)
O6—C8	0. 1301(2)	N3—C11	10. 1323(2)
O7—C11	10. 1247(2)	N3—C12	20. 1466(3)
O5—C8—O6	125. 5(2)	O7—C11—C10	121. 7(2)
O6—C8—C7	112. 3(2)	N3—C12—C13	109. 9(2)
O1—N1—C1—C2	-47. 5(3)	C12—N3—C11—O7	-0. 8(3)
O2—N1—C1—C2	130. 8(2)	C9—C10—C11—O7	111. 2(2)
O1—N1—C1—C6	132. 6(2)	C9—C10—C11—N3	-71. 6(2)
O2—N1—C1—C6	-49. 1(3)	C11—N3—C12—C14	-79. 5(2)

Table 3　Hydrogen - bond distances(nm) of the title compound

D	H	A	Symmetry	D—H	H···A	D···A	D—H···A
O6	H6	O7	$-x+1, y-1/2, -z+1$	0. 082(3)	0. 173(3)	0. 2547(2)	173(3)
N3	H3	O5	$x-1, y, z$	0. 079(2)	0. 224(2)	0. 3009(2)	164(2)
N2	H2	O5		0. 075(2)	0. 225(2)	0. 2646(2)	114(2)
N2	H2	O2		0. 075(2)	0. 244(2)	0. 2998(2)	133(2)

In the title compound, the bond lengths and bond angles show normal values. The O3—S1—O4 angle[120. 65(10)°] deviates significantly from the ideal tetrahedral value. The O—N—C—C torsion angles indicate that the nitro group is twisted away from the plane of the benzene ring(Fig. 1). The glutamine residue adopts a folded conformation. In the glutamine residue, the ψ^1(N2—C7—C8—O5), ψ^2(N2—C7—C8—O6), χ^1(N2—C7—C9—C10), and χ^2(C7—C9—C10—C11) torsion angles are 10. 8(3)°, -170. 59(15)°, 70. 9(2)°, and -66. 5(2)°, respectively; the two planar groups in the residue, C7/C8/O5/O6 and C10/C11/O7/N3, form a dihedral angle of 148. 2(1)°. The molecular structure is stabilized by intramolecular N—H···O hydrogen bonds(Table 3). The crystal structure involves intermolecular O—H···O and N—H···O hydrogen bonds, which link the molecules into a layer parallel to the *ab* plane(Fig. 2).

References

[1] Shaner D. L., Anderson P. C., Stidham M. A., *Plant physiol*, 1984, 76, 545.

[2] LaRossa R. A., Sohloss J. V., *J. Biol. Chem.*, 1984, 259, 8753.

[3] Pang S. S., Guddat L W., Duggleby R. G., *J. Biol. Chem*, 2003, 278, 7639.

[4] De A. U., Pandey J., Majumdar A., *Indian J. Chem. Sect. B*, 1982, 21(5), 481.

[5] Srikanth K., Kumar C. A., Ghoah B., *et al.*, *Bioorg. Med. Chem.*, 2002, 10, 2119.

[6] March J. (Ed.), *Advanced Organic Chemistry, Reactions, Mechanism and Structure*, John Wiley & Sons, New York, 1992, 417.

[7] Sheldrick G. M., *SHELXS97 and SHELXL97*, University of Göttingen, Germany, 1997.

[8] Bruker., *SAINT and SHELXTL*, Bruker AXS Inc., Medison, USA, 1999.

Synthesis, Dimeric Crystal Structure, and Biological Activities of *N* – (4 – Methyl – 6 – oxo – 1,6 – dinydro – pyrimidin – 2 – yl) – *N′* – (2 – trifluoromethyl – phenyl) – guanidine*

Fengqi He, Baolei Wang, Zhengming Li

(Elemento – Organic Chemistry Institute, State – Key Laboratory of Elemento – Organic Chemistry, Nankai University, Tianjin, 300071, China)

Abstract The title compound, *N′* – (4 – methyl – 6 – oxo – 1,6 – dihydro – pyrimidin – 2 – yl) – *N′* – (2 – trifluoromethyl – phenyl) – guanidine, was synthesized and its structure was confirmed by using IR, MS, ^{1}H NMR, and elemental analysis. The single crystal structure of the title compound was determined by X – ray diffraction. The preliminary biological test showed that the synthesized compound has a weak herbicidal activity.

Key words Ketol – acid reductoisomerase, synthesis, X – ray diffraction, crystal structure, biological activity

Introduction

Plants have a more extensive biosynthetic capacity than animals, which is attributed to the fact that animals rely on their diet for the supply of many biochemical compounds and their precursors. Owing to this difference, plants contain numerous enzymes that are potential targets of herbicides. The enzymes involved in the biosynthesis of the branched – chain amino acids (leucine, isoleucine, and valine) are the examples of such a mechanism. Valine and isoleucine can be synthesized by a parallel set of four reactions, whereas the synthetic route to leucine is an extension of the valine pathway.

The first batch of successful herbicides that targeted this pathway were the sulfonylureas[1] and the imidazolinones[2], both of which inhibited the first enzyme, acetohydroxyacid synthase (AHAS). Ever since the discovery of these herbicides, a variety of other compounds targeting this enzyme have been discovered[3]. The successful application of these herbicides has stimulated researches on inhibitors of other enzymes in the pathway, including two of the enzymes of the leucine branch[4,5] and the second enzyme in the common pathway, ketol – acid reductoisomerase (KARI). New inhibitors of KARI act as the potential novel herbicidal compounds.

Based on the reported high – resolution crystal structure of a spinach KARI complex that is 0.165nm high[6], 279 molecules with low binding energies toward KARI were obtained from MDL/ACD 3D database by probing with the DOCK 4.0 program[7]. These potential stuctures provide further information for the design of novel KARI herbicidal molecules. One novel derivative of the KARI herbicidal molecules, namely, the title compound was synthesized. The study

* Reprinted from *Chemical Research in Chinese Universities*, 2006, 22(6): 768 – 771. Supported by the National Basic Research Program of China (No. 2003CB114406).

of the crystal structure of the title compound will provide more information in the study of structure activity relationship of potential KARI inhibitors.

Experimental

1 Materials and Instrument

The melting points were determined by using a Yanaco MP – 241 apparatus and were uncorrected. The infrared spectra were recorded on a Bruker Equinox 55 FTIR spectrophotometer with potassium bromide tablets. The ^{1}H NMR spectra were measured on a Bruker AC – P500 instrument (300MHz) with tetramethylsilane as the internal standard and dimethyl sulfoxide – d_6 as the solvent. The mass spectra were recorded on a Thermo Finnigan LCQ Advantage LC/mass detector instrument. The elemental analyses were performed on a Yanaco MT – 3 CHN elemental analyzer. The single crystal structure of compound **3** was determined on a Bruker SMART 1000 CCD diffractometer.

2 Synthesis of the Title Compound

As shown in Scheme 1, intermediate **1** was prepared according to the procedure reported by Buu – hoγ & Jacquignon *et al.* [8]. 2 – Trifluoromethylaniline hydrochloride and freshly recrystallized cyanoguanidine in hot water were heated under reflux for 2h, and 1 – (2 – trifluoromethylphenyl) biguanidine hydrochloride was formed by precipitation upon cooling. The hydrochlorate was basified by the addition of aqueous sodium hydroxide, which resulted in intermediate **1**. The procedure for the synthesis of the title compound is described as follows.

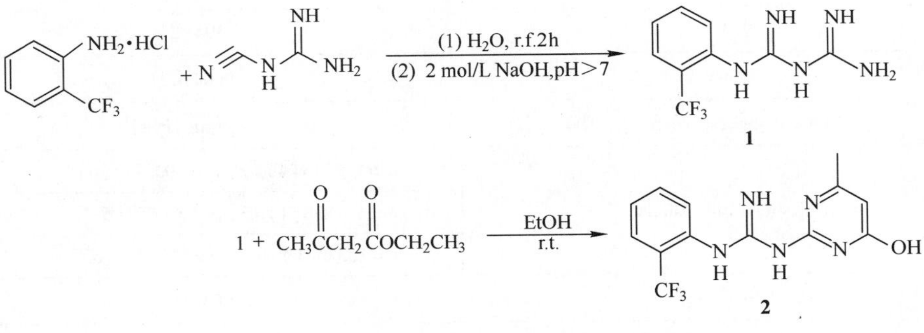

Scheme 1 Synthesis route of the title compound

To a 25 – mL flask, 10mmol of intermediate **1** and 10mL of ethanol were added. Ethyl acetoacetate (10mmol) was added in drops to the above mixture, and the mixture was stirred for 6h at room temperature. The solution was filtered and the solid was collected, which was recrystallized from ethanol and water as white crystals. Yield: 42%; m. p.: 243 – 244℃. ^{1}H NMR (DMSO, 300 MHz), δ: 7.72 (d, 1H, J = 8.4Hz, Ph—H); 7.61 (t, 1H, J = 16.2Hz, Ph—H); 7.41 – 7.53 (m, 2H, Ph—H); 5.81 (s, 1H, Py—H); 2.11 (s, 3H, Py—CH_3). IR (KBr), $\tilde{\nu}$/cm^{-1}: 3473, 3298, 1644, 1554, 1316, 764. MS (ESI), m/z: 623 (2M + 1, 100), 396 (47), 312 (M + 1, 7), 275 (40), 191 (31). Elemental analysis (%), calculated: C 50.16, H 3.89, N 22.5; found: C 49.95, H 4.08, N 22.31.

3 X - ray Crystallographic Analysis

A crystal of the title compound was obtained from the methanol solution by slow evaporation at room temperature. The colorless crystal with dimensions of 0. 24mm × 0. 22mm × 0. 20mm was selected for the X - ray diffraction analysis. The data were collected by a Bruker SMART[9] 1000 CCD area - detector diffractometer with graphite monochromated radiation(λ = 0. 071073nm). In the range from 2. 63° to 22. 09°, 5780 independent reflections were collected.

The structure was resolved by using the SHELXS97 program[10,11]. All the nonhydrogen atoms were anisotropically refined on F^2 by the full - matrix least squares method. Hydrogen atoms were located by using the difference Fourier map and added to the calculation with regard to the structure, but their positions were not refined. The contributions of these hydrogen atoms were included in the structure - factor calculations. The final least - square cycle gave R_1 = 0. 0769 and wR_2 = 0. 2012 for 2629 reflections with $I > 2\sigma(I)$; the weighting scheme is $w = 1/[\sigma^2(F_o^2) + (0.1094P)^2 + 1.1074P]$, where $P = (F_o^2 + 2F_c^2)/3$. The maximum and the minimum differences in peaks and holes are 373 and - 256e/nm^3, respectively. S = 1. 02 and $(\Delta/\sigma)_{max}$ = 0. 002. Atomic scattering factors and anomalous dispersion corrections were from the International Table for X - Ray Crystallography[12]. A summary of the key crystallographic information is given in Table 1.

Table 1 Crystallographic data of the title compound

Empirical Formula	$C_{14}H_{16}F_3N_5O_2$
Molecular weight	343. 32
T/K	294(2)
λ/nm	0. 071073
Crystal system, space group	Triclinic, P - 1
Unit - cell dimensions	a = 1. 1515(4) nm, α = 100. 774(6)°
	b = 1. 1838(4) nm, β = 99. 657(6)°
	c = 1. 3737(4) nm, γ = 110. 745(6)°
V/nm^3, Z	1. 6634(9), 4
Calculated density(mg · m^{-3})	1. 371
Absorption coefficient(mm^{-1})	0. 117
F(000)	712

Results and Discussion

1 Description of the Structure

Figs. 1 and 2 show the molecular structure and the view of crystal packing in a unit cell of the title compound. The atomic coordinates and the equivalent isotropic displacement parameters are given in Table 2. The selected bond lengths, bond angles, and hydrogen - bonding geometry of the title compound are listed in Tables 3 and 4.

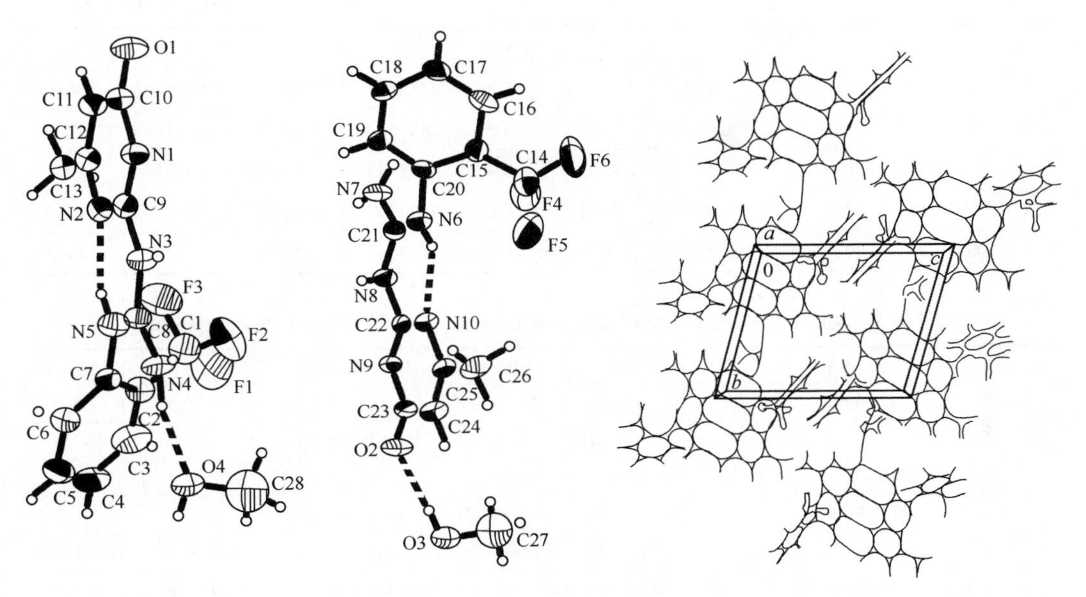

Fig. 1 Molecular structure of the title compound, showing displacement ellipsoids drawn at the 30% probability level

Fig. 2 Crystal packing of the title compound, showing the hydrogen - bonded chains (dashed lines)

Table 2 Atomic coordinates($\times 10^4$) and equivalent isotropic displacement parameters($\times 10nm^2$) of the title compound

Atom	x	y	z	U_{eq}	Atom	x	y	z	U_{eq}
F1	9117(5)	7285(5)	4568(3)	146(2)	C2	10091(6)	6945(5)	3229(4)	58(2)
F2	8146(5)	7218(5)	3093(4)	128(2)	C3	10565(7)	6213(6)	3711(5)	85(2)
F3	9863(5)	8800(4)	3917(3)	114(2)	C4	11260(7)	5622(6)	3266(7)	93(2)
F4	2805(4)	9236(4)	7582(3)	107(1)	C5	11468(6)	5749(5)	2343(6)	77(2)
F5	2480(4)	7907(4)	6188(3)	101(1)	C6	11013(5)	6481(5)	1848(5)	60(2)
F6	1394(4)	9045(5)	6278(4)	132(2)	C7	10311(5)	7068(4)	2294(4)	45(1)
O1	8529(4)	11990(3)	196(3)	65(1)	C8	8973(5)	7417(4)	912(4)	46(1)
O2	6523(4)	5255(3)	8664(3)	71(1)	C9	9101(5)	9529(4)	898(4)	41(1)
N1	8572(4)	10123(3)	340(3)	42(1)	C10	9011(5)	11395(4)	706(4)	46(1)
N2	10013(4)	10060(3)	1776(3)	44(1)	C11	9983(5)	11996(4)	1632(4)	51(1)
N3	8636(4)	8252(3)	494(3)	47(1)	C12	10447(5)	11326(4)	2148(3)	43(1)
N4	8393(5)	6230(4)	439(4)	58(1)	C13	11435(5)	11876(5)	3146(4)	58(1)
N5	9871(4)	7843(4)	1789(3)	54(1)	C14	2568(6)	9074(7)	6579(6)	75(2)
N6	5209(4)	9423(3)	7078(3)	49(1)	C15	3561(5)	10046(4)	6268(4)	50(1)
N7	6568(5)	10992(4)	8507(4)	59(1)	C16	3204(6)	10771(6)	5689(4)	66(2)
N8	6463(4)	8999(3)	8356(3)	44(1)	C17	4118(7)	11677(5)	5414(5)	69(2)
N9	6504(4)	7125(3)	8501(3)	43(1)	C18	5380(6)	11849(5)	5685(4)	61(2)
N10	5129(4)	7194(3)	7027(3)	46(1)	C19	5744(5)	11128(4)	6255(4)	52(1)
C1	9305(8)	7550(8)	3682(5)	89(2)	C20	4838(5)	10215(4)	6542(4)	45(1)

Table 2(continued)

Atom	x	y	x	U_{eq}	Atom	x	y	x	U_{eq}
C21	6070(5)	9816(4)	7968(4)	41(1)	C26	3713(6)	5373(5)	5664(4)	74(2)
C22	5996(4)	7702(4)	7922(4)	39(1)	O3	5401(4)	2743(3)	8415(3)	80(1)
C23	6071(5)	5849(4)	8127(4)	48(1)	O4	8223(4)	4061(3)	1080(3)	83(1)
C24	5147(5)	5253(4)	7184(4)	56(2)	C27	4389(9)	2588(8)	8887(7)	134(3)
C25	4696(5)	5931(4)	6654(4)	50(1)	C28	7260(9)	3641(9)	1572(7)	134(3)

Table 3 Selected bond lengths(nm) and bond angles(°) of the title compound

O1 - C10	0.1275(5)	N4—C8	0.1300(6)	N9—C22	0.1331(5)	N6—C21	0.1323(6)
N1 - C9	0.1342(6)	N5—C8	0.1326(6)	N10—C22	0.1326(6)	N6—C20	0.1428(6)
N2 - C9	0.1333(6)	N5—C7	0.1425(6)	N8—C21	0.1357(5)	O3—C27	0.1400(10)
N3 - C8	0.1366(6)	O4—C28	0.1386(9)	N8—C22	0.1403(5)		
N3 - C9	0.1376(5)	O2—C23	0.1279(5)	N7—C21	0.1313(6)		
N2—C9—N1	126.3(4)	N5—C8—N3	119.0(4)	N10—C22—N9	127.8(4)	N6—C21—N8	120.4(4)
N2—C9—N3	118.9(4)	N4—C8—N5	122.3(5)	N10—C22—N8	119.1(4)	N7—C21—N6	121.7(4)
N1—C9—N3	114.8(4)	C8—N5—C7	124.2(4)	N9—C22—N8	113.1(4)	C21—N6—C20	125.2(4)
C8—N3—C9	127.2(4)	O1—C10—N1	118.7(4)	C21—N8—C22	126.2(4)	O2—C23—N9	118.0(4)
N4—C8—N3	118.7(5)	N2—C12—C13	115.4(4)	N7—C21—N8	117.8(5)	N10—C25—C26	115.9(4)

Table 4 Hydrogen - bond distances(nm) and bong angles(°) of the title compound

D—H···A	Symmetry	D—H	H···A	D···A	D—H···A
N8—H8···N1	$x,y,z+1$	0.086	0.218	0.3039(6)	178.6
N7—H7A···O1	$x,y,z+1$	0.099(6)	0.172(6)	0.2685(6)	166(4)
O4—H4A···O1	$x,y-1,z$	0.082	0.189	0.2693(5)	166.9
N5—H5···N2	x,y,z	0.086	0.188	0.2574(5)	136.5
O3—H3B···O2	x,y,z	0.082	0.190	0.2716(5)	172.2
N6—H6···N10	x,y,z	0.086	0.191	0.2594(5)	135.6
N4—H4C···O4	x,y,z	0.100(6)	0.185(6)	0.2820(6)	162(5)
N3—H3···N9	$x,y,z-1$	0.086	0.220	0.3060(6)	175.4
N4—H4B···O2	$x,y,z-1$	0.086(5)	0.186(5)	0.2709(6)	170(5)

There are two different planes in the molecule, and each of them has a conjugated system. One is benzene ring, the other is composed of pyrimidine ring and guanidine group, in which the hydrogen atom of the hydroxyl group transfers to the nitrogen atom(N4 or N7). The crystal has two different configurations; the intermolecular hydrogen bonds and $\pi-\pi$ interactions result in the dimeric crystal structure.

In the title compound, the bond lengths and bond angles are generally normal in phenyl and pyrimidine rings[13]. The pyrimidine ring(N1, N2, C9, C10, C11, and C12) with atoms N3, N4, and N5 [pyrimidine ring (N9, N10, C22, C23, C24, and C25) with atoms N6, N7, and N8] is fairly planar with plane equation $9.472x-0.622y-9.449z=7.1709$ ($9.595x-0.301y-9.118z=-1.7350$), and the largest deviation from the least squares plane is 0.00289nm(-

0.00511nm). Table 3shows that the bond lengths of N3—C8, N4—C8, and N5—C8 (N8—C21, N7—C21, and N6—C21) are shorter than the normal C—N single - bond distance of 0.147nm[14], which shows that N3, N4, N5(N6, N7, N8) and the pyrimidine ring form a conjugated system. The N5—H5 ··· N2 (N6—H6 ··· N10) intramolecular hydrogen bond makes N2, N3, N5, H5, C8, and C9(N6, N7, N8, H6, C21, and C22) form a hexatomic ring. Four intermolecular hydrogen bonds (N8—H8 ··· N1, N7—H7A ··· O1, N3—H3 ··· N9, N4—H4B ··· O2) are formed between two molecules with different configurations.

The intramolecular and intermolecular hydrogen bonds are shown in Table 4. In the solid state, these hydrogen bonds stabilize these structures.

2 Biological Activities

2.1 The Herbicidal Activity Test

The compounds to be evaluated were made into an emulsion to aid dissolution. Ten barnyard grass seeds were placed in a 50 - mL cup covered with a layer of glass beads and a piece of filter paper at the bottom, to which 5mL of the inhibitor solution was added in advance. The cup was then placed in a well - lit room and allowed to germinate for 65h at (28 ± 1)℃. The heights of seedlings of the above - ground parts from each cup were measured, and their means were calculated. The test was carried out only by using distilled water. The percentage of the inhibition was calculated[15]

2.2 Results

The primary bioassay showed that the title compound exhibits a weak inhibiting activity on the weeds. Its inhibition rates to seedling growth of barnyard grass reach 8.6% and 16.7% at 10μg/mL and 100μg/mL, respectively.

References

[1] Chaleff R. S., Mauvais C. J., *Science*, 1984, 224, 1443.

[2] Shaner D. L., Anderson P. C., Stidham M. A., *Plant Physiol.*, 1984, 76, 545.

[3] Duggleby R. G., Pang S. S., *J. Biochem. Molec. Biol.*, 2000, 33, 1.

[4] Wittenbach V. A., Teaney P. W., Rayner D. R., *et al.*, *Plant Physiol.*, 1993, 102, 50.

[5] Hawkes T. R., Cox J. M., Fraser T. E. M., *et al.*, *Z. Naturforsch. C*, 1993, 48, 364.

[6] Biou V., Dumas R., Cohen - Addad C., *EMBO J.*, 1997, 16(12), 3405.

[7] Wang Bao - lei, Li Zheng - ming, Ma Yi, *et al.*, *Chin. J. Org. Chem.*, 2004, 24(8), 973.

[8] Buu - hoγ N. P., Jacquignon P., Béranger S., *et al.*, *J. Chem. Soc. Perkin Trans.* 1, 1972, 278.

[9] Bruker, *SMART.*, Bruker AXS Inc., Wisconsin, USA, 1998.

[10] Sheldrick G. M., *SHELXS97 and SHELXL97.*, University of Göttingen, Germany, 1997.

[11] Bruker, *SAINT and SHELXTL*, Bruker AXS Inc., Wisconsin, USA, 1999.

[12] Wilson A. J., *International Table for X - ray Crystallograghy*, *Vol. C*, *Tables* 6.1.1.4 *and* 4.2.6.8, Kluwer Academic Publisher, Dordrecht, 1992, 500, 219.

[13] He Feng - qi, Wang Bao - lei, Li Zheng - ming, *et al.*, *Acta Cryst. E*, 2005, 61(8), o2602.

[14] Carey F. A., *Organic Chemistry*, *4th ed.*, McGraw Hill Press, New York, 2000, 861.

[15] Wang B. L., Duggleby R. G., Li Z. M., *et al.*, *Pest Manag. Sci.*, 2005, 61(4), 407.

Microwave and Ultrasound Irradiation – Assisted Synthesis of Novel Disaccharide – Derived Arylsulfonyl Thiosemicarbazides*

Yuxin Li, Weiguang Zhao, Zhengming Li, Suhua Wang, Weili Dong

(State Key Laboratory of Elemento – Organic Chemistry,
Nankai University, Tianjin, 300071, China)

Abstract The synthesis of 1 – arylsulfonyl – 4 – (1′ – N – hepta – *O* – acetyl – *β* – lactosyl) thiosemicarbazides by reaction of hepta – *O* – acetyl – *α* – D – lactosyl isothiocyanate with substituted phenylsulfonyl hydrazines has been shown to occur in less than 1 min under microwave activation and 8 min under ultrasound irradiation at room temperature. It is noteworthy that when ultrasound and microwaves (MW) were utilized, a cleaner reaction accompanied with higher yields was observed.

Key words lactose; maltose; microwave; thiosemicarbazide; ultrasound irradiation

INTRODUCTION

Some thiosemicarbazides have been reported as fungicidal,[1] herbicidal,[2] and pesticidal[3] agents. Recently, it has been well documented that thiosemicarbazides are a novel class of non-peptide antagonists for the bradykinin B_2[4] which has a therapeutic effect in the symptomatic treatment of chronic pain and various inflammatory disorders. In addition, thiosemicarbazide is an useful intermediate in the synthesis of important heterocyclic molecules such as thiazol – 4 – one[5] and thiadiazol – 2 – one.[6] Its derivatives also exhibit herbicidal anti – inflammatory, and analgesis activities.

Over the past two decades the biological properties of carbohydrates have been reported. They have played key function in cell recognition, cell growth, cell differentiation, and signal transduction.[7] On the other hand, carbohydrates also compose important structural subunits in some natural pesticides such as spinosads.[8] A survey of the literature revealed that carbohydrates linked to phenylsulfonyl hydrazines have seldom been reported. As part of our program aimed at developing a new class of agrochemicals, we designed and synthesized the title compounds and studied their bioactivity.

Previously we had synthesized novel 1 – arylsulfonyl – 4 – (1′ – N – 2′,3′,4′,6′ – tetra – *O* – acetyl – *β* – D – glucopyranosyl) thiosemicarbazides via condensation of 2,3,4,6′ – tetra – *O* – actetyl – *β* – glucopyranosyl isothiocyanote with swastituted phenylsulfonyl hydrazines by the conventional method.[9] In the synthesis of 1 – arylsulfonyl – 4 – (1′ – N – hepta – *O* – acetyl – *β* – lactosyl) thiosemicarbazides, we found that a longer reaction time (at least 24h at rt or

* Reprinted from *Synthetic Communications*, 2006, 36: 1471 – 1477. The project was supported by National Natural Science Foundation of China (NNSFC) (No. 20432010) and Tianjin Natural Science Foundation (No. 033803411).

30min under reflux) was needed with the conventional method. At the same time, the by – product accompained by lower yields was observed under reflux.

It is well documented that ultrasound irradiation and microwave activation have been widely used in chemical synthesis. The microwave – irradiation technique has been used to assist in nuclephilic reaction, [10] aromatic substitutions, [11] cycloaddition reactions, [12] and so on. [13] Under certain circumstances, ultrasonic irradiation can replace the heating procedure. [14] Our target compounds were synthesized by applying microwave activation, ultrasound irradiation, or both. The synthesis of novel 1 – arylsulfonyl – 4 – (1′ – Nhepta – *O* – acetyl – β – lactosyl) thiosemicarbazides **3a – e** and 1 – arylsulfonyl – 4 – (1′ – hepta – *O* – acetyl – β – maltosyl) thiosemicarbazides **3f – j** are described as in Scheme 1.

RESULTS AND DISCUSSION

Hepta – *O* – acetyl – β – D – lactosyl isothiocyanate **2a** and hepta – *O* – acetyl – α – Dmaltosyl isothiocyanate **2b** were prepared according to the literature. [15] The reaction between **2a, b** and substituted phenylsulfonyl hydrazines under the conventional method was investigated. Under conventional conditions, the reaction between **2a** with phenylsulfonyl hydrazines did not proceed within 20h at room temperature and 30min under reflux. For example, if the synthesis of **3a** and **3h** was carried out at room temperature, 24h was needed; if refluxed, 30min was needed. Obviously, temperature is one decisive factor for this reaction. However, although the reaction can proceed within 30min under conventional heating conditions, lower yields were obtained because **2a** or **2b** was decomposed. In general, the common method for the synthesis of 1 – arylsulfonyl thiosemicarbazide compounds involves the reaction of isothiocyanate with hydrazine in ethonal, [16] but this reaction did not proceed well because the addition product N – (hepta – *O* – acetyl – β – D – maltosyl) thiocarbamic acid *O* – ethyl ester was obtained as the major compound. If acetonitrile is used as the solvent, target compound could be obtained with reasonable yields (70% – 80%).

1a-b → (i) → 2a-b → (ii) → 3a-j

Scheme 1 Reagents and conditions: (i) $Pb(NCS)_2$, toluene, reflux; (ii) $p-R_3-Ph-SO_2NHNH_2$, acetonitrile, ultrasonic, rt

The same reaction under ultrasound required only 5 – 8min to complete with excellent yields (Table 1). Ultrasonic irradiation was carried out with KQ – 218 ultrasonic cleaner 20kHz/50W. It is reported that when an ultrasonic wave passes through a liquid medium, a large number of microbubbles form, grow, and collapse in a very short time (a few microseconds), an effect that is called the cavitation effect. [14] Sonochemical theory calculations suggested that ultrasonic cavitation could generate local temperatures as high as 5000K and local pressures as 500 atm, with heating and cooling rates greater than 10^8 K/s, a very rigorous environment. [7]

Compared with the conventional method, a shorter time, lower temperature, cleaner reaction accompained with higher yields were observed(from 70% -80% to 92% -99% compared with conventional method at rt). Therefore, the bound breakdown was avoided by using ultrasound irradiation at rt. After the ultrasonication, the solvent was evporated in vaccum. The crude product was purified by a silicon - gel column eluted with ethyl acetate/petroleum ether(1∶2). Generally, the reaction proceeded quickly with excellent yield as shown in Table 2. The excellent yield of **3a - j** is shown in Tables 1 and 2.

Table 1 Effect of different reaction conditions

Compound	Reaction time			Yield(%)		
	Reflux	MW	Ultrasound	Reflux	MW	Ultrasound
3a	30min	1min	8min	71.8	100	96.8
3b	30min	1min	8min	73.6	96.9	97.2
3d	20min	30s	6min	80.2	96.2	98.2
3f	30min	1min	8min	74.0	95.5	94.3
3h	30min	30s	5min	76.5	98.3	99.0

Table 2 Physical data of title compounds and elemental analysis

Compound	R_1	R_2	R_3	Yield(%)	mp(℃)	Elemental analysis, % (calcd., %)		
						C	H	N
3a	OAc	H	H	96.8	218 - 221	46.40(46.64)	4.97(5.10)	5.07(4.94)
3b	OAc	H	CH_3	97.2	212 - 215	47.04(47.27)	5.19(5.25)	4.74(4.86)
3c	OAc	H	F	91.5	214 - 216	45.44(45.67)	4.72(4.88)	5.07(4.84)
3d	OAc	H	Cl	98.2	213 - 215	44.72(44.82)	4.79(4.74)	5.04(4.75)
3e	OAc	H	Br	99.1	213 - 215	42.40(42.68)	4.54(4.56)	4.68(4.52)
3f	H	OAc	H	94.3	180 - 183	46.50(46.64)	5.10(5.10)	4.99(4.94)
3g	H	OAc	CH_3	92.7	187 - 188	46.52(47.27)	4.94(5.25)	5.08(4.86)
3h	H	OAc	Cl	98.4	186 - 190	44.69(44.82)	4.77(4.74)	4.83(4.75)
3i	H	OAc	Br	98.6	191 - 193	42.90(42.68)	4.53(4.56)	4.45(4.52)
3j	H	OAc	F	93.4	171 - 173	45.50(45.67)	4.95(4.88)	4.96(4.84)

The reaction under microwave(MW) heating conditions was also investigated to demonstrate the specific microwave effect. As we all know, microwave - assisted reactions have several advantages over conventional heating by not only significantly reducing the reaction time but also improving the reaction yield dramatically and, in the process, greatly suppressing the side reactions. This increase in reaction rate is due in large part to the vast amount of microwave energy being absorbed by the system compared with the required energy necessary to attain the requisite activation energy.[18] The same reaction under MW heating conditions for only 30 - 60s afforded excellent product yield. The yield of synthesis of **3a - h** increased from 70% - 80% to 95% - 100% respectively, compared with refluxing under conventional conditions. The effects of different reaction conditions are shown in Table 1.

In conclusion, we have described a straightforward and high - yielding method for the synthesis of **3a - h** from their corresponding **2a - b** under mild conditions using microwave irradiation or ultrasonic irradiation. It is noteworthy that when microwave and ultrasound were utilized, a cleaner reaction accompanied with higher yields was observed. The application of microwave activation and ultrasonic irradiation to this reaction has been shown significantly enhance the rate.

EXPERIMENTAL

Melting points were conducted on a Yanaco MP - 500 micro melting - point apparatus ^{1}H NMR and ^{13}CNMR spectra were recorded in $CDCl_3$ as solvent on Bruker AC - 400 instrument using TMS as an internal standard. Elemental analysis was performed on a Yanaco CHN Corder MF - 3 automatic elemental analyzer. Mass spectra were recorded with VG ZAB - HS, 8kV, 1mA using the Fast Atom Bombardment (FAB) method Ultrasonic irradiation was carried out with Ultrasonic Cleaning 20kHz/50W. Microwave activation was carried out with LWMC - 201.

General Procedure of 1 - phenylsulfonyl - 4 - (1′ - N - hepta - *O* - acetyl - β - lactosyl) Thiosemicarbazide 3a

2a (0. 68g, 10mmol) and benzenesulfonyl hydrazide (0. 18g, 10mmol) were dissolved in 10mL of acetonitrile. The reaction mixture was heated for 1min under MW heating conditions. After the microwave heating, the solvent was evaporated in vacuum. The crude product obtained was purified by silicon - gel column eluted with ethyl acetate/petroleum ether (1:2) and **3a** (0. 82g, 96. 8%) was obtained. Mp 218 - 221℃. ^{1}H NMR ($CDCl_3$): δ 1. 95, 1. 99, 2. 04, 2. 05, 2. 07, 2. 08, 215 (s, 21H, CH_3), 3. 68 - 5. 53 (m, 14H, lactosyl ring), 7. 04 (bs, 1H, NH), 7. 55 - 7. 90 (m, 7H, Ar—H, 2 × NH). MS (FAB): 850. 0 (849. 83 calcd. for $C_{33}H_{43}N_3O_{19}S_2$, M).

3b. Yield: 97. 2% Mp 212 - 215℃. ^{1}H NMR ($CDCl_3$): δ 1. 96, 2. 00, 2. 04, 2. 06, 2. 07, 2. 09, 2. 15 (s, 21H, CH_3), 2. 44 (s, 3H, CH_3), 3. 69 - 5. 54 (m, 14H, lactosyl ring), 6. 90 (bs, 1H, NH), 7. 35 - 7. 37 (m, 2H, Ar—H), 7. 45 (bs, 1H, NH), 7. 75 - 7. 77 (m, 2H, Ar—H), 7. 92 (bs, 1H, NH).

3c. Yield: 91. 5% Mp 214 - 216℃. ^{1}H NMR ($CDCl_3$): δ 1. 96, 1. 99, 2. 04, 2. 06, 2. 07, 2. 09, 2. 16 (s, 21H, CH_3), 3. 67 - 5. 50 (m, 14H, lactosyl ring), 7. 14 (bs, 1H, NH), 7. 23 - 7. 27 (m, 3H, Ar—H, NH), 7. 78 (bs, 1H, NH), 7. 90 - 7. 92 (bs, 2H, NH).

3d. Yield: 98. 2%. Mp 213 - 214℃. ^{1}H NMR ($CDCl_3$): δ 1. 96, 1. 98, 2. 03, 2. 04, 2. 07, 2. 08, 2. 16 (s, 21H, CH_3), 3. 66 - 5. 49 (m, 14H, lactosyl ring), 7. 24 (bs, 1H, NH), 7. 53 - 7. 55 (m, 2H, Ar—H), 7. 82 - 7. 84 (m, 2H, Ar—H), 7. 90 (bs, 2H, NH).

3e. Yield: 99. 1%. Mp 213 - 215℃. ^{1}H NMR ($CDCl_3$): δ 1. 96, 1. 99, 2. 04, 2. 04, 2. 07, 2. 08, 2. 16 (s, 21H, CH_3), 3. 67 - 5. 47 (m, 14H, lactosyl ring), 7. 15 (bs, 1H, NH), 7. 70 - 7. 76 (m, 4H, Ar—H), 7. 84 (bs, 1H, NH), 7. 90 (bs, 1H, NH).

3f. Yield: 94. 3% Mp 180 - 183℃. ^{1}H NMR ($CDCl_3$); δ 1. 99, 2. 00, 2. 02, 2. 06, 2. 10, 2. 11 (6s, 21H, CH_3), 3. 79 - 5. 61 (m, 14H, maltosyl ring), 7. 09 (bs, 1H, NH), 7. 56 - 7. 67 (m, 5H, 2 × Ar—H, NHNH), 7. 87 - 7. 89 (m, 3H, 3 × Ar—H).

3g. Yield: 92. 7% Mp 187 - 188℃. ^{1}H NMR ($CDCl_3$): δ 2. 00, 2. 01, 2. 02, 2. 03, 2. 07,

2. 10,2. 12(7s,21H,CH_3),2. 45(s,3H,CH_3),3. 77 – 5. 54(m,14H,CH_2),7. 50 – 7. 83 (m,6h,4 × Ar—H,NHNH).

3h. Yield:98. 4%. Mp 186 – 190℃ ^{1}H NMR($CDCl_3$)^{1}HNMR($CDCl_3$):δ 1. 98,2. 01, 2. 02,2. 04,2. 06,3. 08,2. 10(7s,21H,CH_3),3. 77 – 5. 54(m,14H,CH_2),7. 50 – 7. 83(m, 6H,4 × Ar—H,NHNH).

3i. Yield:98. 6%. Mp 191 – 193℃. ^{1}H NMR($CDCl_3$):δ 1. 98,1. 99,2. 01,2. 02,2. 05, 2. 09,2. 10(7s,21H,CH_3),3. 72 – 5. 54(m,14H,CH_2),7. 56 – 7. 74(m,4H,4 × Ar—H), 7. 87(bs,1H,NH),8. 07(bs,1H,NH),8. 12(bs,1H,NH).

3j. Yield:93. 4%. Mp 171 – 173℃. ^{1}H NMR($CDCl_3$):δ 1. 92,1. 98,1. 99,2. 02,2. 09, 2. 11(s,21H,CH_3),3. 77 – 5. 57(m,14H,maltosyl ring),7. 15(bs,1H,NH),7. 22 – 7. 25 (m,2H,Ar—H,NH),7. 34(bs,1H,NH). 7. 84(bs,1H,NH),7. 89 – 7. 92(m,2H,Ar—H). ^{13}C NMR($CDCl_3$):δ 20. 81,21. 02,21. 17(7 × CH_3),60. 61,61. 25(2 × CH_2),62. 50, 68. 20, 68. 72, 69. 49, 70. 30, 70. 77, 72. 72, 74. 47, 75. 11, 82. 20, 95. 75 (sugar ring), 116. 97,117. 21,131. 67,131. 76,165. 08,167. 63(6 × Ar—H),169. 67,169. 93,169. 94, 170. 07,170. 79,171. 10,171. 67(7 × C ═O),184. 10(C ═S). Anal. calcd. for $C_{33}H_{42}FN_3O_{19}S_2$: C,45. 67;H,4. 88;N,4. 84. Found: C,45. 50;H,4. 95;N,4. 96.

References

[1] Van Der Lans, R. G. J. M. *Proceedings of the Fourth British Insecticide and Fungicide Conference*, Brighton, England, 1967, 2, 562 – 569.

[2] Achgill, R. K.; Call, L. W. Preparation of 4 – methyl – 3 – thiosemicarbazide as herbicide intermediate. EP 339964, 1989.

[3] Heuer, L.; Kugler, M.; Paulus, W.; Lorentzen, J. P.; Dehne, H. W.; Erdelen, C. EP 571857, 1993.

[4] Dziadulewicz, E. K.; Ritchie, T. J.; Hallett, A.; Snell, C. R.; Ko, S. Y. 1 – (2 – Nitrophenyl) thiosemicarbazides: A novel class of potent, orally active non – peptide antagonist of the bradykinin B_2 receptor. *J. Med. Chem.* 2000, 43, 769 – 771.

[5] Schenone, S.; Bruno, O.; Ranise, A.; Bondavalli, F.; Filippelli, W.; Falcone, G. 3 – Arylsulphonyl – 5 – arylamino – 1, 3, 4 – thiadiazol – 2 (3*H*) ones as anti – inflammatory and analgesic agents. *Bioorg. Med. Chem.* 2001, 9, 2149 – 2153.

[6] Suzuki, M.; Morita, K. Synthesis and herbicidal activity of 4 – thiazolone derivatives and their effect on plant secretory pathway. *J. Pest. Sci.* 2003, 28(1), 37 – 43.

[7] Wells, L.; Vosseller, K.; Hart, G. W. Glycosylation of nucleocytoplasmic proteins: Signal transduction and *O* – GlcNAc. *Science* 2001, 291(23), 2376 – 2378.

[8] Thompson, G. D.; Dutton, R.; Sparks, T. C. Spinosad—a case study: An example from a natural products discovery programme. *Pest. Manag. Sci.* 2000, 56, 696 – 702.

[9] Li, Y. X.; Li, Z. M.; Zhao, W. G.; Wang, S. H.; Dong, W. L. Synthesis of novel 1 – arylsulfonyl – 4 – (1′ – N – 2′,3′,4′,6 – tetra – *O* – acetyl – *β* – D – glucopyranosyl) thiosemicarbazides. *Chin. Chem. Lett.* 2006, 2, 152 – 155.

[10] Hamilton, S. K.; Wilkinson, D. E.; Hamilton, G. S.; Wu, Y. Q. Microwave assisted synthesis of N, N′ – diaryl cyanoguanidines. *Org. Lett.* 2005, 7(12), 2429 – 2431.

[11] Laurent, R.; Laporteire, A.; Dubac, J.; Berlan, J. Microwave – assisted Lewis acid catalysis: Application to the synthesis of alkyl – or arylhalogermanes. *Organometallics* 1994, 2493 – 2495.

[12] Alan, R. K.; Sandeep, K. S. Synthesis of C – carbamoyl – 1,2,3 – triazoles by microwave – induced 1,3 –

dipolar cycloaddition of organic azides to acetylenic amides. *J. Org. Chem.* 2002,67,9077 – 9079.

[13] Lin,M. J. ;Sun,C. M. Microwave – assisted traceless synthesis of thiohydantion. *Tetrahedron. Lett.* 2003,44,8739 – 8742.

[14] Rathke,M. W. ;Weipert,P. In *Comprehensive Organic Synthesis*; Trost,B. M. , Fleming,K. ,Eds. ;Pergamon:New York,1991;Vol. 2,p. 277.

[15] Lemieux,R. U. *Methods in Carbohydrate Chemistry II*;Academic Press:New York,1963;Vol. 2,pp. 221 – 222.

[16] Yu,J. X. ;Yi,Y. X. Stereoselective synthesis of 1 – [1′ – aroyl thiosemicarbazide – 4′ – YL] – N – β – D – glycopyranosides. *Chin. J. Syn. Chem.* 2000,8(2),137 – 141.

[17] Suslick,K. S. Sonochemistry. *Science* 1990,247,1439 – 1445.

[18] Hayes,B. L. *Microwave Synthesis,Chemistry at the Speed of Light*; CEM Publishing:Matthews,NC,2002; Chap. 1,p. 16.

Synthesis of New Plant Growth Regulator: *N* – (Fatty acid) *O* – aryloxyacetyl Ethanolamine*

Liang Han[1], Jianrong Gao[1], Zhengming Li[2], Yun Zhang[3], Weiming Guo[3]

([1]College of Chemical Engineering and Material Science, Zhejiang University of Technology, Hangzhou, 310014, China;
[2]National Key Laboratory of Elemento – Organic Chemistry, Nankai University, Tianjin, 300071, China;
[3]College of Horticulture, Nanjing Agricultural University, Nanjing, 210095, China)

Abstract *N* – (Fatty acyl) *O* – aryloxyacetyl ethanolamines, prepared from *N* – acylethanolamine (NAE) and aryloxyacetic acid, were tested for plant growth regulating activity. Compared with *N* – stearoylethanolamine, most compounds exhibit improved plant growth stimulating activity. In particular, those with chlorine on aryl ring show better activity than 2,4 – dichlorophenyloxyacetic acid in stimulating hypocotyls elongation of rape which indicates that chlorine on aryl ring appears significant. Moreover, these derivatives display improved solubility.

Key words *N* – acylethanolamines; aryloxyacetic acids; plant growth regulator; synthesis

N – Acylethanolamines (NAEs), a family of endogenous fatty acid amides, are known in mammals since some decades. They are proved to be an important bioactive substance in microgram level existing in the animal cells and play biochemical and pharmacological role. [1] However, their widespread occurrence, metabolism, and physiological significance in plants have only recently begun to be appreciated. [2] More investigation supported the fact that NAEs play a lipid mediator role in plants and are implicated as transducers in plant defense signaling, [2d,f,3] and at elevated levels, interfered with normal seedling root development. [4] The identification and active metabolism of NAEs in these physiological situations supports the emerging concept that NAEs are endogenous signaling compounds in plant systems. [5] Furthermore, LPE (lysophosphatidylethanolamine), capable of inhibiting plant PLD α (phospholipase α), has a profound effect on the physiological symptoms associated with postharvest senescence of flowers and fruits[6]; however, NAE12:0 and NAE14:0 are at least a 100 – fold more potent than LPE in inhibiting PLD α activity, so these NAEs may found novel agrochemical applications in the future. [7,8] The applications of exogenous NAEs further confirmed the recognition of their important role in plant. [2f,9]

Though the above reports are exciting, NAEs have very low solubility in water and limited solubility in most common organic solvents, which discourages further study and application. It ur-

---

* Reprinted from *Bioorganic & Medicinal chemistry Letters*, 2007, 17 (11): 3231 – 3234. Supported by National Natural Science Foundation NNSFC#20432010.

ges the molecular modification of NAEs to improve the solubility, which has not been reported to date. On the other hand, it is well known that substituted aryloxyacetic acids exhibit high plant growth regulating activity. [10] Yet, sometimes they induce unfavorable responses. For example, when substituted aryloxyacetic acids were used to improve fruit set, some suppression of leaf growth happened. [10] The report that the ester of aryloxyacetic acids can not only improve the translocatibility but also reduce the extent of side effects encourages us to incorporate the aryloxyacetic acid type regulator into NAEs. A series of *N* – (fatty acyl) *O* – aryloxyacetyl ethanolamines were synthesized and their bioactivity is reported here. The preparation of title compounds followed the reaction sequence depicted in Scheme 1.

$$NH_2CH_2CH_2OH + CH_3(CH_2)_nCOOH \longrightarrow \underset{\textbf{NAE}}{HOCH_2CH_2NHCO(CH_2)_nCH_3}$$

$$R\text{-}C_6H_4\text{-}OCH_2COOH + SOCl_2 \longrightarrow R\text{-}C_6H_4\text{-}OCH_2COCl$$

$$\xrightarrow[Et_3N/CHCl_3]{\textbf{NAE}} \underset{\textbf{1}}{R\text{-}C_6H_4\text{-}O-CH_2COO\frown NHCO(CH_2)_nCH_3}$$

Scheme 1

Several NAE types have been identified in a variety of plant species. These NAE types identified contain acyl chains of 12 – 18℃ in length and up to three double bonds, reflecting the typical acyl moieties prevalent in higher plants. [2a] In terms of bioactivity, plants are particularly responsive to low concentration of medium chain saturated NAEs; whereas animal physiology is largely regulated by low concentration of long chain polyunsaturated NAEs(e. g. , NAE 20:4). [5] In our work, four kinds of NAEs, namely NAE12:0, NAE14:0, NAE16:0, NAE18:0, were synthesized to incorporate into the structure of several aryloxyacetic acids. NAE18:0 was chosen to couple with five kinds of aryloxyacetic acids to study the influence of substituent group on bioactivity. And the 2,4 – dichlorophenoxyacetic acid(2,4 – D) was used to react with all synthesized NAEs to investigate the bioactivity variety along with chain length.

NAEs were prepared by refluxing 1 mol of the fatty acid with 1.5 mol of ethanolamine for 6h in 60% – 90% yield. [11] The products were purified by recrystallization from the solution of trichloromethane and acetone instead of 95% ethanol of Ref. [11] Aryloxyacetic acids were prepared according to Ref. [12] and then transformed to the acid chlorides by refluxing with $SOCl_2$ for 5h. NAEs, suspended in $CHCl_3$ in the presence of Et_3N, were dissolved gradually with the addition of the acid chlorides, which indicated the reaction happened and the products had better solubility in $CHCl_3$ than NAEs. After the usual workup, title compounds **1** were obtained through the purification by column chromatography. Structures of title compounds were confirmed by elemental analyses and ^1H NMR. [13]

On the other hand, the solubility of title compounds and NAEs was detected in $CHCl_3$, EtOAc, and EtOH. The solubility data of some title compounds with NAE12:0 and NAE18:0 are listed in Table 1. As compared with respective NAE, title compounds show evident solubility improvement.

Table 1 Solubility of some title compounds and parent NAEs

Compound	n	R	Solubility(g)		
			$CHCl_3$	EtOAc	EtOH
NAE12:0	10	—	1.76	0.38	9.14
1j	10	2,4-2Cl	15.10	11.0	10.42
NAE18:0	16	—	0.16	0.03	0.34
1a	16	$m-CH_3O$	7.76	2.67	1.99
1b	16	$o-CH_3$	53.97	8.27	3.08
1e	16	p-Cl	4.37	1.54	1.88
1g	16	2,4-2Cl	10.22	3.54	1.03

Since NAE was reported to prolong the shelf life of cut flowers, delay deterioration and leaf drop, and extend the overall appearance and quality of the plant cutting,[9] title compounds were tested for their plant growth stimulating activity in hypocotyls elongation of rape, cotyledon expansion of cucumber, and coleoptiles growth of common wheat.[14-16] The increase percent in these assays over control samples with no regulator is listed in Table 2 and also are the data of NAE18:0 and the conventional plant growth regulator 2,4-D for comparison. Just like NAE 18:0 and 2,4-D, title compounds exhibit good stimulating effect on hypocotyls elongation of rape. In hypocotyls elongation test, improved stimulating effect was observed for all title compounds except compound **1a** when compared with NAE18:0, and compounds **1e-1j** with different chain length and substituent group even have better growth stimulating activity than 2,4-D. It is interesting to note that the hypocotyls elongation of rape appears strongly associated with the substituent group on the benzene ring. The compounds with substituent chlorine are obviously superior to those with substituent methyl and methoxy group, which suggests that chlorine on aryl ring appears significant. For example, compound **1g** has an increase percent of 73.8% in stimulating hypocotyls elongation of rape, while the increase percent of compound **1b** is only 57.0%. However, introduction of additional chlorine does not affect the elongation of rape hypocotyls and the activity of 4-chloro-substituent **1e** is comparable to that of 2,4-dichloro-substituent **1g**. In coleoptiles growth test of common wheat, all title compounds also exhibit better stimulating effect than NAE18:0 except compound **1b**, unfortunately, only compound **1c** shows a little improved stimulating activity in cotyledon expansion test of cucumber. The change of fatty chain length seems to have intricate effect on the bioactivity. Compound **1f** has the same bioactivity in stimulating hypocotyls elongation of rape as **1e**, though there is a difference of six methylenes in their fatty chain structure. But when stimulating cotyledon expansion of cucumber and coleoptile growth of common wheat, compounds with fatty chain of 16 carbons and 12 carbons have better activity than compounds with fatty chain of 18 carbons and 14 carbons.

In conclusion, a series of *N*-(fatty acyl) *O*-aryloxyacetyl ethanolamines were synthesized by reaction of NAEs of different chain length with substituted aryloxyacetic acids. All new compounds have improved solubility. The preliminary bioactivity data show that most of them exhibit better plant growth stimulating effect than NAE18:0, moreover, those with chlorine on benzene ring have better stimulation of hypocotyls elongation than conventional 2,4-D.

Table 2 Plant growth regulating bioactivity in vitro

Compound	n	R	Increased bioactivity(% ,10mg/L)		
			Hypocotyls elongation of rape	Cotyledon expansion of cucumber	Coleoptiles growth of common wheat
1a	16	m – CH_3O	-1.4	-0.1	5.0
1b	16	*o* – CH_3	57.0	5.7	0.7
1c	16	*p* – CH_3	38.7	9.8	3.3
1d	10	*p* – CH_3	54.9	5.9	4.1
1e	16	*p* – Cl	71.8	6.3	5.0
1f	10	*p* – Cl	71.8	2.9	16.0
1g	16	2,4 – 2Cl	73.8	0.7	7.5
1h	14	2,4 – 2Cl	67.3	3.5	17.0
1i	12	2,4 – 2Cl	68.1	-0.1	8.9
1j	10	2,4 – 2Cl	69.0	4.0	17.9
NAE(18)	16	—	35.3	6.2	2.1
2,4 – D	—	2,4 – 2Cl	65.7	0.8	28.3

Supplementary data

Supplementary data associated with this article can be found, in the online version, at doi: 10.1016/j.bmcl.2007.03.013.

References and notes

[1] (a) Kuehl, F. F.; Jacob, T. A.; Ganley, O. H.; Ormond, R. E.; Meisinger, M. A. P. *J. Am. Chem. Soc.* 1957, 79, 5577; (b) Parinandi, N. L.; Schmid, H. H. O. *FEBS Lett.* 1988, 237, 49; (c) Gulaya, N. M.; Kuzmenko, A. I.; Margitich, V. M.; Govseeva, N. M.; Melnichuk, S. D.; Goridko, T. M.; Zhukov, A. D. *Chem. Phys. Lipids* 1998, 97, 49; (d) Skaper, S. D.; Buriani, A.; Dal Toso, R.; Petrelli, L.; Romanello, S.; Facci, L.; Leon, A. *Proc. Natl. Acad. Sci. U. S. A.* 1996, 93, 3984; (e) Facci, L.; Dal Toso, R. *Proc. Natl. Acad. Sci. U. S. A.* 1995, 92, 3376; (f) Mazzari, S.; Canella, R.; Petrelli, L.; Marcolongo, G.; Leon, A. *Eur. J. Pharmacol.* 1996, 300, 227; (g) Molina – Holgado, F.; Lledo′, A.; Guaza, C. *Neuroreport* 1997, 8, 1929.

[2] (a) Chapman, K. D.; Venables, B.; Blair, R., Jr.; Berringer, C. *Plant Physiol.* 1999, 120, 1157; (b) Schmid, H. H.; Berdyshev, E. V. *Prostaglandins Leukot Essent Fatty Acids* 2002, 66, 363; (c) Di Marzo, V.; De Petrocellis, L.; Fezza, F.; Ligresti, A.; Bisogno, T. *Prostaglandins Leukot Essent Fatty Acids* 2002, 66, 377; (d) Chapman, K. D.; Tripathy, S.; Venables, B.; Desouza, A. *Plant Physiol.* 1998, 116, 1163; (e) Stella, N.; Schweitzer, P.; Piomelli, D. *Nature (London)* 1997, 388, 773; (f) Tripathy, S.; Venables, B. J.; Chapman, K. D. *Plant Physiol.* 1999, 121, 1299; (g) Chapman, K. D.; Venables, B. J.; Dian, E. E.; Gross, G. W. *J. Am. Oil. Chem. Soc. (JAOCS)* 2003, 80, 223; (h) Giuffrida, A.; Rodriguez de Fonseca, F.; Piomelli, D. *Anal. Biochem.* 2000, 280, 87; (i) Schmid, P. C.; Schwartz, K. D.; Smith, C. N.; Krebsbach, R. J.; Berdyshev, E. V.; Schmid, H. H. Chem. *Phys. Lipids* 2000, 104, 185.

[3] (a) Chapman, K. D. *Chem. Phys. Lipids* 2000, 108, 221; (b) Tripathy, S.; Kleppinger – Sparace, K.; Dixon, R. A.; Chapman, K. D. *Plant Physiol.* 2003, 131, 1781.

[4] Blancaflor, E. B.; Hou, G.; Chapman, K. D. *Planta* 2003, 217, 206.

[5] Chapman, K. D. *Prog. Lipid Res.* 2004, 43, 302.

[6] Ryu, S. ; Bjourn, K. ; Ozgen, M. ; Plata, J. *Proc. Natl. Acad. Sci. U. S. A.* 1997, 94, 12717.

[7] Austin – Brown, S. ; Chapman, K. D. *Plant Physiol.* 2002, 129, 1892.

[8] Sang, Y. ; Zheng, S. ; Li, W. ; Huang, B. ; Wang, X. *Plant J.* 2001, 28, 1.

[9] Chapman, K. D. PCT Int. Appl. WO01, 301, 43 A2, 2001.

[10] Krewson, C. F. ; Saggese, J. E. ; Carmichael, J. F. ; Ard, J. S. ; Drake, T. F. *J. Agri. Food Chem.* 1959, 7, 118.

[11] Roe, E. T. ; Miles, T. D. ; Swern, D. *J. Am. Chem. Soc.* 1952, 74, 3442.

[12] Bao, M. ; He, Q. L. ; He, X. – Zh. ; Liu, B. – D. *J. Appl. Chem.* (*in Chinese*) 1997, 14, 90.

[13] *General method for preparing title compounds.* A mixture of NAE (1mmol) and Et_3N (1.2mmol) in dry $CHCl_3$ (5mL) was stirred at room temperature and then the solution of aryloxyacetic acid chloride (1.2mmol) in dry $CHCl_3$ (2mL) was added drop by drop. During the addition, NAE was dissolved gradually and the solution turned yellow. After stirred for 12h, the solution was washed with water and then dried with $MgSO_4$. The solvent was removed and column chromatography of the residue with PE/EtOAc (1.5:1, v/v) gave the product as a white solid. Compound **1b**, yield 75% : mp 63 – 64℃; 1H NMR ($CDCl_3$, 300MHz): δ 7.20 – 7.12 (dd, 2H, Ar$\underline{H}$), 6.94 – 6.90 (t, 1H, Ar$\underline{H}$), 6.72 – 6.69 (d, 1H, Ar$\underline{H}$), 5.45 (br s, 1H, N$\underline{H}$CO), 4.70 (s, 2H, ArO$\underline{CH_2}$), 4.28 – 4.24 (t, 2H, O$\underline{CH_2}$$CH_2$N), 3.51 – 3.46 (dd, 2H, O$CH_2$$\underline{CH_2}$N), 2.30 (s, 3H, Ar$\underline{CH_3}$), 2.08 – 2.02 (t, 2H, $\underline{CH_2}$CONH), 1.58 – 1.53 (t, 2H, $\underline{CH_2}$$CH_2$CO), 1.25 (br s, 28H, $(\underline{CH_2})_{14}$), 0.90 – 0.86 (t, 3H, $\underline{CH_3}(CH_2)_{14}$) ppm; Anal. calcd for $C_{29}H_{49}NO_4$: C 73.22, H 10.38, N 2.94; found: C73.24, H 10.40, N 2.85.

[14] *Hypocotyls elongation test of rape.* After soaking, the seeds with similar magnitude were chosen to use. Tested compounds were dissolved in DMF and later dropped evenly on 6 – cm – diameter fitter paper. After air – volatilization of solvent, the filter paper was placed in 6 – cmdiameter glass utensil with distilled water to give 10mg/L compound solutions, and then ten seeds were added. The seeds were cultured at 25℃ in dark. The length of hypocotyls was measured after 3 days and compared with those treated with distilled water to estimate the activity. Two replicates were included in the evaluation.

[15] *Cotyledon expansion test of cucumber.* After soaking, the seeds were germinated in covered enamelware containing 0.7% agar and cultured for 3 days in dark at 26℃, and then the cotyledons of similar magnitude were chosen to use. Tested compounds were dissolved in DMF and later dropped evenly on 6 – cm – diameter fitter paper. After air – volatilization of solvent, the filter paper was placed in 6 – cm – diameter glass utensil with distilled water to give 10mg/L compound solutions and ten pieces of cotyledons were added. The cotyledons were cultured in light (300Lux, 26℃). After 3 days, the total weight of cotyledon was measured and compared with those treated with distilled water to estimate the activity. Two replicates were included in the evaluation.

[16] *Coleoptiles growth test of common wheat.* After soaking, the seeds were germinated in covered enamelware containing 0.7% agar and cultured for 3 days in dark at 25℃. When the seedling grew to 2.5 – 3.0cm tall, the first 3 mm of coleoptile top was rejected. Coleoptile (5mm) was truncated and dunked in distilled water for 1h to remove endogenous hormone, then it was chopped into 10 segments. Tested compounds were dissolved in DMF and later dropped evenly on 6 – cm – diameter fitter paper. After air – volatilization of solvent, the filter paper was placed in 10mL beaker with 0.01mol/L phosphoric acidcitric acid buffer solution (pH5) containing 2% sucrose to give 10mg/L compound solutions and ten coleoptile segments were added afterward. The coleoptiles were cultured at 25℃ in dark for 18 – 20h and then the length of coleoptiles was measured and compared with those treated without tested compounds to estimate the activity. Two replicates were included in the evaluation.

Synthesis, Bioactivity, Theoretical and Molecular Docking Study of 1 – Cyano – *N* – substituted – cyclopropanecarboxamide as Ketol – acid Reductoisomerase Inhibitor*

Xinghai Liu, Peiquan Chen, Baolei Wang, Yonghong Li, Suhua Wang, Zhengming Li

(The Research Institute of Elemento – Organic Chemistry, The State – Key Laboratory of Elemento – Organic Chemistry, National Pesticide Engineering Research Center, Nankai University, Tianjin, 300071, China)

Abstract Ketol – acid reductoisomerase (KARI; EC 1. 1. 1. 86) catalyzes the second common step in branched – chain amino acid biosynthesis. The catalyzed process consists of two stages, the first of which is an alkyl migration from one carbon atom to its neighbouring atom. The likely transition state is a cyclopropane derivative, thus a series of new cyclopropane derivatives, such as 1 – cyano – *N* – substituted – cyclopropanecarboxamide, were designed and synthesized. Their structures were verified by ^1H NMR, FTIR spectrum, MS and elemental analysis. The K_i values of active compounds **2**, **4b** against rice KARI were 95. 30 ± 13. 71, 207. 9 ± 21. 99 μM, respectively. The X – ray crystal structure of compound **4a** was also determined. Auto – Dock was used to predict the binding mode of **4a**. This was done by analyzing the interaction of the compounds **4a** with the active sites of spinach KARI. This result was in accord with the result analyzed by the frontier molecular orbital theory.

Key words cyclopropanecarboxamide; synthesis; structure; moleculardocking; theoretical study; KARI

Plants and most micro – organisms have biosynthetic ability which allows them to survive on relatively simple nutrients. For this reason, plants and microorganisms contain numerous enzymes that are potential targets for designing bioactive compounds such as herbicides and antibiotics. Enzymes involved in the biosynthesis of the branched chain amino acids are one such example. Isoleucine and valine are synthesized in a parallel set of four reactions while an extension of the valine pathway results in leucine.[1]

This pathway is the target for the sulfonylureas,[2] the imidazolinones[3] and a variety of other herbicides,[1] which all inhibit the first enzyme, acetohydroxyacid synthase. The success of these herbicides has stimulated research into inhibitors of other enzymes in the pathway, including the second enzyme in the common pathway,[4] ketol – acid reductoisomerase (KARI; EC 1. 1. 1. 86), and two enzymes in the leucine extension.[5,6] The reaction catalyzed by KARI is shown in Scheme 1 which consists of two steps,[7,8] an alkyl migration followed by a NADPH dependent reduction. Both steps require a divalent metal ion, such as Mg^{2+}, Mn^{2+} or Co^{2+}, but

* Reprinted from *Bioorganic & Medicinal chemistry Letters*, 2007, 17: 3784 – 3788. This work was supported by the National Basic Research Key Program of China (Grant No. 2003 CB114406), the National Natural Science Foundation Key Project of China (Grant No. 20432010) and the High Performance Computing Project of Tianjin Ministry of Science and Technology of China (Grant No. 043185111 – 5).

the alkyl migration is highly specific for Mg^{2+}. HOE 704[9] and IpOHA[10] are potent competitive inhibitors of the enzyme (Scheme 1).

A transition state being a cyclopropane is postulated and mimicked by Gerwick et al. [11] They showed that cyclopropane-1,1-dicarboxylate (CPD) can inhibit *Escherichia coli* KARI. They also showed that application of CPD to various plant tissues caused the accumulation of the substrate 2-acetolactate; in vivo data strongly suggest that the CPD can inhibit the activity of KARI Scheme 1. [12]

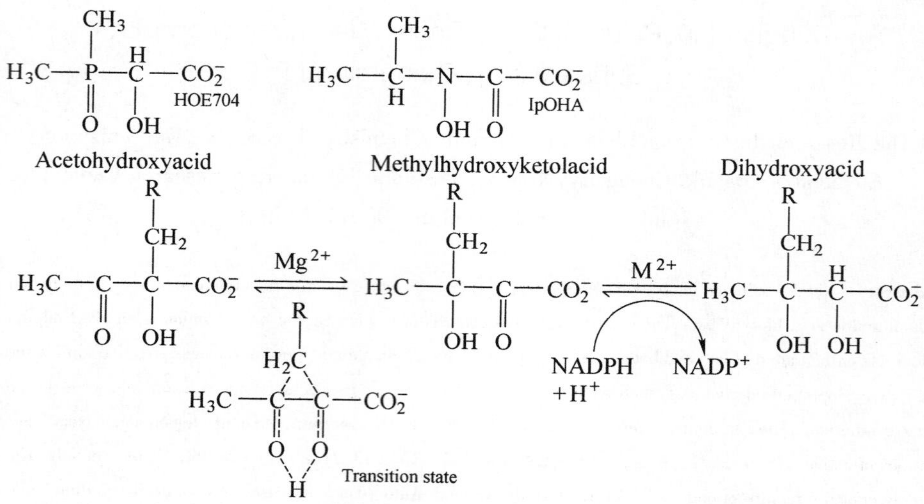

HOE704 IpOHA

Acetohydroxyacid Methylhydroxyketolacid Dihydroxyacid

Mg^{2+} M^{2+} NADPH $+H^+$ $NADP^+$

Transition state

Scheme 1 Reaction catalyzed by KARI

The first step in the KARI catalyzed process involves an alkyl migration from one carbon atom to its neighboring atom. The likely transition state is a cyclopropane derivative. For this reason, some new cyclopropane derivatives were synthesized in our laboratory (Scheme 2).

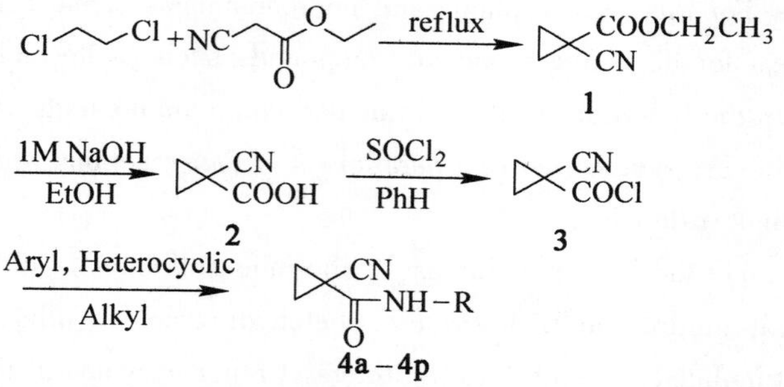

reflux $COOCH_2CH_3$ CN **1**

1M NaOH / EtOH → CN, COOH **2**; $SOCl_2$ / PhH → CN, COCl **3**

Aryl, Heterocyclic / Alkyl → CN, NH-R **4a-4p**

R = p-$CH_3C_6H_4$, 2-$CHCl_2C_2N_2S$, p-BrC_6H_4, 2-CH_3C_3HNS, 2,4,5-$Cl_3C_6H_2$, m-BrC_6H_4, C_6H_5, 2,4-$Cl_2C_6H_3$, o-$CH_3C_6H_4$ p-ClC_6H_4, $OHCH_2CH_2$, p-$CF_3C_6H_4$, m-ClC_6H_4 o-$CF_3C_6H_4$, m-$CF_3C_6H_4$, p-$OCH_3C_6H_4$

Scheme 2 Synthesis route for compounds **4a-p**

Biological studies revealed that some of these compounds inhibit ketol-acid reductoisomerase in vivo effectively.

The 1-cyan-1-cyclopropane carboxylic acid, prepared from 1,2-dichloroethane and ethyl cyanacetate was cyclized for 16h at refluxing temperature. In order to optimize the reaction time, microwave assisted irradiation was applied which shortened the reaction time to 40min. Compound **3** reacted with substituted anilines, heterocyclic amine or alkyl amines in the presence of inorganic base to yield substituted cyclopropanecarboxamides as shown in Scheme 2. [13]

The KARI activities in vitro of these compounds were determined. [14] The results for compound, 1, 2 and compounds **4a – p** are summarized in Table 1.

Table 1 Inhibition rate (%) of compounds 4a – p against rice KARI at 200ppm in vitro

Compound	R	KARI activity
1		0
2		100
4a	$p-CH_3C_6H_4-$	61.21
4b	$2-CHCl_2C_2N_2S-$	100
4c	$p-BrC_6H_4-$	32.23
4d	$2-CH_3C_3HNS-$	69.81
4e	$2,4,5-Cl_3C_6H_2-$	0
4f	$m-BrC_6H_4-$	17.25
4g	C_6H_5-	77.23
4h	$2,4-Cl_2C_6H_3-$	97.04
4i	$o-CH_3C_6H_4-$	100
4j	$p-ClC_6H_4-$	93.92
4k	$OHCH_2CH_2-$	98.92
4l	$p-CF_3C_6H_4-$	0
4m	$m-ClC_6H_4-$	0
4n	$o-CF_3C_6H_4-$	0
4o	$m-CF_3C_6H_4-$	0
4p	$p-OCH_3C_6H_4-$	3.95

It was found from Table 1 that compounds **2, 4b, 4h, 4i, 4j** and **4k** have favourable inhibitory activity against KARI. The data given in Table 1 indicated that the change of substituent at phenyl ring affects the KARI activity. When the benzene ring is substituted by CF_3 group, the compounds generally have no KARI bioactivity, as **4l, 4n, 4o**. While for heterocyclic and alkane substituents, their inhibitory activities increase for **4b** and **4k**. For the compounds, **2, 4a** and **4b**, further bioassay was conducted and their K_i values against KARI were 95.30 ± 13.71, >300 and $207.9 \pm 21.99\mu M$, respectively. Hence, these identified cyclopropane derivatives could be useful for further optimization work in finding the potential KARI inhibitors.

In order to study the structure – activity relationship, the single – crystal structure of **4a** was determined[15] by X – ray crystallography[16] as illustrated in Fig. 1 in which three C – N bond lengths C(5) – N(2), C(6) – N(2) and N(1) – C(1) are 0.135, 0.142, and 0.114 nm, respectively, which are all longer than that of 0.134 nm in the single heterocycle ring. [17] In **4a**, the bond length of C(1) – N(1) is 0.1396 nm, which is longer than the double C—N bond.[18] Based on the computal results by Gaussian,[19,20] it was seen that DFT, HF and MP2 have good coherence with the crystal diffraction, for example it can be observed that C(2) – C(5) > C(1) – C(2) > N(2) – C(6) > N(2) – C(5) > O(1) – C(5) > N(1) – C(1) in crystal structure, which is accordance with the order of C(2) – C(5) > C(1) – C(2) > N(2) – C(6) > N(2) – C(5) > O(1) – C(5) > N(1) – C(1) in all calculation structures.

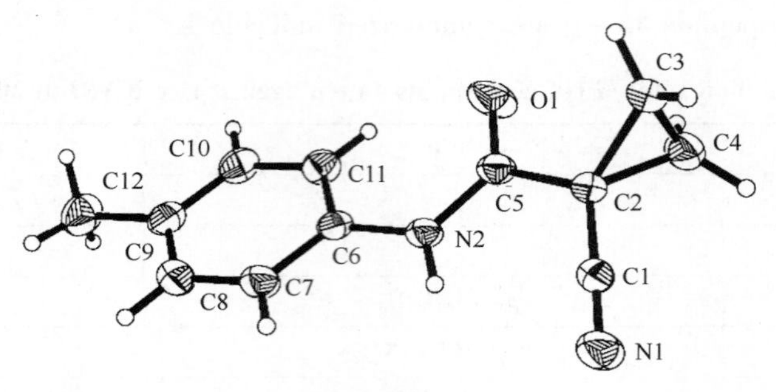

Fig. 1 Molecular structure of **4a**

According to the frontier molecular orbital theory, HOMO and LUMO are two most important factors which affect the bioactivities of compounds. HOMO has the priority to provide electrons, while LUMO accepts electrons first.[21,22] Thus, study on the frontier orbital energy can provide some useful information for the active mechanism. Taking HF (Hatree – Fork) results, the geometry of the frame of **4a** is hardly influenced by the introduction of either the cyano group or the cyclopropane ring from Fig. 2. The HOMO of **4a** is mainly located on aromatic ring and the amide group. On the other hand, the LUMO of **4a** contains aromatic ring, the amido group, the cyano group and the cyclopropane ring. The fact that **4a** has strong affinity suggests the importance of the frontier molecular orbital in the $\pi - \pi$ stacking or hydrophobic interactions. This also implies that the orbital interaction between **4a** and the rice KARI amino acid residues is dominated by $\pi - \pi$ of hydrophobic interaction between the frontier molecular orbitals.

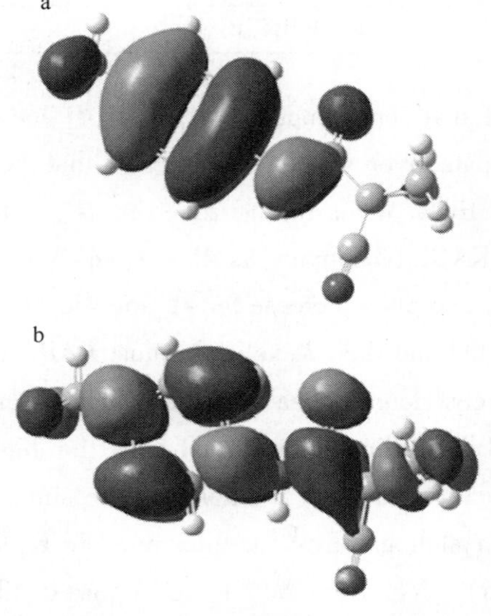

Fig. 2 Frontier molecular orbitals of compound **4a**:
(a) HOMO of compound **4a**; (b) LUMO of compound **4a**

To make the results predicted by our frontier molecular orbital model more relevant to the active sites of the enzyme and to further explore a probable binding site in the KARI, the compound **4a** was docked[21,22] into the active sites of KARI.[23]

Visual inspection of the conformation of **4a** docked into the KARI binding site revealed that the phenyl rings are hosted in the pocket of KARI and are oriented to establish π - π stacking interactions with the His 226 side chains (Fig. 3a). Moreover, two hydrogen bonds between the amino groups of **4a** and the carbonyl oxygen of Glu 319 and Asp 315 side chain are also observed. Furthermore, the cyclopropane ring and aromatic ring are embedded in a large hydrophobic pocket formed by Ser 225, His 226, Glu 496, Leu 323, Ser 518, Glu 319 and Asp 315 (Fig. 3b).

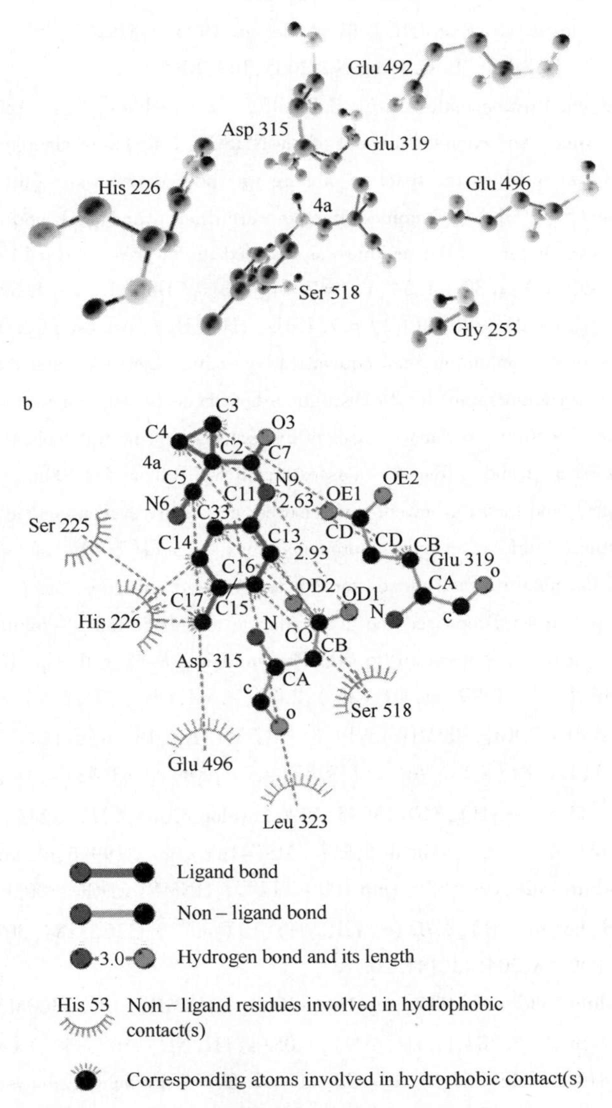

Fig. 3 Binding modes of compound **4a** in the active sites of spinach KARI:
(a) π - π stacking interaction between the His 226 side chain and phenyl ring;
(b) hydrogen bond and hydrophobic interaction between **4a** and the rice KARI amino acid residues

References and notes

[1] Duggleby, R. G.; Pang, S. S. *J. Biochem. Molec. Biol.* 2000, 33, 1.

[2] Chaleff, R. S.; Mauvais, C. J. *Science* 1984, 224, 1443.

[3] Shaner, D. L.; Anderson, P. C.; Stidham, M. A. *Plant Physiol.* 1984, 76, 545.

[4] Dumas, R.; Biou, V. F.; Douce, H. R.; Duggleby, R. G. *Acc. Chem. Res.* 2001, 34, 399.

[5] Wittenbach, V. A.; Teaney, P. W.; Rayner, D. R.; Schl oss, J. V. *Plant Physiol.* 1993, 102, 50.

[6] Hawkes, T. R.; Cox, J. M.; Fraser, T. E. M.; Lewis, T. *Z. Naturforsch.* 1993, C 48, 364.

[7] Dumas, R.; Butikofer, M. C.; Job, D.; Douce, R. *Biochemistry* 1995, 34, 6026.

[8] Chunduru, S. K.; Mrachko, G. T.; Calvo, K. C. *Biochemistry* 1989, 28, 486.

[9] Schulz, A.; Sponemann, P.; Kocher, H.; Wengenmayer, F. *FEBS Lett.* 1988, 238, 375.

[10] Aulabaugh, A.; Schloss, J. V. *Biochemistry* 1990, 29, 2824.

[11] Gerwick, B. C.; Mireles, L. C.; Eilers, R. J. *Weed Technol.* 1993, 7, 519.

[12] Lee, Y. T.; Ta, H. T.; Duggleby, R. G. *Plant Sci.* 2005, 168, 1035.

[13] General procedure: ethyl cyanacetate (22.6g, 0.2mol), 1,2 – dichloroethane (160g, 0.2mol), potassium carbonate (220g, 1.6mol) and catalytic amount of Bu4NHSO4 (1.0g) were vigorously refluxed in 1,2 – dichloroethane for 6 h after which the reaction mixture was poured into water (800mL). The product was extracted with ether (5×100mL), combined extracts were dried over $MgSO_4$ and then the solvent was removed on a rotary evaporator and the residue was distilled under pressure: bp 115 – 118/15mmHg. Yield 85%. ^{1}H NMR (δ, $CDCl_3$): 1.30 – 1.34 (t, $J = 7.14$Hz, 3H, CH3), 1.59 – 1.69 (m, $J = 3.27$Hz, 4H, cyclopropane – CH_2), 4.21 – 4.27 (q, $J = 7.13$Hz, 2H, CH_2). An ester (0.03mol) was added to a ca. 15% aqueous solution containing 3mol equivalents of sodium hydroxide and the suspension was vigorously stirred at ambient temperature for 2 days until a homogeneous solution was formed. The solution was extracted with ether (2×50mL) to remove traces of unreacted ester, the water phase was acidified with concentrated hydrochloric acid and a free acid was extracted with ether (3×100mL). The combined extracts were dried over $MgSO_4$ and then the solvent was removed on a rotary evaporator. Yields 51%. Mp 88 – 90℃. To a benzene solution (25mL) of cyanocyclopropanecarboxylic acid (7.50mmol) was added thionyl chloride (30mmol) and the mixture was refluxed for 2 h to give acid chloride.

Preparation of compound **4a**. Dropwised the acid chloride to substituted *p* – toluidine (7.50mmol), then vigorously stirred at ambient temperature for 4h. The yield was 84% with mp (106 – 108) ℃; ^{1}H NMR ($CDCl_3$, 300MHz) δ: 1.58 – 1.79 (m, 4H, CH_2), 2.92 (s, 3H, CH_3), 7.15 – 7.39 (m, 4H, ArH), 7.96 (s, 1H, NH); ^{13}C NMR ($CDCl_3$, 300MHz) δ: 14.35 (s, 2C, CH_2), 18.44 (s, 1C, C), 21.13 (s, 1C, CH_3), 120.30 (s, 1C, CN), 120.81 (s, 2C, ArC), 129.82 (s, 2C, ArC), 163.48 (s, 1C, CO), 135.40 (d, 1C, ArC); IR (KBr) ν: 3338 (—NH), 3107, 3035, 2920 (cyclopropane, CH), 2243 ($C \equiv N$), 1898 ($C = O$), 1689, 1602, 1523 (Ar C—C), 810cm^{-1}, ESI – MS (41eV): m/z: 199.6, 65.96.

Compound **4b**. A white solid, yield 82%; mp 109 – 117℃; ^{1}HNMR ($CDCl_3$, 300M) 1.73 – 1.80 (m, 4H, CH_2), 6.26 (s, 1H, hetero – H), 7.02 (s, 1H, NH); IR (cm^{-1}) 3336, 3184, 3077, 2991, 2268, 1731. ESI – MS: 273.48, 230.08, 204.43, 141.20.

Compound **4c**. A white solid, yield 93%; mp 106 – 108℃; ^{1}HNMR ($CDCl_3$, 300M) 1.59 – 1.72 (m, 4H, CH_2), 7.15 – 7.39 (m, $J = 8.784$Hz, 4H, ArH), 8.05 (s, 1H, NH); IR (cm^{-1}) 3347, 3093, 2236, 1692, 1598, 1525, 1489, 814. ESIMS: 264.98, 197.95, 116.02, 80.99. Elemental analysis: C, 49.80; H, 3.28; N, 10.58; calculated from $C_{12}H_9BrN_2O$. Observed: C, 49.84; H, 3.42; N, 10.57.

Compound **4d**. A white solid, yield 93%; mp 106 – 108℃; ^{1}HNMR ($CDCl_3$, 300M) 1.26 (s, 3H, CH_3), 1.73 – 1.94 (m, 4H, CH_2), 7.81 (d, 1H, Heterocycle); IR (cm^{-1}) 3450, 3077, 2991, 2855, 2268,

1731. ESI – MS:206. 37.

Compound **4e**. A white solid, yield 93%; mp 106 – 108℃; ^{1}HNMR(CDCl$_3$, 300M) 1. 62 – 1. 84(m, 4H, CH$_2$), 7. 52(s, 1H, ArH), 8. 50(s, 1H, ArH), 8. 68(s, 1H, NH); IR(cm^{-1}) 3369, 3122, 3020, 2229, 1692, 1576, 1496, 1456, 879. ESIMS: 287. 22, 219. 95, 253. 84, 222. 10, 183. 72. Elemental analysis: C, 45. 82; H, 2. 21; N, 9. 41; calculated from $C_{11}H_7Cl_3N_2O$. Observed: C, 45. 63; H, 2. 44; N, 9. 67.

Compound **4f**. A white solid, yield 93%; mp 106 – 108℃; ^{1}HNMR(CDCl$_3$, 300M) 1. 61 – 1. 83(m, 4H, CH$_2$), 7. 15(d, J = 7. 573Hz, 1H, ArH), 7. 31(d, J = 3. 893Hz, 1H, ArH), 7. 66(d, J = 1. 938Hz, 1H, ArH), 8. 50(s, 1H, ArH), 8. 68(s, 1H, NH); IR(cm^{-1}) 3320, 3195, 3115, 2244, 1680, 1590, 1527, 1476, 883, 810, 766, 656.

Compound **4g**. A white solid, yield 93%; mp 106 – 108℃; ^{1}HNMR(CDCl$_3$, 300M) 1. 59 – 1. 81(m, 4H, CH$_2$), 7. 18(t, J = 6. 848Hz, 1H, ArH), 7. 36(t, J = 7. 933Hz, 2H, ArH), 7. 50(d, J = 8. 293Hz, 2H, ArH), 8. 03(s, 1H, NH); IR(cm^{-1}) 3249, 3191, 3127, 2239, 1680, 1595, 1536, 1488, 751, 693. ESIMS: 185. 41, 65. 97. Elemental analysis: C, 70. 94; H, 5. 41; N, 15. 39; calculated from $C_{11}H_{10}N_2O$. Observed: C, 70. 95; H, 5. 41; N, 15. 04.

Compound **4h**. A white solid, yield 93%; mp 106 – 108℃; ^{1}HNMR(CDCl$_3$, 300M) 1. 63 – 1. 82(m, 4H, CH$_2$), 7. 28(d, J = 2. 270Hz, 1H, ArH), 7. 44(d, J = 2. 297Hz, 1H, ArH), 7. 28(d, J = 8. 888Hz, 1H, ArH), 8. 67(s, 1H, NH); IR(cm^{-1}) 3398, 3115, 2236, 1699, 1583, 1510, 959, 923, 821, 727. ESIMS: 253. 29, 185. 98, 149. 81, 114. 05. Elemental analysis: C, 51. 70; H, 3. 21; N, 10. 79; calculated from $C_{11}H_8Cl_2N_2O$. Observed: C, 51. 79; H, 3. 16; N, 10. 98.

Compound **4i**. A white solid, yield 93%; mp 106 – 108℃; ^{1}HNMR(CDCl$_3$, 300M) 1. 59 – 1. 81(m, 4H, CH$_2$), 7. 08(t, J = 6. 815Hz, 2H, ArH), 7. 18(d, J = 5. 187Hz, 1H, ArH), 7. 77(d, J = 8. 998Hz, 1H, ArH), 7. 99(s, 1H, NH); IR(cm^{-1}) 3435, 3105, 3019, 2225, 1695, 1588, 1531, 1459, 758. ESIMS: 200. 15, 171. 18, 106. 16, 77. 11.

Compound **4j**. A white solid, yield 93%; mp 106 – 108℃; ^{1}HNMR(CDCl$_3$, 300M) 1. 62 – 1. 81(m, 4H, CH$_2$), 7. 32(d, J = 8. 794Hz, 2H, ArH), 7. 47(d, J = 8. 856Hz, 2H, ArH), 8. 03(s, 1H, NH); IR(cm^{-1}) 3331, 3120, 2941, 2247, 1667, 1593, 1534, 1490, 832. ESIMS: 219. 42, 152. 06, 116. 04. Elemental analysis: C, 59. 66; H, 3. 96; N, 12. 58; calculated from $C_{11}H_9ClN_2O$. Observed: C, 59. 88; H, 4. 11; N, 12. 70.

Compound **4k**. A white solid, yield 93%; mp 106 – 108℃; ^{1}HNMR(CDCl$_3$, 300M) 1. 24(t, 1H, OH), 1. 51 – 1. 71(m, 4H, CH$_2$), 3. 62(t, 2H, J = 5. 380Hz, NH—CH$_2$), 4. 17(q, 2H, J = 5. 239Hz, OH—CH$_2$), 6. 88(s, 1H, NH); IR(cm^{-1}) 3356, 3195, 3115, 2251, 1698, 1588, 1534, 1486, 832.

Compound **4l**. A white solid, yield 93%; mp 106 – 108℃; ^{1}HNMR(CDCl$_3$, 300M) 1. 64 – 1. 84(m, 4H, CH$_2$), 7. 60(d, J = 9. 086Hz, 2H, ArH), 7. 65(d, J = 9. 071Hz, 2H, ArH), 8. 17(s, 1H, NH); IR(cm^{-1}) 3500, 3327, 3120, 2239, 1674, 1617, 1567, 1495, 808. ESI – MS: 287. 34(M + Na), 160. 33.

Compound **4m**. A white solid, yield 93%; mp 106 – 108℃; ^{1}HNMR(CDCl$_3$, 300M) 1. 63 – 1. 83(m, 4H, CH$_2$), 7. 47(dd, J = 7. 530Hz, 2H, ArH), 7. 65(d, J = 7. 631Hz, 1H, ArH), 7. 89(s, 1H, ArH), 8. 12(s, 1H, NH); IR(cm^{-1}) 3313, 3198, 3120, 2261, 1681, 1602, 1538, 1474, 872, 765, 672. ESIMS: 219. 33, 152. 02.

Compound **4n**. A white solid, yield 93%; mp106 – 108℃; ^{1}HNMR(CDCl$_3$, 300M) 1. 61 – 1. 82(m, 4H, CH$_2$), 7. 55(d, J = 7. 361Hz, 1H, ArH), 7. 65(dd, J = 7. 793Hz, 1H, ArH), 7. 94(d, J = 8. 150Hz, 1H, ArH), 8. 48(d, J = 8. 065Hz, 1H, ArH), 6. 64(s, 1H, NH); IR(cm^{-1}) 3312, 3188, 3115, 2244, 1680, 1593, 1527, 1476, 766. ESIMS: 285. 63(M + Na), 194. 94, 117. 05.

Compound **4o**. A white solid, yield 93%; mp106 – 108℃; ^{1}HNMR(CDCl$_3$, 300M) 1. 61 – 1. 81(m, 4H, CH$_2$), 7. 15(m, J = 7. 626Hz, 1H, ArH), 7. 30(dd, J = 7. 894Hz, 1H, ArH), 7. 51(d, J = 8. 841Hz, 1H,

ArH),7.66(d,J=1.914Hz,1H,ArH),8.05(s,1H,NH);IR(cm^{-1})3334,3127,3062,2247,1688,1602,1531,1476,844. ESIMS:253.22,186.20,166.02,145.82. Elemental analysis:C,59.41;H,4.12;N,12.13;calculated from $C_{11}H_9ClN_2O$. Observed:C,59.88;H,4.11;N,12.70.

Compound **4p**. A white solid,93% yield;mp 106 – 108℃;^{1}HNMR($CDCl_3$,300M)1.57 – 1.81(m,4H,CH_2),3.80(s,3H,CH_3),6.88(d,J=9.015Hz,2H,ArH),7.04(d,J=9.001Hz,2H,ArH),7.94(s,1H,NH);IR(cm^{-1})3334,3195,3115,2251,1680,1600,1534,1490,Elemental analysis:C,66.59;H,5.51;N,12.96;calculated from $C_{12}H_{12}N_2O_2$. Observed:C,66.65;H,5.59;N,12.96.

[14] KARI activity was measured by following the decrease in A340 at 30℃ in solutions containing 0.2mM NADPH,1mM $MgCl_2$,substrate(2 – acetolactate or hydroxypyruvate) and CPD or other inhibitors as required,in 0.1M Tris – HCl,pH8.0. The reaction was started by adding the enzyme except for inhibitor preincubation experiments,where the substrate was added last.

[15] Crystal data of **4a**. $C_{12}H_{12}N_2O$,M=200.24,Monoclinic,a=7.109(4),b=13.758(7),c=11.505(6)Å,β=102.731(8)°,V=1097.6(9)$Å^3$,T=294(2)K,space group P2(1)/c,Z=4,Dc=1.212g/cm_3,μ(Mo – $K\alpha$)=0.71073mm^{-1},F(000)=424.6109 reflections measured,2290 unique(R_{int}=0.0294),which were used in all calculation. Fine R_1 = 0.0490,$wR(F^2)$=0.1218(all data). Full crystallographic details of **4a** have been deposited at the Cambridge Crystallographic Data Center and allocated the deposition number CCDC 612888.

[16] (a) Bruker, *SMART.*, Bruker AXS Inc., Madison, Wisconsin, USA, 1998; (b) Sheldrick G. M., *SHELXS97 and SHELXL97.*, University of Göttingen, Germany, 1997; (c) Bruker, *SAINT and SHELXTL.* Bruker AXS Inc., Madison, Wisconsin, USA, 1999.

[17] Dmitry,S.;Yufit,M. J. T.;Judith,A. K. H. *Acta Cryst.* 2006,*E*,62,o1237.

[18] Zhao,G. L.;Feng,Y. L.;Liu,X. H. *Chin. J. Inorg. Chem.* 2005,21,598.

[19] According to the above crystal structure,a crystal unit was selected as the initial structure,while HF/6 – 31G(d,p),DFT – B3LYP/6 – 31G(d,p) and MP2/6 – 31G(d,p) methods in *Gaussian* 03 package were used to optimize the structure of the title compound. Vibration analysis showed that the optimized structures were in accordance with the minimum points on the potential energy surfaces,which means no virtual frequencies,proving that the obtained optimized structures were stable. All the convergent precisions were the system default values,and all the calculations were carried out on the Nankai Stars supercomputer at Nankai University.

[20] Frisch,M. J.;Trucks,G. W.;Schlegel,H. B.;Scuseria,G. E.;Robb,M. A.;Cheeseman,J. R.;Montgomery,J. A. Jr.;Vreven,T.;Kudin,K. N.;Burant,J. C.;Millam,J. M.;Iyengar,S. S.;Tomasi,J.;Barone,V.;Mennucci,B.;Cossi,M.;Scalmani,G.;Rega,N.;Petersson,G. A.;Nakatsuji,H.;Hada,M.;Ehara,M.;Toyota,K.;Fukuda,R.;Hasegawa,J.;Ishida,M.;Nakajima,T.;Honda,Y.;Kitao,O.;Nakai,H.;Klene,M.;Li,X.;Knox,J. E.;Hratchian,H. P.;Cross,J. B.;Adamo,C.;Jaramillo,J.;Gomperts,R.;Stratmann,R. E.;Yazyev,O.;Austin,A. J.;Cammi,R.;Pomelli,C.;Ochterski,J. W.;Ayala,P. Y.;Morokuma,K.;Voth,G. A.;Salvador,P.;Dannenberg,J. J.;Zakrzewski,V. G.;Dapprich,S.;Daniels,A. D.;Strain,M. C.;Farkas,O.;Malick,D. K.;Rabuck,A. D.;Raghavachari,K.;Foresman,J. B.;Ortiz,J. V.;Cui,Q.;Baboul,A. G.;Clifford,S.;Cioslowski,J.;Stefanov,B. B.;Liu,G.;Liashenko,A.;Piskorz,P.;Komaromi,I.;Martin,R. L.;Fox,D. J.;Keith,T.;Al – Laham,M. A.;Peng,C. Y.;Nanayakkara,A.;Challacombe,M.;Gill,P. – M. W.;Johnson,B.;Chen,W.;Wong,M. W.;Gonzalez,C.;Pople,J. A. *Gaussian* 03,Revision C.01,Gaussian,Inc.,Wallingford CT,2004.

[21] (a) Ma,H. X.;Song,J. R.;Xu,K. Z.;Hu,R. Z.;Zhai,G. H.;Wen,Z. Y.;Yu,K. B. *Acta Chem. Sinica* 2003,61,1819;(b) Ma,H. X.;Song,J. R.;Hu,R. Z.;Li,J. *Chin. J. Chem.* 2003,21,1558.

[22] All docking procedures were done in NanKai Stars supercomputer at Nankai University. The automated molecular docking calculations were carried out using AutoDock 3.05. The AUTOTORS module of

AutoDock defined the active torsions for each docked compound. The active sites of the protein were defined using AutoGrid centred on the IpOHA in the crystal structure. The grid map with 60 × 60 × 60 points centred at the center of mass of the KARI and a grid spacing of 0. 375Å was calculated using the AutoGrid program to evaluate the binding energies between the inhibitors and the protein. The Lamarckian genetic algorithm(LGA) was used as a searching method. Each LGA job consisted of 50 runs, and the number of generation in each run was 27000 with an initial population of 100 individuals. The step size was set to 0. 2Å for translation and 5° for orientation and torsion. The maximum number of energy evaluations was set to 15,00,000. Operator weights for cross – over, mutation and elitism were 0. 80, 0. 02 and 1, respectively. The docked complexes of the inhibitor – enzyme were selected according to the criterion of interaction energy combined with geometrical and electronic matching quality.

[23] Biou, V. ; Dumas, R. ; Cohen – Addad, C. ; Douce, R. ; Job, D. ; Pebay – Peyroula, E. *EMBO J.* 1997, 16, 3405.

单嘧磺隆晶体－活性构象转换的分子动力学模拟*

陈沛全[1,2,4]　孙宏伟[3,4]　李正名[1,2,4]　王建国[1,2]　马　翼[1,2]　赖城明[2,3]

（[1]南开大学元素有机化学研究所；[2]元素有机化学国家重点实验室；
[3]化学系；[4]科学计算研究所，天津　300071）

摘　要　应用分子动力学模拟方法对单嘧磺隆在水、正辛醇和正辛烷3种不同溶剂中的构象行为、单嘧磺隆与3种溶剂之间的相互作用能及氢键相互作用进行了计算研究。计算结果表明，在3种不同的溶剂中，单嘧磺隆的优势构象不同；其构象转换过程，特别是转换成活性构象的过程主要发生在水溶液中；与溶剂分子间的相互作用是分子构象行为的决定因素；单嘧磺隆的脲桥部分可以和含氢键接受体的溶剂形成氢键，分子间与分子内氢键的竞争可能是从晶体构象转换成活性构象的主要驱动力。

关键词　单嘧磺隆　分子动力学模拟　溶液　构象　氢键

磺酰脲类除草剂是近20多年来开发出来的超高效除草剂，有着重要的用途[1]。研究结果表明[2,3]，磺酰脲类除草剂的靶酶是乙酰羟基酸合成酶（AHAS）。Levitt等[4]认为磺酰脲类化合物的杂环上面必须有两个取代基才能具有较好的活性，本课题组的研究发现，杂环单取代磺酰脲化合物单嘧磺隆[5]同样具有很好的除草活性。

药物分子的活性构象往往与其最低能量构象或晶体构象不同[6]，因而研究农药分子各种构象之间的转换过程对于了解药物药效性能及药物如何与靶酶分子作用有着重要的意义。最近的研究分别获得了单嘧磺隆和单嘧磺隆－拟南芥AHAS复合物结晶，对其结构测定的结果表明[7]，单嘧磺隆的晶体构象（图1（A））与其活性构象（图1（B））在脲桥部分存在着较大的差异。这表明单嘧磺隆与靶酶作用的过程中必然经历由晶体构象向活性构象转换的过程。

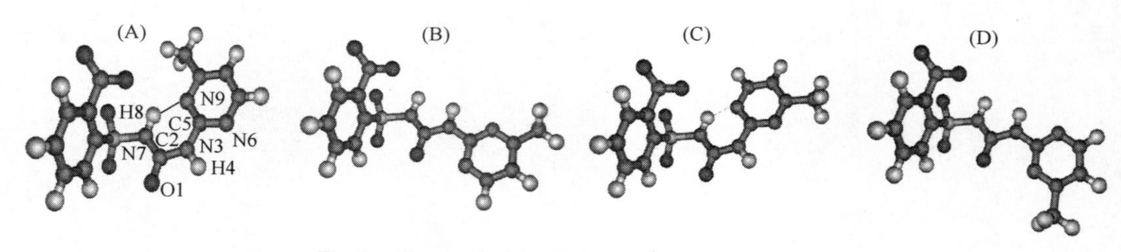

Fig. 1　Four typical conformations of monosulfuron

（A）and（C）the conformations with intramolecular hydrogen bond of N—H···N type;
（B）and（D）the conformations with intramolecular hydrogen bond destroyed due to intramolecular rotation

农药分子从施药位置到达靶酶的作用位点的过程一般可认为是在溶液中进行的，

* 原发表于《高等学校化学学报》，2007，28（2）：278－282。基金项目：国家“973”计划（批准号：2003CB114406）、国家自然科学基金重点项目（批准号：20432010）和天津市科委高性能计算项目（批准号：043185111－5）资助。

因而研究溶液条件下单嘧磺隆的构象转换具有重要的意义。为了探讨单嘧磺隆的晶体构象-活性构象之间转换过程及不同溶液环境对构象转换过程的影响，我们在前文[8~10]工作的基础上，进一步利用分子动力学模拟方法，对单嘧磺隆在水、正辛醇、正辛烷3种不同溶剂中的行为进行模拟，从分子结构水平上理解单嘧磺隆晶体构象-活性构象转换的具体过程，从而为设计新型除草剂提供理论信息。

1 计算方法

药物从施药位置到达药效作用位置的过程中，需要多次通过细胞膜。一般认为，连续的脂双分子层组成膜的主体，膜的表面是极性亲水的，膜的内部是非极性疏水的。为了模拟单嘧磺隆在植物细胞中的传输过程，选取了水、正辛醇和正辛烷3种溶剂分别模拟植物体的细胞液、细胞膜表面和膜内部，应用分子动力学模拟方法，研究单嘧磺隆在3种溶剂中的运动状态及构象转换过程。

在模拟过程中，单嘧磺隆分子采用OPLS-AA力场，水分子采用SPC模型，正辛烷和正辛醇分子采用OPLS-UA力场；所有模拟均采用GROMACS 3.2.1软件包，在等温等压（NPT）系统下进行，通过BERENDSEN算法来保持恒定的温度（298.15K）和恒定的压力（1.01325×10^5Pa），温度的弛豫时间为0.1ps，压力弛豫时间为1ps；范德华（VDW）相互作用采用双截断（cut-off）半径（0.9nm和1.4nm）。为了提高静电相互作用，尤其是长程静电相互作用的计算精度，模拟过程中采用PME方法；分子中的共价键用LINCS算法进行刚性约束。

根据文献［7］构建单嘧磺隆的晶体构象及活性构象，将两种不同的构象分别置于预先用分子动力学模拟平衡过后的分别含有水、正辛烷及正辛醇溶剂分子的立方溶剂盒子中，盒子的边长为4.5nm。对这6个不同的体系都进行了20ns的长时间分子动力学模拟，模拟的时间步长为0.002ps，模拟过程中每隔1ps记录1次体系的能量和轨迹。

2 结果与讨论

2.1 单嘧磺隆在不同溶液中的构象转换

比较晶体构象和活性构象可知，单嘧磺隆的构象转换可以通过脲桥部分的C2—N3键与N3—C5键作内旋转而实现（原子编号见图1（A））。密度泛函计算结果[11]表明，单嘧磺隆分子由于C2—N3键和N3—C5键的内旋转，可以产生4类典型的构象形式(图1)；这4类典型构象又可分为两大类，一类是存在N—H…N型分子内氢键的A类构象和C类构象，另一类是分子内氢键由于分子内旋转而被破坏的B类构象和D类构象。A与C类构象之间，以及B与D类构象之间的差别在嘧啶环上甲基的空间取向上。因此，分子构象多样性除了与能否形成N—H…N型氢键有关外。还与嘧啶环上CH_3的空间取向有关。为监控分子动力学模拟过程中以上4类构象的变化，定义了二面角θ_1（O1—C2—N3—H4）及θ_2（H4—N3—C5—N6）。

在模拟过程中，θ_1和θ_2基本上在0°或180°附近波动。对比3种不同溶剂中特征二面角的变化可以看出，模拟时间内水溶液中4类构象形式均能出现，而在正辛醇及正辛烷溶液中，只出现其中1种或2种类型特定的构象形式。这说明单嘧磺隆在水、正辛醇及正辛烷3种不同溶剂中的构象变化存在着明显的区别，在水溶液中的构象灵活程度最大，而在正辛醇及正辛烷中构象的灵活程度相对较小。

在水溶液中，无论以晶体构象（A 类）还是活性构象（B 类）为初始构象，单嘧磺隆的构象都能发生转换。从 A 出发，在约 2.7ns 附近，分子首先绕 C2—N3 旋转约 180°转换成 D；在约 6.5ns 附近，分子进一步绕 N3—C5 旋转约 180°转换成有活性的 B；B 稳定存在一段时间后，分子绕 C2—N3 和 N3—C5 协同旋转回到 A，进而又进行 D 向 B 的转换。从 B 出发，在模拟时间内，分子主要发生了 B 与 D 的互相转换。在 1.12 和 1.26ns 附近，分别发生 B 向 C 和 C 向 B 的转换。

从水溶液中的模拟可见，4 类构象在水溶液中可以通过 C2—N3 和 N3—C5 的内旋转而相互转换；在模拟时间内，构象出现的概率大小顺序为 B > D > A > C，即不形成分子内 N—H…N 氢键的 B 和 D 出现的概率较形成分子内 N—H…N 氢键的 A 和 C 大；从模拟结果还可以看出，A 与 B 之间的转换可以通过分子绕 C2—N3 和 N3—C5 协同内旋转一步完成，也可以通过先绕 C2—N3 或 N3—C5 作内旋转转换成 C 或 D，然后进一步转换成有活性的 B。在模拟时间内并未出现 A→C→B 的转换，这可能是模拟时间未能对分子的构象空间充分取样的缘故，延长模拟时间可能得到更好的结果。

在正辛醇及正辛烷溶液中，无论是以 A 还是 B 作初始构象，特征二面角在模拟过程中基本保持不变。从 A 出发，在正辛醇溶液中模拟约 19.9ns，分子绕 C2—N3 内旋转约 180°转换为 D；而在随后进一步以 D 为起始构象的 20ns 模拟中，并没发生由 D 向其他类型的构象转换；B 在正辛烷溶液中模拟的最初 100ps 中，C2—N3 内旋转约 180°转换为 C。模拟结果暗示了单嘧磺隆在正辛醇溶液中主要以不形成分子内 N—H…N 氢键的 B 和 D 类构象存在，而在正辛烷溶液中主要以形成分子内 N—H…N 氢键的 A 和 C 类构象存在。

从以上的模拟结果可以得出如下结论：（1）分子在不同溶剂中的构象行为不同。在水溶液中，4 类构象之间可以通过内旋转进行转换，在水和正辛醇中倾向以不形成分子内 N—H…N 氢键的 B 和 D 类构象存在，而在正辛烷中倾向以形成分子内 N—H…N 氢键的 A 和 C 构象存在。这可能是由于以 B 类和 D 类构象存在时，有利于增加分子与水及正辛醇溶剂间的相互作用，特别有利于与溶剂间氢键的形成。而正辛烷由于不存在氢键的供体或受体，与药物分子难以形成氢键，因而分子以形成分子内氢键的 A 及 B 类构象存在；（2）农药分子在植物体内传输的过程中，构象的转换主要发生在水相（即细胞液）中，在细胞膜表面分子主要以 B 或 D 类构象存在，在细胞膜内部主要以 A 或 C 类构象存在。

2.2 单嘧磺隆与不同溶剂的相互作用能

单嘧磺隆在不同溶剂中的构象行为不同，说明了不同溶剂效应对于分子的构象变化影响不同，溶质分子与溶剂分子之间的相互作用能是溶剂效应中的一个重要因素。因而研究不同溶剂与单嘧磺隆的相互作用能具有重要的意义。相互作用能的计算方法如下：

$$E_{\text{drug-solvent}} = \sum E_{\text{drug-solvent}}^{\text{vdw}} + \sum E_{\text{drug-solvent}}^{\text{electrostatic}}$$

计算结果表明，相互作用能的变化与构象变化表现了时间上的相关性，即出现不同类构象对应不同的相互作用能。表 1 列出了各类构象与不同溶剂间的相互作用能的统计结果。由于在正辛醇和正辛烷溶液的模拟中没出现 C 和 D 类构象，因此表 1 中并未列出其相应数据。

Table 1 Interaction energies (kJ/mol) between monosulfuron and different solvents

Energy	Water				n - Octanol			n - Octane		
	A	B	C	D	A	B	D	A	B	C
$\sum E_{\text{drug - solvent}}^{\text{vdw}}$	-107.01	-107.45	-109.07	-108.68	-181.71	-193.78	-190.90	-176.55	-176.58	-176.54
$\sum E_{\text{drug - solvent}}^{\text{electrostatic}}$	-177.45	-229.01	-175.59	-225.21	-107.84	-151.28	-150.10	0.00	0.00	0.00
$E_{\text{drug - solvent}}$	-284.46	-336.56	-284.66	-333.89	-289.55	-345.06	-341.00	-176.55	-176.58	-176.54

表1数据表明，不同的溶剂与单嘧磺隆之间的相互作用能不同，强弱顺序为正辛醇≈水>正辛烷，这说明药物分子与极性溶剂（水和正辛醇）间的相互作用远远强于非极性溶剂，这是药物分子与水或正辛醇之间存在着强的静电相互作用，而药物分子与正辛烷之间基本上不存在静电相互作用的缘故。另一方面，水溶液中药物分子与溶剂的静电相互作用的贡献大于范德华相互作用，正辛醇及正辛烷溶液中范德华相互作用的贡献大于静电相互作用。

由表1数据可见，水及正辛醇溶液中药物分子不同的构象形式在同一溶剂中的相互作用能不同。一般情况下，A与C和B与D与相同的溶剂间相互作用能相近，且C或D类构象与溶剂间的相互作用能比A或C类构象低约50kJ/mol。说明分子的B或D类构象与溶剂间的相互作用较A或C类构象强，且相互作用的强弱主要取决于脲桥部分两个NH基团的相对取向（即是否形成分子内氢键）。嘧啶环上甲基的空间取向对相互作用能影响不大。分析相互作用能的能量组成可见，相同的极性溶剂（水和正辛醇）与不同构象的药物分子间的范德华相互作用相近，静电相互作用决定了药物分子与溶剂间的相互作用能。另外，在模拟力场条件下，非极性溶剂正辛烷与药物分子之间的静电相互作用为0，且溶剂分子与不同构象药物分子之间的范德华相互作用基本相同，从而导致正辛烷与不同构象药物分子间的相互作用能基本相同。

前文[10]的密度泛函计算结果表明，单嘧磺隆气态条件下的构象稳定性为A≈C>B≈D，即从A（晶体）或C类构象转换成B（活性）或D类构象的过程是一个吸热过程。在水和正辛醇溶液中，药物分子从A与C类构象转换成B与D类构象的过程中药物分子与溶剂间的相互作用增强，从而有效地补偿了构象转换过程中所需要的能量，因此，在水及正辛醇极性溶剂中药物分子主要以B和D类构象存在，即主要以不形成分子内氢键的构象存在；而在正辛烷溶液中，由于不同形式的构象与溶剂间的相互作用相差不大，药物分子主要以较稳定的A和C类构象存在。不同形式的构象与溶剂之间的相互作用能决定了单嘧磺隆在溶液状态下主要以何种形式构象存在。

2.3 单嘧磺隆与溶剂间形成的分子间氢键

在水及正辛醇溶液中，不同构象与同种溶剂之间的相互作用能取决于不同构象的药物分子与溶剂之间的静电相互作用，反映在结构上就是分子的不同构象与溶剂分子所形成的氢键模式不同。径向分布函数是反映液体微观状态的物理量，它能给出溶液中的局部结构特征。为了探讨与构象转换过程有关的药物分子-溶剂分子氢键相互作用，计算了分子脲桥部分两个NH基团上的H原子（H4和H8）和水及正辛醇溶剂分子中氢键受体O原子之间的径向分布函数（图2）。

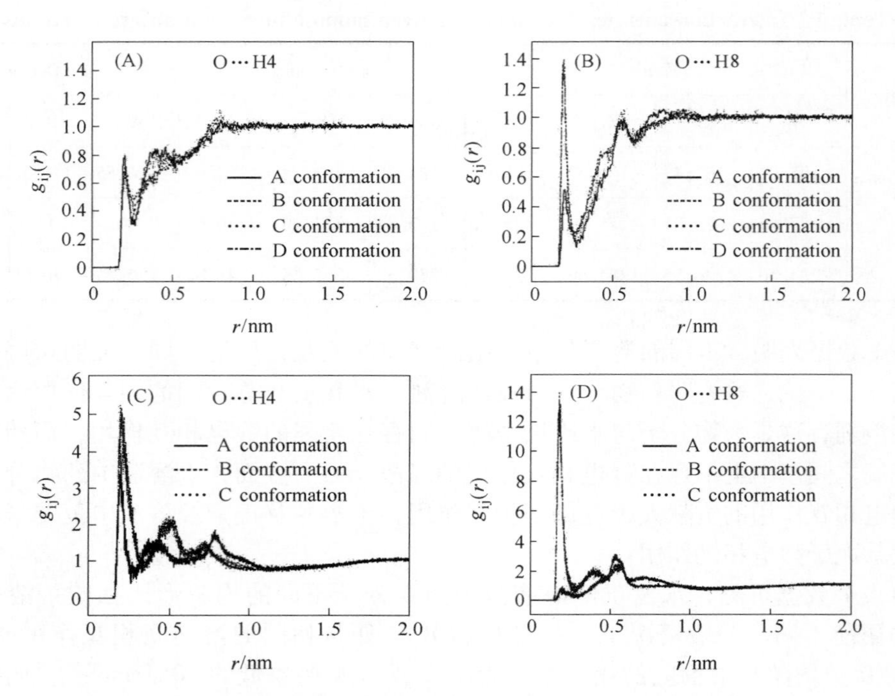

Fig. 2 Radial distribution functions of hydrogen bond in water（A，B）and n－octanol solutions（C，D）

从图2可以看出，无论在水溶液还是正辛醇溶液中，不同构象状态下，H4原子与溶剂中的氢键接受体O原子之间的径向分布函数基本相似，都在0.24，0.50及0.76nm处出现峰值。不同的是，在水溶液中，径向分布函数第一峰峰值最小，为0.8左右，而正辛醇溶液中径向分布函数第一峰峰值最大，为3～5附近。由于0.24nm比H原子与O原子的范德华半径之和（0.26nm）小，因而在此位置可能形成N—H···O型分子间氢键。

在不同构象状态下，H8原子与溶剂中的氢键接受体O原子之间的径向分布函数不同，但出现峰值的位置与H4原子基本类似，也在0.24，0.50及0.76nm处出现峰值。一般情况下，A及C类构象的径向分布函数相似，B及D类构象的径向分布函数相似，这是因为A及C类构象和B及D类构象在脲桥部分的结构基本上相同。对比A及C类构象和B及D类构象的径向分布函数可以看出，分子从A或C类构象转换成B或D类构象，溶剂氢键接受体O原子和药物分子的H8原子之间的径向分布函数的第一峰峰值在水溶液中从0.5增加到1.4，在正辛醇溶液中从0.8增加到14，而径向分布函数的其它位置无明显的变化。

不同构象的H8与溶剂中O原子的径向分布函数变化的原因如下：A或C类构象H8与嘧啶环上的N原子形成分子内氢键，因而不易再与溶剂分子的O原子形成分子间氢键；而在B或D类构象中，H8倾向与溶剂分子中O原子形成分子间氢键，且这种分子间氢键的形成增强了药物分子与溶剂之间的相互作用，使B或D类构象在水及正辛醇溶液中占优势。因此在水及正辛醇溶液中，H8存在着与嘧啶环上的N形成分子内氢键及与溶剂形成分子间氢键的竞争，这种竞争作用可能是单嘧磺隆发生构象转换的驱动力。

溶剂分子与药物分子之间的氢键相互作用对于溶液条件下的构象行为起着重要的作用。图3给出的是模拟过程中单嘧磺隆D类构象与溶剂形成分子间氢键作用的模式图（在正辛醇溶液中作用方式类似）。B或D类构象与溶剂形成分子间氢键的模式有两种（A）和（B），一种是脲桥部分的两个NH基团分别与一个水分子形成分子间氢键，两个水分子间形成一个氢键，从而形成一个八元环结构，另一种是两个NH基团都与同一个水分子形成氢键，从而形成一个六元环结构。

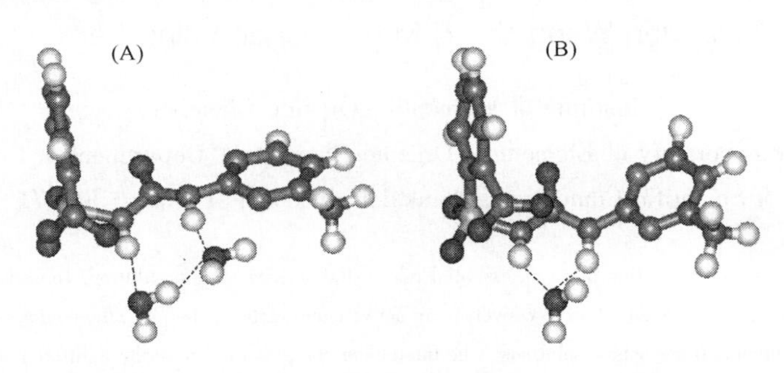

Fig. 3 Two types of hydrogen bonding modes in water solution from the snapshoot
（A） Eight – membered ring structure；（B） six – membered ring structure.

参 考 文 献

[1] LI Zheng – Ming（李正名），LAI Cheng – Ming（赖城明）. Chinese J. Org. Chem.（有机化学）[J]，2001，21（11）：810 – 815.

[2] Pang S. S.，Duggleby R. G.，Guddat L. W.. J. Mol. Biol. [J]，2002，17：249 – 262.

[3] Pang S. S.，Guddat L. W.，Duggleby R. G.. J. Biol. Chem. [J]，2003，278：7639 – 7644.

[4] Levitt G.. Synthesis and Chemistry of Agrochemicals II，ACS Symposium SeriesNo. 443 [M]，Washington DC：American Chemistry Society，1991：16 – 31.

[5] LI Zheng – Ming（李正名），JIA Guo – Feng（贾国锋），WANG Ling – Xiu（王玲秀），*et al.*. Conposite for Preventing and Killing Weeds in the Corn Field [P]，CN 1080116，1994.

[6] CHEN Kai – Xian（陈凯先），JIANG Hua – Liang（蒋华良），JI Ru – Yun（嵇汝运）. Computer – Aided Drug Design（计算机辅助药物设计）[M]，Shanghai：Shanghai Scientific & Technical Publishers，2000：205 – 206.

[7] WANG Jian – Guo（王建国），The Biological Activities，Molecular Basis and Three Dimensional QSAR Investigation on Novel Sulfonylurea Herbicedes [D]，Tianjin：Elem. Org. Chem. Ins.，Nankai Univ.，2004.

[8] SHEN Rong – Xin（沈荣欣），FANG Ya – Yin（方亚寅），MA Yi（马翼），*et al.* Chem. J. Chinese Universities（高等学校化学学报）[J]，2001，22（6）：952 – 954.

[9] MA Yi（马翼），LI Zheng – Ming（李正名），LAI Cheng – Ming（赖城明）. Chinese J. Pestic. Sci.（农药学学报）[J]，2004，6（1）：71 – 73.

[10] CHEN Pei – Quan（陈沛全），SUN Hong – Wei（孙宏伟），LI Zheng – Ming（李正名），*et al.*. Acta Chim. Sinica（化学学报）[J]，2006，64（13）：1341 – 1348.

[11] Wang J. G.，Li Z. M.，Ma L.，*et al.*. J. Comput. Aid. Mol. Des. [J]，2005，19：801 – 820.

Molecular Dynamics Simulation of Conformational Conversion Between Crystal Conformation and Active Conformation of Herbicidal Monosulfuron

Peiquan Chen[1,2,4], Hongwei Sun[3,4], Zhengming Li[1,2,4], Jianguo Wang[1,2], Yi Ma[1,2], Chengming Lai[2,3]

([1]Institute of Elemento - Organic Chemistry;
[2]State Key Laboratory of Elemento - Organic Chemistry; [3]Department of Chemistry;
[4]Institute of Scientific Computing, Nankai University, Tianjin, 300071, China)

Abstract Molecular conformation plays an essential role in the activity of monosulfuron. To understand the conformational conversion, especially how to convert to an active conformation, herein, the conformational behavior in water, *n* - octanol and *n* - octane solutions, the interaction energies with these three different solvents and hydrogen bond interaction with solvents of monosulfuron were investigated by a series of molecular dynamics simulations. The simulation results indicate that dominative conformations were different in different solvents, and the conformational conversion, especially converting to an active conformation, was mainly occurs in the aqueous solution in the plant. The calculation results also implies that the interaction between the monosulfuron and solvents was main factor which determined the conformational behavior of monosulfuron. The NH group in the sulfonylurea bridge may interact with the solution which has hydrogen bond acceptor atoms through hydrogen bond, and the competition between forming internal N - H⋯N hydrogen bond and forming hydrogen bond with solution was probably the driven force from crystal conformation to active conformation conversion.

Key words monosulfuron; molecular dynamics simulation; solution; conformation; hydrogen bond

基于受体结构的 AHAS 抑制剂的设计、合成及生物活性*

肖勇军　王建国　刘幸海　李永红　李正名

（南开大学元素有机化学研究所，元素有机化学国家重点实验室，天津　300071）

摘　要　在 AHAS 与磺酰脲类除草剂复合物的晶体结构基础上，利用分子对接程序 DOCK4.0，通过 MDL/ACD 三维数据库虚拟筛选，得到了 296 个与 AHAS 结合能较低的小分子化合物结构信息，从中选取了部分小分子进行化学合成，并且测试了其生物活性。部分化合物的体内和体外活性表现出一定的一致性。

关键词　乙酰乳酸合成酶（AHAS）　分子对接　有机合成　生物活性

乙酰乳酸合成酶（AHAS，EC2.2.2.6）是催化支链氨基酸生物合成的关键酶，通过抑制酶的活性中断支链氨基酸的生物合成而使植物死亡，达到除草目的[1]。人畜等温血动物自身不能合成支链氨基酸，因此设计和研制结构新颖的 AHAS 抑制剂具有一定的理论和现实意义。生物合成药物分子设计直接从作用对象出发，研究靶酶的活性位点，探索药物分子与靶酶之间的相互作用关系，高效率地进行新药设计。Duggleby 等[2~4]首次报道了来自酵母 AHAS 纯酶的晶体结构及其与氯嘧磺隆复合物以及拟南芥 AHAS 与氯嘧磺隆复合物的晶体结构，为基于受体结构的除草剂分子设计提供了新的机遇。在已报道的 AHAS 与磺酰脲类除草剂复合物的晶体结构基础上，利用分子对接程序 DOCK4.0[5,6]，对 MDL/ACD－3D 有机化合物数据库进行虚拟筛选，得到 296 个与 AHAS 结合能较低的小分子结构信息。本文从中选取部分化合物（图 1）进行合成，用生物等排等原则合成了部分类似物，测试了其除草活性（*in vivo*）以及抑制酶的离体活性（*in vitro*），以期望从中找到活性较好的先导结构化合物。

1　实验部分

1.1　仪器、试剂及目标化合物合成

Bruker AV－300MHz 核磁共振仪，Varian 400MHz，TMS 为内标；Carlo Erba 1106 型元素分析仪；北京泰克 X－4 数字显示显微熔点仪。所用试剂和原料均为分析纯或化学纯，经常规方法处理。

目标化合物按照文献［7～15］方法合成，部分新化合物的物化性质及核磁共振数据见表 1 和表 2。

1.2　化合物离体和活体活性的测定

用油菜平皿法和稗草小杯法[16]测定了所设计化合物的除草活性，按照 Singh 等[17]

* 原发表于《高等学校化学学报》，2007，28（7）：1280－1282。基金项目：国家“973”计划项目（批准号：2003CB114406）和国家自然科学基金（批准号：20432010，20602021）资助。

方法测定了化合物抑制拟南芥 AHAS 酶的离体抑制活性。

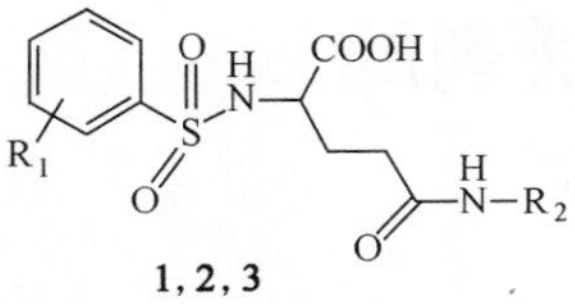

Fig. 1 Structures of the designed and syn thesized compounds

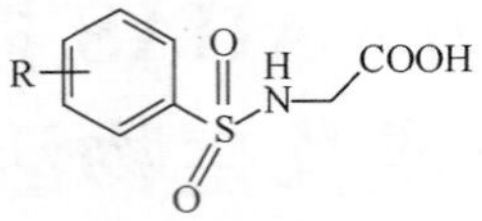

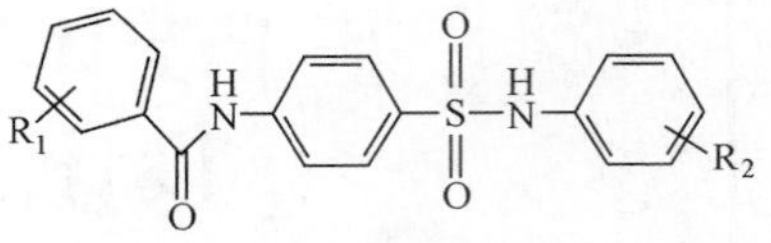

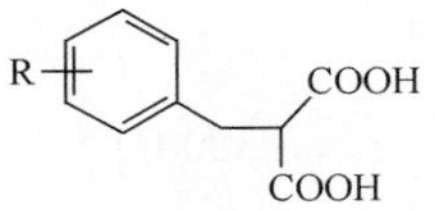

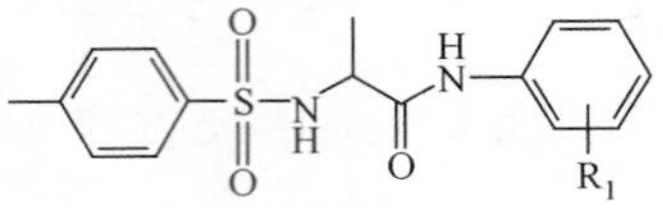

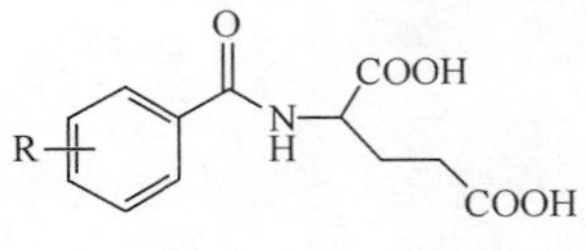

1. R_1: $p-CH_3$, R_2: C_2H_5; **2**. R_1: H, R_2CH_3; **3**. R_1: $o-NO_2$, R_2: CH $(CH_3)_2$; **4**. R: $p-CH_3$; **5**. R: H; **6**. R: $m-NO_2$; **7**. R_1: $p-NO_2$, R_2: $p-CH_3$; **8**. R: $m-Cl$; **9**. R: H; **10**. R: $p-CH_3$; **11**. R_1: $m-Cl$; **12**. R_1: $p-CH_3$; **13**. R: H; **14**. R: 2, 4-di-Cl; **15**. R: $o-CH_3$

Table 1 Physical data of the compounds syn thesized

Compd.	m. p. /℃ (Ref)	Yeild(%)	Appearance	Compd.	m. p. /℃ (Ref)	Yeild(%)	Appearance
1	162 ~ 163(160 ~ 162)[8]	71.3	White crystal	**9**	118 ~ 119(120)[12]	69.8	White crystal
2	165 ~ 167(161 ~ 163)[8]	87.3	White crystal	**10**	121 ~ 123(162.5 ~ 163.5)[13]	58.6	White crystal
3	158 ~ 160	47.9	White crystal	**11**	144 ~ 146	76.5	White crystal
4	135 ~ 137(149 ~ 150)[9]	56.3	White crystal	**12**	224 ~ 226(174 ~ 175)[14]	72.9	White crystal
5	167 ~ 169(167 ~ 169)[9]	70.6	White crystal	**13**	138 ~ 140(130 ~ 132)[15]	74.6	White crystal
6	136 ~ 138(147 ~ 148)[10]	42.2	White crystal	**14**	134 ~ 135	67.8	White crystal
7	271 ~ 273	47.3	White crystal	**15**	107 ~ 109	46.7	White crystal
8	102 ~ 103(98 ~ 99)[11]	49.9	White crystal				

Table 2 ^{1}H NMR data of compounds 3, 11, 15 (DMSO − d_6)

Compd.	^{1}H NMR, δ
3	8.04 ~ 7.81 (m, 4H, ArH), 3.90 (m, 1H, $NHCHCH_2$), 3.84 ~ 3.74 [m, 1H, $NHCH(CH_3)_2$], 2.12 (m, 2H, CH_2CH_2CO), 2.04 ~ 1.91 (m, 1H, $NHCHCH_2$), 1.85 ~ 1.70 (m, 1H, $NHCHCH_2$), 1.01 [d, 6H, J = 6.0Hz, $CH(CH_3)_2$]
11	8.26 (br, 1H, CONH), 7.76 (d, 2H, J = 8.1Hz, $ArHCH_3$), 7.58 (s, 1H, ArHCl), 7.35 (d, 2H, J = 8.1Hz, $ArHCH_3$), 7.31 ~ 7.10 (m, 3H, ArHCl), 5.46 (s, 1H, SO_2NH), 3.92 ~ 3.88 (m, 1H, $CHCH_3$), 2.39 (s, 3H, $ArCH_3$), 1.30 (d, 3H, J = 7.1Hz, $CHCH_3$)
15	7.70 (d, 1H, J = 8.8Hz, CONH), 7.42 ~ 7.17 (m, 4H, ArHCO), 4.53 ~ 4.48 (m, 1H, $NHCHCH_2$), 2.49 (t, 2H, J = 7.3Hz, CH_2CO_2H), 2.26 (s, 3H, CH_3Ar), 2.20 ~ 2.15 (m, 1H, $NHCHCH_2$), 2.04 ~ 1.96 (m, 1H, $NHCHCH_2$)

2 结果与讨论

2.1 合成及波谱数据

在化合物**1**的^{1}H NMR（表2）中，1.87～1.74和1.67～1.57两处的多重峰是与手性碳相连的亚甲基上2个H的吸收峰，受手性碳上H原子影响，它们的化学位移不同，裂分成多重峰。我们用*L*－谷氨酸二乙酯与取代苯甲酰氯的方法合成化合物**13～15**，该反应在无水条件下进行，避免了酰氯的水解。

2.2 生物活性

生物活性结果（表3）表明，部分化合物的活体和离体活性一致。如化合物**6**在100μg/mL和10μg/mL浓度下，对稗草地上部分的抑制率分别为79.6%和70.4%；在100μg/mL浓度下，对AHAS抑制率为57.8%。化合物**11**在100μg/mL和10μg/mL浓度下，对油菜胚根生长抑制率分别为82.6%和63.2%；在100μg/mL浓度下，对AHAS抑制率为51.3%。部分化合物具有较好的除草活性，同时对AHAS也有一定的抑制作用。如化合物**14**在100和10μg/mL浓度下，对稗草地上部分的抑制率分别为99.0%和43.3%；在100μg/mL浓度下，对AHAS抑制率为24.5%。化合物**8**在100和10μg/mL浓度下，对稗草地上部分的抑制率分别为77.0%和74.5%；在100μg/mL浓度下，对AHAS抑制率为33%。个别化合物对AHAS具有较好的抑制作用，但除草活性一般。如化合物**7**在100μg/mL浓度下，对AHAS抑制率为74.1%；在100μg/mL浓度下，对稗草地上部分的抑制率为40.2%，在10μg/mL浓度下，活性更差。个别化合物具有较好的除草活性，对AHAS无抑制作用。如在100μg/mL浓度下，化合物**15**对油菜胚根生长抑制率为83.4%，对AHAS的抑制率却为零。

Table 3 Bioactiv ities of compounds in vivo and in vitro

Compd.	*Echinochloa crusg－alli* cup test		*Brassica cam pest－ris* root test		AHAS	Compd.	*Echinochloa crusg－alli* cup test		*Brassica cam pest－ris* root test		AHAS
	10μg/mL	100μg/mL	10μg/mL	100μg/mL	100μg/mL		10μg/mL	100μg/mL	10μg/mL	100μg/mL	100μg/mL
1	0	0	0	0	0	**9**	15.0	0	15.1	0	25.0
2	3.12	7.4	15.9	5.0	0	**10**	29.1	0	59.6	3.4	0
3	0	0	3.0	0	30.3	**11**	0	0	82.6	63.2	51.3
4	18.8	0	0	0	0	**12**	22.0	15.1	41.0	11.6	0
5	16.6	0	7.1	0	38.6	**13**	0	0	38.4	0	28.1
6	79.6	70.4	32.0	0	57.8	**14**	0	0	99.0	43.3	26.5
7	40.2	21.5	1.5	0	74.1	**15**	32.8	18.8	83.4	18.1	0
8	28.1	0	63.3	0	10.3						

通过化合物活体和离体活性的比较，可以发现基于受体结构的分子设计具有一定的合理性。在此基础上完善分子对接模型，提高分子设计的成功率；还要考虑小分子从植物表皮到作用靶酶的传输、代谢等相关因素，化合物离体和活体活性表现出更好的一致性。

参考文献

[1] Hofgen R., Laber B., Schuttke I., *et al.*. Plant Physiol. [J], 1995, 107: 469-477.
[2] Pang S. S., Duggelby R. G., Guddat L. W. J.. Mol. Biol. [J], 2002, 317: 249-262.
[3] Pang S. S., Guddat L. W., Duggelby R. G.. J. Biol. Chem. [J], 2003, 278 (9): 7639-7644.
[4] McCourt J. A., Pang S. S., Scott J. K., *et al.*. PNAS [J], 2006, 103 (3): 569-573.
[5] Ewing T. J. A., Kuntz I. D.. J. Comput. Chem. [J], 1997, 18 (9): 1175-1189.
[6] Wang J. G., Xiao Y. J., Li Y. H., *et al.*. Bioorg. Med. Chem. [J], 2007, 15 (1): 374-380.
[7] Srikanth K., Kumar C. A., Ghosh B., *et al.*. Bioorg. Med. Chem. [J], 2002, 10 (7): 2119-2131.
[8] De A. U., Pandey J., MajumdarA.. Indian J. Chem. [J], 1982, 21B: 481-483.
[9] Mahajan P. K., Patial V. P., Sharma P.. Indian J. Chem [J], 2002, 41B: 2635-2641.
[10] Gosh N. N., Majumder M. N.. J. Indian. Chem. Soc. [J], 1967, 44: 115-117.
[11] Kenner J., Witham G.. J. Chem. Soc. [J], 1921, 119: 1452-1461.
[12] Fexier F., Marchand E., Carrie R.. Tetrahedron [J], 1974, 30: 3185-3192.
[13] Westfahl J. C., Gresham T. L.. J. Am. Chem. Soc. [J], 1954, 76: 1076-1080.
[14] EI-Naggar A. M., Ismail I. M., Gomoa A. M.. Egyption J. Chem. [J], 1980, 21 (4): 309-313.
[15] Fischer E.. Chem. Ber. [J], 1899, 32: 2451.
[16] Wang B. L., Duggleby R. G., Li Z. M., *et al.*. Pest. Manag. Sci. [J], 2005, 61: 407-412.
[17] Singh B. K., Stidham M. A., Shaner D. L.. Anal. Biochem. [J], 1988, 171: 173-179.

Molecular Design, Synthesis and Biological Activity Evaluation of Novel AHAS Inhibitors Basedon Receptor Structure

Yongjun Xiao, Jianguo Wang, Xinghai Liu, Yonghong Li, Zhengming Li

(Elemento-Organic Chemistry Institute, State-Key Laboratory of Elemento-Organic Chemistry, Nankai University, Tianjin, 300071, China)

Abstract Based on the crystal structure of AHAS/sulfonylurea complex, 296 molecules were obtained with low binding energy towards AHAS from MDL/ACD 3D database *via* virtual screening with program DOCK 4.0, from which some compounds were synthesized. The biological activities of the synthesized compounds were measured *in vitro* and *in vivo*. The preliminary bioassay indicates some compounds displayed a good herbicidal activity on rape and barnygrass and had AHAS inhibition to some extent. These studies indicate the rationality of molecular design based on the crystal structure of AHAS complex.

Key words acetohydroxyacid synthase (AHAS); molecular docking; organic synthesis; biological activity

The Design, Synthesis, and Biological Evaluation of Novel Substituted Purines as HIV - 1 Tat - TAR Inhibitors*

Dekai Yuan[1,3], Meizi He[1], Ruifang Pang[1], Shrongshi Lin[2], Zhengming Li[3], Ming Yang[1]

([1]National Research Laboratory of Natural and Biomimetic Drugs, Peking University Health Science Center, Beijing, 100083, China;
[2]College of Chemistry and Molecular Engineering, Peking University, Beijing, 100871, China;
[3]The Institute of Elemento - Organic Chemistry, The State Key Laboratory of Elemento - Organic Chemistry, Nankai University, Tianjin, 300071, China)

Abstract A series of novel substituted purines containing a side chain with a terminal amino or guanidyl group were designed and synthesized as HIV - 1 Tat - TAR inhibitors. All the compounds could effectively block the TAR transactivation in human 293T cells with the CAT expression percentage ranging from 34.4% to 65.7% and showed high antiviral effects with low cytotoxicities in inhibiting the formation of SIV - induced syncytium in CEM174 cells. Molecular modeling studies by Auto - dock process suggest that the compounds bind to TAR RNA in two different modes.

Key words substituted purines; HIV Tat - TAR inhibitors; molecular modeling

1 Introduction

Trans - activator of transcription (Tat) protein plays a determinant role in HIV replication by specific interaction with the trans - activation response element region (TAR), a 59 - nucleotide stem - loop structure containing a six - nucleotide loop, a three - nucleotide bulge, and two single - nucleotide bulges at the 5' - end of all nascent transcribed HIV - 1 mRNAs. [1,2] The tri - nucleotide bulge (U23, C24, U25) of HIV TAR is essential for high - affinity and specific binding to the basic domain (RKKRRQRRR, 49 - 57) of Tat protein. The arginine 52 residue of Tat directly binds the tri - nucleotide bulge and its guanidyl group is largely responsible for the Tat - TAR interaction. [3-6] The arginine residue (or arginineamide) can bind to TAR and induce a change in RNA conformation largely mimicking a portion of Tat - TAR complex. [7] It has also been proved that the high - affinity binding of some small molecules with amino, guanidyl group or arginine residue to TAR RNA is governed by electrostatic interaction between these groups and the region near the UCU bulge. [8,9] Due to the specificity of Tat - TAR interaction and the high conservation of TAR sequence, the interaction of Tat - TAR is an attractive target for designing novel anti - HIV drugs. [10,11]

Our previous work has demonstrated that some substituted β - carboline, isoquinoline, α, α -

* Reprinted from *Bioorganic&Medicinal Chemistry*, 2007, 15: 265 - 272. This project was supported by the National Natural Science Foundation of China (No. 30570387 and No. 20332010) and the Doctoral Program Foundation of China (No. 20030001041).

trehalose derivatives with a flexible side – chain with a terminal amino or guanidyl group could interact with TAR and inhibit the replication of HIV – 1. [12] These studies above provide us an idea for designing one type of new Tat – TAR inhibitors containing an 'activator', an 'anchor' and a 'linker' with suitable length. An 'activator' is defined as a group that could recognize and bind to the tribased bulge of TAR, usually an amino or guanidyl group. An 'anchor' is a functional group that could interact with TAR in different ways from the 'activator', such as stacking into tri – based bulge, forming hydrogen bond with unpaired base, intercalating into the upper or lower stem, or falling into the major groove near the bulge. Its structure may vary in a wide range, such as single, linked or fused aromatic cycles, glucosides, etc. A suitable 'anchor' can not only reinforce the affinity but also improve the solubility and lyphohydrophilic character of the compound which would affect the final activity. A 'linker' is the structure linking the 'activator' and the 'anchor' with optimal length. If the linker could form additional interactions such as hydrogen bond or electrostatic interaction with TAR, the interaction between the small molecule and TAR will be reinforced. The β – carboline, isoquinoline, and α, α – trehalose derivatives as well as several other kinds of small molecular inhibitors of Tat – TAR fall into the model we proposed. [7,8,11 – 17]

In order to obtain more potential Tat – TAR inhibitors and make further study of the TAR RNA binding property of them, we have designed and synthesized a novel series of substituted purines as HIV – 1 Tat – TAR inhibitors, as shown in Figure 1, on the basis of the molecular model we proposed above. Multi – substituted purine is selected as an 'anchor' here for the following reasons: (1) Purine and its analogues attract much attention for serving as key recognition and anchoring elements in a variety of cofactors and signaling molecules in bio – systems and are usually studied as anti – cancer and antivirus agents. [18 – 21] (2) The purine would bind to TAR RNA with higher affinity than β – carboline or isoquinoline for it could mimic Watson – Crick interactions with pyrimidine bases of the tri – based bulge, or form hydrogen bonds between its N – atom and the nucleotides of TAR. (3) The amphiphlic 'anchor' would give the compounds higher solubility than β – carbolines and isoquinolines. The 'linker' here is located at C8 position of the purine and is somewhat longer but more flexible than our former compounds. The 'activator' is an amino or guanidyl group for its specific binding with TAR at the tri – nucleotide bulge.

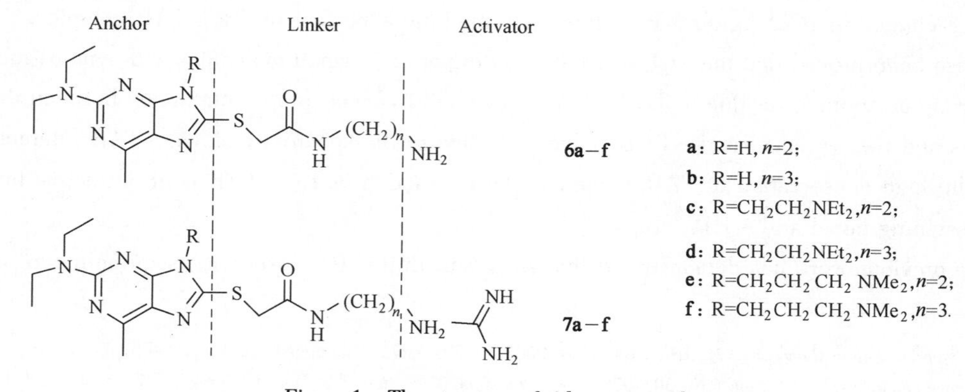

Figure 1 The structure of title compounds

All the title compounds are synthesized for the first time and biological evaluation proves that all of them possess inhibitory activity to Tat - TAR interaction with low cytotoxicity as compared with the substituted β - carboline, isoquinoline, α, α - trehalose derivatives we reported before.

2 Results and discussion

The synthetic work is illustrated in Scheme 1 and the synthetic procedure is explained in Section 4. Twelve title compounds were obtained and their MS, ^{1}H and ^{13}C NMR spectroscopy data are provided in Section 4.

1 → (i) 2a～c → (ii) 3a～c → (iii) 4a～c → (iv) 5a～c

a: R=H;
b: $R=CH_2CH_2NEt_2$;
c: $R=CH_2CH_2CH_2NMe_2$.

→ (v) 6a～f → (vi) 7a～f

a: R=H, n=2;
b: R=H, n=3;
c: $R=CH_2CH_2NEt_2$, n=2;
d: $R=CH_2CH_2NEt_2$, n=3;
e: $R=CH_2CH_2CH_2NMe_2$, n=2;
f: $R=CH_2CH_2CH_2NMe_2$, n=3.

Scheme 1 Synthesis for substituted purines. Reagents and conditions: (i) amine, ethanol, 24h, rt; (ii) H_2 (1 atm.), Pd/C, methanol, 36h, rt; (iii) CS_2, NaOH, ethanol, 5h, reflux; (iv) $ClCH_2CO_2Et$, KI, K_2CO_3, THF, 12h, rt; (v) ω, ω - diamino alkane (5 equiv), methanol, 8h, reflux; (vi) $AIMSO_3H \cdot H_2O$ (1.1 equiv), anhydrous ethanol, 4h, 35 - 45℃

All the 12 title compounds (**6a - 6f** and **7a - 7f**) were evaluated for inhibiting HIV - 1 Tat - TAR interaction in human 293T cells using Tat dependent HIV - 1 LTR - driven CAT gene expression colorimetric enzyme assays at a concentration of 30μM. As a report gene, the depressed CAT expression indicated the high inhibitory activity of the compound. As shown in Figure 2, the decreased CAT activities in the presence of the title compounds suggested that all of them could effectively block the interaction of Tat - TAR RNA in vivo.

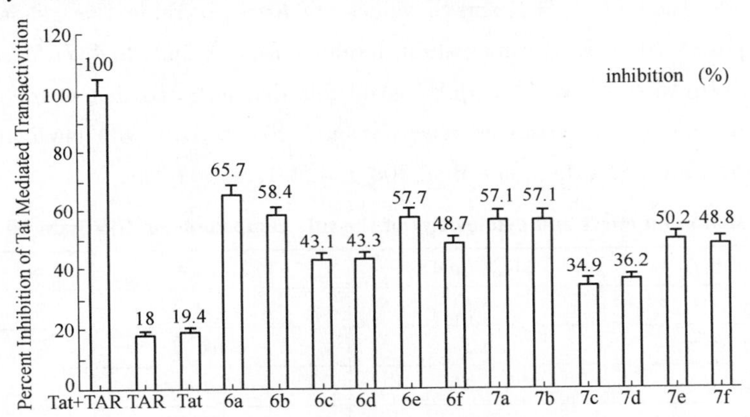

Figure 2 Effects of title compounds on Tat Trans - activation in 293T cells

The range of inhibited CAT expression induced by the title compounds is from 34.9% to 65.7%, and the lowest data, 34.9% and 36.2%, appeared in the presence of **7c** and **7d**, respectively, while **6a** showed the highest (65.7%). As we reported before, at the same concentration, only two of the twelve β-carbolines depressed CAT expression below 60% (Refs. 12a, d), and for isoquinoline derivatives and α,α-trehalose-arginine conjugates, CAT expressions were all above 40% (Refs. 12e, b, c). It is very clear that the purine derivatives contain the best activity in the CAT assay. This might come from the ability of purine to mimic Watson-Crick interactions with pyrimidine bases.

Taking a detailed look at the data in Figure 2, it is found as follows: (a) when the amino group at the end of the side chain was converted to the guanidyl group, the inhibitory activities of the compounds would be reinforced, such as **6a** to **7a**, **6c** to **7c**, **6e** to **7e**, etc.; (b) with a side chain terminated with a trisubstituted amino group at N9 of the purine ring, the inhibition of CAT expression in 293T cells is increased by **6c**, **6d**, **6e**, and **6f** compared to **6a** and **6b** as well as **7c**, **7d**, **7e**, and **7f** compared to **7a** and **7b**. Among them, **6c**, **6d**, **7c**, and **7d** appeared to be the most effective compounds in restraining the CAT expression; (c) the length of the side chain at C8 had little effect on the activities of the title compounds for there was no significant difference in activity between compounds **6c** and **6d**, **7a** and **7b**, **7c** and **7d**, **7e** and **7f**, respectively (Figure. 3).

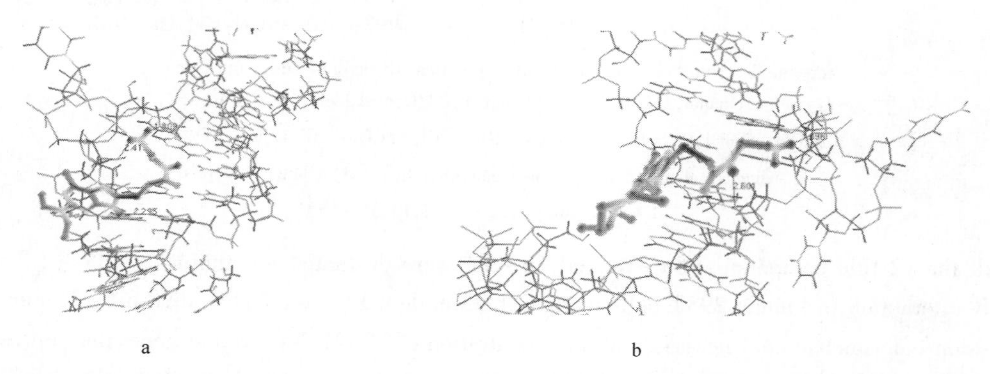

a b

Figure 3 (a) Interaction of compound **6a** to TAR RNA; (b) Interaction of compound **6c** to TAR RNA

We also evaluated the biological activities of the title compounds using SIV-induced syncytium in CEM cells. Their EC_{50}, TC_{50} and SI values are listed in Table 1. As shown in Table 1, each of them possessed an EC_{50} value within the range from 0.2μM to 4.7μM, and a TC_{50} value more than 100μM except **6c** (52.5μM). And this demonstrated that nearly all compounds we mentioned in this article possessed effective anti-SIV activity with low cytotoxicity, especially **6b** and **6f** with a SI value more than 200 and 500, respectively.

Table 1 Inhibition effect and cytotoxicity of the title compounds on SIV induced syncytium

Compound①	EC_{50}② (μM)	TC_{50}③ (μM)	SI④ (TC_{50}/EC_{50})
6a	2.0	>100	>50
6b	0.5	>100	>200
6c	1.3	52.2	40.2
6d	1.4	>100	>71.4

Table 1 (continued)

Compound①	EC_{50}② (μM)	TC_{50}③ (μM)	SI④ (TC_{50}/EC_{50})
6e	3.4	>100	>29.4
6f	0.2	>100	>500
7a	4.2	>100	>23.8
7b	0.7	>100	>142.9
7c	0.7	>100	>142.9
7d	4.7	>100	>21.3
7e	7.2	>100	>13.9
7f	3.2	>100	>31.25

① AZT was used as the positive control at a concentration of 10μM here. Its EC_{50} is 0.0122μM and TC_{50} is above 100μM in this system.

② EC_{50}, concentration required to protect cells against the cytopathogenicity of SIV by 50%.

③ TC_{50}, concentration required to inhibit uninfected cells proliferation by 50%.

④ SI, selective index.

Compounds with an amino group as the 'activator' possessed smaller EC_{50} than those with an guanidyl group which might be due to the lyphohydrophilic characters of the compounds. Under the physiological condition, compounds with an 'activator' of amino group would possess higher lyphophilicity and could be easily absorbed by the cells. But with a guanidyl group as a cation, it was difficult for the compounds to be absorbed by the cells. We used 293T cell line in CAT assay while CEM174 cell line in anti - SIV assay, which might account for some differences of activities inside the cells. These two different biological systems in which the compounds might present different solubilities, ionic forms and especially different absorptions by cells would affect the final activities.

Here using Auto - dock 3.0 process,[22] we studied the interaction between title compounds and HIV - 1 TAR RNA and the energetic data of the interaction and the figures of the compounds binding to TAR were obtained.

In Table 2, the free energy and K_i values of the title compounds reflect the binding affinity of the compounds and the TAR RNA element. The negative free energy and low K_i values suggest that the title compounds could bind to TAR RNA element in theory. A terminal guanidyl group on the side chain could raise the K_i value and these results are consistent with those of the SIV assay. However, the substitution at N9 seems to depress the binding affinity, which is different from the CAT assay. Considering that the model we used here came from the binding complex of rbt203 and TAR, we assume that **6a** and **6b** with similar conformation to rbt203 could fit the model better than those with a side - chain at N9 of the ring.

Table 2 Energy data of the title compounds in Autodock process

Compound	Docked energy (kcal/mol)	Free energy (kcal/mol)	K_i	Intermolecular energy (kcal/mol)	Internal energy (kcal/mol)
6a	-14.38	-11.35	9.78e-09	-14.15	-0.23
6b	-15.01	-11.7	2.67e-09	-14.81	-0.2
6c	-15.29	-10.23	3.7e-08	-14.59	-0.7

Table 2(continued)

Compound	Docked energy (kcal/mol)	Free energy (kcal/mol)	K_i	Intermolecular energy (kcal/mol)	Internal energy (kcal/mol)
6d	-15.83	-10.34	2.63e-08	-15.01	-0.82
6e	-15.58	-11.14	6.84e-09	-15.19	-0.4
6f	-15.47	-10.57	1.78e-08	-14.93	-0.54
7a	-14.44	-11.04	8.08e-09	-14.15	-0.29
7b	-13.82	-10.03	4.49e-08	-13.45	-0.37
7c	-14.26	-8.96	2.71e-07	-13.63	-0.63
7d	-15.2	-9.52	1.06e-07	-14.5	-0.7
7e	-13.53	-8.72	4.04e-07	-13.08	-0.45
7f	-15.71	-10.48	2.09e-08	-15.15	-0.56

Compounds **6a**,**6b**,**7a**,and **7b** inserted into the enlarged minor groove with the rings underneath U23 of the tribased bugle and the amino or guanidyl groups binding to the backbone of the bulge or the lower stem region near A22:U40 and G21:C41. Hydrogen bonds were formed by the amino group with the backbone of the bulge and by the N9 or N7 of the ring with the amino group of the base. While compounds **6c**,**6d**,**6e**,**6f**,**7c**,**7e**,and **7f**,each with an additional side chain at N9,presented a different binding mode:they bound to the major groove of TAR RNA with their rings and to the bulge with their amino or guanidyl groups inserted between the base A22 and U23. In this pattern,the binding was stabilized by hydrogen bonds formed by the N7 or N9 atom with the backbone or the base in the groove. Although the side chain at N9 did not form hydrogen bond,it might contribute to the affinity of the anchor by electrostatic attraction. The number of hydrogen bond,formed by compounds **7a-7f** with TAR was fewer than compounds **6a-6f** and it was due to the fact that under physiologic conditions guanidyl group was fully protonated.

The fact that more potent HIV-1 Tat-TAR inhibitory activities of the 12 title compounds than those of our previous β-carboline and isoquinoline derivatives might suggest the increased interaction with TAR RNA of the purine ring. The substitution at N9 of the purine ring would affect the interaction between the compound and TAR. Unlike the previous work,there was no significant relationship between the biological activity of the title compounds and the length of the 'linker'. It was indicated that our purine derivatives might interact with TAR in a mode somewhat different from β-carboline and isoquinoline derivatives.

From the molecular modeling results,we found that the interaction pattern to TAR RNA of the title compounds is related to the structure of 'anchor'. Without a side chain at N9,the ring of compounds(**6a**,**6b**,**7a**,and **7b**)inserted into the minor groove near the tri-based bulge. In the other case,with a side chain at N9 of the ring,compounds(**6c**,**6d**,**6e**,**6f**,**7c**,**7e**,and **7f**) presented a different binding mode:the rings stayed in the major groove of TAR RNA and the amino or guanidly groups bound to the bulge by inserting between the bases A22 and U23. This is some different from the newly reported TAR RNA binding reagents such as rbt 203 and rbt 550.[14,15] Both of the binding models would efficiently block the Tat-TAR interaction. as compared with the TAR RNA binding reagents we reported previously.

3 Conclusion

In this work, we proposed a general molecular design for Tat – TAR inhibitors and synthesized a novel series of substituted purines as HIV – 1 Tat – TAR inhibitors. Compared with our previous β – carboline and isoquinoline derivatives, the title compounds were more active in inhibiting HIV – 1 Tat – TAR interaction in 239T cells and depressed SIV – induced syncytium in CEM174 cells, which might be due to the increased interaction with TAR RNA of the purine ring. Molecular modeling studies suggested the compounds could bind to TAR in two different modes according to the purine structures. All the findings provide us with new ideas in designing novel Tat – TAR inhibitors.

4 Experimental

4.1 Chemistry

All reactions were performed with commercially available reagents and they were used without further purification. Solvents were dried by standard methods and stored over molecular sieves. All reactions were monitored by thin – layer chromatography (TLC) and viewed with UV light. Melting points were determined on a XA – 4 instrument and are uncorrected. All the title compounds were characterized by ^1H NMR spectra on a Brucker 300MHz photometer using the solvents described. Chemical shifts were reported in δ ppm (parts per million) relative to Tetramethyl Silane (TMS) for deuteriorated water (D_2O). Signals were quoted as s (singlet), d (doublet), t (triplet), q (quartet), and m (multiplet). The mass spectra (EI and FAB +) were recorded on JEOL – JMS – SX – PNBA, (*p* – nitrobenzyl alcohol) was used as matrix (M^+) which showed M + 1 peak at 154, 2M + 1 peak at 307.

4.2 General synthetic procedure

The synthetic work was carried out starting from 4 – chloro – *N*,*N* – diethyl – 6 – methyl – 5 – nitro – pyrimidine – 2 – amine **1** and 5 – nitro – pyrimidine – 2,4 – diamine **2** was easily prepared from it by amination.[23] Hydrogen reduction was preferred to prepare pyrimidine – 2,4,5 – triamine **3**. Without further purification, cyclization of **3** with CS_2 was performed in alkali ethanol and substituted purine – 8 – thiol **4** was obtained in a high yield after acidization of the reaction system. Under the catalysis of K_2CO_3/KI in THF, (purin – 8 – ylsulfanyl) – acetic acid ethyl ester **5** was obtained from **4** in a moderate yield at reflux temperature. By reactions with **5** equiv or more 1,2 – ethylenediamine or 1,3 – propylenediamine in methanol, **5** was conveniently converted to **6** in good yields. As a result the side chain with full length was introduced into the purine ring at C8. In the last step, amino iminomethane sulfonic acid ($AIMSO_3H \cdot H_2O$) was selected as guanidylation reagent in anhydrous ethanol considering the amphiphilicity or hydrophilicity of target compounds. $AIMSO_3H \cdot H_2O$ was chosen to construct the guanidyl group at the end of the side chain for its high reactivity, mild condition, simple procedure, and easy final treatment,[24] and which could be efficiently used in preparing other amphiphilic or hydrophilic compounds with guanidyl groups as well.

4.2.1 Aminoiminomethane sulfonic acid($AIMSO_3H \cdot H_2O$)

$AIMSO_3H \cdot H_2O$ was prepared from thiourea by the method reported in reference and obtained as a colorless crystal, mp 132 – 134℃ (dec).[24]

4.2.2 4 – Chloro – *N,N* – diethyl – 6 – methyl – 5 – nitro – pyrimidine – 2 – amine(1)

This intermediate was obtained as yellow crystal mp 69 – 71℃.[23]

4.2.3 N^2, N^2 – Diethyl – 6 – methyl – 5 – nitro – pyrimidine – 2,4 – diamine(**2a**)

Compound **1** (1.5g; 0.06mol) was dissolved in anhydrous ethanol (20mL) and $NH_3 \cdot H_2O$ (26%, 10mL) was added dropwise under stirring. After being reacted for 24h at room temperature, the mixture was poured into water(100mL) and(**2a**) was precipitated as yellow solid and then purified by column chromatography(silica gel, petroleum ether/ethyl acetate = 5:1, v/v), 1.35g, mp 89 – 92℃, yield = 98%.

4.2.4 N^4 – (2 – Diethylaminoethyl) – N^2, N^2 – diethyl – 6 – methyl – 5 – nitro – pyrimidine – 2,4 – diamine(**2b**)

Intermediate compounds(**2b**) were obtained from(**1**) and N^1, N^1 – diethyl – ethane – 1,2 – diamine, After purified by column chromatography (silica gel, organic layer of $CHCl_3/CH_3OH/NH_3 \cdot H_2O$ = 40:2:1, v/v/v), **2b** was obtained as yellow jelly in nearly stoichiometric yield, respectively. Without further purification, it was used in the next step directly.

4.2.5 N^4 – (3 – Dimethylaminopropyl) – N^2, N^2 – diethyl – 6 – methyl – 5 – nitro – pyrimidine – 2,4 – diamine(**2c**)

Compound **2c** was also obtained from N^1, N^1 – dimethyl – propane – 1,3 – diamine and(**1**) by the same procedure and used in the next step directly.

4.2.6 N^2, N^2 – Dimethyl – 6 – methyl – pyrimidine – 2,4,5 – triamine(**3a**)

Compound **2a**(2.0g; 0.0089mol) was dissolved in anhydrous methanol(30mL) containing Pd/C(10%, 0.1g) and H_2 gas was bubbled into the solution under stirring for 36h. Compound **3a** was obtained as colorless needle – like solid in stoichiometric yield when methanol was removed under reduced pressure. Without further purification, it was used in the next step directly.

4.2.7 N^4 – (2 – Diethylaminoethyl) – N^2, N^2 – diethyl – 6 – methylpyrimidine – 2,4,5 – triamine(**3b**)

Compound **3b** was obtained in the same procedure for **3a** after reaction for 8h as pale – green jelly in stoichiometric yield, Without further purification, **3b** was directly used in the next step.

4.2.8 N^4 – (3 – Dimethylaminopropyl) – N^2, N^2 – diethyl – 6 – methyl – pyrimidine – 2,4,5 – triamine(**3c**)

Compound **3c** was obtained in the same procedure for **3a** after reaction for 8h as pale – green jelly in stoichiometric yield. Without further purification, **3c** was directly used in the next step.

4.2.9 2 – Diethylamino – 6 – methyl – 7,9 – dihydro – purine – 8 – thione(**4a**)

Compound **3a**(2.0g; 0.0102mol) was dissolved in ethanol(20mL) containing KOH(0.58g; 0.0102mol) and CS_2 (0.5mL) was added dropwise under stirring. After being reacted at room temperature for 1.5h, the mixture was then heated to 78℃ and refluxed for 5h. After ethanol was removed, the residue was dissolved in water (100mL) and neutralized with dilute HCl (10%) to pH5 - 6, then (**4a**) precipitated as palepink solid. After being purified by column chromatography (silica gel, petroleum ether/ethyl acetate = 3:1, v/v), **4a** was obtained as white flocculent solid, 2.16g, mp 285℃ (dec), yield = 88.5%.

4.2.10 2 - Diethylamino - 9 - (2 - diethylaminoethyl) - 6 - methyl - 9*H* - purine - 8 - thiol (**4b**)

The compound was obtained from **3b** in the same procedure for (**4a**). After being purified by column chromatography (silica gel, petroleum ether/ethyl acetate = 1:2, v/v), **4b** was obtained as white flocculent solid, mp 211 - 214℃, yield = 87.1%. ^{1}H NMR (300MHz, CD_3OD) δ 1.12 - 1.21 (m, 12H), 2.43 (s, 3H), 2.70 (q, 4H), 2.90 (m, 2H), 3.64 (q, 4H), 4.25 (m, 2H). MS (EI) *m/z* calcd: 336.21, found: 336 (M^+).

4.2.11 2 - Diethylamino - 9 - (3 - dimethylaminopropyl) - 6 - methyl - 9*H* - purine - 8 - thiol (**4c**)

Compound **4c** was also prepared from **3c** and purified by column chromatography (silica gel, ethyl acetate), white solid, mp 190 - 193℃, yield = 83.3%. ^{1}H NMR (300MHz, CD_3OD) δ 1.17 (t, 6H), 2.01 - 2.12 (m, 2H), 2.28 (s, 6H), 2.44 (s, 3H), 2.42 - 2.50 (m, 2H) 3.62 (q, 4H), 4.23 ~ 4.30 (m, 2H). MS (EI) *m/z* calcd: 322.19, found: 322 (M^+).

4.2.12 (2 - Diethylamino - 6 - methyl - 9*H* - purin - 8 - ylsulfanyl) - acetic acid ethyl ester (**5a**)

Compound **4a** (1.5g; 0.0063mol) was dissolved in HTF (50mL) containing ethyl chloroacetate (1.16g; 0.010mol), K_2CO_3 (1.3g; 0.0095mol) KI (0.05g) and reacted at room temperature for 12h. After the residue was purified by column chromatography (silica gel, petroleum ether/ethyl acetate = 3:1, v/v), **5a** was obtained as white solid, 1.3g, mp 84 - 88℃, yield = 78.6%.

4.2.13 Ethyl [2 - diethylamino - 9 - (2 - diethylaminoethyl) - 6 - methyl - 9*H* - purin - 8 - ylsulfanyl] acetate (**5b**)

The compound was prepared from **4b** in the same way as (**5a**). After column chromatography (silica gel, petroleum ether/ethyl acetate = 1:1, v/v), **5b** was obtained as yellow oil, yield = 73.5%. ^{1}H NMR (300MHz, CD_3Cl_3) δ 1.00 - 1.26 (t t, 15H), 2.52 (s, 3H), 2.57 - 2.61 (q, 4H), 2.73 - 2.76 (m, 2H), 3.59 - 3.63 (q, 4H), 4.06 (s, 2H), 4.06 - 4.10 (m, 2H), 4.17 - 4.19 (q, 2H). MS (EI) *m/z* calcd: 422.25, found: 422 (M^+).

4.2.14 Ethyl [2 - diethylamino - 9 - (3 - dimethylaminopropyl) - 6 - methyl - 9*H* - purin - 8 - ylsulfanyl] - acetate (**5c**)

Compound **5c** was prepared from (**4c**) and column chromatography (silica gel, methanol/ethyl acetate = 1:1, v/v), pale - pink solid, yield = 78.9%. ^{1}H NMR (300MHz, CD_3Cl_3) δ 1.16 - 1.20 (m, 9H), 2.09 - 2.11 (m, 2H), 2.35 (s, 6H), 2.51 - 2.56 (m, 2H), 2.56 (s, 3H), 3.25 - 3.67 (m, 6H), 3.76 (s, 2H), 4.12 - 4.14 (m, 2H). MS (EI) *m/z* calcd: 408.23. found: 408 (M^+).

4.2.15 *N* - (2 - Aminoethyl) - 2 - (2 - diethylamino - 6 - methyl - 9*H* - purin - 8 - ylsulfanyl) acetamide (**6a**)

Compound **5a** (0.45g; 0.0014mol) was dissolved in anhydrous methanol (10mL) containing 1, 2 - ethylenediamine (0.42g; 0.0070mol) and refluxed for 8h. After methanol was removed under reduced pressure, the residue was purified by column chromatography (silica gel, the organic layer of MeOH/$CHCl_3$/$NH_3 \cdot H_2O$(26%) = 5:3:1, v/v/v). Compound **6a** was obtained as pale-yellow sticky jelly, yield = 85.2%. ^{1}H NMR (300MHz, CD_3OD) δ 1.01 (t, 6H, J = 6.54Hz), 2.43 (s, 3H), 2.67 (m, 2H), 3.23 (m, 2H), 3.44 (dq, 4H, J = 6.54Hz), 3.49 (s, 2H); MS (EI) m/z calcd: 337.17, found: 337 (M^+).

4.2.16 *N*-(3-Aminopropyl)-2-(2-diethylamino-6-methyl-9*H*-purin-8-ylsulfanyl) acetamide (**6b**)

Compound **6b** was prepared in the same way as (**6a**), pale-yellow sticky jelly, yield = 73.6%. ^{1}H NMR (300MHz, CD_3OD) δ 0.98 (m, 6H), 1.20, 1.50 (d t, 2H, J = 6.5Hz), 2.20, 2.50 (d t, 2H, J = 6.51Hz), 2.41 (s, 3H), 3.00, 3.15 (dt, 2H, J = 6.5Hz), 3.21 (s, 2H), 3.49 (m, 4H); MS (EI) m/z calcd: 351.18, found: 351 (M^+).

4.2.17 *N*-(2-Aminoethyl)-2-[2-diethylamino-9-(2-diethylaminoethyl)-6-methyl-9*H*-purin-8-ylsulfanyl] acetamide (**6c**)

Compound **6c** was obtained from **5b**. After purified by column chromatography (silica gel, the organic layer of $CHCl_3$/MeOH/$NH_3 \cdot H_2O$(26%) = 20:3:1, v/v/v). Compound **6c** was obtained as yellow oil, yield = 94.9%. ^{1}H NMR (300MHz, CD_3OD) δ1.10-1.27 (m, 12H), 2.56-2.63 (m, 4H), 2.63 (s, 3H), 2.75-2.82 (dm, 4H), 3.29-3.35 (m, 2H), 3.62-3.69 (q, 4H), 3.83 (s, 2H), 4.06 (t, 2H), 8.59 (s, 1H); MS (FAB+) m/z calcd: 436.27, found: 437.2 [$(M+1)^+$].

4.2.18 *N*-(3-Aminopropyl)-2-[2-diethylamino-9-(2-diethylaminoethyl)-6-methyl-9*H*-purin-8-ylsulfanyl] acetamide (**6d**)

Compound **6d** was obtained from **5b** and purified by column chromatography as (**6c**). Compound **6d**: yellow oil, yield = 95.7%. ^{1}H NMR (300MHz, CD_3OD) δ 1.00-1.20 (d t, 12H), 1.56-1.65 (m, 2H), 2.55 (s, 3H), 2.60-2.79 (m, 4H), 2.52-2.58 (m, 4H), 3.30-3.37 (m, 2H), 3.66 (q, 4H), 3.82 (s, 2H), 4.03-4.08 (t, 2H), 8.57 (s, 1H). MS (FAB+) m/z calcd: 450.29, 450.7 (M^+).

4.2.19 *N*-(2-Aminoethyl)-2-[2-diethylamino-9-(3-dimethylaminopropyl)-6-methyl-9*H*-purin-8-ylsulfanyl] acetamide (**6e**)

Compound **6e** was prepared from **5c** and purified by column chromatography (silica gel, the organic layer of $CHCl_3$/MeOH/$NH_3 \cdot H_2O$(26%) = 40:5:1, v/v/v). Compound **6e**: yellow oil, yield = 96.7%. ^{1}H NMR (300MHz, CD_3OD) δ 1.20 (t, 6H), 1.93-1.98 (m, 2H), 2.24 (s, 6H), 2.32 (t, 2H), 2.57 (s, 3H), 2.78-2.82 (m, 2H), 3.29-3.35 (m, 2H), 3.66 (q, 4H), 3.84 (s, 2H), 4.04-4.09 (m, 2H), 8.54 (s, 1H). MS (FAB+) m/z calcd: 426.26, found: 426.0 (M^+).

4.2.20 *N*-(3-Aminopropyl)-2-[2-diethylamino-9-(3-dimethylaminopropyl)-6-methyl-9*H*-purin-8-ylsulfanyl] acetamide (**6f**)

Compound **6f** was prepared from **5c** and purified by column chromatography as (**6e**). Compound **6f**: yellow oil, yield = 93.6%. ^{1}H NMR (300MHz, CD_3OD) δ 1.19 (t, 6H), 1.56-1.63 (m, 2H), 1.92-1.97 (m, 2H), 2.24 (s, 6H), 2.30-2.35 (m, 2H), 2.55 (s, 3H), 2.68-2.71 (m, 2H), 3.30-3.34 (m, 2H) 3.66 (q, 4H), 3.83 (s, 2H), 4.03-4.08 (m, 2H). MS (FAB

+) m/z calcd:436. 27, found:436. 8(M^+).

4.2.21 *N*-(2-Guanidinoethyl)-2-(2-diethylamino-6-methyl-9*H*-purin-8-ylsulfanyl) acetamide hydrosulfite(**7a**)

$AIMSO_3H \cdot H_2O$ (0. 15g, 0. 001mol) was added by portion to the solution of **6a** (0. 4g: 0. 00091mol) in 15mL anhydrous ethanol in water bath at 35 – 45℃ in 1h. After being stirred for 2h, the solvent was moved under reduced pressure and a pale – yellow jelly was obtained. After recrystallized in isopropanol, **7a** was obtained as moisture regained jelly, 0. 42g, yield = 87. 7%. ^1H NMR(300MHz, CD_3OD) δ 1. 02(m,6H), 2. 62(s,3H), 3. 20(m,2H), 3. 32(dq,4H, J = 6. 54Hz), 3. 90(m,2H), 4. 10(s,2H); MS(FAB +) m/z calcd:379. 19, found:379. 8[$(M+1)^+$].

4.2.22 *N*-(3-Guanidinopropyl)-2-(2-diethylamino-6-methyl-9*H*-purin-8-ylsulfanyl) acetamide hydrosulfite(**7b**)

Compound **7b** was prepared from **6b** moisture regained jelly, 0. 45g, yield = 83. 1%. ^1H NMR (300MHz, CD_3OD) δ 1. 01(t,6H, J = 6. 18Hz), 1. 20, 1. 50(dt,2H, J = 6. 9Hz), 2. 43(s, 3H), 2. 63, 3. 20(dt,2H, J = 6. 8Hz), 3. 08(m,4H, J = 5. 8Hz), 3. 48(m,2H, J = 6. 8Hz); MS(FAB +) m/z calcd:393. 21, found:393. 4(M^+).

4.2.23 2-[2-Diethylamino-9-(2-diethylaminoethyl)-6-methyl-9*H*-purin-8-ylsulfanyl] -*N*-(2-guanidinoethyl)acetamide(**7c**).

Compound **7c** was prepared from **6c** in the same way as(**7a**)and(**7b**). After being purified by column chromatography(silica gel, the organic layer of $CHCl_3/MeOH/NH_3 \cdot H_2O$(26%) = 40:60:7, v/v/v), **7c** was obtained as pale – yellow jelly, yield = 91. 2%. ^1H NMR(300MHz, D_2O) δ 0. 97, 1. 13(d t,12H), 2. 34(s,3H), 2. 99 – 3. 06(m,2H), 3. 11 – 3. 23(m,4H), 3. 21 – 3. 24(m,4H), 3. 43(q,4H), 3. 81(s,1H), 4. 28, 4. 63(d t,2H). MS(FAB +) m/z calcd:478. 30, found:479. 1[$(M+1)^+$].

4.2.24 2-[2-Diethylamino-9-(2-diethylaminoethyl)-6-methyl-9*H*-purin-8-ylsulfanyl] -*N*-(3-guanidino-propyl)acetamide(**7d**).

Compound **7d** was prepared from(**6d**)and purified by column chromatography as **7c**: pale – yellow jelly, yield = 91. 5%. ^1H NMR(300MHz, D_2O) δ 1. 00, 1. 13(dt,12H), 1. 40 – 1. 70 (m,2H), 2. 41(s,3H), 3. 00 – 3. 10(m,2H), 3. 12 – 3. 17(m,4H), 3. 28 – 3. 33(m,2H), 3. 76(s,2H), 4. 59(q,2H); MS(FAB +) m/z calcd:492. 31, found:493. 4 [$(M+1)^+$].

4.2.25 2-[2-Diethylamino-9-(3-dimethylaminopropyl)-6-methyl-9*H*-purin-8-ylsulfanyl]-*N*-(2-guanidino-ethyl)acetamide(**7e**)

Compound **7e**: pale – yellow jelly, yield = 90. 9%. ^1H NMR(300MHz, D_2O) δ 0. 99(t,6H), 2. 16 – 2. 21(m,2H), 2. 40(s,3H), 2. 68(s,6H), 3. 01 – 3. 07(m,2H), 3. 18 – 3. 29(d m, 4H), 3. 46(q,4H), 3. 61(s,2H), 4. 26 – 4. 31(m,2H). MS(FAB +) m/z calcd:464. 28, found:464. 9(M^+).

4.2.26 2-[2-Diethylamino-9-(3-dimethylaminopropyl)-6-methyl-9*H*-purin-8-ylsulfanyl]-*N*-(3-guanidino-propyl)acetamide(**7f**)

Compound **7f**: pale – yellow jelly, yield = 91. 2%. ^1H NMR(300MHz, D_2O) δ 0. 99(t,6H),

1.62(m,2H),2.19(m,2H),2.41(s,3H),2.68(s,6H),3.04 – 3.17(d m,4H),3.37(t, 2H),3.48(q,4H),3.61(s,2H),4.30(t,2H). MS(FAB +) *m/z* calcd:478.30, found: 477.8[(M – 1)$^+$].

4.3 Biological evaluation

Transient transfection and CAT assays:293T cells were grown as monolayer in Dulbecco's modified Eagle's medium(DMEM)(Gibco – BRL) supplemented with 10%(v/v) fetal calf serum, penicillin(100 U · mL^{-1}), and streptomycin(100 U · mL^{-1}) at 37℃ in 5% CO_2 containing humidified air. The cells were seeded at a six – well plate 24h prior to transfection which was performed by standard calcium phosphate coprecipitation techniques with optimum amounts of the plasmids pLTRCAT and pSVCMVTAT. Twenty four hours later, the culture medium was removed and the cells were washed twice with phosphate – buffered saline(PBS). Then the transfected cells were added to fresh medium together with diluted compounds at final concentration of 30μM, respectively, and incubated for another 24h. After 48h post – transfection, the cells were harvested and analyzed for CAT activity using a commercial CAT ELISA kit(Roche Molecular Biochemicals) in accordance with the manufacturer's protocol. All data were reported as a percentage of CAT activity(± SD). Results shown were representative of three independent experiments.

Inhibition of SIV – induced syncytium in CEM174 cell cultures was measured in a 96 – well microplate containing 2 × 10^5 CEM cells/mL infected with 100 $TCID_{50}$ of SIV per well and containing appropriate dilutions of the tested compounds. After 5 days of incubation at 37℃ in 5% CO_2 containing humidified air, CEM giant(syncytium) cell formation was examined microscopically. The EC_{50} was defined as the compound concentration required to protect cells against the cytopathogenicity of SIV by 50%. AZT was used as the positive control at a concentration of 10μM here.

4.4 Molecular modeling

The initial structures of our compounds were subjected to minimization using MOPAC in Chemoffice 2002 and the 3D structure of HIV – 1 TAR RNA in complex with its inhibitor rbt 203 was recovered from the Protein Database(http://www.PDB.org) with the code as 1UUI. The advanced docking program Auto – dock 3.0 was used to remove the small molecule and perform the automatic molecular docking with our compounds. The number of generations, energy evaluation, and docking runs were set to 370000, 1500000 and 30, respectively, and the kinds of atomic charges were taken as Kollman – all – atom for HIV – 1 TAR RNA and Gasteiger – Hücel for the compounds.

References and notes

[1] Kingsman, S. M.; Kingsman, A. *Eur. J. Biochem.* 1996, 240, 491.

[2] Yang, M. *Curr. Drug. Targets – Infect. Dis.* 2005, 5, 433.

[3] Weeks, K. M.; Ampe, C.; Schultz, S. C.; Steitz, T. A.; Crothers, D. M. *Science* 1990, 249, 1281.

[4] Rana, T. M.; Jeang, K. T. *Arch. Biochem. Biophys.* 1999, 365, 175.

[5] Calnan, B. J. ; Tidor, B. ; Biancalana, S. ; Hudson, D. ; Frankel, A. D. *Science* 1991, 252, 1167.
[6] Wang, X. ; Huq, I. ; Rana, T. M. *J. Am. Chem. Soc.* 1997, 119, 6444.
[7] Nifosi, R. ; Reyes, C. M. ; Kollman, P. A. *Nucleic Acids Res.* 2000, 28, 4944.
[8] Peytou, V. ; Condom, R. ; Patino, N. ; Guedj, R. ; Aubertin, A. M. ; Gelus, N. ; Bailly, C. ; Terreux, R. ; Cabrol - Bass, D. *J. Med. Chem.* 1999, 42, 4042.
[9] Mei, H. Y. ; Cui, M. ; Heldsinger, A. ; Lemrow, S. M. ; Loo, J. A. ; Sannes - Lowery, K. A. ; Sharmeen, L. ; Czarnik, A. W. *Biochemistry* 1998, 37, 14204.
[10] Gelus, N. ; Bailly, C. ; Hamy, F. ; Klimkait, T. ; Wilson, W. D. ; Boykin, D. W. *Bioorg. Med. Chem.* 1999, 7, 1089.
[11] Daelemans, D. ; Vandamme, A. M. ; De Clercq, E. *Antiviral Chem. Chemother.* 1999, 10, 1.
[12] (a) Yu, X. L. ; Lin, W. ; Li, J. Y. ; Yang, M. *Bioorg. Med. Chem. Lett.* 2004, 14, 3127; (b) Wang, M. ; Xu, Z. D. ; Tu, P. F. ; Yu, X. L. ; Xiao, S. L. ; Yang, M. *Bioorg. Med. Chem. Lett.* 2004, 14, 2585; (c) Wang, M. ; Tu, P. F. ; Xu, Z. D. ; Yang, M. *Helv. Chim. Acta* 2003, 86, 2637; (d) Yu, X. L. ; Lin, W. ; Pang, R. F. ; Yang, M. *Eur. J. Med. Chem.* 2005, 40, 831; (e) He, M. Z. ; Yuan, D. K. ; Lin, W. ; Pang, R. F. ; Yu, X. L. ; Yang, M. *Bioorg. Med. Chem. Lett.* 2005, 15, 3978.
[13] Hamasaki, H. ; Ueno, A. *Bioorg. Med. Chem.* 2001, 11, 591.
[14] Davis, B. ; Afshar, M. ; Varani, G. ; Murchie, A. I. H. ; Karn, J. ; Lentzen, G. ; Drysdale, M. ; Bower, J. ; Potter, A. J. ; Starkey, I. D. ; Swarbrick, T. ; Aboul - ela, F. *J. Mol. Biol.* 2004, 336, 343.
[15] Murchie, A. I. H. ; Davis, B. ; Isel, C. ; Afshar, M. ; Drysdale, M. J. ; Bower, J. ; Potter, A. J. ; Starkey, I. D. ; Swarbrick, T. M. ; Mirza, S. ; Prescott, C. D. ; Vaglio, P. ; Aboul - ela, F. ; Karn, J. *J. Mol. Biol.* 2004, 336, 625.
[16] Gelus, N. ; Hamy, F. ; Bailly, C. *Bioorg. Med. Chem.* 1999, 7, 1075.
[17] Patino, N. ; Di - Giorgio, C. ; Dan - covalciuc, C. ; Peyou, V. ; Terreux, R. ; Cabrol - Bass, D. ; Bailly, C. ; Condom, C. *Eur. J. Med. Chem.* 2002, 37, 573.
[18] Belmont, P. ; Jourdan, M. ; Demeunynck, M. ; Constant, J. F. ; Garcia, J. ; Lhomme, J. *J. Med. Chem.* 1999, 42, 5153.
[19] Haruhiko, M. ; Noriyuki, A. ; Shinji, M. ; Mikari, E. ; Kohei, Y. ; Kenji, K. ; Yuichi, Y. ; Shinji, S. ; Osamu, I. ; Yoshito, E. *Antiviral Res.* 1998, 39, 129.
[20] Baraldi, P. G. ; Broceta, A. U. ; de las Infantas, M. J. P. ; Mochun, J. J. D. ; Espinosab, A. ; Romagnoli, R. *Tetrahedron* 2002, 58, 7607.
[21] Arris, C. E. ; Boyle, F. T. ; Calvert, A. H. ; Curtin, N. J. ; Endicott, J. A. ; Garman, E. F. ; Gibson, A. E. ; Golding, B. T. ; Grant, S. ; Griffin, R. J. ; Jewsbury, P. ; Johnson, L. N. ; Lawrie, A. M. ; Newell, D. R. ; Noble, M. E. M. ; Sausville, E. A. ; Schultz, R. ; Yu, W. J. *Med. Chem.* 2000, 43, 2797.
[22] Morris, G. M. ; Goodsell, D. S. ; Halliday, R. S. ; Huey, R. ; Hart, W. E. ; Belew, R. K. ; Olson, A. J. *J. Comput. Chem.* 1998, 19, 1639.
[23] Yuan, D. K. ; Li, Z. M. ; Zhao, W. G. *Chin. J. Org. Chem.* 2003, 23, 1155.
[24] Pan, Z. X. Chin. *J. Appl. Chem.* 2001, 18, 62.

酵母 AHAS 酶与磺酰脲类抑制剂作用模型的分子对接研究*

李 琼[1] 陈沛全[1,2] 陈 兰[1] 孙宏伟[1,2] 沈荣欣[1] 赖城明[1] 李正名[1]

([1] 南开大学化学学院，天津 300071；[2] 南开大学科学计算研究所，天津 300071)

摘 要 基于酵母乙酰羟酸合成酶（AHAS）与磺酰脲类抑制剂复合物的晶体结构，用分子对接方法对 AHAS 与 5 个磺酰脲类抑制剂相互作用的方式进行了系统的分子对接研究。晶体复合物对接和假复合物对接两种模式对接的结果基本相同，并与实验结果吻合。在进一步的对接中逐级考虑了辅酶 FAD 和 TPP 的影响，结果表明，辅酶 FAD 和 TPP 的加入，对 AHAS 酶与磺酰脲类抑制剂的结合顺序基本没有影响。其中 FAD 的加入使 AHAS 与抑制剂的结合更加稳定，这主要是由于抑制剂的 R_2 取代基与 FAD 中的平面基团 Flavin 环间存在的范德华相互作用所致；抑制剂与 TPP 间存在的静电相互作用可能是加速 TPP 降解的原因。

关键词 酵母乙酰羟酸合成酶（AHAS） 对接 磺酰脲 作用模型

乙酰羟酸合成酶（AHAS）是植物和微生物中支链氨基酸合成的关键酶。AHAS 抑制剂的研究已成为新型、高效、低毒除草剂开发的主流。AHAS 抑制剂主要有磺酰脲类、咪唑啉酮类、磺酰胺类和嘧啶水杨酸类[1]，研究主要集中在实验[2~4]和虚拟筛选[5]上。酵母 AHAS 的晶体结构（1JSC）[2]中含 AHAS 二聚体、2 分子辅酶 FAD、2 分子 TPP（thiamine pyrophosphate）和 2 个 Mg^{2+}。2 个活性位点分别位于酶相对的两侧，每个活性位点都与 2 个单体氨基酸残基相接。Dugg leby 等[4]制得 5 种磺酰脲类抑制剂与酵母 AHAS 的复合晶体[6]，提供了抑制剂与 AHAS 结合位点与结合模式的信息。本文用分子对接方法系统研究了 AHAS 与磺酰脲抑制剂的作用方式，建立了一个 AHAS 与磺酰脲抑制剂作用的模型。

1 计算方法

用 Jackal 软件[7]添加 5 个复合物晶体结构（1NOH，1T9A，1T9B，1T9C，1T9D）中缺失的残基和侧链。5 个配体小分子均为磺酰脲类抑制剂，其典型结构由芳基、磺酰脲桥及杂环 3 部分组成，如图 1 所示。配体小分子的结构均取自复合物的晶体数据，采用 Quacpac 软件[8]添加氢原子，并赋予 Gasteiger - Marsilli 电荷，将其作为分子对接中配体的初始构型。

对上述体系进行了两种模式的对接：（1）复合物分子对接，即对接中的受体和配体均直接取自晶体结构；（2）假复合物分子对接，即受体选用 1NOH 的蛋白质，而配体取自其他晶体中相应部分。除这两种模式外，还将受体分子加入了辅酶 FAD 和 TPP。由于复合物晶体结构中只有 1NOH 的蛋白质结构含有一分子完整的 TPP（理论上应有

* 原发表于《高等学校化学学报》，2007，28（8）：1552 - 1555。基金项目：国家“863”计划（批准号：2006AA10A 213）和南开大学“南开之星”高性能计算项目资助。

两分子 TPP），故在第一种对接模式中，没有考虑 TPP 的影响。

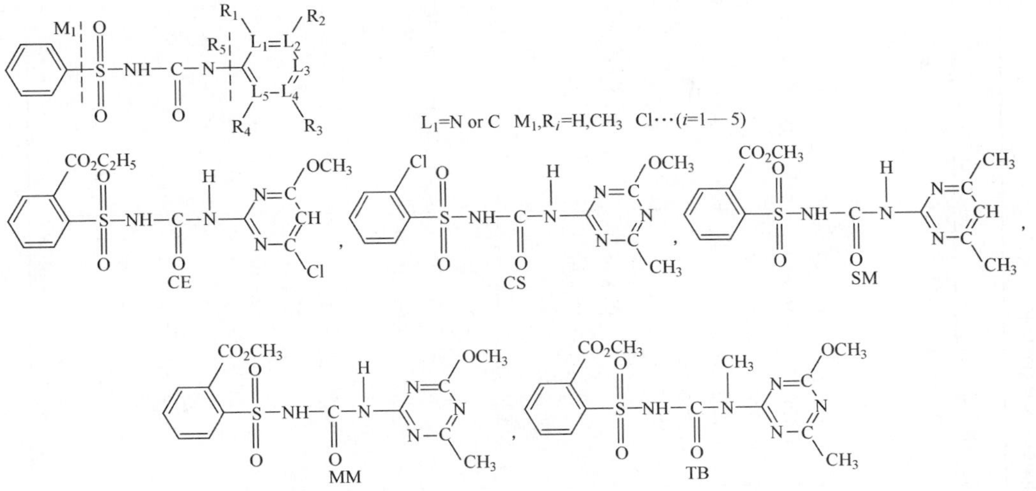

L_1=N or C　M_1,R_i=H,CH_3　Cl…(i=1—5)

Fig. 1　Structures of sulfonylureas inhibitors

分子对接采用 Autodock 3.0 程序进行，受体格点盒子为边长 2.25nm 的立方体，格点间距 0.375nm。复合物分子对接时，盒子中心位于抑制剂的中心；假复合物分子对接时，盒子中心位于 1NOH 复合物晶体中抑制剂 CE 的中心。氢键和范德华相互作用分别采用12－10 和 12－6 Lennard－Jones 参数形式。运用 Lamarckian 遗传算法 LGA，将局部能量搜索与遗传算法相结合，以半经验势函数作为能量打分函数，对小分子构象和位置进行全局搜索，Population 为 150，ga_ num_ evals 设置为 10^6。进行 100 次独立对接，选择均方根偏差 RMSD＜0.05nm 作为评价标准聚类。计算在“南开之星”超级计算机上完成。

2　结果与讨论

2.1　复合物分子对接

受体和配体小分子的结构分别取自 5 个补齐蛋白质链的复合物晶体中的相应部分，并考虑了辅酶 FAD 在对接过程中的影响。对接模式、对接结果列于表 1，加入辅酶 FAD 前后的两种对接结果与晶体中小分子的叠合图见图 2。从表 1 的计算结果可以看出，对接能量最低的构型与晶体结构的 RMSD 值在 0.36～0.082nm 之间。因此，使用 Autodock 3.0 的分子对接结果，可以很好地重现这 5 个磺酰脲分子与 AHAS 的作用方式。加入辅酶 FAD 对整个对接结果影响不大，两种情况下，小分子构型基本没有发生变化；但是加入辅酶 FAD，使 5 个复合物对接更加稳定。

2.2　假复合物分子对接

2.2.1　两种对接模式的比较

假复合物分子对接的受体取自 1NOH 复合物晶体，并逐级考虑辅酶 FAD 和 TPP 的影响。5 个配体小分子初始构型分别来自 5 个晶体复合物中的相应部分。对比表 1 中计算结果可知，在对接过程中采用何种受体模型，对对接结果基本上没有影响。分析这 5 个复合物的对接结果可知，5 个结构中，磺酰脲抑制剂的结合位点的残基在空间构象上

Table 1 Molecule docking results obtained from the calculation with autodock 3.0 package

Ligand	Dock	Mode	No of cluster	No in the best cluster	Docked energy① /kJ · mol^{-1}	Internolecular energy /kJ · mol^{-1}	Intemal energy of ligand /kJ · mol^{-1}	RMSD /nm	K_i	
									Calcd	Exp
CE	I		1/9	77/100	−48.28(−49.41)	−47.20	−2.22	0.0528		
		FAD	1/2	98/100	−50.42(−50.79)	−47.28	−3.51	0.0368		
	II		1/9	77/100	−48.28(−49.41)	−47.20	−2.22	0.0528	3.63×10^{-7}	
		FAD	1/2	98/100	−50.42(−50.79)	−47.28	−3.51	0.0368	3.50×10^{-7}	$(3.25 \pm 0.28) \times 10^{-9}$
		FAD + TPP	1/1	100/100	−50.38(−50.79)	−47.37	−3.43	0.0422	3.63×10^{-7}	
CS	I		1/1	100/100	−38.74(−38.79)	−37.78	−1.00	0.0803		
		FAD	1/1	100/100	−40.08(−40.12)	−39.16	−0.96	0.0816		
	II		1/14	26/100	−40.38(−41.53)	−39.63	−0.96		1.59×10^{-6}	
		FAD	1/13	42/100	−41.38(−41.63)	−40.63	−1.00		1.06×10^{-6}	$(1.27 \pm 0.17) \times 10^{-7}$
		FAD + TPP	1/10	33/100	−41.38(−41.63)	−40.71	−0.92		1.02×10^{-6}	
SM	I		1/2	98/100	−45.90(−46.07)	−43.35	−2.72	0.0658		
		FAD	1/1	100/100	−46.65(−46.82)	−44.02	−2.80	0.0642		
	II		1/10	86/100	−46.02(−46.61)	−43.85	−2.76		4.88×10^{-7}	
		FAD	1/10	87/100	−46.49(−47.07)	−44.27	−2.80		4.08×10^{-7}	$(5.08 \pm 0.21) \times 10^{-8}$
		FAD + TPP	1/8	79/100	−46.44(−46.99)	−44.31	−2.68		4.01×10^{-7}	
MM	I		1/1	100/100	−45.65(−45.77)	−43.18	−2.60	0.0535		
		FAD	1/1	100/100	−46.82(−46.99)	−44.52	−2.43	0.0644		
	II		1/30	40/100	−46.28(−46.86)	−44.15	−2.72		7.27×10^{-7}	
		FAD	1/30	47/100	−47.28(−47.82)	−45.27	−2.51		4.62×10^{-7}	$(9.40 \pm 1.30) \times 10^{-9}$
		FAD + TPP	1/27	41/100	−47.28(−47.99)	−45.48	−2.51		4.25×10^{-7}	
TB	I		1/2	99/100	−51.17(−51.34)	−47.53	−3.81	0.0496		
		FAD	1/1	100/100	−52.38(−52.55)	−48.87	−3.68	0.0482		
	II		1/13	79/100	−49.33(−50.04)	−46.11	−3.98		5.63×10^{-7}	
		FAD	1/12	78/100	−50.41(−51.30)	−47.28	−4.02		3.52×10^{-7}	$(1.14 \pm 0.07) \times 10^{-7}$
		FAD + TPP	2/13	63/100	−50.46(−51.00)	−47.12	−3.90		3.70×10^{-7}	

①Average energy of the best cluster (the low est docked energy).

Table 2　Interaction energies of yeast AHAS and sulfonylureas inhibitors in different substituents and different accep tor manners

	Inhibitor	VDW/kJ/mol^{-1}			ELEC/kJ · mol^{-1}			Substituent
			FAD	FAD + TPP		FAD	FAD + TPP	
R_2	CE	-2.97	-3.93	-4.02	-0.33	-0.29	-1.13	OCH_3
	CS	-3.39	-3.85	-3.85	-0.29	-0.29	-0.75	OCH_3
	SM	-2.26	-2.68	-2.72	0.08	0.04	0.59	CH_3
	MM	-2.85	-3.72	-3.68	-0.33	-0.25	-1.46	OCH_3
	TB	-3.05	-4.14	-4.14	-0.38	-0.25	-1.00	OCH_3
R_3	CE	-3.64	-3.60	-3.93	0.00	0.00	-0.59	C1
	CS	-2.80	-2.80	-3.22	0.00	0.00	0.96	CH_3
	SM	-3.35	-3.35	-3.51	0.00	0.00	0.54	CH_3
	MM	-3.22	-3.26	-3.56	0.00	0.00	0.75	CH_3
	TB	-2.97	-3.05	-3.22	0.00	0.04	0.84	CH_3

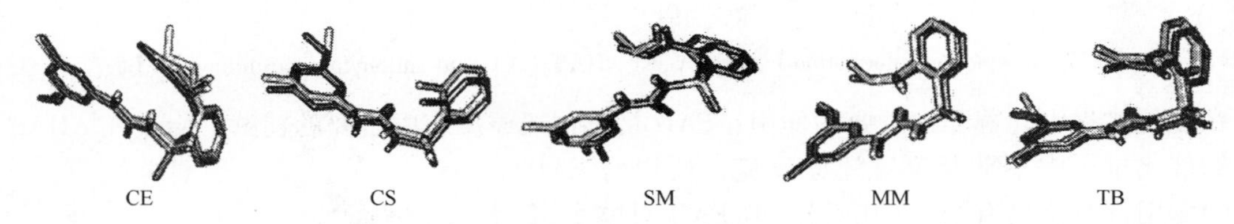

Fig. 2　Overlay of five sulfonylurea inhibitors

非常相似，表明 AHAS 在结合不同的磺酰脲类抑制剂时，其结合位点的构型未发生太大的变化，即磺酰脲上的取代基不影响其与 AHAS 的结合模式。表 1 同时列出了实验测定的抑制常数 $K_i^{[4]}$ 及分子对接预测的 K_i 值。实验测定的结合能力顺序为 CE > MM > SM > TB≈CS，假复合物的 3 种对接计算的结合能力大小顺序为 CE > SM > TB > MM > CS，CE > TB > SM≈MM > CS（受体中含 FAD）及 CE > TB > SM≈MM > CS（受体中含 FAD 和 TPP），在结合能力预测方面，分子对接和实验测定结果基本吻合。进一步说明应用 Autodock 3.0 的 LGA 算法及其评价函数，能有效地重现磺酰脲类抑制剂与 AHAS 的作用模式和结合强弱程度。

2.2.2　抑制剂与 AHAS 的相互作用

图 3（A）示出用 LigPlot 程序[9]分析得到的 5 个抑制剂与 AHAS 之间存在氢键和疏水作用的残基。可以看出，抑制剂与 AHAS 结合位点附近的残基产生很强的疏水作用。从作用模式［图 3（B）］可看出，抑制剂正好堵住通向 AHAS 活性位点的通道，杂环插入到结合位点疏水空腔的底部（活性位点所处的位置），而芳香环则处于结合位点与溶剂的界面，且芳环平面基本上与活性位点通道相垂直，有效地阻碍了底物进入酶的活性位点。杂环基团与磺酰脲桥接近于一个平面，形成一个较大的共轭体系，有利于电子传递。抑制剂 CE 的脲桥 O 及磺基 O 分别可与酶中的 ARG 380 及 LYS251 形成 2 个氢键，其他 4 个分子脲桥部分则只能与 ARG380 形成氢键，与 LYS251 形成氢键可能是由于 CE 与 AHAS 结合能力较强所致，这与晶体结构结果一致[5]。

2.2.3　FAD 和 TPP 的作用

实验结果表明[1]，AHAS 催化过程的第一步就是 TPP 中 C2 的去质子化，因而 TPP

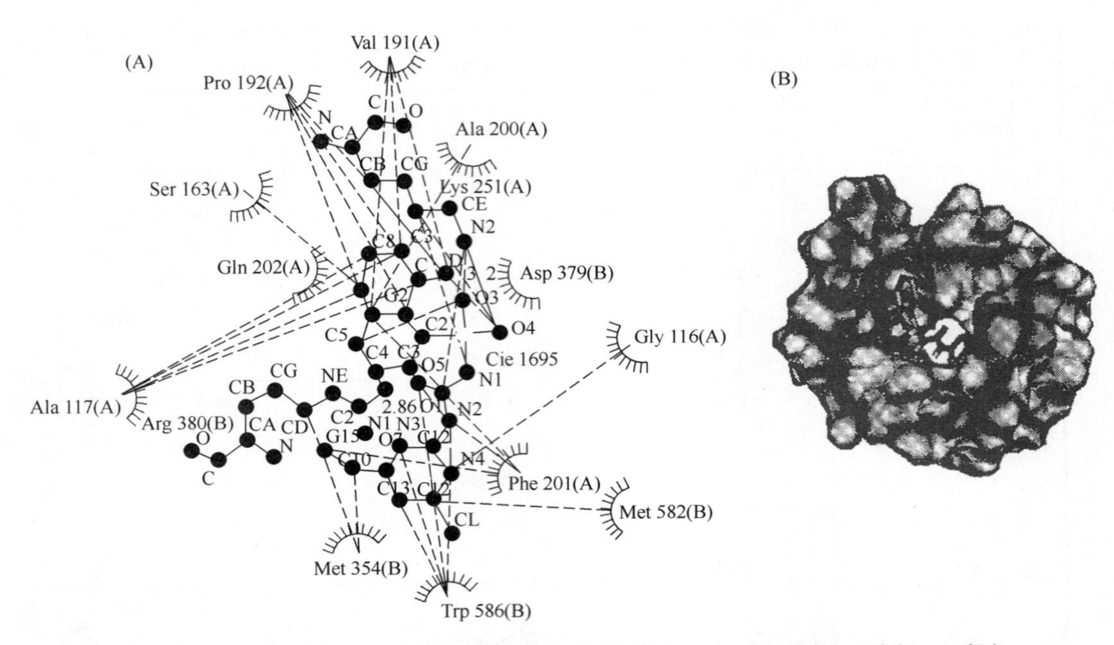

Fig. 3 Interaction information between yeast AHAS（A）and sulfonylureas inhibitors（B）

在催化过程中起着非常重要的作用；FAD 是氧化还原反应中重要的辅酶，由于在 AHAS 的催化过程中不涉及氧化还原反应，所以对 FAD 的作用目前还没有统一的解释，但 FAD 对整个蛋白质起结构支撑作用已有共识。表 2 列出了依次加入 FAD 和 TPP 后，5 个配体小分子各部分的范德华相互作用和静电相互作用能，图 4 给出了 5 个配体小分子与 FAD 和 TPP 的空间位置。对接结果表明，加入 FAD 对整个体系起稳定作用，这主要是由于抑制剂 R_2 取代基与 FAD 中平面基团 Flavin 环形成较强的范德华相互作用。加入 FAD 后，

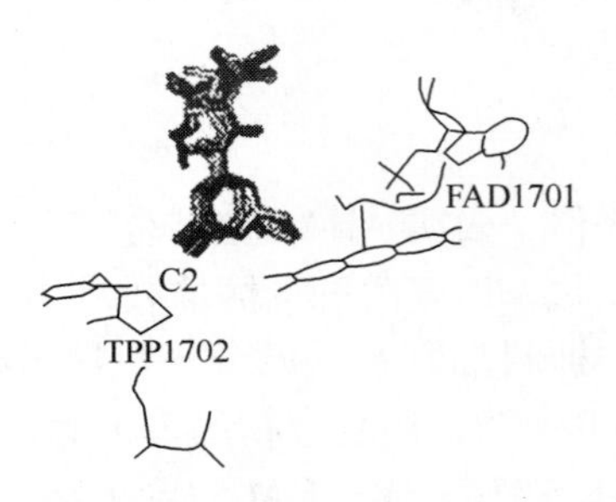

Fig. 4 Relative position of moleculek

SM 和 CS 的增加了 0.4kJ/mol，其他 3 个分子增加 1.0kJ/mol 从抑制剂结构可知，除 SM 的 R_2 为甲基外，其余均为甲氧基，甲基疏水性大于甲氧基，因此造成范德华相互作用增加的原因并非疏水作用，而应是 R_2 的空间取向，即甲氧基在空间上能更接近 FAD 的 Flavin 环，进而有利于形成较强的范德华相互作用。TPP 加入与否对总对接能影响不大。TPP 加入后主要与 R_3 取代基发生作用，使 CE 与酶静电相互作用加强，即其静电作用能从 0 变化到 -0.59kJ/mol，对其他 4 种抑制剂则减弱，即从 0 变化到 0.80kJ/mol，由于 TPP 本身不稳定[5]，因此不利的静电作用可能是加速 TPP 降解的原因。也可能正是这种原因，只有在 AHAS 与 CE 的复合晶体中可以找到完整的 TPP 分子，而 AHAS 与其他 4 种抑制剂的复合物晶体中得到的只是 TPP 降解的产物。

参考文献

[1] Duggleby R. G. , Pang S. S. J Biochem. Mol. Biol. [J] . 2000, 33: 1 -36.

[2] Pang S. S, Duggleby R. G. , Guddat L. W. . J Mol. Biol. [J] . 2002, 317: 249 -262.

[3] Pang S. S, Guddat L. W. , Duggleby R. G. . J Biol. Chem. [J] . 2003, 278: 7639 -7644.

[4] Duggleby R. G., Pang S. S., Yu H., *et al.* Eur. J. Biochem. [J]. 2003: 270: 2895-2904.
[5] ZHANG Tao (张涛), DONG Xi-Cheng (董喜成), CHEN Hai-Feng (陈海峰), *et al.* Acta Chin. Sinica (化学学报) [J]. 2006, 64: 899-905.
[6] Jenn ifer A. McCourt, Pang S. S., Guddat L. W., *et al. Biochemistry* [J]. 2005, 44: 2330-2338.
[7] Petrey D., Xiang X., Tang C. L. Protein Stric Func Genet [J]. 2003, 53: 430-435.
[8] QuACPAC; OpenEye Scientific Sofware [CP/OL]. http//www.eyesopen.com.
[9] Wallace A. C., Laskowski R. A., Thomton J M., Prot. Eng. [J], 1995, 8: 127-134.

Molecular Docking Study on Interaction Mode between Yeast AHAS and Sulfonylureas Inhibitors

Qiong Li[1], Peiquan Chen[1,2], Lan Chen[1], Hongwei Sun[1,2], Rongxin Shen[1], Chengming Lai[1], Zhengming Li[1]

([1]Department of Chemistry;
[2]Institute of Scientific Computing, Nankai University, Tianjin, 300071, China)

Abstract On the basis of the complex structures of AHAS and sulfonylureas inhibitors, systematic molecule docking study of five sulfonylureas inhibitors to AHAS were performed with autodock 3.0 package. The systematic docking results indicate that two kinds of docking modes are consistent basically each other, and closely correlated with expermental results as well. The further research reveals that the sequence of docking results were not affected *via* the appearance of FAD and TPP. Due to the VDW interaction between R_2 substituent and flav in ring of FAD, the docking complex of inhibitors to AHAS became more stable. It was assumed that the unfavorable electrostatic interaction between the inhibitors and TPP may be the factor which accelerates the degradation of TPP.

Key words yeast AHAS; docking; sulfonylureas; interaction mode

High Throughput Screening Under Zinc – Database and Synthesis a Dialkylphosphinic Acid as a Potential Kari Inhibitor*

Xinghai Liu, Peiquan Chen, Wancheng Guo, Suhua Wang, Zhengming Li

(The Research Institute of Elemento – Organic Chemistry, The State – Key Laboratory of Elemento – Organic Chemistry, National Pesticide Engineering Research Center, Nankai University, Tianjin, China)

Abstract A novel phosphorus derivative was synthesized through HTVS. The title compounds were confirmed by MS, ^{1}H NMR, ^{31}P NMR. The DOCK was also studied.

Key words DOCK; KARI; phosphorus derivatives; synthesis

1 INTRODUCTION

Ketol – acid reductoisomerase (KARI; EC 1. 1. 1. 86) [1] is an attractive target for argochemical discovery because it catalyzes the second important step in the biosynthesis of the branched chain amino acids that exist in higher plants only and not in animals. Thus, it is an ideal target from which to design nontoxic KARI – inhibitors as potential novel herbicides. The reaction catalyzed by KARI is shown in Figure 1.

In particular, as high throughput screening considerably increased the numbers of new chemical entities to be studied, the expectations to find new bio – active molecules among these compounds were high at the beginning. However, high throughput screenings (HTS) yielded so far the expected success, and therefore virtual screening approaches emerged and largely evolved. Today, high throughput virtual screening (HTVS) is useful in argochemical innovation.

2 METHODS AND EXPERIMENTS

2.1 HTVS

Based on the reported crystal structure of complexes of the enzyme KARI, 1000 new molecules were predicted with high affinity for KARI from ZINC – database (Drug – Like) searching, using program eHiTS. [5] The computational flow chart for this study was shown as in Figure 2. Among them, many compounds contain phosphorus structures. A dialkylphosphinic acid was noticed which is analogy of HOE 704 (Figure 1).

* Reprinted from *Phosphorus, sulfur and silicon and the Related Elements*, 2008, 183 (2 – 3): 775 – 778. This work was funded by the National Basic Research Program of China (No. 2003CB114406), National Natural Science Foundation Key Project of China (No. 20432010), and The High Performance Computing Project of Tianjin Ministry of Science and Technology of China (No. 043185111 – 5).

Figure 1 Reaction catalyzed by KARI

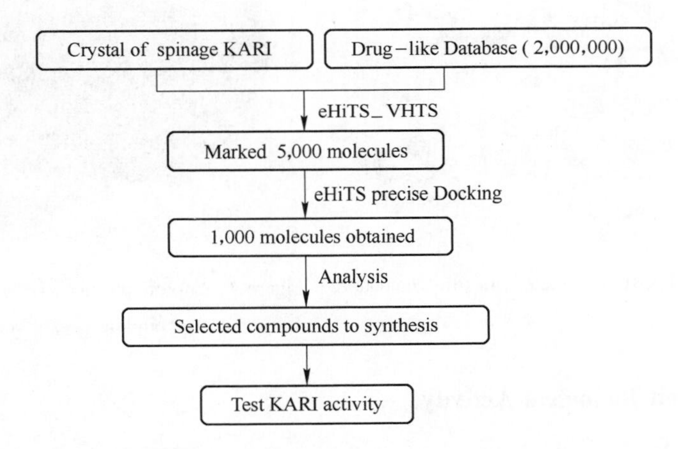

Figure 2 Computational and experimental flow chart for this study

2.2 Synthesis

The dialkylphosphinic acid was synthesized according the reference[6] (Scheme 1). The physical chemistry data were accordance with the reference.

Scheme 1 The synthesis route of the title compound

3 RESULTS AND DISCUSSION

3.1 HTVS

A dialkylphosphinic acid was docked with KARI and estimated the pKi = −5. 235. The interaction pattern(Figures 3 and 4) between the titled compound and KARI was same as IpOHA(Figure 1). The Mg^{2+} ions and the crystal water molecules, inhibitor in the active site are necessary for constructing and maintaining the conformation of the active site and thus the binding affinity of ligand or the enzymatic activity of title compound and IpOHA in complex with spinage KARI.

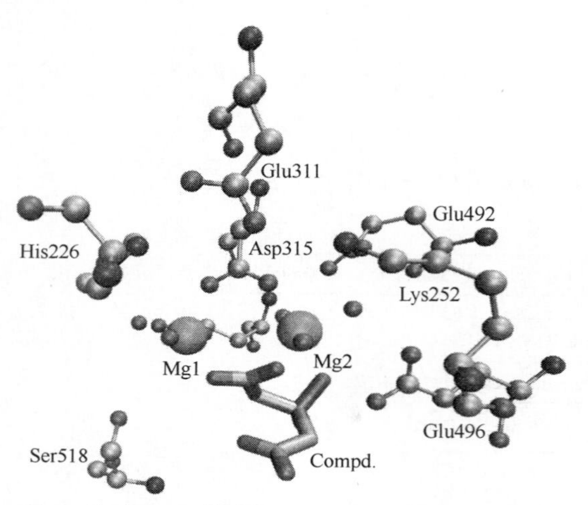

Figure 3 Spinage KARI active site with title compound

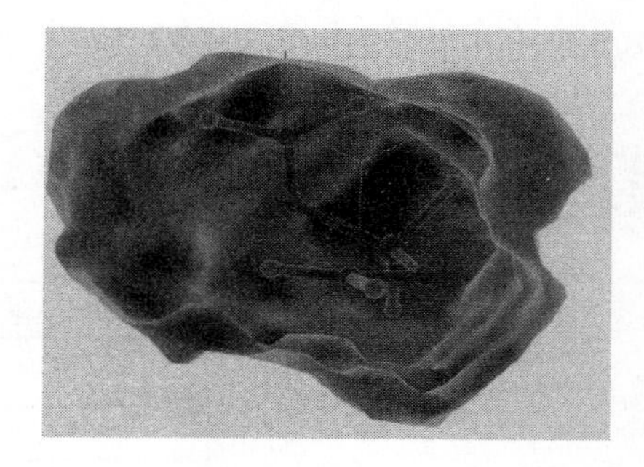

Figure 4 Superimposition of the docked modeled into the binding pocket conformation

3.2 Synthesis and Biological Activity

The synthesis route of the title compound is shown in the Scheme 1. The method was not modified. The dialkylphosphinic acid was synthesized for further bioassay experiments.

4 CONCLUSION

A new potential lead compound of KARI inhibitor was found by virtual and laboratory screening for further optimization.

REFERENCES

[1] R. Dumas, V. Biou, F. Halgand, R. Douce, and R. G. Duggleby, *Acc. Chem. Res.* ,34,399(2001).

[2] A. Schulz, P. Spönemann, H. Köcher, and F. Wengenmayer FEBS, *Lett.* ,238,375(1988).

[3] A. Aulabaugh and J. V. Schloss *Biochemistry*,29,2824(1990).

[4] Y. T. Lee, H. T. Ta, and R. G . Duggleby, *Plant Science*,168,1035(2005).

[5] SimBioSys Inc. ,http://www. simbiosys. ca/ehits/.

[6] P. Majewski, *Phosphorus*, *Sulfur*, *Silicon*,45,151(1989).

N－(4′－取代嘧啶－2′－基) －2－甲氧羰基－5－(取代)苯甲酰胺基苯磺酰脲化合物的合成及除草活性研究*

王美怡　郭万成　兰　峰　李永红　李正名

（南开大学元素有机化学研究所，元素有机化学国家重点实验室，天津　300071）

摘　要　为进一步寻找高效、安全和对环境更加友好的除草剂，以商品化除草剂单嘧磺酯为研究基础，对其结构中的苯环5－位取代基作了结构修饰，合成了26个未见文献报道的新型 *N*－(4′－取代嘧啶－2′－基)－2－甲氧羰基－5－苯甲酰胺基苯磺酰脲化合物，通过 ^{1}H NMR、质谱及元素分析确定了化合物的结构。经油菜平皿法及盆栽法测试了所有化合物的除草活性，结果表明，当苯环5－位取代基为苯甲酰胺时，活性较好，其对双子叶植物的除草活性与商品化的甲嘧磺隆相当。

关键词　苯磺酰脲　5－位取代苯环　合成　除草活性

磺酰脲类除草剂自20世纪80年代问世以来，以其高效、低毒、高选择性等卓越特性在世界各地得到了广泛应用。但由于其活性极高（用量一般为2～75g/ha）、选择性极强，因此在使用不当的情况下，某些化合物在土壤中的少量残留会对后茬敏感作物产生药害[1,2]，特别是对于不同作物连作的地区，此问题更加突出。

磺酰脲类除草剂的药效和残留药害不仅与土壤的pH值、温度、湿度和土壤的有机质含量等外部因素有关，更主要的是与化合物本身的结构密切相关[3]。在1995年的布莱顿植保会议上介绍了5－取代的氟啶嘧磺隆（5－取代是指在芳环的5位取代），它继承了磺酰脲类除草剂用量小、杀草谱广、选择性高、对动物毒性低等优点，并且以降解速度快的特点得到了广泛关注[4～6]。为考察芳环5位取代基变化对此类磺酰脲化合物除草活性的影响，进一步寻找更加高效、安全和对环境友好的除草剂，本文在高效除草剂品种单嘧磺酯的结构基础上[7～9]，对苯环5位取代基作了结构修饰，合成了26个未见文献报道的新型 *N*－（4′－取代嘧啶－2′－基）－2－甲氧羰基－5－取代苯甲酰胺基苯磺酰脲化合物，并对其除草活性进行了研究。目标化合物的合成路线如Scheme 1所示。

1　实验部分

1.1　试剂与仪器

X－4数字显示显微熔点测定仪（北京泰克仪器有限公司），温度计未校正；Yanaco CHN CORDER MT－3元素分析仪；BRUKER AVANCE－300 MHz和Varian Mercury Vx400 MHz核磁共振仪，DMSO－d_6 为溶剂，TMS为内标；Thermo Finigan LCQ Advantage液－质联用仪。

* 原发表于《有机化学》，2008，28（4）：649－656。国家自然科学基金（No. 20432010）、国家“973”（No. 2003CB114406）资助项目。

R^1 = H (**a**), 4-Cl(**b**), 2-F(**c**), 4-F(**d**), 2-Cl(**e**), 2-NO_2(**f**), 4-NO_2(**g**), 2-OCH_3(**h**) 4-OCH_3(**i**), 4-Br(**j**), 3-F(**k**), 2,4-Cl_2(**l**), 3, 4-Cl_2(**m**), 4-CH_3(**n**), 4-C_2H_5(**o**)

3a R^1=H, R^2=CH_3; **3b** R^1=H, R^2=OCH_3; **3c** R^1=4-Cl, R^2=CH_3; **3d** R^1=4-Cl, R^2=OCH_3;
3e R^1=2-F, R^2=CH_3; **3f** R^1=2-F, R^2=OCH_3; **3g** R^1=4-F, R^2=CH_3; **3h** R^1=4-F, R^2=OCH_3;
3i R^1=2-Cl, R^2=CH_3; **3j** R^1=2-Cl, R^2=OCH_3; **3k** R^1=2-NO_2, R^2=CH_3; **3l** R^1=2-NO_2, R^2=OCH_3;
3m R^1=4-NO_2, R^2=CH_3; **3n** R^1=4-NO_2, R^2=OCH_3; **3o** R^1=2-OCH_3, R^2=CH_3; **3p** R^1=2-OCH_3, R^2=OCH_3;
3q R^1=4-OCH_3, R^2=CH_3; **3r** R^1=4-OCH_3, R^2=OCH_3; **3s** R^1=4-Br, R^2=CH_3; **3t** R^1=4-Br, R^2=OCH_3;
3u R^1=3-F, R^2=CH_3; **3v** R^1=3-F, R^2=OCH_3; **3w** R^1=2,4-Cl_2, R^2=CH_3; **3x** R^1=3,4-Cl_2, R^2=CH_3;
3y R^1=4-CH_3, R^2=CH_3; **3z** R^1=4-C_2H_5, R^2=CH_3

Scheme 1

所有试剂均为分析纯，溶剂均经过常规处理后使用。参照文献［10，11］制备了2－甲氧酯基5－氨基苯磺酰胺（1），参照文献［12］制备了4－取代嘧啶－2－基氨基甲酸苯酯。

1.2　2－甲氧酯基－5－取代苯甲酰胺基苯磺酰胺（2）的合成通法

将2mmol化合物1溶于15mL四氢呋喃中，冷却至适当的温度，保持此温度搅拌下缓慢滴加2mmol取代苯甲酰氯，滴毕升温至回流4～8h，薄层色谱（TLC）监测反应终点。过滤反应液，将滤液旋干后用丙酮和乙酸乙酯重结晶得产物**2**；滤饼用蒸馏水洗涤后过滤，干燥，亦可得到化合物**2**。合并两部分产物即为总收率。

2a：苯甲酰氯加入时温度低于0℃，产物为白色晶体，产率72%，m. p. 238～240℃。**2b**：对氯苯甲酰氯加入时温度低于5℃，产物为白色晶体，产率65%，m. p. 244～246℃。**2c**：邻氟苯甲酰氯加入时温度低于0℃，产物为黄色晶体，产率82%，m. p. 188～189℃。**2d**：对氟苯甲酰氯加入时温度低于0℃，产物为黄色晶体，产率89%，m. p. 249～251℃。**2e**：邻氯苯甲酰氯加入时温度低于5℃，产物为黄色晶体，产率83%，m. p. 209～210℃。**2f**：邻硝基苯甲酰氯加入时温度低于0℃，产物为黄色晶体，产率62%，m. p. 215～216℃。**2g**：对硝基苯甲酰氯加入时温度低于0℃，产物为白色晶体，产率61%，m. p. 264～265℃。**2h**：邻甲氧基苯甲酰氯加入时温度低于10℃，产物为白色晶体，产率88%，m. p. 231～232℃。**2i**：对甲氧基苯甲酰氯加入时温度低于10℃，产物为白色晶体，产率85%，m. p. 261～262℃。**2j**：对溴苯甲酰氯加入时温度低于5℃，产物为白色晶体，产率78%，m. p. 244～246℃。**2k**：间氟苯甲酰氯加入时温度低于0℃，产物为白色晶体，产率77%，m. p. 248～249℃。**2l**：2，4－二氯苯甲酰氯加入时温度低于0℃，产物为白色晶体，产率83%，m. p. 219～220℃。**2m**：3，4－二氯苯甲酰氯加入时温度低于0℃，产物为白色晶体，产率82%，m. p. 252～253℃。**2n**：对甲基苯甲酰氯加入时温度低于10℃，产物为白色晶体，产率86%，m. p. 245～246℃。**2o**：对乙基苯甲酰氯加入时温度低于10℃，产物为白色晶体，产率85%，m. p. 236～237℃。

1.3 N-(4′-取代嘧啶-2′-基)-2-甲氧羰基-5-取代苯甲酰胺基苯磺酰脲(3)的合成通法

在50mL圆底烧瓶中加入1mmol化合物**2**，12mL无水乙腈，磁力搅拌下加入1mmol（4-取代嘧啶-2-基）-氨基甲酸苯酯，溶解后滴入由2mL乙腈稀释的1.1mmol DBU（1，8-二氮杂双环［5，4，0］十一-7-烯）。加入时可以见到颜色变深。在室温下继续搅拌4h后，加入20mL水使反应混合物清亮，滤去不溶物，滤液用20%的盐酸酸化至pH=2~3，析出固体，过滤，依次用水、乙醚、乙腈洗涤，抽干，收集固体，干燥，得目标化合物**3**。如果酸化后不能析出沉淀，则用适量二氯甲烷萃取，合并有机相，无水硫酸钠干燥，脱溶，用乙腈重结晶可得到最后目标产物**3**。

3a：白色晶体，产率56.6%，m.p.227~228℃；^{1}H NMR（DMSO-d_6，300MHz）δ：2.50（s，3H，pyrim-CH_3），3.81（s，3H，CO_2CH_3），7.14（d，J=5.4Hz，1H，pyrim-H^5），7.58（d，J=8.4Hz，2H，Ar—H），7.63（d，J=8.4Hz，1H，Ar—H），7.81（d，J=8.4Hz，1H，Ar—H），8.00~8.03（m，2H，Ar—H），8.35（dd，J=8.4Hz，1.8Hz，1H，Ar—H），8.59（d，J=5.4Hz，1H，pyrim-H^6），8.68（s，1H，Ar—H），10.71（s，1H，NHCOAr），10.89（s，1H，CON**H**Het），12.98（brs，1H，SO_2NH）；ESI-MS（positive ion）m/z：470（M^++1）。Anal. calcd for $C_{21}H_{19}N_5O_6S$：C 53.73，H 4.08，N 14.92；found C53.61，H 4.11，N 15.04。

3b（DBU盐）：白色固体，产率62.4%，m.p.151~152℃；^{1}H NMR（DMSO-d_6，400MHz）δ：1.58~1.64（m，6H，DBU-H^3，H^4，H^5），1.85~1.89（m，2H，DBU-H^{10}），2.61（d，J=8.7Hz，2H，DBU-H^6），3.23（t，J=5.2Hz，2H，DBU-H^9），3.45（t，J=5.5Hz，2H，DBU-H^{11}），3.53（d，J=7.6Hz，2H，DBU-H^2），3.73（s，3H，pyrim-OCH_3），3.81（s，3H，CO_2CH_3），6.30（d，J=5.6Hz，1H，pyrim-H^5），7.32（d，J=8.4Hz，1H，Ar—H），7.53~7.60（m，3H，Ar—H），8.01（d，J=8.4Hz，2H，Ar—H），8.13（d，J=5.6Hz，1H，pyrim-H^6），8.29（s，2H，Ar—H），10.59（s，1H，NHCOAr）；ESI-MS（positive ion）m/z：638（M^++1）。Anal. calcd for $C_{30}H_{35}N_7O_7S$：C 56.50，H 5.53，N 15.37；found C 56.34，H 5.67，N 15.11。

3c：黄色晶体，产率81.6%，m.p.234~236℃；^{1}H NMR（DMSO-d_6，300MHz）δ：2.50（s，3H，pyrim-CH_3），3.81（s，3H，CO_2CH_3），7.14（d，J=5.4Hz，1H，pyrim-H^5），7.64（d，J=8.4Hz，2H，Ar—H），7.81（d，J=8.4Hz，1H，Ar—H），8.05（d，J=8.4Hz，2H，Ar—H），8.35（dd，J=8.4，1.8Hz，1H，Ar—H），8.58（d，J=5.4Hz，1H，pyrim-H^6），8.64（d，J=1.8Hz，1H，Ar—H），10.70（s，1H，NHCOAr），10.93（s，1H，CON**H**Het），12.98（brs，1H，SO_2NH）；ESI-MS（positive ion）m/z：504（M^++1）。Anal. calcd for $C_{21}H_{18}ClN_5O_6S$：C 50.05，H 3.60，N 13.90；found C 49.80，H 3.64，N 13.72。

3d：白色晶体，产率76.7%，m.p.230~231℃；^{1}H NMR（DMSO-d_6，300MHz）δ：3.81（s，3H，pyrim-OCH_3），3.97（s，3H，CO_2CH_3），6.68（d，J=5.4Hz，1H，pyrim-H^5），7.65（d，J=8.4Hz，2H，Ar—H），7.78（d，J=8.4Hz，1H，Ar—H），8.05（d，J=8.4Hz，2H，Ar—H），8.31（dd，J=8.4，1.8Hz，1H，Ar—H），8.43（d，J=5.4Hz，1H，Pyrim-H^6），8.63（d，J=1.8Hz，1H，Ar—H），

10.72 (s, 1H, NHCOAr), 10.92 (s, 1H, CON**H**Het); ESI - MS (positive ion) *m/z*: 520 (M^+ + 1)。Anal. calcd for $C_{21}H_{18}ClN_5O_7S$: C 48.51, H 3.49, N 13.47; found C 48.26, H 3.79, N 13.59。

3e：白色晶体，产率 61.0%，m. p. 178 ~ 180℃；^{1}H NMR (DMSO - d_6, 300MHz) δ: 2.50 (s, 3H, pyrim - CH_3), 3.81 (s, 3H, CO_2CH_3), 7.13 (d, *J* = 5.4Hz, 1H, pyrim - H^5), 7.33 ~ 7.41 (m, 2H, Ar—H), 7.58 ~ 7.66 (m, 1H, Ar—H), 7.71 ~ 7.76 (m, 1H, Ar—H), 7.81 (d, *J* = 8.4Hz, 1H, Ar—H), 8.22 (dd, *J* = 8.4Hz, 1.8Hz, 1H, Ar—H), 8.57 ~ 8.60 (m, 2H, Ar—H, Pyrim—H^6), 10.68 (s, 1H, NHCOAr), 11.08 (s, 1H, CON**H**Het), 12.98 (brs, 1H, SO_2NH); ESI - MS (positiveion) *m/z*: 488 (M^+ + 1)。Anal. calcd for $C_{21}H_{18}FN_5O_6S$: C 51.74, H 3.72, N 14.37; found C 51.52, H 3.75, N 14.33。

3f：白色晶体，产率 58.2%，m. p. 195 ~ 197℃；^{1}H NMR (DMSO - d_6, 300MHz) δ: 3.81 (s, 3H, pyrim - OCH_3), 3.97 (s, 3H, CO_2CH_3), 6.68 (d, *J* = 5.4Hz, 1H, pyrim - H^5), 7.37 (d, *J* = 8.4Hz, 2H, Ar—H), 7.59 ~ 7.67 (m, 1H, Ar—H), 7.72 (d, *J* = 8.4, 1H, Ar—H), 7.79 (d, *J* = 8.4Hz, 1H, Ar—H), 8.21 (dd, *J* = 8.4, 1.8Hz, 1H, Ar—H), 8.43 (d, *J* = 5.4Hz, 1H, pyrim - H^6), 8.58 (s, 1H, Ar—H) 10.75 (s, 1H, NHCOAr), 11.08 (s, 1H, CON**H**Het), 12.98 (brs, 1H, SO_2NH); ESI - MS (negative ion) *m/z*: 502 (M^+ - 1)。Anal. calcd for $C_{21}H_{18}FN_5O_7S$: C 50.10, H 3.60, N 13.91; found C 49.86, H 3.70, N 13.71。

3g：黄色晶体，产率 62.3%，m. p. 226 ~ 227℃；^{1}H NMR (DMSO - d_6, 300MHz) δ: 2.23 (s, 3H, pyrim - CH_3), 3.86 (s, 3H, CO_2CH_3), 7.14 (d, *J* = 5.4Hz, 1H, pyrim - H^5), 7.40 (d, *J* = 8.4Hz, 2H, Ar—H), 7.80 (d, *J* = 8.4Hz, 1H, Ar—H), 8.05 (d, *J* = 8.4Hz, 2H, Ar—H), 8.33 (dd, *J* = 8.4, 1.8Hz, 1H, Ar—H), 8.59 (d, *J* = 5.4Hz, 1H, pyrim - H^6), 8.66 (d, *J* = 1.8Hz, 1H, Ar—H), 10.71 (s, 1H, NHCOAr), 10.90 (s, 1H, CON**H**Het); ESI - MS (negative ion) *m/z*: 486 (M^+ - 1)。Anal. calcd for $C_{21}H_{18}FN_5O_6S$: C 51.74, H 3.72, N 14.37; found C 51.62, H 3.66, N 14.32。

3h：白色晶体，产率 71.6%，m. p. 220 ~ 221℃；^{1}H NMR (DMSO - d_6, 300MHz) δ: 3.81 (s, 3H, pyrim - OCH_3), 3.97 (s, 3H, CO_2CH_3), 6.67 (d, *J* = 5.4Hz, 1H, pyrim - H^5), 7.40 (d, *J* = 8.4Hz, 2H, Ar—H), 7.79 (d, *J* = 8.4Hz, 1H, Ar—H), 8.11 (d, *J* = 8.4Hz, 2H, Ar—H), 8.32 (dd, *J* = 8.4Hz, 1.8Hz, 1H, Ar—H), 8.44 (d, *J* = 5.4Hz, 1H, pyrim - H^6), 8.64 (d, *J* = 1.8Hz, 1H, Ar—H), 10.73 (s, 1H, NHCOAr), 10.88 (s, 1H, CON**H**Het), 12.98 (brs, 1H, SO_2NH); ESIMS (positive ion) *m/z*: 504 (M^+ + 1)。Anal. calcd for $C_{21}H_{18}FN_5O_7S$: C 50.10, H 3.60, N 13.91; found C 49.87, H 3.74, N 13.83。

3i：黄色晶体，产率 75.2%，m. p. 175 ~ 177℃；^{1}H NMR (DMSO - d_6, 400MHz) δ: 2.26 (s, 3H, pyrim - CH_3), 3.85 (s, 3H, CO_2CH_3), 6.53 (d, *J* = 5.4Hz, 1H, pyrim - H^5), 7.50 ~ 7.55 (m, 2H, Ar—H), 7.61 ~ 7.65 (m, 2H, Ar—H), 7.78 (d, *J* = 8.4Hz, 1H, Ar—H), 7.97 (dd, *J* = 8.4, 1.8Hz, 1H, Ar—H), 8.12 (d, *J* = 5.4Hz, 1H, pyrim - H^6), 8.50 ~ 8.51 (d, *J* = 1.8Hz, 1H, Ar—H), 10.71 (s,

1H，NHCOAr)，11.18（s，1H，CONHHet)；ESI - MS（positive ion）*m/z*：504（M^+ +1)。Anal. calcd for $C_{21}H_{18}ClN_5O_6S$：C 50.05，H 3.60，N 13.90；found C 49.84，H 3.80，N 14.14。

3j：白色晶体，产率 76.7%，m. p. 195 ~ 197℃；^{1}H NMR（DMSO - d_6，400MHz）δ：3.81（s，3H，pyrim—OCH_3)，3.97（s，3H，CO_2CH_3)，6.69（d，*J* = 5.4Hz，1H，pyrim - H^5)，7.55 ~ 7.57（m，2H，Ar—H)，7.79 ~ 7.82（m，2H，Ar—H)，7.98（d，*J* = 8.4Hz，1H，Ar—H)，8.19（d，*J* = 8.4Hz，1H，Ar—H)，8.44（d，*J* = 5.4Hz，1H，pyrim - H^6)，8.59（s，1H，Ar—H)，10.77（s，1H，NHCOAr)，11.18（s，1H，CONHHet)，12.99（brs，1H，SO_2NH)；ESI - MS（positive ion）*m/z*：520（M^+ +1)。Anal. calcd for $C_{21}H_{18}ClN_5O_7S$：C 48.51，H 3.49，N 13.47；found C 48.62，H 3.31，N 13.40。

3k（DBU 盐)：白色晶体，产率 65.7%，m. p. 163 ~ 165℃；^{1}H NMR（DMSO - d_6，400MHz）δ：1.58 ~ 1.63（m，6H，DBU - H^3，H^4，H^5)，1.85 ~ 1.89（m，2H，DBU - H^{10})，2.50（s，3H，pyrim - CH_3)，2.61（d，*J* = 8.7Hz，2H，DBU - H^6)，3.23（t，*J* = 5.2Hz，2H，DBU - H^9)，3.45（t，*J* = 5.5Hz，2H，DBU - H^{11})，3.53（d，*J* = 7.6Hz，2H，DBU - H^2)，3.73（s，3H，CO_2CH_3)，6.76（d，*J* = 5.4Hz，1H，pyrim - H^5)，7.35（d，*J* = 8.4Hz，1H，Ar—H)，7.76 ~ 7.94（m，3H，Ar—H)，8.13 ~ 8.16（m，2H，Ar—H)，8.27（d，*J* = 5.4Hz，1H，pyrim - H^6)，8.30（s，1H，Ar—H)；ESI - MS（positive ion）*m/z*：667（M^+ +1)。Anal. calcd for $C_{30}H_{34}N_8O_8S$：C 54.05，H 5.14，N 16.81；found C 54.15，H 5.33，N 16.68。

3l（DBU 盐)：白色晶体，产率 71.2%，m. p. 164 ~ 166℃；^{1}H NMR（DMSO - d_6，400MHz）δ：1.59 ~ 1.64（m，6H，DBU - H^3，H^4，H^5)，1.85 ~ 1.89（m，2H，DBU - H^{10})，2.61（d，*J* = 8.7Hz，2H，DBU - H^6)，3.23（t，*J* = 5.2Hz，2H，DBU - H^9)，3.45（t，*J* = 5.5Hz，2H，DBU - H^{11})，3.52（d，*J* = 7.6Hz，2H，DBU - H^2)，3.73（s，3H，pyrim - OCH_3)，3.82（s，3H，CO_2CH_3)，6.30（d，*J* = 5.4Hz，1H，pyrim - H^5)，7.35（d，*J* = 8.4Hz，1H，Ar—H)，7.74 ~ 7.94（m，3H，Ar—H)，8.11 ~ 8.16（m，2H，Ar—H)，8.28（d，*J* = 5.4Hz，1H，pyrim—H^6)，8.31（s，1H，Ar—H)；ESI - MS（positive ion）*m/z*：683（M^+ +1)。Anal. calcd for $C_{30}H_{34}N_8O_9S$：C 52.78，H 5.02，N 16.41；found C 52.81，H 5.07，N 16.32。

3m（DBU 盐)：黄色晶体，产率 66.8%，m. p. 189 ~ 191℃；^{1}H NMR（DMSO - d_6，400MHz）δ：1.58 ~ 1.63（m，6H，DBU - H^3，H^4，H^5)，1.85 ~ 1.89（m，2H，DBU - H^{10})，2.50（s，3H，pyrim - CH_3)，2.61（d，*J* = 8.7Hz，2H，DBU - H^6)，3.23（t，*J* = 5.2Hz，2H，DBU - H^9)，3.45（t，*J* = 5.5Hz，2H，DBU - H^{11})，3.53（d，*J* = 7.6Hz，2H，DBU - H^2)，3.66（s，3H，CO_2CH_3)，7.39（d，*J* = 5.4Hz，1H，pyrim - H^5)，8.08（d，*J* = 8.4Hz，1H，Ar—H)，8.12（d，*J* = 8.4Hz，1H，Ar—H)，8.18 ~ 8.25（m，4H，Ar—H)，8.36 ~ 8.38（m，2H，Ar—H，pyrim - H^6)，10.92（s，1H，NHCOAr)；ESI - MS（positive ion）*m/z*：667（M^+ +1)。Anal. calcd for $C_{30}H_{34}N_8O_8S$：C 54.05，H 5.14，N 16.81；found C 54.13，H 5.28，N 16.99。

3n（DBU 盐)：白色晶体，产率 64.1%，m. p. 193 ~ 195℃；^{1}H NMR（DMSO - d_6，400MHz）δ：1.60 ~ 1.65（m，6H，DBU - H^3，H^4，H^5)，1.86 ~ 1.90（m，2H，DBU

$-H^{10}$), 2.62 (d, $J=8.7Hz$, 2H, DBU $-H^6$), 3.24 (t, $J=5.2Hz$, 2H, DBU $-H^9$), 3.46 (t, $J=5.5Hz$, 2H, DBU $-H^{11}$), 3.53 (d, $J=7.6Hz$, 2H, DBU $-H^2$), 3.74 (s, 3H, pyrim $-OCH_3$), 3.81 (s, 3H, CO_2CH_3), 6.30 (d, $J=5.4Hz$, 1H, pyrim $-H^5$), 7.37 (d, $J=8.4Hz$, 1H, Ar—H), 8.13 (d, $J=8.4Hz$, 1H, Ar—H), 8.21 ~ 8.40 (m, 5H, Ar—H, pyrim—H^6), 8.54 (s, 1H, Ar—H); ESI-MS (positive ion) m/z: 683 ($M^+ +1$)。Anal. calcd for $C_{30}H_{34}N_8O_9S$: C 52.78, H 5.02, N 16.41; found C 53.05, H 4.92, N 16.11。

3o：黄色晶体，产率 78.2%，m. p. 198 ~ 199℃；^1H NMR (DMSO $-d_6$, 400MHz) δ: 2.50 (s, 3H, pyrim $-CH_3$), 3.81 (s, 3H, CO_2CH_3), 3.88 (s, 3H, Ar—OCH_3), 7.08 (d, $J=5.4Hz$, 1H, pyrim $-H^5$), 7.14 ~ 7.20 (m, 2H, Ar—H), 7.51 ~ 7.54 (m, 1H, Ar—H), 7.59 ~ 7.62 (m, 1H, Ar—H), 7.78 (d, $J=8.4Hz$, 1H, Ar—H), 8.16 (d, $J=5.4Hz$, 1H, Pyrim $-H^6$), 8.60 (d, $J=8.4Hz$, 1H, Ar—H), 8.65 (s, 1H, Ar—H), 10.72 (s, 1H, NHCOAr), 10.82 (s, 1H, CON**H**Het), 12.94 (brs, 1H, SO_2NH); ESI-MS (positive ion) m/z: 500 ($M^+ +1$)。Anal. calcd for $C_{22}H_{21}N_5O_7S$: C 52.90, H 4.24, N 14.02; found C 52.81, H 4.35, N 14.04。

3p：白色晶体，产率 80.5%，m. p. 202 ~ 204℃；^1H NMR (DMSO $-d_6$, 400MHz) δ: 3.80 (s, 3H, Pyrim—OCH_3), 3.85 (s, 3H, $ArOCH_3$), 3.97 (s, 3H, CO_2CH_3), 6.69 (d, $J=5.4Hz$, 1H, pyrim—H^5), 7.58 ~ 7.66 (m, 4H, Ar—H), 7.74 ~ 7.78 (m, 1H, Ar-H), 8.08 (d, $J=8.4Hz$, 1H, Ar—H), 8.12 ~ 8.16 (m, 1H, Ar-H), 8.43 (d, $J=5.4Hz$, 1H, pyrim $-H^6$), 8.64 (s, 1H, Ar—H) 10.75 (s, 1H, NHCOAr), 10.82 (s, 1H, CON**H**Het), 12.93 (brs, 1H, SO_2NH); ESI-MS (positiveion) m/z: 516 ($M^+ +1$)。Anal. calcd for $C_{22}H_{21}N_5O_8S$: C 51.26, H 4.11, N 13.59; found C 50.98, H 4.13, N 13.61。

3q：黄色晶体，产率 58.8%，m. p. 230℃ (dec)；^1H NMR (DMSO $-d_6$, 400MHz) δ: 2.50 (s, 3H, pyrim $-CH_3$), 3.51 (s, 3H, $ArOCH_3$), 3.84 (s, 3H, CO_2CH_3), 6.45 (d, $J=5.4Hz$, 1H, pyrim $-H^5$), 7.08 (d, $J=8.4Hz$, 2H, Ar—H), 7.23 (s, 1H, Ar—H), 7.74 (d, $J=8.4Hz$, 2H, Ar—H), 7.89 (dd, $J=8.4$, 1.8Hz, 1H, Ar—H), 8.16 (d, $J=5.4Hz$, 1H, pyrim $-H^6$), 8.63 (d, $J=1.8Hz$, 1H, Ar—H), 10.60 (s, 1H, NHCOAr), 10.72 (s, 1H, CON**H**Het), 12.93 (brs, 1H, SO_2NH); ESI-MS (positive ion) m/z: 500 ($M^+ +1$)。Anal. calcd for $C_{22}H_{21}N_5O_7S$: C 52.90, H 4.24, N 14.02; found C 52.81, H 3.96, N 14.24。

3r：白色晶体，产率 62.2%，m. p. 243 ~ 245℃；^1H NMR (DMSO $-d_6$, 300MHz) δ: 3.83 (s, 3H, Pyrim—OCH_3), 3.85 (s, 6H, $ArOCH_3$, CO_2CH_3), 6.12 (d, $J=5.4Hz$, 1H, Pyrim—H^5), 7.09 (d, $J=8.4Hz$, 2H, Ar—H), 7.23 (s, 1H, Ar—H), 7.76 (d, $J=8.4Hz$, 2H, Ar—H), 7.98 ~ 8.02 (m, 2H, Ar—H), 8.11 ~ 8.15 (m, 1H, Ar—H), 8.54 (d, $J=5.4Hz$, 1H, pyrim $-H^6$), 10.56 (s, 1H, NHCOAr), 10.62 (s, 1H, CON**H**Het), 12.93 (brs, 1H, SO_2NH); ESI-MS (positiveion) m/z: 516 ($M^+ +1$)。Anal. calcd for $C_{22}H_{21}N_5O_8S$: C 51.26, H 4.11, N 13.59; found C 51.50, H 4.21, N 13.77。

3s：白色晶体，产率 85.5%，m. p. 226 ~ 227℃；^{1}H NMR（DMSO - d_6，400MHz）δ：2.50（s，3H，pyrim—CH_3），3.82（s，3H，CO_2CH_3），7.15（d，J = 5.4Hz，1H，pyrim—H^5），7.49 ~ 7.50（m，1H，Ar—H），7.62 ~ 7.64（m，1H，Ar—H），7.98（d，J = 8.4Hz，1H，Ar—H），8.16（d，J = 8.4Hz，1H，Ar—H），8.35（dd，J = 8.4Hz，1.8Hz，1H，Ar—H），8.47（s，1H，Ar—H），8.59（d，J = 5.4Hz，1H，pyrim—H^6），8.66（s，1H，Ar—H），10.74（s，1H，NHCOAr），10.95（s，1H，CON**H**Het），13.00（brs，1H，SO_2NH）；ESI - MS（positive ion）m/z：549（M^+ + 1）。Anal. calcd for $C_{21}H_{18}BrN_5O_6S$：C 46.00，H 3.31，N 12.77；found C 45.82，H 3.30，N 12.78。

3t：白色晶体，产率 67.7%，m. p. 218 ~ 220℃；^{1}H NMR（DMSO - d_6，300MHz）δ：3.81（s，3H，pyrim - OCH_3），3.97（s，3H，CO_2CH_3），6.68（d，J = 5.4Hz，1H，pyrim - H^5），7.65（d，J = 8.4Hz，2H，Ar—H），7.78（d，J = 8.4Hz，1H，Ar—H），8.06（d，J = 8.4Hz，2H，Ar—H），8.32（dd，J = 8.4，1.8Hz，1H，Ar—H），8.43（d，J = 5.4Hz，1H，pyrim - H^6），8.63（d，J = 1.8Hz，1H，Ar—H），10.71（s，1H，NHCOAr），10.92（s，1H，CON**H**Het），12.98（brs，1H，SO_2NH）；ESI - MS（positive ion）m/z：565（M^+ + 1）。Anal. calcd for $C_{21}H_{18}BrN_5O_7S$：C 44.69，H 3.21，N 12.41；found C 44.76，H 3.40，N 12.18。

3u：白色晶体，产率 85.7%，m. p. 195 ~ 196℃；^{1}H NMR（DMSO - d_6，400MHz）δ：2.50（s，3H，pyrim - CH_3），3.82（s，3H，CO_2CH_3），7.14（d，J = 5.4Hz，1H，pyrim - H^5），7.49 ~ 7.50（m，1H，Ar—H），7.62 ~ 7.64（m，1H，Ar—H），7.98（d，J = 8.4Hz，1H，Ar—H），8.16（dd，J = 8.4，1.8Hz，1H，Ar—H），8.35（d，J = 8.4Hz，1H，Ar—H），8.47（d，J = 1.8Hz，1H，Ar—H），8.59（d，J = 5.4Hz，1H，pyrim - H^6），8.66（d，J = 1.8Hz，1H，Ar—H），10.74（s，1H，N**H**COAr），10.95（s，1H，CON**H**Het），13.00（brs，1H，SO_2NH）；ESI - MS（positive ion）m/z：488（M^+ + 1）。Anal. calcd for $C_{21}H_{18}FN_5O_6S$：C 51.74，H 3.72，N 14.37；found C 51.60，H 3.90，N 14.38。

3v：白色晶体，产率 72.2%，m. p. 201 ~ 203℃；^{1}H NMR（DMSO - d_6，400MHz）δ：3.81（s，3H，pyrim - OCH_3），3.97（s，3H，CO_2CH_3），6.68（d，J = 5.4Hz，1H，pyrim - H^5），7.53 ~ 7.56（m，2H，Ar—H），7.89（d，J = 8.4Hz，1H，Ar—H），8.05（d，J = 8.4Hz，2H，Ar—H），8.35（d，J = 8.4Hz，1H，Ar—H），8.42（d，J = 5.4Hz，1H，pyrim - H^6），8.63（s，1H，Ar—H），10.71（s，1H，NH-COAr），10.92（s，1H，CON**H**Het），12.99（brs，1H，SO_2NH）；ESI - MS（positive ion）m/z：488（M^+ + 1）。Anal. calcd for $C_{21}H_{18}FN_5O_7S$：C 50.10，H 3.60，N 13.91；found C 49.92，H 3.57，N 13.65。

3w：白色晶体，产率 88.7%，m. p. 207 ~ 209℃；^{1}H NMR（DMSO - d_6，400MHz）δ：2.50（s，3H，pyrim - CH_3），3.82（s，3H，CO_2CH_3），7.15（d，J = 5.4Hz，1H，pyrim - H^5），7.60（d，J = 8.4Hz，1H，Ar—H），7.74（d，J = 8.4Hz，1H，Ar—H），7.83（d，J = 8.4Hz，1H，Ar—H），7.98（dd，J = 8.4，1.8Hz，1H，Ar—H），8.20（d，J = 8.4Hz，1H，Ar—H），8.56（s，1H，Ar—H），8.59（d，J = 5.4Hz，1H，pyrim - H^6），10.75（s，1H，NHCOAr），11.23（s，1H，CON**H**Het），13.00（brs，

1H, SO_2NH); ESI-MS (positive ion) *m/z*: 539 (M^+ +1)。Anal. calcd for $C_{21}H_{17}Cl_2N_5O_6S$: C 46.85, H 3.18, N 13.01; found C 46.62, H 3.14, N 12.84。

3x：白色晶体，产率 81.2%，m.p. 197~198℃；^1H NMR（DMSO-d_6, 300MHz）δ: 2.50 (s, 3H, pyrim-CH_3), 3.82 (s, 3H, CO_2CH_3), 7.15 (d, *J*=5.4Hz, 1H, pyrim-H^5), 7.81~7.88 (m, 2H, Ar—H), 8.00 (d, *J*=8.4Hz, 1H, Ar—H), 8.14 (d, *J*=8.4Hz, 1H, Ar—H), 8.35 (d, *J*=8.4Hz, 1H, Ar—H), 8.59 (d, *J*=5.4Hz, 1H, pyrim-H^6), 8.63 (s, 1H, Ar—H), 10.74 (s, 1H, NHCOAr), 11.02 (s, 1H, CON**H**Het), 13.00 (brs, 1H, SO_2NH); ESI-MS (positive ion) *m/z*: 539 (M^+ +1)。Anal. calcd for $C_{21}H_{17}Cl_2N_5O_6S$: C 46.85, H 3.18, N 13.01; found C 46.61, H 3.29, N 12.84。

3y：白色晶体，产率 78.6%，m.p. 189~191℃；^1H NMR（DMSO-d_6, 400MHz）δ: 2.41 (s, 3H, CH_3—Ar), 2.50 (s, 3H, pyrim-CH_3), 3.81 (s, 3H, CO_2CH_3), 7.14 (d, *J*=5.4Hz, 1H, pyrim-H^5), 7.37 (d, *J*=8.4Hz, 2H, Ar—H), 7.80 (d, *J*=8.4Hz, 1H, Ar—H), 7.93 (d, *J*=8.4Hz, 2H, Ar—H), 8.35 (d, *J*=8.4Hz, 1H, Ar—H), 8.59 (d, *J*=5.4Hz, 1H, pyrim—H^6), 8.67 (s, 1H, Ar—H), 10.73 (s, 1H, NHCOAr), 10.90 (s, 1H, CON**H**Het), 12.99 (brs, 1H, SO_2NH); ESIMS (positive ion) *m/z*: 484 (M^+ +1)。Anal. calcd for $C_{22}H_{21}N_5O_6S$: C 54.65, H 4.38, N 14.48; found C 54.51, H 4.27, N 14.61。

3z：白色晶体，产率 81.5%，m.p. 199~200℃；^1H NMR（DMSO-d_6, 400MHz）δ: 1.23 (t, *J*=7.2Hz, 3H, $CH_2C\mathbf{H}_3$), 2.50 (s, 3H, pyrim-CH_3), 2.70 (q, 2H, $C\mathbf{H}_2CH_3$), 3.81 (s, 3H, CO_2CH_3), 7.14 (d, *J*=5.4Hz, 1H, pyrim-H^5), 7.40 (d, *J*=8.4Hz, 2H, Ar—H), 7.80 (d, *J*=8.4Hz, 1H, Ar—H), 7.94 (d, *J*=8.4Hz, 2H, Ar—H), 8.35 (d, *J*=8.4Hz, 1H, Ar—H), 8.59 (d, *J*=5.4Hz, 1H, pyrim-H^6), 8.68 (s, 1H, Ar—H), 10.72 (s, 1H, NHCOAr), 10.82 (s, 1H, CON**H**Het), 12.98 (brs, 1H, SO_2NH); ESI-MS (positive ion) *m/z*: 498 (M^+ +1)。Anal. calcd for $C_{23}H_{23}N_5O_6S$: C 55.52, H 4.66, N 14.08; found C 55.38, H 4.42, N 14.11。

2 结果与讨论

2.1 ^1H NMR 谱图解析

该系列磺酰脲化合物的核磁共振氢谱中，典型的化学位移有有机盐中 DBU 环上八个亚甲基氢的化学位移、嘧啶环的 H^5，H^6，脲桥上面的 2 个活泼氢和苯环 5 位酰胺的活泼 H。参照文献［13］详细归属了 DBU 环上 8 个亚甲基氢的化学位移，结果与文献报道的一致。嘧啶环 H^5 的化学位移一般为 δ 6.30~7.20 之间的双峰，H^6 的化学位移一般为 δ 8.40~8.60 之间的双峰，它们的耦合常数都为 5.4Hz。苯环 B 上不同取代基的影响使得嘧啶环 H^6 有时与苯环上的 H 峰混杂在一起而形成多重峰。苯环 A 上 H^6 的化学位移 δ 在 8.30~8.68 之间，部分化合物显示的是独立的单峰，有时也会与苯环 A 上 H^4 耦合而裂分为双峰，耦合常数 *J* 为 1.8Hz。苯环 5 位酰胺活泼氢的化学位移 δ 在 10.56~10.92 之间；CON**H**Het 的化学位移 δ 为 10.82~11.23 范围内的鼓包峰；$SO_2N\mathbf{H}CO$ 中的氢原子由于与嘧啶环上的氮原子形成六元环分子内氢键，从而使得共振吸收移向低场，化学位移 δ 为 12.92 ~ 13.00 范围内的鼓包峰。核磁谱图中脲桥上的两个活泼氢和苯环 5 位酰胺的活泼 H 有时会与氘代试剂中水的氢交换而不出峰。

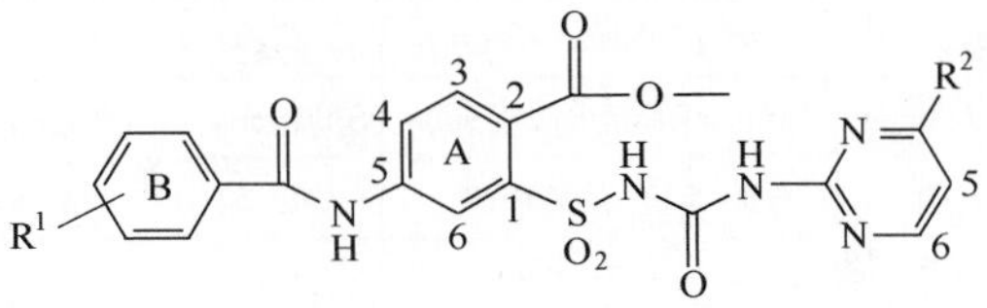

图 1　目标化合物的结构示意图

Figure 1　Structure of target compounds

2.2　除草活性

对所合成的 26 个新型磺酰脲化合物分别进行了油菜平皿法和盆栽法的除草活性测试。测试方法如下：

油菜平皿法：直径 6cm 的培养皿中铺好一张直径 5.6cm 的滤纸，加入 2mL 一定浓度的供试化合物溶液，播种浸种 4 ~ 6h 的油菜种子 15 粒，(28 ± 1)℃下，黑暗培养 48h 后测定胚根长度。通过黑暗条件下化合物对油菜胚根的生长抑制来检测化合物的除草活性。结果见表 1。

表 1　平皿法测定目标化合物 3 的除草活性（抑制率/%）

Table 1　Herbicidal activity (inhibition ratio/%) of the target compounds 3 by rape disc assay

Compd.	100 μg/mL	10 μg/mL	Compd.	100 μg/mL	10 μg/mL	Compd.	100 μg/mL	10 μg/mL
3a	71.4	70.7	**3j**	72.7	66.6	**3s**	71.9	69.2
3b	66.5	66.0	**3k**	63.5	62.1	**3t**	57.1	32.1
3c	72.7	68.1	**3l**	68.8	64.0	**3u**	73.5	70.6
3d	62.5	39.3	**3m**	66.5	65.6	**3v**	65.1	59.4
3e	73.5	72.0	**3n**	71.0	63.7	**3w**	71.7	69.9
3f	69.0	68.6	**3o**	75.0	66.6	**3x**	73.8	67.2
3g	52.5	18.2	**3p**	58.7	50.9	**3y**	63.3	60.1
3h	39.9	30.2	**3q**	70.1	61.9	**3z**	59.4	52.3
3i	71.6	71.6	**3r**	59.5	58.2			
Sulfometuron - methyl	74.9	72.4						

盆栽法：在直径 8cm 的塑料小杯中放入一定量的土，加入一定量的水，播种后覆盖一定厚度的土壤，于花房中培养，幼苗出土前以塑料覆盖。每天加以一定量的清水以保持正常生长。处理剂量为 25g/ha，处理包括土壤处理（出苗前）和茎叶（幼苗一叶一心期）处理两种。试验靶标为油菜（*Brassica napus*）、稗草（*Echinochloacrusgalli*）、马唐（*Digitaria adscendens*）和反枝苋（*Amaranthus retroflexus*）。20d 后调查结果，测定地上部鲜重，以鲜重抑制百分数来表示药效。试验结果见表 2。

表2 盆栽法测定目标化合物3的除草活性（抑制率/%）

Table 2 Herbicidal activity (inhibition ratio/%) of the target compounds 3 by pot bioassay experiments

Compd.	*Brassica napus*		*Amaranthus retroflexus*		*Digitaria adscendens*		*Echinochloa crusgalli*	
	Soil treatment	Foliage spray	Soil treatment	Foliage spray	Soil treatment	Foliage spray	Soil treatment	Foliage spray
3a	82.6	92.0	39.5	91.8	23.5	31.8	5.7	18.2
3b	39.2	68.7	44.6	72.4	9.4	0	14.7	0
3c	68.6	68.9	19.5	49.2	5.2	7.4	23.3	0
3d	39.9	34.2	20.3	82.2	15.6	3.3	21.2	2.7
3e	78.4	92.8	44.1	100	14.6	1.5	12.3	3.9
3f	66.3	59.4	8.4	98.4	14.6	9.8	5.0	0
3g	50.9	38.2	31.6	57.0	6.8	9.8	7.8	11.6
3h	48.3	59.9	9.4	96.7	3.9	4.5	4.8	0
3i	77.4	36.1	57.6	66.4	9.4	0	0	0
3j	69.8	46.2	34.2	54.1	18.4	34.7	0	0
3k	7.4	36.1	27.6	46.4	9.4	0	0	0
3l	6.5	48.5	16.8	35.1	17.0	0	30.8	0
3m	7.2	32.1	14.2	24.3	11.3	0	2.8	1.7
3n	7.8	41.7	3.4	25.7	17.0	17.6	35.7	4.4
3o	51.9	41.5	16.2	49.2	0	14.5	22.8	4.5
3p	48.9	44.2	3.8	67.2	0	0	14.2	3.9
3q	60.1	17.3	0.3	52.4	0	3.3	25.3	3.9
3r	23.3	50.7	34.2	93.0	18.7	0	9.8	0
3s	25.5	23.7	24.1	17.6	11.3	0	0	2.2
3t	0	10.4	0	12.2	9.4	0	4.9	0
3u	27.5	34.6	48.3	39.2	7.5	1.8	7.7	0
3v	26.8	30.1	44.8	100	0	0	27.3	31.7
3w	10.8	23.7	31.0	25.7	0	0	0	0
3x	0	54.1	20.7	24.3	0	0	37.8	0
3y	20.6	9.3	3.4	20.3	0	0	9.8	0
3z	21.1	4.8	6.9	6.8	9.4	9.1	0	0
Sulfometuron - methyl	81.8	81.7	100	100	81.1	100	100	100

结果表明，该系列化合物对双子叶作物的除草活性明显高于单子叶，茎叶处理的活性比土壤处理的活性要好。嘧啶环为单甲基取代的化合物比单甲氧基取代的化合物的除草活性高。在平皿法的测试中，该系列化合物均显示出较好的生物活性，然而盆栽的活性均有不同程度的降低。这可能是由于当苯环B上连有取代基时，整个磺酰脲分子的体积较大，使得化合物在植物体内不能被很好的吸收和传导。例如，当苯环B上没有任何取代基时（**3a**），其在平皿法和盆栽法中均对双子叶植物油菜和反枝苋显示出较高的活性，与商品化品种甲嘧磺隆相当；当苯环B上邻位连有体积较小的氟原子

时（**3e**，**3f**），同样也对双子叶植物表现出较高的活性。总体来看，该系列化合物活性的高低主要受苯环 B 上取代基的影响：苯环 B 上没有取代基时活性最高，随着取代基体积的增大，盆栽中显示出的抑制活性明显降低；苯环 B 邻位连有体积较小的吸电子基团时比对位连有相同吸电子基团的化合物的活性要高（如：**3e**，**3f**，**3i**，**3j** 的盆栽活性高于**3c**，**3d**，**3g**，**3h**）；对于相同位置的取代基，有供电子作用的化合物的活性比吸电子的要高（如：**3o～3r** 的盆栽活性高于 **3k～3n**）；苯环 B 的对位连有取代基时，体积增大，活性降低。

References

[1] Wang, J. –P.; Sun, Y. –Q.; Wu, W. –Z. *Pesticides* 1992, 31 (1), 3 (in Chinese).
（王建平，孙永泉，吴维中，农药，1992，31（1），3.）

[2] Sabadie, J. *Weed Res.* 1995, 35, 33.

[3] Ajit, K. S.; Jean S. *J. Agric. Food Chem.* 2002, 50, 6253.

[4] Teaney, S. R.; Armstrong, L.; Bentley, K.; Cotterman, D.; Leep, D.; Liang, P. H.; Powley, C.; Summers, J.; Cranwell, S.; Lichtner, F.; Stichburg, R. *Brighton Crop Prot. Conf. – Weeds* 1995, pp. 49～56.

[5] Suzanne, K. S.; Gary, M. D.; David, M. K.; Bruce, C. M.; Anne, D. L. – P.; Aldos, C. B.; Frederick, Q. B. *Pestic. Sci.* 1999, 55, 288.

[6] Rouchaud, J.; Neus, O.; Cools, K.; Bulcke, R. *J. Agric. Food Chem.* 1999, 47, 3872.

[7] Li, Z. – M.; Jia, G. – F.; Wang, L. – X.; Yang, Z. – Q.; Lai, C. – M. *CN* 1080116, 1994 [*Chem. Abstr.* 1994, 121, 29277.].

[8] Li, Z. – M.; Jia, G. – F.; Wang, L. – X.; Fan, C. – W.; Yang, Z. CN1106393, 1995 [*Chem. Abstr.* 1995, 124, 261066.].

[9] Li, Z. – M.; Mu, X. – L.; Fan, Z. – J.; Li, Y. – H.; Liu, B.; Zhao, W. – G.; Wang, J. – G.; Wang, S. – H.; Wang, B. – L. CN 1702064, 2005 [*Chem. Abstr.* 2005, 145, 271802.].

[10] Hiroyoshi, K.; Shuji, Y.; Masato, N. *Bull. Chem. Soc. Jpn.* 1982, 55, 3824.

[11] Dalelio, G. F.; Fessler, W. A.; Feigl, D. M. *J. Macromol. Sci.*, *Chem.* 1969, *A3* (5), 941

[12] Ma, N. *Ph. D. Disertation*, Nankai University, Tianjin, 2004 (in Chinese).
（马宁，博士论文，南开大学，天津，2004.）

[13] Li, P. – F.; Wang, B. – L.; Ma, N.; Wang, S. – H.; Li, Z. – M. *Chin. J. Org. Chem.* 2005, 25 (9), 1057 (in Chinese).
（李鹏飞，王宝雷，马宁，王素华，李正名，有机化学，2005，25（9），1057.）

Synthesis and Herbicidal Activity of *N* –（4′ – Substituted pyrimidin – 2′ – yl） – 2 – methoxycarbonyl – 5 – [（un） substituted benzamido] Phenylsulfonylurea Derivatives

Meiyi Wang, Wancheng Guo, Feng Lan, Yonghong Li, Zhengming Li

（Research Institute of Elemento – Organic Chemistry, State Key Laboratory of Elemento – organic Chemistry, Nankai University, Tianjin, 300071）

Abstract In order to search for ecologically – safer and environmentally – benign sulfonylurea herbicides, on

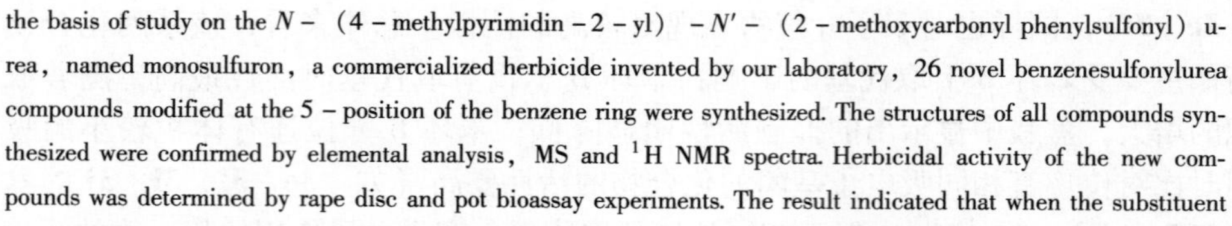

the basis of study on the N – (4 – methylpyrimidin – 2 – yl) – N' – (2 – methoxycarbonyl phenylsulfonyl) urea, named monosulfuron, a commercialized herbicide invented by our laboratory, 26 novel benzenesulfonylurea compounds modified at the 5 – position of the benzene ring were synthesized. The structures of all compounds synthesized were confirmed by elemental analysis, MS and ^{1}H NMR spectra. Herbicidal activity of the new compounds was determined by rape disc and pot bioassay experiments. The result indicated that when the substituent group at 5 – position of the benzene ring was benzamide, the herbicidal activity was good compared to sulfometuron – methyl.

Key words benzene sulfonylurea; 5 – substituted phenyl ring; synthesis; herbicidal activity

Synthesis and Herbicidal Activity of Novel Sulfonylureas Containing Thiadiazol Moiety*

Wancheng Guo, Xinghai Liu, Yonghong Li, Suhua Wang, Zhengming Li

(Institute of Elemento – Organic Chemistry, State Key Laboratory of Elemento – Organic Chemistry, Nankai University, Tianjin 300071, China)

Abstract Thirteen novel sulfonylureas containing thiadiazole moiety were synthesized in a two – step reaction. Their structures were determined using IR, ^{1}H NMR, HRFTMS, and elemental analysis. Herbicidal activities of these compounds were determined in the green house bio – assay. The results show that four compounds among them exhibit some activity toward four tested herbs.

Key words sulfonylurea; synthesis; herbicidal activity

1 Introduction

Since the synthesis and application of chlorsulfuron, a sulfonylurea herbicide, for the control of weed in 1982, various novel sulfonylurea herbicides, such as nicosulfuron, primisulfuron, foramsulfuron *etc.* [1-3], have been rapidly commercialized worldwide. Many scientists have devoted their large efforts to this field becasue of their high activity at low application rates (in the range of 2 to 75g/ha) and low mammal toxicity. Considering that 2,5 – disubstituted – 1,3,4 – thiadiazole derivatives have biological activity[4-6], a novel series of sulfonylureas containing 1,3,4 – thiadiazole moiety was designed and synthesized as potential herbicidal agents (Scheme 1). Herbicidal activity of these compounds was tested in the green house on four types of herbs. The results show that four compounds among them exhibit some activity against barnyardgrass.

R—COOH + H_2NC(=S)$NHNH_2$ —Reflux→ H_2N–(1,3,4-thiadiazol-2-yl)–R (1a—1m) —(2-NO_2-C_6H_4-SO_2NCO)→ 2-NO_2-C_6H_4-$SO_2NHCONH$–(1,3,4-thiadiazol-2-yl)–R (2a—2m)

Scheme 1 Synthesis route of compounds **2a – 2m**

R = H, Me, Et, *n* – propyl, cyclopropyl, phenyl, 2 – methylphenyl, 2 – chlorophenyl, 2 – fluorophenyl, 3 – methylphenyl, 4 – methylphenyl, furyl, pyridyl

2 Experimental

2.1 Reagents and Instruments

Melting points were measured using Yanaco MP – 500 micromelting point apparatus. Infrared spectra were recorded on a Bruker Equinox55 spectrophotometer with KBr pellets. ^{1}H NMR spectra were recorded on a Bruker AC – P500 instrument (300MHz) with tetramethylsilane as

---

* Reprinted from *Chemical Research in Chinese Universities*, 2008, 24(1): 32 – 35. Supported by the National Basic Research Program of China (No. 2003CB11406) and Key Project of National Natural Science Foucndation of China (No. 20432010).

internal standard and DMSO – d_6 as solvent. Elemental analyses were performed using Yanaco MT – 3CHN elemental analyzer. HRFTMS were performed using FT – ICR 7.0 – T mass spectrometer.

2.2 Syntheses of Title Compounds

2.2.1 General Procedure for the Synthesis of Compounds 1a – 1d

As the method shown in ref. [7], a mixture of thiosemicarbazide (2.73g, 30mmol), carboxylic acid (30 mmol), and hydrochloric acid (15.21g, 75mmol) was refluxed for 3h. Having been cooled, the reaction mixture was neutralized to pH 8 – 9 with aqueous sodium hydroxide. The products **1**a – **1**d were obtained by filtration and recrystallization from water.

2.2.2 General Procedure for the Synthesis of Compounds 1e – 1m

As the method shown in ref. [8], a mixture of thiosemicarbazide (2.73g, 30mmol), substituted benzoic acid (50 mmol), and phosphorus oxychloride (12.42g, 81mmol) was refluxed for 0.5h at 75℃. After the mixture was cooled to room temperature, 30mL of water was added to it and the mixture was refluxed for 4h at 110℃. Then the reaction mixture was neutralized to pH 8 with aqueous sodium hydroxide. The products **1**e – **1**m were obtained by filtration and recrystallization from alcohol.

2.2.3 General Procedure for the Synthesis of Compounds 2a – 2m

As the method shown in ref. [9], to a stirred suspension of 2 – amino – 5 – substituted thiadiazole (3mmol) in 12mL of anhydrous acetonitrile at room temperature, 2 – nitrobenzenesulfonyl isocyanate (0.68g, 3mmol) was added. The mixture was stirred for 12 – 20h. Then the products **2**a – **2**m were separated by filtration, washed with acetonitrile, and recrystallized from acetonitrile.

2.3 Data for Title Compounds

2.3.1 1 – (2 – Nitrobenzensulfonyl) – 3 – ([1,3,4]thiadiazol – 2 – yl) – urea (2a)

Yield 92.4%; m.p. 244 – 246℃; ^{1}H NMR (300MHz, DMSO – d_6), δ: 7.86 – 7.90 (m, 2H, benzo—H), 7.93 – 7.97 (m, ^{1}H, benzo—H), 8.18 – 8.23 (m, ^{1}H, benzo—H), 8.97 (s, ^{1}H, Het—H); IR (KBr), $\tilde{\nu}$/cm^{-1}: 3376, 3277 (N—H), 1712 (C=O), 1365 and 1160 (S=O); HRFTMS (m/z): 327.9821 (M^-—H for $C_9H_6N_5O_5S_2$, 327.9816).

2.3.2 1 – (2 – Nitrobenzensulfonyl) – 3 – (5 – methyl – [1,3,4]thiadiazol – 2 – yl) – urea (2b)

Yield 94.5%; m.p. 254 – 256℃; ^{1}H NMR (300MHz, DMSO – d_6), δ: 2.50 (s, 3H, CH_3), 7.86 – 7.89 (m, 2H, benzo—H), 7.93 ~ 7.97 (m, ^{1}H, benzo—H), 8.20 ~ 8.23 (m, ^{1}H, benzo—H); IR (KBr), $\tilde{\nu}$/cm^{-1}: 3375, 3262 (N—H), 1702 (C=O), 1363 and 1157 (S=O); HRFTMS (m/z): 341.9970 (M^-—H for $C_{10}H_8N_5O_5S_2$, 341.9972); elemental anal. (%) calcd. for $C_{10}H_9N_5O_5S_2$: C 34.98, H 2.64, N 20.40; found: C 35.06, H 2.65, N 20.21.

2.3.3 1 – (2 – Nitrobenzensulfonyl) – 3 – (5 – ethyl – [1,3,4]thiadiazol – 2 – yl) – urea (2c)

Yield 79.6%; m.p. 272 – 274℃; ^{1}H NMR (300MHz, DMSO – d_6), δ: 1.22 (t, J = 7.5Hz, 3H,

CH_3),2.86(q,J=7.5Hz,2H,CH_2),7.84~7.90(m,2H,benzo-H),7.93-7.97(m,[1]H,benzo—H),8.18~8.23(m,[1]H,benzo—H);IR(KBr),$\tilde{\nu}$/cm^{-1}:3223,3129(N—H),1728(C=O),1361 and 1162(S=O);HRFTMS(m/z):356.0127(M^- -H for $C_{11}H_{10}N_5O_5S_2$, 356.0129); elemental anal. (%) calcd. for $C_{11}H_{11}N_5O_5S_2$: C 36.97, H 3.10, N 19.60; found: C 36.78, H 3.20, N 19.45.

2.3.4 1-(2-Nitrobenzensulfonyl)-3-(5-propyl-[1,3,4]thiadiazol-2-yl)-urea(**2d**)

Yield 75.2%;m.p. 233-235℃;^{1}H NMR(300MHz,DMSO-d_6),δ:0.89(t,J=7.5Hz,3H,CH_3),1.64(m,J=7.5Hz,2H,CH_2),2.80(t,J=7.2Hz,2H,CH_2),7.73-7.78(m,3H,benzo—H),8.10-8.13(m,1H,benzo—H);IR(KBr),$\tilde{\nu}$/cm^{-1}: 3375,3262(N—H),1717(C=O),1363 and 1150(S=O);HRFTMS(m/z):370.0281(M^- -H for $C_{12}H_{12}N_5O_5S_2$, 370.0285).

2.3.5 1-(2-nitrobenzensulfonyl)-3-(5-cyclopropyl-[1,3,4]thiadiazol-2-yl)-urea(**2e**)

Yield 49.1%;m.p. 237-239℃;^{1}H NMR(300MHz,DMSO-d_6),δ: 0.82-0.86(m,2H,cyclopropyl—H),1.02-1.05(m,2H,cyclopropyl—H),2.19-2.26(m,[1]H,CH),7.65-7.75(m,3H,benzo—H),8.07(d,J=7.6Hz,1H,benzo—H);IR(KBr),$\tilde{\nu}$/cm^{-1}: 3376,3267(N—H),1702(C=O),1365 and 1162(S=O);HRFTMS(m/z): 368.0140(M^- -H for $C_{12}H_{10}N_5O_5S_2$,368.0129).

2.3.6 1-(2-Nitrobenzensulfonyl)-3-(5-phenyl-[1,3,4]thiadiazol-2-yl)-urea(**2f**)

Yield 55.5%;m.p. 147-149℃;^{1}H NMR(300MHz,DMSO-d_6),δ:7.48-7.51(m,4H,benzo—H),7.54-7.58(m,1H,benzo—H),7.84-7.86(m,2H,benzo—H),7.91-7.93(m,1H,benzo—H),8.19-8.22(m,1H,benzo—H);IR(KBr),$\tilde{\nu}$/cm^{-1}: 3441,3231(N—H),1725(C=O),1358 and 1155(S=O);HRFTMS(m/z): 404.0121(M^- -H for $C_{15}H_{10}N_5O_5S_2$,404.0129); elemental anal. (%) calcd. for $C_{15}H_{11}N_5O_5S_2$: C 44.44, H 2.73, N 17.27;found:C 44.60,H 2.78,N 17.30.

2.3.7 1-(2-Nitrobenzensulfonyl)-3-(5-[2-chlorophenyl]-[1,3,4]thiadiazol-2-yl)-urea(**2g**)

Yield 32.8%;m.p. 195-197℃;^{1}H NMR(300MHz,DMSO-d_6),δ: 7.43-7.50(m,3H,benzo—H),7.57-7.61(m,[1]H,benzo—H),7.69-7.77(m,2H,benzo—H),7.95-8.00(m,[1]H,benzo—H),8.09-8.11(m,[1]H,benzo—H);IR(KBr),$\tilde{\nu}$/cm^{-1}: 3311,3205(N—H),1724(C=O),1363 and 1164(S=O); HRFTMS(m/z): 437.9740(M^- -H for $C_{15}H_9ClN_5O_5S_2$,437.9739); elemental anal. (%) calcd. for $C_{15}H_{11}ClN_5O_5S_2$: C 40.96, H 2.29, N 15.92;found:C 41.32,H 2.18,N 15.55.

2.3.8 1-(2-Nitrobenzensulfonyl)-3-(5-[2-fluorophenyl]-[1,3,4]thiadiazol-2-yl)-urea(**2h**)

Yield 48.9%;m.p. 287-289℃;^{1}H NMR(300MHz,DMSO-d_6),δ: 7.32-7.42(m,2H,Ar′—H),7.49-7.53(m,2H,Ar′—H),7.60-7.70(m,2H,benzo—H),8.06-8.17(m,2H,benzo—H),10.98(s,[1]H,NH);IR(KBr),$\tilde{\nu}$/cm^{-1}: 3411,3311(N—H),1724(C=O),1363 and 1164(S=O); HRFTMS(m/z): 422.0027(M^- -H for $C_{15}H_9FN_5O_5S_2$, 422.0035); elemental anal. (%) calcd. for $C_{15}H_{10}N_5O_5S_2$: C 42.55, H 2.38, N 16.54;

found: C 42.43, H 2.52, N 16.37.

2.3.9 1-(2-nitrobenzensulfonyl)-3-(5-[2-methylphenyl]-[1,3,4]thiadiazol-2-yl)-urea(**2i**)

Yield 44.8%; m.p. 293 - 295℃; ^{1}H NMR (300MHz, DMSO - d_6), δ: 2.46 (s, 3H, CH_3), 7.35 (m, 3H, benzo—H), 7.57 (m, 1H, benzo—H), 7.65 (m, 3H, benzo—H), 8.06 (m, ^{1}H, benzo—H), 10.86 (s, ^{1}H, NH); IR (KBr), $\tilde{\nu}$/cm^{-1}: 3441, 3231 (N—H), 1700 (C=O), 1360 and 1169 (S=O); HRFTMS (m/z): 418.0296 (M^- - H for $C_{16}H_{12}N_5O_5S_2$, 418.0285).

2.3.10 1-(2-nitrobenzensulfonyl)-3-(5-[4-methoxyphenyl]-[1,3,4]thiadiazol-2-yl)-urea(**2j**)

Yield 69.0%; m.p. 165 - 167℃; ^{1}H NMR (300MHz, DMSO - d_6), δ: 3.80 (s, 3H, OCH_3), 7.02 (d, J=8.4Hz, 2H, benzo—H), 7.10 (q, 1H, benzo—H), 7.66 (q, ^{1}H, benzo—H), 7.76 (d, J = 8.0Hz, 1H, benzo—H), 7.90 - 7.95 (m, 1H, benzo—H), 8.08 (d, J = 7.6Hz, 1H, benzo—H); IR (KBr), $\tilde{\nu}$/cm^{-1}: 3376, 3267 (N—H), 1750 (C=O), 1365 and 1162 (S=O); HRFTMS (m/z): 434.0240 (M^- - H for $C_{16}H_{12}N_5O_6S_2$, 434.0235).

2.3.11 1-(2-Nitrobenzensulfonyl)-3-(5-[3-methylphenyl]-[1,3,4]thiadiazol-2-yl)-urea(**2k**)

Yield 42.3%; m.p. 141 - 143℃; ^{1}H NMR (300MHz, DMSO - d_6), δ: 2.36 (s, 3H, CH_3), 7.27 (m, 1H, benzo - H), 7.36 (m, 1H, benzo—H), 7.56 (m, 1H, benzo—H), 7.64 (m, 1H, benzo - H), 7.76 - 7.83 (m, 3H, benzo - H), 8.16 (d, J=6.8Hz, 1H, benzo—H); IR (KBr), $\tilde{\nu}$/cm^{-1}: 3382, 3248 (N—H), 1752 (C=O), 1355 and 1165 (S=O); HRFTMS (m/z): 418.0278 (M^- - H for $C_{16}H_{12}N_5O_5S_2$, 418.0285).

2.3.12 1-(2-Nitrobenzensulfonyl)-3-(5-furyl-[1,3,4]thiadiazol-2-yl)-urea(**2l**)

Yield 41.2%; m.p. 233 - 234℃; ^{1}H NMR (300MHz, DMSO - d_6), δ: 6.63 (d, J=2.8Hz, 1H, furyl—H), 6.95 (d, J = 2.8Hz, 1H, furyl—H), 7.43 (s, 1H, furyl—H), 7.57 - 7.67 (m, 2H, benzo—H), 7.81 (d, J=8.0Hz, ^{1}H, benzo—H), 8.04 (d, J=7.6Hz, 1H, benzo—H), 10.98 (s, 1H, NH); IR (KBr), $\tilde{\nu}$/cm^{-1}: 3310, 3173 (N—H), 1702 (C=O), 1358 and 1169 (S=O); HRFTMS (m/z): 393.9916 (M^- - H for $C_{13}H_9N_5O_6S_2$, 393.9922); elemental anal. (%) calcd. for $C_{13}H_{10}N_5O_6S_2$: C 39.49, H 2.29, N 17.71; found: C 39.22, H 2.32, N 17.52.

2.3.13 1-(2-Nitrobenzensulfonyl)-3-(5-pyridyl-[1,3,4]thiadiazol-2-yl)-urea(**2m**)

Yield 42.1; m.p. 214 - 216℃; ^{1}H NMR (300MHz, DMSO - d_6), δ: 7.50 (q, J = 4.8Hz, 1H, Benzo - RH), 7.57 (s, 4H, Py—H), 8.14 (q, 1H, benzo—H), 8.61 (d, J = 4.8Hz, 1H, benzo—H), 8.95 (s, 1H, benzo—H); IR (KBr), $\tilde{\nu}$/cm^{-1}: 3376, 3274 (N—H), 1689 (C=O), 1373 and 1162 (S=O); HRFTMS (m/z): 405.0087 (M^- - H for $C_{14}H_9N_6O_5S_2$, 405.0081).

2.4 Tests of Herbicidal Activity

The tested herbs were two monocotyledons: *Echinochloa Crusgalli L* (*barnyardgrass*) and *Digitaria sanguinalis Scop* (Common crabgrass) and two dicotyledons: Brassica napus *L.* (winter rape) and *Amaranthus retroflexus L.* (Red Amaranth).

2.4.1 Pre-emergence

Sandy clay(100g) in a 6-cm-diameter plastic box was wetted with water. The box was covered with plastic film to keep moist. Sprouting seeds(15) of the weed under test were planted in fine earth(0.6cm depth) in the green house and sprayed with the test compound dissolved in a suitable solvent at 1.5kg/ha.

2.4.2 Post-emergence

Seedlings(one leaf and one stem) of the weed were sprayed with the tested compounds at the same rate as in pre-emergence test.

For both the methods, the fresh weights were determined after 20days, and the inhibition percent was used to describe the control efficiency of the compounds.

3 Results and Discussion

3.1 Synthesis

The title compounds were synthesized in a two-step reaction as shown in Scheme 1. Thiadiazoles **1a-1m** were synthesized according to literatures [7,8]. Compounds **2a-2m** were identified using IR, ^{1}H NMR, HRFTMS, and elemental analysis.

3.2 Herbicidal Activity

All the target compounds were tested in the green house on four herbs. The results are listed in Table 1. In comparison with the tested tribenuron-methyl that, it is found that the compounds **2g**, **2i**, **2l**, **2m** showed weaker herbicidal activity on barnyardgrass, whereas others showed lower herbicidal activity.

Table 1 Herbicidal activities of compounds 2a-2m(% inhibition)①

Compound②	*Echinochloa Crusgalli L.*		*Digitaria sanguinalis Scop.*		*Brassica napus L.*		*Amaranthus retroflexus L.*	
	Pre	Post	Pre	Post	Pre	Post	Pre	Post
2a	1.3	16.4	23.6	20.0	31.2	47.8	0	16.2
2b	2.9	0	6.4	0	3.4	56.5	6.1	0
2c	5.3	0	0	0	2.0	37.2	2.7	13.9
2d	11.2	0	0	0	0.4	1.3	0	0
2e	36.8	16.1	3.7	0	0	17.8	38.8	0
2f	38.6	33.2	20.4	0	0	19.8	0	42.3
2g	42.2	35.8	0	0	0	12.3	0	0
2h	37.2	38.9	3.7	0	0	0	0	0
2i	41.7	29.5	5.6	0	0	31.9	10.2	0
2j	36.8	16.6	1.9	0	0	25.3	0	0.9
2k	37.7	17.1	11.1	0	1.7	27.5	0	0
2l	40.4	18.7	0	0	0	4.2	18.4	0
2m	48.4	24.4	0	3.8	0	25.5	27.4	11.7
Tribenuron-methyl	50.9	39.4	62.0	46.8	100.0	99.5	94.0	90.3

①Pre: pre-emergence; Post: post-emergence; ②Rates: 2a-2m 1.5kg/hm^2, tribenuron-methyl 30g/hm^2.

References

[1] Russell M. H. ,Saladini J. L. ,Lichtner F. ,*Pesticide Outlook*,2002,(8),166.
[2] Liu C. L. ,*Pesticides*,1999,38(3),39.
[3] Liu C. L. ,*Fine and Speciality Chemicals*,2000,17,6.
[4] Sengupta A. K. ,Chandra U. ,*Journal of the Indian Chemical Society*,1979,56(6),645.
[5] Foroumadi A. ,Soltani F. ,Moshafi M. H. ,*et al.* ,*Farmaco*,2003,58(10),1023.
[6] Foroumadi A. ,Emami S. ,Hassanzadeh A. ,*et al.* ,*Bioorganic & Medicinal Chemistry Letters*,2005,15(20),4488.
[7] Le Z. G. ,Ding J. H. ,Yang S. J. ,*Chemical World*,2002,43(7),s366.
[8] Shuai X. ,Sun X. ,Shao R. Q. ,*et al.* ,*Journal of Shandong University*(*Health Sciences*),2002,40(2),182.
[9] Levitt G. ,Wilmington D. ,US 4394506,1983.

The Outset Innovation of Agrochemicals in China*

Zhengming Li[1], Yibin Zhang[2]

([1]National Pesticide Engineering Research Center, Nankai University, Tianjin, 300071, China;
[2]Shanghai Pesticide Research Institute, 2354 Xie - Tu Road, Shanghai, 200032, China)

Abstract This paper outlines some of the recent discoveries in agrochemical research in China.

As a result of the Chinese Government's successful "Reformation and Open" policy, all branches of industry in China have made great strides over the last decades. During the last few years, pesticide production in China has attained the level of the leading countries in the world. In 2006, China exported 398,000 tonnes of both chemical and biological generic pesticides (CPCIA). In recent years, it has been emphasized in China to uphold the innovation principle in the advancement of science and technology. As well as in other fields, agrochemical research has also started its innovation program recently. The aim of this article is to briefly introduce a taste of the new agrochemical compounds which have been invented and developed to a commercial stage by Chinese researchers. At present in China every new agrochemical has to go through the required legal process to obtain the official licenses before being allowed to go into the market, the so called "3 licences" system for official registration that all new products are required to follow. These 3 licences are as follows:

1. One from the Ministry of Agriculture (implemented by its subordinated Institute for Control of Agrochemicals, abbreviated as ICAMA) to carefully scrutinize all the related chemical data, related greenhouse and field biological tests, environmental evaluation, toxicological data, residue data etc.

2. Another one from State Development and Reformation Commission for a special permit to produce the new product with its manufacturing technology (including waste disposal) in an officially registered plant.

3. The last one from State Bureau of Inspection and Assay for approval of the reliability of the quality control of products, intermediates, formulations etc.

Only after all 3 official licenses have been obtained, a new agrochemical can then be cleared to enter the market.

Most of the products listed below are in various stages of development and production, although most of them have already been cleared for 3 licenses and some have reached to the commercial stage. With support from different governmental agencies, scores of other newly discovered bioactive structures are in their early stages of development, but these are not mentioned in this article. It is expected that agrochemical innovation R&D in China once started will receive more input and will upscale its level of research, while local industry will gradually

* Reprinted from *Dutlooks on Pest Management*, 2008, 19(3): 135 - 138.

participate in the stream of innovation of new agrochemicals in the future.

Ⅰ. Insecticides (including acaricides)

1. JS – 118

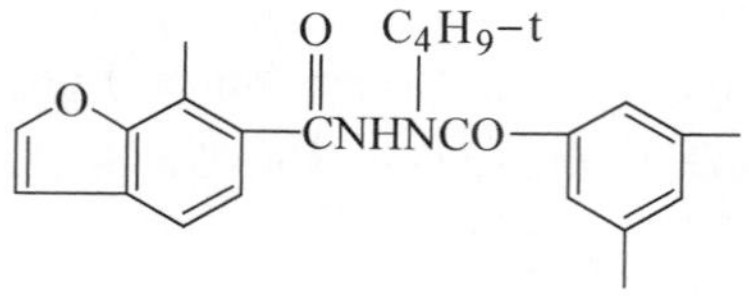

Developed by Jiangsu Pesticide Research Institute as a diacyl – hydrazine insect growth regulator. It acts as an insect moulting inhibitor. It is effective against a wide variety of lepidopteran larvae including diamond – back moth (*Plutella xylostella*), beet armyworm (*Spodoptera exigua*), corn borer (*Ostrinia nubilalis*), cotton bollworm (*Helicoverpa zea* and *Helicoverpa armigera*) etc.

2. D – *trans* – chloropropinylchrysanthemate (Beisu – juezhi)

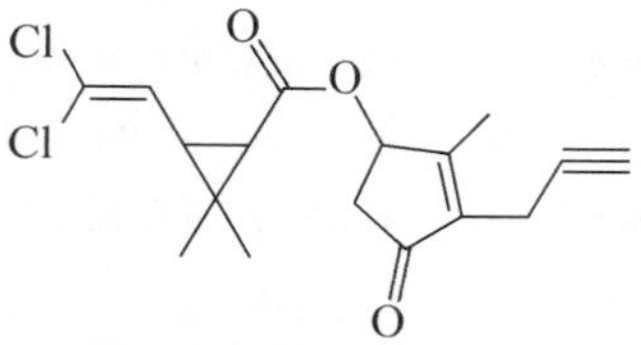

Developed by Yangnong Chemical Industry Company as a public hygiene insecticide in aerosol formulations with high knock – down properties against mosquitoes, flies and cockroaches, with results better then furamethrin and tetramethrin.

3. Xiaochong – liulin

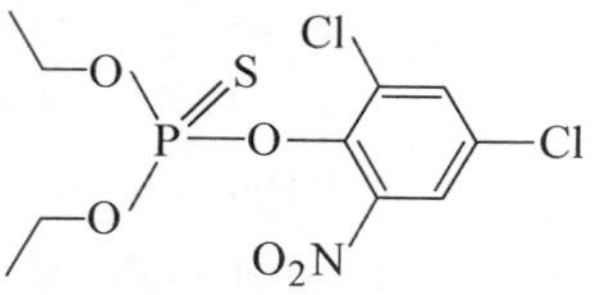

Developed by Sichuan Chemical Industry Research Institute. An organo – phosphorus insecticide and acaricide, this is a new analogue derived from diclorfenthion. Effective against orange red – mite (*Panonychus citri*), orange bollworm (*Heliothis armigera*), oriental tobacco budworm (*Heliothis assulta*), diamond back moth, rice planthopper (*Nilaparvata lugens*) etc.

4. Dingxi – fuchongqing

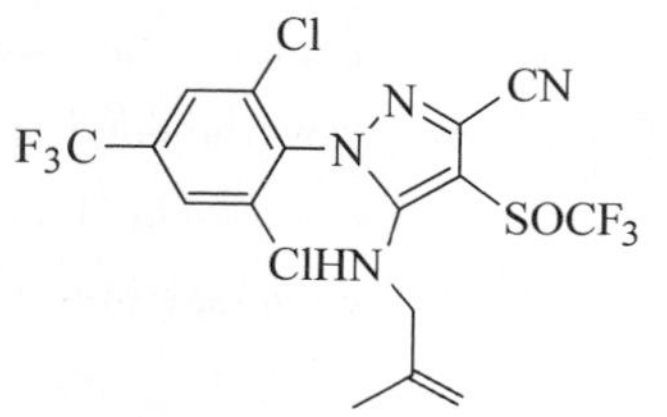

Developed by Dalian Pesticide Company it is an optimized structure derived from fipronil, effective against diamond – back moth, oriental armyworm (*Mythimna separata*), rice planthopper etc. Its toxicity to fish is lower than fipronil.

Ⅱ. Fungicides

5. Flumorph(SYP - L 190)

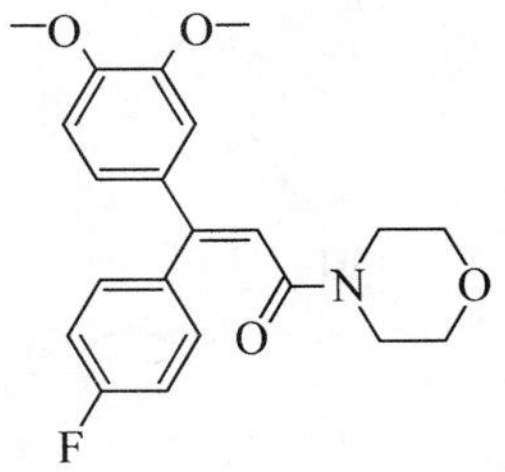

Developed by Shenyang Chemical Industry Institute it is structurally optimized from dimethomorph. It is effective against oomycetes fungi such as downy mildew, *Phytophthora* blight, black rot etc.

6. Enestroburin(SYP - 2071)

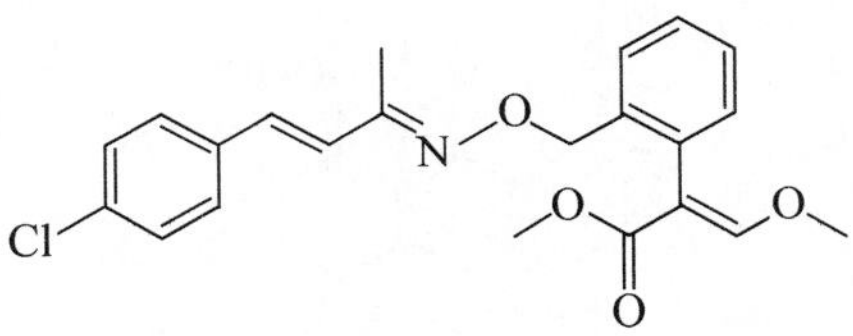

Developed by Shenyang Chemical Industry Institute. This is a methoxyacrylic ester analogue, used to control cucumber downy mildew (*Pseudoperonospora cubensis*), *Fusarium* blight of wheat, pear blight spot, wheat powdery mildew (*Blumeria* spp), pear black spot(*Botryosphaeria* spp), tomato late blight(*phytophthora infestans*), grape downy mildew(*Plasmopara viticola*) etc.

7. SYP - Z048

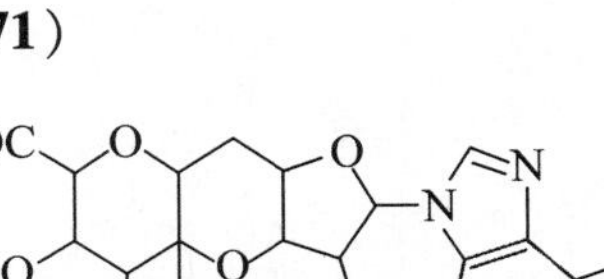

Developed by Shenyang Chemistry Industry Institute. This is a pyridinyl oxazole fungicide used to control gray mold (*Botrytis* spp) of tomato, leaf mold of tomato, gray mold of cucumber etc.

8. Aureonucleomycin(SPRI - 371)

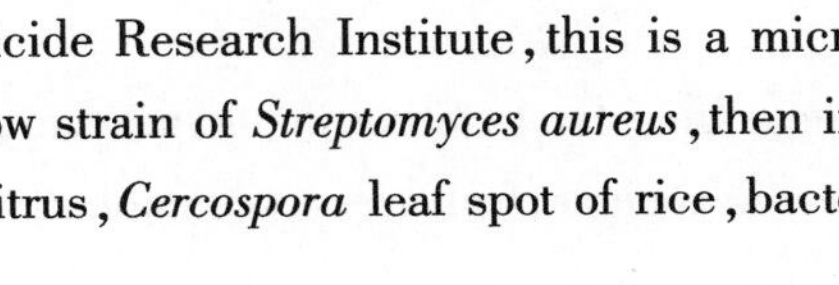

Developed by shanghai Pesticide Research Institute, this is a microbial origin fungicide, which was isolated from the Soochow strain of *Streptomyces aureus*, then incubated and fermented, It is effective against canker of citrus, *Cercospora* leaf spot of rice, bacterial leaf blight of rice etc.

9. SPRI – 2098

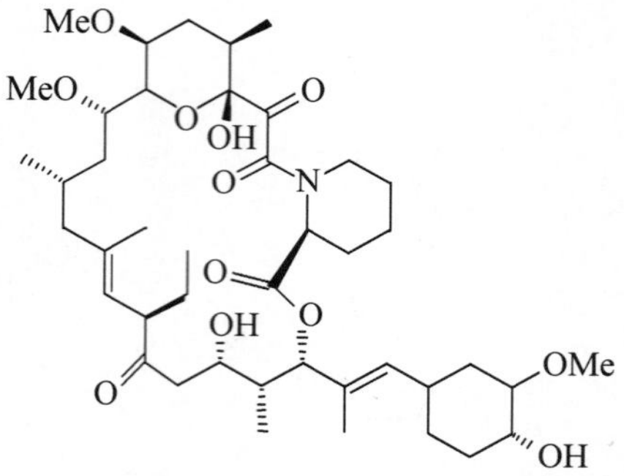

Developed by Shanghai Pesticide Research Institute this is a microbial fungicide which was isolated from a new isolate of *Streptomyces hygroscopicus*. It inhibits powdery mildew, early blight and black spot of fruits and vegetables.

10. M18

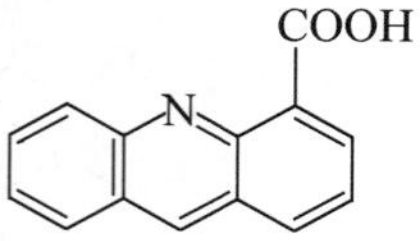

Developed by Shanghai Jiaotong University. It was isolated from a fluorescent *Pseudomonas* strain originated from the soil of Shanghai suburb. It is active against *Fusariwm* wilt of watermelon, *Phytophthora* blight of pepper, gummy stem blight of melon etc.

11. JS 399 – 19

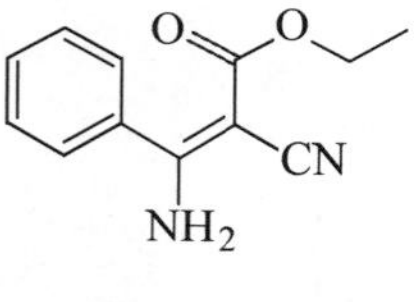

Developed by Jiangsu Pesticide Research Institute. This cyanoacrylic fungicide is active against *Fusarium* blight, *Rhizoctonia* of wheat, *Fusarium* wilt of cotton, *phytophthora* blight of pepper, gray mold of cucumber(*Botrytis*), Bakanae disease of rice(*Gibberella*) etc.

12. ZJ – 0712

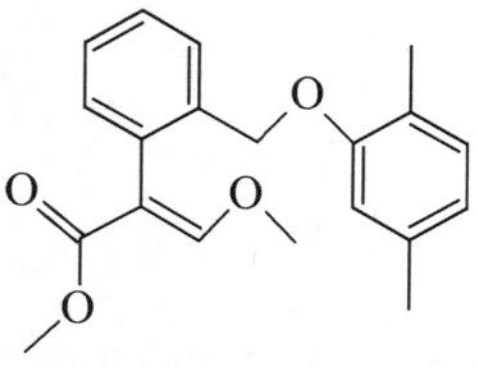

Developed by Zhejiang Chemical Industry Institute. A methoxyacrylic fungicide used to control powdery mildew, downy mildew in melons, fruit trees, vegetables, wheat, tobacco, flowers etc.

Ⅲ. Herbicides and plant growth regulators

13. Monosulfuron and monosulfuron – ester

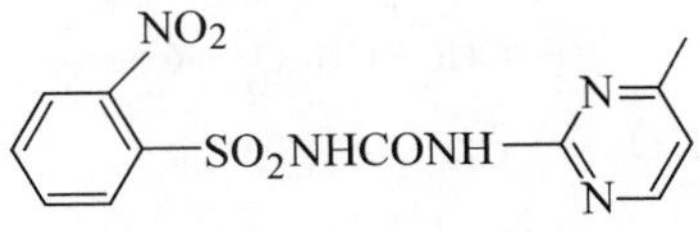

Monosulfuron

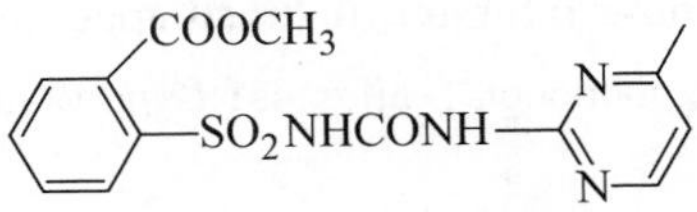

Monosulfuron–ester

Developed by Nankai University. These compounds belong to a new type of sulfonylurea herbicides, structurally unique because they have a mono – substituted pyridine ring which differs from the classical sulfonylurea herbicides.

Monosulfuron shows its exclusive control against weeds selectively in millet and against noxious weeping alkali weeds in wheat fields. Monosulfuron – ester is effective against lambsquarter and goosefoot (*Chenopodium* spp), knotweed (*Polygonum* spp), wild indian mustard (*Brassica juncea*) etc. Both herbicides are used in ultra – low dosages.

14. H –9201

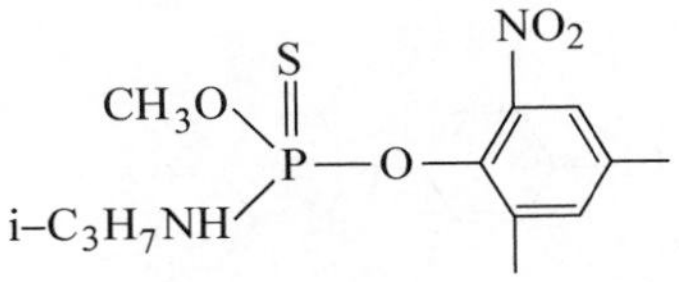

Developed by Nankai University. An organic phosphorus herbicide that may be used in both dry and wet land. It has systemic properties, and is selective in soybean, rice, wheat, carrot crops for annual monocot and dicot weeds.

15. ZJ –0273 and ZJ –0702

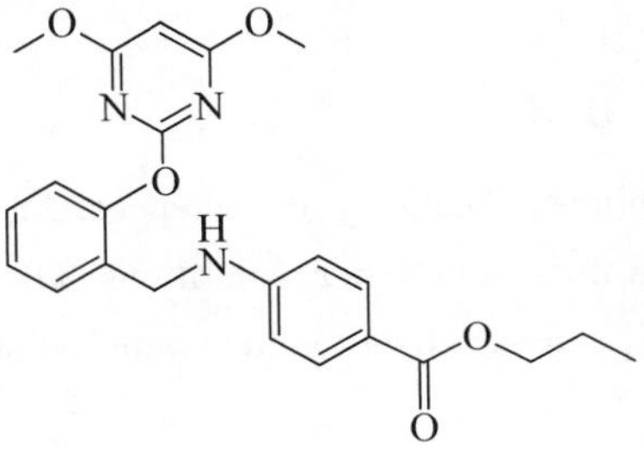

ZJ–0273

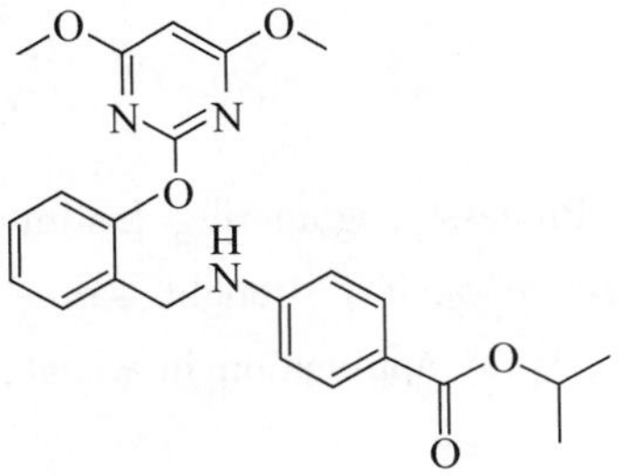

ZJ–0702

Developed by Shanghai Institute of Organic Chemistry and Zhejiang Chemical Industry Institute. Unusual benzylpyrimidine herbicides used to control monocot and dicotweeds, such as blackgrass (*Alopecurus myosuroides*), hairy bittercress (*Cardamine hirsuta*), common chickweeds (*Stellaria media*) etc. in rape fields.

16. HW－02

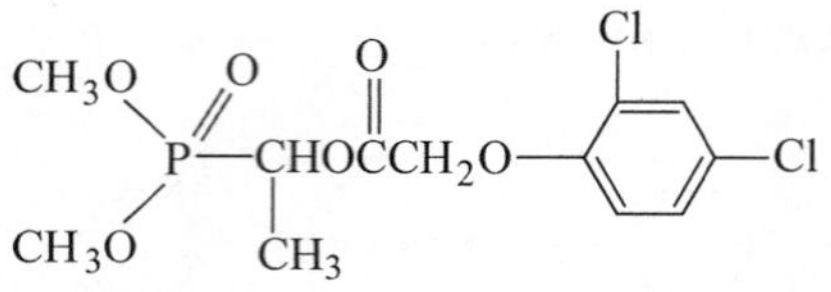

Developed by Central China Normal University. An organic phosphorus herbicide with a mode of action as a pyruvic dehydrogenase inhibitor. A broad spectrum herbicide by pre－ or post－emergence application to control monocots, nutgrass (*Cyperus* spp), bracken fern (*pteridium*) weeds etc.

17. HNPC－C9908

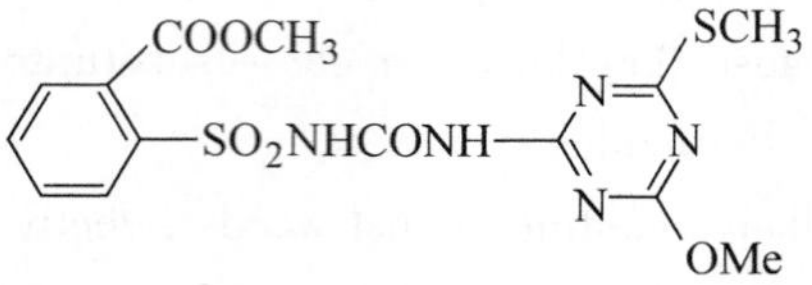

Developed by Hunan Chemical Industry Institute. Structurally it introduced a methylthio group into the heterocylic moiety instead of the classical methoxy group common in other sulfonylurea herbicides. It is applied in wheat fields to control broad leaf and graminaceous weeds.

18. BAU－9403

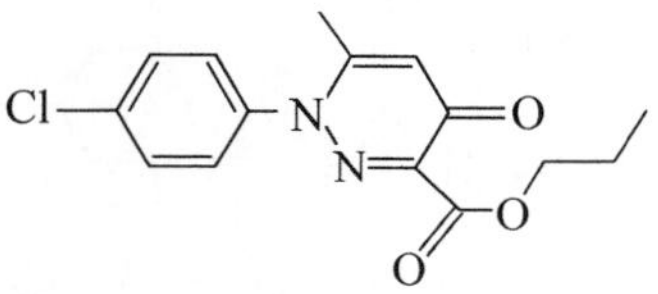

Developed by China Agriculture University. A wheat hybridization chemical, applied in wheat hybridized seed treatments.

19. RAD

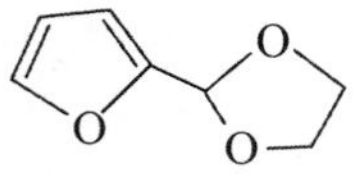

Developed by Process Engineering Institute, Chinese Academy of Sciences. An anti－stress compound effective against drought, saline conditions and other conditions when applied by seed soaking or spray application in wheat, corn, soybean, beet, cotton, watermelon, cucumber, flowers etc.

Reference

CPCIA (2007) Statistics from the China Petroleum & Chemistry Industry Association (CPCIA).

Quantitive – Structure Activity Relationship(QSAR) Study of a New Heterocyclic Insecticides Using CoMFA and CoMSIA*

Weili Dong, Xinghai Liu, Yi Ma, Zhengming Li

(State – Key Laboratory of Elemento – Organic Chemistry, National Pesticide Engineering Research Center, Institute of Elemento – Organic Chemistry. Nankai University, Tianjin, 300071, China)

Abstract Three – dimensional quantitative structure – activity relationship (3D – QSAR) studies were carried out on a series of 33 anthranilic diamides related to their insecticide activity as ryanodine receptor activation using CoMFA and CoMSIA. All models were carried out over a training set including 29 anthranilic diamides. For CoMFA model, cross – validated correlation coefficient q^2 is 0.720, non – cross – validated correlation coefficient r^2 is 0.894, F values is 81.449 and standard error of estimate (SE) values is 0.465. For CoMSIA model, they are 0.732, 0.850, 0.554 and 54.706 respectively. The predictive ability of the models was validated by four compounds that were not included in the training set. The deviation between prediction and experiment is small. These research results can provide valuable information for designing new potential insecticides interacting with ryanodine receptor.

Key words anthranilic diamides; 3D – QSAR; insecticidal activity

Introduction

The safe and effective use of insecticides to fight serious crop damage from harmful pests is an essential element in both guarantee of food supply and prevention of disease transmission. Due to the ability of insects to rapidly develop resistance, the discovery of agents that act on new biochemical targets is an important tool for effective pest management. A new class of insecticides has been discovered, the anthranilic diamides, that provides exceptional control through action on a novel target, the ryanodine receptor. Rynaxypry™ (Dupont's) is the first new insecticide from the anthranilic diamides, characterized by its high levels of insecticidal activity and low toxicity to mammals attributed to a high selectivity for insect over mammalian ryanodine receptors[1]. Owing to their prominent insecticidal activity, unique modes of action and good environmental profiles, anthranilic diamides and their chemical synthesis have recently attracted considerable attention in the field of novel agricultural insecticides[2-4].

In addition, the mechanism of insecticidal action of anthranilic diamides has not yet been clearly established, and there is so far no report about the binding model of anthranilic diamides with the receptor, which is crucial for the design of novel molecules. [5,6] Due to the limited reports on structure – activity approach, 3D – QSAR study were performed applying comparative molecular field analysis (CoMFA) and comparative molecular similarity indices analysis (CoM-

* Reprinted from *Heterocgclic Communications*, 2009, 15(1): 17 – 21.

SIA). Both 3D – QSAR techniques compare molecular interaction fields In CoMFA, interaction fields are represented as steric and electrostatic interaction energies calculated using Lennard – Jones potential and Coulombic potential for a molecule in the data set at the intersections of a grid embedding that molecule. [7] Another molecular interaction field in CoMSIA, [8], which uses Gaussian functions to describe the similarities of steric, electrostatic, hydrophobic, and hydrogen bond donor and acceptor properties[9]. The outcome of the present work can provide valuable information for designing potential ryanodine receptor activator with high insecticidal activities.

Materials and methods

Selection of compounds and activities

The title compounds (1 – 33 in Tables 1) and activities studied in this work were taken from the literature[1,10]. The insecticidal activity reported are against fall armyworm (*Spodoptera frugiperda*, *Sf*), insecticidal potency as LC_{50} in ppm, which were collected and transformed into log ($10^6/LC_{50}$) values. For a stronger evaluation of model applicability for prediction on new chemicals, the data set was divided into two subdata sets. Four compounds were chosen randomly as a test set and were used for external validation of the 3D – QSAR models; the training sets included all the remaining 29 compounds. The structures and insecticidal activities of anthranilic diamides are summarized in Tables 1.

Superposition of molecules

All molecular modeling studies, CoMFA and CoMSIA, were performed using the Sybyl 6.91 software of Tripos running on a SGI (Silicon Graphics, Inc.) workstation. Compound16 (Rynaxypyr™) was used in the systematic conformational search. First, all of the rotatable bonds in compound 16 were varied by using a step of 10°. Then, the lowest energy conformation identified in this conformational search was used as a template to build the other molecular structures. Each structure was fully geometry – optimized using a conjugate gradient minimization algorithm based on the Tripos force field and Gasteiger – Hückel charges and then aligned by an atom – by – atom least – square fit. We used the backbone of the 16 in its optimized conformation as a template, the atoms marked with an asterisk were used for rms – fitting onto the corresponding atoms of the template structure as shown in Tables 1 (Series Ⅰ), the superposition of all 29 compounds as shown in Fig 1.

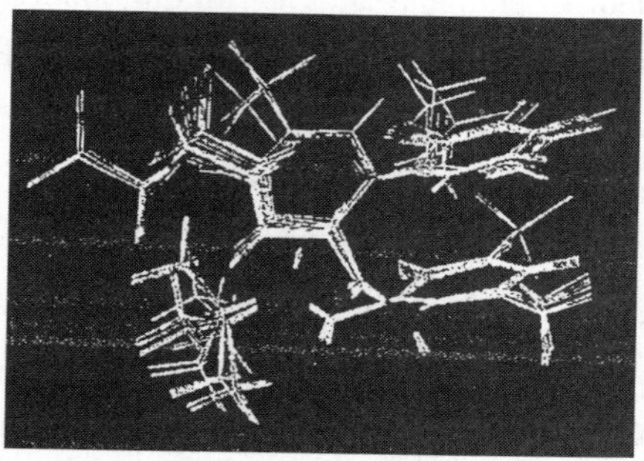

Fig. 1 Super position of 29 anthranilic diamides in the training and test sets

Table 1 Structures, experimental activities and predicted activities by the 3D-QSAR model from CoMFA and CoMSIA analysis of anthranilic diamides

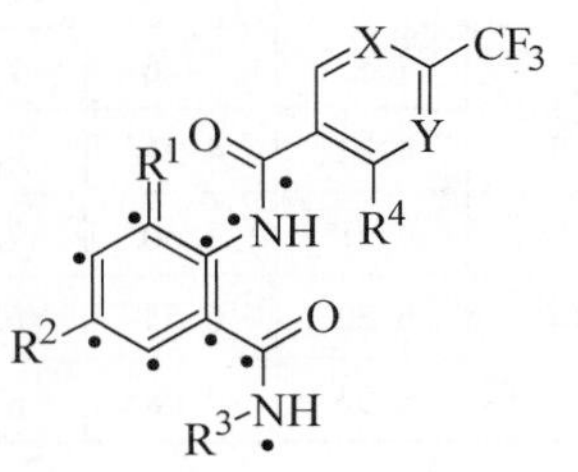

I 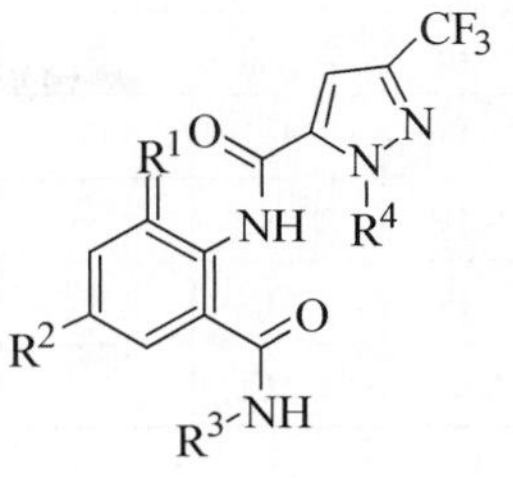　　Ⅱ 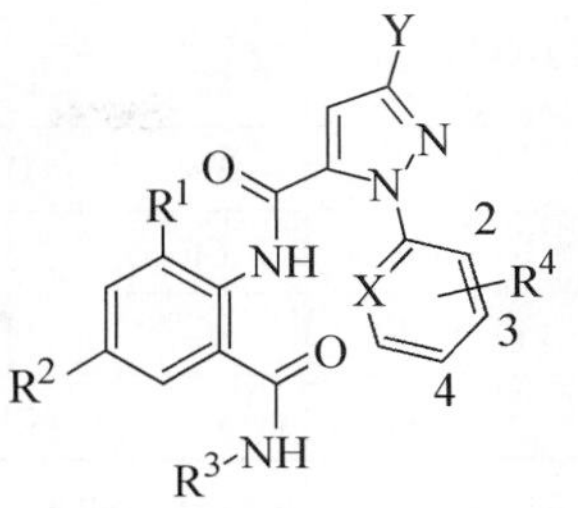　　Ⅲ

No.	Series	R^1	R^2	R^3	R^4	X	Y	pLC_{50}	Predicted	
									CoMFA	CoMSIA
1	I	Me	H	i-Pr	Me	C	C	4.70	4.54	4.65
2	I	Me	H	i-Pr	Me	N	N	4.15	4.25	4.51
3	Ⅱ	Me	H	i-Pr	Me	—	—	4.31	4.26	4.42
4	Ⅱ	Me	H	i-Pr	Et	—	—	4.05	4.18	4.41
5	Ⅱ	Me	H	i-Pr	i-Pr	—	—	4.64	4.87	4.53
6	Ⅲ	Me	H	i-Pr	H	C	CF_3	4.32	4.15	4.31
7	Ⅲ	Me	H	i-Pr	2-Cl	C	CF_3	6.70	5.32	5.10
8	Ⅲ	Cl	H	i-Pr	2-Cl	C	CF_3	6.40	6.25	5.58
9	Ⅲ	Me	H	i-Pr	H	C	CF_3	5.54	5.45	5.16
10	Ⅲ	Me	H	i-Pr	2-F	C	CF_3	6.28	5.13	4.80
11	Ⅲ	Me	H	i-Pr	3-Cl	C	CF_3	3.30	4.13	4.55
12	Ⅲ	Me	H	i-Pr	4-Cl	C	CF_3	3.30	4.28	4.41
13	Ⅲ	Me	H	i-Pr	2-Cl	N	CF_3	7.0	7.25	7.04
14	Ⅲ	Me	Cl	Me	2-Cl	N	CF_3	7.70	7.46	7.14
15	Ⅲ	Me	Cl	i-Pr	2-Cl	N	CF_3	7.52	7.73	7.32
16	Ⅲ	Me	Cl	Me	2-Cl	N	Br	7.70	7.72	7.14
17	Ⅲ	Me	Cl	i-Pr	2-Cl	N	Br	7.40	7.84	7.34
18	Ⅲ	Me	Cl	Me	2-Cl	N	Cl	7.52	7.26	7.13
19	Ⅲ	Me	Cl	i-Pr	2-Cl	N	Cl	7.30	7.07	7.09
20	Ⅲ	Me	Br	i-Pr	2-Cl	N	CF_3	7.52	6.74	7.03
21	Ⅲ	Me	Br	Me	2-Cl	N	Br	6.74	7.18	7.23
22	Ⅲ	Me	1	Me	2-Cl	N	CF_3	6.59	6.67	7.27
23	Ⅲ	Me	1	Me	2-Cl	N	Br	6.89	6.88	7.32
24	Ⅲ	Me	CF_3	Me	2-Cl	N	CF_3	6.28	6.46	7.15
25	Ⅲ	Me	CF_3	i-Pr	2-Cl	N	CF_3	6.41	6.87	7.20
26	Ⅲ	Me	Cl	Me	2-Cl	N	OCH_3	6.71	6.46	6.62
27	Ⅲ	Me	Cl	Me	2-Cl	N	OCF_2H	6.68	6.53	6.73
28	Ⅲ	Me	Cl	i-Pr	2-Cl	N	OCF_2H	6.85	6.85	6.86

Table 1(continued)

No.	Series	R^1	R^2	R^3	R^4	X	Y	pLC_{50}	Predicted	
									CoMFA	CoMSIA
29	Ⅲ	Me	Cl	*i* – Pr	2 – Cl	N	OCH_2CF_3	7. 52	7. 26	7. 00
30①	I	Me	H	*i* – Pr	Me	C	N	4. 34	4. 34	4. 61
31①	Ⅲ	Cl	H	*i* – Pr	2 – Cl	N	CF_3	7. 0	7. 30	7. 11
32①	Ⅲ	Me	Cl	*i* – Pr	2 – Cl	N	OCH_3	6. 52	6. 53	6. 64
33①	Ⅲ	Me	Cl	Me	2 – Cl	N	OCH_2CF_3	6. 96	7. 08	6. 90

①These compounds were used as a test set.

CoMFA and CoMSIA modeling

The steric and electrostatic potential fields for CoMFA were calculated at each lattice intersection of a regularly spaced grid of 2. 0Å. The lattice was definde automatically and was extended 4Å units past Van der Waals volume of all molecules in X, Y, and Z directions. An sp^3 carbon atom with Van der Waals radius of 1. 52A and +1. 0 charge served as the probe atom to calculate steric(Lennard – Jones 6 – 12 potential) field energies and electrostatic (Columbic potential) fields with a distance – dependent dielectric at each lattice point. The steric and electrostatic contributions were truncated to 30. 0kcal/mol, and electrostatic contributions were ignored at lattice intersections with maximum steric interactions. The CoMFA steric and electrostatic fields generated were scaled by CoMFA standard option given in SYBYL. In the CoMSIA analyses, similarity is expressed in terms of steric occupancy, electrostatic interactions, local hydrophobicity, and H – bond donor and acceptor properties, as the same method of CoMFA. The experimental and predicted activity values for the training and test set molecules by the 3D – QSAR model from CoMFA and CoMSIA analysis are given in Table 1. Both the CoMFA and CoMSIA models obtained exhibited a good predictability on these compounds.

Partial Least – Square(PLS) calcualations and validations

PLS methodology was used for all 3D – QSAR analyses[11,12], in which the CoMFA and CoMSIA descriptors were used as independent variables and LC_{50} values were used as dependent variables. The cross validation with leave – one – out(LOO) option and the SAMPLS program[13], rather than column filtering, were carried out to obtain the optimum number of components to be used in the final analysis. After the optimum number of components(n) was determined, a non – cross – validated analysis was performed without column filtering. The cross – validated correlation coefficient q^2, non – cross – validated correlation coeffcient r^2, and F values and standard error of estimate(SE) valued were computed according to the definitions in SYBYL.

Result and discussion

CoMFA and CoMSIA analysis results

The results of CoMFA and CoMSIA studies are summarized in Table 2. The number of components in the PLS models was three. The two models had a high cross – validated correlation coefficient(q^2 >0. 7) and non – cross – validated correlation coefficient(r^2 >0. 85) , a low standard error of estimate(SE) and a high Fischer ratio(F). The CoMFA analyses revealed that contributions of steric field and electrostatic field was 67. 4% and 32. 6% , respectively, the steric field had the major contribution in the model. In CoMSIA model, contributions of steric field, electrostaticfield, hydrophobic field and hydrogen bond donor and acceptor field was 7. 1% , 23. 2% ,42. 7% ,26. 4% and 0. 5% , respectively, the hydrophobic field had the major contribution in the model.

Table 2 CoMFA and CoMSIA analysis results

Model	Cross – validated			Conventional		Relative contributions(%)				
	q^2	n	r^2	SE	F	Steric	Electro – static	Hydro – phobic	Hydrogen bond	Hydrogen receptor
CoMFA	0. 720	3	0. 894	0. 465	81. 449	67. 4	32. 6			
CoMSIA	0. 732	3	0. 850	0. 554	54. 706	7. 1	23. 2	42. 7	26. 4	0. 5

CoMFA and CoMSIA coefficient contour plots

The coefficient contour plots are helpful to identify important regions where any change in the steric, electrostatic, hydrophobic fields and hydrogen bond donor and acceptor field may affect the biological activity. The CoMFA and CoMSIA coefficient contour plots are shown in Fig. 2 and Fig. 3. According to the CoMFA steric maps(Fig. 2(a)) , which had the major contribution

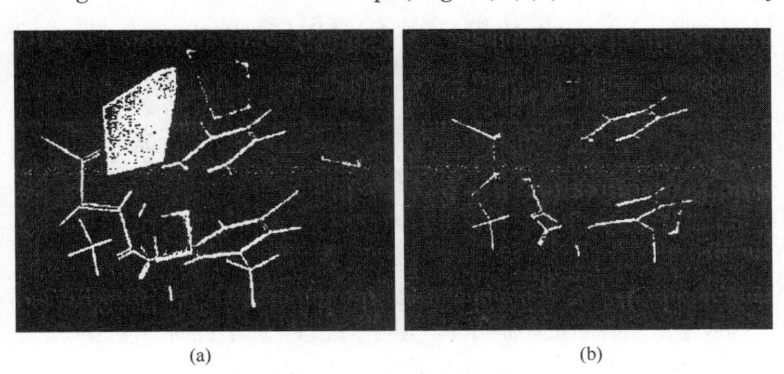

(a) (b)

Fig. 2 CoMFA model

(a) steric field; (b) electrostatic field

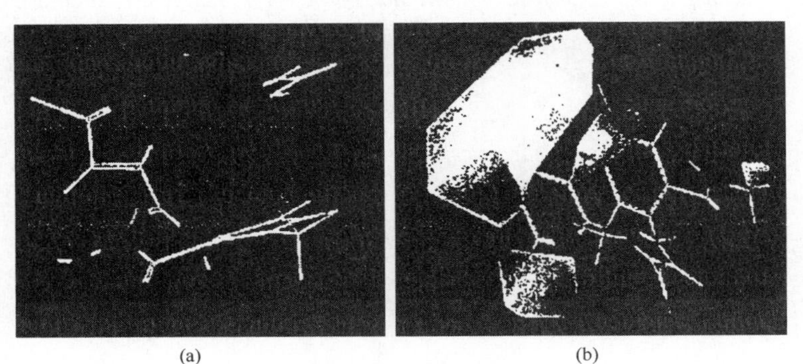

(a) (b)

Fig. 3 CoMSIA model

(a) electrostatic field; (b) hydrophobic filed

in the CoMFA model, the green contour defines a region where bulkier substituents at pyridyl position may give compounds with improved activity. For the electrostatic maps, CoMFA and CoMSIA analyses reveal essentially similar results here. The blue contour defines a region where increasing positive charge will result in increasing the activity, whereas the red contour defines a region of space where increasing electron density is favorable. A predominant feature of the electrostatic plot is the presence of a red contour surrounding the pyridylpyrazole ring. It could be reasonably presumed that there is a significant electrostatic interaction between the pyridylpyrazole ring and the possible receptor, and it may be assumed that the faction of receptor around the red region is electropositive. Gray and yellow contours of the currently reported CoMSIA model in Fig. 3(b) indicated the areas where hydrophilic and hydrophobic properties were preferred, respectively, and will be useful in selecting specific areas of the molecules to be utilized for adjusting the lipophilicity and hydrophilicity to improve insecticidal activity. According to the CoMSIA hydrophobic map, hydrophobic residues at 2 – substitution position of phenylpyrazole or pyridylpyrazole (Series Ⅲ) are preferred for increasing their insecticidal activity. This can be seen from the activities of compounds 7, 8 and 13 that possess hydrophobic groups at the 2 – substituent(R^4) of phenylpyrazole or pyridylpyrazole.

References

[1] Lahm G. P., Stevenson T. M., Sellby T. P., *et al.*, *Bioorg. Med. Chem. Lett.*, 2007, 17, 6274.

[2] Lahm G. P., Selby T. P., W. O. 2005118552, 2005[*Chem, Abstr.* 2005, 144, 5157 4].

[3] Jeanguenat A., O'sullivan A. C., W. O. 2006061200, 2006[*Chem Abstr.* 2006, 145, 62886].

[4] Hughes K. A., Lahm G. P., Selby T. P., *et al.*, W. O. 2004067528, 2004 [*Chem. Abstr.* 2004, 141, 190786].

[5] Ebbinghaus – Kintscher U., Luemmen P., Lobitz. N., *et al.*, *Cell Calcium*, 2006, 39, 21.

[6] Cordova D., Benner E. A., Sacher M. D., *et al.*, *Pest. Biochem. Physiol*, 2006, 84, 196.

[7] Cramer R. D., Patterson D. E., Bunce J. D., *J. Am. Chem. Soc.* 1988, 110, 5959.

[8] Klebe G., Abraham U., Mietzner T., *J. Med. Chem.* 1994, 37, 4130.

[9] Klebe G., Abraham U., *J. Comput. – Aided Mol. Des.* 1999, 13, 1.

[10] Lahm G. P., Selby T. P., Freudenberger J. H., *et al.*, *Bioorg. Med. Chem. Lett*, 2005, 15, 4898.

[11] Wold S., Rhue A., Wold H., *et al.*, *SIAM J. Sci. Stat. Comput.* 1984, 5, 735.

[12] Clark M., Cramer R. D., *Quant. Struct. – Act. Relat.* 1993, 12, 137.

[13] Bush B. L., Nachbar R. B., *J. Comput. – Aided Mol. Des.* 1993, 7, 587.

Received on September 1, 2008.

Design, Synthesis, and Fungicidal Activity of Novel Analogues of Pyrrolnitrin*

Mingzhong Wang, Han Xu, Qi Feng, Lizhong Wang, Suhua Wang, Zhengming Li

(State Key Laboratory of Elemento – organic Chemistry, Research Institute of Elemento – organic Chemistry, Nankai University, Tianjin, 300071, China)

Abstract A series of novel analogues of pyrrolnitrin containing a thiophene moiety were designed and synthesized by a facile method, and their structures were characterized by ^1H nuclear magnetic resonance (NMR) and high – resolution mass spectrometry. The isomers **IV – h** and **V – h** were isolated, and their structures were identified by 2D NMR, including heteronuclear multiple – quantum coherence (HMQC), heteronuclear multiple – bond correlation (HMBC), and nuclear Overhauser effect spectrometry (NOESY) spectra. Their fungicidal activities against five fungi were evaluated, and the results indicated that some of the title compounds showed excellent fungicidal activities *in vitro* against *Alternaria solani*, *Gibberella zeae*, *Physalospora piricola*, *Fusarium omysporum*, and *Cercospora arachidicola* at the dosage of 50μg · mL^{-1}. Some compounds shown moderate activity at low dosage. Compound **V – h** could be considered as a leading structure for further design of agricultural fungicides.

Key words Pyrrolnitrin; thiophene; fungicidal activities; leading structure; synthesized

Introduction

Pyrrolnitrin (A, Figure 1) is an antibiotic first isolated from the bacterial cells of *Pseudomonas* by Arima et al. [1], and the structure was established by the same author [1]. The antibiotic was totally synthesized by Nakano et al. in 1965 [1]. It was also isolated from a non – obligate predator bacterial strain 679 – 2, which has growth – inhibitory properties against bacteria, fungi, and other predator bacteria and yet with an ability to thrive in the absence of prey microorganisms [2]. Pyrrolnitrin is one of the most important natural products inhibiting *Mycobacterium tuberculosis* in early days [3], and this structure has been studied in crop protection. Fenpiclonil (**B**, Figure 1) and fludioxonil (**C**, Figure 1) were discovered as a novel class of agrochemical fungicides on the basis of the synthetic optimization of the natural structure [4]. Both compounds have broad – spectrum activity across many classes of fungi and are safe to mammals and the environment [4].

In the design of new fungicides, when the pyrrole group in pyrrolnitrin was replaced by furanone, pyrazole, and isoxazolinol, respectively, all of the analogues reported were found to be devoid of biological activity [5]. The thiophene and pyrrole rings are considered to be bioisosteric analogues, and there are many fungicides that have the thiophene group, such as bethoxazin [6], penthiopyrad [7], ethaboxam [8], etc. Therefore, in this paper, the pyrrole ring was replaced by the thiophene ring. To study the structure – activity relationship (SAR) of those analogues, the thiophene and aromatic rings of some analogues were nitrified to know whether that might im-

* Reprinted from *Journal of Agricultural and Food Chemistry*, 2009, 57(17): 7912 – 7918. Supported by the National Key Project for Basic Research and the National Natural Science Foundation of China.

prove or decrease the fungicide activity [9]. In addition, adding some halogen atoms to the aromatic ring might enhance the lipophilic ability and fungicidal activity of those compounds [4]. A total of 24 analogues (**D**, Figure 1) were synthesized with a facile synthetic method. The target compounds were evaluated for fungicide activity *in vitro* against five fungi, and some compounds showed moderate activity at low dosage.

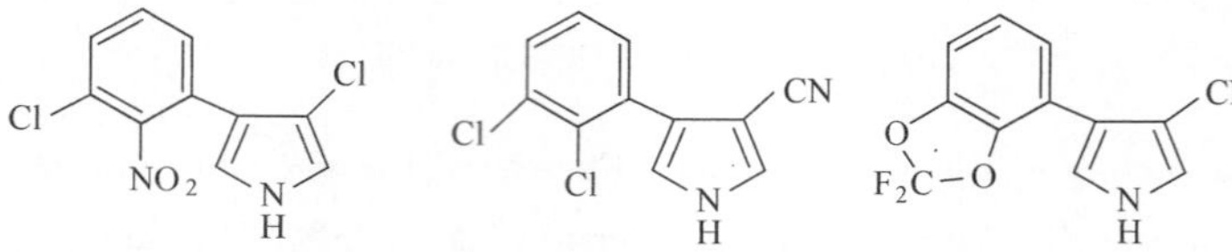

Pyrrolnitrin (**A**) Fenpiclonil (**B**) Fludioxonil (**C**)

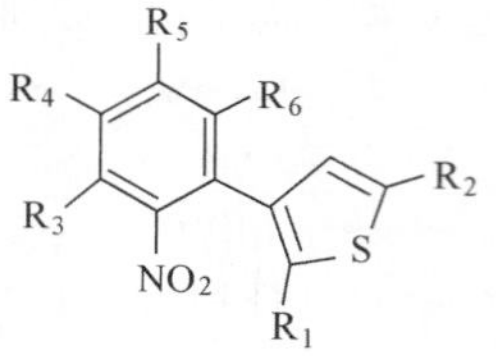

$R_1, R_2 = H, NO_2$
$R_3, R_4, R_5, R_6 = H, F, Cl, NO_2$

D

Figure 1 Chemical structures of compounds **A** – **D**

Materials and methods

Instruments

^1H nuclear magnetic resonance (NMR), ^{13}C NMR, heteronuclear multiple – quantum coherence (HMQC), heteronuclear multiplebond correlation (HMBC), and nuclear Overhauser effect spectrometry (NOESY) spectra were obtained at 300MHz using a Bruker AV300 spectrometer or at 400MHz using a Varian Mercury Plus 400 spectrometer in $CDCl_3$ solution, with tetramethylsilane as the internal standard. Chemical – shift values (δ) were given in parts per million (ppm). Highresolution mass spectrometry (HRMS) data were obtained on a VGZABHS instrument. The melting points were determined on an X – 4 binocular microscope melting point apparatus (Beijing Tech Instruments Co., Beijing, China) and were uncorrected. Yields were not optimized. The reagents were all analytically or chemically pure. All solvents and liquid reagents were dried by standard methods in advance and distilled before use. 1 – Bromo – 2 – nitro – benzene (**Ⅱ – a**) and all halogen – benzene were bought from the Alfa Aesar Company (Tianjin, China). Fenpiclonil (**B**) was bought from the laboratory of Dr. Ehrenstorfer – Schäfers (Augsburg, Germany).

General Synthetic Procedures for Ⅱ – b – Ⅱ – h

A mixture of 65% nitric acid (1.5mL) and 1 – bromo – 4 – fluorobenzene (**Ⅰ – b**, 3.4mL, 30mmol) in 98% sulfuric acid (12.0mL) was kept 15min at 0℃, then poured onto crushed ice (20g), and extracted with dichloromethane (2 × 20mL), and the solvent was removed. The residue was purified by flash chromatography on silica gel eluting with petroleum ether (60 – 90℃) to provide **Ⅱ – b** as a light yellow crystal[10]. Compounds **Ⅱ – c**, **Ⅱ – d**, **Ⅱ – e**, **Ⅱ – f**,

Ⅱ－g, and Ⅱ－h were prepared through the same process. In compounds Ⅱ－e and Ⅱ－f preparation procedures, the mixture was kept 40min at 0℃ and then poured onto crushed ice. In compounds Ⅱ－g preparation procedures, the mixture was kept 60min at 60℃ and then poured onto crushed ice. Compound Ⅱ－h was obtained from Ⅱ－c, and the mixture was kept 30min at 60℃ and then poured onto crushed ice. All of the substituents at the aromatic rings were listed in Scheme 1.

Data for **Ⅱ－b**. Yield, 61.1%. mp, 37－39℃ (acetone/petroleum ether). ^{1}H NMR (300MHz, $CDCl_3$) δ: 7.19－7.25 (m, 1H, Ar—H), 7.62 (dd, $^3J_{HH}$ = 7.8Hz, $^4J_{HF}$ = 3.0Hz, 1H, Ar—H), 7.74 (dd, $^3J_{HF}$ = 9.0Hz, ^4J_HH = 5.1Hz, 1H, Ar—H).

Data for **Ⅱ－c**. Yield, 59.2%. mp, 39－41℃ (acetone/petroleumether). ^{1}H NMR (300MHz, $CDCl_3$) δ: 7.15－7.19 (m, 1H, Ar—H), 7.39 (dd, $^3J_{HF}$ = 8.1Hz, $^4J_{HH}$ = 2.4Hz, 1H, Ar—H), 7.89 (dd, $^4J_{HF}$ = 5.4Hz, $^3J_{HH}$ = 9.0Hz, 1H, Ar—H).

Data for **Ⅱ－d**. Yield, 48.7%. mp, 65－67℃ (acetone/petroleum ether). ^{1}H NMR (300MHz, $CDCl_3$) δ: 7.42 (dd, $^3J_{HH}$ = 8.4Hz, $^4J_{HH}$ = 2.4Hz, 1H, Ar—H), 7.68 (d, $^3J_{HH}$ = 8.7Hz, 1H, Ar—H), 7.85 (s, 1H, Ar—H).

Data for **Ⅱ－e**. Yield, 90.0%. mp, 63－65℃ (acetone/petroleumether). ^{1}H NMR (300MHz, $CDCl_3$) δ: 7.61 (d, $^4J_{HH}$ = 2.4Hz, 1H, Ar—H), 7.66 (d, $^4J_{HH}$ = 2.4Hz, 1H, Ar—H).

Data for **Ⅱ－f**. Yield, 49.5%. mp, 24－26℃ (acetone/petroleum ether). ^{1}H NMR (400MHz, $CDCl_3$) δ: 7.76 (d, $^3J_{HF}$ = 8.0Hz, 1H, Ar—H), 7.83 (d, $^4J_{HF}$ = 6.8Hz, 1H, Ar—H).

Data for **Ⅱ－g**. Yield, 67.8%. mp, 100－102℃ (acetone/petroleum ether). ^{1}H NMR (300MHz, $CDCl_3$) δ: 7.73 (s, 1H, Ar—H).

Data for **Ⅱ－h**. Yield, 55.0%. mp, 95－97℃ (acetone/petroleum ether). ^{1}H NMR (300MHz, $CDCl_3$) δ: 7.78 (d, $^3J_{HF}$ = 9.6Hz, 1H, Ar—H), 8.72 (d, $^4J_{HF}$ = 6.9Hz, 1H, Ar—H).

Synthesis of 3－Bromo－thiophene

3－Bromo－thiophene was synthesized through an improved procedure referenced in the literature [11,12]: A solution of thiophene (93.0g, 1.1mol) in 130mL of chloroform was cooled to 0℃. Bromine (150mL, 2.9mol) was added dropwise over 4.5h and then refluxed for 3.5h, after which 200mL of 2M sodium hydroxide was added and stirred for 0.5h, then extracted with chloroform (150mL), washed with water, and dried, and then the solvent was removed to afford a light reddish oil, which was used in the subsequent step without further purification. In a 2L flask equipped with a distillation apparatus, zinc powder (220g, 3.3mol) was added to 500mL of water; glacial acetic acid was then added (210mL, 3.64mol). The crude product obtained above was added. The mixture was stirred and refluxed for 8h. The subsequent steps of distillation are the same as in the literature. Distillation afforded 3－bromo－thiophene as a colorless liquid. Yield, 80.8%. ^{1}H NMR (300MHz, $CDCl_3$) δ: 6.90 (dd, $^4J_{HH}$ = 2.1Hz, $^3J_{HH}$ = 4.2Hz, 1H), 7.08 (d, $^3J_{HH}$ = 4.2Hz, 1H), 7.10 (s, 1H).

General Synthetic Procedures for Target Compounds Ⅲ－a－Ⅲ－h

Compound **Ⅲ－a** was synthesized by a facile method according to the literature [13]. Cuprous iodide (7.6g, 39.8mmol) and 3 – bromo – thiophene (8.2g, 50.0mmol) were mixed together and heated to 160℃ for 1h. Then, the bath temperature was raised and maintained at 200℃. With stirring, compound **Ⅱ – a** (8.0g, 39.8mmol) was dissolved in 3 – bromo – thiophene (16.4g, 100.0mmol) before it was added dropwise into the above mixture. The process of adding would need 50min. During the above adding process, copper powder (24.0g, 375.0mmol) was added in five portions. After **Ⅱ – a** was added up, the mixture reacted for 15h while keeping the bath temperature between 200 and 210℃. Upon cooling, the products were extracted with dichloromethane (2 × 50mL) and the extraction was filtered through celite and evaporated to give a brown oil, which was purified by flash chromatography on silica gel eluting with petroleumether (60 – 90℃) to provide an orange oil **Ⅲ – a** (3.27g; yield, 40.1%). Compounds **Ⅲ – b**, **Ⅲ – c**, **Ⅲ – d**, **Ⅲ – e**, **Ⅲ – f**, **Ⅲ – g**, and **Ⅲ – h** were prepared according to the same process. All of the substituents at the aromatic rings were listed in Scheme 1.

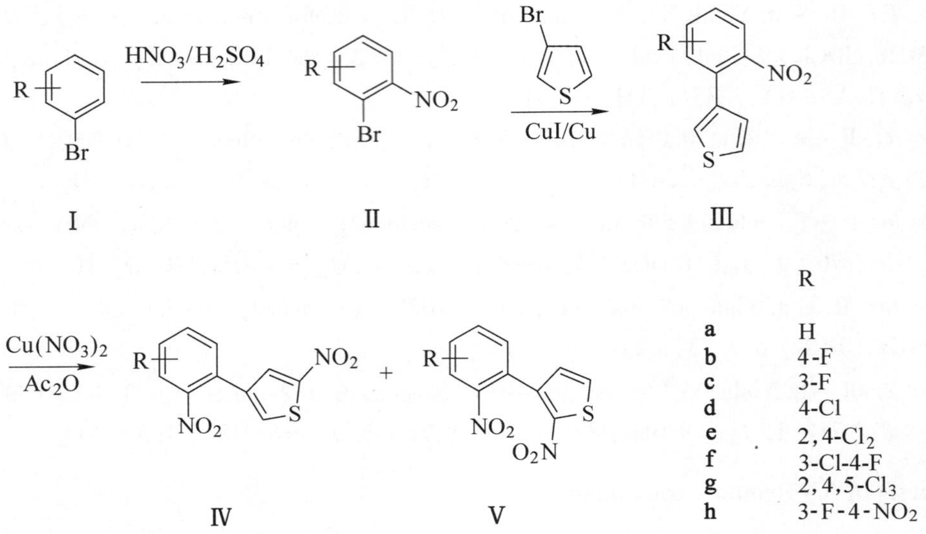

HNO_3/H_2SO_4 ; CuI/Cu ; $Cu(NO_3)_2$ / Ac_2O

I II III IV V

	R
a	H
b	4-F
c	3-F
d	4-Cl
e	2,4-Cl_2
f	3-Cl-4-F
g	2,4,5-Cl_3
h	3-F-4-NO_2

Scheme 1 Synthetic Routes to the Title Compounds **Ⅲ**, **Ⅳ**, and **Ⅴ**

Data for **Ⅲ – a**. Yield, 40.1%; oil. ^1H NMR (300MHz, $CDCl_3$) δ: 6.96 (dd, $^4J_{HH}$ = 1.2Hz, $^3J_{HH}$ = 5.1Hz, 1H, thiophene), 7.21 (d, $^4J_{HH}$ = 1.2Hz, 1H, thiophene), 7.25 (dd, $^4J_{HH}$ = 3Hz, $^3J_{HH}$ = 5.1Hz, 1H, thiophene), 7.32 (t, $^3J_{HH}$ = 7.5Hz, 1H, Ar—H), 7.36 (t, $^3J_{HH}$ = 7.8Hz, 1H, Ar—H), 7.44 (dd, $^3J_{HH}$ = 7.5Hz, $^3J_{HH}$ = 7.5Hz, 1H, Ar—H), 7.66 (d, $^3J_{HH}$ = 7.8Hz, 1H, Ar—H). HRMS, *m/z* 205.0156. Calcd for $C_{10}H_7NO_2S$, 205.0197. All of the data are the same as the known compound (13).

Data for **Ⅲ – b**. Yield, 39.8%; oil. ^1H NMR (400MHz, $CDCl_3$) δ: 7.06 (d, $^3J_{HH}$ = 4.8Hz, 1H, thiophene), 7.31 (d, $^4J_{HH}$ = 2.4Hz, 1H, thiophene), 7.33 (dt, $^4J_{HH}$ = 2.4Hz, $^3J_{HF}$ = 8.0Hz, 1H, Ar—H), 7.40 (dd, $^4J_{HH}$ = 3.2 Hz, $^3J_{HH}$ = 4.8Hz, 1H, thiophene), 7.49 (dd, $^4J_{HF}$ = 5.6Hz, $^3J_{HH}$ = 8.4Hz, 1H, Ar—H), 7.55 (dd, $^4J_{HH}$ = 2.4Hz, 3J_{HF} = 8.0Hz, 1H, Ar—H). HRMS, *m/z* 223.0103. Calcd for $C_{10}H_6FNO_2S$, 223.0109.

Data for **Ⅲ – c**. Yield, 42.2%; oil. ^1H NMR (400MHz, $CDCl_3$) δ: 7.07 (dd, $^3J_{HH}$ = 5.2Hz, $^4J_{HH}$ = 1.2Hz, 1H, thiophene), 7.12 – 7.16 (m, 1H, Ar—H), 7.19 (dd, $^4J_{HH}$ =

2.4Hz, $^3J_{HF}$ = 8.8Hz, 1H, Ar—H), 7.36 – 7.37 (m, 1H, thiophene), 7.41 (dd, $^4J_{HH}$ = 2.8Hz, $^3J_{HH}$ = 5.2Hz, 1H, thiophene), 7.88 (dd, $^4J_{HF}$ = 5.2Hz, $^3J_{HH}$ = 8.8Hz, 1H, Ar—H). HRMS, *m/z* 223.0103. Calcd for $C_{10}H_6FNO_2S$, 223.0113.

Data for **Ⅲ – d**. Yield, 34.8%; oil. ^{1}H NMR (400MHz, $CDCl_3$) δ: 7.07 (dd, $^4J_{HH}$ = 1.2Hz, $^3J_{HH}$ = 4.8Hz, 1H, thiophene), 7.33 – 7.34 (m, 1H, thiophene), 7.40 (dd, $^4J_{HH}$ = 2.8Hz, $^3J_{HH}$ = 4.8Hz, 1H, thiophene), 7.45 (d, $^3J_{HH}$ = 8.0Hz, 1H, Ar—H), 7.57 (dd, $^4J_{HH}$ = 2.0Hz, $^3J_{HH}$ = 8.0Hz, 1H, Ar—H), 7.81 (d, $^4J_{HH}$ = 2.0Hz, 1H, Ar—H). HRMS, *m/z* 238.9807. Calcd for $C_{10}H_6ClNO_2S$, 238.9806.

Data for **Ⅲ – e**. Yield, 45.3%. mp, 70 – 72℃ (acetone/petroleum ether). ^{1}H NMR (400MHz, $CDCl_3$) δ: 7.07 (dd, $^4J_{HH}$ = 1.2Hz, $^3J_{HH}$ = 4.8Hz, 1H, thiophene), 7.28 (dd, $^4J_{HH}$ = 1.2Hz, $^4J_{HH}$ = 2.8Hz, 1H, thiophene), 7.45 (dd, $^4J_{HH}$ = 2.8Hz, $^3J_{HH}$ = 5.1 Hz, 1H, thiophene), 7.69 (d, $^4J_{HH}$ = 2.0Hz, 1H, Ar—H), 7.71 (d, $^4J_{HH}$ = 2.4Hz, 1H, Ar—H). HRMS, *m/z* 272.9418. Calcd for $C_{10}H_5Cl_2NO_2S$, 272.9414.

Data for **Ⅲ – f**. Yield, 35.0%; oil. ^{1}H NMR (400MHz, $CDCl_3$) δ: 7.06 (dd, $^4J_{HH}$ = 1.2Hz, $^3J_{HH}$ = 4.8Hz, 1H, thiophene), 7.34 (dd, $^4J_{HH}$ = 1.2Hz, $^4J_{HH}$ = 2.8Hz, 1H, thiophene), 7.41 (dd, $^4J_{HH}$ = 2.8Hz, $^3J_{HH}$ = 4.8Hz, 1H, thiophene), 7.57 (d, $^4J_{HF}$ = 7.2Hz, 1H, Ar—H), 7.71 (d, $^3J_{HF}$ = 8.0Hz, 1H, Ar—H). HRMS, *m/z* 256.9713. Calcd for $C_{10}H_5ClFNO_2S$, 256.9747.

Data for **Ⅲ – g**. Yield, 45.8%; mp, 66 – 68℃ (acetone/petroleum ether). ^{1}H NMR (400MHz, $CDCl_3$) δ: 7.10 (d, $^3J_{HH}$ = 4.8Hz, 1H, thiophene), 7.37 (s, 1H, thiophene), 7.43 (dd, $^4J_{HH}$ = 2.8Hz, $^3J_{HH}$ = 5.1Hz, 1H, thiophene), 7.74 (s, 1H, Ar—H). HRMS, *m/z* 306.9028. Calcd for $C_{10}H_4Cl_3NO_2S$, 306.9027.

Data for **Ⅲ – h**. Yield, 35.8%; viscosity. ^{1}H NMR (400MHz, $CDCl_3$) δ: 6.98 (d, $^3J_{HH}$ = 5.2Hz, 1H, thiophene), 7.34 (t, $^3J_{HH}$ = 5.2Hz, 1H, thiophene), 7.37 (s, 1H, thiophene), 7.38 (s, 1H, Ar—H), 8.47 (d, $^3J_{HF}$ = 6.8Hz, 1H, Ar—H). HRMS, *m/z* 267.9955. Calcd for $C_{10}H_5FN_2O_4S$, 267.9954.

General Synthetic Procedures for Target Compounds **Ⅳ – a – Ⅴ – h**. Nitration of compound **Ⅲ – a** (2.0g, 9.8mmol) in 10mL of acetic anhydride was added dropwise to a solution of $Cu(NO_3)_2 \cdot 3H_2O$ (2.4g, 9.9mmol) in 10mL of acetic anhydride. The solution was held at 10 – 12℃ for 2h. Then, the copper salts were removed by filtration, and the residue was poured into ice water. Continuous extraction with dichloromethane (2 × 20mL) gave a thick oil, which was evaporated and purified by flash chromatography on silica gel eluting with petroleum ether (60 – 90℃) to provide a brown mixture, which consisted of 60% **Ⅳ – a** ($^4J_{HH}$ = 2.0Hz, thiophene) and 40% **Ⅴ – a** ($^3J_{HH}$ = 5.2Hz, thiophene), as determined by ^{1}H NMR [14,15]. Compounds **Ⅳ – a** and **Ⅴ – a** were isolated with preparative thin – layer chromatography [elution solvent: ethyl acetate/petroleum ether (60 – 90℃), 1∶9 (v/v)]. Compounds **Ⅳ – b – Ⅴ – h** were prepared according to the same process, except compounds **Ⅳ – g**, **Ⅴ – g**, **Ⅳ – h**, and **Ⅴ – h**, which were isolated with preparative thin – layer chromatography [elution solvent: chloroform/petroleum ether (60 – 90℃), 1∶3 (v/v)]. All of the substituents at the aromatic rings were listed in Scheme 1.

Data for **Ⅳ – a**. Yield, 44.6% . mp, 118 – 120℃ (acetone/petroleum ether). ^{1}H NMR (400MHz, $CDCl_3$) δ:7.48 (d, $^{3}J_{HH}$ = 7.6Hz, 1H, Ar—H), 7.49 (d, $^{4}J_{HH}$ = 2.0Hz, 1H, thiophene), 7.59 (t, $^{3}J_{HH}$ = 7.2Hz, 1H, Ar—H), 7.69 (t, $^{3}J_{HH}$ = 7.6Hz, 1H, Ar—H), 7.93 (d, $^{4}J_{HH}$ = 1.6Hz, 1H, thiophene), 7.98 (d, $^{3}J_{HH}$ = 8.0Hz, 1H, Ar—H). HRMS, *m/z* 250.0047. Calcd for $C_{10}H_6NO_2S$, 250.0048. All of the data are the same as the known compound (15).

Data for **Ⅳ – b**. Yield, 49.3% . mp, 97 – 99℃ (acetone/petroleum ether). ^{1}H NMR (400MHz, $CDCl_3$) δ:6.99 (d, $^{4}J_{HH}$ = 4.0Hz, 1H, thiophene), 7.43 (dt, $^{3}J_{HF}$ = 7.6Hz, $^{4}J_{HH}$ = 2.4Hz, 1H, Ar—H), 7.56 (dd, $^{3}J_{HH}$ = 8.4Hz, $^{4}J_{HF}$ = 5.6Hz, 1H, Ar—H), 7.71 (dd, $^{3}J_{HF}$ = 7.6, $^{4}J_{HH}$ = 2.4Hz, 1H, Ar—H), 7.89 (d, $^{4}J_{HH}$ = 4.0Hz, 1H, thiophene). HRMS, *m/z* 267.9936. Calcd for $C_{10}H_5FNO_2S$, 267.9954.

Data for **Ⅳ – c**. Yield, 49.0% . mp, 99 – 101℃ (acetone/petroleum ether). ^{1}H NMR (400MHz, $CDCl_3$) δ:7.02 (d, $^{4}J_{HH}$ = 4.0Hz, 1H, thiophene), 7.25 (dt, $^{3}J_{HF}$ = 7.6Hz, $^{4}J_{HH}$ = 2.4Hz, 1H, Ar—H), 7.33 (ddt, $^{3}J_{HF}$ = 9.6Hz, $^{3}J_{HH}$ = 9.2Hz, $^{4}J_{HH}$ = 2.4Hz, 1H, Ar—H), 7.89 (d, $^{4}J_{HH}$ = 4.0 Hz, 1H, thiophene), 8.06 (dd, $^{3}J_{HH}$ = 8.8Hz, $^{4}J_{HF}$ = 4.8, 1H, Ar—H). HRMS, *m/z* 267.9955. Calcd for $C_{10}H_5FNO_2S$, 267.9954.

Data for **Ⅳ – d**. Yield, 42.2% . mp, 136 – 138℃ (acetone/petroleum ether). ^{1}H NMR (400MHz, $CDCl_3$) δ:7.35 (d, $^{3}J_{HH}$ = 8.4Hz, 1H, Ar—H), 7.41 (d, $^{4}J_{HH}$ = 2.0Hz, 1H, thiophene), 7.58 (dd, $^{3}J_{HH}$ = 8.4Hz, $^{4}J_{HH}$ = 2.0Hz, 1H, Ar—H), 7.83 (d, $^{4}J_{HH}$ = 2.0Hz, 1H, thiophene), 7.90 (d, $^{4}J_{HH}$ = 2.0, 1H, Ar—H). HRMS, *m/z* 283.9651. Calcd for $C_{10}H_5ClNO_2S$, 283.9658.

Data for **Ⅳ – e**. Yield, 41.7% . mp, 96 – 98℃ (acetone/petroleum ether). ^{1}H NMR (400MHz, $CDCl_3$) δ:7.45 (d, $^{4}J_{HH}$ = 1.6Hz, 1H, thiophene), 7.50 (s, 1H, Ar—H), 7.82 (d, $^{4}J_{HH}$ = 2.0Hz, 1H, thiophene), 8.20 (s, 1H, Ar—H). HRMS, *m/z* 317.9266. Calcd for $C_{10}H_4Cl_2NO_2S$, 317.9268.

Data for **Ⅳ – f**. Yield, 46.1% . mp, 107 – 109℃ (acetone/petroleum ether). ^{1}H NMR (400MHz, $CDCl_3$) δ:7.01 (d, $^{4}J_{HH}$ = 4.0Hz, 1H, thiophene), 7.63 (d, $^{4}J_{HF}$ = 7.2Hz, 1H, Ar—H), 8.20 (d, $^{3}J_{HF}$ = 7.6Hz, 1H, Ar—H), 7.89 (d, $^{4}J_{HH}$ = 4.0Hz, 1H, thiophene). HRMS, *m/z* 301.9569. Calcd for $C_{10}H_4ClFNO_2S$, 301.9564.

Data for **Ⅳ – g**. Yield, 42.1% . mp, 105 – 107℃ (acetone/petroleum ether). ^{1}H NMR (400MHz, $CDCl_3$) δ:7.54 (d, $^{4}J_{HH}$ = 1.6Hz, 1H, thiophene), 7.79 (s, 1H, Ar—H), 7.91 (d, $^{4}J_{HH}$ = 2.0Hz, 1H, thiophene). HRMS, *m/z* 351.8868. Calcd for $C_{10}H_3Cl_3NO_2S$, 351.8879.

Data for **Ⅳ – h**. Yield, 34.2% . mp, 113 – 115℃ (acetone/petroleum ether). ^{1}H NMR (400MHz, $CDCl_3$) δ:7.48 (d, $^{3}J_{HF}$ = 10.0Hz, H – 6′), 7.64 (d, $^{4}J_{HH}$ = 1.6Hz, H – 2), 7.93 (d, $^{4}J_{HH}$ = 1.6Hz, H – 4), 8.79 (d, $^{4}J_{HF}$ = 6.8Hz, H – 3′). ^{13}C NMR (400MHz, $CDCl_3$) δ: 120.9, 121.2 (dd, C – 6′), 122.8 (d, C – 3′), 126.5 (d, C – 2), 129.8 (d, C – 4), 132.6 (s, C – 3), 135.4 (s, C – 1′), 142.6 (s, C – 5′), 152.2 (s, C – 5), 154.2 (s, C – 2′), 156.9 (s, C – 4′). HRMS, *m/z* 312.9802. Calcd for $C_{10}H_4FN_3O_6S$, 312.9804.

Data for **Ⅴ – a**. Yield, 29.4% . mp, 108 – 110℃ (acetone/petroleum ether). ^{1}H NMR (400MHz, $CDCl_3$) δ:7.01 (d, $^{3}J_{HH}$ = 5.2Hz, 1H, thiophene), 7.38 (dd, $^{3}J_{HH}$ = 7.6Hz, $^{4}J_{HH}$ =

1.2Hz,1H,Ar—H),7.59(d, $^3J_{HH}$ = 5.2Hz,1H,thiophene),7.64(dt, $^3J_{HH}$ = 8.0Hz, $^4J_{HH}$ = 1.6Hz,1H, Ar—H),7.72(dt, $^3J_{HH}$ = 7.6Hz, $^4J_{HH}$ = 1.2Hz,1H, Ar—H),8.25(dt, $^3J_{HH}$ = 8.4Hz, $^4J_{HH}$ = 1.2Hz,1H,Ar—H). HRMS, *m/z* 249.0057 [M − 1]$^+$. Calcd for $C_{10}H_6NO_2S$, 250.0048. All of the data are the same as the known compound (15).

Data for **V − b**. Yield, 24.0%. mp, 132 − 134℃ (acetone/petroleum ether). ^1H NMR (400MHz, $CDCl_3$) δ: 6.92(d, $^3J_{HH}$ = 5.2Hz,1H,thiophene),7.30(dt, $^3J_{HH}$ = 8.4Hz, $^4J_{HF}$ = 5.6Hz,1H, Ar—H),7.37(dt, $^3J_{HH}$ = 8.4Hz, $^3J_{HF}$ = 7.2Hz,1H, Ar—H),7.53(d, $^3J_{HH}$ = 5.2Hz,1H,thiophene),7.88(d, $^3J_{HF}$ = 8.0Hz,1H,Ar—H). HRMS, *m/z* 267.9947. Calcd for $C_{10}H_5FNO_2S$, 267.9954.

Data for **V − c**. Yield, 20.9%. mp, 150 − 152℃ (acetone/petroleum ether). ^1H NMR (400MHz, $CDCl_3$) δ: 6.92 (d, $^3J_{HH}$ = 5.6Hz,1H,thiophene),7.01(dd, $^3J_{HF}$ = 8.4Hz, $^4J_{HH}$ = 2.4Hz,1H, Ar—H),7.37(ddt, $^3J_{HF}$ = 9.2Hz, $^3J_{HH}$ = 7.2Hz, $^4J_{HH}$ = 2.4Hz,1H, Ar—H), 7.54(d, $^3J_{HH}$ = 5.2Hz,1H, thiophene),8.24(dd, $^3J_{HH}$ = 8.4Hz, $^4J_{HF}$ = 4.8,1H, Ar—H). HRMS, *m/z* 267.9966. Calcd for $C_{10}H_5FNO_2S$, 267.9954.

Data for **V − d**. Yield, 26.0%. mp, 108 − 110℃ (acetone/petroleum ether). ^1H NMR (400MHz, $CDCl_3$) δ: 6.91 (d, $^3J_{HH}$ = 5.2Hz,1H,thiophene),7.25(d, $^3J_{HH}$ = 8.0Hz,1H, Ar—H),7.54 (d, $^3J_{HH}$ = 5.6Hz,1H,thiophene),7.62 (dd, $^3J_{HH}$ = 8.0Hz, $^4J_{HH}$ = 2.0Hz,1H, Ar—H), 8.16 (d, $^4J_{HH}$ = 2.0, 1H, Ar—H). HRMS, *m/z* 283.9639. Calcd for $C_{10}H_5ClNO_2S$, 283.9658. All of the data are the same as the known compound (6).

Data for **V − e**. Yield, 24.5%. mp, 118 − 120℃ (acetone/petroleum ether). ^1H NMR (400MHz, $CDCl_3$) δ: 7.01(d, $^3J_{HH}$ = 5.2Hz,1H, thiophene),7.47(s,1H, Ar—H),7.63 (d, $^3J_{HH}$ = 5.2Hz,1H,thiophene),8.52(s,1H,Ar—H). HRMS, *m/z* 317.9265. Calcd for $C_{10}H_4Cl_2NO_2S$, 317.9268.

Data for **V − f**. Yield, 20.2%. mp, 125 − 127℃ (acetone/petroleum ether). ^1H NMR (400MHz, $CDCl_3$) δ: 6.99(d, $^3J_{HH}$ = 5.6Hz,1H,thiophene),7.47(d, $^4J_{HF}$ = 7.2Hz,1H,Ar—H),7.63(d, $^3J_{HH}$ = 5.6Hz,1H,thiophene),8.08(d, $^3J_{HF}$ = 8.0Hz,1H,Ar—H). HRMS, *m/z* 301.9569. Calcd for $C_{10}H_4ClFNO_2S$, 301.9564.

Data for **V − g**. Yield, 20.4%; oil. ^1H NMR(400MHz, $CDCl_3$) δ: 7.03(d, $^3J_{HH}$ = 4.0Hz, 1H,thiophene),7.81(s,1H, Ar—H),7.89(d, $^3J_{HH}$ = 4.0Hz,1H, thiophene). HRMS, *m/z* 351.8876. Calcd for $C_{10}H_3Cl_3NO_2S$, 351.8879.

Data for **V − h**. Yield, 22.1%; viscosity. ^1H NMR (400MHz, $CDCl_3$) δ: 7.06(d, $^3J_{HH}$ = 4.0Hz,H − 4),7.49(d, $^3J_{HF}$ = 10.0Hz,1H,H − 6′),7.85 (d, $^3J_{HH}$ = 4.0Hz,H − 5),8.68 (d, $^4J_{HF}$ = 6.8Hz,1H,H − 3′). ^{13}C NMR (400MHz, $CDCl_3$) δ: 121.6,121.8 (dd,C − 6′), 122.8 (d,C − 3′),127.2 (d,C − 4),127.4 (d,C − 5),133.0 (s,C − 3),135.8 (s,C − 1′),139.1 (s,C − 2),143.1 (s,C − 5′),153.9 (s,C − 2′),156.6 (s,C − 4′). HRMS, *m/z* 312.9817. Calcd for $C_{10}H_4FN_3O_6S$, 312.9804.

Bioassays

The fungicidal activity of the compounds **Ⅲ − a − Ⅲ − h** and **Ⅳ − a − V − h** were tested *in vitro* against *Alternaria solani*, *Gibberella zeae*, *Physalospora piricola*, *Fusarium omysporum*, and

Cercospora arachidicola, and their relative inhibitory ratio (%) has been determined using the mycelium growth rate method [16]. Fenpiclonil was used as a control. After the mycelia grew completely, the diameters of the mycelia were measured, and the inhibition rate is calculated according to the formula:

$$I = (D1 - D2)/D1 \times 100\%$$

In the formula, I is the inhibition rate, $D1$ is the average diameter of mycelia in the blank test, and $D2$ is the average diameter of mycelia in the presence of those compounds. The inhibition ratios of those compounds at the dose of 50, 20, and 10μg · mL^{-1} have been determined, and the experimental results are summarized in Table 1. The EC_{50} values of the high fungicidal activity compounds **Ⅲ-g**, **Ⅲ-h**, **Ⅳ-h**, and **Ⅴ-h** and fenpiclonil have been calculated by the Scatchard method. The results are summarized in Table 2.

Table 1 Fungicidal Activity of the Compounds Ⅲ-a – Ⅴ-h①

compound	concentration (μg · mL^{-1})	*A. solani* inhibition (%)	*G. zeae* inhibition (%)	*P. piricola* inhibition (%)	*F. omysporum* inhibition (%)	*C. arachidicola* inhibition (%)
Ⅲ-a	50	75	65	88	68	18
	20	30	43	68	31	0
	10	0	13	20	12	0
Ⅲ-b	50	55	39	80	75	0
	20	35	30	20	50	0
	10	10	8	0	0	0
Ⅲ-c	50	75	65	72	87	25
	20	25	43	36	50	12
	10	5	21	0	18	0
Ⅲ-d	50	55	39	80	75	25
	20	20	21	38	43	6
	10	0	13	14	25	0
Ⅲ-e	50	60	39	82	87	25
	20	55	26	72	81	18
	10	20	8	42	56	0
Ⅲ-f	50	60	47	80	87	25
	20	35	30	74	43	18
	10	0	0	64	18	0
Ⅲ-g	50	49	47	90	36	50
Ⅲ-h	50	68	100	100	100	100
Ⅳ-a	50	22	25	20	0	20
	20	16	7	9	0	16
	10	6	1	0	0	8
Ⅳ-b	50	41	12	46	25	44
	20	25	0	10	17	27
	10	6	0	4	7	11
Ⅳ-c	50	35	16	27	28	33
	20	16	0	8	10	22
	10	0	0	0	0	0

Table 1(continued)

compound	concentration (μg · mL⁻¹)	*A. solani* inhibition (%)	*G. zeae* inhibition (%)	*P. piricola* inhibition (%)	*F. omysporum* inhibition (%)	*C. arachidicola* inhibition (%)
Ⅳ-d	50	19	4	2	17	2
	20	0	0	0	0	0
	10	0	0	0	0	0
Ⅳ-e	50	41	24	25	39	27
	20	32	16	17	32	16
	10	25	8	10	21	5
Ⅳ-f	50	25	0	14	10	16
	20	16	0	2	7	11
	10	0	0	0	0	0
Ⅳ-g	50	34	0	24	4	35
	20	26	0	10	0	20
	10	17	0	0	0	5
Ⅳ-h	50	58	54	100	60	100
Ⅴ-a	50	25	23	30	22	25
	20	16	14	25	13	16
	10	9	6	0	0	8
Ⅴ-b	50	48	0	14	17	16
	20	25	0	4	7	0
	10	9	0	0	0	0
Ⅴ-c	50	29	0	10	21	11
	20	16	0	0	0	0
	10	0	0	0	0	0
Ⅴ-d	50	9	0	25	17	0
	20	0	0	10	0	0
	10	0	0	0	0	0
Ⅴ-e	50	38	0	27	28	27
	20	32	0	17	14	11
	10	16	0	2	10	0
Ⅴ-f	50	61	20	42	39	18
	20	54	4	36	28	5
	10	48	0	12	14	0
Ⅴ-g	50	34	0	40	22	40
	20	26	0	13	0	25
	10	0	0	0	0	5
Ⅴ-h	50	100	100	100	100	85
fenpiclonil	50	100	100	100	18	100

①The data are the average of three duplicate results.

Table 2 EC_{50} Values of the Compounds **Ⅲ-g**, **Ⅲ-h**, **Ⅳ-h**, **Ⅴ-h** and Fenpiclonil

Ⅲ-g	$y=a+bx$	EC_{50}	R	Ⅲ-h	$y=a+bx$	EC_{50}	R
A. solani		>50		*A. solani*	$y=2.392+1.860x$	25.20	0.99
G. zeae		>50		*G. zeae*	$y=0.559+3.084x$	27.53	0.99
P. piricola	$y=3.399+1.023x$	36.67	0.96	*P. piricola*	$y=4.044+1.481x$	4.41	0.99
F. omysporum		>50		*F. omysporum*	$y=1.855+2.628x$	4.42	0.99
C. arachidicola		>50		*C. arachidicola*	$y=3.127+2.190x$	7.16	0.99
Ⅳ-h				Ⅴ-h			
A. solani	$y=2.868+1.398x$	33.49	0.99	*A. solani*	$y=3.461+1.045x$	29.62	0.94
G. zeae	$y=1.458+2.795x$	18.50	0.97	*G. zeae*	$y=3.104+2.059x$	8.33	0.95
P. piricola	$y=4.048+1.481x$	4.39	0.99	*P. piricola*	$y=3.854+2.070x$	3.57	0.99
F. omysporum	$y=3.584+0.893x$	38.44	0.94	*F. omysporum*	$y=3.932+1.434x$	5.55	0.99
C. arachidicola	$y=3.443+1.602x$	9.36	0.96	*C. arachidicola*	$y=3.798+1.487x$	6.42	0.99
fenpiclonil							
A. solani	$y=5.340+1.322x$	0.55	0.96				
G. zeae	$y=4.584+1.780x$	1.71	0.97				
P. piricola	$y=5.226+2.352x$	0.80	0.99				
F. omysporum		>50					
C. arachidicola	$y=5.477+1.153x$	0.38	0.95				

Results and discussion

Synthesis and Structure Elucidation. In the previous paper, compound **Ⅲ-a** was synthesized using 1-bromo-2-nitrobenzene and 3-iodothiophene by the Ullmann coupling reaction but the yield was not reported [13]. Under the same conditions, **Ⅲ-a** was synthesized using 1-bromo-2-nitrobenzene and 3-bromo-thiophene, which reacted for 20h with 20.0% yield and obtained the byproduct 2,2′-dinitrobiphenyl (yield, 80.0%). One moiecular equivalent of cuprous iodide was added, mixed with 3-bromothiophene, and then heated to 160℃ for 1h before 1-bromo-2-nitrobenzene was added. Compound **Ⅲ-a** (yield, 40.1%) and 2,2′-dinitrobiphenyl (yield, 59.9%) were obtained after 15h. The other target compounds were all synthesized in this novel way. Regrettably, 2-bromo-1-nitro-4-(trifluoromethyl) benzene, 1-bromo-3-nitrobenzene, and 2-bromo-1-nitro-3-(trifluoromethyl) benzene were not successful in obtaining the corresponding target compounds. Additionally, in the process of the synthesis of 3-bromo-thiophene, the production of bromization of thiophene in the first step was used in the subsequent step without further purification and had a satisfactory yield (80.8%).

Because the compounds **Ⅳ-a-Ⅴ-h** were not easily separated, they were finally isolated with the preparative layer chromatography. The structures of two important compounds (**Ⅳ-h** and **Ⅴ-h**) were elucidated by 1D and 2D NMR.

The molecular formula of **Ⅳ-h** was revealed as $C_{10}H_4FN_3O_6S$ by HRMS data $[M]^+$ (calcd., 312.9804; found, 312.9802). The ^1H and ^{13}C NMR (Data for **Ⅳ-h**) spectra showed the signals of six quaternary, four CH, carbon atoms. Considering the reagents, the

HMQC spectra showed that δ_H = 7.48 (d, H-6′) and δ_H = 8.79 (d, H-3′) belong to the Ar—H, which was confirmed by $^3J_{HF}$ = 10.0Hz and $^4J_{HF}$ = 6.8Hz; δ_H = 7.64 (d, H-2) and δ_H = 7.93 (d, H-4) belong to the thiophene ring, which was confirmed by the $^4J_{HH}$ = 1.6Hz. The chemical-shift value of C-6′ in the ^{13}C NMR spectra was split by Ar—F into two signals δ = 120.9 and 121.2, which was confirmed by the HMQC spectra.

On the basis of the HMBC spectra, the correlations between H-6′ (δ_H = 7.48) and C-1′, C-5′, C-2′, and C-4′ [δ_C = 135.4 (s, C-1′), 142.6 (s, C-5′), 154.2 (s, C-2′), and 156.9 (s, C-4′)], the correlations between H-3′ (δ_H = 8.79) and C-1′, C-5′, C-2′, and C-4′ [δ_C = 135.4 (s, C-1′), 142.6 (s, C-5′), 154.2 (s, C-2′), and 156.9 (s, C-4′)], the correlations between H-2 (δ_H = 7.64) and C-3 and C-4 [δ_C = 132.6 (s, C-3) and 129.8 (d, C-4)], the correlations between H-4 (δ_H = 7.93) and C-1′, C-2, C-3, and C-5 [δ_C = 135.4 (s, C-1′), 126.5 (d, C-2), 132.6 (s, C-3), and 152.2 (s, C-5)] indicated that the structure of **Ⅳ-h** should be as follows in Figure 2.

In the NOESY spectra, the correlations between H-6′ (δ_H = 7.48) and H-4 (δ_H = 7.93) indicated that H-6′ and H-4 are on the same side.

The ^{1}H and ^{13}C NMR (Data for **Ⅴ-h**) spectra of **Ⅴ-h** are the same as those of **Ⅳ-h** on the whole. However, there are evident differences in the 2D NMR as follows: In the HMBC spectra, the correlations between H-6′ (δ_H = 7.49) and C-1′, C-5′, C-2′, and C-4′ [δ_C = 135.8 (s, C-1′), 143.1 (s, C-5′), 153.9 (s, C-2′), and 156.6 (s, C-4′)], the correlations between H-3′ (δ_H = 8.68) and C-1′, C-5′, C-2′, and C-4′ [δ_C = 135.8 (s, C-1′), 143.1 (s, C-5′), 153.9 (s, C-2′), and 156.6 (s, C-4′)], the correlations between H-4 (δ_H = 7.06) and C-5, C-3, C-2, and C-1′ [δ_C = 127.4 (d, C-5), 133.0 (s, C-3), 139.1 (s, C-2), and 135.8 (s, C-1′)], the correlations between H-5 (δ_H = 7.85) and C-4 and C-2 [δ_C = 127.2 (d, C-4) and 139.1 (s, C-2)] indicated that the structure of Ⅴ-h should be as follows in Figure 2.

In the NOESY spectra, the correlations between H-4 (δ_H = 7.06) and H-5 (δ_H = 7.85) indicated that H-4 and H-5 are nearby.

Biological Assay and SAR. The data in Table 1 show the fungicidal activities against *A. solani*, *G. zeae*, *P. piricola*, *F. omysporum*, *and C. arachidicola* of the title compounds **Ⅲ-a-Ⅴ-h**. The data in Table 2 show the EC_{50} values of the high fungicidal activity compounds **Ⅲ-g**, **Ⅲ-h**, **Ⅳ-h**, and **Ⅴ-h** and the commercial fungicide fenpiclonil.

Fungicidal Activity against A. solani. The screening data of Table 1 indicated that, at the dosage of 50μg · mL^{-1}, some compounds of **Ⅲ-a-Ⅴ-h** exhibited excellent activity against *A. solani*. For instance, the inhibition activity of **Ⅲ-a-Ⅲ-h** were more than 50%, and the inhibition activity of **Ⅴ-h** was equal to the commercialized fenpiclonil. At the dosage of 20 and 10μg · mL^{-1}, the fungicidal activities evidently decreased and the fungicidal activities of compounds **Ⅳ-a-Ⅴ-h** are lower than those of **Ⅲ-a-Ⅲ-h**. It may be owing to the nitration of the thiophene ring that resulted in the activity to be decreased.

Fungicidal Activity against G. zeae. At the dosage of 50μg · mL^{-1}, compounds of **Ⅲ-a**, **Ⅲ-c**, and **Ⅳ-h** exhibited moderate activity and **Ⅳ-f**, **Ⅳ-g**, **Ⅴ-b-Ⅴ-e**, and **Ⅴ-g** exhibited no activity. However, the fungicidal activity of **Ⅲ-h** and **Ⅴ-h** were equal to the fenpiclonil at this dosage. As the concentration of compounds **Ⅲ－a－Ⅴ－h** declining, its biological activity decreased.

Fungicidal Activity against P. piricola. At the dosage of 50μg · mL^{-1}, compounds of **Ⅲ－a－Ⅲ－h** exhibited excellent activity, while **Ⅳ－a－Ⅳ－g** and **Ⅴ－a－Ⅴ－g** exhibited low activity against *P. piricola.* Compound **Ⅲ－g** showed 90% inhibition, which might be owing to the biological activity of adding halogen atoms to the aromatic ring. The fungicidal activities of **Ⅲ－h**, **Ⅳ－h**, and **Ⅴ－h** were the same as fenpiclonil at this dosage. Although at the low concentrations the fungicidal activities evidently declined, **Ⅲ－h**, **Ⅳ－h**, and **Ⅴ－h** still exhibited moderate activity, which can be shown from their EC_{50} values (Table 2). This might be owing to the biological activity of adding nitro to the aromatic ring.

Fungicidal Activity against F. omysporum. The biological activity rules of **Ⅲ－a－Ⅴ－h** against *F. omysporumare* generally the same as the test against *P. piricola.* The EC_{50} value of **Ⅴ－h** against *F. omysporum* is 5.55μg · mL^{-1}, while that of fenpiclonil is more than 50μg · mL^{-1} (Table 2).

Fungicidal Activity against C. arachidicola. The screening data of Table 1 indicated that most target compounds exhibited low activity against *C. arachidicola*, except for **Ⅲ－h**, **Ⅳ－h**, and **Ⅴ－h**. At the dosage of 50μg · mL^{-1}, **Ⅲ－h**, **Ⅳ－h**, and fenpiclonil exhibited 100% inhibition against *C. arachidicola* and **Ⅴ－h** showed 85% inhibition. Owning to adding nitro to the aromatic ring might be essential for high fungicidal activity.

The screening data of Tables 1 and 2 indicated that, athough the fungicidal activities of most target compounds are lower than that of fenpiclonil, compound **Ⅴ－h** has higher fungicidal activity than fenpiclonil against *F. omysporum.* Therefore, the **Ⅴ－h** could be developed as a leading compound for further structural optimization.

In conclusion, a novel and facile procedure for preparation analogues of pyrrolnitrin from 3－bromo－thiophene with corresponding substituted nitrobenzene was described, and 24 3－(2－nitrophenyl) thiophene derivatives were synthesized. The results of the bioassay showed that some of these title compounds exhibited favorable fungicidal activities against *A. solani*, *G. zeae*, *P. piricola*, *F. omysporum*, and *C. arachidicola* at the dosage of 50μg · mL^{-1}; **Ⅲ－h**, **Ⅳ－h**, and **Ⅴ－h** still exhibited moderate activity against *P. piricola* at low dosage. Although the fungicidal activities of most target compounds are lower than that of fenpiclonil, the EC_{50} value of **Ⅴ－h** against *F. omysporum* is 5.55μg · mL^{-1}, while that of fenpiclonil is more than 50μg · mL^{-1} (Table 2). Therefore, the **Ⅴ－h** could be developed as a leading compound for further structural optimization. The possible SAR is as follows: The nitration of the thiophene ring may result from the activity to be decreased. The adding of halogen atoms and nitro to the aromatic ring might be essential for high fungicidal activity.

Literature cited

[1] Nakano, H.; Umio, S.; Kariyone, K.; Tanaka, K.; Kishimoto, T.; Noguchi, H.; Ueda, I.; Nakamura, H.; Morimoto, Y. Total synthesis of pyrrolnitrin, a new antibiotic. *Tetrahedron Lett.* 1966, 7, 737－740.

[2] Cain, C. C.; Lee, D.; Waldo, R. H.; Henry, A. T.; Casida, E. J. J.; Wani, M. C.; Wall, M. E.; Oberlies, N. H.; Falkinham, J. O. Synergistic antimicrobial activity of metabolites produced by a nonobligatory bacterial predator. *Antimicrob. Agents Chemother.* 2003, 7, 2113－2117.

[3] Copp, B. R.; Pearce, A. N. Natural product growth inhibitors of *Mycobacterium tuberculosis*. *Nat. Prod. Rep.* 2007, 24, 278 – 297.

[4] Ackermann, P.; Beiter, K. G.; Zeun, R. Chemistry and biology of fludioxonil, fenpiclonil, and quinoxyfen. In *Modern Crop Protection Compounds*; Krämer, W., Schirmer, U., Eds.; Wiley: New York, 2007; pp 568 – 575.

[5] Garcia, E. E.; Benjamin, L. E.; Ian Fryer, R. Enamine intermediates in the synthesis of analogs of pyrrolnitrin. *J. Heterocycl. Chem.* 1974, 11 (2), 275 – 277.

[6] Davis, R. A.; Valcke, A. R. A.; Brouwer, W. G. Wood preservatives oxathiazines. W. O. Patent 9,506,043, 1995.

[7] Yoshikma, Y.; Tomiya, K.; Katsuta, H.; Kawashima, H.; Takahashi, O.; Inami, S.; Yanase, Y. Substituted thiophene derivative and agricultural and horticulture fungicide containing the same as active ingredient. U. S. Patent 5,747,518, 1998.

[8] Rew, Y. S.; Cho, J.; Ra, C. S.; Ahn, S. C.; Kim, S. K.; Lee, Y. H.; Jung, B. Y.; Choi, W. B.; Rhee, Y. H.; Yoon, M. Y.; Chun, S. W. 2 – Aminothiazolecarboxaminde derivatives, processes for preparing the same and use thereof for controlling phytopathogenic organisms. U. S. Patent 5,514,643, 1996.

[9] Shlley, R. H.; Forsberg, J. L.; Perry, R. S.; Dickerson, D. R.; Finger, G. C. Fungicidal activity of some fluoroaromatic compounds. *J. Fluorine Chem.* 1975, 5, 371 – 376.

[10] Manfred, S.; Assunta, G.; Frédéric, L. In search of simplicity and flexibility: Arational access to twelve fluoroindolecarboxylic acids. *Eur. J. Org. Chem.* 2006, 2956 – 2969.

[11] Nielsen, C. B.; Bjornholm, T. New regiosymmetrical dioxopyrroloand dihydropyrrolo – functionalized polythiophenes. *Org. Lett.* 2004, 6 (9), 3381 – 3384.

[12] Guo, H. C. The study of synthesis of 3 – bromo – thiophene. *Anhui Chem. Ind.* 2008, 1, 33 – 34.

[13] Barton, J. W.; Lapham, D. J.; Rowe, D. J. Intramolecular diazo coupling of 2 – aminophenylthiophenes. The formation of isomeric thieno[c]cinnolines. *J. Chem. Soc.*, *Perkin Trans.* I 1985, 131 – 133.

[14] Kellogg, R. M.; Buter, J. Cyclopropylthiophenes. Syntheses, reactions, and ultraviolet spectra. *J. Org. Chem.* 1971, 36 (16), 2236 – 2244.

[15] Gronowitz, S.; Gjos, N. Nitration of 2 – and 3 – phenylthiophene. Lund University, Lund, Sweden. *Acta Chem. Scand.* 1967, 21, 2823 – 2833.

[16] Chen, N. C. *Bioassay of Pesticides*; Beijing Agricultural University Press: Beijing, China, 1991; pp 161 – 162.

新型苯环5－（取代）苯甲酰胺基苯磺酰脲类化合物的比较定量构效关系研究*

王美怡[1,2]　马　翼[2]　李正名[2]　王素华[2]

（[1] 北方民族大学化学与化学工程学院，国家民委化工技术重点实验室，银川　750021；
[2] 南开大学元素有机化学研究所，元素有机化学国家重点实验室，天津　300071）

摘　要　采用比较分子力场分析（CoMFA）方法，对26个新型苯环5－（取代）苯甲酰胺基苯磺酰脲类化合物的除草活性进行了三维定量构效关系（3D－QSAR）研究，建立了三维定量构效关系CoMFA模型（$R^2=0.948$，$F=91.364$，SE＝0.141）。结果表明，此类磺酰脲类化合物的除草活性与苯环5位取代基的立体结构和电场性质密切相关。根据CoMFA模型的立体场和静电场三维等值线图不仅直观地解释了结构与活性的关系，而且为进一步设计高活性的目标化合物提供理论依据。

关键词　比较分子力场分析（CoMFA）　三维定量构效关系（3D－QSAR）　苯环5－（取代）苯甲酰胺基苯磺酰脲类化合物

磺酰脲类除草剂自20世纪80年代问世以来，便以其高效、低毒和高选择性等卓越特性在世界各地得到了广泛应用[1~4]。随着人们环境保护意识的不断加强，新农药不仅要研究开发具有高效、高选择性的品种，更重要的是低残留、无污染和环境相容性好。在目前已经商品化的磺酰脲类除草剂品种中，氟啶嘧磺隆（Flupyrsulfuron－Methyl）、碘甲磺隆（Iodosulfuron－Methyl）、甲磺胺磺隆（Mesosulfuron－Methyl）和甲酰胺磺隆（Foramsulfuron）这4个化合物的苯环5位均连有取代基，它们不但高效而且降解快速，是一类很受欢迎的除草剂品种[5,6]。

三维定量构效关系（3D－QSAR）研究作为计算机辅助药物设计的重要手段，日益受到人们的广泛关注，并将其应用到新药研发的领域中。比较分子力场（CoMFA）方法是研究药物－受体三维定量构效关系最为通用和标准的方法[7~11]，它能够充分考虑分子的三维结构信息，通过计算药物分子周围一定范围内的三维空间中假想网状格点上的立体场和静电场参数，再用偏最小二乘法（PLS）寻找这些三维特征信息与药物活性的联系。使用该方法建立的模型可以将药物分子周围的作用力场与药物分子的活性定量地联系起来，为进一步设计及合成此类高效化合物提供了一定的理论基础[12,13]。

本文采用比较分子力场分析方法对本课题组研发的26个新型苯环5－（取代）苯甲酰胺基苯磺酰脲类化合物进行了三维定量构效关系研究，讨论了其三维静电场与立体场空间结构与除草活性之间的关系，为进一步设计合成此类苯环5位含有取代基的磺酰脲类化合物提供有益参考。

1　实验部分

1.1　材料与方法

采用美国Tripos公司的SYBYL/6.91软件进行计算，在SGI350服务器和SGI Feul

＊　原发表于《高等学校化学学报》，2009，30（7）：1361－1364。基金项目：国家重点基础研究发展计划（批准号：2003CB114406）和天津市应用基础研究计划（批准号：07JCYBJC00200）资助。

图形工作站上完成。实验所用的26个新型苯环5－（取代）苯甲酰胺基苯磺酰脲类化合物为本实验室合成[14]。在定量构效关系模型中使用油菜盆栽法的茎叶处理数据表示活性，处理剂量为375g/hm^2，施药方式为喷施，处理20d后，测定地上部分鲜重，以鲜重抑制百分数来表示化合物的除草活性。

1.2 生物活性分子构象的优化及分子叠合

确定化合物的构象是建立定量构效关系（QSAR）的关键步骤之一。采用分子的最低能量构象作为药效构象。本文所研究的磺酰脲类化合物是在单嘧磺酯（代号NK94827）的基础上，对苯环5位进行取代基的结构修饰后所得到的一系列化合物。因此，对于26个化合物的能量最低构象的确定，是在单嘧磺酯晶体结构的基础上用分子力学的方法优化得到的。选用Tripos力场优化，Powell方法优化500轮，加载Gasteiger－Huckel电荷，得到各分子的低能构象。

在CoMFA分析中，构象分子的叠合是非常重要的步骤。应使所有分子按一定规则在空间叠合，以减小其空间结构的差别。考虑到本系列化合物的特征，所有化合物经构象优化后，以生物活性最高的化合物**1**为模板分子，将分子中有共同原子特征的骨架结构进行叠合（图1中“＊”所示原子）。

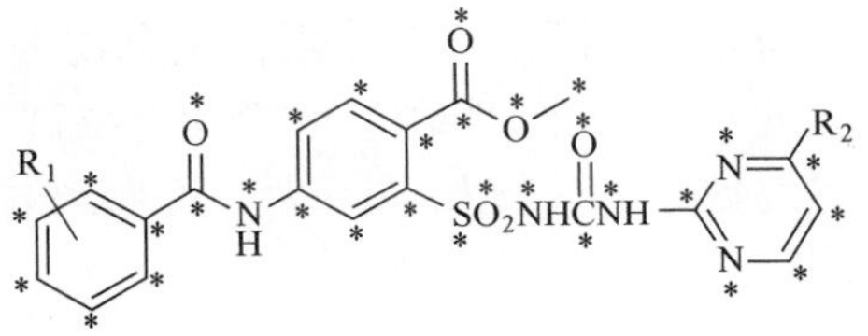

Fig. 1 Structure of sufonylurea compounds

采用Alignment database模块使余下化合物分子中的公共部分均与模板分子中的基本骨架相互重叠。采用此方法叠合保证每个分子力场的取向具有一致性。

1.3 CoMFA模型建立

选定三维网格边距为1nm×1nm×1nm，以带一个单位正电荷的sp^3杂化C原子为探针，步长为0.2nm，计算分子与网格点上探针原子的相互作用，建立CoMFA场。在偏最小二乘法（PLS）分析的第一步采用Leave－one－out方法进行交叉验证，得到最佳组分数（n）为4，交叉验证系数（q^2）为0.510（一般认为，交叉验证系数q^2值大于0.5时，模型具有可信预报能力）；第二步采用非交叉验证进行回归计算，获得CoMFA模型。

2 结果与讨论

由最佳组分数得到的CoMFA模型，$R^2=0.948$，$F=91.364$，标准偏差SE＝0.141，立体场的贡献为56.6%，静电场的贡献为43.4%。用CoMFA方法得到的3D－QSAR结果以三维等值线图表示。图2为26个化合物的分子叠合图。图3为化合物的相关性示意图。图4为用模板分子化合物**1**作为代表、采取stdev＊coeff方法表示的三维等值线图。用不同颜色表示此系列化合物分子的主要空间和静电能。红色和蓝色区域代表静电场效应，绿色和黄色区域代表立体场效应。应用CoMFA模型进行计算，对所研究的化合物进行活性预测所得的结果见表1。其中利用训练集所构建的CoMFA模型对测试集中的分子进行了活性预测，以验证模型的准确性。结果表明，预测值与实验值吻合度较高，差值在误差允许的范围内。表1中EPI[15]的数据处理方程为$EPI=\lg\{I/[(100-I)\times M_w]\}$，式中，$I$为实验测定的活性数据，$M_w$为化合物的相对分子质量。

PPI_{CoMFA}为应用 CoMFA 模型进行计算预测所得的活性数据。

Fig. 2 Overlay of all 26 compounds

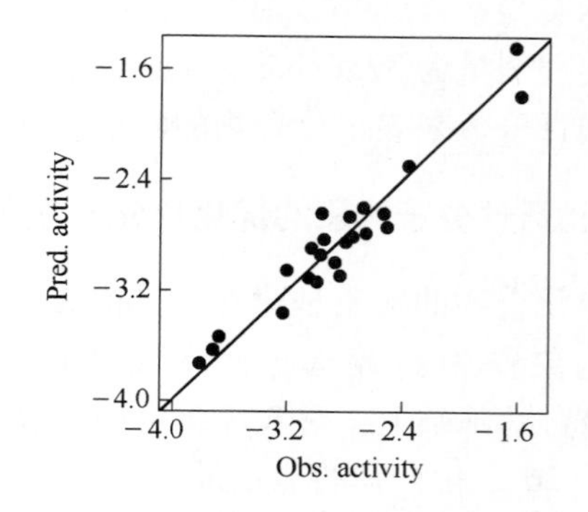

Fig. 3 Observed versus predicted activities in the training set and the testing set

由表 1 和图 3 可以看出，化合物的预测活性与实验值有良好的相关性。用该方法建立的 CoMFA 模型对这 26 个化合物进行的活性预报值 PPI_{CoMFA} 与由实验测定值 *I* 计算所得的数据 EPI 值相差不大，说明该模型能够对此系列化合物的活性进行了很好的预测。

Table 1 Structures and quantum chemical parameters of the compounds

Compound	R_1	R_2	*I*（Inhibit ratio）（%）	EPI	PPI_{CoMFA}	Residual
1	H	Me	92. 0	−1. 61	−1. 45	−0. 16
2	H	OMe	68. 7	−2. 34	−2. 34	0
3	*p*−Cl	Me	68. 9	−2. 36	−3. 25	5. 45
4	*p*−Cl	OMe	34. 2	−3. 00	−2. 89	−0. 11
5①	*o*−F	Me	92. 8	−1. 58	−1. 79	0. 21
6	*o*−F	OMe	59. 4	−2. 54	−2. 66	0. 12
7	*p*−F	Me	38. 2	−2. 90	−2. 99	0. 09
8	*p*−F	OMe	59. 9	−2. 53	−2. 70	0. 17
9①	*o*−Cl	Me	36. 1	−2. 95	−2. 66	−0. 29
10	*o*−Cl	OMe	46. 2	−2. 78	−2. 81	0. 03
11	*o*−NO_2	Me	36. 1	−2. 96	−2. 85	−0. 11
12	*o*−NO_2	OMe	48. 5	−2. 75	−2. 67	−0. 08
13	*p*−NO_2	Me	32. 1	−3. 04	−3. 11	0. 07
14	*p*−NO_2	OMe	4. 17	−2. 87	−2. 99	0. 12
15	*o*−OCH_3	Me	4. 15	−2. 85	−3. 08	0. 23
16	*o*−OCH_3	OMe	44. 2	−2. 81	−2. 82	0. 01
17	*p*−OCH_3	Me	17. 3	−3. 77	−3. 75	−0. 02
18	*p*−OCH_3	OMe	50. 7	−2. 70	−2. 63	−0. 07
19	*p*−Br	Me	23. 7	−3. 25	−3. 35	0. 10
20	*p*−Br	OMe	10. 4	−3. 68	−3. 62	−0. 06
21	*m*−F	Me	34. 6	−2. 96	−2. 93	−0. 03

续表 1

Compound	R_1	R_2	*I* (Inhibit ratio) (%)	EPI	PPI_{CoMFA}	Residual
22	*m* - F	OMe	30.1	-3.07	-3.08	0.01
23	2, 4 - Cl_2	Me	23.7	-3.23	-3.05	-0.18
24①	3, 4 - Cl_2	Me	54.1	-2.66	-2.76	0.10
25①	*p* - Me	Me	9.3	-3.67	-3.56	-0.11
26	*p* - Et	Me	4.8	-3.99	-4.01	0.02

①Compounds 5, 9, 24, 25 are the testing set, and all the other compounds are the training set.

从静电场等值线图（图4）可以看出，在红色区域，即5位取代苯甲酰胺基的邻位区域内引入负电荷基团；在蓝色区域，即5位取代苯甲酰胺基的间位和对位区域内引入正电荷的基团有助于增加化合物的除草活性。从立体场等值线图可以看出，减小黄色区域，即5位取代苯甲酰胺基与苯环共平面的对位区域的空间体积或增大绿色区域，即5位取代苯甲酰胺基的间位和对位区域的空间体积，均有助于增加化合物的除草活性。这说明并不是取代基碳链越长或体积越大，其活性就越好，其中应当有个最优条件，经过如此改造的分子与受体的几何匹配程度更好。模型结果还表明，分子体积和形状对于活性的影响稍大于静电效应对于活性的影响，这对于进一步设计合成此类化合物具有一定的指导意义。

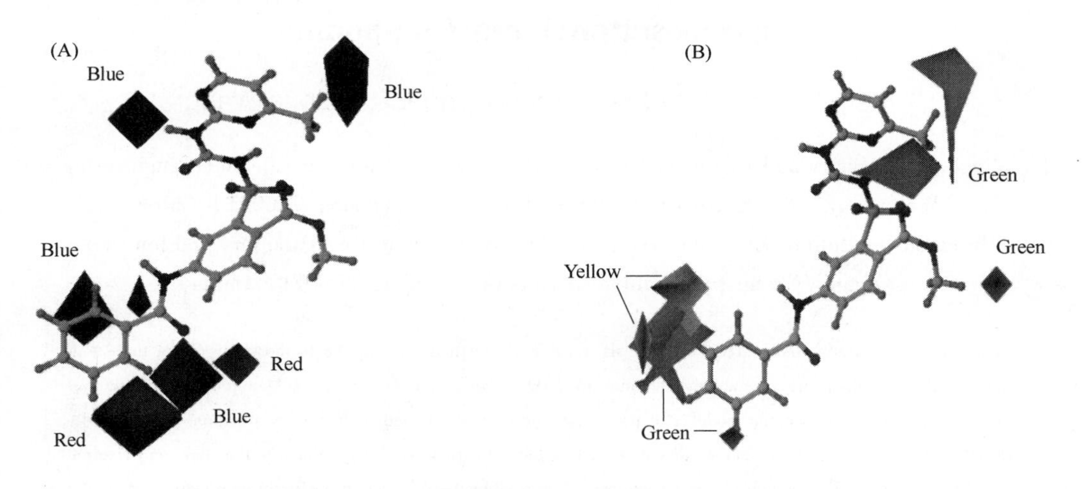

Fig. 4 3D con tour plots of the CoMFA model electrostatic field (A) and stericfield (B)

参 考 文 献

[1] Levitt G., Ploeg H. L., Hans L., *et al.*. J. Agric. Food Chem. [J], 1981, 29: 416-418.

[2] Levitt G., BakerD. R., Fenyes J. G., *et al.*. Synthesis and Chemistry of Agrochemicals Ⅱ, ACS Symposium Series No. 443 [M], Washington DC: American Chemistry Society, 1991: 16-17.

[3] LI Zheng - Ming (李正名), JIA Guo - Feng (贾国峰), WANG Ling - Xiu (王玲秀), *et al.*. Chem. J. Chinese Universities (高等学校化学学报) [J], 1994, 15 (3): 391-395.

[4] JIANG Lin (姜林), LI Zheng - Ming (李正名), GAO Fa - Wang (高发旺). Chin. J. Appl. Chem. (应用化学) [J], 2002, 19 (5): 416-419.

[5] Suzanne K. S., Gary M. D., DavidM. K., *et al.*. Pestic. Sci. [J], 1999, 55 (3): 288-300.

[6] Rouchaud J., NeusO., Cools K., *et al.*. J. Agric. Food Chem. [J], 1999, 47 (9): 3872-3878.

[7] MA Ning（马宁），LI Peng－Fei（李鹏飞），LI Yong－Hong（李永红）. Chem. J. Chinese Universities（高等学校化学学报）[J]，2004，25（12）：2259－2263.

[8] WANG Bao－Lei（王宝雷），MA Ning（马宁），WANG Jian－Guo（王建国），*et al.*. Acta Phys. Chim. Sin.（物理化学学报）[J]，2004，20（6）：577－581.

[9] WANG Xiao－Dong（王晓栋），LIN Zhi－Fen（林志芬），YIN Da－Qiang（尹大强）. Science in China，Series B（中国科学，B辑）[J]，2004，6：498－503.

[10] WANG Jian－Guo（王建国），CHEN Han－Song（陈寒松），ZHAO Wei－Guang（赵卫光）. Acta Chim. Sinica（化学学报）[J]，2002，60（11）：2043－2048.

[11] WU Bin（吴斌），LI Min－Yong（李敏勇），JIANG Zhen－Zhou（江振洲）. Chin. J. Org. Chem.（有机化学）[J]，2004，24（12）：1587－1594.

[12] Julio C.，Mario S.，Michael F.，*et al.*. J. Agric. Food Chem. [J]，2007，55：8101－8104.

[13] Dalelio G. F.，FesslerW. A.，Feigl D. M.，*et al.*. J. Macromol. Sci. Chem. [J]，1969，A3（5）：941－958.

[14] WANG Mei－Yi（王美怡），GUO Wan－Cheng（郭万成），LAN Feng（兰峰），*et al.*. Chin. J. Org. Chem.（有机化学）[J]，2008，28（4）：649－656.

[15] YANG Hua－Zheng（杨华铮），WU Ye（吴烨），REN Kang－Tai（任康太），*et al.*. Chem. J. Chinese Universities（高等学校化学学报）[J]，1995，16（5）：710－714.

3D－QSAR Study of Novel 5－Substituted Benzenesulfonylurea Compounds

Meiyi Wang[1,2]，Yi Ma[2]，Zhengming Li[2]，Suhua Wang[2]

（[1]College of Chemistry and Chemical Engineering，Key Laboratory for Chemical Engineering and Technology，The North University for Ethnics，Yinchuan，750021，China；
[2]Research Institute of Elemento－Organic Chemistry，State Key Labortary of Elemento－Organic Chemistry，Nankai University，Tianjin，300071，China）

Abstract Comparative molecular field analysis（CoMFA）method was applied to the study of the three－dimensional quantitative structure activity relationship（3D－QSAR）on 26 novel 5－substituted benzenesulfony－lurea compounds. A reasonable model with predictive ability was obtained from the investigation（$R^2=0.948$，$F=91.364$，SE＝0.141）. The results show that herbicidal activity is closely related to the steric and electronic properties of the 5－position of the benzene ring. The contour maps based on the analysis of steric and electrostatic CoMFA coefficients can not only explain the relationship between the structures and biological activity，but also lead to insight into the further design of highly active title compounds.

Key words comparative molecular field analysis（CoMFA）；three－dimensional quantitative structure activity relationship（3D－QSAR）；5－substituted benzenesulfonylurea compounds

Regioselective Synthesis of Novel 3 – Alkoxy(phenyl)thiophosphorylamido – 2 – (per – *O* – acetylglycosyl – 1′ – imino) Thiazolidine – 4 – one Derivatives from *O* – alkyl *N*⁴ – glycosyl(thiosemicarbazido) Phosphonothioates*

Yuxin Li, Haoan Wang, Xiaoping Yang, Haiying Cheng, Zhihong Wang, Yiming Li, Zhengming Li, Suhua Wang, Dongwen Yan

(State Key laboratory of Elemento – Organic Chemistry, Nankai University, Tianjin, 300071, China)

Abstract A series of 3 – alkoxy(phenyl)thiophosphorylamido – 2 – (per – *O* – acetylglycosyl – 1′ – imino) thiazolidine – 4 – one derivatives were prepared by the reaction of 1 – alkoxy(phenyl)thiophosphoryl – 4 – (per – *O* – acetylglycosyl) thiosemicarbazides with ethyl bromoacetate. $^{1}H/^{13}C$ HMBC measurements corroborated by X – ray crystallographic results revealed the exclusive regioselectivity of these ring closures toward the N – 2 position of the thiosemicarbazide moiety. The bioactivity data of **3a – k** suggest that the thiazolidine – 4 – one ring is critical for the herbicidal and fungicidal activities.

Key words 2 – iminothiazolidine – 4 – ones; thiosemicarbazidophosphonothioates; $^{1}H/^{13}C$ HMBC correlation spectra; regioselectivity

Thiazolidine – 4 – one compounds have been shown to display antimicrobial,[1] antimycobacterial,[2] anticonvulsant,[3] anti – inflammatory,[4] anti – HIV – 1,[5] analgesic,[6] antituberculosis,[7] immunosuppressant,[8] and antimalarial[9] activities. Recently, it was reported that thiazolidinones displayed antidegenerative activity on human chondrocyte cultures.[10] Previously, we have described the synthesis of bioactive 2 – arylsulfonylhydrazono – 3 – (per – *O* – acetylglycosyl) thiazolidine – 4 – ones using 1 – arylsulfonyl – 4 – (per – *O* – acetylglycosyl) thiosemicarbazides as precursor.[11] Since organo phosphorus compounds are known as highly effective agrochemicals,[12] we planned to prepare a series of 2 – alkoxy(phenyl) thiophosphorylhydrazono – 3 – (per – *O* – acetylglycosyl) thiazolidine – 4 – one derivatives.

We started from 1 – alkoxy(phenyl) thiophosphoryl – 4 – (per – *O* – acetylglycosyl) thiosemicarbazides **2a – k**, obtained from the reaction of glycosyl isothiocyanates[13,14] and thiophosphoryl hydrazines.[15] Reaction of **2a – k** with ethyl bromoacetate was investigated (Scheme 1). We first tried to use chloroform as the reaction medium. Unfortunately, we failed to obtain any cyclization product. However, we successfully separated a moderately stable S – alkylated intermediate in the presence of sodium acetate. FABMS (*m/z* 948 [M + H]$^{+}$ for **2k** with ethyl bromoacetate) indicates that it is an uncyclized product. It is likely that this S – alkylated intermediate was unstable when

* Reprinted from *Carbohydrate Research*, 2009, 344(10): 1248 – 1253. The project was supported by NNSFC #20432010 and the Science and Technology Innovation Foundation of Nankai University.

the reaction was refluxed in chloroform. Finally, we obtained cyclization products when we used dichloromethane as solvent. A reaction time of 12h at reflux in dichloromethane was maintained since a longer reaction time did not improve the yield.

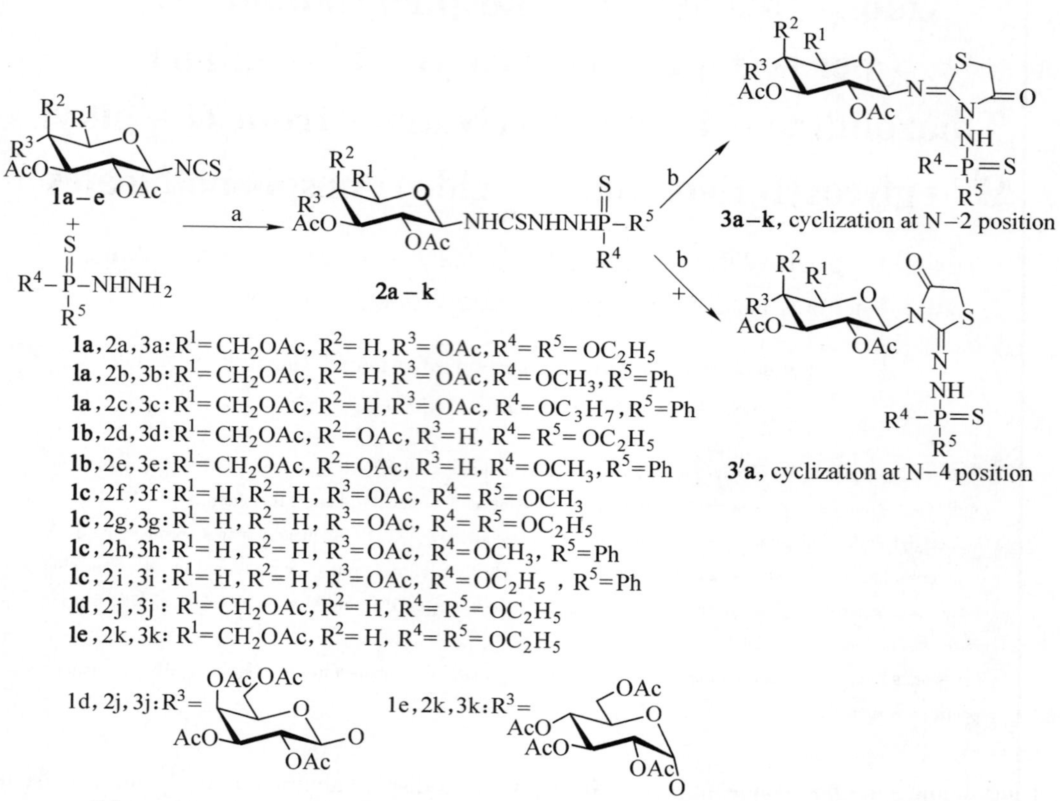

1a,2a,3a: R^1= CH_2OAc, R^2= H, R^3= OAc, R^4= R^5= OC_2H_5
1a,2b,3b: R^1= CH_2OAc, R^2= H, R^3= OAc, R^4= OCH_3, R^5=Ph
1a,2c,3c: R^1= CH_2OAc, R^2= H, R^3= OAc, R^4= OC_3H_7, R^5=Ph
1b,2d,3d: R^1= CH_2OAc, R^2=OAc, R^3=H, R^4= R^5= OC_2H_5
1b,2e,3e: R^1= CH_2OAc, R^2=OAc, R^3=H, R^4= OCH_3, R^5=Ph
1c,2f,3f: R^1=H, R^2=H, R^3=OAc, R^4= R^5= OCH_3
1c,2g,3g: R^1=H, R^2=H, R^3=OAc, R^4= R^5= OC_2H_5
1c,2h,3h: R^1=H, R^2=H, R^3=OAc, R^4= OCH_3, R^5=Ph
1c,2i,3i: R^1=H, R^2=H, R^3=OAc, R^4= OC_2H_5, R^5=Ph
1d,2j,3j: R^1= CH_2OAc, R^2= H, R^4= R^5= OC_2H_5
1e,2k,3k: R^1= CH_2OAc, R^2= H, R^4= R^5= OC_2H_5

3a–k, cyclization at N–2 position

3′a, cyclization at N–4 position

Scheme 1 Reagents and conditions: (a) MeCN, rt and (b) CH_2Cl_2, $BrCH_2CO_2Et$, reflux

Several reports regarding the condensation of thiosemicarbazides with alkyl halides have pointed out that main products are of type **3′a**,[16,17] resulting from cyclization at N – 4. 2 – (Dimethylhydrazono) – 3 – phenylthiazol – 4 – one was obtained by reflux of thiosemicarbazide with ethyl bromoacetate for 7 days in dichloromethane.[18] In the presence of a weak base, Saleh et al. cyclized substituted thioureas with chloroacetic acid, giving thiazolidine – 4 – one compounds substituted at N – 4.[19] Conversely, the cyclization at N – 2 has been more scarcely reported. Moghaddam and Hojabri[20] and Rajanarendar et al.[21] have reported cyclizations at N – 2 position. A detailed study of the condensation of thiosemicarbazides with ethyl bromoacetate to give N – 2 isomers as major products has been reported.[22]

It is of interest that the cyclization of thiosemicarbazides **2a – k**, which includes a relatively bulky substituent, with ethyl bromoacetate leads to the formation of N – 2 isomers **3a – k**. When these reactions were monitored by TLC, we found that the polarity of **3a – k** was smaller than that of **2a – k** and the polarity of previously reported 2 – arylsulfonylhydrazono – 3 – (per – *O* – acetylglycosyl) thiazolidine – 4 – ones was higher than that of 1 – arylsulfonyl – 4 – (per – *O* – acetylglycosyl) thiosemicarbazides. ^{1}H and ^{13}C NMR spectra of final products (after work – up) suggest absence of any trace of the isomers, indicating that the final products are isomerically pure.

Due to the presence of the imino group, the chemical shift value of H – 1′ for **3f** (4. 76ppm) is smaller than that of H - 1′(5.42 ppm), as we previously reported in the case of 2 - phenylsulfonylhydrazono - 3 - (2′,3′,4′ - tri - *O* - acetyl - β - D - xylopyranosyl) thiazolidine - 4 - one, and the chemical shift value of C - 1′ for **3f**(88.90 ppm) is larger than that of C - 1′ (81.63ppm).[11] These differences in chemical shifts may indicate different types of ring closures. Formation of the N - 4 vs N - 2 type closure for the sulfonylhydrazono versus glycosylimino derivatives has now been proven by HMBC experiments. Figure 1 shows the HMBC spectra of 2 - phenylsulfonylhydrazono - 3 - (2′,3′,4′ - tri - *O* - acetyl - β - D - xylopyranosyl) thiazolidine - 4 - one[11] (Fig. 1A), **3a**(Fig. 1C) and **3f**(Fig. 1 E). Based on the HMBC correlations in Figure 1, we can unambiguously assign all chemical shift values of carbon and proton atoms in the sugar ring. On the other hand, the three - bond connectivity of H - 1′ at δ 5.42 ppm with C - 2 at δ 164ppm and with C - 4 at δ172 ppm in the HMBC spectrum confirmed the cyclization of 1 - arylsulfonyl - 4 - (per - *O* - acetylglycosyl) thiosemicarbazides with ethyl bromoacetate at N - 4(Figure 1B); the three - bond connectivity of H - 1′ at δ 4.76ppm with C - 2 at δ 155ppm in the HMBC spectrum proved the cyclization of 1 - alkoxy(phenyl) thiophosphoryl - 4 - (per - *O* - acetyl glycosyl) thiosemicarbazides with ethyl bromoacetate at N - 2(Fig. 1F). This type of ring closure was confirmed by X - ray crystallography.[23]

Compounds **2a - k** and **3a - k** were tested in soil treatment against several herbs such as *Brassica campestris*, *Echinochloa crus - galli*, *Amaranthus retroflexus L.*, and *Digitaria sanguinalis*(*L.*) *Scop* at 1.5 kg/ha.[24] The bioassay results showed that **3a - k** have a rather moderate herbicidal activity. The inhibitory rate of **3b** against Brassica campestris was 62.7%. Compounds **2a - k** and **3a - k** were also tested for their growth inhibition and fungicidal activity in vitro against *Gibberella zeae*, *Alternaria solani*, *Cercospora arachidicola*, *Physalospora piricola*, *Phoma asparagi* at 50ppm. The inhibitory rates of **3a - k** against *Physalospora piricola*, *Phoma asparagi* were between 40% and 50%. Higher inhibitory rate of **3a - k** compared to that of **2a - k** indicates that the thiazolidine - 4 - one ring is critical for the herbicidal and fungicidal activities.

1 Experimental

1.1 General methods

Melting points were obtained on a Yanaco MP - 500 micro - melting point apparatus. ^1H and ^{13}C NMR spectra were recorded on a Bruker AC - 400 instrument using $CDCl_3$ as solvent and Me_4Si as internal standard. Elemental analysis was performed on a Yanaco CHN Corder MF - 3 automatic elemental analyzer. Mass spectra were recorded with a VG ZAB - HS instrument using the FAB ionization mode.

1.2 1 - Alkoxy(phenyl) thiophosphoryl - 4 - (per - *O* - acetylglycosyl) - thiosemicarbazide derivatives(2a - k), general method

To a soln of 1a - e(4mmol) in MeCN(20mL), the corresponding thiophosphoryl hydrazine (6mmol) was added with stirring. The mixture was heated at reflux for 2h and the solvent was removed under diminished pressure. Concentration resulted in a crude product which was purified on a silica gel column(1:2 EtOAcpetroleum ether).

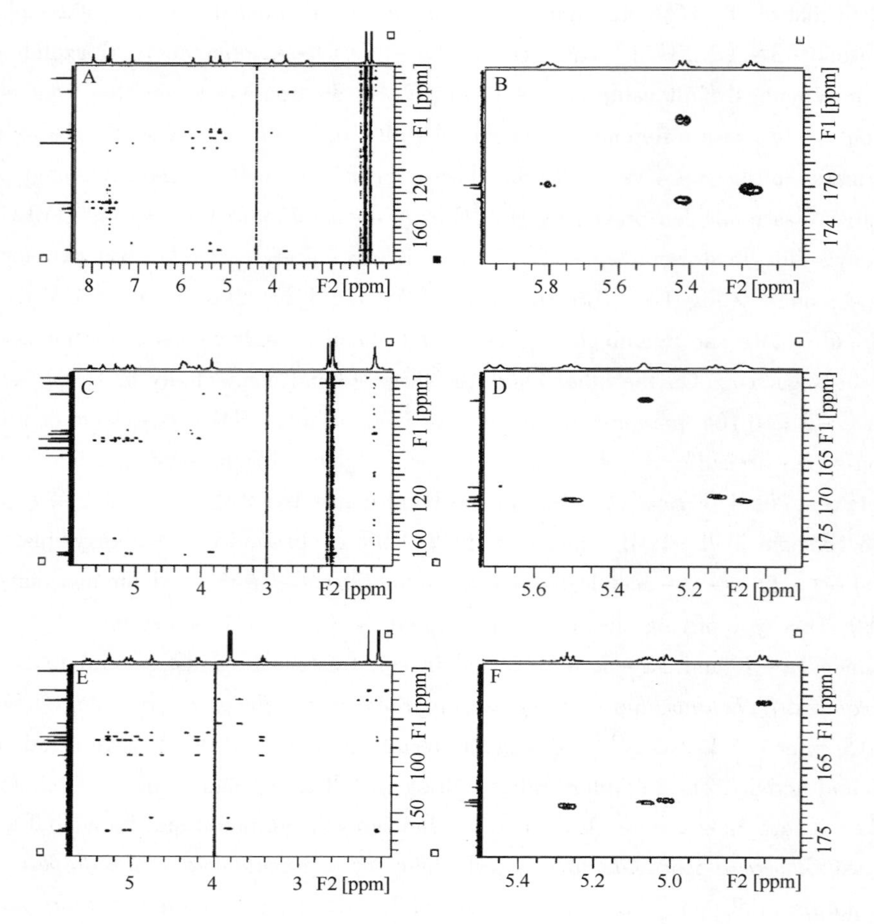

Figure 1 (A) $^{1}H-^{13}C$ HMBC spectra of 2 - phenylsulfonylhydrazono - 3 - (2′,3′,4′ - tri - *O* - acetyl - β - *D* - xylopyranosyl) thiazolidine - 4 - one[11]; (B) $^{1}H-^{13}C$ HMBC spectra of H - 1′/C - 2, C - 4 in 2 - phenylsulfonylhydrazono - 3 - (2′,3′,4′ - tri - *O* - acetyl - β - *D* - xylopyranosyl) thiazolidine - 4 - one; (C) $^{1}H-^{13}C$ HMBC spectra **3a**; (D) $^{1}H-^{13}C$ HMBC spectra of H - 1′ with C - 2 in **3a**; (E) $^{1}H-^{13}C$ HMBC spectra **3f**; and (F) $^{1}H-^{13}C$ HMBC spectra of H - 1′ with C - 2 in **3f**

1.2.1 1 - Diethoxythiophosphoryl - 4 - (2,3,4,6 - tetra - *O* - acetyl - β - D - glucopyranosyl) thiosemicarbazide (**2a**)

From 1.56g (4mmol) of **1a** (1.97g, 87%); white crystals from EtOAc - petroleum ether; mp 164 ~ 165℃. ^{1}H NMR ($CDCl_3$, ppm): δ 7.64 (d, *J* 8.4Hz, 1H, NH), 7.60 (br s, 1H, NH), 5.78 (t, 1H, H - 1′), 5.37 (m, 2H, H - 2′, H - 3′), 5.06 (m, 2H, H - 4′, H - 6′), 4.31 (dd, J_1 4.2Hz, J_2 3.6Hz, 1H, H - 5′), 4.13 (m, 5H, H - 6′, 2 × $POCH_2$), 3.86 (d, *J* 8.4Hz, 1H, NH), 2.10, 2.06, 2.04, 2.02 (4s, 12H, 4 × OAc), 1.32 (t, 6H, 2 × $POCH_2CH_3$); ^{13}C NMR ($CDCl_3$, ppm): δ 183.67 (C ═S), 171.20, 170.83, 169.99, 169.74 (4C, 4 × CO), 82.28 (C - 6′), 73.59 (C - 3′), 72.91 (C - 2′), 68.30 (C - 1′), 64.66, 63.89 (2C, 2 × $POCH_2$), 63.50 (C - 5′), 61.82 (C - 4′), 20.82, 20.75, 20.70, 20.67 (4C, 4 × OAc), 15.83 (2C, 2 × $POCH_2CH_3$); ^{31}P NMR ($CDCl_3$, ppm): δ 79.16. Anal. Calcd for $C_{19}H_{32}N_3O_{11}PS_2$: C, 41.78; H, 5.32; N, 5.04. Found: C, 41.58; H, 5.61; N, 4.98.

1.2.2 1-Methoxy(phenyl)thiophosphoryl-4-(2,3,4,6-tetra-*O*-acetyl-β-D-glucopyranosyl)thiosemicarbazide(**2b**)

From 1.56g(4mmol) of **1a**(1.80g,76%); white crystals from EtOAc - petroleum ether; mp 169-171℃. ^{1}H NMR($CDCl_3$, ppm): δ 7.83(m,2H,Ar—H), 7.70(d,*J* 9.2Hz,1H,NH), 7.54-7.46(m,4H,Ar—H,NH), 5.89(br s,1H,NH), 5.71-3.69(m,10H,OCH_3,H-1′, H-2′,H-3′,H-4′,H-5′,H-6′), 2.08, 2.01, 1.98, 1.96, 1.98, (4s,12H,4×OAc); ^{13}C NMR($CDCl_3$, ppm): δ 183.65(C═S), 171.02, 170.89, 170.04, 169.75, 170.04, 170.89, 171.02(4C,4×CO), 133.34, 131.94, 131.86, 131.74, 129.12, 128.98(6C, Ar—C), 82.35, 73.74, 73.00, 70.78, 68.18, 61.75(6C,C-1′,C-2′,C-3′,C-4′,C-5′,C-6′) 53.36(OCH_3), 21.01, 20.81(4C, 4×OAc); ^{31}P NMR($CDCl_3$, ppm): δ 81.94, 81.76. Anal. Calcd for $C_{22}H_{30}N_3O_{10}PS_2$: C,44.67; H,5.11; N,7.10. Found: C,44.38; H, 5.24; N,7.01.

1.2.3 1-Phenyl(propoxy)thiophosphoryl-4-(2,3,4,6-tetra-*O*-acetyl-β-D-glucopyranosyl)thiosemicarbazide(**2c**)

From 1.56g(4mmol) of **1a**(1.70g,69%); white crystals from EtOAc - petroleum ether; mp 166-167℃. ^{1}H NMR($CDCl_3$, ppm): δ 7.91(br s,1H,NH), 7.79-7.41(m,6H,Ar—H, NH), 5.89(br s,1H,NH), 5.68-3.75(m,9H,OCH_2,H-1′,H-2′,H-3′,H-4′,H-5′, H-6′), 2.09, 2.03, 1.97, 1.93(4s,12H,4×OAc), 1.29(m,2H,$POCH_2CH_2$), 0.91(t,3H, CH_3); ^{13}C NMR($CDCl_3$): δ 184.08(C═S), 170.33, 170.13, 170.04, 169.84(4C,4×CO), 140.33, 132.65, 131.87, 131.29, 129.43, 138.96(6H, Ar—C), 82.89, 73.79, 69.34, 63.85, 63.52, 63.43, 61.45(7C,C-1′,C-2′,C-3′,C-4′,C-5′,C-6′,OCH_2), 33.68 ($POCH_2CH_2$), 20.81, 20.54, 20.36, 20.12(4C,4×OAc), 16.29(CH_3); ^{31}P NMR($CDCl_3$, ppm): δ 79.27, 79.01. Anal. Calcd for $C_{24}H_{34}N_3O_{10}PS_2$: C,46.52; H,5.53; N,6.78. Found: C,46.23; H,5.60; N,6.95.

1.2.4 1-Diethoxythiophosphoryl-4-(2,3,4,6-tetra-*O*-acetyl-β-D-galactopyranosyl)thiosemicarbazide(**2d**)

From 1.56g(4 mmol) of **1b**(1.56g,68%); white crystals from EtOAc - petroleum ether; mp 167-168℃. ^{1}H NMR($CDCl_3$, ppm): δ 7.59(m,2H,NHNH), 5.71(t,1H,H-1′), 5.43 (m,2H,H-2′,H-3′), 5.17(m,2H,H-4′,H-6′), 4.01(m,7H,H-5′,H-6′,2× $POCH_2$,NH), 2.14, 2.04, 2.01, 1.97(4s,12H,4×OAc), 1.31(t,6H,2×$POCH_2CH_3$); ^{13}C NMR($CDCl_3$, ppm): δ 183.76(C═S), 171.41, 170.60, 170.26, 169.95(4C,4×CO), 82.96(C-6′), 72.63(C-3′), 71.01(C-2′), 68.79(C-1′), 67.38(C-5′), 65.09, 64.92(2C,2×$POCH_2$), 61.24(C-4′), 21.08, 20.83, 20.73, 20.56(4C,4×OAc), 16.10, 16.04(2C,2×$POCH_2CH_3$); ^{31}P NMR($CDCl_3$, ppm): δ 79.16. Anal. Calcd for $C_{19}H_{32}N_3O_{11}PS_2$: C,41.78; H,5.32; N,5.04. Found: C,41.35; H,5.11; N,4.98.

1.2.5 1-Methoxy(phenyl)thiophosphoryl-4-(2,3,4,6-tetra-*O*-acetyl-β-D-galactopyranosyl)thiosemicarbazide(**2e**)

From 1.56g(4mmol) of **1b**(1.7g,72%); white crystals from EtOAc - petroleum ether; mp 178-179℃. ^{1}H NMR($CDCl_3$, ppm): δ 8.10(br s,1H,NH), 7.87(m,6H,Ar—H,NH), 5.84(d,*J* 8.4Hz,1H,NH), 5.78-3.57(m,10H,H-1′,H-2′,H-3′,H-4′,H-5′,H-

6′,OCH_3),2.15,2.04,1.98,1.96(4s,12H,4 × OAc);^{13}C NMR($CDCl_3$,ppm):δ 184.98 (C=S),171.32,171.03,170.27,169.67(4C,4 × CO),139.76,134.02,133.10,132.68, 130.64,128.96(6C,Ar—C),81.98,73.85,72.69,71.78,67.56,61.71,61.55(7C,C-1′, C-2′,C-3′,C-4′,C-5′,C-6′,CH_3O),21.00,20.89,20.81,20.79(4C,4 × OAc);^{31}P NMR($CDCl_3$,ppm):δ 79.86,79.65. Anal. Calcd for $C_{22}H_{30}N_3O_{10}PS_2$:C,44.67;H,5.11;N, 7.10. Found:C,44.34;H,4.89;N,7.28.

1.2.6 1-Dimethoxythiophosphoryl-4-(2,3,4-tri-*O*-acetyl-β-D-xylopyranosyl) thiosemicarbazide(**2f**)

From 1.26g(4mmol) of **1c**(1.48g,78%);white crystals from EtOAc - petroleum ether; mp 181-182℃. ^{1}H NMR($CDCl_3$,ppm):δ 8.02(br s,1H,NH),7.84(d,*J* 8.4Hz,1H,NH), 5.63(br s,1H,NH),5.57-3.54(m,12H,H-1′,H-2′,H-3′,H-4′,H-5′,2 × OCH_3),2.04,2.03,1.89(3s,9H,3 × OAc);^{13}C NMR($CDCl_3$,ppm):δ 185.13(C=S), 170.56,170.03,169.84(3C,3 × CO),80.97,72.36,68.72,67.88,64.66,66.68,59.89 (7C,C-1′,C-2′,C-3′,C-4′,C-5′,2 × OCH_3),20.98,20.24,20.02(3C,3 × OAc); ^{31}P NMR($CDCl_3$,ppm):δ 69.63. Anal. Calcd for $C_{14}H_{24}N_3O_9PS_2$:C,35.52;H,5.11;N, 8.88. Found:C,35.50;H,5.21;N,8.39.

1.2.7 1-Diethoxythiophosphoryl-4-(2,3,4-tri-*O*-acetyl-β-D-xylopyranosyl) thiosemicarbazide(**2g**)

From 1.26g(4mmol) of **1c**(1.38g,69%);white crystals from EtOAc - petroleum ether; mp 179-181℃. ^{1}H NMR($CDCl_3$,ppm):δ 8.01(br s,1H,NH),7.72(d,*J* 8.4Hz,1H,NH), 5.68(br s,1H,NH),5.58-3.44(m,10H,H-1′,H-2′,H-3′,H-4′,H-5′,2 × OCH_2),2.01,2.00,1.98(3s,9H,3 × OAc),1.25(t,6H,2 × $POCH_2CH_3$);^{13}C NMR($CDCl_3$, ppm):δ 184.56(C=S),171.25,170.32,169.98(3C,3 × CO),81.32,71.67,68.77, 67.82,64.65,63.25,61.04(7C,C-1′,C-2′,C-3′,C-4′,C-5′,2 × OCH_2),21.33, 20.72,20.54(3C,3 × OAc),16.16,16.11(2C,2 × $POCH_2CH_3$);^{31}P NMR($CDCl_3$,ppm): δ 69.74,69.61. Anal. Calcd for $C_{16}H_{28}N_3O_9PS_2$:C,38.33;H,5.63;N,8.38. Found:C, 38.78;H,5.38;N,7.98.

1.2.8 1-Methoxy(phenyl) thiophosphoryl-4-(2,3,4-tri-*O*-acetyl-β-D-xylopyranosyl) thiosemicarbazide(**2h**)

From 1.26g(4mmol) of **1c**(1.48g,72%);white crystals from EtOAc - petroleum ether; mp 163-164℃. ^{1}H NMR($CDCl_3$,ppm):δ 7.86(m,7H,NHNH,Ar—H),5.71-3.52(m,10H, H-1′,H-2′,H-3′,H-4′,H-5′,OCH_3,NH),2.09,2.04,1.98(3s,9H,3 × OAc);^{13}C NMR($CDCl_3$,ppm):δ 183.89(C=S),171.22,170.96,170.38(3C,3 × CO),140.82, 133.22,131.97,130.65,128.39,127.56(6H,Ar—C),82.80,72.19,69.17,64.12,63.03, 61.25(6C,C-1′,C-2′,C-3′,C-4′,C-5′,OCH_3),20.86,20.43,19.87(3C,3 × OAc). Anal. Calcd for $C_{19}H_{26}N_3O_8PS_2$:C,43.93;H,5.04;N,8.09. Found:C,43.60;H,4.69; N,7.93.

1.2.9 1-Ethoxy(phenyl) thiophosphoryl-4-(2,3,4-tri-*O*-acetyl-β-D-xylopyranosyl) thiosemicarbazide(**2i**)

From 1.26g(4mmol) of **1c**(1.56g,73%);white crystals from EtOAc - petroleum ether; mp

163 – 164℃. ^{1}H NMR(CDCl$_3$, ppm): δ 8.02(br s, 1H, NH), 7.45(d, *J* 8.4Hz, 1H, NH), 7.86 – 3.49(m, 14H, H – 1′, H – 2′, H – 3′, H – 4′, H – 5′, OCH$_2$, NH, Ar—H), 2.11, 2.03, 2.00(3s, 9H, 3 × OAc), 1.56(t, 3H, CH$_3$); ^{13}C NMR(CDCl$_3$, ppm): δ 183.66(C =S), 170.82, 170.49, 170.36(3C, 3 × CO), 149.87, 132.65, 131.87, 131.39, 128.34, 126.97 (6C, Ar—C), 83.60, 82.55, 72.16, 69.46, 65.64, 63.56, 63.01(6C, C – 1′, C – 2′, C – 3′, C – 4′, C – 5′, OCH$_2$), 20.96, 20.89, 20.43(3C, 3 × OAc), 16.33(POCH$_2$CH$_3$). Anal. Calcd for $C_{20}H_{28}N_3O_8PS_2$: C, 45.02; H, 5.29; N, 7.88. Found: C, 44.73; H, 5.00; N, 7.66.

1.2.10 1 – Diethoxythiophosphoryl – 4 – [2,3,4,6 – tetra – *O* – acetyl – β – D – galactopyranosyl – (1 →4) – 2,3,6 – tri – *O* – acetyl – β – D – glucopyranosyl] thiosemicarbazide(**2j**)

From 2.7g(4mmol) of **1d**(2.3g, 67%); white crystals from EtOAc – petroleum ether; mp 140 – 141 ℃. ^{1}H NMR(CDCl$_3$, ppm): δ 7.88(br s, 1H, NH), 7.43(br s, 1H, NH), 5.76 – 3.56 (m, 18H, 2 × POCH$_2$, sugar ring), 2.04, 2.01, 1.98, 1.97, 1.95, 1.92, 1.91(7s, 21H, 7 × OAc), 1.26(t, 6H, 2C, 2 × POCH$_2$CH$_3$); ^{13}C NMR(CDCl$_3$, ppm): δ 184.86(C =S), 170.96, 170.48, 170.26, 170.12, 169.81, 169.38, 169.27(7C, 7 × CO), 89.37, 79.69, 76.33, 74.39, 72.93, 71.76, 70.73, 68.28, 67.83, 62.96, 62.34, 61.97, 61.36, 58.12(14C, sugar ring, 2 × P × POCH$_2$), 21.32, 21.13, 21.09, 21.03, 21.01, 20.99, 20.94, 20.88(7C, 7 × OAc), 15.97, 15.88(2C, 2 × POCH$_2$CH$_3$); ^{31}P NMR(CDCl$_3$, ppm): δ 79.82. Anal. Calcd for $C_{31}H_{48}N_3O_{19}PS_2$: C, 43.20; H, 5.61; N, 4.88. Found: C, 42.90; H, 5.60; N, 5.12.

1.2.11 1 – Diethoxythiophosphoryl – 4 – [2,3,4,6 – tetra – *O* – acetyl – α – D – glucopyranosyl – (1 →4) – 2,3,6 – tri – *O* – acetyl – β – D – glucopyranosyl] – thiosemicarbazide(**2k**)

From 2.7g(4mmol) of **1e**(1.58g, 75%); white crystals from EtOAc – petroleum ether; mp 160 – 163℃. ^{1}H NMR(CDCl$_3$, ppm): δ 7.59(br s, 1H, NH), 7.48(d, *J* 8.8Hz, 1H, NH), 5.75(t, 1H, H – 1′), 5.38 – 3.81(m, 18H, 2 × POCH$_2$, sugar ring), 2.10, 2.06, 2.03, 2.01, 1.99, 1.98, 1.96(7s, 21H, 7 × OAc), 1.29(t, 6H, 2C, 2 × POCH$_2$CH$_3$); ^{13}C NMR(CDCl$_3$, ppm): δ 183.84(C =S), 171.19, 170.85, 170.72, 170.58, 170.00, 169.87, 169.66(7C, 7 × CO), 95.78, 82.18, 75.17, 74.20, 72.88, 71.61, 70.23, 68.72, 68.48, 68.15, 65.11, 64.98, 62.79, 61.16(14C, sugar ring, 2 × POCH$_2$), 21.06, 20.96, 20.81(7C, 7 × OAc), 16.14, 16.02(2C, 2 × POCH$_2$CH$_3$); ^{31}P NMR(CDCl$_3$, ppm): δ 69.47. Anal. Calcd for $C_{31}H_{48}N_3O_{19}PS_2$: C, 43.20; H, 5.61; N, 4.88. Found: C, 43.18; H, 5.78; N, 4.67.

1.3 3 – Alkoxy(phenyl)thiophosphorylamido – 2 – (per – *O* – acetylglycosyl – 1′ – imino)thiazolidine – 4 – one derivatives(3a – k), general method

To a soln of **2a – k**(2mmol) in CH_2Cl_2(20mL), ethyl bromoacetate(0.5g, 3mmol) was added dropwise with stirring. The mixture was heated at reflux for 12h and the solvent was removed under diminished pressure. Concentration resulted in a crude product which was purified on a silica gel column(1:2 EtOAc – petroleumether).

1.3.1 3 – Diethoxythiophosphorylamido – 2 – (2′,3′,4′,6′ – tetra – *O* – acetyl – β – D – glucopyranosyl – 1′ – imino) thiazolidine – 4 – one(**3a**)

From 1.14g of **2a**(0.59g, 48%); white crystals from EtOAc – petroleum ether; mp 140 – 141℃; ^{1}H NMR(CDCl$_3$, ppm): δ 5.71(d, *J* 8.74Hz, 1H, NH), 5.50(m, 1H, H – 2′), 5.32

(d,1H,*J* 3.6Hz,H－1′),5.13(m,1H,H－3′),5.06(t,1H,H－4′),4.28(m,7H,H－5′,H－6′,2×$POCH_2$),3.84(s,2H,H－5),2.06,2.02,1.99,1.98(4s,12H,4×OAc),1.28(t,6H,2×$POCH_2CH_3$);^{13}C NMR($CDCl_3$,ppm):δ 170.71,170.11,170.01,169.53(4C,4×CO),168.18(C－4),156.70(C－2),83.41(C－6′),71.33(C－3′),70.68(C－2′),68.69(C－1′),64.59,64.56(2C,2×$POCH_2$),63.75(C－5′),61.89(C－4′),29.93(C－5),20.84,20.77,20.68,20.59(4C,4×OAc),15.90,15.85(2C,2×$POCH_2CH_3$);^{31}P NMR($CDCl_3$,ppm):δ 71.17. FABMS:*m/z* 613[M＋H]$^+$. Anal. Calcd for $C_{21}H_{32}N_3O_{12}PS_2$:C,41.11;H,5.26;N,6.85. Found:C,41.24;H,5.27;N,6.76.

1.3.2 3－Methoxy(phenyl)thiophosphorylamido－2－(2′,3′,4′,6′－tetra－*O*－acetyl－β－D－glucopyranosyl－1′－imino)thiazolidine－4－one(**3b**)

From 1.18g of **2b**(0.79g,64%);white crystals from EtOAc－petroleum ether;mp 87－89℃;^1H NMR($CDCl_3$,ppm):δ 8.07－7.36(m,5H,Ar—H),5.70(d,*J* 8.4Hz,1H,NH),5.48(t,1H,H－2′),5.27(d,*J* 4.4Hz,1H,H－1′),5.22(t,1H,H－3′),5.11(t,1H,H－4′),5.02(m,2H,H－6′,H－5′),3.90(s,3H,CH_3O),3.72(m,3H,H－6′,H－5),2.06,2.02,2.00,1.96(4s,12H,4×OAc);^{13}C NMR($CDCl_3$,ppm):δ 170.93,170.81,170.30,169.73(4C,4×CO),168.68(C－4),157.40(C－2),132.74,131.83,131.69,131.57,128.86,128.71(6C,Ar—C),88.30(CH_3O),83.73(C－6′),71.52(C－3′),70.87(C－2′),68.92(C－1′),62.01(C－5′),53.23(C－4′),29.98(C－5),21.15,20.99,20.94,20.86(4C,4×OAc);^{31}P NMR($CDCl_3$,ppm):δ 81.75,81.69. FABMS:*m/z* 631[M＋H]$^+$. Anal. Calcd for $C_{24}H_{30}N_3O_{11}PS_2$:C,45.64;H,4.79;N,6.65. Found:C,45.91;H,4.96;N,6.92.

1.3.3 3－Phenyl(propoxy)thiophosphorylamido－2－(2′,3′,4′,6′－tetra－*O*－acetyl－β－D－glucopyranosyl－1′－imino)thiazolidine－4－one(**3c**)

From 1.2g of **2c**(0.55g,42%);white crystals from EtOAc－petroleum ether;mp 107－108℃;^1H NMR($CDCl_3$,ppm):δ 7.79－7.41(m,5H,Ar—H),5.68(d,*J* 8.4Hz,1H,NH),5.50(t,1H,H－2′),5.22(d,*J* 4.4Hz,1H,H－1′),5.11(t,1H,H－3′),5.09(m,3H,H－4′,H－5′,H－6′),3.94(m,2H,CH_2O),3.75(m,3H,H－6′,H－5),2.06,1.98,1.97,1.95(4s,12H,4×OAc),1.28(t,2H,CH_2),0.93(t,3H,CH_3);^{13}C NMR($CDCl_3$,ppm)δ 170.27,169.94,169.58,169.20(4C,4×CO),168.88(C－4),154.92(C－2),132.34,131.65,131.54,128.69,128.54,128.38(6C,Ar—C),86.74,74.71,73.66,68.23,67.34,62.79,62.43(7C,C－1′,C－2′,C－3′,C－4′,C－5′,C－6,$POCH_2$),32.81(C－5),23.67($POCH_2CH_2$),20.92,20.79,20.55(4C,4×OAc),10.45($POCH_2CH_2CH_3$);^{31}P NMR($CDCl_3$,ppm):δ 82.37,82.26. FABMS:*m/z* 659[M＋H]$^+$. Anal. Calcd for $C_{26}H_{34}N_3O_{11}PS_2$:C,47.34;H,5.20;N,6.37. Found:C,47.58;H,5.39,N,5.53.

1.3.4 3－Diethoxythiophosphorylamido－2－(2′,3′,4′,6′－tetra－*O*－acetyl－β－D－galactopyranosyl－1′－imino)thiazolidine－4－one(**3d**)

From 1.14g of **2d**(0.34g,24%);syrup;^1H NMR($CDCl_3$,ppm):δ 5.65(d,*J* 8.4Hz,1H,NH),5.41(t,1H,H－2′),5.22(t,1H,H－3′),5.10(t,1H,H－4′),4.70(d,*J* 8.8Hz,1H,H－1′),4.22－4.10(m,6H,H－6′,H－5′,2×$POCH_2$),3.60(m,3H,H－6′,H－5),2.13,2.07,2.02,1.98(4s,12H,4×OAc),1.30(t,6H,2×$POCH_2CH_3$);^{13}C NMR($CDCl_3$, ppm): δ 170. 61, 170. 41, 170. 17, 168. 54 (4C, 4 × CO), 167. 80 (C – 4), 156. 46 (C – 2), 84. 23 (C – 1′), 72. 01 (C – 5′), 67. 79 (C – 3′), 66. 69 (C – 2′), 64. 74 (C – 4′), 64. 01, 63. 77 (2C, 2 × $POCH_2$), 61. 56 (C – 6′), 33. 47 (C – 5), 21. 16, 20. 92, 20. 84, 20. 77 (4C, 4 × OAc), 16. 17, 16. 09 (2C, 2 × $POCH_2CH_3$); ^{31}P NMR ($CDCl_3$, ppm): δ 71. 74. FABMS: m/z 613 [M + H]$^+$. Anal. Calcd for $C_{21}H_{32}N_3O_{12}PS_2$: C, 41. 11; H, 5. 26; N, 6. 85. Found: C, 41. 00; H, 5. 43; N, 6. 78.

1. 3. 5 3 – Methoxy (phenyl) thiophosphorylamido – 2 – (2′, 3′, 4′, 6′ – tetra – *O* – acetyl – β – D – galactopyranosyl – 1′ – imino) thiazolidine – 4 – one (**3e**)

From 1. 18g of **2e** (0. 77g, 61%); white crystals from EtOAc – petroleum ether; mp 98 – 100℃; ^1H NMR ($CDCl_3$, ppm): δ 7. 93 – 7. 37 (m, 5H, Ar—H), δ 5. 70 (d, *J* 8. 4Hz, 1H, NH), 5. 44 (t, 1H, H – 5′), 5. 25 (t, 1H, H – 3′), 5. 16 (t, 1H, H – 2′), 4. 82 (d, *J* 8. 8Hz, 1H, H – 1′), 4. 33 – 4. 19 (m, 5H, H – 6′, H – 4′, $POCH_3$), 3. 73 (m, 3H, H – 6′, H – 5), 2. 05, 2. 03, 1. 98, 1. 95 (4s, 12H, 4 × OAc); ^{13}C NMR ($CDCl_3$, ppm): δ 170. 69, 170. 59, 170. 32, 170. 10 (4C, 4 × CO), 169. 55 (C – 4), 156. 75 (C – 2), 132. 56, 131. 75, 131. 65, 131. 52, 128. 52, 128. 37 (6C, Ar—C), 95. 88 (C – 1′), 76. 37 (C – 5′), 76. 14 (C – 3′), 70. 18 (C – 2′), 69. 49 (C – 4′), 68. 73 ($POCH_3$), 68. 22 (C – 6′), 30. 07 (C – 5), 30. 07, 21. 04, 20. 84, 20. 75 (4C, 4 × OAc); ^{31}P NMR ($CDCl_3$, ppm): δ 83. 87, 83. 62. FABMS: m/z 631 [M + H]$^+$. Anal. Calcd for $C_{24}H_{30}N_3O_{11}PS_2$: C, 47. 59; H, 5. 18; N, 4. 50. Found: C, 47. 44; H, 5. 58; N, 4. 70.

1. 3. 6 3 – Dimethoxythiophosphorylamido – 2 – (2′, 3′, 4′ – tri – *O* – acetyl – β – D – xylopyranosyl – 1′ – imino) thiazolidine – 4 – one (**3f**)

From 0. 84g of **2f** (0. 44g, 43%); white crystals from EtOAc – petroleum ether; mp 146 – 148℃; ^1H NMR ($CDCl_3$, ppm): δ 5. 53 (br s, 1H, NH), 5. 27 (m, 1H, H – 3′), 5. 06 (m, 1H, H – 4′), 5. 01 (m, 1H, H – 2′), 4. 76 (d, *J* 8. 4Hz, 1H, H – 1′), 4. 20 (m, 1H, H – 5′), 3. 83 (m, 8H, 2 × CH_3O, 2 × H – 5), 3. 43 (m, 1H, H – 5′), 2. 07, 2. 06, 2. 05 (3s, 9H, 3 × OAc); ^{13}CNMR ($CDCl_3$, ppm): δ 170. 30, 169. 8, 169. 5 (3C, 3 × CO), 167. 76 (C – 4), 155. 8 (C – 2), 89. 73 (C – 1′), 81. 67 (C – 5′), 72. 8 (C – 3′), 72. 43 (C – 2′), 68. 97 (C – 4′), 54. 97, 54. 76 (2C, 2 × $POCH_3$), 29. 98 (C – 5), 21. 00, 20. 95, 20. 84 (3C, 3 × OAc); ^{31}P NMR ($CDCl_3$, ppm): δ 70. 65. FABMS: m/z 514. 2 [M + H]$^+$. Anal. Calcd for $C_{16}H_{24}N_3O_{10}PS_2$: C, 37. 43; H, 4. 71; N, 8. 18. Found: C, 37. 35; H, 5. 02; N, 8. 32.

1. 3. 7 3 – Diethoxythiophosphorylamido – 2 – (2′, 3′, 4′ – tri – *O* – acetyl – β – D – xylopyranosyl – 1′ – imino) thiazolidine – 4 – one (**3g**)

From 1. 02g of **2g** (0. 60g, 58%); white crystals from EtOAc – petroleum ether; mp 101 – 103℃; ^1H NMR ($CDCl_3$, ppm): δ 7. 96 – 7. 52 (m, 5H, Ar—H), 5. 54 (br s, 1H, NH), 5. 30 (m, 2H, H – 3′), 5. 01 (m, 1H, H – 4′), 4. 92 (m, 1H, H – 2′), 4. 39 (d, *J* 8. 4Hz, 1H, H – 1′), 4. 03 (s, 2H, H – 5′), 3. 92 (m, 6H, 2 × $POCH_2$, 2 × H – 5), 3. 70 (m, 1H, H – 5′), 2. 06, 2. 05, 2. 03 (3s, 9H, 3 × OAc), 1. 40 (t, 6H, 2 × $POCH_2CH_3$); ^{13}C NMR ($CDCl_3$, ppm): δ 170. 87, 170. 33, 169. 42 (3C, 3 × CO), 168. 70 (C – 4), 157. 11 (C – 2), 89. 84 (C – 1′), 82. 37 (C – 5′), 71. 99 (C – 2′), 71. 01, 67. 45, 60. 29, 54. 96 (4C, C – 3′, C – 4′, 2 × $POCH_2$), 30. 47 (C – 5), 21. 01, 20. 68, 20. 43 (3C, 3 × OAc), 16. 23, 16. 17 (2C, 2 ×

$POCH_2CH_3$);^{31}P NMR($CDCl_3$,ppm):δ 70.66. FABMS:*m/z* 541[M+H]$^+$. Anal. Calcd for $C_{18}H_{28}N_3O_{10}PS_2$:C,39.92;H,5.21;N,7.76. Found:C,39.74;H,5.30;N,7.88.

1.3.8 3-Methoxy(phenyl)thiophosphorylamido-2-(2′,3′,4′-tri-*O*-acetyl-β-D-xylopyranosyl-1′-imino)thiazolidine-4-one(**3h**)

From 1.03g of **2h** (0.52g, 47%); white crystals from EtOAc - petroleum ether; mp95-96℃;^{1}H NMR($CDCl_3$,ppm):δ 8.06-7.51(m,5H,Ar—H),5.56(br s,1H,NH),5.29(m,2H,H-3′,H-4′),5.00(m,1H,H-2′),4.36(d,*J* 8.4Hz,1H,H-1′),4.02(m,1H,H-5′),3.91(m,5H,CH_3O,2×H-5),3.79(m,1H,H-5′),2.07,2.06,2.05(3s,9H,3×OAc);^{13}C NMR($CDCl_3$,ppm):δ 170.50,170.46,169.35(3C,3×CO),168.76(C-4),156.40(C-2),132.82,131.79,131.67,131.50,128.66,128.45(6C,Ar—C),93.74,83.46,73.42,70.46,68.82,62.45,53.43(6C,C-1′,C-2′,C-3′,C-4′,C-5′,CH_3O),31.54(C-5),21.00,20.89,20.46(3C,3×OAc);^{31}P NMR($CDCl_3$,ppm):δ 81.70,81.52. FABMS:*m/z* 559[M+H]$^+$. Anal. Calcd for $C_{21}H_{26}N_3O_9PS_2$:C,45.08;H,4.68;N,7.51. Found:C,45.11;H,4.96;N,7.22.

1.3.9 3-Ethoxy(phenyl)thiophosphorylamido-2-(2′,3′,4′-tri-*O*-acetyl-β-D-xylopyranosyl-1′-imino)thiazolidine-4-one(**3i**)

From 1.06g of **2i**(0.71g,63%);syrup;^{1}H NMR($CDCl_3$,ppm):δ 8.01-7.66(m,5H,Ar—H),5.57(br s,1H,NH),5.31(m,2H,H-3′),5.12(m,1H,H-4′),5.02(m,1H,H-2′),4.33(d,*J* 8.4Hz,1H,H-1′),3.99(m,4H,2×H-5,CH_2O),3.83(m,1H,H-5′),2.07,2.04,2.03(3s,9H,3×OAc),1.36(t,3H,CH_3);^{13}C NMR($CDCl_3$,ppm):δ 170.52,170.12(3C,3×CO),168.99(C-4),159.52(C-2),135.29,134.36,134.64,133.58,130.66,129.87(6C,Ar—C),89.45,80.21,73.26,69.37,62.55,53.60(6C,C-1′,C-2′,C-3′,C-4′,C-5′,CH_3O),30.57(C-5),20.94,20.83,20.32(3C,3×OAc),16.25($POCH_2CH_3$),^{31}P NMR($CDCl_3$,ppm):δ 81.85,81.70. FABMS:*m/z* 573[M+H]$^+$. Anal. Calcd for $C_{22}H_{28}N_3O_9PS_2$:C,46.07;H,4.92;N,7.33. Found:C,46.39;H,5.18;N,7.70.

1.3.10 3-Diethoxythiophosphorylamido-2-[2,3,4,6-tetra-*O*-acetyl-β-D-galactopyranosyl-(1→4)-2,3,6-tri-*O*-acetyl-β-D-glucopyranosyl-1′-imino]thiazolidine-4-one(**3j**)

From 1.72g of **2j** (0.6g, 35%); white crystals from EtOAc - petroleum ether; mp 88-91℃;^{1}H NMR($CDCl_3$,ppm):δ 5.53-3.48(m,20H, lactosyl ring,2×$POCH_2$,5-H),2.14,2.10,2.09,2.04,2.03,1.94(m,21H,7×OAc),1.27(t,6H,2×$POCH_2CH_3$);^{13}C NMR($CDCl_3$,ppm):δ 170.55,170.49,170.33,170.27,169.67(7C,7×CO),169.22(C-4),156.13(C-2),91.36,88.30,74.22,72.77,71.18,70.92,69.23,66.80,63.76,62.04,61.02(14C, lactosyl ring,2×CH_2O),33.69(C-5),21.08,20.97,20.85,20.72(7C,7×$COCH_3$),16.19,16.11(2C,2×$POCH_2CH_3$);^{31}P NMR($CDCl_3$,ppm):δ 71.05. FABMS:*m/z* 901[M+H]$^+$. Anal. Calcd for $C_{33}H_{48}N_3O_{20}PS_2$:C,43.95;H,5.36;N,4.66. Found:C,44.16;H,5.54;N,4.97.

1.3.11 3-Diethoxythiophosphorylamido-2-[2,3,4,6-tetra-O-acetyl-α-D-glucopyranosyl-(1→4)-2,3,6-tri-*O*-acetyl-β-D-glucopyranosyl-1′-imino] thiazolidine-4-one(**3k**)

From 0.26g of **2k**(0.56g,33%);syrup;^{1}H NMR($CDCl_3$,ppm):δ 5.58-3.48(m,20H,maltosyl ring, 2 × CH_2O, 5 - H), 2.19, 2.13, 2.10, 2.07, 2.03, 2.00, 1.97 (s, 21H, $COCH_3$), 1.22 (t, 6H, 2 × $POCH_2CH_3$); ^{13}C NMR ($CDCl_3$, ppm): δ 168.17, 168.01, 167.81, 167.56, 167.35, 167.05, 166.40, 165.26 (8C, 7 × CO, C - 4), 160.17 (C - 2), 93.09, 85.18, 80.83, 76.47, 73.50, 70.48, 70.43, 69.55, 67.65, 67.01, 66.04, 65.67, 60.57, 59.58 (14C, sugar ring, 2 × CH_2O), 30.57 (C - 5), 18.55, 18.49, 18.30, 18.21 (7C, 7 × $COCH_3$), 14.26, 13.99 (2C, 2 × $POCH_2CH_3$); ^{31}P NMR ($CDCl_3$, ppm): δ 68.99. FABMS: *m/z* 901 [M + H]$^+$. Anal. Calcd for $C_{33}H_{48}N_3O_{20}PS_2$: C, 43.95; H, 5.36; N, 4.66. Found: C, 43.88; H, 5.76; N, 4.38.

References

[1] Rida, S. M.; Ashour, F. A.; El - Hawash, S. A. M.; El - Semary, M. M.; Badr, M. *Arch. Pharm.* 2007, 340, 185 - 194.

[2] Ulusoy, N.; Forsch, A. *Drug Res.* 2002, 52, 565 - 571.

[3] Shiradkar, M. R.; Ghodake, M.; Bothara, K. G.; Bhandari, S. V.; Nikalje, A.; Akula, K. C.; Desai, N. C. *ARKIVOC* 2007, 14, 58 - 74.

[4] Ottana, R.; Maccari, R.; Barreca, M. L.; Bruno, G.; Rotondo, A.; Rossi, A.; Chiricosta, G.; Di Paola, R.; Sautebin, L.; Cuzzocrea, S.; Vigorita, M. G. *Bioorg. Med. Chem.* 2005, 13, 4243 - 4252.

[5] Rao, A.; Balzarini, J.; Carbone, A.; Chimirri, A.; Clercq, E.; Luca, De L. D.; Monforte, A. M.; Monforte, P. C.; Zappalà, P. M. *Farmaco.* 2004, 59, 33 - 39.

[6] Schenone, S.; Bruno, O.; Ranise, A.; Bondavalli, F.; Filippelli, W.; Falcone, G.; Giordano, L.; Redenta Vitelli, M. *Bioorg. Med. Chem.* 2001, 9, 2149 - 2153.

[7] Babaoglu, K.; Page, M. A.; Jones, V. C.; McNeil, M. R.; Dong, C.; Naismith, J. H.; Lee, R. E. *Bioorg. Med. Chem. Lett.* 2003, 13, 3227 - 3230.

[8] Bolli, M.; Scherz, M.; Mueller, C.; Mathys, B.; Binkert, C. WO 2005 054215; *Chem. Abstr.* 143: 59967.

[9] Takasu, K.; Pudhom, K.; Kaiser, M.; Brun, R.; Ihara, M. *J. Med. Chem.* 2006, 49, 4795 - 4798.

[10] Ottana, R.; Maccari, R.; Ciurleo, R.; Vigorita, M. G.; Panico, A. M.; Cardile, V.; Garufi, F.; Ronsisvalle, S. *Bioorg. Med. Chem.* 2007, 15, 7618 - 7625.

[11] Li, Y. X.; Wang, S. H.; Li, Z. M.; Su, N.; Zhao, W. G. *Carbohydr. Res.* 2006, 341, 2867 - 2870.

[12] Eto, M. *Organophosphorous Pesticides. Organic and Biological Chemistry*; CRC Press: Cleveland, OH, 1974. pp 123 - 158.

[13] Lemieux, R. U. *Methods Carbohydr. Chem.* 1963, 2, 221 - 222.

[14] Ortiz Mellet, C.; Garcia Fernandez, J. M. *Adv. Carbohydr. Chem. Biochem.* 1999, 55, 35 - 135.

[15] Tsao, L.; Chzhou, Ch.; Sun, Ts.; Koroteev, A. M. *Russ. J. Org. Chem.* 2003, 39, 1608 - 1612.

[16] Kücükgüzel, G.; Kocatepe, A.; De Clercq, E.; Sahin, F.; Güllüce, M. Eur. *J. Med. Chem.* 2006, 41, 353 - 359.

[17] Demirbas, A.; Ceylan, S.; Demirbas, N. *J. Heterocycl. Chem.* 2007, 44, 1271 - 1280.

[18] Cooley, J. H.; Sarker, S.; Vij, A. *J. Org. Chem.* 1995, 60, 1464 - 1465.

[19] Saleh, M. A.; Hafez, Y. A.; Abdel - Hay, F. E.; Gad, W. I. *J. Heterocycl. Chem.* 2003, 40, 973 - 978.

[20] Moghaddam, F. M.; Hojabri, L. *J. Heterocycl. Chem.* 2007, 44, 35 - 38.

[21] Rajanarendar, E.; Karunakar, D.; Ramu, K. *Heterocycl. Commun.* 2006, 12, 123 - 128.

[22] Guersoy, A.; Terzioglu, N. *Turkish J. Chem.* 2005, 29, 247 - 254.

[23] Cheng, H. Y.; Li, Y. X.; Wang, X. A.; Li, Z. M.; Song, H. B. *Acta Crystallogr.*, *Sect. E* 2007, 63, o2861 - o2862.

[24] Fan, Z. J.; Liu, B.; Liu, X. F.; Zhong, B.; Liu, C. L.; Li, Z. M. Chem. *J. Chinese Univ.* 2004, 25, 663 - 666.

Synthesis, Herbicidal Activities and Comparative Molecular Field Analysis Study of Some Novel Triazolinone Derivatives*

Lei Wang, Yi Ma, Xinghai Liu, Yonghong Li,
Haibin Song, Zhengming Li

(The Research Institute of Elemento – Organic Chemistry, State – Key Laboratory of Elemento – Organic Chemistry, National Pesticide Engineering Research Center (Tianjin), Nankai University, 300071. Tianjin, China)

Abstract A series of novel triazolinones were synthesized and their structures were characterized by ^1H NMR, elemental analysis and single – crystal X – ray diffraction analysis. The herbicidal activities were evaluated against *Echinochloa crusgalli* (L.) Beauv. , *Digitaria adscendens*, *Brassica napus and Amaranthus retroflexus*. The herbicidal activity data indicated that the title compounds had higher activities with substituted benzyl group moieties than with other groups such as sulfonyl, alkyl, etc. To further investigate the structure – activity relationship, comparative molecular field analysis was performed on the basis of herbicidal activity data. Both the steric and electronic field distributions of comparative molecular field analysis are in good agreement in this work. The results showed that a bulky and electronegative group around the ortho – or para – positions of the benzene ring would possibly lead to higher activity. Based on the comparative molecular field analysis, compound Ⅰ – 23 was designed and synthesized, which display as good herbicidal activities as the commercial herbicide, carfentrazone – ethyl. The activity against Digitaria adscendens is 66.1% under pre – emergence at 300 g of a. i. /ha.

Key words herbicidal activities; synthesis; three – dimensional quantitative structure – activity relationship; triazolinone

Protoporphyrinogen oxidase (Protox, EC 1.3.3.4) is an attractive target for argochemical discovery[1-6] because of its important role in a key step toward biosynthesis of chloroph yll[7-9]. Protoporphyrinogen oxidase is a potent photosensitizing agent, which induces peroxidation of cell membranes with consequent membrane leakage, pigment breakdown and eventually necrosis of the leaf. Since the first triazolinone herbicide, azafenidin was found by Dupont during the 1970s[10], a large variety of triazolinone derivatives have been synthesized and lots of them, such as amicarbazone[11], azafenidin[10], sulfentrazone[12] and carfentrazone – ethyl, are commercially available[13,14]. The action mode of these commercial herbicides is the inhibition of Protox, which causes the accumulation of protoporphyrin Ⅸ (Proto Ⅸ), which is involved in the light – dependent formation of singlet oxygen responsible for membrane peroxidation.

In our search for compounds with herbicidal activities[15-17], a series of 3 – methyl – 1 – (2,4 – dichlorophenyl) – 1,2,4 – triazol – 5 – one derivatives have been designed and synthe-

* Reprinted from *Chemical Biology & Drug Design*, 2009, 73(6):674 – 681. This work was funded by National Key Basic Research Program of China (No. 2003 CB114406), National Natural Science Foundation Key Project of China (No. 20432010), National Key Project of Scientific and Technical Supporting Programs (2006BAE01A01) and Natural Science Foundation Project of Tianjin Ministry of Science and Technology of China (Grant No. 07JCYBJC00200).

sized. The bioassay results showed that some compounds displayed good herbicidal activity against monocotyledon.

Experimental

Materials and methods

2,4 – Dichloroaniline was purchased from Nanjing Tianzun Zezhong Chemical Co. , Ltd. (Nanjing, Jiangsu, China) and shown to be >95% purity by high – performance liquid chromatography. Other reagents are analytical grade. Melting points were determined using an X – 4 apparatus and uncorrected. ^{1}H NMR spectra were measured on a Bruker AC – P500 instrument (300MHz; Bruker, Fallanden, Switzerland) using TMS as an internal standard and $CDCl_3$ as solvent. Elemental analyses were performed on a Vario EL elemental analyzer (Elementar, Hanau, Germany). Crystallographic data of the compound were collected on a Bruker SMART 1000 CCD diffractometer(Bruker).

General procedure(Scheme 1)

Synthesis of 3a – 3c and 4

To an ice – bath, cooled solution of 2,4 – dichloroaniline – hydrochloride(15g, 92. 5mmol) in water(3mL) and concentrated hydrochloric acid(100mL) was added a solution of sodium nitrite(7. 038g, 102mmol) in water(50mL) dropwise for 15min. The reaction mixture was stirred for 1 h and a solution of tin(Ⅱ) chloride(52g, 231mmol) in concentrated hydrochloric acid (50mL) was added dropwise. After 1. 5h, a bulky precipitate was collected and recrystallized in ethanol to give 2,4 – dichlorophenylhydrazine hydrochloride. Then, diluted NaOH solution was added dropwise to the solution of 2,4 – dichlorophenylhydrazine hydrochloride (pH 9 – 10) to give white solid, and dried after filtration. Melting point 93 – 94℃ (ref. 94℃ [18]). ^{1}H NMR(δ, $CDCl_3$) :7. 07 – 7. 24(m, PhH, 3H) ,5. 70(br, 1H, NH) ,3. 61(br, 2H, NH_2).

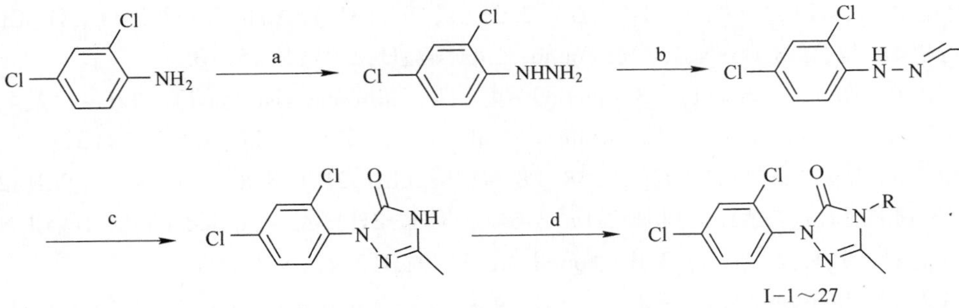

a. 1. $NaNO_2$, 2.$SnCl_2$/HCl; b. CH_3CHO; c. 1. NaOCN, HAc, 2. NaOCl; d. DMF, RCl, r. t. or reflux

Scheme 1 The Synthetic route of title compounds

To an ice – water bath, cooled solution of 2,4 – dichlorophenylhydrazine(2g, 0. 019mol) in *tert* – butyl alcohol and water(v/v = 9: 1) ,40% actaldehyde (17g, 160mmol) was added dropwise, after stirring for 5min, the sodium cyanate(9. 2g, 140. 9mmol) was added. Then acetic acid (9. 3g, 155. 0mmol) was added dropwise for 15min, and the resulting mixture was stirred at 20℃ for 3h. A 10% aqueous solution of sodium hypochlorite(11. 8g, 159. 0mmol) was added for 40min and then the reaction was stirred for another 2h. Then *tert* – butyl alcohol and water were evaporated to give a red residue, which was recrystallized in acetic ether and petroleum ether to afford a white crystal. Melting point 188 – 190℃ (ref. 190 – 192℃[18]). ^{1}H NMR(δ, $CDCl_3$): 11.49(br, 1H, NH), 7.35 – 7.55(m, PhH, 3H), 2.28(s, 3H, CH_3).

To a mixture of 2 – (2,4 – dichlorophenyl) – 5 – methyl – 2,4 – dihydro – 1,2,4 – triazol – 3 – one(0.50g, 2.0mmol) and anhydrous potassium carbonate (0.35g, 2.5mmol) in dimethylformamide (DMF, 10mL), 2 – nitrobenzenesulfonyl chloride (0.46g, 2.1mmol) in DMF (2mL) was added dropwise. The resulting mixture was stirred at room temperature for 2h, and then CH_2Cl_2 (30mL) was added and the organic layer was separated, washed with water and the solvent was then evaporated *in vacuo* to afford compound (Ⅰ) (yield 0.79g, 90%, mp 157 – 158℃).

4 – Acetyl – 1 – (2,4 – dichlorophenyl) – 3 – methyl – 1*H* – 1,2,4 – triazol – 5(4*H*) – one (**Ⅰ – 1**): white crystal, yield, 96.0%; mp, 109 – 110℃; ^{1}H NMR($CDCl_3$), δ 7.56(m, PhH, 1H), 7.37(m, PhH, 2H), 2.73(s, CH_3, 3H), 2.56(s, CH_3, 3H). Anal. Calcd for $C_{11}H_9Cl_2N_3O_2$(%): C, 46.18; H, 3.17; N, 14.69. Found: C, 46.24; H, 3.18; N, 14.70.

4 – Benzyl – 1 – (2,4 – dichlorophenyl) – 3 – methyl – 1*H* – 1,2,4 – triazol – 5(4*H*) – one (**Ⅰ – 2**): white crystal, yield, 87.1%; mp, 76 – 77℃; ^{1}H NMR($CDCl_3$), δ 7.56(m, PhH, 1H), 7.37(m, PhH, 2H), 2.73(s, CH_3, 3H), 2.56(s, CH_3, 3H). Anal. Calcd for $C_{16}H_{13}Cl_2N_3O$(%): C, 57.50; H, 3.92; N, 12.57. Found: C, 57.58; H, 3.89; N, 12.62.

1 – (2,4 – Dichlorophenyl) – 4 – (2 – fluorobenzyl) – 3 – methyl – 1*H* – 1,2,4 – triazol – 5(4*H*) – one(**Ⅰ – 3**): white crystal, yield, 93.2%; mp, 64 – 65℃; ^{1}H NMR($CDCl_3$), δ 7.07 – 7.53(m, PhH, 7H), 4.96(s, CH_2, 2H), 2.24(s, CH_3, 3H). Anal. Calcd for $C_{16}H_{12}Cl_2FN_3O$(%): C, 54.56; H, 3.43; N, 11.93. Found: C, 54.31; H, 3.41; N, 11.71.

4 – ((6 – Chloropyridin – 3 – yl)methyl) – 1 – (2,4 – dichlorophenyl) – 3 – methyl – 1*H* – 1,2,4 – triazol – 5(4*H*) – one(**Ⅰ – 4**): white crystal, yield, 86.7%; mp, 168 – 169℃; ^{1}H NMR($CDCl_3$), δ 8.40(m, ArH, 1H), 7.72 – 7.74(m, ArH, 1H), 7.55(m, ArH, 1H), 7.34 – 7.42(m, ArH, 3H), 4.89(s, CH_2, 2H), 2.25(s, CH_3, 3H). Anal. Calcd for $C_{15}H_{11}Cl_3N_4O$(%): C, 48.47; H, 3.00; N, 15.16. Found: C, 48.67; H, 2.91; N, 15.10.

1 – (2,4 – Dichlorophenyl) – 3 – methyl – 4 – (4 – nitrophenylsulfonyl) – 1*H* – 1,2,4 – triazol – 5(4*H*) – one(**Ⅰ – 5**): white crystal, yield, 90.1%; mp, 188 – 191℃; ^{1}H NMR($CDCl_3$), δ 8.95(m, PhH, 1H), 8.58 – 8.60(m, PhH, 2H), 8.83 – 8.88(m, PhH, 2H), 7.49(s, PhH, 1H), 7.30(s, PhH, 2H), 2.64(s, CH_3, 3H). Anal. Calcd for $C_{15}H_{10}Cl_2N_4O_5S$(%): C, 41.97; H, 2.35; N, 13.05. Found: C, 41.79; H, 2.47; N, 13.05.

1 – (2,4 – Dichlorophenyl) – 3 – methyl – 4 – (2 – nitrophenylsulfonyl) – 1*H* – 1,2,4 – triazol – 5(4*H*) – one(Ⅰ – 6): white crystal, yield, 88.7%; mp, 157 – 158℃; ^{1}H NMR($CDCl_3$), δ 8.58(d, PhH, 1H), 7.81 – 7.91(m, PhH, 3H), 7.49(m, PhH, 1H), 7.28 – 7.36(m, PhH, 2H), 2.61(s, CH_3, 3H). Anal. Calcd for $C_{15}H_{10}Cl_2N_4O_5S$(%): C, 41.97; H, 2.35; N, 13.05. Found: C, 42.01; H, 2.31; N, 12.90.

1 – (2,4 – Dichlorophenyl) – 3 – methyl – 4 – (methylsulfonyl) – 1*H* – 1,2,4 – triazol – 5(4*H*) – one(**Ⅰ – 7**): white crystal, yield, 80.2%; mp, 173 – 174℃; ^{1}H NMR($CDCl_3$), δ 7.55(m, PhH, 1H), 7.37 – 7.42(m, PhH, 2H), 3.57(s, CH_3, 3H), 2.52(s, CH_3, 3H). Anal.

Calcd for $C_{10}H_9Cl_2N_3O_3S$ (%): C, 37.28; H, 2.82; N, 13.04. Found: C, 37.27; H, 2.69; N, 13.09.

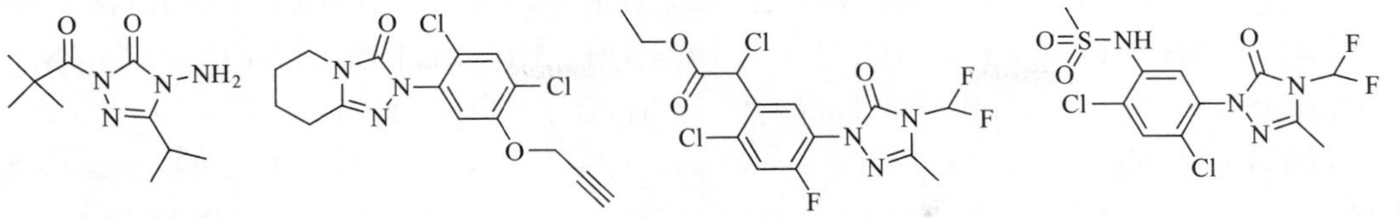

Armicarbazone　　Azafenidin　　Carfentrazone-ethyl　　Sulfentrazone

Figure 1　Some commercial triazolinone herbicides

Methyl 2-(1-(2,4-dichlorophenyl)-3-methyl-5-oxo-1*H*-1,2,4-triazol-4(5*H*)-yl)acetate(**I-8**): white crystal, yield, 88.2%; mp, 90-91℃; ^{1}H NMR($CDCl_3$), δ 7.32-7.54(m, PhH, 3H), 4.48(s, CH_2, 2H), 3.83(s, CH_3, 3H), 2.27(s, CH_3, 3 H). Anal. Calcd for $C_{12}H_{11}Cl_2N_3O_3$ (%): C, 45.59; H, 3.51; N, 13.29. Found: C, 45.55; H, 3.41; N, 13.31.

4-(3-Chlorobenzyl)-1-(2,4-dichlorophenyl)-3-methyl-1*H*-1,2,4-triazol-5(4*H*)-one(**I-9**): white crystal, yield, 86.2%; mp, 98-99℃; ^{1}H NMR($CDCl_3$), δ 7.20-7.54(m, PhH, 7H), 4.88(s, CH_2, 2H), 2.20(s, CH_3, 3H). Anal. Calcd for $C_{16}H_{12}Cl_3N_3O$ (%): C, 52.13; H, 3.28; N, 11.40. Found: C, 52.07; H, 3.21; N, 11.45.

1-(2,4-Dichlorophenyl)-3-methyl-4-octyl-1*H*-1,2,4-triazol-5(4*H*)-one (**I-10**): oil, yield, 75.4%; ^{1}H NMR($CDCl_3$), δ 7.30-7.52(m, PhH, 3H), 3.67(t, *J* = 4.5Hz, CH_2, 2H), 2.31(s, CH_3, 3H), 1.72(m, CH_2, 2H), 1.28-1.34(m, CH_2, 10H), 0.88 (t, *J* = 6.0Hz, CH_3, 3H). Anal. Calcd for $C_{17}H_{23}Cl_2N_3O$ (%): C, 57.31; H, 6.51; N, 11.79. Found: C, 57.21; H, 6.41; N, 11.97.

1-(2,4-Dichlorophenyl)-3-methyl-4-(4-methylbenzyl)-1*H*-1,2,4-triazol-5(4*H*)-one(**I-11**): white crystal, yield, 80.2%; mp, 109-110℃; ^{1}H NMR ($CDCl_3$), δ 7.18-7.55(m, PhH, 7H), 4.87(s, CH_2, 2H), 2.37(s, CH_3, 3H), 2.20(s, CH_3, 3H). Anal. Calcd for $C_{17}H_{15}Cl_2N_3O$ (%): C, 58.63; H, 4.34; N, 12.07. Found: C, 58.59; H, 4.40; N, 11.98.

1-(2,4-Dichlorophenyl)-4-(4-methoxybenzyl)-3-methyl-1*H*-1,2,4-triazol-5(4*H*)-one(**I-12**): white crystal, yield, 79.6%; mp, 115-116℃; ^{1}H NMR ($CDCl_3$), δ 6.90-7.55(m, PhH, 7H), 4.85(s, CH_2, 2H), 3.83(s, CH_3, 3H), 2.21(s, CH_3, 3H). Anal. Calcd for $C_{17}H_{15}Cl_2N_3O_2$ (%): C, 56.06; H, 4.14; N, 11.54. Found: C, 56.04; H, 4.17; N, 11.53.

1-(2,4-Dichlorophenyl)-3-methyl-4-(2-methylbenzyl)-1*H*-1,2,4-triazol-5(4*H*)-one(**I-13**): white crystal, yield, 81.1%; mp, 78-79℃; ^{1}H NMR($CDCl_3$), δ 7.02-7.54(m, PhH, 7H), 4.91(s, CH_2, 2H), 2.37(s, 3H), 2.10(s, CH_3, 3H). Anal. Calcd for $C_{17}H_{15}Cl_2N_3O$ (%): C, 58.63; H, 4.34; N, 12.07. Found: C, 58.60; H, 4.45; N, 12.01.

4-(4-*tert*-Butylbenzyl)-1-(2,4-dichlorophenyl)-3-methyl-1*H*-1,2,4-triazol-5(4H)-one(**I-14**): white crystal, yield, 78.6%; mp, 115-116℃; ^{1}H NMR ($CDCl_3$), δ 7.23-7.53(m, PhH, 7H), 4.86(s, CH_2, 2H), 2.20(s, CH_3, 3H), 1.31(s, CH_3, 9H). Anal. Calcd for $C_{20}H_{21}Cl_2N_3O$ (%): C, 61.55; H, 5.42; N, 10.77. Found: C,

61.55;H,5.47;N,10.89.

1-(2,4-Dichlorophenyl)-4-(4-fluorobenzyl)-3-methyl-1*H*-1,2,4-triazol-5(4*H*)-one(**Ⅰ-15**):white crystal,yield,85.2%;mp,60-61℃;^{1}H NMR(CDCl$_3$),δ 7.04-7.54(m,PhH,7H),4.86(s,CH$_2$,2H),2.19(s,CH$_3$,3H). Anal. Calcd for $C_{16}H_{12}Cl_2FN_3O$(%):C,54.56;H,3.43;N,11.93. Found:C,54.31;H,3.41;N,11.71.

1-(2,4-Dichlorophenyl)-4-(4-ethylbenzyl)-3-methyl-1*H*-1,2,4-triazol-5(4*H*)-one(**Ⅰ-16**):white crystal,yield,79.2%;mp,99-100℃;^{1}H NMR(CDCl$_3$),δ 7.18-7.53(m,PhH,7H),4.86(s,CH$_2$,2H),2.65(q,J=4.5Hz,CH$_2$,2H),2.18(s,CH$_3$,3H),1.23(t,J=4.5Hz,CH$_3$,3H). Anal. Calcd for $C_{18}H_{17}Cl_2N_3O$(%):C,59.68;H,4.73;N,11.60. Found:C,59.60;H,4.85;N,11.45.

4-(3,4-Dichlorobenzyl)-1-(2,4-dichlorophenyl)-3-methyl-1*H*-1,2,4-triazol-5(4*H*)-one(**Ⅰ-17**):white crystal,yield,86.7%;mp,80-81℃;^{1}H NMR(CDCl$_3$),δ 7.16-7.54(m,PhH,6H),4.85(s,CH$_2$,2H),2.21(s,CH$_3$,3H). Anal. Calcd for $C_{16}H_{11}Cl_4N_3O$(%):C,47.67;H,2.75;N,10.42. Found:C,47.60;H,2.71;N,10.47.

4-(2-Chlorobenzyl)-1-(2,4-dichlorophenyl)-3-methyl-1*H*-1,2,4-triazol-5(4*H*)-one(**Ⅰ-18**):white crystal,yield,84.3%;mp,111-112℃;^{1}H NMR(CDCl$_3$),δ 7.17-7.54(m,PhH,7H),5.04(s,CH$_2$,2H),2.18(s,CH$_3$,3H). Anal. Calcd for $C_{16}H_{12}Cl_3N_3O$(%):C,52.13;H,3.28;N,11.40. Found:C,52.05;H,3.12;N,11.40.

4-(2-(Chloromethyl)benzyl)-1-(2,4-dichlorophenyl)-3-methyl-1*H*-1,2,4-triazol-5(4*H*)-one(**Ⅰ-19**):white crystal,yield,83.3%;mp,104-105℃;^{1}H NMR(CDCl$_3$),δ 7.12-7.64(m,PhH,6H),5.09(s,CH$_2$,2H),4.72(s,CH$_2$,2H),2.15(s,CH$_3$,3H). Anal. Calcd for $C_{17}H_{14}Cl_3N_3O$(%):C,53.36;H,3.69;N,10.98. Found:C,53.25;H,3.57;N,10.81.

4-((1-(2,4-Dichlorophenyl)-3-methyl-5-oxo-1*H*-1,2,4-triazol-4(5*H*)-yl)methyl)benzonitrile(**Ⅰ-20**):white crystal,yield,87.8%;mp,122-123℃;^{1}H NMR(CDCl$_3$),δ 7.34-7.71(m,PhH,7H),4.95(s,CH$_2$,2H),2.20(s,CH$_3$,3H). Anal. Calcd for $C_{17}H_{12}Cl_2N_4O$(%):C,56.84;H,3.37;N,15.60. Found:C,56.71;H,3.43;N,15.43.

1-(2,4-Dichlorophenyl)-3-methyl-4-(2-nitrobenzoyl)-1*H*-1,2,4-triazol-5(4*H*)-one(**Ⅰ-21**):white crystal,yield,77.8%;mp,113-114℃;^{1}H NMR(CDCl$_3$),δ 8.27(d,J=8.4Hz,PhH,1H),7.70-7.81(m,PhH,2H),7.47-7.55(m,PhH,2H),7.33(m,PhH,2H),2.75(s,CH$_3$,3H). Anal. Calcd for $C_{16}H_{10}Cl_2N_4O_4$(%):C,48.88;H,2.56;N,14.25. Found:C,48.74;H,2.65;N,14.13.

1-(2,4-Dichlorophenyl)-3-methyl-4-(2-methylbenzoyl)-1*H*-1,2,4-triazol-5(4*H*)-one(**Ⅰ-22**):white crystal,yield,79.9%;mp,116-118℃;^{1}H NMR(CDCl$_3$),δ 7.29-7.51(m,PhH,7H),2.59(s,CH$_3$,3H),2.50(s,CH$_3$,3H). Anal. Calcd for $C_{17}H_{13}Cl_2N_3O_2$(%):C,56.37;H,3.62;N,11.60. Found:C,56.16;H,3.40;N,11.88.

2-((1-(2,4-Dichlorophenyl)-3-methyl-5-oxo-1*H*-1,2,4-triazol-4(5*H*)-yl)methyl)benzonitrile(**Ⅰ-23**):white crystal,yield,93.1%;mp,105-106℃;^{1}H NMR(CDCl$_3$),δ 7.35-7.74(m,PhH,7H),5.14(s,CH$_2$,2H),2.25(s,CH$_3$,3H). Anal. Calcd for $C_{17}H_{12}Cl_2N_4O$(%):C,56.84;H,3.37;N,15.60. Found:C,56.71;H,3.41;N,15.45.

Herbicidal activities

Method of the pot culture(glasshouse tests)

Pre – emergence: Sandy clay (100g) in a plastic box (11 × 7.5 × 6cm) was wetted by water. Then the sprouting seeds (15 grains) of the weed were planted with fine earth (0.6cm depth) covered in glasshouse. Spraying the solution of the compound evaluated in the suitable solvent was carried out at the rate of 1500g/ha.

Postemergence: Spraying the same solution of the compound evaluated on the seedlings (one leaf and one core) of the weed was carried out at the same rate of pre – emergence.

For both methods, 15 days later, the fresh weights were weighed and the percentage of the inhibition was calculated. The herbicidal tests results were listed in Table 1 and 2.

Crystal structure determination

The crystal of title compound with dimensions of 0.26mm × 0.22mm × 0.10mm was mounted on a Bruker SMART[a] 1000CCD area – detector diffractometer with a graphite – monochromated Mo*K*α radiation(λ = 0.71073Å) by using a phi and scan modes at 294(2) K in the range of 1.72° ≤ θ ≤ 25.01°. The crystal belongs to monoclinic system with space group P_1 and crystal parameters of a = 7.416(2) Å, b = 9.153(3) Å, c = 12.255(4) Å, α = 103.918(5)°, β = 91.059(5)°, γ = 104.428(5)°, V = 779.3(4) A^3 Dc = 1.501g/cm^3. The absorption coefficient μ = 0.4300mm^{-1} and Z = 2. The structure was solved by direct methods with SHELXS – 97[b[19]] and refined by the full – matrix least – squares method on F^2 data using SHELXL – 97. The empirical absorption corrections were applied to all intensity data. The H atom of N—H was initially located in a difference Fourier map and were refined with the restraint Uiso(H) = 1.2Ueq(N). Other H atoms were positioned geometrically and refined using a riding model, with d(C—H) = 0.93 – 0.97Å and Uiso(H) = 1.2Ueq(C) or 1.5Ueq(Cmethyl). The final fullmatrix least – squares refinement gave R = 0.0467 and wR = 0.1070 $w = 1/s^2(F_0^2) + (0.0625P)^2 + 0.1002P$]} where $P = (F_0^2) + 2F_c^2$, S = 1.04, $(\Delta/\sigma)_{max}$ < 0.0001, $\Delta_{\rho max}$ = 0.2900e$Å^3$ and $\Delta\rho_{min}$ = –0.38e$Å^3$.

Table 1 Herbicidal activity of compounds at 1500g /ha(% inhibition)

No.	R	*Echinochloa crusgalli* (L.) Beauv.		*Digitaria adscendens*		*Brassica napus*		*Amaranthus retroflexus*	
		A	B	A	B	A	B	A	B
I –1	H_3C–C(=O)–	2.8	27.8	18.8	24.8	0	29.9	9.4	28.0
I –2	benzyl	58.5	93.7	17.6	100	46.3	28.8	37.1	100
I –3	2-fluorobenzyl (F)	100	100	100	100	88.4	48.2	100	100

Table 1 (continued)

No.	R	*Echinochloa crusgalli* (L.) Beauv.		*Digitaria adscendens*		*Brassica napus*		*Amaranthus retroflexus*	
		A	B	A	B	A	B	A	B
I -4	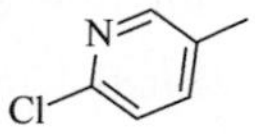 N, Cl	26.2	4.9	30.9	18.2	38.0	22.2	24.3	18.9
I -5	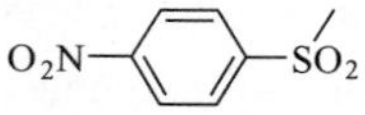 O_2N, SO_2	4.5	11.4	0	6.7	0	69.7	13.7	15.7
I -6	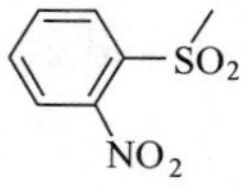 SO_2, NO_2	24.8	18.6	31.4	15.0	0	72.7	15.8	4.6
I -7	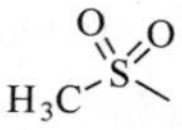 H_3C, S, O, O	25.0	15.5	7.4	21.2	33.8	17.2	5.7	29.3
I -8	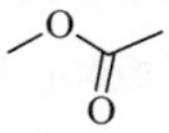 O, O	0	9.0	0	3.3	40.7	12.0	14.7	10.1
I -9	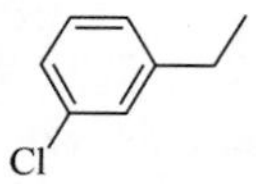 Cl	33.5	66.7	19.1	40.9	21.3	32.5	11.4	87.2
I -10	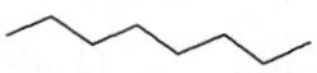	29.8	25.8	11.8	36.6	24.7	0	22.6	19.3
I -11	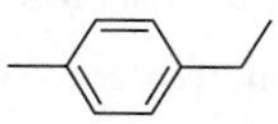	8.3	26.2	24.1	24.8	19.5	24.9	18.9	44.1
I -12	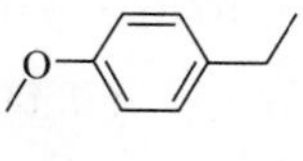 O	20.0	27.8	21.4	21.6	5.3	27.4	37.7	10.9
I -13	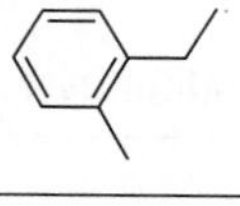	11.7	53.7	40.7	36.0	0	25.4	11.3	0
I -14		13.8	3.0	18.6	10.4	14.6	25.4	0	0
I -15	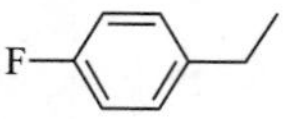 F	11.7	78.5	46.2	100	0	8.9	92.4	82.9
I -16	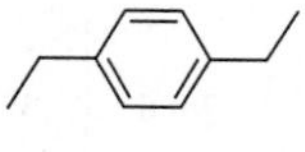	14.0	0	0	24.8	15.9	0.3	30.6	27.1
I -17	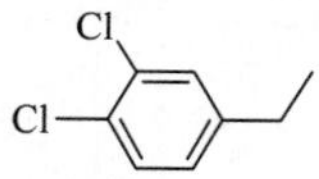 Cl, Cl	6.2	35.0	6.2	36.0	9.3	48.3	15.1	100

Table 1 (continued)

No.	R	*Echinochloa crusgalli* (L.) Beauv.		*Digitaria adscendens*		*Brassica napus*		*Amaranthus retroflexus*	
		A	B	A	B	A	B	A	B
I -18	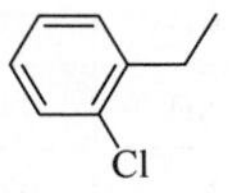 Cl	60.7	79.6	62.8	80.8	12.8	23.2	32.1	91.5
I -19	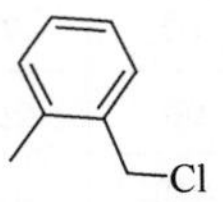 Cl	2.8	14.6	14.5	15.2	7.5	12.3	34.0	15.6
I -20	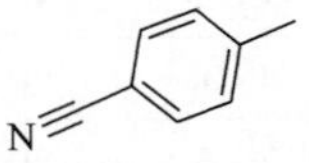 N	0	21.2	7.6	16.8	0	31.8	30.1	10.9
I -21	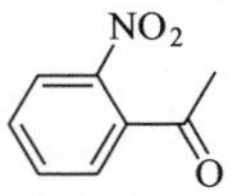 NO_2, O	20.0	36.8	0	17.9	7.0	0	2.9	0
I -22	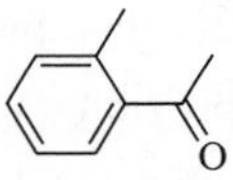 O	11.9	30.8	2.5	30.4	12.3	0	2.3	14.1
I -23	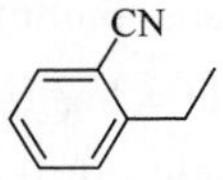 CN	83.8	95.8	100	84.8	74.1	83.5	15.8	47.6

A, postemergence; B, pre - emergence.

Table 2 Herbicidal activity of some compounds (% inhibition)

No.	Dosage (g of a. i. /ha)	*Echinochloa crusgalli* (L.) Beauv.		*Digitaria adscendens*		*Brassica napus*		*Amaranthus retroflexus*	
		A	B	A	B	A	B	A	B
Carfentra - zone - ethyl	150.0	96.7	7.9	8.7	0	100	100	100	100
	300.0	100	18.5	8.7	18.8	100	100	100	100
	600.0	100	86.0	21.7	54.5	100	100	100	100
I -2	150.0	30.7	ND	8.7	ND	ND	ND	19.9	ND
	300.0	68.5	ND	34.8	ND	ND	ND	23.7	ND
	600.0	72.6	ND	41.3	ND	ND	ND	25.7	ND
I -3	150.0	83.8	0	34.8	0	0	0	36.6	3.3
	300.0	89.5	14.7	47.8	0	0	0	83.6	7.1
	600.0	98.9	84.9	87.0	71.4	17.3	5.9	94.2	39.3
I -15	150.0	25.4	ND	28.3	ND	ND	ND	16.5	ND
	300.0	58.0	ND	34.8	ND	ND	ND	27.1	ND
	600.0	62.2	ND	54.3	ND	ND	ND	32.4	ND

Table 2(continued)

No.	Dosage (g of a. i. /ha)	*Echinochloa crusgalli* (L.) Beauv.		*Digitaria adscendens*		*Brassica napus*		*Amaranthus retroflexus*	
		A	B	A	B	A	B	A	B
I -17	150. 0	ND	ND	ND	ND	ND	ND	33. 4	ND
	300. 0	ND	ND	ND	ND	ND	ND	74. 9	ND
	600. 0	ND	ND	ND	ND	ND	ND	91. 3	ND
I -18	150. 0	11. 3	ND	6. 3	ND	ND	ND	7. 8	ND
	300. 0	54. 7	ND	21. 7	ND	ND	ND	26. 2	ND
	600. 0	67. 7	ND	34. 8	ND	ND	ND	42. 6	ND
I -23	150. 0	14. 6	0	15. 2	4. 5	0	23. 5	ND	ND
	300. 0	45. 8	20. 9	15. 2	66. 1	0	45. 1	ND	ND
	600. 0	68. 8	66. 1	34. 8	92. 0	0	70. 6	ND	ND

A, postemergence; B, pre - emergence; ND, the bioactivity was not tested.

Structure - activity relationships

Molecular modeling was performed using SYBYL 6. 91 software[c] and the comparative molecular field analysis (CoMFA) method was been done according our previous work[20,21]. The fungicidal activities of the 22 compounds against *Digitaria adscendens* data(%I) at 1500g/ha under pre - emergence condition used to derive the CoMFA analyses model are listed in Table 1. The activity was expressed in terms of ED by the formula $ED = \log[I/(100 - I) \times Mw]$, where I is percent inhibition and Mw is the molecular weight of the tested compounds. The compound **I -3** was used as a template to build the other molecular structures. Each structure was fully geometry - optimized using a conjugate gradient procedure based on the TRIPOS force field and Gasteiger and Hückel charges. Because these compounds share a common skeleton, seven atoms marked with an asterisk were used for RMS - fitting onto the corresponding atoms of the template structure. CoMFA steric and electrostatic interaction fields were calculated at each lattice intersection on a regularly spaced grid of 2. 0Å. The grid pattern was generated automatically by the SYBYL/CoMFA routine, and an sp^3 carbon atom with a van der Waals radius of 1. 52Å and a + 1. 0 charge was used as the probe to calculate the steric(Lennard. Jones 6 - 12 potential) field energies and electrostatic(Coulombic potential) fields with a distance - dependent dielectric at each lattice point. Values of the steric and electrostatic fields were truncated at 30. 0 kcal/mol. The CoMFA steric and electrostatic fields generated were scaled by the CoMFA - STD method in SYBYL. The electrostatic fields were ignored at the lattice points with maximal steric interactions. A partial least - squares approach was used to derive the 3D - QSAR, in which the CoMFA descriptors were used as independent variables, and ED values were used as dependent variables. The cross - validation with the leave - one - out option and the SAMPLS program, rather than column filtering, was carried out to obtain the optimal number of components to be used in the final analysis. After the optimal number of components was determined, a non - cross - validated analysis was performed without column filtering. The modeling capability(goodness of fit) was judged by the correlation coefficient squared, r^2, and the prediction capability (goodness of prediction) was indicated by the cross - validated r^2 (q^2).

Results and Discussion

Chemistry

Compound **3** was synthesized at 0 - 5℃. Compound **4** was prepared according to the references (18). When the 1 - bromooctane was reacted with compound **4**, NaH was used instead of K_2CO_3, while the temperature must be below 90℃, as higher temperature decreased the yield of product. The structures of all intermediates and title compounds were confirmed by elemental analysis and ^1H NMR. In addition, the crystal structures of **I -3** and **I -8**[22] were determined by X - ray diffraction analysis. As shown in Figure 2, the *o* - flurophenyl ring and the 2,4 - dichlorophenyl ring of the compound **I -3** are parallel, but the two phenyl rings of the compound **I -8** are not. While there is a π - π interaction between the two aromatic rings of compound **I -3**, there is no such interaction for compound **I -8**, which might be able to explain the herbicidal activity differences between the two compounds.

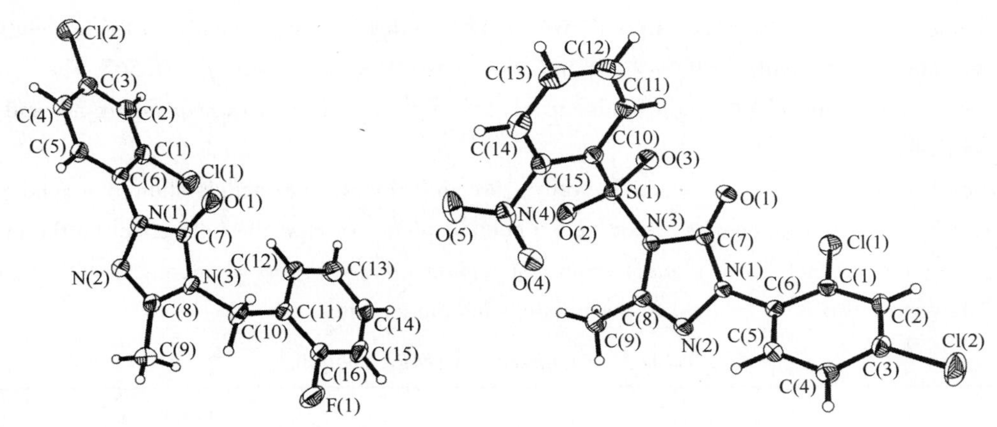

Figure 2 Crystal structural of **I -3** and **I -8**

Bioassay

The postemergence and pre - emergence herbicidal activities of the title compounds were tested in a greenhouse at the concentration of 1500g of a. i. /ha, with carfentrazone - ethyl, a triazolinone - type commercial product, as the control. As shown in Table 1, some of the compounds (**I -2, I -3, I -15, I -17, I -18** and **I -23**) were found to display good herbicidal activities. For example, compound **I -2** displayed >90% inhibition activities against *Digitaria adscendens*, *Amaranthus retroflexus* and *Echinochloa crusgalli* (L.) Beauv. under the postemergence condition, but did not show herbicidal activities against *Brassica napus*. Unfortunately, the compound had low herbicidal acitivities against *Digitaria adscendens*, *Amaranthus retroflexus*, *Brassica napus* and *Echinochloa crusgalli* (L.) Beauv under the pre - emergence condition. Compound **I -3** has excellent herbicidal activities *against Digitaria adscendens*,

Amaranthus retroflexus, *Brassica napus* and *Echinochloa crusgalli*(L.) Beauv(100%) under the pre – emergence condition. Under the postemergence, however, same herbicidal acitivities were observed for compounds **I – 3** and **I – 2**, neither of which has effects on the *Brassica napus*. The comparison(Tables 1 and 2) clearly showed that compound **I – 3** had a higher level of herbicidal activities against *Digitaria adscendens* in postemergence treatment than the commercial herbicide carfentrazone – ethyl at the dosage of 600 g of a. i. /ha. In particular, compound **I – 3** displayed comparable herbicidal activity against *Echinochloa crusgalli*(L.) Beauv. with carfentrazone – ethyl at the dosage of 150g of a. i. /ha.

The data given in Table 1 indicated that the change of substituent affects the herbicidal activity. These compounds have lower herbicidal activities with the SO_2 group than with the CH_2 group, such as **I – 5**, **I – 6** and **I – 7**. The compounds with the benzyl group have higher herbicidal activities than those with alkyl groups. 2 – or 4 – halogen, cyano – substituted benzyl groups lead to higher inhibitory activities, such as **I – 3**, **I – 15**, **I – 17**, **I – 18** and **I – 23**.

Structure – activity relationships

Comparative molecular field analysis method is widely used in drug design, because they allow rapidly generate prediction information of QSAR from the biological activity of newly designed molecules. As listed in Table 3, a predictive CoMFA model was established with the conventional correlation coefficient $r^2 = 0.949$ and the cross – validated coefficient $q^2 = 0.565$. The contributions of steric and electrostatic fields are 61.6% and 38.4% as shown in Figure 3A and B, respectively.

The observed and calculated activity values are in Table 4. The models showed a good predictability on these compounds. Stdev* coeff contour plots can view the field effect on the target property; they are helpful to identify important regions change in the steric, electrostatic fields, and they may also help to identify the possible interaction sites.

Table 3 Summary of CoMFA analysis

	q^2	r^2	Compound	Contribution(%)	
				Steric	Electrostatic
CoMFA	0.565	0.949	I – 3	61.6	38.4

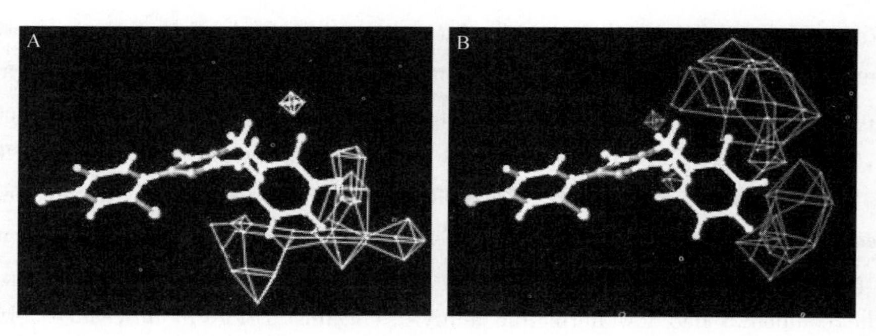

Figure 3 (A) Steric maps from the CoMFA model.
(B) Electrostatic maps from the CoMFA model

Table 4 Experimental and calculated activity of compounds in training set and test set

No.	I	ED	PD_{COMFA}	Residue
1	24.80	-2.938	-3.063	0.125
2	99.90	0.476	0.294	0.182
3	99.90	0.453	0.620	-0.168
4	18.20	-3.220	-3.026	-0.194
7	6.70	-3.776	-3.156	-0.620
8	15.00	-3.386	-3.137	-0.249
9①	21.20	-3.078	-3.306	0.228
10	3.30	-3.967	-4.185	0.218
11①	40.90	-2.726	-2.406	-0.320
12	36.60	-2.792	-2.661	-0.131
13	24.80	-3.024	-2.811	-0.213
14	21.60	-3.121	-3.367	0.246
15	36.00	-2.792	-2.437	-0.355
16	10.40	-3.527	-3.279	-0.248
17①	99.90	0.453	0.227	0.226
18	24.80	-3.041	-3.591	0.550
19	36.00	-2.855	-3.450	0.595
20	80.80	-1.943	-2.158	0.215
21	15.20	-3.345	-3.124	-0.221
23	16.80	-3.267	-3.166	-0.101
25	17.90	-3.256	-3.263	0.007
26	30.40	-2.919	-2.884	-0.035

ED, experimental value of I; PD_{CoMFA}, predictive value of I by CoMFA.

①Represent the compounds in test set.

The steric field contour map is plotted in Figure 3A. The green displays a position where a bulky group would be favorable for higher herbicidal activity. As shown in Figure 3, the CoMFA steric contour plots obviously indicated that a green region is located around the 3 - and 4 - positions of the benzene ring. This means that the bulky substituents at 3 - and 4 - positions will increase the herbicidal activity. The electrostatic contour plot is shown in Figure 3B. The blue contour defines a region where an increase in the positive charge will result in an increase in the activity, whereas the red contour defines a region of space where increasing electron density is favorable. As shown in Figure 3B, the target compounds bearing an electron - withdrawing group at the 3 - or 4 - position of the benzene ring and an electron - donating group at the 2 - position displayed higher activity. These results provide useful information for further optimization of the compounds.

According to the above CoMFA analysis, compound **I -23** (R = CN) was designed and synthesized, and its herbicidal activities against *Echinochloa crusgalli* (L.) Beauv. and *Digitaria adscenden.* was tested. The results indicated that the inhibition of compound **I -23** against

Digitaria adscendens is 66. 1% under pre – emergence at 300g of a. i. /ha, better than the commercial herbicide carfentrazone – ethyl. Moreover, compund **I –23** showed moderate postemergence herbicidal activity against *Echinochloa crusgalli* (L.) Beauv. , which was comparable with that of compound **I –2**, **I –3**, **I –15**, **I –17** and **I –18** at 300g of a. i. /ha.

References

[1] Theodoridis G. (1989) Herbicidal aryl triazolinones. US 4 818 275. Chem Abstr; 111: 153817.

[2] Meazza G. , Bettarini F. , La Porta P. , Piccardi P. , Signorini E. , Portoso D. , Fornara L. (2004) Synthesis and herbicidal activity of novel heterocyclic protoporphyrinogen oxidase inhibitors. Pest Manag Sci; 60: 1178 – 1188.

[3] Hiraki M. , Ohki S. , Sato Y. , Jablonkai I. , Boger P. , Wakabayashi K. (2001) Protoporphyrinogen – IX oxidase inhibitors: bioactivation of thiadiazolidines. Pestic Biochem Physiol; 70: 159 – 167.

[4] Patzoldt W. L. , Hager A. G. , McCormick J. S. , Tranel P. J. (2006) A codon deletion confers resistance to herbicides inhibiting protoporphyrinogen oxidase. Proc Natl Acad Sci USA; 103: 12329 – 12334.

[5] Luo Y. P. , Jiang L. L. , Wang G. D. , Chen Q. , Yang G. F. (2008) Syntheses and herbicidal activities of novel triazolinone derivatives. J Agric Food Chem; 56: 2118 – 2124.

[6] Corradi H. R. , Corrigall A. V. , Boix E. , Mohan C. G. , Sturrock E. D. , Meissner P. N. , Acharya K. R. (2006) Crystal structure of protoporphyrinogen oxidase from *Myxococcus xanthus* and its complex with the inhibitor acifluorfen. J Biol Chem; 281: 38625 – 38633.

[7] Koch M. , Breithaupt C. , Kiefersauer R. , Freigang J. , Huber R. , Messerschmidt A. (2004) Crystal structure of protoporphyrinogen IX oxidase: a key enzyme in haem and chlorophyll biosynthesis. EMBO J; 23: 1720 – 1728.

[8] Hwang I. T. , Hong K. S. , Choi J. S. , Kim H. R. , Joen D. J. , Cho K. Y. (2004) Protoporphyrinogen IX – oxidizing activities involved in the mode of action of a new compound *N* – [4 – chloro – 2 – fluoro – 5 – {3 – (2 – fluorophenyl) – 5 – methyl – 4,5 – ihydroisoxazol – 5 – yl – methoxy} – phenyl] – 3,4,5,6 – tetrahydrophthalimide. Pestic Biochem Physiol; 80: 123 – 130.

[9] Shapiro R. , DiCosimo R. , Hennessey S. M. , Stieglitz B. , Campopiano O. , Chiang G. C. (2001) Discovery and development of a commercial synthesis of azafenidin. Org Process Res Dev; 5: 593 – 598.

[10] Muller K. H. , Lindig M. , Findeisen K. , Konig K. , Lurssen K. , Santel H. J. , Schmidt R. R. , Strang H. (1990) Preparation of 1 – carbamoyl – 3 – (cyclo) alkyl – 4 – amino – 1,2,4 – triazolin – 5 – ones as herbicides. DE 3839 206. Chem Abstr; 113: 172023.

[11] Dayan F. E. , Green H. M. , Weete J. D. , Hancock H. G. (1996) Postemergence activity of sulfentrazone: effects of surfactants and leaf surfaces. Weed Sci; 44: 797 – 803.

[12] Dayan F. E. , Duke S. O. , Weete J. D. , Hancock H. G. (1997) Selectivity and mode of action of carfentrazone – ethyl, a novel phenyl triazolinone herbicide. Pestic Sci; 51: 65 – 73.

[13] Mei X. Y. (2003) Recent advances of research and development of protox inhibitor. Hubei Ind; (Suppl.): 4 – 9.

[14] Pilgram K. H. (1985) An efficient method for the conversion of arylhydrazines into 2 – aryl – 4,4, – dialkylsemicarbazides. Synth Commun; 15: 697 – 703.

[15] Liu X. H. , Chen P. Q. , Wang B. L. , Li Y. H. , Wang S. H. , Li Z. M. (2007) Synthesis bioactivity theoretical and molecular docking study of 1 – cyano – *N* – substituted – cyclopropanecarboxamide as ketol – acid reductoisomerase inhibitor. Bioorg Med Chem Lett; 17: 3784 – 3788.

[16] Liu X. H. , Chen P. Q. , He F. Q. , Li Y. H. , Wang S. H. , Li Z. M. (2007) Structure bioactivity and theoretical study of 1 – cyano – *N* – p – tolylcyclo – propanecarboxamide. Struct Chem; 5: 563 – 568.

[17] Liu X. H. ,Zhang C. Y. ,Guo W. C. ,Li Y. H. ,Chen P. Q. ,Wang T. ,Dong W. L. ,Sun H. W. ,Li Z. M. (2009) Synthesis bioactivity and SAR study of *N'* – (5 – substituted – 1,3,4 – thiadiazol – 2 – yl) – *N* – cyclopropyformyl – thiourea as ketol – acid reductoisomerase inhibitor. J Enzym Inhib Med Chem;24:545 – 552.

[18] Bailey Allan R. ,Bailey M. ,Sortore Eric W. (1993) Triazolinone ring formation in tertiary – butanol. US 5 256 793. Chem Abstr;121:83343.

[19] Sheldrick G. M. SHELXS97 and SHELXL97(1997) Germany:University of Göttingen.

[20] Liu X. H. ,Shi Y. X. ,Ma Y. ,He G. R. ,Dong W. L. ,Zhang C. Y. ,Wang B. L. ,Wang S. H. ,Li B. J. ,Li Z. M. (2009) Synthesis of some *N*,*N'* – diacylhydrazine derivatives with radical scavenging and antifungal activity. Chem Biol Drug Des;73:320 – 327.

[21] Liu X. H. ,Shi Y. X. ,Ma Y. ,Zhang C. Y. ,Dong W. L. ,Li P. ,Wang B. L. ,Li B. J. ,Li Z. M. (2009) Synthesis,antifungal activities and 3D – QSAR study of *N* – (5 – substituted – 1,3,4 – thiadiazol – 2 – yl)cyclopropanecarboxamides. Eur J Med Chem;DOI:10. 1016/j. ejmech. 2009. 01. 012.

[22] Wang L. ,Wang B. L. ,Li Z. M. ,Song H. B. (2006) 2 – (2,4 – Dichlorophenyl) – 5 – methyl – 4 – (2 – nitrophenylsulfonyl) – 2H – 1,2,4 – triazol – 3(4H) – one. Acta Crystallogr E;62:O935.

Notes

[a]Bruker SMART(1998),Madison,WI,USA:Bruker AXS Inc.

[b]Bruker SAINT SHELXTL(1999),Madison,WI,USA:Bruker AXS Inc.

[c]SYBYL version 6. 91(2004),St Louis,MO,USA:Tripos Inc.

Synthesis, Antifungal Activities and 3D - QSAR Study of *N* - (5 - substituted - 1,3,4 - thiadiazol - 2 - yl) Cyclopropanecarboxamides*

Xinghai Liu[1,**], Yanxia Shi[2,**], Yi Ma[1,**], Chuanyu Zhang[1], Weili Dong[1], Li Pan[1], Baolei Wang[1], Baoju Li[2], Zhengming Li[1]

([1]State - Key Laboratory of Elemento - Organic Chemistry, National Pesticidal Engineering Centre (Tianjin), Tianjin Key Laboratory of Pesticide Science, Nankai University, Tianjin, 300071, China;
[2]Institute of Vegetables and Flowers, Chinese Academy of Agricultural Sciences, 300071, China)

Abstract A series of cyclopropanecarboxamide were prepared and tested for antifungal activity *in vivo*. The preliminary bioassays indicated that some compounds are comparable to the commercial fungicides. To further explore the comprehensive structure - activity relationship on the basis of fungicidal activity data, comparative molecular field analysis (CoMFA) was performed, and a statistically reliable model with good predictive power (r^2 = 0. 8, q^2 = 0. 516) was achieved. Based on the CoMFA, compound **7p** was designed and synthesized, which was found to display a good antifungal activity (79. 38%) as **7g** and **7h**.

Key words cyclopropanecarboxamide; 1,3,4 - thiadiazole; antifungal activity; 3D - QSAR; synthesis

1 Introduction

Cyclopropane derivatives, often as bioactive compounds, have been studied for years. At the end of 1960s, some cyclopropane compounds, such as pyrethroids[1], were marketed as low toxic pesticides. Also some pharmaceuticals contain cyclopropane group, such as ciprofloxacin monohydrochloride[2]. So synthesis of broader spectrum and highly bioactive substituted cyclopropane compounds, especially heterocycle substituted ones which are bioactive themselves, becomes the hot spot in the agricultural and medicinal chemistry field. Additionally, sulfur and nitrogen linked heterocyclic compounds received considerable attention in recent times because of their pharmacological and pesticidal importance [3-6]. 2 - Amino - 5 - substituted - 1,3,4 - thiadiazoles are very useful starting materials for the synthesis of various bioactive molecules and applied in medicine and agriculture[7-10] (Fig. 1).

In our previous paper, we reported the synthesis of some cyclopropane derivatives which target herbicidal target KARI (ketol - acid reductoisomerase) [11-13]. As continued our work, a series

* Reprinted from *European Journal of Medicinal Chemistry*, 2009, 44(7): 2782 - 2786. This work was funded by National Key Basic Research Program of China (No. 2003 CB114406), National Natural Science Foundation Key Project of China (No. 20432010), High Performance Computing Project of Tianjin Ministry of Science and Technology of China (No. 043185111 - 5), Natural Science Foundation Project of Tianjin Ministry of Science and Technology of China (No. 07JCYBJC00200 & 08JCYBJC00800), Specialized Research Fund for the Doctoral Program of Higher Education (No. 20070055044).

** These authors contributed equally to this work.

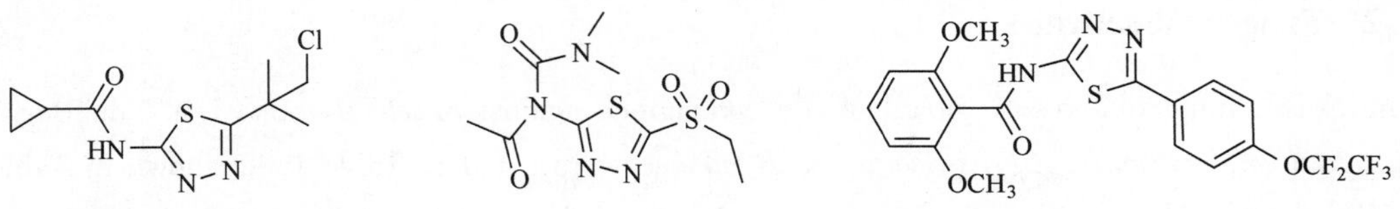

Fig. 1 The pesticides containing 1,3,4 - thiadizole ring

of cyclopropanecarboxamide compounds were prepared, and their fungicidal activities were tested. The preliminary biological tests showed that some compounds exhibit good activity to *Sclerotinia sclerotiorum*(Lib.) de Bary, *Corynespora cassiicola*, *Botrytis cinerea*, *Fusarium oxysporum f.* sp. *cucumerinum*, *Cercospora arachidicola*, and *Rhizoctonia solanii.* The structure - activity relationship was also studied.

2 Results and discussion

2.1 Chemistry

The cyclopropane - 1,1 - dicarboxylic acid, prepared from 1,2 - dichlorethane and diethyl malonate was cyclized for 16h at refluxing temperature. In order to optimize the reaction time, microwave assistant irradiation was applied which shortened the reaction time to 40min. The cyclopropane - 1,1 - dicarboxylic acid was obtained from the hydrolysis of diethyl cyclopropane - 1,1 - dicarboxylate, but the yield of this step is low, about 50%. Cyclopropanecarbonyl chloride was prepared from the cyclopropane dicarboxylic acid and $SOCl_2$, without isolation further reacted with 5 - substituted - 2 - amino - 1,3,4 - thiadizoles at room temperature[12] as shown in Scheme 1. Several procedures are available for the one - step synthesis of 2 - amino - 5 - substituted - 1,3,4 - thiadiazoles derivative. Yet the reaction of substituted aryl and alkyl acid with thiosemicarbazide in the presence of dehydrating agent $POCl_3$ affords a series of 2 - amino - 5 - substituted - 1,3,4 - thiadiazoles under microwave irradiation.

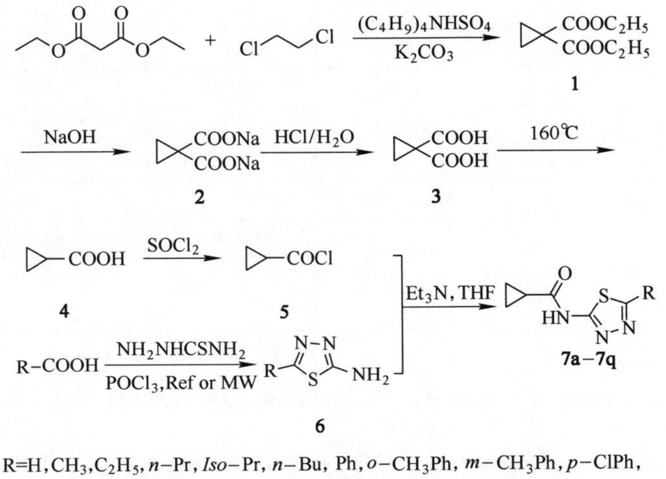

R=H, CH_3, C_2H_5, *n*-Pr, *Iso*-Pr, *n*-Bu, Ph, *o*-CH_3Ph, *m*-CH_3Ph, *p*-ClPh, *o*-ClPh, *o*-FPh, *p*-NO_2Ph, Py, furan, *p*-OCH_3Ph, cyclopropane

Scheme 1 The synthesis route of title compounds

2.2 Fungicidal activities

The in *vivo* fungicidal results of all of the compounds against *S. sclerotiorum* (Lib.) de Bary, *R. solanii*, *F. oxysporum*, *C. cassiicola*, and *B. cinerea* were listed in Table 1. As shown in Table 1, Compounds **7b** and **7l** were found to display good fungicidal activities against *R. solanii*, *F. oxysporum*, *C. cassiicola*, and *B. cinerea*, compounds **7a**, **7c – f**, **7m – o** did not display obvious fungicidal activities against *S. sclerotiorum* (Lib.) de Bary, *R. solanii*, *F. oxysporum*, *C. cassiicola*, and *B. cinerea*. Compound **7k** have fair to good fungicidal activity with the commercial fungicide pyrimethanil against *F. oxysporum*. Among them, these compounds displayed the highest fungicidal activity against *B. cinerea*. All compounds did not exhibit good fungicidal activity against *S. sclerotiorum* (Lib.) de Bary at the concentration of $500\mu g \cdot mL^{-1}$.

Table 1 The antifungal activities of compounds 7a – 7q *in vivo* at $500\mu g \cdot mL^{-1}$

No.	*Sclerotinia sclerotiorum* (Lib.) de Bary	*Rhizoctonia solanii*	*Fusarium oxysporum*	*Corynespora cassiicola*	*Botrytis cinerea*
7a	24.00	37.00	17.00	22.52	83.73
7b	23.00	5.00	84.00	52.43	67.54
7c	51.70	23.00	10.00	49.66	32.09
7d	40.00	35.00	1.00	46.63	57.87
7e	47.70	11.00	20.00	29.30	41.98
7f	10.30	44.00	2.00	21.92	26.02
7g	24.00	9.00	16.00	17.43	80.53
7h	37.50	36.00	21.00	50.79	75.09
7i	20.00	39.00	11.00	8.38	90.67
7j	23.90	62.00	28.00	19.69	13.49
7k	25.50	44.00	91.00	21.45	35.11
7l	23.00	5.00	84.00	68.27	67.54
7m	43.06	16.00	13.00	18.56	30.27
7n	14.00	8.00	14.00	34.06	29.66
7o	5.00	0	0	15.73	38.76
7p	33.0	9.00	12.00	33.26	79.38
7q	26.70	21.00	37.00	13.28	66.59
dimehachlon	96.70				
Jinggangmycin		92.10			
Thiophanate methyl			97.00		
Chlorothalonil				86.43	
Pyrimethanil					99.17

2.3 Quantitative structure – activity relationship (3D – QSAR)

Molecular modeling was performed using SYBYL 6.91 software (Tripos, Inc.)[14]. Each structure was fully geometry – optimized using a conjugate gradient procedure based on the TRIPOS force field and Gasteiger and Hückel charges. Because these compounds share a common skeleton, 10 atoms marked with an asterisk were used for rms – fitting onto the corresponding atoms of the template structure (Figs. 2 and 3).

Fig. 2 The asterisk skeleton of title compounds

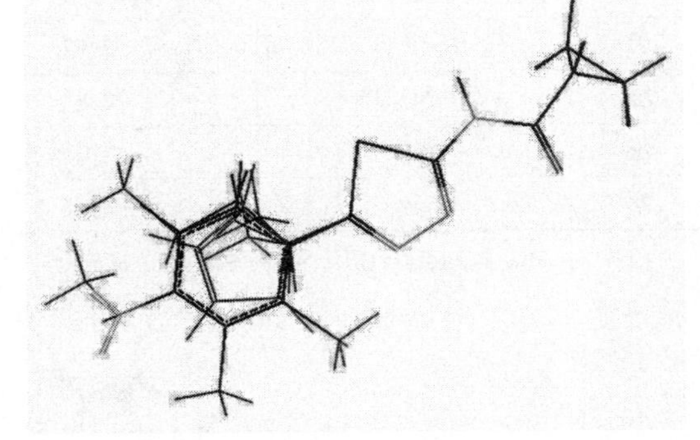

Fig. 3 Superposition modes of compounds

CoMFA steric and electrostatic interaction fields were calculated at each lattice intersection on a regularly spaced grid of 2.0Å. The grid pattern was generated automatically by the SYBYL/CoMFA routine, and an sp^3 carbon atom with a van der Waals radius of 1.52Å and a +1.0 charge was used as the probe to calculate the steric (Lennard – Jones 6 – 12 potential) field energies and electrostatic (Coulombic potential) fields with a distance – dependent dielectric at each lattice point. Values of the steric and electrostatic fields were truncated at 30.0kcal/mol. The CoMFA steric and electrostatic fields generated were scaled by the CoMFA – STD method in SYBYL. The electrostatic fields were ignored at the lattice points with maximal steric interactions. A partial least – squares (PLS) approach was used to derive the 3D – QSAR, in which the CoMFA descriptors were used as independent variables, and ED values were used as dependent variables. The data were analyzed by CoMFA method and fungicidal activity against *B. cinerea* data (% I) at 500μg · mL^{-1} being converted to $ED = \log(I/((100 - I) \times MW))$ [15] as a dependent variable. The observed and calculated activity values for all the compounds are shown in Table 2, and the plots of the predicted versus the actual activity values for all the compounds are shown in Fig. 5.

Table 2 The structures, activities and total score of compounds

No.	R^1	ED	ED''	Residue
7a	H	-1.51	-1.89	0.38
7b	Me	-1.94	-1.92	-0.02
7c	Et	-2.62	-2.23	-0.39
7d	Pr	-2.19	-2.27	0.08
7e	Iso – Pr	-2.46	-2.4	-0.06
7f	Bu	-2.8	-2.35	-0.45
7g	Ph	-1.77	-2.46	0.69
7h	*o* – Me Ph	-1.93	-1.9	-0.03
7i	*m* – Me Ph	-1.42	-1.54	0.12
7j	*p* – Cl Ph	-2.71	-2.75	0.04

Table 2(continued)

No.	R^1	ED	ED″	Residue
7k	*o* – Cl Ph	–2. 1	–2. 13	0. 03
7l	o – F Ph	–2. 82	–2. 46	–0. 36
7m	*p* – NO_2 Ph	–2. 83	–2. 67	–0. 16
7n	Py	–2. 77	–2. 48	–0. 29
7o	Furan	–2. 57	–2. 57	0

ED = experimental value, ED″ = predictive value of ED.

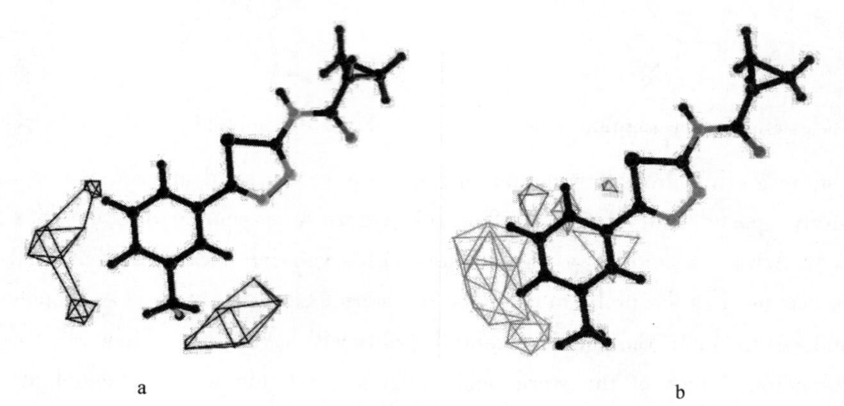

Fig. 4 Steric and electrostatic contribution contour maps of CoMFA

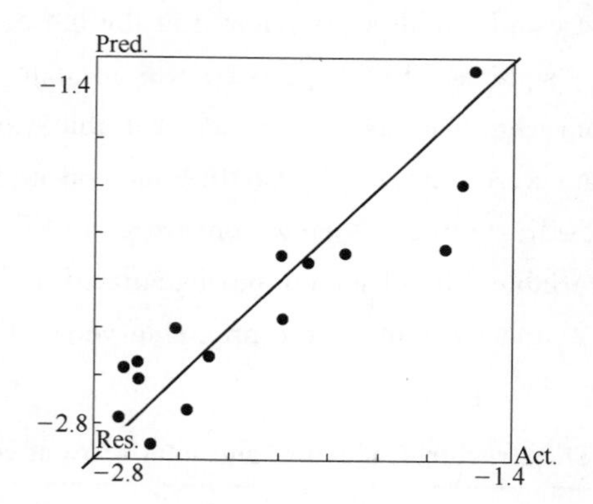

Fig. 5 CoMFA predicted as experimental *p*ED values

The cross – validation with the leave – one – out (LOO) option and the SAMPLS program, rather than column filtering, was carried out to obtain the optimal number of components to be used in the final analysis. After the optimal number of components was determined, a non – cross – validated analysis was performed without column filtering. The modeling capability (goodness of fit) was judged by the correlation coefficient squared, r^2, and the prediction capability (goodness of prediction) was indicated by the cross – validated r^2 (q^2). The 3D – QSAR models gave a good q^2 (cross – validated r^2) = 0. 516 and r^2 (non – cross – validated r^2) = 0. 800, two components. The compound **7i** was illustrated to explain the field contributions of different properties obtained from the CoMFA analyses. The steric and electrostatic contribution contour maps of CoMFA are plotted in Fig. 4. As shown in Fig. 4a, green displays 2 – positions or 3 – position of benzene ring where a bulky group would be favorable for higher antifungal activity. *In contrast, yellow indicates 5 – position of benzene ring where a decrease in the bulk of the target molecules is favored. For example, some compounds bearing 2 – methyl, 3 – methyl of benzene ring, such as **7h**, and **7i**, displayed higher antifungal activity against *B. cinerea.* As shown in Fig. 4b, the title compounds bearing an electron – donating group at the 2 – position, 3 – position or 4 – position of benzene ring can improve the antifungal activity, such as **7h** and **7i**.

According to the above CoMFA analysis, compound **7p**(R = *p* – OMe Ph) was designed, synthesized and tested its antifungal activity against *B. cinerea.* The results indicated that the inhibition of compound **7p** is 79.38%, whose inhibition is as good as **7g** and **7h**.

3 Conclusion

Using easily obtainable compounds **5**, we have prepared a new series of cyclopropanecarboxamide analogues **7** containing 1,3,4 – thiadiazoles in good yields. Some of these compounds **7g**, **7h**, **7i**, **7p** exhibited excellent activity as displayed in Table 1. According to the CoMFA model, when R is substituted benzene groups, substituents at 2 – position, 3 – position or 4 – position of the benzene ring are favored with electron – donating and bulky groups. Meanwhile, electron – withdrawing group is disfavored on these positions, such as NO_2 group.

4 Experimental section

4.1 Materials and methods

All reagents are analytical grade. Melting points were determined using a X – 4 apparatus and were uncorrected. ^1H NMR spectra were measured on a Bruker AC – P500 instrument (300MHz) using TMS as an internal standard and DMSO – d_6 as solvent. HRMS data was obtained on a FTICR – MS instrument (Ionspec 7.0T).

4.2 Synthesis

4.2.1 General procedure

4.2.1.1 Preparation of 7a

The acid chloride was prepared according the reference[12]. Dropwised the acid chloride to substituted 2 – amino – 5 – substituted – 1,3,4 – thiadiazoles (7.50 mmol), then vigorously stirred at ambient temperature for 4h. The corresponding amide **7** precipitated immediately. The product was filtered, washed with THF, dried, and recrystallized from EtOH – H_2O to give the title compounds **7**.

7a: white crystal, yield 84.6%, m. p. 254 – 255℃; ^1H NMR ($CDCl_3$) δ: 1.08 – 1.25 (m, 4H, cyclopropane – CH_2), 1.59 (s, 1H, Het – H), 2.29 – 2.35 (m, 1H, cyclopropane – CH), 8.78 (s, 1H, NH); FTICR – MS for $C_6H_7N_3OS$: found 168.0229, calcd. 168.0237.

7b: white crystal, yield 85.1%, m. p. >300℃; ^1H NMR ($CDCl_3$) δ: 1.05 – 1.19 (m, 4H, cy-

* For interpretation of the references to color in this text, the reader is referred to the web version of this article.

clopropane – CH_2), 1.55(t, 3H, CH_3), 2.16 – 2.26(m, 1H, cyclopropane – CH), 12.85(s, 1H, NH); FTICR – MS for $C_7H_9N_3OS$: found 182.0395, calcd. 182.0394.

7c: white crystal, yield 81.2%, m. p. 197 – 198℃; ^{1}H NMR($CDCl_3$) δ: 1.04 – 1.19(m, 4H, cyclopropane – CH_2), 1.38(t, 3H, CH_3), 2.21 – 2.26(m, 1H, cyclopropane – CH), 3.02 (q, 2H, CH_2), 13.14 (s, 1H, NH); FTICR – MS for $C_8H_{11}N_3OS$: found 196.0556, calcd. 196.0550.

7d: white crystal, yield 84.5%, m. p. 175 – 176℃; ^{1}H NMR($CDCl_3$) δ: 1.02(t, 3H, CH_3), 1.05 – 1.21(m, 4H, cyclopropane – CH_2), 1.81(m, 2H, CH_2), 2.25 – 2.31(m, 1H, cyclopropane – CH), 2.98 (t, 2H, CH_2), 13.43 (s, 1H, NH); FTICR – MS for $C_9H_{13}N_3OS$: found 210.0704, calcd. 210.0707.

7e: white crystal, yield 84.5%, m. p. 127 – 128℃; ^{1}H NMR($CDCl_3$) δ: 0.92 – 0.96(t, 3H, CH_3), 1.04 – 1.19(m, 4H, cyclopropane – CH_2), 1.43(m, 2H, CH_2), 1.56(m, 2H, CH_2), 1.75(m, 2H, CH_2), 2.19 – 2.22(m, 1H, cyclopropane – CH), 13.00(s, 1H, NH); FTICR – MS for $C_{10}H_{15}N_3OS$: found 210.0702, calcd. 210.0707.

7f: white crystal, yield 90.1%, m. p. 183 – 184℃; ^{1}H NMR($CDCl_3$) δ: 1.03 – 1.29(m, 4H, cyclopropane – CH_2), 1.40(d, 6H, CH_3), 2.26 – 2.28(m, 1H, cyclopropane – CH), 3.33 – 3.40(m, 1H, CH), 7.26 – 7.71 (m, 4H, Ar—H), 13.27 (s, 1H, NH); FTICR – MS for $C_9H_{13}N_3OS$: found 224.0865, calcd. 224.0863.

7g: white crystal, yield 86.5%, m. p. 249 – 251℃; ^{1}H NMR($CDCl_3$) δ: 1.11 – 1.26(m, 4H, cyclopropane – CH_2), 2.35(m, 1H, cyclopropane – CH), 7.47 – 7.92(m, 5H, Ar—H), 13.33 (bs, 1H, NH); FTICR – MS for $C_{12}H_{11}N_3OS$: found 244.0550, calcd. 244.0550.

7h: white crystal, yield 82.6%, m. p. 216 – 218℃; ^{1}H NMR($CDCl_3$) δ: 1.04 – 1.25(m, 4H, cyclopropane – CH_2), 2.39 – 2.45(m, 1H, cyclopropane – CH), 2.56(s, 3H, CH_3), 7.29 – 7.65(m, 4H, Ar—H), 13.07(bs, 1H, NH); FTICR – MS for $C_{13}H_{13}N_3OS$: found 260.0841, calcd. 260.0852.

7i: white crystal, yield 80.6%, m. p. 262 – 265℃; ^{1}H NMR($CDCl_3$) δ: 1.11 – 1.26(m, 4H, cyclopropane – CH_2), 2.31 – 2.36(m, 1H, cyclopropane – CH), 2.43(d, 3H, CH_3), 7.26 – 7.71(m, 4H, Ar—H), 13.19(s, 1H, NH); FTICR – MS for $C_{13}H_{13}N_3OS$: found 260.0841, calcd. 260.0852.

7j: white crystal, yield 84.6%, m. p. > 300℃; ^{1}H NMR(DMSO) δ: 0.93 – 1.02(m, 4H, cyclopropane – CH_2), 1.98 – 2.05(m, 1H, cyclopropane – CH), 7.60(d, J = 8.55 Hz, 2H, Ar—H), 7.95(d, J = 8.54 Hz, 2H, Ar—H), 12.99(bs, 1H, NH); FTICR – MS for $C_{12}H_{10}ClN_3OS$: found 280.0296, calcd. 280.0306.

7k: white crystal, yield 88.4%, m. p. 253 – 255℃; ^{1}H NMR(DMSO) δ: 0.93 – 1.02(m, 4H, cyclopropane – CH_2), 1.98 – 2.05(m, 1H, cyclopropane – CH), 7.68(d, J = 7.29Hz, 2H, Ar—H), 8.08(d, J = 7.73 Hz, 2H, Ar—H), 12.99(bs, 1H, NH); FTICR – MS for $C_{12}H_{10}ClN_3OS$: found 280.0296, calcd. 280.0306.

7l: white crystal, yield 83.6%, m. p. > 300℃; ^{1}H NMR(DMSO) δ: 0.73 – 1.16(m, 4H, cyclopropane – CH_2), 1.94 – 2.03(m, 1H, cyclopropane – CH), 7.39 – 7.69(m, 3H, Ar—H), 8.25(m, 1H, Ar—H), 13.00(bs, 1H, NH); FTICR – MS for $C_{12}H_{10}FN_3OS$: found 264.0597, calcd. 264. 0601.

7m: white crystal, yield 85. 4%, m. p. > 300℃; ^{1}H NMR (DMSO) δ: 0. 94 – 0. 99 (m, 4H, cyclopropane – CH_2), 1. 98 – 2. 02 (m, 1H, cyclopropane – CH), 8. 18 (d, J = 8. 83 Hz, 2H, Ar—H), 8. 32 (d, J = 8. 81Hz, 2H, Ar—H), 13. 11 (bs, 1H, NH); FTICR – MS for $C_{12}H_{10}N_4O_3S$: found 289. 0392, calcd. 289. 0401.

7n: white crystal, yield 89. 7%, m. p. 238 – 239℃; ^{1}H NMR (DMSO) δ: 0. 92 – 0. 98 (m, 4H, cyclopropane – CH_2), 1. 98 – 2. 01 (m, 1H, cyclopropane – CH), 7. 54 (m, 1H, Py—H), 8. 29 (m, 1H, Py—H), 8. 66 (m, 1H, Py—H), 9. 01 (m, 1H, Py—H), 13. 03 (bs, 1H, NH); FTICR – MS for $C_{11}H_{10}N_4OS$: found 247. 0637, calcd. 247. 0648.

7o: white crystal, yield 86. 5%, m. p. 272 – 273℃; ^{1}H NMR ($CDCl_3$) δ: 1. 01 – 1. 25 (m, 4H, cyclopropane – CH_2), 2. 24 – 2. 26 (m, 1H, cyclopropane – CH), 6. 56 (d, J = 1. 68 Hz, 1H, furan – H), 7. 03 (d, J = 3. 43Hz, 1H, furan – H), 7. 58 (s, 1H, furan – H), 13. 00 (s, 1H, NH); FTICR – MS for $C_{10}H_9N_3O_2S$: found 236. 0483, calcd. 236. 0488.

7p: white crystal, yield 84. 5%, m. p. 264 – 267℃; ^{1}H NMR (DMSO) δ: 0. 95 – 0. 99 (m, 4H, cyclopropane – CH_2), 1. 98 – 2. 02 (m, 1H, cyclopropane – CH), 3. 84 (s, 3H, CH_3), 7. 15 (d, 2H, Ar—H), 7. 90 (d, 2H, Ar—H), 12. 88 (bs, 1H, NH); FTICR – MS for $C_{13}H_{13}N_3O_2S$: found 276. 0781, calcd. 276. 0801.

7q: white crystal, yield 84. 4%, m. p. 194 – 196℃; ^{1}H NMR ($CDCl_3$) δ: 1. 02 – 1. 18 (m, 8H, cyclopropane – CH_2), 2. 17 – 2. 02 (m, 1H, cyclopropane – CH), 2. 29 – 2. 32 (m, 1H, cyclopropane – CH), 12. 97 (s, 1H, NH); FTICR – MS for $C_9H_{11}N_3OS$: found 210. 0690, calcd. 210. 0696.

4. 3 Antifungal activity

Fungicidal activities of compounds of series 7 against *S. sclerotiorum* (Lib.) de Bary, *R. solanii*, *F. oxysporum*, *C. cassiicola*, and *B. cinerea* were evaluated using pot culture test according to reference[16]. The culture plates were cultivated at (24 ± 1) ℃. Fungicidal activities of commercial fungicides dimehachlon, jinggangmycin, Thiophanate methyl, chlorothalonil, pyrimethanil as a control against above mentioned five fungi were evaluated at the same condition. The relative inhibition rate of the circle mycelium compared to blank assay was calculated via the following equation:

$$\text{Relative inhibition rate}(\%) = \frac{d_{ex} - d'_{ex}}{d_{ex}} \times 100\%$$

where d_{ex} is the extended diameter of the circle mycelium during the blank assay; and d'_{ex} is the extended diameter of the circle mycelium during testing.

References

[1] N. Ohno, K. Fujimoto, Y. Mizutani, T. Mizutani, M. Hirano, N. Itaya, T. Honda, H. Yoshioka, Agric. Biol. Chem. 38 (1974) 881 – 883.

[2] M. P. Yovani, M. M. Ricardo, T. Francisco, M. Yamilen, R. Z. Vicente, C. Eduardo, Bioorg. Med. Chem. 13 (2005) 2881 – 2899.

[3] Y. X. Li, Y. M. Wang, X. P. Yang, S. H. Wang, Z. M. Li, Heteroat. Chem. 14 (2003) 342 – 344.

[4] F. Q. He, X. H. Liu, B. L. Wang, Z. M. Li, J. Chem. Res. 12(2006)809 - 811.
[5] F. Q. He, X. H. Liu, B. L. Wang, Z. M. Li, Heteroat. Chem. 19(2008)21 - 27.
[6] W. G. Zhao, J. G. Wang, Z. M. Li, Z. Yang, Bioorg. Med. Chem. Lett. 16(2006)6107 - 6111.
[7] L. Zhang, A. Zhang, G. Zhou, Z. Zhang, Chin. J. Org. Chem. 22(2002)663 - 666.
[8] Z. Y. Zhang, H. S. Dong, S. Y. Yang, Chin. J. Org. Chem. 16(1996)430 - 435.
[9] F. Liu, X. Q. Luo, B. A. Song, P. S. Bhadury, S. Yang, L. H. Jin, W. Xue, D. Y. Hu, Bioorg. Med. Chem. 16 (2008)3632 - 3640.
[10] W. C. Guo, X. H. Liu, Y. H. Li, Y. H. Li, S. H. Wang, Z. M. Li, Chem. Res. Chin. Univ. 24(2008)32 - 35.
[11] X. H. Liu, P. Q. Chen, B. L. Wang, Y. H. Li, Z. M. Li, Bioorg. Med. Chem. Lett. 17(2007)3784 - 3788.
[12] X. H. Liu, P. Q. Chen, F. Q. He, Y. H. Li, S. H. Wang, Z. M. Li, Struct. Chem. 5(2007)563 - 568.
[13] X. H. Liu, C. Y. Zhang, W. C. Guo, Y. H. Li, P. Q. Chen, T. Wang, W. L. Dong, B. L. Wang, H. W. Sun, Z. M. Li, J. Enzyme Inhib. Med. Chem. (2008). doi:10.1080/14756360802234943.
[14] SYBYL, Version 6.91, Tripos Inc., St. Louis, 2004.
[15] Y. Q. Zhu, C. Wu, H. B. Li, X. M. Zou, X. K. Si, F. Z. Hu, H. Z. Yang, J. Agric. Food Chem. 55 (2007) 1364 - 1369.
[16] Y. X. Shi, L. P. Yuan, Y. B. Zhang, B. J. Li, Chin. J. Pestic. Sci. 9(2007)126 - 130.

Synthesis, Bioactivity and SAR Study of N' – (5 – substituted – 1,3,4 – thiadiazol – 2 – yl) – N – cyclopropylformyl – thioureas as Ketol – acid Reductoisomerase Inhibitors*

Xinghai Liu[1], Chuanyu Zhang[1], Wancheng Guo[1],
Yonghong Li[1], Peiquan Chen[1], Teng Wang[2],
Weili Dong[1], Baolei Wang[1], Hongwei Sun[2], Zhengming Li[1,3]

([1]State – Key Laboratory of Elemento – Organic Chemistry, Nankai University, Tianjin, 300071, China;
[2]Department of Chemistry, College of Chemistry, Nankai University, Tianjin, 300071, China;
[3]National Pesticide Engineering Research Center, Nankai University, Tianjin, 300071, China)

Abstract Ketol – acid reductoisomerase (KARI; EC 1.1.1.86) catalyzes the second common step in branched – chain amino acid biosynthesis. The catalyzed process consists of two steps, the first of which is an alkyl migration from one carbon atom to its neighboring atom. The likely transition state is a cyclopropane derivative, thus a new series of cyclopropanecarbonyl thiourea derivatives were designed and synthesized involving a one – pot phase transfer catalyzed reaction. Rice KARI inhibitory activity of these compounds were evaluated and the 5 – butyl substituted (**3e**) and 3 – pyridinyl substituted (**3n**) compounds reached 100% at 100μg · mL^{-1}. Structure – activity relationship shows that longer chain derivatives had higher KARI inhibitory activity. Meanwhile substitution of the 4 – position of the benzene ring had higher KARI inhibitory activity than that of the 2 and 3 – position. Auto – Dock was used to predict the binding mode of **3n**. This was done by analyzing the interaction of compound **3n** with the active sites of the available spinach KARI. This was in accord with the results analyzed by the frontier molecular orbital theory.

Key words cyclopropane derivatives; Auto – Dock; KARI activity; thiourea; inhibition; ketol – acid reductoisomerase

Introduction

Microorganisms and plants contain numerous enzymes that some of which are potential targets for designing bioactive compounds such as antibiotics and herbicides. Enzymes involved in the biosynthesis of the branched chain amino acids are one such example. The success of these herbicides (sulfonylureas[1], imidazolinones[2], and so on) which target the first enzyme (acetohydroxyacid synthase) has stimulated research into inhibitors of other enzymes in the pathway, including the second enzyme in the common pathway[3], ketol – acid reductoisomerase (KARI; EC 1.1.1.86). But there are no commercial herbicides targeting KARI yet, only HOE 704[4],

* Reprinted from *Journal of Enzyme Inhibition and Medicinal Chemistry*, 2009, 24(2): 545 – 552. This work was supported by the National Basic Research Key Program of China (No. 2003CB114406), the National Natural Science Foundation Key Project of China (No. 20432010), the High Performance Computing Project of Tianjin Ministry of Science and Technology of China (No. 043185111 – 5), Specialized Research Fund for the Doctoral Program of Higher Education (No. 20070055044) and Tianjin Natural Science Foundation (No. 08JCYBJC00800).

IpOHA[5] and CPD analogs[6-7] are reported as potent competitive inhibitors of the enzyme *in vitro* (Scheme 1). Additionally, Grandoni et al. [8] found HOE 704[4], IpOHA[5] can inhibit *tuberculosis* much better than ATCC35801, so these branched chain amino acids inhibitors became novel anti – *tuberculosis* medicine hopefully.

The reaction catalyzed by KARI is shown in Scheme 1 which consists of two steps[9-10], an alkyl migration followed by a NADPH dependent reduction. Both steps require a divalentmetal-ion, such as Mg^{2+}, Mn^{2+} or Co^{2+}, but the alkyl migration is highly specific for Mg^{2+}.

A transition state being a cyclopropane is postulated and mimicked by Gerwick et al. [11] They showed that cyclopropane – 1,1 – dicarboxylate (CPD) can inhibit *Escherichia coli* KARI. They also showed that application of CPD to various plant tissues caused the accumulation of the substrate 2 – acetolactate; which data strongly suggest that the CPD can inhibit the activity of KARI *in vivo*[6].

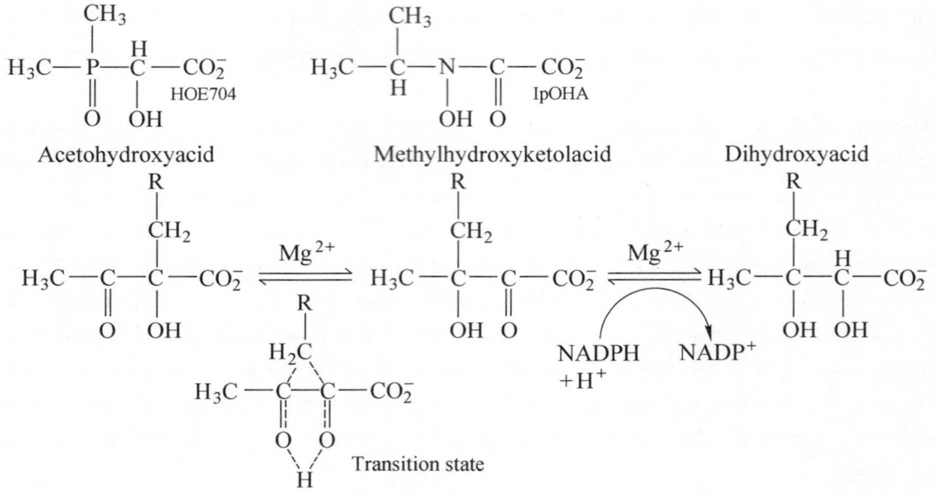

Scheme 1 Reaction catalyzed by KARI, the known inhibitors of HOE704 and IpOHA are the analogies of Acetohydroxyacid and Methylhydroxyketolacid respectively

The first step in the KARI catalyzing process involves an alkyl migration from one carbon atom to its neighboring atom. The likely transition state is a cyclopropane derivative. Also Halgand et al. [12] found that 1,2,3 – thiadiazole can inhibit KARI effectively using high throughput screening. By the way, all these inhibitors contain C =O, P =O, S =O and other groups. For this reason, some new cyclopropane derivatives contain C =O, C =S and 1,3,4 – thiadiazole were synthesized in our laboratory (Scheme 2).

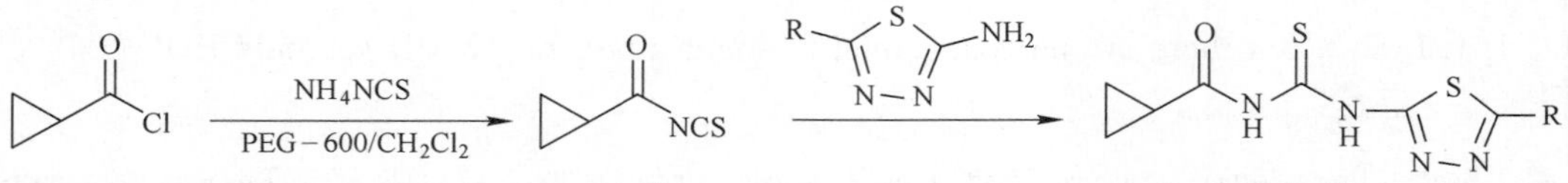

R= CH_3, C_2H_5, *n*–Pr, *iso*–Pr, *n*–Bu, Ph, *o*–CH_3Ph, *m*–CH_3Ph, *o*–ClPh, *p*–ClPh, *o*–FPh, *p*–NO_2Ph, *p*–OCH_3Ph, 3–Py, furan

Scheme 2 Synthetic route for compounds **3a – 3o**

Experimental

Instruments

Melting points were determined using an X – 4 melting apparatus and were uncorrected. Infrared spectra were recorded on a Bruker Equinox55 spectrophotometer as potassium bromide tablets. ^{1}H NMR spectra were measured on a Bruker AC – P500 instrument (300MHz) using tetramethylsilane as an internal standard and deuterochloroform as solvent. Mass spectra were recorded on a Thermo Finnigan LCQ Advantage LC/mass detector instrument. FTMS were determined by Ionspec FT – MS 7.0T.

Synthesis of compounds

The title compounds were synthesized according to the route shown in Scheme 2, and the yields were not optimized. To a solution (25mL) of cyclopropanecarboxylic acid (7.50mmol) was added thionyl chloride (30mmol) and the mixture was refluxed for 2h to give acid chloride. Powdered ammonium thiocyanate (1.14g, 15mmol), cyclopropanecarbonyl chloride (1.04g, 10mmol), PEG – 600 (0.18g, 3% with respect to ammonium thiocyanate) and methylene chloride (25mL) were placed in a dried roundbottomed flask containing a magnetic stirrer bar and stirred at room temperature for 1h. Then 2 – amino – 5 – substituted – 1,3,4 – thiadiazoles (4.5mmol) in methylene dichloride (10mL) was added dropwise over 0.5h, and the mixture was stirred for 1 – 2h while monitored by TLC. The corresponding products precipitated immediately. The product was filtered, washed with water to remove inorganic salts, dried, and recrystallized from DMF – EtOH – H_2O, afforded a light yellow solid.

N′ – (5 – methyl – 1,3,4 – thiadiazol – 2 – yl) – N – cyclopropyformyl – thiourea (***3a***). Light yellow crystals, yield 84.3%, m. p. 248 – 250℃; ^{1}HNMR ($CDCl_3$) δ: 1.04 – 1.19 (m, 4H, cyclopropane – CH_2), 2.17 (s, 1H, cyclopropane – CH), 2.677 (s, 3H, CH_3); IR/cm^{-1}: 3161 (N—H), 1680 (C ═O), 1296 (C ═S); ESI – MS: 182.08 [M – H_2NCS]$^-$; FT – MS for $C_8H_{10}N_4OS_2$: found 182.0395, calcd. 182.0394.

N′ – (5 – ethyl – 1,3,4 – thiadiazol – 2 – yl) – N – cyclopropyformylthiourea (***3b***). Light yellow crystals, yield 88.9%, m. p. 199 – 200℃; ^{1}HNMR ($CDCl_3$) δ: 1.05 – 1.19 (m, 4H, cyclopropane – CH_2), 1.39 (t, 2H, CH_2), 2.26 (s, 1H, cyclopropane – CH), 3.03 (d, 3H, CH_3); IR/cm^{-1}: 3161 (N—H), 1687 (C ═O), 1304 (C ═S); ESI – MS: 198.13 [M – H_2NCS]$^-$; FT – MS for $C_9H_{12}N_4OS_2$: found 255.0384, calcd. 255.0380.

N′ – (5 – n – propyl – 1,3,4 – thiadiazol – 2 – yl) – N – cyclopropyformyl – thiourea (***3c***). Light yellow crystals, yield 91.2%, m. p. 176 – 178℃; ^{1}HNMR ($CDCl_3$) δ: 1.03 – 1.06 (m, 4H, cyclopropane – CH_2), 1.18 (t, 3H, CH_3), 1.72 – 1.83 (m, 2H, CH_2), 2.17 (s, 1H, cyclopropane – CH), 2.98 (t, 3H, CH_2); IR/cm^{-1}: 3169 (N—H), 1687 (C ═O), 1304 (C ═S); ESI – MS: 212.19 [M – H_2NCS]$^-$; FT – MS for $C_{10}H_{14}N_4OS_2$: found 210.0704, calcd. 210.0707.

N′ – (5 – iso – propyl – 1,3,4 – thiadiazol – 2 – yl) – N – cyclopropyformyl – thiourea (***3d***). Light yellow crystals, yield 88.6%, m. p. 173 – 174℃; ^{1}HNMR ($CDCl_3$) δ: 1.02 – 1.29 (m,

4H, cyclopropane − CH_2), 1.41 (d, 6H, CH_3), 2.17 (s, 1H, cyclopropane − CH), 4.37 (m, 1H, CH); IR/cm^{-1}: 3169 (N—H), 1687 (C=O), 1311 (C=S); ESI − MS: 210.16 [M − H_2NCS]; FT − MS for $C_{10}H_{14}N_4OS_2$: found 210.0704, calcd. 210.0707.

N′ − (5 − butyl − 1,3,4 − thiadiazol − 2 − yl) − N − cyclopropyformylthiourea (***3e***). Light yellow crystals, yield 82.6%, m. p. 115 − 117℃; ^{1}HNMR ($CDCl_3$) δ: 0.95 (t, 3H, CH_3), 1.03 − 1.18 (m, 4H, cyclopropane − CH_2), 1.37 − 1.46 (m, 2H, CH_2), 1.68 − 1.78 (m, 2H, CH_2), 2.17 (s, 1H, cyclopropane − CH), 2.87 − 3.03 (m, 2H, CH_2); IR/cm^{-1}: 3161 (N—H), 1680 (C=O), 1304 (C=S); ESI − MS: 284.19 $[M - H]^-$; FT − MS for $C_{10}H_{14}N_4OS_2$: found 283.0693, calcd. 283.0693.

N′ − (5 − phenyl − 1,3,4 − thiadiazol − 2 − yl) − N − cyclopropyformyl − thiourea (***3f***). Light yellow crystals, yield 93.1%, m. p. >270℃; ^{1}HNMR ($CDCl_3$) δ: 1.11 − 1.27 (m, 4H, cyclopropane − CH_2), 2.17 (s, 1H, cyclopropane − CH), 7.96 − 8.45 (m, 5H, C_6H_5); IR/cm^{-1}: 3161 (N—H), 1680 (C=O), 1304 (C=S); ESI − MS: 305.18 $[M + H]^+$; FT − MS for $C_{13}H_{12}N_4OS_2$: found 303.0383, calcd. 303.0380.

N′ − (5 − (2 − methyl − phenyl) − 1,3,4 − thiadiazol − 2 − yl) − N − cyclopropyformyl − thiourea (***3g***). Light yellow crystals, yield 90.5%, m. p. 201 − 203℃; ^{1}HNMR ($CDCl_3$) δ: 1.07 − 1.25 (m, 4H, cyclopropane − CH_2), 2.21 (s, 1H, cyclopropane − CH), 2.57 (s, 3H, CH_3), 7.30 (t, J = 7.00Hz, 2H, C_6H_4), 7.37 (d, J = 7.84Hz, 1H, C_6H_4), 7.63 (d, J = 7.84Hz, 1H, C_6H_4); IR/cm^{-1}: 3161 (N—H), 1687 (C=O), 1289 (C=S); ESI − MS: 317.10 $[M - H]^-$; FT − MS for $C_{14}H_{14}N_4OS_2$: found 318.0609, calcd. 318.0604.

N′ − (5 − (3 − methyl − phenyl) − 1,3,4 − thiadiazol − 2 − yl) − N − cyclopropyformyl − thiourea (***3h***). Light yellow crystals, yield 84.8%, m. p. 219 − 220℃; ^{1}HNMR ($CDCl_3$) δ: 1.10 − 1.25 (m, 4H, cyclopropane − CH_2), 2.31 (s, 1H, cyclopropane − CH), 2.42 (s, 3H, CH_3), 7.28 (d, J = 6.32Hz, 1H, C_6H_4), 7.36 (t, J = 7.36Hz, 1H, C_6H_4), 7.70 (d, J = 8.67Hz, 2H, C_6H_4); IR/cm^{-1}: 3169 (N—H), 1680 (C=O), 1304 (C=S); ESI − MS: 317.04 $[M - H]^-$; FT − MS for $C_{14}H_{14}N_4OS_2$: found 318.0608, calcd. 318.0604.

N′ − (5 − (4 − chlorophenyl) − 1,3,4 − thiadiazol − 2 − yl) − N − cyclopropyformyl − thiourea (***3i***). Light yellow crystals, yield 88.2%, m. p. 171 − 173℃; ^{1}HNMR ($CDCl_3$) δ: 1.08 − 1.26 (m, 4H, cyclopropane − CH_2), 1.99 (s, 1H, cyclopropane − CH), 7.45 (d, J = 8.36Hz, 2H, C_6H_4), 7.86 (d, J = 7.36Hz, 2H, C_6H_4); IR/cm^{-1}: 3169 (N—H), 1680 (C=O), 1296 (C=S); ESI − MS: 336.98 $[M - H]^-$; FT − MS for $C_{13}H_{11}ClN_4OS_2$: found 339.0130, calcd. 339.0136.

N′ − (5 − (2 − chlorophenyl) − 1,3,4 − thiadiazol − 2 − yl) − N − cyclopropyformyl − thiourea (***3j***). Light yellow crystals, yield 91.5%, m. p. 198 − 199℃; ^{1}HNMR ($CDCl_3$) δ: 1.09 − 1.25 (m, 4H, cyclopropane − CH_2), 2.27 (s, 1H, cyclopropane − CH), 7.40 (d, J = 5.31Hz, 2H, C_6H_4), 7.52 (t, J = 7.29Hz, 1H, C_6H_4), 8.11 (d, J = 7.71Hz, 1H, C_6H_4); IR/cm^{-1}: 3154 (N—H), 1680 (C=O), 1318 (C=S); ESI − MS: 336.96 $[M - H]^-$; FT − MS for $C_{13}H_{11}ClN_4OS_2$: found 339.0140, calcd. 339.0136.

N′ − (5 − (2 − florophenyl) − 1,3,4 − thiadiazol − 2 − yl) − N − cyclopropyformyl − thiourea (***3k***). Light yellow crystals, yield 89.6%, m. p. >270℃; ^{1}HNMR ($CDCl_3$) δ: 1.10 − 1.26 (m,

4H, cyclopropane – CH_2), 2.30 (s, 1H, cyclopropane – CH), 7.29 (d, J = 7.08Hz, 2H, C_6H_4), 7.46 (d, J = 6.6Hz, 1H, C_6H_4), 8.23 (t, J = 7.43Hz, 1H, C_6H_4); IR/cm^{-1}: 3255 (N—H), 1687 (C═O), 1296 (C═S); ESI – MS: 321.00 [M – H]$^-$; FT – MS for $C_{13}H_{11}FN_4OS_2$: found 323.0426, calcd. 323.0438.

N′ – (5 – (4 – nitrophenyl) – 1,3,4 – thiadiazol – 2 – yl) – N – cyclopropyformyl – thiourea (***3l***). Light yellow crystals, yield 92.6%, m. p. 174 – 175℃; ^{1}HNMR ($CDCl_3$) δ: 1.07 – 1.14 (m, 4H, cyclopropane – CH_2), 2.17 (s, 1H, cyclopropane – CH), 7.48 (d, J = 3.53Hz, 2H, C_6H_4), 7.91 (d, J = 2.35Hz, 1H, C_6H_4); IR/cm^{-1}: 3198 (N—H), 1689 (C═O), 1302 (C═S); ESI – MS: 348.00 [M – H]$^-$; FT – MS for $C_{13}H_{11}N_5O_3S_2$: found 348.0225, calcd. 348.0231.

N′ – (5 – (4 – methoxphenyl) – 1,3,4 – thiadiazol – 2 – yl) – N – cyclopropyformyl – thiourea (***3m***). Light yellow crystals, yield 91.1%, m. p. 235 – 238℃; ^{1}HNMR ($CDCl_3$) δ: 1.07 – 1.24 (m, 4H, cyclopropane – CH_2), 2.24 (s, 1H, cyclopropane – CH), 3.87 (s, 3H, CH_3), 7.00 (d, J = 8.84Hz, 2H, C_6H_4), 7.84 (d, J = 8.81Hz, 2H, C_6H_4); IR/cm^{-1}: 3161 (N—H), 1694 (C═O), 1304 (C═S); ESI – MS: 332.95 [M – H]$^-$; FT – MS for $C_{14}H_{14}N_4O_2S_2$: found 334.0588, calcd. 334.0552.

N′ – (5 – (3 – pyridinyl) – 1,3,4 – thiadiazol – 2 – yl) – N – cyclopropyformyl – thiourea (***3n***). Light yellow crystals, yield 89.9%, m. p. 216 – 218℃; ^{1}HNMR ($CDCl_3$) δ: 1.09 – 1.25 (m, 4H, cyclopropane – CH_2), 2.21 (s, 1H, cyclopropane – CH), 6.55 (d, J = 1.30Hz, 2H, C_5H_4N), 7.04 (d, J = 3.17Hz, 1H, C_5H_4N), 7.56 (s, 1H, C_5H_4N); IR/cm^{-1}: 3155 (N—H), 1682 (C═O), 1302 (C═S); ESI – MS: 304.00 [M – H]$^-$; FT – MS for $C_{12}H_{11}N_5OS_2$: found 306.0462, calcd. 306.0477.

N′ – (5 – furan – 1,3,4 – thiadiazol – 2 – yl) – N – cyclopropyformylthiourea (***3o***). Light yellow crystals, yield 94.5%, m. p. > 270℃; ^{1}HNMR ($CDCl_3$) δ: 1.09 – 1.24 (m, 4H, cyclopropane – CH_2), 2.17 (s, 1H, cyclopropane – CH), 6.56 (dd, J_{ab} = 1.75Hz, J_{ac} = 1.71Hz, 1H, C_4H_3O), 7.03 (d, J = 3.31Hz, 1H, C_4H_3O), 7.58 (d, J = 0.97Hz, 1H, C_4H_3O); IR/cm^{-1}: 3161 (N—H), 1680 (C═O), 1302 (C═S); ESI – MS: 293.75 [M – H]$^-$; FT – MS for $C_{11}H_{10}N_4O_2S_2$: found 294.0270, calcd. 294.0240.

Theoretical Calculations and DOCK

The structure of N′ – (5 – (3 – pyridinyl) – 1,3,4 – thiadiazol – 2 – yl) – N – cyclopropyformyl – thiourea (**3n**) was selected as the initial structure, while HF/6 – 31G(d,p)[13], DFT – B3LYP/6 – 31G(d,p)[14–15] and MP2/6 – 31G(d,p)[16–18] methods in Gaussian 03 package[19] were used to optimize the structure of **3n**. Vibration analysis showed that the optimized structures were in accordance with the minimum points on the potential energy surfaces. All the convergent precisions were the system default values, and all the calculations were carried out on the Nankai Stars supercomputer at Nankai University.

All docking procedures were done in Nankai Stars supercomputer at Nankai University. The automated molecular docking calculations were carried out using AutoDock 3.05. The AUTOTORS module of AutoDock defined the active torsions for each docked compound. The active sites of the protein were defined using AutoGrid centered on the IpOHA in the crystal structure. The grid map with 60 × 60 × 60 points centered at the center of mass of the KARI and a grid spacing of 0. 375Å was calculated using the AutoGrid program to evaluate the binding energies between the inhibitors and the protein. The Lamarckian genetic algorithm (LGA) was used as a searching method. Each LGA job consisted of 50 runs, and the number of generation in each run was 27000 with an initial population of 100 individuals. The step size was set to 0. 2Å for translation and 5° for orientation and torsion. The maximum number of energy evaluations was set to 1500000. Operator weights for cross - over, mutation, and elitism were 0. 80, 0. 02, and 1, respectively. The docked complexes of the inhibitorenzyme were selected according to the criterion of interaction energy combined with geometrical and electronic matching quality.

KARI assay

Cloning, expression and purification of rice KARI. The KARI resultant expression plasmid was obtained from Professor Ronald G. Duggleby's lab, and was used to transform *Escherichia coli* BL21 (DE3) cells. The methods of expression and purification of rice KARI are according to the reference[6].

Enzyme and protein assays. Gerwick et al. [11] reported that the inhibition of *Escherichia coli* KARI is timedependent. KARI activity was measured by following the decrease in A_{340} at 30℃ in solutions containing 0. 2mM NADPH, 1mM $MgCl_2$, substrate 2 - acetolactate and inhibitors as required, in 0. 1M phosphate buffer, pH 8. 0. Inhibitors was preincubated with enzyme in phosphate buffer at 30℃ for 10min before the reaction was started by adding the substrate combining with NADPH and $MgCl_2$. Protein concentrations were estimated using the bicinchoninic acid method[20] and protein purity was assessed by SDS - PAGE [21]. The yield of recombinant rice KARI from a 30 culture was 50mg with a specific activity (measured with saturating 2 - acetolactate) of 1. 17U/mg. The 2 - acetolactate was prepared according to reference[22].

Results and discussion

Synthesis

One - pot synthesis method was used in this process. Cyclopropanecarbonyl chloride was treated with ammonium thiocyanate, 3% PEG - 600 as the solid - liquid phase - transfer catalyst to afford the intermediate **2**, which was not isolated but reacted with the 2 - amino - 5 - substituted - 1, 3, 4 - thiadiazoles to give the target compounds (Scheme 2). It can easily react with NH_4SCN to form complex [PEG - 600 - NH_4^+] SCN^-, which makes it possible for SCN^- to readily react with cyclopropanecarbonyl chloride. With the enhancement of the ion exchange between inorganic salt and organic solution, PEG - 600 efficiently facilitated this heterogeneous solid - liquid two - phase reaction. As a result, PEG - 600 can fasten the NH_4^+ effectively[23]. Besides, the catalyst PEG - 600 is inexpensive, relatively nontoxic, highly stable and easily available, making this method more applicable.

In addition, the method of synthesis of 2 - amino - 5 - substituted - 1, 3, 4 - thiadiazoles was studied. Several procedures are available for the one - step synthesis of 2 - amino - 5 - substituted – 1,3,4 – thiadiazole derivatives[24].

KARI activity

The KARI inhibitory activities of the title compounds were tested at 100μg · mL^{-1}; a known inhibitor, cyclopropane – 1, 1 – dicarboxylic acid (CPD), was selected as a control. The results are shown in Table Ⅰ where it is seen that some of these compounds inhibit ketol – acid reductoisomerase *in vitro* effectively, such as **3e** and **3n**. The KARI activities of these two compounds are similar to those of other cyclopropane compounds which were synthesized in our lab[7a]. For example, compound **3n** can inhibit KARI to reach 100% at 100μg · mL^{-1}, also compound 1 – cyano – N – *o* – tolylcyclopropanecarboxamide can inhibit KARI effectively at the same level[7a]. Meanwhile the two compounds displayed as good activity as the known inhibitor CPD at 100μg · mL^{-1}.

Table Ⅰ Inhibition(%) of compounds 3a – 3o against rice KARI at 100ppm *in vitro*

No.	R	KARI
3a	CH_3	0
3b	C_2H_5	0
3c	*n* – Pr	0
3d	*Iso* – Pr	66. 18
3e	*n* – Bu	100
3f	C_6H_5	44. 91
3g	*o* – $CH_3C_6H_4$	44. 96
3h	*m* – $CH_3C_6H_4$	37. 30
3i	*p* – ClC_6H_4	82. 27
3j	*o* – ClC_6H_4	22. 77
3k	*o* – FC_6H_4	38. 78
3l	*p* – $NO_2C_6H_4$	32. 09
3m	*p* – $OCH_3C_6H_4$	70. 34
3n	3 – pyridyl	100
3o	furan	18. 08
	CPD	100

Structure – activity relationship

The structure – activity relationship can be summarized from the data given in Table Ⅰ which indicate that the change of substituent affects the KARI activity. The compounds that were substituted at the 4 – position of the phenyl ring had higher potency against KARI than that of the 2 – and 3 – substituted position. With the longer chain compound for alkane substituted, their inhibitory activities increased up to **3e**. The heterocyclic substituent also can enhance the activity, such as **3m**. Hence, these identified cyclopropane derivatives could be useful for further optimization work in finding potential KARI inhibitors.

Theoretical and DOCK

According to the frontier molecular orbital theory, HOMO and LUMO are the two most important factors which affect the bioactivities of compounds. HOMO has the priority to provide electrons, while LUMO accept electrons firstly[25]. Thus a study of the frontier orbital energy can provide some useful information for the active mechanism. Taking HF results, the HOMO of **3n** is mainly located on the pyridine ring, thiadiazole ring and the thiourea group (Figure 1(A)). On the other hand, the HOMO − 1 of **3n** contains the pyridine ring, thiadiazole ring, the thiourea group and the cyclopropane ring (Figure 1(B)). The fact that **3n** has strong affinity suggests the importance of the frontier molecular orbital in the hydrophobic interactions. Meanwhile, the frontier molecular orbital are located on the main groups whose atoms can easily bind with the receptor KARI. This implies that the orbital interaction between **3n** and the rice KARI amino acid residues are dominated by hydrophobic interaction between the frontier molecular orbital.

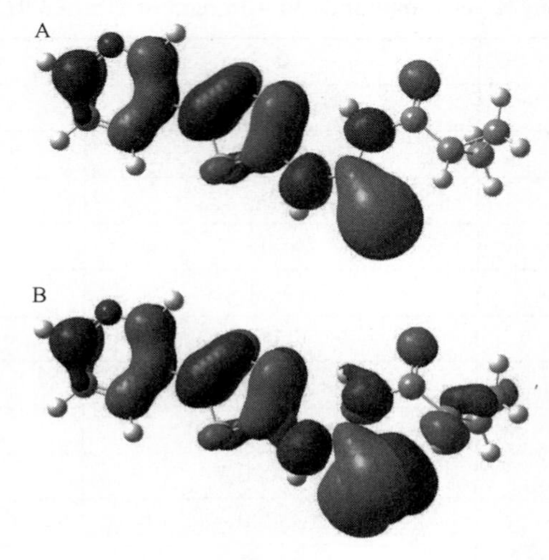

Figure 1　Frontier molecular orbitals of compound **3n**: A. HOMO of compound **3n**; B. HOMO − 1 of compound **3n**

The energies of HOMO and HOMO − 1 of **3n** and CPD are listed in Table Ⅱ which surprisingly shows that compounds **3n** have similar energies with CPD. This probably is the reason for the good activity of the compound **3n** and CPD.

Table Ⅱ　Energies of HOMO, HOMO − 1 of 3n and CPD (Hartree)

	3n	CPD
HOMO	−0.33672	−0.32068
HOMO − 1	−0.33989	−0.34159

To make prediction by our frontier molecular orbital model more relevant to the active sites of the enzyme and to describe a probable binding site in the KARI, the compound **3n** was docked into the active sites of spinach KARI.

Visual inspection of the conformation of **3n** docked into the KARI binding site revealed that the phenyl rings are hosted in the pocket of KARI and three hydrogen bonds between the amino groups of **3n** and the carbonyl oxygen of Glu 311, Pro 251 side chain and the NADPH are also observed. Furthermore, the cyclopropane ring and aromatic ring are embedded in a large hydrophobic pocket formed by His 226, Cys 250, Pro 251, Lys 252, Glu 311, Glu 319, Asp 315, Leu 323, Leu 324, Glu 496, Leu 501, Cys 517, Ser 518 and NADPH (Figure 2).

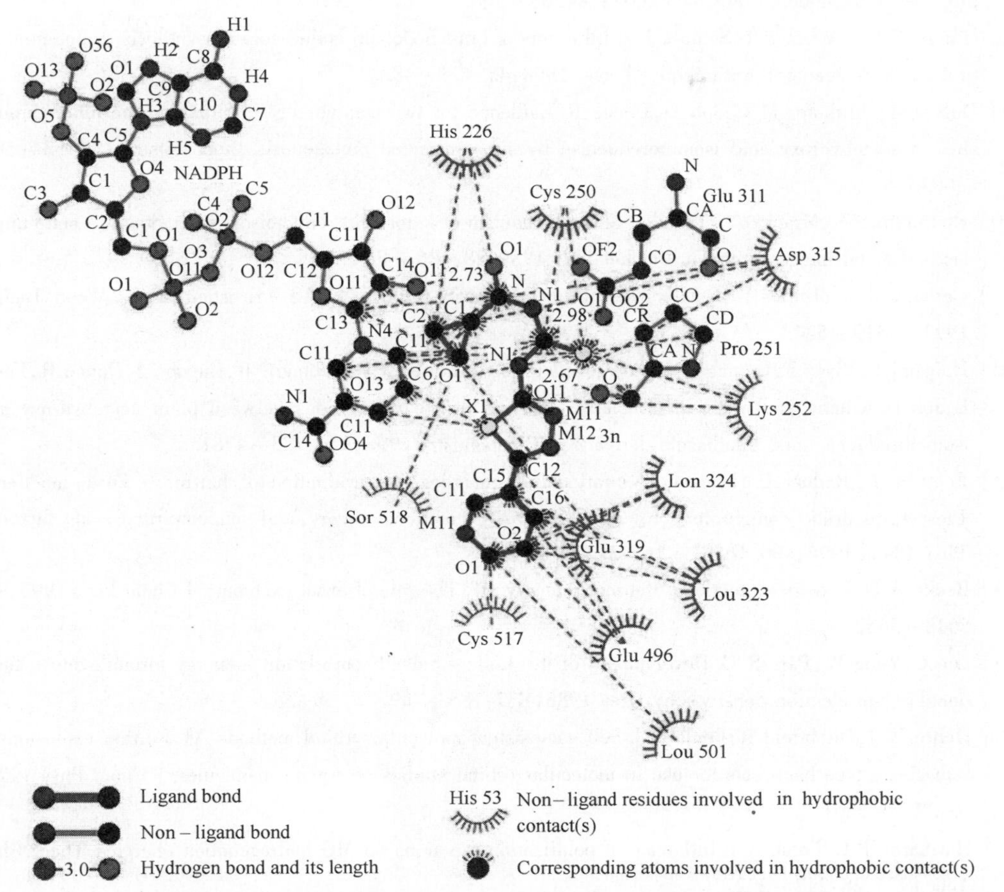

Figure 2 PDB code: 1YVE Binding modes of compound **3n** in the active sites of spinach KARI: hydrogen bond and hydrophobic interaction between **3n** and the rice KARI amino acid residues

References

[1] Chaleff R S, Mauvais C J. Acetolactate synthase is the site of action of two sulfonylurea herbicides in higher plants. Science 1984; 224: 1443 – 1445.

[2] Shaner D L, Anderson P C, Stidham MA. Imidazolinones. Potent inhibitors of acetohydroxy acid synthase. Plant Physiol 1984; 76: 545 – 546.

[3] Dumas R, Biou V F, Douce H R, Duggleby R G. Enzymology, structure, and dynamics of acetohydroxy acid isomeroreductase. Acc Chem Res 2001; 34: 399 – 408.

[4] Schulz A, Sponemann P, Kocher H, Wengenmayer F. The herbicidally active experimental compound Hoe 704 is a potent inhibitor of the enzyme acetolactate reductoisomerase. FEBS Lett 1988; 238: 375 – 378.

[5] Aulabaugh A, Schloss J V. Oxalyl hydroxamates as reactionintermediate analogs for ketol – acid reductoisomerase. Biochemistry 1990; 29: 2824 – 2840.

[6] Lee Y T, Ta H T, Duggleby R G. Cyclopropane – 1, 1 – dicarboxylate is a slow – , tight – binding inhibitor of rice ketol – acid reductoisomerase. Plant Science 2005; 168: 1035 – 1040.

[7] (a)Liu X H, Chen P Q, Wang B L, Li Y H, Wang S H, Li Z M. Synthesis, bioactivity, theoretical and molecular docking study of 1 - cyano - *N* - substituted - cyclopropanecarboxamide as ketolacid reductoisomerase inhibitor. Bioorg Med Chem Lett 2007;17:3784 - 3788. (b)Liu X H, Chen P Q, He F Q, Wang S H, Song H B, Li Z M, Structure, Bioactivity and Theoretical Study of 1 - cyano - *N* - p - Tolylcyclo - propanecarboxamide. Struc Chem 2007;18:563 - 568.

[8] Grandoni J A, Marta P T, Schloss J V. Inhibitors of branchedchain amino acid biosynthesis as potential antituberculosis agents. J Antimicrob Chemot 1998;42:475 - 482.

[9] Dumas R, Butikofer M C, Job D, Douce R. Evidence for two catalytically different magnesium - binding sites in acetohydroxy acid isomeroreductase by site - directed mutagenesis. Biochemistry 1995;34:6026 - 6036.

[10] Chunduru S K, Mrachko G T, Calvo K C. Mechanism of ketol acid reductoisomerase. Steady - state analysis and metal ion requirement. Biochemistry 1989;28:486 - 493.

[11] Gerwick B C, Mireles L C, Eilers R J. Rapid diagnosis of ALS/AHAS - resistant weeds. Weed Technol 1993;7:519 - 524.

[12] Halgand F, Vives F, Dumas R, Biou V, Andersen J, Andrieu J P, Cantegril R, Gagnon J, Douce R, Forest E, Job D. Kinetic and mass spectrometric analyses of the interactions between plant acetohydroxy acid isomeroreductase and thiadiazole derivatives. Biochemistry 1998;37:4773 - 4781.

[13] Scott A P, Radom L. Harmonic Vibrational frequencies: An evaluation of hartree - Fock, moeller - Plesset, quadratic configuration interaction, density functional theory, and semiempirical scale factors. J Phys Chem 1996;100:16502 - 16513.

[14] Becke A D. Density - functional thermochemistry. Ⅲ. The role of exact exchange. J Chem Phys 1993;98:5648 - 5652.

[15] Lee C, Yang W, Parr R G. Development of the Colle - Salvetti correlation - energy formula into a functional of the electron density. Phys Rev 1988;B37:785 - 789.

[16] Hehre W J, Ditchfield R, Pople A J. Self - consistent molecular orbital methods. Ⅻ. Further extensions of Gaussian - type basis sets for use in molecular orbital studies of organic molecules. J Chem Phys 1972;56:2257 - 2261.

[17] Hariharan P C, Pople A J. Influence of polarization functions on MO hydrogenation energies. Theor Chim Acta 1973;28:213 - 222.

[18] Gordon M S. Excited states and photochemistry of saturated molecules. The 1B1 (1T2) surface in silane. Chem Phys Lett 1980;70:343 - 349.

[19] Frisch M J, Trucks G W, Schlegel H B, Scuseria G E, Robb M A, Cheeseman J R, Montgomery J A Jr, Vreven T, Kudin K N, Burant J C, Millam J M, Iyengar S S, Tomasi J, Barone V, Mennucci B, Cossi M, Scalmani G, Rega N, Petersson G A, Nakatsuji H, Hada M, Ehara M, Toyota K, Fukuda R, Hasegawa J, Ishida M, Nakajima T, Honda Y, Kitao O, Nakai H, Klene M, Li X, Knox J E, Hratchian HP, Cross J B, Adamo C, Jaramillo J, Gomperts R, Stratmann R E, Yazyev O, Austin A J, Cammi R, Pomelli C, Ochterski J W, Ayala P Y, Morokuma K, Voth G A, Salvador P, Dannenberg J J, Zakrzewski V G, Dapprich S, Daniels A D, Strain M C, Farkas O, Malick D K, Rabuck A D, Raghavachari K, Foresman J B, Ortiz J V, Cui Q, Baboul A G, Clifford S, Cioslowski J, Stefanov B B, Liu G, Liashenko A, Piskorz P, Komaromi I, Martin R L, Fox D J, Keith T, AlLaham M A, Peng C Y, Nanayakkara A, Challacombe M, Gill PMW, Johnson B, Chen W, Wong M W, Gonzalez C, Pople J A. Gaussian 03 Revision C 01 Gaussian Inc Wallingford CT 2004.

[20] Smith P K, Krohn R I, Hermanson G T, Mallia A K, Gartner F H, Provenzano M D, Fujimoto E K, Goeke N M, Olson BJ, Klenk DC. Measurement of protein using bicinchoninic acid. Analyt Biochem 1985;150:76 - 85.

[21] Laemmli U K. Cleavage of structural proteins during the assembly of the head of bacteriophage T4. Nature 1970;227:680 - 685.

[22] Hans W M. α - Hydroxyacetoacetic acid. Biochem J 1932;6:1033 - 1053.

[23] (a) Ke S Y, Wei T B, Xue S J, Duan L P, Li J Z. Phase transfer catalyzed synthesis under ultrasonic irradiation and bioactivity of N′ - (4,6 - disubstituted - pyrimidin - 2 - yl) - N - (5 - aryl - 2 - furoyl) thiourea derivatives. Indian J Chem B 2005;44:1957 - 1960. (b) Yang XD, Phase transfer catalysts promoting the one - pot synthesis under ultrasonic irradiation and biological activity of N - (5 - substituted - 1, 3,4 - thiadiazole - 2 - yl) - N′ - (5 - methylisoxazoyl) - thiourea derivatives. Heterocycl Commun 2007,13:387 - 392.

[24] (a) Carvalho S A, de Silva E F, SantaRita R M, de Castrod S L, Fragaa C A M. Synthesis and antitrypanosomal profile of new functionalized 1,3,4 - thiadiazole - 2 - arylhydrazone derivatives, designed as non - mutagenic megazol analogues. Bioorg Med Chem Lett 2004;14:5967 - 5970. (b) Malbec F, Milcent R, Barbier G, Synthesis of new derivatives of 4 - amino - 2,4 - dihydro - 1,2,4 - triazol - 3 - one as potential antibacterial agents. J Heterocyclic Chem 1984;21:1689 - 1698. (c) Jung K Y, Kim S K, Gao ZG, Gross A S, Melman N, Jacobson K A, Kim Y C, Structure - activity relationships of thiazole and thiadiazole derivatives as potent and selective human adenosine A3 receptor antagonists. Bioorg Med Chem 2004;12:613 - 623.

[25] (a) Ma H X, Song J R, Xu K Z, Hu R Z, Zhai G H, Wen Z Y, Yu K B. Preparation, crystal structure and theoretical calculation of $(CH_3)_2NH_2^+ C_2N_4O_3H^-$. Acta Chem Sinica 2003;61:1819 - 1823. (b) Chen PQ, Liu X H, Sun H W, Wang B L, Li Z M, Lai C M, Molecular Simulation Studies of Interactions between ketol - acid reductoisomerase and Its Inhibitors. Acta Chem Sinica 2007;65:1693 - 1701.

Synthesis and Biological Activity of Novel 2,3 – Dihydro – 2 – phenylsulfonyl-hydrazono – 3 – (2′,3′,4′,6′ – tetra – *O* – acetyl – *β* – *D* – glucopyranosyl) thiazoles*

Haoan Wang, Yuxin Li, Zhengming Li,
Haiying Cheng, Suhua Wang, Liu Bin

(Elemento – Organic Chemistry Institute, State Key Laboratory of Elemento – Organic Chemistry, National Pesticide Engineering Research Center, Nankai University, Tianjin, 300071, China)

Abstract A series of novel 2,3 – dihydro – 2 – phenylsulfonylhydrazono – 3 – (2′,3′,4′,6′ – tetra – *O* – acetyl – *β* – *D* – glucopyranosyl) thiazoles were designed and synthesized *via* the reaction of thiosemicarbazide with chloroaldehyde. Their chemical structures were characterized by ^1H and ^{13}C NMR spectroscopy and elemental analysis. The bioassay results indicate that some of these compounds exhibit moderate fungicidal and herbicidal activities.

Key words synthesis; 2,3 – dihydrothiazole; biological activity

1 Introduction

Thiazoles have become the focus in the search of new drugs in recent years by exhibiting a broad range of biological activities, such as antitumor, antibiotic, antibacterial, anti – inflammatory, and antifungal activities[1-4]. With the introduction of thiazole, some compounds were introduced as potent inhibitors of inosine monophosphate dehydrogenase (IMPDH)[5], cyclin – dependent kinase – 2 (CDK2)[6], photosynthetic electron transport (PET)[7], metabotropic glutamate receptors (mGluRs)[8], and Itk (Interleukin – 2 – inducible T cell kinases)[9]. Thiazoles have also played an increasingly important role in the development of highly active pesticides[10]. Within the past decades, the biological properties of carbohydrates have been reported. They play a key role in cell recognition, cell growth, cell differentiation, and signal transduction. On the other hand, carbohydrates also compose important structural subunits in numerous natural bioactive products such as nucleocidin[11] *etc.*

A series of glycosyl phenylsulfonyl substituted thiazolidine – 4 – ones were synthesized and the bioassay results indicate that some of these compounds exhibit moderate fungicidal and herbicidal activities[12]. As part of our program aimed at developing ecologically safer and more environmentally agrochemicals, the title compounds were designed and synthesized and their bioactivities were studied.

2 Results and Discussion

In our experiment, glycosyl phenylsulfonyl thiosemicarbazides were prepared according to the

* Reprinted from *Chemical Research in Chinese Universities*, 2009, 25(1): 52 – 55. Supported by the National Basic Research Program of China (No. 2003CB114406) and the Science and Technology Innovation Foundation of Nankai University, China.

method reported in the literature[13]. The condensation reaction of glycosyl isothiocyanate with phenylsulfonyl hydrazine requires 6 – 8h heating to reflux.

The cyclization of suitable open chain organic molecules is a common approach for the synthesis of heterocyclic derivatives. A survey of the literature revealed that thiazole ring can be obtained by condensation of α – halo ketone[14] or chloroaldehyde[15-17] with thiocarbamide. As seen in Scheme 1, glycosyl phenylsulfonyl thiosemicarbazides were allowed to react with chloroaldehyde in *N*, *N'* – dimethyl formamide (DMF) containing a small amount of sodium acetate to give the title compounds. At first, we attempted to use tetrahydrofuran as the solvent, but no title compounds were obtained. It was postulated that the polarity of the solvent may play an important role in the synthesis of the title compounds. We found that the yield of compound **4**a from compound **3**a was low. To optimize the reaction conditions at this step, the reaction time was investigated. A black oily material was obtained when compound **3**a and chloroaldehyde were heated in DMF for 7h. Based on the above finding, the reaction time should be less than 7h.

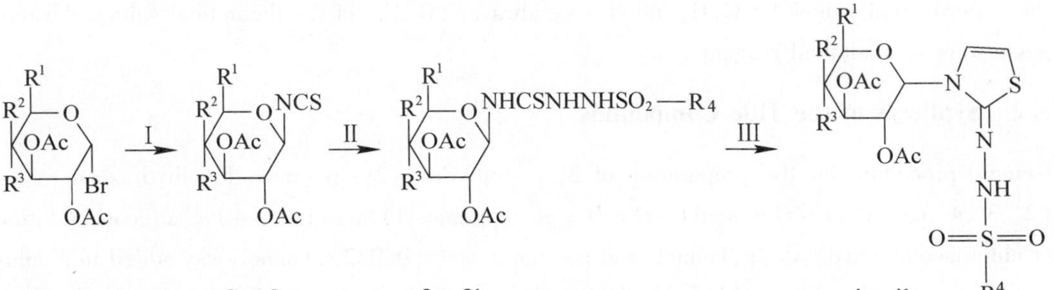

1a – 1c　　2a – 2c　　3a – 3l　　4a – 4l

1a, **2a**: $R^1 = R^2 = H, R^3 = OAc$; **1b**, **2b**: $R^1 = CH_2OAc, R^2 = H, R^3 = OAc$; **1c**, **2c**: $R^1 = CH_2OAc, R^2 = OAc, R^3 = H$; **3a**, **4a**: $R^1 = R^2 = H, R^3 = OAc, R^4 = Ph$; **3b**, **4b**: $R^1 = R^2 = H, R^3 = OAc, R^4 = p-PhCH_3$; **3c**, **4c**: $R^1 = R^2 = H, R^3 = OAc, R^4 = p-PhCl$; **3d**, **4d**: $R^1 = R^2 = H, R^3 = OAc, R^4 = CH_2Ph$; **3e**, **4e**: $R^1 = CH_2OAc, R^2 = H, R^3 = OAc, R^4 = Ph$; **3f**, **4f**: $R^1 = CH_2OAc, R^2 = H, R^3 = OAc, R^4 = p-PhCH_3$; **3g**, **4g**: $R^1 = CH_2OAc, R^2 = H, R^3 = OAc, R^4 = p-PhCl$; **3h**, **4h**: $R^1 = CH_2OAc, R^2 = H, R^3 = OAc, R^4 = CH_2Ph$; **3i**, **4i**: $R^1 = CH_2OAc, R^2 = OAc, R^3 = H, R^4 = Ph$; **3j**, **4j**: $R^1 = CH_2OAc, R^2 = OAc, R^3 = H, R^4 = p-PhCH_3$; **3k**, **4k**: $R^1 = CH_2OAc, R^2 = OAc, R^3 = H, R^4 = p-PhCl$; **3l**, **4l**: $R^1 = CH_2OAc, R^2 = OAc, R^3 = H, R^4 = CH_2Ph$

Scheme 1　Synthesis routes of target compounds

Ⅰ. $Pb(NCS)_2$, toluene, reflux; Ⅱ. $Ph-SO_2NHNH_2$, acetonitrile, r. t.; Ⅲ. $ClCH_2CHO$, NaAc, DMF, 70℃

In the ^1H NMR spectra of the title compounds, there are two sets of double peaks in the field of δ 6.04 – 7.52 assigned to the two protons of the 2,3 – dihydrothiazole ring. All the title compounds exhibit a multiplet in a region of δ 7.98 – 7.30 owing to aromatic protons. In the ^{13}C NMR spectra of compounds **3a** – **3l**, the carbon of C=S resonate nearing δ 184. ^{13}C NMR spectra of compounds **4a** – **4l** show a singlet in the region nearing δ 173 owing to C2 of the 2,3 – dihydrothiazole ring.

All compounds were tested in foliage treatment against several herbs such as *Brassica campestris*, *Echinochloa crus – galli*, *Amaranthus retroflexus* L., and *Digitaria Sanguinalis* (L.) Scop at 1.5kg/ha. The bioassay results show that most of them have rather weak herbicidal activities. The inhibitory rates of compounds **4**c, **4**h, and **4**j against B. *campestris* are 39.2%, 38.0%, and 37.7%, respectively. The new compounds were also tested for their *in vitro* fungicidal activity against *Gibberella zeae*, *Alternaria solani*, *Cercospora arachidicola*, *Physalospora piricola*, and *Phoma asparagi* at 50mg/L. The inhibitory rates of all the new compounds against *P. piricola* and *P. asparagi* range between 30% and 50%. The result indicates that the introduction of the glycosyl into the thiazoles is not good for the enhancement of the activity.

3 Experimental

3.1 Materials and Instruments

All melting points were conducted on a Yanaco MP-500 Micro-melting point apparatus and were uncorrected. ^1H NMR and ^{13}C NMR spectra were recorded in $CDCl_3$ on a Bruker AC-400 instrument with Me_4Si as an internal standard. Elemental analyses of new compounds were performed on a Yanaco CHN Corder MF-3 automatic elemental analyzer for C, H, and N, and the experimental values for C, H, and N were always ±0.3% of the theoretical values. All reagents were of analytical reagent grade.

3.2 Synthesis of the Title Compounds

General procedure for the preparation of 2,3-dihydro-2-phenylsulfonylhydrazono-3-(2′,3′,4′,6′-tetra-*O*-acetyl-*β*-*D*-glucopyranosyl) thiazoles: a 40% aqueous solution of chloroacetaldehyde (0.2g, 1mmol) and sodium acetate (0.082g, 1mmol) was added to a solution of phenylsulfonyl-4-(1′-*N*-2′,3′,4′-tri-*O*-acetyl-*β*-*D*-pyranosyl) thiosemicarbazide (**3**a-**3**l, 1mmol) in DMF (5mL). The mixture was heated at 70℃. The progress was monitored by thin layer chromatography on F254 silica gel developed with petroleum ether and ethyl acetate. When the reaction was complete, 20mL of H_2O was added and the product was extracted with ethyl acetate. Having been dried with Na_2SO_4, the mixture was filtered. The solvent was evaporated under reduced pressure and the concentration resulted in a crude product. The product was isolated on a silica gel chromatographic column with petroleum ether and ethyl acetate as eluent.

Compound **4**a: from 0.489g of compound **3**a (0.267g, 52%); m. p. 170-172℃ (recrystallized from EtOAc-petroleum ether to give white crystals); ^1H NMR ($CDCl_3$), δ: 2.03, 2.04, 2.05 (3s, 9H, 3OAc), 3.36-5.49 (m, 6H, xylosyl-H), 6.06, 6.65 (2d, J = 4.8Hz, 2H, CH═CH), 6.80 (s, 1H, NH), 7.56-7.93 (m, 5H, Ph—H); ^{13}C NMR ($CDCl_3$), δ: 20.5, 20.7, 20.9 ($3CH_3$), 65.6, 68.8, 69.9, 72.7, 82.1 (5sugar ring), 102.7 (C-5), 123.4 (C-4), 128.6, 128.9, 133.5, 137.9 (6Ar—C), 169.8, 170.1, 170.2 (3C═O), 173.5 (C2); ESI-MS: M-H^+ peak at *m/z* 512.3 (513.1). Elemental anal. (%) calcd. for $C_{20}H_{23}N_3O_9S_2$: C 46.89, H 4.67, N 8.22; found: C 46.78, H 4.51, N 8.18.

Compound **4**b: from 0.503g of compound **3**b (0.269g, 51%); m. p. 174-176℃ (recrystallized from EtOAc-petroleum ether to give white crystals); ^1H NMR ($CDCl_3$), δ: 2.07, 2.08, 2.19 (3s, 9H, 3OAc), 2.50 (s, 3H, Ar—CH_3), 3.41-5.55 (m, 6H, xylosyl-H), 6.09, 6.66 (2d, J = 4.5Hz, 2H, CH═CH), 6.38 (s, 1H, NH), 7.39 (d, J = 7.8Hz, 2H, Ph—H), 7.850

(d, J = 8.1Hz, 2H, Ph—H); ^{13}C NMR ($CDCl_3$), δ: 20.5, 20.7, 20.9 ($3CH_3$), 21.9 (Ar—CH_3), 65.6, 68.8, 69.9, 72.7, 82.2 (5sugar ring), 103.1 (C5), 123.4 (C4), 128.6, 129.6, 141.1, 144.4 (6Ar—C), 169.8, 170.0, 170.2 (3C =O), 173.7 (C2); ESI - MS: M - H^+ peak at m/z 526.1 (527.1). Elemental anal. (%) calcd. for $C_{21}H_{25}N_3O_9S_2$: C 47.73, H 4.96, N 7.75; found: C 47.81, H 4.78, N 7.96.

Compound **4c**: from 0.523g of compound **3c** (0.328g, 60%); m. p. 173 - 175℃ (recrystallized from EtOAc - petroleum ether to give white crystals); ^{1}H NMR ($CDCl_3$), δ: 1.97, 1.98, 2.00 (3s, 9H, 3OAc), 3.34 - 6.04 (m, 6H, xylosyl - H), 6.03, 6.60 (2d, J = 4.8Hz, 2H, CH =CH), 6.81 (s, 1H, NH), 7.47 (d, J = 8.4Hz, 2H, Ph—H), 7.80 (d, J = 8.4Hz, 2H, Ph—H); ^{13}C NMR ($CDCl_3$), δ: 20.5, 20.6, 20.8 ($3CH_3$), 65.6, 68.7, 69.9, 72.6, 82.3 (5sugar ring), 103.0 (C5), 123.6 (C4), 129.2, 130.0, 136.3, 140.0 (6Ar—C), 169.8, 170.0, 170.2 (3C =O), 174.0 (C2); ESI - MS: M - H^+ peak at m/z 546.3 (547.1). Elemental anal. (%) calcd. for $C_{20}H_{22}ClN_3O_9S_2$: C 43.60, H 4.21, N 7.42; found: C 43.84, H 4.05, N 7.67.

Compound **4d**: from 0.503g of compound **3d** (0.290g, 55%); m. p. 166 - 168℃ (recrystallized from EtOAc - petroleum ether to give white crystals); ^{1}H NMR ($CDCl_3$), δ: 2.03, 2.04, 2.05 (3s, 9H, 3OAc), 2.15 - 5.43 (m, 8H, xylosyl - H, Ar—CH_2), 6.07, 6.69 (2d, J = 4.8Hz, 2H, CH =CH), 6.59 (s, 1H, NH), 7.39 - 7.47 (m, 5H, Ph—H); ^{13}C NMR ($CDCl_3$), δ: 20.5, 20.7, 20.8 ($3CH_3$), 54.5 (Ar—CH_2), 65.7, 68.9, 70.3, 72.7, 82.2 (5sugar ring), 102.8 (C5), 123.5 (C4), 129.0, 129.2, 139.4, 131.1 (6Ar—C), 169.9, 170.1, 170.2 (3 C =O), 173.8 (C2); ESI - MS: M - H^+ peak at m/z 526.1 (527.1). Elemental anal. (%) calcd. for $C_{21}H_{25}N_3O_9S_2$: C 47.77, H 4.92, N 7.69; found: C 47.81, H 4.78, N 7.96.

Compound **4e**: from 0.561g of compound **3e** (0.415g, 71%); m. p. 193 - 195℃ (recrystallized from EtOAc - petroleum ether to give white crystals); ^{1}H NMR ($CDCl_3$), δ: 2.03, 2.04, 2.05, 2.06 (4s, 12H, 4OAc), 4.09 - 5.59 (m, 7H, glucosyl - H), 6.10, 6.68 (2d, J = 3.6Hz, 2H, CH =CH), 7.25 (s, 1H, NH), 7.59 - 7.94 (m, 5H, Ph—H). ^{13}C NMR ($CDCl_3$), δ: 20.0, 20.4, 20.7, 20.9 ($4CH_3$), 62.9 (CH_2O), 67.9, 69.0, 72.1, 73.9, 81.1 (5sugar ring), 102.9 (C5), 123.3 (C4), 127.9, 129.4, 134.9, 144.3 (6Ar—C), 169.6, 169.8, 170.2, 170.8 (4C =O), 173.6 (C2); ESI - MS: M - H^+ peak at m/z 584.1 (585.1). Elemental anal. (%) calcd. for $C_{23}H_{27}N_3O_{11}S_2$: C 46.95, H 4.68, N 6.94; found: C 47.17, H 4.65, N 7.18.

Compound **4f**: from 0.575g of compound **3f** (0.311g, 52%); m. p. 190 - 192℃ (recrystallized from EtOAc - petroleum ether to give white crystals); ^{1}H NMR ($CDCl_3$), δ: 2.06, 2.07, 2.08, 2.09 (4s, 12H, 4OAc), 2.51 (s, 3H, CH_3), 3.82 - 5.65 (m, 7H, glucosyl - H), 6.10, 6.69 (2d, J = 3.6Hz, 2H, CH =CH), 6.33 (s, 1H, NH), 7.40 (d, J = 6.6Hz, 2H, Ph—H), 7.85 (d, J = 7.8Hz, 2H, Ph—H); ^{13}C NMR ($CDCl_3$), δ: 20.5, 20.7, 20.8, 20.9 ($4CH_3$), 21.9 (Ar—CH_3), 62.0 (CH_2O), 68.1, 69.7, 73.1, 74.8, 81.3 (5sugar ring), 102.9 (C5), 123.4 (C4), 128.6, 129.5, 134.9, 144.4 (6Ar—C), 169.6, 169.7, 170.2, 170.8 (4C =O), 173.6 (C2); ESI - MS: M - H^+ peak at m/z 598.2 (599.1). Elemental anal. (%) calcd. for $C_{24}H_{29}N_3O_{11}S_2$: C 47.96, H 4.71, N 6.90; found: C 48.07, H 4.87, N 7.01.

Compound **4g**: from 0.596g of compound **3g** (0.316g, 51%); m. p. 182 - 183℃ (recrystallized from EtOAc - petroleum ether to give white crystals); ^{1}H NMR ($CDCl_3$), δ: 2.03, 2.04,

2. 05, 2. 07(4s, 12H, 4OAc), 3. 84 – 5. 62(m, 7H, glucosyl – H), 6. 10, 6. 70(2d, J = 4. 5Hz, 2H, CH=CH), 6. 82(s, 1H, NH), 7. 54(d, J = 8. 1Hz, 2H, Ph—H), 7. 86(d, J = 8. 4Hz, 2H, Ph—H); ^{13}C NMR($CDCl_3$), δ: 20. 4, 20. 7, 20. 8, 20. 9($4CH_3$), 62. 1(CH_2O), 68. 1, 69. 8, 73. 0, 74. 9, 81. 3(5sugar ring), 102. 9(C5), 123. 5(C4), 129. 2, 130. 0, 136. 4, 134. 0 (6Ar—C), 169. 7, 170. 2, 170. 8, 171. 1(4C=O), 173. 8(C2); ESI – MS: M – H^+ peak at m/z 618. 0(619. 0). Elemental anal. (%) calcd. for $C_{23}H_{26}ClN_3O_{11}S_2$: C 44. 70, H 4. 20, N 6. 57; found: C 44. 55, H 4. 23, N 6. 78.

Compound **4**h: From 575g of compound **3**h(0. 467g, 78%); m. p. 201 – 202℃ (recrystallized from EtOAc – petroleum ether to give white crystals); ^{1}H NMR($CDCl_3$), δ: 1. 97, 1. 99, 2. 00, 2. 03(4s, 12H, 4OAc), 3. 92 – 5. 38(m, 7H, glucosyl – H, Ar—CH_2), 6. 02, 6. 65(2d, J = 4. 8Hz, 2H, CH=CH), 6. 31(s, 1H, NH), 7. 34 – 7. 43(m, 5H, Ph—H); ^{13}C NMR($CDCl_3$), δ: 20. 5, 20. 6, 20. 8, 21. 0($4CH_3$), 54. 4(Ar—CH_2), 62. 0(CH_2O), 68. 2, 70. 2, 73. 1, 75. 0, 81. 5(5sugar ring), 102. 9(C5), 123. 5(C4), 129. 1, 129. 2, 129. 4, 131. 0(6Ar—C), 169. 6, 169. 8, 170. 2, 170. 8(4C=O), 173. 7(C2); ESI – MS: M – H^+ peak at m/z 598. 16 (599. 1). Elemental anal. (%) calcd. for $C_{24}H_{29}N_3O_{11}S_2$: C 48. 12, H 4. 72, N 7. 12; found: C 48. 07, H 4. 87, N 7. 01.

Compound **4**i: from 0. 561g of compound **3**i(0. 439g, 75%); m. p. 144 – 146℃ (recrystallized from EtOAc – petroleum ether to give white crystals); ^{1}H NMR($CDCl_3$), δ: 1. 97, 1. 98, 2. 02, 2. 15(4s, 12H, 4OAc), 4. 02 – 5. 57(m, 7H, galactosyl – H), 6. 06, 6. 73(2d, J = 4. 8Hz, 2H, CH=CH), 6. 79(s, 1H, NH), 7. 53 – 7. 89(m, 5H, Ph—H); ^{13}C NMR($CDCl_3$), δ: 20. 5, 20. 6, 20. 7, 20. 8($4CH_3$), 61. 5(CH_2O), 67. 3, 67. 5, 71. 3, 73. 7, 81. 6(5sugar ring), 102. 5 (C5), 123. 6(C4), 128. 6, 128. 9, 133. 5, 138. 0(6Ar—C), 169. 9, 170. 2, 170. 3, 170. 6 (4C=O), 173. 5(C2); ESI – MS: M – H^+ peak at m/z 584. 2(585. 1). Elemental anal. (%) calcd. for $C_{23}H_{27}N_3O_{11}S_2$: C 47. 13, H 4. 82, N 7. 03; found: C 47. 17, H 4. 65, N 7. 18.

Compound **4**j: from 0. 575g of compound **3**j(0. 443g, 74%); m. p. 153 – 155℃ (recrystallized from EtOAc – petroleum ether to give white crystals); ^{1}H NMR($CDCl_3$), δ: 1. 97, 1. 99, 2. 02, 2. 15(4s, 12H, 4OAc), 2. 41(s, 3H, CH_3), 3. 81 – 5. 54(m, 7H, galactosyl – H), 6. 05, 6. 73 (2d, J = 4. 8Hz, 2H, CH=CH), 6. 85(s, 1H, NH), 7. 28(d, J = 7. 6Hz, 2H, Ph—H), 7. 74 (d, J = 8. 0Hz, 2H, Ph—H); ^{13}C NMR($CDCl_3$), δ: 20. 6, 20. 7, 20. 8, 20. 9($4CH_3$), 21. 9 (Ar—CH_3), 61. 6(CH_2O), 67. 3, 67. 5, 71. 3, 73. 7, 81. 6(5sugar ring), 102. 5(C5), 123. 6 (C4), 128. 5, 129. 5, 135. 0, 144. 3(6Ar—C), 170. 0, 170. 1, 170. 2, 170. 6(4C=O), 173. 6 (C2); ESI – MS: M – H^+ peak at m/z 598. 3(599. 1). Elemental anal. (%) calcd. for $C_{24}H_{29}N_3O_{11}S_2$: C 48. 04, H 4. 85, N 6. 86; found: C 48. 07, H 4. 87, N 7. 01.

Compound **4**k: from 0. 596g of compound **3**k(0. 360g, 62%); m. p. 149 – 151℃ (recrystallized from EtOAc – petroleum ether to give white crystals); ^{1}H NMR($CDCl_3$), δ: 1. 99, 2. 00, 2. 03, 2. 12(4s, 12H, 4OAc), 4. 09 – 5. 58(m, 7H, galactosyl – H), 6. 09, 6. 75(2d, J = 4. 8Hz, 2H, CH=CH), 6. 79(s, 1H, NH), 7. 50(d, J = 8. 4Hz, 2H, Ph—H), 7. 83(d, J = 8. 8Hz, 2H, Ph—H); ^{13}C NMR($CDCl_3$), δ: 20. 6, 20. 7, 20. 8, 20. 9($4CH_3$), 61. 5(CH_2O), 67. 3, 67. 6, 71. 2, 73. 8, 81. 6(5sugar ring), 102. 7(C5), 123. 7(C4), 129. 2, 130. 0, 136. 4, 140. 1(6Ar—C), 169. 9, 170. 2, 170. 6(4C=O), 174. 0(C2); ESI – MS: M – H^+ peak at m/z 618. 2(619. 0). Elemental anal. (%) calcd. for $C_{23}H_{26}ClN_3O_{11}S_2$: C 44. 38, H 4. 42, N 6. 70;

found: C 44.55, H 4.23, N 6.78.

Compound **4l**: from 0.575g of compound **3l**(0.347g, 58%); m. p. 158 - 160℃ (recrystallized from EtOAc - petroleum ether to give white crystals); ^{1}H NMR($CDCl_3$), δ: 1.99, 2.02, 2.04, 2.18(4s, 12H, 4OAc), 4.16 - 5.49(m, 7H, galactosyl - H, Ar—CH_2), 6.07, 6.77(2d, J = 4.8Hz, 2H, CH =CH), 6.45(s, 1H, NH), 7.38 - 7.46(m, 5H, Ph—H); ^{13}C NMR($CDCl_3$), δ: 20.6, 20.8, 20.9, 21.0 (4CH_3), 54.3 (Ar—CH_2), 61.4 (CH_2O), 67.2, 68.0, 71.2, 73.8, 81.8 (5sugar ring), 102.6(C5), 123.7(C4), 129.1, 129.2, 129.4, 131.0(6Ar—C), 167.0, 170.2, 170.7, 171.3(4C =O), 173.9 (C2); 598.2 (599.1). Elemental anal. (%) calcd. for $C_{24}H_{29}N_3O_{11}S_2$: C 47.96, H 4.96, N 6.80; found: C 48.07, H 4.87, N 7.01.

References

[1] Michael J. G., Tachel M. L., Susan, L. M., *et al.*, *Bioorg. Med. Chem.*, 2004, 12, 1029.

[2] Jung K. Y., Kim S. K., Gao Z. G., *et al.*, *Bioorg. Med. Chem.*, 2004, 12, 613.

[3] Andreani A., Burnelli S., Granaiola M., *et al.*, *J. Med. Chem.*, 2006, 49, 7897.

[4] Matsuya Y., Kawaguchi T., Ishihara K., *et al.*, *Org. Lett.*, 2006, 8, 4609.

[5] Lesiak K., Watanabe K. A., Majumdar A., *et al.*, *J. Med. Chem.*, 1997, 40, 2533.

[6] Wang S. D., Meade's C., Wood G., *et al.*, *J. Med. Chem.*, 2004, 47, 1662.

[7] Dayan F. E., Vincent A. C., Romagni J. G., *et al.*, *J. Agric. Food Chem.*, 2000, 48, 3689.

[8] Iso Y., Grajkowska E., Wroblewski J. T., *et al.*, *J. Med. Chem.*, 2006, 49, 1080.

[9] Das J., Furch J. A., Liu C. J., *et al.*, *Bioorg. Med. Chem. Lett.*, 2006, 16, 3706.

[10] Sawada Y., Yanai T., Nakagawa H., *et al.*, *Pest Manag Sci.*, 2003, 59, 25.

[11] Wells L., Vosseller K., Hart G. W., *Science*, 2001, 291, 2376.

[12] Li Y. X., Wang S. H., Li Z. M., *et al.*, *Carbon Res.*, 2006, 341, 2867.

[13] Li Y. X., Li Z. M., Zhao W. G., *et al.*, *Chin. Chem. Lett.*, 2006, 17, 153.

[14] NuBbaumer T., Krieger C., Neidlein R., *Eur. J. Org. Chem.*, 2000, 2449.

[15] Nguyen V. A., Willis C. L., Gerwick W. H., *Chem. Commun.*, 2001, 1934.

[16] Chambers M. S., Atack J. R., Broughton H. B., *et al.*, *J. Med. Chem.*, 2003, 46, 2227.

[17] Easmon J., Heinisch G., Hofmann J., *et al.*, *Eur. J. Med. Chem.*, 1997, 32, 397.

4，5，6－三取代嘧啶苯磺酰脲类化合物的生物活性、分子对接与3D－QSAR关系研究*

郭万成　马　翼　李永红　王素华　李正名

（南开大学元素有机化学研究所，元素有机化学国家重点实验室，天津　300071）

摘　要　用柔性分子对接方法（FlexX）将15个4，5，6－三取代嘧啶苯磺酰脲化合物以及3个不含5－位取代嘧啶苯磺酰脲化合物（分别为4，6－双取代嘧啶和4－取代嘧啶）和乙酰羟酸合成酶（AHAS）活性口袋进行了对接，对接程序预测的抑制剂和酶之间的相互作用能与抑制活性之间有一定的相关性，相关系数为0.660。然后采用比较分子相似性指数分析（CoMSIA）对27个新型4，5，6－三取代嘧啶苯磺酰脲类化合物的除草活性进行三维定量构效关系（3D－QSAR）研究。建立了三维定量构效关系CoMSIA模型，立体场、静电场和氢键的贡献分别为47.3%，32.8%，19.9%。交叉验证系数q^2值为0.520。根据CoMSIA模型的立体场、静电场、氢键给体场三维等值线图不仅直观地解释了结构与活性的关系，并且与用FlexX预测的结合模式相一致，证明了我们预测的结合模式是可靠的，为进一步设计高活性的标题化合物提供较好的理论指导。

关键词　分子对接　乙酰羟酸合成酶　比较分子相似性指数分析　三维定量构效关系　三取代嘧啶苯磺酰脲　除草活性

乙酰羟酸合成酶（AHAS，EC 2.2.1.6）是催化生物体内支链氨基酸——缬氨酸、亮氨酸和异亮氨酸生物合成过程中的第一个酶，高等动物自身不能合成该酶，需要从植物中摄取，因而该酶是理想的除草剂作用靶点[1]。1982年，第一个商品化的磺酰脲（SU）类除草剂——氯磺隆（chlorsulfuron）问世[2]。在随后几年里，人们逐渐认识到磺酰脲类除草剂的作用靶点是植物体内的乙酰羟酸合成酶，并且由于磺酰脲类除草剂具有高效、低毒、高选择性和对环境友好等特点，所以不断地有商品化品种上市[3~5]。在商品化的品种中，杂环部分通常为4－位和6－位双取代的嘧啶环或三嗪环。文献［2］中认为，磺酰脲类化合物具有较高的除草活性必须满足的条件之一为在脲桥的对位是没有取代基的。为了更充分地理解结构－活性关系，我们合成了具有不同取代的4，5，6－三取代嘧啶的磺酰脲类化合物[6,7]，并测定了其平皿除草活性，结果表明在1μg/mL剂量下，部分化合物显示了较好的除草活性。在此基础上，采用分子对接方法预测了该类化合物与AHAS酶的结合模式，并对该类化合物用CoMSIA方法进行了三维定量构效关系研究，以便对该类化合物进一步的合理设计做出理论指导。

1　材料与方法

1.1　分子对接

分子对接采用Sybyl 6.9[8]中的FlexX模块进行，FlexX是一种半柔性对接方法，是德国国家信息技术研究中心生物信息学算法和科学算法计算研究室的Rarey等[9~12]发

* 原发表于《化学学报》，2009，67（6）：569－574。国家自然科学基金（No. 20432010）和天津市应用基础研究计划（No. 07JCYBJC00200）资助项目。

展的分子对接方法。在 FlexX 中，分子对接的流程主要分为几个步骤[13]：核心结构的确定，核心结构的放置和配体结构的生长。首先，把能对配体和受体之间相互起决定作用的基团作为核心结构，然后把它考虑为刚体放置于活性部位中，最后在此基础上，把配体的其它部分分为小的片断，依次“生长”在核心结构上。在对接过程中考虑了抑制剂的可旋转键的所有柔性以确定抑制剂与 AHAS 酶结合的最佳构象。配体和受体之间结合情况的评价采用了类似 Böhm[14,15] 提出的半经验的自由能得分方程。我们选取 18 个该类化合物进行对接研究（化合物编号为 **F-1 ~ F-18**），使其抑制百分率（油菜平皿法测得，测试剂量为 1μg/mL）在所测出数值中的各个数量级都有分布，对这些化合物的抑制百分率 a 进行 $\log[a/(100-a)]+\log M$ 的变换，变换后的结果不影响内在规律，表示为 D[16]，结构和抑制常数见表 1。所有化合物分子在对接之前先加载 Gasteiger-Hückel 电荷，采用 Tripos 力场进行优化，采用共轭梯度法进行优化，收敛精度为 0.209kJ/mol。计算中选用的各项参数除非特别指明外，均采用缺省值。拟南芥 AHAS（EC 2.2.1.6）晶体结构来源于蛋白质数据库 Protein Data Bank（www.rcsb.org/pdb），PDB 编号为 1YBH。

表 1　所研究化合物的结构、活性数据及对接打分

Table 1　The structures, activities and total score of compounds

No.	R^1	R^2	R^3	Energy score (kJ·mol^{-1})	D	D''	Residue
F-1	*i*-Propyl	Cl	$COOCH_3$	-15.2	2.907	2.941	-0.034
F-2	*i*-Bu	Cl	$COOCH_3$	-15.5	2.999	2.737	0.262
F-3*	*n*-Penta	Cl	$COOCH_3$	-13.4	2.661	2.608	0.053
F-4	*n*-Bu	OEt	$COOCH_3$	-11.2	2.610		
F-5	*n*-Propyl	Cl	NO_2	-16.4	2.649	2.459	0.190
F-6	*n*-Bu	Cl	NO_2	-15.9	1.842	2.339	-0.497
F-7	*n*-Octa	Cl	NO_2	-13.4	2.087	1.953	0.134
F-8	Benzyl	Cl	$COOCH_3$	-16.6	2.601	2.674	-0.073
F-9	*p*-Methylbenzyl	Cl	$COOCH_3$	-16.5	2.948		
F-10	*o*-Florobenzyl	OEt	$COOCH_3$	-13.8	1.958		
F-11	Benzyl	Cl	NO_2	-17.6	2.460	2.377	0.083
F-12	*p*-Methylbenzyl	Cl	NO_2	-17.7	1.727		
F-13	Benzyl	OEt	NO_2	-17.4	3.168	2.906	0.262
F-14	*o*-Florobenzyl	OEt	NO_2	-17.4	3.073	2.905	0.168
F-15	*p*-Methoxybenzyl	OEt	NO_2	-17.9	3.168		
F-16	H	Cl	$COOCH_3$	-17.7	2.946	2.874	0.072
F-17	H	OEt	$COOCH_3$	-17.9	3.019		
F-18	H	H	$COOCH_3$	-17.7	2.805	2.847	-0.042

D = Experimental value；D'' = predictive value of D；**F-18**：NK92825.

将晶体结构中抑制剂所在位置周围 0.65nm 的范围定义为酶的活性口袋。将 18 个抑制剂分子与其进行一一对接，对接中采用晶体结构中的抑制剂构象作为参考。分子对接预测的抑制剂分子和靶酶的结合能见表 1。对结合能和活性因子进行作图（见图 2），得到线性相关系数为 0.660，对接程序预测的抑制剂和酶之间的相互作用能与抑制活性之间有一定的相关性，对接结果可以预测抑制剂与 AHAS 酶之间的结合模式。

图 1　所研究化合物的基本结构

Figure 1　The structures of the title compounds

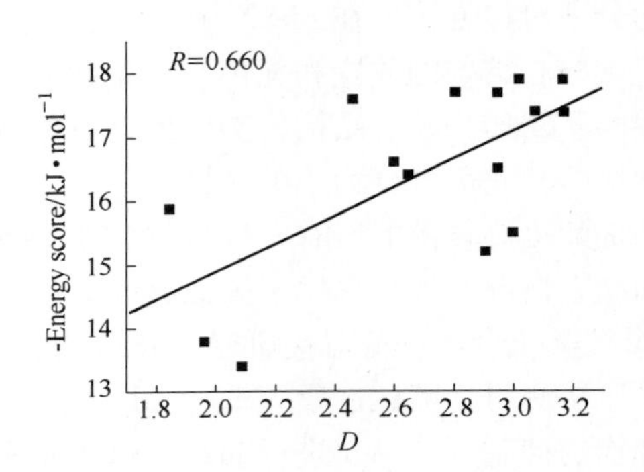

图 2　结合能与活性因子相关性图

Figure 2　Linear regression of the FlexX total score and experimental activities

1.2　三维定量构效关系研究

我们把表 1 中预测的 12 个分子连同表 2 中其它同类分子共 27 个化合物，选取 22 个磺酰脲分子组成训练集，其余 5 个组成测试集（如表 1 和表 2 中标 * 的分子），对这一组构象进行三维定量构效关系研究。采用了比较分子相似性指数分析方法（CoMSIA），该方法是现在最通用的三维定量构效关系研究的一种方法。该软件是 Tripos 公司推出的 Sybyl 分子模拟软件包中的模块，其基本原理认为：在药物分子与受体之间的可逆相互作用主要是通过非共价键结合，如范德华相互作用、静电相互作用、氢键相互作用和疏水相互作用实现的。作用于同一个受体的一系列药物分子，它们与受体之间的各种作用力场应该有一定的相似性。这样，在不了解受体三维结构的情况下，研究这些药物分子周围的力场分布，并把它们与药物分子活性定量地联系起来，既可以预测受体的某些性质，又可依此建立一个定量模型设计新化合物，并定量地预测化合物的活性。

我们选择化合物 **F-18** 作为模板，采用其晶体结构数据，用 Tripos 力场进行优化，优化时采用 Powell 方法优化 1000 轮，电荷计算采用 Gasteiger-Hückel 法，直至达到能量收敛标准为 0.209kJ/mol。分子力场计算以 +1 价 sp^3 杂化的碳原子作为探针，步长为 0.2nm，其余参数均为缺省值。优化后的构象作为活性构象，也就是模板分子，其它化合物在此基础之上进行取代基变换再优化，采用分子力学方法，选用 Tripos 力场优化，得到各分子的低能构象。叠合时选取图 1 中 * 标记的原子作为公共结构，将所有化合物的公共部分均与模板分子的基本骨架相重叠。采用此方法叠合保证每个分子力场的取向具有一致性。化合物分子的叠合如图 3 所示。叠合 RMS 值 <0.05。

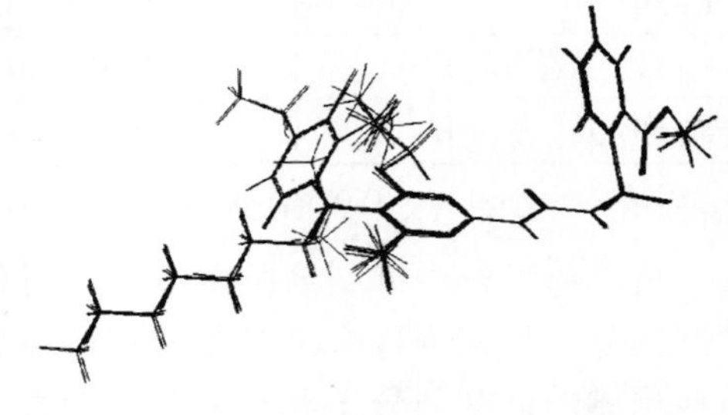

图 3　化合物分子的叠合方式

Figure 3　Superposition modes of compounds

在 CoMSIA 模型中，选取立体场（S）、静电场（E）、疏水场（H）、氢键供体场（D）和氢键受体场（A）等场类型来考虑化合物与受体相互作用，介电常数与距离有关，阈值为 30kJ/mol，其余各参数均用系统默认值得到分子场。首先分析了考虑全部五种场贡献的

CoMSIA 模型，然后又考察了各种场组合来找到能得到较好交叉验证系数的分子场。交叉验证采用留一法（leave - one - out，LOO），得到最佳组分数和交叉验证系数。为使模型具有可信预报能力，交叉验证系数 q^2 值最终大于 0.5。接下来通过非交叉验证进行回归计算，建立相关的 CoMSIA 模型，用模板分子 **F - 18** 作为代表，用三维等值线图显示，结果见图 5。用此模型进行活性预测，预测结果见表 2。

表 2　所研究化合物的结构、活性数据及预测结果

Table 2　The structure, experimental and calculated activity of studied compounds

Compd.	R^1	R^2	R^3	D	D''	Residue
Q - 1	*n* - Penta	Cl	NO_2	2.053	2.210	-0.157
Q - 2	*n* - Bu	OEt	NO_2	2.700	2.527	0.173
Q - 3	*n* - Hepta	OEt	NO_2	2.210	2.200	0.010
Q - 4	*o* - Florobenzyl	Cl	$COOCH_3$	2.931	2.765	0.166
Q - 5	*o* - Methylbenzyl	Cl	$COOCH_3$	2.923	2.643	0.280
Q - 6	Benzyl	OEt	$COOCH_3$	2.931	3.082	-0.151
Q - 7	*m* - Chlorobenzyl	OEt	$COOCH_3$	2.872	3.077	-0.205
Q - 8	*o* - Methylbenzyl	OEt	$COOCH_3$	2.985	3.052	-0.067
Q - 9	2，4 - Dichlorobenzyl	OEt	$COOCH_3$	3.085	3.282	-0.197
Q - 10	*m* - Chlorobenzyl	OEt	NO_2	3.073	2.905	0.168
Q - 11	*p* - Methylbenzyl	OEt	NO_2	3.075	2.909	0.166
Q - 12*	*m* - Chlorobenzyl	Cl	$COOCH_3$	2.537	2.758	-0.221
Q - 13*	*o* - Methylbenzyl	Cl	NO_2	2.003	2.348	0.345
Q - 14*	*o* - Florobenzyl	Cl	NO_2	2.770	2.868	-0.098
Q - 15*	2，4 - Dichlorobenzyl	Cl	NO_2	2.998	3.137	-0.139

D = Experimental value；D'' = Predictive value of D；* test - set compounds.

2　结果与讨论

2.1　SU 类抑制剂与 AHAS 的相互作用方式

对预测结合能和活性因子作图，发现二者有很好的相关性，如图 2 所示，线性回归系数 R = 0.660。分子对接结果表明，化合物小分子可以很好地对接到酶的活性位点，如图 3 所示。对接打分函数结果见表 1，从表中可以看出，用 FlexX 程序预测的最低对接能构象的对接能为 - 17.9kJ/mol。从整个对接能看出，4，5，6 - 三取代嘧啶磺酰脲化合物的活性与其对接能体现了很好的相关性，化合物活性较好则对接能较低，如化合物 **F - 11 ~ F - 15** 在 1μg/mL 剂量下对油菜的根长抑制较好，则对接能也较低。化合物 **F - 16**，**F - 17** 和 **F - 18** 分别为 4，6 - 双取代嘧啶磺酰脲化合物和 4 - 单取代嘧啶商品磺酰脲，从对接能来看，该三个化合物与化合物 **F - 11 ~ F - 15** 相比显示了相近的对接能，并且与活性具有相同的趋势。这也验证了 FlexX 程序在研究磺酰脲类抑制剂与靶酶 AHAS 相互作用模式的适用性和所采用对接参数的合理性。

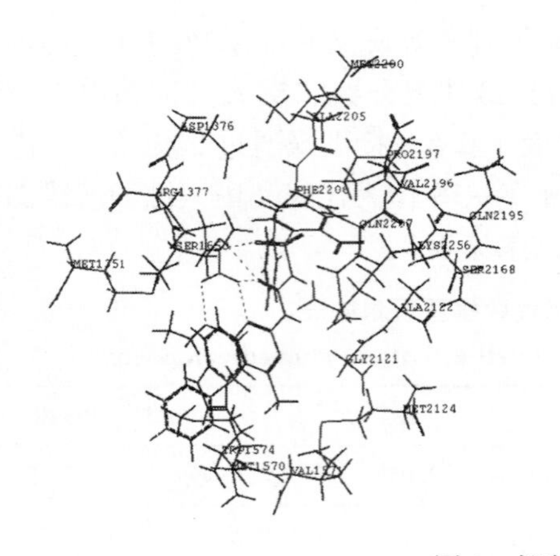

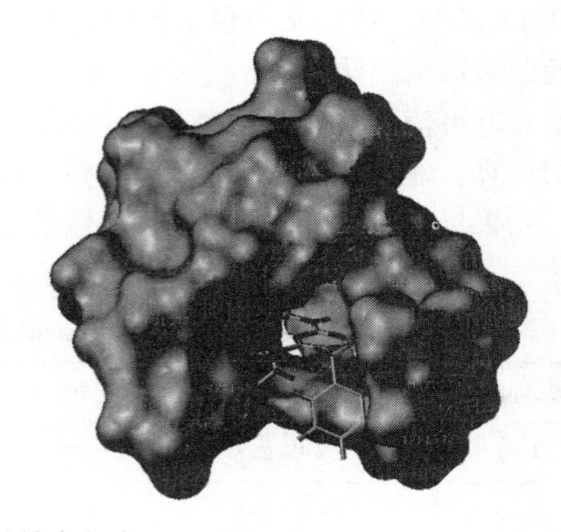

图4　抑制剂与酶的结合方式

Figure 4The binding mode of inhibitor with enzyme

分子对接结果表明（如图4 所示），三取代嘧啶磺酰脲类抑制剂与靶酶 AHAS 的主要作用模式是：（1）抑制剂磺酰基上的 O 原子可与 Ser1653 形成氢键；（2）脲桥上 NH—CO—NH 上的 O 原子与 Ser1653 形成氢键；（3）嘧啶环 1 - 位 N 原子和 6 - 位乙氧基的 O 原子与精氨酸残基 Arg1377 均可形成氢键；（4）在苯环位置形成一个疏水中心。该作用模式与双取代和单取代磺酰脲化合物相一致。

因此，4，5，6 - 三取代嘧啶磺酰脲与 4，6 - 双取代嘧啶磺酰脲化合物和 4 - 位单取代嘧啶磺酰脲化合物相比较，5 - 位取代基的引入对于化合物与酶的结合并不产生很明显的影响，保持了与单取代嘧啶和双取代嘧啶磺酰脲与酶同样的结合方式，并且 5 - 位取代基没有明显的与酶作用的位点。

2.2　CoMSIA 模型对结合模式的验证和对抑制剂结构改造的指导

CoMSIA 在移动步长为 0.20nm 时所得 3D - QSAR 模型的统计结果列于表 3。其中 CoMSIA6 是所有场组合中 q^2 最高的分子场（S，E，D）分析。由最佳组分得到的 CoMSIA，$r^2 = 0.739$，$F = 19.776$，标准偏差 SE = 0.208，立体场的贡献为 $S = 47.3\%$，静电场的贡献为 $E = 32.8\%$，氢键给体场的贡献为 $D = 19.9\%$。从表 3 可以看出，在步长为 0.2nm 时，CoMSIA6 中，综合来看考虑了立体场、静电场和氢键给体场的贡献，而疏水场和氢键受体场贡献不大，在所有不同场组合当中所得交叉验证系数最好，所得模型相对最为可靠。

表 3　所得的 3D - QSAR 模型统计结果

Table 3　The results of the 3D - QSAR model

	CoMSIA1	CoMSIA2	CoMSIA3	CoMSIA4	CoMSIA5	CoMSIA6
q^2	0.51	0.519	0.439	0.485	0.088	0.52
N	4	3	3	4	3	3
r^2	0.806	0.738	0.702	0.744	0.536	0.739
SE	0.184	0.209	0.222	0.211	0.277	0.208
F	20.753	19.734	16.500	14.531	8.102	19.776
Steric（S）	0.582	0.470	0.261	0.317	0.627	0.473
Electrostatic（E）	0.418	0.325	0.287	0.271	0.085	0.328

续表 3

	CoMSIA1	CoMSIA2	CoMSIA3	CoMSIA4	CoMSIA5	CoMSIA6
Hydrophobic（H）	—	—	0.452	0.383	—	—
H－bond donor（D）	—	0.198	—	0.024	0.008	0.199
H－bond acceptor（A）	—	0.007	—	0.005	0.28	—

q^2：Cross－validated correlational coefficient；r^2：non－validated correlational coefficient；SE：standard error of estimate；F：the Fischer ratio；N：the number of compounds used in the correlation.

CoMSIA 模型的三维等值线图，可从分子周围空间看到化合物的各种不同场对活性带来的影响。用不同颜色表示该系列分子的主要空间和静电性能。红色和蓝色区域代表静电效应，绿色和黄色区域代表立体效应。图 5a 为立体场分布图，可以看出，当在绿色区域引入大体积取代基或在黄色区域减小取代基的体积有利于化合物活性增加，经过这样改造的分子能够更加实现与受体的几何匹配，如化合物 **Q－3** 比 **Q－1** 活性好，**Q－2** 比 **Q－3** 活性好。图 5b 为静电场分布图，如要提高化合物的活性，可以在蓝色区域增加化合物的正电性，在红色区域引入电负性较大的基团如化合物 **Q－9** 比 **Q－8** 活性好。图 5c 为氢键受体场三维等值线分布图，从中可以看出在紫色区域增加氢键给体场有利于化合物活性的提高。

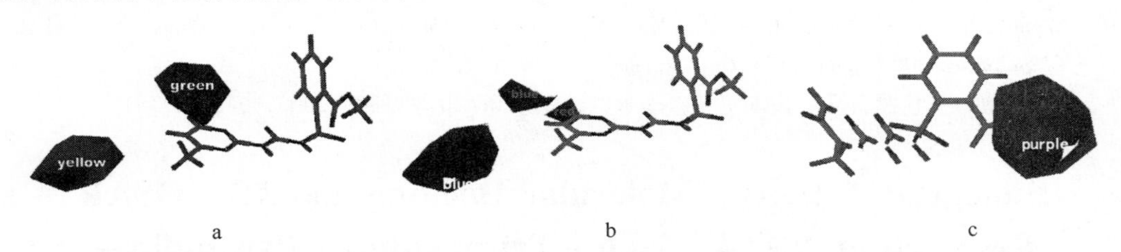

图 5　化合物 CoMSIA 的立体场（a）、静电场（b）和氢键给体场（c）三维等值线图

Figure 5　The steric maps（a），electrostatic maps（b）and hydrogen－bond donor maps（c）from the CoMSIA model

3　结论

采用分子对接方法预测了一组含 5－位取代基的 4，5，6－三取代嘧啶苯磺酰脲化合物、不含 5－位取代基的 4，6－二取代嘧啶苯磺酰脲化合物和 4－单取代嘧啶苯磺酰脲化合物与 AHAS 酶间的相互作用模式，得到的复合物结构表明，该类抑制剂与 AHAS 之间的主要结合方式均为氢键和疏水相互作用，并且嘧啶环 5－位取代基与酶没有明显的结合方式，因此 5－位取代基的引入对整个分子的活性不产生太大的影响。对该类化合物进行的三维定量构效关系研究进一步验证了我们预测的结合模式，所得的构效关系模型表明，应从不同的氢键给体场和空间体积进行改造，以期提高活性。

References

[1] Duggleby，R. G.；Pang，S. S. *J. Biochem. Mol. Biol.* 2000，33，1.

[2] Levitt，G. *Synthesis and Chemistry of Agrochemicals II*，ACS Symposium Series No. 443，American Chemical Society，Washington，DC，1991，pp. 16－33.

[3] Ray，T. B. *Plant Physiol.* 1984，75，827.

[4] Chaleff，R. S.；Mauvais，C. J. *Science* 1984，224，1443.

[5] LaRossa, R. A.; Schloss, J. V. *J. Biol. Chem.* 1984, 259, 8753.

[6] Guo, W. – C.; Wang, M. – Y.; Liu, X. – H.; Li, Y. – H.; Wang, S. – H.; Li, Z. – M. *Chem. J. Chin. Univ.* 2007, 28, 1666 (in Chinese).
(郭万成，王美怡，刘幸海，李永红，王素华，李正名，高等学校化学学报，2007，28，1666.)

[7] Guo, W. – C.; Tan, H. – Z.; Liu, X. – H.; Li, Y. – H.; Wang, S. – H.; Li, Z. – M. *Chem. J. Chin. Univ.* 2008, 29, 319 (in Chinese).
(郭万成，谭海忠，刘幸海，李永红，王素华，李正名，高等学校化学学报，2008，29，319.)

[8] *Tripos 6. 91 User Guide*, *Tripos*, Inc. St. louis. USA, 2003.

[9] Rarey, M.; Kramer, B.; Lengauer, T.; Klebe, G. *J. Mol. Biol.* 1996, 261, 470.

[10] Zhou, M.; Zhang, W.; Cheng, Y. – H.; Ji, M. – J.; Xu, X. – J. *Acta Chim. Sinica* 2005, 63, 2131 (in Chinese).
(周梅，章威，成元华，计明娟，徐筱杰，化学学报，2005，63，2131.)

[11] Böhm, H. J. *J. Comput. – Aided Mol. Des.* 1994, 8, 243.

[12] Klebe, G. *J. Mol. Biol.* 1994, 237, 212.

[13] Li, J. – T., *M. S. Thesis*, Tianjin University, Tianjin, 2006 (in Chinese).
(李金涛，硕士论文，天津大学，天津，2006.)

[14] Böhm, H. J. *J. Comput. – Aided Mol. Des.* 1992, 6, 61.

[15] Böhm, H. J. *J. Comput. – Aided Mol. Des.* 1992, 6, 593.

[16] Wang, B. – L.; Wang, J. – G.; Ma, Y.; Li, Z. – M.; Li, Y. – H.; Wang, S. – H. *Acta Chim. Sinica* 2006, 64, 1373 (in Chinese).
(王宝雷，王建国，马翼，李正名，李永红，王素华，化学学报，2006，64，1373.)

Biological Activity, Molecular Docking and 3D – QSAR Research of *N* –(4, 5, 6 – Trisubstituted Pyrimidin – 2 – yl) – *N′* – benzenesulfonylureas

Wancheng Guo, Yi Ma, Yonghong Li, Suhua Wang, Zhengming Li

(State Key Laboratory of Elemento – Organic Chemistry, Nankai University, Tianjin, 300071)

Abstract Fifteen *N* – (4, 5, 6 – trisubstituted pyrimidin – 2 – yl) – *N′* – benzenesulfonylureas and three analogous benzenesulfonylureas without 5 – substituent at pyrimidine moiety were studied on a flexing molecular docking method (FlexX) according to their biological activity to the active site of acetohydroxyacid synthase (AHAS). The predicted binding affinities of the molecules were found to be linearly relevant to their experimental activities ($R = 0.660$). Then 27 *N* – (4, 5, 6 – trisubstituted pyrimidin – 2 – yl) – *N′* – benzenesulfonylureas were studied on comparative molecular similarity index analysis. The results showed that the contribution of steric, electrostatic and H – bond donor was 47.3%, 32.8% and 19.9%, respectively. The cross – validated q^2 was 0.520. The results indicate that the 3D – QSAR model is significant and has good predictability, providing useful information for designing new good activity AHAS inhibitors prior to their synthesis.

Key words molecular docking; AHAS; CoMSIA; 3D – QSAR; *N* – (trisubstituted pyrimidin – 2 – yl) – *N′* – benzenesulfonylurea; herbicidal activity

Synthesis, Crystal Structure and Insecticidal Activities of Novel Neonicotinoid Derivatives*

Yu Zhao, Gang Wang, Yongqiang Li, Suhua Wang, Zhengming Li

(State Key Laboratory of Elemento – Organic Chemistry, Research Institute of Elemento – Organic Chemistry, National Pesticide Engineering Research Center (Tianjin), Nankai University, Tianjin, 300071, China)

Abstract Ten new *N* – oxalyl derivatives of neonicotinoid compound were designed and synthesized. Their structures were confirmed by ^{1}H NMR spectra, elemental analysis and X – ray diffraction. The preliminary insecticidal activities of the new compounds were evaluated. The results of bioassays indicate that the title compounds exhibit moderate insecticidal activities. Surprisingly, the insecticidal activity of compound **7**b against bean aphids at 200mg/kg is 100%, which is comparable to that of the commercialized imidacloprid.

Key words neonicotinoid; *N* – Oxalyl derivative; insecticidal activity; bean aphid; imidacloprid

1 Introduction

Imidacloprid (**1**, Scheme 1), the first neonicotinoid insecticide acting on a nicotinic acetylcholine receptor (nAChR), has been widely used to control not only various plant pest, but also fleas on cats and dogs, and termites[1-4]. And then, the extensive research and development have led to the discovery of some insecticides from this class of chemicals[5]. Following imidacloprid, six new market products were developed by replacing the pyridine ring with a thiazole ring or a saturated heterocyclic ring, changing the nitroimino group to an isoelectronic nitromethylene or cyanoimine group, or reconstructing the imidazolidine ring with bioisosteric cyclic or acyclic moiety[6,7]. All of these compounds are characterized by their high insecticidal activities against insects and relative safety toward mammals and aquatic life[8,9].

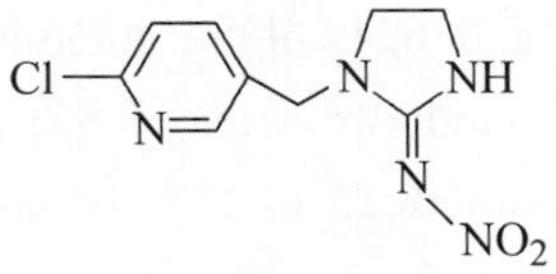

Cl N NH N N NO_2

1

Scheme 1 Chemical structure of imidacloprid

It was reported that *N* – oxalyl derivatives of carbofuran containing a carboxylic acid or ester substituent (**2**) displayed an insecticidal activity comparable or superior to that of carbofuran[10]. In another paper, the synthesis and insecticidal evaluation of novel *N* – oxalyl derivatives of tebufenozide (**3**) have been reported (Scheme 2), and the results of bioassay show that

* Reprinted from *Chemical Research in Chinese Universities*, 2010, 26(3): 380 – 383.

they exhibit excellent larvicidal activity[11]. Encouraged by these reports, an idea was developed that the introduction of an oxalyl substituent into a neonicotinoid molecule by substituting the hydrogen on the nitrogen atom could improve biological properties and decrease resistance. And the thiazole group can be taken as a bioisostere of pyridine. Hence, using thiazolyl group to replace pyridyl, ten novel *N* - oxalyl derivatives of neonicotinoid compound were designed and synthesized. And the preliminary insecticidal activities of the new compounds were evaluated.

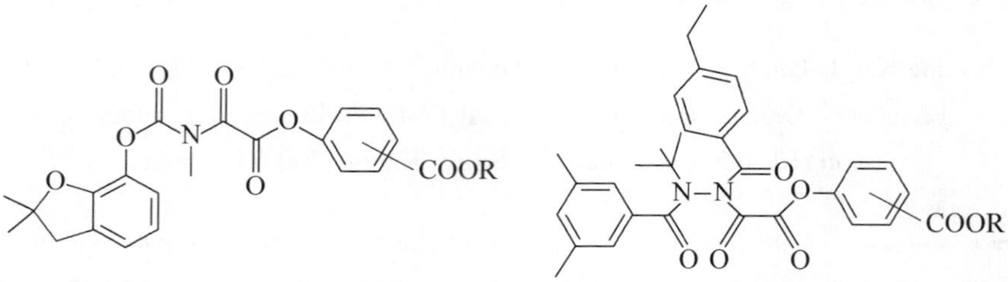

2　　　　3

Scheme 2　Structures of the two reported *N* - oxalyl derivatives

2　Experimental

2.1　Instruments

^{1}H NMR spectra were obtained at 400MHz on a Varian Mercury Plus 400 spectrometer in $CDCl_3$ solution with tetramethylsilane as the internal standard. Elemental analyses were carried out on a Yanaca CHN Corder MT - 3 elemental analyzer. The melting points were determined on an X - 4 binocular microscope melting point apparatus (Beijing Tech Instruments Co., Beijing, China) and were uncorrected. The reagents were all analytically or chemically pure. All the solvents and liquid reagents were dried by standard methods in advance and distilled before use. Imidacloprid (**1**) was prepared according to the literature[12-14].

2.2　Synthesis of the Title Compounds

2.2.1　Synthetic Procedure for Compound 4

A solution of ethylenediamine (1.44g, 0.024mol) in chloroform (2mL) was added with stirring to a solution of dimethyl cyanodithioimidocarbonate (2.92g, 0.02mol) in chloroform (15mL), whilst the reaction temperature was kept at 25 to 27℃ by cooling with ice - water. The reaction mixture was stirred at room temperature for 3h. The solvent was evaporated under reduced pressure and the residue was washed with diethyl ether (5mL × 2) and dried to give *N* - (imidazolidin - 2 - ylidene) cyanamide (**4**) (2.17g, yield 98.9%). m. p. 205 - 206℃. ^{1}H NMR (400MHz, DMSO), δ: 7.68 (s, br, 2H, NH), 3.27 - 3.47 (m, 4H, CH_2CH_2).

2.2.2　Synthetic Procedure for Compound 5

Synthetic route of compound **5** was shown in Scheme 3.

A solution of 2 - chloro - 5 - chloromethylthiazole (1.68g, 0.01mol) in acetonitrile (20mL) was added dropwise to a mixture of compound **4** (1.32g, 0.012mol) in acetonitrile (10mL) and anhydrous potassium carbonate (2.07g, 0.015mol). The mixture was heated at refluxed temperature

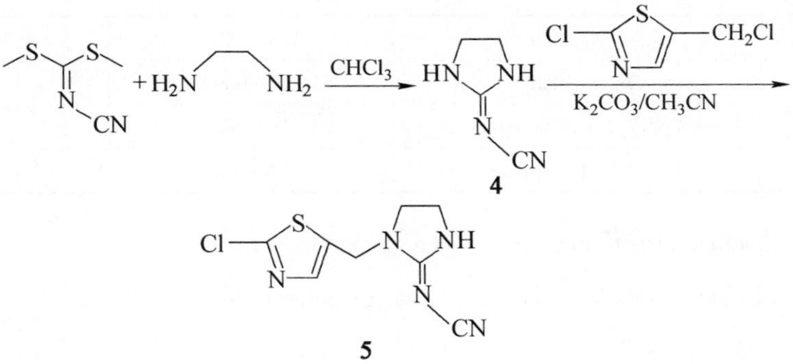

Scheme 3 Synthetic route of compound **5**

for 3h. After the reaction, acetonitrile was distilled off under reduced pressure, and dichloromethane(60mL) was added to the residue. The mixture was washed successively with water(20mL × 3) and brine(20mL), and then dried over anhydrous sodium sulfate. After removal of the solvent, the residue was purified by column chromatography on a silica gel using petroleum ether(60 – 90℃) and ethyl acetate as the eluent to afford *N* – {1 – [(2 – chlorothiazol – 5 – yl) methyl] imidzolidin – 2 – yli – dene} cyanamide (**5**). The yield was 92.1%. m. p. 150 – 152℃. ^1H NMR (400MHz, $CDCl_3$), δ: 7.44 (s, 1H, thiazole – H); 6.91 (s, 1H, NH); 4.51 (s, 2H, thiazole – CH_2); 3.61 (t, $^3J_{HH}$ = 8.4Hz, 2H, CH_2—CH_2); 3.52 (t, $^3J_{HH}$ = 8.8Hz, 2H, CH_2—CH_2).

2.2.3 Synthetic Procedure for Alkyloxyoxalyl Chlorides 6a – 6j

Synthetic route of compounds **6**a – **6**j was shown in Scheme 4.

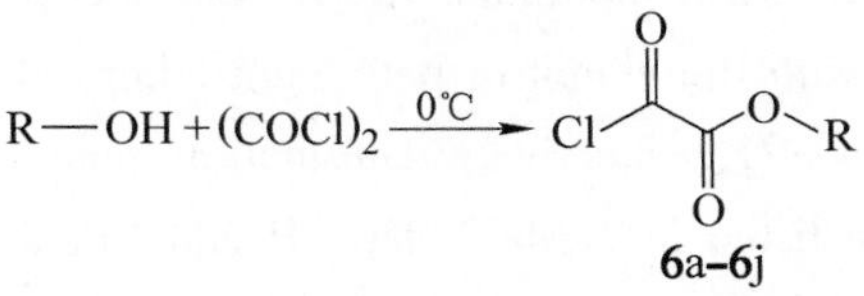

Scheme 4 Synthetic route of compounds **6**a – **6**j

The appropriate alcohol(0.1mol) was added dropwise over 20min to an excess of oxalyl chloride(0.2mol) at 0℃. When the addition was complete, the mixture was allowed to warm to room temperature for 2h. Excess oxalyl chloride was removed by vacuum distillation. Further distillation afforded alkyloxyoxalyl chlorides **6**a – **6**j. The boiling points and yields of compounds **6**a – **6**j are listed in Table 1.

Table 1 Boiling points and yields of compounds 6a – 6j

Compd.	R	b. p. (℃)	Yield(%)
6a	Methyl	118 – 120	75.3
6b	Ethyl	132 – 135	69.6
6c	*n* – Propyl	150 – 152	77.2
6d	*i* – Propyl	149 – 151	71.0
6e	*n* – Butyl	77 – 79(3mmHg)	66.4
6f	*s* – Butyl	76 – 78(3mmHg)	63.7
6g	Cyclohexyl	95 – 97(3mmHg)	43.6

Table 1(continued)

Compd.	R	b. p. (℃)	Yield(%)
6h	Allyl	63 - 66(3mmHg)	84. 8
6i	2 - Methoxyethyl	97 - 99(3mmHg)	60. 3
6j	2,2,2 - Trifluoroethyl	115 - 117	53. 8

2. 2. 4 General Synthetic Procedure for the Title Compounds 7a - 7j

Synthetic route of compounds **7**a - **7**j was shown in Scheme 5.

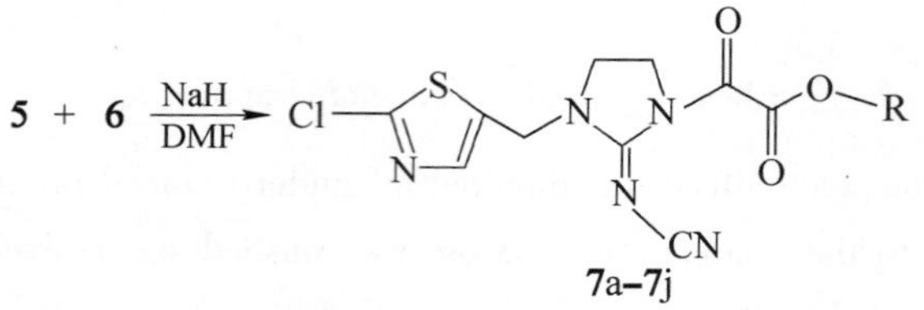

Scheme 5 Synthetic route of the title compounds **7**a - **7**j

Compound **5**(0. 01mol) was dissolved in dry dimethylformamide(30mL), and sodium hydride (0. 011mol) was added to it at 10℃. The mixture was stirred at room temperature until the generation of hydrogen ceased. Then, appropriate alkyloxyoxalyl chloride(**6**, 0. 011mol) was added to the mixture, and the mixture was stirred at 30℃ for 5h, and then poured into ice water (50mL). The aqueous layer was extracted with dichloromethane (40mL × 3). The dichloromethane layer was washed with water (40mL × 3) and dried over anhydrous sodium sulfate. Then the dichloromethane was evaporated. The residue was purified by column chromatography on a silica gel eluted with the eluent of petroleum ether(60 - 90℃) and ethyl acetate to afford the title compounds **7**a - **7**j, respectively. The melting points, yields and elemental analyses of compounds **7**a - **7**j are listed in Table 2. The ^{1}H NMR data are listed in Table 3.

Table 2 Melting points, yields and elemental analyses of the title compounds 7a - 7j

Compd.	R	m. p. (℃)	Yield(%)	Elemental analysis(%), calcd. (found)		
				C	H	N
7a	Methyl	111 - 113	62. 7	40. 31(40. 04)	3. 08(3. 05)	21. 37(21. 32)
7b	Ethyl	150 - 151	58. 9	42. 17(42. 16)	3. 54(3. 52)	20. 49(20. 49)
7c	*n* - Propyl	114 - 116	65. 1	43. 88(43. 81)	3. 97(4. 02)	19. 68(19. 77)
7d	*i* - Propyl	121 - 123	70. 3	43. 88(43. 62)	3. 97(3. 89)	19. 68(19. 71)
7e	*n* - Butyl	133 - 135	56. 8	45. 47(45. 60)	4. 36(4. 44)	18. 94(18. 82)
7f	*s* - Butyl	139 - 140	64. 9	45. 47(45. 48)	4. 36(4. 51)	18. 94(18. 99)
7g	Cyclohexyl	55 - 57	66. 1	48. 54(48. 61)	4. 58(4. 33)	17. 69(17. 70)
7h	Allyl	165 - 167	68. 5	44. 13(44. 09)	3. 42(3. 58)	19. 80(19. 77)
7i	2 - Methoxyethyl	120 - 122	69. 6	42. 00(41. 86)	3. 80(3. 93)	18. 84(18. 86)
7j	2,2,2 - Trifluoroethyl	129 - 131	66. 2	36. 42(36. 40)	2. 29(2. 35)	17. 70(17. 61)

Table 3 ^{1}H NMR of the title compounds 7a – 7j

Compd.	^{1}H NMR(400MHz, $CDCl_3$), δ
7a	7.63(s,1H,thiazole – H),5.20(s,2H,thiazole – CH_2),3.98(t,$^3J_{HH}$ = 7.8Hz,2H,CH_2—CH_2),3.93(s,3H,CH_3);3.75(t,$^3J_{HH}$ = 7.8Hz,2H,CH_2—CH_2)
7b	7.62(s,1H, thiazole – H),5.19(s,2H, thiazole – CH_2),4.39(q,$^3J_{HH}$ = 7.8Hz,2H,OCH_2),3.97(t,$^3J_{HH}$ = 7.8Hz,2H,CH_2—CH_2),3.74(t,$^3J_{HH}$ = 7.8Hz,2H,CH_2—CH_2),1.38(t,$^3J_{HH}$ = 7.1Hz,3H,CH_3)
7c	7.63(s,1H,thiazole – H),5.20(s,2H,thiazole – CH_2),4.27(t,$^3J_{HH}$ = 6.8Hz,2H,OCH_2),4.07(t,$^3J_{HH}$ = 7.8Hz,2H,CH_2—CH_2),3.77(t,$^3J_{HH}$ = 7.8Hz,2H,CH_2—CH_2),1.75(m,2H,CH_2CH_3),0.98(t,$^3J_{HH}$ = 7.5Hz,3H,CH_3)
7d	7.62(s,1H,thiazole – H),5.23(m,1H,OCH),5.20(s,2H,thiazole – CH_2),3.96(t,$^3J_{HH}$ = 8.0Hz,2H,CH_2—CH_2),3.74(t,$^3J_{HH}$ = 8.0Hz,2H,CH_2—CH_2),1.35[d,$^3J_{HH}$ = 6.4Hz,6H,$CH(CH_3)_2$]
7e	7.62(s,1H,thiazole – H),5.20(s,2H,thiazole – CH_2),4.34(t,$^3J_{HH}$ = 6.7Hz,2H,OCH_2),3.97(t,$^3J_{HH}$ = 8.0Hz,2H,CH_2—CH_2),3.74(t,$^3J_{HH}$ = 8.0Hz,2H,CH_2—CH_2),1.72(m,2H,OCH_2CH_2),1.42(m,2H,CH_2CH_3),0.94(t,$^3J_{HH}$ = 7.4Hz,3H,CH_3)
7f	7.63(s,1H,thiazole – H),5.20(s,2H,thiazole – CH_2),5.08(m,1H,OCH),3.97(t,$^3J_{HH}$ = 7.8Hz,2H,CH_2—CH_2),3.74(t,$^3J_{HH}$ = 7.8Hz,2H,CH_2—CH_2),1.76(m,2H,CH_2CH_3),1.35(d,$^3J_{HH}$ = 6.3Hz,3H,$CHCH_3$),0.98(t,$^3J_{HH}$ = 7.2Hz,3H,CH_2CH_3)
7g	7.62(s,1H,thiazole – H),5.19(s,2H,thiazole – CH_2),4.99(m,1H,OCH),3.96(t,$^3J_{HH}$ = 8.0Hz,2H,CH_2—CH_2),3.74(t,$^3J_{HH}$ = 8.0Hz,2H,CH_2—CH_2),1.96(m,2H,CH_2),1.77(m,2H,CH_2),1.55(m,2H,CH_2),1.39(m,2H,CH_2),1.30(m,2H,CH_2)
7h	7.63(s,1H, thiazole – H),5.98(m,1H,CH ═),5.43(dd,m,2H,═CH_2),5.20(s,2H, thiazole – CH_2),4.80(d,$^3J_{HH}$ = 6.2Hz,2H,CH_2CH ═),3.97(t,$^3J_{HH}$ = 8.0Hz,2H,CH_2—CH_2),3.74(t,$^3J_{HH}$ = 8.0Hz,2H,CH_2—CH_2)
7i	7.63(s,1H, thiazole – H),5.20(s,2H,thiazole – CH_2),4.47(t,$^3J_{HH}$ = 4.5Hz,2H,$COOCH_2$),3.96(t,$^3J_{HH}$ = 8.0Hz,2H,CH_2—CH_2),3.74(t,$^3J_{HH}$ = 8.0Hz,2H,CH_2—CH_2),3.69(t,$^3J_{HH}$ = 4.5Hz,2H,OCH_2CH_2),3.42(s,3H,CH_3)
7j	7.63(s,1H, thiazole – H),5.20(s,2H,thiazole – CH_2),4.67(q,2H,CF_3CH_2,$^3J_{HF}$ = 8.4Hz),3.97(t,$^3J_{HH}$ = 8.0Hz,2H,CH_2—CH_2),3.76(t,$^3J_{HH}$ = 8.0Hz,2H,CH_2—CH_2)

3 Results and Discussion

3.1 Synthesis

In the present work, the syntheses of some new *N* – oxalyl derivatives of neonicotinoid compound as well as their insecticidal activities against bean aphids were studied. The target *N* – oxalyl derivatives of neonicotinoid compounds **7**a – **7**j were synthesized by a simple and convenient four – step procedure started from dimethyl cyanodithioimidocarbonate and ethylenediamine. The reaction of these two compounds in chloroform at 25 – 27℃ afforded *N* – (imidazolidin – 2 – ylidene) cyanamide(**4**) in a yield of 98.9%. Compound **4** was reacted with 2 – chloro

–5 – chloromethylthiazole in acetonitrile in the presence of potassium carbonate as a base to give *N* – {1 – [(2 – chlorothiazol – 5 – yl) methyl] imidazolidin – 2 – ylidene} cyanamide(**5**) in a good yield. Alkyloxyoxalyl chlorides **6**a – **6**j were obtained from the appropriate alcohol and oxalyl chloride as shown in Scheme 4. It was convenient to purify the alkyloxyoxalyl chloride by distillation. Then compound **6** was reacted with the key intermediate (**5**) in dry dimethylformamide in the presence of sodium hydride as a base to yield *N* – oxalyl derivatives **7**a – **7**j. The title compounds **7**a – **7**j could be purified by column chromatography on a silica gel with petroleum ether(60 – 90℃) and ethyl acetate as the eluent.

3.2 Crystal Structure Analysis

Compound **7**a was recrystallized from ethyl acetate to give colorless crystals suitable for X – ray single – crystal diffraction with the following crystallographic parameters: $C_{11}H_{10}ClN_5O_3S$, M = 327.75, a = 1.5322(3) nm, b = 1.1902(2) nm, c = 1.6047(3) nm, $\alpha = 90°$, $\beta = 90°$, $\gamma = 90°$, μ = 0.421 mm^{-1}, V = 2.9264(10) nm^3, Z = 8, D_x = 1.488 mg/m^3, $F(000)$ = 1344, T = 113(2) K, $2.51° \leqslant \theta \leqslant 25.02°$, and the final R factor, R_1 = 0.0361, ωR_2 = 0.1012. The crystal is orthorhombic.

It could be seen from the X – ray single – crystal analysis that the molecule consists of one thiazole ring and one imidazolidine ring. The bond lengths of C7 – N2[0.1332(2) nm] and C7 – N3[0.1393(2) nm] are shorter than that of the normal C – N single bond (0.149 nm), which suggests that the electron density is delocalized among N4 – C7 – N2 and N3. The atoms N4 – C7 – N2 and N3 are close to a plane. The molecular structure of compound **7**a is shown in Fig. 1, and the packed structure of this compound is shown in Fig. 2.

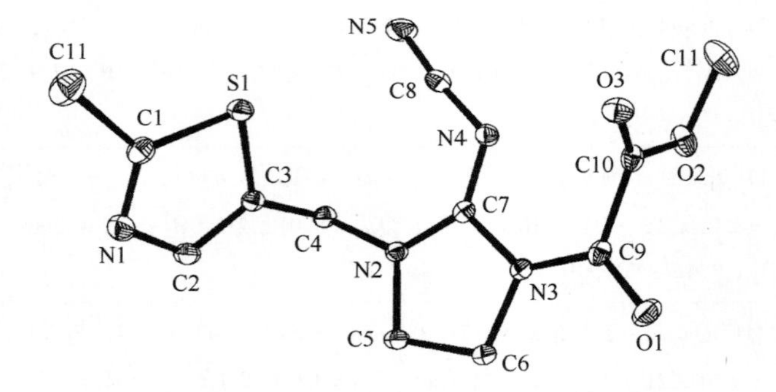

Fig. 1 Crystal structure of the compound **7**a

3.3 Insecticidal Activity

The insecticidal activities of the title compounds **7**a – **7**j and imidacloprid against bean aphids at 200mg/kg were evaluated. Bean aphids were dipped according to a slightly modified FAO dip test[15]. The results of the assay are summarized in Table 4.

The results of insecticidal activities given in Table 4 indicate that some of the title compounds exhibit a good activity against bean aphids, which are better than the commercialized imidacloprid. For instance, the insecticidal activities of compounds **7**f and **7**g against bean aphids at 200mg/kg are 85% and 96%, respectively. Surprisingly, the results indicate that the activity of

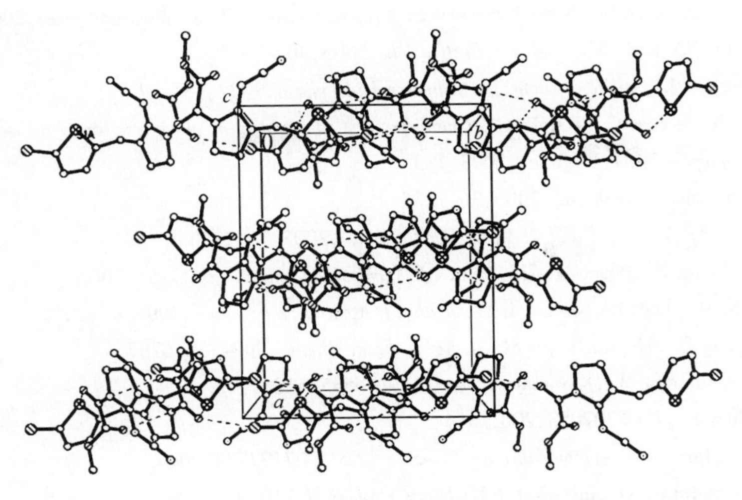

Fig. 2 Packed diagram of compound **7**a

compound **7**b against bean aphids at 200mg/kg is 100% ,which is equal to that of the commercial imidacloprid.

Table 4 Insecticidal activities of compounds 7a – 7j against bean aphids

Compd.	Larvicidal activity(%) at 200mg/kg
7a	70
7b	100
7c	30
7d	60
7e	40
7f	85
7g	96
7h	50
7i	47
7j	30
Imidacloprid	100

4 Conclusions

Ten novel *N* – oxalyl neonicotinoids were designed and synthesized with structures characterized by ^{1}H NMR spectroscopy, single crystal X – ray diffraction analysis and elemental analysis. The insecticidal activities of the new compounds were evaluated. The results of bioassays show that these title compounds exhibit moderate insecticidal activities. Surprisingly, the results indicate that the activity of compound **7**b against bean aphids at 200mg/kg is 100% ,which is equal to that of the commercial imidacloprid.

References

[1] Tomizawa M. ,Casida J. E. ,*Annu. Rev. Pharmacol. Toxicol.* ,2005,45,247.

[2] Maienfish P. ,Brandl F. ,Kobel W. ,*et al.* ,Eds. :Yamamoto I. ,Casida J. E. ,*Nicotinoid Insecticides and the Nicotinic Acetylcholine Receptor*;Springer – Verlag,Tokyo,Japan,1999,177.

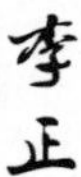

[3] Matsuda K. ,Sattelle D. B. ,*New Discoveries in Agrochemicals*,ACS Symposium Series,2005,892,172.
[4] Kagabu S. ,Ito N. ,Imai R. ,*et al.* ,*J. Pestic. Sci.* ,2005,30,409.
[5] Harry R. H. ,Wu J. ,Xu W. ,*Chem. Res. Chinese Universities*,2002,18,481.
[6] Kagabu,S. ,Eds. :Voss G. ,Ramos G. ,*Chemistry of Crop Protection,Progress and Prospects in Science and Regulation*,Wiley - VCH,Weiheim,2003,193.
[7] Kagabu S. ,*Synthetic Commun.* ,2006,36,1235.
[8] Tian Z. ,Shao X. ,Li Z. ,*et al.* ,*J. Agric. Food Chem.* ,2007,55,2288.
[9] Yokota T. ,Mikata K. ,Nagasaki H. ,*et al.* ,*J. Agric. Food Chem.* ,2003,51,7066.
[10] Mallipudi N. M. ,Lee A. ,Kapoor I. P. ,*et al.* ,*J. Agric. Food Chem.* ,1994,42,1019.
[11] Mao C. ,Wang Q. ,Huang R. ,*et al.* ,*J. Agric. Food Chem.* ,2004,52,6737.
[12] Kojima S. ,Funabora M. ,Kawahara N. ,*et al.* ,*US* 5453529,1995.
[13] Yeh C. ,Chen C. ,*US* 6307053,2001.
[14] Shroff D. K. ,Jain A. K. ,Chaudhari R. P. ,*et al.* ,*US* 2007197792,2007.
[15] Food and Agricluture Organization,*FAO Plant Prot. Bull.* ,1979,27,29.

区域选择性合成2-取代磺酰基亚肼基-3-全乙酰糖基-2，3-二氢噻唑及其表征、生物活性研究*

王浩安　李玉新　李一鸣　王素华　李正名

（南开大学元素有机化学国家重点实验室，天津　300071）

摘　要　以取代磺酰肼和糖基异硫氰酸酯反应，生成8种1-取代磺酰基-4-全乙酰糖基氨基硫脲，然后在$ClCH_2CHO/CH_3COONa/DMF$条件下关环，区域选择性地合成了一系列新型2-取代磺酰基亚肼基-3-全乙酰糖基-2，3-二氢噻唑。化合物**3e**的$^1H-^{13}C$ HMBC谱图表明1-取代磺酰基-4-全乙酰糖基氨基硫脲与氯乙醛的缩合反应关环位置为氨基硫脲的N-4，而非N-2。所有新化合物的结构均经过^1H NMR，^{13}C NMR和元素分析确证，生物活性初步测定结果表明该类二氢噻唑化合物具有一定的除草活性。

关键词　$^1H-^{13}C$ HMBC相关谱图　区域选择性　二氢噻唑

噻唑类化合物作为具有广泛生物活性的一类含氮杂环，因其独特的结构特征而在农药和医药领域受到化学家们的普遍关注[1]，近年来一些商品化杀虫（杀螨）剂、杀菌剂和除草剂均含有噻唑杂环[1~4]。

糖类化合物是生物体内的信号分子，参与了生命体内几乎所有的生物过程，并且糖苷类化合物也表现出抗病毒、抗肿瘤等生理活性[5]。糖基异硫氰酸酯作为常用的有机合成中间体，被用于制备多种具有生物活性及药用价值的碳水化合物的衍生物[6]。

由于糖基修饰的先导化合物能延长药物作用时间、增加药效和降低毒性[7]，将糖基和磺酰肼基同时引入二氢噻唑环中，对于开发二氢噻唑潜在的生物活性和降低化合物对动物的毒性很有意义。

此前，本课题组报道了一系列具有一定的杀菌及除草活性的2-取代磺酰基亚肼基-3-全乙酰糖基噻唑-4-酮[8]。为了寻找活性较好的化合物，以1-取代磺酰基-4-全乙酰糖基氨基硫脲为原料，与氯乙醛缩合合成了一系列双糖取代的2-取代磺酰基亚肼基-3-全乙酰糖基-2，3-二氢噻唑类化合物（Scheme 1），并通过化合物的$^1H-^{13}C$ HMBC谱研究其区域选择性。

1　实验部分

1.1　仪器与试剂

核磁采用Bruker AV-300MHz核磁共振仪（TMS为内标）；元素分析用Yanaco CHN CORDER MT-3元素分析仪；熔点测定用北京泰克X-4数字显示显微熔点测定仪（温度未校正）；质谱采用Finnigan LCQ Advantage MAX质谱仪；薄层色谱采用薄层

* 原发表于《有机化学》，2010，30（5）：703-706。天津市自然科学基金重点（No. 09JCZDJC21300）资助项目。

层析硅胶板（烟台市芝罘黄务硅胶开发试剂厂）。1－取代磺酰基－4－全乙酰糖基氨基硫脲按文献［9］方法合成；所用试剂和原料均为分析纯，并经常规处理。

1.2　目标化合物的合成

以2－苯磺酰基亚肼基－3－（七－*O*－乙酰基－*β*－麦芽糖基）－2，3－二氢噻唑（**3a**）为例，将0.874g（0.001mol）1－苯磺酰基－4－七－*O*－乙酰基－*β*－麦芽糖基氨基硫脲溶于20mL DMF中，加入0.082g CH_3COONa（0.001mol），在冰浴下滴加0.078g $ClCH_2CHO$（0.001mol），然后加热至60～70℃。TLC监测反应，约4h反应完毕，停止反应并加入20mL水，乙酸乙酯萃取，无水硫酸镁干燥，减压脱除乙酸乙酯，以乙酸乙酯和石油醚作为洗脱剂（*V*:*V*＝1:3），柱层析分离得到产物2－苯磺酰基亚肼基－3－（七－*O*－乙酰基－*β*－麦芽糖基）－2，3－二氢噻唑（**3a**）：收率70%，m. p. 123～125℃；^{1}H NMR（$CDCl_3$）*δ*：1.96～2.07（m，21H，7×OAc），3.93～5.59（m，14H，maltosyl－H），6.02（d，*J*＝4.8Hz，H－5），6.57（d，*J*＝4.8Hz，H－4），6.79（s，1H，NH），7.55～7.58（m，2H，ArH），7.62～7.64（m，1H，ArH），7.89～7.92（m，2H，ArH）；^{13}C NMR（$CDCl_3$）*δ*：20.44，20.79，20.86，21.01（7C，7×CH_3），61.65，62.91（2C，2×CH_2O），68.14，68.87，69.46，70.19，70.35，72.85，75.24，75.43，96.03（9C，9×sugar ring），81.00（C′－1），102.79（C－5），122.98（C－4），128.65，128.90，137.88，133.52（6C，6×Ar—C），169.61，169.98，170.04，170.17，170.59，170.71（7C，7×C＝O），173.38（C－2）。Anal. calcd for $C_{35}H_{43}N_3O_{19}S_2$：C 48.11，H 4.96，N 4.81；found C 48.03，H 4.95，N 4.59。

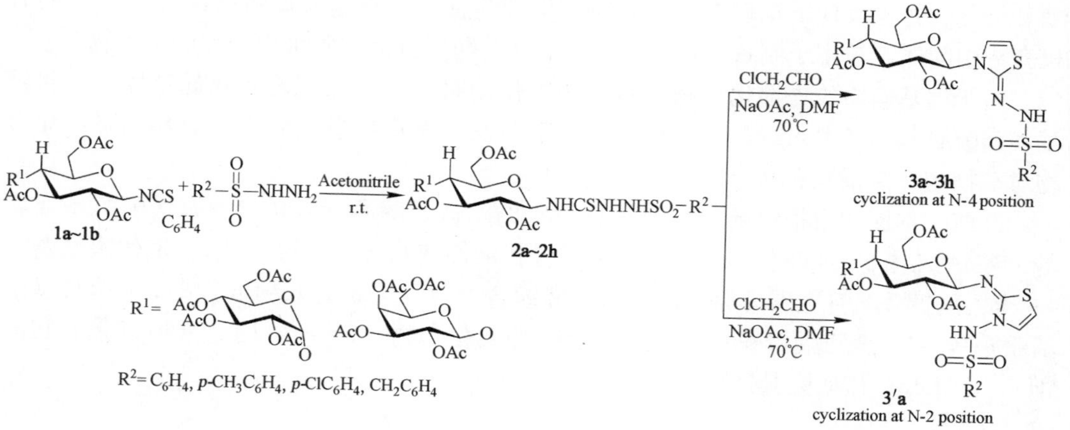

Scheme　1

2－对甲苯磺酰基亚肼基－3－（七－*O*－乙酰基－*β*－麦芽糖基）－2，3－二氢噻唑（**3b**）：收率66%，m. p. 133～135℃；^{1}H NMR（$CDCl_3$）*δ*：1.99～2.09（m，21H，7×OAc），2.45（s，3H，CH_3），3.94～5.62（m，14H，maltosyl－H），6.04（d，*J*＝4.8Hz，H－5），6.58（d，*J*＝4.8Hz，H－4），6.64（s，1H，NH），7.36（d，*J*＝8.0Hz，2H，Ar—H），7.80（d，*J*＝8.0Hz，2H，Ar—H）；^{13}C NMR（$CDCl_3$）*δ*：20.45，20.79，20.88，20.99，21.03（7C，7×CH_3），21.91（$ArCH_3$），61.62，62.87（2C，2×CH_2O），68.11，68.86，69.45，70.21，70.40，72.72，75.23，

75.41，95.97（9C，9 × sugar ring），81.08（C′ - 1），103.10（C - 5），123.06（C - 4），128.65，129.60，134.84，144.38（6C，6 × Ar—C），169.32，169.68，170.02，170.25，170.32，170.56（7C，7 × C ═O），173.44（C - 2）。Anal. calcd for $C_{36}H_{45}N_3O_{19}S_2$：C 48.70，H 5.11，N 4.73；found C 48.74，H 5.34，N 4.97。

2 - 对氯苯磺酰基亚肼基 - 3 - （七 - *O* - 乙酰基 - *β* - 麦芽糖基） - 2，3 - 二氢噻唑（**3c**）：收率 76%，m. p. 139 ~ 141℃；^{1}H NMR（$CDCl_3$）*δ*：1.94，1.96，1.97，1.98，2.03，2.04，2.06（7s，21H，7 × OAc），3.81 ~ 5.59（m，14H，maltosyl - H），6.02（d，*J* = 4.8Hz，H - 5），6.57（d，*J* = 4.8Hz，H - 4），7.02（s，1H，NH），7.52（d，*J* = 8.8Hz，2H，ArH），7.83（d，*J* = 8.8Hz，2H，ArH）；^{13}C NMR（$CDCl_3$）*δ*：20.38，20.73，20.93，20.98，21.17（7C，7 × CH_3），60.53，61.65（2C，2 × CH_2O），62.87，68.16，68.84，69.43，70.20，70.42，72.65，75.36，95.87（9C，9 × sugar ring），81.03（C′ - 1），102.85（C - 5），123.02（C - 4），129.17，130.14，136.42，139.91（6C，6 × Ar—C），169.25，169.69，169.95，170.28，170.52（7C，7 × C ═O），173.79（C - 2）。Anal. calcd for $C_{35}H_{42}ClN_3O_{19}S_2$：C 46.28，H 4.66，N 4.63；found C 46.19，H 4.76，N 4.78。

2 - 苯甲磺酰基亚肼基 - 3 - （七 - *O* - 乙酰基 - *β* - 麦芽糖基） - 2，3 - 二氢噻唑（**3d**）：收率 62%，m. p. 156 ~ 158℃；^{1}H NMR（$CDCl_3$）*δ*：1.94，1.95，1.97，1.98，2.01，2.04，2.09（7s，21H，7 × OAc），4.49（s，2H，CH_2），4.07 ~ 5.76（m，14H，maltosyl - H），6.01（d，*J* = 4.8Hz，H - 5），6.56（s，1H，NH），6.59（d，*J* = 4.8Hz，H - 4），7.35 ~ 7.37（m，2H，ArH），7.38 ~ 7.40（m，1H，Ar—H），7.44 ~ 7.46（m，2H，ArH）；^{13}C NMR（$CDCl_3$）*δ*：20.50，20.74，20.85，20.98（7C，7 × CH_3），54.30（$ArCH_2$），61.64，62.91（2C，2 × CH_2O），68.13，68.87，69.37，70.21，70.78，72.99，75.23，75.49，96.06（9C，9 × sugar ring），81.25（C′ - 1），102.76（C - 5），123.17（C - 4），128.96，129.19，129.48，131.06（6C，6 × Ar—C），169.58，170.03，170.08，170.56，170.67，170.73（7C，7 × C ═O），173.55（C - 2）。Anal. calcd for $C_{36}H_{45}N_3O_{19}S_2$：C 48.70，H 5.11，N 4.73；found C 48.71，H 4.97，N 4.88。

2 - 苯磺酰基亚肼基 - 3 - （七 - *O* - 乙酰基 - *β* - 乳糖基） - 2，3 - 二氢噻唑（**3e**）：收率 71%，m. p. 121 ~ 123℃；^{1}H NMR（$CDCl_3$）*δ*：1.89 ~ 2.08（m，21H，7 × OAc），3.67 ~ 5.51（m，14H，lactosyl - H），6.08（d，*J* = 4.8Hz，H - 5），6.63（d，*J* = 4.8Hz，H - 4），6.95（s，1H，NH），7.50 ~ 7.52（m，2H，ArH），7.58 ~ 7.60（m，1H，ArH），7.85 ~ 7.88（m，2H，ArH）；^{13}C NMR（$CDCl_3$）*δ*：20.47，20.67，20.79，20.90，20.98（7C，7 × CH_3），60.99，62.18（2C，2 × CH_2O），66.85，69.28，70.05，70.85，71.12，73.05，75.77，76.21，101.30（9C，9 × sugar ring），81.12（C′ - 1），102.50（C - 5），122.71（C - 4），128.58，128.91，133.54，137.82（6C，6 × Ar—C），169.31，169.68，169.99，170.26，170.32，170.56（7C，7 × C ═O），173.30（C - 2）。Anal. calcd for $C_{35}H_{43}N_3O_{19}S_2$：C 48.11，H 4.96，N 4.81；found C 48.12，H 5.18，N 4.63。

2 - 对甲苯磺酰基亚肼基 - 3 - （七 - *O* - 乙酰基 - *β* - 乳糖基） - 2，3 - 二氢噻唑（**3f**）：收率 64%，m. p. 130 ~ 132℃；^{1}H NMR（$CDCl_3$）*δ*：1.90，1.98，1.99，2.00，2.01，2.02，2.10（7s，21H，7 × OAc），2.42（s，3H，CH_3），3.81 ~ 5.54（m，

14H, lactosyl－H), 6.01 (d, J=4.8Hz, H－5), 6.57 (d, J=4.8Hz, H－4), 6.79 (s, 1H, NH), 7.30 (d, J=8.0Hz, 2H, ArH), 7.75 (d, J=8.0Hz, 2H, ArH); ^{13}C NMR ($CDCl_3$) δ: 20.49, 20.68, 20.80, 20.92, 21.00 (7C, 7 × CH_3), 21.87 ($ArCH_3$), 60.98, 62.20 (2C, 2 × CH_2O), 66.84, 69.30, 70.06, 70.88, 71.12, 73.12, 75.75, 76.29, 101.39 (9C, 9×sugar ring), 81.06 (C′－1), 102.71 (C－5), 123.01 (C－4), 128.56, 129.56, 134.90, 144.39 (6C, 6×Ar—C), 169.32, 169.68, 170.02, 170.25, 170.32, 170.56 (7C, 7×C ═O), 173.44 (C－2)。Anal. calcd for $C_{36}H_{45}N_3O_{19}S_2$: C 48.70, H 5.11, N 4.73; found C 48.52, H 5.02, N 4.51。

2－对氯苯磺酰基亚肼基－3－(七－O－乙酰基－β－乳糖基)－2, 3－二氢噻唑(**3g**)：收率77%，m.p. 141～143℃；^1H NMR ($CDCl_3$) δ: 1.92, 1.99, 2.00, 2.01, 2.03, 2.04, 2.12 (7s, 21H, 7×OAc), 3.82～5.54 (m, 14H, lactosyl－H), 6.04 (d, J=4.8Hz, H－5), 6.59 (d, J=4.8Hz, H－4), 6.88 (s, 1H, NH), 7.51 (d, J=8.8Hz, 2H, ArH), 7.83 (d, J=8.8Hz, 2H, ArH); ^{13}C NMR ($CDCl_3$) δ: 20.50, 20.71, 20.83, 20.92, 21.02 (7C, 7 × CH_3), 61.01, 62.15 (2C, 2 × CH_2O), 66.83, 69.28, 70.08, 70.92, 71.12, 73.01, 75.85, 76.17, 101.35 (9C, 9×sugar ring), 81.17 (C′－1), 102.85 (C－5), 123.18 (C－4), 129.25, 130.07, 136.34, 140.05 (6C, 6 × ArC), 169.25, 169.69, 169.95, 170.28, 170.56 (7C, 7×C ═O), 173.79 (C－2)。Anal. calcd for $C_{35}H_{42}ClN_3O_{19}S_2$: C 46.28, H 4.66, N 4.63; found C 46.12, H 4.83, N 4.38。

2－苯甲磺酰基亚肼基－3－(七－O－乙酰基－β－乳糖基)－2, 3－二氢噻唑(**3h**)：收率61%，m.p. 155～157℃；^1H NMR ($CDCl_3$) δ: 1.89, 1.98, 1.96, 1.99, 2.01, 2.06, 2.08 (7s, 21H, 7 × OAc), 4.44 (s, 2H, CH_2), 3.83～5.36 (m, 14H, lactosyl－H), 6.00 (d, J=4.8Hz, H－5), 6.59 (d, J=4.8Hz, H－4), 6.63 (s, 1H, NH), 7.32～7.34 (m, 2H, ArH), 7.37～7.39 (m, 1H, ArH), 7.41～7.43 (m, 2H, ArH); ^{13}C NMR ($CDCl_3$) δ: 20.53, 20.68, 20.82, 20.93, 21.03 (7C, 7 × CH_3), 54.27 ($ArCH_2$), 61.02, 62.15 (2C, 2 × CH_2O), 66.84, 69.28, 70.41, 70.87, 71.07, 72.99, 75.82, 76.05, 101.24 (9C, 9 × sugar ring), 81.35 (C′－1), 102.68 (C－5), 123.29 (C－4), 128.95, 129.13, 131.09, 129.42, 138.93 (6C, 6×Ar—C), 169.31, 169.76, 170.00, 170.23, 170.31, 170.55 (7C, 7×C ═O), 173.62 (C－2)。Anal. calcd for $C_{36}H_{45}N_3O_{19}S_2$: C 48.70, H 5.11, N 4.73; found C 48.45, H 5.34, N 4.48。

2 结果与讨论

2.1 化合物的合成

中间体1－芳磺酰基－4－全乙酰糖基氨基硫脲的合成以乙腈为溶剂，室温反应3～6h即可以完成，且产率较高（>95%）。

在目标化合物的合成过程中，最初分别以四氢呋喃和氯仿为溶剂，没有得到关环产物。以DMF为溶剂，在乙酸钠的作用下得到了目标化合物。

2.2 图谱解析及反应区域选择性

根据化合物**3e**的^1H－^{13}C HMBC（图1），二氢噻唑环上的H－4及H－5化学位移分别在δ 6.63 及 6.08；C－4 及 C－5 的化学位移分别在δ 122.7 及 102.5。中间体氨基硫脲中 C－3 的化学位移在δ 184；关环后，二氢噻唑 C－2 化学位移移向高场，大部分在δ 173 左右。

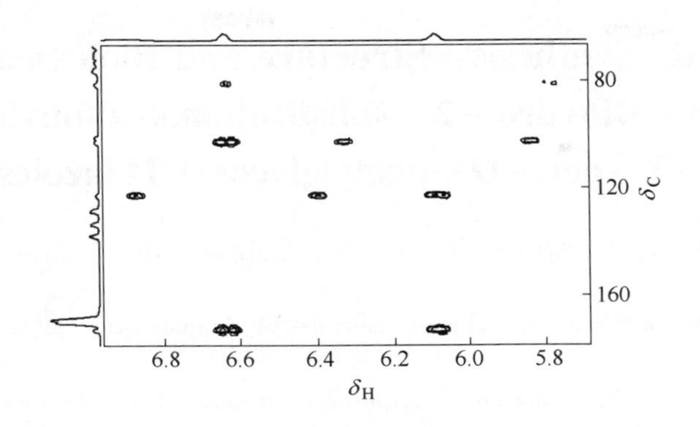

图 1　化合物 **3e** 的 HMBC 谱图

Figure 1　HMBC spectrum of compound **3e**

此前，本课题组曾报道了 1－取代磺酰基－4－全乙酰糖基氨基硫脲与溴乙酸乙酯关环的反应，所得到的目标化合物是氨基硫脲 N－4 关环的产物[8]；1－硫代磷酰肼基－4－全乙酰糖基氨基硫脲与溴乙酸乙酯关环的反应，得到的目标化合物是氨基硫脲 N－2 关环的产物[10]。而 1－取代磺酰基－4－全乙酰糖基氨基硫脲与氯乙醛缩合在理论上也可能得到两种不同的目标化合物，既可以在氨基硫脲 N－2 位形成噻唑环，也可以在 N－4 位形成噻唑环。这两种不同的化合物是同分异构体，具有相同的分子式，也很难通过 ^{1}H NMR 和 ^{13}C NMR 来区分。在反应过程中，TLC 监控没有其它新化合物出现。根据化合物 **3e** 的 ^{1}H－^{13}C HMBC（图 1），存在 H－4 与糖环异头碳原子 C′－1 相关，不存在 H－5 与糖环异头碳原子 C′－1 相关，谱图中 H－4，H－5 与 C－2 存在明显相关，由此可以确定 1－芳磺酰基－4－全乙酰基糖基氨基硫脲与氯乙醛缩合反应，其关环位置为氨基硫脲的 N－4，而非 N－2。

除草活性测试结果表明，在 1.5kg/ha 的剂量下，大多数化合物未表现出明显的除草活性，但化合物 **4b**，**4d** 和 **4h** 对双子叶杂草油菜茎叶处理分别表现出 35.4%，33.2% 和 38.5% 的抑制活性，且对双子叶植物的抑制活性普遍高于对单子叶植物的抑制活性。但总的说来，该类化合物的除草活性仍不太理想。而中间体 1－取代磺酰基－4－全乙酰基糖基氨基硫脲在 1.5kg/ha 的剂量下没有除草活性。

References

[1] Michael, J. G.; Tachel, M. L.; Susan, L. M. *Bioorg. Med. Chem.* 2004, 12, 1029.

[2] Jung, K. Y.; Kim, S. K.; Gao, Z. G. *Bioorg. Med. Chem.* 2004, 12, 613.

[3] Andreani, A.; Burnelli, S.; Granaiola, M. *J. Med. Chem.* 2006, 49, 7897.

[4] Matsuya, Y.; Kawaguchi, T.; Ishihara, K. *Org. Lett.* 2006, 8, 4609.

[5] Wells, L.; Vosseller, K.; Hart, G. W. *Science* 2001, 291, 2376.

[6] Qian, Z. －S.; Zhou, C. －J.; Cao, L. －H. *Prog. Chem.* 2006, 18, 429 (in Chinese).
（钱兆生，周传健，曹玲华，化学进展，2006，18，429.）

[7] Wu, P.; Cao, L. －H. *Chin. J. Org. Chem.* 2005, 25, 1121 (in Chinese).
（吴鹏，曹玲华，有机化学，2005，25，1121.）

[8] Li, Y. -X. ; Wang, S. -H. ; Li, Z. -M. *Carbon Res.* 2006, 341, 2867.
[9] Li, Y. -X. ; Li, Z. -M. ; Zhao, W. -G. *Chin. Chem. Lett.* 2006, 17, 153.
[10] Li, Y. -X. ; Wang, H. -A. ; Li, Z. -M. *Carbohydr. Res.* 2009, 344, 1248.

Regioselectivite Synthesis, Structure and Biological Activity of Novel 2, 3 - Dihydro - 2 - substitutionalsulfonylhydrazono - 3 - per - *O* - *a*cetylglycosyl Thiazoles

Haoan Wang, Yuxin Li, Yiming Li, Suhua Wang, Zhengming Li

(State Key Laboratory of Elemento - Organic Chemistry, Nankai University, Tianjin, 300071)

Abstract A series of novel 2, 3 - dihydro - 2 - phenylsulfonylhydrazono - 3 - per - *O* - acetylglycosyl thiazoles were designed and synthesized via reaction of the chloroacetaldehyde with 1 - arylsulfonyl - 4 - per - *O* - acetylglycosylthiosemicarbazides, which were synthesized via condensation of per - *O* - acetylglycosylisothiocyanate with substituted phenylsulfonylhydrazines. Their chemical structures were characterized by ^{1}H NMR, ^{13}C NMR and elemental analysis. ^{1}H - ^{13}C HMBC of compound **3e** revealed the exclusive regioselectivity of these ring closures toward the N - 4 position of the thiosemicarbazide moiety. The bioassaay data indicated that this type of novel compounds showed some herbicidal activity.

Key words ^{1}H - ^{13}C HMBC correlation spectra; regioselectivity; 2, 3 - dihydrothiazoles

3位苯甲酰腙和亚氨基取代硫脲取代的吲哚满二酮衍生物的合成及对AHAS的抑制活性*

李慧东　商建丽　谭海忠　王建国　李永红　李正名

（南开大学元素有机化学研究所，元素有机化学国家重点实验室，天津　300071）

摘　要　基于前期生物设计AHAS抑制剂的研究，设计合成了15个吲哚满二酮类衍生物，其中9个为新化合物，其结构均经过^{1}H NMR，MS和元素分析确证，并对所有化合物进行了离体和活体活性测试。实验结果表明，这类化合物在体内和体外均具有一定的生物活性，在离体活性测试中，所有化合物在100μg/mL浓度下对拟南芥AHAS均表现出明确的抑制活性，其中化合物**4d**，**4e**，**5a**和**5f**在10μg/mL浓度下仍然对AHAS表现出45%以上的抑制率，但此类化合物除草活性普遍较差。

关键词　AHAS抑制剂　吲哚满二酮类化合物　生物活性

乙酰乳酸合成酶（Acetohydroxyacid synthase，AHAS，EC 2.2.1.6）是催化植物和微生物体内支链氨基酸生物合成过程的第一个关键酶，通过抑制该酶的活性可以阻断支链氨基酸的生物合成[1]，从而达到除草目的，并且对人畜无害，因而AHAS成为设计新颖绿色除草剂的理想作用靶点。

本课题组利用生物合理分子设计方法，基于AHAS的晶体结构，通过虚拟筛选，发现了一些具有新结构特征的AHAS抑制剂[2]，并对部分化合物的类似结构进行了生物活性验证[3,4]。其中吲哚满二酮类化合物具有抗衰老、抗老年性震颤麻痹、抗氧化、抗癌和抗菌等广泛的生物活性[5]。本课题组前期通过购买和简单合成，初步得到一些吲哚满二酮类化合物，并对其进行了离体AHAS抑制活性和除草生物活性研究。结果表明，吲哚满二酮类化合物作为一类新型AHAS抑制剂，其在体外和体内均具有一定的生物活性[4]。为了寻找新的具有高活性的AHAS抑制剂，本文对已有的吲哚满二酮类衍生物的构效关系进行分析，并对其结构进行优化，设计并合成了15个吲哚满二酮类衍生物，通过^{1}H NMR，MS和元素分析对其结构进行了确证，生物活性测定结果表明，大部分化合物在离体测试条件下具有中等的生物活性。

1　实验部分

1.1　仪器及试剂

X－4数字显示显微熔点测定仪（北京泰克仪器有限公司）；Bruker AV－300MHz和Varian 400MHz核磁共振仪（DMSO－d_6为溶剂，TMS为内标）；Yanaco CHN CORDER MT－3型元素分析仪；Thermo Finigan LCQ Advantage液－质联用仪。

所用试剂均为市售分析纯或化学纯，并均经过常规方法处理。

* 原发表于《高等学校化学学报》，2010，30（5）：703－706。基金项目：国家“973”计划项目（批准号：2010CB126103）、国家自然科学青年基金（批准号：20602021）和国家基金委中澳合作特别项目（批准号：20911120022）资助。

1.2 目标化合物的合成

化合物**4**和**5**的合成路线见 Scheme 1。中间体**3**按照文献［6～15］方法合成，所有化合物的物理化学性质及部分未见文献报道化合物的元素分析、^{1}H NMR 和 MS 数据分别列于表 1 和表 2。

a—$Cl_3CCH(OH)_2$，$NH_2OH \cdot HCl$，Na_2SO_4，H_2O；b—H_2SO_4，H_2O；c—Br_2，AcOH；d—thiosemicarbazide or phenylhydrazine，EtOH

Scheme 1　Synthetic routes of compounds **4**a－**4**i and **5**a－**5**f

Table 1　Physical data of compounds 4a－4i and 5a－5f

Compd.	R^1	R^2	m. p. (℃)	Yield (%)	Elemental analysis(%, calcd.)			Appearance
					C	H	N	
4a	NO_2	OH	＞300	64	43.99(44.47)	2.57(2.22)	13.93(13.83)	Bright－yellow crystal
4b	F	F	299～300	60	47.20(47.39)	2.26(2.12)	11.08(11.05)	Bright－yellow crystal
4c①	Br	H	＞300	95				Brown－yellow crystal
4d	NO_2	F	＞300	83	43.78(44.25)	2.46(1.98)	13.49(13.76)	Bright－yellow crystal
4e①	F	OH	＞300	66				Yellow crystal
4f①	Br	OH	＞300	94				Light－grey crystal
4g	NO_2	H	＞300	88	46.29(46.29)	2.52(2.33)	14.07(14.40)	Bright－yellow crystal
4h①	F	H	289～290	56				Yellow crystal
4i①	Br	F	297～299	88				Bright－yellow crystal
5a	NO_2	H	＞300	57	31.80(31.41)	2.22(2.19)	20.21(20.35)	Yellow crystal
5b	F	*p*－Fluorophenyl	＞300	61	43.50(43.81)	2.50(2.21)	13.41(13.62)	Red－brown crystal
5c①	Br	H	271～273	82				Bright－yellow crystal
5d	NO_2	*p*－Fluorophenyl	268～269	60	41.12(41.11)	2.32(2.07)	15.62(15.98)	Brown－yellow crystal
5e	F	H	289～292	58	33.80(34.08)	2.18(1.91)	17.27(17.67)	Yellow crystal
5f	Br	*p*－Fluorophenyl	264～265	65	38.47(38.16)	2.12(1.92)	12.05(11.87)	Brown－yellow crystal

①Corresponding compounds are known ones which have been reported elsewhere.

Table 2 ^{1}H NMR and MS data of new compounds

Compd.	^{1}H NMR and MS
4a	^{1}H NMR (DMSO－d_6, 400MHz), δ: 14.38 (s, 1H, NNH), 12.05 (overlapping, 2H, NH and OH), 8.44 (s, 1H, Ar_1－H6), 8.24 (s, 1H, Ar_1－H4), 8.03 (d, 1H, J=7.6Hz, Ar_2－H6), 7.51 (t, 1H, J=7.6Hz, Ar_2－H4), 7.05～6.99 (m, 2H, Ar_2－H3 and Ar_2－H5); ESI－MS, m/z: 404 ([M－H]$^-$)
4b	^{1}H NMR (DMSO－d_6, 400MHz), δ: 13.78 (s, 1H, NNH), 11.64 (s, 1H, NH), 8.02 (s, 1H, Ar_1－H4), 7.77～7.72 (m, 1H, Ar_2－H4), 7.62～7.60 (d, 1H, J=9.2Hz, Ar_2－H6), 7.50～7.43 (m, 3H, Ar_2－H3, Ar_2－H5 and Ar_1－H6); ESI－MS, m/z: 379 ([M－H]$^-$)
4d	^{1}H NMR (DMSO－d_6, 400MHz), δ: 13.43 (s, 1H, NNH), 12.10 (s, 1H, NH), 8.31 (s, 1H, Ar_1－H6), 8.04 (s, 1H, Ar_1－H4), 7.87 (s, 1H, Ar_2－H6), 7.62～7.57 (m, 1H, Ar_2－H4), 7.36～7.28 (m, 2H, Ar_2－H3 and Ar_2－H5); ESI－MS, m/z: 406 ([M－H]$^-$)
4g	^{1}H NMR (DMSO－d_6, 400MHz), δ: 13.67 (s, 1H, NNH), 12.32 (s, 1H, NH), 8.47 (s, 1H, Ar_1－H6), 8.24 (s, 1H, Ar_1－H4), 7.93 (d, 2H, J=7.6Hz, Ar_2－H2 and Ar_2－H6), 7.75 (t, 1H, J=7.6Hz, Ar_2－H4), 7.66 (t, 2H, J=7.6Hz, Ar_2－H3 and Ar_2－H5); ESI－MS, m/z: 387 ([M－H]$^-$)
5a	^{1}H NMR (DMSO－d_6, 400MHz), δ: 12.17 (s, 1H, NNH), 12.14 (s, 1H, NH), 9.27 (s, 1H, NH_2), 9.11 (s, 1H, NH_2), 8.62 (d, 1H, J=2.4Hz, Ar_1－H6), 8.42 (d, 1H, J=2.4Hz, Ar_1－H4); ESI－MS, m/z: 343 ([M－H]$^-$)
5b	^{1}H NMR (DMSO－d_6, 400MHz), δ: 12.62 (s, 1H, NNH), 11.60 (s, 1H, NH), 10.91 (s, 1H, SCNH), 7.67 (dd, 1H, J=8, 2.4Hz, Ar_1－H4), 7.64～7.55 (m, 3H, Ar_2－H3, Ar_2－H5 and Ar_1－H6), 7.28 (t, 2H, J=8.8Hz, Ar_2－H2 and Ar_2－H6); ESI－MS, m/z: 410 ([M－H]$^-$)
5d	^{1}H NMR (DMSO－d_6, 300MHz), δ: 12.53 (s, 1H, NNH), 12.21 (s, 1H, NH), 11.14 (s, 1H, SCNH), 8.69 (d, 1H, J=2.1Hz, Ar_1－H6), 8.46 (d, 1H, J=2.1Hz, Ar_1－H4), 7.60～7.56 (m, 2H, Ar_2－H3 and Ar_2－H5), 7.30 (t, 2H, J=7.8Hz, Ar_2－H2 and Ar_2－H6); ESI－MS, m/z: 336 ([M－H]$^-$)
5e	^{1}H NMR (DMSO－d_6, 400MHz), δ: 12.29 (s, 1H, NNH), 11.53 (s, 1H, NH), 9.20 (s, 1H, NH_2), 8.84 (s, 1H, NH_2), 7.56～7.52 (m, 2H, Ar_1－H4 and Ar_1－H6); ESI－MS, m/z: 316 ([M－H]$^-$)
5f	^{1}H NMR (DMSO－d_6, 300MHz), δ: 12.56 (s, 1H, NNH), 11.70 (s, 1H, NH), 10.95 (s, 1H, SCNH), 8.01 (d, 1H, J=1.8Hz, Ar_1－H6), 7.82 (d, 1H, J=1.8Hz, Ar_1－H4), 7.61～7.56 (m, 2H, Ar_2－H3 and Ar_2－H5), 7.29 (t, 2H, J=7.8Hz, Ar_2－H2 and Ar_2－H6); ESI－MS, m/z: 471 ([M－H]$^-$)

1.3 目标化合物离体和活体生物活性的测定

用油菜平皿法和稗草小杯法[16]先测定了化合物的除草活性，然后按照文献［17］方法测定了化合物抑制拟南芥 AHAS 酶的离体抑制活性。

2 结果与讨论

基于本课题组前期对吲哚满二酮类化合物的构效关系研究，本文在苯环5位和7位引入硝基、氟及溴等电负性较大的基团，同时利用邻位羟基、氟取代的苯甲酰肼或氨基硫脲修饰3位的羰基，进一步考察3位附近对应区域的静电场和立体场对该类化合物活性的影响，合成了15个3位苯甲酰腙和硫代酰腙取代的吲哚满二酮类衍生物，并以本课题组创制研发的超高效除草剂单嘧磺隆（Monosulfuron）为对照药对其进行了生物活性测试，结果见表3。

Table 3 Bioactivities（inhibition,%）of compounds 4a－4i and 5a－5f in vivo and in vitro

Compd.	AHAS		*Brassica campestris* root		*Echinochloa crusgalli* cup①	
	100μg/mL	10μg/mL	100μg/mL	10μg/mL	100μg/mL	10μg/mL
4a	85	21	16.7	0	0	0
4b	73	3	0	0	10.0	0
4c	89	25	13.1	0	5.0	0
4d	86	47	0	0	0	0
4e	98	47	0	0	0	0
4f	82	38	0	0	0	0
4g	51	0	0	0	0	0
4h	97	21	5.7	0	5.0	0
4i	53	47	0	0	0	0
5a	97	55	0	0	0	0
5b	67	36	13.1	0	5.0	0
5c	85	17	0	0	5.0	0
5d	78	35	0	0	5.0	0
5e	99	33	0	0	0	0
5f	81	49	0	0	10.0	0
C②	97	71	77.4	65.2	—	—

①Not sensitive to *E. crusgalli*, the test has not been done;

②Control herbicide, commercial sulfonylurea herbicide Monosulfuron.

离体活性测试表明，所有化合物在100μg/mL浓度下对AHAS的抑制率均在50%以上，表现出该类化合物对AHAS明确的抑制作用，其中化合物**4**e，**4**h，**5**a和**5**e在100μg/mL浓度下对AHAS的抑制率均在95%以上；当抑制剂浓度降低到10μg/mL时，化合物**4**d，**4**e，**5**a和**5**f对AHAS仍能表现出45%以上的抑制率。单从活性数据直观分析并不能很好地诠释3位附近对应区域的静电场和立体场的直接影响，因此进一步建立CoMFA/CoMSIA等模型来定量描述该类化合物的3D－QSAR，结果将另文发表。活体除草活性测试结果表明，此类化合物的生物活性较低，与离体活性不一致。分析其原因可能是该类化合物水溶性不好，疏水常数过大，造成在配制抑制液溶液时溶解困难，同时在药物作用活体上的吸收及传导效果并不是很好。

参考文献

[1] Duggleby R. G., Pang S. S. J. Biochem. Molec. Biol.［J］, 2000, 33: 1－36.

[2] Wang J. G., Xiao Y. J., Li Y. H., *et al.* Bioorg. Med. Chem.［J］, 2007, 15: 374－380.

[3] LI Wen - Ming（李文明），TAN Hai - Zhong（谭海忠），WANG Jian - Guo（王建国），*et al*. Chem. J. Chinese Universities（高等学校化学学报）[J]，2009，30（4）：728 - 730.

[4] TAN Hai - Zhong（谭海忠），LI Hui - Dong（李慧东），WANG Jian - Guo（王建国），*et al*. Chem. J. Chinese Universities（高等学校化学学报）[J]，2009，30（3）：510 - 512.

[5] Surendra N. P.，Sivakumar S.，Mayank J.，*et al*. Acta Pharm. [J]，2005，55：27 - 46.

[6] Milind R.，Frank D. P. J. Med. Chem. [J]，1988，31：1001 - 1005.

[7] Vincent L.，Max R.，Sylvain R. J. Org. Chem. [J]，2000，65：4193 - 4194.

[8] LI Wen（李雯），YOU Qi - Dong（尤启冬）. Journal of Zhengzhou University（郑州大学学报）[J]，2004，25（3）：29 - 32.

[9] LIU Cui - Ying（柳翠英），GE Wei - Ying（葛蔚颖），QIAN Jun（钱俊），*et al*. Chemical Reagent（化学试剂）[J]，2003，25（3）：160 - 162.

[10] LI Ying - Chun（李迎春），XU Li - Jun（徐丽君）. Acta Pharm. Sinica（药学学报）[J]，1990，25（8）：593 - 597.

[11] ZHAO Quan - Qin（赵全芹），WANG Xing - Po（王兴坡），LIU Cui - Ying（柳翠英）. Chemical Reagent（化学试剂）[J]，2001，23（4）：224 - 225.

[12] Kara L. V.，Julie M. L.，Marie R.，*et al*. Bioorg. Med. Chem. [J]，2007，15：931 - 938.

[13] LI Zai - Guo（李在国），WANG Qing - Min（汪清民），HUANG Jun - Min（黄君珉）. Organic Intermediate Preparation，Second Edition（有机中间体制备，第二版）[M]，Beijing：Chemical Industry Press，2001：99，105.

[14] Michael C. P.，Sunil V. P.，Koushik D. S.，*et al*. J. Med. Chem. [J]，2005，48：3045 - 3050.

[15] James R. M.，Dan A. W.，Alan C. W.，*et al*. J. Med. Chem. [J]，1984，27：1565 - 1570.

[16] Wang B. L.，Duggleby R. G.，Li Z. M.，*et al*. Pest. Manag. Sci. [J]，2005，61：407 - 412.

[17] Singh B. K.，Stidham M. A.，Shaner D. L. Anal. Biochem. [J]，1988，171：173 - 179.

Design，Synthesis of 3 - Substituted Benzoyl Hydrazone and 3 - (Substituted Thiourea)yl Isatin Derivatives and Their Inhibitory Ativities of AHAS

Huidong Li，Jianli Shang，Haizhong Tan，Jianguo Wang，Yonghong Li，Zhengming Li

(State Key Laboratory of Elemento - Organic Chemistry，Elemento Organic Chemistry Institute，Nankai University，Tianjin，300071，China)

Abstract Based on the results of our previous bio - rational design of AHAS inhibitors research，15 isatin derivatives were designed and synthesized，anong which 9 compounds are new. Their structures were confirmed *via* ^{1}H NMR，MS and elemental analysis. Preliminary bioassay showed that most of the isatin derivatives exhibited some inhibition aginast AHAS *in vivo* and *in vitro*. All the compounds exhibited good inhibition of *Arabidopsis thaliana* AHAS at the concentration of 100μg/mL. Compounds **4d**，**4e**，**5a** and **5f** showed over 45% inhibition even at 10μg/mL concentration. However，most compounds are week inhibitors *in vivo*，which means that other factors such as hydrophobicity should be considered in the future to enhance the *in vivo* activity.

Key words novel AHAS inhibitor；isatin derivative；biological activity

Synthesis and Biological Activity of Some Novel Trifluoromethyl - Substituted 1,2,4 - Triazole and Bis(1,2,4 - Triazole) Mannich Bases Containing Piperazine Rings*

Baolei Wang,[1,**] Yanxia Shi,[2,**] Yi Ma[1],Xinghai Liu[1],
Yonghong Li[1],Haibin Song[1],Baoju Li[2],Zhengming Li[1]

([1]State Key Laboratory of Elemento - Organic Chemistry,Tianjin Key Laboratory of Pesticide Science,Elemento - Organic Chemistry Institute,Nankai University,Tianjin,300071,China;
[2]Institute of Vegetables and Flowers,Chinese Academy of Agricultural Sciences,
Beijing,100081,China)

Abstract A series of trifluoromethyl - substituted 1,2,4 - triazole Mannich base **6** and bis(1,2,4 - triazole) Mannich base **7** containing pyrimidinylpiperazine rings via the Mannich reaction were synthesized and characterized by infrared(IR), ^{1}H nuclear magnetic resonance(NMR), and elemental analysis. The fungicidal tests indicated that most of compounds **6** and **7** possessed excellent fungicidal activity. Among 19 novel compounds, some showed superiority over commercial fungicides Dimethomorph, Thiophanate - methyl, Iprodione, and Zhongshengmycin. Some compounds also exhibited favorable herbicidal activity in the preliminary studies. On the basis of the comparative molecular field analysis(CoMFA), five novel compounds were subsequently synthesized, their activities were estimated fairly accurately, and compounds **6 - A1** and **7 - A2** displayed good fungicidal activity against *Pseudoperonospora cubensis*(96.9 and 84.9%) as **6h** and **7c**, respectively.

Key words 1,2,4 - triazole; pyrimidinylpiperazine; Mannich base; fungicidal activity; herbicidal activity

INTRODUCTION

The application of agrochemicals to protect vegetable and cereal crops is an established part of conventional agriculture. This has provided healthy crops and increased yields as well as economic benefits for over many years. The main purpose of the research for agrochemicals is to develop novel active compounds with lower application doses and high selectivity and that are environmentally friendly[1,2]. Compounds containing a 1,2,4 - triazole ring are often associated with various useful pesticidal activity[3-6]. In the early 1970s, the azole fungicides were synthesized and used for the protection of various crops. Since the discovery of these fungicides, a variety of other 1,2,4 - triazole compounds have been discovered[7-9]. The introduction of these 1,2,4 - triazole sterol biosynthesis inhibitors represented significant progress in the chemical control of fungal diseases. This class of pesticides includes several excellent systemic fungicides

* Reprinted from *Journal of Agriculture and Food Chemistry*, 2010, 58(9): 5515 - 5522. This work was supported by the National Basic Research Key Program of China(2003CB114406), the Tianjin Natural Science Foundation(08JCYBJC00800 and 07JCYBJC00200), and the Specialized Research Fund for the Doctoral Program of Higher Education(20070055044).

** These authors contributed equally to this work.

with long, protective, and curative activity against a broad spectrum of foliar, root, and seedling diseases caused by many ascomycetes, basidiomycetes, and imperfect fungi[10,11]. On the other hand, some structures with a 1,2,4 – triazole ring also exhibited outstanding herbicidal activity. For example, 3 – amino – 1,2,4 – triazole was introduced as a selective herbicide and has been used successfully in the control of several woody species during the mid – 1950s[12]. Also, in the early 1990s, 3 – trifluoromethyl – 4 – aryl – 1,2,4 – triazole – 5(4*H*) – thiones were reported to possess good herbicidal activity[13]. Furthermore, bioactive compounds possess a trifluoromethyl moiety that might be expected to enhance their biological activity[14-17], which is often considered an important point by researchers for designing various biological molecules. All of these results encouraged us to synthesize some other novel compounds containing such structural moiety.

In our previous work, we reported some interesting Mannich bases derived from 1,2,4 – triazole Schiff base[18] and benzotriazoles containing a pyrimidinylpiperzine ring, which are associated with various useful biological activities[19]. As a continuation of our work, a series of novel Mannich base and bis – Mannich base with trifluoromethyl – 1,2,4 – triazole and pyrimidinylpiperazine moiety were synthesized and their fungicidal and herbicidal activities were investigated in this paper.

MATERIALS AND METHODS

Instruments and Materials. The melting points were determined on a X – 4 binocular microscope melting point apparatus (Beijing Tech Instrument Co., Beijing, China) and were uncorrected. Infrared spectra were recorded on a Nicolet MAGNA – 560 spectrophotometer as KBr tablets. ^{1}H nuclear magnetic resonance (NMR) spectra were measured on a Bruker AC – P500 instrument (400MHz) using tetramethylsilane (TMS) as the internal standard and dimethylsulfoxide (DMSO) – d_6 or $CDCl_3$ as the solvent. Elemental analyses were performed on a Vario EL elemental analyzer. Crystallographic data of the compound were collected on a Rigaku MM – 07 Saturn 724 charge coupled device (CCD) diffractometer. All of the solvents and materials were analytical – grade.

Synthetic Procedures. 4 – Amino – 5 – trifluoromethyl – 4*H* – 1,2,4 – triazole – 3 – thiol **4** was prepared according to the literature[20].

General Synthetic Procedures for 4 – (4,6 – **Disubstituted – pyrimidin** – 2 – **yl) piperazine 2.** 2 – Chloro – 4,6 – disubstituted – pyrimidines **1** were prepared by the reaction of the diazonium salts of 4,6 – disubstituted – pyrimidin – 2 – amines with concentrated HCl and $ZnCl_2$[21]. Compound **2** was prepared according to ref[22], and the method was improved. To a stirred solution of piperazine (45mmol) and K_2CO_3 (16.5mmol) in water (20mL) was added chloropyrimidine **1** (18mmol) in small portions at 50 – 65℃. The mixture was stirred for 1h at 60 – 65℃ and cooled to 35℃. The yellow solid, 1,4 – bis(4,6 – disubstituted – pyrimidin – 2 – yl) piperazine **3**, was filtered off, and the filtrate was then extracted 3 times with chloroform, dried over Na_2SO_4, and evaporated in vacuum to give compound **2**, which was used for the following reactions without further purification.

2a ($R^1 = R^2 = H$): yellow oil, yield 88%. **2b** ($R^1 = H; R^2 = Me$): yellow solid, yield 81%, mp

45 – 48℃. **2c**($R^1 = R^2$ = Me): yellow solid, yield 79%, mp 82 – 84℃.

3a($R^1 = R^2$ = H): yellow solid, yield 4%, mp 273 – 274℃ [literature mp 275 – 278℃[23]]. **3b**(R^1 = H, R^2 = Me): yellow crystal, yield 14%, mp180 – 182℃. **3c**($R^1 = R^2$ = Me): yellow crystal, yield 19%, mp 216 – 218℃.

General Synthetic Procedures for 4 – (Substituted) benzylideneamino – 5 – trifluoromethyl – 4*H* – 1,2,4 – triazole – 3 – thiol 5. Compound **4** (10mmol) and aromatic aldehyde (10.5mmol) were mixed in acetic acid (15mL). After the reaction mixture was stirred and refluxed for 15min, it was cooled to room temperature. The resulting crystals were filtered and washed with ethanol to give Schiff base **5**.

5a(R^3 = H): white crystal, yield 74%, mp 198 – 199℃ [literature mp 200 – 201℃[18]]. ^{1}H NMR (DMSO – d_6, 400MHz) δ: 14.90 (s, 1H, SH), 9.97 (s, 1H, CH), 7.58 – 7.92 (m, 5H, Ph—H).

5b(R^3 = 2 – F): white crystals, yield 82%, mp 192 – 193℃. ^{1}H NMR (DMSO – d_6, 400MHz) δ: 14.93 (s, 1H, SH), 10.44 (s, 1H, CH), 7.41 – 8.03 (m, 4H, Ph—H).

5c(R^3 = 4 – MeO): white crystals, yield 71%, mp 195 – 196℃. ^{1}H NMR (DMSO – d_6, 400MHz) δ: 14.84 (s, 1H, SH), 9.73 (s, 1H, CH), 7.86 (d, *J* = 8.8Hz, 2H, Ph—H), 7.13 (d, *J* = 8.8Hz, 2H, Ph—H), 3.86 (s, 3H, OCH_3).

5d(R^3 = 3,4 – Me_2): white crystals, yield 73%, mp 205 – 206℃. ^{1}H NMR (DMSO – d_6, 400MHz) δ: 14.86 (s, 1H, SH), 9.79 (s, 1H, CH), 7.66 (s, 1H, Ph—H), 7.62 (d, *J* = 8.0Hz, 1H, Ph—H), 7.35 (d, *J* = 8.0Hz, 1H, Ph—H), 2.31 (s, 6H, CH_3).

5e(R^3 = 4 – Cl): white crystals, yield 72%, mp 208 – 209℃. ^{1}H NMR (DMSO – d_6, 400MHz) δ: 14.92 (s, 1H, SH), 10.05 (s, 1H, CH), 7.93 (d, *J* = 8.4Hz, 2H, Ph—H), 7.67 (d, *J* = 8.4Hz, 2H, Ph—H).

5f(R^3 = 2 – NO_2): yellow crystals, yield 89%, mp 201 – 202℃. ^{1}H NMR (DMSO – d_6, 400MHz) δ: 14.96 (s, 1H, SH), 10.88 (s, 1H, CH), 7.87 – 8.23 (m, 4H, Ph—H).

General Synthetic Procedures for 1 – [4 – (4,6 – Disubstituted – pyrimidin – 2 – yl) piperazin – 1 – yl) methyl] – 4 – (substituted) benzylideneamino – 3 – trifluoromethyl – 1*H* – 1, 2, 4 – triazole – 5 (4*H*) – thione 6. Schiff base **5** (1mmol) and 40% formalin (1.2mmol) were dissolved in ethanol (15mL), and the mixture was stirred at room temperature for 10min. A solution of pyrimidylpiperazine **2** (1mmol) in ethanol (2mL) was slowly added dropwise. Then, the reaction mixture was stirred for 2 – 3h and placed in a refrigerator overnight. The resulting precipitate was filtered and recrystallized from ethanol to give Mannich base **6**.

General Synthetic Procedures for 1,1′ – [Piperazin – 1,4 – diylbis – (methylene)] bis[4 – (substituted) benzylideneamino – 3 – trifluoromethyl – 1*H* – 1,2,4 – triazole – 5(4*H*) – thione] 7. Schiff base **5** (1.8mmol) and 40% formalin (2mmol) were dissolved in ethanol (30mL), and the mixture was stirred at room temperature for 10min. Asolution of piperazine (0.9mmol) in ethanol (2mL) was slowly added dropwise. Then, the reaction mixture was stirred for 2 – 3h and placed in a refrigerator overnight. The resulting precipitate was filtered and recrystallized from ethanol to give bis – Mannich base **7**.

Fungicidal Activity Tests. Fungicidal activity of compounds **6** and **7** against *Corynespora cassiicola*, *Pseudomonas syringae* pv. *lachrymans*, *Ascochyta citrallina* Smith, *Pseudoperonospora cubensis*, and *Sclerotinia sclerotiorum* were evaluated according to ref [24], and a potted plant test method was adopted. Four commericial fungicides, Dimethomorph, Thiophanate – methyl, Iprodione, and Zhongshengmycin, were evaluated as controls at the same condition. Germination was conducted by soaking cucumber seeds in water for 2h at 50℃ and then keeping the seeds moist for 24h at 28℃ in an incubator. When the radicles were 0. 5cm, the seeds were grown in plastic pots containing a 1∶1 (v/v) mixture of vermiculite and peat. Cucumber plants used for inoculations were at the stage of two seed leaves. Tested compounds and commercial fungicides were sprayed with a hand spray on the surface of the seed leaves on a fine morning, at the standard concentration of 500μg/mL. After 2h, inoculations of *C. cassiicola*, *A. citrullina* Smith, and *P. cubensis* were carried out by spraying a conidial suspension, inoculation of *P. syringae* pv. *lachrymans* was carried out by spraying a suspension, and inoculation of *S. sclerotiorum* was carried out by spraying a mycelial suspension. The experiment was repeated 4 times. After inoculation, the plants were maintained at 18 – 30℃ [mean temperature of 24℃ and above 80% relative humidity (RH)]. The fungicidal activity were evaluated when the nontreated cucumber plant (blank) fully developed symptoms. The area of inoculated treated leaves covered by disease symptoms was assessed and compared to that of nontreated ones to determine the average disease index[24]. The relative control efficacy of compounds compared to the blank assay was calculated via the following equation:

$$\text{relative control efficacy}(\%) = [(CK - PT)/CK] \times 100\%$$

where CK is the average disease index during the blank assay and PT is the average disease index after treatment during testing.

Herbicidal Activity Tests. *Inhibition of the Root Growth of Rape (Brassica campestris)*. The evaluated compounds were dissolved in water and emulsified if necessary. Rape seeds were soaked in distilled water for 4h before being placed on a filter paper in a 6cm Petri plate, to which 2mL of inhibitor solution had been added in advance. Usually, 15 seeds were used on each plate. The plate was placed in a dark room and allowed to germinate for 65h at 28 ± 1℃. The lengths of 10 rape roots selected from each plate were measured, and the means were calculated. The check test was carried out in distilled water only. The percentage of the inhibition was calculated.

Inhibition of the Seedling Growth of Barnyardgrass (Echinochloa crusgalli). The evaluated compounds were dissolved in water and emulsified if necessary. A total of 10 barnyardgrass seeds were placed into a 50mL cup covered with a layer of glass beads and a piece of filter paper at the bottom, to which 5mL of inhibitor solution had been added in advance. The cup was placed in a bright room and allowed to germinate for 65h at 28 ± 1℃. The heights of seedlings of above – ground plant parts from each cup were measured, and the means were calculated. The check test was carried out in distilled water only. The percentage of the inhibition was calculated.

Crystal Structure Determination. Compound **6i** was dissolved in hot alcohol, and the resulting solution was allowed to stand in air at room temperature to give a single crystal of compound **6i**. A yellow crystal of compound **6i** suitable for X – ray diffraction with dimensions of 0. 16 × 0. 14 × 0. 10mm was mounted on a Rigaku MM – 07 Saturn 724 CCD diffractometer with Mo $K\alpha$ radiation (λ = 0. 71073Å) for data collection. A total of 17003 reflections were collected in the range of 2. 01 < θ < 25. 02 using phi and scan modes at 113 (2) K, of which 4363 were independent with R_{int} = 0.0614. All calculations were refined anisotropically (SHELXS - 97). All hydrogen atoms were located from a difference Fourier map, placed at calculated positions, and included in the refinements in the riding mode with isotropic thermal parameters. The compound crystallizes in space group $Pca2_1$ of the orthorhombic system with cell parameters: a = 20.312(4) Å, b = 14.395(3) Å, c = 8.8173(18) Å, α = 90°, β = 90°, γ = 90°, V = 2578.2(9) $Å^3$, Z = 4, D_c = 1.367mg/m^3, μ = 0.186mm^{-1}, and $F(000)$ = 1100. The final refinement converged at R = 0.0603 and wR = 0.1494 for 3552 observed reflections with $I > 2\sigma(I)$, where $W = 1/[\sigma^2(F_o^2) + (0.0980P)^2 + 0.0000P]$ with $P = (F_o^2 + 2F_c^2)/3$, S = 1.073, $(\Delta/\sigma)_{max} <$ 0.0001, $(\Delta\rho)_{max}$ = 0.434 e/$Å^3$, and $(\Delta\rho)_{min}$ = −0.265e/$Å^3$.

Three - Dimensional Quantitative Structure - Activity Relationship (3D QSAR) Analysis. Molecular modeling was performed using SYBYL 6.91 software, and the comparative molecular field analysis (CoMFA) method has been performed according to our previous papers[25,26]. The fungicidal activities of 19 compounds (**6a - 6o** and **7a - 7d**, for training sets compounds) against *P. cubensis* (% *I*) at 500μg/mL used to derive the CoMFA analysis model were listed in Table 7. The activity was expressed in terms of *D* by the formula $D = \log\{[I/(100 - I)] \times \text{MW}\}$, where *I* is the percent control efficacy and MW is the molecular weight of the tested compounds. The compound **6i**, owing to the determination of the crystal structure, was used as a template to build the other molecular structures. Each structure was fully geometry - optimized using a conjugate gradient procedure based on the TRIPOS force field and Gasteiger and Hückel charges. Because these compounds share a common skeleton, 19 atoms marked with an asterisk were used for root - mean - square (rms) fitting onto the corresponding atoms of the template structure. CoMFA steric and electrostatic interaction fields were calculated at each lattice intersection on a regularly spaced grid of 2.0Å. The grid pattern was generated automatically by the SYBYL/CoMFA routine, and an sp^3 carbon atom with a van der Waals radius of 1.52Å and a +1.0 charge was used as the probe to calculate the steric (Lennard - Jones 6 - 12 potential) field energies and electrostatic (Coulombic potential) fields with a distance - dependent dielectric at each lattice point. Values of the steric and electrostatic fields were truncated at 30.0kcal/mol. The CoMFA steric and electrostatic fields generated were scaled by the CoMFA - STD method in SYBYL. The electrostatic fields were ignored at the lattice points with maximal steric interactions. A partial least - squares approach was used to derive the 3D QSAR, in which the CoMFA descriptors were used as independent variables and ED values were used as dependent variables. The cross - validation with the leave - one - out option and the SAMPLS program[27] rather than column filtering was carried out to obtain the optimal number of components to be used in the final analysis. After the optimal number of components was determined, a non - cross - validated analysis was performed without column filtering. The modeling capability (goodness of fit) was judged by the correlation coefficient squared, r^2, and the prediction capability (goodness of prediction) was indicated by the cross - validated r^2 (q^2).

RESULTS AND DISCUSSION

Synthesis. The synthesis procedures for compounds **6** were shown in Scheme 1, and the synthesis procedures for compounds **7** were shown in Scheme 2. According to the method described in ref[22] for the synthesis of required 4 – (4,6 – disubstituted – pyrimidin – 2 – yl) piperazine **2**, a small amount of 1,4 – bis(4,6 – disubstituted – pyrimidin – 2 – yl) piperazine **3** was also obtained, which was not mentioned in the literature. Compound **3** was further confirmed by ^{1}H NMR spectra. In reference to an acetic acid solvent method reported by Wu et al. for the condensation of amine and aldehyde[28], Schiff base **5**, namely, 4 – amino – 5 – trifluoromethyl – 4*H* – 1,2,4 – trizole – 3 – thiol, was prepared successfully. The Mannich reaction of compound **5** with formaldehyde and pyrimidyl – piperazine **2** in ethanol at room temperature led to novel trifluoromethyl – substituted 1,2,4 – triazole Mannich base **6**. In a 2∶1 molar ratio of Schiff base **5** and piperazine, novel trifluoromethyl – substituted bis(1,2,4 – triazole) Mannich base **7** was prepared conveniently in 60% – 88% yield using the same procedures for compound **6**. Compounds **6** and **7** were identified by ^{1}H NMR and infrared(IR) spectra(Tables 2 and 3). The measured elemental analyses were also consistent with the corresponding calculated ones(Table 1).

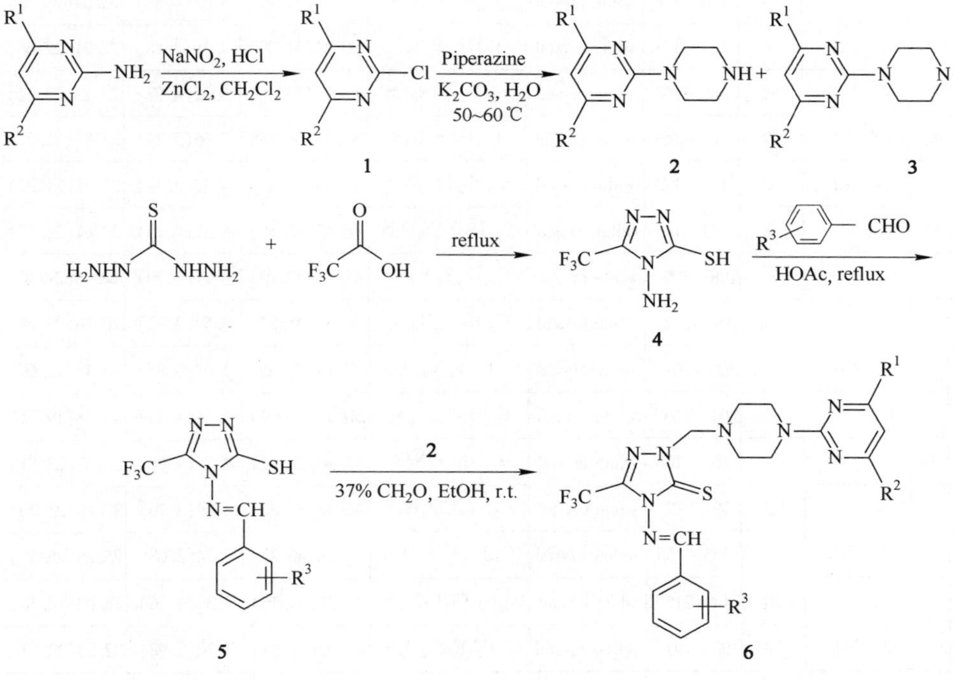

Scheme 1

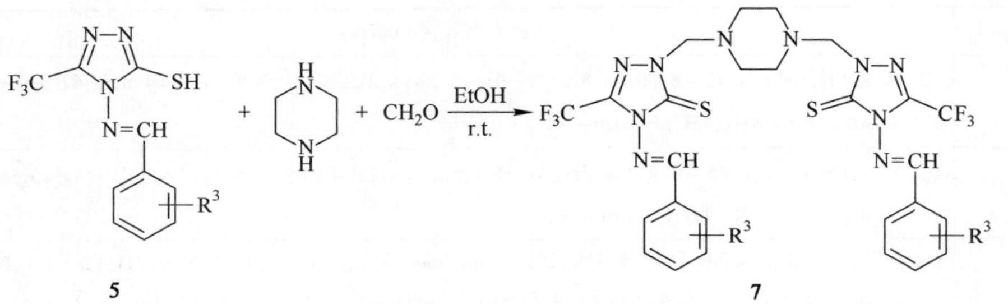

Scheme 2

Table 1 Analytical Data for Compounds 6 and 7

compound	R^1	R^2	R^3	yield (%)	mp(℃)	appearance	formula	elemental analysis (calculated %)		
								C	H	N
6a	H	H	H	48	156 – 158	white crystal	$C_{19}H_{19}F_3N_8S$	50.68(50.89)	4.53(4.27)	25.10(24.99)
6b	H	H	2 – F	65	150 – 152	white crystal	$C_{19}H_{18}F_4N_8S$	48.64(48.92)	4.21(3.89)	23.88(24.02)
6c	H	H	4 – MeO	64	154 – 156	white crystal	$C_{20}H_{21}F_3N_8OS$	49.96(50.20)	4.65(4.42)	23.27(23.42)
6d	H	Me	H	44	138 – 140	white crystal	$C_{20}H_{21}F_3N_8S$	51.67(51.94)	4.94(4.58)	24.38(24.23)
6e	H	Me	2 – F	42	113 – 115	white crystal	$C_{20}H_{20}F_4N_8S$	49.96(49.99)	4.49(4.20)	23.56(23.32)
6f	H	Me	4 – MeO	72	163 – 164	white crystal	$C_{21}H_{23}F_3N_8OS$	50.97(51.21)	4.98(4.71)	23.02(22.75)
6g	H	Me	3,4 – Me_2	58	158 – 159	white crystal	$C_{22}H_{25}F_3N_8S$	53.87(53.87)	5.28(5.14)	23.24(22.84)
6h	H	Me	4 – Cl	48	160 – 162	white crystal	$C_{20}H_{20}ClF_3N_8S$	48.05(48.34)	4.29(4.06)	22.69(22.55)
6i	H	Me	2 – NO_2	58	88 – 89	yellow crystal	$C_{20}H_{20}F_3N_9O_2S$	47.08(47.33)	3.81(3.97)	24.53(24.84)
6j	Me	Me	H	54	154 – 156	white crystal	$C_{21}H_{23}F_3N_8S$	52.91(52.93)	4.65(4.86)	23.54(23.51)
6k	Me	Me	2 – F	46	153 – 154	white crystal	$C_{21}H_{22}F_4N_8S$	50.26(51.00)	4.31(4.48)	22.71(22.66)
6l	Me	Me	4 – MeO	52	160 – 161	white crystal	$C_{22}H_{25}F_3N_8OS$	52.12(52.16)	5.02(4.97)	22.23(22.12)
6m	Me	Me	3,4 – Me_2	47	162 – 164	white crystal	$C_{23}H_{27}F_3N_8S$	54.33(54.75)	5.46(5.39)	21.81(22.21)
6n	Me	Me	4 – Cl	63	171 – 173	white crystal	$C_{21}H_{22}ClF_3N_8S$	48.95(49.36)	4.72(4.34)	21.61(21.93)
6o	Me	Me	2 – NO_2	48	143 – 145	yellow crystal	$C_{21}H_{22}F_3N_9O_2S$	48.43(48.36)	4.32(4.25)	23.87(24.17)
7a			H	60	198 – 199①	white crystal	$C_{26}H_{24}F_6N_{10}S_2$	47.64(47.70)	4.01(3.70)	20.96(21.40)
7b			2 – F	63	201 – 202①	white crystal	$C_{26}H_{22}F_8N_{10}S_2$	45.14(45.22)	3.50(3.21)	20.16(20.28)
7c			4 – MeO	75	200 – 201①	white crystal	$C_{28}H_{28}F_6N_{10}O_2S_2$	47.13(47.05)	4.08(3.95)	19.45(19.60)
7d			3,4 – Me_2	80	204 – 206①	white crystal	$C_{30}H_{32}F_6N_{10}S_2$	50.42(50.69)	4.38(4.54)	19.38(19.71)
6 – A1	H	H	3,4 – Me_2	52	166 – 168	white crystal	$C_{21}H_{23}F_3N_8S$	52.90(52.93)	4.78(4.86)	23.67(23.51)
6 – A2	H	H	4 – Cl	72	191 – 192	white crystal	$C_{19}H_{18}ClF_3N_8S$	46.92(47.26)	3.84(3.76)	22.88(23.20)
6 – A3	H	H	2 – NO_2	63	153 – 155	yellow crystal	$C_{19}H_{18}F_3N_9O_2S$	45.98(46.25)	3.74(3.68)	25.26(25.55)
7 – A1			4 – Cl	88	214 – 215①	white crystal	$C_{26}H_{22}Cl_2F_6N_{10}S_2$	42.91(43.16)	3.30(3.06)	18.80(19.36)
7 – A2			2 – NO_2	72	196 – 197①	yellow crystal	$C_{26}H_{22}F_6N_{12}O_4S_2$	41.92(41.94)	3.20(2.98)	22.23(22.57)

①Decomposed.

Table 2 ^{1}H NMR Spectral Data of Compounds 6 and 7

compound	δ($CDCl_3$,400MHz)
6a	10.34(s,1H,CH),6.46 – 8.29(m,8H,Ar—H),5.28(s,2H,CH_2),3.87(t,J=4.8Hz,4H,piperazine – H),2.90(t,J=4.8Hz,4H,piperazine – H)
6b	10.70(s,1H,CH),6.45 – 8.29(m,7H,Ar—H),5.28(s,2H,CH_2),3.86(t,J=4.8Hz,4H,piperazine – H),2.90(t,J=4.8Hz,4H,piperazine – H)
6c	10.07(s,1H,CH),8.28(d,J=4.8Hz,2H,pyrimidine – H),7.82(d,J=8.8Hz,2H,Ph—H),6.98(d,J=8.8Hz,2H,Ph—H),6.47(t,J=4.8Hz,1H,pyrimidine – H),5.28(s,2H,CH_2),3.88(s,3H,OCH_3),3.86(t,J=4.8Hz,4H,piperazine – H),2.90(t,J=4.8Hz,4H,piperazine – H)

Table 2 (continued)

compound	δ($CDCl_3$,400MHz)
6d	10. 35(s,1H,CH),8. 14(d,J = 4. 4Hz,1H,pyrimidine – H),7. 47 – 7. 88(m,5H,Ph—H),6. 35(d,J = 4. 4Hz,1H,pyrimidine – H),5. 29(s,2H,CH_2),3. 87(t,J = 4. 0Hz,4H,piperazine – H),2. 89(t,J = 4. 0Hz,4H,piperazine – H),2. 31(s,3H,CH_3)
6e	10. 70(s,1H,CH),8. 13(d,J = 4. 4Hz,1H,pyrimidine – H),7. 15 – 8. 09(m,4H,Ph—H),6. 35(d,J = 4. 4Hz,1H,pyrimidine – H),5. 28(s,2H,CH_2),3. 88(bs,4H,piperazine – H),2. 89(bs,4H,piperazine – H),2. 31(s,3H,CH_3)
6f	10. 08(s,1H,CH),8. 13(d,J = 4. 8Hz,1H,pyrimidine – H),7. 82(d,J = 8. 0Hz,2H,Ph—H),6. 99(d,J = 8. 0Hz,2H,Ph—H),6. 35(d,J = 4. 8Hz,1H,pyrimidine – H),5. 28(s,2H,CH_2),3. 88(bs,7H,CH_3O + piperazine – H),2. 89(bs,4H,piperazine – H),2. 31(s,3H,CH_3)
6g	10. 11(s,1H,CH),8. 13(d,J = 4. 8Hz,1H,pyrimidine – H),7. 63(s,1H,Ph—H),7. 60(d,J = 8. 0Hz,1H,Ph—H),7. 24(d,J = 8. 0Hz,1H,Ph—H),6. 35(d,J = 4. 8Hz,1H,pyrimidine – H),5. 28(s,2H,CH_2),3. 87(bs,4H,piperazine – H),2. 89(bs,4H,piperazine – H),2. 33(s,3H,Ph—CH_3),2. 32(s,3H,Ph—CH_3),2. 31(s,3H,pyrimidine – CH_3)
6h	10. 44(s,1H,CH),8. 13(d,J = 4. 8Hz,1H,pyrimidine – H),7. 80(d,J = 8. 0Hz,2H,Ph—H),7. 47(d,J = 8. 0Hz,2H,Ph—H),6. 35(d,J = 4. 8Hz,1H,pyrimidine – H),5. 28(s,2H,CH_2),3. 87(bs,4H,piperazine – H),2. 89(bs,4H,piperazine – H),2. 31(s,3H,CH_3)
6i	11. 21(s,1H,CH),7. 70 – 8. 18(m,5H,Ph—H + pyrimidine – H),6. 36(d,J = 4. 8Hz,1H,pyrimidine – H),5. 28(s,2H,CH_2),3. 88(bs,4H,piperazine – H),2. 90(bs,4H,piperazine – H),2. 31(s,3H,CH_3)
6j	10. 35(s,1H,CH),7. 47 – 7. 88(m,5H,Ph—H),6. 25(s,1H,pyrimidine – H),5. 29(s,2H,CH_2),3. 89(t,J = 4. 4Hz,4H,piperazine – H),2. 89(t,J = 4. 4Hz,4H,piperazine – H),2. 26(s,6H,CH_3)
6k	10. 70(s,1H,CH),7. 15 – 8. 09(m,4H,Ph—H),6. 25(s,1H,pyrimidine – H),5. 29(s,2H,CH_2),3. 88(t,J = 4. 0Hz,4H,piperazine – H),2. 89(t,J = 4. 0Hz,4H,piperazine – H),2. 26(s,6H,CH_3)
6l	10. 08(s,1H,CH),7. 83(d,J = 8. 0Hz,2H,Ph—H),6. 99(d,J = 8. 0Hz,2H,Ph—H),6. 25(s,1H,pyrimidine – H),5. 29(s,2H,CH_2),3. 89(bs,7H,CH_3O + piperazine – H),2. 88(bs,4H,piperazine – H),2. 26(s,6H,CH_3)
6m	10. 11(s,1H,CH),7. 63(s,1H,Ph—H),7. 60(d,J = 8. 0Hz,1H,Ph—H),7. 25(d,J = 8. 0Hz,1H,Ph—H),6. 25(s,1H,pyrimidine – H),5. 28(s,2H,CH_2),3. 88(t,J = 4. 4Hz,4H,piperazine – H),2. 88(t,J = 4. 4Hz,4H,piperazine – H),2. 34(s,3H,Ph—CH_3),2. 33(s,3H,Ph—CH_3),2. 26(s,6H,pyrimidine – CH_3)
6n	10. 44(s,1H,CH),7. 80(d,J = 8. 0Hz,2H,Ph—H),7. 47(d,J = 8. 0Hz,2H,Ph—H),6. 25(s,1H,pyrimidine – H),5. 28(s,2H,CH_2),3. 88(bs,4H,piperazine – H),2. 88(bs,4H,piperazine – H),2. 26(s,6H,CH_3)
6o	11. 21(s,1H,CH),7. 70 – 8. 18(m,4H,Ph—H),6. 25(s,1H,pyrimidine – H),5. 29(s,2H,CH_2),3. 89(bs,4H,piperazine – H),2. 89(bs,4H,piperazine – H),2. 26(s,6H,CH_3)
7a	10. 35(s,2H,CH),7. 48 – 7. 59(m,10H,Ph—H),5. 19(s,4H,CH_2),2. 88(s,8H,piperazine – H)
7b	10. 71(s,2H,CH),7. 15 – 8. 10(m,8H,Ph—H),5. 18(s,4H,CH_2),2. 88(s,8H,piperazine – H)
7c	10. 08(s,2H,CH),7. 83(d,J = 8. 8Hz,4H,Ph—H),6. 99(d,J = 8. 8Hz,4H,Ph—H),5. 18(s,4H,CH_2),3. 89(s,6H,OCH_3),2. 87(s,8H,piperazine – H)

Table 2 (continued)

compound	δ($CDCl_3$, 400MHz)
7d	10.12(s,2H,CH), 7.24 – 7.64(m,6H,Ph—H), 5.18(s,4H,CH_2), 2.88(s,8H,piperazine – H), 2.34(s,6H,CH_3), 2.33(s,6H,CH_3)
6 – A1	10.11(s,1H,CH), 8.28(d,J = 4.8Hz,2H,pyrimidine – H), 7.63(s,1H,Ph—H), 7.60(d,J = 8.0Hz,1H,Ph—H), 7.24(d,J = 8.0Hz,1H,Ph—H), 6.47(t,J = 4.8Hz,1H,pyrimidine – H), 5.28(s,2H,CH_2), 3.87(t,J = 4.8Hz,4H,piperazine – H), 2.90(t,J = 4.8Hz,4H,piperazine – H), 2.33(s,3H,CH_3), 2.32(s,3H,CH_3)
6 – A2	10.44(s,1H,CH), 8.28(d,J = 4.8Hz,2H,pyrimidine – H), 7.80(d,J = 8.8Hz,2H,Ph—H), 7.46(d,J = 8.8Hz,2H,Ph—H), 6.47(t,J = 4.8Hz,1H,pyrimidine – H), 5.28(s,2H,CH_2), 3.87(t,J = 4.4Hz,4H,piperazine – H), 2.90(t,J = 4.4Hz,4H,piperazine – H)
6 – A3	11.22(s,1H,CH), 8.29(d,J = 4.8Hz,2H,pyrimidine – H), 7.70 – 8.19(m,4H,Ph—H), 6.47(t,J = 4.8Hz,1H,pyrimidine – H), 5.28(s,2H,CH_2), 3.87(t,J = 4.4Hz,4H,piperazine – H), 2.91(t,J = 4.4Hz,4H,piperazine – H)
7 – A1	10.45(s,2H,CH), 7.81(d,J = 8.4Hz,4H,Ph—H), 7.47(d,J = 8.4Hz,4H,Ph—H), 5.17(s,4H,CH_2), 2.87(s,8H,piperazine – H)
7 – A2	11.21(s,2H,CH), 7.71 – 8.19(m,8H,Ph—H), 5.18(s,4H,CH_2), 2.89(s,8H,piperazine – H)

Table 3 IR Spectral Data of Compounds 6 and 7

compound	ν(cm^{-1})(KBr)
6a	2860, 2825(C – H), 1587(C=N), 1550, 1506, 1483, 1465(Ar), 1310, 1205(C – F), 1161(C=S)
6d	2980, 2830(C – H), 1578(C=N), 1566, 1489, 1467, 1449(Ar), 1316, 1194(C – F), 1161(C=S)
6j	2941, 2853(C – H), 1577(C=N), 1504, 1467, 1449(Ar), 1311, 1194(C – F), 1156(C=S)
7a	2935, 2822(C – H), 1599(C=N), 1583, 1463, 1452(Ar), 1317, 1193(C – F), 1156(C=S)

Crystal Structure. The structure of compound **6i** was further confirmed by single – crystal X – ray diffraction analysis (Figure 1). From the molecular structure, it can be seen that both groups on the N atoms of the piperazine ring (triazole – CH_2 and pyrimidine) are in the e – bond positions of chair conformation in the sixmembered ring. The dihedral angle between the 2 – nitrobenzene ring and the triazole ring is 32°, which indicates that the two rings are not coplanar in the molecular structure. The X – ray analysis also reveals that, in this typical compound **6i**, the substituted benzene ring and the triazole ring are on the opposite sides of the C=N double bond (Figure 1). The torsion angle of C(6) – C(7) – N(2) – N(3) is 176.52°, which indicates that the C=N double bond is in the *E* configuration.

Fungicidal Activity. The *in vivo* fungicidal results of the Mannich base **6a – 6o** and bis – Mannich base **7a – 7d** against *C. cassiicola*, *P. syringae* pv. *lachrymans*, *A. citrallina* Smith, *P. cubensis*, and *S. sclerotiorum* were listed in Table 4. Most of the compounds showed promising results in inhibiting the mycelial growth of all of the test fungi at a concentration of 500μg/mL. Meanwhile, all of these tested compounds were found safe for the cucumber plants. The comparison of the fungicidal activity of compounds **6** and **7** for five test fungi to those of com-

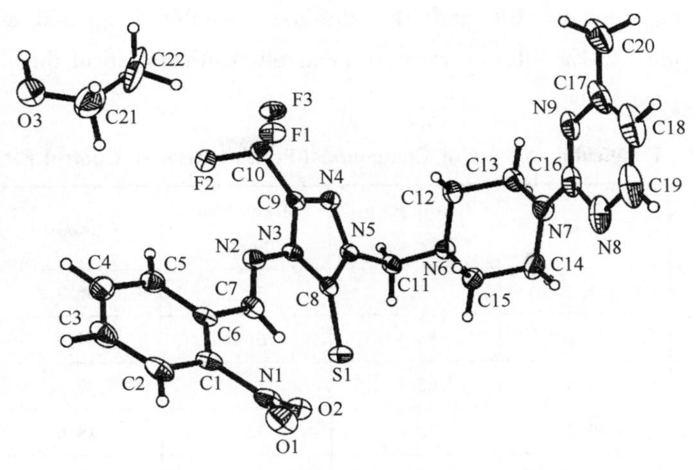

Figure 1 Molecular structure of compound **6i**

mercial fungicides leads to the following conclusions: (a) Compounds **6l**, **6m**, **6n**, **6o**, and **7b** exhibited a significant inhibition effect against *C. cassiicola*, and the fungicidal activities (control efficacy of 83.7% – 94.4%) were higher than those of Thiophanate – methyl and Iprodione. Especially, compound **6n** showed a control efficacy of 94.4%, which was similar to that of the most active fungicide Dimethomorph (95.3%). It can be seen that variances among R^1, R^2, and R^3 can greatly effect the fungicidal activity of compounds against *C. cassiicola*. When $R^1 = R^2 =$ Me and a bulky group at the 4 position of the benzene ring in Mannich base **6** (e. g., compound **6n**, $R^3 = Cl$), there is an apparent increase of fungicidal activity, while the introduction of the electron – withdrawing group in the benzene ring of bis – Mannich base **7** is favorable to the improvement of activity. (b) Compounds **6c**, **6e**, and **6n** possessed efficacy rates 68.1%, 62.3%, and 61.5% against *P. syringae* pv. *lachrymans*, respectively. All of the three compounds were more effective than Dimethomorph, Thiophanate – methyl, and Iprodione. Among them, compound **6c** had almost the same activity as that of Zhongshengmycin against *P. syringae* pv. *lachrymans*. For this fungi, when $R^1 = H$ and $R^2 = Me$ (Mannich base **6**), compounds with an electron – donating group in the benzene ring will affect lower fungicidal activities than those of others. However, the small and electron – donating group in the benzene ring of bis – Mannich base **7** is favorable. (c) Compounds **6l**, **6o**, **7b**, and **7c** held 86.5% – 89.7% efficacy rates against *A. citrallina* Smith, while all of them were less effective than all of the contrasts. It was found that compounds with two methyl groupds in the pyrimidine ring ($R^1 = R^2 = Me$) have a higher level of fungicidal activity than others ($R^1 = R^2 = H$ and $R^1 = H$ and $R^2 = Me$), which indicates that the introduction of hydrophobic groups at 4 and 6 positions of the pyrimidine ring could be an activation process for this group of compounds. (d) All of the compounds exhibited good control efficacy against *P. cubensis* (exceeding 64% efficacy rate), and most of them had higher activity than Thiophanate – methyl, Iprodione, and Zhongshengmycin. It was worthy to note that compounds **6h**, **6l**, **6n**, and **7c**, whose efficacy rates were 91.2%, 87.3%, 91.6%, and 84.5%, respectively, were found to be more effective compared to the most active fungicide Dimethomorph (83.8%) against *P. cubensis*. To further explore the comprehensive structure – activity relationship on the basis of these data, CoMFA was subsequently performed. (e) For

S. sclerotiorum, compounds **6c**, **6h**, and **6k** exhibited favorable fungicidal activity and held 75.3%, 70.3%, and 72.0% efficacy rates, respectively. However, all of them were less effective than all of the contrasts.

Table 4 Fungicidal Activity of Compounds (Percent Relative Control Efficacy)

compound	*C. cassiicola*	*P. syringae* pv *lachrymans*	*A. citrullina* Smith	*P. cubensis*	*S. sclerotiorum*
6a	56.2	55.8	37.1	66.3	41.7
6b	40.3	45.4	61.3	64.9	47.3
6c	73.4	68.1	73.4	79.5	75.3
6d	49.2	20.4	62.4	78.6	36.8
6e	-5.3	62.3	57.8	75.5	52.2
6f	52.8	30.6	67.7	72.5	50.9
6g	38.7	22.0	70.0	69.5	51.9
6h	71.1	21.1	61.3	91.2	70.3
6i	73.8	44.5	44.3	76.8	32.8
6j	22.1	55.0	61.5	74.1	48.0
6k	78.7	56.1	83.5	79.2	72.0
6l	83.7	50.3	87.6	87.3	62.6
6m	85.9	59.5	82.9	67.8	48.9
6n	94.4	61.5	62.8	91.6	38.0
6o	87.8	46.5	88.4	77.2	39.3
7a	70.9	56.4	25.6	78.9	67.6
7b	84.7	6.8	86.5	79.9	25.9
7c	44.1	55.6	89.7	84.5	62.0
7d	67.4	36.7	26.0	78.7	32.7
Dimethomorph	95.3	14.0	98.4	83.8	98.4
Thiophanate - methyl	78.8	54.2	100	68.7	83.4
Iprodione	53.5	38.5	98.4	55.9	97.8
Zhongshengmycin	90.2	69.7	100	62.6	84.4

Herbicidal Activity. As shown in Table 5, most compounds **6** and **7** displayed herbicidal activity based on the rape (*B. campestris*) root and barnyardgrass (*E. crusgalli*) cup tests at a concentration of 100μg/mL. Compounds **6a** - **6c**, **6h**, **6j** - **6l**, and **6n** showed inhibition rates of 66.8% - 88.5% to the root growth of dicotyledonous rape at 100μg/mL. Especially, compounds **6b** and **6k** held by inhibition rates 70.8% and 67.1% at the concentration of 10μg/mL, respectively, which were similar to that of contrast Chlorsulfuron. Against the seedling growth of monocotyledonous barnyardgrass, all of the compounds showed little inhibitory activitity at 10μg/mL, but compounds **6h** exhibited 96.7% inhibition at 100μg/mL. On the whole, these compounds might exhibit less of an effect in comparison to the commercial sulfonylureas against rape root at 1μg/mL [e. g., Chlorsulfuron, 64.2% inhibition; Metsulfuron – methyl, 81.0% inhibition[29]]; however, some showed favorable herbicidal activity in these preliminary studies and might be further optimized to enhance their activity.

Table 5　Herbicidal Activity of Compounds (Percent Inhibition)

compound	rape (*B. campestris*) root test		barnyardgrass (*E. crusgalli*) cup test	
	100μg/mL	10μg/mL	100μg/mL	10μg/mL
6a	66.8	49.2	0	0
6b	74.5	70.8	15.3	9.9
6c	78.8	53.5	18.6	0
6d	12.5	0	15.3	8.7
6e	24.4	0	8.0	0
6f	30.3	0	34.3	15.3
6g	11.2	0	27.8	0
6h	68.6	0	96.7	10.2
6i	0	0	33.9	11.5
6j	71.1	36.9	9.8	0
6k	72.4	67.1	33.2	17.1
6l	67.7	30.6	29.5	0
6m	20.5	6.0	8.7	0
6n	88.5	9.5	36.6	12.3
6o	11.2	0	10.2	0
7a	35.1	0	25.1	16.2
7b	4.3	0	26.6	10.7
7c	29.5	0	25.5	5.5
7d	27.4	0	28	1.5
Chlorsulfuron	80.4	76.0	29.9	8.2

CoMFA Analysis. The CoMFA method is widely used in drug design, because it allows for rapid generation of predictive information of QSAR from the biological activity of newly designed molecules[30]. Starting from the structural alignment of Figure 2, comprehensive CoMFA analyses were performed. The results of these computations were summarized in Table 6. A comparison of experimental and predicted activity by CoMFA for all of the compounds in this study was presented in Table 7. The cross – validated results were assessed by their q^2 value (see the Materials and Methods), where a value above 0.3 indicates that the probability of chance correlation is less than 5% and a value over 0.5 is highly significant. As listed in Table 6, a predictive CoMFA model was established with the cross – validated coefficient $q^2 = 0.526$ and the conventional correlation coefficient $r^2 = 0.913$. The contributions of steric and electrostatic fields are 74.6% and 25.4% as shown in panels a and b of Figure 3, respectively.

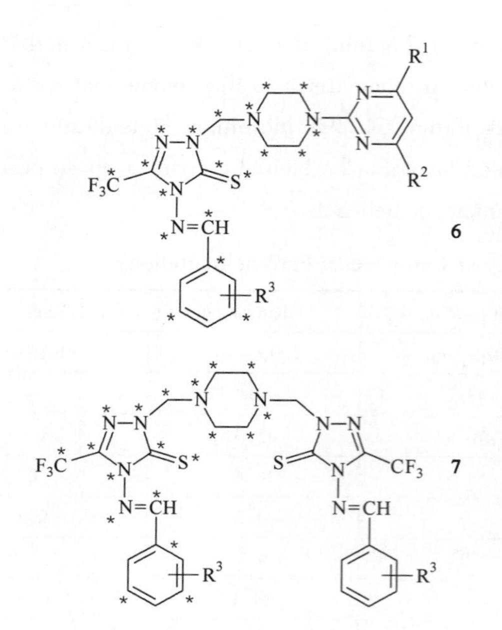

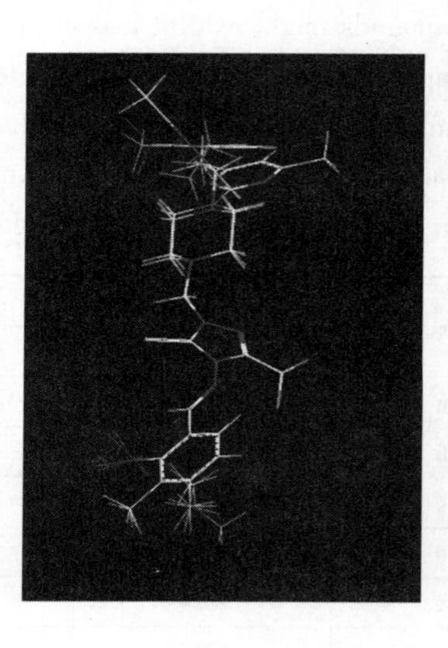

Figure 2 Superimposition of compounds **6** and **7** for 3D QSAR studies

Table 6 Summary of CoMFA Analysis

	q^2	r^2	S	F	compound	contribution(%)	
						steric	electrostatic
CoMFA	0.526	0.913	0.085	24.448	**6h**	74.6	25.4

Table 7 Experimental and Predicted Fungicidal Activity(*P. cubensis*) of 3D QSAR

compound	MW	I	ED①	PD②	residual
6a	448	66.3	2.945	3.002	-0.057
6b	466	64.9	2.935	3.003	-0.068
6c	478	79.5	3.268	3.162	0.106
6d	462	78.6	3.230	3.249	-0.019
6e	480	75.5	3.170	3.139	0.031
6f	492	72.5	3.113	3.272	-0.159
6g	490	69.5	3.048	3.072	-0.024
6h	496	91.2	3.711	3.701	0.010
6i	507	76.8	3.225	3.265	-0.040
6j	476	74.1	3.134	3.250	-0.116
6k	494	79.2	3.274	3.250	0.024
6l	506	87.3	3.541	3.605	-0.064
6m	504	67.8	3.026	3.028	-0.002
6n	510	91.6	3.745	3.599	0.146
6o	521	77.2	3.247	3.259	-0.012
7a	654	78.9	3.388	3.296	0.092
7b	690	79.9	3.438	3.394	0.044

Table 7(continued)

compound	MW	I	ED①	PD②	residual
7c	714	84.5	3.590	3.606	-0.016
7d	710	78.7	3.419	3.458	-0.039
6-A1③	476	96.9	4.173	4.002	0.171
6-A2③	482	74.2	3.142	3.341	-0.199
6-A3③	493	76.2	3.198	2.903	0.295
7-A1③	722	81.0	3.488	3.568	-0.080
7-A2③	744	84.9	3.622	3.677	-0.055

①ED = experimental D value; ②PD = predicted D value; ③Test set compounds.

In Figure 3, the isocontour diagrams of the steric and electrostatic field contributions ("standard deviation × coefficient") obtained from the CoMFA analysis are illustrated. The steric field contour map is plotted in Figure 3a. The green displays 2 positions, where a bulky group would be favorable for higher fungicidal activity. In contrast, yellow indicates positions where a decrease in the bulk of the target molecules is favored. As shown in Figure 3a, the CoMFA steric contour plots obviously indicated that a yellow region is located around the 2 and 3 position of the benzene ring. This means that the bulky substituents at the 2 and 3 position will decrease the fungicidal activity. For example, some compounds bearing substituents at the 2 position of the benzene ring, such as **6b**, **6e**, **6i**, **6j**, and **6o**, displayed lower fungicidal activity. The electrostatic contour plot is shown in Figure 3b. The blue contour defines a region where an increase in the positive charge will result in an increase in the activity, whereas the red contour defines a region of space where increasing electron density is favorable. As shown in Figure 3b, the target compounds bearing an electron-donating group at the 4 or 6 position of the pyrimidine ring or an electron-withdrawing group at the 3 or 4 position of the benzene ring, such as **6h**, **6n**, and **7c**, displayed higher activity. According to the 3D QSAR model obtained, five novel Mannich bases (**6-A1-6-A3**, **7-A1**, and **7-A2**) as test set compounds were subsequently synthesized and bioassayed. As shown in Table 7, their activities were estimated fairly accurately from the analysis, indicating that the alignment of the molecules was valid. For example, compounds

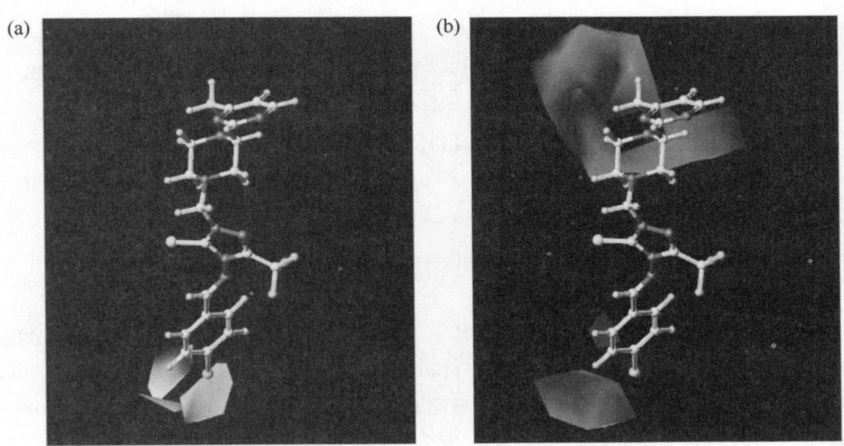

Figure 3 (a) Steric map from the CoMFA model. (b) Electrostatic map from the CoMFA model

6 – **A1** and 7 – **A2** were found to display good fungicidal activity against *P. cubensis* (96.9 and 84.9%) as compounds **6h** and **7c**, respectively. These results provide useful information for further optimization of the compounds.

In summary, we have conveniently synthesized a series of trifluoromethyl – substituted 1,2,4 – triazole Mannich base **6** and bis(1,2,4 – triazole) Mannich base **7** containing pyrimidinylpiperazine rings via the Mannich reaction in good yields. The fungicidal tests indicated that most compounds **6** and **7** possessed excellent fungicidal activity. Among 19 novel compounds, some showed superiority over the commercial fungicides Dimethomorph, Thiophanate – methyl, Iprodione, and Zhongshengmycin during the present studies and could be further developed as fungicides. Some compounds also exhibited favorable herbicidal activity in the preliminary studies. In addition, the 3D QSAR results provide useful information for guiding optimization of such structures to accelerate the discovery of compounds with high fungicidal activity.

LITERATURE CITED

[1] Kramer, W.; Schirmer, U. *Modern Crop Protection*; Wiley – VCH: Weinheim, Germany, 2007.

[2] Tanaka, K.; Tsukamoto, Y.; Sawada, Y.; Kasuya, A.; Hotta, H.; Ichinose, R.; Watanabe, T.; Toya, T.; Yokoi, S.; Kawagishi, A.; Ando, M.; Sadakane, S.; Katsumi, S.; Masui, A. Chromafenozide: A novel lepidopteran insect control agent. *Annu. Rep. Sankyo Res. Lab.* 2001, 53, 1 – 49.

[3] Bucchel, K. H.; Draber, W. Ger. Offen. 1940628, 1971. *Chem. Abstr.* 1971, 74, No. 125698t.

[4] Greenfield, S. A.; Seldel, M. C.; Von Meyer, W. C. Ger. Offen. 1943915, 1970. *Chem. Abstr.* 1970, 72, No. 100713q.

[5] Okano, S.; Yasujaga, K. Jpn. Pat. 7017191, 1970. *Chem. Abstr.* 1970, 73, No. 77251x.

[6] Reisser, F. Ger. Offen. 1918795, 1969. *Chem. Abstr.* 1970, 72 No. 43692f.

[7] Bianchi, D.; Cesti, P.; Spezia, S.; Garavaglia, C.; Mirenna, L. Chemoenzymic synthesis and biological activity of both enantiomeric forms of tetraconazole, a new antifungal triazole. *J. Agric. Food Chem.* 1991, 39, 197 – 201.

[8] Chen, T.; Dwyre – Gygax, C.; Hadfield, S. T.; Willetts, C.; Breuil, C. Development of an enzyme – linked immunosorbent assay for a broad spectrum triazole fungicide: Hexaconazole. *J. Agric. Food Chem.* 1996, 44, 1352 – 1356.

[9] Garca Ruano, J. L.; Cifuentes Garca, M.; Martn Castro, A. M.; Rodrguez Ramos, J. H. First highly stereoselective synthesis of fungicide Systhane. *Org. Lett.* 2002, 4, 55 – 57.

[10] Song, Z.; Nes, W. D. Sterol biosynthesis inhibitors: Potential for transition state analogs and mechanism – based inactivators targeted at sterol methyltransferase. *Lipids* 2007, 42, 15 – 33.

[11] Cao, X.; Li, F.; Hu, M.; Lu, W.; Yu, G. A.; Liu, S. H. Chiral γ – aryl – 1*H* – 1,2,4 – triazole derivatives as highly potential antifungal agents: Design, synthesis, structure, and in vitro fungicidal activities. *J. Agric. Food Chem.* 2008, 56, 11367 – 11375.

[12] Naylor, A. W. Metabolism of herbicides, complexes of 3 – amino – 1,2,4 – triazole in plant metabolism. *J. Agric. Food Chem.* 1964, 12, 21 – 25.

[13] Simmons, K. A.; Dixson, J. A.; Halling, B. P.; Plummer, E. L.; Plummer, M. J.; Tymonko, J. M.; Schmidt, R. J.; Wyle, M. J.; Webster, C. A.; Baver, W. A.; Witkowski, D. A.; Peters, G. R.; Gravelle, W. D. Synthesis and activity optimization of herbicidal substituted 4 – aryl – 1,2,4 – triazole – 5(1*H*) – thiones. *J. Agric. Food Chem.* 1992, 40, 297 – 305.

[14] Filler, R.; Kobayashi, Y. *Biomedicinal Aspects of Fluorine Chemistry*; Kodansha and Elsevier Biomedical:

Amsterdam, The Netherlands, 1983.

[15] Clark, R. D. Synthesis and QSAR of herbicidal 3 - pyrazolyl α, α, α - trifluorotolyl ethers. *J. Agric. Food Chem.* 1996, 44, 3643 - 3652.

[16] Liu, A.; Wang, X.; Ou, X.; Huang, M.; Chen, C.; Liu, S.; Huang, L.; Liu, X.; Zhang, C.; Zheng, Y.; Ren, Y.; He, L.; Yao, J. Synthesis and fungicidal activities of novel bis (trifluoromethyl) phenyl - based strobilurins. *J. Agric. Food Chem.* 2008, 56, 6562 - 6566.

[17] Liu, A.; Wang, X.; Chen, C.; Pei, H.; Mao, C.; Wang, Y.; He, H.; Huang, L.; Liu, X.; Hu, Z.; Ou, X.; Huang, M.; Yao, J. The discovery of HNPC - A3066: A novel strobilurin acaricide. *Pest Manage. Sci.* 2009, 65, 229 - 234.

[18] Liu, F. M.; Wang, B. L.; Lu, W. J. Synthesis of trifluoromethyl - substituted 1, 2, 4 - triazole mannich bases. *Chin. J. Org. Chem.* 2000, 20, 738 - 742.

[19] He, F. Q.; Liu, X. H.; Wang, B. L.; Li, Z. M. Synthesis and biological activity of new 1 - [4 - (substituted) - piperazin - 1 - ylmethyl] - 1*H* - benzotriazole. *J. Chem. Res.* 2006, 12, 809 - 811.

[20] George, T.; Mehta, D. V.; Tahilramani, R.; David, J.; Talwalker, P. K. Synthesis of some s - triazoles with potential analgesic and antiinflammatory activities. *J. Med. Chem.* 1971, 14, 335 - 338.

[21] Adams, R. R.; Whitmore, F. C. Heterocyclic basic compounds. Ⅳ. 2 - Aminoalkylamino - pyrimidines. *J. Am. Chem.* Soc. 1945, 67, 735 - 738.

[22] Zlatoidsky, P.; Maliar, T. Synthesis of 1 - (4 - acyloxybenzoyloxyacetyl) - 4 - alkylpiperazines and 1 - (4 - acyloxybenzoyl) - 4 - alkylpiperazines as inhibitors of chymotrypsin. *Eur. J. Med. Chem.* 1996, 31, 669 - 673.

[23] Matsumoto, K.; Hashimoto, S.; Minatogawa, K.; Munakata, M.; Otani, S. Synthesis of linked heterocycles mediated by high pressure SNAr reactions. *Chem. Express* 1990, 5, 473 - 476.

[24] Shi, Y. X.; Yuan, L. P.; Zhang, Y. B.; Li, B. J. Fungicidal activity of novel fungicide chlorphenomizole and efficacy in controlling main diseases on vegetable in field. *Chin. J. Pestic. Sci.* 2007, 9, 126 - 130.

[25] Zhao, W. G.; Wang, J. G.; Li, Z. M.; Yang, Z. Synthesis and antiviral activity against tobacco mosaic virus and 3D - QSAR of α - substituted - 1, 2, 3 - thiadiazoleacetamides. *Bioorg. Med. Chem.* 2006, 16, 6107 - 6111.

[26] Liu, X. H.; Shi, Y. X.; Ma, Y.; He, G. R.; Dong, W. L.; Zhang, C. Y.; Wang, B. L.; Wang, S. H.; Li, B. J.; Li, Z. M. Synthesis of some N, N′ - diacylhydrazine derivatives with radical scavenging and antifungal activity. *Chem. Biol. Drug Des.* 2009, 73, 320 - 327.

[27] Bush, B. L.; Nachbar, R. B. Sample - distance partial least squares PLS optimized for many variables, with application to CoMFA. *J. Comput. - Aided Mol. Des.* 1993, 7, 587 - 619.

[28] Wu, T. X.; Li, Z. J.; Zhao, J. C. A facile method for the synthesis of 4 - amino - 5 - hydrocarbon - 2, 4 - dihydro - 3*H* - 1, 2, 4 - triazole - 3 - thione schiff bases. *Chem. J. Chin. Univ.* 1998, 19, 1617 - 1619.

[29] Wang, B. L.; Duggleby, R. G.; Li, Z. M.; Wang, J. G.; Li, Y. H.; Wang, S. H.; Song, H. B. Synthesis, crystal structure and hebricidal activity of mimics of intermediates of the KARI reaction. *Pest Manage. Sci.* 2005, 61, 407 - 412.

[30] Chen, Q.; Zhu, X. L.; Jiang, L. L.; Liu, Z. M.; Yang, G. F. Synthesis antifungal activity and CoMFA analysis of novel 1, 2, 4 - triazolo[1, 5 - *a*] pyrimidine derivatives. *Eur. J. Med. Chem.* 2008, 43, 595 - 603.

Synthesis and Fungicidal Activity of Novel Aminophenazine - 1 - carboxylate Derivatives*

Mingzhong Wang, Han Xu, Shujing Yu,
Qi Feng, Suhua Wang, Zhengming Li

(State Key Laboratory of Elemento - Organic Chemistry, Research Institute of Elemento - organic Chemistry, Nankai University, Tianjin, 300071, China)

Abstract A series of novel 6 - aminophenazine - 1 - , 7 - aminophenazine - 1 - and 8 - aminophenazine - 1 - carboxylate derivatives were synthesized by a facile method, and their structures were characterized by ^{1}H NMR, ^{13}C NMR and high - resolution mass spectrometry. Some unexpected byproducts **V - 7b**—**V - 8d** were noticed and isolated, and their structures were identified by 2D NMR spectra including heteronuclear multiple - quantum coherence (HMQC), heteronuclear multiple - bond correlation (Hmbc) and H—H correlation spectrometry (H—H COSY) approach. Their fungicidal activities against five fungi were evaluated, which indicated that most of the title compounds showed low fungicidal activities in vitro against *Alternaria solani*, *Cercospora arachidicola*, *Fusarium omysporum*, *Gibberella zeae*, and *Physalospora piricola* at a dosage of 50μg · mL^{-1}, while compounds **IV - 6a** and **IV - 6b** exhibited excellent activities against *P. piricola* at that dosage. Compound **IV - 6a** could be considered as a leading structure for further design of fungicides.

Key words aminophenazine - 1 - carboxylate; derivatives; synthesized; fungicidal activities; leading structure

INTRODUCTION

In the past century, many potential antibiotics containing the planar tricyclic heteroaromatic phenazine were isolated from the marine microorganism *Streptomyces antibioticus* strain Tü 2706[1]. The derivatives included the diphenazine antibiotics esmeraldin A and B as well as simpler monomeric structures containing 6 - (1 - hydroxyethyl) - 1 - phenazine carboxylic acid (**A**, Figure 1) have shown extensive antimicrobial activity toward a broad range of bacteria[2]. Phenazine - 1 - carboxylic acid (**A**, Figure 1) isolated from *Pseudomonas* sp. had high antimicrobial activity against nine bacterial strains, inhibiting settlement of barnacle larvae, and reducing *Ulva lactuca* spore settlement and percent cover of germlings[3]. Recently, a series of substituted (—Cl, —OCH_3, —CH_3) phenazine - 1 - carboxylic acids and the corresponding carboxamides (**B**, Figure 1) were prepared as antitumor drugs acting on electron - deficient DNA - intercalating ligands and evaluated against L1210 leukemia in vitro and against P388 leukemia and Lewis lung carcinoma *in vivo*[4].

The total synthesis of aminophenazine - 1 - carboxylic acid was completed in the 1960s[5-7], but few published studies reported their biological activity. Considering the other potential biological activity of phenazine structures, the derivatives of aminophenazine - 1 - carboxylic acid with the phenazine skeleton were synthesized and tested against fungi in vitro. In this paper, for-

---

* Reprinted from *Journal of Agriculture and Food Chemistry*, 2010, 58(6): 3651 - 3660. Supported by the National Key Project for Basic Research and the National Natural Science Foundation of China.

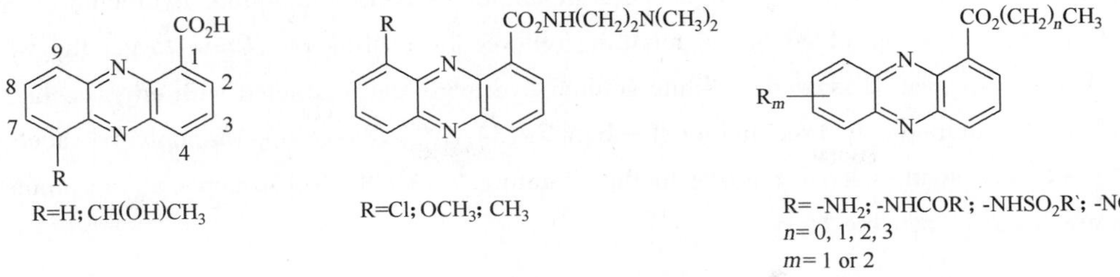

CO_2H 9 1 8 2 7 3 4 R

R=H; $CH(OH)CH_3$

A

R $CO_2NH(CH_2)_2N(CH_3)_2$

R=Cl; OCH_3; CH_3

B

$CO_2(CH_2)_nCH_3$ R_m

R= $-NH_2$; -NHCOR`; $-NHSO_2R`$; $-NO_2$
n= 0, 1, 2, 3
m= 1 or 2

C

Figure 1 Chemical structure of compounds **A** – **C**

ty – six 6 – aminophenazine – 1 – ,7 – aminophenazine – 1 – and 8 – aminophenazine – 1 – carboxylate derivatives, including six known compounds (**C**, Figure 1), were synthesized with a facile synthetic method.

Those compounds' fungicidal activities against five fungi were evaluated, and their possible structure – activity relationships (SAR) were discussed. In order to enhance the lipophilic properties, aminophenazine – 1 – carboxylic acid was derivatized to its ester with corresponding $CH_3(CH_2)_nOH$ (n = 0, 1, 2, 3); The —NH_2 group was substituted at different positions of the phenazine ring; Different carboxamide and sulfonamide derivatives were optimized with acyl chloride or sulfonyl chloride; the phenazine rings of some carboxamide derivatives were further nitrified. All the new derivatives were designed to explore whether that might improve or decrease the fungicidal activities.

MATERIALS AND METHODS

Instruments. ^1H NMR, ^{13}C NMR, HMQC, HMBC, H—H COSY spectra were obtained at 300MHz using a Bruker AV300 spectrometer or at 400MHz using a Bruker AV400 spectrometer in $CDCl_3$ or d_6 – DMSO solution with tetramethylsilane as the internal standard. Chemical shift values (δ) are given in ppm. High – resolution mass spectrometry (HRMS) data were obtained on a Varian QFT – ESI instrument. The melting points were determined on an X – 4 binocular microscope melting point apparatus (Beijing Tech Instruments Co., Beijing, China) and are uncorrected. Yields were not optimized. The reagents were all analytically or chemically pure. All solvents and liquid reagents were dried by standard methods in advance and distilled before use. Silica gel (200 – 300mesh) was obtained from Qingdao Marine Chemical Factory, Qingdao, P. R. China. Three kinds of benzene – diamine, 2 – bromo – 3 – nitrobenzoic acid and $NaBH_4$ were bought from the Alfa Aesar Company (Tianjin, China). Phenazine – 1 – carboxylic acid was synthesized according to the literature[4].

General Synthetic Procedures for I – a, I – b and I – c. To a solution of *o* – phenylenediamine (8.0g, 74.0mmol) in dry THF (50.0mL) was added dropwise a solution of Ac_2O (8.0mL) in dry THF (20.0mL) under 15℃ within 2h. The mixture was stirred for 5h at 15℃, and the solvent was removed, followed by cooling to get the crude product and recrystallization with ethyl acetate to give *N* – (2 – aminophenyl) acetamide (**I – a**, 5.3 g, 47.7%).

To a solution of *m* – phenylenediamine (12.0g, 111.0mmol) in dry THF (30.0mL) was added dropwise a solution of Ac_2O (9.0mL) in dry THF (35.0mL) under – 10℃ within 2h. The mixture was stirred for 3h keeping the temperature under -10℃, and dilute hydrochloric acid (20.0mL, 18.5%) was added to the mixture, followed by cooling for 30min to get the white salt. Then the salt was dissolved in dilute sodium hydroxide and extracted with ethyl acetate to give *N*-(3-aminophenyl) acetamide(**Ⅰ-b**, 5.7g, 34.2%). *N*-(4-Aminophenyl) acetamide(**Ⅰ-c**) was synthesized according to the literature[8]. All the substituents at the aromatic rings are listed in Scheme 1.

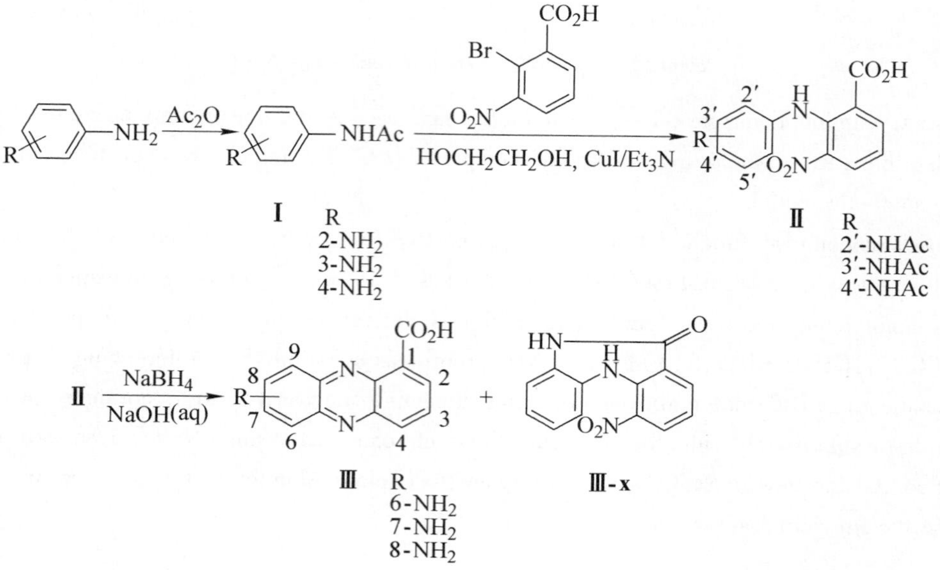

Scheme 1 Synthetic Routes to Compounds Ⅲ

Data for Ⅰ-a. Yield: 47.7%; white crystal; mp, 129-131℃ (ethyl acetate/petroleum ether). ^{1}H NMR(400MHz, d_6-DMSO) δ: 9.11 (s, 1H, NHAc), 7.14 (d, $^3J_{HH}$ = 7.6Hz, 1H, Ar—H), 6.88 (t, $^3J_{HH}$ = 7.6Hz, 1H, Ar—H), 6.70 (d, $^3J_{HH}$ = 7.6Hz, 1H, Ar—H), 6.52 (t, $^3J_{HH}$ = 7.6Hz, 1H, Ar—H), 4.84 (s, 2H, NH_2), 2.02 (s, 3H, $COCH_3$).

Data for Ⅰ-b. Yield: 34.2%; white crystal; mp, 82-84℃ (ethyl acetate/petroleumether). ^{1}H NMR(400MHz, d_6-DMSO) δ: 9.60 (s, 1H, NHAc), 6.92 (s, 1H, Ar—H), 6.89 (t, $^3J_{HH}$ = 8.0Hz, 1H, Ar—H), 6.65 (d, $^3J_{HH}$ = 8.0Hz, 1H, Ar—H), 6.23 (dd, $^3J_{HH}$ = 8.0Hz, $^3J_{HH}$ = 1.2Hz, 1H, Ar—H), 5.02 (s, 2H, NH_2), 1.99 (s, 3H, $COCH_3$).

Data for Ⅰ-c. Yield: 72.0%; white crystal; mp, 162-164℃ (ethyl acetate/petroleum ether). ^{1}H NMR(300MHz, d_6-DMSO) δ: 9.46 (s, 1H, NHAc), 7.18 (d, $^3J_{HH}$ = 8.4Hz, 2H, Ar—H), 6.48 (d, $^3J_{HH}$ = 8.4Hz, 2H, Ar—H), 4.80 (s, 2H, NH_2), 1.94 (s, 3H, $COCH_3$).

General Synthetic Procedures for Ⅱ-a, Ⅰ-b and Ⅱ-c. A mixture of *N*-(2-aminophenyl) acetamide (Ⅰ-a) (4.0g, 26.6mmol), 2-bromo-3-nitrobenzoic acid (6.3g, 26.6mmol), CuI (1.3g, 6.8mmol) in triethylamine (12.5mL) and ethane-1,2-diol (50.0mL) were stirred at 80℃ for 3h. The cooled mixture was diluted with 0.2N aqueous NaOH, clarified with charcoal, and filtered through Celite. The resulting clear solution was acidified with dilute HCl to give 2-(2-acetamidophenylamino)-3-nitrobenzoic acid(**Ⅱ-a**) (3.0g, 35.8%)[9]. 2-(3-acetamidophenylamino)-3-nitrobenzoic acid(**Ⅱ-b**) and 2-

(4 - acetamidophenylamino) - 3 - nitrobenzoic acid(**Ⅱ - c**) were given through the same process. All the substituents at the aromatic rings were listed in the Scheme 1.

Data for Ⅱ - a. Yield: 35.8%; orange solid; mp, 202 - 204℃ (acetone/petroleum ether). ^1H NMR(400MHz, d_6 - DMSO) δ: 13.56(s, 1H, COOH), 9.80(s, 1H, NHAc), 9.63(s, 1H, Ar—NH—Ar), 8.14(d, $^3J_{HH}$ = 7.2Hz, 1H, Ar—H), 8.01(d, $^3J_{HH}$ = 7.6Hz, 1H, Ar—H), 7.39(d, $^3J_{HH}$ = 6.8Hz, 1H, Ar—H), 7.02(t, $^3J_{HH}$ = 8.4Hz, 3H, Ar—H), 6.82(d, $^3J_{HH}$ = 7.2Hz, 1H, Ar—H), 2.07(s, 3H, $COCH_3$).

Data for Ⅱ - b. Yield: 62.6%; orange solid; mp, 222 - 224℃ (acetone/petroleum ether). ^1H NMR(400MHz, d_6 - DMSO) δ: 9.88(s, 1H, NHAc), 9.84(s, 1H, Ar—NH—Ar), 8.21(d, $^3J_{HH}$ = 7.6Hz, 1H, Ar—H), 8.08(d, $^3J_{HH}$ = 8.0Hz, 1H, Ar—H), 7.25(s, 1H, Ar—H), 7.18(d, $^3J_{HH}$ = 7.6Hz, 1H, Ar—H), 7.15(d, $^3J_{HH}$ = 7.6Hz, 1H, Ar—H), 7.11(d, $^3J_{HH}$ = 7.6Hz, 1H, Ar—H), 6.56(d, $^3J_{HH}$ = 7.6Hz, 1H, Ar—H), 2.01(s, 3H, $COCH_3$).

Data for Ⅱ - c. Yield: 50.6%; orange solid; mp, 218 - 220℃ (acetone/petroleum ether). ^1H NMR(400MHz, d_6 - DMSO) δ: 9.93(s, 1H, Ar—NH—Ar), 9.87(s, 1H, NHAc), 8.19(d, $^3J_{HH}$ = 7.6Hz, 1H, Ar—H), 8.04(d, $^3J_{HH}$ = 8.0Hz, 1H, Ar—H), 7.44(d, $^3J_{HH}$ = 8.4Hz, 2H, Ar—H), 7.02(t, $^3J_{HH}$ = 8.0Hz, 1H, Ar—H), 6.88(d, $^3J_{HH}$ = 8.4Hz, 2H, Ar—H), 2.01(s, 3H, $COCH_3$).

General Synthetic Procedures for Ⅲ - a, Ⅲ - b and Ⅲ - c. A solution of 2 - (3 - acetamido phenylamino) - 3 - nitrobenzoic acid(**Ⅱ - b**) (6.0g, 19.0mmol) and $NaBH_4$ (4.8g, 126.0mmol) in 2 N NaOH(500.0mL) was refluxed for 4h. Cooling gave the sodium salt of the phenazine acid, which was acidified to give a mixture of 6 - aminophenazine - 1 - carboxylic acid(**Ⅲ - a**) and 8 - aminophenazine - 1 - carboxylic acid(**Ⅲ - c**). The mixture consisted of 50% **Ⅲ - a** and 50% **Ⅲ - c**, as determined by ^1H NMR. A small amount of the mixture was purified by flash chromatography on silica gel [elution solvent: ethyl acetate/petroleum ether (60 - 90℃), 1:4, v/v]. Similar procedures were used to prepare 7 - aminophenazine - 1 - carboxylic acid(**Ⅲ - b**) by reductive cyclization of 2 - (4 - acetamido phenylamino) - 3 - nitrobenzoic acid(**Ⅱ - c**). The compound **Ⅲ - x** was the only product by reductive cyclization of 2 - (2 - acetamido phenylamino) - 3 - nitrobenzoic acid(**Ⅱ - a**). All the substituents at the aromatic rings are listed in Scheme 1.

Data for Ⅲ - a. Yield: 59.0%; purple solid; mp, 306 - 308℃ (acetone/petroleum ether)[5]. ^1H NMR(300MHz, d_6 - DMSO) δ: 15.27(s, 1H, COOH), 8.62(d, $^3J_{HH}$ = 6.6Hz, 1H, Ar—H), 8.48(d, $^3J_{HH}$ = 8.4Hz, 1H, Ar—H), 8.03(t, $^3J_{HH}$ = 7.5Hz, 1H, Ar—H), 7.82(t, $^3J_{HH}$ = 7.8Hz, 1H, Ar—H), 7.40(d, $^3J_{HH}$ = 8.4Hz, 1H, Ar—H), 6.98(d, $^3J_{HH}$ = 7.5Hz, 1H, Ar—H), 6.74(s, 2H, NH_2).

Data for Ⅲ - b. Yield: 69.0%; red solid; mp, beyond 340℃ (acetone/petroleum ether)[4]. ^1H NMR(400MHz, d_6 - DMSO) δ: 8.33(d, $^3J_{HH}$ = 6.0Hz, 1H, Ar—H), 8.23(d, $^3J_{HH}$ = 8.0Hz, 1H, Ar—H), 8.05(d, $^3J_{HH}$ = 8.0Hz, 1H, Ar—H), 7.91(t, $^3J_{HH}$ = 7.2Hz, 1H, Ar—H), 7.58(d, $^3J_{HH}$ = 8.0Hz, 1H, Ar—H), 6.95(s, 1H, Ar—H), 6.81(s, 2H, NH_2).

Data for Ⅲ - c. Yield: 59.0%; red solid; mp, beyond 340℃ (acetone/petroleumether)[5]. ^1H NMR(300MHz, d_6 - DMSO) δ: 8.41(d, $^3J_{HH}$ = 6.6Hz, 1H, Ar—H), 8.34(d, $^3J_{HH}$ =

8.7Hz, 1H, Ar—H), 7.98 (d, $^{3}J_{HH}$ = 9.6Hz, 1H, Ar—H), 7.78 (d, $^{3}J_{HH}$ = 7.8Hz, 1H, Ar—H), 7.57 (d, $^{3}J_{HH}$ = 9.0Hz, 1H, Ar—H), 7.25 (s, 2H, NH_2), 6.86 (s, 1H, Ar—H).

CO_2H 1 2 3 4 6 7 8 9 R N N — $CH_3(CH_2)_nOH$ / H_2SO_4 → $CO_2(CH_2)_nCH_3$ R N N + $CO_2(CH_2)_nCH_3$ R N N

Ⅲ R: 6-NH_2, 7-NH_2, 8-NH_2

Ⅳ R: 6-NH_2, 7-NH_2, 8-NH_2; n = 0, 1, 2, 3

Ⅴ R=7-NH_2, 8-$O(CH_2)_nCH_3$; n=1, 2, 3; R=7-$O(CH_2)_nCH_3$, 8-NH_2; n= 3

$CO_2(CH_2)_nCH_3$ 1 2 3 4 6 7 8 9 R N N — R′COCl or R′SO_2Cl / pyridine or CH_2Cl_2/pyridine → $CO_2(CH_2)_nCH_3$ RHN N N

Ⅳ R= 6-NH_2 or 7-NH_2; n= 0, 3

Ⅵ R= 6-COR′ or -SO_2R′; 7-COR′ or -SO_2R′; n= 0, 3

$CO_2(CH_2)_nCH_3$ RHN N N — HNO_3 / H_2SO_4 → $CO_2(CH_2)_nCH_3$ R N N

Ⅵ R=6-$CO(CH_2)_3Cl$; n= 0, 3; R=7-$CO(CH_2)_3Cl$; n= 3

Ⅶ R
a 6-$NHCO(CH_2)_3Cl$, 7-NO_2; n= 0
b 6-NH_2, 7-NO_2, 9-NO_2; n= 0
c 6-$NHCO(CH_2)_3Cl$, 7-NO_2; n= 3
d 6-NH_2, 7-NO_2, 9-NO_2; n= 3
e 7-NH_2, 6-NO_2; n= 3

Scheme 2 Synthetic Routes to the Title Compounds Ⅳ, Ⅵ and Ⅶ

Data for Ⅲ – x. Yield: 20.0%; red solid; mp, 305 – 307℃ (acetone/petroleum ether)[10]. ^{1}H NMR (400MHz, d_6 – DMSO) δ: 10.41 (s, 1H, NHCO), 8.79 (s, 1H, Ar—NH—Ar), 8.20 (d, $^{3}J_{HH}$ = 7.2Hz, 1H, Ar—H), 8.05 (d, $^{3}J_{HH}$ = 6.8Hz, 1H, Ar—H), 7.13 (t, $^{3}J_{HH}$ = 8.0Hz, 1H, Ar—H), 7.02 – 7.06 (m, 4H, Ar—H).

General Synthetic Procedures for Compounds **Ⅳ – 6a—Ⅴ – 8d.** To a solution of the mixture 6 – aminophenazine – 1 – carboxylic acid (**Ⅲ – a**) and 8 – aminophenazine – 1 – carboxylic acid (**Ⅲ – c**) (2.0g) in methanol (500.0mL), H_2SO_4 (0.5mL) was added dropwise and refluxed for 7h. Most of the methanol was then removed by distillation, and the residue was diluted with water. The black solution was made alkaline with $NH_3 \cdot H_2O$[5] and extracted with dichloromethane (2 × 50.0mL). The extraction was evaporated and purified by flash chromatography on silica gel eluting [elution solvent: ethyl acetate/petroleum ether (60 – 90℃), 1∶4, v/v] to provide red solid **Ⅳ – 6a** (0.52g, Yield, 25.1%) and **Ⅳ – 8a** (0.25g, yield, 11.8%). Compounds **Ⅳ – 6b – Ⅳ – 8d** were prepared according to the same process. **Ⅴ – 7b, Ⅴ – 7c, Ⅴ – 7d**, and **Ⅴ – 8d** were the byproducts of **Ⅳ – 7b, Ⅳ – 7c, Ⅳ – 7d** and **Ⅳ – 8d** respectively. All the substituents at the phenazine rings are listed in Scheme 2 and Table 1.

Table 1 The R Group and the Value of *n* in Compounds Ⅳ − 6a − Ⅶ − e

Compd.	R	*n*	Compd.	R	*n*
Ⅳ − 6a	6 − NH_2	0	**Ⅳ − 6b**	6 − NH_2	1
Ⅳ − 6c	6 − NH_2	2	**Ⅳ − 6d**	6 − NH_2	3
Ⅳ − 7a	7 − NH_2	0	**Ⅳ − 7b**	7 − NH_2	1
Ⅳ − 7c	7 − NH_2	2	**Ⅳ − 7d**	7 − NH_2	3
Ⅳ − 8a	8 − NH_2	0	**Ⅳ − 8b**	8 − NH_2	1
Ⅳ − 8c	8 − NH_2	2	**Ⅳ − 8d**	8 − NH_2	3
Ⅴ − 7b	7 − NH_2	1	**Ⅴ − 7c**	7 − NH_2	2
Ⅴ − 7d	7 − NH_2	3	**Ⅴ − 8d**	8 − NH_2	3
Ⅵ − 6a	6 − $COCH_3$	0	**Ⅵ − 6b**	6 − $COCH_2OOCCH_3$	0
Ⅵ − 6c	6 − $CO(CH_2)_3Cl$	0	**Ⅵ − 6d**	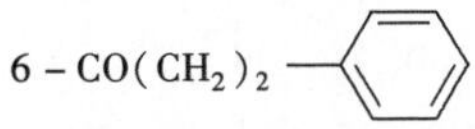 6 − $CO(CH_2)_2$ —	0
Ⅵ − 6e	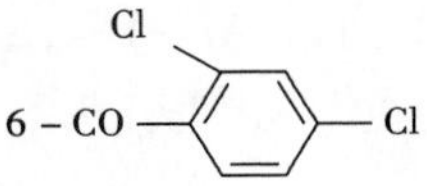 Cl 6 − CO — — Cl	0	**Ⅵ − 6f**	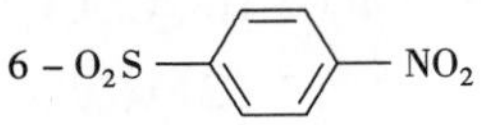 6 − O_2S — — NO_2	0
Ⅵ − 6g	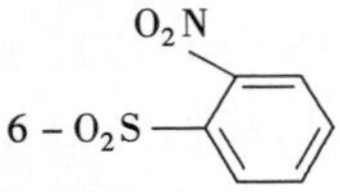 O_2N 6 − O_2S —	0	**Ⅵ − 6h**	6 − $CO(CH_2)_3Cl$	3
Ⅵ − 7a	7 − $COCH_3$	0	**Ⅵ − 7b**	7 − $COCH_2OOCCH_3$	0
Ⅵ − 7c	7 − $CO(CH_2)_3Cl$	0	Ⅵ − 7d	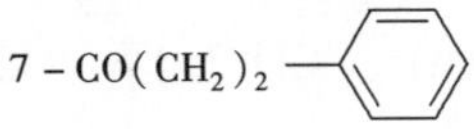 7 − $CO(CH_2)_2$ —	0
Ⅵ − 7e	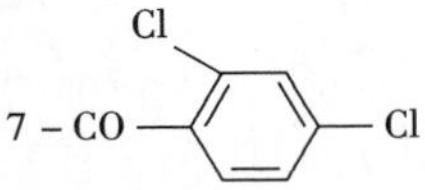 Cl 7 − CO — — Cl	0	**Ⅵ − 7f**	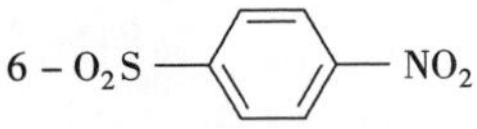 6 − O_2S — — NO_2	0
Ⅵ − 7g	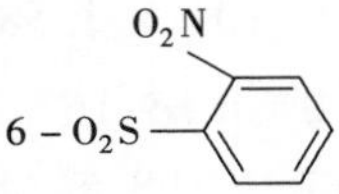 O_2N 6 − O_2S —	0	**Ⅵ − 7h**	7 − $CO(CH_2)_3Cl$	3
Ⅵ − 7i	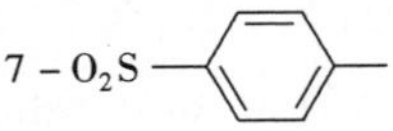 7 − O_2S — —	0	**Ⅵ − 7j**	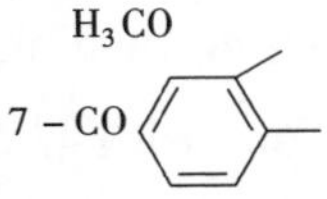 H_3CO 7 − CO —	0
Ⅵ − 7k	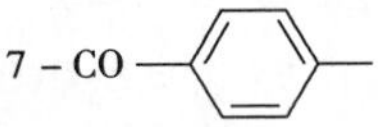 7 − CO — —	0	**Ⅵ − 7l**	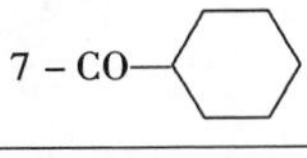 7 − CO—	0
Ⅵ − 7m	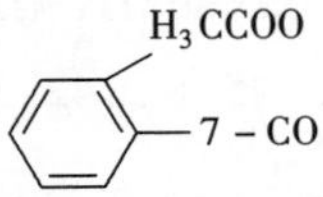 H_3CCOO — 7 − CO	0	**Ⅵ − 7n**	7 − $COCl_3$	0
Ⅶ − a	6 − $NHCO(CH_2)_3Cl$, 7 − NO_2	0	**Ⅶ − b**	6 − NH_2, 7 − NO_2, 9 − NO_2	0
Ⅶ − c	6 − $NHCO(CH_2)_3Cl$, 7 − NO_2	3	**Ⅶ − d**	6 − NH_2, 7 − NO_2, 9 − NO_2	3
Ⅶ − e	7 − NH_2, 6 − NO_2	3			

Data for Ⅳ－6a. Yield: 25.1%; red solid; mp, 142－144℃ (ethyl acetate/petroleum ether)[6]. ^{1}H NMR(400MHz, $CDCl_3$) δ: 8.14 (d, $^3J_{HH}$ = 8.8Hz, 1H, Ar—H), 8.06 (d, $^3J_{HH}$ = 6.8Hz, 1H, Ar—H), 7.60 (t, $^3J_{HH}$ = 8.0Hz, 1H, Ar—H), 7.51 (d, $^3J_{HH}$ = 8.4Hz, 1H, Ar—H), 7.47 (d, $^3J_{HH}$ = 8.4Hz, 1H, Ar—H), 6.74 (d, $^3J_{HH}$ = 6.0Hz, 1H, Ar—H), 5.16 (s, 2H, NH_2), 3.98 (s, 3H, OCH_3). ^{13}C NMR (100MHz, $CDCl_3$) δ: 167.3, 144.3, 143.9, 141.2, 140.4, 134.9, 133.4, 132.6, 131.9, 131.0, 127.8, 117.7, 108.0, 52.6.

Data for Ⅳ－6b. Yield: 41.7%; red solid; mp, 108－110℃ (ethyl acetate/petroleum ether). ^{1}H NMR(400MHz, $CDCl_3$) δ: 8.20 (dd, $^3J_{HH}$ = 8.8Hz, $^4J_{HH}$ = 1.2Hz, 1H, Ar—H), 8.09 (dd, $^3J_{HH}$ = 7.2Hz, $^4J_{HH}$ = 1.2Hz, 1H, Ar—H), 7.65 (dd, $^3J_{HH}$ = 8.8Hz, $^3J_{HH}$ = 7.2Hz, 1H, Ar—H), 7.52 (d, $^3J_{HH}$ = 4.4Hz, 2H, Ar—H), 6.80 (d, $^3J_{HH}$ = 4.4Hz, 1H, Ar—H), 5.16 (s, 2H, NH_2), 4.50 (q, $^3J_{HH}$ = 7.2Hz, 2H, OCH_2), 1.42 (t, $^3J_{HH}$ = 7.2Hz, 3H, CH_3). ^{13}C NMR (100MHz, $CDCl_3$) δ: 165.8, 143.3, 142.8, 140.2, 139.4, 133.9, 132.1, 131.4, 130.5 (2C), 126.9, 116.9, 107.0, 60.5, 13.3. HRMS: *m/z* 268.1082. Calcd for $C_{15}H_{13}N_3O_2$: 268.1081 $[M+H]^+$.

Data for Ⅳ－6c. Yield: 33.1%; red solid; mp, 98－100℃ (ethyl acetate/petroleum ether). ^{1}H NMR(400MHz, $CDCl_3$) δ: 8.16 (d, $^3J_{HH}$ = 8.4Hz, 1H, Ar—H), 8.07 (d, $^3J_{HH}$ = 6.4Hz, 1H, Ar—H), 7.62 (t, $^3J_{HH}$ = 7.6Hz, 1H, Ar—H), 7.50 (brd, $^4J_{HH}$ = 1.2Hz, 2H, Ar—H), 6.76 (brs, 1H, Ar—H), 5.15 (s, 2H, NH_2), 4.41 (t, $^3J_{HH}$ = 6.0Hz, 2H, OCH_2), 1.77－1.82 (m, 2H, CH_2), 1.04 (t, $^3J_{HH}$ = 7.2Hz, 3H, CH_3). ^{13}C NMR (100MHz, $CDCl_3$) δ: 167.2, 144.3, 143.8, 141.2, 140.4, 134.9, 133.2, 132.5, 131.6 (2C), 127.9, 117.8, 108.0, 67.2, 22.2, 10.7. HRMS: *m/z* 282.1241. Calcd for $C_{16}H_{15}N_3O_2$: 282.1237 $[M+H]^+$.

Data for Ⅳ－6d. Yield: 60.1%; brown solid; mp, 110－112℃ (ethyl acetate/petroleum ether). ^{1}H NMR(400MHz, $CDCl_3$) δ: 8.30 (dd, $^3J_{HH}$ = 8.8Hz, $^4J_{HH}$ = 1.2Hz, 1H, Ar—H), 8.17 (dd, $^3J_{HH}$ = 6.8Hz, $^4J_{HH}$ = 1.2Hz, 1H, Ar—H), 7.75 (dd, $^3J_{HH}$ = 8.8Hz, $^4J_{HH}$ = 2.8Hz, 1H, Ar—H), 7.63 (dt, $^3J_{HH}$ = 8.8Hz, $^3J_{HH}$ = 6.8Hz, 1H, Ar—H), 7.61 (dt, $^3J_{HH}$ = 8.8Hz, $^3J_{HH}$ = 6.8Hz, 1H, Ar—H), 6.91 (dd, $^3J_{HH}$ = 6.8Hz, $^3J_{HH}$ = 1.6Hz, 1H, Ar—H), 5.24 (s, 2H, NH_2), 4.53 (t, $^3J_{HH}$ = 6.4Hz, 2H, OCH_2), 1.83－1.87 (m, 2H, CH_2), 1.58－1.64 (m, 2H, CH_2), 1.03 (t, $^3J_{HH}$ = 7.2Hz, 3H, CH_3). ^{13}C NMR (100MHz, $CDCl_3$) δ: 167.1, 144.3, 143.8, 141.3, 140.5, 135.0, 133.2, 132.5, 131.7, 131.6, 128.0, 117.9, 108.1, 65.5, 30.8, 19.3, 13.8. HRMS: *m/z* 296.1394. Calcd for $C_{17}H_{17}N_3O_2$: 296.1393 $[M+H]^+$.

Data for Ⅳ－7a. Yield: 47.6%; red solid; mp, 201－203℃ (acetone/petroleum ether)[5]. ^{1}H NMR(400MHz, d_6－DMSO) δ: 8.11 (d, $^3J_{HH}$ = 8.4Hz, 1H, Ar—H), 7.90 (d, $^3J_{HH}$ = 8.8Hz, 1H, Ar—H), 7.86－7.87 (brd, 1H, Ar—H), 7.79 (d, $^3J_{HH}$ = 8.0Hz, 1H, Ar—H), 7.49 (d, $^3J_{HH}$ = 8.4Hz, 1H, Ar—H), 6.91 (s, 1H, Ar—H), 6.65 (s, 2H, NH_2), 3.96 (s, 3H, OCH_3). ^{13}C NMR (100MHz, d_6－DMSO) δ: 167.4, 151.5, 145.8, 142.5, 139.2, 136.5, 132.0, 130.9, 130.4, 128.9, 127.4, 126.9, 100.7, 52.3.

Data for Ⅳ－7b. Yield: 59.7%; red solid; mp, 190－192℃ (ethyl acetate/petroleumether). ^{1}H NMR(300MHz, $CDCl_3$) δ: 8.20 (dd, $^3J_{HH}$ = 7.5Hz, $^4J_{HH}$ = 1.2Hz, 1H, Ar—H), 8.07 (d, $^3J_{HH}$ = 9.3Hz, 1H, Ar—H), 8.03 (dd, $^3J_{HH}$ = 7.2Hz, $^4J_{HH}$ = 1.2Hz, 1H, Ar—H), 7.74

(dd, $^3J_{HH}$ = 8.7Hz, $^3J_{HH}$ = 7.2Hz, 1H, Ar—H), 7.32 (dd, $^3J_{HH}$ = 9.3Hz, $^4J_{HH}$ = 2.4Hz, 1H, Ar—H), 7.13 (s, 1H, Ar—H), 4.58 (q, $^3J_{HH}$ = 6.9Hz, 2H, OCH_2), 4.53 (s, 2H, NH_2), 1.49 (t, $^3J_{HH}$ = 6.9Hz, 3H, CH_3). ^{13}C NMR (100MHz, $CDCl_3$) δ: 167.1, 148.8, 145.4, 143.1, 140.4, 138.7, 132.2, 131.7 (2C), 129.2, 128.8, 125.7, 104.5, 61.5, 14.4. HRMS: *m/z* 290.0899. Calcd for $C_{15}H_{13}N_3O_2$: 290.0900 $[M+Na]^+$.

Data for Ⅳ - 7c. Yield: 29.9%; red solid; mp, 170 - 172℃ (ethyl acetate/petroleum ether). ^1H NMR (400MHz, $CDCl_3$) δ: 8.20 (d, $^3J_{HH}$ = 8.4Hz, 1H, Ar—H), 8.03 (brd, $^3J_{HH}$ = 6.8Hz, 2H, Ar—H), 7.73 (t, $^3J_{HH}$ = 7.6Hz, 1H, Ar—H), 7.29 (d, $^3J_{HH}$ = 9.6Hz, 1H, Ar—H), 7.10 (s, 1H, Ar—H), 4.61 (s, 2H, NH_2), 4.47 (t, $^3J_{HH}$ = 6.0Hz, 2H, OCH_2), 1.87 - 1.89 (m, 2H, CH_2), 1.12 (t, $^3J_{HH}$ = 6.8Hz, 3H, CH_3). ^{13}C NMR (100MHz, $CDCl_3$) δ: 167.3, 148.9, 145.4, 143.1, 140.3, 138.6, 132.1, 131.8, 131.5, 129.1, 128.8, 125.8, 104.3, 67.1, 22.1, 10.6. HRMS: *m/z* 304.1059. Calcd for $C_{16}H_{15}N_3O_2$: 304.1056 $[M+Na]^+$.

Data for Ⅳ - 7d. Yield: 31.7%; red solid; mp, 202 - 204℃ (ethylacetate/petroleum ether). ^1H NMR (400MHz, $CDCl_3$) δ: 8.20 (d, $^3J_{HH}$ = 8.4Hz, 1H, Ar—H), 8.07 (d, $^3J_{HH}$ = 9.6Hz, 1H, Ar—H), 8.03 (t, $^3J_{HH}$ = 6.8Hz, 1H, Ar—H), 7.75 (t, $^3J_{HH}$ = 8.0Hz, $^3J_{HH}$ = 6.8Hz, 1H, Ar—H), 7.33 (dd, $^3J_{HH}$ = 9.6Hz, $^4J_{HH}$ = 2.0Hz, 1H, Ar—H), 7.14 (d, $^4J_{HH}$ = 2.4Hz, 1H, Ar—H), 4.51 (t, $^3J_{HH}$ = 6.4Hz, 2H, OCH_2), 4.46 (s, 2H, NH_2), 1.81 - 1.88 (m, 2H, CH_2), 1.57 - 1.62 (m, 2H, CH_2), 1.02 (t, $^3J_{HH}$ = 7.2Hz, 3H, CH_3). ^{13}C NMR (100MHz, $CDCl_3$) δ: 167.3, 148.7, 145.4, 143.1, 140.4, 138.8, 132.2, 131.9, 131.7, 129.2, 128.8, 125.7, 104.6, 65.4, 30.8, 19.3, 13.8. HRMS: *m/z* 296.1391. Calcd for $C_{17}H_{17}N_3O_2$: 296.1393 $[M+H]^+$.

Data for Ⅳ - 8a. Yield: 11.8%; red solid; mp, 185 - 187℃ (ethylacetate/petroleum ether)[6]. ^1H NMR (400MHz, $CDCl_3$) δ: 8.27 (d, $^3J_{HH}$ = 8.4Hz, 1H, Ar—H), 8.16 (d, $^3J_{HH}$ = 6.8Hz, 1H, Ar—H), 7.99 (d, $^3J_{HH}$ = 9.2Hz, 1H, Ar—H), 7.67 (dt, $^3J_{HH}$ = 8.4Hz, $^3J_{HH}$ = 7.2Hz, 1H, Ar—H), 7.34 (d, $^3J_{HH}$ = 9.2Hz, 1H, Ar—H), 7.25 (s, 1H, Ar—H), 4.58 (s, 2H, NH_2), 4.08 (s, 3H, OCH_3).

Data for Ⅳ - 8b. Yield: 19.0%; red solid; mp, 136 - 138℃ (ethylacetate/petroleum ether). ^1H NMR (400MHz, $CDCl_3$) δ: 8.18 (d, $^3J_{HH}$ = 8.0Hz, 1H, Ar—H), 8.06 (d, $^3J_{HH}$ = 7.2Hz, 1H, Ar—H), 7.93 (d, $^3J_{HH}$ = 9.2Hz, 1H, Ar—H), 7.60 (dd, $^3J_{HH}$ = 8.4Hz, $^3J_{HH}$ = 7.2Hz, 1H, Ar—H), 7.25 (dd, $^3J_{HH}$ = 9.3Hz, $^4J_{HH}$ = 2.4Hz, 1H, Ar—H), 7.17 (d, $^4J_{HH}$ = 2.4Hz, 1H, Ar—H), 4.49 (q, $^3J_{HH}$ = 7.2Hz, 2H, OCH_2), 4.37 (s, 2H, NH_2), 1.42 (t, $^3J_{HH}$ = 7.2Hz, 3H, CH_3). ^{13}C NMR (100MHz, $CDCl_3$) δ: 166.0, 147.5, 144.7, 140.5, 139.6, 138.9, 132.2, 130.5, 129.7 (2C), 125.5, 124.9, 104.7, 60.4, 13.3. HRMS: *m/z* 268.1082. Calcd for $C_{15}H_{13}N_3O_2$: 268.1081 $[M+H]^+$.

Data for Ⅳ - 8c. Yield: 21.2%; red solid; mp, 144 - 146℃ (ethyl acetate/petroleum ether). ^1H NMR (400MHz, $CDCl_3$) δ: 8.16 (d, $^3J_{HH}$ = 8.4Hz, 1H, Ar—H), 8.05 (d, $^3J_{HH}$ = 6.8Hz, 1H, Ar—H), 7.88 (d, $^3J_{HH}$ = 9.2Hz, 1H, Ar—H), 7.58 (dd, $^3J_{HH}$ = 8.0Hz, $^3J_{HH}$ = 7.2Hz, 1H, Ar—H), 7.21 (dd, $^3J_{HH}$ = 9.2Hz, $^3J_{HH}$ = 8.0Hz, 1H, Ar—H), 7.13 (s, 1H, Ar—H), 4.50 (s, 2H, NH_2), 4.39 (t, $^3J_{HH}$ = 6.0Hz, 2H, OCH_2), 1.77 - 1.82 (m, 2H, CH_2), 1.04 (t, $^3J_{HH}$ = 6.8Hz, 3H, CH_3). ^{13}C NMR (100MHz, $CDCl_3$) δ: 166.3, 147.7, 144.7, 140.5,

139.5, 138.8, 132.2, 130.5, 129.7, 129.6, 125.4, 125.0, 104.4, 66.0, 21.1, 9.6. HRMS: *m/z* 282.1243. Calcd for $C_{16}H_{15}N_3O_2$: 282.1237[M + H]$^+$.

Data for Ⅳ - 8d. Yield: 50.5%; red solid; mp, 141 - 143℃ (ethyl acetate/petroleum ether). ^{1}H NMR (400MHz, $CDCl_3$) δ: 8.10 (d, $^3J_{HH}$ = 8.4Hz, 1H, Ar—H), 7.99 (d, $^3J_{HH}$ = 6.8Hz, 1H, Ar—H), 7.74 (d, $^3J_{HH}$ = 9.2Hz, 1H, Ar—H), 7.51 (t, $^3J_{HH}$ = 7.6Hz, 1H, Ar—H), 7.13 (d, $^3J_{HH}$ = 9.2Hz, 1H, Ar—H), 7.06 (s, 1H, Ar—H), 4.78 (s, 2H, NH_2), 4.37 (t, $^3J_{HH}$ = 6.4Hz, 2H, OCH_2), 1.67 - 1.70 (m, 2H, CH_2), 1.39 - 1.44 (m, 2H, CH_2), 0.85 (t, $^3J_{HH}$ = 7.2Hz, 3H, CH_3). ^{13}C NMR (100MHz, $CDCl_3$) δ: 166.2, 148.2, 144.8, 140.3, 139.3, 138.7, 132.1, 130.4, 129.5, 129.4, 125.2 (2C), 103.7, 64.2, 29.7, 18.2, 12.7. HRMS: *m/z* 296.1392. Calcd for $C_{17}H_{17}N_3O_2$: 296.1393[M + H]$^+$.

Data for Ⅴ - 7b. Yield: 5.1%; red solid; mp, 159 - 161℃ (ethyl acetate/petroleum ether). ^{1}H NMR (400MHz, $CDCl_3$) δ: 8.22 (d, $^3J_{HH}$ = 8.4Hz, 1H, H - 4), 8.04 (d, $^3J_{HH}$ = 6.8Hz, 1H, H - 2), 7.66 - 7.72 (m, 1H, H - 3), 7.35 (s, 1H, H - 9), 7.13 (s, 1H, H - 6), 4.89 (s, 2H, NH_2), 4.56 (t, $^3J_{HH}$ = 7.2Hz, 2H, CH_2 - 2′), 4.29 (t, $^3J_{HH}$ = 7.2Hz, 2H, CH_2 - 1″), 1.55 (q, $^3J_{HH}$ = 7.2Hz, 3H, CH_3 - 2″), 1.48 (q, $^3J_{HH}$ = 7.2Hz, 3H, CH_3 - 3′). ^{13}C NMR (100MHz, $CDCl_3$) δ: 167.0 (s, C - 1′), 152.8 (s, C - 8), 143.6 (s, C - 7), 142.2 (s, C - 9a), 142.2 (s, C - 5a), 141.2 (s, C - 4a), 138.6 (s, C - 10a), 131.9 (d, C - 4), 130.5 (s, C - 1), 129.1 (d, C - 2), 127.3 (d, C - 3), 105.7 (d, C - 9), 103.4 (d, C - 6), 64.9 (t, C - 1″), 61.3 (t, C - 2′), 14.5 (q, C - 2″), 14.4 (q, C - 3′). HRMS: *m/z* 334.1163. Calcd for $C_{17}H_{17}N_3O_3$: 334.1162[M + Na]$^+$.

Data for Ⅴ - 7c. Yield: 10.6%; red solid; mp, 140 - 142℃ (ethyl acetate/petroleum ether). ^{1}H NMR (300MHz, $CDCl_3$) δ: 8.14 (dd, $^3J_{HH}$ = 7.2Hz, $^4J_{HH}$ = 1.2Hz, 1H, Ar—H), 7.97 (dd, $^3J_{HH}$ = 6.9Hz, $^3J_{HH}$ = 1.2Hz, 1H, Ar—H), 7.61 (dd, $^3J_{HH}$ = 7.2Hz, $^4J_{HH}$ = 1.2Hz, 1H, Ar—H), 7.28 (s, 1H, Ar—H), 7.04 (s, 1H, Ar—H), 4.72 (s, 2H, NH_2), 4.39 (t, $^3J_{HH}$ = 6.6Hz, 2H, OCH_2), 4.13 (t, $^3J_{HH}$ = 6.6Hz, 2H, OCH_2), 1.85 - 1.92 (m, 2H, CH_2), 1.77 - 1.85 (m, 2H, CH_2), 1.04 - 1.06 (m, 6H, $2CH_3$). ^{13}C NMR (100MHz, $CDCl_3$) δ: 166.2, 151.8, 142.3, 141.6, 141.0, 140.6, 137.7, 131.2, 129.6, 128.1, 126.1, 104.6, 102.7, 69.6, 65.9, 21.2 (2C), 9.6, 9.5. HRMS: *m/z* 340.1653. Calcd for $C_{19}H_{21}N_3O_3$: 340.1656[M + H]$^+$.

Data for Ⅴ - 7d. Yield: 30.2%; red solid; mp, 134 - 136℃ (ethyl acetate/petroleum ether). ^{1}H NMR (400MHz, $CDCl_3$) δ: 8.20 (d, $^3J_{HH}$ = 8.0Hz, 1H, Ar—H), 8.04 (d, $^3J_{HH}$ = 7.2Hz, 1H, Ar—H), 7.68 (dd, $^3J_{HH}$ = 8.4Hz, $^3J_{HH}$ = 7.2Hz, 1H, Ar—H), 7.35 (s, 1H, Ar—H), 7.11 (s, 1H, Ar—H), 4.83 (s, 2H, NH_2), 4.51 (t, $^3J_{HH}$ = 6.8Hz, 2H, OCH_2), 4.23 (t, $^3J_{HH}$ = 6.4Hz, 2H, OCH_2), 1.82 - 1.92 (m, 4H, $2CH_2$), 1.54 - 1.60 (m, 4H, $2CH_2$), 1.00 - 1.03 (m, 6H, $2CH_3$). ^{13}C NMR (100MHz, $CDCl_3$) δ: 167.3, 152.9 143.4, 142.5, 142.0, 141.6, 138.7, 132.2, 130.7, 129.1, 127.2, 105.6, 103.7, 68.9, 65.2, 30.8 (2C), 19.3 (2C), 13.8, 13.7. HRMS: *m/z* 390.1786. Calcd for $C_{21}H_{25}N_3O_3$: 390.1788[M + Na]$^+$.

Data for Ⅴ - 8d. Yield: 14.6%; red solid; mp, 198 - 200℃ (ethyl acetate/petroleum ether). ^{1}H NMR (400MHz, $CDCl_3$) δ: 8.13 (d, $^3J_{HH}$ = 8.4Hz, 1H, Ar—H), 7.99 (d, $^3J_{HH}$ = 6.8Hz, 1H, Ar—H), 7.57 (t, $^3J_{HH}$ = 7.2Hz, 1H, Ar—H), 7.18 (s, 1H, Ar—H), 7.13 (s, 1H, Ar—H), 4.70 (s, 2H, NH_2), 4.43 (t, $^3J_{HH}$ = 6.4Hz, 2H, OCH_2), 4.17 (t, $^3J_{HH}$ = 6.4Hz, 2H,

OCH_2),1.83-1.86(m,2H,CH_2),1.75-1.78(m,2H,CH_2),1.48-1.54(m,4H,2CH_2), 0.94-0.99(m,6H,2CH_3). ^{13}C NMR(100MHz, $CDCl_3$) δ: 166.4, 152.0, 142.1, 142.0, 140.4, 139.4, 138.9, 131.3, 129.6, 128.9, 125.2, 103.9, 103.5, 67.8, 64.2, 29.8(2C), 18.3(2C), 12.8(2C). HRMS: m/z 368.1970. Calcd for $C_{21}H_{25}N_3O_3$: 368.1969[M+H]$^+$.

General Synthetic Procedures for Target Compounds Ⅵ-6a—Ⅵ-7n. Methyl-6-aminophenazine-1-carboxylate(Ⅳ-**6a**,50.0mg,0.2mmol) was dissolved in the solution of dry CH_2Cl_2(4.0mL) and anhydrous pyridine(0.2mL). Acetyl chloride(0.2mL) diluted in CH_2Cl_2(6.0mL) was added dropwise to the above solution. The mixture was stirred under 15℃ for 0.5h, and then 20mL of CH_2Cl_2 was added, followed by washing with dilute 2N hydrochloric acid, saturated sodium bicarbonate and water respectively. Separating the organic layer, the solvent was removed by distillation. The residue was recrystallized with ethyl acetate/petroleum ether(60-90℃) to give methyl-6-acetamidophenazine-1-carboxylate(Ⅵ-**6a**,23.0mg, 40.0%. Compounds Ⅵ-**6b**-Ⅵ-**6e**, Ⅵ-**6h** and Ⅵ-**7h** were prepared with corresponding acyl chloride through the same process. Compounds Ⅵ-**6f** and Ⅵ-**6g** were prepared as follows: Methyl-6-aminophenazine-1-carboxylate(Ⅳ-**6a**,50.0mg,0.2mmol) and 4-nitrobenzene-1-sulfonyl chloride(88mg,0.4mmol) were dissolved in a solution of CH_2Cl_2 (20.0mL) and pyridine(0.2mL). The mixture was refluxed for 1h, and then 20mL of CH_2Cl_2 was added, followed by washing with dilute 2N hydrochloric acid, saturated sodium bicarbonate and water respectively. Leaving the organic layer, the solvent was removed by distillation. The residue was recrystallized with ethyl acetate/petroleum ether(60-90℃) to give methyl 6-(4-nitrophenylsulfonamido) phenazine-1-carboxylate(Ⅵ-**6f**, 52.0mg, 60%). To be pointed out, in the process of preparing compounds Ⅵ-**7a**—Ⅵ-**7e** and Ⅵ-**7j**—Ⅵ-**7n**, methyl-7-aminophenazine-1-carboxylate(Ⅳ-**7a**) was dissolved only in the anhydrous pyridine. The rest of the procedures were the same as preparing for compound Ⅵ-**6a**. In the process of preparing compounds Ⅵ-**7f**, Ⅵ-**7g** and Ⅵ-**7i**, compound Ⅳ-**7a** and the corresponding sulfonyl chloride were refluxed in anhydrous pyridine[11]. The other procedures were the same as preparing for compound Ⅵ-**6f**. All the substituents at the phenazine rings are listed in Scheme 2 and Table 1.

Data for Ⅵ-6a. Yield: 79.3%; yellow solid; mp, 212-214℃ (ethyl acetate/petroleum ether). ^{1}H NMR(400MHz, $CDCl_3$) δ: 9.66(s,1H,NHCO), 8.81(d, $^3J_{HH}$=7.6Hz,1H,Ar—H), 8.33(dd, $^3J_{HH}$=8.8Hz, $^4J_{HH}$=1.2Hz, 1H, Ar—H), 8.23(dd, $^3J_{HH}$=6.8Hz, $^4J_{HH}$=1.2Hz,1H,Ar—H), 7.95(d, $^3J_{HH}$=8.0Hz,1H,Ar—H), 7.81(dt, $^3J_{HH}$=8.8Hz, $^4J_{HH}$=1.6Hz, 2H,Ar—H), 4.11(s,3H,OCH_3), 2.43(s,3H,$COCH_3$). ^{13}C NMR(100MHz, $CDCl_3$) δ: 168.8, 166.9, 143.3, 141.2, 140.3, 134.3, 133.9, 132.7, 132.2, 132.1, 131.4, 129.1, 123.7, 116.4, 52.7, 25.1. HRMS: m/z 296.1031. Calcd for $C_{16}H_{13}N_3O_3$: 296.1030[M+H]$^+$.

Data for Ⅵ-6b. Yield: 35.7%; yellow solid; mp, 165-167℃ (ethyl acetate/petroleum ether). ^{1}H NMR(400MHz, $CDCl_3$) δ: 10.03(s,1H,NHCO), 8.77(d, $^3J_{HH}$=7.6Hz,1H,Ar—H), 8.22(d, $^3J_{HH}$=6.8Hz,1H,Ar—H), 8.18(d, $^3J_{HH}$=8.4Hz,1H,Ar—H), 7.99(d, $^3J_{HH}$=8.8Hz, 1H, Ar—H), 7.82(dt, $^3J_{HH}$=7.2Hz, $^4J_{HH}$=2.4Hz, 2H, Ar—H), 4.89(s,2H, OCH_2), 4.11(s,3H,OCH_3), 2.42(s,3H,$COCH_3$). ^{13}C NMR(100MHz, $CDCl_3$) δ: 169.3,

166.8, 165.4, 143.2, 141.2, 140.3, 134.3, 132.8, 132.6, 132.2, 132.0, 131.5, 129.3, 124.5, 116.8, 63.3, 52.7, 20.8. HRMS: *m/z* 376.0899. Calcd for $C_{18}H_{15}N_3O_5$: 376.0904 [M + Na]$^+$.

Data for Ⅵ - 6c. Yield: 80.3%; orange solid; mp, 158 - 160℃ (ethyl acetate/petroleum ether). ^{1}H NMR (400MHz, $CDCl_3$) δ: 9.70 (s, 1H, NHCO), 8.78 (d, $^3J_{HH}$ = 7.6Hz, 1H, Ar—H), 8.32 (d, $^3J_{HH}$ = 8.4Hz, 1H, Ar—H), 8.22 (d, $^3J_{HH}$ = 6.8Hz, 1H, Ar—H), 7.96 (d, $^3J_{HH}$ = 8.8Hz, 1H, Ar—H), 7.82 (t, $^3J_{HH}$ = 8.0Hz, 2H, Ar—H), 4.11 (s, 3H, OCH_3), 3.75 (t, $^3J_{HH}$ = 6.4Hz, 2H, CH_2Cl), 2.85 (t, $^3J_{HH}$ = 6.8Hz, 2H, CH_2CO), 2.29 - 2.35 (m, 2H, CH_2). ^{13}C NMR (100MHz, $CDCl_3$) δ: 170.4, 166.9, 143.3, 141.2, 140.3, 134.3, 133.7, 132.8, 132.1 (2C), 131.4, 129.1, 123.9, 116.4, 52.7, 44.4, 34.5, 27.8. HRMS: *m/z* 380.0774. Calcd for $C_{18}H_{16}ClN_3O_3$: 380.0772 [M + Na]$^+$.

Data for Ⅵ - 6d. Yield: 90.1%; yellow solid; mp, 191 - 193℃ (ethyl acetate/petroleum ether). ^{1}H NMR (400MHz, $CDCl_3$) δ: 9.60 (s, 1H, NHCO), 8.82 (d, $^3J_{HH}$ = 6.8Hz, 1H, Ar—H), 8.25 (d, $^3J_{HH}$ = 8.4Hz, 1H, Ar—H), 8.20 (d, $^3J_{HH}$ = 6.0Hz, 1H, Ar—H), 7.96 (d, $^3J_{HH}$ = 8.4Hz, 1H, Ar—H), 7.82 (t, $^3J_{HH}$ = 8.0Hz, 2H, Ar—H), 7.31 (brs, 4H, Ar—H), 7.19 (brs, 1H, Ar—H), 4.10 (s, 3H, OCH_3), 3.17 (t, $^3J_{HH}$ = 7.2Hz, 2H, CH_2CO), 2.96 (t, $^3J_{HH}$ = 7.2Hz, 2H, Ar—CH_2). ^{13}C NMR (100MHz, $CDCl_3$) δ: 170.8, 166.9, 143.3, 141.1, 140.6, 140.3, 134.3, 133.8, 132.8, 132.3, 132.2, 131.3, 129.1, 128.6 (2C), 128.4 (2C), 126.4, 123.7, 116.4, 52.8, 39.7, 31.3. HRMS: *m/z* 386.1501. Calcd for $C_{23}H_{19}N_3O_3$: 386.1499 [M + H]$^+$.

Data for Ⅵ - 6e. Yield: 92.0%; yellow solid; mp, 226 - 228℃ (ethyl acetate/petroleum ether). ^{1}H NMR (400MHz, $CDCl_3$) δ: 10.63 (s, 1H, NHCO), 8.98 (d, $^3J_{HH}$ = 7.2Hz, 1H, Ar—H), 8.33 (d, $^3J_{HH}$ = 8.4Hz, 1H, Ar—H), 8.25 (d, $^3J_{HH}$ = 6.4Hz, 1H, Ar—H), 8.06 (d, $^3J_{HH}$ = 8.8Hz, 1H, Ar—H), 7.92 (t, $^3J_{HH}$ = 7.2Hz, 2H, Ar—H), 7.85 (d, $^3J_{HH}$ = 8.0Hz, 1H, Ar—H), 7.56 (s, 1H, Ar—H), 7.43 (d, $^3J_{HH}$ = 8.0Hz, 1H, Ar—H), 4.11 (s, 3H, OCH_3). ^{13}C NMR (100MHz, $CDCl_3$) δ: 166.9, 163.6, 143.3, 141.4, 140.5, 137.6, 134.6, 133.8, 133.2, 132.9, 132.3, 132.1, 131.9 (2C), 131.4, 130.5, 129.3, 127.8, 124.6, 117.1, 52.7. HRMS: *m/z* 426.0408. Calcd for $C_{21}H_{13}Cl_2N_3O_3$: 426.0407 [M + H]$^+$.

Data for Ⅵ - 6f. Yield: 54.2%; yellow solid; mp, 226 - 228℃ (ethyl acetate/petroleum ether). ^{1}H NMR (400MHz, d_6 - DMSO) δ: 11.12 (s, 1H, $NHSO_2$), 8.29 - 8.31 (brd, $^3J_{HH}$ = 8.4Hz, 3H, Ar—H), 8.25 (d, $^3J_{HH}$ = 6.8Hz, 1H, Ar—H), 8.20 (d, $^3J_{HH}$ = 8.8Hz, 2H, Ar—H), 8.00 - 8.04 (m, 2H, Ar—H), 7.96 (t, $^3J_{HH}$ = 7.2Hz, 1H, Ar—H), 7.87 (d, $^3J_{HH}$ = 6.8Hz, 1H, Ar—H), 3.99 (s, 3H, OCH_3). ^{13}C NMR (100MHz, d_6 - DMSO) δ: 166.5, 149.7, 145.2, 142.6, 140.6, 139.9, 136.4, 133.2, 132.1, 131.8, 131.6 (2C), 130.2, 128.5 (2C), 125.9, 124.3 (2C), 121.5, 52.5. HRMS: *m/z* 439.0708. Calcd for $C_{20}H_{14}N_4O_6S$: 439.0707 [M + H]$^+$.

Data for Ⅵ - 6g. Yield: 83.3%; brown solid; mp, 192 - 194℃ (ethyl acetate/petroleum ether). ^{1}H NMR (400MHz, $CDCl_3$) δ: 10.00 (s, 1H, $NHSO_2$), 8.37 (d, $^3J_{HH}$ = 8.8Hz, 1H, Ar—H), 8.23 (d, $^3J_{HH}$ = 9.2Hz, 1H, Ar—H), 8.10 - 8.13 (m, 1H, Ar—H), 8.08 (d, $^3J_{HH}$ = 7.2Hz, 1H, Ar—H), 8.01 (d, $^3J_{HH}$ = 9.2Hz, 1H, Ar—H), 7.85 (d, $^3J_{HH}$ = 8.0Hz, 1H, Ar—

H), 7.82 (d, $^{3}J_{HH}$ = 8.0Hz, 2H, Ar—H), 7.59 - 7.61 (m, 2H, Ar—H), 4.07 (s, 3H, OCH_3). ^{13}C NMR (100MHz, $CDCl_3$) δ: 166.7, 147.9, 143.3, 141.5, 140.9, 134.8, 134.1, 133.3, 132.9, 132.7, 132.6, 132.5, 131.3, 131.2, 131.1, 129.6, 125.7, 125.4, 116.3, 52.7. HRMS: *m/z* 439.0705. Calcd for $C_{20}H_{14}N_4O_6S$: 439.0707[M + H]$^+$.

Data for Ⅵ - 6h. Yield: 37.0%; yellow solid; mp, 101 - 103℃ (ethyl acetate/petroleum ether). ^{1}H NMR(400MHz, $CDCl_3$) δ: 9.76 (s, 1H, NHCO), 8.81 (d, $^{3}J_{HH}$ = 7.2Hz, 1H, Ar—H), 8.36 (d, $^{3}J_{HH}$ = 8.8Hz, 1H, Ar—H), 8.22 (d, $^{3}J_{HH}$ = 6.8Hz, 1H, Ar—H), 7.96 (d, $^{3}J_{HH}$ = 8.8Hz, 1H, Ar—H), 7.87 (d, $^{3}J_{HH}$ = 6.8Hz, 1H, Ar—H), 7.83 (d, $^{3}J_{HH}$ = 7.6Hz, 1H, Ar—H), 4.54 (t, $^{3}J_{HH}$ = 6.8Hz, 2H, OCH_2), 3.75 (t, $^{3}J_{HH}$ = 6.4Hz, 2H, CH_2Cl), 2.84 - 2.88 (m, 2H, CH_2CO), 2.31 - 2.34 (m, 2H, CH_2), 1.84 - 1.88 (m, 2H, CH_2), 1.58 - 1.64 (m, 2H, CH_2), 1.04 (t, $^{3}J_{HH}$ = 7.2Hz, 3H, CH_3). ^{13}C NMR(100MHz, $CDCl_3$) δ: 170.4, 166.8, 143.3, 141.3, 140.4, 134.3, 133.8, 132.6, 132.1 (2C), 131.8, 129.3, 123.8, 116.4, 65.6, 44.4, 34.5, 30.8, 27.8. 19.3, 13.7. HRMS: *m/z* 400.1419. Calcd for $C_{21}H_{22}ClN_3O_3$: 400.1422[M + H]$^+$.

Data for Ⅵ - 7a. Yield: 34.5%; yellow solid; mp, 234 - 236℃ (ethyl acetate/petroleum ether). ^{1}H NMR(400MHz, $CDCl_3$) δ: 8.40 (s, 1H, NHCO), 8.25 (d, $^{3}J_{HH}$ = 8.8Hz, 1H, Ar—H), 8.12 - 8.14 (brd, $^{3}J_{HH}$ = 8.4Hz, 2H, Ar—H), 7.95 (s, 1H, Ar—H), 7.87 (d, $^{3}J_{HH}$ = 8.8Hz, 1H, Ar—H), 7.75 (t, $^{3}J_{HH}$ = 7.2Hz, 1H, Ar—H), 4.03 (s, 3H, OCH_3), 2.22 (s, 3H, $COCH_3$). ^{13}C NMR (100MHz, $CDCl_3$) δ: 167.8, 166.2, 143.0, 142.1, 140.3, 139.1 (2C), 132.0, 130.3, 130.2, 130.0, 128.1, 125.2, 113.9, 51.7, 23.8. HRMS: *m/z* 318.0847. Calcd for $C_{16}H_{13}N_3O_3$: 318.0849[M + Na]$^+$.

Data for Ⅵ - 7b. Yield: 50.1%; yellow solid; mp, 206 - 208℃ (ethyl acetate/petroleum ether). ^{1}H NMR(400MHz, $CDCl_3$) δ: 8.56 (s, 1H, NHCO), 8.28 (brs, 4H, Ar—H), 7.85 - 7.96 (brd, 2H, Ar—H), 4.80 (s, 2H, OCH_2), 4.11 (s, 3H, OCH_3), 2.28 (s, 3H, $COCH_3$). ^{13}C NMR(100MHz, $CDCl_3$) δ: 169.3, 167.1, 165.5, 150.9, 143.9, 143.2, 141.5, 140.4, 138.7, 133.2, 131.6, 131.4, 129.3, 126.0, 116.1, 63.4, 52.7, 20.8. HRMS: *m/z* 376.0902. Calcd for $C_{18}H_{15}N_3O_5$: 376.0904[M + Na]$^+$.

Data for Ⅵ - 7c. Yield: 47.5%; brown solid; mp, 198 - 200℃ (ethyl acetate/petroleum ether). ^{1}H NMR (400MHz, $CDCl_3$) δ: 8.49 (s, 1H, NHCO), 8.45 (s, 1H, Ar—H), 8.31 (d, $^{3}J_{HH}$ = 8.4Hz, 1H, Ar—H), 8.12 (d, $^{3}J_{HH}$ = 7.2Hz, 1H, Ar—H), 8.10 (d, $^{3}J_{HH}$ = 9.2Hz, 1H, Ar—H), 8.00 (d, $^{3}J_{HH}$ = 9.2Hz, 1H, Ar—H), 7.79 (dd, $^{3}J_{HH}$ = 8.4Hz, $^{3}J_{HH}$ = 7.2Hz, 1H, Ar—H), 4.03 (s, 3H, OCH_3), 3.58 (t, $^{3}J_{HH}$ = 6.0Hz, 2H, CH_2Cl), 2.62 (t, $^{3}J_{HH}$ = 6.8Hz, 2H, CH_2CO), 2.12 - 2.15 (m, 2H, CH_2). ^{13}C NMR(100MHz, $CDCl_3$) δ: 171.0, 166.9, 142.3, 141.7, 141.4, 141.1, 140.0, 131.6, 131.4, 131.3, 131.1, 130.1, 126.7, 113.3, 52.8, 44.3, 34.3, 27.6. HRMS: *m/z* 380.0777. Calcd for $C_{18}H_{16}ClN_3O_3$: 380.0772[M + Na]$^+$.

Data for Ⅵ - 7d. Yield: 65.7%; yellow solid; mp, 215 - 217℃ (ethyl acetate/petroleum ether). ^{1}H NMR(400MHz, $CDCl_3$) δ: 8.44 (s, 1H, NHCO), 8.33 (d, $^{3}J_{HH}$ = 8.0Hz, 1H, Ar—H), 8.28 (s, 1H, Ar—H), 8.12 - 8.17 (m, 2H, Ar—H), 7.95 (d, $^{3}J_{HH}$ = 8.0Hz, 1H, Ar—H), 7.81 (t, $^{3}J_{HH}$ = 7.2Hz, 1H, Ar—H), 7.20 - 7.26 (m, 5H, Ar—H), 4.08 (s, 3H, OCH_3), 3.06 (t, $^{3}J_{HH}$ = 6.4Hz, 2H, CH_2CO), 2.80 (t, $^{3}J_{HH}$ = 6.4Hz, 2H, Ar—CH_2). ^{13}C NMR

(100MHz, $CDCl_3$) δ: 171.2, 167.1, 143.3, 142.2, 141.5, 140.7, 140.3, 140.0, 132.5, 131.4, 131.2, 131.0, 129.5, 128.6 (2C), 128.4 (2C), 126.7, 126.4, 114.4, 52.8, 39.6, 31.3. HRMS: m/z 408.1322. Calcd for $C_{23}H_{19}N_3O_3$: 408.1319 $[M+Na]^+$.

Data for Ⅵ-7e. Yield: 30.1%; yellow solid; mp, 246-248℃ (ethyl acetate/petroleum ether). ^1H NMR (400MHz, $CDCl_3$) δ: 11.27 (s, 1H, NHCO), 8.83 (s, 1H, Ar—H), 8.38 (d, $^3J_{HH}$ = 8.4Hz, 1H, Ar—H), 8.25 (d, $^3J_{HH}$ = 9.6Hz, 1H, Ar—H), 8.16 (d, $^3J_{HH}$ = 6.0Hz, 1H, Ar—H), 8.12 (d, $^3J_{HH}$ = 8.8Hz, 1H, Ar—H), 7.99 (dd, $^3J_{HH}$ = 7.6Hz, $^3J_{HH}$ = 6.8Hz, 1H, Ar—H), 7.85 (s, 1H, Ar—H), 7.79 (d, $^3J_{HH}$ = 8.0Hz, 1H, Ar—H), 7.64 (d, $^3J_{HH}$ = 7.2Hz, 1H, Ar—H), 4.01 (s, 3H, OCH_3). ^{13}C NMR (100MHz, $CDCl_3$) δ: 166.8, 165.0, 143.7, 142.5, 140.8, 140.5, 139.0, 135.3, 135.1, 132.0 (2C), 131.2, 130.5, 130.3 (2C), 129.9, 129.3, 127.5, 126.9, 114.4, 52.5. HRMS: m/z 426.0408. Calcd for $C_{21}H_{13}Cl_2N_3O_3$: 426.0407 $[M+H]^+$.

Data for Ⅵ-7f. Yield: 38.0%; green solid; mp, 235-237℃ (ethyl acetate/petroleum ether). ^1H NMR (400MHz, d_6-DMSO) δ: 11.68 (s, 1H, $NHSO_2$), 8.38 (d, $^3J_{HH}$ = 8.4Hz, 2H, Ar—H), 8.29 (d, $^3J_{HH}$ = 8.4Hz, 1H, Ar—H), 8.19 (d, $^3J_{HH}$ = 8.4Hz, 2H, Ar—H), 8.16 (d, $^3J_{HH}$ = 6.8Hz, 1H, Ar—H), 8.12 (d, $^3J_{HH}$ = 6.8Hz, 1H, Ar—H), 7.94 (t, $^3J_{HH}$ = 8.0Hz, 1H, Ar—H), 7.82 (s, 1H, Ar—H), 7.78 (d, $^3J_{HH}$ = 8.8Hz, 1H, Ar—H), 3.95 (s, 3H, OCH_3). ^{13}C NMR (100MHz, d_6-DMSO) δ: 166.7, 150.1, 144.2, 143.1, 142.3, 140.2, 139.7, 139.0, 131.9, 131.2, 130.5, 130.1, 128.3 (2C), 126.2, 124.9 (3C), 112.9, 52.5. HRMS: m/z 461.0528. Calcd for $C_{20}H_{14}N_4O_6S$: 461.0526 $[M+Na]^+$.

Data for Ⅵ-7g. Yield: 37.1%; orange solid; mp, 199-201℃ (ethyl acetate/petroleum ether). ^1H NMR (400MHz, $CDCl_3$) δ: 8.30 (d, $^3J_{HH}$ = 8.4Hz, 1H, Ar—H), 8.25 (d, $^3J_{HH}$ = 9.6Hz, 1H, Ar—H), 8.20 (d, $^3J_{HH}$ = 7.2Hz, 1H, Ar—H), 7.99-8.01 (brd, 2H, Ar—H), 7.85-7.87 (brd, 2H, Ar—H), 7.82 (d, $^3J_{HH}$ = 10.0Hz, 1H, Ar—H), 7.67 (t, $^3J_{HH}$ = 7.6Hz, 1H, Ar—H), 7.56 (t, $^3J_{HH}$ = 7.6Hz, 1H, Ar—H), 4.09 (s, 3H, OCH_3). ^{13}C NMR (100MHz, $CDCl_3$) δ: 166.9, 148.1, 143.3, 143.1, 141.7, 140.6, 138.1, 134.5, 133.0, 132.8, 132.1, 131.9, 131.8, 131.7, 131.5, 129.7, 127.0, 125.6, 118.1, 52.8. HRMS: m/z 461.0531. Calcd for $C_{20}H_{14}N_4O_6S$: 461.0526 $[M+Na]^+$.

Data for Ⅵ-7h. Yield: 40.1%; yellow solid; mp, 200-202℃ (ethyl acetate/petroleum ether). ^1H NMR (400MHz, $CDCl_3$) δ: 8.44 (s, 1H, NHCO), 8.29 (d, $^3J_{HH}$ = 8.8Hz, 1H, Ar—H), 8.15 (d, $^3J_{HH}$ = 7.2Hz, 1H, Ar—H), 8.13 (d, $^3J_{HH}$ = 9.2Hz, 1H, Ar—H), 8.11 (d, $^3J_{HH}$ = 6.0Hz, 1H, Ar—H), 7.94 (dd, $^3J_{HH}$ = 8.0Hz, $^4J_{HH}$ = 1.2Hz, 1H, Ar—H), 7.82 (dd, $^3J_{HH}$ = 8.0Hz, $^3J_{HH}$ = 7.2Hz, 1H, Ar—H), 4.53 (t, $^3J_{HH}$ = 6.8Hz, 2H, OCH_2), 3.67 (t, $^3J_{HH}$ = 6.0Hz, 2H, CH_2Cl), 2.65-2.68 (m, 2H, CH_2CO), 2.22-2.25 (m, 2H, CH_2), 1.83-1.87 (m, 2H, CH_2), 1.57-1.62 (m, 2H, CH_3), 1.03 (t, $^3J_{HH}$ = 7.2Hz, 3H, CH_3). ^{13}C NMR (100MHz, $CDCl_3$) δ: 170.6, 167.1, 143.9, 143.1, 141.3, 140.2, 139.8, 132.9, 131.8, 131.0, 130.9, 129.2, 126.2, 115.2, 65.6, 44.3, 34.2, 30.8, 27.7, 19.3, 13.8. HRMS: m/z 400.1423. Calcd for $C_{21}H_{22}ClN_3O_3$: 400.1422 $[M+H]^+$.

Data for Ⅵ-7i. Yield: 35.5%; yellow solid; mp, 183-185℃ (ethyl acetate/petroleum ether). ^1H NMR (300MHz, $CDCl_3$) δ: 8.22 (d, $^3J_{HH}$ = 8.7Hz, 1H, Ar—H), 8.09 (d, $^3J_{HH}$ =

6.3Hz, 1H, Ar—H), 8.07 (d, $^3J_{HH}$ = 6.9Hz, 1H, Ar—H), 7.75 – 7.78 (brd, 4H, Ar—H), 7.62 (d, $^3J_{HH}$ = 8.7Hz, 1H, Ar—H), 7.13 (d, $^3J_{HH}$ = 8.1Hz, 2H, Ar—H), 4.01 (s, 3H, OCH_3), 2.23 (s, 3H, Ar—CH_3). ^{13}C NMR (100MHz, $CDCl_3$) δ: 166.0, 143.5, 141.7, 141.0, 139.1, 134.5, 131.2, 130.8, 130.5, 130.3, 128.9 (2C), 126.4 (2C), 125.0, 123.4, 122.9, 118.0, 112.4, 51.7, 20.5. HRMS: *m/z* 430.0839. Calcd for $C_{21}H_{17}N_3O_4S$: 430.0832 [M + Na]$^+$.

Data for Ⅵ – 7j. Yield: 45.7%; yellow solid; mp, 202 – 204℃ (ethyl acetate/petroleum ether). ^{1}H NMR (400MHz, $CDCl_3$) δ: 10.22 (s, 1H, Ar—H), 8.57 (s, 1H, Ar—H), 8.21 – 8.26 (m, 3H, Ar—H), 8.10 (d, $^3J_{HH}$ = 6.8Hz, 1H, Ar—H), 8.07 (d, $^3J_{HH}$ = 8.8Hz, 1H, Ar—H), 7.76 (t, $^3J_{HH}$ = 7.2Hz, 1H, Ar—H), 7.41 (brs, 1H, Ar—H), 7.07 (brs, 1H, 1H, Ar—H), 6.95 (d, $^3J_{HH}$ = 7.2Hz, 1H, Ar—H), 4.09 (s, 3H, OCH_3), 4.05 (s, 3H, OCH_3). ^{13}C NMR (100MHz, $CDCl_3$) δ: 167.1, 163.5, 157.1, 144.2, 143.1, 141.4, 140.5, 140.0, 133.7, 133.1, 132.5, 131.3, 131.1, 130.9, 128.9, 127.0, 121.6, 120.9, 115.3, 111.4, 56.2, 52.6. HRMS: *m/z* 410.1111. Calcd for $C_{22}H_{17}N_3O_4$: 410.1111 [M + Na]$^+$.

Data for Ⅵ – 7k. Yield: 20.1%; yellow solid; mp, 238 – 240℃ (ethyl acetate/petroleum ether). ^{1}H NMR (400MHz, d_6 – DMSO) δ: 10.88 (s, 1H, NHCO), 8.95 (s, 1H, Ar—H), 8.43 (d, $^3J_{HH}$ = 8.4Hz, 1H, Ar—H), 8.37 (d, $^3J_{HH}$ = 8.0Hz, 1H, Ar—H), 8.29 (d, $^3J_{HH}$ = 8.4Hz, 1H, Ar—H), 8.21 (d, $^3J_{HH}$ = 6.0Hz, 1H, Ar—H), 8.02 – 8.04 (brd, 3H, Ar—H), 7.47 (d, $^3J_{HH}$ = 6.4Hz, 2H, Ar—H), 4.06 (s, 3H, OCH_3), 2.48 (s, 3H, CH_3). ^{13}C NMR (100MHz, d_6 – DMSO) δ: 166.9, 166.4, 155.0, 143.8, 142.4, 142.2, 141.5, 140.4, 138.8, 132.0, 131.5, 130.1, 129.8, 129.7, 129.0 (2C), 127.9 (2C), 127.8, 114.4, 52.5, 21.0. HRMS: *m/z* 394.1159. Calcd for $C_{22}H_{17}N_3O_3$: 394.1162 [M + Na]$^+$.

Data for Ⅵ – 7l. Yield: 41.9%; yellow solid; mp, 100 – 102℃ (ethyl acetate/petroleum ether). ^{1}H NMR (400MHz, $CDCl_3$) δ: 8.53 (s, 1H, NHCO), 8.42 (s, 1H, Ar—H), 8.16 (d, $^3J_{HH}$ = 8.8Hz, 1H, Ar—H), 8.06 (d, $^3J_{HH}$ = 6.8Hz, 1H, Ar—H), 8.00 (d, $^3J_{HH}$ = 8.8Hz, 1H, Ar—H), 7.92 (d, $^3J_{HH}$ = 8.8Hz, 1H, Ar—H), 7.69 (t, $^3J_{HH}$ = 8.0Hz, 1H, Ar—H), 4.00 (s, 3H, OCH_3), 2.21 – 2.27 (m, 1H, cyclohexyl – CH), 1.81 – 1.87 (m, 2H, cyclohexyl – CH_2), 1.65 – 1.67 (m, 2H, cyclohexyl – CH_2), 1.41 – 1.52 (m, 3H, cyclohexyl – CH_2), 1.07 – 1.09 (m, 3H, cyclohexyl – CH_2). ^{13}C NMR (100MHz, $CDCl_3$) δ: 174.5, 166.2, 142.8, 141.7, 140.2, 139.7, 138.8, 131.9, 130.2, 130.0, 129.5, 128.0, 125.9, 113.8, 51.7, 45.4, 28.4 (2C), 24.5 (3C). HRMS: *m/z* 386.1482. Calcd for $C_{21}H_{21}N_3O_3$: 386.1475 [M + Na]$^+$.

Data for Ⅵ – 7m. Yield: 21.7%; orange solid; mp, 60 – 62℃ (ethyl acetate/petroleum ether). ^{1}H NMR (400MHz, $CDCl_3$) δ: 8.63 (s, 1H, NHCO), 8.51 (d, $^4J_{HH}$ = 2.0Hz, 1H, Ar—H), 8.25 (d, $^3J_{HH}$ = 8.8Hz, 1H, Ar—H), 8.18 (d, $^3J_{HH}$ = 9.2Hz, 1H, Ar—H), 8.10 (d, $^3J_{HH}$ = 6.8Hz, 1H, Ar—H), 7.99 (dd, $^3J_{HH}$ = 9.2Hz, $^4J_{HH}$ = 2.0Hz, 1H, Ar—H), 7.79 (d, $^3J_{HH}$ = 8.0Hz, 1H, Ar—H), 7.75 (dd, $^3J_{HH}$ = 7.2Hz, $^3J_{HH}$ = 6.8Hz, 1H, Ar—H), 7.44 (d, $^3J_{HH}$ = 8.0Hz, 1H, Ar—H), 7.25 (t, $^3J_{HH}$ = 8.0Hz, 1H, Ar—H), 7.10 (d, $^3J_{HH}$ = 8.8Hz, 1H, Ar—H), 4.02 (s, 3H, OCH_3), 2.28 (s, 3H, $COCH_3$). ^{13}C NMR (100MHz, $CDCl_3$) δ: 168.3, 166.1, 163.1, 146.9, 143.0, 142.1, 140.4, 139.2, 138.9, 132.1, 131.6, 130.4, 130.3, 129.9, 128.8, 128.2, 127.0, 125.5, 125.1, 122.4, 114.4, 51.7, 20.1. HRMS: *m/z* 416.1248. Calcd for $C_{23}H_{17}N_3O_5$:416.1241[M+H]$^+$.

Data for Ⅵ-7n. Yield: 47.7%; yellow solid; mp, 246-248℃ (ethyl acetate/petroleum ether). ^{1}H NMR(400MHz, d_6-DMSO) δ: 11.48 (s, 1H, NHCO), 8.67 (s, 1H, Ar—H), 8.38 (d, $^3J_{HH}$=8.8Hz, 1H, Ar—H), 8.23-8.29 (m, 2H, Ar—H), 8.18 (d, $^3J_{HH}$=6.4Hz, 1H, Ar—H), 7.99 (dd, $^3J_{HH}$=8.0Hz, $^3J_{HH}$=7.2Hz, 1H, Ar—H), 4.00 (s, 3H, OCH_3). ^{13}C NMR (100MHz, d_6-DMSO) δ: 166.7, 160.2, 143.1, 142.4, 140.6, 139.6, 139.3, 132.1, 131.9, 130.8, 130.2, 130.0, 127.2, 116.8, 51.5. HRMS: *m/z* 419.9688. Calcd for $C_{16}H_{10}Cl_3N_3O_3$: 419.9680[M+Na]$^+$.

General Synthetic Procedures for Target Compounds Ⅶ-a—Ⅶ-e. At 0℃, 65% nitric acid (0.2mL) was dropped to the solution of Ⅵ-**6c** (80mg, 0.22mmol) in 98% sulfuric acid (10.0mL). The mixture was stirred and kept at 0℃ for 2h, then poured onto crushed ice (20g), and extracted with dichloromethane (2×20mL), followed by washing with saturated sodium bicarbonate and water respectively. The solvent of the organic layer was removed[12]. The residue was recrystallized with ethyl acetate/petroleum ether (60-90℃) to give methyl 6-amino-7,9-dinitrophenazine-1-carboxylate (Ⅶ-**b**, 40.0mg, 53.3%). Two recrystallizations from ethyl acetate/petroleum ether (60-90℃) gave methyl-6-(4-chlorobutanamido)-7-nitrophenazine-1-carboxylate (Ⅶ-**a**, 5.0mg, 5.5%). Compounds **Ⅶ-c** and **Ⅶ-d** were gained from **Ⅵ-6h** with the same process as above. Compound **Ⅶ-e** was the only product from **Ⅵ-7h**. The structures of **Ⅶ-c**, **Ⅶ-d** and **Ⅶ-e** were testified by the 2D-NMR spectra. All the substituents at the phenazine rings are listed in Scheme 2 and Table 1.

Data for Ⅶ-a. Yield: 5.5%; orange solid; mp, beyond 340℃ (ethyl acetate/petroleum ether). ^{1}H NMR(400MHz, $CDCl_3$) δ: 9.95 (s, 1H, NHCO), 8.90 (d, $^3J_{HH}$=8.4Hz, 1H, Ar—H), 8.56 (d, $^3J_{HH}$=8.4Hz, 1H, Ar—H), 8.44 (d, $^3J_{HH}$=8.0Hz, 1H, Ar—H), 8.40 (d, $^3J_{HH}$=6.0Hz, 1H, Ar—H), 8.02 (dd, $^3J_{HH}$=8.0Hz, $^3J_{HH}$=7.2Hz, 1H, Ar—H), 4.16 (s, 3H, OCH_3), 3.77 (brs, 2H, CH_2Cl), 2.93 (brs, 2H, CH_2CO), 2.34 (brs, 2H, CH_2). ^{13}C NMR (100MHz, $CDCl_3$) δ: 170.9, 166.9, 141.8, 140.4 (2C), 138.5, 135.6, 134.4, 133.9, 133.6, 132.5, 131.3, 130.0, 113.1, 53.0, 44.2, 34.5, 27.5. HRMS: *m/z* 426.0620. Calcd for $C_{18}H_{15}ClN_4O_5$: 425.0623[M+Na]$^+$.

Data for Ⅶ-b. Yield: 53.3%; orange solid; mp, 257-259℃ (ethyl acetate/petroleum ether). ^{1}H NMR(400MHz, d_6-DMSO) δ: 9.66-9.77 (brd, 2H, NH_2) 9.14 (s, 1H, Ar—H), 8.44 (d, $^3J_{HH}$=8.4Hz, 1H, Ar—H), 8.37 (d, $^3J_{HH}$=6.8Hz, 1H, Ar—H), 8.11 (t, $^3J_{HH}$=7.2Hz, 1H, Ar—H), 3.97 (s, 3H, OCH_3). ^{13}C NMR (100MHz, d_6-DMSO) δ: 166.4, 148.7, 140.7, 138.9, 137.9, 135.3, 133.9, 132.3, 132.0, 131.7, 131.4, 127.8, 121.4, 52.4. HRMS: *m/z* 366.0442. Calcd for $C_{14}H_9N_5O_6$: 366.0445[M+Na]$^+$.

Data for Ⅶ-c. Yield: 18.0%; yellow solid; mp, 141-143℃ (ethyl acetate/petroleum ether). ^{1}H NMR(400MHz, $CDCl_3$) δ: 9.85 (s, 1H, NHCO), 8.80 (d, $^3J_{HH}$=8.8Hz, 1H, Ar—H), 8.41 (d, $^3J_{HH}$=8.4Hz, 1H, Ar—H), 8.33 (d, $^3J_{HH}$=8.8Hz, 1H, Ar—H), 8.27 (d, $^3J_{HH}$=6.8Hz, 1H, Ar—H), 7.92 (t, $^3J_{HH}$=8.0Hz, 1H, Ar—H), 4.48 (t, $^3J_{HH}$=7.2Hz, 2H, OCH_2), 3.69 (t, $^3J_{HH}$=6.0Hz, 2H, CH_2Cl), 2.85 (t, $^3J_{HH}$=7.2Hz, 2H, CH_2CO), 2.25-2.28 (m, 2H, CH_2), 1.79-1.82 (m, 2H, CH_2), 1.41-1.46 (m, 2H, CH_2), 0.91 (t, $^3J_{HH}$=

7.2Hz,3H,CH_3). ^{13}C NMR(100MHz,$CDCl_3$)δ:169.8,165.6,140.8,139.7,139.4,137.3, 134.6,132.5, 132.0, 131.6, 131.2, 130.2, 128.4, 112.0, 65.3, 43.2, 33.5, 29.6, 26.5, 18.2,12.7. HRMS:*m/z* 467.1100. Calcd for $C_{21}H_{21}ClN_4O_5$:467.1093[M + Na]$^+$.

Data for Ⅶ - d. Yield:51.9%; orange solid; mp, 176 - 178℃ (ethyl acetate/petroleum ether). ^1H NMR(400MHz,$CDCl_3$)δ:9.30(s,1H,H-8),9.01(s,1H,NH-α),8.40(brs, 1H,NH-β),8.40(brs,1H,H-4),8.40(brs,1H,H-2),8.02(dd,$^3J_{HH}$ = 7.2Hz,$^3J_{HH}$ = 6.8Hz,1H,H-3),4.54(t,$^3J_{HH}$ = 5.6Hz,2H,CH_2-2'),1.85-1.88(m,2H,CH_2-3'), 1.48-1.51(m,2H,CH_2-4'),0.99(t,$^3J_{HH}$ = 6.4Hz,3H,CH_3-5'). ^{13}C NMR(100MHz, $CDCl_3$)δ:166.2(s,C-1'),147.2(s,C-9),142.5(s,C-5a),139.9(s,C-4a),138.1 (s,C-7),135.0(d,C-4),134.1(s,C-6),133.7(s,C-10a),132.9(s,C-1),132.5 (d,C-2),131.6(d,C-3),126.5(d,C-8),122.6(s,C-9a),66.4(t,C-2'),30.6(t, C-3'),19.2(t,C-4'),13.7(q,C-5'). HRMS:*m/z* 386.1102. Calcd for $C_{17}H_{15}N_5O_6$: 386.1095[M + H]$^+$.

Data for Ⅶ - e. Yield:50.8%; yellow solid; mp, 181 - 183℃ (ethyl acetate/petroleum ether). ^1H NMR(400MHz,$CDCl_3$)δ:8.33(d,$^3J_{HH}$ = 8.4Hz,1H,Ar—H),8.16(d,$^3J_{HH}$ = 7.2Hz,1H,Ar—H),8.08(d,$^3J_{HH}$ = 9.2Hz,1H,Ar—H),7.84(dd,$^3J_{HH}$ = 8.4Hz,$^3J_{HH}$ = 7.2Hz,1H,Ar—H),7.35(d,$^3J_{HH}$ = 9.2Hz,1H,Ar—H),6.91(s,2H,NH_2),4.50(t,$^3J_{HH}$ = 6.4Hz,2H,OCH_2),1.82-1.85(m,2H,CH_2),1.56-1.58(m,2H,CH_2),1.02(t,$^3J_{HH}$ = 7.2Hz,3H,CH_3). ^{13}C NMR(100MHz,$CDCl_3$)δ:166.6,146.8,142.7,139.5,138.7,138.5, 136.3, 133.1, 131.2, 131.1, 130.1, 125.8, 123.8, 65.5, 30.7, 19.3, 13.7. HRMS: *m/z* 341.1236. Calcd for $C_{17}H_{16}N_4O_4$,:341.1244[M + H]$^+$.

Bioassays. The fungicidal activities of the compounds **Ⅲ - a—Ⅶ - e** were tested in vitro against *Alternaria solani*, *Cercospora arachidicola*, *Fusarium omysporum*, *Gibberella zeae*, *Physalospora piricola*, and their relative inhibitory ratio(%) had been determined by using the mycelium growth rate method[13]. Phenazine-1-carboxylic acid was used as a control. After the mycelia grew completely, the diameters of the mycelia were measured and the inhibition rate was calculated according to the formula

$$I = (D1 - D2)/D1 \times 100\%$$

In the formula, I is the inhibition rate, D1 is the average diameter of mycelia in the blank test, and D2 is the average diameter of mycelia in the presence of those compounds. The inhibition ratio of those compounds at the dose of 50μg·mL^{-1} is summarized in Table 2. The EC_{50} of compounds **Ⅳ-6a**, **Ⅳ-6b** and phenazine-1-carboxylic acid had been experimented and calculated by the Scatchard method. The results are summarized in Table 3.

Table 2 Fungicidal Activity of the Compounds Ⅲ - a—Ⅶ - e at Dosage of 50μg·mL^{-1}①

compd.	inhibition(%)				
	A. solani	*C. arachidicola*	*F. omysporum*	*G. zeae*	*P. piricola*
Ⅲ - a	0	7.7	13.8	12.9	19.2
Ⅲ - b	5.1	11.1	2.7	1.6	4.9
Ⅲ - c	5.9	11.5	6.9	33.9	9.6

Table 2 (continued)

compd.	inhibition (%)				
	A. solani	*C. arachidicola*	*F. omysporum*	*G. zeae*	*P. piricola*
Ⅳ-6a	100	26.9	40.0	36.3	100
Ⅳ-6b	51.0	32.0	33.3	41.8	100
Ⅳ-6c	46.7	30.0	36.7	43.6	57.1
Ⅳ-6d	24.2	8.3	21.4	2.7	15.6
Ⅳ-7a	33.3	0	10.8	15.1	25.0
Ⅳ-7b	29.4	7.7	31.0	11.3	44.2
Ⅳ-7c	16.7	26.1	10.0	25.5	19.6
Ⅳ-7d	43.3	17.4	13.3	7.3	29.6
Ⅳ-8a	10.0	4.0	23.3	18.2	37.1
Ⅳ-8b	20.0	12.0	16.7	20.0	51.4
Ⅳ-8c	20.2	20.0	33.3	27.3	48.6
Ⅳ-8d	23.3	24.0	26.7	23.6	34.3
Ⅴ-7b	17.6	11.5	6.9	14.5	26.9
Ⅴ-7c	20.0	21.7	0	20.0	26.5
Ⅴ-7d	13.3	13.0	0	23.6	25.5
Ⅴ-8d	17.4	16.7	20.0	42.4	34.3
Ⅵ-6a	13.8	8.3	14.3	18.2	46.4
Ⅵ-6b	10.3	0	21.4	22.7	32.1
Ⅵ-6c	13.8	0	0	25.0	41.1
Ⅵ-6d	4.3	0	28.0	30.3	31.4
Ⅵ-6e	3.4	0	7.1	27.3	10.7
Ⅵ-6f	13.8	25.0	17.9	31.8	17.9
Ⅵ-6g	3.4	0	0	43.2	28.6
Ⅵ-6h	6.1	8.3	10.7	0	0
Ⅵ-7a	3.3	0	0	12.7	3.9
Ⅵ-7b	6.5	4.0	5.3	4.8	1.8
Ⅵ-7c	37.7	7.7	0	36.8	0
Ⅵ-7d	19.1	2.6	11.1	2.4	0
Ⅵ-7e	3.0	0	7.1	0	0
Ⅵ-7f	6.5	8.0	15.8	0	18.2
Ⅵ-7g	13.3	10.9	6.7	12.7	2.0
Ⅵ-7h	6.1	8.3	3.6	0	0
Ⅵ-7i	33.3	22.2	13.5	17.5	8.0
Ⅵ-7j	6.5	24.0	2.6	13.1	23.6
Ⅵ-7k	19.1	0	0	14.6	0
Ⅵ-7l	23.4	12.8	11.1	9.8	7.4
Ⅵ-7m	17.2	4.2	10.7	36.4	33.9

Table 2 (continued)

compd.	inhibition (%)				
	A. solani	*C. arachidicola*	*F. omysporum*	*G. zeae*	*P. piricola*
Ⅵ -7n	0	12.0	5.3	4.8	21.8
Ⅶ - a	20.5	35.0	16.3	40.4	34.4
Ⅶ - b	21.7	22.2	32.0	57.6	50.0
Ⅶ - c	17.4	11.1	16.0	51.5	37.1
Ⅶ - d	17.4	5.6	24.0	53.0	42.9
Ⅶ - e	17.2	12.5	17.9	25.0	16.1
phenazine - 1 - carboxylic acid	100	92.3	65.4	43.1	100

①The data is the average of three duplicate results.

Table 3 EC_{50} of the Compounds Ⅳ -6a, Ⅳ -6b, and Phenazine - 1 - carboxylic Acid

	$Y = a + bx$	EC_{50}	*R*
Ⅳ -6a			
A. solani	$Y = 3.913 + 0.827x$	20.53	0.96
C. arachidicola		>50	
F. omysporum		>50	
G. zeae		>50	
P. piricola	$Y = 4.329 + 1.402x$	3.00	0.96
Ⅳ -6b			
A. solani	$Y = 3.250 + 1.0908x$	40.13	0.94
C. arachidicola		>50	
F. omysporum		>50	
G. zeae		>50	
P. piricola	$Y = 4.149 + 1.312x$	4.44	0.98
phenazine - 1 - carboxylic acid			
A. solani	$Y = 4.400 + 0.472x$	18.5	0.99
C. arachidicola	$Y = 3.185 + 1.615x$	13.2	0.96
F. omysporum	$Y = 4.108 + 0.758x$	14.9	0.98
G. zeae		>50	
P. piricola	$Y = 3.445 + 2.803x$	3.58	0.99

RESULTS AND DISCUSSION

Synthesis and Structural Elucidation. In the previous papers[5,6], compounds **Ⅲ - a—Ⅲ - c** and **Ⅳ -6a—Ⅳ -8a** were synthesized as follows: methyl - 3 - nitro - 2 - (3 - nitrophenylamino) benzoate was hydrogenated over PtO_2 to give methyl 3 - amino - 2 - (3 - aminophenylamino) benzoate. The above product was refluxed in nitrobenzene for 30h to give a mixture of **Ⅳ - 6a** and **Ⅳ -8a** (1: 11). Compound **Ⅲ - a** was the hydrolysis product of **Ⅳ -6a. Ⅲ - c** was the product from 3 – nitro – 2 – (3 – nitrophenylamino) benzoic acid[6]. **Ⅲ – b** and **Ⅳ – 7a** were synthesized by the same author with the same process in principle[5]. The whole procedure was rather tedious, and the total yield was low.

Compounds **Ⅲ – a**—**Ⅳ – 8d** were synthesized through an improved procedure referencing to the literature[4–7,14–16]: Compound **Ⅱ – b** with $NaBH_4$ in 2 N NaOH was refluxed for 4 – 5h, cooled, and acidified to give a mixture of **Ⅲ – a** and **Ⅲ – c**, which consisted of 50% **Ⅲ – a** and 50% **Ⅲ – c**, as determined by ^1H NMR. The mixture of **Ⅲ – a** and **Ⅲ – c** was refluxed in $CH_3(CH_2)_nOH$ (n = 0, 1, 2, 3) with a catalytic amount of H_2SO_4 to give **Ⅳ – 6a**—**Ⅳ – 6d** and **Ⅳ – 8a**—**Ⅳ – 8d**, which were easily purified by flash chromatography on silica gel [elution solvent: ethyl acetate/petroleum ether (60 – 90℃)]. Similar procedures were used to prepare **Ⅲ – b** and **Ⅳ – 7a**—**Ⅳ – 7d** from **Ⅱ – c**. Regrettably, compound **Ⅲ – x** was the only product from **Ⅱ – a**. So the compound 9 – aminophenazine – 1 – carboxylic acid did not come into being.

In the preparation of **Ⅳ – 7a**—**Ⅳ – 7d**, some unexpected byproducts (**Ⅴ – 7b**—**Ⅴ – 7d**) were isolated by flash chromatography on silica gel, and the yields increased with increasing value of n in $CH_3(CH_2)_nOH$. However, no corresponding byproduct was isolated when the n was equal to 0. **Ⅴ – 8d** was the only byproduct from **Ⅳ – 8d**. No corresponding byproduct was isolated when n was equal to 0, 1, or 2 in the series of **Ⅳ – 8**. The structures of compounds **Ⅴ – 7b**—**Ⅴ – 8d** were testified by 1D NMR and 2D NMR spectra. One typical compound (**Ⅴ – 7b**) was elucidated as follows: The molecular formula of **Ⅴ – 7b** was revealed as $C_{17}H_{17}N_3O_3$ by HRMS data $[M + Na]^+$ (found 334.1163, calcd 334.1162). The ^1H and ^{13}C NMR (Data for **Ⅴ – 7b**) spectra showed the signals of eight quaternary, five CH, two CH_2, two CH_3 carbon atoms. Considering the reagents, the HMQC spectra showed as follows: 8.22 (d, H – 4), 8.04 (d, H – 2), 7.66 – 7.72 (m, H – 3), 7.35 (s, H – 9) and 7.13 (s, H – 6) belonged to the Ar—H, which was confirmed by $^3J_{HH}$ = 8.4Hz, $^3J_{HH}$ = 6.8Hz and HMBC spectra; 4.89 (s, 2H) belonged to NH_2; 4.56 (t, CH_2 – 2′), 4.29 (t, CH_2 – 1″), 1.55 (q, CH_3 – 2″), 1.48 (q, CH_3 – 3′) belonged to two side chains, which was confirmed by the $^3J_{HH}$ = 7.2Hz and HMBC spectra. The chemical shift values of C – 9a and C – 5a in ^{13}C NMR spectra overlapped; the chemical shift values of C – 4a and C – 1 in ^{13}C NMR spectra were not easily differentiated by HMBC spectra, which were mentioned by the reference data of **Ⅴ – 7c**, **Ⅴ – 7d** and known compound[17].

Based on the HMBC spectra, the correlations between H – 4 (δ_H = 8.22) and C – 2, C – 10a [δ_C = 129.1 (d, C – 2), 138.6 (s, C – 10a)], the correlations between H – 2 (δ_H = 8.04) and C – 1′, C – 4, C – 10a [δ_C = 167.0 (s, C – 1′), 131.9 (d, C – 4), 138.6 (s, C – 10a)], the correlations between H – 3 (δ_H = 7.66 – 7.72) and C – 1, C – 4a [δ_C = 130.5 (s, C – 1), 141.2 (s, C – 4a)], the correlations between H – 9 (δ_H = 7.35) and C – 5a, C – 7, C – 8, C – 9a [δ_C = 142.2 (s, C – 5a), 143.6 (s, C – 7), 152.8 (s, C – 8), 142.2 (s, C – 9a)], the correlations between H – 6 (δ_H = 7.13) and C – 8, C – 9a [δ_C = 152.8 (s, C – 8), 142.2 (s, C – 9a)], the correlations between H – 2′ (δ_H = 4.56) and C – 1′, C – 3′ [δ_C = 167.0 (s, C – 1′), 14.4 (q, C – 3′)], the correlations between H – 3′ (δ_H = 1.48) and C – 2′ [δ_C = 61.3 (t, C – 2′)], the correlations between H – 1″ (δ_H = 4.29) and C – 8, C – 2″ [δ_C = 152.8 (s, C – 8), 14.5 (q, C – 2″)], and the correlations between H – 2″ (δ_H = 1.55) and C – 1″ [δ_C = 64.9 (t, C – 1″)] indicated the structure of **Ⅴ – 7b** should be as shown in Figure 2.

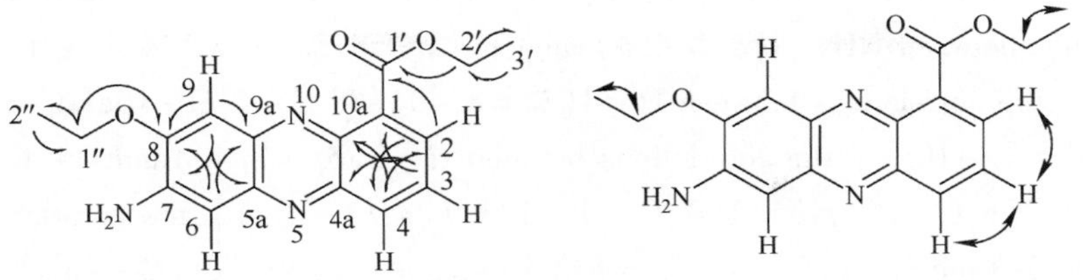

Figure 2 Key HMBC correlations(left) and key H—H COSY(right) correlations of **V -7b**

In the H—H COSY spectra, the correlations between H -3(δ_H =7.66 -7.72) and H -4(δ_H =8.22), the correlations between H -3(δ_H =7.66 -7.72) and H -2(δ_H =8.04) indicated H -4, H -3 and H -2 were in nearby positions. The correlations between H -2′(δ_H =4.56) and H -3′(δ_H =1.48) and the correlations between H -1″(δ_H =4.29) and H -2″(δ_H =1.55) further verified the positions of two side chains were right in the structure of **V -7b**. As the possible synthesis mechanisms to those unexpected byproducts would be further studied in the future.

Compounds **VI -7a**—**VI -7g** and **VI -7i**—**VI -7n** were given using **IV -7a** with corresponding acyl chloride or sulfonyl chloride. As the amount of **IV -6a** was limited, compounds **VI -6a**—**VI -6g** were synthesized using **IV -6a** with acyl chloride or sulfonyl chloride which were selected based on the SAR of the **VI -7** series. Considering the high activity of the **IV -6** series and the activity contribution of butan -1 -ol to **IV -7d** and 4 -chlorobutanoyl chloride to **VI -7c**, compounds **VI -6h** and **VI -7h** were designed and synthesized. In order to know whether the nitryl group on the phenazine ring improved the fungicidal activity or not, some nitration derivatives were given from **VI -6c**, **VI -6h** and **VI -7h**. In the nitration process of **VI -6c** and **VI -6h**, both compounds had two nitration derivatives at 0℃. One had a single nitryl in the 7 -position, and the other had two nitryls in the 7,9 -position with the carboxamide bond hydrolyzed. Compound **IIV -e** was the only nitration product from **IV -7h** whenever the reaction's temperature was kept at 0 or 20℃. The structures of compounds **VII -a**—**VII -d** were proved by 1D NMR and 2D NMR spectra. One typical compound(**VII -d**) was elucidated as follows: The molecular formula of **VII -d** was revealed as $C_{17}H_{17}N_3O_3$ by HRMS data$[M + Na]^+$ (found 334.1163, calcd 334.1162). The ^1H and ^{13}C NMR(Data for **VII -d**) spectra showed the signals of nine quaternary, four CH, three CH_2, one CH_3 carbon atoms. Considering the reagents(**VI -6h**), the HMQC spectra showed as follows: 9.30(s, H -8), 8.40(brs, H -4), 8.40(brs, H -2), 8.02(dd, H -3) belonged to the Ar—H, which was confirmed by $^3J_{HH}$ =7.2Hz, $^3J_{HH}$ =6.8Hz(H -3) and HMBC spectra; 9.01(s, H -α), 8.40(brs, H -β) belonged to NH_2; 4.54(t, CH_2 -2′), 1.85 -1.88(m, CH_2 -3′), 1.48 -1.51(m, CH_2 -4′), 0.99(t, CH_3 -5′) belonged to a side chain, which was confirmed by the $^3J_{HH}$ value and HMBC spectra. The chemical shift values of H -4, H -2 and NH -β in ^1H NMR spectra overlapped; the chemical shift values of C -4a and C -1 in ^{13}C NMR spectra were not easily differentiated by HMBC spectra, which were mentioned by the reference data of **VII -a**—**VII -c** and known compound[17].

Based on the HMBC spectra, the correlations between H -8(δ_H =9.03) and C -6, C -7,

C-9, C-9a[δ_C = 134.1(s, C-6), 138.1(s, C-7), 147.2(s, C-9), 122.6(s, C-9a)], the correlations between NH-β(δ_H = 8.40) and C-6, C-5a[δ_C = 134.1(s, C-6), 142.5(s, C-5a)], the correlations between H-4(δ_H = 8.40) and C-2, C-10a[δ_C = 132.5(d, C-2), 133.7(s, C-10a)], the correlations between H-2(δ_H = 8.40) and C-1′, C-4, C-10a[δ_C = 166.2(s, C-1′), 135.0(d, C-4), 133.7(s, C-10a)], the correlations between H-3(δ_H = 8.02) and C-1, C-4a[δ_C = 132.9(s, C-1), 139.9(s, C-4a)], the correlations between H-2′(δ_H = 4.54) and C-1′, C-3′, C-4′[δ_C = 166.2(s, C-1′), 30.6(t, C-3′), 19.2(t, C-4′)], the correlations between H-3′(δ_H = 1.85 - 1.88) and C-2′, C-4′, C-5′[δ_C = 66.4(t, C-2′), 19.2(t, C-4′), 13.7(q, C-5′)], the correlations between H-4′(δ_H = 1.48 - 1.51) and C-2′, C-3′, C-5′[δ_C = 66.4(t, C-2′), 30.6(t, C-3′), 13.7(q, C-5′)], and the correlations between H-5′(δ_H = 0.99) and C-3′, C-4′[δ_C = 30.6(t, C-3′), 19.2(t, C-4′)] indicated the structure of **VII-d** should be as shown in Figure 3.

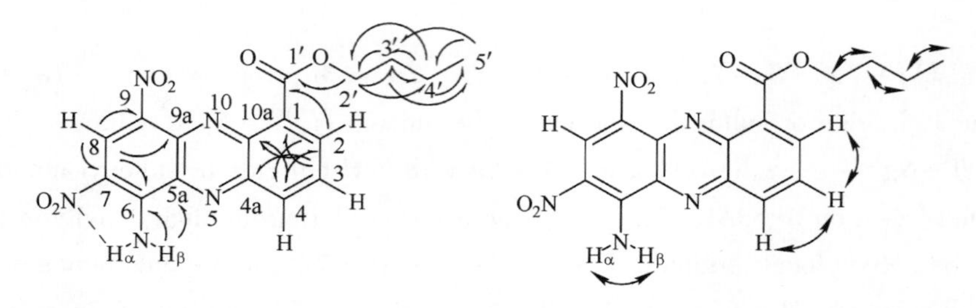

Figure 3 Key HMBC correlations(left) and key H—H COSY(right) correlations of **VII-d**

In the H—H COSY spectra, the correlations between H-3(δ_H = 8.02) and H-4(δ_H = 8.40) and the correlations between H-3(δ_H = 8.02) and H-2(δ_H = 8.40) indicated H-4, H-3 and H-2 were nearby. The correlations between H-α(δ_H = 9.01) and H-β(δ_H = 8.40) further proved H-α and H-β belonged to NH_2-6. The chemical shift value of H-α in ^1H NMR spectra might be influenced by the nitryl in the 7-position. The correlations between H-2′(δ_H = 4.54) and H-3′(δ_H = 1.85 - 1.88), the correlations between H-3′(δ_H = 1.85 - 1.88) and H-4′(δ_H = 1.48 - 1.51), and the correlations between H-4′(δ_H = 1.48 - 1.51) and H-5′(δ_H = 0.99) indicated H-2′, H-3′, H-4′ and H-5′ were adjacent.

Biological Assay and Structure - Activity Relationship. Table 2 showed the fungicidal activities against *A. solani*, *C. arachidicola*, *F. omysporum*, *G. zeae*, and *P. piricola* of the title compounds **III-a**—**VII-e**. Table 3 showed the EC_{50} of the high fungicidal activity compounds **IV-6a**, **IV-6b** and phenazine-1-carboxylic acid.

Fungicidal Activity against A. solani. The screening data of Table 2 indicated that, at a dosage of 50μg · mL^{-1}, most compounds of **III-a**—**VII-e** exhibited low activities against *A. solani* except the series of compounds **IV-6**. The inhibition activity of **IV-6a** was 100%, which was equal to that of the phenazine-1-carboxylic acid. The inhibition activity of **IV-6b** was 51.0%; as the side chain prolonged, the fungicidal activities of compounds **IV-6c** and **IV-6d** evidently decreased. The series of compounds **IV-7** and **IV-8** did not show satisfactory fungicidal activity, and their activity rules seemed to be contrary to the **IV-6** series, such as the typical compounds **Ⅳ -7d** and **Ⅳ -8d** only exhibited 43.4% and 23.3% inhibition respectively. That might be due to the side chain extension. To be mentioned, compared with the biological activities of **Ⅲ - a—Ⅲ - c**, the carboxyl group substituted by the ester group in aminophenazine derivatives had a vital function for improving the fungicidal activity.

Fungicidal Activity against C. arachidicola. At a dosage of 50μg · mL^{-1}, only compounds **Ⅳ -6a**, **Ⅳ -6b** and **Ⅳ -6c** exhibited moderate activity. The fungicidal activity of **Ⅳ -6a**, **Ⅳ -6b** and **Ⅳ -6c** were 26.9%, 32.3% and 30.0% respectively, which might be attributed to the —NH_2 group in the 6 - position of phenazine ring. Most compounds of **Ⅲ - a—Ⅶ - e** exhibited low activities against *C. arachidicola.*

Fungicidal Activity against F. omysporum. The screening data of Table 2 indicated that, at a dosage of 50μg · mL^{-1}, most compounds of **Ⅲ - a—Ⅶ - e** exhibited low activities against *F. omysporum* except the compounds **Ⅳ -6a**, **Ⅳ -6b** and **Ⅳ -6c**. The fungicidal activities of **Ⅳ -6a**, **Ⅳ -6b** and **Ⅳ -6c** were 40.0%, 33.3% and 36.7% respectively, which further explained that the —NH_2 group in the 6 - position of the phenazine ring was vital for improving their biological activity. Regrettably, the fungicidal activities of **Ⅵ - 6a—Ⅵ - 7n** against *F. omysporum* were as low as those against *C. arachidicola.* It might be concluded that the —NH_2 group being substituted by—NHCO—or—$NHSO_2$—group would result in the fungicidal activity being decreased.

Fungicidal Activity against G. zeae. The biological activity rules of **Ⅲ - a—Ⅵ - 7n** against *G. zeae* were generally the same as the test against *C. arachidicola* and *F. omysporum* showed. To be pointed out, compounds **Ⅶ - a—Ⅶ - d** were the nitration derivatives of **Ⅵ -6c** and **Ⅵ -6h**. The fungicidal activities of **Ⅶ - a—Ⅶ - d** were 40.4%, 57.6%, 51.5% and 53.0% respectively. The fungicidal activities of **Ⅶ - a—Ⅶ - d** were higher than those of **Ⅳ -6a—Ⅳ -6d** or **Ⅵ -6c**, **Ⅵ -6h** (Table 2). It might be explained by that the nitration of the phenazine ring resulted in the activity against *G. zeae* being increased.

Fungicidal Activity against P. piricola. The screening data of Table 2 indicated that most target compounds exhibited low activities against *P. piricola* except for **Ⅳ -6a** and **Ⅳ -6b**. At a dosage of 50μg · mL^{-1}, compounds **Ⅳ -6a**, **Ⅳ -6b** and phenazine - 1 - carboxylic acid exhibited 100% inhibition. The EC_{50} values of **Ⅳ -6a**, **Ⅳ -6b** and phenazine - 1 - carboxylic acid against *F. omysporum* were 3.00μg · mL^{-1}, 4.44μg · mL^{-1}, 3.58μg · mL^{-1} (Table 3), which fully explained the rules as follows: The side chain of C - 1 should not be too long. The —NH_2 group should be in the 6 - position of the phenazine ring and not form the —NHCO— or —$NHSO_2$— group. The above terms were essential for high fungicidal activity of the aminophenazine derivatives.

Though the screening data of Table 2 and Table 3 indicated that the fungicidal activities of most target compounds were lower than that of phenazine - 1 - carboxylic acid, compound **Ⅳ -6a** had higher fungicidal activity than that of phenazine - 1 - carboxylic acid against *P. piricola*. **Ⅳ -6a** also showed excellent activity against *A. solani* at low dosage. So **Ⅳ -6a** could be developed as a leading compound for further structural optimization. In conclusion, a novel and facile procedure for preparation of derivatives of aminophenazine - 1 - carboxylate was developed from 2 - bromo - 3 - nitrobenzoic acid with corresponding substituted benzene -

diamine. Forty – six aminophenazine – 1 – carboxylate derivatives were synthesized. The results of bioassay showed that the fungicidal activities of most target compounds were lower than that of phenazine – 1 – carboxylic acid; a few of the title compounds exhibited moderate activities against *A. solani*, *C. arachidicola*, *F. omysporum*, *G. zeae*, *P. piricola* at a dosage of 50μg · mL^{-1}; **Ⅳ – 6a** and **Ⅳ – 6b** exhibited excellent activity against *A. solani* and *P. piricola* at that dosage; the EC_{50} of **Ⅳ – 6a** against *P. piricola* was 3.00μg · mL^{-1}, which was lower than that of phenazine – 1 – carboxylic acid (Table 3). To our knowledge, **Ⅳ – 6a** was a known compound; however, no biological activity or feasible synthesis method was reported in any literature. So **Ⅳ – 6a** could be developed as a leading compound for further structural optimization. The possible SAR of aminophenazine – 1 – carboxylate derivatives were as follows: The —NH_2 group should be in the 6 – position of the phenazine ring and not be substituted by the —NHCO— or —$NHSO_2$— group. The aminophenazine – 1 – carboxylic acid should be esterified to aminophenazine – 1 – carboxylate, and the side chain of C – 1 should not be too long. The above terms were essential for high fungicidal activity of the aminophenazine derivatives. The nitration of the phenazine ring could increase the fungicidal activity against *G. zeae* but not for *A. solani* or *P. piricola*. The byproducts **Ⅴ – 7b—Ⅴ – 8d** did not show noticeable fungicidal activities.

LITERATURE CITED

[1] McDonald, M.; Wilkinson, B.; Van't Land, C. W.; Mocek, U.; Lee, S.; Floss, H. G. Biosynthesis of Phenazine Antibiotics in Streptomyces antibioticus: Stereochemistry of Methyl Transfer from Carbon – 2 of Acetate. *J. Am. Chem. Soc.* 1999, 24(121), 5619 – 5624.

[2] Laursen, J. B.; Visser, P. C.; Nielsen, H. K.; Jensen, K. J.; Nielsen, J. Solid – Phase Synthesis of New Saphenamycin Analogues with Antimicrobial Activity. *Bioorg. Med. Chem. Lett.* 2002, 12, 171 – 175.

[3] Paul, V. J.; Puglisi, M. P.; Williams, R. R. Marine Chemical Ecology. *Nat. Prod. Rep.* 2006, 23, 153 – 180.

[4] Rewcastle, G. W.; Denny, W. A.; Baguley, B. C. Potential Antitumor Agents. 51. Synthesis and Antitumor Activity of Substituted Phenazine – 1 – carboxamides. *J. Med. Chem.* 1987, 30, 843 – 851.

[5] Holliman, F. G.; Jeffery, B. A.; Brock, D. J. H. The Synthesis of 7 – Aminophenazine – 1 –, 7 – Aminophenazine – 2 and 8 – Aminophenazine – 2 – carboxylic acid. *Tetrahedron* 1963, 19, 1841 – 1848.

[6] Brock, D. J. H.; Holliman, F. G. The Synthesis of 3 – Aminophenazine – 1 –, 3 – Aminophenazine – 2 and 8 – Aminophenazine – 1 – carboxylic acid. *Tetrahedron* 1963, 19, 1903 – 1909.

[7] Brock, D. J. H.; Holliman, F. G. The Synthesis of 2 – Aminophenazine – 1 – carboxylic acid. *Tetrahedron* 1963, 19, 1911 – 1917.

[8] Troisi, F.; Russo, A.; Gaeta, C.; Bifulcob, G.; Neria, P. Aramidocalix[4]arenes as new anion receptors. *Tetrahedron Lett.* 2007, 48, 7986 – 7989.

[9] Spicer, J. A.; Gamage, S. A.; Rewcastle, G. W.; Finlay, G. J.; Bridewell, D. J. A.; Baguley, B. C.; Denny, W. A. Bis(phenazine – 1 – carboxamides): Structure – Activity Relationships for a New Class of Dual Topoisomerase Ⅰ/Ⅱ – Directed Anticancer Drugs. *J. Med. Chem.* 2000, 43, 1350 – 1358.

[10] Breslin, H. J.; Kukla, M. J.; Ludovici, D. W.; Mohrbacher, R.; Ho, W.; Miranda, M.; Rodgers, J. D.; Hitchens, T. K.; Leo, G.; Gauthier, D. A.; Ho, C. Y.; Scott, M. K.; Clercq, E. D.; Pauwels, R.; Andries, K.; Janssen, M. A. C.; Janssene, P. A. J. Synthesis and Anti – HIV – 1 Activity of 4,5,6,7 – Tetrahydro – 5 – methylimidazo – [4, 5, 1 – jk] – [1, 4] benzodiazepin – 2 (1H) – one (TIBO) Derivatives. 3. *J. Med. Chem.* 1995, 38, 771 – 793.

[11] Purushottamachar, P.; Khandelwal, A.; Vasaitis, T. S.; Bruno, R. D.; Gediyaa, L. K.; Njar, V. C. O. Po-

tent anti - prostate cancer agents derived from a novel androgen receptor down - regulating agent. *Bioorg. Med. Chem.* 2008,16,3519 - 3529.

[12] Wang, M. Z. ; Xu, H. ; Feng, Q. ; Wang, L. Z. ; Wang, S. H. ; Li, Z. M. Design, Synthesis, and Fungicidal Activity of Novel Analogues of Pyrrolnitrin. *J. Agric. Food Chem.* 2009,57,7912 - 7918.

[13] Chen, N. C. *Bioassay of Pesticides*; Beijing Agricultural University Press: Beijing, China, 1991; pp 161 - 162.

[14] Gamage, S. A. ; Rewcastle, G. W. ; Baguley, B. C. ; Charltonb, P. A. ; Dennya, W. A. Phenazine - 1 - carboxamides: Structure - cytotoxicity relationships for 9 - substituents and changes in the H - bonding pattern of the cationic side chain. *Bioorg. Med. Chem.* 2006,14,1160 - 1168.

[15] Vicker, N. ; Burgess, L. ; Chuckowree, I. S. ; Dodd, R. ; Folkes, A. J. ; Hardick, D. J. ; Hancox, T. C. ; Miller, W. ; Milton, J. ; Sohal, S. ; Wang, S. M. ; Wren, S. P. ; Charlton, P. A. ; Dangerfield, W. ; Liddle, C. ; Mistry, P. ; Stewart, A. J. ; Denny, W. A. Novel Angular Benzophenazines: Dual Topoisomerase I and Topoisomerase Ⅱ Inhibitors as Potential Anticancer Agents. *J. Med. Chem.* 2002,45,721 - 739.

[16] Rewcastle, G. W. ; Denny, W. A. Unequivocal Synthesis of Phenazine - 1 - carboxylic acids: Selective Displacement of Fluorine during Alkaline Borohydride reduction of N - (2 - fluorophenyl) - 3 - nitroanthranilic acids. *Synth. Commun.* 1987,1(10),1171 - 1179.

[17] Breitmaier, E. ; Hollstein, U. Carbon - 13 Nuclear Magnetic Resonance Chemical Shifts of Substituted Phenazines. *J. Org. Chem.* 1976,12(41),2104 - 2108.

High Throughput Receptor - based Virtual Screening under ZINC Database, Synthesis, and Biological Evaluation of Ketol - acid Reductoisomerase Inhibitors*

Xinghai Liu[1,**], Peiquan Chen[1,**], Baolei Wang[1], Weili Dong[1], Yonghong Li[1], Xingqiao Xie[2], Zhengming Li[1,3]

([1]State Key Laboratory of Elemento - Organic Chemistry, Nankai University, Tianjin, 300071, China;
[2]College of Life Science, Nankai University, Tianjin, 300071, China;
[3]National Pesticide Engineering Research Center (Tianjin), Nankai University, Tianjin, 300071, China)

Abstract Ketol - acid reductoisomerase (KARI; EC 1. 1. 1. 86) catalyzes the second common step in branched-chain amino acid biosynthesis. This enzyme is an important target for drug design. Based on the crystal structure of ketol - acid reductoisomerase/N - hydroxy - N - isopropyloxamate (IpOHA) complex, we have carried out high throughput receptorbased virtual screening of the ZINC/drug like database (2000000 compounds) to look for novel inhibitors of KARI for the first time. Some novel compounds were found to inhibit rice KARI *in vitro* among 15 procured compounds. This method can provide useful information for further design and discovery of KARI inhibitors.

Key words docking; high throughput virtual screening; KARI; KARI activity

Ketol - acid reductoisomerase (KARI; EC 1. 1. 1. 86) is an attractive target for agro - chemical and medicinal discovery, because it catalyzes the second important step in the biosynthesis of the branched - chain amino acid[1]. The KARI exists in microorganisms and plants, not in mammals. Thus, it is an ideal target from which to design nontoxic KARI inhibitors as potential novel drugs.

Enlightened from the biosynthesis of the branched - chain amino acid pathway, some bioactive compounds, such as sulfonylureas[2], imidazolinones[3], sulfonamide[4], and pyrimidinylbenzoic acids[5], which inhibit the first enzyme as acetohydroxyacid synthase have been successfully developed. The success of these inhibitors has stimulated the research into inhibitors of other enzymes in the pathway, including the second enzyme[6], KARI (EC 1. 1. 1. 86), and two other enzymes in the leucine extension[7]. The reaction catalyzed by KARI is shown in Figure 1 which consists of two steps[8,9], an alkyl migration followed by a NADPH - dependent reduc-

* Reprinted from *Chemical Biology & Drug Design*, 2010, 75(2): 228 - 232. This work was supported by National Basic Research Key Program of China (No. 2003CB114406), National Natural Science Foundation Key Project of China (No. 20432010), High Performance Computing Project of Tianjin Ministry of Science and Technology of China (No. 043185111 - 5) and Specialized Research Fund for the Doctoral Program of Higher Education (No. 20070055044) and Tianjin Natural Science Foundation (No. 08JCYBJC00800).

** X. H. Liu and P. Q. Chen contributed equally to this work.

tion. Both steps require a divalent metal ion, such as Mg^{2+}, Mn^{2+} or Co^{2+}, but the alkyl migration is highly specific for Mg^{2+} interaction.

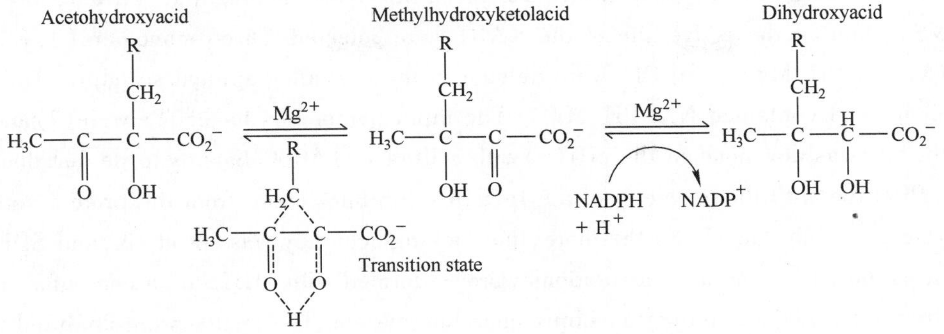

Acetohydroxyacid ⇌ (Mg^{2+}) Methylhydroxyketolacid ⇌ (Mg^{2+}; NADPH + H^+ → $NADP^+$) Dihydroxyacid

Transition state

Figure 1 Reaction catalyzed by ketol – acid reductoisomerase

No commercial herbicides and drugs targeting KARI have been developed. Until now, only 2 – dimethylphosphinoyl – 2 – hydroxyacetate (HOE 704)[10], IpOHA[11], 1, 2, 3 – thiadiazoles[12], and cyclopropane – 1, 1 – dicarboxylic acid (CPD) derivatives[13-16] were shown to be potential inhibitors targeting KARI (Figure 2). In addition, it is found HOE 704[10], IpOHA[11] can inhibit tuberculosis much better than ATCC35801[17], so these branched – chain amino acids inhibitors became novel anti – tuberculosis medicine, and CPD[18,19] had good antifungal activities.

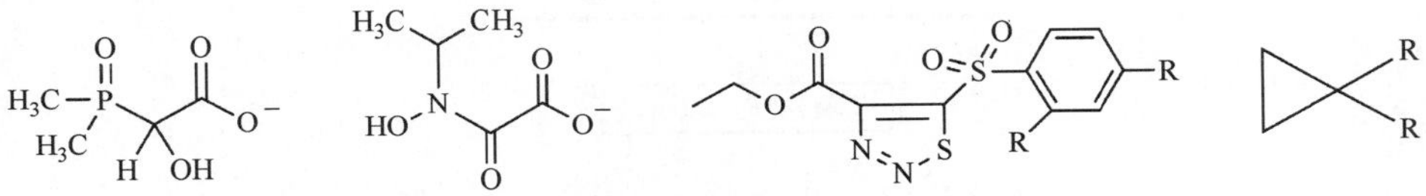

HOE 704   IpOHA   Thiadiazoles   CPD

Figure 2 Known inhibitors of HOE 704, IpOHA, thiadiazole, and CPD analogies

Structure – based drug design method is widely used in drug design, because the potential inhibitors are designed on the basis of the receptor geometry. Virtual screening could reduce the cost and present diversified structures in drug discovery. In our previous work, based on this crystal structure of the KARI complex 1YVE, MDL/ACD 3D database searching was carried out using program DOCK4.0 DOCK (San Francisco, USA) without Mg^{2+} ions and crystal waters[20], but the result was not satisfactory. Mg^{2+} ions and crystal waters in the active site of plant KARI could provide valuable information in the design of more competitive inhibitors which could block the enzyme's action at certain stage. On the basis of a three – dimensional model of the spinach KARI, molecular docking – based virtual screening was employed to search from the ZINC database/Drug like.

Methods and Material

Virtual screening method

The experiment was conducted using eHiTS version 5.3 software[a]. All the programs were adopted the default. 2000000 molecules which are 3D structures were docked to the KARI enzyme (Figure 3). Enzyme geometries were taken from the Protein Data Bank (codes 1YVE) which is homology as the rice KARI. To speed up High Throughput Virtual Screening (HTVS), data on the active site of the KARI were selected. Three monomers (J, K, L), IpOHA, NADPH, Mg^{2+}, and Cl^- were deleted in the crystallographical structure. The residue monomer I contained NADPH, Mg^{2+}. The input parameters for eHiTS were (i) an SDF file (the file was download in the eHiTS) with a library of 5000 ligands to be fast docked; (ii) a PDB file with the geometry of the IpOHA – inhibitor taken from the protein complex (enzyme active site data). Furthermore, the docking accuracy was set at six, and SDF was chosen as the output format. Calculations were performed with eHiTS on a computer cluster built from eight nodes, running on Linux operation system. The results were analyzed using CheVi 6. 1[b].

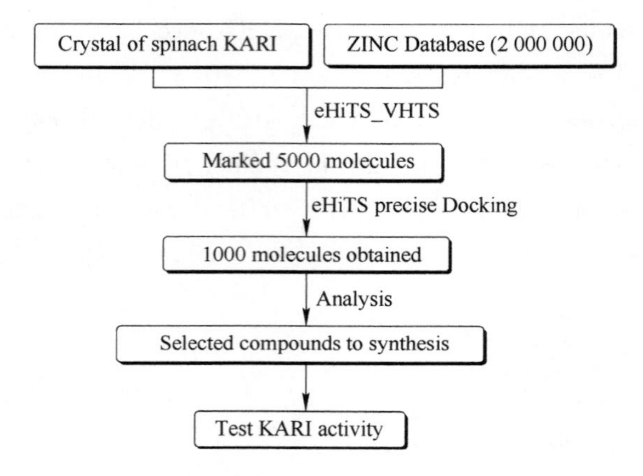

Figure 3 High throughput screening flow chart

Experimental

Instruments

Melting points were determined using an X – 4 melting apparatus and were uncorrected. Infrared spectra were recorded on a Bruker Equinox55 spectrophotometer (Bruker, Fallanden, Switzerland) as KBr tablets. ^1H NMR spectra were measured on a Bruker AC – P500 instrument (300MHz; Bruker) using tetramethylsilane (TMS) as an internal standard and $CDCl_3$ or DMSO – d_6 as solvent. Mass spectra were recorded on a Thermo Finnigan LCQ Advantage LC/mass detector instrument (Thermo – Finnigan, MA, USA).

All the compounds were synthesized according the references, and their structure were confirmed by ^1H MNR, M. P.[21-35] (Table 1), and MS (Figure 4).

Table 1 The melting point (m. p.) or boiling point (b. p.) of title compounds

No.	m. p. or b. p. /℃	Ref. m. p. or b. p. /℃
1	138/760mmHg	135 – 140/760mmHg
2	98 – 99	96 – 99

Table 1 (continued)

No.	m. p. or b. p. /℃	Ref. m. p. or b. p. /℃
3	132 – 133	135 – 136
4	135 – 136	140
5	188 – 189	190
6	160 – 161	162 – 163
7	Liquid	Liquid
8	88 – 89	86 – 87
9	73 – 74	74 – 76
10	72 – 73	74
11	183 – 184	186 – 187
12	177 – 178	178
13	215 – 216	218 – 219
14	183 – 184	185
15	74 – 75	78 – 79

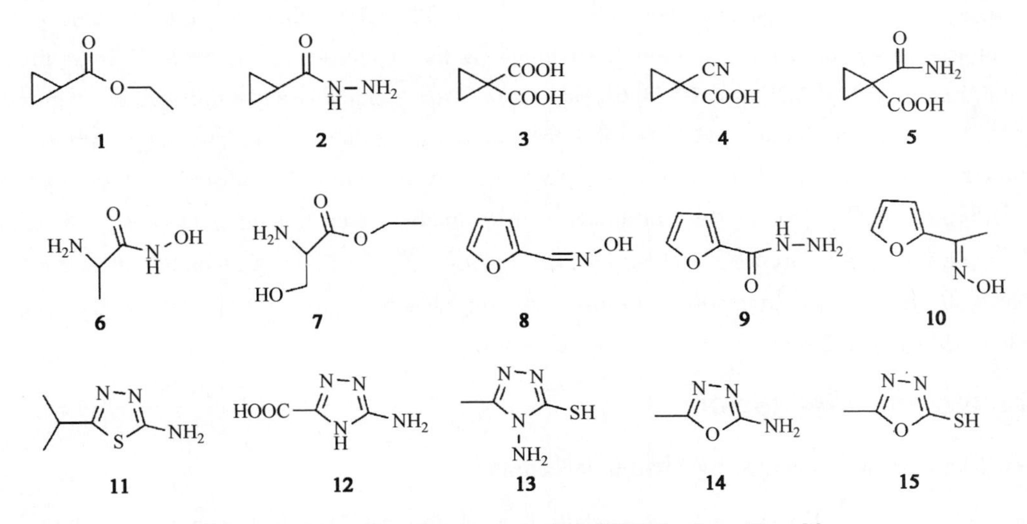

Figure 4 The selected compounds from ZINC database/Drug like

Biochemistry of KARI

Cloning, expression, and purification of rice KARI

The KARI resultant expression plasmid was obtained from the Prof. Ronald G. Duggleby's lab[13] and was used to transform *Escherichia coli* BL21 (DE3) cells. A single colony of these cells was inoculated into 20mL of Luria – Bertani (LB) medium containing 50mg/mL kanamycin. The culture was incubated overnight at 37℃ and was used to inoculate each of two 1000mL volumes of LB medium containing 50mg/mL kanamycin; the cultures were incubated at 37℃

with shaking. When an OD_{600} of 0. 6 was reached, expression was induced by adding 1μL isopropyl β-D-thiogalactoside to each culture; these were then incubated at room temperature (37℃) for a further 2h with shaking, and the cells were harvested by centrifugation and were kept in -30℃.

The frozen cell pellet was thawed, suspended in ice-cold purification buffer [50mM Tris-HCl (pH 7.9)/500mM NaCl] containing 5mM imidazole and then treated with lysozyme (10mg/g of cells for 30min at 0℃). The cells were disrupted by sonication, insoluble material was removed by centrifugation, and the supernatant was passed through a 0. 45mm filter. The cell extract was applied to a 7mL column of His Bind resin (Novagen, Wisconsin, USA) that had been charged using 50mM $NiSO_4$ then equilibrated with purification buffer containing 5mM imidazole. The loaded column was washed with 23mL of the same buffer, followed by 30mL of purification buffer containing 25mM imidazole, and then KARI was eluted with 30mL of purification buffer containing 1M imidazole. Fractions containing the enzyme were pooled, concentrated to 2. 5mL by ultrafiltration, and exchanged into 20mM Na-Hepes buffer, pH 8. 0 using a Pharmacia PD-10 (North Peapack, NJ, USA) column. The eluate was snap-frozen in low-temperature refrigerator and stored at -78℃.

Enzyme and protein assays (*in vitro*)

Gerwick et al. [36] reported that the inhibition of *E. coli* KARI is time dependent. Ketol-acid reductoisomerase activity was measured by following the decrease in A_{340} at 30℃ in solutions containing 0. 2mM NADPH, 1mM $MgCl_2$, substrate 2-acetolactate, and inhibitors as required, in 0. 1M phosphate buffer, pH 8. 0. Inhibitors were preincubated with enzyme in phosphate buffer at 30℃ for 10min before the reaction was started by adding the substrate combining with NADPH and $MgCl_2$. Protein concentrations were estimated using the bicinchoninic acid method[37], and protein purity was assessed by SDS-PAGE[38]. The yield of recombinant rice KARI from a 30 culture was 50mg with a specific activity (measured with saturating 2-acetolactate) of 1. 17U/mg. The 2-acetolactate was prepared by us.

Results and Discussion

Identification of molecules by virtual screening

Based on the plant KARI structural models, Drug-like database with approximate 2000000 small compounds was searched using the program eHiTS. From the virtual screening, the top 1000 candidate molecules with the highest energy scores were considered as potential KARI inhibitors for further study. For pharmaceutical or agrochemical compounds, hydrophobic property is an important factor to reach its site of action. Their log p values were calculated in the relating database.

On the basis of virtual screening and log p prediction, 15 candidate compounds were selected to test their biologic activity. The compounds containing one or more negative charge groups, such as NH_2, COOH, and OH, could approach Mg^{2+} ion. The results are in accord with the log p prediction. Table 2 shows the calculated log p values.

Table 2 *In vitro* ketol - acid reductoisomerase (KARI) activity data of compounds (% inhibition) and log p value

No.	KARI/100μg/mL	K_i/μM	Log p
1	100	—	1.02
2	33.03	—	-0.48
3	100	76.56 ± 14.49	0.19
4	100	95.30 ± 13.71	0.67
5	100	31.16 ± 9.26	-0.46
6	—	—	-1.38
7	43.77	—	-1.15
8	—	—	0.78
9	19.85	—	-0.69
10	32.33	—	0.35
11	30.80	—	2.44
12	1.09	—	-0.52
13	100	—	0.68
14	23.23	—	0.01
15	100	—	1
IpOHA	100	2.75 ± 0.72	0

- indicates the compound can not resolve in our test system, so no data obtained.

KARI activity

The 15 compounds were synthesized and tested for rice KARI assay *in vitro*. Among them, six compounds exhibited good inhibition against rice KARI. It also included three cyclopropane compounds (Figure 3). Compound **3** was reported before in the literature (11), whose K_i value was also tested here (Table 2). The K_i values of **3**, **4**, and **5** are 76.56 ± 14.49, 95.30 ± 13.71, and 31.16 ± 9.26μM. These compounds are less active than reported IpOHA. However, this preliminary result has some significance, because these structures are completely different from any known KARI inhibitors. Further modification of these novel structures will be helpful to further discover new structures with enhanced affinity toward KARI enzyme.

Binding modes of the new inhibitors

The binding conformation of compound **3** is shown in Figure **5** (the binding mode of compound **4** is shown in Supporting Information). The inhibitors appear to fit well in the active site pocket showing several hydrogen bonds and Van Der Waals interactions with the Mg^{2+} ion and amino residues. The O atom of the compound **3** has interaction with the KARI via Mg^{2+} (Figure 5).

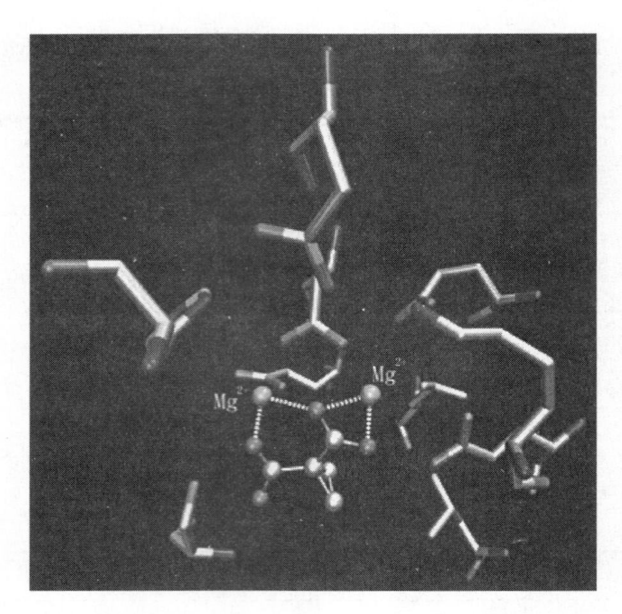

Figure 5 The binding modes of compound **3** in the active sites of spinach ketol – acid reductoisomerase

References

[1] Duggleby R. G. ,Pang S. S. (2000) Acetohydroxyacid synthase. J Biochem Mol Biol;33:1 – 36.

[2] Chaleff R. S. ,Mauvais C. J. (1984) Acetolactate synthase is the site of action of two sulfonylurea herbicides in higher plants. Science;224:1443 – 1445.

[3] Shaner D. L. , Anderson P. C. , Stidham M. A. (1984) Imidazolinones. Potent inhibitors of acetohydroxy acid synthase. Plant Physiol;76:545 – 546.

[4] Hung S. (1983) Analytical investigation of the process for manufacturing 1,2,4 – 1*H* – triazole. Ind Chem; 11:291 – 299.

[5] Duggleby R. G. ,Pang S. S. (2000) Acetohydroxyacid synthase. J Biochem Mol Biol;33:1 – 36.

[6] Dumas R. ,Biou V. F. ,Douce H. R. ,Duggleby R. G. (2001) Enzymology, structure, and dynamics of acetohydroxy acid isomeroreductase. Acc Chem Res;34:399 – 408.

[7] Wittenbach V. A. ,Teaney P. W. ,Rayner D. R. ,Schloss J. V. (1993) Herbicidal activity of an isopropylmalate dehydrogenase inhibitor. Plant Physiol;106:321 – 328.

[8] Dumas R. ,Butikofer M. C. ,Job D. ,Douce R. (1995) Evidence for two catalytically different magnesium – binding sites in acetohydroxy acid isomeroreductase by site – directed mutagenesis. Biochemistry; 34: 6026 – 6036.

[9] Chunduru S. K. ,Mrachko G. T. ,Calvo K. C. (1989) Mechanism of ketol acid reductoisomerase. Steady – state analysis and metal ion requirement. Biochemistry;28:486 – 493.

[10] Schulz A. , Sponemann P. , Kocher H. , Wengenmayer F. (1988) The herbicidally active experimental compound Hoe 704 is a potent inhibitor of the enzyme acetolactate reductoisomerase. FEBS Lett; 238: 375 – 378.

[11] Aulabaugh A. ,Schloss J. V. (1990) Oxalyl hydroxamates as reaction – intermediate analogs for ketol – acid reductoisomerase. Biochemistry;29:2824 – 2840.

[12] Halgand F. ,Vives F. ,Dumas R. ,Biou V. ,Andersen J. ,Andrieu J. P. ,Cantegril R. ,Gagnon J. ,Douce R. ,Forest E. ,Job D. (1998) Kinetic and mass spectrometric analyses of the interactions between plant acetohydroxy acid isomeroreductase and thiadiazole derivatives. Biochemistry;37:4773 – 4781.

[13] Lee Y. T. ,Ta H. T. ,Duggleby R. G. (2005) Cyclopropane – 1,1 – dicarboxylate is a slow – ,tight – binding inhibitor of rice ketol – acid reductoisomerase. Plant Sci;168:1035 – 1040.

[14] Liu X. H. ,Chen P. Q. ,Wang B. L. ,Li Y. H. ,Wang S. H. ,Li Z. M. (2007) Synthesis bioactivity theoretical and molecular docking study of 1 – cyano – N – substituted – cyclopropanecarboxamide as ketolacid reductoisomerase inhibitor. Bioorg Med Chem Lett;17:3784 – 3788.

[15] Liu X. H. ,Chen P. Q. ,He F. Q. ,Li Y. H. ,Wang S. H. ,Li Z. M. (2007) Structure bioactivity and theoretical study of 1 – cyano – N – p – tolylcyclo – propanecarboxamide. Struct Chem;5:563 – 568.

[16] Liu X. H. ,Zhang C. Y. ,Guo W. C. ,Li Y. H. ,Chen P. Q. ,Wang T. ,Dong W. L. ,Sun H. W. ,Li Z. M. (2009) Synthesis bioactivity and SAR study of N′ – (5 – substituted – 1,3,4 – thiadiazol – 2 – yl) – N – cyclopropyformyl – thiourea as ketol – acid reductoisomerase inhibitor. J Enzyme Inhib Med Chem;24: 545 – 552.

[17] Grandoni J. A. ,Marta P. T. ,Schloss J. V. (1998) Inhibitors of branched – chain amino acid biosynthesis as potential antituberculosis agents. J Antimicrob Chemother;42:475 – 482.

[18] Liu X. H. ,Shi Y. X. ,Ma Y. ,He G. R. ,Dong W. L. ,Zhang C. Y. ,Wang B. L. ,Wang S. H. ,Li B. J. ,Li Z. M. (2009) Synthesis of some N,N′ – diacylhydrazine derivatives with radical scavenging and antifungal activity. Chem Biol Drug Des;73:320 – 327.

[19] Liu X. H. ,Shi Y. X. ,Ma Y. ,Zhang C. Y. ,Dong W. L. ,Li P. ,Wang B. L. ,Li B. J. ,Li Z. M. (2009) Synthesis, antifungal activities and 3D – QSAR study of *N* – (5 – substituted – 1,3,4 – thiadiazol – 2 – yl) cyclopropanecarboxamides. Eur J Med Chem;44:2782 – 2786.

[20] Wang B. L. ,Li Z. M. ,Ma Y. ,Wang J. G. ,Luo X. M. (2004) 3D – Database searching based on the crystal structure of ketol – acid reductoisomerase (KARI) complex. Chin J Org Chem;24:973 – 976.

[21] Razus A. C. ,Bartha E. ,Arvay Z. ,Glatz A. M. (1984) Synthesis of para – substituted benzyl cyclopropyl ketones. Rev Roum Chim;29:719 – 725.

[22] Vogel A. I. (1934) Physical properties and chemical constitution. I. Esters of normal dibasic acids and substituted malonic acids. J Chem Soc;333 – 341.

[23] Hell Z. ,Finta Z. ,Dmowski W. ,Faigl F. ,Pustovit Y. M. ,Toke L. ,Harmat V. (2000) reaction of cyclopropane carboxylic acid derivatives with sulphur tetrafluoride – an example of a diastereoselective ring opening. J Fluor Chem;104:297 – 301.

[24] Brenner M. ,Huber W. (1953) Preparation of – amino acid esters by alcoholysis of the methyl esters. Helv Chim Acta;36:1109 – 1115.

[25] Allen C. F. H. ,Boyer R. (1933) Action of sulfuric acid on certain derivatives of cyclopropane. Can J Res;9:159 – 168.

[26] Roberts J. D. (1951) Small – ring compounds. V. Synthesis of cyclopropanecarboxaldehyde by the MacFadyen – Stevens reduction. J Am Chem Soc;73:2959.

[27] Knobler Y. ,Bittner S. ,Frankel M. (1964) Reaction of N – carboxyamino acid anhydrides with hydrochlorides of hydroxylamine, O – alkylhydroxylamines, and amines; syntheses of aminohydroxamic acids, amidooxy peptides, and – amino acid amides. J Chem Soc;3941 – 3951.

[28] Datta A. ,Walia S. ,Parma B. S. (2001) Some furfural derivatives as nitrification inhibitors. J Agric Food Chem;49:4726 – 4731.

[29] Sy M. ,de Malleray B. (1963) Synthesis of 3 – aminothenoyl – and 3 – aminofuroylrhodanines. Bull Soc Chim Fr;3:1278 – 1279.

[30] Ocskay G. ,Vargha L. (1958) Furan compounds. V. Preparation and configuration of furyl ketoximes. Tetrahedron;2:140 – 150.

[31] Wojahn H. ,Wuckel H. (1951) Synthesis of aminothiadiazoles and the corresponding sulfonamide derivatives. Arch Pharm;284:53 – 62.

[32] Henkel C. (1959) Heterocyclic carboxylic acids. ,GB 816531. Chem Abstr;54:7325.

[33] Chande M. S. ,Karnik B. M. ,Inamdar A. N. ,Ganguly N. (1990) Design,synthesis and biological screening of new s – triazolothiadiazine derivatives. J Indian Chem Soc;67:220 – 222.

[34] Zinner G. ,Neitzel M. ,Holdt I. (1978) Reaction of cyanates with carboxylic acid hydrazides. Arch Pharm (Weinheim);311:1050 – 1055.

[35] Crast L. ,Leonard B. Jr,(1972) Antibacterial cephalosporins. DE 2145928. Chem Abstr;77:48487.

[36] Gerwick B. C. ,Mireles L. C. ,Eilers R. J. (1993) Rapid diagnosis of ALS/AHAS – resistant weeds. Weed Technol;7:519 – 524.

[37] Smith P. K. ,Krohn R. I. ,Hermanson G. T. ,Mallia A. K. ,Gartner F. H. ,Provenzano M. D. ,Fujimoto E. K. ,Goeke N. M. , Olson B. J. , Klenk D. C. (1985) Measurement of protein using bicinchoninic acid. Anal Biochem;150:76 – 85.

[38] Laemmli U. K. (1970) Cleavage of structural proteins during the assembly of the head of bacteriophage T4. Nature;227:680 – 685.

Notes

[a]http://www. simbiosys. ca/ehits/.

[b]http://www. simbiosys. ca/chevi/.

Supporting Information

Additional Supporting Information may be found in the online version of this article:

Figure S1. The Binding modes of compound **4** in the active sites of spinach KARI.

Please note: Wiley – Blackwell is not responsible for the content or functionality of any supporting materials supplied by the authors. Any queries (other than missing material) should be directed to the corresponding author for the article.

Synthesis and Insecticidal Activities of Novel Analogues of Chlorantraniliprole Containing Nitro Group*

Qi Feng, Mingzhong Wang, Lixia Xiong, Zhili Liu, Zhengming Li

(State Key Laboratory of Elemento – Organic Chemistry, Institute of Elemento – Organic Chemistry, Nankai University, Tianjin, 300071, China)

Abstract Twelve novel analogues of chlorantraniliprole containing nitro group were synthesized, and their structures were characterized by ^{1}H NMR and high – resolution mass spectrometry (HRMS). Their evaluated insecticidal activities against oriental armyworm (*Mythimna separata*) indicate that the nitro – containing analogues showed favorable insecticidal activities, while the activity of compounds **5**g at 0.25mg/L was 40%, but still lower than chlorantraniliprole.

Key words chlorantraniliprole; nitro; insecticidal activity; anthranilic diamide

1 Introduction

Chlorantraniliprole (Rynaxypyr™; DPX – E2Y45, **A**, Fig. 1) is a highly potent and selective activator of insect ryanodine receptor, which regulates the release of intracellular stored calcium critical for muscle contraction[1-3]. Two classes of synthetic chemicals exhibit their action on this target: one includes anthranilic diamides, with chlorantraniliprole as its representive compound, and the other contains phthalic acid diamides, firstly discovered by Nihon Nohyaku[4,5], its representive compound is flubendiamide (**B**, Fig. 1). Both the compounds show exceptional insecticidal activity on a broad range of Lepidoptera and chlorantraniliprole has more advantages[6]. Besides Lepidoptera, other insect species such as Coleoptera, Diptera and Isoptera have been effectively controlled. In addition to larvicidal activity, significant ovicidal activity was observed to some Lepidoptera pests[7].

Since the discovery of chlorantraniliprole, structural modification of this kind of compounds has attracted considerable attention in the field of insecticidal researches. Most modification focused on the anthranilic amide moiety (**A**)[8-10], such as the case in cyantraniliprole (Cyazypyr™, **C**, Fig. 1)[9], which was introduced a cyano group instead of 4 – halo substituent of the former anthranilic diamides. Improved plant mobility and increased spectrum were reported on this new compound[6].

It is known that nitration is the most important method for the introduction of nitrogen functionality on aromatic ring while the nitro group is the key pharmacophore in neonicotinoids researches[11]. In this work, twelve novel anthranilic diamides containing nitro group were de-

* Reprinted from *Chemical Research in Chinese Universities*, 2011, 27(4): 610 – 613. Supported by the National Basic Research Program of China (No. 2010CB126106).

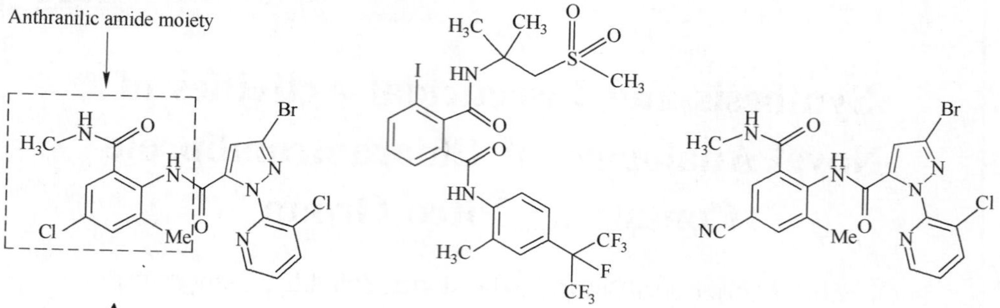

A B C

Fig. 1 Chemical structures of compounds **A** – **C**

signed and synthesized, the bioassay was tested accordingly.

2 Experimental

2.1 Materials and Instruments

^1H NMR spectra were recorded at 400MHz on a Bruker AV – 400 spectrometer in $CDCl_3$ or DMSO – d_6 solution with tetramethylsilane as the internal standard. High – resolution mass spectrometry (HRMS) data were obtained on a Varian QFT – ESI instrument. The melting points were determined on an X – 4 binocular microscope melting point apparatus (Beijing Tech. Instruments Co., Beijing, China) and were uncorrected. Yields were not optimized. Reagents were all analytically or chemically pure. All the solvents and liquid reagents were dried by standard methods in advance and distilled before use.

2.2 General Procedure

2 – Amino – 3 – methylbenzamide derivatives (**2**) were synthesized according to the published procedure[12]. 3 – Bromo (or chloro) – 1 – (3 – chloro – 2 – pyridinyl) – 1*H* – pyrazole – 5 – carboxylic acid (**3**) was prepared according to the literature[2,12]. Referring to known method[12], intermediate anthranilic diamides (**4**) were prepared by the reaction of compound **3** with compound **2**. Nitration of intermediate **4** afforded the title compounds **5**a – **5**l as shown in Scheme 1.

2.3 General Synthetic Procedure of Intermediate Anthranilic Diamides (4)

Carboxylic acid (**3**, 1.0mmol) and oxalyl chloride (1.5mmol) were added in 20mL of anhydrous CH_2Cl_2, followed with a drop of anhydrous DMF. The mixture was stirred at room temperature for 2h and the solvent was removed under vacuum. The resulting yellow solid was dissovled in 20mL of anhydrous tetrahydrofuran, then added dropwise to a solution of compound **2** (1.0mmol) and pyridine (1.0mmol) in 30mL anhydrous tetrahydrofuran in an ice bath. After stirring for 1h at room temperature, the solvent was evaporated under reduced pressure. The residual solid was purified by chromatography on silica gel with petroleum ether – ethyl acetate as eluent to afford intermediate anthranilic diamides (**4**).

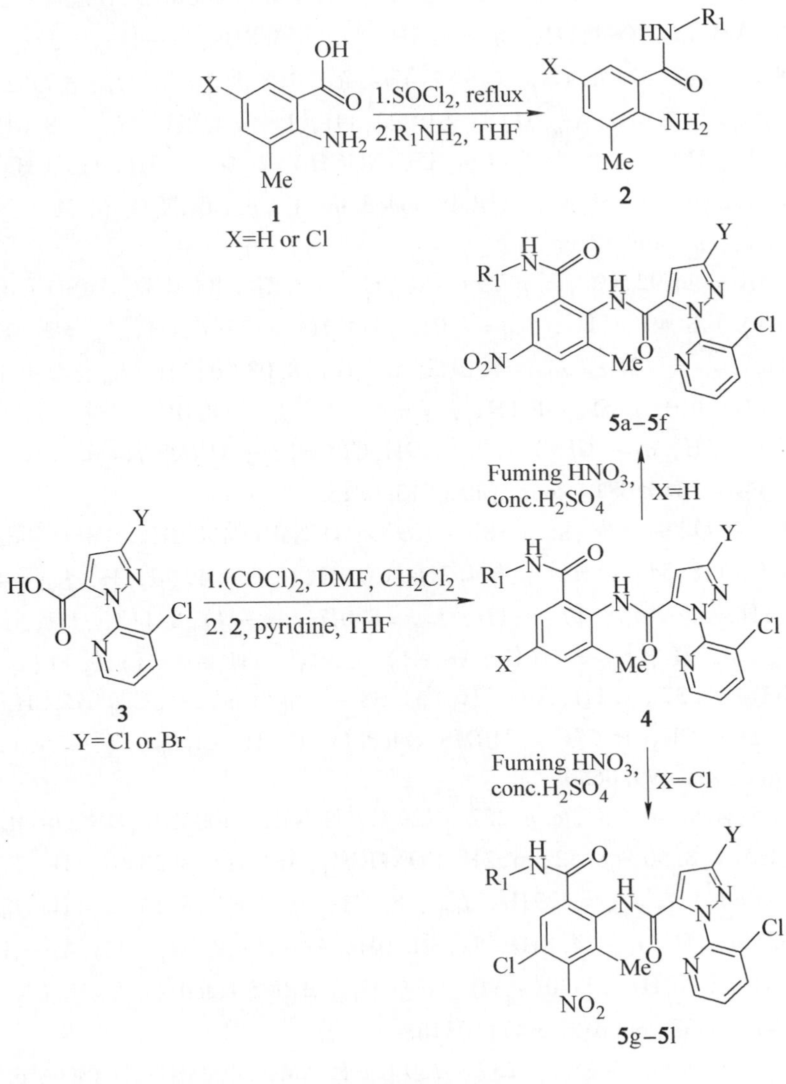

$R_1 = CH_3, i-Pr, n-Pr, t-Bu, n-Bu$, cyclohexyl

Scheme 1 General synthetic procedure of compounds **5a – 5l**

2.4 General Synthetic Procedure of Title Compounds 5a – 5l

The procedure was similar to the one described by Montoya – Pelaez[13]. Intermediate anthanilic diamide(**4**, 1.0mmol) was dissolved in concentrated H_2SO_4 (8mL), and the resulting mixture was cooled to 0℃. Fuming HNO_3 (6.0mmol) was added dropwise to the mixture. Once the addition was completed, the resulting mixture was allowed to stir in an ice bath for another 8h. Then the resulting yellow solution was poured into a flask containing 50g of crushed ice. The mixture was stirred vigorously for 5min, resulting in a white precipitate. The solid was collected by suction filtration, washed throughly with cold water, and further dried. After recrystallization with mixed solvent of petroleum ether – ethyl acetate(1:2, volume ratio), the title compounds **5a – 5l** were obtained as white solid.

Compound **5**a: yield 95.4%; m.p. 207 – 208℃; ^{1}H NMR(400MHz, DMSO – d_6), δ: 10.58 (s, 1H, CONHAr), 8.49 (dd, 1H, $^4J_{HH}$ = 1.2Hz, $^3J_{HH}$ = 4.8Hz, Ar—H), 8.37 (d, 1H, $^3J_{HH}$ = 7.6Hz, CONHCH), 8.27 (d, 1H, $^3J_{HH}$ = 2.4Hz, Ar—H), 8.16 (d, $^3J_{HH}$ = 7.2Hz, Ar—H), 8.07 (d, 1H, $^3J_{HH}$ = 2.4Hz, Ar—H), 7.61 (dd, 1H, $^3J_{HH}$ = 4.8Hz, $^3J_{HH}$ = 8.0Hz, Ar—H), 7.40 (s, 1H, Ar—H), 3.88 – 3.96 (m, 1H, NHC**H**), 2.31 (s, 3H, Ar—CH_3), 1.05 [d, 6H, $^3J_{HH}$ = 6.4Hz, CH(C$\mathbf{H}_3$)$_2$]. HRMS calcd. for $C_{20}H_{18}Cl_2N_6O_4$ ([M + Na]$^+$): *m/z* 499.0658; found: *m/z* 499.0659.

Compound **5**b: yield 92.1%; m.p. 259 – 261℃; ^{1}H NMR(400MHz, DMSO – d_6), δ: 10.53 (s, 1H, CONHAr), 8.49 (d, 1H, $^3J_{HH}$ = 4.0Hz, Ar—H), 8.25 (d, 1H, $^3J_{HH}$ = 2.4Hz, Ar—H), 8.18 (dd, 1H, $^4J_{HH}$ = 1.2Hz, $^3J_{HH}$ = 8.0Hz, Ar—H), 8.03 (d, 1H, $^3J_{HH}$ = 2.0Hz, Ar—H), 7.99 (s, 1H, CONHC), 7.61 (dd, 1H, $^3J_{HH}$ = 4.4Hz, $^3J_{HH}$ = 8.0Hz, Ar—H), 7.39 (s, 1H, Ar—H), 2.30 (s, 3H, Ar – CH_3), 1.26 [s, 9H, C(CH_3)$_3$]. HRMS calcd. for $C_{21}H_{20}Cl_2N_6O_4$ ([M + Na]$^+$): *m/z* 513.0819; found: *m/z* 513.0815.

Compound **5**c: yield 94.8%; m.p. 187 – 189℃; ^{1}H NMR(400MHz, DMSO – d_6), δ: 10.60 (s, 1H, CONHAr), 8.54 (t, 1H, $^3J_{HH}$ = 4.2Hz, CON**H**CH_2), 8.49 (d, 1H, $^3J_{HH}$ = 4.4Hz, Ar—H), 8.28 (s, 1H, Ar—H), 8.17 (d, 1H, $^3J_{HH}$ = 8.0Hz, Ar—H), 8.11 (s, 1H, Ar—H), 7.62 (dd, 1H, $^3J_{HH}$ = 4.8Hz, $^3J_{HH}$ = 7.6Hz, Ar—H), 7.38 (s, 1H, Ar—H), 3.11 (q, 2H, $^3J_{HH}$ = 6.4Hz, NHC$\mathbf{H}_2$), 2.31 (s, 1H, Ar—CH_3), 1.38 – 1.47 (m, 2H, CH_2C$\mathbf{H}_2$$CH_3$), 0.84 (t, 3H, $^3J_{HH}$ = 7.2Hz, CH_2CH_2C$\mathbf{H}_3$). HRMS calcd. for $C_{20}H_{18}Cl_2N_6O_4$ ([M + Na]$^+$): *m/z* 499.0652; found: *m/z* 499.0659.

Compound **5**d: yield 96.5%; m.p. 227 – 229℃; ^{1}H NMR(400MHz, DMSO – d_6), δ: 10.63 (s, 1H, CONHAr), 8.50 – 8.52 (m, 2H, CON**H**CH_3, Ar—H), 8.28 (d, 1H, $^3J_{HH}$ = 2.4Hz, Ar—H), 8.19 (dd, 1H, $^4J_{HH}$ = 1.2Hz, $^3J_{HH}$ = 8.4Hz, Ar— H), 8.13 (d, 1H, $^3J_{HH}$ = 2.4Hz, Ar—H), 7.62 (dd, 1H, $^3J_{HH}$ = 4.8Hz, $^3J_{HH}$ = 8.0Hz, Ar—H), 7.38 (s, 1H, Ar—H), 2.69 (d, 3H, $^3J_{HH}$ = 4.4Hz, NHC$\mathbf{H}_3$), 2.30 (s, 1H, Ar—CH_3). HRMS calcd. for $C_{18}H_{14}Cl_2N_6O_4$ ([M + Na]$^+$): *m/z* 471.0338; found: *m/z* 471.0346.

Compound **5**e: yield 90.2%; m.p. 237 – 239℃; ^{1}H NMR(400MHz, DMSO – d_6), δ: 10.58 (s, 1H, CONHAr), 8.48 (d, 1H, $^3J_{HH}$ = 4.4Hz, Ar—H), 8.36 (d, 1H, $^3J_{HH}$ = 8.0Hz, CON**H**CH), 8.28 (d, 1H, $^3J_{HH}$ = 2.4Hz, Ar—H), 8.16 (d, 1H, $^3J_{HH}$ = 8.0Hz, Ar—H), 8.06 (d, 1H, $^3J_{HH}$ = 2.4Hz, Ar—H), 7.61 (dd, 1H, $^3J_{HH}$ = 4.8Hz, $^3J_{HH}$ = 8.0Hz, Ar—H), 7.42 (s, 1H, Ar—H), 3.58 – 3.61 (m, 1H, cyclohexyl), 2.32 (s, 1H, Ar—CH_3), 1.56 – 1.71 (m, 5H, cyclohexyl), 1.07 – 1.28 (m, 6H, cyclohexyl). HRMS calcd. for $C_{23}H_{22}Cl_2N_6O_4$ ([M + Na]$^+$): *m/z* 539.0965; found: *m/z* 539.0972.

Compound **5**f: yield 95.0%; m.p. 204 – 205℃; ^{1}H NMR(400MHz, DMSO – d_6), δ: 10.55 (s, 1H, CONHAr), 8.49 (d, 1H, $^3J_{HH}$ = 4.0, Ar—H), 8.36 (d, 1H, $^3J_{HH}$ = 7.6Hz, CON**H**CH), 8.27 (s, 1H, Ar—H), 8.16 (d, 1H, $^3J_{HH}$ = 8.0, Ar— H), 8.07 (d, 1H, $^3J_{HH}$ = 1.6Hz, Ar—H), 7.61 (dd, 1H, $^3J_{HH}$ = 4.8Hz, $^3J_{HH}$ = 8.0Hz, Ar—H), 7.45 (s, 1H, Ar—H), 3.88 – 3.96 (m, 1H, NHC**H**), 2.31 (s, 1H, Ar—CH_3), 1.05 [d, 6H, $^3J_{HH}$ = 6.8Hz, CH(C$\mathbf{H}_3$)$_2$]. HRMS calcd. for $C_{20}H_{18}Cl_2N_6O_4$ ([M + Na]$^+$): *m/z* 543.0159, found: *m/z* 543.0154.

Compound **5**g: yield 85.8%; m.p. 233 – 235℃; ^{1}H NMR(400MHz, DMSO – d_6), δ: 10.56

(s, 1H, CONHAr), 8.56 (d, 1H, $^3J_{HH}$ = 4.8Hz, Ar—H), 8.43 (d, 1H, $^3J_{HH}$ = 4.8Hz, CONHCH_3), 8.27 (d, 1H, $^3J_{HH}$ = 8.0Hz, Ar—H), 7.88 (s, 1H, Ar—H), 7.69 (dd, 1H, $^3J_{HH}$ = 4.8Hz, $^3J_{HH}$ = 8.0Hz, Ar—H), 7.42 (s, 1H, Ar—H), 2.68 (d, 1H, $^3J_{HH}$ = 4.4Hz, NHCH_3), 2.29 (s, 1H, Ar—CH_3). HRMS calcd. for $C_{18}H_{13}BrCl_2N_6O_4$ ([M + Na]$^+$): *m/z* 548.9444; found: *m/z* 548.9451.

Compound **5**h: yield 90.3%; m. p. 187 – 189℃; ^{1}H NMR (400MHz, DMSO – d_6), δ: 10.43 (s, 1H, CONHAr), 8.49 (d, 1H, $^3J_{HH}$ = 4.4Hz, Ar—H), 8.30 (d, 1H, $^3J_{HH}$ = 7.6Hz, CON**H**CH), 8.18 (d, 1H, $^3J_{HH}$ = 8.0Hz, Ar—H), 7.81 (s, 1H, Ar—H), 7.62 (dd, 1H, $^3J_{HH}$ = 4.8Hz, $^3J_{HH}$ = 7.6Hz, Ar—H), 7.38 (s, 1H, Ar—H), 3.81 – 3.89 (m, 1H, NHC**H**), 2.23 (s, 1H, Ar—CH_3), 0.97 [d, 6H, $^3J_{HH}$ = 6.8Hz, CH(C**H**$_3$)$_2$]. HRMS calcd. for $C_{20}H_{17}BrCl_2N_6O_4$ ([M + Na]$^+$): *m/z* 576.9770; found: *m/z* 576.9764.

Compound **5**i: yield 92.5%; m. p. 286 – 288℃; ^{1}H NMR (400MHz, DMSO – d_6), δ: 10.52 (s, 1H, CONHAr), 8.55 (d, 1H, $^3J_{HH}$ = 4.4Hz, Ar—H), 8.25 (d, 1H, $^3J_{HH}$ = 8.0Hz, Ar—H), 8.02 (s, 1H, CONHC), 7.84 (s, 1H, Ar—H), 7.69 (dd, 1H, $^3J_{HH}$ = 5.2Hz, $^3J_{HH}$ = 7.6Hz, Ar—H), 7.45 (s, 1H, Ar—H), 2.28 (s, 1H, Ar—CH_3), 1.25 [s, 9H, C(CH_3)$_3$]. HRMS calcd. for $C_{21}H_{19}BrCl_2N_6O_4$ ([M + Na]$^+$): *m/z* 569.0104; found: *m/z* 569.0101.

Compound **5**j: yield 91.2%; m. p. 238 – 240℃; ^{1}H NMR (400MHz, DMSO – d_6), δ: 10.47 (s, 1H, CONHAr), 8.49 (d, 1H, $^3J_{HH}$ = 4.4Hz, Ar—H), 8.46 (t, 1H, $^3J_{HH}$ = 5.2Hz, CON**H**CH_2), 8.19 (d, 1H, $^3J_{HH}$ = 8.0Hz, Ar—H), 7.81 (s, 1H, Ar—H), 7.61 – 7.64 (m, 1H, Ar—H), 7.36 (s, 1H, Ar—H), 3.03 (q, 1H, $^3J_{HH}$ = 6.4Hz, C**H**$_2$$CH_2CH_3$), 2.22 (s, 1H, Ar—$CH_3$), 1.30 – 1.39 (m, 2H, CH_2C**H**$_2$$CH_3$), 0.78 (t, 3H, $^3J_{HH}$ = 7.2Hz, CH_2CH_2C**H**$_3$). HRMS calcd. for $C_{20}H_{17}BrCl_2N_6O_4$ ([M + Na]$^+$): *m/z* 576.9763; found: *m/z* 576.9764.

Compound **5**k: yield 90.6%; m. p. 237 – 239℃; ^{1}H NMR (400MHz, DMSO – d_6), δ: 10.47 (s, 1H, CONHAr), 8.50 (d, 1H, $^3J_{HH}$ = 4.4Hz, Ar—H), 8.44 (t, 1H, $^3J_{HH}$ = 5.4Hz, CON**H**CH_2), 8.19 (d, 1H, $^3J_{HH}$ = 8.4Hz, Ar—H), 7.81 (s, 1H, Ar—H), 7.60 – 7.64 (m, 1H, Ar—H), 7.36 (s, 1H, Ar—H), 3.06 (q, 2H, $^3J_{HH}$ = 6.0Hz, C**H**$_2$$CH_2CH_2CH_3$), 2.22 (s, 3H, Ar—$CH_3$), 1.27 – 1.33 (m, 2H, CH_2C**H**$_2$$CH_2CH_3$), 1.17 – 1.24 (m, 2H, CH_2CH_2C**H**$_2$$CH_3$), 0.79 (t, 3H, $^3J_{HH}$ = 7.2Hz, $CH_2CH_2CH_2$C**H**$_3$). HRMS calcd. for $C_{21}H_{19}BrCl_2N_6O_4$ ([M + Na]$^+$): *m/z* 590.9925; found: *m/z* 590.9920.

Compound **5**l: yield 88.3%; m. p. 201 – 203℃; ^{1}H NMR (400MHz, DMSO – d_6), δ: 10.46 (s, 1H, CONHAr), 8.49 (d, 1H, $^3J_{HH}$ = 4.8Hz, Ar—H), 8.32 (d, 1H, $^3J_{HH}$ = 8.0Hz, CON**H**CH_3), 8.19 (d, 1H, $^3J_{HH}$ = 8.0Hz, Ar—H), 7.81 (s, 1H, Ar—H), 7.62 (dd, 1H, $^3J_{HH}$ = 4.8Hz, $^3J_{HH}$ = 8.0Hz, Ar—H), 7.33 (s, 1H, Ar—H), 3.81 – 3.89 (m, 1H, NHC**H**), 2.22 (s, 1H, Ar—CH_3), 0.97 [d, 6H, $^3J_{HH}$ = 6.8Hz, CH(C**H**$_3$)$_2$]. HRMS calcd. for $C_{20}H_{17}Cl_3N_6O_4$ ([M + Na]$^+$): *m/z* 533.0268; found: *m/z* 533.0269.

3 Results and Discussion

3.1 Synthesis

The nitro – containing analogues **5**a – **5**l were synthesized as shown in Scheme 1. The key reaction of this procedure was the nitration. In the beginning, corresponding nitro – containing anlines were tried to react with pyrazole carbonyl chloride under different conditions. Unfortunately, these reactions were unsuccessful, probably due to the strong electron – withdrawing nitro group resulting in a poor reactivity of anline. A different approach for the synthesis of nitro – containing analogues was adopted, namely, the nitration of the intermediate anthranilic diamides (**4**) directly with fuming HNO_3 in concentrated H_2SO_4. After examing reaction conditions, this reaction was found to be of high regioselectivity with excellent yield. It was postulated that an orientation effect of substituents on the anthraniloyl moiety existed while the rest part of the molecular structure was stable enough under the severe nitration condition.

3.2 Insecticidal Activity

The insecticidal activity against oriental armyworm (*Mythimna separata*) was tested as the literature[14,15], with the data shown in Table 1. The results indicate that the nitrocontaining analogues have excellent toxicities against oriental armyworm. Both *p* – substituted (**5a** – **5f**) and m – substituted (**5g** – **5l**) nitro – containing analogues show similar insecticidal effect. Different alkylamines on the aliphatic amide moiety have slight influence on toxicity, such as **5a**, **5b**, **5c** and **5d**. However, the cyclohexyl (**5e**) and *n* – butyl (**5k**) with more steric effect decreased the activity. Additionally, the bromo and chloro pyrazole structures had the same effect on the toxicities, such as **5a** and **5f**. Finally, the methyl amide (**5g**) showed much higher lethal activity (40%) at a concentration of 0.25mg/L than the other analogues, but still lower than chlorantraniliprole.

Table 1 Insecticidal activities of compounds 5a – 5l and chlorantraniliprole against oriental armyworm

Compd.	Y	R_1	Larvicidal activity (%)		
			20mg/L	10mg/L	1mg/L
5a	Cl	*i* – Pr	100	100	40
5b	Cl	*t* – Bu	100	100	50
5c	Cl	*n* – Pr	100	100	50
5d	Cl	CH_3	100	100	60
5e	Cl	Cyclohexyl	100	20	nt②
5f	Br	*i* – Pr	100	100	40
5g	Br	CH_3	100	100	40③
5h	Br	*i* – Pr	100	100	60
5i	Br	*t* – Bu	100	100	40
5j	Br	*n* – Pr	100	100	20
5k	Br	*n* – Bu	100	100	40④
5l	Cl	*i* – Pr	100	100	40
Control①			100	100	100③

①Chlorantraniliprole; ②nt, not test; ③at a concentration of 0.25mg/L; ④at a concentration of 5mg/L.

In summary, twelve novel analogues of chlorantraniliprole containing nitro group were designed and synthesized according to the nitration of intermediate anthranilic diamides(**4**). The results of bioassays indicate that the series of nitrocontaining analogues under the assigned concentration show favourable/good insecticidal activities.

References

[1] Lahm G. P. ,Selby T. P. ,Freudenberger J. H. ,Stevenson T. M. ,Myers B. J. ,Seburyamo G. ,Smith B. K. , Flexner L. ,Clark C. E. ,Cordova D. ,*Bioorg. Med. Chem. Lett.* ,2005,15,4898.

[2] Lahm G. P. ,Stevenson T. M. ,Selby T. P. ,Freudenberger J. H. ,Cordova D. ,Flexner L. ,Bellin C. A. , Dubas C. M. ,Smith B. K. ,Hughs K. A. ,et al. ,*Bioorg. Med. Chem. Lett.* ,2007,17,6274.

[3] Sattelle D. B. ,Cordova D. ,Cheek T. R. ,*Invert Neurosci.* ,2008,8,107.

[4] Tsubata K. ,Tohnishi M. ,Kodama H. ,Seo A. ,*Pflanzenschutz – Nachrichten Bayer.* ,2007,60,105.

[5] Ebbinghaus – Kintscher U. ,Lummen P. ,Raming K. ,Masaki T. ,Yasokawa N. ,*Pflanzenschutz – Nachrichten Bayer.* ,2007,60,117.

[6] Lahm G. P. ,Cordova D. ,Barry J. D. ,*Bioorg. Med. Chem.* ,2009,17,4127.

[7] *DuPont Rynaxypyr_ Insect Control Technical Bulletin*, http://www2. dupont. com/Production_Agriculture/en_US/assets/downloads/pdfs/Rynaxypyr_Tech_Bulletin. pdf.

[8] Finkelstein B. L. ,Lahm G. P. ,Mccann S. F. ,Song Y. ,Stevenson T. M. ,*Substituted Anthranilamides for Controlling Invertebrate Pests*, *WO*03016284*A*1,2003.

[9] Hughes K. A. ,Lahm G. P. ,Selby T. P. ,Stevenson T. M. ,*Cyano Anthranilamides Insecticides*, *WO*04067-528*A*1,2004.

[10] Hall R. G. ,Loiseleur O. ,Pabba J. ,Pal S. ,Jeanguenat A. ,Edmunds A. ,Stoller A. ,*Novel Insecticides*, *WO*2009010260*A*2,2009.

[11] Jeschke P. ;Eds. :Wolfgang K. ,Ulrich S. ,*Modern Crop Protection Compounds*, WILEY – VCH Verlag GmbH & Co. KGaA,Weinheim,2007,958.

[12] Dong W. L. ,Xu J. Y. ,Xiong L. X. ,Liu X. H. ,Li Z. M. ,*Chin. J. Chem.* ,2009,27,579.

[13] Montoya – Pelaez P. J. ,Uh Y. S. ,Lata C. ,Thompson M. P. ,Lemieux R. P. ,Crudden C. M. ,*J. Org. Chem.* ,2006,71,5921.

[14] Abbott W. S. ,*J. Econ. Entomol.* ,1925,18,265.

[15] Zhao Q. Q. ,Shang J. ,Liu Y. X. ,Wang K. Y. ,Bi F. C. ,Huang R. N. ,Wang Q. M. ,*J. Agric.* Food Chem. ,2007,55,9614.

Modulations of High – voltage Activated Ca^{2+} Channels in the Central Neurones of Spodoptera Exigua by Chlorantraniliprole*

Yuxin Li, Mingzhen Mao, Yiming Li, Lixia Xiong,
Zhengming Li, Junying Xu

(State Key Laboratory of Elemento – Organic Chemistry, Nankai University, Tianjin, China)

Abstract The modulation of voltage – gated calcium channels by chlorantraniliprole in the central neurones isolated from third – instar larvae of *Spodoptera exigua is studied* by the whole – cell patch – clamp technique. The current of calcium in the third – instar larvae of *S. exigua* is identified as a high – voltage activated Ca^{2+} current. During the 10 – min recording, the current – voltage relationship curves of whole – cell calcium channels are shifted in a negative direction by 10 mV compared with the control group. The fact that the gravity rundown of calcium current in the treated group is more apparent than in the control group demonstrates that the open channels are constantly inactivated. In addition, chlorantraniliprole inhibits the recorded calcium currents in a concentration – dependent manner, which is irreversible on washout.

Key words calcium channel; chlorantraniliprole; neurones; *Spodoptera exigua*; whole – cell patch – clamp

Introduction

Spodoptera exigua Hübner (Lepidoptera: Noctuidae) is a serious pest of a variety of crops in China, including sugarbeet, cotton, sesame, corn, hemp, tobacco, green pepper and cucumber (Wen & Zhang, 2010). The areas in China affected by this pest have gradually expanded subsequent to 1986, and *S. exigua* currently poses a severe threat to vegetable production in the country. Thus, the development of effective insecticides against *S. exigua* has become a major interest in China's agrochemical industry.

Chlorantraniliprole, a novel anthranilic diamide insecticide discovered by DuPont for the control of Lepidopteran and Coleopteran pests, was introduced recently into the Asian market (Lahm *et al.*, 2003, 2007). Chlorantraniliprole shares the same mode of action with flubendiamide (Tohnishi *et al.*, 2005; Ebbinghaus – Kintscher *et al.*, 2007). It interacts with the ryanodine receptor (RyR) in muscle cells and causes muscle contraction, paralysis and death (DuPont, 2007). However, the impact of chlorantraniliprole on the calcium current of *S. exigua* central neurones remains unreported. Furthermore, the details of the mechanism of action of chlorantraniliprole are largely unknown.

The whole – cell (extracellular) patch – clamp technique facilitates the study of currents through single ionic channels from a wide variety of cells (Neher, 1981). This technique uses one electrode continuously for voltage recording and the passage of current. In the present stud-

* Reprinted from *Physiological Entomology*, 2011, 36(3): 230 – 234. The project was supported by the National Basic Research Programme of China (2010CB126106) and the Fundamental Research Funds for the Central Universities.

y, the whole – cell patch – clamp technique is used to investigate the effects of chlorantraniliprole on calcium channels in the central neurones of *S. exigua*. We aim to obtain new evidence to explain the mechanism of action of chlorantraniliprole on *S. exigua* and provide useful information for designing novel insecticides.

Materials and methods

Isolation of neural cells

Spodoptera exigua were initially obtained from shallot fields in Tianjin City, China and reared indoors in climatic chambers on an agar – based semisynthetic diet under an LD 16∶8h photocycle at 27 ± 1℃ and 75 ± 5% relative humidity (Huang *et al.*, 2002). The insects were reared for two generations before the experiment. Third – instar larvae of *S. exigua* were first anaesthetised with 70% ethanol and their thoracic and abdomen ganglia were removed and placed in saline. The thoracic and abdomen ganglia were transferred to a solution containing 0.3% trypsin for 6 min at 28℃, plated into a 35 – mm culture dish containing 1mL of improved L – 15 Leibovitz culture medium supplemented with fetal calf serum (15%, v/v) (He *et al.*, 2001) and then mechanically dissociated by repeated trituration using a fire – polished Pasteur pipette. The cultures were maintained at 27℃ for 2 h to allow the cell to adhere to the dish. All procedures were carried out under sterile conditions.

Reagents

The extracellular solution contained (mmol · L^{-1}) NaCl 100, CsCl 4, $BaCl_2$ 3, $MgCl_2$ 2, Hepes 10, glucose 10, TEA – Cl 30, tetrodotoxin 0.001, pH adjusted to 6.85 with 1mmol · L^{-1} CsOH. The intracellular solution contained (mmol · L^{-1}) CsCl 120, $MgCl_2$ 2, Na_2 – ATP 5, ethyleneglycol tetraacetic acid (EGTA) 11, Hepes 5, pH adjusted to 6.85 with 1mmol · L^{-1} CsOH. CsCl, CsOH and Hepes were purchased from Gibco (Gaithersburg, Maryland). Chlorantraniliprole was synthesised as described previously (Lahm *et al.*, 2007).

Electrophysiological recordings

Currents of calcium (I_{Ca}) were recorded from single neurone with diameter of 20 – 25μm (the neurone size was measured under the microscope) using the whole – cell patchclamp technique at room temperature (Hamill *et al.*, 1981). Micropipettes (diameter 1 – 2μm) made from borosilicated glass capillary tubings were pulled in a two – step vertical puller (Narishige, PP – 830; Japan) and fire polished. The resistance of the micropipettes was 2 – 3MΩ after being filled with intracellular solution. Neurones were clamped using a patch – clamp amplifier (EPC – 10; HEKA Electronik, Germany). Capacitive current was compensated by a certain cancellation routine. A series of resistance was compensated electronically by 70% – 85%. The voltage clamp protocols were generated on a computer using pulse software, version 8.52 (HEKA Electronik).

After a giga – seal, the membrane was ruptured by applying a slight negative suction. The current traces were elicited with 90 – ms depolarising voltage steps from. – 40up to 50mV in 10 – mV increments from a holding potential of – 50 mV. The calcium current trace became relatively stable 4 – 10min after patch rupture (27 cells were tested). We recorded the current of calcium at 5 min as the control group, and then chlorantraniliprole was injected in the surroundings of the cell.

Statistical analysis

Each experiment was repeated at least three times. The data were analysed using SPSS, version 17.0 (SPSS Inc., Chicago, Illinois) and MICROCAL ORIGIN, version 8.0 (Origin Lab Corp., Northampton, Massachusetts). Results are expressed as the mean ± SD (n, number of cells). Statistical significance was determined by using Student's paired or unpaired t – tests. $P < 0.05$ and $P < 0.01$ were considered statistically significant as indicated.

Results

The typical inward calcium currents recorded from one central neurone of third – instar larvae are shown in Fig. 1. With barium as current carrier, the current of Ca^{2+} channels was recorded by blocking Na^+ and K^+ currents. The neurone was held at –50 or –70 mV and the test voltage depolarised from –90 to +50mV or from –50 to +70 mV for 90ms with an interval of 10mV. Currents of calcium under relatively steady – state during 4 min after the whole – cell patch – clamp configuration were established. Currents of calcium were recorded every 1, 2 or 3 min under steady – state. The current. voltage ($I - V$) relationship curves of Ca^{2+} channels (n = 6) showed that there was no significant change in the threshold of activation, maximal value and peak current when the neurones were held at –50 or –70mV. The current of calcium was activated at approximately –30mV and reached maximal value at approximately 0 – 10mV. Less than 100μmol · L^{-1}. $CdCl_2$ or 20μmol · L^{-1} nifedipine in extracellular solution can cause the blockade of the current (Sontheimer & Olsen, 2007). Taken together, these data indicate that the neurones of *S. exigua* larvae express a high – voltage activated L – type calcium channel [HVA $Ca^{2+}_{(L)}$].

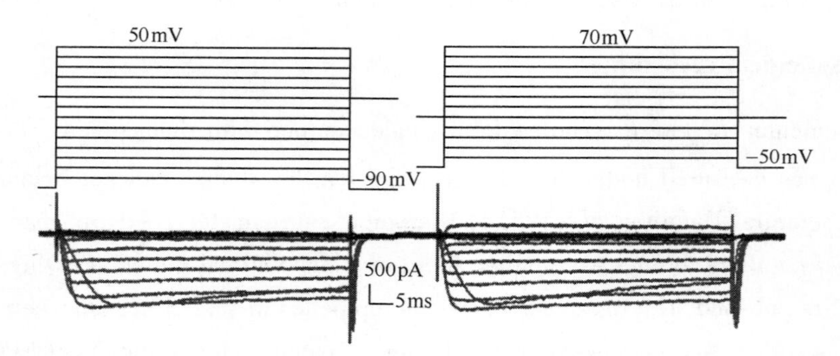

Fig. 1 Typical Ca^{2+} current in the central neurones of *Spodoptera exigua*. Test voltage depolarised from –90 to +50mV and –50 to +70mV for 90ms with an interval of 10mV

In the subsequent experiments, the neurones were all held at –50mV and the test voltage depolarised from –40 to +50mV for 90 ms with an interval of 10mV. Addition of EGTA to the intracellular recording solution has been shown to impede rundown (Horn & Korn, 1992). By the end of the 15 – min recording, 85.44 ± 1.145% (n = 6) of the initial peak current was remained(Fig. 2 and Table 1). The "rundown" in the current was lower than 20% (compared with the original data). The time course to activate $Ca^{2+}_{(L)}$ current and to reach maximal value was not affected by the run – down in peak current.

Figure 2 shows the change rate of peak current amplitude versus recording time in the 0. 1mmol · L^{-1}, 0. 01mmol · L^{-1}, 10nmol · L^{-1}, 1nmol · L^{-1}, 500pmol · L^{-1}, 250pmol · L^{-1}, 100pmol · L^{-1} chlorantraniliprole treated and control neurones. The peak currents were reduced to 83. 27 ± 3. 28% (n = 3), 75. 19 ± 1. 95% (n = 4), 57. 12 ± 0. 87% (n = 3), 48. 69 ± 5. 07% (n = 3), 45. 43 ± 3. 76% (n = 5), 45. 38 ± 1. 00% (n = 3) and 39. 43 ± 5. 02% (n = 8) of the initial value by the end of the 10 – min recording, when the neurones were treated with 100pmol · L^{-1}, 250pmol · L^{-1}, 500pmol · L^{-1}, 1 nmol · L^{-1}, 10nmol · L^{-1}, 0. 01mmol · L^{-1} and 0. 1mmol · L^{-1} of chlorantraniliprole, respectively. Compared with the control (85. 44 ± 1. 14%), chlorantraniliprole at 100pmol · L^{-1} has no effect on $Ca^{2+}_{(L)}$ currents. By contrast, the peak current was reduced to 39. 43 ± 5. 02% (n = 8) when the neurones were treated with 0. 1mmol · L^{-1} chlorantraniliprole. The recorded peak currents of calcium channels treated with the chlorantraniliprole are in a concentration – dependent manner(Fig. 2 and Table 1).

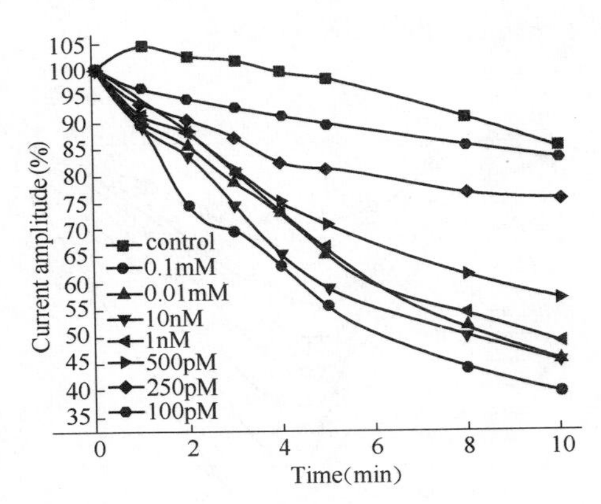

Fig. 2 Variation of peak current for the whole – cell calcium channels in control and different concentration chlorantraniliprole – treated neurones at different times during patch – clamp recording compared with the peak value at 0 min

Table 1 Analysis of the peak currents change rate with time in control and chlorantraniliprole – treated neurones

Time (min)	Control	0. 1mmol · L^{-1} (%)	0. 01mmol · L^{-1} (%)	10nmol · L^{-1} (%)	1nmol · L^{-1} (%)	500pmol · L^{-1} (%)	250pmol · L^{-1} (%)	100pmol · L^{-1} (%)
0	1	1	1	1	1	1	1	1
1	104. 47 ± 2. 33	89. 62 ± 2. 16②	90. 34 ± 2. 55②	89. 36 ± 2. 58②	94. 79 ± 1. 23②	91. 28 ± 3. 97②	93. 66 ± 3. 67②	96. 27 ± 3. 76②
2	102. 75 ± 2. 45	74. 01 ± 8. 01②	85. 55 ± 2. 13②	84. 00 ± 1. 16②	88. 10 ± 0. 83②	88. 11 ± 3. 68②	90. 52 ± 4. 75②	94. 42 ± 5. 18②
3	101. 76 ± 2. 7	69. 57 ± 9. 43②	78. 69 ± 3. 13②	74. 57 ± 3. 63②	80. 78 ± 1. 38②	81. 21 ± 5. 16②	87. 10 ± 6. 97②	92. 86 ± 5. 2②
4	99. 66 ± 1. 64	63. 36 ± 7. 71②	73. 13 ± 2. 73②	64. 92 ± 22. 77②	72. 87 ± 0. 70②	74. 93 ± 4. 16②	82. 55 ± 5. 92②	91. 01 ± 4. 56②
5	98. 06 ± 1. 01	. 55. 79 ± 4. 73②	65. 94 ± 1. 79②	59. 05 ± 0. 88②	65. 00 ± 3. 14②	70. 77 ± 2. 68②	80. 92 ± 2. 56②	89. 62 ± 7. 6②
8	90. 85 ± 0. 65	44. 31 ± 2. 62②	48. 58 ± 2. 68②	50. 00 ± 3. 30②	54. 21 ± 4. 61②	61. 34 ± 0. 82②	76. 61 ± 2. 96②	85. 5 ± 3. 343②
10	85. 44 ± 1. 14	39. 43 ± 5. 02②	45. 38 ± 1. 00②	45. 43 ± 3. 76②	48. 69 ± 5. 07②	57. 12 ± 0. 87②	75. 19 ± 1. 95②	83. 27 ± 3. 28②

①Significant difference at $P < 0.05$. ②Significant difference at $P < 0.01$. Values are the mean ± SD.

There is a significant change in maximal value of peak current when the neurones were held at – 50mV with the treatment of 0. 1mmol · L^{-1} chlorantraniliprole (Fig. 3A). The calcium current decreased when chlorantraniliprole was injected in the surrounding of the neurone cell. The maximal value of I_{Ca} (– 2. 0902 ± 0. 0574nA) shifted to the negative direction by approximately 10mV after the neurones were treated with chlorantraniliprole for 1min. By the end of the 5 - and 10 - min recordings, I_{Ca} decreased to - 1.2988 ± 0.1085 and - 0.8755 ± 0.1532nA, respectively, and the $I-V$ curves were also shifted to the negative direction by approximately 10 mV during the recording.

Figure 3B shows the $I-V$ relationship curves of calcium channels recorded in 100pmol · L^{-1} chlorantraniliprole - treated neurones at different times. Unlike the results for I_{Ca} described above, the maximal value of I_{Ca} did not shift to a negative direction. Combined with Fig. 2, at the concentration ≤100 pmol · L^{-1}, there is no effect on calcium channels in the central neurones of *S. exigua* third larvae.

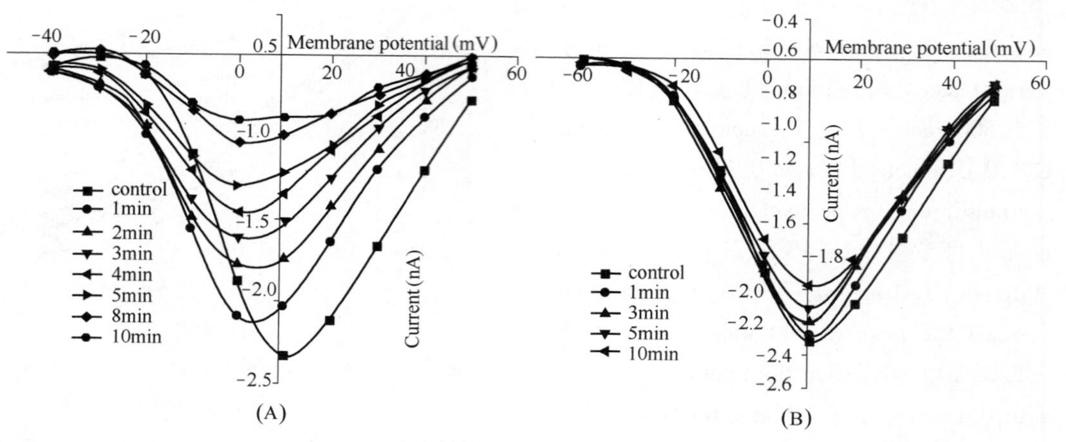

Fig. 3 The current - voltage relationship curves of whole - cell calcium channels recorded in 0.1mmol · L^{-1} (A) and 100 pmol · L^{-1} (B) chlorantraniliprole - treated neurones in *Spodoptera exigua* at different times

Discussion

Representing the most important second messenger, calcium ion has a unique property and the universal ability to transmit diverse signals that trigger primary physiological actions in cells (Fill & Copello, 2002). Voltage - sensitive (or voltagegated) calcium channels are transmembrane proteins that allow Ca^{2+} influx to occur when the open configuration is produced during a depolarisation signal (Bertolino & Llinas, 1992). By allowing Ca^{2+} flux, calcium channels play pivotal roles in the regulation of a vast array of cellular processes, including axonal growth, enzyme modulation, membrane excitability, muscle contraction and neurotransmitter release. Pharmacological, electrophysiological and ligand - binding studies indicate the presence of diverse voltage - sensitive calcium channels in insects that appear to be pharmacologically distinct from those of vertebrates (Olivera *et al.*, 1994).

Phthalic acid diamides and anthranilic diamides are confirmed to control insects via the activation of RyR, which leads to uncontrolled calcium release in muscle (Ebbinghaus - Kintscher *et al.*, 2006, 2007; Cordova *et al.*, 2006). Flubendiamide has a low solubility and can therefore be used only at nanomolar concentrations. Compared with flubendiamide, chlorantraniliprole has extremely broad spectrum activity and higher aqueous solubility. Chlorantraniliprole stabilises insect RyR channels to the open state, evokes massive calcium release from intracellular stores and then disrupts calcium homeostasis, and possesses distinct pharmacological characteristics (Lahm *et al.*, 2009). The high level of mammalian safety of chlorantraniliprole is attributed to a strong selectivity for insect over mammalian receptors (DuPont, 2007). In a recent study, chlorantraniliprole is reported to impair *S. exigua* larvae with poisoning symptoms, which include rapid feeding cessation (Hannig *et al.*, 2009) regurgitation, lethargy, and contractile paralysis ($10nmol \cdot L^{-1}$). At the very least, the insecticidal effects can be explained by the neuronal effect.

Cordova *et al.* (2006) report that anthranilamide stimulates the release of RyR – mediated Ca^{2+} stores in *Periplaneta americana* Linnaeus embryonic neurones, whereas voltagegated Ca^{2+} channels are unaffected. Interestingly, no significant effects on the central nervous system function are reported. The present study describes an inhibitory effect of chlorantraniliprole on high – voltage activated L – type calcium channel in *S. exigua* larvae central neurones. With a treatment concentration $> 100pmol \cdot L^{-1}$, chlorantraniliprole significantly decreases the peak current of the calcium channel, shifts the *I* – *V* relationship curves in a negative direction by approximately 10mV and leads to the inhibition of the peak current of calcium channels in a highly concentration – dependent manner. These results indicate that $Ca^{2+}_{(L)}$ channels of *S. exigua* neurones are modulated by chlorantraniliprole and partly closed after the action of chlorantraniliprole. Although there is no notable change on activation voltage, the negatively shifted *I* – *V* relationship curves indicate that voltage – dependence is influenced by chlorantraniliprole. The gravity rundown in calcium current is more apparent than that shown in the control group, demonstrating that the opened channels are constantly decreased and the channels closed quickly, resulting in inactivation. All of these results suggest that the $Ca^{2+}_{(L)}$ channels of *S. exigua* neurones are the possible target of chlorantraniliprole. After exposure to chlorantraniliprole, washout does not reverse the calcium current to the control level as long as the recording continues. This suggests that the action site may be chemically modified on either the transmembrane domain or the intramembrane of the calcium channel.

Anthranilamides hold great promise for pest management strategies, although further experiments are required to provide more information on the mechanism of action of chlorantraniliprole in the central neurones of insects.

Acknowledgements

We sincerely thank Professor AnXi Liu for his kind support and advice.

References

[1] Bertolino, M. & Llinas, R. R. (1992) The central role of voltage – activated and receptor – operated calcium channels in neuronal cells. *Annual Review of Pharmacology and Toxicology*, 32, 399 – 421.

[2] Cordova, D., Benner, E. A., Sacher, M. D. *et al.* (2006) Anthranilic diamides: a new class of insecticides with a novel mode of action, ryanodine receptor activation. *Pesticide Biochemistry and Physiology*, 84, 196 – 214.

[3] DuPont (2007) *Rynaxypyr® insect control technical bulletin* [WWW document]. URL http://www2dupont.com/Production_Agriculture/en_US/assets/downloads/pdfs/Rynaxypyr_Tech_Bulletin.pdf [accessed on

October 2007].

[4] Ebbinghaus－Kintscher, U., Lüemmena, P., Lobitz, N. *et al.* (2006) Phthalic acid diamides activate ryanodine－sensitive Ca^{2+} release channels in insects. *Cell Calcium*, 39, 21－33.

[5] Ebbinghaus－Kintscher, U., Lümmen, P., Raming, K. *et al.* (2007) Flubendiamide, the first insecticide with a novel mode of action on insect ryanodine receptors. *Pflanzenschutz－Nachrichten Bayer*, 60, 117－140.

[6] Fill, M. & Copello, J. A. (2002) Ryanodine receptor calcium release channels. *Physiological Reviews*, 82, 893－922.

[7] Hamill, O. P., Marty, A., Neher, E. et al. (1981) Improved patch－clamp techniques for high－ resolution current recording from cells and cellfree membrane patches. *Pflugers Archiv*, 391, 85－100.

[8] Hannig, G. T., Ziegler, M. & Marcon, P. G. (2009) Feeding cessation effects of chlorantraniliprole, a new anthranilic diamide insecticide, in comparison with several insecticides in distinct chemical classes and mode－of－action groups. *Pest Management Science*, 65, 969－974.

[9] He, B. J., Liu, A. X., Chen, J. T. *et al.* (2001) Acute isolation and culture of nerve cell from cotton bollworm and patchclamp study on the voltage－gated ion channels in the cultured neurons. *Acta Entomologica Sinica*, 44, 422－427.

[10] Horn, R. & Korn, S. J. (1992) Prevention of rundown in electrophysiological recording. *Methods of Enzymology* (ed. by B. Rudy and L. E. Iverson), pp. 149－155. Academic Press, NewYork, NewYork.

[11] Huang, C. X., Zhu, L. M., Ni, J. P. & Cao, X. Y. (2002) A method of rearing the beet armyworm *Spodoptera exigua*. *Entomological Knowledge*, 39, 229－231.

[12] Lahm, G. P., Selby, T. P. & Stevenson, T. M. (2003) *Arthropodicidal Anthranilamides*. WO Patent 03/015519.

[13] Lahm, G. P., Stevenson, T. M., Selby, T. P. *et al.* (2007) Rynaxypyr™ a new insecticidal anthranilic diamide that acts as a potent and selective ryanodine receptor activator. *Bioorganic and Medicinal Chemistry Letters*, 17, 6274－6279.

[14] Lahm, G. P., Cordova, D. & Barry, J. D. (2009) New and selective ryanodine receptor activators for insect control. *Bioorganic and Medicinal Chemistry*, 17, 4127－4133.

[15] Neher, E. (1981) Unit conductance studies in biological membranes. *Techniques in Cellular Physiology* (ed. by P. F. Baker), Elsevier, The Netherlands.

[16] Olivera, B. M., Miljanich, G. P., Ramachandran, J. & Adams, M. E. (1994) Calcium channel diversity and neurotransmitter release: the ω－conotoxins and ω－agatoxins. *Annual Review of Biochemistry*, 63, 823－867.

[17] Sontheimer, H. & Olsen, M. (2007) Whole－cell patch－clamp recording. *Patch－Clamp Analysis: Advanced Techniques Neuromethods*, 2nd edn (ed. by W. Walz), pp. 35－68. Humana Press, Canada.

[18] Tohnishi, M., Nakao, H., Furuya, T. *et al.* (2005) Flubendiamide, a novel insecticide highly active against lepidopterous insect pests. *Journal of Pesticide Science*, 30, 354－360.

[19] Wen, L. Z. & Zhang, Y. J. (2010) Modelling of the relationship between the firequency of large－scale outbreak of the beet armyworm, *Spodoptera exigua* (Lepidoptera Noctuidae) and the wide－area temperature and rainfall trends in China. *Acta Entomologica Sinica*, 53, 1367－381.

Synthesis, Structure and Biological Activity of Novel 1,2,4 – Triazole Mannich Bases Containing a Substituted Benzylpiperazine Moiety*

Baolei Wang, Xinghai Liu, Xiulan Zhang, Jifeng Zhang, Haibin Song, Zhengming Li

(The Research Institute of Elemento – Organic Chemistry, State – Key Laboratory of Elemento – Organic Chemistry, Nankai University, Tianjin, 300071, China)

Abstract A series of novel Mannich bases with trifluoromethyl – 1,2,4 – triazole and substituted benzylpiperazine moieties were synthesized. Their structures were confirmed by IR, ^{1}H NMR and elemental analysis. The single crystal structure of compound **4r** was also determined. The preliminary bioassays showed that most of the lead compounds had low herbicidal activity against *Brassica campestris*, *Echinochloa crusgalli*, and KARI enzyme. However, most of them exhibited significant fungicidal activity at the dosage of 50 μg/mL toward five test fungi. Among the 18 novel compounds, several showed superiority over the commercial fungicide Triadimefon against *Cercospora arachidicola* and *Fusarium oxysporum* f. sp. *cucumerinum* during this study. Meanwhile, some compounds displayed plant growth regulatory activity at the dosage of 10 μg/mL.

Key words 1,2,4 – triazole; benzylpiperazine; fungicidal activity; herbicidal activity; Mannich base; plant growth regulatory activity

Sulfur – and nitrogen – linked heterocyclic compounds have received considerable attention in recent times because of their medicinal and pesticidal importance[1-4]. Triazoles, like many other five – membered heterocyclic compounds are used very often in pharmacological, medicinal, and agricultural applications. The compounds containing a triazole ring, such as 1 – (substituted phenyl) – 3 – (1 – alkoxy carboxyl alkoxy) – 1,2,4 – 1*H* – triazoles have shown a versatility and useful biological properties, and they have been developed as fungicides, herbicides, or plant growth regulators (PGRS)[5]. The incorporation of different active functional groups into triazole ring was proved to be a good way to produce novel active pesticides[6].

In recent years, many 1,2,4 – triazole derivatives such as Diniconazole, Triadimefon, Triadimenol, Flusilazole, Fluconazole, Itraconazole, and so on (Figure 1) have been found and developed as fungicides. These compounds represent the most important category of fungicides todate and have excellent protective, curative, and eradicant power toward a wide spectrum of crop diseases[7-9]. Meanwhile some structures with a 1,2,4 – triazole ring also exhibited outstanding herbicidal activity[10,11]. Because of their diverse properties, 1,2,4 – triazole pesticides may become one of the focuses in the research and development of agrochemicals. Thus, the synthesis

* Reprinted from *Chemical Biology & Drug Design*, 2011, 78(1): 42 – 49. This work was supported by National Basic Research Key Program of China (No. 2010CB126106), the Fundamental Research Funds for the Central Universities, Specialized and Research Fund for the Doctoral Program of Higher Education (No. 20070055044), Tianjin Natural Science Foundation (No. 08JCYBJC00800).

of broader spectrum and highly bioactive 1,2,4 – triazole compounds has become an active research in agricultural chemistry. Further, piperazine derivatives were also found to possess various biological activities[12-14]. Many literature references have shown that Mannich bases possess potent biological activity such as antibacterial, antifungal, anti – inflammatory, antimalarial, and pesticide properties[15-18].

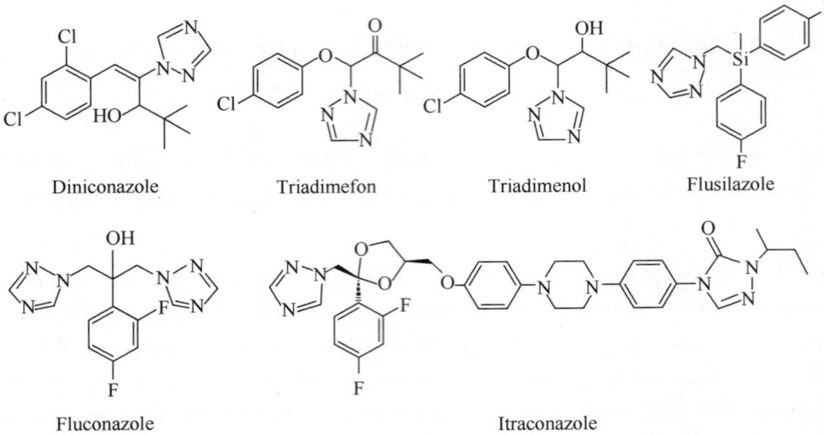

Diniconazole Triadimefon Triadimenol Flusilazole

Fluconazole Itraconazole

Figure 1 Some representative structures of triazole fungicides

In our previous work, we reported some interesting Mannich bases derived from 1,2,4 – triazole Schiff bases[19], and benzotriazoles containing benzylpiperzine ring, which are associated with various useful herbicidal activity[12]. In view of all these facts and as continuation of our research on pesticidal important heterocycles, hereby a series of novel Mannich bases with trifluoromethyl – 1,2,4 – triazole and substituted benzylpiperazine moieties were synthesized, and their herbicidal, fungicidal, and PGR activity were investigated. It was worthy of note that the structures we synthesized here are similar to those of benzotriazoles Mannich bases we reported earlier, during our search for novel inhibitors of KARI (one of the key herbicidal target enzymes involved in the biosynthesis of chain amino acids)[12]. So the *in vitro* and *in vivo* herbicidal activity of lead compounds were both investigated. The preliminary biological tests showed that some compounds exhibit significant fungicidal activity and PGR activity.

Experimental

Instruments and materials

The melting points were determined on an X – 4 binocular microscope melting point apparatus (Beijing Tech Instrument Co., Beijing, China) and are uncorrected. Infrared spectra were recorded on a Nicolet MAGNA – 560 spectrophotometer (Nicolet Instrument Corp., Madison, WI, USA) as KBr tablets. ^{1}H NMR spectra were measured on a Bruker AC – P500 instrument (400MHz) (Bruker, Fallanden, Switzerland) using TMS as an internal standard and DMSO – d_6 or $CDCl_3$ as solvent. Elemental analysis was performed on a Vario EL elemental analyzer (Elementar, Hanau, Germany). Crystallographic data of the compound were collected on a Rigaku MM – 07 Saturn 724 CCD diffractometer (Rigaku International Corp., Tokyo, Japan). All of the solvents and materials are of analytical grade.

Synthesis

The lead compounds were synthesized according to the route shown in Scheme 1.

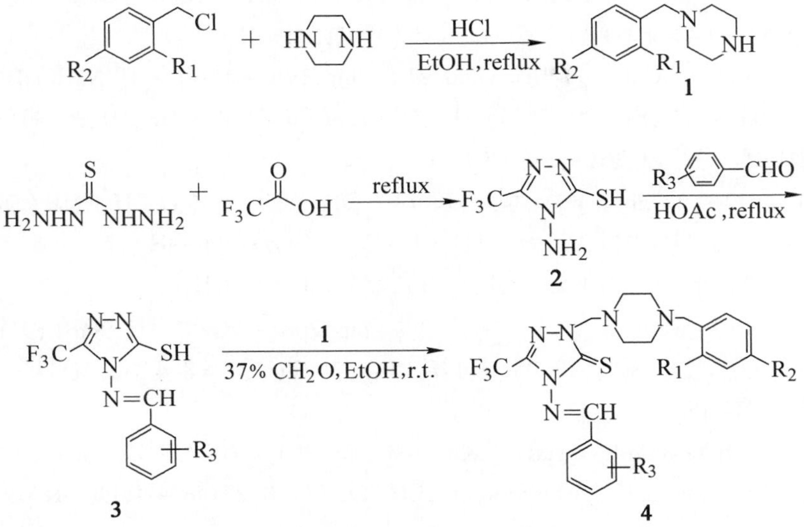

Scheme 1 Synthetic route of title compounds

Synthesis of compound triazolthiol(2)

4 – Amino – 5 – trifluoromethyl – 4*H* – 1,2,4 – triazole – 3 – thiol(**2**) was prepared according to the literature[20].

Preparation of 4 – (substituted) benzylpiperazine(1)

The procedures are similar to that described previously[21]. To a solution of anhydrous piperazine(50mmol) in 20mL 96% of ethanol was added conc. HCl (25mmol). The mixture was stirred under reflux, and substituted benzyl chlorine (25 mmol) was added dropwise for over 5min. The mixture was refluxed for 4 – 8h with TLC monitoring, then left to stand overnight at room temperature. The solid precipitated was filtered and washed with ethanol, the filtrate was evaporated in vacuum, and the residue was dissolved in 30mL of saturated K_2CO_3 aq. , extracted with chloroform(8mL × 5). The chloroform solution was dried with anhydrous Na_2SO_4 and evaporated in vacuum. The residue was then distilled under reduced pressure to give compound **1** as a colorless liquid.

1a(R_1 = R_2 = H): yield 58%, bp 131 – 134℃/10mmHg [Lit. bp 154 – 160℃/18mmHg[22]]; **1b**(R_1 = H, R_2 = Cl): yield 41%, bp 138 – 141℃/10mmHg [Lit. bp 92 – 96℃/0. 1mmHg[23]]; **1c**(R_1 = R_2 = Cl): yield 36%, bp 147 – 150℃/6mmHg [Lit. bp 106 – 108℃/0. 05mmHg[23]].

General synthetic procedures for 4 – (substituted) benzylideneamino – 5 – trifluoromethyl – 4*H* – 1,2,4 – triazole – 3 – thiol(3)

Compound **2**(10mmol) and aromatic aldehyde(10. 5mmol) were mixed in acetic acid(15mL). After having been stirred and refluxed for 15min, the reaction mixture was cooled to room temperature. The resulting crystals were filtered and washed with ethanol to give Schiff base **3**.

3a(R_3 = H): white crystal, yield 74%, mp 198 – 199℃ [Lit. mp 200 – 201℃ (19)]; ^{1}H NMR(DMSO – d_6, 400MHz) δ 14.90(s, 1H, SH), 9.97(s, 1H, CH), 7.58 – 7.92(m, 5H, Ph—H).

3b(R_3 = 2 – F): white crystals, yield 82%, mp 192 – 193℃; ^{1}H NMR(DMSO – d_6, 400MHz) δ 14.93(s, 1H, SH), 10.44(s, 1H, CH), 7.41 – 8.03(m, 4H, Ph—H).

3c(R_3 = 4 – MeO): white crystals, yield 71%, mp 195 – 196℃; ^{1}H NMR(DMSO – d_6, 400MHz) δ 14.84(s, 1H, SH), 9.73(s, 1H, CH), 7.86(d, J = 8.8 Hz, 2H, Ph—H), 7.13(d, J = 8.8Hz, 2H, Ph—H), 3.86(s, 3H, OCH_3).

3d(R_3 = 3,4 – Me_2): white crystals, yield 73%, mp 205 – 206℃; ^{1}H NMR(DMSO – d_6, 400MHz) δ 14.86(s, 1H, SH), 9.79(s, 1H, CH), 7.66(s, 1H, Ph—H), 7.62(d, J = 8.0 Hz, 1H, Ph—H), 7.35(d, J = 8.0 Hz, 1H, Ph—H), 2.31(ds, 6H, CH_3).

3e(R_3 = 4 – Cl): white crystals, yield 72%, mp 208 – 209℃; ^{1}H NMR(DMSO – d_6, 400MHz) δ 14.92(s, 1H, SH), 10.05(s, 1H, CH), 7.93(d, J = 8.4 Hz, 2H, Ph—H), 7.67(d, J = 8.4 Hz, 2H, Ph—H).

3f(R_3 = 2 – NO_2): yellow crystals, yield 89%, mp 201 – 202℃; ^{1}H NMR (DMSO – d_6, 400MHz) δ 14.96(s, 1H, SH), 10.88(s, 1H, CH), 7.87 – 8.23(m, 4H, Ph—H).

General synthetic procedures for 1 – [(4 – substituted – benzylpiperazin – 1 – yl) methyl] – 4 – (substituted) benzylideneamino – 3 – trifluoromethyl – 1*H* – 1,2,4 – triazole – 5(4*H*) – thione(4)

Schiff base **3**(1mmol) and 40% formalin(1.2mmol) were dissolved in 15mL of ethanol, and the mixture was stirred at room temperature for 10min. A solution of substituted benzylpiperazine **1** (1mmol) in 2mL ethanol was slowly added to. Then, the mixture was stirred for 2 – 3h and placed in a refrigerator overnight. The resulting precipitate was filtered and recrystallized from ethanol to give 1,2,4 – triazole Mannich base **4**.

Crystal structure determination

Compound **4r** was dissolved in hot ethyl alcohol, and the resulting solution was allowed to stand in air at room temperature to give single crystal of **4r**. A yellow crystal of **4r** suitable for X – ray diffraction with dimensions of 0.30 × 0.28 × 0.16mm was mounted on a Rigaku MM – 07 Saturn 724 CCD diffractometer with *Mo – Kα* radiation(λ = 0.71073Å) for data collection. A total of 13449 reflections were collected in the range of $2.01 < \theta < 25.02$ by using a phi and scan modes at 113(2)K, of which 4345 were independent with R_{int} = 0.0277. All calculations were refined anisotropically (SHELXS – 97). All hydrogen atoms were located from a difference Fourier map, placed at calculated positions, and included in the refinements in the riding mode with isotropic thermal parameters. The compound crystallizes in space group $P2_1/n$ of the monoclinic system with cell parameters: a = 11.884(2) Å, b = 8.6972(17) Å, c = 24.051(5) Å, $\alpha = 90°$, $\beta = 96.33(3)°$, $\gamma = 90°$, V = 2470.7(9) Å^3, Z = 4, D_c = 1.544mg/m^3, μ = 0.406per mm, and $F(000)$ = 1176. The final refinement converged at R = 0.0315, wR = 0.0814 for 3852 observed reflections with $I > 2\sigma(I)$, where $W = 1/[\sigma^2(F_o^2) + (0.0554P)^2 + 0.2627P]$ with $P = (F_o^2 + 2F_c^2)/3$, S = 1.046, $(\Delta/\sigma)_{max} < 0.0001$, $(\Delta\rho)_{max}$ = 0.252e/Å^3 and $(\Delta\rho)_{min}$ =

−0. 312e / $Å^3$.

The herbicidal activity

In vivo herbicidal activity of compounds **4a – r** was determined by rape root and barnyardgrass cup tests according to the reported method[24].

Inhibition of the root growth of Rape(*Brassica campestris*)

The evaluated compounds were dissolved in water and emulsified if necessary. Rape seeds were soaked in distilled water for 4h before being placed on a filter paper in a 6cm Petri plate, to which 2mL of inhibitor solution had been added in advance. Usually, 15seeds were used on each plate. The plate was placed in a dark room and allowed to germinate for 65 h at 28 ± 1℃. The lengths of 10 rape roots selected from each plate were measured and the means were calculated. The check test was carried out only in distilled water. The inhibitive rates were calculated from the root length using the following equation:

$$\text{Relative inhibition rate}(\%) = [(CK - PT)/CK] \times 100\%$$

where CK is the average root length during the blank assay and PT is the average root length after treatment during testing.

Inhibition of the seedling growth of Barnyardgrass(*Echinochloa crusgalli*)

The evaluated compounds were dissolved in water and emulsified if necessary. Ten barnyardgrass seeds were placed into a 50mL cup covered with a layer of glass beads and a piece of filter paper at the bottom, to which 5mL of inhibitor solution had been added in advance. The cup was placed in a bright room and allowed to germinate for 65 h at 28 ± 1℃. The heights of seedlings of the above – ground plant parts from each cup were measured and the means were calculated. The check test was carried out only in distilled water. The inhibitive rates were calculated from the plant heights using the following equation:

$$\text{Relative inhibition rate}(\%) = [(CK - PT)/CK] \times 100\%$$

where CK is the average plant height during the blank assay and PT is the average plant height after treatment during testing.

KARI activity test

The cloning of rice KARI has been described previously[25], and enzyme expression and purification followed that protocol. Protein concentrations were estimated *via* the bicinchoninic acid method and protein purity was assessed by SDS – PAGE[26,27]. KARI activity was measured with a continuous assay method[25], following the consumption of NADPH at 340nm and 30℃. Assay solutions contained 0. 2mM NADPH, 1 mM $MgCl_2$, 0. 1 mM substrate(2 – acetolactate), and inhibitors(**4a – r** and IpOHA), in 0. 1 M phosphate buffer(pH 8. 0). Inhibitors were preincubated with the enzyme, NADPH, and $MgCl_2$ in phosphate buffer at 30℃ for 10min. The reaction was then started by adding the substrate. The percentage of the inhibition was calculated using the following equation:

$$\text{Relative inhibition rate}(\%) = [(Abs_{blank} - Abs_{sample})/Abs_{blank}] \times 100\%$$

where Abs_{blank} and Abs_{sample} are the absorption curve slopes of NADPH at 340 nm with blank and inhibitor in testing solution, respectively.

The fungicidal activity

Fungicidal activity of **4a – r** against *G. zeae* Petch, *Phytophthora infestans* (Mont.) de Bary, *Cercospora arachidicola*, *Botryosphaeria berengeriana* f. sp. piricola (Nose) koganezawa et Sakuma, and *Fusarium oxysporum* f. sp. *cucumerinum* were evaluated using the mycelium growth rate test[28].

The method for testing the primary biological activity was performed in an isolated culture. Under a sterile condition, 1mL of sample was added to the culture plates, followed by the addition of 9mL of culture medium. The final mass concentration was 50μg/mL. The blank assay was performed with 1mL of sterile water. Circle mycelium with a diameter of 4 mm was cut using a drill. The culture plates were cultivated at 24 ± 1℃. The extended diameters of the circle mycelium were measured after 72h. The relative inhibition rate of the circle mycelium compared to blank assay was calculated via the following equation:

$$\text{Relative inhibition rate}(\%) = [(d_{ex} - d_{ex})/d'_{ex}] \times 100\%$$

where d_{ex} is the extended diameter of the circle mycelium during the blank assay; and d'_{ex} is the extended diameter of the circle mycelium during testing.

The blank test was made using acetone. Three replicates were performed. The fungicidal tests results were given in Table 1.

Table 1 The fungicidal and plant growth regulatory activity of compounds 4a – r

Compound	Growth inhibition (%, 50μg/mL)					Rhizogenesis (%, 10μg/mL)
	G. zeae Petch	*Phytophthora infestans* (Mont.) de Bary	*Cercospora arachidicola*	*Botryosphaeria berengeriana* f. sp. *piricola* (Nose) koganezawa et Sakuma	*Fusarium oxysporum* f. sp. *cucumerinum*	
4a	33.2 ± 2.7	21.0 ± 1.8	70.3 ± 1.9	10.8 ± 2.4	10.1 ± 1.1	4.6 ± 1.8
4b	50.1 ± 2.1	60.4 ± 1.2	70.4 ± 2.4	70.1 ± 1.4	80.4 ± 2.8	68.6 ± 3.2
4c	0	0	0	0	18.7 ± 2.2	10.4 ± 2.0
4d	0	0	19.5 ± 1.3	0	0	109.3 ± 3.4
4e	0	10.3 ± 0.8	50.3 ± 2.7	0	0	97.6 ± 1.7
4f	50.2 ± 4.1	53.1 ± 3.5	72.3 ± 3.2	20.1 ± 3.3	40.2 ± 1.6	62.7 ± 2.8
4g	0	24.3 ± 2.0	50.4 ± 2.3	19.6 ± 1.4	50.9 ± 3.2	−12.7 ± 1.1
4h	50.3 ± 2.9	51.7 ± 2.1	70.6 ± 3.4	52.6 ± 4.0	71.9 ± 3.1	51.1 ± 2.3
4i	11.2 ± 2.0	19.7 ± 2.3	50.7 ± 2.9	10.0 ± 0.9	60.4 ± 3.4	16.2 ± 1.7
4j	0	19.6 ± 2.0	44.5 ± 3.3	0	40.1 ± 3.1	27.9 ± 2.8
4k	30.6 ± 2.5	30.1 ± 3.2	60.7 ± 3.0	0	30.3 ± 1.8	39.5 ± 2.6
4l	18.8 ± 2.3	19.8 ± 2.1	70.2 ± 3.4	30.0 ± 1.9	50.2 ± 2.4	16.4 ± 1.3
4m	50.3 ± 2.6	51.8 ± 4.5	70.4 ± 4.1	30.5 ± 2.2	70.8 ± 3.1	39.5 ± 2.3
4n	50.0 ± 2.7	50.4 ± 2.3	60.2 ± 2.9	41.5 ± 1.2	82.6 ± 1.8	45.3 ± 1.6
4o	51.8 ± 2.4	36.0 ± 1.3	50.2 ± 3.3	0	50.4 ± 3.0	22.0 ± 1.8
4p	50.7 ± 3.6	50.9 ± 3.9	80.8 ± 2.0	70.3 ± 3.2	70.5 ± 2.8	62.9 ± 3.0
4q	30.0 ± 2.1	50.5 ± 2.8	70.4 ± 3.4	70.6 ± 3.7	60.7 ± 2.9	10.2 ± 0.7
4r	30.5 ± 1.8	30.2 ± 2.4	63.6 ± 3.8	41.6 ± 2.8	60.8 ± 3.1	45.3 ± 3.3
Triadimefon	72.6 ± 3.6	66.1 ± 2.7	73.4 ± 2.3	86.7 ± 3.2	81.5 ± 2.6	50.2 ± 3.4

The plant growth regulatory activity

The PGR activity of compounds **4a – r** was evaluated by means of cucumber cotyledon test according to a reported procedure[29]. The cucumber seeds (JINKE, No. 4) were supplied by the Biological Assay Center, Nankai University, China. These seeds were incubated at 24℃ in a dark room for 3 days and 10 pieces of cotyledons of the same size were selected. The test samples were dissolved in *N*, *N* – dimethylformamide (DMF) at a concentration of 10μg/mL. A sample solution (0.3mL) was sprayed over a filter paper (6cm diameter), and solvent was volatilized to dryness on air. The filter paper thus prepared was placed into an incubation vessel (6cm diameter) and soaked with distilled water (3mL). Finally, 10pieces of cotyledons were added. These cotyledons were incubated at 24℃ in a dark room for 3 days. The rhizogenesis numbers of every 10pieces of hypocotyls were measured. Each treatment was performed three times. In contrast, the distilled water was used as a control. The relative ratios of cucumber cotyledon rhizogenesis were calculated according to the following formula:

$$\text{Relative ratio}(\%) = (N_S - N_C)/N_C \times 100\%$$

where N_S and N_C are the numbers of cucumber cotyledon rhizogenesis of tested compound and control experiment, respectively.

Results and Discussion

Synthesis

The synthesis procedures for compound **4** were shown in Scheme 1. Based on an acetic acid. solvent method reported by Wu *et al.* for the condensation of amine and aldehyde[30], Schiff base **3**, namely 4 – amino – 5 – trifluoromethyl – 4*H* – 1,2,4 – trizole – 3 – thiol, was prepared successfully with high yield and short reaction time. The Mannich reaction of **3** with formaldehyde and substituted benzylpiperazine **1** in ethanol at room temperature led to novel trifluoromethyl – substituted 1,2,4 – triazole Mannich base **4** in 44% – 83% yield. Compounds **4** were identified by melting point, ^{1}H NMR and IR spectra. The measured elemental analyses were also consistent with the corresponding calculated ones (see Supporting Information).

Proton magnetic resonance spectra

The proton magnetic resonance spectra of the Mannich bases have been recorded in $CDCl_3$ and those of the Schiff bases in DMSO – d_6. A comparison of the spectra of the products with the intermediates leads to the following conclusions: The —CH ═N proton appeared at δ 9.73 – 10.88 as a singlet in the Schiff base **3**, which shifted downfield to δ 10.07 – 11.21 in Mannich bases **4**. The Schiff bases **3** can exist either as a thione or the thiol tautomeric forms or as an equilibrium mixture of both forms, because they have a thioamide, —NH—C(═S) function. The chemical shift at δ 14.84 – 14.96 as a singlet in **3** may be because of SH proton, indicating that **3** existed not as thione but as the thiol tautomeric forms in solution[31]. In the Mannich bases **4**, neither a NH signal nor a thiol SH signal is visible. The signal of CH_2 protons neighboring to the triazole ring was observed at δ approximately 5.22 as a singlet, and the substituted benzyl CH_2 proton appeared at δ 3.45 – 3.56 as a singlet in Mannich bases **4**. The chemical shifts at 2.86 – 2.88 ppm and 2.46 – 2.54 ppm may be because of piperazine ring protons, which were appeared as two broad singlets, respectively. The signal because of benzene ring protons appears in the region δ 6.98 – 8.23 in the spectra of all products and intermediates.

Infrared spectra

The spectra of Mannich bases **4a**, **4g**, and **4m** showed bands at 1601 – 1600 per cm for C ═N stretching. The characteristic stretching vibrations ν(C—F) and ν(C ═S) appears at 1320 – 1318 per cm, 1197 – 1194 per cm and 1166 – 1154 per cm, respectively.

Crystal structure

The structure of compound **4r** was further confirmed by single – crystal X – ray diffraction analysis (Figure 2). From the molecular structure, it can be seen that both groups on the N atoms of piperazine ring (triazole – CH_2 and 2,4 – dichlorophenyl – CH_2) are in the e – bond positions of chair conformation in the six – member ring. The dihedral angel between 2 – nitrobenzene ring and triazole ring is 2.6°, which indicates the two rings are almost coplanar in the molecular structure. The X – ray analysis also reveals that in this typical compounds **4r**, the substituted benzene ring and triazole ring are on the opposite sides of the C ═N double bond (Figure 2). The torsion angle of C(6) – C(7) – N(2) – N(3) is 177.33°, which indicates that the C ═N double bond is in the (*E*) – configuration.

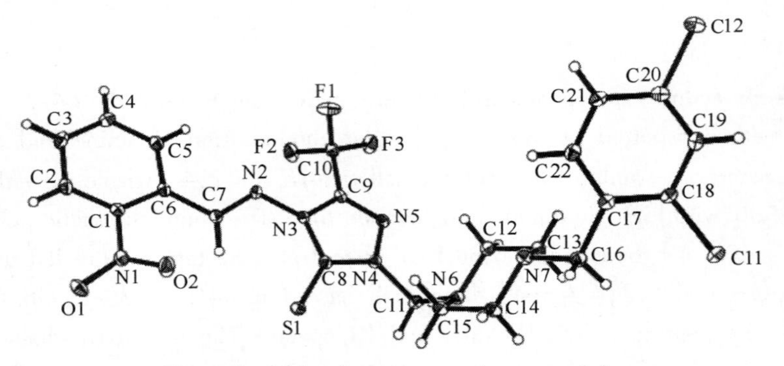

Figure 2 Molecular structure of compound **4r**

Herbicidal activity

To investigate the KARI inhibitory activity and herbicidal activity of the compounds referring to those of benzotriazoles Mannich bases, we reported before during our search for novel KARI inhibitors (12), *N* – hydroxy – *N* – isopropyloxamate (IpOHA)[32], a potent inhibitor of KARI *in vitro*, was used as a control. As shown in Table 2, compounds **4m** and **4r** showed higher inhibition abilities of rape root at a concentration of 100 μg/mL. Compound **4a** showed obvious inhibitory activity to the rice KARI enzyme with an inhibition of 76.6%. Except that, most of compounds exhibited weak herbicidal activity *in vivo* and *in vitro* against rape and barnyardgrass and KARI enzyme.

Fungicidal activity

The *in vitro* fungicidal results of the Mannich bases **4a – r** in inhibiting the mycelial growth of five test fungi were listed in Table 1. The results of preliminary bioassays were compared with

Table 2 Herbicidal activities data of title compounds(% inhibition)

Compound	R_1	R_2	R_3	Rape root test(*Brassica campestris*)	Barnyardgrass cup test(*Echinochloa crusgalli*)	KARI test
				100μg/mL (10μg/mL)	100μg/mL (10μg/mL)	100μg/mL
4a	H	H	H	27. 8	10. 4	76. 6
4b	H	H	2 - F	10. 7	15. 2	21. 3
4c	H	H	4 - MeO	17. 0	5. 3	9. 7
4d	H	H	3,4 - Me_2	0	5. 0	0
4e	H	H	4 - Cl	2. 2	30. 2	1. 1
4f	H	H	2 - NO_2	17. 4	10. 1	1. 9
4g	H	Cl	H	31. 4	15. 2(5. 1)	0
4h	H	Cl	2 - F	30. 1(10. 3)	20. 4	0
4i	H	Cl	4 - MeO	0. 9	30. 2	—
4j	H	Cl	3,4 - Me_2	7. 6	10. 1(5. 0)	21. 6
4k	H	Cl	4 - Cl	39. 3	15. 3(5. 2)	15. 9
4l	H	Cl	2 - NO_2	27. 8	15. 4	17. 5
4m	Cl	Cl	H	50. 2	5. 1	22. 8
4n	Cl	Cl	2 - F	38. 1	15. 4	12. 1
4o	Cl	Cl	4 - MeO	33. 4	20. 4(10. 2)	9. 57
4p	Cl	Cl	3,4 - Me_2	13. 9	25. 4(5. 3)	28. 8
4q	Cl	Cl	4 - Cl	7. 2	15. 1	28. 0
4r	Cl	Cl	2 - NO_2	48. 0	20. 4(5. 2)	30. 0
IpOHA				71. 6(50. 2)	14. 4	99. 5

– ,Indicates the compound cannot be dissolved in our test system,so no data obtained.

that of a commercial fungicide Triadimefon. As indicated in Table 1,some compounds exhibited potential fungicidal activity. For example,compounds **4b**,**4f**,**4h**,and **4m** – **4p** possess approximately 50% inhibitory rates against *G. zeae* Petch. Also compound **4b** held a 60. 4% inhibitory rate against *Phytophthora infestans*(Mont.)de Bary. It was observed that most of the compounds exhibited significant inhibition effect against *Cercospora arachidicola* (inhibition rates of 50. 2% – 80. 8%)and the inhibitory effect of **4p**(80. 8%)on *Cercospora arachidicola* was better than that of the control Triadimefon (73. 4%). For *Botryosphaeria berengeriana* f. sp. *piricola*(Nose) koganezawa et Sakuma, compounds **4b**, **4h**, **4p**, and **4q** showed favorable activity and held 70. 1% , 52. 6% ,70. 3% ,and 70. 6% inhibitory rates, respectively. In addition, almost half of the compounds exhibited good inhibition effect against *Fusarium oxysporum* f. sp. *cucumerinum*(inhibition rates of 60. 4% – 82. 6%). Especially, compound **4b** and **4n** showed 80. 4% and 82. 6% inhibitory activity against *Fusarium oxysporum* f. sp. *cucumerinum*, respectively, and the effects of which were comparable to that of the control(81. 5%). It was worthy of note that in these

Mannich bases, compounds with two chlorine atoms on benzene ring of benzyl group (at 4 – position of piperazine ring), exhibited more obvious activity than others.

Plant growth regulatory activity

Plant growth regulatory (PGR) activities are defined here as those effects produced by the compound that leaves the numbers of cucumber cotyledon rhizogenesis indistinguishable from the untreated control in the same testing condition. The positive value of activity data means the compound exhibits potential promoting activity and can be an activator for the plant growth, while the negative value means the compound exhibits potential inhibitory activity and can be an inhibitor for the plant growth. The PGR activity of the lead compounds **4a – r** was screened by cucumber cotyledon test at a concentration of 10μg/mL. As shown in Table 1, some compounds, such as **4b**, **4d**, **4e**, **4f**, **4h**, and **4p**, exhibited better promoting activity than Triadimefon, with values of 68.6%, 109.3%, 97.6%, 62.7%, 51.1%, and 62.9%, respectively. These results were consistent with those of their low *in vivo* herbicidal activity based on *Brassica campestris* and *Echinochloa crusgalli* tests at a dose of 10μg/mL (almost no activity), referring to the herbicidal activity data listed in Table 2. Compound **4g** displayed weak inhibitory activity for the plant growth, with values of –12.7%, while the same effect could also be observed in its herbicidal activity assays.

It was found that compounds without substituent in the benzene ring of the benzyl group ($R_1 = R_2 = H$) almost have a higher level of promoting activity for plant growth than others ($R_1 = H, R_2 = Cl$ and $R_1 = R_2 = Cl$). For herbicidal activity of the lead compound, compounds with one or two chloro group(s) in the benzene ring of the benzyl group showed relatively favorable activity than others; however, the extent of the herbicidal effect was varying with concentration. On the whole, the promoting activity of the lead compounds is mainly for plant growth. Among these compounds, 1 – [(4 – benzylpiperazin – 1 – yl) methyl] – 4 – (3,4 – dimethyl) benzylideneamino – 3 – trifluoromethyl – 1*H* – 1,2,4 – triazole – 5(4*H*) – thione (**4d**) was the most effective and least inhibitory to plant and could be a favorable activator of plants growth at a low dose of concentration.

Based on these preliminary studies, the fungicidal activity was outstanding for this series of compounds. It appeared prior to the synthesis and testing of substituted derivatives that the fungicidal activity might be influenced by steric factors. Other novel Mannich bases derive from Schiff bases containing various alkyl substituents, or other aryl groups, such as heterocycles, may result in novel fungicides. Anyhow, it is important for their chemical structure that both two chloro groups are in the benzene ring of the benzyl group of such kind of compounds.

In summary, we have conveniently synthesized a series of trifluoromethyl – substituted 1,2,4 – triazole Mannich bases containing substituted benzylpiperazine ring *via* Mannich reaction in good yields. The preliminary bioassays showed that most of the compounds had low herbicidal activity against *Brassica campestris*, *Echinochloa crusgalli*, and KARI enzyme. However, most of them exhibited significant fungicidal activity at the dosage of 50μg/mL toward five test fungi. Among the 18 novel compounds, several showed superiority over the commercial fungicide Triadimefon against *Cercospora arachidicola* and *Fusarium oxysporum* f. sp. *cucumerinum* during this study and could be further developed as fungicides. Meanwhile, some compounds displayed

PGR activity at the dosage of 10μg/mL in the preliminary studies.

Acknowledgments

We thank the teachers of the Biological Assay Center, Nankai University, for kind bioassay assistance of compounds.

References

[1] Singh H., Misra A. R., Yadav L. D. S. (1987) Synthesis and fungitoxicity of new 1,3,4 - oxadiazolo [3,2 - a] - s - triazine - 5,7 - diones and their thione analogues. Indian J Chem;26B:1000 - 1002.

[2] Patel H. V., Fernandes P. S., Vyas K. A. (1990) A novel synthesis of a few substituted s - triazolo[3,4 - b] 1,3,4 - thiadiazoles and evaluation of their antibacterial activity. Indian J Chem;29B:135 - 137.

[3] Khalil Z. H., Abdel H., Ali A., Ahmed A. (1989) New pyrimidine derivatives. Synthesis and application of thiazolo(3,2 - a) - and triazolo[4,3 - a] pyrimidine as bactericides, fungicides and bioregulators. Phosphorus Sulfur Silicon Relat Elem;45:81 - 93.

[4] Demirbas N., Karaoglu S. A., Demirbas A., Sancak K. (2004) Synthesis and antimicrobial activities of some new 1 - (5 - phenylamino - [1,3,4] thiadiazol - 2 - yl) methyl - 5 - oxo - [1,2,4] triazole and 1 - (4 - phenyl - 5 - thioxo - [1,2,4] triazol - 3 - yl) methyl - 5 - oxo - [1,2,4] triazole derivatives. Eur J Med Chem;39:793 - 804.

[5] Li W., Wu Q., Ye Y., Luo M., Hu L., Gu Y., Niu F., Hu J. (2004) Density functional theory and ab initio studies of geometry, electronic structure and vibrational spectra of novel benzothiazole and benzotriazole herbicides. Spectrochim Acta A Mol Biomol Spectrosc;60:2343 - 2354.

[6] He H. W., Meng L. P., Hu L. M., Liu Z. J. (2002) Synthesis and plant growth regulating activity of 1 - (1 - phenyl - 1,2,4 - triazole - 3 - oxy acetoxy) alkyl phosphonates. Chin J Pest Sci;4:14 - 18.

[7] Crofton K. M. (1996) A structure - activity relationship for the neurotoxicity of triazole fungicides. Toxicol Lett;84:155 - 159.

[8] Cao X., Li F., Hu M., Lu W., Yu G. - A., Liu S. H. (2008) Chiral γ - aryl - 1H - 1,2,4 - triazole derivatives as highly potential antifungal agents: design, synthesis, structure, and in vitro fungicidal activities. J Agric Food Chem;56:11367 - 11375.

[9] Menegola E., Broccia M. L., Renzo F. D., Giavini E. (2001) Antifungal triazoles induce malformations in vitro. Reprod Toxicol;15:421 - 427.

[10] Simmons K. A., Dixson J. A., Halling B. P., Plummer E. L., Plummer M. J., Tymonko J. M., Schmidt R. J., Wyle M. J., Webster C. A., Baver W. A., Witkowski D. A., Peters G. R., Gravelle W. D. (1992) Synthesis and activity optimization of herbicidal substituted 4 - aryl - 1,2,4 - triazole - 5(1H) - thiones. J Agric Food Chem;40:297 - 305.

[11] Li X. - Z., Si Z. - X. (2003) Research advances on triazole compounds bearing biological activities. Pesticides;42:4 - 5.

[12] He F. Q., Liu X. H., Wang B. L., Li Z. M. (2006) Synthesis and biological activity of new 1 - [4 - (substituted) - piperazin - 1 - ylmethyl] - 1H - benzotriazole. J Chem Res;12:809 - 811.

[13] Chaudhary P., Kumar R., Verma A. K., Singh D., Yadav V., Chhillar A. K., Sharmab G. L., Chandra R. (2006) Synthesis and antimicrobial activity of N - alkyl and N - aryl piperazine derivatives. Bioorg Med Chem;14:1819 - 1826.

[14] Upadhayaya R. S., Sinha N., Jain S., Kishore N., Chandra R., Arora S. K. (2004) Optically active antifungal azoles: synthesis and antifungal activity of (2R,3S) - 2 - (2,4 - difluorophenyl) - 3 - (5 - {2 - [4 - arylpiperazin - 1 - yl] - ethyl} - tetrazol - 2 - yl/1 - yl) - 1 - [1,2,4] - triazol - 1 - yl - butan

-2-ol. Bioorg Med Chem;12:2225-2238.

[15] Karthikeyan M. S., Prasad D. J., Poojary B., Bhat K. S., Holla B. S., Kumari N. S. (2006) Synthesis and biological activity of Schiff and Mannich bases bearing 2,4-dichloro-5-fluorophenyl moiety. Bioorg Med Chem;14:7482-7489.

[16] Satyanarayana D., Kalluraya B., George N. (2002) Synthesis and biological evaluation of some Mannich bases derived from 1,2,4-triazole derivatives. J Saudi Chem Soc;6:459-464.

[17] Li Y., Yang Z. S., Zhang H., Cao B. J., Wang F. D., Zhang Y., Shi Y. L., Yang J. D., Wu B. A. (2003) Artemisinin derivatives bearing Mannich base group: synthesis and antimalarial activity. Bioorg Med Chem;11:4363-4368.

[18] Ram V. J., Mishra L., Pandey H. N., Vlietnick A. J. (1986) Pesticidal mannich bases derived from isatinimines. J Heterocycl Chem;23:1367-1369.

[19] Liu F. M., Wang B. L., Lu W. J. (2000) Synthesis of trifluoromethylsubstituted 1,2,4-triazole mannich bases. Chin J Org Chem;20:738-742.

[20] George T., Mehta D. V., Tahilramani R., David J., Talwalker P. K. (1971) Synthesis of some s-triazoles with potential analgesic and antiinflammatory activities. J Med Chem;14:335-338.

[21] Zlatoidsky P., Maliar T. (1996) Synthesis of 1-(4-acyloxybenzoyloxyacetyl)-4-alkylpiperazines and 1-(4-acyloxybenzoyl)-4-alkylpiperazines as inhibitors of chymotrypsin. Eur J Med Chem;31: 669-674.

[22] Baltzly R., Buck J. S., Lorz E., Schon W. (1944) Preparation of Nmonosubstituted and unsymmetrically disubstituted piperazines. J Am Chem Soc;66:263-266.

[23] Shapiro S. L., Friedman L., Soloway H. (1962) Piperazine and homopiperazine bronchodilators. U. S. Vit. Pharm. Patent; BE 617599.

[24] Wang B.-L., Duggleby R. G., Li Z.-M., Wang J.-G., Li Y.-H., Wang S.-H., Song H.-B. (2005) Synthesis, crystal structure and herbicidal activity of mimics of intermediates of the KARI reaction. Pest Manag Sci;61:407-412.

[25] Lee Y. T., Ta H. T., Duggleby R. G. (2005) Cyclopropane-1,1-dicarboxylate is a slow-, tight-binding inhibitor of rice ketol-acid reductoisomerase. Plant Sci;168:1035-1040.

[26] Smith P. K., Krohn R. I., Hermanson G. T., Mallia A. K., Gartner F. H., Provenzano M. D., Fujimoto E. K., Goeke N. M., Olson B. J., Klenk D. C. (1985) Measurement of protein using bicinchoninic acid. Anal Biochem;150:76-85.

[27] Laemmli U. K. (1970) Cleavage of structural proteins during the assembly of the head of bacteriophage T4. Nature;227:680-685.

[28] Liu Z. M., Yang G. F., Qin X. H. (2001) Syntheses and biological activities of novel diheterocyclic compounds containing 1,2,4-triazolo[1,5-a]pyrimidine and 1,3,4-oxadiazole. J Chem Technol Biotechnol;76:1154-1158.

[29] Demuner A. J., Barbosa L. C. A., Veloso D. P. (1998) New 8-oxabicyclo[3,2,1]oct-6-en-3-one derivatives with plant growth regulatory activity. J Agric Food Chem;46:1173-1176.

[30] Wu T. X., Li Z. J., Zhao J. C. (1998) A facile method for the synthesis of 4-amino-5-hydrocarbon-2,4-dihydro-3H-1,2,4-triazole-3-thione schiff bases. Chem J Chin Universities; 19: 1617-1619.

[31] Ma C. L., Wang Q. F., Zhang R. F. (2008) Syntheses and crystal structures of triorganotin(Ⅳ) complexes of schiff base derived from 4-amino-5-phenyl-4H-1,2,4-trizole-3-thiol. Heteroatom Chemistry;19:583-591.

[32] Aulabaugh A., Schloss J. V. (1990) Oxalyl hydroxamates as reaction-intermediate analogues for ketol-acid reductoisomerase. Biochemistry;29:2824-2830.

Supporting Information

Additional Supporting Information may be found in the online version of this article:

Table S1. Analytical data for compounds **4**.

Table S2. ^{1}H NMR spectral data of compounds **4**.

Table S3. IR spectral data of compounds **4**.

Please note: Wiley – Blackwell is not responsible for the content or functionality of any supporting materials supplied by the authors. Any queries (other than missing material) should be directed to the corresponding author for the article.

Design, Synthesis and Antifungal Activities of Novel Pyrrole Alkaloid Analogs*

Mingzhong Wang, Han Xu, Tuanwei Liu, Qi Feng,
Shujing Yu, Suhua Wang, Zhengming Li

(State Key Laboratory of Elemento - Organic Chemistry, Research Institute of Elemento - Organic Chemistry, Nankai University, Tianjin, 300071, China)

Abstract A series of novel analogs of pyrrole alkaloid were designed and synthesized by a facile method and their structures were characterized by ^1H NMR, ^{13}C NMR and high - resolution mass spectrometry (HRMS). The structure of compound **2a** was identified by 2D NMR including heteronuclear multiple - quantum coherence (HMQC), heteronuclear multiple - bond correlation (HMBC) and H - H correlation spectrometry (H - H COSY) spectra. Their antifungal activities against five fungi were evaluated, and the results indicated that some of the title compounds showed moderate fungicidal activities in vitro against *Alternaria solani*, *Cercospora arachidicola*, *Fusarium omysporum*, *Gibberella zeae* and *Physalospora piricola* at the dosage of 50 μg · mL^{-1}. Compound **2a** and **3a** exhibited good activities against *P. piricola* at low dosage.

Key words analogues; pyrrole alkaloid; 2D NMR; antifungal activities

1 Introduction

In the past century, a lot of bromopyrrole alkaloids had been isolated from marine organism and exhibited good biological activities[1-3]. For instance, the pyrrolomycin (**A**, Fig. 1) had distinguished antibiotic activity[4]. It was commonly assumed that many bromopyrrole alkaloid metabolites were served as a chemical protection role for the organism including antifeedant, antibacterial and antifungal agent, etc[5-7]. For example, natural product dispacamide (**B**, Fig. 1) and its derivatives isolated from sponge had evident antifeedant and antibacterial activities[7-9]. Anthranilamide (**C**, Fig. 1) isolated from a culture of marine *Streptomyces* sp. strain B7747 showed remarkable antimicroalgal activity[10]. Comparing to dispacamide, there were two similar four - membered carboxamide moieties among them (—N (CH_3) $COCH_2CH_2$— and —CONHCH_2CH_2—). Additionally, it is reported that *N* - (benzyloxy) benzamide (**D**, Fig. 1) and its analogs with the *O* - benzylhydroxylamine moiety (—CONHOCH_2—Ar) possessed antibacterial, herbicidal and enzyme inhibiting activities[11-14]. The pharmacophore (—CONHOCH_2—) was generally considered to be the bioisosteric analog (—CONHCH_2CH_2—) for drug' s design[15]. Considering the potential antifungal activity of the bromopyrrole alkaloid and the activity contribution of the *O* - benzylhydroxylamine group to the *N* - (benzyloxy) benzamide derivatives[11-14], a series of new dibromopyrrole alkaloid analogs (**E**, Fig. 1) were designed, synthesized and tested against fungi in vitro. Their antifungal activities against five fungi were evaluated and the possible structure - activity relationships (SAR) were discussed. In order to study

* *Reprinted from European Journal of Medicinal Chemistry*, 2011, 46(5): 1463 - 1472. This work was supported by the National Key Project for Basic Research (2010CB126106) and the Tianjin S&T Commission.

the SAR, these analogs' benzene ring was replaced with different groups. Based on the SAR, some analogs with the pyrrole ring substituted by —Cl($m=2$), —Br($m=3$) or unsubstituted by any halogen atom were also synthesized. And the pyrrole ring of some derivatives was further methylated. All the new derivatives were designed to learn whether those might increase or decrease their biological activities.

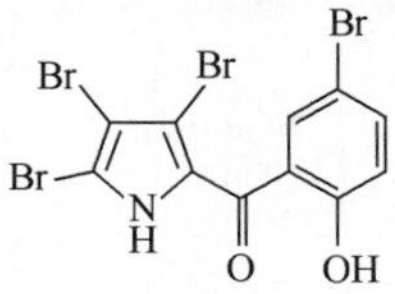

Pyrrolomycin **A**

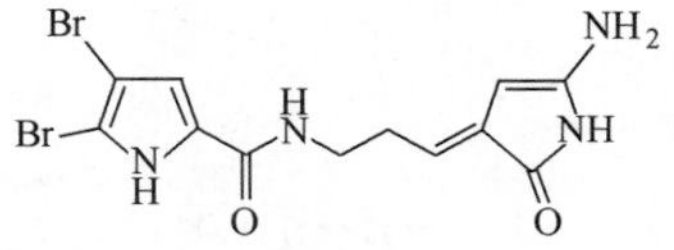

Dispacamide **B**

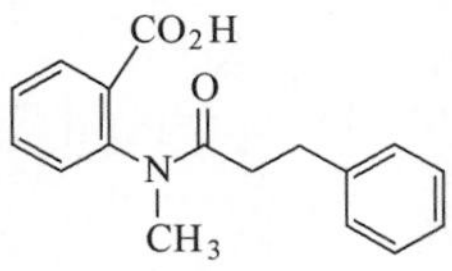

Anthranilamide **C**

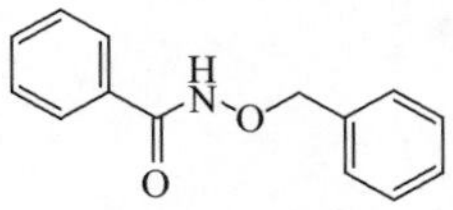

Compound **D**

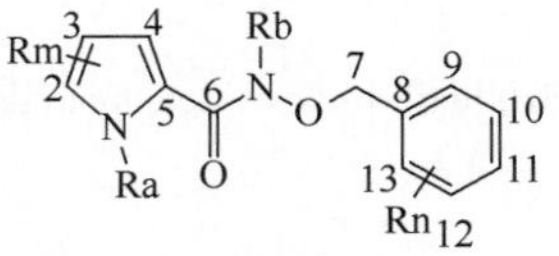

Compound **E**

Ra=H; CH_3

Rb=H; CH_3

Rm= Cl, m=2; Br, m=0,2,3

Rn=F; Cl; Br; CN; NO_2 etc.

n=0,1,2

Target compoanels in present work

Fig. 1 Chemical structure of compounds **A** – **E**

2 Chemistry

2.1 Synthesis

In this paper, 1*H* – pyrrole – 2 – carboxylic acid **2** was successfully synthesized through an improved procedure (see experimental Section 5.2.1 and Scheme 1). Compound **2** was failed to be given when the mixture was stirred for 10h at room temperature according to the literature[16]. 4,5 – Dichloro – 1*H* – pyrrole – 2 – carboxylic acid **5** was also successfully synthesized using amixture solution (see experimental Section 5.2.1 and Scheme 1). It is failed to get compound **5** when the solution was only dichloromethane and stirred for 6h at room temperature according to the literature[17]. Compounds **3a – r, 2a, 2g, 2r, 4a, 4g, 4r, 5a, 5g, 5r, 2aa, 2ab** were all synthesized by a facile method referencing to the literature (see experimental Section 5.2.1.4 and 5.2.1.5, Scheme 2 and Table 1)[18]. Especially, Compound **7r** was fully synthesized by a novel method described in the synthesis section. Since the reaction of bromine with 1*H* – pyrrole – 2 – carboxylic acid was complicated and the byproduct was difficult to separate from each other[19-23], it is unfortunately to obtain the monobromine – 1*H* – pyrrole – 2 – carboxylic acid and corresponding target product.

2.2 Structure elucidation

The structure of the potential compound **2a** was elucidated by 1D NMR and 2D NMR as follows: The molecular formula of **2a** was revealed as $C_{16}H_{16}N_2O_4$ by HRMS data $[M+Na]^+$ (found 323.1004, calcd 323.1002). The ^1H and ^{13}CNMR (**2a**) spectra showed the signals of six quaternary, seven CH, two CH_2 and one CH_3 carbon atoms. Considered the reagents, in the HMQC spectra showed $\delta_H = 11.58$ (s, 1H) and $\delta_H = 11.24$ (s, 1H) belonged to the Pyrrole—NH

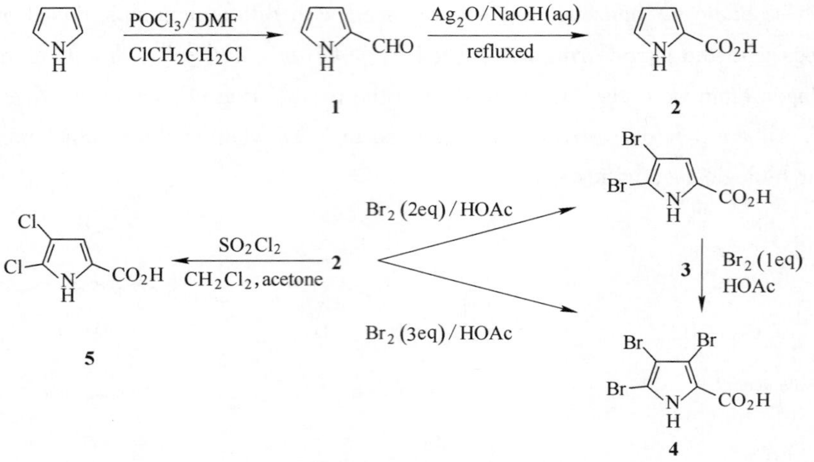

POCl$_3$/DMF
ClCH$_2$CH$_2$Cl
CHO
1
Ag$_2$O/NaOH(aq)
refluxed
CO$_2$H
2
Br$_2$ (2eq)/HOAc
SO$_2$Cl$_2$
CH$_2$Cl$_2$, acetone
5
3
Br$_2$ (1eq)
HOAc
Br$_2$ (3eq)/HOAc
4

Scheme 1 Synthetic routes to compounds **1 – 5**

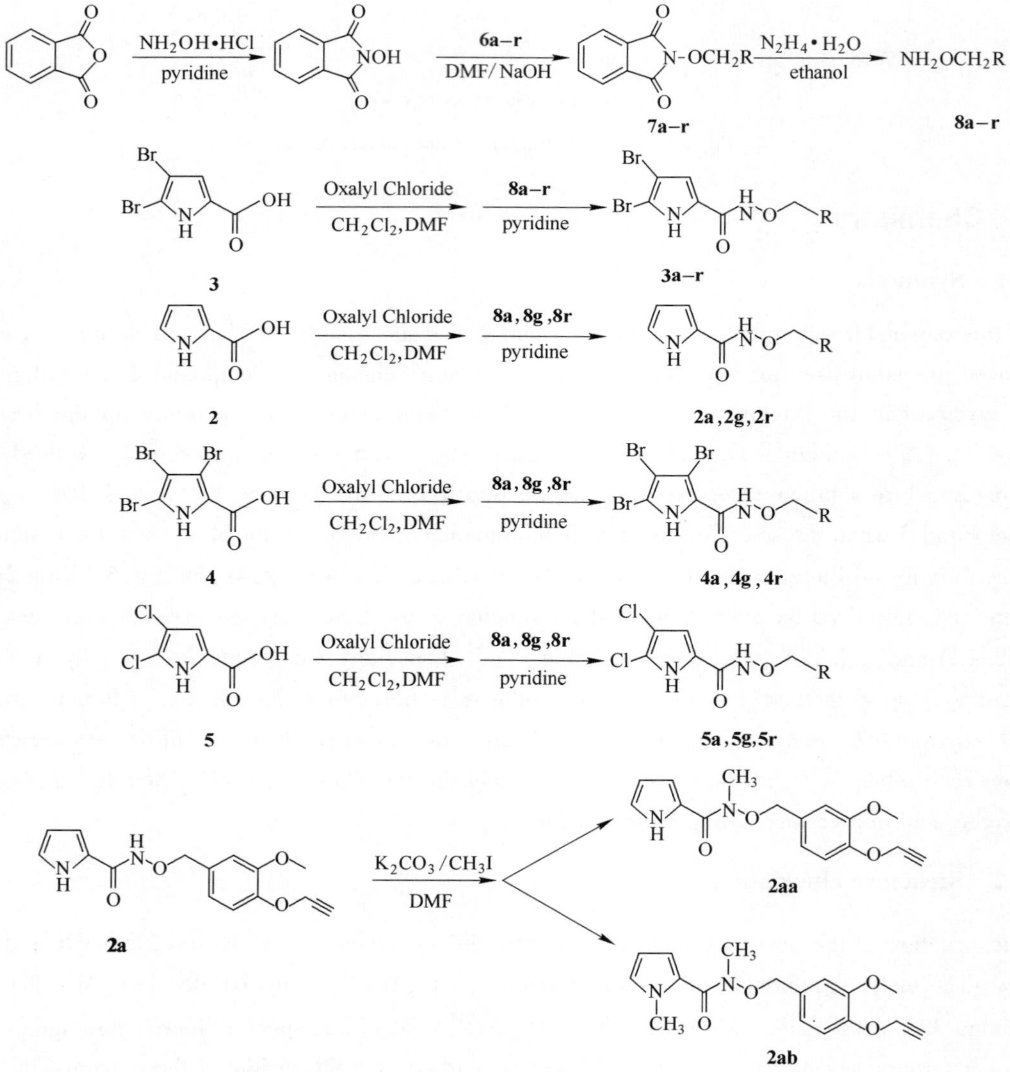

NH$_2$OH•HCl
pyridine
6a–r
DMF/NaOH
N-OCH$_2$R
7a–r
N$_2$H$_4$•H$_2$O
ethanol
NH$_2$OCH$_2$R
8a–r
Oxalyl Chloride
CH$_2$Cl$_2$,DMF
8a–r
pyridine
3
3a–r
8a, 8g, 8r
2
2a, 2g, 2r
4
4a, 4g, 4r
5
5a, 5g, 5r
K$_2$CO$_3$/CH$_3$I
DMF
2a
2aa
2ab

Scheme 2 Synthetic routes to compounds **7a – 2ab**

Table 1 The R group in compounds of 6(Br—CH_2R or Cl—CH_2R) –5r

Compd.	R	Compd.	R	Compd.	R
6a	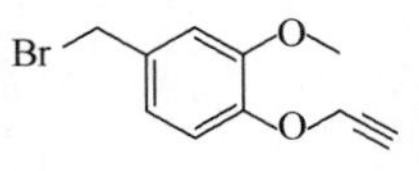 Br O O	**6g**	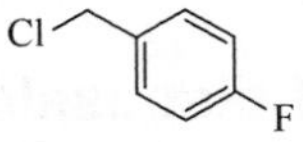 Cl F	**6m**	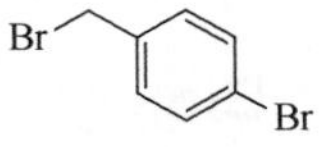 Br Br
6b	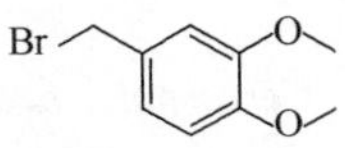 Br O O	**6h**	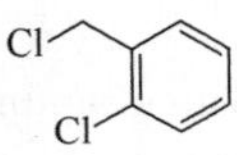 Cl Cl	**6n**	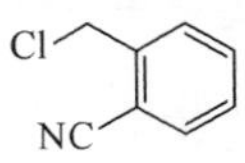 Cl NC
6c	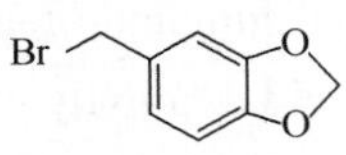 Br O O	**6i**	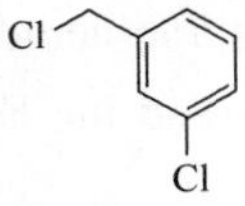 Cl Cl	**6o**	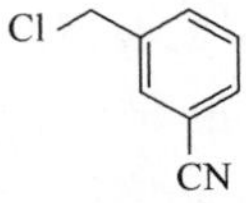 Cl CN
6d	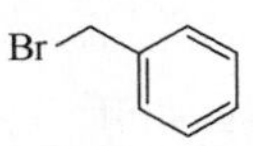 Br	**6j**	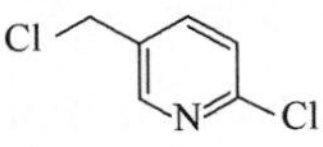 Cl N Cl	**6p**	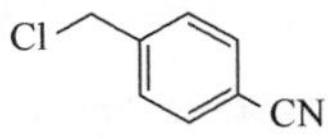 Cl CN
6e	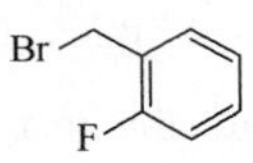 Br F	**6k**	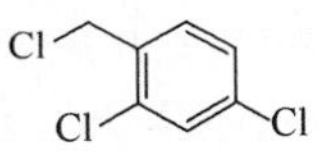 Cl Cl Cl	**6q**	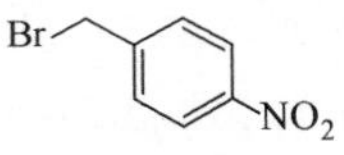 Br NO_2
6f	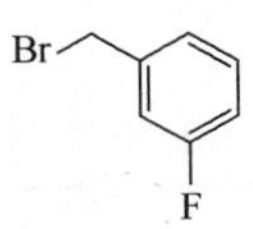 Br F	**6l**	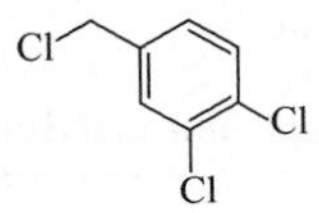 Cl Cl Cl	**6r**	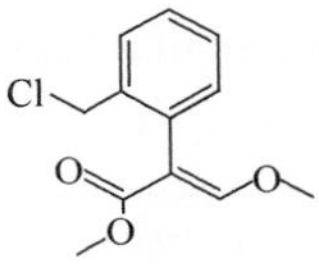 Cl O O O

or —CONH—; δ_H = 7.02 (d, 1H) and δ_H = 6.95 (d, 1H) belonged to the Ar—H, which was confirmed by their J value (J = 8.0Hz). δ_H = 7.10 (s, 1H), 6.92 (s, 1H), 6.71 (s, 1H) and 6.09 (s, 1H) belonged to the Ar—H or Pyrrole—H.

Based on the HMBC and ^{13}C Dept spectra, the correlations between δ_H = 11.24 (s, 1H)/δ_C = 159.7 (s), δ_H = 4.83 (s, 2H)/δ_C = 129.6 (s), 4.79 (s, 2H)/δ_C = 146.4 (s) and δ_H = 3.78 (s, 3H)/δ_C = 148.9 (s) indicated δ_H = 11.24, 4.83, 4.79, 3.78 belonged to —CONH—, H–7, H–14, H–17 and δ_C = 159.7, 129.6, 146.4, 148.9 belonged to C–6, C–8, C–11, C–10 respectively. So δ_H = 11.58 (s, 1H) belonged to the Pyrrole—NH. The correlations between δ_H = 7.10 (s, 1H)/C–10 (δ_C = 148.9), δ_H = 6.92 (s, 1H), 6.09 (s, 1H)/C–6 (δ_C = 159.7) indicated δ_H = 7.10 belonged to H–9 (Ar—H); δ_H = 6.92, 6.09 belonged to Pyrrole—H. So δ_H = 6.71 (s, 1H) belonged to the H–2. The correlations between δ_H = 6.95 (d, 1H)/C–7 (δ_C = 77.1) explained δ_H = 6.95 belonged to H–13; so δ_H = 7.02 (d, 1H) belonged to H–12. The correlations between δ_H = 4.79 (H–14)/δ_C = 79.2 (s), 78.2 (d); δ_H = 3.55 (s, 1H)/C–14 (δ_C = 55.9) indicated δ_H = 3.55 and δ_C = 79.2 belonged to H–16 and C–15 respectively. All the data indicated the structure of **2a** should be as follow (Fig. 2).

In the H–H COSY spectra, the correlations between δ_H = 6.09 (s, 1H)/δ_H = 6.92 (s, 1H) and δ_H = 6.09 (s, 1H)/δ_H = 6.71 (H–2) indicated δ_H = 6.09, 6.92 belonged to H–3 and H–4 respectively. The correlations between δ_H = 11.58 (Pyrrole —NH)/δ_H = 6.09 (H–3), 6.71 (H–2), 6.92 (H–4), the correlations between δ_H = 7.02 (H–12)/δ_H = 6.95 (H–13)

李正名文集

and the correlations between δ_H = 4.79(H-14)/δ_H = 3.55(H-16) further verified the positions of all the protons. Additionally, the structures of compounds **2aa** and **2ab** were also testified by 1D NMR and 2D NMR spectra.

3 Biological results and discussion

3.1 Fungicidal activities

Table 2 showed the fungicidal activities against *Alternaria solani*, *Cercospora arachidicola*, *Fusarium omysporum*, *Gibberella zeae*, *Physalospora piricola* of the compounds **3a-r**, **2a**, **2g**, **2r**, **4a**, **4g**, **4r**, **5a**, **5g**, **5r**, **2aa**, **2ab**, anthranilamide and the commercial fungicide fenpiclonil at the dosage of 50μg · mL^{-1}. Table 3 showed the EC_{50} of the high fungicidal activity compounds **2a**, **3a**, and fenpiclonil.

3.1.1 Fungicidal activity against A. solani

The screening data of Table 2 indicated that at the dosage of 50 μg · mL^{-1}, most compounds exhibited low activities against *A. solani except* the compounds **3a**, **3b**, **3d** - **3g** and **3i**. Comparing the biological activities with the rest of the compounds', the inhibition activities of **3a**, **3b**, **3d-g** and **3i** were all over 60.0%. In comparison with **3d**, the results indicated that H-10 and H-11 substituted by $-OCH_3$ (**3b**) or H-13 substituted by -F(**3e**) in the aroma ring would improve its fungicidal activity.

Table 2 Fungicidal activities at dosage of 50μg · mL^{-1}

Compd. Inhibition(%)	*A. solani*	*C. arachidicola*	*F. omysporum*	*G. zeae*	*P. piricola*
3a	64.9	56.3	47.1	8.7	90.0
3b	73.3	60.0	41.2	4.6	50.0
3c	31.6	75.0	23.8	14.7	18.5
3d	66.7	53.3	40.0	9.1	40.9
3e	76.2	50.3	41.2	18.2	27.3
3f	61.9	54.0	40.8	4.5	40.9
3g	66.5	62.5	58.0	21.7	55.1
3h	19.0	20.0	0	13.6	13.4
3i	66.7	80.0	47.0	17.4	38.9
3j	5.3	28.6	23.8	0	37.0
3k	21.1	35.7	33.3	11.8	33.3
3l	40.0	43.8	23.5	17.4	27.8
3m	31.6	50.0	4.8	2.9	33.3
3n	10.5	14.3	14.0	14.7	14.8
3o	21.1	35.7	18.0	8.8	25.3
3p	31.6	35.0	19.0	20.6	48.1
3q	26.7	18.8	17.6	13.0	16.7
3r	52.4	26.7	6.7	0	13.6

Table 2(continued)

Compd. Inhibition(%)	*A. solani*	*C. arachidicola*	*F. omysporum*	*G. zeae*	*P. piricola*
2a	35.3	20.0	16.7	9.1	100.0
2g	0	12.0	0	7.0	6.2
2r	0	13.3	0	8.5	16.1
4a	11.8	33.3	22.2	9.1	25.8
4g	19.2	28.0	8.3	15.5	18.5
4r	5.9	26.7	27.8	13.6	25.0
5a	42.3	44.0	0	21.1	6.2
5g	30.8	16.0	22.2	19.5	18.5
5r	42.3	40.0	8.3	21.1	24.7
2aa	13.3	19.2	31.6	0	5.8
2ab	16.7	7.7	31.6	20.0	6.0
Anthranilamide	42.3	52.0	27.8	38.0	43.2
Fenpiclonil	100	100	18.0	100	100

The data is the average of three duplicate results.

3.1.2 Fungicidal activity against C. arachidicola

The biological activity rules of **3a – 2ab** against *C. arachidicola* were generally the same as the test against *A. solani* showed. To be pointed out, the inhibition activities of **3a – g** and **3i** were all over 50.0%. Comparing with **3d** and anthranilamide, the results indicated that H – 10 and H – 11 substitutedby —OCH_2O—(**3c**) or H – 12 substitutedby —Cl(**3i**) in the aroma ring would evidently increase its fungicidal activity.

3.1.3 Fungicidal activity against F. omysporum

The screening data of Table 2 indicated that at the dosage of 50 μg · mL^{-1}, most compounds exhibited low activities against *F. omysporum* except the compounds **3a**, **3b**, **3d – g** and **3i**. The fungicidal activities of **3a**, **3b**, **3e – g** and **3i** were all over 40.0%. Comparing with **3d**, the test indicated that H – 10 substituted by —OCH_3 and H – 11 substituted by propargyl(**3a**) or H – 11 substituted by —F(**3g**), H – 12 substituted by —Cl(**3i**) in the aroma ring would enhance fungicidal activity.

3.1.4 Fungicidal activity against G. zeae

The screening data of Table 2 indicated that nearly all of the compounds(**3a – 2ab**) exhibited low activities against *G. zeae*.

3.1.5 Fungicidal activity against P. piricola

The screening data of Table 2 indicated that the target compounds exhibited low activities against P. piricola except for **3a**, **3b**, **3g**, **3p** and **2a**. At the dosage of 50 μg · mL^{-1}, the fungicidal activities of compounds **3a**, **3b**, **3g**, **3p** and **2a** were 90.0%, 50.0%, 55.1%, 48.1%, 100% respectively. Comparing with 3d, the antifungal activities of compounds **3a**, **3b** and **3g** further testified their rules exhibited in other fungi. To be pointed out, the activity of compound

3p showed that H – 11 in the aroma ring was substituted by —CN would enhance its activity. The EC_{50} value of **2a** and **3a** against *P. piricola* was 4. 50 μg · mL^{-1} and 9. 78 μg · mL^{-1} respectively (Table 3), which implied that the pyrrole ring substituted by few halogen atoms and the **6a** moiety were vital for high activity against *P. piricola*.

Table 3 EC_{50} (μg · mL^{-1}) values of 2a, 3a and fenpiclonil

2a	$Y = a + bx$	EC_{50}	R	3a	$Y = a + bx$	EC_{50}	R
A. solani		>50		*A. solani*	$Y = 3.311 + 1.296x$	20. 03	0. 99
C. arachidicola		>50		*C. arachidicola*	$Y = 2.735 + 1.632x$	24. 39	0. 99
F. omysporum		>50		*F. omysporum*		>50	
G. zeae		>50		*G. zeae*		>50	
P. piricola	$Y = 4.283 + 1.095x$	4. 50	0. 94	*P. piricola*	$Y = 3.380 + 1.634x$	9. 78	0. 98

Fenpiclonil	$Y + a = bx$	EC_{50}	R
A. solani	$Y = 5.340 + 1.322x$	0. 55	0. 96
C. arachidicola	$Y = 5.477 + 1.153x$	0. 38	0. 95
F. omysporum		>50	
G. zeae	$Y = 4.584 + 1.780x$	1. 71	0. 97
P. piricola	$Y = 5.226 + 2.352x$	0. 80	0. 99

Comparing the fungicidal activities of the series of compounds **2**, **3**, **4**, **5** against all the fungi except for **2a** against *P. piricola*, it would be found that the series of compounds **3** had a little more activities and spectrums than the **2**, **4** and **5** series, which indicated that the addition of two bromine atoms to the pyrrole ring 2 and 3 position might improve its antifungal activity. Comparing with the series of **3**, the addition of two chlorine atoms (**5a**, **5g** and **5r**) or three bromine atoms (**4a**, **4g** and **4r**) to the pyrrole ring did not increase its fungicidal activity. The biological activities of **2aa** and **2ab** against all the fungi implied that the methylated pyrrole ring might decrease their activities.

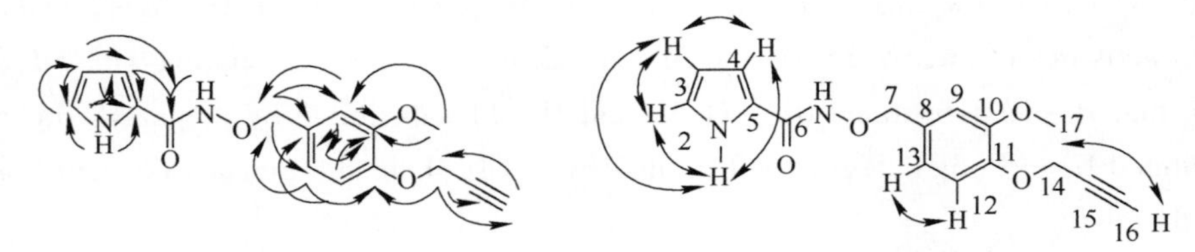

Fig. 2 Key HMBC correlations (left) and key H—H correlations (right) of **2a**

4 Conclusion

Twenty – nine pyrrole alkaloid analogs were designed and synthesized with a novel and facile procedure. The results of bioassay showed that the fungicidal activities of most target compounds exhibited moderate activities against *A. solani*, *C. arachidicola*, *F. omysporum*, *G. zeae and P. piricola* at the dosage of 50μg · mL^{-1}. The possible SAR of those targetcompounds were as follows: H – 10 and H – 11 substituted by—OCH_3; or H – 10 and H – 11 substituted by —OCH_3 and propargyl; H – 11 substituted by —F or —CN; H – 12 substituted by—Cl; H – 13 substituted by − F in the aroma ring might improve their fungicidal activities. The addition of bromine atoms to the pyrrole ring 2 and 3 position might enhance its antifungal activity and spectrum. The above terms were essential for high fungicidal activity in this type of pyrrole alkaloid analogs. The addition of chlorine atoms at 2 and 3 position or three bromine atoms to the pyrrole ring did not increase its activity. And the methylated pyrrole ring would weaken its fungicidal activity. The **2a** could be considered as a potential compound for further structural optimization.

5 Experimental section

5.1 Materials and methods

^{1}H NMR, ^{13}C NMR, HMQC, HMBC and H − H COSY spectra were obtained at 300MHz using a Bruker AV300 spectrometer or at 400MHz using a Bruker AV400 spectrometer in $CDCl_3$ or DMSO − d_6 solution with tetramethylsilane (TMS) as the internal standard. Chemical shift values (δ) were given in ppm. High − resolution mass spectrometry (HRMS) data were obtained on a Varian QFT − ESI instrument. The melting points were determined on an X − 4 binocular microscope melting point apparatus (Beijing Tech Instruments Co. , Beijing, China) and were uncorrected. Yields were not optimized. The reagents were all analytically or chemically pure. All solvents and liquid reagents were dried by standard methods in advance and distilled before use. Pyrrole and benzyl halogen (**6d − q**) were purchased from the Alfa Aesar Company (Tianjin, China). **6a − c**[24,25], **6r**[26] and anthranilamide[10] were synthesized according to literatures. Fenpiclonil was bought from labor Dr. Ehrenstorfer − Schäers (Augsburg, Germany).

5.2 Synthesis

5.2.1 General synthetic

5.2.1.1 Synthesis procedures for compounds **1 − 5**.

^{1}*H* − Pyrrole − 2 − carbaldehyde **1** was synthesized referencing to the literatures [27,28]. 1*H* − Pyrrole − 2 − carboxylic acid **2** was synthesized through an improved procedure referencing to literature[16]: To a suspension of sliver oxide was added **1** in ethanol. After stirring for 1h at reflux (It is failed to get compound **2** when the mixture was stirred for 10 h at room temperature according to the literature[16]), the precipitate was filtered out and washed with hot water. The combined filtrate and washings were acidified with concentrated hydrochloride acid at room temperature and extracted with ethyl acetate, drying ($MgSO_4$) and removal the ethyl acetate under vacuum to give the crude **2**. 4,5 − Dibromo − 1*H* − pyrrole − 2 − carboxylic acid **3** and 3,4,5 − tribromo − 1*H* − pyrrole − 2 − carboxylic acid **4** were synthesized referencing to the literature[7,9]. When the bromine was 3 equivalents of the **2** in HOAc and the reaction time was prolonged. Compound **2** could be fully converted to the crude product **4** which could be also obtained from compound **3** reacting with the equivalent of bromine in HOAc. The crude **4** was then recrystallized in the solution of ethanol/water to give the pure product. 4,5 − Dichloro − 1*H* − pyrrole − 2 − carboxylic acid **5** was synthesized through an improved procedure referencing to literature[17]: To pyrrole − 2 − carboxylic acid **2** in the mixture solution of dichloromethane and acetone (v/v = 4∶1, It is failed to get compound **5** when the solution was only dichloromethane

李正名文集

and stirred for 6h at room temperature according to the literature[17]), sulfuryl chloride was added dropwise in dichloromethane. The reaction continued for 0.5h at room temperature; then the reaction mixture was poured slowly into water, followed by extraction with ethyl acetate and washed with water, evaporation the solvent to give the crude product **5** (Scheme 1).

5.2.1.1.1 *1H-Pyrrole-2-carbaldehyde* (**1**). Light yellow solid; yield 58.8%; mp 40–42℃; ^{1}H NMR (400 MHz, $CDCl_3$) δ: 10.97 (brs, 1H, Pyrrole—NH), 9.49 (s, 1H, —CHO), 7.18 (brs, 1H, Pyrrole—H), 7.01 (brs, 1H, Pyrrole—H), 6.33 (brs, 1H, Pyrrole—H). ^{13}C NMR (100MHz, $CDCl_3$) δ: 179.6, 132.8, 127.4, 122.3, 111.3.

5.2.1.1.2 *1H-Pyrrole-2-carboxylic acid* (**2**). White solid; yield 67.9%; mp 193–195℃; ^{1}H NMR (400MHz, DMSO-d_6) δ: 12.24 (brs, 1H, —COOH), 11.72 (brs, 1H, Pyrrole—NH), 6.95 (brs, 1H, Pyrrole—H), 6.72 (brs, 1H, Pyrrole—H), 6.13 (brs, 1H, Pyrrole—H). ^{13}C NMR (100MHz, DMSO-d_6) δ: 161.8, 123.3, 122.8, 114.6, 109.2.

5.2.1.1.3 *4,5-Dibromo-1H-Pyrrole-2-carboxylic acid* (**3**). Light red solid; yield 67.2%; mp 180–182℃; ^{1}H NMR (400 MHz, DMSO-d_6) δ: 12.96 (brs, 1H, —COOH), 6.83 (brs, 1H, Pyrrole—H). ^{13}C NMR (100 MHz, DMSO-d_6) δ: 160.2, 124.9, 116.8, 106.6, 98.7.

5.2.1.1.4 *3,4,5-Tribromo-1H-Pyrrole-2-carboxylic acid* (**4**). Red solid; yield 50.1%; mp 202–204℃; ^{1}H NMR (400 MHz, DMSO-d_6) δ: 13.36 (brs, 1H, —COOH), 13.16 (brs, 1H, Pyrrole—NH). ^{13}C NMR (100 MHz, DMSO-d_6) δ: 159.3, 122.5, 106.6, 104.2, 103.3.

5.2.1.1.5 *4,5-Dichloro-1H-Pyrrole-2-carboxylic acid* (**5**). Purple solid; yield 75.0%; mp 169–171℃; ^{1}H NMR (400 MHz, DMSO-d_6) δ: 13.02 (brs, 1H, —COOH), 12.84 (brs, 1H, Pyrrole—NH), 6.81 (brs, 1H, Pyrrole—H). ^{13}C NMR (100 MHz, DMSO-d_6) δ: 160.3, 121.6, 116.6, 113.9, 108.8.

5.2.1.2 Synthetic procedures for compounds **7a–r**.

Compound **7a** was synthesized referencing to the literatures[18,29]. To be pointed out, the crude products of **7b**, **7i**, **7k**, **7l**, **7n–p** required towash with the solution of ethyl acetate/petroleum ether (v/v = 1:5) again; In the preparation of **7r**, the whole mixture of *N*-hydroxyphthalimide, sodium hydroxide and **6r** in DMF needed to be stirred at 80℃ for 2h, cooled, then poured into water, and stirred for 1h to afford a mucilaginous solid which was collected by vacuum filtration and washed with water. The above mucilaginous solid was then dissolved in the solution of dichloromethane/water (v/v = 1:1), leaving the organic layer and removal the solvent under vacuum to give the crude **7r** which was then recrystallized in the solution of ethyl acetate/petroleum ether (v/v = 1:4). All the substituents at the benzene ring were listed in Scheme 2 and Table 1.

5.2.1.2.1 *2-(3-Methoxy-4-prop-2-ynyloxy-benzyloxy)-isoindole-1,3-dione* (**7a**). White solid; yield 79.8%; mp 151–153℃; ^{1}H NMR (400 MHz, $CDCl_3$) δ: 7.80 (brd, *J* = 2.4Hz, 2H, Ar—H), 7.73 (brs, 2H, Ar—H), 7.17 (s, 1H, Ar—H), 6.99–7.01 (m, 2H, Ar—H), 5.17 (s, 2H, Ar—CH_2O—), 4.75 (s, 2H, Ar—OCH_2—), 3.91 (s, 3H, Ar—OCH_3), 2.50 (s, 1H, alkynyl—H). ^{13}C NMR (100 MHz, $CDCl_3$) δ: 163.5 (2C), 149.5, 147.5, 134.4 (2C), 128.8 (2C), 127.5, 123.4 (2C), 122.3, 113.5, 113.2, 79.6, 78.2, 75.9, 56.6, 55.9.

5.2.1.2.2 *2-(3,4-Dimethoxy-benzyloxy)-isoindole-1,3-dione* (**7b**). White solid;

yield 49.3%; mp 143 – 145℃; ^{1}H NMR(400MHz, $CDCl_3$) δ: 7.80(dt, J = 8.0, 3.2Hz, 2H, Ar—H), 7.73(dt, J = 8.0, 3.2Hz, 2H, Ar—H), 7.13(s, 1H, Ar—H), 7.03(d, J = 8.0Hz, 1H, Ar—H), 6.82(d, J = 8.0Hz, 1H, Ar—H), 5.17(s, 2H, Ar—CH_2O—), 3.91(s, 3H, Ar—OCH_3), 3.87(s, 3H, Ar—OCH_3). ^{13}C NMR(100 MHz, $CDCl_3$) δ: 163.5(2C), 149.8, 148.9, 134.4(2C), 128.8(2C), 126.1, 123.4(2C), 122.7, 112.7, 110.6, 79.7, 55.9, 55.8.

5.2.1.2.3 2 – (Benzo[1,3]dioxol – 5 – ylmethoxy) – isoindole – 1,3 – dione (**7c**). White solid; yield 83.3%; mp 172 – 174℃; ^{1}H NMR(400 MHz, $CDCl_3$) δ: 7.80(dt, J = 8.0, 3.2Hz, 2H, Ar—H), 7.73(dt, J = 8.0, 3.2Hz, 2H, Ar—H), 7.06(s, 1H, Ar—H), 6.95(d, J = 8.0Hz, 1H, Ar—H), 6.77(d, J = 8.0Hz, 1H, Ar—H), 5.97(s, 2H, —OCH_2O—), 5.10(s, 2H, Ar—CH_2O—). ^{13}CNMR(100 MHz, $CDCl_3$) δ: 163.5(2C), 148.5, 147.8, 134.4(2C), 128.8(2C), 127.4, 124.1, 123.5(2C), 110.3, 108.1, 101.2, 79.7.

5.2.1.2.4 2 – Benzyloxy – isoindole – 1,3 – dione (**7d**). White solid; yield 67.5%; mp 141 – 143℃; ^{1}H NMR(400MHz, $CDCl_3$) δ: 7.79(dt, J = 8.8, 3.2Hz, 2H, Ar—H), 7.71(dt, J = 8.8, 3.2Hz, 2H, Ar—H), 7.51 – 7.54(m, 2H, Ar—H), 7.36 – 7.37(m, 3H, Ar—H), 5.20(s, 2H, Ar—CH_2O—). ^{13}CNMR(100MHz, $CDCl_3$) δ: 163.5(2C), 134.4(2C), 133.6, 129.9(2C), 129.3, 128.8(2C), 128.5(2C), 123.5(2C), 79.8.

5.2.1.2.5 2 – (2 – Fluoro – benzyloxy) – isoindole – 1,3 – dione (**7e**). White solid; yield 79.2%; mp 153 – 155℃; ^{1}H NMR(400 MHz, $CDCl_3$) δ: 7.80(dt, J = 8.4, 3.2Hz, 2H, Ar—H), 7.73(dt, J = 8.4, 3.2Hz, 2H, Ar—H), 7.53 – 7.57(m, 1H, Ar—H), 7.34 – 7.37(m, 1H, Ar—H), 7.14 – 7.18(m, 1H, Ar—H), 7.05 – 7.09(m, 1H, Ar—H), 5.29(s, 2H, Ar—CH_2O—). ^{13}C NMR(100 MHz, $CDCl_3$) δ: 163.3(2C), 162.8(Ar – F), 160.3(Ar – F), 134.4(2C), 132.2, 131.5, 128.8(2C), 124.3, 123.5(2C), 121.2, 115.5, 73.0.

5.2.1.2.6 2 – (3 – Fluoro – benzyloxy) – isoindole – 1,3 – dione (**7f**). White solid; yield 54.8%; mp 129 – 131℃; ^{1}H NMR(400 MHz, $CDCl_3$) δ: 7.74(dt, J = 8.4, 3.2Hz, 2H, Ar—H), 7.67(dt, J = 8.4, 3.2Hz, 2H, Ar—H), 7.24 – 7.28(m, 1H, Ar—H), 7.18 – 7.24(m, 2H, Ar—H), 6.98 – 7.01(m, 1H, Ar—H), 5.12(s, 2H, Ar—CH_2O—). ^{13}C NMR(100 MHz, $CDCl_3$) δ: 163.9(Ar—F), 161.4(Ar—F), 163.4(2C), 136.0, 134.5(2C), 130.1, 128.8(2C), 125.2, 123.6(2C), 116.5, 116.2, 78.9.

5.2.1.2.7 2 – (4 – Fluoro – benzyloxy) – isoindole – 1,3 – dione (**7g**). White solid; yield 43.7%; mp 148 – 150℃; ^{1}H NMR(400 MHz, $CDCl_3$) δ: 7.81(dt, J = 8.4, 3.2Hz, 2H, Ar—H), 7.74(dt, J = 8.4, 3.2Hz, 2H, Ar—H), 7.50 – 7.54(m, 2H, Ar—H), 7.04 – 7.08(m, 2H, Ar—H), 5.17(s, 2H, Ar—CH_2O—). ^{13}C NMR(100 MHz, $CDCl_3$) δ: 164.6(Ar—F), 162.1(Ar – F), 163.4(2C), 134.5(2C), 131.9(2C), 129.6, 128.8(2C), 123.5(2C), 115.6(2C), 79.0.

5.2.1.2.8 2 – (2 – Chloro – benzyloxy) – isoindole – 1,3 – dione (**7h**). White solid; yield 29.0%; mp 154 – 156℃; ^{1}H NMR(400 MHz, $CDCl_3$) δ: 7.81(dt, J = 8.4, 3.2Hz, 2H, Ar—H), 7.74(dt, J = 8.4, 3.2Hz, 2H, Ar—H), 7.62 – 7.64(m, 1H, Ar—H), 7.37 – 7.39(m, 1H, Ar—H), 7.29 – 7.31(m, 2H, Ar—H), 5.36(s, 2H, Ar—CH_2O—). ^{13}C NMR(100 MHz, $CDCl_3$) δ: 163.3(2C), 134.7, 134.4(2C), 131.9, 131.7, 130.5, 129.6, 128.8(2C), 127.0, 123.5(2C), 76.4.

5. 2. 1. 2. 9　2 – (3 – Chloro – benzyloxy) – isoindole – 1,3 – dione(**7i**). White solid; yield 29. 2%; mp 134 – 136℃; ^{1}H NMR (400 MHz, $CDCl_3$) δ: 7. 82 (dt, J = 8. 4, 3. 2Hz, 2H, Ar—H), 7. 75 (dt, J = 8. 4, 3. 2Hz, 2H, Ar—H), 7. 54 (s, 1H, Ar—H), 7. 43 – 7. 44 (m, 1H, Ar—H), 7. 32 – 7. 34 (m, 2H, Ar—H), 5. 17 (s, 2H, Ar—CH_2O—). ^{13}C NMR (100 MHz, $CDCl_3$) δ: 163. 4 (2C), 135. 6, 134. 5 (2C), 134. 3, 129. 8, 129. 7, 129. 4, 128. 8 (2C), 127. 7, 123. 6 (2C), 78. 9.

5. 2. 1. 2. 10　2 – (6 – Chloro – pyridin – 3 – ylmethoxy) – isoindole – 1,3 – dione(**7j**). White solid; yield 67. 4%; mp 151 – 153℃; ^{1}H NMR (400MHz, DMSO – d_6) δ: 8. 54 (s, 1H, Pyridine—H), 8. 05 (d, J = 8. 0Hz, 1H, Pyridine—H), 7. 86 (brs, 4H, Ar—H), 7. 58 (d, J = 8. 0Hz, 1H, Pyridine—H), 5. 24 (s, 2H, Pyridine—CH_2O—). ^{13}C NMR (100MHz, DMSO – d_6) δ: 163. 0 (2C), 150. 8 (2C), 141. 1, 134. 8 (2C), 129. 9, 128. 4 (2C), 124. 1, 123. 3 (2C), 75. 6.

5. 2. 1. 2. 11　2 – (2,4 – Dichloro – benzyloxy) – isoindole – 1,3 – dione (**7k**). White solid; yield 51. 8%; mp 154 – 156℃; ^{1}H NMR (400MHz, $CDCl_3$) δ: 7. 82 (dt, J = 8. 4, 3. 2Hz, 2H, Ar—H), 7. 75 (dt, J = 8. 4, 3. 2Hz, 2H, Ar—H), 7. 59 (d, J = 8. 4Hz, 1H, Ar—H), 7. 41 (d, J = 1. 6Hz, 1H, Ar—H), 7. 27 (dt, J = 8. 4, 1. 6Hz, 1H, Ar—H), 5. 32 (s, 2H, Ar—CH_2O—). ^{13}C NMR (100MHz, $CDCl_3$) δ: 163. 3 (2C), 135. 8, 135. 3, 134. 5 (2C), 132. 4, 130. 5, 129. 5, 128. 7 (2C), 127. 4, 123. 6 (2C), 75. 7.

5. 2. 1. 2. 12　2 – (3,4 – Dichloro – benzyloxy) – isoindole – 1,3 – dione(**7l**). Purple solid; yield 51. 0%; mp 182 – 184℃; ^{1}H NMR (400MHz, $CDCl_3$) δ: 7. 82 (dt, J = 8. 4, 3. 2Hz, 2H, Ar—H), 7. 75 (dt, J = 8. 4, 3. 2Hz, 2H, Ar—H), 7. 64 (d, J = 1. 6Hz, 1H, Ar—H), 7. 46 (d, J = 8. 4Hz, 1H, Ar—H), 7. 39 (dt, J = 8. 4, 1. 6Hz, 1H, Ar—H), 5. 15 (s, 2H, Ar—CH_2O—). ^{13}C NMR (100MHz, $CDCl_3$) δ: 163. 4 (2C), 134. 6 (2C), 133. 9, 133. 5, 132. 7, 131. 4, 130. 6, 128. 8 (2C), 128. 7, 123. 6 (2C), 78. 2.

5. 2. 1. 2. 13　2 – (4 – Bromo – benzyloxy) – isoindole – 1,3 – dione(**7m**). White solid; yield 60. 6%; mp 143 – 145℃; ^{1}H NMR (400MHz, $CDCl_3$) δ: 7. 81 (dt, J = 8. 0, 3. 2Hz, 2H, Ar—H), 7. 74 (dt, J = 8. 0, 3. 2Hz, 2H, Ar—H), 7. 50 (d, J = 8. 4Hz, 2H, Ar—H), 7. 41 (d, J = 8. 4Hz, 2H, Ar—H), 5. 16 (s, 2H, Ar—CH_2O—). ^{13}C NMR (100MHz, $CDCl_3$) δ: 163. 4 (2C), 134. 5 (2C), 132. 7, 131. 7 (2C), 131. 4 (3C), 128. 7 (2C), 123. 6 (2C), 78. 9.

5. 2. 1. 2. 14　2 – (1,3 – Dioxo – 1,3 – dihydro – isoindol – 2 – yloxymethyl) – benzonitrile (**7n**). Purple solid; yield 54. 3%; mp 195 – 197℃; ^{1}H NMR (300MHz, $CDCl_3$) δ: 7. 28 – 7. 81 (m, 8H, Ar—H), 5. 42 (s, 2H, Ar—CH_2O—). ^{13}C NMR (75MHz, $CDCl_3$) δ: 163. 2 (2C), 137. 3, 134. 6 (2C), 133. 0, 132. 9, 130. 8, 129. 6, 128. 7 (2C), 123. 6 (2C), 116. 8, 113. 0, 77. 3.

5. 2. 1. 2. 15　3 – (1,3 – Dioxo – 1,3 – dihydro – isoindol – 2 – yloxymethyl) – benzonitrile (**7o**). White solid; yield 54. 5%; mp 173 – 175℃; ^{1}H NMR (300MHz, $CDCl_3$) δ: 7. 79 – 7. 84 (m, 4H, Ar—H), 7. 75 – 7. 78 (m, 2H, Ar—H), 7. 60 – 7. 66 (m, 1H, Ar—H), 7. 50 – 7. 55 (m, 1H, Ar—H), 5. 23 (s, 2H, Ar—CH_2O—). ^{13}C NMR (75MHz, $CDCl_3$) δ: 163. 3 (2C), 135. 4, 134. 6 (2C), 133. 8, 132. 9, 132. 8, 129. 4, 128. 7 (2C), 123. 6 (2C), 118. 2, 112. 8, 78. 4.

5. 2. 1. 2. 16　4 – (1,3 – Dioxo – 1,3 – dihydro – isoindol – 2 – yloxymethyl) – benzonitrile

(**7p**). White solid; yield 54.0%; mp 194 – 196℃; ^{1}H NMR(400MHz, DMSO – d_6)δ: 7.89(d, J = 8.4Hz, 2H, Ar—H), 7.86(brs, 4H, Ar—H), 7.74(d, J = 8.4Hz, 2H, Ar—H), 5.28(s, 2H, Ar—CH_2O—). ^{13}C NMR(100MHz, DMSO – d_6)δ: 163.0(2C), 139.8, 134.8(2C), 132.3(2C), 130.0(2C), 128.4(2C), 123.2(2C), 118.5, 111.5, 78.1.

5.2.1.2.17 2 – (4 – Nitro – benzyloxy) – isoindole – 1,3 – dione (**7q**). White solid; yield 61.7%; mp 191 – 193℃; 1H NMR(400MHz, $CDCl_3$)δ: 8.25(d, J = 8.0Hz, 2H, Ar—H), 7.80(d, J = 8.0Hz, 2H, Ar—H), 7.74(brs, 4H, Ar—H), 5.31(s, 2H, Ar—CH_2O—). ^{13}C NMR(100MHz, $CDCl_3$)δ: 163.3(2C), 148.3, 140.8, 134.7(2C), 130.0(2C), 128.6(2C), 123.7(4C), 78.3.

5.2.1.2.18 2 – [2 – (1,3 – Dioxo – 1,3 – dihydro – isoindol – 2 – yloxymethyl) – phenyl] – 3 – methoxy – acrylic acid methyl ester (**7r**). Gray solid; yield 32.6%; mp 162 – 164℃; ^{1}H NMR(400MHz, $CDCl_3$)δ: 7.79(brs, 3H, Ar—H), 7.71(brs, 2H, Ar—H), 7.62(s, 1H, —$CHOCH_3$), 7.38(brs, 2H, Ar—H), 7.15(brs, 1H, Ar—H), 5.12(s, 2H, Ar—CH_2O—), 3.75(s, 3H, – OCH_3), 3.61(s, 2H, —OCH_3). ^{13}C NMR(100MHz, $CDCl_3$)δ: 167.8, 163.3(2C), 160.6, 134.3(2C), 133.1, 133.0, 130.9, 130.5, 129.0, 128.9(2C), 128.0, 123.3(2C), 109.5, 77.3, 61.9, 51.6.

5.2.1.3 Synthetic procedures for compounds **8a – r**.

Compound **8a** was synthesized by a facile method improving on the literature[18]: To a solution of compound **7a** in anhydrous ethanol, hydrazine monohydrate was added. The mixture was refluxed for 1h. The resulting solid was removed by filtration, and the filtrate was evaporated on a rotavap. The residue was slurried in anhydrous ether and filtered, and the filtrate was washed with water again. The ether layer was separated, dried with $CaCl_2$ and evaporated to afford Compound **8a**. According to the above process, Compounds **8b – q** were prepared from **7b – q** respectively. To be pointed out, as the product **8r** was unstable, the mixture of **7r**, ethanol and hydrazine monohydrate was stirred overnight at room temperature. The final ether layer was dried with $CaCl_2$ and used rapidly in the next step of synthesis. All the substituents at the benzene ring were listed in Scheme 2 and Table 1. As substance **8b – q** was unstable, the boiling points of them were not detected.

5.2.1.3.1 O – (3 – Methoxy – 4 – prop – 2 – ynyloxy – benzyl) – hydroxylamine (**8a**). White solid; yield 68.9%; mp 30 – 32℃; ^{1}H NMR(400 MHz, $CDCl_3$)δ: 7.00(d, J = 8.0Hz, 1H, Ar—H), 6.93(s, 1H, Ar—H), 6.90(d, J = 8.0Hz, 1H, Ar—H), 5.38(s, 2H, —NH_2), 4.75(d, J = 2.0Hz, 2H, Ar—OCH_2—), 4.62(s, 2H, Ar—CH_2O—), 3.88(s, 3H, Ar—OCH_3), 2.50(s, 1H, alkynyl—H). ^{13}C NMR(100 MHz, $CDCl_3$)δ: 149.7, 146.6, 131.4, 120.8, 114.1, 112.0, 78.5, 77.8, 75.8, 56.7, 55.8.

5.2.1.3.2 O – (3,4 – Dimethoxy – benzyl) – hydroxylamine (**8b**). Colorless oil; yield 20.5%; ^{1}H NMR(400MHz, $CDCl_3$)δ: 6.89 – 6.91(m, 2H, Ar—H), 6.82 – 6.85(m, 1H, Ar—H), 4.61(s, 2H, Ar—CH_2O—), 3.88(s, 3H, Ar—OCH_3), 3.86(s, 3H, Ar—OCH_3). ^{13}C NMR(100MHz, $CDCl_3$)δ: 149.0, 148.9, 129.9, 121.0, 111.6, 110.0, 77.8, 55.8(2C).

5.2.1.3.3 O – Benzo[1,3] dioxol – 5 – ylmethyl – hydroxylamine (**8c**). Colorless oil; yield 69.6%; ^{1}H NMR(400MHz, $CDCl_3$)δ: 6.82(s, 1H, Ar—H), 6.73 – 6.75(m, 2H, Ar—H),

5.86(s,2H,—OCH_2O—),5.32(brs,2H,—NH_2),4.51(s,2H,Ar—CH_2O—). ^{13}C NMR (100MHz,$CDCl_3$)δ:147.7,147.3,131.4,122.0,108.9,108.0,101.0,77.5.

5.2.1.3.4 O – Benzylhydroxylamine(**8d**). Colorless oil;yield 58.8%;^{1}H NMR(400MHz, $CDCl_3$) δ: 7.29 – 7.31 (m, 5H, Ar—H), 5.25 (brs, 2H,—NH_2), 4.61 (s, 2H, Ar—CH_2O—). ^{13}C NMR(100MHz,$CDCl_3$)δ:137.6,128.5(2C),128.4(2C),127.9,77.9.

5.2.1.3.5 O – (2 – Fluoro – benzyl) – hydroxylamine(**8e**). Colorless oil;yield 49.3%;^{1}H NMR(400MHz, $CDCl_3$) δ: 7.36 – 7.41 (m, 1H, Ar—H), 7.26 – 7.28 (m, 1H, Ar—H), 7.01 – 7.14(m,2H,Ar—H),5.42(brs,2H,—NH_2),4.74(s,2H,Ar—CH_2O—). ^{13}C NMR (100MHz, $CDCl_3$) δ: 162.8 (Ar—F), 159.5 (Ar—F), 130.8, 129.7, 124.5, 123.9, 115.4,71.3.

5.2.1.3.6 O – (3 – Fluoro – benzyl) – hydroxylamine(**8f**). Colorless oil;yield 44.1%;^{1}H NMR(400MHz, $CDCl_3$) δ: 7.26 – 7.31 (m, 1H, Ar—H), 7.04 – 7.10 (m, 2H, Ar—H), 6.95 – 6.99(m,1H,Ar—H),5.45(brs,2H,—NH_2),4.63(s,2H,Ar—CH_2O—). ^{13}C NMR (100MHz, $CDCl_3$) δ: 164.1 (Ar—F), 161.6 (Ar—F), 140.3, 130.0, 123.7, 115.0, 114.7,76.9.

5.2.1.3.7 O – (4 – Fluoro – benzyl) – hydroxylamine (**8g**). Colorless oil; yield 66.7%;^{1}H NMR(400MHz,$CDCl_3$)δ:7.29 – 7.32(m,2H,Ar—H),7.00 – 7.04(m,2H, Ar—H),5.38(brs,2H,—NH_2),4.61(s,2H,Ar—CH_2O—). ^{13}C NMR(100MHz,$CDCl_3$)δ: 163.7(Ar—F),161.3(Ar—F),133.3,130.2(2C),115.3(2C),77.1.

5.2.1.3.8 O – (2 – Chloro – benzyl) – hydroxylamine (**8h**). Colorless oil; yield 90.9%;^{1}H NMR(400MHz,$CDCl_3$)δ:7.38 – 7.40(m,1H,Ar—H),7.31 – 7.33(m,1H, Ar—H),7.17 – 7.21(m,2H,Ar—H),5.26(brs,2H,—NH_2),4.76(s,2H,Ar—CH_2O—). ^{13}C NMR(100MHz,$CDCl_3$)δ:135.2,133.6,129.9,129.4,129.0,126.7,74.9.

5.2.1.3.9 O – (3 – Chloro – benzyl) – hydroxylamine(**8i**). Colorless oil;yield 88.0%;^{1}H NMR(400MHz,$CDCl_3$)δ:7.34(s,1H,Ar—H),7.25 – 7.26(m,2H,Ar—H),7.20 – 7.21 (m,1H,Ar—H),5.43(brs,2H,—NH_2),4.62(s,2H,Ar—CH_2O—). ^{13}C NMR(100MHz, $CDCl_3$)δ:139.8,134.3,129.7,128.2,128.0,126.2,76.9.

5.2.1.3.10 O – (6 – Chloro – pyridin – 3 – ylmethyl) – hydroxylamine(**8j**). White solid; yield 57.8%;mp 58 – 60℃;^{1}H NMR(400MHz,DMSO – d_6)δ:8.34(d,J = 2.1Hz,1H,Pyridine – H),7.68(dt,J = 8.1,2.1Hz,1H,Pyridine – H),7.30(d,J = 8.1Hz,1H,Pyridine – H),5.60(brs,2H,—NH_2),4.65(s,2H,Pyridine—CH_2O—). ^{13}C NMR(100MHz,DMSO – d_6)δ:150.6,149.5,138.9,132.2,123.9,74.0.

5.2.1.3.11 O – (2,4 – Dichloro – benzyl) – hydroxylamine (**8k**). Colorless oil; yield 90.3%;^{1}H NMR(400MHz,$CDCl_3$)δ:7.34(s,1H,Ar—H),7.32(d,J = 8.0Hz,1H,Ar—H),7.20(d,J = 8.0Hz,1H,Ar—H),5.50(brs,2H,—NH_2),4.72(s,2H,Ar—CH_2O—). ^{13}C NMR(100MHz,$CDCl_3$)δ:134.2,134.0(2C),130.6,129.2,126.9,74.2.

5.2.1.3.12 O – (3,4 – Dichloro – benzyl) – hydroxylamine (**8l**). Colorless oil; yield 94.3%;^{1}H NMR(400MHz,$CDCl_3$)δ:7.40(s,1H,Ar—H),7.36(d,J = 8.0Hz,1H,Ar—H),7.13(d,J = 8.0Hz,1H,Ar—H),5.45(brs,2H,—NH_2),4.57(s,2H,Ar—CH_2O—). ^{13}C NMR(100MHz,$CDCl_3$)δ:138.1,132.3,131.6,130.3,130.0,127.4,76.1.

5. 2. 1. 3. 13 O – (4 – Bromo – benzyl) – hydroxylamine (**8m**). Colorless oil; yield 94. 0%; ^{1}H NMR(400MHz, $CDCl_3$)δ: 7. 44(d, J = 8. 1Hz, 2H, Ar—H), 7. 19(d, J = 8. 1Hz, 2H, Ar—H), 5. 39 (brs, 2H, —NH_2), 4. 58 (s, 2H, Ar—CH_2O—). ^{13}C NMR (100MHz, $CDCl_3$)δ: 136. 6, 131. 5(2C), 130. 0(2C), 121. 8, 77. 0.

5. 2. 1. 3. 14 2 – Aminooxymethyl – benzonitrile (**8n**). Colorless oil; yield 70. 2%; ^{1}H NMR (300MHz, $CDCl_3$)δ: 7. 52 – 7. 57 (m, 1H, Ar—H), 7. 43 – 7. 51 (m, 2H, Ar—H), 7. 31 – 7. 34 (m, 1H, Ar—H), 5. 50 (brs, 2H, —NH_2), 4. 76 (s, 2H, Ar—CH_2O—). ^{13}C NMR (75MHz, $CDCl_3$)δ: 141. 2, 132. 8, 132. 7, 129. 4, 128. 3, 117. 4, 112. 2, 75. 2.

5. 2. 1. 3. 15 3 – Aminooxymethyl – benzonitrile (**8o**). Colorless oil; yield 80. 2%; 1H NMR (300MHz, $CDCl_3$)δ: 7. 63 (s, 1H, Ar—H), 7. 56 – 7. 58 (m, 2H, Ar—H), 7. 43 – 7. 47 (m, 1H, Ar—H), 5. 57 (brs, 2H, —NH_2), 4. 69 (s, 2H, Ar—CH_2O—). ^{13}C NMR (75MHz, $CDCl_3$) δ: 139. 5, 132. 4, 131. 4, 131. 2, 129. 1, 118. 7, 112. 1, 76. 2.

5. 2. 1. 3. 16 4 – Aminooxymethyl – benzonitrile (**8p**). Colorless oil; yield 81. 7%; ^{1}H NMR (400MHz, $CDCl_3$)δ: 7. 50 (d, J = 6. 8Hz, 2H, Ar—H), 7. 35 (d, J = 6. 8Hz, 2H, Ar—H), 5. 48 (brs, 2H, —NH_2), 4. 61 (s, 2H, Ar—CH_2O—). ^{13}C NMR (100MHz, $CDCl_3$)δ: 143. 5, 132. 1 (2C), 128. 4(2C), 118. 8, 111. 2, 76. 5.

5. 2. 1. 3. 17 O – (4 – Nitro – benzyl) – hydroxylamine (**8q**). Colorless oil; yield 86. 5%; ^{1}H NMR(400MHz, $CDCl_3$)δ: 8. 14 (d, J = 8. 4Hz, 2H, Ar—H), 7. 50 (d, J = 8. 4Hz, 2H, Ar—H), 5. 70 (brs, 2H, —NH_2), 4. 77 (s, 2H, Ar—CH_2O—). ^{13}C NMR (100MHz, $CDCl_3$)δ: 147. 2, 145. 7, 128. 3(2C), 123. 4(2C), 76. 2.

5. 2. 1. 4 Synthetic procedures for target compounds **3a – r, 2a, 2g, 2r, 4a, 4g, 4r, 5a, 5g, 5r.**

Compound **3a** was synthesized as follows: To a solution of compound **3** (0. 40g, 1. 5mmol) in dry CH_2Cl_2 (15. 0mL) 0. 5mL oxalyl choride was added and stirred. 10min later, two drops of dry DMF was added to the above solution. The mixture was stirred overnight at room temperature. Then the dark solution was evaporated under vacuum to get the 4, 5 – dibromo – 1H – Pyrrole – 2 – carbonyl chloride. Compounds **8a** (0. 31g, 1. 5mmol) was dissolved in the solution of dry CH_2Cl_2 (20. 0mL) and anhydrous pyridine (0. 3mL). The 4, 5 – dibromo – 1H – Pyrrole – 2 – carbonyl chloride diluted in dry CH_2Cl_2 (10. 0mL) was added dropwise to the above solution of **8a**. The whole mixture continued to stir for 1h at room temperature after the carbonyl chloride dripped off, then 50. 0mL water and 100. 0mL ethyl acetate were added, washed with 2 N dilute hydrochloric acid, saturated sodium bicarbonate and water respectively. Separating the organic layer, the solvent was removed by distillation. The residue was recrystallized with ethyl acetate/petroleum ether (60 – 90℃, v/v = 1 : 4) to give 4, 5 – dibromo – *N* – (3 – methoxy – 4 – (prop – 2 – ynyloxy) benzyloxy) – 1H – Pyrrole – 2 – carboxamide (**3a**, 0. 38 g, 56. 5%). Using the same procedure, compounds **2a – 5r** were prepared from **8a – r** with corresponding carbonyl chloride respectively. To be pointed out, compound **3r** need to be isolated from its crude residue by flash chromatography on silica gel eluting with ethyl acetate/petroleum [elution solvent: ethyl acetate/petroleum ether (60 – 90℃), v/v = 1 : 4]. All the substituents at the benzene ring were listed in Scheme 2 and Table 1.

5. 2. 1. 4. 1 4, 5 – Dibromo – 1H – Pyrrole – 2 – carboxylic acid (3 – methoxy – 4 – prop –

2 – ynyloxy – benzyloxy) – amide (**3a**). Gray solid; yield 56.5%; mp 164 – 166℃; ^{1}H NMR (400MHz, DMSO – d_6) δ: 12.84 (s, 1H, Pyrrole—NH), 11.40 (brs, 1H, —CONH—), 7.08 (s, 1H, Ar—H), 7.02 (d, J = 8.0Hz, 1H, Ar—H), 6.93 (d, J = 8.0Hz, 1H, Ar—H), 6.80 (d, J = 2.0Hz, 1H, Pyrrole – H), 4.82 (s, 2H, Ar—CH_2O—), 4.78 (d, J = 2.0Hz, 2H, Ar—OCH_2—), 3.78 (s, 3H, Ar—OCH_3), 3.54 (s, 1H, alkynyl—H). ^{13}C NMR (100MHz, DMSO – d_6) δ: 157.6, 148.9, 146.5, 129.3, 125.2, 121.2, 113.6, 112.9 (2C), 105.4, 97.9, 79.2, 78.2, 77.2, 55.9, 55.4. HRMS, *m/z* 456.9223. Calcd for $C_{16}H_{14}Br_2N_2O_4$, 456.9227.

5.2.1.4.2 4,5 – Dibromo – 1H – Pyrrole – 2 – carboxylic acid (3,4 – dimethoxy – benzyloxy) – amide (**3b**). Gray solid; yield 30.1%; mp 197 – 199℃; ^{1}H NMR (300MHz, DMSO – d_6) δ: 12.83 (s, 1H, Pyrrole – NH), 11.38 (brs, 1H, —CONH—), 7.03 (s, 1H, Ar—H), 6.93 (brs, 2H, Ar—H), 6.79 (s, 1H, Pyrrole—H), 4.80 (s, 2H, Ar—CH_2O—), 3.76 (s, 3H, Ar—OCH_3), 3.75 (s, 3H, Ar—OCH_3). ^{13}C NMR (75MHz, DMSO – d_6) δ: 157.6, 148.9, 148.5, 128.1, 125.3, 121.6, 112.7 (2C), 111.3, 105.4, 97.8, 77.3, 55.4, 55.3. HRMS, *m/z* 432.9219. Calcd for $C_{14}H_{14}Br_2N_2O_4$, 432.9227.

5.2.1.4.3 4,5 – Dibromo – 1H – Pyrrole – 2 – carboxylic acid (benzo[1,3] dioxol – 5 – ylmethoxy) – amide (**3c**). Gray solid; yield 61.8%; mp 99 – 101℃; ^{1}H NMR (400MHz, DMSO – d_6) δ: 12.83 (s, 1H, Pyrrole—NH), 11.39 (brs, 1H, —CONH—), 7.00 (s, 1H, Ar—H), 6.89 – 6.90 (m, 2H, Ar—H), 6.79 (d, J = 1.8Hz, 1H, Pyrrole—H), 6.02 (s, 2H, —OCH_2O—), 4.77 (s, 2H, Ar—CH_2O—). ^{13}C NMR (100MHz, DMSO – d_6) δ: 157.6, 147.2, 147.1, 129.6, 125.2, 122.8, 112.7, 109.3, 107.9, 105.4, 101.0, 97.8, 77.1. HRMS, *m/z* 416.8906. Calcd for $C_{13}H_{10}Br_2N_2O_4$, 416.8914.

5.2.1.4.4 4,5 – Dibromo – 1H – Pyrrole – 2 – carboxylic acid benzyloxyamide (**3d**). Gray solid; yield 34.0%; mp 113 – 115℃; ^{1}H NMR (400MHz, $CDCl_3$) δ: 10.94 (s, 1H, Pyrrole—NH), 9.56 (brs, 1H, —CONH—), 7.13 – 7.21 (m, 5H, Ar—H), 6.67 (s, 1H, Pyrrole—H), 4.82 (s, 2H, Ar—CH_2O—). ^{13}C NMR (100MHz, $CDCl_3$) δ: 159.9, 134.5, 129.3 (2C), 129.0, 128.7 (2C), 123.7, 115.3, 107.3, 100.2, 79.1. HRMS, *m/z* 372.9022. Calcd for $C_{12}H_{10}Br_2N_2O_2$, 372.9016.

5.2.1.4.5 4,5 – Dibromo – 1H – Pyrrole – 2 – carboxylic acid (2 – fluorobenzyloxy) – amide (**3e**). Gray solid; yield 57.6%; mp 82 – 84℃; ^{1}H NMR (400MHz, DMSO – d_6) δ: 12.86 (s, 1H, Pyrrole – NH), 11.47 (brs, 1H, —CONH—), 7.43 – 7.50 (m, 2H, Ar—H), 7.21 – 7.23 (m, 2H, Ar—H), 6.79 (s, 1H, Pyrrole—H), 4.95 (s, 2H, Ar—CH_2O—). ^{13}C NMR (100MHz, DMSO – d_6) δ: 162.0 (Ar—F), 159.6 (Ar—F), 157.7, 132.1, 130.8, 125.1, 124.3, 122.6, 115.3, 112.8, 105.5, 97.9, 70.7. HRMS, *m/z* 390.8922. Calcd for $C_{12}H_9Br_2FN_2O_2$, 390.8922.

5.2.1.4.6 4,5 – Dibromo – 1H – Pyrrole – 2 – carboxylic acid (3 – fluorobenzyloxy) – amide (**3f**). Gray solid; yield 63.8%; mp 99 – 101℃; ^{1}H NMR (400MHz, DMSO – d_6) δ: 12.87 (s, 1H, Pyrrole—NH), 11.50 (brs, 1H, —CONH—), 7.42 – 7.44 (m, 1H, Ar—H), 7.28 – 7.30 (m, 2H, Ar—H), 7.19 – 7.26 (m, 1H, Ar—H), 6.79 (s, 1H, Pyrrole—H), 4.91 (s, 2H, Ar—CH_2O—). ^{13}C NMR (100MHz, DMSO – d_6) δ: 163.3 (Ar—F), 160.8 (Ar—F), 157.8, 138.8, 130.3, 125.1, 124.6, 115.3, 114.9, 112.9, 105.7, 97.9, 76.5. HRMS, *m/z*

390.8925. Calcd for $C_{12}H_9Br_2FN_2O_2$, 390.8922.

5.2.1.4.7 4,5 - Dibromo - 1H - Pyrrole - 2 - carboxylic acid(4 - fluorobenzyloxy) - amide(**3g**). Gray solid; yield 58.3%; mp 140 - 142℃; ^1H NMR(400MHz, DMSO - d_6) δ: 12.85 (s, 1H, Pyrrole - NH), 11.44 (brs, 1H, —CONH—), 7.47 - 7.50 (m, 2H, Ar—H), 7.19 - 7.24 (m, 2H, Ar—H), 6.80 (s, 1H, Pyrrole - H), 4.87 (s, 2H, Ar—CH_2O—). ^{13}C NMR (100MHz, DMSO - d_6) δ: 163.2 (Ar—F), 160.8 (Ar—F), 157.6, 132.1, 131.2 (2C), 125.1, 115.1 (2C), 112.7, 105.5, 97.9, 76.5. HRMS, *m/z* 390.8923. Calcd for $C_{12}H_9Br_2FN_2O_2$, 390.8922.

5.2.1.4.8 4,5 - Dibromo - 1H - Pyrrole - 2 - carboxylic acid(2 - chlorobenzyloxy) - amide(**3h**). Gray solid; yield 44.5%; mp 124 - 126℃; ^1H NMR(400MHz, DMSO - d_6) δ: 12.90 (s, 1H, Pyrrole - NH), 11.54 (brs, 1H, —CONH—), 7.59 - 7.60 (m, 1H, Ar—H), 7.50 - 7.52 (m, 1H, Ar—H), 7.40 - 7.42 (m, 2H, Ar—H), 6.84 (s, 1H, Pyrrole—H), 5.05 (s, 2H, Ar—CH_2O—). ^{13}C NMR (100MHz, DMSO - d_6) δ: 157.8, 133.3, 133.2, 131.3, 130.2, 129.2, 127.2, 125.0, 112.9, 105.6, 97.9, 74.1. HRMS, *m/z* 406.8619. Calcd for $C_{12}H_9Br_2ClN_2O_2$, 406.8625.

5.2.1.4.9 4,5 - Dibromo - 1H - Pyrrole - 2 - carboxylic acid(3 - chlorobenzyloxy) - amide(**3i**). Gray solid; yield 80.0%; mp 121 - 123℃; ^1H NMR(400MHz, $CDCl_3$) δ: 10.87 (s, 1H, Pyrrole—NH), 9.79 (brs, 1H, —CONH—), 7.21 (s, 1H, Ar—H), 7.08 - 7.13 (m, 3H, Ar—H), 6.69 (s, 1H, Pyrrole—H), 4.79 (s, 2H, Ar—CH_2O—). ^{13}C NMR(100MHz, $CDCl_3$) δ: 159.9, 136.7, 134.4, 129.9, 129.0 (2C), 127.1, 123.5, 115.5, 107.4, 100.3, 78.1. HRMS, *m/z* 406.8625. Calcd for $C_{12}H_9Br_2ClN_2O_2$, 406.8625.

5.2.1.4.10 4,5 - Dibromo - 1H - Pyrrole - 2 - carboxylic acid(6 - chloropyridin - 3 - ylmethoxy) - amide(**3j**). Gray solid; yield 58.8%; mp 214 - 216℃; ^1H NMR(400MHz, DMSO - d_6) δ: 12.87 (s, 1H, Pyrrole - NH), 11.47 (brs, 1H, —CONH—), 8.47 (s, 1H, Pyridine—H), 7.93 (d, J = 8.0Hz, 1H, Pyridine—H), 7.56 (d, J = 8.0Hz, 1H, Pyridine—H), 6.78 (s, 1H, Pyrrole—H), 4.93 (s, 2H, Pyridine —CH_2O—). ^{13}C NMR (100MHz, DMSO - d_6) δ: 157.8, 150.1 (2C), 140.4, 131.1, 124.9, 124.0, 112.9, 105.6, 97.9, 73.8. HRMS, *m/z* 407.8578. Calcd for $C_{11}H_8Br_2ClN_3O_2$, 407.8578.

5.2.1.4.11 4,5 - Dibromo - 1H - Pyrrole - 2 - carboxylic acid(2,4 - dichloro - benzyloxy) - amide(**3k**). Gray solid; yield 50.0%; mp 202 - 204℃; ^1H NMR(400MHz, DMSO - d_{63}) δ: 12.86 (s, 1H, Pyrrole—NH), 11.47 (s, 1H, —CONH—), 7.66 (s, 1H, Ar—H), 7.58 (d, J = 8.4Hz, 1H, Ar—H), 7.47 (d, J = 8.4Hz, 1H, Ar—H), 6.79 (s, 1H, Pyrrole—H), 4.98 (s, 2H, Ar—CH_2O—). ^{13}C NMR(100MHz, DMSO - d_6) δ: 157.8, 134.3, 133.9, 132.7, 132.6, 128.7, 127.3, 125.0, 112.9, 105.6, 97.9, 73.5. HRMS, *m/z* 440.8227. Calcd for $C_{12}H_8Br_2Cl_2N_2O_2$, 440.8234.

5.2.1.4.12 4,5 - Dibromo - 1H - Pyrrole - 2 - carboxylic acid(3,4 - dichloro - benzyloxy) - amide(**3l**). Gray solid; yield 54.5%; mp 94 - 96℃; ^1H NMR(400MHz, DMSO - d_6) δ: 12.87 (s, 1H, Pyrrole - NH), 11.48 (brs, 1H, —CONH—), 7.74 (s, 1H, Ar—H), 7.65 (d, J = 8.0Hz, 1H, Ar—H), 7.42 (d, J = 8.0Hz, 1H, Ar—H), 6.78 (s, 1H, Pyrrole—H), 4.89 (s, 2H, Ar—CH_2O—). ^{13}C NMR(100MHz, DMSO - d_6) δ: 157.8, 137.2, 130.9, 130.8, 130.6, 130.4, 128.9, 124.9, 112.8, 105.7, 97.9, 75.7. HRMS, *m/z*

440.8228. Calcd for $C_{12}H_8Br_2Cl_2N_2O_2$, 440.8234.

5.2.1.4.13 4,5-Dibromo-1H-Pyrrole-2-carboxylic acid(4-bromobenzyloxy)-amide(**3m**). Gray solid; yield 80.1%; mp 160-162℃; ^{1}H NMR(400MHz, DMSO-d_6)δ: 12.84(s,1H, Pyrrole-NH), 11.44(brs,1H, —CONH—), 7.59(d, J=8.0Hz, 2H, Ar—H), 7.39(d, J=8.0Hz, 2H, Ar—H), 6.78(d, J=2.4Hz, 1H, Pyrrole-H), 4.86(s, 2H, Ar—CH_2O—). ^{13}C NMR(100MHz, DMSO-d_6)δ: 157.7, 135.3, 131.2(2C), 131.0(2C), 125.1, 121.5, 112.8, 105.5, 97.9, 76.4. HRMS, m/z 450.8114. Calcd for $C_{12}H_9Br_3N_2O_2$, 450.8121.

5.2.1.4.14 4,5-Dibromo-1H-Pyrrole-2-carboxylic acid(2-cyanobenzyloxy)-amide(**3n**). Gray solid; yield 46.1%; mp 193-195℃; ^{1}H NMR(400MHz, DMSO-d_6)δ: 12.86(s, 1H, Pyrrole—NH), 11.51(brs, 1H, —CONH—), 7.87(d, J=6.8Hz, 1H, Ar—H), 7.70-7.74(m, 2H, Ar—H), 7.58(d, J=6.8Hz, 1H, Ar—H), 6.78(s, 1H, Pyrrole—H), 5.06(s, 2H, Ar—CH_2O—). ^{13}C NMR(100MHz, DMSO-d_6)δ: 157.8, 138.9, 133.2, 132.9, 130.8, 129.3, 124.9, 117.2, 112.9, 112.1, 105.6, 97.9, 74.7. HRMS, m/z 397.8971. Calcd for $C_{13}H_9Br_2N_3O_2$, 397.8968.

5.2.1.4.15 4,5-Dibromo-1H-Pyrrole-2-carboxylic acid(3-cyanobenzyloxy)-amide(**3o**). Gray solid; yield 40.1%; mp 190-192℃; ^{1}H NMR(300MHz, DMSO-d_6)δ: 12.87(s, 1H, Pyrrole—NH), 11.50(brs, 1H, —CONH—), 7.92(s, 1H, Ar—H), 7.76e7.83(m, 2H, Ar—H), 7.59-7.62(m, 1H, Ar—H), 6.79(s, 1H, Pyrrole—H), 4.95(s, 2H, Ar—CH_2O—). ^{13}C NMR(75MHz, DMSO-d_6)δ: 157.8, 137.7, 133.4, 132.0, 131.9, 129.5, 125.0, 118.6, 112.9, 111.3, 105.6, 97.9, 76.1. HRMS, m/z 397.8971. Calcd for $C^{13}H_9Br_2N_3O_2$, 397.8968.

5.2.1.4.16 4,5-Dibromo-1H-Pyrrole-2-carboxylic acid(4-cyanobenzyloxy)-amide(**3p**). Gray solid; yield 35.2%; mp 212-214℃; ^{1}H NMR(400MHz, DMSO-d_6)δ: 12.87(s, 1H, Pyrrole—NH), 11.52(brs, 1H, —CONH—), 7.86(d, J=8.0Hz, 2H, Ar—H), 7.64(d, J=8.0Hz, 2H, Ar—H), 6.78(s, 1H, Pyrrole—H), 4.98(s, 2H, Ar—CH_2O—). ^{13}C NMR(100MHz, DMSO-d_6)δ: 157.8, 141.6, 132.2(2C), 129.2(2C), 124.9, 118.7, 112.9, 110.8, 105.7, 97.9, 76.3. HRMS, m/z 397.8968. Calcd for $C_{13}H_9Br_2N_3O_2$, 397.8968.

5.2.1.4.17 4,5-Dibromo-1H-Pyrrole-2-carboxylic acid(4-nitrobenzyloxy)-amide(**3q**). Gray solid; yield 60.6%; mp 191-193℃; ^{1}H NMR(400MHz, DMSO-d_6)δ: 12.88(s, 1H, Pyrrole—NH), 11.56(s, 1H, —CONH—), 8.25(d, J=8.4Hz, 2H, Ar—H), 7.73(d, J=8.4Hz, 2H, Ar—H), 6.79(d, J=2.4Hz, 1H, Pyrrole—H), 5.04(s, 2H, Ar—CH_2O—). ^{13}C NMR(100MHz, DMSO-d_6)δ: 157.8, 147.2, 143.7, 129.5(2C), 124.9, 123.3(2C), 112.9, 105.7, 97.9, 76.0. HRMS, m/z 417.8859. Calcd for $C_{12}H_9Br_2N_3O_4$, 417.8867.

5.2.1.4.18 2-{2-[(4,5-Dibromo-1H-Pyrrole-2-carbonyl)-aminooxymethyl]-phenyl}-3-methoxy-acrylic acid methyl ester(**3r**). Light yellow oil; yield 32.6%; ^{1}H NMR(400MHz, DMSO-d_6)δ: 12.85(s, 1H, Pyrrole—NH), 11.43(brs, 1H, —CONH—), 7.65(s, 1H, =$CHOCH_3$), 7.56(d, J=7.2Hz, 1H, Ar—H), 7.30-7.35(m, 2H, Ar—H), 7.09(d, J=7.2Hz, 1H, Ar—H), 6.79(s, 1H, Pyrrole—H), 4.73(s, 2H, Ar—CH_2O—), 3.77(s, 3H, —OCH_3), 3.57(s, 3H, —OCH_3). ^{13}C NMR(100MHz, DMSO-d_6)δ: 166.9,

160.8, 157.6, 134.9, 132.1, 130.8, 128.2, 127.5, 127.4, 125.2, 112.8, 108.4, 105.4, 97.8, 74.7, 61.7, 51.2. HRMS, *m/z* 486.9330. Calcd for $C_{17}H_{16}Br_2N_2O_5$, 486.9333.

5.2.1.4.19 1H-Pyrrole-2-carboxylic acid (3-methoxy-4-prop-2-ynyloxy-benzyloxy)-amide (**2a**). White solid; yield 58.8%; mp 129-131℃; ^{1}H NMR (400MHz, DMSO-d_6) δ: 11.58 (s, 1H, Pyrrole—NH), 11.24 (brs, 1H, —CONH—), 7.10 (s, 1H, H-9), 7.02 (d, J=8.0Hz, 1H, H-12), 6.95 (d, J=8.0Hz, 1H, H-13), 6.92 (s, 1H, H-4), 6.71 (s, 1H, H-2), 6.09 (s, 1H, H-3), 4.83 (s, 2H, H-7), 4.79 (s, 2H, H-14), 3.78 (s, 3H, H-17), 3.55 (s, 1H, H-16). ^{13}C NMR (100MHz, DMSOd_6) δ: 159.7 (s, C-6), 148.9 (s, C-10), 146.4 (s, C-11), 129.6 (s, C-8), 123.1 (s, C-5), 122.0 (d, C-4), 121.1 (d, C-13), 113.6 (d, C-12), 112.8 (d, C-9), 110.3 (d, C-2), 108.6 (d, C-3), 79.2 (s, C-15), 78.2 (d, C-16), 77.1 (t, C-7), 55.9 (t, C-14), 55.4 (q, C-17). HRMS, *m/z* 323.1004. Calcd for $C_{16}H_{16}N_2O_4$, 323.1002 [M+Na]$^+$.

5.2.1.4.20 1H-Pyrrole-2-carboxylic acid (4-fluoro-benzyloxy)-amide (**2g**). Gray solid; yield 76.9%; mp 144-146℃; ^{1}H NMR (300MHz, DMSO-d_6) δ: 11.58 (s, 1H, Pyrrole—NH), 11.27 (brs, 1H, —CONH—), 7.48-7.51 (m, 2H, Ar—H), 7.19-7.24 (m, 2H, Ar—H), 6.91 (s, 1H, Pyrrole—H), 6.68 (s, 1H, Pyrrole—H), 6.08 (d, J=2.4Hz, 1H, Pyrrole—H), 4.87 (s, 2H, Ar—CH_2O—). ^{13}C NMR (75MHz, DMSO-d_6) δ: 163.2 (Ar—F), 160.7 (Ar—F), 159.7, 132.4, 131.0 (2C), 123.0, 122.1, 115.0 (2C), 110.2, 108.6, 76.4. HRMS, *m/z* 257.0701. Calcd for $C_{12}H_{11}FN_2O_2$, 257.0697 [M+Na]$^+$.

5.2.1.4.21 3-Methoxy-2-{2-[(1H-Pyrrole-2-carbonyl)-aminooxymethyl]-phenyl}-acrylic acid methyl ester (**2r**). Light yellow oil; yield 56.6%; ^{1}H NMR (400MHz, DMSO-d_6) δ: 11.59 (s, 1H, Pyrrole—NH), 11.25 (brs, 1H, —CONH—), 7.67 (s, 1H, =$CHOCH_3$), 7.64 (d, J=7.2Hz, 1H, Ar—H), 7.31-7.36 (m, 2H, Ar—H), 7.12 (d, J=7.2Hz, 1H, Ar—H), 6.93 (s, 1H, Pyrrole—H), 6.72 (s, 1H, Pyrrole—H), 6.11 (s, 1H, Pyrrole—H), 4.79 (s, 2H, Ar—CH_2O—), 3.78 (s, 3H, -OCH_3), 3.60 (s, 3H, -OCH_3). ^{13}C NMR (100MHz, DMSO-d_6) δ: 166.9, 160.8, 159.7, 135.3, 131.8, 130.8, 128.0, 127.3 (2C), 123.1, 122.0, 110.3, 108.6, 108.5, 74.7, 61.7, 51.1. HRMS, *m/z* 353.1101. Calcd for $C_{17}H_{18}N_2O_5$, 353.1108 [M+Na]$^+$.

5.2.1.4.22 3,4,5-Tribromo-1H-Pyrrole-2-carboxylic acid (3-methoxy-4-prop-2-ynyloxy-benzyloxy)-amide (**4a**). White solid; yield 39.0%; mp 142-144℃; ^{1}H NMR (400MHz, DMSO-d_6) δ: 13.13 (s, 1H, Pyrrole—NH), 11.08 (brs, 1H, —CONH—), 7.08 (s, 1H, Ar—H), 7.01 (d, J=8.0Hz, 1H, Ar—H), 6.97 (d, J=8.0Hz, 1H, Ar—H), 4.82 (s, 2H, Ar—CH_2O—), 4.78 (d, J=2.0Hz, 2H, Ar—OCH_2—), 3.78 (s, 3H, Ar—OCH_3), 3.54 (s, 1H, alkynyl-H). ^{13}C NMR (100MHz, DMSO-d_6) δ: 157.3, 148.8, 146.5, 129.1, 124.2, 121.3, 113.5, 112.8, 104.7, 102.0, 100.1, 79.2, 78.2, 77.1, 55.9, 55.4. HRMS, *m/z* 558.8296. Calcd for $C_{16}H_{13}Br_3N_2O_4$, 558.8297 [M+Na]$^+$.

5.2.1.4.23 3,4,5-Tribromo-1H-Pyrrole-2-carboxylic acid (4-fluoro-benzyloxy)-amide (**4g**). Gray solid; yield 45.4%; mp 172-174℃; ^{1}H NMR (400MHz, DMSO-d_6) δ: 13.16 (brs, 1H, Pyrrole—NH), 11.14 (s, 1H, —CONH—), 7.49-7.52 (m, 2H, Ar—H), 7.19-7.24 (m, 2H, Ar—H), 4.87 (s, 2H, Ar—CH_2O—). ^{13}C NMR (100MHz, DMSO-d_6) δ: 163.2 (Ar—F), 160.8 (Ar—F), 157.4, 132.0, 131.2 (2C), 124.1, 115.0 (2C), 104.9, 102.0,

100.1,76.3. HRMS, m/z 492.7999. Calcd for $C_{12}H_8Br_3FN_2O_2$, 492.7992 $[M+Na]^+$.

5.2.1.4.24 3-Methoxy-2-{2-[(3,4,5-tribromo-1H-Pyrrole-2-carbonyl)-aminooxymethyl]-phenyl}-acrylic acid methyl ester(**4r**). White solid; yield 34.7%; mp 179-181 ℃; ^{1}H NMR(400MHz, DMSO-d_6)δ: 13.15(s,1H,Pyrrole—NH), 11.02(brs,1H,—CONH—), 7.68(s,1H,=$CHOCH_3$), 7.56(d,J=7.2Hz,1H,Ar—H), 7.31-7.34(m,2H,Ar—H), 7.10(d,J=7.2Hz,1H,Ar—H), 4.75(s,2H,Ar—CH_2O—), 3.80(s,3H,—OCH_3), 3.60(s,3H,—OCH_3). ^{13}C NMR(100MHz, DMSO-d_6)δ: 166.9, 160.9, 157.1, 134.7, 132.0, 130.9, 128.2, 127.6, 127.4, 124.0, 108.4, 104.8, 102.1, 100.2, 74.7, 61.7, 51.2. HRMS, m/z 588.8397. Calcd for $C_{17}H_{15}Br_3N_2O_5$, 588.8403 $[M+Na]^+$.

5.2.1.4.25 4,5-Dichloro-1H-Pyrrole-2-carboxylic acid(3-methoxy-4-prop-2-ynyloxy-benzyloxy)-amide(**5a**). Yellow solid; yield 41.8%; mp 169-171℃; ^{1}H NMR(400MHz, DMSO-d_6)δ: 12.90(s,1H,Pyrrole—NH), 11.45(brs,1H,—ONH—), 7.08(s,1H,Ar—H), 7.02(d,J=8.4Hz,1H,Ar—H), 6.93(d,J=8.4Hz,1H,Ar—H), 6.77(s,1H,Pyrrole—H), 4.81(s,2H,Ar—CH_2O—), 4.78(s,2H,Ar—OCH_2—), 3.77(s,3H,Ar—OCH_3), 3.55(s,1H,alkynyl—H). ^{13}C NMR(100MHz, DMSO-d_6)δ: 157.7, 148.9, 146.5, 129.2, 121.9, 121.2, 115.6, 113.5, 112.8, 109.7, 108.0, 79.2, 78.2, 77.2, 55.9, 55.3. HRMS, m/z 391.0232. Calcd for $C_{16}H_{14}Cl_2N_2O_4$, 391.0223 $[M+Na]^+$.

5.2.1.4.26 4,5-Dichloro-1H-Pyrrole-2-carboxylic acid(4-fluorobenzyloxy)-amide(**5g**). Yellow solid; yield 69.5%; mp 123-125℃; ^{1}H NMR(400MHz, DMSO-d_6)δ: 12.96(brs,1H,Pyrrole—NH), 11.55(brs,1H,—CONH—), 7.51-7.54(m,2H,Ar—H), 7.23-7.28(m,2H,Ar—H), 6.81(s,1H,Pyrrole—H), 4.91(s,2H,Ar—CH_2O—). ^{13}C NMR(100MHz, DMSO-d_6)δ: 163.2(Ar—F), 160.8(Ar—F), 157.7, 132.1, 131.2(2C), 121.8, 115.7, 115.1(2C), 109.8, 108.0, 76.5. HRMS, m/z 324.9921. Calcd for $C_{12}H_9Cl_2FN_2O_2$, 324.9917 $[M+Na]^+$.

5.2.1.4.27 2-{2-[(4,5-Dichloro-1H-Pyrrole-2-carbonyl)-aminooxymethyl]-phenyl}-3-methoxy-acrylic acid methyl ester(**5r**). Light yellow oil; yield 53.5%; ^{1}H NMR(400MHz, DMSO-d_6)δ: 12.89(s,1H,Pyrrole—NH), 11.45(brs,1H,—CONH—), 7.64(s,1H,=$CHOCH_3$), 7.55-7.58(m,1H,Ar—H), 7.30-7.37(m,2H,Ar—H), 7.07-7.10(m,1H,Ar—H), 6.75(s,1H,Pyrrole—H), 4.73(s,2H,Ar—CH_2O—), 3.77(s,3H,—OCH_3), 3.57(s,3H,—OCH_3). ^{13}C NMR(100MHz, DMSO-d_6)δ: 166.9, 160.8, 157.7, 134.9, 132.0, 130.8, 128.2, 127.5, 127.4, 121.9, 115.6, 109.9, 108.5, 108.0, 74.7, 61.7, 51.1. HRMS, m/z 421.0327. Calcd for $C_{17}H_{16}Cl_2N_2O_5$, 421.0328 $[M+Na]^+$.

5.2.1.5 Synthetic procedures for compounds **2aa**, **2ab**.

Compound **2aa** and **2ab** were synthesized as follows: To a solution of compound **2a**(0.1g, 0.33mmol) in dry DMF(20.0mL) potassium carbonate(1.0g, 7.2mmol) was added and stirred. 10min later, two drops of CH_3I was added to the above solution. The mixture was stirred for 5h at room temperature[5]. Then the suspension was extracted with ethyl acetate/water system. Separating the organic layer, the solvent was removed by distillation to give the crude residue. **2aa** and **2ab** were isolated from the residue by flash chromatography on silica gel eluting with ethyl acetate/petroleum [elution solvent: ethyl acetate/petroleum ether(60-90℃), v/v =1:8]. Respective substituents at the benzene ring were listed in Scheme 2 and Table 1.

5. 2. 1. 5. 1 *1H - Pyrrole - 2 - carboxylic acid(3 - methoxy - 4 - prop - 2 - ynyloxy - benzyloxy) - methyl - amide* (**2aa**). Light yellow solid; yield 60. 1% ; mp 99 - 101℃ ; ^{1}H NMR (400MHz, $CDCl_3$) δ: 9. 86 (brs, 1H, Pyrrole—NH), 6. 98 (d, *J* = 8. 0Hz, 1H, Ar—H), 6. 87e6. 89(m, 2H, Ar—H), 6. 87 - 6. 89 (m, 2H, Pyrrole—H), 6. 16 - 6. 19 (m, 1H, Pyrrole—H), 4. 79 (s, 2H, Ar—CH_2O—), 4. 70 (d, *J* = 2. 4Hz, 2H, Ar—OCH_2—), 3. 80 (s, 3H, Ar—OCH_3), 3. 28 (s, 3H, —$CONCH_3$), 2. 45 (s, 1H, alkynyl - H). ^{13}C NMR (100MHz, $CDCl_3$) δ: 160. 7, 148. 6, 146. 2, 127. 1, 122. 5, 121. 0, 120. 9, 114. 3, 112. 9, 111. 8, 109. 3, 77. 2, 75. 2, 75. 0, 55. 6, 54. 9, 33. 7. HRMS, *m/z* 337. 1165. Calcd for $C_{17}H_{18}N_2O_4$, 337. 1159[M - Na]$^+$.

5. 2. 1. 5. 2 *1 - Methyl - 1H - Pyrrole - 2 - carboxylic acid (3 - methoxy - 4 - prop - 2 - ynyloxy - benzyloxy) - methyl - amide* (***2ab***). Light yellow solid; yield 20. 0% ; mp 94 - 96℃ ; ^{1}H NMR(400MHz, $CDCl_3$) δ: 6. 99 (d, *J* = 8. 0Hz, 1H, Ar—H), 6. 88 (s, 1H, Ar—H), 6. 86 (d, *J* = 8. 0Hz, 1H, Ar—H), 6. 83 (s, 1H, Pyrrole—H), 6. 71 (s, 1H, Pyrrole - H), 6. 08 - 6. 09 (m, 1H, Pyrrole—H), 4. 76 (s, 2H, Ar—CH_2O—), 4. 76 (s, 2H, Ar—OCH_2—), 3. 84 (s, 3H, Ar—OCH_3), 3. 81 (s, 3H, Pyrrole—NCH_3), 3. 34 (s, 3H, —$CONCH_3$), 2. 51 (s, 1H, alkynyl—H). ^{13}C NMR (100MHz, $CDCl_3$) δ: 162. 9, 149. 5, 147. 1, 128. 4, 127. 8, 123. 6, 122. 0, 116. 5, 113. 7, 113. 1, 107. 1, 78. 3, 75. 9, 75. 8, 56. 6, 55. 8, 36. 8, 35. 1. HRMS, *m/z* 351. 1316. Calcd for $C_{18}H_{20}N_2O_4$, 351. 1315 [M + Na]$^+$.

5. 3 Antifungal activity

The fungicidal activities of the compounds **3a - r, 2a, 2g, 2r, 4a, 4g, 4r, 5a, 5g, 5r, 2aa, 2ab** were tested in vitro against *A. solani*, *C. arachidicola*, *F. omysporum*, *G. zeae* and *P. piricola* and their relative inhibitory ratio (%) had been determined by using the mycelium growth rate method[30]. Anthranilamide and fenpiclonil were used as control. After the mycelia grew completely, the diameter of the mycelia was measured and the inhibition rate was calculated according to the formula:

$$I = (D1 - D2)/D1 \times 100\%$$

where I was the inhibition rate, *D*1 was the average diameter of mycelia in the blank test, and *D*2 was the average diameter of mycelia in the presence of those compounds. The inhibition ratio of those compounds at the dose of 50μg · mL^{-1} was summarized in Table 2. The EC_{50} values of compounds **2a**, **3a** and fenpiclonil had been experimented and calculated by the Scatchard method. And the results were summarized in Table 3.

References

[1] G. W. Gribble, Chem. Soc. Rev. 28(1999)335 - 346.

[2] J. F. Liu, S. P. Guo, B. Jiang, Chin. J. Org. Chem. 25(2005)788 - 799.

[3] G. W. Gribble, J. Nat. Prod. 55(1992)1353 - 1395.

[4] K. H. Pée, J. M. Ligon, Nat. Prod. Rep. 17(2000)157 - 164.

[5] P. M. Fresneda, P. Molina, M. A. Sanz, Tetrahedron Lett. 42(2001)851 - 854.

[6] T. Lindel, H. Hoffmann, M. Hochgurtel, J. Pawlik, J. Chem. Ecol. 26(2000)1477 - 1496.

[7] J. A. Ponasik, D. J. Kassab, B. Ganem, Tetrahedron Lett. 37(1996)6041 - 6044.

[8] B. Chanas, J. R. Pawlik, T. Lindelb, W. Fenicalb, J. Exp. Mar. Biol. Eco. 208(1996)185 - 196.

[9] J. A. Ponasik, S. Conova, D. Kinghorn, W. A. Kinney, D. Rittschof, B. Ganem, Tetrahedron 54 (1998) 6977 – 6986.

[10] M. A. F. Biabani, M. Baake, B. Lovisetto, H. Laatsch, E. Helmke, H. Weyland, J. Antibiot. 51 (1998) 333 – 340.

[11] B. J. Ludwig, F. Dursch, M. Auerbach, K. Tomeczek, F. M. Berger, J. Med. Chem. 10(1967)556 – 563.

[12] S. A. Glover, G. P. A. Hammond, J. Org. Chem. 63(1998)9684 – 9689.

[13] M. A. Bailén, R. Chinchilla, D. J. Dodsworth, C. Nájera, Tetrahedron Lett. 42(2001)5013 – 5016.

[14] N. D. Kokare, R. R. Nagawade, V. P. Ranea, D. B. Shindea, Tetrahedron Lett. 48(2007)4437 – 4440.

[15] H. Sauter, Strobilurinsand other complex III inhibitors. in: W. Krämer, U. Schirmer(Eds.), Modern Crop Protection Compounds. Wiley, 2007, pp. 457 – 491.

[16] B. P. Hodge, R. W. Rickards, J. Chem. Soc. Perkin. Trans. 1(1963)2543 – 2545.

[17] H. Kikuchi, M. Sekiya, Y. Katou, K. Ueda, T. Kabeya, S. Kurata, Y. Oshima, Org. Lett. 11 (2009) 1693 – 1695.

[18] Y. Li, H. Q. Zhang, J. Liu, X. P. Yang, Z. J. Liu, J. Agric, Food Chem. 54(2006)3636 – 3640.

[19] C. X. Zeng, S. H. Xu, Y. Q. Li, Y. F. Wang, Chin. J. Org. Chem. 25(2005)954 – 958.

[20] H. J. Anderson, S. F. Lee, Canad. J. Chem. 43(1965)409 – 414.

[21] B. P. Hodge, R. W. Rickards, J. Chem. Soc. (1965)459 – 470.

[22] B. R. Clark, C. D. Murphy, Org. Biomol. Chem. 7(2009)111 – 116.

[23] D. O. A. Garrido, G. Buldain, M. I. Ojea, B. Frydman, J. Org. Chem. 53(1988)403 – 407.

[24] H. T. Wu, P. L. Micca, M. S. Makar, M. Miura, Bioorg. Med. Chem. 14(2006)5083 – 5092.

[25] A. V. Oeveren, J. F. G. A. Jansen, B. L. Feringa, J. Org. Chem. 59(1994)5999 – 6007.

[26] M. Yasuyuki, S. Takahiro, I. Yutaka, Y. Hiroyuki, F. Makoto, T. Mitsuru, I. Yoshiyuki, Y. Satoru, K. Noriaki. US 2004152894(2004).

[27] D. M. Wallace, S. H. Leung, M. O. Senge, K. M. Smith, J. Org Chem 58(1993)7245 – 7257.

[28] A. Mai, S. Massa, R. Ragno, I. Cerbara, F. Jesacher, P. Loidl, G. Brosch, J. Med. Chem. 46(2003)512 – 524.

[29] L. F. Fieser, M. Fieser, Regent for Organic Synthesis. John Wiley and Sons, Inc., New York, 1967, pp. 485 – 486.

[30] N. C. Chen, Bioassay of Pesticides. Beijing Agricultural University Press, Beijing, China, 1991, pp. 161 – 162.

Synthesis and Insecticidal Evaluation of Novel *N* – Pyridylpyrazolecarboxamides Containing Different Substituents in the *ortho* – Position*

Qi Feng(冯启),Guanping Yu(于观平),Lixia Xiong(熊丽霞)
Mingzhong Wang(王明忠),Zhengming Li(李正名)

(State Key Laboratory of Elemento – Organic Chemistry,Institute of Elemento – Organic Chemistry,Nankai University,Tianjin,300071,China)

Abstract In order to look for novel insecticides containing *N* – pyridylpyrazole, ten novel pyrazolecarboxamides containing different *ortho* – substituents in the aniline part were synthesized, and their structures were characterized by ^{1}H NMR, ^{13}C NMR and HRMS. The single crystal structure of **10b** was determined by X – ray diffraction. Their evaluated insecticidal activity against oriental armyworm(*Mythimna separata*) indicated that all the compounds exhibited moderate insecticidal activities.

Key words insecticidal activity; *N* – pyridylpyrazole; pyrazolecarboxamide, *ortho* – substituents

Introduction

Since the introduction of chlorantraniliprole(Figure 1, **A**) as an insecticide for crop protection in recent years,[1] several anthranilic diamides insecticides have been developed due to their high potency, low mammalian toxicity, broad insecticidal spectra, and new mode of action.[2,3] From their chemical structures, two features are evident: one is the *N* – pyridylpyrazole moiety, the other is the anthranilamide moiety. Previous research indicated that some modification focused on the anthranilamide while preserving the *N* – pyridylpyrazole.[4] However, the amides in the *ortho* – position have been changed a lot in the reported patents, such as heterocyclic groups,[5,6] *N* – alkoxy amides,[7] cyanocontaining amides[8], *etc.* Recently a novel pyrazolecarboxamide with thiadiazol group in the *ortho* – position(Figure 1, **B**, JS 9117) was reported with high insecticidal activity,[9,10] indicating the *ortho* – position is still attractive in further modification. In this article, novel pyrazolecarboxamides containing different *ortho* – substituents in the aniline part, including methylene acohols, methylene amines, aldehydes, oximes and thiazolidines, were synthesized as shown in Scheme 1 and their insecticidal activity against oriental armyworm was tested accordingly.

Experimental

Materials and instruments

NMR spectra were recorded at 400MHz using a Bruker AV – 400 spectrometer in $CDCl_3$ or DMSO – d_6 solution with tetramethylsilane as the internal standard, High – resolution mass spec-

* Reprinted from *Chinese Journal of Chemistry*, 2011, 29: 1651 – 1655. Project supported by the National Basic Research Program of China(No. 2010CB126106) and the National Natural Science Foundation of China(No. 20872069).

trometry (HRMS) data were obtained on a Varian QFT – ESI instrument. The melting points were determined on an X – 4 binocular microscope melting point apparatus (Beijing Tech Instruments Co., Beijing, China) and were uncorrected. Manganese dioxide (activated, tech.) was purchased from Alfa Aesar. Reagents were all analytically pure. All solvents and liquid reagents were dried by standard methods in advance and distilled before use. Chlorantraniliprole was synthesized according to the literature[11] and used as control.

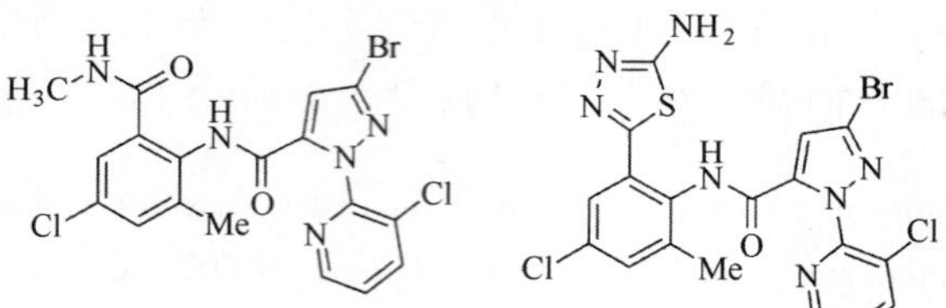

Figure 1 Chemical structures of compounds **A** and **B**

General procedure

The intermediate 3 – bromo – 1 – (3 – chloro – 2 – pyridyl) – 1*H* – pyrazole – 5 – caboxylic acid (**4**) was synthesized according to procedure reported by Dong *et al.*[11] The title compounds were prepared as shown in Scheme 1.

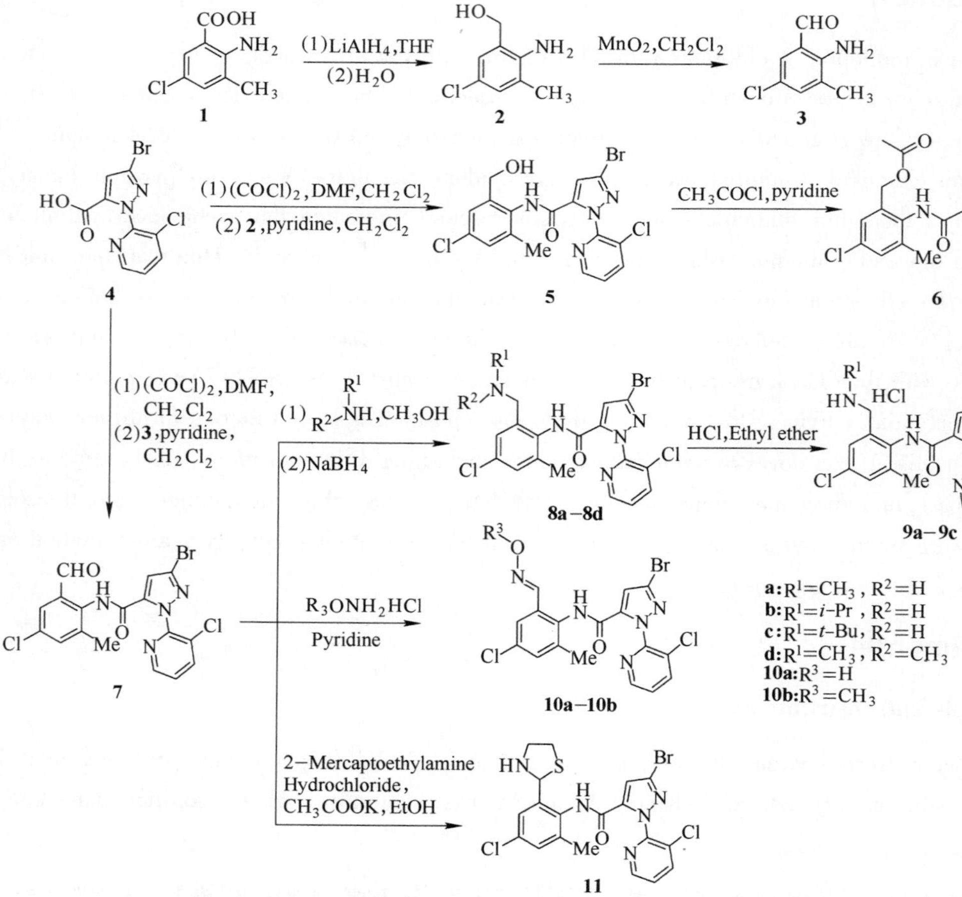

a: R^1=CH_3, R^2=H
b: R^1=*i*-Pr, R^2=H
c: R^1=*t*-Bu, R^2=H
d: R^1=CH_3, R^2=CH_3
10a: R^3=H
10b: R^3=CH_3

Scheme 1 General synthetic procedure of title compounds **5 – 11**

(2 - Amino - 5 - chloro - 3 - methylphenyl) methanol (2) The solution of compound **1** (60. 0mmol) in 20mL of THF was added dropwise to an ice - cold mixture of lithium aluminium hydride (0. 120mol) in 60mL of THF with the temperature maintained below 5℃ throughout. After stirring at 5℃ for 3h, 60mL of 5% sodium hydroxide solution was added to quench the reaction. The resulting mixture was extracted with ethyl acetate ($40mL \times 3$). The combined extracts were washed with brine, dried with anhydrous sodium sulfate and evaporated to give the title compound as a dust - color solid. Yield 97. 0%, m. p. 83 - 85℃; ^1H NMR ($CDCl_3$, 400MHz) δ: 7. 02 (d, J = 2. 4Hz, 1H, Ar—H), 6. 93 (d, J = 2. 4Hz, 1H, Ar—H), 4. 63 (s, 2H, Ar—CH_2), 2. 16 (s, 3H, Ar—CH_3); ^{13}C NMR ($CDCl_3$, 100MHz) δ: 142. 6, 129. 9, 126. 5, 125. 5, 124. 4, 121. 9, 63. 9, 17. 2.

2 - Amino - 5 - chloro - 3 - methylbenzaldehyde (3) Manganese dioxide (4. 0mmol) was added to a solution of compound **2** (2. 0mmol) in 20mL of dichloromethane in one portion. The resulting mixture was stirred at room temperature for 6 h, then an additional portion of manganese dioxide (2. 0mmol) was added and the stirring was continued overnight. Silica gel (2. 0g) was added and the mixture was evaporated, the resulting black powder was applied to flash chromatography on silica gel with petroleum ether/ethyl acetate (4/1, V/V) to give the title compound as a yellow solid. Yield 77. 0%, m. p. 114 - 116℃; ^1H NMR ($CDCl_3$, 400MHz) δ: 9. 81 (s, 1H, CHO), 7. 35 (s, 1H, Ar—H), 7. 21 (s, 1H, Ar—H), 6. 23 (br s, 2H, Ar—NH_2), 2. 16 (s, 3H, Ar—CH_3); ^{13}C NMR ($CDCl_3$, 100MHz) δ: 193. 1, 147. 0, 135. 6, 132. 2, 124. 9, 120. 3, 117. 8, 16. 5.

3 - Bromo - *N* - (4 - chloro - 2 - (hydroxymethyl) - 6 - methylphenyl) - 1 - (3 - chloropyridin - 2 - yl) - 1*H* - pyrazole - 5 - carboxamide (5) Carboxylic acid **4** (1. 0mmol) was dissolved in 20mL of dichloromethane under stirring; oxalyl chloride (1. 5mmol) and a drop of DMF were added successively. After stirring for 2h at room temperature, dichloromethane was removed in vacuum. The residue was dissolved in 20mL of dichloromethane, and then added dropwise to an ice - cold solution of compound **2** (1. 0mmol), pyridine (1. 0mmol) in 20mL of dichloromethane. The mixture was stirred at room temperature for 2h before it was washed with 2mol/L hydrochloric acid, saturated sodium bicarbonate solution and brine successively. The organic layer was dried with sodium sulfate and evaporated. The residue was subjected to flash chromatography on silica gel with petroleum ether/ethyl acetate (2/1, V/V) to give the title compound as white solid. Yield 50. 0%, m. p. 92 - 94℃; ^1H NMR ($CDCl_3$, 400MHz) δ: 8. 51 (s, 1H, CONH), 8. 42 (dd, J = 1. 4, 4. 6Hz, 1H, Ar—H), 7. 87 (dd, J = 1. 6, 8. 0Hz, 1H, Ar—H), 7. 37 (dd, J = 4. 4, 8. 0Hz, 1H, Ar—H), 7. 15 (d, J = 2. 0Hz, 1H, Ar—H), 7. 10 (d, J = 2. 0Hz, 1H, Ar—H), 6. 90 (s, 1H, Ar—H), 4. 48 (s, 2H, Ar—CH_2), 2. 15 (s, 3H, Ar—CH_3); ^{13}C NMR ($CDCl_3$, 100MHz) δ: 156. 1, 148. 6, 146. 8, 139. 3, 139. 1, 137. 6, 137. 3, 132. 8, 131. 8, 130. 6, 128. 9, 128. 2, 126. 8, 125. 9, 110. 4, 63. 1, 18. 5; HRMS calcd for $C_{17}H_{13}BrCl_2N_4O_2Na$ ($[M + Na]^+$) 476. 9494, found 496. 9491.

2 - (3 - Bromo - 1 - (3 - chloropyridin - 2 - yl) - 1*H* - pyrazole - 5 - carboxamido) - 5 - chloro - 3 - methylbenzyl acetate (6) Acetyl chloride (0. 52mmol) was added dropwise to an ice - cold solution of compound **5** (0. 50mmol) in 10mL of pyridine. After stirring for 2h, 40mL of water was added to quench the reaction and the solution was extracted with ethyl acetate(40mL × 3). The extracts were combined, washed with 2mol/L hydrochloride acid and brine, dried, then evaporated. The resulting residue was applied to flash chromatography on silica gel with petroleum ether/ethyl acetate (2/1, *V/V*) to give the title compound as white solid. Yield 95.8%, m.p. 77 - 79 ℃; ^{1}H NMR ($CDCl_3$, 400MHz) δ: 9.29 (s, 1H, CONH), 8.46 (dd, *J* = 1.4, 4.6Hz, 1H, Ar—H), 7.86 (dd, *J* = 1.2, 8.2Hz, 1H, Ar—H), 7.38 (dd, *J* = 4.8, 8.0Hz, 1H, Ar—H), 7.25 (d, *J* = 2.0Hz, 1H, Ar—H), 7.22 (d, *J* = 2.4Hz, 1H, Ar—H), 7.10 (s, 1H, Ar—H), 4.99 (s, 2H, CH_2), 2.19 (s, 3H, CH_3CO), 2.11 (s, 3H, Ar—CH_3); ^{13}C NMR ($CDCl_3$, 100MHz) δ: 172.4, 155.8, 149.0, 146.9, 139.0, 138.8, 138.5, 133.0, 132.7, 131.9, 131.6, 129.0, 128.6, 128.2, 125.8, 110.3, 63.1, 21.0, 18.7; HRMS calcd for $C_{19}H_{15}BrCl_2N_4O_3Na$ ($[M+Na]^+$) 474.9335, found 474.9335.

3 - Bromo - *N* - (4 - chloro - 2 - formyl - 6 - methylphenyl) - 1 - (3 - chloropyridin - 2 - yl) - 1*H* - pyrazole - 5 - carboxamide (7) This compound was prepared following the same procedure described for compound **5** using 2 - amino - 5 - chloro - 3 - methylbenzaldehyde (**3**) instead of compound **2**. Yellow solid, yield 60.0%, m.p. 86 - 88℃; ^{1}H NMR ($CDCl_3$, 400MHz) δ: 9.95 (s, 1H, CHO), 9.91 (s, 1H, CONH), 8.46 (d, *J* = 4.4Hz, 1H, Ar—H), 7.86 (d, *J* = 8.0Hz, 1H, Ar—H), 7.58 (s, 1H, Ar—H), 7.46 (s, 1H, Ar—H), 7.39 (dd, *J* = 4.4, 8.0Hz, 1H, Ar—H), 7.09 (s, 1H, Ar—H), 2.21 (s, 3H, Ar—CH_3); ^{13}C NMR ($CDCl_3$, 100MHz) δ: 193.0, 155.7, 148.7, 146.9, 139.1, 138.9, 137.8, 137.0, 134.2, 132.1, 132.0, 129.1, 128.9, 128.3, 125.8, 110.9, 19.1; HRMS calcd for $C_{17}H_{11}BrCl_2N_4O_2Na$ ($[M+Na]^+$) 518.9591, found 518.9597.

The general synthetic procedure for the compounds 8a - 8d

Alkylamine (0.60mmol) was added to a solution of compound **7** (0.60mmol) in 20mL ethanol. The resulting solution was stirred at room temperature for 2h, and then the solvent was evaporated. The residue was dissolved in methanol (20mL), and sodium borohydride (0.80mmol) was added. The reaction was stirred for 2h, and then most of the solvent was evaporated. Ethyl acetate (50mL) was added to the residue, which was washed with potassium carbonate solution (10%, 40mL) and brine successively. The resulting organic phase was concentrated and subjected to flash chromatography on silica gel with petroleum ether: ethyl acetate (1/1, *V/V*) to give the title compounds **8a - 8d**.

3 - Bromo - *N* - (4 - chloro - 2 - methyl - 6 - ((methylamino) - methyl) phenyl) - 1 - (3 - chloropyridin - 2 - yl) - 1*H* - pyrazole - 5 - carboxamide (8a) White solid, yield 50.2%, m.p. 134 - 136℃; ^{1}H NMR ($CDCl_3$, 400MHz) δ: 10.80 (brs, 1H, CONH), 8.50 (d, *J* = 4.4Hz, 1H, Ar—H), 7.88 (d, *J* = 8.0Hz, 1H, Ar—H), 7.41 (dd, *J* = 4.8, 8.0Hz, 1H, Ar—H), 7.18 (s, 1H, Ar—H), 7.06 (s, 1H, Ar—H), 6.94 (s, 1H, Ar—H), 3.76 (s, 2H, CH_2), 2.48 (s, 3H, **CH_3**NH), 2.20 (s, 3H, Ar—CH_3).

3 - Bromo - *N* - (4 - chloro - 2 - ((isopropylamino) methyl) - 6 - methylphenyl) - 1 - (3 - chloropyridin - 2 - yl) - 1*H* - pyrazole - 5 - carboxamide (8b) White solid, yield 85.0%, m.p. 161 - 163 ℃; ^{1}H NMR ($CDCl_3$, 400MHz) δ: 10.96 (brs, 1H, CONH), 8.50 (dd, *J* = 1.4, 4.6Hz, 1H, Ar—H), 7.87 (dd, *J* = 1.2, 8.0Hz, 1H, Ar—H), 7.40 (dd, *J* = 4.8, 8.0Hz, 1H, Ar—H), 7.16 (s, 1H, Ar—H), 7.06 (s, 1H, Ar—H), 6.92 (s, 1H, Ar—H), 3.78

(s,1H,CH_2),2.88 – 2.82(m,1H,CH),2.19(s,3H,Ar—CH_3),1.15(d,J=6.4Hz,6H,CH_3).

N – (2 – ((*tert* – butylamino) methyl) – 4 – chloro – 6 – methylphenyl) – 3 – bromo – 1 – (3 – chloropyridin – 2 – yl) – 1*H* – pyrazole – 5 – carboxamide (8c) White solid, yield 97.5%, m.p. 60 – 62℃; ^1H NMR($CDCl_3$,400MHz)δ:10.96(br s,1H,CONH),8.48(dd,J=1.2,4.8Hz,1H,Ar—H),7.82(dd,J=1.2,4.8Hz,1H,Ar—H),7.36(dd,J=4.8,8.0Hz,1H,Ar—H),7.12(s,1H,Ar—H),7.05(s,1H,Ar—H),6.91(s,1H,Ar—H),3.71(s,2H,CH_2),2.15(s,3H,Ar—CH_3),1.15(s,9H,C(CH_3)$_3$).

3 – Bromo – N – (4 – chloro – 2 – ((dimethylamino) methyl) – 6 – methylphenyl) – 1 – (3 – chloropyridin – 2 – yl) – 1*H* – pyrazole – 5 – carboxamide (8d) White solid, yield 45.0%, m.p. 218 – 220℃; ^1H NMR($CDCl_3$,400MHz)δ:10.82(br s,1H,CONH),8.49(d,J=6.0Hz,1H,Ar—H),7.86(dd,J=1.6,10.8Hz,1H,Ar—H),7.39(dd,J=6.2,10.8Hz,1H,Ar—H),7.16(s,1H,Ar—H),7.00(s,1H,Ar—H),6.86(s,1H,Ar—H),3.41(s,2H,CH_2),2.29(s,6H,CH_3),2.19(s,3H,Ar—CH_3); ^{13}C NMR($CDCl_3$,100MHz)δ:154.9,149.1,146.9,139.7,138.9,134.0,131.3,131.1,130.4,129.0,128.9,128.1,127.1,125.8,109.7,62.8,45.0,19.0; HRMS calcd for $C_{19}H_{19}BrCl_2N_5O$ ([M + H]$^+$) 482.0150, found 482.0145.

Compounds **8a – 8c** were further treated with hydrochloric acid gas in ethyl ether to give compounds **9a – 9c**.

3 – Bromo – N – (4 – chloro – 2 – methyl – 6 – ((methylamino) – methyl) phenyl) – 1 – (3 – chloropyridin – 2 – yl) – 1*H* – pyrazole – 5 – carboxamide hydrochloride (9a) White solid, quantitative yield, m.p. 168 – 170 ℃; ^1H NMR($CDCl_3$,400MHz)δ:10.80(s,1H,CONH),9.19(brs,2H,NH_2^+ Cl^-),8.52(d,J=4.8Hz,1H,Ar—H),8.21(d,J=8.4Hz,1H,Ar—H),7.73(s,1H,Ar—H),7.63(dd,1H,J=4.8,8.0Hz,Ar—H),7.57(s,1H,Ar—H),7.48(s,1H,Ar—H),4.01(t,J=5.8Hz,2H,CH_2),2.55(t,J=5.0Hz,3H,CH_3),2.17(s,3H,Ar—CH_3); ^{13}C NMR($CDCl_3$,100MHz)δ:155.9,148.3,147.2,139.4,138.9,133.1,131.7,131.6,130.8,128.2,128.1,127.7,126.6,111.4,47.1,32.4,17.8; HRMS calcd for $C_{18}H_{17}BrCl_2N_5O$ ([M – Cl]$^+$) 467.9984, found 467.9988.

3 – Bromo – N – (4 – chloro – 2 – ((isopropylamino) methyl) 6 – methylphenyl) – 1 – (3 – chloropyridin – 2 – yl) – 1*H* – pyrazole – 5 – carboxamide hydrochloride (9b) White solid, quantitative yield, m.p. 163 – 165℃; ^1H NMR($CDCl_3$,400MHz)δ:9.97(s,1H,CONH),9.15(br s,2H,NH_2^+ Cl^-),8.45(d,J=3.6Hz,1H,Ar—H),7.84(d,J=7.6Hz,1H,Ar—H),7.55(s,1H,Ar—H),7.38 – 7.34(m,2H,Ar—H),7.27(s,1H,Ar—H),3.82(s,2H,CH_2),3.10(s,1H,CH),2.23(s,3H,Ar—CH_3),1.18(d,J=6.4Hz,6H,CH(CH_3)$_2$); ^{13}C NMR($CDCl_3$,100MHz)δ:155.8,148.4,147.2,139.4,139.3,138.8,133.2,131.8,131.6,130.8,128.8,127.8,126.9,126.7,111.3,50.4,43.4,18.5,17.8; HRMS calcd for $C_{20}H_{21}BrCl_2N_5O$ ([M – Cl]$^+$) 496.0305, found 496.0301.

N – (2 – ((*tert* – Butylamino) methyl) – 4 – chloro – 6 – methylphenyl) – 3 – bromo – 1 – (3 – chloropyridin – 2 – yl) – 1*H* – pyrazole – 5 – carboxamide hydrochloride (9c) White solid, quantitative yield, m.p. 268 – 271℃; ^1H NMR($CDCl_3$,400MHz)δ:10.94(s,1H,

CONNH), 9.01(br s, 2H, NH_2^+ Cl^-), 8.51(d, J = 4.0Hz, 1H, Ar—H), 8.20(d, J = 8.0Hz, 1H, Ar—H), 7.77(s, 1H, Ar—H), 7.63(dd, J = 4.8, 8.0Hz, 1H, Ar—H), 7.56(s, 1H, Ar—H), 7.49(s, 1H, Ar—H), 3.97(s, 2H, CH_2), 2.16(s, 3H, Ar—CH_3), 1.37(s, 9H, C$(CH_3)_3$); ^{13}C NMR($CDCl_3$, 100MHz)δ: 155.7, 148.4, 147.1, 139.5, 139.3, 138.7, 133.5, 131.7, 131.6, 130.9, 129.5, 127.8, 126.9, 126.7, 111.1, 57.4, 40.5, 25.0, 17.7; HRMS calcd for $C_{21}H_{23}BrCl_2N_5O$([M − Cl]$^+$) 510.0454, found 510.0458.

(*Z*) −3 − Bromo − *N* − (4 − chloro −2 − ((hydroxyimino) methyl) −6 − methylphenyl) −1 − (3 − chloropyridin −2 − yl) −1*H* − pyrazole −5 − carboxamide(10a) Hydroxylamine hydrochloride(0.30mmol) was added to a solution of compound **7**(0.15mmol) in 10mL pyridine. After stirring for 12h at room temperature, ethyl acetate(50mL) was added. The mixture was washed with 2mol/L hydrochloric acid, saturated sodium bicarbonate solution and brine successively. After dried and evaporated, the title compound was obtained as white solid. Yield 98%, m. p. 172 − 174℃; ^{1}H NMR($CDCl_3$, 400MHz)δ: 9.15(s, 1H, CONH), 8.44(dd, J = 1.2, 4.8Hz, 1H, Ar—H), 8.32(s, 1H, OH), 8.05(s, 1H, CH), 7.87(dd, J = 1.0, 8.2Hz, 1H, Ar—H), 7.38(dd, J = 4.4, 8.0Hz, 1H, Ar—H), 7.21(s, 1H, Ar—H), 7.19(s, 1H, Ar—H), 6.92(s, 1H, Ar—H), 2.18(s, 3H, CH_3); ^{13}C NMR($CDCl_3$, 100MHz)δ: 156.3, 148.0, 147.1, 144.2, 139.5, 139.1(2C), 132.0, 131.9, 131.8, 130.6, 127.4, 127.1, 126.6, 122.4, 110.7, 17.4; HRMS calcd for $C_{17}H_{12}BrCl_2N_5O_2Na$ ([M + Na]$^+$) 489.9445, found 489.9444.

3 − Bromo − *N* − (4 − chloro −2 − ((*Z*) − methoxyiminomethyl) −6 − methylphenyl) −1 − (3 − chloropyridin −2 − yl) −1*H* pyrazole −5 − carboxamide(10b) This compound was prepared following the same procedure described for compound **10a** using methoxamine hydrochloride instead of hydroxylamine hydrochloride. White solid, yield 90%, m. p. 219 − 220℃; ^{1}H NMR($CDCl_3$, 400MHz)δ: 9.37(s, 1H, CONH), 8.48(d, J = 4.0Hz, 1H, Ar—H), 8.05(s, 1H, CH), 7.85(d, J = 8.0Hz, 1H, Ar—H), 7.39(dd, J = 4.8, 8.0Hz, 1H, Ar—H), 7.22(s, 2H, Ar—H), 6.96(s, 1H, Ar—H), 3.97(s, 3H, OCH_3), 2.20(s, 3H, Ar—CH_3); ^{13}C NMR($CDCl_3$, 100MHz)δ: 155.7, 148.7, 148.1, 147.0, 139.3, 139.1, 137.6, 132.3, 132.0, 131.4, 128.8, 128.0, 127.8, 127.2, 125.8, 110.3, 62.6, 19.0; HRMS calcd for $C_{18}H_{14}BrCl_2N_5O_2Na$([M + Na]$^+$) 503.9595, found 503.9600.

3 − Bromo − *N* − (4 − chloro −2 − methyl −6 − (thiazolidin −2 − yl) phenyl) −1 − (3 − chloropyridin −2 − yl) −1*H* − pyrazole −5 − carboxamide(11) To a solution of compound 7 (0.17g, 0.37mmol) and 2 − mercaptoethylamine hydrochloride (0.084g, 0.74mmol) in 20mL of ethanol, potassium acetate(0.073g, 0.74mmol) was added in one portion. The resulting suspension was stirred at room temperature for 12 h, and then most of the solvent was removed in vacuum. The residue was dissolved in 50mL of ethyl acetate, washed with water and brine. The ethyl acetate solution was evaporated and then subjected to flash chromatography using petroleum ether/ethyl acetate(2/1, *V/V*) as the eluant to give the title compound as white solid, yield 67.7%, m. p. 177 − 179℃; ^{1}H NMR($CDCl_3$, 400MHz)δ: 9.33(s, 1H, CONH), 8.44(d, J = 4.4Hz, 1H, Ar—H), 7.87(d, J = 8.4Hz, Ar—H), 7.51(s, 1H, Ar—H), 7.37(dd, J = 4.8, 8.0Hz, 1H, Ar—H), 7.19(s, 1H, Ar—H), 6.92(s, 1H, Ar—H), 5.56(s, 1H, CH), 3.39 − 3.33(m, 1H, CH), 3.24 − 3.18(m, 1H, CH), 3.11 − 3.06(m, 1H, CH), 3.01 − 2.96(m,

1H, CH), 2.19 (s, 3H, Ar—CH_3); ^{13}C NMR ($CDCl_3$, 100MHz) δ: 155.4, 148.7, 146.8, 139.5, 139.2, 137.9, 135.7, 132.5, 131.9, 130.8, 128.8, 128.1, 126.5, 125.8, 110.2, 70.2, 51.6, 35.5, 18.6; HRMS calcd for $C_{19}H_{17}BrCl_2N_5OS$ ($[M+H]^+$) 511.9718, found 511.9709.

Biological activity

The insecticidal activities of compounds **5 - 11** and chlorantraniliprole were evaluated using the reported procedure.[11] The insecticidal activity against oriental armyworm was tested by foliar application, individual corn leaves were placed on moistened pieces of filter paper in Petri dishes. The leaves were then sprayed with the test solution and allowed to dry. Then every 10 fourth - instar oriental armyworm larvae were put into each dish. Percent mortalities were evaluated 2d after treatment. Each treatment was replicated for three times. For comparative purpose, chlorantraniliprole was tested as control under the same conditions.

Results and discussion

Structure of compound 10b

In order to verify the configuration of oxime ether, compound **10b** was recrystallized from ethanol to give a colorless crystal for X - ray single - crystal diffraction.[12] The crystal with dimensions of 0.22mm × 0.20mm × 0.18mm was mounted and analyzed on a Rigaku Saturn 724 CCD diffractometer with a graphite monochromatized MoKα radiation (λ = 0.71073 Å) using the ω scan mode (1.82° ≤ θ ≤ 27.89°) at 113 K. The crystal belongs to a orthorhombic system with space group PbCa and crystal parameters of a = 21.315(3) Å, b = 8.3900(10) Å, c = 22.375(3) Å, V = 4001.4(9) A^3, D_c = 1.604g/cm^3, μ = 2.346mm^{-1}, $F(000)$ = 1936, R = 0.0328, wR = 0.0736 and Z = 8. Crystal structure of **10b** was shown in Figure 2. The average bond lengths and bond angles of the pyridine ring, the phenyl ring and the pyrazole ring were normal.[13] The pyrazole and the pyridine rings make a dihedral angle of 58.3(3)°. The torsion angle of C(17) - N(5) - O(2) - C(18) indicated that the C = N double bond was in the (Z) - configuration.

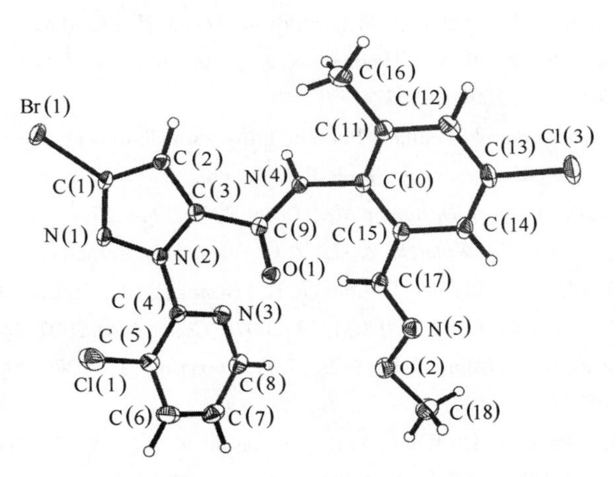

Figure 2 Crystal structure of **10b**

Insecticidal activity

The insecticidal activity of compounds **5** – **11** against oriental armyworm was listed in Table 1. Compound **5** and its acetylated product **6** showed the same activity, while compounds **10a** and **10b** containing oxime groups had the similar trend. The activity of compounds **9a** – **9c** and **8d** with methylene amines substituted indicated that the change of substituent affected the activity with the trend $CH(CH_3)_2 > C(CH_3)_3 > CH_3 >$ two – methyl substituted, and had a decrease in activity compared with their amides analogues. In addition, the thiazolidines **11** did not lead to higher activity compared to its precursor compound **7**.

Table 1 Insecticidal activities of compounds 5 – 11 and chlorantraniliprole against oriental armyworm

Comp.	Larvicidal activity/%			
	Concentration of/mg · L^{-1}			
	200	100	50	20
5	100	40		
6	100	40		
7	100	20		
8d	50			
9a	80			
9b	100	100	100	60
9c	100	100	100	40
10a	100	40		
10b	100	60		
11	100	10		
Chlorantraniliprole				100

References and note

[1] Lahm, G. P. ; Stevenson, T. M. ; Selby, T. P. ; Freudenberger, J. H. ; Cordova, D. ; Flexner, L. ; Bellin, C. A. ; Dubas, C. M. ; Smith, B. K. ; Hughes, K. A. ; Hollingshaus, J. G. ; Clark, C. E. ; Benner, E. A. *Bioorg. Med. Chem. Lett.* 2007, 17, 6274.

[2] DuPont Rynaxypyr_ Insect Control Technical Bulletin: http://www2. dupont. com/Production_Agriculture/en_US/assets/downloads/pdfs/Rynaxypyr_Tech_Bulletin. pdf.

[3] Lahm, G. P. ; Cordova, D. ; Barry, J. D. *Bioorg. Med. Chem.* 2009, 17, 4127.

[4] Feng, Q. ; Liu, Z. L. ; Xiong, L. X. ; Wang, M. Z. ; Li, Y. Q. ; Li, Z. M. *J. Agric. Food Chem.* 2010, 58, 12327.

[5] Kruger, B. ; Hense, A. ; Alig, B. ; Fischer, R. ; Funke, C. ; Gesing, E. R. ; Malsam, O. ; Drewes, M. W. ; Arnold, C. ; Lummen, P. ; Sanwald, E. *WO* 2007031213, 2007 [*Chem. Abstr.* 2007, 146, 358885].

[6] Clark, D. A. ; Finkelstein, B. L. ; Lahm, G. P. ; Selby, T. P. ; Stevenson, T. M. *WO* 2003016304, 2003 [*Chem. Abstr.* 2003, 138, 205057].

[7] Liu, A. P. ; Hu, Z. B. ; Wang, Y. J. ; Wang, X. G. ; Huang, M. Z. ; Ou, X. M. ; Mao, C. H. ; Pang, H. L. ; Huang, L. *CN* 101337959, 2009 [*Chem. Abstr.* 2009, 150, 168342].

[8] Muehlebach, M. ; Craig, G. W. *WO* 2008064891, 2008 [*Chem. Abstr.* 2008, 149, 10005].

[9] Zhang, X. N. ; Zhu, H. J. ; Tan, H. J. ; Li, Y. H. ; Ni, J. P. ; Shi, J. J. ; Zeng, X. ; Liu, L. ; Zhang, Y. N. ; Zhou, Y. L. ; He, H. B. ; Feng, H. M. ; Wang, N. *CN* 101845043, 2010 [*Chem. Abstr.* 2010, 153, 530554].

[10] Ni, J. P. ; Zhang, Y. N. ; Zhang, X. N. ; Tan, H. J. ; Zeng, X. *Modern Agrochem.* 2010, 9, 21 (in Chinese).

[11] Dong, W. L. ; Xu, J. Y. ; Xiong, L. X. ; Liu, X. H. ; Li, Z. M. *Chin. J. Chem.* 2009, 27, 579.

[12] Supplementary material CCDC – 799316 contains the supplementary crystallographic data for this compound. These data can be obtained free of charge at www. ccdc. cam. ac. uk/conts/retrieving or from the Cambridge Crystallographic Data Centre, 12 Union Road, Cambridge CB 21EZ, UK; fax: +44 – 1223 – 336033; email: deposit@ ccdc. cam. ac. uk.

[13] Liu, X. F. ; Liu, X. H. *Acta Cryst.* 2011, E67, O202.

Synthesis, Crystal Structure, Bioactivity and DFT Calculation of New Oxime Ester Derivatives Containing Cyclopropane Moiety*

Xinghai Liu[1], Li Pan[2], Chengxia Tan[1], Jianquan Weng[1], Baolei Wang[2], Zhengming Li[2]

([1]College of Chemical Engineering and Materials Science, Zhejiang University of Technology, Hangzhou, 310014, China; [2]State - Key Laboratory of Elemento - Organic Chemistry, National Pesticidal Engineering Centre (Tianjin), Nankai University, Tianjin, 300071, China)

Abstract A new series of oxime esters containing cyclopropane moiety were designed and synthesized. Their structures were confirmed by ^1H NMR, MS and elemental analysis. The single crystal structure of compound **7b** was determined to further elucidate the structure. The KARI activity indicated that compound **7k** exhibits favorable inhibition rate; the herbicidal assay showed that most of them have moderate activity against *Echinochloa frumentacea*, some of which have moderate activity against *Brassica campestris*.

Key words cyclopropane; oxime ester; transit structure; KARI activity; herbicidal activity

1 Introduction

Ketol - acid reductoisomerase (KARI; EC 1. 1. 1. 86) is an attractive target for agro - chemical and medicinal discovery because it catalyzes the second important step in the biosynthesis of the branched chain amino acid[1]. The KARI only exists in microorganisms and plants, not in mammals. Thus it is an ideal target from which to design non - toxic KARI - inhibitors as potential novel drugs.

In the process of biosynthesis of the branched chain amino acid, some commercial compounds, such as sulfonylureas[2], imidazolinones[3], sulfonamide[4], and pyrimidinylbenzoic acids[5], which inhibit the first enzyme as acetohydroxyacid synthase (AHAS). The success of these inhibitors has stimulated the research into inhibitors of other enzymes in the pathway. The reaction catalyzed by KARI is shown in Fig. 1 which consists of two steps[6]. Both steps require a divalent metal ion, such as Mg^{2+}, Mn^{2+} or Co^{2+}, but the alkyl migration is highly specific for Mg^{2+} interaction.

Until now, no commercial herbicides and drugs targeting KARI have been developed, only HOE 704[7], IpOHA[8], 1, 2, 3 - thiadiazoles[9] and CPD derivatives [10-13] were shown to be potential inhibitors targeting KARI (Fig. 2).

Herein, a series of oxime ester derivatives were reported. Some of these compounds had herbicidal and KARI activity.

---

* Reprinted from *Pesticide Biochemistry and physiology*, 2011, 101: 143 - 147. This work was funded by Natural Science Foundation of China (No. 21002090) and Scientific Research Fund of Zhejiang Education Department (Y201018479).

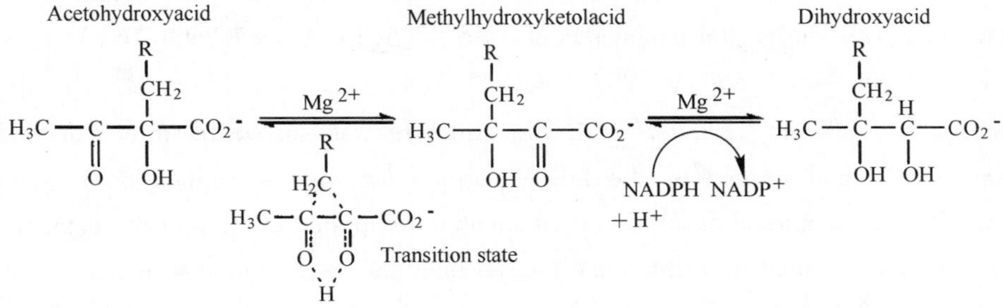

Acetohydroxyacid

Methylhydroxyketolacid

Dihydroxyacid

Transition state

Fig. 1 Reaction catalyzed by KARI

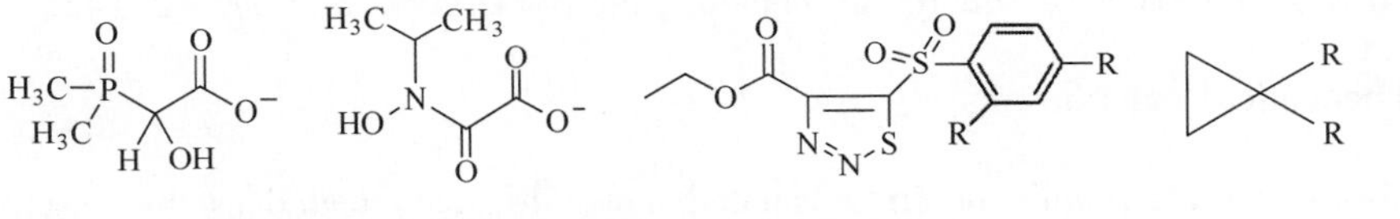

Hoe 704 IpOHA Thiadiazoles CPD

Fig. 2 The structures of inhibitors targeting KARI

2 Experiments

2.1 Materials and methods

All reagents are analytical grade. Melting points were determined using an X－4 apparatus and were uncorrected. ^1H NMR spectra were measured on a Bruker AV400 instrument(400MHz)using TMS as an internal standard and DMSO－d_6 as solvent. Mass spectra were recorded on a Thermo Finnigan LCQ Advantage LC/mass detector instrument. Elemental analyses were performed on a Vario EL elemental analyzer. Crystallographic data of the compound were collected on a Rigaku MM－07 Saturn 724 CCD diffractometer.

2.2 Synthesis

2.2.1 *General procedure: preparation of* 7a

The acid chloride and oximes were prepared according to the references[14,15]. Dropwised the acid chloride to substituted oximes (7.50mmol in 25mL THF) and 7.5mmol Et_3N, then vigorously stirred at ambient temperature for overnight. The corresponding products **7** precipitated immediately. The product was washed with saturated $NaHCO_3$, water, and then CH_2Cl_2 (30mL) was added and the organic layer was separated, dried, and the solvent was then evaporated *in vacuo* to afford crude compound, recrystallized from ethyl acetate－petroleum ether to give the title compounds 7.

2.3 Crystal structure determination

Compound **7b** was dissolved in hot alcohol and the resulting solution was allowed to stand in air at room temperature to give single crystal of **7b**. The crystal of title compound with dimensions of 0.18mm 0.16mm ×0.12mm was mounted on a Rigaku MM－07 Saturn 724 CCD diffractometer with a graphite－monochromated Mo$K\alpha$ radiation(λ =0.71073Å) by using a phi and scan modes at 294(2) K in the range of 1.96° ≤ θ ≤ 25.00°. The crystal belongs to Monoclinic system with space group P2_1/n and crystal parameters of a = 6.4705(13) Å b = 7.7509(16) Å, c = 20.877 (4) Å, α = 90°, β = 94.67(3)°, γ = 90°, V = 1043.5(4) A^3, D_c = 1.294g/cm^3. The absorption coefficient μ = 0.088mm^{-1}, and Z = 4. The structure was solved by direct methods with SHELXS - 97[16] and refined by the full - matrix least squares method on F^2 data using SHELXL - 97. The empirical absorption corrections were applied to all intensity data. H atom of N - H was initially located in a difference Fourier map and were refined with the restraint Uiso (H) = 1.2Ueq(N). Other H atoms were positioned geometrically and refined using a riding model, with d(C···H) = 0.93 - 0.97Å and Uiso(H) = 1.2 Ueq(C) or 1.5Ueq(Cmethyl). The final full - matrix least squares refinement gave R = 0.0502 and wR = 0.1465.

2.4 Theoretical calculations

On the basis of the structure of **7n**, a isolated molecule was selected as the initial structure, while DFT - B3LYP/6 - 31G(d,p)[17] methods in Gaussian 03 package[18] were used to optimize the structure of **7n**. Vibration analysis showed that the optimized structures were in accordance with the minimum points on the potential energy surfaces. All the convergent precisions were the system default values, and all the calculations were carried out on the Nankai Stars supercomputer at Nankai University.

3 KARI assay

3.1 Cloning, expression and purification of rice KARI

The DNA sequence corresponding to mature KARI was amplified by PCR using the oligonucleotide primers 5′ - aaaggatCCATGGTCGCGGCGC - 3′ and 5′ - cccAaaTTtgaagcttCTACG ATGACTGCCGGAG - 3′. In these sequences, lower case represents mismatched bases, underlining indicates the location of introduced *Bam*HI and *Hind*III restriction sites, and italics show the Met - 54 codon or the reverse complement of the TAG stop codon. The PCR product was digested with *Bam*HI and *Hind*III and ligated into the pET - 30a plasmid that had been digested with the same enzymes. The resultant expression plasmid was used to transform *Escherichia coli* BL21(DE3) cells.

A single colony of these cells was inoculated into 20mL of LB medium containing 50mg/mL kanamycin. The culture was incubated overnight at 37℃ and was used to inoculate each of two 500mL volumes of LB medium containing 50mg/mL kanamycin; the cultures were incubated at 37℃ with shaking. When an OD_{600} of 0.8 was reached, expression was induced by adding 0.5mM isopropyl β - *D* - thiogalactoside to each culture; these were then incubated at room temperature (~22℃) for a further 4h with shaking and the cells were harvested by centrifugation.

The frozen cell pellet was thawed, suspended in ice - cold purification buffer [20mM Tris - HCl(pH 7.9)/500mM NaCl] containing 5mM imidazole and then treated with lysozyme (10mg/g of cells for 30min at 0℃). The cells were disrupted by sonication, insoluble material was removed by centrifugation and the supernatant was passed through a 0.45mm filter. The cell extract was applied to a 7mL column of His Bind resin (Novagen) that had been charged by using 50mM $NiSO_4$ then equilibrated with purification buffer containing 5mM imidazole. The loaded column was washed with 23mL of the same buffer, followed by 30mL of purification buffer containing 25mM imidazole, and then KARI was eluted with 30mL of purification buffer containing 400mM imidazole. Fractions containing the enzyme were pooled, concentrated to 2.5mL by ultrafiltration and exchanged into 20mM Na – Hepes buffer, pH 8.0 using a Pharmacia PD – 10 column. The eluate was snapfrozen in liquid nitrogen and stored at –70℃[19].

3.2 Enzyme and protein assays

Gerwick et al.[20] reported that the inhibition of *E. coli* KARI is time – dependent. To characterise the steady – state inhibition constant, *E. coli* KARI was preincubated for 10min with NADPH, Mg^{2+} and the title compound, then the reaction was initiated with hydroxypyruvate. Under these conditions, the change in A_{340} was found to be linear with time.

3.3 Herbicidal activities assay

Inhibition of the root growth of rape(*Brassica campestris*).

The evaluated compounds were dissolved in water and emulsified if necessary. Rape seeds were soaked in distilled water for 4h before being placed on a filter paper in a 6cm Petri plate, to which 2mL of inhibitor solution had been added in advance. Usually, 15 seeds were used on each plate. The plate was placed in a dark room and allowed to germinate for 65 h at 28 ±1℃. The lengths of 10 rape roots selected from each plate were measured, and the means were calculated. The check test was carried out in distilled water only. The percentage of the inhibition was calculated.

3.4 Inhibition of the seedling growth of barnyard grass(*Echinochloa crusgalli*)

The evaluated compounds were dissolved in water and emulsified if necessary. A total of 10 barnyard grass seeds were placed into a 50mL cup covered with a layer of glass beads and a piece of filter paper at the bottom, to which 5mL of inhibitor solution had been added in advance. The cup was placed in a bright room and allowed to germinate for 65 h at 28 ±1℃. The heights of seedlings of aboveground plant parts from each cup were measured, and the means were calculated. The check test was carried out in distilled water only. The percentage of the inhibition was calculated.

4 Results and discussion

4.1 Synthesis and spectroscopy

The synthesis procedures for title compounds were shown in Scheme 1. Cyclopropanecarboxylic acid and oximes was synthesized according the reference. Microwave assistant synthesis method was used in this process. Referring to this method for diethyl cyclopropane – 1, 1 – dicarboxylate was prepared successfully with high yield and short reaction time. Compound **7** were identified by melting point, ^{1}H NMR and MS. The measured elemental analyses were also consistent with the corresponding calculated ones(Tables 1 and 2).

The cyclopropane proton was appeared at δ 1. 18 – 1. 97 as multiplet. The chemical shift the =CH proton of these compounds is around δ 8. 36. The FTIR spectra of oxime **7d** showed bands at 1750cm^{-1} for C =O stretching. The characteristic stretching vibrations ν C =N appears at 1608cm^{-1}. All the title compounds of mass spectra are 2M + Na or M + H peak.

$$\text{diethyl malonate} + ClCH_2CH_2Cl \xrightarrow[K_2CO_3]{(C_4H_9)_4NHSO_4} \text{1 (cyclopropane-}COOC_2H_5, COOC_2H_5)$$

$$\xrightarrow{NaOH} \text{2 (}COONa, COONa) \xrightarrow{HCl/H_2O} \text{3 (}COOH, COOH) \xrightarrow{160℃}$$

$$\text{4 (cyclopropyl-}COOH) \xrightarrow{SOCl_2} \text{5 (cyclopropyl-}COCl) \xrightarrow[Et_3N, THF]{oximes} \text{7}$$

a: R^1=*o*-Meph, R^2=H; **b**: R^1=*p*-Me ph, R^2=H; **c**: R^1=*p*-Br ph, R^2=H; **d**: R^1=*m*-NO_2ph, R^2=H; **e**: R^1=2,4-Cl_2, R^2=H; **f**: R^1=ph, R^2=CN; **g**: R^1=*p*-OMe, R^2=H; **h**: R^1=*o*-NO_2, R^2=H; **i**: R^1=3, 4-$(OMe)_2$, R^2=H; **j**: R^1= Furan, R^2=H; **k**: R^1=*p*-Cl ph, R^2=H

Scheme 1 The synthesitic route of the title compounds

Table 1 The physiochemical data of title compounds

No.	m. p.	Color	Yield(%)	Elemental analysis(% found)		
				C	H	N
7a	61 – 62	White crystal	59. 4	70. 92(70. 82)	6. 45(6. 06)	6. 89(7. 02)
7b	191 – 193	White crystal	63. 4	70. 92(70. 24)	6. 45(6. 82)	6. 89(7. 09)
7c	90 – 91	White crystal	78. 1	49. 28(48. 82)	3. 76(3. 70)	5. 22(50. 32)
7d	86 – 87	White crystal	70. 6	56. 41(56. 51)	4. 30(4. 41)	11. 96(11. 72)
7e	91 – 92	White crystal	66. 7	51. 19(51. 32)	3. 51(3. 46)	5. 43(5. 67)
7f	99 – 100	White crystal	58. 5	67. 28(67. 31)	4. 71(4. 92)	13. 08(13. 00)
7g	74	White crystal	61. 6	65. 74(65. 60)	5. 98(6. 08)	6. 39(6. 55)
7h	62 – 63	White crystal	56. 9	56. 41(56. 30)	4. 30(4. 60)	11. 96(12. 38)
7i	—	Ceraceous solid	65. 8	62. 64(62. 45)	6. 07(5. 89)	5. 62(5. 55)
7j	86 – 87	White crystal	77. 8	60. 33(59. 88)	5. 06(5. 59)	7. 82(8. 02)
7k	89 – 90	White crystal	74. 9	59. 07(58. 55)	4. 51(5. 01)	6. 26(6. 39)

Table 2 The ^{1}H NMR and MS data of compounds 7a – 7k

No.	ESI – MS	^{1}H NMR(400M, $CDCl_3$)
7a	428. 94[2M + Na]$^+$, 204. 01[M + H]$^+$	0. 96 – 1. 00(m, 2H, cycloprane – CH_2), 1. 14 – 1. 18(m, 2H, cycloprane – CH_2), 1. 74 – 1. 80(m, 1H, cyclopropane – CH), 2. 39(s, 3H, CH_3), 7. 22(d, *J* = 7. 94Hz, 2H, Ph—H), 7. 64(d, *J* = 8. 08Hz, 2H, Ph—H), 8. 35(s, 1H, CH)
7b	428. 88[2M + Na]$^+$, 204. 00[M + H]$^+$	1. 02 – 1. 08(m, 2H, cycloprane – CH_2), 1. 13 – 1. 20(m, 2H, cycloprane – CH_2), 1. 74 – 1. 81(m, 1H, cyclopropane – CH), 2. 42(s, 3H, CH_3), 7. 93(d, *J* = 8. 64Hz, 2H, Ph—H), 8. 28(d, *J* = 8. 64Hz, 2H, Ph—H), 8. 47(s, 1H, CH)

Table 2 (continued)

No.	ESI－MS	^1H NMR(400M, $CDCl_3$)
7c	588.70[2M + Na]$^+$, 269.08[M + H]$^+$	0.98 – 1.00 (m, 2H, cycloprane – CH_2), 1.14 – 1.15 (m, 2H, cycloprane – CH_2), 1.72 – 1.77 (m, 1H, cyclopropane – CH), 7.55 (d, J = 8.22Hz, 2H, Ph—H), 7.61 (d, J = 8.22Hz, 2H, Ph—H), 8.34 (s, 1H, CH)
7d	490.69[2M + Na]$^+$, 235.03[M + H]$^+$	1.01 – 1.05 (m, 2H, cycloprane – CH_2), 1.17 – 1.19 (m, 2H, cycloprane – CH_2), 1.74 – 1.81 (m, 1H, cyclopropane – CH), 7.64 – 7.74 (m, 2H, Ph—H), 8.09 (d, J = 7.64Hz, 1H, Ph—H), 8.16 (d, J = 8.02Hz, 1H, Ph—H), 9.00 (s, 1H, CH)
7e	538.67[2M + Na]$^+$, 257.00[M + H]$^+$	0.98 – 1.00 (m, 2H, cycloprane – CH_2), 1.14 – 1.18 (m, 2H, cycloprane – CH_2), 1.73 – 1.79 (m, 1H, cyclopropane – CH), 7.28 (dd, J = 2.03, 1.99Hz, 1H, Ph—H), 7.43 (d, J = 2.08Hz, 1H, Ph—H), 8.03 (d, J = 8.52Hz, 1H, Ph—H), 8.76 (s, 1H, CH)
7f	428.89[2M + H]$^+$, 214.79[M + H]$^+$	1.11 – 1.14 (m, 2H, cycloprane – CH_2), 1.25 – 1.29 (m, 2H, cycloprane – CH_2), 1.88 – 1.93 (m, 1H, cyclopropane – CH), 7.46 – 7.61 (m, 3H, Ph—H), 7.97 (d, J = 7.75Hz, 2H, Ph—H)
7g	438.78[2M + H]$^+$, 220.02[M + H]$^+$	1.01 – 1.08 (m, 2H, cycloprane – CH_2), 1.14 – 1.18 (m, 2H, cycloprane – CH_2), 1.66 – 1.72 (m, 1H, cyclopropane – CH), 7.29 – 7.33 (m, 2H, Ph—H), 7.40 – 7.43 (m, 2H, Ph—H), 8.40 (s, 1H, CH)
7h	252.07[M + H_2O]$^+$, 2235.08[M + H]$^+$	1.00 – 1.04 (m, 2H, cycloprane – CH_2), 1.06 – 1.21 (m, 2H, cycloprane – CH_2), 1.75 – 1.83 (m, 1H, cyclopropane – CH), 7.64 – 7.43 (m, 2H, Ph—H), 8.10 (d, J = 8.11Hz, 2H, Ph—H), 8.170 (d, J = 7.52Hz, 2H, Ph—H), 9.01 (s, 1H, CH)
7i	481.85[2M + H]$^+$	0.96 – 1.01 (m, 2H, cycloprane – CH_2), 1.14 – 1.18 (m, 2H, cycloprane – CH_2), 1.73 – 1.79 (m, 1H, cyclopropane – CH), 6.52 (s, 1H, Furan – H), 6.92 (d, J = 3.43Hz, 1H, Furan – H), 7.58 (s, 1H, Furan – H), 8.26 (s, 1H, CH)
7j	380.77[2M + Na]$^+$, 179.96[M + H]$^+$	0.97 – 1.02 (m, 2H, cycloprane – CH_2), 1.14 – 1.18 (m, 2H, cycloprane – CH_2), 1.73 – 1.79 (m, 1H, cyclopropane – CH), 7.40 (d, J = 8.34Hz, 2H, Ph—H), 7.68 (d, J = 8.35Hz, 2H, Ph—H), 8.36 (s, 1H, CH)
7k	468.73[2M + Na]$^+$	0.99 – 1.06 (m, 2H, cycloprane – CH_2), 1.16 – 1.20 (m, 2H, cycloprane – CH_2), 1.75 – 1.81 (m, 1H, cyclopropane – CH), 7.29 – 7.33 (m, 2H, Ph—H), 7.40 – 7.43 (m, 2H, Ph—H), 8.40 (s, 1H, CH)

4.2 Crystal structure

The structure of compound **7b** was further confirmed by single crystal X – ray diffraction analysis (Fig. 3). From the molecular structure, it can be seen that the dihedral angel between benzene ring and omixe ester bond is 5.4(3)° which indicates the two groups are almost coplanar in the molecular structure; meanwhile, the benzene ring is vertical with cyclopropane ring. The X – ray analysis also reveals that in this typical compound **7b**, the benzene ring and cyclopropane ring are of the opposite sides of the C ═N double bond (Fig. 3). The torsion angle of C(6) – C(5) – N(1) – O(1) is – 174.91(16)°, which indicates that the C ═N double bond is in the(E) – configuration.

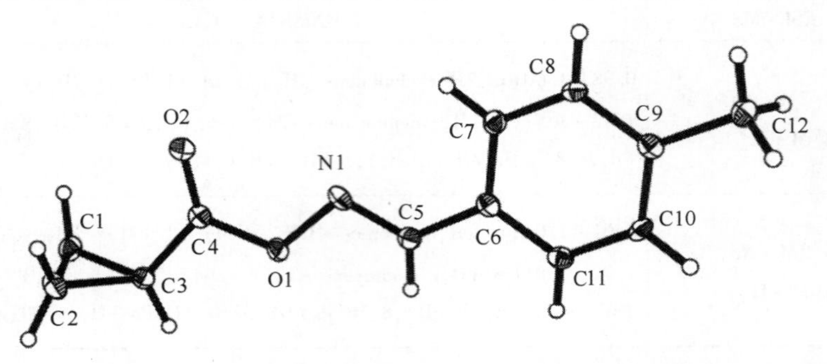

Fig. 3 The crystal structure of compound **7b**

4.3 Biological activity and structure – activity relationship

The herbicidal activities of these compounds were determined *in vivo* and *in vitro*. The results for these compounds **7a – k** are summarized in Table 3. As shown in Table 3, Compound **7k** showed obvious inhibitory activity to the rice KARI enzyme with an inhibition of 73.48%. From the Table 3, some compounds exhibited potential herbicidal activity. For example, compounds possess about 50% inhibitory rates against dicotyledonous(*Brassica napus*) rape and monocotyledon(*Echinochloa crusgalli*). In general, the *in vivo* activity of compounds **7** against B. napus indicated that the change of substituent affects the activity with the trend *p* – Cl Ph > furan > o – CH_3 Ph > two substituted Ph(**7e** and **7i**) > *o* – NO_2 Ph. In addition, the compound with *o* – Me or *m* – NO_2 lead to higher activity against *E. crusgalli*. Their herbicidal activity is higher than that of control CPD. But the many of these compounds revealed that they inhibit ketol – acid reductoisomerase *in vitro* ineffectively. Notably, compound **7k** exhibited good herbicidal activity against rape and barnyard grass and KARI enzyme *in vivo* and *in vitro*.

Table 3 Inhibition rate(%) of compounds 7a – 7o *in vitro* and *in vivo* at 100ppm

No.	KARI	*Echinochloa crusgalli*	*Brassica campestris*
7a	11.97	57.4	62.9
7b	34.65	31.6	32.8
7c	0	48.7	24.7
7d	5.23	48.6	69.2
7e	0	56.7	52.9
7f	—	41.0	59.2
7g	—	53.6	0
7h	3.99	50.2	23.5
7i	0	59.1	0
7j	20.25	72.9	47.1
7k	73.48	73.6	41.1
CPD	100	17.2	27.7

Note: – indicate the compound cannot resolve in our test system, so no data obtain.

4.4 Theoretical calculation

According to the frontier molecular orbital theory, HOMO and LUMO are two most important factors which affect the bioactivities of compounds. HOMO has the priority to provide electrons, while LUMO accept electrons firstly. Thus study on the frontier orbital energy can provide some useful information for the active mechanism. Taking B3LYP results, the HOMO of **7k** is mainly located on phenyl ring, oxime group (Fig. 4**A**). On the other hand, the LUMO of **7k** contains phenyl ring, oxime group and the cyclopropane ring (Fig. 4**B**). The red and green parts represent the cloud density of frontier orbital. The frontier molecular orbital are located on the main groups which atoms can easily bind with the receptor KARI. This implies that the orbital interaction between **7k** and the rice KARI amino acid residues be dominated by hydrophobic interaction between the frontier molecular orbital.

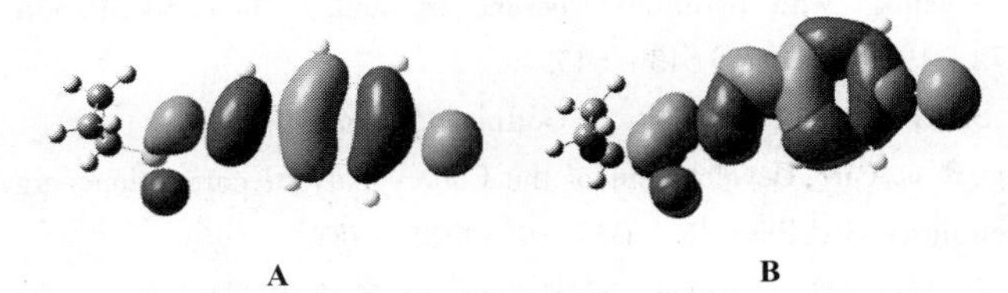

Fig. 4 Frontier molecular orbitals of compound **7b**
(A) HOMO of compound **7b**; (B) LUMO of compound **7b**

References

[1] R. G. Duggleby, S. S. Pang, Acetohydroxyacid synthase, J. Biochem. Mol. Biol. 33 (2000) 1 – 36.

[2] R. S. Chaleff, C. J. Mauvais, Acetolactate synthase is the site of action of two sulfonylurea herbicides in higher plants, Science 224 (1984) 1443 – 1445.

[3] D. L. Shaner, P. C. Anderson, M. A. Stidham, Imidazolinones. Potent inhibitors of acetohydroxy acid synthase, Plant Physiol. 76 (1984) 545 – 546.

[4] S. Hung, Analytical investigation of the process for manufacturing 1, 2, 4 – ^{1}Htriazole, Ind. Chem. 11 (1983) 291 – 299.

[5] S. J. Koo, S. C. Ahn, J. S. Lim, S. H. Chae, J. S. Kim, J. H. Lee, J. H. Cho, Biological activity of the new herbicide LGC – 40863 [benzophenone – [2,6 – bis [(4,6 – dimethoxy – 2 – pyrimidinyl) oxy] benzoyl] oxime], Pestic. Sci. 51 (1997) 109 – 114.

[6] S. K. Chunduru, G. T. Mrachko, K. C. Calvo, Mechanism of ketol acid reductoisomerase. Steady – state analysis and metal ion requirement, Biochemistry 28 (1989) 486 – 493.

[7] A. Schulz, P. Sponemann, H. Kocher, F. Wengenmayer, The herbicidally active experimental compound Hoe 704 is a potent inhibitor of the enzyme acetolactate reductoisomerase, FEBS Lett. 238 (1988) 375 – 378.

[8] A. Aulabaugh, J. V. Schloss, Oxalyl hydroxamates as reaction – intermediate analogs for ketol – acid reductoisomerase, Biochemistry 29 (1990) 2824 – 2840.

[9] F. Halgand, F. Vives, R. Dumas, V. Biou, J. Andersen, J. P. Andrieu, R. Cantegril, J. Gagnon, R. Douce, E. Forest, D. Job, Kinetic and mass spectrometric analyses of the interactions between plant acetohydroxy acid isomeroreductase and thiadiazole derivatives, Biochemistry 37 (1998) 4773 – 4781.

[10] Y. T. Lee, H. T. Ta, R. G. Duggleby, Cyclopropane – 1,1 – dicarboxylate is a slowtight – binding inhibitor of rice ketol – acid reductoisomerase, Plant Sci. 168 (2005) 1035 – 1040.

[11] X. H. Liu, P. Q. Chen, B. L. Wang, Y. H. Li, Y. H. Li, Z. M. Li, Synthesis bioactivity theoretical and mo-

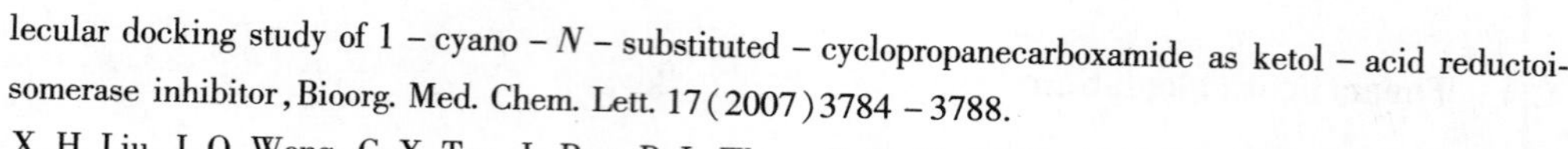
lecular docking study of 1 – cyano – *N* – substituted – cyclopropanecarboxamide as ketol – acid reductoisomerase inhibitor, Bioorg. Med. Chem. Lett. 17(2007)3784 – 3788.

[12] X. H. Liu, J. Q. Weng, C. X. Tan, L. Pan, B. L. Wang, Z. M. Li, Synthesis. Biological activities and DFT calculation of α – aminophosphonate containing cyclopropane moiety, Asian J. Chem. 23 (2011) 4031 – 4036.

[13] X. H. Liu, C. Y. Zhang, W. C. Guo, Y. H. Li, P. Q. Chen, T. Wang, W. L. Dong, B. L. Wang, H. W. Sun, Z. M. Li, Synthesis bioactivity and SAR study of *N'* – (5 – substituted – 1,3,4 – thiadiazol – 2 – yl) – *N* – cyclopropyformyl – thiourea as ketol – acid reductoisomerase inhibitor, J. Enzym. Inhib. Med. Chem. 24(2009)545 – 552.

[14] X. H. Liu, Y. X. Shi, Y. Ma, C. Y. Zhang, W. L. Dong, P. Li, B. L. Wang, B. J. Li, Z. M. Li, Synthesis, antifungal activities and 3D – QSAR study of *N* – (5 – substituted – 1,3,4 – thiadiazol – 2 – yl) cyclopropanecarboxamides, Eur. J. Med. Chem. 44(2009)2782 – 2786.

[15] M. Kawase, Y. Kikugawa, Chemistry of amine – boranes. Part 5. Reduction of oximes, O – acyl – oximes, and O – alkyl – oximes with pyridine – borane in acid. J. Chem. Soc. Perkin Trans. 1: Org. Bio – Org. Chem. (1972 – 1999), (1979)643 – 645.

[16] G. M. Sheldrick, SHELXS – 97, University of Gottingen, Germany, 1997.

[17] C. Lee, W. Wang, R. G. Parr, Development of the Colle – Salvetti correlationenergy formula into a functional of the electron density, Phys. Rev. B37(1988)785 – 789.

[18] M. J. Frisch, G. W. Trucks, H. B. Schlegel, G. E. Scuseria, M. A. Robb, J. R. Cheeseman, J. A. Jr. Montgomery, T. Vreven, K. N. Kudin, J. C. Burant, J. M. Millam, S. S. Iyengar, J. Tomasi, V. Barone, B. Mennucci, M. Cossi, G. Scalmani, N. Rega, G. A. Petersson, H. Nakatsuji, M. Hada, M. Ehara, K. Toyota, R. Fukuda, J. Hasegawa, M. Ishida, T. Nakajima, Y. Honda, O. Kitao, H. Nakai, M. Klene, X. Li, J. E. Knox, H. P. Hratchian, J. B. Cross, C. Adamo, J. Jaramillo, R. Gomperts, R. E. Stratmann, O. Yazyev, A. J. Austin, R. Cammi, C. Pomelli, J. W. Ochterski, P. Y. Ayala, K. Morokuma, G. A. Voth, P. Salvador, J. J. Dannenberg, V. G. Zakrzewski, S. Dapprich, A. D. Daniels, M. C. Strain, O. Farkas, D. K. Malick, A. D. Rabuck, K. Raghavachari, J. B. Foresman, J. V. Ortiz, Q. Cui, A. G. Baboul, S. Clifford, J. Cioslowski, B. B. Stefanov, G. Liu, A. Liashenko, P. Piskorz, I. Komaromi, R. L. Martin, D. J. Fox, T. Keith, M. A. AlLaham, C. Y. Peng, A. Nanayakkara, M. Challacombe, P. M. W. Gill, B. Johnson, W. Chen, M. W. Wong, C. Gonzalez, J. A. Pople, Gaussian 03, Revision C. 01, Gaussian, Inc. , Wallingford CT, 2004.

[19] X. H. Liu, P. Q. Chen, B. L. Wang, W. L. Dong, Y. H. Li, X. Q. Xie, Z. M. Li, High throughput receptor based virtual screening under zinc – database. synthesis and biological evaluation of ketol – acid reductoisomerase inhibitors, Chem. Biol. Drug Des. 75(2010)228 – 232.

[20] B. C. Gerwick, L. C. Mireles, R. J. Eilers, Rapid diagnosis of ALS/AHAS – resistant weeds, Weed Technol. 7(1993)519 – 524.

The Structure - activity Relationship in Herbicidal Monosubstituted Sulfonylureas*

Zhengming Li[1], Yi Ma[1], Luke Guddat[2], Peiquan Chen[1], Jianguo Wang[1], Siew S Pang[2], Yuhui Dong[3], Chengming Lai[1], Lingxiu Wang[1], Guofeng Jia[1], Yonghong Li[1], Suhua Wang[1], Jie Liu[1], Weiguang Zhao[1], Baolei Wang[1]

([1]National Key Laboratory of Elemento - Organic Chemistry, Nankai University, Tianjin, China; [2]School of Molecular and Microbial Sciences, Queensland University, Brisbane, QLD, Australia; [3]Beijing Synchrotron Radiation Facility, Institute of High Energy Physics, China Academy of Sciences, Beijing, China)

Abstract BACKGROUND: The herbicide sulfonylurea (SU) belongs to one of the most important class of herbicides worldwide. It is well known for its ecofriendly, extreme low toxicity towards mammals and ultralow dosage application. The original inventor, G Levitt, set out structure - activity relationship (SAR) guidelines for SU structural design to attain superhigh bioactivity. A new approach to SU molecular design has been developed.
RESULTS: After the analysis of scores of SU products by X - ray diffraction methodology and after greenhouse herbicidal screening of 900 novel SU structures synthesised in the authors' laboratory, it was found that several SU structures containing a monosubstituted pyrimidine moiety retain excellent herbicidal characteristics, which has led to partial revision of the Levitt guidelines.
CONCLUSIONS: Among the novel SU molecules, monosulfuron and monosulfuron - ester have been developed into two new herbicides that have been officially approved for field application and applied in millet and wheat fields in China. A systematic structural study of the new substrate - target complex and the relative mode of action in comparison with conventional SU has been carried out. A new mode of action has been postulated.
Key words sulfonylurea herbicide; acetolactic synthase; acetohydroxyacid synthase; docking; substrate - target enzyme complex; Levitt model; Nankai model

1 INTRODUCTION

The innovative research by Dupont's GLevitt on sulfonylureas (SUs) has been acknowledged as a milestone in herbicidal chemistry owing to the ultralow dosages and ecofriendly characteristics of SUs. Thousands of new SU structures have since been designed and bioassayed, and more than 400 patents have been applied for worldwide (Figs. 1 and 2). Levitt set out the structure - activity relationship (SAR) for superactive SUs as follows:[1]

(a) a guanidine system involving a sulfonylureido bridge;

* Reprinted from Pest Mangement Science, 2012, 68: 618 - 628. Supported by the National Science Foundation of China (Nos 2923010 and 29832050), the Ministry of Science and Technology (No. 2004DFA01500), the National Basic Research Programme of China (Nos 2003CB114400 and 2010CB126106), the Ministry of Education, the Tianjin Science and Technology Commission and the National Key Laboratory of Elemento - Organic Chemistry.

Fig. 1 The general formula according to Levitt's guidelines(X = nitrogen or carbon atom)

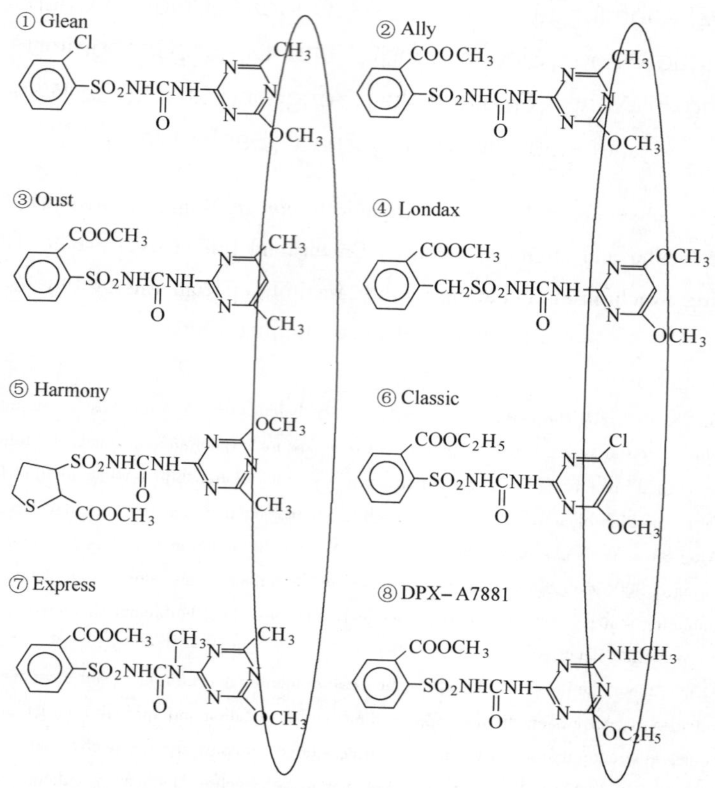

Fig. 2 Dupont's main commercial SU herbicides

(b) substituents at both meta positions to the bridge;

(c) no substituent at the para position to the bridge;

(d) an aromatic heteroatom ring.

Since then all SU innovation research and development has followed Levitt's guidelines, as shown in Figs 3 and 4, where all compounds have two active groups substituted at the heterocyclicring within the bioactive molecule (Figs. 3 and 4).

It is of great interest to investigate further the structure – activity relationship in sulfonylurea chemistry, which is critical to the pursuit of the development of ecofriendly herbicides.

2 MATERIALS AND METHODS

2.1 Structural identification

Infrared spectra were recorded on a Nicolet MAYNA – 560 spectrophotometer as KBr tablets. ^{1}H nuclear magnetic resonance (NMR) spectra were measured on a Bruker AC – P500 in-

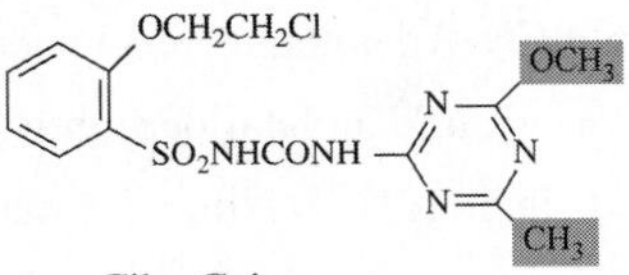

Ciba-Geigy

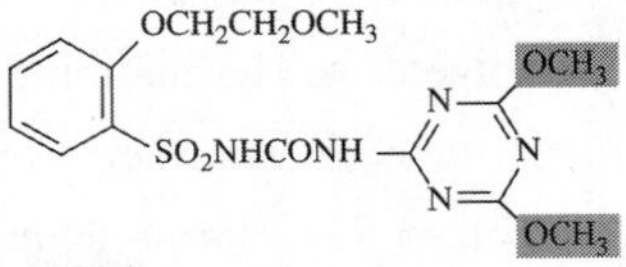

Ciba-Geigy

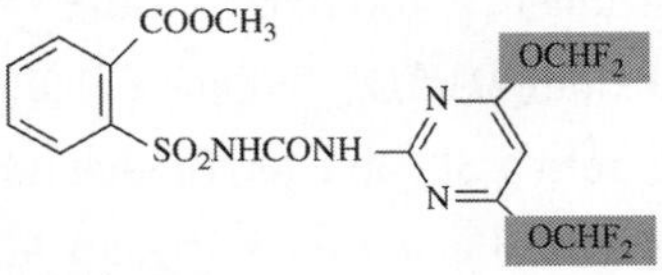

iba-Geigy

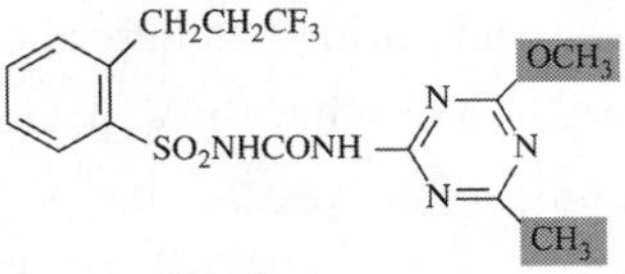

Ciba-Geigy

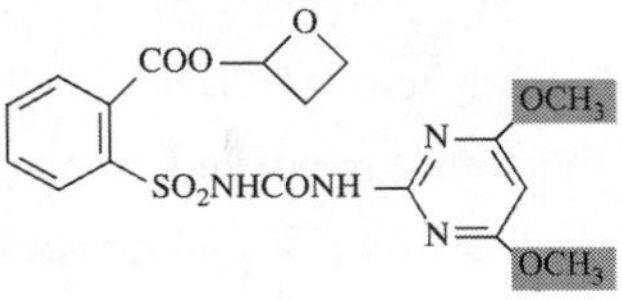

Norvatis

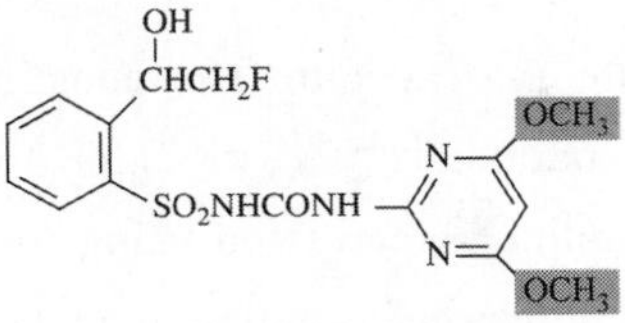

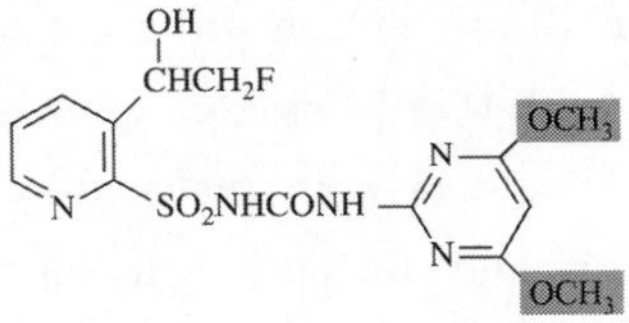

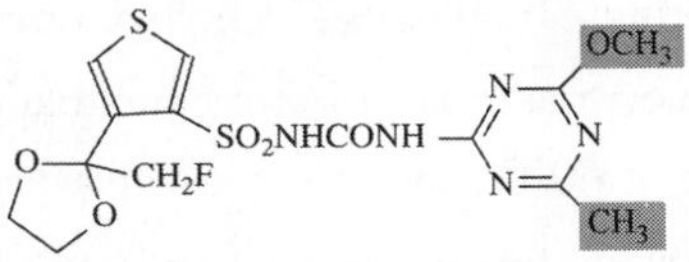

Korean Chemical Institute

Fig. 3 New SUs discovered in Switzerland and South Korea

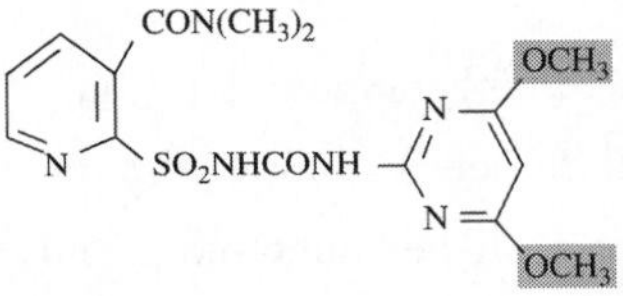

Ishihar Sangyo Kaisha Ltd

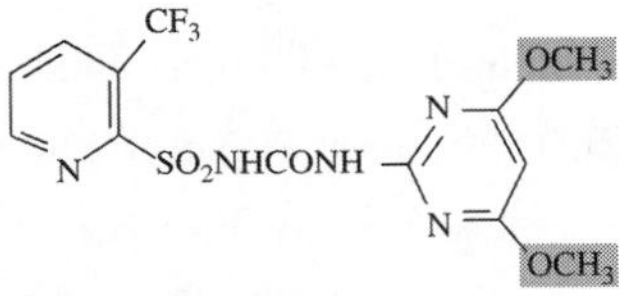

Ishihar Sangyo Kaisha Ltc

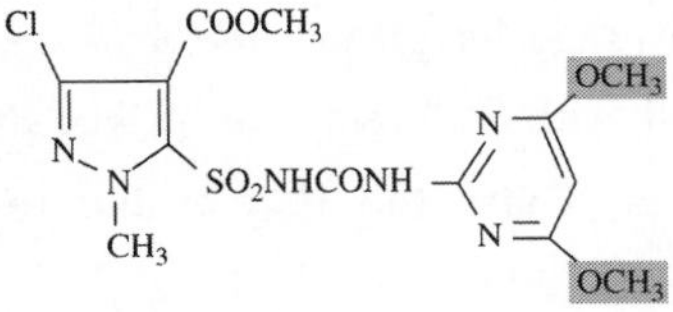

Nippon Kayaku Co

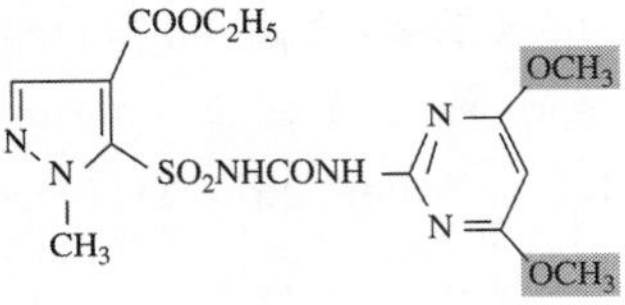

Nippon Kayaku Co

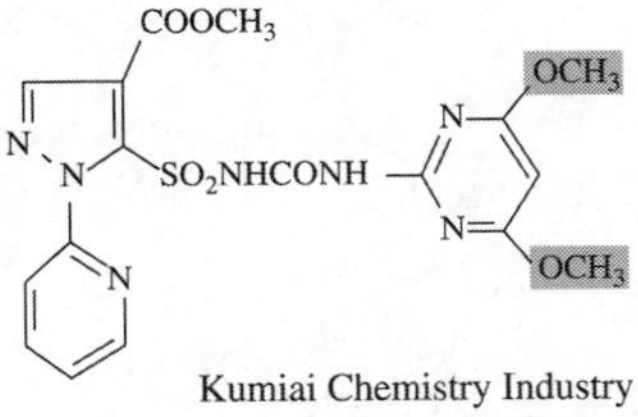

Kumiai Chemistry Industry Co

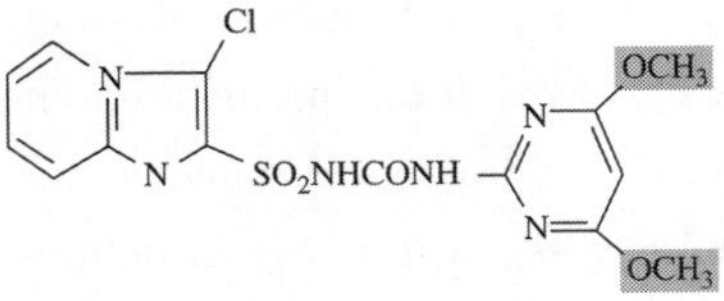

Tajeda Chemical Industries Ltd

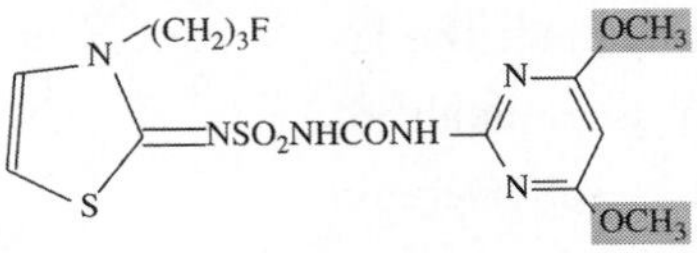

Nawakami Co

Fig. 4 New SUs discovered in Japan

strument (400MHz), using tetramethylsilane (TMS) as the internal standard and dimethylsulfoxide (DMSO) – d_6 or $CDCl_3$ as the solvent. A Thermofinnigan LCQ Advantage mass spectrograph (ESI) and an ElementarVario EL type Ⅲ analyser were used, and crystallographic data were collected on a Rigaku MM – 07 Saturn 724 charge coupled device (CCD) diffractometer.

2. 1. 1 *Biological assessment in vitro*

The plant acetohydroxyacid synthase (AHAS) was expressed and purified as described previously.[2] AHAS activity was measured by using the colorimetric assay in 50mM potassium phosphate (pH 7.0) containing 50mM pyruvate, 1mM thiamine diphosphate, 10mM $MgCl_2$ and 10μM FAD. The test compounds were made into emulsions to aid dissolution. The mixture was incubated at 30℃ for 30min, the reaction was stopped with 25μL of 10% H_2SO_4, and then heating at 60℃ was carried out for 15min to convert acetolactate into acetoin. The acetoin formed was quantified by incubation with 0.5% creatine and α – naphthol (5%, w/v) for 15 min at 60℃, and A_{525} was measured. The data were analysed by nonlinear regression using the following equation to estimate the values and standard errors for the apparent inhibition constant (K_i^{app}) and the uninhibited rate (V_0):

$$V = \frac{V_0}{1 + [I]/K_i^{app}}$$

2. 1. 2 *Biological assessment in vivo*

Solutions were prepared to dilute the stock solutions of test compounds (30mg · mL^{-1}, dissolved in DMSO) with 0.1% Tween 80 solutions. Groups of 25 seeds of rape (*Brassica campestris*) were placed on two pieces of 5.6cm filter paper in 6cm petri dishes containing 3mL of compound solutions. An equal volume of distilled water was used as control. Petri dishes were placed in darkness at 28 ± 1℃ for 45h. The radicle length of seedlings was measured. All experiments had two replications. The inhibition percentage of average length to control wasused to describe the activity of compounds. The data were calculated with DPS software (Data Processing System, v. 8.01; Refine Information Tech. Co., Ltd) to obtain the IC_{50} of the respective compounds.

3 RESULTS

Since the early 1990s, the authors' laboratory has carried out basic research on sulfonylurea chemistry. A systematic study of novel mono – and disubstituted heterocycles within SU structures was carried out to determine the structure – activity relationship (SAR). A series of X – ray diffraction spectra of bioactive structures was studied. The laboratory was the first to report that there is an inner hydrogen bond within Dupont's sulfometuron molecule[3] (Fig. 5).

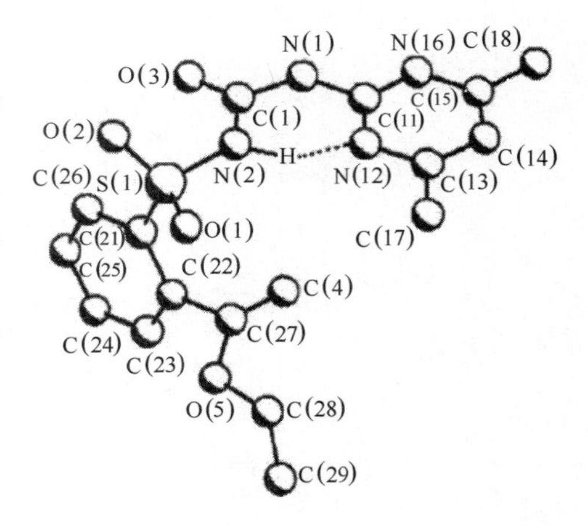

Fig. 5 X – ray diffraction graph of sulfometuron

The X – ray data showed this inner hydrogen bond between N_2 and N_{12}. The IR spectra with a broad band at 3395cm^{-1} and a shoulder at 3265cm^{-1} confirmed this inner hydrogen bond between the H linked to N(2) and N(12) in the pyrimidine ring, within an atomic distance of 0.186nm, which favourably forms a six – member ring. All bonds between C and N atoms are shorter than the normal single C – N bond (0.137nm) and longer than the C ═N double bond (0.127nm). The S – N bond (0.165nm) is also shorter than the normal value 0.178nm. This factor keeps the sulfonylureido – pyrimidine molecule in a planar configuration.

It was confirmed that the new bioactive SUs also have this type of inner hydrogen bond configuration (Fig. 6).

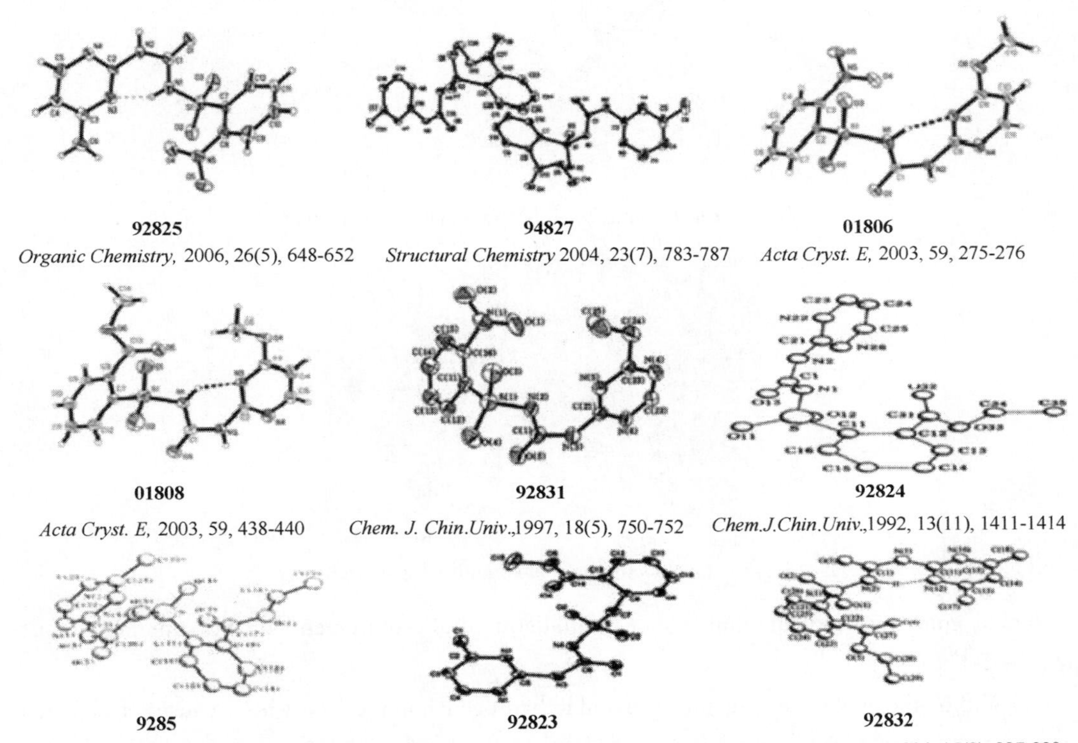

Fig. 6 X – ray diffraction graphs of novel monosubstituted SUs

It was somewhat surprising to discover that, among the 900 novel SUs synthesised in the authors' laboratory, there were five structures (92825, 9285, 94827, 01806 and 01808), containing a peculiar monosubstituent on the pyrimidine ring, that exhibited superactive herbicidal properties. Quantum mechanics calculations showed the distribution pattern of electrostatic potentials in these structures[4,5] (Fig. 7). CoMFA was applied to obtain a 3D contour map:[6] the red part denotes a negative charge and the blue part denotes a positive charge (Fig. 8a); the yellow part denotes spatial limitation and the green part denotes spatial allowance for a larger substituent (Fig. 8b).

The substitution pattern in these new SU herbicides contradicts one of Levitt's guidelines: "to attain superactivity, BOTH substituents should be meta to the guanidine bridge". Data from the present authors' laboratory has shown clearly that Levitt's guidelines do work when the heterocycle is a triazine heterocycle, but do not work when the SU contains a pyrimidine ring.[7]

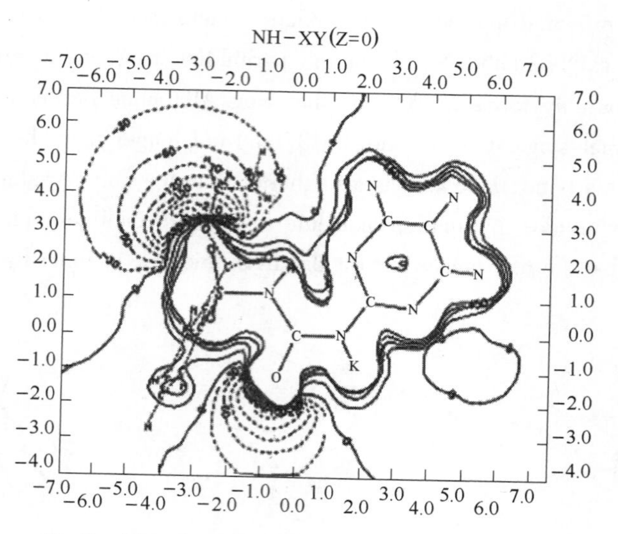

Fig. 7 Electrostatic potential(kJ · mol^{-1}) calculation

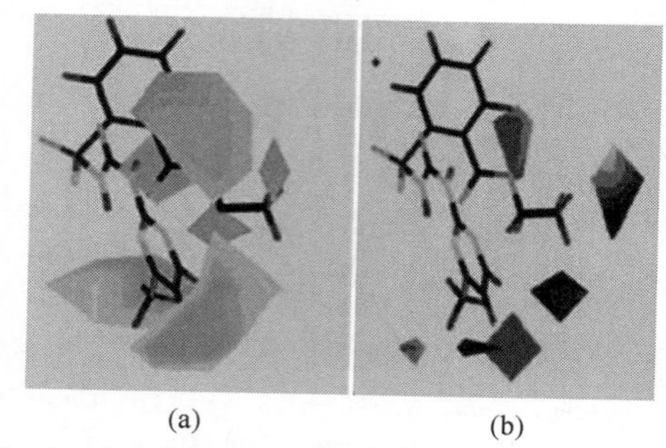

(a) (b)

Fig. 8 (a) Electrostatic effect and (b) spatial effect

Nankai guidelines indicate that three essential structural requirements for SU superactivity are required:[8]

(a) Within the molecules, an intramolecular hydrogen bond exists, which causes a coplanar conjugated system between the heterocycle and ureido moiety.

(b) Carbonyl oxygen, sulfonyl oxygen and heterocyclic nitrogen collectively form a three - negative - centre system.

(c) An electron - withdrawing substituent in an ortho postion to the sulfonylureido chain is crucial for bioactivity.

The ortho substituent as a critical factor for bioactivity, which regrettably was not mentioned in Levitt's SAR guidelines, should be emphasised (in a private discussion with Dr George Levitt, this viewpoint was mutually agreed with the present authors).

A caliper model was also proposed for bioactive SUs[9] (Fig. 9).

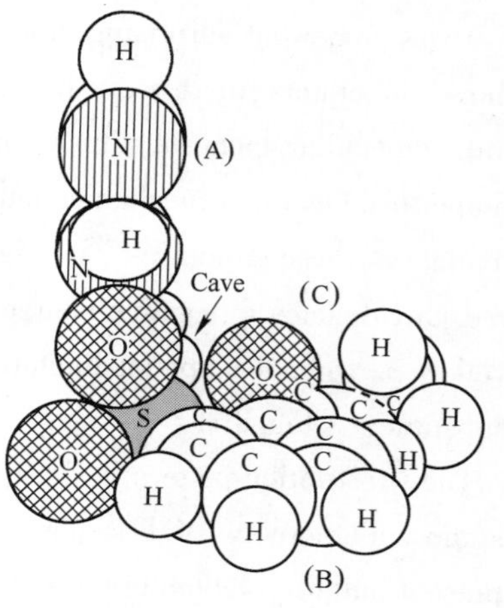

Fig. 9 A proposed caliper model

The Levitt model and Nankai model are schematically described as follows (Fig. 10). From the above - mentioned compounds, five monosubsituted SU candidates [NK 9285,

92825,94827,01806 and 01808(Fig. 6)],monousulfuron(92825,CA registry number 155860 – 63 – 2) and monosulfuron – ester(94827,CA registry number 175076 – 90 – 1)(Figs. 11 and 12). Exhibited excellent herbicidal performance in glasshouse tests and were patented for further development. [10] Following successive field tests and acute and chronic toxicological [acute toxicology data: LD_{50} > 4640mg · kg^{-1}(92825); >10000 mg · kg^{-1}(94827)] and environmental evaluation, monosulfuron was granted a first state registration license(PD – 20070369) as a novel herbicide in China. It has become the exclusive herbicide to be applied in millet fields. Monosulfuron – ester has also been granted a state registration temporary license(LS20041087) to be applied in wheat fields. The state certificates for monosulfuron are: LS – 991690(China Ministry of Agriculture), HNP12049 – C1876(Production License from China State Planning and Development Commission) and Q/12NY0218 – 2002(China State Bureau of Standards); those for monosulfuronester are: LS – 20041087, HNP12049 – C2685 and Q/12NY0394 – 2008. As novel patented herbicides, they have been applied in fields covering over 150000 hectares in northern China.

M, R1, $SO_2NHCONH$, N, X, N, R2 — Levitt's model

M, CH_3, $SO_2NHCONH$, N, N — Nankai model

Fig. 10 Comparison between the two SU templates

Fig. 11 Monosulfuron(92825)

The target enzyme of SU herbicides is acetolactate synthase (ALS), also known as acetohydroxyacid synthase (AHAS, EC 4.1.3.18). [11] It occurs in microorganisms and higher plants, but not in mammals, and catalyses the first step of three crucial branched chain amino acid biosyntheses. These amino acids cannot be synthesised in animals, and the strategy to design new herbicides that inhibit ALS has the advantage of their certainly not affecting mammals from a toxicological viewpoint. Schloss *et al.* [12] were the first to report that the target enzyme is the ALS enzyme. Regrettably, all early attempts to elucidate the structure of the target enzyme were unsuccessful. In 2002, Duggleby's group[13] was the first to report the absolute structure of the target enzyme ALS(AHAS) purified from the yeast *Saccharomyces cerevisiae*, Meyen ex EC Hansen(Fig. 13).

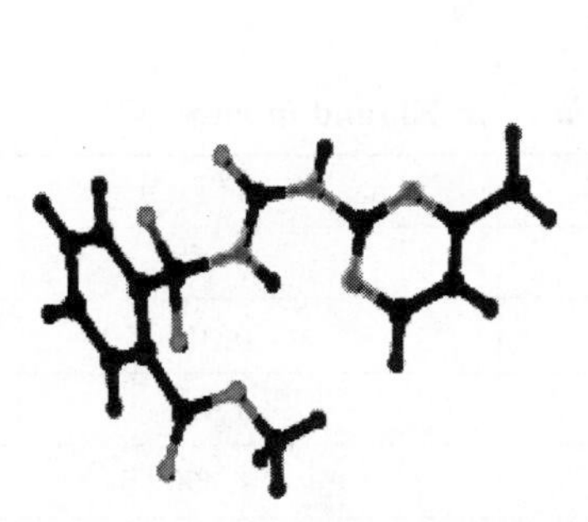

Fig. 12 Monosulfuron – ester(94827)

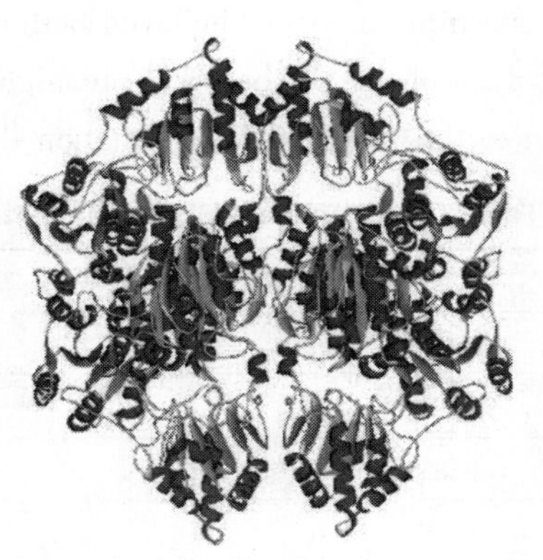

Fig. 13 AHAS(ALS) composed of four subunits[13]

By systemic bioscreening in both *in vivo* and in vitro tests it was confirmed that the five novel SU structures discovered in the present authors' laboratory have shown superactivity against weeds in glasshouse testsin *vivo* as well as towards ALS enzyme [from *Aradibopsis thaliana* (L Heynh)] *in vitro* (Figs. 14 and 15).

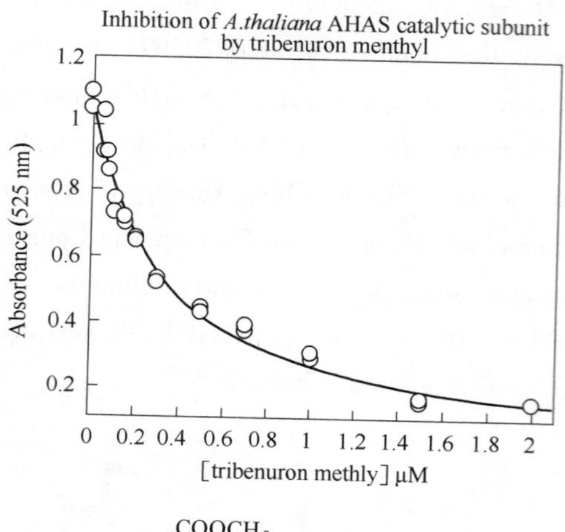

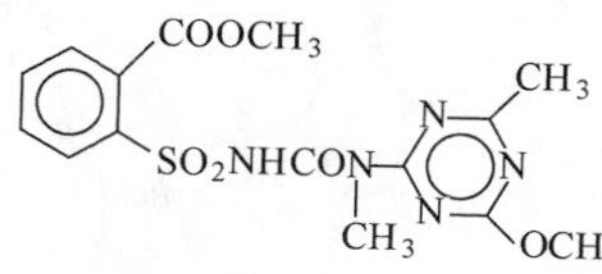

Simple inhibition curve
Simple weighting
reduced Chi squared = 0. 001696

Variable	Value	Std. Err.
Uninhibited rate	1. 0686	0. 0214
Ki(app)	0. 3162	0. 0210

Fig. 14 Tribenuron inhibition, Ki, on AHAS(ALS)

Inhibition of *A thaliana* AHAS catalytic subunit by NK92825

Absorbance (525 nm)

[NK92825] μM

simple inhibition curve
simple weighting
reduced Chi squared = 0. 0005348

Variable	Value	Std. Err.
Uninhibited rate	1. 3979	0. 0129
Ki(app)	0. 2453	0. 0071

Fig. 15 Monosulfuron inhibition, Ki, on AHAS(ALS)

Table 1 shows that the bioactivity *in vitro* and *in vivo* are in parallel. Both monosulfuron (92825) and monosulfuron - ester (92847) showed good inhibition of ALS *in vitro*. [14] A high inhibition rate in the case of chlorimuron *in vitro* was also noted, yet in *in vivo* tests monosulfuron and monosulfuron - ester behaved better. Greenhouse tests involved absorption, translocation and metabolism of the SU herbicide in higher plants, which usually indicated that *in vivo* tests were closer to the practical field situation than *in vitro* tests.

Table 1 A summary of the evaluation of novel SUs *in vitro* (Ki) and *in vivo* (IC)

	Ki_{50} ① (μM)	IC_{50} ② (μM)
92825③	0. 2453	0. 489
94827④	0. 3626	0. 315
9285	0. 2661	0. 522
01806	0. 3447	1. 490
01808	0. 5602	0. 845

Table 1 (continued)

	Ki_{50}① (μM)	IC_{50}② (μM)
Tribenuron	0.3612	0.807
Chlorsulfuron	0.014	0.787

① Ki_{50} is the concentration of SU that inhibits 50% of the target enzyme AHAS.

② IC_{50} is the concentration of SU that inhibits 50% on dicotelydon rape.

③ 92825 is the code for monosulfuron.

④ 94827 is the code for monosulfuron – ester.

As monosulfuron and monosulfuron – ester did not comply structurally with the conventional SU structures, it was of interest to study their mode of action. With reference to Duggleby et al., by the hanging – drop diffusion method,[15] monosulfuron was successfully docked onto AHAS (from *A. Thaliana*) to obtain a substrate – enzyme complex in crystal form[16] (Fig. 16).

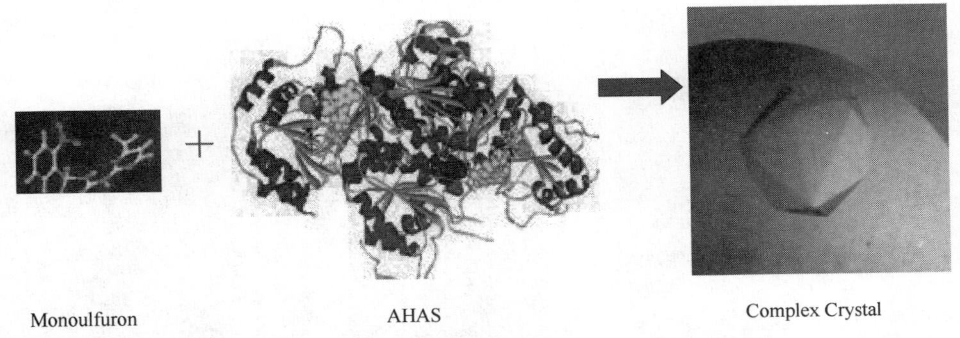

Fig. 16 Docking of monosulfuron on AHAS (the same results were obtained with monosulfuron – ester)

The crystalline complexes have been annotated by the International PDB and PCSB (Research Collaboratory for Structural Bio – Informatics) Office with new ID numbers PDB 3E9Y and 3EA4. The successful docking of the two new SUs on *Arabidopsis thaliana* acetohydroyacid synthase (AHAS) and their complex crystals have been given the following ID code numbers: (a) monosulfuron – AHAS complex conferred: RCSB rcsb049060, PDB 3E9Y; (b) monosulfuron – ester – AHAS complex conferred: RCSB rcsb049066, PDB 3EA4. The crystalline complexes were sent to Advance Photon Source at the Argonne National Laboratory, Chicago University, Chicago, Illinois, for study by the application of the DENZO and SCALEPACK programs. From the data collected from synchrotron radiation (BioCARS) at 2.8Å the crystalline monosulfuron – AHAS complex was further analysed, from which it was elucidated that the complex includes 18720 atoms (no hydrogen atom showed up) divided by four subunits, while each subunit could be divided into 582 amino acids. The monosulfuron – ester – AHAS complex gave similar information. These experimental data provided reliable information for the further study of the interaction between related substrate and target enzyme.

Recently, Duggleby's group was the first to describe the chlorimuron molecule at the site of AHAS[15,17] (Fig. 17).

By means of software O[18] a schematic picture of the substrate – receptor at the site of action of monosulfuron was depicted. For convenience in further discussion, the bioactive SU structure was artificially separated into three parts: the benzene ring; sulfonylureido; the heterocycle. Duggleby's group[17] provided valuable information concerning the site of action relating to the benzene and sulfonylureido parts, but regrettably the crucial heterocyclic moiety was not described. Firstly, at the site of action of monosulfuron, it was noted that amino acid residues surrounding the benzene moiety and sulfonylureido moiety were identical to Dupont's chlorimuron, with common residues W574, P197′, D376, S653, K256 and S168′ (the substrate is considered to intrude into the crevice between the two subunits of the ALS enzyme),[19] which validated the present approach and methodology. The site of action was then viewed from the opposite side, which made it possible to explore the pyrimidine domain from the reverse direction (Fig. 18). It was clear that the pyrimidine surrounding amino acids M570, W574, S653, F206′, F207, G121′ and V571 were in an entirely different pattern from the front view.[20] It was very interesting to explore further the difference, if any, between the two types of SU in the substrate – enzyme environment.

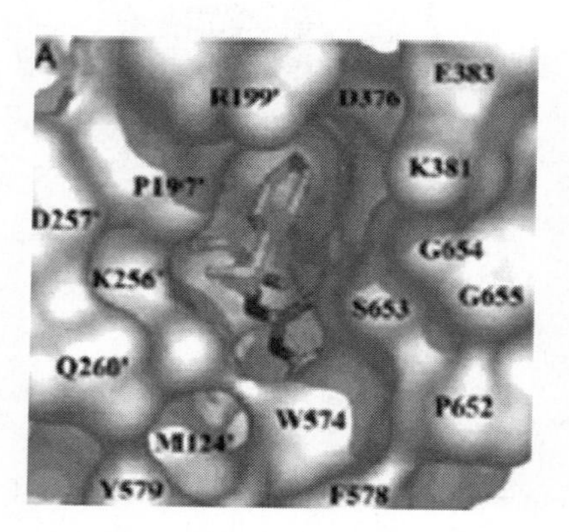

Fig. 17 Chlorimuron interacts with AHAS[17]

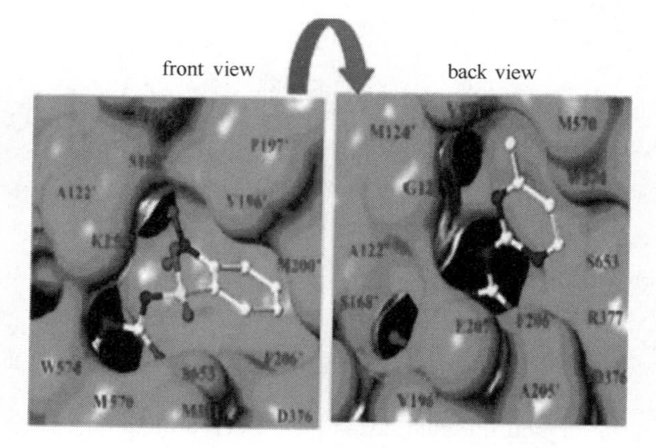

Fig. 18 Monosulfuron at the site of action: benzene moiety (front view), pyrimidine moeity (back view)

Monosulfuron was compared with chlorimuron as a conventional SU within the pyrimidine domain, as both types of SU are structurally similar (Fig. 19).

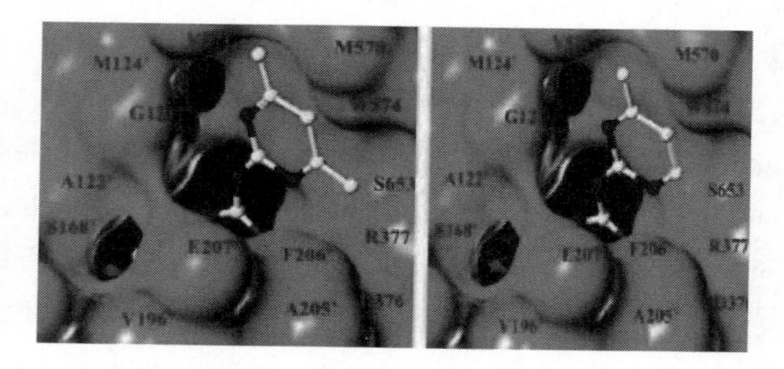

Fig. 19 Conventional SU (left) and monosulfuron (right) at the site of action (pyrimidine moiety environment)

Another point of interest was to compare monosulfuron (92825) with another non – substituted pyrimidine – containing SU (code number 92826); the latter was synthesised specially for the purpose of comparison (Fig. 20). The results indicated that the unsubstituted 92826 did not interact with the amino residues W574, G121′, V196 and V571 and fell way out of the particular site of action (Fig. 20).

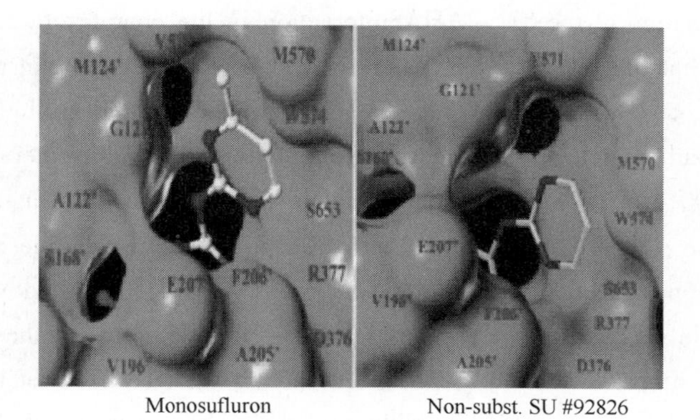

Fig. 20　Monosulfuron 92826 and a mimic that has no substituted group on the pyrimidine ring

When the similar herbicidal potency between monosulfuron and conventional SU was considered, it was postulated that, during the susbstrate – receptor docking interaction, there might exist two cavities (Fig. 20) surrounding the pyrimidine moiety. The 'smaller cavity' seemed to be more important, while the other 'larger cavity' accommodated the second substituent on the pyrimidine ring with supplementary interaction. This might partly explain why some of the present monosubstituted SUs exhibited superactive herbicidal activity similar to that of conventional SUs (Figs. 21 and 22).

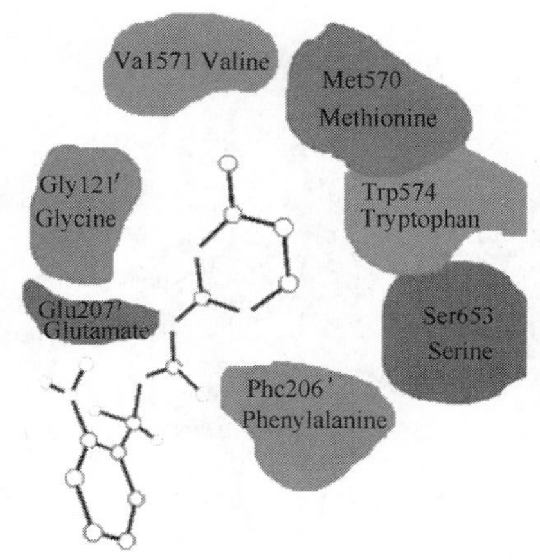

Fig. 21　A schematic graph depicting the environ – ment around the monosubstituted pyrimidine moiety, which is different from the benzene environment

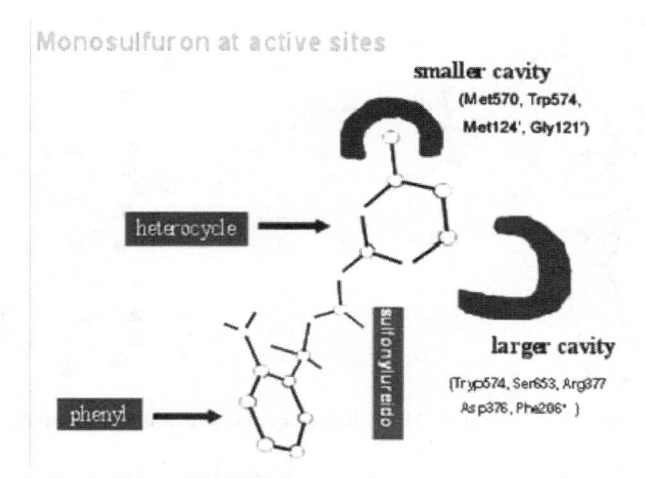

Fig. 22　Schematic display of the environment of the two substituents

There are usually four kinds of non – covalent force involved in the substrate – enzyme interaction, which need to be considered as ionic, hydrogen bond, VDW force and $\pi - \pi$ interaction. Further examination of the SU – AHASinteraction in the complex by free energy calculation demonstrated that the disubstituted SU showed lower free energy than monosubstituted SU ($-46 kcal \cdot mol^{-2} < -34 kcal \cdot mol^{-2}$). From conformational analysis by Logplot,[22] it was observed that in disubstituted SUs the benzene ring had a hydrophobic interaction with Arg377, Pro197′ and Met200′. The two nitrogen atoms in the guanidine moiety in the arginine residue coordinated with the α – carbonyl group in both SUs, while the pyrimidine ring, forming a fundamental coplanar configuration with the sulfonylureido chain, interacted with a hydrophobic force with the indole ring within Trp574 through a $\pi - \pi$ interaction. In the case of monosulfuron, the pyrimidine ring also formed a $\pi - \pi$ interaction with Trp574, yet the pyrimidine ring could rotate around the C – N chain, possibly causing a shift in the methyl conformation in a more spatial environment[21-23] (Figs. 23 and 24).

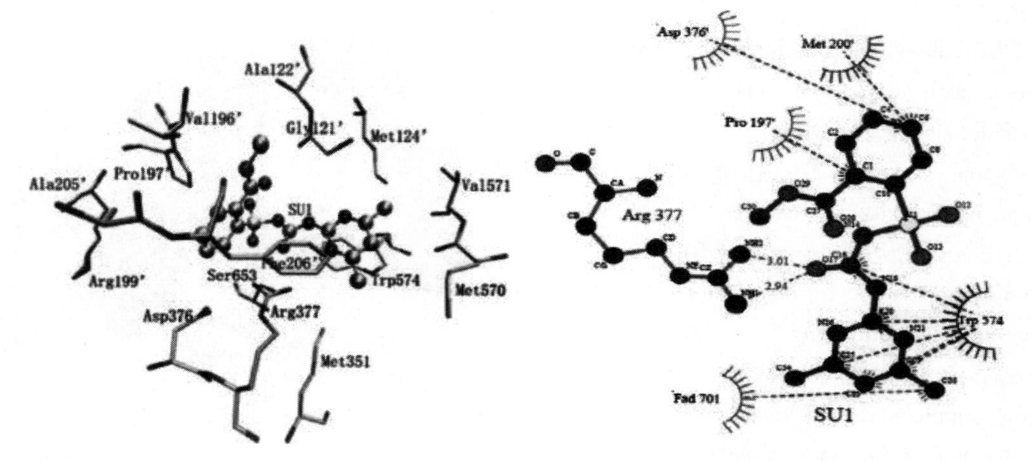

Fig. 23 Substrate – AHAS interaction of conventional SU (sulfometuron): green dotted line shows hydrogen or coordination bond; red dotted line shows hydrophobic interaction

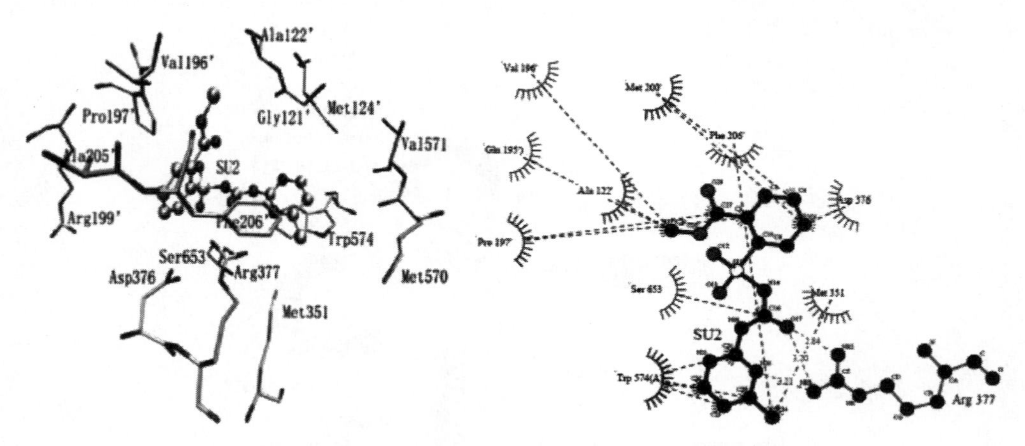

Fig. 24 Substrate – AHAS interaction for monosulfuron – ester

MM – PBSA calculations of free energy in the interaction between SUs and AHAS showed that VDW force and unpolar solvation energy are factors to be considered. It was surprising to note that, in sulfometuron, when considering the contribution of each amino residue at the site of action, the top five amino acids are R377, W574, G121′, F206′ and V196′(arginine, tryptophane, glycine, phenyalanine and valine), and in monosulfuron the top five are G121′, W574′, V196′, F206′ and A122′(glycine, trytophane, valine, pheylalanine and alanine) in an entirely different pattern(Fig. 25). In other words, in sulfometuron the contribution of the Arg 377 residue ranks highest in interaction free energy($-14kJ \cdot mol^{-1}$), which indicates that Arg 377 forms a hydrogen bond with the carbonyl oxygen atom in the urea moiety. Tryptophane interacts with the pyrimidine ring of the substrate to form a $\pi-\pi$ interaction($-12kJ \cdot mol^{-1}$) insulfometuron, while in monosulfuron the $\pi-\pi$ interaction is also working, although somewhat more weakly($-8kJ \cdot mol^{-1}$). Nevertheless, as shown in Figs. 19, 23 and 24, although monosulfuron has no substituent in the 6-position, tryptophane 574 and G121′ might still play a role in interaction, which strengthens the present authors' view that the 'smaller cavity' is more important, as shown in Fig. 22.

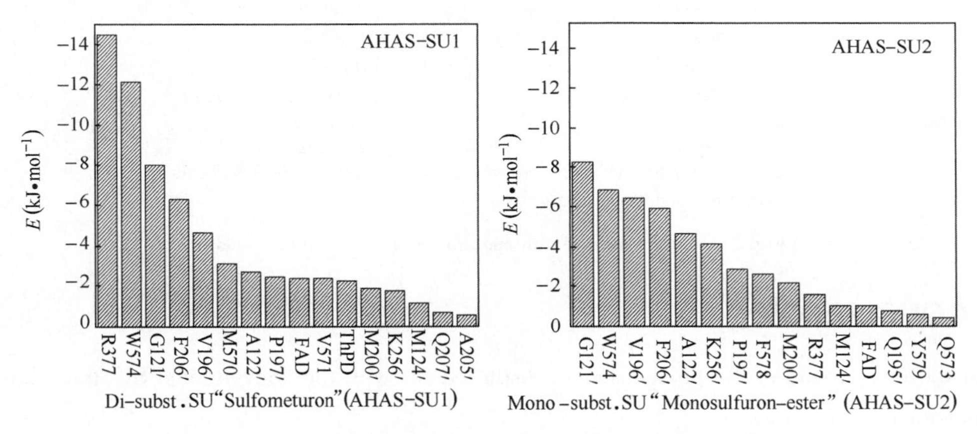

Fig. 25 MM-GBSA energy analysis of main residue contribution

According to greenhouse and field tests over the past 5 years, monosulfuron is the only herbicide available for application in millet fields in northern China, as the millet seedlings are exceptional fragile to other commercial SU herbicides. Fig. 26 shows schematically the survival rate of millet seedlings in field tests with different SUs at a dosage of $15g \cdot ha^{-1}$; monosulfuron gave the best results among the SUs in the tests.

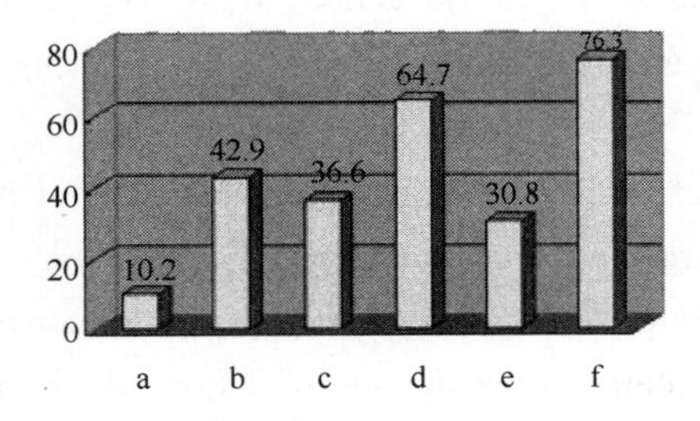

Fig. 26 Inhibition of millet seedlings(%) by SUs(in dosage of $15g\ ha^{-1}$, field tests): a, monosulfuron; b, chlorimuron; c, tribemuron; d, metsulfuron; e, flazasulfuron; f, chlorsulfuron. b to d, classical disubstituted SU products

Further, in a comparison between sulfometuron (OUST) and the present monosulfuron - ester - the two structures are very much alike except that in the latter there is one methyl group less than in the former (Fig. 27) - an interesting phenomenon was observed from bioassay data: their biological characteristics are strikingly different, although both of them have superactivity towards herbs and very low toxicity (both have an acute oral toxicity of $>5000\text{mg} \cdot \text{kg}^{-1}$). Sulfometuron as a non - selective herbicide is restricted to application only in forests and non - crop land,[19] while monosulfuron - ester (with one methyl group less than the former) can be safely applied in wheat and barley fields.

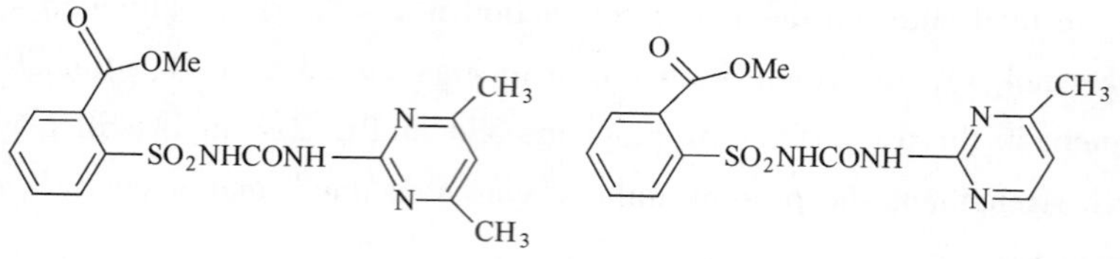

O, OMe, CH_3, N, $SO_2NHCONH$, N, CH_3

O, OMe, CH_3, N, $SO_2NHCONH$, N

Sulfometuron

Oust

LD_{50} >5000mg/kg

Ki = 0.0396 (*in vitro*) *

appl. dosage

(70 - 840g/ha)

non - selective herbicide

* our own experimental data in vitro

Monosulfuron

NK94827

LD_{50} >10000mg/kg

Ki = 0.3626 (*in vitro*) *

appl. dosage

(20 - 25g/ha)

wheat (barley) herbicide

Fig. 27 Comparison of sulfometuron and monosulfuron - ester

4 CONCLUSION

From the comparison of the present monosubstituted SU with the conventional SU, the following conclusions can be drawn:

1. Tests, both *in vivo* and *in vitro*, indicated that the two types of SU have a similar bioactivity potency at the level of ultralow dosage.

2. Both are AHAS inhibitors, and their sites of action are similar.

3. At the active sites of the target enzyme, it seems that a smaller cavity could be more important than a larger cavity.

4. In the case of monosulfuron, amino residues W574, G121′, V196, etc., make an important contribution to the interaction at the site of action, where Tryp 574 could take part in critical $\pi - \pi$ interaction on the monosubstituted pyrimidine ring, although a second methyl group at the 6 - position is lacking.

5. Between the two types of SU, there are likely to be small but noticeable differences in the interaction pattern on the target enzyme, which might explain why, under some circumstances, monosubstituted SUs manifest preference in selectivity and tolerance to certain crops, probably owing to their more flexible interaction at the site of action, which are factors that are of utmost concern during any herbicidal innovation programme, along with efficacy and toxicology.

ACKNOWLEDGEMENTS

A special tribute and sincere thanks to Professor Ronald G Duggleby, Queensland University,

Australia, who has given valuable and generous assistance to this project. Sincere thanks also to Advance Photon Source at the Argonne National Laboratory, Chicago University, Chicago, Illinois, for important technical support.

References

[1] Levitt G, *Synthesis and Chemistry of Agrochemicals*, ed. by Baker DR, Fenyes JG and Moberg WK. *ACS Symposium Series* 443, American Chemical Society, Washington, DC, pp. 17 – 47(1991).

[2] Wang J – G, Li Z – M, Ma N, Wang B – L, Jiang L, Pang SS, *et al*, Structure – activity relationships for a new family of sulfonylurea herbicides. *J Comput – Aided Mol Des* 19:801 – 820(2005).

[3] Li Z – M, Jia G – F, Wang L – X, Lai C – M, Wang R – J and Wang H – G, Synthesis, structure and SAR study on new sulfonylurea compounds(II). *Chem J Chin Univ* 13:1411 – 1414(1992).

[4] Lai C – M, Luo H, Yuan M and Li Z, Investigation of the relationship of applied molecular graphics, molecular mechanics, quantum chemistry and static potential methods to properties(II). *Acta Scientiarum Naturalium Universitatis Nankaiersis*(1):56 – 60(1993).

[5] Li Z – M, Jia G – F and Wang L – X, Research on the synthesis, structure and SAR study on new sulfonylureas(IV). *Chem J Chin Univ* 15:281 – 284(1994).

[6] Liu J, Li Z – M, Wang X, Ma Y, Lai C – M, Jia G – F, *et al*, Application of the CoMFA method to study the 3D SAR relationship of sulfonylureas. Sci China(Ser B)41:50 – 53(1998).

[7] Liu J, Doctoral Dissertation, Nankai University, Nankai, Tianjin, China (1998).

[8] Li Z – M, The structure – activity relationship study on herbicidal sulfonylureas, *in Proceedings of the Second International Conference on Crop Protection Chemicals*(Nantong, China), ed. by Copping LG and Suganam B. China Agriculture Press, Beijing, China, pp. 30 – 36(1999).

[9] Lai C – M, Yuan M – X, Li Z – M, Jia G – F and Jia L – X, The primitive model of sulfonylurea molecules interaction with receptor. *Chem J Chin Univ* 15:286 – 287(1994).

[10] Li Z – M, Wang L – X, *et al*, Chinese Patent ZL94118793.4, 10/06/98; ZL – 94 – 118793.4, 23/4/98; ZL – 96 – 1 – 06731.4, 17/2/03.

[11] Duggleby R G and Pang S S, J Biochem Molec Biol 1 – 36(2000).

[12] Schloss JV, Ciskanik LM and Van Dyk DE, Nature(Lond)331:360 – 362(1988).

[13] Pang S S, Duggleby R G and Guddat L W, JMol Biol 317:249 – 262(2002).

[14] Wang J – G, Li Z – M, Ma N, Wang B – L, Jiang L, Pang SS, *et al*, *J Comput – AidedMol Des* 19:801 – 820(2005).

[15] Pang S S, Guddat L W and Duggleby R G, Acta Cryst D57:1321 – 1323(2001).

[16] Wang J – G, Lee PK – M, Dong Y – H, Pang SS, Duggleby RG, Li Z – M, *et al*, *FEBS J* 276:1282 – 1290(2009).

[17] Duggleby R G and Pang S S, PNAS 103:569 – 573(2006).

[18] Jones T A, Zou J Y, Cowan S W and Kjeldgard M, *Acta Crystallogr A* 47:110 – 119(1991).

[19] Tomlin CDS, *A World Compendium – The Pesticide Manual*, 15th edition. BCPC, Farnham, Surrey, UK, pp. 1060 – 1061(2009).

[20] Wallace A C, Laskowski R A and Thornton J M, Logplot: a program to generate schematic diagrams of protein – ligand interaction. *Protein Eng Des Select* 8:127 – 134(1955).

[21] Chen P – Q, Sun H – W, Li Z – M, Wang J – G, Ma Yi and Lai C – M, *J Chin Chem* 64:1341 – 1348(2006).

[22] Chen P – Q, Sun H – W, Li Z – M, Wang J – G, Ma Y and Lai C – M, *Chem J Chin Univ* 28:278 – 282(2007).

[23] Chen P – Q, *Doctoral Dissertation*, in Pesticide Chemistry, Nankai University, Nankai, Tianjin, China(2007).

Synthesis, Crystal Structure and Biological Activities of Novel Anthranilic (Isophthalic) Acid Esters*

Tao Yan, Guanping Yu, Pengfei Liu, Lixia Xiong, Shujing Yu, Zhengming Li

(State Key Laboratory of Elemento – Organic Chemistry, Institute of Elemento – Organic Chemistry, Nankai University, Tianjin, 300071, China)

Abstract In search of environmentally benign insecticides with high activity, low toxicity and low resistance, a series of novel anthranilic (isophthalic) acid esters was designed and synthesized based on the structure of ryanodine modulating agent. All the compounds were characterized by ^1H NMR spectra, elemental analysis or high resolution mass spectrometry (HRMS). The preliminary results of biological activity assessment indicate that some of the title compounds exhibit certain but unremarkable insecticidal activity against *Mythimna separata* Walker at 200mg/L and fungicidal activities against five funguses at 50mg/L.

Key words anthranilic acid; isophthalic acid; crystal structure; biological activity

1 Introduction

The ryanodine receptor (RyR) derives its name from the plant metabolite ryanodine, a natural insecticide from Ryania speciosa, known to modify calcium channels[1-4]. It has been conjectured that RyRs provide an excellent target for insect control[5]. The phthalic diamides (– *Flubendiamide* –)[6] from Nihon Nohyaku Co., Ltd. and the anthranilic diamides (– *Chlorantraniliprole* –)[7-9] from DuPont Co., Ltd. are the first two synthetic classes of potent activators of RyRs (Fig. 1).

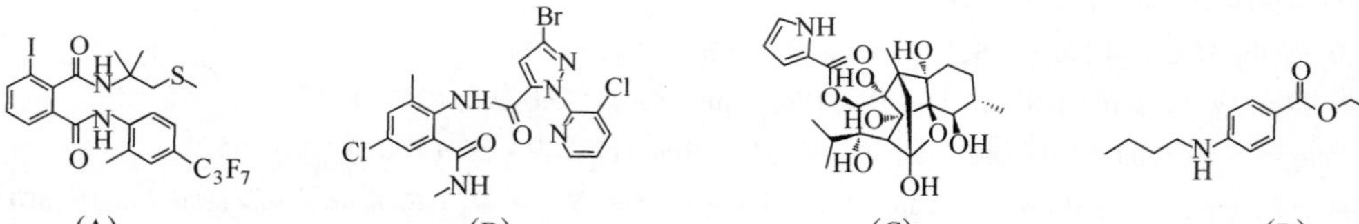

(A) (B) (C) (D)

Fig. 1 Structures of flubendiamide (A), chlorantraniliprole (RynaxypyrTM; DPX – E2Y45) (B), ryanodine (C) and tetracaine (D)

These two commercial insecticides are significant in the field of crop protection, particularly important in light of their ability to control harmful insects, which have developed the resistance to other traditional insecticides[5]. The structures of these two insecticides contain the important amido moiety and an ester moiety contained in many RyR modulating agents, such as ryanodine and tetracaine[10] (Fig. 1), which indicates that the importance of the ester moiety might be similar to the amide moiety for insecticidal activity. Besides, ester moiety could play an important role in the improvement of hydrophobic solubility. In consideration of the above viewpoints, the title compounds were synthesized and characterized by ^1H NMR, elemental analysis or HRMS, single

* Reprinted from *Chemical Research in Chinese Universities*, 2012, 28(1): 53 – 56. Supported by the National Natural Science Foundation of China (No. 20872069) and the National Basic Research Program of China (No. 2010CB126106).

crystal X - ray crystallographic analysis and biological activity test.

2 Experimental

2.1 Materials and Instruments

^1H NMR spectra were recorded on a Bruker 400MHz nuclear magnetic resonance spectrometer with $CDCl_3$ as the solvent and tetramethylsilane(TMS) as internal standard. Elemental analyses were carried out on a Yanaca CHN Corder MT - 3 elemental analyzer. High resolution mass spectrometry (HRMS) data were obtained on a Varian quantum field theoryelectrospray ionization (QFT - ESI) instrument. The melting points were determined on an X - 4 melting point apparatus. All the reagents were of analytical grade.

2.2 Syntheses of Target Compounds

2.2.1 Syntheses of Intermediates

Synthetic routes of intermediates (compounds **2 - 7**) is shown in Scheme **1**. They were prepared according to the reported method[11-15]. The melting points, yields and elemental analysis or HRMS of the intermediates are listed in Table 1; ^1H NMR data are listed in Table 2.

Scheme 1 Synthetic routes of the title compounds

5a: X^1 = Cl; **5b**: X^1 = Br; **6a**: X^1 = Cl, R = CH_3; **6b**: X^1 = Br, R = CH_3; **6c**: X^1 = Cl, R = C_2H_5; **6d**: X^1 = Br, R = C_2H_5; **6e**: X^1 = Cl, R = $n-C_3H_7$; **6f**: X^1 = Br, R = $n-C_3H_7$; **6g**: X^1 = H, R = CH_3; **7a**: X^2 = Cl; **7b**: X^2 = Br; **8a**: X^1 = Cl, X^2 = Cl, R = CH_3; **8b**: X^1 = Cl, X^2 = Br, R = CH_3; **8c**: X^1 = Br, X^2 = Cl, R = CH_3; **8d**: X^1 = Br, X^2 = Br, R = CH_3; **8e**: X^1 = Cl, X^2 = Cl, R = C_2H_5; **8f**: X^1 = Cl, X^2 = Br, R = C_2H_5; **8g**: X^1 = Br, X^2 = Cl, R = C_2H_5; **8h**: X^1 = Br, X^2 = Br, R = C_2H_5; **8i**: X^1 = Cl, X^2 = Cl, R = $n-C_3H_7$; **8j**: X^1 = Cl, X^2 = Br, R = $n-C_3H_7$; **8k**: X^1 = Br, X^2 = Cl, R = $n-C_3H_7$; **8l**: X^1 = Br, X^2 = Br, R = $n-C_3H_7$; **8m**: X^1 = H, X^2 = Cl, R = CH_3; **8n**: X^1 = H, X^2 = Br, R = CH_3

Table 1 Physical data and HRMS of the synthesized compounds①

Compd.	m. p. (℃)	Yield(%)	Appearance	Elemental analysis found(%, calcd.) or HRMS(m/z, calcd.)		
				C	H	N
2	215 - 217	74.6	White solid		192.0666(192.0661 [M - H]⁻)	
3	224 - 225	48.6	White solid		222.0398(222.0402 [M - H]⁻)	
4	260(dec.)	90.0	White solid		180.0293(180.0297 [M - H]⁻)	
5a	210 - 211	93.2	Yellow solid		213.9913(213.9907 [M - H]⁻)	
5b	250(dec.)	95.5	Yellow solid		257.9407(257.9502 [M - H]⁻)	

Table 1 (continued)

Compd.	m. p. (℃)	Yield (%)	Appearance	Elemental analysis found (%, calcd.) or HRMS (*m/z*, calcd.)		
				C	H	N
6a	150 - 152	66. 3	Yellow solid	48. 97 (49. 30)	4. 31 (4. 14)	5. 58 (5. 75)
6b	162 - 164	58. 0	Yellow solid	41. 83 (41. 69)	3. 62 (3. 50)	4. 73 (4. 86)
6c	118 - 120	55. 7	Yellow solid	53. 06 (53. 05)	5. 41 (5. 19)	5. 03 (5. 16)
6d	118 - 120	79. 2	Yellow solid	45. 38 (45. 59)	4. 62 (4. 46)	4. 38 (4. 43)
6e	62 - 64	68. 3	Yellow solid	56. 38 (56. 10)	5. 80 (6. 05)	4. 94 (4. 67)
6f	63 - 65	71. 4	Yellow solid		344. 0489 (340. 0942 [M + H]$^+$)	
6g	102 - 104	65. 6	Yellow solid		232. 0583 (232. 0580 [M + Na]$^+$)	
8a	217 - 219	35. 7	White solid		504. 9848 (504. 9844 [M + Na]$^+$)	
8b	184 - 186	24. 2	White solid	43. 07 (43. 21)	2. 45 (2. 48)	10. 48 (10. 61)
8c	213 - 214	34. 3	White solid		548. 9345 (548. 9344 [M + Na]$^+$)	
8d	212 - 213	25. 8	White solid		592. 8833 (592. 8831 [M + Na]$^+$	
8e	179 - 181	39. 5	White solid		533. 0165 (533. 0157 [M + Na]$^+$)	
8f	179 - 181	38. 7	White solid		578. 9618 (578. 9629 [M + Na]$^+$)	
8g	178 - 180	26. 8	White solid		576. 9655 (576. 9657 [M + Na]$^+$)	
8h	205 - 206	31. 3	White solid		620. 9147 (620. 9146 [M + Na]$^+$)	
8i	135 - 136	30. 6	White solid		561. 0475 (561. 0470 [M + Na]$^+$)	
8j	136 - 137	33. 6	White solid		606. 9951 (606. 9942 [M + Na]$^+$)	
8k	146 - 147	35. 9	White solid	47. 46 (47. 28)	3. 72 (3. 62)	9. 81 (9. 59)
8l	136 - 137	29. 0	White solid		648. 9459 (648. 9462 [M + Na]$^+$)	
8m	198 - 199	37. 2	White solid		471. 0229 (471. 0234 [M + Na]$^+$)	
8n	181 - 182	36. 0	White solid		514. 9726 (514. 9728 [M + Na]$^+$)	

①**7**a: m. p. 198 - 200℃ (lit. [14]: 200 - 201℃); **7**b: m. p. 198 - 200℃ (lit. [14]: 197 - 200℃).

2. 2. 2 Syntheses of Title Compounds

Synthetic routes of title compounds is shown in Scheme 1.

The synthetic procedure is described below. To a suspension of intermediate **7** (1mmol) in dichloromethane (10mL) were added oxalyl chloride (3mmol) and dimethylformamide (1 drop). The solution was stirred at room temperature for 3h, and then the mixture was concentrated *in vacuo* to obtain the crude acid chloride. The crude acyl chloride in dichloromethane (10mL) was added dropwise slowly to a stirred solution of intermediate **6** (1. 2mmol) in dichloromethane (20mL) in an ice bath. After 20min, triethylamine (1mmol) was added. The solution was warmed to room temperature and stirred for 12h, then diluted with CH_2Cl_2 (20mL), and washed with 1mol/L aqueous HCl solution (10mL), saturated aqueous $NaHCO_3$ solution (10mL), and brine (10mL), respectively. The organic extract was separated, dried, filtered, concentrated and purified by silica gel chromatography to afford the desired compound **8**. The melting points, yields and elemental analysis or HRMS of the title compounds are listed in Table 1, and ^1H NMR data are listed in Table 2.

Table 2 ^{1}H NMR data of the synthesized compounds①

Compd.	^{1}H NMR($CDCl_3$), δ
2	12.74(s, 1H, —COOH), 9.49(s, 1H, —NH—), 7.58(d, J = 7.34Hz, 1H, Ar—H), 7.44(d, J = 11.32Hz, 1H, Ar—H), 7.22(t, J = 6.72Hz, 1H, Ar—H), 2.19(s, 3H, Ar—CH_3), 2.00(s, 3H, —$COCH_3$)
3	8.97(s,1H,—NH—),7.90(t,1H,Ar—H),7.52(d,J=20.96Hz,2H,Ar—H),2.02(s,3H,—$COCH_3$)
4	8.03(d,J=7.84Hz,2H,Ar—H),6.59(t,J=7.84Hz,1H,Ar—H)
5a	7.95(s,2H,Ar—H)
5b	8.06(s,2H,Ar—H)
6a	8.12(s,2H,Ar—NH_2),8.06(s,2H,Ar—H),3.89(s,6H,—CH_3)
6b	8.18(s,2H,Ar—H),8.14(s,2H,Ar—NH_2),3.88(s,6H,—CH_3)
6c	8.13(s,2H,Ar—NH_2),8.05(s,2H,Ar—H),4.34(q,4H,—$\mathbf{CH_2}CH_3$),1.39(t,6H,—$CH_2\mathbf{CH_3}$)
6d	8.18(s,2H,Ar—NH_2),8.15(s,2H,Ar—H),4.33(q,4H,—$\mathbf{CH_2}CH_3$),1.39(t,6H,—$CH_2\mathbf{CH_3}$)
6e	8.10(s,2H, Ar—NH_2), 8.04(s, 2H, Ar—H), 4.24(t, 4H, —$\mathbf{CH_2}$ CH_2CH_3), 1.86 - 1.73(m, 4H, —CH_2 $\mathbf{CH_2}CH_3$), 1.03(t,6H,—CH_2 $CH_2\mathbf{CH_3}$)
6f	8.16(s,2H, Ar—NH_2), 8.12(s, 2H, Ar—H), 4.23(t, 4H, —$\mathbf{CH_2}$ CH_2CH_3), 1.85 - 1.72(m, 4H, —CH_2 $\mathbf{CH_2}CH_3$), 1.04(t,6H,—CH_2 $CH_2\mathbf{CH_3}$)
6g	8.15(s,2H,Ar—NH_2),8.09(d,J=7.82Hz,2H,Ar—H),6.56(t,1H,Ar—H),3.88(s,6H,—CH_3)
8a	11.36(s,1H, Ar—NH), 8.44(d, J = 4.70Hz, 1H, pyridyl—H), 8.01(s, 2H, Ar—H), 7.86(d, J = 8.00Hz,1H,pyridyl—H),7.39(dd,J=4.70,8.00Hz,1H,pyridyl—H),7.02(s,1H,pyrazolyl—H),3.85(s,6H,—CH_3)
8b	11.36(s,1H, Ar—NH), 8.44(d, J = 4.70Hz, 1H, pyridyl—H), 8.01(s, 2H, Ar—H), 7.86(d, J = 8.00Hz,1H,pyridyl—H),7.39(dd,J=4.70,8.00Hz,1H,pyridyl—H),7.10(s,1H,pyrazolyl—H),3.85(s,6H,—CH_3)
8c	11.37(s,1H, Ar—NH), 8.44(d, J = 4.70Hz, 1H, pyridyl—H), 8.16(s, 2H, Ar—H), 7.86(d, J = 8.00Hz,1H,pyridyl—H),7.39(dd,J=4.70,8.00Hz,1H,pyridyl—H),7.02(s,1H,pyrazolyl—H),3.85(s,6H,—CH_3)
8d	11.38(s,1H, Ar—NH), 8.44(d, J = 4.70Hz, 1H, pyridyl—H), 8.16(s, 2H, Ar—H), 7.86(d, J = 8.00Hz,1H,pyridyl—H),7.39(dd,J=4.70,8.00Hz,1H,pyridyl—H),7.10(s,1H,pyrazolyl—H),3.85(s,6H,—CH_3)
8e	11.43(s,1H, Ar—NH), 8.42(d, J = 4.70Hz, 1H, pyridyl—H), 8.00(s, 2H, Ar—H), 7.85(d, J = 8.00Hz,1H,pyridyl—H),7.37(dd,J=4.70,8.00Hz,1H,pyridyl—H),7.00(s,1H,pyrazolyl—H),4.31(q,4H,—$\mathbf{CH_2}CH_3$),1.32(t,6H,—$CH_2\mathbf{CH_3}$)
8f	11.43(s,1H, Ar—NH), 8.42(d, J = 4.70Hz, 1H, pyridyl—H), 8.00(s, 2H, Ar—H), 7.85(d, J = 8.00Hz,1H,pyridyl—H),7.38(dd,J=4.70,8.00Hz,1H,pyridyl—H),7.09(s,1H,pyrazolyl—H),4.30(q,4H,—$\mathbf{CH_2}CH_3$),1.32(t,6H,—$CH_2\mathbf{CH_3}$)
8g	11.43(s,1H, Ar—NH), 8.42(d, J = 4.70Hz, 1H, pyridyl—H), 8.14(s, 2H, Ar—H), 7.85(d, J = 8.00Hz,1H,pyridyl—H),7.38(dd,J=4.70,8.00Hz,1H,pyridyl—H),7.00(s,1H,pyrazolyl—H),4.32(q,4H,—$\mathbf{CH_2}CH_3$),1.32(t,6H,—$CH_2\mathbf{CH_3}$)

Table 2(continued)

Compd.	^1H NMR($CDCl_3$),δ
8h	11.44(s,1H,Ar—NH),8.42(d,J = 4.70Hz,1H,pyridyl—H),8.14(s,2H,Ar—H),7.85(d,J = 8.00Hz,1H,pyridyl—H),7.38(dd,J = 4.70,8.00Hz,1H,pyridyl—H),7.09(s,1H,pyrazolyl—H),4.30(q,4H,—$\mathbf{CH_2}CH_3$),1.32(t,6H,—$CH_2\mathbf{CH_3}$)
8i	11.43(s,1H,Ar—NH),8.42(d,J = 4.80Hz,1H,pyridyl—H),7.99(s,2H,Ar—H),7.85(d,J = 7.40Hz,1H,pyridyl—H),7.38(dd,J = 4.80,7.40Hz,1H,pyridyl—H),7.00(s,1H,pyrazolyl—H),4.20(t,4H,—$\mathbf{CH_2}CH_2CH_3$),1.78 – 1.65(m,4H,—$CH_2\mathbf{CH_2}CH_3$),0.97(t,6H,—$CH_2CH_2\mathbf{CH_3}$)
8j	11.45(s,1H,Ar—NH),8.42(d,J = 4.60Hz,1H,pyridyl—H),8.13(s,2H,Ar—H),7.85(d,J = 8.00Hz,1H,pyridyl—H),7.38(dd,J = 4.60,8.00Hz,1H,pyridyl—H),7.00(s,1H,pyrazolyl—H),4.20(t,4H,—$\mathbf{CH_2}CH_2CH_3$),1.80 – 1.64(m,4H,—$CH_2\mathbf{CH_2}CH_3$),0.97(t,6H,—$CH_2CH_2\mathbf{CH_3}$)
8k	11.45(s,1H,Ar—NH),8.43(d,J = 4.70Hz,1H,pyridyl—H),8.13(s,2H,Ar—H),7.85(d,J = 8.00Hz,1H,pyridyl—H),7.38(dd,J = 4.70,8.00Hz,1H,pyridyl—H),7.00(s,1H,pyrazolyl—H),4.20(t,4H,—$\mathbf{CH_2}CH_2CH_3$),1.78 – 1.65(m,4H,—$CH_2\mathbf{CH_2}CH_3$),0.97(t,6H,—$CH_2CH_2\mathbf{CH_3}$)
8l	11.45(s,1H,Ar—NH),8.43(d,J = 4.80Hz,1H,pyridyl—H),8.13(s,2H,Ar—H),7.85(d,J = 7.40Hz,1H,pyridyl—H),7.38(dd,J = 4.80,7.40Hz,1H,pyridyl—H),7.08(s,1H,pyrazolyl—H),4.20(t,4H,—$\mathbf{CH_2}CH_2CH_3$),1.78 – 1.64(m,4H,—$CH_2\mathbf{CH_2}CH_3$),0.97(t,6H,—$CH_2CH_2\mathbf{CH_3}$)
8m	11.52(s,1H,Ar—NH),8.44(d,J = 4.70Hz,1H,pyridyl—H),8.04(d,J = 7.80Hz,2H,Ar—H),7.86(d,J = 8.00Hz,1H,pyridyl—H),7.39(dd,J = 4.70,8.00Hz,1H,pyridyl—H),7.25(t,1H,Ar—H),7.05(s,1H,pyrazolyl—H),3.84(s,6H,—CH_3)
8n	11.52(s,1H,Ar—NH),8.44(d,J = 4.70Hz,1H,pyridyl—H),8.04(d,J = 7.80Hz,2H,Ar—H),7.86(d,J = 8.00Hz,1H,pyridyl—H),7.39(dd,J = 4.70,8.00Hz,1H,pyridyl—H),7.25(t,1H,Ar—H),7.13(s,1H,pyrazolyl—H),3.84(s,6H,—CH_3)

①compounds **2**,**4**,**5**a,**5**b,solvent:DMSO – d_6.

2.3 Biological Assay

Insecticidal activity against *Mythimna separate*(*MS*) and antifungal activities against funguses[*Fusarium omysporum* (*FO*),*Physalospora piricola*(*PP*),*Alternaria solani*(*AS*),*Cercospora arachidicola* (*CA*),and *Gibberella sanbinetti*(*GS*)]were tested according to the reported method[14,16].

2.4 X – ray Crystallographic Analysis

The single crystals of compound **8**e were obtained from a methanol solution by slow evaporation at room temperature. A white crystal with a dimension of 0.22mm × 0.14mm × 0.12mm was selected for X – ray diffraction analysis. The data were collected on a Rigaku Raxis – IV diffractometer with graphite monochromated Mo $K\alpha$ radiation(λ = 0.071075nm). The structure was solved by direct methods and expanded *via* difference Fourier techniques with SHELXS – 97[17].

3 Results and Discussion

Fig. 2 and Fig. 3 show the molecular structure and the perspective view of the crystal packed in the unit cell of compound **8**e. The X – ray analysis reveals that it crystallized in a monoclinic system with space group $P2_1/n$, $a = 1.0555(1)$ nm, $b = 0.90390(8)$ nm, $c = 2.3662(3)$ nm, $\alpha = 90°$, $\beta = 93.266(5)°$, $\gamma = 90°$, $C_{21}H_{17}Cl_3N_4O_5$, $M_r = 511.74$, $V = 2.2538(4)$ nm^3, $Z = 4$, $D_c = 1.508$ g/cm^3, $F(000) = 1048.0$, $\mu = 0.448$ mm^{-1}, the final $R = 0.0457$ and $wR = 0.1289$ for 5337 unique reflections. The dihedral angles made by the pyrazole ring with the other pyridine ring and benzene ring are 80.54° and 18.31°, respectively. The bond lengths of C1 – N3 [0.1329(3) nm] and C6 – N1 [0.1362(3) nm] are shorter than that of the normal C – N single bond (0.1490nm), which suggests that the electron is delocalized in pyridine and pyrazole rings. From Fig. 3, we also observe that there are weak hydrogen bonds and C – H···π supramolecular interactions in the lattice, which form a threedimensional hydrogen – bonded network to stabilize the crystal structure. Meanwhile, the intra – and intermolecular hydrogen bonds could have a great impact on the biological activity.

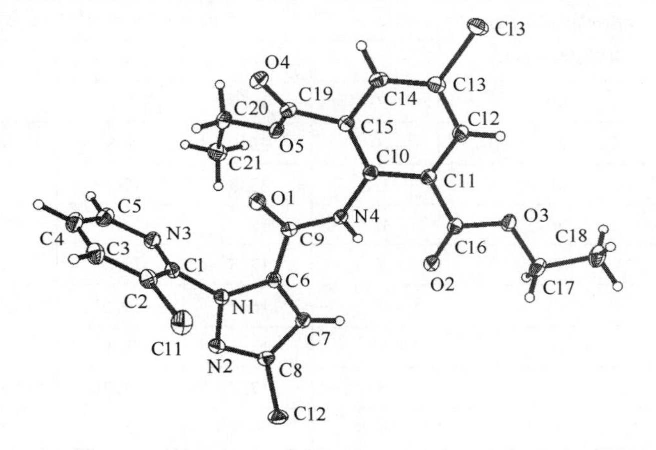

Fig. 2 Molecular structure of compound 8e

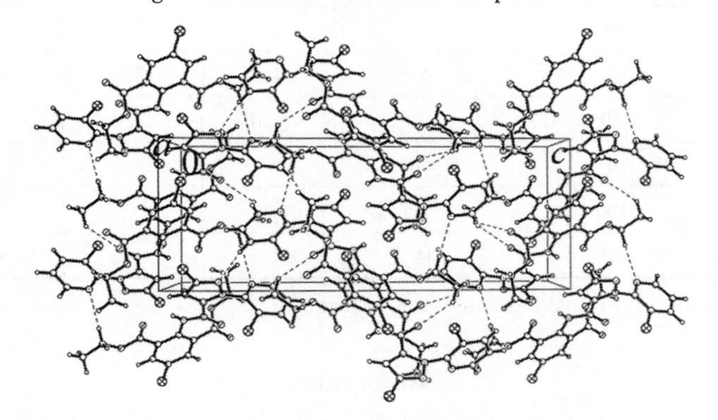

Fig. 3 View of the crystal packing down the axis a for compound 8e

The biological activities of the title compounds are listed in Table 3, which exhibit certain but unremarkable insecticide activities against *MS* at 200mg/L (compared with *rynaxypyr*™) and antifungal activities against five funguses at 50mg/L (compared with *chlorothalonil*). All the insecticide activities of compounds **8**a, **8**f, **8**m (at 200mg/L) against *MS* were 40%, which are inferior in contrast to *rynaxypyr*™. The preliminary structure – activity relationship of the title compounds indicate that the methyl on the benzene or the amido moiety is essential to the insecticide activity. However, the steric effect of the title compounds maybe have a great impact on the insecticide activities. The antifungal activities (at 50mg/L) of compounds **8**g (against *GS*) and **8**c (against *PP*) are 35.6% and 33.3%, which indicate that some of the title compounds exhibit favorable antifungal activities to certain fungus.

In conclusion, fourteen new anthranilic acid esters were designed and synthesized with the structures characterized by ^{1}H NMR spectroscopy, single crystal X-ray diffraction analysis and elemental analysis or HRMS. The biological activities of the new compounds were evaluated. The results of bioassays show that these title compounds exhibit certain insecticide and antifungal activities. The title compounds **8**a, **8**f and **8**m exhibit an insecticidal activity of 40% against *MS* at 200mg/L; compound **8**c exhibits a fungicidal activity of 33.3% against *PP* and compound **8**g exhibits a fungicidal activity of 35.6% against *GS* at 50mg/L.

Table 3 Biological activities (%) of the title compounds①

Compd.	Insecticidal activity *MS* (200 mg/L)	Antifungal activity (50mg/L)				
		FO	*PP*	*AS*	*CA*	*GS*
8a	40	0	16.7	5.3	0	32.4
8b	20	0	25.0	5.3	0	8.8
8c	0	4.8	33.3	10.5	0	14.7
8d	20	4.8	12.5	0	0	35.3
8e	0	0	14.8	12.9	2.8	24.4
8f	40	0	16.7	10.5	7.1	17.6
8g	0	0	14.6	9.8	0	35.6
8h	0	0	17.8	11.7	6.4	14.7
8i	20	19.5	20.8	16.1	0	23.5
8j	20	19.0	29.2	5.3	0	14.7
8k	0	0	0	10.5	0	26.5
8l	20	4.8	29.2	15.8	7.1	8.8
8m	40	0	25.0	15.8	0	20.6
8n	0	0	15.7	17.6	0	25.9
Control-1	100	NT	NT	NT	NT	NT
Control-2	NT	84.0	88.5	63.6	58.8	73.1

①NT: not tested; control-1: c = 1mg/L; control-2, c = 50mg/L.

References

[1] Sattelle B. D., Cordova D., Cheek R. T., *Invert Neurosci.*, 2008, 8, 107.

[2] Fill M., Copello A. J., *Physiol. Rev.*, 2002, 82, 893.

[3] Gyorke S., Fill M., *Science*, 1993, 260, 807.

[4] Coronado R., Morrissette J., Sukhareva M., Vaughan D. M., *Am. J. Physiol.*, 1994, 266, C1485.

[5] Lahm G. P., Cordova D., Barry J. D., *Bioorg. Med. Chem.*, 2009, 17, 4127.

[6] Tohnishi M., Nakao H., Furuya T., Seo A., Kodama H., Tsubata K., Fujioka S., Kodama H., Hirooka T., Nishimatsu T., *J. Pestic. Sci.*, 2005, 30, 354.

[7] Lahm G. P., Selby T. P., Freudenberger J. H., Stevenson T. M., Myers B. J., Seburyamo G., Smith B. K., Flexner L., Clark C. E., Cordova D., *Bioorg. Med. Chem. Lett.*, 2005, 15, 4898.

[8] Pszczolkowski A. M. ,Olson E. ,Rhine C. ,Ramaswamy B. S. ,*J. Insect Physiol.* ,2008,54,358.
[9] Caboni P. ,Sarais G. ,Angioni A. ,Vargiu S. ,Pagnozzi D. ,Cabras P. ,Casida E. J. ,*J. Agric. Food Chem.* , 2008,56,7696.
[10] Santonastasi M. ,Wehrens X. H. T. ,*Acta Pharmacol. Sin.* ,2007,28,937.
[11] Montoya - pelaez P. J. , Uh Y. , Lata C. , Thompson M. P. , Lemieux R. P. , Crudden C. M. , *J. Org. Chem.* ,2006,71,5921.
[12] Kremer B. C. ,J. *Chem. Educ.* ,1956,33,71.
[13] Shapiro R. ,Taylor E. G. ,Zimmerman W. T. ,*Method for Preparing. N - Phenylpyrazole - 1 - carboxamides*, *WO* 2006062978,2006 - 06 - 15.
[14] Dong W. L. ,Xu J. Y. ,Xiong L. X,Liu X. H. ,Li Z. M. ,*Chin. J. Chem.* ,2009,27,579.
[15] Feng Q. ,Liu Z. L. ,Wang M. Z. ,Xiong L. X. ,Yu S. J. ,Li Z. M. ,*Chem. J. Chinese Universities*,2011,32(1),74.
[16] Dong W. L. ,Xu J. Y. ,Liu X. H. ,Li Z. M. ,Li B. J. ,Shi Y. X. ,*Chem. J. Chinese Universities*,2008,29(10),1990.
[17] Sheldrick G. M. , *SHELXTL - 97*, *Program for Crystal Structure Refinement*, University of Göttingen, Göttingen,1997.

李正名文集

Synthesis and Evaluation of Novel Monosubstituted Sulfonylurea Derivatives as Antituberculosis Agents*

Li Pan[1,**], Ying Jiang[2,**], Zhen Liu[3,**], Xinghai Liu[1], Zhuo Liu[1], Gang Wang[1], Zhengming Li[1], Di Wang[2]

([1]State-Key Laboratory of Elemento-Organic Chemistry, National Pesticide Engineering Research Center (Tianjin), Nankai University, Tianjin, 300071, China;
[2]Department of Clinical Laboratory, the 309 Hospital of Chinese People's Liberation Army, Beijing, 100091, China;
[3]The 309 Hospital of Chinese People's Liberation Army, Beijing, 100091, China)

Abstract A series of novel monosubstituted sulfonylurea derivatives **10a-y** were synthesized and characterized by ^1H NMR, ^{13}C NMR and HRMS. These compounds were evaluated against *Mycobacterium tuberculosis* H37Rv *in vitro*. The results showed compounds **10f**, **10k** and **10s** exhibited moderate antituberculosis activities with MIC values in the range of 20-100mg/L. Compounds **10b** and **10o** displayed good antituberculosis activities (MIC 10 mg/L), which were comparable with that of the sulfometuron methyl. Both of the two compounds showed little cytotoxicities, with an IC_{50} against THP-1 cells greater than 100mg/L.

Key words sulfonylurea; C_5-Substituted benzene ring; antimycobacterial activity; mycobacterium tuberculosis; cytotoxicity

1 Introduction

Tuberculosis (TB) remains the international public health concern. More than 2 billion people, roughly one-third of the world's population, are infected with TB bacilli, the microbes that cause TB[1]. Multidrug-resistant TB (MDR TB) refers to strains that are resistant to at least isoniazid (INH) and rifampin (RFP). Extensively drug-resistant TB (XDR TB) refers to strains that are MDR-plus resistance to at least one of the three injectable second-line drugs, amikacin, kanamycin or capreomycin and a fluoroquinolone. MDR TB and XDR TB make the problem increasingly more complex for which little effective strategy was confirmed[2]. Therefore, it appears to be a very urgent need for discovery and development of new anti-TB agents that act via novel mechanisms, especially for MDR and XDR TB cases[3].

One effective strategy for the development of new therapies against TB is to target essential biosynthetic pathways of the microorganism that are absent in humans. Recently, aceto-

* Reprinted from *European Journal of Medicinal Chemistry*, 2012, 50: 18-26. The project was supported by the National Natural Science Foundation of China (project number NNSFC 20872069), National Basic Research Program of China (2010CB126106), the key project of National Natural Science Foundation of China (No. 30970419), the National Natural Science Foundation of China (No. 81000001) and the Project of PLA 309 Hospital Foundation (No. 309-09-14 and No. 309-09-02).

** Co-first authors Li Pan, Ying Jiang and Zhen Liu contributed equally to this work.

hydroxyacid synthase (AHAS, EC 2.2.1.6, also referred as acetolactate synthase, ALS) has been identified as an attractive target for design new generation of anti – TB agents[4,5]. AHAS catalyses the conversion of two molecules of pyruvate to 2 – acetolactate and CO_2, which plays an important part in biosynthesis of the branched – chain amino acids (leucine, isoleucine and valine) by higher plants, algae, fungi and bacteria, and its homologous protein was not found in humans and animals[6]. Identified inhibitors of plant AHAS, such as sulfonylurea (Fig. 1) and imidazolinone, have already been widely and safely used as commercial herbicides[7,8]. Previous researches have studied the bacteriostasis of plant AHAS inhibitors against TB, founding that some of sulfonylurea compounds, such as sulfometuron methyl (SM), chlorimuron ethyl (CE) and metsulfuron methyl (MM) displayed potent efficacies against TB strains *in vitro*[4,9,10]. Grandoni et al. demonstrated that high doses of sulfometuron methyl significantly prevented the growth of the standard TB strain H37Rv in the lungs of a mouse model, proving its anti – TB activity *in vivo*[4]. However, imidazolinone compounds did not show any inhibition against TB[10].

Chlorsulfuron   Pyrazosulfuron Ethyl

Metsulfuron Methyl   Sulfometuron Methyl

Triasulfuron   Nicosulfuron

Fig. 1 Commercial sulfonylureas as inhibitors of plant AHAS

The difficulty in developing new anti – TB drugs using sulfonylurea compounds is that they are not efficient enough. MIC values suggested that even the most powerful sulfonylurea compound, to our best knowledge, demonstrated much lower efficacy than the current TB treatment regimen (for example, NIH and RFP). Therefore, synthesis of more effective sulfonylurea derivatives will be required.

The general features of the sulfonylureas are a central sulfonylurea bridge with an *o* – substituted aromatic ring attached to the sulfur atom and a heterocyclic ring attached to the nitrogen atom. Sohn et al. discussed the relationship between the structure of sulfonylureas and anti – TB activities, and it was presumed that the aromatic backbone and the heterocyclic ring tail structure of these compounds determine the differences in anti – TB activity[9]. However, the heterocyclic ring was substituted in both meta – positions in all the compounds tested by them. Our previous studies showed that some sulfonylurea compounds with only one substituent at the metaposition of the heterocyclic ring exhibited significant and peculiar AHAS inhibitory activities, and we also found monosulfuron and monosulfuron - ester[17] (both with 4 - monosubstituted pyrimidine) had observable activities against TB[18].

Herein, 25 novel sulfonylurea compounds containing 4 - monosubstituted pyrimidine modified at C_5 - substituted acyl aniline were designed, synthesized and tested for antimycobacterial activities against standard TB strain H37Rv. Five new compounds were identified to have antimycobacterial activities with the MIC values less than 100mg/L, among of which 2 compounds showed the activities equal to sulfometuron methyl. We then evaluated their activities for clinical isolates, including MDR and XDR cases. In order to understand the possibility of drug development based these compounds, the antimicrobial activities against nontuberculosis mycobacteria (NTM) and the cytotoxicity were also evaluated.

2 Results and discussion

2.1 Chemistry

The synthetic route of intermediates (**6a - 6m**) is outlined in Scheme 1. We tried to prepare compound **3** from the reaction of 4 - chloronitrobenzene with chlorosulfonic acid. However, when the reaction was quenched by water, black sticky solid was obtained with low yield. So the synthetic route of compound **3** was optimized according another route. The substituted benzenesulfonic acid **2** was synthesized by electrophilic substitution of 4 - chloronitrobenzene with fuming sulfuric acid in good yield. Subsequent reacting with chlorosulfonic acid by heating at 130℃ afforded the desired benzenesulfonyl chloride **3**. The substituted benzenesulfonyl chloride was subsequently condensed with ammonia water to give corresponding substitude benzenesulfonic amide **4**. Compound **5** was obtained by reduction of **4** at hydrogen atmosphere in the presence of

$ClSO_3H$, 120℃, 22h

4-Chloronitrobenzene 1 2 3

4 5 6a–k

Acyl chloride

6a: R = Me; **6b**: R = CH_2Cl; **6c**: R = *i* - Pr; **6d**: R = cyclo - Pr; **6e**: R = *t* - Bu;
6f: R = vinyl; **6g**: R = $CHClCH_3$; **6h**: R = acetoxymethyl;
6i: R = mono - Ethyl oxalyl; **6j**: R = *n* - C_3F_7; **6k**: R = Ph.

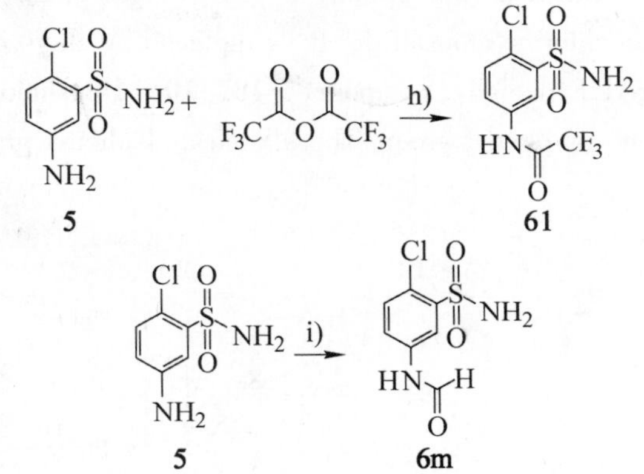

a) oleum, 115℃, 6h; b) NaOH, H_2O, rt, 3h; c) $ClSO_3H$, 130℃, 4h;

d) NH_3 (aq), rt, overnight; e) Pd/C, H_2, CH_3OH, 50℃, 4h;

f) $(COCl)_2$, DMF, rt, 4h; g) Et_3N, THF, 0℃→rt, 2h;

h) THF, 0℃→rt, 2h; i) HCOOH, HCOONa, rt, 10h

Scheme 1 The synthesis route of the substituted benzenesulfonic amides **6a – 6m**

Pd/C as a catalyst under mild condition. A mixture of **5**, Et_3N, and THF was sirred at 0℃. Then, acyl chloride was added, and the mixture was stirred at room temperature for 2h, then corresponding acylated products **6a – 6j** were obtained in excellent yield. Similarly, the reaction of **5** with trifluoroacetic anhydride gave **6l** in 84% yield. According to the literature[19], formylation of 5 gave the intermediate **6m** in excellent yield by using catalytic amount of sodium formate in formic acid under solvent – free conditions.

As indicated in Scheme 2 the title compounds (**10a – 10y**) were prepared via two steps or four steps. Treatment of isocytosine with phosphorus oxychloride followed by water indeed resulted in the chlorinated product **7**. Surprisingly, if the reaction added 10% chlorosulfonic acid as catalyst, the yield is higher (from 59% to 73%). Methylation with sodium methoxide gave the product **8a** in good yield. The pyrimidinamines (**8a**, **8b**) was converted to the carbamates (**9a**, **9b**) via the nucleophilic substitution. The condensation of the substituted benzenesulfonic amides **6a – 6m** with the phenylcarbamates (**9a**, **9b**) under DBU as catalyst in acetonitrile afforded the final products (**10a – 10y**).

2.2 Biological activity

The minimum inhibitory concentrations (MIC) of the title compounds **10a – 10y** against H37Rv were listed in Table 1. According to the references[9,10,12], sulfonylurea compounds with good antituberculosis activities contain 4,6 – disubstituted pyrimidine. But as shown in Table 1, we found that some of 4 – monosubstituted pyrimidine sulfonylureas also showed high antituberculosis activities. The most active agents were **10b** and **10o**, with the MIC values at 10mg/L, and they demonstrated the same efficacies with sulfometuron methyl tested in our study. Another three compounds, **10f**, **10s** and **10k**, also exhibited visible antituberculosis activities, with the MIC values at 20mg/L, 40mg/L and 100mg/L, respectively. However, other compounds did not exhibit significant inhibition against H37Rv at the MIC values above 100mg/L. Surprisingly, it was found that compounds **10b** and **10o** contain the same CH_2Cl group, possibly because of the inductive effect of a single chlorine atom. If the R_1 is replaced by alkyl, ester or perfluoro alkyl, the activity is disappeared. Especially, compounds **10f**, **10s**, **10k** held 20 – 100mg/L MICs against H37Rv with alkene or phenyl group. The alkene and phenyl group may play a role of electron transport ability.

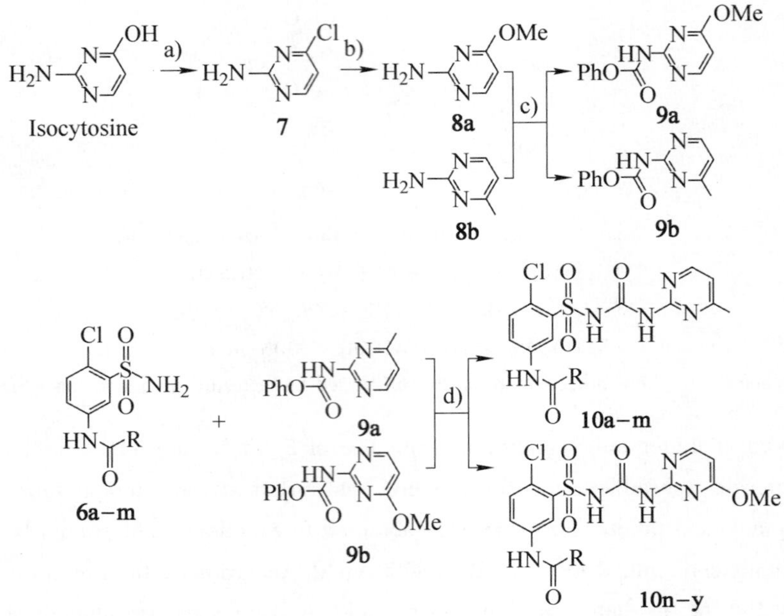

Isocytosine → a) → 7 → b) → 8a; 8a, 8b → c) → 9a, 9b

6a–m + 9a / 9b → d) → 10a–m, 10n–y

a) $POCl_3$, $ClSO_3H$, −5℃→45℃→90℃, 5h; b) CH_3ONa, CH_3OH, reflux, 3h; c) ClCOOPh, acetone, K_2CO_3, rt, 2h; d) CH_3CN, DBU, rt, overnight

Scheme 2 The synthesis route of title compounds

Table 1 Antimycobacterial activities of novel sulfonylurea derivatives

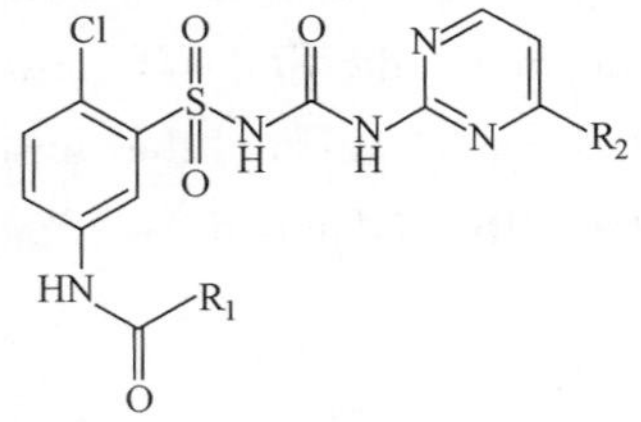

Compounds	R_1	R_2	MIC for H37Rv (mg/L)
10a	CH_3	CH_3	>100
10b	CH_2Cl	CH_3	10
10c	$CH(CH_3)_2$	CH_3	>100
10d	▷—	CH_3	>100
10e	$C(CH_3)_3$	CH_3	>100
10f	$CH=CH_2$	CH_3	20
10g	$CHClCH_3$	CH_3	>100
10h	CH_2OOCCH_3	CH_3	>100
10i	$COOC_2H_5$	CH_3	>100

Table 1 (continued)

Compounds	R_1	R_2	MIC for H37Rv (mg/L)
10j	$n-C_3F_7$	CH_3	>100
10k	C_6H_5	CH_3	100
10l	CF_3	CH_3	>100
10m	H	CH_3	>100
10n	CH_3	OCH_3	>100
10o	CH_2Cl	OCH_3	10
10p	$CH(CH_3)_2$	OCH_3	>100
10q	▷—	OCH_3	>100
10r	$C(CH_3)_3$	OCH_3	>100
10s	$CH=CH_2$	OCH_3	40
10t	$CHClCH_3$	OCH_3	>100
10u	CH_2OOCCH_3	OCH_3	>100
10v	$COOC_2H_5$	OCH_3	>100
10w	$n-C_3F_7$	OCH_3	>100
10x	CF_3	OCH_3	>100
10y	H	OCH_3	>100
SM	—	—	10

The most active compounds, **10b** and **10o**, were further evaluated for their activities against clinical TB strains isolated from the 309 Hospital of Chinese People's Liberation Army, which is the center for tuberculosis research and treatment, located in Beijing. Characterization of the drug resistance phenotype of the clinical TB strains against first-line and second-line anti-TB drugs was described in another report[11]. A to J strains were randomly selected from these isolates. The drug-resistant phenotype of certain isolates is shown in Table 2. As expected, the two compounds, exhibited the same degree of activities against all the isolates. This suggested that anti-TB agents based on **10b** and **10o** would be effective for the clinical isolates, even MDR or XDR strains.

Table 2 Antimycobacterial activities of compound 10b and 10o against clinical isolates of TB and non-tuberculosis mycobacteria

No. of isolates	Isolates	Resistance phenotypes	MIC (mg/L)	
			10b	**10o**
	H37Rv	Susceptible	10	10
A	All-susceptible	Susceptible	10	10
B	Single-resistant	INH	10	10
C	Single-resistant	INH	10	10
D	MDR	RFP, INH	10	10
E	MDR	RFP, INH, SM	10	10
F	MDR	RFP, INH, EMB	10	10

Table 2(continued)

No. of isolates	Isolates	Resistance phenotypes	MIC(mg/L)	
			10b	**10o**
G	MDR	RFP,INH,EMB	10	10
H	XDR	RFP,INH,AM,OLF	10	10
I	XDR	RFP,INH,KN,OLF	10	10
J	XDR	RFP,INH,AM,OLF	10	10
	M. bovis		10	10
	M. avium		>100	>100
	M. kansasii		20	20

In many countries, non – tuberculous mycobacteria lung disease due to *Mycobacterium avium* and *Mycobacterium kansasii* were more commonly than *Mycobacterium tuberculosis*. Sohn et al. found that SM demonstrated the activity against *M. kansasii* approximately two – to four – fold higher than the MIC against *M. tuberculosis*, and did not significantly inhibit the growth of *M. avium*[9]. Therefore, in our study, *Mycobacterium bovis*, *M. avium* and *M. kansasii* were also selected to test the activities of the compounds (Table 1). Similar with Sohn's report, we also found that **10b** and **10o** were not as effective against these NTM strains as they showed to TB.

2.3 Cytotoxicity

In order to determine whether **10b** and **10o** have deleterious effects in mammalian cells, we evaluated their inhibition of cell viability against THP – 1 cells using the MTT assay (see Experimental Section). Both of the two compounds showed little cytotoxicities, with an IC_{50} against THP – 1 cells greater than 100mg/L (**10b**: 137.92mg/L; **10o**: 179.70mg/L) tested 72h after compounds were added, which is 10 fold higher than the MICs of these compounds for the *M. tuberculosis strains.*

3 Conclusion

To our best knowledge, this is the first report to discuss the antituberculosis activities of sulfonylurea compounds with 4 – monosubstituted pyrimidine. In this work, 25 novel structures of sulfonylurea compounds were synthesized and their activities against *M. tuberculosis* were evaluated. Experimental data demonstrated that newly identified compounds are promising lead compounds for development of novel antimycobacterial agents for the treatment of *M. tuberculosis*. Further structural modification is necessary for improving efficacy against MTB strains. Furthermore, it is need to evaluate the *in vivo* efficacies of these agents for the treatment of MDR and XDR TB.

4 Experimental

4.1 Materials and methods

^{1}H NMR spectra and ^{13}C NMR spectra were recorded at 300MHz using a Bruker AV 300 spectrometer or 400MHz using a Bruker AV 400 spectrometer (Bruker CO., Switzerland) in $CDCl_3$ or DMSO – d_6 solution with tetramethylsilane as the internal standard, and chemical shift values (δ) were given in ppm. High – resolutionmass spectrometry (HRMS) data were obtained on a high resolution FTICR – MS (Varian 7.0T). Flash chromatography was performed with silica gel (200 – 300mesh). The melting points were determined on an X – 4 binocular microscope melting point apparatus (Beijing Tech Instruments Co., Beijing, China) and were uncorrected. Reagents were all analytically or chemically pure. All solvents and liquid reagents were dried by standard methods in advance and distilled before use.

4.2 Synthesis

4.2.1 2 – Chloro – 5 – nitrobenzenesulfonic acid (**1**)[13]

Parachloronitrobenzene (60g, 0.381mol) was dissolved in 180g of 20% oleum with good agitation at 0℃. The mixture was heated cautiously at 110 – 115℃ with vigorous stirring for 6h. After completion of the reaction, the resulting mixture was then cooled and poured onto 2000mL of sodium chloride solution (15% by weight) at 5 – 10℃. After seeding and stirring for 0.5h, the crystalline precipitate of 2 – chloro – 5 – nitrobenzenesulfonic acid **1** which formed was filtered off and dried at about 60℃. The yield was 80%, m. p. 169℃ (decomposed).

4.2.2 Sodium 2 – chloro – 5 – nitrobenzenesulfonate (**2**)

To a solution of 2 – chloro – 5 – nitrobenzenesulfonic acid (25g, 0.105mol) in 15mL of water, was added a solution of sodium hydroxide (4.21g, 0.105mol) in 15mL of water with vigorous stirring. The resulting mixture was stirred at room temperature for 1h. The resulting suspension was filtered, washed with water, dried to offer, compound **2** as a gray solid powder, 21.72g, yield, 80%; m. p. >300℃.

4.2.3 2 – Chloro – 5 – nitrobenzenesulfonyl chloride (**3**)

To chlorosulphonic acid (14.96g, 8.45mL) was added sodium 2 – chloro – 5 – nitrobenzenesulfonate (5.56g, 21.4mmol) with stirring at 5 – 10℃. The reaction mixture was stirred at 130℃ for 4h and then poured onto 120g of ice with vigorous stirring. After stirring for 10min, the gray precipitate was filtered and pressed dry. The solid was dried overnight in a desiccator under vacuum (4.43g, 81% yield, m. p. 88 – 90℃).

4.2.4 2 – Chloro – 5 – nitrobenzenesulfonamide (**4**)

To 28% ammonia water (4.71g, 69.2mmol), compound **3** (4.43g, 17.3mmol) was added in small portions at 10℃. When the addition was complete, the reaction was stirred at room temperature overnight. The suspension was acidified with 5mol/L hydrochloric acid to pH 7. The white precipitate was collected by filtration to give benzenesulfonamide **4**, 3.68g, yield 90%, m. p. 184 – 185℃ ^1H NMR (400MHz, DMSO – d_6) δ (ppm): 7.969 (d, 1H, J_H = 8.8Hz, Ar—H), 8.017 (s, 2H, SO_2NH_2), 8.414 – 8.442 (m, 1H, Ar—H), 8.671 – 8.677 (m, 1H, Ar—H).

4.2.5 5 – Amino – 2 – chlorobenzenesulfonamide (**5**)[14]

The compound **4** (14.15g, 59.8mmol) and 300mL of methanol were placed into a flask and stirred at 50℃ until dissolved completely and then 10% Pd/C (1.4g) was added. The mixture was intensively stirred under H_2 at 50℃ and monitored with TLC. After 4h, the reaction was complete, and the suspension was filtered. The resulting solution was evaporated to give the crude product. It was dissolved in 30mL of 9mol/L hydrochloric acid solution and filtered. The filtrate was alkalified with 10mol/L sodium hydroxide solution to pH 7. The suspension was filtered to afford the compound **5**(10.27g, 83% yield, m.p. 166 – 167℃). ^{1}H NMR(400MHz, DMSO – d_6)δ(ppm): 7.969(d, 1H, J_H = 8.8Hz, Ar—H), 8.017(s, 2H, SO_2NH_2), 8.414 – 8.442(m, 1H, Ar—H), 8.671 – 8.677(m, 1H, Ar—H).

4.2.6 General synthetic procedure for intermediates 6a – 6k

To a mixture of 5 – amino – 2 – chlorobenzenesulfonamide(1.24g, 6.0mmol) and Et_3N(0.67g, 6.6mmol) in 60mL of THF at 0℃, was added dropwise acyl chloride(6.5mmol). The reaction was monitored with TLC and complete after 2h of stirring at room temperature. Then the resulting suspension was filtered, washed with THF(10mL). The filtrate was evaporated, and 30mL of water was added to the residue. The mixture was intensively stirred for 20min, and the resulting suspension was filtered to give the intermediates **6a – 6k**.

4.2.7 *N* – (4 – chloro – 3 – sulfamoylphenyl) – 2,2,2 – trifluoroacetamide(6l)

To a solution of 5 – amino – 2 – chlorobenzenesulfonamide(1.24g, 6.0mmol) in 60mL of THF at 0℃, was added dropwise trifluoroacetic anhydride(1.37g, 6.5mmol). The reaction was monitored with TLC and complete after 3h of stirring at room temperature. Then the solvent was evaporated, and 30mL of water was added to the residue. The mixture was intensively stirred for 10min, and the resulting suspension was filtered to afford the intermediate **6l**.

4.2.8 *N* – (4 – chloro – 3 – sulfamoylphenyl) formamide(6m)[19]

The compound **5** (1.24g, 6.0mmol), formic acid (2.23g, 48.0mmol) and sodium formate (82mg, 1.2mmol) were placed into a flask. The reaction was monitored with TLC and complete after 10h of stirring at room temperature. The resulting mixture was poured onto 50mL of water, and the suspension was stirred for 10min. After stewing at 10℃ for 10min, the white precipitate was filtered to give the intermediate **6m**.

N – (4 – chloro – 3 – sulfamoylphenyl) acetamide(**6a**)

White solid, yield 92%, m.p. 245 – 247℃; ^{1}H NMR(400MHz, DMSO – d_6)δ(ppm): 2.073 (s, 3H, $COCH_3$), 7.549(d, 1H, J_H = 8.8Hz, Ar—H), 7.611(s, 2H, SO_2NH_2), 7.816 – 7.843(m, 1H, Ar—H), 8.298(d, 1H, J_H = 2.4Hz, Ar—H), 10.398(s, 1H, CONH—Ar).

2 – Chloro – *N* – (4 – chloro – 3 – sulfamoylphenyl) acetamide(**6b**)

Light yellow solid, yield 94%, m.p. 190 – 192℃; ^{1}H NMR(400MHz, DMSO – d_6)δ(ppm): 4.294(s, 2H, $COCH_2Cl$), 7.600(d, 1H, J_H = 8.8Hz, Ar—H), 7.658(s, 2H, SO_2NH_2), 7.826 – 7.853(m, 1H, Ar—H), 8.311(d, 1H, J_H = 2.0Hz, Ar—H), 10.750(s, 1H, CONH—Ar).

N – (4 – chloro – 3 – sulfamoylphenyl) isobutyramide(**6c**)

White solid, yield 93%, m.p. 215 – 217℃; ^{1}H NMR(400MHz, DMSO – d_6)δ(ppm): 1.107 (d, 6H, J_H = 6.8Hz, isopropyl – CH_3), 2.543 – 2.672(m, 1H, isopropyl – CH), 7.547(d, 1H, J_H = 8.8Hz, Ar—H), 7.588(s, 2H, SO_2NH_2), 7.841 – 7.867(m, 1H, Ar—H), 8.337 (d, 1H, J_H = 1.6Hz, Ar—H), 10.243(s, 1H, CONH—Ar).

N – (4 – chloro – 3 – sulfamoylphenyl) cyclopropanecarboxamide(**6d**)

White solid, yield 84%, m. p. 259 - 260℃; ^1H NMR(400MHz, DMSO - d_6) δ(ppm): 0.814 - 0.840(m, ^1H, cyclopropyl - CH_2), 1.751 - 1.782(m, 1H, cyclopropyl - CH), 7.532 - 7.554 (m, 1H, Ar—H), 7.586(s, 2H, SO_2NH_2), 7.820 - 7.839(m, 1H, Ar—H), 8.311(s, 1H, Ar—H), 10.602(s, 1H, CONH—Ar).

N-(4-chloro-3-sulfamoylphenyl) pivalamide(**6e**)

White solid, yield 90%, m. p. 236 - 237℃; ^1H NMR(400MHz, DMSO - d_6) δ(ppm): 1.225(s, 9H, CO - neopentyl), 7.540(d, 1H, J_H = 8.8Hz, Ar—H), 7.556(s, 2H, SO_2NH_2), 7.932 - 7.960(m, 1H, Ar—H), 8.385(d, 1H, J_H = 2.8Hz, Ar—H), 9.621(s, 1H, CONH—Ar).

N-(4-chloro-3-sulfamoylphenyl) acrylamide(**6f**)

White solid, yield 78%, m. p. 231 - 233℃; ^1H NMR(400MHz, DMSO - d_6) δ(ppm): 5.832 (s, 1H, CH $=CH_2$), 6.297 - 6.452(m, 2H, CH $=CH_2$), 7.633(s, 3H, Ar—H and SO_2NH_2), 7.946(s, 1H, Ar—H), 8.372(s, 1H, Ar—H), 10.566(s, 1H, CONH—Ar).

2-Chloro-*N*-(4-chloro-3-sulfamoylphenyl) propanamide(**6g**)

White solid, yield 89%, m. p. 208 - 210℃; ^1H NMR(400MHz, DMSO - d_6) δ(ppm): 1.635 (d, 3H, J_H = 6.0Hz, $COCHClCH_3$), 4.656 - 4.704(m, 1H, $COCHClCH_3$), 7.617(d, 1H, J_H = 8.8Hz, Ar—H), 7.649(s, 2H, SO_2NH_2), 7.860(d, 1H, J_H = 8.4Hz, Ar—H), 8.338(s, 1H, Ar—H), 10.746(s, 1H, CONH—Ar).

2-(4-Chloro-3-sulfamoylphenylamino)-2-oxoethyl acetate(**6h**)

White solid, yield 86%, m. p. 203 - 204℃; ^1H NMR(400MHz, DMSO - d_6) δ(ppm): 2.132 (s, 3H, $COCH_2OCOCH_3$), 4.671(s, 2H, $COCH_2OCOCH_3$), 7.586(d, 1H, J_H = 8.4Hz, Ar—H), 7.634(s, 2H, SO_2NH_2), 7.792 - 7.817(m, 1H, Ar—H), 8.313(s, 1H, Ar—H), 10.510 (s, 1H, CONH—Ar).

Ethyl 2-(4-chloro-3-sulfamoylphenylamino)-2-oxoacetate(**6i**)

White solid, yield 89%, m. p. 210 - 212℃; ^1H NMR(400MHz, DMSO - d_6) δ(ppm): 1.317 (t, 3H, J_{12} = 6.4Hz, OCH_2CH_3), 4.315(q, 2H, J_{12} = 6.8Hz, J_{13} = 13.2Hz, OCH_2CH_3), 7.622 - 7.644(m, 3H, Ar—H and SO_2NH_2), 7.951(d, 1H, J_H = 8.8Hz, Ar—H), 8.528(s, 1H, Ar—H), 11.183(s, 1H, CONH—Ar).

N-(4-chloro-3-sulfamoylphenyl)-2,2,3,3,4,4,4-heptafluorobutanamide(**6j**)

White solid, yield 88%, m. p. 188 - 190℃; ^1H NMR(400MHz, DMSO - d_6) δ(ppm): 7.702 (d, 1H, J_H = 8.8Hz, Ar—H), 7.714(s, 2H, SO_2NH_2), 7.931 - 7.958(m, 1H, Ar—H), 8.434(d, 1H, J_H = 2.0Hz, Ar—H), 11.691(s, 1H, CONH—Ar).

N-(4-chloro-3-sulfamoylphenyl) benzamide(**6k**)

White solid, yield 78%, m. p. 271 - 272℃; ^1H NMR(400MHz, DMSO - d_6) δ(ppm): 7.544 - 7.634(m, 6H, Ar—H and SO_2NH_2), 7.986 - 8.083(m, 3H, Ar—H), 8.561(s, 1H, Ar—H), 10.644(s, 1H, CONH—Ar).

N-(4-chloro-3-sulfamoylphenyl)-2,2,2-trifluoroacetamide(**6l**)

Light yellow solid, yield 84%, m. p. 258 - 259℃; ^1H NMR(400MHz, DMSO - d_6) δ(ppm): 7.691(d, 1H, J_H = 8.8Hz, Ar—H), 7.705(s, 2H, SO_2NH_2), 7.915(d, 1H, J_H = 8.0Hz, Ar—H), 8.405(s, 1H, Ar—H), 11.648(s, 1H, CONH—Ar).

N-(4-chloro-3-sulfamoylphenyl) formamide(**6m**)

White solid, yield 82%, m. p. 227 - 229℃; ^1H NMR(400MHz, DMSO - d_6) δ(ppm): 7.592 (d, 1H, J_H = 8.8Hz, Ar—H), 7.630(s, 2H, SO_2NH_2), 7.789 - 7.816(m, 1H, Ar—H),

8.336(d,1H,J_H=2.0Hz,Ar—H),8.347(s,1H,CHO),10.591(s,1H,CONH—Ar).

4.2.9 4-Chloro-2-pyrimidinamine(**7**)

In a three neck 250mL round bottom flask equipped with reflux condenser was added isocytosine(20g,0.18mol) followed by phosphorus oxychloride(40mL,0.43mol) at -5℃ and under an atmosphere of nitrogen. To this was cautiously added chlorosulfonic acid(4mL,0.06mol). The solution was permitted to heat spontaneously to room temperature and then heated at 45℃ for 1h. It was then heated cautiously at 90℃ for 4h, before it was cooled to room temperature. The mixture was then cooled in an ice bath before it was pour onto 400mL of ice water with vigorous stirring. Adjusted the pH to 7 with ammonia water(28% by weight)(temperature was held below 20℃). A tan colored solid was collected by filtration. The solid was dried to afford 4-chloro-2-pyrimidinamine(17g,73% yield) as an off white solid. m. p. 166-167℃.

4.2.10 4-Methoxy-2-pyrimidinamine(**8a**)

Na(2.2g,92.6mmol) was dissolved in 30mL of methanol. To this solution was added **7**(6.0g, 46.3mmol) in small portions at 5℃. When the addition was complete, this mixture was heated to reflux for 4h. After cooling, then the solvent was evaporated. The residue was dissolved in 30mL of hydrochloric acid solution(20% by weight) and filtered. The filtrate was alkalified with 10mol/L sodium hydroxide solution to pH 9. After seeding and stirring for 1h, the crystalline precipitate of the pyrimidinamine **8a** which formed was filtered off and dried at about 60℃. The yield was 84%, m. p. 117-119℃.

4.2.11 Phenyl 4-methoxypyrimidin-2-ylcarbamate(**9a**)[15]

The compound **8a**(12.5g,0.1mol), potassium carbonate(16.6g,0.12mol) and 150 mL of acetone were placed into a flask. To this mixture was added phenyl chloroformate(15.7g,0.1mol) for 20-30min. After stirring at room temperature for 2h, the suspension was filtered. The precipitate was washed with sodium carbonate solution(10% by weight), hydrochloric acid solution (5% by weight), and water. The crude carbamate **9a** was obtained as a white powder(18.6g, 76% crude yield), which was used directly in the next step without further purification. m. p. 147-149℃(decomposed). ^{1}H NMR(400MHz, DMSO-d_6)δ:3.903(s,3H,pyrim-OCH_3),6.616 (d,1H,J_H=5.6Hz, pyrim-H_5),7.208(d,2H,J_H=7.6Hz, Ar—H),7.271(t,1H,J_{12}= 7.6Hz,Ar—H),7.437(t,2H,J_{12}=7.6Hz,Ar—H),8.361(d,1H,J_H=5.6Hz,pyrim-H_6), 10.864(s,1H,CONH-pyrim).

4.2.12 Phenyl 4-methylpyrimidin-2-ylcarbamate(**9b**)[15]

The pyrimidinamine **8b**(10.9g,0.1mol), potassium carbonate(16.6g,0.12mol) and 250mL of acetone were placed into a flask. To this mixture was added phenyl chloroformate(15.7g, 0.1mol) for 20-30min. After stirring at room temperature for 2h, the suspension was filtered. The filtrate was collected, and the solvent was evaporated to give the yellow oil. The residue was washed with sodium carbonate solution(10% by weight), hydrochloric acid solution (5% by weight), and water to give the crude product **9b** as a light yellow powder(16.7g,73% crude yield), which was used directly in the next step without further purification. m. p. 118-122℃ ^{1}H NMR(300MHz, $CDCl_3$)δ:2.502(s,3H,pyrim-CH_3),6.892(d,1H,J_H=6.0Hz, pyrim - H_5), 7. 167 - 7. 242(m, 3H, Ar—H), 7. 328 - 7. 409(m, 2H, Ar—H), 8. 501(d, 1H, J_H = 6. 0Hz, pyrim - H_6), 8. 882(s, 1H, CONH - pyrim).

4. 2. 13 General synthetic procedure for title compounds **10a - 10y**

At room temperature, 5 - acylamino - 2 - chlorobenzenesulfonamide (2. 0mmol) and phenyl *N* - (4 - substitutedpyrimidine - 2 - yl) carbamate (2. 0mmol) were suspended in 15mL of acetonitrile and thereafter 1, 8 - diazabicyclo[5. 4. 0]undec - 7 - ene (DBU) in 3mL of acetonitrile was added dropwise to this suspension. After 8 - 24h of stirring at room temperature the solution was slowly adjusted to pH = 1 using 2mol/L hydrochloric acid. The resulting precipitate was collected by filtration, washed with water and dried in air to obtain the final product **10**. If there was not any precipitate appearance, the mixture was extracted with dichloromethane (15mL × 3). The extract was dried over sodium sulfate and evaporated to yield compound **10**. The crude product was purified by recrystallization from the mixture of petroleumether and acetone or by flash chromatography on silica gel with petroleum ether/acetone (2∶1).

N - (4 - chloro - 3 - (*N* - (4 - methylpyrimidin - 2 ylcarbamoyl) sulfamoyl) phenyl) acetamide (**10a**)

Light yellow solid, yield 75%, m. p. 192 - 193℃; ^{1}H NMR (300MHz, DMSO - d_6) δ (ppm): 2. 086(s, 3H, $COCH_3$), 2. 458(s, 3H, pyrim - CH_3), 7. 144(d, 1H, J_H = 5. 4Hz, pyrim - H_5), 7. 600(d, 1H, J_H = 8. 7Hz, Ar—H), 7. 968 - 8. 006 (m, 1H, Ar—H), 8. 384 (d, 1H, J_H = 2. 7Hz, Ar—H), 8. 573 (d, 1H, J_H = 5. 4Hz, pyrim - H_6), 10. 465 (s, 1H, CONH—Ar), 10. 745(br, 1H, CONH - pyrim), 13. 388(br, 1H, SO_2NH). ^{13}C NMR (300MHz, DMSO - d_6) δ (ppm): 22. 368, 22. 860, 108. 338, 114. 537, 120. 798, 122. 053, 123. 369, 130. 978, 137. 417, 147. 697, 155. 419, 156. 265, 167. 637, 167. 816. ESI - FTICR - MS for $C_{14}H_{14}ClN_5O_4S$: found 384. 0529[M + H]$^+$, calcd 384. 0528 [M + H]$^+$ (error 0. 260 ppm).

2 - Chloro - *N* - (4 - chloro - 3 - (*N* - (4 - methylpyrimidin - 2 - ylcarbamoyl) sulfamoyl) phenyl) acetamide (**10b**)

Light yellow solid, yield 86%, m. p. 194 - 195℃; ^{1}H NMR (400MHz, DMSO - d_6) δ (ppm): 2. 465(s, 3H, pyrim - CH_3), 4. 310(s, 2H, $COCH_2Cl$), 7. 151 (d, 1H, J_H = 5. 2Hz, pyrim - H_5), 7. 658(d, 1H, J_H = 8. 8Hz, Ar—H), 7. 985 - 8. 013 (m, 1H, Ar—H), 8. 405 (d, 1H, J_H = 2. 4Hz, Ar—H), 8. 576 (d, 1H, J_H = 5. 2Hz, pyrim - H_6), 10. 773 (br, 1H, CONH - pyrim), 10. 844(s, 1H, CONH—Ar), 13. 433 (br, 1H, SO_2NH). ^{13}C NMR (400MHz, DMSO - d_6) δ (ppm): 23. 502, 43. 433, 109. 434, 115. 670, 122. 336, 124. 127, 124. 914, 132. 332, 137. 733, 148. 892, 156. 440, 157. 277, 165. 280, 168. 834. ESI - FTICR - MS for $C_{14}H_{13}C_{12}N_5O_4S$: found 439. 9951[M + Na]$^+$, calcd 439. 9957[M + Na]$^+$ (error 1. 364ppm).

N - (4 - chloro - 3 - (*N* - (4 - methylpyrimidin - 2 - ylcarbamoyl) sulfamoyl) phenyl) isobutyramide (**10c**)

Light yellow solid, yield 80%, m. p. 152 - 154℃; ^{1}H NMR (400MHz, DMSO - d_6) δ (ppm): 1. 123(d, 6H, J_H = 6. 8Hz, isopropyl - CH_3), 2. 460(s, 3H, pyrim - CH_3), 2. 578 - 2. 645(m, 1H, isopropyl - CH), 7. 146(d, 1H, J_H = 4. 8Hz, pyrim - H_5), 7. 505(d, 1H, J_H = 8. 8Hz, Ar—H), 8. 016(d, 1H, J_H = 8. 8Hz, Ar—H), 8. 457(s, 1H, Ar—H), 8. 576(d, 1H, J_H = 4. 8Hz, pyrim - H_6), 10. 369(s, 1H, CONH—Ar), 10. 764(br, 1H, CONH - pyrim), 13. 377(br, 1H, SO_2NH). ^{13}C NMR (400MHz, DMSO - d_6) δ (ppm): 19. 274, 23. 481, 35. 059, 109. 436, 115. 436, 122. 055, 123. 093, 124. 653, 132. 065, 138. 667, 148. 744, 156. 485, 157. 401,

168.758,175.824. ESI-FTICR-MS for $C_{16}H_{18}ClN_5O_4S$: found 434.0654[M+Na]$^+$ calcd 434.0660[M+Na]$^+$(error 1.382 ppm).

N-(4-chloro-3-(*N*-(4-methylpyrimidin-2-ylcarbamoyl)sulfamoyl)phenyl)cyclopropanecarboxamide(**10d**)

White solid, yield 77%, m.p. 259-261℃; ^{1}H NMR(400MHz, DMSO-d_6)δ(ppm): 0.848 (d,4H, J_H=5.2Hz, cyclopropyl-CH_2), 1.756-1.799(m,1H, cyclopropyl-CH), 2.456 (s,3H, pyrim-CH_3), 7.139(d,1H, J_H=3.6Hz, pyrim-H_5), 7.597(d,1H, J_H=8.0Hz, Ar—H), 7.971(d,1H, J_H=8.8Hz, Ar—H), 8.433(s,1H, Ar—H), 8.569(d,1H, J_H=3.6Hz, pyrim-H_6), 10.734(s,2H, CONH—Ar and CONH-pyrim), 13.396(br,1H, SO_2NH). ^{13}C NMR(300MHz, DMSO-d_6)δ(ppm): 7.567, 14.618, 23.462, 109.435, 115.610, 121.887, 123.074, 124.439, 132.066, 138.500, 148.847, 156.546, 157.364, 168.699, 172.251. MALDI-FTICR-MS for $C_{16}H_{16}ClN_5O_4S$: found 432.0509[M+Na]$^+$ calcd 432.0504 [M+Na]$^+$(error 1.157ppm).

N-(4-chloro-3-(*N*-(4-methylpyrimidin-2-ylcarbamoyl)sulfamoyl)phenyl)pivalamide(**10e**)

Light yellow solid, yield 83%, m.p. 154-156℃; ^{1}H NMR(400MHz, DMSO-d_6)δ(ppm): 1.245(s,9H, CO-neopentyl), 2.462(s,3H, pyrim-CH_3), 7.149(d,1H, J_H=5.2Hz, pyrim-H_5), 7.603(d,1H, J_H=8.8Hz, Ar—H), 8.127(d,1H, J_H=8.4Hz, Ar—H), 8.529(s, 1H, Ar—H), 8.580(d,1H, J_H=4.8Hz, pyrim-H_6), 9.765(s,1H, CONH—Ar), 10.768 (br,1H, CONH-pyrim), 13.379(br, 1H, SO_2NH). ^{13}C NMR(400MHz, DMSO-d_6)δ (ppm): 23.477, 26.939, 39.356, 109.446, 115.675, 122.981, 123.270, 125.549, 131.795, 138.776, 148.759, 156.497, 157.387, 168.759, 177.102. ESI-FTICR-MS for $C_{17}H_{20}ClN_5O_4S$: found 448.0810[M+Na]$^+$, calcd 448.0817 [M+Na]$^+$(error 1.562 ppm).

N-(4-chloro-3-(*N*-(4-methylpyrimidin-2-ylcarbamoyl)sulfamoyl)phenyl)acrylamide(**10f**)

White solid, yield 76%, m.p. 142-144℃; ^{1}H NMR(400MHz, DMSO-d_6)δ(ppm): 2.463 (s,3H, pyrim-CH_3), 5.842(d,1H, J_H=10.0Hz, CH=CH_2), 6.305-6.467(m,2H, CH=CH_2), 7.149(d,1H, J_H=5.2Hz, pyrim-H_5), 7.647(d,1H, J_H=8.8Hz, Ar—H), 8.100(d,1H, J_H=8.8Hz, Ar—H), 8.470(s,1H, Ar—H), 8.577(d,1H, J_H=4.8Hz, pyrim-H_6), 10.686(s, 1H, CONH—Ar), 10.770(s, 1H, CONH-pyrim), 13.416(br, 1H, SO_2NH). ^{13}C NMR(300MHz, DMSO-d_6)δ(ppm): 23.470, 109.443, 115.573, 122.295, 123.743, 124.747, 127.930, 131.247, 132.139, 138.234, 149.037, 156.571, 157.315, 163.592, 168.703. ESI-FTICR-MS for $C_{15}H_{14}ClN_5O_4S$: found 418.0338[M+Na]$^+$ calcd 418.0347 [M+Na]$^+$(error 2.153ppm).

2-Chloro-*N*-(4-chloro-3-(N-(4-methylpyrimidin-2-ylcarbamoyl)sulfamoyl)phenyl)propanamide(**10g**)

White solid, yield 72%, m.p. 175-176℃; ^{1}H NMR(400MHz, DMSO-d_6)δ(ppm): 1.636 (d,3H, J_H=6.4Hz, $COCHClCH_3$), 2.451(s,3H, pyrim-CH_3), 4.678(q,1H, J_{12}=6.4Hz, J_{13}=12.8Hz, $COCHClCH_3$), 7.129(d,1H, J_H=2.8Hz, pyrim-H_5), 7.645(d,1H, J_H=8.0Hz, Ar—H), 8.003(d,1H, J_H=7.6Hz, Ar—H), 8.425(s,1H, Ar—H), 8.560(d,1H, J_H=1.6Hz, pyrim-H_6), 10.737(br,1H, CONH-pyrim), 10.855(s,1H, CONH—Ar), 13.439(br, 1H, SO_2NH). ^{13}C NMR(300MHz, DMSO-d_6)δ(ppm): 20.781, 23.474, 54.577, 109.434, 115.601, 122.491, 124.236, 124.938, 132.249, 137.724, 148.961,

156.519, 157.266, 167.851, 168.767. MALDI - FTICR - MS for $C_{15}H_{15}Cl_2N_5O_4S$: found 454.0111[M + Na]$^+$ calcd 454.0114 [M + Na]$^+$ (error 0.661 ppm).

2-(4-Chloro-3-(*N*-(4-methylpyrimidin-2-ylcarbamoyl) sulfamoyl) phenylamino) -2-oxoethyl acetate(**10h**)

Yellow solid, yield 80%, m. p. 185 - 186℃; ^{1}H NMR(400MHz, DMSO - d_6) δ(ppm): 2.145 (s, 3H, $COCH_2OCOCH_3$), 2.465 (s, 3H, pyrim - CH_3), 4.691 (s, 2H, $COCH_2OCOCH_3$), 7.151 (d, 1H, J_H = 5.2Hz, pyrim - H_5), 7.644 (d, 1H, J_H = 8.4Hz, Ar—H), 7.965 - 7.992 (m, 1H, Ar—H), 8.430 (d, 1H, J_H = 2.4Hz, Ar—H), 8.578 (d, 1H, J_H = 5.2Hz, pyrim - H_6), 10.639 (s, 1H, CONH—Ar), 10.797 (br, 1H, CONH - pyrim), 13.429 (br, 1H, SO_2NH). ^{13}C NMR (400MHz, DMSO - d_6) δ (ppm): 20.405, 23.500, 62.469, 109.457, 115.675, 122.309, 123.839, 124.909, 132.264, 137.678, 148.795, 156.430, 157.295, 166.226, 168.847, 170.037. MALDI - FTICR - MS for $C_{16}H_{16}ClN_5O_6S$: found 442.0581 [M + H]$^+$, calcd 442.0583[M + H]$^+$ (error 0.452 ppm).

Ethyl 2-(4-chloro-3-(*N*-(4-methylpyrimidin-2-ylcarbamoyl) sulfamoyl) phenylamino) -2-oxoacetate(**10i**)

White solid, yield 75%, m. p. 193 - 194℃; ^{1}H NMR(400MHz, DMSO - d_6) δ(ppm): 1.328 (t, 3H, J_{12} = 6.8Hz, OCH_2CH_3), 2.464 (s, 3H, pyrim - CH_3), 4.327 (q, 2H, J_{12} = 6.8Hz, J_{13} = 13.6Hz, OCH_2CH_3), 7.151 (d, 1H, J_H = 5.2Hz, pyrim - H_5), 7.694 (d, 1H, J_H = 8.8Hz, Ar—H), 8.130 (d, 1H, J_H = 8.4Hz, Ar—H), 8.577 (d, 1H, J_H = 4.8Hz, pyrim - H_6), 8.503 (s, 1H, Ar—H), 10.787 (s, 1H, CONH - pyrim), 11.287 (s, 1H, CONH—Ar), 13.433 (br, 1H, SO_2NH). ^{13}C NMR (400MHz, DMSO - d_6) δ (ppm): 13.800, 23.501, 62.593, 109.504, 115.687, 123.555, 125.200, 126.116, 132.208, 136.912, 148.825, 155.814, 156.453, 157.284, 159.883, 168.863. ESI - FTICR - MS for $C_{16}H_{16}ClN_5O_6S$: found 464.0397 [M + Na]$^+$, calcd 464.0402[M + Na]$^+$ (error 1.078ppm).

N-(4-chloro-3-(N-(4-methylpyrimidin-2-ylcarbamoyl) sulfamoyl) phenyl) -2,2,3,3,4,4,4-heptafluorobutanamide(**10j**)

White solid, yield 80%, m. p. 181 - 182℃; ^{1}H NMR(400MHz, DMSO - d_6) δ(ppm): 2.469 (s, 3H, pyrim - CH_3), 7.160 (d, 1H, J_H = 5.2Hz, pyrim - H_5), 7.756 (d, 1H, J_H = 8.8Hz, Ar—H), 8.112 (d, 1H, J_H = 8.0Hz, Ar—H), 8.555 (s, 1H, Ar—H), 8.578 (d, 1H, J_H = 5.2Hz, pyrim - H_6), 10.834 (s, 1H, CONH - pyrim), 11.810 (s, 1H, CONH—Ar), 13.507 (br, 1H, SO_2NH). ^{13}C NMR (400MHz, DMSO - d_6) δ (ppm): 23.490, 107.865, 109.427, 115.175, 115.666, 121.609, 124.250, 126.601, 126.758, 132.495, 135.578, 148.971, 155.461, 156.369, 157.126, 168.994. MALDI - FTICR - MS for $C_{16}H_{11}ClF_7N_5O_4S$: found 538.0183[M + H]$^+$, calcd 538.0181[M + H]$^+$ (error 0.372 ppm).

N-(4-chloro-3-(N-(4-methylpyrimidin-2-ylcarbamoyl) sulfamoyl) phenyl) benzamide (**10k**)

White solid, yield 85%, m. p. 262 - 263℃; ^{1}H NMR(400MHz, DMSO - d_6) δ(ppm): 2.468 (s, 3H, pyrim - CH_3), 7.151 (d, 1H, J_H = 4.8Hz, pyrim - H_5), 7.547 - 7.691 (m, 4H, Ar—H), 8.005 - 8.024 (m, 2H, Ar—H), 8.232 (d, 1H, J_H = 8.4Hz, Ar—H), 8.583 (d, 1H, J_H = 4.8Hz, pyrim - H_6), 8.687 (s, 1H, Ar—H), 10.758 (s, 2H, CONH - Ar and CONH - pyrim), 13.421 (br, 1H, SO_2NH). ^{13}C NMR (400MHz, DMSO - d_6) δ (ppm): 23.498, 109.433,

115.667, 123.192, 123.879, 125.762, 127.734, 127.774, 128.449, 132.019, 134.080, 138.550, 148.822, 156.514, 157.396, 165.889, 168.795. ESI - FTICR - MS for $C_{19}H_{16}ClN_5O_4S$: found 468.0500[M + Na]$^+$ calcd 468.0504[M + Na]$^+$ (error 0.855 ppm).

N-(4-chloro-3-(N-(4-methylpyrimidin-2-ylcarbamoyl)sulfamoyl)phenyl)-2,2,2-trifluoroacetamide(**10l**)

Light yellow solid, yield 72%, m. p. 191 - 192℃; ^1H NMR(400MHz, DMSO - d_6) δ(ppm): 2.475(s, 3H, pyrim - CH_3), 7.165(d, 1H, J_H = 4.8Hz, pyrim - H_5), 7.755(d, 1H, J_H = 8.8Hz, Ar—H), 8.098(d, 1H, J_H = 8.8Hz, Ar—H), 8.548(s, 1H, Ar—H), 8.585(d, 1H, J_H = 4.4Hz, pyrim - H_6), 10.828(br, 1H, CONH - pyrim), 11.775(s, 1H, CONH—Ar), 13.512(br, 1H, SO_2NH). ^{13}C NMR(300MHz, DMSO - d_6) δ(ppm): 23.485, 109.448, 113.562, 115.663, 124.014, 126.280, 126.557, 132.452, 135.746, 148.950, 154.115, 156.396, 157.161, 168.936. ESI - FTICR - MS for $C_{14}H_{11}ClF_3N_5O_4S$: found 460.0056[M + Na]$^+$, calcd 460.0064 [M + Na]$^+$ (error 1.739 ppm).

N-(4-chloro-3-(N-(4-methylpyrimidin-2-ylcarbamoyl)sulfamoyl)phenyl)formamide (**10m**)

Yellow solid, yield 46%, m. p. 180 - 181℃; ^1H NMR(400MHz, DMSO - d_6) δ(ppm): 2.468 (s, 3H, pyrim - CH_3), 7.153(d, 1H, J_H = 4.8Hz, pyrim - H_5), 7.647(d, 1H, J_H = 8.8Hz, Ar—H), 7.940 - 7.967(m, 1H, Ar—H), 8.385(s, 1H, CHO), 8.459(d, 1H, J_H = 2.0Hz, Ar—H), 8.581(d, 1H, J_H = 5.2Hz, pyrim - H_6), 10.714(s, 1H, CONH—Ar), 10.799(s, 1H, CONH - pyrim), 13.425(br, 1H, SO_2NH). ^{13}C NMR(400MHz, DMSO - d_6) δ(ppm): 23.500, 109.455, 115.683, 122.095, 123.907, 124.889, 132.331, 137.427, 148.846, 156.424, 157.339, 160.236, 168.874. MALDI - FTICR - MS for $C_{13}H_{12}ClN_5O_4S$: found 392.0192 [M + Na]$^+$, calcd 392.0191 [M + Na]$^+$ (error 0.255 ppm).

N-(4-chloro-3-(N-(4-methoxypyrimidin-2-ylcarbamoyl)sulfamoyl)phenyl)acetamide (**10n**)

White solid, yield 82%, m. p. 189 - 190℃; ^1H NMR(400MHz, DMSO - d_6) δ(ppm): 2.088 (s, 3H, $COCH_3$), 3.953(s, 3H, pyrim - OCH_3), 6.683(d, 1H, J_H = 5.6Hz, pyrim - H_5), 7.570(d, 1H, J_H = 8.0Hz, Ar—H), 7.957(d, 1H, J_H = 8.4Hz, Ar—H), 8.366 - 8.386(m, 2H, Ar—H and pyrim - H_6), 10.434(s, 1H, CONH—Ar), 10.873(br, 1H, CONH - pyrim), 13.472(br, 1H, SO_2NH). ^{13}C NMR(300MHz, DMSO - d_6) δ(ppm): 23.941, 54.431, 102.652, 119.273, 121.850, 123.287, 124.085, 131.883, 138.385, 150.018, 155.212, 156.310, 168.869, 170.169. ESI - FTICR - MS for $C_{14}H_{14}ClN_5O_5S$: found 422.0297 [M + Na]$^+$, calcd 422.0296[M + Na]$^+$ (error 0.237 ppm).

2-Chloro-N-(4-chloro-3-(N-(4-methoxypyrimidin-2-ylcarbamoyl)sulfamoyl)phenyl)acetamide(**10o**)

White solid, yield 91%, m. p. 188 - 189℃; ^1H NMR(400MHz, DMSO - d_6) δ(ppm): 3.995 (s, 3H, pyrim - OCH_3), 4.305(s, 2H, $COCH_2Cl$), 6.685(d, 1H, J_H = 6.0Hz, pyrim - H_5), 7.607(d, 1H, J_H = 8.0Hz, Ar—H), 7.953(d, 1H, J_H = 8.4Hz, Ar—H), 8.346 - 8.371(m, 2H, Ar—H and pyrim - H_6), 10.816(s, 1H, CONH—Ar), 10.931(br, 1H, CONH - pyrim), 13.499(br, 1H, SO_2NH). ^{13}C NMR(400MHz, DMSO - d_6) δ(ppm): 43.442, 54.570, 102.690, 119.653, 122.224, 123.291, 124.281, 132.045, 137.534, 150.421, 156.046,

156.542, 165.196, 170.338. ESI - FTICR - MS for $C_{14}H_{13}Cl_2N_5O_5S$: found 455.9899 [M + Na]$^+$, calcd 455.9907 [M + Na]$^+$ (error 1.754 ppm).

N-(4-chloro-3-(N-(4-methoxypyrimidin-2-ylcarbamoyl)sulfamoyl)phenyl)isobutyramide (**10p**)

Yellow solid, yield 68%, m.p. 137 - 139℃; ^{1}H NMR (400MHz, DMSO - d_6) δ (ppm): 1.124 (d, 6H, J_H = 6.4Hz, isopropyl - CH_3), 2.602 - 2.667 (m, 1H, isopropyl - CH), 3.959 (s, 3H, pyrim - OCH_3), 6.696 (d, 1H, J_H = 6.0Hz, pyrim - H_5), 7.578 (d, 1H, J_H = 8.8Hz, Ar—H), 7.987 (d, 1H, J_H = 7.6Hz, Ar—H), 8.377 (d, 1H, J_H = 6.0Hz, pyrim - H_6), 8.460 (s, 1H, Ar—H), 10.408 (s, 1H, CONH—Ar), 10.922 (br, 1H, CONH - pyrim). ^{13}C NMR (400MHz, DMSO - d_6) δ (ppm): 19.297, 35.015, 54.492, 102.720, 119.412, 121.988, 123.232, 124.251, 131.851, 138.528, 150.006, 154.985, 156.223, 170.216, 175.801. MALDI - FTICR - MS for $C_{16}H_{18}ClN_5O_5S$: found 428.0793 [M + H]$^+$, calcd 428.0790 [M + H]$^+$ (error 0.701 ppm).

N-(4-chloro-3-(N-(4-methoxypyrimidin-2-ylcarbamoyl)sulfamoyl)phenyl)cyclopropanecarboxamide (**10q**)

White solid, yield 86%, m.p. 140 - 141℃; ^{1}H NMR (400MHz, DMSO - d_6) δ (ppm): 0.841 (d, 4H, J_H = 6.4Hz, cyclopropyl - CH_2), 1.785 - 1.846 (m, 1H, cyclopropyl - CH), 3.951 (s, 3H, pyrim - OCH_3), 6.690 (d, 1H, J_H = 6.0Hz, pyrim - H_5), 7.565 (d, 1H, J_H = 8.4Hz, Ar—H), 7.927 - 7.955 (m, 1H, Ar—H), 8.367 (d, 1H, J_H = 6.4Hz, pyrim - H_6), 8.437 (d, 1H, J_H = 2.4Hz, Ar—H), 10.778 (s, 1H, CONH—Ar), 10.918 (br, 1H, CONH - pyrim). ^{13}C NMR (400MHz, DMSO - d_6) δ (ppm): 7.566, 14.595, 54.532, 102.790, 119.217, 121.850, 123.212, 124.149, 131.910, 138.436, 149.991, 154.806, 156.111, 170.275, 172.273. MALDI - FTICR - MS for $C_{16}H_{16}ClN_5O_5S$: found 426.0629 [M + H]$^+$, calcd 426.0633 [M + H]$^+$ (error 0.939 ppm).

N-(4-chloro-3-(N-(4-methoxypyrimidin-2-ylcarbamoyl)sulfamoyl)phenyl)pivalamide (**10r**)

White solid, yield 88%, m.p. 130 - 132℃; ^{1}H NMR (400MHz, DMSO - d_6) δ (ppm): 1.245 (s, 9H, CO - neopentyl), 3.953 (s, 3H, pyrim - OCH_3), 6.681 (d, 1H, J_H = 6.0Hz, pyrim - H_5), 7.571 (d, 1H, J_H = 8.8Hz, Ar—H), 8.088 (d, 1H, J_H = 8.4Hz, Ar—H), 8.388 (d, 1H, J_H = 4.0Hz, pyrim - H_6), 8.503 (s, 1H, Ar—H), 9.728 (s, 1H, CONH—Ar), 10.846 (br, 1H, CONH - pyrim), 13.446 (br, 1H, SO_2NH). ^{13}C NMR (400MHz, $CDCl_3$) δ (ppm): 27.479, 39.834, 54.729, 103.353, 123.509, 125.710, 126.125, 132.287, 136.068, 137.540, 149.346, 156.502, 157.609, 170.602, 177.215. ESI - FTICR - MS for $C_{17}H_{20}ClN_5O_5S$: found 464.0759 [M + Na]$^+$, calcd 464.0766 [M + Na]$^+$ (error 1.508 ppm).

N-(4-chloro-3-(N-(4-methoxypyrimidin-2-ylcarbamoyl)sulfamoyl)phenyl)acrylamide (**10s**)

White solid, yield 83%, m.p. 168 - 170℃; ^{1}H NMR (400MHz, DMSO - d_6) δ (ppm): 3.964 (s, 3H, pyrim - OCH_3), 5.843 (d, 1H, J_H = 9.6Hz, CH ═CH_2), 6.308 - 6.476 (m, 2H, CH═CH_2), 6.691 (d, 1H, J_H = 4.0Hz, pyrim - H_5), 7.615 (d, 1H, J_H = 8.4Hz, Ar—H), 8.075 (d, 1H, J_H = 8.4Hz, Ar—H), 8.380 (d, 1H, J_H = 4.0Hz, pyrim - H_6), 8.444 (s, 1H, Ar—H), 10.643 (s, 1H, CONH—Ar), 10.888 (br, 1H, CONH - pyrim), 13.508 (br, 1H,

SO_2NH). ^{13}C NMR (300MHz, DMSO − d_6) δ (ppm): 54.449, 102.636, 119.646, 122.256, 123.865, 124.380, 127.871, 131.274, 131.975, 138.109, 150.234, 154.959, 156.276, 163.545, 170.209. MALDI − FTICR − MS for $C_{15}H_{14}ClN_5O_5S$: found 434.0290 [M + Na]$^+$, calcd 434.0296 [M + Na]$^+$ (error 1.382 ppm).

2 − Chloro − N − (4 − chloro − 3 − (N − (4 − methoxypyrimidin − 2 − ylcarbamoyl) sulfamoyl) phenyl) propanamide (**10t**)

White solid, yield 82%, m. p. 127 − 129℃; 1H NMR (400MHz, DMSO − d_6) δ (ppm): 1.637 (d, 3H, J_H = 6.4Hz, $COCHClCH_3$), 3.956 (s, 3H, pyrim − OCH_3), 4.681 (q, 1H, J_{12} = 6.4Hz, J_{13} = 12.8Hz, $COCHClCH_3$), 6.685 (d, 1H, J_H = 6.0Hz, pyrim − H_5), 7.613 (d, 1H, J_H = 8.8Hz, Ar—H), 7.969 (d, 1H, J_H = 7.6Hz, Ar—H), 8.357 (d, 1H, J_H = 4.4Hz, pyrim − H_6), 8.408 (s, 1H, Ar—H), 10.820 (s, 1H, CONH—Ar), 10.928 (br, 1H, CONH − pyrim), 13.539 (br, 1H, SO_2NH). ^{13}C NMR (400MHz, DMSO − d_6) δ (ppm): 20.799, 54.545, 54.610, 102.680, 119.805, 122.384, 123.394, 124.378, 132.049, 137.569, 150.611, 154.386, 156.142, 167.797, 170.319. MALDI − FTICR − MS for $C_{15}H_{15}Cl_2N_5O_5S$: found 448.0239 [M + H]$^+$, calcd 448.0244 [M + H]$^+$ (error 1.116 ppm).

2 − (4 − Chloro − 3 − (N − (4 − methoxypyrimidin − 2 − ylcarbamoyl) sulfamoyl) phenylamino) − 2 − oxoethyl acetate (**10u**)

White solid, yield 83%, m. p. 130 − 131℃; ^1H NMR (400MHz, DMSO − d_6) δ (ppm): 2.138 (s, 3H, $COCH_2OCOCH_3$), 3.959 (s, 3H, pyrim − OCH_3), 4.690 (s, 2H, $COCH_2OCOCH_3$), 6.702 (d, 1H, J_H = 6.4Hz, pyrim − H_5), 7.595 (d, 1H, J_H = 8.8Hz, Ar—H), 7.912 − 7.939 (m, 1H, Ar—H), 8.348 (d, 1H, J_H = 6.0Hz, pyrim − H_6), 8.420 (d, 1H, J_H = 2.4Hz, Ar—H), 10.698 (s, 1H, CONH—Ar), 11.005 (s, 1H, CONH − pyrim). ^{13}C NMR (400MHz, DMSO − d_6) δ (ppm): 20.406, 54.709, 62.488, 102.916, 119.597, 122.182, 124.005, 124.354, 131.973, 137.553, 150.576, 153.581, 155.625, 166.155, 170.068, 170.504. MALDI − FTICR − MS for $C_{16}H_{16}ClN_5O_7S$: found 458.0539 [M + H]$^+$, calcd 458.0532 [M + H]$^+$ (error 1.528 ppm).

Ethyl 2 − (4 − chloro − 3 − (N − (4 − methoxypyrimidin − 2 − ylcarbamoyl) sulfamoyl) phenylamino) − 2 − oxoacetate (**10v**)

White solid, yield 53%, m. p. 199 − 201℃; ^1H NMR (400MHz, DMSO − d_6) δ (ppm): 1.332 (t, 3H, J_{12} = 6.8Hz, OCH_2CH_3), 3.959 (s, 3H, pyrim − OCH_3), 4.331 (q, 2H, J_{12} = 6.8Hz, J_{13} = 14.0Hz, OCH_2CH_3), 6.682 (d, 1H, J_H = 6.0Hz, pyrim − H_5), 7.646 (d, 1H, J_H = 8.8Hz, Ar—H), 8.066 (d, 1H, J_H = 7.6Hz, Ar—H), 8.363 (d, 1H, J_H = 5.2Hz, pyrim − H_6), 8.634 (s, 1H, Ar—H), 10.889 (br, 1H, CONH − pyrim), 11.230 (s, 1H, CONH—Ar), 13.487 (br, 1H, SO_2NH). ^{13}C NMR (400MHz, DMSO − d_6) δ (ppm): 13.794, 54.537, 62.556, 102.661, 120.902, 123.419, 125.355, 125.468, 131.919, 136.668, 150.652, 154.443, 155.805, 156.128, 159.971, 170.321. MALDI − FTICR − MS for $C_{16}H_{16}ClN_5O_7S$: found 458.0537 [M + H]$^+$, calcd 458.0532 [M + H]$^+$ (error 1.097 ppm).

N − (4 − chloro − 3 − (N − (4 − methoxypyrimidin − 2 − ylcarbamoyl) sulfamoyl) phenyl) − 2,2,3,3,4,4,4 − heptafluorobutanamide (**10w**)

White solid, yield 83%, m. p. 225 − 226℃; ^1H NMR (400MHz, DMSO − d_6) δ (ppm): 3.965 (s, 3H, pyrim − OCH_3), 6.693 (d, 1H, J_H = 6.4Hz, pyrim − H_5), 7.689 (d, 1H, J_H = 8.4Hz,

Ar—H), 8.030(d, 1H, J_H = 7.6Hz, Ar—H), 8.340(d, 1H, J_H = 3.2Hz, pyrim - H_6), 8.523 (s, 1H, Ar—H), 11.009(br, 1H, CONH - pyrim), 11.744(s, 1H, CONH—Ar), 13.639(br, 1H, SO_2NH). ^{13}C NMR (300MHz, DMSO - d_6) δ (ppm): 54.586, 102.643, 107.901, 115.195, 119.015, 121.657, 124.092, 126.007, 126.772, 132.138, 135.310, 151.047, 153.719, 155.399, 155.967, 170.449. MALDI - FTICR - MS for $C_{16}H_{11}ClF_7N_5O_5S$: found 554.0139[M + H]$^+$, calcd 554.0130 [M + H]$^+$ (error 1.625 ppm).

N-(4-chloro-3-(N-(4-methoxypyrimidin-2-ylcarbamoyl) sulfamoyl) phenyl)-2,2,2-trifluoroacetamide(**10x**)

Light yellow solid, yield 77%, m. p. 183 - 184℃; ^1H NMR(400MHz, DMSO - d_6) δ(ppm): 3.970(s, 3H, pyrim - OCH_3), 6.698(d, 1H, J_H = 6.0Hz, pyrim - H_5), 7.691(d, 1H, J_H = 8.4Hz, Ar—H), 8.022(d, 1H, J_H = 8.0Hz, Ar—H), 8.349(d, 1H, J_H = 4.4Hz, pyrim - H_6), 8.513(s, 1H, Ar—H), 10.003(br, 1H, CONH - pyrim), 11.713(s, 1H, CONH—Ar), 13.635(br, 1H, SO_2NH). ^{13}C NMR (400MHz, DMSO - d_6) δ (ppm): 54.667, 102.692, 111.210, 121.332, 123.788, 125.724, 126.426, 132.111, 135.457, 151.452, 153.552, 154.227, 155.847, 170.501. MALDI - FTICR - MS for $C_{14}H_{11}ClF_3N_5O_5S$: found 454.0196 [M + H]$^+$, calcd 454.0194 [M + H]$^+$ (error 0.441 ppm).

N-(4-chloro-3-(N-(4-methoxypyrimidin-2-ylcarbamoyl) sulfamoyl) phenyl) formamide (**10y**)

White solid, yield 73%, m. p. 192 - 193℃; ^1H NMR(400MHz, DMSO - d_6) δ(ppm): 3.955 (s, 3H, pyrim - OCH_3), 6.681(d, 1H, J_H = 5.6Hz, pyrim - H_5), 7.593(d, 1H, J_H = 7.6Hz, Ar—H), 7.906(d, 1H, J_H = 8.0Hz, Ar—H), 8.351 - 8.366(m, 2H, CHO and pyrim - H_6), 8.409(s, 1H, Ar—H), 10.652(s, 1H, CONH—Ar), 10.923(br, 1H, CONH - pyrim), 13.536(br, 1H, SO_2NH). ^{13}C NMR (400MHz, DMSO - d_6) δ (ppm): 54.549, 102.682, 119.456, 121.991, 124.063, 124.229, 132.043, 137.234, 150.588, 154.568, 156.105, 160.135, 170.315. MALDI - FTICR - MS for $C_{13}H_{12}ClN_5O_5S$: found 386.0323 [M + H]$^+$, calcd 386.0320[M + H]$^+$ (error 0.777 ppm).

5 MIC determination

M. tuberculosis H37Rv (ATCC 27294), *M. bovis* (ATCC 25523), *M. avium* (ATCC 25291), *M. kansasii* (ATCC 12478) were purchased from Beijing Institute for Tuberculosis Control. Ten clinical isolates, which were identified from PLA 309 hospital in China, and characterized for their susceptibility to the antibiotics by our team, were randomly selected in this work and numbered from A to J. All the isolates were cultured by LJ medium before used.

A total of 25 sulfonylurea derivates were tested for the activities against TB *in vitro* assay. Compounds were diluted two - fold on Middlebrook 7H10 agar media supplemented with OADC (oleic acid, albumin, dextrose, and catalase). Strains were harvested from LJ medium and collected by the solution with 0.05% tween - 80. An inoculum of 5×10^4 CFU of each prepared strains was added to each medicated media. The medias were incubated at 36.5℃ and the examination for bacterial growth was recorded four weeks later. Minimum inhibitory concentration (MIC) was defined as the lowest concentration of a compound that will inhibit the visible bacterial growth of a microorganism after incubation.

6 Cytotoxity assay

The 3 – (4,5 – dimethylthiazol – 2 – yl) – 2,5 – diphenyl tetrazolium bromide(MTT) assay developed by Mosmann[16] which test for cell proliferation and survival was used in this study to assay the cytotoxity of the compounds. THP – 1 cells were incubated in 1640 medium plus 10% calf serum in 6cm plate before used. MTT reduction by cultured cells was performed in 96 – well plates containing 100μL of medium per well. When the cells were subcultured to 50% – 60% concentrations, compounds were added at different concentrations and 72h later, a total of 50μL MTT solution was added to each well and incubated at 37℃ temperature and 5% CO_2 and 95% relative humidity for 4h. After incubation, MTT was aspirated and 150μL per well of DMSO was added to each well to dissolve the foramazan precipitate. Subsequently, ELISA Reader read the optical densities of the plates at 570nm.

Supplementary data

Supplementary data associated with this article can be found in the online version, at doi: 10.1016/j.ejmech.2012.01.011. These data include MOL files and InChiKeys of the most important compounds described in this article.

References

[1] World Health Organization, Global Tuberculosis Control – Surveillance, Planning, Financing, World Health Organization, Geneva, Switzerland, 2007.

[2] World Health Organization, Anti – tuberculosis Drug Resistance in the World. In report No. 4, World Health Organization, Geneva, Switzerland, 2008.

[3] A. Koul, et al., The challenge of new drug discovery for tuberculosis, Nature 469(2011)483 – 490.

[4] J. A. Grandoni, P. T. Marta, J. V. Schloss, Inhibitors of branched – chain amino acid biosynthesis as potential antituberculosis agents, J. Antimicrob. Chemother. 4(1998)475 – 482.

[5] V. Singh, et al., Biochemical and transcription analysis of acetohydroxyacid synthase isoforms in Mycobacterium tuberculosis identifies these enzymes as potential targets for drug development, Microbiology 157(Pt 1)(2011)29 – 37.

[6] D. Chipman, Z. Barak, J. V. Schloss, Biosynthesis of 2 – aceto – 2 – hydroxy acids: acetolactate synthases and acetohydroxyacid synthases, Biochim. Biophys. Acta 2(1998)401 – 419.

[7] R. S. Chaleff, C. J. Mauvais, Acetolactate synthase is the site of action of two sulfonylurea herbicides in higher plants, Science 224(1984)1443 – 1445.

[8] D. L. Shaner, P. C. Anderson, M. A. Stidham, Imidazolinones: potent inhibitors of acetohydroxyacid synthase, Plant Physiol. 2(1984)545 – 546.

[9] H. Sohn, et al., In vitro and ex vivo activity of new derivatives of acetohydroxyacid synthase inhibitors against Mycobacterium tuberculosis and nontuberculous mycobacteria, Int. J. Antimicrob. Agents 6(2008) 567 – 571.

[10] K. J. Choi, et al., Characterization of acetohydroxyacid synthase from Mycobacterium tuberculosis and the identification of its new inhibitor from the screening of a chemical library, FEBS Lett. 21 (2005) 4903 – 4910.

[11] D. Wang, et al., Prevalence of multidrug and extensively drug – resistant tuberculosis in Beijing, China: a hospital – based retrospective study, Jpn. J. Infect. Dis. 5(2010)368 – 371.

[12] Y. Zohar, et al. , Acetohydroxyacid synthase from Mycobacterium avium and its inhibition by sulfonylureas and imidazolinones, Biochim. Biophysic. Acta, Proteins and Proteomics 1(2003)97 – 105.
[13] B. Buckman, et al. , Novel Inhibitors of hepatitis C virus replication. PCT Int. Appl. WO 2011075607 A1, June 23, 2011.
[14] M. Weitman, et al. , Structure – activity relationship studies of 1 – (4 – chloro – 2, 5 – dimethoxyphenyl) – 3 – (3 – propoxypropyl) thiourea, a non – nucleoside reverse transcriptase inhibitor of human immunodeficiency virus type – 1, Eur. J. Med. Chem. 46(2011)447 – 467.
[15] N. Ma, et al. , Synthesis and herbicidal activities of pyridyl sulfonylureas. More convenient preparation process of phenyl pyrimidylcarbamates, Chin. Chem. Lett. 11(2008)1268 – 1270.
[16] T. Mosmann, Rapid colorimetric assay for cellular growth and survival: application to proliferation and cytotoxicity assays, J. Immunol. Methods 65(1983)55 – 63.
[17] J. G. Wang, et al. , Crystal structures of two novel sulfonylurea herbicides in complex with *Arabidopsis thaliana* acetohydroxyacid synthase, FEBS J. 5(2009)1282 – 1290.
[18] M. Dong, et al. , In vitro efficacy of acetohydroxyacid synthase inhibitors against clinical strains of Mycobacterium tuberculosis isolated from PLA 309 hospital in Beijing, China, Saudi Med. J. 32 (2011) 1122 – 1126.
[19] G. Brahmachari, S. Laskar, A very simple and highly efficient procedure for N – formylation of primary and secondary amines at room temperature under solvent – free conditions, Tetrahedron Lett. 51 (2010) 2319 – 2322.

基于Ugi反应的新型鱼尼丁受体杀虫剂的设计、合成及生物活性*

刘鹏飞　周　莎　熊丽霞　于淑晶　张　晓　李正名

（南开大学元素有机化学国家重点实验室，元素有机化学研究所，天津　300071）

摘　要　利用Ugi反应设计合成了一系列未见文献报道的α-苯基-α-酰胺基-酰胺类化合物，所有化合物均通过^{1}H NMR谱、元素分析和高分辨质谱表征确定。初步的生物活性测试结果表明，在浓度为200mg/L时，化合物7h对粘虫有一定抑制活性；在浓度为50mg/L时，化合物7q对苹果轮纹病菌、化合物7e对小麦赤霉菌有一定的抑菌活性。

关键词　Ugi反应　氯虫酰胺　杀虫活性　抑菌活性

1998年日本农药公司发现的邻苯二甲酰胺类［氟虫酰胺（Flubendiamide）］杀虫剂[1]和2001年杜邦公司[2]发现的邻甲酰胺基苯甲酰胺类［氯虫酰胺（Chlorantraniliprole）］杀虫剂是一类新型的作用于鱼尼丁受体（Ryanodine receptor，RyR）的杀虫剂。这两种杀虫剂结构独特，作用方式新颖，对鳞翅目害虫效果好，杀虫谱广，对各种益虫和天敌安全，并与现用的杀虫剂无交互抗性[3]。氯虫酰胺具有比氟虫酰胺更宽的杀虫谱和独特的杀卵活性[1]，创制此类新型抑制剂是当今农用化学品研究的热点[4~10]。

氟虫酰胺和氯虫酰胺均含有2个酰胺基结构，在不同位置的双肽键结构对杀虫活性可能具有重要意义。本文利用Ugi反应，在尽可能保持分子结构多样性的基础上，导入不同位置的双酰胺结构，以探讨此类新型结构的生物活性及其构效关系。将氯虫酰胺中β-氨基酸结构变为α-氨基酸结构，设计合成了一系列未见文献报道的α-苯基-α-酰胺基-酰胺类化合物（Scheme 1），以期获得更高活性的化合物，并通过^{1}H NMR谱、元素分析和高分辨质谱等方法对目标化合物的结构进行了确证。初步的生物活性测试结果表明，部分化合物具有一定的杀虫和抑菌活性。

Chlorantraniliprole
(Rynaxypyr™;DPX-E2Y45)

Scheme 1　Synthetic routes of novel α-phenyl-α-amide-amide compounds

* 原发表于《高等学校化学学报》，2012，33（4）：738-743。国家“973”计划项目（批准号：2010CB126106）、国家自然科学基金（批准号：20872069）和“十二五”国家科技支撑计划项目（批准号：2011BAE06B05）资助。

1 实验部分

1.1 试剂与仪器

所有试剂均为国产分析纯，均经过常规方法处理。

Bruker Avance 400 型核磁共振仪（TMS 为内标，$CDCl_3$ 或 DMSO－d_3 为溶剂）；Thermofinnigan LCQ Advantage 型质谱仪（ESI）；Elementar Vario EL Ⅲ型元素分析仪（德国）；X－24 型数字显示显微熔点仪（北京泰克仪器有限公司）；H 型柱层析硅胶（青岛海洋化工厂）。

1.2 中间产物的合成

1.2.1 4－取代－2－硝基－1－甲酰苯胺（**2**）的合成

化合物**2**的合成参照文献[11]方法，选择新的催化剂制备。在 100mL 圆底烧瓶中加入 10.0mmol 4－取代邻硝基苯胺（**1**）、30.0mmol 无水甲酸和 1.0mmol $AlCl_3$，加热回流 4h，将反应液趁热倒入分液漏斗中，微冷后用 50mL 二氯甲烷稀释，依次用水、饱和 Na_2CO_3 溶液和饱和食盐水洗涤，干燥，过滤，得产品 4－取代－2－硝基－1－甲酰苯胺（**2**）的二氯甲烷溶液，直接用于下一步合成。

1.2.2 化合物 4－取代－2－硝基苯基异腈（**3**）的合成

化合物**3**的合成参考文献［12］方法，产品可在乙醚中重结晶纯化。

1.2.3 化合物**5**的合成

化合物**5**的合成参考文献［13］方法。

1.2.4 化合物**6**的合成

化合物**6**的合成参考文献［14］方法，化合物**6a**：m. p. 198～200℃（文献值[14] 197～200℃）；化合物**6b**：m. p. 198－200℃（文献值[14]200～201℃）。

1.3 目标产物 7 的合成

目标化合物的合成路线见 Scheme 2。

参照文献［15］方法，在 10mL 单口圆底烧瓶中加入化合物**3**（1.0mmol），**5**（1.0mmol），**6**（1.0mmol）和无水甲醇（3.0mL），室温下搅拌反应 12～24h，反应完毕，析出大量固体，抽滤，用少许甲醇和乙醚洗涤滤饼，烘干得目标化合物**7**。如果需要进一步提纯，可以用乙酸乙酯/石油醚重结晶或柱层析纯化。

1.4 生物活性测试

对所合成的目标化合物进行杀虫和抑菌生物活性测试。杀虫活性采用文献［14］的方法对东方粘虫（*Mythimna separate*）进行测试。抑菌测试采用离体平皿法[16]，在 50mg/L 浓度下，对黄瓜枯萎病菌（*Fusarium omysporum*）、苹果轮纹病菌（*Physalospora piricola*）、番茄早疫病菌（*Alternaria solani*）、花生褐斑病菌（*Cercospora arachidicola*）、小麦赤霉病菌（*Gibberella sanbinetti*）和辣椒疫霉（*Phytophthora capsici*）等 6 种常见病菌进行测试。

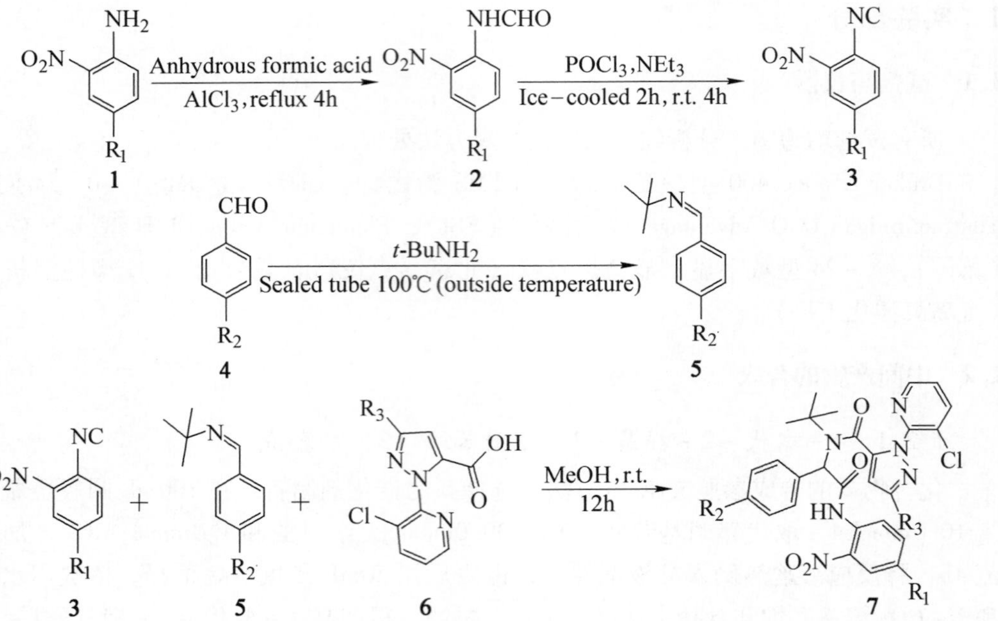

Anhydrous formic acid, $AlCl_3$, reflux 4h; $POCl_3$, NEt_3, Ice-cooled 2h, r.t. 4h; t-$BuNH_2$, Sealed tube 100℃ (outside temperature); MeOH, r.t. 12h

Scheme 2 Synthetic routes of compounds 7

7a: R_1 = Me, R_2 = F, R_3 = Br; 7b: R_1 = Me, R_2 = F, R_3 = Cl; 7c: R_1 = Me, R_2 = Cl, R_3 = Br; 7d: R_1 = Me, R_2 = Cl, R_3 = Cl; 7e: R_1 = Me, R_2 = NO_2, R_3 = Br; 7f: R_1 = Me, R_2 = NO_2, R_3 = Cl; 7g: R_1 = Cl, R_2 = F, R_3 = Br; 7h: R_1 = Cl, R_2 = F, R_3 = Cl; 7i: R_1 = Cl, R_2 = Cl, R_3 = Br: 7j: R_1 = Cl, R_2 = Cl, R_3 = Cl; 7k: R_1 = Cl, R_2 = NO_2, R_3 = Br; 7l: R_1 = Cl, R_2 = NO_2, R_3 = Cl; 7m: R_1 = OMe, R_2 = F, R_3 = Br; 7n: R_1 = OMe, R_2 = F, R_3 = Cl; 7o: R_1 = OMe, R_2 = Cl, R_3 = Br; 7p: R_1 = OMe, R_2 = Cl, R_3 = Cl; 7q: R_1 = OMe, R_2 = NO_2, R_3 = Br; 7r: R_1 = OMe, R_2 = NO_2, R_3 = Cl.

2 结果与讨论

2.1 化合物的合成

在化合物**2**的合成过程中，实验尝试了其他几种甲酰化方法（如与甲酸乙酯回流、DMF中回流及甲乙混酐等方法），但由于反应物**1**的活性较低，导致收率低，产物需要柱层析分离。在Shekhar等[11]报道的方法中，虽提到了$AlCl_3$可以催化甲酰化反应，但并未报道反应细节。在本文的反应条件下，反应物**1**在反应时间4h以上时，反应基本完全，收率较高，无需进行后处理，即可进行异腈的制备。

在目标化合物**7**的合成中，如果使用对位硝基或者氯取代的异腈，由于异腈活性不高，不预先制备亚胺**5**，而直接使用化合物**4**和叔丁胺，产物的收率会降低，甚至不能发生反应，得不到产物。溶剂中微量水的存在会阻碍反应的进行，如果反应溶剂为未处理的甲醇，则不能发生反应。溶剂的选取也是关键，通常Ugi反应使用的溶剂是甲醇，而以二氯甲烷为溶剂时，产物的收率会降低；当选用DMF为溶剂时，在本实验中未得到产物。因此在本实验中，预先制备了亚胺中间体，采用除水甲醇为溶剂进行反应。反应在尽可能少的甲醇中进行，以利于转化，提高收率，产物在甲醇中也有一定的溶解性，如果对后处理的滤液进行柱层析，也可以提高产物收率。

2.2 化合物的表征

中间产物和目标化合物的物化参数和^{1}H NMR 数据列于表 1 和表 2 中。

Table 1 Physical data, HRMS and elemental analysis results of the synthesized compounds①

Compd.	m. p. /℃	Yield(%)	Appearance	HRMS(calcd.) [M+Na]$^+$	Elemental analysis(%, calcd.)		
					C	H	N
3a	66 ~ 67(dec)	90.15(two step)	Yellow solid	—	—	—	—
3b	90 ~ 91(dec)	88.33(two step)	Yellow solid	—	—	—	—
3c	106 ~ 108(dec)	92.17(two step)	Yellow solid	—	—	—	—
5a	—	95.01	Yellow oil	—	—	—	—
5b	33 ~ 34	94.00	White solid	—	—	—	—
5c	69 ~ 70	95.00	Yellow solid	—	—	—	—
7a	211 ~ 212	8.54	Yellow solid	665.0683 (665.0685)	—	—	—
7b	223 ~ 224	11.69	Yellow solid	—	56.03(56.10)	4.14(4.20)	14.06(14.02)
7c	210 ~ 211	26.97	Yellow solid	—	50.91(50.93)	3.73(3.82)	12.81(12.73)
7d	224 ~ 225	35.88	Yellow solid	—	54.52(54.60)	3.98(4.09)	13.73(13.65)
7e	202 ~ 203	23.70	Yellow solid	—	50.25(50.13)	3.86(3.76)	14.65(14.61)
7f	214 ~ 215	29.84	Yellow solid	—	53.60(53.68)	3.89(4.02)	15.85(15.65)
7g	210 ~ 211	9.53	Yellow solid	685.0136 (685.0139)	—	—	—
7h	210 ~ 212	9.98	Yellow solid	—	52.12(52.32)	3.76(3.58)	13.54(13.56)
7i	210 ~ 211	29.96	Yellow solid	700.9850 (700.9844)	—	—	—
7j	199 ~ 200	24.69	Yellow solid	—	50.69(50.96)	3.65(3.48)	13.29(13.21)
7k	201 ~ 202	16.35	Yellow solid	712.0080 (712.0084)	—	—	—
7l	205 ~ 206	14.68	Yellow solid	668.0592 (668.0589)	—	—	—
7m	205 ~ 206	62.58	Yellow solid	—	50.95(50.96)	3.87(3.82)	13.90(13.74)
7n	205 ~ 206	54.63	Yellow solid	—	54.58(54.64)	4.34(4.09)	13.85(13.66)
7o	207 ~ 208	60.80	Yellow solid	697.0339 (697.0339)	—	—	—
7p	214 ~ 215	63.13	Yellow solid	—	53.39(53.22)	3.71(3.99)	13.39(13.30)
7q	206 ~ 207	43.09	Yellow solid	—	49.03(48.96)	3.55(3.67)	14.33(14.27)
7r	215 ~ 216	44.08	Yellow solid	—	52.50(52.35)	3.90(3.92)	15.45(15.26)

①—not tested.

Table 2 ¹H NMR data of the synthesized compounds

Compd.	Frequency /MHz	¹H NMR ($CDCl_3$), δ
3a	300	7.91 (s, 1H, Ar—H), 7, 49 (s, 2H, Ar—H), 2.49 (s, 3H, Me)
3b	400	8.12 (d, j = 2.1Hz, 1H, Ar—H), 7.69 (dd, J = 8.5, 2.2Hz, 1H, Ar—H), 7.57 (d, J = 8.5Hz, 1H, Ar—H)
3c	400	7.60 (d, J = 2.8Hz, 1H, Ar—H), 7.54 (d, J = 8.9Hz, 1H, A—H), 7.20 (dd, J = 8.9, 2.8Hz, 1H Ar—H), 3.95 (s, 3H, Me)
5a	300	8.23 (s, 1H), 7.79 (brs, 2H, Ar—H), 7.09 (t, J = 8.6Hz, 2H, Ar—H), 1.31 (s, 9H, ᵗBu)
5b	300	8.22 (s, 1H, CHN), 7.87 ~ 7.73 (m, 2H, Ar—H), 7.39 (d, J = 8.4Hz, 2H, Ar—H), 1.34 (s, 9H, ᵗBu)
5c	300	8.33 (s, 1H, CHN), 8.26 (d, J = 8.8Hz, 2H, Ar—H), 7.93 (d, J = 8.5Hz, 2H, Ar—H), 1.32 (s, 9H, ᵗBu)
7a	400	10.07 (s, 1H, NH), 8.52 (d, J = 8.7Hz, 1H, Ar—H), 8.36 (d, J = 4.6Hz, 1H, pyridyl—H), 7.96 ~ 7.90 (m, 1H, pyridyl—H), 7.87 (s, 1H, Ar—H), 7.46—7.32 (m, 4H, Ar—H), 7.09 (t, J = 8.5Hz, 2H, pyridyl—H, Ar—H), 6.71 (s, 1H, pyrazolyl—H), 5.24 (s, 1H, NCHCO), 2.31 (s, 3H, Me), 1.49 (s, 9H, ᵗBu)
7b	400	10.08 (s, 1H, NH), 8.53 (d, J = 8.5Hz, 1H, Ar—H), 8.36 (d, J = 3.8Hz, 1H, pyridyl—H), 7.93 (d, J = 7.9Hz, 1H, pyridyl—H), 7.87 (s, 1H, Ar—H), 7.50 ~ 7.39 (m, 2H, Ar—H, pyridyl—H), 7.35 (t, J = 6.5Hz, 2H, Ar—H), 7.09 (t, J = 8.4Hz, 2H, Ar—H), 6.63 (s, 1H, pyrazolyl—H), 5.24 (s, 1H, NCHCO), 2.31 (s, 3H, Me), 1.49 (s, 9H, ᵗBu)
7c	400	10.09 (s, 1H, NH), 8.50 (d, J = 8.7Hz, 1H, Ar—H), 8.37 (dd, J = 4.6, 1.3Hz, 1H, pyridyl—H), 7.93 (dd, J = 8.0, 1.4Hz, 1H, pyridyl—H), 7.88 (s, 1H, Ar—H), 7.42 ~ 7.32 (m, 6H; 5H, Ar—H, 1H, pyridyl—H), 6.71 (s, 1H, pyrazolyl—H), 5.23 (s, 1H, NCHCO), 2.32 (s, 3H, Me), 1.49 (s, 9H, ᵗBu)
7d	400	10.09 (s, 1H, NH), 8.50 (d, J = 8.6Hz, 1H, Ar—H), 8.37 (d, J = 4.1Hz, 1H, pyridyl—H), 7.92 (d, J = 7.9Hz, 1H, pyridyl—H), 7.84 (d, J = 22.5Hz, 1H, Ar—H), 7.46 ~ 7.31 (m, 5H; 4H, Ar—H, 1H, pyridyl—H), 6.63 (s, 1H, pyrazolyl—H), 5.24 (s, 1H, NCHCO), 2.36 (s, 3H, Me), 1.49 (s, 9H, ᵗBu)
7e	400	10.20 (s, 1H, NH), 8.54 (d, J = 8.6Hz, 1H, Ar—H), 8.50 ~ 8.36 (m, 1H, pyridyl—H), 8.28 (d, J = 8.7Hz, 2H, Ar—H), 7.98 (dd, J = 8.0, 1.3Hz, 1H, pyridyl—H), 7.92 (s, 1H, Ar—H), 7.65 (d, J = 8.6Hz, 2H, Ar—H), 7.51—7.37 (m, 2H; 1H, Ar—H, 1H, pyridyl—H), 6.78 (s, 1H, pyrazolyl—H), 5.40 (s, 1H, NCHCO), 2.36 (s, 3H, Me), 1.55 (s, 9H, ᵗBu)
7f	400	10.17 (s, 1H, NH), 8.51 (d, J = 8.6Hz, 1H, Ar—H), 8.41 (d, J = 4.0Hz, 1H, pyridyl—H), 8.25 (d, J = 8.3Hz, 2H, Ar—H), 7.95 (d, J = 7.9Hz, 1H, pyridyl—H), 7.89 (s, 1H, Ar—H), 7.63 (d, J = 8.2Hz, 2H, Ar—H), 7.40 (dd, J = 14.4, 8.6Hz, 2H, pyridyl—H), 6.67 (s, 1H, pyrazolyl—H), 5.36 (s, 1H, NCHCO), 2.33 (s, 3H, Me), 1.52 (s, 9H, ᵗBu)

Compd.	Frequency /MHz	^{1}H NMR ($CDCl_3$), δ
7g	400	10.12 (s, 1H, NH), 8.71 (d, J = 9.2Hz, 1H, Ar—H), 8.39 (dd, J = 4.6, 1.5Hz, 1H, pyridyl—H), 8.09 (d, J = 2.5Hz, 1H, Ar—H), 7.96 (dd, J = 8.0, 1.5Hz, 1H, pyridyl—H), 7.53 (dd, J = 9.2, 2.5Hz, 1H, Ar—H), 7.44—7.37 (m, 3H, Ar—H), 7.12 (t, J = 8.6Hz, 2H, Ar—H), 6.73 (s, 1H, pyrazolyl—H), 5.24 (s, 1H, NCHCO), 1.51 (s, 9H, ^{t}Bu)
7h	400	10.13 (s, 1H, NH), 8.72 (d, J = 9.1Hz, 1H, Ar—H), 8.39 (dd, J = 1.2, 4.4Hz, 1H, pyridyl—H), 8.09 (d, J = 2.4Hz, 1H, Ar—H), 7.96 (dd, J = 1.2, 8Hz, 1H, pyridyl—H), 7.53 (dd, J = 9.1, 2.4Hz, 1H, Ar—H), 7.43 (dd, J = 8.5, 5.2Hz, 2H, Ar—H), 7.39 (dd, J = 8.0, 4.7Hz, 1H, pyridyl—H), 7.12 (t, J = 8.5Hz, 2H, Ar—H), 6.65 (s, 1H, pyrazolyl—H), 5.24 (s, 1H, NCHCO), 1.51 (s, 9H, ^{t}Bu)
7i	400	10.13 (s, 1H, NH), 8.71 (d, J = 8.9Hz, 1H, Ar—H), 8.41 (d, J = 4.8Hz, 1H, pyridyl—H), 8.11 (d, J = 7.3Hz, 1H, Ar—H), 7.98 (d, J = 7.9Hz, 1H, pyridyl—H), 7.57 ~ 7.35 (m, 5H; 4H, Ar—H, 1H, pyridyl—H), 7.30 (d, J = 9.5Hz, 1H, Ar—H), 6.74 (s, 1H, pyrazolyl—H), 5.25 (s, 1H, NCHCO), 1.53 (s, 9H, ^{t}Bu)
7j	400	10.13 (s, 1H, NH), 8.68 (d, J = 9.1Hz, 1H, Ar—H), 8.40 (d, J = 4.6Hz, 1H, pyridyl—H), 8.09 (d, J = 2.2Hz, 1H, Ar—H), 7.96 (d, J = 8.0Hz, 1H, pyridyl—H), 7.53 (dd, J = 9.1, 2.2Hz, 1H, Ar—H), 7.46 ~ 7.34 (m, 5H; 4H, Ar—H, 1H, pyridyl—H), 6.65 (s, 1H, pyrazolyl—H), 5.24 (s, 1H, NCHCO), 1.51 (s, 9H, ^{t}Bu)
7k	400	10.18 (s, 1H, NH), 8.67 (d, J = 9.2Hz, 1H, Ar—H), 8.42 (d, J = 4.5Hz, 1H, pyridyl—H), 8.27 (t, J = 7.4Hz, 2H, Ar—H), 8.09 (d, J = 2.2Hz, 1H, Ar—H), 7.97 (d, J = 7.9Hz, 1H, pyridyl—H), 7.64 (dd, J = 19.2, 9.9Hz, 2H, Ar—H), 7.53 (dd, J = 9.1, 2.1Hz, 1H, Ar—H), 7.49 ~ 7.38 (m, 1H, pyridyl—H), 6.74 (s, 1H, pyrazolyl—H), 5.35 (s, 1H, NCHCO), 1.54 (d, J = 20.1Hz, 9H, ^{t}Bu)
7l	400	10.17 (s, 1H), 8.51 (d, J = 8.6Hz, 1H, Ar—H), 8.41 (d, J = 4.0Hz, 1H, pyridyl—H), 8.25 (d, J = 8.3Hz, 2H, Ar—H), 7.95 (d, J = 7.9Hz, 1H, pyridyl—H), 7.89 (s, 1H, Ar—H), 7.63 (d, J = 8.2Hz, 2H, Ar—H), 7.47 ~ 7.33 (m, 2H; 1H, pyridyl—H, 1H, Ar—H), 6.67 (s, 1H, pyrazolyl—H), 5.36 (s, 1H, NCHCO), 2.33 (s, 3H, Me), 1.52 (s, 9H, ^{t}Bu)
7m	400	9.92 (s, 1H, NH), 8.52 (d, J = 9.3Hz, 1H, Ar—H), 8.36 (d, J = 4.5Hz, 1H, pyridyl—H), 7.92 (d, J = 8.0Hz, 1H, pyridyl—H), 7.53 (d, J = 3.0Hz, 1H, Ar—H), 7.42 (dd, J = 8.4, 5.3Hz, 2H, pyridyl—H), 7.35 (dd, J = 8.0, 4.7Hz, 1H, Ar—H), 7.11 (dt, J = 17.3, 5.7Hz, 3H, Ar—H), 6.71 (s, 1H, pyrazolyl—H), 5.23 (s, 1H, NCHCO), 3.79 (s, 3H, Me), 1.49 (s, 9H, ^{t}Bu)

续表 2

Compd.	Frequency /MHz	^{1}H NMR (CDCl$_3$), δ
7n	400	9.95 (s, 1H, NH), 8.55 (d, J=9.3Hz, 1H, Ar—H), 8.39 (dd, J=4.6, 1.5Hz, 1H, pyridyl—H), 7.95 (dd, J = 8.0, 1.5Hz, 1H, pyridyl—H), 7.55 (d, J = 3.0Hz, 1H, Ar—H), 7.45 (dd, J=8.6, 5.2Hz, 2H, pyridyl—H), 7.37 (dd, J= 8.0, 4.6Hz, 1H, Ar—H), 7.14 (dt, J = 17.2, 5.8Hz, 3H, Ar—H), 6.65 (s, 1H, pyrazolyl—H), 5.26 (s, 1H, NCHCO), 3.81 (s, 3H, Me), 1.51 (s, 9H, ^{t}Bu)
7o	400	9.97 (s, 1H, NH), 8.53 (d, J=9.3Hz, 1H, Ar—H), 8.39 (dd, J=4.6, 1.4Hz, 1H, pyridyl—H), 7.94 (dd, J = 8.0, 1.4Hz, 1H, pyridyl—H), 7.55 (d, J = 3.0Hz, 1H, Ar—H), 7.47 ~ 7.35 (m, 4H, Ar—H, pyridyl—H), 7.15 (dd, J = 9.3, 3.0Hz, 1H, Ar—H), 6.73 (s, 1H, pyrazolyl—H), 5.24 (s, 1H, NCHCO), 3.81 (s, 3H, Me), 1.51 (s, 9H, ^{t}Bu)
7p	400	9.95 (s, 1H, NH), 8.52 (d, J=9.3Hz, 1H, Ar—H), 8.40 (d, J=4.0Hz, 1H, pyridyl—H), 7.96 (d, J=7.5Hz, 1H, pyridyl—H), 7.57 (d, J=2.8Hz, 1H, Ar—H), 7.46 ~ 7.35 (m, 4H, Ar—H, pyridyl—H), 7.16 (dd, J = 9.3, 2.8Hz, 1H, Ar—H), 6.65 (s, 1H, pyrazolyl—H), 5.27 (s, 1H, NCHCO), 3.82 (s, 3H, Me), 1.52 (s, 9H, ^{t}Bu).
7q	400	10.03 (s, 1H, NH), 8.54 (d, J=9.2Hz, 1H, Ar—H), 8.44 (d, J=3.4Hz, 1H, pyridyl—H), 8.28 (d, J=8.2Hz, 2H, Ar—H), 7.98 (d, J=7.7Hz, 1H, pyridyl—H), 7.66 (d, J=8.1Hz, 2H, Ar—H), 7.57 (d, J=2.0Hz, 1H, Ar—H), 7.49 ~ 7.41 (m, 1H, pyridyl—H), 7.17 (dd, J=7.2Hz, J=2.0Hz, 1H, Ar—H), 6.77 (s, 1H, pyrazolyl—H), 5.40 (s, 1H, NCHCO), 3.86 (s, 3H, Me), 1.54 (s, 9H, ^{t}Bu)
7r	400	10.01 (s, 1H, NH), 8.52 (d, J=9.3Hz, 1H, Ar—H), 8.41 (d, J=3.8Hz, 1H, pyridyl—H), 8.25 (d, J=8.7Hz, 2H, Ar—H), 7.95 (d, J=7.1Hz, 1H, pyridyl—H), 7.63 (d, J = 8.6Hz, 2H, Ar—H), 7.55 (d, J = 3.0Hz, 1H, Ar—H), 7.41 (dd, J=8.0, 4.6Hz, 1H, pyridyl—H), 7.15 (dd, J=9.3, 3.0Hz, 1H, Ar—H), 6.66 (s, 1H, pyrazolyl—H), 5.36 (s, 1H, NCHCO), 3.80 (s, 3H, Me), 1.52 (s, 9H, ^{t}Bu)

2.3 目标化合物的生物活性

生物活性测定结果（表3）表明，目标化合物 **7a –7o** 总体对东方粘虫的活性一般，化合物 **7h** 在 200mg/L 时杀虫活性为 30%，与氯虫酰胺相差较大。

Table 3 Biological activities of the synthesized compounds (50μg/mL)

Compd.	Insecticidal activities *Mythimna separata* (200mg/L)	Inhibition ratio (%)					
		Fusarium omysporum	*Cercospora arachidicola*	*Physalospor apiricola*	*Alternaria solani*	*Gibberella sanbinetti*	*Phytophthora capsici*
7a	0	7.5	3.1	23.1	25.0	27.9	8.8
7b	20	0.0	12.5	28.8	15.0	18.6	8.8

续表 3

Compd.	Insecticidal activities *Mythimna separata* (200mg/L)	Inhibition ratio (%)					
		Fusarium omysporum	*Cercospora arachidicola*	*Physalospor apiricola*	*Alternaria solani*	*Gibberella sanbinetti*	*Phytophthora capsici*
7c	0	10.0	6.3	28.8	15.0	11.6	17.6
7d	10	2.5	6.3	38.5	20.0	20.9	11.8
7e	0	7.5	12.5	26.9	40.0	16.3	5.9
7f	0	7.5	9.4	21.2	30.0	16.3	14.7
7g	10	2.5	0.0	25.0	15.0	11.6	14.7
7h	30	5.0	31.3	30.8	20.0	16.3	8.8
7i	20	5.0	15.6	26.9	20.0	16.3	8.8
7j	0	0.0	6.3	42.3	30.0	16.3	11.8
7k	10	10.0	6.3	7.7	30.0	18.6	17.6
7l	10	7.5	12.5	25.0	30.0	25.6	14.7
7m	20	0.0	21.9	26.9	20.0	23.3	11.8
7n	20	7.5	3.1	23.1	20.0	20.9	23.5
7o	20	5.0	6.3	32.7	25.0	16.3	8.8
7p	0	2.5	6.3	13.5	10.0	16.3	17.6
7q	20	5.0	3.1	46.2	20.0	14.0	14.7
7r	0	0.0	3.1	17.3	20.0	16.3	17.6
Contrast①	100	NT③	NT③	NT③	NT③	NT③	NT③
Contrast②	NT③	84	58.8	88.5	63.6	73.1	82.6

①Chlorantraniliprole, $\rho = 1$mg/L; ②chlorothalonil, $\rho = 50$mg/L; ③NT: not tested.

目标化合物 **7** 的抑菌活性测试结果（表 3）表明，在 50mg/L 浓度下，大部分化合物对黄瓜枯萎病菌、花生褐斑病菌、苹果轮纹病菌、番茄早疫病菌、小麦赤霉病菌和辣椒疫霉表现出一定的抑菌活性。其中，化合物 **7**q 对苹果轮纹病菌的防效较高，为 46.2%；化合物 **7**e 对番茄早疫病菌的防效为 40.0%，但与对照药百菌清（*Chlorothalonil*）的抑菌活性仍有差距。

3 结论

本文设计合成了 18 个新型 α－苯基－α－酰胺基－酰胺类化合物。由于双肽键的移位导致生物活性不够理想，但个别化合物仍显示出一定的生物活性。这可能是由于邻氨基苯甲酸的刚性结构是杀虫剂的关键结构片段，不宜改变；吡唑联吡啶片段与苯环是通过酰胺桥连接的，仍需要保持 2 个活性片段间固定的相对空间位置。而在系列化合物 **7** 中，吡唑联吡啶片段与苯环片段是通过酰胺键和 1 个 sp^3 碳连接的，还引入了 1 个大基团叔丁胺，导致与靶标结合能力变弱，活性降低；另外，新引入的取代硝基苯胺代替原来的甲酰胺部分，空间位阻增大，导致与靶标结合能力变弱，活性降低。

参考文献

[1] Tohnishi M., Nakao H., Kohno E., Nishida T., Furuya T., Shimizu T., Seo A., Sakata K., Fujioka S., Kanno H.. Preparation of Phthalic Acid Diamides as Agricultural and Horticultural Insecticides, EP 919542 [P], 1999-06-02.

[2] Lahm G. P., Myers B. J., Selby T. P., Stevenson T. M.. Preparation of Insecticidal Anthranilamides, WO2001070671 [P], 2001-09-27.

[3] Tang Z. H., Tao L. M.. Acta Entomologica Sinica [J], 2008, 51 (6): 646~651.

[4] Feng M. L., Li Y. F., Zhu H. J., Zhao L., Xi B. B., Ni J. P.. J. Agric. Food Chem. [J], 2010, 58 (20): 10999~11006.

[5] Tohnishi M., Nakao H., Furuya T., Seo A., Kodama H., Tsubata K., Fujioka S., Kodama H., Hirooka T., Nishimatsu T.. J. Pestic. Sci. [J], 2005, 30 (4): 354~360.

[6] Lahm G. P., Selby T. P., Freudenberger J. H., Stevenson T. M., Myers B. J., Seburyamo G., Smith B. K., Flexner L., Clark C. E., Cordova D.. Bioorg. Med. Chem. Lett. [J], 2005, 15: 4898~4906.

[7] Zhang J., Tang X. H., Isaac I., Cao S., Wu J. J., Yu J. L., Li H., Qian X. H.. J. Agric. Food Chem. [J], 2010, 58: 2736~2740.

[8] George P. L., Daniel C., James D. B.. Bioorg. Med. Chem. [J], 2009, 17: 4127~4133.

[9] Maciej A. P., Emily O., Crystal R., Sonny B. R.. J. Insect. Physiol. [J], 2008, 54: 358~366.

[10] Pierluigi C., Giorgia S., Alberto A., Simona V., Daniela P., Paolo C., John E. C.. J. Agric. Food Chem. [J], 2008, 56: 7696~7699.

[11] Shekhar A. C., Kumar A. R., Sathaiah G., Paul V. L., Sridhar M., Rao P. S.. Tetrahed. Lett. [J], 2009, 50: 7099~7101.

[12] Lacerda R. B., Cleverton K. F., Leandro L., Romeiro N. C., Miranda A. P., Barreiro E. J., Fraga C. A. M.. Bioorg. Med. Chem. [J] 2009, 17: 74~84.

[13] Guimond N., Fagnou K.. J. Am. Chem. Soc. [J], 2009, 131: 12050~12051.

[14] Dong W. L., Xu J. Y., Xiong L. H., Liu X. H., Li Z. M.. Chin. J. Chem. [J], 2009, 27: 579~586.

[15] Marcaccini S., Torroba T.. Nature Protocols [J], 2007, 2 (3): 632~639.

[16] DONG Wei-Li (董卫莉), XU Jun-Ying (徐俊英), LIU Xing-Hai (刘幸海), LI Zheng-Ming (李正名), Li Bao-Ju (李宝聚), SHI Yan-Xia (石延霞). Chem. J. Chinese Universities (高等学校化学学报) [J], 2008, 29 (10): 1990~1994.

Design, Synthesis and Biological Activities of New Ryanodine Receptor Pesticides Based on Ugi Reaction

Pengfei Liu, Sha Zhou, Lixia Xiong, Shujing Yu, Xiao Zhang, Zhengming Li

(State Key Laboratory of Elemento-Organic Chemistry, Institute of Elemento-Organic Chemistry, Nankai University, Tianjin, 300071, China)

Abstract Chlorantraniliprole, invented by DuPont company in 2001, is a new type of insecticide with high efficiency, low toxicity, broad-spectrum insecticidal activity which act at Ryanodine receptor of target insects. Both flubendiamide and chlorantraniliprole contain two amide groups in different locations of the structure, and this could be of great significance in insecticidal activity. Referring to their structural composition, a series of novel α – phenyl – α – amide – amide compounds was designed and synthesized. We changed chlorantraniliprole's β – amino – acid – amide to the α – phenyl – α – amide – amide. The Ugi reaction was carried on to uphold diversity in the molecular. A series of compounds was obtained and bio – assayed and their structure – activity relationship was discussed. Also, their structures were characterized by ^{1}H NMR, elemental analysis and HRMS. The preliminary results of biological activity experiment show that the compounds at 200mg/L exhibit some insecticidal activity against *Mythimna separate* (7h); the compounds also exhibit some fungicidal activity against *Physalospora piricola* (7q) and *Alternaria solani* (7e).

Key words Ugi reaction; Chlorantraniliprole; insecticidal activity; fungicidal activity

李正名文集

Synthesis and Insecticidal Activities of Novel Anthranilic Diamides Containing Acylthiourea and Acylurea*

Jifeng Zhang[1], Junying Xu[2], Baolei Wang[1], Yuxin Li[1], Lixia Xiong[1], Yongqiang Li[1], Yi Ma[1], Zhengming Li[1]

([1]State Key Laboratory of Elemento - Organic Chemistry, College of Chemistry, Nankai University, Tianjin, 300071, China;
[2]CNOOC Tianjin Chemical Research and Design Institute, Tianjin, 300131, China)

Abstract Two series of anthranilic diamides containing acylthiourea and acylurea linkers were designed and synthesized, with changed length and flexibility of the linkers to compare to known anthranilic diamide insecticides. In total, 26 novel compounds were synthesized, and all compounds were characterized by ^1H nuclear magnetic resonance and high - resolution mass spectrometry. Their insecticidal activities against oriental armyworm (*Mythimna separata*), mosquito larvae (*Culex pipiens pallens*), and diamondback moth (*Plutella xylostella*) were evaluated. The larvicidal activities against oriental armyworm indicated that the introduction of acylthiourea into some structures could retain their insecticidal activity; 8 of the 15 compounds (**13a - 13e**, **14a - 14e**, and **15a - 15e**) exhibited 100% larvicidal activity at 10 mg/L. However, the introduction of acylurea decreased the insecticidal activity; only 3 of the 11 compounds (**17a - 17k**) exhibited 100% larvicidal activity at 200mg/L. The whole - cell patch - clamp technique indicated that compound **13b** and chlorantraniliprole exhibited similar effects on the voltage - gated calcium channel. The calcium - imaging technique was also applied to investigate the effects of compounds **13b** and **15a** on the intracellular calcium ion concentration ($[Ca^{2+}]_i$), which indicated that they released stored calcium ions from endoplasmic reticulum. Experimental results denoted that several new compounds are potential activators of the insect ryanodine receptor (RyR).

Key words anthranilic diamide; insecticidal activity; acylthiourea; acylurea; calcium channel

INTRODUCTION

For years, scientists have been dedicated in searching for insecticides with new mechanisms to cope with the global food shortage. Recently, the anthranilic diamides chlorantraniliprole (Rynaxypyr; DPX - E2Y45) (compound **A** in Figure 1) and cyantraniliprole (Cyazypyr) (compound **B**) were marketed by Dupont, targeting at the insect ryanodine receptor,[1,2] which exhibit their action by activating the uncontrolled release of the calcium store.[1] Their broad insecticidal spectra, high efficiency, and low toxicity aroused interests worldwide. Some modified structures (compound **C**) have been reported, which mainly focused on the 1 - (3 - chloropyridyl) pyrazole moiety (Ⅰ), the aliphatic amide moiety (Ⅱ), and the anthraniloyl moiety

* Reprinted from *Journal of Agricultural and Food Chemistry*, 2012, 60: 7565 - 7572. This work has been supported by the National Basic Research Program of China (973 Project 2010CB126106) and the National Natural Science Foundation of China (NNSFC) (31000861). We are also grateful for the financial support from the National Key Technologies Research and Development Program (2011BAE06B05), the Fundamental Research Funds for the Central Universities, and the Natural Science Foundation Project of Tianjin Ministry of Science and Technology of China (11JCYBJC04500).

(Ⅲ).[2-11] Nevertheless, the modifications about the amide bridge part as a linker of two aryl rings were seldom reported.[12]

In crop protection and bioactive chemicals, acylthioureas and acylureas have been reported to display a variety of biological activities, such as insecticidal, fungicidal, herbicidal, antimicrobial, antitumor, etc.[13-18] With this in mind, two series of novel anthranilic diamide derivatives (compound **D**) were designed and synthesized in this paper by introducing acylthiourea and acylurea moieties to increase the linker length and flexibility. Their synthetic routes were shown in Schemes 1 - 5. Their insecticidal activities against oriental armyworms, mosquito larvae, and diamondback moths were tested accordingly. The preliminary structure - activity relationship (SAR) was discussed. To provide insight to further study the biological effect of target compounds, the whole - cell patch - clamp and calcium imaging techniques were used to investigate the effects of compounds **13b** and **15a** on calcium channels in the central neurons of *Spodoptera exigua*.

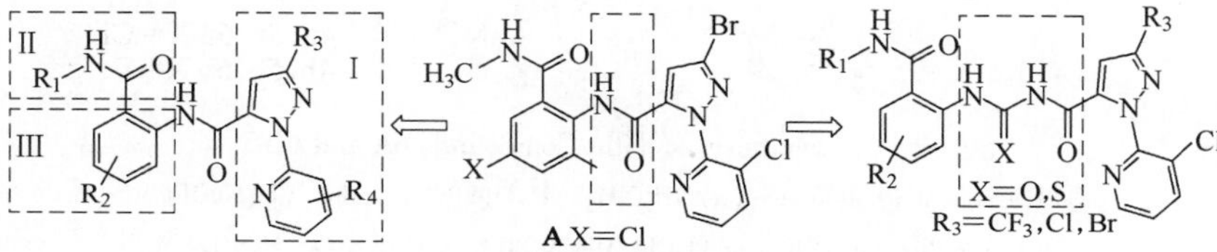

Figure 1 Chemical structures of compounds A - D.

MATERIALS AND METHODS

Instruments. ^1H nuclear magnetic resonance (NMR) spectra were recorded at 300MHz using a Bruker AC - P300 spectrometer or 400MHz using a Bruker AV 400 spectrometer (Bruker Co., Switzerland) in $CDCl_3$ or DMSO - d_6 solution with tetramethylsilane as the internal standard, and chemical - shift values (δ) were given in parts per million (ppm). High - resolution mass spectrometry (HRMS) data were obtained on a Varian QFT - ESI instrument. Flash chromatography was performed on Combi Flash Companion (Teledyne Isco, Inc., Lincoln, NE) with silica gel (300 - 400mesh). The melting points were determined on a X - 4 binocular microscope melting point apparatus (Beijing Tech Instruments Co., Beijing, China) and were uncorrected. The whole - cell patch clamp was performed using a patch - clamp amplifier (EPC - 10, HEKA Electronik, Lambrecht, Germany). Reagents were all analytically or chemically pure. All solvents and liquid reagents were dried by standard methods in advance and distilled before use. Chlorantraniliprole used in this work was synthesized according to the literature only for bioassay reference.[19]

General Synthetic Procedure for Compounds 6a and 6b. The title compounds 6a and 6b were prepared in our laboratory according to the methods reported by Dong et al. (Scheme 1), with some modifications. The melting points and ^1H NMR data of all of the compounds were consistent with the literature.[19,20]

1,1,1 - Trifluoro - 4 - (furan - 2 - yl) - 4 - hydroxybut - 3 - en - 2 - one (8). Sodium

(1.38g,60mmol) was dissolved in 30mL of anhydrous methanol slowly. Methanol was evaporated to dryness under reduced pressure to give the white solid sodium methylate. Then, diethyl ether (50mL), ethyl trifluoroacetate (8.52g, 60mmol), and 2-acetylfuran (6.60g, 60mmol) were successively added dropwise, stirred at room temperature for 5h, and filtered, which was acidized with 1M sulfuric acid to give a red liquid product. Yield = 64.9%. ^{1}H NMR (400MHz, $CDCl_3$) δ: 13.67 (br s, 1H, OH), 7.67 (s, 1H, Ar—H), 7.33 (d, 1H, J = 3.6Hz, Ar—H), 6.63 (d, 1H, J = 3.6Hz, Ar—H), 6.48 (s, 1H, CH).

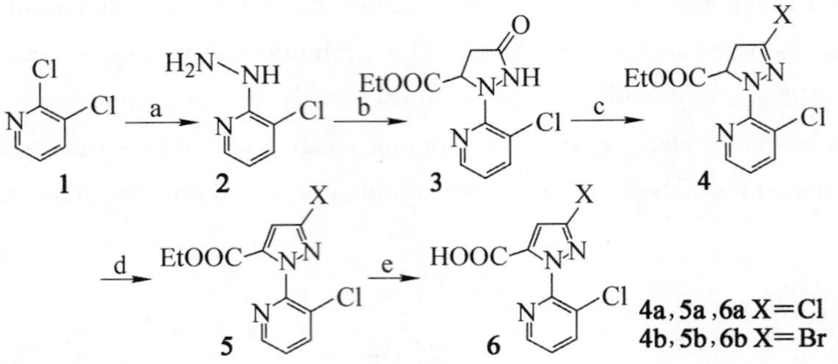

Scheme 1 Synthesis of Title Compounds **6a** and **6b**[①]

①Reagents and conditions: (a) $NH_2NH_2 \cdot H_2O$ (80%), reflux, (b) NaOEt, EtOH, reflux, (c) $POBr_3$ or $POCl_3$, MeCN, 80℃, (d) $K_2S_2O_8$, H_2SO_4, MeCN, reflux, and (e) (i) aqueous NaOH, MeOH and (ii) aqueous HCl

3-Chloro-2-(5-(furan-2-yl)-3-(trifluoromethyl)-1*H*-pyrazol-1-yl) pyridine (9). In a three-neck 1000mL round-bottomed flask, 1,1,1-trifluoro-4-(furan-2-yl)-4-hydroxybut-3-en-2-one (8, 27.1g, 130mmol) and 3-chloro-2-hydrazinylpyridine (18.9g, 130mmol) were added to 300mL of acetic acid. The mixture was refluxed for 8h, and then the excess acetic acid was removed under reduced pressure. The residue was further purified by flash column chromatography on silica gel with petroleum ether/ethyl acetate (5:1) to give intermediate **9** as a red oil. Yield = 42.1%. ^{1}H NMR (400MHz, $CDCl_3$) δ: 8.58 (d, J = 4.8Hz, 1H, pyridyl-H), 7.96 (d, J = 8.0Hz, 1H, pyridyl-H), 7.51 (dd, J = 8.0, 4.8Hz, 1H, pyridyl-H), 7.36 (s, 1H, pyrazolyl-H), 6.93 (s, 1H, furanyl-H), 6.41-6.30 (s, 1H, furanyl-H), 6.01 (d, J = 3.4Hz, 1H, furanyl-H).

1-(3-Chloropyridin-2-yl)-3-(trifluoromethyl)-1*H*-pyrazole-5-carboxylic acid (6c). To a solution of 3-chloro-2-(5-(furan-2-yl)-3-(trifluoromethyl)-1*H*-pyrazol-1-yl) pyridine (**9**, 3.88g, 12.4mmol) in acetone (35mL) and water (35mL), potassium permanganate (9.80g, 62mmol) was added in portion. When the addition was complete, the reaction was refluxed for 30min. The resulting mixture was filtered. The filtrate was acidified with 2MHCl and extracted with ethyl acetate (50mL × 3). The extracts were combined, washed with water (30mL × 2) and brine (30mL), then dried with anhydrous sodium sulfate, and evaporated to give compound **6c** as a white solid (1.85g). Yield = 51.3%. Melting point (mp) = 176-179℃. ^{1}H NMR (400MHz, DMSO-d_6) δ: 14.14 (s, 1H), 8.61 (dd, J = 4.8, 1.5Hz, 1H, pyridyl-H), 8.31 (dd, J = 8.0, 1.5Hz, 1H, pyridyl-H), 7.74 (dd, J = 8.0, 4.8Hz, 1H, pyridyl-H), 7.61 (s, 1H, pyrazolyl-H) (Scheme 2).

General Synthetic Procedure for Compounds 11a – 11m. The compounds **11a – 11m** were prepared in our laboratory according to the methods reported by Dong et al. (Scheme 3 and Table 1). The melting points and ^{1}H NMR data of all of the compounds were consistent with the literature.[19,20]

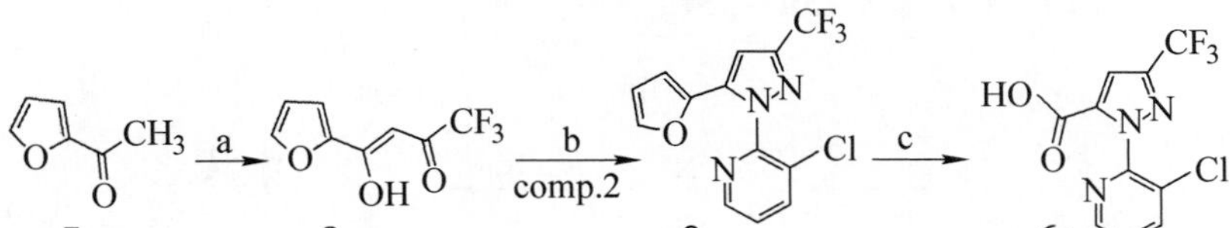

Scheme 2 Synthesis of Compound **6c**①

①Reagents and conditions: (a) $NaOCH_3$, ethyl trifluoroacetate, (b) AcOH, reflux, and (c) $KMnO_4$, acetone/H_2O, reflux.

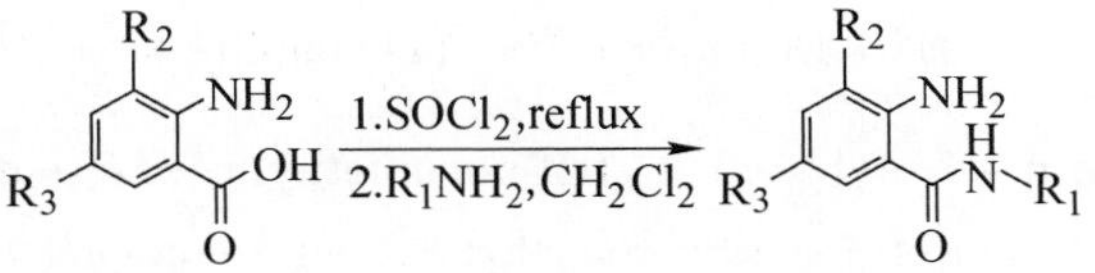

Scheme 3 Synthesis of Compounds **11a – 11m**

Table 1 Structure of Compounds 11a – 11m

compound	R_1	R_2	R_3
11a	Me	Me	Cl
11b	Et	Me	Cl
11c	*i* – Pr	Me	Cl
11d	*i* – Pr	H	Cl
11e	*n* – Pr	H	Cl
11f	*n* – Bu	Me	Cl
11g	*t* – Bu	Me	Cl
11h	cyclopropyl	Me	Cl
11i	cyclohexyl	Me	Cl
11j	*i* – Pr	Me	Br
11k	benzyl	Me	Br
11l	*i* – Pr	Me	H
11m	*i* – amyl	Me	H

General Synthetic Procedure for Compounds 13a – 13e, 14a – 14e, and 15a – 15e. To a suspension of 1 – (3 – chloropyridin – 2 – yl) – 1*H* – pyrazole – 5 – carboxylic acid derivative (**6a – 6c**) (1mmol) in 25mL of dichloromethane, oxalyl chloride (2.0mmol) and a drop of *N*, *N* – dimethylformamide (DMF) were added. After stirring at room temperature for 3h, the solution was evaporated. The resulting acyl chloride was dissolved in 20mL of acetonitrile and added to a solution of potassium thiocyanate in 15mL of acetonitrile with two drops of polyethylene glycol – 400 (PEG – 400). After stirring at room temperature for 30min, the mixture was filtered to give the acyl isothiocyanate derivatives **12a – 12c**, which were used without further purification, to which the 2 – amino – 3 – methylbenzamide derivatives (**11a – 11m**) (1.0mmol) were added and stirred overnight, then filtered, and further purified by recrystallization using methanol or by a silica – gel column eluted with petroleum ether/ethyl acetate (4:1) to obtain the title compounds **13a – 13e**, **14a – 14e**, and **15a – 15e**.

6a R_4=Cl
6b R_4=Br
6c R_4=CF_3
12a R_4=Cl
12b R_4=Br
12c R_4=CF_3
13a–e R_4=Cl
14a–e R_4=Br
15a–e R_4=CF_3

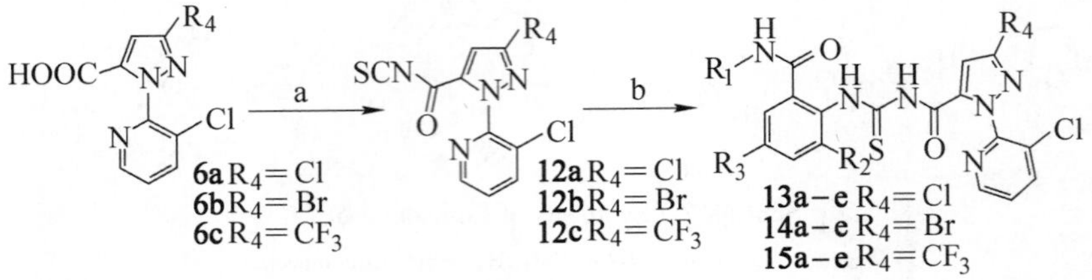

Scheme 4 Synthesis of Compounds **13a – 13e**, **14a – 14e** and **15a – 15e**①

①Reagents and conditions: (a) (i) oxalyl chloride, CH_2Cl_2 and (ii) KSCN, PEG – 400, CH_3CN and (b) **11a – 11m**, CH_3CN

1 – (3 – Chloropyridin – 2 – yl) – 3 – (trifluoromethyl) – 1*H* – pyrazole – 5 – carboxamide (16). To a suspension of 3 – trifluoromethyl – 1 – (3 – chloropyridin – 2 – yl) – 1*H* – pyrazole – 5 – carboxylic acid (**6c**) (2.40g, 8.23mmol) in 50mL of dichloromethane, oxalyl chloride (17.0mmol) and a drop of DMF were added. After stirring at room temperature for 3h, the solution was evaporated. The resulting acyl chloride was dissolved in 50mL of tetrahydrofuran (THF) and added to aqueous ammonia (25%) (5.76g, 41.15mmol) at 0℃. After stirring overnight, a large amount of solid formed and was filtered, which was further purified by a silica – gel column eluted with petroleum ether/ethyl acetate (2:1) to give the title compound as a white solid (1.85g). Yield = 77.3%. mp = 183 – 185℃. ^{1}H NMR (400MHz, $CDCl_3$) δ: 8.56 (d, J = 4.8Hz, 1H, pyridyl – H), 8.00 (d, J = 8.0Hz, 1H, pyridyl – H), 7.53 (dd, J = 8.0, 4.8Hz, 1H, pyridyl – H), 7.38 (s, 1H, pyrazolyl – H), 6.78 (s, 1H, NH), 5.77 (s, 1H, NH).

General Synthetic Procedure for Compounds 17a – 17k. A suspension of compound **16** (0.29g, 1mmol) in 20mL of 1,2 – dichloroethane and oxalyl chloride (3.0mmol) was heated to reflux and kept for 5h. Then, the mixture was evaporated, and the residue was dissolved in acetonitrile, to which the 2 – amino – 3 – methylbenzamide derivatives (**11a – 11m**) (1mmol) were added and stirred overnight. Then, the produced precipitate was filtered and washed with acetonitrile (5mL × 3) to give the title compounds **17a – 17k**.

Larvicidal Activity against Oriental Armyworm (*M. separata*). The larvicidal activities of compounds **13a – 13e**, **14a – 14e**, **15a – 15e**, **17a – 17k**, and chlorantraniliprole were evaluated using the reported procedure.[19] The insecticidal activity against oriental armyworms was tested by foliar application; individual corn (Tangyu 10, *Zea mays* L.) leaves were placed on moistened pieces of filter paper in Petri dishes. The leaves were then sprayed with the test solution and allowed to dry. The dishes were infested with 10 fourth – instar oriental armyworm larvae. Percentage mortalities were evaluated 2 days after treatment. Each treatment was replicated 3 times.

Larvicidal Activity against Mosquito Larvae (*C. pipiens pallens*). The larvicidal activities of compounds **13a – 13e**, **14a – 14e**, and chlorantraniliprole against mosquito larvae were evaluated by the reported procedure. [22] The compounds **13a – 13e**, **14a – 14e**, and chlorantraniliprole were prepared to different concentrations by dissolving compounds in acetone and adding distilled water. Then, 20 fourth – instar mosquito larvae were placed in 10mL of test solution and raised for 3 days. Each treatment was performed 3 times, and the average value of the three tested values was calculated. The results were expressed by death percentage.

Larvicidal Activity against Diamondback Moth (*P. xylostella*). The larvicidal activity of compounds **15a – 15e** and chlorantraniliprole was tested by the leaf – dip method. At first, a solution of each test sample in DMF (AR, purchased from Alfa Aesar) at a concentration of 200mg/L was prepared and then diluted to the required concentration with water (distilled). Leaf disks (6×2cm) were cut from fresh cabbage leaves and then sprayed with the test solution for 3s and allowed to dry. The resulting leaf disks were placed individually into glass tubes. Each disk was infested with 30 s – instar diamondback moth larvae. Percentage mortalities were evaluated 2 days after treatment. Each treatment was performed 3 times.

Electrophysiological Recording and Calcium Imaging. The currents of calcium (I_{Ca}) were recorded using the reported procedure. [23] Calibration of the fluorescence signal was achieved using the method by Iakahashi et al., with some modifications. [24] The attached neurons were rinsed in standard physiological saline [150mM NaCl, 4mM KCl, 2mM $MgCl_2$, 2mM $CaCl_2$, and 10mM *N* – 2 – hydroxyethylpiperazine – *N′* – 2 – ethanesulfonic acid (HEPES), buffered to pH 6.9] and then incubated in the dark at 28℃ in standard external saline containing the dye fluo – 3 – AM (Sigma, 10μM) for 45min or incubated in the dark at 28℃ in external saline [150mM NaCl, 4mM KCl, 2mM $MgCl_2$, 2mM ethylene glycol bis (2 – aminoethyl ether) – *N*, *N*, *N′*, *N′* – tetraacetic acid (EGTA), and 10mM HEPES, buffered to pH 6.9] containing the dye fluo – 5N (Invitrogen, 5μM) for 5h. After dye loading, cells were again rinsed in physiological saline twice.

For full depletion of thapsigargin – sensitive stores, cells were incubated with 1μM thapsigargin for 10min. Calcium ratio imaging studies were conducted using the imaging system coupled to an inverted fluorescence microscope with a Fluor 40 × oil immersion objective (Olympus IX71). Cells which excited at 488nm and 530nm under fluorescence emission were acquired using a charge – coupled device (CCD) camera (Image Pri – 6.0).

Each experiment was repeated at least 6 times. The data were analyzed using SPSS, Inc., version 17.0, and Microcal Origin, version 8.0 (Origin Lab Corp., Northampton, MA). Results were expressed as the mean ± standard deviation (SD) (n = number of cells). Statistical significance was determined by Student's paired or unpaired t tests. Fluorescence values were expressed as F/F_0, with F_0 being the resting (or baseline) fluorescence and F being the change in fluorescence from baseline after the drug application.

RESULTS AND DISCUSSION

Synthesis. The carboxylic acids **6a** and **6b** were prepared using the procedure reported by Dong et al. (Scheme 1), [19,20] with some modifications, for the synthesis of intermediates. With the reported procedure, intermediate **2** was prepared in 69% yield by refluxing 2,3 - dichloropyridine(1) and hydrazine hydrate(50%) in ethanol for 36h. In our experiments, hydrazine hydrate (80%) was used to directly reflux with compound 1 for 5 - 6h without solvent ethanol. By the improvement, compound 2 could be obtained in a 93% high yield.

The title compounds **13a - 13e**, **14a - 14e** and **15a - 15e** were synthesized, as shown in Scheme 4. Carboxylic acid(**6a - 6c**) was treated with oxalyl chloride and then added to potassium thiocyanate in acetonitrile with two drops of PEG - 400. The corresponding acyl isothiocyanate derivative was filtered after stirring at room temperature for 30min and reacted with 1 equiv of amines **11a - 11m** to afford the title compounds **13a - 13e**, **14a - 14e** and **15a - 15e**, without further purification. Unlike the reported procedure that a reflux condition was usually needed,[14] this reaction was worked out at room temperature under the catalysis of a small amount of PEG - 400 with high yield.

Scheme 5 Synthesis of Compounds **17a - 17k**①

①Reagents and conditions: (a) (i) oxalyl chloride, CH_2Cl_2 and (ii) $NH_3 \cdot H_2O$ and (b) (i) oxalyl chloride, $ClCH_2CH_2Cl$, reflux and (ii) **11a - 11m**, CH_3CN

Table 2 Insecticidal Activities of Compounds 13a - 13e, 14a - 14e, 15a - 15e, 17a - 17k, and Chlorantraniliprole against Oriental Armyworms

compound	R_1	R_2	R_3	R_4	larvicidal activity (%) at a concentration of mg/L					
					200	100	50	20	10	5
13a	Me	Me	Cl	Cl	100	100	100	100	40	
13b	*i* - Pr	Me	Cl	Cl	100	100	100	100	100	40
13c	*i* - Pr	Me	Br	Cl	100	80	40			
13d	*i* - Pr	Me	H	Cl	100	100	100	100	100	0
13e	*i* - amyl	Me	H	Cl	60					
14a	Me	Me	Cl	Br	100	100	100	100	100	50
14b	*i* - Pr	Me	Br	Br	100	100	100	100	100	0
14c	*t* - Bu	Me	Cl	Br	100	100	100	100	100	40
14d	cyclohexyl	Me	Cl	Br	0					
14e	*i* - amyl	Me	H	Br	100	30				
15a	Me	Me	Cl	CF_3	100	100	100	100	100	50
15b	Et	Me	Cl	CF_3	100	100	100	100	100	0
15c	cyclohexyl	Me	Cl	CF_3	40					
15d	cyclopropyl	Me	Cl	CF_3	100	100	100	100	100	30
15e	benzyl	Me	Br	CF_3	100	100	100	100	0	
17a	Me	Me	Cl	CF_3	100	20				

Table 2(continued)

compound	R_1	R_2	R_3	R_4	larvicidal activity(%) at a concentration of mg/L					
					200	100	50	20	10	5
17b	cyclopropyl	Me	Cl	CF_3	20					
17c	*n* - Bu	Me	Cl	CF_3	40					
17d	*t* - Bu	Me	Cl	CF_3	60					
17e	cyclohexyl	Me	Cl	CF_3	0					
17f	*i* - Pr	Me	Cl	CF_3	70					
17g	cyclopropyl	Me	Br	CF_3	100	0				
17h	*n* - Pr	H	Cl	CF_3	0					
17i	cyclopropyl	Me	H	CF_3	100	80				
17j	cyclopropyl	H	Cl	CF_3	30					
17k	*n* - Pr	Me	H	CF_3	60					
chlorantraniliprole					100	100	100	100	100	100

The title compounds **17a – 17k** were synthesized, as shown in Scheme 5. Carboxylic acid 6c was converted to amide **16** by reacting with oxalyl chloride and then aqueous ammonia. Compound **16** was then refluxed with oxalyl chloride in 1,2 – dichloroethane to give its corresponding acyl isocyanate derivative, which coupled with amines **11a – 11m** to afford the title compounds **17a – 17k**.

The ^1H NMR data of acylthiourea and acylurea derivatives were characteristic in all of the target compounds. In compounds **13a – 13e** and **14a – 14e**, the two active proton signals of NHCS and NHCO on carbonyl thiourea moiety were observed in DMSO – d_6 at 12.15 – 12.04 and 11.48 – 11.37 ppm, respectively. In compounds **15a – 15e**, the introduction of a trifluoromethyl group increased the hydrophobicity of the whole molecule; therefore, the ^1H NMR determination was carried out in $CDCl_3$. The chemical shifts of these two active protons were at 11.34 – 11.21 and 10.10 – 9.77 ppm. In compounds **17a – 17k**, when R_2 = Me, the two active proton signals of CONHCO and NHCO on the carbonyl acylurea moiety appeared at 11.55 – 11.46 and 9.93 – 9.72 ppm, respectively. Whereas their values shifted to 11.54 – 11.52 and 11.16 – 11.11 ppm, respectively, when R_2 = H(**17h** and **17j**) in DMSO – d_6.

SAR. *Larvicidal Activities against Oriental Armyworms* (*M. separata*). The larvicidal activities of compounds **13a – 13e**, **14a – 14e**, **15a – 15e**, **17a – 17k**, and commercial chlorantraniliprole against oriental armyworms were summarized in Table 2. In general, most of the compounds exist in moderate to good pesticidal activities. Particularly, eight compounds (**13a – 13e**, **14a – 14e**, and **15a – 15e**) exhibited 100% larvicidal activities at 10mg/L against oriental armyworm. From the activities against oriental armyworms, we could obviously find that the compounds containing the acylthiourea moiety (**13a – 13e**, **14a – 14e** and **15a – 15e**) gave higher insecticidal activities than the corresponding acylurea derivatives (**17a – 17k**). When compounds **15a** and **15d** are compared to compounds 17a and 17b, we could find that one atom difference between acylthiourea and acylurea can cause a great change in its insecticidal activity. When different substitutions in the aliphatic amide moiety of title compounds **13a – 13e**, **14a – 14e** and **15a – 15e** are compared, it was found that large substituents, such as i – amyl and cyclohexyl (**13e – 14d**, **14e**, and **15c**) decreased the insecticidal activity against oriental armyworms. Furthermore, compounds with the i – propyl and *t* – butyl substituents showed the best larvicidal activity against ariental armyworms in our research, which was consistent with the results reported in previous SARs. [2,25]

Biological Assay. Insecticidal activities against oriental armyworms (*Mythimna separata*), mosquito larvae (*Culex pipiens pallens*), and diamondback moths (*Plutella xylostella*) were performed on test organisms reared in a greenhouse. The bioassay was replicated at 25 ± 1℃ according to statistical requirements. Assessments were made on a dead/alive basis, and mortality rates were corrected applying Abbott's formula. [21] Evaluation was based on a percentage scale of 0 – 100, where 0 equals no activity and 100 equals total kill. Error of the experiments was 5%. For comparative purposes, chlorantraniliprole was tested as a reference. The insecticidal activity was summarized in Tables 2 – 4.

Table 3 Insecticidal Activities of Compounds 13a – 13e, 14a – 14e, and Chlorantraniliprole against Mosquito Larvae

compound	larvicidal activity (%) at a concentration of mg/L		
	2	1	0.5
13a	70	30	
13b	40		
13c	90	60	20
13d	20		
13e	10		
14a	60	20	
14b	90	70	30
14c	100	90	30
14d	40		
14e	20		
chlorantraniliprole	100	100	100

Table 4 Insecticidal Activities of Compounds 15a – 15e and Chlorantraniliprole against Diamondback Moths

compound	larvicidal activity (%) at a concentration of mg/L				
	20	1	0.1	0.01	0.001
15a	100	100	100	71	43
15b	100	100	100	86	29
15c	30	0			
15d	100	100	100	100	71
15e	100	100	100	100	57
chlorantraniliprole	100	100	100	100	100

Larvicidal Activities against Mosquito Larvae (*C. pipiens pallens*). The larvicidal activities of compounds **13a – 13e** and **14a – 14e** against mosquito larvae were shown in Table 3. Most of the compounds tested showed good larvicidal activities, while compounds with big substituents i – amyl and cyclohexyl in the aliphatic amide moiety revealed somewhat less activity.

Larvicidal Activity against Diamondback Moths (*P. xylostella*). The larvicidal activities against diamondback moths of compounds **15a – 15e** and chlorantraniliprole were shown in Table 4. From it, we can see that most of the compounds tested showed excellent larvicidal activities. Compounds **15d** and **15e** showed the best activities (both 100% at 0.01mg/L). Similarly, compound 15e with a large substituent (R_2 = benzyl) also showed good larvicidal activity against diamondback moths.

The toxicity profile (LC_{50} values) of compounds **15d** and **15e**, which were found to be the most active insecticides of these two series of compounds against diamondback moths (100% at 0.01mg/L), was shown in Table 5. The LC_{50} values of compounds **15d** and **15e** were 0.00025 and 0.00065mg/L, respectively, but still higher than that of chlorantraniliprole (0.0000123mg/mL; Table 5).

Table 5 LC_{50} Values of Compounds 15d, 15e, and Chlorantraniliprole against Diamondback Moths

compound	$y = a + bx$	LC_{50} (mg/L)	R	95% confidence interval
15d	$y = 11.04 + 1.68x$	0.00025	0.9974	0.00023 – 0.00028
15e	$y = 12.90 + 2.48x$	0.00065	0.9723	0.00053 – 0.00081
chlorantraniliprole	$y = 11.50 + 1.32x$	0.0000123	0.9995	$1.16 \times 10^{-5} - 1.31 \times 10^{-5}$

Effects of Compounds 13b and 15a on Calcium Channels of Neurons from *S. exigua*. Figure 2 shows the change rate of peak current amplitude versus recording time in 0.1 and 0.001mmol/L compound **13b** – and 0.1mmol/L chlorantraniliprole – treated and control neurons. The peak currents were reduced to 54.71 ± 4.95% (n = 6), 74.63 ± 5.37% (n = 5), 39.43 ± 5.02% (n = 8), and 85.44 ± 1.14% (n = 6) of the initial value by the end of the 10min recording, when the neurons were treated with 0.1 and 0.001mmol/L compound **13b**, 0.1mmol/L chlorantraniliprole, and the control, respectively. The peak current was reduced to 54.71 ± 4.95% (n = 6) when the neurons were treated with 0.1 mmol/L compound **13b**. In comparison to the control (85.44 ± 1.14%), compound **13b** at 0.001mmol/L has a weak effect on $Ca^{2+}_{(L)}$ currents. The recorded peak currents of calcium channels treated with compound **13b** were in a concentrationdependent manner (Figure 2 and Table 6).

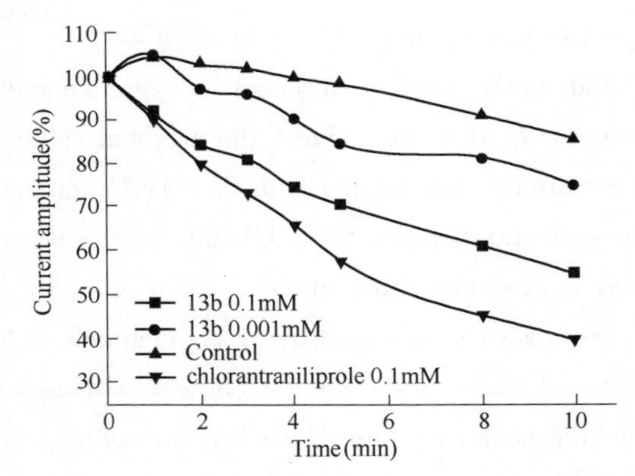

Figure 2 Variation of the peak current for the whole – cell calcium channels in neurons (treated with different concentrations of compound **13b** and chlorantraniliprole) at different times during patch – clamp recording in comparison to the peak value at 0min

Table 6 **Analysis of the Peak Current Change Rate with Time in the Neurons of *S. exigua*(Treated by Compound 13b and Chlorantraniliprole)** ①

time(min)	0.1mM compound **13b**	0.001mM compound **3b**	0.1mM chlorantraniliprole	control
0	100	100	100	100
1	91.492 ±3.47②	104.964 ±1.34③	89.62 ±2.16②	104.47 ±2.33②
2	84.128 ±2.16②	97.215 ±7.58②	74.01 ±8.01②	102.75 ±2.45③
3	80.727 ±4.77②	95.803 ±7.26②	69.57 ±9.43②	101.76 ±2.7③
4	74.437 ±5.31②	90.134 ±6.41②	63.36 ±7.71②	99.66 ±1.64③
5	70.252 ±8.48②	84.442 ±5.48②	55.79 ±4.73②	98.06 ±1.01③
8	61.013 ±3.28②	81.209 ±3.69②	44.31 ±2.62②	90.85 ±0.65②
10	54.715 ±4.95②	74.632 ±6.12②	39.685 ±5.02②	85.44 ±1.14②

①Values are the mean ± SD. ②Significant difference at $p<0.01$. ③Significant difference at $p<0.05$.

There was a significant change in maximal value of the peak current when the neurons were held at −70mV with the treatment of 0.1mmol/L compound **13b** (Figure 3). The calcium current decreased when compound **13b** was injected in the surrounding of the neuron cells. The maximal value of I_{Ca} (−2.1303 ± 0.0237nA) shifted to the negative direction by approximately 10mV after the neurons were treated with compound **13b** for 1min. By the end of the 5 and 10 min recordings, I_{Ca} decreased to −1.6356 ± 0.0641 and −1.2738 ± 0.882nA, respectively, and the $I-V$ curves were also shifted to the negative direction by approximately 10 mV during the recording. Unlike the results for I_{Ca} described above, the maximal value of I_{Ca} did not shift to a negative direction when the neurons were treated with 0.001mM compound **13b**, from which we concluded that, when the concentration was ≤0.001mM, there was no effect on calcium channels in the central neurons of *S. exigua* third larvae.

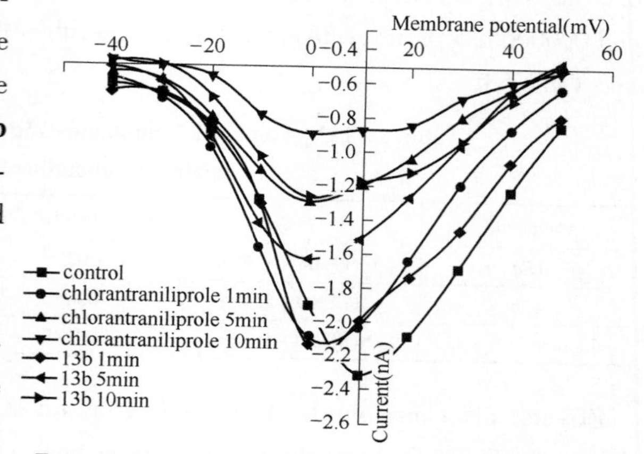

Figure 3 Current – voltage relationship curves of whole – cell calcium channels recorded in neurons of *S. exigua* (treated with 0.1 mmol/L compound **13b** and chlorantraniliprole) at different intervals (in minutes)

These results indicated that $Ca^{2+}_{(L)}$ channels of *S. exigua* neurons were modulated by compound **13b** and partly closed after the action. Although there was no notable change on activation voltage, the negatively shifted $I-V$ relationship curves indicate that voltage dependence was influenced by compound **13b**. All of these results suggest that the $Ca^{2+}_{(L)}$ channels of *S. exigua* neurons were the possible target of compound **13b**.

Figure 4 illustrated the change of $[Ca^{2+}]_i$ versus the recording time when the neurons were treated with compounds **13b**, **15a**, and chlorantraniliprole. The peak of $[Ca^{2+}]_i$ was 116.863 ± 2.33% ($n=10$), 112.408 ±1.26% ($n=8$), and 123.292 ±2.17% ($n=21$) of the initial value by the end of the 10min recording when the cells were treated with 1000mg/L compound **13b**, 1000mg/L compound **15a**, and 1000mg/L chlorantraniliprole, respectively. In comparison to the control (99.91 ± 2.56%), compounds **13b** and **15a** induced a $[Ca^{2+}]_i$ increase without extracellular Ca^{2+}. It indicated that compounds **13b** and **15a** could activate the calcium release channel in the endoplasmic reticulum(ER) membrane. Figure 3 also indicated that the recorded $[Ca^{2+}]_i$ (F/F_0) had a good positive correlation with bioactivities.

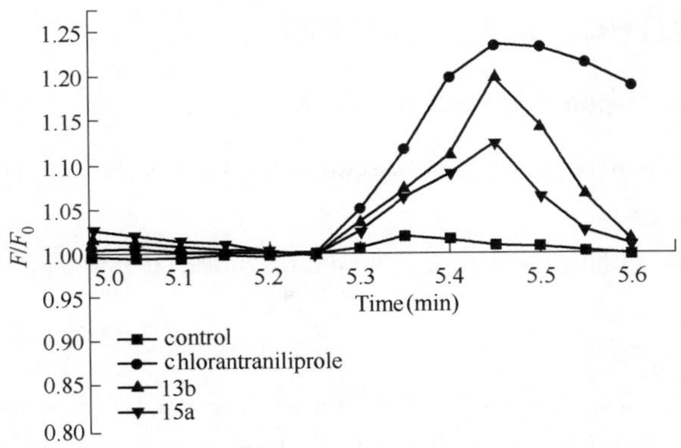

Figure 4 Change of $[Ca^{2+}]_i$ versus recording time when the neurons were treated with 1000 mg/L compounds **13b**, **15a**, and chlorantraniliprole

There were two kinds of calcium release channels in the ER membrane, namely, RyR and IP_3R Ca^{2+} channels.[26] To test which pathway was involved in the elevation of $[Ca^{2+}]_i$, the primary cultured neurons were dyed, loading with fluo-5N, then treated with heparin (10mg/mL, a competitive antagonist of IP_3) for 20min, and incubated with 1μM thapsigargin for 10min. When external Ca^{2+} was free, IP_3 receptors were blocked using heparin and the intracellular calcium store was depleted with thapsigargin; the decrease of $[Ca^{2+}]_i$ was only attributed to compound **13b** (1000mg/L). These data indicated that RyRs would be the possible action target of this series of novel compounds.

In summary, two novel series of anthranilic diamides containing acylthiourea and acylurea moieties were synthesized, and their larvicidal activities against oriental armyworm, mosquito larvae, and diamondback moth were evaluated. The results indicated that the introduction of acylthiourea moiety into some structures could retain their insecticidal activity; 8 of the 15 compounds (**13a – 13e**, **14a – 14e**, and **15a – 15e**) exhibited 100% larvicidal activity at 10 mg/L against oriental armyworm. However, the introduction of acylurea moiety decreased the insecticidal activity; only 3 of the 11 compounds (**17a – 17k**) exhibited 100% larvicidal activity at 200mg/L against oriental armyworm. We speculated that it might be the solubility factor involved because compounds containing acylthiourea moiety exhibited better solubility in organic solvents during our experiments. The effects on calcium channels of neurons from *S. exigua* indicated that the title compounds influenced the same target RyRs as chlorantraniliprole. Also, the experiment of intracellular calcium of neurons provided us a rapid detection for the activity of the target compound.

ASSOCIATED CONTENT

Supporting Information

^{1}H NMR, HRMS, and melting point data for compounds **13a – 13e**, **14a – 14e**, **15a – 15e**, and **17a – 17k**. This material is available free of charge via the Internet at http://pubs.acs.org.

AUTHOR INFORMATION

Corresponding Author

E – mail: liyx128@ nankai. edu. cn(Y. – X. L.); nkzml@ vip. 163. com(Z. – M. L.).

Notes

The authors declare no competing financial interest.

References

[1] Lahm, G. P.; Selby, T. P.; Frendenberger, J. H.; Stevenson, T. M.; Myers, B. J.; Seburyamo, G.; Smith, B. K.; Flexner, L.; Clark, C. E.; Cordova, D. Insecticidal anthranilicdiamides: A new class of potent ryanodine receptor activators. *Bioorg. Med. Chem. Lett.* 2005, 15, 4898 – 4906.

[2] Feng, Q.; Liu, Z. L.; Xiong, L. X.; Wang, M. Z.; Li, Y. Q.; Li, Z. M. Synthesis and insecticidal activities of novel anthranilicdiamides containing modified *n* – pyridylpyrazoles. *J. Agric. Food Chem.* 2010, 58, 12327 – 12336.

[3] Clark, D. A.; Lahm, G. P.; Smith, B. K.; Barry, J. D.; Clagg, D. G. Synthesis of insecticidal fluorinated anthranilicdiamides. *Bioorg. Med. Chem.* 2008, 16, 3163 – 3170.

[4] Lahm, G. P.; Cordova, D.; Barry, J. D. New and selective ryanodine receptor activators for insect control. *Bioorg. Med. Chem.* 2009, 17, 4127 – 4133.

[5] Lahm, G. P.; Stevenson, T. M.; Selby, T. P.; Freudenberger, J. H.; Cordova, D.; Flexner, L.; Bellin, C. A.; Dubas, C. M.; Smith, B. K.; Hughes, K. A.; Hollingshaus, J. G.; Clark, C. E.; Benner, E. A. Rynaxypyr: A new insecticidal anthranilicdiamide that acts as a potent and selective ryanodine receptor activator. *Bioorg. Med. Chem. Lett.* 2007, 17, 6274 – 6279.

[6] Li, B.; Yang, H. B.; Wang, J. F.; Yu, H. B.; Zhang, H.; Li, Z. N.; Song, Y. Q. 1 – Substituted pyridyl – pyrazolyl amide compounds and uses thereof. U. S. Patent 201146186 A1, 2011.

[7] Alig, B.; Fischer, R.; Funke, C.; Gesing, E. F.; Hense, A.; Malsam, O.; Drewes, M. W.; Gorgens, U.; Murata, T.; Wada, K.; Arnold, C.; Sanwald, E. Anthranilic acid diamide derivative with hetero – aromatic and hetero – cyclic substituents. WO Patent 2007144100 A1, 2007.

[8] Loiseleur, O.; Hall, R. G.; Stoller, A. D.; Graig, G. W.; Jeanguenat, A.; Edmunds, A. Novel insecticides. WO Patent 2009024341 A2, 2009.

[9] Loiseleur, O.; Durieux, P.; Trah, S.; Edmunds, A.; Jeanguenat, A.; Stoller, A.; Hughes, D. J. Pesticides containing a bicyclic bisamide structure. WO Patent 2007093402 A1, 2007.

[10] Kruger, B.; Hense, A.; Alig, B.; Fischer, R.; Funke, C.; Gesing, E. R.; Malsam, A.; Drewes, M. W.; Arnold, C.; Lummen, P.; Sanwald, E. Dioxazine – and oxadiazine substituted arylamides. WO Patent 2007031213 A1, 2007.

[11] Dumas, D. J. Process for preparing 2 – amido – 5 – cyanobenzoic acid derivatives. WO Patent 2009111553 A1, 2009.

[12] Hughes, D. J.; Peace, J. E.; Riley, S.; Russell, S.; Swanborough, J. J.; Jeanguenat, A.; Renold, P.; Hall, R. G.; Loiseleur, O.; Trah, S.; Wenger, J. Synergistic pesticidal mixtures with nitrogen – containing component. WO Patent 2007009661 A2, 2007.

[13] Sun, R. F.; Zhang, Y. L.; Chen, L.; Li, Y. Q.; Li, Q. S.; Song, H. B.; Huang, R. Q.; Bi, F. C.; Wang, Q. M. Design, synthesis, bioactivity, and structure – activity relationship (SAR) studies of novel benzoylphenylureas containing oxime ether group. *J. Agric. Food Chem.* 2008, 56, 11376 – 11391.

[14] Saeed, A.; Batool, M. Synthesis and bioactivity of some new 1 – tolyl – 3 – aryl – 4 – methylimidazole – 2 – thiones. *Med. Chem. Res.* 2007, 16, 143 – 154.

[15] Cesarini, S. ; Spallarossa, A. ; Ranise, A. ; Schenone, S. ; Rosano, C. ; La Colla, P. ; Sanna, G. ; Busonera, B. ; Loddo, R. N – Acylated and N, N′ – diacylated imidazolidine – 2 – thione derivatives and N, N′ – diacylated tetrahydropyrimidine – 2 (*1H*) – thione analogues: Synthesis and antiproliferative activity. *Eur. J. Med. Chem.* 2009, 44, 1106 – 1118.

[16] Rao, X. P. ; Wu, Y. ; Song, Z. Q. ; Shang, S. B. ; Wang, Z. D. Synthesis and antitumor activities of unsymmetrically disubstitutedacylthioureas fused with hydrophenanthrene structure. *Med. Chem. Res.* 2011, 20, 333 – 338.

[17] Sun, C. W; Huang, H. ; Feng, M. Q. ; Shi, X. L. ; Zhang, X. D. ; Zhou, P. A novel class of potent influenza virus inhibitors: Polysubstitutedacylthiourea and its fused heterocycle derivatives. *Bioorg. Med. Chem. Lett.* 2006, 16, 162 – 166.

[18] Hallur, G. ; Jimeno, A. ; Dalrymple, S. ; Zhu, T. ; Jung, M. K. ; Hidalgo, M. ; Isaacs, J. T. ; Sukumar, S. ; Hamel, E. ; Khan, S. R. Benzoylphenylurea sulfur analogues with potent antitumor activity. *J. Med. Chem.* 2006, 49, 2357 – 2360.

[19] Dong, W. L. ; Xu, J. Y. ; Xiong, L. X. ; Liu, X. H. ; Li, Z. M. Synthesis, structure and biological activities of some novel anthranilic acid esters containing *N* – pyridylpyrazole. *Chin. J. Chem.* 2009, 27, 579 – 586.

[20] Dong, W. L. ; Liu, X. H. ; Xu, J. Y. ; Li, Z. M. Design and synthesis of novel anthranilicdiamides containing 5,7 – dimethyl[1,2,4]triazolo – [1,5 – *a*]pyrimidine. *J. Chem. Res.* 2008, 9, 530 – 533.

[21] Abbott, W. S. A method of computing the effectiveness of an insecticide. *J. Econ. Entomol.* 1925, 18, 265 – 267.

[22] Chen, L. ; Huang, Z. Q. ; Wang, Q. M. ; Shang, J. ; Huang, R. Q. ; Bi, F. C. Insecticidal benzoylphenylurea – *S* – carbamate: A new propesticide with two effects of both benzoylphenylureas and carbamates. *J. Agric. Food Chem.* 2007, 55, 2659 – 2663.

[23] Li, Y. X. ; Mao, M. Z. ; Li, Y. M. ; Xiong, L. X. ; Li, Z. M. ; Xu, J. Y. Modulations of high – voltage activated Ca^{2+} channels in the central neurones of *Spodoptera exigua* by chlorantraniliprole. *Physiol. Entomol.* 2011, 36, 230 – 234.

[24] Takahashi, A. ; Camacho, P. ; Lechleiter, J. D. ; Herman, B. Measurement of intracellular calcium. *Physiol. Rev.* 1999, 79, 1089 – 1125.

[25] Liu, A. P. ; Hu, Z. B. ; Wang, Y. J. ; Wang, X. G. ; Huang, M. Z. ; Ou, X. M. ; Mao, C. H. ; Pang, H. L. ; Huang, L. Preparation of pyrazole containing 2 – amino – *N* – oxybenzamide compounds as pesticides. CN Patent 101337959 A, 2009.

[26] Striggow, F. ; Ehrlich, B. E. Ligand – gated calcium channels inside and out. *Curr. Opin. Cell. Biol.* 1996, 8, 490 – 495.

Design, Synthesis and Biological Activities of Novel Anthranilic Diamide Insecticide Containing Trifluoroethyl Ether*

Yu Zhao(赵毓), Yongqiang Li(李永强), Lixia Xiong(熊丽霞), Hongxue Wang(王红学), Zhengming Li(李正名)

(National Pesticidal Engineering Centre(Tianjin), State Key Laboratory of Elemento - organic Chemistry, Nankai University, Tianjin, 300071, China)

Abstract Two series of novel anthranilic diamide insecticide containing trifluoroethyl ether were designed and synthesized, and their structures were characterized by ^{1}H NMR spectroscopy, elemental analysis and single crystal X - ray diffraction analysis. The insecticidal activities of the new compounds were evaluated. The results of bioassays indicated that some of these title compounds exhibited excellent insecticidal activities. The insecticidal activities of compounds **19a**, **19b**, **19d**, **19g**, **19k** and **19m** against oriental armyworm at 2. 5 mg. kg^{-1} were 100%. The larvicidal activities of **19a**, **19b**, **19c**, **19d**, **19e**, **19g** and **19n** against diamond - back moth were 100% at 0. 1mg · kg^{-1}. Surprisingly, most of them still exhibited perfect insecticidal activity against diamond - back moth when the concentration was reduced to 0. 05mg · kg^{-1}, which was higher than the commercialized Chlorantraniliprole.

Key words anthranilic diamide; ryanodine receptor; trifluoroethyl ether; insecticidal activity

Introduction

Resistance has often been a problem or a potential problem for insecticide and is one of the most important reasons why insecticides with a new mode of action have been desired. [1] Recently, two new classes of insecticidal phthalic acid diamides and anthranilic diamides have been discovered with exceptional insecticidal activity on a range of Lepidoptera, which exhibit their action by binding to ryanodine receptors and activating the uncontrolled release of calcium stores. [2] Since then diamides have be the focus of synthesis activities within the agrochemical industry. Anthranilic diamides and their chemistry have recently attracted considerable attention in the field of novel agricultural insecticides, owing to their prominent insecticidal activity, unique modes of action and good environmental profiles. [3,4] Anthranilic diamides act on the sarcosplasmic reticulum of cardiac and skeletal muscle cells to open internal calcium stores causing muscle contraction, paralysis, and death. [5,6]

Anthranilic diamide insecticide is characterized by a three - part chemical structure as shown in Figure 1: (A) an anthraniloyl moiety, (B) an aromatic acyl moiety and (C) an aliphatic amide moiety. Notably, anthranilic diamides containing *N* - pyridylpyrazole in the second section (B) showed significantly better activity than other heterocyclic derivatives. Work in this area has led to the discovery of Chlorantraniliprole, a highly potent and selective activator of insect ryanodine

---

* Reprinted from *Chinese Journal of Chemistry*, 2012, 30(8): 1748 - 1758. This work was supported by the National Basic Research Program of China (No. 2010CB126106), the Fundamental Research Funds for the Central Universities and the Specialized Research Fund for the Doctoral Program of Higher Education (No. 20110031120011).

receptors with exceptional activity on a broad range of Lepidoptera. As the first new insecticide from this class (Figure 1),[7] Chlorantraniliprole demonstrates field use – rates that are significantly less than current commercial standards, varying from 50g/ha to less than 1g/ha and with good safety toward beneficial insects. In addition to possessing extremely high levels of potency on insects, Chlorantraniliprole shows remarkable safety to mammals as a result of poor intrinsic activity on mammalian ryanodine receptors with a margin of selectivity of the order of 10^3.

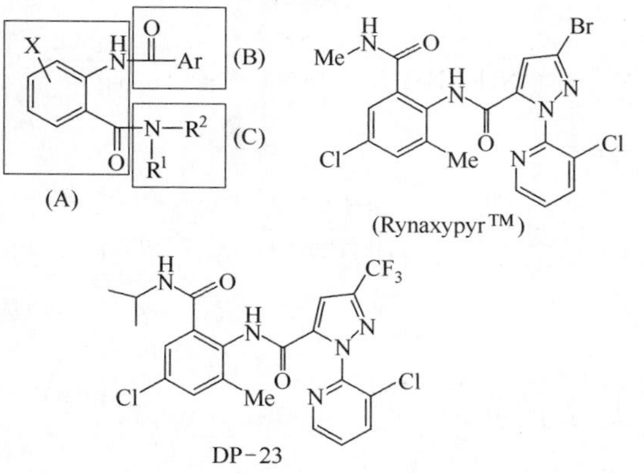

Figure 1 Chemical structures of anthranilic diamides insecticides

Recently, anthranilic diamides derivatives have drawn much attention in insecticidal research due to their significant bioactivity. In addition, many investigations have indicated that introducing the F or CF_3 group into heterocyclic molecules mostly results in the improvement of physical, chemical and biological properties.[8-11] It was reported that a series of fluorinated derivatives of anthranilic diamides displayed an insecticidal activity comparable or superior to that of Chlorantraniliprole. The synthesis and insecticidal evaluation of DP – 23 have been reported and the results of bioassay showed that it exhibit excellent larvicidal activity (Figure 1).[12]

Encouraged by these reports, an idea was developed that the introduction of a trifluoroethyl ether substituent into the Chlorantraniliprole molecules by substituting the halogen atoms on the pyrazole ring could improve biological properties. Therefore, in a search for new anthranilic diamide insecticides with improved profiles, two series of anthranilic diamide derivatives containing trifluoroethyl ether were designed and synthesized.

Experimental

Materials and methods

^{1}H NMR spectra were obtained at 400MHz using a Bruker AV400 spectrometer or Varian Mercury Plus400 spectrometer in $CDCl_3$ solution with tetramethylsilane as the internal standard. Elemental analyses were determined on a Yanaco CHN Corder MT – 3 elemental analyzer. The melting points were determined on an X – 4 binocular microscope melting point apparatus (Beijing Tech Instruments Co., Beijing, China) and were uncorrected. All solvents and liquid reagents were dried by standard methods and distilled before use.

General procedures

Chlorantraniliprole was prepared according to the route shown in Scheme 1. The title compounds **19** and **20** were synthesized from compound **18** and the appropriate intermediate **14** or **15** (obtained from the intermediate **13** and corresponding alcohol or amine – see Table 1) in dry tetrahydrofuran using triethylamine as base as shown in Scheme 2.

Scheme 1

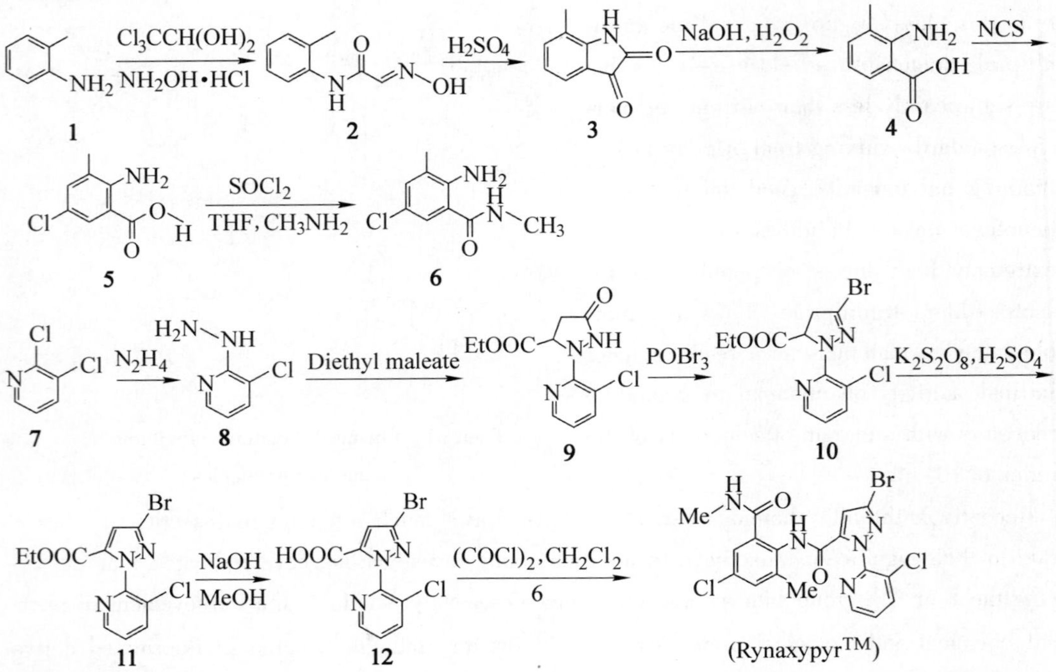

1 (NH_2) → ($Cl_3CCH(OH)_2$, $NH_2OH \cdot HCl$) → 2 (O, N, H, N, OH) → (H_2SO_4) → 3 (H, N, O, O) → ($NaOH$, H_2O_2) → 4 (NH_2, OH, O) → (NCS) →

5 (Cl, NH_2, O, H, O) → ($SOCl_2$; THF, CH_3NH_2) → 6 (Cl, NH_2, H, N, CH_3, O)

7 (Cl, N, Cl) → (N_2H_4) → 8 (H_2N—NH, N, Cl) → (Diethyl maleate) → 9 (EtOOC, O, N, NH, N, Cl) → ($POBr_3$) → 10 (EtOOC, Br, N, N, N, Cl) → ($K_2S_2O_8$, H_2SO_4) →

11 (EtOOC, Br, N, N, N, Cl) → (NaOH, MeOH) → 12 (HOOC, Br, N, N, N, Cl) → ($(COCl)_2$, CH_2Cl_2; 6) → (Me, H, N, O, H, N, O, Br, N, N, N, Cl, Cl, Me)

(Rynaxypyr[TM])

Scheme 2

13 (R^1, NH_2, R^2, OH, O) → (R^3—NH_2) → 14a—14n (R^1, NH_2, R^2, H, N, R^3, O)

13 → (R^3—OH) → 15a—15b (R^1, NH_2, R^2, O, R^3, O)

9 (EtOOC, O, N, NH, N, Cl) → ($K_2S_2O_8$, H_2SO_4) → 16 (EtOOC, O, N, NH, N, Cl) → (CF_3CH_2I) → 17 (EtOOC, O—CH_2CF_3, N, N, N, Cl) → (NaOH) →

18 (HOOC, O—CH_2CF_3, N, N, N, Cl) → (14) → 19a-19n (R^3, H, N, O, H, N, O, O—CH_2CF_3, N, N, N, Cl, R^2, R^1)

18 → (15) → 20a-20b (R^3, O, O, H, N, O, O—CH_2CF_3, N, N, N, Cl, R^2, R^1)

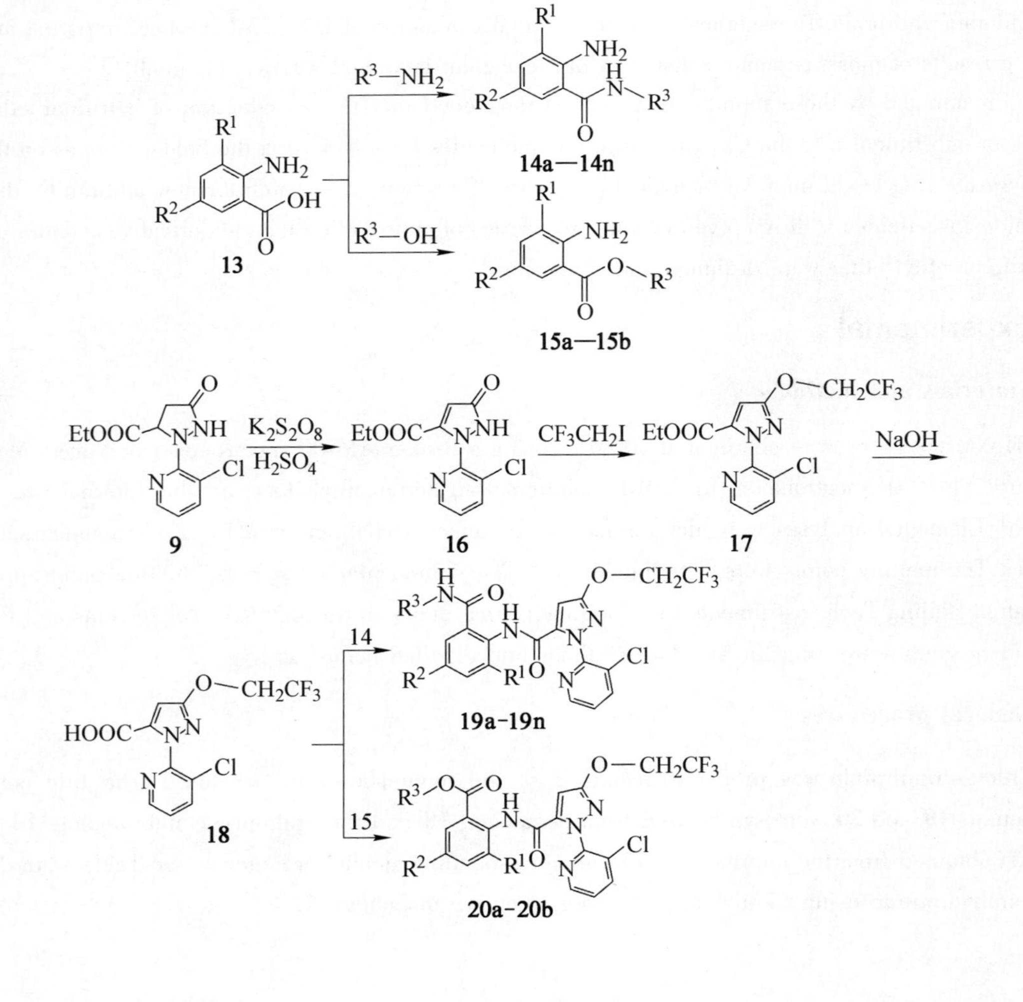

Synthetic procedure for 2 - amino - 3 - methyl - benzoic acid(4)

Compounds **4** was prepared according to the literature.[13] Chloralhydrate(8.1g,55mmol,1.1 equiv.) and Na_2SO_4(71.0g,0.5mol,10equiv.) were dissolved in water(200mL) in a three - neck 500mL round - bottom flask. The solution was stirred with a mechanical stirrer and heated to 40℃ until the mixture became clear. A warm solution of the commercial *o* - toluidine **1** (5.4g,50mmol) in water(50mL) and an aqueous solution of concentrated HCl(5.32g,4.5mL, 52.5mmol,1.05equiv.) were added, followed by a warm solution of hydroxylamine hydrochloride (10.4g,0.15mol,3.0equiv.) in water(45mL). The mixture was heated to reflux under vigorous stirring, allowed to reflux for 10min, and then cooled to room temperature. The product precipitated out of solution, and after standing overnight, the solid were collected and dried to obtain 2 - hydroxyimino - *N* - *o* - tolyl - acetamide.

Sulfuric acid(60mL) was heated in a three - neck 250mL round - bottom flask to 60℃ and then removed. The dry 2 - hydroxyimino - *N* - *o* - tolyl - acetamide(**2**) was added in portions with stirring over 30 min so that the temperature did not exceed 70℃. The mixture was then heated to 80℃ for 20 min, then allowed to cool to room temperature. The reaction mixture was poured over crushed ice(100g) and left to stand for 1h, yielding a crude precipitate that was collected by suction filtration. The product was washed with water(50mL × 2) and filtered to give crude 7 - methyl - 1*H* - indole - 2,3 - dione, which was directly used for the next step without further purification.

To a stirred suspension of compound **3** in a 5% aqueous sodium hydroxide solution(150mL), this mixture was cooled to 0℃, and added dropwise a 30% aqueous hydrogen peroxide solution (150mL). The reaction mixture was stirred at 50℃ for 30 min and then allowed to reach room temperature. The filtered solution was acidified to pH 4 with an aqueous 1mol · L^{-1} hydrochloric acid solution, and a tan precipitate was collected by filtration, washed thoroughly with cold water, and dried under vacuum to afford 2 - amino - 3 - methyl - benzoic acid(**4**). The overall yield of compound **4** was 25.6%, m.p. 173 - 174℃. ^1H NMR(DMSO - d_6, 400MHz) δ: 7.61(d, J = 8.0Hz, 1H, Ph - H), 7.15(d, J = 7.0Hz, 1H, Ph - H), 6.48 - 6.51(m, 1H, Ph - H), 2.09(s, 3H, CH_3).

Synthetic procedure for 2 - amino - 5 - chloro - 3 - methylbenzoic acid(5)

2 - amino - 5 - chloro - 3 - methylbenzoic acid(**5**) was prepared according to the literature.[14] To a solution of 2 - amino - 3 - methylbenzoic acid(10g, 66mmol) in DMF (40mL) was added *N* - chlorosuccinimide(8.8g, 66mmol) and the reaction mixture was heated to 100℃ for 40min. The reaction was cooled to room temperature and let stand overnight. The reaction mixture was then slowly poured into ice - water(150mL) to precipitate a white solid. The solid was filtered and washed with water(50mL × 3) and then taken up in ethyl acetate (600mL). The ethyl acetate solution was dried over magnesium sulfate, evaporated under reduced pressure and the residual solid was washed with ether(30mL × 3) to afford intermediate 2 - amino - 5 - chloro - 3 - methylbenzoic acid(**5**): White solid, m.p. 196 - 197℃ (dec.), yield 76.0%; ^1H NMR(DMSO, 400MHz) δ: 7.53(s, 1H, Ph - H), 7.21(s, 1H,

Ph – H),2.09(s,3H,CH_3).

Synthesis of intermediates *N* – methyl 2 – amino – 5 – chloro – 3 – methyl – benzamide(6)

N – Methyl 2 – amino – 5 – chloro – 3 – methyl – benzamide(**6**) was prepared according to the literature.[15] To a 100mL round – bottomed flask was placed 2 – amino – 5 – chloro – 3 – methylbenzoic acid(**5**)(3.7g,20mmol) and then was added 50mL of thionyl chloride. The resulting mixture was refluxed for 3h. The mixture was evaporated *in vacuo* to dryness and then 60mL of THF was added. To this solution was added dropwise 50g of 25% aqueous methylamine solution under an ice bath. The resulting solution was allowed to stir at room temperature for 12h and then water(200mL) was added. The yellow precipitate was collected by filtration and dried to give 2.36g(59.3%) of compound **6**, m. p. 130 – 132℃; ^{1}H NMR ($CDCl_3$, 400MHz) δ: 7.16(d, J = 2.2Hz, 1H, Ph – H), 7.09(d, J = 1.6Hz, 1H, Ph – H), 6.01(br s, 1H, NH), 5.52(br s, 2H, NH_2), 2.95(d, J = 4.8Hz, 3H, $NHCH_3$), 2.13(s, 3H, CH_3).

Synthetic procedure for(3 – chloro – pyridin – 2 – yl) – hydrazine(8)

To a suspension of 2,3 – dichloropyridine **7**(100.0g,0.676mol) in anhydrous ethanol(420mL) was added 50% hydrazine hydrate(280mL,2.884mol). The resulting mixture was refluxed for 36h, and then cooled to room temperature. The product precipitated out of solution, the white crystal was collected by filtration, washed thoroughly with cold ethanol and dried to give white crystals(74.4g, 76.8%), m. p. 163 – 164℃; ^{1}H NMR ($CDCl_3$, 300MHz) δ: 8.09(d, J = 3.9Hz, 1H, pyridyl – H), 7.47(d, J = 8.1Hz, 1H, pyridyl – H), 6.64(dd, J = 3.9, 8.1Hz, 1H, pyridyl – H), 6.21(s, 1H, NH), 3.97(br s, 2H, NH_2).

Synthetic procedure for 2 – (3 – chloro – pyridin – 2 – yl) – 5 – oxo – pyrazolidine – 3 – carboxylic acid ethyl ester(9)

To 200mL of absolute ethanol in a 500mL three – necked round – bottomed flask was added 6.9 g(0.3mol) of sodium cut in pieces of suitable size. When all the sodium has reacted, the mixture was heated to reflux and (3 – chloro – pyridin – 2 – yl) – hydrazine(**8**)(39.82 g, 0.277mol) was added. The mixture was refluxed for 10min, then diethyl maleate(51.65g, 0.3mol) was added dropwise. The resulting orange – red solution was held at reflux for 30min. After being cooled to 65℃, the reaction mixture was treated with glacial acetic acid (30g,0.51mol). The mixture was diluted with water(30mL). After removal of most solvent, the residue was treated with water(300mL). The slurry formed was dissolved in aqueous ethanol (70%, 200mL) and was stirred thoroughly. The solid was collected by filtration, washed with aqueous ethanol(50%, 50mL × 3) to give 2 – (3 – chloro – pyridin – 2 – yl) – 5 – oxo – pyrazolidine – 3 – carboxylic acid ethyl ester(**9**)(36.6 g, 49.0%), m. p. 132 – 134℃; ^{1}H NMR (DMSO – d_6, 400MHz) δ: 10.18(s, 1H, NH), 8.25(d, J = 4.8Hz, 1H, pyridyl – H), 7.91(d, J = 7.4Hz, 1H, pyridyl – H), 7.18(dd, J = 4.8, 7.4Hz, 1H, pyridyl – H), 4.81(d, J = 9.8Hz, 1H, CH), 4.17(q, J = 7.0Hz, 2H, OCH_2), 2.89(dd, J = 9.8, 16.8Hz, 1H, CH_2 – H), 2.34 (d, J = 16.8Hz, 1H, CH_2 – H), 1.20(t, J = 7.0Hz, 3H, CH_3).

5 – Bromo – 2 – (3 – chloro – pyridin – 2 – yl) – 3,4 – dihydro – 2*H* – pyrazole – 3 – carboxylic acid ethyl ester (10)

To a solution of 2 – (3 – chloro – pyridin – 2 – yl) – 5 – oxopyrazolidine – 3 – carboxylic acid ethyl ester (**9**) (27g, 0.1mol) in acetonitrile (300mL) was added phosphorous oxybromide (34.4g, 0.12mmol). The reaction mixture was refluxed for 5h, then 250mL of solvent was removed by distillation. The concentrated reaction mixture was slowly poured into saturated aq. Na_2CO_3 (250mL) and was stirred vigorously for 30min. The resulting mixture was extracted with CH_2Cl_2 (250mL × 2), the organic extract was separated, dried, filtered, concentrated and purified by silica gel chromatography to afford 5 – bromo – 2 – (3 – chloro – pyridin – 2 – yl) – 3,4 – dihydro – 2*H* – pyrazole – 3 – carboxylic acid ethyl ester (**10**) (31.0g, 93.0%), m. p. 59 – 60℃; ^{1}H NMR (DMSO – d_6, 400MHz) δ: 8.10 (d, J = 4.4Hz, 1H, pyridyl – H), 7.83 (d, J = 7.7Hz, 1H, pyridyl – H), 6.98 (dd, J = 4.4, 7.7Hz, 1H, pyridyl – H), 5.17 (dd, J = 8.7, 11.8Hz, 1H, CH), 4.08 (q, J = 7.0Hz, 2H, OCH_2), 3.27 (dd, J = 8.7, 17.6Hz, 1H, CH_2 – H), 3.57 (dd, J = 11.8, 17.6Hz, , 1H, CH_2 – H), 1.12 (t, J = 7.0Hz, 3H, CH_3).

Synthetic procedure for 5 – bromo – 2 – (3 – chloro – pyridin – 2 – yl) – 2*H* – pyrazole – 3 – carboxylic acid ethyl ester (11)

To a solution of 5 – bromo – 2 – (3 – chloro – pyridin – 2 – yl) – 3,4 – dihydro – 2*H* – pyrazole – 3 – carboxylic acid ethyl ester (**10**) (17g, 51mmol) in acetonitrile (250mL) was added sulfuric acid (98%, 10g, 102mmol). After being stirred for several minutes, the reaction mixture was treated with $K_2S_2O_8$ (21g, 76.5mmol) and was refluxed for 4.5h. After being cooled to 60℃, the mixture was filtered, the filter cake was washed with acetonitrile (30mL). The filtrate was concentrated to 100mL, then was added slowly to water (250mL) under stirring. The solid was collected by filtration, washed with acetonitrile (30mL × 3), water (30mL), and then dried to give 5 – bromo – 2 – (3 – chloro – pyridin – 2 – yl) – 2*H* – pyrazole – 3 – carboxylic acid ethyl ester (**11**) (15.6g, 92.7%), m. p. 117 – 118℃; ^{1}H NMR ($CDCl_3$, 300MHz) δ: 8.52 (d, J = 4.8Hz, 1H, pyridyl – H), 7.92 (d, J = 8.1Hz, 1H, pyridyl – H), 7.45 (dd, J = 4.8, 8.1Hz, 1H, pyridyl – H), 6.95 (s, 1H, pyrazolyl – H), 4.24 (q, J = 7.2Hz, 2H, CH_2), 1.21 (t, J = 7.2Hz, 3H, CH_3).

Synthetic procedure for 5 – bromo – 2 – (3 – chloropyridin – 2 – yl) – 2*H* – pyrazole – 3 – carboxylic acid (12)

To a mixture of the ethyl 5 – bromo – 2 – (3 – chloro – pyridin – 2 – yl) – 2*H* – pyrazole – 3 – carboxylic acid ethyl ester (**11**) (15.6g, 47.2mmol) in methanol (120mL) was added aqueous sodium hydroxide solution (60mL, 1 mol · L^{-1}). The solution was stirred at room temperature for 6h, then was concentrated *in vacuo* to about 50mL. The concentrated mixture was diluted with H_2O (150mL), and washed with ethyl acetate (150mL). The aqueous solution was acidified using concentrated hydrochloric acid to pH = 2. The solid was collected by filtration, washed with ether (30mL), and then dried to give 5 – bromo – 2 – (3 – chloro – pyridin – 2 – yl) – 2*H* – pyrazole – 3 – carboxylic acid (**12**) (12.75g, 89.3%), m. p. 197 – 200℃; ^{1}H NMR ($CDCl_3$, 300MHz) δ: 8.52 (dd, J = 1.5, 4.8Hz, 1H, pyridyl – H), 7.94 (dd, J = 1.5, 8.1Hz,

1H, pyridyl – H), 7.48(dd, J = 4.8, 8.1Hz, 1H, pyridyl – H), 7.10(s, 1H, pyrazolyl – H).

Synthetic procedure for Chlorantraniliprole

Chlorantraniliprole was prepared according to the literatures.[16,17] To a suspension of N – pyridylpyrazole acid **12** (0.30g, 1mmol) in dichloromethane (20mL) was added oxalyl chloride (0.38g, 3mmol), followed by dimethylformamide (2drops). The solution was stirred at room temperature. After 6h the mixture was concentrated *in vacuo* to obtain the crude acid chloride. The crude acid chloride in dichloromethane (20mL) was added slowly to a stirred solution of 2 – amino – 5 – chloro – 3 – N – dimethyl – benzamide (**6**) (0.24g, 1.2mmol) in dichloromethane (20mL) in an ice bath. After 20min, ethyl – diisopropyl – amine (0.13g, 1mmol) was added dropwise. The solution was warmed to room temperature and stirred for 12h. The solution was diluted with CH_2Cl_2 (20mL), and washed with 1 mol · L^{-1} aq. HCl solution (10mL), saturated aq. $NaHCO_3$ (10mL), and brine (10mL). The organic extract was separated, dried, filtered, and concentrated and purified by silica gel chromatography to afford the Chlorantraniliprole. (0.43g, 89.3%), m. p. 197 – 200℃; ^1H NMR ($CDCl_3$, 400MHz) δ: 10.10 (br s, 1H, NH), 8.46 (dd, J = 1.6, 4.8Hz, 1H, pyridyl – H), 7.85 (dd, J = 1.6, 8.0Hz, 1H, pyridyl – H), 7.38 (dd, J = 4.8, 8.0Hz, 1H, pyridyl – H), 7.24 (d, J = 2.0Hz, 1H, Ph – H), 7.21 (d, J = 2.0Hz, 1H, Ph – H), 7.11 (s, 1H, pyrazolyl – H), 6.15 – 6.18 (m, 1H, NHCO), 2.95 (d, J = 4.9Hz, 2H, $NHCH_3$), 2.17 (s, 3H, CH_3).

Synthetic procedure for 2 – amino – 5 – chloro – 3 – methyl – N – propyl – benzamide (14a)

To a 100mL round – bottomed flask was placed 2 – amino – 5 – chloro – 3 – methylbenzoic acid (**5**) (5.0 g, 27mmol) and then was added 50mL of thionyl chloride. The resulting mixture was refluxed for 3h. The mixture was evaporated *in vacuo* to dryness and then 40mL of THF was added. The solution was added slowly to a stirred solution of propylamine (15.8g, 270mmol) in tetrahydrofuran (40mL) in an ice bath. The resulting solution was allowed to stir at room temperature for 12h. Then the solution was concentrated *in vacuo* and diluted with ethyl acetate (150mL), and washed with water (50mL × 3). The organic extract was separated, dried, filtered, and concentrated and purified by silica gel chromatography to afford the desired title compound **14a**.

Compounds **14b – 14n** and **15a – 15b** were prepared by similar method above using the appropriate substrates. The melting points and yields of compounds **14** and **15** are listed in Table 1. The ^1H NMR data are listed in Table 2.

Table 1 Melting points and yields of the compounds 14a – 14n and 15a – 15b

Compd.	R^1	R^2	R^3	m. p(℃)	Yield(%)
14a	CH_3	Cl	n – propyl	119 – 121	88.0
14b	CH_3	Cl	cyclopropyl	122 – 124	92.4
14c	CH_3	Cl	n – butyl	87 – 88	77.9
14d	CH_3	Cl	i – butyl	117 – 122	68.1
14e	CH_3	Cl	cyclohexyl	167 – 168	89.2

Table 1 (continued)

Compd.	R^1	R^2	R^3	m. p(℃)	Yield(%)
14f	H	Cl	*n* – propyl	120 – 122	79.3
14g	H	Cl	*i* – propyl	161 – 162	58.0
14h	H	Cl	cyclopropyl	143 – 145	61.9
14i	H	Cl	*n* – butyl	108 – 110	66.4
14j	H	Cl	cyclohexyl	179 – 181	56.9
14k	CH_3	H	*n* – propyl	88 – 90	70.2
14l	CH_3	H	*i* – propyl	137 – 139	70.2
14m	CH_3	H	cyclopropyl	118 – 120	80.0
14n	CH_3	H	cyclohexyl	158 – 159	69.6
15a	CH_3	Cl	methyl	33 – 35	52.9
15b	CH_3	Br	methyl	50 – 53	47.6

Table 2 ^{1}H NMR of the compounds 14a – 14n and 15a – 15b

Compd.	^{1}H NMR δ
14a	(400MHz, DMSO – d_6) δ: 8.37 (br s, 1H, CONH), 7.41 (d, J = 1.8Hz, 1H, Ph – H), 7.13 (d, J = 1.8Hz, 1H, Ph – H), 6.32 (s, 2H, $PhNH_2$), 3.14 – 3.17 (m, 2H, $NHCH_2$), 2.08 (s, 3H, $PhCH_3$), 1.50 – 1.52 (m, 2H, CH_2CH_3), 0.88 (t, J = 7.4Hz, 3H, CH_2CH_3)
14b	(400MHz, $CDCl_3$) δ: 7.08 – 7.10 (m, 2H, Ph – H), 6.10 (br s, 1H, NH), 5.60 (br s, 2H, NH_2), 2.80 – 2.86 (m, 1H, cyclopropyl – H), 2.13 (s, 3H, CH_3), 0.84 – 0.87 (m, 2H, cyclopropyl – H), 0.58 – 0.62 (m, 2H, cyclopropyl – H)
14c	(400MHz, DMSO – d_6) δ: 8.34 (br s, 1H, CONH), 7.41 (d, J = 1.8Hz, 1H, Ph – H), 7.12 (d, J = 1.8Hz, 1H, Ph – H), 6.32 (s, 2H, $PhNH_2$), 3.19 – 3.22 (m, 2H, $NHCH_2$), 2.08 (s, 3H, $PhCH_3$), 1.45 – 1.52 (m, 2H, CH_2), 1.29 – 1.36 (m, 2H, CH_2CH_3), 0.89 (t, J = 7.3Hz, 3H, CH_2CH_3)
14d	(400MHz, DMSO – d_6) δ: 8.38 (br s, 1H, CONH), 7.42 (d, J = 2.0Hz, 1H, Ph – H), 7.14 (d, J = 2.0Hz, 1H, Ph – H), 6.29 (s, 2H, $PhNH_2$), 3.01 – 3.03 (m, 2H, $NHCH_2$), 2.08 (s, 3H, $PhCH_3$), 1.77 – 1.88 (m, 1H, $CH(CH_3)_2$), 0.88 (d, J = 6.6Hz, 6H, $CH(CH_3)_2$)
14e	(400MHz, $CDCl_3$) δ: 7.05 – 7.14 (m, 2H, Ph – H), 5.81 (br s, 1H, NH), 5.46 (br s, 2H, NH_2), 3.82 – 3.94 (m, 1H, cyclohexyl – H), 2.13 (s, 3H, CH_3), 1.18 – 2.04 (m, 10H, cyclohexyl – H)
14f	(400MHz, DMSO – d_6) δ: 8.34 – 8.36 (m, 1H, CONH), 7.53 (d, J = 2.4Hz, 1H, Ph – H), 7.15 (dd, J = 8.7, 2.4Hz, 1H, Ph – H), 6.70 (d, J = 8.7Hz, 1H, Ph – H), 6.54 (s, 2H, $PhNH_2$), 3.13 – 3.16 (m, 2H, $NHCH_2$), 1.46 – 1.53 (m, 2H, CH_2CH_3), 0.87 (t, J = 7.2Hz, 3H, CH_2CH_3)
14g	(400MHz, DMSO – d_6) δ: 8.11 – 8.13 (m, 1H, CONH), 7.54 (d, J = 2.4Hz, 1H, Ph – H), 7.15 (dd, J = 8.8, 2.4Hz, 1H, Ph – H), 6.70 (d, J = 8.8Hz, 1H, Ph – H), 6.50 (s, 2H, $PhNH_2$), 4.01 – 4.09 (m, 1H, CH), 1.14 (d, J = 6.6Hz, 6H, $CH(CH_3)_2$)
14h	(400MHz, DMSO – d_6) δ: 8.35 – 8.37 (m, 1H, CONH), 7.53 (d, J = 2.4Hz, 1H, Ph – H), 7.20 (dd, J = 8.8, 4.8Hz, 1H, Ph – H), 6.76 (d, J = 8.8Hz, 1H, Ph – H), 6.61 (s, 2H, $PhNH_2$), 2.82 – 2.88 (m, 1H, cyclopropyl – H), 0.69 – 0.74 (m, 2H, cyclopropyl – H), 0.58 – 0.62 (m, 2H, cyclopropyl – H)

Compd.	^{1}H NMR δ
14i	(400MHz, DMSO - d_6) δ: 8.30 - 8.32 (m, 1H, CONH), 7.51 (d, J = 2.0Hz, 1H, Ph - H), 7.15 (dd, J = 8.7, 2.0Hz, 1H, Ph - H), 6.70 (d, J = 8.7Hz, 1H, Ph - H), 6.53 (s, 2H, $PhNH_2$), 3.17 - 3.19 (m, 2H, $NHCH_2$), 1.42 - 1.51 (m, 2H, CH_2CH_2), 1.26 - 1.35 (m, 2H, CH_2CH_3), 0.89 (t, J = 7.2Hz, 3H, CH_2CH_3)
14j	(400MHz, DMSO - d_6) δ: 8.10 - 8.12 (m, 1H, CONH), 7.52 (d, J = 2.0Hz, 1H, Ph - H), 7.14 (dd, J = 8.7, 2.0Hz, 1H, Ph - H), 6.69 (d, J = 8.7Hz, 1H, Ph - H), 6.47 (s, 2H, $PhNH_2$), 3.67 - 3.70 (m, 1H, NHCH), 1.08 - 1.79 (m, 10H, cyclohexyl - H)
14k	(400MHz, DMSO - d_6) δ: 8.20 - 8.22 (m, 1H, CONH), 7.35 (d, J = 7.6Hz, 1H, Ph - H), 7.06 (d, J = 7.2Hz, 1H, Ph - H), 6.45 - 6.49 (m, 1H, Ph - H), 6.18 (s, 2H, $PhNH_2$), 3.14 - 3.19 (m, 2H, $NHCH_2$), 2.07 (s, 3H, $PhCH_3$), 1.46 - 1.55 (m, 2H, CH_2CH_3), 0.88 (t, J = 7.4Hz, 3H, CH_2CH_3)
14l	(400MHz, DMSO - d_6) δ: 8.19 - 8.21 (m, 1H, CONH), 7.33 (d, J = 7.6Hz, 1H, Ph - H), 7.05 (d, J = 7.0Hz, 1H, Ph - H), 6.46 - 6.49 (m, 1H, Ph - H), 6.11 (s, 2H, $PhNH_2$), 4.22 - 4.30 (m, 1H, CH), 2.18 (s, 3H, CH_3), 1.25 (d, J = 6.6Hz, 6H, $CH(CH_3)_2$)
14m	(400MHz, DMSO - d_6) δ: 8.18 (s, 1H, CONH), 7.30 (d, J = 7.6Hz, 1H, Ph - H), 7.05 (d, J = 6.8Hz, 1H, Ph - H), 6.43 - 6.47 (m, 1H, Ph - H), 6.21 (s, 2H, $PhNH_2$), 2.78 - 2.80 (m, 1H, cyclopropyl - H), 2.07 (s, 3H, $PhCH_3$), 0.65 - 0.67 (m, 2H, CH_2CH_2, cyclopropyl - H), 0.53 - 0.55 (m, 2H, CH_2CH_2, cyclopropyl - H)
14n	(400MHz, $CDCl_3$) δ: 6.56 - 7.19 (m, 3H, Ph - H), 5.90 (br s, 1H, NH), 5.53 (br s, 2H, NH_2), 3.88 - 3.97 (m, 1H, cyclohexyl - H), 2.15 (s, 3H, CH_3), 1.15 - 2.03 (m, 10H, cyclohexyl - H)
15a	(400MHz, DMSO - d_6) δ: 7.56 (s, 1H, Ph - H), 7.26 (s, 1H, Ph - H), 6.33 (br s, 2H, NH_2), 3.79 (s, 3H, OCH_3), 2.12 (s, 3H, CH_3)
15b	(400 MHz, DMSO - d_6) δ: 7.80 (s, 1H, Ph - H), 7.21 (s, 1H, Ph - H), 5.78 (br s, 2H, NH_2), 3.84 (s, 3H, OCH_3), 2.29 (s, 3H, CH_3)

Synthetic procedure for 2 - (3 - chloro - pyridin - 2 - yl) - 5 - oxo - 2,5 - dihydro - 1*H* - pyrazole - 3 - carboxylic acid ethyl ester (16)

To a solution of 2 - (3 - chloro - pyridin - 2 - yl) - 5 - oxopyrazolidine - 3 - carboxylic acid ethyl ester (**9**) (10 g, 37mmol) in acetonitrile (150mL) was added sulfuric acid (98%, 7.2g, 74mmol). After being stirred for several min, the reaction mixture was treated with $K_2S_2O_8$ (15g, 56mmol) and was refluxed for 4.5h. After being cooled to 60℃, the mixture was filtered, the filter cake was washed with acetonitrile (30mL). The filtrate was concentrated and poured into ice water (200mL). The aqueous layer was extracted with dichloromethane (150mL × 3). The organic layer was washed with water (100mL × 3) and dried over anhydrous sodium sulfate. Then the ethyl acetate was concentrated. The residue was purified by column chromatography over silica gel using petroleum ether (60 - 90℃) and ethyl acetate as the eluent to afford the 2 - (3 - chloro - pyridin - 2 - yl) - 5 - oxo - 2,5 - dihydro - 1*H* - pyrazole - 3 - carboxylic acid ethyl ester (**16**). (6.2g,62.4%), m. p. 136 – 138℃; ^{1}H NMR ($CDCl_3$, 400MHz)δ: 9.35(s, 1H, NH), 8.52(d, J = 4.4Hz, 1H, pyridyl – H), 7.90(d, J = 8.0Hz, 1H, pyridyl – H), 7.43(dd, J = 4.4, 8.0Hz, 1H, pyridyl – H), 6.36(s, 1H, pyrazolyl – H), 4.19(q, J = 7.2Hz, 2H, CH_2), 1.19(t, J = 7.2Hz, 3H, CH_3).

Synthetic procedure for 2 – (3 – chloro – pyridin – 2 – yl) – 5 – (2,2,2 – trifluoro – ethoxy) – 2,5 – dihydro – 1*H* – pyrazole – 3 – carboxylic acid ethyl ester (17)

The ester **17** was prepared according to the literature.[7] Compound **16** (1.0 g, 3.7mmol) was dissolved in 30mL of dry dimethylformamide, and potassium carbonate (0.76g, 5.5mmol) was added. The mixture was heated to 100℃. The 2,2,2 – trifluoroiodoethane (0.94g, 4.4mmol) in dry dimethylformamide (5mL) was added slowly to the mixture. The solution was warmed at 100℃ and stirred for 3h and poured into ice water (50mL). The aqueous layer was extracted with ethyl acetate (40mL × 3). The organic layer was washed with water (40mL × 3) and dried over anhydrous sodium sulfate. Then the ethyl acetate was concentrated. The residue was purified by column chromatography on a silica gel using petroleum ether (60 – 90℃) and ethyl acetate as the eluent to afford the 2 – (3 – chloro – pyridin – 2 – yl) – 5 – (2,2,2 – trifluoro – ethoxy) – 2,5 – dihydro – 1*H* – pyrazole – 3 – carboxylic acid ethyl ester (**17**). (1.28g, 99%), m. p. 63 – 65℃; ^{1}H NMR ($CDCl_3$, 400MHz)δ: 8.52(dd, J = 1.6, 4.8Hz, 1H, pyridyl – H), 7.91(dd, J = 1.6, 8.0Hz, 1H, pyridyl – H), 7.43(dd, 1H, J = 4.8, 8.0Hz, 1H, pyridyl – H), 6.54(s, 1H, pyrazolyl – H), 4.66(q, J = 16.4Hz, 2H, CH_2CF_3), 4.20(q, J = 7.2Hz, 2H, CH_2), 1.22(t, J = 7.2Hz, 3H, CH_3).

Synthetic procedure for 2 – (3 – chloro – pyridin – 2 – yl) – 5 – (2,2,2 – trifluoro – ethoxy) – 2,5 – dihydro – 1*H* – pyrazole – 3 – carboxylic acid (18)

To a mixture of the compound **17** (1.28g, 3.6mmol) in methanol (20mL) was added aqueous sodium hydroxide solution (5mL, 1mol · L^{-1}). The solution was stirred at room temperature for 6h, then was concentrated *in vacuo* to about 5mL. The concentrated mixture was diluted with H_2O (40mL), and washed with ethyl acetate (20mL). The aqueous solution was acidified using concentrated hydrochloric acid to pH = 2. The solid was collected by filtration, washed with ether (10mL), and then dried to give 2 – (3 – chloro – pyridin – 2 – yl) – 5 – (2,2,2 – trifluoro – ethoxy) – 2,5 – dihydro – 1*H* – pyrazole – 3 – carboxylic acid (**18**) (0.84 g, 71.3%), m. p. 165 – 167℃. ^{1}H NMR ($CDCl_3$, 300MHz)δ: 8.51(dd, J = 1.6, 4.8Hz, 1H, pyridyl – H); 7.92(dd, J = 1.6, 8.0Hz, 1H, pyridyl – H); 7.44(dd, 1H, J = 4.8, 8.0Hz, pyridyl – H); 6.59 (s, 1H, pyrazolyl – H); 4.65(q, J = 16.4Hz, 2H, CH_2CF_3).

Synthetic procedure for the title compounds 19 and 20

To a suspension of *N* – pyridylpyrazole acid **18** (1mmol) in dichloromethane (20mL) was added oxalyl chloride (3mmol) and dimethylformamide (2 drops). The solution was stirred at ambient temperature for 4 h. Then the mixture was concentrated *in vacuo* to give the crude acid chloride. The crude acid chloride in tetrahydrofuran (25mL) was added slowly to a stirred solution of **14** or **15** (1.2mmol) and triethylamine (1.2mmol) in tetrahydrofuran (15mL). The mixture was stirred at ambient temperature for 8h. Then the solution was concentrated in vacuo and diluted with CH_2Cl_2 (60mL), and washed with 1mol · L^{-1} aq. HCl solution (15mL), saturated aq. $NaHCO_3$ (15mL), and brine (15mL). The organic extract was separated, dried, filtered, and concentrated and purified by silica gel chromatography to afford the desired title compounds **19** and **20**. The melting points, yields, and elemental analyses of compounds **19** and **20** are listed in Table 3. The ^1H NMR data are listed in Table 4.

Table 3 The melting points, yields and elemental analyses of the title compounds 19 and 20

Compd.	R^1	R^2	R^3	m. p(℃)	Yield (%)	Elemental analysis(%) calcd. (found)		
						C	H	N
19a	CH_3	Cl	*n* – propyl	200 – 201	59. 9	49. 82(49. 66)	3. 80(3. 92)	13. 21(13. 22)
19b	CH_3	Cl	cyclopropyl	211 – 213	62. 7	50. 02(49. 92)	3. 43(3. 78)	13. 26(13. 21)
19c	CH_3	Cl	*n* – butyl	173 – 175	67. 6	50. 75(50. 56)	4. 07(3. 99)	12. 87(12. 98)
19d	CH_3	Cl	*i* – butyl	220 – 221	63. 9	50. 75(50. 81)	4. 07(4. 11)	12. 87(12. 81)
19e	CH_3	Cl	cyclohexyl	160 – 162	58. 7	52. 64(52. 54)	4. 24(4. 41)	12. 28(12. 24)
19f	H	Cl	*n* – propyl	102 – 104	60. 5	48. 85(48. 77)	3. 51(3. 55)	13. 56(13. 29)
19g	H	Cl	*i* – propyl	186 – 188	70. 1	48. 85(48. 79)	3. 51(3. 46)	13. 56(13. 40)
19h	H	Cl	cyclopropyl	198 – 199	56. 9	49. 04(49. 40)	3. 14(3. 47)	13. 62(13. 07)
19i	H	Cl	*n* – butyl	96 – 98	63. 2	49. 82(49. 70)	3. 80(3. 84)	13. 21(13. 10)
19j	H	Cl	cyclohexyl	145 – 146	66. 7	51. 81(51. 93)	3. 99(4. 06)	12. 59(13. 45)
19k	CH_3	H	*n* – propyl	185 – 187	67. 0	53. 29(53. 41)	4. 27(4. 11)	14. 12(13. 97)
19l	CH_3	H	*i* – propyl	177 – 179	69. 3	53. 29(53. 22)	4. 27(4. 35)	14. 12(14. 40)
19m	CH_3	H	cyclopropyl	172 – 173	54. 8	53. 50(53. 25)	3. 88(4. 09)	14. 18(13. 94)
19n	CH_3	H	cyclohexyl	147 – 149	64. 7	56. 03(56. 07)	4. 70(4. 60)	13. 07(13. 08)
20a	CH_3	Cl	methyl	50 – 52	69. 1	47. 73(47. 99)	3. 00(2. 85)	11. 13(11. 01)
20b	CH_3	Br	methyl	40 – 42	65. 4	43. 86(43. 90)	2. 76(3. 02)	10. 23(10. 45)

Table 4 ^1H NMR of the title compounds 19a – 19n and 20a – 20b

Compd.	^1H NMR(400MHz, $CDCl_3$) δ
19a	9. 97(s, 1H, CONH), 8. 38(dd, *J* = 4. 8, 1. 6Hz, 1H, pyridyl – H), 7. 75(dd, *J* = 8. 0, 1. 6Hz, 1H, pyridyl – H), 7. 27(dd, *J* = 8. 0, 4. 8Hz, 1H, pyridyl – H, 1H), 7. 14(s, 1H, Ph – H), 7. 10(s, 1H, Ph – H), 6. 58(s, 1H, pyrazolyl – H), 6. 09 – 6. 11(m, 1H, $NHCH_2$), 4. 60(q, *J* = 8. 2Hz, 2H, CH_2CF_3), 3. 23 – 3. 25(m, 2H, CH_2NH), 2. 10(s, 3H, $PhCH_3$), 1. 47 – 1. 48(m, 2H, CH_2CH_3), 0. 87(t, *J* = 7. 0Hz, 3H, CH_2CH_3)
19b	10. 02(s, 1H, CONH), 8. 45(dd, *J* = 4. 4, 1. 6Hz, 1H, pyridyl – H), 7. 82(dd, *J* = 8. 0, 1. 6Hz, 1H, pyridyl – H, 1H), 7. 34(dd, *J* = 8. 0, 4. 4Hz, 1H, pyridyl – H, 1H), 7. 17(s, 1H, Ph – H), 7. 10(s, 1H, Ph – H), 6. 74(s, 1H, pyrazolyl – H), 6. 42 – 6. 44(m, 1H, PhCONH), 4. 67(q, *J* = 8. 2Hz, 2H, CH_2CF_3), 2. 77 – 2. 79(m, 1H, CH), 2. 15(s, 3H, $PhCH_3$), 0. 82 – 0. 84(m, 2H, CH_2CH_2), 0. 53 – 0. 55(m, 2H, CH_2CH_2)
19c	9. 99(s, 1H, CONH), 8. 39(dd, *J* = 4. 4, 1. 2Hz, 1H, pyridyl – H), 7. 77(dd, *J* = 8. 0, 1. 6Hz, 1H, pyridyl – H), 7. 29(dd, *J* = 8. 0, 4. 4Hz, 1H, pyridyl – H), 7. 27(s, 1H, Ph – H), 7. 15(s, 1H, Ph – H), 6. 55(s, 1H, pyrazolyl – H), 6. 10 – 6. 12(m, 1H, PhCONH), 4. 60(q, *J* = 8. 2Hz, 2H, CH_2CF_3), 3. 30 – 3. 32(m, 2H, $NHCH_2$), 2. 14(s, 3H, $PhCH_3$), 1. 53 – 1. 55(m, 2H, CH_2CH_2), 1. 32 – 1. 34(m, 2H, CH_2CH_3), 0. 89(t, *J* = 7. 2Hz, 3H, CH_2CH_3)

Table 4 (continued)

Compd.	^1H NMR (400MHz, $CDCl_3$) δ
19d	10.04 (s, 1H, CONH), 8.47 (dd, J = 4.8, 1.6Hz, 1H, pyridyl – H), 7.84 (dd, J = 8.0, 1.6Hz, 1H, pyridyl – H), 7.36 (dd, J = 8.0, 4.8Hz, 1H, pyridyl – H), 7.27 (s, 1H, Ph – H), 7.22 (s, 1H, Ph – H), 6.59 (s, 1H, pyrazolyl – H), 6.19 – 6.21 (m, 1H, PhCONH), 4.68 (q, J = 8.2Hz, 2H, CH_2CF_3), 3.21 – 3.23 (m, 2H, CH_2NH), 2.19 (s, 3H, $PhCH_3$), 1.84 – 1.86 (m, 1H, CH), 0.95 (d, J = 6.7Hz, 6H, $CH(CH_3)_2$)
19e	10.06 (s, 1H, CONH), 8.47 (dd, J = 4.8, 1.6Hz, 1H, pyridyl – H), 7.84 (dd, J = 8.0, 1.6Hz, 1H, pyridyl – H), 7.35 (dd, J = 8.0, 4.8Hz, 1H, pyridyl – H), 7.26 (s, 1H, Ph – H), 7.20 (s, 1H, Ph – H), 6.60 (s, 1H, pyrazolyl – H), 5.95 – 5.97 (m, 1H, PhCONH), 4.69 (q, J = 8.2Hz, 2H, CH_2CF_3), 3.85 – 3.87 (m, 1H, CH), 2.19 (s, 3H, $PhCH_3$), 1.96 – 1.98 (m, 2H, cyclohexanyl – H), 1.73 – 1.75 (m, 2H, cyclohexanyl – H), 1.63 – 1.65 (m, 2H, cyclohexanyl – H), 1.38 – 1.40 (m, 2H, cyclohexanyl – H), 1.18 – 1.21 (m, 2H, cyclohexanyl – H)
19f	12.24 (s, 1H, CONHPh), 8.52 (d, J = 4.0Hz, 1H, pyridyl – H), 8.46 – 8.48 (m, 1H, Ph – H), 7.92 (d, J = 6.8Hz, 1H, pyridyl – H), 7.37 – 7.48 (m, 3H, pyridyl – H, Ph – H), 6.61 (s, 1H, pyrazolyl – H), 6.33 – 6.35 (m, 1H, PhCONH), 4.69 (q, J = 8.2Hz, 2H, CH_2CF_3), 3.44 – 3.46 (m, 2H, $NHCH_2$), 1.66 – 1.68 (m, 2H, CH_2CH_2), 1.01 (t, J = 6.8Hz, 3H, CH_2CH_3)
19g	12.26 (s, 1H, CONHPh), 8.50 (dd, J = 4.8, 1.6Hz, 1H, pyridyl – H), 8.47 (d, J = 8.9Hz, 1H, Ph – H), 7.92 (d, J = 8.0Hz, 1H, pyridyl – H), 7.42 – 7.45 (m, 2H, pyridyl – H, Ph – H), 7.34 (d, J = 9.0Hz, 1H, Ph – H), 6.58 (s, 1H, pyrazolyl – H), 6.05 – 6.06 (m, 1H, PhCONH), 4.66 (q, J = 8.2Hz, 2H, CH_2CF_3), 4.27 – 4.29 (m, 1H, CH), 1.30 (d, J = 6.4Hz, 6H, $CH(CH_3)_2$)
19h	12.27 (s, 1H, CONHPh), 8.52 (dd, J = 4.8, 1.6Hz, 1H, pyridyl – H), 8.46 (d, J = 8.9Hz, 1H, Ph – H), 7.93 (d, J = 8.0Hz, 1H, pyridyl – H), 7.44 – 7.48 (m, 2H, pyridyl – H, Ph – H), 7.35 (d, J = 9.0Hz, 1H, Ph – H), 6.62 (s, 1H, pyrazolyl – H), 6.37 – 6.39 (m, 1H, PhCONH), 4.68 (q, J = 8.2Hz, 2H, CH_2CF_3), 4.27 – 4.29 (m, 1H, CH), 0.95 – 0.99 (m, 2H, CH_2CH_2), 0.66 – 0.69 (m, 2H, CH_2CH_2)
19i	12.15 (s, 1H, CONHPh), 8.52 (dd, J = 4.8, 1.6Hz, 1H, pyridyl – H), 8.45 (d, J = 8.9Hz, 1H, Ph – H), 7.93 (d, J = 8.0Hz, 1H, pyridyl – H), 7.42 – 7.46 (m, 2H, pyridyl – H, Ph – H), 7.33 – 7.35 (m, 1H, Ph – H), 6.51 (s, 1H, pyrazolyl – H), 6.14 – 6.16 (m, 1H, PhCONH), 4.60 (q, J = 8.2Hz, 2H, CH_2CF_3), 3.31 – 3.33 (m, 2H, $NHCH_2$), 1.53 – 1.56 (m, 2H, CH_2CH_2), 1.32 – 1.37 (m, 2H, CH_2CH_3), 0.90 (t, J = 7.2Hz, 3H, CH_2CH_3)
19j	12.26 (s, 1H, CONH), 8.52 (dd, J = 4.8, 1.6Hz, 1H, pyridyl – H), 8.48 (d, J = 9.0Hz, 1H, Ph – H), 7.89 (d, J = 1.2Hz, 1H, pyridyl – H), 7.40 – 7.45 (m, 2H, pyridyl – H, Ph – H), 7.33 – 7.35 (m, 1H, Ph – H), 6.60 (s, 1H, pyrazolyl – H), 6.20 – 6.23 (m, 1H, PhCONH), 4.68 (q, J = 8.2Hz, 2H, CH_2CF_3), 3.95 – 3.97 (m, 1H, CH), 1.78 – 1.81 (m, 2H, cyclohexanyl – H), 1.68 – 1.70 (m, 2H, cyclohexanyl – H), 1.43 – 1.45 (m, 2H, cyclohexanyl – H), 1.28 – 1.30 (m, 2H, cyclohexanyl – H), 1.19 – 1.22 (m, 2H, cyclohexanyl – H)
19k	10.10 (s, 1H, CONH), 8.39 (dd, J = 4.7, 1.5Hz, 1H, pyridyl – H), 7.75 (dd, J = 8.0, 1.5Hz, 1H, pyridyl – H), 7.36 (dd, J = 8.0, 4.7Hz, 1H, pyridyl – H), 7.25 – 7.28 (m, 1H, Ph – H), 7.17 – 7.20 (m, 1H, Ph – H), 7.07 – 7.10 (m, 1H, Ph – H), 6.48 (s, 1H, pyrazolyl – H), 6.05 – 6.08 (m, 1H, PhCONH), 4.61 (q, J = 8.2Hz, 2H, CH_2CF_3), 3.26 – 3.28 (m, 2H, CH_2NH), 2.15 (s, 3H, $PhCH_3$), 1.51 – 1.55 (m, 2H, CH_2CH_3), 0.89 (t, J = 7.2Hz, 3H, CH_2CH_3)

Table 4(continued)

Compd.	^1H NMR(400MHz, $CDCl_3$)δ
19l	8.54(s,1H,CONH),8.38(dd,J=4.7,1.4Hz,1H,pyridyl-H),7.92(dd,J=8.0,1.4Hz,1H,pyridyl-H),7.53-7.55(m,1H,Ph-H),7.42-7.45(m,1H,Ph-H),7.39(dd,J=8.0,4.7Hz,1H,pyridyl-H),7.21-7.23(m,1H,Ph-H),6.54(s,1H,pyrazolyl-H),6.03-6.06(m,1H,PhCONH),4.69(q,J=8.3Hz,2H,CH_2CF_3),4.20-4.22(m,1H,CH),2.30(s,3H,$PhCH_3$),1.16(d,J=6.6Hz,6H,CH(CH_3)$_2$)
19m	10.16(s,1H,CONH),8.38(d,J=4.8Hz,1H,pyridyl-H),7.76(d,J=8.0Hz,1H,pyridyl-H),7.27(dd,J=4.8,8.0Hz,1H,pyridyl-H),7.13-7.15(m,1H,Ph-H),7.00-7.05(m,2H,Ph-H),6.57(s,1H,pyrazolyl-H),6.44-6.46(m,1H,PhCONH),4.61(q,J=8.2Hz,2H,CH_2CF_3),2.64-2.65(m,1H,CH),2.10(s,3H,$PhCH_3$),0.72-0.74(m,2H,CH_2CH_2),0.40-0.43(m,2H,CH_2CH_2)
19n	10.22(s,1H,CONH),8.47(dd,J=4.7,1.5Hz,1H,pyridyl-H),7.83(dd,J=8.0,1.5Hz,1H,pyridyl-H),7.34(dd,J=8.0,4.7Hz,1H,pyridyl-H),7.26-7.27(m,1H),7.22-7.25(m,1H),7.14-7.15(m,1H),6.61(s,1H,pyrazolyl-H),6.21-6.23(m,1H,PhCONH),4.69(q,J=8.2Hz,2H,CH_2CF_3),3.86-3.88(m,1H CH),2.23(s,3H,$PhCH_3$),1.94-1.96(m,2H,cyclohexanyl-H),1.73-1.75(m,2H,cyclohexanyl-H),1.63-1.66(m,2H,cyclohexanyl-H),1.38-1.40(m,2H,cyclohexanyl-H),1.18-1.21(m,2H,cyclohexanyl-H)
20a	10.02(s,1H,CONH),8.45(d,J=4.5Hz,1H,pyridyl-H),7.80(d,J=1.7Hz,1H,pyridyl-H),7.34-7.37(m,3H,pyridyl-H,Ph-H),6.56(s,1H,pyrazolyl-H),6.54-6.57(m,1H,PhCONH),4.68(q,J=8.2Hz,2H,CH_2CF_3),3.91(s,3H,OCH_3),2.21(s,3H,$PhCH_3$)
20b	10.03(s,1H,CONH),8.45(dd,J=4.6,1.4Hz,1H,pyridyl-H),7.92-7.94(m,1H,Ph-H),7.84(dd,J=7.8,1.4Hz,1H,pyridyl-H),7.51-7.53(m,1H,Ph-H),7.36(dd,J=4.6,1.4Hz,1H,pyridyl-H),6.56(s,1H,pyrazolyl-H),4.69(q,J=8.2Hz,2H,CH_2CF_3),3.91(s,3H,OCH_3),2.20(s,3H,$PhCH_3$)

Biological assay

All bioassays were performed on representative test organisms reared in the laboratory. The bioassay was repeated at(25 ±1)℃ according to statistical requirements. Assessments were made on a dead/alive basis, and mortality rates were corrected using Abbott's formula. Evaluations are based on a percentage scale of 0-100 in which 0 equals no activity and 100 equals total kill.

Insecticidal activity against oriental armyworm(*Mythimna separata*) The insecticidal activities of the title compounds **19a-19n** and **20a-20b** against oriental armyworm were evaluated using the reported procedure.[18,19] The insecticidal activity against Oriental armyworm was tested by foliar application, individual corn leaves were placed on moistened pieces of filter paper in Petri dishes. The leaves were then sprayed with the test solution and allowed to dry. The dishes were infested 10 fourth-instar Oriental armyworm larvae. Percentage mortalities were evaluated 2 d after treatment. Each treatment was performed three times. For comparative purposes, Chlorantraniliprole was tested under the same conditions. The results were summarized in Table 5.

Table 5 Insecticidal activities against oriental armyworm of the title compounds 19a – 19n, 20a – 20b and Chlorantraniliprole

Compd.	larvicidal activity(%) at conc(mg · kg^{-1})									
	200	100	50	25	10	5	2.5	1	0.5	0.25
19a	100	100	100	100	100	100	100	40		
19b	100	100	100	100	100	100	100	100	100	40
19c	100	100	100	100	100	100	70	0		
19d	100	100	100	100	100	100	100	50		
19e	100	100	100	100	100	100	30			
19f	100	100	100	40						
19g	100	100	100	100	100	100	100	80	0	
19h	100	100	100	100	100	80	0			
19i	80	20								
19j	20									
19k	100	100	100	100	100	100	100	40		
19l	100	100	100	100	0					
19m	100	100	100	100	100	100	100	100	20	
19n	100	100	100	100	100	50				
20a	100	90	30							
20b	90	10								
Chlorantraniliprole	100	100	100	100	100	100	100	100	100	100

Insecticidal activity against diamond – back moth (*Plutella xylostella Linnaeus*) The insecticidal activities of the title compounds **19a – 19n** against diamondback moth were evaluated using the leaf disc assay.[20] The leaf discs (5cm × 3cm) were cut from fresh cabbage leaves and then dipped into the test solution for 15 s. After air – drying, the treated leaf discs were placed individually into boxes (80cm^3), and then the second – instar diamondback moth larvae were transferred to the Petri dish. Three replicates (seven larvae per replicate) were carried out. The commercial insecticide Chlorantraniliprole was used as a standard. The results were summarized in Table 6.

Table 6 Insecticidal activities against diamond – back moth of the title compounds 19a – 19n and Chlorantraniliprole

Compd.	larvicidal activity(%) at conc(mg · kg^{-1})								
	50	20	10	5	1	0.5	0.25	0.1	0.05
19a	100	100	100	100	100	100	100	100	100
19b	100	100	100	100	100	100	100	100	100
19c	100	100	100	100	100	100	100	100	100
19d	100	100	100	100	100	100	100	100	100
19e	100	100	100	100	100	100	100	100	100
19f	100	100	100	100	40		100	100	57

Table 6 (continued)

Compd.	larvicidal activity (%) at conc (mg · kg^{-1})								
	50	20	10	5	1	0.5	0.25	0.1	0.05
19g	100	100	100	100	100	100			
19h	100	100	100	100	0				
19i	100	100	100	0	0				
19j	100	100	60	80	0				
19k	100	100	100	100	100	100	43		
19l	100	100	100	0	0				
19m	100	100	100	100	100	100	100	43	
19n	100	100	100	100	100	100	100	100	100
Chlorantraniliprole	100	100	100	100	100	100	100	43	0

Insecticidal activity against beet armyworm (*Laphygma exigua Hubner*) The insecticidal activities of the title compounds **19a – 19n** against beet armyworm were tested by the leaf – dip method using the reported procedure.[21] Leaf discs (1.8 cm diameter) were cut from fresh cabbage leaves and then were dipped into the test solution for 15 s. After air – drying, the treated leaf discs were placed in a Petri dish (9 cm diameter). Each dried treated leaf disc was infested with seven third – instar beet armyworm larvae. Percentage mortalities were evaluated 3 d after treatment. Leaves treated with water and acetone were provided as controls. Each treatment was performed three times. For comparative purposes, Chlorantraniliprole was tested under the same conditions. The results were summarized in Table 7.

Table 7 Insecticidal activities against beet armyworm of the title compounds 19a – 19n and Chlorantraniliprole

Compd.	larvicidal activity (%) at conc (mg · kg^{-1})				
	20	5	2	1	0.5
19a	100	100	100	100	57
19b	100	100	100	100	86
19c	100	100	28		
19d	100	83			
19e	100	83			
19f	100	50			
19g	100	100	57		
19h	100	100	71		
19i	0				
19j	0				
19k	67				
19l	0				
19m	100	100	86		
19n	50				
Chlorantraniliprole	100	100	100	100	100

Results and Discussion

Synthesis

In the present work, the synthesis of two series of novel anthranilic diamide derivatives as well as their insecticidal activities against three lepidopterous pests were studied. The target trifluoroethyl ether compounds **19** and **20** were synthesized by a simple and convenient four – step procedure starting from the key intermediate 2 – (3 – chloro – pyridin – 2 – yl) – 5 – oxo – pyrazolidine – 3 – carboxylic acid ethyl ester(**9**). Compound **9** was oxidized to give pyrazolone **16** in low yield. The compound **16** was reacted with 2, 2, 2 – trifluoroiodoethane in dry dimethylformamide to yield 1 – (2 – chloro – 5 – pyridylmethyl) – 2 – cyanoiminoimidazolidine(**17**). Then compound **17** was hydrolyzed to give the key intermediate 2 – (3 – chloro – pyridin – 2 – yl) – 5 – (2,2,2 – trifluoro – ethoxy) – 2,5 – dihydro – 1*H* – pyrazole – 3 – carboxylic acid (**18**). The title compounds **19** and **20** were synthesized from compound **18** and the appropriate intermediate **14** or **15** (obtained from the intermediate **13** and corresponding alcohol or amine – see Table 1) in dry tetrahydrofuran using triethylamine as base.

Crystal structure analysis

Compound **19g** was recrystallized from ethyl acetate/petroleum ether to give colorless crystal suitable for X – ray single – crystal diffraction with the following crystallographic parameters: $a = 11.976(2)$ Å, $b = 14.274(3)$ Å, $c = 14.350(3)$ Å, $\alpha = 90°$, $\beta = 111.73(3)°$, $\gamma = 90°$, $\mu = 0.344 mm^{-1}$, $V = 2278.6(8)$ Å^3, $Z = 4$, $D_x = 1.505 mg \cdot m^{-3}$, $F(000) = 1056$, $T = 113(2)$ K, $3.06° \leqslant \theta \leqslant 25.02°$, and the final R factor, $R_1 = 0.0431$, $wR_2 = 0.1048$. The crystal is monoclinic.

The molecular structure of **19g** contains the following three – plane subunit: the benzene ring C(1) – C(6) (*p*1), the pyridine ring C(17) – C(21) – N(5) (*p*2), and the pyrazole ring(*p*3). The dihedral angel between the plane of the pyridine ring p2 and the plane of the pyrazole ring *p*3 is about 44.0°. The crystal packing structure of this compound is shown in Figure 3.

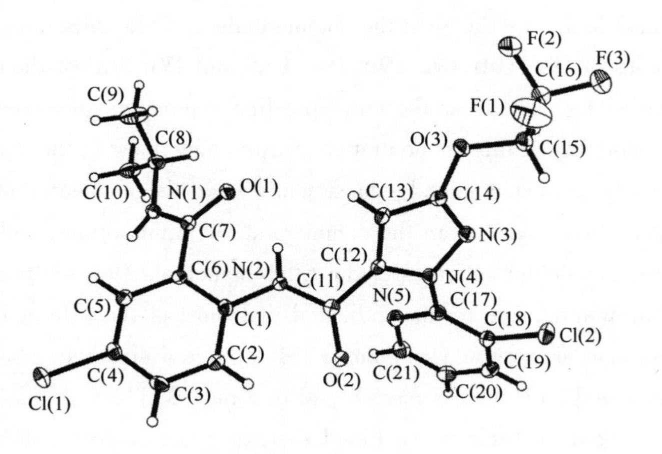

Figure 2 Molecular structure of the compound **19g**

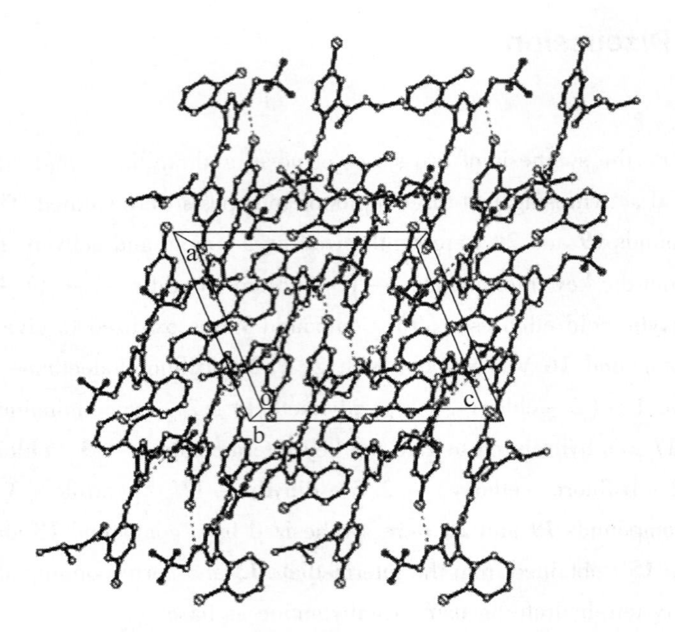

Figure 3 Packing diagram of the compound **19g**

Biological activity

Table 5 shows the insecticidal activities of the title compounds **19a – 19n** and **20a – 20b** and Chlorantraniliprole against oriental armyworm. The results of insecticidal activities given in Table 5 indicated that most of the title compounds exhibited excellent activity against oriental armyworm comparable to the commercialized Chlorantraniliprole. For instance, the insecticidal activities of compounds **19a**, **19b**, **19d**, **19g**, **19k** and **19m** against oriental armyworm at 2.5mg · kg^{-1} were 100%. Moreover, some of them still exhibited good insecticidal activity against oriental armyworm when the concentration was reduced to 1mg · kg^{-1}.

Table 6 shows the insecticidal activities of the title compounds **19a – 19n** and Chlorantraniliprole against diamond – back moth. The results indicate that the title compounds **19** have excellent insecticidal activities against diamond – back moth and that some of the title compounds **19** exhibit higher larvicidal activities than the commercialized Chlorantraniliprole. For example, the larvicidal activities of **19a**, **19b**, **19c**, **19d**, **19e**, **19g**, and **19n** against diamond – back moth were 100% at 0.1mg · kg^{-1}, whereas the corresponding commercial insecticide Chlorantraniliprole caused 43% mortality at this concentration. Surprisingly, most of them still exhibited perfect insecticidal activity against diamond – back moth when the concentration was reduced to 0.05 mg · kg^{-1}, which was higher than the commercialized Chlorantraniliprole.

Table 7 shows the insecticidal activities of the title compounds **19a – 19n** and Chlorantraniliprole against beet armyworm. The results indicated that most of the title compounds exhibited good activity against beet armyworm. Compounds **19b** exhibited moderate insecticide activity against beet armyworm and had > 80% mortality at 0.5 mg · kg^{-1}.

From the data presented in Table 5, we found that the bioactivities of the second series **20** were weaker than that of the first series **19**. Therefore the amide – substituted analogue showed a higher insecticidal activity than did the corresponding ester – substituted analogue. Among those compounds, replacing the nitrogen atom with oxygen atom resulted in decreased insecticidal activity. From Tables 5 – 7, we can also see that the larvicidal activities of the title compounds appeared to be strongly associated with the substituent R and its position on the benzene. Methyl – substituted at *ortho* and chloro – substituted at *para* are very important for increasing activity. Further studies on structural optimization and structure – activity relationships of these anthranilic diamide derivatives are in progress.

Conclusions

In summary, two series of novel anthranilic diamide insecticide containing trifluoroethyl ether were designed and synthesized with structures characterized by ^1H NMR spectroscopy, single crystal X – ray diffraction analysis and elemental analysis. The insecticidal activities of the new compounds were evaluated. The results of bioassays indicated that some of these title compounds exhibited excellent insecticidal activities. Surprisingly, most of them still exhibited perfect insecticidal activity against diamond – back moth when the concentration was reduced to $0.05 mg \cdot kg^{-1}$, which was higher than the commercialized Chlorantraniliprole.

References

[1] Tohnishi, M.; Nakao, H.; Furuya, T.; Seo, A.; Kodama, H.; Tsubata, K.; Fujioka, S.; Kodama, H.; Hirooka, T.; Nishimatsu, T. *J. Pestic. Sci.* 2005, 30, 354.

[2] Ebbinghaus – Kintscher, U.; Luemmena, P.; Lobitz, N.; Schulte, T.; Funke, C.; Fischer, R.; Masaki, T.; Yasokawa, N.; Tohnishi, M. *Cell Calcium* 2006, 39, 21.

[3] Gewehr, M.; Puhl, M.; Dickhaut, J.; Bastiaans, H. M. M.; Anspaugh, D. D.; Kuhn, D. G.; Oloumi – Sadeghi, H.; Armes, N. *WO* 2007082841, 2007 [*Chem. Abstr.* 2007, 147, 159937].

[4] Muehlebach, M.; Jeanguenat, A.; Hall, R. G. *WO* 2007080131, 2007 [*Chem. Abstr.* 2007, 147, 553327].

[5] Cordova, D.; Benner, E. A.; Sacher, M. A.; Rauh, J. J.; Sopa, J. S.; Lahm, G. P.; Selby, T. P.; Stevenson, T. M.; Flexner, L.; Gutteridge, S.; Rhoades, D. F.; Wu, L.; Smith, R. M.; Tao, Y. *Pestic. Biochem. Physiol.* 2006, 84, 196.

[6] Clark, D. A.; Lahm, G. P.; Smith, B. K.; Berry, J. D.; Clagg, D. G. *Bioorg. Med. Chem.* 2008, 16, 3163.

[7] Lahm, G. P.; Stevenson, T. M.; Selby, T. P.; Freudenberger, J. H.; Cordova, D.; Flexner, L.; Bellin, C. A.; Dubas, C. M.; Smith, B. K.; Hughes, K. A.; Hollingshaus, J. G.; Clark, C. E.; Benner, E. A. *Bioorg. Med. Chem. Lett.* 2007, 17, 6274.

[8] Sun, L.; Wang, T.; Ye, S. *Chin. J. Chem.* 2012, 30, 190.

[9] Smart, B. E. *J. Fluorine Chem.* 2001, 109, 3.

[10] Gong, Y. F.; Kato, K. *Curr. Org. Chem.* 2004, 8, 1659.

[11] Begue, J. P.; Bonnet – Delpon, D.; Crousse, B.; Legros, J. *Chem. Soc. Rev.* 2005, 34, 562.

[12] Lahm, G. P.; Selby, T. P.; Freudenberger, J. H.; Stevenson, T. M.; Myers, B. J.; Seburyamo, G. S.; Smith, B. K.; Flexner, L.; Clark, C. E.; Cordova, D. *Bioorg. Med. Chem. Lett.* 2005, 15, 4898.

[13] Montoya – Pelaez, P. J.; Uh, Y. – S.; Lata, C.; Thompson, M. P.; Lemieux, R. P.; Crudden, C. M. *J. Org. Chem.* 2006, 71, 5921.

[14] Shapiro, R.; Taylor, E. G.; Zimmerman, W. T. *WO* 2006062978, 2006 [*Chem. Abstr.* 2006, 145, 62887].

[15] Zhou, Z. L.; Kher, S. M.; Cai, S. X.; Whittemore, E. R.; Espitia, S. A.; Hawkinson, J. E.; Tran, M.; Woodward, R. M.; Weberb, E.; Keana, J. F. W. *Bioorg. Med. Chem.* 2003, 11, 1769.

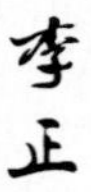

[16] Li, B. ; Wu, H. ; Yu, H. ; Yang, H. *WO* 2009121288, 2009 [*Chem. Abstr.* 2009, 151, 425742].
[17] Davis, R. F. ; Shapiro, R. ; Taylor, E. G. *WO* 2008010897, 2008 [*Chem. Abstr.* 2008, 148, 191926].
[18] Wang, B. ; Ma, Y. ; Xiong, L. ; Li, Z. *Chin. J. Chem.* 2012, 30, 815.
[19] Dong, W. ; Xu, J. ; Xiong, L. ; Liu, X. ; Li, Z. *Chin. J. Chem.* 2009, 27, 579.
[20] Wang, Y. ; Ou, X. ; Pei, H. ; Lin, X. ; Yu, K. *Agrochem. Res. Appl.* 2006, 10, 20.
[21] Busvine, J. R. *FAO Plant Production and Protection Paper*, Rome, Italy, 1980, No. 21, pp. 3 - 13, 119 - 122.

Evaluation of the in Vitro and Intracellular Efficacy of New Monosubstituted Sulfonylureas against Extensively Drug – resistant Tuberculosis*

Di Wang[1,**], Li Pan[2,**], Gang Cao[2], Hong Lei[1], Xianghong Meng[1], Jufang He[1], Mei Dong[1], Zhengming Li[2], Zhen Liu[3]

([1]Department of Clinical Laboratory, The 309th Hospital of Chinese People's Liberation Army, Beijing, 100091, China;
[2]State Key Laboratory of Elemento – Organic Chemistry, National Pesticide Engineering Research Center (Tianjin), Nankai University, Tianjin, 300071, China;
[3]The 309th Hospital of Chinese People's Liberation Army, Beijing, 100091, China)

Abstract Acetohydroxyacid synthase (AHAS) has been regarded as a potential drug target against *Mycobacterium tuberculosis* as it catalyses the first step in the pathway for biosynthesis of branched – chain amino acids. In our previous work, several monosubstituted sulfonylureas that are inhibitors of AHAS showed obvious in vitro activity against *M. tuberculosis.* In this study, further exploration of the antitubercular activity of newly synthesised monosubstituted sulfonylureas was conducted. A series of new compounds were identified that exhibit significant activity against in vitro and intracellular extensively drug – resistant *M. tuberculosis.* These results provide a further insight into the structural requirements for targeting AHAS to develop potential new agents to combat tuberculosis.

Key words Sulfonylurea; mycobacterium tuberculosis; antitubercular activity; extensively drug – resistant

1 Introduction

The emergence of multidrug – resistant tuberculosis (MDR – TB) and extensively drug – resistant tuberculosis (XDR – TB) seriously threatens the control of TB globally[1-4]. To combat these MDR/XDR – TB infections, new and effective pharmaceuticals are urgently needed[5,6]. Recently, acetohydroxyacid synthase (AHAS; EC 2. 2. 1. 6) has been identified as an attractive target for designing a new generation of anti – TB agents [7-9]. AHAS is a key enzyme in the biosynthesis of branched – chain amino acids (leucine, isoleucine and valine) by higher plants, algae, fungi and bacteria, and no homologous enzyme has been observed in humans or animals[10]. Sulfonylureas, with the general features of a central sulfonylurea bridge with an *o* – substituted aromatic ring attached to the sulphur atom and a heterocyclic ring attached to the nitrogen atom, have been recorded as inhibitors of AHAS[8]. Preliminary studies indicated that some sulfonylurea compounds, such as sulfometuron – methyl (SM), chlorimuron – methyl and metsulfuron – methyl, displayed antitubercular activity. A common feature of all the above com-

* Reprinted from *International Journal of Antimicrobial Agents*, 2012, 40 (5): 463 – 466. This project was supported by the National Natural Science Foundation of China (project no. NNSFC 20872069), the National Basic Research Program of China (2010CB126106), the Key Project of the National Natural Science Foundation of China (No. 30970419), the National Natural Science Foundation of China (No. 81000001) and the Project of PLA309 Hospital Foundation (No. 309 – 09 – 14).

** These two authors contributed equally to this work.

pounds is that they are meta – substituted in both meta – positions[7,8].

Our team have reported that some monosubstituted sulfonylureas in which the heterocyclic ring is substituted with only one substituent at the meta – position also exhibited potent activity against TB in vitro, even against clinical MDR and XDR cases[11,12]. Experimental data demonstrated that these compounds are potential lead structures for the development of novel antimycobacterial agents. Further structural optimisation is required to find more effective anti – TB agents. In addition, because TB is an intracellular infection, demonstration of in vitro activity must be followed by evaluation of the compound's killing ability against intracellular organisms. In previous studies, a branched – chain amino acid auxotrophic mycobacterial strain failed to proliferate because of its inability to use amino acids from the host[13], indicating that inhibitors of branched – chain amino acid biosynthesis could kill the infectious *Mycobacterium* despite amino acids being freely available from the host. Several reports have confirmed that sulfonylurea compounds are effective against TB in cultured macrophages[8] and in mice[7]. Here we report a series of new monosubstituted sulfonylureas and determined their inhibitory activity against intracellular XDR – TB isolates.

2 Materials and methods

2.1 Compounds and strains

Synthesis and determination of the in vitro activity of monosubstituted sulfonylurea derivative compounds **30 – 33** were as described in our previous report[11]. Synthesis of compounds **1 – 12** was as reported in the literature[14]. Other compounds reported in this paper were first synthesised as follows: 1,8 – diazabicyclo [5,4,0] undec – 7 – ene was added to 10 mL of acetonitrile containing sulfonamide (2 mmol) and phenyl 4 – substituted pyrimidin – 2 – ylcarbamate (2 mmol). After 8 – 24h of stirring at room temperature, the solution was slowly adjusted to pH 1. The resulting precipitate was collected, washed with water and dried to obtain the final product, which was recrystallised (petroleum ether and acetone) or purified by flash chromatography on silica gel (petroleum ether/acetone).

Standard *M. tuberculosis* strain H37Rv (ATCC 27294), which is susceptible to all the anti – TB drugs, was purchased from the Beijing Institute for Tuberculosis Control. XDR – TB refers to strains that are resistant to at least: (i) isoniazid (INH) and rifampicin (RIF); (ii) one of the three injectable second – line drugs amikacin (AMK), kanamycin (KAN) or capreomycin; and (iii) one of the fluoroquinolones. In total, 16 clinical XDR – TB isolates identified from The 309th Hospital of Chinese People's Liberation Army (PLA) (Beijing, China) were used in this work[15].

2.2 Determination of in vitro activity

The detailed protocol to determine the in vitro activity of the compounds against *M. tuberculosis* has been described previously [11,12]. Briefly, each compound was diluted two – fold on Middlebrook 7H10 agar media supplemented with OADC (oleic acid – albumin – dextrose – catalase). Prepared *M. tuberculosis strains* were added to the plates, incubated at 36.5℃ and cultures were examined for bacterial growth 4 weeks later. The minimum inhibitory concentration(MIC) was defined as the lowest concentration of a compound that inhibited visible bacterial growth of *M. tuberculosis*.

2.3 Intracellular activity assay

The 3 – [4,5 – dimethylthiazol – 2 – yl] – 2,5 – dipheny tetrazolium bromide (MTT) assay was used to test the cytotoxicity of the active compounds in the human acute monocytic leukaemia cell line (THP – 1) as previously described[11,12]. Then, the intracellular activity assay was performed as described below. THP – 1 cells were distributed in the wells of a 24 – well plate and were incubated in RPMI 1640 medium plus 10% foetal bovine serum and 1% L – glutamine. Cells were differentiated into macrophages using 5 nM phorbol myristate acetate(PMA). Then, 3 days later, three washes were performed with RPMI medium to remove all of the nonadherent cells and new medium without PMA was added and the cells were subsequently incubated for another 2 days.

Differentiated THP – 1 cells were infected with *M. tuberculosis* for 4h at a ratio of 10 CFU/macrophage. Infected THP – 1 cells were then washed three times with RPMI to remove all of the nonphagocytosed bacteria. Cells were treated with 50mg/L of each tested compound(including SM as the positive control) in RPMI medium. Medium without compound was used as a negative control. After 5 days, the supernatant was discarded, the infected cells were lysed with 0.1% Triton – X 100, and CFU counts of the cell lysates were determined.

3 Results

3.1 In vitro activity assay

By in vitro evaluation of the inhibitory efficacy of the monosubstituted sulfonylurea compounds against reference strain H37Rv, eight new compounds were found to exhibit observable activity, with MIC values ranging from 10mg/L to 80 mg/L(Table 1), as well as four compounds that have been reported elsewhere[12]. In particular, compounds **3** and **15** exhibited excellent activities, with MIC values of 10mg/L. The efficacy of compounds **3**, **15**, **30**, **31** and SM was also observed in medium containing additional branched – chain amino acids. Interestingly, *M. tuberculosis* was rescued from growth inhibition by the compounds in the presence of 300μM each of leucine, isoleucine and valine as seen by the four – fold increase in the MIC of SM and **3** (40mg/L) and the two – fold increase in the MIC of **15**, **30** and **31**(20mg/L). This result indicated that growth inhibition was indeed due to blocking of branched – chain amino acid biosynthesis.

The activities are comparable with that of control SM and our previously identified compounds **30** and **31**. Compounds **2**, **5**, **20**, **21**, **23** and **26** showed MICs in the range 20 – 80mg/L. The efficacy of the compounds against clinical XDR – TB strains isolated from Chinese PLA 309 Hospital were further determined(Table 1). All 16 XDR – TB isolates used in this work were resistant to INH, RFP and ofloxacin, 5 were resistant to KAN, whilst another 11 were resistant to AMK. Compounds retained their potency against all of the clinical XDR – TB strains, indicating that there was no cross - resistance between these compounds and conventional anti - TB drugs.

Table 1 Antimycobacterial activities of title sulfonylurea compounds

Compound	R_1	R_2	R_3	MIC for H37Rv (mg/L)	MIC range for XDR - TB isolates (mg/L)[1]
1	$COOC_2H_5$	NO_2	CH_3	>80	N/D
2	$COOCH_3$	Cl	CH_3	40	40
3	$COOC_2H_5$	Cl	CH_3	10	5 - 10
4	$COOC_2H_5$	Cl	OCH_3	>80	N/D
5	$COOCH_3$	Br	CH_3	40	40 - 80
6	$COOCH_3$	Br	OCH_3	>80	N/D
7	$COOCH_3$	$-NH-N=C(CN)CO_2C_2H_5$	CH_3	>80	N/D
8	$COOCH_3$	$-NH-N=C(CN)CO_2C_2H_5$	OCH_3	>80	N/D
9	$COOC_2H_5$	$-NH-N=C(CN)CO_2C_2H_5$	CH_3	>80	N/D
10	$COOC_2H_5$	$-NH-N=C(CN)CO_2C_2H_5$	OCH_3	>80	N/D
11	$COOCH_3$	$NHSO_2CH_3$	CH_3	>80	N/D
12	$COOCH_3$	$NHPO(OC_2H_5)_2$	CH_3	>80	N/D
13	$COOCH_3$	$COOC_2H_5$	CH_3	>80	N/D
14	Br	$COOCH_3$	OCH_3	>80	N/D
15	Br	$COOCH(CH_3)_2$	CH_3	10	10 - 20
16	Br	$CONH_2$	CH_3	>80	N/D
17	Br	$CONHC_2H_5$	OCH_3	>80	N/D
18	Br	$CONHCH(CH_3)_2$	OCH_3	>80	N/D
19	$COOCH_3$	$NHCOCH_2Cl$	CH_3	>80	N/D
20	$COOCH_3$	$NHCOCHCl_2$	CH_3	20	10 - 20
21	$COOCH_3$	$NHCOCCl_3$	CH_3	80	40 - 80
22	$COOCH_3$	$NHCOCH_2Cl$	OCH_3	>80	N/D
23	$COOCH_3$	$NHCOCHCl_2$	OCH_3	40	20 - 40
24	$COOCH_3$	$NHCOCCl_3$	OCH_3	>80	N/D
25	$COOC_2H_5$	$NHCOCH_2Cl$	CH_3	>80	N/D
26	$COOC_2H_5$	$NHCOCHCl_2$	CH_3	80	80
27	$COOC_2H_5$	$NHCOCH_2Cl$	OCH_3	>80	N/D
28	$COOCH(CH_3)_2$	$NHCOCH_2Cl$	CH_3	>80	N/D

Table 1 (continued)

Compound	R_1	R_2	R_3	MIC for H37Rv (mg/L)	MIC range for XDR - TB isolates (mg/L) ①
29	$COOCH(CH_3)_2$	$NHCOCH_2Cl$	OCH_3	>80	N/D
30②	Cl	$NHCOCH_2Cl$	CH_3	10	5 - 10
31②	Cl	$NHCOCH_2Cl$	OCH_3	10	5 - 10
32②	Cl	$NHCOCH{=}CH_2$	CH_3	20	20
33②	Cl	$NHCOCH{=}CH_2$	OCH_3	40	20 - 40

MIC, minimum inhibitory concentration; XDR - TB, extensively drug - resistant *Mycobacterium tuberculosis*; N/D, not determined.

①In total, 16 XDR - TB strains isolated from The 309th Hospital of Chinese People's Liberation Army (Beijing, China).

②Reference[12].

3.2 Intracellular activity assay

THP - 1 cells were used to test the intracellular activity of the compounds according to previous reports. Before that, in order to determine whether the compounds have deleterious effects, their inhibition of cell viability against THP - 1 cells was evaluated by the MTT assay. No compounds were found to be toxic at 100mg/L (data not shown). Intracellular inhibition of the compounds against 16 XDR - TB is shown in Table 2. Amongst the tested compounds, **30** and **31** displayed the most significant inhibition activity, with kill rates of >60% at a concentration of 50mg/L, followed by **15**, **32**, **3** and **20**. The other compounds also showed visible efficacy but with lower activities. The ex vivo activities of the compounds were much lower than those shown in the in vitro study.

Table 2 Intracellular antimycobacterial activities of screened sulfonylurea compounds at 50mg/L

Compound	Intracellular inhibition (%) (mean ± S. D.)①
2	22.4 ± 11.2
3	43.9 ± 13.3
5	26.5 ± 7.3
15	52.3 ± 10.3
20	40.4 ± 10.4
21	N/D
23	31.9 ± 11.5
26	N/D
30	66.6 ± 9.5
31	64.4 ± 8.8
32	44.3 ± 14.2
33	20.8 ± 14.6
SM	70.1 ± 6.6

S. D., standard deviation; N/D, not determined; SM, sulfometuron - methyl.

①In total, 16 extensively drug - resistant *Mycobacterium tuberculosis* strains isolated from The 309th Hospital of Chinese People's Liberation Army (Beijing, China).

4 Discussion

Treatment of serious infections with *M. tuberculosis* remains a major threat to the human population and new drugs are urgently needed, especially for MDR/XDR – TB cases[5]. Following our previous work[11,12], this paper described a further evaluation of monosubstituted sulfonylureas as potential anti – TB agents.

In general, in vitro activity testing of the title compounds indicated that a change of substituents influences their activities (Table 1). When R_1 = ester, R_2 = halogen and R_3 = methyl, there is an apparent increase in anti – TB activity. Introduction of ethoxycarbonyl and chlorine in the 2 – and 5 – positions of the benzene ring is favourable to increase activity (MIC = 10mg/L). When R_1 = bromine, R_2 = isopropoxycarbonyl and R_3 = methyl, compound **15** retained good efficacy against H37Rv (MIC = 10mg/L). When R_1 = chlorine, the potency of sulfonylureas with vinyl in the benzene ring is less than that of chloroacetyl (MIC = 10mg/L). When R_1 = ester, compounds containing dichloroacetyl have a higher level of antitubercular activity than others such as chloroacetyl and trichloroacetyl. Compounds with methyl in the pyrimidine ring are more effective compared with methoxyl.

Although hundreds of compounds have activity in vitro against *M. tuberculosis* strains, few are efficient where macrophages reside in the mammalian lung. Therefore, it is of far greater significance that these compounds are capable of killing intracellular *M. tuberculosis* in human macrophages. The THP – 1 cells used here have been shown to be a good cell – based model to provide an understanding of the bacteriostatic ability of agents against intracellular *M. tuberculosis* because the cell itself possesses slight killing activities against bacteria. Inhibition is evident by the end of 5 days post infection by most compounds at a concentration of 50 mg/L in the cultured cells. The inhibitory activities are basically consistent with, but lower than, their inhibitory activities in vitro. The lower efficacy in cells than that in vitro may be due to the following reasons: (i) penetration of compounds into cells is not good enough; (ii) the availability of certain amino acids within mycobacteria – containing vacuoles is not completely limited; and (iii) there may be some breakdown of sulfonylureas in the cultures[7,8]. To generate complete inhibition in intracellular *M. tuberculosis*, compounds at concentrations >50mg/L, or combination with other anti – TB agents, must be used.

It is particularly exciting that screened compounds were tested against clinical XDR – TB with some success. The XDR – TB strains were isolated from PLA 309 Hospital in China, which has the second largest number of TB patients in the world and where resistance to anti – TB drugs is a significant problem. Monosubstituted sulfonylurea compounds have the potential to become part of new anti – TB drugs if they can retain their activity against resistant isolates. In this work, compounds showed similar activities against all XDR – TB strains tested in this study, both in vitro and in THP – 1 cells, regardless of the antibiotic susceptibility of the strains.

Overall, the results suggest that monosubstituted sulfonylurea derivatives exhibit strong antimycobacterial potential and are excellent candidates for further evaluation in the treatment of TB infections in human. These observations provide us with an impetus for designing a new series in the near future based on the active compounds identified in this work and our previous report.

References

[1] World Health Organization. Global tuberculosis control – surveillance, planning, financing. Geneva, Switzerland: WHO; 2007.

[2] World Health Organization. Anti – tuberculosis drug resistance in the world. Report No. 4. Geneva, Switzerland: WHO; 2008.

[3] Centers for Disease Control Prevention (CDC). Emergence of *Mycobacterium tuberculosis* with extensive resistance to second – line drugs – worldwide, 2000 – 2004. Morb Mortal Wkly Rep 2006; 55: 301 – 5.

[4] World Health Organization. Multidrug and extensively drug – resistant tuberculosis: 2010 global report on surveillance and response. Geneva, Switzerland: WHO; 2010.

[5] Koul A, Arnoult E, Lounis N, Guillemont J, Andries K. The challenge of new drug discovery for tuberculosis. Nature 2011: 469, 483 – 90.

[6] Sacks L V, Behrman R E. Developing new drugs for the treatment of drugresistant tuberculosis: a regulatory perspective. Tuberculosis 2008; 88 (Suppl. 1): S93 – 100.

[7] Grandoni J A, Marta P T, Schloss J V. Inhibitors of branched – chain amino acid biosynthesis as potential antituberculosis agents. J Antimicrob Chemother 1998; 42: 475 – 82.

[8] Sohn H, Lee K S, Ko Y K, Ryu J W, Woo J C, Koo D W, et al. In vitro and ex vivo activity of new derivatives of acetohydroxyacid synthase inhibitors against *Mycobacterium tuberculosis* and non – tuberculous mycobacteria. Int J Antimicrob Agents 2008; 31: 567 – 71.

[9] Zohar Y, Einav M, Chipman D M, Barak Z. Acetohydroxyacid synthase from *Mycobacterium avium* and its inhibition by sulfonylureas and imidazolinones. Biochim Biophys Acta 2003; 1649: 97 – 105.

[10] Chipman D, Barak Z, Schloss J V. Biosynthesis of 2 – aceto – 2 – hydroxy acids: acetolactate synthases and acetohydroxyacid synthases. Biochim Biophys Acta 1998; 1385: 401 – 19.

[11] Pan L, Jiang Y, Liu Z, Liu X H, Liu Z, Wang G, et al. Synthesis and evaluation of novel monosubstituted sulfonylurea derivatives as antituberculosis agents. Eur J Med Chem 2012; 50: 18 – 26.

[12] Dong M, Wang D, Jiang Y, Zhao L, Yang C E, Wu C. In vitro efficacy of acetohydroxyacid synthase inhibitors against clinical strains of *Mycobacterium tuberculosis* isolated from a hospital in Beijing, China. Saudi Med J 2011; 32: 1122 – 6.

[13] Guleria I, Teitelbaum R, McAdam R A, Kalpana G, Jacobs Jr W R, Bloom B R. Auxotrophic vaccines for tuberculosis. Nat Med 1996; 2: 334 – 7.

[14] Cao G, Wang M Y, Wang M Z, Wang S H, Li Y H, Li Z M. Synthesis and herbicidal activity of novel sulfonylurea derivatives. Chem Res Chinese Universities 2011; 27: 60 – 5.

[15] Wang D, Yang C, Kuang T, Lei H, Meng X, Tong A, et al. Prevalence of multidrug and extensively drug – resistant tuberculosis in Beijing China: a hospital – based retrospective study. Jpn J Infect Dis 2010; 63: 368 – 71.

Synthesis and Insecticidal Evaluation of Novel N – Pyridylpyrazolecarboxamides Containing Cyano Substituent in the *ortho* – Position*

Mingzhen Mao, Yuxin Li, Qiaoxia Liu, Yunyun Zhou, Xiulan Zhang, Lixia Xiong, Yongqiang Li, Zhengming Li

(State Key Laboratory of Elemento – Organic Chemistry, Institute of Elemento – Organic Chemistry, Nankai University, Tianjin, 300071, China)

Abstract In an attempt to search for potent insecticides targeting the ryanodine receptor (RyR), a series of novel *N* – pyridylpyrazolecarboxamides containing cyano substituent in the *ortho* – position were designed and synthesized. Their insecticidal activities of target compounds against oriental armyworm (*Mythimna separata*) and diamondback moth (*Plutella xylostella*) indicated that most of the compounds showed moderate to high activities at the tested concentrations. In particular, compound **6l** and **6o** showed 86% larvicidal activities against *Plutella xylostella* at the concentration of 0.1mg/L, while the activity of compound **6h** against *Mythimna separate* was 80% at 1mg/L. The calcium imaging technique was applied to investigate the effects of some title compounds on the intracellular calcium ion concentration ($[Ca^{2+}]_i$), experimental results demonstrated that compound **6h** stimulates a transient elevation in $[Ca^{2+}]_i$ in the absence of external calcium after the central neurons dye loading with fluo – 3 AM. However, when the central neurons were dyed with fluo – 5 N and incubated with 2 – APB, $[Ca^{2+}]_i$ decreased transiently by treated of compound **6h**. All of the calcium imaging technique experiments demonstrated that these novel compounds deliver calcium from endoplasmic reticulum to cytoplasm, which proved that the title compounds were the possible activators of insect RyR.

Key words *N* – pyridylpyrazolecarboxamides; cyano; insecticidal activity; calcium channel

In order to overcome resistance and ecobiological problems associated with conventional insecticides, there is an urgent need to discover novel potent insecticides with a new mode of action. In recently years, Dupont discovered chlorantraniliprole[1] (Fig. 1**A**), which has an anthranilic diamide structure, exhibits exceptional broad – spectrum activity, high potency and low mammalian toxicity, and proves itself to be selective activators of the insect ryanodine receptor.[2] Due to its unique modes of action and good environmental profiles, anthranilic diamides have attracted considerable attention.

Most modifications in chlorantraniliprole structure in following researches preserve the anthranilic amide moiety, indicating that anthranilic amide is a key pharmacophore in this kind of compounds.[3-5] The introduction of a cyano group to replace the 4 – halo substituent led to the discovery of cyantraniliprole[4] (Fig. 1**B**), which had improved plant mobility and in-

* Reprinted from *Bioorganic & Medicinal Chemistry Letters*, 2013, 23(1): 42 – 46. The project was supported by the National Basic Research Programm of China (973 project # 2010CB126106), the Fundamental Research Funds for the Central Universities, Natural Science Foundation of China (No. 31000861) and the National Key Technologies R&D Program (2011BAE06B05).

creased spectra of insect control. However, there were also reported for the structural modification of the amides in the *ortho* – position, such as hydrazone,[6] heterocyclic groups[7,8] and cyano – containing amides.[9,10] The compound **C** reported by Li et al, showed excellent larvicidal activity against beet armyworm (*Spodoptera exigua*).[9] In view of the above information, introducing the cyano group into the structure of chlorantraniliprole skeleton would improve plant mobility and insecticidal activities. In order to obtain compounds with higher larvicidal activity and study the structure – activity relationship, a series of novel *N* – pyridylpyrazolecarboxamides (Fig. 1, **D**) containing cyano at the *ortho* – position were designed and synthesized. The larvicidal activities against oriental armyworms and diamondback moths were evaluated and the relating structure – activity relationships were also discussed. To further explore the mode of action for the target compounds, the effect of some target compounds on $[Ca^{2+}]_i$ in the central neurons isolated from the third instar of *Spodoptera exigua* was studied by calcium imaging techniques.

2 – Amino – 5 – substituited – 3 – methylbenzoic acid (**1b – d**) were synthesized by referring to the known procedure.[11,12] Compounds **2a – d** were prepared according to the reported method with minor improvements as shown in Scheme 1,[13] the pure product were easily obtained after filtration instead of extraction using sodium sulfate decahydrate as the quencher. Subsequent reaction with 8 equiv of manganese dioxide yielded compounds **3a – d** in excellent yields. 2 – Amino – 5 – cyano – 3 – methylbenzaldehyde was obtained from **3d** in the presence of cuprous cyanide in DMF at 140℃.[12]

Compounds **4a – c** were synthesized by the method reported by literature.[11,12,14] The key intermediates **5a – i** were achieved according to our previous work (Scheme 2).[13,14] Nevertheless, attempts to synthesize compound **5m** with same procedure failed, probably due to the strong electron – withdrawing cyano group resulting in a poor reactivity of amino moiety. Instead, **5m** was obtained via reflux in acetonitrile (Scheme 3).[15] The nitro – containing intermediates (**5l**, **5j**, **5k**) were synthesized as shown in Scheme 4, and the compounds (**5a**, **5c**, **5g**) were treated with fuming HNO_3 in concentrated H_2SO_4 to yield corresponding nitro – containing products in good yields and high regioselectivity.[16]

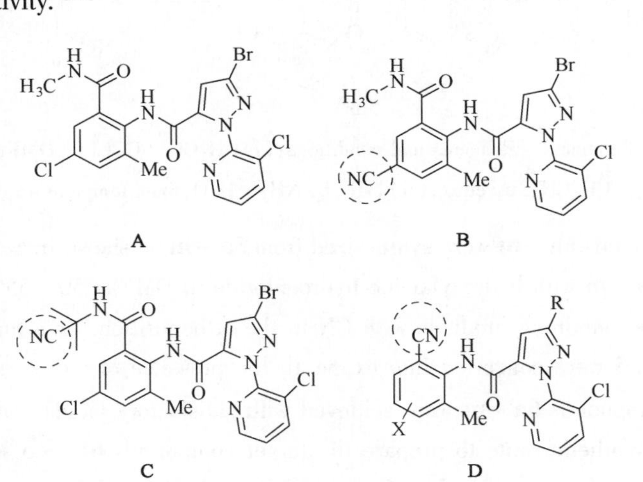

Figure 1 Chemical structures of compounds **A – D**

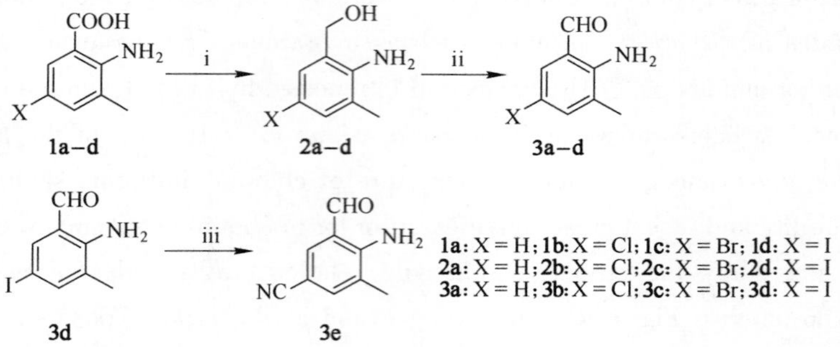

Scheme 1　Reagents and conditions: (i) $LiAlH_4$, THF, 0℃, then $NaSO_4 \cdot 10H_2O$, room temperature; (ii) MnO_2, CH_2Cl_2, room temperature; (iii) CuCN, DMF, 140℃

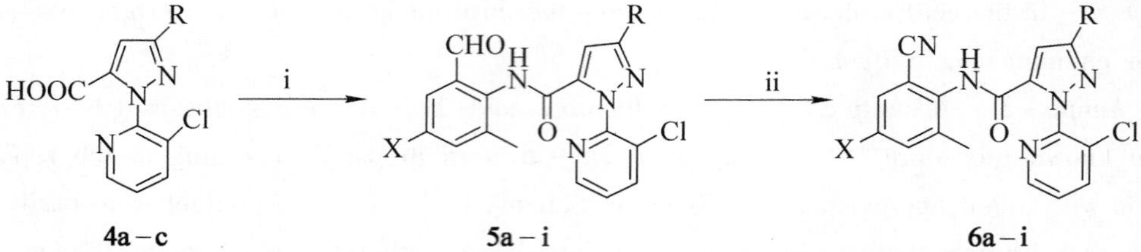

1a: X = H; 1b: X = Cl; 1c: X = Br; 1d: X = I
2a: X = H; 2b: X = Cl; 2c: X = Br; 2d: X = I
3a: X = H; 3b: X = Cl; 3c: X = Br; 3d: X = I

4a: R = Br; 4b: R = CF_3; 4c: R = OCH_2CF_3; 5a: R = Br X = Cl; 5b: R = Br X = Br;
5c: R = Br X = H; 5d: R = Br X = I; 5e: R = CF_3 X = Cl; 5f: R = CF_3 X = Br;
5g: R = CF_3 X = H; 5h: R = OCH_2CF_3 X = Cl; 5i: R = OCH_2CF_3 X = Br;
6a: R = Br X = Cl; 6b: R = Br X = Br; 6c: R = Br X = H; 6d: R = Br X = I;
6e: R = CF_3 X = Cl; 6f: R = CF_3 X = Br; 6g: R = CF_3 X = H;
6h: R = OCH_2CF_3 X = Cl; 6i: R = OCH_2CF_3 X = Br

Scheme 2　Reagents and conditions: (i) CH_2Cl_2, $(COCl)_2$, DMF; CH_2Cl_2, pyridine, **3a – d**, 0℃, then room temperature; (ii) THF, $I_2/NH_3 \cdot H_2O$, room temperature

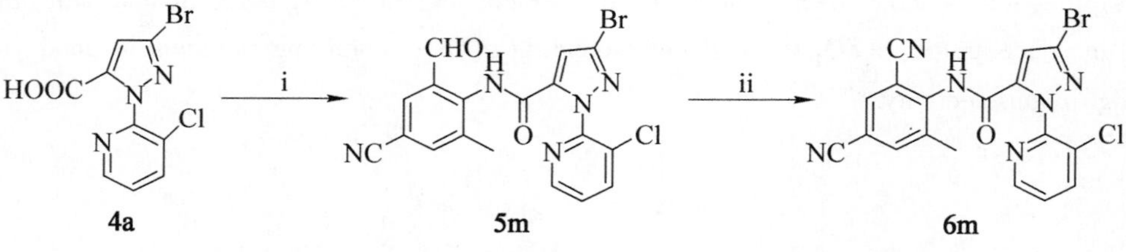

Scheme 3　Reagents and conditions: (i) CH_2Cl_2, $(COCl)_2$, DMF; CH_3CN, **3e**, reflux; (ii) THF, $I_2/NH_3 \cdot H_2O$, room temperature

The target compounds **6a – m** were synthesized from **5a – m** as shown in Schemes 2 – 4. We attempted to treat **5a – m** with hydroxylamine hydrochloride in DMF at 50 – 55℃ to afford the corresponding cyano – containing products with CN in the orthoposition. Unfortunately, the aldehyde group of compound **5** were converted into oxime. In the presence of iodine and aqueous NH_3 in THF, the target compounds **6a – m** were achieved with satisfactory yields and purity.[17]

The alternative synthetic route to prepare the target compounds **6m – o** with cyano group in the 4 – position of the benzene ring from **8a – c** as shown in Scheme 5. Intermediates **7a – c** and **8a – c** were synthesized following the previously reported procedure with minor improvements.[4,18] No reaction occurred using 2,4,6 – trichloro – 1,3,5 – triazine as dehydrant to synthesize **6m – o** from **8a – c**.[19] However, the dehydration reaction proceeded smoothly with thionyl dichloride and the products were obtained in excellent yields.[20]

5c; 5g
c: R = Br
g: R = CF_3

5j; 5k
j: R = Br
k: R = CF_3

6j; 6k
j: R = Br
k: R = CF_3

5a **5l** **6l**

Scheme 4 Reagents and conditions: (i) concentrated sulfuric acid, fuming nitric acid, 0℃, then room temperature; (ii) THF, $I_2/NH_3 \cdot H_2O$, room temperature

1e **4a – c** **7a – c** **8a – c**

7a: R = Br; **7b**: R = CF_3; **7c**: R = OCH_2CF_3;
8a: R = Br; **8b**: R = CF_3; **8c**: R = OCH_2CF_3;
6m: R = Br; **6n**: R = CF_3; **6o**: R = OCH_2CF_3;

6m – o

Scheme 5 Reagents and conditions: (i) CH_3CN, pyridine, CH_3SO_2Cl, room temperature; (ii) CH_3CN, $NH_3 \cdot H_2O$, room temperature; (iii) DMF, $SOCl_2$, 0℃, then room temperature

Most of the intermediates were determined by ^1H NMR, and all new target compounds were characterized with ^1H NMR, ^{13}C NMR and elemental analysis (or HRMS) (see Supplementary data). Compounds **6b** was selected to further investigate the IR spectrum characterization of this kind of compounds. The characteristic stretching vibration $\nu(C\equiv N)$ appears at $2235cm^{-1}$.

The larvicidal activity of compounds **6a – o** against oriental armyworms is summarized in Table 1. The bioassay results indicated that most compounds have excellent larvicidal activities against oriental armyworm. For example, the larvicidal activities of **6a**, **6e**, **6h** and **6i** against oriental armyworm at 1.0mg/L were 60%, 80%, 50%, 20%, respectively. Activities varied significantly depending upon the types of substituents on the 3 – position pyrazole. Compared with 3 – Br and 3 – CF_3 in pyrazole, compounds with 3 – OCH_2CF_3 substituents showed higher insecticidal activities against oriental armyworm, with the sequence of **6h** > **6a** > **6e**, **6i** > **6f** > **6b** and **6o** > **6m** > **6n**, which suggests that the introduction of the 2,2,2 – trifluoroethoxy groups in the 3 – position of pyrazole has a positive effect on the larvicidal activities. Furthermore, different substituents in benzene ring had various influence on activity. When R was fixed as Br, the bioactivity of compounds with different X indicated the sequence of Cl > Br > CN > I > NO_2 > H, while compounds with 3 – CF_3 and 3 – OCH_2CF_3 in pyrazole showed a similar trend. However, the compounds with cyano group in 4 – position of the benzene ring did not exhibited higher activities as we expected. For example, the larvicidal activities of **6m** and **6n** at a concentration of 10mg/L were 70% and 20%, respectively. In addition, the introduction of nitro group at the 5 – position of the benzene ring led to a significant decrease in activity, such as **6l**.

The larvicidal activity of compounds **6a – o** against diamondback moth were evaluated as shown in Table 2. Most of them had excellent larvicidal activity against diamondback moth. In particular, compounds **6l** and **6o** had around 86% mortality at the concentration of 0.1mg/L, approaching closer to chlorantraniliprole. Surprisingly, compound **6d**(X = 4 – I) showed good activity against diamondback moth (86% death rate at 1mg/L).

Table 1 Insecticidal activities of compounds 6a – o and chlorantraniliprole against oriental armyworms

Compound	Larvicidal activity (%) at a concentration of (mg/L)				
	25	10	5	2.5	1
6a	100	100	100	100	60
6b	100	80			
6c	40				
6d	100	60			
6e	100	100	100	100	50
6f	100	100	100	80	
6g	100	100	40		
6h	100	100	100	100	80
6i	100	100	100	100	20
6j	100	20			
6k	100	100	60		
6l	100	100	60		
6m	100	70			
6n	100	20			
6o	100	100	60		
Control①	100	100	100	100	100

①Chlorantraniliprole.

Table 2 Insecticidal activities of compounds 6a – o and chlorantraniliprole against diamondback moth

Compound	Larvicidal activity(%) at a concentration of(mg/L)		
	10	1	0.1
6a	100	43	0
6b	86	43	0
6c	43	29	0
6d	100	86	14
6e	100	29	0
6f	57	29	0
6g	71	29	0
6h	100	71	14
6i	100	43	0
6j	100	57	14
6k	71	29	0
6l	100	100	86
6m	100	71	29
6n	100	57	14
6o	100	100	86
Control①	100	100	100

①Chlorantraniliprole.

Figure 2 illustrated the change of $[Ca^{2+}]_i$ versus recording time when the neurons were treated with **6b**,**6c**,**6e**,**6h**,**6k**,**6l**,**6n** and chlorantraniliprole. The peak of $[Ca^{2+}]_i$ were elevated to 117.38 ±4.21% (n = 18), 111.71 ±3.29% (n = 13), 119.29 ±3.47% (n = 13), 114.63 ± 4.11% (n = 9), 114.43 ±3.78% (n = 9), 109.23 ±2.37% (n = 9) and 122.06 ±2.54% (n = 18) of the initial value when the cells were treated with 1000mg/L of **6b**,**6c**,**6e**,**6h**,**6k**,**6l**,**6n** and chlorantraniliprole, respectively. Compared with the control (99.91 ± 2.56%), these compounds induced $[Ca^{2+}]_i$ increase without extracellular Ca^{2+}. It indicated that compounds could activate the calcium release channel in the endoplasmic reticulum (ER) membrane. Figure 2 also indicated that the recorded $[Ca^{2+}]_i(F/F_0)$ had a good positive correlation with bioactivities.

As shown in Figure 3, brief application of compound **6h** continued to stimulate a transient elevation in $[Ca^{2+}]_i$ in the absence of external calcium. Reintroduction of standard saline allowed depleted calcium stores to become refilled and thereby available for the next **6h** challenge but resulted in an attenuated response.

To test why compound **6h** and chlorantraniliprole can cause $[Ca^{2+}]_i$ elevation, the primary cultured neurones were dyed loading with fluo – 5 N. Figure 4 illustrated the change of $[Ca^{2+}]_i$ versus recording time when the neurons were treated with **6h** and chlorantraniliprole. Compound **6h** and chlorantraniliprole decrease $[Ca^{2+}]_i$ to 95.12 ±2.06% (n = 12) and 90.34 ±3.64%

($n = 18$), respectively. These data indicated that $[Ca^{2+}]_i$ decreased by 1000mg/L of **6h** and chlorantraniliprole. It means that compound **6h** and chlorantraniliprole could deliver calcium from endoplasmic reticulum(ER) to cytoplasm.

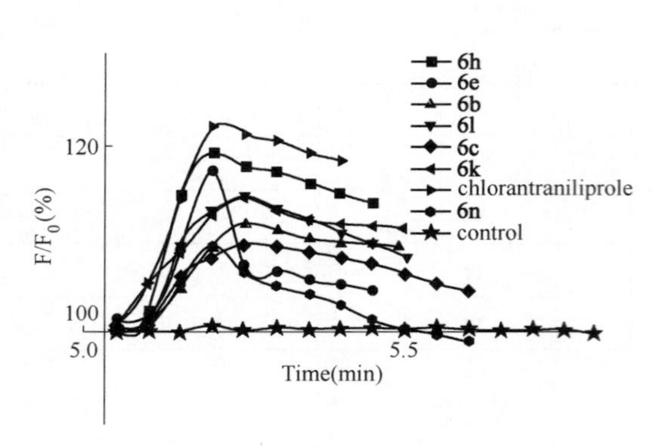

Figure 2 The change of $[Ca^{2+}]_i$ versus recording time when the neurons treated with 0.1 mg/L **6b**, **6c**, **6e**, **6h**, **6k**, **6l**, **6n** and chlorantraniliprole

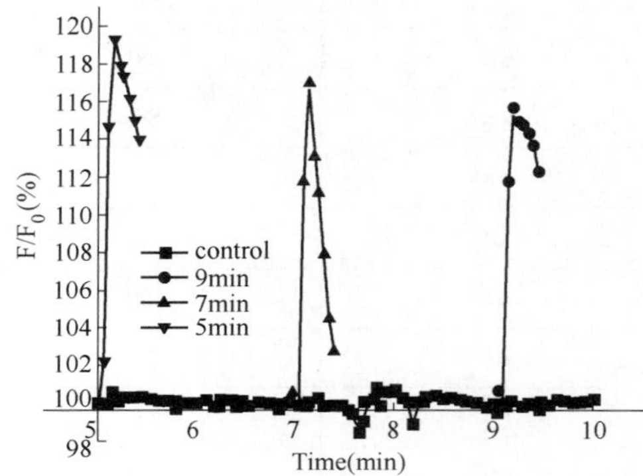

Figure 3 Characterization of compound **6h** stimulated calcium responses in the central neurons of *S. exgua* third larvae. Repeated challenges with compound **6h** in calcium – free saline(1mM EGTA with $CaCl_2$ omitted)

There were two kinds of calcium release channels in the ER membrane, namely RyR and IP_3R Ca^{2+} channels. To test which pathway was involved in the elevation of $[Ca^{2+}]_i$, the primary cultured neurones were dyed loading with fluo – 5 N(low – affinity calcium indicator, accurately tracks the dynamic changes in calcium in the ER and SR), and then incubated with 2 – aminoethoxydiphenyl borate(2 – APB 50μM, a chemical that acts to inhibit both IP_3 receptors and TRP channels) for 20min. As shown in Figure 5, when external Ca^{2+} was free, IP_3 receptors were blocked using 2 – APB, the decrease of $[Ca^{2+}]_i$ was only attributed to compound **6h** (1000mg/L). After the central neurons were incubated with 2 – APB, compound **6h** decrease $[Ca^{2+}]_i$ to 94.23 ± 3.48% ($n = 12$) and there was no statistically difference in the calcium response. It means that compound **6h** only has weak influence on TRP channels. More importantly, these data demonstrated that compound **6h** delivers calcium from endoplasmic reticulum (ER) to cytoplasm by RyRs.

In conclusion, a series of novel *N* – pyridylpyrazolecarboxamides containing cyano in the *ortho* – position were designed and synthesized. The bioassays showed that some of the compounds exhibited excellent insecticidal activities against oriental armyworm (*Mythimna separata*) and diamondback moth (*Plutella xylostella*). In particular, compound **6l** and **6o** showed 86% larvicidal activities against Plutella *xylostella* at the concentration of 0.1mg/L, while the activity of compound **6h** against Mythimna separate was 80% at 1mg/L. The calcium imaging techniques were used to investigate the effects of some title compounds on the $[Ca^{2+}]_i$, which indicated that the title compounds were the possible activators of the RyR. The results of the present study provide useful information for further structural optimization of these compounds and a rapid detection for the activity of the target compounds.

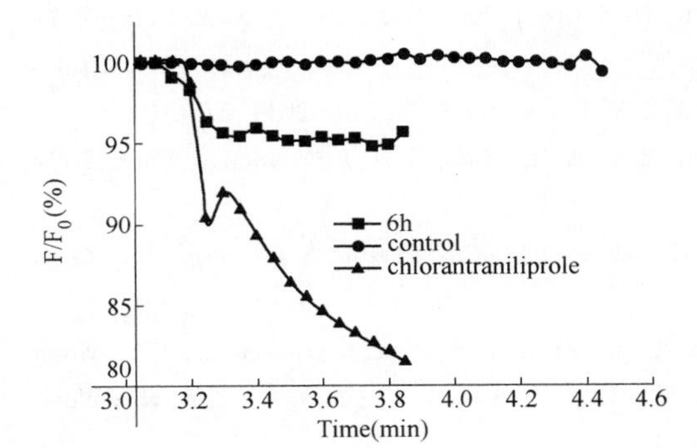

Figure 4 Effect of treatments with of **6h** and chlorantraniliprole on intracellular Ca^{2+} at different time when extracellular Ca^{2+} was in absence (EGTA replace Ca^{2+}). The central neurons of *S. exigua* third larvae dye loading with fluo - 5 N

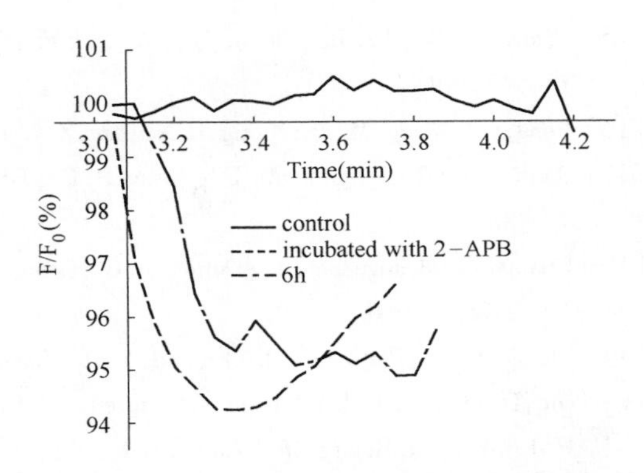

Figure 5 Effect of treatments with of **6h** on intracellular Ca^{2+} at different time when extracellular Ca^{2+} was in absence (EGTA replace Ca^{2+}). The central neurons of *S. exigua* third larvae dye loading with fluo - 5 N, then incubated with 2 - aminoethoxydiphenyl borate for 20 min

Supplementary data

Supplementary data associated with this article can be found, in the online version, at http://dx. doi. org/10. 1016/j. bmcl. 2012. 11. 045.

References and Notes

[1] DuPont Rynaxypyr_insect ontrol Technical Bulletin: http://www2. dupont. com/Production_Agriculture/en_US/assets/downloads/pdfs/Rynaxypyr_Tech_Bulletin. pdf.

[2] Lahm, G. P.; Cordova, D.; Barry, J. D. *Bioorg. Med. Chem.* 2009, 17, 4127.

[3] Finkelstein, B. L.; Lahm, G. P.; Mccann, S. F.; Song, Y.; Stevenson, T. M. WO2009/03016284A1; *Chem. Abstr.* 2003, 138, 205054.

[4] Hughes, K. A.; Lahm, G. P.; Selby, T. P.; Stevenson, T. M. WO2004/067528A1; *Chem. Abstr.* 2004, 141, 190786.

[5] Hall, R. G.; Loiseleur, O.; Pabba, J.; Pal, S.; Jeanguenat, A.; Edmunds, A.; Stoller, A. WO2009/010260A2; *Chem. Abstr.* 2009, 150, 14469.

[6] Wu, J.; Song, B. A.; Hu, D. Y.; Yue, M.; Yang, S. *Pest Manag. Sci.* 2012, 68, 801.

[7] Clark, D. A.; Finkelstein, B. L.; Lahm, G. P.; Selby, T. P.; Stevenson, T. M. WO2003/016304 A1; *Chem. Abstr.* 2003, 138, 205057.

[8] Zhang, X. N.; Zhu, H. J.; Tan, H. J.; Li, Y. H.; Ni, J. P.; Shi, J. J.; Zeng, X.; Liu, L.; Zhang, Y. N.; Zhou, Y. L.; He, H. B.; Feng, H. M.; Wang, N. CN 101845043, 2010; *Chem. Abstr.* 2010, 153, 530554.

[9] Li, B.; Xiang, D.; Chai, B. S.; Yuan, J.; Yang, H. B.; Zhang, H.; Wu, H. F.; Yu, H. B. WO 2008/134969 A1; *Chem. Abstr.* 2008, 149, 534209.

[10] Muchlebach, M.; Caring, G. W. WO 2008/064891 A1; *Chem. Abstr.* 2008, 149, 10005.

[11] Dong, W. L.; Xu, J. Y.; Xiong, L. X.; Liu, X. H.; Li, Z. M. *Chin. J. Chem.* 2009, 27, 579.

[12] Chai, B. S.; He, X. M.; Wang, J. F.; Li, Z. N.; Liu, C. L. *Agrochemicals* 2010, 49, 167 (in Chinese).

[13] Feng, Q.; Yu, G. P.; Xiong, L. X.; Wang, M. Z.; Li, Z. M. *Chin. J. Chem.* 2011, 29, 1651.

[14] Li, Z. M.; Wang, B. L.; Zhang, J. F.; Xu, J. Y.; Xiong, L. X.; Zhao, Y.; Wang, G. CN102276580A, 2011; *Chem. Abstr.* 2011, 156, 74423.

[15] Yang, H. B. ; Li, B. ; Chen, H. ; Wu, H. F. ; Yu, H. B. CN 102020633A, 2011; *Chem. Abstr.* 2011, 154,486349.

[16] Feng,Q. ;Wang,M. Z. ;Xiong,L. X. ;Liu,Z. L. ;Li,Z. M. *Chem. Res. Chin. Univ.* 2011,27,610.

[17] Jadhav, G. R. ; Shaikh, M. U. ; Kale, R. P. ; Ghawalkar, A. R. ; *Gill, C. H. J. Heterocycl. Chem.* 2009, 46,980.

[18] Gnamm, C. ; Jeanguenat, A. ; Dutton, A. C. ; Grimm, C. ; Kloer, D. P. ; Crossthwaite, A. J. *Bioorg. Med. Chem. Lett.* 2012,22,3800.

[19] Zhang,Y. K. ;Plattner,J. J. ;Freund,Y. R. ;Easom,E. E. ;Zhou,Y. S. ;Gut,J. ;Rosenthal,P. J. ;Waterson,D. ;Gamo,F. J. ;Angulo – Barturen,I. ;Ge,M. ;Li,Z. Y. ;Li,L. C. ;Jian,Y. ;Cui,H. ;Wang, H. L. ;Yang,J. Bioorg. *Med. Chem. Lett.* 2011,15,644.

[20] Liu,Z. L. ;Feng,Q. ;Xiong,L. X. ;Wang,M. Z. ;Li,Z. M. *Chin. J. Chem.* 2010,28,1757.

N－[4－氯－2－取代氨基甲酰基－6－甲基苯基]－1－芳基－5－氯－3－三氟甲基－1H－吡唑－4－甲酰胺的合成及生物活性*

张秀兰　王宝雷　毛明珍　熊丽霞　于淑晶　李正名

（南开大学元素有机化学国家重点实验室，元素有机化学研究所，天津　300071）

摘　要　通过改变传统氯虫酰胺中吡唑环上氨基甲酰基与吡啶环之间的相对位置，或以其它芳环取代原分子中的吡啶环，设计合成了24个结构新颖的 N－[4－氯－2－取代氨基甲酰基－6－甲基苯基]－1－芳基－5－氯－3－三氟甲基－1H－吡唑－4－甲酰胺类化合物。所有目标化合物的结构均通过 ^1H NMR谱、元素分析或高分辨质谱表征确定。初步的生物活性测试结果表明，部分化合物对东方粘虫具有较好的杀虫活性，其中化合物**6m**在浓度为50mg/L时具有80%的杀虫活性。同时，在浓度为50mg/L时目标化合物对5种常见病菌具有明显的抑制作用，其中化合物6n和6x对苹果轮纹菌的抑菌率达62.1%。

关键词　Vilsmeier反应　氯虫酰胺　邻甲酰胺基苯甲酰胺　杀虫活性　抑菌活性

21世纪人类面临着各种各样的挑战，其中最严峻的问题之一就是粮食问题。随着害虫对现存杀虫剂抗性的提高，如何应对日益严重的昆虫抗药性问题成为植物保护成功的关键。寻求具有新作用机制的高效、绿色杀虫剂对于人类的生存和发展具有重要作用。新型的邻甲酰胺基苯甲酰胺类化合物在此方面取得了突破性进展，其作用机理是激活昆虫体内的钙离子通道———鱼尼丁受体，引起钙离子持续释放，从而导致昆虫死亡。该类杀虫剂因靶标新颖、作用机理独特、高效、对非靶标生物安全以及与传统农药无交互抗性而引起人们的关注。自2008年上市以来，氯虫酰胺[1,2]以其杀虫谱广和杀虫杀卵活性高的优势[3,4]，成为农药研究领域的热点[5,6]。前文[7,8]对氯虫酰胺的结构进行了修饰，但都保留了其中的 N－吡啶基吡唑基团，并保持吡唑环上氨基甲酰基与吡啶基处于邻位。鉴于底物与鱼尼丁受体结合区域的三维结构尚不明确，若改变吡唑环上氨基甲酰基与吡啶基之间的相对位置，或以其它芳基取代吡啶基是否有助于提高活性呢？本文通过Vilsmeier反应合成关键中间体，设计合成了一系列氨基甲酰基与吡啶基在吡唑环上处于间位的新型邻甲酰胺基苯甲酰胺类化合物（Scheme 1），并研究了其杀虫和抑菌活性，以期获得更高生物活性的新化合物。

1　实验部分

1.1　仪器与试剂

Bruker Avance 400型核磁共振仪（TMS为内标，$CDCl_3$ 或 DMSO－d_6 为溶剂）；

* 原发表于《高等学校化学学报》，2013，34（1）：96－102。国家“973”计划项目（批准号：2010CB126106）、国家自然科学基金（批准号：20872069）和“十二五”国家科技支撑计划项目（批准号：2011BAE06B05）资助。

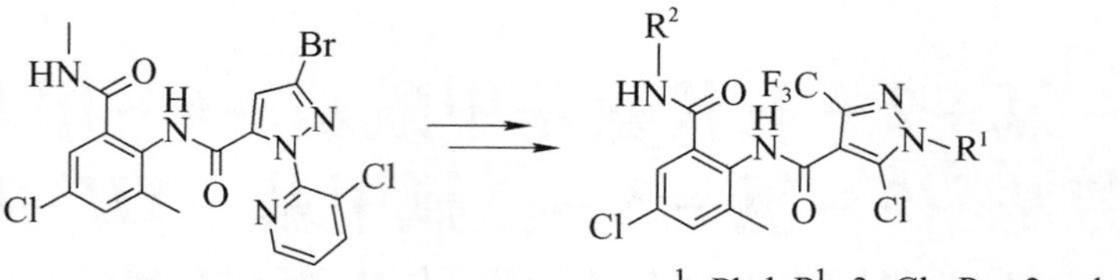

Chlorantraniliprole
Rynaxypyr™, DPX－E2Y45

a. R^1=Ph; b. R^1=3－Cl－Py－2－yl;
c. R^1=2,4－Cl_2－Ph
R^2=C_1－C_6 alkyl

Scheme 1　Design strategy of the title compounds

Thermofinnigan LCQ Advantage 型质谱仪（ESI）；Elementar Vario EL Ⅲ型元素分析仪（德国）；X－4 型数字显示显微熔点仪（北京泰克仪器有限公司）。

H 型柱层析硅胶（青岛海洋化工厂）；所用试剂均为国产分析纯，均经过常规方法处理。

1.2　中间体的合成

中间体吡唑醇（**1a～1c**）、吡唑醛（**2a～2c**）、吡唑酸（**3a～3c**）和吡唑酰氯（**4a～4c**）参照文献［9～17］方法制备。合成路线见 Scheme 2。

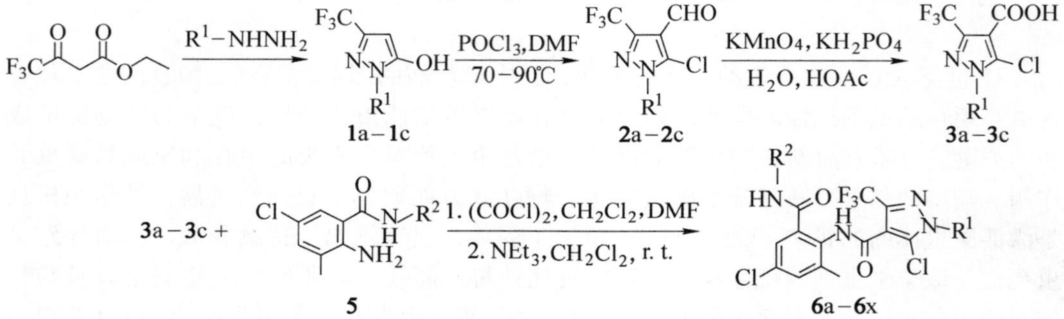

R^1－$NHNH_2$；$POCl_3$, DMF, 70－90℃；$KMnO_4$, KH_2PO_4, H_2O, HOAc

1a－1c　**2a－2c**　**3a－3c**

3a－3c + **5**　1. $(COCl)_2$, CH_2Cl_2, DMF　2. NEt_3, CH_2Cl_2, r. t.　**6a－6x**

Scheme 2　Synthetic routes of the title compounds

1a，**2a**，**3a**：R^1 = Ph；**1b**，**2b**，**3b**：R^1 = 2，4 － Cl_2 － Ph；**1c**，**2c**，**3c**：R^1 = 3 － Cl － Py － 2 － yl；
6a－6h：R^1 = Ph；**6i－6p**：R^1 = 3 － Cl － Py － 2 － yl；**6q－6x**：R^1 = 2，4 － Cl_2 － Ph；**6a**，**6i**，
6q：R^2 = CH_3；**6b**，**6j**，**6r**：R^2 = C_2H_5；**6c**，**6k**，**6s**：R^2 = n － C_3H_7；**6d**，**6l**，
6t：R^2 = i － C_3H_7；**6e**，**6m**，**6u**：R^2 = cycloproyl；**6f**，**6n**，**6v**：R^2 = n － C_4H_9；
6g，**6o**，**6w**：R^2 = t － C_4H_9；**6h**，**6p**，**6x**：R^2 = cyclohexyl

1.3　目标化合物 6a～6x 的合成

在 50mL 单口圆底烧瓶中加入化合物 **3**（1.5mmol）、草酰氯（0.6mL）和二氯甲烷（20mL），搅拌下反应 10min，滴加 2 滴二甲基甲酰胺（DMF），待溶液变澄清后继续搅拌反应 3h，减压蒸除溶剂得酰氯粗品。用 15mL 干燥的四氢呋喃溶解酰氯粗品，在冰盐浴下缓慢滴加到化合物 5 的 DMF 溶液中，滴加完毕继续搅拌反应 20min，加入三乙胺（2.25mmol）作缚酸剂，将反应液缓慢升温至回流，2～3h 后减压蒸除溶剂，用乙酸乙酯将其溶解，然后分别用稀盐酸（20mL）、饱和碳酸氢钠水溶液（20mL）和饱和氯化钠水溶液（20mL）洗涤有机层，无水硫酸钠干燥，减压蒸除溶剂后残余物经柱层

析分离得化合物**6**（当 $R^1=2,4-Cl_2-Ph$ 时直接用乙酸乙酯洗涤得到目标产物）。

1.4 目标化合物的生物活性测试

对目标化合物进行了杀虫和抑菌生物活性测试。杀虫活性参照文献［18］方法对3~4龄东方粘虫（*Mythimna separate*）进行测试。抑菌测试采用离体平皿法[19~21]，在50mg/L浓度下，分别对苹果轮纹病菌（*Physalospora piricola*）、番茄早疫病菌（*Alternaria solani*）、花生褐斑病菌（*Cercospora arachidicola*）、小麦赤霉病菌（*Gibberella sanbinetti*）和辣椒疫霉（*Phytophthora capsici*）5种常见病菌进行了测试。

2 结果与讨论

2.1 化合物的合成

在中间体**1b**的合成过程中，参照文献［11］方法进行类似合成时，以3-氯-2-肼基吡啶与三氟乙酰乙酸乙酯为原料，只能得到脱去1分子 H_2O 的未合环的中间体。实验中曾尝试多种反应条件如乙酸催化回流、采用乙醇/NaOH或乙醇/乙醇钠体系等，但这些反应或者不能得到产物，或者产率低、条件繁琐，最终采用对甲苯磺酸/甲苯体系反应，并用分水器回流不断除去反应生成的乙醇，高产率得到了中间体**3**。

在中间体**1c**的合成过程中，参照文献［11］方法以2，4-二氯苯肼与三氟乙酰乙酸乙酯为原料进行反应未得到目标产物，而是同样得到脱去一分子 H_2O 的未合环产物。以对甲苯磺酸/甲苯体系并用分水器回流进行的反应虽可进行，但产率较低。当采用HCl/EtOH体系回流时，以原料2，4-二氯苯肼盐酸盐（无需中和处理）直接与三氟乙酰乙酸乙酯反应，反应过程中固体慢慢溶解消失，最后经处理以高产率得到中间体**1c**。中间体**1c**的溶解性较差，可以用 CH_2Cl_2 洗涤或乙醇/H_2O 重结晶。

2.2 化合物的表征

中间产物和目标化合物的物化参数和 ^1H NMR数据分别列于表1和表2。

Table 1 Physical data, HRMS and elemental analysis results of the synthesized compounds①

Compd.	Appearance	m. p. /℃	Yield(%)	Elemental analysis (%, calcd.) C	H	N	[M-H]⁻ HRMS (calcd.), *m/z*
1b	Yellow solid	128-130	90	NT	NT	NT	NT
2b	White solid	108-110	90	NT	NT	NT	NT
3b	White solid	170-171	88	NT	NT	NT	NT
6a	White solid	255-256	43.2	51.00(50.97)	3.10(3.21)	11.76(11.89)	NT
6b	White solid	265-266	41.5	51.80(51.97)	3.85(3.85)	11.34(11.35)	NT
6c	White solid	250-251	39.4	52.91(52.92)	4.07(3.84)	11.13(11.22)	NT
6d	White solid	249-250	43.2	52.82(52.92)	3.92(3.84)	11.25(11.22)	NT
6e	White solid	277-278	39.5	NT	NT	NT	519.0573(519.0573)②
6f	White solid	249-250	44.6	53.89(53.81)	4.16(4.16)	10.88(10.91)	NT
6g	White solid	149-151	48.3	53.81(53.86)	4.12(4.68)	10.82(10.91)	NT
6h	White solid	217-218	32.8	55.67(55.67)	4.24(4.30)	10.32(10.39)	NT

续表 1

Compd.	Appearance	m. p. /℃	Yield(%)	Elemental analysis (%, calcd.)			$[M-H]^-$ HRMS (calcd.), m/z
				C	H	N	
6i	White solid	243 - 244	40.7	45.03(45.04)	2.10(2.59)	13.88(13.82)	NT
6j	White solid	228 - 229	42.1	46.40(46.13)	2.77(2.90)	13.04(13.45)	NT
6k	White solid	223 - 224	39.2	47.17(47.21)	3.22(3.20)	12.92(13.10)	NT
6l	White solid	211 - 212	35.5	47.20(47.17)	3.26(3.20)	12.67(13.10)	NT
9m	White solid	237 - 238	38.7	47.15(47.35)	3.03(2.84)	13.24(13.15)	NT
6n	White solid	212 - 213	32.6	48.20(48.15)	3.58(3.49)	12.51(12.76)	NT
6o	White solid	235 - 236	39.8	47.92(48.15)	3.73(3.49)	12.63(12.76)	NT
6p	White solid	241 - 242	40.2	49.88(50.15)	3.28(3.68)	11.72(12.18)	NT
6q	White solid	256 - 257	31.8	NT	NT	NT	538.9641(538.9637)
6r	White solid	257 - 258	48.5	NT	NT	NT	552.9795(552.9794)
6s	White solid	232 - 233	47.2	NT	NT	NT	566.9962(566.9950)
6t	White solid	207 - 208	36.6	NT	NT	NT	566.9950(566.9950)
6u	White solid	273 - 274	54.2	NT	NT	NT	564.9798(564.9794)
6v	White solid	240 - 242	48.5	47.43(47.25)	3.35(3.29)	9.88(9.62)	NT
6w	White solid	245 - 246	55.0	NT	NT	NT	605.0078(605.0082)
6x	White solid	240 - 241	50.3	49.47(49.36)	3.52(3.48)	9.16(9.21)	NT

①NT: not tested; ② $[M+Na]^+$。

Table 2 ^{1}H NMR data of the synthesized compounds

Compd.	^{1}H NMR (400MHz), δ
1b①	5.91 (s, 1H, Pyrazolyl—H), 7.30 (dd, J = 8.1, 4.6Hz, 1H, Pyrazolyl—H), 8.04 (dd, J = 8.1, 1.4Hz, 1H, Pyrazolyl—H), 8.32 (dd, J = 4.6, 1.4Hz, 1H, Pyrazolyl—H)
2b①	7.58 (dd, J = 8.1, 4.6Hz, 1H, Pyrazolyl—H), 8.03 (dd, J = 8.1, 1.4Hz, 1H, Pyrazolyl—H), 8.60 (dd, J = 4.6, 1.4Hz, 1H, Pyrazolyl—H), 10.07 (s, 1H, —CHO)
3b①	7.57 (dd, J = 8.1, 4.6Hz, 1H, Pyrazolyl—H), 8.01 (d, J = 8.1Hz, 1H, Pyrazolyl—H), 8.61 (d, J = 4.6Hz, 1H, Pyrazolyl—H)
6a②	2.28 (s, 3H, Ar—CH_3), 2.72 (d, J = 4.0Hz, 3H, —$CONHCH_3$), 7.38 (s, 1H, Ar—H), 7.53 (s, 1H, Ar—H), 7.67 - 7.68 (m, 5H, Ar—H), 8.35 (q, J = 4.0Hz, 1H, Ar—CONH—), 10.21 (s, 1H, Ar—NHCO—)
6b②	1.08 (t, J = 7.2Hz, 3H, —CH_2CH_3), 2.26 (s, 3H, Ar—CH_3), 3.17 - 3.24 (m, 2H, —CH_2CH_3), 7.37 (s, 1H, Ar—H), 7.52 (s, 1H, Ar—H), 7.66 - 7.67 (m, 5H, Ar—H), 8.43 (t, J = 4.8Hz, 1H, Ar—CONH—), 10.16 (s, 1H, Ar—NHCO—)
6c①	0.96 (t, J = 7.4Hz, 3H, —$CH_2CH_2CH_3$), 1.58 - 1.64 (m, 2H, —$CH_2CH_2CH_3$), 2.34 (s, 3H, Ar—CH_3), 3.32 - 3.37 (m, 2H, —$CH_2CH_2CH_3$), 6.11 (t, J = 5.4Hz, 1H, Ar—CONH—), 7.29 (d, J = 2.0Hz, 1H, Ar—H), 7.38 (d, J = 2.0Hz, Ar—H), 7.56 (m, 5H, Ar—H), 9.51 (s, 1H, Ar—NHCO—)

Compd.	^1H NMR (400MHz), δ
6d①	1.24 [d, J=6.5Hz, 6H, —CH (CH_3)$_2$], 2.34 (s, 3H, Ar—CH_3), 4.13 -4.20 [m, 1H, —CH (CH_3)$_2$], 5.91 (d, J=7.6Hz, 1H, Ar—CONH—), 7.27 (d, J=2.0Hz, 1H, Ar—H), 7.38 (d, J=2.0Hz, 1H, Ar—H), 7.56 (m, 5H, Ar—H), 9.55 (s, 1H, Ar—NHCO—)
6e②	0.53 -0.54 [m, 2H, —CH (CH_2)$_2$], 0.62 -0.67 [m, 2H, —CH (CH_2)$_2$], 2.26 (s, 3H, Ar—CH_3), 2.74 -2.79 [m, 1H, —CH (CH_2)$_2$], 7.33 (d, J=2.0Hz, 1H, Ar—H), 7.51 (d, J=2.0Hz, 1H, Ar—H), 7.65 -7.67 (m, 5H, Ar—H), 8.47 (d, J=4.0Hz, 1H, Ar—CONH—), 10.17 (s, 1H, Ar—NHCO—)
6f②	0.85 (t, J=7.2Hz, 3H, —$CH_2CH_2CH_2CH_3$), 1.27 -1.32 (m, 2H, —$CH_2CH_2CH_2CH_3$), 1.42 -1.47 (m, 2H, —$CH_2CH_2CH_2CH_3$), 2.26 (s, 3H, Ar—CH_3), 3.15 - 3.20 (m, 2H, —$CH_2CH_2CH_2CH_3$), 7.36 (d, J=2.0Hz, 1H, Ar—H), 7.53 (d, J=2.0Hz, 1H, Ar—H), 7.65 -7.66 (m, 5H, Ar—H), 8.43 (d, J=5.4Hz, 1H, Ar—CONH—), 10.12 (s, 1H, Ar—NHCO—)
6g①	1.42 [s, 9H, —C (CH_3)$_3$], 2.33 (s, 3H, Ar—CH_3), 5.89 (s, 1H, Ar—CONH—), 7.25 (d, J=2.0Hz, 1H, Ar—H), 7.36 (d, J=2.0Hz, 1H, Ar—H), 7.56 (m, 5H, Ar—H), 9.34 (s, 1H, Ar—NHCO—)
6h①	1.18 -1.27 [m, 2H, —CH (CH_2)$_5$], 1.34 -1.43 [m, 2H, —CH (CH_2)$_5$], 1.62 -1.66 [m, 2H, —CH (CH_2)$_5$], 1.73 - 1.76 [m, 2H, —CH (CH_2)$_5$], 1.95 - 2.02 [m, 2H, —CH (CH_2)$_5$], 3.82 -3.90 [m, 1H, —CH (CH_2)$_5$], 5.95 (d, J=8.0Hz, 1H, Ar—CONH—), 7.38 (d, J=2.0Hz, 1H, Ar—H), 7.53 (d, J=2.0Hz, 1H, Ar—H), 7.56 (m, 5H, Ar—H), 9.60 (s, 1H, Ar—NHCO—)
6i①	2.33 (s, 3H, Ar—CH_3), 2.95 (d, J=4.8Hz, 3H, —$CONHCH_3$), 6.20 (q, J=2.0Hz, 1H, Ar—CONH—), 7.30 (s, 1H, Ar—H), 7.37 (s, 1H, Ar—H), 7.56 (dd, J=8.1, 4.6Hz, 1H, Pyridyl—H), 8.02 (d, J=8.1Hz, 1H, Pyridyl—H), 8.60 (d, J=4.6Hz, 1H, Pyridyl—H), 9.51 (s, 1H, Ar—NHCO—)
6j①	1.22 (t, J=7.2Hz, 3H, —CH_2CH_3), 2.33 (s, 3H, Ar—CH_3), 3.42 (m, 2H, —CH_2CH_3), 6.11 (t, J=4.8Hz, 1H, Ar—CONH—), 7.29 (d, J=2.0Hz, 1H, Ar—H), 7.37 (d, J=2.0Hz, 1H, Ar—H), 7.56 (dd, J=8.1, 4.6Hz, 1H, Pyridyl—H), 8.02 (d, J=8.1Hz, 1H, Pyridyl—H), 8.60 (d, J=4.6Hz, 1H, Pyridyl—H), 9.53 (s, 1H, Ar—NHCO—)
6k②	0.85 (t, J=7.2Hz, 3H, —$CH_2CH_2CH_3$), 1.48 (m, 2H, —$CH_2CH_2CH_3$), 2.27 (s, 3H, Ar—CH_3), 3.13 (m, 2H, —$CH_2CH_2CH_3$), 7.35 (d, J=2.0Hz, 1H, Ar—H), 7.52 (d, J=2.0Hz, 1H, Ar—H), 7.87 (dd, J=8.1, 4.6Hz, 1H, Pyridyl—H), 8.38 (t, J=5.6Hz, 1H, Ar—CONH—), 8.45 (dd, J=8.1, 1.4Hz, 1H, Pyridyl—H), 8.73 (dd, J=4.6, 1.4Hz, 1H, Pyridyl—H), 10.30 (s, 1H, Ar—NHCO—)
6l①	1.22 [d, J=6.6Hz, 6H, —CH (CH_3)$_2$], 2.33 (s, 3H, Ar—CH_3), 4.18 [m, 1H, —CH (CH_3)$_2$], 5.91 (d, J=7.5Hz, 1H, Ar—CONH—), 7.26 (d, J=2.0Hz, 1H, Ar—H), 7.37 (d, J=2.0Hz, 1H, Ar—H), 7.56 (dd, J=8.1, 4.6Hz, 1H, Pyridyl—H), 8.02 (dd, J=8.1, 1.4Hz, 1H, Pyridyl—H), 8.60 (dd, J=4.6, 1.4Hz, 1H, Pyridyl—H), 9.54 (s, 1H, Ar—NHCO—)

续表2

Compd.	^{1}H NMR (400MHz), δ
6m①	0.61 [m, 2H, —CH (CH_2)$_2$], 0.86 [m, 2H, —CH (CH_2)$_2$], 2.33 (s, 3H, Ar—CH_3), 2.82 [m, 1H, —CH (CH_2)$_2$], 6.27 (d, J=4.8Hz, 1H, Ar—CONH—), 7.24 (d, J=2.0Hz, 1H, Ar—H), 7.37 (d, J=2.0Hz, 1H, Ar—H), 7.56 (dd, J=8.1, 4.6Hz, 1H, Pyridyl—H), 8.01 (d, J=8.1Hz, 1H, Pyridyl—H), 8.60 (d, J=4.6Hz, 1H, Pyridyl—H), 9.54 (s, 1H, Ar—NHCO—)
6n①	0.93 (t, J=7.2Hz, 3H, —$CH_2CH_2CH_2CH_3$), 1.37 (m, 2H, —$CH_2CH_2CH_2CH_3$), 1.56 (m, 2H, —$CH_2CH_2CH_2CH_3$), 2.33 (s, 3H, Ar—CH_3), 3.37 (m, 2H, —$CH_2CH_2CH_2CH_3$), 6.14 (t, J=5.2Hz, 1H, Ar—CONH—), 7.27 (d, J=2.0Hz, 1H, Ar—H), 7.36 (d, J=2.0Hz, 1H, Ar—H), 7.56 (dd, J=8.1, 4.6Hz, 1H, Pyridyl—H), 8.02 (d, J=8.1Hz, 1H, Pyridyl—H), 8.60 (d, J=4.6Hz, 1H, Pyridyl—H), 9.55 (s, 1H, Ar—NHCO—)
6o①	1.40 [s, 9H, —CH (CH_3)$_3$], 2.33 (s, 3H, Ar—CH_3), 5.89 (s, 1H, Ar—CONH—), 7.24 (d, J=2.0Hz, 1H, Ar—H), 7.35 (d, J=2.0Hz, 1H, Ar—H), 7.56 (dd, J=8.1, 4.6Hz, 1H, Pyridyl—H), 8.02 (dd, J=8.1, 1.4Hz, 1H, Pyridyl—H), 8.61 (dd, J=4.6, 1.4Hz, 1H, Pyridyl—H), 9.31 (s, 1H, Ar—NHCO—)
6p①	1.21-1.97 [m, 10H, —CH (CH_2)$_5$], 2.33 (s, 3H, Ar—CH_3), 3.87 [m, 1H, —CH (CH_2)$_5$], 5.93 (d, J=8.0Hz, 1H, Ar—CONH—), 7.26 (d, J=2.0Hz, 1H, Ar—H), 7.37 (d, J=2.0Hz, 1H, Ar—H), 7.56 (dd, J=8.1, 4.6Hz, 1H, Pyridyl—H), 8.02 (d, J=8.1Hz, 1H, Pyridyl—H), 8.60 (dd, J=4.6Hz, 1H, Pyridyl—H), 9.57 (s, 1H, Ar—NHCO—)
6q②	2.25 (s, 3H, Ar—CH_3), 2.70 (d, J=4.5Hz, 3H, —$CONHCH_3$), 7.37 (d, J=2.0Hz, 1H, Ar—H), 7.52 (d, J=2.0Hz, 1H, Ar—H), 7.76 (dd, J=8.5, 2.2Hz, 1H, Ar—H), 7.87 (d, J=8.5Hz, 1H, Ar—H), 8.09 (d, J=2.2Hz, 1H, Ar—H), 8.35 (q, J=4.5Hz, 1H, Ar—CONH—), 10.28 (s, 1H, Ar—NHCO—)
6r②	1.07 (t, J=7.2Hz, 3H, —CH_2CH_3), 2.26 (s, 3H, Ar—CH_3), 3.20 (m, 2H, —CH_2CH_3), 7.37 (d, J=2.0Hz, 1H, Ar—H), 7.52 (d, J=2.0Hz, 1H, Ar—H), 7.76 (dd, J=8.5, 2.2Hz, 1H, Ar—H), 7.88 (d, J=8.5Hz, 1H, Ar—H), 8.09 (d, J=2.2Hz, 1H, Ar—H), 8.42 (t, J=4.5Hz, 1H, Ar—CONH—), 10.24 (s, 1H, Ar—NHCO—)
6s②	0.85 (t, J=7.4Hz, 3H, —$CH_2CH_2CH_3$), 1.48 (m, 2H, —$CH_2CH_2CH_3$), 2.26 (s, 3H, Ar—CH_3), 3.13 (m, 2H, —$CH_2CH_2CH_3$), 7.36 (d, J=2.0Hz, 1H, Ar—H), 7.53 (d, J=2.0Hz, 1H, Ar—H), 7.76 (dd, J=8.5, 2.2Hz, 1H, Ar—H), 7.88 (d, J=8.5Hz, 1H, Ar—H), 8.10 (d, J=2.2Hz, 1H, Ar—H), 8.42 (t, J=5.4Hz, 1H, Ar—CONH—), 10.18 (s, 1H, Ar—NHCO—)
6t②	1.10 [d, J=6.6Hz, 6H, —CH (CH_3)$_2$], 2.25 (s, 3H, Ar—CH_3), 3.98 [m, 1H, —CH (CH_3)$_2$], 7.34 (d, J=2.0Hz, 1H, Ar—H), 7.51 (d, J=2.0Hz, 1H, Ar—H), 7.76 (dd, J=8.5, 2.2Hz, 1H, Ar—H), 7.88 (d, J=8.5Hz, 1H, Ar—H), 8.08 (d, J=2.2Hz, 1H, Ar—H), 8.29 (d, J=7.8Hz, 1H, Ar—CONH—), 10.15 (s, 1H, Ar—NHCO—)
6u②	0.52 [m, 2H, —CH (CH_2)$_2$], 0.64 [m, 2H, —CH (CH_2)$_2$], 2.25 (s, 3H, Ar—CH_3), 2.76 [m, 1H, —CH (CH_2)$_2$], 7.33 (d, J=2.0Hz, 1H, Ar—H), 7.51 (d, J=2.0Hz, 1H, Ar—H), 7.77 (dd, J=8.5, 2.2Hz, 1H, Ar—H), 7.88 (d, J=8.5Hz, 1H, Ar—H), 8.10 (d, J=2.2Hz, 1H, Ar—H), 8.45 (d, J=4.1Hz, 1H, Ar—CONH—), 10.23 (s, 1H, Ar—NHCO—)

Compd.	^{1}H NMR (400MHz), δ
6v②	0.85 (t, J=7.6Hz, 3H, —$CH_2CH_2CH_2CH_3$), 1.29 (m, 2H, —$CH_2CH_2CH_2CH_3$), 1.45 (m, 2H, —$CH_2CH_2CH_2CH_3$), 2.26 (s, 3H, Ar—CH_3), 3.17 (m, 2H, —$CH_2CH_2CH_2CH_3$), 7.36 (d, J=2.0Hz, 1H, Ar—H), 7.52 (d, J=2.0Hz, 1H, Ar—H), 7.77 (dd, J=8.5, 2.2Hz, 1H, Ar—H), 7.88 (d, J=8.5Hz, 1H, Ar—H), 8.10 (d, J=2.2Hz, 1H, Ar—H), 8.40 (t, J=4.8Hz, 1H, Ar—CONH—), 10.17 (s, 1H, Ar—NHCO—)
6w②	1.31 [s, 9H, —C (CH_3)$_3$], 2.25 (s, 3H, Ar—CH_3), 7.33 (d, J=2.0Hz, 1H, Ar—H), 7.50 (d, J=2.0Hz, 1H, Ar—H), 7.77 (dd, J=8.5, 2.2Hz, 1H, Ar—H), 7.89 (d, J=8.5Hz, 1H, Ar—H), 7.98 (s, 1H, Ar—CONH—), 8.09 (d, J=2.2Hz, 1H, Ar—H), 10.00 (s, 1H, Ar—NHCO—)
6x②	1.11-1.29 [m, 4H, —CH (CH_2)$_5$], 1.52-1.60 [m, 2H, —CH (CH_2)$_5$], 1.64-1.71 [m, 2H, —CH (CH_2)$_5$], 1.74-1.81 [m, 2H, —CH (CH_2)$_5$], 2.25 (s, 3H, Ar—CH_3), 3.65 [m, 1H, —CH (CH_2)$_5$], 7.33 (d, J=2.0Hz, 1H, Ar—H), 7.52 (d, J=2.0Hz, 1H, Ar—H), 7.77 (dd, J=8.5, 2.2Hz, 1H, Ar—H), 7.89 (d, J=8.5Hz, 1H, Ar—H), 8.09 (d, J=2.2Hz, 1H, Ar—H), 8.28 (d, J=7.1Hz, 1H, Ar—CONH—), 10.13 (s, 1H, Ar—NHCO—)

①$CDCl_3$；②DMSO-d_6。

2.3 目标化合物的生物活性

生物活性测定结果（表3）表明，大部分目标化合物在200mg/L浓度下对东方粘虫具有一定的杀虫活性，其中化合物**6m**和**6o**相对较好，在200和100mg/L浓度时杀虫活性均为100%，在50mg/L时杀虫活性分别为80%和50%，但不及对照药氯虫酰胺。通过数据分析得出：（1）当目标产物中R^1为3-Cl-Py-2-yl时，活性较好，活性顺序为R^2=cycloproyl>$t-C_4H_9$>$i-C_3H_7$，CH_3>其它取代基，说明在目标化合物结构中R^2具有较大空间位阻有利于保持活性，同时也说明先导氯虫酰胺中吡啶基是与受体结合的重要基团，此取代基的存在有助于保持活性。（2）当将吡唑上的氨基甲酰基与吡啶基的相对位置调至间位，并在吡唑的5位上引入氯原子之后，活性下降，这可能是由于氯原子的空间位阻影响了吡唑环与所连接芳环之间的二面角大小，导致与受体结合受阻，说明在吡唑环上氨基甲酰基与吡啶基处于邻位是活性保持的重要条件。

Table 3 Biological activities of the synthesized compounds①

Compd.	Insecticidal activity (%)			Antifungal activity (%)				
	Mythimna separate			*Physalospora piricola*	*Alternaria solani*	*Cercospora arachidicola*	*Gibberella sanbinettim*	*Phytophthora capsici*
	200mg/L	100mg/L	50mg/L					
6a	10	NT	NT	15.9	21.1	34.6	16.2	32.1
6b	20	NT	NT	17.8	15.8	19.2	13.5	28.6
6c	10	NT	NT	47.3	31.6	15.4	16.2	32.1
6d	30	NT	NT	31.6	15.8	19.2	21.6	28.6
6e	10	NT	NT	49.2	15.8	15.4	16.2	28.6

续表3

Compd.	Insecticidal activity (%)			Antifungal activity (%)				
	Mythimna separate			*Physalospora piricola*	*Alternaria solani*	*Cercospora arachidicola*	*Gibberella sanbinettim*	*Phytophthora capsici*
	200mg/L	100mg/L	50mg/L					
6f	20	NT	NT	39.6	15.8	15.4	21.6	28.6
6g	0	NT	NT	19.8	26.3	19.2	21.6	35.7
6h	20	NT	NT	10.0	21.1	15.4	18.9	32.1
6i	100	20	NT	39.4	26.3	7.7	21.6	32.1
6j	100	20	NT	57.1	26.3	11.5	16.2	28.6
6k	23.3	NT	NT	56.5	26.3	15.6	16.2	32.1
6l	60	NT	NT	47.5	26.3	15.4	16.2	32.1
6m	100	100	80	51.2	31.6	19.2	21.6	35.7
6n	30	NT	NT	62.1	26.3	30.8	16.2	32.1
6o	100	100	50	55.1	31.6	11.5	16.2	32.1
6p	20	NT	NT	10.0	15.8	3.8	16.2	28.6
6q	20	NT	NT	45.3	10.5	11.5	13.5	28.6
6r	10	NT	NT	39.4	15.8	26.9	16.2	32.1
6s	10	NT	NT	39.4	26.3	30.8	24.3	28.6
6t	30	NT	NT	29.6	21.1	3.8	18.9	25.0
6u	10	NT	NT	45.3	15.8	11.5	18.9	32.1
6v	10	NT	NT	53.1	15.8	11.5	16.2	32.1
6w	10	NT	NT	49.2	26.3	15.4	16.2	32.1
6x	60	NT	NT	62.1	15.8	11.5	16.2	28.6
Chlorantraniliprole	100	100	100	NT	NT	NT	NT	NT
Chlorothalonil	NT	NT	NT	88.5	63.6	58.8	73.1	82.6

①NT：Not tested。

目标化合物的抑菌活性测试结果（表3）表明，在50mg/L浓度下，目标化合物对5种测试病菌均表现出一定的抑菌活性，尤其对于苹果轮纹病菌的抑制性普遍较好。其中，化合物**6**n和**6**x对苹果轮纹病菌的抑菌率均为62.1%；化合物**6**c，**6**m和**6**o对番茄早疫病菌的抑菌率为31.6%；化合物**6**a对花生褐斑病菌的抑菌率为34.6%，但与对照药百菌清（*Chlorothalonil*）的抑菌活性相比有一定差距。

3 结论

设计合成了24个未见文献报道的*N*－［4－氯－2－取代氨基甲酰基－6－甲基苯基］－1－芳基－5－氯－3－三氟甲基－1*H*－吡唑－4－甲酰胺类化合物，由于吡唑环上氨基甲酰基位置的改变以及氯原子的引入导致生物活性不够理想，但个别保留氯代吡啶环的化合物对东方粘虫仍表现出中等活性，这可能是由于*N*－吡啶基吡唑基团的存在以及吡唑环与吡啶环之间的特定二面角的保持是化合物与靶标充分结合的关键因素，不宜改变。本文结果对今后鱼尼丁类杀虫剂的开发具有一定的借鉴意义。

参考文献

[1] Lahm G. P., Myers B. J., Selby T. P., Stevenson T. M., *Preparation of Insecticidal Anthranilamides*, *WO* 2001070671, 2001-09-27.

[2] Lahm G. P., Steven son T. M., Selby T. P., Freudenberger J. H., Cordova D., Flexner L., Bell in C. A., Dubas C. M., Smith B. K., Hughes K. A., *Bioorg. Med. Chem. Lett.*, 2007, 17 (22), 6274-6279.

[3] Lahm G. P., Cordova D., Barry J. D., *Bioorg. Med. Chem.*, 2009, 17 (12), 4127-4133.

[4] Tohnishi M., Nakao H., Kohno E., Nishida T., Furuya T., Shimizu T., Seo A., Sakata K., Fujioka S., Kanno H., *Preparation of Phthalic Acid Diamides as Agricultural and Horticultural Insecticides*, *EP* 919542, 1999-06-02.

[5] Lahm G. P., Thomas P. S., John H. F., Thomas M. S., Brian J. M., Gilles S., Smith K. B., Lindsey F., Christopher E. C., Cordova D., *Bioorg. Med. Chem. Lett.*, 2005, 15 (22), 4898-4906.

[6] Pierluigi C., Giorgia S., Alberto A., Simona V., Daniela P., Paolo C., John E. C., *J. Agric. Food Chem.*, 2008, 56 (17), 7696-7699.

[7] Feng Q., Liu Z. L., Wang M. Z., Xiong L. X., Yu S. J., Li Z. M., *Chem. J. Chinese Universities*, 2011, 32 (1), 74-78（冯启，刘智力，王明忠，熊丽霞，于淑晶，李正名. 高等学校化学学报，2011，32（1），74-78）.

[8] Yan T., Yu G. P., Xiong L. X., Yu S. J., Wang S. H., Li Z. M., *Chem. J. Chinese Universities*, 2011, 32 (8), 1750-1754.（闫涛，于观平，熊丽霞，于淑晶，王素华，李正名. 高等学校化学学报，2011，32（8），1750-1754）.

[9] Xue Y. L., Liu X. H., Zhang Y. G., *Asian J. Chem.*, 2012, 24, 1571-1574.

[10] Gilbert A. M., Bursavich M. G., Lombardi S., Georgiadis K. E., Reifenberg E., Flannery C. R., Morris E. A., *Bioorg. Med. Chem. Lett.*, 2007, 17 (5), 1189-1192.

[11] Mao C. H., Lu Y. F., Huang M. Z., Huang L., Chen C., Ou X. M., *Chin. J. Syn. Chem.*, 2002, 10 (2), 167-169, 171.（毛春晖，卢艳芬，黄明智，黄路，陈灿，欧晓明. 合成化学，2002，10（2），167-169，171）.

[12] Datterl B., Troestner N., Kucharski Do., Holzer W., *Molecules*, 2010, 15 (9), 6106-6126.

[13] Zhou Z. L., Kher S. M., Cai S. X., Whittemore E. R., Espitia S. A., Hawkinson J. E., Tran M., Woodward R. M., Weber E., Keana J. F. W., *Bioorg. Med. Chem.*, 2003, 11 (8), 1769-1780.

[14] Liu W. D., Li J. S., Li Z. Y., Wang X. G., Gao B. D., *Chin. J. Pestic. Sci.*, 2004, 6 (1), 17-21（刘卫东，李江胜，李仲英，王晓光，高必达. 农药学学报，2004，6（1），17-21）.

[15] LiuX. H., Tan C. X., Weng J. Q., Liu H. J., *Acta Cryst.*, 2012, E68, o493.

[16] Tan C. X., Weng J. Q., Liu Z. X., Liu X. H., Zhao W. G., *Phosphorus Sulfur Silicon Relat. Elem.*, 2011, 187, 990-996.

[17] Xue Y. L., Zhang Y. G., Liu X. H., *Asian J. Chem.*, 2012, 24, 3016-3018.

[18] Xu J. Y., Dong W. L., Xiong L. X., Li Z. M., *Chem. J. Chinese Universities*, 2012, 33 (2), 298-302.（徐俊英，董卫莉，熊丽霞，李正名. 高等学校化学学报，2012，33（2），298-302）.

[19] Liu X. H., Tan C. X., Weng J. Q., *Phosphorus Sulfur Silicon Relat. Elem.*, 2011, 186, 552-557.

[20] Liu X. H., Tan C. X., Weng J. Q., *Asian J. Chem.*, 2011, 23, 4064-4066.

[21] Liu X. H., Tan C. X., Weng J. Q., *Phosphorus Sulfur Silicon Relat. Elem.*, 2011, 186, 558-564.

Synthesis and Insecticidal Activity of 5 – Chloro – *N* – [4 – chloro – 2 – (substitutedcarbamoyl) – 6 – methylphenyl] – 1 – aryl – 3 – (trifluoromethyl) – 1*H* – pyrazole – 4 – carboxamide

Xiulan Zhang, Baolei Wang , Mingzhen Mao, Lixia Xiong, Shujing Yu, Zhengming Li

(State Key Laboratory of Elemento – Organic Chemistry, Institute of Elemento – Organic Chemistry, Nankai University, Tianjin, 300071, China)

Abstract In the 21st century, one of the most serious challenges to mankind is food which plays a critical role to their survival and development. Seeking novel pesticides with new mechanism is an important approach in plant protection science. Anthranilic diamides recently appeared are of new type of potent insecticides which acts at ryanodine receptor of target insects. DuPont' s Chlorantraniliprole has become one of the focuses in the innovation of agrochemicals due to its high efficiency, low toxicity, broader insecticidal spectra and significant ovicidal activity. Since it came to the world market from 2008, there have been numerous patents relating to new structural optimization reports. Most of them retained the *N* – pyridylpyrazole moiety, and kept carbamoyl in the *ortho* – position with pyridyl moiety in pyrazole ring. It was postulated that the activity could be improved by changing the distance between carbamoyl and pyridyl in pyrazole ring or displacing the pyridyl with other aryls. 24 novel anthranilic diamides containing 1 – aryl – 5 – chloro – 3 – (trifluoromethyl) – 1*H* – pyrazole moieties were designed and synthesized. Their structures were characterized by ^1H NMR and elemental analysis or HRMS. The preliminary bioassay results indicated that all compounds exhibit moderate insecticidal activity against *Mythimna separate* at 200mg /L and some of them exhibited good insecticidal activity at a concentration of 50mg /L; most compounds exhibited unexpectedly fungicidal activity against five funguses at a concentration of 50mg /L.

Key words Vilsmeier reaction; Chlorantraniliprole; anthranilic diamides; insecticidal activity; fungicidal activity

Synthesis, Insecticidal Activities, and SAR Studies of Novel Pyridylpyrazole Acid Derivatives Based on Amide Bridge Modification of Anthranilic Diamide Insecticides*

Baolei Wang, Hongwei Zhu, Yi Ma, Lixia Xiong, Yongqiang Li, Yu Zhao, Jifeng Zhang, Youwei Chen, Sha Zhou, Zhengming Li

(State - Key Laboratory of Elemento - Organic Chemistry, Institute of Elemento - Organic Chemistry, Nankai University, Tianjin, 300071, China)

Abstract Anthranilic diamides are one of the most important classes of modern agricultural insecticides. To discover new structure - modified compounds with high activity, series of novel carbonyl thioureas, carbonyl ureas, oxadiazoles, carbonyl thiophosphorylureas, oxadiazole - containing amides, and thiazoline - containing amides were designed through the modification of the amide bridge based on the structure of chlorantraniliprole and were synthesized, and bioassays were carried out. The compounds were characterized and confirmed by melting point, IR, ^{1}H NMR, and elemental analyses or HRMS. Preliminary bioassays indicated that some compounds exhibited significant insecticidal activities against oriental armyworm, diamondback moth, beet armyworm, corn borer, and mosquito. Among them, trifluoroethoxyl - containing carbonyl thiourea **20a** showed best larvicidal activity against oriental armyworm, with LC_{50} and LC_{95} values of 0. 1812 and 0. 7767mg/L, respectively. Meanwhile, **20c** and **20e** showed 86 and 57% death rates against diamondback moth at 0. 005mg/L, and the LC_{50} values of the two compounds were 0. 0017 and 0. 0023mg/L, respectively, which were lower than that of the control chlorantraniliprole. The relationship between structure and insecticidal activity was discussed, and the HF calculation results indicated that the carbonyl thiourea moiety plays an important role in the insecticidal activity. The present work demonstrated that the trifluoroethoxylcontaining carbonyl thioureas can be used as lead compounds for further development of novel insecticides.

Key words pyridylpyrazole acid derivative; synthesis; insecticidal activity

INTRODUCTION

Synthetic pesticides play important roles for controlling pests harmful to crop growth in the current agricultural system. To overcome resistance and ecobiological problems associated with conventional insecticides, there is an urgent need to discover novel potent insecticides with new modes of action and ecofriendly properties such as easy degradability to nontoxic residues, harmlessness to human beings, and benefits in meeting the demands of crop protection.

The ryanodine receptor (RyR), also known as the calcium ion channel receptor, has been regarded as one of the potential targets for novel insecticide discovery since it was found that natural ryanodine possessed insecticidal activity.[1-3] In the past decade, Nippon Kayaku Co.,

* Reprinted from *Journal of Agricultural and Food Chemistry*, 2013, 61: 5483 - 5493. This work was supported by the National Basic Research Program of China (No. 2010CB126106), the National Key Technologies Research and Development Program (2011BAE06B05 - 3), the National Natural Science Foundation of China (No. 20872069), and the Tianjin Natural Science Foundation (No. 11JCYBJC04500).

Ltd., found the phthalic diamide insecticide flubendiamide targeting insect RyR.[4] Also, DuPont discovered the anthranilic diamides,[5] which originated from the insecticidal phthalic diamides as highly potent and selective activators of the insect RyR.[6] Until now, two representive phthalic diamides, chlorantraniliprole (Rynaxypyr; DPX – E2Y45) and cyantraniliprole (Cyazypyr) (Figure 1) have been marketed.[7,8] These compounds show exceptional insecticidal activity on a broad range of Lepidoptera, Coleoptera, Diptera, and Isoptera insects.[6] In addition to larvicidal activity, chlorantraniliprole has been found to have significant ovicidal activity among some Lepidopteran pests.[9]

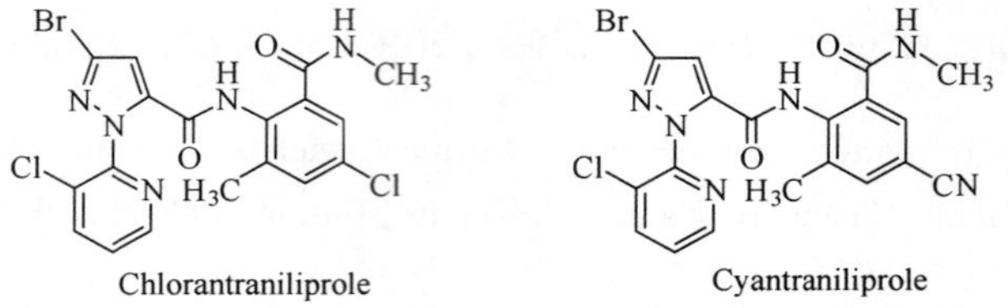

Chlorantraniliprole   Cyantraniliprole

Figure 1 Chemical structures of anthranilic diamide insecticides

Since the discovery of phthalic diamides and anthranilic diamides, compounds with such structural features and their chemical synthesis have attracted considerable attention.[10,11] There are two amide moieties in the structures of flubendiamide and chlorantraniliprole, respectively. In previous work, most modifications for chlorantraniliprole were related to the benzene moiety and the *N* – pyridylpyrazole heterocycle moiety. The most successful example is cyantraniliprole (Cyazypyr),[8] which replaced a cyano group of the 4 – Cl substituent of the former chlorantraniliprole. However, the modification for the two amide moieties was relatively seldom reported, especially the alteration of the amide bridge, a key pharmacophore that links the benzene ring and the *N* – pyridylpyrazole heterocycle, was not much reported in previous patents. It is known that acylurea or acylthiourea compounds, representing an important class of biologically active molecules, such as benzoylphenylureas (e. g., diflubenzuron), are a familiar type of insect growth regulators (IGRs), which have attracted considerable attention for decades because of their unique mode of action and low toxicity to nontarget organisms (beneficial arthropods).[12-14] We also found some *N* – pyridylpyrazole acylthioureas showed favorable insecticidal activities in our preliminary research, which encouraged us to further synthesize novel compounds with bridge – modified structure.[15] Heterocycles such as oxadiazole, thiazoline, and other heterocycles are important pharmacophores for insecticidal molecular design. Given these considerations, series of novel compounds were designed by changing the amide bridge to carbonyl thiourea, carbonyl urea, oxadiazole, carbonyl thiophosphorylurea, and thiazoline amide moieties with a substituted *N* – pyridylpyrazole ring based on the structure of chlorantraniliprole and were synthesized successfully as shown in Schemes 1 – 9. Their insecticidal activities against oriental armyworm, diamondback moth, beet armyworm, corn borer, and mosquito were tested accordingly. The preliminary structure – activity relationship (SAR) was also discussed.

MATERIALS AND METHODS

Instruments and Materials. The melting points were determined on an X – 4 binocular micro-

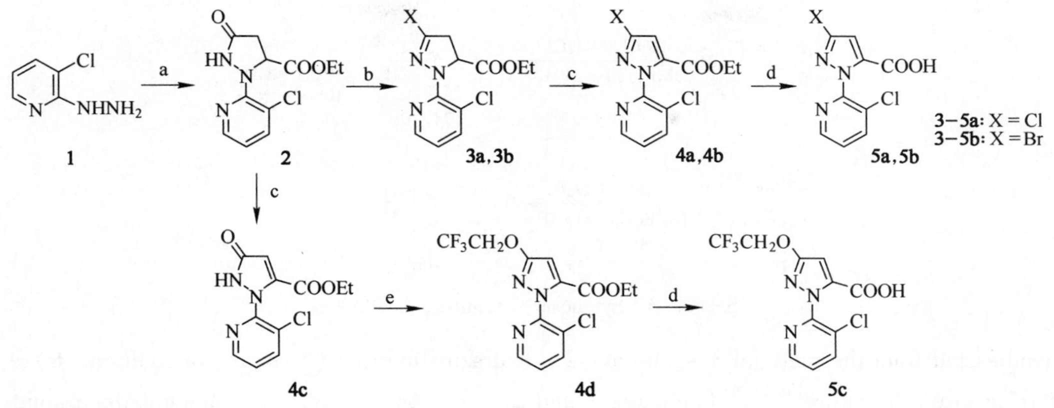

Scheme 1 Synthesis of Compounds **5a – c**①

①Reagents and conditions: (a) diethyl maleate, NaOEt, EtOH, reflux; (b) $POCl_3$ or $POBr_3$, MeCN, **80℃**; (c) $K_2S_2O_8$, H_2SO_4, MeCN, reflux; (d) (i) aqueous NaOH, MeOH and (ii) aqueous HCl; and (e) CF_3CH_2I, K_2CO_3, DMF, **100℃**

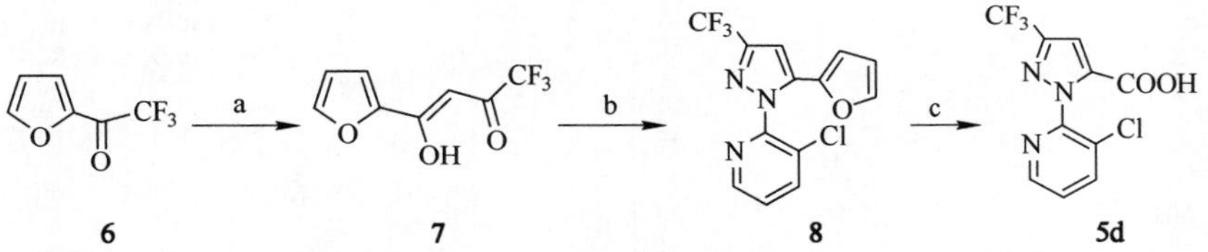

Scheme 2 Synthesis of Compound **5d**①

①Reagents and conditions: (a) $NaOCH_3$, ethyl trifluoroacetate; (b) **3** – chloro – **2** – hydrazinylpyridine (**1**), AcOH, reflux; and (c) $KMnO_4$, acetone/H_2O, reflux

scope melting point apparatus (Beijing Tech Instrument Co., Beijing, China) and were uncorrected. Infrared spectra were recorded on a Nicolet MAGNA – 560 spectrophotometer as KBr tablets. ^1H NMR spectra were recorded at 300MHz using a Bruker ACP300 spectrometer or 400MHz using a Bruker AV 400 spectrometer (Bruker Co., Switzerland) in $CDCl_3$ or DMSO – d_6 solution with tetramethylsilane as the internal standard, and chemical shift values (δ) were given in parts per million (ppm). Elemental analyses were performed on a Vario EL elemental analyzer. High – resolution mass spectrometry (HRMS) data were obtained on a Varian QFT – ESI instrument.

Flash chromatography was performed with silica gel (200 – 300mesh). Reagents were all analytically pure. All solvents and liquid reagents were dried by standard methods in advance and distilled before use. 5 – Methyl – 1*H* – tetrazole was purchased from Adamas Reagent Co., Ltd. 2 – Aminoethanol and 1 – amino – 2 – propanol were purchased from Aladdin Reagent Database Inc. Commercial insecticides chlorantraniliprole and diflubenzuron were used only as contrast compounds and synthesized according to the literature. [16,17]

Synthetic Procedures. 4 – Amino – 5 – methyl – 4*H* – 1,2,4 – triazole – 3 – thiol was prepared according to the literature. [18] The key intermediate pyrazole carboxylic acid (**5a – d**) was

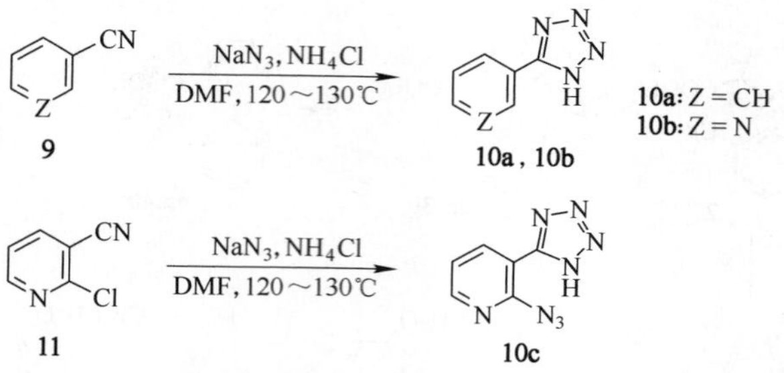

Scheme 3 Synthesis of Compounds **10a－c**

synthesized from the material 3－chloro－2－hydrazinylpyridine(**1**) or 2－acetylfuran(**6**) referring to the literature[15,19,20] (Schemes 1 and 2). 2－Amino－3,5－disubstitutedbenzamide derivatives(**14a－g**) were synthesized with 2－amino－3,5－disubstituted－benzoic acids (**12**) as materials in moderate yield according to the method reported by Dong et al.[21] (Scheme 4).

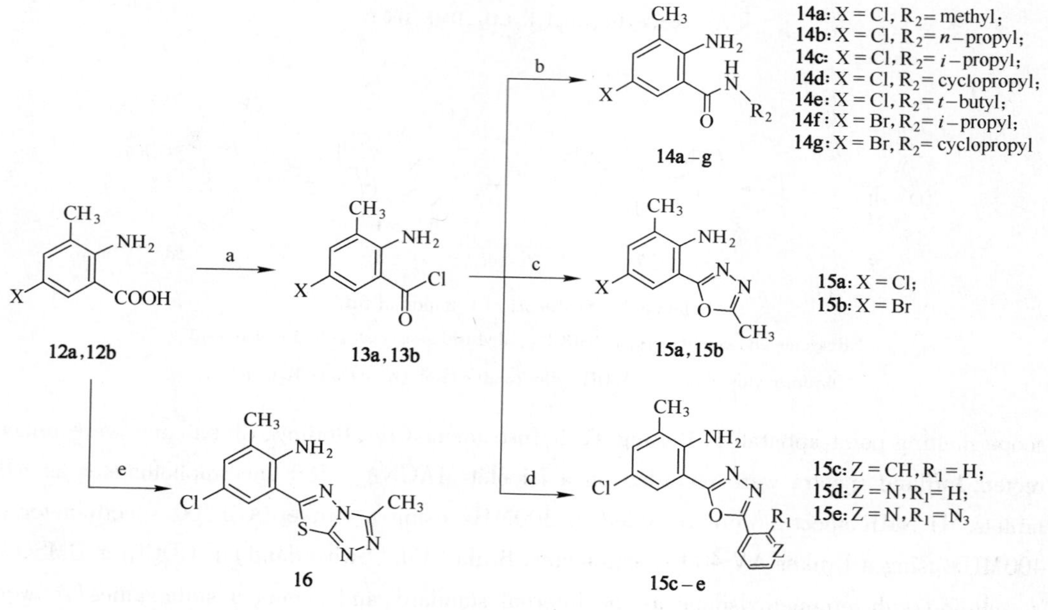

Scheme 4 Synthesis of Compounds **14a－g**, **15a－e** and **16**①

①Reagents and conditions: (a) $SOCl_2$, reflux; (b) R_2NH_2, CH_2Cl_2; (c) **5－methyl－1*H*－tetrazole**, Py, **100－110℃**; (d) compounds **10a－c**, Py, 1,4－dioxane, 100－110℃; and (e) 4－amino－5－methyl－4*H*－1,2,4－triazole－3－thiol, $POCl_3$, reflux

Synthesis of Intermediate Tetrazole Derivatives(10a－c). [22] To a solution of benzonitrile (20mmol) in *N*, *N*－dimethylformamide (DMF)(25mL) was added ammonium chloride (40mmol) and sodium azide(40mmol), and the resultant slurry was vigorously stirred at 120℃ for 2－3h. After cooling to room temperature, the mixture was filtered, and the solid was washed with DMF. The filtrates were combined and the DMF was removed in vacuo. To the residue was added water(30mL), the mixture was adjusted to pH 2－3 with concentrated HCl, and the solid was collected by filtration and washed with water and acetone successively to afford 5－phenyl－

1*H* – tetrazole(**10a**)as white crystals:yield 68% ,mp 213 – 215℃(lit.[23] mp 215 – 216℃).

3 – Cyanopyridine was used as material to afford 3 – (1*H* – tetrazol – 5 – yl)pyridine(**10b**)as white crystals:yield 54% ,mp 240 – 242℃(lit.[22] mp 239 – 241 ℃).

2 – Chloro – 3 – cyanopyridine was used as material to afford 2 – azido – 3 – (1*H* – tetrazol – 5 – yl)pyridine(**10c**)as a light yellow solid:yield 52% ;mp 245 – 247℃;^{1}H NMR($CDCl_3$, 400MHz)δ 7.68(t,J = 7.2Hz,1H,Py – H),8.60(d,J = 7.2Hz,1H,Py – H),9.55(d,J = 7.2Hz,1H,Py – H). HRMS calcd for $C_6H_4N_8$([M – H]$^-$.),187.0481;found,187.0490.

Synthesis of Oxadiazole – Containing Substituted Aniline(15a,15b). A mixture of 2 – amino – 5 – halo – 3 – methylbenzoic acid(**12**,10mmol)with thionyl chloride($SOCl_2$)(15mL) was refluxed for 3 – 5h and condensed under reduced pressure to give the corresponding crude carbonyl chloride(**13**),which was immediately dissolved in dry toluene(10mL)and added to a solution of 5 – methyl – 1*H* – tetrazole(9.5mmol)in dry pyridine(20mL)with stirring at room temperature. The mixture was then stirred at 100 – 110℃ for 2 – 3h,the solvent was removed in vacuo,the residue was added to a 10% Na_2CO_3 solution(30mL)and stirred for 5 min,and the solid was filtered,washed with water,and recrystallized with 95% EtOH to give compound **15a** or **15b**.

4 – Chloro – 2 – methyl – 6 – (5 – methyl – 1,3,4 – oxadiazol – 2 – yl)aniline(**15a**):brown solid;yield 69% ;mp 150 – 153℃;^{1}H NMR(400MHz,DMSO – d_6)δ 2.19(s,3H,CH_3),2.58 (s,3H,CH_3),6.64(br,0.67H,NH_2),7.24(d,J = 1.6Hz,1H,Ph – H),7.49(d,J = 1.6Hz, 1H,Ph – H).

4 – Bromo – 2 – methyl – 6 – (5 – methyl – 1,3,4 – oxadiazol – 2 – yl)aniline(**15b**):yellow solid;yield 72% ;mp 133 – 136℃;^{1}H NMR(400MHz,$CDCl_3$)δ 2.22(s,3H,CH_3),2.62(s,3H, CH_3),5.91(br,2H,NH_2),7.27(d,J = 2.0Hz,1H,Ph – H),7.75(d,J = 2.0Hz,1H,Ph – H).

Synthesis of Oxadiazole – Containing Substituted Aniline(15c – e). A mixture of 2 – amino – 5 – chloro – 3 – methylbenzoic acid(**12a**,5 mmol)with $SOCl_2$(7 mL)was refluxed for 3 h and condensed under reduced pressure to give a crude carbonyl chloride(**13a**),which was immediately dissolved in 1,4 – dioxane(5mL)and added to a solution of tetrazole derivative (**10a – c**,4 mmol)in dry pyridine(10mL)with stirring. The mixture was further stirred at 100℃ for 4 – 6h,the solvent was removed in vacuo,the residue was added to a 10% Na_2CO_3 solution(20mL)and stirred for 5min,and the solid was filtered,washed with water,and recrystallized with a mixed solvent of EtOH and DMF to give the product.

4 – Chloro – 2 – methyl – 6 – (5 – phenyl – 1,3,4 – oxadiazol – 2 – yl)aniline(**15c**):yellow crystal;yield 51% ;mp 205 – 208℃;^{1}H NMR(400MHz,DMSO – d_6)δ 2.21(s,3H,CH_3),6.75 (br,0.71H,NH_2),7.29(s,1H,Ph – H),7.62 – 7.68(m,3H,Ph – H),7.83(d,J = 2.0Hz,1H, Ph – H),8.19(t,J = 2.0Hz,2H,Ph – H).

4 – Chloro – 2 – methyl – 6 – (5 – (pyridin – 3 – yl) – 1,3,4 – oxadiazol – 2 – yl)aniline (**15d**):yellow crystal;yield 84% ;mp 202 – 205℃;^{1}H NMR(400MHz,DMSO – d_6)δ 2.22(s, 3H,CH_3),6.75(br,0.85H,NH_2),7.30(s,1H,Ph – H),7.67(dd,J_1 = 8.0Hz,J_2 = 4.4Hz, 1H,Py – H),7.89(s,1H,Ph – H),8.57(d,J = 7.6Hz,1H,Py – H),8.83(d,J = 4.4Hz, 1H,Py – H),9.37(s,1H,Py – H).

2 – (5 – (2 – Azidopyridin – 3 – yl) – 1,3,4 – oxadiazol – 2 – yl) – 4 – chloro – 6 – methyla-

niline(**15e**):yellow solid;yield 78%;mp 254 - 256℃;^{1}H NMR(400MHz,DMSO - d_6)δ 2.24 (s,3H,CH_3),6.82(br,2H,NH_2),7.34(d,J=2.0Hz,1H,Ph - H),7.67(t,J=7.2Hz,1H, Py - H),7.85(d,J=2.0Hz,1H,Ph - H),8.79(d,J=7.2Hz,1H,Py - H),9.59(d,J= 7.2Hz,1H, Py - H). HRMS calcd for $C_{14}H_{10}ClN_7O$ ([M + Na]$^+$), 350.0533; found,350.0530.

Synthesis of 4 - Chloro - 2 - methyl - 6 - (3 - methyl - [1,2,4] triazolo - [3,4 - b] [1,3,4] thiadiazol - 6 - yl) aniline(16). A mixture of 2 - amino - 5 - chloro - 3 - methylbenzoic acid(**12a**,2 mmol),4 - amino - 5 - methyl - 4*H* - 1,2,4 - triazole - 3 - thiol (2mmol),and phosphorus oxychloride($POCl_3$)(7mL) was refluxed for 5h and condensed under reduced pressure. The residue was poured into ice - water,and the mixture was adjusted to pH 10 - 11 with 30% sodium hydroxide solution. The solid was collected by filtration, washed with water, and recrystallized with DMF - H_2O to afford compound **16** as a yellow solid: yield 47%;mp 266 - 268℃;^{1}H NMR(400MHz,DMSO - d_6)δ 2.20(s,3H,CH_3),2.67(s,3H, CH_3),6.56(br,2H,NH_2),7.31(s,1H,Ph - H),7.36(s,1H,Ph - H).

Synthesis of 1 - (3 - Chloro - 2 - pyridyl) - 3 - (substituted) - 1*H* - pyrazole - 5 - carbonyl chloride(17a - d). To a suspension of pyrazole carboxylic acid(**5**)(1mmol) in 25 mL of dichloromethane were added successively oxalyl chloride(4mmol) and two drops of DMF under vigorous stirring. After stirring at room temperature for 3h, the solution was evaporated to dryness in vacuo, and the residue, that is, the crude pyrazole carbonyl chloride, was obtained as a yellow powder(100%), which was used directly in the next step without further purification.

General Synthetic Procedure for *N'* - (2,4,6 - Trisubstitutedphenyl) - *N* - (1 - (3 - chloro - 2 - pyridyl) - 3 - halo - 1*H* - prazole - 5 - carbonyl) thiourea(19a - n). To a mixture of 0.24g(2.5mmol) of potassium thiocyanate(KSCN) in 15mL of dry acetonitrile were added two drops of polyethylene glycol - 400(PEG - 400), and the mixture was stirred at room temperature for 5min to give a homogeneous solution; then, the solution of crude pyrazole carbonyl chloride(**17a** or **17b**)(1mmol) in 5mL of dry acetonitrile was added. After stirring at room temperature for 40 min, the mixture was filtered to give the acetonitrile solution of pyrazole carbonyl isothiocyanate(**18a** or **18b**), and then the 2,4,6 - trisubstituted aniline(0.85mmol) was added. After stirring for a further 3 - 4h at room temperature, the reaction mixture was evaporated in vacuo, and the residue was subjected to column chromatography on silica gel with petroleum ether and ethyl acetate as solvents to give product(**19a - n**).

General Synthetic Procedure for *N* - (4 - Halo - 2 - methyl - 6 - (substituted - carbamoyl) phenylcarbamothioyl) - 1 - (3 - chloropyridin - 2 - yl) - 3 - (2,2,2 - trifluoroethoxy) - 1*H* - pyrazole - 5 - carboxamide(20a - g). To a mixture of 0.24g(2.5mmol) of KSCN in 15mL of dry acetonitrile were added two drops of PEG - 400, and the mixture was stirred at room temperature for 5 min to give a homogeneous solution; then, the solution of crude pyrazole carbonyl chloride(**17c**)(1mmol) in 5mL of dry acetonitrile was added. After stirring at room temperature for 40min, the reaction was filtered to give the acetonitrile solution of pyrazole carbonyl isothiocyanate(**18c**), and then the amino compound (**14a - g**)(0.85mmol) was added; after stirring at room temperature for another 3 - 4h, the reaction mixture was evaporated in vacuo, and the residue was subjected to column chromatography on silica gel with petroleum ether and ethyl acetate as solvents to give product.

General Synthetic Procedure for 3 – Substituted – *N* – (4 – halo – 2 – methyl – 6 – (5 – methyl – 1,3,4 – oxadiazol – 2 – yl) phenyl) – 1 – (3 – chloropyridin – 2 – yl) – 1*H* – pyrazole – 5 – carboxamide(21a – c). To a mixture of oxadiazole – containing substituted aniline(**15a** or **15b**)(0.95mmol) and triethylamine(1mmol) in dichloromethane(15mL) was added the solution of pyrazole carbonyl chloride(**17b** or **17c**, 1mmol) in dry dichloromethane (7mL) at 0 – 5℃. The reaction mixture was then stirred at room temperature for 3 – 4h and washed with 5% $NaHCO_3$ solution and water successively. The organic layer was dried over anhydrous sodium sulfate. The solvent was removed in vacuo, and the residue was subjected to column chromatography on silica gel with petroleum ether and ethyl acetate as solvents to give product.

General Synthetic Procedure for *N* – (2 – (5 – (2 – Azidopyridin – 3 – yl) – 1,3,4 – oxadiazol – 2 – yl) – 4 – chloro – 6 – methylphenyl) – 3 – bromo – 1 – (3 – chloropyridin – 2 – yl) – 1*H* – pyrazole – 5 – carboxamide(22). The synthesis was similar to that of compounds **21a – c**. Using oxadiazolecontaining aniline(**15e**) and pyrazole carbonyl chloride(**17b**) as materials, compound **22** was obtained as a yellow solid.

General Synthetic Procedure for *N′* – (4 – Chloro – 2 – methyl – 6 – (5 – methyl – 1,3,4 – oxadiazol – 2 – yl) phenyl) – *N* – (1 – (3 – chloro – 2 – pyridyl) – 3 – bromo – 1H – prazole – 5 – carbonyl) thiourea(23). The procedure was similar to those of **19** and **20**. Using **15a** as substituted aniline material, compound **23** was obtained as a yellow solid.

General Synthetic Procedure for 3 – Halo – *N* – (4 – chloro – 2 – methyl – 6 – (5 – aryl – 1,3,4 – oxadiazol – 2 – yl) phenylcarbamothioyl) – 1 – (3 – chloropyridin – 2 – yl) – 1*H* – pyrazole – 5 – carboxamide(24a – c). The synthesis was similar with those of **19** and **20**. Using **15c** or **15d** as substituted aniline material, compounds **24a – c** were obtained.

Synthesis of 3 – Substituted – 1 – (3 – chloropyridin – 2 – yl) – 1*H* – pyrazole – 5 – carboxamide(25a, 25b). To a suspension of pyrazole carboxylic acid(**5b** or **5d**)(8mmol) in 50mL of dichloromethane were added successively oxalyl chloride(17mmol) and two drops of DMF. After stirring at room temperature for 3 – 4h, the solution was evaporated. The resulting carbonyl chloride was dissolved in 50mL of dry tetrahydrofuran(THF) and added to a mixture of aqueous ammonia (25%)(40mmol) and water(100mL) at 0℃. After stirring overnight, a large amount of solid formed and was filtered, which was further purified by a silica gel column eluted with petroleum ether and ethyl acetate to give the pyrazole caboxamide. **25a**: white solid; yield 84%; mp 200 – 202℃. **25b**: white solid; yield 77%; mp 183 – 185℃.

General Synthetic Procedure for 3 – Bromo – *N* – (4 – chloro – 2 – methyl – 6 – (5 – substituted – 1,3,4 – oxa diazol – 2 – yl) – phenylcarbamoyl) – 1 – (3 – chloropyridin – 2 – yl) – 1*H* – pyrazole – 5 – carboxamide(26a, 26b). A suspension of compound **25** (1mmol) in 20mL of 1,2 – dichloroethane and oxalyl chloride(1.1mmol) was heated to reflux and kept for 2h. After the mixture had cooled slightly, oxadiazolecontaining aniline(**15a** or **15d**) was added, and the mixture was further refluxed for 1 – 2h. After cooling, the produced precipitate was filtered and washed with 1,2 – dichloroethane and ethyl acetate successively to give product.

General Synthetic Procedure for *O*,*O* – Dimethyl 3 – Substituted – 1 – (3 – chloropyridin – 2 – yl) – 1H – pyrazole – 5 – carbonylcarbamoylphosphoramidothioate (27a,27b). The procedure was similar to that of **26** using 93% spermine as amine material, and the solid filtered was washed with 1,2 – dichloroethane to give pure product.

General Synthetic Procedure for 2 – (3 – Bromo – 1 – (3 – chloropyridin – 2 – yl) – 1*H* – pyrazol – 5 – yl) – 5 – substituted – 1,3,4 – oxadiazole (28,29a – c). A mixture of pyrazole carboxylic acid (**5b**, 1.5mmmol) with $SOCl_2$ (10mL) was refluxed for 4h and condensed under reduced pressure to give carbonyl chloride (**17b**), which was immediately dissolved in dry toluene (7mL) and added to a solution of 5 – methyl – 1*H* – tetrazole or compound 10 (1.2mmol) in dry pyridine (10mL) with stirring. The mixture was further stirred at 100 – 110℃ for 2h and cooled to room temperature, and water (50mL) was added. The mixture was then extracted with ethyl acetate (10mL ×4), and the extracts were dried with anhydrous sodium sulfate. The solvent was removed in vacuo, and the residue was recrystallized with 95% EtOH to give product.

Synthesis of 6 – (3 – Bromo – 1 – (3 – chloropyridin – 2 – yl) – 1*H* – pyrazol – 5 – yl) – 3 – methyl – [1,2,4] triazolo[3,4 – *b*][1,3,4]thiadiazole (30). A mixture of pyrazole carboxylic acid (**5b**, 2mmol) and 4 – amino – 5 – methyl – 4*H* – 1,2,4 – triazole – 3 – thiol (2mmol) in $POCl_3$ (7mL) was refluxed for 5h and condensed under reduced pressure. The residue was poured into ice – water, and the mixture was adjusted to pH 10 – 11 with 30% sodium hydroxide solution. The solid was collected by filtration, washed with water, and recrystallized with DMF – H_2O to afford compound **30** as pink crystals.

Synthesis of 1 – (2 – Hydroxyethyl/hyroxypropyl) – 3 – (2,4,6 – trisubstituted – phenyl) thiourea (32a – d). According to the method in the literature,[24] to a solution of 1,3,5 – trisubstituted – 2 – isothiocyanatobenzene (**31a – c**, 4mmol) in diethyl ether (15mL) was added 2 – aminoethanol or 1 – amino – 2 – propanol (4.5mmol), and the mixture was refluxed for 1 – 2h. The produced precipitate was filtered and washed with diethyl ether to give compound **32** as a white solid, which was used for following reaction without purification.

Synthesis of 5 – Substituted – *N* – (2,4,6 – trisubstituted – phenyl) – 4,5 – dihydrothiazol – 2 – amine (33a – d). The intermediate thiourea (**32**, 3mmol) was mixed with concentrated HCl (8mL) and refluxed for 2h. The reaction solution was diluted with water (15mL) and decolorized by activated charcoal. After cooling to room temperature, the mixture was adjusted to pH ~ 10 with aqueous sodium hydroxide (6mol/L), and the solid was collected by filtration, washed with water and recrystallized with EtOH – H_2O to afford compound **33**.

33a: white crystal; yield 40%; mp 117 – 119℃ (lit.[25] mp 118 – 119℃).

33b: white crystal; yield 46%; mp 124 – 126℃ (lit.[24] mp 124 – 125℃).

33c: white crystal; yield 65%; mp 165 – 166℃ (lit.[24] mp 165.5 – 166℃).

33d: white crystal; yield 42%; mp 122 – 123℃; ^1H NMR ($CDCl_3$, 400MHz) δ 1.44 (d, J = 6.8Hz, 3H, CH_3), 3.34 (dd, J_1 = 10.0Hz, J_2 = 6.4Hz, 1H, CH_2), 3.78 (dd, J_1 = 10.0Hz, J_2 = 6.4Hz, 1H, CH_2), 3.88 (m, 1H, CH), 6.80 (s, 1H, NH), 7.30 (s, 2H, Ph – H).

General Synthetic Procedure for 3 – Substituted – 1 – (3 – chloropyridin – 2 – yl) – *N* – (5 – substituted – 4,5 – dihydrothiazol – 2 – yl) – *N* – (2,4,6 – trisubstituted – phen-

yl) – 1*H* – pyrazole – 5 – carboxamide(34a – e). The synthesis was similar to the synthesis of compounds **21a – c**. Using thiazoline – containing aniline(**33**) and pyrazole carbonyl chloride (compound **17b** or **17c**) as materials, compound **34** was obtained.

Biological Assay. All bioassays were performed on representative test organisms reared in the laboratory. The bioassay was repeated at 25 ± 1℃ according to statistical requirements. Assessments were made on a dead/alive basis, and mortality rates were corrected using Abbott's formula.[26] Evaluations were based on a percentage scale of 0 – 100, in which 0 = no activity and 100 = total kill. The standard deviations of the tested biological values were ±5%. LC_{50} and LC_{95} values were calculated by probit analysis.[27]

Larvicidal Activity against Oriental Armyworm(*Mythimna separata* Walker) and Corn Borer(*Ostrinia nubilalis*). The larvicidal activity of the title compounds and contrast compounds chlorantraniliprole and diflubenzuron against oriental armyworm and corn borer was tested according to the leaf – dip method using the reported procedure.[28] Leaf disks(about 5cm) were cut from fresh corn leaves and then were dipped into the test solution for 3 – 5s. After air drying, the treated leaf disks were placed individually into a glass – surface vessel(7cm). Each dried treated leaf disk was infested with 10 thirdinstar oriental armyworm or corn borer larvae. Percentage mortalities were evaluated 4 days after treatment. Leaves treated with acetone were provided as controls. Each treatment was performed three times. The insecticidal activity is summarized in Tables 1, 2, and 4.

Table 1 Larvicidal Activity of Compounds and Chlorantraniliprole against Oriental Armyworm(*Mythimna separata* Walker) at 200 mg/L Concentration

compd	larvicidal activity(%)	compd	larvicidal activity(%)	compd	larvicidal activity(%)	compd	larvicidal activity(%)	compd	larvicidal activity(%)
15a	20	**19e**	10	**20a**	100	**22**	100	**29a**	0
15b	16.7	**19f**	20	**20b**	100	**23**	100	**29b**	10
15c	36.7	**19g**	10	**20c**	100	**24a**	0	**29c**	0
15d	23.3	**19h**	0	**20d**	100	**24b**	0	**30**	13.3
15e	6.67	**19i**	20	**20e**	100	**24c**	0	**34a**	30
16	10	**19j**	30	**20f**	100	**26a**	60	**34b**	16.7
19a	20	**19k**	40	**20g**	100	**26b**	10	**34c**	50
19b	20	**19l**	10	**21a**	100	**27a**	10	**34d**	6.67
19c	100	**19m**	20	**21b**	100	**27b**	10	**34e**	10
19d	100	**19n**	10	**21c**	43.3	**28**	33.3	control①	100

①chlorantraniliprole.

Table 2 Larvicidal Activity of Compounds, Chlorantraniliprole, and Diflubenzuron against Oriental Armyworm(*Mythimna separata* Walker)

compd	larvicidal activity(%) at								
	100mg/L	50mg/L	25mg/L	10mg/L	5mg/L	2.5mg/L	1mg/L	0.5mg/L	0.25mg/L
19c	100	100	20	0					
19d	100	100	100	15					

Table 2(continued)

compd	larvicidal activity(%)at								
	100mg/L	50mg/L	25mg/L	10mg/L	5mg/L	2.5mg/L	1mg/L	0.5mg/L	0.25mg/L
20a	100	100	100	100	100	100	100	100	40
20b	100	100	100	100	0				
20c	100	100	100	100	40				
20d	100	100	100	100	0				
20e	100	100	100	100	100	40			
20f	100	100	100	100	0				
20g	100	100	100	100	100	100	80		
21a	100	60		0					
21b	30			0					
22	100	60		0					
23	100	40		0					
26a	0								
34c	10								
diflubenzuron	100	100	100	100	100	40			
chlorantraniliprole	100	100	100	100	100	100	100	100	70①

①At a concentration of 0.1mg/L.

Larvicidal Activity against Diamondback Moth(*Plutella xylostella* L.)and Beet Armyworm(*Laphygma exigua* Hübner). The larvicidal activity of the title compounds and contrast compounds chlorantraniliprole and diflubenzuron against diamondback moth and beet armyworm was tested by the leaf-dip method using the reported procedure.[29,30] Leaf disks(5cm×3cm) were cut from fresh cabbage leaves and then dipped into the test solution for 3s. After air-drying, the treated leaf disks were placed individually into boxes(80cm). Each dried treated leaf disk was infested with 30 third-instar beet armyworm or 30 second-instar diamondback moth larvae. Percentage mortalities were evaluated 3 days after treatment. Leaves treated with water and acetone were provided as controls. Each treatment was performed three times. The insecticidal activity is summarized in Tables 3 and 4.

Table 3 Larvicidal Activity of Compounds, Diflubenzuron, and Chlorantraniliprole against Diamondback Moth(*Plutella xylostella* L.)

compd	larvicidal activity(%)at								
	50mg/L	25mg/L	10mg/L	2.5mg/L	1.25mg/L	0.125mg/L	0.1mg/L	0.01mg/L	0.005mg/L
19c	90		57						
19d	43		0						
20a		100		100	100	100	71	43	29
20c		100		100	100	100	100	100	86

Table 3 (continued)

compd	larvicidal activity (%) at								
	50mg/L	25mg/L	10mg/L	2.5mg/L	1.25mg/L	0.125mg/L	0.1mg/L	0.01mg/L	0.005mg/L
20e		100		100	100	100	100	71	57
20g		100		100	100	86	71	30	0
22	100		60						
23	100		60						
26a	29								
34c	29								
diflubenzuron		29		0					
chlorantraniliprole		100		100	100	86	100	86	57

Table 4 Larvicidal Activity of Compounds 20a, 20c, 20e, 20g, Diflubenzuron, and Chlorantr-aniliprole against Beet Armyworm (*Laphygma exigua* Hübner) and Corn Borer (*O. nubilalis*)

compd	*L. exigua* Hübner		*O. nubilalis*	
	concn (mg/L)	larvicidal activity (%)	concn (mg/L)	larvicidal activity (%)
20a	10	0	25	100
	1	0	10	60
20c	10	70	25	100
	1	60	10	60
	0.1	40		
20e	10	80	25	100
	1	60	10	60
	0.1	40		
20g	10	20	25	30
	1	0		
diflubenzuron	10	0	25	20
	1	0		
chlorantraniliprole	10	80	25	100
	1	80	10	80
	0.1	40		

Larvicidal Activity against Mosquito (*Culex pipiens pallens*). The larvicidal activity of the title compounds and contrast compounds against mosquito were evaluated by using the reported procedure.[31] The compounds were prepared to different concentrations by dissolving the compounds in acetone and adding distilled water. Then 20 fourth-instar mosquito larvae were put into 10mL of the test solution and raised for 8 days. Each treatment was performed three times. The biological data in Table 5 were the average value of the three tested values. The results were expressed by death percentage. For comparative purposes, chlorantraniliprole and diflubenzuron were tested under the same conditions.

Table 5 Larvicidal Activity of Compounds 19c,19d,20a – g,Diflubenzuron, and Chlorantraniliprole against Mosquito(*Culex pipiens pallens*)

compd	larvicidal activity(%) at			
	2mg/L	1mg/L	0.5mg/L	0.25mg/L
19c	20			
19d	0			
20a	100	100	40	
20b	100	60	20	
20c	60			
20d	50			
20e	60			
20f	30			
20g	50			
diflubenzuron	100	100	30	
chlorantraniliprole	100	100	100	100

Hartree – Fock(HF) Calculation. The structures of compounds **20a**,**20c** and chlorantraniliprole were selected as the initial structures,whereas the HF/6 – 31G(d,p)[32] method in the Gaussian 03 package[33] was used to optimize their structures. Vibration analysis showed that the optimized structures were in accordance with the minimum points on the potential energy surfaces. All of the convergent precisions were the system default values,and all of the calculations were carried out on the NanKai – Star supercomputer.

RESULTS AND DISCUSSION

Chemistry. The tetrazole compounds(**10a**,**10b**)were prepared according to the literature[22] by heating cyanocontaining compound **9** with 2 equiv of sodium azide and ammonium chloride in DMF with satisfactory yields. When using 2 – chloro – 3 – cyanopyridine as material,we did not obtain the desired 2 – chloro – 3 – (1*H* – tetrazol – 5 – yl) pyridine;however,2 – azido – 3 – (1*H* – tetrazol – 5 – yl) pyridine(**10c**),which is a novel biheterocyclic compound containing eight nitrogen atoms,was obtained. This result indicated that the chloro atom in the 2 – position of pyridine can be substituted by nucleophilic reagent N_3^-. during cyclization of the cyano group with hydrazoic acid,and the reactivity of the —Cl group may be increased by its neighboring —CN group of the material or tetrazole group of its product,so the —Cl group could be easier to leave. Tetrazole compound(5 – methyl – 1*H* – tetrazole or **10a**,**10b**)reacted with 3,5 – disubstituted – 2 – amino benzoyl chloride(**13**)in pyridine,undergoing a Huisgen reaction to give oxadiazole – containing substituted aniline(**15a – d**)with the amino group conserved. Similarly, azido – containing tetrazole compound(**10c**)can also undergo such reaction to afford corresponding oxadizolecontaining substituted aniline(**15e**). The azido group is not affected in the following reactions,too,which can be confirmed by ^{1}H NMR and HRMS of their products(**22** and **29c**). Moreover,2 – amino – 5 – chloro – 3 – methylbenzoic acid(**12a**)and 4 – amino – 5 – methyl – 4*H* – 1,2,4 – triazole – 3 – thiol refluxed in $POCl_3$ to successfully give corresponding fused heterocyclic substituted aniline(**16**)via a cyclization reaction. The reaction of oxadiazole-containing aniline(**15a**,**15b**,**15e**)with pyrazole carbonyl chloride can give oxadiazole – containing amide(**21a – c**,**22**), whereas it is difficult for compound **16** to undergo this reaction to afford acylation product, which may be because of the very weak solubility and weak reactivity of fused heterocyclic aniline **16**.

There are some common methods for synthesizing the carbonyl thioureas, for example, carbonyl chloride compound refluxes with thiocyanate in acetone[34] or reacts with thiocyanate catalyzed by PEG – 600 in methane dichloride at room temperature[35] to produce carbonyl isothiocyanate, which further reacts with amine to give carbonyl thiourea. In our experiments for the syntheses of carbonyl thioureas(**19**,**20**,**23**, and **24a – c**), a PEG – 400 PTC method was adopted(illustrated in Scheme 5 and Scheme 6). Pyrazole carbonyl chloride(**5**) was prepared from the reaction of pyrazole acid with oxalyl chloride, and it was treated with potassium thiocyanate under the conditions of solid – liquid phase transfer catalysis using a small amount of PEG – 400 as the catalyst to give pyrazole carbonyl isothiocyanate (**18**) at room temperature. This compound does not need to be isolated, and it was treated immediately with 2,4,6 – trisubstituted anilines or 2 – amino – 3 – methylbenzamide derivatives(**14a – g**) to give compound **19** or **20** in good to excellent yields.

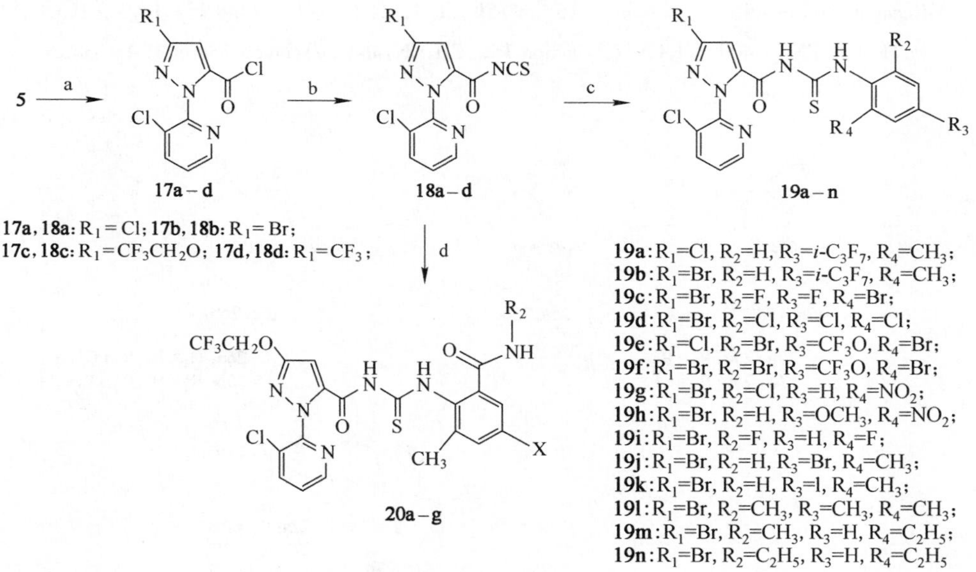

17a – d **18a – d** **19a – n**

17a, **18a**: R_1 = Cl; **17b**, **18b**: R_1 = Br;
17c, **18c**: R_1 = CF_3CH_2O; **17d**, **18d**: R_1 = CF_3;

19a: R_1=Cl, R_2=H, R_3=i-C_3F_7, R_4=CH_3;
19b: R_1=Br, R_2=H, R_3=i-C_3F_7, R_4=CH_3;
19c: R_1=Br, R_2=F, R_3=F, R_4=Br;
19d: R_1=Br, R_2=Cl, R_3=Cl, R_4=Cl;
19e: R_1=Cl, R_2=Br, R_3=CF_3O, R_4=Br;
19f: R_1=Br, R_2=Br, R_3=CF_3O, R_4=Br;
19g: R_1=Br, R_2=Cl, R_3=H, R_4=NO_2;
19h: R_1=Br, R_2=H, R_3=OCH_3, R_4=NO_2;
19i: R_1=Br, R_2=F, R_3=H, R_4=F;
19j: R_1=Br, R_2=H, R_3=Br, R_4=CH_3;
19k: R_1=Br, R_2=H, R_3=I, R_4=CH_3;
19l: R_1=Br, R_2=CH_3, R_3=CH_3, R_4=CH_3;
19m: R_1=Br, R_2=CH_3, R_3=H, R_4=C_2H_5;
19n: R_1=Br, R_2=C_2H_5, R_3=H, R_4=C_2H_5

20a – g

20a: X=Cl, R_2=methyl; **20b**: X=Cl, R_2=n-propyl;
20c: X=Cl, R_2=i-proply; **20d**: X=Cl, R_2=cyclopropyl;
20e: X=Cl, R_2=t-butyl; **20f**: X=Br, R_2=i-propyl; **20g**: X=Br, R_2=cyclopropyl

Scheme 5 Synthesis of Compounds **19a – n** and **20a – g**①

①Reagents and conditions: (a) $(COCl)_2$, DMF, CH_2Cl_2; (b) KSCN, PEG – 400, CH_3CN; (c) substituted aniline, CH_3CN; and (d) compd **14a – g**, CH_3CN

Compounds **26** were synthesized, as shown in Scheme 7. Pyrazole carboxylic acid **5** was converted to amide **25** by reaction with oxalyl chloride and then aqueous ammonia. Compound **25** was then refluxed with oxalyl chloride in 1,2 – dichloroethane to give its corresponding pyrazole carbonyl isocyanate derivative, which coupled with amines **15** to afford the compounds **26**. Similarly, using 93% spermine as amine material the carbonyl thiophosphorylureas **27** were obtained.

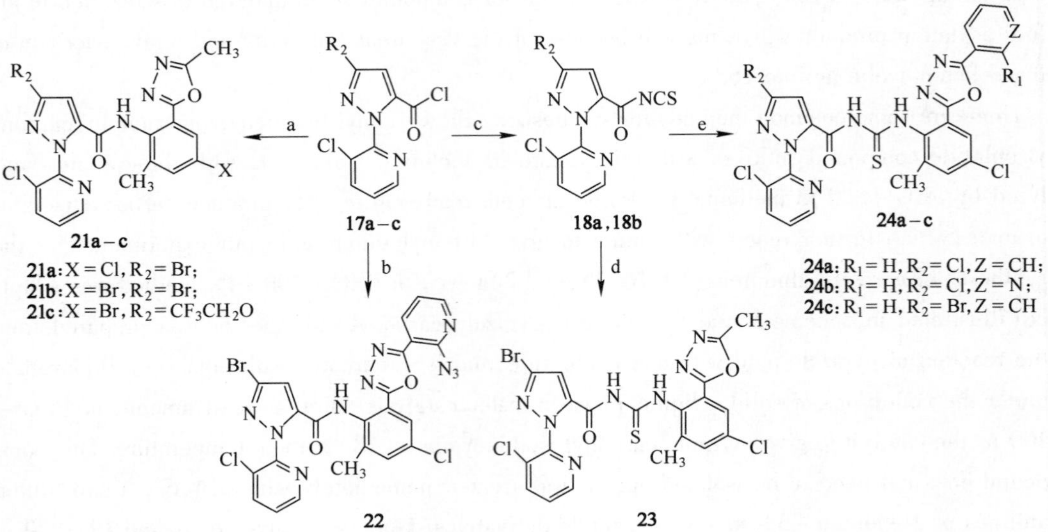

Scheme 6 Synthesis of Compounds **21a－c**,**22**,**23**,and **24a－c**①

①Reagents and conditions:(a) compd **15a** or **15b**, Et_3N, CH_2Cl_2; (b) compd **15e**, Et_3N, CH_2Cl_2; (c) KSCN, PEG－400, CH_3CN; (d) compd **15a**, CH_3CN; and (e) compd **15c** or **15d**, CH_3CN

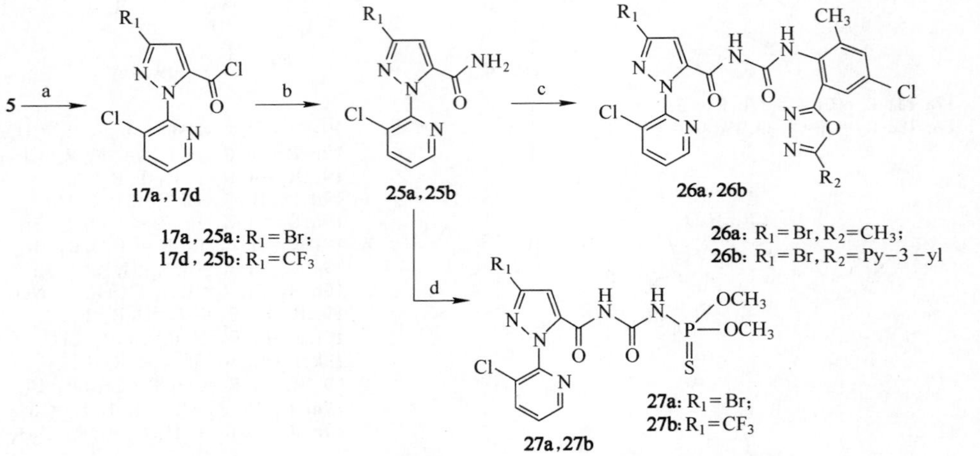

Scheme 7 Synthesis of Compounds **26a**,**26b**,**27a**,and **27b**①

①Reagents and conditions:(a) $(COCl)_2$, DMF, CH_2Cl_2; (b) $NH_3 \cdot H_2O$; (c) (i) $(COCl)_2$, $ClCH_2CH_2Cl$, reflux, and (ii) compd **15a** or **15d**; and (d) (i) $(COCl)_2$, $ClCH_2CH_2Cl$, reflux, and (ii) 93% spermine

The reaction of tetrazole compound (5－methyl－1*H*－tetrazole or **10**) with pyrazole carbonyl chloride **17b** in pyridine led to oxadiazole compound (**28**,**29**; Scheme 8) by a Huisgen reaction, whereas pyrazole carboxylic acid **5b** directly reacted with 4－amino－5－methyl－4*H*－1,2,4－triazol－3－thiol in $POCl_3$ to afford the fused heterocyclic compound **30**.

Referring to the literature method reported by He et al.,[24] thiazoline amines (**33**) were prepared by a cyclization reaction in a concentrated HCl solution of hydroxyethylthiourea or hydroxypropylthiourea (**32**), which derived from the condensation of isothiocyanate (**31**) with 2

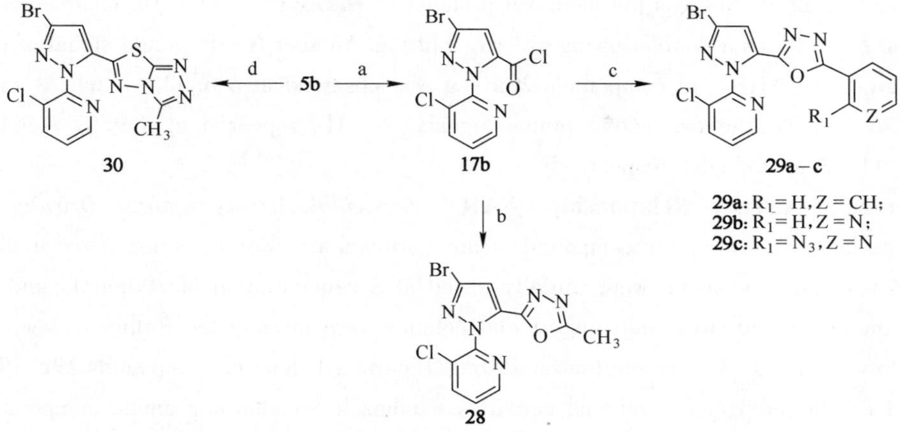

Scheme 8 Synthesis of Compounds **28**, **29a – c**, and **30**①

①Reagents and conditions: (a) $SOCl_2$, benzene, reflux; (b) 5 – methyl – 1*H* – tetrazole, Py, 100 – 110℃; (c) compd **10a – c**, Py, 1,4 – dioxane, 100 – 110℃; and (d) 4 – amino – 5 – methyl – 4*H* – 1,2,4 – triazole – 3 – thiol, $POCl_3$, reflux

– aminoethanol or 1 – amino – 2 – propanol. Then compound **33** reacted with pyrazole acyl chloride (**17**) using Et_3N as deacid reagent to give acylation product (**34**; Scheme 9).

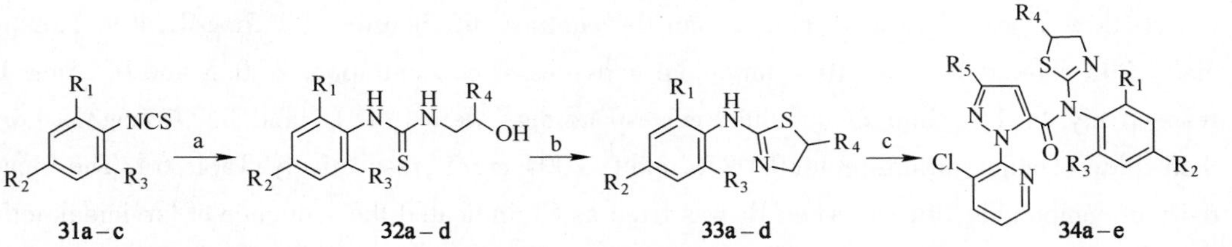

31a: R_1=H, R_2=Cl, R_3=CH_3;
31b: R_1=R_2=R_3=CH_3;
31c: R_1=R_2=R_3=Cl

32a,33a,34a: R_1=R_4=H, R_2=Cl, R_3=CH_3, R_5=Br;
32b,33b,34b: R_1=R_2=R_3=CH_3, R_4=H, R_5=Br;
32c,33c,34c: R_1=R_2=R_3=Cl, R_4=H, R_5=Br;
32d,33d,34d: R_1=R_2=R_3=Cl, R_4=CH_3, R_5=Br;
34e: R_1=R_2=R_3=Cl, R_4=CH_3, R_5=CF_3CH_2O

Scheme 9 Synthesis of Compounds **34a – e**①

①Reagents and conditions: (a) 2 – aminoethanol or 1 – amino – 2 – propanol, Et_2O; (b) (i) concentrated HCl, reflux, and (ii) NaOH; and (c) compd **17b** or **17c**, Et_3N, dry THF

The novel structures of these title compounds have been characterized by melting point, ^1H NMR, and elemental analyses or HRMS; several compounds with typical structures were also characterized by IR spectra. All spectral and analytical data were consistent with the assigned structures. The infrared spectrum of carbonyl thiourea compounds **19a**, **20a – g**, **23**, and **24a – c** showed absorption bands at 3143 – 3446 cm^{-1} for N – H stretching. The strong peaks at 1742 – 1579 cm^{-1} are ascribed to the C=O group and C=N group of heterocyclic ring skeleton as well. The characteristic stretching vibrations ν(C=S) and ν(C—N) appeared at 1273 – 1369 and 1152 – 1167 cm^{-1}, respectively. In the ^1H NMR spectra of compounds **19a – n** and **20a – g**, the proton signals of —NHCO— and —NHCS— on the carbonyl thiourea bridge were observed at δ 9.73 – 10.42 and 11.08 – 12.48 as a singlet, respectively. The trifluoroethoxyl protons ($—OCH_2CF_3$) of compounds **20a – g**, **21c**, and **34e** appeared at δ 4.65 – 4.92 as a quartet due to "F" splitting, whereas the methoxyl protons($(CH_3O)_2P(=S)-$)of compound **27** appeared at $\delta \sim 3.85$ as a doublet owing to "P" splitting. Another N—H proton signal of the amide moiety($—CONHR_2$) in compounds **20a – g** was observed at δ 6.32 – 6.64. As for compounds **26** and **27**, the two active proton signals (N—H) appeared at δ 9.67 – 9.93 and 10.75 – 11.53 as a singlet, respectively.

Structure – Activity Relationship (SAR) . *Larvicidal Activity against Oriental Armyworm.* The larvicidal activity of compounds against oriental armyworm is summarized in Tables 1 and 2. All of the compounds were initially tested at a concentration of 200mg/L, and consequently the compounds with high insecticidal potency were investigated further at low concentration. From Table 1, we can see that at 200mg/L carbonyl thiourea compounds **19c**, **19d**, **20a – g**, and **23** showed 100% larvicidal activities, oxadiazole – containing amide compounds **21a – c** and **22** exhibited 43.3% – 100% activities, and oxadiazole – containing carbonyl urea **26a** and thiazoline – containing amide **34c** possessed 60 and 50% activities, respectively. Other compounds including novel carbonyl thiophosphorylureas **27** and intermediates oxadiazole – containing or fused heterocyclic substituted aniline **15** and **16** showed comparably lower activity and were not tested further. From Table 2, we can see that compounds **20a – g** also showed 100% larvicidal activity against oriental armyworm even at the concentration of 10mg/L. At lower concentration, for example, 2.5mg/L, **20a** and **20g** still exhibited excellent larvicidal activity(100%), and were more potent than the contrast diflubenzuron(2.5mg/L, 40%). Especially, **20a** showed 100 and 40% larvicidal activities at concentrations of 0.5 and 0.25mg/L, respectively. Its LC_{50} and LC_{95} values correspondingly were 0.1812 and 0.7767mg/L, lower than those of chlorantraniliprole(0.0276 and 0.2004 mg/L, respectively, Table 6). The bioactivity of compounds **20a – e**, when R_1 was fixed as Cl, indicated the sequence of larvicidal activity was $CH_3 > C(CH_3)_3 > CH(CH_3)_2 > CH_3CH_2CH_2$, cyclopropyl in the aliphatic amide moiety(R_2).

Table 6 LC_{50} and LC_{95} Values of Compound 20a and Chlorantraniliprole against Oriental Armyworm

compd	$y = a + bx$	R	LC_{50} (mg/L)	LC_{95} (mg/L)
20a	$y = 6.93 + 2.60x$	0.98	0.1812	0.7767
chlorantraniliprole	$y = 7.98 + 1.91x$	0.91	0.0276	0.2004

Larvicidal Activity against Diamondback Moth. Table 3 shows that the bioassay results against diamondback moth at 1.25mg/L, compounds **20a**, **20c**, **20e**, **20g** were more effective than diflubenzuron(2.5mg/L, 0%). At 0.125mg/L the four compounds have the same or better larvicidal activity than chlorantraniliprole(86% – 100%). Even at 0.005mg/L, **20e** showed 57% activity and reached the same larvicidal level as chlorantraniliprole. It was worth noting that **20c** showed a death rate of 86% at 0.005mg/L, which is more effective than chlorantraniliprole (57%) against diamondback moth. The LC_{50} value of **20a** was 0.0054mg/L, higher than that of chlorantraniliprole(0.0045mg/L, Table 7). The LC_{50} values of **20c** and **20e** were 0.0017 and 0.0023mg/L, respectively, lower than that of chlorantraniliprole (0.0045mg/mL, Table 7), which were in accordance with the results in Table 3. The activities of compounds **20a**, **20c**, and

20e, where R_1 was fixed as Cl, indicated the trend $CH(CH_3)_2 > C(CH_3)_3 > CH_3$ in the aliphatic amide moiety(R_2).

Table 7 LC_{50} Values of Compounds 20a, 20c, 20e, and Chlorantraniliprole against Diamondback Moth

compd	$y = a + bx$	R	LC_{50} (mg/L)
20a	$y = 15.11 + 4.46x$	0.99	0.0054
20c	$y = 10.79 + 2.09x$	0.97	0.0017
20e	$y = 15.22 + 3.87x$	0.92	0.0023
chlorantraniliprole	$y = 18.06 + 5.57x$	0.97	0.0045

Larvicidal Activity against Beet Armyworm. The larvicidal activity of **20a**, **20c**, **20e**, and **20g** against beet armyworm at different concentrations is summarized in Table 4. It was found that only **20c** and **20e** exhibited a 40% death rate at 0.1mg/L, similar to that of chlorantraniliprole. Surprisingly, **20a** and **20g**, with excellent activity against oriental armyworm (100% at 2.5mg/L), showed no activity against beet armyworm at 1mg/L.

Larvicidal Activity against Corn Borer. From Table 4, it was found that **20a**, **20c**, and **20e** possessed 100% death rate against corn borer at 25mg/L, more effective than 20g (30%) and diflubenzuron (20%) as well. All three compounds exhibited a 60% death rate at 10mg/L, which was somewhat less effective than chlorantraniliprole (80%).

Larvicidal Activity against Mosquito. The larvicidal activity of **19c**, **19d**, and **20a - g** against mosquitoes is summarized in Table 5, from which we can see that **20a - g** exhibited significant larvicidal activity and showed a 30% - 100% death rate at 2mg/L. **20a** and **20b** showed 40% and 20% activity against mosquitoes at 0.5mg/L, respectively, lower than that of chlorantraniliprole but similar to that of diflubenzuron (0.5mg/L, 30%).

HF Calculation. According to the frontier molecular orbital theory, HOMO and LUMO are the two most important factors that affect the bioactivities of compounds. HOMO has the priority to provide electrons, whereas LUMO accepts electrons first.[36-38] Thus, a study of the frontier orbital energy can provide some useful information for the active mechanism. We therefore calculated the frontier molecular orbital of representative compounds **20a** and **20c** that have good insecticidal activity against oriental armyworm and diamondback moth with chlorantraniliprole as contrast by means of HF (Hartree - Fock).

The energies of LUMO, HOMO, and HOMO - 1 of **20a**, **20c** and chlorantraniliprole are listed in Table 8 and the LUMO and HOMO maps of **20a**, **20c** and chlorantraniliprole are shown in Figure 2. Taking HF calculation results, there is something in common between **20a** and **20c** in the LUMO and HOMO maps. In the HOMO of **20a** and **20c**, electrons are mainly delocalized on the benzene ring and the thiourea group, especially the latter. However, when electron transitions take place, some electrons in the HOMO will enter into the LUMO;[37] then, in the LUMO of **20a** and **20c**, the electrons will mainly be delocalized on the aromatic rings [benzene ring, pyrazole ring (including O atom of trifluoroethyloxyl group in the 3 - position of pyrazole) and pyridine ring] and the carbonyl thiourea group. It was observed that the carbonyl thiourea moiety in both of the molecules (**20a**, **20c**) makes a major contribution to the activity. These are largely through hydrophobic interaction and are most obvious in the HOMO maps (especially the S atom of the bridge). Comparing with those of chlorantraniliprole, we can see that the LUMO of chlorantraniliprole mainly contains three aromatic rings and the amide bridge group, which is similar to that of **20a** and **20c**, whereas the HOMO contains the pyrazole ring, the pyridine ring, and a small part of the amide bridge (mainly the N atom). The frontier molecular orbitals are located on the main groups, the atoms of which can easily bind with the receptor. [38] Thus, it can be concluded that the three aromatic rings of compounds **20a** and **20c** and chlorantraniliprole as well may contribute activity through $\pi-\pi$ and hydrophobic interactions; meanwhile, the carbonyl thiourea bridge that links the benzene and pyrazole rings of the compounds also plays an important role in the insecticidal activity, like the amide bridge group of chlorantraniliprole, which may contribute activity through hydrophobic interaction.

Table 8 Energies of LUMO, HOMO, and HOMO - 1 of 20a, 20c and Chlorantraniliprole (Hartree)

	20a	**20c**	Chlorantraniliprole
LUMO	-0.20895	-0.20902	-0.09190
HOMO	-0.29959	-0.29961	-0.25188
HOMO - 1	-0.30856	-0.308448	-0.25569

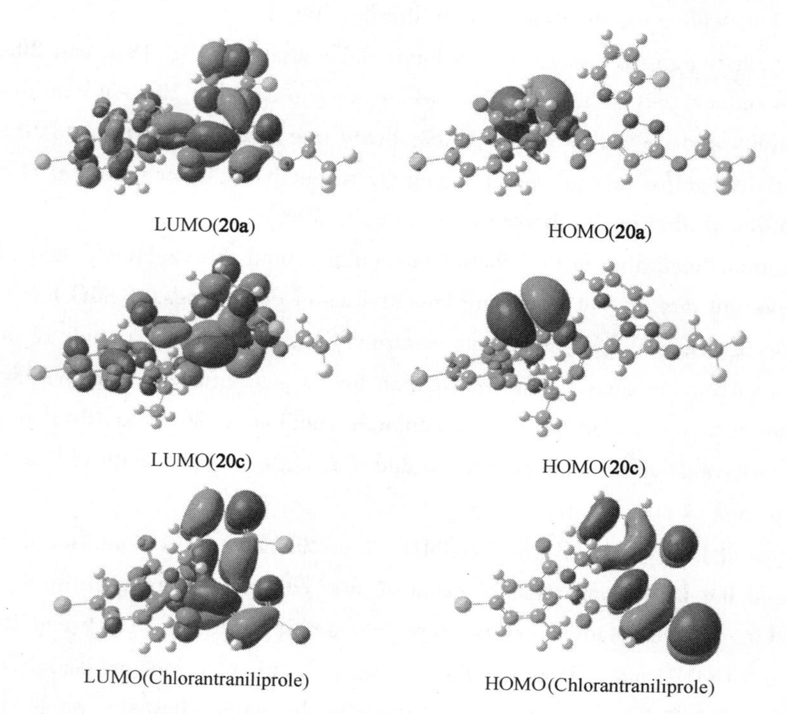

Figure 2 LUMO and HOMO maps for compounds **20a**, **20c**, and chlorantraniliprole from HF calculations. The green parts represent positive molecular orbital, and the red parts represent negative molecular orbital

In summary, series of novel amide bridge modified compounds containing pyridylpyrazole were conveniently synthesized on the basis of the structures of anthranilic diamide insecticides, and their structures were characterized and confirmed by melting point, IR, ^{1}H NMR, and elemental analyses or HRMS. The insecticidal activities against oriental armyworm, diamondback moth, beet armyworm, corn borer, and mosquito of these heterocycle compounds were evaluated. The bioassay results indicated that carbonyl urea, oxadiazole, carbonyl thiophosphorylurea, and thiazoline - containing amide compounds showed weak insecticidal activity against oriental armyworm, whereas part of the carbonyl thiourea and oxadiazole - containing amide compounds exhibited significant larvicidal activity against oriental armyworm, and some of them showed favorable activity against mosquito, diamondback moth, beet armyworm, and corn borer. In general, carbonyl thiourea compounds **20a**, **20c**, and **20e** exhibited comparably outstanding activity compared with others, even prior to the contrasts. In some cases, **20c** and **20e** had better larvicidal effects against diamondback moth than chlorantraniliprole. The preliminary structure - activity relationship of compounds was discussed, and the HF calculation results indicated that the carbonyl thiourea moiety of the compounds **20a** and **20c** plays an important role in the insecticidal activity. It is worth noting that the change of the amide bridge in chlorantraniliprole to carbonyl thiourea can keep the insecticidal activities at a low concentration in our earlier studies, [15] whereas the introduction of both carbonyl thiourea and trifluoroethoxyl moieties to such a structure can increase the insecticidal activities in this study, such as in the case of **20c** and **20e**. The present work demonstrated that the trifluoroethoxycontaining carbonyl thioureas can be used as lead compounds for the development of new insecticidal structures.

ASSOCIATED CONTENT

Supporting Information

^{1}H NMR, HRMS, or elemental analyses and melting point data for compounds **19a - n**, **20a - g**, **21a - c**, **22**, **23**, **24a - c**, **26a**, **26b**, **27a**, **27b**, **28**, **29a - c**, **30**, and **34a - e** (^{31}P NMR and IR data for part compounds). This material is available free of charge via the Internet at http://pubs. acs. org.

AUTHOR INFORMATION

Corresponding Author

* Phone: +86 - 22 - 23503732. E - mail: nkzml@ vip. 163. com.

Notes

The authors declare no competing financial interest.

References

[1] Rogers, E. F.; Koniuszy, F. R.; Shavel, J., Jr.; Folkers, K. Plant insecticides. I. Ryanodine, a new alkaloid from Ryania speciosa. *J. Am. Chem. Soc.* 1948, 70, 3086 - 3088.

[2] Kuna, S.; Heal, R. E. Toxicological and pharmacological studies on the powdered stem of Ryania speciosa, a plant insecticide. *J. Pharmacol. Exp. Ther.* 1948, 93, 407 - 413.

[3] Procita, L. Some pharmacological actions of ryanodine in the mammal. *J. Pharmacol. Exp. Ther.* 1958, 123, 296 - 305.

[4] Tohnishi, M. ; Nakao, H. ; Furuya, T. ; Seo, A. ; Kodama, H. ; Tsubata, K. Flubendiamide, a new insecticide characterized by its novel chemistry and biology. *J. Pestic. Sci.* 2005, 30, 354 – 360.

[5] Lahm, G. P. ; Selby, T. P. ; Frendenberger, J. H. ; Stevenson, T. M. ; Myers, B. J. ; Seburyamo, G. ; Smith, B. K. ; Flexner, L. ; Clark, C. E. ; Cordova, D. Insecticidal anthranilicdiamides: a new class of potent ryanodine receptor activators. *Bioorg. Med. Chem. Lett.* 2005, 15, 4898 – 4906.

[6] Lahm, G. P. ; Cordova, D. ; Barry, J. D. New and selective ryanodine receptor activators for insect control. *Bioorg. Med. Chem.* 2009, 17, 4127 – 4133.

[7] Lahm, G. P. ; Stevenson, T. M. ; Selby, T. P. ; et al. Rynaxypyr™: a new insecticidal anthranilic diamide that acts as a potent and selective ryanodine receptor activator. *Bioorg. Med. Chem. Lett.* 2007, 17, 6274 – 6279.

[8] Hughes, K. A. ; Lahm, G. P. ; Selby, T. P. ; Stevenson, T. M. Cyano *anthranilamide insecticides.* WO 2004067528 A1, 2004.

[9] DuPont Rynaxypyr Insect Control Technical Bulletin; http://www2. dupont. com/Production_Agriculture/en_US/assets/downloads/pdfs/Rynaxypyr_Tech_Bulletin. pdf.

[10] Gewehr, M. ; Puhl, M. ; Dickhaut, J. ; Bastiaans, H. M. M. ; Anspaugh, D. D. ; Kuhn, D. G. ; Oloumi – Sadeghi, H. ; Armes, N. ; Armers, N. *Pesticidal mixtures.* WO 2007082841 A2, 2007.

[11] Muehlebach, M. ; Jeanguenat, A. ; Hall, R. G. *Novel insecticides.* WO 2007080131 A2, 2007.

[12] Xu, X. Y. ; Qian, X. H. ; Li, Z. ; Huang, Q. C. ; Chen, G. Synthesis and insecticidal activity of new substituted *N* – aryl – *N'* – benzoylthiourea compounds. *J. Fluorine Chem.* 2003, 121, 51 – 54.

[13] Yoon, C. J. ; Yang, O. S. ; Kang, H. ; Kim, G. H. Insecticidal properties of bistrifluron against sycamore lace bug, Corythucha ciliate (Hemiptera: Tingidae). *J. Pestic. Sci.* 2008, 33, 44 – 50.

[14] Sun, R. F. ; Li, Y. Q. ; Lü. , M. Y. ; Xiong, L. X. ; Wang, Q. M. Synthesis, larvicidal activity, and SAR studies of new benzoylphenylureas containing oxime ether and oxime ester group. *Bioorg. Med. Chem. Lett.* 2010, 20, 4693 – 4699.

[15] Zhang, J. – F. ; Xu, J. – Y. ; Wang, B. – L. ; Li, Y. – X. ; Xiong, L. – X. ; Li, Y. – Q. ; Ma, Y. ; Li, Z. – M. Synthesis and insecticidal activities of novel anthranilic diamides containing acylthiourea and acylurea. *J. Agric.* Food Chem. 2012, 60, 7565 – 7572.

[16] Dong, W. L. ; Xu, J. Y. ; Xiong, L. X. ; Liu, X. H. ; Li, Z. M. Synthesis, structure and biological activities of some novel anthranilic acid esters containing N – pyridylpyrazole. *Chin. J. Chem.* 2009, 27, 579 – 586.

[17] Schneider, L. ; Graham, D. E. Method of making *N* – {[(4 – *chlorophenyl*) *arnino*] *carbonyl*} – 2, 6 – *difluorobenzamide.* U. S. 4117009 – A, 1978.

[18] George, T. ; Mehta, D. V. ; Tahilramani, R. ; David, J. ; Talwalker, P. K. Synthesis of some striazoles with potential analgesic and antiinflammatory activities. *J. Med. Chem.* 1971, 14, 335 – 338.

[19] Xu, J. ; Dong, W. ; Xiong, L. ; Li, Y. ; Li, Z. Design, synthesis and biological activities of novel amides (sulfonamides) containing Npyridylpyrazole. *Chin. J. Chem.* 2009, 27, 2007 – 2012.

[20] Lahm, G. P. ; Selby, T. P. ; Stevenson, T. M. ; George, P. L. ; Philip, L. G. ; Paul, S. T. ; Martin, S. T. *Arthropodicidal anthranilamides.* WO 2003015519 A1, 2003.

[21] Dong, W. L. ; Liu, X. H. ; Xu, J. Y. ; Li, Z. M. Design and synthesis of novel anthranilic diamides containing 5, 7 – dimethyl[1, 2, 4]triazolo – [1, 5 – a]pyrimidine. *J. Chem. Res.* 2008, 9, 530 – 533.

[22] Denton, T. T. ; Zhang, X. ; Cashman, J. R. 5 – Substituted, 6 – substituted, and unsubstituted 3 – heteroaromatic pyridine analogues of nicotine as selective inhibitors of cytochrome P – 450 2A6. *J. Med. Chem.* 2005, 48(1), 224 – 239.

[23] David, A. ; Romina, B. ; Francesco, F. ; Ferdinando, P. ; Luigi, V. TBAF – catalyzed synthesis of 5 – substituted 1*H* – tetrazoles under solventless conditions. *J. Org. Chem.* 2004, 69(8), 2896 – 2898.

[24] He, H. – D. ; Tu, S. – Z. Synthesis and structure – activity relationship of analogs of intravenous anesthetic – xylazine. *Chinese J. Pharm.* 1992, 23(5), 208 – 211.

[25] Hirashima, A.; Yoshii, Y.; Eto, M. Synthesis and biological activity of 2 – aminothiazolines and 2 – mercaptothiazolines as octopaminergic agonists. *Agric. Biol. Chem.* 1991, 55(10), 2537 – 2545.

[26] Abbott, W. S. A method of computing the effectiveness of an insecticide. *J. Econ. Entomol.* 1925, 18, 265 – 267.

[27] Raymond, M. Presentation d' um programme basicd' analyse logprobit pour miero – ordinateur. *Cah. ORSTOM Ser. Ent. Med. Parasitol.* 1985, 23, 117 – 121.

[28] Wu, Y. D.; Shen, J. L.; Chen, J.; Lin, X. W.; Li, A. M. Evaluation of two resistance monitoring methods in *Helicoverpa armigera*: topical application method and leaf dipping method. *Zhiwu* Baohu 1996, 22(5), 3 – 6.

[29] Ma, H.; Wang, K. Y.; Xia, X. M.; Zhang, Y.; Guo, Q. L. The toxicity testing of five insecticides to different instar larvae of Spodoptera exigua. *Xiandai Nongyao* 2006, 5, 44 – 46.

[30] Busivine, J. R. *Recommended Methods for Measurement of Pest Resistance to Pesticides*; FAO Plant Production and Protection Paper 21; FAO: Rome, Italy, 1980; pp 3 – 13 and 119 – 122.

[31] Chen, L.; Huang, Z. Q.; Wang, Q. M.; Shang, J.; Huang, R. Q.; Bi, F. C. Insecticidal benzoylphenylurea – S – carbamate: a new propesticide with two effects of both benzoylphenylureas and carbamates. *J. Agric. Food Chem.* 2007, 55, 2659 – 2663.

[32] Scott, A. P.; Radom, L. Harmonic vibrational frequencies: an evaluation of Hartree – Fock, Moeller – Plesset, quadratic configuration interaction, density functional theory, and semiempirical scale factors. *J. Phys. Chem.* 1996, 100, 16502 – 16513.

[33] Frisch, M. J.; Trucks, G. W.; Schlegel, H. B.; Scuseria, G. E.; Robb, M. A.; Cheeseman, J. R.; et al. *Gaussian* 03, revision C01; Gaussian, Inc.: Wallingford, UK, 2004.

[34] Nitulescu, G. M.; Draghici, C.; Missir, A. V. Synthesis of new pyrazole derivatives and their anticancer evaluation. Eur. *J. Med. Chem.* 2010, 45, 4914 – 4919.

[35] Liu, X. – H.; Zhang, C. – Y.; Guo, W. – C.; Li, Y. – H.; Chen, P. – Q.; Wang, T.; Dong, W. – L.; Wang, B. – L.; Sun, H. – W.; Li, Z. – M. Synthesis, bioactivity and SAR study of *N'* – (5 – substituted – 1, 3, 4 – thiadiazol – 2 – yl) – Ncyclopropylformyl – thioureas as ketol – acid reductoisomerase inhibitors. *J. Enzym. Inhib. Med. Chem.* 2009, 24(2), 545 – 552.

[36] Chen, P. Q.; Liu, X. H.; Sun, H. W.; Wang, B. L.; Li, Z. M.; Lai, C. M. Molecular simulation studies of interactions between ketol – acid reductoisomerase and its inhibitors. *Acta Chem. Sinica* 2007, 65, 1693 – 1701.

[37] Liu, X. – H.; Zhao, W. – G.; Wang, B. – L.; Li, Z. – M. Synthesis, bioactivity and DFT structure. activity relationship study of novel 1, 2, 3 – thiadiazole derivatives. Res. *Chem. Intermed.* 2012, 38(8), 1999 – 2008.

[38] Liu, X. – H.; Pan, L.; Tan, C. – X.; Weng, J. – Q.; Wang, B. – L.; Li, Z. – M. Synthesis, crystal structure, bioactivity and DFT calculation of new oxime ester derivatives containing cyclopropane moiety. Pestic. Biochem. *Physiol.* 2011, 101, 143 – 147.

Design, Synthesis and Structure - activity of *N* - Glycosyl - 1 - pyridyl - 1*H* - pyrazole - 5 - carboxamide as Inhibitors of Calcium Channels*

Yunyun Zhou[1], Yuxin Li[1], Yiming Li[1], Xiaoping Yang[2]
Mingzhen Mao[1], Zhengming Li[1]

([1]State Key Laboratory of Elemento - Organic Chemistry, Institute of Elemento - Organic Chemistry, Nankai University, Tianjin, 300071, China;
[2]School of Pharmacy, University of Wyoming, Laramie WY 82071, USA)

Abstract Carbohydrates, with broad - spectrum structures and biological functions, are key organic compounds in nature, along with nucleic acids and proteins. As part of our ongoing efforts to develop a new class of pesticides with novel mechanism of action, a series of novel *N* - glycosyl - 1 - pyridyl - 1*H* - pyrazole - 5 - carboxamide was designed and synthesized *via* the reactions of glycosyl methanamides and pyridyl - pyrazole acid. The compounds were characterized by ^{1}H NMR and ^{13}C NMR. The bioassay results indicate that some of these compounds exhibit moderate insecticidal activities and assessed as potential inhibitors of calcium channels. The modulation of voltage - gated calcium channels by compounds **4**a and **5**a in the central neurons isolated from the third instar larvae of Spodoptera exigua was studied by whole - cell patch - clamp technique. In addition, compound **5**a inhibits the recorded calcium currents reversible on washout. Experimental results also indicate that compound **5**a did not release stored calcium from the Endoplasmic Reticulum. The present work demonstrates that *N* - glycosyl - 1 - pyridyl - 1*H* - pyrazole - 5 - carboxamides cannot be used as possible inhibitors of calcium channels for developing novel pesticides.

Key words *N* - glycosyl; pyridyl; pyrazole; insecticidal activity; calcium channel

1 Introduction

Voltage - sensitive (or voltage - gated) calcium channels are transmembrane proteins that allow Ca^{2+} influx to occur when the open configuration is produced during a depolarization signal[1]. Pharmacological, electrophysiological, and ligandbinding studies have indicated the presence of diverse voltage - sensitive calcium channels in insects which appear to be pharmacologically distinct from those of vertebrates[2].

Chlorantraniliprole[3-6] (Fig. 1), a novel anthranilic diamide insecticide discovered by DuPont for the control of Lepidopteran and Coleopteran pests, shares the mode of action with flubendiamide[7,8]. In our previous study[9-12], the modulation of voltage - gated calcium channels by chlorantraniliprole in the central neurons isolated from the third instar larvae of Spodoptera ...a was investigated by whole - cell patch - clamp technique[13]. We found that chlorantra-

...n *Chemical Research in Chinese Universities*, 2013, 29(2): 249 - 255. Supported by the National Basic Re... No. 2010CB126106), the National Natural Science Foundation of China (No. 31000861), the Tianjin ..., China (No. 11JCYBJC08600) and the National Key Technologies R&D Program of China (No.

niliprole inhibited the recorded calcium currents in a concentration – dependent manner and the current – voltage ($I-V$) relationship curves of whole – cell calcium channels shifted in a negative direction by 10mV compared with the control group. It has been shown that modification of these three substructures has benefits to obtaining better bioactivity. In the previous paper[14], N – pyridylpyrazole moiety was described as an important pharmacophore in this series of compounds.

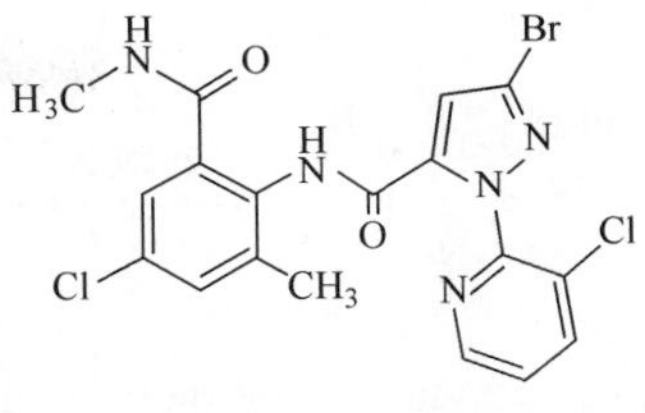

Fig. 1 Chemical structure of chlorantraniliprole

Carbohydrates participate in various vital processes, showing important physiological and biological activities *in vivo*. Yet carbohydrates linked with N – pyridylpyrazole moiety have not been reported. As part of our efforts[15,16] to develop a novel class of pesticides with new mechanism of action, we synthesized a series of novel compounds by replacing the substructures with glycosides. Additionally, we also replaced N – pyridylpyrazole by cyclopropyl to test their insecticidal activities. Their pesticidal activities *in vitro* against *S. exigua* were evaluated and the relationships between structure and bioactivity were also explored. The effect of these compounds on voltage – gated calcium channels in the central neurons was studied by whole – cell patch – clamp technique as well.

2 Experimental

2.1 General Methods

Melting points were measured on a Yanaco MP – 500 micro – melting point apparatus. ^1H NMR and ^{13}C NMR spectra were recorded in $CDCl_3$ or DMSO – d_6 on a Bruker AC – 400 instrument with Me_4Si as an internal standard. Mass spectral or high – resolution mass spectral (HRMS) data were recorded on a VG ZAB – HS mass spectrometer with the fast atom bombardment (FAB) method or on a Varian QFT – ESI instrument. All the chemical reagents and solvents used were of analytical grade that were obtained from commercial sources unless otherwise indicated.

2.2 Syntheses of Intermediates

The synthetic route is shown in Scheme 1. The substituents are listed in Table 1. Glycosyl azides[17] (**1**a – **1**d) and 3 – chloro(bromo) – 1 – (3 – chloropyridin – 2 – yl) – 1H – pyrazole – 5 – carboxylic acid (**3**a, **3**b) were prepared according to the literature procedures[4,18]. Glycosyl amines (**2**a – **2**d) were synthesized *via* the efficient reduction of glycosyl azides with hydrogen and palladium in ethanol.

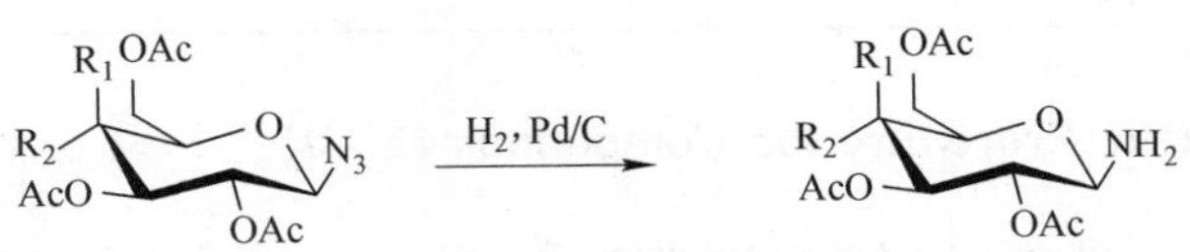

H_2, Pd/C

1a—1d → 2a—2d

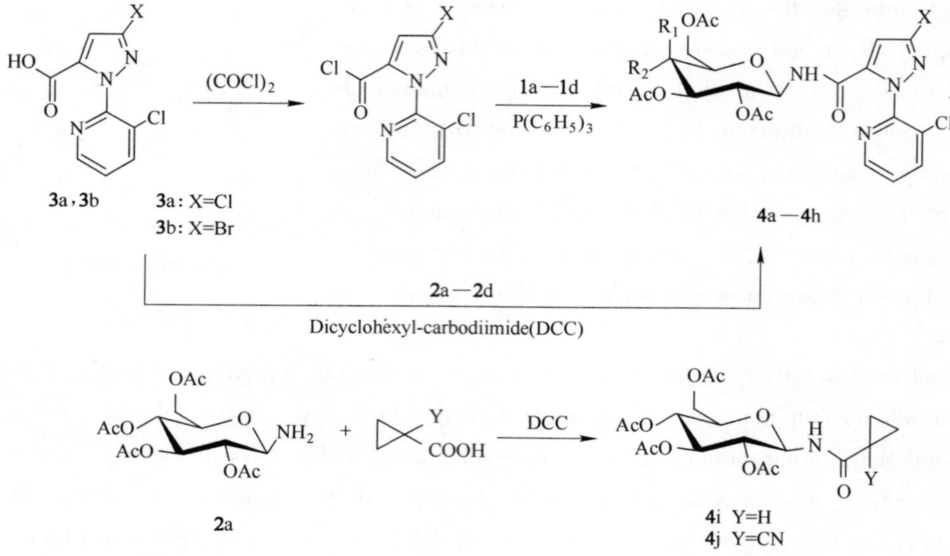

HO, X, N, N, O, N, Cl

3a, 3b　3a: X=Cl　3b: X=Br

$(COCl)_2$

Cl, X, N, N, O, N, Cl

1a—1d $P(C_6H_5)_3$

R_1, OAc, R_2, AcO, O, OAc, NH, X, N, N, O, N, Cl

4a—4h

2a—2d

Dicyclohexyl-carbodiimide(DCC)

OAc, AcO, AcO, O, NH_2, OAc　+　Y, COOH

2a

DCC

OAc, AcO, AcO, O, H, N, OAc, O, Y

4i Y=H
4j Y=CN

Scheme 1　Syntheses of compounds **4a – 4j**

Table 1　Substituents of the compounds

Compound	R_1	R_2	X
1a,2a,4a	H	OAc	Cl
1a,2a,4b	H	OAc	Br
1b,2b,4c	OAc	H	Cl
1b,2b,4d	OAc	H	Br
1c,2c,4e	H	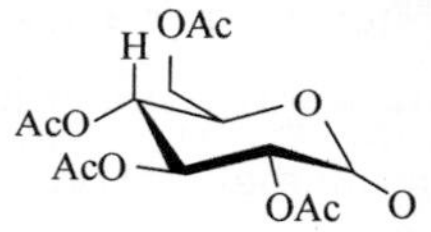 H, OAc, AcO, O, AcO, OAc, O	Cl
1c,2c,4f	H	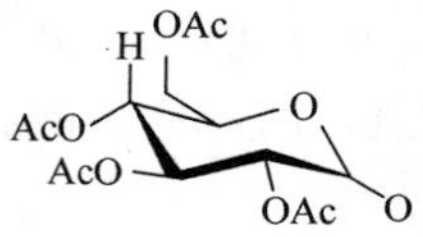 H, OAc, AcO, O, AcO, OAc, O	Br
1d,2d,4g	H	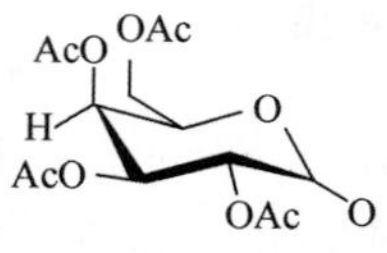 AcO, OAc, H, O, AcO, OAc, O	Cl
1d,2d,4h	H	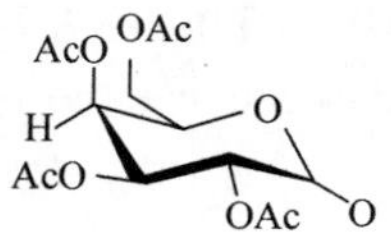 AcO, OAc, H, O, AcO, OAc, O	Br

2.3　General Synthetic Procedure for Compounds 4a – 4j

2.3.1　Synthesis of 3 – Chloro – 1 – (chloropyridin – 2 – yl) – *N* – (2,3,4,6 – tetra – O – acetyl – β – D – glucopyranosyl) – 1H – pyra zole – 5 – carboxamide (4a) via Staudinger Reaction

The synthesis of compound **4**a is based on lit. [19]. To a suspension of 3 – chloro – 1 – (3 – chloropyridin – 2 – yl) – 1*H* – pyrazole – 5 – carboxylic acid 3a(0. 258g, 1mmol) in anhydrous dichloromethane(30mL) were added oxalyl chloride(0. 381g, 3mmol) and 0. 05mL of dimethylformamide until the solution became clear. After stirring overnight, the solvent was evaporated. 2,3,4,6 – Tetra – *O* – acetyl – *β* – *D* – glucopyranosyl azide(**1**a, 0. 373g, 1mmol) was added to the solution of acyl chloride in 10mL of dichloromethane and then the solution of triphenylphosphine(0. 262g, 1mmol) in 10mL of dichloromethane was added to the mixture. After 2h agitation, 30mL of dichloromethane was added to it, neutralized by addition of aqueous sodium bicarbonate and washed with water and brine. The evaporated residue was subjected to flash chromatography on a silica gel column eluted with petroleum ether/ethyl acetate(2∶1, volume ratio) to give compound **4**a as a white solid(0. 05g, 16%).

2. 3. 2 General Synthetic Procedure for Compounds 4a – 4j

To a solution of 2,3,4,6 – tetra – *O* – acetyl – *β* – *D* – glucopyranosyl amine(**2**a, 0. 347g, 1mmol) and 3 – chloro(bromo) – 1 – (3 – chloropyridin – 2 – yl) – 1*H* – pyrazole – 5 – carboxylic acid(**3**a) in anhydrous dichloromethane(30mL) was added DCC(0. 206g, 1mmol) in small portions at 0℃. The reaction mixture was allowed to warm up to room temperature and stirred overnight. After the stirring, the mixture was filtered, the organic phase was washed with water and brine and dried with anhydrous magnesium sulfate. The crude product was evaporated and then purified on a silica – gel column eluted with petroleum ether/ethyl acetate(2∶1, volume ratio) to obtain compound **4**a(0. 23g) as a white solid: yield 40%, m. p. 102 – 106℃. ^{1}H NMR (400MHz, $CDCl_3$), *δ*: 8. 48(d, 1H, *J* = 4. 5Hz, NH), 7. 90(d, 1H, *J* = 8. 0Hz, pyridyl – H), 7. 43(dd, 1H, *J* = 3. 2, 12. 8Hz, pyridyl – H), 7. 14(d, 1H, *J* = 8. 8Hz, pyridyl – H), 6. 71(s, 1H, pyrazolyl – H), 5. 32 – 4. 97(m, 4H, H – 1,2,3,6), 4. 32(dd, 1H, *J* = 8. 4 16. 7Hz, H – 4), 4. 04(d, 1H, *J* = 12. 3Hz, H – 6), 3. 76(d, 1H, *J* = 9. 7Hz, H – 5), 2. 03, 2. 04, 2. 07(3s, 12H, 4OAc). ^{13}C NMR(100MHz, $CDCl_3$), *δ*: 171. 68, 170. 57, 169. 83, 169. 53(4C, 4CO), 157. 35(CONH), 148. 79, 146. 90, 141. 66, 139. 12, 137. 54, 129. 17, 125. 98, 107. 17(8C, 3pyrazolyl – C, 5pyridyl – C), 78. 44, 73. 69, 72. 34, 70. 68, 68. 06, 61. 49(6C, 6glucosyl – C), 20. 71, 20. 57(4C, $4CH_3CO$). FABMS, *m/z* calcd. for [$C_{23}H_{24}Cl_2N_4O_{10}$ + H]$^+$: 586; found: 587.

3 – Bromo – 1 – (3 – chloropyridin – 2 – yl) – *N* – (2,3,4,6 – tetra – *O* – acetyl – *β* – *D* – glucopyranosyl) – 1*H* – pyrazole – 5 – carboxamide(**4**b): yield 33%, a white solid. m. p. 231 – 233℃. ^{1}H NMR(400MHz, $CDCl_3$), *δ*: 8. 49(d, *J* = 4. 6Hz, 1H, NH), 7. 90(d, 1H, *J* = 8. 0Hz, pyridyl – H), 7. 43 – 7. 37(m, 2H, pyridyl – H), 6. 84(s, 1H, pyrazolyl – H), 5. 29 – 4. 98(m, 4H, 4glucosyl – H), 4. 28(m, 1H, glucosyl – H), 4. 04(d, 1H, *J* = 12. 2Hz, glucosyl – H), 3. 76(d, 1H, *J* = 8. 2Hz, glucosyl – H), 2. 03, 2. 06, 2. 07(3s, 12H, 4OAc). ^{13}C NMR (100MHz, $CDCl_3$), *δ*: 171. 48, 170. 63, 169. 88, 169. 60(4C, 4CO), 157. 29(CONH), 157. 20, 148. 77, 146. 90, 139. 10, 137. 76, 129. 13, 128. 08, 126. 00, 110. 59(8C, 3pyrazolyl – C, 5pyridyl – C), 78. 32, 73. 66, 72. 45, 70. 62, 68. 07, 61. 57(6C, 6glucosyl – C), 20. 72, 20. 57(4C, $4CH_3CO$). FABMS, *m/z* calcd for [$C_{23}H_{24}BrClN_4O_{10}$ + H]$^+$: 630; found: 631.

3 – Chloro – 1 – (3 – chloropyridin – 2 – yl) – *N* – (2,3,4,6 – tetra – *O* – acetyl – *β* – *D* – galactopyranosyl) – 1*H* – pyrazole – 5 – carboxamide(**4**c): yield 42%, a white solid. m. p.

197 – 199℃. ^{1}H NMR(400MHz, $CDCl_3$), δ: 8.49(d, J = 4.6Hz, 1H, NH), 7.90(d, 1H, J = 7.7Hz, pyridyl – H), 7.43(dd, 1H, J = 4.7, 8.0Hz, pyridyl – H), 7.09(d, 1H, J = 7.7Hz, pyridyl – H), 6.71(s, 1H, pyrazolyl – H), 5.44 – 5.15(m, 4H, 4galactosyl – H), 4.15 – 3.97 (m, 3H, 3galactosyl – H), 2.17, 2.08, 2.02(3s, 12H, 4OAc). ^{13}C NMR(100MHz, $CDCl_3$), δ: 172.05, 170.34, 170.00, 169.76 (4C, 4CO), 157.28 (CONH), 148.78, 146.93, 141.66, 139.06, 137.54, 129.13, 125.97, 106.97 (8C, 3pyrazolyl – C, 5pyridyl – C), 78.74, 72.40, 70.50, 68.37, 66.95, 60.83 (6C, 6galactosyl – C), 20.84, 20.62, 20.56 (4C, $4CH_3CO$). FABMS, m/z calcd. for [$C_{23}H_{24}Cl_2N_4O_{10}$ + H]$^+$: 586; found: 587.

3 – Bromo – 1 – (3 – chloropyridin – 2 – yl) – N – (2,3,4,6 – tetra – O – acetyl – β – D – galactopyranosyl) – 1H – pyrazole – 5 – carboxamide(**4**d): yield 48%, a white solid. m. p. 192 – 194℃. ^{1}H NMR (400MHz, $CDCl_3$), δ: 8.49 (d, 1H, J = 4.7Hz, NH), 7.90 (d, 1H, J = 8.0Hz, pyridyl – H), 7.44(dd, 1H, J = 4.7, 7.9Hz, pyridyl – H), 7.10(d, 1H, J = 7.3Hz, pyridyl – H), 6.79(s, 1H, pyrazolyl – H), 5.44 – 5.13(m, 4H, 4galactosyl – H), 4.10 – 3.97 (m, 3H, 3galactosyl – H), 2.17, 2.08, 2.01(3s, 12H, 4OAc). ^{13}C NMR(100MHz, $CDCl_3$), δ: 172.03, 170.35, 170.00, 169.77 (4C, 4CO), 157.15 (CONH), 148.76, 146.93, 139.06, 137.75, 129.12, 128.13, 125.98, 110.33 (8C, 3pyrazolyl – C, 5pyridyl – C), 78.72, 72.40, 70.52, 68.35, 66.96, 60.85 (6C, 6galactosyl – C), 20.84, 20.65, 20.62, 20.56 (4C, $4CH_3CO$). FABMS, m/z calcd. for [$C_{23}H_{24}BrClN_4O_{10}$ + H]$^+$: 630; found: 631.

3 – Chloro – 1 – (3 – chloropyridin – 2 – yl) – N – [2,3,4,6 – tetra – O – acetyl – α – D – glucopyranosyl – (1→4) – 2,3,6 – tri – O – acetyl – β – D – glucopyranosyl] – 1H – pyrazole – 5 – carboxamide (**4**e): yield 50%, a white solid. m. p. 213 – 215℃. ^{1}H NMR (400MHz, $CDCl_3$), δ: 8.48(d, 1H, J = 4.5Hz, NH), 7.90(d, 1H, J = 5.9Hz, pyridyl – H), 7.44(t, 1H, pyridyl – H), 7.01(d, 1H, J = 8.8Hz, pyridyl – H), 6.67(s, 1H, pyrazolyl – H), 5.38 – 4.08 (m, 7H, 7maltosyl – H), 4.40 – 3.71(m, 7H, 7maltosyl – H), 2.12, 2.10, 2.06, 2.04, 2.02 (6s, 21H, 7OAc). ^{13}C NMR(100MHz, $CDCl_3$), δ: 171.70, 170.65, 170.52, 170.40, 169.87, 169.62, 169.45 (7C, 7CO), 157.24 (CONH), 148.73, 146.91, 141.61, 139.15, 137.64, 129.13, 125.99, 107.10 (8C, 3pyrazolyl – C, 5pyridyl – C), 95.64, 78.01, 74.55, 74.03, 72.54, 71.41, 69.98, 69.25, 68.61, 67.93, 62.55, 61.41 (12C, 12maltosyl – C), 20.86, 20.81, 20.68, 20.61, 20.58(7, $7CH_3CO$). FABMS, m/z calcd. for [$C_{35}H_{40}Cl_2N_4O_{18}$ + H]$^+$: 874; found: 875.

3 – Bromo – 1 – (3 – chloropyridin – 2 – yl) – N – [2,3,4,6 – tetra – O – acetyl – α – D – glucopyranosyl – (1→4) – 2,3,6 – tri – O – acetyl – β – D – glucopyranosyl] – 1H – pyrazole – 5 – carboxamide (**4**f): yield 47%, a white solid. m. p. 216 – 218℃. ^{1}H NMR (400MHz, $CDCl_3$), δ: 8.49(d, 1H, J = 4.4Hz, NH), 7.91 (d, 1H, J = 8.0Hz, pyridyl – H), 7.44(dd, 1H, J = 6.2 7.7Hz, pyridyl – H), 7.15(d, 1H, J = 9.0Hz, pyridyl – H), 6.78(s, 1H, pyrazolyl – H), 5.38 – 4.82(m, 7H, 7maltosyl – H), 4.37 – 3.72(m, 7H, 7maltosyl – H), 2.12, 2.10, 2.06, 2.04, 2.03(6s, 21H, 7OAc). ^{13}C NMR(100MHz, $CDCl_3$), δ: 171.58, 170.64, 170.53, 170.40, 169.87, 169.65, 169.45 (7C, 7CO), 157.14 (CONH), 148.72, 146.92, 139.13, 137.87, 129.10, 128.04, 126.00, 110.49 (8C, 3pyrazolyl – C, 5pyridyl – C), 95.69, 77.96, 74.68, 74.02, 72.58, 71.40, 70.00, 69.28, 68.60, 67.93, 62.60, 61.42 (12C, 12maltosyl – C), 20.87, 20.81, 20.67, 20.58(7, $7CH_3CO$). FABMS, m/z calcd. for [$C_{35}H_{40}BrClN_4O_{18}$ +

H] $^{+}$:918;found:919.

3 – Chloro – 1 – (3 – chloropyridin – 2 – yl) – *N* – [2,3,4,6 – tetra – *O* – acetyl – *β* – *D* – galactopyranosyl – (1→4) – 2,3,6 – tri – *O* – acetyl – *β* – D – glucopyranosyl] – 1*H* – pyrazole – 5 – carboxamide (**4**g): yield 53%, a white solid. m. p. 211 – 213℃. ^{1}H NMR (400MHz, $CDCl_3$), *δ*: 8. 48 (d, *J* = 4. 5Hz, 1H, NH), 7. 89 (d, 1H, *J* = 8. 0Hz, pyridyl – H), 7. 44 (dd, 1H, *J* = 8. 6, 4. 9Hz, pyridyl – H), 7. 09 (d, 1H, *J* = 8. 7Hz, pyridyl – H), 6. 68 (s, 1H, pyrazolyl – H), 5. 36 – 4. 87 (m, 6H, 6lactosyl – H), 4. 47 – 3. 67 (m, 8H, 8lactosyl – H), 2. 15, 2. 10, 2. 06, 2. 02, 1. 96 (5s, 21H, 7OAc). ^{13}C NMR (100MHz, $CDCl_3$), *δ*: 171. 84, 170. 36, 170. 28, 170. 13, 170. 07, 169. 28, 168. 97 (7C, 7CO), 157. 30 (CONH), 148. 76, 146. 89, 141. 58, 139. 12, 137. 68, 129. 13, 125. 96, 107. 07 (9C, 3pyrazolyl – C, 5pyridyl – C, CONH), 100. 96, 78. 26, 75. 93, 74. 59, 72. 07, 71. 01, 70. 96, 70. 76, 69. 02, 66. 60, 61. 78, 60. 82 (12C, 12lactosyl – C), 20. 76, 20. 64, 20. 50 (7, 7CH_3CO). FABMS, *m/z* calcd. for [$C_{35}H_{40}Cl_2N_4O_{18}$ + H] $^{+}$:874;found:875.

3 – Bromo – 1 – (3 – chloropyridin – 2 – yl) – *N* – [2,3,4,6 – tetra – *O* – acetyl – *β* – *D* – galactopyranosyl – (1→4) – 2,3,6 – tri – *O* – acetyl – *β* – *D* – glucopyranosyl] – 1*H* – pyrazole – 5 – carboxamide (**4h**): yield 57%, a white solid. m. p. 170 – 171℃. ^{1}H NMR (400MHz, $CDCl_3$), *δ*: 8. 48 (d, *J* = 4. 6Hz, 1H, NH), 7. 89 (d, 1H, *J* = 8. 0Hz, pyridyl – H), 7. 43 (dd, 1H, *J* = 4. 8, 8. 0Hz, pyridyl – H), 7. 17 (d, 1H, *J* = 8. 8Hz, pyridyl – H), 6. 78 (s, 1H, pyrazolyl – H), 5. 36 – 4. 88 (m, 6H, 6lactosyl – H), 4. 47 – 3. 66 (m, 8H, 8lactosyl – H), 2. 16, 2. 10, 2. 08, 2. 06, 2. 02, 1. 98 (6s, 21H, 7OAc). ^{13}C NMR (100MHz, $CDCl_3$), *δ*: 171. 77, 170. 39, 170. 32, 170. 15, 170. 09, 169. 32, 169. 00 (7C, 7CO), 157. 18 (CONH), 148. 72, 146. 89, 139. 12, 137. 89, 129. 09, 128. 02, 125. 98, 110. 44 (9C, 3pyrazolyl – C, 5pyridyl – C, CONH), 100. 92, 78. 20, 75. 92, 74. 59, 72. 14, 70. 96, 70. 73, 69. 02, 66. 62, 61. 81, 60. 83 (12C, 12lactosyl – C), 20. 84, 20. 76, 20. 64, 20. 58, 20. 50 (7C, 7CH_3CO). FABMS, *m/z* calcd. for [$C_{35}H_{40}BrClN_4O_{18}$ + H] $^{+}$:918;found:919.

N – (2,3,4,6 – *tetra* – *O* – *acetyl* – *β* – *D* – *glucopyranosyl*) – *cyclopropanecarboxamide* (**4i**): yield 56%, a white solid. m. p. 183 – 186℃. ^{1}H NMR (400MHz, $CDCl_3$), *δ*: 6. 43 (d, 1H, *J* = 9. 3Hz, glucosyl – H, NH), 5. 34 – 5. 25 (m, 2H, glucosyl – H), 5. 08 (t, 1H, *J* = 9. 6Hz, glucosyl – H), 4. 96 (t, 1H, *J* = 9. 6Hz, glucosyl – H), 5. 33 (dd, 1H, *J* = 4. 0, 12. 4Hz, glucosyl – H), 4. 09 (d, 1H, *J* = 12. 5Hz, glucosyl – H), 3. 82 (d, 1H, *J* = 9. 0Hz, glucosyl – H), 2. 09, 2. 07, 2. 03 (3s, 12H, CH_3), 1. 34 (m, 1H, cyclopropyl – CH), 0. 99 (d, 2H, *J* = 3. 9Hz, cyclopropyl – CH_2), 0. 81 (d, 2H, *J* = 7. 0Hz, cyclopropyl – CH_2). ^{13}C NMR (100MHz, $CDCl_3$), *δ*: 174. 13, 171. 13, 170. 64, 169. 89, 169. 60 (5C, CONH, $COCH_3$), 78. 29, 73. 53, 72. 75, 70. 65, 68. 17, 61. 64 (6C, glucosyl – C), 20. 75, 20. 70, 20. 60 (4C, CH_3CO), 14. 87 (1C, cyclopropyl—CH), 8. 34, 8. 15 (2C, cyclopropyl – CH_2). FABMS, *m/z* calcd. for [$C_{18}H_{25}NO_{10}$ + H] $^{+}$:415;found:416.

1 – Cyano – *N* – (2,3,4,6 – tetra – *O* – acetyl – *β* – D – glucopyranosyl) – cyclopropanecarboxamide (4j): yield 60%, a white solid. m. p. 177 – 179℃. ^{1}H NMR (400MHz, $CDCl_3$), *δ*: 8. 02 (d, 1H, *J* = 9. 4Hz, glucosyl – H, NH), 5. 35 – 5. 31 (m, 2H, glucosyl – H), 5. 13 – 5. 05 (m, 2H, glucosyl – H), 4. 30 (dd, 1H, *J* = 4. 5, 12. 5Hz, glucosyl – H), 4. 10 (1H, *J* = 12. 2Hz,

glucosyl – H), 3.89 (dd, 1H, J = 2.6Hz, glucosyl – H), 3.80 (t, J = 6.6Hz, cyclopropyl—CH_2), 3.15 (t, J = 6.6Hz, cyclopropyl – CH_2), 2.09, 2.05, 2.04 (3s, 12H, CH_3). ^{13}C NMR (100MHz, $CDCl_3$), δ: 170.55, 169.85, 169.49, 156.27 (4C, $COCH_3$), 155.51 (1C, CONH), 151.95 (1C, CN), 121.24 (1C, C – CN), 78.44, 73.97, 72.58, 70.41, 67.90, 61.55 (6C, glucosyl – C), 40.32, 30.79 (2C, cyclopropyl – CH_2), 20.70, 20.62, 20.56 (4C, CH_3CO). FABMS, m/z calcd. for $[C_{19}H_{26}N_2O_{10} + H]^+$: 440; found: 441.

2.4 General Synthetic Procedure for Compounds 5a – 5h

The synthetic route of compounds **5a** – **5h** is shown in Scheme 2. The substituents of compounds **5a** – **5h** are listed in Table 2.

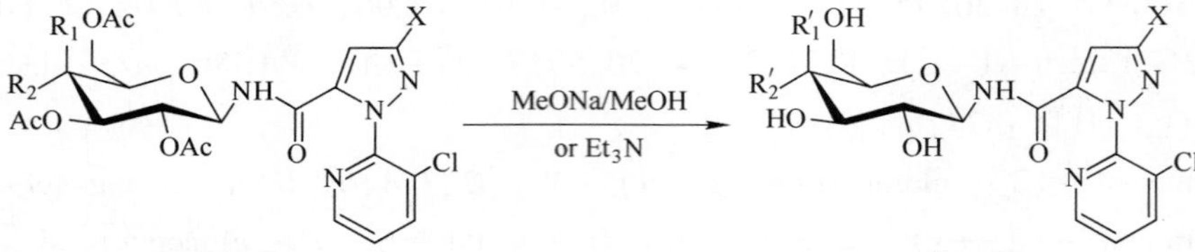

MeONa/MeOH or Et_3N

4a—4h → 5a—5h

Scheme 2 General synthetic procedure of compounds 5a – 5h

Table 2 Substituents of compounds 5a – 5h

Compound	R'_1	R'_2	X
5a	H	OH	Cl
5b	H	OH	Br
5c	OH	H	Cl
5d	OH	H	Br
5e	H	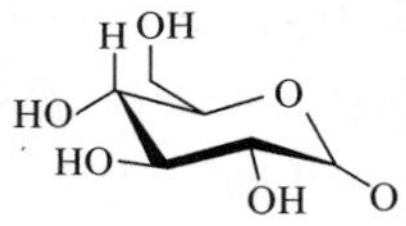 (glycosyloxy structure: H, OH, HO, HO, OH, O)	Cl
5f	H	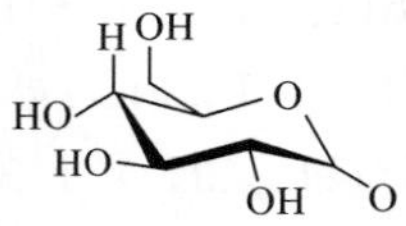 (glycosyloxy structure: H, OH, HO, HO, OH, O)	Br
5g	H	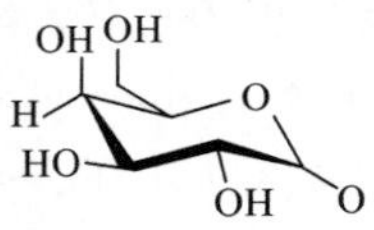 (glycosyloxy structure: OH, OH, H, HO, OH, O)	Cl
5h	H	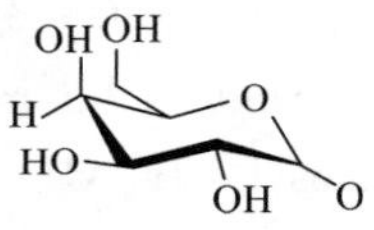 (glycosyloxy structure: OH, OH, H, HO, OH, O)	Br

To an ice - cooled solution of sodium(28 mg, 1.22 mmol) in 28mL of anhydrous methanol, 3 - chloro - 1 - (3 - chloropyridin - 2 - yl) - *N* - (2,3,4,6 - tetra - *O* - acetyl - *β* - *D* - glucopyranosyl) - 1*H* - pyrazole - 5 - carboxamide(**4**a, 0.176g, 0.3mmol) was added portion - wise, and then stirred at 0℃ until the reaction completed(thin layer chromatogram, TLC). After the mixture was neutralized by strongly acid cation exchange resin, the result suspension was filtered and the filtrate was removed under vaccum. The residue was subjected to flash chromatography on silica gel with ethyl acetate/methanol(10: 1, volume ratio) to give the title compound as a white solid.

3 - Chloro - 1 - (3 - chloropyridin - 2 - yl) - *N* - glycosyl - 1*H* - pyrazole - 5 - carboxamide(**5**a): yield 77%, a white solid. m. p. 148 - 151℃. ^1H NMR(400MHz, DMSO - d_6), δ: 9.18(d, 1H, *J* = 8.9Hz, NH), 8.20(d, 1H, *J* = 8.1Hz, pyridyl - H), 7.90(d, 1H, *J* = 4.6Hz, pyridyl - H), 7.65(dd, 1H, *J* = 8.0, 4.8Hz, pyridyl - H), 7.34(s, 1H, pyrazolyl - H), 5.08 - 3.08(m, 11H, 7glucosyl - H, 4OH). ^{13}C NMR(100MHz, DMSO - d_6), δ: 156.86, 148.81, 147.03, 139.37, 139.17, 138.76, 128.23, 126.70, 107.061(9C, 3pyrazolyl - C, 5pyridyl - C, CONH), 79.65, 78.80, 77.32, 72.01, 69.78, 60.73 (6C, 6glucosyl - C). HRMS, *m/z* calcd. for $[C_{15}H_{16}Cl_2N_4O_6 + H]^+$: 419; found: 419.05.

3 - Bromo - 1 - (3 - chloropyridin - 2 - yl) - *N* - (*β* - *D* - glucopyranosyl) - 1*H* - pyrazole - 5 - carboxamide(**5**b): yield 80%, a white solid. m. p. 160 - 162℃. ^1H NMR(400MHz, DMSO - d_6), δ: 9.16(d, 1H, *J* = 7.6Hz, NH), 8.51(s, 1H, pyridyl - H), 8.20(d, 1H, *J* = 6.8Hz, pyridyl - H), 7.65(s, 1H, pyridyl - H), 7.43(s, 1H, pyrazolyl - H), 5.06 - 3.08(m, 11H, 7glucosyl - H, 4OH). ^{13}C NMR(100MHz, DMSO - d_6), δ: 156.74, 148.83, 147.02, 139.15, 138.91, 128.20, 126.66, 126.50, 110.471 (9C, 3pyrazolyl - C, 5pyridyl - C, CONH), 79.70, 78.79, 77.30, 71.96, 69.78, 60.72 (6C, 6glucosyl - C). FABMS, *m/z* calcd. for $[C_{15}H_{16}BrClN_4O_6 + H]^+$: 462; found: 463.

3 - Chloro - 1 - (3 - chloropyridin - 2 - yl) - *N* - (*β* - *D* - galactopyranosyl) - 1*H* - pyrazole - 5 - carboxamide (**5**c): yield 76%, a white solid. m. p. 148 - 151℃. ^1H NMR (400MHz, DMSO - d_6), δ: 9.18(d, 1H, *J* = 8.9Hz, NH), 8.50(d, 1H, *J* = 4.6Hz, pyridyl - H), 8.20(d, 1H, *J* = 8.0Hz, pyridyl - H), 7.64(dd, 1H, *J* = 4.9, 8.0Hz, pyridyl - H), 7.39 (s, 1H, pyrazolyl - H), 4.86 - 3.40(m, 11H, 7galactosyl - H, 4OH). ^{13}C NMR(100MHz, DMSO - d_6), δ: 156.84, 148.89, 146.99, 139.31, 139.13, 138.87, 128.20, 126.61, 107.11(9C, 3pyrazolyl - C, 5pyridyl - C, CONH), 80.14, 77.05, 74.02, 69.24, 68.18, 60.46 (6C, 6galactosyl - C). FABMS, *m/z* calcd. for $[C_{15}H_{16}Cl_2N_4O_6 + H]^+$: 418; found: 419.

3 - Bromo - 1 - (3 - chloropyridin - 2 - yl) - *N* - (*β* - *D* - galactopyranosyl) - 1*H* - pyrazole - 5 - carboxamide (**5**d): yield 78%, a white solid. m. p. 153 - 157℃. ^1H NMR (400MHz, DMSO - d_6), δ: 9.18(d, *J* = 8.9Hz, 1H, NH), 8.51(d, 1H, *J* = 3.8Hz, pyridyl - H), 8.20(d, 1H, *J* = 8.0Hz, pyridyl - H), 7.64(dd, 1H, *J* = 7.9, 4.8Hz, pyridyl - H), 7.45 (s, 1H, pyrazolyl - H), 4.88 - 3.38(m, 11H, 7galactosyl - H, 4OH). ^{13}C NMR(100MHz, DMSO - d_6), δ: 156.75, 148.87, 146.98, 139.11, 138.98, 128.18, 126.60, 126.50, 110.50(9C, 3pyrazolyl - C, 5pyridyl - C, CONH), 80.15, 77.03, 74.01, 69.22, 68.18, 60.46(6galactosyl - C). FABMS, *m/z* calcd. for $[C_{15}H_{16}BrClN_4O_6 + H]^+$: 462; found: 463.

3 - Chloro - 1 - (3 - chloropyridin - 2 - yl) - *N* - [*α* - *D* - glucopyranosyl - (1→4) - *β* -

D – glucopyranosyl] – 1*H* – pyrazole – 5 – carboxamide (**5**e): yield 83%, a white solid. m. p. 155 – 157℃. ^{1}H NMR(400MHz, DMSO – d_6), δ: 9.23(d, J = 6.8Hz, 1H, NH), 8.51 (s, 1H, pyridyl – H), 8.21(d, 1H, J = 6.0Hz, pyridyl – H), 7.65(s, 1H, pyridyl – H), 7.36 (s, 1H, pyrazolyl – H), 5.62 – 4.54, 3.62 – 3.06(m, 21H, 14maltosyl – H, 7OH). ^{13}C NMR (100MHz, DMSO – d_6), δ: 156.87, 148.79, 147.05, 139.36, 139.18, 138.68, 128.21, 126.71, 107.09(9C, 3pyrazolyl – C, 5pyridyl – C, CONH), 100.83, 79.58, 79.30, 77.00, 77.00, 73.44, 73.20, 72.40, 71.52, 69.83, 60.71, 60.23(12C, 12maltosyl – C). FABMS, *m/z* calcd. for[$C_{21}H_{26}Cl_2N_4O_{11}$ + H]$^+$: 580; found: 581.

3 – Bromo – 1 – (3 – chloropyridin – 2 – yl) – *N* – [α – *D* – glucopyranosyl – (1→4) – β – *D* – glucopyranosyl] – 1*H* – pyrazole – 5 – carboxamide (**5**f): yield 85%, a white solid. m. p. 179 – 182℃. ^{1}H NMR(400MHz, DMSO – d_6), δ: 9.20(d, J = 8.6Hz, 1H, NH), 8.51 (d, 1H, J = 4.4Hz, pyridyl – H), 8.21(d 1H, J = 8.0Hz, pyridyl – H), 7.65(m, 1H, pyridyl – H), 7.42(s, 1H, pyrazolyl – H), 5.61 – 4.52, 3.64 – 3.06(m, 21H, 14maltosyl – H, 7OH). ^{13}C NMR (100MHz, DMSO – d_6), δ: 156.78, 148.77, 147.03, 139.16, 138.82, 128.19, 126.68, 126.53, 110.49(9C, 3pyrazolyl – C, 5pyridyl – C, CONH), 100.81, 79.61, 79.30, 77.00, 73.44, 73.22, 72.41, 71.52, 69.86, 60.76, 60.25(12C, 12maltosyl – C). FABMS, *m/z* calcd. for[$C_{21}H_{26}BrClN_4O_{11}$ + H]$^+$: 624; found: 625.

3 – Chloro – 1 – (3 – chloropyridin – 2 – yl) – *N* – [β – *D* – galactopyranosyl – (1→4) – β – *D* – glucopyranosyl] – 1*H* – pyrazole – 5 – carboxamide (**5**g): yield 67%, a white solid. m. p. 166 – 170℃. ^{1}H NMR(400MHz, DMSO – d_6), δ: 9.20(d, J = 8.7Hz, 1H, NH), 8.51 (d, 1H, J = 4.4Hz, pyridyl – H), 8.20(d, 1H, J = 8.0Hz, pyridyl – H), 7.65(m, 1H, pyridyl – H), 7.42(s, 1H, pyrazolyl – H), 5.61 – 4.52, 3.64 – 3.06(m, 21H, 14lactosyl – H, 7OH). ^{13}C NMR (100MHz, DMSO – d_6), δ: 156.98, 148.79, 147.03, 139.16, 138.70, 128.62, 128.20, 126.68, 107.09(9C, 3pyrazolyl – C, 5pyridyl – C, CONH), 103.80, 80.41, 79.41, 76.72, 75.60, 75.50, 73.17, 71.58, 70.53, 68.12, 64.98, 60.40(12C, 12lactosyl – C). FABMS, *m/z* calcd. for[$C_{21}H_{26}Cl_2N_4O_{11}$ + H]$^+$: 580; found: 581.

3 – Bromo – 1 – (3 – chloropyridin – 2 – yl) – *N* – [β – *D* – galactopyranosyl – (1→4) – β – *D* – glucopyranosyl] – 1*H* – pyrazole – 5 – carboxamide (**5**h): yield 66%, a white solid. m. p. 194 – 197℃. ^{1}H NMR(400MHz, DMSO – d_6), δ: 9.22(d, J = 8.8Hz, 1H, NH), 8.51 (dd, 1H, J = 1.3, 4.7Hz, pyridyl – H), 8.20(dd, 1H, J = 1.2, 8.1Hz, pyridyl – H), 7.65(dd, 1H, J = 4.7, 8.1Hz, pyridyl – H), 7.41(s, 1H, pyrazolyl – H), 5.21 – 4.21, 3.61 – 3.28(m, 21H, 14lactosyl – H, 7OH). ^{13}C NMR(100MHz, DMSO – d_6), δ: 156.87, 148.77, 147.04, 139.17, 138.81, 128.18, 126.68, 126.54, 110.46 (9C, 9lactosyl – C), 103.79, 80.38, 79.39, 76.69, 75.58, 75.49, 73.15, 71.55, 70.52, 68.10, 60.38, 60.06 (12C, 12lactosyl – C). FABMS, *m/z* calcd. for [$C_{21}H_{26}BrClN_4O_{11}$ + H]$^+$: 624; found: 625.

2.5 Biological Assay

2.5.1 Electrophysiological Recording

The current of calcium(I_{Ca}) was recorded via the method by Li *et al*[13].

Neurons were clamped with a patch – clamp amplifier (EPC – 10, HEKA Electronik, Lambrecht, Germany). The voltage clamp protocols were generated on a computer via Pulse software (version 8.52; HEKA Electronic).

Each experiment was repeated at least three times. The data were analyzed with the help of Spss Inc, version 17.0 and Microcal Origin, version 8.0 (Origin Lab Corp., Northampton, MA). Results were expressed as mean ± SD (n = number of cells). Statistical significance was determined by Student's paired or unpaired t – tests and the differences were considered significant at $P < 0.05$ or at $P < 0.01$.

2.5.2 Reagents

The extracellular solution contained (mmol/L) NaCl (100), CsCl (4), $BaCl_2$ (3), $MgCl_2$ (2), Hepes (10), Glucose (10), TEA – Cl (30), TTX (0.001), pH adjusted to 6.85 with 1mmol/L CsOH. The intracellular solution contained (mmol/L) CsCl (120), $MgCl_2$ (2), Na_2 – ATP (5), EGTA (11), Hepes (5), pH adjusted to 6.85 with 1mmol/L CsOH.

CsCl, CsOH and Hepes were purchased from Gibco. Chlorantraniliprole was s ynthesized as reported previously[4].

3 Results and Discussion

3.1 Effect of Minor Structural Modifications Around Compound Methanamide

The reaction of acyl chloride with azide is commonly used for the synthesis of amide. Plus triphenylphosphine was used to reduce azide via a stable iminophosphorane intermediate in the Staudinger reaction. We found that this Staudinger reaction gave us fairly low yield with many different by – products in the synthesis of N – glycosyl – 1 – pyridyl – 1H – pyrazole – 5 – carboxamide. Only a slight increase of yield was found when toluene and acetonitrile were used as solvents in the reaction systems.

Fairly good yields with no by – products were achieved in the hydrogenolysis (H_2, Pd/C, 5%) of glycosyl azide and the reaction of per – O – acetylglycosyl amine with pyrazole – 5 – carboxylic acid.

Deprotection is the key step to obtain the desired target compound. In our experiments, the deprotection reaction proceeded rapidly when 3 – 9 mmol/L sodium in methanol was used. When the concentration of sodium in methanol was greater than 15mmol/L, the protected compounds would decompose. Therefore, low concentration of sodium in methanol is able to prevent the decomposition of compounds **4**a – **4**h.

3.2 Insecticidal Activity and the Effect of *N* – Glycosyl – 1 – pyridyl – 1*H* – pyrazole – 5 – carboxamide on Voltage – gated Calcium Channel

The larvicidal activity against *S. exigua* is summarized in Table 3. In general, most of the compounds **5**a – **5**g show moderate potency against *S. exigua* compared with compounds **4**a – **4**j. Compounds **4**i and **4**j exhibit no larvacidal activity. Higher pesticidal activities of compounds **5**a – **5**g compared to the corresponding ones of compounds **4**a – **4**j indicate that the deprotection of the protected glycosyl is critical for increasing the pesticidal activity. The activities of compounds **5**a – **5**d are higher than that of the disaccharide substituted amide, indicating that monosaccharide substituted amide is liable to protein binding.

Table 3 Insecticidal activities of compounds 4a – 4j and 5a – 5g and chlorantraniliprole against *S. exigua*

Compound	Larvicidal activity(%)		
	200mg/L	100mg/L	50mg/L
4a	30	0	0
4b	40	0	0
4e	20	0	0
4f	30	0	0
4i	0	0	0
4j	0	0	0
5a	100	100	80
5b	100	70	0
5c	100	100	70
5d	100	60	0
5e	40	0	0
5f	30	0	0
5g	40	0	0
5h	0	0	0
Chlorantraniliprole	100	100	100

Chlorantraniliprole stabilizes insect RyR channels to open state, evokes massive calcium released from intracellular stores and then disrupts the calcium homeostasis, possesses distinct pharmacological characteristics[20]. In our previous work[21], we found that the $Ca^{2+}_{(L)}$ channels of *S. exigua* neurons were inhibited and the voltage – dependent calcium channel was influenced by chlorantraniliprole. In order to provide useful information for designing novel insecticides and to study the effect of glycosylation, the whole – cell patch – clamp technique was used to investigate the effect of compound **5**a on calcium channels in the central neurons of *S. exigua*.

Fig. 2 and Table 4 show the change rate of peak current amplitude versus recording time in the 0. 1mmol/L compound **5**a, 0. 1mmol/L chlorantraniliprole treated neurons and control neu-

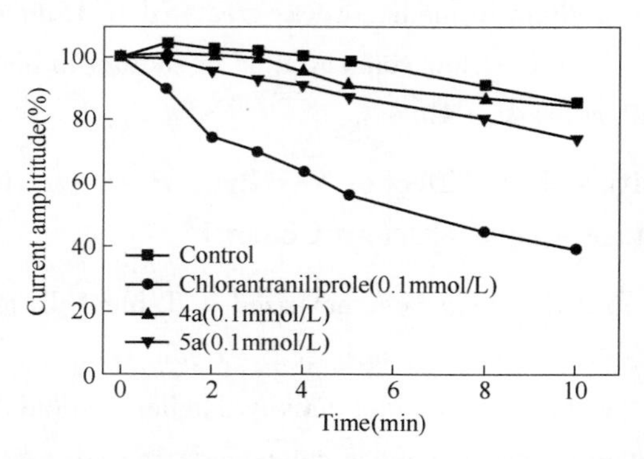

Fig. 2 Variation percentages of peak current for the whole – cell calcium channels in control, chlorantraniliprole and compound **4**a, **5**a – treated neurone at different time during patch – clamp recording compared with those at 0 – min

rons. The peak currents were reduced to (73.85 ± 2.31)% ($n = 6$) of the initial value at the end of the 10 - min recording when the neurons was treated in 0.1 mmol/L compound **5**a. Compared with the chlorantraniliprole (39.43 ± 2.56)% compound **5**a at 0.1 mmol/L has weak inhibitory effect on $Ca^{2+}_{(L)}$ currents.

Table 4 Analysis of peak currents change rate (%) with time in 0.1mmol/L compounds 4a, 5a, control and chlorantraniliprole - treated neurons

Time/min	**4**a	**5**a	Control	Compound chlorantraniliprole
0	100	100	100	100
1	101.26 ± 5.38②	99.42 ± 2.58①	104.47 ± 2.33②	89.62 ± 2.16②
2	100.42 ± 6.27②	95.33 ± 6.39②	102.75 ± 2.45①	74.01 ± 8.01②
3	99.89 ± 3.52②	92.67 ± 8.65②	101.76 ± 2.7①	69.57 ± 9.43②
4	95.41 ± 7.36②	90.42 ± 5.32②	99.66 ± 1.64①	63.36 ± 7.71②
5	90.38 ± 4.51②	87.26 ± 4.38②	98.06 ± 1.01①	55.79 ± 4.73②
8	86.41 ± 5.24②	80.07 ± 5.46②	90.85 ± 0.65②	44.31 ± 2.62②
10	83.95 ± 6.73②	73.85 ± 2.31②	85.44 ± 1.14②	39.43 ± 2.56②

①Significant difference at $P < 0.05$; ②significant difference at $P < 0.01$. Values are the mean ± SD.

There was no significant change in maximal value of peak current when the neuron was held at −50 mV with the treatment of 0.1 mmol/L **5**a (Fig. 3). Though calcium current decreased when compound **5**a was injected in the surrounding of the neuron cell, the maximal value of I_{Ca} did not shift to the negative direction.

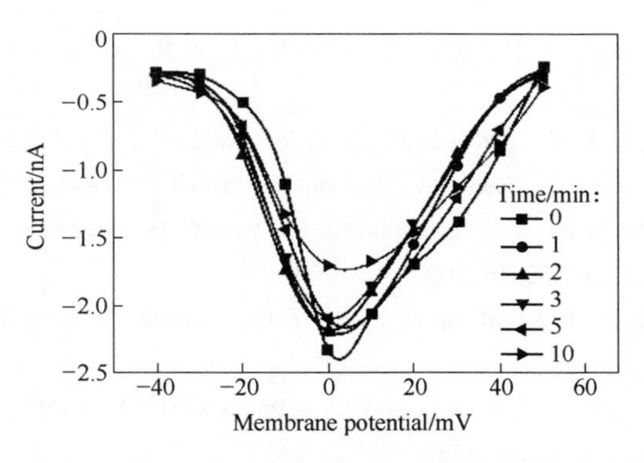

Fig. 3 *I* − *V* curves of whole − cell calcium channels recorded in compound **5**a − treated neurons in *S. exigua* at different time

Unaffected *I* − *V* relationship curves indicate that voltagedependence is not influenced by compound **5**a. After exposure to compound **5**a, washout could reverse the calcium current to the control level as long as the recording continued. This suggests that the action site may have been physically modified on either the transmembrane domain or extramembrane of calcium channel. Compound **4**a did not affect calcium channel of the central neurons.

Calcium imaging technique (Fur − 3) was also used to investigate the effects of compound **5**a on calcium channels in the central neurons of *S. exigua*. Experimental results indicate that compound **5**a did not release stored calcium from the Endoplasmic Reticulum.

In summary, the preliminary structure − activity relationship of the target compounds indicates that the unprotected monosaccharide amides can be easily combined with receptors. More importantly, for the development of novel pest management agents, it is not appropriate to use glycosyl groups instead of anthraniloyl moiety to modify N – pyridylpyrazole moiety.

References

[1] Bertolino M., Llinas R. R., *Annul. Rev. Pharmacool Toxicol*, 1992, 32, 399.

[2] Olivera B. M., Miljanich G. P., Ramachandran J., Adams M. E., *Annul. Rev. Biochem.*, 1994, 63, 823.

[3] Lahm G. P., Selby T. P., Stevenson T. M., *Arthropodicidal Anthranilamides*, *US* 0225615, 2003.

[4] Lahm G. P., Selby T. P., Freudenberger J. H., Stevenson T. M., Myers B. J., Seburyamo G., Smith B. K., Flexner L., Clark C. E., Cordova. D., *Bioorg. Med. Chem. Lett.*, 2005, 15, 4898.

[5] Lahm G. P., Stevenson T. M., Selby T. P., Freudenberger J. H., Cordova D., Flexner L., Bellin C. A., Dubas C. M., Smith B. K., Hughes K. A., Hollingshaus J. G., Clark C. E., Benner E. A., *Bioorg. Med. Chem. Lett.*, 2007, 17, 6274.

[6] Peter J., *Modern Crop Protection Compounds*, Wileg – VCH Publishers, Weinheim, 2007, 1212.

[7] Tohnishi M., Nakao H., Furuya T., Seo A., Kodama H., Tsubata K., Fujioka S., Kodama H., Hirooka T., Nishimatsu T., *J. Pestic. Sci.*,. 2005, 30, 354.

[8] Ebbinghaus – Kintscher U., Lümmen P., Raming K., Masaki T., Yasokawa N., *Pflanzenschutz – Nachrichten Bayer*, 2007, 60, 117.

[9] Feng Q., Wang M. Z., Xiong L. X., Liu Z. L., Li Z. M., *Chem. Res. Chinese Universities*, 2011, 27(4), 610.

[10] Yan T., Yu G. P., Liu P. F., Xiong L. X., Yu S. J., Li Z. M., *Chem. Res. Chinese Universities*, 2012, 28(1), 53.

[11] Liu P. F., Zhang J. F., Yan T., Xiong L. X., Li Z. M., *Chem. Res. Chinese Universities*, 2012, 28(3), 430.

[12] Yan T., Liu P. F., Zhang J. F., Yu S. J., Xiong L. X., Li Z. M., *Chem. J. Chinese Universities*, 2012, 33(8), 1745.

[13] Li Y. X., Mao M. Z., Li Y. M., Xiong L. X., Li Z. M., Xu J. Y.,. *Physiol. Entomol.*, 2011, 36, 231.

[14] Alig B., Fischer R., Funke C., Hense A., Murata T., Arnold C., Gorgens U., Gesing E., Malsam O., Fischer R., Drewes M., Wada K.,. *Anthranilic Acid Diamide Derivative with Hetero – aromatic and Hetero – cyclic Substituents*, *EP* 005016, 2007.

[15] Feng Q., Liu Z. L., Xiong L. X., Wang M. Z., Li Y. Q., Li Z. M., *J. Agric. Food Chem.*, 2010, 58, 12327.

[16] Li Y. X., Wang H. A., Yang X. P., Cheng H. Y., Wang Z. H., Li Y. M., Li Z. M., Wang S. H., Yan D. W., *Carbohydr. Res.*, 2009, 344, 1248.

[17] Tropper F. D., Andersson F. O., Braun S., Roy R., *Synth.*, 1992, 7, 618.

[18] Shapiro R., Taylor E. D., Zimmerman W. T., *Method for Preparing* N – *Phenylpyrazole* – 1 – *carboxamides*, *US* 2005044131, 2005.

[19] Boullanger P., Maunier V., Lafont D., *Carbohydr. Res.*, 2000, 324, 97.

[20] Lahm G. P., Cordova D., Barry J. D., Bioorg. *Med. Chem.*, 2009, 17, 4127.

[21] Li Y. X., Mao M. Z., Li Y. M., Xiong L. X., Li Z. M., Xu J. Y., *Physiol Entomol.*, 2011, 36, 230.

新型含三唑啉酮的磺酰脲类化合物的合成、晶体结构及除草活性研究*

潘　里　陈有为　刘　卓　李永红　李正名

（南开大学元素有机化学研究所，元素有机化学国家重点实验室，天津　300071）

摘　要　为了进一步寻找高效的磺酰脲类除草剂，以商品化磺酰脲类除草剂为基础，将三唑啉酮杂环引入到分子中，合成了23个未见文献报道的新型磺酰脲化合物，通过 ^{1}H NMR、高分辨质谱以及X射线单晶衍射确定了其结构。经盆栽法和平皿法测试了所有化合物的除草活性以及部分化合物对油菜的 IC_{50} 值。结果表明，化合物**5h**具有优秀的除草活性，其对油菜的 IC_{50} 值与对照药醚苯磺隆和醚磺隆相近，在3.75g/ha浓度下对单子叶杂草稗草具有优异的盆栽抑制活性。

关键词　磺酰脲　三唑啉酮　合成　晶体结构　除草活性

乙酰羟基酸合成酶（AHAS）也称乙酰乳酸合成酶（ALS）。它是高等植物体内支链氨基酸——缬氨酸、亮氨酸和异亮氨酸生物合成第一步的催化酶[1]，能催化一分子丙酮酸脱羧并与另一分子丙酮酸缩合成一分子乙酰乳酸。由于高等动物自身不能合成这些支链氨基酸，需要从外界食物中摄取，因此该酶是理想的除草剂作用靶点[2~5]。磺酰脲是以AHAS酶为靶标的一类重要除草剂[6,7]，从而使得磺酰脲类除草剂对哺乳动物具有很低的毒性。正因为这个特点，磺酰脲类除草剂得到了广泛应用[8~10]。

杂环化合物[11]由于其多变的结构和广泛的生物活性使得在农药的开发中受到广泛的关注，尤其是三唑类含氮杂环[12~14]。三唑啉酮类化合物因具有很好的除草活性[15~18]已成为农药创制中的热点。自从杜邦公司发现了第一个三唑啉酮类除草剂唑啶草酮后，很多三唑啉酮类除草剂相继被开发出来，例如：氨唑草酮、甲磺草胺、唑草酮等。同时，近年来出现的氟酮磺隆、丙苯磺隆也具有三唑啉酮结构。因此新三唑啉酮类化合物具有很好的研究前景。

综上所述，我们将三唑啉酮引入到已知商品化除草剂（如醚苯磺隆、醚磺隆等）分子中，设计出结构新颖的含三唑啉酮的苯基磺酰脲或杂环磺酰脲结构，以期得到除草活性更好的磺酰脲类化合物。目标化合物合成路线见Scheme 1。

1　结果与讨论

1.1　合成讨论

在三唑甲酸苯酯**3**的合成中，曾尝试使用氢氧化钠、碳酸钾、三乙胺、吡啶等碱作为缚酸剂，但是除三乙胺以外，其他几种缚酸剂在反应中均得不到较好的收率，所以将此反应的缚酸剂确定为三乙胺，这可能与三乙胺碱性适中有关，同时过量的三乙胺可以通过旋转蒸发的方式除去，不会残留在体系中。同时在合成**3**的过程中，为

* 原发表于《有机化学》，2013，33（3）：542－550。国家自然科学基金（No. 21272129）、国家重点基础研究发展计划（“973”，No. 2010CB126106）、国家科技攻关计划（No. 2011BAE06B05）资助项目。

了防止反应过于剧烈，用乙腈将氯甲酸苯酯稀释再滴入到反应体系中，然而仅仅得到少量目标产物，大多数为氯甲酸苯酯缩合的副产物——碳酸二苯酯。当用纯的氯甲酸苯酯滴加时，时间应控制在 20 ~ 30min 内，如果滴加时间过短会导致反应剧烈放热，从而使反应失败，过长则会使氯甲酸苯酯缩合成为碳酸二苯酯而影响目标产物的收率。

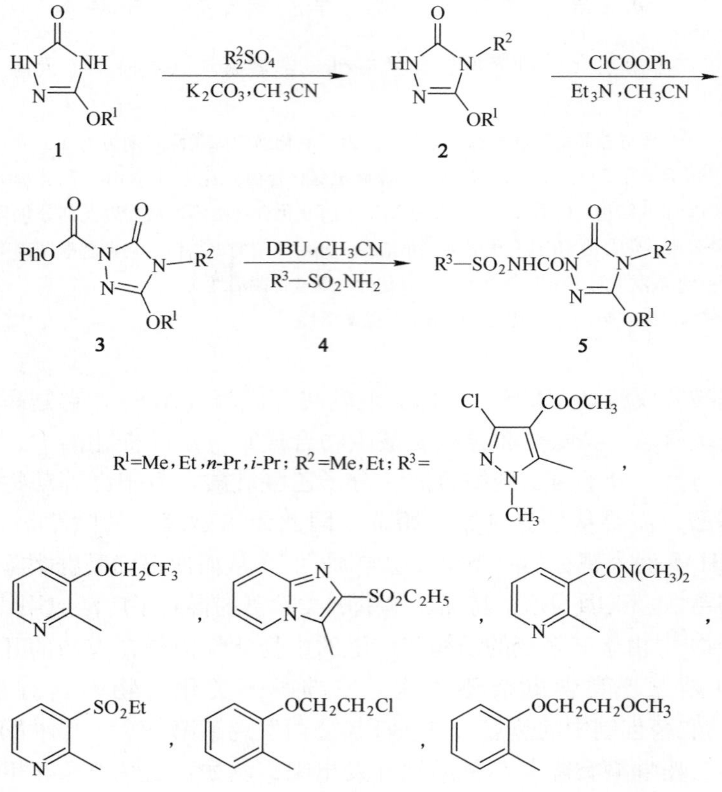

$R^2_2SO_4$; K_2CO_3, CH_3CN; ClCOOPh; Et_3N, CH_3CN; DBU, CH_3CN; $R^3—SO_2NH_2$

R^1=Me, Et, *n*-Pr, *i*-Pr; R^2=Me, Et; R^3=

Scheme 1

在目标化合物 **5** 的合成中，曾尝试使用三乙胺作为催化剂，但是用于碱性不够强，从而导致缩合反应不能彻底进行，降低产品收率因此换用常用有机强碱——DBU 作为反应催化剂，使得反应的收率得到较大提高。

1.2 晶体结构

晶体结构分析表明（图 1）：证明了三唑啉酮环中 1 位的氮原子与脲桥上的羰基相连，4 位的氮原子与烷基相连。吡唑环、磺酰基、酰胺基、酯基的键长键角都与文献［19 ~ 21］报道的一致。C(2)—N(2)—C(1)—O(1) 之间二面角为 - 176.6 (2)°，N (3)—N (2) —C (1) —O (1) 之间二面角为 - 0.8 (3)°，说明了羰基与三唑啉环基本在一个平面上，存在比较强的共轭作用。另外 C (6) —C (8) —C (10) —O (6) 之间的二面角为 10.4 (4)°，C (7) —C (8) —C (10) —O (6) 之间的二面角为 - 167.8 (2)°，也说明羰基与吡唑环基本在同一平面。三唑啉酮［N (2) —N (3) —C (3) —N (4) —C (2)］的五元环为一个平面，与吡唑环平面［N (5) —N (6) —C (7) —C (8) —C (6)］形成的二面角为 93.5°，几乎垂直。脲桥上 N1 相连的 H1 与三唑啉酮羰基上的 O_2 形成了氢键作用；同时三唑酮的甲基的 C—H 与吡唑环存在 C—H - π 的作用，其距离为 3.547Å，这对整个分子起到了稳定作用。

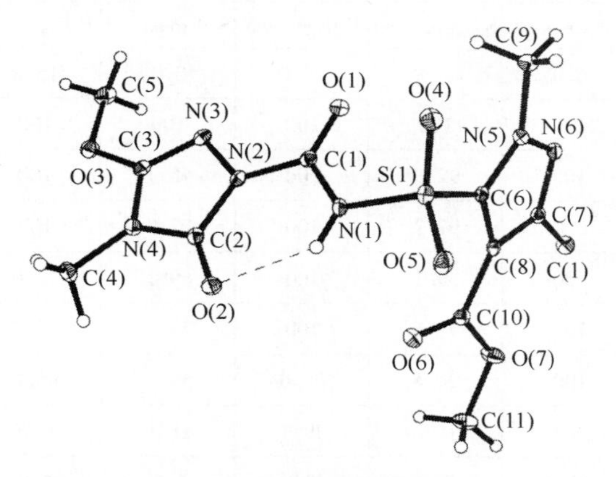

图 1　化合物 **5a** 的晶体结构图

Figure 1　Crystal structure of compound **5a**

1.3　生物活性

目标化合物 **5** 对双子叶杂草油菜（*Brassica napus*）和反枝苋（*Amaranthus retroflexus*）及单子叶杂草稗草（*Echinochloa crusgalli*）和马唐（*Digitaria adscendens*）的生物活性测定结果如表 1 ~3 所示。

从表 1 的 375g/ha 剂量的初筛结果可以看出，大部分化合物对双子叶作物的除草活性明显高于单子叶作物。当 R^3 是 2 -(甲氧基乙氧基）苯基或 2 -(氯乙氧基）苯基时，除草活性优于其它取代基。R^2 是甲基的化合物普遍除草活性高于乙基。表 2 的降低浓度复筛结果表明，大部分化合物在降低浓度后，活性有所下降。但 **5h** 在浓度变成初筛的 1/100 后，对油菜仍具有良好的除草活性，抑制率高于对照药醚磺隆，与醚苯磺隆和单嘧磺酯相当。同时 **5h** 在 3.75g/ha 浓度下对稗草的茎叶处理抑制活性也在 90% 以上，优于三种对照药的活性。化合物 **5j** 在 15g/ha 的浓度下，对双子叶杂草具有一定的选择性（对油菜具有良好的活性，而对反枝苋则没有活性），此化合物值得进一步进行结构改造，寻找具有选择性更好的磺酰脲类除草剂。我们对一些降低浓度后仍具有一定除草活性的化合物进行了 IC_{50} 值测定（表 3），结果显示 **5h** 具有和商品化磺酰脲除草剂醚磺隆、醚苯磺隆相近的超高效除草活性，此化合物可作为结构进一步优化的先导结构。

表 1　目标化合物 5 在 375g/ha 剂量下对杂草的抑制率（盆栽法）(抑制率/%)

Table 1　Inhibitory rate（%） of the target compounds 5 by pot - culture method against herbs at 375g/ha

Compd.	*Brassica napus*		*Amaranthus retroflexus*		*Echinochloa crusgalli*		*Digitaria adscendens*	
	Soil treatment	Foliage spray	Soil treatment	Foliage spray	Soil treatment	Foliage spray	Soil treatment	Foliage spray
5a	0.0	0.0	0.0	15.0	0.0	10.0	0.0	0.0
5b	0.0	0.0	29.4	53.4	0.0	0.0	0.0	15.0
5c	0.0	0.0	0.0	15.0	0.0	10.0	35.0	15.0
5d	0.0	0.0	0.0	52.8	0.0	0.0	40.0	10.0

续表1

Compd.	Brassica napus		Amaranthus retroflexus		Echinochloa crusgalli		Digitaria adscendens	
	Soil treatment	Foliage spray	Soil treatment	Foliage spray	Soil treatment	Foliage spray	Soil treatment	Foliage spray
5e	0. 0	0. 0	45. 9	15. 0	0. 0	15. 0	10. 0	10. 0
5f	100	100	100	100	100	100	100	100
5g	90. 8	100	92. 9	100	45. 1	100	100	100
5h	93. 3	100	97. 2	100	77. 3	100	78. 0	67. 9
5i	95. 0	100	99. 1	100	69. 1	73. 0	65. 0	75. 0
5j	95. 0	100	92. 6	100	73. 2	82. 8	35. 0	70. 0
5k	78. 7	100	94. 4	95. 0	59. 8	95. 0	30. 0	35. 0
5l	0. 0	20. 0	0. 0	29. 1	0. 0	0. 0	10. 0	5. 0
5m	0. 0	40. 0	0. 0	23. 0	0. 0	5. 0	30. 0	5. 0
5n	15. 0	15. 0	0. 0	5. 0	10. 0	0. 0	15. 0	0. 0
5o	0. 0	5. 0	0. 0	10. 0	0. 0	10. 0	10. 0	5. 0
5p	86. 6	100	95. 4	97. 0	64. 9	82. 8	60. 0	46. 2
5q	90. 5	100	94. 4	100	41. 2	97. 0	60. 0	10. 0
5r	83. 8	100	98. 1	100	74. 2	100	77. 0	70. 0
5s	10. 0	89. 5	0. 0	90. 0	26. 8	56. 5	15. 0	34. 6
5t	93. 8	100	98. 1	100	73. 2	100	75. 0	65. 0
5u	81. 0	100	93. 5	99. 0	34. 0	81. 3	10. 0	40. 0
5v	78. 2	100	61. 1	55. 0	19. 6	72. 3	15. 0	5. 0
5w	29. 5	100	76. 9	85. 0	10. 0	10. 0	10. 0	25. 0
Monosulfuron – ester	100	100	100	100	100	100	100	100
Triasulfuron	100	100	100	100	100	100	100	100
Cinosulfuron	100	100	100	100	100	100	100	100

表2　部分目标化合物在15，7.5和3.75g/ha剂量下对杂草的抑制率（盆栽法）（抑制率/%）

Table 2　Inhibitory rate（%）of some target compounds by pot – culture method against herbs at 15，7.5 and 3.75g/ha

Compd.	Dosage(g · ha^{-1})	Brassica napus		Amaranthus retroflexus		Echinochloa crusgalli
		Soil treatment	Foliage spray	Soil treatment	Foliage spray	Foliage spray
5f	15	100	100	99. 4	80. 7	97. 1
	7. 5	85. 6	65. 9	61. 0	55. 4	66. 0
	37. 5	41. 3	29. 8	34. 8	18. 1	36. 4
5g	15	20. 0	100	0. 0	22. 8	0. 0
	7. 5	—	34. 1	—	—	—
	3. 75	—	21. 9	—	—	—
5h	15	98. 3	94. 1	77. 2	22. 2	100

Compd.	Dosage(g·ha^{-1})	Brassica napus		Amaranthus retroflexus		Echinochloa crusgalli
		Soil treatment	Foliage spray	Soil treatment	Foliage spray	Foliage spray
5h	7.5	94.8	100	76.6	—	100
	3.75	91.5	92.0	28.7	—	90.1
5i	15	71.1	77.4	25.1	12.3	0.0
	7.5	60.5	23.7	—	—	—
	3.75	30.3	0.0	—	—	—
5j	15	92.7	73.2	0.0	0.0	0.0
	7.5	77.5	31.8	—	—	—
	37.5	49.1	0.0	—	—	—
5k	15	0.0	63.1	0.0	0.0	21.2
	7.5	—	42.0	—	—	—
	3.75	—	12.5	—	—	—
5p	15	0.0	80.7	0.0	0.0	0.0
	7.5	—	70.5	—	—	—
	3.75	—	9.1	—	—	—
5q	15	66.5	19.1	21.5	12.3	5.5
	7.5	64.9	—	—	—	—
	3.75	57.9	—	—	—	—
5t	15	81.7	74.8	16.9	29.2	21.2
	7.5	69.7	37.8	—	—	—
	37.5	62.4	26.0	—	—	—
Monosulfur-on-ester	7.5	96.3	100	86.1	53.4	23.5
	3.75	81.2	92.6	27.6	26.5	16.7
Triasulfuron	7.5	100	100	100	83.6	46.9
	3.75	98.2	100	94.4	29.4	0.0
Cinosulfuron	7.5	88.2	56.8	99.4	48.0	29.6
	3.75	79.7	52.8	75.5	27.0	8.6

表 3　部分目标化合物对油菜的 IC_{50} 值（平皿法）

Table 3　IC_{50} values of some target compounds by culture dish method against *Brassica napus*

Compd.	IC_{50}(mg·L^{-1})	Compd.	IC_{50}(mg·L^{-1})
5f	0.1019	**5p**	0.3040
5g	1.3380	**5q**	0.8442
5h	0.0761	**5t**	0.3516
5i	0.4473	Monosulfuron-ester	0.0078
5j	0.2649	Triasulfuron	0.0102
5k	0.9837	Cinosulfuron	0.0265

2 结论

合成了23个含有三唑啉酮结构的新型磺酰脲类化合物，发现部分化合物具有良好的除草活性以及选择性。其中化合物**5h**对双子叶杂草油菜具有与商品化磺酰脲除草剂相当的抑制活性，对单子叶杂草稗草的抑制活性高于商品化对照药。因此，化合物**5h**可以作为先导化合物进一步进行结构优化，寻找活性更好的磺酰脲类除草剂。

3 实验部分

3.1 仪器与试剂

X－4数字显示显微熔点测定仪（北京泰克仪器有限公司）；FTICR－MS（Varian 7.0T）型高分辨率质谱仪；Bruker Avance－300MHz和Bruker Avance－400MHz核磁共振仪，$CDCl_3$或DMSO－d_6为溶剂，TMS为内标。所用试剂均为市售分析纯或化学纯。

3.2 合成

3.2.1 4，5－二氢－3－烷氧基－4－烷基－5－氧代－1*H*－1，2，4－三唑甲酸苯酯（**3**）的合成

中间体**2**参考文献［22］合成，不经分离纯化，直接用于下一步反应。

在250mL圆底烧瓶中加入100mL乙腈、60mmol中间体**2**和70mmol三乙胺，搅拌溶解。在冰浴冷却下，向其中滴加66mmol（10.33g）氯甲酸苯酯，25min左右滴加完毕。于室温搅拌2h后，减压脱溶，残余物用200mL水洗涤，有固体产生，抽滤，干燥后得到中间体**3**。

4，5－二氢－3－甲氧基－4－甲基－5－氧代－1*H*－1，2，4－三唑甲酸苯酯（**3ae**）：白色固体，产率74%。m.p.150～151℃；^{1}H NMR（400MHz，DMSO－d_6）δ：3.060（s，3H，NCH_3），4.022（s，3H，triazolone－OCH_3），7.261～7.499（m，5H，ArH）。

4，5－二氢－3－乙氧基－4－甲基－5－氧代－1*H*－1，2，4－三唑甲酸苯酯（**3be**）：白色固体，产率69%。m.p.110～112℃；^{1}H NMR（400MHz，DMSO－d_6）δ：1.385（t，*J*＝6.8Hz，3H，triazolone－OCH_2CH_3），3.056（s，3H，NCH_3），4.405（q，*J*＝6.8Hz，2H，triazolone－OCH_2CH_3），7.255～7.503（m，5H，ArH）。

4，5－二氢－3－正丙氧基－4－甲基－5－氧代－1*H*－1，2，4－三唑甲酸苯酯（**3ce**）：白色固体，产率70%。m.p.92～93℃；^{1}H NMR（400MHz，DMSO－d_6）δ：0.976（t，*J*＝7.2Hz，3H，CH_2CH_3），1.735～1.822（m，2H，CH_2CH_3），3.065（s，3H，triazolone－NCH_3），4.309（t，*J*＝6.4Hz，2H，triazolone－OCH_2），7.257～7.496（m，5H，ArH）。

4，5－二氢－3－异丙氧基－4－甲基－5－氧代－1*H*－1，2，4－三唑甲酸苯酯（**3de**）：白色固体，产率65%。m.p.80～81℃；^{1}H NMR（400MHz，DMSO－d_6）δ：1.390（d，*J*＝6.0Hz，6H，isopropyl—CH_3），3.036（s，3H，NCH_3），5.035～5.096（m，1H，isopropyl—CH），7.259～7.496（m，5H，ArH）。

4，5－二氢－3－甲氧基－4－乙基－5－氧代－1*H*－1，2，4－三唑甲酸苯酯（**3af**）：白色固体，产率62%。m.p.51～53℃；^{1}H NMR（400MHz，DMSO－d_6）δ：

1.301（t，J = 7.2Hz，3H，NCH_2CH_3），3.684（q，J = 7.2Hz，2H，NCH_2CH_3），4.116（s，3H，triazolone – OCH_3），7.261 ~7.443（m，5H，ArH）。

4，5 – 二氢 – 3 – 乙氧基 – 4 – 乙基 – 5 – 氧代 – 1H – 1，2，4 – 三唑甲酸苯酯（**3bf**）：白色固体，产率 55%。m. p. 27 ~ 29℃；^1H NMR（400MHz，DMSO – d_6）δ：1.193（t，J = 7.2Hz，3H，NCH_2CH_3），1.385（t，J = 6.8Hz，3H，triazolone – OCH_2—CH_3），3.559（q，J = 7.2Hz，2H，NCH_2CH_3），4.414（q，J = 7.2Hz，2H，triazolone – OCH_2CH_3），7.270 ~7.506（m，5H，ArH）。

4，5 – 二氢 – 3 – 正丙氧基 – 4 – 乙基 – 5 – 氧代 – 1H – 1，2，4 – 三唑甲酸苯酯（**3cf**）：白色固体，产率 60%。m. p. 22 ~ 23℃；^1H NMR（400MHz，$CDCl_3$）δ：1.051（t，J =7.2Hz，3H，CH_2CH_3），1.322（t，J =7.2Hz，3H，NCH_2CH_3），1.814 ~1.911（m，2H，CH_2CH_3），3.707（q，J =7.2，14.4Hz，2H，NCH_2CH_3），4.418（t，J = 6.4Hz，2H，triazolone – OCH_2），7.263 ~7.454（m，5H，ArH）。

3.2.2　目标化合物 **5** 的合成

15mL 乙腈中加入 2mmol 磺胺 **4** 和 2mmol 三唑甲酸苯酯 3，搅拌均匀后滴加 1.5mmol 1，8 – 二氮杂二环［5.4.0］ –7 – 十一烯（DBU）的 2mL 乙腈溶液，室温搅拌过夜。将反应液减压脱溶，向残余物中加入 20mL 质量分数为 5% 的盐酸溶液，充分搅拌。用 60mL 二氯甲烷分 3 次对水相进行萃取，有机相干燥，减压脱溶，残余物经柱层析分离［V（石油醚）：V（丙酮） =8:1］得到目标化合物 **5**。

4，5 – 二氢 – 3 – 甲氧基 – 4 – 甲基 – 5 – 氧代 – N – （1 – 甲基 – 3 – 氯 – 4 – 甲氧基甲酰基 – 5 – 吡唑磺酰基） – 1H – 1，2，4 – 三唑 – 1 – 甲酰胺（**5a**）：白色固体，产率 72%。m. p. 152 ~153℃；^1H NMR（400MHz，DMSO – d_6）δ：3.008（s，3H，triazolone – NCH_3），3.772（s，3H，pyrazole—CH_3），3.968（s，3H，triazolone – OCH_3），4.114（s，3H，$COOCH_3$）；ESI – FTICR – MS calcd for $C_{11}H_{12}ClN_6O_7S$［M – H］$^-$ 407.0180，found，407.0182。

4，5 – 二氢 – 3 – 甲氧基 – 4 – 甲基 – 5 – 氧代 – N – ［3 – （2，2，2 – 三氟乙氧基） – 2 – 吡啶磺酰基］ – 1H – 1，2，4 – 三唑 – 1 – 甲酰胺（**5b**）：白色固体，产率 64%。m. p. 132 ~ 133℃；^1H NMR（400MHz，$CDCl_3$）δ：3.219（s，3H，NCH_3），4.093（s，3H，triazolone – OCH_3），4.573（q，J =8.0Hz，2H，OCH_2CF_3），7.500 ~7.572（m，2H，Py – H），8.385（d，J =3.6Hz，1H，Py – H），10.723（br，1H，SO_2NH）。ESI – FTICR – MS calcd for $C_{12}H_{11}F_3N_5O_6S$［M – H］$^-$ 410.0388，found 410.0382。

4，5 – 二氢 – 3 – 甲氧基 – 4 – 甲基 – 5 – 氧代 – N – ［2 – （乙基磺酰）咪唑并［1，2 – a］吡啶 – 3 – 磺酰基］ – 1H – 1，2，4 – 三唑 – 1 – 甲酰胺（**5c**）：白色固体，产率 75%。m. p. 198 ~ 199℃；^1H NMR（400MHz，DMSO – d_6）δ：1.150（t，J = 7.2Hz，3H，$SO_2CH_2CH_3$），2.970（s，3H，NCH_3），3.684（q，J = 7.2Hz，2H，$SO_2CH_2CH_3$），3.920（s，3H，triazolone – OCH_3），7.350（t，J = 6.8Hz，1H，Py – H），7.670（t，J =6.0Hz，1H，Py – H），7.890（d，J =8.8Hz，1H，Py – H），9.091（d，J =7.2Hz，1H，Py – H）。ESI – FTICR – MS calcd 443.0449 for $C_{14}H_{15}N_6O_7S_2$［M – H］$^-$ 443.0449，found 443.0447。

4，5 – 二氢 – 3 – 甲氧基 – 4 – 甲基 – 5 – 氧代 – N – （3 – 二甲基氨基甲酰基 – 2 – 吡啶磺酰基） – 1H – 1，2，4 – 三唑 – 1 – 甲酰胺（**5d**）：白色固体，产率 62%。

m. p. 165 ~ 166℃; ^{1}H NMR（400MHz，DMSO－d_6）δ：2.835（s，3H，$CONCH_3$），2.995（s，3H，$CONCH_3$），3.053（s，3H，NCH_3），4.005（s，3H，triazolone－OCH_3），7.791 ~ 7.818（m，1H，Py－H），8.043 ~ 8.063（m，1H，Py－H），8.740 ~ 8.754（m，1H，Py－H）。ESI－FTICR－MS calcd for $C_{13}H_{15}N_6O_6S$［M－H］$^-$ 383.0779，found 383.0780。

4，5－二氢－3－甲氧基－4－甲基－5－氧代－*N*－（3－乙基磺酰基－2－吡啶磺酰基）－1*H*－1，2，4－三唑－1－甲酰胺（**5e**）：白色固体，产率68%。m. p. 164 ~ 165℃; ^{1}H NMR（400MHz，DMSO－d_6）δ：1.204（t，*J*＝7.6Hz，3H，$SO_2CH_2CH_3$），3.050（s，3H，NCH_3），3.741（q，*J*＝7.2Hz，2H，$SO_2CH_2CH_3$），4.020（s，3H，triazolone－OCH_3），8.026（dd，*J*＝4.8，8.0Hz，1H，Py－H），8.615（dd，*J*＝1.2，6.8Hz，1H，Py－H），8.979（dd，*J*＝1.2，4.8Hz，1H，Py－H）。ESI－FTICR－MS calcd for $C_{12}H_{14}N_5O_7S_2$［M－H］$^-$ 404.0340，found 404.0345。

4，5－二氢－3－甲氧基－4－甲基－5－氧代－*N*－［2－（2－氯乙氧基）苯基磺酰基］－1*H*－1，2，4－三唑－1－甲酰胺（**5f**）：白色固体，产率62%。m. p. 154 ~ 155℃; ^{1}H NMR（400MHz，DMSO－d_6）δ：3.052（s，3H，NCH_3），3.953 ~ 3.981（m，5H，CH_2Cl，triazolone－OCH_3），4.420（t，*J*＝4.8Hz，2H，Ar－OCH_2），7.190（t，*J*＝7.6Hz，1H，ArH），7.314（d，*J*＝8.4Hz，1H，ArH），7.707（t，*J*＝7.6Hz，1H，ArH），7.923（d，*J*＝7.6Hz，1H，ArH），10.890（br，1H，SO_2NH）。ESI－FT-ICR－MS calcd for $C_{13}H_{14}ClN_4O_6S$［M－H］$^-$ 389.0328，found 389.0326。

4，5－二氢－3－甲氧基－4－甲基－5－氧代－*N*－［2－（2－甲氧基乙氧基）苯基磺酰基］－1*H*－1，2，4－三唑－1－甲酰胺（**5g**）：白色固体，产率71%。m. p. 128 ~ 129℃; ^{1}H NMR（400MHz，DMSO－d_6）δ：3.073（s，3H，NCH_3），3.255（s，3H，CH_2OCH_3），3.682（t，*J*＝4.8Hz，2H，CH_2OCH_3），3.977（s，3H，triazolone－OCH_3），4.260（t，*J*＝4.8Hz，2H，Ar－OCH_2），7.167（t，*J*＝7.6Hz，1H，ArH），7.307（d，*J*＝8.4Hz，1H，ArH），7.704（t，*J*＝6.8Hz，1H，ArH），7.908（d，*J*＝7.2Hz，1H，ArH），10.865（br，1H，SO_2NH）。ESI－FTICR－MS calcd for $C_{14}H_{17}N_4O_7S$［M－H］$^-$ 385.0823，found 385.0820。

4，5－二氢－3－乙氧基－4－甲基－5－氧代－*N*－［2－（2－氯乙氧基）苯基磺酰基］－1*H*－1，2，4－三唑－1－甲酰胺（**5h**）：白色固体，产率60%。m. p. 137 ~ 138℃; ^{1}H NMR（400MHz，DMSO－d_6）δ：1.346（t，*J*＝7.2Hz，3H，triazolone－OCH_2—CH_3），3.052（s，3H，NCH_3），3.973（t，*J*＝4.8Hz，2H，CH_2Cl），4.358（q，*J*＝7.2Hz，2H，triazolone－OCH_2CH_3），4.432（t，*J*＝4.8Hz，2H，Ar－OCH_2），7.199（t，*J*＝7.6Hz，1H，ArH），7.323（d，*J*＝8.4Hz，1H，ArH），7.722（t，*J*＝7.6Hz，1H，ArH），7.930（d，*J*＝8.0Hz，1H，ArH），10.907（br，1H，SO_2NH）。MALDI－FTICR－MS calcd for $C_{14}H_{17}Cl－N_4O_6SNa$［M＋Na］$^+$ 427.0450，found 427.0457。

4，5－二氢－3－乙氧基－4－甲基－5－氧代－*N*－［2－（2－甲氧基乙氧基）苯基磺酰基］－1*H*－1，2，4－三唑－1－甲酰胺（**5i**）：白色固体，产率56%。m. p. 131 ~ 132℃; ^{1}H NMR（400MHz，DMSO－d_6）δ：1.347（t，*J*＝6.8Hz，3H，triazolone－OCH_2—CH_3），3.071（s，3H，NCH_3），3.254（s，3H，CH_2OCH_3），3.679（t，

$J=4.8Hz$，2H，CH_2OCH_3），4.239（t，$J=5.2Hz$，2H，Ar－OCH_2），4.359（q，$J=6.8Hz$，2H，triazolone－OCH_2CH_3），7.170（t，$J=76Hz$，1H，ArH），7.310（d，$J=8.4Hz$，1H，ArH），7.707（t，$J=7.6Hz$，1H，ArH），7.911（d，$J=8.0Hz$，1H，ArH），10.889（br，1H，SO_2NH）。MALDIFTICR－MS calcd for $C_{15}H_{20}N_4O_7SNa$［M＋Na］$^+$ 423.0945，found 423.0948。

4，5－二氢－3－正丙氧基－4－甲基－5－氧代－*N*－［2－（2－氯乙氧基）苯基磺酰基］－1*H*－1，2，4－三唑－1－甲酰胺（**5j**）：白色固体，产率68%。m.p. 136～138℃；^1H NMR（400MHz，DMSO－d_6）δ：0.941（t，$J=7.2Hz$，3H，CH_2CH_3），1.698～1.787（m，2H，CH_2CH_3），3.060（s，3H，NCH_3），3.974（t，$J=4.8Hz$，2H，CH_2Cl），4.265（t，$J=6.4Hz$，2H，triazolone－OCH_2），4.431（t，$J=4.8Hz$，2H，Ar－OCH_2），7.198（t，$J=7.6Hz$，1H，ArH），7.323（d，$J=8.0Hz$，1H，ArH），7.720（t，$J=7.6Hz$，1H，ArH），7.929（d，$J=7.6Hz$，1H，ArH），10.894（br，1H，SO_2NH）。MALDI－FTICR－MS calcd for $C_{15}H_{19}ClN_4O_6SNa$［M＋Na］$^+$ 441.0606，found 441.0610。

4，5－二氢－3－正丙氧基－4－甲基－5－氧代－*N*－［2－（2－甲氧基乙氧基）苯基磺酰基］－1*H*－1，2，4－三唑－1－甲酰胺（**5k**）：白色固体，产率59%。m.p. 122～123℃；^1H NMR（400MHz，DMSO－d_6）δ：0.941（t，$J=7.2Hz$，3H，CH_2CH_3），1.699～1.786（m，2H，CH_2CH_3），3.078（s，3H，NCH_3），3.254（s，3H，CH_2OCH_3），3.682（t，$J=4.4Hz$，2H，CH_2OCH_3），4.250～4.281（m，4H，triazolone－OCH_2，Ar－OCH_2），7.169（t，$J=7.6Hz$，1H，ArH），7.307（d，$J=8.4Hz$，1H，ArH），7.704（t，$J=7.6Hz$，1H，ArH），7.908（d，$J=7.2Hz$，1H，ArH），10.865（br，1H，SO_2NH）。MALDI－FTICR－MS calcd for $C_{16}H_{22}N_4O_7SNa$［M＋Na］$^+$ 437.1101，found 437.1100。

4，5－二氢－3－正丙氧基－4－甲基－5－氧代－*N*－（1－甲基－3－氯－4－甲氧基甲酰基－5－吡唑磺酰基）－1*H*－1，2，4－三唑－1－甲酰胺（**5l**）：白色固体，产率48%。m.p. 135～136℃；^1H NMR（400MHz，$CDCl_3$）δ：1.006（t，$J=7.2Hz$，3H，CH_2CH_3），1.791～1.880（m，2H，CH_2CH_3），3.219（s，3H，triazolone－NCH_3），3.924（s，3H，pyrazole－CH_3），4.311（s，3H，$COOCH_3$），4.376（t，$J=6.8Hz$，2H，triazolone－OCH_2），11.097（br，1H，SO_2NH）。MALDI－FTICR－MS calcd for $C_{13}H_{17}ClN_6O_7SNa$［M＋Na］$^+$ 459.0460，found 459.0462。

4，5－二氢－3－正丙氧基－4－甲基－5－氧代－*N*－［3－（2，2，2－三氟乙氧基）－2－吡啶磺酰基］－1*H*－1，2，4－三唑－1－甲酰胺（**5m**）：白色固体，产率50%。m.p. 138～139℃；^1H NMR（400MHz，DMSO－d_6）δ：0.967（t，$J=7.2Hz$，3H，CH_2CH_3），1.728～1.816（m，2H，CH_2CH_3），3.087（s，3H，NCH_3），4.312（t，$J=6.8Hz$，2H，triazolone－OCH_2），5.067（q，$J=8.4Hz$，2H，OCH_2CF_3），7.794～7.826（m，1H，Py－H），7.983（d，$J=8.8Hz$，1H，Py－H），8.338（d，$J=4.4Hz$，1H，Py－H）。MALDI－FTICR－MS calcd for $C_{14}H_{16}F_3N_5O_6SNa$［M＋Na］$^+$ 462.0666，found 462.0665。

4，5－二氢－3－正丙氧基－4－甲基－5－氧代－*N*－［2－（乙基磺酰）咪唑并［1，2－*a*］吡啶－3－磺酰基］－1*H*－1，2，4－三唑－1－甲酰胺（**5n**）：白色固体，产率58%。m. p. 168 ~ 169℃；^{1}H NMR（400MHz，$CDCl_3$）δ：0.973（t，J = 7.2Hz，3H，CH_2CH_3），1.430（t，J = 7.6Hz，3H，$SO_2CH_2CH_3$），1.750 ~ 1.838（m，2H，CH_2CH_3），3.181（s，3H，NCH_3），3.588（q，J = 7.6Hz，2H，$SO_2CH_2CH_3$），4.309（t，J = 6.8Hz，2H，triazolone − OCH_2），7.246（t，J = 6.8Hz，1H，Py − H），7.606（t，J = 7.6Hz，1H，Py − H），7.860（d，J = 8.8Hz，1H，Py − H），9.350（d，J = 7.2Hz，1H，Py − H），11.008（br，1H，SO_2NH）。MALDI − FTICR − MS calcd for $C_{16}H_{20}N_6O_7S_2Na$ [M + Na]$^+$ 495.0727，found 495.0723。

4，5 − 二氢 − 3 − 正丙氧基 − 4 − 甲基 − 5 − 氧代 − N − （3 − 乙基磺酰基 − 2 − 吡啶磺酰基） − 1H − 1，2，4 − 三唑 − 1 − 甲酰胺（**5o**）：白色固体，产率53%。m. p. 153 ~ 154℃；^{1}H NMR（400MHz，$CDCl_3$）δ：0997（t，J = 7.6Hz，3H，CH_2CH_3），1.346（t，J = 7.2Hz，3H，$SO_2CH_2CH_3$），1.776 ~ 1.864（m，2H，CH_2CH_3），3.220（s，3H，NCH_3），3.760（q，J = 7.2Hz，2H，$SO_2CH_2CH_3$），4.348（t，J = 6.8Hz，2H，triazolone − OCH_2），7.753（dd，J = 4.4，7.2Hz，1H，Py − H），8.597（d，J = 8.0Hz，1H，Py − H），8.850（d，J = 3.6Hz，1H，Py − H），10.865（br，1H，SO_2NH）。MALDI − FTICR − MS calcd for $C_{14}H_{19}N_5O_7-S_2Na$ [M + Na]$^+$ 456.0618，found 456.0617。

4，5 − 二氢 − 3 − 异丙氧基 − 4 − 甲基 − 5 − 氧代 − N − [2 − （2 − 氯乙氧基）苯基磺酰基] − 1H − 1，2，4 − 三唑 − 1 − 甲酰胺（**5p**）：白色固体，产率70%。m. p. 149 ~ 150℃；^{1}H NMR（400MHz，DMSO − d_6）δ：1.346（d，J = 6.0Hz，6H，isopropyl—CH_3），3.030（s，3H，NCH_3），3.971（t，J = 4.8Hz，2H，CH_2Cl），4.434（t，J = 4.8Hz，2H，ArOCH_2），4.977 ~ 5.038（m，1H，isopropyl—CH），7.198（t，J = 7.6Hz，1H，ArH），7.323（d，J = 8.0Hz，1H，ArH），7.719（t，J = 7.6Hz，1H，ArH），7.929（d，J = 7.6Hz，1H，ArH），10.928（br，1H，SO_2NH）。MALDI − FTICR − MS calcd for $C_{15}H_{19}ClN_4O_6SNa$ [M + Na]$^+$ 441.0606，found 441.0603。

4，5 − 二氢 − 3 − 异丙氧基 − 4 − 甲基 − 5 − 氧代 − N − [2 − （2 − 甲氧基乙氧基）苯基磺酰基] − 1H − 1，2，4 − 三唑 − 1 − 甲酰胺（**5q**）：白色固体，产率66%。m. p. 190 ~ 120℃；^{1}H NMR（400MHz，DMSO − d_6）δ：1.349（d，J = 6.0Hz，6H，isopropyl—CH_3），3.050（s，3H，NCH_3），3.252（s，3H，CH_2OCH_3），3.685（t，J = 4.4Hz，2H，CH_2OCH_3），4.269（t，J = 4.4Hz，2H，ArOCH_2），4.972 ~ 5.032（m，1H，isopropyl—CH），7.174（t，J = 7.6Hz，1H，ArH），7.313（d，J = 8.4Hz，1H，ArH），7.710（t，J = 7.6Hz，1H，ArH），7.902（d，J = 7.6Hz，1H，ArH），10.921（br，1H，SO_2NH）。MALDI − FTICR − MS calcd for $C_{16}H_{22}N_4O_7SNa$ [M + Na]$^+$ 437.1101，found 437.1103。

4，5 − 二氢 − 3 − 甲氧基 − 4 − 乙基 − 5 − 氧代 − N − [2 − （2 − 氯乙氧基）苯基磺酰基] − 1H − 1，2，4 − 三唑 − 1 − 甲酰胺（**5r**）：白色固体，产率77%。m. p. 147 ~ 149℃；^{1}H NMR（400MHz，$CDCl_3$）δ：1.281（t，J = 7.2Hz，3H，NCH_2CH_3），3.661（q，J = 7.2Hz，2H，NCH_2CH_3），3.904（t，J = 5.6Hz，2H，CH_2Cl），4.054（s，3H，triazolone − OCH_3），4.356（t，J = 5.6Hz，2H，Ar − OCH_2），6.984（d，J = 8.4Hz，1H，ArH），7.126（t，J = 7.6Hz，1H，ArH），7.569（t，J = 7.2Hz，1H，ArH），8.136（d，J = 8.0Hz，1H，ArH），10.673（s，1H，SO_2NH）。MALDI − FTICR − MS calcd for $C_{14}H_{17}ClN_4O_6SNa$ [M + Na]$^+$ 427.0450，found 427.0450。

4，5－二氢－3－甲氧基－4－乙基－5－氧代－*N*－［2－（2－甲氧基乙氧基）苯基磺酰基］－1*H*－1，2，4－三唑－1－甲酰胺（**5s**）：白色固体，产率69%。m. p. 107～108℃；^{1}H NMR（400MHz，$CDCl_3$）δ：1.287（t，*J*＝7.2Hz，3H，NCH_2CH_3），3.414（s，3H，CH_2OCH_3），3.664（q，*J*＝7.2Hz，2H，NCH_2CH_3），3.806（t，*J*＝4.8Hz，2H，CH_2OCH_3），4.056（s，3H，triazolone－OCH_3），4.233（t，*J*＝4.8Hz，2H，$ArOCH_2$），7.016（d，*J*＝8.4Hz，1H，ArH），7.087（t，*J*＝7.6Hz，1H，ArH），7.553（t，*J*＝8.0Hz，1H，ArH），8.108（d，*J*＝7.6Hz，1H，ArH），10.681（s，1H，SO_2NH）。MALDI－FTICR－MS calcd for $C_{15}H_{20}N_4O_7SNa$［M＋Na］$^+$ 423.0945，found 423.0947。

4，5－二氢－3－乙氧基－4－乙基－5－氧代－*N*－［2－（2－氯乙氧基）苯基磺酰基］－1*H*－1，2，4－三唑－1－甲酰胺（**5t**）：白色固体，产率55%。m. p. 130～131℃；^{1}H NMR（400MHz，DMSO－d_6）δ：1.175（t，*J*＝6.8Hz，3H，NCH_2CH_3），1.349（t，*J*＝7.2Hz，3H，triazolone－OCH_2CH_3），3.555（q，*J*＝6.4Hz，2H，NCH_2CH_3），3.977（t，*J*＝4.8Hz，2H，CH_2Cl），4.370（q，*J*＝6.4Hz，2H，triazolone－OCH_2CH_3），4.443（t，*J*＝4.8Hz，2H，$ArOCH_2$），7.200（t，*J*＝7.6Hz，1H，ArH），7.326（d，*J*＝8.4Hz，1H，ArH），7.724（t，*J*＝8.0Hz，1H，ArH），7.933（d，*J*＝7.6Hz，1H，ArH），10.932（br，1H，SO_2NH）。MALDI－FTICR－MS calcd for $C_{15}H_{19}ClN_4O_6SNa$［M＋Na］$^+$ 441.0606，found 441.0606。

4，5－二氢－3－乙氧基－4－乙基－5－氧代－*N*－［2－（2－甲氧基乙氧基）苯基磺酰基］－1*H*－1，2，4－三唑－1－甲酰胺（**5u**）：白色固体，产率59%。m. p. 105～106℃；^{1}H NMR（400MHz，DMSO－d_6）δ：1.192（t，*J*＝6.8Hz，3H，NCH_2CH_3），1.350（t，*J*＝6.8Hz，3H，triazolone－OCH_2CH_3），3.251（s，3H，CH_2OCH_3），3.572（q，*J*＝6.8Hz，2H，NCH_2CH_3），3.685（t，*J*＝4.8Hz，2H，CH_2OCH_3），4.273（t，*J*＝4.8Hz，2H，$ArOCH_2$），4.369（q，*J*＝6.8Hz，2H，triazolone－OCH_2—CH_3），7.172（t，*J*＝7.6Hz，1H，ArH），7.314（d，*J*＝8.4Hz，1H，ArH），7.710（t，*J*＝7.6Hz，1H，ArH），7.913（d，*J*＝7.6Hz，1H，ArH），10.910（br，1H，SO_2NH）。MALDI－FTICRMS calcd for $C_{16}H_{22}N_4O_7SNa$［M＋Na］$^+$ 437.1101，found 437.1100。

4，5－二氢－3－正丙氧基－4－乙基－5－氧代－*N*－［2－（2－氯乙氧基）苯基磺酰基］－1*H*－1，2，4－三唑－1－甲酰胺（**5v**）：白色固体，产率64%。m. p. 136～137℃；^{1}H NMR（400MHz，DMSO－d_6）δ：0.941（t，*J*＝7.2Hz，3H，CH_2CH_3），1.179（t，*J*＝6.8Hz，3H，NCH_2CH_3），1.703～1.788（m，2H，CH_2CH_3），3.564（q，*J*＝6.8Hz，2H，NCH_2CH_3），3.978（t，*J*＝4.4Hz，2H，CH_2Cl），4.279（t，*J*＝6.4Hz，2H，triazolone－OCH_2），4.442（t，*J*＝4.8Hz，2H，$ArOCH_2$），7.200（t，*J*＝7.6Hz，1H，ArH），7.325（d，*J*＝8.4Hz，1H，ArH），7.723（t，*J*＝7.6Hz，1H，ArH），7.933（d，*J*＝7.6Hz，1H，ArH），10.927（br，1H，SO_2NH）。MALDI－FTICR－MS calcd for $C_{16}H_{21}ClN_4O_6SNa$［M＋Na］$^+$ 455.0763，found 455.0767。

4，5－二氢－3－正丙氧基－4－乙基－5－氧代－*N*－［2－（2－甲氧基乙氧基）苯基磺酰基］－1*H*－1，2，4－三唑－1－甲酰胺（**5w**）：白色固体，产率73%。m. p. 126～127℃；^{1}H NMR（400MHz，DMSO－d_6）δ：0.943（t，*J*＝7.2Hz，3H，CH_2CH_3），1.196（t，*J*＝7.2Hz，3H，NCH_2CH_3），1.705～1.790（m，2H，

CH_2CH_3), 3.251 (s, 3H, CH_2OCH_3), 3.581 (q, J = 7.2Hz, 2H, NCH_2CH_3), 3.685 (t, J = 4.4Hz, 2H, CH_2OCH_3), 4.262 ~ 4.293 (m, 4H, triazolone - OCH_2, $ArOCH_2$), 7.172 (t, J = 7.6Hz, 1H, ArH), 7.313 (d, J = 8.4Hz, 1H, ArH), 7.710 (t, J = 8.0Hz, 1H, ArH), 7.911 (d, J = 7.6Hz, 1H, ArH), 10.907 (br, 1H, SO_2NH)。MALDI - FTICR - MS calcd for $C_{17}H_{24}N_4O_7SNa$ $[M + Na]^+$ 451.1258, found 451.1261。

3.3 晶体结构测定

取约 50mg 化合物 **5a** 溶于适量丙酮中，室温下缓慢挥发溶剂得无色透明块状晶体。选取大小为 0.22mm × 0.18mm × 0.16mm 的晶体，在 Rigaku Saturn CCD 衍射仪上，于 113 (2) K 下用 Mo Kα 射线 (λ = 0.71075Å) 以 ω 扫描方式在 2.00° ≤ θ ≤ 27.96° 范围内共收集到 13322 个衍射点，其中 $I > 2\sigma(I)$ 的可观测衍射点 3803 (R_{int} = 0.0956)。所有数据经过半经验方法进行吸收校正。晶体结构用 Bruker SHELXTL 程序包求解，非氢原子坐标用直接法解出，最小二乘法对非氢原子进行各种异性温度因子修正。化合物 **5a** 属于单斜晶系，晶胞参数为 a = 9.1370 (9) Å，b = 20.384 (2) Å，c = 8.7700 (7) Å，α = 90°，β = 103.595 (6)°，γ = 90°，V = 1587.6 (3) $Å^3$，Z = 4，D_c = 1.710mg/m^3，μ = 0.426mm^{-1}，F (000) = 840。晶体结构数据存于英国剑桥数据中心，CCDC 号为 903774。

3.4 除草活性测试

用油菜平皿法、稗草小杯法和盆栽法[23~25]对 1 ~ 23 号化合物进行了除草活性测试，并以商品除草剂醚磺隆，醚苯磺隆和单嘧磺酯作为对照药剂。

盆栽法：在直径 8cm 的塑料小杯中放入一定量的土，加入一定量的水，播种后覆盖一定厚度的土壤，于花房中培养，幼苗出土前以塑料布覆盖。每天加以定量的清水以保持正常生长。测试试材：油菜 (*Brassica napus*)、反枝苋 (*Amaranthus retroflexus*)、稗草 (*Echinochloa crusgalli*) 和马唐 (*Digitaria adscendens*)。施药方法为喷施，处理 21d 后调查结果，测定地上部鲜重，以鲜重抑制百分数来表示药效。

平皿法：试验靶标为油菜 (*Brassica napus*)。直径 6cm 的培养皿中铺好一张直径 5.6cm 的滤纸，加入 2mL 一定浓度的供试化合物溶液，待测药剂至少 4 ~ 5 个浓度，每个浓度重复 4 次，设清水为对照。播种浸种 4 ~ 6h 的油菜种子 15 粒，(30 ± 1)℃下，黑暗培养 48h 后测定胚胎长度。与空白对照比较，计算抑制百分率。之后采用 DPS 数据处理系统 (8.01 版) 数量型数据机值分析法计算得到 IC_{50}值。

References

[1] Guo, W. C.; Ma, Y.; Li, Y. H.; Wang, S. H.; Li, Z. M. *Acta Chim. Sinica* 2009, 67, 569 (in Chinese).
(郭万成，马翼，李永红，王素华，李正名，化学学报，2009，67，569.)
[2] Xue, Y. L.; Liu, X. H.; Zhang, Y. G. *Asian J. Chem.* 2012, 24, 1571.
[3] Duggleby, R. G.; Pang, S. S. *J. Biochem. Mol. Biol.* 2000, 33, 1.
[4] Xue, Y. L.; Zhang, Y. G.; Liu, X. H. *Asian J. Chem.* 2012, 24, 3016.
[5] Xue, Y. L.; Zhang, Y. G.; Liu, X. H. *Asian J. Chem.* 2012, 24, 5087.

[6] Wang, J. G.; Ma, N.; Wang, B. L.; Wang, S. H.; Song, H. B.; Li, Z. M. *Chin. J. Org. Chem.* 2006, 26, 648 (in Chinese).
(王建国，马宁，王宝雷，王素华，宋海斌，李正名，有机化学，2006，26，648.)
[7] Wang, H. X.; Li, F.; Xu, L. P.; Li, Y. H.; Wang, S. H.; Li, Z. M. *Chem. J. Chin. Univ.* 2010, 31, 64 (in Chinese).
(王红学，李芳，许丽萍，李永红，王素华，李正名，高等学校化学学报，2010，31，64.)
[8] Chen, P. Q.; Tan, C. X.; Weng, J. Q.; Liu, X. H. *Asian J. Chem.* 2012, 24, 2808.
[9] Wang, M. Y.; Li, Z. M.; Li, Y. X. *Chin. J. Org. Chem.* 2010, 30, 877 (in Chinese).
(王美怡，李正名，李玉新，有机化学，2010，30，877.)
[10] Ban, S. R.; Niu, C. W.; Chen, W. B.; Li, Q. S.; Xi, Z. *Chin. J. Org. Chem.* 2010, 30, 564 (in Chinese).
(班树荣，牛聪伟，陈文彬，李青山，席真，有机化学，2010，30，564.)
[11] Wang, M. Y.; Guo, W. C.; Lan, F.; Li, Y. H.; Li, Z. M. *Chin. J. Org. Chem.* 2008, 28, 649 (in Chinese).
(王美怡，郭万成，兰峰，李永红，李正名，有机化学，2008，28，649.)
[12] Liu, X. H.; Tan, C. X.; Weng, J. Q. *Phosphorus, Sulfur Silicon Relat. Elem.* 2011, 186, 552.
[13] Tan, C. X.; Shi, Y. X.; Weng, J. Q.; Liu, X. H.; Li, B. J.; Zhao, W. G. *Lett. Drug Des. Discovery* 2012, 9, 431.
[14] Tong, J. Y.; Shi, Y. X.; Liu, X. H.; Sun, N. B.; Li, B. J. *Chin. J. Org. Chem.* 2012, 32, 2373 (in Chinese).
(童建颖，石延霞，刘幸海，孙娜波，李宝聚，有机化学，2012，32，2373.)
[15] Tan, C. X.; Shi, Y. X.; Weng, J. Q.; Liu, X. H.; Li, B. J.; Zhao, W. G. *J. Heterocycl. Chem.* 2013, DOI: 10.1002/jhet.1656.
[16] Lan, F.; Xu, J. Y.; Li, Y. H.; Wang, S. H.; Li, Z. M. *Chem. J. Chin. Univ.* 2009, 30, 712 (in Chinese).
(兰峰，徐俊英，李永红，王素华，李正名，高等学校化学学报，2009，30，712.)
[17] Wang, L.; Ma, Y.; Liu, X. H.; Li, Y. H.; Song, H. B.; Li, Z. M. *Chem. Biol. Drug Des.* 2009, 73, 674.
[18] Wang, L.; Tang, M.; Li, W. M.; Li, Y. H.; Wang, S. H.; Li, Z. M. *Chem. J. Chin. Univ.* 2008, 29, 1371 (in Chinese).
(王雷，唐蜜，李文明，李永红，王素华，李正名，高等学校化学学报，2008，29，1371.)
[19] Weng, J. Q.; Wang, L.; Liu, X. H. *J. Chem. Soc. Pakistan* 2012, 34, 1248.
[20] Liu, X. H.; Tan, C. X.; Weng, J. Q. *Phosphorus, Sulfur Silicon Relat. Elem.* 2011, 186, 558.
[21] Liu, X. F.; Liu, X. H. *Acta Crystallogr.* 2011, E67, O202.
[22] Liu, H. J.; Weng, J. Q.; Tan, C. X.; Liu, X. H. *Acta Crystallogr.* 2011, E67, O1940.
[23] Xu, Q. G.; Liu, T. Y.; Tian, R.; Ma, D. Y.; Li, Q. G. *Chin. J. Org. Chem.* 2008, 28, 234 (in Chinese).
(徐启贵，刘天渝，田睿，马德银，李勤耕，有机化学，2008，28，234.)
[24] Liu, X. H.; Pan, L.; Weng, J. Q.; Tan, C. X.; Li, Y. H.; Wang, B. L.; Li, Z. M. *Mol. Diversity* 2012, 16, 251.
[25] Liu, X. H.; Pan, L.; Tan, C. X.; Weng, J. Q.; Wang, B. L.; Li, Z. M. *Pestic. Biochem. Phys.* 2011, 101, 143.
[26] Liu, X. H.; Zhao, W. G.; Wang, B. L.; Li, Z. M. *Res. Chem. Intermed.* 2012, 38, 1999.
[27] Liu, X. H.; Pan, L.; Ma, Y.; Weng, J. Q.; Tan, C. X.; Li, Y. H.; Shi, Y. X.; Li, B. J.; Li, Z. M.; Zhang, Y. G. *Chem. Biol. Drug Des.* 2011, 78, 689.

Synthesis, Crystal Structure and Herbicidal Activity of Novel Sulfonylureas Containing Triazolinone Moiety

Li Pan, Youwei Chen, Zhuo Liu, Yonghong Li, Zhengming Li

(Research Institute of Elemento - Organic Chemistry, State Key Laboratory of Elemento - organic Chemistry, Nankai University, Tianjin, 300071)

Abstract In order to search for efficient sulfonylurea herbicides, the title compounds were designed by introducing triazolinones into some known sulfonylurea structures. 23 novel sulfonylurea compounds had been synthesized and characterized by ^{1}H NMR, HRMS and X - ray diffraction. Herbicidal activities of title compounds were determined by pot - culture method. IC_{50} values of some effective sulfonylurea compounds were tested by culture dish method. It was found that compound **5h** attained the super - active level which was comparable to the controls as triasulfuron and cinosulfuron. It was worthy to note that compound **5h** had potent inhibitory activity against *Echinochloa crusgalli* by pot - culture method at 3. 75g/ha.

Key words sulfonylurea; triazolinone; synthesis; crystal structure; herbicidal activity

Design, Syntheses and Biological Activities of Novel Anthranilic Diamide Insecticides Containing *N* – Pyridylpyrazole*

Yu Zhao, Yongqiang Li, Lixia Xiong, Liping Xu,
Lina Peng, Fang Li, Zhengming Li

(National Pesticidal Engineering Centre (Tianjin), State Key Laboratory of Elemento – organic Chemistry, Nankai University, Tianjin, 300071, China)

Abstract In search of environmentally benign insecticides with high activity, low toxicity and low residue, a series of novel anthranilic diamide derivatives containing *N* – pyridylpyrazole was designed and synthesized. All the compounds were characterized by ^1H NMR spectroscopy and elemental analysis. The single crystal structure of compound **8**j was determined by X – ray diffraction. The insecticidal activities of the new compounds were evaluated. The results show that some compounds exhibited moderate insecticidal activities against Lepidoptera pests. Among this series of compounds, compounds **8**o and **8**p showed 100% larvicidal activity against *Mythimna separate* Walker, *Plutella xylostella Linnaeus* and *Laphygma exigua Hubner* at a test concentration of 200mg/kg, which is equal to the commercial chlorantraniliprole.

Key words anthranilic diamide; ryanodine receptor; insecticidal activity

1 Introduction

Resistance has often been a problem or a potential problem for insecticide and is one of the most important reasons why insecticides with a new mode of action have been desired[1]. The ryanodine receptor (RyR) derives its name from the plant metabolite ryanodine [Fig. 1 (A)], a natural insecticide from *Ryania speciosa*, known to modify calcium channels[2-5]. As ryanodine is a potent natural insecticide, it has been conjectured that RyRs would provide an excellent target for insect control. The phthalic dimides[6-8] from Nihon Nohyaku and the anthranilic diamides[9-12] from Dupont are the first synthetic classes of potent activators of insect RyRs. The recent commercial introduction of RyR insecticides flubendiamide [Fig. 1 (B)] and chlorantraniliprole [Fig. 1 (C)], is significant in the field of crop protection, particularly important in light of ability of insects to rapidly develop resistance and the need for safe and effective pesticides that act at new biochemical targets[13,14].

Owing to their prominent insecticidal activity, unique modes of action and good environmental profiles, anthranilic diamides and their chemical synthesis have recently attracted considerable attention in the field of novel agricultural insecticides. There were lots of literatures reported for the modification of the anthranilic diamides[15-18]. Most modification was related to a variation

* Resrinted from *Chemical Research in Chinese Universities*, 2013, 29 (1): 51 – 56. Supported by the Fundamental Research Funds for the Central Universities and the Specialized Research Fund for the Doctoral Program of Higher Education (No. 20110031120011).

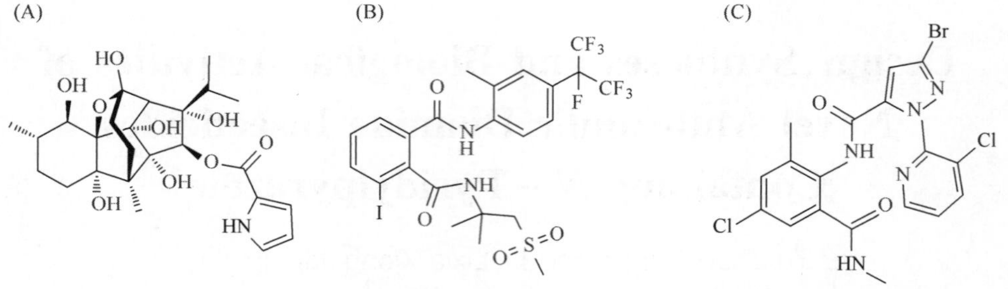

(A) (B) (C)

Fig. 1 Structures of ryanodine(A), flubendiamide(B) and chlorantraniliprole(C)

of the substitution pattern in part of the aliphatic amide moiety. Although less researches have been devoted to the modification of the anthraniloyl skeleton, it has been reported that the biological activity of such compounds can be affected by changing the anthraniloyl skeleton to a large extent[19]. In continuation of our researches on biologically active heterocycles[20], a series of novel anthranilic diamide derivatives containing *N* – pyridylpyrazole was designed and synthesized. And their insecticidal activities were tested. The results show that some compounds exhibited moderate insecticidal activities against *Mythimna separate* Walker, *Plutella xylostella Linnaeus* and *Laphygma exigua Hubner*.

2 Experimental

2.1 Instruments

^1H NMR spectra were recorded in $CDCl_3$ or dimethyl sulfoxide(DMSO – d_6) on a Bruker AC – 400 instrument with tetramethylsilane(TMS) as an internal standard. Elemental analyses were performed on a Yanaca CHN Corder MT – 3 elemental analyzer. The melting points were determined on an X – 4 binocular microscope melting point apparatus (Beijing Tech Instruments Co., Beijing, China) and were uncorrected. Crystallographic data of compound **8**j were collected on a Bruker Smart 1000 CCD diffractometer at 113(2) K($1.47° \leqslant \theta \leqslant 25.02°$). The reagents were all analytically or chemically pure. All the solvents and liquid reagents were dried by standard methods in advance and distilled before use.

2.2 Synthesis of Title Compounds

2.2.1 Synthetic Procedure for Compounds 3a – 3p

The synthetic route of compounds **3**a – **3**p is illustrated in Scheme 1.

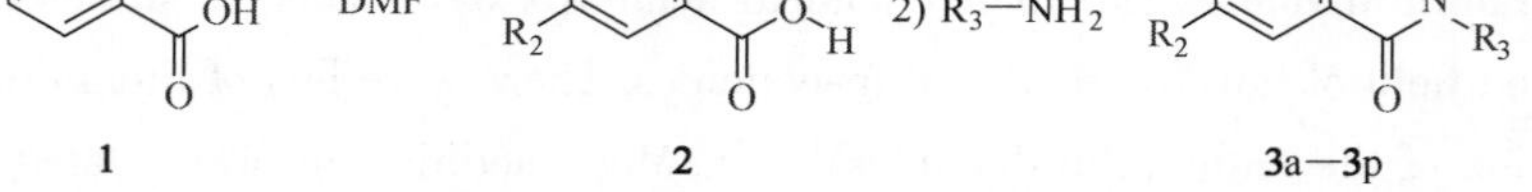

1 → NCS(NBS), DMF → 2 → 1) $SOCl_2$ 2) R_3—NH_2 → 3a—3p

Scheme 1 Synthetic route of compounds **3**a – **3**p

To a solution of compound **1**a(10g, 66mmol) in *N*, *N* – dimethyl formamide(DMF, 40mL) was added *N* – halosuccinimide(66mmol) and the reaction mixture was heated at 100℃ for 40 min, cooled to room temperature, left overnight, and then slowly poured into ice – water(150mL) to precipitate a white solid. The solid was filtered, washed with water(50mL × 3), then taken up in ethyl acetate(600mL). The ethyl acetate solution was dried over magnesium sulfate, evaporated under reduced pressure and the residual solid was washed with ether(30mL × 3) to afford intermediate **2**.

To a 100mL round – bottomed flask was placed 2 – amino – 5 – chloro – 3 – methylbenzoic acid(**2**a) (5.0g, 27mmol) and then was added 50mL of thionyl chloride. The resulting mixture was refluxed for 3h. The mixture was evaporated *in vacuo* to dryness and then 40mL of tetrahydrofuran(THF) was added to it. The solution was added slowly to a stirred solution of propylamine(15.8g, 270mmol) in THF(40mL) in an ice bath. The resulting solution was allowed to stir at room temperature for 12h. Then the solution was concentrated in vacuo and diluted with ethyl acetate(150mL), and washed with water(50mL × 3). The organic extract was separated, dried, filtered, concentrated and purified by silica gel chromatography to afford desired title compound **3**a. Compounds **3**b – **3**p were prepared by similar method above mentioned from the appropriate substrates. The melting points and yields of compounds **3**a – **3**p are listed in Table 1. The ^{1}H NMR data are listed in Table 2.

Table 1 Melting points and yields of compounds 3a – 3p

Compd.	R_1	R_2	R_3	m. p(℃)	Yield(%)
3a	CH_3	Cl	Methyl	130 – 132	50.9
3b	CH_3	Cl	*n* – Propyl	119 – 121	88.0
3c	CH_3	Cl	*i* – Propyl	161 – 162	81.2
3d	CH_3	Cl	Cyclopropyl	122 – 124	92.4
3e	CH_3	Cl	*n* – Butyl	87 – 88	77.9
3f	CH_3	Cl	*i* – Butyl	117 – 122	68.1
3g	CH_3	Cl	*t* – Butyl	109 – 110	65.4
3h	CH_3	Cl	Cyclohexyl	167 – 168	89.2
3i	H	Cl	*n* – Propyl	120 – 122	79.3
3j	H	Cl	*i* – Propyl	161 – 162	58.0
3k	H	Cl	Cyclopropyl	143 – 145	61.9
3l	CH_3	H	*n* – Propyl	88 – 90	70.2
3m	CH_3	H	*i* – Propyl	137 – 139	70.2
3n	CH_3	H	Cyclopropyl	118 – 120	80.0
3o	CH_3	Br	*n* – Propyl	102 – 104	66.3
3p	CH_3	Br	Cyclopropyl	124 – 126	52.8

Table 2 ^{1}H NMR data of compounds 3a—3p

Compd.	^{1}H NMR(400MHz, $CDCl_3$), δ
3a	7.16(d, *J* = 22Hz, 1H, Ph—H), 7.09(d, *J* = 16Hz, 1H, Ph—H), 6.01(br, 1H, CONH), 5.52(br, 2H, NH_2), 2.95(d, *J* = 48Hz, 3H, $NHCH_3$), 2.13(s, 3H, CH_3)
3b	8.37(br, 1H, CONH), 7.41(d, *J* = 1.8Hz, 1H, Ph—H), 7.13(d, *J* = 1.8Hz, 1H, Ph—H), 6.32(s, 2H, $PhNH_2$), 3.14 – 3.17(m, 2H, $NHC\mathbf{H_2}$), 2.08(s, 3H, $PhCH_3$), 1.50 – 1.52(m, 2H, $C\mathbf{H_2}CH_3$), 0.88(t, *J* = 7.4Hz, 3H, $CH_2C\mathbf{H_3}$)

Table 2 (continued)

Compd.	^{1}H NMR(400MHz, $CDCl_3$), δ
3c	7.10 – 7.17(m,2H,Ph—H),5.78(br,1H,CONH),5.52(br,2H,$PhNH_2$),4.15 – 4.26(m,1H,CH),2.14(s,3H,$PhCH_3$),1.25[d,J=6.4Hz,6H,$CH(CH_3)_2$]
3d	7.08 – 7.10(m,2H,Ph—H),6.10(br,1H,NH),5.60(br,2H,NH_2),2.80 – 2.86(m,1H,cyclopropyl – H),2.13(s,3H,CH_3),0.84 – 0.87(m,2H,cyclopropyl – H),0.58 – 0.62(m,2H,cyclopropyl – H)
3e	8.34(br,1H,CONH),7.41(d,J = 18Hz,1H,Ph—H),7.12(d,J = 1.8Hz,1H,Ph—H),6.32(s,2H,$PhNH_2$),3.19 – 3.22(m,2H,$NHC\mathbf{H}_2$),2.08(s,3H,$PhCH_3$),1.45 – 1.52(m,2H,CH_2),1.29 – 1.36(m,2H,$C\mathbf{H}_2CH_3$),0.89(t,J=7.3Hz,3H,$CH_2C\mathbf{H}_3$)
3f	8.38(br,1H,CONH),7.42(d,J = 2.0Hz,1H,Ph—H),7.14(d,J = 2.0Hz,1H,Ph—H),6.29(s,2H,$PhNH_2$),3.01 – 3.03(m,2H,$NHCH_2$),2.08(s,3H,$PhCH_3$),1.77 – 1.88[m,1H,$CH(CH_3)_2$],0.88[d,J=6.6Hz,6H,$CH(C\mathbf{H}_3)_2$]
3g	8.35(br,1H,CONH),7.43(d,J = 2.0Hz,1H,Ph—H),7.12(d,J = 2.0Hz,1H,Ph—H),6.25(s,2H,$PhNH_2$),2.09(s,3H,$PhCH_3$),0.92[s,9H,$CH(C\mathbf{H}_3)_3$]
3h	7.05 – 7.14(m,2H,Ph—H),5.81(br,1H,NH),5.46(br,2H,NH_2),3.82 – 3.94(m,1H,cyclohexyl – H),2.13(s,3H,CH_3),1.18 – 2.04(m,10H,cyclohexyl – H)
3i	8.34 – 8.36(m,1H,CONH),7.53(d,J = 2.4Hz,1H,Ph—H),7.15(dd,J = 8.7,2.4Hz,1H,Ph—H),6.70(d,J = 8.7Hz,1H,Ph—H),6.54(s,2H,$PhNH_2$),3.13 – 3.16(m,2H,$NHC\mathbf{H}_2$),1.46 – 1.53(m,2H,$C\mathbf{H}_2CH_3$),0.87(t,J=7.2Hz,3H,$CH_2C\mathbf{H}_3$)
3j	8.11 – 8.13(m,1H,CONH),7.54(d,J = 2.4Hz,1H,Ph—H),7.15(dd,J = 8.8,2.4Hz,1H,Ph—H),6.70(d,J=8.8Hz,1H,Ph—H),6.50(s,2H,$PhNH_2$),4.01 – 4.09(m,1H,CH),1.14[d,J=6.6Hz,6H,$CH(C\mathbf{H}_3)_2$]
3k	8.35 – 8.37(m,1H,CONH),7.53(d,J=2.4Hz,1H,Ph—H),7.20(dd,J=8.8,4.8Hz,1H,Ph—H),6.76(d,J=8.8Hz,1H,Ph—H),6.61(s,2H,$PhNH_2$),2.82 – 2.88(m,1H,cyclopropyl – H),0.69 – 0.74(m,2H,cyclopropyl – H),0.58 – 0.62(m,2H,cyclopropyl – H)
3l	8.20 – 8.22(m,1H,CONH),7.35(d,J=7.6Hz,1H,Ph—H),7.06(d,J=7.2Hz,1H,Ph—H),6.45 – 6.49(m,1H,Ph—H),6.18(s,2H,$PhNH_2$),3.14 – 3.19(m,2H,$NHC\mathbf{H}_2$),2.07(s,3H,$PhCH_3$),1.46 – 1.55(m,2H,$C\mathbf{H}_2CH_3$),0.88(t,J=7.4Hz,3H,$CH_2C\mathbf{H}_3$)
3m	8.19 – 8.21(m,1H,CONH),7.33(d,J = 7.6Hz,1H,Ph—H),7.05(d,J = 7.0Hz,1H,Ph—H),6.46 – 6.49(m,1H,Ph—H),6.11(s,2H,$PhNH_2$),4.22 – 4.30(m,1H,CH),2.18(s,3H,CH_3),1.25[m,6H,J=6.6Hz,$CH(C\mathbf{H}_3)_2$]
3n	8.18(s,1H,CONH),7.30(d,J = 7.6Hz,1H,Ph—H),7.05(d,J = 6.8Hz,1H,Ph—H),6.43 – 6.47(m,1H,Ph—H),6.21(s,2H,$PhNH_2$),2.78 – 2.80(m,1H,cyclopropyl – H),2.07(s,3H,$PhCH_3$),0.65 – 0.67(m,2H,CH_2CH_2,cyclopropyl – H),0.53 – 0.55(m,2H,CH_2CH_2,cyclopropyl – H)
3o	8.24(br,1H,CONH),7.39(d,J = 1.8Hz,1H,Ph—H),7.11(d,J = 1.8Hz,1H,Ph—H),6.35(s,2H,$PhNH_2$),3.11 – 3.14(m,2H,$NHC\mathbf{H}_2$),2.03(s,3H,$PhCH_3$),1.51 – 1.53(m,2H,$C\mathbf{H}_2CH_3$),0.89(t,J =7.4Hz,3H,$CH_2C\mathbf{H}_3$)
3p	7.02 – 7.06(m,2H,Ph—H),6.13(br,1H,NH),5.55(br,2H,NH_2),2.82 – 2.87(m,1H,cyclopropyl – H),2.15(s,3H,CH_3),0.85 – 0.89(m,2H,cyclopropyl – H),0.57 – 0.61(m,2H,cyclopropyl – H)

2.2.2 Synthetic Procedure for Key Intermediate 7

The synthetic route of intermediate **7** is illustrated in Scheme 2.

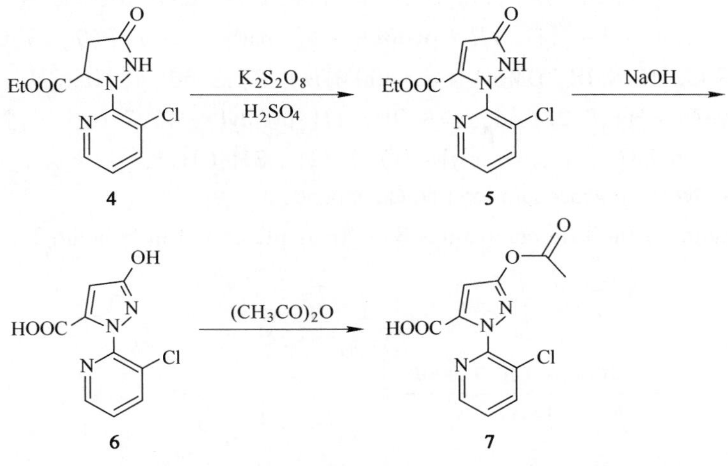

Scheme 2 Synthetic route of compound 7

Intermediate 2 - (3 - chloro - pyridin - 2 - yl) - 5 - oxopyrazolidine - 3 - carboxylic acid ethyl ester(**4**) was prepared according to the literatures[10,11].

To a solution of compound **4**(13.5g, 0.05mol) in acetonitrile(250mL) was added sulfuric acid(98%, 10g, 102mmol). Having been stirred for several minutes, the reaction mixture was treated with $K_2S_2O_8$(21g, 76.5mmol) and was refluxed for 4.5h. Having been cooled to 60℃, the mixture was filtered, the filter cake was washed with acetonitrile(30mL). The filtrate was concentrated to 100mL, then was added slowly to water(250mL) under stirring. The solid was collected by filtration, washed with acetonitrile(30mL × 3), water(30mL), and then dried to give ethyl 2 - (3 - chloropyridin - 2 - yl) - 5 - oxo - 2,5 - dihydro - 1*H* - pyrazole - 3 - carboxylate(**5**) (11.6 g, 86.9%), m. p. 112 - 113℃. ^1H NMR($CDCl_3$, 400MHz), δ: 8.51(d, J = 4.8Hz, 1H, pyridyl - H), 7.94(d, J = 8.0Hz, 1H, pyridyl - H), 7.45(dd, J = 4.8, 8.0Hz, 1H, pyridyl - H), 6.73(s, 1H, pyrazolyl - H), 4.22(q, J = 7.2Hz, 2H, CH_2), 1.21(t, J = 7.2Hz, 3H, CH_3).

To a solution of compound **5**(10.7g, 0.04mmol) in methanol(100mL) was added an aqueous sodium hydroxide solution(50mL, 1mol/L). The solution was stirred at room temperature for 6h, then was concentrated *in vacuo* to a volume of about 50mL. The concentrated mixture was diluted with H_2O(150mL), and washed with ethyl acetate(150mL). The aqueous solution was acidified with concentrated hydrochloric acid to pH = 2. The solid was collected by filtration, washed with ether(50mL), and then dried to give 1 - (3 - chloropyridin - 2 - yl) - 3 - hydroxy - 1*H* - pyrazole - 5 - carboxylic acid(**6**) (8.87g, 92.5%), m. p. 232 - 232℃. ^1H NMR ($CDCl_3$, 300MHz), δ: 8.52(dd, J = 1.5, 4.8Hz, 1H, pyridyl - H), 7.94(dd, J = 1.5, 8.1Hz, 1H, pyridyl - H), 7.48(dd, 1H, J = 4.8, 8.1Hz, pyridyl - H), 7.10(s, 1H, pyrazolyl - H).

To a solution of compound **6**(6.5g, 0.027mol) in THF(100mL) was added pyridine(5.3g, 0.068mol) and acetic anhydride(6.9g, 0.068mol). The reaction mixture was refluxed for 2h, then 250mL of solvent was removed by distillation. The concentrated reaction mixture was slowly poured into 100mL of water. The aqueous solution was acidified with concentrated hydrochloric acid to pH = 2. The resulting mixture was extracted with ethyl acetate(250mL × 2), the organic extract was separated, dried, filtered, concentrated and then dried to give 3 – acetoxy – 1 – (3 – chloropyridin – 2 – yl) – 1*H* – pyrazole – 5 – carboxylic acid(**7**) (7.01 g, 92.5%), m. p. 206 – 207℃. ^{1}H NMR (DMSO – d_6, 400MHz), δ: 13.50 (s, 3H, CH_3), 8.54 (d, J = 4.6Hz, 1H, pyridyl – H), 8.22 (dd, J = 8.0Hz, 1H, pyridyl – H), 7.66 (dd, J = 4.6, 8.0Hz, 1H, pyridyl – H), 6.88 (s, 1H, pyrazolyl – H), 2.32 (s, 3H, CH_3).

2.2.3 General Synthetic Procedure for Title Compounds 8a – 8p

The synthetic route of the title compounds **8**a – **8**p is illustrated in Scheme 3.

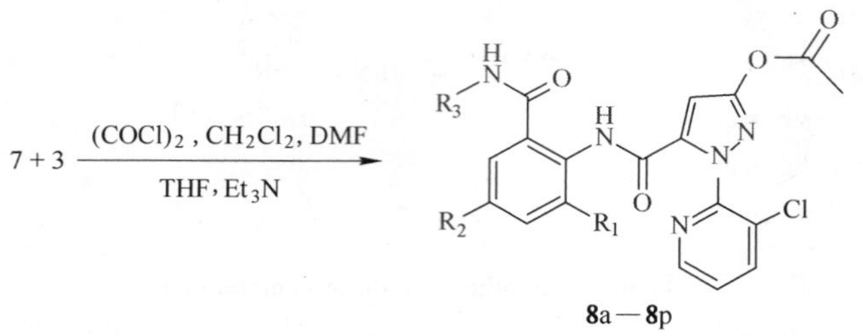

7 + 3 —— $(COCl)_2$, CH_2Cl_2, DMF / THF, Et_3N ——→ **8**a—**8**p

Scheme **3** Synthetic route of title compounds **8**a – **8**p

To a suspension of *N* – pyridylpyrazole acid **6**(1mmol) in dichloromethane(20mL) was added oxalyl chloride(3mmol) and DMF(0.1mL). The solution was stirred at ambient temperature for 4h. Then the mixture was concentrated in vacuo to give the crude acid chloride. The crude acid chloride in THF(25mL) was added slowly to a stirred solution of compound **3**(1.2mmol) and triethylamine(1.2mmol) in THF(15mL). The mixture was stirred at ambient temperature for 8h. Then the solution was concentrated *in vacuo* and diluted with CH_2Cl_2(60mL), and washed with saturated aqueous $NaHCO_3$(15mL) and brine(15mL). The organic extract was separated, dried, filtered, and concentrated and purified by silica gel chromatography to afford desired title compounds **8**a – **8**p. The melting points, yields, and elemental analyses of compounds **8**a – **8**p are listed in Table 3. The ^{1}H NMR data are listed in Table 4.

Table 3 Melting points and yields of compounds 8a – 8p

Compd.	R_1	R_2	R_3	m. p. /℃	Yield (%)	Elemental analysis calcd. (%, found)		
						C	H	N
8a	CH_3	Cl	Methyl	188 – 189	63.0	51.96(51.71)	3.71(3.70)	15.15(15.24)
8b	CH_3	Cl	*n* – Propyl	108 – 110	58.3	53.89(53.63)	4.32(4.29)	14.28(14.13)
8c	CH_3	Cl	*i* – Propyl	164 – 166	52.8	53.89(53.91)	4.32(4.24)	14.28(14.25)
8d	CH_3	Cl	Cyclopropyl	208 – 210	62.4	54.11(54.09)	3.92(4.07)	14.34(14.14)
8e	CH_3	Cl	*n* – Butyl	195 – 197	56.9	54.77(54.71)	4.60(4.59)	13.89(14.07)
8f	CH_3	Cl	*i* – Butyl	202 – 204	54.1	54.77(54.54)	4.60(4.42)	13.89(14.02)
8g	CH_3	Cl	*t* – Butyl	161 – 163	55.4	54.77(54.78)	4.60(4.50)	13.89(13.88)
8h	CH_3	Cl	Cyclohexyl	159 – 161	59.3	56.61(56.57)	4.75(4.55)	13.20(13.24)
8i	H	Cl	*n* – Propyl	181 – 182	61.9	52.95(52.84)	4.02(3.99)	14.70(14.93)
8j	H	Cl	*i* – Propyl	185 – 187	50.6	52.95(52.85)	4.02(3.97)	14.70(14.82)
8k	H	Cl	Cyclopropyl	215 – 216	60.4	53.13(53.18)	3.61(3.51)	14.77(14.76)

Table 3 (continued)

Compd.	R_1	R_2	R_3	m. p. /℃	Yield (%)	Elemental analysis calcd. (%, found) C	H	N
8l	CH_3	H	*n* - Propyl	227 - 229	53. 7	57. 96(57. 84)	4. 86(4. 60)	15. 36(15. 44)
8m	CH_3	H	*i* - Propyl	196 - 198	56. 5	57. 96(58. 01)	4. 86(4. 62)	15. 36(15. 42)
8n	CH_3	H	Cyclopropyl	235 - 237	69. 1	58. 32(58. 51)	4. 44(4. 43)	15. 43(15. 22)
8o	CH_3	Br	*n* - Propyl	214 - 215	54. 4	49. 41(49. 32)	3. 96(3. 78)	13. 10(12. 95)
8p	CH_3	Br	Cyclopropyl	244 - 246	58. 2	49. 60(49. 71)	3. 59(3. 55)	13. 15(12. 90)

Table 4 ^{1}H NMR data of title compounds 8a - 8p

Compd.	^{1}H NMR(400MHz, $CDCl_3$), δ
8a	9. 98(s, 1H, CONH), 8. 46(d, *J* = 4. 5Hz, 1H, pyridyl - H), 7. 86(d, *J* = 8. 0Hz, 1H, pyridyl - H), 7. 38(dd, *J* = 4. 5, 8. 0Hz, 1H, pyridyl - H), 7. 26(s, 1H, Ph—H), 7. 23(s, 1H, Ph—H), 7. 09(s, 1H, pyrazolyl - H), 6. 16 - 6. 19(m, 1H, N**H**CH_2), 2. 97(d, *J* = 4. 5Hz, 3H, NHC$\mathbf{H}_3$), 2. 37(s, 3H, CH_3CO), 2. 22(s, 3H, PhCH_3)
8b	10. 17(s, 1H, CONH), 8. 43(d, *J* = 4. 1Hz, 1H, pyridyl - H), 7. 82(d, *J* = 7. 9Hz, 1H, pyridyl - H), 7. 40(dd, *J* = 4. 1, 7. 9Hz, 1H, pyridyl - H), 7. 28(s, 1H, Ph—H), 7. 12(s, 1H, Ph—H), 7. 11(s, 1H, pyrazolyl - H), 6. 43 - 6. 45(m, 1H, N**H**CH_2), 3. 30 - 3. 32(m, 2H, C$\mathbf{H}_2$NH), 2. 35(s, 3H, CH_3CO), 2. 14(s, 3H, PhCH_3), 1. 48 - 1. 50(m, 2H, C$\mathbf{H}_2$$CH_3$), 0. 89(t, *J* = 7. 2Hz, 3H, CH_2C$\mathbf{H}_3$)
8c	10. 02(s, 1H, CONH), 8. 46(d, *J* = 4. 5Hz, 1H, pyridyl - H), 7. 85(d, *J* = 8. 0Hz, 1H, pyridyl - H), 7. 75(dd, *J* = 4. 5, 8. 0Hz, 1H, pyridyl - H), 7. 27(s, 1H, Ph—H), 7. 22(s, 1H, Ph—H), 7. 08(s, 1H, pyrazolyl - H), 5. 91 - 5. 93(m, 1H, N**H**CH), 4. 19 - 4. 21(m, 1H, CH), 2. 37(s, 3H, CH_3CO), 2. 22(s, 3H, PhCH_3), 1. 23[d, *J* = 6. 5Hz, 6H, $(CH_3)_2$]
8d	9. 90(s, 1H, CONH), 8. 37(d, *J* = 4. 5Hz, 1H, pyridyl - H), 7. 76(d, *J* = 8. 0Hz, 1H, pyridyl - H), 7. 28(dd, *J* = 4. 5, 8. 0Hz, 1H, pyridyl - H), 7. 16(s, 1H, Ph—H), 7. 08(s, 1H, Ph—H), 7. 03(s, 1H, pyrazolyl - H), 6. 18 - 6. 20(m, 1H, N**H**CH), 2. 73 - 2. 76(m, 1H, N**H**CH), 2. 27(s, 3H, CH_3CO), 2. 11(s, 3H, PhCH_3), 0. 77 - 0. 79(m, 2H, cyclopropyl - H), 0. 48 - 0. 50(m, 2H, cyclopropyl - H)
8e	9. 97(s, 1H, CONH), 8. 46(d, *J* = 4. 8Hz, 1H, pyridyl - H), 7. 86(d, *J* = 8. 0Hz, 1H, pyridyl - H), 7. 38(dd, *J* = 4. 8, 8. 0Hz, 1H, pyridyl - H), 7. 28(s, 1H, Ph—H), 7. 22(s, 1H, Ph—H), 7. 05(s, 1H, pyrazolyl - H), 6. 13 - 6. 15(m, 1H, N**H**CH_2), 3. 38 - 3. 40(m, 2H, NHC$\mathbf{H}_2$), 2. 37(s, 3H, CH_3CO), 2. 21(s, 3H, PhCH_3), 1. 54 - 1. 57(m, 2H, C$\mathbf{H}_2$$CH_2$), 1. 38 - 1. 40(m, 2H, CH_2C$\mathbf{H}_2$), 0. 95(t, *J* = 7. 2Hz, 3H, CH_2C$\mathbf{H}_3$)
8f	9. 98(s, 1H, CONH), 8. 46(d, *J* = 4. 5Hz, 1H, pyridyl - H), 7. 85(d, *J* = 7. 9Hz, 1H, pyridyl - H), 7. 38(dd, *J* = 4. 5, 7. 9Hz, 1H, pyridyl - H), 7. 28(s, 1H, Ph—H), 7. 24(s, 1H, Ph—H), 7. 05(s, 1H, pyrazolyl - H), 6. 13 - 6. 15(m, 1H, N**H**CH_2), 3. 23 - 3. 25(m, 2H, NHC$\mathbf{H}_2$), 2. 37(s, 3H, CH_3CO), 2. 22(s, 3H, PhCH_3), 1. 83 - 1. 85(m, 1H, C**H**CH_2), 0. 97[d, *J* = 6. 6Hz, 6H, $(CH_3)_2$]
8g	10. 10(s, 1H, CONH), 8. 45(d, *J* = 4. 5Hz, 1H, pyridyl - H), 7. 84(d, *J* = 7. 8Hz, 1H, pyridyl - H), 7. 37(dd, *J* = 4. 5, 7. 8Hz, 1H, pyridyl - H), 7. 25(s, 1H, Ph—H), 7. 17(s, 1H, Ph—H), 7. 15(s, 1H, pyrazolyl - H), 5. 93 - 5. 95(m, 1H, NH), 2. 36(s, 3H, CH_3CO), 2. 18(s, 3H, PhCH_3), 1. 38[s, 9H, $(CH_3)_3$]

Table 4 (continued)

Compd.	^1H NMR (400MHz, $CDCl_3$), δ
8h	9.98 (s, 1H, CONH), 8.48 (d, J = 4.8Hz, 1H, pyridyl – H), 7.89 (d, J = 8.0Hz, 1H, pyridyl – H), 7.39 (dd, J = 4.8, 8.0Hz, 1H, pyridyl – H), 7.24 (s, 1H, Ph—H), 7.00 (s, 1H, Ph—H), 6.73 (s, 1H, pyrazolyl – H), 6.00 – 6.03 (m, 1H, N**H**CH_2), 3.71 – 3.73 (m, 1H, cyclohexyl – H), 2.37 (s, 3H, CH_3CO), 2.22 (s, 3H, $PhCH_3$), 1.95 – 1.98 (m, 2H, cyclohexyl – H), 1.73 – 1.76 (m, 2H, cyclohexyl – H), 1.21 – 1.23 (m, 2H, cyclohexyl – H), 1.19 – 1.21 (m, 2H, CH_2), 1.16 – 1.19 (m, 2H, CH_2)
8i	12.22 (s, 1H, CONH), 8.48 – 8.52 (m, 2H, Ph—H, pyridyl – H), 7.92 (d, J = 7.9Hz, 1H, pyridyl – H), 7.37 – 7.46 (m, 3H, Ph—H, pyridyl – H), 7.00 (s, 1H, pyrazolyl – H), 6.26 – 6.29 (m, 1H, N**H**CH_2), 3.44 – 3.46 (m, 2H, NHC**H**$_2$), 2.37 (s, 3H, CH_3CO), 1.69 – 1.72 (m, 2H, CH_2CH_3), 1.05 (t, J = 7.3Hz, 3H, CH_2CH_3)
8j	12.24 (s, 1H, CONH), 8.47 – 8.50 (m, 2H, Ph—H, pyridyl – H), 7.92 (d, J = 7.5Hz, 1H, pyridyl – H), 7.37 – 7.46 (m, 3H, Ph—H, pyridyl – H), 7.00 (s, 1H, pyrazolyl – H), 6.09 – 6.11 (m, 1H, N**H**CH), 4.31 – 4.32 (m, 1H, NHC**H**), 2.37 (s, 3H, CH_3CO), 1.32 [d, J = 6.5Hz, 6H, $(CH_3)_2$]
8k	12.23 (s, 1H, CONH), 8.47 – 8.50 (m, 2H, Ph—H, pyridyl – H), 7.91 (d, J = 7.7Hz, 1H, pyridyl – H), 7.34 – 7.44 (m, 3H, Ph—H, pyridyl – H), 7.00 (s, 1H, pyrazolyl – H), 6.55 – 6.57 (m, 1H, N**H**CH), 2.90 – 2.92 (m, 1H, NHC**H**), 2.37 (s, 3H, CH_3CO), 0.92 – 0.94 (m, 2H, cyclopropyl – H), 0.66 – 0.68 (m, 2H, cyclopropyl – H)
8l	10.18 (s, 1H, CONH), 8.45 (d, J = 4.6Hz, 1H, pyridyl – H), 7.84 (d, J = 7.6Hz, 1H, pyridyl – H), 7.36 (dd, J = 4.6, 7.6Hz, 1H, pyridyl – H), 7.05 – 7.25 (m, 3H, Ph—H), 7.05 (s, 1H, pyrazolyl – H), 6.29 – 6.31 (m, 1H, N**H**CH), 3.32 – 3.34 (m, 2H, CH_2CH_2), 2.36 (s, 3H, CH_3CO), 2.23 (s, 3H, $PhCH_3$), 1.56 – 1.58 (m, 2H, **C**H$_2$CH$_3$), 0.95 (t, J = 6.8Hz, 3H, CH_2**C**H$_3$)
8m	10.18 (s, 1H, CONH), 8.45 (d, J = 4.5Hz, 1H, pyridyl – H), 7.84 (d, J = 7.8Hz, 1H, pyridyl – H), 7.36 (dd, J = 4.5, 7.8Hz, 1H, pyridyl – H), 7.08 – 7.25 (m, 3H, Ph—H), 7.09 (s, 1H, pyrazolyl – H), 6.08 – 6.11 (m, 1H, N**H**CH), 4.19 – 4.21 (m, 1H, CH), 2.37 (s, 3H, CH_3CO), 2.230 (s, 3H, $PhCH_3$), 1.20 [d, J = 6.2Hz, 6H, $(CH_3)_2$]
8n	9.90 (s, 1H, CONH), 8.46 (d, J = 4.5Hz, 1H, pyridyl – H), 7.86 (d, J = 8.0Hz, 1H, pyridyl – H), 7.36 (dd, J = 4.5, 8.0Hz, 1H, pyridyl – H), 7.11 – 7.27 (m, 3H, Ph—H), 7.04 (s, 1H, pyrazolyl – H), 6.34 – 6.36 (m, 1H, N**H**CH), 2.79 – 2.82 (m, 1H, CH), 2.37 (s, 3H, CH_3CO), 2.23 (s, 3H, $PhCH_3$), 0.84 – 0.86 (m, 2H, cyclopropyl – H), 0.53 – 0.55 (m, 2H, cyclopropyl – H)
8o	9.94 (s, 1H, CONH), 8.47 (s, 1H, pyridyl – H), 7.81 (d, J = 7.3Hz, 1H, pyridyl – H), 7.40 – 7.42 (m, 1H, Ph—H), 7.32 – 7.35 (m, 2H, pyridyl – H, Ph—H), 6.54 (s, 1H, pyrazolyl – H), 5.93 – 5.95 (m, 1H, N**H**CH), 4.18 – 4.20 (m, 1H, CH), 4.02 (s, 3H, CH_3CO), 2.22 (s, 3H, $PhCH_3$), 1.24 [d, J = 5.6Hz, 6H, $(CH_3)_2$]
8p	9.65 (s, 1H, CONH), 8.26 (s, 1H, pyridyl – H), 7.62 (d, J = 7.1Hz, 1H, pyridyl – H), 7.19 – 7.21 (m, 1H, Ph—H), 7.08 – 7.12 (m, 2H, pyridyl – H, Ph—H), 6.32 (s, 1H, pyrazolyl – H), 6.03 – 6.05 (m, 1H, N**H**CH), 3.80 (s, 3H, CH_3CO), 2.60 – 2.62 (m, 1H, NHC**H**), 1.99 (s, 3H, $PhCH_3$), 0.66 – 0.68 (m, 2H, cyclopropyl – H), 0.37 – 0.39 (m, 2H, cyclopropyl – H)

3 Results and Discussion

3.1 Synthesis

In present work, the synthesis of a series of novel anthranilic diamide derivatives as well as their insecticidal activities against three lepidopterous pests were studied. Target compounds **8**a – **8**p were synthesized via a simple and convenient four – step procedure from the key intermediate **4**. Compound **4** was oxidized to give pyrazolone **5** in a low yield. Compound **5** was hydrolyzed to give compound **6**. Intermediate **6** reacted with acetic anhydride in dry THF to yield the key intermediate **7**. The title compounds were synthesized from compound **7** and the appropriate intermediates **3**a – **3**p. Chlorantraniliprole (C) was prepared according to the literatures[10,11].

3.2 Crystal Structure Analysis

Compound **8**j was recrystallized from ethyl acetate to give colorless crystals suitable for X – ray single – crystal diffraction with the following crystallographic parameters: monoclinic system, space group $P2_1/c$, molecular formula $C_{21}H_{19}Cl_2N_5O_4$, $M_r = 476.31$, $a = 0.75735(15)$ nm, $b = 1.0297(2)$ nm, $c = 2.7637(6)$ nm, $\alpha = 90°$, $\beta = 91.79(3)°$, $\gamma = 90°$, $V = 2.1542(7)$ nm^3, $Z = 4$, $\mu = 0.341$ mm^{-1}, $D_c = 1.469$ Mg/m^3, $F(000) = 984$, and the final R factor, $R_1 = 0.0496$, $\omega R_2 = 0.1491$

The molecular structure of compound **8**j (Fig. 2) contains the following three – plane subunits: the benzene ring C5 – C10 (plane 1), the pyridine ring C17 – C21 – N5 (plane 2), and the pyrazole ring (plane 3) The dihedral angel between the plane of the pyridine ring (plane 2) and the plane of the pyrazole ring (plane 3) is about 83.5°. The crystal packing structure of compound **8**j is shown in Fig. 3.

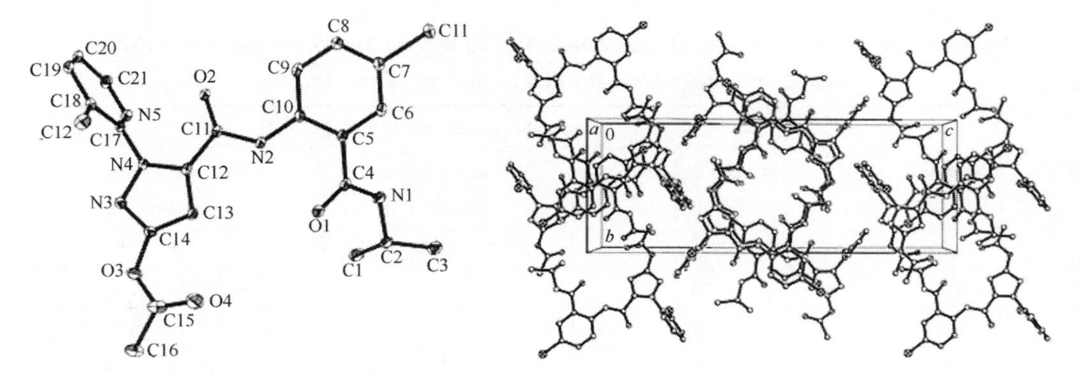

Fig. 2 Crystal structure of compound **8**j

Fig. 3 Packing diagram of compound **8**j

3.3 Insecticidal Activity

All the bioassays were performed on representative test organisms reared in the laboratory. The bioassay was repeated at (25 ± 1)℃ according to statistical requirements. Assessments were made on a dead/alive basis, and mortality rates were corrected *via* Abbott's formula Evaluations were based on a percentage scale of 0 – 100% of which 0 means no activity and 100% means total kill.

3.3.1 Insecticidal activity against oriental armyworm(*Mythimna separata*)

The insecticidal activities of title compounds **8**a – **8**p against oriental armyworm were evaluated via the reported procedure[21]. The insecticidal activity against Oriental armyworm was tested by foliar application, individual corn leaves were placed on moistened pieces of filter paper in Petri dishes. The leaves were then sprayed with the test solution and allowed to dry. The dishes were infested 10 fourth – instar Oriental armyworm larvae. Percentage mortalities were evaluated 2d after treatment. Each treatment was performed three times.

3.3.2 Insecticidal activity against diamond – backmoth(*Plutella xylostella Linnaeus*)

The insecticidal activities of title compounds **8**a – **8**p against diamondback moth were evaluated *via* the leaf disc assay[22]. The leaf discs (5cm × 3cm) were cut from fresh cabbage leaves and then dipped into the test solution for 15s. After air – drying, the treated leaf discs were placed individually into boxes($80cm^3$), and then the second – instar diamondback moth larvae were transferred to the Petri dish. Three replicates(seven larvae per replicate) were carried out.

3.3.3 Insecticidal activity against beet armyworm(*Laphygma exigua Hubner*)

The insecticidal activities of title compounds **8**a – **8**p against beet armyworm were tested by the leaf – dip method *via* the reported procedure[23]. Leaf discs(1.8cm dia – meter) were cut from fresh cabbage leaves and then were dipped into the test solution for 15s. After air – drying, the treated leaf discs were placed in a Petri dish(9cm diameter). Each dried treated leaf disc was infested with seven third – instar beet armyworm larvae. Percentage mortalities were evaluated 3d after treatment. Leaves treated with water and acetone were taken as controls. Each treatment was performed three times.

For comparative purposes, chlorantraniliprole was tested under the same conditions. The results are summarized in Table 5.

Table 5 Insecticidal activities of compounds 8a – 8p against *Mythimna separate Walker*, *Plutella xylostella Linnaeus* and *Laphygma exigua Hubner*

Compd.	Larvicidal activity(%) at $200mg \cdot kg^{-1}$		
	Mythimna separate Walker	*Plutella xylostella Linnaeus*	*Laphygma exigua Hubner*
8a	0	100	0
8b	20	0	100
8c	100	0	100
8d	20	0	0
8e	20	0	0
8f	60	0	0
8g	40	100	70
8h	0	0	0
8i	0	0	0
8j	0	0	0
8k	0	0	0
8l	20	0	0

Table 5 (continued)

Compd.	Larvicidal activity (%) at 200mg · kg^{-1}		
	Mythimna separate Walker	*Plutella xylostella Linnaeus*	*Laphygma exigua Hubner*
8m	100	0	0
8n	20	0	0
8o	100	100	100
8p	100	100	100
Chlorantraniliprole	100	100	100

The results of insecticidal activities given in Table 5 indicate that some of the title compounds exhibited good activity against *Mythimna separate* Walker, *Plutella xylostella Linnaeus* and *Laphygma exigua Hubner*, which were good compared to the commercialized chlorantraniliprole. For instance, the insecticidal activities of compounds **8**c and **8**m against *Mythimna separate* Walker at 200mg/kg were 100%, compounds **8**a and **8**g against *Plutella xylostella Linnaeus* at 200mg/kg were 100%, compounds **8**b and **8**c against *Laphygma exigua Hubner* at 200mg/kg were 100%. Surprisingly, the results indicate that compounds **8**o and **8**p against *Mythimna separate* Walker, *Plutella xylostella Linnaeus* and *Laphygma exigua Hubner* at 200mg/kg exhibited an insecticidal activity of 100% respectively, which is equal to that of the commercial chlorantraniliprole. Therefore the bromine – substituted analogue showed a higher insecticidal activity than did the corresponding chlorine – substituted analogue. Further studies on structural optimization and structure – activity relationships of these anthranilic diamide derivatives are in progress.

4 Conclusions

In summary, sixteen novel anthranilic diamide insecticides containing *N* – pyridylpyrazole were designed and synthesized with structures characterized by ^{1}H NMR spectroscopy, single crystal X – ray diffraction analysis and elemental analysis. The insecticidal activities of the new compounds were evaluated. The results of bioassays show that these title compounds exhibited moderate insecticidal activities. The bromine – substituted analogue showed a higher insecticidal activity than did the corresponding chlorine – substituted analogue. Compounds **8**o and **8**p exhibited excellent insecticidal activities against some Lepidoptera pests at 200mg/kg respectively, which is equal to that of the commercial chlorantraniliprole.

References

[1] Tohnishi M., Nakao H., Furuya T., Seo A., Kodama H., Tsubata K., Fujioka S., Kodama H., Hirooka T., Nishimatsu T., *J. Pestic. Sci.*, 2005, 30, 354.

[2] Thany S. H., *Adv. Exp. Med. Biol.*, 2010, 683, 75.

[3] Legocki J., Polec I., Zelechowski K., *Pestycydy. Pesticides*, 2010, 59.

[4] Coronado R., Morrissette J., Sukhareva M., Vaughan D. M., *Am. J. Physiol.*, 1994, 266, C1485.

[5] Sattelle D. B., Cordova D., Cheek T. R., *Invert. Neurosci.*, 2008, 8, 107.

[6] Tohnishi M. ,Nishimatsu T. ,Motoba K. ,Hirooka T. ,Akira S. ,*J. Pestic. Sci.* ,2010,35,490,508.

[7] Tohnishi M. ,Nakao H. ,Kohno E. ,Nishida T. ,Furuya T. ,Shimizu T. ,Seo A. ,Sakata K. ,Fujioka S. , Kanno H. , *Preparation of Phthalic Acid Diamides as Agricultural and Horticultural Insecticides*, *EP* 919542,1999.

[8] Tohnishi M. ,Nakao H. ,Kohno E. ,Nishida T. ,Furuya T. ,Shimizu T. ,Seo A. ,Sakata K. ,Fujioka S. , Kanno H. ,*Preparation of Phthalamides as Agrohorticultural Insecticides*,*EP* 1006107,2000.

[9] Lahm G. P. ,Selby T. P. ,Freudenberger J. H. ,Stevenson T. M. ,Myers B. J. ,Seburyamo G. ,Smith B. K. , Flexner L. ,Clark C. E. ,Cordova D. ,*Bioorg. Med. Chem. Lett.* ,2005,15,4898.

[10] Lahm G. P. ,Stevenson T. M. ,Selby T. P. ,Freudenberger J. H. ,Cordova D. ,Flexner L. ,Bellin C. A. , Dubas C. M. , Smith B. K. , Hughes K. A. , Hollingshaus J. G. , Clark C. E. , Benner E. A. , *Bioorg. Med. Chem. Lett.* ,2007,17,6274.

[11] Lahm G. P. ,Myers B. J. ,Selby T. P. ,Stevenson T. M. ,*Preparation of Insecticidal Anthranilamides*,*WO* 2001070671,2001.

[12] Lahm G. P. ,Selby T. P. ,Stevenson T. M. ,*Arthropodicidal Anthranilamides*,*WO* 2003015519,2003.

[13] Tohnishi M. , Nishimatsu T. , Motoba K. , Hirooka T. , Akira S. , *J. Pestic. Sci.* (*Tokyo*, *Jpn.*), 2010, 35,490.

[14] Lahm G. P. ,Cordova D. ,Barry J. D. ,*Bioorg. Med. Chem. Lett.* ,2009,17,4127.

[15] Liu P. F. ,Zhou S. ,Xiong L. X. ,Yu S. J. ,Zhang X. ,Li Z. M. ,Chem. *J. Chinese Universities*,2012,33 (4),738.

[16] Hughes K. A. , Lahm G. P. , Selby T. P. ,*Novel Pyrazole – based Anthranilamide Insecticides and Their Preparation*,*Compositions*,*and Use*,*WO* 2004046129,2004.

[17] Lahm G. P. ,Selby T. P. ,Stevenson T. M. ,*Preparation of Anthranilamide Derivatives for Controlling Invertebrate Pests*,*WO* 2004033468,2004.

[18] Yan T. ,Yu G. P. ,Xiong L. X. ,Yu S. J. ,Wang S. H. ,Li Z. M. ,*Chem. J. Chinese Universities*,2011,32 (8),1750.

[19] George P. L. ,Thomas P. S. ,*Insecticidal Compositions Containing Diamides*,*WO* 2003026415,2003.

[20] Liu P. F. , Zhang J. F. , Yan T. , Xiong L. X. , Li Z. M. , *Chem. Res. Chinese Universities*, 2012, 28 (3),430.

[21] Luo Y. ,Yang G. ,*Bioorg. Med. Chem.* ,2007,15,1716.

[22] Wang Y. ,Ou X. ,Pei H. ,Lin X. ,Yu K. ,*Agrochem. Res. Appl.* ,2006,10,20.

[23] Busvine J. R. ,*Recommended Methods for Measurement of Pest Resistance to Pesticides*,The Food and Agriculture Organization of the United Nations(FAO),Rome,1980,21,3,119.

Synthesis and Biological Activities of Novel Anthranilic Diamides Analogues Containing Benzo[*b*]thiophene*

Jifeng Zhang, Chen Liu, Pengfei Liu, Tao Yan,
Baolei Wang, Lixia Xiong, Zhengming Li

(State Key Laboratory of Elemento – Organic Chemistry, Institute of Elemento – Organic Chemistry, Nankai University, Tianjin, 300071, China)

Abstract A series of novel anthranilic diamides analogues containing benzo[*b*] thiophenyl ring was designed and synthesized. Their structures were characterized by melting points, ^{1}H nuclear magnetic resonance(^{1}H NMR) and high – resolution mass spectrometry (HRMS). The bioassay tests indicate that their insecticidal activities were weak to moderate. Antibacterial tests indicate that some of the compounds showed favourable activity in vitro against *Physalospora piricola*, *Alternaria solani*, *Cercospora arachidicola*, *Gibberella sanbinetti* and *Phytophthora infestans* at a dosage of 50mg/L.

Key words anthranilic diamide; benzo[*b*] thiophene; insecticidal activity; antibacterial activity

1 Introduction

With the continuous expansion of the world's population, food production and supply are under great challenge. The application of pesticides has been an irreplaceable way to overcome the food crisis. However, along with the environmental pollution of the traditional pesticide and the emergence of pesticide resistance, the development of novel pesticides has become increasingly urgent. Chlorantraniliprole (A in Fig. 1) and cyantraniliprole (B in Fig. 1) were discovered as novel insect ryanodine receptor insecticide which had aroused great interests[1-7]. We also reported some of our research work concerning the modification of chlorantraniliprole[8-11].

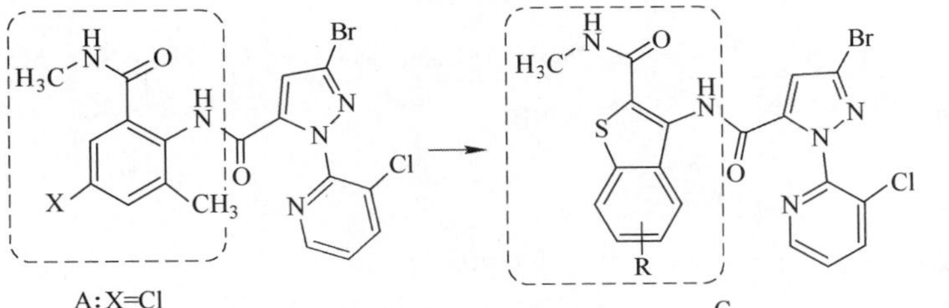

A: X=Cl
B: X=CN

C

Fig. 1 Design of title compounds

Benzothiophene moiety is usually present in other bioactive chemicals as pharmacophore, such as pesticide[12,13], antiproliferative[14], antifungal agents[15], and anti – inflammatory agents[16]

* Reprinted from *Chemical Research in Chinese Universities*, 2013, 29(4): 714 – 720. Supported by the National Natural Science Foundation of China (No. 20872069), the National Basic Research Program of China (No. 2010CB126106), the National Key Technologies R&D Program of China (No. 2011BAE06B05), the Tianjin Natural Science Foundation, China (No. 11JCYBJC08600) and the Fund of State Key Laboratory of Elemento – Organic Chemistry, China.

and so on. In this work it was introduced into the anthranilic diamide structure to replace the phenyl ring. A new series of 3 – bromo – *N* – (2 – carbamoylbenzo[*b*]thiophen – 3 – yl) – 1 – (3 – chloropyridin – 2 – yl) – 1*H* – pyrazole – 5 – carboxamide derivatives(C in Fig. 1) was designed and synthesized, with the two amides in the *ortho* position of the thiophene ring. Totally 14 novel compounds were synthesized and their insecticidal activities against oriental armyworm (*Mythimna separata*) and their fungicidal activities against five fungi were evaluated.

2 Experimental

2.1 Materials and Instruments

Reagents were all analytically or chemically pure. All the solvents and liquid reagents were dried by standard methods in advance and distilled before use.

^{1}H NMR spectra were recorded at 300MHz on a Bruker AC – P300 spectrometer or 400MHz on a Bruker AV 400 spectrometer (Bruker Co., Switzerland) in $CDCl_3$ or DMSO – d_6 solution with tetramethylsilane (TMS) as the internal standard. High – resolution mass spectrometry (HRMS) data were obtained on a Varian QFT – ESI instrument. Flash chromatography was performed on CombiFlash Companion (Teledyne Isco, Inc., America) with silica gel (300 – 400 mesh). The melting points were determined on an X – 4 binocular microscope melting point apparatus (Beijing Tech. Instruments Co., Beijing, China) and were uncorrected.

2.2 General Synthetic Procedure

The synthetic routes of compounds **6**a – **6**n are illustrated in Scheme 1. $SOCl_2$ (70mL) was added to 2 – nitrobenzoic acid (**1**a, 16.72g, 0.1mol) and the mixture was refluxed for 3h and evaporated in vacuo, the residue was dissolved in 50mL of tetrahydrofuran (THF) to which 50mL of aqueous ammonia (25%, volume fraction) was added at 0℃. After stirring overnight, a large amount of solid was formed and filtered to give 15.85g of compound **2**a: a white solid, yield 95.4%, m. p. 175 – 176℃ (174 – 176℃[17]).

1a–5a, 6a–6d: R_1=H, R_2=H; 1b–5b, 6e–6i: R_1=Me, R_2=H; 1c–5c, 6j–6n: R_1=H, R_2=Cl

Scheme 1 Synthetic routes of compounds **6**a – **6**n

With the same procedure, compounds **2**b and **2**c were prepared. Compound **2**b: a white solid, yield 93.6%, m. p. 190 – 191℃ (189 – 191℃[18]); Compound **2**c: white solid, yield 97.7%, m. p. 170 – 172℃ (170℃[19]).

To a solution of compound **2**a(13.2g,80mmol) in 100mL of dimethylformamide(DMF) was added cyanuric chloride (18.4g, 0.1mol) in an ice bath and the mixture was stirred for 4h. After the reaction finished, 500mL of H_2O was added to it and extracted with ethyl acetate, organic layer was collected, dried and evaporated in vacuo to give 2 – nitrobenzonitrile(**3**a) as a white solid: 10.3 g, yield 86.9%. ^1H NMR(400MHz, $CDCl_3$), δ: 8.41 – 8.35 (m, 1H, Ph—H), 7.99 – 7.94 (m, 1H, Ph—H), 7.90 – 7.84 (m, 2H, Ph—H).

With the same procedure, compounds **3**b and **3**c were prepared. Compound **3**b: ^1H NMR (400MHz, $CDCl_3$), δ: 7.551 – 7.43 (m, 3H, Ph—H), 2.42 (s, 3H, Ph—CH_3); compound **3**c: ^1H NMR(400MHz, $CDCl_3$) δ: 8.34 (d, J = 2.0Hz, 1H, Ph—H), 7.88 (d, J = 8.3Hz, 1H, Ph—H), 7.81 (dd, J = 8.3, 2.0Hz, 1H, Ph—H).

Methyl 3 – aminobenzo[*b*]thiophene – 2 – carboxylate(**4**a), methyl 3 – amino – 7 – methylbenzo[*b*]thiophene – 2 – carboxylate(**4**b) and methyl 3 – amino – 6 – chlorobenzo[*b*]thiophene – 2 – carboxylate(**4**c) were synthesized with the similar procedure reported[20]. To a solution of compound **3**a(4.5g, 30mmol), methyl thioglycolate(3.2g, 30mmol) and 60mL of DMF was dropwise added KOH(5.0g, 90mmol in 20mL of H_2O) in an ice bath. After 30 min the reaction was finished, the mixture was added to 200mL ice water and stirred for 30min. The solid precipitated was filtered and collected to give compound **4**a as a pale – yellow solid: 4.95g, yield 79.7%, m. p. 120 – 121℃; ^1H NMR(400MHz, $CDCl_3$), δ: 7.73 (d, J = 8.1Hz, 1H, Ph—H), 7.63 (d, J = 8.1Hz, 1H, Ph—H), 7.47 (t, J = 7.6Hz, 1H, Ph—H), 7.36 (t, J = 7.6Hz, 1H, Ph—H), 5.92 (s, 2H, NH_2), 3.89 (s, 3H, CO_2CH_3).

With the same procedure, compounds **4**b and **4**c were prepared. Compound **4**b: yellow solid, m. p. 124 – 125℃; ^1H NMR(400MHz, $CDCl_3$), δ: 7.52 (d, J = 7.4Hz, 1H, Ph—H), 7.36 – 7.29 (m, 2H, Ph—H), 5.94 (s, 2H, NH_2), 3.93 (s, 3H, CO_2CH_3), 2.52 (s, 3H, Ph—CH_3); HRMS, *m/z*: 244.0402 [M + Na]$^+$, calcd. for $C_{11}H_{11}NO_2S$: 244.0403. Compound **4**c: yellow solid, m. p. 138 – 139℃; ^1H NMR(400MHz, $CDCl_3$), δ: 7.74 (s, 1H, Ph—H), 7.57 (d, J = 8.6Hz, 1H, Ph—H), 7.36 (d, J = 8.6Hz, 1H, Ph—H), 5.91 (s, 2H, NH_2), 3.91 (s, 3H, Ph—CH_3).

To a solution of compound **4**a(1.3g, 6.3mmol) was added 20mL of aqueous NaOH(5.0g, 90mmol in 20mL of H_2O) and the mixture was refluxed for 40 min, and then cooled and acidified with concentrated HCl. The solid precipitated was filtered to give 3 – aminobenzo[*b*]thiophene – 2 – carboxylic acid(**5**a) as a pale – yellow solid and was used without further purification: 1.09 g, yield 89.3%; m. p. 143 – 145℃. ^1H NMR(400MHz, DMSO – d_6), δ: 8.11 (d, J = 8.0Hz, 1H, Ph—H), 7.82 (d, J = 8.0Hz, 1H, Ph—H), 7.49 (t, J = 7.6Hz, 1H, Ph—H), 7.39 (t, J = 7.6Hz, 1H, Ph—H).

With the same procedure, 3 – amino – 7 – methylbenzo[*b*] – thiophene – 2 – carboxylic acid (**5**b) and 3 – amino – 6 – chlorobenzo[*b*]thiophene – 2 – carboxylic acid(**5**c) were prepared.

Compound **5**b: light red solid, m. p. 151 – 152℃; ^1H NMR(400MHz, DMSO – d_6), δ: 7.98 – 7.93 (m, 1H, Ph—H), 7.36 – 7.30 (m, 2H, Ph—H), 3.35 (s, 2H, NH_2), 2.41 (s, 3H, Ph—CH_3).

Compound **5**c: yellow solid, m. p. 178 – 180℃; ^1H NMR(400MHz, DMSO – d_6), δ: 8.13 (d,

J =8.7Hz,1H,Ph—H),8.00(d,J = 1.8Hz,1H,Ph—H),7.45(dd,J = 8.7,1.9Hz,1H, Ph—H).

$SOCl_2$(15mL)was added to compound **5**a(0.96g,5mmol)and the mixture was refluxed for 3 h then evaporated under reduced pressure. The residue was dissolved in 15mL of THF and added to a solution of methylamine aqueous solution (5mL in 10mL THF) dropwise in an ice bath. After 3 h,the mixture was evaporated then 50mL of H_2O and 70mL of ethyl acetate were added to the residue,the organic layer was collected,dried and evaporated to give a brown oil, which was further purified by flash chromatography to give 3 – amino – N – methylbenzo[b]thiophene – 2 – carboxamide (**6**a) as a brown solid:0.64g, yield 62.1%, m. p. 160 – 161℃ (163 – 164℃[20]). ^{1}H NMR(400MHz,$CDCl_3$),δ:7.71(d,J =8.0Hz,1H,Ph—H),7.64(d, J =7.9Hz,1H,Ph—H),7.48 – 7.42(m,1H,Ph—H),7.41 – 7.35(m,1H,Ph—H),6.05 (s,2H,NH_2),5.53(s,1H,CONH),2.99(t,J =5.3Hz,3H,NHC**H_3**).

With the same procedure,3 – amino – N – isopropylbenzo[b] – thiophene – 2 – carboxamide (**6**b),3 – amino – N – cyclopropylbenzo – [b]thiophene – 2 – carboxamide(**6**c),3 – amino – N – butylbenzo[b] – thiophene – 2 – carboxamide(**6**d),3 – amino – N,7 – dimethylbenzo[b] – thiophene – 2 – carboxamide(**6**e),3 – amino – N – isopropyl – 7 – methylbenzo[b]thiophene – 2 – carboxamide(**6**f),3 – amino – N – cyclopropyl – 7 – methylbenzo[b]thiophene – 2 – carboxamide(**6**g),3 – amino – N – butyl – 7 – methylbenzo[b]thiophene – 2 – carboxamide(**6**h), 3 – amino – N – (tert – butyl) – 7 – methylbenzo[b]thiophene – 2 – carboxamide(**6**i),3 – amino – 6 – chloro – N – methylbenzo[b]thiophene – 2 – carboxamide(**6**j),3 – amino – 6 – chloro – N – isopropylbenzo[b] – thiophene – 2 – carboxamide(**6**k),3 – amino – 6 – chloro – N – cyclopro – pylbenzo – [b]thiophene – 2 – carboxamide(**6**l),3 – amino – N – butyl – 6 – chlorobenzo[b]thiophene – 2 – carboxamide(**6**m) and 3 – amino – N – (tertbutyl) – 6 – chlorobenzo[b]thiophene – 2 – carboxamide(**6**n)were prepared.

Compound **6**b: yield 63%; brown solid, m. p. 105 – 106℃; ^{1}H NMR(400MHz, $CDCl_3$), δ: 7.73(d,J = 8.0Hz,1H,Ph—H),7.65(d,J = 7.9Hz,1H,Ph—H),7.49 – 7.44(m,1H, Ph—H),7.43 – 7.37(m,1H,Ph—H),6.06(s,2H,NH_2),5.32(d,J =6.8Hz,1H,CONH), 4.35 – 4.23(m,1H,NHC**H**),1.29(d,J =6.5Hz,6H,i – Pr—CH_3);HRMS,m/z:257.0715 $[M+Na]^+$,calcd. for $C_{12}H_{14}N_2OS$:257.0719.

Compound **6**c:yield 68%;brown solid,m. p. 80 – 81℃;^{1}H NMR(400MHz,$CDCl_3$),δ:7.72 (d,J =8.0Hz,1H,Ph—H),7.66(d,J =8.0Hz,1H,Ph—H),7.50 – 7.45(m,1H,Ph—H), 7.43 – 7.38(m,1H,Ph—H),6.12(s,2H,NH_2),5.68(s,1H,CONH),2.93 – 2.89(m,1H, NHC**H**),0.91 – 0.85(m,2H,cyclopropyl – CH_2),0.69 – 0.63(m,2H,cyclopropyl – CH_2); HRMS,m/z:255.0568$[M+Na]^+$,calcd. for $C_{12}H_{12}N_2OS$:255.0563.

Compound **6**d:yield 81%;brown solid,m. p. 81 – 82℃;^{1}H NMR(400MHz,$CDCl_3$),δ:7.71 (d,J =8.0Hz,1H,Ph—H),7.63(d,J =8.0Hz,1H,Ph—H),7.48 – 7.42(m,1H,Ph—H), 7.42 – 7.36(m,1H,Ph—H),6.03(s,2H,NH_2),5.50(s,1H,CONH),3.43(dd,J = 13.0, 7.1Hz,2H,n – Bu – CH_2),1.65 – 1.55(m,2H,n – Bu – CH_2),1.45 – 1.39(m,2H,n – Bu – CH_2),0.96(t,J = 7.3Hz,3H,n – Bu – CH_3);HRMS,m/z:271.0876$[M+Na]^+$, calcd. for $C_{13}H_{16}N_2OS$:271.0877.

Compound **6**e: yield 74%; yellow solid, m. p. 124 – 126℃; ^1H NMR (400MHz, $CDCl_3$), δ: 7.49(d, J=7.9Hz, 1H, Ph—H), 7.32(t, J=7.5Hz, 1H, Ph—H), 7.26(t, J=3.5Hz, 1H, Ph—H), 6.04(s, 2H, NH_2), 5.58(s, 1H, CONH), 2.99(t, J=5.3Hz, 3H, NHC**H**$_3$), 2.48 (s, 3H, Ph – CH_3); HRMS, *m/z*: 243.0563[M + Na]$^+$, calcd. for $C_{11}H_{12}N_2OS$: 243.0565.

Compound **6**f: yield 64%; yellow solid, m. p. 121 – 123℃; ^1H NMR (400MHz, $CDCl_3$), δ: 7.51(d, J=7.8Hz, 1H, Ph—H), 7.34(t, J=7.5Hz, 1H, Ph—H), 7.28(t, J=3.5Hz, 1H, Ph—H), 6.05(s, 2H, NH_2), 5.39(d, J=7.0Hz, 1H, CONH), 4.30(dq, J=13.2, 6.6Hz, 1H, NHC**H**), 2.51(s, 3H, Ph – CH_3), 1.29(d, J=6.6Hz, 6H, i – Pr – CH_3); HRMS, *m/z*: 271.0878[M + Na]$^+$, calcd. for $C_{13}H_{16}N_2OS$: 271.0877.

Compound **6**g: yield 83%; yellow solid, m. p. 101 – 103℃; ^1H NMR (400MHz, $CDCl_3$), δ: 7.49(d, J=7.9Hz, 1H, Ph—H), 7.32(t, J=7.5Hz, 1H, Ph—H), 7.27(d, J=3.7Hz, 1H, Ph—H), 6.09(s, 2H, NH_2), 5.72(s, 1H, CONH), 2.89 – 2.81(m, 1H, NHC**H**), 2.47(s, 3H, Ph – CH_3), 0.90 – 0.83(m, 2H, cyclopropyl – CH_2), 0.68 – 0.60(m, 2H, cyclopropyl – CH_2); HRMS, *m/z*: 269.0715[M + Na]$^+$, calcd. for $C_{13}H_{14}N_2OS$: 269.0719.

Compound **6**h: yield 59%; brown solid, m. p. 90 – 92℃; ^1H NMR (400MHz, $CDCl_3$), δ: 7.51(d, J=7.9Hz, 1H, Ph—H), 7.34(t, J=7.5Hz, 1H, Ph—H), 7.28(t, J=3.5Hz, 1H, Ph—H), 6.05(s, 2H, NH_2), 5.58(s, 1HCONH), 3.50 – 3.41(m, 2H, $NHCH_2$), 2.51(s, 3H, Ph – CH_3), 1.68 – 1.58(m, 2H, *n* – Bu – CH_2), 1.49 – 1.41(m, 2H, *n* – Bu – CH_2), 0.99(t, J=7.3Hz, 3H, *n* – Bu – CH_3); HRMS, *m/z*: 285.1031[M + Na]$^+$, calcd. for $C_{14}H_{18}N_2OS$: 285.1032.

Compound **6**i: yield 67%; yellow solid, m. p. 183 – 185℃; ^1H NMR (400MHz, $CDCl_3$), δ: 7.50(d, J=7.9Hz, 1H, Ph—H), 7.34(t, J=7.5Hz, 1H, Ph—H), 7.27(t, J=3.5Hz, 1H, Ph—H), 6.01(s, 2H, NH_2), 5.42(s, 1H, CONH), 2.51(s, 3H, Ph – CH_3), 1.50(s, 9H, t – Bu – CH_3); HRMS, *m/z*: 285.1033[M + Na]$^+$, calcd. for $C_{14}H_{18}N_2OS$: 285.1032.

Compound **6**j: yield 81%; yellow solid, m. p. 146 – 148℃; ^1H NMR (400MHz, $CDCl_3$), δ: 7.70(s, 1H, Ph—H), 7.57(d, J=8.6Hz, 1H, Ph—H), 7.35(dd, J=8.6, 1.2Hz, 1H, Ph—H), 6.05(s, 2H, NH_2), 5.53(s, 1H, CONH), 3.00(d, J=4.8Hz, 3H, NHC**H**$_3$); HRMS, *m/z*: 263.0015[M + Na]$^+$, calcd. for $C_{10}H_9ClN_2OS$: 263.0016.

Compound **6**k: yield 71%; yellow solid, m. p. 126 – 128℃; ^1H NMR (400MHz, $CDCl_3$), δ: 7.70(d, J=1.8Hz, 1H, Ph—H), 7.56(d, J=8.6Hz, 1H, Ph—H), 7.36(dd, J=8.6, 1.8Hz, 1H, Ph—H), 6.04(s, 2H, NH_2), 5.29(d, J=7.0Hz, 1H, N**H**CH), 4.31 – 3.25(m, 1H, NHC**H**), 1.28(d, J=6.5Hz, 6H, i – Pr – CH_3); HRMS, *m/z*: 291.0328[M + Na]$^+$, calcd. for $C_{12}H_{13}ClN_2OS$: 291.0329.

Compund **6**l: yield 83%; brown solid, m. p. 150 – 152℃; ^1H NMR (400MHz, $CDCl_3$), δ: 7.70(d, J=1.4Hz, 1H, Ph—H), 7.57(d, J=8.6Hz, 1H, Ph—H), 7.36(dd, J=8.6, 1.4Hz, 1H, Ph—h), 6.10(s, 2H, NH_2), 5.65(s, 1H, CONH), 2.86(dt, J=6.8, 3.2Hz, 1H, NHC**H**), 0.89(q, J=6.7Hz, 2H, cyclopropyl – CH_2), 0.66(t, J=7.9Hz, 2H, cyclopropyl – CH_2); HRMS, *m/z*: 289.0175[M + Na]$^+$, calcd. for $C_{12}H_{11}ClN_2OS$: 289.0173.

Compound **6**m: yield 64%; yellow solid, m. p. 93 – 95℃; ^1H NMR (400MHz, $CDCl_3$), δ: 7.69 (d, J=1.7Hz, 1H, Ph—H), 7.55(d, J=8.6Hz, 1H, Ph—H), 7.35(dd, J=8.6, 1.8Hz, 1H,

Ph—H), 6.01(s, 2H, NH_2), 5.46(s, 1H, CONH), 3.46 – 3.38(m, 2H, NHC**H_2**), 1.65 – 1.55 (m, 2H, *n* – Bu – CH_2), 1.42(m, 2H, *n* – Bu – CH_2), 0.96(t, *J* = 7.3Hz, 3H, *n* – Bu – CH_3); HRMS, *m/z*: 305.0485[M + Na]$^+$, calcd. for $C_{13}H_{15}ClN_2OS$: 305.0486.

Compound **6n**: yield 80%; yellow solid, m. p. 142 – 144℃; ^{1}H NMR(400MHz, $CDCl_3$), δ: 7.69(d, *J* = 1.6Hz, 1H, Ph—H), 7.55 (d, *J* = 8.6Hz, 1H, Ph—H), 7.35 (dd, *J* = 8.6, 1.6Hz, 1H, Ph—H), 5.98(s, 2H, NH_2), 5.31(s, 1H, CONH), 1.49(s, 9H, t – Bu – CH_3); HRMS, *m/z*: 305.0488[M + Na]$^+$, calcd. for $C_{13}H_{15}ClN_2OS$: 305.0486.

Compounds **7 – 11** were prepared in our laboratory *via* the reported procedure[21].

General synthetic procedure of title compounds **12** are illustrated in Scheme 2. To a suspension of acid **11**(1mmol) in 25mL of CH_2Cl_2 were added oxalyl chloride(3.0mmol) and 0.05mL of DMF. After stirring at room temperature for 3h, the solution was evaporated. The resulting acyl chloride was dissolved in 25mL of CH_2Cl_2 and added to a solution of compound **6** in 20mL of CH_2Cl_2 at room temperature. The mixture was stirred overnight and washed with H_2O, brine and dried with anhydrous Na_2SO_4. After evaporation, the residue was further purified by flash chromatography to give the title compounds.

3 – Bromo – 1 – (3 – chloropyridin – 2 – yl) – *N* – [2 – (methylcarbamoyl) benzo[*b*]thiophen – 3 – yl] – 1*H* – pyrazole – 5 – carboxamide(**12a**): yield 68%; white solid, m. p. 211 – 212℃; ^{1}H NMR(400MHz, $CDCl_3$), δ: 11.35(s, 1H, CONH), 8.49(dd, *J* = 4.7, 1.4Hz, 1H, pyridyl – H), 7.97(d, *J* = 8.2Hz, 1H, Ph—H), 7.87(dd, *J* = 8.0, 1.4Hz, 1H, pyridyl – H), 7.71(d, *J* = 8.2Hz, 1H, Ph—H), 7.46 – 7.32(m, 3H, pyridyl – H and Ph—H), 7.16(s, 1H, pyrazolyl – H), 6.00(d, *J* = 4.4Hz, 1H, N**H**CH_3), 2.98(d, *J* = 4.8Hz, 3H, NHC**H_3**); HRMS, *m/z*: 511.9550[M + Na]$^+$, calcd. for $C_{19}H_{13}BrClN_5NaO_2S$: 511.9554.

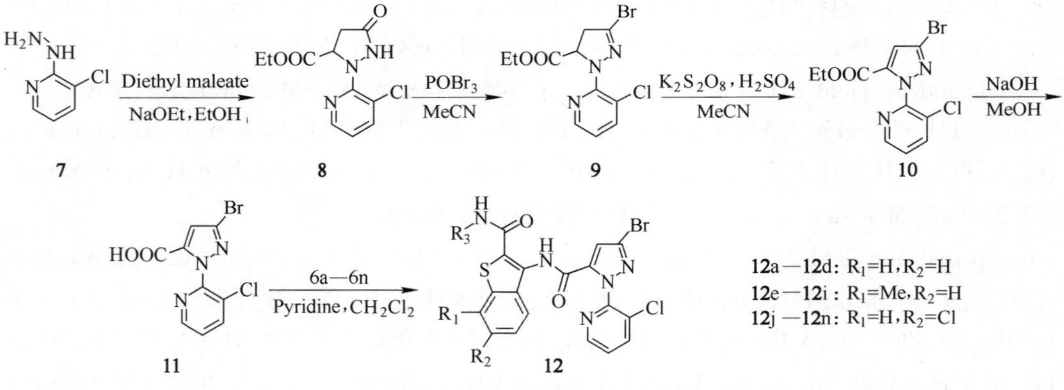

Scheme 2 Synthetic routes of title compounds **12**a – **12**n

3 – Bromo – 1 – (3 – chloropyridin – 2 – yl) – *N* – [2 – (isopropylcarbamoyl) benzo[*b*]thiophen – 3 – yl] – 1*H* – pyrazole – 5 – carboxamide(**12b**): yield 64%; white solid, m. p. 202 – 204℃; ^{1}H NMR(400MHz, $CDCl_3$), δ: 11.42(s, 1H, CONH), 8.51(dd, *J* = 4.7, 1.5Hz, 1H, pyridyl – H), 7.99(d, *J* = 8.2Hz, 1H, Ph—H), 7.89(dd, *J* = 8.0, 1.5Hz, 1H, pyridyl – H), 7.73(d, *J* = 8.1Hz, 1H, Ph—H), 7.47 – 7.34(m, 3H, pyridyl – H and Ph—H), 7.18(s, 1H, pyrazolyl – H), 5.73 (d, *J* = 7.3Hz, 1H, CON**H**CH), 4.30 (dq, *J* = 13.3, 6.6Hz, 1H, NHC**H**), 1.32(d, *J* = 6.6Hz, 6H, i – Pr – CH_3); HRMS, *m/z*: 539.9867[M + Na]$^+$, calcd. for

$C_{21}H_{17}BrClN_5NaO_2S$: 539.9867.

3 - Bromo - 1 - (3 - chloropyridin - 2 - yl) - *N* - [2 - (cyclopropylcarbamoyl) benzo[*b*] thiophen - 3 - yl] - 1*H* - pyrazole - 5 - carboxamide (**12c**): yield 56%; white solid, m. p. 217 - 218℃; ^1H NMR (400MHz, $CDCl_3$), δ: 11.30 (s, 1H, CONH), 8.48 (dd, *J* = 4.7, 1.5Hz, 1H, pyridyl - H), 7.94 (d, *J* = 8.2Hz, 1H, Ph—H), 7.87 (dd, *J* = 8.0, 1.5Hz, 1H, pyridyl - H), 7.69 (d, *J* = 8.0Hz, 1H, Ph—H), 7.45 - 7.32 (m, 3H, pyridyl - H and Ph—H), 7.18 (s, 1H, pyrazolyl - H), 6.18 (s, 1H, CON**H**CH), 2.77 (dt, *J* = 10.1, 3.3Hz, 1H, NHC**H**), 0.93 - 0.84 (m, 2H, cyclopropyl - CH_2), 0.67 - 0.59 (m, 2H, cyclopropyl - CH_2); HRMS, *m/z*: 537.9714 [M + Na]$^+$, calcd. for $C_{21}H_{15}BrClN_5NaO_2S$: 537.9711.

3 - Bromo - *N* - [2 - (butylcarbamoyl) benzo[*b*] thiophen - 3 - yl] - 1 - (3 - chloropyridin - 2 - yl) - 1*H* - pyrazole - 5 - carboxamide (**12d**): yield 70%; pale solid, m. p. 202 - 204℃; ^1H NMR (400MHz, $CDCl_3$), δ: 11.38 (s, 1H, CONH), 8.51 (dd, *J* = 4.7, 1.4Hz, 1H, pyridyl - H), 7.98 (d, *J* = 8.2Hz, 1H, Ph—H), 7.89 (dd, *J* = 8.0, 1.4Hz, 1H, pyridyl - H), 7.73 (d, *J* = 8.0Hz, 1H, Ph—H), 7.48 - 7.34 (m, 3H, pyridyl - H and Ph—H), 7.19 (s, 1H, pyrazolyl - H), 6.01 (s, 1H, CON**H**CH_2), 3.43 (dd, *J* = 13.1, 7.0Hz, 2H, *n* - Bu - CH_2), 1.62 (dt, *J* = 14.9, 7.4Hz, 3H, *n* - Bu - CH_2), 1.43 (dq, *J* = 14.5, 7.3Hz, 3H, *n* - Bu - CH_2), 0.99 (t, *J* = 7.4Hz, 3H, *n* - Bu - CH_3); HRMS, *m/z*: 554.0020 [M + Na]$^+$, calcd. for $C_{22}H_{19}BrClN_5NaO_2S$: 554.0024.

3 - Bromo - 1 - (3 - chloropyridin - 2 - yl) - *N* - [7 - methyl - 2 - (methylcarbamoyl) benzo[*b*] thiophen - 3 - yl] - 1*H* - pyrazole - 5 - carboxamide (**12e**): yield 59%; white solid, m. p. 201 - 202℃; ^1H NMR (400MHz, $CDCl_3$), δ: 11.45 (s, 1H, CONH), 8.53 (d, *J* = 4.7Hz, 1H, pyridyl - H), 7.91 (d, *J* = 8.0Hz, 1H, pyridyl - H), 7.84 (d, *J* = 8.1Hz, 1H, Ph—H), 7.44 (dd, *J* = 8.0, 4.7Hz, 1H, pyridyl - H), 7.41 (s, 1H, pyrazolyl - H), 7.35 - 7.29 (m, 1H, Ph—H), 7.28 - 7.24 (m, 1H, Ph—H), 6.05 (s, 1H, CONH), 3.04 (d, *J* = 4.8Hz, 3H, NHC$\mathbf{H}_3$), 2.51 (s, 3H, Ph - CH_3); HRMS, *m/z*: 525.9713 [M + Na]$^+$, calcd. for $C_{20}H_{15}BrClN_5NaO_2S$: 525.9716.

3 - Bromo - 1 - (3 - chloropyridin - 2 - yl) - *N* - [2 - (isopropylcarbamoyl) - 7 - methylbenzo[*b*] thiophen - 3 - yl] - 1*H* - pyrazole - 5 - carboxamide (**12f**): yield 75%; white solid, m. p. 224 - 225℃; ^1H NMR (400MHz, DMSO - d_6), δ: 10.77 (s, 1H, CONH), 8.51 (dd, *J* = 4.7, 1.5Hz, 1H, pyridyl - H), 8.19 (dd, *J* = 8.1, 1.4Hz, 1H, pyridyl - H), 8.02 (d, *J* = 5.6Hz, 1H, CON**H**CH), 7.62 (dd, *J* = 8.1, 4.7Hz, 1H, pyridyl - H), 7.58 (d, *J* = 7.9Hz, 1H, Ph—H), 7.53 (s, 1H, pyrazolyl - H), 7.39 (t, *J* = 7.6Hz, 1H, Ph—H), 7.33 (d, *J* = 7.1Hz, 1H, Ph—H), 4.02 (dq, *J* = 13.3, 6.6Hz, 1H, NHC**H**), 2.50 (s, 3H, Ph - CH_3), 1.10 (d, *J* = 6.6Hz, 6H, i - Pr - CH_3); HRMS, *m/z*: 554.0023 [M + Na]$^+$, calcd. for $C_{22}H_{19}BrClN_5NaO_2S$: 554.0024.

3 - Bromo - 1 - (3 - chloropyridin - 2 - yl) - *N* - [2 - (cyclopropylcarbamoyl) - 7 - methylbenzo[*b*] thiophen - 3 - yl] - 1*H* - pyrazole - 5 - carboxamide (**12g**): yield 53%; white solid, m. p. 223 - 224℃; ^1H NMR (400MHz, $CDCl_3$), δ: 11.33 (s, 1H, CONH), 8.49 (dd, *J* = 4.7, 1.4Hz, 1H, pyridyl - H), 7.87 (dd, *J* = 8.0, 1.4Hz, 1H, pyridyl - H), 7.82 (d, *J* = 8.0Hz, 1H, Ph—H), 7.38 (dd, *J* = 8.0, 4.7Hz, 1H, pyridyl - H), 7.31 - 7.27 (m, 1H, Ph—H), 7.23 (d, *J* = 7.1Hz, 1H, Ph—H), 7.18 (s, 1H, pyrazolyl - H), 6.16 (s, 1H, N**H**CH),

2.84(dt, J = 7.0, 3.3Hz, 1H, NHC**H**), 2.47(s, 3H, Ph − CH_3), 0.92(q, J = 6.9Hz, 2H, cyclopropyl − CH_2), 0.66(q, J = 7.0Hz, 2H, cyclopropyl − CH_2); HRMS, m/z: 551.9865 [M + Na]$^+$, calcd. for $C_{22}H_{17}BrClN_5NaO_2S$: 551.9867.

3 − Bromo − N − [2 − (butylcarbamoyl) − 7 − methylbenzo[b]thiophen − 3 − yl] − 1 − (3 − chloropyridin − 2 − yl) − 1H − pyrazole − 5 − carboxamide (**12**h): yield 69%; white solid, m. p. 209 − 210℃; ^1H NMR(400MHz, DMSO − d_6), δ: 10.80(s, 1H, CONH), 8.52(d, J = 4.7Hz, 1H, pyridyl − H), 8.20(d, J = 8.1Hz, 1H, pyridyl − H), 8.16(t, J = 5.6Hz, 1H, N**H**CH_2), 7.62(dd, J = 8.1, 4.7Hz, 1H, pyridyl − H), 7.56(d, J = 7.9Hz, 1H, Ph—H), 7.49(s, 1H, pyrazolyl − H), 7.39(t, J = 7.6Hz, 1H, Ph—H), 7.34(d, J = 7.1Hz, 1H, Ph—H), 3.26 − 3.20(m, 2H, NHC$\mathbf{H_2}$), 2.50(s, 3H, Ph − CH_3), 1.50 − 1.39(m, 2H, n − Bu − CH_2), 1.34 − 1.23(m, 2H, n − Bu − CH_2), 0.86(t, J = 7.3Hz, 3H, n − Bu − CH_3); HRMS, m/z: 568.0186 [M + Na]$^+$, calcd. for $C_{23}H_{21}BrClN_5NaO_2S$: 568.0180.

3 − Bromo − N − [2 − (tert − butylcarbamoyl) − 7 − methylbenzo[b] − thiophen − 3 − yl] − 1 − (3 − chloropyridin − 2 − yl) − 1H − pyrazole − 5 − carboxamide(**12**i): yield 83%; white solid, m. p. 219 − 221℃; ^1H NMR(400MHz, $CDCl_3$), δ: 11.30(s, 1H, CONH), 8.51(dd, J = 4.7, 1.5Hz, 1H, pyridyl − H), 7.88(dd, J = 8.0, 1.5Hz, 1H, pyridyl − H), 7.79(d, J = 8.1Hz, 1H, Ph—H), 7.40(dd, J = 8.0, 4.7Hz, 1H, pyridyl − H), 7.30(t, J = 6.2Hz, 1H, Ph—H), 7.24(d, J = 7.1Hz, 1H, Ph—H), 7.19(s, 1H, pyrazolyl − H), 5.82(s, 1H, CONH), 2.50(s, 3H, Ph − CH_3), 1.51(s, 9H, t − Bu); HRMS, m/z: 568.0184 [M + Na]$^+$, calcd. for $C_{23}H_{21}BrClN_5NaO_2S$: 568.0180.

3 − Bromo − N − [6 − chloro − 2 − (methylcarbamoyl)benzo[b]thiophen − 3 − yl] − 1 − (3 − chloropyridin − 2 − yl) − 1H − pyrazole − 5 − carboxamide(**12**j): yield 72%; white solid, m. p. 199 − 201℃; ^1H NMR(400MHz, $CDCl_3$), δ: 11.45(s, 1H, CONH), 8.51(d, J = 4.6Hz, 1H, pyridyl − H), 7.95(d, J = 8.8Hz, 1H, Ph—H), 7.90(d, J = 7.9Hz, 1H, pyridyl − H), 7.72(s, 1H, Ph—H), 7.42(dd, J = 8.0, 4.6Hz, 1H, pyridyl − H), 7.33(d, J = 8.9Hz, 1H, Ph—H), 7.18(s, 1H, pyrazolyl − H), 5.98(s, 1H, N**H**CH_3), 3.04(d, J = 4.7Hz, 3H, NHC$\mathbf{H_3}$); HRMS, m/z: 545.9161 [M + Na]$^+$, calcd. for $C_{19}H_{12}BrCl_2N_5NaO_2S$: 545.9164.

3 − Bromo − N − [6 − chloro − 2 − (isopropylcarbamoyl)benzo[b] − thiophen − 3 − yl] − 1 − (3 − chloropyridin − 2 − yl) − 1H − pyrazole − 5 − carboxamide(**12**k): yield 67%; white solid, m. p. 229 − 231℃; ^1H NMR(400MHz, DMSO − d_6), δ: 10.81(s, 1H, CONH), 8.52(dd, J = 4.7, 1.2Hz, 1H, pyridyl − H), 8.20(dd, J = 7.8, 1.2Hz, 2H, pyridyl − H and Ph—H), 8.07(d, J = 7.6Hz, 1H, N**H**CH), 7.72(d, J = 8.7Hz, 1H, Ph—H), 7.62(dd, J = 8.1, 4.7Hz, 1H, pyridyl − H), 7.56 − 7.46(m, 2H, Ph—H and pyrazolyl − H), 4.07 − 3.92(m, 1H, NHC**H**), 1.10(d, J = 6.6Hz, 6H, i − Pr − CH_3); HRMS, m/z: 573.9485 [M + Na]$^+$, calcd. for $C_{21}H_{16}BrCl_2N_5NaO_2S$: 573.9477.

3 − Bromo − N − [6 − chloro − 2 − (cyclopropylcarbamoyl)benzo[b] − thiophen − 3 − yl] − 1 − (3 − chloropyridin − 2 − yl) − 1H − pyrazole − 5 − carboxamide(**12**l): yield 53%; white solid, m. p. 226 − 227℃; ^1H NMR(400MHz, DMSO − d_6), δ: 10.79(s, 1H, CONH), 8.53(d, J = 4.7Hz, 1H, pyridyl − H), 8.36(d, J = 3.0Hz, 1H, CONH), 8.24 − 8.16(m, 2H, pyridyl − H and Ph—H), 7.72(d, J = 8.7Hz, 1H, Ph—H), 7.63(dd, J = 8.0, 4.7Hz, 1H, pyridyl − H), 7.54 − 7.46(m, 2H, Ph—H and pyrazolyl − H), 2.78 − 2.70(m, 1H, NHC**H**), 0.71 − 0.63

(m,2H,cycloproprl – CH_2),0.51 – 0.44(m,2H,cycloproprl – CH_2);HRMS,m/z:571.9329 [M + Na]$^+$,calcd. for $C_{21}H_{14}BrCl_2N_5NaO_2S$:571.9321.

3 – Bromo – N – [2 – (butylcarbamoyl) – 6 – chlorobenzo[b]thiophen – 3 – yl] – 1 – (3 – chloropyridin – 2 – yl) – 1H – pyrazole – 5 – carboxamide (**12m**): yield 74%; white solid, m. p. 205 – 206℃; ^{1}H NMR(400MHz, DMSO – d_6), δ: 10.82(s,1H,CONH),8.53(d,J = 4.7Hz,1H,pyridyl – H),8.26 – 8.15(m,3H,pyridyl – H,Ph—H and N**H**CH_2),7.69(d,J = 8.7Hz,1H,Ph—H),7.63(dd,J = 8.1,4.7Hz,1H,pyridyl – H),7.54(s,1H,pyrazolyl – H),7.51(dd,J = 8.7,1.6Hz,1H,Ph—H),3.21(dd,J = 12.9,6.7Hz,2H,NHC**H_2**),1.49 – 1.38(m,2H,n – Bu – CH_2),1.34 – 1.22(m,2H,n – Bu – CH_2),0.85(t,J = 7.3Hz,3H, n – Bu – CH_3);HRMS,m/z:587.9626[M + Na]$^+$,calcd. for $C_{22}H_{18}BrCl_2N_5NaO_2S$:587.9634.

3 – Bromo – N – [2 – (tert – butylcarbamoyl) – 6 – chlorobenzo[b] – thiophen – 3 – yl] – 1 – (3 – chloropyridin – 2 – yl) – 1H – pyrazole – 5 – carboxamide (**12n**): yield 81%; white solid, m. p. 216 – 217℃; ^{1}H NMR(400MHz, DMSO – d_6), δ: 10.77(s,1H,CONH),8.51(d, J = 4.7Hz,1H,pyridyl – H),8.23 – 8.14(m,2H,CONH and pyridyl – H),7.74(d,J = 8.7Hz, 1H,Ph—H),7.62(dd,J = 8.0,4.7Hz,1H, pyridyl – H),7.57 – 7.48(m,3H, Ph—H and pyrazolyl – H),1.28(s,9H,t – Bu – CH_3);HRMS,m/z:587.9630[M + Na]$^+$,calcd. for $C_{22}H_{18}BrCl_2N_5NaO_2S$:587.9634.

2.3 Biological Assay

Insecticidal activities against oriental armyworms (*Mythimna separata*) were performed on test organisms reared in a greenhouse. The bioassay was replicated at (25 ± 1)℃ according to statistical requirements. Assessments were made on a dead/alive basis, and mortality rates were corrected *via* Abbott's formula[22]. Evaluation was based on a percentage scale of 0 – 100, where 0 means no activity, and 100 means total kill. Error of the experiments was 5%.

The fungicidal activities of compounds **12**a – **12**n were tested in vitro against *Physalospora piricola* (*P. piricola*), *Alternaria solani* (*A. solani*), *Cercospora arachidicola* (*C. arachidicola*), *Gibberella sanbinetti* (*G. sanbinetti*) *and Phytophthora infestans* (*P. infestans*), and their relative inhibitory ratio (%) was determined with the mycelium growth rate method[23]. Chlorothalonil and carbendazim were used as controls. After the mycelia grew completely, the diameters of the mycelia were measured, and the inhibition rate was calculated according to the formula

$$I(\%) = (D_1 - D_2) \times 100\%$$

where $I(\%)$ is the inhibition rate, D_1 is the average diameter of mycelia in the blank test, and D_2 is the average diameter of mycelia in the presence of each of those compounds. The inhibition ratios of those compounds at a dose of 50mg/L were determined.

3 Results and Discussion

3.1 Synthesis

The syntheses of title compounds are illustrated in Scheme 1 and Scheme 2. Compound **2** was dehydrated by cyanuric chloride to give compound **3** in a high yield. Compound **4** was synthesized from o – nitrobenzonitrile analogues **3** and methyl thioglycolate in the presence of 20mL

4.5mmol/L KOH referring to the former report[20]. A sufficient amount of KOH is critical to this reaction, but with a less amount of KOH, compound **3** would not be completely converted. Compound **5** was obtained by the hydrolysis of ester **4** by aqueous sodium hydroxide, and the reaction was carried out under reflux condition.

Title compounds **12** were synthesized from acid **11** and aniline **6** at room temperature. In the reactions CH_2Cl_2, CH_3CN, THF were used as solvents, among which CH_2Cl_2 was proved to be the best. Triethylamine (TEA), *N*, *N* – diisopropyl ethyl amine (DIPEA) and pyridine were tested as acid binding reagents, and pyridine was preferential.

3.2 Biological Activity

Biological evaluation results of title compounds are listed in Table 1. The insecticidal activities of them against oriental armyworms are only moderate (10% – 70%) at 200mg/L. Comparing the insecticidal activities of all the title compounds, we could conclude that when R_3 is isopropyl or cyclopropyl moiety, the insecticidal activities are better than those of the ones with *n* – butyl or *t* – butyl moiety.

Table 1 Insecticidal activities of compounds 12a – 12n and chlorantraniliprole against oriental armyworms at a concentration of 200mg/L

Compd.	R_1	R_2	R_3	Larvicidal activity (%)
12a	H	H	Me	30.0
12b	H	H	*i* – Pr	50.0
12c	H	H	Cyclopropyl	16.7
12d	H	H	*n* – Bu	16.7
12e	Me	H	Me	60.0
12f	Me	H	*i* – Pr	40.0
12g	Me	H	Cyclopropyl	70.0
12h	Me	H	*n* – Bu	10.0
12i	Me	H	*t* – Bu	50.0
12j	H	Cl	Me	33.3
12k	H	Cl	*i* – Pr	56.7
12l	H	Cl	Cyclopropyl	30.0
12m	H	Cl	*n* – Bu	30.0
12n	H	Cl	*t* – Bu	40.0
Chlorantraniliprole				100.0

The fungicidal activities against *P. piricola*, *A. Solani*, *C. Arachidicola*, *G. Sanbinetti* and *P. infestans* were also evaluated with the results recorded in Table 2. All the compounds tested showed moderate antibacterial activities, among which **12**c showed an inhibition rate of 66.7% at 50mg/L against *C. Arachidicola*, which is better than control carbendazim (8.3%), and close to chlorothalonil (75%). Compound **12**d showed an inhibition rate of 62.1% at 50mg/L against *P. piricola*, which is the best of all the compounds tested. The fungicidal activity data of all the compounds indicate that these compounds have intrinsic fungicidal activities.

Table 2 Antifungal activities(relative inhibitory ratio, %) of title compounds 12a – 12n, chlorothalonil and carbendazim in vitro(50mg/L)

Compd.	*P. piricola*	*A. solani*	*C. arachidicola*	*G. sanbinetti*	*P. infestans*
12a	44. 8	40. 7	26. 7	21. 2	18. 8
12b	27. 6	18. 5	13. 3	30. 3	28. 1
12c	34. 5	37. 0	66. 7	36. 4	28. 1
12d	62. 1	25. 9	26. 7	15. 2	18. 8
12e	31. 0	33. 3	13. 3	27. 3	18. 8
12f	26. 7	18. 8	15. 2	34. 5	25. 0
12g	20. 7	25. 9	26. 7	21. 2	28. 1
12h	24. 1	37. 0	26. 7	27. 3	37. 5
12i	31. 0	25. 9	26. 7	30. 3	26. 7
12j	44. 8	25. 9	40. 0	21. 2	37. 5
12k	20. 7	25. 9	20. 0	21. 2	18. 8
12l	31. 0	37. 0	20. 0	30. 3	25. 0
12m	27. 6	40. 7	0	39. 4	43. 8
12n	10. 3	40. 7	6. 7	30. 3	18. 8
Chlorothalonil	92. 3	73. 9	75. 0	66. 7	69. 5
Carbendazim	97. 4	43. 5	8. 3	100. 0	74. 6

In summary, a series of novel anthranilic diamides analogues containing benzo[*b*]thiophenyl ring has been synthesized and characterized. All the 14 compounds show moderate larvicidal activity against *Mythimna separate* at 200mg/L and moderate antibacterial activities against five fungi tested. Though the most active compound **12**g shows a less effect against *Mythimna separate* than the control chlorantraniliprole, its novel structure makes it an interesting compound for further studies. Also, this work provides us some insights into these novel heterocyclic structures.

References

[1] Lahm G. P., Selby T. P., Frendenberger J. H., Stevenson T. M., Myers B. J., Seburyamo G., Smith B. K., Flexner L., Clark C. E., Cordova D., *Bioorg. Med. Chem. Lett.*, 2005, 15, 4898.

[2] Alig B., Fischer R., Funke C., Gesing E. F., Hense A., Malsam O., Drewes M. W., Gorgens U., Murata T., Wada K., Arnold C., Sanwald E., *Anthranilic Acid Diamide Derivative with Hetero – aromatic and Hetero – cyclic Substituents*, *WO 2007144100*, 2007.

[3] Loiseleur O., Hall R. G., Stoller A. D., Graig G. W., Jeanguenat A., Edmunds A., *Novel Insecticides*, *WO 2009024341*, 2009.

[4] Loiseleur O., Durieux P., Trah S., Edmunds A., Jeanguenat A., Stoller A., Hughes D. J., *Pesticides Containing a Bicyclic Bisamide Structure*, *WO* 2007093402, 2007.

[5] Kruger B., Hense A., Alig B., Fischer R., Funke C., Gesing E. R., Malsam A., Drewes M. W., Arnold C., Lummen P., Sanwald E., *Dioxazine – and Oxadiazine – Substituted Arylamides*, *WO 2007031213*, 2007.

[6] Dumas D. J., Process for Preparing 2 – Amino – 5 – cyanobenzoic *Acid Derivatives*, *WO* 2009111553, 2009.

[7] Hughes D. J. , Peace J. E. , Riley S. , Russell S. , Swanborough J. J. , Jeanguenat A. , Renold P. , Hall R. G. , Loiseleur O. , Trah S. , Wenger J. , *Pesticidal Mixtures*, *WO 2007009661*, 2007.

[8] Zhang J. F. , Xu J. Y. , Wang B. L. , Li Y. X. , Xiong L. X. , Li Y. Q. , Ma Y. , Li Z. M. , *J. Agric. Food Chem.* , 2012, 60, 7565.

[9] Liu P. F. , Zhou S. , Xiong L. X. , Yu S. J. , Zhang X. , Li Z. M. , *Chem. J. Chinese Universities*, 2012, 33(4), 738.

[10] Liu P. F. , Zhang J. F. , Yan T. , Xiong L. X. , Li Z. M. , Chem. Res. *Chinese Universities*, 2012, 28(2), 430.

[11] Yan T. , Yu G. P. , Liu P. F. , Xiong L. X. , Yu S. J. , Li Z. M. , *Chem. Res. Chinese Universities*, 2012, 28(1), 53.

[12] Meyer M. D. , Altenbach R. J. , Basha F. Z. , Carroll W. A. , Condon S. , Elmore S. W. , Kerwin J. F. , Sippy K. B. , Tietje K. , Wendt M. D. , Hancock A. A. , Brune M. E. , Buckner S. A. , Drizin I. , *J. Med. Chem.* , 2000, 43, 1586.

[13] Kilsheimer J. R. , Kaufman H. A. , Foster H. M. , Driscoll P. R. , GlickL. A. , Napier R. P. , *J. Agric. Food Chem.* , 1969, 17, 91.

[14] Romagnoli R. , Baraldi P. G. , Cara C. L. , Hamel E. , Basso G. , Bortolozzi R. , Viola G. , *Eur. J. Med. Chem.* , 2010, 45, 5781.

[15] Boateng C. A. , Eyunni S. V. K. , Zhu X. Y. , Etukala J. R. , Bricker B. A. , Ashfaq M. K. , Jacob M. R. , Khan S. I. , Walker L. A. , Ablordeppey S. Y. , *Bioorg. Med. Chem.* , 2011, 19, 458.

[16] Chen Z. D. , Ginn J. D. , Hickey E. R. , Liu W. M. , Mao C. , Morwick T. M. , Nemoto P. A. , Spero D. , Sun S. X. , *Substituted Benzothiophene Compounds and Uses Thereof*, *WO 2005012283*, 2005.

[17] Owston N. A. , Parker A. J. , Williams J. M. J. , *Org. Lett.* , 2007, 9, 3599.

[18] Campbell J. B. , Warawa E. J. , *Certain 1H-pyrrold[3,4-b]quinolin-1-one-9-amino-2,3-dihydro Derivatives Useful for Treating Anxiety*, *US 4975435*, 1990.

[19] Kreimeyer A. , Laube B. , Sturgess M. , Goeldner M. , Foucaud B. , *J. Med. Chem.* , 1999, 42, 4394.

[20] Beck J. R. , *J. Org. Chem.* , 1972, 37, 3224.

[21] Yan T. , Yu G. P. , Xiong L. X. , Yu S. J. , Wang S. H. , Li Z. M. , *Chem. J. Chinese Universities*, 2011, 32(8), 1750.

[22] Abbott W. S. , *J. Econ. Entomol.* , 1925, 18, 265.

[23] Chen N. C. , *Bioassay of Pesticides*, Beijing Agricultural University Press, Beijing, 1991, 161.

附录

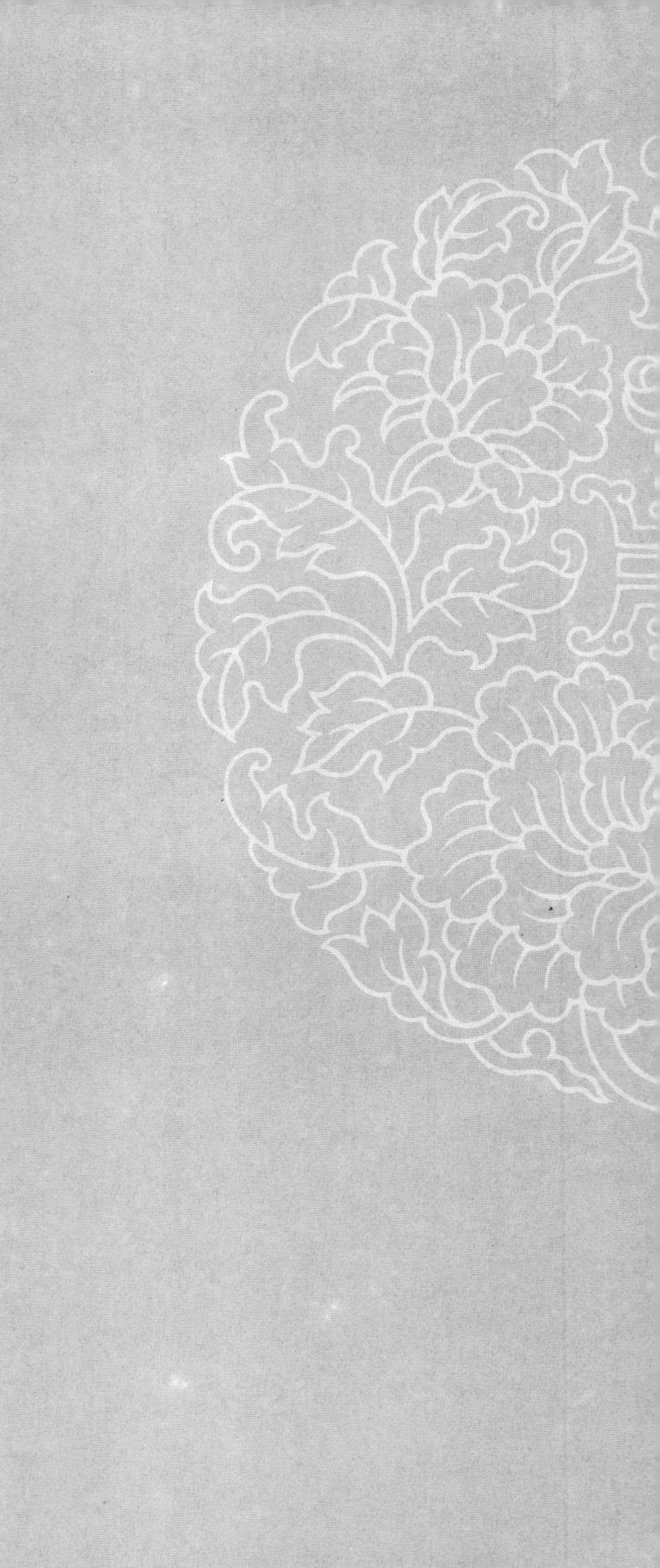

附录1 论文总目录*

1957 年

1. 萘和苯的衍生物类生长素对植物插枝生根作用的初步报告．杨石先，叶超然，姚珍，李正名，崔徵．南开大学学报（自然科学），1957（04）：132－148.

1959 年

2. 有机磷杀虫剂的研究 Ⅰ．杨石先，陈天池，李正名，李毓桂，王琴荪，颜茂恭，董希阳．化学学报，1959，25（6）：402－408；1960，7：897－906.

1962 年

3. 有机磷杀虫剂的研究Ⅱ．杨石先，陈天池，李正名，李毓桂，董希阳，高绍仪，董松琦．化学学报，1962，28（3）：187－190.

1963 年

4. 有机磷杀虫剂的研究Ⅲ．杨石先，陈天池，王琴荪，李正名．化学学报，1963，29（3）：153－158.
5. 折射度法在测定有机磷化合物的结构上的应用．李正名．化学通报，1963，12：724.

1964 年

6. 有机磷杀虫剂的研究Ⅴ．杨石先，陈天池，李正名，唐除痴，黄润秋，王惠林．南开大学学报，1964，5（2）：59－63.
7. 有机磷杀虫剂的研究Ⅵ．杨石先，陈天池，李正名，么恩云，刘天麟．南开大学学报，1964，5(3)：79－88.
8. 近年来有机磷化学及其应用的发展．陈天池，李正名．化学通报，1964，6：321.

1965 年

9. 有机磷杀虫剂的研究Ⅶ．杨石先，陈天池，李正名，王惠林，黄润秋，唐除痴，刘天麟，张金碚．化学学报，1965，31(5)：399－406.

1972 年

10. 新内吸性杀菌剂抑枯灵的合成研究初报．南开大学杀菌组．农药，1972，3：8－12.
11. 国外内吸性杀菌剂（化学治疗剂）进展．南开大学元素化学研究所杀菌剂组．农药，1972，3：39.

1974 年

12. 防治水稻白叶枯病新杀菌剂 Tf－114 的合成研究．南开大学杀菌组．南开大学学报（自然科学），1974，1：91－94.

* 第210、244、417、498 篇是应邀所写的论文。序号标灰的论文已收录在本书中。

13. 近年来国内外水稻白叶枯病化学防治的进展．李正名．南开大学学报（自然科学学报），1974，1：77－90.
14. 红外光谱在测定有机磷化合物结构上的应用．李正名．燃化科技资料，1974，6：17－46.

1975 年

15. 水稻白叶枯病的化学防治．李正名．化学通报，1975，5：21.

1979 年

16. 近年来农药科研的进展和趋势．杨石先，李正名．世界农业（第四辑），1979，13－17.

1983 年

17. UV－Ozonation and Land Disposal of Aqueous Pesticide Wastes. *P. C. Kearney*, *J. R. Plimmer*, *Zhengming Li*. IUPAC Pesticide Chemistry Monographs Published by Pergamon Press. 1983，397－400.
18. 内吸杀菌剂三唑醇立体异构体研究初报．邵瑞链，李正名，陈宗庭，王笃祜，谢龙观．农药，1983，5：2－3，14.
19. 美国昆虫激素研究概况．李正名．农药译丛，1983，5（1）：23.

1984 年

20. 棉铃虫拟信息素正十四烷基甲酸酯及有关化合物的触角电位反应．任自立，王银淑，尚稚珍，刘天麟，李正名，郭虎森．昆虫激素，1984，1：43－46.
21. 毛细管气相色谱仪分离柱及进样方式的若干进展．李正名．分析仪器，1984，1：64－68.
22. 一种合成亚洲玉米螟性外激素的新方法．李正名，*M. Schwarz*. 中国科学（B 辑），1984，1：38－43.
23. A Convenient Synthetic route for the Sex Pheromone of the Asian Corn Borer Moth（*Ostrinia furnacalis Guenee*）. *Zhengming Li*, *M. Schwarz*. Scientia Sinica（Series B），1984，1：38－43.
24. 改良的 Wittig 反应．李正名，董丽雯．有机化学，1984，5：334.

1985 年

25. 内吸性杀菌剂三唑醇（酮）立体异构体的研究Ⅱ．李正名，董丽雯，李国炜，张祖新．有机化学，1985，2：188－192.
26. Stereochemistry of New Systemic Fungicides Triadimenol and Triadimefon Ⅱ. *Li Zhengming*, *Dong Liwen*, *Li Guowei*, *Zhang Zuxin*. 有机化学，1985，1：190－192.
27. 1984 年 IUPAC 农药化学委员会会议追记．李正名．农药译丛，1985，7(4)：52.
28. 手性化合物的气相色谱分析．李正名，么恩云．有机化学，1985，6：443－449.
29. 双键立体构型转化用有机试剂．李正名，王天生．化学试剂，1985，7(6)：340.

1986 年

30. 水稻二化螟拟信息素的合成．李正名，刘天麟，刘子平，郭虎森，么恩云．高等学校化学学报，1986，7（3）：228－232［转载于 The synthesis of parapheromone of rice stem borer moth. *Li Zhengming*, *Liu Tianlin*, *Liu Ziping*, *Guo Husen*, *Yao Enyun*. Chemical Journal of Chinese Universities（英文版），1985，2（2）：439］.

1987 年

31. 内吸性杀菌剂三唑醇（酮）立体异构体的研究（Ⅱ）．李正名，董丽雯，李国炜，张祖新，曹秋文，王素华，窦士琦．高等学校化学学报，1987，8（3）：235－239.

32. 1－(4－氯苯氧基)－3，3－二甲基－1－（1，2，4－三氮唑－1基）－丁醇2右旋A体的晶体结构和绝对构型．窦士琦，姚家星，李正名，董丽雯．物理化学学报，1987，3（1）：74－77.

33. 昆虫信息素研究Ⅲ．李正名，王天生，么恩云，陈学仁，朱兰惠，王素华．化学学报，1987，45：1124－1128.

34. 第六届国际农药化学会议概况介绍（上）．李正名．农药译丛，1987，9（5）：1－6.

35. 第六届国际农药化学会议概况介绍（下）．李正名．农药译丛，1987，9（6）：2－6.

36. 中国昆虫信息素的研究与开发．李正名，刘孟英．化学生态物质，1987，1：1－6.

1988 年

37. 家蝇性诱剂对舍蝇诱引作用的研究．毕富春，王文丽，王玲秀，刘天麟，李正名．化学生态物质，1988，1：46－48.

38. 昆虫信息素的研究Ⅳ—印度谷螟性信息素的合成研究．郭海生，王素华，李正名．农药，1988，27(3)：19－21.

39. 昆虫信息素的研究Ⅴ—家蝇性诱剂（Z）－9－二十三碳烯的合成．刘天麟，李正名，王立坤，梁格．高等学校化学学报，1988，9（5）：522－524.

40. 昆虫性信息素的研究Ⅱ—顺－11－十六碳烯醛的合成．刘天麟，徐宝元，李正名．有机化学，1988，8：156－158.

41. 有机硼试剂在碳－碳键生成中的应用．王天生，李正名．化学试剂，1988，10（6）：333－342.

42. 有机硅试剂在碳－碳键合成上的应用．李正名，郭海生．昆虫激素，1988，2：1.

1989 年

43. 叶醇及其衍生物的立体有择合成．李正名，蒋益民．化学通报，1989，12：23－24.

44. 新农药的创制研究Ⅰ．李正名，陈林．农药，1989，28（6）：5－6.

45. A Stereoselective Synthesis of Pear Ester via Arsenic Ylide. *Zhengming Li*，*Tiansheng Wang*，*Diankun Zhang*，*Zhenheng Guo*. Synthetic Communication，1989，19：91－96.

46. 有机铜试剂在定向合成烯烃中的应用．李正名，柴生勇．化学通报，1989，3：20－25.

1990 年

47. The Chemical Characterization of the Cephalic Secretion of the Australian Colletid bee，*Hylaeus albonitens*（Gnathopro－sopis Cockerell）. *Zhengming Li*，*Batra S. W. T.*，*Plimmer J. R.* Chinese Journal of Chemistry，1990，2：160－168.

48. The Wittig Sigmatropic Rearrangement in a Conjugated Diene System I. *Zhengming Li*，*Tiansheng Wang*，*EnyunYao*，*Zhengheng Gao*. Chinese Journal of Chemistry，1990，3：265－270.

49. 山梨酸的合成及其生物活性研究．李正名，王天生，张祖新，张殿坤，么恩云．有机化学，1990，10：117－125.

50. 已二烯基烯丙基醚的Wittig迁移重排反应．李正名，王天生，高振衡．有机化学，1990，10：427－431.

51. 三唑醇及三唑酮的不同光学异构体在温室中对小麦叶锈病的生物活性测定．张祖新，曹秋文，董丽雯，李正名．植物保护学报，1990，17（3）：262－263.

52. 新农药的创制研究Ⅱ．李正名，陈林，王立坤．农药，1990，29（2）：2－3.

1991 年

53. 含氟咪唑新杀菌剂的合成与抑菌活性．李正名，宋宝安．农药，1991，30（5）：14－16.

54. 新农药研究（Ⅲ）咪唑啉酮类化合物的合成．王立坤，李正名，刘同柱，蒋益民．高等学校化学学报，1991，12（7）：912－914.
55. Report on Structural Elucidation of Sex Pheromone Components of a Tea Pest. *Yao Enyun*, *Li Zhengming*, *Luo Zhiqiang*, *Shang Zhizhen*. Natural Science, 1991, 1 (6): 556－559.
56. 茶尺蠖性信息素化学结构研究初报．么恩云，李正名，罗志强，尚稚珍，殷坤山，洪北边．自然科学进展，1991，5：452－454.
57. 1，3－二氧戊环－4－甲醇酯－1H－咪唑相转移催化合成法的研究．李正名，宋宝安，廖仁安．化学通报，1991，8：44－50.
58. 昆虫信息素的研究（Ⅵ）．刘天麟，柴生勇，李正名．高等学校化学学报，1991，12（10）：1347－1349.
59. Synthesis and Study of Structure－Activity Relationship in a Class of 4－Methoxycarbonyl－1, 3－dioxolane－2－ylmethylimidazoles. *Song Baoan*, *Li Zhengming*, *Li Shuzheng*, *Liao Renan*, *Zhang Suhua*. Progress in Natural Science, 1991, 2 (6): 533－538.
60. 农药化学进展．化学通讯，1991，2：10－24.
61. 第四届亚洲化学会农业化学学术报告简介．李正名．化学通讯，1991，112（5）：6－7.
62. 农药化学的新进展．李正名．Science and Technology Review（科技导报），1991，3：45－48.
63. 农药化学又一次黄金时代．李正名．国际学术动态，1991，1：84－86.

1992 年

64. 新磺酰脲类化合物的合成、结构及构效关系研究（Ⅰ）．李正名，贾国锋，王玲秀，赖城明，王如骥，王宏根．高等学校化学学报，1992，13（11）：1411－1414.
65. 应用分子图形学、分子力学、量子化学及静电势研究农药分子结构与性能的关系（Ⅲ）．严波，赖城明，林少凡，李正名．高等学校化学学报，1992，13（12）：1555－1557.
66. 具有生物活性的脲类化合物的合成研究．王立坤，李正名．化学通报，1992，1：32－37.
67. 槐尺蠖性信息素的不对称合成．王胜新，么恩云，李正名．化学通报，1992，10：22－25.
68. 4－羧酸甲酯－1，3－二氧戊环－2－基甲基－1H－咪唑类杀菌剂合成与构效关系研究．宋宝安，李正名，李树正，廖仁安，张素华．自然科学进展—国家重点实验室通讯，1992，2（5）：441－446.
69. 我国农药化学进展．李正名．化学通报，1992，8：31.
70. 有机金属化合物在立体有择合成上的新进展．乔立新，李正名．化学通报，1992，7：1.
71. 新杀菌剂氟硅唑的创制途径与经验．李正名，贺峥杰．农药，1992，31（5）：36－39.
72. 分子设计中的手性元方法．乔立新，李正名．化学通报，1992，11：12－17.
73. 美国农药公司访后感．李正名．农药译丛，1992，14（1）：1－5.

1993 年

74. 新磺酰脲类化合物的合成、结构及构效关系研究（Ⅱ）．李正名，贾国锋，王玲秀，赖城明，王宏根，王如骥．高等学校化学学报，1993，14（3）：349－352.
75. 应用分子图形学、分子力学、量子化学及静电势研究农药分子结构与性能关系（Ⅳ）．严波，赖城明，林少凡，李正名．高等学校化学学报，1993，14（11）：1534－1537.
76. 茶尺蠖性信息素生物学综合研究．殷坤山，洪北边，尚稚珍，么恩云，李正名．自然科学进展—国家重点实验室通讯，1993，3（4）：332.
77. 柠檬醛的非均相催化氢化二氢香茅醛的制备．刘天麟，谢文权，李正名．应用化学，1993，10（2）：107－108.
78. 1－取代苯基－1，4－二氢－6－甲基－4－哒嗪酮－3－酰胺的合成．李正名，邹霞娟，么恩云，王素华. 应用化学，1993，10（6）：86－88.

79. 应用分子图形学、分子力学、量子化学及静电势研究农药分子结构与性能关系Ⅱ. 研究农药分子的微机静电势软件包. 赖城明，骆辉红，袁满雪，李正名. 南开大学学报（自然科学），1993，1：56－60.

80. Structural Elucidation of Sex Pheromone Components of the Geometridae *Semiothisa cinerearia*（Bremer et Grey） in China. *Zhengming Li*，*Enyun Yao*，*Tianlin Liu*，*Ziping Liu*，*Suhua Wang*，*Haiqing Zhu*，*Gang Zhao*，*Zili Ren*. Chinese Journal of Chemistry，1993，11（3）：251－256.

81. Reaction of a－（1H－1，2，4－triazol－1－yl）acetophenone with Phenyl isothiocyanate. *Zheng Ming LI*，*Zhen Nian HUANG*. Chinese Chemical Letters，1993，4（9）：763－766.

82. 农药化学现状和发展动向. 李正名. 应用化学，1993，10（5）：14－21.

1994 年

83. 稠杂环化合物的研究（X）. 张自义，李明，赵岚，李正名，廖仁安. 高等学校化学学报，1994，15（2）：220－223.

84. 新磺酰脲类化合物的合成、结构及构效关系研究（Ⅲ）. 李正名，贾国锋，王玲秀，赖城明. 高等学校化学学报，1994，15（2）：227－229.

85. 经由胂叶立德立体有择合成仿保幼激素——增丝素（ZR－512）及其（2Z，4E）－异构体. 刘天麟，倪赤友，谢文权，李正名. 高等学校化学学报，1994，15（3）：387－390.

86. 新磺酰脲类化合物的合成、结构及其构效关系研究（Ⅳ）. 李正名，贾国锋，王玲秀，赖城明. 高等学校化学学报，1994，15（3）：391－395.

87. 磺酰脲除草剂分子与受体作用的初级模型. 赖城明，袁满雪，李正名，贾国锋，王玲秀. 高等学校化学学报，1994，15（5）：693－694.

88. 应用分子图形学、分子力学、量子化学及静电势研究农药分子结构与性能的关系（V）. 赖城明，袁满雪，李正名，贾国锋. 高等学校化学学报，1994，15（7）：1004－1008.

89. 3－烷基－6－芳氧亚甲基均三唑并［3，4，－b］－1，3，4－噻二唑的合成. 张自义，赵岚，李明，李正名，廖仁安. 有机化学，1994，14：74－80.

90. 增产胺及其类似物的合成. 刘天麟，王中文，李正名. 应用化学，1994，11（2）：102－104.

91. 3－［取代苯胺基甲叉］－6－甲基－α－吡喃酮的酸性成环反应. 李正名，邹霞娟，王素华. 应用化学，1994，11（6）：101－103.

92. 碧冬茄的挥发性化学成分鉴定及其驱蚊活性研究初报. 么恩云，李正名，平霄飞，施红林，毕富春，王文丽，杨瑞华. 化学通报，1994，2：28－29.

93. 昆虫生长调节剂3，7，11－三甲基－2，4－十二碳二烯酸乙酯及其衍生物的合成. 刘天麟，谢文权，毕富春，李正名. 化学通报，1994，1：41－44.

94. 相转移催化合成增产胺及其类似物. 刘天麟，王中文，李正名. 化学试剂，1994，16（3）：171－173.

95. 茶尺蠖性信息素几种活性成分的合成. 刘天麟，李正名，罗志强，蒋益民，么恩云. 南开大学学报（自然科学），1994，1：82－86.

96. 呔嗪酮类衍生物的合成研究. 王家喜，王立坤，么恩云，李正名. 南开大学学报（自然科学），1994，2：97－98.

97. The Reaction of a－oxo－a－Triazolylketene Dithioacetal with Amines，Hydrazine and Guanidine. *Zhengming Li*，*Huang Zhennian*. Chinese Chemical Letters，1994，5(1)：31－34.

98. Studies on Sex Pheromone of Ectropis obliqua Prout. *Kunshan Yin*，*Beibian Hong*，*Zhizhen Shang*，*Enyun Yao*，*Zhengming Li*. Progress in Natural Science，1994，4(6)：732－740.

1995 年

99. 3－芳脘－6－甲基－α，γ－吡喃酮的结构测定及其关环反应研究．李正名，邹霞娟，么恩云，张殿坤，王如骥．高等学校化学学报，1995，16（2）：220－224.
100. α－羰基－α－（1H－1，2，4－三唑－1－基）乙烯基芳基酮衍生物的合成及生物活性研究．李正名，黄震年．高等学校化学学报，1995，16（10）：1555－1558.
101. 1－（取代异噁唑基）－1，2，4－三唑和1－（取代嘧啶基）－1，2，4－三唑的合成及生物活性．黄震年，李正名．高等学校化学学报，1995，16（11）：1740－1743.
102. 磺酰脲类除草剂的三维药效团模型．孙红梅，谢前，谢桂荣，周家驹，许志宏，李正名，贾国锋，王玲秀．物理化学学报，1995，11（9）：773－776.
103. 含磷增产胺类似物的合成与生物活性．刘天麟，王中文，李正名．精细化工，1995，12（5）：41－45.
104. Synthesis of 5 – Mercaptoalkylamino and 5 – Anilinoalkylthio – pyrazolyl – 1，2，4 – triazoles by Ring Chain Transformation. *Zhennian Huang*，*Zhengming Li*. Heterocycles，1995，41（8）：1653－1658.
105. Synthesis of（3 – Mercaptoalkylamino）isoxazolyl – 1，2，4 – triazoles by Ring Chain Transformation. *Zhen – Nian Huang*，*Zheng – Ming Li*. Synthetic Communications，1995，25（20）：3219－3224.
106. Synthesis of（ω – Functionalized alkylheteroatomic pyrazolyl）－1，2，4 – triazoles by Ring Chain Transformation. *Zhennian Huang*，*Zhengming Li*. Synthetic Communications，1995，25（22）：3603－3609.
107. Two Novel Cell Division Type PGRs. *Li Zheng – ming*，*Qiao Li – xin*，*Zhao Zhong – ren*，*Liu Tian – lin*. Chemical Research in Chinese Universities，1995，11（1）：28－31.
108. Quantitative Studies on Structure – Activity Relationships（QSAR）of Cytokinin – Active Phenyl urea Derivatives（PUD）. *QIAO Li – xin*，*Li Zheng – ming*，*Yang Hua – zheng*，*Zhao Zhong – ren*，*Zhang Dian – kun*，*Lai Cheng – ming*. Chemical Research in Chinese Universities，1995，11（4）：291－298.
109. α－（1，2，4－三唑－1－基）－α－苯甲酰基烯酮N，S；N，N；N，O和O，S－缩醛的合成及生物活性研究．李正名，黄震年．高等学校化学学报，1995，16（12）：1874－1877.
110. 某些生物调控物质的化学研究．李正名．自然科学进展，1995，5（3）：267－276.
111. 含吡啶农药的近年发展概况及最新动向．余申义，李正名．农药译丛，1995，17（4）：4－9.
112. Synthesis of 5 – mercapto alkyl amino – and 5 – anilino alkyl thiopyrazolyl – 1，2，4 – triazoles by ring chain transformation. *Huang Zhen – Nian*，*Li Zheng – Ming*. Heterocycles，1995，41（8）：1633－1638.

1996 年

113. 斯德酮类化合物的合成及生物活性研究．李正名，范传文，王素华．高等学校化学学报，1996，17（6）：923－924.
114. 含硅氨基甲酸酯衍生物的合成、结构及生物活性．贺峥杰，李正名，王宏根，王如冀．应用化学，1996，13（2）：1－6.
115. 磺化二氯苯醚菊酯的结构鉴定．廖云，李正名，金星．南开大学学报（自然科学），1996，29（4）：95－98.
116. 美洲大蠊初孵若虫在杀虫活性研究中的应用初探．尚稚珍，徐建华，宁黔冀，苏胜忠，杨淑华，贺铮杰，李正名．医学动物防制，1996，12（3）：5－8.
117. Synthesis and Base – Catalyzed Protodesilylation of 5 –（Silylmethylthio）－3（2H）－pyridazi-

none Derivatives. *Zhengjie He*, *Zhengming Li*. Phosphorus, Sulfur and silicon, 1996, 117 (1): 1 – 9.

118. Syntheses of Some New 4 – Amino – 5 – (N – methyl – arylsulfonamido) methyl – 1, 2, 4 – triazole – 3 – thiones and Their Derivatives. *Guofeng Jia*, *Zhengming Li*. Heteroatom Chemistry, 1996, 7 (4): 263 – 267.
119. Synthesis of 5 – Mercaptoalkylthiopyrazolyl and 3 – Mercaptoalkylthioisoxazolyl – 1, 2, 4 – triazoles by Ring Chain Transformation. *Zhennian Huang*, *Zhengming Li*. Synthetic Communications, 1996, 26 (16): 3115 – 3120.
120. Asymmetric Michael – Type Alkylation of Chiral Imines Diastereoselective Synthesis of (+) – α – Cyperone and (–) – 10 – Epi – α – Cyperone. *Zhao Ming XIONG*, *Jiong YANG*, *Yu Lin LI*, *Ren An LIAO*, *Zheng Ming LI*. Chinese Chemical Letters, 1996, 7 (8): 695 – 696.
121. Base – Catalyzed Protodesilylation of (2H) – Pyridazinon – 5 – yl Silylmethyl Sulfides. *Zheng Jie HE*, *Zheng Ming LI*. Chinese Chemical Letters, 1996, 7 (7): 603 – 604.
122. Syntheses and Herbicidal Activity of New Pyrimidinyl (triazinyl) oxy Substituted. Carboxylic Derivatives. *LI Zheng – ming*, *LIAO Yun*, *WANG Ling – Xiu*, *YANG Zhao*. Chemical Research in Chinese Universities, 1996, 12 (4): 360 – 367.
123. 应用分子图形学、分子力学、量子化学及静电势研究农药分子结构与性能关系（Ⅷ）. 表面积及构象差异对磺酰脲分子活性影响的研究. 王霞，孙莹，袁满雪，赖城明，李正名. 高等学校化学学报，1996，17（12）：1874 – 1877.
124. 酶催化反应在立体有机合成中的应用. 李纪敏，李正名. 合成化学，1996，4（4）：352.
125. Diastereoselective Synthesis of (–) – 10 – Epi – α – cyperone. *Zhaoming Xiong*, *Yulin Li*, *Renan Liao*, *Zhengming Li*. Journal of Chemical Research (S), 1996, 477.
126. 氯磺酰异氰酸酯在杂环化合物合成中的应用. 杨小平，李正名. 合成化学，1996，4（1）：31 – 39.
127. 利用 DAST 试剂形成碳 – 氟键的新进展. 王中文，刘天麟，李正名. 化学试剂，1996，18（5）：276 – 278.
128. Chemistry of Novel Bio – Regulating Substances. *Li Zheng – Ming*. Journal of Pesticide Science, 1996, 21 (1): 124 – 128.

1997 年

129. 应用分子图形学、分子力学、量子化学及静电势研究农药分子结构与性能关系（Ⅸ）. 结构参数及计算方法的选择对提高磺酰脲类除草剂活性预报准确性的影响. 王霞，袁满雪，赖城明，刘洁，李正名. 高等学校化学学报，1997，18（1）：60 – 63.
130. Comparative Molecular Field Analysis on a Set of New Herbicidal Sulfonylurea Compounds. *Jie Liu*, *Xia Wang*, *Yi Ma*, *Zheng Ming Li*, *Cheng Ming Lai*, *Guo Feng Jia*, *Ling Xiu Wang*. Chinese Chemical Letters, 1997, 8 (6): 503 – 504.
131. 新磺酰脲类化合物的合成、结构及构效关系研究（Ⅴ）. 李正名，刘洁，王霞，袁满雪，赖城明. 高等学校化学学报，1997，18（5）：750 – 752.
132. 1 –（取代硅基甲硫基）甲基 – 2，8，9 – 三氧杂 – 5 – 氮杂 – 1 – 硅杂双环［3.3.3］十一烷的合成与生物活性. 贺峥杰，李正名，杨志强. 高等学校化学学报，1997，18（1）：76 – 78.
133. 含硅二硫代磷酸酯的合成及生物活性. 贺峥杰，李正名，毕富春，王文丽. 高等学校化学学报，1997，18（5）：739 – 743.
134. α – 羰基 – α – 咪唑基烯酮缩醛类化合物的合成、反应及生物活性. 李纪敏，李正名，贾国锋，王素华. 应用化学，1997，14（5）：1 – 4.

135. 应用MOPAC方法研究磺酰脲类化合物氮杂环结构与其除草活性的关系．刘洁，李正名，王霞，马翼，赖城明，贾国锋，王玲秀．计算机与应用化学，1997（增刊）：155－156.
136. 应用分子图形学、分子力学、量子化学及静电势研究农药分子结构与性能关系（Ⅶ）．磺酰脲分子构象差异对活性影响的ANN研究．王霞光，孙莹，袁满雪，马淑英，赖城明，李正名．南开大学学报（自然科学），1997，30（4）：92－95.
137. 应用分子图形学、分子力学、量子化学及静电势研究农药分子结构与性能关系（Ⅹ）．磺酰脲分子内旋转通道的分子力学研究．刘艾林，曹炜，赖城明，袁满雪，张金碚，林少凡，李正名．高等学校化学学报，1997，18（4）：574－576.
138. 1，2，3－噻二唑衍生物的合成及其生物活性．赵卫光，李正名，陈寒松．高等学校化学学报，1997，18（10）：1651－1653.
139. 吡唑衍生物的合成及生物活性．李正名，陈寒松，赵卫光，张锴，黄兴盛．高等学校化学学报，1997，18（11）：1794－1799.
140. A Facile One Pot Synthesis of 2，5－Dipyrazolyl－1，3，4－oxadiazole. *Zheng Ming Li*，*Han Song Chen.* 合成化学，1997（增刊）：170.
141. 新磺酰脲类除草剂的分子设计，合成及构效关系．李正名．合成化学，1997，5（A10）：1.
142. 3－吡唑基－6－取代均三唑并（3，4－b）－1，3，4－噻二唑的合成及生物活性．陈寒松，李正名．合成化学，1997，5（A10）：169.
143. 一种合成2，6－二芳基吡嗪的新方法．贾国锋，李正名．合成化学，1997，5（A10）：171.
144. 新型杀菌剂β－甲氧基丙烯酸甲酯类化合物的合成方法概述．王忠文，李正名，刘天麟．合成化学，1997，5（3）：241－245.

1998年

145. 4－二甲氨基吡啶催化硫代磷酰化反应及其在有机磷杀虫剂合成中的应用．廖联安，李正名．化工进展，1998，6：43－45.
146. 3－吡唑基－6－取代均三唑并［3，4－b］－1，3，4－噻二唑的合成及生物活性．陈寒松，李正名．应用化学，1998，15（3）：59－62.
147. β－芳氧乙氧基－α－苯氧基丙烯酸乙酯类化合物的合成和生物活性的研究．王忠文，李正名，刘天麟，李树正．合成化学，1998，6（3）：311－314.
148. Polymer－Supported Regioselective Synthesis of Methyl 2－O－and 2，3－di－O－（4，6－dimethylpyrimidin－2yl）－α－D－glucopyranosides. *Yun Liao*，*Zhengming Li.* Synthetic Communications，1998，28（19）：3539－3547.
149. 含硅不对称硫代磷酸酯及其碳类似物的合成与杀虫活性．贺峥杰，李正名，王有名，杨志强，钱宝英．应用化学，1998，15（6）：29－32.
150. Comparative Molecular Field Analysis（CoMFA）of New Herbicidal Sulfonylurea Compounds. *Jie Liu*，*Zhengming Li*，*Xia Wang*，*Yi Ma*，*Chengming Lai*，*Guofeng Jia*，*Lingxiu Wang.* Science in China（Series B），1998，14（1）：39－42.
151. Syntheses of Di－heterocyclic Compounds；Pyrazolylimidazoles and Isoxazolylimidazoles. *Zhengming Li*，*Jimin Li*，*Guofeng Jia.* Heteroatom Chemistry，1998，9（3）：317－320.
152. 3－芳基磺酰氧基（取代）异噻唑的合成及除草活性．杨小平，李正名，王玲秀，李永红．高等学校化学学报，1998，19（2）：228－231.
153. 应用CoMFA研究磺酰脲类化合物的三维构效关系．刘洁，李正名，王霞，马翼，赖城明，贾国锋，王玲秀．中国科学（B辑），1998，28（1）：60－64.
154. Crystal Structure of 3－（4－nitro－）phenylsulfonyloxyisothiazole $NO_2C_6H_4SO_3C_3H_2NS$. *Yang Xiao－Ping*，*Li Zheng－Ming*，*Wang Hong－Gen*，*Yao Xin－Kan.* Chinese J. Struct. Chem.，1998，

17（2）：81－84.

155. A Facile Preparation of 2，6－Diarylpyrazines. *Guofeng Jia*，*Zhengming Li*，*Yaoshu Zhang*. Heteroatom Chemistry，1998，9（3）：341－345.

156. 2，5－二吡唑基－1，3，4－噁二唑的一锅煮法合成及生物活性．陈寒松，李正名．高等学校化学学报，1998，19（4）：572－573.

157. 新型含硅氨基甲酸酯衍生物的合成、杀虫活性及抗乙酰胆碱酯酶活性研究．贺峥杰，李正名，徐建华，尚稚珍．昆虫学报，1998，41（2）：130－134.

158. 三价手性氮化合物的研究进展．余申义，李正名．大学化学，1998，13（5）：26－30.

159. 2，6－二取代苯胺衍生物生物活性的研究进展．王有名，李正名，李佳凤．化学研究，1998，9（4）：1－11.

1999 年

160. 2－芳甲酰基硫羰基－3－异噻唑酮及4－氰基－5－甲硫基－2－芳甲酰氨基硫碳基－3－异噻唑酮的合成与生物活性．杨小平，李正名，陈寒松，刘洁，李树正．高等学校化学学报，1999，20（3）：395－398.

161. β－取代苯（氧）丙烯酸酯（酰胺）类化合物的合成及其杀菌活性的研究．王忠文，李正名，陈寒松，刘天麟，李树正．高等学校化学学报，1999，20（8）：1248－1253.

162. 除草剂安全剂应用研究近况．姜林，李正名．农药学学报，1999，1（2）：1－8.

163. 一种新型可立体控制的酰基合成方法及其在具有生物活性物质合成中的应用．高喜麟，王有名，李正名．化学通报网络版，1999，99089.

164. 3－芳胺甲烯基－6－烷（芳）基－5，6－2H－吡喃－2，4－二酮化合物的3D－QSAR研究．刘洁，王有名，李正名．农药学学报，1999，1（1）：78－80.

165. 新农药创制的现状和发展趋势．李正名．世界农药，1999，21（6）：1－4.

166. The Design and Synthesis of ALS Inhibitors from Pharmacophore Models. *Jie Liu*，*Zhengming Li*，*Han Yan*，*Lingxiu Wang*，*Junpeng Chen*. Bioorganic & Medicinal Chemistry Letters，1999，9：1927－1932.

167. 2－取代苯氧乙硫基－5－吡唑基－1，3，4－噁二唑及1，ω－双（5－吡唑基－1，3，4－噁二唑－硫代）烷烃．陈寒松，李正名，王忠文．合成化学，1999，7（2）：164－169.

168. A New Method for the Synthesis of Thieno［2，3－C］Pyrazole. *Zhongwen Wang*，*Jun Ren*，*Hansong Chen*，*Zhengming Li*. Heteroatom Chem.，1999，10（4）：303－305.

169. 3－氯－4－氰基－5－异噻唑基脲的合成及生物活性．杨小平，李正名，刘洁，陈寒松．应用化学，1999，16（2）：115－116.

170. 1－杂环基吡唑的合成及生物活性研究．陈寒松，李佳凤，李正名．厦门大学学报（自然科学版），1999，38（增刊）：507.

171. 依维菌素的合成改进．廖联安，李正名．厦门大学学报（自然科学版），1999，38（10）（增刊）：530.

172. 新型杀菌剂β－取代丙烯酸酯类化合物的合成和生物活性的研究．王忠文，李正名，刘天麟，李树正，张祖新．合成化学，1999，7（1）：62－67.

173. 3－芳胺甲烯基－6－烷基（芳基）－5，6－二氢－2H－吡喃－2，4－二酮的合成及其杀菌活性的研究．王有名，李正名，李佳凤，李树正，张素华．高等学校化学学报，1999，20（10）：1559－1563.

174. A Novel Tandem Reaction for the Synthesis of Thieno［2，3－C］Pyrazole. *Wang Zhongwen*，*Ren Jun*，*Chen Hansong*，*Li Zhengming*. Chinese Chemical Letters，1999，10（3）：189－190.

175. A new condensation reaction of b－keto－d－valerolactone with substituted aniline. *You－Ming*

Wang, *Jia – Feng Li*, *Zheng – Ming Li.* Chinese Chemical Letters, 1999, 10 (4): 269 – 270.

176. Condensation products from b – keto – d – valerolactones. *You – Ming Wang*, *Zheng – Ming Li*, *Jia – Feng Li.* Chinese Chemical Letters, 1999, 10 (5): 345 – 346.
177. New Fungicidally Active Pyrazol – Substifuted 1, 3, 4 – Thiadiazole Compounds and Their Preparation. *Chen Hansong*, *Li Zhengming*, *Han Yufeng*, *Wang Zhongwen.* Chinese Chem. Letters, 1999, 10 (5): 365 – 366.
178. A Convenient Synthesis of Pyrazolyl pyrazoles using 2 – Oxo – Ketene S, S – and N, S – Acetals. *Chen Hansong*, *Li Zhengming*, *Wang Zhongwen.* Chinese Chemical Letters, 1999, 10 (8): 643 – 646.
179. A new synthesis for methyl 2 – benzyloxylphenylacetate. *Zhong – Wen Wang*, *Zheng – Ming Li*, *Tian Lin*, *Jun Ren.* Synthetic Communications, 1999, 29 (13): 2361 – 2364.
180. A Novel Synthesis of Thieno [2, 3 – C] Pyrazle. *Wang Zhongwen*, *Li Zhengming*, *Ren Jun*, *Chen Hansong.* Synthetic Commumications, 1999, 29 (13): 2355 – 2359.

2000 年

181. 水分子对磺酰脲类分子构效关系影响的研究．马翼，刘洁，李正名．高等学校化学学报，2000，21（1）：85 – 87.
182. 麦谷宁生测方法及其对玉米的安合性研究．范志金，钱传范，党宏斌，李正名，王玲秀．农药学学报，2000，2（1）：63 – 70.
183. 新磺酰脲类化合物的合成及构效关系研究（Ⅵ）．姜林，李正名，翁林红，冷雪冰．结构化学，2000，19（2）：149 – 152.
184. 在水介质中进行的 Barbier – Grignard 反应．廖联安，李正名．有机化学，2000，20（3）：306 – 318.
185. 含嘧啶环的苄基磺酰脲，吡唑磺酰脲的合成及生物活性．姜林，李正名，陈寒松，高发旺，王玲秀．应用化学，2000，17（4）：349 – 352.
186. Synthesis of some heteroaryl pyrazole dterivatves and their biolosical activities. *Chen Han – Song*, *Li Zheng – Ming.* Chinese Journal of Chemisty, 2000, 18 (14): 596 – 602.
187. 关于联合国开发总署（NNDP）医药农药工业在全球竞争中的技术升级研讨会．李正名．中国工程科学，2000，2（7）：92 – 93.
188. 具有杀菌活性的吡唑联杂环类化合物．陈寒松，赵卫光，李正名，韩玉芬，颜寒，赖俊英，王素华．浙江化工（增刊），2000，31：27 – 29.
189. Synthesis and Fungicidal Activity against *Rhizoctania Solani* of 2 – Alkyl (Alkylthio) – 5 – Pyrazolyl – 1, 3, 4 – Oxadiazoles (Thiadia – zoles). *Chen Han – Song*, *Li Zheng – Ming*, *Han Yu – Feng.* Journal of Agricultural & Food Chemisty, 2000, 48 (11): 5312 – 5315.
190. 2 – 取代 – 5 – 吡唑基 – 1，3，4 – 噁二唑类化合物的合成及生物活性．陈寒松，李正名，李佳凤．高等学校化学学报，2000，21（10）：1520 – 1523.
191. Three – Dimensional Quantative Structure – Activity Relationship Analysis of the New Potent Sulfonylureas using Comparative Molecular Similarity Indices Analysis. *Hou T. J.*, *Li Z. M.*, *Li Z.*, *Liu J.*, *Xu X. J.* J. Chem. Inf. Comput. Sci., 2000, 40: 1002 – 1009.
192. AM1 study of catalytic hydrogenation of 3 – anilinomethylidene – 6 – aryl – 5, 6 – 2H – dihydropyran – 2, 4 – diones. *Liu Jie*, *Fang Yayin*, *Li Zhengming*, *Lai Chengming.* Journal of Molecular (Theochem), 2000, 532: 103 – 107.
193. 二硫缩醛合成方法的改进．赵卫光，王素华，王文艳，李正名．化学试剂，2000，22（6）：376.
194. 应用分子图形学、分子力学、量子化学及静电势研究农药分子结构与性能关系（Ⅻ）——

化学模式识别对磺酰脲类除草剂杂环结构与活性关系的分类研究．王霞，李正名．南开大学学报（自然科学版），2000，33（2）：11－14.

195. Synthesis，Crystal Structure and Fungicidal Activity of 3－（4－Chloro－3－ehytl－1－methyl－1H－pyrazol－5－yl）－6（E）－hexylvinyl－triazol［3，4－6］－1，3，4－thiadiazole. *Chen Han－Song*，*Li Zheng－Ming*，*Yang Xiao－Ping*. Chinese J. Structural Chemisty，2000，10（5）：317－321.

196. A novel method for the synthesis of pyrazolo［5，1－b］thiazole. *Wang Z W*，*Ren J*，*Li Z M*. Synthetic Communications，2000，30（4）：763－769.

2001 年

197. 用分子动力学模拟方法研究磺酰脲化合物在溶液中构象的变化．沈荣欣，方亚寅，马翼，孙宏伟，赖城明，李正名．高等学校化学学报，2001，22（6）：952－954.

198. 应用 AM1 方法研究 3－芳胺甲烯基－2－烷基－5，6－二氢－二氢吡喃－2，4－二酮的催化氢化反应机理．方亚寅，沈荣欣，孙宏伟，袁满雪，刘洁，李正名，赖城明．高等学校化学学报，2001，22（9）：1506－1510.

199. 2－氰基－3－（6－氯－3－吡啶甲基）胺基－3－脂肪胺基丙烯腈的合成与生物活性．余申义，李正名．农药学学报，2001，3（3）：18－22.

200. 3－芳胺甲叉－6－苯基－5，6－二氢－2H－吡喃 2，4－二酮类新化合物的设计与合成．贾强，李正名，姜林，赵卫光，王素华．应用化学，2001，18（1）：84－86.

201. 通过碳－碳键形成杂环的合成方法及其应用．罗铁军，李正名．有机化学，2001，21（7）：505－513.

202. 2，2，6－三甲基－4H－1，3－二噁烷－4－酮在有机合成中的应用．赵卫光，陈寒松，王素华，李正名．化学通报网络版，2001，01009.

203. 1－磺酰基－3，5－二氨基－1H－吡唑－4－腈的合成及其生物活性．赵卫光，曹一兵，李正名，高发旺，王素华，王建国．应用化学，2001，18（6）：423－427.

204. 3－杂环胺基甲烯基－6－烷基（芳基）－5，6－二氢－2H－吡喃－2，4－二酮的合成及生物活性研究．王有名，李正名，韩玉芬，贾保军，王玉林．应用化学，2001，18（7）：524－527.

205. N－（取代嘧啶－2′－基）－2－三氟乙酰氨基苯磺酰脲的合成及除草活性．姜林，刘洁，高发旺，李正名．应用化学，2001，18（3）：225－227.

206. 含吡唑的双杂环化合物的合成及其生物活性．赵卫光，陈寒松，李正名，韩玉芬，颜寒，赖俊英，王素华．高等学校化学学报，2001，22（6）：939－942.

207. 双乙烯酮在杂环化学中的应用．赵卫光，李正名，贾强．合成化学，2001，9（4）：302－309.

208. 3－甲基－4－乙氧基（二乙氧基）羰基－1H－吡唑衍生物的合成及生物活性．赵卫光，李正名，袁平伟，袁德凯，贾强，王文艳，王素华．有机化学，2001，21（8）：593－598.

209. No－solvent Condensation Reaction of Amino－Acids and their Derivatives with Pyrandiones . *Jia Qiang*，*Li Zheng－Ming*，*Wang Su－Hua*，*Zhao Wei－Guang*，*Wang You－Ming*. Chinese Chemical Letters，2001，12（6）：475－476.

210. 新磺酰脲类除草活性构效关系研究．李正名，赖城明．有机化学，2001，21（11）：810－815.

211. 现代化学农药研究中值得注意的课题．罗铁军，李正名，赵卫光．农药，2001，40（10）：1－6.

212. Synthesis and Biological Activity of 3－methyl－1H－Pyraxole－4－Carbozylic Eester Derivatives. *Zhao Weiguang*，*Li Zhengming*，*Yuan Pingwei*，*Wang Wenyen*. Chinese Journal of Chemis-

try, 2001, 19 (2): 184 - 188.

213. A Concise strategy for Polymer - supported regio - oriented Introduction of Various Building Blocks onto Glucopyranoside Scaffold. *Liao Yun*, *Li Zheng - Ming*, *Henry N. C. Wong*. Chinese Journal of Chemistry, 2001, 19: 1119 - 1129.
214. Mechanism analysis of catalytic hydrogenation of 3 - anilinomethylidene - 6 - alkyl - 5, 6 - 2H - dihydropyran - 2, 4 - diones. *Fang Ya - Yin*, *Shen Rong - Xin*, *Sun Hong - Wei*, *Yuan Man - Xue*, *Liu Jie*, *Li Zheng - Ming*, *Lai Cheng - Ming*. Journal of molecular structure (Theochem), 2001, 578: 71 - 78.
215. 吡啶类植物保护剂研究进展和创制思路. 赵卫光, 王建国, 罗铁军, 李正名. 新农药, 2001, 18 (2): 13 - 17.

2002 年

216. 超临界流体在有机合成中的应用. 野国中, 李正名. 化学通报, 2002, (4): 221 - 226.
217. 卤代 - 2 - (3 - 甲基 - 5 - 取代 - 4H - 1, 2, 4 - 三唑 - 4 - 基) - 苯甲酸的合成. 罗铁军, 李正名, 王素华, 廖任安, 李之春. 应用化学, 2002, 19 (6): 594 - 596.
218. 吡啶类农药研究趋势及新发现的含吡啶环的天然活性物质. 赵卫光, 王建国, 袁德凯, 罗铁军, 李正名. 农药, 2002, 41 (7): 8 - 11.
219. 生物农药的研究进展 (一) ——天然源农药来源、类型及其生物活性. 钟滨, 范志金, 李正名. 新农药, 2002, (5 - 6): 19 - 32.
220. 5 - 阿维菌素 B1a 酯的合成及生物活性. 廖联安, 李正名, 方红云, 赵卫光, 陈明德, 范志金, 刘桂荣. 高等学校化学学报, 2002, 23 (9): 1709 - 1714.
221. 5 - 依维菌素 B1a 酯的合成和生物活性. 廖联安, 李正名, 方红云, 赵卫光, 范志金, 刘桂龙. 化学学报, 2002, 60 (3): 468 - 474.
222. 一类吡唑衍生物的 3D - QSAR 研究. 王建国, 陈寒松, 赵卫光, 马翼, 李正名, 韩玉芬. 化学学报, 2002, 60 (11): 2043 - 2048.
223. 原卟啉原氧化酶 (PPO) 及其抑制剂的近期研究. 王建国, 赵卫光, 王素华, 李正名. 农药, 2002, 43 (8): 1 - 4.
224. 2 - 氰基 - 3 - [(6 - 氯) - 3 - 吡啶甲基] 氨基 - 3 - 脂肪氨丙烯酸乙酯的合成. 余申义, 李正名. 农药学学报, 2002, 4 (3): 79 - 82.
225. HPPD 抑制剂的研究进展. 罗铁军, 李正名. 新农药, 2002, (2 - 3): 19 - 24.
226. 2 - 甲基 - 3 - 芳基 - 7 - (5, 5 - 二甲基 - 3 - 酮 - 1 - 环己烯 - 1 - 基) 甲酸酯 - 4 (3H) - 喹唑啉酮的合成. 罗铁军, 李正名, 赵卫光, 范志金. 有机化学, 2002, 22 (10): 741 - 745.
227. 氯磺隆的化学行为. 李伟, 范志金, 陈建宇, 王海英, 李正名, 艾应伟. 四川师范大学学报 (自然科学版), 2002, 25 (5): 521 - 524.
228. 几种除草剂靶酶及其抑制剂的研究进展. 王建国, 赵卫光, 范志金, 王素华, 李正名. 植物保护学报, 2002, 29 (3): 279 - 284.
229. (Z) - 2 - (1H - 1, 2, 4 - 三唑 - 1 - 基) - 1 - (2, 3, 4 - 三甲氧基苯) 乙酮肟酯的合成和结构表征. 杨松, 宋宝安, 李正名, 廖仁安. 农药学学报, 2002, 4 (2): 23 - 28.
230. 1, 3 - 氮硫杂环化合物防治芦笋茎枯病的 QSAR 研究. 王建国, 马翼, 赵卫光, 李正名, 韩嘉祥. 农药学学报, 2002, 4 (2): 35 - 39.
231. 单嘧磺隆稳定性的研究. 陈建宇, 王海英, 范志金, 陈俊鹏, 李正名, 钱传范. 四川师范大学学报 (自然科学版), 2002, 25 (3): 313 - 315.
232. 2 - (1H - 咪唑 - 1 - 基) - 1 - (2, 3, 4 - 三甲氧基) 苯乙酮肟酯新化合物合成与生物

活性研究．杨松，宋宝安，李正名，廖仁安，刘刚，胡德禹．有机化学，2002，22（5）：345－349.

233. 3－芳胺甲烯基－5，6－二氢－二氢吡喃－2，4－二酮类化合物催化氢化反应选择性的理论研究．方亚寅，沈荣欣，孙宏伟，袁满雪，刘洁，李正名，赖城明．高等学校化学学报，2002，23（6）：1056－1059.
234. 2－吡啶氨基磺酰脲的合成及除草活性．姜林，李正名．山东农业大学学报（自然科学版），2002，33（3）：384－385.
235. N－（4'－取代嘧啶－2'－基）－2－取代苯氧基磺酰脲的合成及除草活性．姜林，李正名，高发旺，王素华．应用化学，2002，19（5）：416－419.
236. 5－O－三苯硅基－4′－O－酰基阿维菌素 B1a 的合成和生物活性．廖联安，李正名，方红云，赵卫光，范志金，陈明德，刘桂龙．应用化学，2002，19（6）：521－526.
237. 2－（1H－唑基）－1－（2，3，4－三甲氧基）苯乙酮衍生物的合成与结构表征．杨松，宋宝安，李正名，廖仁安．化学通报，2002，（3）：198－200.
238. 3－氯－4－氰基－5－取代苯氧基异噻唑的合成及生物活性．王昕，杨小平，梁鑫森，李正名．化学研究与应用，2002，14（6）：677－679.
239. 2－(1H－咪唑－1－基)－1－（2，3，4－三甲氧基)乙酮的合成．杨松，宋宝安，李正名，廖仁安．应用化学，2002，19（5）：491－493.
240. （Z）－2－（1H－咪唑－1－基）－1－（2，3，4－三甲氧基）苯乙酮肟酯的设计与合成．杨松，宋宝安，李正名，廖仁安，刘刚．应用化学，2002，19（3）：259－262.
241. A novel and facile synthesis of pyrazolo［3，4－b］pyridines. *Zhao Wei－Guang*，*Li Zheng－Ming*，*Yuan De－Kai*. J. Chem. Research（s），2002，454－455.
242. Mechanism Analysis of Catalytic Hydrogenation of 3－Anilino－methylidene－6－alkyl－5，6－2H－dihydropyran－2，4－diones. *Yayin Fang*，*Rongxin Shen*，*Hongwei Sun*，*Manxue Yuan*，*Jie Liu*，*Zhengming Li*，*Chengming Lai*. Journal of Molecular Structure（THEOCHEM），2002，578：71－78.
243. Crystal and Molecular Structure of N－（Pyridin－2－yl－carbonyl）－2－ethoxycarbonyl－benzene sulfonamide. *Jiang Lin*，*Li Zheng－Ming*，*Weng Lin－Hong*，*Leng Xue－Bing*. Chinese J. Struct. Chem.，2002，21（3）：284－287.
244. 迎接新农药创制研究所面临的挑战．李正名，杨华铮，杜灿屏，刘鲁生，张恒主编．21 世纪有机化学发展战略，北京：化学工业出版社，2002，371－380.
245. 乙酰乳酸合成酶（ALS）除草剂抗性的近期研究进展．王建国，赵卫光，王素华，李正名．农药（增刊），2002，41：1－2.

2003 年

246. 1，3－二甲基－5－甲硫基－4－苯脲基羰基吡唑的合成及抑菌活性．范志金，钟滨，王素华，李正名．应用化学，2003，20（4）：365－367.
247. 固态有机合成反应进展．臧洪俊，李正名，王宝雷．有机化学，2003，23（10）：1058－1063.
248. 10% 单嘧磺隆可湿性粉剂防除谷子地杂草田间药效试验．范志金，李香菊，吕德兹，范仁俊，申日升，王玲秀，李正名．农药，2003，42（3）：34－36.
249. 10% 单嘧磺酯 WP 防除冬小麦田杂草试验．孟和生，王玲秀，刘亦学，李正名，陈俊鹏，高发旺．杂草科学，2003，（4）：37－38.
250. 以天然产物为先导化合物开发的农药品种（Ⅱ）——杀虫杀螨剂．刘长令，钟滨，李正名．农药，2003，42（12）：1－8.
251. 以天然产物为先导化合物开发的农药品种（Ⅰ）——杀菌剂．刘长令，李正名．农药，

2003，42（11）：1－4.

252. 2－取代氨基－5－吡唑基－1，3，4－噁二唑的合成及生物活性．袁德凯，李正名，赵卫光，陈寒松．应用化学，2003，20（7）：624－628.

253. 2－取代苯并噁唑衍生物的合成及生物活性．钟滨，范志金，李正名．应用化学，2003，20（7）：684－686.

254. 单嘧磺隆对小麦的安全性及在麦田除草效果的研究．范志金，钱传范，陈俊鹏，李正名，王玲秀，党宏斌．中国农学通报，2003，19（3）：4－8.

255. 新型稻田杀菌剂噻酰菌胺．赵卫光，刘桂龙，王素华，李正名．农药，2003，42（10）：47－48.

256. 2－巯基噻唑［5，4－d］并嘧啶类化合物及其衍生物的合成．袁德凯，李正名，赵卫光．有机化学，2003，23（10）：1155－1158.

257. 氯磺隆和苯磺隆对玉米乙酰乳酸合成酶抑制作用的研究．范志金，钱传范，于维强，陈俊鹏，李正名，王玲秀．中国农业科学，2003，36（2）：173－178.

258. 1－（3，5－二氯－4－异噻唑甲酰基）－3－芳基脲及1－（3，5－二甲硫基－4－异噻唑甲酰基）－3－芳基脲的合成．李玉新，谭凤娇，王昕，杨小平，李正名．湘潭大学自然科学学报，2003，25（1）：32－35.

259. 单嘧磺隆对靶标乙酰乳酸合成酶活性的影响．范志金，陈俊鹏，党宏斌，王玲秀，钱传范，李正名．现代农药，2003，2（2）：15－17.

260. 玉米乙酰乳酸合成酶活性的测定及其性质初探．范志金，于维强，艾应伟，陈俊鹏，胡继业，钱传范，李正名．安全与环境学报，2003，3（2）：19－23.

261. Strobin 类杀菌剂的创制经纬．刘长令，李正名．农药，2003，42（3）：43－46.

262. 3－N－乙酰基－2－取代芳基－5－［5'－甲基－异噁唑－3'］－Δ3－1，3，4－噁唑啉类化合物的合成．钟滨，赵卫光，李正名，王素华．应用化学，2003，20（8）：719－722.

263. Facile Synthesis of Nicotinic Acid Derivatives with Unsymmetrical Substitution Patterns. *Zhao Wei－Guang*，*Liu Zheng－Xiao*，*Li Zheng－Ming*，*Wang Bao－Lei.* Synthetic Communications，2003，33（24）：4229－4234.

264. 2（1H）－喹啉－2，4－二酮类化合物抗小麦锈病的3D－QSAR研究．王建国，符新亮，王有名，马翼，李正名，张祖新．高等学校化学学报，2003，24（11）：2010－2013.

265. Metal Ion Interactions with Sugars. The crystal Structure and FT－IR Study of the $NdCl_3$－ribose complex. *Yan Lu*，*Guocai Deng*，*Fangming Miao*，*Zhengming Li.* Carbohydrate Research，2003，338（24）：2913－2919.

266. 新磺酰脲类化合物的合成及生物活性．野国中，范志金，李正名，李永红，高发旺，王素华．高等学校化学学报，2003，24（9）：1599－1603.

267. Sugar complexation with calcium ion. Crystal structure and FT－IR study of a hydrated calcium chloride complex of D－ribose. *Lu Yan*，*Deng Guocai*，*Miao Fangming*，*Li Zhengming.* J. Inorg. Biochem.，2003，96：487－492.

268. 1－（4－Methoxypyrimidin－2－yl）－3－（2－nitorphenylsulfonyl）urea. *Ma Ning*，*Li Zhengming*，*Wang Jianguo*，*Song Haibin.* Acta Crystallographica，2003，E59，o275－276.

269. Methy 2－（4－methoxypyrimidin－2－yl－carbamoylsulfamoyl）Benzoate. *Ma Ning*，*Wang Baolei*，*Wang Jianguo*，*Song Haibin*，*Wang Suhua*，*Li Zhengming.* Acta Crystallographica，2003，E59，o438－440.

270. N－（吡啶－2－基羰基）取代芳基磺酰胺的合成及除草活性．姜林，李正名，高发旺，王素华．应用化学，2003，20（1）：77－79.

271. Ugi 反应的研究新进展．马宁，李正名，赵卫光．化学进展，2003，15（3）：186－193.

272. Milbemycins 的结构修饰．袁德凯，赵卫光，李正名．农药学学报，2003，5（1）：1－11.

273. 3 - Methylthio - pyrano [4, 3 - c] pyrazol - 4 (2H) - ones from 3 - (Bis - methylthio) methylene - 2H - pyran - 2, 4 - diones and Hydrazines. *Yuxin Li, Youming Wang, Xiaoping Yang, Suhua Wang, Zhengming Li.* Heteroatom Chemistry, 2003, 14 (4): 342 - 344.

274. A facile synthesis of novel 1, 7 - dihydropyrazolo [3, 4 - b] pyridine - 4, 6 - diones. *Yuan De - Kai, Li Zheng - Ming, Zhao Wei - Guang.* Journal of Chemical Research (S), 2003, 782 - 783.

275. Condensation Reaction of 5, 6 - Dihydro - 6 - methyl - 6 - piperonyl - 2Hpyran - 2, 4 - dione, Ethyl Orthoformate and Substituted Anilines. *WANG You Ming, HE Ke, ZHAO Guo Feng, LI Zheng Ming.* Chinese Chemical Letters, 2003, 14 (3): 221 - 224.

276. 基因组学对基于结构的药物设计的影响．王宝雷，李正名，臧洪俊．化学进展，2003，15 (6)：505 - 511.

277. Synthesis, X - ray Crystal Structure and Biological Activities of alpha - Phenoxyl - 1, 2, 3 - thiadiazoleacetamide. *Zhao Wei - Guang, Li Zheng - Ming, Yang Zhao.* Journal of Heterocyclic Chemistry, 2003, 40 (5): 925 - 928.

2004 年

278. 新型先导化合物 4 - [4 - (3, 4 - 二甲氧基苯基) - 2 - 甲基噻唑 - 5 - 甲酰基] 吗啉的设计、合成与生物活性．刘长令，李正名．农药，2004，43 (4)：157 - 159.

279. 1, 8 - 萘二甲酸酐对高浓度单嘧磺隆协迫下玉米的解毒作用．范志金，钱传范，陈俊鹏，王玲秀，李正名．农药学学报，2004，6 (4)：55 - 61.

280. 新单取代苯磺酰脲衍生物的合成及生物活性．马宁，李鹏飞，李永红，李正名，王玲秀，王素华．高等学校化学学报，2004，25 (12)：2259 - 2262.

281. 新磺酰脲类化合物的合成及除草活性．马宁，李正名，李永红，郝静君，范传文，王玲秀，王素华．应用化学，2004，21 (10)：989 - 992.

282. 新多取代吡啶类化合物的合成及生物活性．钟滨，李正名，刘长令，赵卫光．有机化学，2004，24 (10)：1304 - 1306.

283. N - (4 - 取代嘧啶 - 2 - 基) 苄基磺酰脲和苯氧基磺酰脲的 3D - QSAR 研究．马翼，姜林，李正名，赖城明．高等学校化学学报，2004，25 (11)：2031 - 2033.

284. N - 杂环卡宾的反应及应用．柳清湘，李正名．化学通报，2004，10：715 - 722，749.

285. 基于酮醇酸还原异构酶 KARI 复合物晶体结构的三维数据库搜寻．王宝雷，李正名，马翼，王建国，罗小民，左之利．有机化学，2004，24 (8)：973 - 976.

286. 新磺酰脲类化合物除草活性的 3D - QSAR 分析．王宝雷，马宁，王建国，马翼，李正名，李永红．物理化学学报，2004，20 (6)：577 - 581.

287. 10% 单嘧磺酯可湿性粉剂的 HPLC 分析．范志金，陈建宇，王海英，寇俊杰，陈俊鹏，李正名．农药，2004，43 (7)：321 - 322.

288. 新磺酰脲类化合物 N - [2 - (4 - 甲基) 嘧啶基] - N′ - 2 - 甲氧羰基苯磺酰脲的合成及其二聚体结构．王宝雷，马宁，王建国，马翼，李正名，冷雪冰．结构化学，2004，23 (7)：783 - 787.

289. 单嘧磺隆正辛醇 - 水分配系数的测定．范志金，陈俊鹏，李正名，艾应伟，钱传范．环境化学，2004，23 (4)：431 - 434.

290. N - 杂环 - 3 - N′ - 苄氧羰基 - β - 氨基丁酰胺的合成和结构表征．臧洪俊，李正名，韩亮，王宝雷，赵卫光．有机化学，2004，24 (6)：669 - 672.

291. 含吡啶环的 1, 3, 4 - 噁二唑衍生物的合成及生物活性研究．范志金，刘斌，刘秀峰，钟滨，刘长令，李正名．高等学校化学学报，2004，25 (4)：663 - 666.

292. 新型除草剂 1 - (4 - 甲基嘧啶 - 2 - 基) - 3 - (2 - 硝基苯磺酰) 脲的溶液构象研究．马翼，李正名，赖城明．农药学学报，2004，6（1）：71－73.

293. 除草剂靶标酮醇酸还原异构酶（KARI）研究进展．王宝雷，李正名．农药学学报，2004，6（1）：11－16.

294. N－芳基－3－N′－苄氧羰基－β－氨基丁酰胺的合成和结构表征．臧洪俊，李正名，赵卫光，王宝雷．应用化学，2004，21(3)：313－315.

295. 新磺酰脲类除草剂 NK#94827 的除草活性．范志金，王玲秀，陈俊鹏，李正名．中国农学通报，2004，20(1)：198－200.

296. 动态组合化学研究进展及其在药物设计中的应用．刘征骁，赵卫光，李正名．有机化学，2004，24（1）：1－6.

297. 单嘧磺隆防除小麦田难除杂草碱茅．孟和生，王玲秀，李正名，陈俊鹏，高发旺．农药，2004，43（1）：20－21.

298. 以天然产物为先导化合物开发的农药品种（3）——除草剂．刘长令，韩亮，李正名．农药，2004，43（1）：1－4.

299. 单嘧磺酯的除草活性及其对玉米的安全性初探．范志金，陈俊鹏，艾应伟．安全与环境学报，2004，14（1）：22－25.

300. 二［4－羟基－5，6－二氢－6－烷基（芳基）－2H－吡喃－2－酮－3－］烃的合成及生物活性的研究．李玉新，王有名，王素华，杨小平，李正名．高等学校化学学报，2004，25（2）：281－283.

301. 高通量蛋白质结晶及其在药物设计中的应用．赵卫光，李正名，王宝雷，王建国．化学进展，2004，16（1）：105－109.

302. Synthesis of 2－Alkyl（aryl）－3－methylthio－6－methyl－6－arylpyrano－［4，3－c］pyrazol－4（2H）－ones. *LI Yu－Xin*，*Wang You－Ming*，*Yang Xiao－Ping*，*Wang Su－Hua*，*Li Zheng－Ming*. Chinese Chemical Letters，2004，15（1）：14－16.

303. 9，10－Bis（propylammoniomethyl）anthracene dichloride. *Liu Qingxiang*，*Song Haibin*，*Li Zhengming*. Acta Cryst，2004，E60：o777－778.

304. Design，Synthesis，and Biological Activity of Novel 4－（3，4－dimethoxyphenyl）－2－methylthiazole－5－carboxylic Acid Derivatives. *Changling Liu*，*Lin Li*，*Zhengming Li*. Bioorganic & Medicinal Chemistry，2004，12：2825－2830.

305. Residue analysis and dissipation of monosulfuron in soil and wheat. *Fan Zhi－Jin*，*Hu Ji－ye*，*Ai Ying－wei*，*Qian Chuan－fan*，*Yu Wei－qiang*，*Li Zheng－Ming*. Journal of Environmrntal Science，2004，16（5）：717－721.

306. Syntheses and Bioactivity of 4"－Sulfonate－5－triphenyl silyl Avermectin B1a and Ivermectin B_{1a} Derivatives. *Lianan Liao*，*Hongyun Fang*，*Zhengming Li*，*Weiguang Zhao*，*Zhijin Fan*. Chemical research in Chinese University，2004，20（5）：551－557.

307. Synthesis and Biological Activity of Novel 2－Methyl－4－trifluoromethyl－thiazole－5－carboxamide Derivatives. *Changling Liu*，*Zhengming Li*，*Bin Zhong*. Journal of Fluorine Chemistry，2004，125：1287－1290.

308. N－（9－anthrylmethyl）propylaminium diphenylphos－phinate monohydrate. *Qing－Xiang Liu*，*Hai－Bin Song*，*Zheng－Ming Li*. Acta Crystallogrphica Section E－Structure Reports，2004，60，o1759－o1761.

309. Metal－ion Interactions with Sugars Crystal Structures and FT－IR Studies of the $LaCl_3$－ribopyranose and $CeCl_3$－ribopyranose Complexes. *Lu Yan*，*Guocai Deng*，*Fangming Miao*，*Zhengming Li*. Carbohydrate Reasearch，2004，339（10）：1689－1696.

310. Syntheses of 1，3，4－Thia（oxa）diazole－substituted Pyrazole Derivatives and Their Fungicidal Activities. *Wenyan Wang*，*Weiguang Zhao*，*Zhengming Li*. Chemical research in Chinese Universi-

ty, 2004, 20 (5): 543 – 547.

311. New condensation reaction of b – keto – d – valerolactones, carbon disulfide and alkyl halides. *You Ming WANG, Yu Xin LI, Su Hua WANG, Zheng Ming LI.* Chinese Chemical Letters, 2004, 15 (10): 1135 – 1136.

312. N – (1 – Cyano – 1, 2 – dimethylpropyl) – 5 – methylisoxazole – 4 – carboxamide. *Bin Zhong, Xiao – Li Mu, Zheng – Ming Li, Hai – Bin Song.* Acta Crystallogrphica Section E – Structure Reports, 2004, 60, o1797 – o1799.

313. 2, 2, 6 – 三甲基 – 4H – 1, 3 – 二噁烷 – 4 – 酮在有机合成中的应用. 赵卫光, 陈寒松, 王素华, 李正名. 精细化工原料及中间体, 2004, 12: 11 – 15.

314. 新型杀虫杀螨剂的研究进展. 钟滨, 刘长令, 李正名. 新农药, 2004, 2: 7 – 12.

315. 2, 2, 6 – 三甲基 – 4H – 1, 3 – 二噁烷 – 4 – 酮在有机合成中的应用. 赵卫光, 陈寒松, 王素华, 李正名. 精细化工原料及中间体, 2004, 11: 12 – 15.

316. 1 – 芳酰基 – 4 – 取代异噁唑甲酰基氨基硫脲和环化产物的合成. 钟滨, 韩亮, 王素华, 李正名. 有机化学 (增刊), 2004, 24 (suppl): 205.

2005 年

317. 多取代吡啶类化合物的合成及生物活性. 钟滨, 李正名, 韩亮, 王素华. 应用化学, 2005, 22 (12): 1354 – 1356.

318. 一类新型嘧啶苯氧(硫)醚的合成及生物活性. 袁德凯, 李正名, 赵卫光, 范志金, 王素华. 应用化学, 2005, 22 (10): 1045 – 1049.

319. 咪唑甘油磷酸酯脱水酶及其抑制剂的研究进展. 肖勇军, 王建国, 何凤琦, 李正名. 农药, 2005, 44 (10): 433 – 436.

320. 植物 N – 酰基乙醇胺 (NAEs) 的代谢机制及其生理功能. 张云, 林凡, 郭维明, 韩亮, 李正名. 西北植物学报, 2005, 25 (10): 2134 – 2138.

321. 4 – (取代嘧啶 – 2 – 基) – 1 – 芳磺酰基氨基脲化合物的合成与表征. 李鹏飞, 王宝雷, 马宁, 王素华, 李正名. 有机化学, 2005, 25 (9): 1057 – 1061.

322. (R) – 3 – O – 2 – 酰基 – 2 – O – 苄基 – 甘油二烷基磷酸酯的合成和结构表征. 韩亮, 李正名, 杨娜, 张云, 郭维明. 应用化学, 2005, 22 (6): 630 – 633.

323. β – 卤素取代对杯[4]吡咯构象及主 – 客体相互作用的影响Ⅱ—密度泛函理论研究. 陈沛全, 孙宏伟, 陈兰, 沈荣欣, 袁满雪, 赖城明, 李正名. 化学学报, 2005, 63 (13): 1175 – 1181.

324. 2 – 氯磺酰基苯甲酰氯的合成. 李鹏飞, 王宝雷, 马宁, 李正名, 王素华. 化学试剂, 2005, 27 (7): 445.

325. 苯环 2 – 位不同酯基取代的磺酰脲类化合物的合成及其除草活性. 李鹏飞, 马宁, 王宝雷, 李正名, 王素华, 李永红. 高等学校化学学报, 2005, 26 (8): 1459 – 1462.

326. 2 – 硬脂酰胺乙基乙基取代苯基磷酸酯的合成和生物活性. 韩亮, 李正名, 张云, 杨娜, 郭维明. 农药, 2005, 44 (4): 165 – 167.

327. N – 杂环卡宾及其金属络合物的性质和合成方法. 柳清湘, 李正名. 化学研究与应用, 2005, 17 (2): 147 – 150, 208.

328. 磷酯的合成进展. 韩亮, 李正名, 郭维明, 陈素梅. 化学通报, 2005, 6: 408 – 416.

329. 要用科学发展观来推动农药学科的持续进步. 李正名. 华中师范大学学报 (自然科学版), 2005, 39 (1): 1.

330. 单嘧磺隆原药组成的定性和定量分析. 范志金, 钱传范, 艾应伟, 陈俊鹏, 李正名. 高等学校化学学报, 2005, 26 (2): 235 – 237.

331. Structure – activity Relationships for a New Family of Sulfonylurea Herbicides. *Jianguo Wang,*

Zhengming Li, Ning Ma, Baolei Wang, Lin Jiang, Siew Siew Pang, Yu – Ting Lee, Guddat Luke W., *Duggleby Ronald G.* Journal of Computer – Aided Molecular Design, 2005, 19: 801 – 820.

332. 1 – Benzoyl – 6 – methyl – 3 – methylsulfanyl – 6, 7 – dihydro – 1H – pyrano [4, 3 – c] pyrazol – 4 – one. *Hao Jing – Jun, Gao Feng, Wang You – Ming, Li Zheng – Ming.* Acta Cryst., 2005, E61, o3470 – o3471.

333. 1 – [4 – (4 – Methylpyrimidin – 2 – yl) piperazin – 1 – ylmethyl] – 1H – benzotriazole. *He Feng – Qi, Wang Bao – Lei, Li Zheng – Ming, Song Hai – Bin.* Acta Cryst., 2005, E61, o2602 – o2604.

334. 3 – (2 – Chlorophenyl) – N – (2 – cyano – 4 – methyl – 2 – pentyl) – 5 – methylisoxazole – 4 – carboxamide. *Zhong Bin, Li Zheng – Ming, Song Hai – Bin.* Acta Cryst., 2005, E61, o2621 – o2622.

335. 6, 7 – Dihydro – 6 – methyl – 3 – (methylsulfanyl) – 1 – (p – tolyl) pyrano [4, 3 – c] pyrazol – 4 (1H) – one. *Hao Jing – Jun, Wang You – Ming, Li Zheng – Ming, Song Hai – Bing.* Acta Cryst., 2005, E61, o1424 – o1425.

336. 3 – (Imidazolidin – 2 – ylidene) – 6 – methyl – 3, 4, 5, 6 – tetrahydro – 2H – pyran – 2, 4 – dione. *Hao Jing – Jun, Xu Hu, Song Hai – Bing, Wang You – Ming, Li Zheng – Ming.* Acta Cryst., 2005, E61, o749 – o750.

337. N, N – Bis (anthracen – 9 – ylmethyl) propylamine. *Liu Qing – Xiang, Song Hai – Bin, Li Zheng – Ming.* Acta Cryst., 2005, E61, o385 – o386.

338. Synthesis and Fungicidal Activity of Novel 2 – Oxocycloalkylsulfonylureas. *Li Xinghai, Yang Xinling, Ling Yun, Fan Zhijin, Liang Xiaomei, Wang Daoquan, Chen Fuheng, Li Zhengming.* Journal of Agricultural and Food Chemistry, 2005, 53, 2202 – 2206.

339. Synthesis, Crystal Structure and Herbicidal Activity of Mimics of Intermediates of the KARI Reaction. *Baolei Wang, Ronald G. Dugglegy, Zhengming Li, Jianguo Wang, Yonghong Li, Suhua Wang, Haibin Song.* Pest Management Science, 2005, 61: 407 – 412.

340. Synthesis and Herbicidal Activity of 2 – Alkyl (aryl) – 3 – methylsulfonyl (sulfinyl) pyrano – [4, 3 – c] pyrazol – 4 (2H) – ones. *Yuxin Li, Youming Wang, Bin Liu, Suhua Wang, Zhengming Li.* Heteroatom Chemistry, 2005, 16 (4): 255 – 258.

341. Herbicide activity of monosulfuron and its mode of action. *Fan Zhi – Jin, Ai Ying – wei, Qian Chuan – fan, Li Zheng – Ming.* Journal of Environmental Science – China, 2005, 17 (3): 399 – 403.

342. N2 – (2 – Nitorphenylsulfonyl) – N5 – n – propylgulutamine. *Xiao Yong – Jun, Wang Jian – guo, Li Zheng – Ming, Song Hai – bin.* Acta Crystallogrphica Section E – Structure Reports, 2005, 61 (10): o3461 – o3463.

2006 年

343. 一类新型嘧啶苯氧（硫）醚的合成及生物活性．袁德凯，李正名．农化新世纪，2006，(10)：28.

344. 吡唑衍生物与碘甲烷反应的密度泛函理论研究．沈福刚，段文勇，孙宏伟，陈兰，陈沛全，赖城明，李正名．高等学校化学学报，2006，27 (11)：2175 – 2178.

345. 44% 单嘧·扑灭 W G 防除夏谷子田杂草．寇俊杰，鞠国栋，王贵启，李正名，苏立军．农药，2006，45 (9)：643 – 645.

346. 水杨酸类糖酯化合物的合成及其生物活性．臧洪俊，李正名，倪长春，沈宙，范志金，刘秀峰．高等学校化学学报，2006，27（10）：1877－1880.
347. 新型除草剂单嘧磺酯亚慢性毒性研究．王静，张静，姜树卿，王晓军，杨友润，姜文玲，李正名，寇俊杰，赵卫光．中国公共卫生，2006，22（8）：988－989.
348. 脒类 KARI 酶抑制剂的分子对接和 3D－QSAR 研究．王宝雷，王建国，马翼，李正名，李永红，王素华．化学学报，2006，64（13）：1373－1378.
349. 单嘧磺隆除草剂的晶体构象－活性构象转换的密度泛函理论研究．陈沛全，孙宏伟，李正名，王建国，马翼，赖城明．化学学报，2006，64（13）：1341－1348.
350. 低聚氨基葡萄糖的化学合成及修饰研究进展．刘幸海，王宝雷，李正名．化学通报，2006，7：484－492.
351. N－月桂酰乙醇胺对香石竹切花瓶插效应的影响．张云，郭维明，韩亮，李正名．园艺学报，2006，33（2）：422－425.
352. N－［2－（4－甲基）嘧啶基］－N′－2－硝基苯磺酰脲的合成、晶体结构、生物活性及其与酵母 AHAS 的分子对接．王建国，马宁，王宝雷，王素华，宋海斌，李正名．有机化学，2006，26（5）：648－652.
353. “链接”化学及其应用．董卫莉，赵卫光，李玉新，刘征骁，李正名．有机化学，2006，26（3）：271－277.
354. 新型先导化合物 2－（2－氯－4－三氟甲基苯氧基）－4－三氟甲基噻唑－5－甲酸乙酯的设计、合成与生物活性．刘长令，李正名．农药，2006，45（4）：246－247.
355. β－氨基丁酸诱导黄瓜抗黄瓜炭疽病筛选体系的构建．张永刚，范志金，刘秀峰，苑建勋，李正名．农药，2006，45（4）：239－242，245.
356. N－（2－甲氧羰基－1－甲基）－乙基－（N′－D－乙酰基吡喃型糖）硫脲类化合物的合成及生物活性．臧洪俊，李正名，杨宇红，赵卫光，王素华．应用化学，2006，23（3）：242－245.
357. O－取代苯基－O－（2－硬脂酰胺基）乙基－N，N－二（2－氯乙基）磷酰胺的合成和生物活性．韩亮，李正名，张云，郭维明．有机化学，2006，26（2）：242－246.
358. 壳寡糖的制备工艺最新研究进展．刘幸海，李正名，王宝雷．天津化工，2006，20（1）：1－6.
359. 酮醇酸还原异构酶抑制剂的设计、合成及除草活性．王宝雷，李正名，赵卫光，王素华，李永红．农药学学报，2006，8（1）：14－19.
360. 具有农业生物活性壳寡糖的研究进展．刘幸海，李正名，王宝雷．农药学学报，2006，8（1）：1－7.
361. 3－N－苄氧羰基－β－氨基丁酸水杨酸酯类化合物的合成及其生物活性．臧洪俊，李正名，倪长春，沈宙，范志金，刘秀峰．高等学校化学学报，2006，27（3）：468－471.
362. 杯［4］二吡咯与卤素阴离子相互作用的密度泛函理论研究．陈东辉，陈沛全，孙宏伟，陈兰，沈荣欣，袁静，袁满雪，赖城明，李正名．高等学校化学学报，2006，27（2）：332－335.
363. 一类新型多取代嘌呤类化合物的合成及生物活性．袁德凯，李正名，赵卫光，范志金，王素华．高等学校化学学报，2006，27（2）：258－262.
364. 重排反应在精细化工中间体合成中的应用．孙小军，赵卫光，董卫莉，李正名．精细化工中间体，2006，36（1）：7－10.
365. 药物分子设计中的 Lipinski 规则．杨二冰，李正名．化学通报，2006，1：16－19.
366. Design，synthesis and biological activity of substituted－N－［4－（substituted－phenylsufamoyl）phenyl］benzamide as potential AHAS Inhaibitors. *Wang Jian Guo*，*Xiao Yong Jun*，*Li Yong Hong*，*Liu Xing Hai*，*Li Zheng Ming*. Chinese Chemical Letters，2006，17（12）：1555－1558.

367. Synthesis and Crystal Structure of 5 – N – i – Propyl – 2 – (2′ – nitrobenzenesulfonyl) – glutamine. *Yongjun Xiao, Jianguo Wang, Baolei Wang, Zhengming Li, Haibin Song*. Chemical Journal of Chinese Universities, 2006, 22 (6): 760 – 762.

368. Synthesis, Dimeric Crystal Structure, and Biological Activities of N – (4 – Methyl – 6 – oxo – 1, 6 – dihydro – pyrimidin – 2 – yl) – N′ – (2 – trifluromethyl – phenyl) – guanidine. *Fengqi He, Baolei Wang, Zhengming Li*. Chemical Journal of Chinese Universities, 2006, 22 (6): 768 – 771.

369. Synthesis and Crystal Structure of a Sodium Monosulfuron – ester (N – [2′ – (4 – Methyl) pyrimidinyl] – 2 – carbomethoxy Benzyl Sulfonylurea Sodium). *Kou Jun – jie, Li Zheng – ming, Song Hai – bin*. Chinese J. Struct. Chem., 2006, 25 (11): 1414 – 1417.

370. Synthesis and Crystal Structure of 1 – [4 – (pyrimidin – 2 – yl) piperazin – 1 – ylmethyl] – 1H – benzotriazole. *He Feng – qi, Wang Bao – lei, Li Zheng – ming, Song Hai – bin*. Chinese J. Struct. Chem., 2006, 25 (5): 543 – 546.

371. Synthesis of Novel 1 – Arylsulfonyl – 4 – (1′ – N – 2′, 3′, 4′, 6′ – tetra – O – acetyl – β – D – glucopyranosyl) thiosemicarbazides. *Li Yu – xin, Li Zheng – ming, Zhao Wei – guang, Dong Wen – li, Wang Su – hua*. Chinese Chemical Letters, 2006, 17 (2): 153 – 155.

372. 2 – (2, 4 – Dichlorophenyl) – 5 – methyl – 4 – (2 – nitrophenylsulfonyl) – 2H – 1, 2, 4 – triazol – 3 (4H) – one. *Wang Lei, Wang Bao – Lei, Li Zheng – Ming, Song Hai – Bin*. Acta Crystallographica Section E (Structure Reports, Online), 2006, E62, o935 – o937.

373. 4 – Bromobenzaldehyde O – (2 – ethoxybenzyl) oxime. *Yang Er – Bing, Wang Lei, Li Zheng – Ming, Song Hai – Bin*. Acta Crystallographica Section E (Structure Reports, Online), 2006, E62, o1657 – o1658.

374. 4 – {2 – [(4 – Methoxyphenyl) sulfanyl] – 2 – (1, 2, 3 – thiadiazol – 4 – yl) acetyl} morpholine. *Dong Wei – Li, Zhao Wei – Guang, Li Zheng – Ming, Song Hai – Bin*. Acta Crystallographica Section E (Structure Reports, Online), 2006, E62, o1645 – o1646.

375. 2 – Amino – 4 – (2, 2, 2 – trifluoroethoxy) pyrimidine. *Sun Fei – Fei, Ma Ning, Li Zheng – Ming, Song Hai – Bin*. Acta Cryst., 2006, E62, o3864 – o3865.

376. Synthesis and antiviral activity against tobacco mosaic virus and 3D – QSAR of substituted – 1, 2, 3 – thiadiazoleacetamides. *Zhao Wei – Guang, Wang Jian – Guo, Li Zheng – Ming, Yang Zhao*. Bioorganic & Medicinal Chemistry Letters, 2006, 16: 6107 – 6111.

377. Synthesis of novel 2 – phenylsulfonylhydrazono – 3 – (2′, 3′, 4′, 6′ – tetra – O – acetyl – β – D – glucopyranosyl) thiazolidine – 4 – ones from thiosemicarbazide precursors. *Li Yu – Xin, Wang Su – Hua, Li Zheng – Ming, Su Na, Zhao Wei – Guang*. Carbohydrate Research, 2006, 341: 2867 – 2870.

378. Synthesis and Biological Activity of Novel 2 – (3 – trifluoromethylphenoxy) – 4 – trifluoromethylthiazole – 5 – Carboxamide Derivatives. *Changling Liu, Miao Li, Huiwei Chi, Chunqing Hou, Zhengming Li*. Journal of Fluorine Chemistry, 2006, 127: 796 – 799.

379. Silver (Ⅰ) Complexes in Coordination Supramolecular System with Bulky Acridine – Based Ligands: Syntheses, Crystal Structures, and Theoretical Investigations on C – H⋯Ag Close Interaction. *Liu Chun – Sen, Chen Pei – Quan, Yang En – Cui, Tian Jin – Lei, Bu Xian – He, Li Zheng – Ming, Sun Hong – Wei, Lin Zhen – Yang*. Inorg. Chem., 2006, 45 (15): 5812 – 5821.

380. Microwave and Ultrasound Irradiation – Assisted Synthesis of Novel Disaccharide – Derived Arylsulfonyl Thiosemicarbazides. *Yuxin Li, Weiguang Zhao, Zhengming Li, Suhua Wang, Weili Dong*. Synthetic Communications, 2006, 36: 1471 – 1477.

381. Synthesis and Biological Activity of New 1 – [4 – (substituted) – piperazin – 1 – ylmethyl] – 1H – benzotriazole. *He Fengqi*, *Liu Xinghai*, *Wang Baolei*, *Li Zhengming*. J. Chemical Research, 2006, 12: 809 – 811.

2007 年

382. Synthesis of New Plant Growth Regulator – N – (fatty acid) *O* – aryloxyacetyl ethanol. *Liang Han*, *Jianrong Gao*, *Zhengming Li*, *Yun Zhang*, *Weiming Guo*. Bioorganic & Medicinal Chemistry Letters, 2007, 17 (11): 3231 – 3234.

383. Identification of Some Novel AHAS Inhibitors via Molecular Docking and Virtual Screening Approach. *Jianguo Wang*, *Yongjun Xiao*, *Yonghong Li*, *Yi Ma*, *Zhengming Li*. Bioorganic & Medicinal Chemistry, 2007, 15 (1): 374 – 380.

384. Synthesis and Crystal structure of N – [4 – (4 – Methyl – phenylsulfamoyl) – phenyl] – 4 – methyl – 1, 2, 3 – thiadiazole – 5 – carboxamide. *Wang Jian – Guo*, *Xiao Yong – Jun*, *Li Zheng – Ming*, *Song Hai – Bin*. Chinese J. Struct. Chem., 2007, 26 (5): 501 – 504.

385. Synthesis, Bioactivity, Theoretical and Molecular Docking Study of 1 – Cyano – N – substituted – cyclopropanecarboxamide as Ketol – acid Reductoisomerase Inhibitor. *Xinghai Liu*, *Peiquan Chen*, *Baolei Wang*, *Yonghong Li*, *Suhua Wang*, *Zhengming Li*. Bioorganic & Medicianl Chemistry Letters, 2007, 17: 3784 – 3788.

386. 3 – N – 苄氧羰基 – β – 氨基丁酸糖酯的合成基生物活性. 臧洪俊, 李正名, 范志金, 刘秀峰, 王素华. 高等学校化学学报, 2007, 28 (8): 1512 – 1515.

387. 酮醇酸还原异构酶 (KARI) 与其抑制剂间相互作用的分子模拟研究. 陈沛全, 刘幸海, 孙宏伟, 王宝雷, 李正名, 赖城明. 化学学报, 2007, 65 (16): 1693 – 1701.

388. 单嘧磺隆晶体—构象活性构象转换的分子动力学模拟. 陈沛全, 孙宏伟, 李正名, 王建国, 马翼, 赖城明. 高等学校化学学报, 2007, 28 (2): 278 – 282.

389. 单嘧磺隆对 3 种鱼腥藻的毒性. 罗伟, 沈健英, 李正名. 农药, 2007, 46 (5): 345 – 348.

390. 基于受体结构的 AHAS 抑制剂的设计、合成及生物活性. 肖勇军, 王建国, 刘幸海, 李永红, 李正名. 高等学校化学学报, 2007, 28 (7): 1280 – 1282.

391. α – 芳氧基乙酸 – (2 – 取代氧基) – 苄酯的合成与生物活性. 杨二冰, 李永红, 刘秀峰, 李正名. 高等学校化学学报, 2007, 28 (6): 1077 – 1079.

392. 1 – (2, 4 – Dimethylphenyl) – 3 – methyl – 4 – (2 – nitrophenylsulfonyl) 1H – 1, 2, 4 – triazol – 5 (4H) – one. *Mi Tang*, *Lei Wang*, *Zheng – Ming Li*, *Hai – Bin Song*. Acta Cryst., 2007, E63, o2513 – o2514.

393. (S) – N – 5 – Ethyl – N – 2 – (2 – nitrophenylsulfonyl) glutamine. *Wang Jian – Guo*, *Li Wen – Ming*, *Li Zheng – Ming*, *Song Hai – Bin*. Acat Crystallographica Section E – Structure Reports Online, 2007, 63: O2668 – O2669 Part 5.

394. 3 – (Dimethoxythiophosphorylamido) – 2 – (2′, 3′, 4′ – tri – O – acetyl – beta – D – xylopyranos – 1′ – ylimino) – 1, 3 – thiazolidin – 4 – one. *Cheng Hai – Ying*, *Li Yu – Xin*, *Wang Xian – An*, *Li Zheng – Ming*, *Song Hai – Bin*. Acat Crystallographica Section E – Structure Reports Online, 2007, 63: O2861 – O2862 Part 6.

395. Ethyl 2 – [4 – (5 – methyl – 1, 3, 4 – oxadiazol – 2 – yl) phenoxy] propanoate. *Ma Rui – Dian*, *Wang Bao – Lei*, *Li Zheng – Ming*, *Song Hai – Bin*. Acat Crystallographica Section E – Structure Reports Online, 2007, 63: O2897 – U2395 Part 6.

396. Ethyl 2 – (N – cyclohexyl – 2 – nitrophenyl – sulfonamido) – 2 – oxoacetate. *Wang Bao – Lei*, *He Feng – Qi*, *Li Zheng – Ming*, *Song Hai – Bin*. Acat Crystallographica Section E – Structure Reports

Online, 2007, 63: O3264 - U3973 Part 7.

397. Syntheses and Biological Activities of Ethyl N - hydroxy - N - (substituted) phenyloxamates as KARI Inhibitors. *Baolei Wang*, *Zhengming Li*, *Yonghong Li*, *Suhua Wang*. Chemical Research in Chinese Universities, 2007, 23 (3): 280 - 283.

398. New triphenylamine - based organic dyes for efficient dye - sensitized solar cells. *Liang Mao*, *Xu Wei*, *Cai Fengshi*, *Chen Peiquan*, *Peng Bo*, *Chen Jun*, *Li Zhengming*. Journal of Physical Chemistry C, 2007, 111 (11): 4465 - 4472.

399. The Design, Synthesis, and Biological Evaluation of Novel Substituted Purines as HIV - 1 Tat - TAR Inhibitors. *Dekai Yuan*, *Meizi He*, *Ruifang Pang*, *Shrongshi Lin*, *Zhengming Li*, *Ming Yang*. Bioorganic & Medicinal Chemistry, 2007, 15: 265 - 272.

400. Synthesis and Crystal Structure of 2 - Amino - 4 - chloro - 5 - (4′ - methylbenzyl) - 6 - methylpyrimidine. 郭万成，刘幸海，陈沛全，宋海斌，李正名. 结构化学，2007，26 (9): 1005 - 1008.

401. 4，5，6 - 三取代嘧啶磺酰脲化合物的合成和除草活性. 郭万成，王美怡，刘幸海，李永红，王素华，李正名. 高等学校化学学报，2007，28 (9): 1666 - 1670.

402. 1，2，3 - 噻二唑 - 4 - 乙酰胺（吗啉）类衍生物的合成与生物活性. 董卫莉，姚红伟，王凤龙，李正名，申丽丽，钱玉梅，赵卫光. 高等学校化学学报，2007，28 (6): 1671 - 1676.

403. 酵母 AHAS 酶与磺酰脲类抑制剂作用模型的分子对接研究. 李琼，陈沛全，陈兰，孙宏伟，沈荣欣，赖城明，李正名. 高等学校化学学报，2007，28 (8): 1552 - 1555.

404. N - 月桂酰乙醇胺对香石竹开放和衰老进程中花瓣微粒体膜组分和功能的调节. 张云，郭维明，陈发棣，韩亮，李正名. 中国农业科学，2007，10: 2303 - 2308.

405. 2，4 - 二氯苯戊酮的工艺改进. 王红学，李正名. 天津化工，2007，5: 19 - 20.

2008 年

406. Molecular design, synthesis and biological activities of amidines as new ketol - acid reductoisomerase inhibitors. *Wang Bao Lei*, *Li Yong Hong*, *Wang Jian Guo*, *Ma Yi*, *Li Zheng Ming*. Chinese Chemical Letters, 2008, 19 (6): 651 - 654.

407. Syntheses and crystal structures of (S) - methyl 2 - (4 - R - phenylsulfonamido) - 3 - (1H - indol - 3 - yl) propanoate (R = H (1) and Cl (2)). *Li Wen - Ming*, *Wang Jian - Guo*, *Guo Wan - Cheng*, *Li Zheng - Ming*, *Song Hai - Bin*. Chinese Journal of Structural Chemistry, 2008, 27 (6): 691 - 696.

408. Synthesis, Crystal Structure and Biological Activity of 2 - Chloro - N - {2 - fluoro - 5 - [N - (phenylsulfonyl) phenylsulfonamido] phenyl} benzamide. *Wenming Li*, *Jianguo Wang*, *Zhengming Li*, *Haibin Song*. Chemical Research in Chinese Universities, 2008, 24 (3): 295 - 298.

409. High Throughput Screening under Zinc - database and Synthesis a Dialkylphosphinic Acid as a Potential Kari Inhibitor. *Xinghai Liu*, *Peiquan Chen*, *Wancheng Guo*, *Suhua Wang*, *Zhengming Li*. Phosphorus, Sulfur and Silicon and the Related Elements, 2008, 183 (2 - 3): 775 - 778.

410. Design, synthesis and biological activities of novel 1, 2, 3 - thiadiazole derivatives containing oxime ether. *Dong Wei - li*, *Yao Hong - wei*, *Li Zheng - ming*, *Zhao Wei - guang*. Journal of Chemical Research, 2008, (3): 145 - 147.

411. N - (4′ - 取代嘧啶 - 2′ - 基) - 2 - 甲氧羰基 - 5 - (取代) 苯甲酰胺基苯磺酰脲化合物的合成及除草活性研究. 王美怡，郭万成，兰峰，李永红，李正名. 有机化学，2008，28 (4): 649 - 656.

412. The research progress in green chemistry. *Li Wen - ming*, *Wang Jian - guo*, *Li Zheng - ming*. Tianjin Huagong, 2008, 22 (2): 1 -4, 60.

413. 酰胺类 KARI 酶抑制剂的设计、合成和生物活性．王宝雷，李正名，李永红，王素华．高等学校化学学报，2008，29（3）：523 -527.

414. Synthesis and Herbicidal Activity of Novel Sulfonylureas Containing Thiadiazol Moiety. *Wancheng Guo*, *Xinghai Liu*, *Yonghong Li*, *Suhua Wang*, *Zhengming Li*. Chemical Research in Chinese Universities, 2008, 24 (1): 32 -35.

415. Synthesis and Biological Activities of Novel Bis - heterocyclic Pyrrodiazole Derivatives. *Fengqi He*, *Xinghai Liu*, *Baolei Wang*, *Zhengming Li*. Heteroatom Chemistry, 2008, 19 (1), 21 -27.

416. A photoluminescent hexanuclear silver (Ⅰ) complex exhibiting C - H Ag close interactions. *Liu Chun - Sen*, *Chen Pei - Quan*, *Chang Ze*, *Wang Jun - Jie*, *Yan Li - Fen*, *Sun Hong - Wei*, *Bu Xian - He*, *Lin Zhenyang*, *Li Zheng - Ming*, *Batten*, *Stuart R*. Inorganic Chemistry Communications, 2008, 11 (2): 159 -163.

417. The Outset Innovation of Agrchemicals in China. *Zhengming Li*, *Yibin Zhang*. Outlooks on Pest Management, 2008, 19 (3): 135 -138.

418. 含 4 -噻唑啉酮环的新烟碱类化合物的合成及生物活性．孙小军，苏娜，刘幸海，董卫莉，李正名，赵卫光．高等学校化学学报，2008，29（7）：1359 -1362.

419. 4 -烷氧/苄氧基苯基四唑和 1，3，4 -噁二唑类化合物的合成及除草活性．王宝雷，李正名，李永红，王素华．高等学校化学学报，2008，29（1）：90 -94.

420. 新三取代嘧啶苯磺酰脲衍生物的合成与生物活性．郭万成，谭海忠，刘幸海，李永红，王素华，李正名．高等学校化学学报，2008，29（2）：319 -323.

421. 新型 1，2，4 -三唑啉酮类化合物的合成及生物活性．王雷，唐蜜，李文明，李永红，王素华，李正名．高等学校化学学报，2008，29（7）：1371 -1375.

422. Synthesis and herbicidal activities of pyridy - sulfonylureas: More convenient preparation process of phenyl pyrimidylcarbamates. *Ning Ma*, *Zhi - Jin Fan*, *Bao - Lei Wang*, *Yong - Hong Li*, *Zheng - Ming Li*. Chinese Chemical Letters, 2008, 19: 1268 -1270.

423. Facile Synthesis and Herbicidal Activities of New Isoxazole Derivatives via 1, 3 - Dipolar Cycloadditions Reaction. *Chuan - Ye Zhang*, *Bao - Lei Wang*, *Xing - Hai Liu*, *Yong - Hong Li*, *Su - Hua Wang*, *Zheng - Ming Li*. Heterocyclic Communications, 2008, 14 (6): 397 -403.

2009 年

424. Quantitive - structure activity relationship (QSAR) study of a new heterocyclic insecticides using CoMFA and CoMSIA. *Weili Dong*, *Xinghai Liu*, *Yi Ma*, *Zhengming Li*. Heterocyclic Communications, 2009, 15 (1): 17 -21.

425. A Convenient Method for the Aerobic Oxidation of Thiols to Disulfides. *Weili Dong*, *Guangying Huang*, *Zhengming Li*, *Weiguang Zhao*. Phosphorus, Sulfur and Silicon and the Related Elements, 2009, 184 (8): 2058 -2065.

426. Design, Synthesis, and Fungicidal Activity of Novel Analogues of Pyrrolnitrin. *Mingzhong Wang*, *Han Xu*, *Qi Feng*, *Lizhong Wang*, *Suhua Wang*, *Zhengming Li*. Journal of Agricultural and Food Chemistry, 2009, 57 (17), 7912 -7918.

427. 新型苯环 5 -（取代）苯甲酰胺基苯磺酰脲类化合物的比较定量构效关系研究．王美怡，马翼，李正名，王素华．高等学校化学学报，2009，30（7）：1361 -1364.

428. Design, synthesis and bioactivity of new N - methoxycarbamate containing pyrazole. *Li Miao*, *Zhang Jin - Bo*, *Yang Ji - Chun*, *Li Zhi - Nian*, *Liu Chang - Ling*, *Li Zheng - Ming*. Chemical

Journal of Chinese Universities – Chinese, 2009, 30 (7): 1348 – 1352.

429. Regioselective synthesis of novel 3 – alkoxy (phenyl) thiophosphorylamido – 2 – (per – O – acetylglycosyl – 1′ – imino) thiazolidine – 4 – one derivatives from O – alkyl N^4 – glycosyl (thiosemicarbazido) phosphonothioates. *Yuxin Li, Haoan Wang, Xiaoping Yang, Haiying Cheng, Zhihong Wang, Yiming Li, Zhengming Li, Suhua Wang, Dongwen Yan.* Carbohydrate Research, 2009, 344 (10): 1248 – 1253.

430. Synthesis, Herbicidal Activities and Comparative Molecular Field Analysis Study of Some Novel Triazolinone Derivatives. *Lei Wang, Yi Ma, Xinghai Liu, Yonghong Li, Haibin Song, Zhengming Li.* Chemical Biology & Drug Design, 2009, 73 (6): 674 – 681.

431. 3 – 取代 – 5 – 芳基噁唑 – 4 – 酮类化合物的合成与生物活性. 高等学校化学学报, 2009, 30 (5): 908 – 912.

432. N – 取代 – 4 – 三氟甲基嘧啶 – 2 – 胺的合成. 王红学, 许丽萍, 李芳, 王宝雷, 李正名. 化学试剂, 2009, 31 (5): 365 – 366, 372.

433. Syntheses and herbicidal activities of novel 1, 2, 4 – triazolinone derivatives. *Lan Feng, Xu Jun – Ying, Li Yong – Hong, Wang Su – Hua, Li Zheng – Ming.* Chemical Journal of Chinese Universities – Chinese, 2009, 30 (4): 712 – 715.

434. Synthesis, Antifungal Activities and 3D – QSAR Study of N – (5 – substituted – 1, 3, 4 – thiadiazol – 2 – yl) Cyclopropanecarboxamides. *Xinghai Liu, Yanxia Shi, Yi Ma, Chuanyu Zhang, Weili Dong, Li Pan, Baolei Wang, Baoju Li, Zhengming Li.* European Journal of Medicinal Chemistry, 2009: 44 (7): 2782 – 2786.

435. 新型芳磺酰基色氨酸酯以及芳磺酰基谷氨酸二酯类化合物的合成与生物活性研究. 李文明, 谭海忠, 王建国, 李永红, 李正名. 高等学校化学学报, 2009, 30 (4): 728 – 730.

436. Synthesis, Bioactivity and SAR Study of N′ – (5 – substituted – 1, 3, 4 – thiadiazol – 2 – yl) – N – cyclopropylformyl – thioureas as Ketol – acid Reductoisomerase Inhibitors. *Xinghai Liu, Chuanyu Zhang, Wancheng Guo, Yonghong Li, Peiquan Chen, Teng Wang, Weili Dong, Baolei Wang, Hongwei Sun, Zhengming Li.* Journal of Enzyme Inhibition and Medicinal Chemistry, 2009, 24 (2): 545 – 552.

437. Synthesis of isatin derivatives and their inhibition against AHAS. *Tan Hai – Zhong, Li Hui – Dong, Wang Jian – Guo, Li Wen – Ming, Li Yong – Hong, Li Zheng – Ming.* Chemical Journal of Chinese Universities – Chinese, 2009, 30 (3): 510 – 512.

438. Synthesis and Characterization of O, O – Dimethyl – N – (2, 2, 2 – trichloro – 1 – arylaminoethyl) Phosphoramidothioates. *Baolei Wang, Xinghai Liu, Zhengming Li.* Phosphorus Sulfur and Silicon and the Related Elements. 2009, 184 (9): 2281 – 2287.

439. Synthesis and Biological Activity of Novel 2, 3 – Dihydro – 2 – phenylsulfonylhydrazono – 3 – (2′, 3′, 4′, 6′ – tetra – O – acetyl – D – glucopyranosyl) thiazoles. *Haoan Wang, Yuxin Li, Zhengming Li, Haiying Cheng, Suhua Wang, Bin Liu.* Chemical Research in Chinese Universities, 2009, 25 (1): 52 – 55.

440. Synthesis of Some N, N′ – Diacylhydrazine Derivatives with Radical – scavenging and Antifungal Activity. *Xinghai Liu, Yanxia Shi, Yi Ma, Guorong He, Weili Dong, Chuanyu Zhang, Baolei Wang, Suhua Wang, Baoju Li, Zhengming Li.* Chemical Biology & Drug Design, 2009, 73 (3): 320 – 327.

441. Crystal structures of two novel sulfonylurea herbicides in complex with Arabidopsis thaliana acetohydroxyacid synthase. *Wang Jian – Guo, Lee Patrick K. – M., Dong Yu – Hui, Pang Siew Siew, Duggleby Ronald G., Li Zheng – Ming, Guddat Luke W.* FEBS Journal, 2009, 276 (5): 1282 –

1290.

442. Studies on the virtual screening, synthesis and biological activity of KARI inhibitors. *Wang Bao－lei*, *Chen Pei－quan*, *Liu Xing－hai*, *Li Yong－hong*, *Li Zheng－ming*. Zhongguo Keji Lunwen Zaixian, 2009, 4 (9): 638－643.

443. 含4－正庚/正辛氧基苯基杂环化合物的设计、合成及生物活性．王宝雷，马瑞典，赵卫光，李正名，李永红．农药学学报，2009，11（1）：13－18.

444. Design, Synthesis, and Insecticidal Activity of Novel Neonicotinoid Derivatives Containing N－oxalyl groups. *Yu Zhao*, *Yongqiang Li*, *Suhua Wang*, *Zhengming Li*. ARKIVOC (Gainesville, FL, United States), 2009, (11): 152－164.

445. Synthesis, Structure and Biological Activities of Some Novel Anthranilic Acid Esters Containing N－Pyridylpyrazole. *Weili Dong*, *Junying Xu*, *Lixia Xiong*, *Xinghai Liu*, *Zhengming Li*. Chinese Journal of Chemistry, 2009, 27 (3): 579－586.

446. 4，5，6－三取代嘧啶苯磺酰脲类化合物的生物活性、分子对接与3D－QSAR关系研究．郭万成，马翼，李永红，王素华，李正名．化学学报，2009，67（6）：569－574.

447. 玉嘧磺隆新合成方法研究．童军，郑占英，严东文，王素华，李正名．化工中间体，2009，05：29－31.

448. 新型含嘧啶环的甲氧基丙烯酸酯类化合物的设计、合成及杀菌活性．李淼，刘若霖，李志念，刘长令，李正名．农药，2009，06：405－408.

449. 妇炎灵泡腾片质量标准的研究．陈建宇，周福军，李正名．天津中医药大学学报，2009，03：138－141.

450. 单嘧磺酯的HPLC/MS/MS研究．蔡飞，陈建宇，王海英，李正名．农药学学报，2009，03：388－391.

451. 30%单嘧·氯氟水分散粒剂防除冬小麦田杂草田间试验．鞠国栋，寇俊杰，王满意，王贵启，李正名，王建平．农药，2009，10：765－766，770.

452. 78%单嘧·扑草净WG防除麦田杂草碱茅药效试验．寇俊杰，王满意，鞠国栋，段美生，李正名，王贵启．现代农药，2009，06：21－22，25.

2010年

453. 新（4′－三氟甲基嘧啶基）－2－苯磺酰脲衍生物的合成及生物活性．王红学，李芳，许丽萍，李永红，王素华，李正名．高等学校化学学报，2010，31（1）：64－67.

454. Synthesis of some new N－pyridylpyrazoles and determination of their fungicidal activity. *Zhao Yu*, *Wang Gang*, *Dong Wei－li*, *Shi Yan－xia*, *Li Bao－ju*, *Wang Su－hua*, *Li Zheng－ming*. ARKIVOC, 2010, Part 2: 16－30.

455. 微波及超声波辅助合成新型苯环5位Schiff碱取代苯磺酰脲类化合物及除草活性研究．王美怡，李正名，李永红．有机化学，2010，30（6）：877－883.

456. Synthesis, Crystal Structure and Insecticidal Activities of Novel Neonicotinoid Derivatives. *Yu Zhao*, *Gang Wang*, *Yongqiang Li*, *Suhua Wang*, *Zhengming Li*. Chemical Research in Chinese Universities, 2010, 26 (3): 380－383.

457. Solvent－and Catalyst－free Synthesis and Antifungal Activities of alpha－Aminophosphonate Containing Cyclopropane Moiety. *Pan Li*, *Liu Xing－hai*, *Shi Yan－xia*, *Wang Bao－lei*, *Wang Su－hua*, *Li Bao－ju*, *Li Zheng－ming*. Chemical Research in Chinese Universities, 2010, 26 (3): 389－393.

458. 区域选择性合成2－取代磺酰基亚肼基－3－全乙酰糖基－2，3－二氢噻唑及其表征、生物活性研究．王浩安，李玉新，李一鸣，王素华，李正名．有机化学，2010，30（5）：703

–706.

459. 3位苯甲酰腙和亚氨基取代硫脲取代的吲哚满二酮衍生物的合成及对AHAS的抑制活性．李慧东，商建丽，谭海忠，王建国，李永红，李正名．高等学校化学学报，2010，31（5）：953–956.

460. Synthesis and Biological Activity of Some Novel Trifluoromethyl – Substituted 1，2，4 – Triazole and Bis（1，2，4 – Triazole）Mannich Bases Containing Piperazine Rings. *Baolei Wang*，*Yanxia Shi*，*Yi Ma*，*Xinghai Liu*，*Yonghong Li*，*Haibin Song*，*Baoju Li*，*Zhengming Li*. Journal of Agricultural and Food Chemistry，2010，58（9）：5515–5522.

461. Synthesis and Antiviral Activity of New Acrylamide Derivatives Containing 1，2，3 – Thiadiazole as Inhibitors of Hepatitis B Virus Replication. *Weili Dong*，*Zhengxiao Liu*，*Xinghai Liu*，*Zhengming Li*，*Weiguang Zhao*. European Journal of Medicinal Chemistry，2010，45（5）：1919–1926.

462. Design，Synthesis and Insecticidal Activities of Novel N – Oxalyl Derivatives of Neonicotinoid Compound. *Yu Zhao*，*Gang Wang*，*Yongqiang Li*，*Suhua Wang*，*Zhengming Li*. Chinese Journal of Chemistry，2010，28（3）：475–479.

463. Synthesis and Antifungal Activities of New Pyrazole Derivatives via 1，3 – dipolar Cycloaddition Reaction. *Chuanyu Zhang*，*Xinghai Liu*，*Baolei Wang*，*Suhua Wang*，*Zhengming Li*. Chemical Biology & Drug Design，2010，75（5）：489–493.

464. Synthesis and Fungicidal Activity of Novel Aminophenazine – 1 – carboxylate Derivatives. *Mingzhong Wang*，*Han Xu*，*Shujing Yu*，*Qi Feng*，*Suhua Wang*，*Zhengming Li*. Journal of Agricultural and Food Chemistry，2010，58（6）：3651–3660.

465. Synthesis and Biological Activity of New（E）– alpha –（Methoxyimino）benzeneacetate Derivatives Containing a Substituted Pyrazole Ring. *Miao Li*，*Changling Liu*，*Jichun Yang*，*Jinbo Zhang*，*Zhinian Li*，*Hong Zhang*，*Zhengming Li*. Journal of Agricultural and Food Chemistry，2010，58（5）：2664–2667.

466. Synthesis，Structure and Biological Activity of 2 –［4 –（5 – Substituted – 1，3，4 – oxadiazol – 2 – yl）phenoxy］Propanoate/Acetate. *Wang Baolei*，*Ma Ruidian*，*Li Yonghong*，*Song Haibin*，*Li Zhengming*. Chinese Journal of Organic Chemisty，2010，30（1）：92–97.

467. 新（4′–三氟甲基嘧啶基）–2–苯磺酰脲衍生物的合成及生物活性．王红学，李芳，许丽萍，李永红，王素华，李正名．高等学校化学学报，2010，31（1）：64–67.

468. Design，Synthesis and biological activities of new strobilurin derivatives containing substituted pyrazoles. *Miao Li*，*Changling Liu*，*Lin Li*，*Hao Yang*，*Zhinian Li*，*Hong Zhang*，*Zhengming Li*. Pest Management Science，2010，66（1）：107–112.

469. High Throughput Receptor – based Virtual Screening under ZINC Database，Synthesis，and Biological Evaluation of Ketol – Acid Reductoisomerase Inhibitors. *Xinghai Liu*，*Peiquan Chen*，*Baolei Wang*，*Weili Dong*，*Yonghong Li*，*Xingqiao Xie*，*Zhengming Li*. Chemical Biology & Drug Design，2010，75（2）：228–232.

470. Synthesis and Insecticidal Activities of Novel Anthranilic Diamides Containing Modified N – Pyridylpyrazoles. *Qi Feng*，*Zhili Liu*，*Lixia Xiong*，*Mingzhong Wang*，*Yongqiang Li*，*Zhengming Li*. Journal of Agricultural and Food Chemistry，2010，58（23）：12327–12336.

471. 4，6–二甲氧基嘧啶衍生物的合成及生物活性测定．李洋，张志国，朱敏娜，李志念，刘长令，李正名．农药，2010，（04）：246–249.

472. 烟酸糖酯的合成、晶体结构及生物活性．李一鸣，李玉新，王浩安，徐俊英，王素华，李正名．化学试剂，2010，（06）：481–483.

473. 2，3–二氢吡喃自动控制工艺的初步设计研究．张树军，刘桂龙，严冬文，王素华，李正名．化工中间体，2010，（01）：42－45.

474. 2－酰基－氰基乙酸衍生物的合成及生物活性．李洋，张志国，迟会伟，罗艳梅，刘长令，李正名．高等学校化学学报，2010，31（9）：1798－1804.

475. 三环己基氯化锡对烟草花叶病毒的抑制作用．李祝，周蕴赟，陈伟，毛明珍，李一鸣，李玉新，李正名，唐蜜，宋宝安．农药学学报，2010，12（3）：279－282.

2011 年

476. Synthesis and Insecticidal Activities of Novel Analogues of Chlorantraniliprole Containing Nitro Group. *Qi Feng*，*Mingzhong Wang*，*Lixia Xiong*，*Zhili Liu*，*Zhengming Li*. Chemical Research in Chinese Universities，2011，27（4）：610－613.

477. Modulations of High－voltage Activated Ca^{2+} Channels in the Central Neurones of *Spodoptera exigua* by Chlorantraniliprole. *Yuxin Li*，*Mingzhen Mao*，*Yiming Li*，*Lixia Xiong*，*Zhengming Li*，*Junying Xu*. Phyaiological Entomology，2011，36（3）：230－234.

478. Integrating molecular docking，DFT and CoMFA/CoMSIA approaches for a series of naphthoquinone fused cyclic alpha－aminophosphonates that act as novel topoisomerase Ⅱ inhibitors. *Ma Yi*，*Wang Jian－Guo*，*Wang Bin*，*Li Zheng－Ming*. Journal of Molecular Modeling，2011，17（8）：1899－1909.

479. Syntheses，Crystal Structures and Bioactivities of Two Novel Isatin Derivatives. *Shang Jian－li*，*Li Hui－dong*，*Shang Jun*，*Song Hai－bin*，*Li Zheng－ming*，*Wang Jian－guo*. Chemical Research in Chinese Universities，2011，27（4）：595－598.

480. Synthesis and Biological Evaluation of Isosteric Analogs of Mandipropamid for the Control of Oomycete pathogens. *Su Na*，*Wang Zhen－Jun*，*Wang Li－Zhong*，*Zhang Xiao*，*Dong Wei－Li*，*Wang Hong－Xue*，*Li Zheng－Ming*，*Zhao Wei－Guang*. Chemical Biology & Drug Design，2011，78（1）：101－111.

481. Synthesis，Structure and Biological Activity of Novel 1，2，4－Triazole Mannich Bases Containing a Substituted Benzylpiperazine Moiety. *Baolei Wang*，*Xinghai Liu*，*Xiulan Zhang*，*Jifeng Zhang*，*Haibin Song*，*Zhengming Li*. Chemical Biology & Drug Design，2011，78（1）：42－49.

482. Design，Synthesis and Structure－activity Relationship of Novel Coumarin Derivatives. *Aiying Guan*，*Changling Liu*，*Miao Li*，*Hong Zhang*，*Zhinian Li*，*Zhengming Li*. Pest Management Science，2011，67（6）：647－655.

483. Design，Synthesis and Antifungal Activities of Novel Pyrrole Alkaloid Analogs. *Mingzhong Wang*，*Han Xu*，*Tuanwei Liu*，*Qi Feng*，*Shujing Yu*，*Suhua Wang*，*Zhengming Li*. European Journal of Medicinal Chemistry，2011，46（5）：1463－1472.

484. Synthesis and Herbicidal Activity of Novel Sulfonylurea Derivatives. *Gang Cao*，*Meiyi Wang*，*Mingzhong Wang*，*Suhua Wang*，*Yonghong Li*，*Zhengming Li*. Chem. Res. Chinese Universities，2011，27（1）：60－65.

485. Synthesis and Crystal Structure of（S）－Ethyl－2－（2－methoxy－phenylsulfenamido）－3－（1H－indol－3－yl）propanoate. *Shang Jian－Li*，*Song Hai－Bin*，*Li Zheng－Ming*，*Wang Jian－Guo*. Chinese Journal of Structural Chemistry，2011，30（2）：252－256.

486. Syntheses，Crystal Structures and Bioactivities of Two Isatin Derivatives. *Tan Hai－Zhong*，*Wang Wei－Min*，*Shang Jian－Li*，*Song Hai－Bin*，*Li Zheng－Ming*，*Wang Jian－Guo*. Chinese Journal of Structural Chemistry，2011，30（4）：502－507.

487. Chemical Synthesis，in Vitro Acetohydroxyacid Synthase（AHAS）Inhibition，Herbicidal Activity，and Computational Studies of Isatin Derivatives. *Jianguo Wang*，*Haizhong Tan*，*Yonghong Li*，*Yi*

Ma, Zhengming Li, Luke W. Guddat. Journal of Agricultural and Food Chemistry, 2011, 59: 9892 – 9900.

488. Synthesis and Insecticidal Evaluation of Novel N – Pyridylpyrazolecarboxamides Containing Different Substituents in the ortho – Position. *Qi Feng, Guanping Yu, Lixia Xiong, Mingzhong Wang, Zhengming Li.* Chinese Journal of Chemistry, 2011, 29, 1651 – 1655.

489. Regioselective Synthesis of Novel 3 – Thiazolidine Acetic Acid Derivatives from Glycosido Ureides. *Yuxin Li, Wei Chen, Xiaoping Yang, Guanping Yu, Mingzhen Mao, Yunyun Zhou, Tuanwei Liu, Zhengming Li.* Chemical Biology & Drug Design, 2011, 78 (6): 969 – 978.

490. Design, Synthesis, Biological Activities, and 3D – QSAR of New N, N′ – Diacylhydrazines Containing 2 – (2, 4 – dichlorophenoxy) propane Moiety. *Xinghai Liu, Li Pan, Yi Ma, Jianquan Weng, Chengxia Tan, Yonghong Li, Yanxia Shi, Baoju Li, Zhengming Li, Yonggang Zhang.* Chemical Biology & Drug Design, 2011, 78 (4): 689 – 694.

491. Synthesis and Biological Activity of Novel 1 – Substituted Phenyl – 4 – [N – [(2′ – morpholinothoxy) phenyl] aminomethyl] – 1H – 1, 2, 3 – Triazoles. *Mingzhen Mao, Yuxin Li, Yunyun Zhou, Wei Chen, Tuanwei Liu, Shujing Yu, Suhua Wang, Zhengming Li.* Chemical Biology & Drug Design, 2011, 78 (4): 695 – 699.

492. Synthesis, Crystal Structure, Bioactivity and DFT Calculation of New Oxime Ester Derivatives Containing Cyclopropane Moiety. *Xinghai Liu, Li Pan, Chengxi Tan, Jianquan Weng, Baolei Wang, Zhengming Li.* Pesticide Biochemistry and Physiology, 2011, 101: 143 – 147.

493. 含七氟异丙基的氯虫酰胺类似物的设计合成及生物活性．冯启，刘智力，王明忠，熊丽霞，于淑晶，李正名．高等学校化学学报，2011，32 (1)：74 – 78.

494. α – 取代苯乙酰胺类化合物的合成及生物活性．李欢欢，王振军，王力钟，李正名，赵卫光．高等学校化学学报，2011，32 (1)：79 – 83.

495. 杀螟丹中间体与异构体互变异构理论研究．于观平，马翼，刘鹏飞，闫涛，李正名．高等学校化学学报，2011，32 (11)：2539 – 2543.

496. 新型邻甲酰胺基间苯二甲酰胺类化合物的设计合成及生物活性．闫涛，于观平，熊丽霞，于淑晶，王素华，李正名．高等学校化学学报，2011，32 (8)：1750 – 1754.

497. 沙蚕毒素类杀虫剂研究进展．于观平，王刚，王素华，李正名．农药学学报，2011，13 (2)：103 – 109.

498. 杨石先对农药化学学科的重要贡献及其学术思想．李正名．化学进展，2011，23 (1)：14 – 18.

499. Synthesis, Biological Activities and DFT Calculation of alpha – Aminophosphonate Containing Cyclopropane Moiety. *Liu Xing – Hai, Weng Jian – Quan, Tan Cheng – Xia, Pan Li, Wang Bao – Lei, Li Zheng – Ming.* Asian Journal of Chemistry, 2011, 23 (9): 4031 – 4036.

500. Synthesis, Crystal Structure and Theoretically Study of 2 – (Dimethylamino) – 1, 3 – dithiocyanatopropane and Its Isomer. *Guanping Yu, Yi Ma, Zhuo Liu, Gang Wang, Zhengming Li.* Chemical Research in Chinese Universities, 2011, 27 (6): 963 – 967.

501. 杀螟丹中间体硫氰化物及其异构体的高效液相色谱分析．于观平，闫涛，王素华，李正名．农药，2011，50 (3)：204 – 205.

2012 年

502. The Structure – activity Relationship in Herbicidal Mono – substituted Sulfonylureas. *Zhengming Li, Yi Ma, Luke Guddat, Peiquan Cheng, Jianguo Wang, Siew S Pang, Yuhui Dong, Chengming Lai, Lingxiu Wang, Guofeng Jia, Yonghong Li, Suhua Wang, Jie Liu, Weiguang Zhao, Bao-*

lei Wang. Pest Management Science, 2012, 68, 618 – 628.

503. Synthesis and Insecticidal Activity of Novel N – Pyridylpyrazole Carbonyl Thioureas. *Wang Baolei*, *Ma Yi*, *Xiong Lixia*, *Li Zhengming*. Chinese Journal of Chemistry, 2012, 30: 815 – 821.

504. Design, Synthesis and Biological Activities of Novel Benzoyl Hydrazines Containing Pyrazole. *Tao Yan*, *Shujing Yu*, *Pengfei Liu*, *Zhuo Liu*, *Baolei Wang*, *Lixia Xiong*, *Zhengming Li*. Chinese Journal of Chemistry, 2012, 30: 919 – 923.

505. Synthesis, Crystal Structure and Biological Activities of Novel Anthranilic (Isophthalic) Acid Esters. *Tao Yan*, *Guanping Yu*, *Pengfei Liu*, *Lixia Xiong*, *Shujing Yu*, *Zhengming Li*. Chemical Research in Chinese Universities, 2012, 28 (1): 53 – 56.

506. 含N – 吡啶联吡唑杂环的（磺）酰胺类化合物的设计、合成及生物活性．徐俊英，董卫莉，熊丽霞，李正名．高等学校化学学报，2012，33（2）：298 – 302.

507. Synthesis and Evaluation of Novel Monosubstituted Sulfonylurea Derivatives as Antituberculosis Agents. *Li Pan*, *Ying Jiang*, *Zhen Liu*, *Xinghai Liu*, *Zhuo Liu*, *Gang Wang*, *Zhengming Li*, *Di Wang*. European Journal of Memicinal Chemistry, 2012, 50: 18 – 26.

508. 基于Ugi反应的新型鱼尼丁受体杀虫剂的设计、合成及生物活性．刘鹏飞，周莎，熊丽霞，于淑晶，张晓，李正名．高等学校化学学报，2012，33（4）：738 – 743.

509. Synthesis, bioactivity and DFT structure – activity relationship study of novel 1, 2, 3 – thiadiazole derivatives. *Xing – Hai Liu*, *Wei – Guang Zhao*, *Bao – Lei Wang*, *Zheng – Ming Li*. Res. Chem. Intermed., 2012, 38 (8): 1999 – 2008.

510. Synthesis, Structure, and Biological Activity of Novel (Oxdi/Tri) azoles Derivatives Containing 1, 2, 3 – Thiadiazole or Methyl Moiety. *Xinghai Liu*, *Li Pan*, *Jianquan Weng*, *Chengxia Tan*, *Yonghong Li*, *Baolei Wang*, *Zhengming Li*. Molecular Diversity, 2012, 16 (2): 251 – 260.

511. Design, Synthesis and Insecticidal Activities of Novel Phenyl Substituted Isoxazolecarboxamides. *Pengfei Liu*, *Jifeng Zhang*, *Tao Yan*, *Lixia Xiong*; *Zhengming Li*. Chemical Research in Chinese Universities, 2012, 28 (3): 430 – 433.

512. 新型不对称草酰二胺类化合物的设计、合成及生物活性．闫涛，刘鹏飞，张吉凤，于淑晶，熊丽霞，李正名．高等学校化学学报，2012，33（8）：1745 – 1750.

513. Synthesis and Insecticidal Activities of Novel Phthalic Acid Diamides. *Tao Yan*, *Yuxin Li*, *Yongqiang Li*, *Duoyi Wang*, *Wei Chen*, *Zhuo Liu*, *Zhengming Li*. Chinese Journal of Chemistry, 2012, 30 (7): 1445 – 1452.

514. Synthesis and Insecticidal Activities of Novel Anthranilic Diamides Containing Acylthiourea and Acylurea. *Jifeng Zhang*, *Junying Xu*, *Baolei Wang*, *Yuxin Li*, *Lixia Xiong*, *Yongqiang Li*, *Yi Ma*, *Zhengming Li*. Journal of Agricultural and Food Chemistry, 2012, 60: 7565 – 7572.

515. Design, Synthesis and Biological Activities of Novel Anthranilic Diamide Insecticide Containing Trifluoroethyl Ether. *Yu Zhao*, *Yongqiang Li*, *Lixia Xiong*, *Hongxue Wang*, *Zhengming Li*. Chinese Journal of Chemistry, 2012, 30 (8): 1748 – 1758.

516. Synthesis, Structure and Insecticidal Activities of Some Novel Amides Containing N – Pyridylpyrazole Moeities. *Weili Dong*, *Jingying Xu*, *Lixia Xiong*, *Zhengming Li*. Molecules, 2012, 17 (9): 10414 – 10428.

517. 单取代嘧啶基吡啶磺酰脲化合物的合成和生物活性．童军，郑占英，李永红，王素华，李正名．农药，2012，51（9）：638 – 641.

518. 4 – 取代嘧啶基苯磺酰脲类化合物的合成与除草活性．郑占英，陈建宇，刘桂龙，李永红，李正名．农药学学报，2012，14（6）：607 – 611.

519. 新型2 – 甲氧羰基 – 5 – 芳甲亚胺基苯磺酰胺的合成及其生物活性．王美怡，李正名．合成化学，2012，20（1）：65－68.

520. Evaluation of the in Vitro and Intracellular Efficacy of New Monosubstituted Sulfonylureas against Extensively Drug－resistant Tuberculosis. *Di Wang*，*Li Pan*，*Gang Cao*，*Hong Lei*，*Xianghong Meng*，*Jufang He*，*Mei Dong*，*Zhengming Li*，*Zhen Liu*. International Journal of Antimicrobial Agents，2012，40（5）：463－466.

521. Synthesis，Crystal Structure and Biological Activity of Novel Anthranilic Diamide Insecticide Containing Alkyl Ether Group. *Yu Zhao*，*Liping Xu*，*Jun Tong*，*Yongqiang Li*，*Lixia Xiong*，*Fang Li*，*Lina Peng*，*Zhengming Li*. Molecular Diversity，2012，16：711－725.

522. 25%单嘧·2甲4氯钠盐水剂防除冬小麦田杂草田间试验．鞠国栋，李正名．农药，2012，51（12）：924－926.

2013年

523. Synthesis and insecticidal evaluation of novel N－pyridylpyrazolecarboxamides containing cyano Substituent in the ortho－position. *Mingzhen Mao*，*Yuxin Li*，*Qiaoxia Liu*，*Yunyun Zhou*，*Xiulan Zhang*，*Lixia Xiong*，*Yongqiang Li*，*Zhengming Li*. Bioorganic & Medicinal Chemistry Letters，2013，23（1）：42－46.

524. N－［4－氯－2－取代氨基甲酰基－6－甲基苯基］－1－芳基－5－氯－3－三氟甲基－1H－吡唑－4－甲酰胺的合成及生物活性．张秀兰，王宝雷，毛明珍，熊丽霞，于淑晶，李正名．高等学校化学学报，2013，34（1）：96－102.

525. Synthesis，Insecticidal Activities，and SAR Studies of Novel Pyridylpyrazole Acid Derivatives Based on Amide Bridge Modification of Anthranilic Diamide Insecticides. *Baolei Wang*，*Hongwei Zhu*，*Yi Ma*，*Lixia Xiong*，*Yongqiang Li*，*Yu Zhao*，*Jifeng Zhang*，*Youwei Chen*，*Sha Zhou*，*Zhengming Li*. Journal of Agricultural and Food Chemistry，2013，61：5483－5493.

526. Synthesis of Novel 3－Chloropyridin－2－yl－Pyrazole Derivatives and their Insecticidal，Fungicidal Activities and QSAR Study. *Zhang Ji－Feng*，*Liu Chen*，*Ma Yi*，*Wang Bao－Lei*，*Xiong Li－Xia*，*Yu Shu－Jing*，*Li Zheng－Ming*. Lett. Drug Des. Discov.，2013，10（6）：497－506.

527. Synthesis，structure and insecticidal activity of some novel amides containing N－pyridylpyrazole. *Dong Wei－Li*，*Xu Jun－Ying*，*Xiong Li－Xia*，*Li Zheng－Ming*. Journal of the Iranian Chemical Society，2013，10（3）：429－437.

528. Design，Synthesis and Structure－activity of N－Glycosyl－1－pyridyl－1H－pyrazole－5－carboxamide as Inhibitors of Calcium Channels. *Yunyun Zhou*，*Yuxin Li*，*Yiming Li*，*Xiaoping Yang*，*Mingzhen Mao*，*Zhengming Li*. Chemical Research in Chinese Universities，2013，29（2）：249－255.

529. 新型含三唑啉酮的磺酰脲类化合物的合成、晶体结构及除草活性研究．潘里，陈有为，刘卓，李永红，李正名．有机化学，2013，33（3）：542－550.

530. Design，Syntheses and Biological Activities of Novel Anthranilic Diamide Insecticides Containing N－Pyridylpyrazole. *Yu Zhao*，*Yongqiang Li*，*Lixia Xiong*，*Liping Xu*，*Lina Peng*，*Fang Li*，*Zhengming Li*. Chemical Research in Chinese Universities，2013，29（1）：51－56.

531. Synthesis and Biological Evaluation of Nonsymmetrical Aromatic Disulfides as Novel Inhibitors of Acetohydroxyacid Synthase. *Zaishun Li*，*Weimin Wang*，*Wei Lu*，*Congwei Niu*，*Yonghong Li*，*Zhengming Li*，*Jianguo Wang*. Bioorganic & Medicinal Chemistry Letters，2013，23（13）：3723－3727.

532. Recent advances in 4－hydroxyphenylpyruvate dioxygenase（HPPD）inhibitors. *Zhou Yun－yun*，*Li Zheng－ming*. Shijie Nongyao，2013，35（1）：1－7.

533. Synthesis and Biological Activities of Novel Anthranilicdiamides Analogues Containing Benzo [b] thiophene. *Jifeng Zhang, Chen Liu, Peng - fei Liu, Tao Yan, Baolei Wang, Lixia Xiong, Zhengming Li*. Chemical Research in Chinese Universities, 2013, 29 (4): 714 - 720.
534. Synthesis and herbicidal activity of novel sulfonylureas containing 1, 2, 4 - triazolinone moiety. *Liu Zhuo, Pan Li, Li Yong - hong, Wang Su - hua, Li Zheng - ming*. Chemical Research in Chinese Universities, 2013, 29 (3): 466 - 472.
535. Larvicidal Activity and Click Synthesis of 2 - Alkoxyl - 2 - (1, 2, 3 - Triazole - 1 - yl) Acetamide Library. *Su Na - Na, Xiong Li - Xia, Yu Shu - Jing, Zhang Xiao, Cui Can, Li Zheng - Ming, Zhao Wei - Guang*. Combinatorial Chemistry & High Throughput Screening, 2013, 16 (6): 484 - 493.
536. Synthesis and antiviral activity of hydrogenated ferulic acid derivatives. *Cui Can, Wang Zhi - Peng, Du Xiu - Jiang, Wang Li - Zhong, Yu Shu - Jing, Liu Xing - Hai, Li Zheng - Ming, Zhao Wei - Guang*. Journal of Chemistry, 2013, 269434.
537. Synthesis and evaluation of novel sulfenamides as novel anti Methicillin - resistant Staphylococcus aureus agents. *Shang Jian - Li, Guo Hui, Li Zai - Shun, Ren Biao, Li Zheng - Ming, Dai Huan - Qin, Zhang Li - Xin, Wang Jian - Guo*. Bioorganic & Medicinal Chemistry Letters, 2013, 23 (3): 724 - 727.
538. Synthesis, Crystal Structure and Biological Activity of a Novel Anthranilicdiamide Insecticide Containing Allyl Ether. *Yu Zhao, Lixia Xiong, Liping Xu, Hongxue Wang, Han Xu, Huabin Li, Jun Tong, Zhengming Li*. Research on Chemical Intermediates, 2013, 39 (7): 3071 - 3088.
539. 单嘧磺隆土壤残留 12 个月对主要后茬作物的安全性 [J]. 王满意，王宇，边强，鞠国栋，黄元炬，刘桂龙，李正名，黄春艳. 农药，2013，04：278 - 280.
540. 含有单取代嘧啶的新型磺酰脲类化合物的设计、合成及除草活性 [J]. 潘里，刘卓，陈有为，李永红，李正名. 高等学校化学学报，2013，06：1416 - 1422.
541. 单嘧磺酯钠盐的合成及除草活性 [J]. 寇俊杰，鞠国栋，李正名. 农药学学报，2013，03：356 - 358.
542. 纤维素生物合成抑制剂 (CBI) 类除草剂研究进展 [J]. 张秀兰，李正名. 世界农药，2013，02：10 - 15.
543. N - (4′ - 芳环取代嘧啶基 - 2′ - 基) - 2 - 乙氧羰基苯磺酰脲衍生物的合成及抑菌活性 [J]. 刘卓，潘里，于淑晶，李正名. 高等学校化学学报，2013，08：1868 - 1872.
544. Synthesis and biological activity of novel 1 - substituted phenyl (glycosyl) - 4 - {4 - [4, 6 - dimethoxy pyrimidin - 2 - yl] piperazin - 1 - yl} methyl - 1H - 1, 2, 3 - triazoles; *Mao, MZ (Mao Ming - zhen), Li YX (Li Yu - xin), Zhou YY (Zhou Yun - yun), Yang XP (Yang Xiao - ping), Zhang XL (Zhang Xiu - lan), Zhang X (Zhang Xiao), Li ZM (Li Zheng - ming)*. Chemical Research in Chinese Universities, 2013, 29 (5): 900 - 905.
545. Synthesis and insecticidal activities of 2, 3 - dihydroquinazolin - 4 (1H) - one derivatives targeting calcium channel; *Zhou YY (Zhou Yunyun), Feng Q (Feng Qi), Di FJ (Di Fengjuan), Liu QX (Liu Qiaoxiao), Wang DY (Wang Duoyi), Chen YW (Chen Youwei), Xiong LX (Xiong Lixia), Song HB (Song Haibin), Li YX (Li Yuxin), Li ZM (Li Zhengming)*. Bioorganic & Medicinal Chemistry, 2013, 21 (17): 4968 - 4975.
546. Three - level regular designs with general minimum lower - order confounding. *Li ZM (Li, Zhi ming), Zhang TF (Zhang, Tianfang), Zhang RC (Zhang, Runchu)*. Canadian Journal of Statistics - revue Canadienne de Statistique, 2013, 41 (1): 192 - 210.
547. Synthesis and characteristics of (Hydrogenated) ferulic acid derivatives as potential antiviral agents with insecticidal activity. *Huang GY (Huang Guang - Ying), Cui C (Cui Can), Wang ZP*

(*Wang Zhi – Peng*), *Li YQ* (*Li Yong – Qiang*), *Xiong LX* (*Xiong Li – Xia*), *Wang LZ* (*Wang Li – Zhong*), *Yu SJ* (*Yu Shu – Jing*), *Li ZM* (*Li Zheng – Ming*), *Zhao WG* (*Zhao Wei – Guang*). Chemistry Central Journal, 2013, 7: 33.

548. Synthesis and Herbicidal Activities of Sulfonylureas Bearing 1, 3, 4 – Thiadiazole Moiety. *Song XH* (*Song Xiang – Hai*), *Ma N* (*Ma Ning*), *Wang JG* (*Wang Jian – Guo*), *Li YH* (*Li Yong – Hong*), *Wang SH* (*Wang Su – Hua*), *Li ZM* (*Li Zheng – Ming*). Journal of Heterocyclic Chemistry, 2013, 50 (zengl): E67 – E72.

549. Discovery of Novel Acetohydroxyacid Synthase Inhibitors as Active Agents against Mycobacterium tuberculosis by Virtual Screening and Bioassay. *Wang D* (*Wang Di*), *Zhu XL* (*Zhu Xuelian*), *Cui CJ* (*Cui Changjun*), *Dong M* (*Dong Mei*), *Jiang HL* (*Jiang Hualiang*), *Li ZM* (*Li Zhengming*), *Liu Z* (*Liu Zhen*), *Zhu WL* (*Zhu Weiliang*), *Wang JG* (*Wang Jianguo*). Journal of Chemical Information and Modeling, 2013, 53 (2): 343 – 353.

550. Synthesis and Biological Activities of [5 – (4 – Substituted phenylsulfinyl/sulfonyl) – 1, 3 – dimethyl – 1H – pyrazol – 4 – yl] – arylmethanones. *Wang Bao – Lei*, *Li Qing – Nan*, *Zhan Yi – Zhao*, *Xiong Li – Xia*, *Yu Shu – Jing*, *Li Zheng – Ming*. Phosphorus, Sulfur and Silicon and the Related Elements (2013).

附录2 专利及著作目录

参与申请的专利

序号	国别	申请号	专利号	项目名称	发明人
1	中国		ZL200710151155.5	新结构磺酰脲化合物水溶盐的除草剂组合物	李正名、郑占英、李永红、王素华、童军、严冬文、陈建宇、寇俊杰、刘幸海、张树军
2	中国		ZL200810052445.9	乙酰乳酸合成酶 AHAS 抑制剂组合物	王建国、李正名、李永红、谭海忠
3	中国		ZL200810052812.5	含1，2，3－噻二唑环并具有抗乙肝病毒活性的丙烯酰胺类化合物	赵卫光、李正名、刘征骁、李玉新
4	中国		ZL93101976.1	防治玉米田杂草组合物	李正名、贾国锋、王玲秀、杨志强、赖城明
5	中国		ZL94118793.4	新型磺酰脲类化合物除草剂	李正名、贾国锋、王玲秀、范传文、杨炤
6	中国		ZL200310106640.2	嘧啶衍生物以及其制备方法	席真、班树荣、李正名、崔东亮、张弘、罗丁、牛聪伟、李志念、李峰、吴丽欢
7	中国		ZL200410018791.7	大豆田的田间除草方法	李正名、马宁、李永红、王建国、王玲秀、王素华、范传文
8	中国		ZL200410019224.3	磺酰脲类化合物及其制备方法和用途	席真、班树荣、李正名、崔东亮、张弘、罗丁、牛聪伟、陈文彬、吴丽欢
9	中国		ZL200310105079.6	具有杀虫、杀菌活性的苯并吡喃酮类化合物及制备与应用	刘长令、关爱莹、李志念、李林、李正名、李淼、张明星、张弘
10	中国		ZL200510013223.2	磺酰脲化合物水溶盐的除草剂组合物	李正名、寇俊杰、陈俊鹏、王玲秀、王素华
11	中国		ZL200510013913.8	磺酰脲化合物及除草活性	李正名、穆小丽、范志金、李永红、刘斌、赵卫光、王建国、王素华、王宝雷
12	中国		ZL200480020125.5	苯并吡喃酮类化合物及其制备与应用	刘长令、关爱莹、张弘、张明星、李正名、李淼、李林、李志念、侯春青

续表

序号	国别	申请号	专利号	项目名称	发明人
13	中国		ZL00131571.4	一种植物生长调节剂	李正名、贾强、黄桂琴
14	中国		ZL01136684.2	1，2，3－噻二唑类化合物及其制备方法和生物活性	李正名、赵卫光、杨炤
15	中国		ZL200410019058.7	用于制备N－（D）－脱氧核糖醇基－3，4－二甲基苯胺的催化剂	李正名、张津枫、邓国才
16	中国		ZL200710150519.8	单嘧磺酯类化合物的复配除草剂组合物	李正名、寇俊杰、王满意、鞠国栋、王素华、张晓光、李永红、王建明
17	中国		ZL201110147505.7	吡唑甲酰基硫脲衍生物与制备方法和应用	李正名、王宝雷、张吉凤、徐俊英、熊丽霞、赵毓、王刚
18	中国	200810053750.X		一种苯基三唑啉酮类化合物及其应用	李正名、唐蜜、王雷、兰峰、李永红、王素华
19	中国	85103016		多霉净的制备及应用	成俊然、石素娥、李正名、方仁慈、彭永冰、张素华、李树正、贺水济、张森、李振山
20	中国	93101416.6		细胞分裂因子类植物生长调节剂	李正名、乔立新、赵仲仁
21	中国	96115429.2		磺酰脲类除草混剂及其应用	郭武棣、张瑞亭、李正名、王丕业、姜斌、贾国峰、黑中一、周惠中、王玲秀、詹福康
22	中国	96106731.4		一种新的除草剂复配制剂	王玲秀、李正名、程穆如、贾国锋、李永红
23	中国	98100257.9		独角隆在谷田中的应用	王玲秀、李正名、贾国锋、陈俊鹏
24	中国	00134157.X		5－阿（依）维菌素B_{1a}酯的合成及生物活性	廖联安、李正名、方红云、范志金
25	中国	200310106702.X		小麦田的田间除草方法	李正名、马宁、李永红、王建国、王玲秀、王素华
26	中国	200410072054.5		DL－β－氨基丁酸衍生物及其制备方法和应用	李正名、臧洪俊、倪长春、范志金、王素华
27	中国	200510013111.7		苯并噻二唑衍生物及其合成方法和诱导抗病活性的筛选	范志金、刘凤丽、刘秀峰、聂开晟、范志银、鲍丽丽、张永刚、李正名

续表

序号	国别	申请号	专利号	项目名称	发明人
28	中国	200610013488.7		三唑啉酮类化合物和制备方法及其应用	李正名、王雷、李永红、马翼、王利、王素华
29	中国	200910070429.7		一类吡唑酰胺衍生物及其应用	李正名、赵毓、冯启、王宝雷、李玉新、李永强、熊丽霞、王素华
30	中国	201110147522.0		2，3－二氢喹唑啉酮类衍生物及其制备方法和应用	李正名、冯启、李玉新、王宝雷、熊丽霞、李永强、王素华
31	中国	201110075018.4		环丙烷肟酯衍生物及其制备方法和用途	刘幸海、李正名、张传玉、王宝雷、谭成侠、翁建全
32	中国	201110170304.9		一种噻二唑类化合物及其制备与应用	刘幸海、李正名、谭成侠、翁建全、李宝聚、石延霞、刘会君、曹耀艳
33	中国	201110289907.0		双酰胺衍生物及其制备和应用	李正名、闫涛、刘团伟、刘鹏飞、周莎、赵毓、李永强、熊丽霞
34	中国	201210548870.3		旱田除草剂组合物及其田间除草方法	李正名、寇俊杰、鞠国栋、边强、王满意、陈建宇
35	中国	201310183112.0		具有光学活性和几何异构的双酰胺衍生物与制备及应用	李正名、周莎、闫涛、周蕴赟、熊丽霞、李永强
36	中国	201310456588.7		一类取代苯基吡唑酰胺衍生物及其制备方法和应用	李正名、张秀兰、马金龙、周莎、熊丽霞、李永强、王宝雷
37	中国	201310398053.9		新型酰肼衍生物及其制备方法和应用	李正名、周蕴赟、刘辰、周莎、狄凤娟、熊丽霞、王宝雷、李玉新、赵毓

著 作 目 录

（1）李正名．第二部分 杀菌剂//南开大学元素有机化学研究所．国外农药进展．北京：化学工业出版社，1979：49－85.

（2）Zhengming Li. *Pesticide Chemistry and Regulation in People's Republic of China*// S. K. Bandal，G. J. Marco，L. Golberg，M. L. Leng. *The Pesticide Chemist and Modern Toxicology*. American Chemical Society，1981：522－536.

（3）李正名．农药化学//惠永正，戴立信，林国强，李正名，胡振元编辑．有机化学．北京：科学出版社，1994：262－268.

（4）李正名．有机立体化学进展．北京：中国轻工业出版社，1994.

（5）Zhengming Li. *The Structure－activity Relationship study on Herbicidal Sulfonylureas*// L. G. Copping，B. Sugavanam. *Crop Protection Chemicals—Present Developments and Future Prospects in the 21st Century*. Beijing：China Agriculture Press，1999：30－36.

（6）李正名，杨华铮．迎接新农药创制研究所面临的挑战//杜灿屏，刘鲁生，张恒编辑．21世纪有机化学发展战略．北京：化学工业出版社，2002：371－381.

附录3　历年指导的学生名单

年　级	姓　名	学　历	年　级	姓　名	学　历
81 级	董丽雯（女）	硕　士		王建国	硕　士　博　士
82 级	王天生	硕　士　博　士		马　宁	博　士
83 级	郭海生	硕　士		罗铁军	博　士
84 级	陈　林	硕　士	00 级	袁德凯	博　士
85 级	蒋益民	硕　士	01 级	王宝雷	博　士
86 级	刘同柱	硕　士		臧洪俊（女）	博　士
88 级	王胜新	硕　士		鲁　彦（女）	博　士
89 级	贾国锋	硕　士　博　士		张津枫	博　士
	邹霞娟（女）	硕　士		刘征骁	硕　士
90 级	乔立新	硕　士		李鹏飞	硕　士
91 级	范传文	硕　士		陈俊鹏	硕　士
	黄震年	博　士	02 级	韩　亮（女）	博　士
	贺铮杰	博　士		刘长令	博　士
92 级	廖　云	硕　士　博　士		郝静君	硕　士
	刘　洁（女）	硕　士　博　士		张建颖（女）	硕　士
93 级	李纪敏（女）	硕　士		穆小丽（女）	硕　士
	王忠文	博　士		王红学	硕　士
	余申义	博　士	03 级	李玉新（女）	博　士
94 级	赵卫光	硕　士		杨二冰（女）	硕　士
	杨小平	博　士		何凤琦	硕　士
	陈寒松	博　士		肖勇军	硕　士
95 级	曹一兵	硕　士	04 级	陈沛全	博　士
96 级	颜　寒	硕　士		马瑞典	硕　士
	王有名	博　士		程海英（女）	硕　士
97 级	高喜麟	硕　士		刘幸海	硕　士　博　士
	廖联安	博　士		王　雷	硕　士
	姜　林	博　士		孙　蕊（女）	硕　士
	朱向前	博士后		寇俊杰	硕　士
98 级	袁平伟	硕　士		严东文	硕　士
	王文艳（女）	硕　士	05 级	王美怡（女）	博　士
	赵卫光	博　士		郭万成	博　士
	贾　强	博　士		王浩安	硕　士　博　士
99 级	野国中	硕　士		李文明	硕　士
	符新亮	硕　士		唐　蜜（女）	硕　士
	钟　滨	硕　士　博　士		乌　婧（女）	硕　士

续表

年 级	姓 名	学 历	年 级	姓 名	学 历
	童 军	硕 士		张秀兰（女）	硕 士 博 士
	陈建宇	硕 士		陈 伟	硕 士 博 士
	王海英（女）	硕 士		刘团伟	硕 士
	郑占英	硕 士		于观平	博士后
	张树军	硕 士		鞠国栋	硕 士
06 级	董卫莉（女）	博 士	10 级	刘 卓（女）	博 士
	李 淼（女）	博 士		周 莎（女）	硕 士 博 士
	兰 峰	硕 士		陈有为（女）	硕 士 博 士
	谭海忠	硕 士		王多义	硕 士
	张传玉	硕 士	11 级	狄凤娟（女）	硕 士
	徐俊英（女）	硕 士		刘巧霞（女）	硕 士
07 级	赵 毓	博 士		刘 辰	硕 士
	李 洋	博 士		郭延军	博 士
	曹 刚	硕 士		朱洪伟（女）	硕 士
	李慧东	硕 士		郑 洁（女）	博 士
	李一鸣	硕 士	12 级	马金龙	硕 士
08 级	潘 里	硕 士 博 士		魏 巍	硕 士
	冯 启	博 士		刘敬波	硕 士
	王明忠	博 士		詹益周	硕 士
	王 刚	硕 士	13 级	华学文	博 士
	刘智力	硕 士		刘 明	博 士
	张吉凤	硕 士 博 士		张冬凯	硕 士
	毛明珍	硕 士 博 士		张丽媛（女）	硕 士
	李 芳（女）	硕 士		周 沙（女）	硕 士
09 级	刘鹏飞	博 士	14 级	吴长春	博 士
	闫 涛	博 士		魏 巍	博 士
	周蕴赟（女）	硕 士 博 士		刘敬波	博 士